In Memory Of

PETER GODFREY BARTON

Ecology and Classification of North American Freshwater Invertebrates

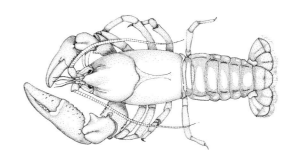

Ecology and Classification of North American Freshwater Invertebrates

Edited by

James H. Thorp
*Water Resources Laboratory
and Biology Department
University of Louisville
Louisville, Kentucky*

and

Alan P. Covich
*Department of Zoology
University of Oklahoma
Norman, Oklahoma*

Academic Press, Inc.
Harcourt Brace Jovanovich, Publishers
San Diego   New York   Boston
London   Sydney   Tokyo   Toronto

Front cover photograph:
Scanning electron micrograph of a heterotardigrade, *Echiniscus spiniger*.
See Figure 15.1 for details. Courtesy of Diane R. Nelson.

Academic Press, Inc.
San Diego, California 92101

United Kingdom Edition published by
Academic Press Limited
24–28 Oval Road, London NW1 7DX

Library of Congress Cataloging-in-Publication Data

Ecology and classification of North American Freshwater invertebrates
    / edited by James H. Thorp,   Alan P. Covich.
        p.   cm.
      ISBN   0-12-690645-9
        1. Freshwater invertebrates--North America--Ecology.
    2. Freshwater invertebrates--North America--Classification.
    I. Thorp, James H.   II. Covich, Alan P.
    QL151.E36   1191
    592.092'97--dc20   90-46694
                                                                  CIP

Printed in the United States of America
91   92   93   94   9   8   7   6   5   4   3   2   1

To my parents and wife
who kindled my scientific curiosity and kept the faith
JHT

To my family and to Professor G.E. Hutchinson
for their inspiration and patience
APC

Contents

Preface   xi

1 Introduction to Freshwater Invertebrates   1
James H. Thorp and Alan P. Covich

I. Introduction   1
II. Approaches to Taxonomic Classification   1
III. Synopses of the North American Freshwater Invertebrates   2
Literature Cited   15

2 An Overview of Freshwater Habitats   17
James H. Thorp and Alan P. Covich

I. Introduction   17
II. Lotic Environments   17
III. Underground Aquatic Habitats   26
IV. Lentic Ecosystems   28
Literature Cited   35

3 Protozoa   37
William D. Taylor and Robert W. Sanders

I. Introduction   37
II. Anatomy and Physiology   38
III. Ecology and Evolution   52
IV. Collecting, Rearing, and Preparation for Identification   61
V. Identification of Protozoa   62
Literature Cited   86

4 Porifera   95
Thomas M. Frost

I. Introduction   95
II. Anatomy and Physiology   96
III. Ecology and Evolution   103
IV. Collection, Culture, and Preparation for Identification   112
V. Classification   113
Literature Cited   121

5 The Freshwater Cnidaria—or Coelenterates   125
Lawrence E. Slobodkin and Patricia E. Bossert

I. Introduction   125
II. General Biology of Cnidaria   126
III. Descriptive Ecology of Freshwater Cnidaria   130
IV. Collection and Maintenance of Freshwater Cnidaria   138
V. Classification of Freshwater Cnidaria   140
Literature Cited   142

6 Flatworms: Turbellaria and Nemertea   145
Jerzy Kolasa

Turbellaria
I. Introduction: Status in the Animal Kingdom   146
II. Anatomy and Physiology   146
III. Ecology   148
IV. Current and Future Research Problems   151
V. Collecting, Rearing, and Identification Techniques   152
VI. Identification of North American Genera of Microturbellaria   153

Nemertea
VII. General Characteristics, External and Internal Anatomical Features   164
VIII. Ecology   164
IX. Current and Future Research Problems   166
X. Collection, Culturing, and Preservation   166
XI. Identification   166
Literature Cited   167
Appendix 6.1. List of North American Species of Microturbellaria   170

7   Gastrotricha   173

David L. Strayer and William D. Hummon

   I. Introduction   173
   II. Anatomy and Physiology   174
   III. Ecology and Evolution   175
   IV. Collecting, Rearing, and Preparation for
      Identification   180
   V. Identification of the Gastrotrichs of
      North America   180
      Literature Cited   183

8   Rotifera   187

Robert L. Wallace and Terry W. Snell

   I. Introduction   187
   II. Anatomy and Physiology   189
   III. Ecology and Evolution   199
   IV. Collecting, Rearing, and Preparation for
      Identification   222
   V. Classification and Systematics   224
      Literature Cited   240

**9   Nematoda and
Nematomorpha   249**

George O. Poinar, Jr.

Nematoda
   I. Introduction   249
   II. Morphology and Physiology   250
   III. Development and Life History   254
   IV. Ecology   255
   V. Collecting and Rearing
      Techniques   259
   VI. Identification   263

Nematomorpha
   VII. Introduction   273
   VIII. Morphology and Physiology   273
   IX. Development and Life History   275
   X. Sampling   277
   XI. Identification   277
      Literature Cited   280
      Appendix 9.1. Systematic
      Arrangement of the Nematode
      Genera   282

10   Mollusca: Gastropoda   285

Kenneth M. Brown

   I. Introduction   285
   II. Anatomy and Physiology   286
   III. Ecology and Evolution   290
   IV. Collecting and Culturing Freshwater
      Gastropods   300
   V. Identification of the Freshwater
      Gastropods of North America   301
      Literature Cited   309

11   Mollusca: Bivalvia   315

Robert F. McMahon

   I. Introduction   315
   II. Anatomy and Physiology   316
   III. Ecology and Evolution   335
   IV. Collecting, Preparation for
      Identification, and Rearing   370
   V. Identification of the Freshwater
      Bivalves of North America   373
      Literature Cited   390

**12   Annelida: Oligochaeta and
Branchiobdellida   401**

Ralph O. Brinkhurst and Stuart R. Gelder

Oligochaeta
   I. Introduction to Oligochaeta   401
   II. Oligochaete Anatomy and
      Physiology   402
   III. Ecology and Evolution of
      Oligochaeta   408
   IV. Collecting, Rearing, and Preparation
      of Oligochaetes for
      Identification   414
   V. Taxonomic Keys for
      Oligochaeta   416

Branchiobdellida
   VI. Introduction to
      Branchiobdellida   428
   VII. Anatomy and Physiology of
      Branchiobdellidans   428
   VIII. Ecology and Distribution   429
   IX. Identification of
      Branchiobdellida   430
      Literature Cited   433

**13   Annelida: Leeches, Polychaetes, and
Acanthobdellids   437**

Ronald W. Davies

   I. Introduction   437
   II. Taxonomic Status and
      Characteristics   438
   III. Leeches and Acanthobdellidae   439
   IV. Ecology   446
   V. Collection and Rearing   454
   VI. Identification   455
   VII. Polychaeta   457
   VIII. Taxonomic Keys   459
      Literature Cited   469
      Appendix 13.1. Distribution of Leech
      and Acanthobdellid Taxa in North
      America   476
      Appendix 13.2. Distribution of
      Polychaete Taxa in North
      America   478

14 Bryozoans   481
Timothy S. Wood

   I. Introduction   481
  II. Anatomy and Physiology   481
 III. Ecology and Evolution   487
 IV. Entoprocta   490
  V. Study Methods   491
     Literature Cited   497

15 Tardigrada   501
Diane R. Nelson

   I. Introduction   501
  II. Anatomy   502
 III. Physiology (Latent States)   508
 IV. Reproduction and Development   509
  V. Ecology   511
 VI. Techniques for Collection, Extraction, and Microscopy   514
 VII. Identification   515
     Literature Cited   518

16 Water Mites   523
Ian M. Smith and David R. Cook

   I. Introduction   523
  II. External Morphology and Internal Anatomy   527
 III. Ecology   537
 IV. Collecting, Rearing, and Preparation for Study   544
  V. Taxonomic Keys to Subfamilies of Water Mites in North America   546
     Literature Cited   589

17 Diversity and Classification of Insects and Colembola   593
William L. Hilsenhoff

   I. Introduction   594
  II. Orders of Aquatic Insects   597
 III. Partially Aquatic Orders of Insects   613
 IV. Semiaquatic Collembola—Springtails   631
  V. Identification of the Freshwater Insects and Collembola   632
     Literature Cited   653

18 Crustacea: Introduction and Peracarida   665
Alan P. Covich and James H. Thorp

   I. Introduction to the Subphylum Crustacea   665

  II. Anatomy and Physiology of Crustacea   668
 III. Ecology and Evolution of Selected Crustacea   673
 IV. Collecting, Rearing, and Preparation for Identification   681
  V. Classification of Peracarida and Branchiura   682
     Literature Cited   683

19 Ostracoda   691
L. Denis Delorme

   I. Introduction   691
  II. Anatomy and Physiology   692
 III. Life History   697
 IV. Distribution   699
  V. Chemical Habitats   699
 VI. Physical Habitat   701
 VII. Physiological and Morphological Adaptations   703
VIII. Behavioral Ecology   703
 IX. Foraging Relationships   704
  X. Population Regulation   704
 XI. Collecting and Rearing Techniques   704
 XII. Current and Future Research Problems   706
XIII. Classification of Ostracodes   706
     Literature Cited   716

20 Cladocera and Other Branchiopoda   723
Stanley I. Dodson and David G. Frey

Cladocera
   I. Anatomy and Physiology   725
  II. Ecology of Cladocerans   730
 III. Current and Future Research Problems   743
 IV. Collecting and Rearing Techniques   744
  V. Taxonomic Keys for Cladoceran Genera   745

Other Branchiopods
 VI. Anatomy and Physiology of the Other Branchiopods   764
 VII. Ecology of the Other Branchiopods   765
VIII. Current and Future Research Problems   770
 IX. Collecting and Rearing Techniques   770
  X. Taxonomic Keys for Non-Cladoceran Genera of Branchiopods   770
     Literature Cited   776

21 *Copepoda*   787
Craig E. Williamson

    I. Introduction   787
   II. Anatomy and Physiology   788
  III. Ecology and Evolution   790
  IV. Current and Future Research
      Problems   803
    V. Collecting and Rearing
      Techniques   804
   VI. Identification Techniques   805
  VII. Identification of the Order
      Calanoida   806
VIII. Identification of the Order
      Cyclopoida   807
  IX. Identification of the Order
      Harpacticoida   810
      Literature Cited   813

22 *Decapoda*   823
H. H. Hobbs III

    I. Introduction   823
   II. Anatomy and Physiology   824
  III. Ecology and Evolution   829
  IV. Current and Future Research
      Problems   847
    V. Collecting and Rearing
      Techniques   848
   VI. Identification   850
      Literature Cited   852

Glossary   859

Index   875

Preface

An exhaustive coverage of the diverse structural, physiological, behavioral, and general ecological adaptations of freshwater animals is obviously beyond the scope of a single book. The more focused theme of the present book—ecology and classification—reflects our evaluation of the needs of the discipline and the personal scientific interests of the chapter authors. Even within those two areas, the chapters of this book are meant to serve only as an introduction. Authors of each chapter briefly review the general biology of their group and then present a detailed summary of the most notable ecological aspects. Every chapter includes a broad introduction to the more recent ecological and taxonomic literature.

We have also constrained the breadth of this text by limiting the taxonomic keys in most cases to the generic level (occasionally the keys are to species; rarely is the coverage less detailed than to the family tier). Although we appreciate that significant ecological differences exist among species, we felt that taxonomic keys at the species level in all chapters would be impractical. Furthermore, as this book was designed for use by a general audience, we did not intend for individual chapters to be employed extensively by scientific experts in those fields. We anticipate that a continuing series of specialized monographs will treat these species-level relationships. Most authors refer the readers to other current publications for regional and species-level taxonomic keys.

We have also severely reduced the coverage of freshwater insects to avoid unnecessary duplication of the significant contributions in freshwater entomology currently available to the reader (e.g., broad treatments by McCafferty 1981, Merritt and Cummins 1984, as well as several books on specific insect orders; for references, see Chapter 1). Rather than entirely eliminate the insects from this book, however, we included a single, overview chapter at the family level (Chapter 17) so that students taking a freshwater invertebrate course (other than aquatic entomology) would have a general text to guide their studies, with more specific keys supplemented by the instructor.

The organisms discussed in this book are almost exclusively "true" freshwater species; that is, estu-arine taxa are rarely treated. Generally, only species with at least one free-living life stage are included here; hence, parasites of aquatic hosts are either ignored or given scant attention. We have also limited the coverage to species in Canada and the United States to avoid the task of handling the enormous tropical fauna in the southern portion of North America. Freshwater invertebrates within this north temperate to arctic region certainly number in excess of 10,000 species but probably encompass much fewer than 15,000 taxa.

While reading this book, you are bound to notice differences among chapters in such features as organization, focal points, breadth and detail of coverage, number of references, and writing style. Some variations should be expected, because the nature and scope of scientific research has not been the same for each major taxon covered in this text. We anticipated some of these differences and instituted an editorial process of external and internal reviews. Given the options of more consistency among chapters than is possible with a single or coauthored book versus the greater expertise that is gained from a multi-authored, edited volume, we chose the latter approach. We believe that our success in enlisting the help of outstanding biologists for each invertebrate group had made this choice the correct one.

Editing a book of this magnitude has proved to be a stimulating and demanding task. Although finishing this volume was a great relief, we continue to seek ways to improve it. We encourage you to send us your comments so that we may incorporate those suggestions in any future editions.

Just as money does not grow on trees, books do not appear in print without the help of a large cast of characters whose names are absent from the book cover. We have been consistently pleased with the staff at Academic Press and wish to acknowledge in particular the assistance of Chuck Arthur, Barb Heiman, and Jean Thompson Black who aided us in the initial steps of publication. Two of our clerical staff, Mary Jenkins (University of Louisville) and Paula Goldberg (Fordham University), also deserve special thanks for their work.

The present chapters were improved significantly by comments made by many reviewers in the United States, Canada, and Europe. Without their contributions, we would not have attempted to publish this book. Among all reviewers and authors, we would particularly like to express our appreciation to Dave Strayer for his many excellent suggestions in the early stages of putting this book together and for his consistently high quality reviews. Several of the following reviewers commented on more than one chapter; their names are followed by an asterisk: Denton Belk, Ralph O. Brinkhurst, Joseph C. Britton, Kenneth M. Brown, John H. Bushnell, Richard D. Campbell, John O. Corliss, David C. Culver, Ronald W. Davies, George M. Davis, Stanley I. Doson*, Richard M. Forester, John J. Gilbert, Al Grigarick, B. J. Hann, John E. Havel, William Hilsenhoff, John Holsinger, Dan M. Johnson*, Eileen H. Jokinen, R. D. Kathman*, Jacek Kisielewski, Donald J. Klemm, Jerzy Kolasa, Janet L. Leonard, Denis H. Lynn, Gerry Mackie, Robert F. McMahon, Rodger Mitchell, Thomas Nogrady, Michael L. Pace, James F. Payne, C. Patterson, Barbara L. Peckarsky*, Michael A. Poirrier, Vincent H. Resh, Robert O. Schuster, Robert A. Short, Thomas W. Simmons, Tracy L. Simpson, Bruce P. Smith, Douglas G. Smith, Peter L. Starkweather, Roy A. Stein, Richard S. Stemberger, David L. Strayer*, John H. Tietjen, Alan J. Tessier, Bryn H. Tracy, Seth Tyler, Carl M. Way, Mitchell J. Weiss, and David S. White.

James H. Thorp
Alan P. Covich

Introduction to Freshwater Invertebrates

1

James H. Thorp
Department of Biology and Water Resources Laboratory
University of Louisville
Louisville, Kentucky 40292

Alan P. Covich
Department of Zoology
University of Oklahoma
Norman, Oklahoma 73019

Chapter Outline

I. INTRODUCTION
II. APPROACHES TO TAXONOMIC CLASSIFICATION
III. SYNOPSES OF THE NORTH AMERICAN FRESHWATER INVERTEBRATES
   A. Protozoa
   B. Porifera
   C. Cnidaria
   D. Flatworms: Turbellaria and Nemertea
   E. Gastrotricha
   F. Rotifera
   G. Nematoda and Nematomorpha
   H. Mollusca
      1. Gastropoda
      2. Bivalvia
   I. Annelida
      1. Oligochaeta and Branchiobdellida
      2. Leeches, Polychaetes, and Acanthobdellids
   J. Bryozoans
   K. Tardigrada
   L. Arthropoda
      1. Water Mites
      2. Aquatic Insects and Collembola
      3. Crustaceans
   M. Taxonomic Key to Major Taxa of Freshwater Invertebrates
   Literature Cited

I. INTRODUCTION

Next to their more vibrant and often larger marine relatives, freshwater invertebrates may initially seem drab and uninspiring. Yet once students of freshwater biology penetrate beyond these superficial appearances and thoroughly examine the diverse structural, physiological, behavioral, and general ecological adaptations of freshwater invertebrates, few fail to be impressed by the enticing nature of these fascinating animals.

This chapter serves as a very fleeting introduction to the freshwater invertebrates of North America. Following a short discussion of procedures for naming and classifying organisms, we briefly describe the groups covered in Chapters 3–22. We also include a taxonomic key to help you begin the process of identifying the major taxonomic classification of a freshwater invertebrate you may have found or been asked to identify. This key will lead you to the appropriate chapter for more detailed ecological and taxonomic information.

II. APPROACHES TO TAXONOMIC CLASSIFICATION

No phylum is totally immune to an occasional reshuffling of species, renaming of constituent taxa, and addition to, or substraction from, the total number of species. This dynamic feature of the discipline is, in part, a necessary response to the acquisition of new knowledge, but it also results from the ubiquitous and almost inherent disagreements that usually typify groups of two or more scientists! Taxonomists are frequently labeled as splitters or lumpers according to whether they tend to acknowledge more or fewer species within an assemblage, respectively. Classifications based on diverse scientific approaches, such as cladistics and numerical taxonomy, can produce dramatically different views of relationships among taxa. Although arguments at the species level are especially rampant, debates occasionally break out regarding the categorization of higher tiers, including classes and phyla. While reading through this book, you will often encounter references to current taxonomic debates. The final classification of taxa has been left to the authors of individual chapters. As editors, we have allowed the authors a great amount of leeway in this area, insisting only that they inform the reader about the principal areas of dispute.

The final arbiter for naming taxa is the International Commission on Zoological Nomenclature, a judicial body elected periodically by the International Congress of Zoology. The commission bases its findings on rules covering the priority of names and designation of "type" taxa. As the name implies, "priority" specifies that the first name assigned in a publication is generally the official name of that taxon. The commission has, however, plenary powers to set aside the "rightful" name of a species if by doing so the cause of science would be further advanced. For example, alteration of a familiar, long-standing name could cause a major confusion in the scientific literature. Occasionally, two distinct taxa are inadvertently assigned the same binomial name by scientists working in different laboratories. In that case, the species named last loses its name as a junior homonym. To lessen the state of confusion and diminish the possibilities for future arguments, one individual of each species is designated as the prime example of that taxon. Such "type" specimens are deposited in one of the numerous, internationally recognized museums and made available for study by responsible scientists.

Information on systematic zoology is published in various journals, such as the *Zoological Record*, *Zeitschrift für Wissenschaftliche Zoologie* (Abteilung B), and *Bulletin Signaletique* (CNRS), as well as in numerous society publications and journals dealing with a broad range of subjects including classification. Readers interested in knowing where best to search for papers on a given taxon should consult the Literature Cited section in the appropriate chapter of this book or see Pennak (1989).

The International Code of Zoological Nomenclature passes judgement on the names of genera and species but does not strictly control higher classification. For that reason, some disagreements exist about the numbers and names of higher classification tiers, especially at the ordinal level and above

Table 1.1 Principal Zoological Ranks and Suffixes Where Designated

Kingdom
  Phylum
    Class
      Cohort
        Order
          Superfamily (-oidea [a])
           Family (-idae)
             Subfamily (-inae)
               Tribe (-ini [a])
                 Subtribe (-ina [b])
             Genus
               Species

[a] An ending recommended but not mandatory according to the International Code of Zoological Nomenclature.
[b] An ending customary but not cited in the code.

(e.g., the phyla Cnidaria vs. Coelenterata, Bryozoa vs. Ectoprocta, and disagreements on whether Crustacea is a phylum, subphylum, or class). The system for adding suffixes to taxonomic names as a means of identifying the classification level has been confusing in the past but is beginning to change for the better. A nonexclusive list of the principal zoological ranks and suffixes is shown in Table 1.1.

The importance of correct identification and consistent approaches to classification soon becomes apparent to all people interested in learning about general biotic relationships. Vast amounts of information are catalogued in research libraries and museums that use the Linnean system of binomial classification. Once a specimen is correctly identified, a large number of pertinent references in specialized journals can be accessed, not only in systematics but also in ecology, evolution, animal behavior, and medical sciences. Given the rapid development of computer-based information retrieval at most larger libraries, students are encouraged to use examples of genera to learn what others have discovered. This trail of "information networking" can lead to many more new ideas on how various taxa live together.

III. SYNOPSES OF THE NORTH AMERICAN FRESHWATER INVERTEBRATES

All major phyla of invertebrates, other than Echinodermata, have some freshwater representatives, but only a few are more diverse in freshwaters than in the world's oceans. Some salient features of these freshwater taxa, as discussed in Chapters 3–22, are summarized in the remainder of this section.

A. *Protozoa*

Protozoa are unicellular (or acellular), eukaryotic organisms existing either as individuals or members of a loose-knit colony (Fig. 1.1). The term protozoa is a taxon of convenience, having no significance in classification schemes within the kingdom Protista (or Protoctista) other than as a traditional functional grouping of all heterotrophic and motile species. Modern taxonomic treatments divide the kingdom Protista into as many as 27 phyla, of which two contain significant numbers of free-living, animal-like species. These two phyla, Ciliophora and Sarcomastigophora, are the subject of Chapter 3, by William D. Taylor and Robert W. Sanders.

Protozoa play important, though often overlooked, roles in freshwater ecosystems. For example, flagellate protozoans are the predominant predators of picoplankton (primarily bacteria and small cyanobacteria), and thus are essential components of the "microbial loop" within the epilimnion of many lakes. Many genera consume bacteria within the surficial layer of sediments, and larger taxa capture and eat algae or other protozoans.

They, in turn, are preyed upon by some species of oligochaetes, chironomids, and rotifers.

B. *Porifera*

About 25 species of sponges colonize freshwater habitats in Canada and the United States. Although less diverse than protozoans, freshwater members of the phylum Porifera—the sponges—nonetheless are commonly found on hard substrata within many lakes and streams (Fig. 1.2). Except for the marine mesozoans and placozoans, sponges have the simplest structure of all multicellular phyla, showing little variation in external form and having a tissue-level organization internally. Freshwater taxa routinely contain symbiotic algae, which provide nutrition to supplement the food gained by filter feeding on bacteria and algae in the water column.

The annual life cycle of a freshwater sponge is characterized by shifts between periods of active growth and dormancy in the gemmule stage. Availability of suitable hard substrata and adequate food

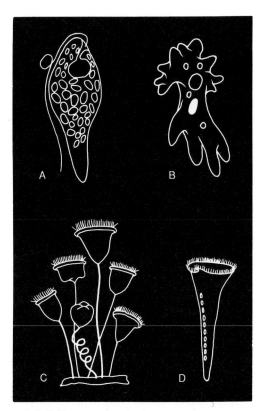

Figure 1.1 Solitary and colonial protozoa: (A) the zooflagellate *Khawkinea*; (B) *Amoeba*; (C) the colonial ciliate *Vorticella*; and, (D) the solitary ciliate *Stentor*. [See Chapter 3.] [These and subsequent figures in this chapter were redrawn and modified from illustrations in Chapters 3–22.]

Figure 1.2 Freshwater sponges (phylum Porifera): (A) drawing of a living sponge on a stick; (B) representative sponge spicules (all 25–150 μm in length). [See Chapter 4.]

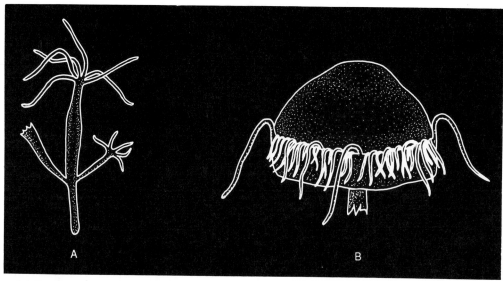

Figure 1.3 Freshwater coelenterates (phylum Cnidaria): (A) a polypoid *Hydra*; (B) a medusoid *Craspedacusta*. [See Chapter 5.]

seem to be the critical factors regulating the spread and growth of sponges, as described by Thomas M. Frost in Chapter 4.

C. Cnidaria

Freshwater cnidarians, or coelenterates, have been relatively unsuccessful in adapting to freshwater, being represented only by one class (Hydrozoa) with five purely freshwater genera: the common hydras

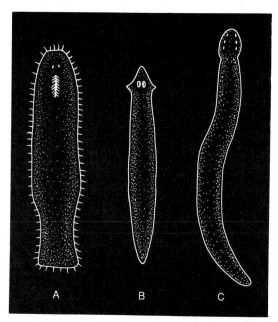

Figure 1.4 Turbellarian flatworms (phylum Platyhelminthes) and a ribbon worm (phylum Nemertea): (A) *Macrostomum*; (B) *Dugesia*; and (C) the nemertean *Prosotoma*. [See Chapter 6.]

(brown *Hydra* and green *Chlorohydra*) and the rare medusoid *Craspedacusta* and polypoid *Calposoma* and *Polypodium* (Fig. 1.3). Despite their low diversity, cnidarians—or at least hydras—are routinely present on almost any reasonably hard substratum in permanent ponds and lakes and in streams ranging from headwater streams to large rivers.

The cnidarian body consists of a generally thin, acellular layer of mesoglea sandwiched between two thicker cellular layers (ecto- and endoderm), all surrounding a central body cavity (the coelenteron). Coelenterates dispatch their mostly invertebrate prey with ectodermal batteries of stinging or sticky nematocysts. Green hydra are named for the presence of endosymbiotic *Chlorella*, algae that provide a significant source of nutrition for the cnidarian. The biology of freshwater cnidarians is described in Chapter 5, by Lawrence B. Slobodkin and Patricia E. Bossert.

D. Flatworms: Turbellaria and Nemertea

In Chapter 6 on unsegmented flatworms, Jerzy Kolasa describes the biology and classification of the more than 200 species of turbellarians (phylum Platyhelminthes) and 3 species of ribbon worms (phylum Nemertea) found in North America (Fig. 1.4). He discusses the acoelomate Turbellaria, the simplest bilaterally symmetrical eumetazoan, and the more complex, coelomate nemerteans, which have an anus and a closed circulatory system. These latter flatworms possess an eversible, muscular proboscis resting in a rhynchocoel.

Flatworms are well represented in the benthos of lakes and streams. Ribbon worms of the single

North American genus *Prostoma* are rarely reported but can be common in their typical shallow, littoral zone habitats; all are carnivores that capture their prey with a sticky proboscis. The much more diverse turbellarians are distributed more widely than ribbon worms; but they are not much better known, in part because they are very difficult to identify with the collection and preservation techniques normally employed in field studies. The great majority of these flattened or cylindrical worms are free-living denizens of lakes and streams from shallow to deep waters. Most are small (< 1 mm), especially the more diverse microturbellarians. Turbellarians consume prey in accordance with their size—variously eating bacteria, algae, protozoans, and small or large invertebrates.

E. Gastrotricha

The pseudocoelomate gastrotrichs are common residents of epigean waters throughout North America, numbering as high as 100,000 animals per m^2. Although fewer than 100 freshwater species have been reported from North America, the actual diversity is probably much higher (Fig. 1.5). Possibly because gastrotrichs are so small (50–800 μm), our ecological knowledge of this aschelminth group has not advanced very far. They colonize sediments and are members of the aufwuchs assemblage on plants. Gastrotrichs thrive on a diet of bacteria, and genera occur in both aerobic and anoxic sediments. Their ecology and classification are examined in Chapter 7 by David L. Strayer and William D. Hummon.

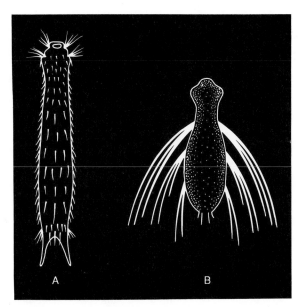

Figure 1.5 Representative freshwater Gastrotricha: (A) *Chaetonotus*; (B) *Stylochaeta*. [See Chapter 7.]

F. Rotifera

Perhaps no other phylum is as clearly associated with freshwater as is Rotifera (Fig. 1.6). Nearly 2000 species of rotifers, or "wheel animals," inhabit freshwaters throughout the world, whereas only about 50 species are exclusively marine. These unsegmented, bilaterally symmetrical pseudocoelomates are distinguished by two principal anatomical features: an apical, ciliated region known as the corona and a muscular pharynx, termed the mastax, with its complex set of hard jaws.

As Robert L. Wallace and Terry W. Snell point out in Chapter 8, rotifers are one of the three principal animal taxa in the plankton (along with protozoans and microcrustaceans). Because they are more efficient than cladocerans when feeding on minute algae, rotifers can exert a greater grazing pressure on the small picoplankton. Other rotifers are important predators on prokaryotes, protozoans, and small metazoans in the plankton, and many motile and sessile species play significant roles as herbivores and carnivores in the littoral zone assemblages.

G. Nematoda and Nematomorpha

Nematodes, which are best known for their damaging effects on crops, are also abundant in freshwater and are represented in North America by nearly 70 genera of roundworms that spend all or a large part of their life cycle in lakes or streams (Fig. 1.7a). Nematodes are significant components of the micro- and macrobenthos of many aquatic ecosystems. For example, in Mirror Lake, New Hampshire, they have constituted 60% of all the individuals of benthic metazoans, although only about 1% of the total biomass (see citation for Strayer 1985 in Chapter 9). Aside from the parasitic species, which are not extensively covered by George O. Poinar in Chapter 9, free-living roundworms obtain nourishment from bacteria, algae, and protozoans. Nematodes are unsegmented worms with a body cavity (pseudocoel) and complete alimentary tract.

Another phylum of worms that, in a very rough way, resembles nematodes is Nematomorpha (Fig. 1.7b). These horsehair worms are free living as adults but parasitic as larvae in various invertebrates. They are also unsegmented pseudocoelomates, but their intestine is nonfunctional and they are generally much larger than roundworms.

H. Mollusca

Two classes within the phylum Mollusca contain large numbers of freshwater species: the classes

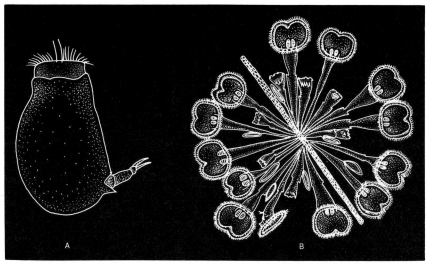

Figure 1.6 "Wheel animals" of the phylum Rotifera: (A) a solitary *Gastropus*; (B) a colony of *Sinantherina*. [See Chapter 8.]

Gastropoda (snails) and Bivalvia (mussels and clams). These ubiquitous and abundant molluscs are discussed separately in Chapters 10 (snails) and 11 (mussels).

1. Gastropoda

Snails are one of the more diverse freshwater groups (having about 350 species in North America) and are certainly among the most easily recognized by the public. These soft-bodied, unsegmented coelomates have a body organized into a muscular foot, distinct head region, and visceral mass; a fleshy mantle covers the viscera and secretes a spiraled, univalve shell (Fig. 1.8). They are commonly found in the littoral region of lentic and lotic habitats, where they collect bottom detritus, graze on the algal film covering rocks, wood snags, and plants, or even float upside down at the water surface consuming trapped and neustonic algae. Recent ecological work on freshwater gastropods, as described by Kenneth M. Brown in Chapter 10, has focused on their important roles as grazers on periphyton and as prey of benthic-feeding fish and invertebrates.

2. Bivalvia

The North American fauna of freshwater bivalves is the richest in the world, with over 250 species of mussels and clams (Fig. 1.9; see also Chapter 11, by Robert F. McMahon). Bivalves, or pelecypods, are

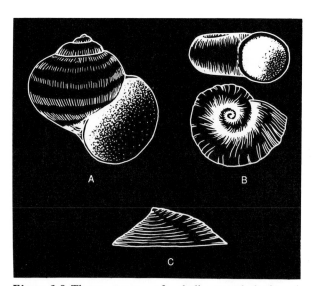

Figure 1.8 Three types of shell morphologies in freshwater snails and limpets (phylum Mollusca, class Gastropoda): (A) the most common, spiral shell, as shown by *Pomacea*; (B) two views of the planospiral snail *Planorbella* (*Helisoma*); and (C) the conical freshwater limpet *Ferrissia*. [See Chapter 10.]

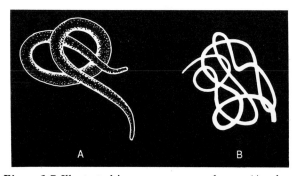

Figure 1.7 Illustrated is a common roundworm (A; phylum Nematoda) and a less frequently observed, adult horsehair worm (B; phylum Nematomorpha). [See Chapter 9.]

major deposit and filter feeders within permanent lakes, streams, and large rivers, where they are often the largest invertebrates in body mass. Except for the smaller pea clams (Sphaeriidae), native mussels (Unionacea) are uniquely capable of growing large enough to escape predation by almost all aquatic vertebrates and invertebrates. Bivalves are much less motile than gastropods, and their body is enclosed in two hinged, calcareous shell valves; a radula, which is employed by snails for grazing, is absent in clams and mussels. Unlike the vast majority of North American freshwater invertebrates, bivalves have been of direct economic importance to humans; they were formerly used for button production and are now harvested for freshwater pearls or for the core elements needed to grow marine-cultured pearls.

I. Annelida

The phylum Annelida consists of two relatively important freshwater groups (the oligochaete worms and the leeches) and several minor taxa composed of two exclusively freshwater groups (Branchiobdellida and Acanthobdellida) and one class that is predominantly a marine taxon (Polychaeta). The taxonomy of Annelida is in great dispute—an observation reflected by the diverse views of the authors of Chapters 12 and 13. The principal areas of controversy involve their status as either a class or lower taxonomic level and their placement either in the phylum Annelida (the traditional approach and the one used here) or in the arthropod subphylum Uniramia (a view which has yet to garner more than a few adherents). Annelids are legless, segmented coelomates with serially arranged organs; they occur in both lentic and lotic environments.

1. Oligochaeta and Branchiobdellida

In Chapter 12, Ralph O. Brinkhurst and Stuart R. Gelder discuss the free-living oligochaete worms and the odd branchiobdellids, a small group of 15 genera that are ectocommensals of astacid crayfish (Fig. 1.10). Oligochaetes are among the more widely distributed freshwater taxa and one of the very few invertebrates to colonize moist terrestrial environments. In addition to the characteristics of the phylum as a whole, oligochaetes lack suckers and generally have four bundles of setae on every segment but the first. Oligochaetes are best known for their ingestion of sediments in the littoral and profundal zones (a process prominent in the recycling of nutrients), but some species feed on benthic algae and other epiphytic organisms. Their great abundance and ease of consumption and assimilation make them favorite prey items for many invertebrate and vertebrate predators.

2. Leeches, Polychaetes, and Acanthobdellids

The remainder of the phylum Annelida is comprised of fewer than 100 species, over 80% of which are

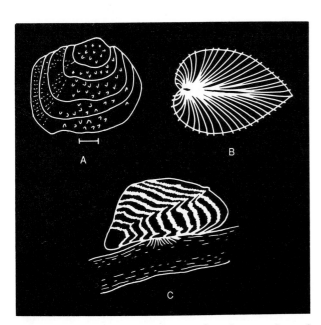

Figure 1.9 One native and two invader species of freshwater bivalves (phylum Mollusca, class Bivalvia): (A) the native mussel *Quadrula*; (B) the ubiquitous Asiatic clam, *Corbicula*; and, (C) a newcomer to North America, the rapidly spreading zebra mussel, *Dreissena*. [See Chapter 11.]

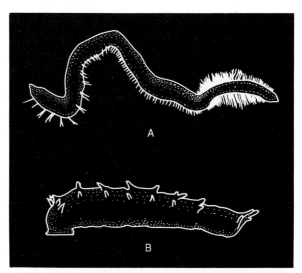

Figure 1.10 Two freshwater annelids with distinctive lifestyles: (A) the free-living oligochaete *Branchiura*, showing its posterior gill filaments; (B) *Ceratodrilus*, a branchiobdellidan which is an ectocommensal of crayfish. [See Chapter 12.]

leeches. Although leeches were traditionally listed as a separate class (Hirudinea) within Annelida, recent treatments have tended to group them within Oligochaeta, perhaps as the superfamily Hirudinoidea (for a discussion of the controversy, see Chapter 13, by Ronald W. Davies). Unlike most oligochaetes, however, leeches have suckers (anterior and posterior) and lack setae (Fig. 1.11). Another major distinction between these two annelids is their feeding niches. Leeches are either predators of invertebrates or temporary ectoparasites feeding on the blood of vertebrates (in contrast to the detritivorous nature of most oligochaetes).

J. Bryozoans

Bryozoans, or "moss animals," are generally sessile, colonial invertebrates with ciliated tentacles—the lophophore—for capturing suspended food particles (Fig. 1.12). All but one North American species are in the phylum Ectoprocta (or Bryozoa, according to your preference) and are composed of animals whose mouth, but not anus, is located within the whorl of feeding tentacles. The sole exception, *Urnatella gracilis*, is classified in the phylum Entoprocta, a taxon with dubious relation to the ectoprocts. Both the mouth and anus of entoprocts are placed within the whorl of feeding tentacles; other important structural differences also exist. Although the diversity of freshwater bryozoans is not high (22 species in North America), they are actually quite common occupants of hard, relatively stationary substrata (see Chapter 14, by Timothy S. Wood). They occur in lakes but thrive better in flowing water, where a constant source of suspended food passes their tentacles. Our ecological knowledge of bryozoans is relatively sparse despite the long-standing familiarity of scientists with this group.

K. Tardigrada

Tardigrades, or "water bears," are easy to overlook because of their diminutive nature (adults average 250–500 μm), but they are widely dispersed in freshwater and terrestrial environments, and a few of the 600 species worldwide occur in the ocean. They are segmented micrometazoans with four fleshy legs terminating in claws (Fig. 1.13). Tardigrades have piercing/sucking mouthparts and obtain nutrition from plants, animals, and detritus. An important part of the water bear life cycle is the latent stage (cryptobiosis) entered into when environmental conditions become inhospitable. The biology and systematics of these little-known creatures are described by Diane R. Nelson in Chapter 15.

L. Arthropoda

The most successful terrestrial phylum and one of the most prominent freshwater taxa is Arthropoda. Its three subphyla with freshwater members—Chelicerata (water mites and aquatic spiders), Uniramia (aquatic insects), and Crustacea (crayfish, fairy shrimp, copepods, etc.) —are all diverse and important components of lakes and streams. Arthropods occupy every heterotrophic niche in benthic and pelagic habitats of most permanent and temporary aquatic systems. These metameric coelomates are characterized by a chitinous exoskeleton and stiff, jointed appendages modified as legs, mouthparts, and antennae (except in water mites).

1. Water Mites

The class Arachnida (subphylum Chelicerata) is represented in freshwater by a few semiaquatic, true spiders and a colossal number of water mites (subclass Acari; Fig. 1.14). In Chapter 16, the biology and classification of five distantly related groups of mites are discussed by Ian M. Smith and David R. Cook, the most diverse of which is Hydrachnida. As they point out, a single square meter of substratum from a littoral weed bed in a eutrophic lake may contain 2000 mites representing up to 75 species! Because of their prodigious numbers and wide distribution, mites probably play a major, albeit often ignored role in aquatic ecosystems as predators, ec-

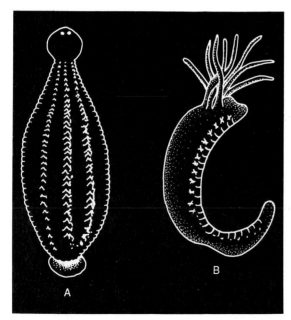

Figure 1.11 Schematic view of two other annelids: (A) the leech *Placobdella*; (B) the polychaete *Hypaniola*. [See Chapter 13.]

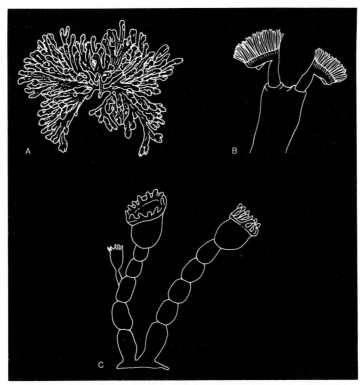

Figure 1.12 Freshwater bryozoans: (A) *Plumatella* colony (phylum Bryozoa or Ectoprocta); (B) close-up of two individuals in an ectoproct colony; and (C) the ubiquitous entoproct *Urnatella*. [See Chapter 14.]

toparasites, and prey. For example, mites interact intimately with all life stages of another major group of arthropods, the aquatic insects. Mites differ from members of other arthropod subphyla in lacking antennae; adults generally have four pairs of jointed legs and segmentation is greatly reduced or not immediately apparent.

2. Aquatic Insects and Collembola

Perhaps the most diverse and best studied groups of freshwater invertebrates are the ten insect orders containing aquatic species (Fig. 1.15). William L. Hilsenhoff accepted the daunting task of summarizing some basic aspects of the biology and classification of aquatic insects in Chapter 17, leaving the more detailed discussions to books devoted to all

or specific group of aquatic insects (e.g., Ward and Whipple 1959, McCafferty 1981, Merritt and Cummins 1984). Like the water mites, insects occupy every freshwater habitat and heterotrophic niche and are present at tremendous densities and diversities. The majority spend most of their lives as larvae, only briefly departing the freshwater milieu to mate. Aquatic insects feature mandibles and one pair of antennae; adults have three pairs of legs and generally possess functional wings.

Chapter 17 also covers the semi-aquatic springtails, or Collembola. These wingless arthropods were once classified as insects but are now generally regarded as members of the separate class Entog-

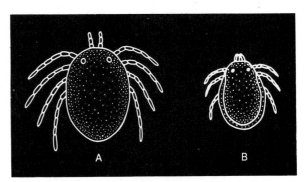

Figure 1.13 Lateral view of a heterotardigrade (phylum Tardigrada). [See Chapter 15.]

Figure 1.14 Generalized larval (A) and adult (B) water mites (phylum Arthropoda, Acari). [See Chapter 16.]

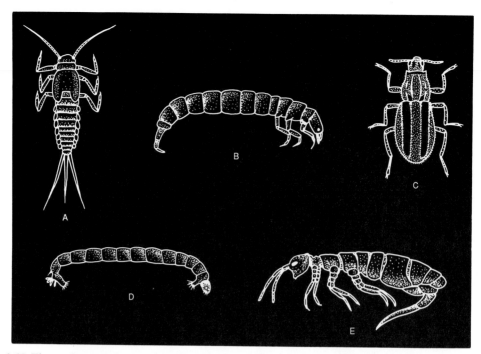

Figure 1.15 Five arthropods in the superclass Hexapoda: (A) *Baetis*, a mayfly nymph (animals drawn in Fig. 1.15 a–d are all in the class Insecta); (B) the caddisfly larva *Polycentropus*; (C) an adult elmid beetle *Stenelmis*; (D) the midge larva *Chironomus*; and (E) a semi-aquatic springtail (class Entognatha, order Collembola). [See Chapter 17.]

natha within the superclass Hexapoda, which also includes insects. They are usually associated with the water surface (neuston) or riparian habitats.

3. Crustaceans

Nearly 4000 species of crustaceans inhabit freshwaters around the world, occupying a great diversity of habitats and feeding niches. Within pelagic and littoral zones, water fleas and copepods are the principal macrozooplankton, and benthic littoral areas shelter vast numbers of seed shrimps, scuds, and other crustaceans. An omnivorous feeding habit is typical of crustaceans, although there are many strict herbivores, carnivores, and detritivores. Members of the subphylum Crustacea are characterized by a head with paired mandibular jaws, a pair of maxillae, and two pairs of antennae. Their appendages are often biramous.

a. Peracarida and Branchiura

There are several groups of crustaceans, most in the class Malacostraca, superorder Peracarida, that have a relatively small number of freshwater representatives. Two of these, Amphipoda and Isopoda, are important taxa within marine systems. Chapter 18 (by Alan P. Covich and James H. Thorp) includes both an introduction to the subphylum Crustacea

and coverage of the ecology and systematics of three malacostracan orders: Amphipoda (scuds), Isopoda (aquatic sow bugs), and Mysidacea (opossum shrimp) (Fig. 1.16). Of these three taxa, the scuds are the most widely distributed and diverse within North American freshwaters. They occur in ecosystems ranging from large rivers to small ponds and are also well represented among subterranean fauna. The subclass Branchiura, or "fishlike," attach themselves to the gills of fish.

b. Ostracoda

The over 400 species of mussel and seed shrimps (class Ostracoda) found on this continent are easily distinguished from other crustaceans by the typical, bivalved carapace enveloping their bodies (Fig. 1.17). Ostracods occur within the benthos of nearly every conceivable aquatic system but rarely enter the plankton. Almost all are free living and the majority are herbivores or detritivores, as discussed by L. Denis Delorme in Chapter 19.

c. Cladocera and Other Branchiopoda

The class Branchiopoda includes the common *Daphnia* and other cladocerans frequently studied in the laboratory component of introductory biology courses, along with poorer known representatives of

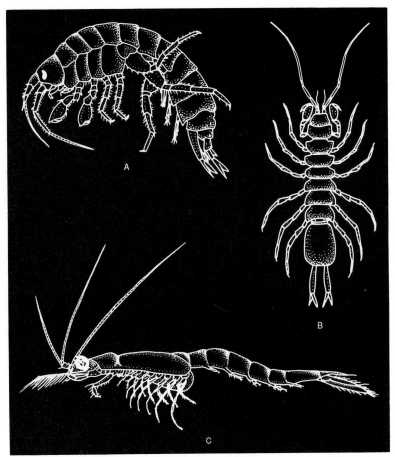

Figure 1.16 Representative freshwater crustaceans: (A) *Gammarus*, a scud (order Amphipoda); (B) *Caecidotea*, an aquatic sow bug (order Isopoda); and (C) *Mysis*, an opossum shrimp (order Mysidacea). [See Chapter 18.]

the class such as the elusive fairy shrimp of ephemeral pools and saline lakes (Fig. 1.18). Branchiopods are a heterogeneous group linked by similar mouthparts and leaf-like thoracic legs (phyllopods). Virtually all species within the eight extant orders of Branchiopoda are limited to freshwaters (mostly lentic systems). As pointed out by Stanley I. Dodson and David G. Frey in Chapter 20, branchiopods occupy key positions in aquatic communities both as important herbivores, eating algae and bacteria, and as major prey items in the diets of fish, waterfowl, and certain other vertebrate and invertebrate predators.

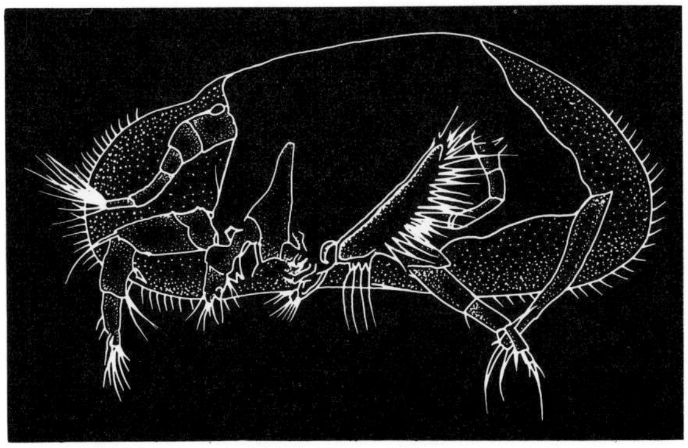

d. *Copepoda*

Copepods are common in aquatic and semiaquatic habitats, spanning the gamut from groundwaters and wetlands to the benthos and plankton of large lakes; they are also major constituents of marine zooplankton. Most are free living, but two of the seven orders contain species that are parasites on freshwater fish. As was the case for cladocerans, copepods have an integral role in aquatic food webs both as primary and secondary consumers (most are omnivorous) and as a major source of food for many fish and larger invertebrates (see discussion by Craig E. Williamson in Chapter 21). Copepods can be distinguished from other small (0.5–2.0 mm) crustaceans

Figure 1.17 Female ostracode *Candona* (crustacean class Ostracoda) with left valve of carapace removed. [See Chapter 19.]

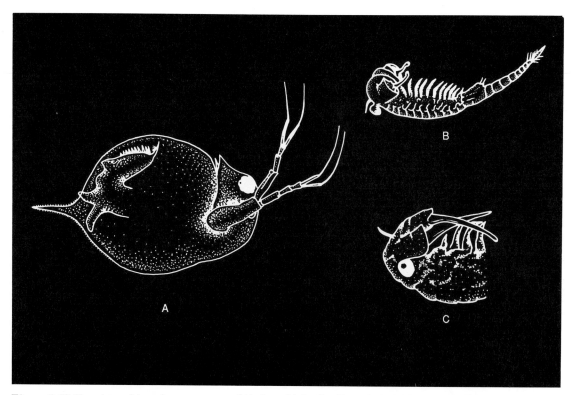

Figure 1.18 Two branchiopod crustaceans: (A) the widely distributed "water flea" *Daphnia*; (B) the more localized fairy shrimp *Eubranchipus*, showing an adult male; and, (C) an expanded view of the head of the fairy shrimp *Linderiella*. [See Chapter 20.]

by their possession of (1) a cylindrical, segmented body; (2) two setose, caudal rami on the posterior end of the abdomen; and (3) a single, simple anterior eye (Fig. 1.19).

e. *Decapoda*

Decapoda is the most diverse order of the class Malacostraca in marine and freshwater ecosystems. It encompasses 315 species of crayfish and nearly 20

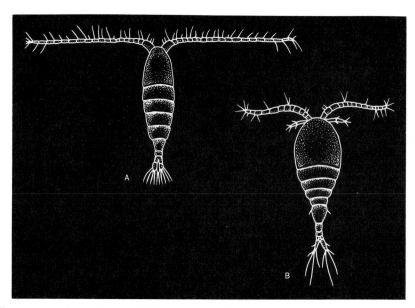

Figure 1.19 Two copepod crustaceans: (A) a generalized calanoid species; (B) a representative cyclopoid copepod. [See Chapter 21.]

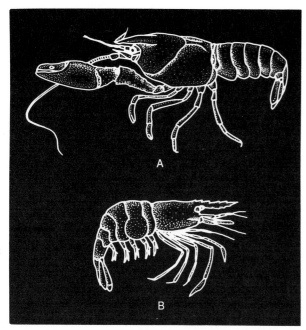

Figure 1.20 Crayfish (A) are extremely common decapod crustaceans, while palaemonid shrimp (B) have a more restricted distribution (although occasionally abundant). [See Chapter 22.]

species of shrimp in epigean and subterranean freshwaters north of Mexico (Fig. 1.20). All are

characterized by terminal claws on the first three of five pairs of thoracic appendages and a branchial chamber enclosed within the carapace (see Chapter 22 by H. H. Hobbs III). Decapods are the largest motile invertebrates in North American freshwaters. Their size, activity, and omnivorous diet make them key players in the benthos of many ecosystems. They can influence the species distribution of snails within a community by selectively preying on certain size classes or shell morphologies, and they can be voracious herbivores that devastate the vascular plant flora in a pond. Decapods, unlike practically all other North American invertebrates, are increasingly consumed by humans, especially in the southern regions of the United States.

M. Taxonomic Key to Major Taxa of Freshwater Invertebrates

The following dichotomous, taxonomic key is designed specifically to lead the reader to appropriate chapters within this book. It can be used for the larval and adult stages of aquatic insects, but it generally applies to the adults of other taxa.

1a.	Unicellular (or acellular) organisms present as individuals or colonies; heterotrophic and/or autotrophic; protozoans (Fig.1.1) .... kingdom Protista (Chapter 3)
1b.	Multicellular, heterotrophic organisms (sometimes with symbiotic autotrophs) 2
2a(1b).	Multicellular animals without discrete organs and with a tissue–level construction; sponges (Fig. 1.2) .... phylum Porifera (Chapter 4)
2b.	Multicellular animals with organ or organ-system construction and 2–3 embryonic cell layers 3
3a(2b).	Two embryonic cell layers; adults with a central body cavity opening to the exterior and surrounded by cellular endoderm, an acellular mesoglea, and a cellular ectoderm; polyps (e.g., hydra) and medusae (e.g., *Craspedacusta*) (Fig. 1.3) .... phylum Cnidaria (Chapter 5)
3b.	Three embryonic cell layers; adults with cellular ectoderm, mesoderm, and endoderm 4
4a(3b).	Flattened or cylindrical, bilateral, acoelomate worms with only one opening to the digestive tract; flatworms (some groups occasionally called planarians) (Fig. 1.4) .... phylum Platyhelminthes: class Turbellaria (Chapter 6)
4b.	Pseudocoelomate or coelomate eumetazoans with a complete digestive tract 5
5a(4b).	Flattened, unsegmented worms with an eversible proboscis in a rhynchocoel and a closed circulatory system; ribbon worms (Fig. 1.5) .... phylum Nemertea (Chapter 6)
5b.	Not with above characteristics 6

6a(5b).   Pseudocoelomates; legs, lophophorate tentacles, and segmentation absent; gastrotrichs (Fig. 1.5), rotifers (Fig. 1.6), roundworms (Nematoda; Fig. 1.7a), or hairworms (Nematomorpha; Fig. 1.7b) 7

6b.   Coelomates; legs, lophophorate tentacles, and/or metameres present; snails and mussels (Mollusca; Figs. 1.8 and 1.9), segmented worms (Annelida; Figs. 1.10 and 1.11), bryozoans (Fig. 1.12), water bears (Tardigrada; Fig. 1.13), and arthropods (mites, insects, and crustaceans; Figs. 1.14–1.20) 10

7a(6a).   Small (50–800 μm), spindle- or tenpin-shaped, ventrally flattened pseudocoelomates; more or less distinct head, bearing sensory cilia; cuticle usually ornamented with spines or scales of various shapes; posterior of body normally formed into a furca with distal adhesive tubes; gastrotrichs (Fig. 1.5) .... phylum Gastrotricha (Chapter 7)

7b.   Not with above characteristics 8

8a(7b).   Nonvermiform pseudocoelomates with apical (and usually ciliated) corona and muscular pharynx (mastax) with complex set of jaws; wheel animals, or rotifers (Fig. 1.6) .... phylum Rotifera (Chapter 8)

8b.   Not with above characteristics; nonsegmented worms 9

9a(8b).   Alimentary tract present; cylindrical body usually tapering at both ends; noncellular cuticle without cilia, often with striations, punctuations, minute bristles, etc.; most under 1 cm long (except family Mermithidae); roundworms (Fig. 1.7a) .... phylum Nematoda (Chapter 9)

9b.   Degenerate intestine; anterior and posterior tips normally are obtusely rounded or blunt; cuticle opaque and epicuticle usually crisscrossed by minute grooves; pseudocoel frequently filled with mesenchymal cells with large clear vacuoles, giving tissue a foamy appearance; length several cm to 1 m, width 0.25–3 mm; only adults with free-living stage; hairworms or horsehair worms (Fig. 1.7b) .... phylum Nematomorpha (Chapter 9)

10a(6b).   Soft-bodied coelomates, usually with a hard calcareous shell; most with ciliated gills, a ventral muscular foot, and a fleshy mantle covering internal organs; snails and mussels (Figs. 1.8 and 1.9) .... phylum Mollusca   11

10b.   Body not enclosed in a single, spiraled shell or in a hinged, bivalved shell, or if a bivalved shell is present then animal has jointed legs 12

11a(10a).   Body enclosed in a single shell, which usually has obvious spiral coils (limpets are the exception); body basically partitioned into head, foot, and visceral mass; radula present; snails and limpets (Fig. 1.8) .... phylum Mollusca: class Gastropoda (Chapter 10)

11b.   Body enclosed in two hinged shells; head and radula absent; mussels and clams (Fig. 1.9) .... phylum Mollusca: class Bivalvia (Chapter 11)

12a(10b).   Legs absent in all life stages 13

12b.   Adults and most larval stages with legs; if larvae without legs or prolegs (some insects), then cephalic region with paired mandibles .... phylum Arthropoda   16

13a(12a).   Lophophorate tentacles absent; segmented worms 14

13b.   Ciliated, lophophorate feeding tentacles present 15

14a(13a).   No suckers and four bundles of chaetae on each segment except the first (Oligochaeta and Aphanoneura; Fig. 1.10a) "or" posterior, disc-shaped sucker on segment 11 and head composed of peristomium and three segments (Branchiobdellida; (Fig. 1.10b) .... phylum Annelida (in part) (Chapter 12)

14b.   Body with anterior and posterior suckers and no chaetae; leeches (Hirudinoidea, or Hirudinea; Fig. 1.11a) "or" no suckers and usually each segment after head with paired parapodia (fleshy lateral outgrowths) bearing bundles of setae (Polychaeta; Fig. 1.11b) .... phylum Annelida (in part) (Chapter 13)

15a(13b).   Anus opens outside of a generally U-shaped lophophore; bryozoans (Fig. 1.12a) .... phylum Ectoprocta (Bryozoa) (Chapter 14)

15b.   Both mouth and anus open within lophophore; calyx with single whorl of 8–16 ciliated tentacles (Fig. 1.12b) .... phylum Entoprocta (Chapter 14)

16a(12b).   Minute adults with four pairs of clawed, nonjointed legs; water bears (Fig. 1.13) .... phylum Tardigarda (Chapter 15)

16b.   Adults and most larvae with jointed legs; arthropods 17

17a(16b).   Antennae absent; body divided into cephalothorax (prosoma) and abdomen (opisthosoma); adults generally with four pairs of legs; water mites (Fig. 1.14) (water spiders are also included in this class but are not treated in this book) .... subphylum Chelicerata, class Arachnida, subclass Acari (Chapter 16)

17b.   Antennae present; insects, springtails, and crustaceans 18

18a(17b).   Mandibles and one pair of antennae; adults (most of which are terrestrial) generally with functional wings, larvae with wing pads or without any indication of wings; larvae and adults normally with three pairs of legs (some larvae without legs); wingless adults in class Entognatha (order Collembola, the springtails) and winged adults are in class Insecta (aquatic insects) (Fig. 1.15) .... subphylum Uniramia (Chapter 17)

18b.   Two pairs of antennae and mandibles (at some life stage); adults with variable numbers of legs, never with wings subphylum .... Crustacea   19

19a(18b).   Body with 19 segments (head: 5; thorax: 8; abdomen: 6); abdomen with pleopods; carapace covers head and all or part of thorax .... class Malacostraca   20

19b.   Body enveloped in a bivalved carapace (ostracodes), "or" thoracic legs flattened and leaf-like (water fleas and other branchiopods), "or" eleven postcephalic segments and a single cephalic eye (copepods) 21

20a(19a).   Malacostracans with first three of five pairs of thoracic appendages modified as maxillipeds (including usually large chelae on first pair in crayfish) and carapace encloses branchial chamber; crayfish and atyid and palaemonid shrimp (Fig. 1.20) .... order Decapoda (Chapter 22)

20b.   Malacostracan crustaceans with other characteristics, including scuds (order Amphipoda), aquatic sow bugs (order Isopoda), and opossum shrimp (order Mysidacea) (Fig 1.16) .... (Chapter 18)

21a(19b).   Body enveloped in laterally compressed, bivalved carapace; mussel and seed shrimp (Fig. 1.17) .... class Ostracoda (Chapter 19)

21b.   Body not enclosed within a bivalved carapace 22

22a(21b).   Thoracic legs flattened and leaf-like (phyllopod), mandibles are simple unsegmented rods with inner corrugated grinding surfaces; last body segment with a pair of spines or claws; water fleas (cladocerans) and other branchiopods (Fig. 1.18) .... class Branchiopoda (Chapter 20)

22b.   Cylindrical, segmented body, two setose caudal rami on posterior end of abdomen, and a single, simple anterior eye; copepods (Fig. 1.19) .... class Copepoda (Chapter 21)

LITERATURE CITED

McCafferty, W. P. 1981. Aquatic entomology. Science Books International, Boston, Massachusetts.

Merritt, R. W., and K. W. Cummins. 1984. An introduction to the aquatic insects of North America. 2nd Edition. Kendall/Hunt, Dubuque, Iowa.

Pennak, R. W. 1989. Fresh-Water Invertebrates of the United States: Protozoa to Mollusca. 3rd Edition. Wiley, New York.

Ward, H. B., and G. C. Whipple, editors. 1959. Freshwater biology (revised by W. T. Edmondson). Wiley, New York.

An Overview of Freshwater Habitats

2

James H. Thorp
Department of Biology and Water Resources Laboratory
University of Louisville
Louisville, Kentucky 40292

Alan P. Covich
Department of Zoology
University of Oklahoma
Norman, Oklahoma 73019

Chapter Outline

I. INTRODUCTION
II. LOTIC ENVIRONMENTS
  A. The Physical–Chemical Milieu
    1. Basin Morphometry and Characteristics of Flowing Water
    2. The Importance of Substratum Type
    3. Thermal and Chemical Constraints on Distribution
  B. Ecosystem Changes Along the Stream Course
    1. Patterns Along the Continuum
    2. Nature of the River's Source
    3. Large Rivers
III. UNDERGROUND AQUATIC HABITATS
  A. Hyporheic and Phreatic Zones
  B. Aquatic Habitats Within Caves and Other Karst Topography
IV. LENTIC ECOSYSTEMS
  A. Geomorphology and Abiotic Zonation of Lakes
  B. Biotic Zonation of Lakes
    1. Neuston and Zooplankton of Pelagic and Littoral Habitats
    2. Littoral and Profundal Benthos
  C. Wetlands, Ephemeral Ponds, and Swamps
  D. Hypersaline Lakes
  Literature Cited

I. INTRODUCTION

The contribution of inland waters to the total biospheric water content is insignificant in terms of percentage (< 1% according to Wetzel 1983) but absolutely crucial from the perspective of terrestrial and freshwater life. Although inland lakes contain 100 times as much water as surface rivers, most of the volume of the former is held within massive basins, such as the Laurentian Great Lakes of North America, Lake Baikal of Siberia, and Lake Tanganyika of East Africa. Because the great majority of freshwater invertebrates are clustered within the shallow, well-lit zones of these and other lakes, the relative importance of small ponds, creeks, and rivers as habitats for aquatic invertebrates is much greater than their volume percentages would otherwise indicate. The composition, species richness, and total density of invertebrates vary considerably among the inland water habitats, as discussed in this chapter.

Distributions of invertebrates are influenced by an interaction of physical, chemical, and biological characteristics. The general importance of these abiotic and biotic factors are examined in this chapter. Detailed information on the ecology of individual taxa in inland water ecosystems can be gleaned from Chapters 3–22.

II. LOTIC ENVIRONMENTS

Flowing waters, or lotic environments, were a principal pathway for the evolutionary movement of animals from the sea to lakes and land. Even today, many taxa of freshwater invertebrates are restricted to brooks, creeks, and rivers by the unique environmental characteristics of these ecosystems. In comparison to nonflowing waters, or lentic ecosystems, streams are more turbulent on the whole than lakes and, therefore, stratification of the water mass is rare. High turbulence generally maintains high oxygen concentrations and reduces within-stream temperature differences. Temperatures in streams fluctuate over a smaller range than is typical of the shallow littoral zone of lentic ecosystems, where most lake-dwelling animals reside (Hynes 1970). Except in the most northern rivers, ice is less commonly encountered and is generally not as thick as in lakes. Flowing water habitats frequently possess more habitat heterogeneity, and the food web in forested drainage basins is more dependent on allochthonous production (input of externally produced plant matter). Lotic ecosystems are also more permanent on both evolutionary and ecological time frames than most lentic habitats. Both heterogeneity and permanence increase diversity within these ecosystems.

In this section, we review features of epigean, or surface waters. The characteristics of aquatic environments where water flows underground (hyporheic, phreatic, and cave environments) are discussed in Section III.

A. The Physical–Chemical Milieu

1. Basin Morphometry and Characteristics of Flowing Water

Although stream invertebrates are influenced directly by the physical features of the local habitat and only indirectly by the geological forces shaping the morphometry of the entire ecosystem, ecologists and invertebrate zoologists must consider all those factors that control community composition. Some of the elements of stream morphometry that influence aquatic communities are stream order, channel patterns (e.g., meandering and braiding), erosion, deposition, and the formation and characteristics of pools and riffles.

Classifying streams by their order, or pattern of tributary connections, is a simple technique that helps in the analysis of longitudinal (i.e., upstream–downstream) changes in stream characteristics within a single catchment; however, it is less useful, and indeed often misleading, when used to compare streams in different catchments. For example, communities within second-order streams in the southwest desert bear no close resemblance to communities within second-order streams in the eastern coastal plain. According to this scheme, as refined

Figure 2.1 Bifurcation of streams illustrating the system for classifying lotic ecosystems according to stream order. The maximum order of the hypothetical stream shown here is sixth.

by Strahler (1952) and illustrated in Figure 2.1, a continuously running, headwater creek with no permanent tributaries is classified as a first-order stream (permanency is defined here as not being ephemeral on a seasonal or other regular basis). Two first-order streams combine to form a second-order segment; two second-order creeks join as a third-order stream, etc. Stream order increases only when two streams of equivalent rank come together; hence, a third-order stream that flows into a fourth-order system has no effect on the subsequent numbering scale of the larger stream. The Mississippi River, the largest river in North America, attains a stream order of 11–13 in its lower reaches (depending on which maps are used as the basis for analysis).

Viewed from the air, streams clearly do not flow in a simple, straight direction for significant distances (Fig. 2.2), nor do stream beds maintain constant depths for long stretches. Instead, the channel traces a meandering pattern of gentle and sharp bends through the basin with alternating bars, pools, and riffles (Fig. 2.3) as the water travels downstream along a roughly helical path. The distance between meanders is a function of several factors, especially stream width. The river channel also has a tendency to undergo braiding (i.e., the division of the channel into a network of branches) in response to the presence of erodible banks, sediment transport, and large, rapid, and frequent variations in stream discharge (Ritter 1986).

Figure 2.2 A river in Alaska, showing a pattern of meanders and straight reaches. (photograph courtesy of Nina Hemphill).

Water flows in a somewhat helical manner along the surface toward an undercutting bank where the deepest pool is usually located and then passes along the bottom toward a "point" or "alternate" sediment bar on the opposite bank from the pool. Sediment is continually being eroded from some areas (e.g., undercut banks and the head of islands) and deposited in other regions (such as on bars and the foot of islands). Successive deepwater pools are separated by shallow riffle regions near the midpoint of the two pools. This sequence of bars, pools, and riffles is characteristic of both straight and meandering segments of most rivers, especially those with poorly sorted loads. [See Ritter (1986) and refer-

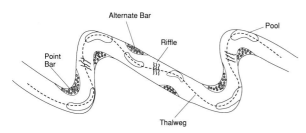

Figure 2.3 Schematic drawing of a stream showing straight reaches and meanders. In both sections of a river, a sequence of deep (pools) and shallow (riffle) areas is manifested. Point bars and alternate bars develop as bed material is deposited in areas opposite from the pools. The thalweg, a line connecting the deepest part of the channel, migrates back and forth across the river.

ences cited therein for additional details on fluvial geomorphology.]

In contrast to our knowledge of the biological significance of stream meandering and braiding, scientists have appreciated the influence of alternating pools, riffles, and (to a much lesser extent) bars for some time. In comparison to pools, riffles are characterized by greater habitat heterogeneity, sediment size, stream velocity, and sometimes oxygen content. In general, well-flushed, stony riffles support more species and individuals than silty reaches and pools (Hynes 1970). This richness is a consequence, in part, of the different adaptations required to live in these physically distinct areas, especially with regard to dissolved oxygen.

Our initial separation of aquatic ecosystems into flowing and standing water environments recognizes the fundamental importance of this character to the biotic composition of inland water environments. Scientists measure the movement of water in a stream in several ways, including stream discharge, turbulence, and velocity.

The mean and peak values of stream discharge influence various ecosystem characteristics such as substratum particle size, bed movement, nutrient spiraling, and rate of animal drift. Stream discharge is calculated by the following equation:

$$Q = wdv$$

where Q is discharge in m³/sec (cms) or ft³/sec (cfs),

w is width, d is depth, and v is velocity. As stream flow increases, shear stress is magnified; when the stress reaches a critical point, the substratum is picked up, rolled, or generally moved downstream. Various aspects of stream discharge, including mean annual discharge and yearly and historical extremes, are described annually for most rivers in the United States in the Water-Data Reports published by the U.S. Geological Survey. This information is derived from multiple gaging stations located near the mouth of each river and often extending into the headwaters; water quality data are also available for a few gaging stations in some of these rivers. These reports are available in the government documents section of most major libraries. For more details on hydrologic processes, consult articles or books on this subject, such as Newbury (1984), Ritter (1986), and Statzner et al. (1988).

Although freshwater invertebrates are certainly affected by the mean and range of stream discharge, they are probably influenced at least as significantly by one component of that equation—water velocity. The ability of an invertebrate to maintain its position on a rock and the amount of food passing its way (for filter-feeders in particular) are both affected by current velocity. The mean current velocity of a stream at any given time occurs at a depth of about 0.6 times the maximum depth. The current velocity diminishes progressively toward the bottom and drops dramatically near the substratum in a 1–3 mm thick area known as the boundary layer. For this reason, the velocities experienced by two invertebrates living within boundary layers of separate streams— one with high and the other with low average current velocity—may not differ appreciably. Many stream invertebrates are partially protected from dislodgment by having a low profile that keeps their bodies mostly within the boundary layer. Other possible adaptations for life in fast flow include streamlining; reduction of projecting structures; development of suckers, friction pads, hooks, or grapples; and secretions of adhesive products.

Under normal flow conditions, the average current velocity varies from the headwaters to the mouth of a river. Although people may believe that headwater streams flow faster than large order rivers, this popular idea has been repeatedly disproved (e.g., Leopold 1953, Ledger 1981). This erroneous view of streams may have arisen because the waters of a turbulent headwater stream (Fig. 2.4) can "appear" to the casual observer to be moving faster than the waters of a higher order, laminar flow river (Fig. 2.5). It is important to note, however, that microhabitats in any river order can exceed significantly the average current velocity of the stream reach.

Fluctuating current velocity rather than constancy is typical of all lotic systems. Except in the largest rivers, zero or near-zero flow rates have been recorded at least once in most lotic environments since the U.S. Geological Survey began recording stream discharges in the nineteenth century. The smaller the stream and the drier the surrounding area (e.g., desert streams), the more likely that a lotic system will be intermittent on at least a seasonal basis. At the other extreme, all streams have peak discharges on seasonal and multiyear bases. Flood events, or spates, may be very predictable, such as those produced by seasonal snow melt in mountainous streams, or they may occur dramatically over unpredictable time periods (Resh et al. 1988). Current velocities during floods and normal periods are related to the width, depth, and roughness of the stream bed but rarely exceed 3 m/sec even during floods (falling water seldom surpasses 6 m/sec) (Hynes 1970). Above 2 m/sec, most rivers begin to enlarge their beds by erosion, unless constrained by rock banks or human construction.

Invertebrates are often adapted morphologically or in life-history traits to moderate, seasonally predictable spates, but large unpredictable floods may eliminate all or most of the fauna of a stream by causing bed movement and severely scouring the bottom. When these scouring events occur, the sources for recolonization may be the assemblages found in the adjacent hyporheic zone, in surface waters located upstream or downstream, or in separate, nearby stream catchments (Wallace 1990).

Figure 2.4 Photograph of a turbulent headwater stream (photograph by J. H. Thorp).

Figure 2.5 Photograph of a large river (the Ohio River upstream of Louisville, Kentucky), showing the laminar flow nature of its waters (photograph by J. H. Thorp).

2. *The Importance of Substratum Type*

Because the overwhelming majority of lotic invertebrates are benthic, the nature and provision of substrata are of prime importance to their survival (see reviews in Cummins 1966, Hynes 1970, Minshall 1984). Substrata provide sites for resting, food acquisition, reproduction, and development (e.g., places for pupal case attachment) as well as refuges from predators and inhospitable physical conditions. Hard substrata are generally of value only for their surface qualities (except for species that can bore into wood), whereas relatively soft substrata (ranging from sand to silt) provide a three-dimensional environment.

The desirability of a given substratum type can be evaluated for a specific taxon using a combination of at least the following interrelated characteristics (not necessarily in order of importance):

1. Whether it is mineral (ranging from large boulders to fine clay particles), living (algae, mosses, vascular plant stems and roots, and, rarely, animal hosts), or formerly living (e.g., abscised leaves and snags from tree limbs and trunks)
2. The silt and organic content of, or on, the substratum, including whether it is bare or covered by significant amounts of algae, mosses, or vascular plants
3. Its particle size, surface texture, and porosity
4. Its stability, as judged by its retention in the system (e.g., leaves are a short-lived

substratum, whereas boulders are retained almost indefinitely) and its tendency to shift position (ranging from almost constantly shifting sand to boulders that rarely turn)
5. The physical–chemical milieu in which the substratum resides, such as the water depth in which it is located, the ambient oxygen content, and the range of current velocities surrounding it.

Even if a particular site meets all the physical and chemical criteria for a desirable substratum, it may not be appropriate for an individual invertebrate if a competitor has preempted the space or if it is located in an area that would make a resident especially susceptible to predators or abiotic disturbances. For these reasons, the distribution of a species across a spectrum of substratum types may not reflect so much the selectivity of the invertebrate species as the effects of substratum limitation. While a particular taxon may not benefit greatly by the heterogeneity of the substratum, the species diversity of the entire community tends to rise with an increase in habitat heterogeneity. In a like manner, an individual may benefit from maximum stability of the substratum (and perhaps from other characteristics of the ecosystem), but the community as a whole may flourish better with some disturbances. This latter concept, embodied in part by the "intermediate disturbance hypothesis" (Connell 1978, Hutchinson 1961, Ward and Stanford 1983, Thorp and Cothran 1984), suggests that intermediate levels of biotic or abiotic disturbances (e.g., frequency of substratum shifting) can promote maximum species diversity under certain circumstances. For a more detailed discussion of the general effects of substratum characteristics (with examples drawn from the aquatic insect world), the reader should initially consult the review by Minshall (1984).

3. Thermal and Chemical Constraints on Distribution

a. The Thermal Environment

Because temperatures vary on a seasonal basis and a gradient exists from headwaters to mouth, it is not surprising that the distribution and abundance of freshwater invertebrates are correlated with the annual mean and range of temperatures (e.g., Hynes 1970, Ward 1986). A general stream temperature pattern on an annual basis is a progressive increase from headwaters to mouth, especially in streams originating in the mountains and fed primarily by terrestrial runoff. [This general pattern may not hold for low-gradient, north-flowing streams.] Streams originating from coldwater springs, however, with their relatively constant temperatures at the headwaters, can be warmer in the winter near their source than somewhat farther downstream; in the summer, the opposite pattern exists (Minckley 1963). In many temperate zone rivers, the annual fluctuation in temperature increases downstream, while the diel temperature change peaks in mid-order reaches (Statzner and Higler 1985). In contrast, the diel and annual temperature amplitudes are very low in the middle and lower reaches of tropical streams, although high elevation, tropical streams do have significant seasonal temperature gradients (Covich 1988).

Predictions of thermal patterns must be modified to account for the effects of tributaries (often colder) and dams on downstream reaches. Likely differences in thermal regimes between natural and regulated rivers are described by the "serial discontinuity concept" (Stanford *et al.* 1988), which predicts that the thermal character of the stream below a dam will be "reset" toward that typical of reaches above the dam. This pattern is evident if you calculate the number of degree days (sum of the average daily temperatures over a period) in stream reaches above and below the dam. The timing of emergence of insects, the ability of an invertebrate to tolerate specific habitats, and the annual productivity of individual species and communities are all influenced directly or indirectly by the number of degree days in those habitats (Sweeney and Vannote 1978).

Species richness and the nature of the invertebrate fauna vary along a continuum from headwaters to mouth. Crustaceans (benthic and planktonic), molluscs, and fish increase in abundance and diversity with stream order, whereas insects generally decrease downstream. Because of previously noted patterns for temperature means and fluctuations with stream order, it was not surprising that aquatic ecologists would see a link between thermal regimes and species diversity. Temperature may influence organisms directly or have an indirect effect through changes in oxygen saturation levels (the oxygen-carrying capacity of water decreases at higher temperatures). While it is true that a great many lotic invertebrates are more stenothermal (i.e., able to survive in only a narrow range of temperatures) than their lentic counterparts (Hynes 1970), it has not been demonstrated that the relationship between invertebrate diversity and temperature is causative rather than merely coincidental. The problem in distinguishing causative factors is that many biotic and abiotic parameters other than temperature also change along the river gradient. For example, habitat complexity, which has often been demonstrated to have a significant influence on diversity, varies downstream, often reaching a maximum value in mid-order sections of the river.

b. The Chemical Environment

Certain natural chemical features of streams and lakes significantly affect invertebrate abundance and diversity by influencing habitat quality. Of these, the most crucial are probably dissolved oxygen and salinity (water hardness). Hydrogen ion concentration, or pH, is occasionally important in natural systems. Anthropogenic pollution has had a severe impact on the integrity of aquatic ecosystems by affecting these chemical parameters as well as by introducing organic and inorganic toxicants.

As a general rule, dissolved oxygen decreases downstream. This pattern occurs because the upper reaches of a river are usually more turbulent, the surface to volume ratio is more favorable for diffusion, and temperatures are lower. In the last case, oxygen solubility is related in an inverse, nonlinear manner to temperature; this gas is markedly more abundant in cold waters. Pure water in equilibrium with air at standard pressure holds 12.770, 10.084, 8.263, and 6.949 mg O_2/liter at $5°$, $15°$, $25°$, and $35°C$, respectively (Mortimer 1981 in Wetzel 1983, p. 158). Because oxygen is produced and respired by plants, it may be locally more abundant or even supersaturated during the day in regions with heavy plant growth. However, oxygen potentially decreases at night to lethal levels in weed-choked streams because of high oxygen demands for decomposition and plant respiration. The probability of lethal oxygen concentrations occurring for the invertebrate species typical of a stream reach rises as currents diminish; hence, the problem is more serious in lakes and seasonally sluggish streams than in fast-flowing lotic habitats.

Invertebrates differ both in their respiratory needs and in their anatomical and behavioral adaptations for obtaining oxygen, as described in subsequent chapters. The respiratory rates of lotic species are generally dependent on external oxygen concentration (or tension) at medium to low levels, while the oxygen consumption rates of animals in lentic or typically sluggish lotic environments tend to be independent of ambient tensions, at least at medium to high concentrations (Hynes 1970). Currents provide a steady renewal of oxygen to the respiratory surfaces of invertebrates. Where velocities are too low, the animal may need to ventilate its gills or obtain atmospheric oxygen. For example, crayfish normally move water over their respiratory surfaces with ''gill balers'' but some species will climb partially or completely out of water if oxygen tensions drop too low (returning when necessary to keep the gills moist). Because invertebrates have different oxygen requirements, it is tempting to assume that interspecific differences in distribution are partially caused by oxygen limitations (e.g., contrasting assemblages of caddisflies in mountain streams and floodplain rivers). Certainly many stream animals, such as the blackfly larva *Simulium*, cannot obtain adequate oxygen in still water (Wu 1931 in Hynes 1970, p. 163), but this explanation for the restriction from lentic habitats is less effective in resolving species differences in distribution along the entire reach of a stream. Even if one demonstrated in the laboratory that the oxygen tensions in high-order rivers were inadequate to support a caddisfly from a mountain stream, this is not evidence that oxygen was the principal factor leading to the present pattern of distribution. Other factors, such as interspecific competition, resource type (e.g., particle size of the seston), substratum form, or predators, may have been initially responsible, with fine tuning of respiratory adaptations coming later.

The ionic content of inland waters is generally extremely dilute in comparison to seawater, with four cations (Ca^{2+}, Mg^{2+}, Na^+, and K^+) and four anions (HCO_3, CO_3^{2-}, SO_4^{2-}, and Cl^-) representing the total ionic salinity in most freshwater systems, for all practical purposes (Wetzel 1983). Soft waters are generally found in catchments with acidic igneous rocks, whereas hard waters (i.e., relatively high salinities for freshwater) occur in drainage regions with alkaline earths, usually derived from calcareous rocks (Hutchinson 1957). The global mean salinity of river water is 120 mg/liter; North American waters are slightly harder, averaging 142 mg/liter

(from Livingstone 1963 and Benoit 1969 in Wetzel 1983, p. 180). Most of the soft waters in North America have salinities of less than 50 mg/liter. Salinities are slightly elevated during periods of low flow, and they generally increase downstream (Hynes 1970).

Except for a few hypersaline environments, such as mineral springs and alkaline and brine lakes, inland waters are hypotonic to the internal fluids of freshwater species and, therefore, all freshwater animals need some ability either to osmoregulate (almost all species) or, at the minimum, to control the relative abundances of certain ions (necessary even for freshwater Cnidaria; see Chapter 5). Because animals with either partially calcareous exoskeletons (e.g., crayfish, ostracodes, and cladocerans) or hard calcareous shells (snails and mussels) have an added need for calcium ions, one might expect them to be particularly excluded from soft-water habitats (see Chapter 18). Indeed, molluscs are rare in soft-water lakes and less abundant in soft-water streams, but this limitation is not as severe as in lentic ecosystems. The degree to which a mollusc can penetrate soft-water streams is made possible by its ability to tolerate extremely dilute tissue fluids (the lowest of any metazoan; see Chapter 11). The minimum concentration of Ca^{2+} tolerated by mussels (2–2.5 mg Ca/liter) is just adequate to prevent the rate of shell dissolution from exceeding the rate of shell deposition. Despite their heightened need for calcium, crayfish can be abundant in many soft-water streams of the Southeast where these ions are difficult to obtain. The problem of ionic loss at the critical period when the old exoskeleton is molted is partially overcome by the crayfish habit of consuming its shed exuvia as a way of extracting additional calcium ions. This behavior and the efficient osmoregulatory system of the crayfish (with gastrolith formation) allow this decapod to colonize a greater chemical range of habitats (see Chapter 22).

In the last decade, the catastrophic effects of acid precipitation and mine drainage on freshwater ecosystems have drawn great attention from scientists and politicians alike, but relatively unpolluted rainwater and the runoff from thick forest litter are also slightly acidic (though not at lethal levels). Bogs and the streams draining them are among the most naturally acidic ecosystems; *Sphagnum* bogs usually have a pH below 4.5 and occasionally below 3.0 (Wetzel 1983). Streams draining limestone catchments are well buffered by carbonates and, therefore, have moderate to high pH values as well as plentiful supplies of calcium. In contrast, acidic waters are normally low in calcium, causing potential osmotic problems for their residents. Despite probable ionic and pH difficulties which may have eliminated certain taxa (as covered in many of the following chapters), naturally acidic ecosystems often support a surprisingly diverse invertebrate assemblage.

B. Ecosystem Changes Along the Stream Course

1. Patterns Along the Continuum

Physical and biological characteristics clearly change from the headwaters to the mouth of a river. The river widens, deepens, and becomes more turbid downstream, often removing part of the bottom from the photic zone. Stream shading from riparian trees lessens, with a corresponding increase in the percentage of the surface receiving direct solar radiation. Concurrently, the relative importance of autochthonous (internally produced) and allochthonous carbon to the total energy budget alters along with the ratio of production to respiration. As a result of these and other habitat changes, the invertebrate fauna undergoes dramatic shifts in species composition, relative abundances, and functional feeding groups.

This holistic view of the river was not synthesized until formation of the river continuum concept (RCC) by a group of aquatic ecologists working in temperate deciduous streams of North America (Vannote et al. 1980, Minshall et al. 1983, Minshall et al. 1985). This body of hypotheses has received both support and criticism (as examples of the latter, see Winterbourn et al. 1981, Winterbourn 1982, Statzner and Higler 1985). It is not our purpose to defend or assail this holistic, ecosystem concept; instead, we wish to acquaint you with some of its predictions as they could influence your understanding of the habitat and ecology of lotic invertebrates. Because the RCC was originally developed from studies of relatively undisturbed, North American streams with forested headwaters, the following comments may not be as applicable to stream invertebrate communities in other biomes or regions of the world.

The river continuum concept relies on three theoretical pillars of support. The first is the concept that stream communities originate in response to a continuous gradient of physical variables present from the headwaters to the mouth; this view contrasts with a previous portrayal of natural streams as a series of disjunct communities proceeding downstream in a stepwise fashion. The second canon is that the biotic community within the "wetted channel" of the river cannot be divorced from either its adjacent riparian zone (the trees and shrubs on the nearby banks) or the surrounding biotic and geomorphic catchment area that funnels water, nutrients, and other material into the aquatic ecosystem. The final tenet is that the nature of a downstream assemblage is inextricably linked with processes occurring upstream.

Of the many physical, chemical, and biological predictions made by the RCC, the one most relevant to understanding invertebrate ecology concerns longitudinal changes in the relative abundance of functional feeding groups and their food resources. The proposed relationships between stream size and progressive alterations in community composition are illustrated in Figure 2.6. As heterotrophs, invertebrates consume living and/or dead organic matter, processing it in ways characteristic of several "functional feeding groups." These groups include: "shredders," which break up coarse particulate organic matter (CPOM) such as abscised leaves; "collectors," which gather or filter CPOM, fine particulate organic matter (FPOM), and/or live, drifting organisms; "grazers" or "scrapers," which remove algae or other aufwuchs growing on rocks or other substrata; and "predators," which may include carnivores and some herbivores.

The best means of assigning a functional feeding label to a life stage of a particular invertebrate is to observe the feeding behavior of the organism (preferably in the field) and analyze its gut contents on a seasonal basis. This task is time consuming and may be impractical for some taxa. References to the food resources and feeding behavior of species or larger taxonomic categories are reported in many research papers (for examples, see chapters in this book and the tables and references on aquatic insects in Merritt and Cummins 1984). The danger in relying on general sources, however, is the well-established observation that functional feeding classifications can vary significantly with life-history stage, season, and other ecological features. It is also difficult to assign omnivores to any single or consistent functional feeding group.

The RCC predicts that the invertebrate assemblages of headwater streams, with their heavy canopy of riparian trees, will be dominated by shredders and collectors. As the stream widens and more light reaches the water surface, autochthonous production from benthic algae and rooted macrophytes increases, allowing the development of a significant grazer fauna; shredders should diminish dramatically in these mid-order reaches of the river. Farther downstream, the river deepens so that much of the bottom is below the photic zone, and the channel is sufficiently wide to preclude any relatively significant input of allochthonous production from the riparian zone. In this high-order section of

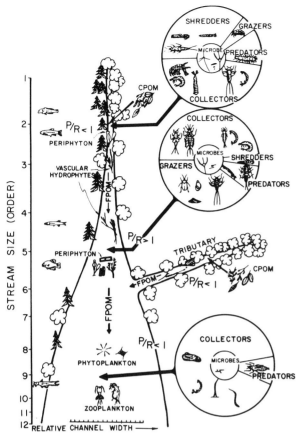

Figure 2.6 Predicted relationships between stream size and structural and functional attributes of a freshwater ecosystem as embodied in the river continuum concept. P/R refers to the ratio of production to respiration. (From Fig. 1 in Vannote *et al.* 1980.)

the river, collector–filterers are expected to dominate the ecosystem.

2. Nature of the River's Source

One weakness in the original formulation of the RCC was its reliance on data collected from a narrow range of stream sizes (mostly stream orders 1–3). Even today, our knowledge of the communities living at the opposite ends of the continuum—springs and intermittent streams versus large rivers—is woefully inadequate (Meyer 1990).

Research on intermittent, or ephemeral, streams is more common in arid regions of North America (e.g., Matthews 1988, Grimm and Fisher 1989), but has recently expanded into headwater streams in more humid regions of eastern North America (e.g., Delucchi and Peckarsky 1989). Almost all streams are subject to seasonal and/or aperiodic disturbances from unusually high current velocities and stream discharges, but only intermittent streams fluctuate regularly from dry to wetted basin. When a stream stops flowing, a resident invertebrate may

have several options for its survival or that of its future progeny: it may move downstream, seek shelter in deep pools, enter the hyporheic zone (i.e., the wetted substratum below the surface waters, as described in Section III.A), emerge as adults (some aquatic insects), develop a resistant stage, or lay eggs and then die. Although the composition of the invertebrate fauna in permanent and intermittent streams may differ, life-history patterns of invertebrates in ephemeral streams are not particularly unique (Delucchi and Peckarsky 1989), perhaps because the options for migrating downstream or entering the hyporheic zone are readily available for many species. The ecology of that portion of the ecosystem, however, may be significantly different from community functioning within permanent streams (e.g., Grimm and Fisher 1989).

There is a tendency to think of streams as originating exclusively from terrestrial runoff, but many headwater systems are derived from springs. Near their source, spring-fed streams often vary in physical, chemical, and biological features from other types of first-order streams (e.g., Covich 1988). Temperatures within spring-fed streams are relatively more constant, and oxygen concentrations are occasionally lower, sometimes to near-lethal levels for metazoa. In other cases, hot springs may provide habitats that range into the upper lethal limits for eukaryotes (see Chapters 18 and 19) and even prokaryotes. Metazoan life near the upwelling region is rarely possible; but, as ambient temperatures approach 40°C or lower in side channels and pools and in areas farther downstream, conditions become suitable for some invertebrates (e.g., Barnby and Resh 1988). These include a very restricted number of species of nematodes, oligochaetes, mites, dipterans (such as brine flies), and ostracodes and a few other crustaceans. Geothermal springs are often extremely vigorous environments not only because temperatures may approach the boiling point but because their waters are frequently laden with high concentrations of sulfur. The ionic content of springs can differ from nearby streams supplied by runoff from land. Springs in limestone regions are well buffered and are frequently highly charged with calcium bicarbonate (Hynes 1970). Saline springs exist where the ionic content is too high for many invertebrates.

The biota of springs often vary from other first-order streams as a result of their connection to underground rivers and their aquatic constancy. A net placed near the spring source can capture unusual subsurface species that have been washed out of their subterranean environments. In this way, blind troglobitic shrimp have been retrieved from surface waters in the springs of the San Marcos River of

Texas (Longley 1986, Strenth *et al.* 1988). Because springs are relatively uniform environments over long ecological periods, it is not unusual to find relict species in these environments that have survived the retreat of the glaciers only in these limited habitats (Hynes 1970).

3. Large Rivers

At the opposite extreme of the continuum, the invertebrate biota and ecological processes in large rivers are virtually unknown despite their importance to humans (but see studies by, for example, Anderson and Day 1986, Beckett and Miller 1986). Consequently, certain misconceptions about the nature of large rivers have arisen, such as the descriptions of the "typical" large river as a slow, meandering, soft-bottomed stream. In contrast, observations by one of us (JHT) in the Mississippi, Ohio, and Tennessee Rivers suggest that the bottom is more likely to be sand, gravel, or cobble except in the shallow, slow-moving water near banks and in the mouths of tributaries where silt often accumulates. Furthermore, evidence for differences in meandering patterns along a river continuum has not yet been generated, and in comparison to upstream reaches, the current velocity is generally higher in large rivers. Because governments have been removing fallen trees and regulating both flow and channel depth in navigable rivers since the last century, the present nature of these ecosystems may be considerably different from the original state (Minshall 1988). For example, removal of snags has greatly reduced the abundance of shallow water hard substrata, thereby influencing both the composition of the insect assemblage and the fish that normally fed upon these insects (cf. Benke *et al.* 1984).

Many aquatic ecologists lacking experience in large rivers often think of them as "lake-like," perhaps expecting similar physical and chemical conditions. Unlike lentic environments, however, riverine currents prevent formation of thermoclines and major zones of oxygen depletion within the water column. The greater depth and turbidity of sediment-laden rivers reduce the amount of benthos receiving light so that only a narrow photic zone exists. In combination, these factors blur or eliminate biotic patterns of distribution with depth below the level of quasi-predictable fluctuations in water level (Haag and Thorp, manuscript submitted).

The "typical" large river assemblage may also prove to be different from that proposed by the RCC. For example, although filterers (e.g., mussels and midges) predominate in the Ohio River, as would be anticipated from Figure 2.6, the predicted absence of grazers is certainly incorrect. Ongoing studies have revealed large numbers of snails and other grazers in the Ohio River (K. Greenwood and JHT, unpublished data). Rather than feeding strictly on the perilithic algae found in mid-order streams, these riverine snails seem to be consuming algae, other aufwuchs, and detritus (depending on their depth distribution).

III. UNDERGROUND AQUATIC HABITATS

A. Hyporheic and Phreatic Zones

Until recently, aquatic ecologists considered the boundaries of rivers to extend only from the air–water interface down to the silty, sandy, or rocky "bottom" of the river. We now realize that this definition is overly limiting and view streams as four-dimensional ecosystems with additional longitudinal (upstream–downstream), vertical (deeper into the sediments), lateral (floodplain), and temporal components (cf. Ward 1989).

Early in this century, scientists recognized the potential importance of interstitial spaces within the sediments of lakes and streams as refuges for small animals. This interstitial habitat is termed the hyporheic zone within lotic systems and the psammon in lentic environments. The depth of the hyporheic zone varies greatly according to the porosity of the substratum, being greater, for example, in rivers having a gravel bed rather than a mud bottom (the habitable interstitial space is influenced by, but is not directly related to, sediment porosity). The overlying surface, or epigean, waters strongly influence the physical and chemical characteristics of this zone and, consequently, the nature of its fauna, or "hyporheos" (Williams and Hynes 1974).

Hyporheic organisms are usually abundant in streams unless the substratum is bedrock, clay, or other material containing pore sizes under 50 μm (Williams 1984). Below a few centimeters, the density of hyporheos is generally inversely related to substratum depth, with very few individuals occurring much below one meter into the sediment.

The hyporheic zone also extends laterally into the stream bank, where it was assumed to penetrate a similar distance horizontally from the river proper as vertically into the river bed. While this assumption may be valid for many if not most streams, recent data (Stanford and Ward 1988) from the Flathead River, a gravelly stream of the northern Rocky Mountains, have shown that the hyporheos permeates the groundwater as far as 2 km from the surface water of the river. It has been known for some time that a unique fauna can occupy the phreatic zone—a saturated underground area where the water moves slowly laterally in the absence of direct influence from surface water rivers. The study by Stanford and Ward, however, demonstrated that the ground-water community within up to 2 km from the Flat-head River was hyporheic rather than phreatic because there were physical (flow), nutrient, and biotic connections between the two areas. The biota of the adjacent phreatic zone was dominated by sub-terranean crustaceans, while the hyporheic zone supported stoneflies and other typically riverine taxa.

The hyporheic food web is based on detrital inputs from surface waters, which in turn are derived from allochthonous inputs of leaves and autochthonous production from vascular plants and algae. Indeed, Schwoerbel (1961) showed a direct relationship between the amount of detritus in the hyporheic zone and the density of hyporheos. Almost all freshwater invertebrate phyla have representatives in hyporheos, with the possible exceptions of Cnidaria and Porifera. They include several orders of insects (e.g., midges, elmid beetles, and stoneflies), mites, crustaceans (including cladocerans, ostracodes, and amphipods), oligochaetes, turbellarians, rotifers, snails, tardigrades, and protozoa.

B. Aquatic Habitats Within Caves and Other Karst Topography

Regions of the world containing large amounts of soluble limestone and adequate underground water often form complex subterranean cavities; such karst topography is widespread in North America (Fig. 2.7). Karsts develop when rocks are fractured and especially susceptible to solution (e.g., crystalline, high calcite limestone). Over time, small cavities can enlarge to produce huge underground caverns. The public generally becomes aware of the

solution processes only if an above-ground entrance exists or if a sinkhole develops (these are also called sinks or, more properly, dolines).

Caves can be defined as karsts with an entrance, passageways, rooms, and a terminal blockage (which prevents humans but not water from passing through). Cave environments provide habitats (e.g., Fig. 2.8) for many types of invertebrates. These habitats differ in several significant ways from surface, or epigean, aquatic habitats. They have very constant temperatures that are equal to the average yearly surface temperature of the local geographic region. Light is dim or absent, and abiotic disturbances are probably less frequent and severe (although floods do occur in caves).

Food is quite scarce in caves, as it usually arrives only through the groundwater or the periodic movement of animals that seek refuge within an open cave after having fed outside (e.g., bats which deposit organic guano in the cave). The principal forms of organic matter supporting the food web are particulate (POM) and dissolved organic matter (DOM) along with the bacteria growing in the water or on the suspended and benthic detritus. Consequently, the food web is rather simple, consisting mostly of detritivores and their predators; the density and species diversity of a single cave community are usually quite low.

In contrast to the simplicity of a cave community, the composition of one cave can be markedly different from an adjacent but unconnected neighboring cave. Endemism is especially prominent in caves because of the geographic isolation. An example of this endemism is the species radiation of amphipods in North America, which has been much greater in karsts than in surface waters (see Chapter 18).

Obligate cave species, or troglobites (as opposed to the facultative troglophiles), usually have low metabolic rates and a much longer lifespan in comparison to their surface relatives; they are also frequently blind and lack pigmentation. The reproductive yield is spread out over longer periods with fewer but more yolky eggs formed during a given reproductive bout. For example, a female crayfish of the troglobitic species *Orconectes inermis* in Indiana caves may carry only a few dozen large eggs while nearby epigean, congeneric individuals can bear hundreds of small eggs (JHT, personal observation and H.H. Hobbs, III, personal communication; see also ecologically similar species in Fig. 2.8).

Information on karst communities can be gleaned from journals such as the *National Speleological Society Bulletin* and from various reviews (e.g., Culver 1970). See also Chapters 18 and 22 in this book for details about cave amphipod and decapod crustaceans, respectively.

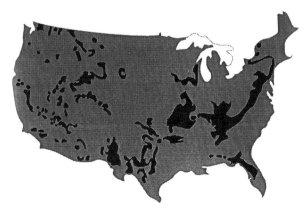

Figure 2.7 Karst topography in the 48 contiguous states of the United States (from Ritter 1986).

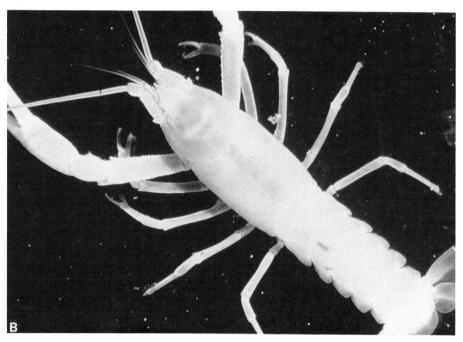

Figure 2.8 (A) A cave habitat in southern Indiana; (B) the blind, troglobitic crayfish *Orconectes testii* (both photographs courtesy of H. H. Hobbs, III).

IV. LENTIC ECOSYSTEMS

Unlike the situation for lotic habitats, the chemistry, physics, geology, and biology of lentic (or lacustrine) environments are well covered in several standard limnology textbooks (e.g., Cole 1983, Goldman and Horne 1983, Wetzel 1983), most of which devote very little space to stream studies. Lake limnology is also discussed in more depth in various specialty texts (e.g., Taub 1984, Carpenter 1988) and in three volumes on limnology by Hutchinson (1957, 1967, 1975; a fourth volume on the benthos will soon be released). For this reason, our discussion of lake habitats is briefer than accorded to stream ecosystems.

A. *Geomorphology and Abiotic Zonation of Lakes*

Although rivers and streams are important biologically as well as being fascinating environments to study, more than 99% of the inland surface waters are in fresh and saline lakes. These standing water, or lentic, ecosystems are present in every continent but are concentrated in the formerly glaciated regions of the Northern Hemisphere. Except for about 20 lakes with depths exceeding 400 m, the majority of natural and human-constructed lentic systems have average depths of less than 20 m (Wetzel 1983). Most are geologically young, dating from the last glacial period in the case of natural lakes or from the last century for manmade lakes. According to Hutchinson (1957), natural lakes are formed by over 70 distinct processes; most are the result of catastrophic phenomena (such as landslides and glacial, tectonic, and volcanic activities), but some form less violently from the action of rivers (e.g., oxbow lakes), waves, and rock solution. Before the advent of extensive, commercial fur trapping in North America, ponds built by beavers were a pervasive feature of the headwaters of many North American streams (Naiman *et al.* 1986).

Lentic ecosystems can be divided into several abiotic zones, primarily on the bases of distance from shore, light penetration, and temperature change (Fig. 2.9). The photic zone is bounded by the water surface and below by the depth of 1% light penetration; all primary producers and most heterotrophic animals live within this zone. The areas below these depths are variously called the aphotic or profundal zone (the latter often in reference to the benthos). The shallow, nearshore region of the photic zone where rooted macrophytes exist is termed the littoral zone. The entire mass of open water located away from both the shore and the littoral zone is known as the limnetic or pelagic zone.

On a seasonal basis, most lentic ecosystems in North America become stratified with a layer of lighter water, the epilimnion, floating over the denser hypolimnion (Fig. 2.9). During at least the summer, the epilimnion is considerably warmer than the hypolimnion. These two zones are separated by a layer of rapid temperature change called the metalimnion. The boundary between the epilimnion and the hypolimnion where temperature changes occur most rapidly with depth is defined as the thermocline (it is typically within the metalimnion). Because there is minimal exchange of water between upper and lower zones during stratification, the hypolimnion is frequently lower in oxygen, higher in nutrients, and different in chemical concentrations and pH. When lake stratification breaks down for a short period during one or more seasons, much or all of

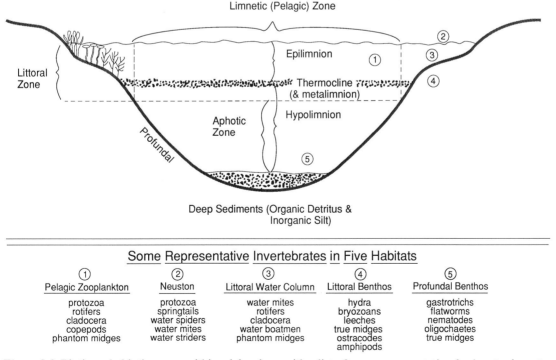

Some Representative Invertebrates in Five Habitats				
① Pelagic Zooplankton	② Neuston	③ Littoral Water Column	④ Littoral Benthos	⑤ Profundal Benthos
protozoa	protozoa	water mites	hydra	gastrotrichs
rotifers	springtails	rotifers	bryozoans	flatworms
cladocera	water spiders	cladocera	leeches	nematodes
copepods	water mites	water boatmen	true midges	oligochaetes
phantom midges	water striders	phantom midges	ostracodes	true midges
			amphipods	

Figure 2.9 Biotic and abiotic zones within a lake along with a list of some representative freshwater invertebrates found within these zones.

the water mass recirculates in a process referred to as lake turnover. If a lake mixes completely twice a year, it is referred to as dimictic; if it turns over only once a year it is called monomictic. In other lentic systems, complete mixing either rarely takes place (in oligomictic lakes) or occurs more than twice per year (polymictic lakes). In addition to thermal zones, some lakes are chemically stratified (commonly with a heavier layer of saline water in the hypolimnion). Some of these lakes never mix completely and are termed meromictic; they only circulate in the upper zones. As a consequence, they are typically devoid of oxygen in the deeper zones, where chemical concentrations of nutrients can accumulate over time.

The benthic zone of lakes and ponds can be further subdivided into somewhat arbitrary categories based on other characteristics of the ecosystem, such as substratum type, wave action, and type of vegetation. The importance of substratum type for lentic environments is similar to that described earlier for lotic ecosystems.

B. Biotic Zonation of Lakes

The invertebrate assemblages within a lake or pond can be subdivided into "zooplankton," which occupy the water column and "benthos," which live in, on, or just above the bottom (principally on macrophytes). A third, smaller group encompasses the "neuston," which live at the air–water interface. These first two categories remain quite distinct except in three general cases. A few species, such as the phantom midge *Chaoborus*, migrate upward at dusk into the plankton and return to the benthos near dawn. Meroplankton, in contrast to the much more diverse holoplankton, spend only a portion of their life cycle in the water column. For example, true midges (Diptera: Chironomidae) are benthic for the majority of their larval lives (the short-lived adults are aerial) but swim within the water column during the earliest stages of their larval existence. The largest meroplanktonic group of "exceptions" to the planktonic–benthic dichotomy thrives within the littoral zone. In this shallow, nearshore region, many species, such as littoral zone water fleas (cladoceran crustaceans; see Chapter 20), alternate periods of resting or foraging on the bottom with intervals of swimming and foraging in the water column.

1. Neuston and Zooplankton of Pelagic and Littoral Habitats

The term neuston refers to the assemblage of organisms associated with the surface film of a water body. It generally includes species that live just underneath the water surface (the hyponeuston), individuals that are above but immersed in the water (epineuston), and taxa that travel over the surface on hydrophobic structures (superneuston or, more properly, a form of epineuston). This name is similar to, or a subset of the older name, pleuston (sometimes neuston is used in reference to the microscopic components of the more encompassing pleuston). Quiet waters of lake embayments generally support a greater concentration of neuston than do more turbulent open-water areas.

The food web within the neuston is supported by a thin bacterial film on the upper surface of the water, a concentration of phytoplankton near the surface, and allochthonous inputs from trapped terrestrial and aquatic organisms. Protozoa are common in this assemblage (which also includes algae and floating macrophytes), whereas other typically planktonic taxa are rare (one exception is the cladoceran *Scapholeberis*). Moving over the water surface are springtails (Collembola), some arachnids (mites and water spiders), and various families of heteropteran bugs (e.g., water striders, Gerridae) (refer to Chapters 16 for information on arachnids and to Chapter 17 for details about insects and springtails).

Zooplankton include all animals in the water column that float, drift, or swim weakly (i.e., they are at the mercy of currents). Fish, which are powerful swimmers, are the principal components of the freshwater nekton in the littoral and pelagic environments. Some predatory zooplankton are active swimmers and comprise the "nektoplankton"; examples are the large cladoceran *Leptodora kindtii* (see Chapter 20) and the mysid "shrimp" *Mysis relicta* (see Chapter 18). Zooplankton thrive within two distinct habitats: the open water epilimnion and the nearshore littoral zone. The littoral habitat is the more heterogeneous of the two in terms of spatial and temporal contexts. The seasonal presence of macrophytic plants (algae and vascular plants) and the proximity of the planktonic and benthic habitats provide important aspects of physical complexity to the littoral zone. Additionally, temperatures fluctuate more rapidly in shallow water, and wave turbulence is greater. Oxygen is rarely limiting nearshore (except somewhat in eutrophic lakes), but the effects of ice formation can be more severe. The majority of lentic research has dealt with the pelagic zone.

Biological interactions in the two planktonic habitats are different in response to those spatial and temporal differences and to the variance in the bases of the two food webs. Pelagic plankton rely on nutrients regenerated by lake turnover, offshore transport, and internal recycling; photosynthesis is exclusively by phytoplankton. Littoral plankton derive nutrients from the same three sources, but they have first access to allochthonous carbon and nutrients. They can also gain energy and recycled nutrients from the benthos on a continual basis (not possible for pelagic zooplankton except during lake turnover). Phytoplankton, periphyton, and macrophytes provide photosynthetic products to support the invertebrate assemblage of the productive littoral zone.

Predator–prey relationships and zooplankton behaviors vary between pelagic and littoral zones. Open-water copepods, cladocerans, and a few rotifers migrate vertically on a diel basis. Fewer littoral zooplankton migrate, and, among those that do, the typical pattern is for offshore movement at dusk and inshore migration at dawn. Ecologists believe that an important cause for both horizontal and vertical migration is predator avoidance. The types of predators in the two habitats are considerably different. Aquatic insects in general are of minor importance in the pelagic zone, but insect predators, such as dragonflies and predaceous beetles, are extremely abundant nearshore. Although predaceous fish are usually more numerous in shallow, vegetated regions, the greater habitat complexity of the littoral habitat reduces foraging efficiency relative to open-water regions. Microcrustacean predators are present in both habitats.

2. Littoral and Profundal Benthos

Benthic invertebrates are often divided into three arbitrary size classes based on the mesh dimensions of the sieve used to process the samples. Macroinvertebrates are retained by coarse sieves ($\geq 200\ \mu$m mesh) and are usually sorted with No. 60 or 35 U.S. Geological Sieves (250 and 500 μm mesh, respectively). Meiofauna, which are plentiful but rarely identified, need to be collected with fine sieves ranging from 40–200 μm mesh (usually $\leq 100\ \mu$m); microbenthos pass even these minute openings. The effort required to process benthic samples, especially from vegetated and/or silty habitats, increases tremendously with diminishing mesh size; consequently, species smaller than macrofauna are often ignored. While these size categories in themselves have no natural ecological or taxonomic significance, the frequent exclusion of the smaller forms from ecological analyses may have severely influenced our perception of how benthic assemblages function. For example, Strayer (1985) estimated that the micro- and meiobenthic animals contributed 68% of the species, 98% of the individuals, 25% of the biomass, and 35% of the production of the zoobenthos in Mirror Lake, New Hampshire. Studies in

Mirror Lake revealed that over half of the benthic micro- and meiofauna live in the top centimeter of sediments (Strayer 1985, 1986).

Most freshwater benthos reach their maximum densities and diversity in shallow water and decline perceptibly with increasing depth in the profundal zone (Thorp and Diggins 1982, Diggins and Thorp 1985); few macroinvertebrates tolerate conditions in the profundal zone beneath the seasonal thermocline, but micro- and meiofauna can be abundant in deep water. This pattern probably reflects gradients of oxygen availability, habitat heterogeneity, and food resources—all of which are greater in the littoral zone. Studies of vegetated and nonvegetated regions of the littoral zone demonstrate the great value of macrophytes in reducing predation rates on benthic macrofauna (e.g., Hershey 1985). This refuge is especially crucial because benthic animals are generally poor swimmers and have difficulty escaping motile predators.

A variety of distinctive habitats are available for benthic invertebrates. Nematodes, gastrotrichs, flatworms, and other members of the meiobenthos frequent the infaunal habitats where they move between sediment particles (see Chapters 6–9). Those that live on or among sand grains are called psammon. Macroinvertebrates, such as freshwater mussels, oligochaetes, and some crayfish, regularly burrow in the substratum (see Chapters 11, 12, and 22). Many motile and sedentary species live on the surface of the mud, rocks, or submerged plant debris. Smaller animals and algae that live on firmer substrata are sometimes called aufwuchs; they include sessile rotifers, midges, and nematodes (see Chapters 8, 17, and 9, respectively). Sponges and bryozoans also attach to solid substrata, such as dead branches or rocks, but they are often not classified among the aufwuchs (see Chapters 4 and 14, respectively). In addition to the attached aufwuchs, macrophytes support some larger, motile invertebrates that either feed on the aufwuchs (e.g., grazing snails; Chapter 10) or use the plant stems and leaves as perching sites for resting or foraging (e.g., sprawling dragonfly nymphs; Chapter 17).

C. Wetlands, Ephemeral Ponds, and Swamps

For many years the public has ignored or labeled as undesirable the vast acreage of wetlands, ephemeral ponds, and swamps in North America, considering them breeding grounds for mosquitoes and snakes. Many have been drained, filled, and bulldozed for housing and commercial development without regard to their intrinsic value to wildlife and the environment. Our knowledge of these fascinating aquatic ecosystems is relatively limited in comparison to information available about lakes and streams. Interestingly enough, one factor that has started to reverse this trend has been the institution of strict environmental laws to protect the broad category of wetlands; such laws have also led to an influx of research funds to study these ecosystems.

In its broader definition, the term freshwater wetlands refers to nontidal ecosystems whose soils are saturated with water on a permanent or seasonal basis. Emergent aquatic vegetation is prominent and may alternate with annual terrestrial plants in these marshy, or palustrine, habitats. The extensive biomass of trees, herbaceous vegetation, grasses, and other plants in many wetlands is responsible for their recently recognized abilities to contribute to the "natural filtration" of pollutants in the environment. The distinctions among the many types of freshwater wetlands are not always clear to the nonspecialist (but see definitions in Cowardin *et al.* 1979). Wetlands vary from shallow, seasonally ephemeral ponds (such as Carolina bays and other pocosins, Sharitz and Gibbons 1982) to relatively permanent vegetation-choked marshes to semilotic alluvial swamps (Fig. 2.10).

Because wetlands are generally more ephemeral than other lentic ecosystems, their biotic communities are strongly influenced by the water cycle. Wetlands invertebrates are affected by ecosystem permanency (e.g., years since last exposure), predictability of drying, and season of drying (if at all). The nature of the community also reflects the volume and depth of open water and sometimes the current velocity (for alluvial swamps).

Invertebrates of seasonal wetlands, such as the Carolina bays of the southeastern United States, have evolved various characteristics that have directly or indirectly adapted them for ecosystems that alternate between aquatic and terrestrial states. Wiggins *et al.* (1980) divided animals of temporary pools into four groups based on their adaptations for tolerating or avoiding drought and their period of recruitment. "Group 1 are year-round residents incapable of active dispersal, which avoid desiccation either as resistant stages or by burrowing into pool sediments. Group 2 are spring recruits which must oviposit on water but subsequently aestivate and overwinter in the dry basin in various stages. Group 3 are summer recruits ovipositing in the dry basin and overwintering as eggs or larvae. Group 4 are nonwintering migrants leaving the pool before the dry phase, which is spent in permanent water; they return in spring to breed." When a temporary pool refills after a drought, the first colonizers are usually detritivores which exploit the abundant biomass of recently dead and decaying terrestrial vege-

tation; these species provide the animal tissue that supports the later arrival of predaceous invertebrates (Wiggins *et al.* 1980). Many species in ephemeral environments are ecological generalists because they live in both temporary and permanent aquatic ecosystems.

Alluvial swamps are at an opposite continuum from temporary wetland pools. Contrasting with the typical picture of a swamp as a stagnant marsh, alluvial (or riverine) swamps contain many areas of slowly to rapidly moving waters; portions of these wetlands somewhat resemble a highly braided stream. While it is true that southeastern alluvial swamps, such as those bordering the Savannah River between South Carolina and Georgia, are infested with alligators and poisonous snakes, they are otherwise extremely beautiful ecosystems with very few aerial insect pests once you have moved inward from the surrounding brush and terrestrial forest. Alluvial swamps are extremely heterogeneous, organically rich environments; oxygen does not appear to be limiting, at least in the shaded, flowing water regions.

The many soft- and hard-sediment habitats of swamps are home to a great diversity and density of aquatic invertebrates. In a study of an alluvial swamp in South Carolina, Thorp *et al.* (1985) showed that invertebrates rapidly colonized wood snags, reaching a rough steady state in numbers within the first week of a eight-week study. Peak densities were equivalent to nearly 18,000 animals/m^2! "Filter-feeding taxa were numerically dominant early but soon were subordinate to gatherer and scraper functional feeding groups. Current velocity, seston particle size [as a food resource], dispersal capacity and competition for space may be important factors affecting community structure and colonization patterns in these aquatic ecosystems (p. 56)."

D. Hypersaline Lakes

Natural saline lakes are common worldwide but generally have neither the average size nor abundance of freshwater lakes (Eugster and Hardie 1978, Hammer 1984). They are frequently encountered in the prairies and Great Basin of the western United States and in the provinces of Alberta, Saskatchewan, and British Columbia of western Canada (Hammer 1984). The Great Salt Lake of Utah is the largest saline lake in North America. Saline lakes usually occur in more arid regions where closed basins promote the concentrations of salts in these lentic ecosystems (Fig. 2.11). They include: "saltern lakes," which are high in sodium

Figure 2.10 (A) Photograph of a low-flow portion of a cypress-tapelo, alluvial swamp in South Carolina; water is covered by floating duckweed. Floating platforms in center of photograph are suspending wood snags as part of colonization study by Thorp *et al*. 1985. (B) Photograph of a weed-clicked marsh in South Carolina. Despite its name, Dry Bay, this Carolina bay contained fish, amphibians, crayfish, and many species of aquatic insects (both photographs by J. H. Thorp).

chloride; lakes with large concentrations of sulfates and borates; and "soda" lakes characterized by abundant sodium carbonates and bicarbonates. Salinities in the Great Salt Lake have been recorded as high as 200 g/liter (versus an average of 35 g/liter in the oceans of the world). In addition to salinities that often surpass those of full-strength seawater, hyper-

saline lakes differ in pH, metal concentrations, and alkalinity from more typical inland lakes.

The abiotic characteristics of saline lakes make them extremely rigorous environments. Cyanobacteria colonize saline lakes as do some sedges, but the submerged macrophyte flora is relatively sparse in comparison to that in freshwater lakes. Because in-

Figure 2.11 Photographs of hypersaline Mono Lake in California showing CaCO₃ mounds known as "tufa" (photograph by J. H. Thorp).

vertebrate density and diversity are enhanced by abundant littoral plants, the depauperate flora of saline lakes and the lack of certain other microhabitats undoubtedly depress the invertebrate fauna of briny pools. Few invertebrates of inland waters have the ability to regulate the osmotic and ionic concentrations of their tissues in a hypertonic environment. As a result, hypersaline environments contain very few species, although their densities may be high. For example, the only permanent metazoans in Mono Lake, California (Fig. 2.11), are the ubiquitous brine shrimp, *Artemia salina*, and a species of brine fly. High densities of the brine shrimp in Mono Lake are crucial to the survival of the migratory eared grebe, *Podiceps nigricollis*, which molts its flight feathers while at Mono Lake, depending on these grazing crustaceans as its sole food source (Cooper *et al.* 1984). [This simple but important food chain is threatened by the possible demise of the brine shrimp as a result of rising salinities brought on by the removal of freshwater by Los Angeles from the catchment of Mono Lake.] Other saline lakes support more species, including the rotifer *Brachionus plicatilis*, the calanoid copepod *Diaptomus nevadensis*, and several families of flies, beetles, and true bugs.

As saline lakes become less salty, the diversity and density of their flora and fauna are enhanced. For example, the meromictic Waldsea Lake of Saskatchewan, which is one of the better studied saline (≥ 3 ppt salts) inland lakes in Canada, has a much higher species diversity than the much saltier Mono Lake. It contains two macrophyte species, some filamentous algae, a sparse phytoplankton population, and a dense but rather uniform population of zooplankton, dominated by the calanoid copepod *Diaptomus connexus*, the rotifers *Hexarthra fennica* and *Brachionus plicatilis*, and the water flea *Daphnia similis* (Hammer 1984). The littoral zone supports high numbers of relatively few species (in comparison to freshwater lakes but not to Mono Lake). These include several species of true bugs, nine genera of beetles, midges (mostly *Cricotopus*), a few caddisflies, the damselfly *Enallagma clausum*, one snail species (*Lymnaea stagnalis*), and several genera of crustaceans (see references in Hammer 1984). Although Waldsea Lake is certainly diverse in comparison to the hypersaline Mono Lake, its community is depauperate in comparison to most permanent, freshwater lentic environments.

The inland waters of North America are sufficiently diverse and abundant to prevent the valid formation of any simple, all-encompassing theory on how the biotic communities of these aquatic ecosystems are regulated. This does not imply, however, that we should abandon the task of assimilating and evaluating data from a diversity of ecosystems in order to understand better the effects of a few identifiable factors, such as the influence of habitat characteristics. A comparative approach, combined with intensive descriptive and experimental studies, will, in the long run, enhance our ability to develop a holistic view of stream and lake ecosystems. As we have shown in this chapter and as you will learn in subsequent chapters of this book, the habitat characteristics of an organism are extremely crucial in determining its distribution, survival, and reproductive output, but the nature of the habitat is only one of a large suite of factors controlling the success of a freshwater invertebrate.

LITERATURE CITED

Anderson, R. V., and D. M. Day. 1986. Predictive quality of macroinvertebrate-habitat associations in lower navigation pools of the Mississippi River. Hydrobiologia 136:101–112.

Barnby, M. A., and V. H. Resh. 1988. Factors affecting the distribution of an endemic and a widespread species of brinefly (Diptera: Ephydridae) in a northern California thermal saline spring. Annals of the Entomological Society of America 81:437–446.

Beckett, D. C., and M. C. Miller. 1986. Macroinvertebrate colonization of multiplate samplers in the Ohio River: the effects of dams. Canadian Journal of Fisheries and Aquatic Sciences 39:1622–1627.

Benke, A. C., T. C. Van Arsdall, Jr., D. M. Gillespie, and F. K. Parrish. 1984. Invertebrate productivity in a subtropical blackwater river: the importance of habitat and life history. Ecological Monographs 54:25–63.

Carpenter, S. R. (editor). 1988. Complex interactions in lake communities. Springer-Verlag, New York. 283 p.

Cole, G. A. 1983. Textbook of limnology. Third edition. Mosby, St. Louis, Missouri. 401 p.

Connell, J. H. 1978. Diversity in tropical rain forests and coral reefs. Science 199:1302–1310.

Cooper, S. D., D. W. Winkler, and P. H. Lenz. 1984. The effects of grebe predation on a brine shrimp population. Journal of Animal Ecology 53:51–64.

Covich, A. P. 1988. Geographical and historical comparisons of neotropical streams: biotic diversity and detrital processing in highly variable habitats. Journal of the North American Benthological Society 7:361–386.

Cowardin, L. M., V. Carter, F. C. Golet, and E. T. LaRoe. 1979. Classification of wetlands and deepwater habitats of the United States. FWS/OBS-79/31, U.S. Fish and Wildlife Service, Office of Biological Services, Habitat Preservation Program, Washington, D.C. 103 p.

Culver, D. C. 1970. Analysis of simple cave communities. I. Caves as islands. Evolution 24:463–474.

Cummins, K. W. 1966. A review of stream ecology with special emphasis on organism-substrate relationships. Pymatuning Laboratory of Ecology Special Publication 4:2–51.

Delucchi, C. M. and B. A. Peckarsky. 1989. Life history patterns of insects in an intermittent and a permanent stream. Journal of the North American Benthological Society 8:308–321.

Diggins, M. R. and J. H. Thorp. 1985. Winter-spring depth distribution of Chironomidae in a southeastern reservoir. Freshwater Invertebrate Biology 4:8–21.

Eugster, H. P. and L. A. Hardie. 1978. Saline lakes. Pages 237-293 in: A. Lerman (ed.), Lakes: chemistry, geology, and physics. Springer-Verlag, New York.

Goldman, C. R., and A. J. Horne. 1983. Limnology. McGraw-Hill, New York. 464 p.

Grimm, N. B., and S. G. Fisher. 1989. Stability of periphyton and macroinvertebrates to disturbance by flash floods in a desert stream. Journal of the North American Benthological Society 8:293–307.

Haag, K. H., and J. H. Thorp. Cross-channel distribution patterns of invertebrate benthos in a regulated reach of the Tennessee River. (Submitted).

Hammer, U. T. 1984. The saline lakes of canada. Pages 521-540 in: F. B. Taub (ed.), Ecosystems of the world, No. 23: lakes and reservoirs. Elsevier, New York.

Hershey, A. E. 1985. Effects of predatory sculpin on the chironomid communities in an arctic lake. Ecology 66:1131–1138.

Hutchinson, G. E. 1957. A treatise on limnology. I. Geography, physics, and chemistry. John Wiley, New York. 1015 p.

Hutchinson, G. E. 1961. The paradox of the plankton. American Naturalist 95:137–145.

Hutchinson, G. E. 1967. A treatise on limnology. II. Introduction to lake biology and limnoplankton. John Wiley, New York. 1115 p.

Hutchinson, G. E. 1975. A treatise on limnology. III. Limnological botany. John Wiley, New York. 660 p.

Hynes, H. B. N. 1970. The ecology of running waters. University of Toronto Press, Toronto, Ontario. 555 p.

Ledger, D. C. 1981. The velocity of the River Tweed and its tributaries. Freshwater Biology 11:1–10.

Leopold, L. B. 1953. Downstream change of velocity in rivers. American Journal of Science 251:606–624.

Longley, G. 1986. The biota of the Edwards aquifer and the implications for paleozoogeography. Pages 51–54 in: P. L. Abbott and C. M. Woodruff, Jr. (eds.), The Balcones Escarpmernt, Central Texas. Geological Society of America.

Matthews, W. J. 1988. North American prairie streams as systems for ecological study. Journal of the North American Benthological Society 7:387–409.

Merritt, R. W., and K. W. Cummins (eds.). 1984. An introduction to the aquatic insects of North America. Second Edition. Kendall/Hunt, Dubuque, Iowa. 722 p.

Meyer, J. L. 1990. A blackwater perspective on riverine ecosystems. BioScience 40:643–651.

Minckley, W. L. 1963. The ecology of a spring stream Doe Run, Meade County, Kentucky. Wildlife Monographs No. 11 (A publication of the Wildlife Society). 124 p.

Minshall, G. W. 1984. Aquatic insect-substratum relationships. Pages 358–399 in: V. H. Resh and D. M. Rosenberg (eds.), The ecology of aquatic insects. Praeger, New York.

Minshall, G. W. 1988. Stream ecosystem theory: a global perspective. Journal of the North American Benthological Society 7:263–288.

Minshall, G. W., K. W. Cummins, R. C. Petersen, C. E. Cushing, D. A. Bruns, J. R. Sedell, and R. L. Vannote. 1985. Developments in stream ecosystem theory. Canadian Journal of Fisheries and Aquatic Sciences 42:1045–1055.

Minshall, G. W., R. C. Petersen, K. W. Cummins, T. L. Bott, J. R. Sedell, C. E. Cushing, and R. L. Vannote. 1983. Interbiome comparisons of stream ecosystem dynamics. Ecological Monographs 53:1–25.

Naiman, R. J., J. M. Melillo, and J. E. Hobbie. 1986. Ecosystem alteration of boreal forest streams by beaver (*Castor canadensis*). Ecology 67:1254–1269.

Newbury, R. W. 1984. Hydrologic determinants of aquatic insect habitats. Pages 323–357 in: V. H. Resh and D. M. Rosenberg (eds.), The ecology of aquatic insects. Praeger, New York.

Resh, V. H., A. V. Brown, A. P. Covich, M. E. Gurtz, H. W. Li, G. W. Minshall, S. R. Reice, A. L. Sheldon, J. B. Wallace, and R. C. Wissmar. 1988. The role of disturbance in stream ecology. Journal of the North American Benthological Society 7:433–455.

Ritter, D. F. 1986. Process geomorphology. Second edition. Wm. C. Brown Publishers, Dubuque, Iowa. 579 p.

Schwoerbel, J. 1961. Über die Lebensbedingungen und die Besiedlung des hyporheischen Lebensraumes. Archiv für Hydrobiologie Supplement 25:182–214.

Sharitz, R. R., and J. W. Gibbons. 1982. The ecology of southeastern shrub bogs (pocosins) and Carolina bays: a community profile. FWS/OBS-82/04. U.S. Fish and Wildlife Service, Division of Biological Services, Washington, D.C. 93 p.

Stanford, J. A., F. R. Hauer, and J. V. Ward. 1988. Serial discontinuity in a large river system. Verhandlungen Internationale Vereinigung für theoretische und angewandte Limnologie 23:114–118.

Stanford, J. A., and J. V. Ward. 1988. The hyporheic habitat of river ecosystems. Nature 335:64–66.

Statzner, B., J. A. Gore, and V. H. Resh. 1988. Hydraulic stream ecology: observed patterns and potential applications. Journal of the North American Benthological Society 7:307–360.

Statzner, B., and B. Higler. 1985. Questions and comments on the River Continuum Concept. Canadian Journal of Fisheries and Aquatic Sciences 42:1038–1044.

Strayer, D. 1985. The benthic micrometazoans of Mirror Lake, New Hampshire. Archive für Hydrobiologie/Suppl. 72 3:287–426.

Strayer, D. 1986. The size structure of a lacustrine zoobenthic community. Oecologia 69:513–516.

Strahler, A. N. 1952. Dynamic basis of geomorphology. Geological Society of America Bulletin 63:1117–1142.

Strenth, N. E., J. D. Norton, and G. Longley. 1988. The larval development of the subterranean shrimp *Palaemonetes antrorum* Benedict (Decapoda, Palaemonidae) from central Texas. Stygologia 4:363–370.

Sweeney, B.W., and R.L. Vannote. 1978. Size variation and the distribution of hemimetabolous aquatic insects: two thermal equilibrial hypotheses. Science 200: 444–446.

Taub, F. B. (ed.). 1984. Lakes and reservoirs. Ecosystems of the world, Volume 23. Elsevier, New York 643 p.

Thorp, J. H., and M. L. Cothran. 1984. Regulation of freshwater community structure at multiple intensities of dragonfly predation. Ecology 65:1546–1555.

Thorp, J. H., and M. R. Diggins. 1982. Factors affecting depth distribution of dragonflies and other benthic insects in a thermally destabilized reservoir. Hydrobiologia 87:33–44.

Thorp, J. H., E. M. McEwan, M. F. Flynn, and F. R. Hauer. 1985. Invertebrate colonization of submerged wood in a cypress-tupelo swamp and blackwater stream. American Midland Naturalist 113:56–68.

Vannote, R. L., G. W. Minshall, K. W. Cummins, J. R. Sedell, and C. E. Cushing. 1980. The river continuum concept. Canadian Journal of Fisheries and Aquatic Sciences 37:130–137.

Wallace, J. B. 1990. Recovery of lotic macroinvertebrate communities from disturbance. Environmental Management 14:605–620.

Ward, J. V. 1986. Altitudinal zonation in a Rocky Mountain stream. Archiv für Hydrobiologie, Suppl. 74: 133–199.

Ward, J. V. 1989. The four-dimensional nature of lotic ecosystems. Journal of the North American Benthological Society 8:2–8.

Ward, J. V., and J. A. Standford. 1983. The intermediate disturbance hypothesis: an explanation for biotic diversity patterns in lotic ecosystems. Pages 347–356 in: T. D. Fontaine and S. M. Bartell (eds.), Dynamics of lotic ecosystems. Ann Arbor Science Publishers, Ann Arbor, Michigan.

Wetzel, R. G. 1983. Limnology. Second Edition. Saunders, Philadelphia. 857 p.

Wiggins, G. B., R. J. Mackay, and I. M. Smith. 1980. Evolutionary and ecological strategies of animals in annual temporary pools. Archive für Hydrobiologie/Suppl. 58 1:97–206.

Williams, D. D. 1984. The hyporheic zone as a habitat for aquatic insects and associated arthropods. Pages 430–455 in: Resh, V. H. and D. M. Rosenberg (eds.), The ecology of aquatic insects. Praeger, New York.

Williams, D. D., and H. B. N. Hynes. 1974. The occurrence of benthos deep in the substratum of a stream. Freshwater Biology 4:233–256.

Winterbourn, M. J. 1982. The River Continuum Concept–reply to Barmuta and Lake, New Zealand Journal of Marine and Freshwater Research 16:229–231.

Winterbourn, M. J., J. S. Rounick, and B. Cowie. 1981. Are New Zealand stream ecosystems really different? New Zealand Journal of Marine and Freshwater Research 15:321–328.

Protozoa

3

William D. Taylor
Department of Biology
University of Waterloo
Waterloo, Ontario N2L 3G1
Canada

Robert W. Sanders
Patrick Center for Environmental Research
Academy of Natural Sciences of Philadelphia
Philadelphia, Pennsylvania 19103

Chapter Outline

I. INTRODUCTION
II. ANATOMY AND PHYSIOLOGY
  A. Protozoa as Cells and Organisms
  B. Protozoan Organelles
    1. Food Vacuoles
    2. Contractile Vacuoles
    3. Cilia and Flagella
    4. Extrusomes
  C. Anatomy of Zoomastigophora
  D. Anatomy of Phytomastigophora
  E. Anatomy of Sarcodina
  F. Anatomy of Heliozoa
  G. Anatomy of Ciliophora
  H. Environmental Physiology
    1. Factors Affecting the Distribution of Protozoa
    2. Temperature
    3. Oxygen
    4. Environmental Contaminants
    5. Energetics of Protozoa
III. ECOLOGY AND EVOLUTION
  A. Diversity and Distribution
  B. Reproduction and Life History
  C. Ecological Interactions
    1. Intraspecific Interactions Among Protozoa
    2. Interspecific Competition Among Protozoa
    3. Predatory Interactions Among Protozoa
    4. Symbionts of Protozoa
    5. Epizooic and Epiphytic Protozoa
    6. Protozoa as Bacterivores
    7. Algivorous Protozoa
    8. Histophagous Ciliates
    9. Metazoan Predators of Protozoa
    10. Protozoa and Nutrient Cycling
  D. Evolutionary Relationships
IV. COLLECTING, REARING, AND PREPARATION FOR IDENTIFICATION
V. IDENTIFICATION OF PROTOZOA
  A. Notes about the Key
  B. Taxonomic Key to the Phyla or Subphyla of Protozoa
  C. Taxonomic Key to Classes and Orders in the Subphylum Mastigophora
  D. Taxonomic Key to Orders and Families in the Subphylum Sarcodina
  E. Taxonomic Key to Orders and Families in the Phylum Ciliophora
  Literature Cited

I. INTRODUCTION

Protozoa are ubiquitous; they are present in an active state in any aquatic or moist environment, and cysts are present everywhere in the biosphere, ready to give rise to active populations. Although readily overlooked, they play an important role in many communities. However, for biologists, protozoa have most often played the role of model organisms. Cell biologists, physiologists, geneticists, and even developmental biologists frequently turn to protozoa to address questions that would be more difficult to answer with metazoa, but in the hope that the answers are relevant to metazoa. The result is a surprising amount of information about a relatively few species which have filled the role of laboratory model. Accordingly, detailed monographs are available on several taxa, including *Paramecium* (e.g., Wichterman 1986), *Tetrahymena* (Elliott 1973), and *Amoeba* (Jeon 1973). Each contains thousands of references. Ecologists also have used protozoan populations and communities as model systems to investigate processes such as competition and predation or to test models, such as island biogeography, that would be much more difficult to study using larger organisms. Again, the result has been an extensive body of literature that is largely restricted to a few species and/or to unusual or artificial habitats.

At the same time, we know relatively little about most freshwater protozoa and their roles in natural communities. This is changing; recent interest in microzooplankton (zooplankton 20–200 μm) and the protozoan grazers of picoplankton (plankton 0.2–2.0 μm), is generating a great deal of research on or including protozoa. As aquatic ecologists come to realize that protozoa are not only present, but play a major quantitative role in material and energy flow, this interest should extend to other freshwater environments.

The goal of this chapter is to provide an introduction to the ecology and classification of protozoa for ecologists who wish to include them in studies of freshwater environments. The chapter focuses on the biology of the animal-like protozoa, largely ignoring the related pigmented organisms, which are adequately covered in phycological sources. The chapter is oriented toward the organisms, not the freshwater habitats themselves. Readers wishing to learn about the fauna of particular habitats might consult the bibliography on freshwater protozoa prepared by Finlay and Ochsenbein-Gattlen (1982) or some more recent papers on particular habitats: lake plankton (Beaver and Crisman 1989, Carrick and Fahnenstiel 1989, 1990); stream sediments (Bott and Kaplan 1989); stream aufwuchs (Baldock *et al.* 1983, Taylor 1983a); wetlands (Pratt and Cairns 1985); or the general outline by Bamforth (1985).

What are protozoa? The answers to this question are both simple and complex. Protozoa are unicellular or colonial eukaryotes, including all of the heterotrophic and motile ones. It is a taxon of convenience. Although the protozoa were considered a phylum in early classifications, modern treatments usually split the protozoa among several phyla. This reflects the immense diversity of this assemblage; differences among these unicellular organisms are at least as profound as those that separate plant and animal phyla (see Section III.D). Margulis and Schwartz (1988), in their five-kingdom classification of living organisms, reserve the kingdom Animalia, along with the kingdoms Plantae and Fungi, for the major lines of multicellular organisms. [Note that

the cells of some "multicellular" organisms are no more specialized than are those of "colonial" protozoa.] The unicellular eukaryotes and some of their multicellular descendants, 27 phyla in all, are placed in the kingdom Protoctista—or as favored by most authors and as used in our chapter, Protista. Other modern classifications partition the protists into more or fewer phyla. The classification of unicellular eukaryotes is in a state of flux, with much work to be done. In any event, it is among these groups that the organisms we know as protozoa are scattered.

For the purpose of this chapter, we define the term protozoa as those unicellular or colonial eukaryotes that are heterotrophic, and we will restrict ourselves to a discussion of free-living, phagotrophic forms in freshwater. Among the protozoa we will not discuss are the marine Foraminifera and Radiolaria; the soil-dwelling Labyrinthulamycota; Plasmodiophoromycota, and Acrasida; the saprophytic Hyphochytridiomycota, Oomycota, and Chytridiomycota; and the parasitic Apicomplexa, Microspora, and Myxospora. The classification we use and the citations to figures illustrating the major taxa are outlined in Table 3.1. This classification is conservative; many of the classes and orders of Sarcomastigophora have been raised to the level of phylum in recent treatments. However, it was chosen to concur as much as possible with the most useful references for identification.

Despite the fact that the term protozoa is perhaps not a proper taxonomic group, it remains a useful, functional one. For ecologists conducting studies involving aquatic animals, including those of microscopic proportions, it is natural to include the animal-like protozoa in those studies. Similarly, botanists may wish to include phototrophic Protista (algae) in their work. After all, it is functional similarity that generally concerns the practicing zoologist or ecologist.

II. ANATOMY AND PHYSIOLOGY

A. Protozoa as Cells and Organisms

Protozoa have been extensively studied by cell biologists as model cells. It is a matter of definition as to whether large protozoa that have many nuclei, or large, amitotic, polyploid nuclei, are actually acellular rather than unicellular. But the zoologist or ecologist can glean much useful information from the labors of cell biologists who have chosen protozoa as model eukaryotic cells. Therefore, this section unavoidably relies heavily on information from a few taxa. We first describe some of the organelles that are important and widely distributed among protozoa and then discuss major taxa of protozoa with respect to their more unique organelles. Lastly,

Table 3.1 Major Taxonomic Categories of Protists as Used in This Chapter

Phylum
  Subphylum
    Class
      Subclass
        Order
	Figure Reference

Sarcomastigophora	
Mastigophora	
Phytomastigophora	
Cryptomonadida	(3.13J)
Dinoflagellida	(3.13A–C)
Euglenida	(3.13D, F–I)
Chrysomonadida	(3.13K–N; 3.14A)
Volvocida	(3.13E)
Zoomastigophora	
Choanoflagellida	(3.14B–H)
Bicosoecids	(3.14I)
Cercomonadida	(3.14K)
Kinetoplastida	(3.14J,L,N,O,P)
Diplomonadida	(3.14M)
Sarcodina	
Heterolobosea	
Schizopyrenida	(3.15A–C)
Acrasida	
(terrestrial)	
Lobosea	
Gymnamoebia	
Euamoebida	(3.15D–M,O,P)
Acanthopodida	(3.15N)
Testacealobosia	
Arcellinida	(3.16B–Q)
Himatismenida	(3.16A)
Caryoblastea	
Pelobiontida	(3.16V)
Filosea	
Aconchulinida	(3.16R; 3.17A)
Gromiida	(3.16S–U,W; 3.19B–K)
Granuloreticulosea	
Athalamida	(3.17P,Q)
Monothalamida	(3.17L–O)
Promycetozoida	(3.17R)
Heliozoea	
Actinophryida	(3.18B,C)
Desmothoracida	(3.18A)
Ciliophrida	(3.18E)
Centrohelida	(3.18D,F,H)
Rotosphaerida	(3.18G)
Ciliophora	
Postciliodesmatophora	
Karyorelictea	(3.23J)
Spirotrichea	
Heterotrichia	(3.21J–M; 3.22A,B,D–F)
Choreotrichia	(3.21S–X)
Stichotrichia	(3.21N–Q)
Rhabdophora	
Prostomatea	(3.22H,J,M,O; 3.23 A–C)
Litostomatea	(3.22G,I,K,L,N,P–W; 3.23D–I)

(continued)

Table 3.1 *(Continued)*

Phylum
  Subphylum
    Class
      Subclass
        Order
	Figure Reference

Cyrtophora	
Phyllopharyngea	
Phyllopharyngia	(3.24T)
Chonotrichia	
(marine)	
Suctoria	(3.19A–P)
Nassophorea	
Nassophoria	(3.20V,W; 3.24A,B,D,E,G,I,J)
Hypotrichia	(3.20X,Y)
Oligohymenophorea	
Hymenostomatia	(3.23L–X; 3.24C,F,H,O)
Peritrichia	(3.20A–U)
Colpodea	(3.22C; 3.23K; 3.24K,N,P–S)

we revert to an overall view in discussing environmental physiology of protozoa. In all sections, we emphasize those organelles that relate to feeding, locomotion, ecology, and morphology at the light microscope level.

B. Protozoan Organelles

1. Food Vacuoles

Phagocytosis in protozoa leads to the formation of food vacuoles (Fig. 3.1). A food vacuole is essentially a membrane-bound vesicle surrounding the ingested material. Digestion is, as one would expect, primarily intracellular in free-living protozoa. The formation and behavior of food vacuoles is best known in the ciliates *Tetrahymena* (see Rasmussen 1976, Nilsson 1976, 1987) and *Paramecium* (Allen 1984). A general review is provided by Nisbet (1984).

The formation of food vacuoles is stimulated by the presence of particulate food. In protein broth medium, *Tetrahymena* shows very slow rates of food vacuole formation, ingestion, and growth unless particles are present (Ricketts 1972), even though it can grow without forming food vacuoles in complex medium via carrier-mediated uptake of dissolved compounds across the plasma membrane. During starvation, vacuoles resembling food vacuoles may be created to digest cellular components. Formation of such "cytolosomes" is also known from *Amoeba* (Chapman-Andresen 1973).

Many protozoa have a fixed site of food-vacuole formation—a cytostome with accompanying struc-

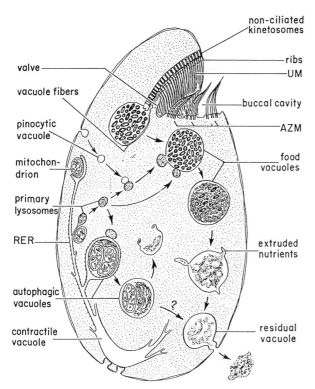

Figure 3.1 The formation and fate of food vacuoles, autophagic vacuoles, and pinocytotic vacuoles in *Tetrahymena pyriformis*. UM, undalating membrane; AZM, adoral zone of membranes (polykinetids).

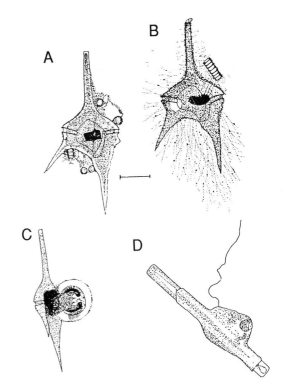

Figure 3.2 Phagocytosis of large food items by protoplasmic extensions in the dinoflagellate *Ceratium hirudinella* (A–C), and the consumption of a large diatom by a *Paraphysomonas*-like chrysophyte (D). (A–C from Hofeneder 1930; D after Suttle *et al.* 1986.)

tures to aid with ingestion. In others, including many Sarcomastigophora, the site of ingestion is flexible, perhaps only limited to a characteristic region of the cell. The Suctoria have many fixed sites for ingestion. The contents of a single vacuole may vary from a single large food item in macrophagous species to thousands of small items, such as bacteria, in microphagous forms. Similarly, individual protozoa may have one to many food vacuoles. The vacuoles of microphagous forms may be perfectly spherical, while larger vacuoles containing single prey may be quite irregular, conforming to the contours of the prey. Macrophagous protozoa may distend themselves to surround large food items, leave food items protruding from their cytoplasm, or even share a prey item among several individuals. Large food items may be digested by cytoplasmic extensions in dinoflagellates and, possibly, in other protozoa (Fig. 3.2). In some amebas (Schizopyrenida), food vacuoles containing bacteria become surrounded by concentric membranous whorls (Schuster 1979).

Digestion may be initiated by the toxicysts of some predatory ciliates. In others, it begins when granules (acidosomes) fuse with the developing food vacuole. The pH of the new vacuole drops precipitously, then increases minutes later. Hydrolytic enzymes, such as acid phosphatase, are introduced as the food vacuole fuses with lysosomes. The size of

the food vacuole may diminish, reflecting the absorption of water and nutrients, and vacuoles may coalesce around the undigested residue. Some protozoa, such as ciliates, have a fixed site of egestion, or cytoproct. When the contents of the food vacuole are released, the membrane surrounding the vacuole is retained and recycled (Allen 1984). It is unlikely that a cellular equivalent of a digestive tract exists in protozoa. Nevertheless, it has been possible to estimate the feeding rate of protozoa in the field from the rate at which food vacuoles are eliminated (Goulder 1972).

2. *Contractile Vacuoles*

These organelles are widely distributed in protists and are almost universal in freshwater protozoa and sponges. They are absent in dinoflagellates and are generally lacking in those protists with rigid cell walls and/or those from marine or endosymbiotic habitats. A large body of evidence suggests that contractile vacuoles eliminate water and, thereby, counteract the tendency of naked protists to swell in freshwater (Patterson 1980). The activity of the contractile vacuole is correlated with changes in the osmotic balance between the cell and its medium and a loss of contractile function results in swelling of the cell. Of course, both the occurrence of contractile vacuoles in naked, freshwater protists and their absence in most rigid-walled, marine or endo-symbiotic ones also support this theory.

Associated with the contractile vacuole is a specialized cytoplasm, the spongiome, which collects the solute to be expelled by the contractile vacuole. The spongiome may contain vesicles or canals visible with the light microscope, and therefore be of diagnostic value (Fig. 3.3). Patterson (1980) recognized four types of contractile vacuole complexes in protozoa, with six different patterns of behavior. Permanent contractile vacuole pores are present in ciliates and are readily visible in silver-stained specimens. The contractile vacuole may have a fixed location in some other groups, for example in the gullet or reservoir of cryptomonad or euglenid flagellates, but the fixed pores, if they exist, are not visible.

Several important functional aspects of the contractile vacuole complex remain unknown; for example, the composition of the fluid expelled and, therefore, the role of the contractile vacuole complex in salt balance and excretion. It is also debated whether expulsion of contractile vacuole contents is active or passive and how the activity of the contractile vacuole is controlled.

Dinoflagellates possess a functionally homologous organelle, the pusule. This organelle consists of a single chamber or tubule associated with a flagellar opening and is lined with vesicles (Dodge and Greuet 1987).

3. Cilia and Flagella

Flagella, cilia, and derived structures are widely distributed in eukaryotes, including protozoa. Their ultrastructural similarities, including the basic 9 + 2 pattern of microtubules, indicates homology despite the great diversity in form and function. The sense organs of many animals contain cilia, which emphasizes that they are probably intrinsically sensory as well as motile. Cilia or flagella that are thought to be sensory occur in many protozoa, e.g., the "brosse" kinetids of prostomateans and the flagella of phototactic euglenids.

It is well established that the motile force in cilia and flagella is the sliding of microtubules in the shaft against each other. Therefore, the active movement is the bending of the shaft and not the movement of the shaft relative to the base. The bending of the shaft takes on the appearance of a beat or stroke in short cilia or a wave in long flagella.

Flagella usually occur in pairs and are relatively long (often longer than the cell that carries them, and reaching lengths of up to 50 μm). A major functional and morphologic dichotomy among flagella is between those with and without mastigonemes (Fig. 3.4). Mastigonemes are thin, hair-like projections perpendicular to the shaft of the flagellum. They are not visible with light microscopy, but are functionally important. A distally directed wave of bending in a smooth flagellum will push the cell in the opposite direction from which the flagellum is pointed, as in animal sperm; whereas, the same wave form in a flagellum with mastigonemes will pull the cell in the direction to which the flagellum is pointed. To understand the motility of protozoa, especially those using cilia and flagella, it is necessary to appreciate the viscous nature of water at the scale (i.e., at the low Reynold's number) at which protozoa operate (Holwill 1974, Fenchel 1987).

The difference between cilia and flagella is one of form and function. Cilia are generally short and

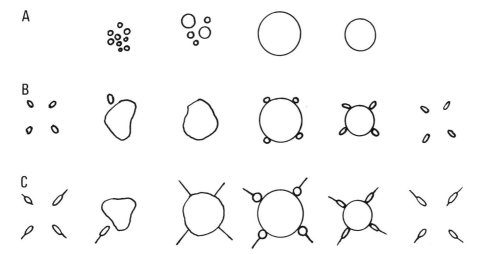

Figure 3.3 Top views of three morphologies of contractile vacuoles found in protozoa, showing their cycles of activity from left to right as seen with a light microscope. (A) A vacuole that forms by the coalescing of smaller vesicles, as in common in Sarcomastigophora. Vacuole (B) has ampullae in fixed locations, while vacuole (C) also has visible canals. (After Patterson 1980.)

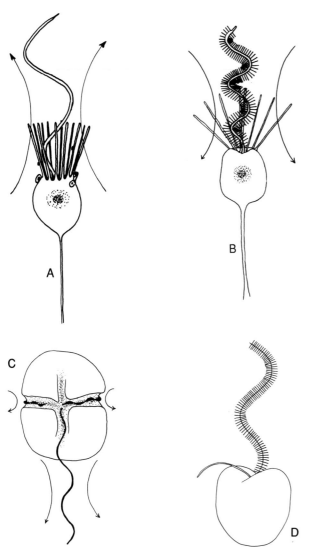

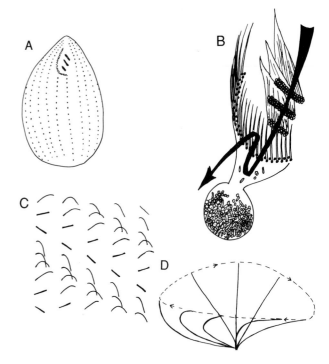

Figure 3.5 The distribution of cilia on *Tetrahymena* (A). The compound ciliary organelles associated with the mouth (B) are used to collect bacteria into the forming food vacuole. Somatic cilia beat metachronously to propel the cell (C). (D) The path of a single cilium viewed from the side, showing the power stroke and the recovery stroke in different planes.

Figure 3.4 Different uses of flagella in two sessile bacterivores, and two swimming forms. Arrows indicate movement of water relative to the cells. Note that distally directed waves of beating produce opposite forces in smooth flagella and flagella with mastigonemes. (A) The choanoflagellate *Monosiga,* modified from Fenchel (1982a); (B) *Actinomonas,* which is probably a heliozoan, also modified from Fenchel; (C) a dinoflagellate; (D) a chrysophyte.

densely packed. They are organized in rows (kineties) that are typically aligned in parallel, to produce fields of cilia whose activities are coordinated (Fig. 3.5A). The movement of adjacent somatic cilia is not quite synchronous; rather, each is slightly out of phase with its neighbors, or metachronous. A field of cilia, therefore, moves in waves that may or may not correspond in direction to the actual power stroke of the cilia. Compound ciliary organelles (Fig. 3.5B) or polykinetids are restricted fields of cilia acting as a single unit in locomotion, as in the cirri of stichotrichs, or feeding, as in the buccal apparatus of

spirotrichs. They may be relatively long compared to other cilia and fused so that their compound nature is not obvious under light microscopy except as indicated by their diameter.

Ciliary movement is faster than flagellar movement, even accounting for ciliates generally being larger than flagellates (Sleigh and Blake 1977). Indeed, within both groups there is relatively little increase in absolute swimming speed with body size; flagellates average about 0.2 mm/sec, while ciliates average about 1 mm/sec.

4. Extrusomes

Extrusive organelles or extrusomes are membrane-bound organelles associated with the pellicles of protists and contain material that can be ejected or extruded from the cells. Most are visible at the light microscopic level, at least in the extruded state. They are widespread in occurrence and diverse in structure and function; they are probably not homologous. The function of many is in doubt. However, following Hausmann (1978), we group them by functional and structural similarities for discussion, omitting those restricted to marine groups.

Spindle trichocysts of nassophorean ciliates, especially those of *Paramecium*, are well known. They are abundantly distributed over the cortex (e.g., 8000 per cell in *Paramecium*) and each may discharge a pointed projectile on a proteinaceous shaft in response to mechanical or physical stimulation. These are probably effective deterrents to some potential predators, although specialists, such as the predatory ciliate *Didinium*, are undaunted (Fig. 3.6). Expulsion of the trichocyst is extremely

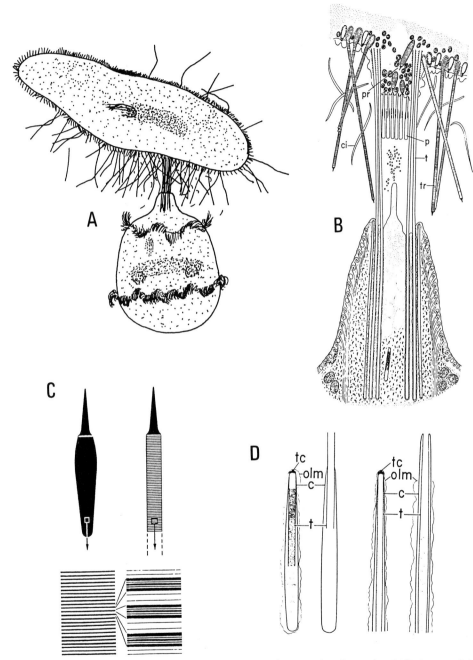

Figure 3.6 (A) Capture of *Paramecium* by *Didinium*. Note the use of toxicysts by *Didinium,* and the spindle trichocysts released by *Paramecium* in response. (B) The oral apparatus and zone of contact, as revealed by electron microscopy. Note that there are two varieties of extrusomes involved in the attack by *Didinium* and that the deciliation of *Paramecium* is part of the result. (C) The ultrastructure of spindle trichocysts of *Paramecium* in the resting (left) and discharged (right) state. (D) Ultrastructure of *Didinium* pexicysts (left) and toxicysts (right). Both are shown in resting and discharged states. c, Capsule; ci, cilia; olm, outer limiting membrane; p, discharged pexicysts; pr, protuberance of *Paramecium* cytoplasm; t, discharged toxicyst; tc, terminal cap; tr, discharged trichocyst. (B and D are from Wessenberg and Antipa 1970; C is from Hausmann 1978.)

rapid and is driven by a change in the paracrystalline structure of protein in the shaft of the trichocyst. The increase in length of the shaft following expulsion is approximately eight-fold.

Extrusomes are found in several flagellate groups; most dinoflagellates have spindle trichocysts similar to those of ciliates. Until recently, these groups were not considered to be even remotely related. However, recent studies of ribosomal RNA suggest that they are less distant than one might suppose, based on gross morphology (Gunderson *et al.* 1987). Ejectisomes are found in most cryptomonads and a few other phytoflagellates (Fig. 3.7*c*). They are typically dimorphic, although of similar morphology; larger ejectisomes are associated with the gullet, while smaller ones are external to the gullet and near the anterior of the cell. Because some cryptomonads are heterotrophic and photosynthetic ones may be phagotrophic, it is tempting to speculate that the ejectisomes are used in food capture. However, the gullet is not the site of phagotrophy, so an offensive role of these extrusomes is doubtful. The mechanism of expulsion of ejectisomes is very different from that of spindle trichocysts; the unexpelled ejectisome is a tight coil of ribbon-like material, which uncoils and rolls into a tube on expulsion. A similarly coiled secondary structure forms a pointed tip.

Discobolocysts are also extrusomes of flagellates, in this case primarily of chrysomonads. They are almost spherical, rather than spindle-shaped, but their expulsion is reminiscent of spindle trichocysts; the shaft of the discobolocyst elongates, apparently through a conformational change in the material in the shaft (Fig. 3.7B). Although their function is unknown, it is possible that they discourage phagotrophs.

Extrusomes are used for capturing food by predatory ciliates in the classes Prostomatea and Litostomatea, which have toxicysts, and Phyllopharyngea, which have haptocysts. Toxicysts are extruded by eversion, and in doing so release material that appears to be toxic to the prey. Several types of toxicysts are known, and up to three kinds may be found within a species. The use of toxicysts by *Didinium* in capturing *Paramecium* is illustrated in Fig. 3.6. Although the toxicysts of *Didinium* are concentrated on the proboscis, they may be elsewhere in other ciliates. *Loxophyllum* (Fig. 3.23I) has them in clusters along its lateral margin, while *Actinobolina* (Fig. 3.22G) has toxicysts on tentacles that are widely distributed over the body.

Haptocysts are found in the tips of the feeding tentacles of Suctoria (Fig. 3.19). Prey (mostly other ciliates) that touch the feeding tentacles are caught and held by the extrusion of the haptocysts (Fig. 3.7D), which penetrate the prey, inject substances (probably including hydrolytic enzymes), and aid in the fusion of the membranes of the predator and prey. The cytoplasm of the prey is then drawn into the suctorian through the feeding tentacles.

Some Heliozoa possess extrusomes of various types, which may give the elongate pseudopodia (axopodia) a bumpy appearance (Fig. 3.7A). Kinetocysts resemble haptocysts or small trichocysts. Extrusomes in Heliozoa also include mucocysts and electron dense bodies that also expel amorphous material onto the cell surface. All probably facilitate prey adhesion and capture.

Mucocysts are widespread in ciliates and flagellates. They secrete an amorphous material onto the cell surface that, at least in many species, is involved in cyst formation. *Tetrahymena* can use mucocysts to create a temporary capsule in response to introduction of a dye; this is probably a defensive reaction, although it is not known which other stimuli cause this response. Wolfe (1988) suggested that mucocysts may be the first line of defense of protozoans. Mucocysts may also be involved in nutrition; the mucous secreted may be reingested after attracting dissolved compounds or flocculating small food items (see Section III.C).

Mucocysts occur in many flagellates. Euglenoid flagellates have larger muciferous bodies in addition to mucocysts, although some authors do not differentiate between them. These provide a continuous coat of mucous on the surface of the cell and, in some species, participate in cyst formation, adhesion to the substratum, or stalk formation.

C. Anatomy of Zoomastigophora

This diverse, probably polyphyletic group contains heterotrophic protists with one to many flagella. It is likely that many of these protists could be assigned to Phytomastigophora if only they resembled a pigmented form closely enough for their affinity to be recognized. In free-living taxa, as opposed to parasitic species, the number of flagella is limited; *Paramastix* has two rows of 8–12 flagella, but others have 1–4 (usually 2). Among the Zoomastigophora, there are four orders containing freshwater species: Choanoflagellida, Kinetoplastida, Diplomonadida and Cercomonadida. We include the bicoecids here, although their affinities are uncertain. Among these orders, the choanoflagellates and kinetoplastids are raised to phyla by some authors, while bicoecids are occasionally put with chrysomonads. Schizopyrenids, also called amoeboflagellates, are sometimes included.

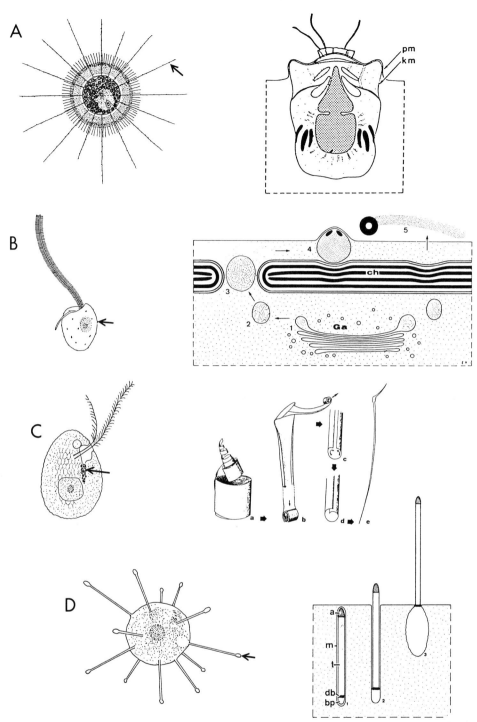

Figure 3.7 Examples of extrusomes in other protozoa. (A) Kinetocysts from the axopodia of the heliozoan *Hetrophrys* (from Febvre-Chevalier 1985.) (B) Discobolocysts from the cell surface of the chrysophyte *Ochromonas,* and the steps from formation (1) to discharge (5). (C) Ejectisomes from the gullet of *Crytomonas*. a, Undischarged; b–d, discharging, showing uncoiling in two directions and rolling of the uncoiled ejectisome into a tube; e, discharged ejectisome. (D) Haptocysts from the tentacles of a suctorian. a, Apex; bp, proximal part that swells on extrusion; db, dark band; m, membrane. Arrows indicate the location of the extrusomes to the right. (Extrusomes are from Hausmann 1978.)

Choanoflagellates, or collared flagellates, are distinctive for the collar that surrounds the single flagellum (Figs. 3.4A, 3.14B–H). They bear a strong resemblance to choanocytes of Porifera. The collar may be difficult to distinguish with light microscopy, but when distinguished by electron microscopy, is evidently composed of microvilli. These intercept bacteria drawn past the cell by the flagellum. Because most choanoflagellates are attached or colonial, the distally directed waves of the smooth flagellum serve to push water past the protozoan. Bacteria caught on microvilli are transported to the cell and ingested by pseudopodia at the base of the collar (Leadbeater and Morton 1974). Many choanoflagellates attach to the substratum, and many have an external, loose-fitting covering or lorica (although, again, it may be difficult to see with the light microscope). In one marine family (Acanthoecidae) the lorica is basket-like.

Bicoecids (Fig. 3.14I) resemble choanoflagellates, although they lack a collar. Like choanoflagellates, they are enclosed in a lorica and have a flagellum that is used to create a feeding current. A second flagellum lies along the cell and continues posteriorly to become an attachment to the base of the lorica.

Kinetoplastida (Fig. 3.14J, L, N–P) are known mostly as parasites, especially *Trypanosoma* and its relatives, but many members of the suborder Bodina live in freshwater. They have a unique single mitochondrion, which runs the length of the cell, and one or more DNA-rich organelles, kinetoplasts, associated with this mitochondrion near the flagella. The best known genus is *Bodo*, which, like other bodonids, has two flagella (Fig. 3.14J). One trails, often in contact with the substratum, while the other extends ahead. The trailing flagellum may attach temporarily to the substratum. In the related, but sessile and colonial genus *Cephalothamnium* (Fig. 3.14L), the attachment is permanent.

Diplomonadida and Cercomonadida are relatively unimportant groups to the freshwater ecologist; diplomonads are largely parasitic, with a few free-living forms that may be found in organically enriched water (Fig. 3.14M). The order Cercomonadida was created for one genus *Cercomonas* (Fig. 3.14K).

D. Anatomy of Phytomastigophora

Autotrophic protists, including flagellates, are classified by botanists. Nevertheless, in recognition of their motility and the ubiquity of heterotrophy in these groups, the flagellated forms are given a place in zoological nomenclature as Phytomastigophora. Indeed, many of the pigmented, autotrophic taxa are capable of phagotrophy, a condition referred to as mixotrophy. In keeping with our desire to cover the animal-like protists, we survey heterotrophic and mixotrophic flagellates from the Phytomastigophora. Pigmentation and chloroplast morphology are generally important taxonomic characters, but we emphasize features to be found in heterotrophs.

The Cryptomonadida include many common heterotrophs, autotrophs, and mixotrophs. The two flagella are unequal in length and differ in appearance (Fig. 3.7C). The longer flagella has two rows of mastigonemes while the shorter has only one. These flagella arise from a subapical invagination commonly referred to as a gullet, although it does not appear to be the site of ingestion in heterotrophic forms. The contractile vacuole empties into this vestibule. Ejectisomes (see earlier) may not be visible to the light microscopist unless they are discharged, which they sometimes do on fixation. The pellicle is covered with plates, although these also are not generally visible.

The Dinoflagellida are a very large and unique group, probably more important in marine than freshwater environments. Certainly our knowledge of this group is mostly based on marine studies. Their unique arrangement of flagella—one spiral flagellum around the cell in a groove (girdle) and a second distally directed in another groove (sulcus) —makes them distinctive. Again, heterotrophy and mixotrophy are common (Fig. 3.2). Because dinoflagellates sometimes ingest large prey, their phagotrophic habits have been more often recognized. A covering of plates may or may not be present (hence ''armored'' and ''naked'' dinoflagellates). Ingestion in some species is assisted by an appendage, the peduncle, which emerges from the sulcus and allows them to subdue prey many times their size (Gaines and Elbrächter 1987; Jacobson and Anderson 1986). The trichocysts also assist in prey capture in some forms. Very large prey may be digested by protoplasmic extensions and this tendency intergrades with ectoparasitism, which is common in this group.

Chrysomonadida also contain both colorless heterotrophs and pigmented mixotrophs. Even among the mixotrophs, the degree to which different taxa rely on phagotrophy varies greatly. Chrysomonads are generally small, and their prey are picoplankton. They have two unequal flagella, one long and directed anteriorly with rows of mastigonemes and the other short, smooth, and directed laterally (Figs. 3.4D, 3.13K–M). They are naked or covered in fine siliceous scales which are not always visible with light microscopy; many are amoeboid. Their carbohydrate storage product, chrysolaminarin, occurs in liquid globules and may be useful in recognizing members of this group.

Euglenida are generally large flagellates with two flagella, although in many taxa only one flagellum emerges from the gullet (Fig. 3.13D). Several heterotrophic species creep over the substrate with the second flagellum trailing and hidden beneath the cell (Fig. 3.13F–H), as in some bodonids. Indeed, recent molecular genetic studies indicate that the Euglenida may be related to the Kinetoplastida, and both may be rather remote from other flagellates (Gunderson *et al.* 1987). Heterotrophic species may have an ingestatory apparatus comprised of two rods, allowing them to swallow relatively large items. Again, the gullet is not used in ingestion.

E. Anatomy of Sarcodina

These are the protozoa we generally refer to as amebas, in reference to the well known *Amoeba proteus*. Locomotion is mostly by pseudopodia. Flagella are restricted to temporary swimming stages in the order Schizopyrenida, and in some terrestrial forms (e.g., Acrasida). Lobosea, Heterolobosia, and Karyoblastea mostly have blunt pseudopodia called lobopodia. Sarcodines are up to 2 mm in diameter, while the smallest reach only a few micrometers. Although many lack a fixed external morphology, the characteristic morphologies shown by the various taxa are surprisingly distinctive, even if difficult to quantify (Figs. 3.14 and 3.15). By using also the number, size and structure of organelles, and characteristics of tests (where present), identification is not as difficult for living specimens as might be imagined.

The morphology of amebas is plastic. Many adopt a stellate morphology (Fig. 3.9) if suspended in water. Few are truly planktonic, but rather they live on surfaces or in sediments. Most show an inner granular cytoplasm and an outer hyaline cytoplasm, or hyaloplasm, with a characteristic thickness and distribution around the cell. Locomotion may be achieved by extending many pseudopodia simultaneously, as in *Amoeba* (Fig. 3.15F), or by moving as a single mass on a broad front (e.g., Figs. 3.8F, 3.15E, P) or as a cylinder (limax amebas, Figs. 3.8A, 3.15I, K, L). Not only do pseudopodia have characteristic shapes, but the tail end or uroid may be distinctive (Fig. 3.15J, L) and the cell surface may be distinctly sculptured, as in *Thecamoeba* (Fig. 3.15D). The dynamics of movement also vary; some amebas move continuously and smoothly, while the cytoplasm of others seems to burst forward intermittently; this motion is called eruptive.

Members of the subclass Testacealobosia are distinguished by their external coverings, called tests (Fig. 3.16A–Q). These are generally vase-shaped,

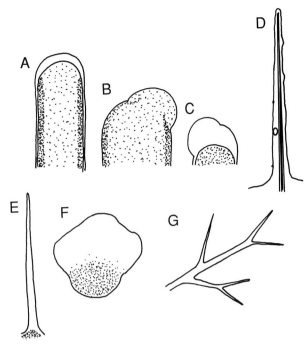

Figure 3.8 Types of pseudopodia. (A) The cylindrical pseudopodia of Amoebidae; (B) the cylindrical, granular, eruptive pseudopodia of Pelomyxidae; (C) the flattened, hyaline, eruptive pseudopodia of some Entamoebidae; (D) the axopodia of Heliozoa; (E) the filiform pseudopodia of some Mayorellidae; (F) the broad, flattened pharopodia of some Mayorellidae; (G) the filopodia of athalamids. (Modified from Bovee and Jahn 1973.)

with a single opening through which pseudopodia emerge. The test may be proteinaceous but is often covered with secreted plates or inorganic particles. Many are terrestrial; however, benthic forms are common and a few are planktonic. The latter have gas-filled vacuoles to compensate for the weight of the test.

Schizopyrenida are small amebas, many of which are capable of transforming into flagellates (Fig. 3.9C). The flagellated form may have two or four flagella. As amebas, they are small, uninucleate, and usually simple in form. The amoeboid form may be a soil-dweller, transforming into a flagellate on being suspended in water. The best known ameboflagellate of this kind is *Naegleria*. *Naegleria gruberi* has been referred to by Page (1976) as possibly the most common ameba in freshwater. At one time, *Naegleria* was best known as a model system for microtubule assembly, since it could be induced to assemble its flagellum on command in the laboratory. The similar *Naegleria fowleri* was discovered as a cause of fatal encephalitis in humans in 1958, and the genus is now better known in infamy as an unwanted resident of air conditioners and swimming pools. Pathogenic strains are able to survive at 37°C, which, fortunately, most free-living amebas cannot

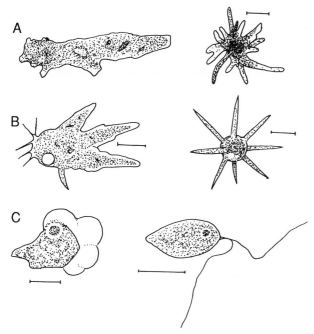

Figure 3.9 Some amebas and their "floating forms" on the right. (A) *Ameba proteus*, scale = 100 μm.; (B) *Polychaos timidum*, scale = 20 μm; (C) *Naegleria gruberi*, scale = 10 μm. (After Page 1976.)

tolerate. Increased abundance of pathogenic *N. fowleri* is a dangerous consequence of thermal pollution (Tyndall *et al.* 1989). The interesting history of this discovery, as well as other aspects of this group can be found in Carter (1968) and Singh (1975).

Filosea and Granuloreticulosida are amebas with fine, pointed pseudopodia. Filosea (Figs. 3.16.R–U, W, 3.17.A–K) have pointed filopodia (Fig. 3.8G), while Granuloreticulosida have a network of branching and anastomosing reticulopodia (3.17L–R). Most have tests. They are among the least studied amebas, except for the marine Granuloreticulosida (Foraminiferida).

Amebas feed by phagocytosis, the smaller being bacterivores and the larger being predators of algae, other protozoa, and small metazoa. Particulate food is surrounded by ectoplasm, sometimes engulfed by a pseudopod, and enveloped in a food vacuole. In contrast to most other phagotrophic protozoa, there is no fixed site of ingestion or cytostome. Testacealobosia are largely algivores, and many feed on filamentous algae, drawing in filaments through the opening in their test.

Pinocytosis or "cell drinking" can be induced in *Amoeba* by organic substances. Channels appear in the ectoplasm through which dissolved materials, concentrated on the cell surface, are imbibed. Although *Amoeba* does not have mucocysts, a mucous coat on the cell surface appears to be involved (Chapman-Andresen 1973). The significance of this

form of nutrition in the field is unknown, but one must conclude that *Amoeba* uses this ability somewhere other than on microscope slides in undergraduate laboratories.

F. Anatomy of Heliozoa

The Heliozoa are primarily freshwater counterparts of the better known marine Radiolaria and are placed with the Radiolaria in a separate phylum in some classifications. Both have unique pseudopodia, called axopodia (Fig. 3.8D). Axopodia are strengthened by a microtubular support or axoneme. The origin and ultrastructure of axonemes is diverse; indeed, the Heliozoa may well be polyphyletic. Most Heliozoa lack the skeleton that is so characteristic of Radiolaria, although some are covered in siliceous or organic scales (Fig. 3.18H, F), and some have a perforated shell or capsule (order Desmothoracida, Fig. 3.18A).

As diverse as axonemes may be, the axopodia they support are similar in function. The surface is sticky, possibly because of extrusomes on the axopodia or near their bases (Fig. 3.7A). The cytoplasm surrounding the axonemes appears to be in constant motion, streaming out and returning on opposite sides of each axopod. Food items adhere and are transported with the flow of cytoplasm until they reach the surface of the cell body and are engulfed. In this way they are functionally analogous to the microvilli of choanoflagellates. Food items may range from picoplankton to mesozooplankton in different Heliozoa. Large prey items, such as copepods in the case of *Actinosphaerium*, may be entangled in several axopodia and engulfed by pseudopods. Several individuals may participate in the capture of one prey.

Although Heliozoa are frequently planktonic, many sessile forms with stalks are known. In sessile forms, cell division is likely to be unequal, producing a dispersal stage that may be flagellated or amoeboid.

G. Anatomy of Ciliophora

Unlike many of the other protozoan groups, the phylum Ciliophora is certainly monophyletic; despite their diversity, ciliates share many features. After discussing these common features, we discuss the important differences among the subphyla and classes, emphasizing organelles involved in feeding.

Their having cilia is obvious and distinctive. The cilia may be reduced in number, especially in sessile forms, or organized into larger compound ciliary organelles such as cirri. The only large group that does not always possess cilia is the Suctoria; sessile predators whose dispersal stages are, however, ciliated. The novice should easily recognize this distinctive group by their feeding tentacles. The novice should also take care not to confuse other small, ciliated organisms with ciliates; the size of ciliates overlaps that of metazoans such as turbellaria, rotifers, and gastrotrichs.

The cortex of Ciliophora is a wonderful example of the complex cell ultrastructure to be found among protozoa. We refer the reader to Lynn (1981) and Lynn and Corliss (1990). This ultrastructure is important in the definition of higher taxa, while the arrangement of both somatic cilia and the compound ciliary organelles, including those associated with the cytostome, are important at all taxonomic levels.

The other distinctive characteristic of ciliates is their unique nuclear organization, or nuclear dualism. One or more large, amitotic, highly polyploid macronuclei are functional in RNA synthesis, while one or more smaller, diploid micronuclei function as gametic nuclei during conjugation.

The ciliates are divisible into three subphyla containing a total of seven classes (Small and Lynn 1985). The first subphylum is the Postciliodesmatophora, whose members possess a relatively simple cytostome, unsupported by the obvious microtubular or microfibrillar structures seen in other ciliates. However, compound ciliary organelles associated with the cytostome and used in feeding may be prominent in the class Spirotrichea. Spirotrichs are abundant in many freshwater habitats, from plankton (subclass Choreotrichia) to anaerobic benthos (family Metopidae), and most consume algae and other nannoplankton-sized food particles. The subclass Stichotricha contains dorsoventrally flattened crawlers, whose somatic cilia are all compound cirri. Large spirotrichs in the order Heterotricha, such as *Stentor* and *Spirostomum* (Fig. 3.22F, A) are familiar as teaching material. The other class in this subphylum, the Karyorelictea, is thought to be primitive for the phylum. Their macronuclei, rather than being polyploid as in other ciliates, are diploid and sometimes numerous. The one freshwater genus, *Loxodes* (Fig. 3.23J), is perhaps better known in its environmental physiology and field ecology than any other heterotrophic protozoan (reviewed in Finlay and Fenchel 1986, Goulder 1980).

The subphylum Rhabdophora contains classes Prostomatea and Litostomatea, which are largely predators, often of other ciliates. Prostomes generally have apical cytostomes, while many litostomes have subapical, sometimes slit-like cytostomes. The mouth is encircled by a crown of cilia from whose bases (kinetosomes) arise the "rhabdos," a cylinder

of microtubules surrounding and supporting the cytopharynx. Toxicysts are found in most species, and are used to subdue active prey. Toxicysts may be found around the cytostome, on a proboscis, on tentacles, or elsewhere on the body. A number of short, specialized kineties are often found near the anterior of the protozoan. This brosse (brush) probably assists in prey recognition.

The subphylum Cyrtophora is more diverse with respect to its feeding organelles and nutrition. The cytopharynx again is supported by a basket-like structure, the cyrtos. As in Spirotrichea, oral polykinetids are common (except in Suctoria), although polykinetids differ in composition and in appearance among classes. Cyrtophoran classes Phyllopharyngea and Nassophorea contain many species with relatively simple oral polykinetids and a prominent cyrtos; they are largely algivores. Several are capable of ingesting algal filaments that are surprisingly large relative to their own size. Also included in the Phyllopharyngea are the Suctoria, which, as previously mentioned, are sessile predators without cilia. A few are free-floating in the plankton. Rather than a single mouth, Suctoria are unusual in that most have several feeding tentacles. Prey ciliates stick to these tentacles because of the firing of haptocysts (see extrusomes, earlier), and their cytoplasm is withdrawn through the tentacles. Suctoria reproduce by unequal binary fission (budding), which yields a ciliated dispersal stage or "swarmer."

The class Nassophorea contains the peniculine ciliates, including the much studied *Paramecium*. *Paramecium*, unlike most other nassophoreans, is a bacterivore, using compound ciliary organelles associated with a buccal cavity to collect small prey. This microphagous habit is also common in the nassophorean subclass Hypotrichia. In hypotrichs, the compound ciliary organelles associated with the mouth are numerous and prominent, such that they are reminiscent of spirotrichs, particularly stichotrichs. Indeed, these groups were formerly classified together (e.g., by Corliss 1979); and, if their division into separate subphyla is valid, they represent a striking case of convergent evolution. As in stichotrichs, the somatic cilia are all compound, forming limb-like cirri.

The third class of Cyrtophora is the Oligohymenophorea. These are named for the compound ciliary organelles that are found in a buccal cavity surrounding the cytostome. The most common pattern (subclass Hymenostomatia) is three polykinetids on the left side of the buccal cavity and an undulating (paroral) membrane on the right. The net result is three brushes, the polykinetids, working against a curved wall, the undulating membrane, to deliver small particles to the cytostome. Most are effective bacterivores, although some are histophagous (consuming tissue of injured or recently killed animals) or parasitic. The large subclass Peritrichia contains mostly sessile bacterivores in which the buccal cavity is deepened as an infundibulum, and the polykinetids wind down it to the cytostome after encircling a prominent peristome. Somatic ciliature is absent in most species. Many are attached to the substrate by a contractile stalk, as in the common *Vorticella* (Fig. 3.22K). Many are colonial, and a few species are secondarily free-swimming.

H. Environmental Physiology

1. Factors Affecting the Distribution of Protozoa

Protozoa are found over an extremely broad range of environmental conditions, but individual species, especially those adapted to unusual circumstances, may have relatively narrow limits to their distribution. Several microbial communities with characteristic fauna have been described. Biological, physical, and chemical factors must interact to govern the distribution of these assemblages in a manner that defies simple description. Nonetheless, several physical–chemical factors are important to the growth and/or distribution of protozoa. Among these, temperature and oxygen are probably the most important and most well known. Light is probably underestimated in its importance to heterotrophic forms.

An extensive and still rapidly expanding literature concerns the importance of environmental contaminants. The distribution of protozoa, with respect to organic pollution and its associated environmental conditions, has been used as part of a system for classifying the state of aquatic habitats according to their fauna (Sládecek 1973). Recently, Foissner (1988) clarified the taxonomy of ciliates used in Sládecek's saprobity system and provided a summary of the saprobity values of ciliates. Similarly, Bick (1972) provided a guide to ciliates that are useful as biological indicators in freshwater, along with their ecologic distribution with respect to several parameters.

An earlier list of species occurrences versus environmental parameters was assembled by Noland (1925). His conclusion that "the wide range of tolerance that most ciliates show toward the physicochemical factors studied, together with the evident correlation of several of these factors with the food habits of the species, suggests that the nature and amount of available food has more to do with distribution of the freshwater ciliates than any other one factor" still appears to be warranted. The physical

nature of the substratum, including the biological substratum of periphyton, is also important to the community of surface-oriented protozoans that develops on it (Picken 1937, Ricci 1989). Indeed, the relative importance of biotic components of the environment to the distribution of species, versus the structural and chemical–physical components, is difficult to ascertain from survey data.

2. Temperature

As in all organisms, metabolic processes in protozoa and, ultimately, the rate of production of new protozoan biomass, are dependent on temperature. Because of the difficulty of measuring growth rates *in situ* for protozoa, several workers have investigated the relationship of temperature and growth rate in the laboratory to facilitate the extrapolation of laboratory data on growth rate to field situations (Finlay 1977, Baldock *et al.* 1980). Generally, Q_{10} values for protozoa are about 2.0, as they are for many biologic processes, but important deviations occur (Baldock and Berger 1984).

Active protozoa are found in freshwater environments at temperatures between 0° and 50°C. For example, ciliates were abundant throughout the 11-month, ice-covered season in Char Lake, 74° N (Rigler *et al.* 1974). Some protozoa can be cultured in the laboratory at or near 0°C. However, it is equivocal whether all protozoa can function at such low temperatures, or whether those that do function at low temperatures can perform any better than one would expect from extrapolating the metabolic rate versus temperature curve for species collected from more moderate temperatures. Lee and Fenchel (1972) found that *Euplotes* collected from Antarctic, temperate, and subtropical waters had similar growth–temperature relationships; but the Antarctic strains did not multiply above about 12°C, while the subtropical strain multiplied even above 35°C, but could not reproduce below 6°C.

A similar situation is to be expected with respect to high temperatures; most species continue to increase their rates of cell division as the temperature increases to 25°C or more and tolerate at least 30°C. But there are numerous records of species from hot environments, up to 50°C (Noland and Godjics 1967). Tolerance to extremes of temperature and some other variables have been induced by gradual acclimation (see Fauré-Fremiet 1967).

3. Oxygen

Protozoa are sufficiently small that organelles to increase the surface area for respiration are unnecessary. Aerobic forms appear to function at extremely low oxygen tensions and thereby avoid competition and predation from larger animals; it is in these circumstances, especially when they are generated by high densities of bacteria in response to organic enrichment, that protozoa are most abundant. Aerobic protozoa that are typically found under such conditions are called microaerophilic. The succession of species following organic enrichments and resulting changes in oxygen tension has been a focus for research in protozoan ecology.

More recently, a great deal of research has been done on the autecology of *Loxodes*, a microaerophilic ciliate living in the benthos and plankton of hypereutrophic waters where it consumes nannoplanktonic algae (Goulder 1980, Finlay and Fenchel 1986). One of the most surprising outcomes of this research is the discovery that *Loxodes* can use nitrate rather than oxygen as an electron acceptor. It undergoes diel vertical migrations between the anoxic hypolimnion and oxic hypolimnion, but cannot survive in either if constrained. It aggregates at low oxygen concentrations guided, in part, by unique organelles (Müller's vesicles) now known to be responsible for geotaxis. The direction of the geotactic response is determined by the oxygen tension, probably detected by cytochromes, and by light, detected by pigment granules (pigmentocysts) in the pellicle. Although other ciliates have been intensively studied in the laboratory, the work on *Loxodes* is unusual in the degree to which laboratory studies have been guided by field observations and suggests that autecological work on other species could be very rewarding.

Some microaerophilic ciliates contain endosymbiotic green algae (see Section III). Despite their photosynthetic symbionts, they aggregate at depths where oxygen and light are low (Finlay *et al.* 1987). At very low oxygen tensions they move into dim light, suggesting a role for the symbionts not only as producers of carbon for their hosts but as a means for maintaining a low intracellular level of oxygen.

Many free-living protozoa are tolerant of anaerobic conditions (Noland and Godjics 1967, Bick 1957). Among the ciliates, this diverse assemblage of species has been referred to as sapropelobionts (e.g., Jankowski 1964a,b), although some of the species embraced by this term are microaerophilic rather than anaerobic. Fenchel (1969) suggested that the term "sulfide ciliates" is more appropriate for the truly anaerobic fauna of reducing sediments. These are intolerant of oxygen, and those examined by Fenchel *et al.* (1977) from estuarine sediment lacked mitochondria and cytochrome oxidase. Many sulfide ciliates have epi- and endozooic bacteria, but the significance of these associations to either party is unknown. Related freshwater species are common, but have not been examined.

Sarcomastigophora that can probably exist under anaerobic conditions include the giant ameba *Pelomyxa palustris*, which has been said to consume at least some oxygen (Bovee and Jahn 1973) although it has no mitochondria (Daniels 1973). The Diplomonadida, which have two free-living genera (*Hexamita* and *Trepomonas*), also lack mitochondria. Many of the true anaerobes among the protozoa belong to taxa that include gut-dwelling endosymbionts.

4. Environmental Contaminants

The relative abundance and diversity of protozoa has been proposed as a useful indicator of organic and toxic pollution (Cairns *et al.* 1972), and the response of protozoa to contaminants has been reviewed in that light (Cairns 1974). Others have advocated the use of protozoa in single-species bioassays (Dive and Leclerc 1975). Slabbert and Morgan (1982) proposed that respiration of *Tetrahymena* cultures be used as a rapid bioassay, as respiration rate was found to respond quickly to toxicants. Responses included both decreases and transient increases in respiration. Protozoan predators and prey have been used to explore complex effects of contaminants on communities (Doucet and Maly 1990).

Among environmental contaminants, the response of protozoa to oil and other hydrocarbons is relatively well known. They are generally resistant (Rogerson and Berger 1981). Scott *et al.* (1984) found that freshwater ponds treated with oil alone or oil and oil-dispersant mixtures had similar numbers and biomass of protozoa to those of control ponds, although some taxa were reduced while others increased. The resistant ciliate *Colpidium colpoda* was found to accumulate hydrocarbons in intracellular inclusions, up to about 20% of the cell volume. The number of mucocysts increased, implicating these organelles as having a protective function (see Section II.B.4).

The toxicity of various hydrocarbons to *Tetrahymena* was similar when differences in solubility and partitioning between the water and ciliates were considered, indicating a similar mode of action once they were in the cell (Rogerson *et al.* 1983). Biomagnification of chlorinated hydrocarbons may be substantial. For the pesticide Mirex, the concentration in *Tetrahymena* was 193 times that present in the medium, and 82% of the available contaminant was in the cells (Cooley *et al.* 1972). For Arochlor 1254, the biomagnification was 60-fold. A wide variety of insecticides have been shown to cause dose-dependent inhibition of growth and biomagnification in *Tetrahymena* (Dhanaraj *et al.* 1989, Lal *et al.* 1987). Some were toxic at higher concentrations.

The toxicity of heavy metals to protozoa also has been explored in some detail (Ruthven and Cairns 1973). Intraspecific variation in heavy metal tolerance occurs and appears to be related to environmental variation in contamination (Nyberg and Bishop 1983). To some degree, tolerance to metals is specific, but is also generally greater in outbreeding species (Nyberg 1974; see Section III.B for discussion of breeding systems). Free ion concentration, rather than total amount of metal, appears to determine toxicity (Stoecker *et al.* 1986).

5. Energetics of Protozoa

Because protozoa are fast growing and many are easily cultivated in the laboratory, their energetics have been studied extensively. Ingestion, growth, respiration, egestion, and other parameters have been measured or estimated, and quotients catalogued. We deal with some of these parameters elsewhere; for example, maximum growth rates are discussed in Section II.B, field estimates of growth rate are mentioned in Section III.B, and assimilation efficiencies are referred to in Section III.C.10. Laybourn-Parry (1984) summarized much of the existing information on protozoan energetics, and Calow (1977) integrated it into his wider review of animal energetics.

Fenchel's (1974) classic paper drew attention to the relationship between body size and rate of population increase for organisms, and the more rapid growth rate of smaller compared to greater ones. His analysis also suggested a cost associated with multicellularity; both net growth efficiency (growth/ assimilation) and growth are generally greater for protozoa than for metazoa of similar size. Calow's summary of conversion efficiencies confirms that protozoa are generally efficient converters of energy, with many species showing gross growth efficiencies of 50% or more. However, efficiencies for some taxa are much lower (e.g., Scott 1985 and references therein) and it is difficult to generalize.

Given the high growth rates of protozoa, we expect that field populations of protozoa will be found to have high production to biomass (P/B) ratios relative to metazoa, on the order of one per day. However, few field estimates of growth rate (P/B for protozoa) are available. Estimates for benthos include Schönborn (1977) and Taylor (1983b). With the current interest in planktonic protozoa, methods are being developed for the estimation of population growth rate *in situ*, and rates for marine populations are appearing (e.g., Coats and Heinbokel 1982, Stoecker *et al.* 1983, Heinbokel 1988). Freshwater studies of a similar nature should be forthcoming soon. Some radiotracer studies of nutrient cycling yielded rough estimates for freshwater planktonic protozoa (e.g., Taylor 1984).

III. ECOLOGY AND EVOLUTION

A. Diversity and Distribution

The concept of species within the protozoa has always been problematic. The definition of species as interbreeding or potentially interbreeding populations is irrelevant to the many asexual forms. On the other hand, studies of the genetics of what were once believed to be species, such as *Paramecium aurelia*, *Tetrahymena pyriformis,* and *Euplotes patella*, have revealed groups of interfertile mating types (Sonneborn 1975; Corliss and Daggett 1983). These groups of mating types are called syngens; but, of course, they are sibling species. In some cases, new species names have been created to recognize them (e.g., *P. monaurelia*, *P. biaurelia*, and so on). Careful morphological analysis has uncovered differences among these species; but in practice they are not morphologically separable. They must be identified by mating reactions with known lines, or by using electrophoretic differences in some cytoplasmic enzymes. At the same time, the phenotypic plasticity of protozoa has also led to the description of invalid species names (Gates 1978). With that said, it remains that some protozoan groups have as many species as more familiar animal taxa. For example, ciliates comprise some 7000 species (Corliss 1979) of the tens of thousands of protozoan species.

At this point, it should be admitted that the taxonomy of free-living protozoa, at least as practiced by ecologists, is in a primitive state. A textbook on protozoa will advise the student to collect protozoa at the margin of a pond or in some other small and protected waters rich in organic material. Collections from these environments are likely to to yield a rich assortment of large protozoa, readily identifiable at least to genus under the dissecting or compound-phase microscope. The working freshwater ecologist will more likely be interested in lake plankton, littoral or profundal benthos, or stream aufwuchs. Samples from these situations will more likely yield minute forms too small to be seen under a dissecting microscope and too few to be examined under a compound microscope without concentration. They may not be figured in the textbooks. Many will be undescribed species, or species descriptions will either be in specialist journals not in most libraries, or so outdated as to be of dubious value. Techniques of sample preservation, concentration, and enumeration (see Section IV) if they exist for a particular biotope, may not always allow for identification. Largely because of these problems, most ecologists who include protozoa in their studies of aquatic habitats do not identify protozoa, even if they do count them and measure them for biomass estimates.

The distribution of most free-living protozoa is probably broad. Certainly, some morphologic species are worldwide in distribution and are essentially ubiquitous (Cairns and Ruthven 1972). Protozoa quickly colonize new environments, as was shown by Maguire's study of new volcanic islands (Maguire 1977). On the other hand, sibling species of the *Paramecium aurelia* complex have limited distributions on a world-wide scale (Sonneborn 1975) even though, as a species complex, they are ubiquitous.

Certainly, some species appear to have restricted distributions. For example, Lake Baikal in Siberia contains several planktonic protozoa that are probably endemic (Dogiel 1965). Even a protozoan taxonomist expects to find new species when collecting in an unsurveyed or novel habitat.

B. Reproduction and Life History

Most protozoa reproduce by equal binary fission and therefore their life histories are simple. Multiple fission is more common among endosymbionts but does also occur in free-living forms. Among free-living protozoa, multiple fission appears to be related to intense temporal heterogeneity in the environment. For example, it is found in some terrestrial forms, which are active only for brief periods when their environment is moist, or in histophages whose food resources are very patchy and ephemeral (Fig. 3.10). Therefore, some analogy can be drawn between the significance of multiple fission for some free-living protozoa and some endosymbionts using the common theme of dispersal (Taylor 1981).

A phenomenon perhaps related to multiple fission in that it is related to dispersal is unequal fission or budding. Binary fission is only approximately equal, and differences in the partitioning of cytoplasm occur and are adjusted in the next cell cycle. But consistent asymmetry of daughter cells is widely distributed among sessile forms; the larger daughter cell remains attached, while the smaller is motile. This pattern undoubtedly has evolved independently a number of times. It is perhaps a reflection of the probability of survival of the two daughter cells not being equal and resources being partitioned according to the odds. The smaller daughter cell is usually a propagule, capable of establishing itself in a new location. In some cases (e.g., *Zoothamnium*, Fig. 3.20H, I), it may function as a gamete.

Generation times are highly variable among pro-

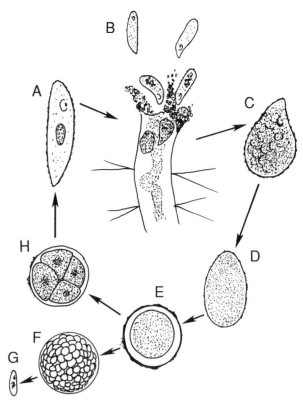

Figure 3.10 Life cycle of the histophagous ciliate *Ophryoglena*. Change from the theront (hunting morphology) (A) to the trophont (feeding) morphology (B, C) occurs when food is found. The prototomont stage (D) seeks a suitable place to settle, and leads to the cyst or tomont stage (E, H). The tomont usually produces 4–8 new theronts, or may produce a much larger number of microtheronts (F, G). The function of microtheronts is unknown. Theronts may encyst if they do not find food, and re-emerge as small secondary theronts. (Modified from Canella and Rocchi-Canella 1976.)

tozoa, but the maximum rates of population increase attainable when food is not a constraint generally increase with temperature and decrease with body size. A multiple regression equation summarizing these relationships for ciliates has been produced and used to estimate potential production of field populations (Montagnes *et al.* 1988). However, even for protozoa of the same size growing at the same temperature, maximum rates of cell division vary widely among taxa. Some of this variation is correlated with genome size and may reflect a causative relationship.

Among most eukaryotic cells with diploid, mitotic nuclei, there is a relationship among DNA content, cell size and, negatively, fission rate. It appears that among such cells, a large cell must have a proportionately large DNA content and nucleus, and a slow division rate. This relationship holds among most plant and animal cells and among most protozoa. However, it has also been established in plants that polyploidy allows a large DNA content and large cell size, but without a concomitant decrease in fission rate. Fission rate is apparently related to the size of the genome, not the number of copies.

The diverse nuclear arrangements of protozoa can be viewed as various solutions to the genome size–cell size–fission rate constraint (Cavalier-Smith 1980). Multinucleate cells, nuclear dualism, amitotic nuclei, and colony formation may represent ways to get larger without slowing growth. For example, in ciliates, cell size is correlated with macronuclear DNA content, but cells divide as frequently as one would expect from the size of their micronucleus (Shuter *et al.* 1983). In this context, both multinuclearity and multicellularity can be seen as means of avoiding constraints imposed by the basic eukaryotic plan.

Encystment is a feature of many protozoan life cycles, especially those that live in environments that dry out (Corliss and Esser 1974). Many protozoa also produce cysts at times when environmental conditions are unsuitable for other reasons, such as lack of food. Encystment under these conditions can be considered as a means of coping with temporal heterogeneity. Still other protozoa incorporate encystment into the life cycle following feeding (the "digestive cyst") or during cell division (the "reproductive cyst"). Digestive cysts are common in ciliates that rapidly consume large amounts of food, possibly after a long period of searching, e.g., histophages and predators. Some species enter a cyst after feeding and emerge after cell division, combining these functions. The encysted stage may reduce energy expenditure and/or the risk of predation when the acquisition of food is either not possible or unnecessary. Nonencysting protozoa may persist days or weeks in the absence of food; this survival time is not closely related to cell size, is density-dependent, and in ciliates may reflect differences in energy costs associated with locomotory behavior (Jackson and Berger 1985). Survival time was negatively correlated with the intrinsic rate of population increase (see Section III.C.2) among eleven species, but there were exceptions to this trend.

Sex is widespread, but far from universal among protozoa. For example, it is common among the Ciliophora, is found in some Dinoflagellida, is limited to perhaps a few genera of Heliozoa, is absent in most Zoomastigophora, and is apparently absent among the Gymnamoebia. Grell (1973) and Margulis and Sagan (1986) discuss the occurrence and variety of sexual phenomena among the protists. Grell organizes sexual phenomena into three basic types: ga-

metogamy, in which gametes fuse as free-swimming cells; autogamy, where gametes or gametic nuclei of the same individual fuse; and gamontogamy, where mates (gamonts) unite to exchange gametes or gametic nuclei. Gametogamy is found in Phytomastigophora where, typically, isogametes are produced by multiple fission. However, anisogamy also occurs. Autogamy is present in Heliozoa and some Ciliophora. Gamontogamy is the usual form of sex in Ciliophora.

As in many small metazoa, sexuality interrupts asexual reproduction and is associated with environmental change. It should be emphasized that sexuality is not necessarily associated with reproduction in protozoa and is best understood as a separate phenomenon (Margulis and Sagan 1986). Indeed, sex usually marks the end of the existence of an individual protist; it becomes the gamete or gametes and its genome ceases to exist. We believe that the restricted occurrence of sex among protozoa and the separation of sex from reproduction in many taxa make protozoa a promising group for further study concerning the evolution and significance of sex.

The relationship between sexual and asexual reproduction has been carefully examined in some Ciliophora, especially *Paramecium* (see Wichterman 1986 for a review of this topic). Exchange of haploid nuclei between two conjugants (gamontogamy) leads to two individuals with new and different genotypes. Each is the founding member of a new clone, expanding by asexual binary fission. Individuals of a new clone will undergo a period of "sexual immaturity" when sex does not occur. The duration of this immature period varies among species; species with short immature periods are considered to be "inbreeders," as mating with close relatives is favored, whereas species with long immature periods are considered to be "outbreeders."

The immature period is followed by a period of sexual maturity when conjugation will occur under the following circumstances: (1) a partner of a complementary mating type is available; (2) both are at the right stage of the cell cycle; and (3) both are slightly starved. If conjugation does not occur before the end of the period of maturity, a period of sexual senescence follows. The rate of binary fission declines; cells are less likely to survive conjugation if a mate is found; and autogamy (self-fertilization) may occur. Although autogamy may prolong vigor, the clones ultimately die out. An analogy can be drawn between the life of a clone of ciliates and the development of a metazoan from a zygote—a large population of cells develops asexually from the single cell resulting from sexual fusion; but, ultimately, the size of that population is limited by senescence, which must be circumvented by another sexual episode.

As some ciliates do not undergo conjugation and others vary greatly in the frequency necessary to prevent senescence (a strain of *Tetrahymena* "*pyriformis*" without a micronucleus and, therefore, not capable of sex was isolated in 1923 and is still going strong) one is tempted to speculate on the function of both sex and senescence. Is sex required to postpone senescence? Or is senescence a mechanism to eliminate micronuclear or gametic genes that have not been expressed for many generations? Or is the need for sex a function of the unique amitosis of the ciliate macronucleus? The finite life span of animal cells in tissue culture versus the immortality of cultures of cancer cells begs similar questions. That autogamy in ciliates is a stop-gap measure to postpone senescence contrasts with sex in the heliozoan *Actinophrys sol*, where fusion of the products of meiosis within the same cell is the normal pattern. Bell (1988) recently reviewed this fascinating subject and concluded that senescence of both sexual and asexual protists can be viewed as a common phenomenon, explicable in terms of the genetic deterioration of finite populations.

Although sex is a necessary event in the life cycle of some ciliates, it is a dangerous undertaking; genetic load is high in many ciliates, as is the mortality following conjugation. As noted previously, some species avoid close matings by having a long immaturity period. As if conjugation were not dangerous enough, prokaryotic endosymbionts render some ciliates "mate-killers;" their partners in conjugation pass on their genes but are lethally affected (see Section III.C.4).

C. Ecological Interactions

1. Intraspecific Interactions Among Protozoa

Protozoa have been used extensively in the study of competition and predator–prey interactions. Their obvious advantages for this type of study include small size, short generation time, and for some species, ease of maintenance in the laboratory. These advantages do not carry over to field experiments and little of the work has been done *in situ*.

Intraspecific competition in protozoa is evident in the dynamics of batch cultures, where a few individuals are introduced into a large, but finite environment. If inoculated as starved individuals, some ciliates will respond to this situation by growing rapidly and increasing in size past the size at which the cell might be expected to divide. This has been interpreted as a strategy to sequester the maximum amount of food should the supply become ex-

hausted. However, if food is still available, these cells enter into a phase of exponential growth where they divide at a similar size during each cycle. When the food supply is depleted, some individuals will continue to divide at the expense of cell size. Again, this can be interpreted as an adaptation for an r-selected or colonizing species to produce as many propagules as possible (Taylor 1981, Fenchel 1987).

Competition between individuals as outlined above is purely through exploitation of common resources. However, the exhaustion of food in some species induces cannibalism, sometimes with abrupt changes in morphology for the cannibals (see Giese 1973 for a thorough discussion of this phenomenon in the ciliate *Blepharisma*). Another form of interference is the "killer" phenomenon. Some protozoa harbor prokaryotic endosymbionts that render them lethal to conspecifics (mate-killers were mentioned previously). On a more subtle level, Curds and Cockburn (1971) found that in order to model the population dynamics of *Tetrahymena* in batch cultures it was necessary to allow for negative density dependence or interference in their equations. This was not necessary for continuous cultures.

2. Interspecific Competition Among Protozoa

Interspecific competition in laboratory cultures of ciliates has been studied by Gause (1934) and by numerous authors since this work. Gause provided clear evidence that environmental factors could influence the outcome of competition.

The Lotka-Volterra equations predict that the winner of exploitative competition for resources in stable environments will be the species with the greater K or carrying capacity, that is, the more efficient user of the resource. However, K is defined as numbers, not biomass, so smaller species will tend to have a higher K. In his studies of competitive interactions of bacterivorous ciliates, Luckinbill (1979) concluded that K did not predict the outcome of competition. But, in accordance with part of the r/K selection hypothesis, the intrinsic rate of increase, r, was negatively related to competitive ability. [The r/K selection hypothesis holds that r, which measures fitness in density-independent or stochastic environments, is negatively correlated among species with K, which measures fitness in stable environments.] Thus the theorem seems to hold for ciliates, except that K, being highly correlated to body size, does not measure competitive ability in stable environments. As noted earlier, r is also correlated with body size.

To factor out this correlation, Taylor (1978b) expressed the intrinsic rate of increase as r/r_e, where r_e is the expected r for a ciliate, based on a regression of *r* on body volume. Therefore, r/r_e is a measure of how fast-growing or r-selected a ciliate is for its body size. This measure was found to be negatively correlated with frequency of occurrence in the field and positively correlated with macronuclear ploidy (Taylor and Shuter 1981). This research suggests that protozoan communities contain fast-growing but ephemeral opportunists as well as more conservative, specialist species that have more modest growth rates but stable populations. A similar picture emerges from a comparison of the population dynamics of a relatively fast-growing species and a relatively slow-growing species in the laboratory (Luckinbill and Fenton 1978).

Field studies of exploitative competition in protozoa are few, as are field measurements of growth rates for comparison to laboratory estimates under conditions of excess food. Accordingly, we know little about the relative importance of density dependence or independence. However, production estimates are often based on the extrapolation of laboratory growth rates under conditions of excess food to field situations.

Maguire (1963a,b) found strong evidence for the exclusion of colonizing species in the genus *Colpoda* by *Paramecium* and associated species. Hairston and Kellerman (1965) and Hairston (1967) found that one member of the *Paramecium aurelia* species complex dominated sympatric sibling species in competition experiments and predominated in field collections. The field population that they studied appeared to be food limited. In contrast, Taylor (1983b) found that sessile ciliates on an artificial substrate in a stream had high growth and mortality rates, implying little competition. The extensive studies on colonization by Cairns and co-workers (reviewed in Cairns and Yongue 1977) provided indirect evidence for species interactions, including competition.

3. Predatory Interactions Among Protozoa

Many protozoa are predators of other protozoa. For example, Chardez (1985) listed a diverse assortment of protozoa consuming testaceans. Protozoa have been used extensively in laboratory studies of predator–prey interactions, including such topics as selectivity, stability of predator–prey interactions, and the behavioral and energetic determinants of functional and numerical response. Much of this research has been done with *Didinium nasutum*. Hewett (1980a,b) and Berger (1979) showed that *Didinium* changes its body size, feeding behavior, and feeding and growth rate when fed different prey and may encounter some difficulty handling large prey after adjusting to small ones. The extrusomes, which enable *Didinium* and other predatory ciliates, dino-

flagellates, and Heliozoa to handle their prey were mentioned in Section II. Although extrusomes are important weapons of many other ciliates, especially in the Litostomatea and Suctoria, other predators are simply swallowers, including the giant cannibal forms of *Blepharisma* and *Tetrahymena*, large ciliates such as *Bursaria*, and some other predators such as *Peranema* among the Euglenida and the large amebas.

Recent interest in the "microbial loop" (the route by which picoplankton production reaches larger consumers) has focused attention on the contribution of protozoa, as consumers of picoplankton, to production at higher trophic levels (see Section III.C.9). Although some crustaceans can efficiently ingest protozoa and even reproduce on a diet of heterotrophic flagellates (Sanders and Porter 1990), a microbial food chain leading from picoplankton to fish via protozoa and crustacean zooplankton could be inefficient because of the number of steps involved. This is especially true if predatory protozoa process some of that production. Unfortunately, nothing is known about the quantitative significance of predatory protozoa in planktonic food webs except that known predators of other protozoa, such as *Dileptus* and *Actinosphaerium*, are commonly observed in the plankton. The feeding biology of many planktonic rhabdophorines, which are even more common, is not known. Because this predation within the protozoan microplankton is an unknown loss in the transfer of energy to higher trophic levels, and therefore bears on the current debate concerning the importance of the microbial loop, it is a hiatus in our knowledge of microbial food webs which should be filled.

4. Symbionts of Protozoa

Symbionts of protozoa include both ecto- and endosymbiotic prokaryotes and eukaryotes, including a wide variety of parasites (Curds 1977, Jeon 1983, Lee and Corliss 1985). The most obvious and widespread eukaryote symbionts of freshwater protozoa are algae of the genus *Chlorella*, often called zoochlorellae. These are most frequently studied in the ciliate *Paramecium bursaria*, but are also known from a wide variety of other ciliates as well as flagellates, amebas, and many metazoans. In some cases, the symbiosis appears to benefit the protozoan because of the photosynthates released by the algae; *Paramecium bursaria*, for example, probably receives maltose from its symbionts and can grow better at low food concentrations when they are present.

Although some algae-bearing protozoa display positive phototaxis and aggregate at the surface of the water (Berninger *et al.* 1986, Taylor and Heynen

1987), others aggregate deeper, in dim light. Some of these microaerophilic forms are negatively phototactic unless the medium is anaerobic (Finlay *et al.* 1987), and zoochlorellae may provide low internal oxygen tensions in anaerobic external environments. Therefore, there may be two functional types of protozoan–algal symbioses; one in which the primary benefit for the host is to receive photosynthate, characterized by positive phototaxis, and one in which the primary benefit for the host is oxygen production, characterized by oxygen-dependent phototaxis.

Some marine choreotrichs (e.g., *Strombidium* spp.) harbor isolated chloroplasts which remain functional (Stoecker *et al.* 1989). The chloroplasts may be from more than one species of alga. In some cases, whole cells may be retained as symbionts. The host ciliate becomes functionally mixotrophic, able to fix significant amounts of carbon. The freshwater species, *Strombidium viride*, also appears to sequester chloroplasts, perhaps of diatoms (Rogerson *et al.* 1989).

Prokaryotic endosymbionts are widely distributed in protozoa. Some are infectious parasites that may kill their hosts, while others are necessary for the survival of their hosts. Perhaps the most interesting from an ecological point of view are the intensively studied "killer" particles of ciliates. Some transform their hosts into "killers" that cause the death of conspecific "sensitives" simply by their proximity. Others transform their hosts into "mate-killers," whose partners in conjugation are lethally affected. The significance of these endosymbionts to field populations of *Paramecium* has been investigated by Landis (1981). He concluded that the infection must be maintained by natural selection, because killers rarely have a chance to mate and transmit their infection to new hosts. Mating and cell division are the only known means of transmission.

5. Epizooic and Epiphytic Protozoa

Many freshwater protozoa have developed epizooic associations with animals. Some associations are likely to be rather nonspecific, involving species that can also be found attached to or crawling on nonliving substrata as part of the aufwuchs community. Other associations (for example, peritrichs of the order Mobilida that are epizooic on fish) are perhaps better left to the parasitologists. However, some cases deserve brief mention here because they are common and likely to be noticed.

One common association is between planktonic, colonial cyanobacteria and peritrich ciliates (Pratt and Rosen 1983, Kerr 1983). Peritrichs are frequently so abundant as to rival the biomass of their

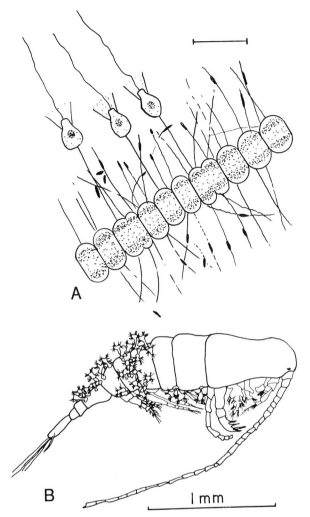

Figure 3.11 (A) Choanoflagellates epiphytic on filamentous cyanobacteria from Georgian Bay, Lake Huron. Scale bar = 10 μm. (B) *Tokophrya quadripartita* (Suctoria) epizooic on *Limnocalanus* from Lake Michigan. (From Evans *et al.* 1979.)

host colony and likely affect its nutrient environment and motility. Smaller peritrichs, choanoflagellates, and bicoecids are common epizooics on diatoms and filamentous cyanobacteria (Fig. 3.11.A).

Peritrichs and Suctoria colonize the exoskeletons of planktonic crustacea. Kankaala and Eloranta (1987) found that the clearance rates of *Vorticella* epizooic on *Daphnia* can rival those of their host, suggesting that competition for food with *Vorticella* could be significant to the *Daphnia*. Epizooic suctoria (*Tokophrya* spp., Fig. 3.11.B) on mysid and calanoid crustacea in the Great Lakes may be heavy enough to impair their hosts (Evans *et al.* 1981) and may also compete with them for food. Henebry and Ridgeway (1980) suggested that epizooic protozoa on crustacea may be useful pollution indicators.

Benthic invertebrates may also be infested with peritrichs (see Baldock 1986 for a brief summary).

The biomass of these may be a significant portion of the protozoan biomass in streams.

6. Protozoa as Bacterivores

Protozoa are most widely known as bacterivores. Bacterivorous protozoa are abundant in activated sludge sewage treatment plants and, indeed, are necessary for their proper functioning (Curds and Cockburn 1970). At the opposite extreme, there is a great deal of evidence that in plankton communities, especially oligotrophic ones, heterotrophic flagellates are the dominant bacterivores and much of the carbon flow passes through them (e.g., Sanders *et al.* 1989, Sherr *et al.* 1989). Although ciliates may be abundant bacterivores in enriched environments (hence the old name "infusorians" or animals from organic infusions), they are thought to be relatively unimportant in oligotrophic ones. The bacterivorous ciliates that have been studied in the laboratory cannot grow at the bacterial concentrations observed in oligotrophic plankton systems (Fenchel 1980b). However, recent field measurements suggest that this may not be true of ciliates isolated from those environments (Sherr and Sherr 1987); the relative importance of flagellates and ciliates as bacterivores in different habitats is, therefore, an open question. The question should probably be phrased in terms of the size distribution of grazers of picoplankton and the answer probably involves species interactions and food web structure as well as trophic status.

The small Heliozoa are an overlooked group of largely planktonic bacterivores, and small amebas are bacterivorous also. Many of the Phytomastigophora are phagotrophic, and it has recently come to light that they may play a very significant role as bacterivores in freshwater plankton (Bird and Kalff 1986, Sanders and Porter 1988). Light, temperature, and food levels may determine the relative importance of phagotrophy versus photosynthesis in these organisms (Bird and Kalff 1986, Sanders *et al.* 1989). Many protozoa glean bacteria from surfaces or attached to particles (Albright *et al.* 1987, Caron 1987), but the quantitative role of bacterivorous protozoa in benthic habitats is relatively unknown. Pioneering studies in this area include Baldock and Sleigh (1988) and Bott and Kaplan (1989).

Small protozoa or "heterotrophic nannoplankton" (flagellates, amebas, and perhaps some small ciliates) capture and consume bacteria individually, with rates of ingestion of tens to hundreds per hour (e.g., Fenchel 1982b). Most ciliates, in contrast, use their complex oral ciliature to collect hundreds into each food vacuole, producing feeding rates of up to thousands per hour (e.g., Taylor 1986). Although the former are likely to be selective in their feeding, the larger ciliates are probably very limited in their

ability to discriminate. Currently, the use of fluorescent beads to measure bacterivory (McManus and Fuhrman 1986) has resulted in renewed interest in the ability of protozoa to discriminate against inorganic particles in favor of bacteria (Nygaard *et al.* 1988) and in the influence of chemical coatings and surface charge (Sanders 1988). Both ciliate and flagellate bacterivores discriminate among bacteria on the basis of size (Fenchel 1980b, Andersson *et al.* 1986, Sanders 1988). Several studies have documented that ciliates differ in their abilities to use different bacteria as food; some may be unable to consume filamentous bacteria, while some bacteria, especially pigmented ones, produce toxins that render them unsuitable to support the growth of some ciliates (Dive 1973, Taylor and Berger 1976).

A poorly understood and usually ignored facet of bacterivory in protozoa is the role of extracellular products. The presence of some ciliates leads to the flocculation of bacteria (Hardin 1943, Curds 1963, Nilsson 1976). Mucous-like extracellular materials are involved in uptake of dissolved material and possibly small particles by pinocytosis in *Amoeba*. Small amebas secrete various hydrolases into their medium (Schuster 1979), as do flagellates. Porter (1988) suggested that flocculation and external digestion may be important in bacterivorous flagellates. The involvement of mucocysts or other extrusomes is likely but not confirmed.

As one might expect, the feeding rate (functional response) and growth rate (numerical response) of bacterivores increases with bacterial density and the nature of these response curves has been studied because of their relevance to interspecific competition and the control of bacterial numbers. In addition to the expected hyperbolic functions, thresholds or sigmoid responses have also been observed and attributed to changes from feeding to searching behavior at low particle concentrations (e.g., Taylor 1978a, Sanders 1988, Sanders *et al.* 1990). Accordingly, low concentrations of bacteria (10^4–10^7 per ml, usually 10^5–10^6) can persist even in stationary phase cultures of bacterivorous protozoa. The behavioral and physiological mechanisms responsible for this phenomenon are still in doubt (Sambanis and Fredrickson 1988); however, it is probably relevant to the densities of picoplankton that we observe in field situations.

7. Algivorous Protozoa

Larger protozoa are less likely to be bacterivores, and most consume algae or other protozoa. Suspension-feeding ciliates each ingest a distinct size spectrum of particles, related to the form and function of the oral apparatus (Fenchel 1980a), and this spectrum generally shifts in larger ciliates towards larger particles (i.e., algae and other protozoa). Larger flagellates, amebas, and Heliozoa also consume other algae and protozoa.

Protozoa feeding on larger items are more likely to be selective, as the rate at which particles are consumed is much reduced. For example, Rapport *et al.* (1972) documented selective feeding by the heterotrich ciliate *Stentor* consuming other protozoa at up to about five prey per minute. Stentors consistently biased their feeding towards *Tetrahymena* relative to three phytomastigophorans when mixed versus single species feeding suspensions were compared. The behavioral response of algivorous choreotrich ciliates appears to involve both mechanical and chemical stimuli and suggests that although ciliates may ingest inert particles, they may discriminate among them as well as among food organisms (Buskey and Stoecker 1989). Selective feeding has been noted for *Amoeba* (Mast and Hahnert 1935) and *Thecamoeba* (Page 1977).

Although most phagotrophic flagellates are bacterivores, larger species may be effective algivores. Dinoflagellates, for example, may consume diatoms and other large phytoplankton (see Section II.B.1). Zoomastigophora are probably underrated as consumers of algae (Canter 1973, Goldman and Caron 1985, Suttle *et al.* 1986). The colorless chrysomonad *Paraphysomonas* (Fig. 3.2D) was observed to feed selectively on diatoms, phagocytosing cells that were nearly their own volume (Grover 1989, Goldman and Dennet 1990). The phagotrophic flagellate *Peranema* (Fig. 3.13H) feeds on related autotrophic euglenids.

The consumption of filamentous algae (including cyanobacteria) would appear to be mechanically difficult for protozoa, but many have specialized to this role. Large amebas such as *Pelomyxa* include filamentous algae in their omnivorous diets and some Testacealobosia, such as *Arcella* and *Difflugia* are readily cultured on filamentous algae, which they engulf through the aperture of their tests. The nassophorine ciliates contain many forms that feed on filamentous algae, including *Nassula*, *Pseudomicrothorax,* and *Frontonia*. *Nassula* spp.(Fig. 3.12) appear to limit their diet to *Oscillatoria*. Detailed ultrastructural studies have revealed the role of the

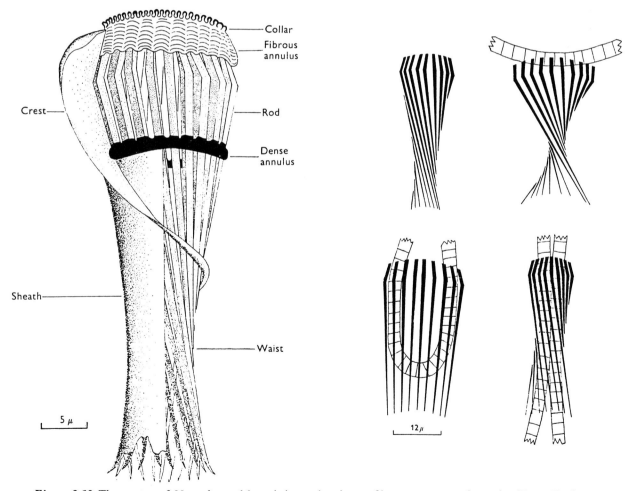

Figure 3.12 The cyrtos of *Nassula*, and how it is used to ingest filamentous cyanobacteria. (From Tucker 1968.)

cyrtos in feeding, especially how it is used to break the algal filament so that ingestion can begin (Tucker 1968, Peck 1985).

Abundant epibenthic diatoms, especially noncolonial and unattached pennate diatoms such as *Navicula*, usually attract abundant herbivorous ciliates and amebas. Many dorsoventrally flattened and thigmotactic ciliates in the order Cyrtophorida and class Nassophorea are diatom feeders.

8. Histophagous Ciliates

Several genera and species of ciliates feed on metazoan tissue. Some, such as *Coleps hirtus* and *Prorodon* spp., are omnivores and only opportunistic histophages. Strict histophages, feeding only on fresh tissue, include all hymenostomes of the genus *Ophryoglena* (Fig. 3.10) and some *Tetrahymena* spp. (e.g., Corliss 1973, Lynn 1975). *Deltopylum* is also a histophage; but in our experience it is relatively rare. *Philaster* appears to be marine only. These forms feature complex life cycles with reproductive cysts and multiple fission, which are probably adaptations to an extremely patchy food resource (see Section III.A). Like other carrion feeders, they increase the efficiency of food webs by bypassing the detrital pathway. *Ophryoglena* and the related fish parasite *Ichthyophthirius* share the unique organelle of Lieberkühn (Canella and Rocchi-Canella 1976), or watchglass organelle, whose function is unknown but probably related to their feeding mode.

Related hymenostomes are insect parasites; *Lambornella* is a parasite of the cuticle of mosquito larvae (Corliss and Coats 1976); it and some *Tetrahymena* species may have significant impacts on aquatic Diptera (Golini and Corliss 1981, Egerter *et al.* 1986). *Espejoia* feeds on the egg masses of aquatic insects.

9. Metazoan Predators of Protozoa

Protozoa, as quantitatively important components of plankton, aufwuchs, or sediment communities, probably contribute accordingly to the nutrition of many animal feeding guilds, such as filter feeders, scrapers, and deposit feeders. However, because most protozoa are fragile and without structures that resist digestion, they are seldom recorded among the gut contents of animals despite their importance. Among benthic animals, minute predators such as Naididae (Oligochaeta) and Orthocladinae (Diptera, Chironomidae) might consume protozoa as a large fraction of their diet (Taylor 1980, 1983b and references therein).

Slightly more is known about the feeding of planktonic animals on protozoa (Sorokin and Pavel-jeva 1972, Porter *et al.* 1985), reflecting the recent interest in the microbial loop (see Section III.C.3) and how picoplankton production reaches higher trophic levels (e.g., Pomeroy 1974, Sherr *et al.* 1987). Cyclopoda especially are known to be consumers of protozoa and recent marine studies suggest that copepods may actively select protozoa from among similarly sized plankton (see Wiadnyana and Rassoulzadegan 1989 and references therein). Cladocera can ingest ciliates as well as both photosynthetic and nonphotosynthetic flagellates (Porter *et al.* 1979, Sanders and Porter 1988, 1990). There is recent evidence, however, that ciliates may be less suitable food for Cladocera (DeBiase *et al.* 1990.) Many planktonic protozoa are capable of rapid or saltatory movements that might reduce their probability of being captured, although these are not entirely effective against some predators (Archbold and Berger 1985).

Fish would seem to be unlikely predators of protozoa, but protozoa may be important food for some larval fishes (Kornienko 1971, Kentouri and Divanach 1986).

10. Protozoa and Nutrient Cycling

Regeneration of the inorganic constituents of living organisms occurs through a process involving ingestion, digestion, and egestion by consumers, catabolism and autolysis of all organisms, and the enzyme-mediated breakdown of nonliving organic matter by decomposers. Autotrophs are net consumers of these inorganic constituents, whereas phagotrophs regenerate them. Decomposers, the heterotrophic bacteria and fungi, regenerate carbon; but they may or may not be net producers of other inorganic constituents, depending on the composition of the organic matter they decompose. If it is poor in other constituents such as organic nitrogen and phosphorus (as when it is mostly carbohydrate), then decomposers will compete with autotrophs for these other constituents. In the final analysis, the incorporation and regeneration of inorganic constituents will be in balance unless organic matter is stored or exported.

Heterotrophic protozoa fit into this picture as both typical consumers, eating other organisms and releasing unassimilated organic material and end products of catabolism, and as major consumers of the decomposers. The latter role is manifested in the ability of bacterivorous protozoa to accelerate the breakdown of organic material by bacteria (Stout 1980). This property of protozoa may be due to their keeping the bacteria in an actively growing state at moderate abundances, rather than directly due to their contribution to the remineralization process.

Protozoa also contribute directly to mineralization, and some early studies suggested protozoa were particularly active in releasing inorganic nutrients. However, the basis of this view has been challenged (Taylor 1986). Indeed, the relatively independent literature on energetics of protozoa (see Section II) suggested that they are, if anything, more rather than less efficient in converting the food they ingest into cytoplasm. Recent studies on nutrient recycling by flagellates suggest that mineralization of nitrogen (N) and phosphorus (P) by protozoa may be particularly low when it is most needed, that is, when their food organisms are themselves N or P limited (Goldman *et al.* 1985, Caron *et al.* 1988). *In situ* measurements of nutrient release relative to ingestion are few, but release of P for planktonic ciliates appears to be close to the 50% or so we expect, based on conversion efficiencies (Taylor 1984). Protozoa probably make an important contribution to the regeneration of nutrients in many environments, but no more so than one would expect from their contribution to consumption and production.

D. Evolutionary Relationships

The relationship of protozoa to plants and animals, as well as the evolutionary relationships among the protozoa, are controversial areas of current research. New tools for the study of phylogeny, developing from molecular biology, are contradicting morphology-based classifications, but the number of organisms examined is still small. Here, we briefly introduce some of the major theories and controversies.

One change in view which has occurred over recent years is that the major dichotomy among living things is not between plants and animals, but between prokaryotes and eukaryotes. The evolution of eukaryotes from the prokaryotes is perhaps the major event in organic evolution on earth and may have occurred by essentially Lamarckian means; the inheritance of acquired characteristics (Taylor 1987). That is, the eukaryotic cell represents the symbiotic amalgamation of several distantly related prokaryotes (Margulis 1981). Although this concept is controversial, the controversy is largely one of degree in that the symbiotic origin of some organelles, such as mitochondria and chloroplasts, is widely accepted.

Within the unicellular eukaryotes, or Protista, the most primitive forms are unanimously believed to be among the Sarcomastigophora. Margulis and Schwartz (1988) concluded that the giant amoeba *Pelomyxa*, which lacks mitochondria and has amitotic nuclear fission, is a primitive species. Many consider the flagellum to be as old as the eukaryotic cell,

but disagree as to whether colorless flagellates are primitive or derived from pigmented forms. Recent analyses of ribosomal RNA strongly support the former argument; the colorless (and parasitic) flagellate *Giardia*, which also lacks mitochondria, is clearly very close to the prokaryotes relative to other protozoa, strongly suggesting that its features are primitive rather than derived (Sogin *et al.* 1989). The same analysis reveals that the volvocid *Chlamydomonas* is more closely aligned with higher plants than other protozoa, including other phytoastigophorans.

Although our view of the relationship between unicellular eukaryotes is in a state of flux, the molecular techniques currently being applied reveal that these groups have had long and independent evolutionary histories. For example, the genetic distances within ciliates or phytoastigophorans are greater than the distances between volvocids and higher plants, or between plants and animals (Sogin and Elwood 1986, Gunderson *et al.* 1987).

There are two major schools of thought with respect to the relationship of protozoa and metazoa (Hanson 1977). One considers that the protozoan ancestor of metazoa is a colonial flagellate, probably a choanoflagellate, and that Porifera, Placozoa, Cnidaria, some gastrotrichs, and other animals with monociliated epidermal cells are among the most primitive animals. Others believe that primitive animals were derived by the compartmentalization of a ciliated, multinucleate protozoan. The early development of animals best supports the former hypothesis, as do the data collected thus far on ribosomal RNA; the protozoa least distant from animals are flagellates, not ciliates, which seem to have had a very long independent history.

Perhaps the most important conclusions an ecologist can draw from the dynamic field of protist systematics is that although the protozoa are all unicellular or colonial and comprise one to many phyla in different classification schemes, they are more evolutionarily distant among themselves than multicellular animals, and at least as morphologically and physiologically diverse.

IV. COLLECTING, REARING, AND PREPARATION FOR IDENTIFICATION

The small size, delicacy, and lack of hard parts in most protozoa makes the quantitative collection and enumeration of protozoa particularly difficult. The methods for sample collection and examination vary with the group of interest and the habitat and there is room for improvement in almost all of the following procedures. General techniques for the beginner are provided in Lee *et al.* (1985) and Finlay *et al.* (1988).

For some groups, especially amebas and heterotrophic flagellates when identification is required or when the sample contains sediment or some other material, examination of live organisms in fresh collections under a dissecting and/or compound microscope is the only method available. With care, this can be done in a quantitative manner (e.g., Bott and Kaplan 1989). The "most probable number" method of enumeration, based on inoculating culture medium with small aliquots of sample and scoring the fraction of such inocula producing cultures, is common in the study of soil protozoa and has been applied to freshwater. However, the resulting estimates sometimes have been very low relative to direct counts (Bott and Kaplan 1989). Given the difficulty in culturing many freshwater protozoa, this method should be reserved for estimating species known to be able to grow from small inocula on the medium used. Extraction procedures have been devised for benthic protozoa but are not widely used (Hairston and Kellerman 1964; Fenchel 1967). Fixed samples are not useful in benthic studies, as the task of enumerating fixed but motionless protozoa in sediment is virtually impossible.

Studies on epibenthic or aufwuchs protozoa have generally resorted to artificial substrata. Those that allow for microscopic examination are particularly useful; glass slides (Schönborn 1977), petri dishes (Spoon and Burbank 1967; Taylor 1983b), and strips of nylon mesh (Taylor 1983a) have all proven useful. Polyurethane foam has been extensively used as a substrate in studying protozoan community interactions and response to pollution (Cairns *et al.* 1979).

Planktonic samples present fewer problems. Large protozoa can be settled from samples and counted by inverted microscopy as for phytoplankton. Useful fixatives are Lugol's Iodine (1% volume/volume, e.g., Taylor and Heynen 1987) or mercuric chloride (5% saturated mercuric chloride plus a drop of 0.04% bromophenol blue, Pace and Orcutt 1981). These procedures do not lend themselves to the identification of ciliates, nor to the detection of chromatophores in small flagellates. Filtration methods may aid with both of these problems. The quantitative protargol or QPS method of Montagnes and Lynn (1987a,b) produces permanent, quantitative, stained preparations for identification of ciliates and flagellates, although it requires that samples be fixed in a concentrated Bouin's fixative. The use of epifluorescence microscopy on filtered samples (Caron 1983) can facilitate the separation of heterotrophic and phototrophic protists and is generally the method of choice for enumerating very small planktonic protozoa, even though it does not lend itself to identification.

A culture of most protozoa is best begun by maintaining samples collected in the field. Samples kept in the laboratory will usually undergo succession, with different species waxing and waning in abundance. Screening out metazoan predators such as copepods may encourage protozoa to increase in number. Supplementing the organic or nutrient content of the sample may help. Cereal grains (rice or wheat, boiled for quicker effect or plain), broth from boiled vegetable matter (hay, lettuce) or nutrient broth are commonly used. Remember that adding too much is easy and results in foul conditions that may encourage different species from those important in the original sample; or one may simply kill all the aerobic protozoa. Generally, additions of organic matter to cultures should be no more than about 0.1% weight to volume for whole vegetable matter and no more than 10% as much (100 mg/liter) for more labile material.

The first step in isolating species of interest is to prepare some protozoan-free medium by filtering or boiling water from the site where the protozoans were collected, and inoculating it with individuals of the desired species. Mineral water is frequently used for culturing heterotrophic forms. Again, rice or wheat grains or organic infusions are useful in promoting the moderate growth of bacteria for bacterivorous species. Algae or other protozoans may be required as food for others. White worms (enchytreids) are convenient food for histophagous forms.

Ultimately, you may wish to establish monoxenic (the organism of interest and its prey only) or axenic cultures, possibly cloned from a single individual. Artificial freshwater media are described in Finlay *et al.* (1988) and Thompson *et al.* (1988). Although involving marine protozoa, the recent efforts by Gifford (1985) might prove a useful template for attempts to culture difficult freshwater forms. A manual of elementary techniques has recently been prepared by Finlay *et al.* (1988).

V. IDENTIFICATION OF PROTOZOA

A. Notes about the Key

It is often difficult and time consuming to identify protozoa to the level of species. In many cases, unambiguous identification requires specialized staining techniques or the use of electron microscopy. The taxonomic grouping used in this key is an amalgamation of recent publications by specialists on the different groups (Lee *et al.* 1985, Page 1987, 1988). As such, the taxonomy is based on ultrastructural features and morphometric attributes that are not included in the key. This key is designed to make possible the identification of many protozoa to a family or genus level using light microscopy alone. Although observation of living organisms is important for identification, the key should still be useful for many fixed samples. Most of the specialized terms used in the key are defined in the general glossary. If one becomes more interested in identifying protozoa to a species level, specialized techniques and keys in the publications cited earlier can be used.

The genera listed in the key are representative of the families and not all inclusive. Likewise, the illustrations used as examples here are of one or more species considered typical of a genus, but do not necessarily represent all of the species in the genus.

Advances in the systematics of the protozoa are continuous. Due to the heterogeneity of the group, it is often difficult to remain abreast of the changes. For example, the ciliate classification of Small and Lynn (1985) was followed for this key, but has been significantly changed by Lynn and Corliss (1990). Classification below the ordinal level in the subphylum Mastigophora and the class Heliozoa of the subphylum Sarcodina are not conclusively agreed upon or are under revision. Thus, the use of familial names was avoided for most orders of these groups in the key.

Orders of Phytomastigophora marked with an asterisk (*) contain many or mostly photosynthetic members; genera listed have one or more species that lack chloroplasts and are thus heterotrophic, or contain chloroplasts but also exhibit heterotrophic nutrition.

B. Taxonomic Key to the Phyla or Subphyla of Protozoa

1a.	Single type of nucleus; flagella, pseudopodia, or both types of locomotor organelles; sexuality, if present, syngamy .... phylum Sarcomastigophora	2
1b.	Two types of nuclei (macronucleus and micronucleus), with rare exceptions; simple cilia or ciliary organelles typical in at least one part of life cycle, with subpellicular infraciliature present even when cilia are not; transverse binary fission, but budding and multiple fission also occur; sexuality involves conjugation, autogamy, and autogamy-in-pairs; typically with contractile vacuole .... phylum Ciliophora (see Section V.E)	
2a(1a).	Commonly 1–4 flagella in trophozoites, 16–24 in one free-living genus (*Paramastix*); binary fission .... subphylum Mastigophora (see Section V.C)	
2b.	Pseudopodia, or amoeboid movement (protoplasmic flow) without discrete pseudopodia; flagella, if present, usually restricted to temporary stage; cell naked or with external or internal test or skeleton; asexual reproduction by fission, sexual reproduction rare or absent in most groups, if present usually associated with flagellated gamete .... subphylum Sarcodina (see Section V.D)	

C. Taxonomic Key to Classes and Orders in the Subphylum Mastigophora

1a.	With median groove (annulus and sulcus); two flagella, one extending transversely around the cell .... order Dinoflagellida*	
	(*Peridinium, Gymnodinium, Gyrodinium,* Fig. 3.13A–C)	
1b.	Without a median groove	2
2a(1b).	Cells with pharyngeal rods (sometimes difficult to see) or containing the reserve material paramylon; one or usually two flagella arise from within an anterior invagination (reservoir); contractile vacuole associated with reservoir; elongate; body shape often plastic when living .... order Euglenida*	
	(*Khawkinea, Entosiphon, Petalomonas, Peranema, Urceolus,* Figs. 3.13D, F–I)	
2b.	Cells not as above	3
3a(2b).	Cells with two or four equal flagella .... order Volvocida*	
	(*Polytoma, Polytomella,* Fig. 3.13E)	
3b.	Cells with two unequal flagella	4

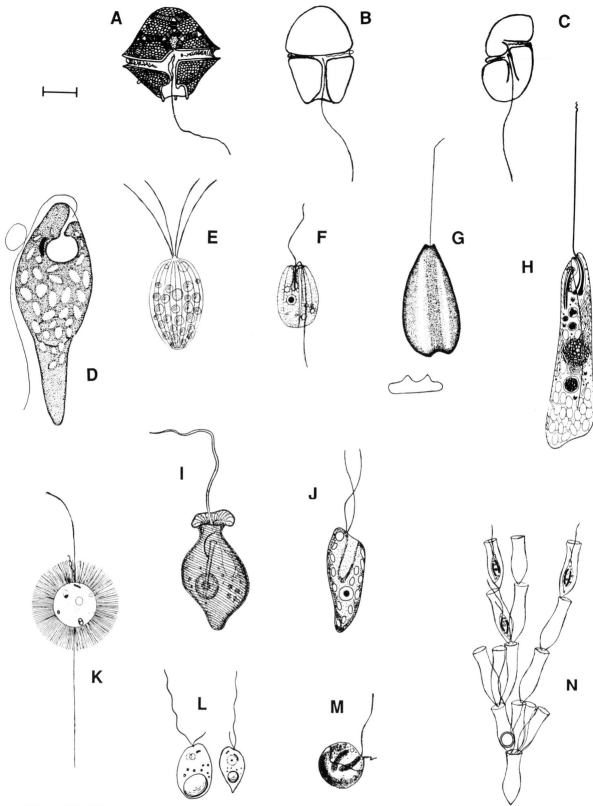

Figure 3.13 (A) *Peridinium*; (B) *Gymnodinium*; (C) *Gyrodinium*; (D) *Khawkinea halli*; (E) *Polytomella citri*; (F) *Entosiphon sulcatum*; (G) *Petalomonas abcissa*; (H) *Peranema trichophorum*; (I) *Urceolus cyclostomas*; (J) *Chilomonas paramecium*; (K) *Paraphysomonas vestita*; (L) *Spumella* (= *Monas*) *vivapara*, two cell shapes; (M) *Ochromonas variabilis*[a]; (N) *Dinobryon sertularia*[a]. (After: Bourrelly 1968, L; Calaway and Lackey 1962, E, F, J, N; Eddy 1930, A; Jahn and McKibben 1937, D; Leedale 1985, H; Pascher 1913, M; Lemmermann 1914, K; Shawhan and Jahn 1947, G; Smith 1950, I.) Scale = 5 μm for E; 10 μm for A, B, C, G, I, J, K, L, M; 20 μm for D, F, H, N. [a]Mixotrophic.

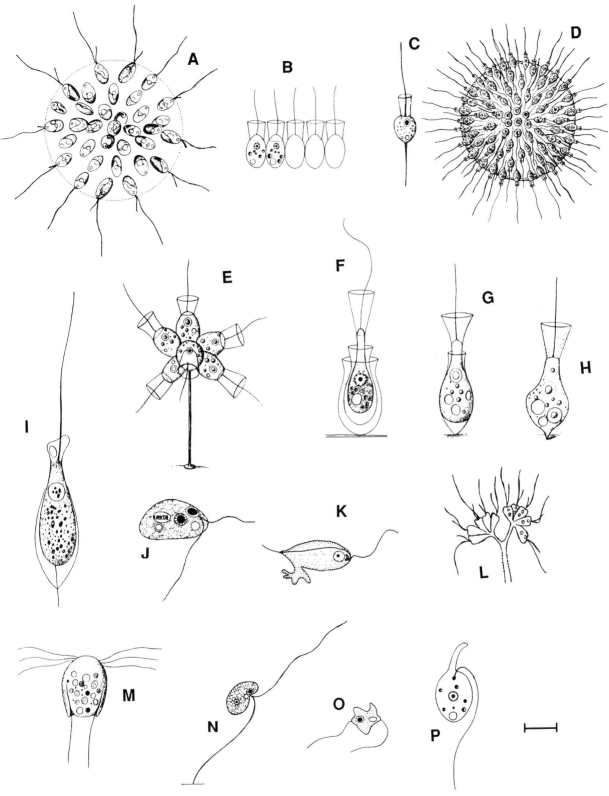

Figure 3.14 (A) *Uroglena americana*[a]; (B) *Desmarella moniliformis*; (C,D) *Sphaeroeca volvox*, individual cell and colony; (E) *Codosiga botrys*; (F) *Diploeca placita*; (G) *Salpingoeca fusiformis*; (H) *Monosiga ovata*; (I) *Bicoeca lacustris*; (J) *Bodo caudatus*; (K) *Cercomonas* sp.; (L) *Cephalothamnion cyclopum*; (M) *Hexamita inflata*; (N,O) *Pleuromonas jaculans*, attached and amoeboflagellate forms; (P) *Rhynchomonas nasuta*. (After: Bourrelly 1968; L; Calaway and Lackey 1962, N, O, P; Lackey 1959, B, F; Lee 1985, K; Pascher 1913, A; Pascher 1914, C, D, E, G, H, I, J, M.) Scale = 2.5 μm for P; 5 μm for F, G, H, I, K, L; 10 μm for A, B, C, J, M, N, O; and 20 μm for E. [a]Mixotrophic.

4a(3b).   Small cells with a deep, subapical gullet; two nearly equal length flagella arise from gullet .... order Cryptomonadida*
           (*Chilomonas*, Fig. 3.13J)

4b.   Cells without deep gullet 5

5a(4b).   Typically with two unequal flagella; long flagellum usually directed forward during swimming, with short flagellum directed backward if emergent; compact Golgi body frequently visible anterior to nucleus of larger cells; chrysolaminarin vesicle(s) often fill posterior of cell; contractile vacuole usually present in extreme anterior end of cell; many species capable of simultaneous photosynthesis and phagocytosis .... order Chrysomonadida*
           (*Dinobryon, Monas, Ochromonas, Paraphysomonas, Uroglena*, Figs. 3.7, 3.13K–N, 3.14A)

5b.   Cell not as above .... class Zoomastigophora   6

6a(5b).   Single anterior flagellum encircled laterally by a tentacular, funnel-shaped collar; solitary or colonial; with or without theca .... order Choanoflagellida
           (*Codosiga, Desmarella, Sphaeroeca, Diploeca, Monosiga, Salpingoeca*, Figs. 3.4, 3.14B–H)

6b.   Without collar, or with indistinct collar 7

7a(6b).   Like Choanoflagellida, but with second trailing flagellum attached to base of lorica; protoplasmic collar rudimentary .... bicosoecid flagellates, affinities uncertain
           (*Bicoeca*, Fig. 3.14I)

7b.   Without protoplasmic collar 8

8a(7b).   Amoeboflagellate with two unequal length flagella, one trailing; ingestion by pseudopodia .... order Cercomonadida
           (*Cercomonas*, Fig. 3.14K)

8b.   Not as above 9

9a(8b).   With one or usually two flagella arising from a flagellar pocket (depression); characteristically elongate or bean-shaped ...... order Kinetoplastida, family Bodonidae
           (*Bodo, Cephalothamnion, Pleuromonas, Rhynchomonas*, Fig. 3.14J, L, N–P)

9b.   Cells with one or two karyomastigonts, each with 1–4 flagella; no Golgi apparatus; when two karyomastigonts, mirrored symmetry of nuclei and flagella .... order Diplomonadida
           (*Hexamita*, Fig. 3.14M)

D. Taxonomic Key to Orders and Families in the Subphylum Sarcodina

1a.   Trophic stages usually amoeboid, often with transitory flagellate stage; no flagellate stage known in some genera; usually cylindrical, monopodial, and eruptive; usually uninucleate .... class Heterolobosea   2

1b.   Pseudopodia lobose to filiform and produced from a broader, usually hyaline lobopodium; typically uninucleate; flagellate stages unknown .... class Lobosea   4

1c.   Cylindrical or ovoid multinucleate amoeba; bacterial symbionts; usually containing mineral particles; nonmotile flagellar-like extensions; oxygen-poor habitats .... class Karyoblastea, order Pelobiontida, family Pelomyxidae
           (*Pelomyxa*, Fig. 3.16V)

1d.   Hyaline, filiform pseudopodia, often branching, not produced from lobopodia; no known flagellate stages or spores .... class Filosea   21

1e.   Delicate pseudopodia with fine, granular appearance; pseudopodia often branching and anastomosing to form complex reticulum .... class Granuloreticulosea   27

1f.   Long slender axopods and short pseudopodia radiating from cell; with or without skeletal elements .... class Heliozoa   32

2a(1a).   Class Heterolobosea; No fruiting bodies formed .... order Schizopyreniida   3

2b.   Fruiting bodies .... order Acrasida
           [Typically occur as fruiting bodies on bark or other dead plant parts,

not free in running or standing water, thus not considered further in this key.]

3a(2a).　　Amoeboid form cylindrical, monopodial, eruptive; temporary flagellate stages common; nucleolus divides to form polar masses in mitosis .... family Vahlkampfiidae
　　　　　　(*Vahlkampfia, Naegleria*, Fig. 3.15A–B)

3b.　　　Nucleolus disintegrates during mitosis, nuclear membrane does not; single known freshwater species; commonly flattened; no flagellate stage known .... family Gruberellidae
　　　　　　(*Stachyamoeba*, Fig. 3.15C)

4a(1b).　Class Lobosea; lacking external test .... subclass Gymnamoebia　　5

4b.　　　Incompletely enclosed in a test or other flexible cuticle of microscales .... subclass Testacealobosia　　13

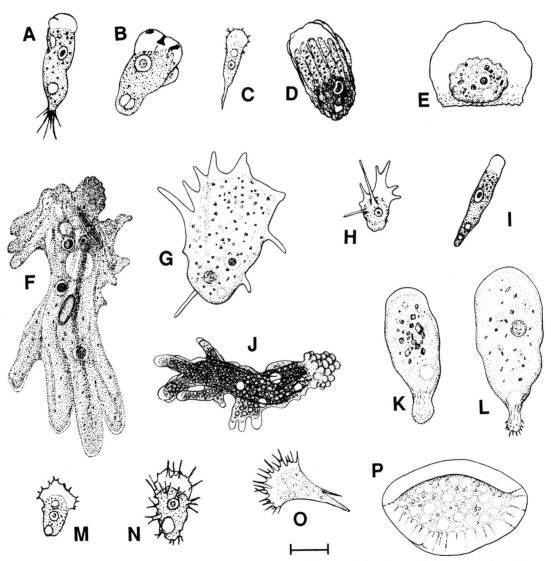

Figure 3.15 (A) *Vahlkampfia avaria*; (B) *Naegleria*; (C) *Stachyamoeba lipophora*; (D) *Thecamoeba sphaeronucleolus*; (E) *Vanella miroides*; (F) *Amoeba proteus*; (G) *Mayorella bigemma*; (H) *Vexillifera telemathalassa*; (I) *Hartmanella vermiformis*; (J) *Chaos illinoisense*; (K) *Saccamoeba lucens*; (L) *Trichamoeba cloaca*; (M) *Echinamoeba exudans*; (N) *Acanthamoeba*; (O) *Filamoeba nolandi*; (P) *Hylodiscus rubicundus*. (After: Bovee 1985, A, B, C, D, H, I, J, M, N, O; Kudo 1966, F; Page 1988, E, G, K, L, P.) Scale = 10 μm for A, B, C, E, I, M; 15 μm for H, N, O, P; 30 μm for D, G, K, L; 50 μm for F; 100 μm for J.

5a(4a).   Cylindrical or flattened; flattened forms with regular outline; no trailing
           uroidal filaments, with rare exceptions; not strikingly eruptive . . . . order Euamoebida   6
5b.   Usually flattened; frequent changes in shape typical; sometimes
           eruptive; subpseudopodia usually present, often furcate   11

6a(5a).   Without subpseudopodia   7
6b.   With subpseudopodia   9

7a(6a).   Cell body flattened   8
7b.   Cell body subcylindrical   10

8a(7a).   Cell usually oblong; cresent-shaped hyaline margin at anterior end;
           pellicle-like layer, with dorsum often wrinkled and/or ridged; usually
           uninucleate; no cytoplasmic crystals . . . . family Thecamoebidae
              (*Thecamoeba*, Fig. 3.15D)
8b.   Body usually fan-shaped, oval, or spoon-shaped, with hyaline margin
           occupying up to half of length . . . . family Vannellidae
              (*Vannella*, Fig. 3.15E)

9a(6b).   Subpseudopodia hyaline, blunt, digitiform, usually from anterior hyaline
           margin; uninucleate; nucleolar material in central body . . . . family Paramoebidae
              (*Mayorella*, Fig. 3.15G)
9b.   Few slender, conical, or linear subpseudopodia, from anterior hyaline
           margin or cell surface; uninucleate . . . . family Vexilliferidae
              (*Vexillifera*, Fig. 3.15H)

10a(7b).   Most species polypodial; length usually more than 75 μm; uni- or
           multinucleate; numerous cytoplasmic crystals . . . . family Amoebidae
              (*Amoeba, Chaos, Trichamoeba*, Fig. 3.15F, J, L)
10b.   Cell monopodial, pseudopods rare; uninucleate with central nucleolus;
           cytoplasmic crystals in some; cysts common . . . . family Hartmannellidae
              (*Hartmannella, Saccamoeba*, Fig. 3.15I, K)

11a(5b).   Cell regularly discoid, flattened ovoid, or fan-shaped; usually broader
           than wide; postcentral granular mass, usually surrounded, sometimes
           completely, by hyaline border with short subpseudopodia . . . . family Hyalodiscidae
              (*Hyalodiscus*, Fig. 3.15P)
11b.   Not as above   12

12a(11b). Cell flattened, broad, and irregular in outline, though sometimes
           elongate during locomotion; slender, tapering subpseudopodia,
           sometimes furcate, produced from broad, hyaline lobopodium; often
           with small lipid globules; uninucleate . . . . order Acanthopodida, family Acanthamoebidae
              (*Acanthamoeba*, Fig. 3.15N)
12b.   Several to many fine, sometimes furcate subpseudopodia, finer than in
           Acanthamoebidae . . . . order (*incertae sedis*) family Echinamoebidae
              (*Echinamoeba, Filamoeba*, Fig. 3.15M, O)

13a(4b).   Test more or less rigid with distinct aperture . . . . order Arcellinida   14
13b.   Discoid or sometimes globose amoeba incompletely closed in a flexible
           tectum, no well-defined aperture . . . . order Himatismenida, family Cochliopodiidae
              (*Cochliopodium*, Fig. 3.16A)

14a(13a). Pseudopods conical, clear, sometimes anastomosing . . . . suborder Phryganellina,
                                                                                                  family Phryganellidae
              (*Phryganella*, Fig. 3.16B–C)
14b.   Pseudopods digitate and finely granular   15

15a(14b). Test membranous or chitinoid, pliable or rigid; no plates or scales, but
           may have attached debris . . . . suborder Arcellina   16
15b.   Test chitinoid or not, rigid, with embedded and/or attached plates,
           scales, siliceous granules . . . . suborder Difflugina   18

16a(15a). Test flexible to semirigid; aperture ventral with variable shape . . . . family Microcoryciidae
              (*Penardochlamys*, Fig. 3.16H–I)
16b.   Test rigid, chitinoid   17

17a(16b). Test often areolar, smooth; aperture ventral, round . . . . family Arcellidae
              (*Arcella, Pyxidicula*, Fig. 3.16D, F–G)

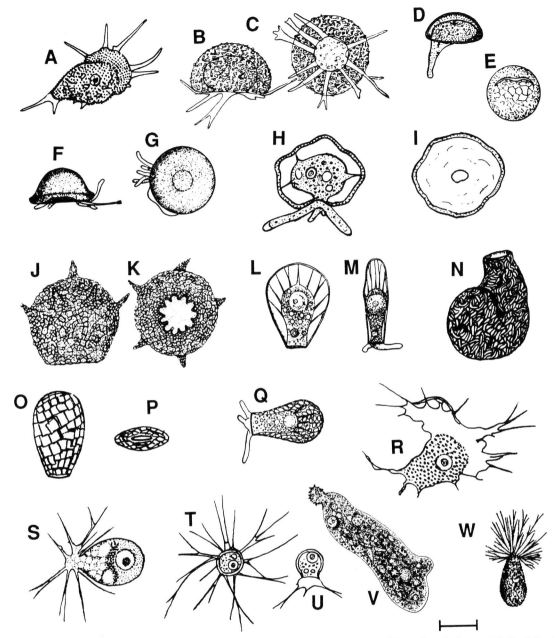

Figure 3.16 (A) *Cochliopodium bilimbosum*; (B, C) *Phryganella nidulus*; side and oral views; (D) *Pyxidicula operculata*; (E) *Plagiopyxis callida*; (F,G) *Arcella vulgaris*, side and dorsal views; (H, I) *Penardochlamys arcelloides*, side and oral views; (J, K) *Difflugia corona*, side and oral views; (L, M) *Hyalosphenia cuneata*; (N) *Lesquereusia spiralis*; (O, P) *Quadrulella symmetrica*; (Q) *Nebela collaris*; (R) *Penardia granulosa*; (S) *Chlamydophrys minor*; (T, U) *Lecythium hyalinum*, dorsal and side; (V) *Pelomyxa palustrus*; (W) *Pseudo-difflugia gracilis*; (After: Bovee 1985, A, B, C, H, I, J, K, N, O, P, R, T, U; Deflandre 1959, D, E, F, G, L, M, Q, S, W; Kudo 1966, V.) Scale = 10 μm for C, R, S; 25 μm for A, H; 30 μm for L, T, W; 45 μm for N, O; 60 μm for Q; 90 μm for B, E, F, J; 500 μm for V.

17b.	Test oval to flask-shaped, nonareolar, clear; aperture terminal .... family Hyalospheniidae
	(*Hyalosphenia*, Fig. 3.16L–M)
18a(15b).	Aperture round, broadly oval or wavy .... family Difflugiidae
	(*Difflugia*, *Lesquereusia*, Fig. 3.16J–K, N)
18b.	Aperture slit-like or narrowly oval 19
19a(18b).	Aperture terminal 20

19b.	Aperture anteroventral, invaginated, slit-like with overhanging lip .... family Plagiopyxidae
	(*Plagiopyxis*, Fig. 3.16E)
20a(19a).	Test particles rectangular .... family Paraquadrulidae
	(*Quadrulella*, Fig. 3.16O–P)
20b.	Test particles not rectangular .... family Nebelidae
	(*Nebela*, Fig. 3.16Q)
21a(1d).	Class Filosca, without distinct test, sometimes with scales .... order Aconchulinida 22
21b.	With test .... order Gromiida 23
22a(21a).	Filopodia nonanastomosing and more or less radiate .... family Vampyrellidae
	(*Vampyrella*, Fig. 3.17A)

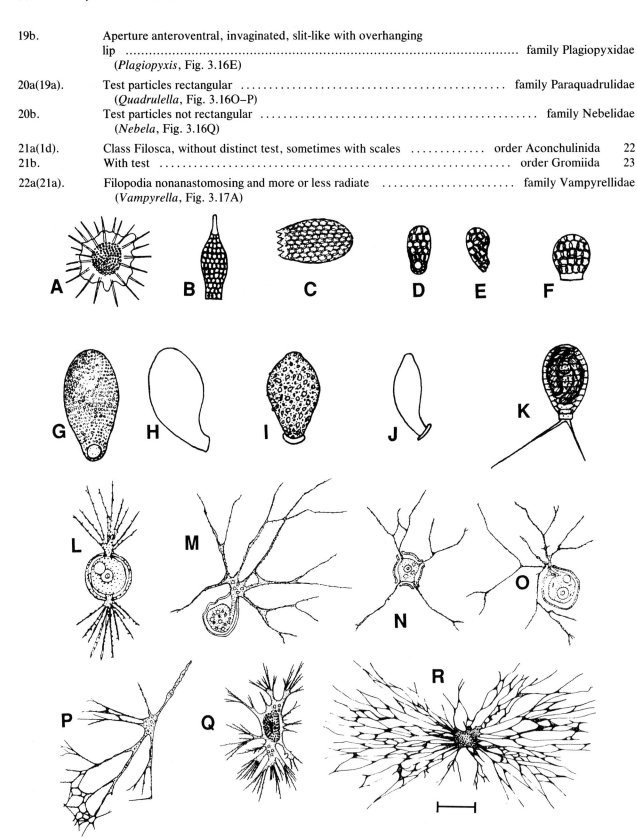

Figure 3.17 (A) *Vampyrella lateritia*; (B) *Paraeuglypha reticulata*; (C) *Euglypha tuberculata*; (D, E) *Trinema enchelys*, oral and side views; (F) *Sphenoderia lenta*; (G, H) *Cyphoderia ampulla*; (I, J) *Campascus triqueter*; (K) *Paulinella chromataphora*; (L) *Diplophyrys archeri*; (M) *Lieberkuehnia wagnerella*; (N) *Micrometes paludosa*; (O) *Microgromia haeckeliana*; (P) *Biomyxa vagans*; (Q) *Chlamydomyxa montana*; (R) *Reticulomyxa filosa*. (After: Bovee 1985, B, D, E, F, G, H, I, J, K, L, M, N, O, P, Q, R; Deflandre 1959, A, C.) Scale = 10 μm for L, N, O; 15 μm for K; 25 μm for A, B, C; 40 μm for D, F, G; 50 μm for I, M, Q; 80 μm for P; 10,000 μm for R.

22b.   Filopodia distally branching and anastomosing .... family Reticulosidae
            (*Penardia*, Fig. 3.16R)

23a(21b).   Test without scales; may have spines, attached debris .... suborder Gromiina   24
23b.   Test with secreted, siliceous scales arranged in definitive
            patterns .... suborder Euglyphina   25

24a(23a).   Test round, thin; may have spines or spicules .... family Chlamydophryidae
            (*Chlamydophrys, Lecythium*, Fig. 3.16S–U)
24b.   Test not round .... family Pseudodifflugiidae
            (*Pseudodifflugia*, Fig. 3.16W)

25a(23b).   Scales round to elliptical, thin, overlapping, adjacent or
            scattered .... family Euglyphidae
            (*Paraeuglypha, Euglypha, Trinema, Sphenoderia*, Fig. 3.17B–F)
25b.   Scales not as above 26

26a(25b).   Scales thick, cylindrical; test usually with aperture at end of a neck bent
            to one side .... family Cyphoderiidae
            (*Cyphoderia, Campascus*, Fig. 3.17G–I)
26b.   Scales long, with long axes perpendicular to pseudostome; with two
            symbiotic cyanobacterial filaments .... family Paulinellidae
            (*Paulinella*, Fig. 3.17K)

27a(1e).   Class Granuloreticulosea; with a chambered test .... order Monothalamida   28
27b.   Without test 30

28a(27a).   With one pseudostome 29
28b.   With more than one pseudostome .... family Amphitremidae
            (*Diplophyrys, Microcometes*, Fig. 3.17L, N)

29a(28a).   Test flattened on one side .... family Microgromiidae
            (*Microgromia*, Fig. 3.17O)
29b.   Test not flattened on one side .... family Lieberkuehnidae
            (*Lieberkuehnia*, Fig. 3.17M)

30a(27b).   Naked amoebae with or without anastomosing pseudopodia; one or a
            few nuclei .... order Athalamida   31
30b.   Naked, multinucleate and highly reticulate plasmodia .... order Promycetozoida,
                                                                                  family Reticulomyxidae

            (*Reticulomyxa*, Fig. 3.17R)

31a(30a).   Body mass changeable in form .... family Biomyxidae
            (*Biomyxa*, Fig. 3.17P)
31b.   Body mass more or less round and central .... family Chlamydomyxidae
            (*Chlamydomyxa*, Fig. 3.17Q)

32a(1f).   Class Heliozoa[1]; axopodia without visible axonemes 33
32b.   Axonemes usually visible with light microscopy 34

33a(32a).   Cell enclosed in latticed organic capsule (skeleton); generally stalked; no
            centroplast; cell body spherical in adults .... order Desmothoracida
                (representative genus: *Clathrulina*, Fig.3.18A)
33b.   Lack centroplast and apparently lack axonemes, although lack of
            microtubules (axonemes) in axopods has not been confirmed and this
            order may be abandoned .... order Rotosphaerida
                (representative genus: *Lithocolla*, Fig. 3.18G)

34a(32b).   Stalked or not; long, thin granule-studded axopods that usually arise
            from centroplast; axonemes frequently discernible with light
            microscopy; usually with skeleton of siliceous or organic plates and/or
            spicules; highly contractile axopods or stalk .... order Centrohelida
                (representative genera: *Heterophrys, Acanthocystis, Raphidiophrys*,
                Figs. 3.7, 3.18.D, F, H)
34b.   No skeleton 35

[1] Ordinal, familial and generic distinctions of the Heliozoa are in a state of flux due to the advance of electron microscopic studies. In lieu of a key based on ultrastructural criteria (e.g., Febvre-Chevalier 1985), descriptions of characters visible with light microscopy are given for the currently recognized orders (Febvre-Chevalier 1985).

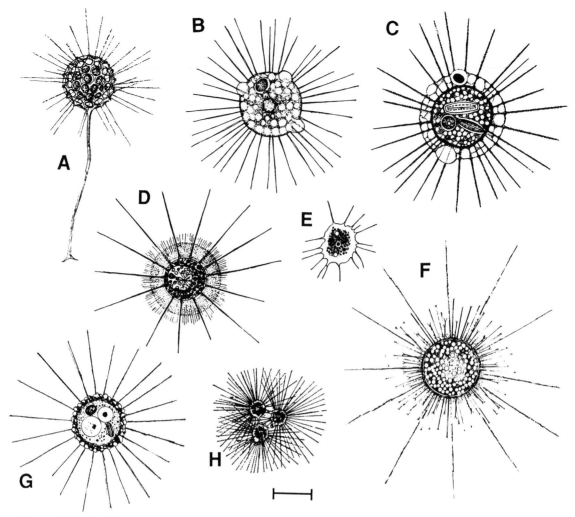

Figure 3.18 (A) *Clathrulina elegans*; (B) *Actinophrys sol*; (C) *Actinosphaerium eichhorni*; (D) *Heterophrys myriopoda*; (E) *Ciliophrys infusionum*; (F) *Acanthocystis turfacea*; (G) *Lithocolla globosa*; (H) *Raphidiophrys elegans*. (After: Deflandre 1959, H; Kudo 1966, B, C, D, E; Rainer 1968, A, F, G.) Scale = 15 μm for E; 30 μm for B, D, G; 50 μm for A; 75 μm for F, H; 160 μm for C.

35a(34b).	No centroplast or axoplast; axopods granule-studded and thicker at bases; large central nucleus surrounded by lacunar ectoplasm or several nuclei at periphery of central area with vesicular ectoplasm .... order Actinophryida
	(representative genera: *Actinophrys*, *Actinosphaerium*, Fig. 3.18B–C)
35b.	Microtubule organizing center of dense plaques from the nuclear membrane or from centroplast; may be confused with Actiniphryida without knowledge of fine structure (origin and pattern of axopod microtubules) .... order Ciliophyrida
	(representative genus: *Ciliophrys*, Fig. 3.18E)

E. Taxonomic Key to Orders and Families in the Phylum Ciliophora

1a.	Suctorial tentacles present; no true cytostome or cytopharynx; adults (trophonts) usually sessile, many species ectosymbiotic, some planktonic species; cilia absent except during free-swimming juvenile (larval) stages; with or without lorica .... subclass Suctoria 9
	[Note: Patterns of division and release of the larvae form the basis for dividing the subclass into orders. Other characters are useful in identification, but knowledge of the full life cycle is often required before suctorians can be confidently assigned to a genus.]

1b.	Cilia present in unencysted stages; no suctorial tentacles	2
2a(1b).	Conspicuous buccal ciliature at apical pole; buccal ciliature winds clockwise toward the center when viewed from oral end; somatic ciliature reduced or absent; mobile or sessile; solitary, gregarious, or colonial; some species loricate; oral region can contract and withdraw in most species .... class Oligohymenophora, subclass Peritrichia	17
2b.	Not as above	3
3a(2b).	Oral area usually bordered by a well-developed adoral zone of membranelles (AZM) consisting of more than three membranelles (polykineties); with or without ventral cirri	26
3b.	Oral area not as above; no ventral cirri	4
4a(3b).	Cytostome at or near surface; buccal cavity, if present, without cilia or paroral membranes	5
4b.	Cytostome at end of a buccal cavity with cilia or paroral membrane(s) associated	6
5a(4a).	Cytostome at or near anterior end, or continuing down side as a slit	52
5b.	Cytostome lateral or ventral	7
6a(4b).	Paroral membrane(s) within or leading into buccal cavity	65
	(Sometimes difficult to see; follow both alternatives in key when in doubt.)	
6b.	No paroral membrane; simple cilia in buccal cavity	82
7a(5b).	Cytostome a barely visible, lateral slit on the convex side of a tapering front end, or a lateral opening at the base of an anterior proboscis	63
	(See also Spathidiidae 55)	
7b.	Circular mouth located midventrally; no proboscis; large cyrtos	8
8a(7b).	Body nearly ellipsoid, rounded in cross-section, sometimes flattened ventrally; medium to large (some > 100 μm); densely ciliated all over; oral depression present .... class Nassophorea, order Nassulida, family Nassulidae	
	(*Nassula*, Fig. 3.20Z)	
8b.	Body usually flattened; ventrum ciliated, dorsum bare or with a few cilia; anterior preoral arcs of right ventral ciliary rows continuous with more posterior parts .... class Phyllopharyngea, order Cyrtophorida, family Chiliodonellidae	
	(*Chilodonella*, Fig. 3.24T)	
9a(1a).	Budding and cytokinesis on surface of trophont .... order Exogenida	10
9b.	Budding begun in a pouch	12
10a(9a).	Trophont basally attached to bottom of lorica near stalk .... family Thecacinetidae	
	(*Thecacineta*, Fig. 3.19A)	
10b.	Not as above	11
11a(10b).	With vase-like lorica, sometimes with stalk-like base; tentacles extend through one or more slits .... family Urnulidae	
	(*Metacineta, Paracineta*, Fig. 3.19B–D)	
11b.	Trophont small, pyriform to spherical; usually stalked; tentacles apical or evenly distributed; often attached to other ciliates .... family Podophryidae	
	(*Podophrya, Sphaerophyra*, Fig. 3.19E, G)	
12a(9b).	Budding and cytokinesis completed in brood pouch .... order Endogenida	13
12b.	Cytokinesis begun in pouch, completed exogenously .... order Evaginogenida	15
13a(12a).	Trophont without stalk or stalk-like process; attached to substrate by broad part of body .... family Dendrosomatidae	
	(*Dendrosoma, Trichophrya*, Fig. 3.19H, P)	
13b.	Not as above	14
14a(13b).	With lorica, usually stalked; tentacles in a few fascicles or distributed over body .... family Acinetidae	
	(*Acineta*, Fig. 3.19F)	
14b.	Lacks lorica; tentacles single, fascicles, or evenly distributed .... family Tokophryidae	
	(*Tokophyra, Multifasciculatum*, Fig. 3.11; 3.19K, L)	
15a(12b).	Conical tentacles; body often hemispherical .... family Dendrocometidae	
	(*Dendrocometes, Stylocometes*, Fig. 3.19I, O)	

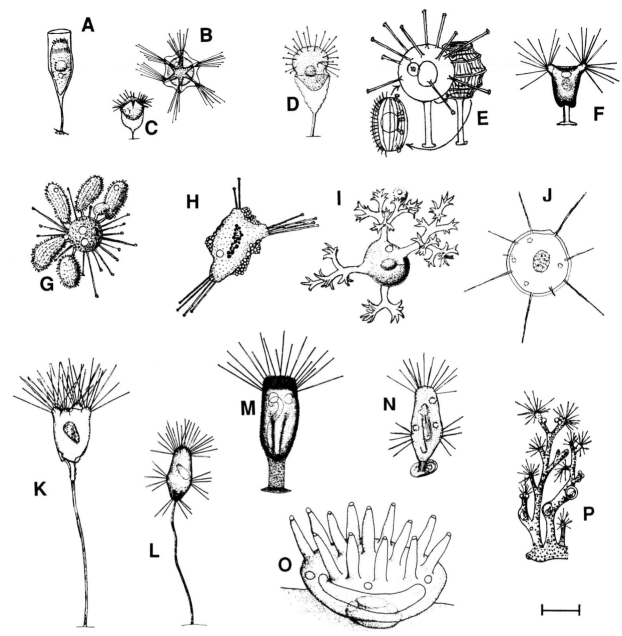

Figure 3.19 (A) *Thecacineta cothurniodes*; (B, C) *Metacineta mystacina*, top and side views; (D) *Paracineta crenata*; (E) *Podophyra fixa*, showing trophont, encysted form, and swarmer; (F) *Acineta limnetis*; (G) *Sphaerophyra magna*; (H) *Trichophyra epistylidis*; (I) *Dendrocometes paradoxus*; (J) *Heliophyra reideri*; (K) *Tokophyra quadripartita*; (L) *Multifasciculatum elegans*; (M) *Squalorophyra macrostyla*; (N) *Discophyra elongata*; (O) *Stylocometes digitalis*; (P) *Dendrosoma radians*. (After: Corliss 1979, P; Goodrich and Jahn 1943, F, K, L, M; Kent 1880–1882, G, I; Matthes 1972, J, O; Noland 1959, A, B, C, D, N; Small and Lynn 1985, E, H.) Scale = 15 μm for E, H, J, O; 30 μm for A, D, F, G; 50 μm for I, L, M, N; 75 μm for B, K; 200 μm for P.

15b.	Not as above, capitate tentacles	16
16a(15b).	Body discoidal, flattened against substrate, with peripheral attachment ring .... family Heliophryidae (*Heliophrya*, Fig. 3.19J)	
16b.	Thin tentacles; body spherical to ovoid; some species loricate and/or stalked .... family Discophryidae (*Discophrya*, *Squalorophyra*, Fig. 3.19M, N)	
17a(2a).	Trophont mobile; symbiotic on other organisms; with complex, aboral holdfast .... order Mobilida	18

17b.	Trophont usually sessile, attached with stalk; often attached to other organisms; many gregarious or colonial species .... order Sessilida	19

18a(17a). Denticles of holdfast with hooks and spines; ectosymbionts on *Hydra*, fishes, amphibia .... family Trichodinidae
        (*Trichodina*, Fig. 3.20D)

18b. Denticles of holdfast simple, toothed; elongated macronucleus, ectosymbionts on turbellarians .... family Urceolariidae
        (*Urceolaria*, Fig. 3.20C)

19a(17b). Free-swimming; aboral end drawn to point with one or two rigid bristles or cilia; swim with oral end forward .... family Astylozoidae
        (*Astylozoon, Hastatella*, Fig. 3.20A, B)

19b. Aborally attached to substratum directly or by stalk 20

20a(19b). Colonial, with zooids in gelatinous matrix .... family Ophrydiidae
        (*Ophrydium*, Fig. 3.20J)

20b. Not in gelatinous matrix 21

21a(20b). Stalk absent, tapering body may resemble short stalk .... family Scyphidiidae
        (*Scyphidia*, Fig. 3.20E)

21b. Stalked; if stalkless, in lorica 22

22a(21b). With contractile stalk; solitary or colonial 23
22b. If stalked, stalk not contractile; solitary or colonial 24

23a(22a). Each zooid independently contractile; if solitary, entire stalk contractile .... family Vorticellidae
        (*Carchesium, Vorticella*, Fig. 3.20K, Q–R)

23b. Entire colony contracts in unison due to shared, continuous myonemes .... family Zoothamniidae
        (*Zoothamnium*, Fig. 3.20H–I)

24a(22b). With vase-shaped lorica, stalkless or short stalked; slender body attached to lorica aborally .... family Vaginicolidae
        (*Cothurnia, Platycola, Pyxicola, Thuricola, Vaginicola*, Fig. 3.20F, G, L–N)

24b. Not as above 25

25a(24b). Stalked; peristomal area raised on short neck with furrow separating it from margin of zooid .... family Operculariidae
        (*Opercularia*, Fig. 3.20S)

25b. Stalked; solitary or colonial; peristomal area not on neck, no furrow .... family Epistylididae
        (*Campanella, Epistylis, Rhabdostyla*, Fig. 3.20O, P, T, U)

26a(3a). Locomotor cirri present on ventral surface; generally dorsoventrally flattened 27
26b. Not as above 36

27a(26a). Adoral zone of membranelles (AZM) well developed, prominent 28
27b. AZM reduced; no marginal or caudal cirri, ventral cirri conspicuous .... class Nassophorea, subclass Hypotrichia, order Euplotida, family Aspidiscidae

        (*Aspidisca*, Fig. 3.20X)

28a(27a). Right marginal row of cirri absent, one or more left marginal cirri; large caudal cirri; paroral membrane only as a field on right of oral area .... class Nassophorea, subclass Hypotrichia, order Euplotida, family Euplotidae

        (*Euplotes*, Fig. 3.20Y)

28b. Left oral cilia as a "collar" and "lapel," right oral cilia variable with one or more paroral membranes .... class Spirotrichea, subclass Stichotrichia, order Stichotrichida 29

29a(28b). Ventral cirri not in distinct file from anterior to posterior (exception: *Gastrostyla*) .... family Oxytrichidae
        (*Gastrostyla, Oxytricha, Stylonychia*, Fig. 3.21A, C, E)

29b. Ventral cirri in distinct linear or zig-zag file(s) 30

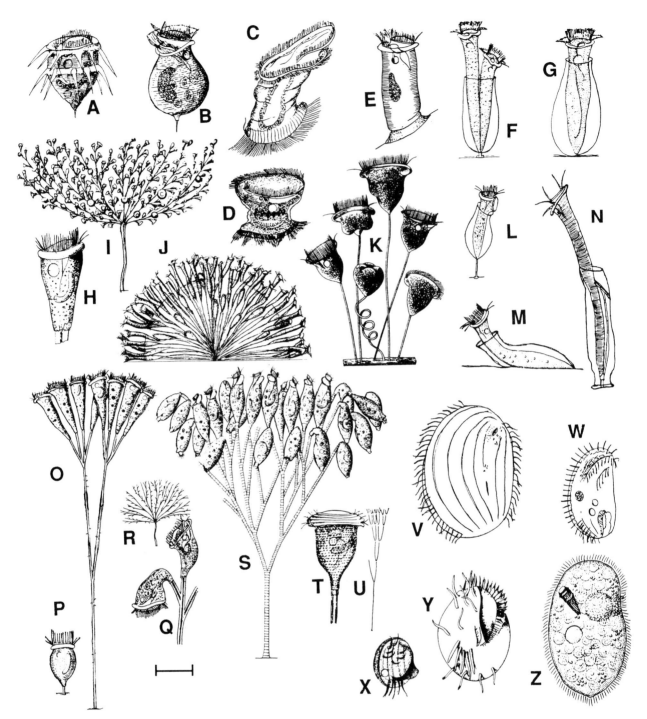

Figure 3.20 (A) *Hastatella radians*; (B) *Astylozoon faurei*; (C) *Urceolaria mitra*; (D) *Trichodina pediculis*; (E) *Scyphidia physarum*; (F) *Cothurnia imberbis*; (G) *Vaginicola ingenita*; (H, I) *Zoothamnium arbuscula*, individual and colony; (J) *Ophrydium eichhorni*; (K) *Vorticella campanula*; (L) *Pyxicola affinis*; (M) *Platycola longicollis*; (N) *Thuricola folliculata*; (O) *Epistylis plicatilis*; (P) *Rhabdostyla pyriformis* (Q, R) *Carchesium polypinum*, individual and colony; (S) *Opercularia nutans*; (T, U) *Campanella umbellaria*, individual and colony; (V) *Pseudomicrothorax agilis*; (W) *Microthorax pusillus*; (X) *Aspidisca costata*; (Y) *Euplotes patella*; (Z) *Nassula ornata*. (After: Corliss 1959, V, Y; Kahl 1930–1935, A, B, C, D, E, H, L, N, Q, R, T, U, W; Kent 1880–1882, I, J, K, O, S, X; Noland 1959, F, G, M, P.) Scale = 15 μm for V, W; 20 μm for A, B, G, P; 25 μm for D, H, F, X; 30 μm for C, Z; 40 μm for E, L, M, N, S; 50 μm for O, Y; 75 μm for K, Q, U; 200 μm for J.

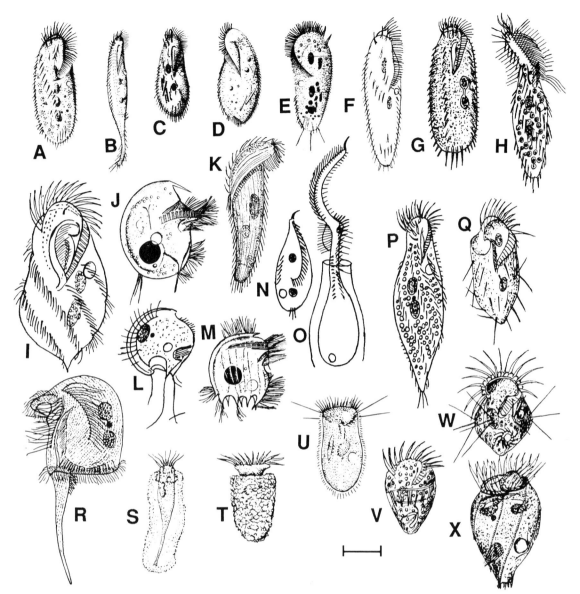

Figure 3.21 (A) *Gastrostyla steini*; (B) *Uroleptus piscis*; (C) *Oxytricha fallax*; (D) *Urostyla grandis*; (E) *Stylonychia mytilus*; (F) *Gonostomum affine*; (G) *Amphisiella oblonga*; (H) *Stichotricha aculeata*; (I) *Hypotrichidium conicum*; (J) *Discomorphella pectinata*; (K) *Metopus es*; (L) *Myelostoma flagellatum*; (M) *Saprodinium dentatum*; (N, O) *Chaetospira mülleri*, contracted and extended forms; (P) *Strongylidium crassum*; (Q) *Psilotricha acuminata*; (R) *Caenomorpha medusula*; (S) *Tintinnidium fluviatile*; (T) *Tintinnopsis cylindricum*; (U) *Strombidinopsis setigera*; (V) *Strombidium viride*; (W) *Halteria grandinella*; (X) *Strombilidium gyrans*. (After: Jankowski 1964, J, M; Kahl 1930–1935, F, G, H, I, K, L, N, O, P, Q, R, V, W, X; Kent 1880–1882, A, B, C, D, E; Noland 1959, S, T, U.) Scale = 15 μm for L; 25 μm for H, W, X; 30 μm for F, I, J, P, Q, R, T; 40 μm for G, K, M, N, O, S, U, V; 60 μm for B, E; 80 μm for A, C; 140 μm for D.

30a(29b).	Frontoventral cirri run length of ventrum as zig-zag files	family Urostylidae	
	(*Urostyla, Uroleptus*, Fig. 3.21B, D)		
30b.	Ventral cirri in one or more linear files of varied length	suborder Stichotrichina	31
31a(30b).	Body in lorica; without marginal cirri; anterior end long, narrow, contractile	family Chaetospiridae	
	(*Chaetospira*, Fig. 3.21N, O)		
31b.	Not as above		32
32a(31b).	Files of cirri curved or spiralled along body		33
32b.	Files of cirri straight or oblique, ventral only		35

33a(32a).   Files of cirri end on dorsum; oral region extends over 25% of body
            length .... family Strongylidiidae
            (*Strongylidium*, Fig. 3.21P)

33b.   Not as above 34

34a(33b).   Ventral files of cirri curve to left rear .... family Psilotrichidae
            (*Psilotricha*, Fig. 3.21Q)

34b.   Ventral files of cirri spiral obliquely around body .... family Spirofilidae
            (*Hypotrichidium*, *Stichotricha*, Fig. 3.21H, I)

35a(32b).   Files of cirri parallel to long axis of oral region along its right
            border .... family Gonostomatidae
            (*Gonostomum*, Fig. 3.21F)

35b.   Files of cirri not parallel to oral region, one or more of the ventral cirral
            files extends from anterior to well past midventrum .... family Amphisiellidae
            (*Amphisiella*, Fig. 3.21G)

36a(26b).   Body surface ciliated 46

36b.   Somatic ciliature sparse or absent 37

37a(36b).   Body flattened, rigid, often with spines; body ciliature present as short,
            generally obvious rows; oral ciliature relatively inconspicuous; mainly
            anaerobic .... class Spirotrichea, subclass Heterotrichia, order
                                                                                       Odontostomatida   38

37b.   Not as above 40

38a(37a).   Body box-shaped, generally with short posterior spines; short rows of
            body cilia at front and rear, those in front parallel to and forward of
            oral opening .... family Epalxellidae
            (*Saprodinium*, Fig. 3.21M)

38b.   Body round to discoidal 39

39a(38b).   Distinct anterior spine; band of cilia on ridge overhanging buccal
            cavity .... family Discomorphellidae
            (*Discomorphella*, Fig. 3.21J)

39b.   Body ciliature very sparse; lacks spines; pair of cirri at posterior
            end .... family Myelostomatidae
            (*Myelostoma*, Fig. 3.21L)

40a(37b).   Membranelles numerous in circle or almost complete circle at oral end;
            body generally conical or bell-shaped; mostly
            planktonic .... class Spirotrichea, subclass Choreotrichia   41

40b.   Body rounded or conical, spirally twisted to left; twisting of body less
            prominent than Metopidae; often with long caudal spine; oral region
            spiralled; dense preoral cilia along anterior border of oral cavity; body
            cilia absent except for caudal tuft and anterior cirrus-like tufts;
            anaerobic .... class Spirotrichea, subclass Heterotrichia, order
                                                                         Armophorida, family Caenomorphidae
            (*Caenomorpha*, Fig. 3.21R)

41a(40a).   Oral ciliature forms a closed circle .... order Choreotrichida   42

41b.   Oral ciliature forms an open anterior collar with an anterioventral
            "lapel" .... order Oligotrichida   45

42a(41a).   Loricate, attached to inner wall of lorica .... suborder Tintinnina   43

42b.   Not loricate 44

43a(42a).   Delicate, gelatinous or mucoid tubular lorica, with attached debris,
            often translucent .... family Tintinnididae
            (*Tintinnidium*, Fig. 3.21S)

43b.   Lorica rigid, agglomerate of mineral grains and diatom
            fragments .... family Codonellidae
            (*Tintinnopsis*, Fig. 3.21T)

44a(42b).   Rows of somatic ciliature equally distributed around body, extending
            length of body; body cilia may be long .... suborder Strombidinopsina, family Strombidinopsidae
            (*Strombidinopsis*, Fig. 3.21U)

44b.   One or more rows of body cilia shorter than body; body ciliature short .... suborder Strobilidiina, family Strobilidiidae
         (*Strobilidium*, Fig. 3.21X)

45a(41b).   Somatic ciliature reduced to a few cirrus-like bristles arranged circumferentially around the equator .... family Halteriidae
         (*Halteria*, Fig. 3.21W)

45b.   Somatic ciliature greatly reduced or absent, no bristles; sometimes with prominent girdle close to middle of cell .... family Strombidiidae
         (*Strombidium*, Fig. 3.21V)

46a(36a).   Uniform somatic ciliation; often with conspicuous tuft of caudal cilia; buccal membranelles large, but not always obvious; anterior end of body twisted left; oral region spiralled with dense preoral ciliation; in richly organic environments with low oxygen .... class Spirotrichea, subclass Heterotrichia, order Armophorida, family Metopidae

         (*Metopus*, Fig. 3.21K)

46b.   Not as above 47

47a(46b).   Somatic ciliature well developed, uniform, usually inserts on oral region; body size large; many contractile species; rarely loricate; some species pigmented .... class Spirotrichea, subclass Heterotrichia, order Heterotrichida, suborder Heterotrichina   48

47b.   Somatic ciliature well developed; body large (500–1000 μm), broad; buccal cavity prominent, funnel-like; macronucleus elongate; carnivorous .... class Colpodea, order Bursariomorphida, family Bursariidae
         (*Bursaria*, Fig. 3.22C)

48a(47a).   Body trumpet-shaped; highly contractile; oral ciliature spirals clockwise around flared anterior end; often pigmented and/or with algal endosymbionts .... family Stentoridae
         (*Stentor*, Fig. 3.22F)

48b.   Body not trumpet-shaped 49

49a(48b).   Body laterally compressed, pyriform to ellipsoid; often pigmented pink, red, or purple; long, narrow peristome on left margin; noncontractile .... family Blepharismidae
         (*Blepharisma*, Fig. 3.22B)

49b.   Body shape not as above 50

50a(49b).   Body large, very long; oral cavity shallow; peristomal area long, narrow, oral ciliature sometimes inconspicuous; contractile .... family Spirostomidae
         (*Spirostomum*, Fig. 3.22A)

50b.   Buccal ciliature prominent; oral cavity deeper than above 51

51a(50b).   Buccal cavity occupying much of anterior part of body; paroral membrane inconspicuous; somatic ciliation dense; with or without endosymbiotic algae .... family Climacostomidae
         (*Climacostomum*, Fig. 3.22D)

51b.   Paroral membrane prominent; contractile; somatic ciliation dense .... family Condylostomatidae
         (*Condylostoma*, Fig. 3.22E)

52a(5a).   Body ovoid; with tentacles extending from the cell in all directions when at rest, but retracted and hardly visible when swimming .... class Litostomatea, order Pharyngophorida, family Actinobolinidae

         (*Actinobolina*, Fig. 3.22G)

52b.   Tentacles absent 53

53a(52b).   Translucent $CaCO_3$ plates (armor) in cortex; typically barrel-shaped; body frequently spiny, often with prominent anterior and caudal thorns; long caudal cilium common; brosse present, but inconspicuous .... class Prostomatea, order Prorodontida, family Colepidae

         (*Coleps*, Fig. 3.22H)

53b.   No such armor, not as above 54

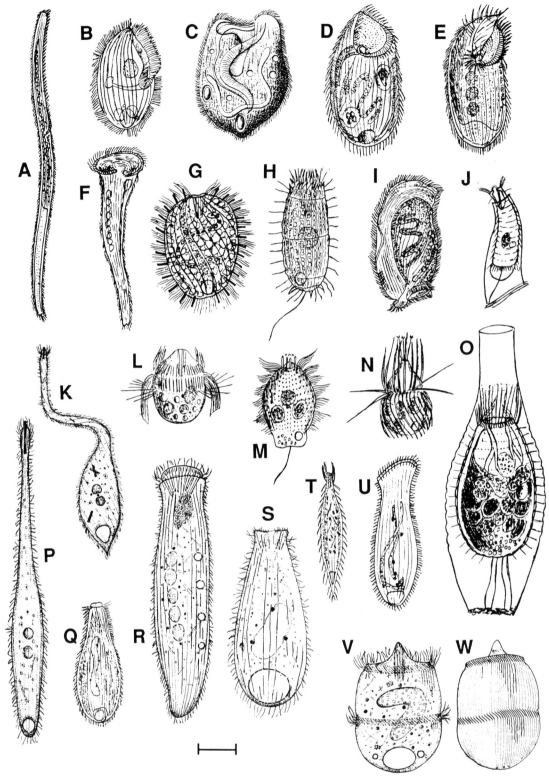

Figure 3.22 (A) *Spirostomum minus*; (B) *Blepharisma lateritium*; (C) *Bursaria truncatella*; (D) *Climacosto-mum virens*; (E) *Condylostoma tardum*; (F) *Stentor polymorphus*, half extended; (G) *Actinobolina radians*; (H) *Coleps hirtus*; (I) *Bryophyllum lieberkühni*; (J) *Metacystis recurva*; (K) *Lacrymaria olor*; (L) *Askensia volvox*; (M) *Urotricha farcta*; (N) *Mesodinium pulex*; (O) *Vasicola ciliata*; (P) *Trachelophyllum apiculatum*; (Q) *Enchelyodon elegans*; (R) *Homalozoon vermiculare*; (S) *Enchelys simplex*; (T) *Chaena teres*; (U) *Spathi-dium spathula*; (V, W) *Didinium nasutum*, live and silver stained. (After: Dragesco 1966a, K, S, V, W; Dragesco 1966b, P, R; Kahl 1930–1935, A, B, D, E, F, G, H, I, J, L, M, N, O, Q, T, U; Kent 1880–1882: C.) Scale = 10 μm for M, N; 20 μm for H, J, L, P, S; 30 μm for G, O, U; 40 μm for B, K, T; 60 μm for E, Q, R; 80 μm for D, V, W; 100 μm for A, F, I; 200 μm for C.

54a(53b). Sessile in pseudochitinous lorica; one or more caudal cilia; paratenes
obvious .... class Prostomatea, order Prostomatida, family Metacystidae
  (*Metacystis, Vasicola*, Fig. 3.22J, O)

54b. Not as above 55

55a(54b). Mouth at anterior end, rounded or only slightly elongated 56

55b. Apex fan-shaped to varying degree; mouth a long slit, beginning at
anterior end; slit may extend down side of cell .... class Litostomatea, order Haptorida,
family Spathidiidae

  (*Byrophyllum, Spathidium*, Fig. 3.22I, U)

56a(55a). Mouth at anterior pole, surrounded by an unciliated area and often
raised as a blunt cone; circumferential ciliary girdle of closely apposed
cilia, otherwise naked or with short cilia .... class Litostomatea, order Haptorida   57

56b. Uniform ciliation around mouth, and typically over the entire cell 58

57a(56a). Girdle ciliature of one type, one or two girdles .... family Didiniidae
  (*Didinium*, Figs. 3.6; 3.22V, W)

57b. Body ciliature of two types, one cirrus-like, as a single girdle .... family Mesodiniidae
  (*Askenasia, Mesodinium*, Fig. 3.22L, N)

58a(56b). Rear 1/3 to 1/5 of body unciliated, except for one or more long caudal
cilia; other somatic ciliation evenly distributed .... class Prostomatea, order Prorodontida,
family Urotrichidae

  (*Urotricha*, Fig. 3.22M)

58b. Not as above 59

59a(58b). Body uniformly ciliated at rear; mouth with oral dome or bulb;
cytopharynx not permanently inverted, but inverts during
ingestion .... class Litostomatea, order Haptorida   60

59b. Not as above; lack oral dome 62

60a(59a). Anterior tapering to a ciliated neck; neck set off by groove and a circle
of longer cilia; contractile .... family Lacrymariidae
  (*Lacrymaria*, Fig. 3.22K)

60b. Not as above 61

61a(60b). Body long, generally greater than 4 times width; ovoid or flask-shaped;
oral region simple dome, sometimes pointed .... family Trachelophyllidae
  (*Chaenea, Trachelophyllum*, Fig. 3.22P, T)

61b. Oral region flattened at apex; cytostome circular to ovoid; body usually
shorter than four times width .... family Enchelyidae
  (*Homalozoon, Enchelys, Enchelyodon*, Fig. 3.22Q–S)

62a(59b). Short dorsal brush extends backward from apical mouth on one side
(may not be obvious without silver staining); body ellipsoid .... class Prostomatea,
order Prorodontida, family Prorodontidae

  (*Prorodon, Pseudoprorodon*, Fig. 3.23A, B)

62b. Dorsal brush lacking, may have reduced brosse; similar to and easily
confused with *Prorodon* .... class Prostomatea, order Prorodontida, family Holophryidae
  (*Holophyra*, Fig. 3.23A, C)

63a(7a). Body long, flattened; mouth on convex side of tapering anterior end;
ciliated only on right side of body .... class Karyorelictea, order Loxodida, family Loxodidae
  (*Loxodes*, Fig. 3.23J)

63b. Not as above 64

64a(63b). Cytopharynx permanent, round, located at base of proboscis or
tapering front end; ciliature along proboscis; toxicysts on ventrum
of proboscis .... class Litostomatea, order Pharyngophorida,
family Tracheliidae

  (*Dileptus, Paradileptus, Trachelius*, Fig. 3.23D, E, H)

64b. Mouth slit-like, located along convex side of tapering anterior end,
evident only when feeding; body laterally compressed; both sides
ciliated, cilia may be only bristles on one side .... class Litostomatea, order Pleurostomatida,
family Amphileptidae

  (*Amphileptus, Litonotus, Loxophyllum*, Fig. 3.23F, G, I)

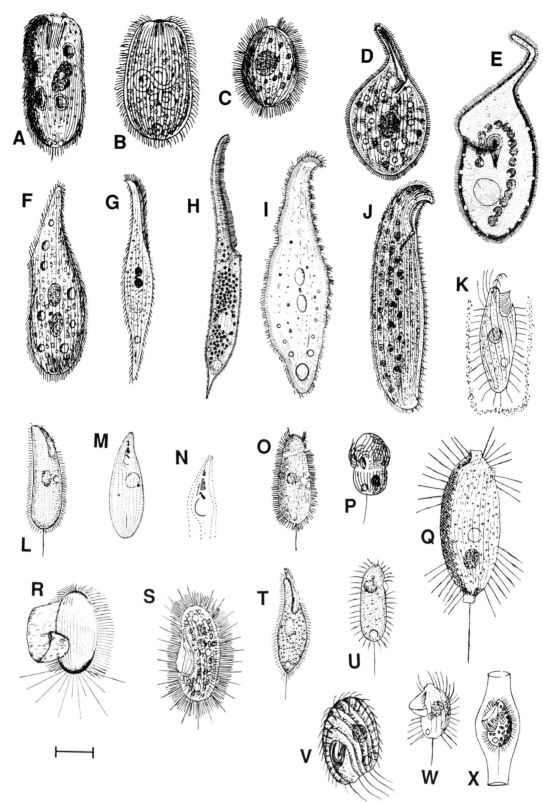

Figure 3.23 (A) *Prorodon teres*; (B) *Pseudoprorodon ellipticus*; (C) *Holophyra simplex*; (D) *Trachelius ovum*; (E) *Paradileptus robustus*; (F) *Amphileptus claparedi*; (G) *Litonotus fasciola*; (H) *Dileptus anser*; (I) *Loxophyllum helus*; (J) *Loxodes magnus*; (K) *Cyrtolophosis mucicola*; (L,M,N) *Philasterides armata*, live, silver stained, and oral detail of silver-stained specimen; (O) *Loxocephalus plagius*; (P) *Urozona bütschlii*; (Q) *Balanonema biceps*; (R) *Pleuronema coronatum*; (S) *Histobalantium natans*; (T) *Cohnilembus pusillus*; (U) *Uronema griseolum*; (V) *Cinetochilum margaritaceum*; (W) *Cyclidium glaucoma*; (X) *Calyptotricha pleuronemoides*. (After: Corliss 1959, R; Dragesco 1966a, I; Grolière 1980, M, N; Kahl 1930–1935, A, B, C, F, G, J, K, O, P, Q, S, V, W, X; Kudo 1966, D, E, H; Noland 1959, L, T, U.) Scale = 10 μm for K, Q; 15 μm for P, V; 20 μm for T, U, W, X; 25 μm for G, H, L, M; 30 μm for C, I, S; 40 μm for B, R; 50 μm for F; 60 μm for A, O; 75 μm for D, E, J.

65a(6a). With linear oral furrow or groove leading from anterior end to mouth, bordered on right by paroral membrane 66

65b. Without linear preoral groove bordered by membranes, but generally with membranes inside oral cavity 71

66a(65a). With an apparent double paroral membrane on the right side of the furrow (actually the paroral membrane plus a row of somatic cilia) .... class Oligohymenophora, order Scuticociliatida, family Cohnilembidae

(*Cohnilembus*, Fig. 3.23T)
[Many genera of the order Scuticociliatida (e.g., lines 66–70 and lines 75–76) are difficult to identify without special preparation techniques such as Protargol or other silver staining.]

66b. Paroral membrane along furrow not "double" 67

67a(66b). Preoral furrow long, shallow, ciliated with three ciliary fields; generally a single caudal cilium; body long, ovoid .... class Oligohymenophora, order Scuticociliatida, family Philasteridae

(*Philasterides*, Fig. 3.23L–N)

67b. Bottom of preoral furrow not ciliated, paroral membrane curves around rear of mouth 68

68a(67b). Dwells in lorica that is open at both ends; resembles *Pleuronema* or *Cylidium* .... class Oligohymenophora, order Scuticociliatida, family Calyptotrichidae

(*Calyptotricha*, Fig. 3.23X)

68b. Not in a lorica as above 69

69a(68b). Small (15–60 μm long); with few somatic cilia; distinct caudal cilium .... class Oligohymenophora, order Scuticociliatida, family Cyclidiidae

(*Cyclidium*, Fig. 3.23W)

69b. Larger than above 70

70a(69b). Paroral membrane a prominent velum, extends from anterior to well past equator of cell; one to many stiff, long caudal cilia; a single contractile vacuole .... class Oligohymenophora, order Scuticociliatida, family Pleuronematidae

(*Pleuronema*, Fig. 3.23R)

70b. Somatic ciliation uniform; long stiff cilia distributed over body surface; paroral membrane less prominent than above .... class Oligohymenophora, order Scuticociliatida, family Histobalantiidae

(*Histobalantium*, Fig. 3.23S)

71a(65b). Oral cavity huge, covers most of ventrum; dorsal side convex; cilia of paroral membrane long .... class Nassophorea, order Peniculida, family Lembadionidae

(*Lembadion*, Fig. 3.24G)

71b. Mouth smaller than above, not over 50% of body length 72

72a(71b). With one or more long caudal cilia 73

72b. Without caudal cilia 77

73a(72a). With one or two girdles of cilia 74

73b. Cilia not in girdles, usually in longitudinal rows 75

74a(73a). Oral cavity equatorial; distinct constriction in the ciliated girdle; ends bare except for caudal cilium .... class Oligohymenophora, order Scuticociliatida, family Urozonidae

(*Urozona*, Fig. 3.23P)

74b. Oral cavity deep, equatorial; 2 distinct ciliary girdles; ciliary tuft on posterior end .... class Nassophorea, order Peniculida, family Urocentridae

(*Urocentrum*, Fig. 3.24D)

75a(73b). Mouth at, or anterior to, middle of cell 76

75b. Oral area large, toward posterior half of cell ($\approx$ midventral); body usually flattened; cilia denser ventrally .... class Oligohymenophora, order Scuticociliatida, family Cinetochilidae

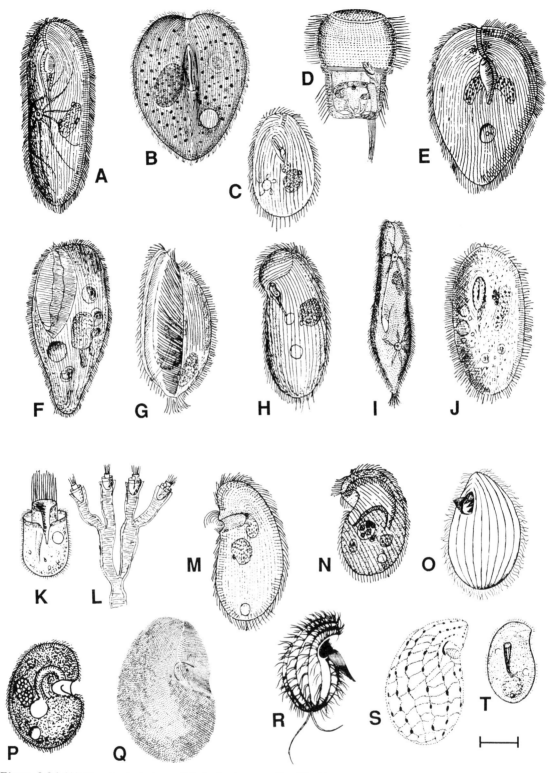

Figure 3.24 (A) *Frontonia leucas*; (B) *Stokesia vernalis*; (C) *Glaucoma scintillans*; (D) *Urocentrum turbo*; (E) *Disematostoma bütschlii*; (F) *Turaniella vitrea*; (G) *Lembadion magnum*; (H) *Colpidium colpoda*; (I) *Paramecium caudatum*; (J) *Clathrostoma viminale*; (K, L) *Maryna socialis*, individual and colony; (M) *Plagiopyla nasuta*; (N) *Bresslaua vorax*; (O) *Tetrahymena pyriformis*; (P,Q) *Tillina magna*, live and line drawing of silver-stained specimen; (R, S) *Colpoda steinii*, live and silver stained; (T) *Chilodonella uncinata*. (After: Corliss 1959, O, R; Dragesco 1966b, B; Kahl 1930–1935, A, C, D, E, F, G, H, I, J, K, L, M, P; Kudo 1966, N; Lynn 1976, S; Lynn 1977, Q; Noland 1959: T.) Scale = 15 μm for G, O, R; 25 μm for C, H; 30 μm for D, F; 40 μm for B, E, J, M; 60 μm for I, N; 75 μm for A, K, Q.

(*Cinetochilum*, Fig. 3.23V)

76a(75a). Small, < 50 μm; anterior pole flat, unciliated .... class Oligohymenophora,
order Scuticociliatida, family Uronematidae

(*Uronema*, Fig. 3.23U)

76b. Somatic ciliation even; body long ovoid; oral area small, closer to
anterior end .... class Oligohymenophora, order Scuticociliatida,
family Loxocephalidae

(*Balanonema, Loxocephalus*, Fig. 3.23O, Q)

77a(72b). Body small, ovoid to ellipsoid; in mucilaginous case from which it can
emerge freely; cytostome near anterior end .... class Colpodea, order Cyrtolophosidida,
family Cyrtolophosidae

(*Cyrtolophosis*, Fig. 3.23K)

77b. Not associated with mucilaginous envelope 78

78a(77b). Body distinctly heart- or cone-shaped; oral region relatively large,
covers most of ventral surface .... class Nassophorea, order Peniculida, family Stokesiidae
(*Stokesia*, Fig. 3.24B)

78b. Body shaped otherwise 79

79a(78b). Body ovoid to ellipsoid; mouth in anterior half of body, pointed in front,
with preoral and/or postoral sutures, sometimes hard to see;
nematodesmata prominent to side and rear of mouth; contractile
vacuoles with long collecting canals .... class Nassophorea, order Peniculida, family Frontoniidae
(*Disematostoma, Frontonia*, Fig. 3.24A, E)

79b. Mouth typically rounded or blunt in front; body ciliature with preoral
suture, no postoral suture; one paroral membrane and three
membranelles, which are typically inconspicuous .... order Hymenostomatida, suborder
Tetrahymenina 80

[Genera of this suborder, lines 80–81, are difficult to identify without
special preparation techniques, i.e., silver staining.]

80a(79b). Body pyriform to cylindrical; bases of membranelles of uniform width;
mouth roughly triangular .... family Tetrahymenidae
(*Tetrahymena*, Figs. 3.1; 3.24O)

80b. Not as above 81

81a(80b). Right ventral cilia curve left and twist forward parallel to suture .... family Turaniellidae
(*Colpidium, Turaniella*, Fig. 3.24F, H)

81b. Right ventral cilia curve left, but do not twist forward toward
mouth .... family Glaucomidae
(*Glaucoma*, Fig. 3.24C)

82a(6b). In a gelatinous tube-like lorica, which may have dichotomous branches;
with a terminal cone-like protuberance bearing long cilia and surrounded
by a circular adoral groove leading to the buccal
cavity .... class Colpodea, order Colpodida, family Marynidae
(*Maryna*, Fig. 3.24K, L)

82b. No gelatinous case; mouth area not as described above 83

83a(82b). Body small (< 100 μm), laterally flattened; ciliation relatively sparse, on
pellicular ridges 84

83b. Not as above 85

84a(83a). Cytostome in rear half of body; small cyrtos hidden by ventral pellicular
fold .... class Nassophorea, order Microthoracida,
family Microthoracidae

(*Microthorax*, Fig. 3.20W)

84b. Cytostome opens laterally in anterior 1/3 of body, long tubular cyrtos;
body nearly oval .... class Nassophorea, order Propeniculida
family Leptophrygidae

(*Pseudomicrothorax*, Fig. 3.20V)

85a(83b). Body ciliation uniform; ventral cytostome in an oral vestibule;
nematodesmata form a ring or "basket" around mouth; similar to
Frontoniidae .... class Nassophorea, order Peniculida,
family Clathrostomatidae

(*Clathrostoma*, Fig. 3.24J)

85b.   Mouth area sunken in, with oral groove leading to it, or with ciliated
           pharyngeal tube, or both 86

86a(85b). Medium to large (≥ 150 μm in some species); body foot-shaped or
           ellipsoidal, with pointed or rounded caudal end; uniform body ciliation;
           long, broad, ciliated oral groove leads into buccal cavity; two contractile
           vacuoles; one species with algal
           symbiont .... class Nassophorea, order Peniculida, family Parameciidae
           (*Paramecium*, Figs. 3.6; 3.24I)

86b.   Not as above; with single contractile vacuole 87

87a(86b). Oral groove transverse, in anterior part of body; cytostome at base of
           deep tubular pocket; densely ciliated; common in anaerobic
           habitats .... class Oligohymenophorea, order Plagiopylida,
           family Plagyopylidae
           (*Plagiopyla*, Fig. 3.24M)

87b.   Body reniform, indented in middle on one side; oral groove passes
           around groove toward mouth; division in a cyst .... class Colpodea, order Colpodida,
           family Colpodidae

(*Bresslaua*, *Tillina*, *Colpoda*, Fig. 3.24N, P–S)

LITERATURE CITED

Albright, L. J., E. B. Sherr, B. F. Sherr, and R. D. Fallon. 1987. Grazing of ciliated protozoa on free and particle-associated bacteria. Marine Ecology Progress Series 38:125–129.

Allen, R. D. 1984. *Paramecium* phagosome membrane: from oral region to cytoproct and back again. Journal of Protozoology 31:1–6.

Andersson, A., U. Larsson, and Å. Hagström. 1986. Size-selective grazing by a microflagellate on pelagic bacteria. Marine Ecology Progress Series 33:51–57.

Archbold, J. H. G., and J. Berger. 1985. A qualitative assessment of some metazoan predators of *Halteria grandinella*, a common freshwater ciliate. Hydrobiologia 126:97–102.

Baldock, B. M. 1986. Peritrich ciliates epizoic on larvae of *Brachycentrus subnilus* (Trichoptera): importance in relation to the total protozoan population in streams. Hydrobiologia 131:125–131.

Baldock, B. M., and J. Berger. 1984. Effects of low temperature on the growth of four fresh-water amoebae (Protozoa: Gymnamoebia). Transactions of the American Microscopical Society 103:233–239.

Baldock, B. M., and M. A. Sleigh. 1988. The ecology of benthic protozoa in rivers—seasonal variation in numerical abundance in fine sediments. Archiv für Hydrobiologie 111:409–422.

Baldock, B. M., J. H. Baker, and M. A. Sleigh. 1980. Laboratory growth rates of six species of freshwater Gymnamoebia. Oecologia (Berlin) 47:156–159.

Baldock, B. M., J. H. Baker, and M. A. Sleigh. 1983. Abundance and productivity of protozoa in a chalk stream. Holarctic Ecology 6:238–246.

Bamforth, S. S. 1985. Ecology of protozoa. Pages 8–15 *in:* J. J. Lee, S. H. Hutner, and E. C. Bovee, editors. An illustrated guide to the protozoa. Society of Protozoologists, Lawrence, Kansas.

Beaver, J. R., and T. L. Crisman. 1989. The role of ciliated protozoa in pelagic freshwater ecosystems. Microbial Ecology 17:111–136.

Bell, G. 1988. Sex and death in Protozoa. Cambridge University Press, London.

Berger, J. 1979. The feeding behavior of *Didinium nasutum* on an atypical prey ciliate (*Colpidium campylum*). Transactions of the American Microscopical Society 141:261–275.

Berninger, U.-G., B. J. Finlay, and H. M. Canter. 1986. The spatial distribution and ecology of zoochlorellae-bearing ciliates in a productive pond. Journal of Protozoology 33:557–563.

Bick, H. 1957. Beiträge zur Ökologie einiger Ciliaten des saprobiensystems. Vom Wasser 24:224–246.

Bick, H. 1972. Ciliated protozoa. An illustrated guide to the species used as biological indicators in freshwater biology. World Health Organization, Geneva. 198 pp.

Bird, D. F., and J. Kalff. 1986. Bacterial grazing by planktonic algae. Science 231:493–495.

Bird, D. F., and J. Kalff. 1987. Algal phagotrophy: regulating factors and importance relative to photosynthesis in *Dinobryon* (Chrysophyceae). Limnology and Oceanography 32:277–284.

Bott, T. L., and L. A. Kaplan. 1989. Densities of benthic protozoa and nematodes in a Piedmont stream. Journal of the North American Benthological Society 8:187–196

Bourelly, P. 1968. Les algues d'eau douce, Vol. 2. N. Boubée, Paris.

Bovee, E. C. 1985. Classes Lobosea Carpenter; Filosea Leidy; Granuloreticulosa De Saedeleer, Order Athalamida Haeckel. Pages.158–252 *in:* J. J. Lee, S. H. Hutner, and E. C. Bovee, editors. An illustrated guide to the Protozoa. Society of Protozoologists.Lawrence, Kansas.

Bovee, E. C., and T. L. Jahn. 1973. Taxonomy and phylogeny. Pages 37–82 *in:* K. W. Jeon, editor. The biology of amoeba. Academic Press, New York.

Buskey, E. J., and D. K. Stoecker. 1989. Behavioral responses of the marine tintinnid *Favella* sp. to phytoplankton: influence of chemical, mechanical and photic stimuli. Journal of Experimental Marine Biology and Ecology 132:1–16.

Cairns, J., Jr. 1974. Protozoans (Protozoa). Pages 1–18 *in:* C. W. Hart, Jr. and S. L. H. Fuller, editors. Pollution ecology of freshwater invertebrates. Academic Press, New York.

Cairns, J., Jr., and J. A. Ruthven. 1972. A test of the cosmopolitan distribution of fresh-water protozoans. Hydrobiologia 39:405–427.

Cairns, J., Jr., and W. H. Yongue. 1977. Factors affecting the number of species in freshwater protozoan communities. Pages 257–303 *in:* J. Cairns, Jr., editor. Aquatic microbial communities. Garland, New York.

Cairns, J., Jr., G. R. Lanza, and B. C. Parker. 1972. Pollution related structural and functional changes in aquatic communities with emphasis on freshwater algae and protozoa. Proceedings of the Academy of Natural Sciences of Philadelphia 124:79–127.

Cairns, J., Jr., D. L. Kuhn, and J. L. Plafkin. 1979. Protozoan colonization of artificial substrates. Pages 34–57 *in:* R. L. Wetzel, editor. Methods and measurements of periphyton communities: a review. Special Technical Publication No. 690. American Society for Testing and Materials, Philadelphia, Pennsylvania.

Calaway, W. T., and J. B. Lackey. 1962. Waste treatment protozoa. Florida Engineering and Industrial Experiment Station, Gainesville, Florida.

Calow, P. 1977. Conversion efficiencies in heterotrophic organisms. Biological Reviews of the Cambridge Philosophical Society 52:385–409.

Canella, M. F., and I. Rocchi-Canella. 1976. Biologie des Ophryoglenina. Contributions à la connaissance des Ciliés VIII. Annali dell Universitià Università di Ferrara. (Nuova Serie) 3:Suppl. 2.

Canter, H. M. 1973. A new primitive protozoan devouring centric diatoms in the plankton. Zoological Journal of the Linnean Society 52:63–83.

Caron, D. A. 1983. Technique for enumeration of heterotrophic and phototrophic nannoplankton, using epifluorescence microscopy, and comparison with other procedures. Applied and Environmental Microbiology 46:491–498.

Caron, D. A. 1987. Grazing of attached bacteria by heterotrophic microflagellates. Microbial Ecology 13:203–218.

Caron, D. A., J. C. Goldman, and M. R. Dennett. 1988. Experimental demonstration of the roles of bacteria and bacterivorous protozoa in plankton nutrient cycles. Hydrobiologia 159:27–40.

Carrick, H. J., and G. L. Fahnenstiel. 1989. Biomass, size-structure, and composition of phototrophic and heterotrophic nanoflagellate communities in lakes Huron and Michigan. Canadian Journal of Fisheries and Aquatic Sciences 46:1922–1928.

Carrick, H. J., and G. L. Fahnenstiel. 1990. Protozoa in lakes Huron and Michigan: seasonal abundance and composition of ciliates and dinoflagellates. Journal of Great Lakes Research 16:319–329

Carter, R. F. 1968. Primary amoebic meningoencephalitis: clinical, pathological and epidemiological features of six fatal cases. Journal of Pathology and Bacteriology 96:1–25.

Cavalier-Smith, T. 1980. *r*- and *K*-tactics in the evolution of protist developmental systems: genome size, phenotype diversifying selection, and cell cycle patterns. BioSystems 12:43–59.

Chapman-Andresen, C. 1973. Endocytic processes. Pages 319–48 *in:* K. W. Jeon, editor. The biology of *Amoeba*. Academic Press, New York.

Chardez, D. 1985. Protozoaires prédateurs de Thécamoebiens. Protistologica 21:187–194.

Coats, D. W., and J. F. Heinbokel. 1982. A study of reproduction and other life history phenomena in planktonic protists using an acridine orange fluorescence technique. Marine Biology 67:71–79.

Cooley, N. R., J. M. Keltner, Jr., and J. Forester. 1972. Mirex and Aroclorr 1254: effect on and accumulation by *Tetrahymena pyriformis*. Journal of Protozoology 19:636–638.

Corliss, J. O. 1959. An illustrated key to the higher groups of the ciliated protozoa, with definition of terms. Journal of Protozoology 6:265–284.

Corliss, J. O. 1973. History, taxonomy, and evolution of species of *Tetrahymena*. Pages 1–55 *in:* A. M. Elliott, editor. Biology of *Tetrahymena*. Dowden, Hutchinson & Ross, Stroudsburg, Pennsylvania.

Corliss, J. O. 1979. The ciliated protozoa. Characterization, classification and guide to the literature. Pergamon, Oxford.

Corliss, J. O., and D. W. Coats. 1976. A new cuticular cyst-producing tetrahymenid ciliate, *Lambornella clarki* n. sp., and the current status of ciliatosis in culicine mosquitoes. Transactions of the American Microscopical Society 95:725–739.

Corliss, J. O., and P.-M. Daggett. 1983. "*Paramecium aurelia*" and "*Tetrahymena pyriformis*": current status of the taxonomy and nomenclature of these popularly known and widely used ciliates. Protistologica 19:307–322.

Corliss, J. O., and S. C. Esser. 1974. Comments on the role of the cyst in the life cycle and survival of free-living protozoa. Transactions of the American Microscopical Society 93:578–593.

Curds, C. R. 1963. The flocculation of suspended matter by *Paramecium caudatum*. Journal of General Microbiology 33:357–363.

Curds, C. R. 1977. Microbial interactions involving protozoa. Pages 69–105 *in:* F. A. Skinner and J. M. Shewan, editors. Aquatic microbiology. Academic Press, New York.

Curds, C. R., and A. Cockburn. 1970. Protozoa in biological sewage-treatment processes. II. Protozoa as indicators in the activated sludge process. Water Research 4:237–249.

Curds, C. R., and A. Cockburn. 1971. Continuous monoxenic culture of *Tetrahymena pyriformis*. Journal of General Microbiology 66:95–108.

Daniels, E. W. 1973. Ultrastructure. Pages 125–169 *in:* K. W. Jeon, editor. The biology of *Amoeba*. Academic Press, New York.

DeBiase, A. E., R. W. Sanders, and K. G. Porter. 1990. Relative nutritional value of ciliate protozoa and algae as food for *Daphnia*. Microbial Ecology 19:199–210.

Deflandre, G. 1959. Rhizopoda and Actinopoda. Pages 232–264 *in:* W. T. Edmondson, editor. Freshwater biology. Wiley, New York.

Dhanarj, P. S., B. R. Caushal, and R. Lal. 1989. Effects on uptake by the metabolism of aldrin and phorate in a protozoan, *Tetrahymena pyriformis*. Acta Protozoologica 28:157–163.

Dive, D. 1973. Nutrition holozoïque de *Colpidium campylum* aux dépens de bactéries pigmentées ou synthétisant des toxines. Protistologica 10:517–525.

Dive, D., and H. Leclerc. 1975. Standardized test method using protozoa for measuring water pollutant toxicity. Progress in Water Technology 7:67–72.

Dodge, J. D., and C. Greuet. 1987. Dinoflagellate ultrastructure and complex organelles. Pages 92–142 *in:* F. J. R. Taylor, editor. The biology of dinoflagellates. Academic Press, New York.

Dogiel, V. A. 1965. General Protozoology. 2nd Edition, revised by J. I. Poljansij and E.M. Chejsin. Oxford University Press, London.

Doucet, C. M., and E. J. Maly. 1990. Effects of copper on the interaction between the predator *Didinium nasutum* and its prey *Paramecium caudatum*. Canadian Journal of Fisheries and Aquatic Sciences 47:1122–1127.

Dragesco, J. 1966a. Observations sur quelques ciliés libres. Archiv für Protistenkunde 109:155–206.

Dragesco, J. 1966b. Ciliés libres de thonon et sus environs. Protistologica 2:59–95.

Eddy, S. 1930. The freshwater armored or thecate dinoflagellates. Transactions of the American Microscopical Society 49:277–321.

Egerter, D. E., J. R. Anderson, and J. O. Washburn. 1986. Dispersal of the parasitic ciliate *Lambornella clarkii*: implications for ciliates in the biological control of mosquitoes. Proceedings of the National Academy of Sciences U.S.A. 83:7335–7339.

Elliott, A. M., editor. 1973. Biology of *Tetrahymena*. Dowden, Hutchinson & Ross, Stroudsburg, Pennsylvania.

Elliott, A. M., and G. L. Clemmons. 1966. An ultrastructural study of ingestion and digestion in *Tetrahymena pyriformis*. Journal of Protozoology 13:311–323.

Evans, M. S., D. W. Sell, and A. M. Beeton. 1981. *Tokophrya quadripartita* and *Tokophrya* sp. (Suctoria) associations with crustacean zooplankton in the Great Lakes region. Transactions of the American Microscopical Society 100:384–391.

Evans, M. S., L. M. Sicko-Goad, and M. Omair. 1981. Seasonal occurrence of *Tokophrya quadripartita* (Suctoria) as epibionts on adult *Limnocalanus macrurus* (Copepoda: Calanoida) in southeastern Lake Michigan. Transactions of the American Microscopical Society 98:102–109.

Fauré-Fremiet, E. 1967. Chemical aspects of ecology. Pages 21–54 *in:* M. Florkin and B. T. Scheer, editors. Chemical ecology. Vol. 1: Protozoa. Academic Press, New York.

Febvre-Chevalier, C. 1985. Subphylum Sarcodina Schmarda. IV. Class Heliozoea Haeckel. Pages 302–317 *in:* J. J. Lee, S. H. Hutner, and E. C. Bovee, editors. An illustrated guide to the Protozoa. Society of Protozoologists. Lawrence, Kansas.

Fenchel, T. 1967. The ecology of the marine microbenthos. I. The quantitative importance of ciliates as compared with metazoans in various types of sediments. Ophelia 4:121–137.

Fenchel, T. 1969. The ecology of the marine microbenthos. IV. Structure and function of the benthic ecosystem, its chemical and physical factors and the microfauna communities with special reference to the ciliated protozoa. Ophelia 6:1–182.

Fenchel, T. 1974. Intrinsic rate of increase: the relationship with body size. Oecologia (Berlin) 14:317–326.

Fenchel, T. 1980a. Suspension feeding in ciliated protozoa: functional response and particle size selection. Microbial Ecology 6:1–11.

Fenchel, T. 1980b. Suspension feeding in ciliated protozoa: feeding rates and their ecological significance. Microbial Ecology 6:13–25.

Fenchel, T. 1982a. Ecology of heterotrophic microflagellates. I. Some important forms and their functional morphology. Marine Ecology Progress Series 8:211–223.

Fenchel, T. 1982b. Ecology of heterotrophic microflagellates. II. Bioenergetics and growth. Marine Ecology Progress Series 8:225–231.

Fenchel, T. 1987. Ecology of Protozoa. Springer-Verlag, Berlin.

Fenchel, T., T. Perry, and A. Thane. 1977. Anaerobiosis and symbiosis with bacteria in free-living ciliates. Journal of Protozoology 24:154–163.

Finlay, B. J. 1977. The dependence of reproductive rate on cell size and temperature in freshwater ciliated protozoa. Oecologia (Berlin) 30:75–81.

Finlay, B. J., and T. Fenchel. 1986. Physiological ecology of the ciliated protozoan *Loxodes*. Reports of the Freshwater Biological Association 54:73–96.

Finlay, B. J., and C. Ochsenbein-Gattlen. 1982. Ecology of Free-Living Protozoa. A Bibliography. Occasional Paper No. 17. Freshwater Biological Association, Ambleside, England.

Finlay, B. J., U.-G. Berninger, L. J. Stewart, R. M. Hindle, and W. Davison. 1987. Some factors controlling the distribution of two pond-dwelling ciliates with algal symbionts (*Frontonia vernalis* and *Euplotes daideleos*). Journal of Protozoology 34:349–356.

Finlay, B. J., A. Rogerson, and A. J. Cowling. 1988. A Beginners Guide to the Collection, Isolation, Cultivation and Identification of Freshwater Protozoa. Culture Collection of Algae and Protozoa. Freshwater Biological Association, Ambleside, England.

Foissner, W. 1988. Taxonomic and nomenclatural revision of Sládecek's list of ciliates (Protozoa: Ciliophora) as indicators of water quality. Hydrobiologia 166:1–64.

Gaines, G., and M. Elbrächter. 1987. Heterotrophic nutrition. Pages 224–268 *in:* F. J. R. Taylor, editor. The biology of dinoflagellates. Academic Press, New York.

Gates, M. A. 1978. An essay on the principles of ciliate systematics. Transactions of the American Microscopical Society 97:221–235.

Gause, G. F. 1934. The struggle for existence. Williams & Wilkins, Baltimore, Maryland.

Giese, A. C. 1973. *Blepharisma*. Stanford University Press, Stanford, California.

Gifford, D. J. 1985. Laboratory culture of marine planktonic oligotrichs (Ciliophora; Oligotrichida). Marine Ecology Progress Series 23:257–267.

Goldman, J. C., and D. A. Caron. 1985. Experimental studies on an omnivorous microflagellate: implications for grazing and nutrient recycling in the marine microbial food chain. Deep-Sea Research 32:899–915.

Goldman, J. P., and M. R. Dennet. 1990. Dynamics of prey selection by an omnivorous flagellate. Marine Ecology Progress Series 59:183–194.

Goldman, J.C., D.A. Caron, O.K. Andersen, and M.R. Dennett. 1985. Nutrient cycling in a microflagellate food chain. Marine Ecology Progress Series 24:231–242.

Golini, V.I., and J.O Corliss. 1981. A note on the occurrence of the hymenostome ciliate *Tetrahymena* in chironomid larvae. Transactions of the American Microscopical Society 100:89–93.

Goodrich, J.P., and T.L. Jahn. 1943. Epizoic suctoria (Protozoa) from turtles. Transactions of the American Microscopial Society 62:245–253.

Goulder, R. 1972. Grazing by the ciliated protozoon *Loxodes magnus* on the alga *Scenedesmus* in a eutrophic pond. Oecologia (Berlin) 13:177–182.

Goulder, R. 1980. The ecology of two species of primitive ciliated protozoa commonly found in standing freshwaters (*Loxodes magnus* Stokes and *L. striatus* Penard). Hydrobiologia 72:131-158.

Grell, K.G. 1973. Protozoology. Springer-Verlag, Berlin.

Grolière, C.-A. 1980. Morphologie et stomatogenèse chez deux ciliés Scuticociliatida des genres *Philasterides* Kahl, 1926 et *Cyclidium* O.F. Müller, 1786. Acta Protozoolgica 19:195–206.

Grover, J.P. 1989. Effects of Si:P supply ratio, supply variability, and selective grazing in the plankton: an experiment with a natural algal and protistan assemblage. Limnology and Oceanography 34:349–367.

Gunderson, J.H., H. Elwood, A. Ingold, K. Kindle, and M.L. Sogin. 1987. Phylogenetic relationships between chlorophytes, chrysophytes, and oomycetes. Proceedings of the National Academy of Sciences 84:5823–5827.

Hairston, N.G. 1967. Studies on the limitation of a natural population of *Paramecium aurelia*. Ecology 48:904–910.

Hairston, N.G., and S.L. Kellerman. 1964. *Paramecium* ecology: electromigration of field samples and observations on density. Ecology 45:373–376.

Hairston, N.G., and S.L. Kellerman. 1965. Competition between varieties 2 and 3 of *Paramecium aurelia*: the influence of temperature in a food-limited system. Ecology 46:134–139.

Hanson, E.D. 1977. The origin and early evolution of animals. Wesleyan Univ. Press, Middletown, Connecticut.

Hardin, G. 1943. Flocculation of bacteria by protozoa. Nature (London) 151:642.

Hausmann, K. 1978. Extrusive organelles of protozoa. International Review of Cytology 52:197–276.

Heinbokel, J.F. 1988. Reproductive rates and periodicities of oceanic tintinnine ciliates. Marine Ecology Progress Series 47:239–248.

Henebry, M.S., and R.T. Ridgeway. 1980. Epizooic ciliated protozoa of planktonic copepods and cladocerans and their possible use as indicators of organic water pollution. Transactions of the American Microscopical Society 98:495–508.

Hewett, S.W. 1980a. Prey-dependent cell size in a protozoan predator. Journal of Protozoology 27:311–313.

Hewett, S.W. 1980b. The effect of prey size on the functional response and numerical response of a protozoan predator to its prey. Ecology 61:1075–1081.

Hofeneder, H. 1930 Über die animalische Ernährung von *Ceratium hirudinella* O.F. Müller und über die Rolle des Kernes bei dieser Zellfunkton. Archiv für Protistenkunde 71:1–32.

Holwill, M.E.J. 1974. Hydrodynamic aspects of ciliary and flagellar movement. Pages 143–175 *in:* M.A. Sleigh, editor. Cilia and flagella. Academic Press, New York.

Jackson, K.M., and J. Berger. 1985. Life history attributes of some ciliated protozoa. Transactions of the American Microscopical Society 104:53–62.

Jacobson, D.M., and D.M. Anderson. 1986. Thecate heterotrophic dinoflagellates: feeding behavior and mechanisms. Journal of Phycology 22:249–258.

Jahn, T.L., and W.R. McKibben. 1937. A colorless euglenoid flagellate, *Khawkkinea halli* n. gen., n. sp. Transactions of the American Microscopical Society 56:48–54.

Jankowski, A.W. 1964a. Morphology and evolution of Ciliophora. III. Diagnoses and phylogenesis of 53 sapropelebionts, mainly of the order Heterotrichida. Archiv für Protistenkunde 107:185–294.

Jankowski, A.W. 1964b. Morphology and evolution of Ciliophora. IV. Sapropelebionts of the family Loxocephalidae fam. nova, their taxonomy and evolutionary history. Acta Protozoologica 2:33–58.

Jeon, K.W. editor. 1973. The Biology of *Amoeba*. Academic Press, New York.

Jeon, K.W. editor. 1983. Intracellular Symbiosis. International Review of Cytology. Suppl. No.14.

Kahl, A. 1930–1935. Urtiere oder Protozoa. I: Wimpertiere oder Cilata(Infusoria). pp. 1–886 *in:* Dahl, editor. Die Tierwelt Deutschlands 18 (1930), 21 (1931), 25 (1932),30 (1935). Fischer-Verlag, Jena.

Kankaala, P., and P. Eloranta. 1987. Epizooic ciliates (*Vorticella* sp.) compete for food with their host *Daphnia longispina* in a small polyhumic lake. Oecologia (Berlin) 73:203–206.

Kent, W.S. 1881–1882. A manual of the infusoria, Vol. 2. David Bogue,London.

Kentouri, M., and P. Divanach. 1986. Sur l'mportance des ciliés pélagiques dans l'alimentation des stades larvaires de poissons. (On the importance of pelagic ciliates in the feeding of larval stages of fish.) Année Biologique 25:307–318.

Kerr, S.J. 1983. Colonization of blue-green algae by *Vorticella* (Ciliata, Peritrichida). Transactions of the American Microscopical Society 102:38–47.

Kornienko, G.S. 1971. The role of infusoria in the feeding of larvae of herbivorous fish. Journal of Ichthyology 11:241–246.

Kudo, R.R. 1966. Protozoology, 5th edition. C.C. Thomas, Springfield Illinois.

Lackey, J.B. 1959. Zooflagellates. Pages 190–231 *in:* W.T. Edmondson, editor. Fresh-water biology. Wiley, New York.

Lal, S., B.M. Saxena, and R. Lal. 1987. Uptake, metabolism and effects of DDT, Fenthrothion and chlorphyrifos on *Tetrahymena pyriformis*. Pesticide Science 21:181–191.

Landis, W.G. 1981. The ecology of the killer trait, and interactions of five species of the *Paramecium aurelia* complex inhabiting the littoral zone. Canadian Journal of Zoology 59:1734–1743.

Laybourn-Parry, J. 1984. A Functional Biology of Free-Living Protozoa. University of California Press, Berkeley.

Leadbeater, B.S.C., and C. Morton. 1974. A microscopical study of a marine species of *Codosiga* James-Clark (Choanoflagellata) with special reference to the ingestion of bacteria. Biological Journal of the Linnean Society 6:337–347.

Lee, C.C., and T. Fenchel. 1972. Studies on ciliates associated with sea ice from Antarctica. II. Temperature responses and tolerances in ciliates from Antarctic, temperate and tropical habitats. Archiv für Protistenkunde 114:237–244.

Lee, J.J. 1985. Order 2. Cercomonadida Vickerman. Page 117 *in:* J.J. Lee, S.H. Hutner, and E.C. Bovee, editors, An illustrated guide to the Protozoa. Society of Protozoologists. Lawrence, Kansas.

Lee, J.J., and J.O. Corliss. 1985. Symposium on Symbiosis in Protozoa: introductory remarks. Journal of Protozoology 32:371–372.

Lee, J.J., E.B. Small, D.H. Lynn, and E.C. Bovee. 1985. Some techniques for collecting, cultivating, and observing protozoa. pages 1–7 *in:* J.J. Lee, S.H. Hutner, and E.C. Bovee, editors. An illustrated guide to the Protozoa. Society of Protozoologists, Lawrence, Kansas.

Leedale, G.F. 1985. Order 3. Euglenida Bütschli, 1984. Pages 41–54*in:* J.J. Lee, S.H. Hutner, and E.C. Bovee, editors. An illustrated guide to the Protozoa. Society of Protozoologists. Lawrence, Kansas.

Lemmermann, E. 1914. Protomastiginae. Pages 30–121 *in:* A. Pascher, editor. Die süsswasser-flora deutschlands, österreichs und der schweiz, 1. Flagellatae I. Fischer, Jena.

Luckinbill, L.S. 1979. Selection and the r/K continuum in experimental populations of protozoa. American Naturalist 113:427–437.

Luckinbill, L.S., and M.M. Fenton. 1978. Regulation and environmental variability in experimental populations of protozoa. Ecology 59:1271–1276.

Lynn, D.H. 1975. The life cycle of the histophagous ciliate *Tetrahymena corlissi* Thompson, 1955. Journal of Protozoology 22:188–195.

Lynn, D.H. 1976. Comparative ultrastructure and systematics of the Colpodida. Structural conservatism hypothesis and a description of *Colpoda Steinii* Maupas. Journal of Protozoology 23:302–314.

Lynn, D.H. 1977. Comparative ultrastructure and systematics of the Colpodida. Fine structural specializations associated with large body size in *Tillina magna* Gruber, 1880. Protistologica 12:629–648.

Lynn, D.H. 1981. The organization and evolution of the microtubular organelles in ciliated protozoa. Biological Reviews of the Cambridge Philosophical Society 56:243–292.

Lynn, D.H., and J.O Corliss. 1990. Phylum Ciliophora (Ciliates). Pages 331–465 *In:* F.W. Harrison, and J.O. Corliss, editors. Microscopic Anatomy of Invertebrates. Vol. 1: The Protozoa. Wiley-Liss, New York. In press.

Maguire, B., Jr. 1963a. The passive dispersal of small aquatic organisms and their colonization of isolated bodies of water. Ecological Monographs 33:161–185.

Maguire, B., Jr. 1963b. The exclusion of *Colpoda* (Ciliata) from superficially favorable habitats. Ecology 44:781–784.

Maguire, B., Jr. 1977. Community structure of protozoans and algae with particular emphasis on recently colonized bodies of water. Pages 355–397 *in:* J. Cairns Jr., editor. Aquatic microbial communities. Garland, New York.

Margulis, L. 1981. Symbiosis in cell evolution. Freeman, San Francisco, California.

Margulis, L., and D. Sagan. 1986. Origins of sex. Yale University Press, New Haven, Connecticut.

Margulis, L., and K.V. Schwartz. 1988. Five kingdoms: an illustrated guide to life on earth. 2nd Edition. Freeman, San Francisco, California.

Mast, S.O., and W.F. Hahnert. 1935. Feeding, digestion, and starvation in *Amoeba proteus* Leidy. Physiological Zoology 8:255–272.

Matthes, D. 1954. Suktorienstudien. VI. Die gattung Heliophrya Saedeleer and Tellier 1929. Archiv für Protistenkunde 100:143–152.

McManus, G.B., and J.A. Fuhrman. 1986. Bacterivory in seawater studied with the use of inert fluorescent particles. Limnology and Oceanography 31:420–426.

Montagnes, D.J.S., and D.H. Lynn. 1987a. A quantitative protargol stain (QPS) for ciliates: a method description and test of its quantitative nature. Marine Microbial Food Webs 2:83–93.

Montagnes, D.J.S., and D.H. Lynn. 1987b. Agar embedding on cellulose filters: an improved method of mounting protists for protargol and Chatton-Lwoff staining. Transactions of the American Microscopical Society 106:183–186.

Montagnes, D.J.S., D.H. Lynn, J.C. Roff, and W.D. Taylor. 1988. The annual cycle of heterotrophic planktonic ciliates in the waters surrounding the Isles of Shoals, Gulf of Maine: an assessment of their trophic role. Marine Biology 99:21–30.

Nilsson, J.R. 1976. Phagotrophy in *Tetrahymena*. Pages 339–379 *in:* M. Levandowsky and S.H. Hutner, editors. Biochemistry and Physiology of Protozoa. Vol. 2. Academic Press, New York.

Nilsson, J.R. 1987. Structural aspects of digestion of *Escherichia coli* in *Tetrahymena*. Journal of Protozoology 34:1–6.

Nisbet, B. 1984. Nutrition and feeding strategies in Protozoa. Croom Helm, London.

Noland, L.E. 1925. The factors affecting the distribution of fresh water ciliates. Ecology 6:437–452.

Noland, L.E. 1959. Ciliophora. Pages 265–297 *in:* W.T. Edmondson, editor. Fresh-water biology. Wiley, New York.

Noland, L.E., and M. Godjics. 1967. Ecology of free-living Protozoa. Pages 215–266 *in:* T.-T. Chen, editor. Research in protozoology. Vol. 2. Pergamon, Oxford.

Nyberg, D. 1974. Breeding systems and resistance to environmental stress in ciliates. Evolution 28:367–380.

Nyberg, D., and P. Bishop. 1983. High levels of phenotypic variability of metal and temperature tolerance in *Paramecium*. Evolution 37:341–357.

Nygaard, K., K.Y. Borsheim, and T.F. Thingstad. 1988. Grazing rates on bacteria by marine heterotrophic microflagellates compared to uptake rates of bacteria-sized monodisperse fluorescent latex beads. Marine Ecology Progress Series 44:159–165.

Pace, M.L., and J.D. Orcutt, Jr. 1981. The relative importance of protozoans, rotifers, and crustaceans in a freshwater zooplankton community. Limnology and Oceanography 26:822–830.

Page, F.C. 1976. An illustrated key to freshwater and soil amoebae, with notes on cultivation and ecology. Scientific Publication Number 34, Freshwater Biological Association, Ambleside, England.

Page, F.C. 1977. The genus *Thecamoeba* (Protozoa, Gymnamoebia). Species distinctions, locomotive morphology, and protozoan prey. Journal of Natural History 11:25–63.

Page, F.C. 1987. The classification of 'naked' amoebae (Phylum Rhizopoda). Archiv für Protistenkunde 133:199–217.

Page, F.C. 1988. A New Key to Freshwater and Soil Gymnamoebae. Freshwater Biological Association, Ambleside, England.

Pascher, A. 1913. Chrysomonadinae. Pages 7–95 *in:* A. Pascher, editor. Die süswasser-flora deutschlands, österreichs und der schweiz, 2. Flagellatae II. Fischer, Jena.

Patterson, D.J. 1980. Contractile vacuoles and associated structures: their organization and function. Biological Reviews of the Cambridge Philosophical Society 55:1–46.

Peck, R.K. 1985. Feeding behavior in the ciliate *Pseudomicrothorax dubius* is a series of morphologically and physiologically distinct events. Journal of Protozoology 32:492–501.

Picken, L.E.R. 1937. The structure of some protozoan communities. Journal of Ecology 25:368–384.

Pomeroy, L.R. 1974. The ocean's food web: a changing paradigm. BioScience 24: 499-504.

Porter, K.G. 1988. Phagotrophic phytoflagellates in microbial food webs. Hydrobiologia 159: 89-97.

Porter, K.G., M.L. Pace, and J.F. Battey. 1979. Ciliate protozoans as links in freshwater planktonic food chains. Nature (London) 277:563–565.

Porter, K.G., E.B. Sherr, B.F. Sherr, M. Pace, and R.W. Sanders. 1985. Protozoa in planktonic food webs. Journal of Protozoology 32:409–415.

Pratt, J.R., and B.H. Rosen. 1983. Associations of species of *Vorticella* (Peritrichida) and planktonic algae. Transactions of the American Microscopical Society 102:48–54.

Rainier, H. 1968. Heliozoa. *in:* F. Dahl, editor. Die tierwelt deutschlands 56:3–174. Fischer-Verlag, Jena.

Rapport, D., J. Berger, and D.B. Reid. 1972. Determination of food preference of *Stentor coeruleus*. Biological Bulletin 142:103–109.

Rasmussen, L. 1976. Nutrient uptake in *Tetrahymena*. Carlsberg Research Communications 41:143–167.

Ricci, N. 1989. Microhabitats of ciliates: specific adaptations to different substrates. Limnology and Oceanography 34:1089–1097.

Ricketts, T.R. 1972. The interaction of particulate material and dissolved foodstuffs in food uptake by *Tetrahymena pyriformis*. Archive für Mikrobiologie 81:344–349.

Rigler, F.H., M.E. McCallum, and J.C. Roff. 1974. Production of zooplankton in Char Lake. Journal of the Fisheries Research Board of Canada 31:637–646.

Rogerson, A., and J. Berger. 1981. Effect of crude oil and petroleum-degrading microorganisms on the growth of freshwater and soil protozoa. Journal of General Microbiology 124:53–59.

Rogerson, A., W.Y. Shiu, G.L. Huang, D. Mackay, and J. Berger. 1983. Determination and interpretation of hydrocarbon toxicity to ciliate protozoa. Aquatic Toxicology 3:215–228.

Rogerson, A., B.J. Finlay, and U.-G. Berninger. 1989. Sequestered chloroplasts in the freshwater ciliate *Strombidium viride* (Ciliophora: Oligotrichina). Transactions of the American Microscopical Society 108:117–126.

Ruthven, J.A., and J. Cairns Jr. 1973. Response of freshwater protozoan artificial communities to metals. Journal of Protozoology 20:127–135.

Sambanis, A., and A.G. Fredrickson. 1988. Persistence of bacteria in the presence of viable, nonencysting bacterivorous ciliates. Microbial Ecology 16:197–212.

Sanders, R.W. 1988. Feeding by *Cyclidium* sp. (Ciliophora, Scuticociliatida) on particles of different sizes and surface properties. Bulletin of Marine Science 43:446–457.

Sanders, R.W., and K.G. Porter. 1988. Phagotrophic phytoflagellates. Advances in Microbial Ecology 10:167–192.

Sanders, R.W., and K.G. Porter. 1990. Bacterivorous flagellates as food resources for the freshwater cladoceran zooplankter *Daphnia ambigua*. Limnology and Oceanography 35:188–191.

Sanders, R.W., K.G. Porter, S.J. Bennett, and A.E. DeBiase. 1989. Seasonal patterns of bacterivory by flagellates, ciliates, rotifers and cladocerans in a freshwater planktonic community. Limnology and Oceanography 34:673–687.

Sanders, R.W., K.G. Porter, and D.A. Caron. 1990. Relationship between phototrophy and phagotrophy in the mixotrophic chrysophyte *Poterioochromonas mahlhamensis*. Microbial Ecology 19:97–109.

Schönborn, W. 1977. Production studies on Protozoa. Oecologia (Berlin) 27:171–184.

Schuster, F.L. 1979. Small amebas and ameboflagellates. Pages 215–285, *in:* M. Levandowsky and S.H. Hutner, editors. Biochemistry and physiology of protozoa. Vol. 1. Academic Press, New York.

Scott, B.F., P.J. Wade, and W.D. Taylor. 1984. Impact of oil and oil-dispersant mixtures on the fauna of freshwater ponds. The Science of the Total Environment 35:191–206.

Scott, J.M. 1985. The feeding rates and efficiencies of a marine ciliate, *Strombidium* sp., grown under chemostat steady-state conditions. Journal of Experimental Marine Biology and Ecology 90:81–95.

Shawhan, F.M., and T.L. Jahn. 1947. A survey of the genus *Petalomonas* Stein (Protozoa: Euglenida). Transactions of the American Society 66:182–189.

Sherr, B.F., and E.B. Sherr. 1987. High rates of consumption of bacteria by pelagic ciliates. Nature (London) 325:710–711.

Sherr, B.F., E.B. Sherr, and C. Pedrós-Alió. 1989. Simultaneous measurement of bacterioplankton production and protozoan bacterivory in estuarine water. Marine Ecology Progress Series 54:209–219.

Sherr, E.B., B.F. Sherr, and L.J. Albright. 1987. Bacteria: Link or sink? Science 235:88–89.

Shuter, B.J., J.E. Thomas, W.D. Taylor, and A.M. Zimmerman. 1983. Phenotypic correlates of genomic DNA content in unicellular eukaryotes and other cells. American Naturalist 122:26–44.

Singh, B.N. 1975. Pathogenic and non-pathogenic amoebae. Macmillan, London.

Slabbert, J.L., and W.S.G. Morgan. 1982. A bioassay technique using *Tetrahymena pyriformis* for the rapid assessment of toxicants in water. Water Research 16:517–523.

Sládecek, V. 1973. System of water quality from the biological point of view. Ergebnisse der Limnologie 7:1–218.

Sleigh, M.A., and J.R. Blake. 1977. Methods of ciliary propulsion and their size limitations. Pages 243–256 *in:* T.J. Pedley, editor. Scale effects in animal locomotion. Academic Press, New York.

Small, E.B., and D.H. Lynn. 1985. Phylum Ciliophora Doflein, 1901. Pages 393–575 *in:* J.J. Lee, S.H. Hutner, and E.C. Bovee, editors. An illustrated guide to the protozoa. Society of Protozoologists, Lawrence, Kansas.

Smith, G.M. 1950. The fresh-water algae of the United States, 2nd ed. McGraw-Hill, New York.

Sogin, M.L., and H.J. Elwood. 1986. Primary structure of the *Paramecium aurelia* small-subunit rRNA coding region: phylogenetic relationships within the Ciliophora. J. Molecular Evolution 23:53–60.

Sogin, M.L., J.H. Gunderson, H.J. Elwood, R.A. Olonso, and D.A. Peattie. 1989. Phylogenetic meaning of the Kingdom concept: an unusual ribosomal RNA from *Giardia lamblia*. Science 243:75–77.

Sonneborn, T.M. 1975. The *Paramecium aurelia* complex of fourteen sibling species. Transactions of the American Microscopical Society 94:155–178.

Sorokin, Y.I., and E.B. Paveljeva. 1972. On the quantitative characteristics of the pelagic ecosystem of Dalnee lake (Kamchatka). Hydrobiologia 40:519–552.

Spoon, D.M., and W.D. Burbank. 1967. A new method for collecting sessile ciliates in plastic petri dishes with tight-fitting lids. Journal of Protozoology 14:735–744.

Stoecker, D.K., L.H. Davis, and A. Provan. 1983. Growth of *Favella* sp. (Ciliata: Tintinnina) and other microzooplankters in cages incubated *in situ* and comparison to growth *in vitro*. Marine Biology 75:293–302.

Stoecker, D.K., W.G. Sunda, and L.H. Davis. 1986. Effects of copper and zinc on two planktonic ciliates. Marine Biology 92:21–29.

Stoecker, D.K., M.W. Silver, A.E. Michaels, and L.H. Davis. 1989. Enslavement of algal chloroplasts by four *Strombidium* spp. (Ciliophora, Oligotrichida). Marine Microbial Food Webs 3:79–100.

Stout, J.D. 1980. The role of protozoa in nutrient cycling and energy flow. Advances in Microbial Ecology 4:1–50.

Suttle, C.A., A.M. Chan, W.D. Taylor, and P.J. Harrison. 1986. Grazing of planktonic diatoms by microflagellates. Journal of Plankton Research 8:393–398.

Taylor, F.J.R. 1987. Some eco-evolutionary aspects of intracellular symbiosis. Pages 1–28 *in:* K.W. Jeon, editor. Intracellular symbiosis. International Review of Cytology. Suppl. No. 14.

Taylor, W.D. 1978a. Growth responses of ciliate protozoa to the abundance of their bacterial prey. Microbial Ecology 4:207–214.

Taylor, W.D. 1978b. Maximum growth rate, size and commonness in a community of bacterivorous ciliates. Oecologia (Berlin) 36:263–272.

Taylor, W.D. 1980. Observations on the feeding and growth of the predaceous oligochaete *Chaetogaster langi* on ciliated protozoa. Transactions of the American Microscopical Society 99:360–368.

Taylor, W.D. 1981. Temporal heterogeneity and the ecology of lotic ciliates. Pages 209–224 *in:* M.A. Locke and D.D. Williams, editors. Perspectives in running water ecology. Plenum, New York.

Taylor, W.D. 1983a. A comparative study of the sessile ciliates of several streams. Hydrobiologia 98:125–133.

Taylor, W.D. 1983b. Rate of population increase and mortality for sessile ciliates in a stream riffle. Canadian Journal of Zoology 61:2023–2028.

Taylor, W.D. 1984. Phosphorus flux through epilimnetic zooplankton from Lake Ontario: relationship with body size and significance to phytoplankton. Canadian Journal of Fisheries and Aquatic Sciences 41:1702–1712.

Taylor, W.D. 1986. The effect of grazing by a ciliated protozoan on phosphorus limitation in batch culture. J. Protozoology 33:47–52.

Taylor, W.D., and J. Berger. 1976. Growth responses of cohabiting ciliate protozoa to various prey bacteria. Canadian Journal of Zoology 54:1111–1114.

Taylor, W.D., and M.L. Heynen. 1987. Seasonal and vertical distribution of Ciliophora in Lake Ontario. Canadian Journal of Fisheries and Aquatic Sciences 44:2185–2191.

Taylor, W.D., and B.J. Shuter. 1981. Body size, genome size, and intrinsic rate of increase in ciliated protozoa. The American Naturalist 118:160–172.

Thompson, A.S., J.C. Rhodes, and I. Pettman. 1988. Catalogue of strains. Culture collection of algae and protozoa, Freshwater Biological Association. Ambleside, England.

Tucker, J.B. 1968. Fine structure and function of the pharyngeal basket in the ciliate *Nassula*. Journal of Cell Science 3:493–514.

Tyndall, R.L., K.S. Ironside, P.L. Metler, E.L. Tan, T.C. Hazen, and C.B. Fliermanns. 1989. Effect of thermal additions on the density and distribution of thermophilic amoebae and pathogenic *Naegleria fowleri* in a newly created cooling lake. Applied and Environmental Microbiology 55:722–732.

Weisse, T. 1990. Trophic interactions among heterotrophic microplankton, nanoplankton, and bacteria in Lake Constance. Hydrobiologia 191:111–112.

Wessenberg, H., and G. Antipa. 1970. Capture and digestion of *Paramecium* by *Didinium nasutum*. Journal of Protozoology 17:250–270.

Wiadnyana, N.N., and F. Rassoulzadegan. 1989. Selective feeding of *Acartia clausi* and *Centropages typicus* on microzooplankton. Marine Ecology Progress Series 53:37–45.

Wichterman, R. 1986. The biology of *Paramecium*. 2nd Edition. Plenum, New York.

Wolfe, J. 1988. Analysis of *Tetrahymena* mucocyst material with lectins and Alcian blue. Journal of Protozoology 35:46–51.

Porifera

4

Thomas M. Frost
Trout Lake Station
Center for Limnology
University of Wisconsin
Madison, Wisconsin 53706

Chapter Outline

I. INTRODUCTION
II. ANATOMY AND PHYSIOLOGY
  A. External Morphology
  B. Internal Anatomy and Physiology
    1. Water Processing
    2. Digestion
    3. Skeleton
    4. Reproduction
III. ECOLOGY AND EVOLUTION
  A. Diversity and Distribution
    1. Diversity
    2. Distribution
  B. Reproduction and Life History
    1. Dormancy
    2. Growth, Sexual Reproduction, and Dispersal within Habitats
    3. Dispersal among Habitats
  C. Ecologic Interactions
    1. Availability of Substrata
    2. Nutritional Ecology
    3. Biotic Interactions
    4. Functional Role in Ecosystems
    5. Paleolimnology
  D. Evolutionary Relationships
IV. COLLECTION, CULTURE, AND PREPARATION FOR IDENTIFICATION
V. CLASSIFICATION
  A. Taxonomic Key to Species of Freshwater Porifera
  B. Descriptions of Freshwater Sponge Species in Canada and the United States
  Literature Cited

I. INTRODUCTION

The presence of sponges in freshwater comes as a surprise to many who think that the members of the phylum Porifera dwell only in the ocean. Yet sponges are common and sometimes abundant inhabitants of a broad diversity of freshwater communities. In some habitats they comprise a major component of the benthic fauna and may play important roles in ecosystem processes.

Sponges are the simplest of the multicellular phyla. Lacking organs, their highest degree of organization is at the tissue level, and specialized cells accomplish many basic biological functions. Despite their simplicity, however, sponges display a variety of elegant adaptations to freshwater habitats including complex life cycles, a capacity to feed selectively on a broad range of particulate resources, and, in most species, an intimate association with algal symbionts. This chapter will introduce the structure, function, and diversity of freshwater sponges with an emphasis on their basic ecology. For those whose interests are whetted by this introduction, Berquist (1978) and Simpson (1984) provide detailed summaries on the biology of the phylum in general.

Within the zoological community, there has been substantial debate whether sponges should be treated as colonies or individuals (Hartman and Reiswig 1973, Simpson 1973). This conflict hinges on whether a particular, continuous growth of sponge tissue should be considered to either function as a single unit or as a set of units that function independently. To an ecologist, the critical element of this debate involves the capability of sponges to continue growth after they have been separated into pieces from an originally intact body. In some large marine species that exhibit body forms suggesting a single, integrated unit, only intact sponges can survive. In many other species, however, sponges can be broken down into components that can function independently when they are separated. This is the case for freshwater sponges, which can be separated into large numbers of independently growing units (Frost *et al.* 1982). As such, freshwater sponges, as well as most marine sponges (Jackson 1977) can be considered functionally as colonial organisms. Simply stated, sponges are much more like corals or bryozoans in terms of the ecological function of their growth form than they are like arthropods or vertebrates. From an ecological perspective, the "colony versus individual" debate for sponges can be traced to attempts to apply concepts derived from metazoans to sponges, where, because of their level of organization, such concepts are not appropriate.

II. ANATOMY AND PHYSIOLOGY

There is a striking contrast between the simplicity of the external features of a sponge and the complexity of its internal structure and function. Viewed macroscopically, freshwater sponges exhibit a limited range of body forms, but a microscopic examination

reveals a variety of internal features. This contrast derives from the basic organization of sponges, in which tissues rather than organs represent the highest level of morphological complexity. In fact, specialized cells, operating independently or in association with related cells, accomplish such primary functions as food gathering, digestion, and reproduction. An appreciation of sponge structure, therefore, is possible only from microscopic examinations. Likewise, it is difficult to discuss the physiological functions of sponges independent of their microscopic structure and these topics are presented jointly, in a general overview.

A. External Morphology

Freshwater sponges display a variety of morphologies which range from encrusting (Fig. 4.1) to rounded (Fig. 4.2) to fingerlike (Fig. 4.3) growth forms. Because these forms can vary substantially within a species (Penney and Racek 1968), general morphology is of little use in sponge taxonomy. Variation in growth form is influenced by environ-

Figure 4.1 Gemmulated colonies of *Spongilla lacustris* and *Eunapius mackayi* exhibiting an encrusted growth form. The distinct zone between the two species illustrates the typical nonmerging front that occurs when two different freshwater species grow into contact with each other on a substratum.

Figure 4.2 A colony of *Ephydatia muelleri* exhibiting a rounded growth form encrusted on a branch from a fallen tree. (From Neidhoefer 1940.)

mental conditions such as water current (Jewell 1935) and the availability of light (Frost and Williamson 1980). Differences can be particularly dramatic between lentic and lotic habitats (Neidhoefer 1940).

The surface structure of sponges varies from flat (Fig. 4.1) to strongly convoluted (Fig. 4.2). Although there are some species-associated differences in surface characteristics, this feature, too, is strongly influenced by habitat conditions. The more detailed

Figure 4.3 A colony of the common freshwater sponge *Spongilla lacustris*, exhibiting a characteristic, finger-like growth form.

characteristics of external sponge anatomy are closely related to internal features. Examples, discussed later, include the incurrent and excurrent portions of the feeding canals and extensions of the skeletal system.

B. Internal Anatomy and Physiology

The basic organization of a sponge consists of a surface epithelium made up of pinacocytes, surrounding an organic matrix, termed the mesohyl, which contains a broad variety of specialized cells. Cells within the mesohyl interact with the epidermal cells to accomplish basic functions, including the processing of water for food uptake and gas exchange, digestion, structural support, and reproduction. Some sponge cells are highly plastic in their behavior and can shift their function with changing environmental conditions.

It also appears that all sponge cells, including those in both the epidermis and the mesohyl are involved in osmoregulation (Weissenfels 1975). This capacity has been critical in allowing the invasion of freshwater by sponges from their original marine habitats.

1. Water Processing

A sponge can be considered, essentially, as a series of progressively finer filters with a mechanism that circulates water through them. The sponge filters are capable of removing particles ranging from large algae to bacterial cells less than 1 μm in diameter and possibly colloidal organic matter (Frost 1987). In addition, the movement of water through the canal system, with its extensive surface area, fosters the transport of gases and excreted materials. The main features of the sponge water processing system (Figs. 4.4 and 4.5) have been identified through the efforts of numerous investigators (reviewed in Frost 1976, 1987), with several papers by Weissenfels and co-workers (Weissenfels 1975, 1976, 1980, 1981, 1982, Langenbruch and Weissenfels 1987) providing particularly important details on structure. Although there are some subtle differences in structures among species, the following summary provides a general overview of the structures associated with sponge water processing.

Water first enters a sponge through a series of 50-μm diameter openings termed ostia that are spread across its surface epithelium. Passing through the surface, inflowing water next enters a

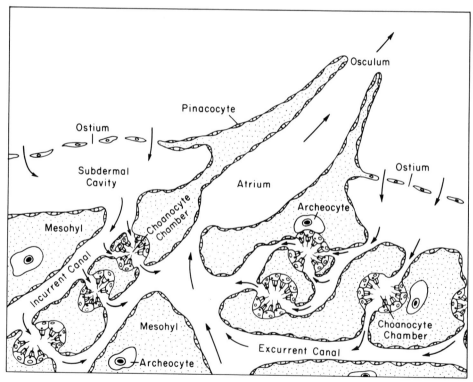

Figure 4.4 A schematic diagram indicating the primary components of the sponge feeding canal system. The diagram is not drawn to scale but Fig. 4.5 provides an indication of the size and arrangement of the feeding system in an actual sponge. Water flow patterns are indicated by arrows.

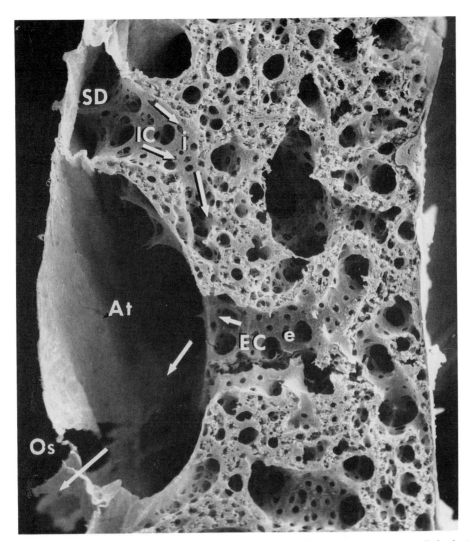

Figure 4.5 A scanning electron micrograph of a cross-section of the freshwater sponge *Ephydatia fluviatilis*, indicating the form, size, and arrangement of the feeding canal system. Illustrated are a subdermal cavity (SD), an incurrent canal (IC) with lateral branches into the sponge (i), and excurrent canal (EC) and its lateral branches (e), an atrium (At), and an osculum (Os). Arrows indicate the flow path of water and numerous choanocyte chambers are visible throughout the sponge. Magnification, approximately 100X. (From Weissenfels l982.)

large subdermal cavity. From there, water flows into incurrent canals that branch repeatedly into canals of progressively narrower diameters until they reach a choanocyte chamber (Fig. 4.4).

Choanocyte chambers are the keystones of the sponge feeding system. Each chamber consists of numerous choanocytes, which are characterized by the presence of a flagellum and a collar of microvilli. Flagellar beating within these chambers is primarily responsible for setting up the flow of water through the sponge. Water flow is also facilitated by the basic hydrodynamic structure of the sponge. Vogel (1974) has demonstrated that water will actually flow pas-

sively, albeit at a reduced rate, through the feeding canals of killed sponges because of this structure. Choanocytes also serve as the final filters in the feeding system. The microvilli that make up the collars of the choanocytes are spaced to form openings less than 0.1 μm in diameter.

Once water has passed the choanocyte chambers, it enters the excurrent portion of the water processing system. Exiting the choanocyte chambers, the canals join together in an anastamosing fashion, forming larger and larger diameter vessels, and finally entering a broad chamber termed an atrium. From this chamber, water exits the sponge body through a specialized structure termed an osculum. Such oscula channel water from large portions of a sponge, and each sponge is likely to have several oscula. Water is forced from the oscula with sufficient force that materials that have passed through the sponge are transported far enough away that they are unlikely to be recaptured by the flow of water entering that sponge.

Although this description may suggest that the structure of the sponge canals is fixed, the system is actually dynamic. The positions of both the incurrent and excurrent canals, as well as their components, can shift as the sponge grows or as portions of the feeding canals are occluded by ingested materials (Mergner 1966).

2. Digestion

A second key feature of sponge feeding involves the capacity of the cells within the canal network to engulf particles through phagocytosis. Cells ranging from the porocytes, which form the ostia on the sponge surface, to the pinacocytes lining the incurrent canals and the choanocytes all exhibit this capacity for phagocytosis. Digestion and transport of nutrients within the sponge body also involve phagocytic activities in which materials are exchanged between cells. Information on activities by cells in sponge digestion has been gained by several detailed studies employing light and electron microscopy (van Weel 1949, Schmidt 1970, Willenz 1980).

Phagocytosis can occur whenever a particle makes contact with a surface within the canal system. Food cells large enough to occlude the ostia of a sponge are taken up directly at the epithelium. Smaller particles are taken up after contact with either the cells lining the canal walls or the collars of the choanocytes. Following initial uptake, food particles are transferred to digestive cells, termed archaeocytes, which move freely within the mesohyl. As digestion proceeds, archaeocytes bearing ingested materials move through the sponge interior and transfer nutritional materials to other cells, such as those involved with reproduction or skeleton formation.

After the digestion of ingested materials is completed, archaeocytes move to the excurrent portion of the canal network. Nondigested material is released into the water flowing out of the sponge by a reverse phagocytic action. Egested particles are then carried by excurrent flow out through oscula at the sponge surface.

Detailed analyses of cell activities in sponge digestion have revealed a surprising degree of complexity. The time necessary for a particle to traverse the digestive system completely, from initial uptake to release into an excurrent canal, can vary markedly with the nature of the food particle and with environmental conditions. Ordinarily, digestible food particles may be held within archaeocytes for more than 12 hr prior to release into the sponge excurrent. Under certain circumstances, sponges can sense indigestible particles and release them after much shorter time periods (Frost 1980a). The capacity to speed the processing time for ingested materials is most evident when a sponge is faced with a dense suspension of particles. At such times, food particles may be taken up, cycled, and released almost immediately (Schmidt 1970).

The sophistication of the sponge feeding system is particularly evident in digestive activities because the rapid cycling of ingested particles can be highly selective. While overabundant particles are being quickly cycled through the sponge, other food particles, present in lower concentrations in the feeding suspension, are held within the sponge for periods that are sufficiently long to allow normal digestion (Frost 1980a, 1981). Sponges, therefore, are selective feeders, not in their initial uptake of particles as selective feeding is typically defined, but in their ultimate use of these resources (Frost 1980b).

The capacity of a sponge to vary particle transit time is linked to the large number of cells that are operating within it. Digestive cells are sufficiently numerous that, under normal feeding conditions, each cell handles only one ingested particle at a time (Willenz 1980). Each individual archaeocyte can then be considered as a separately functioning digestive unit with the capacity to tune its actions to the characteristics of the food particle that it is processing.

3. Skeleton

The gross structure of freshwater sponges derives from an interplay between two fundamentally different components, a mineral skeleton made up of siliceous structures termed spicules and an organic skeleton made up of collagen. Collagen binds spicules together into rigid structures yielding the basic framework of the sponge. All freshwater sponges exhibit skeletal systems comprised of siliceous spicules and collagen. Marine sponges exhibit a broader diversity of supporting structures (Berquist 1978).

a. Siliceous Spicules

Sponges are unique among multicellular animals in the use of silica as a primary component of their skeletons. Freshwater sponge spicules are made up of opalescent silica, which has been laid down along an axial, organic filament by specialized cells termed sclerocytes. Considerable diversity exists in spicule morphology and, because much of this variability is species specific, their structure plays a critical role in sponge taxonomy.

Three general classes of spicules are recognized: megascleres, which make up the main framework of a sponge; microscleres, which appear to add structural reinforcement for sponge tissues; and gemmoscleres, which form part of the resistant coat of gemmules. Megascleres are needle-shaped structures (Figs. 4.6 and 4.7) that range in length from 150–450 μm. Microscleres are similar in form to megascleres but are usually less than 20% of their length (Fig. 4.6). Gemmoscleres are similar in size to microscleres and both exhibit a variety of forms ranging from needle like to dumbbell- or star-shaped. Any of these three spicule forms may be smooth or spined depending upon the species.

b. *Collagen*

The predominant form of collagen in sponges is spongin, which serves primarily to bind the spicules of the inorganic skeleton together. A second form of collagen provides smaller-scale structure within the sponge mesohyl. Finally, collagen, often combined with gemmoscleres, forms the resistant coat of gemmules.

4. Reproduction

Both sexual and asexual processes can play major roles in sponge reproduction. Sexual reproduction involves the activities of numerous, isolated reproductive cells functioning throughout the body of an active sponge. Asexual reproduction often involves major changes in all of the cells within a sponge.

a. *Sexual Reproduction*

Although detailed observations have been made for only a few freshwater species, it appears that the mode of sexual reproduction is gonochoristic, with each separate sponge being entirely male or female. In at least some cases however, the gender of a particular sponge is not fixed but appears to depend upon environmental conditions. In at least one species, *Spongilla lacustris*, sponges have been shown to switch sex from one year to the next (Gilbert and Simpson 1976a).

Sexual reproduction is accomplished by specialized cells that develop within the mesohyl during restricted portions of the year. Reproductive cells are derived from other, more common, sponge cells. Spermatogenesis occurs within distinct spermatic cysts, which appear to be formed from choanocytes. Egg cells (oocytes) appear to develop either from archaeocytes or choanocytes. Oocyte growth is dependent upon a variety of nurse cells that are phagocytized as the egg develops (Gilbert 1974).

Fertilization occurs when sperm that have been released into the open water by other sponges are brought into the canal systems of female sponges by their feeding currents. Sperm are then taken up by the sponge, probably by choanocytes, and conveyed to an egg cell where fertilization takes place.

As is the general case within the Porifera, freshwater sponges are viviparous. Larvae undergo extensive development prior to their release from the mother sponge. Material for growth is provided by nurse cells. In the final stages of development, the sponge larva, termed a parenchymula, contains archaeocytes, pinacocytes, sclerocytes, and choanocyte chambers. In addition, its surface is covered by flagellated epithelial cells that allow the larva to swim upon release.

Once development is completed, larvae are released through the excurrent portion of the feeding canal system. The larvae swim until they settle onto a substratum, where they undergo metamorphosis and quickly develop the structures typical of adult sponges.

b. *Asexual Reproduction*

Nonsexual reproduction in sponges may be simple or complex. On one extreme, clonal development of separate, independently functioning sponges can take place through fragmentation. The plastic nature of sponge cells makes it possible for even very small pieces to develop into active, fully functional sponges. Propagation in this fashion occurs commonly in some habitats (Frost *et al.* 1982).

On the other extreme, most freshwater sponges typically form gemmules in a complex process of dedifferentiation, in which the structures of the active sponge regress into a mass of cells surrounded by a resistant coat (Fig. 4.3). This process, which often occurs prior to the onset of rigorous environmental conditions, permits sponges to weather periods of stress and to disperse into other habitats.

The coat of a mature gemmule usually consists of three collagen layers. In most species, specialized spicules, gemmoscleres, which are formed only within gemmules, are also embedded within the gemmule coat. The coat of each gemmule is continuous except for the presence of a single structure

Figure 4.6 A scanning electron micrograph of megascleres from *Spongilla lacustris* collected in a lake with a high concentration of dissolved silica. A spined microsclere is also present. The bar is 10 μm in length.

termed a micropyle or foraminal aperture, through which cells emerge during germination. Within the gemmule are large numbers of thesocytes, which contain numerous yolk inclusions. Thesocytes have the capacity to store energy-rich materials that are used for growth following germination. The aggregation of cells to form gemmules takes place within the mesohyl. Gemmules are often present in large numbers within the vestigial skeleton of a previously active sponge.

Completely formed gemmules exhibit low metabolic rates until germination is induced. Upon germination, the thesocytes increase their metabolic rate, exit from the gemmule coat through the micropyle, and differentiate into the varied cells typical of an active sponge.

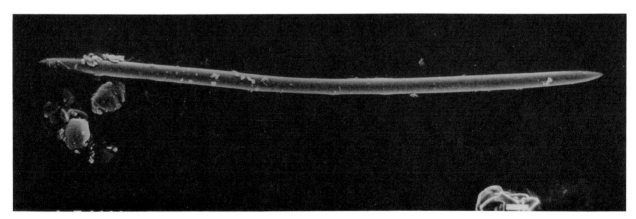

Figure 4.7 A scanning electron micrograph of a megasclere of *Spongilla lacustris* collected in a lake with a low concentration of dissolved silica. The bar is 10 μm in length.

III. ECOLOGY AND EVOLUTION

A. Diversity and Distribution

1. Diversity

The diversity of sponges in freshwater is low in comparison with those in the marine habitats from which they have evolved. Although the entire phylum Porifera consists of more than 5000 species (Berquist 1978), worldwide there are probably fewer than 300 freshwater species. In the most recent taxonomic revision of the Spongillidae, the largest family of freshwater sponges, Penney and Racek (1968) listed a total of 95 species in 18 genera. While it is certain that there are additional, undescribed species, particularly in tropical regions that have not been thoroughly investigated (e.g., the Amazon basin; Volkmer-Ribeiro 1981), it is clear that freshwater forms represent a small group within the sponge phylum.

In turn, North American freshwater sponges represent only a restricted subset of the freshwater species in the world. Until recently, all North American species were traced to a monophyletic group in the Spongillidae (Penney and Racek 1968). In 1986, Volkmer-Ribeiro proposed that two species in the genus *Corvomeyenia* should be grouped along with several South American species into the family Metaniidae, with an evolutionary history in freshwater distinct from the spongillids. Overall, within both the Spongillidae and Metaniidae, my current best estimate is that a total of 27 species in 11 genera occur in North America north of Mexico and the Caribbean.

2. Distribution

Factors affecting the distribution of sponges can be considered to operate on three different scales. From the broadest perspective, the distribution of sponge species is influenced by biogeographic factors. On an intermediate level, the general environmental conditions that exist within a lake or stream will control the presence of a particular sponge species. Finally, conditions such as light, substrata, and wave action will determine sponge distribution within a particular water body.

The freshwater sponge species in North America exhibit a variety of distribution patterns (Penney 1960, Penney and Racek 1968). Several species have been reported throughout the United States and Canada, while others have been observed in only one or a few habitats. In some cases, a broad distribution in North America represents only a subset of

a more extensive range. *Ephydatia fluviatilis* and *Eunapius fragilis* are distributed worldwide, while *Spongilla lacustris*, *Ephydatia muelleri*, and *Trochospongilla horrida* occur throughout temperate regions of the northern hemisphere. Other species are broadly distributed throughout North America but have not been observed on other continents (e.g., *Eunapius mackayi* and *Trochospongilla pennsylvanica*). Of the remaining North American species, most appear to be restricted to limited regions within Canada and the United States and, in the most extreme cases, several species have been reported from single locations.

The apparently limited ranges of some species may result from a lack of observations rather than actual restrictions in sponge distributions. However, with this caveat in mind, there appear to be some general trends in the distribution of sponges across North America. For example, relatively few species of sponges have been reported in western regions and there appears to be a general east-to-west decrease in sponge diversity. A similar trend occurs between northern and southern regions, with more species reported from the northern United States and southern Canada than more southern regions of the continent. In some situations, the limited distribution of a particular species may involve adaptations to climatic conditions. However, in many other cases, biogeographic factors, particularly limitations on dispersal, are likely to control regional distribution patterns (cf. Jones and Rützler 1975).

Within these regional distribution patterns, the factors that regulate the occurrence of a species in a particular lake or stream are less well understood. While *Spongilla alba* appears to be restricted to brackish water habitats (Poirrier 1976), other species are influenced by a variety of environmental factors. Jewell (1935, 1939) has provided the most detailed quantitative information that is available on this topic. In comparing species across lakes in northern Wisconsin, she found considerable variation in the habitat requirements of each species. Some species were restricted to a narrow subset of environmental conditions, while others occurred throughout a broad range of habitats. Chemical factors correlated with sponge distributions included the pH, silica content, overall mineral content, and organic matter content of habitat waters. In addition, some species were favored by flowing water while others were more typically associated with standing water. Because many of the environmental characteristics of the sponge habitats studied by Jewell are correlated (e.g., pH and organic matter content), it is difficult to pinpoint the specific factors that control the distribution of sponges. In general, however, sponges are sufficiently broad in their distribution that many lakes can be expected to contain at least some small specimens. This is particularly true in some regions (e.g., northern Wisconsin) where sponges occur in nearly every lake. In some cases, particularly in smaller lakes, large populations of several sponge species may be common. Aside from cases of geographic isolation, sponges would be excluded only from habitats with frequent physical disturbances, with high pollution levels, or with high loadings of silt or particulates that can clog their feeding systems (Harrison 1974).

Within a lake or stream that is suitable for growth, sponge distribution is controlled by finer-scale environmental features. For most sponges, a hard substratum is essential for growth and this factor will limit the distribution of sponges, even where other conditions are highly favorable. Only a few species, notably *Spongilla lacustris* and to a lesser extent, *Ephydatia muelleri,* can grow in soft sediments. In rivers and streams, current exerts a major influence and sponges are excluded from regions with high flow due to physical disruption. In lakes, sponges can be limited in shallow waters by wave action or ice scour (Bader 1984). In deeper waters, sponges can be limited by low oxygen or by colder temperatures. Some species with symbiotic algae may be limited by light, particularly where water is darkly stained by dissolved organic materials, but species with less dependence on symbionts may thrive in such habitats.

B. *Reproduction and Life History*

The annual life-history pattern for most freshwater sponges involves periods of active growth and dormancy and includes both asexual and sexual reproductive processes. Transitions to and from dormant stages usually involve the asexual processes of gemmule formation and hatching. Asexual reproduction also occurs during active growth periods, both in the formation of gemmules and in the generation of separately functioning sponges through fragmentation. Sexual reproduction occurs only for a limited time during periods of active growth. Key features of the freshwater sponge life cycle are summarized in Fig. 4.8 and will be discussed in detail. Overall, reproduction in sponges involves responses to two distinct problems, propagation within a single habitat and dispersal among habitats. Within-habitat processes are discussed separately, prior to a consideration of dispersal.

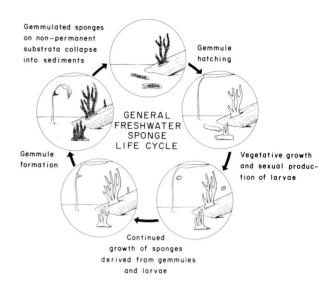

Figure 4.8 General features of the annual life history of freshwater sponges (see text for detailed explanations).

1. *Dormancy*

In most cases, dormant periods for freshwater sponges are characterized by the complete transformation of all active tissue in a sponge into gemmules (Simpson and Fell 1974, Simpson 1984). More rarely, tissue regression occurs, yielding a growth form that functions as a gemmule without its specialized structures. In either situation, sponge feeding ceases and respiratory activities are greatly reduced. Most studies of dormancy in sponges have focused on transitions involving gemmules.

Despite the common role of gemmules in dormancy, their presence does not, in itself, necessarily signify a transition to a dormant life-cycle period. In many cases, particularly where sponges are large, a few gemmules are produced during most of the active season.

Most freshwater sponges undergo a period of dormancy at some time during the year, typically during periods of environmental stress. In temperate habitats, dormant periods occur most commonly during winter. This is the case for the majority of sponges in North America. However, this pattern can be reversed at lower latitudes, with dormancy during hot summer periods and active growth during cooler seasons (Poirrier 1974). In addition, dormant periods have been observed in response to drying in ephemeral habitats. While dormancy is a common feature of freshwater sponge life cycles, it does not always occur; some sponges maintain active growth even under winter ice cover (T. M. Frost personal observation).

The timing of dormancy varies both among species within particular habitats and among habitats for individual species. In Lake Pontchartrain, Louisiana, *Spongilla alba* exhibits a typical pattern of gemmule formation during winter months (Poirrier 1976), while a co-occurring population of *Ephydatia fluviatilis* exhibits a reciprocal pattern of summer dormancy (Harsha *et al.* 1983). At the same time, populations of *E. fluviatilis* in more northern habitats form dormant stages in winter. In Little Rock Lake, Wisconsin, *S. lacustris* undergoes a winter dormancy period while co-occurring populations of *Corvomeyenia everetti* and *Ephydatia muelleri* maintain active growth forms during the entire winter. Populations of *E. muelleri* in nearby Wisconsin lakes undergo typical winter dormancy.

The transition from an active to a dormant state for sponges usually appears to be cued by environmental factors, such as changing temperatures or declining water levels. In some species, this transition may also involve a complex response to previous growth conditions (Gilbert 1975). Within a region, the timing of gemmulation is nearly synchronous across habitats for all populations of a species that exhibit dormancy (T. M. Frost personal observation). In contrast, different species within a region can vary substantially in the seasonal timing of dormancy and in the rate at which the transition from active tissue to gemmules proceeds. In Mud Pond, New Hampshire, *Trochospongilla pennsylvanica* begins forming gemmules in mid-August when water temperatures are at their maximum, while *S. lacustris* does not form gemmules until October when water temperatures reach 10°C (Simpson and Gilbert 1973). The transition to a fully gemmulated state took a longer time in *T. pennsylvanica* than in *S. lacustris* within the same habitat.

Sponge species also vary in the factors that induce their release from dormancy (Simpson and Fell 1974). Temperature appears to be the primary environmental cue that induces gemmule hatching in temperate habitats. However, gemmules of some sponges hatch while water temperatures are at near-winter levels, suggesting that another cue is operating, such as photoperiod or the availability of light.

In some species, gemmules exhibit a form of diapause in which some period at cold temperatures is necessary before hatching can occur (Simpson and Fell 1974). Such a delay prevents hatching during short warming periods prior to the onset of winter. Hatching at such times could lead to major population reductions if time or energy was insufficient for a second formation of gemmules. Not all species form diapausing gemmules and in these cases, hatching will take place during any period of increasing water temperature. The occurrence of diapause appears to be linked to the timing of gemmule formation. Species that form gemmules later in the year are less likely to exhibit diapause. While there are obvious costs to hatching at the wrong time, there may also be advantages to being able to hatch quickly without diapause under some circumstances. Studies of dormancy in freshwater sponges have generally focused on cold-water populations and less is known about the factors that control the transitions to and from dormant stages when they occur in response to hot or dry conditions.

In general, while it is clear that the formation of dormant stages plays a crucial role in the overall life history of most sponges, substantial variability is apparent in: (1) the morphological forms involved in dormant stages; (2) the timing of transitions to and from dormant periods; and (3) the occurrence of diapause. This variability suggests that the overall value of dormancy and the successful strategies by which it can be employed differ markedly among habitats and that sponge responses to such differences involve a diversity of adaptations.

2. Growth, Sexual Reproduction, and Dispersal within Habitats

Nondormant periods in the freshwater sponge life cycle appear to be characterized by continuous growth. Although quantitative studies have been limited, sponge growth in some habitats can be extremely prolific. In a population of *S. lacustris* in Mud Pond, New Hampshire, 10 mg dry mass of gemmule tissue that hatched in the spring produced an average of 18 g of active sponge tissue just prior to gemmule formation the following fall (Frost *et al.* 1982). Growth is largely indeterminate for individual sponges, which may reach surprisingly large sizes. In some extreme habitats for example, a single specimen of *S. lacustris* consisting of intertwined fingers (as in Fig. 4.3) can occupy more than a cubic meter of space (T. M. Frost personal observation).

Sexual reproduction for most freshwater sponge populations occurs synchronously throughout a habitat in a 1–3 month period following hatching from dormant conditions (Simpson and Gilbert 1973). Synchrony across a population is particularly important; otherwise, the sperm released by male sponges would have a low probability of successfully fertilizing an egg (Gilbert 1974).

The motile larvae produced by sexual reproduction in freshwater sponges can play an important role in the establishment of new sponges. In particular, the colonization of new substrata will often depend upon such motile forms. While settling on an appropriate substratum is clearly critical for the successful growth of a new sponge, little information is available on substratum choice by settling sponge larvae. Studies of a variety of marine organisms suggest, however, that substratum selection by freshwater sponge larvae is possible.

Fragmentation of intact specimens into separately functioning units may also play an important role in freshwater sponge life cycles. In some cases, such fragmentation occurs during periods of tissue regression and may not lead to growth on new substrata. In other habitats, however, particularly those with large amounts of aquatic vegetation to which sponges can attach, fragmentation and growth may occur repeatedly during an active season, leading to widespread dispersal throughout a habitat (Frost *et al.* 1982).

Although gemmules are frequently associated with dormant periods, they may also be present in freshwater species during periods of growth. Many sponge species routinely contain a few gemmules within otherwise active tissue, particularly in areas of thick growth. In an extreme case, *S. lacustris* develops specialized summer gemmules, which have much thicker coats than the gemmules formed for over-winter periods (Gilbert and Simpson 1976b). These investigators have suggested that summer gemmules may serve in a "bet-hedging" strategy against some environmental catastrophe, which could occur prior to more extensive gemmule formation during dormant periods. It seems likely that summer gemmules may also serve in dispersal among habitats.

3. Dispersal among Habitats

As is the case for most freshwater organisms, an ability to disperse among different habitats is critical for the continued success of sponges in freshwater environments. Most lakes and streams are ephemeral from an evolutionary perspective. Nearly all lakes in North America are less than 20,000 years old (Frey 1963), a brief period compared to the 100 million year history of freshwater sponges (Racek and Harrison 1975).

While there has been little documentation of dispersal by freshwater sponges among habitats, strong evidence points to gemmules as the primary agents for such movement. The two alternative life-cycle stages that might function in dispersal among habitats, larvae and sponge fragments, possess neither the structural nor the physiological mechanisms necessary for movement across significant dis-

tances. It is conceivable that these forms, particularly the larvae, could function in short-range dispersal among connected water bodies but their fragile structure would preclude any out-of-water transport. Gemmules, on the other hand, with their resistant coats and ability to withstand harsh environmental conditions appear to be well suited for dispersal among habitats.

C. Ecological Interactions

As is the case with all organisms, the size of a freshwater sponge population and its subsequent impact on an ecosystem result from an interplay of growth and loss processes. Growth is regulated by the availability of nutrients and suitable habitats. Losses are influenced by physical environmental conditions and, potentially, by interactions with other organisms. Although quantitative studies are few, it appears that the dynamics of freshwater sponge populations are mediated largely by the availability of suitable microhabitats within a lake or stream. Where substrata for sponge growth are short-lived or where physical disturbances are frequent, growth and loss rates can be extremely rapid and may operate primarily in a density-independent fashion. Under such circumstances, sponge populations would be regulated by an interplay between the physical environment and their own growth and life history processes. Concomitantly, interactions with other organisms would exert only minimal influences on sponge population processes. In contrast, where substrata are permanent, sponge populations appear more likely to be regulated by intra- or interspecific biotic interactions (Jackson and Buss 1975). In general, because freshwater sponges are rarely associated with permanent substrata, such biotic interactions may typically exert only a secondary role in their population dynamics.

1. Availability of Substrata

The development of freshwater sponge populations usually depends upon the availability of hard substrata. Most species must be attached to a substratum in order to grow. In addition, all species would appear to depend upon the presence of some solid surface for the successful settling and growth of their free-swimming larvae. A variety of surfaces are suitable. Examples range from boulders or exposed bedrock to the branches of fallen trees to the leaves and stems of aquatic macrophytes. Potential substrata vary markedly in their permanence relative to the longevity of sponges, and these differences, in turn, have important consequences for sponge population dynamics.

Among North American species, only *S. lacustris* has been routinely observed growing out of soft sediments. A quantitative investigation of its population dynamics (Frost *et al.* 1982), however, illustrated a crucial role for substrata even where sponges did not depend upon such surfaces for their growth. In Mud Pond, New Hampshire, *S. lacustris* grows either directly from soft bottom sediments or attaches to any of several species of aquatic macrophytes that exhibit large summer populations but die back completely before winter. Due to the near absence of any permanent substrata in Mud Pond, gemmulated sponges overwinter almost exclusively in sediments on the pond bottom. Successful hatching in spring depends upon the presence of enough gemmulated tissue to grow out of these sediments, and many sponges are not large enough to make this transition. Most sponges are unsuccessful in hatching and spring populations are sparse. Hatching from sediments, therefore, is the limiting step in the Mud Pond sponge population dynamics. Despite remarkable growth during summer (>1000-fold increase in biomass), the population of *S. lacustris* in Mud Pond was maintained at nearly constant levels throughout five years of observations. Essentially, the population is maintained in a density-independent fashion by a combination of the lack of permanent substrata and the difficulty of gemmule hatching from soft sediments.

Although quantitative data are limited, it appears that sponge population dynamics are substantially different in habitats where more permanent substrata are available. For example, in Mary Lake, Wisconsin, *S. lacustris* grows attached to trees that have fallen into the lake and sponges repeatedly make the transition from active tissue to gemmules and back to active tissue within the same skeleton over the course of several years. In contrast with the large fluctuations in sponge density observed in Mud Pond, it seems that the populations of *S. lacustris* in Mary Lake exhibit higher densities throughout the year and lower growth rates during the course of the summer than those seen in Mud Pond (T. M. Frost personal observation). Under such circumstances, population dynamics are more likely to operate in a density-dependent fashion.

Overall, the availability of substrata exerts a major control over sponge population dynamics. For many species, the lack of substrata limits growth completely. But even where species are not wholly dependent upon substrata, their nature and availability can exert a major influence over sponge population dynamics. The longevity of an individual sponge, including repeated transitions between active and gemmulated states, would appear to be directly related to the permanence of its substratum.

In addition, it seems likely that the importance of both intra- and interspecific biotic interactions in sponge population dynamics will be influenced by the time that a particular substratum has been available for colonization. Only on long-lived substrata would sponges be likely to reach sufficient densities to interact with each other.

2. Nutritional Ecology

When suitable habitats are available, sponges are capable of prolific growth. In some cases, these high growth rates can be traced primarily to the efficient particle-gathering mechanisms of a sponge. However, most freshwater sponges contain large numbers of symbiotic algae and form symbiotic associations with nutritional processes that combine autotrophy and heterotrophy. In addition, because sponges require substantial amounts of silica for their spicules, population growth and maintenance can be influenced by silica availability.

a. Sponge Feeding

Although there have been several detailed studies of both the mechanisms by which sponges feed and the resources they consume (Frost 1987), much less is known about the influence of food availability on sponge growth under natural conditions. The rates at which sponges filter water can be extremely high. During summer, specimens of *S. lacustris* typically filter more than 6 ml/hr/mg dry mass of tissue (Frost 1980b). At this rate, a finger-sized sponge could filter more than 125 liters in a day. Because a feeding sponge effectively removes all particles ranging in size from bacteria to large algae from the water it filters, these rates suggest that sponges may exhaust available food under some circumstances. Food limitation would be particularly likely in still water or where particulate resources are sparse. A comparison of several lakes in northern Wisconsin confirms that sponges vary substantially in their dependence on particulate food for growth under different environmental conditions (T. M. Frost personal observation).

b. Algal Symbionts

In many cases, freshwater sponges are bright green due to the presence of chlorophyll contained within extensive populations of algal symbionts (Gilbert and Allen 1973a, 1973b). This situation is so common that sponges are frequently mistaken for plants. The algal symbiosis in freshwater sponges is representative of a large number of algal–invertebrate associations that are common in marine and freshwater habitats (Reisser 1984). In these mutualistic associations, the nutritional processes of algae and invertebrates are closely coupled. In freshwater sponges, and in many other invertebrates (see Chapters 3 and 5), algae are maintained endosymbiotically within the cells of their host.

Algal–invertebrate symbioses combine autotrophic processes with normal animal heterotrophy. In the resulting mixotrophic nutrition, symbiotic algae provide photosynthetically fixed carbon to the invertebrate host, which, in turn, supplies nutrients such as nitrogen or phosphorus or carbon dioxide to the algae. This combination has proven particularly successful in nutrient poor conditions such as coral reefs (Muscatine and Porter 1977), but occurs over a broad range of conditions.

Algal symbionts play a major role in the growth of some freshwater sponges. Contributions from symbiotic algae have accounted for 50–80% of the growth of *S. lacustris* in Mud Pond, New Hampshire (Frost and Williamson 1980). The contribution of algae was determined by maintaining sponges *in situ* under darkened conditions, which led to the loss of their algae. Comparisons were then made between the growth of such aposymbiotic sponges and that of normal green sponges maintained under lighted but otherwise identical conditions. In similar experiments in Little Rock Lake, Wisconsin, *S. lacustris* and two other sponge species were incapable of any growth and died after a few weeks under darkened conditions (T. M. Frost personal observation). Not all sponges depend upon algal symbionts, however. Some species can live in darkness where their nutrition is based solely on heterotrophic processes.

The factors that control the proportional contribution of autotrophic and heterotrophic processes to overall sponge nutrition are less clear. The observation that sponges failed to grow under darkened conditions in only some habitats indicates that the relationship between a sponge and its algal symbionts can range from facultative to obligate, depending upon environmental conditions. Even for species that sometimes depend to a large extent upon their algae, the amount of algal chlorophyll within a sponge (a rough measure of the potential contribution of its algal symbionts) varies markedly both among habitats for a species and among species within a single habitat. In a survey of northern Wisconsin lakes, the average concentration of algal chlorophyll in *S. lacustris* varied among lakes by more than a factor of three (T. M. Frost personal observation). Similarly, in Little Rock Lake, Wisconsin, three sponges exhibited consistent species-specific differences in their chlorophyll content that were coupled with significantly different rates of both photosynthesis and feeding (Frost and Elias 1990). These differences occurred even though the

three sponge species shared identical environmental conditions.

The symbiotic relationship between sponges and algae differs fundamentally among sponge species. Some species depend to a large extent upon algae for growth while others exhibit little dependence. These differences among species may involve specializations to particular environmental conditions, in that sponges that contain large amounts of algal chlorophyll may only be capable of growth where light is readily available. Comparing habitats, it seems likely that sponges would vary the density of symbionts that they contain in response to the relative availability of light and particulate resources. However, this relationship is not simple. Detailed examinations of several sponge species indicate that increasing particulate resources may favor higher rather than lower concentrations of algae (Frost and Elias 1990).

Until recently, it had been thought that all symbiotic associations of algae and invertebrates in freshwater involved the presence of one group of green algae, termed zoochlorellae (Reisser 1984). Invertebrates hosting these algae include protozoans, hydra, flat worms, clams, and sponges. It has now been determined, however, that at least one freshwater sponge species, *Corvomeyenia everetti,* and likely its congener, *C. carolinensis,* contain a yellow-green alga as their symbiont (T. M. Frost and L. Graham personal observation). The *Corvomeyenia* species are thought to have an evolutionary history that is long distinct from sponges with green algal symbionts (Volkmer-Ribeiro 1986). These distinct histories, combined with numerous overall similarities between sponges with green and yellow-green symbionts indicate that two closely convergent symbiotic associations have evolved separately in the freshwater sponges. Morphologic affinities between marine algae and the yellow-green symbiont suggest that the symbiotic association in *Corvomeyenia* may have developed in an ancestral marine form and has been maintained during a subsequent invasion of freshwater. It is important to note that differences between zoochlorellae and yellow-green algae cannot explain the observed differences in algal contributions to overall sponge nutrition described earlier. Substantial differences in the contribution of algae have been observed in comparisons of sponge species that contain only green algal symbionts.

The variety of ecological interactions between sponges and algae and the complexity of their evolutionary history indicate the overall importance of algal symbiosis in the diversity and distribution of freshwater sponges.

c. *Influence of Silica on Skeletal Growth*

The availability of silica within a habitat may exert both direct and indirect effects on sponge growth. As discussed previously, some sponges are restricted to high- or low-silica habitats. However, even species distributed across habitats exhibiting a broad range of silica conditions show strong responses to its availability. For example, in populations of *S. lacustris* in different lakes in northern Wisconsin, the total amount of biogenic silica and both the number and morphology of megascleres vary directly with lake silica concentrations (T. M. Frost personal observation). Total biogenic silica and spicule width (Figs. 4.6 and 4.7) increase as more silica is available. However, while the size of megascleres decreases under low-silica conditions, their numbers increase substantially—a situation that maximizes the structural support gained per unit of available silica. This suggests a direct and complex response by sponges to different availabilities of silica.

Indirect responses to low-silica conditions are related to changes in the stiffness and growth form of sponges. Specimens of *S. lacustris* from high-silica habitats are much stronger and more resistant to breakage than those growing where silica is low. Sponges in low-silica habitats, therefore, may be limited to growing closely attached to substrata or in waters with little wave action. Also, because spicules may play a role in the resistance of sponges to predation (discussed later), populations in low-silica habitats may be more vulnerable to consumption.

3. *Biotic Interactions*

a. *Competition for Substrata*

The availability of substrata plays a crucial role in the population dynamics of freshwater sponges. As such, competition for surfaces would seem to have a potentially important effect on sponge distribution and abundance. A sponge has the potential to interact for substratum any time that its growth leads to contact with another sponge or any other organism. There are reports of significant interactions for substratum within and between sponge species, as well as between sponges and other organisms. However, such interactions do not appear to be common occurrences and the role of competition for substratum in the regulation of freshwater sponge populations remains largely unknown. In part, this is due to the minimal attention that interactions for substrata have received in freshwater in general. In addition, because freshwater sponge populations are often limited by density-independent processes, the abundance of sponges within a habitat is usually sparse, with little potential for interactions on substrata.

When specimens of the same sponge species grow into contact with each other on a substratum, two distinctly different results have been observed (Van de Vyver 1970). In some cases, the separate sponges will form a single, functionally integrated unit. In other instances, a distinct structural barrier is erected in areas of contact and the two sponges continue to function as completely separate units. The genetic relationship between sponges appears to control whether they will merge or develop a barrier. Separate sponges grown from fragments or gemmules taken from the same original sponge will always merge with each other. In contrast, experimental manipulations have indicated the existence of distinct strains within a sponge species that will not merge with each other (Van de Vyver and Willenz 1975). Studies of marine sponges originally suggested that only sponges that are genetically identical would merge with each other (Niegel and Avise 1983). However, subsequent analyses have indicated that, while the probability of merging appears to be a function of the genetic similarity of sponges, specimens need not be genetically identical in order to join into a single unit (Stoddart *et al.* 1985). Such fusion interactions in sponges operate at a cellular level (Van de Vyver 1979) and have attracted the attention of researchers interested in the basic processes of cell–cell recognition. These studies have direct implications for the understanding of immune response systems in vertebrates and more complex invertebrates (Smith and Hildemann 1984).

Interactions for substrata among different sponge species have been thoroughly documented in marine systems but have received little attention in freshwater systems. Marine studies have demonstrated a complex hierarchy, in which certain sponge species can completely overgrow and kill other species (Jackson and Buss 1975). These interactions have a major influence on overall community ecology (Jackson 1977). Observations of interspecies interactions in freshwater sponges have been limited to instances where sponges have been shown to develop a nonmerging front similar to that described previously for intraspecies interactions (Fig. 4.3). In habitats where substrata are extremely long-lived, it seems likely that some sponge species may be able to completely overgrow other sponges, but the significance of such interactions in freshwater habitats remains to be demonstrated.

Interactions for substrata between sponges and other organisms have likewise received little attention in freshwater. The spreading of a sponge across any surface must involve the overgrowth of microscopic algae and bacteria. Sponges often encrust on aquatic macrophytes but, although it seems likely that plant growth could be reduced by their presence, sponges do not usually kill the plants on which they encrust even when they grow over most of the surface of a plant. In a few cases, freshwater sponges have been observed to overgrow and kill colonies of bryozoans (T. M. Frost personal observation). Overall, however, the role of such interactions remains poorly understood.

b. *Predation and Infaunal Organisms*

Conspicuous predation upon sponges, in which major portions are consumed by a larger organism, is rare. A variety of smaller organisms consume portions of sponges, but they appear to act as grazers or parasites, leaving the sponge that they consume largely intact. Such species that feed on sponges represent a subset of the diverse infaunal community that typically occurs in freshwater sponges. Other infaunal organisms appear only to make use of the structure provided by a sponge in a commensal relationship.

Resistance to predation by larger organisms appears to be characteristic of the Porifera in general (Randall and Hartman 1968, Resh 1976). Mechanical and chemical defenses combine to make sponges resistant to predation. Spicules provide mechanical deterrence to predators; these sharp spines are thought to act as an irritant to their mouths and digestive systems. The effectiveness of spicules in deterring predation was evident in experimental tests, in which snails would not consume *S. lacustris* from habitats where it contained robust spicules, but would consume specimens from habitats in which low-silica availability limited the size of spicules (T. M. Frost and A. P. Covich personal observation). Chemical deterrence to predation has been attributed to a variety of toxic and pharmacologically active compounds, which are well documented in marine sponges (Cecil *et al.* 1976) and are also likely to occur in freshwater species.

Despite the general effectiveness of sponge defense mechanisms, a few specialized organisms prey effectively upon sponges. In many cases, sponge predators have developed a nearly complete dependence upon sponges. The few large predators on freshwater sponges include fishes, crayfishes, and possibly snails. Extensive predation by fish on sponges has been documented in Africa (Dominey 1987) and South America (Volkmer-Ribeiro and Grosser 1981), but does not appear to be important in North America. Crayfish were implicated as causing major reductions in a sponge population in a

Massachusetts stream (Williamson 1979) but widespread consumption of sponges by crayfish has not been reported. Snails will consume and can subsist on freshwater sponges under laboratory conditions (T. M. Frost and A. P. Covich personal observation) but here too, their significance as predators under natural conditions remains unknown.

Numerous small invertebrates reside within, or attached to, freshwater sponges. Examples include protozoans, oligochaetes, nematodes, rotifers, bivalves, water mites, and aquatic insects (Berzins 1950, Resh 1976, Volkmer-Ribeiro and De Rosa-Barbosa 1974). A variety of relationships exist between these "infaunal" organisms and their sponge hosts. In some cases the relationships are obligate, at least for certain life-cycle stages and such species may show a striking degree of specialization to the sponge host. Other organisms are only occasionally associated with sponges.

Several groups of aquatic insects typify organisms with obligate, specialized feeding on freshwater sponges. Numerous species within three insect orders, the Diptera, Neuroptera, and Trichoptera depend upon freshwater sponges during major portions of their life cycle (Resh 1976). One neuropteran family, the Sisyridae, is so commonly associated with sponges that its members are referred to as spongillaflies. Not all relationships between insects and sponges are obligate. Some insect genera that contain obligately sponge-feeding species also contain species that are only occasionally or never associated with sponges (Resh *et al.* 1976). The variety of interactions evident between sponges and related insect species indicates that the evolutionary history of insect–sponge relationships is diverse and complex.

Obligate relationships with freshwater sponges have also been documented for water mites (Crowell and Davids 1979), which apparently depend upon the sponge primarily for structure. The water mites live and lay their eggs within the feeding canal system of a sponge. These and other organisms that live within sponges (e.g., clams, insects, and protozoans) may be taking advantage of the lack of predation on sponges. Spicules and chemical defenses may provide a general refuge against predators on infauna as well as sponges.

Little is known about the specificity of relationships between infaunal organisms and particular sponge species. In a comparison of four sponges from the same habitat, the same major taxonomic groups were observed in each sponge species (J. Elias and T. M. Frost personal observation). More detailed taxonomic data on these infaunal species might have revealed specific associations, but this area remains to be investigated. Differences were observed in the relative abundances of the infaunal groups among sponge species. Similarly, little is known regarding the effects of infauna on the growth of their sponge hosts. Since there can be large numbers of organisms within sponges, it seems likely that infauna could have adverse effects; however, this relationship has not been quantified.

The community ecology of these infaunal organisms is also complex. The presence and abundance of a particular infaunal species result from an interplay of its own life cycle, the life cycle of the host sponge, and, potentially, the abundance of other infaunal organisms. Data from northern Wisconsin sponge populations indicate that most other infaunal species are rare or absent during periods when water mites are abundant within sponges (J. Elias and T. M. Frost personal observation). Again, this is an area that has received little attention.

4. Functional Role in Ecosystems

In most lakes and streams, populations of freshwater sponges are sufficiently sparse that they are unlikely to play a major role in ecosystem function. However, there are marked exceptions to this general pattern. In some habitats, sponges are the dominant component of the benthic community and exert a substantial influence on total nutrient cycling and primary production. Quantitative analyses of freshwater sponge populations are rare, but a detailed study of one population illustrates the potentially major impact of sponges on ecosystems (Frost 1978; Frost *et al.* 1982). At the peak of its population size in early October, *S. lacustris*, in Mud Pond, New Hampshire, reached an average biomass of greater than 3.5 g dry mass per m². At this density, the total sponge population in the pond would filter more than ten million liters of pond water per day, a rate that would cycle a volume equivalent to that of the entire pond every seven days. Since the sponges are capable of removing most phytoplankton and bacteria from the water, their potential impact on Mud Pond is substantial. Because of their algal symbionts, sponges can also account for a substantial portion of the primary production in Mud Pond. During late summer and fall, the chlorophyll within the sponges is approximately equivalent to that of the entire phytoplankton community (Frost 1978; Frost and Williamson 1980).

Sponges also play a major role in flowing water ecosystems. Mann *et al.* (1972), in a detailed study of the River Thames in England, found that sponges accounted for nearly 40% of the total production by benthic animals.

Clearly, sponges can be a major component of the biota of some ecosystems. While detailed quantitative studies are rare, observations indicate that the populations in Mud Pond and the River Thames are representative of sponges in a variety of other lakes and streams. Sponge populations in some northern Wisconsin lakes, for example, appear to be substantially larger than those in Mud Pond (T. M. Frost personal observation). At the same time, sponges in many ecosystems represent only a minor component when compared with other organisms.

Even when sponges account for substantial portions of the primary and secondary production of an ecosystem, they may not interact directly with higher trophic levels. The lack of predation on sponges can lead to an accumulation of material in sponge biomass that is not passed directly up the food chain. Sponge interactions with other ecosystem components may be indirect, operating primarily to limit the availability of materials to other organisms within the food web.

5. Paleolimnology

The siliceous nature of sponge spicules makes them resistant to decomposition under most circumstances. As such, they are usually well preserved in lake sediments and have the potential to serve as a useful tool in indicating historic lake conditions. In situations where specific habitat requirements can be defined for a species (Jewell 1935), the presence of its spicules in the sediments of a lake from which it is currently absent can indicate changes over time in the environmental conditions of that lake. Likewise, even when a sponge species is present throughout an extended period within a lake, quantitative changes in spicule morphology can indicate shifts in the availability of silica within that lake over time (T. M. Frost and T. Kratz personal observation). Although paleolimnological investigations employing sponge spicules have been limited, this is a potentially fruitful area for further study (Harrison 1986).

D. Evolutionary Relationships

All freshwater sponges belong to the class Demospongiae, the most diverse and morphologically complex of the four major groups in the sponge phylum. Within the demosponges, the evolutionary history of the freshwater sponges is long and polyphyletic. Fossil freshwater sponges are reported from at least 100 million years ago (Racek and Harrison 1975) and, although there is still debate in this area, there appears to have been at least four separate, successful invasions of freshwater habitats from marine habitats. The results of these invasions are presently classified into separate families within the freshwater sponges (Volkmer-Ribeiro and De Rosa-Barbosa 1979; Volkmer-Ribeiro 1986).

The phylogenetic diversity of freshwater sponges is not well represented in North America, where species are confined primarily to the most cosmopolitan family, the Spongillidae (Penney and Racek 1968). Only two species are exceptions, *Corvomeyenia everetti* and *C. carolinensis*. These are classified in the recently described Metaniidae, which has a worldwide, primarily tropical distribution (Volkmer-Ribeiro 1986). The two other freshwater groups are absent from North America. Species in the Lumbomerskiidae appear to be confined to a few large and ancient lakes in Europe and Asia (e.g., Lakes Baikal and Ochrid) and the family is characterized by a large degree of endemism. Species in the Potamolepidae are widely distributed throughout Africa and South America (Volkmer-Ribeiro and De Rosa-Barbosa 1979).

IV. COLLECTION, CULTURE, AND PREPARATION FOR IDENTIFICATION

Collecting sponges is usually straightforward. In many cases, sponges grow in shallow water and can be obtained simply by hand or with a long-handled rake. Where sponges are rare, it may be easiest to collect them while snorkeling. This may be the case particularly when collecting in areas with numerous hard substrata, where small stones must be overturned and the undersides of fallen trees and large boulders must be examined. Where sponges grow in deeper waters, scuba techniques afford the most efficient means of collection. Dredges or nets dragged across substrata in deeper waters are likely to miss most sponges and, at the same time, are likely to disrupt substantial bottom areas. In very deep waters, however, these may be the only practical means of collecting.

When searching a lake or stream for sponges, it is best to examine as broad a variety of substrata as possible. Aquatic macrophytes, rocks, and fallen logs are common sites for sponge growth. In bog habitats, the roots of vegetation growing at the edges of bog mats or on the undersides of the mats themselves are likely sites for sponges. Where sponges are rare, it may be necessary to examine a large number of different substrata before finding any specimens. In other habitats, however, sponges will be conspicuous.

Growing or maintaining sponges under controlled environmental conditions is much more difficult than collecting them. A number of investigators have grown small sponges from gemmules in the laboratory for cytologic investigations (Rasmont 1962, Fell 1974). Such specimens can also be particularly useful in examining the finer structural details of sponges. Poirrier *et al.* (1981) have developed an effective, continuous-flow system for sponges using bacteria as a food source. Laboratory investigations of this sort can provide important insights into sponge ecology (e.g., Harsha *et al.* 1983), particularly when controlled conditions are essential. However, sponges appear to be highly sensitive to environmental conditions and the responses of sponges obtained in laboratory studies must be compared carefully with their behavior under field situations. As such, to evaluate sponge behavior under natural conditions, it is often best to work with freshly collected sponges and to conduct experiments *in situ* whenever possible.

The identification of freshwater sponges depends primarily on characteristics of spicules and on features of intact gemmules. Successful species identifications are dependent upon obtaining all of the spicules (megascleres, gemmoscleres, and, if present in a species, microscleres) that can occur in a species. Thus, it is necessary to obtain samples of sponges that include all spicule types. This can be a problem because gemmules may only occur during certain times of the year.

As the primary step in species identification, samples of all of the types of spicules that occur in a species should be prepared for microscopic examination. To obtain spicule preparations, the organic portion of a sponge is digested with acid and the remaining spicules are mounted on a glass microscope slide. Sponge tissue can be digested in a centrifuge tube by boiling for one hour with concentrated nitric acid. The spicules can then be concentrated by gentle centrifugation, washed with ethanol or methanol and recentrifuged at least two times, then mounted on a microscope slide using a coverslip and a permanent medium (e.g., Permount). Reiswig and Browman (1987) have described a more elaborate technique for a quantitative method of processing sponge spicules.

Intact gemmules can be removed from sponge tissue, dried, and mounted directly on microscope slides. Specimens of entire sponges are necessary for museum collections. Whole sponges can be dried or preserved in alcohol. Simpson (1963) has described methods that are appropriate for preparing freshwater sponges for cytological techniques, such as are necessary for examining reproductive cycles.

V. CLASSIFICATION

The classification scheme reported here is derived primarily from Penney and Racek (1968), who have completed the most recent taxonomic revision of the family Spongillidae. Their monograph provides more detailed descriptions of most of the species reported here and should be consulted for rigorous taxonomic investigations. Information for several species is based upon reports other than that of Penney and Racek (1968), most of which have been published since 1968. In the cases where substantial information about a species has been obtained from a source other than or in addition to Penney and Racek, I cite that source along with the species description in the following list of species. In addition, there are a number of cases where species reported for Canada and the United States by Penney and Racek (1968) have been subsequently judged as invalid (e.g., Poirrier 1974, 1977; Harrison and Harrison 1977; Harrison *et al.* 1977; Harrison 1979), and I have not included those species here.

In general, the freshwater sponge species in Canada and the United States can be distinguished by characteristics of their spicules. Taxa are separated primarily by the presence, shape, and spination of microscleres and/or gemmoscleres, although the presence or absence of spines on megascleres can be a useful characteristic in some cases in distinguishing species within a genus. Spicules are usually either needle- to rod-like (Fig. 4.6) or dumbbell-shaped (Fig. 4.9). The latter are termed birotulate and their rounded ends are called rotules. In the descriptions here, a spicule should be assumed to be needle- to rod-like unless it is described as otherwise. Also, the descriptions reported here are based upon examinations with light microscopes. More detailed examinations with scanning electron microscopes (SEM) could reveal fine structures, primarily spines, that are not apparent in light microscopic observations. More detailed observations should be interpreted with caution when employing the descriptions and key that are provided here.

Gemmoscleres are critical in the classification of freshwater sponges and it is important to include gemmules when preparing sponge specimens for identification. Obtaining gemmoscleres may be a problem, particularly in young sponges derived from larvae, where gemmules and their spicules may not be present during certain times of the year. Examine a collected specimen closely to ensure that gemmules are present (Fig. 4.1). It is possible, particularly in larger specimens, that some gemmoscleres may have been left in a sponge body from the gemmules of a previous year. However, it is also possible that a few stray gemmoscleres from another sponge species may have been incorporated into a sponge specimen. Thus, identifications based upon only a few gemmoscleres should be made with caution.

In some species with microscleres, it may be difficult to distinguish between microscleres and gemmoscleres on the basis of structure alone. By separating gemmules from intact sponges and making separate spicule preparations of them, it is possible to obtain samples in which gemmoscleres predominate. Alternatively, by carefully sampling tissue from the surface of a sponge in areas of new growth, it may be feasible to procure material in which gemmoscleres are rare and microscleres and megascleres predominate. Because of the important role that microscleres play in separating species, identifications based on only a few microscleres should also be made with caution.

In using the key presented here, there are only two cases in which characteristics other than the structure of spicules are required to distinguish among taxa. In contrasting the genera *Ephydatia* and *Radiospongilla*, it is necessary to acquire a cross-section of intact gemmules in order to view the arrangement of gemmoscleres within the gemmule coat. Similarly, within the genus *Heteromeyenia*, the size, shape, and structures of the foraminal apertures of entire gemmules must be employed to distinguish among species. Particularly important for *Heteromeyenia* are the forms of structures on the ends of the foraminal tubules, termed cirrous projections.

Note that the successful use of this key depends on obtaining a fully representative sample of all types of spicules that can occur within a sponge species (megascleres, gemmoscleres, and, if they occur in a species, microscleres). If a species ordinarily exhibits gemmoscleres or microscleres and they are not contained within a specimen, it will not be possible to identify even its genus. The one exception to this rule occurs for *Spongilla lacustris*, which is unique among the species of this key in lacking gemmoscleres in its winter gemmules. If gemmules are clearly present in a specimen but gemmoscleres are absent, it is most likely *S. lacustris*. [Caution is necessary here because some foreign bodies in sponges may resemble gemmules superficially, for example, the eggs of water mites. Gemmules can be distinguished by their highly resistant coats.] It is important to recognize that *S. lacustris* may have gemmoscleres; they occur when summer gemmules are present.

Also note that this key is intended only for use

with the species from Canada and the United States, as listed herein. Those examining specimens from other regions will need to consult Penney and Racek (1968) and more recent taxonomic work (e.g., Volkmer-Ribeiro 1986 and the references cited in Section V.B). The key is arranged systematically so that genera are separated prior to species within a genus. In most cases, therefore, a valid identification to genus may be possible even if a species has not previously been reported in Canada or the United States. Even to separate genera, however, it is necessary to have included all of the possible spicule types from a specimen.

A. Taxonomic Key to Species of Freshwater Porifera

1a.	Microscleres present	2
1b.	Microscleres absent	16
2a(1a).	Microscleres star-shaped (Fig. 4.13) or birotulate (Fig. 4.11)	3
2b.	Microscleres rod-shaped to needlelike in structure (Fig. 4.17)	7
3a(2a).	Microscleres star-shaped with several rays extending from central region of the spicules (Fig. 4.13) .... *Dosilia*	4
3b.	Microscleres birotulate (Fig. 4.11)	5
4a(3a).	Gemmoscleres birotulate in two distinctly different size categories (45–82 μm and 120–230 μm in length) .... *Dosilia radiospiculata*	
4b.	Gemmoscleres birotulate in two size categories that are nearly equal in length (55–85 μm) .... *Dosilia palmeri*	
5a(3b).	Gemmoscleres birotulate (Fig. 4.11) .... *Corvomeyenia*	6
5b.	Gemmoscleres rod-shaped (as in Fig. 4.17) .... *Corvospongilla becki*	
6a(5a).	Microscleres straight to slightly curved birotules with a predominance of straight forms (Fig. 4.12) .... *Corvomeyenia everetti*	
6b.	Microscleres a mixture of straight to strongly curved birotules with a predominance of curved forms (Fig. 4.11) .... *Corvomeyenia carolinensis*	
7a(2b).	Gemmoscleres birotulate (as in Fig. 4.11) .... *Heteromeyenia*	8
7b.	Gemmoscleres needlelike (as in Fig. 4.17) or absent	11
8a(7a).	Foraminal aperture of gemmules without terminal cirrous projections (contrast with Figs. 4.18–4.20) .... *Heteromeyenia baileyi*	
8b.	Foraminal aperture of gemmules with distinct, terminal cirrous projections (Figs. 4.18–4.20)	9
9a(8b).	Foraminal aperture of gemmules with one to two very long cirrous projections starting from a flat disk, flat and ribbon-like at the base and cylindrical at the end (Fig. 4.18) .... *Heteromeyenia latitenta*	
9b.	Foraminal aperture of gemmules with three to six cirrous projections that are short when compared with those of *H. latitenta*	10
10a(9b).	Foraminal aperture of gemmules distinctly tubular and very long ranging from 0.5–0.9 times the diameter of the gemmule (Fig. 4.20) .... *Heteromeyenia tubisperma*	
10b.	Foraminal aperture of gemmules distinctly tubular but short, less than 0.4 times the diameter of the gemmule (Fig. 4.19) .... *Heteromeyenia tentasperma*	
11a(7b).	Gemmoscleres absent or covered with robust, conspicuous spines (Fig. 4.22) .... *Spongilla*	12
11b.	Gemmoscleres smooth or covered with very fine spines .... *Stratospongilla penneyi*	
12a(11a).	Megascleres covered with spines (as in Fig. 4.25) .... *Spongilla heterosclerifera*	
12b.	Megascleres smooth (as in Fig. 4.26)	13
13a(12b).	Microscleres smooth .... *Spongilla aspinosa*	
13b.	Microscleres covered with spines (Fig. 4.22)	14
14a(13b).	Gemmoscleres club-like with spines conspicuously more dense on the ends than in the center (Fig. 4.22) .... *Spongilla alba*	

| 14b. | Gemmoscleres absent or with spines distributed evenly across the length of the spicule (Fig. 4.24) .... | 15 |

15a(14b).   Microscleres with conspicuously denser and longer spines in the central region (Fig. 4.23) .... *Spongilla cenota*

15b.   Microscleres with spines distributed evenly across the length of the spicule (Fig. 4.24) .... *Spongilla lacustris*

16a(1b).   Gemmoscleres birotulate (Fig. 4.25) 17
16b.   Gemmoscleres rod-shaped or needlelike (Fig. 4.21) 24

17a(16a).   Margins of rotules completely smooth (Fig. 4.25) .... *Trochospongilla*   18
17b.   Margins of rotules spined or serrated (Figs. 4.14 and 4.15) 20

18a(17a).   Megascleres conspicuously spined (Fig. 4.25) 19
18b.   Megascleres smooth or with very fine spines (Fig. 4.26) .... *Trochospongilla leidii*

19a(18a).   Gemmoscleres with rotules of two distinctly different diameters (Fig. 4.27) .... *Trochospongilla pennsylvanica*
19b.   Gemmoscleres with rotules of nearly equivalent diameters (Fig. 4.25) .... *Trochospongilla horrida*

20a(17b).   Gemmoscleres occurring in two distinct size classes (Fig. 4.9) .... *Anheteromeyenia*   21
20b.   Gemmoscleres occurring in only one size class (Fig. 4.14) .... *Ephydatia*   22

21a(20a).   Gemmoscleres of two classes that are similar in shape but clearly different in length (Fig. 4.9) .... *Anheteromeyenia argyrosperma*
21b.   Gemmoscleres of two classes that are distinct in both shape and size (Fig. 4.10) .... *Anheteromeyenia ryderi*

22a(20b).   Rotules of gemmoscleres with margins that are nearly smooth, bearing numerous, very small incisions (Fig. 4.15) .... *Ephydatia millsii*
22b.   Rotules of gemmoscleres with clearly indented margins (Fig. 4.16) 23

23a(22b).   Gemmoscleres less than 20 μm in length, with rotules bearing fewer that 12 teeth deeply incised into long rays (Fig. 4.16) .... *Ephydatia muelleri*
23b.   Gemmoscleres greater than 25 μm in length, with rotules bearing greater than 20 teeth that are not deeply incised (Fig. 4.14) .... *Ephydatia fluviatilis*

24a(12b).   Most gemmoscleres in intact gemmules arrayed in a distinctly radial fashion within the gemmule coat .... *Radiospongilla*   25
24b.   Gemmoscleres in intact gemmules arrayed only tangentially to the surface of the gemmule .... *Eunapius*   26

25a(24a).   Megascleres sparsely covered with small spines .... *Radiospongilla crateriformis*
25b.   Megascleres entirely smooth .... *Radiospongilla cerebellata*

26a(24b).   Megascleres entirely smooth .... *Eunapius fragilis*
26b.   Megascleres covered with coarse spines (Fig. 4.17) .... *Eunapius mackayi*

B. Descriptions of Freshwater Sponge Species in Canada and the United States.

Summarized here are the spicular characteristics and distribution patterns of freshwater sponges reported from the United States and Canada. Distribution patterns emphasize the occurrence of sponges in the United States and Canada but also describe the overall, reported distributions of each species.

Anheteromeyenia argyrosperma (Potts)

Spicules: megascleres slender, 240–280 μm in length, and sparsely covered with small, sharply pointed spines; microscleres absent; gemmoscleres birotulates of two distinct length groups, 65–80 or 110–125 μm, with both size classes similar in form, exhibiting spines on their entire shaft and conspicuous, recurved, clawlike hooks on their ends. (See Fig. 4.9.)

Distribution: reported in the eastern half of North America from Florida to Canada, but confined to this region.

Anheteromeyenia ryderi (Potts)

Spicules: megascleres 190–220 μm in length, variable in shape, and covered with broadly conical

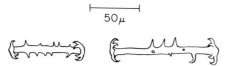

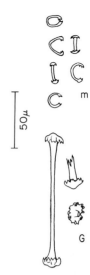

Figure 4.9 Gemmoscleres of *Anheteromeyenia argyrosperma*. (After Penney and Racek l968.)

spines; microscleres absent; gemmoscleres birotulates of two distinct length groups, 30–40 or 50–75 μm, with distinct differences in the shapes of the two length classes; shorter forms have shafts with only one or a few spines and flattened rotules with a large number of small teeth; larger forms are robust with numerous recurved spines on their shafts and with strongly recurved hooks on their ends. (See Fig. 4.10.)

Distribution: reported in the eastern half of North America, from Louisiana to Canada, but confined to this region.

Corvomeyenia carolinensis (Harrison)

Spicules: megascleres slender, straight to slightly curved, and entirely smooth, ranging in length from 194–280 μm; microscleres birotulate with straight to strongly curved (>80%), smooth shafts, 15–25 μm in length, terminating in rotules 4–7 μm in diameter with 4–6 recurved hooks; gemmoscleres birotulates with straight to slightly curved, smooth shafts ranging in length from 60–158 μm and terminating in rotules of 13–22 μm diameters with 5–8 recurved hooks (Harrison 1971). (See Fig. 4.11.)

Distribution: reported from only one pound in South Carolina.

Corvomeyenia everetti (Mills)

Spicules: megascleres slender, slight curved, and entirely smooth, 195–285 μm in length; microscleres birotulates, 16–19 μm in length, terminating in rotules 3–6 μm in diameter with 3–5 small, distinctly recurved spines; gemmoscleres birotulates with straight to slightly curved, smooth shafts, ranging in length from 42–140 μm and terminating in rotules bearing 5–7 recurved hooks. (See Fig. 4.12.)

Distribution: reported only from the eastern half of the northern United States and Canada.

Corvospongilla becki (Poirrier)

Spicules: megascleres stout, 130–218 μm in length, usually curved, covered with spines, the

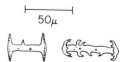

Figure 4.10 Gemmoscleres of *Anheteromeyenia ryderi*. (After Penney and Racek l968.)

Figure 4.11 Microscleres (m) and gemmoscleres (G) of *Corvomeyenia carolinensis*. (After Harrison l971.)

spines being larger near the ends of the spicule; microscleres birotulates, 25–44 μm in length, rotules 9–17 μm in diameter usually with four recurved hooks; gemmoscleres of two distinct length classes, the smaller class is 28–56 μm in length, slightly to strongly curved and spined except in the inner curved region, the larger class is 71–139 μm in length, straight to slightly curved and completely spined (somewhat similar in form to the gemmoscleres of *S. lacustris* (Fig. 4.24) (Poirrier 1978).

Distribution: reported from one lake in Louisiana.

Dosilia palmeri (Potts)

Spicules: megascleres slender, 370–450 μm in length, slightly curved to nearly straight, covered with sparse spines in their central portion; microscleres stellate with 8–12 usually smooth rays arising from a central nodule, length is extremely variable; gemmoscleres birotulates, 55–85 μm in length, occurring in two subtly distinct size classes, with

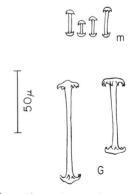

Figure 4.12 Microscleres (m) and gemmoscleres (G) of *Corvomeyenia everetti*. (After Harrison l971.)

strong spines on their central shaft and with equally sized rotules, 23–25 μm in diameter, bearing numerous blunt, recurved teeth (Penney 1960).

Distribution: reported from Florida in North America (north of Mexico) and from other locations in Central America.

Dosilia radiospiculata (Mills)

Spicules: megascleres slender, 290–400 μm in length, entirely smooth or covered with minute spines; microscleres stellate with 6–8 microspined rays projecting from their center, length is extremely variable; gemmoscleres birotulates of two distinctly different size classes with longer forms exhibiting nonspined shafts ranging in length from 120–230 μm in length and shorter forms exhibiting strongly spined shafts ranging in length from 45–82 μm. [There is some questions as to whether the two species of *Dosilia* are distinct or simply ecomorphic variants of a single species.] (See Fig. 4.13.)

Distribution: reported from the Canadian border south to Mexico, but only from this region.

Ephydatia fluviatilis (Linneaus)

Spicules: megascleres usually slightly curved, 210–400 μm in length, and entirely smooth; microscleres absent; gemmoscleres birotulates of one class, 26–30 μm in length with a slender and smooth shaft and with flat irregularly shaped rotules that have equal, 18–21 μm diameters and that usually have more than 20 teeth that are not deeply incised. (See Fig. 4.14.)

Distribution: appears to be truly cosmopolitan with more frequent occurrences in temperate rather than in tropical zones.

Ephydatia millsii (Potts)

Spicules: megascleres slightly curved to nearly straight, 180–270 μm in length, with numerous small spines except at the tips; microscleres absent; gemmoscleres birotulates of one size class, 36–48 μm in length, with smooth shafts that are clearly broader near the rotules and with distinctly flat, circular disk-shaped rotules that have equal 22–28 μm

Figure 4.14 Gemmoscleres of *Ephydatia fluviatilis*. (After Penney and Racek 1968.)

diameters and that have only very small incisions at their margins. (See Fig. 4.15.)

Distribution: reported only from Florida.

Ephydatia muelleri (Lieberkühn)

Spicules: megascleres straight to slightly curved, 200–350 μm in length, usually with small spines but smooth in rare cases; microscleres absent; gemmoscleres birotulates of one class, 12–20 μm in length, with thick, smooth shafts and with flat, irregularly shaped rotules that have equal, 20–25 μm diameters and usually fewer than 12 teeth that are deeply incised into long rays. (See Fig. 4.16.)

Distribution: widely distributed throughout the northern hemisphere with a preference for temperate regions.

Eunapius fragilis (Leidy)

Spicules: megascleres entirely smooth, 180–270 μm in length; microscleres absent; gemmoscleres straight to slightly curved, covered with conspicuous spines, which are often more dense near the tips, 75–140 μm in length. Gemmoscleres in intact gemmules arrayed tangentially to the surface of the gemmule.

Distribution: truly cosmopolitan.

Eunapius mackayi (Potts)

Spicules: megascleres straight or slightly curved, 190–265 μm in length, covered with coarse spines; microscleres absent; gemmoscleres stout and slightly curved, covered with coarse spines, 190–265 μm in length. Gemmoscleres in intact gemmules

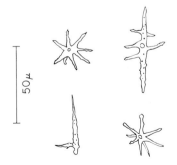

Figure 4.13 Microscleres of *Dosilia radiospiculata*. (After Penney and Racek 1968.)

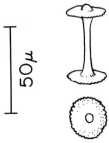

Figure 4.15 Gemmoscleres of *Ephydatia millsii*. (After Penney and Racek 1968.)

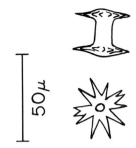

Figure 4.16 Gemmoscleres of *Ephydatia muelleri*. (After Penney and Racek 1968.)

arrayed tangentially to the surface of the gemmule. (This species is reported as *Eunapius igloviformis* in Penney and Racek, 1968.) (See Fig. 4.17.)

Distribution: throughout the United States and Canada, but confined to these regions.

Heteromeyenia baileyi (Bowerbank)

Spicules: megascleres slender, 255–315 μm in length, with sparse microspines except near the tips; microscleres delicate, slightly curved to almost straight, with spines that occur throughout their length but which increase in length towards the central region, where they are often knobbed and distinctly perpendicular to the main axis, 75–85 μm in length; gemmoscleres birotulates of two distinct length groups, 50–60 or 80–85 μm, with conspicuous conical spines on their shafts and with both rotules of 22-μm diameters with conspicuous, strongly recurved spines (the shorter gemmoscleres are much more abundant than the larger forms). Foraminal apertures of gemmules lack terminal cirrous projections.

Distribution: widely distributed throughout the eastern United States and Canada, with a few reports from Europe.

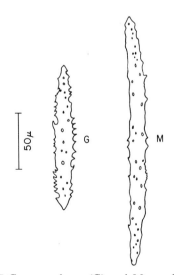

Figure 4.17 Gemmoscleres (G) and Megascleres (M) of *Eunapius mackayi*. (After Neidhoefer 1940.)

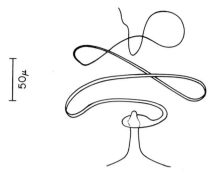

Figure 4.18 Foraminal aperture of *Heteromeyenia latitenta*. (After Neidhoefer 1940.)

Heteromeyenia latitenta (Potts)

Spicules: megascleres straight, 265–285 μm in length, smooth to sparsely microspined; microscleres slender and entirely spined with those in the central region only slightly larger than those on the ends, 85–100 μm in length; gemmoscleres birotulates of two length groups, 50–55 or 60–78 μm with shafts bearing numerous stout and pointed spines and with both rotules of equal, 16–18 μm diameters, with margins forming numerous conspicuous, recurved teeth. Foraminal apertures of gemmules with one or two very long cirrous projections starting from a disk initially flat but rounded in its terminal regions. (See Fig. 4.18.)

Distribution: reported only from the northeastern United States.

Heteromeyenia tentasperma (Potts)

Spicules: megascleres very slender, 260–280 μm in length, with sparse microspines; microscleres slender with sparse microspines, 75–80 μm in length; gemmoscleres birotulates of two length groups, 50–55 or 65–72 μm with stout shafts bearing a small number of acute spines and with both rotules of equal, 15–18 μm diameters and consisting of an arrangement of lateral spines. Foraminal apertures of gemmules distinctly tubular and relatively short with 3–6 long and irregular cirrous projections. (See Fig. 4.19.)

Distribution: reported only from the northeastern to the midwestern United States.

Heteromeyenia tubisperma (Potts)

Spicules: megascleres slender, 190–230 μm in length, with sparse microspines; microscleres slen-

Figure 4.19 Foraminal aperture of *Heteromeyenia tentasperma*. (After Neidhoefer 1940.)

der and entirely spined, with spines in the central portion distinctly larger, 85–90 μm in length; gemmoscleres birotulates of two length groups, 40–48 or 60–70 μm with stout shafts bearing a small number of acute spines and with both rotules of equal, 15–18 μm diameters and consisting of an arrangement of lateral spines. Foraminal apertures of gemmules distinctly tubular, slender, and very long (0.5–0.9 times the diameter of the gemmule) with 5–6 cylindrical cirrous projections. (See Fig. 4.20.)

Distribution: reported only from the eastern half of North America.

Radiospongilla cerebellata (Bowerbank)

Spicules: megascleres straight to slightly curved, 240–330 μm in length, entirely without spines; microscleres absent, although immature gemmoscleres may be abundant in some portions of the dermal membrane; gemmoscleres usually distinctly curved and rarely straight, 72–110 μm in length, covered with abundant spines that are pronouncedly recurved toward the terminal ends (in intact gemmules, gemmoscleres are arrayed in two distinct layers, with those in the inner layer arranged in a radial fashion and those in the outer layer arranged tangentially to the inner layer) (Poirrier 1972).

Distribution: reported only from Texas in the United States, but widely distributed in tropical and subtropical Asia and Africa.

Radiospongilla crateriformis (Potts)

Spicules: megascleres slender and slightly curved, 240–300 μm in length, sparsely covered by

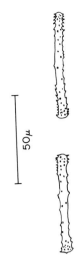

Figure 4.21 Gemmoscleres of *Radiospongilla crateriformis*. (After Penney and Racek 1968.)

very small spines except at their tips; microscleres absent; gemmoscleres slender and covered with small conical spines except at the tips, where slightly recurved spines are sufficiently dense to form pseudorotules, 60–75 μm in length, gemmoscleres in intact gemmules are arranged radially throughout the gemmule coat. (See Fig. 4.21.)

Distribution: reported primarily from the eastern half of the United States but as far west as Wisconsin and Texas; likely to occur in southern Canada, also reported in China, Japan, Southeast Asia, and Australia.

Spongilla alba (Carter)

Spicules: megascleres entirely smooth, 144–420 μm in length; microscleres slender and slightly curved with erect spines that are longer in the central region, 49–124 μm in length; gemmoscleres slightly to moderately curved with stout, sharp, recurved spines that are more dense at the ends, forming distinct heads, 48–130 μm (Poirrier 1976). (See Fig. 4.22.)

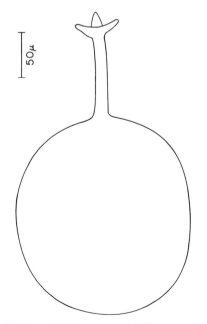

Figure 4.20 Foraminal aperture of *Heteromeyenia tubisperma*. (After Neidhoefer 1940.)

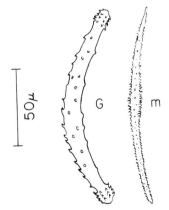

Figure 4.22 Microscleres (m) and gemmoscleres (G) of *Spongilla alba*. (After Penney and Racek 1968.)

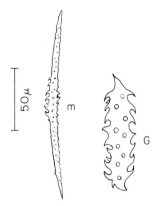

Figure 4.23 Microscleres (m) and gemmoscleres (G) of *Spongilla cenota*. (After Penney and Racek 1968.)

Distribution: occurs in warmer regions world-wide with a strong preference for brackish water, reported from the southeast coastal regions of the United States.

Spongilla aspinosa (Potts)
Spicules: megascleres slender and entirely smooth, 155–215 μm in length; microscleres rare and smooth, 30–42 μm; gemmoscleres unknown but gemmule coats are very tough.
Distribution: reported only in the United States from Florida to Michigan.

Spongilla cenota (Penney and Racek)
Spicules: megascleres stout and completely smooth, 310–410 μm in length; microscleres numerous and slender, covered with small spines at their tips and with a group of clearly larger spines in their centers, 68–123 μm in length; gemmoscleres extremely stout, entirely covered with stout, sharp, recurved spines, 65–86 μm (Poirrier 1976). (See Fig. 4.23.)
Distribution: reported only from Florida and the Yucatan in Mexico.

Spongilla heterosclerifera (Smith)
Spicules: megascleres spined; microscleres elongate and spined, occurring only in the region of the sponge growing near its attachment to a substratum; gemmoscleres robust and blunt, covered

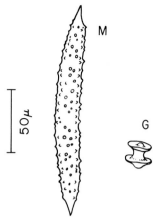

Figure 4.25 Megascleres (M) and gemmoscleres (G) of *Trochospongilla horrida*. (After Penney and Racek 1968.)

with spines, occurring primarily on the portions of the gemmules that bear the foraminal pores (Harrison and Harrison 1979). There appears to be some confusion as to whether the spicules growing near the base of the growing sponge are microscleres or a form of gemmoscleres. If they are not microscleres, this species may not belong in the genus *Spongilla* but rather, perhaps, in the genus *Eunapius*.
Distribution: reported only from Oneida Lake, New York, United States.

Spongilla lacustris (Linneaus)
Spicules: megascleres entirely smooth, 200–350 μm in length; microscleres entirely covered with small spines, 30–130 μm in length; gemmoscleres absent from winter gemmules; thick-coated summer gemmules slightly curved, usually covered with strong, curved spines, 80–130 μm (Poirrier *et al.* 1987). (See Fig. 4.24.)
Distribution: throughout the northern hemisphere including frequent reports from throughout the United States and Canada.

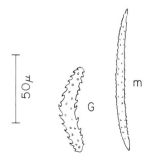

Figure 4.24 Microscleres (m) and gemmoscleres (G) of *Spongilla lacustris*. (After Penney and Racek 1968.)

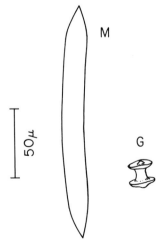

Figure 4.26 Megascleres (M) and gemmoscleres (G) of *Trochospongilla leidii*. (After Penney and Racek 1968.)

Figure 4.27 Gemmoscleres of *Trochospongilla pennsylvanica*. (After Penney and Racek 1968.)

Stratospongilla penneyi (Harrison)

Spicules: megascleres slightly curved, 215–296 μm in length, smooth to very delicately microspined; microscleres slender, 38–75 μm in length, covered with very small spines; gemmoscleres curved to distinctly bent, 48–123 μm in length, smooth to delicately microspined with sharply pointed apices.

Distribution: reported from only one location, a canal in southern Florida (Harrison 1979).

Trochospongilla horrida (Weltner)

Spicules: megascleres straight to slightly curved, 170–235 μm in length, covered with stout, sharp spines; microscleres absent; gemmoscleres small birotulates, 11 μm in length, with stout smooth shafts and rotules with circular margins of equal size, 9–12 μm in diameter. (See Fig. 4.25.)

Distribution: widely dispersed throughout the northern hemisphere.

Trochospongilla leidii (Bowerbank)

Spicules: megascleres straight to slightly curved and entirely smooth, 150–170 μm in length; microscleres absent; gemmoscleres small birotulates, 11 μm in length, terminating in rotules with circular margins and equal, 12–14 μm diameters. (See Fig. 4.26.)

Distribution: reported from limited regions of the eastern United States, with one report of a population from the Panama Canal (Jones and Rützler 1975).

Trochospongilla pennsylvanica (Potts)

Spicules: megascleres slender and slightly curved, 140–210 μm in length, entirely covered with sharp, conical spines; microscleres absent; gemmoscleres small birotulates with slender shafts, 9–11 μm in length, terminating in rotules with circular margins and with two distinctly different diameters, 3–9 μm and 16–20 μm. (See Fig. 4.27.)

Distribution: reported throughout but apparently restricted to the North American continent.

ACKNOWLEDGMENTS

The preparation of this chapter was supported by grants from the National Science Foundation. I also thank Joan Elias and Susan Knight for their assistance.

LITERATURE CITED

Bader, R. B. 1984. Factors affecting the distribution of a freshwater sponge. Freshwater Invertebrate Biology 3:86–95.

Berquist, P. R. 1978. Sponges. University of California Press, Berkeley.

Berzins, B. 1950. Observations on rotifers on sponges. Transactions of the American Microscopical Society 69:189–193.

Cecil, J. T., M. F. Stempien, Jr., G. D. Ruggieri, and R. F. Nigrelli. 1976. Cytological abnormalities in a variety of normal and transformed cell lines treated with extracts from sponges. Pages 171–182 *in* F. E. W. Harrison and R. R. Cowden, editors. Aspects of sponge biology. Academic Press, New York.

Crowell, R. M., and C. Davids. 1979. The developmental cycle of sponge-associated water mites. Recent Advances in Acarology 1:563–566.

Dominey, W. 1987. Sponge-eating by *Pungu maclareni*, an endemic Cichlid fish from Lake Barombi Mbo, Cameroon. National Geographic Research 3:389–393.

Fell, P. E. 1974. Porifera. Pages 51–132 *in:* A. C. Giese and J. S. Pearse, editors. Reproduction of Marine Invertebrates. Volume 1. Academic Press, New York.

Frey, D. G. 1963. Limnology in North America. University of Wisconsin Press, Madison.

Frost, T. M. 1976. Sponge feeding: A review with a discussion of some continuing research. Pages 283–298 in F. W. Harrison and R. R. Cowden, editors. Aspects of sponge biology. Academic Press, New York.

Frost, T. M. 1978. The impact of the freshwater sponge *Spongilla lacustris* on a sphagnum bog pond. Verhandlungen Internationale Vereinigung für Theoretische und Angewandte Limnologie 20:2368–2371.

Frost, T. M. 1980a. Clearance rate determinations for the freshwater sponge *Spongila lacustris*: Effect of temperature, particle type and concentration, and sponge size. Archiv für Hydrobiologie 90:330–356.

Frost, T. M. 1980b. Selection in sponge feeding processes. Pages 33–44 *in:* D. C. Smith and Y. Tiffon, editors. Nutrition in lower Metazoa. Pergamon, Oxford.

Frost, T. M. 1981. Analysis of ingested materials within a freshwater sponge. Transaction of the American Microscopical Society 100:271–277.

Frost, T. M. 1987. Porifera. Pages 27–53 in T. J. Pandian and J. F. Vernberg, editors. Animal Energetics. Volume 1. Academic Press, New York.

Frost, T. M., and J. E. Elias. 1990. The balance of autotrophy and heterotrophy in three freshwater sponges with algal symbionts. In W. D. Hartman and K. Ruetzler, editors. Proceedings of the Third International Conference on Sponge Biology. Smithsonian Inst. Press, Washington, D.C. In press.

Frost, T. M., and C. E. Williamson. 1980. In situ determination of the effect of symbiotic algae on the growth of the freshwater sponge *Spongilla lacustris*. Ecology 61:1361–1370.

Frost, T. M., G. S. de Nagy, and J. J. Gilbert. 1982. Population dynamics and standing biomass of the freshwater sponge, *Spongilla lacustris*. Ecology 63: 1203–1210.

Gilbert, J. J. 1974. Field experiments on sexuality in the freshwater sponge *Spongilla lacustris*. Control of oocyte production and the fate of unfertilized oocytes. The Journal of Experimental Zoology 188:165–178.

Gilbert, J. J. 1975. Field experiments on gemmulation in the freshwater sponge *Spongilla lacustris*. Transactions of the American Microscopical Society 94:347–356.

Gilbert, J. J., and H. L. Allen. 1973a. Studies on the physiology of the green freshwater sponge *Spongilla lacustris*: primary productivity, organic matter, and chlorophyll content. Verhandlungen Internationale Vereinigung für Theoretische und Angewandte Limnologie 20:2368–2371.

Gilbert, J. J., and H. L. Allen. 1973b. Chlorophyll and primary productivity of some green, freshwater sponges. Internationale Revue der Gesamten Hydrobiologie 58:633–658.

Gilbert, J. J., and T. L. Simpson. 1976a. Sex reversal in a freshwater sponge. The Journal of Experimental Zoology 195:145–151.

Gilbert, J. J., and T. L. Simpson. 1976b. Gemmule polymorphism in the freshwater sponge *Spongilla lacustris*. Archiv für Hydrobiologie 78:268–277.

Harrison, F. W. 1971. A taxonomical investigation of the genus *Corvomeyenia* Weltner (Spongillidae) with an introduction of *Corvomeyenia carolinensis sp.nov.* Hydrobiologia 38:123–140.

Harrison, F. W. 1974. Sponges (Porifera: Spongillidae). Pages 29–66 *in*: C. W. Hart, Jr. and S. L. H. Fuller, editors. Pollution Ecology of Freshwater Invertebrates. Academic Press, New York.

Harrison, F. W. 1979. The taxonomic and ecological status of the environmentally restricted spongillid species of North America. V. *Ephydatia subtilis* (Weltner) and *Stratospongilla penneyi sp. nov.* Hydrobiologia 62:99–105.

Harrison, F. W. 1986. Fossil fresh-water sponges (Porifera: Spongillidae) from Western Canada: An overlooked group of Quaternery paleoecological indicators. Transactions of the American Microscopical Society 105:110–120.

Harrison, F. W., and M. B. Harrison. 1977. The taxonomic and ecological status of the environmentally restricted spongillid species of North America. II. *Anheteromeyenia biceps* (Lindenschmidt, 1950). Hydrobiologia 62:99–105.

Harrison, F. W., and M. B. Harrison. 1979. The taxonomic and ecological status of the environmentally restricted spongillid species of North America. IV. *Spongilla heterosclerifera* Smith 1918. Hydrobiologia 62:99–105.

Harrison, F. W., L. Johnston, D. B. Stansell, and W. McAndrew. 1977. The taxonomic and ecological status of the environmentally restricted spongillid species of

North America. I. *Spongilla sponginosa* Penney 1957. Hydrobiologia 53:199–202.

Harsha, R. E., J. C. Francis, and M. A. Poirrier. 1983. Water temperature: A factor in the seasonality of two freshwater sponge species, *Ephydatia fluviatilis* and *Spongilla alba*. Hydrobiologia 102:145–190.

Hartman, W. D., and H. M. Reiswig. 1973. The individuality of sponges. Pages 567–584 *in*: R. S. Boardmann, A. H. Cheetham, and W. A. Oliver, editors. Animal colonies. Dowden, Hutchinson & Ross, Stroudsburg, Pennsylvania.

Jackson, J. B. C. 1977. Competition on marine hard substrata: the adaptive significance of solitary and colonial strategies. American Naturalist 111:743–767.

Jackson, J. B. C., and L. W. Buss. 1975. Allelopathy and spatial competition among coral reef invertebrates. Proceedings of the National Academy of Science, U.S.A. 72:5160–5163.

Jewell, M. E. 1935. An ecological study of the fresh-water sponges of northern Wisconsin. Ecological Monographs 5:461–504.

Jewell, M. E. 1939. An ecological study of the fresh-water sponges of northern Wisconsin, II. The influence of calcium. Ecology 20:11–28.

Jones, M. L. and K. Rützler. 1975. Invertebrates of the Upper Chamber, Gatún Locks, Panama Canal, with emphasis on *Trochospongilla leidii* (Porifera). Marine Biology 33:57–66.

Langenbruch, P.-F., and N. Weissenfels. 1987. Canal systems and choanocyte chambers in freshwater sponges (Porifera, Spongillidae). Zoomorphology 107:11–16.

Mann, K. H., R. H. Britton, A. Kowalczewski, T. J. Lack, C. P. Mathews, and I. McDonald. 1972. Productivity and energy flow at all trophic levels in the River Thames, England. Pages 579–596 *in*: Z. Kajak and A. Hillbricht-IlKowska, editors. Productivity problems of freshwaters. PWN, Polish Scientific Publishers, Warsaw.

Mergner, H. 1966. Zum Nachweis der Artspezifität des Induktionsstoffes bei Oscularrohrneubildungen von Spongilliden. Verhandlunger der Deutschen Zoologischen Gesselschaft in Gettingen 30:522–564.

Muscatine, L., and J. W. Porter. 1977. Reef corals: mutualistic symbioses adapted to nutrient-poor environments. Bioscience 27:454–460.

Neidhoefer, J. R. 1940. The fresh-water sponges of Wisconsin. Transactions of the Wisconsin Academy of Sciences, Arts, and Letters 32:177–197.

Niegel, J. E., and J. C. Avise. 1983. Histocompatability bioassays of population structure in marine sponges. Journal of Heredity 74:134–140.

Penney, J. T., 1960. Distribution and bibliography (1892–1957) of the freshwater sponges. University of South Carolina Publications, Series 3, 3:1–97.

Penney, J. T., and A. A. Racek. 1968. Comprehensive revision of a worldwide collection of freshwater sponges (Porifera: Spongillidae). United States National Museum Bulletin No. 272:1–184.

Poirrier, M. A. 1972. Additional records of Texas freshwater sponges (Spongillidae) with the first record of *Radiospongilla cerebellata* (Bowerbank, 1863) from the Western Hemisphere. The Southwestern Naturalist 16:434–435.

Poirrier, M. A. 1974. Ecomorphic variation in gemmoscleres of *Ephydatia fluviatilis* Linneaus (Porifera: Spongillidae) with comments upon its systematics and ecology. Hydrobiologia 44:337–347.

Poirrier, M. A. 1976. A taxonomic study of the *Spongilla alba, S. cenota, S. wagneri* species group (Porifera: Spongillidae) with ecological observations of *S. Alba*. Pages 203–214 *in:* F. W. Harrison and R. R. Cowden, editors. Aspects of sponge biology. Academic Press, New York.

Poirrier, M. A. 1977. Systematic and ecological studies of *Anheteromeyenia ryderi* (Porifera: Spongillidae) in Louisiana. Transactions of the American Microscopical Society 96:62–67.

Poirrier, M. A. 1978. *Corvospongilla becki N. Sp.*, A freshwater sponge from Louisiana. Transactions of the American Microscopical Society 97:240–243.

Poirrier, M. A., J. C. Francis, and R. A. Labiche. 1981. A continuous-flow system for growing fresh-water sponges in the laboratory. Hydrobiologia 79:225–259.

Poirrier, M. A., P. S. Martin, and R. J. Baerwald. 1987. Comparative morphology of microsclere structure in *Spongilla alba. S. cenota,* and *S. lacustris* (Porifera: Spongillidae). Transactions of the American Microscopical Society 106:302–310.

Racek, A. A., and F. W. Harrison. 1975. The systematic and phylogenetic position of *Palaelospongilla chubutensis* (Porifera: Spongillidae). Proceeding of the Linnean Society of New South Wales 99:157–165.

Randall, J. E., and W. D. Hartman. 1968. Sponge-feeding fishes of the West Indies. Marine Biology 1:216–225.

Rasmont, R. 1962. The physiology of gemmulation in fresh-water sponge. Symposium of the Society for the Study of Development and Growth 20:3–25.

Reisser, W. 1984. The taxonomy of green algae endosymbiotic in ciliates and a sponge. British Phycological Journal 19:309–318.

Reiswig, H. M., and H. I. Browman. 1987. Use of membrane filters for microscopic preparations of sponge spicules. Transactions of the American Microscopical Society 106:10–20.

Resh, V. H. 1976. Life cycles of invertebrate predators of freshwater sponge. Pages 299–314 *in:* F. W. Harrison and R. R. Cowden, editors. Aspects of sponge biology. Academic Press, New York.

Resh, V. H., J. C. Morse, and I. D. Wallace. 1976. The evolution of the sponge feeding habit in the caddisfly genus *Ceraclea* (Trichoptera: Leptoceridae). Annals of the Entomological Society of America 69: 937–941.

Schmidt, I. 1970. Phagocytose et pinocytose cez les Spongillidae. Zeitschrift für Vergleichende Physiologie 66:398–420.

Simpson, T. L. 1963. The biology of the marine sponge *Microciona prolifera* (Ellis and Solander): I. A study of cellular function and differentiation. The Journal of Experimental Zoology 154:135–151.

Simpson, T. L. 1973. Coloniality among the Porifera. Pages 549–565 *in:* R. S. Boardmann, A. H. Cheetham

and W. A. Oliver, editors. Animal colonies. Dowden, Hutchinson & Ross, Stroudsburg, Pennsylvania.

Simpson, T. L. 1984. The cell biology of sponges. Springer-Verlag, New York.

Simpson, T. L., and P. E. Fell. 1974. Dormancy among the Porifera: gemmule formation and hatching in freshwater and marine sponges. Transactions of the American Microscopical Society 93:544–577.

Simpson, T. L., and J. J. Gilbert. 1973. Gemmulation, gemmule hatching, and sexual reproduction in freshwater sponges: I. The life cycle of *Spongilla lacustris* and *Tubella pennsylvanica*. Transactions of the American Microscopical Society 92:422–433.

Smith, L. C., and W. H. Hildemann. 1984. Alloimmune memory is absent in *Hymeniacidon sinapium*, a marine sponge. The Journal of Immunology 133:2351–2355.

Stoddart, J. A., D. J. Ayre, G. Willis, and A. J. Heyward. 1985. Self-recognition in sponges and corals? Evolution 39:461–463.

van Weel, P. B. 1949. On the physiology of the tropical fresh-water sponge *Spongilla proliferens* Annand. I. Ingestion, digestion, and excretion. Physiolgia Comparata et Oecologia 1:110–128.

Van de Vyver, G. 1970. La non confluence intraspécifique cez les spongiaires et la notion d'individu. Extrait des Annales d'Embryologie et de Morphogenèse 3:251–262.

Van de Vyver, G. 1979. Cellular mechanisms of recognitions and rejection among sponges. Pages 195–204 *in:* C. Levi and N. Boury-Esnault, editors. Biologie des Spongiaires. Colloques Internationaux du Centre National de la Recherche Scientifique, No. 291. CNRS, Paris.

Van de Vyver, G., and P. Willenz. 1975. An experimental study of the life cycle of the fresh-water sponge *Ephydatia fluviatilis* in its natural surroundings. Wihelm Roux' Archiv 177:41–52.

Vogel, S. 1974. Current induced flow through the sponge, *Halichondria*. Biological Bulletin (Woods Hole, Massachusetts) 147:443–456.

Volkmer-Ribeiro, C. 1981. Porifera. Pages 86–95 *in:* S. H. Hurlbert, G. Rodriquesz, and N. D. Santos, editors. Aquatic biota of tropical South America. Part 2: Anarthropoda. San Diego State University, San Diego, California.

Volkmer-Ribeiro, C. 1986. Evolutionary study of the freshwater sponge genus *Metania* GRAY, 1867: III. Metaniidae, new family. Amazoniana 9:493–509.

Volkmer-Ribeiro, C., and R. De Rosa-Barbosa. 1974. A freshwater sponge-mollusk association in Amazonian Waters. Amazoniana 2:285–291.

Volkmer-Ribeiro, C., and R. De Rosa-Barbosa. 1979. Neotropical freshwater sponges of the Family Potamolepidae Brien, 1967. Pages 497–502 *in:* C. Levi and N. Boury-Esnault, editors. Biologie des Spongiaires. Colloques Internationaux du Centre National de la Recherche Scientifique, No. 291. CNRS, Paris.

Volkmer-Ribeiro, C., and K. M. Grosser. 1981. Gut contents of *Leporinus obtusidens* ''sensu'' Von Ihering (Pisces, Characoidei) used in a survey for freshwater sponges. Revista Brasileira de Biologia 41:175–183.

Weissenfels, N. 1975. Bau und Funktion des Süsswasserschwamms *Ephydatia fluviatilis* L. (Porifera). II. Anmerkungen zum Körperbau. Zeitschrift für Morphologie der Tiere 81:241–256.

Weissenfels, N. 1976. Bau und Funktion des Süsswasserschwamms *Ephydatia fluviatilis* L. (Porifera). III. Nahrungsaufnahme, Verdauung und Defäkation. Zoomorphologie 85:73–88.

Weissenfels, N. 1980. Bau und Funktion des Süsswasserschwamms *Ephydatia fluviatilis* L. (Porifera). VII. Die Porocyten. Zoomorphologie 95:27–40.

Weissenfels, N. 1981. Bau und Funktion des Süsswasserschwamms *Ephydatia fluviatilis* L. (Porifera). VIII. Die Entstehung und Entwicklung der Kragtengeisssel-kammern und ihre Verbindung mit dem ausführenden Kanalsystem. Zoomorphology 98:35–45.

Weissenfels, N. 1982. Bau und Funktion des Süsswasserschwamms *Ephydatia fluviatilis* L. (Porifera). IX. Rasterelektronmikrosckpische Histologie und Cytologie. Zoomorphology 100:75–87.

Willenz, P. 1980. Kinetic and morphological aspects of particle ingestion by the freshwater sponge *Ephydatia fluviatilis* L. Pages 163–187 *in:* D. C. Smith and Y. Tiffon, editors. Nutrition in lower Metazoa. Pergamon, Oxford.

Williamson, C. W. 1979. Crayfish predation on freshwater sponges. American Midland Naturalist 101:245–246.

The Freshwater Cnidaria— or Coelenterates

5

Lawrence B. Slobodkin
Department of Ecology and Evolution
State University of New York
Stony Brook, New York 11794

Patricia E. Bossert
Science Department
Northport High School
Northport, New York 11768

Chapter Outline

I. INTRODUCTION

II. GENERAL BIOLOGY OF CNIDARIA
 A. Body Plan
 B. Nematocysts
 C. Feeding
 D. Reproduction and Metamorphoses
 E. Ecological Interactions

III. DESCRIPTIVE ECOLOGY OF FRESHWATER CNIDARIA
 A. Hydra
  1. General Biology of Hydra
  2. Reproduction and Mortality
  3. Feeding
  4. Regeneration
  5. Symbiosis
 B. *Craspedacusta*
  1. Life Cycle
  2. Ecology
 C. *Calposoma*
 D. *Cordylophora*
 E. *Polypodium hydriforme*

IV. COLLECTION AND MAINTENANCE OF FRESHWATER CNIDARIA
 A. Collecting Techniques
 B. Maintenance Procedures

V. CLASSIFICATION OF FRESHWATER CNIDARIA
 A. Species Groups of Hydras
 B. Taxonomic Key to Genera of Freshwater Cnidaria
 Literature Cited

I. INTRODUCTION

Pelagic jellyfish and attached anemones, corals, and other polyps compose an ancient and remarkably successful phylum. They occur as fossils in the lithographic stone of the Mid-Cambrian Burgess shale and have not changed very dramatically since then. The group is sometimes referred to as the phylum Cnidaria, from the Greek term for "nail," based on their possession of nematocysts, which look like stiff rods attached to a round capsule. The other name for the phylum is Coelenterata, a term alluding to their sac-like internal space, the coelenteron. A general overview of the phylum and survey of older literature is provided by Hyman (1940). Muscatine and Lenhoff (1974) and Mackie (1976) sample more modern research.

The freshwater representatives are small animals belonging to the class Hydrozoa, with relatively few species and somewhat monotonous anatomy. They consist of the following taxa.

1. The common and familiar hydras, a group of secondarily simple, solitary polyps;
2. The sporadically common *Craspedacusta* and *Limnocodium*, jellyfish with minute, polypoid larvae;
3. *Calposoma* (Fuhrman 1939), a tiny, colonial polyp, which is so small and inconspicuous that it is probably more common than it appears to be;
4. *Polypodium*, which spends part of its life cycle as a parasite in the eggs of sturgeon and part as an ambulatory, predaceous polyp (this has been described primarily from Russian rivers but should also be found in North American sturgeon);

5. Various estuarine coelenterates, which may occasionally occur in relatively fresh water; colonial, sessile animals of the genus *Cordylophora* will serve as an example of these.

Hydra and *Cordylophora* belong to the order Hydroida of the class Hydrozoa. *Craspedacusta*, *Limnocnida*, and *Calposoma* are classified as members of the order Limnomedusae. The parasitic *Polypodium* has been assigned to the order Trachylina of the same class (cf. Hyman 1940). The fact that the other three orders of the class Hydrozoa (Actinulida, Siphonophora, Hydrococorallina) and the other three classes of coelenterates (Scyphozoa, Cubozoa, and Anthozoa) are not represented outside of the sea is curious. There is not even any serious speculation as to why freshwater invasion by coelenterates has been so severely limited. There is no special osmoregulatory organ in the phylum, but this is not an explanation since its absence has not stopped the successful invaders. Because of their small size, soft bodies, and often sessile habits, freshwater Cnidaria are either not collected or not well preserved in most routine collecting procedures. They are, however, widely distributed and can be found in most ponds and streams when specifically searched for. When they are abundant, they can be major predators of small invertebrates and even of tiny fish. In turn, hydra are fed upon by flatworms, and crayfish eat *Craspedacusta*; probably other animals prey on Cnidaria but this has not been carefully studied. Their ecology will be described in more detail in Section III.

II. GENERAL BIOLOGY OF CNIDARIA

A. Body Plan

All cnidarians share a simple body plan of a central cavity surrounded by two cellular layers (Fig. 5.1). The endoderm lines the interior cavity, the coelenteron. Between the endoderm and the ectoderm is an intermediate, noncellular mesoglea. They all have unique endocellular capsules called nematocysts, which aid in feeding and in repelling predators.

The feeding aperture or mouth leads into the coelenteron. By convention, the end with the mouth is termed "oral," the opposite end "aboral." The coelenteron functions as a gut. The oral aperture serves at different times as mouth and anus. When the mouth is closed, the pressure of fluid in the coelenteron can stiffen the body, even in the com-

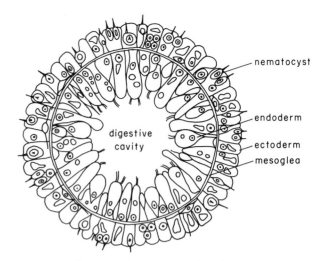

Figure 5.1 Cross-section through digestive cavity (coelenteron) of generalized cnidarian.

plete absence of any hard tissue. The coelenteron, therefore, serves also as a hydrostatic skeleton.

The mesoglea varies enormously in thickness among different members of the group. At its most meager, as in hydra, it is not more than 200 μm thick, containing only wandering cells, nonliving fibrous components, and fibers from neuromuscular cells. It is more fully developed in medusae such as *Craspedacusta* and may contain a great deal of collagenous or gelatinous material.

The ectodermal body wall has neural and contractile properties and is also the location for ripe nematocysts. The presence of definite cell layers with differentiated functions distinguishes these animals as true metazoans. The absence of mesoderm implies that they do not have organs like higher metazoans, but such terms as tentacles and gut are used in a functional sense.

The basic body plan can be manifested as either a polyp or a medusa (Fig. 5.2). Polyps are typically elongated along the oral–aboral axis. Medusae are approximately bell-shaped and usually have their greatest body dimension perpendicular to the oral–aboral axis. The coelenteron of a polyp is usually deeper than its body is wide, while a medusa is usually wider than its coelenteron is deep. Also, medusae generally have relatively thicker mesoglea. In some species, different generations, and in some cases, the same individual organism at different stages of its development can take on the form of either a polyp or a medusa.

Around the feeding aperture of polyps or the edge of the bell of medusae is typically a ring of tentacles. Tentacles are extensions of the two cellular layers into more or less elongated projections. The coelenteron may or may not extend into the tentacles. Tentacular ectoderm is especially rich in nema-

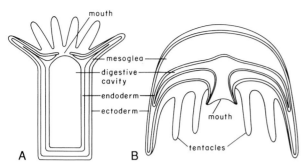

Figure 5.2 Basic body plan of Cnidaria showing (A) polyp and (B) medusa.

tocysts, which may be arranged in rosette or ring-shaped batteries. Tentacles are used in food capture, defense, and in some cases, locomotion.

B. Nematocysts

The entire phylum is characterized by the presence of cnidoblasts, ectodermal cells that produce cell products called cnidaria or nematocysts. Because there are many kinds of elaborately spined nematocysts, these structures are valuable characters for the classification of coelenterates, particularly in such morphologically monotonous groups as hydra.

Cnidoblasts differentiate into a thin, living layer surrounding a capsule containing a thread-like tube (Fig. 5.3a). In fully developed cnidoblasts, the protoplasm is greatly reduced or absent. Near the base of the tube is an outward pointed projection that is reminiscent of a trigger. Mature nematocysts come to lie on the tentacles, often after having moved through the mesoglea from some other region.

After firing, a nematocyst consists of a long thread with a capsule at its base (Fig. 5.3b). Some of the nematocysts, the stenoteles or penetrants, are open at the thread tip, giving the appearance of a hypodermic syringe with a long needle. The stenoteles may have a complex of thornlike structures around the base of the thread. Stenoteles eject a neurotoxin, which partially paralyzes the prey. Desmonemes (or volvonts), another form of nematocyst, do not seem to contain poison but rather eject sticky threads that wind about the spines and hairs on the body of the prey, interfering with movement. The capsules of some volvonts remain fixed to the tentacles after firing so that their exploded

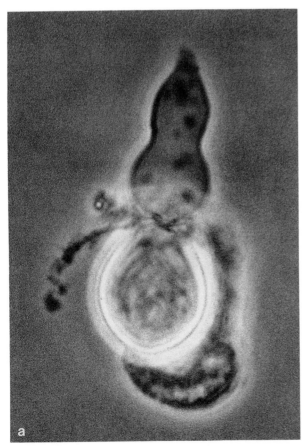

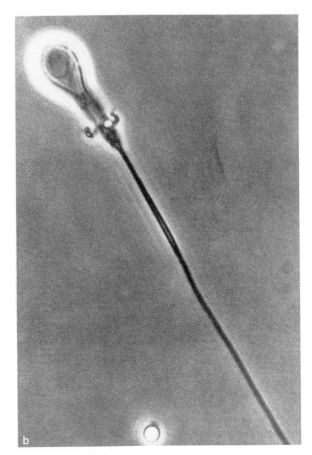

Figure 5.3 Discharged nematocyst from *Hydra*.

threads fasten prey to the tentacles as if by many tiny ropes or grappling hooks.

At least 17 morphologically distinct forms of nematocysts are present in the phylum (Fig. 5.4). The penetrants and volvonts of hydra come in several forms, classified in terms of capsule size, spination of the threads, and shape and distribution of the basal spines. A single animal may have five or more types of nematocysts. Nematocysts of basically the same type may differ among species in how the thread is coiled inside the capsule prior to eversion. Some may appear like a coiled spring, with gyres at right angles to the longest dimension of the capsule, while others are coiled parallel to the long dimension of the capsule. Nematocyst structure was initially considered a central taxonomic character; but recent evidence indicates that details of shape and coiling and also the relative abundance of nematocysts of different types are somewhat variable even within clones of hydra (Campbell 1987).

The microanatomy and function of nematocysts is a rich research area. Hessinger and Lenhoff (1988) have edited a large review volume that cannot be summarized here. The enormous interest of nematocysts and their production arises in part from the following observations. On the level of electron microscopy they are extremely elaborate structures, of interest as examples of complex cell differentiation. Also, the mechanism by which the thin thread of the nematocyst everts from its coiled state within the capsule, like an enormously elongated, inverted finger of a glove suddenly turning itself inside out, is a difficult problem in fluid pressures and their effects. The poison that is secreted by some nematocysts through the open tip of the thread is of medical interest. Furthermore, the enormous diversity of shapes, sizes, and spination of nematocysts, within single organisms and among the different coelenterates, poses a problem in cell differentiation and genetics.

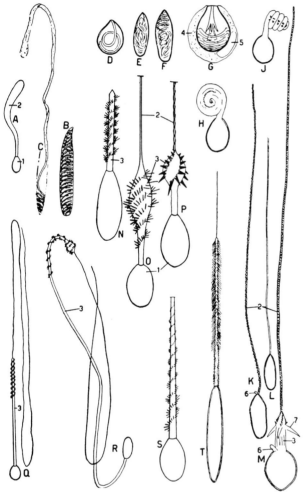

Figure 5.4 Some of the types of nematocysts present in Cnidaria (redrawn from Hyman 1940). (A) rhopaloneme (only in Siphonophora); (B) spirocyst; (C) same as B, unraveling (not discharged); (D) desmoneme [(H) and (J) same as D but not discharged]; (E) atrichous hydrorhiza [(L) same as E but discharged]; (F) holotrichous isorhiza ((K) same as F but discharged); (G) stenotele inside its cnidoblast ((M) same as G but discharged); (N) microbasic amastigophore; (O) homotrichous microbasic eurytele; (P) heterotrichous microbasic eurytele; (Q) macrobasic mastigophore; (R) teleotrichous macrobasic eurytele; (S) heterotrichous anisorhiza; (T) microbasic mastigophore. 1, Capsule; 2, tube; 3, butt; 4, cnidoblast; 5, its nucleus; 6, lid; 7, stylet.

C. Feeding

After a prey has been stung and encumbered by the nematocysts, the tentacles move the victim to the mouth, which opens to admit it into the coelenteron. Often the prey is still alive and active, but activity ceases as digestive juices are secreted by the gastric cells lining the coelenteron.

Since there is no anus, food cannot be passed along the gut while digestion continues, as in higher metazoans. Instead, feeding stops until the digestion process is completed and the indigestible remnants have been regurgitated. In hydra, ingested food decomposes within an hour into a slurry of particles (L. B. Slobodkin personal observation). The free borders of the digestive cells ingest particles by pinocytosis. Individual food particles are enclosed in vacuoles that are moved through the endodermal cells; eventually, their indigestible residues are returned to the coelenteron. The role of food vacuoles is reminiscent of the feeding process in protozoa and sponges (cf. Barnes 1966 p. 165) (see Chapters 3 and 4).

D. Reproduction and Metamorphoses

The coelenterates, like all metazons, can reproduce sexually. The fertilized eggs may produce larvae differing anatomically and ecologically from the adult sexually reproducing stage. In marine coelenterates, there may be an elaborate succession of larval stages, some of which may form colonies or reproduce vegetatively by budding or fragmentation. In any particular species, one or more of these stages may be missing. There are medusae that produce eggs that go through various larval stages to produce new medusae (Fig. 5.5A). Some polyps produce gonads (Fig. 5.5B) and others bud off medusae that either swim away to become sexual (Fig. 5.5C) or remain attached to their parents and become sexual without ever feeding independently.

Many kinds of vegetative or asexual reproduction exist. Polyps or medusae may produce new individuals that may or may not resemble their "parent." A new individual may separate from its parent; but, in

many coelenterates, the asexually produced individuals stay attached and form a colony. The development of colonies is absent in hydra, but occurs among all other freshwater Cnidaria.

In the simplest coelenterate colonies, such as the colonial "microhydra" larvae of the freshwater *Craspedacusta*, all attached individuals are essentially similar in both form and function (Payne 1924). Each has its own mouth and coelenteron and each is capable of budding new polyps. Microhydra larvae can produce medusae by budding. Colonies of many coelenterates are much more elaborate.

In a common modification of this process, certain members of the colony become "gonozooids" that neither feed nor have tentacles but instead, consist of a stalk rising from the common stolon that buds off medusae. These medusae, in turn, either form

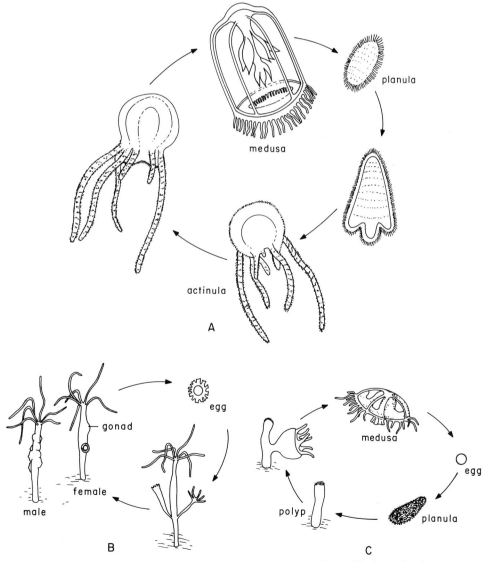

Figure 5.5 (A) Sexual medusa produces larva; larva forms new medusa; (B) Sexual polyp generates new polyp; (C) polyp produces medusa which becomes sexual and forms new polyp. (Redrawn from Barnes 1966).

gonads while still attached to their colony or leave the colony and then produce gonads. This reproductive phenomenon occurs in the brackish water hydrozoan *Cordylophora lacustris* (Roos 1979).

The best known freshwater coelenterates, the hydra, have lost the medusa stage entirely. In hydra, polyps may produce new polyps by asexual budding or may temporarily switch to sexual reproduction using gonads. Fertilized eggs derived directly from polyps may then hatch to produce new polyps. There are several marine examples, but only one freshwater example of medusae budding new medusae. This is in a species presumed endemic to the great African rift lake, Tanganyika (Thiel 1973).

E. Ecologic Interactions

Although some species may gain part of their nourishment from intracellular algal symbionts, all cnidarians are carnivores. Their prey consists primarily of small coelomate animals, organisms that evolved long after the appearance of coelenterates. They generally do not feed on protozoans, nematodes, or sponges, nor will these animals trigger the nematocysts. This raises the curious question—for which there is no evident answer—as to what the first coelenterates ate.

Prey include crustaceans, worms, and insect larvae. Hydra are extremely effective predators. They have been seriously considered as a biologic control organism for mosquitoes. Hydra and *Craspedacusta* can both sting small fishes very badly. Fish that are too big to swallow often cannot survive being stung. For this reason, hydra are sometimes serious pests of fish hatcheries.

Nematocysts are sufficiently unpleasant that relatively few predators attack coelenterates. Turtles, fish, crabs, worms, echinoderms, and flatworms are among the predators on marine coelenterates. Predation on the freshwater coelenterates is not well studied. Crayfish eat *Craspedacusta* (Dodson and Cooper 1983) and flatworms are reported to eat hydra (Hyman 1940, Kanaev 1969), but we have seen small flatworms withdraw from contact with hydra. The cydorid cladoceran *Anchistropus* has been observed to cling to and apparently feed on the body wall of hydra (Griffing 1967, L. B. Slobodkin personal observation). The amoeba *Hydraamoeba hydroxena* feeds on hydra (Stiven 1976) and the hypotrichous ciliate *Kerona* lives on the surface of hydra (Hyman 1940).

Some marine coelomates are not affected by nematocysts and are commensal with coelenterates. On coral reefs, clown fish live among tentacles of large anemones and other fish occur only in close association with the Portuguese man-of-war. In freshwater, although some coelomates seem immune to attack (see Section III.A.3), there is no known commensalism with other animals. *Polypodium hydriforme*, however, is parasitic on a fish for part of its life.

III. DESCRIPTIVE ECOLOGY OF FRESHWATER CNIDARIA

Their anatomical simplicity and the ease with which hydra can be cultured in the laboratory make freshwater Cnidaria very important as experimental material in cellular and developmental biology, as well as in neurobiology (Schaller 1983). Hydra also lends itself to studies in cell growth, morphogenesis, microanatomy, and symbiosis (cf. Lenhoff 1983, Muscatine and Lenhoff 1974). In the present chapter, we will focus on cnidarian ecology and natural history, which have been less well studied.

A. Hydra

1. General Biology of Hydra

Hydra are by far the most thoroughly studied cnidarians, being a favorite object of research from the time of Trembley (1744). They are small polyps from 1–20 mm in body length. The body is crowned by up to ten or twelve tentacles. Usually the tentacles are approximately the same length as the body but may be somewhat shorter, particularly in the green hydra, and can exceed 20 cm in length in hungry brown hydra in quiet water. There is no medusa. Reproduction is by budding or by gametes produced directly from ectoderm. The fertilized eggs may enter a resting stage; but when development proceeds, it immediately develops into a polyp (see Fig. 5.5).

Hydra are found attached to almost any reasonably hard surface. Slight bacterial films may make surfaces more attractive, but heavy growth of microalgae may be avoided. Water lily stems, charophytes, dead leaves, sticks, and stones are favored substrates. They may appear as single animals or as dense fur-like aggregations. There are reports of fishing nets becoming completely covered with brown hydra, resulting in rashes on the hands and arms of the fishermen (cf. Batha 1974). The same stem or leaf may be thick with hydra one day and completely bare the next. If the wait for food is longer than approximately 12 hr, the hydra begin to change their locations on the substrate.

They have two different methods of movement. Small scale movements on a substratum may occur by attaching the stretched tentacles to the substratum or in shallow water, to the surface film, releasing the pedal attachment and contracting the tentacles and reattaching (Ritte 1969). Sufficiently crowded or hungry hydra float off their substratum (Lomnicki and Slobodkin 1966) and may appear in the plankton or be found floating upside down with the pedal disk in the water surface film and their tentacles hanging downward.

Batha (1974) has extensively documented the existence of planktonic hydra in Lake Michigan by use of divers and of suitable attachment surfaces suspended in midwater. In addition, the reports of fishnets covered by hydra and observations by Griffing (1965) of sudden relocations of hydra within a single pond all suggest that hydra are probably much more important components of lake plankton than has been generally realized. Floating animals will sink and settle either from wave and current action or from having just fed. This behavior keeps hydra in areas of abundant food. It is also evolutionarily important as a dispersal mechanism. From the perspective of a naturalist, it has the effect of making it relatively easy to collect hydra at the outflow of lakes and ponds.

Hydra occur in bodies of freshwater from the Amazon to Alaska, at depths from shallow water to 60 m or more. Although not tolerant of heavy metals, they can thrive in even highly eutrophic water at temperatures from near freezing to 25°C. This cosmopolitan distribution may be a result of the portability of the thecate eggs; or perhaps, as has been suggested by Campbell (1987), it might be due to the four species groups (see Section V) having differentiated before the primeval continental masses separated in the Mesozoic era.

Several hydra species may coexist in a pond; often a small green hydra and at least one large brown species will co-occur. Also, strains of hydra may replace each other seasonally. Bossert (1988) showed that green hydra collected several months apart from the same small pond had very different size and growth characteristics when maintained under very similar conditions in the laboratory.

Despite many years of collecting and observation, we have never found a species of green hydra that was consistently larger than any strain of asymbiotic brown hydra, nor have we found any contradictory account in the scientific literature. If it occurs, it certainly seems rare. Some large brown hydra can be caused to become green in the laboratory (Rahat and Reich 1985), but it is not clear that this is of significance in nature.

2. Reproduction and Mortality

Asexual budding is the primary reproductive mechanism during periods of population increase. Under optimal conditions of food supply, temperature, and water quality, each adult hydra polyp can produce two buds per day and each bud can mature and begin reproducing in a week or less. Generally, the smaller strains of hydra bud more rapidly than the larger ones, at equal feeding rates. Buds are from 12–20% of the size of the mother, varying with the strain.

While hydra usually reproduce by producing free-living buds, under deleterious environmental conditions they may develop testes or single egg ovaries and engage in sexual reproduction. External fertilization by free-swimming sperm occurs while the egg is attached to the body wall of the mother. The embryo may enter a resting period of days or even months before proceeding with direct development into new polyps.

During sexual reproduction, gonads develop along the stalk in place of buds (Fig. 5.6). Mature male gonads are a mound of tissue with a distinct apical nipple from which sperm extrude. Mature female gonads consist of a single large egg cell resting on a cushion of smaller cells. The fertilized eggs are surrounded by a theca, which may be smooth or ornamented or may consist of polyhedral plates. The features of the theca are of taxonomic importance.

Sexuality in hydra seems to occur only under deleterious environmental conditions, although the precise cues are unknown. Chemical change, temperature fluctuations, and perhaps sudden nutritional variation can all induce sexuality at some times (Rutherford *et al.* 1983, Loomis 1964). Individual hydra have been reported as unisexual and some bisexual individuals have been noted (Kanaev 1969, Hyman 1940). If conditions improve, sexual hydra can return to asexual reproduction. Buds and gonads may occur concurrently in the same individual.

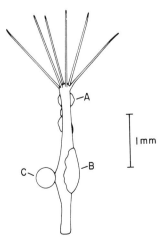

Figure 5.6 Hydra. (A) male gonad; (B) female gonad; (C) egg. (Redrawn from Campbell 1987.)

In most plants and animals, population size and genetic recombination are associated through the process of sexual reproduction, but in hydra, sexual reproduction is not significant in population dynamics. Presumably, it is primarily of importance in maintaining genetic heterogeneity. Most studies of hydra sexuality are made on animals that have been maintained in the laboratory. Batha (1974) has suggested that sexual reproduction is very rare in nature, based on the complete absence of gonads among thousands of brown hydra collected in Lake Michigan. Although we have found eggs in green hydra collected from a small pond and other field reports of gonads do exist (Kanaev 1969), it seems clear that in hydra the numerically most important reproductive process is budding.

In a single pond or stream, most of the animals are likely to belong to a rather small number of vegetative reproductive lines, being genetically identical except for occasional mutations. Even those that have emerged from eggs are likely to be the result of relatively close inbreeding, since sexuality is usually found in crowded local populations whose members are descended from a very small number of clonal lines. There is no direct evidence on sperm survival in nature, but it seems unlikely that sperm can travel great distances.

Hydra that are very small, either because of youth or starvation, will not produce buds. If nutrition is limited, animals may be kept indefinitely in a condition of neither growing nor budding. Both body size and budding rate are proportional to feeding rate up to a point of food saturation.

The capacity of hydra to reduce its body size is of ecologic interest. Animals with hard skeletons are committed to a particular body size in the sense that they cannot shrink below a given point during periods of food shortage. The inability to reduce size may contribute to death by starvation. By contrast, hydra shrink in size when starved but can be restored to full reproductive size by increasing their food supply. At a given temperature, hydra will come to a steady state body size and budding rate if food supply is sufficient and constant. The time for complete size adjustment to either temperature change or feeding level change on the part of an individual brown hydra is approximately three weeks (Hecker and Slobodkin 1976).

Reduction of temperature decreases growth rate but increases body size. Hecker and Slobodkin (1976) have shown that these size differences are due to changes in cell number rather than cell size (Fig. 5.7). Because larger animals have greater food requirements, the smaller green hydra can produce buds when fed one or two *Artemia salina* nauplii per day while the larger brown hydra require from 5–10 per day before they will bud.

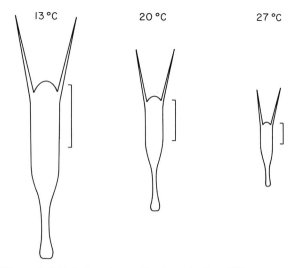

Figure 5.7 Relative sizes of hydra raised at different temperatures. Scale shown is the same size at each temperature. (Redrawn from Hecker and Slobodkin 1976.)

It seems likely that hydra polyps are potentially immortal. All deaths in hydra can be assigned to such things as alterations in water quality or food supply, temperature shocks, excessively severe starvation, or predation. Single polyps have been maintained in the laboratory without budding for at least a year, their lives terminating only from human error. There is no clear evidence for senescence of any kind in hydra.

3. Feeding

The tentacles extend above and lateral to the body. They are generally motionless except in the presence of potential food or during locomotion. In healthy animals, the tentacles are cylindrical or tapered but never clubbed. The length of the tentacles varies somewhat with species but considerably more with environmental circumstances. In quiet water, animals with bodies no longer than 2 cm have been observed to constrict their tentacles into thin threads extending at least 20 cm (L. B. Slobodkin personal observation). This is unusual, however, and tentacles equal to two body lengths or less are more typical.

Tentacles are quickly retracted if the animal is disturbed or if food organisms brush against them. Whole prey or extracts of their body fluids will initiate active waving movements of the tentacles (Lenhoff 1974). If a prey organism brushes against the tentacles, nematocysts will discharge, poisoning the prey and attaching it to the tentacle. Other tentacles and their nematocysts then join the attack. In a matter of 1–4 min, the prey will have been pressed against the mouth surface by the tentacles and will have entered the coelenteron. Meanwhile, other prey may have been caught by the tentacles for subsequent swallowing.

The number of prey swallowed in one feeding encounter varies with the size of the hydra, the size of the prey, and the previous feeding condition of the hydra. One large *Daphnia magna* may fill the coelenteron completely, stretching its walls to give the appearance of a *Daphnia* stuffed into a thin expandable sack of hydra tissue.

During the digestion process, the body may become rounded as the swallowed prey are reduced to a slurry. After approximately one hr, material is suddenly discharged from the coelenteron through the mouth, the body momentarily appearing like a punctured balloon. The columnar shape is then restored, tentacles regain their virulence, and the animal waits for its next meal. There is evidence that several days of starvation will increase the appetite of a brown hydra. Brown hydra can survive more than 40 days without food and green hydra can live 4 months or longer.

In general, hydra eat small, open water plankters but are less effective at capturing animals that normally inhabit underwater surfaces. The common cladoceran genera *Simocephalus*, *Scapholebris*, and *Chydoris* and at least some ostracodes are immune to the activities of hydra (Schwartz *et al.* 1983). Hydra can eat very small fish and insects, but sufficiently large animals with hard skeletons and strong swimming force can escape after being stung. The long bristles on small midge larvae have been found to impede predation by hydra (Hershey and Dodson 1987).

Hydra primarily feed on the kinds of prey that they are least likely to encounter, suggesting that hydra are sufficiently important as natural predators that only those crustaceans that have evolved immunity to hydra can coexist closely with these coelenterate predators.

4. Regeneration

Trembley (1744) demonstrated the capacity of hydra to regenerate perfectly even after most severe mutilations. The regenerative and regulatory powers of hydra make them favorite classroom objects even today. They can be decapitated, bisected, have their tentacles amputated, or even be turned inside out and in a matter of one day regenerate missing parts or regain their proper organization. Within a clone, rings of hydra stalk can be threaded on hairs like quoits and may fuse to form a single tube. Even without operations, different accidental conditions in the field or laboratory will produce hydra with various mutilations, more or fewer tentacles, missing heads and so on. All of these hydra will reorganize themselves neatly in a period of several days,

if water chemistry is not deleterious and if they have been reasonably well nourished before the mutilations occurred [cf. Section VII in Lenhoff (1983) and Kanaev (1969).]

5. Symbiosis

While species of hydra may be difficult to distinguish, there is a very clear distinction between the brilliant green color of some species of hydra and the yellow, brown, and gray colors of those that do not have symbiotic algae. The green hydra are accepted as a taxonomically distinct group and have been assigned their own generic status as *Chlorohydra*, which we use here interchangeably with "green hydra." No certainty exists as to the monophyletic character of this genus.

The green color arises from *Chlorella* cells, unicellular algae each occupying a vacuole in the endodermal cells of their hosts. Each endodermal cell contains 10–35 algae-laden vacuoles (Fig. 5.8), the precise number varying with species but being nearly constant within a hydra strain and a set of environmental circumstances. Algae are also found in the central cells of the tentacles but in somewhat smaller numbers. The endoderm of the buds contains algae like those of the mother.

Ample evidence exists showing that green hydra gain nourishment from their symbionts. Radioactive tracer experiments have demonstrated that maltose, a secondary photosynthetic compound, leaks from the algal symbionts to their hosts (Ciernichiari *et al.* 1969, Muscatine 1965). Also, there is microscopic evidence that algal cells can be attacked by host lysozymes (Dunn 1986). In competition experiments between brown hydra and green hydra populations, light provides a significant advantage to the green hydra (Slobodkin 1964).

The algae can be removed from some strains of green hydra by use of a variety of techniques, including photosynthetic poisons, prolonged darkness or extremely strong light, and dilute glycerine solutions (Pardy 1983). Such hydra are referred to as aposymbiotic and are susceptible to reinvasion by algae. It is possible to develop green patches in brown hydra by injecting their coelenteron with algae or by feeding green hydra to brown ones (L. B. Slobodkin personal observation), but this coloration fades with time.

If the association with algae is advantageous, why aren't big hydra green? Symbiotic relationships involve a delicate interaction between the two partners, particularly when one partner lives within the cells of the other. A primary requirement for stable endosymbiosis is a mechanism providing balanced rates of mitosis of the host and symbiont cells. If there were no control of the algal rate of increase,

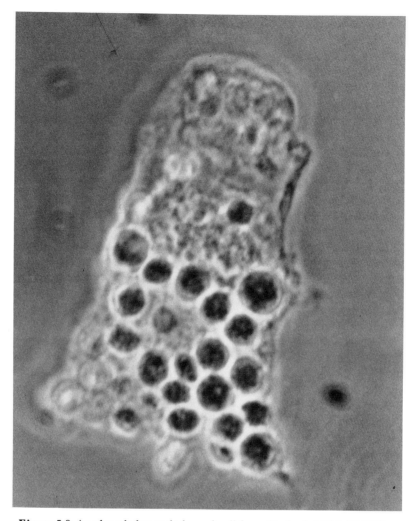

Figure 5.8 An algae-laden endodermal cell from *Hydra* (*viridis*) (1000X).

they would be expected to kill their host cells by filling them with algae; and if the hydra cells excessively limited the algae, they would be eliminated and the hydra would be brown.

While the number of algae per host cell stays constant in all green hydra, maintaining this constancy seems more difficult in the larger green hydra strains. Bossert and Dunn (1986) have shown that algal cells are increasing faster than host cells in green hydra strains of all sizes, but the disparity between algal and animal mitotic rates is greatest in the largest strains. Dunn (1987) suggested that algal cells are being actively expelled or digested by all green hydra, but especially by the larger strains.

Size in green and brown hydra is apparently regulated by the relative amounts of several hormones, some of which activate the formation of buds while others inhibit budding (Schaller *et al.* 1977, Schaller 1983). Bossert (1988) has found that one of the hormones implicated in producing smaller hydra size

also inhibits algal mitosis within green hydra cells. Perhaps the same hormonal mechanism that produces small size aids in maintaining the balance between algal cell and hydra cell increases, while a hormonal balance that permits larger size makes control of algae more difficult and, in the largest strains, impossible. More investigation of this problem is needed.

Green hydra are never found in nature as aposymbionts and seem to have evolved a dependence on these algae. However, *Chlorella* have probably not evolved a dependence on hydra. In fact, how the algae benefit from the association is not at all clear. Certainly the *Chlorella* in a green hydra are immune to being eaten by filter feeders, have an assured source of mineral nutrients, CO_2, and nitrogen, and are moved into light as the hydra move. However, the actual rate of increase of algal cells inside hydra is probably lower than those of algal cells outside, and it is not obvious that the number of *Chlorella* cells contained in the entire green hydra population is a significant fraction of the natural *Chlorella* population.

B. *Craspedacusta*

The first scientific accounts of freshwater medusae were based on specimens found in the giant water-lily tank of Regents Park in London in 1880 (Fig. 5.9A). These were described by two authors separately. Lankester named them *Craspedacusta sowberii* after the discoverer, Mr. Sowber, and Allman named the same organisms *Limnocodium victoria* after the lily. In the same year and the same tanks, a tiny colonial hydroid was found. This animal had no tentacles; each polyp terminated in a bulbous "capitulum" studded with nematocysts. In the center of the capitulum was the mouth. These little polyp colonies (Fig. 5.9B) were initially and correctly assumed to be the larval stage of the medusae (Payne 1924). However, very similar polyps were discovered in a water tank in Philadelphia and were described and named as a separate species. It wasn't until 1928 that it became once again obvious that the polyp named *Microhydra ryderi* was the larva of *Craspedacusta* (Boulenger and Flower 1928). Since the initial description, *Craspedacusta* medusae have been found in many locations around the world, apparently transported with ornamental aquatic plants and with the water hyacinth, a recently spread pest species.

1. Life Cycle

The most conspicuous stage of the life cycle is the small medusa (0.5–1.5 cm). Except in the Yang-tse river system of China, the occurrence of noticeable populations of the medusae of *Craspedacusta* is sporadic and often surprising to local naturalists. As

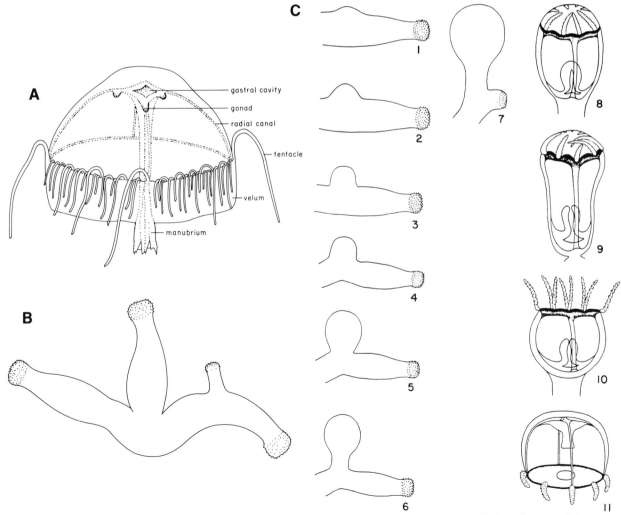

Figure 5.9 (A) *Craspedacusta sowberii*, also named *Limnocodium victoria* (see text); (B) larval stage of A, also named *Microhydra ryderi* (see text); (C) stages in development of *Craspedacusta*. (Redrawn from Payne 1924).

the water warms in a pond or in the slow current of a stream backwater, a swarm of medusae appears, often where it has never been seen before or at least has not been apparent for many years. These medusae feed on zooplankton. As they grow, the number of tentacles increases from 8–12 up to as many as 100. After several weeks of growth, gonads develop in pouches of the radial canals. Fertilized eggs produce a small crawling planula, which then differentiates into a microhydra (Payne 1924). The planula consists of two cell layers forming a double-walled, sausage-shaped sac. The differentiation of the planula into a microhydra is rather simple. It stands on end as a polyp and a mouth and a capitulum develop on the unattached end.

The new polyps may continue to bud off new microhydra or medusae. Buds may remain attached after they have developed a capitulum, producing colonies of up to 12 polyps attached to a common stolon. Starvation and severely abnormal temperatures cause the polyps to shrink to a cellular ball, surrounded by a chitin-like membrane. This can persist through the severe conditions and then redifferentiate as a polyp.

The buds of medusae appear initially as rounded swellings of the polyp wall (Fig. 5.9C). Over a period of several weeks, they enlarge and develop a central manubrium. An endoderm-lined circular canal with four radial canals leads to the base of the manubrium. Eight tentacles emerge from the circular canal. The medusa bud is initially covered by a layer of tissue that eventually perforates centrally, remaining as the vellum of the adult. The mouth opens into the manubrium. The medusa is by now at least as large as the polyp, to which it is still attached at the aboral end. Eventually, it begins locomotory pulsations and separates from the polyp, completing the life cycle (Payne 1924). Medusae usually reproduce sexually; fertilized eggs develop into planulae, which transform into microhydra.

2. Ecology

The microhydra polyps feed in essentially the same fashion as hydra. Their small size limits their prey, but this is partially compensated for by their colonial growth pattern, which permits several polyps to make a simultaneous attack. Free-living medusae feed on various crustacean zooplankters. They can even kill, but apparently not swallow, the large (0.5 cm), predaceous cladoceran, *Leptodora* (Dodson and Cooper 1983). As a medusa feeds, it grows larger, adding additional tentacles at both the vellum margin and its inner edge. Several hundred tentacles are found on mature animals. Medusae

appear sporadically in shallow ponds, natural lakes, and artificial reservoirs throughout the north and south temperate zones. Often entire medusa populations are unisexual. The polyps are so inconspicuous that they probably have a much broader distribution than the literature reports. Polyps, but not medusae, are also found in streams as well as in lakes and ponds. Many locations containing polyps are reported not to have medusae. This bewildering picture has been clarified by Acker (1976) and Kramp (1950). Acker suggests that medusa production by the polyps requires temperatures greater than approximately 20°C during a period of increasing temperature and adequate but not enormous food levels. Other environmental alterations may be significant but have not been tested. [These conditions are reminiscent of those that produce sexuality in hydra.] Acker and Kramp maintain that the original natural habitat of *Craspedacusta sowerbii* is the Yang-tse-kiang region of China. In the upper river valley, two *Craspedacusta* species, *C. sowerbii* and *C. sinensis* coexist while only *C. sowerbii* reaches the down-stream areas. From this habitat, *C. sowerbii* traveled with water hyacinths and other plants to its present worldwide distribution.

Shallow pools exist along the lower Yang-tse valley which are subject to large temperature changes and to sudden flooding from the main river during high water. Plankton populations in these ponds fluctuate strongly. In these ponds, medusae are a recurring phenomenon throughout the year. The occurrence of medusae in the spring is so regular that they are given a common name which translates to "peach blossom fish."

The generic name *Limnocodium* is usually applied to freshwater medusae of the Old World and *Craspedacusta* to those of the New; but it is not clear if this is more than a geographic distinction nor has there been enough investigation to determine how many species of freshwater medusae actually exist. Due to their sporadic occurrence, the changes in size and tentacle number with developmental stage, the simplicity of larval anatomy, and the difficulties of preserving specimens, morphologic studies of freshwater medusae are difficult. As in hydra, many species have been described, but perhaps some of these are based on nutritional or developmental history or preservation artifacts. The strongest evidence for multiple species is that sympatric populations of two species of medusae are known from the Yang-tse (Acker 1976) and that a medusa from Lake Tanganyika has an extraordinarily different life cycle. In this latter case, a mature medusa buds new medusae from its manubrium in addition to reproducing sexually (Bouillon 1957).

C. *Calposoma*

This is a colonial polyp not much bigger than a *Paramecium* and similar in general appearance to the microhydra except that it is considerably smaller and has tentacles rather than a capitulum. As far as is known, there is only one species, *Calposoma dactyloptera* (Fuhrman 1939). Some of its properties have been described by Rahat and Campbell (1974a, 1974b). These organisms are so small that their tentacles consist of a single cell, a tentaculocyte, which contains a row of miniscule nematocysts. The full natural history of these animals and the details of their life cycle are not known. Their general anatomy is reminiscent of a miniaturized hydra with stiff tentacles, but this does not necessarily indicate taxonomic or evolutionary proximity.

D. *Cordylophora*

The genus *Cordylophora* is an athecate member of the primarily brackish water and marine hydrozoan family Clavidae. It grows as a branching colony up to 5 cm high. The feeding polyps have a conical hypostome on which filiform tentacles are irregularly arranged. The colonies also include gonophores, which produce gonads. There is a chitinous periderm (Fig. 5.10). During periods of stress, the animals regress to "metanonts," masses of resting tissue in the hydrorhizae (Naumov 1960). The metanonts appear when the plant stalks on which they grow begin decomposing in the fall (Roos 1979).

Cordylophora was first found in the Caspian and Black Seas. It has been suggested that it evolved during the time that these two bodies of water were connected. It now occurs in rivers of Europe and America, apparently having spread during the last century attached to ship bottoms. The recent expansion of range may be attributed to the relatively greater speed of seagoing vessels during the last hundred years, which would shorten the time of salt water immersion between brackish water ports (Roos 1979).

Industrial and navigational developments have extended the region of brackish water in some estuaries, and the increasing pollution of rivers may have duplicated estuarine conditions. This may be expected to encourage the geographic spread and the upriver movement of not only *Cordylophora*, but also other estuarine Cnidaria (Hubschman and Kishler 1972). The possibility of foreign Cnidaria arising where they have never appeared before is also enhanced by rapid vessels and the use of water ballast discharged in ports of arrival. *Cordylophora* is one example of an expanding range. We expect that

Figure 5.10 *Cordylophora*. (Redrawn from Roos 1979.)

there are many others, which we have not attempted to survey.

E. *Polypodium hydriforme*

The remaining freshwater coelenterate that we will consider is the strange and poorly understood *Polypodium hydriforme*, originally described by Lipin (1911, 1926) and later examined in detail by Raikova (1973, 1980). This tiny hydrozoan was first discovered in eastern Europe as a parasite inside the eggs of the European sterlet (*Acipenser ruthenus*), a small member of the sturgeon family. It is particularly interesting as it is perhaps the only endoparasitic coelenterate. Obviously there are modifications associated with its parasitic habit that make classification difficult, but it is reported to have only one type of nematocyst. These match the type found in the Narcomedusae, an extremely toxic group of hydrozoans.

The development of the larval *Polypodium* (Fig. 5.11) is closely coordinated with that of the sturgeon egg. A binucleate, single cell stage is known from

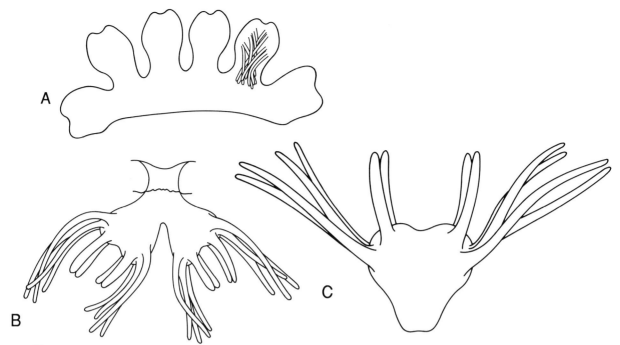

Figure 5.11 (A) *Polypodium* stolon with apex of knobs directed toward center of egg; (B) emerging polyp; (C) mature polyp. (Redrawn from Lipin 1911.)

immature sterlet oocytes. The two nuclei are unequal in size and chromosome number. The smaller, reportedly haploid nucleus is surrounded by the large polyploid nucleus, which develops into a trophic envelope around the embryo formed by the division of the small nucleus and its surrounding cytoplasm. By the time the host oocyte has started to accumulate yolk, the *Polypodium* is a two layered planula approximately 1 mm long. It has a flagellated external layer, which eventually will become the endoderm and an internal layer of ultimate ectoderm, all surrounded by a capsule that serves as a digestive organ for consuming yolk (Raikova 1980).

After a month, the planula has developed into a stolon with internally directed buds and tentacles. At this stage, the *Polypodium* is a colony consisting of a straight stolon from which projects as many as a dozen knobs, arranged linearly with their apices towards the center of the egg (Fig. 5.11A). Each knob develops two indentations. From each indentation, twelve tentacles, two of which are short and stubby, project into the stolon itself (Fig. 5.11B).

As the fish eggs ripen and are released from the fish, the tentacles evert through a slit in the stolon. Simultaneously, the knobs invert, developing a coelenteron lined by what had been the surface exposed to the egg yolk. The stolon breaks up and the knobs now appear as somewhat bifurcated polyps with twelve tentacles on each head and a coelenteron full of fish egg yolk (Fig. 5.11C).

The free-living polyps subdivide by longitudinal fission. They crawl on the bottom using tentacles as

walking legs, aided by nematocysts (isotrichous isorhiza) that hold the substratum. They feed on turbellaria and oligochaetes. Gonads form on the polyps; both single sex and hermaphroditic individuals are known. The genital anatomy is considerably more complex than that of hydra. The gonads and their accessory structures arise from endoderm. The presumed ovule is diploid and is released into the gastric cavity. The presumed male gonads become filled with binucleate cells. One nucleus remains haploid, while the other becomes polyploid. These may fall out of the coelenteron, but there are observations of polyps crawling onto young sterlets and placing these gonads on the fish. The transition from fish surface to immature oocyte has not been observed.

At least five species of *Acipenser* are parasitized by *Polypodium* in all of the major rivers of the Soviet Union. Eighty percent of sterlet (*Acipenser ruthenus*) and 20% of the sturgeon (*A. guldenstadti*) are infested, with sporadic infections in other species of the genus. A careful search of North American sturgeon will probably reveal the presence of this cnidarian.

IV. COLLECTION AND MAINTENANCE OF FRESHWATER CNIDARIA

A. Collecting Techniques

Hydra are so ubiquitous that their presence should be expected in any reasonably unpolluted body of water. Unfortunately, since they are usually sedentary, they are not easily found in plankton tows. Also, lacking hard body parts, they are often badly damaged by preservatives. A careful examination of suitable substrata is required. If rocks, leaves, *Myriophyllum*, *Elodea*, or other submerged vegetation are collected and placed overnight in a glass or enamel pan and the pan is carefully examined under a low power dissection microscope, hydra are usually found. Abundance will depend on the seasonal distribution of zooplankton and may vary among lakes.

Since hydra float when hungry and attach again after they have been fed, the best place to look for them is often the downstream end of a lake or the pools in the stream immediately below the lake. If there is a dam, this should be examined with particular care. In a typical body of water, there may be at least three species of hydra: a green one, a large brown one, and an intermediate-sized brown one.

Collecting of the medusae of *Craspedacusta* must be done with buckets rather than nets. The greater the volume the better, to prevent anoxia and excess temperature change. A compromise must be found between excessive agitation of the water, which will damage the animals and insufficient stirring, which will permit them to settle or to become anoxic. The microhydra larvae can be found on fragmentary organic debris, much like hydra, but we have not collected them.

B. Maintenance Procedures

Craspedacusta and *Cordylophora* can be kept in natural water and fed on field-collected zooplankton. Even the free-living stage of *Polypodium* can be kept in the laboratory and fed on oligochaetes and turbellaria (Raikova 1973). Dodson and Cooper (1983) have fed a range of foods to *Craspedacusta*, including large rotifers and the copepod *Diaptomus*.

Generally, it is extremely difficult to maintain jellyfish in aquaria for any length of time. Their tissues are so fragile that they are battered to pieces by most aquarium aerating or stirring systems. The more or less sedentary polyps are therefore easier to study. However, the microhydra larvae of *Craspedacusta* have not been extensively studied, perhaps due to their small size.

Hydra are laboratory animals par excellence. General directions for most of the things one might want to do with hydra are discussed in detail in the papers collected by Lenhoff (1983). They can be maintained in artificial pond water of simple composition and fed either live natural foods such as daphnia or copepods or larvae of midges and mosquitoes. They will not eat protozoans.

Although the brine shrimp, *Artemia*, do not normally occur in freshwater, many pond animals eat them avidly in the laboratory. Many investigators use *Artemia* nauplii as food for hydra, *Craspedacusta*, planarians, and other small freshwater invertebrates. The *Artemia* eggs are collected at the edges of salt ponds and lakes. They are sold in vacuum-packed cans for use by aquarists and may be stored for long periods at cool temperatures.

The procedure for feeding hydra with *Artemia* consists of following the package directions for hatching the brine shrimp (a process requiring about one day), draining and rinsing the hatched nauplii free of the salt water, and adding them to the hydra. The nauplii must be alive and vigorous or the hydra will not eat them. [Dead and moribund *Artemia* can be fed to flatworms.] *Artemia* are best drained in a net that can be made by inserting a taut sheet of bolting silk in an embroidery hoop. The *Artemia* and their salt solution are poured into the net—wet on both sides to speed drainage. The shimmering mass of *Artemia* and eggs are then rinsed into a new container of hydra culture water. If this is permitted to stand for 15 min, the active nauplii will swim toward a light, leaving the unhatched eggs behind. These active swimmers can be taken with a medicine dropper and fed to the hydra. This keeps salt and hydra separate and also permits elimination of most of the unhatched *Artemia* eggs.

These small prey may be eaten in great numbers. Large brown hydra can eat as many as 100 *Artemia* at a single meal and even the smallest green hydra can consume 1 or 2. No more than 10 min after the first prey is swallowed, the swallowing process stops even if some prey remain on the tentacles. We have seen that for some time after a hydra has been satiated, organisms may brush against the tentacles with impunity. This suggests either that nematocysts are no longer discharging or that the nematocyst supply has been temporarily exhausted.

The primary danger in this feeding process is that water contaminated with either dead *Artemia* or with the regurgitation products of previous feedings is highly detrimental. The live food must therefore be added to the hydra; and after approximately 1 hr, the hydra must be placed in clean water, either by moving them or by discarding the tainted water. Since hydra usually stick to the substratum, the old medium can be simply poured off and new medium added. The glass surface will eventually become coated with bacteria. This can be prevented somewhat by scraping the glass with a rubber spatula before discarding the old medium, but this is not usually effective for more than about one week, at which time it is best to provide new, clean containers. The hydra are remarkably sensitive to heavy metals and to detergents. Even a very short length of copper tubing in a water supply will kill hydra. Therefore, in cleaning the dishes, physical dirt is often less dangerous than the detergents. Dishes must be very thoroughly rinsed in metal-free water. To obtain nontoxic water, one may either use water from the collecting site or pretest the water with hydra to make sure it is innocuous or use glass-distilled or deionized water. When rinsing dishes containing hydra, appropriate salts are added to the distilled water.

Various recipes for culture media are available. We use two stock solutions adapted by K. Dunn (personal communication) from solutions developed by Loomis, Lenhoff, and others. Solution A contains 81 g $NaHCO_3$ in 1 liter of distilled water. Solution B contains 7.46 g KCl, 20.33 g $MgCl_2 \cdot 6H_2O$, and 147.02 g $CaCl_2 \cdot 2 H_2O$ in 1 liter of distilled water. These stock solutions are added at the rate of 1 ml per liter to distilled water to make artificial pond water. Hydra die at temperatures above approximately 30°C, but most seem healthy at temperatures as low as 5°C. The growth and reproductive rates are proportional to temperature, as are the food demands. Stocks are therefore best maintained at low temperatures except when rapid growth is desired.

V. CLASSIFICATION OF FRESHWATER CNIDARIA

The five groups of freshwater Cnidaria are clearly distinct from each other. Except for the hydras, they are relatively rare and sporadic in their distribution. In North America there is no evidence, at present, for more than one species of *Craspedacusta, Polypodium,* or *Calposoma.* There may be several species of *Cordylophora,* but we are considering *Cordylophora* as one example of a brackish water rather than a completely freshwater coelenterate. Our taxonomic key, therefore, consists of five very coarse divisions, one of which, that for the hydras, is then subdivided in somewhat greater detail.

A. Species Groups of Hydras

There are two closely related genera of hydras: the brown genus known as *Hydra* and the green hydras assigned to the genus *Chlorohydra.* By convention the term hydra refers to members of both genera, unless otherwise specified. Hydra tend to look superficially similar, except for the distinction between the asymbiotic and green species. On more careful examination, differences become apparent, not only among apparently unrelated species but even within a clone. Asymbiotic animals from the

same clone change color depending on the color of their food. All hydra will also change body size as a function of the amount of food and of temperature (Hecker and Slobodkin 1976). Despite the within-clone plasticity, unrelated hydra maintained under identical conditions, side by side in the laboratory, retain consistent differences. Batha (1974) noticed that field-collected hydra differ in appearance from their own clonal descendants maintained for long periods in the laboratory. Even animals collected from the same pond at different seasons and kept in the laboratory sometimes show consistent differences in laboratory cultures, despite apparent similarities of the field-collected specimens. The effect of all this is to make taxonomy extremely difficult. Some investigators have tended to describe new species on the basis of rather unstable, variant appearances, while others have despaired of making any precise identifications.

Campbell (1987) has provided a compromise position, in which he has classified the hydra into clearly distinct "species groups," each composed of an uncertain number of more closely related species whose precise identity may have to await new techniques of examination. Color, presence or absence of a body stalk, nematocyst shape, and the order of appearance of tentacles on the new buds are the characters used. The structure of the theca around the fertilized eggs, the appearance of gonads, proportions of different kinds of nematocysts, microscopic details of symbiotic algae, and certain physiological characteristics are all likely to be important in subdividing the various species groups. Unfortunately, these subdivisions are difficult for all but the most serious students and professionals. Since the species group is adequate for most purposes, it seems advisable to quote extensively from

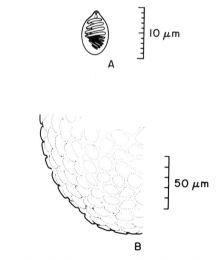

Figure 5.12 (A) Holotrichous isorhiza; (B) nonspiny embryotheca. (Redrawn from Campbell 1987.)

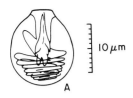

10 μm

Figure 5.13 (A) Stenotele; (B) tentacle formation in *Hydra oligactis* group. (Redrawn from Campbell 1987.)

Campbell's descriptions. The four species groups are given as follows.

1. The easily recognized group "*Hydra viridissima*" (also known as *Chlorohydra*). In addition to the brilliant green color, Campbell lists the following characteristics: "small to moderate body size and tentacles, stalk not distinct from the rest of the column; hermaphroditic; embryotheca spherical, made of polygonal plates; nematocysts very small; tentacles arising simultaneously on buds."
2. The stalked hydra or "*oligactis*" group. "large size; pronounced translucent stalk in large individuals; long tentacles; dioecious; spherical embryo with simple theca; slender stenotele (capsules) and large, blunt, cylindrical holotrichous isorhiza (capsules); distinct golden color in culture due to yellow crystals in the ectoderm; two lateral tentacles arising before the others in the bud."
3. The "*vulgaris*" group or common hydra. "moderate size; tentacles of moderate length; dioecious or monoecious, sometimes switching between the two conditions; spherical

embryotheca ornamented with spines; broad stenotele (capsules); slender holotrichous isorhiza; often slipper-shaped, with the anterior end broadly pointed; tentacles arising on the buds simultaneously or nearly so."
4. The "gracile hydra" or "*braueri*" group. "small to medium size; tentacles moderately short; hermaphroditic; embryotheca and egg flattened, with the embryotheca adherent to the substratum; embryotheca smooth or papillate; tentacles arising simultaneously on the buds; holotrichous isorhiza (capsule) broader than half its length; and stenotele (capsule) plump; body often pale colored during laboratory cultivation."

No single character, not even the nematocysts, can by itself distinguish one group from another. Every group probably contains a large number of species, each of which has been described under a large number of names. Perhaps biochemical procedures may eventually sort them out, but for the moment and for most purposes, strains collected from the field should be classified to a species group and then supplied with enough ancillary description so that other workers can at least know if they have the same or a similar organism. Campbell (1987) provided a key to the four groups of hydra, which we have incorporated in a key to the other freshwater Cnidaria.

B. Taxonomic Key to Genera of Freshwater Cnidaria

The polyp and medusa stages of a single species of coelenterate can differ greatly, both morphologically and ecologically. The key has been arranged to give the same results, whether one starts with polyps or medusae. Also, the key refers to "species groups" of *Hydra* as names in parentheses.

1a.	Parasitic in fish eggs	*Polypodium*
1b.	Nonparasitic	2
2a(1b).	Medusae	*Craspedacusta*
2b.	Polyps	3
3a(2b).	Polyp without basal attachment, with oral surface downward	*Polypodium*
3b.	Polyps with basal disk	4
4a(3b).	Solitary polyps with single circle of multicellular tentacles; no medusa; gonads are body stalk	7
4b.	Colonial polyps	5
5a(4b).	Filiform tentacles, irregularly arranged on conical hypostome; branching colony with gonophores	*Cordylophora*

| 5b. | Colonial; atentacular or tentacles unicellular | 6 |

| 6a(5b). | Atentacular, oral capitulum; polyps from common stolon; may bud medusa from polyp; "microhydra" | *Craspedacusta* |
| 6b. | Minute with unicellular tentacles | *Calposoma* |

| 7a(4a). | Bright green | *Hydra (viridis)* |
| 7b. | Not green | 8 |

| 8a(7b). | Holotrichous isorhiza nematocysts at least half as broad as long; embryotheca without spines | *Hydra (braueri)* |
| 8b. | Holotrichous isorhiza nematocysts slender or if plump, then embryotheca spined | 9 |

| 9a(8b). | Buds acquire two lateral tentacles before others appear; otherwise stenotele nematocysts at least 1.5 times as long as broad | *Hydra (oligactis)* |
| 9b. | Buds acquire tentacles in some other order; otherwise stenotele nematocysts less than 1.5 times as long as broad | *Hydra (vulgaris)* |

LITERATURE CITED

Acker, T. S. 1976. *Craspedacusta sowerbii*: An analysis of an introduced species. Pages 219–226 *in:* G. O. Mackie, editor. Coelenterate biology and behavior. Plenum, New York.

Barnes, R. D. 1966. Invertebrate zoology. Saunders, Philadelphia, Pennsylvania.

Batha, J. 1974. The distribution and ecology of the genus *Hydra* in the Milwaukee area of Lake Michigan. Ph.D. Thesis, Zoology Department, University of Wisconsin, Madison.

Bossert, P. 1988. The effect of hydra strain size on growth of endosymbiotic alga. Ph.D. Thesis, State University of New York, Stony Brook.

Bossert, P., and K. Dunn. 1986. Regulation of intracellular algae by various strains of the symbiotic *Hydra viridissima*. *Journal of Cell Science* 85:187–195.

Bouillon, J. 1957. Etude monographique du genre Limnocnida (Limnomedusae). Annales de la Societe Royale Zoologique de Belgique 87:254–500.

Boulenger, C. L., and W. U. Flower. 1928, The Regent's Park medusa, *Craspedacusta sowerbii* and its identity with *C. (Microhydra) ryderi*. Proceedings of the Zoological Society of London 66:1005–1015.

Campbell, R. D. 1987. A new species of *Hydra* (Cnidaria: Hydrozoa) from North America with comments on species clusters within the genus. Zoological Journal of the Linnean Society 91:243–263.

Cernichiari, E., L. Muscatine, and D. C. Smith. 1969. Maltose excretion by the symbiotic algae of *Hydra viridis*. Proceedings of the Royal Society of London Series B 173:557–576.

Dodson, S. I., and S. D. Cooper. 1983. Trophic relationships of the freshwater jellyfish *Craspedacusta sowerbii* Lankester 1880. Limnology and Oceanography 28:345–351.

Dunn, K. 1986. Adaptations to endosymbiosis in the green hydra, *Hydra viridissima*. Ph.D. Thesis, State Univ. of New York, Stony Brook.

Dunn, K. 1987. Growth of endosymbiotic algae in the green hydra, *Hydra viridissima*. Journal of Cell Science 88:571–578.

Fuhrman, O. 1939. *Craspedacusta sowerbii* Lank. et un noveau coelentere d'eau douce, *Calpasoma dactyloptera*, n.g., n.sp. Revue Suisse de Zoologie 46:363–368.

Griffing, T. 1965. Dynamics and energetics of populations of brown hydra. Ph.D. Thesis, Zoology Department, University of Michigan, Ann Arbor.

Hecker, B., and L. B. Slobodkin. 1976. Responses of *Hydra oligactis* to temperature and feeding rate. Pages 175–186 *in:* Coelenterate ecology and behavior. Plenum, New York.

Hershey, A., and S. Dodson. 1987. Predator avoidance by *Cricotopus*: Cyclomorphosis and the importance of being big and hairy. Ecology 68:913–920.

Hessinger, D., and H. M. Lenhoff, editors. 1988. The biology of nematocysts. Vol. 3. Academic Press, London. 600 pp.

Hubschman, J. H., and W. J. Kishler. 1972. *C. sowerbii* Lankester 1880 and *Cordylophora lacustris* Allman 1871 in Western Lake Erie (Coelenterata). Ohio Journal of Science 72:318–332.

Hyman, L. H. 1940. The invertebrates: Protozoa through Ctenophora. McGraw-Hill, New York. 726 pp.

Kanaev, I. I. 1969. Essays on the biology of freshwater polyps. H. M. Lenhoff, editor. (Translated by E. T. Burrows and H. M. Lenhoff.) Publ. by H. M. Lenhoff, Univ. of Miami, Coral Gables, Florida. (Originally published by Soviet Academy of Sciences, Moscow, 1952.) 453 pp.

Kramp, P. L. 1950. Fresh water medusae in China. Proceedings of the Zoological Society of London 120:165–184.

Lenhoff, H. M. 1974. On the mechanism of action and evolution of receptors associated with feeding and digestion. Pages 211–243 *in:* L. Muscatine and H. M. Lenhoff, editors. Coelenterate biology: reviews and new perspectives. Academic Press, New York.

Lenhoff, H. M., editor. 1983. Hydra: research methods. Plenum, New York. 463 pp.

Lipin, A. 1911. Morphologie und biologie von *Polypodium*. Zoologische Jahrbuecher, Abteilung fuer Anatomie und Ontogenie der Tiere 31.

Lipin, A. 1926. *Polypodium*. Zoologische Jahrbuecher, Abteilung fuer Anatomie und Ontogenie der Tiere

Lomnicki, A., and B. Slobodkin. 1966. Floating in Hydra littoralis. Ecology 47:881–889.

Loomis, W. F. 1964. Microenvironmental control of sexual differentiation in *Hydra*. Journal of Experimental Zoology 156:289–306.

Mackie, G. O., editor. 1976. Coelenterate ecology and behavior. Plenum, New York.

Muscatine, L. 1965. Symbiosis of hydra and algae. III. Extracellular products of the algae. Comparative Biochemistry and Physiology 16:177–192.

Muscatine, L., and H. M. Lenhoff, editors. 1974. Coelenterate biology. Academic Press, New York. 499 pp.

Naumov, D. V. 1960. Hydroids and medusae of the USSR. (Israel Program for Science Translation, 1969.) *Zoological Institute of the Academy of Science USSR* No. 70, p. 112.

Pardy, R. 1983. Preparing aposymbiotic hydra and introducing symbiotic algae into aposymbiotic hydra. Pages 393–401 *in:* H. M. Lenhoff, editor. Hydra: Research methods. Plenum, New York.

Payne, F. 1924. A study of the freshwater medusae, *craspedacusta ryderi*. Journal of Morphology 38:397–430.

Rahat, M., and R. D. Campbell. 1974a. Three forms of the tentacled and non-tentacled fresh water coelenterate polyp genera *Craspedacusta* and *Calpasoma*. *Transactions of the American Microscopical Society* 93: 235–241.

Rahat, M., and R. D. Campbell. 1974b. Nematocyst migration in the polyp and 1 celled tentacles of the minute freshwater coelenterate *Calposoma dactyoptera*. Transactions of the American Microscopical Society 93:379–385.

Rahat, M., and V. Reich. 1985. A new alga-hydra symbiosis: *Hydra magnipappillata* of the "non-symbiotic" vulgaris group hosts a Chlorococcum-like alga. Symbiosis 1:177-184.

Raikova, E. V. 1973. Life cycle and systematic position of *Polypodium hydriforme* Ussov (Coelenterata), a cnidarian parasite of the eggs of *Acipenseridae*. Publications of the Seto Marine Biological Laboratory 20:165–174.

Raikova, E. V. 1980. Morphology, ultrastructure and development of the parasitic larva and its surrounding trophamnion of *Polypodium hydriforme* Coelenterata. Cell Tissue Research. 206:487–500.

Ritte, U. 1969. Floating and sexuality in laboratory populations of *Hydra littoralis*. Ph.D. Thesis, Univ. of Michigan, Ann Arbor.

Roos, P. J. 1979. Two stage life cycle of a *Cordylophora* population in the Netherlands. Hydrobiologia 62:231–239.

Rutherford, C. L., D. Hessinger, and H. M. Lenhoff. 1983. Culture of sexually differentiated hydra. Pages 71–77 *in:* H. M. Lenhoff, editor. Hydra: research methods. Plenum, New York.

Schaller, H. C. 1983. Hormonal regulation of regeneration in hydra. Pages 1–14 *in:* J. L. Barker and J. F. McKelvey, editors. Current methods in Cellular neurobiology. Vol. 4: Model Systems. Wiley, New York.

Schaller, H. C., T. Schmidt, K. Flick, and C. Grimmelikhuijzen. 1977. Analysis of the morphogenetic mutants of Hydra. I. The aberrant III. The maxi and mini. Wilhelm Roux's Archives of Developmental Biology 183:215–222.

Schwartz, S. S., B. C. Hann, and D. N. Hebert. 1983. The feeding ecology of *Hydra* and possible implications in the structuring of pond plankton communities. Biological Bulletin (*Woods Hole, Mass.*) 164:136–142.

Slobodkin, L. B. 1964. Experimental populations of hydrida. *In:* British Ecological Society Jubilee Symposium. Journal of Animal Ecology, 33 (Suppl.):131–148.

Stiven, A. 1976. The quantitative analkysis of the Hydra-Hydramoeba host–parasite system. Pages 389–400 *in:* G. O. Mackie, editor. Coelenterate ecology and behavior. Plenum, New York.

Thiel, H. 1973. *Limnocnida indica* in Africa. *Publ. Seto Mar. Biol. Lab.* 20:73–79

Trembley, M. 1744. Memoire pour servir a l'histoire d'un gente de polypes d'eau douce a bras en forme de cornes. Leiden.

Flatworms: Turbellaria and Nemertea

6

Jerzy Kolasa
Department of Biology
McMaster University
Hamilton, Ontario L8S 4K1 Canada

Chapter Outline

TURBELLARIA

I. INTRODUCTION: STATUS IN THE ANIMAL KINGDOM

II. ANATOMY AND PHYSIOLOGY
  A. General External and Internal Anatomical Features
  B. Environmental Physiology

III. ECOLOGY
  A. Life History
  B. Distribution
  C. Behavioral Ecology
  D. Foraging Relationships, Predators, and Parasites
  E. Population Regulation and Density
  F. Functional Role in the Ecosystem

IV. CURRENT AND FUTURE RESEARCH PROBLEMS
  A. Dispersal
  B. Rhabdoids
  C. Endosymbiosis
  D. Community Ecology

V. COLLECTING, REARING, AND IDENTIFICATION TECHNIQUES

VI. IDENTIFICATION OF NORTH AMERICAN GENERA OF MICROTURBELLARIA
  A. Taxonomic Key to Orders and Suborders of Turbellaria
  B. Taxonomic Key to Catenulida
  C. Taxonomic Key to Macrostomida
  D. Taxonomic Key to Lecithoepitheliata
  E. Taxonomic Key to Prolecithophora, Proseriata, and Tricladida
  F. Taxonomic Key to Typhloplanoida

  G. Taxonomic Key to Kalyptorhynchia and Similar Forms
  H. Taxonomic Key to Dalyellioida

NEMERTEA

VII. GENERAL CHARACTERISTICS, EXTERNAL AND INTERNAL ANATOMICAL FEATURES

VIII. ECOLOGY
  A. Life History
  B. Physiologic Adaptations
  C. Behavioral Ecology
  D. Functional Role in the Ecosystem

IX. CURRENT AND FUTURE RESEARCH PROBLEMS

X. COLLECTION, CULTURING, AND PRESERVATION

XI. IDENTIFICATION
  A. Taxonomic Key to Species of Freshwater Nemertea
  Literature Cited
  Appendix 6.1. List of North American Species of Microturbellaria

Turbellaria and Nemertea are common and often very numerous inhabitants of freshwaters. Even though more than 200 species of Turbellaria and 3 species of Nemertea live in North America, their ecology and systematics have been less studied than that of many other common aquatic invertebrates. An obvious reason for this limited attention is the difficulty posed by preservation. Most turbellarians become unrecognizable after a routine preservation of field samples in alcohol or formalin. Study of live specimens is the best method and many interesting discoveries in the areas of reproductive biology, dispersal, endosymbiosis, community structure, and other areas lie ahead.

Ecology and Classification of North American Freshwater Invertebrates
Copyright © 1991 by Academic Press, Inc.
All rights of reproduction in any form reserved.

TURBELLARIA

I. INTRODUCTION: STATUS IN THE ANIMAL KINGDOM

The turbellarian flatworms are the lowest acoelomate Bilateria with only a single opening to the digestive tract; that is, they lack a definitive anus. Cestoidea (tapeworms), Trematoda (flukes), and Turbellaria constitute the phylum Platyhelminthes. Because of tradition and practical considerations, turbellarians are divided into microturbellarians and macroturbellarians. This division does not reflect phylogenetic relationships, but rather superficial morphological similarities. Also, depending on the methodology chosen, the taxa discussed may have a rank different from the one used here (see Ehlers 1986). Several orders and suborders of Turbellaria are known from freshwaters (Table 6.1, Fig. 6.1).

Of the approximately 400 species of freshwater microturbellaria, about 350 of these species have been recorded in Europe (Lanfranchi and Papi 1978), and approximately 150 species of microturbellaria occur in North America (Appendix 6.1), many of which are the same as those found in Europe. Many species and genera of microturbellarians remain to be discovered in North America. Records from other countries are sparse. Interestingly, Africa, South America, and Papua New Guinea trail closely behind North America in the number of species. Differences in species richness are clearly due in great part to intensity of investigations and, for this reason, the number of species in North America must be highly underestimated. Triclads (often called planarians) are more thoroughly studied. The number of recorded species in North America is approximately 40, but the actual number of taxa is likely to be substantially higher.

II. ANATOMY AND PHYSIOLOGY

A. General External and Internal Anatomical Features

Most freshwater microturbellaria are less than 1 mm in length, although some can reach several millimeters. There are exceptions, however. Triclads are distinctly larger, with most species exceeding 10 mm and the largest being several centimeters long.

The turbellarian body is elongated, relatively soft, and usually tapered at the ends. Sometimes a short tail-like section or lateral flaps are present near the cerebral region. With the exception of triclads, flatworms are generally not flat—despite their com-

Table 6.1 Selected Characteristics of Commonly Recognized Turbellarian Orders and Suborders Occurring in Freshwaters

Taxon[a] Order Suborder Superfamily	Approx. Number of Species	Average Body Size (mm)	Comments and Special Features
Microturbellarians			
Catenulida	60	0.5–1	Thin chains of zooids; common in various habitats
Acoela	3	0.5–1	Most species marine
Macrostomida	50	1.0–3	Common in various habitats
Prolecithophora	5	5.0–10	Many marine species
Lecithoepitheliata	10	3.0–10	Many marine species; some are terrestrial or semiaquatic
Proseriata	4	2.0–5	Many marine species
Rhabdocoela			
Dalyellioida	100	0.8–1	Common in various habitats
Dalyelliida		0.8–1	Common in various habitats
Temnocephalida		1.0–14	Commensals on crustaceans, snails, turtles, one parasitic
Typhloplanoida			
Typhloplanida	150	0.5–6	Common in various habitats including terrestrial and semiaquatic
Kalyptorhynchia	15	1.0–2	Rare; most species marine
Macroturbellarians			
Tricladida	100	5–20	Greatest diversity associated with karst habitats

[a] Taxonomical rank is indicated by indentation.

mon name. Most microturbellaria are cylindrical in cross-section, with some differentiation of shape of the dorsal and ventral surfaces. Moreover, species that reproduce asexually may be composed of several zooids (Figs. 6.3B, 6.4A, 6.5B, 6.12A), giving a chain-like appearance to an individual. Turbellarians may be colorless, white, red, bluish, green, black, brown, or yellowish depending on epidermal and parenchymal pigments, gut content, and symbiotic algae. Some species, such as *Dugesia tigrina* and *Hydrolimax grisea*, develop characteristic patchy patterns of pigments.

Anatomically, the most prominent turbellarian features are a ciliated epidermis, rhabdoids, an intestine without anus, ventral mouth, and complex reproductive system (Figs. 6.1, 6.9, 6.12D). The ciliated epidermis covers the entire body surface. Rhabdoids are rod-like, light-refracting structures produced in epidermis (dermal rhabdoids) or in the parenchyma (adenal rhabdoids). It is often difficult to determine the type of rhabdoids in live specimens. Adenal rhabdoids are most abundant at or near the front extremity. They often aggregate in juxtaposed groups that are known in the taxonomic literature as rod tracts. Wrona (1986) has reviewed hypotheses on the role of rhabdoids in triclads and has provided experimental evidence that rhabdoids contribute to mucus production. Rhabdoids are also poisonous and used in prey immobilization or in deterring predators. Other components of the body walls include a basal membrane and muscle fibers. The fibers are longitudinal, circular, or in larger forms, diagonal. Some fibers go through the parenchyma. The parenchyma is a loose tissue filling the body cavity and containing various cell types and their products, principally secretory, excretory, reproductive, and formative cells that fulfill important physiological functions (Hyman 1951). In larger turbellarians, the parenchyma is usually more compact than in smaller forms where it can be quite loose and vacuolated. Turbellarians possess numerous glands of which the frontal, pharyngeal, shell-producing, and rhabdoid-producing glands are most common.

A vast array of sensory organs enables turbellarians to react to environmental stimuli. These organs include sensory hairs, eyes, statocysts, and chemoreceptors of various kinds. Some species possess aggregations of sensory cells, or ciliated pits, in lateral depressions situated in front of the brain ganglion. Exact functions of the ciliated pits and structures associated with them are unknown but at least some of these structures (i.e., the refractive bodies of *Stenostomum*) have been implicated in phototaxis (Marcus 1951). Some evidence exists that these bodies function as lenses (Tyler and Burt 1988). Thigmotaxis is common and plays a role in a range of behaviors such as choosing substrate, hunting prey, and avoiding predators. Among American freshwater Turbellaria, a statocyst is present only in *Catenula*, *Rhynchoscolex*, *Otomesostoma*, and an unidentified representative of Acoela (Strayer 1985, Kolasa *et al.* 1987). Turbellarians have two types of simple eyes: inverted and direct cup eyes (Bedini *et al.* 1973). The number of eyes is variable, but one pair of eyes is most frequently present.

Locomotion in the Turbellaria is based on the movement of cilia. Some triclads, however, are capable of fast muscular contractions allowing them to crawl in bursts of leech-like behavior. Smaller forms can both swim and glide on submersed objects, while heavier forms are restricted to gliding only. Adhesive organs in the form of pseudosuckers or papillae are often present.

B. Environmental Physiology

Gaseous exchange in turbellarians occurs through their body walls. Such exchange is dependent on the surface/volume ratio and is not very efficient. Consequently, flatworms may have a limited ability to adapt to low oxygen conditions. This conclusion appears to be true for most flatworms and this limitation is often exploited in extraction of specimens from samples (e.g., Young 1970, Schwank 1981). However, the ability to cope with reduced oxygen levels does not seem to differ substantially from that of other aquatic invertebrates. Oxygen adaptability differs more among species adapted to diverse habitats than among major taxa (cf. Heitkamp 1979b). For example, species of *Macrostomum* normally found in stream headwaters are more sensitive to low oxygen levels, and perhaps high temperature, than are representatives of the same genus from physically more fluctuating habitats (J. Kolasa personal observation). Some species, particularly in the genus *Phaenocora*, have developed adaptations to cope with anaerobic conditions (Young and Eaton 1975). They showed that the presence of algae in *Phaenocora typhlops* enhances its ability to survive low oxygen conditions. This same species, as well as a related *P. unipunctata* (Öersted) produce hemoglobin, allowing them to store oxygen during burrowing in the anoxic mud. Many marine species, however, live in microxic sediments but have neither hemoglobin nor symbiotic algae.

Heitkamp (1979a,b) studied respiration rates of *Opistomum pallidum*, two species of *Mesostoma*, and a species hosting symbiotic zoochlorellae—*Dalyellia viridis*. He showed the respiration to be related to the body size (surface in a very flat *Mesostoma ehrenbergi*, or weight in cylindrical *Mesostoma lingua* and other species). Interestingly, *Dalyellia viridis* that are symbiotic with zoochlorellae had the highest respiration rate of all studied species; possibly their symbiotic algae contribute significantly to the observed rate. Heitkamp (1979b) found the respiration rate of microturbellarians to vary depending on the temperature and diurnal rhythm. Moreover, the respiration rate was a function of body weight in several studied species except in *Mesostoma ehrenbergi*, an unusually flat worm, where it was proportional to the body surface.

A considerable effort has been devoted to studying nutrition and digestion in Turbellaria, particularly by Jennings (1977). In this context, the lipid storage capacity of flatworms and its importance to population dynamics and competition is discussed for *Dugesia polychroa* by Boddington and Mettrick (1977).

Triclads clearly avoid light. Microturbellarians appear to be either indifferent or positively phototactic (e.g., *Typhloplana viridata*), although if a sample is suddenly illuminated, many will attempt to hide in sediments.

Desiccation of eggs can be tolerated by some species, and for a few it may be an obligatory factor in releasing them from diapause and stimulating embryonic development. Similarly, some triclads such as *Phagocata* may undergo encystment as entire or fragmented animals.

Although the majority of Turbellaria appear to have some habitat preferences, within a limited range of physical conditions many species have adaptations conferring considerable flexibility. Some members of the Catenulida are found in habitats as different as fast-flowing streams and temporary pools (e.g., *Stenostomum leucops, Catenula lemnae*). *Mesostoma* species have been reported from both warm rice fields in the Sacramento Valley in California as well as from snow melt pools at elevations above 2400 m (Collins and Washino 1979).

Even though microturbellarians are delicate and fragile organisms, they appear to have a significant potential for dispersal. Results of an experiment conducted by Jan Ciborowski (University of Windsor) were surprising. Ciborowski studied colonization of artificial ponds (flexible swimming pools) situated on the roof of a university building for several weeks. About 50% of these ponds were successfully colonized by six different species of microturbellarians (J. Kolasa unpublished). With only one exception, these species are not known to produce resistant disseminules such as eggs or cocoons.

III. ECOLOGY

A. Life History

In most species, miniature replicas of the adult hatch directly from eggs; these juveniles differ from adults chiefly by the absence of reproductive systems. Some freshwater turbellarians, however, may be ovoviviparous (*Mesostoma*) or may have a larval stage distinctly different from the adult (*Rhynchoscolex*). Numerous modifications of the life cycle have evolved among turbellarians. Many seasonally occurring species are univoltine, particularly those associated with temporary habitats or at the extremes of their geographical ranges. Most other species are multivoltine, with the number of generations depending on habitat availability (see Heitkamp 1982 for an excellent account). A similar diversity of life cycles is observed in triclads, but the ecological reasons for this diversity remain hidden. Reproductive patterns may sometimes coincide with major taxonomic subdivisions. Calow and Read (1986) studied European representatives of the three families that are also represented in North America: Dugesiidae, Planariidae, and Dendrocoelidae. These investigators have found that species breeding over several seasons (Dugesiidae, Planariidae) invest less in reproduction than members of Dendrocoelidae, which reproduce once and die.

As a rule, turbellarians are hermaphroditic. Various modifications of both egg formation and development are found and may include self-fertilization, as in *Mesostoma ehrenbergi* (Fiore and Ioalé 1973, Göltenboth and Heitkamp 1977), protandry or progyny in many other species, or loss of either the male or female gonad as in Catenulida (Rieger 1986). All of these modifications have implications for habitat adaptability, clonal diversity, rate of population growth, and ability to colonize.

Asexual reproduction by means of paratomy, that is transverse division of the body, is common in several genera of microturbellaria, particularly in Catenulida and Macrostomida. In macroturbellarians, architomy occurs frequently in Dugesiidae and in some Planariidae (Calow and Read 1986, Rieger 1986). In some species of *Stenostomum* and in some populations of the common triclad *Dugesia tigrina*, sexual organs have never been observed.

B. Distribution

Freshwater turbellarians are largely free-living animals, although a few European freshwater species such as *Varsoviella kozminski* (Gieysztor et Wiszniewski) and *Phaenocora beauchampi* (Sekera) are ectoparasitic on crustaceans. Several other freshwater forms, including triclads, occurring in Europe and Australia are commensal on crustaceans and turtles (e.g., Jennings 1985). The great majority of freshwater turbellarians are free-living and live in various aquatic systems such as ponds, lakes, streams, hyporheic water, ditches, and temporary puddles. However, some may be found in aquatic habitats normally excluded from the domain of freshwater ecology, such as water films among fallen leaves in a mesic forest or in capillary soil water of a grassy meadow (Sayre and Powers 1966).

In North America microturbellarians are known as far north as Igloolik, a small island at 68°N latitude off Melville Peninsula. Two species, one *Mesostoma* and one unidentified mesostomid were collected by P. D. N. Hebert and associates (unpublished). In Alaska and in the area of Hudson Bay in northern Canada, as many as six different species are known (Holmquist 1967, Schwartz and Hebert 1986). Further south, in temperate climatic zones, the species diversity increases and can easily reach 20 to perhaps 60 species in a single lake or pond. For example, there are 23 species of flatworms in the oligotrophic Mirror Lake, New Hampshire (Strayer 1985) and as many as 57 species occur in the polytrophic Lake Zbechy in Poland (Kolasa 1979). Streams also have rich flatworm fauna. In a short section of Wappinger Creek, an eastern tributary of the Hudson River in New York, 15 species were found (Kolasa *et al.* 1987). A survey of several streams, including springs and underground water in the same area revealed 32 species of microturbellarians, some of which are still undescribed. Because of a dearth of systematic studies, few microturbellarians are known from the west coast. Case and Washino (1979), Collins and Washino (1979), and other related papers report *Mesostoma* "*lingua*" from California, together with some other common species in rice fields.

Many species, particularly among the Catenulida and the Macrostomida, are widely distributed worldwide, while species of the "higher" Turbellaria such as the Typhloplanida and Tricladida are geographically more restricted. About 3% of all species known from Europe are designated as cosmopolitan in distribution (Lanfranchi and Papi 1978). Studies (Mead and Kolasa 1984) conducted in the tropics and in the southern hemisphere permit a conclusion that this figure is too low.

Caves and underground waters hold many unique and endemic species, particularly of triclads (e.g., Kenk 1987). Some troglobiotic triclads and microturbellarians have recent or remote marine origins (e.g., Kawakatsu and Mitchell 1984a, Kolasa 1977a). An extensive study of European underground triclads (Gourbault 1972) revealed a suit of distinct distributional, reproductive, and physiologic characteristics. These often include specialized underground microhabitat selection, greater

frequency of sexual reproduction, and slower metabolism as compared to epigean (or surface-dwelling) triclads.

In contrast, a number of interstitial and underground water microturbellarians, e.g., *Bothrioplana semperi*, *Limnoruanis romanae*, *Stenostomum pegephilum*, occur on more than one continent. Distributions of most genera appear to be worldwide among microturbellarians, but not triclads, whose many genera are known exclusively from either North America, Eurasia, South America, or Australia (Ball 1974).

The majority of triclads from North America are described from the eastern and southern United States and Canada (Kenk 1972). Four triclad species inhabit Alaskan waters and ten live in Tennessee (Chandler and Darlington 1986). Several species of triclads with recent marine origins occur in freshwater caves of Mexico (Kawakatsu and Mitchell 1984b, Benazzi and Giannini 1971).

The ecological distribution of both microturbellarians and triclads has been studied more intensively in Europe but most of the results are directly relevant to North American fauna. The distribution of microturbellarians in lakes has been investigated by Strayer (1985), Chodorowski (1959), Kolasa (1977b, 1979), Rixen (1968), Schwank (1976), and Young (1970, 1973b). These studies show that microturbellarians do differentiate among substrate types and depth zones. The flatworm assemblages of sand and gravel bottoms are distinctly different from assemblages in habitats characterized by bottoms of silt, coarse organic matter, or vegetation. Plant successional zones also differ in their species assemblages. Although the littoral zone has the greatest density and diversity of flatworms, there is apparently no limit to how deep some species may be found as long as oxygen is available. For example, *Otomesostoma auditivum* lives at a depth of 15–45 m in Echo Lake in the Sierra Nevada range, but it occurs at a depth of 145 m in some European alpine lakes (Hyman 1955).

Ecological differentiation of microturbellarians in running waters is more pronounced than in lakes. European studies showed distinct assemblages of species associated with springs, headwaters, and lower stretches of streams and rivers (epirhithron, metarhithron, hyporhithron, and potamon) (Schwank 1981, Kolasa 1983). These regularities in zonation of flatworm assemblages are consistent with patterns observed among other lotic taxa. These patterns, however, still need to be related to newer ecological approaches to running waters, i.e., to the river continuum concept (Vannote *et al.* 1980). Among those classical zones, epirhithron shows the greatest species richness. No comparable studies have been published on North American microturbellarians. Zonation of triclads in streams was studied in a greater detail, including experimentation (Reynoldson 1983). Competition, as well as thermal requirements were implicated as important factors separating several species of flatworms along a water course and in standing waters (Claussen and Walters 1982, Pattee 1980, Reynoldson 1983).

Boreo-alpine distribution of many cold water species has been tied to glaciations in Europe, and in North America similar patterns also seem to occur (Hampton 1988). At lower latitudes, similarities between South American and North American microturbellarian composition (cf. Marcus 1945) suggest considerable interchanges of species.

Both microturbellarians and triclads live in habitats where the physical conditions require additional adaptations in comparison to an average pond or stream environment. Wet mosses and leaves on the water edge, leaves of Bromeliaceae, the high Arctic, ice melt ponds at elevations over 4000 m in central Asia, and hot springs are habitats for one or more species of flatworms. Representatives of freshwater families are also commonly found in humid terrestrial habitats.

C. Behavioral Ecology

Laboratory studies have shown that turbellarians have a limited ability to learn simple tasks such as choosing white over dark branches of a maze (McConnell 1967). However, observations of prey handling (personal observation) indicate that their learning ability may be much greater for tasks that might be advantageous under natural conditions, such as prey or predator recognition.

Possible behavior modification or behavior determination associated with symbiosis is a potentially rewarding research area. *Typhloplana viridata*, a common spring pond species, is unique in that it is positively phototactic. This behavior may be associated with the presence of symbiotic algae. It might be of general interest to know control mechanisms of this behavior.

D. Foraging Relationships, Predators, and Parasites

Microturbellarians eat bacteria, algae, protozoans, and invertebrates, while triclads feed predominantly on larger invertebrates. Scavenging is common in both morphological groups. Various *Mesostoma* species have been studied in detail and illustrate the diversity of feeding habits and techniques within one genus. Leaf-shaped *M. ehrenbergi* usually suspend themselves in the water column on a mucous fila-

ment attached to the surface film (Göltenboth and Heitkamp 1977). In that position, the worm waits for a cladoceran or an insect larva to come into contact. Using its front end, *M. ehrenbergi* traps the prey with sticky mucus, wraps around it, and proceeds to feed. According to Dumont and Carels (1987), *Mesostoma lingua* uses neurotoxins to immobilize its prey in addition to mucus. *Mesostoma* can also glide on the surface of underwater objects or swim directly in the water column. Both activities enhance its chances of encountering prey. Unlike *M. ehrenbergi*, many other species are spindle-shaped (e.g., *M. arctica*). These species hunt more often on the bottom or among submersed objects. Although several *Mesostoma* studied spend more than 50% of their time resting on various objects, they can catch prey as soon as it comes close enough to create vibrations in the water (J. Kolasa personal observation). At that point, *Mesostoma* can erect the front portion of its body and actively move it around in search for contact with the prey. Several species of *Mesostoma*, such as *M. vernale* (Hyman) and *M. californicum* (Hyman), as well as species of *Bothromesostoma* have a flat ventral surface with which they cling to the surface film. These species are more agile and much faster than other species of *Mesostoma* and they prefer hunting at the water surface.

Immobilizing the prey and sucking its body fluids is a common feeding behavior in microturbellarians equipped with a strong pharynx rosulatus, or in triclads with a pharynx plicatus. Many other turbellarians, however, swallow the whole victim or bite chunks of prey. For example, species of *Stenostomum* and *Macrostomum* that feed on algae, protozoans, rotifers, and other meiofauna usually swallow the prey whole. Others such as *Stenostomum predatorium* and various *Phaenocora* species may attack and partly consume larger turbellarians (Kepner and Carter 1931), oligochaetes, or other soft invertebrates. Finally, a few microturbellarians (e.g., *Gyratrix hermaphroditus* and *Prorhynchus stagnalis*), use their copulatory stylets to stab the prey.

Chemical detection of prey undoubtedly plays an important role in feeding. Injured invertebrates attract species of both triclads and microturbellarians under laboratory conditions. Such injured or recently killed prey allows scavenging by many individuals that played no role in immobilizing it.

Turbellarians themselves are subject to predation and parasitism. Ciliates and flagellates are frequent parasites of Catenulida and Typhloplanida, and nematodes have been found in Lecithoepitheliata (*Prorhynchus*; J. Kolasa unpublished). Fish occasionally eat triclads (Davies and Reynoldson 1971). Microturbellarians (*Phaenocora typhlops*) may occasionally be eaten by invertebrates such as a chironomid *Anatopynia* (Young 1973a) or other Turbellaria (e.g., *Stenostomum predatorium*).

E. Population Regulation and Density

It is not clear whether there is any single mode of population regulation that applies to flatworms. The few studies conducted in Britain indicate that intraspecific competition for food may be very important for *Phaenocora typhlops* (Young 1975), while interspecific competition may play a role in regulating several planarians (Reynoldson 1983). In European *Polycelis tenuis* and perhaps in American *Cura foremanii* (Girard), this regulation is expressed by a negative relationship between the population size and density. In fact, long term changes in the population size of *Polycelis* suggest that limit cycles or cyclical fluctuations of the density due to intraspecific competition are involved (Reynoldson 1983 and earlier references therein).

Densities of microturbellaria vary according to the habitat, season, and species. No major generalizations have been made, but some patterns are beginning to emerge, particularly along seasonal and successional axes. The latter pertains to both running and standing waters. In small, second- or third-order streams, the greatest densities of microturbellaria are found in the lower reaches, i.e., in the metarhithron and hyporhithron, while the highest richness is found in the epirhithron zone (Kolasa 1983). For example, a mean density of 1280 individuals/m^2 (of all species) was recorded in Wappinger Creek, New York, in artificial substrate and after a week of colonization.

Schwank (1981), using volume instead of surface to quantify turbellarian abundance in submontane streams in Germany, found densities of about 80 microturbellarians per liter, which, assuming an average sample depth into the substrate of 5 cm, translates into over 4000 individuals/m^2. His data suggest a corresponding density of triclads of 32 individuals per liter and 1600/m^2. In oligotrophic Mirror Lake, turbellarian densities varied from 40,000/m^2 in shallow littoral regions to almost zero in profundal zones; the lakewide mean was 27,000/m^2 (Strayer 1985). Other studies reported densities of 800/m^2 (Nalepa and Quigley 1983), 3500/m^2 (annual mean) in shallow littoral areas of Zbechy Lake in Poland (Kolasa 1979), 3100/m^2 in Lake Paajarvi in Finland (Holopainen and Paasivirta 1977), to 9500/m^2 in ponds in Germany (Heitkamp 1982). These values have to be interpreted cautiously, however, in the light of seasonal succession and abundance changes, which ordinarily produce the greatest richness and abundance in early summer (Bauchhenss 1971, Strayer 1985, Chodorowski 1960).

F. Functional Role in the Ecosystem

The functional role of species is typically discussed in terms of their trophic interactions. All triclads are predatory. Several invertebrates and vertebrates may consume triclads and are a significant source of their mortality. However, triclads constitute a relatively minor component of the diet of their predators (Davies and Reynoldson 1971).

Heitkamp (1982) reports intense predation on *Mesostoma* and *Rhynchomesostoma* by *Dalyellia* in temporary pools. He has not determined the frequency of such predation nor established whether this predation results in population regulation.

High densities of both triclads and microturbellarians suggest that their role in biotic interactions of benthic communities may be greater than their contribution to the diet of other organisms. Densities of *Mesostoma* species (*lingua*) in excess of 1000 individuals/m^2 were observed in more than 30% of the rice fields studied by Collins and Washino (1979) in California. In some cases, microturbellaria may regulate population dynamics of zooplankton in ponds, as demonstrated by Maly *et al.* (1980). They found that the feeding rates of *Mesostoma ehrenbergi* were two zooplankters per day (estimated from Fig. 6.1). However, Schwartz and Hebert (1982) reported that this species can consume about ten cladocerans per day. More important, perhaps, is the functional role of microturbellaria as consumers of protozoans, rotifers, and algae, especially by usually abundant *Stenostomum leucops* and similar forms. Unfortunately, no quantitative data on this subject are available.

Although there are no explicit studies of the energy flow through lake meiobenthos, the detailed study of the energy budget of Mirror Lake (Strayer and Likens 1986) indicated this flow to be equivalent to that of zooplankton. Furthermore, Strayer (1985) showed that Turbellaria, and *Rhynchoscolex* in particular, were a substantial component of meiobenthos in Mirror Lake. This statement might be somewhat misleading because meiobenthos itself is a minor component of the whole budget.

IV. CURRENT AND FUTURE RESEARCH PROBLEMS

A. Dispersal

Understanding the geographical and ecological distribution of Turbellaria is important because their distribution has evolutionary and ecological implications. Yet this understanding will suffer seriously unless the dispersal capabilities of flatworms are assessed in terms of distance and rates. Hebert and Payne (1985) found low levels of gene flow in Arctic *Mesostoma*. Their calculated dispersal rates are inconsistent with both the unpublished data from Ciborowski's experiment mentioned earlier and with common observations that even newly created water bodies have a rich complement of species. A series of questions may be posed. Is dispersal different at various latitudes? Do dispersal strategies and efficiency differ along major taxonomic lines? Are habitat generalists better dispersers than specialists? Are species associated with a particular habitat type (e.g., temporary ponds) better dispersers as opposed to cave dwellers? If yes, what in their life cycle makes them so?

B. Rhabdoids

Wrona's (1986) study demonstrates a lack of support for hypotheses that rhabdoids play a major role in contacts with prey or predators in triclads. However, in many microturbellarian rhabdoids, primarily adenal products, but also many other similar products, are channelled toward the anterior of the body. The front end usually plays a crucial role in exploration of the environment and in initiation of attack against prey. This special position is highly suggestive of more active applications of rhabdoids than the mere production of mucus and, therefore, warrants further research.

C. Endosymbiosis

Green algae occur in *Phaenocora, Dalyellia, Typhloplana, Castrada*, and some other microturbellarians. The nature of the association between the algae and turbellarians is not fully understood. However, it could be used as a convenient general model for laboratory manipulation.

D. Community Ecology

Comparative research on the density and richness of turbellarian assemblages in various water types might suggest which factors have relatively greater influence in structuring flatworm communities. Quantification and standardization of sampling procedures may be a major hurdle for comparative ecologists. Different density estimates obtained in quantification of lake Turbellaria may be due to minor differences in techniques (e.g., Strayer 1985).

V. COLLECTING, REARING, AND IDENTIFICATION TECHNIQUES

A variety of techniques can be used to obtain qualitative and quantitative samples of Turbellaria. These techniques may involve scooping with a plankton net over aquatic vegetation or sediments; collecting bottom sediments by means of bottom samplers such as Ekman, Ponar, or multiple corer (Schwank 1981, Strayer 1985); filtering water taken from wells, springs, hyporheic interstitial, or ponds and puddles using a plankton net (Kolasa *et al.* 1987); and baiting (e.g., *Mesostoma*, triclads) into traps using pieces of liver or injured aquatic insects (Case and Washino 1979, Kenk 1972). In streams, a Surber sampler or similar devices can be applied.

Samples should be cooled, topped with water, and transported to the lab as quickly as possible. In the lab, samples can be transferred to jars or beakers and left for a couple of hours for the water to stagnate. When the water clears, animals swim toward the surface or glide on the glass where they can be picked up with a pipette for further examination. Alternatively, turbellarians can be directly separated from the sediments under a stereomicroscope; but, this procedure is more time consuming. Turbellaria can also be removed from mineral substrates having a low organic matter content by gently heating the sample on a hot plate so that the surface of the sediment reaches 30°–32°C (Kolasa 1983). Other methods of extracting flatworms from sediments were developed for marine sands (Martens 1984, Noldt and Wehrenberg 1984). These methods are based on various combinations of anaesthetizing worms (with MgCl$_2$ or ethanol), stirring water to separate mineral particles, and sieving to catch worms. Some of these methods may be adaptable to freshwater fauna.

Many triclads and microturbellarians can be maintained in the lab in small glass containers as long as the water is regularly changed and appropriate food provided. Synthetic pond water is best for common species. Temperatures between 17°–25°C may be adequate for most species except for cold-water forms. Food requirements vary with the species and a variety of easily available aquatic invertebrates must be tested to ensure success. These may include cladocerans, mosquito larvae (for *Mesostoma, Dugesia*), oligochaetes (for *Phaenocora*), protozoans, or other small turbellarians (*Stenostomum*) (e.g., Kolasa 1987, Yu and Legner 1976). Often, pieces of larger invertebrates such as *Asellus, Gammarus*, and *Tubifex* can be successfully used. McConnell (1967) provides many useful hints on rearing triclads.

Most turbellarians can be identified by squash mounts of live animals. Identification is difficult for a novice. Squash mounts are prepared by placing a live individual in a drop of water under a microscope coverslip. Next, excess water is removed by a fine pipette or a strip of filter paper until the specimen is immobilized. Pressure on the specimen can be varied by removing or adding water. Such mounts, with a little bit of practice, reveal the arrangement and appearance of internal organs—information usually necessary for species determination. Whenever further study is desirable, animals can be fixed and preserved as in the following steps.

1. Anaesthetizing with 7% ethanol, 0.1% chloretone, 1% hydroxylamine hydrochloride, or slowing their locomotion by placing them in a small volume of water on ice (this step is optional).
2. Killing with hot fixatives such as Stieve's, Gilson's, Bouin's, room temperature 70% ethanol, or (optional) killing cooled worms with glutaraldehyde (see electron microscopic techniques for details);
3. Rinsing mercury residues of the above fixatives with 50% ethanol solution of iodine (except Bouin');
4. Storing specimens in 70% ethanol.

Animals so prepared can be sectioned and analyzed according to standard histological methods, including staining with Mallory's stain, Delafield's hematoxylin, and other common stains.

VI. IDENTIFICATION OF NORTH AMERICAN GENERA OF MICROTURBELLARIA

The key provided uses a combination of both phylogenetic and superficial morphological and anatomical characters. Although the phylogenetically important characters permit greater confidence, they are often very difficult to use by an inexperienced researcher. It is strongly recommended that an identification be confirmed by comparing the specimen at hand with exact taxonomic descriptions of the species available in other publications, particularly in monographs by Luther (1955, 1960, 1963) (Dalyelliidae, Macrostomidae, Typhloplanidae), Nuttycombe (1956), Nuttycombe and Waters (1956) (*Catenula, Stenostomum*), Ferguson (1940) (*Macrostomum*), and Gilbert (1938) (*Phaenocora*).

A. Taxonomic Key to Orders and Suborders of Turbellaria

As some of the characters may be difficult to determine, it may be advisable to match the general body plan of the individual being identified with one of the pictures in Fig. 6.1A–H.

1a.	Simple female gonad; egg entolecithal (Acoela, Macrostomida, Catenulida; freshwater members of the latter two often reproduce asexually and ovary is absent) (Figs. 6.1A–C, 6.12A, B)	2
1b.	Heterocellular female gonad, with yolk-producing part separate from the oocyte-producing part (all other Turbellaria) (Figs. 6.1D–H, 6.2)	4
2a(1a).	Mouth opens directly or through a pharynx simplex into the body; no protonephridia; no distinct intestine; gonads without clear walls (Fig. 6.1A); rare in freshwater	Acoela
	[Only one species (unidentified) collected in North America (from Mirror Lake in New Hampshire; Strayer 1985).]	
2b.	Mouth opens to a pharynx simplex; protonephridia present; epithelial, ciliated intestine present	3
3a(2b).	Protonephridia with a single, central excretory duct, often two or more zooids; male gonopore, if present, situated on the dorsal side and in the front of the body; ovary without oviducts and supplementary organs; statocyst present in some species (Figs. 6.1B, 6.12A, B)	Catenulida (see Section VI.B)
3b.	Protonephridia with a pair of main excretory ducts; male gonopore on the ventral side of the posterior part of body; female gonopore usually separate and in front of the male one, more than one zooids in the family Microstomidae only; no statocyst (Figs. 6.1C, 6.5A, B)	Macrostomida (see Section VI.C)

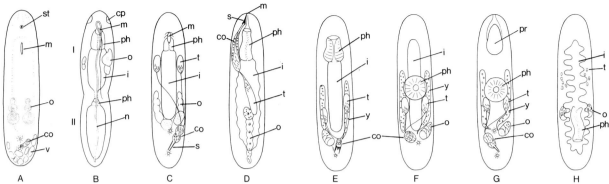

Figure 6.1 Schematic representation of higher turbellarian taxa found in freshwaters with an emphasis on the most useful diagnostic features. (A) Acoela (one unidentified species); (B) Catenulida (e.g., *Catenula, Stenostomum*); (C) Macrostomida (*Macrostomum, Microstomum*); (D) Lecithoepitheliata (e.g., *Prorhynchus, Geocentrophora*); (E) Dalyellioida (*Gieysztoria, Microdalyellia*); (F) Typhloplanoida (e.g., *Mesostoma, Olisthanella, Castrada*); (G) Kalyptorhynchia (e.g., *Gyratrix*; intestine omitted for clarity); (H) Proseriata and Tricladida (e.g., *Bothrioplana, Dendrocoelopsis*). co, Copulatory organ; cp, ciliated pits; i, intestine; m, mouth; n, protonephridial duct; o, ovary; ph, pharynx; pr, proboscis; s, stylet; st, statocyst; t, testes; v, vacuolized tissue; y, yolk glands.

4a(1b).	Single germovitellarium where ova are surrounded by yolk cell epithelium; pharynx variabilis; copulatory organ armed with a sclerotized stylet near the pharynx (Figs. 6.1D, 6.2C) .... Lecithoepitheliata (see Section VI.D)
4b.	Ova not surrounded by yolk epithelium; ovary separate or combined with vitellarium (Figs. 6.1E–H, 6.2A, B) 5
5a(4b).	Pharynx plicatus or variabilis; testicles and ovaries dispersed; germ cells dispersed or aggregated (Fig. 6.1H) .... Proseriata, Prolecithophora, and Tricladida (see Section VI.E)
5b.	Pharynx bulbosus (doliiformis or rosulatus, Fig. 6.1E–G) 6
6a(5b).	Pharynx doliiformis, directed forward (Fig. 6.1E) .... Dalyellioida (see Section VI.H)
6b.	Pharynx rosulatus directed ventrally or, exceptionally as in *Phaenocora*, ventrally and forward 7
7a(6b).	Without a proboscis (Fig. 6.1F) .... Typhloplanoida (see Section VI.F)
7b.	Proboscis present (Fig. 6.1G) .... Kalyptorhynchia (see Section VI.G)

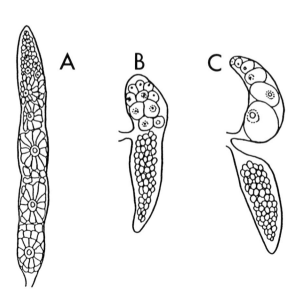

Figure 6.2 Examples of different types of heterocellular female gonads. (A) *Typhloplana, Dalyellia*; (B) *Bothrioplana*; (C) *Prorhynchus, Geocentrophora*.

B. Taxonomic Key to Catenulida

1a.	Brain simple, compact, oval		2
1b.	Brain composed of paired frontal and posterior lobes; a group of sensory cells in front of the brain, often arranged pseudometamerically, i.e., in parallel rows .... family Stenostomidae		5
2a(1a).	Brain situated usually at the basis of but not further than in the middle of prostomium (frontal body section without the intestine); ciliated ventral or lateroventral furrow separating prostomium from the rest of body present .... family Catenulidae		3
2b.	Brain situated in the first 1/3 of prostomium .... family Chordariidae	*Chordarium*	
3a(2a).	A ring of strongly ciliated longitudinal grooves developed on the prostomium (Fig. 6.3A)	*Suomina*	
3b.	No such grooves present		4
4a(3b).	Clearly open intestine extends approximately, to the middle of the postpharyngeal section of body (Figs. 6.3B, C, 6.12A)	*Catenula*	
4b.	Open intestine extends much further, leaving no more than 1/5 to 1/4 of the postpharyngeal section closed; usually the intestinal lumen of individual zooids is connected	*Dasyhormus*	
5a(1b).	Adult individuals composed of two or more zooids; gonads often absent		6
5b.	Adult individuals (with gonads) without signs of asexual divisions; long prostomium, ciliated pits absent (Figs. 6.3D, F, 6.12B)	*Rhynchoscolex*	
6a(5).	A section of the intestine near the pharynx with a strong ring of muscles (Fig. 6.4E)	*Myostenostomum*	
6b.	No such structure, intestine uniformly ciliated (Fig. 6.4A–D)	*Stenostomum*	

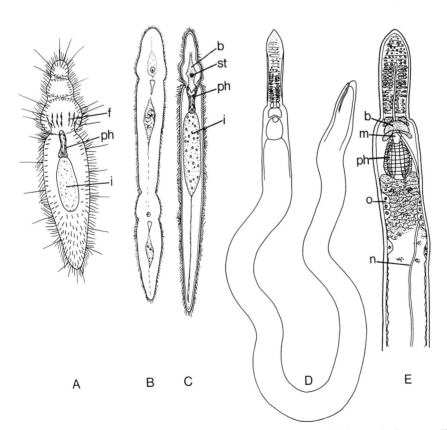

Figure 6.3 Some representatives of Catenulida. (A) *Suomina turgida*; (B) *Catenula lemnae*; (C) *Catenula leptocephala*; (D) and (E) *Rhynchoscolex simplex*. Length of species varies between 0.35 and 5.00 mm. b, Brain; f, furrow; i, intestine; m, mouth; n, protonephridial duct; o, ovary; ph, pharynx; st, statocyst.

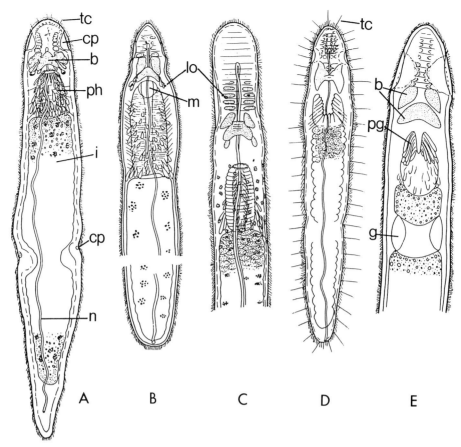

Figure 6.4 Some representatives of Catenulida. (A) *Stenostomum leucops*; (B) *S. beauchampi*; (C) *S. glandulosum*; (D) *S. brevipharyngium*; (E) *Myostenostomum tauricum*. b, Brain; cp, ciliated sensory pits; g, muscular ring of intestine or gizzard; i, intestine; lo, light refracting organs; m, mouth; n, protonephridial duct; pg, pharyngeal glands; ph, pharynx; tc, tactile cilia.

C. Taxonomic Key to Macrostomida

1a.	Ciliated pits present; intestine extends in front of mouth; numerous zooids separated by division planes (Fig. 6.5B, C) . . . . family Microstomidae	*Microstomum*
1b.	No ciliated pits; intestine entirely behind the pharynx; no signs of asexual division (Fig. 6.5A, D, E) . . . . family Macrostomidae	*Macrostomum*

D. Taxonomic Key to Lecithoepitheliata

1a.	Stylet of the copulatory organ straight; no eyes	*Prorhynchus*
1b.	Stylet of the copulatory organ curved; eyes present except in some cave forms	*Geocentrophora*

E. Taxonomic Key to Prolecithophora, Proseriata, and Tricladida

1a.	Mouth situated in front and directed forward	*Hydrolimax*
1b.	Mouth situated elsewhere and directed backward or ventrally	2
2a(1b).	Mouth situated near the middle and directed ventrally; statocyst present	*Otomesostoma*
2b.	Mouth directed backward; no statocyst	3

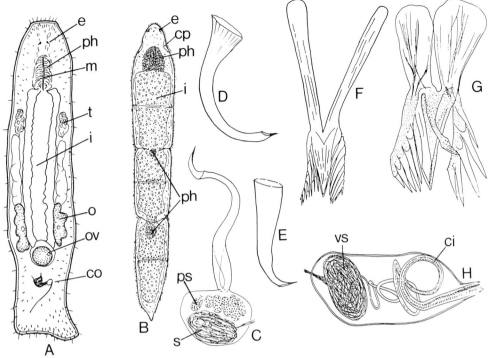

Figure 6.5 Representatives of Macrostomida and copulatory organs of various families. (A) *Macrostomum tuba*; (B) *Microstomum lineare*; (C) copulatory organ of *M. lineare*; (D) stylet of *Macrostomum gilberti*; (E) stylet of an unidentified *Macrostomum* from New Haven; (F) sclerotized copulatory apparatus of *Microdalyellia rossi* (Dalyelliidae); (G) sclerotized copulatory apparatus of *M. tennesseensis*; (H) copulatory organ of *Opistomum pallidum* containing a spiny eversible cirrus (Typhloplanidae). ci, Cirrus; co, copulatory organ; cp, ciliated pits; e, eye; i, intestine; m, mouth; o, ovary; ov, egg; ph, pharynx; ps, prostatic secretions; s, sperm; t, testes; vs, seminal vesicle.

3a(2b).   Intestine split into two branches behind the pharynx (Figs. 6.6,
            6.7) .... Tricladida[1]   4
3b.   Intestine not composed of three branches (Fig. 6.8A) .... *Bothrioplana*

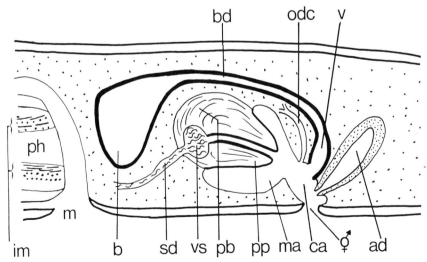

Figure 6.6 Schematic representation of the major anatomic features used in taxonomy of Tricladida. ad, Adenodactyl; b, bursa; bd, bursal duct; ca, common genital atrium; im, internal muscle layer of the pharynx (two possible types shown); m, mouth; ma, male atrium; odc, common oviduct; ph, pharynx; pb, penis bulb; pp, penis papilla; sd, sperm duct; v, vagina; vs, seminal vesicle.

[1] Substantial parts of the key to Tricladida follow Kenk (1972). Separation of *Hymanella* from *Phagocata* is based on Ball *et al.* (1981).

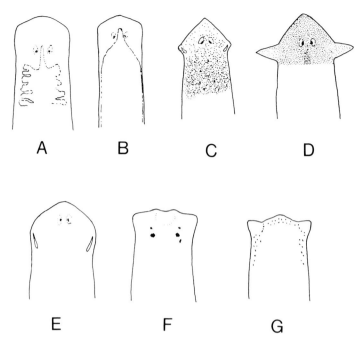

Figure 6.7 Some common representatives of Tricladida. (A) *Hymanella retenuova*; (B) *Phagocata velata*; (C) *Dugesia tigrina*; (D) *D. dorotocephala*; (E) *D. polychroa*; (F) *Procotyla fluviatilis*; (G) *Polycelis coronata* (A and B modified from Ball *et al.* 1981).

4a(3a).	Internal pharyngeal muscles in two distinct layers (Planariidae) (Fig. 6.6)	5
4b.	Internal pharyngeal muscles in one layer of mixed longitudinal and circular fibers (Dendrocoelidae) (Fig. 6.6)	12
5a(4a).	Oviducts, separate or united, open into end part of the bursa stalk	6
5b.	Oviducts united, the common oviduct opens in the genital atrium (Fig. 6.6)	7
6a(5a).	Testes extend to the level of pharynx	*Cura*
6b.	Testes extend to the posterior end of body	*Dugesia*

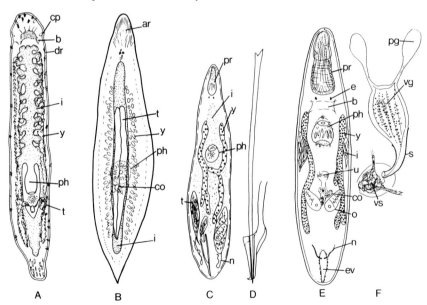

Figure 6.8 Representatives of (A) Proseriata (*Bothrioplana semperi*); (B) Typhloplanoida (*Mesostoma craci*) and Kalyptorhynchia; (C) *Gyratrix hermaphroditus* and (D) its stylet; (E) *Opisthocystis goettei* and (F) its copulatory organs. ar, Adrenal rhabdoids; b, brain; co, copulatory organ; cp, ciliated pits; dr, dermal rhabdoids; e, eye; ev, excretory vesicle; i, intestine; n, protonephridial duct; o, ovary; pg, pharyngeal glands; ph, pharynx; pr, proboscis; s, stylet; t, testes; vg, granular vesicle; vs, seminal vesicle; y, yolk glands.

7a(5b).	Numerous eyes arranged in a band around the anterior end of body (Fig. 6.7G)	*Polycelis*
7b.	Eyes absent or one pair only; if numerous, then not along the head margin	8
8a(7b).	Adenodactyl present	*Planaria*
8b.	Adenodactyl absent	9
9a(8b).	Anterior end with an adhesive organ	10
9b.	Anterior end without an adhesive organ	11
10a(9a).	Body elongated, flat, with a well-developed postpharyngeal section	*Sphalloplana*
10b.	Body turtle-shaped, with reduced postpharyngeal section	*Kenkia*
11a(9b).	Prepharyngeal, fused testes in male phase; large atrium lined by papillose epithelium in female phase; head round and broad; the anterior ramus of the intestine does not reach beyond eyes	*Hymanella*
11b.	Other set of characters	*Phagocata*
12a(4b).	Anterior end of the body with a deeply invaginated adhesive organ	*Macrocotyla*
12b.	Adhesive organ absent, or an adhesive disk present	13
13a(12b).	Penis papilla present	14
13b.	Penis papilla absent, bulb large	*Rectocephala*
14a(13a).	Penis bulb rounded, containing a seminal vesicle	*Dendrocoelopsis*
14b.	Penis bulb elongated, contains a prostatic vesicle	*Procotyla*

F. Taxonomic Key to Typhloplanoida

The basic plan of the freshwater typhloplanoid reproductive system is given in Fig. 6.9.

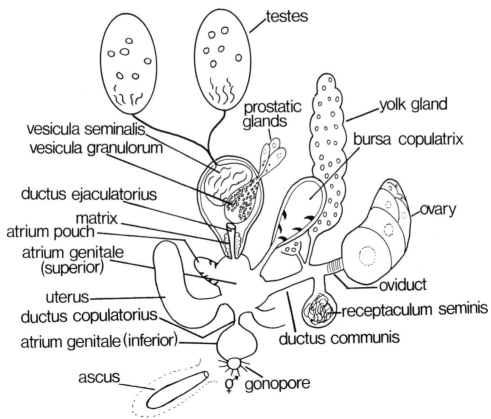

Figure 6.9 A schematic representation of the main reproductive organs and structures of freshwater Typhloplanoida. In some species, additional structures may be present or the relative arrangement among the organs may be slightly different.

1a.	Yolk glands paired, situated on both sides of the intestine	2
1b.	Yolk glands developed as a single strand of cells above the intestine (Fig. 6.10E, F) .... *Limnoruanis*	
2a(1a).	Pharynx developed as a typical, round or slightly elongated, pharynx rosulatus and oriented ventrally or somewhat forward	3
2b.	Pharynx elongated or cylindrical, oriented backward; copulatory organ with a spiny cirrus (Fig. 6.5H) .... *Opistomum*	
3a(2a).	Testes under yolk glands	4
3b.	Testes above yolk glands	11
4a(3a).	Protonephridial ducts open separately on the body surface (Protoplanellinae) (as in Fig. 6.10C)	5
4b.	Protonephridial ducts combined with mouth or gonopore (as in Fig. 6.10A)	7
5a(4a).	Front extremity with a central depression and set slightly apart by two lateral, ciliated pits; pharynx in the first third of body (Fig. 6.10C) .... *Prorhynchella*	
5b.	Front extremity not set apart, rounded; pharynx in the second to third portion of body	6
6a(5b).	Reproductive complex with a bursa copulatrix; ductus ejaculatorius usually surrounded by gelatinous matrix (Fig. 6.11E, F; cf., Fig. 6.9) .... *Krumbachia*	
6b.	Bursa copulatrix absent .... *Amphibolella*	
7a(4b).	Protonephridial ducts joined and combined with mouth; no retractable proboscis (Typhloplaninae)	8
7b.	Protonephridial ducts open into the gonopore; retractable front extremity (Fig. 6.10D) .... *Rhynchomesostoma*	

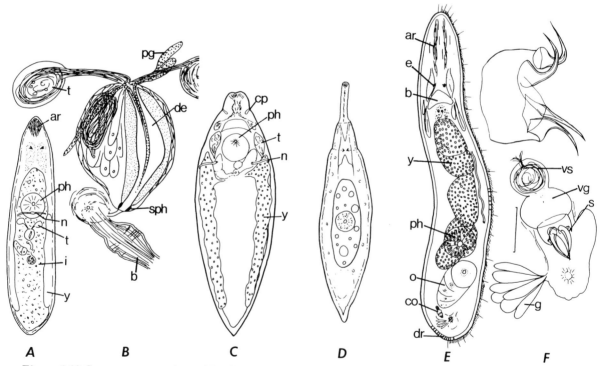

Figure 6.10 Some representatives of Typhloplanoida. (A) *Strongylostoma simplex*; (B) copulatory organ with testes of *S. simplex*; (C) *Prorhynchella minuta*; (D) a juvenile of *Rhynchomesostoma rostratum*; (E) *Limnoruanis romanae*; (F) sclerotized part of the copulatory organ, partly everted (upper picture), and the copulatory organ with the gonopore and glands (lower picture) in *L. romanae* (C modified from Ruebush 1939). ar, adenal rhabdoids; b, brain; co, copulatory organ; cp, ciliated sensory pits; de, ductus ejaculatorius; dr, dermal rhabdoids; e, eye; g, glands; i, intestine; n, protonephridial duct; o, ovary; pg, pharyngeal glands; ph, pharynx; s, stylet; sph, sphincter; t, testes; vg, granular vesicle; vs, seminal vesicle; y, yolk glands.

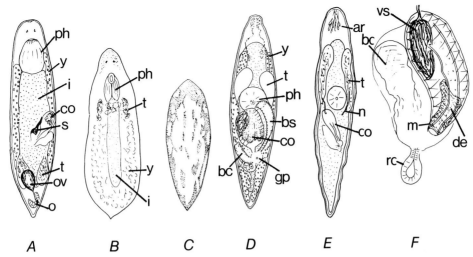

Figure 6.11 Representatives of Dalyellioida. (A) *Castrella pinguis* and Typhloplanida; (B) *Phaenocora* sp. (from Churchill, Manitoba); (C) *Mesostoma vernale* (the habitus of this species is also characteristic of various *Bothromesostoma*); (D) *Castrada virginiana*; (E) *Krumbachia hiemalis*; (F) copulatory complex of *Krumbachia hiemalis*. ar, adenal rhabdoids; bc, copulatory bursa; bs, blind sac of the atrium; co, copulatory organ; de, ductus ejaculatorius; gp, gonopore; i, intestine; m, matrix; n, protonephridial duct; o, ovary; ov, egg capsule; ph, pharynx; r, seminal receptacle; s, stylet; t, testes; vs, seminal vesicle; y, yolk glands.

8a(7a).	With a bursa copulatrix and an atrium copulatorium (Fig. 6.11D)	*Castrada*
8b.	One of the above organs missing	9
9a(8b).	With eyes (Figs. 6.10A, 6.12C)	*Strongylostoma*
9b.	No eyes	10
10a(9b).	Atrium copulatorium absent; bursa copulatrix present	*Typhloplanella*
10b.	Both bursa copulatrix and atrium copulatorium absent; always with green algae; pharynx in the first half of body; copulatory organ pear-shaped	*Typhloplana*
11a(3b).	Pharynx developed as a typical, round pharynx rosulatus	13
11b.	Pharynx barrel-shaped, directed clearly forward, and situated near the front extremity (Fig. 6.11B)	12
12a(11b).	Ductus copulatorius absent; inferior genital atrium inconspicuous (Fig. 6.11B)	*Phaenocora*
12b.	With ductus copulatorius; exceptionally large inferior atrium; occurs in sulfuric springs	*Pseudophaenocora*
13a(11a).	An elongated supplementary organ, the ascus, present in the proximity of the gonopore; or with a well-defined gland complex in its place (Ascophorinae) (cf., Fig. 6.9)	14
13b.	Ascus absent	16
14a(13a).	Pharynx in the middle of the first half of body	*Protoascus*
14b.	Pharynx approximately in the middle of body or slightly behind	15
15a(14b).	Gonopore near the middle of body	*Ascophora*
15b.	Gonopore in the third posterior part of body	*Dochmiotrema*
16a(13b).	Protonephridial ducts open separately on the body surface; eyes often present	*Olisthanella*
16b.	Protonephridial ducts open into mouth	17
17a(16b).	With a ventral pit and a canal connecting the bursa copulatrix to the ovovitellin duct; often flat on the ventral side and strongly convex on the dorsal	*Bothromesostoma*
17b.	Without the above characters (Figs. 6.11C, 6.8B, 6.12D)	*Mesostoma*

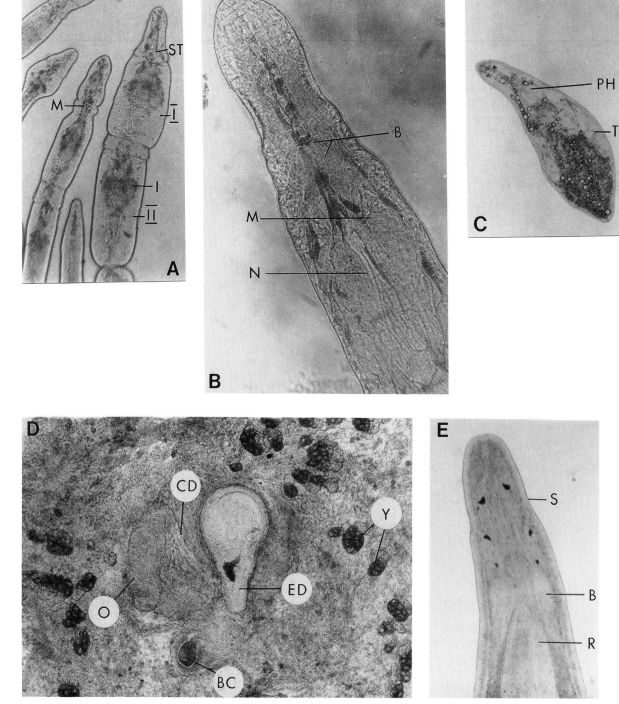

Figure 6.12 Microphotographs of live Turbellaria and Nemertea. (A) Several partly framed individuals of *Catenula sp.* showing chains of zooids; (B) frontal portion of body of *Rhynchoscolex sp.*; (C) *Strongylostoma sp.*, a representative of Typhloplanoida; (D) squash preparation of the portions of the reproductive system of *Mesostoma craci*, with the copulatory organ in the middle; (E) a frontal portion of body of a nemertean *Prostoma sp..* B, brain; BC, bursa copulatrix; CD, common duct; ED, ejaculatory duct, prostatic granulations are visible in its proximity; I, intestine; M, mouth; N, protonephridium; O, ovary; PH, pharynx; R, rhynchocoel; S, sensory pit; ST, statocyst; T, testicle; Y, yolk glands; I, first zooid; II, second zooid.

G. Taxonomic Key to Kalyptorhynchia and Similar Forms

1a. Excretory vesicle absent; protonephridial ducts open directly at the surface of body; single ovary (Fig. 6.8C) 2

1b. Protonephridial ducts open into an excretory vesicle in the rear end of body; paired ovary (Fig. 6.8E, F) .... *Opisthocystis*[2]

2a(1a). Copulatory organ with an almost straight stylet (Fig. 6.8C, D) .... *Gyratrix*

2b. Copulatory organ with a curved stylet .... *Microkalyptorhynchus*[3]

H. Taxonomic Key to Dalyellioida

1a. With a paired ovary; copulatory organ with a straight, funnel-like tube .... *Pilgramilla*

1b. With a single ovary; copulatory organ with a more or less complex sclerotized structure (e.g., Figs. 6.5F, G, 6.11A) 2

2a(1b). Sclerotized structure directly attached to the rest of the copulatory organ 3

2b. Sclerotized structure in a separate pocket adjacent to the bulb of the copulatory organ (Fig. 6.11A) .... *Castrella*

3a(2a). Sclerotized structure typically with two handles and two lateral branches carrying spines, auxiliary parts may be present (Fig. 6.5F, G) 4

3b. Sclerotized structure different; usually in the form of a spiny ring .... *Gieysztoria*

4a(3a). One to several eggs in the parenchyme; normally, older individuals green due to symbiotic zoochlorellae .... *Dalyellia*

4b. Never more than one egg; no zoochlorellae .... *Microdalyellia*

---

[2] An unidentified kalyptorhynchid occurring in a tributary of the Hudson River (Kolasa *et al.* 1987) also meets this criterion, although it may belong to a family other than that containing *Opisthocystis*. *Klattia* is a synonym of *Opisthocystis* (Karling T. G., 1956. *Ark. Zool.* 9:187–279).

[3] Taxonomic placement within Kalyptorhynchia is provisional. Name discussed by W. E. Hazen (1953. Morphology, taxonomy, and distribution of Michigan rhabdocoeles. Ph.D. Thesis, Univ. of Michigan, Ann Arbor. 122 pp.).

NEMERTEA

VII. GENERAL CHARACTERISTICS, EXTERNAL AND INTERNAL ANATOMICAL FEATURES

Nemerteans are coelomate (Turbeville and Ruppert 1985) and unsegmented worms possessing an eversible muscular proboscis resting in the rhynchocoel, a walled, longitudinal, dorsal cavity connected with the mouth (Figs. 6.13, 6.14, see also Fig. 6.12E in Section VI.F.[4] Unlike the Turbellaria, Nemertea have an anus as well as a closed blood circulatory system. The body is enclosed by a monolayered, ciliated epidermis and several layers of muscles. The excretory system is protonephridial. Sensory structures include eyes, statocysts (absent in freshwater nemerteans), and cerebral, frontal, and lateral neural organs. The reproductive system is much simpler than in Turbellaria. Freshwater nemerteans are hermaphroditic and often protandric. Spermatozoa and ova are produced in numerous gonads situated in diverticula of the intestine. As a rule, fertilization is external and self-fertilization can also occur. Development is direct.

It should be emphasized that most of the information provided in this section on nemerteans is based on an excellent review of freshwater and terrestrial nemerteans by Moore and Gibson (1985). That review will be an invaluable source of data and references for anyone working on the ecology and biology of freshwater nemerteans.

Twelve species of freshwater nemerteans are classified into two artificial families and six genera (Moore and Gibson 1985). In the family Heteronemertea, three monotypic genera are recognized: *Planolineus* Beauchamp, 1928, *Siolineus* du Bois-Reymond Marcus 1948, and *Apatronemertes* Wilfert and Gibson, 1974. In the family Hoplonemertea, the genus *Prostoma* Duges 1828 contains most of the species. Two other genera, *Campbellonemertes* Moore and Gibson 1972 and *Potamonemertes* Moore and Gibson 1973 are known from islands in the southern Pacific, Campbell Island and South Island, New Zealand, respectively.

VIII. ECOLOGY

A. Life History and General Ecology

Most nemerteans are marine organisms, although a small minority are terrestrial or freshwater. The

---

[4] Until 1985 Nemertea were considered to be acoelomate. With the new finding (Turbeville and Ruppert 1985), they can no longer be associated with other flatworms. Our presentation is more for convenience than systematic reasons.

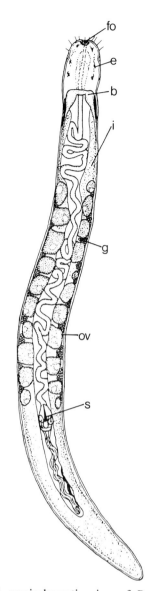

Figure 6.13 A semischematic view of *Prostoma sp.* b, Brain; e, eye; fo, frontal organ; g, gonad; i, intestine; ov, egg; s, stylet.

freshwater forms are benthic, predatory worms 10–40 mm in length and of pink coloration. *Prostoma* feed readily on oligochaete worms and occasionally on crustaceans, nematodes, turbellarians, midge larvae, and other small invertebrates. Feeding activity is most intense at night and prey are captured by the sticky proboscis. Little is known about habitat selection by freshwater nemerteans. North American *Prostoma* species have been found in both streams and lakes, including Lake Huron. In the case of lake sites, *Prostoma* were clearly associated with littoral habitats. Kolasa (1977b) found a high degree of association of *Prostoma* species with filamentous algae in lakes.

Reproduction is exclusively sexual even though some nemerteans have strong regenerative capabilities. Since only limited quantitative studies have

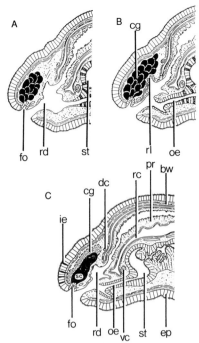

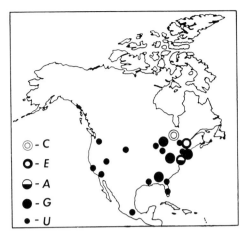

Figure 6.15 Known distribution of *Prostoma* species in North America (after Gibson 1985; modified and updated). C, *P. canadiensis*; E, *P. eilhardi*; A, *P. asensoriatum*; G, *P. graecense*; U, unidentified *Prostoma* .

Figure 6.14 Sections through the anterior extremity of three North American nemerteans. (A) *Prostoma eilhardi*; (B) *P. graecense*; (C) *P. canadiensis* (after Gibson and Moore 1976, 1978). bw, body wall musculature; cg, cephalic gland lobular region; dc, dorsal cerebral commisure; ep, epidermis; fo, frontal organ; id, improvised ducts from cephalic gland lobules; oe, oesophagus; pr, proboscis; rc, rhynchocoel; rd, rhynchodaeum; rl, rhynchodaeal longitudinal muscle fibers; st, stomach; vc, ventral cerebral commisure.

been conducted on the biology of freshwater nemerteans in the field, mechanisms of population regulation remain unknown. Clearly, *Prostoma* can reproduce rapidly with up to 210 eggs per reproductive episode throughout the year as long as the temperature is above 10°C (Young and Gibson 1975).

Distribution of nemerteans in North America is not well known. Strayer (1985) reported *Prostoma rubrum* from Mirror Lake, New Hampshire. However, *P. rubrum* is no longer a valid species (Gibson and Moore 1976) and this observation has to be treated as *Prostoma sp. P. canadiensis* is known from one site in Lake Huron (Gibson and Moore 1978), where it occurs to a depth of 20 m. *Prostoma* species are probably much more common in freshwaters than the infrequent records from eastern North America might indicate. So far, no *Prostoma* have been found in temporary water bodies.

Three established species of nemerteans are known from North America (Fig. 6.14). A fourth, *P. asensoriatum,* is a questionable species. Distribution of these species in North America is summarized in Fig. 6.15, but they are probably present in most standing and running waters. With the exception of the dubious *P. asensoriatum,* all three re-

maining species have been recorded in other parts of the world, including South America, Europe, Africa, and Australia. Zoologists do not know whether the worldwide pattern of distribution is natural or, instead, a result of recent human-mediated dispersal. Analysis of genetic diversity of various populations might shed some light on this question.

B. Physiological Adaptations

Osmoregulation is an ecologically important function in nemerteans as in all other freshwater invertebrates with permeable body walls. It is controlled by the cerebral organs and involves several organs and enzymatic systems associated with blood vessels (Moore and Gibson 1985). Well-developed nephridia may play a role in osmoregulation as well as in the removal of nitrogenous wastes (but the latter has not yet been demonstrated).

Gases, particularly oxygen are exchanged across the surface of the body. The role of blood vessels in oxygen transport is unclear.

Nemerteans, like turbellarians, release copious mucus whose various functions probably include defense, locomotion, physiological barrier, and encystment.

C. Behavioral Ecology

There are no studies devoted exclusively to the behavior of freshwater nemerteans. Nevertheless, some observations on feeding behavior, escape reactions, and locomotion are available. Adult nemerteans can only crawl, while small juveniles also swim. Forward locomotion of *Prostoma* may involve ciliary movements alone, ciliary movements in combination with muscular waves, sinusoidal curves when the animal pushes through the vegetation, or peristalsis when the animal moves backward. The fastest movement forward is provided by the proboscis. This long organ is rapidly everted and attached to substrate, and the body is then pulled toward the point of attachment.

Typically, prey are captured by a sticky proboscis and then swallowed whole by the widely distended mouth (McDermott and Roe 1985). The mechanisms of prey detection are poorly understood. In terrestrial nemerteans, these mechanisms are unusual in that they involve ambushing strategy and reliance on mechanical stimuli.

D. *Functional Role in the Ecosystem*

Necessarily, the functional importance of nemerteans in the ecosystem can only be indirectly inferred from their trophic position and relative densities. I have found densities of *Prostoma* species in Wappinger Creek, a tributary of the Hudson River, to vary between 50 and 590 individuals/m^2, which constitutes a small fraction of all predatory invertebrates identified at the study site. Similarly, low densities of *P. canadiensis* (up to 140/m^2) were observed on gravelly and muddy substrates in Lake Huron.

IX. CURRENT AND FUTURE RESEARCH PROBLEMS

In view of the limited knowledge of nemertean ecology, almost any area of research will provide valuable information. New information on phenology, population dynamics, and habitat selectivity would permit a better evaluation of the role of nemerteans in freshwater communities. Specific studies in the biology of nemerteans appear particularly promising in the areas of reproductive biology and evolutionary ecology. The ecological role of femaleness later

in the life cycle, the production of resting eggs, and the environmental versus genetic cues of life stages all may offer exciting and general models for evolutionary ecology.

X. COLLECTION, CULTURING, AND PRESERVATION

Collection of nemerteans is similar to collection of any soft-bodied, benthic invertebrates such as turbellarians or oligochaetes living in the substrate or on aquatic plants. Sieving and vigorous washing may damage specimens. Placing samples in beakers or jars and topping with water allows nemerteans to crawl toward the edges of water where they can be collected with a pipette.

Freshwater nemerteans are relatively easy to culture. They may be maintained and reproduced in aquaria or even in petri dishes in clean, oxygenated water and on a diet of live aquatic invertebrates. Chopped tubificid oligochaetes provide a good diet.

Worms can be preserved in 80% ethanol. Before preservation, however, worms should be narcotized using 7% ethanol or chloretone until their movement ceases. According to Moore and Gibson (1985), gin can be used in the field if pure ethanol is unavailable, but only as a last resort!

XI. IDENTIFICATION

It should be noted that Gibson and Moore (1976, 1978) provided keys to freshwater nemerteans of the world; those keys are more suitable if a species yet unrecorded in North America is encountered. The present key is an adaptation of the keys by Gibson and Moore (1976, 1978) pertinent to established North American *Prostoma*.

A. *Taxonomic Key to Species of Freshwater Nemertea*

1a.	Cephalic glands open via frontal organs and by improvised ducts (Fig. 6.14C); esophagus distinct but unciliated; rhynchodaeum with isolated longitudinal muscle strands; proboscis with twelve nerves .... *P. canadiensis* Gibson and Moore, 1978	
1b.	Cephalic glands open only via frontal organ, no improvised ducts 2	
2a(1b).	With a distinctive ciliated esophagus; cephalic glands reach back to the brain (Fig. 6.14B); rhynchodaeum with a well-developed layer of longitudinal muscle .... *P. graecense* (Bohmig 1892)	
2b.	Different combination of characters; indistinct and unciliated esophagus (Fig. 6.14A); rhynchodaeum without specifically associated layer of longitudinal muscle fibers; proboscis with 9–10 nerves .... *P. eilhardi* (Montgomery 1894)	

ACKNOWLEDGMENTS

I am very grateful to L. R. G. Cannon, J. Moore, R. Gibson, A. P. Mead, R. Kenk, S. Schwartz, D. Strayer, and S. Tyler for helpful comments on the manuscript and/or permission to reproduce figures, glossary entries, and fragments of keys.

LITERATURE CITED

Ball, I. R. 1974. A contribution to the phylogeny and biogeography of the freshwater triclads (Platyhelminthes: Turbellaria). Pages 339–401 *in:* N. W. Risser, and M. P. Morse, editors. Biology of the Turbellaria. McGraw-Hill, New York.

Ball, I. R., N. Gourbault, and R. Kenk. 1981. The planarians (Turbellaria) of temporary waters in Eastern North America. Life Sciences Contribution, No. 127. Royal Ontario, Mus., Toronto.

Bauchhenss, J. 1971. Die Kleinturbellarien Frankens. Ein Beitrag zur Systematik und Ökologie der Turbellaria excl. Tricladida in Süddeutschland. Internationale Revue des gesamten Hydrobiologie 56:609–666.

Bedini, C., E. Ferrero, and A. Lanfranchi. 1973. Fine structure of the eyes in two species of Dalyelliidae (Turbellaria Rhabdocoela). Monitore Zoologico Italiano 7:51–70.

Benazzi, M., and E. Giannini. 1971. *Cura azteca*, nuova specie di planaria del Messico. Atti della Accademia Nazionale dei Lincei 50:477–481.

Boddington, M. J., and D. F. Mettrick. 1977. A laboratory study of the population dynamics and productivity of *Dugesia polychroa* (Turbellaria: Tricladida). Ecology 58:109–118.

Calow, P., and D. A. Read. 1986. Ontogenetic patterns and phylogenetic trends in freshwater flatworms (Tricladida); constraint or selection? Hydrobiologia 132:263–272.

Cannon, L. R. G. 1986. Turbellaria of the world. A guide to families and genera. Queensland Museum, Brisbane.

Case, T. J., and R. K. Washino. 1979. Flatworm control of mosquito larvae in rice fields. Science 206:1412–1414.

Chandler, C. M., and J. T. Darlington. 1986. Further field studies on freshwater planarians of Tennessee (Turbellaria: Tricladida): III. Western Tennessee. Journal of Freshwater Ecology 3:493–501.

Chodorowski, A. 1959. Ecological differentiation of turbellarians in Harsz Lake. Polskie Archiwum Hydrobiologii 6:33–73.

Chodorowski, A. 1960. Vertical stratification of Turbellaria species in some littoral habitats of Harsz Lake. Polskie Archiwum Hydrobiologii 8:153–163.

Claussen, D. L., and L. M. Walters. 1982. Thermal acclimation in the fresh water planarians, *Dugesia tigrina* and *Dugesia dorotocephala*. Hydrobiologia 94:231–236.

Collins, F. H., and R. K. Washino. 1979. Factors affecting the density of *Culex tarsalis* and *Anopheles freeborni* in northern California rice fields. Proceedings of California Mosquito Control Association 46:97–98.

Davies, R. W., and T. B. Reynoldson. 1971. The incidence and intensity of predation on lake-dwelling triclads in the field. Journal of Animal Ecology 40:191–214.

Dumont, H. J., and I. Carels. 1987. Flatworm predator (*Mesostoma* cf. *lingua*) releases a toxin to catch planktonic prey (*Daphnia magna*). Limnology and Oceanography 32:699–702.

Ehlers, U. 1986. Comments on a phylogenetic system of the Platyhelminthes. Hydrobiologia 132:1–12.

Ferguson, F. F. 1939. A monograph of the genus *Macrostomum* O. Schmidt 1848. Part II. Zoologischer Anzeiger 126:131–144.

Fiore, L., and P. Ioalé. 1973. Regulation of the production of subitaneous and dormant eggs in the turbellarian *Mesostoma ehrenbergii* (Focke). Monitore Zoologico Italiano 7:203–224.

Gibson, R., and J. Moore. 1976. Freshwater nemerteans. Zoological Journal of the Linnean Society 58:177–218.

Gibson, R., and J. Moore. 1978. Freshwater nemerteans: new records of *Prostoma* and a description of *Prostoma canadiensis* sp. nov. Zoologischer Anzeiger. 201:77–85.

Gilbert, C. M. 1938. Two new North American rhabdocoeles—*Phaenocora falciodenticulata* nov. spec. and *Phaenocora kepneri adenticulata* nov. subspec. Zoologischer Anzeiger 122:208–223.

Göltenboth, F., and U. Heitkamp. 1977. *Mesostoma ehrenbergi* (Focke 1836), Platwürmer (Strudelwürmer). Biologie, mikroskopische Anatomie und Cytogenetik. Fischer, Stuttgart.

Gourbault, N. 1972. Recherches sur les triclades paludicoles hypogés. Memoires du Muséum National d'Histoire Naturelle 73:1–249.

Hampton, A.M. 1988. Altitudinal range and habitat of triclads in streams of the Lake Tahoe basin. The American Midland Naturalist 120:302–312.

Hebert, P. D. N., and W. J. Payne. 1985. Genetic variation in populations of the hermaphroditic flatworm *Mesostoma lingua* (Turbellaria, Rhabdocoela). Biological Bulletin (Woods Hole, Mass.) 169:143–151.

Heitkamp, U. 1979a. Der Eifluss endosymbiontischer Zoochlorellen auf die Respiration von *Dalyellia viridis* (G.Shaw, 1791) (Turbellaria Neorhabdocoela). Archiv für Hydrobiologie 86:499–514.

Heitkamp, U. 1979b. Die Respirationsrate neorhabdocoeler Turbellarien mit unterschiedlicher Temperaturvalenz. Archiv für Hydrobiologie 87:95–111.

Heitkamp, U. 1982. Untersuchungen zur Biologie, Ökologie und Systematik limnischer Turbellarien periodischer und perennierender Kleingewässer Südniedersachsens. Archiv für Hydrobiologie, Supplement 64:65–188.

Holmquist, C. 1967. Turbellaria of northern Alaska and northwestern Canada. Internationale Revue des gesamten Hydrobiologie 52:123–139.

Holopainen, I. J., and L. Paasivirta. 1977. Abundance and biomass of the meiozoobenthos in the oligotrophic and mesohumic lake Paajarvi, southern Finland. Annales Zoologici Fennici 14:124–134.

Hyman, L. H. 1951. The invertebrates: Platyhelminthes and Rhynchocoela, the Acoelomate Bilateria. McGraw-Hill, New York.

Hyman, L. H. 1955. Descriptions and records of freshwater Turbellaria from the United States. American Museum Novitates 1714:1–36.

Jennings, J.B. 1977. Patterns of nutrition in free-living and symbiotic Turbellaria and their implications for the evolution of parasitism in the phylum Platyhelminthes. Acta Zoologica Fennica 154:63–79.

Jennings, J. B. 1985. Feeding and digestion in the aberrant planarian *Bdellasimilis barwicki* (Turbellaria: Tricladida: Procerodidae): an ectosymbiote of freshwater turtles in Queensland and New South Wales. Australian Journal of Zoology 33:317–327.

Kawakatsu, M., and R. W. Mitchell. 1984a. A list of retrobursal triclads inhabiting freshwaters and aquatic probursal triclads regarded as marine relicts, with corrective remarks on our 1984 paper published in Zoological Science, Tokyo. Occasional Publications, Biological Laboratory of Fuji Women's College, Sapporo 11:1–8.

Kawakatsu, M., and R. W. Mitchell. 1984b. Redescription of *Dugesia azteca* (Benazzi et Giannini, 1971) based upon the material collected from the type locality in Mexico, with corrective remarks. Bulletin of the National Science Museum, Series A 10:37–50.

Kenk, R. 1972. Freshwater planarians (Turbellaria) of North America. Environmental Protection Agency, Washington, D.C.

Kenk, R. 1987. Freshwater triclads (Turbellaria) of North America. XVI. More on subterranean species of *Phagocata* of the eastern United States. Proceedings of the Biological Society in Washington 100:664–673.

Kepner, W. A., and J. S. Carter. 1931. Ten well-defined new species of *Stenostomum*. Zoologischer Anzeiger 93:108–123.

Kolasa, J. 1977a. Remarks on the marine-originated Turbellaria in the freshwater fauna. Acta Zoologica Fennica 1954:81–87.

Kolasa, J. 1977b. Bottom fauna of the heated Konin Lakes. Turbellaria and Nemertini. Monografie Fauny Polski 7:29–48.

Kolasa, J. 1979. Ecological and faunistical characteristics of Turbellaria in the eutrophic Lake Zbechy. Acta Hydrobiologica 21:435–459.

Kolasa, J. 1983. Formation of the turbellarian fauna in a submontane stream. Acta Zoologica Cracoviensia 26:57–107.

Kolasa, J. 1987. Population growth in some *Mesostoma* species (Turbellaria) predatory on mosquitoes. Freshwater Biology 18:205–212.

Kolasa, J., D. Strayer, and E. Bannon-O'Donnell. 1987. Microturbellarians—from interstitial waters, streams, and springs in southeastern New York. Journal of the North American Benthological Society 6:125–132.

Lanfranchi, A., and F. Papi. 1978. Turbellaria (excl. Tricladida). Pages 5–15 *in:* J. Illies, editor. Limnofauna Europaea. Fisher, Stuttgart.

Luther, A. 1955. Die Dalyelliiden. Acta Zoologica Fennica 87:1–337.

Luther, A. 1960. Die Turbellarien Ostfennoskandiens. I. Acoela, Catenulida, Macrostomida, Lecithoepitheliata, Prolecithophora, und Proseriata. Fauna Fennica 7:1–155.

Luther, A. 1963. Die Turbellarien Ostfennoskandiens. IV. Neorhabdocoela 2. Fauna Fennica 16:1–163.

Maly, E. J., S. Schoenholtz, and M. T. Arts. 1980. The influence of flatworm predation on zooplankton inhabiting small ponds. Hydrobiologia 76:233–240.

Marcus, E. 1945. Sobre microturbellarios do Brasil. Comunicaciones Zoologicas del Museo de Historia Natural de Montevideo 25:1–97.

Marcus, E. 1951. Contributions to the natural history of Brazilian Turbellaria. Comunicaciones Zoologicas del Museo de Historia Natural de Montevideo 63:1–25.

Martens, P. M. 1984. Comparison of three differnt extraction methods for Turbellaria. Marine Ecology Progress Series 14:229–234.

McConnell, J. V. 1967. A manual of psychological experimentation on planarians. A special publication of the Worm Runner's Digest. J.V. McConnell, Ann Arbor, Michigan.

McDermott, J. J., and P. Roe. 1985. Food, feeding behavior and feeding ecology of nemerteans. American Zoologist 25:113–125.

Mead, A. P., and J. Kolasa. 1984. New records of freshwater microturbellaria from Nigeria, West Africa. Zoologischer Anzeiger 212:257–271.

Moore, J., and R. Gibson. 1985. The evolution and comparative physiology of terrestrial and freshwater nemerteans. Biological Reviews. *Cambridge Philos. Soc.* 60:257–312.

Nalepa, T. F., and M. A. Quigley. 1983. Abundance and biomass of meiobenthos in nearshore Lake Michigan with comparisons to macrozoobenthos. Journal of the Great Lakes Research 9:523–529.

Noldt, U., and C. Wehrenberg. 1984. Quantitative extraction of living Plathelminthes from marine sands. Marine Ecology Progress Series 20:193–201.

Nuttycombe, J. W. 1956. The *Catenula* of the eastern United States. The American Midland Naturalist 55:419-433.

Nuttycombe, J. W. and A. J. Waters, 1938. The American species of the genus *Stenostomum*. Proceedings of the American Philosophical Society 79:213–301.

Pattee, E. 1980. Coefficients thermiques et ecologie de quelques planaires d'eau douce VII: leur zonation naturelle. Annales de Limnologie 16:21–41.

Reynoldson, T. B. 1983. The population biology of Turbellaria with special reference to the freshwater triclads of the British Isles. Advances in Ecological Research 13:235–326.

Rieger, R. M. 1986. Asexual reproduction and the turbellarian archetype. Hydrobiologia 132:35–45.

Rixen, J. U. 1968. Beitrag zur Kenntnis der Turbellarienfauna des Bodensees. Archiv für Hydrobiologie 64:335–365.

Ruebush, T. K. 1939. A new North American Rhabdocoel Turbellarian, *Prorhynchella minuta* n. gen., n. sp. Zoologischer Anzeiger 127:204–209.

Sayre, R. M., and E. M. Powers. 1966. A predacious soil turbellarian that feeds on free-living and plant-parasitic nematodes. Nematologica 12:619–629.

Schwank, P. 1976. Quantitative Untersuchungen an litoralen Turbellarien des Bodensees. Jahreshefte der Gesellschaft für Naturkunde in Württemberg 131:163–181.

Schwank, P. 1981. Turbellarien, Oligochaeten und Archianneliden des Breitenbachs und anderer oberhessischer Mittelgebirgsbäche. II. Die Systematik und Autökologie der einzelnen Arten. Archiv für Hydrobiologie, Supplement 62:86–147.

Schwartz, S. S., and P. D. N. Hebert. 1982. A laboratory study of the feeding behavior of the rhabdocoel *Mesostoma ehrenbergii* on pond Cladocera. Canadian Journal of Zoology 60:1305–1307.

Schwartz, S. S., and P. D. N. Hebert. 1986. Prey preference and utilization by *Mesostoma lingua* (Turbellaria, Rhabdocoela) at a low arctic site. Hydrobiologia 135:251–257.

Strayer, D. 1985. The benthic micrometazoans of Mirror Lake, New Hampshire. Archiv für Hydrobiologie, Supplement 72:287–426.

Strayer, D., and G. E. Likens. 1986. An energy budget for the zoobenthos of Mirror Lake, New Hampshire. Ecology 67:303–313.

Turbeville, J. M., and E. E. Ruppert 1985. Comparative ultrastructure and evolution of Nemertines. American Zoologist 25:53–71.

Tyler, S., and M. D. B. Burt. 1988. Lensing by a mitochondrial derivative in the eye of *Urastoma cyprinae* (Turbellaria, Prolecithophora). Fortschritte der Zoologie 36:229–234.

Vannote, R. L., G. W. Minshall, K. W. Cummins, J. R. Sedell, and C. E. Cushing. 1980. The river continuum concept. Canadian Journal of Fisheries and Aquatic Sciences 37:130–137.

Wrona, F. 1986. Distribution, abundance, and size of rhabdoids in *Dugesia polychroa* (Turbellaria: Tricladida). Hydrobiologia 132:287–293.

Young, J. O. 1970. British and Irish freshwater Microturbellaria: historical records, new records and a key for their identification. Archiv für Hydrobiologie 67:210–241.

Young, J. O. 1973a. The prey and predators of *Phaenocora typhlops* (Vejdovsky) (Turbellaria: Neorhabdocoela) living in a small pond. Journal of Animal Ecology 42:637–643.

Young, J. O. 1973b. The occurrence of microturbellaria in some British lakes of diverse chemical content. Archiv für Hydrobiologie 72:202–224.

Young, J. O. 1975. The population dynamics of *Phaenocora typhlops* (Vejdovsky) (Turbellaria: Neorhabdocoela) living in a pond. Journal of Animal Ecology 44:251–262.

Young, J. O., and J. W. Eaton. 1975. Studies on the symbiosis of *Phaenocora typhlops* (Vejdovsky) (Turbellaria; Neorhabdocoela) and *Chlorella vulgaris var. vulgaris*, Fott and Novakova (Chlorococcales). II. An experimental investigation into the survival value of the relationship to host and symbiont. Archiv für Hydrobiologie 75:225–239.

Young, J. O., and R. Gibson. 1975. Some ecological studies on two populations of the freshwater hoplonemertean *Prostoma eilhardi* (Montgomery 1894) from Kenya. Verhandlungen der internationale Vereinigung für Limnologie 19:2803–2810.

Yu, H.-S., and E. F. Legner. 1976. Regulation of aquatic diptera by planaria. Entomophaga 21:3–12.

APPENDIX 6.1. LIST OF NORTH AMERICAN SPECIES OF MICROTURBELLARIA

The arrangements of orders and families follows with some modifications from that of Lanfranchi and Papi (1978).

CATENULIDA
Family Catenulidae
  Catenula confusa
  Catenula lemnae
  Catenula leptocephala
  Catenula sekerai
  Catenula virginia
  Suomina turgida
Family Chordariidae
  Chordarium europaeum
Family Stenostomidae
  Myostenostomum tauricum
  Rhynchoscolex platypus
  Rhynchoscolex simplex
  Rhynchoscolex sp. 1
  Rhychoscolex sp. 2
  Stenostomum anatirostrum
  Stenostomum anops
  Stenostomum arevaloi
  Stenostomum beauchampi
  Stenostomum brevipharyngium
  Stenostomum ciliatum
  Stenostomum cryptops
  Stenostomum glandulosum
  Stenostomum grande
  Stenostomum kepneri
  Stenostomum leucops
  Stenostomum mandibulatum
  Stenostomum membranosum
  Stenostomum occultum
  Stenostomum pegephilum
  Stenostomum predatorium
  Stenostomum pseudoacetabulum
  Stenostomum simplex
  Stenostomum temporaneum
  Stenostomum tuberculosum
  Stenostomum unicolor
  Stenostomum uronephrium
  Stenostomum ventronephrium
  Stenostomum virginianum
    (probably synonymous with S. unicolor)
MACROSTOMIDA
Family Macrostomidae
  Macrostomum collistylum
  Macrostomum curvistylum
  Macrostomum gilberti
  Macrostomum glochostylum
  Macrostomum lewisi
  Macrostomum norfolkense
  Macrostomum ontarioense
  Macrostomum orthostylum
  Macrostomum phillipsi

  Macrostomum reynoldsi
  Macrostomum riedeli
  Macrostomum ruebushi
  Macrostomum sensitivum
  Macrostomum sillimani
  Macrostomum sp. 2
  Macrostomum sp. 3
  Macrostomum stirewalti
  Macrostomum tennesseense
  Macrostomum tuba
  Macrostomum virginianum
  Microstomum lineare
LECITHOEPITHELIATA
Family Prorhynchidae
  Geocentrophora cavernicola
  Geocentrophora sphyrocephala
  Prorhynchus stagnalis
PROSERIATA
Family Plagiostomidae
  Hydrolimax grisea
PROLECITHOPHORA
Family Bothrioplanidae
  Bothrioplana semperi
Family Otomesostomidae
  Otomesostoma auditivum
DALYELLIOIDA
Family Provorticidae
  Genus indet. species indet.
  Pilgramilla virginiensis
Family Dalyelliidae
  Castrella pinguis
  Castrella graffi
  Dalyellia viridis
  Microdalyella abursalis
  Microdalyellia circobursalis
  Microdalyellia fairchildi
  Microdalyellia gilesi sp. 2
  Microdalyellia rheesi
  Microdalyellia rochesteriana
  Microdalyellia rossi
  Microdalyellia ruebushi
  Microdalyellia sillimani
  Microdalyellia tennesseensis
  Microdalyellia virginiana
TYPHLOPLANIDA
Family Typhloplanidae
  Amphibolella spinulosa
  Ascophora elegantissima
  Bothromesostoma personatum
  Castrada hofmanni
  Castrada lutheri
  Castrada virginiana

TYPHLOPLANIDA (*Continued*)
    Castrada sp. indet.
    Krumbachia cf. hiemalis
    Krumbachia minuta
    Krumbachia virginiana
    Limnoruanis romanae
    Mesostoma arctica
    Mesostoma californicum
    Mesostoma columbianum
    Mesostoma craci
    Mesostoma curvipenis
    Mesostoma ehrenbergii
    Mesostoma macroprostatum
    Mesostoma vernale
    Mesostoma virginianum
    Mesostoma platygastricum
    Olisthanella truncula
    Opistomum pallidum
    Phaenocora agassizi

Phaenocora falciodenticulata
Phaenocora highlandense
Phaenocora kepneri
Phaenocora lutheri
Phaenocora virginiana
Prorhynchella minuta
Protoascus wisconsinensis
Pseudophaenocora sulfophila
Rhynchomesostoma rostratum
Strongylostoma cf., elongatum
Strongylostoma radiatum
Typhloplanella halleziana
KALYPTORHYNCHIA
Family Polycystidae
    Gyratrix hermaphroditus
    Opistocystis goettei
Family undetermined
    Microrhynchus virginianus

Gastrotricha

7

David L. Strayer
Institute of Ecosystem Studies
The New York Botanical Garden
Millbrook, N.Y. 12545

William D. Hummon
Department of Zoological and Biomedical Sciences
Ohio University
Athens, Ohio 45701

Chapter Outline

I. INTRODUCTION
II. ANATOMY AND PHYSIOLOGY
  A. External Morphology
  B. Organ System Function
III. ECOLOGY AND EVOLUTION
  A. Diversity and Distribution
  B. Reproduction and Life History
  C. Ecological Interactions
  D. Evolutionary Relationships
IV. COLLECTING, REARING, AND PREPARATION FOR IDENTIFICATION
V. IDENTIFICATION OF THE GASTROTRICHS OF NORTH AMERICA
  A. Taxonomic Key to Genera of Freshwater Gastrotricha
  Literature Cited

I. INTRODUCTION

The gastrotrichs are among the most abundant and poorly known of the freshwater invertebrates. Gastrotrichs are nearly ubiquitous in the benthos and periphyton of freshwater habitats, with densities typically in the range of 10,000–100,000 per m^2. Nonetheless, the remarkable life cycle of freshwater gastrotrichs is only now being worked out, and we know almost nothing about how the distribution and abundance of these animals is controlled in nature. The impact of freshwater gastrotrichs on their food resources and on freshwater ecosystems has not yet been investigated. Ten genera and fewer than 100 species of freshwater gastrotrichs are now known from North America. Because the North American gastrotrich fauna has received so little study, these numbers understate the real diversity of the fauna.

The gastrotrichs are aschelminths most closely allied to the nematodes and are usually considered to constitute a phylum of their own or to be a class of the phylum Aschelminthes or Nemathelminthes. Some important general references on gastrotrichs include Remane (1935–1936), Hyman (1951), Voigt (1958), d'Hondt (1971a), Hummon (1982), and Ruppert (1988). There are two orders of gastrotrichs: the Macrodasyida, which consists almost entirely of marine species, and the Chaetonotida, which contains marine, freshwater, and semiterrestrial species.

Macrodasyids usually are distinguished from chaetonotids by the presence of pharyngeal pores and more numerous adhesive tubules (Fig. 7.1). Macrodasyids are common in marine and estuarine sands, but are barely represented in inland freshwaters. Two species of freshwater gastrotrichs have been placed in the Macrodasyida.

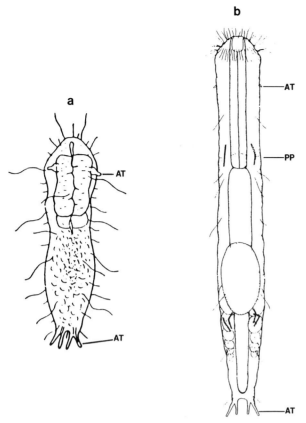

Figure 7.1 Freshwater macrodasyid gastrotrichs: (a) *Marinellina flagellata*, (b) *Redudasys fornerise*. AT, adhesive tube; PP, pharyngeal pore. [From Ruttner-Kolisko (1955) and Kisielewski (1987).]

Ruttner-Kolisko (1955) described an aberrant gastrotrich, *Marinellina flagellata* (Fig. 7.1a), from the hyporheic zone of an Austrian river. Unfortunately, she was able to find only two specimens, both of them apparently immature. Because *Marinellina* has a pair of anterior lateral structures that Ruttner-Kolisko interpreted as adhesive tubules, she placed this species in the Macrodasyida. Remane (1961) rejected the assignment of *Marinellina* to the macrodasyids, and placed it instead in the chaetonotid family Dichaeturidae. Kisielewski (1987) reaffirmed Ruttner-Kolisko's original placement of the species. *Marinellina* has not been found since its original collection. Until *Marinellina* is rediscovered and studied critically, its systematic placement will remain unclear.

Kisielewski (1987) discovered an undoubted macrodasyid, *Redudasys fornerise* (Fig. 7.1b), from the psammon of a Brazilian reservoir. It seems very likely that additional freshwater macrodasyids will be found when appropriate habitats (psammon, hyporheic zone) are explored. The distribution, biology, and evolutionary relationships of any such species will be of great interest.

Unless noted otherwise, the information included in this chapter refers to freshwater members of the Chaetonotida.

II. ANATOMY AND PHYSIOLOGY

The following account of gastrotrich anatomy is brief, and is summarized chiefly from the reviews of Remane (1935–1936), Hyman (1951), Hummon (1982), and Boaden (1985), which should be consulted for greater detail.

A. External Morphology

Gastrotrichs are colorless animals, spindle- or tenpin-shaped, and ventrally flattened (Fig. 7.2). Freshwater gastrotrichs are 50–800 μm long. Conspicuous external features include a more or less distinct head, which bears sensory cilia, and a cuticle, which in most species is ornamented with spines or scales of various shapes. In the most common freshwater family, the Chaetonotidae (as well as in the rare Dichaeturidae and Proichthydidae), the posterior end of the body is formed into a furca, which contains distal adhesive tubes that allow the animal to tenaciously attach itself to surfaces. In other families, these structures are absent but the posterior end of the body bears long spines or sensory bristles. The ventral side of the animal bears longitudinal rows or patches of cilia that provide the forward-gliding locomotion of the animal.

B. Organ System Function

The digestive system begins with a subterminal mouth, which may be surrounded by a ring of short bristles. Between the mouth and the pharynx lies a cuticular buccal capsule, which is often somewhat protrusible. The muscular pharynx is similar to the nematode pharynx, with the triradiate, Y-shaped lumen. Often, there are anterior and posterior swellings, but these lack the valves characteristic of the pharyngeal bulbs of nematodes. Posterior to the pharynx is an undifferentiated gut, which empties through a typically dorsal anus.

The paired reproductive organs lie lateral to the gut in the posterior half of the body. In young animals, large, developing, parthenogenetic eggs are present. As there are no oviducts, the egg is released through a rupture in the ventral body wall. Older animals become hermaphrodites (discussed below) and bear both sperm sacs and developing sexual (i.e., meiotic) eggs lateral to the gut. Also, there

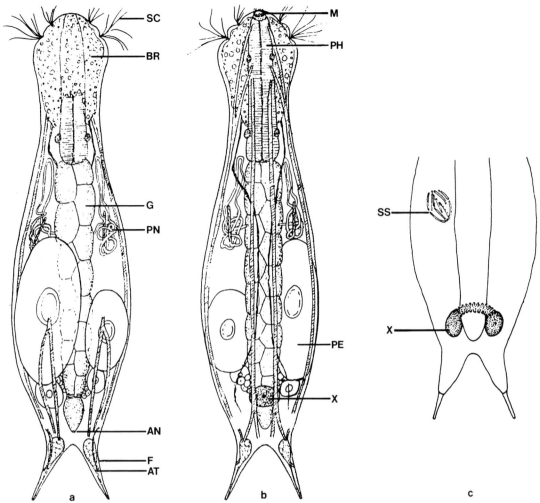

Figure 7.2 Schematic illustration of a typical chaetonotid gastrotrich showing (a) dorsal view, (b) ventral view, and (c) posterior end of a hermaphrodite, showing sexual organs, AT, adhesive tubes; AN, anus; BR, brain; F, furca; G, gut, M, mouth; PH, pharynx; PE, parthenogenetic egg; PH, protonephridium; SC, sensory cilia; SS, sperm sac; X, X-organ. [Modified from Voigt (1958) and Kisielewska (1981).]

is an unpaired organ of unknown function, the X-organ, that lies posterior to the gonads near the anus.

The brain is bilobed, straddling the pharynx. Sensory organs include long cilia and bristles, which presumably are tactile, and a pair of ciliated pits on the head lobes. Balsamo (1980) demonstrated that *Lepidodermella squamata* is photosensitive, although it does not appear to have distinct photoreceptors. Gray and Johnson (1970) found evidence of a tactile chemical sense in a marine macrodasyid; it seems probable that freshwater chaetonotids possess a similar ability.

The excretory system consists of a pair of cystocytic protonephridia (Brandenburg 1962) in the anterior midbody, which empty through pores on the ventral body surface. There is no circulatory or respiratory system per se.

III. ECOLOGY AND EVOLUTION

A. Diversity and Distribution

Gastrotrichs are widely distributed in freshwaters in surface sediments and among vegetation. Some species of the Neogosseidae and Dasydytidae are good swimmers and are occasionally reported from the plankton of shallow, weedy lakes (e.g., Hutchinson 1967, Green 1986). However, none of the gastrotrichs has become as truly planktonic as the daphnid cladocerans or ploimate rotifers.

Although gastrotrichs have been found in all kinds of freshwater habitats, they are apparently scarce in groundwaters other than the hyporheic zone (Renaud-Mornant 1986). Their rarity in underground waters is surprising, because many marine gastrotrichs are interstitial in habit (d'Hondt 1971a,

Renaud-Mornant 1986) and because the small size and bacterial diet of gastrotrichs would seem to preadapt them to the groundwater habitat. Gastrotrichs are common in lakes and ponds. Kisielewski (1981, 1986a) showed that gastrotrich density and species richness are positively correlated with the productivity of the habitat, and several workers have found that gastrotrich density is highest in highly organic sediments (e.g., Strayer 1985). Gastrotrichs also are abundant in unpolluted streams, where they inhabit sand bars, sometimes in great numbers (Hummon *et al.* 1978, Hummon 1987). Apparently, most gastrotrichs live very near the sediment surface (Fig. 7.5).

Gastrotrichs are among the few animals commonly found in anaerobic environments (Moore 1939, Cole 1955, Strayer 1985), remaining abundant even during extended periods (e.g., months) of anoxia. The physiological basis of the anaerobiosis of freshwater gastrotrichs has not yet been studied. It seems likely that freshwater gastrotrichs possess a sulfide detoxification mechanism similar to that demonstrated for marine gastrotrichs (Powell *et al.* 1979, 1980) and freshwater nematodes (Nuss 1984, Nuss and Trimkowski 1984) to deal with the elevated concentrations of H_2S that often accompany extended anoxia.

Little is known of the factors that control the distribution of individual species of gastrotrichs in freshwater. Arguing largely from analogy with marine work (d'Hondt 1971a), we might expect factors of primary importance to include the granulometry, stability, packing, and organic content of the sediment, the amount of dissolved oxygen, and the density and composition of communities of microbes and predators. Also, culture work suggests that the inorganic chemistry of the water and the presence of anthropogenic contaminants can exert a strong influence on gastrotrich populations (Hummon 1974, Faucon and Hummon 1976, Hummon and Hummon 1979).

Most of the genera of freshwater gastrotrichs are known to have intercontinental or cosmopolitan distributions. Exceptions include the macrodasyid *Redudasys*, known in freshwater only from Brazil; *Dichaetura* and *Marinellina*, which are known only from Europe; and some new genera soon to be described from South America (J. Kisielewski personal communication). The geographic distribution of individual species is not well known, because of the primitive state of the species-level taxonomy of freshwater gastrotrichs and the paucity of field studies throughout most of the world. Even in North America, only Michigan (Brunson 1950) and Illinois (Robbins 1965, 1973) have been explored systematically for freshwater gastrotrichs. Some species have been reported to occur over broad ranges, in-

cluding in some cases more than one continent. Until further studies are made, such reported intercontinental distributions should be regarded with suspicion (cf. Frey 1982), especially because "conspecifics" collected from different continents often exhibit marked morphologic differences from one another (e.g., Robbins 1973, Emberton 1981).

Chaetonotus typically dominates gastrotrich faunae everywhere in freshwater, both in terms of numbers of species and numbers of individuals (Table 7.1). Other genera of the Chaetonotidae are common in all kinds of freshwaters, but are usually less abundant than *Chaetonotus*. The Dasydytidae and Neogosseidae are less widespread than the Chaetonotidae, usually living in weedy, productive waters where they may, however, become numerically abundant (e.g., Blinn and Green 1986, Kisielewski 1981, 1986a, Nesteruk 1986). The Dichaeturidae (*Dichaetura* and possibly *Marinellina*) are very rare, the few records coming from cisterns, underground waters, and among moss (Remane 1935–1936, Ruttner-Kolisko 1955).

B. Reproduction and Life History

Until recently, populations of freshwater gastrotrichs were thought to consist entirely of parthenogenetic females (e.g., Hyman 1951, Pennak 1978). However, more detailed recent studies have dispelled this notion, and have revealed a remarkable life cycle among the freshwater gastrotrichs (Fig. 7.3) (Weiss and Levy 1979, Hummon 1984a–c, 1986, Levy 1984, Balsamo and Todaro 1988).

Newly hatched gastrotrichs are relatively large (approximately two thirds the length of adults) (Brunson 1949) and already contain developing parthenogenetic eggs. These eggs develop rapidly under favorable conditions. The first egg may be laid within two days after the mother hatches. Typically, a total of four parthenogenetic eggs is laid over a period of a week. Apparently, parthenogenesis is apomictic, so that offspring are genetically identical to their mother.

There are two kinds of parthenogenetic eggs (Fig. 7.4). The more common kind, tachyblastic eggs, develop immediately and hatch quickly (within a day of being laid, at 20°C). Occasionally, the final parthenogenetic egg laid by a female is not a tachyblastic egg, but a resting, or opsiblastic egg. Opsiblastic eggs are thick-shelled, a little larger than tachyblastic eggs, and are very resistant to freezing and drying (Brunson 1949). The factors that induce the production of opsiblastic eggs are not well known, although such eggs are often produced by animals in crowded cultures. Opsiblastic eggs are almost always the final egg produced by an animal, even if the total number of eggs is fewer than four.

Table 7.1. Number of Species of Gastrotrichs Found In Some Freshwater Habitats

	Chaeto-notus	*Hetero-lepidoderma*	*Aspidio-phorus*	*Ichthy-dium*	*Lepido-dermella*	*Poly-merurus*	*Dasydytes*	*Stylochaeta*	*Neogossea*	Total	Source
Ponds, Poland[a]	10	3	1	2	< 1	2	2	1	< 1	21	Nesteruk (1986), Kisielewski (1986a)
Peat bogs, Poland[b]	16	4	1	2	1	1	1	1	0	27	Kisielewski (1981)
Bog pools, Poland[c]	12	1	2	1	1	3	3	1	0	24	Kisielewska (1982)
Mirror Lake, New Hampshire	8–20	1	2	3	4	2	0	0	0	20–32	Strayer (1985)
Phragmites mats, Romania	16	1	1	2	0	3	4	1		28	Rudescu (1968)
Bog bordering Lake Tschernoe, USSR	26	0	1	2	3	2	1	1	0	36	Preobrajenskaja (1926)
Lake Beloye, USSR	22	0	1	4	2	2	3	1	0	35	Preobrajenskaja (1926)

[a] Mean of seven intensively studied ponds.
[b] Mean of four intensively studied bogs.
[c] Mean of two intensively studied pools.

Following the production of parthenogenetic eggs, animals develop into hermaphrodites (Fig. 7.2c) (Hummon and Hummon 1983, Hummon and Hummon 1983, 1988, 1989). During this time, sperm and meiotic sexual eggs are produced, and the X-body grows. These changes occur slowly, over a period of a week after the last parthenogenetic egg is laid. No one has yet observed sperm transfer or fertilization, although Levy (1984, Levy and Weiss 1980) reported finding a third kind of egg (the "plaque-bearing egg") in cultures of *Lepidodermella squamata*. He suggested that plaque-bearing eggs may be the product of sexual reproduction. Because the sperm are few in number (32-64 per animal) and nonmotile, fertilization probably is internal. Animals reared in isolation do not produce

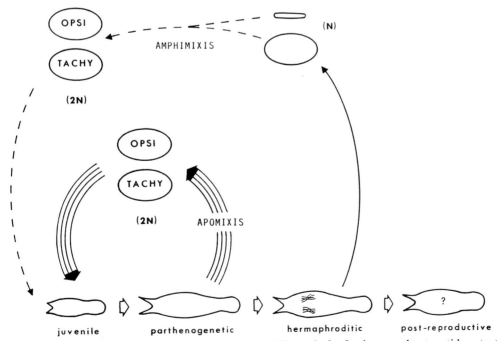

Figure 7.3 Schematic diagram of the proposed generalized life cycle for freshwater chaetonotid gastrotrichs, based predominantly on the study of *Lepidodermella squamata*. Dashed lines show hypothetical events that have not yet been demonstrated. OPSI, opsiblastic egg; TACHY, tachyblastic egg. [From Levy (1984), after ideas presented by Levy and Weiss (1980), with permission of the authors.]

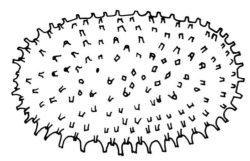

Figure 7.4 Egg of *Chaetonotus maximus*. (From Remane 1935–1936.)

fertilized sexual eggs, so cross-fertilization probably is the rule. The absence of ducts associated with the male or female reproductive system makes it difficult to suggest a mechanism of sperm transfer (Hummon 1986 described one bizarre possibility). Probably the enigmatic X-organ is involved. Much remains to be learned about the postparthenogenetic sexual phase and its importance in nature.

The life cycle just described is unique among invertebrates and offers considerable ecological flexibility to gastrotrich populations. The initial parthenogenetic phase allows for explosive population growth under favorable conditions: workers have commonly reported growth rates (r) of 0.1–0.6 per day in laboratory cultures (e.g., Hummon 1974, Faucon and Hummon 1976, Hummon and Hummon 1979, Hummon 1986, Balsamo and Todaro 1988). Production of parthenogenetic resting eggs (opsiblastic eggs) buffers the population against unfavorable conditions and presumably allows for dispersal among habitats. Finally, the subsequent sexual phase introduces genetic recombination. Sexual reproduction is most likely to occur in populations in which rates of mortality are low enough to allow some gastrotrichs to reach the age required for sexual development.

C. Ecological Interactions

Gastrotrichs feed on bacteria, algae, protozoans, detritus, and small inorganic particles. It is likely that bacteria are of prime importance. Bennett (1979) demonstrated that *Lepidodermella squamata* readily digested bacteria and found that this gastrotrich would not survive in laboratory cultures in the absence of bacteria. He reported that *L. squamata* could digest the green alga *Chlorella* as well, but suggested that algae were of secondary importance in gastrotrich diets. Gray and Johnson (1970) showed that the marine macrodasyid *Turbanella hyalina* could choose among various strains of natural bacteria, apparently on the basis of a tactile

chemical sense. Thus, the quality as well as the quantity of bacterial populations was important to the gastrotrichs. Freshwater gastrotrichs are most likely capable of similar fine discrimination among bacterial prey.

Reported predators of gastrotrichs include heliozoan and sarcodine amoebae, cnidarians, and tanypodine midges (Brunson 1949, Bovee and Cordell 1971, Moore 1979), but many other benthic predators presumably feed on gastrotrichs. Nothing is known about the importance of predation in regulating populations of freshwater gastrotrichs or about the quantitative importance of gastrotrichs as a food item for various predators.

We have no direct information on what regulates gastrotrich populations in nature. Gastrotrichs are capable of enormous population growth (10–60% per day) in laboratory cultures. If potential growth rates are anywhere near this high in nature, as seems likely in some circumstances, then there must be an equally high counterbalancing mortality, perhaps from predation. There are no detailed studies of the population dynamics of freshwater gastrotrichs in nature. The few quantitative studies of the seasonal dynamics of freshwater gastrotrich populations (Fig. 7.6; see also Nesteruk 1986) have shown that population densities are usually (but not always) lowest during the winter. We do not know what drives these seasonal dynamics, but seasonal changes in water temperature, food supply, and predation pressures are obvious possibilities.

Gastrotrichs, along with nematodes and rotifers, are among the most abundant animals in the freshwater benthos, having densities on the order of 10 per cm^2 (100,000 per m^2) (Table 7.2). However, because there have been no direct measurements of the roles of gastrotrichs in freshwater ecosystems, one can only guess at their importance. There is only a single, tentative estimate of gastrotrich metabolism in freshwater: Strayer (1985) estimated secondary production of gastrotrichs in Mir-

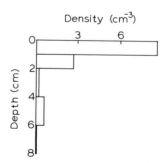

Figure 7.5 Gastrotrich density within the sediments of Mirror Lake, New Hampshire, as a function of depth from the sediment surface. (From Strayer 1985.)

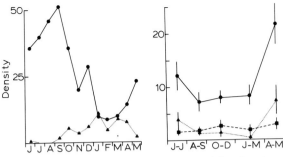

Figure 7.6 Seasonal trends in gastrotrich abundance. *Left*: Density (number/cm³) of two species of *Dasydytes* in the surface sediments of a boggy pool ("Complex B") in Poland: *D. dubius* (●) and *D. ornatus* (▲). (From Kisielewska 1982.) *Right*: Density (number/cm²) of all gastrotrichs (●), an unidentified species (probably of *Heterolepidoderma*) (■), and *Lepidodermella triloba* (▲) on the gyttja sediments of Mirror Lake, New Hampshire (original). Plotted points are means ± standard error (n equals 16).

ror Lake, New Hampshire, to be approximately 100 mg dry mass per m² per yr, which is less than 1% of the total production of the zoobenthic community. This estimate suggests that gastrotrich metabolism and processes correlated with metabolism such as nutrient regeneration by gastrotrichs are of minor importance in freshwater ecosystems. It would be imprudent, however, to dismiss gastrotrichs as quantitatively unimportant to ecosystem functioning without actually measuring gastrotrich activities under defined conditions. The extraordinarily high rates of population turnover and potentially highly

selective feeding behavior of gastrotrichs suggests that they may exert a considerable influence on the composition of natural bacterial communities.

D. *Evolutionary Relationships*

The phylogenetic relationships among the various aschelminth groups are not yet clear. Nonetheless, several points of morphological similarity between nematodes and gastrotrichs (summarized in Lorenzen 1985) show that these two groups are closely allied. Within the Gastrotricha, only the broad outlines of phylogeny have been sketched out. Three major groups of gastrotrichs are widely recognized: the order Macrodasyida and the suborders Multitubulata and Paucitubulata of the order Chaetonotida. The Multitubulata, which includes only the marine *Neodasys*, exhibits many primitive characteristics and is in some ways intermediate between the Macrodasyida and the Paucitubulata, which contains almost all freshwater gastrotrichs. Therefore, Boaden (1985) suggested that modern Macrodasyida and Paucitubulata are descended from a *Neodasys*-like ancestor. Nevertheless, on the basis of digestive tract ciliature and other characteristics, we believe that the ancestral gastrotrich was more likely similar to the dactylopodolid macrodasyids.

Evolutionary relationships among the families, genera, and species of freshwater gastrotrichs are still largely unknown because of a paucity of basic morphological, biochemical, and zoogeographical information about most species. An enormous num-

Table 7.2. Density of Gastrotrichs In Some Freshwater Habitats[a]

Site	Abundance		Source
	Areal (cm⁻²)	Volumetric (cm⁻³)	
Mirror Lake, New Hampshire	13	6[b]	Strayer (1985)
Sandbars of Mississippi River, Minnesota	16–23[c]	8–16[d]	Hummon (1987 and unpublished)
Beaches of Lake Erie, Ohio	5[e]	2[e]	Evans (1982)
Oligotrophic lakes, Poland	–	6[f]	Kisielewski and Kisielewska (1986)
Eutrophic lakes, Poland	–	16[f]	Kisielewski and Kisielewska (1986)
Oligotrophic streams, Poland	–	2[f]	Kisielewski and Kisielewska (1986)
Oligotrophic springs, Poland	–	1[f]	Kisielewski and Kisielewska (1986)
Fish-culture ponds, Poland	–	13[f]	Nesteruk (1986)
Temporary ponds, Poland	–	33[f]	Szkutnik (1986)
Boggy pools, Poland	–	42[g]	Kisielewska (1982)

[a] Data of Nesteruk (1986) and Kisielewska (1982) are annual means; other studies were conducted at various times during the ice-free season.
[b] Top 2 cm of sediment.
[c] Geometric means, top 8 cm of sediments, two sites.
[d] Geometric means, top 1 cm of sediments, two sites.
[e] Top 3 cm of sediments.
[f] Surface sediments, probably an overestimate because authors selected sediments suspected to contain high densities of gastrotrichs.
[g] Mean of two sites.

ber of species have not even been described, let alone studied. Even in North America, perhaps 75–90% of the species of freshwater gastrotrichs have not been described. Certainly there are genera (e.g., *Dichaetura, Marinellina*) whose correct systematic placement will require much additional study. Furthermore, there is a mounting concern (e.g., Remane 1935–1936, Ruppert 1977, Kisielewski 1981, 1987) that some of the genera of the Chaetonotidae (e.g., *Ichthydium*) are polyphyletic and will need to be redefined after more detailed studies are made.

IV. COLLECTING, REARING, AND PREPARATION FOR IDENTIFICATION

Gastrotrichs may be collected by taking samples of sediments or vegetation. For quantitative work, small-diameter (2–5 cm) cores are preferable.

Because living animals are preferable to preserved animals for many purposes, it often is desirable to extract the animals from the sample prior to preservation. If the sample must be preserved immediately, workers recommend narcotizing the animals with 1% $MgCl_2$ for 10 minutes, then fixing them in 10% formalin with rose bengal (e.g., Hummon 1981, 1987).

It is difficult to extract or count the gastrotrichs from a sample. Some workers have handpicked or counted animals under a dissecting microscope (e.g., Hummon 1981, 1987, Evans 1982, Strayer 1985), but this procedure is tedious. Density gradient centrifugation (Nichols 1979, Schwinghamer 1981) may be useful in extracting gastrotrichs from sediment, but this method has not yet been tested on freshwater gastrotrichs. The sample should not be sieved, because gastrotrichs are too small to be retained quantitatively on even very fine mesh sieves. For example, Hummon (1981) found that a 37-μm mesh sieve retained only 31% of the gastrotrichs in a series of samples taken from the upper Mississippi River. For quantitative work, it is important to check the efficiency of whatever extraction or counting method is used, because gastrotrichs are so small and easily overlooked (cf. Strayer 1985, pp. 295–296).

Living gastrotrichs are preferable to dead gastrotrichs for taxonomic work. Living gastrotrichs often are too active for critical observations to be made, so they must be slowed down. This can be accomplished (1) by gently squeezing the animal [either with a rotocompressor (Spoon 1978) or by removing some of the water from beneath a coverslip with a tissue]; (2) by placing it in a viscous medium such as methylcellulose; or (3) by narcotizing it. Cocaine was the traditional narcotic of choice (Brunson 1959), but it is now difficult to obtain for laboratory use. d'Hondt (1967) recommended using MS 222, and we have had very good success with narcotizing gastrotrichs by bleeding 1.8% neosynephrine (available at pharmacies) under the coverslip. Animals may be killed with formalin or fumes of osmium tetroxide (e.g., Brunson 1950) following narcotization. Osmium tetroxide is a superior fixative, but is dangerous and should be used with extreme care. It is sometimes necessary to examine individual scales, which can be isolated from an animal by bleeding 2% acetic acid under the coverslip.

Procedures for culturing gastrotrichs were described by Packard (1936), Brunson (1949), Townes (1968), Hummon (1974), and Bennett (1979). Gastrotrichs have been cultured on 0.1% malted milk, raw egg yolk, wheat grain infusion, baked lettuce infusion, and baker's yeast. Brunson (1949) recommended that animals be acclimated gradually to culture medium when collected from the wild. Hummon (1974) described a procedure for starting individual cultures from eggs that may be especially useful for bioassay work.

V. IDENTIFICATION OF THE GASTROTRICHS OF NORTH AMERICA

It is relatively easy to identify freshwater gastrotrichs to genus and very difficult to identify them to species. Most of the genera of freshwater gastrotrichs are well established and widely recognized (but see Section III.D). Species identification requires a keen eye, careful observation, a cooperative gastrotrich, and some luck because most of the freshwater gastrotrichs of North America undoubtedly are undescribed. The following works are helpful in species identification: Brunson (1950, 1959), who keyed and illustrated species then known from North American freshwaters; Voigt (1958), who provided keys (in German) and drawings for all known gastrotrichs worldwide; Robbins (1965, 1973), who gave additional information and drawings of North American species; d'Hondt (1971b), who gave a key to the species of *Lepidodermella* and defined three subgenera of *Ichthydium*; Kisielewski (1981), who made a critical evaluation of the morphological characters that must be measured to describe (or identify) a species; and Kisielewski (1986b), who gave a recent treatment of *Aspidiophorus*. Schwank and Bartsch's (1990) comprehensive treatment of European freshwater species appeared too late for us to use in preparing this chapter.

A. Taxonomic Key to Genera of Freshwater Gastrotricha

The following key, modified from Voigt (1958) and Brunson (1959), includes all genera of freshwater gastrotrichs known or likely to be found in North America.

1a.	Animal with at least three pairs of adhesive tubules (one anterior and two posterior) and a pair of pharyngeal pores (Fig. 7.1); an almost entirely marine group not yet reported from North American freshwaters .... order Macrodasyida	
1b.	Animal lacking adhesive tubules (Fig. 7.7j–l) or with one pair (very rarely two pairs) of adhesive tubules posteriorly (Fig. 7.7a–i), and no pharyngeal pores (Fig. 7.7); common and widespread in freshwater .... order Chaetonotida	2
2a (1b).	Posterior end with furca bearing adhesive tubules (Fig. 7.7a–i)	3
2b.	Furca absent, although the posterior end may bear spines or rounded protuberances (Fig. 7.7j–l)	9
3a(2a).	Furca doubly branched (Fig. 7.7a); a rare genus not yet reported from North America .... family Dichaeturidae *Dichaetura*	
3b.	Furca singly branched (Fig. 7.7b–i); common and widespread .... family Chaetonotidae	4
4a(3b).	Branches of furca with rings or scales; branches often very long (more than one-half as long as the trunk) (Fig. 7.7b) .... *Polymerurus*	
4b.	Branches of furca without rings or scales; branches short (Fig. 7.7c–i)	5
5a(4b).	Body with cuticular spines or spined scales (Fig. 7.7d–f) .... *Chaetonotus*	
5b.	Body without cuticular spines, or with only a few spines or tactile bristles near the base of the furca (Fig. 7.7c, g–i)	6
6a(5b).	Body without cuticular scales (Fig. 7.7c) .... *Ichthydium*	
6b.	Body with cuticular scales (visible as distinct scales or cuticular roughness at 400×) (Fig. 7.7g–i)	7
7a(6b).	Scales stalked (Fig. 7.7h) .... *Aspidiophorus*	
7b.	Scales not stalked (Fig. 7.7g, i)	8
8a(7b).	Scales keeled (Fig. 7.7g) .... *Heterolepidoderma*	
8b.	Scales without keels (Fig. 7.7i) .... *Lepidodermella*	
9a(2b).	Anterior end with club-shaped tentacles (Fig. 7.7j) .... family Neogosseidae	10
9b.	Anterior end without such tentacles (Fig. 7.7k–l) .... family Dasydytidae	11
10a(9a).	Posterior end of body with a pair of lateral tufts of spines (Fig. 7.7j) .... *Neogossea*	
10b.	Posterior end of body with a median tuft of spines .... *Kijanebalola*	
11a(9b).	Posterior end with pair of peg-shaped protuberances (Fig. 7.7k) .... *Stylochaeta*	
11b.	Posterior end without such protuberances (Fig. 7.7l) .... *Dasydytes*	

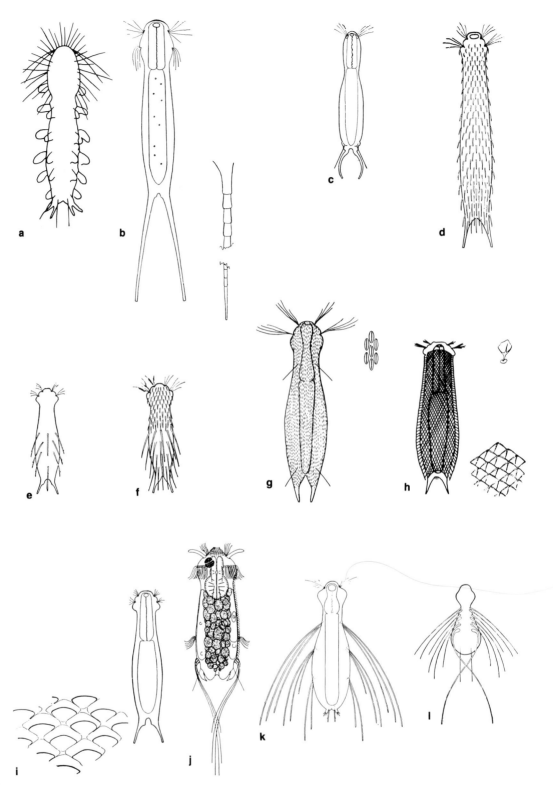

Figure 7.7 Genera of freshwater gastrotrichs. (a) *Dichaetura*; (b) *Polymerurus*, showing detail of ringed branches of furca; (c) *Ichthydium*; (d–f) *Chaetonotus*, showing examples of spination; (g) *Heterolepidoderma*, with detail of scales; (h) *Aspidiophorus*, with detail of coat of scales and a single scale; (i) *Lepidodermella*, with detail of scales; (j) *Neogossea*; (k) *Stylochaeta*; (l) *Dasydytes*. [From Remane (1935–1936), Brunson (1950), Krivanek and Krivanek (1958), Voigt (1958), and Robbins (1965).]

ACKNOWLEDGMENTS

We are grateful to H. Cyr, S. Findlay, J. Kisielewski, J. Kolasa, and M. Weiss for their careful, helpful reviews of this chapter, and to D. Levy and M. Weiss for allowing us to use some of their unpublished work.

LITERATURE CITED

Balsamo, M. 1980. Spectral sensitivity in a fresh-water gastrotrich (*Lepidodermella squammatum* Dujardin). Experientia 36:830–831.

Balsamo, M., and M. A. D. Todaro. 1988. Life history traits of two chaetonotids (Gastrotricha) under different experimental conditions. Invertebrate Reproduction and Development 14:161–176.

Bennett, L. W. 1979. Experimental analysis of the trophic ecology of *Lepidodermella squammata* (Gastrotricha: Chaetonotida) in mixed culture. Transactions of the American Microscopical Society 98:254–260.

Blinn, D.W., and J. Green. 1986. A pump sampler study of microdistribution in Walker Lake, Arizona, U.S.A.: a senescent crater lake. Freshwater Biology 16:175–185.

Boaden, P. J. S. 1985. Why is a gastrotrich? Pages 248-260 *in*: S. Conway Morris, J. D. George, R. Gibson, and H. M. Platt, editors. The origins and relationships of lower invertebrates. Clarendon Press, Oxford.

Bovee, E. C., and D. L. Cordell. 1971. Feeding on gastrotrichs by the heliozoan *Actinophrys sol*. Transactions of the American Microscopical Society 90:365–369.

Brandenburg, J. 1962. Elektronenmikroscopische Untersuchungen des Terminalapparates von *Chaetonotus* sp. (Gastrotrichen) als Beispeil einer Cystocyte bei Aschelminthen. Zeitschrift für Zellforschung und mikroskopische Anatomie 57: 136–144.

Brunson, R. B. 1949. The life history and ecology of two North American gastrotrichs. Transactions of the American Microscopical Society 68:1–20.

Brunson, R. B. 1950. An introduction to the taxonomy of the Gastrotricha with a study of eighteen species from Michigan. Transactions of the American Microscopical Society 69:325–352.

Brunson, R. B. 1959. Gastrotricha. Pages 406-419 *in*: W. T. Edmondson, editor. Fresh-water biology. 2nd Edition. Wiley, New York.

Cole, G. A. 1955. An ecological study of the microbenthic fauna of two Minnesota lakes. American Midland Naturalist 53:213–230.

Emberton, K. C. 1981. First record of *Chaetonotus heideri* (Gastrotricha: Chaetonotidae) in North America. Ohio Journal of Science 81:95–96.

Evans, W. A. 1982. Abundances of micrometazoans in three sandy beaches in the island area of western Lake Erie. Ohio Journal of Science 82:246–251.

Faucon, A. S., and W. D. Hummon. 1976. Effects of mine acid on the longevity and reproductive rate of the Gastrotricha *Lepidodermella squammata* (Dujardin). Hydrobiologia 50:265–269.

Frey, D. G. 1982. Questions concerning cosmopolitanism in Cladocera. Archiv für Hydrobiologie 93:484–502.

Gray, J. S., and R. M. Johnson. 1970. The bacteria of a sandy beach as an ecological factor affecting the interstitial gastrotrich *Turbanella hyalina* Schultze. Journal of Experimental Marine Biology and Ecology 4:119–133.

Green, J. 1986. Associations of zooplankton in six crater lakes in Arizona, Mexico and New Mexico. Journal of Zoology (A) 208:135–159.

d'Hondt, J.-L. 1967. Effets de quelques anaesthétiques sur les Gastrotriches. Experientia 23:1025–1026.

d'Hondt, J.-L. 1971a. Gastrotricha. Oceanography and Marine Biology: Annual Review 9:141–192.

d'Hondt, J.-L. 1971b. Note sur quelques Gastrotriches Chaetonotidae. Bulletin de la Société Zoologique de France 96:215–235.

Hummon, M. R. 1984a. Reproduction and sexual development in a freshwater gastrotrich. 1. Oogenesis of parthenogenic eggs (Gastrotricha). Zoomorphology 104:33–41.

Hummon, M. R. 1984b. Reproduction and sexual development in a freshwater gastrotrich. 2. Kinetics and fine structure of postparthenogenic sperm formation. Cell and Tissue Research 236:619–628.

Hummon, M. R. 1984c. Reproduction and sexual development in a freshwater gastrotrich. 3. Postparthenogenic development of primary oocytes and the X-body. Cell and Tissue Research 236:629–636.

Hummon, M. R. 1986. Reproduction and sexual development in a freshwater gastrotrich. 4. Life history traits and the possibility of sexual reproduction. Transactions of the American Microscopical Society 105:97–109.

Hummon, M. R., and W. D. Hummon. 1979. Reduction in fitness of the gastrotrich *Lepidodermella squammata* by dilute mine acid water and amelioration of the effect by carbonates. International Journal of Invertebrate Reproduction 1:297–306.

Hummon, M. R., and W. D. Hummon. 1983. Gastrotricha. Pages 195–205 *in*: K. G. Adiyodi and R. G. Adiyodi, editors. Reproductive biology of invertebrates. Vol. 2: Spermatogenesis and sperm function. Wiley, London.

Hummon, W. D. 1974. Effects of DDT on longevity and reproductive rate in *Lepidodermella squammata* (Gastrotricha, Chaetonotida). American Midland Naturalist 92:327–339.

Hummon, W. D. 1981. Extraction by sieving: a biased procedure in studies of stream meiobenthos. Transactions of the American Microscopical Society 100:278–284.

Hummon, W. D. 1982. Gastrotricha. Pages 857-863 *in*: S.P. Parker, editor. Synopsis and classification of living organisms. Vol. 1. McGraw-Hill, New York.

Hummon, W. D. 1987. Meiobenthos of the Mississippi headwaters. Pages 125-140 *in*: R. Bertolani, editor. Biology of tardigrades. Selected Symposia and Monographs U.Z.I., 1. Mucchi, Modena, Italy.

Hummon, W. D., and M. R. Hummon. 1983. Gastrotricha. Pages 211–221 *in*: K.G. Adiyodi and R.G.

Adiyodi, editors. Reproductive biology of invertebrates. Vol. 1: Oogenesis, oviposition, and oosorption. Wiley, London.

Hummon, W. D., and M. R. Hummon. 1988. Gastrotricha. Pages 81–85 in: K. G. Adiyodi and R. G. Adiyodi, editors. Reproductive biology of invertebrates. Vol. 3: Accessory sex glands. Oxford and IBH Pub. Co., New Dehli.

Hummon, W. D., and M. R. Hummon. 1989. Gastrotricha. Pages 201–206 in: K. G. Adiyodi and R. G. Adiyodi, editors. Reproductive biology of invertebrates. Volume 4: Fertilization, development, and parental care. Part A. Oxford and IBH Publishing Company, New Delhi.

Hummon, W. D., W. A. Evans, M. R. Hummon, F. G. Doherty, R.H. Wainberg, and W. S. Stanley. 1978. Meiofaunal abundance in sandbars of acid mine polluted, reclaimed, and unpolluted streams in southeastern Ohio. Pages 188–203 in: J.H. Thorp and J.W. Gibbons, editors. Energy and environmental stress in aquatic ecosystems. DOE Symposium Series (CONF-771114). National Technical Information Service, Springfield, Virginia.

Hutchinson, G. E. 1967. A treatise on limnology. Vol. 2: introduction to lake biology and the limnoplankton. Wiley, New York.

Hyman, L. H. 1951. The invertebrates: Acanthocephala, Aschelminthes, and Entoprocta: the pseudocoelomate Bilateria. Vol. 3. McGraw-Hill, New York.

Kisielewska, G. 1981. Hermaphroditism of freshwater gastrotrichs in natural conditions. Bulletin de l'Académie Polonaise des Sciences, Série des Sciences Biologiques 29:167–172.

Kisielewska, G. 1982. Gastrotricha of two complexes of peat hags near Siedlce. Fragmenta Faunistica 27:39–57.

Kisielewski, J. 1979. New and insufficiently known freshwater Gastrotricha from Poland. Annales Zoologici 34:415–435.

Kisielewski, J. 1981. Gastrotricha from raised and transitional peat bogs in Poland. Monografie Fauny Polski 11:1–143.

Kisielewski, J. 1986a. Freshwater Gastrotricha of Poland. VII. Gastrotricha of extremely eutrophicated water bodies. Fragmenta Faunistica 30:267–295.

Kisielewski, J. 1986b. Taxonomic notes on freshwater gastrotrichs of the genus *Aspidiophorus* Voigt (Gastrotricha: Chaetonotidae), with descriptions of four new species. Fragmenta Faunistica 30:139–156.

Kisielewski, J. 1987. Two new interesting genera of Gastrotricha (Macrodasyida and Chaetonotida) from the Brazilian freshwater psammon. Hydrobiologia 153: 23–30.

Kisielewski, J., and G. Kisielewska. 1986. Freshwater Gastrotricha of Poland. I. Gastrotricha from the Tatra and Karkonosze Mountains. Fragmenta Faunistica 30:157–182.

Krivanek, R. C., and J. O. Krivanek. 1958. A new and a redescribed species of *Neogossea* (Gastrotricha) from Louisiana. Transactions of the American Microscopical Society 77:423–428.

Levy, D. P. 1984. Obligate post-parthenogenic hermaphroditism and other evidence for sexuality in the life cycles of freshwater Gastrotricha. Ph.D. Thesis, Rutgers Univ., New Brunswick, New Jersey. 257 pp.

Levy, D. P., and M. J. Weiss. 1980. Sperm in the life cycle of the freshwater gastrotrich, *Lepidodermella squammata*. American Zoologist 20:749. (Abstr.)

Lorenzen, S. 1985. Phylogenetic aspects of pseudocoelomate evolution. Pages 210-223 in: S. Conway Morris, J.D. George, R. Gibson, and H.M. Platt, editors. The origins and relationships of lower invertebrates. Clarendon Press, Oxford.

Martin, L. V. 1981. Gastrotrichs found in Surrey. Microscopy 34:286–300.

Moore, G. M. 1939. A limnological investigation of the microscopic benthic fauna of Douglas Lake, Michigan. Ecological Monographs 9:537–582.

Moore, J. W. 1979. Some factors influencing the distribution, seasonal abundance and feeding of subarctic Chironomidae (Diptera). Archiv für Hydrobiologie 85: 302–325.

Nesteruk, T. 1986. Freshwater Gastrotricha of Poland. IV. Gastrotricha from fish ponds in the vicinity of Siedlce. Fragmenta Faunistica 30:215–233.

Nichols, J. A. 1979. A simple flotation technique for separating meiobenthic nematodes from fine-grained sediments. Transactions of the American Microscopical Society 98:127–130.

Nuss, B. 1984. Ultrastrukturelle und ökophysiologische Untersuchungen an kristalloiden Einschlüssen der Muskeln eines sulfidtoleranten limnischen Nematoden (*Tobrilus gracilis*). Veröeffentlichungen des Instituts für Meeresforschung in Bremerhaven 20:3–15.

Nuss, B., and V. Trimkowski. 1984. Physikalische Mikroanalysen an kristalloiden Einschlüssen bei *Tobrilus gracilis* (Nematoda, Enoplida). Veröeffentlichungen des Instituts für Meeresforschung in Bremerhaven 20:17–27.

Packard, C. E. 1936. Observations on the Gastrotricha indigenous to New Hampshire. Transactions of the American Microscopical Society 55:422–427.

Pennak, R. W. 1978. Fresh-water invertebrates of the United States. 2nd Edition. Wiley (Interscience), New York.

Powell, E. N., M. A. Crenshaw, and R. M. Rieger. 1979. Adaptations to sulfide in the meiofauna of the sulfide system. I. [35]Sulfide accumulation and the presence of a sulfide detoxification system. Journal of Experimental Marine Biology and Ecology 37:57–76.

Powell, E. N., M. A. Crenshaw, and R. M. Rieger. 1980. Adaptations to sulfide in the sulfide-system meiofauna. Endproducts of sulfide detoxification in three turbellarians and a gastrotrich. Marine Ecology Progess Series 2:169–177.

Preobrajenskaja, E. N. 1926. Zur Verbreitung der Gastrotrichen in den Gewässern der Umbegung zu Kossino. Arbeiten der biologischen Station zu Kossino (bei Moskau) 4:3–14. (in Russ.; Ger. summ.)

Remane, A. 1935–1936. Gastrotricha und Kinorhyncha. Klassen und Ordnung das Tierreich 4 (Abt. 2, Buch 1, Teil 2, Lieferung 1–2):1–242.

Remane, A. 1961. *Neodasys uchidai* nov. spec., eine zweite *Neodasys*-Art. Kieler Meeresforschung 17: 85–88.

Renaud-Mornant, J. 1986. Gastrotricha. Pages 86-109 *in*: L. Botosaneaunu, editor. Stygofauna mundi. Brill, Leiden, Netherlands.

Robbins, C. E. 1965. Two new species of Gastrotricha (Aschelminthes) from Illinois. Transactions of the American Microscopical Society 84:260–263.

Robbins, C. E. 1973. Gastrotricha from Illinois. Transactions of the Illinois Academy of Science 66:124–126.

Rudescu, L. 1968. Die Rotatorien, Gastrotrichen und Tardigraden der Schilfrohrgebiete des Donaudeltas. Hidrobiologia (Bucharest) 9:195–202.

Ruppert, E. E. 1977. *Ichthydium hummoni* n. sp., a new marine chaetonotid gastrotrich with a male reproductive system. Cahiers de Biologie Marine 18:1–5.

Ruppert, E. E. 1988. Gastrotricha. Pages 302–311 *in*: R. P. Higgins and H. Thiel, editors. Introduction to the study of meiofauna. Smithsonian Institution Press, Washington, D.C.

Ruttner-Kolisko, A. 1955. *Rheomorpha neiswestnovae* und *Marinellina flagellata*, zwei phylogenetisch interessante Wurmtypen aus dem Süsswasserpsammon. Oesterreichische Zoologische Zeitschrift 6:55–69.

Schwank, P., and I. Bartsch. 1990. Gastrotricha und Nemertini. Süsswasserfauna von Mitteleuropa, Band 3, Teil 1/2. Gustav Fischer Verlag, Stuttgart.

Schwinghamer, P. 1981. Extraction of living meiofauna from marine sediments by centrifugation in a silica sol-sorbitol mixture. Canadian Journal of Fisheries and Aquatic Sciences 38:476–478.

Spoon, D. M. 1978. A new rotary microcompressor. Transactions of the American Microscopical Society 97:412–416.

Strayer, D. 1985. The benthic micrometazoans of Mirror Lake, New Hampshire. Archiv für Hydrobiologie Supplementband 72:287–426.

Szkutnik, A. 1986. Freshwater Gastrotricha of Poland. VI. Gastrotricha of small astatic water bodies with rush vegetation. Fragmenta Faunistica 30:251–266.

Townes, M. M. 1968. The collection, identification, and cultivation of gastrotrichs. Turtox News 46:99–101.

Voigt, M. 1958. Gastrotricha. Die Tierwelt Mitteleuropas, Band I. Lieferung 4a:1–45.

Weiss, M. J., and D. P. Levy. 1979. Sperm in "parthenogenetic" freshwater gastrotrichs. Science 205:302–303.

Rotifera

8

Robert Lee Wallace
Department of Biology
Ripon College
Ripon, Wisconsin 54971

Terry W. Snell
Division of Science and Mathematics
University of Tampa
Tampa, Florida 33606

Chapter Outline

I. INTRODUCTION
II. ANATOMY AND PHYSIOLOGY
  A. External Morphology
  B. Organ System Function
    1. Corona
    2. Trophi and Gut
    3. Organ Systems
  C. Environmental Physiology
    1. Locomotion
    2. Physiological Ecology
    3. Environmental Toxicology
    4. Anhydrobiosis
III. ECOLOGY AND EVOLUTION
  A. Diversity and Distribution
    1. Phenotypic Variation
    2. Distribution and Population Movements
    3. Biogeography
    4. Colonial Rotifers
    5. Sessile Rotifers
  B. Reproduction and Life History
    1. Reproduction
    2. Reproductive Behavior
    3. Aging and Senescence
    4. Population Dynamics
  C. Ecological Interactions
    1. Foraging Behavior
    2. Functional Role in the Ecosystem
    3. Competition with Other Zooplankton
    4. Predator–Prey Interactions
    5. Parasitism on Rotifers
    6. Aquaculture
  D. Evolutionary Relationships

IV. COLLECTING, REARING, AND PREPARATION FOR IDENTIFICATION
  A. Collections
  B. Culture
  C. Preparation for Identification
V. CLASSIFICATION AND SYSTEMATICS
  A. Classification
  B. Systematics
    1. Class Seisonidea
    2. Class Bdelloidea
    3. Class Monogononta
  C. Taxonomic Keys
  D. Taxonomic Key to Families of Freshwater Rotifera
  Literature Cited

I. INTRODUCTION

The phylum Rotifera or Rotatoria is comprised of approximately 2000 species of unsegmented, bilaterally symmetrical pseudocoelomates possessing two distinctive features (Fig. 8.1). First, at the apical end (head) is a ciliated region called the corona, which is used in locomotion and food gathering. In adults of some forms, ciliation is lacking and the corona is a funnel- or bowl-shaped structure at the bottom of which is the mouth. Second, a muscular pharynx, the mastax, possessing a complex set of hard jaws called trophi is present in all rotifers.

When viewing the anterior end of most rotifers, one is struck with the idea of a rotating wheel. This is due to the metachronal beat of cilia on the corona, a structure usually composed of two concentric rings: trochus and cingulum (Fig. 8.2). This same image

Ecology and Classification of North American Freshwater Invertebrates

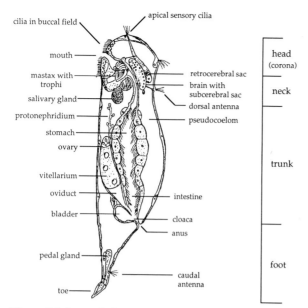

cilia in buccal field

apical sensory cilia

mouth

mastax with
trophi

salivary gland

protonephridium

stomach

ovary

vitellarium

oviduct

bladder

pedal gland

toe

retrocerebral sac

brain with
subcerebral sac

dorsal antenna

pseudocoelom

intestine

cloaca

anus

caudal
antenna

head
(corona)

neck

trunk

foot

Figure 8.1 Lateral view of a generalized rotifer. (Modified from Koste and Shiel 1987, with permission.)

provided early microscopists with the name for the phylum: the etymon is Latin, *rota,* "wheel" and Latin, *ferre,* "to bear" equals "wheel bearers." Although rotifers are often confused with ciliated protozoans and gastrotrichs by beginning students, those organisms do not possess trophi and their ciliation is not distributed in the same way as in rotifers. Rotifers are small organisms, generally ranging from 100–1000 μm long, although a few elongate species may surpass 2000 μm or more. Very few rotifers are parasitic (Rees 1960, 1989); nearly all are free-living herbivores or predators.

Collectively this phylum is widely distributed, being found in all freshwater habitats at densities generally ranging up to about 1000 individuals per liter. However, rotifers occasionally become very abundant if sufficient food is available, and can attain population densities of > 5000 per liter. In some rather unusual water bodies, exceedingly large populations can develop; sewage ponds may contain about 12,000 per liter (Seaman *et al.* 1986), and at certain times in soda-water bodies in Chad, much more than 100,000 per liter may occur (Iltis and

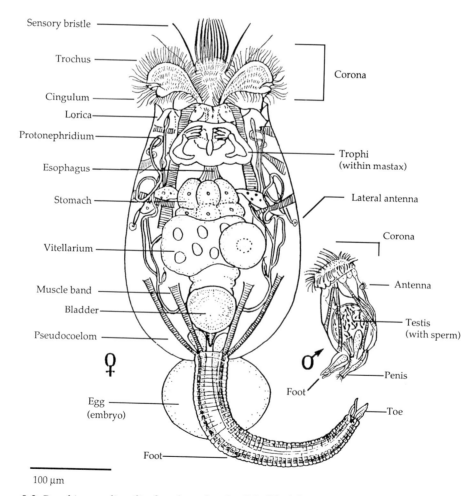

Sensory bristle

Trochus

Cingulum

Lorica

Protonephridium

Esophagus

Stomach

Vitellarium

Muscle band

Bladder

Pseudocoelom

Egg
(embryo)

Foot

Corona

Trophi
(within mastax)

Lateral antenna

Corona

Antenna

Testis
(with sperm)

Penis

Foot

Toe

100 μm

Figure 8.2 *Brachionus plicatilis,* female and male. (Modified from Pourriot 1986, with permission.)

Riou-Duvat 1971)! Although most inhabit fresh-waters, some genera also have members that occur in brackish and marine waters. For example, about 20 of the 32 species comprising the genus *Synchaeta* are described as marine (Nogrady 1982). However, only about 50 species of rotifers are exclusively marine. In general, rotifers are not as diverse or as abundant in marine environments as microcrustaceans, but they occur in many nearshore marine communities (Egloff 1988) and occasionally comprise the dominant portion of the biomass (Schnese 1973, Johansson 1983). One unusual group of rotifers, the bdelloids (Fig. 8.3), may be found inhabiting the film of water covering mosses, lichens, and liverworts. Additionally, they are often abundant in soils (Pourriot 1979); estimates of their densities range from about 32,000 to more than 2 million per m², depending on soil moisture levels. Because of their feeding habits and the fact that they are sometimes more numerous than nematodes, rotifers play an important role in nutrient cycling in soils (Anderson *et al.* 1984, Sohlenius 1982).

Most rotifers are free moving, either by swimming or crawling, but many sessile species live permanently attached to freshwater plants (Edmondson 1944, Wallace 1980). The vast majority of rotifers are solitary, but about 25 species form colonies of various sizes (Wallace 1987). All freshwater rotifers are either exclusively parthenogenetic or produce males for a limited time each year. Therefore, unless collections are made frequently, male rotifers may never be seen. Three very different classes of rotifers are commonly recognized (Seisonidea, Bdelloidea, Monogononta).

Additional accounts of this phylum may be found in most texts of general and invertebrate zoology and in some specialized books about freshwaters (Edmondson 1959, Hutchinson 1967, pp. 506–551, Pennak 1989, pp. 169–225). For detailed reviews of the biology of rotifers, consult the works of de Beauchamp (1965), Hyman (1951, pp. 59–151), Koste (1978), and Ruttner-Kolisko (1974). There is no single scientific journal or set of journals in which researchers publish their work on rotifers; the field is simply too diverse. However, every three years beginning in 1976, a small group of workers (approximately 50-90) have gathered to hold the International Rotifer Symposium. To date, five such meetings have been held and the proceedings for each have been published in a special volume. Some of the papers discussed in this chapter were presented at those meetings.

II. ANATOMY AND PHYSIOLOGY

A. External Morphology

Rotifers are saccate to cylindrical in shape, sometimes appearing worm-like (e.g., many bdelloids, Fig. 8.3). Typically, the body is comprised of four regions: head (with corona), neck, body, and foot (Fig. 8.1). Although these regions may be marked by folds in the body wall, which function like joints, rotifers are not metameric (segmented). Unfortunately, the generalizations noted here do not represent all rotifers well. In some forms, the neck and foot may be quite prominent, while in others, they are absent (Fig. 8.4). In the class Monogononta, male rotifers (Fig. 8.2) and juveniles (larval females) of sessile rotifers are usually much smaller than adult females (Fig. 8.5). In addition, male rotifers are structurally simpler (e.g., the gut commonly functions only as a food reserve) and the larvae of sessile forms generally have a morphology very different from that of the adult (Wallace 1980). When either males or larvae are found in plankton tows, their strikingly different morphologies may lead to improper identification.

The foot is an appendage that extends from the body ventrally (Fig. 8.2). It usually possesses two toes, but the number may vary from 0–4. The foot

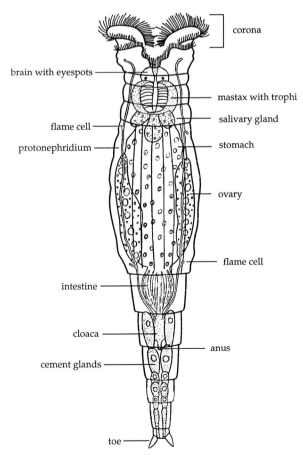

Figure 8.3 Typical bdelloid rotifer (*Philodina*). (Modified from several sources.)

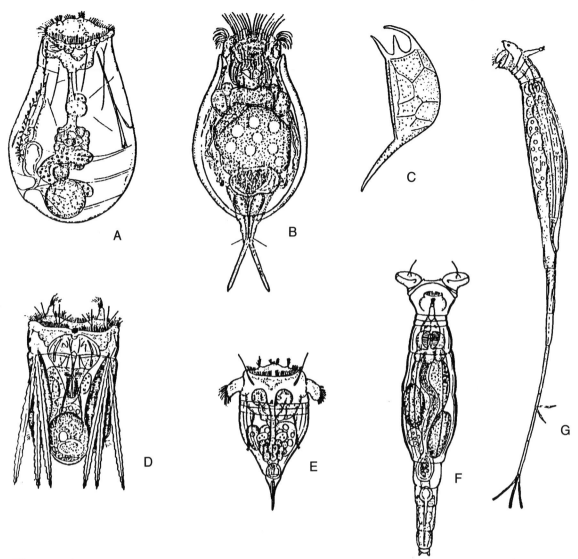

Figure 8.4 Representative rotifers: (a) *Asplanchna;* (b) *Euchlanis;* (c) *Keratella* (lateral view); (d) *Polyarthra;* (e) *Synchaeta* (all monogononts); (f) *Philodina;* (g) *Rotaria* (bdelloids.) (From Koste 1976, with permission.)

also may possess pedal glands whose ducts exit near the toes. These glands secrete a sticky cement for temporarily attaching the rotifer to substrata. In larvae of sessile rotifers, the cement forms a bond with the substratum that is not easily detached; if dislodged, sessile forms do not reattach (Wallace 1980).

Rotifers possess a syncytial integument or body wall containing a filament layer of varying thickness called the intracytoplasmic lamina (Clément 1985, Koehler 1966). This feature apparently is shared with the parasitic pseudocoelomate phylum, Acanthocephala, indicating a phylogenetic relationship (see Section III.D). The integument of a halophile rotifer, *Brachionus plicatilis,* was examined biochemically by Bender and Kleinow (1988) and their work indicates that it contains two filamentous, keratin-like proteins (M_r 39,000 and 47,000) cross-

linked by disulfide bonds. Species in which major portions of the integument are thickened are termed loricate and, if the integument is thin and very flexible, illoricate. However, extremes may be found within a single family or genus (e.g., *Cephalodella*). In general, thickness of the body wall is of little taxonomic significance. Further, even in loricate forms, portions of the integument (lorica) have a less well-developed intracytoplasmic lamina, thus making the lorica flexible in that region. Flexibility is found in the region of the corona, often in the foot, and at articulations between movable spines and the body. In some rotifers, the integument has various projections (e.g., bumps and fixed or movable spines) that serve various functions, chief among these is protection from some predators (see Sections III.A.1 and III.C.4).

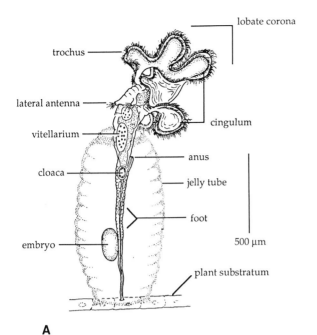

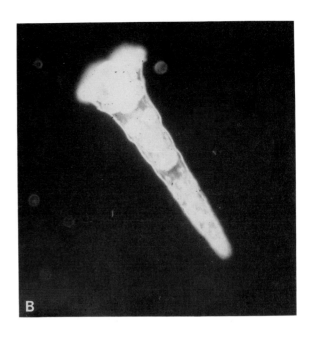

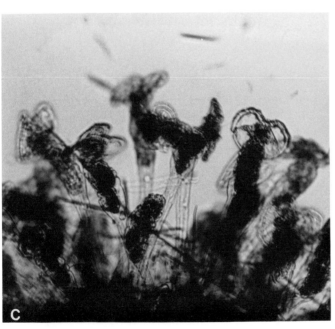

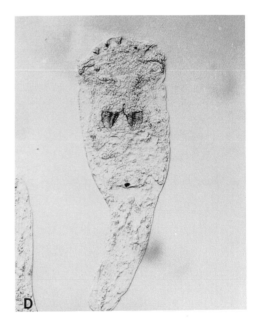

Figure 8.5 Adults and larvae of two sessile rotifers (family Flosculariidae). (A) Adult *Octotrocha speciosa* (from Koste 1989, with permission); (B) larval *Octotrocha speciosa* (dark field illumination, length ~300μm); (C) colony of adult *Lacinularia flosculosa* (bright field illumination, individual length ~1200μm); (D) larval *Lacinularia flosculosa* (bright field illumination, larvae somewhat compressed by the coverglass, length ~450μm).

B. Organ System Function

Internally, rotifers possess a spacious pseudocoelom in which are suspended muscles, nerves, and digestive, reproductive, and protonephridial organs (Figs. 8.1, 8.2, and 8.3). Respiratory and circulatory systems are absent. Rotifers possess two other remarkable features. First, all postembryonic tissues are syncytial. Second, all individuals of a species have a consistent number of nuclei in each organ throughout life. [N.B.: There are approximately 900 nuclei per female (Hyman 1951).] This feature, called eutely, is seen in a few other invertebrates (e.g., nematodes, Chapter 9). Most nuclei may be seen using a standard light microscope, but special

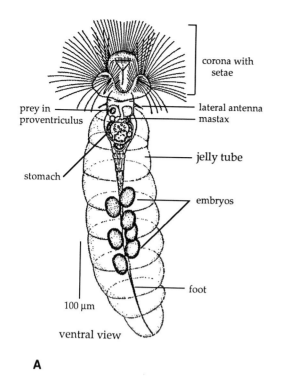

A

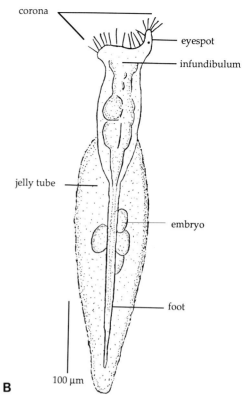

B

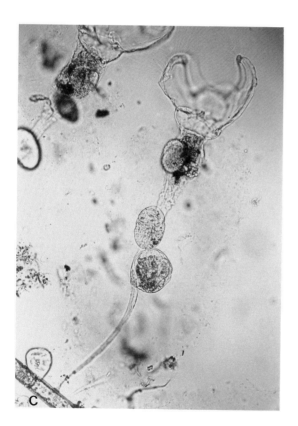

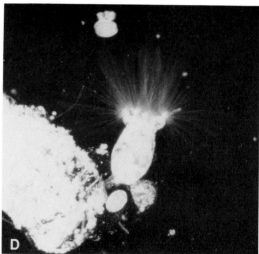

Figure 8.6 Three representatives of the order Collothecacea: (A) *Collotheca trilobata*, a sessile form; (B) *Collotheca mutabilis*, a planktonic form; (C) *Collotheca* sp. photomicrograph with two developing embryos (width of corona = 100 μm); (D) *Collotheca cornuta*, dark field illumination (length of setae ≃ 200 μm). (A, modified from Wallace *et al.* 1989, with permission; B, modified after several sources; C and D, R. L. Wallace original.)

optics such as differential interference contrast (Nomarski) enhance their visibility.

1. Corona

There is considerable variation in the shape of the anterior ends of rotifers and at least seven different types of coronae have been described, based on the placement of the mouth and the distribution of cilia (Koste and Shiel 1987). The corona form of many rotifers is comprised of two ciliated rings called the trochus and cingulum (Figs. 8.2 and 8.5). These structures are responsible for the production of water currents that are used in locomotion and feeding. However, not all rotifers possess this coronal ciliation. Adults of the family Collothecidae exhibit the most extreme variation from the typical plan; in most collothecids, cilia are nearly or completely lacking from the corona and long setae surround the rim of a funnel-shaped structure known as the infundibulum (Latin, a funnel) (Fig. 8.6). These setae prevent escape of prey when the edges of the infundibulum fold over the victim, capturing it in a fashion similar to the Venus flytrap. However, some adult collothecids possess no setae on their corona (e.g. *Cupelopagis*). In other groups, ciliation may be limited to a ventral field or several lobes, as is seen in some creeping forms and some bdelloids. Other structures that may be present on the corona include cirri, sensory antennae, and palps.

2. Trophi and Gut

Once food is captured by the corona, it enters a ventral mouth by passing through a short ciliated tube into a muscular pharynx, termed the mastax. The mastax possesses a chitinous lining on the inside, developed as a set of translucent jaws called trophi. The trophi may work the food in various ways (e.g., grinding) before it is passed to the esophagus and swallowed (Figs. 8.1 and 8.2). In the family Collothecidae, a portion of the mastax is enlarged into a food-storage organ known as the proventriculus (Fig. 8.6). The mastax leads posteriorly to an esophagus, stomach, and in most species, intestine and anus, but the gut ends in a blind stomach in some genera (i.e., *Asplanchna, Asplanchnopus*). The posterior portion of the intestine (cloaca) receives eggs from the oviduct and fluid from either a bladder or directly from paired protonephridia (Fig. 8.1). Often the gut is pigmented, depending on the nature of recently ingested material. Different species found in the same sample may possess guts that vary in color due to differences in diet (Pourriot 1977).

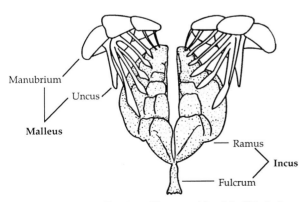

Figure 8.7 Generalized rotifer trophi. (Modified from Wallace *et al.* 1989, with permission.)

Rotifer trophi are composed of several hard parts and associated musculature, which articulate in a specific spacial arrangement. In their basic form, trophi consist of three functional units: an incus (Latin, anvil) and paired mallei (Latin, hammer) (Fig. 8.7). The incus is composed of three pieces: a fulcrum and a pair of rami (Latin, branch) that move like forceps and articulate with the fulcrum at their bases. Each malleus consists of two parts: manubrium and uncus. The manubrium (Latin, handle) resembles a club-shaped structure extended at one end into a cauda (Latin, tail) and flared at the other end (head). The manubrium articulates with a toothed structure called the uncus (Latin, hook). The plane of movement of the pieces comprising the malleus is at right angles to that of the rami. In some species, the trophi may be modified by reduction of the basic parts, addition of accessory structures, or by asymmetrical development of one or more of the pieces.

The trophi of rotifers have been recognized as important taxonomic features; classes, orders, families, and even species may be determined based on the details of the trophi alone (e.g., Salt *et al.* 1978). Nine different types of trophi are recognized based on the size and shape of the seven pieces and the presence of any accessory parts (Figs. 8.7–8.10); transitional and aberrant types also are known. In "malleate" trophi (Fig. 8.8A–C, E–F, P), all parts of the incus and mallei are well developed and functional, but the rami are characteristically massive and may possess teeth along the inner margin. Further, the unci have 4–7 large teeth. This form works by grasping food and grinding it before pumping the crushed material into the esophagus. Malleate trophi are present in such common rotifers as *Keratella* and *Kellicottia*. "Malleoramate" trophi (Fig. 8.8I) are found only in the order Flosculariacea and resemble the malleate form except, in the malleoramate form, the rami are strongly toothed and the unci possess many thin teeth. Similar to the malleoramate form, "ramate" trophi (Fig. 8.9) have large, semicircular shaped rami and unci with many teeth. Ramate trophi generally are considered to be limited to the Bdelloidae. Only members of the order Collothecacea (family Collothecidae) possess "uncinate" trophi (Fig. 8.8H). These trophi are characterized by unci possessing few teeth, usually with one large one and a few small ones. "Virgate" trophi (Fig. 8.8 D, K, L–O, Q, S) are modified for piercing and pumping and generally can be recognized by the long fulcrum and manubria and the presence of a powerful hypopharyngeal muscle. Some trophi of this form are asymmetrical. Virgate trophi are found in the common genera *Notommata, Polyarthra,* and *Synchaeta*. "Forcipate" trophi (Fig. 8.8J) as the name

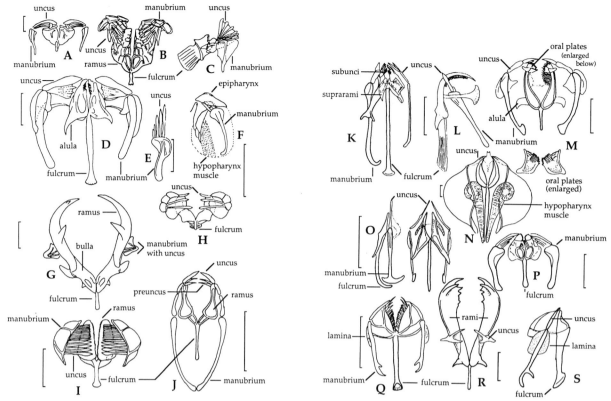

Figure 8.8 Rotifer trophi types: (A) Malleate trophi of *Epiphanes,* ventral; (B) malleate trophi (*Epiphanes*), ventral elevated; (C) malleate trophi (*Epiphanes*), lateral; (D) virgate trophi of *Notommata,* dorsal; (E) malleate trophi of *Proales;* (F) showing hypopharynx (lateral); (G) incudate trophi of *Asplanchna;* (H) uncinate trophi of *Collotheca;* (I) malleoramate of *Ptygura;* (J) forcipate trophi of *Dicranophorus;* (K) asymmetrical virgate trophi of *Trichocera rattus;* (L) virgate trophi of *Eothinia;* (M) virgate trophi of *Itura,* oral plates enlarged; (N) virgate trophi of *Synchaeta* with the powerful hypopharynx muscle; (O) virgate trophi of *Ascomorpha;* (P) malleate trophi of *Proales gigantea,* an intermediate form between malleate and virgate types; (Q) virgate trophi of *Cephalodella,* dorsal; (R) incudate trophi of *Asplanchna;* (S) virgate trophi of *Cephalodella,* lateral. Bars = 20 μm. (From Koste and Shiel 1987, with permission.)

implies, have an action like forceps whereby the trophi are projected from the mouth to grasp prey, which are then brought into the mouth and swallowed. Forcipate trophi are limited to the family Dicranophoridae. "Incudate" trophi (Fig 8.8G, R) function by grasping prey with a forceps-like action, but this form has a different morphology than the forcipate type; the rami are quite large and the mallei

very small. The mastax actually initiates prey capture by creating a suction, drawing prey into the mouth, which is then stuffed into the stomach with the aid of the trophi. Incudate trophi are limited to the family Asplanchnidae. "Cardate" trophi (Fig. 8.10) are found only in the family Lindiidae and function by producing a pumping action, without the hypopharyngeal muscle. The "fulcrate" type of trophi have been described as an aberrant form and are incompletely understood (Edmondson 1959, p. 432). This form is found only in the class Seisonidea, a very small group of marine rotifers.

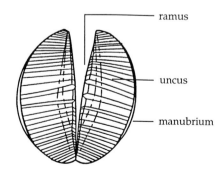

Figure 8.9 Ramate trophi of bdelloid rotifers. (Modified from several sources.)

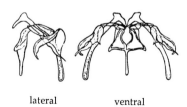

lateral   ventral

Figure 8.10 Cardate trophi of the family Lindiidae. (From Harring and Myers 1922, with permission.)

3. *Organ Systems*

Research on the ultrastructure of rotifers has proceeded rapidly during the past decade and promises to continue to do so. For a consummate overview on the topic of ultrastructural research on rotifers, consult the works of Amsellem and Ricci (1982), Clément (1977, 1980, 1987), Clément *et al.* (1983), Wurdak (1987), Wurdak and Gilbert (1976), and Wurdak *et al.* (1977, 1983). Information as reported in these papers, together with research on population genetics will eventually provide a much better picture of the evolutionary history of the entire phylum.

a. *Muscular*

The muscular system consists of small groups of longitudinal and circular muscles inserted at various points on the integument or between the integument and viscera (Fig. 8.2). In loricate species, the integument provides a firm structure against which the muscles work. Muscle contraction can increase the pressure of the pseudocoel, which then acts as a hydrostatic skeleton. In some species, muscles retract the corona, which increases pressure within the pseudocoel, thus expanding flaccid portions of the integument, known as body-wall outgrowths (*Asplanchna,* Fig. 8.11). In others, this process stiffens spines that articulate with the body in the posterolateral region (*Brachionus,* Fig. 8.12). Muscles are also present in the viscera, particularly in the mastax and stomach. Striated, longitudinal muscles are responsible for retracting the corona and foot and for moving certain articulating spines that are

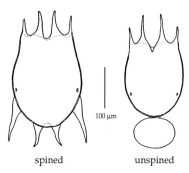

Figure 8.12 Spined and unspined *Brachionus calyciflorus.* (From Koste 1976, with permission.)

not positioned by hydrostatic pressure. Some species possess powerful muscles that control movement of certain locomotory appendages (e.g., *Hexarthra,* Fig. 8.13). Contractions of these muscles cause a swift downward sweep of the appendages which results in a rapid displacement or jump of the rotifer (see Section II.C.1).

b. *Nervous*

The nervous system is simple, consisting of only a cerebral ganglion or brain (Fig. 8.3) located dorsally on the mastax, a few other ganglia present in the mastax and foot, and three types of sensory organs: mechano-, chemo-, and photoreceptors. Mechanoreceptive bristles are situated on the corona, while several antennae are located elsewhere on the body surface, usually laterally and caudally (Fig. 8.2). Chemoreceptive pores are also present on the corona. Many species possess one or more photoreceptive eyespots, sometimes accompanied by a pigmented spot (Fig. 8.3). When present, eyespots are located in the anterior end, usually near the brain (Clément 1980, Clément *et al.* 1983). While most rotifers retain the eyespots throughout life, larvae of sessile rotifers often lose them during metamorphosis. Paired, ventral nerve cords proceed from the brain along the length of the body into the foot. Several other ganglia are usually found in the nerve cords at the exit points for lateral nerves.

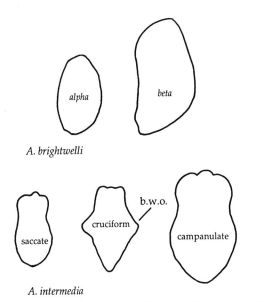

Figure 8.11 Body form variability in the genus *Asplanchna.* (b.w.o. = body-wall outgrowths. (Modified from Gilbert 1980a, with permission.)

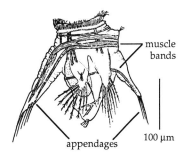

Figure 8.13 *Hexarthra,* showing positioning of muscles that initiate jumps. Lateral view. (Modified from several sources.)

One interesting structure found in the apical region of many bdelloid and monogonont rotifers is the retrocerebral organ. This structure consists of paired subcerebral glands and an unpaired retrocerebral sac, both with ducts that lead to the surface of the corona (Fig. 8.1). While the function of the retrocerebral organ is unknown, Edmondson (1959, p. 422) has noted that there are fewer protonephridia in rotifers that possess well-developed retrocerebral organs. However, Clément (1977) has suggested that it may function as an exocrine gland, perhaps lubricating the anterior part of the body. Although information on neurobiochemistry is very limited, research has shown that acetylcholine functions as a neurotransmitter (i.e., a cholinergic system) in twelve species of rotifers from six families (Nogrady and Alai 1983, Nogrady and Keshmirian 1986a, 1986b). In addition, norepinephrine neuroreceptor sites (i.e., an adrenergic system) have been reported in *Brachionus plicatilis*, and widespread catecholaminergic neuronal systems have been observed in species of *Asplanchna* and *Brachionus* (see Keshmirian and Nogrady 1988 and references therein).

c. Protonephridium

A paired protonephridial system comprised of tubules and flame cells functions in excretion and osmoregulation in all rotifers (Figs. 8.1 and 8.2). Usually there are only a small number (< 6) of flame cells, but large rotifers may possess many more. For example, *Asplanchna sieboldi* may have up to 100 flame cells (Ruttner-Kolisko 1974, p. 11). Normally, the tubules drain into a urinary bladder which leads to a cloaca, but the bladder is absent in some species and a contractile cloaca assumes its function.

d. Reproductive

In addition to their separation by anatomic details of the trophi, the three classes of rotifers are differentiated based on the anatomy of their reproductive systems. The gonads are paired in both the Seisonidea and the Bdelloidea, but in the latter class, males are completely unknown and reproduction is always asexual. Members of the third class (Monogononta) have only one gonad. Although males are present in this group, they have not been described for a large number of species. However, it is generally assumed that most (all) monogononts are capable of producing males given the proper conditions, or at least that the ancestral forms were capable of male production. When males are produced by monogonont populations, they are usually limited to a few days or a week, so that during the year most reproduction is parthenogenetic (see Section III.B.1).

The reproductive organs of female rotifers are comprised of three units, which may be seen with a light microscope: ovary, vitellarium, and follicular layer (Amsellem and Ricci 1982). The ovary of a rotifer is a small, syncytial mass that is closely associated with the yolk-producing vitellarium. At birth, the adult complement of ovocytes already has been formed in the ovary. The vitellarium is also syncytial with a constant number of nuclei, a characteristic useful in the taxonomy of some species. The follicular layer surrounds both the ovary and the vitellarium; in some species, this layer forms the oviduct, which connects with the posterior portion of the gut forming a joint exit, the cloaca (Fig. 8.1).

In general, monogonont males are much smaller (about 100 μm long) than females. In nearly all forms, the gut of males is reduced to a rudiment or is absent entirely (Fig. 8.2). In those forms with a rudimentary gut (e.g., *Asplanchna*), it serves as an energy source for the fast-swimming, nonfeeding male. The single testis is large and saccate, usually containing <50 freely floating mature sperm. A ciliated vas deferens leads from the testis to the penis, usually with one or rarely two pairs of accessory (prostate) glands that discharge into it (see Section III.B.2).

Commonly, rotifers are oviparous, that is, they release their eggs outside the body where the embryos develop. Many planktonic rotifers carry their eggs attached to the body of the mother by a thin thread (*Brachionus*), while others fix them to a substratum (*Epiphanes*) or release them into the plankton (*Notholca*). A few species retain the embryo in the body until the offspring hatches: ovoviviparous (e.g., *Asplanchna* and *Cupelopagis*).

C. Environmental Physiology

1. Locomotion

All rotifers swim during at least a portion of their life cycle and some swim continuously, never attaching even temporarily to surfaces. Swimming influences the acquisition of both food and mates and promotes the dispersal of the larvae of sessile forms. Some rotifers are unusual in that they can swim, but routinely remain in mucous tubes attached to a substratum (e.g., certain *Cephalodella* species, Dodson 1984). Others are free floating in mucous sheaths (*Ascomorpha*, Stemberger 1987).

Most rotifers swim in a helical pattern, so that the actual distance traveled is greater than the linear displacement (Starkweather 1987). However, for practical reasons, most researchers do not attempt to calculate absolute distance when considering the distance traveled.

Although the theoretical power requirements of swimming (i.e., the theoretical energy required to overcome water resistance) is < 1% of total metabolism, the actual energetic cost appears to be much greater. In *Brachionus plicatilis*, Epp and Lewis (1984) calculated this cost to be approximately 38% of the total metabolism.

Swimming speeds of male and female rotifers have been measured for *Brachionus plicatilis* by Luciani *et al.* (1983), Epp and Lewis (1984), and Snell and Garman (1986); for *Asplanchna brightwelli* by Coulon *et al.* (1983); and for two species of *Keratella* by Gilbert and Kirk (1988). These studies have shown that swimming speed is temperature dependent and varies among strains. Values recorded for *B. plicatilis* at 25°C range from 0.6–0.9 mm/sec for females and 1.3–1.5 mm/sec for males, with young and old females swimming about 30% slower than mature females. *Asplanchna brightwelli* and *Keratella* females normally swim at 0.9 and 0.5 mm/sec, respectively.

Little is known about the swimming speeds of the larvae of sessile rotifers, but Wallace (1975) has shown that swimming speeds of the larvae of *Ptygura beauchampi* varied with age. New born (0–2 hr) through mid-aged (ca., 3 hr) larvae swam about 2–2.5 mm/sec, faster than the rates reported for *Brachionus* males. However, older larvae (> 4.5 hr) swam at approximately 1.0 mm/sec. Concomitant with these changes in swimming speed was an increase in turning frequency (Wallace 1980).

Correlations of rotifer ultrastructure with turning frequency, locomotion in general, and other behaviors have been a field of systematic investigation by Clément and co-workers for the past several years (Clément 1987, Clément *et al.* 1983). Their efforts have demonstrated that rotifers are excellent models for comparative neurobehavioral studies, but much more work remains to be done before a complete synthesis of structure, function, and behavior will be possible.

Spines and other appendages influence both swimming speeds and sinking rates in rotifers. For example, Stemberger (1988) has found that unspined *Keratella testudo* generally swim faster and sink more slowly than spined forms. On the other hand, *Polyarthra* normally swims at a much slower velocity (0.24 mm/sec) than any species previously discussed, but is capable of very short bursts (about 0.065 sec long) of rapid movements called jumps. During a jump, *Polyarthra* may attain velocities greater than 50 mm/sec, with a mean velocity of 35 mm/sec, or > 100 times its normal swimming speed (Gilbert 1985a, 1987, Starkweather 1987). Jumps are produced by the movement of 12 appendages, called paddles, which articulate with the body near the head. When *Polyarthra* detects a local disturbance in the water a few body lengths away, powerful striated muscles rapidly flex some of the paddles upward and then during the jump they return to their original positions. During this process, the rotifer is displaced an average of 1.25 mm (ca. 12 body lengths). Jumping helps this rotifer escape invertebrate predators, such as *Asplanchna* (Gilbert 1980b, Gilbert and Williamson 1978) and first instar *Chaoborus* (Moore and Gilbert 1987), as well as the filtering currents of microcrustaceans such as *Daphnia* (Gilbert 1985b, 1987). They are also effective against naive workers attempting to remove *Polyarthra* from plankton samples using micropipets! The genera *Filinia* and *Hexarthra* also can jump rapidly, but movement in these forms has not been well studied (see Section III.C.4). In contrast, *Keratella* cannot jump to avoid the filtering currents of daphnids and, as a result, are severely damaged or killed when swept into the branchial chambers of cladocerans (Burns and Gilbert 1986a, 1986b, Gilbert and Stemberger 1985b). However, *Keratella* is not without some escape abilities; it is capable of increasing its swimming velocity by a factor of about 3.5 when it encounters inhalant currents of *Daphnia* or the predatory rotifer *Asplanchna* (Gilbert and Kirk 1988).

2. Physiological Ecology

Physiological tolerances of organisms prescribe environments where survival and reproduction are possible. Thus an environmental tolerance curve for a species summarizes the range of environments where reproduction occurs and culminates at the upper and lower lethal limits for the species (e.g., temperature range). Within the tolerance curve, the environmental optimum for a species is that environment where survival and reproduction are maximal. Therefore, a set of tolerance curves (including temperature, pH, etc.) indicates the breadth of adaptation of a species and its niche width. Environmental tolerances and niche widths have been characterized for few rotifer species. For example, Epp and Lewis (1980) described the response of *Brachionus plicatilis* to temperature. That study determined respiration rate over temperature ranging from 15°–32°C and recorded Q_{10}s of 1.9–2.4 for broad temperature intervals. Respiration levels off between 20°–28°C, indicating that *B. plicatilis* can maintain a constant metabolic rate over this temperature range. At higher and lower temperatures, respiration rate increased, presumably because of thermal stress which was beyond the homeostatic capability of this rotifer.

Snell (1986) showed that amictic and mictic females have similar temperature tolerance curves. Amictic *B. plicatilis* females reproduced at 20° and 40°C, whereas mictic females did not. In general, amictic females reproduced over a broader environmental range of temperature, salinity, and food level than did mictic females.

The effect of pH on the distribution and abundance of rotifers is a topic that has received a good deal of attention. However, since hydrogen ion concentration is related to other important chemical parameters in freshwaters, studies of rotifer occurrence as a function of pH alone are of limited value. Nevertheless, some extensive early work by F. Myers in the 1930s demonstrated that rotifer species can be classified into a few broad groups based on pH alone: alkaline species, acid species, and those with a broad range. More recently, Berzins and Pejler (1987) concluded that species found in oligotrophic waters had pH optima at or below neutrality (pH = 7.0) and those species common to eutrophic waters had optima at or above neutrality. Further, they noted that acid-water species were often nonplanktonic or semiplanktonic.

Much less is known of the specific metabolic responses of rotifers to pH. The influence of pH on *B. plicatilis* was described by Epp and Winston (1978). They found that swimming activity and respiration rate were not significantly different at pH values of 6.5–8.5. Snell *et al.* (1987) examined the pH from 4.0–9.9 and found swimming activity depressed below pH 5.6 and above pH 8.7. Alkaline waters depressed swimming activity more than acidic conditions.

Osmoregulation has been investigated by Epp and Winston (1977) for *B. plicatilis*, a species that is common in highly alkaline waters and salt lakes. This species was an osmoconformer capable of tolerating osmolarities exceeding 957 mosmol/liter (33.5 ppt). Ito (1960) suggested that the upper osmotic limit for *B. plicatilis* may be as high as 2860 mosmol/liter (97 ppt). Transfer of this rotifer directly from 41 to 957 mosmol/liter caused considerable mortality, but acclimation to high osmolarities was achieved by gradually increasing ionic concentrations. *Brachionus plicatilis* does not tolerate low osmolarities well, and this probably accounts for its restriction to alkaline and brackish waters.

Although most rotifers require oxygen concentrations significantly above 1.0 mg/liter, some can tolerate anaerobic or near-anaerobic conditions for short periods. Other species routinely live in oxygen-poor regions, such as the hypolimnion of eutrophic lakes or in sewage ponds. The physical and chemical factors described, along with food (Bogdan and Gilbert 1987) and predation (Williamson 1983) define the niche boundaries for rotifer species. Niche boundaries for field populations have

been described by Makarewicz and Likens (1975, 1979), Miracle (1974), and Miracle *et al.* (1987). These investigations have generally revealed that rotifer species extensively partition the environment, thus avoiding negative interactions.

3. Environmental Toxicology

Because rotifers fill an important ecological role and are relatively easy to raise in the laboratory, interest is growing in their use for aquatic toxicity testing. The response of rotifers to a variety of toxicants has been characterized in both natural and laboratory populations. For example, effects of various insecticides, herbicides, and wastewater on natural rotifer populations have been investigated (Hurlbert *et al.* 1972, Kaushik *et al.* 1985). In general, rotifers seem to serve as good indicators of environmental water quality (Sladecek 1983) and the use of rotifer population dynamics as sensitive indicators of toxicity has been promoted (Halbach 1984, Halbach *et al.* 1981, 1983). A multispecies approach to toxicity assessment within a laboratory microcosm consisting primarily of rotifers has been attempted by Jenkins and Buikema (1985). Many short-term, acute toxicity tests with a variety of substances have been conducted using rotifers, e.g., insecticides, heavy metals, free ammonia, sodium dodecyl sulfate (Snell and Persoone 1988a, 1988b) crude oil, and petrochemicals (Rogerson *et al.* 1982).

Unfortunately, the results of these studies are not directly comparable, as their methodologies varied considerably. However, testing protocols are comparable from several studies of the halophile *Brachionus plicatilis* and a number of median lethal concentrations (LC_{50}) have been reported. Capuzzo (1979a,b) found *B. plicatilis* to be very sensitive to free chlorine and chloramine. After a 30 min exposure at 25°C, LC_{50} values of 0.09 and < 0.01 mg/liter were recorded for chlorine and chloramine, respectively. *Brachionus* is more sensitive to mercury (LC_{50} of 0.045 mg/liter; Gvozdov 1986) than to chromium (LC_{50} > 500 mg/liter; Persoone *et al.* 1989). Free ammonia is also toxic, with LC_{50} values of 20.4 and 17.7 mg/liter at salinities of 15 and 30 ppt, respectively (Snell and Persoone 1988a).

Pesticides, in contrast, are not very toxic to *B. plicatilis*. The 24 hr LC_{50} values for five organophosphate pesticides and one organochlorine pesticide ranged from 0.9–150 mg/liter for three different strains of *B. plicatilis* (Serrano *et al.* 1986). These same investigators observed that pesticide resistance of this rotifer is about 1000 times greater than that reported for other aquatic organisms. Because pesticides are generally targeted to arthropods, the low sensitivity of rotifers to these compounds is not surprising.

Several median lethal concentrations for a variety of compounds have been reported for freshwater rotifers, mainly in the genus *Brachionus*. Dad and Kant Pandya (1982) recorded 24 hr LC_{50} values for *B. calyciflorus* exposed to two insecticides, and Couillard *et al.* (1989) found the following relative metal toxicities: Hg > Cu > Cd > Zn > Fe > Mn. LC_{50} values of 600 and 0.16 mg/liter for phenol and pentachlorophenol, respectively, were reported by Halbach *et al.* (1983) for *B. rubens*.

A standardized rotifer bioassay for marine and freshwater that employs test animals derived from hatching *B. plicatilis* or *B. rubens* resting eggs has been described (Snell and Persoone 1988a, 1988b). This bioassay has greater precision than *Daphnia* acute toxicity tests and is less expensive because stock cultures are not necessary to obtain test animals. The rotifer bioassay is simple, rapid, and sensitive, so the use of rotifers in aquatic toxicology is expected to expand.

4. Anhydrobiosis

The ability of rotifers to tolerate desiccation and then be revived sometime later has been known since 1702, when Leeuwenhoek observed rehydration of rotifers found in dry sediments of rain gutters (Dobell 1960, pp. 265–267). Anhydrobiosis, which is also termed cryptobiosis and osmobiosis, is limited to bdelloids. The term cryptobiosis refers to the slow or hidden metabolism, which is characteristic of these animals, while anhydrobiosis and osmo-

biosis emphasize the processes whereby loss of water is accomplished through evaporation and external osmotic pressure, respectively. While in the desiccated state, these rotifers resemble a wrinkled barrel (or tun, as in the tardigrades, Chapter 15), with the head and foot retracted into the animal's trunk (Fig. 8.14). The significance of this phenomenon is clear, as many bdelloids inhabit environments that dry completely at irregular intervals. [N.B.: For most bdelloids, a cyst is rarely produced. A few strictly aquatic bdelloids cannot withstand desiccation.]

Anhydrobiosis involves more than simple drying; unless loss of metabolic water proceeds slowly, the rotifer usually dies. During anhydrobiosis, the fine structure of cells is retained, but in a greatly modified state (Ricci 1987, Schramm and Becker 1987). Changes that occur internally include a 50% reduction in the volume of the pseudocoel, a condensation of cells and organs, and a decrease in cytoplasmic volume, so that the entire animal is only about 25–30% of its original size. Nuclei, mitochondria, endoplasmic reticula, and other organelles form a compact mass within cells of the anhydrobiotic animal (Schramm and Becker 1987).

Desiccated bdelloids have been reported to be viable even after more than two decades in the anhydrobiotic state. Recovery from anhydrobiosis may require as little as ten minutes, or it make take several hours, according to prevailing environmental conditions. Survival is negatively affected by starvation before desiccation and by moist environments and high temperatures during the desiccation period (Ricci 1987, Ricci *et al.* 1987, Schramm and Becker 1987).

III. Ecology and Evolution

A. Diversity and Distribution

1. Phenotypic Variation

Phenotypic variation is an important adaptive mechanism in rotifers, but has posed difficult problems for systematists. This variation arises by several mechanisms including cyclomorphosis, dietary- and predator-induced polymorphisms, polymorphisms in hatchlings from resting eggs, and dwarfism. Cyclomorphosis is the seasonal phenotypic change in body size, spine length, pigmentation, or ornamentation found in successive generations of zooplankton. These changes are phenotypic alterations in a single population that are related to physical, chemical, or biologic features of the environment. Each different morphological form is called a morphotype. Specifically excluded from cyclomorphotic change are seasonal successions of sibling spe-

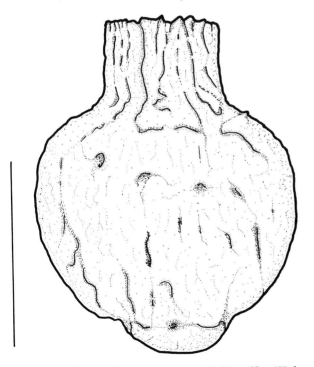

Figure 8.14 Body of a desiccated bdelloid rotifer (*Habrotrocha rosa*); the corona has been withdrawn into the trunk (bar = 50 μm.)

cies and clonal replacements of genotypes, both of which are genetic changes in populations.

A striking phenotypic change in morphology that is associated with a dietary polymorphism was described for three *Asplanchna* species (*brightwelli, intermedia, sieboldi*) by Gilbert (1980a). Diets that include the plant product α-tocopherol (vitamin E) induce saccate females, the smallest morphotype, to produce cruciform daughters. Cruciforms have lateral outgrowths of the body wall (Fig. 8.11) that protect them from cannibalism by conspecifics by making them larger and thus more difficult to ingest if captured. In the presence of α-tocopherol and certain prey types, cruciforms can produce a third morphotype called campanulates (more prevalent in *A. sieboldi* and *A. intermedia*). Campanulates are very large females (> 2000 μm), which heavily cannibalize saccate females. Female polymorphism is much less pronounced in *A. brightwelli* where there is a 50–60% increase in body size, but no campanulates are produced and body wall outgrowths are slight. Dietary polymorphism in *Asplanchna* (gigantism) may have evolved originally as a generalized growth response to larger prey typical of eutrophic waters (Gilbert 1980a, Gilbert and Stemberger 1985c). The tocopherol response probably is adaptive, because it signals the availability of nutritious rotifer and microcrustacean prey.

Another source of phenotypic variation is predator-induced polymorphisms. Spined and unspined forms had been recognized in several rotifer species for many years, but the cause(s) and significance(s) of these variations remained an enigma (Fig. 8.12). However, Gilbert (1966, 1967) was the first to show that spine production could be induced in the offspring of female *B. calyciflorus* if adults were exposed to culture medium which had previously held the predatory species *Asplanchna*. Gilbert (1967) also demonstrated that such spines were strong deterrents to predation by *Asplanchna* (see Section III.C.4). However, Stemberger (1990) has shown that food concentration can dramatically modify the development of spines in *B. calyciflorus*.

Two additional sources of phenotypic variation are polymorphisms called ''aptera generations'' in the hatchlings of resting eggs and dwarfism, both of which have been reported in rotifers of some tropical crater lakes (Green 1977). Aptera morphotypes were initially thought to be different species of *Polyarthra*, but later were shown by Nipkow (1952) to be forms lacking the paddles that are characteristic of this genus. Only the generation hatching from resting eggs lacks paddles; their parthenogenetic offspring develop into typical morphotypes. Similar polymorphisms between resting egg hatchlings and parthenogenetic generations were described for

Keratella quadrata and are suspected for *Notholca acuminata* (Amrén 1964). Dwarfism in *Brachionus caudatus* in Cameroon crater lakes was described by Green (1977) and is characterized by reduced body size and spination as compared to normal morphotypes. Green speculated that high temperature combined with reduced food supply may cause this condition.

2. Distribution and Population Movements

Water bodies are not uniform habitats with respect to concentrations of food and predators and abiotic factors such as dissolved oxygen concentration, light intensity, temperature, and water movements. Therefore, it is not surprising to find that rotifers are not evenly distributed in lakes and ponds; often there is considerable variability with respect to their horizontal and vertical distributions. For example, Hofmann (1982) showed that two *Filinia* species had very different vertical distribution patterns over the course of a year in a small lake in Germany (Fig. 8.15). Striking horizontal variability in rotifer distribution was shown by Green (1985) for a tropical lake and by Nogrady (1988) for a northern temperate one. Thus, we find that some rotifers are strictly littoral, being found in open waters only as occasional migrants, whereas others are pelagic, but the depth at which they are found is a function of season. Consequently, experienced researchers can determine to some degree where and when a water sample has been taken, merely by the composition of the rotifer species it contains. Although these major distribution patterns are the result of differential population growth and water currents and other large-scale water movements, within-lake distribution patterns can be influenced to a lesser degree by locomotory behaviors. One commonly recognized behavior of marine and freshwater zooplankton is a daily (diel) vertical migration in the water column, in which the animals usually come to the surface only during the night. During the day, zooplankton avoid visual predators (fish) that occupy near-surface waters, but when they return to the surface at night they can exploit the rich algal resources present there. In rotifers, diel migrations are never as dramatic as those typical of microcrustaceans; the population maximum usually changes only about 1–2 m over a daily cycle (Fig. 8.16). However, ovigerous (egg-bearing) and nonovigerous females may have different migration patterns and this can cause serious errors in the calculations of birth rates. Errors of nearly an order of magnitude depending on the sampling protocol may occur if the population is sampled at only one depth or at different times during the day (Magnien and Gilbert 1983).

Horizontal movement of rotifers has been investi-

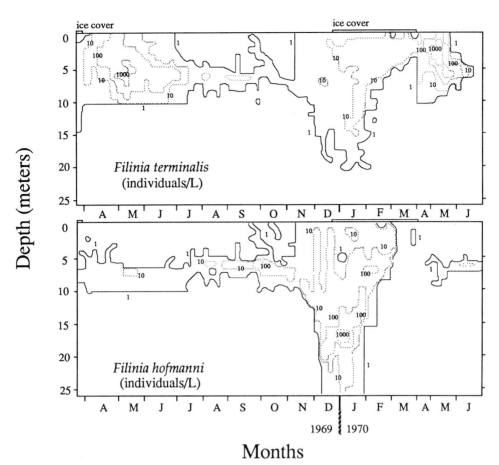

Months

Figure 8.15 Temporal and spatial distribution of two species of *Filinia* in Lake Pluβee, West Germany. Solid line indicates limits of the population at one individual per liter; dotted line indicates boundaries of higher population levels (numbers of individuals per liter.) (Modified from data in Hofmann 1982, with permission.)

gated by Preissler (1980) who showed that some pelagic rotifers apparently avoid the inshore region. Preissler demonstrated this phenomenon, known as "avoidance of the shore," by using a circular plexi-

glass arena in which he simultaneously monitored the swimming direction of several zooplankton species. When he artificially altered the shadow produced by the natural elevation of the shoreline by

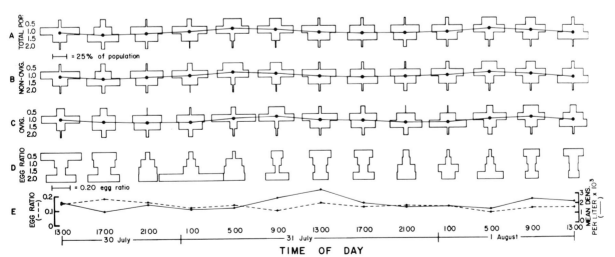

Figure 8.16 Depth distribution of *Keratella crassa* in a small, shallow lake during a 48 hr period in the summer of 1980. (A) Total population; (B) nonovigerous population; (C) ovigerous population; (D) egg ratios calculated for each sampling point; (E) egg ratio and mean density of the entire population throughout the water column. Kite diagrams express the population as relative percentages for each sampling point, while the line plots the mean depth of the entire population. (From Magnien and Gilbert 1983, with permission.)

adding a black collar around the arena, both *Asplanchna priodonta* and *Synchaeta pectinata* swam away from the shadow. The littoral rotifer *Euchlanis dilatata* apparently showed no preference based on light intensity.

3. Biogeography

Early studies of rotifer distributions were dominated by the belief that all rotifers are cosmopolitan, an idea supported by the fact that the resting eggs of monogononts and the anhydrobiotic stages of bdelloids are transported by birds. Such passive dispersal, it was argued, is very effective at ensuring that most species become globally distributed. However, data have accumulated suggesting that this conclusion was premature. Green (1972) showed a latitudinal zonation in certain planktonic rotifers and Pejler (1977b) provided data demonstrating that some species have restricted distributions. In general, endemism seems to be an important theme in several genera, including *Keratella*, *Notholca*, and *Synchaeta* (Dumont 1983).

In comparison to the substantial research effort developed elsewhere, there has been relatively little work on rotifer biogeography in the United States, and most of that work was done prior to the 1950s or is limited to specific geographic areas (e.g., the Great Lakes). Nevertheless, in North America there are several examples of species with restricted distributions in the genera *Anuraeopsis*, *Brachionus*, *Lecane*, and *Lepadella* (Koste and De Paggi 1982). Central America and the southern United States have close affinities to tropical South American fauna, while further north, affinities with Europe are apparent and endemism is strong in *Keratella*, *Notholca*, and *Synchaeta* (Chengalath and Koste 1987). The new view is that while some rotifer species are cosmopolitan, many are not. Dumont (1983) suggests that continental drift and Pleistocene glaciations best explain the current biogeography of rotifers.

Most biogeographic analyses of rotifers have been limited to collating information on the location of populations from descriptions found in the literature. This problem has an additional difficulty because many species undergo periodic polymorphism (cyclomorphosis) (see Section III.A.1), with each population having a slightly different morphology. In the past, this phenomenon confused some workers who considered each morphotype to be a new species. Furthermore, the question of what constitutes a rotifer species becomes problematic as males have never been described for many taxa. Future efforts to clarify these matters probably will be greatly aided by modern techniques in gel electrophoresis (e.g., King 1977, King and Zhao 1987, Snell and Winkler 1984) and by using mating behavior to define species boundaries (Snell and Hawkinson 1983, Snell *et al.* 1988, Snell 1989).

In the future, biogeographic research may play an important role in monitoring environmental conditions by examining the structure of rotifer communities in a variety of lakes subject to environmental perturbations: e.g., acid precipitation and heavy metal input (MacIsaac *et al.* 1987).

4. Colonial Rotifers

Most rotifers are solitary and interact only as potential prey or mates, but about 25 species (in eight genera) of the class Monogononta form permanent colonies (Wallace 1987). All colonial forms are members of two families (Flosculariidae and Conochilidae) of the order Flosculariacea: e.g., *Sinantherina* and *Conochilus*, respectively (Fig. 8.17). None of these taxa are predators and all reproduce in a fashion typical of monogononts. There appears to be a relationship between coloniality and sessile existence. About 70% of colonial forms (18 species) are sessile, but even more striking is the fact that all seven genera of the family Flosculariidae have colonial species. This has led some workers to suggest that sessile species are somehow preadapted for the evolution of coloniality (see Wallace 1987 for a review). The form of colony production has significant implications. Because colonial rotifers do not reproduce by budding or the formation of specialized zooids, colony members are not intimately connected, as are the colonial bryozoans (Chapter 14); therefore, energy resources cannot be shared among colony members. Some colonial rotifers produce tubes from either hardened secretions (e.g., *Limnias*), pellets (e.g., *Floscularia conifera*), or gelatinous secretions (e.g., *Lacinularia* and *Conochilus*). These tubes are important in the overall structure of the colony as they provide a substratum or matrix for the addition of new members to the colony (see Section III.A.5).

The number of individuals in a colony varies greatly among genera (Wallace 1987). *Floscularia ringens* usually builds colonies with fewer than five individuals, as do some members of the genus *Conochilus* (*Conochiloides*). In the latter genus, these small colonies are composed of only one adult and one to several young. Other *Conochilus* species may have up to 25 individuals within a colony. Although some species construct small colonies of fewer than 5 to 24 individuals, they sometimes produce very large colonies of 50 to more than 200 individuals

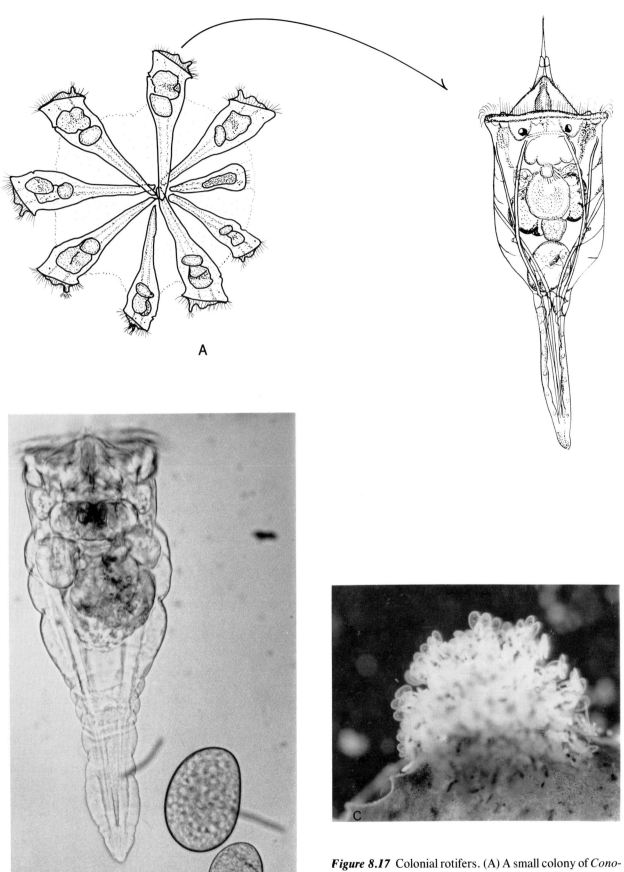

Figure 8.17 Colonial rotifers. (A) A small colony of *Conochilus* (planktonic); (B) An adult *Conochilus* with two embryos within a gelatinous matrix (planktonic); (C) *Sinantherina* (sessile). (A, Modified after several sources; solitary *C. unicornis* by E. Hollowday; others, R. L. Wallace unpublished photomicrographs.)

(e.g., *Floscularia conifera, Sinantherina socialis, Conochilus hippocrepis*). Colonies of truly gargantuan size (> 1000) have been reported in the genus *Lacinularia* (e.g., Vidrine *et al.* 1985).

Colonies are generated by one of two very different methods, known as Type I and Type II (Wallace 1987). In Type I, or allorecruitive colony formation, free-swimming larvae produce colonies by settling on tubes of conspecifics that are attached to some other substratum (e.g., *Floscularia conifera*, Edmondson 1945). Because these larvae may come from females of any colony, genotypic diversity within the colony is probably high. This type of colony is transitory, beginning when larvae attach to previously settled adults and ending when larval recruitment ceases and the adults die. However, in Type II, or autorecruitive colony formation, the young remain in the colony in which they were produced, and genotypic diversity must be low. In some of these colonies, daughter colonies separate from large parent colonies by splitting in two (e.g., *Conochilus unicornis*). Autorecruitive colonies develop continuously throughout the season, increasing in size as new individuals are added and diminishing when the colony divides. Colonies of *Sinantherina socialis* form daughter colonies in a unique way. In this species, young born within a span of a few hours (±3 hr) leave the parent colony as a planktonic aggregate called the larval colony. Members of this young colony subsequently explore and attach to a new substratum together (Wallace 1980). A few species produce interspecific colonies of two or more species (e.g., *Beauchampia crucigera, F. conifera, Ptygura crystallina*).

Two hypotheses have been developed on the adaptive significance of coloniality in rotifers. One hypothesis suggests that colonial animals possess an energetic advantage over solitary individuals of the same species; for example, colonial *F. conifera* apparently live longer and mature faster than solitary individuals (Edmondson 1945). It is argued that juxtaposition of filtering currents produced by two individuals permits an increased filtering rate and/or an enhanced filtering efficiency. Although experiments have not supported the idea that coloniality affects filtration rates, Wallace (1987) has provided some information that supports the view that coloniality increases filtering efficiency. Colonial existence also can protect individuals from certain predators. Gilbert (1980b) showed that large *Conochilus unicornis* colonies are less vulnerable to attack by the predatory rotifer *Asplanchna* because they are too large to be engulfed whole and because individual rotifers can retract into the refuge of the gelatinous matrix of the colony (see also Edmondson and Litt 1987).

5. Sessile Rotifers

Sessile species are found in two families of rotifers: Flosculariidae (seven genera) and Collothecidae (five genera). Although they are often overlooked because their habitats generally are not examined thoroughly, these forms are actually quite common in lakes and rivers in North America, occasionally reaching very high densities on plant surfaces (> 6 individuals/mm^2) (Edmondson 1944, Wallace 1980, Wallace and Edmondson 1986).

The juvenile motile stages of sessile rotifers (Fig. 8.5) are not considered to be larvae by some researchers (Edmondson 1944, Rutner-Kolisko 1974), but others have used this term and there is a conceptual parallel to other sessile invertebrates that undergo an extensive metamorphosis after settlement (Wallace 1980). Not surprisingly, the behavior of larvae changes dramatically once they come into contact with a potential attachment site. These new behaviors have been described using terms such as selection, choice, and preference. However, it should be recognized that the use of such words is not meant to imply cognition on the part of the rotifer; they are merely convenient terms to describe this phenomenon.

Several workers have demonstrated that larvae are capable of selecting a particular substratum among those surfaces available for settlement (Wallace 1980). For example, larvae of *F. conifera* settle with a greater frequency on the tubes of conspecifics than on aquatic plants, although there is substantially more plant surface available (Edmondson 1945). This propensity leads to the formation of intraspecific colonies, each with 50 or more individuals. During the growing season about 75% of the entire population may be colonial (Wallace 1977b).

Some species apparently attach to a surface based on the chemistry of the water in the vicinity of the substratum. Wallace and Edmondson (1986) have shown that greater than 90% of *Collotheca gracilipes* larvae prefer the under surface of *Elodea canadensis* leaves to the upper surface (Fig. 8.18). Choice of the under surface is apparently in response to the way *Elodea* alters the concentration of calcium ions in the water immediately around the leaf. This choice provides a superior habitat in comparison to the upper surface. In short-term laboratory experiments, young *C. gracilipes* that attached to under surfaces of *Elodea* leaves grew significantly taller and produced more eggs per female than those attached to the upper surfaces of the same leaves. Several other species of sessile rotifers exhibit strong preferences for particular substrata, but the significance of these associations has not been fully elucidated (Wallace 1977a, 1978, 1980).

Figure 8.18 *Collotheca gracilipes* colonizing the apical meristem of the aquatic macrophyte *Elodea canadensis*. Many (>25) individuals may be seen attached to the under surfaces of the leaves. The length of the lowest leaf is ~1 cm. (From Wallace and Edmondson 1986, with permission.)

Larval substrate selection behaviors have been described for three species: *Ptygura beauchampi* (Wallace 1975), *Sinantherina socialis* (Wallace 1980), and *C. gracilipes* (Wallace and Edmondson 1986). In general, larvae appear to react to potential surfaces in similar ways. Very young larvae avoid settling for periods of up to several hours after hatching. Once this refractory period is past, all surfaces are explored but some (i.e., the preferred substrata) receive much more attention and elicit different behaviors, including some reminiscent of male mating behavior (see Section III.B.2). A larva will traverse these surfaces with both its corona and foot in contact with the surface, occasionally stopping in this slightly bent position. The larva may continue exploration of the surface for several minutes, but eventually it attaches to the surface using cement from glands in the foot and then undergoes metamorphosis.

Immediately after attachment and metamorphosis have occurred, the young of most sessile rotifers begin to secrete a protective tube. Often this secretion is in the form of a clear, gelatinous material (e.g., *Collotheca* and *Stephanoceros*), but in others the tube becomes somewhat opaque by the adherence of debris and colonizing microorganisms (e.g., several *Ptygura* species and *Lacinularia*). One species of *Limnias* forms a cement tube that looks like a series of rings placed on top of one another. Perhaps the most fascinating example of tube construction is found in the genus *Floscularia*. Some species in this genus possess a small ciliated cup on

the ventral side of the head to which some of the food particles collected by the corona are shunted. The ciliated cup appears to be in constant motion, mixing gelatinous secretions with the particles to form small pellets, either in the shape of bullets (*F. conifera*) or balls (*F. ringens*). Once a pellet is fully formed, the rotifer places it on the top of the tube in a quick motion, much like a bricklayer. In this way, the tube is constantly elongated as the animal grows. Because pellets are manufactured from particles collected from the water, they may be colored a greenish tint, but they are usually brown. When a heavy rain storm temporarily suspends soil in the water the pellets that are produced by *Floscularia* usually turn out to be very dark. Thus the event of the storm is marked as a dark ring in the tubes of all the animals alive at that time. Edmondson (1945) noted this and used suspensions of powdered carmine and carbon black to mark the tubes of *F. conifera* so that he could study the dynamics of population growth and various aspects of rotifer life history in a field population.

B. Reproduction and Life History

1. Reproduction

The type of reproduction varies among the three classes of rotifers. Species in the class Seisonidea reproduce exclusively through bisexual means, with gametogenesis occurring via classical meiosis and the production of two polar bodies. At the other extreme, members of the class Bdelloidea reproduce entirely by asexual parthenogenesis, a process which includes two equational divisions producing two polar bodies. No males have been observed in bdelloids. Species in the class Monogononta exhibit cyclical parthenogenesis, where asexual reproduction predominates but sexual reproduction also occurs occasionally.

a. Cyclical Parthenogenesis

Cyclical parthenogenesis in monogononts (Fig. 8.19) takes place in the absence of males (amictic phase), but periodically males are produced and sexual reproduction takes place (mictic phase). Amictic females are diploid and produce diploid eggs (called amictic eggs) that develop mitotically via a single equational division into females (Birky and Gilbert 1971, Gilbert 1983). The term summer egg has been applied to the eggs produced by amictic females, but this term is misleading because amictic eggs can be

produced at any time of the year, depending on the species in question.

Most of the monogonont rotifer life cycle is spent in the amictic phase, but in certain conditions, sexual reproduction occurs concurrently within the population, presumably initiated by a specific environmental stimulus. Upon receiving the mictic stimulus, amictic females begin producing both mictic and amictic daughters. The proportion of mictic daughters and the duration of their production depends on the strength of the mictic stimulus. Mictic female production is only the initial step in mictic reproduction and is followed by male production, fertilization, and finally formation of a diapausing embryo called a resting egg or cyst (also winter or fertilized egg) (Fig. 8.19). Mictic females produce small haploid eggs via meiosis that, if unfertilized, develop into haploid males. Males are typically smaller, faster swimming, and shorter lived than females. Among sessile species, the male is free swimming while the female is permanently attached to a substratum.

Fertilized mictic eggs become diploid and develop into resting eggs possessing thick, often sculptured, walls (Fig. 8.20). These dormant stages are very resistant to harsh environmental conditions (Gilbert 1974) and may be dispersed over wide areas by the wind, water, or migrating animals. After a period of dormancy, which varies among species, resting eggs respond to species-specific environmental cues and hatch, releasing diploid, amictic females that enter into the asexual phase of the life cycle (Fig. 8.19). The stimuli that induce hatching may include changes in light, temperature, and salinity (Pourriot and Snell 1983).

With a few notable exceptions, the stimulus for initiating sexual reproduction is still poorly understood. Dietary α-tocopherol (vitamin E) controls the shift from amictic to mictic reproduction in most *Asplanchna* species (Gilbert 1980a), and photoperiod plays a similar regulatory role in *Notommata* (Pourriot and Clément 1981). In the genus *Brachionus*, population density has been most frequently cited as a stimulus for mictic female production (e.g., Gilbert 1963a, Pourriot and Snell 1983). In addition to environmental factors, genetic factors also play a major role in determining the sensitivity of particular strains to mictic stimuli (e.g., Snell and Hoff 1985, Lubzens *et al.* 1985).

An unusual variation of the standard monogonont life cycle has been observed in three genera: *Asplanchna*, *Conochilus*, and *Sinantherina*. In some populations, amphoteric females (individuals that produce both female and male offspring) have been recorded (e.g., Ruttner-Kolisko 1977b, King

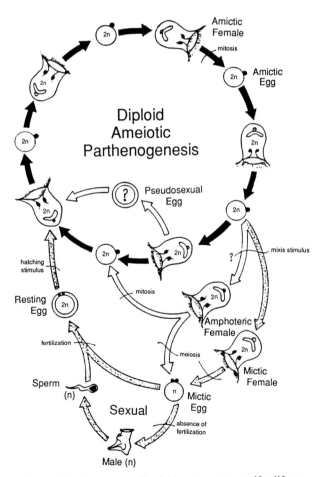

Figure 8.19 The generalized Monogononta rotifer life cycle. The life cycle of bdelloids consists of only the parthenogenetic portion. (From King and Snell 1977a, with permission.)

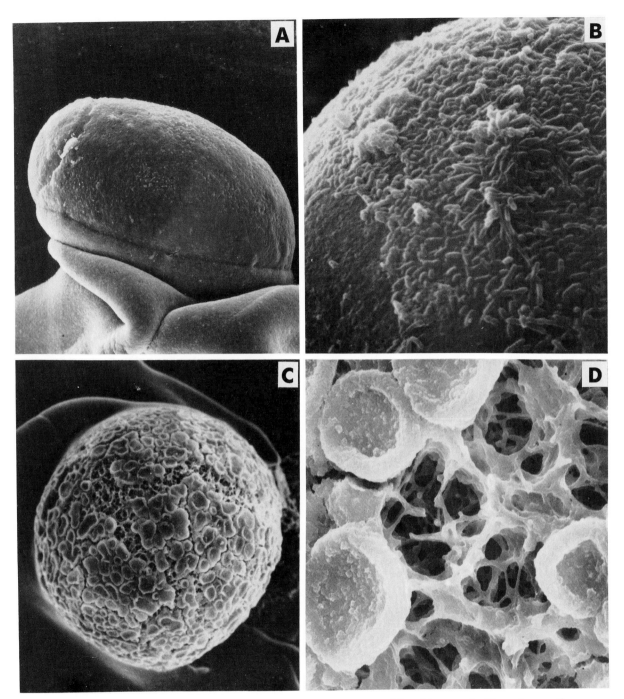

Figure 8.20 (A) Scanning electron photomicrograph of a resting egg of *Brachionus calyciflorus* (≈ μm diameter); (B) Closer view of the surface in A. Rod-shaped bacteria are seen covering the surface; (C) Resting egg of *Asplanchna sieboldi* (≈150 μm diameter); (D) Closer view of the surface in C, showing a highly fenestrate surface with several cup-shaped structures. (From Wurdak *et al.* 1977, with permission.)

and Snell 1977b). Amphoteric rotifers have the ability to produce both diploid (female) and haploid (male) eggs, a condition similar to what is found in cladocerans (see Chapter 20). Some amphoteric females produce both females and resting eggs, or males and resting eggs (Fig. 8.17). The presence of amphoteric reproduction in other rotifers has not been fully investigated, and its significance in the life history of those genera that possess it remains to be determined. There have been reports of the production of pseudosexual eggs; that is, the production of resting eggs via parthenogenesis in the absence of males. However, little is known of this phenomenon.

b. *Resting Eggs*

The density of resting eggs in sediments has not been routinely examined and the few published studies have used different methods for estimating density. Because of their capacity for extended dormancy, resting eggs theoretically could accumulate to high densities in sediments. Nipkow (1961) found resting egg densities for seven species to range from 100–600 eggs/cm^2 in lake sediments, but *Synchaeta oblonga* densities ranged from 3000–4000 eggs/cm^2. Snell *et al.* (1983) found an average of 200 resting eggs/cm^2 for *Brachionus plicatilis* in the surface sediments of a subtropical brackish pond. These workers showed that the highest densities were on the sediment surface and numbers declined exponentially to a depth of 7 cm. May (1987) examined resting egg densities in sediments from Loch Leven, Scotland, by recording the number of animals hatching from sediments incubated at temperatures of 5, 10, and 15°C. This technique is not directly comparable to the others just noted and probably yields lower estimates of resting egg density (see also May 1986). The number of resting eggs in sediments is a function of previous resting egg production, mortality in the sediments, and hatching rates. Certain abiotic features such as sedimentation rates and the uneven deposition of sediments in the benthos (sediment focusing) may affect the final density at particular sites in a lake.

2. Reproductive Behavior

Detailed descriptions of mating behaviors are available only for *Asplanchna brightwelli* (Aloia and Moretti 1973) and three species of *Brachionus* (e.g., Gilbert 1963b, Snell and Hawkinson 1983), but mating is similar in all four species. Males and females show pronounced sexual dimorphism, males being smaller and faster swimmers (Fig. 8.2). Lacking a functional foot, males swim constantly without attaching. Because males and females swim randomly, the probability of male–female encounters can be modeled mathematically (Snell and Garman 1986). Females take no active role in locating a mate or reacting to the male once an encounter occurs. Males, in contrast, display a distinct mating behavior upon encountering conspecific females.

Mating behavior begins only when the corona of the male squarely contacts the female. However, not all head-on encounters result in mating; indeed, the probability of copulation in laboratory cultures generally varies from 10–75%, depending on the strain (Snell and Hawkinson 1983). This requirement for head-on contact by the male is due to the presence of chemoreceptors in his coronal region, which apparently respond only to a species-specific

glycoprotein on the surface of the female (Snell *et al.* 1988, Snell 1989). Mating begins with the male swimming circles around the female, skimming over the surface of her lorica (Fig. 8.21). During this phase, the male maintains contact with the female with both his corona and penis (this requires the male to remain in a slightly bent position). After several seconds of circling, the male attaches his penis to the female, usually in the region of her corona, and loses coronal contact. After about 1.2 min of copulation (Snell and Hoff 1987), sperm transfer is completed and copulation is terminated when the male and female break apart and swim away. In the colonial, sessile rotifer *Sinantherina socialis*, males may copulate with several females of one colony in succession. Newborn males only have about 30 sperm and transfer 2–3 at each insemination (Snell and Childress 1987).

3. Aging and Senescence

Rotifers have long been used as models of aging and senescence. Jennings and Lynch (1928) conducted the first major study of aging in rotifers using the monogonont *Proales sordida*. They found the mean lifespan to be 8 days at 20°C, but a maternal effect

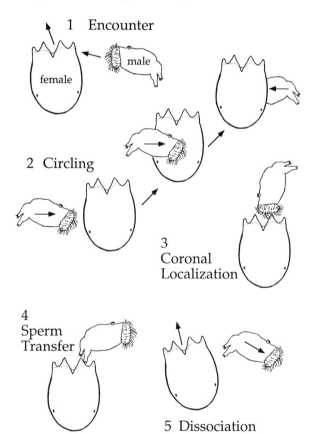

Figure 8.21 Male mating behavior in brachionid rotifers. Arrows indicate general swimming movements by male and female rotifers. (Male not drawn to scale.)

was noted. Offspring from older mothers had shorter lifespans and more variability in their developmental rates and fecundities. Maternal effects were a theme expanded upon by Lansing who, in a series of papers during the 1940s and 1950s, showed that parental age also influenced longevity in the bdelloid *Philodina citrina,* (see King 1969 for a review). Lansing established "orthoclones" by isolating the first offspring of parental females, then the first offspring of F1 females, and so forth, yielding a clone derived exclusively of firstborn females. Orthoclones derived from the offspring of older females also were developed. This protocol is a powerful tool for investigating effects on aging because the only difference among orthoclones is the age of their mothers. In *P. citrina,* mean lifespans of 5, 11, and 16 day orthoclones were 23.8, 18.1, and 16.6 days, respectively. Lansing's general conclusion was that young orthoclones have increasingly longer lifespans, while older orthoclones have progressively shorter lifespans, resulting eventually in clonal extinction. Rate of senescence, Lansing hypothesized, was controlled by a maternal effect that was transmissible and cumulative. However, the effect was reversible since firstborn females from an old orthoclone outlived their parents. Based on these and other studies, Lansing proposed the maternal factor responsible for what is now known as the "Lansing Effect" to be calcium. Rate of calcium accumulation and its importance in rotifer senescence became the reigning hypothesis of the day. Since Lansing's work, researchers have reported similar results in some species, but not in others.

King (1983) re-examined Lansing's data using life table analysis and concluded that Lansing probably did not observe an aging effect. King showed that short-lived lines derived from old parents reproduced earlier and at higher rates in succeeding generations. Long-lived lines derived from young parents delayed initial reproduction to later age classes. King argued that the Lansing Effect actually resulted from shifts in fecundity patterns that concentrated reproduction in a few age classes in short-lived lines. Snell and King (1977) demonstrated that concentrated reproduction in rotifers could shorten lifespan. Until more is known about the physiological basis for the "Lansing Effect," King suggested that it be viewed with skepticism.

The pattern of reproduction also modifies longevity (Snell and King 1977). Female *Asplanchna brightwelli* with high rates of reproduction concentrated early in life have shorter mean lifespans (2.5 days at 25°C) than females with slower reproduction distributed over a greater portion of the lifespan (mean lifespan = 4.5 days). The lengths of prereproductive and postreproductive periods were similar for short- and long-lived individuals. However, length of the reproductive period was about twice as long in long-lived individuals. These data suggest that reproduction influences individual survival negatively, but deleterious effects are minimized when the reproductive period is spread over a greater portion of the lifespan. Further, Rougier and Pourriot (1977) found that maternal age is negatively correlated with mictic female production. When *Brachionus calyciflorus* females began reproducing at age 2 days at 15°C, 80–90% of their daughters were mictic. By age 8 days, only 25% of daughters were mictic, and reproduction by 12 day old females yielded no mictic daughters. This decline in mictic female production with age was linear in 2, 6, and 10 day orthoclones.

Male lifespan has been examined far less thoroughly than that of females, but available data indicate that males live only about half as long as females (Fig. 8.22) (King and Miracle 1980, Snell 1977).

Other experiments on aging in rotifers have indicated that vitamin E (α-tocopherol) can extend mean lifespan about 10% at 23°C in the bdelloid *Philodina* sp., but maximum lifespan is unaffected (Enesco and Verdone-Smith 1980). Vitamin E, at a concentration of 20 μg/ml, also was found to extend the lifespan in *A. brightwelli* (Sawada and Enesco 1984). Of several other antioxidants tested, however, only thiazolidine-4-carboxylic acid (TCA) extended the mean lifespan significantly, presumably by quenching free-radical reactions.

Other factors that affect mean lifespan in *A. brightwelli* are photoperiod (light:dark cycle, L:D), dietary restriction, and ultraviolet (U.V.) radiation. Rotifers exposed to a 6:18 or 0:24 L:D photoperiod showed an 18% and a 22% increased lifespan, respectively, over a control photoperiod of 12:12

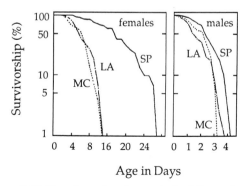

Figure 8.22 Female and male survival of three clones of *Brachionus plicatilis* at 25°C. Clone LA is from La Jolla, California; clone MC is from McKay Bay near Tampa, Florida, and clone SP is from Castellon, Spain. Note that the percentage of survivorship percent is a log scale. (From King and Miracle 1980, with permission.)

(Sawada and Enesco 1984). Dietary restriction has life-extending effects in a wide variety of animals. In *A. brightwelli*, two different modes of dietary restriction yielded longer lifespans. Reduction of food intake increased lifespan by 14.2% whereas intermittent feeding produced an 11.8% longer lifespan. Ultraviolet irradiation in the 200–4800 J/m^2 range shortened lifespan significantly and lifespan declined logarithmically as U.V. dose increased.

4. Population Dynamics

Studies of rotifer population dynamics attempt to explain the causes of changes in population size. By separating and quantifying the relative contributions of reproduction, mortality, and dispersal, ecologists are able to identify the factors regulating population size and the determinants of average abundance. Because some species are easily cultured and experimentally manipulated, rotifers have proven to be useful models for investigating the dynamics of animal populations. Techniques also exist for investigating the dynamics of natural rotifer populations, further broadening their usefulness in ecological studies.

a. Life Tables

The most rigorous approach to studying population dynamics is life table analysis. Life tables are age-specific records of all reproduction and mortality occurring in a population. From these observations, vital population statistics can be calculated such as intrinsic rate of increase (r_m), net reproductive rate (R_o), lifespan, generation time, reproductive value, and stable age distribution. The general protocol for gathering life table data on rotifers is to collect a cohort of newborn females, isolate them in small volumes of culture medium (ca. 1 ml), and make daily observations of their survival and reproduction. Maternal females usually are transferred daily to fresh medium with a specific food level and offspring are counted and removed. A table is then constructed of age, number surviving in each age class, and number of offspring produced in each age class. At least ten replicate females are grouped into each population and the probability of surviving to age x (lx) and age-specific fecundity (mx) are calculated.

The detailed observations required for life table analysis are usually only possible in laboratory populations and a number of genera have been examined in that way, including several species of *Brachionus* (Halbach 1970, King and Miracle 1980, Walz 1987, Pourriot and Rougier 1975), *Asplanchna* (Gilbert 1977, Snell and King 1977), and bdelloids (Ricci 1983). In certain cases, it is possible to perform life table analyses on field populations (i.e., *Floscularia conifera*, Edmondson 1945). Although these and other studies have examined survival and reproduction of parthenogenetic females, survivorship curves for male rotifers also are available for *Asplanchna brightwelli* (Snell 1977) and *Brachionus plicatilis* (King and Miracle 1980).

The major environmental factors affecting age-specific survival and reproduction are temperature, food quantity and quality, genetic strain, and reproductive type (either amictic, or fertilized or unfertilized mictic females). Increased temperature shortens survivorship. In *B. calyciflorus*, for example, 50% of a cohort survive to age 16 days at 15°C, 11 days at 20°C, and 5 days at 25°C (Fig. 8.23). Age-specific fecundity also is compressed at higher temperatures. At 15°C, fecundity occurs at a low rate over 20 days. In contrast, at 25°C reproduction occurs at a high rate and is compressed into 10 days. Females produce a total of 13 offspring at 15°C, 16.6 at 20°C, and 12.9 at 25°C. The intrinsic rates of increase (r_m) calculated from these lx and mx values are 0.34, 0.48, and 0.82 offspring per female per day at the respective temperatures (Halbach 1970).

The effect of food quantity on lx and mx schedules is illustrated in Figure 8.24. *Brachionus calyciflorus* was fed the green alga *Chlorella* at densities from 0.05–5 × 10^6 cells per ml at 20°C. The best survival occurred at algal concentrations of 0.5 ×

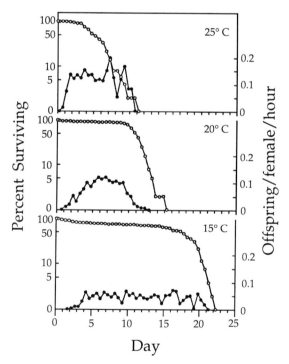

Figure 8.23 Effect of temperature on age-specific survival and fecundity of *Brachionus calyciflorus*. The left Y axis is the percentage of the cohort surviving (open circles) and the right Y axis is the offspring per female per hour (closed circles). (From Halbach 1970, with permission.)

10^6 and 1.0×10^6 cells per ml, where mean lifespan was 9 days (Halbach and Halbach-Keup 1974). Lifespan decreased at lower food levels, reaching 2.5 days at 0.05×10^6 cells per ml. Lower lifespans also were recorded at food concentrations above 1.0×10^6 cells per ml, probably the result of accumulation of algal metabolic products that are toxic to rotifers. Fecundity also peaked at 1.0×10^6 cells per ml, with a mean of 17 offspring per female. In contrast, at 0.05×10^6 cells per ml, lifetime fecundity was only 0.5 offspring per female whereas 3 offspring per female were produced at 5.0×10^6 cells per ml.

Intraspecific differences in survival among strains is illustrated in Figure 8.22. King and Miracle (1980) examined three strains of *B. plicatilis* and found that their mean lifespans at 25°C ranged from 6–13.5

days. Because these strains were acclimated to and tested in a common environment, the observed differences must be genetic. Also shown in the same figure are male survivorship curves; interstrain differences are again apparent with male lifespans.

The three different female types (amictic, and unfertilized and fertilized mictic) differ markedly in their age-specific survival and fecundity. In *Brachionus urceolaris*, for example, amictic females live at 20°C for an average of 9 days and produce 20 female offspring; unfertilized mictic females live 9.5 days and produce 25 male offspring; and fertilized mictic females live 10 days and produce 4 resting eggs (Fig. 8.25). The greatly reduced fecundity of fertilized females is typical of other rotifer species and is likely to be due to higher energy requirements for resting egg formation (Gilbert 1980a). Also reported in Figure 8.25 are development times for each egg (time from egg extrusion to hatching) and the birth intervals between eggs. Birth intervals are shortest in the middle of the reproductive period.

Individuals within a single rotifer population, differing in their growth forms (e.g., spined and unspined) can possess very different intrinsic rates of increase under identical culture conditions. Stemberger (1988) has shown that the r_m value for spined and unspined forms of *Keratella testudo* can differ significantly, but the difference depends on food concentration (Fig. 8.26). At low food levels, there is no significant difference in the r_m for the two forms of this rotifer. At higher food concentrations, however, the unspined form had much greater values of r_m. Presumably, similar phenomena will be found in

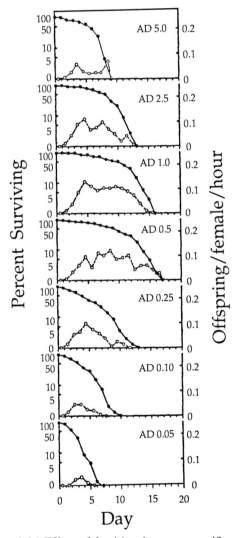

Figure 8.24 Effect of food level on age-specific survival and fecundity of *Brachionus calyciflorus*. The left *Y* axis is the percentage of the cohort surviving (closed circles) and the right *Y* axis is the offspring per female per hour (open circles). The food is *Chlorella pyrenoidosa,* provided at algal doses (AD) ranging from 0.05–5 × 10^6 per ml. (From Halbach and Halbach-Keup 1974, with permission.)

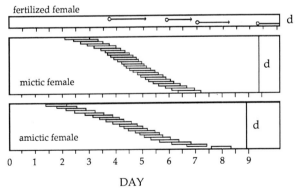

Figure 8.25 The fecundity schedule and egg developmental times of three types of *Brachionus urceolaris* females at 20°C with excess food. Bottom panel: amictic females producing diploid female eggs; middle panel: unfertilized mictic females producing haploid male eggs; top panel: fertilized females producing diploid resting eggs. Each bar represents one egg and the length of the bar indicates the developmental time for that egg. Arrows above the top panel represent the time of resting egg attachment to fertilized mictic females. (From Ruttner-Kolisko 1974, with permission.)

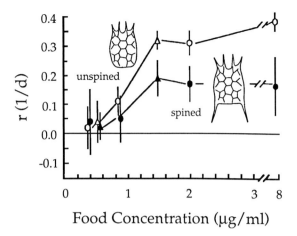

Figure 8.26 Intrinsic rates of population growth of spined and unspined *Keratella testudo*. (Modified from Stemberger 1988, with permission.)

other species in which there are spined and unspined forms (see Sections III.A.1 and III.C.4).

When life-history patterns of zooplankton are compared, rotifers have higher r_m values than either cladocerans or copepods (Allan 1976). The r_m values of these groups range from 0.2–1.6, 0.2–0.6, and 0.1–0.4 offspring per female per day, respectively. Rotifers achieve their high population growth rates by short development times, which more than compensate for their small clutch sizes. Rotifers also show the greatest response to increased temperatures. The high population growth rates of rotifers

may be an adaptive response to predation, against which most rotifer species have few defenses (Stemberger and Gilbert 1987a). As a result of these characteristics, rotifers are regarded as being able to quickly exploit new conditions.

b. *Dynamics of Field Populations*

Annual cycles of natural rotifer populations have been characterized for several species. One particularly thorough multiyear study was conducted by Herzig (1987) in Neusiedlersee, a shallow, well-mixed Austrian lake. Figure 8.27 shows 12 years of data for the abundance of 13 rotifer species. Some species, such as *Keratella quadrata,* have relatively stable population sizes and occupy the lake nearly continuously. In contrast, abundances of other species such as *Rhinoglena fertoensis, Filinia longiseta,* and *Hexarthra fennica* fluctuate much more. Fluctuation of rotifer abundance over two or three orders of magnitude during a seasonal cycle is typical of many natural populations. Techniques have been developed by Edmondson (1960) to characterize the dynamics of natural zooplankton populations. The instantaneous rate of increase of a population (r) is the difference between instantaneous birth (b) and death (d) rates: $r = b - d$ (Eq. 1). The value of r for a given population may be estimated from the equation $r = (\ln N_t = \ln N_o)/T$ (Eq. 2). This procedure requires two successive estimates of population size (N_o and N_t) separated by time interval T.

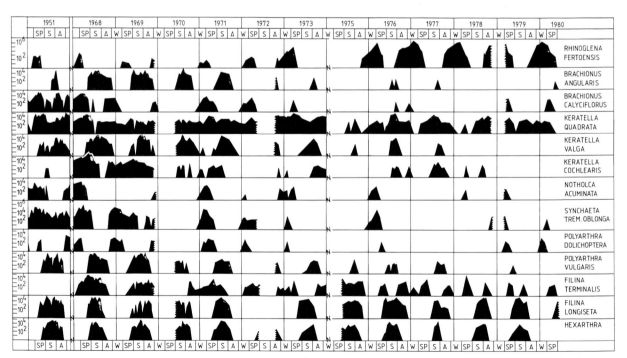

Figure 8.27 Phenologies of several important rotifer species in Neusiedlersee, Austria. The *X* axis is the season (winter, spring, summer, and autumn) and the *Y* axis is rotifer abundance per m³. (From Herzig 1987, with permission.)

The estimate of r is based on several assumptions about the population, including that it is growing exponentially with a stable age distribution. One can also estimate the birth rate in this population by counting the number of eggs carried per female. The finite hatching rate is $B = E/D$ (Eq. 3), where E is the number of eggs per female and D is the developmental rate at a specific temperature. The value E is determined directly from samples of the population and values of D are determined by observing hatching of eggs from a sample brought back to the lab. For example, Herzig (1983) reports values of D for several species. When E and D have been calculated, the instantaneous birth rate, b, can be estimated from the following equation by Paloheimo (1974): $b = \ln(E + 1)/D$ (Eq. 4). With these data, it is possible to estimate death rate, d, by subtraction: $d = b - r$ (Eq. 5).

The egg ratio technique makes it possible to predict growth of natural rotifer populations where reproduction by amictic females is parthenogenetic. A few simple measurements allow ecologists to estimate important population parameters that summarize processes of birth and death occurring in the population. This technique has proven useful for investigating the population dynamics and secondary production of zooplankton. However, before applying the egg ratio method, the works of Edmondson (1960), Paloheimo (1974), and Gabriel *et al.* (1987) should be consulted.

The dynamics of natural rotifer populations are affected by a number of environmental factors including temperature, food quantity and quality, exploitative and interference competition, predation, and parasitism. Temperature is a major factor affecting fertility, mortality, and developmental rates. In general, higher temperatures do not increase the number of offspring produced per female, but instead, shorten birth intervals by decreasing developmental time (Edmondson 1960, Pourriot 1965). Lifespan also is reduced, but the net effect of higher temperatures is increased population growth rate. In laboratory studies, higher temperatures and elevated food levels combine synergistically to increase significantly the mean rate of population increase (r) of *Brachionus calyciflorus* (Fig. 8.28). Other abiotic environmental factors, such as oxygen concentration, light intensity, and pH also influence rotifer population dynamics (see Hofmann 1977 for a review).

Food availability is a major biotic factor regulating rotifer population growth. Species have markedly different food requirements for reproduction. Threshold food concentration, the food level where population growth is zero, was determined for eight species of planktonic rotifers by Stem-

berger and Gilbert (1985). Small species like *Keratella cochlearis* had lower threshold food concentrations than larger species like *Asplanchna priodonta* or *Synchaeta pectinata* (Fig. 8.29). The logarithm of the threshold concentration was positively related to the logarithm of rotifer body mass, so the smallest species had the lowest food thresholds. The food concentration required to support 50% of the maximum population growth rate ($r_{max}/2$) varied 35-fold among the eight rotifer species and also was positively related to body mass. Small rotifer species therefore appear best adapted to food-poor environments because those species possess low food thresholds. Larger species, in contrast, are better adapted to food-rich environments, where they have higher reproductive potentials.

c. Genetic Variation

Our knowledge of rotifer genetics is in its infancy. An early worker, Hertel (1942) observed strong inbreeding depression in five strains of *Epiphanes senta* after one generation of inbreeding by self-fertilization. Net fecundity (R_o) was reduced 70–80% in the F1 generation and continued to decline slowly in the F2 and F3 generations. Birky (1967) found similar results in *Asplanchna brightwelli*, with a 25–45% reduction in the percentage of resting eggs hatching after one generation of selfing. Inbreeding increases homozygosity, thus potentially revealing deleterious recessive alleles previously masked by dominance. Inbreeding depression suggests that substantial amounts of genetic variability exist in natural rotifer populations.

Electrophoretic analysis of allozymes may be used to characterize genetic variation in natural populations. Although application of these techniques to rotifers has been very limited, allozymes have been used to distinguish strains. Snell and Winkler (1984) analyzed 5 enzymes representing 7 loci in 17

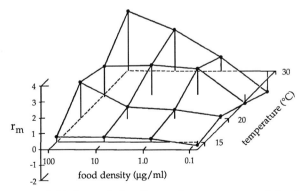

Figure 8.28 Synergistic effects of temperature and food density on mean rate of population increase (r_m) of *Brachionus calyciflorus*. (From Starkweather 1987, with permission.)

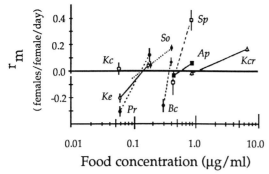

Figure 8.29 Threshold food levels where population growth rate (r_m) = 0. Food concentration is algal dry mass per ml, r_m is offspring per female per day. The solid line indicates food concentrations required to maintain a population growth of zero. Values above the solid line (positive r_m) indicate increasing population densities; values below the solid line (negative r_m) indicate decreasing population densities. Kc = *Keratella cochlearis*, Ke = *Keratella earlinae*, Kcr = *Keratella crassa*, So = *Synchaeta oblonga*, Sp = *Synchaeta pectinata*, Bc = *Brachionus calyciflorus*, Pr = *Polyarthra remata*, Ap = *Asplanchna priodonta*. (From Stemberger and Gilbert 1985, with permission.)

strains of *Brachionus plicatilis* from all over the world. This work showed no relationship between geographic proximity and genetic similarity. In the same species, Serra and Miracle (1985) found low allozyme variation, with only one or two electrophoretic patterns present in each population. These researchers suggested that this low variability resulted from intense inbreeding and interclonal competition, both of which lead to stabilization of just a few highly adapted genotypes. Although genetic variance within populations was low, Serra and Miracle found substantial variance between populations. In contrast, King and Zhao (1987) demonstrated that several electromorphs were present in each collection of *B. plicatilis* from Soda Lake in Nevada, rather than a few dominant types as reported by Serra and Miracle (1985). King and Zhao also observed no seasonal succession of genotypes and concluded that *B. plicatilis* in Soda Lake follow a pattern of incomplete genetic discontinuity. In this pattern, several genotypes persist throughout the year, but their frequencies increase or decrease in response to environmental changes (King 1972).

A variety of perspectives on the genetic structure of rotifer populations have been offered. Ruttner-Kolisko (1963) argued that most rotifer populations are highly homogeneous due to founder effects and strong selection eliminating all but a few genotypes. As a result, bisexual reproduction in rotifers is characterized by extreme inbreeding. Such a genetic structure produces fragmented gene pools where geographically separate populations also are iso-

lated genetically. Birky (1967) disagreed with this view, stating that the substantial heterozygosity revealed by inbreeding experiments suggests that rotifer populations are not genetically homogeneous. The great morphologic similarity among geographically separated populations (Pejler 1977a, Dumont 1983) further supports the idea that rotifer gene pools are better integrated than Ruttner-Kolisko proposed.

Seasonal changes in rotifer population structure have been characterized by King (1972) who proposed three models, one based on physiological adaptation and two based on genetic changes. Working with *Asplanchna girodi* in a small pond in Florida, King (1977, 1980) concluded that the complete genetic discontinuity model best described his observations. *Asplanchna girodi* actually had a succession of genetically distinct populations, each hatching from resting eggs, completing their life cycle, and returning to dormancy as the next population was emerging. Little overlap occurred between populations, making them completely discontinuous. Competition (Snell 1979) and the presence of blue-green bacteria (Snell 1980) were major ecological factors promoting succession among these rotifer populations. This population structure differs from the incomplete genetic discontinuity structure that King and Zhao (1987) reported for *B. plicatilis* in Soda Lake.

Experimental data from interpopulational crosses also illuminate the genetic structure of rotifer populations. Some strains of rotifers can mate successfully with strains from widely separated habitats, while others such as *Asplanchna brightwelli* (Birky 1967, King 1977) are reproductively isolated.

Behavioral reproductive isolating mechanisms are well developed in rotifers and can be used to define species boundaries (Snell 1989). Gilbert (1963b) investigated this phenomenon in *Brachionus calyciflorus* and found that mating behavior was strictly species-specific; *B. calyciflorus* males failed to mate with females of *B. angularis*, *B. quadridentatus*, *Synchaeta* or *Euchlanis*. Although Snell and Hawkinson (1983) reported no behavioral reproductive isolation among temporally separated populations of *B. plicatilis* from the same bay, reduced probabilities of copulation among spatially and geographically isolated populations revealed incipient behavioral isolation.

Reproductive isolating mechanisms are not always effective barriers to gene flow. Pejler (1956) argued that the morphological intermediates of *Polyarthra vulgaris* and *P. dolichoptera* he found in several Swedish tarns were the result of interspecific hybridization. Ruttner-Kolisko (1969) provided some direct evidence of interspecific hybridization when she successfully hybridized *B. urceolaris* and *B. quadridentatus*. Resting eggs were formed from this cross (in both directions), but they could be hatched only when female *B. urceolaris* were crossed with male *B. quadridentatus*. Asymmetry in hybridization is typical of closely related species and suggests that reproductive isolation between *B. urceolaris* and *B. quadridentatus* has yet to be completed.

C. Ecological Interactions

1. Foraging Behavior

Although the diets of some rotifers are highly specialized (Pourriot 1977, Gilbert and Bogdan 1981, Bogdan and Gilbert 1987), many species consume a wide variety of both plant and animal prey and may be described as generalist suspension feeders (Rothhaupt 1990a,b). For example, *Asplanchna* species generally are considered predatory, yet their main food often can be large algae. In contrast, the "herbivorous" *Brachionus* and *Ptygura* will also eat small ciliates. Thus, distinctions between predator and herbivore in rotifers often are not clear; it is more constructive to consider the ways in which rotifers encounter and process food items rather than what they eat.

Research on food selection by rotifers began by correlating the density of natural populations of rotifers to the food available (Edmondson 1965) and by determining food selectivities using laboratory cultures and/or natural populations (Salt 1987). In the latter studies, natural foods (i.e., algae, bacteria, yeast) or artificial materials (e.g., latex microspheres and powdered carmine and carbon black) are used to observe rotifer feeding behavior and to calculate feeding rates (e.g., Rothhaupt 1990a). Natural foods have been labeled using radioisotopes, but loss of label from the food can result in significant errors (Wallace and Starkweather 1985). However, even the simple procedure of adding a few drops of a food suspension to a petri dish with rotifers may lead to interesting observations concerning how rotifers process food. For example, using powered carmine, Wallace (1987) demonstrated that adult *Sinantherina socialis* do not act independently of one another when in a colony. Instead, the coronae of the animals in a large section of the colony all face in the same direction for several minutes. Animals exhibiting this behavior are called an array. Arrays determine where the dominant water currents flow to the colony, how the water currents flow around the colony, and where the currents leave the colony.

Rotifers feed in ways that are directly related to their general life history. Although most planktonic rotifers such as *Asplanchna*, *Brachionus*, *Polyarthra*, and *Rhinoglena* swim at similar rates through comparable areas of the water column, the ways in which they encounter and capture food can be quite different. *Brachionus* uses its swimming currents to sweep small food particles into the buccal region for processing; this is often termed filter or suspension feeding. *Asplanchna*, on the other hand, does not use its swimming currents to gather food. Once this rotifer contacts a potential food item with its corona, it may or may not attempt to ingest the item based on such factors as hunger level and the size and type of prey (Gilbert and Stemberger 1985b, Garreau *et al.* 1988, Salt *et al.* 1978). Some benthic rotifers creep along algal filaments and attempt to feed by piercing the filament and sucking the cytoplasm from the cells; examples are *Notommata copeus* (Clément *et al.* 1983) and *Trichocerca rattus* (Clément 1987). Sessile rotifers capture food in one of two ways, depending on the family. Members of the family Flosculariidae (e.g., *Floscularia*, *Ptygura*, *Sinantherina*) create feeding currents in a manner similar to the planktonic suspension feeders. In contrast, all collothecid rotifers are ambush raptors (e.g., *Collotheca*, *Stephanoceros*). Once a prey enters the infundibular region of the corona of these rotifers, long setae (*Collotheca*) or arms (*Stephanoceros*) fold over the prey, capturing it much like the action of a Venus flytrap. Three members of the family Collothecidae lack setae (e.g., *Cupelopagis*). In these forms, an enlarged, sometimes sheet-like corona folds over prey to capture them. In both cases, prey are pushed through the mouth of the rotifer and into the proventriculus where they are stored until the mastax transfers them into the stomach. When prey are particularly abundant, several live prey may be seen in the proventriculus.

Once food is captured, the process of ingestion begins, but not all captured food items are consumed. By observing individual *Brachionus calyciflorus* in various densities of suspended food particles, Gilbert and Starkweather (1977, 1978) showed that this rotifer regulated its ingestion of food particles by three different mechanisms. First, rotifers used pseudotrochal cirri to screen certain large particles away from the mouth. Second, particles collected by the corona may be rejected later by cilia within the buccal field. Finally, particles that gain entrance to the oral cavity may be returned to the buccal field by action of oral cilia for subsequent rejection. The mastax also may actively reject particles. Microphagus rotifers such as *Brachionus* and *Ptygura* use the mastax to push food (usually larger particles) out of the oral cavity. Members of the genera *Asplanchna* and *Asplanchnopus* also use their trophi to remove materials from their stomachs, but the materials removed by these raptors are empty carapaces of hard-bodied prey such as cladocerans. Occasionally, predatory rotifers such as *Cupelopagis* may become so engorged that attempting to ingest another prey item results in the loss, from the proventriculus, of one previously captured.

2. Functional Role in the Ecosystem

Freshwater zooplankton are dominated by three taxonomic groups (protozoans, rotifers, and microcrustaceans), yet it is the microcrustaceans that usually receive the most attention from researchers. Such a disparity is probably to be expected for two reasons, one significant and one trivial. First, microcrustaceans commonly account for a greater proportion of the total zooplankton biomass than either protozoans or rotifers. Second, microcrustaceans are easier organisms to manipulate and observe in laboratory situations. However, although rotifers usually have a smaller standing biomass than microcrustaceans, recent evidence indicates the importance of rotifers to the trophic dynamics of freshwater planktonic communities (e.g., Bogdan and Gilbert 1982).

Feeding rates of zooplankton generally are referred to as filtration or clearance rates and are measured as microliters of water cleared of a certain food type per animal per unit time (i.e., μl/animal/hr). Usually rotifer clearance rates are lower than that of cladocerans and copepods, although the rates depend heavily on food type, temperature, and animal size (Bogdan and Gilbert 1982). For most rotifers, clearance rates are commonly between 1 and 10 μl/animal/hr, whether determined in the laboratory or in the field. However, a few species can achieve levels exceeding 50 μl/animal/hr (e.g., Bogdan and Gilbert 1987, Bogdan *et al.* 1980, Wallace and Starkweather 1985).

Using estimates of clearance rates, Starkweather (1987) noted that even moderately sized rotifers with body volumes of about 10^{-3} μl process enormous amounts of water with respect to their size: $> 10^3$ times their own body volume each hour! Ingestion rates (biomass consumed per animal per unit time) are also very high for rotifers. An adult rotifer may consume food resources equal to ten times its own dry weight per day. Therefore, by having assimilation efficiences (i.e., assimilation divided by ingestion) of between 20 and 80%, rotifers convert a good deal of their food to animal biomass, which may be passed on to the next trophic level (Starkweather 1980, 1987).

Although microcrustaceans generally have higher clearance rates than rotifers (about 10–150 μl/animal/hr for cladocerans and 100–800 μl/animal/hr for copepods), rotifers can exert greater grazing pressure on phytoplankton than some small cladocerans. Bogdan and Gilbert (1982) report that in a small eutrophic lake, *Keratella cochlearis* accounted for about 80% of the community grazing pressure on small algae during the year. These two workers also showed that *Keratella cochlearis* had filtering rates about 5–13 times higher than the cladoceran *Bosmina longirostris* (Gilbert and Bogdan 1984). Therefore, under certain conditions, rotifers may be important competitors to small, filter-feeding microcrustaceans and are important in nutrient recycling in aquatic systems (Makarewicz and Likens 1979). Furthermore, rotifers can alter the species composition of algae in certain systems. Schlüter *et al.* (1987) showed that intense feeding by *Brachionus rubens* can cause a shift in the dominant algal species from *Scenedesmus* to a spined algae, *Micractinium*. Apparently this shift is based on the inability of *B. rubens* to consume algae with protective spines.

Having the ability to reproduce rapidly, rotifers may account for 50% or more of the zooplankton production, depending on the prevailing conditions (Herzig 1987). This production, in turn, can be an important food source for other rotifers, *Asplanchna* (Gilbert and Stemberger 1985b), cyclopoid and calanoid copepods (Williamson 1983), malacostracans (Mysis, Threlkeld *et al.* 1980), insect larvae (*Chaoborus*, Moore 1988, Moore and Gilbert 1987), and fish (Hewitt and George 1987).

Abundance and species composition of rotifers often reflect the trophic status of lakes (Mäemets 1987, Nogrady 1988). For example, Hillbricht-Ilkowska (1983), Walz *et al.* (1987), and others have reported changes in the maximal total population density of several orders of magnitude when lakes were subject to intense eutrophication. Individual species sometimes undergo dramatic population changes during those periods. Walz *et al.* (1987) showed that the density of *Asplanchna* in Lake Constance increased its maximum population levels 280-fold over a period of 28 yr. Population declines also have been seen. Edmondson and Litt (1982) indicated that *Keratella cochlearis* was abundant during the years when Lake Washington had elevated concentrations of dissolved phosphorus, low water transparency, and high algal densities. However, as these water quality parameters improved, the population of this rotifer declined dramatically. Overall, there was at least a 20-fold increase and then decline during a period of 15 yr.

Studies of the interactions of rotifers with other organisms will probably continue to receive attention in the future, especially predator–prey interactions (Gilbert 1966, 1967, Salt 1987, Williamson 1983), interference competition between other herbivorous zooplankton and rotifers (Burns and Gilbert 1986a, 1986b, Gilbert and Stemberger 1985a), the toxic effects of cyanobacteria (Snell 1980, Starkweather and Kellar 1987), and mate recognition (Snell *et al.* 1988). One under-utilized tool for the study of energetics and trophic interactions is the chemostat, which, unlike batch cultures, maintains constant experimental conditions (Borass 1980, Scott 1988; see also Starkweather 1987 for a review).

3. Competition with Other Zooplankton

Rotifers, cladocerans, and copepods often compete for limited food resources and, in general, rotifers are relatively poor exploitative competitors because their clearance rates are usually many times lower than those of daphnids (see Section III.C.2). Rotifers also have a more limited size range of particles that they can ingest compared to cladocerans. Most rotifers eat algal cells in the 4–17 μm range and are much less efficient at ingesting smaller or larger cells (Gilbert 1985b). In contrast, most cladocerans eat algal and bacterial cells in the 1–17 μm range, and some can ingest much larger cells (Hall *et al.* 1976, Bogdan and Gilbert 1982). Thus, cladocerans generally have broader food niches than rotifers in terms of food type and size (Bogdan and Gilbert 1987).

Exploitative competition between rotifers and daphnids is readily demonstrated when they are grown in single and mixed cultures. *Brachionus calyciflorus* and *Daphnia pulex* both grow well on the alga *Nannochloris oculata* in single-species cultures. When both species are present, however, *Daphnia* removes an increasingly larger proportion of algal cells until the rotifers gradually starve to extinction after 2–3 weeks (Fig. 8.30). *Daphnia* is unaffected by the presence of rotifers.

Population growth of certain rotifers also is inhibited by *Daphnia* through interference competition (Burns and Gilbert 1986a,b, Gilbert and Stemberger 1985b). When in the presence of any one of four *Daphnia* species, *Keratella cochlearis* suffers mechanical damage (i.e., killed, wounded, or loss of eggs) when swept into the branchial chamber of the daphnids. No species differences were detected among the *Daphnia* used, but the rate at which rotifers were killed was directly proportional to daphnid body length. *Keratella* also was found in the guts of some *Daphnia*, documenting a new pathway for trophic interactions. *Daphnia* have a major impact on *Keratella* populations when the cladocerans are larger than 2 mm and are present in densities of 1–5 individuals per liter (cf. May and Jones 1989). Simi-

lar effects have been reported for short-term bottle experiments involving the cladoceran *Scapholeberis kingi* and the rotifer *Synchaeta oblongata* (Gilbert 1989b). Of course, interference and exploitative competition may occur simultaneously (Gilbert 1985b, 1988).

Predation is another important regulatory factor in rotifer population dynamics, as rotifers are prey for several aquatic predators including other rotifers, insect larvae, cladocerans, copepods, and planktivorous fish. Predation affects rotifer population dynamics both directly (by contributing to mortality) and indirectly as a selective force shaping rotifer morphology, physiology, and behavior (see Section III.C.4).

4. Predator–Prey Interactions

Rotifers possess several types of defensive mechanisms that have apparently evolved in response to predation. These mechanisms include morphological features, escape movements, and other ways of avoiding potential predators (see Stemberger and Gilbert 1987a, Williamson 1983, and references therein).

Most rotifers are transparent and quite small, some comparable in size to protozoans (ca. 60–250 μm long). While these features benefit limno-

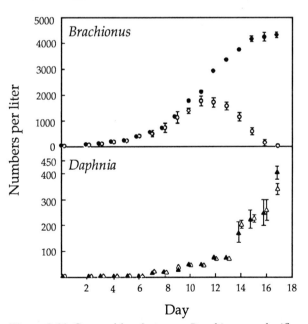

Figure 8.30 Competition between *Brachionus calyciflorus* and *Daphnia pulex*. *Brachionus* and *Daphnia* were grown in single-species (closed symbols) and mixed-species (open symbols) batch cultures at 20°C, daily renewed with 5×10^6 *Nannochloris* cells per ml. Population size (Y axis) is the number of individuals in the 80 ml culture. Error bars (± 1 SE) are visible when they exceed the size of the symbols. (From Gilbert 1985b, with permission.)

planktonic rotifers by reducing their visibility to fish, a small body size renders rotifers more vulnerable to invertebrate predators, which feed by touch rather than by sight. Many rotifers produce a thickened integument (lorica) and/or spines and other projections, or carry their eggs, all of which have been shown to reduce the ability of predatory zooplankton to prey upon them (e.g., Gilbert 1966, 1967, 1980b, Stemberger and Gilbert 1984b). Spines are produced in some rotifers (e.g., *Brachionus calyciflorus*, Figs. 8.12 and 8.31; *Keratella cochlearis*; *Keratella testudo,* Fig. 8.26) in response to a build up of soluble substances released by invertebrate predators such as *Asplanchna* and copepods (Gilbert 1967, Stemberger and Gilbert 1984b, Stemberger 1984). We now know that exposure to a water-soluble factor secreted by the predatory rotifer *Asplanchna* or the copepods *Epischura, Mesocyclops,* and *Tropocyclops* induces *de novo* spine formation or spine lengthening in several planktonic rotifers (Stemberger and Gilbert 1987a, 1987b) (see Section III.A.1). Rotifers for which such polymorphisms have been observed include *Brachionus calyciforus, B. bidentata, B. urceolaris, Filinia longiseta passa, Keratella cochlearis, K. slacki,* and *K. testudo.* The adaptive significance of spined morphotypes is a significant reduction in capture and ingestion by invertebrate predators by making the rotifer more difficult to manipulate and swallow. The presence of spines on *B. calyciflorus* is a good example of the phenomenon (Fig. 8.12). The way that this works is as follows. When disturbed by a potential predator, *B. calyciflorus* will retract its corona and by doing so, increases the hydrostatic pressure within its pseudocoelom. The elevated pressure causes the posterolateral spines to stiffen (Fig. 8.31). Thus, the apparent body volume is larger. For *Asplanchna,* this increase in the size of its prey is sufficient to prevent ingestion after capture. After a period of time during which *Asplanchna* attempts to swallow *B. calyciflorus* (usually >60 sec), the predator will release (reject) the prey, which then swims away unharmed. Thus, the invertebrate predator releases a biochemical cue (an allelochemic) that initiates a developmental change in subsequent generations of the rotifer (spine production), reducing the effect of predation. Some forms of *K. cochlearis* also possess posterior spines, which make them four times more likely to be rejected after capture by *A. girodi* than unspined forms (Stemberger and Gilbert 1984b). However, occasionally spined forms are swallowed by *Asplanchna* or a similar species (*Asplanchnopus*) and the spines become lodged in the pharynx or esophagus of the predator. Presumably, both predator and prey die when that happens. Most rejected prey swim away unharmed.

Spine production has additional consequences beyond predatory avoidance. Stemberger (1988) has demonstrated that in the absence of predators, reproductive effort (survivorship and fecundity) and hydrodynamic characteristics (sinking rates and swimming speeds) vary in spined and unspined *Keratella testudo.* These relationships were further complicated by food concentration (see Section III.B.4.a and Fig. 8.26).

Transparent mucous sheaths produced by some planktonic rotifers (e.g., *Conochilus* and *Lacinularia* species) appear to function in a similar way to spines. The sheaths make the rotifers effectively larger and therefore less vulnerable to invertebrate predators (e.g., *Asplanchna* and predatory copepods) without making them more visible to fish (Gilbert 1980b, Stemberger and Gilbert 1987a, 1987b, Wallace 1987).

Three rotifer genera can often escape predators by making rapid jumps using a variety of appendages: long spines (*Filinia,* Fig. 8.32), setous arm-like appendages (*Hexarthra,* Fig. 8.13), and paddles (*Polyarthra,* Fig. 8.33) (see Section II.C.1). Many rotifers assume a passive posture, displaying what has become known as the "dead-man response," rather than fleeing when predators attack (e.g., *Asplanchna, Brachionus, Keratella, Sinantherina,* and *Synchaeta*). This simple behavior is merely a retraction of the corona into the body and passive sinking. Contraction of the corona stops the animal from swimming, thus eliminating the vibrations it produces in the water that may be detected by the predator. In addition, this behavior may make the rotifer more difficult to grasp in its turgid state. In *Sinantherina spinosa,* this passive posture exposes a group of small spines on its ventral body surface, which may function in defense against planktivorous fish (Wallace 1987).

Defenses against predators like spines, mucous

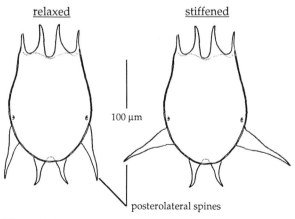

Figure 8.31 Posterolateral spines in *Brachionus calyciflorus.* (Modified from Koste 1976, with permission.)

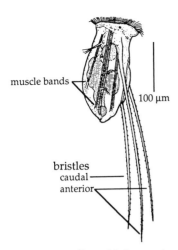

Figure 8.32 *Filinia*, a rotifer with long spines, which are used in making rapid jumps to escape predators. Lateral view. (Modified from several sources.)

sheaths, thickened loricas, and escape movements are energetically demanding. Some species like *Synchaeta pectinata* are not well defended against predators, but have evolved very high maximal population growth rates that offset mortality from predation. Avoiding potential predators in space and/or in time is another simple yet effective defense mechanism against predators. Some rotifers occupy the habitat at a different time of year than an important predator, migrate vertically or horizontally in the habitat, or live in zones with low oxygen concentration, thereby missing predatory pressures altogether.

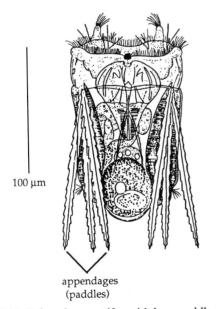

Figure 8.33 *Polyarthra* a rotifer with long paddle-like appendages, which are used in making rapid jumps to escape predators. (Modified from several sources.)

5. *Parasitism on Rotifers*

The importance of parasites in controlling population density in rotifers has not been examined thoroughly, although a few studies have correlated parasitic infection with a decrease in population density of planktonic species (Ruttner-Kolisko 1977a). In certain cases, it appears that parasites can cause the demise of an entire population within a few days (Edmondson 1965).

The sporozoan parasite, *Microsporidium* (*Plistophora aerospora*) frequently infects planktonic rotifers possessing thin loricas (Ruttner-Kolisko 1977a). For example, members of the genera *Asplanchna*, *Brachionus*, *Conochilus*, *Epiphanes*, *Polyarthra*, and *Synchaeta* have been reported to be parasitized. Apparently water temperature is an important mediating factor in the spread of the parasite, as infection rate drops off at water temperatures below 20°C. In infected rotifers, the pseudocoel of the animal becomes nearly filled with sausage-shaped cysts of *Microsporidium* (Fig. 8.34).

Several workers have described endoparasitic fungi that attack soil rotifers of the genera *Adineta* and *Philodina*. These reports describe three avenues of parasitic attack: adhesion, ingestion, and injection. Tzean and Barron (1983) have described one fungus that forms peglike, adhesive appendages on both small conidia (spores, ca. 30 μm long) and long vegetative hyphae. Once the adhesive pegs attach to a rotifer, they germinate and rapidly colonize the pseudocoel (Fig. 8.35A). At least three different genera of fungi produce spores (Fig. 8.35B) that initiate a parasitic attack when ingested (e.g., Barron 1980a, Barron and Tzean 1981). A third avenue of infection occurs via hypodermic injection of a vegetative cell into the rotifer (Barron 1980b, Robb and Barron 1982). Once inside the host, these fungal cells grow into assimilative hyphae, producing more infective cells either inside or outside the rotifer (Fig. 8.35C).

6. *Aquaculture*

In freshwater communities, rotifers serve as food for several invertebrate predators, which in turn are consumed by many economically important fish. They also are used directly as food by many planktivorous adult fish (Evans 1986, O'Brien 1979, 1987). Perhaps the most important ecological link between rotifers and fish is the consumption of rotifers by larval fish. At an early age, most fish larvae feed on microzooplankton, rotifers being one of the most important components of the diet (Torrans 1986). In general, rotifers are highly nutritious and their biochemical composition can be further improved by specialized diets (Watanabe *et al.* 1983). Many fish

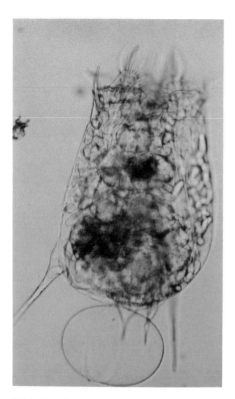

Figure 8.34 *Brachionus* infected with *Microsporidium* in pseudocoel. (R. L. Wallace original photomicrograph.)

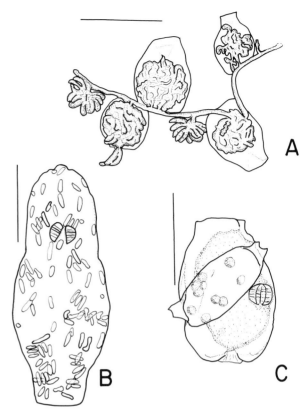

Figure 8.35 Fungal parasites of rotifers. (A) Four rotifers trapped by adhesive pegs on the vegetative hyphae of *Cephaliophora*; (B) Unicellular, assimilative hyphae of *Triaculus*; (C) Zoosporangium of *Haptoglossa* with two evacuation tubes and ten encysted zoospores.

are well adapted to locate, pursue, capture, and ingest rotifers, and rotifers are easy prey because they swim slowly and frequently lack predator defenses (Stemberger and Gilbert 1987a).

Aquaculturists have exploited this important relationship between planktivorous fish and their rotifer prey in intensive aquaculture systems. This field has developed into a major technical discipline in several countries, most notably Israel, Japan, and Kuwait. Most of this work has the practical goal of determining the correct biotic and abiotic factors necessary to maintain mass cultures of rotifers (usually *Brachionus plicatilis*) as food organisms for larval shrimp and fish (Lubzens 1987, Lubzens *et al.* 1989).

Most systems for mass culturing of rotifers are simple batch cultures capable of producing kilogram quantities of rotifer biomass each day (Lubzens 1987). A substantial amount of effort has been devoted to identifying optimal culture conditions for the most commonly grown species (*Brachionus rubens* and *B. calyciflorus* in freshwater and *B. plicatilis* in saltwater). For mass cultures of *B. rubens* raised at 20°C, the optimal pH is 6.0–8.0, with upper and lower lethal limits of 9.5 and 4.5, respectively (Schlüter 1980). Oxygen levels as low as 1.15 mg/liter sustained good growth, but 0.72 mg oxygen per liter was strongly limiting (Schlüter and Groeneweg 1981). Schlüter (1980) also tested eight

algal species and found that *Scenedesmus costatogranulatus*, *Kirchneriella contorta*, and *Chlorella fusca* produced the best rotifer growth. Algal concentrations of 70 mg/liter dry weight of *Scenedesmus* were optimal. *Brachionus rubens* reproduction in mass cultures is inhibited by ammonia levels of 3–5 mg NH_3-N/liter (Schlüter and Groeneweg 1985). Because about 50% of the nitrogenous waste excreted by *Brachionus* is ammonia, waste product removal is a serious consideration in mass culture design (Hirata and Nagata 1982).

Although rotifers have been important in marine fish aquaculture for many years (Lubzens 1987), freshwater rotifers are just beginning to be used. Nauplii of brine shrimp (*Artemia*) have been used extensively as larval feed, but these marine organisms do not live more than about 6 hr in freshwater, and their decomposition contributes to the organic load in culture tanks. A freshwater organism is preferred for freshwater aquaculture, and rotifers fill this role nicely. Rotifers also are being used as food for young fish larvae too small to begin feeding on *Artemia*. For these reasons, the use of rotifers in freshwater aquaculture is expected to grow.

D. Evolutionary Relationships

The phylogenetic position of all eight pseudocoelomate phyla represents a problem that continues to generate debate among workers (Lorenzen 1985). Within this larger question, several opinions have been expressed concerning the evolution of rotifers, including some that propose a polyphyletic origin of the phylum. One theory suggests that rotifers evolved from an ancestor possessing a fluid-filled body cavity (Lorenzen 1985), perhaps a regressed coelom (Remane 1963), while another proposes that rotifers were derived from an ancestral acoel turbellarian (phylum Platyhelminthes, Chapter 6) (Hyman 1951). We believe that the evidence favors the view that rotifers and platyhelminthes are related; these phyla share several common features including protonephridia, a ciliated integument possessing mucous glands, and a cerebral eyespot with a pigment cup. In addition, there is no anatomic evidence to support the view that the body cavity is a regressed coelom (Clément 1985).

Associated with the issue of the origin of the Rotifera is the question of its relationship to another pseudocoelomate phylum, Acanthocephala. The presence of the intracytoplasmic lamina within a syncytial epidermis in both phyla suggests that they are closely related (Clément 1985), even though acanthocephalans are 5–1500 times larger in size than the largest rotifer. Lorenzen (1985) argues that structures identical to the rostrum and lemnisci of acanthocephalans are found in bdelloids, thus making the two monophyletic. Similarities in the ultrastructure of certain muscles that traverse the pseudocoelom of the parasitic rotifer *Embata parasitica* (a bdelloid) and an acanthocephalan of the genus *Polymorphus* (Whitfield 1971) lend support to Lorenzen's view. However, until additional ultrastructural evidence is forthcoming, we believe that these features should not be considered homologous. For now it is probably best to say that these two taxa separated from a common ancestor early in their evolutionary history (Fig. 8.36).

Evolution of the major groups within the phylum Rotifera is less controversial. Using information about eight characters (presence or absence of males, prostate glands, resting eggs, spermatophores, and vitellarium, ability to withstand desiccation, number of gonads, and type of trophi) Wallace and Colburn (1989) analyzed rotifer evolution at the ordinal level and illustrated one of several equally parsimonious cladograms they uncovered (Fig. 8.36). Although other evolutionary trees have been proposed, none have use cladistics for their analysis and they either are (1) very similar to this one (e.g., Epp and Lewis 1979) or (2) require additional evolu-

tionary steps that seem unwarranted (see Wallace and Colburn 1989 for a review).

In constructing their cladogram, Wallace and Colburn (1989) did not use all available information about rotifer morphology. Although information about certain ultrastructural features—degree of structural reduction of males within class Monogononta and presence or absence of sessile and colonial species—was omitted, inclusion of this information does not significantly alter their results. However, they argue that structural differences in members of the orders Collothecacea and Flosculariacea, particularly in the trophi, indicate that the sessile condition evolved independently in each, and that subsequent evolution of some species in both orders led to planktonic existence once again (e.g., *Collotheca mutabilis*, *Ptygura libera*, respectively). Further refinement of the evolutionary tree of rotifers will not be possible until additional ultrastruc-

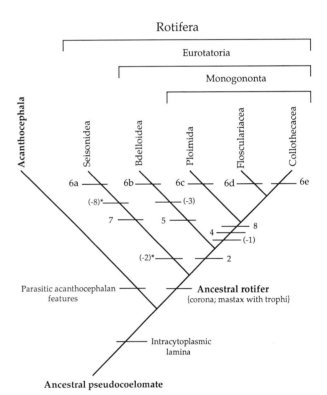

Figure 8.36 Phylogenetic relationships within phylum Rotifera at the ordinal level. Included here is the putative relationship of phylum Acanthocephala to the rotifers. Numbers refer to changes in character state as gains or losses (−). Character states are numbered as follows: 1, paired gonads; 2, vitellarium; 3, male; 4, resting egg formation; 5, desiccation; 6, trophi types, a, aberrant, b, ramate, c, others, d, malleoramate, e, uncinate; 7, spermatophores; 8, prostate glands. Numbers with asterisks (*) indicate alternative positioning of evolutionary events when the status of the ancestral character state has not been ascertained. (Modified from Wallace and Colburn 1989, with permission.)

tural data become available (e.g., Clément 1980, 1985).

IV. COLLECTING, REARING, AND PREPARATION FOR IDENTIFICATION

A. Collections

Collecting rotifers does not require complex or expensive equipment. One can almost always collect several species of planktonic rotifers by towing a fine–mesh (25-50 μm) net through any body of water. [N.B.: Nets with greater mesh sizes tend to miss small-bodied forms.] Productive lakes and ponds usually provide especially good sampling sites. More elaborate equipment such as closing nets and Clarke-Bumpus samplers work well, but are not necessary unless required by a specific experimental protocol. Water collected by discrete sampling devices (e.g., Van Dorn sampler, Kemmerer bottle, plankton trap, and pump) is then filtered through a net of an appropriate size. In weedy areas, a dip net or flexible collecting tube (Pennak 1962) are very useful. Another simple method to collect rotifers is to submerge a 3–4 liter (1 gallon) glass jar in a weedy region and to arrange loosely a few aquatic plants in it. If using this technique, fill the jar to about 2–5 cm from the top and place it near a subdued light source such as a north window. Rotifers that swim to the surface on the lighted side may be removed using a transfer pipette.

Certain aquatic plants such as *Elodea*, *Myriophyllum*, *Utricularia*, and filamentous algae are good substrata to examine for the presence of sessile rotifers (see Wallace 1980 for a review). Plants with highly dissected leaves may be examined in small dishes using a dissecting microscope. Broad-leaved plants must be cut into strips and examined on edge.

The upper few centimeters of moist sand taken just above the water line along a lake shore usually provide several species of rotifers. Unfortunately, very few studies have been conducted on rotifers from this habitat, termed the "psammon" (e.g., Myers 1936, Neel 1948, Pennak 1940, Tzschaschel 1980), in part because of the difficulty in separating the organisms from the sand.

Sediments collected from the bottom using a dredge, coring apparatus, or suction device usually provide several bdelloid and monogonont species. The upper few centimeters of sediments from a core collected in the winter or early spring normally contain resting eggs, which can be induced to hatch within a few days when several milliliters of sediment are incubated at ambient spring or summer temperatures (May 1986, 1987).

Do not overlook laboratory aquaria or holding tanks as potential sources of material. We have found some unusual species in aquaria that had remained almost unattended for months. One might try adding a small amount of sediment from several sites as a way of adding variety to the rotifer community within the aquarium. Sessile rotifers may be present if the aquarium contains aquatic plants recently collected from the field. However, if you attempt to keep sessile forms, be sure first to remove all snails from the aquarium.

B. Culture

King and Snell (1977b), Pourriot (1977), Ricci (1984), and Stemberger (1981) have summarized the general procedures for culturing rotifers. Some species of rotifers may be cultured in the laboratory quite easily, obtaining densities of >100,000 individuals per liter in a few weeks. Others seem impossible to keep even for short periods. Several species are commonly raised under a variety of conditions (Table 8.1). Most of these need regular care a few times per week (changing the medium, feeding, etc.). However, some species require little care. For example, *Habrotrocha rosa* is easily cultured in dilute broths of decaying, powdered baby food and may be nearly ignored for weeks at a time without extinction of the culture (Bateman 1987). Techniques used to culture protozoans (e.g., making extracts and infusions of various grains, manure, soil, etc.) have been adopted for the culture of some rotifer species with great success. Most rotifer cultures can be maintained xenically (i.e., containing several contaminating organisms) without much problem. In fact, sometimes rotifers are found contaminating cultures of protozoans, microcrustaceans, etc., that have been provided by commercial biological supply companies. Rotifers have been maintained under axenic (Dougherty 1963) or monoxenic culture (Gilbert 1970) conditions. Sophisticated culture techniques using single- and two-stage chemostats have been described by several workers (see Starkweather 1987 for a review).

C. Preparation for Identification

Three preservatives are commonly used to preserve rotifers: formalin and Lugol's iodine at concentrations of 5% or less, and ethanol at about 30–50%. Lugol's has two advantages over formalin; it is less toxic to people and stains the specimens slightly, which makes the animals more visible during sorting procedures.

Unless special precautions are taken before fixation, illoricate rotifers (especially bdelloids) will

Table 8.1 Examples of Rotifers That Have Been Cultured[a]

Species	Culture Conditions	Reference
Bdelloids		
Habrotrocha rosa	Batch cultures of simple tap water + powdered baby food, dog food, etc.	Bateman (1987), Schramm and Becker (1987), R. L. Wallace (personal observation)
Philodina acuticornis	Monoxenic culture	Meadow and Barrows (1971)
Several bdelloids	Batch cultures	Ricci (1984), Pourriot (1958)
Monogononts		
Asplanchna various species	Batch cultures	Aloia and Moretti (1973), Gilbert (1967), Stemberger (1981), Stemberger and Gilbert (1984a)
Brachionus angularis	Continuous culture	Walz (1983a)
B. calyciflorus	Batch culture	Gilbert (1963a)
	Monoxenic cultures	Gilbert (1970)
	Chemostat culture	Boraas (1980)
B. patulus	Batch culture	Rao and Sarma (1986)
B. plicatilis	Batch culture	Hirayama and Ogawa (1972)
	Mass culture	Lubzens (1987)
	Chemostat culture	Droop and Scott (1978)
B. rubens	Batch culture	Pilarska (1972)
	Mass culture	Schlüter and Groeneweg (1981)
	Chemostat culture	Rothhaupt (1985)
Cephalodella forficula	Batch culture	Dodson (1984)
Encentrum	Monoxenic cultures	Scott (1983)
E. linnhei	Chemostat culture	Scott (1988)
Euchlanis dilatata	Batch culture	King (1967)
Keratella cochlearis	Continuous and batch cultures	Walz (1983a, 1983b)
K. testudo	Batch culture	Stemberger and Gilbert (1987b)
Lecane inermis	Batch culture	Dougherty (1963)
Notommata copeus	Batch culture	Pourriot (1977)
Polyarthra major	Batch culture	Stemberger (1981)
P. vulgaris	Batch culture	Buikema *et al.* (1977)
Synchaeta cecelia	Batch culture	Egloff (1988)

[a] See also Pourriot (1965) for a list of species that have been cultured prior to about 1959.

contract into a completely unidentifiable lump, making identification difficult if not impossible. Such specimens have lost all value for taxonomic purposes, although the trophi still may be useful (see later discussion). Histological fixatives (e.g., Bouin's) or sugar-formalin solutions (used to prevent osmotic shock in microcrustaceans) are not particularly helpful in preventing contraction in rotifers. Whenever possible, live specimens should be identified first, then fixed to determine the effect of a particular fixative on body shape.

Fortunately, several anesthetics have been found to work well on rotifers (e.g., Edmondson 1959, May 1985, Nogrady and Keshmirian 1986a, 1986b, Pennak 1989, p. 195), but no single anesthetic has been shown to be universally effective. Care should be taken when working with anesthetic agents; some are controlled substances and/or are toxic. An alternative approach is to use the simple hot-water fixation technique described by Edmondson (1959, p. 433), which generally gives excellent results with

a number of species. It is generally a good idea to place live samples in jars over ice for the return trip to the laboratory, although we have found that a few species suffer when cooled (e.g., *Sinantherina socialis*).

Techniques for preservation and mounting of rotifers for examination using light microscopy (Wagstaffe and Fidler 1961, Pennak 1989, p. 196) and transmission and scanning electron microscopy (Amsellem and Clément 1980) have been described. However, the striking beauty of the living rotifer is utterly lost during almost any fixation and mounting procedure.

Observations of live rotifers are not always easily accomplished. The goal is to retard their movements without crushing or distorting the specimen with the cover glass. Both objectives may be accomplished with a compression microscope slide (Edmondson 1959, Spoon 1978). If a compressor is unavailable, then tiny pieces of broken cover glass (not recommended) or clay supports (recommended) work well as corner supports. Clay is superior to glass shards as it does not cut flesh. If the clay supports are too high, a slight pressure from a pencil on each corner will reduce the height of the cover glass to the desired level. With some practice, one can trap a planktonic rotifer sufficiently to prevent swimming without undue constriction. Sessile rotifers are handled more easily. Plant material with attached rotifers can be trimmed with iridectomy scissors to a size suitable for placement on a microscope slide. The animal will remain in place without the need of compression as long as the plant material is large enough to act as an anchor.

Methylcellulose or other viscous agents and fibrous material, such as glass wool or shredded filter paper, also may be used to impede swimming species. Unfortunately, methylcellulose interferes with ciliary function, and fibers reduce observation to a game of hide-and-go-seek. Any lighting conditions may be used, as long as you are careful not to overheat the specimens. Strobe lighting provides a marvelous view of ciliary movements, and darkfield illumination is often spectacular!

The final identification of many rotifer species requires examination of the trophi which, in certain species, may be done by compressing an intact animal. However, Myers (1937) described a method to extract trophi from surrounding soft tissues using a small volume of bleach (sodium hypochlorite) in a depression slide (see also Edmondson 1959 pp. 432–433, Pennak 1989, pp. 195–196, Stemberger 1979). A regular microscope slide may be used, if the cover glass is supported so the arrangements of the pieces of the trophi are not disturbed by compression. Because the trophi are liberated rather quickly when the bleach comes into contact with the rotifer, it is necessary to find the animal rapidly; otherwise it becomes necessary to scan the entire slide for the small trophi. Be aware that bubbles may form and obscure your view when bleach comes into contact with some biological materials.

Methods for preparing trophi for scanning electron microscope (SEM) work were described by Koehler and Hayes (1969) and Salt *et al.* (1978). In all work with trophi, it is important to remember that these structures are very small (< 50 μm long) and they are three-dimensional objects with a particular spatial arrangement among all seven pieces comprising the trophi.

V. CLASSIFICATION AND SYSTEMATICS

A. Classification

Classification schemes differ slightly in the ways that they treat the three basic groups of rotifers. In this key, three classes (Seisonidea, Bdelloidea, Monogononta) are recognized (Edmondson 1959, Nogrady 1982). The first two are comprised of rotifers with two gonads and are sometimes considered orders within the class Digononta, leaving class Monogononta separate (e.g., Pennak 1989, p. 196). Another method of classification emphasizes the unique anatomy and obligatory sexual reproduction of the Seisonidea, separating it from Bdelloidea and Monogononta, which are then placed in the class Eurotatoria (Koste 1978, p. 48). European workers tend to rank rotifers as a class within the phylum Aschelminthes, while North American investigators treat rotifers as a phylum (in which case the group known as aschelminths does not receive taxon status).

B. Systematics

1. Class Seisonidea

This monogeneric taxon (*Seison*) comprises only two recognized species of large (2–3 mm), dioecious, marine rotifers that are epizoic on the gills of a leptostracan crustacean, *Nebalia*. *Seison* often is described as being ectoparasitic. The corona is very reduced and is not used in food gathering or locomotion. Both species have paired gonads and a functional gut in both sexes. The trophi are fulcrate. Females have ovaries without vitellaria. Sexes are of similar size and morphology. These unusual marine rotifers are not included in the following key.

2. Class Bdelloidea

Comprising 18 genera and just over 360 species (Ricci 1987), bdelloids possess a rather uniform morphology and are exclusively females, reproducing by parthenogenesis. Bdelloids are characterized by having paired ovaries with vitellaria, more than two pedal glands, and ramate trophi. Most bdelloids are microphagous with a corona of either two trochal disks or a modified ciliated field. Bdelloids often have a vermiform body with a pseudosegmentation consisting of annuli, which permit shortening and lengthening of the body by telescoping. One order (Bdelloida) comprising four families (Adinetidae, Habrotrochidae, Philodinavidae, Philodinidae) is recognized. Bdelloids generally are not caught in plankton tows, although they may be found in waters with dense vegetation. Bdelloids often occur in sediments or among plant debris or crawling on the surfaces of aquatic plants. Some forms inhabit the capillary water films formed in soils (Pourriot 1979) or covering mosses. Many species are capable of becoming desiccated and then rehydrated (see Section II.C.4). Unfortunately, there has been no systematic review of North American bdelloids. The works of Bartos (1951), Burger (1948), Donner (1965), Koste and Shiel (1986), and Pourriot (1979) should be consulted for more detail on bdelloids.

3. Class Monogononta

The class Monogononta comprises the largest group of rotifers with more than 1600 species in about 95 genera of benthic, free-swimming, and sessile forms. All are assumed to be dioecious with one gonad. In many species, males have never been observed. Females possess one ovary with a vitellarium. Males usually are structurally reduced with a vestigial gut that functions only in energy storage. Males generally are ephemeral, usually being present in the plankton for only a few days or weeks each year. Monogononts are microphagous or raptorial, but a few are parasitic. The corona is modified as broad to narrow disks or ear-like lobes or is vase-shaped with reduced ciliation and long setae used in prey capture. In this key, 3 orders are recognized (Collothecacea, Flosculariacea, Ploimida), collectively comprising 24 families. Other workers include the first two orders as suborders within the order Gnesiotrocha (de Beauchamp 1965, Koste 1978, Koste and Shiel 1987).

Table 8.2 lists the families recognized in the subsequent key and their relevant couplet number.

C. Taxonomic Keys

There are a variety of keys to rotifers including those by Edmondson (1959) and Pennak (1989), covering the North American fauna, that may be followed to the level of genus. However, the specialized keys of Koste and Shiel (1987) for Australian waters, Pontin (1978) for the British Isles, Stemberger (1979) for the Great Lakes, Ruttner-Kolisko (1974) for planktonic rotifers in general, and Koste's (1978) revision of Voigt for the rotifers of central Europe are also important. This second group of works deals nearly

Table 8.2 List of the Rotifer Families Recognized in the Taxonomic Key Presented Here

Order Bdelloidea	Order Ploimida (*continued*)
Adinetidae (4b)	Branchionidae (23b)
Habrotrochidae (2)	Colurellidae (19)
Philodinavidae (4)	Dicranophoridae (12)
Philodinidae (3)	Epiphanidae (22)
Order Collothecacea	Euchlanidae (21)
Collothecidae (5)	Gastropodidae (27b)
Order Flosculariacea	Lecanidae (21b)
Conochilidae (8)	Lindiidae (14)
Filiniidae (11)	Microcodinidae (26)
Flosculariidae (9)	Mytilinidae (20)
Hexarthridae (7)	Notommatidae (26b)
Testudinellidae (11b)	Proalidae (22b)
Trochosphaeridae (10)	Synchaetidae (27)
Order Ploimida	Trichocercidae (24)
Asplanchnidae (15)	Trichotriidae (23)
Birgeidae (14b)	

exclusively with monogonont rotifers. While these keys may not be comprehensive enough to cover all of the variations in size and morphology that are sometimes found within a species, they will probably prove to be more than adequate for most work where detailed descriptions are needed.

One of the fundamental differences among the higher taxonomic levels in phylum Rotifera is the structure of the trophi, with characteristic types of jaws being recognized in each family (see Section II.B.2). [Transitional forms are also known (Koste 1978, Koste and Shiel 1987).] Although the following key is designed with the nonspecialist in mind, commensurate with the central importance of the trophi, we have used both the structure of the trophi and other obvious characters of anatomy and morphology as principal points of separation. In keeping with the nature of this text, the key is taken only to the level of family following, for the monogononts, the more recent taxonomy of Koste (1978) and Koste and Shiel (1987).

D. Taxonomic Key to Families of Freshwater Rotifera

1a.	Rotifers with paired ovaries (Digononta) and ramate trophi (Fig. 8.9) . . . . order Bdelloidea	2
1b.	Rotifers with a single ovary and trophi other than ramate . . . . class Monogononta	5

[Although this step obviously is very important, special care is generally not necessary to resolve this couplet. The ramate trophi of bdelloids are usually identifiable in whole animals without resorting to their isolation (see Section IV.C). If the type of trophi and number of ovaries cannot be determined, there are other clues that may be useful for live organisms. Upon contacting a substratum, many bdelloids will crawl on the surface in a manner reminiscent of a leech, hence the etymon of the name (Greek, *bdella*, leech). Further, the corona of some bdelloids has the appearance of two separate wheels, while there are few monogononts giving this impression.]

2a(1a).   Stomach without lumen, as a syncytial mass of food vacuoles that gives
            the gut a frothy appearance .... family Habrotrochidae
                  [Three genera (e.g., *Habrotrocha*) reported in North America; mainly
                  on moss or benthic, includes about 120 species (e.g., Fig. 8.37).]
2b.   Stomach with lumen 3
3a(2b).   Corona with the appearance of two separate wheels (trochus) extended
            on pedicels .... family Philodinidae
                  [About 10 genera and >150 species, of which *Philodina* and *Rotaria*
                  are common; also *Dissotrocha* and *Macrotrachella* (Fig. 8.38).]
3b   Corona not as above 4
4a(3b).   Corona ciliated lobes or regions near mouth; rostrum ciliated .... family Philodinavidae
                  [One rare species reported from North America (*Abrochta*,
                  Fig. 8.39).]
4b.   Corona as flat, ciliated fields on ventral side (no cingulum) .... family Adinetidae
                  [Two genera (*Adineta, Bradyscela*) of about 15 species (Fig. 8.40).]
5a(1b).   Trophi uncinate (Fig. 8.8H); corona lacking typical ciliated regions, but
            is open as an infundibulum with or without elongate setae around
            the margin .... order Collothecacea
                                                                         family Collothecidae

                  [Five genera comprising >50 species (Figs. 8.6 and 8.41); mostly
                  sessile, 5 species are free swimming, 1 sedentary on bottom or on
                  plants; many produce clear gelatinous tubes; none are colonial. Koste
                  (1978) divides this order: Collothecidae, with 2 genera (*Collotheca* and
                  Stephanoceros) possessing elongate setae and Atrochidae with 3
                  genera (*Acyclus, Atrochus, Cupelopagis*) lacking setae.]
5b.   Trophi other than uncinate; corona with typical ciliated lobes or fields,
            or ciliated bands 6

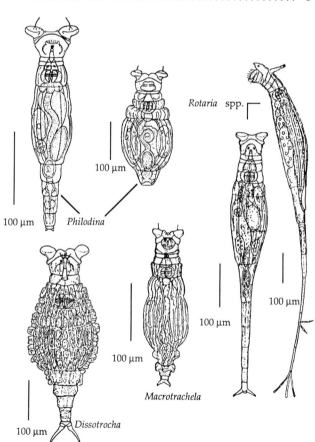

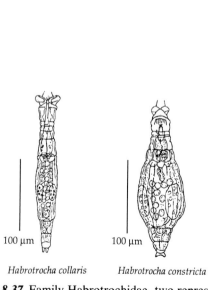

Figure 8.37 Family Habrotrochidae, two representatives of the genus *Habrotrocha*. (From Koste 1976, with permission.)

Figure 8.38 Representatives of the family Philodinidae. (From Koste 1976, with permission.)

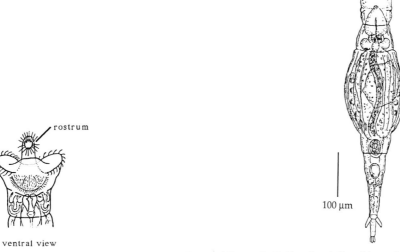

Figure 8.39 Family Philodinavidae (*Abrochta*, ventral view of corona.) (Adapted from several sources.)

Figure 8.40 Family Adinetidae: dorsal view of *Adineta* sp. (From Koste 1976, with permission.)

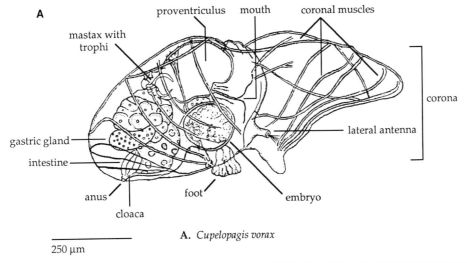

A. *Cupelopagis vorax*

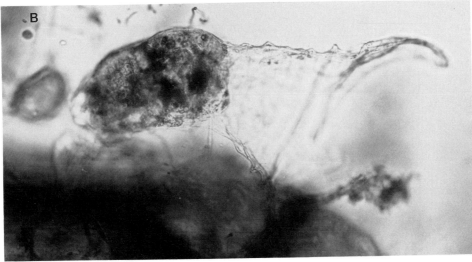

Figure 8.41 Family Collothecidae: (A) *Cupelopagis vorax* (adapted from several sources) and (B) *Collotheca ferox*. (R. L. Wallace, original photomicrograph.)

6a(5b).   Rotifers possessing malleoramate trophi (Fig.8.8I) .... order Flosculariacea   7
               [Malleoramate trophi possess unci with numerous fine teeth, nearly
               completely overlying the rami (unlike the malleate type; see the
               description in couplet 16); teeth close to the fulcrum are usually larger
               than those more distant. In live animals, the nearly constant
               movement of the trophi in a grinding or pounding-like action is
               characteristic.]

6b.   Rotifers with trophi other than malleoramate .... order Ploimida   12

7a(6a).   Conical body with six hollow, arm-like, setose appendages that are
               outgrowths of the integument; appendages inserted with powerful
               muscles .... family Hexarthridae
               [Monogeneric family *Hexarthra* (*Pedalia*) with about eight species
               (Fig. 8.13), some of which inhabit salt or brackish waters. Although
               illoricate, *Hexarthra* spp. preserve well in formalin without serious
               contraction.]

7b.   Body lacking hollow, arm-like, setose appendages   8

8a(7b).   Rotifers with a corona of horseshoe or U-shaped ciliated bands;
               possessing a ventral gap in the coronal ciliation; free-swimming; solitary
               or small to very large colonies (>150 individuals) within a
               gelatinous matrix .... family Conochilidae
               [Two genera recognized, based on the position of antennae
               (*Conochilus* and *Conochiloides*), but Koste (1978) and
               Ruttner-Kolisko (1974) do not consider antennal position to be a
               significant feature and subsume *Conochiloides* within *Conochilus* (Fig.
               8.17).]

8b.   Rotifers with corona other than a horseshoe or U-shaped ciliated band;
               solitary or colonial; with or without a gelatinous matrix   9

9a(8b).   Rotifers typically with elongate bodies and large, circular to lobate,
               ear-like corona (Fig. 8.42); solitary or colonial; with or without a tube or
               gelatinous matrix; mostly sessile, but some free swimming .... family Flosculariidae
               [Seven genera of mainly sessile species (Figs. 8.5 and 8.43), a few
               are free-swimming; includes the genera *Floscularia*, *Lacinularia*,
               Ptygura, and *Sinantherina*.]

9b.   Rotifers lacking a large, circular to lobate, ear-like corona   10

10a(9b).   Spherical rotifers with a corona as a circular band around the equator or
               towards one end .... family Trochosphaeridae
               [Genus *Trochosphaera* (Fig. 8.44) with two species in eutrophic
               waters, rare, but when present may be very abundant. A second rare
               genus (*Horaëlla*) apparently not reported in the United States.]

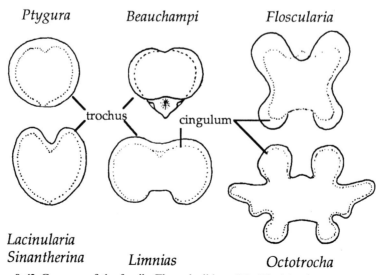

Figure 8.42 Coronae of the family Flosculariidae. (Modified from several sources.)

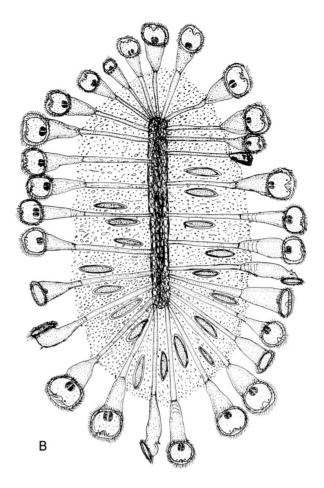

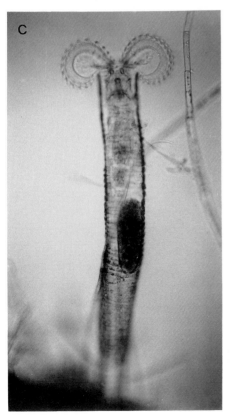

Figure 8.43 Family Flosculariidae: (A) *Floscularia conifera* portion of a sessile colony, with two adults; (B) *Lacinularia elliptica* (plantonic, to 500 individuals, colony diameter to 3000 μm); (C) *Limnias melicerta* (sessile in clear cement tube, individuals to about 900 μm); (D) *Ptygura beauchampi* (sessile. Pb, lateral view of the rotifer; e, egg; gth, plant hair, bmc, plant mucilage; td, substratum; t, tube; bar, 50 μm); (E) *Sinantherina socialis* (sessile colonies may contain over 300 individuals, shorter individuals have contracted due to a disturbance, colony diameter 2000 μm). (B and E from Vidrine *et al.* 1985, with permission; others R. L. Wallace original photomicrographs.) (*Figure continues*)

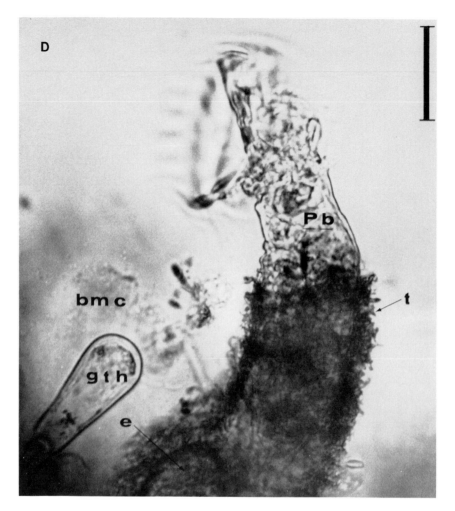

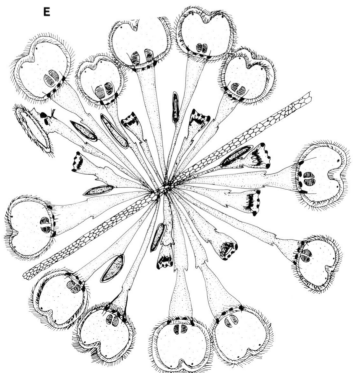

Figure 8.43 (Continued)

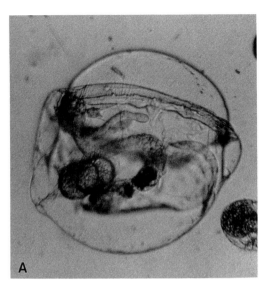

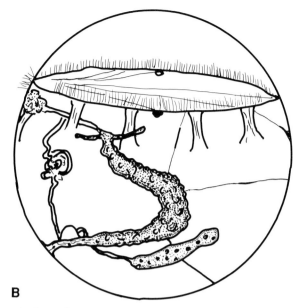

Figure 8.44 Family Trochosphaeridae: (A) photomicrograph of *Trochosphaera* and (B) line drawing interpretation (diameter of the individual in the photomicrograph ca. 800 μm; original photomicrograph by T. Nogrady.)

| 10b. | Rotifers not spherical in shape . . . . | 11 |

11a(10b). Rotifers with two movable anterior spines (bristles) below the corona, of varying lengths often much longer than the body and one (rarely two) rigid caudal spines; foot absent . . . . family Filiniidae

[Genus *Filinia* (Fig. 8.32) with two North American species and apparently many hybrids. The long spines point posteriorly when the animal is swimming, but the anterior ones point forward when the corona is withdrawn.]

11b. Rotifers without long spines as described above . . . . family Testudinellidae

[Two dissimilar genera. (1) Genus *Testudinella*, lorica greatly flattened dorsoventrally, with dorsal and ventral plates fused along the lateral margin; foot annulated and retractile ending in a ciliated cup, which is difficult to see; approximately 12 littoral species up to 250 μm long (Fig. 8.45). (2) Genus *Pompholyx*, body having four nearly equal lobes in cross-section; with a pair of frontal eyespots, (Fig. 8.45) common in lakes and ponds.]

12a(6b). Rotifers with forcipate trophi (Fig. 8.8J) . . . . family Dicranophoridae

[About 12 genera of creeping, littoral rotifers (Fig. 8.46) with symmetrical or asymmetrical forcipate trophi, a feature easily determined by gentle compression of the specimen; no planktonic or semiplanktonic species are present in this family.]

12b. Rotifers with trophi other than forcipate 13

13a(12b). Mostly littoral rotifers possessing cardate trophi (Fig. 8.10) having a sucking action or trophi highly modified and stomach with zoochlorellae 14

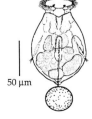

50 μm

Pompholyx

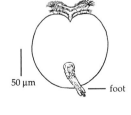

50 μm

foot

Testudinella

◄ **Figure 8.45** Family Testudinellidae: ventral views of *Testudinella* and *Pompholyx*. (Modified from several sources.)

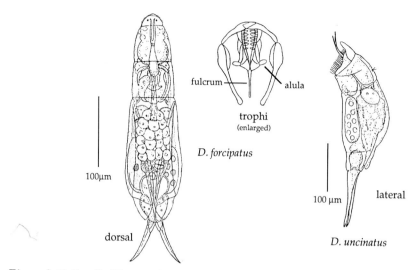

fulcrum

alula

trophi
(enlarged)

D. forcipatus

100μm

dorsal

100 μm

lateral

D. uncinatus

Figure 8.46 Family Dicranophoridae (*Dicranophorus*). (From Koste 1976, with permission.)

13b.	Rotifers with trophi other than cardate and stomach without zoochlorellae 15
14a(13a).	Manubria of the trophi with hooks that may be determined in lateral view of the animal (Fig. 8.10); body spindle-shaped .... family Lindiidae [Monogenetic family, genus *Lindia* (Fig. 8.47); mostly littoral.]
14b.	Highly modified trophi, possessing pseudunci; stomach with zoochlorellae, gastric glands absent (Fig. 8.48) .... family Birgeidae [Monospecific family (*Birgea enantia*, a rare littoral species).]
15a(13b).	Saccate rotifers, with incudate or modified incudate trophi, (Fig. 8.8G, R) some lacking intestine and anus .... family Asplanchnidae [Three genera. Two lack an intestine: *Asplanchna*, foot absent, with six species (Figs. 8.11 and 8.49) and *Asplanchnopus*, foot present, with three species. One rare, benthic genus with intestine and foot, *Harringia*, (Fig. 8.50) with two species.]
15b.	Rotifers not possessing incudate trophi 16
16a(15b).	Rotifers possessing malleate trophi .... 17 [In malleate trophi (Fig. 8.8A, B, C, E, F, P), each uncus has only 4–7 teeth (unlike the condition found in malleoramate trophi; see description couplet 6). The fulcrum may be short (malleate) or long (submalleate).]
16b.	Rotifers possessing virgate trophi .... 24 [Virgate trophi are often asymmetrical (Fig. 8.8D, K, L, M, N, O, Q, S), with a long fulcrum and manubria and generally small rami.]

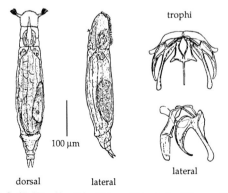

trophi

lateral

100 μm

dorsal   lateral

Figure 8.47 Family Lindiidae (*Lindia*). (From Harring and Myers 1922, with permission.)

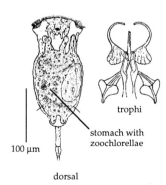

trophi

stomach with
zoochlorellae

100 μm

dorsal

Figure 8.48 Family Birgeidae (*Birgea*). (From Harring and Myers 1922, with permission.)

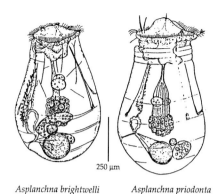

Asplanchna brightwelli   *Asplanchna priodonta*

250 µm

Figure 8.49 Family Asplanchnidae (*Asplanchna*). (From Koste 1976, with permission.)

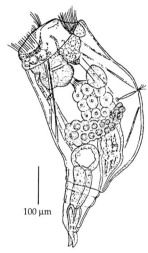

100 µm

Figure 8.50 Family Asplanchnidae (*Harringia*). (From Koste 1976, with permission.)

17a(16a).   Loricate rotifers: body wall thickened and firm 18
17b.   Illoricate rotifers: body wall not thickened or firm 22
[Interpreting whether the body wall is thickened (loricate) or not (illoricate) can be difficult and it does take some experience to judge this characteristic. To get an indication as to how firm the lorica is, follow this procedure. When working with fresh material, gently apply pressure from the point of a pencil or probe onto the cover glass while observing the specimen. If the body puffs out under pressure and returns to its original shape when the pressure is released, the specimen is illoricate. Loricate forms will exhibit much less flexibility of the body wall. In preserved materials, illoricate forms tend to shrivel up, while the body wall of loricate rotifers will retain its shape even if the inner organs separate from the body wall, collapsing into a central mass of tissue.]

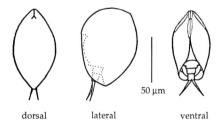

dorsal   lateral   ventral

50 µm

A. *Colurella*

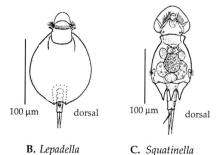

100 µm dorsal   100 µm dorsal

B. *Lepadella*   **C.** *Squatinella*

Figure 8.51 Family Colurellidae. (A) *Colurella;* (B) *Lapedella;* (C) *Squatinella*. (Modified from several sources.)

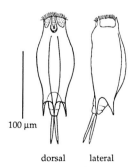

100 µm

dorsal   lateral

Figure 8.52 Family Mytilinidae (*Mytilina*). (Modified from several sources.)

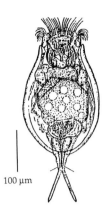

100 μm

◀ ***Figure 8.53*** Family Euchlanidae (*Euchlanis*). (Modified from several sources.)

18a(17a).	Lorica possessing furrows, grooves, or sulci; or with a dorsal, semicircular head shield covering the corona; or with a very strongly developed lorica and a dorsal transverse ridge 19
18b.	Lorica lacking furrows, grooves, sulci or dorsal head shield; without a strongly developed lorica and dorsal transverse ridge 23
19a(18a).	Lorica with medial, ventral furrow (but no lateral furrow or grooves) extending the full length of the animal or with a ventral notch in which the foot lies, or with a dorsal, semicircular head shield covering the corona .... family Colurellidae

[Four genera found in the littoral including *Colurella* (possessing a medial ventral furrow, Fig. 8.51A), a very common genus of some 15 species, which browse on epiphytic microorganisms; *Lapedella* (possessing a ventral notch in which the foot lies, Fig. 8.51B), very common with many species; *Squatinella* (possessing a dorsal, semicircular head shield covering the corona, Fig. 8.51C). [N.B.: *Diplois daviesiae* (family Euchlanidae), a rare monospecific genus present in sphagnum bogs will key to family Colurellidae, if the lateral furrows of this species are missed.]

19b.	Lorica lacking medial, ventral furrow or notch, and lacking a head shield 20

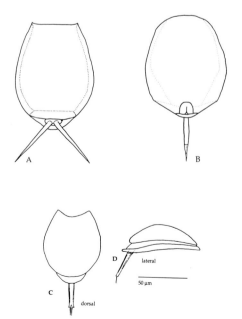

◀ ***Figure 8.54*** Examples of the family Lecanidae: (A) *Lecane;* (B) *Lecane (Monostyla)*; (C) *Lecane lunaris* (dorsal); (D) Lateral view. (A and B from several sources, C and D from Dartnall and Hollowday 1985, with permission.)

20a(19b).   Lorica with dorsal, medial sulcus (double keel) or with a strongly
            developed lorica usually possessing a dorsal transverse ridge .... family Mytilinidae
            [Two littoral genera: (1) *Mytilina* (Fig. 8.52) possessing a laterally
            flattened lorica with a dorsal longitudinal sulcus, spines on all four
            anterior corners, about ten species common and (2) *Lophocharis* with
            a strongly developed lorica, usually having a dorsal transverse ridge.]
20b.   Lorica lacking a dorsal, medial sulcus and lacking a dorsal transverse ridge 21
21a(20b).   Foot projecting from between dorsal and ventral plates at the posterior
            end of the lorica; dorsal and ventral plates separated by a deep furrow
            or groove .... family Euchlanidae
            [Four genera common in the littoral, including *Euchlanis* (Fig. 8.53).]
21b.   Foot projecting through a hole in the ventral plate at the posterior end
            of the lorica; dorsal and ventral plates connected by a weak furrow or
            groove .... family Lecanidae
            [A large number of littoral species in two genera (Fig. 8.54): *Lecane*
            (with two toes that may be partially fused at the base) and *Monostyla*
            (one toe or toe divided distally); Koste (1978) has subsumed
            Monostyla within *Lecane*.]
22a(17b).   Mouth set in a funnel-shaped buccal field .... family Epiphanidae
            [Six genera including *Epiphanes* (Fig. 8.55), *Cyrtonia, Mikrocodides,*
            and *Rhinoglena* all littoral or in small, shallow lakes. *Epiphanes,*
            sometimes incorrectly described as a 'typical rotifer' in textbooks, is
            not common, but may be found in ponds receiving animal wastes.]
22b.   Mouth set in an oblique, ciliated field on ventral side; body wormlike to
            fusiform .... family Proalidae
            [Four genera, including the large genus *Proales;* generally found in
            littoral zones and sandy beaches (Fig. 8.56); *Proales daphnicola* is
            planktonic and may be attached to cladocerans.]

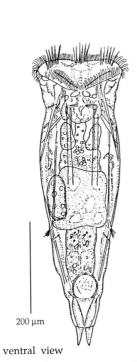

ventral view

Figure 8.55 Family Epiphanidae (*Epiphanes*). (From
Dartnall and Hollowday 1985, with permission.)

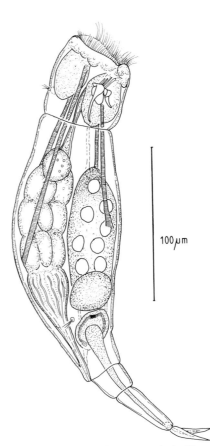

100 μm

Figure 8.56 Family Proalidae (*Proales*). (From Chengalath 1985, with permission.)

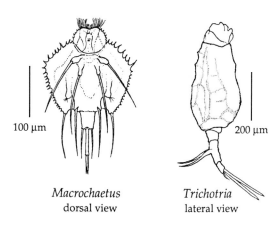

Macrochaetus
dorsal view

Trichotria
lateral view

◀ *Figure 8.57* Family Trichotriidae (*Macrochaetus* and *Trichotria*). (Modified from several sources.)

23a(18b).   Lorica extending beyond body to head, foot, and toes .... family Trichotriidae
            [Three genera, including *Macrochaetus* (Fig. 8.57A), with numerous
            bilaterally placed spines (about seven littoral species) and *Trichotria*
            (Fig. 8.57B), with a pair of heavy spines on the dorsal side of the foot
            (about ten littoral species).]

23b.   Lorica not extending beyond body .... family Brachionidae
            [A large family comprising six genera of common rotifers. *Brachionus*,
            a very common genus with about 25 species of littoral and planktonic
            rotifers (Fig. 8.58). *B. plicatilis* is frequently found in salt and
            brackish waters. *Kellicottia*, with two common planktonic species,
            possessing long anterior spines of unequal length (Fig. 8.59).
            Keratella, with more than 15 species, having the dorsal plate
            decorated with a characteristic facet pattern (Fig. 8.60). *K. cochlearis*
            is very common. *Notholca*, a genus common in cool waters (Fig.
            8.61).]

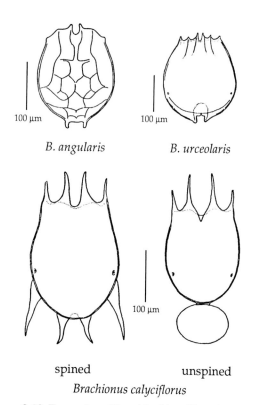

B. angularis

B. urceolaris

spined   unspined
Brachionus calyciflorus

Figure 8.58 Examples of brachionid rotifers, family Brachionidae. (From Koste 1976, with permission.)

Figure 8.59 Kellicottia, a common genus of the family Brachionidae. Dorsal view. (From several sources.)

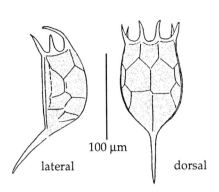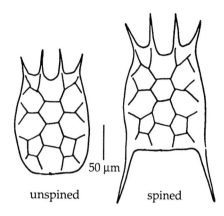

A. *Keratella cochlearis*   **B.** *Keratella testudo*

Figure 8.60 *Keratella*, a common genus of the family Brachionidae: (A) *K. cochlearis* (from Koste 1976, with permission); (B) *K. testudo* (from Stemberger 1988, with permission).

24a(16b).   Body twisted as a partial helix (asymmetrical) and/or trophi asymmetrical; or small saccate animals parasitic in the colonial alga *Volvox* .... family Trichocercidae
              [Three genera: (1) genus *Trichocera* (body twisted as a partical helix; unequal toes; asymmetrical trophi) with some 90 species, may be important in the plankton (Fig. 8.62); (2) genus *Elosa* (body showing in cross section three lobes, two lateral and one dorsal, plus ventrally an inconspicuous lobe; asymmetrical trophi) with two species common in the psammon and with *Sphagnum* moss; (3) *Ascomorphella volvocicola*, parasitic in *Volvox* (Fig. 8.63).]

24b.   Body and trophi not as described above 25

25a(24b).   Free-swimming rotifers 27
25b.   Crawling or creeping rotifers in the littoral (occasional species in the plankton) 26

26a(25b).   Purple plates positioned anterior to the mastax; corona wide, flat, and somewhat circular; foot jointed and long, about 50% of the total length of the animal, with a single toe .... family Microcodonidae
              [Monospecific family of one uncommon species, *Microcodon clavus*, mostly littoral, but may be found in the plankton.]

26b.   Lacking purple plates as described earlier, corona ventral and not circular .... family Notommatidae
              [A varied family with many species in 13 genera including *Cephalodella* (Fig. 8.64), *Notommata* (Fig. 8.65), and *Pleurotrocha*.]

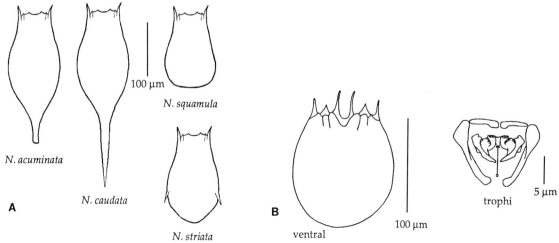

Figure 8.61 *Notholca*, a common genus of the family Brachionidae, usually in cool waters: (A) Variation within the genus (modified from several sources). (B) *N. squamula* (modified from May 1980, with permission).

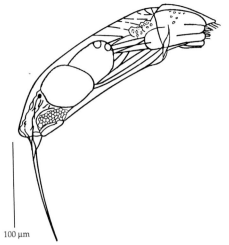

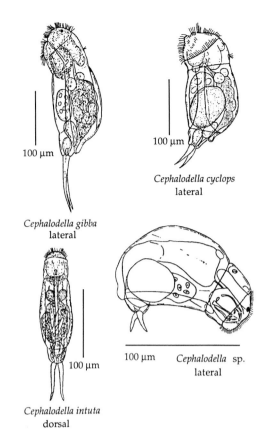

Figure 8.62 *Trichocera*, an important genus of the family Trichocercidae. Lateral view. (From Wallace *et al.* 1989, with permission.)

Figure 8.64 Family Notommatidae (*Cephalodella*). (*Cephalodella* sp. from Dartnall and Hollowday 1985; others from Koste 1976, with permission.)

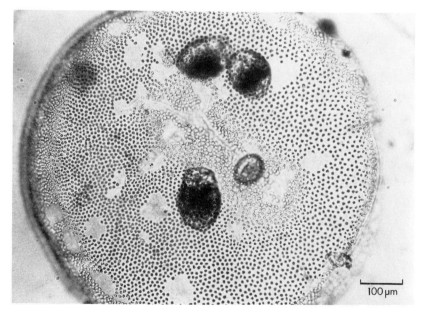

Figure 8.63 Three *Ascomorphella volvocicola* (family Trichocercidae), having done extensive damage to a colony of *Volvox*. (From Ganf *et al.* 1983, with permission.)

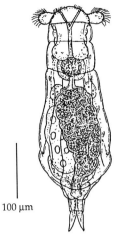

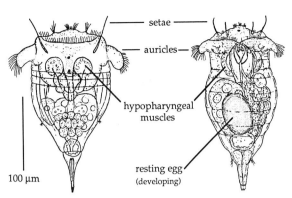

Figure 8.66 Family Synchaetidae (*Synchaeta*). (From Koste 1976, with permission.)

Figure 8.65 Family Notommatidae (*Notommata*). Dorsal view. (From Koste 1976, with permission.)

27a(25a).   Corona with four prominant setae (sensory bristles) and auricles (ear-like structures); or possessing 12 movable, flattened, sword-to paddle- or feather-shaped appendages (paddles); or with a sculptured lorica having ridges, grooves, or areolations .... family Synchaetidae

   [Four genera: with sensory bristles and auricles, *Synchaeta* (Fig. 8.66), comprising some 12 freshwater species plus others of marine and brackish waters, may be important in the plankton; with paddles, *Polyarthra* (Fig. 8.33), fewer than 10 species separated based on paddle morphology, nuclei of vitellarium, and lateral antennae; *Pseudoploesoma* (Fig. 8.67A) and *Ploesoma* (Fig. 8.67B).]

27b.   Lacking prominant setae, auricles, and paddles, and lorica not sculptured .... family Gastropodidae

   [Two genera: *Ascomorpha* (Fig. 8.68A) (incorporating *Chromogaster*) comprising six species, possessing a fingerlike projection (palp) from the corona; *Gastropus* (Fig. 8.68B) with three species, having a laterally compressed body, may be important in the plankton.]

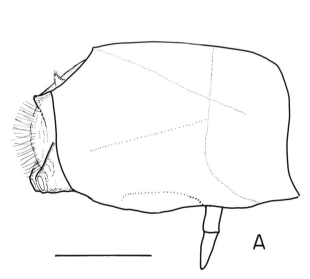

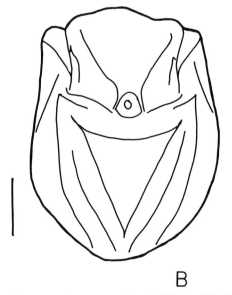

Figure 8.67 Family Synchaetidae. (A) *Pseudoploesoma* and (B) *Ploesoma*. Bars approximately 75 μm. (Modified from several sources.)

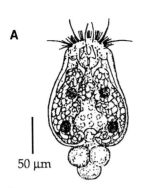
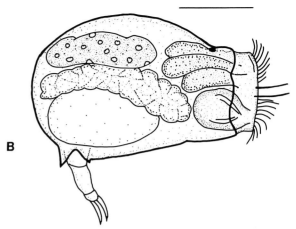

Figure 8.68 Family Gastropodidate. (A) *Ascomorpha* sp; (B) *Gastropus*, (bar = 100 μm). (A from Koste 1976, with permission; B modified from several sources.)

Acknowledgments

We wish to thank our mentors J.J. Gilbert and C.E. King for all of the help and encouragement they have given us over the years. We also acknowledge W. Koste, a good friend and inspiration; a worker who knows more about rotifers than we ever thought possible. T. Nogrady, P.L. Starkweather, R.S. Stemberger, and G. H. Wittler reviewed the manuscript for this chapter, a task very much appreciated by the authors. However, we remain responsible for any errors that may still be present.

Literature Cited

Allan, J. D. 1976. Life history patterns in zooplankton. American Naturalist 110:165–180.

Aloia, R. C., and R. L. Moretti. 1973. Sterile culture techniques for some species of the rotifer *Asplanchna*. Transactions of the American Microscopical Society 92:364–371.

Amsellem, J., and P. Clément. 1980. A simplified method for the preparation of rotifers for transmission and scanning electron microscopy. Hydrobiology 73:119–122.

Amsellem, J., and C. Ricci. 1982. Fine structure of the female genital apparatus of *Philodina* (Rotifera, Bdelloidea). Zoomorphology 100:89–105.

Anderson, R. V., R. E. Ingham, J. A. Trofymow, and D. C. Coleman. 1984. Soil mesofaunal distribution in relation to habitat types in shortgrass prairie. Pedobiologia 26:257–261.

Amrén, H. 1964. Ecological and taxonomic studies on zooplankton from Spitsbergen. Zool. Bidr. Uppsala 36:193–208.

Barron, G. L. 1980a. Fungal parasites of rotifers: *Harposporium*. Canadian Journal of Botany 58:443–446.

Barron, G. L. 1980b. A new *Haptoglossa* attacking roti-fers by rapid injection of an infective sporidium. Mycologia 72:1186–1194.

Barron, G. L., and S. S. Tzean. 1981. A subcuticular endoparasite impaling bdelloid rotifers using three-pronged spores. Canadian Journal of Botany 59:1207–1212.

Bartos, E. 1951. The Czechoslovak Rotatoria of the order Bdelloidea. Vestnik Ceskoslovenské Zoologické Spolecnosti 15:241–500.

Bateman, L. 1987. A bdelloid rotifer living as an inquiline in leaves of the pitcher plant, *Sarracenia purpurea*. Hydrobiologia 147:129–133.

de Beauchamp, P. 1965. Classe des Rotifères. Pages 1225–1379 *in:* P. P. Grassé, editor. Traité de Zoologie IV, 3. Masson, Paris.

Bender, K., and W. Kleinow. 1988. Chemical properties of the lorica and related parts from the integument of *Brachionus plicatilis*. Comparative Biochemistry and Physiology 89B:483–487.

Berzins, B., and B. Pejler. 1987. Rotifer occurrence in relation to pH. Hydrobiologia 147:107–116.

Birky, C.W., Jr. 1967. Studies on the physiology and genetics of the rotifer Asplanchna III. Results of out-crossing, selfing, and selection. Journal of Experimental Zoology 165:104–116.

Birky, C.W., Jr., and J.J. Gilbert. 1971. Parthenogenesis in rotifers: the control of sexuality and asexuality. American Zoologist 11:245–266.

Bogdan, K. G., and J. J. Gilbert. 1982. Seasonal patterns of feeding by natural populations of *Keratella, Polyarthra,* and *Bosmina:* clearance rates, selectivities, and contributions to community grazing. Limnology and Oceanography 27:918–934.

Bogdan, K. G., and J. J. Gilbert. 1987. Quantitative comparison of food niches in some freshwater zooplankton. Oecologia 72:331–340.

Bogdan, K. G., J. J. Gilbert, and P. L. Starkweather. 1980. *In situ* clearance rates of planktonic rotifers. Hydrobiologia 73:73–77.

Boraas, M. E. 1980. A chemostat system for the study of

rotifer–algal–nitrate interactions. Pages 173-182 *in*: W.C. Kerfoot, editor. Evolution and ecology of zooplankton communities. University of New England Press, Hanover, New Hampshire.

Buikema, A. L., Jr., J. Cairns, Jr., P.C. Edmunds, and T.H. Krakauer. 1977. Culturing and ecology studies of the rotifer, *Polyarthra vulgaris*. EPA-600/3-77-051. Environmental Research Laboratory, Office of Research and Development, U.S. Environmental Protection Agency, Duluth, Minnesota.

Burger, A. 1948. Studies on the moss dwelling bdelloids (Rotifera) of eastern Massachusetts. Transactions of the American Microscopical Society 67:111–142.

Burns, C. W., and J. J. Gilbert. 1986a. Effects of daphnid size and density on interference between *Daphnia* and *Keratella cochlearis*. Limnology and Oceanography 31:848–858.

Burns, C. W., and J. J. Gilbert. 1986b. Direct observations of the mechanisms of interference between *Daphnia* and *Keratella cochlearis*. Limnology and Oceanography 31:859–866.

Capuzzo, J. 1979a. The effect of temperature on the toxicity of chlorinated cooling waters to marine animals—a preliminary review. Marine Pollution Bulletin 10:45–47.

Capuzzo, J. 1979b. The effect of halogen toxicants on survival, feeding and egg production of the rotifer *Brachionus plicatilis*. Estuarine and Coastal Marine Science 8:307–316.

Chengalath, R. 1985. The Rotifera of the Canadian Arctic Sea ice, with description of a new species. Canadian Journal of Zoology 63:2212–2218.

Chengalath, R., and W. Koste. 1987. Rotifera from northwestern Canada. Hydrobiologia 147:49–56.

Clément, P. 1977. Ultrastructural research on rotifers. Archiv für Hydrobiologie Beiheft 8:270–297.

Clément, P. 1980. Phylogenetic relationships of rotifers, as derived from photoreceptor morphology and other ultrastructural analysis. Hydrobiologia 73:93–117.

Clément, P. 1985. The relationships of rotifers. Pages 224–247 *in*: S. Conway Morris, J. D. George, and H. M. Platt, editors. The origins and relationships of lower invertebrates. Systematics Association, Vol. 28. Clarendon Press, Oxford.

Clément, P. 1987. Movements in rotifers: correlations of ultrastructure and behavior. Hydrobiologia 147:339–359.

Clément, P., E. Wurdak, and J. Amsellem. 1983. Behavior and ultrastructure of sensory organs in rotifers. Hydrobiologia 104:89–130.

Couillard, Y., P. Ross, and B. Pinel-Alloul. 1989. Acute toxicity of six metals to the rotifer *Brachionus calyciflorus*. *Toxicity Assessment* 4:451–462.

Coulon, P. Y., J. P. Charras, J. L. Chasse, P. Clément, A. Cornillac, A. Luciani, and E. Wurdak. 1983. An experimental system for automatic tracking and analysis of rotifer swimming behavior. Hydrobiologia 104:197–202.

Dad, N. K., and V. Kant Pandya. 1982. Acute toxicity of two insecticides to rotifer *Brachionus calyciflorus*. In-

ternational Journal of Environmental Studies 18:245–246.

Dartnell, H. J. G. and E. D. Hollowday. 1985. Anarctic rotifers. British Antarctic Survey Scientific Reports 100:1–46.

Dobell, C. 1960. Antony van Leeuwenhoek and his "little animals." Dover, New York.

Dodson, S.I. 1984. Ecology and behaviour of a free-swimming tube-dwelling rotifer *Cephalodella forficula*. Freshwater Biology 14:329–334.

Donner, J. 1965. Ordnung Bdelloidea. Bestimmungsbücher zur Bodenfauna Europas. Vol. 6. Akademie-Verlag, Berlin. 267 pp.

Dougherty, E. C. 1963. Cultivation and nutrition of micrometazoa III. The minute rotifer *Lecane inermis*. Journal of Experimental Zoology 153:183–186.

Droop, M. R., and J. M. Scott. 1978. Steady-state energetics of a planktonic herbivore. Journal of the Marine Biological Association of the United Kingdom 58:749–772.

Dumont, H. 1983. Biogeography of rotifers. Hydrobiologia 104:19–30.

Edmondson, W. T. 1944. Ecological studies of sessile Rotatoria, Part I. Factors affecting distribution. Ecological Monographs 4:32–66.

Edmondson, W. T. 1945. Ecological studies of sessile Rotatoria, Part II. Dynamics of populations and social structure. Ecological Monographs 15:141–172.

Edmondson, W. T. 1959. Rotifera. Pages 420-494 *in*: W. T. Edmondson, editor. Fresh-water Biology. 2nd Edition. Wiley, New York.

Edmondson, W. T. 1960. Reproductive rates of rotifers in natural populations. Memorie dell'Instituto Idrobiologia 12:21–77.

Edmondson, W. T. 1965. Reproductive rate of planktonic rotifers as related to food and temperature in nature. Ecological Monographs 35:61–111.

Edmondson, W. T., and A.H. Litt. 1982. Daphnia in Lake Washington. Limnology and Oceanography 30:180–188.

Edmondson, W. T., and A. H. Litt. 1987. *Conochilus* in Lake Washington. Hydrobiologia 147:157–162.

Egloff, D. A. 1988. Food and growth relations of the marine microzooplankter, *Synchaeta cecelia* (Rotifera). Hydrobiologia 157:129–141.

Enesco, H. E., and C. Verdone-Smith. 1980. α-Tocopherol increases lifespan in the rotifer *Philodina*. Experimental Gerontology 15:335–338.

Epp, R. W., and W. M. Lewis. 1979. Sexual dimorphism in *Brachionus plicatilis* (Rotifera): evolutionary and adaptive significance. Evolution 33:919–928.

Epp, R. W., and W. M. Lewis. 1980. Metabolic uniformity over the environmental temperature range in *Brachionus plicatilis* (Rotifera). Hydrobiologia 73:145–147.

Epp, R. W., and W. M. Lewis. 1984. Cost and speed of locomotion for rotifers. Oecologia 61:289–292.

Epp, R. W., and P. W. Winston. 1977. Osmotic regulation in the brackish-water rotifer *Brachionus plicatilis* (Muller). Journal of Experimental Biology 68:151–156.

Epp, R. W., and P. W. Winston. 1978. The effects of

salinity and pH on the activity and oxygen consumption of *Brachionus plicatilis* (Rotatoria). Comparative Biochemistry and Physiology 59A:9–12.

Evans, M. 1986. Biological principles of pond aquaculture: zooplankton. Pages 27-37 *in*: J.E. Lannan, R.O. Smitherman, and G. Tchobanoglous, editors. Principles and Practices of Pond Aquaculture. Oregon State University Press, Corvallis.

Gabriel, W., B. E. Taylor, and S. Kirsch-Prokosch. 1987. Cladoceran birth and death rates: experimental comparisons of egg ratio methods. Freshwater Biology 18:361–372.

Ganf, G. G., R. J. Shiel, and C. J. Merrick. 1983. Parisitism: the possible collapse of a *Volvox* population in Mount Bold Reservoir, South Australia. Australian Journal of Marine and Freshwater Research 34:489–494.

Garreau, F., C. Rougier, and R. Pourriot. 1988. Exploitation des resources alimentaires par le prédateur planctonique *Asplanchna girodi* de Guerne 1888 (Rotifères) dan un lac de sablére. Archiv für Hydrobiologie 112:91–106.

Gilbert, J. J. 1963a. Mictic female production in the rotifer *Brachionus calyciflorus*. Journal of Experimental Zoology 153:113–123.

Gilbert, J. J. 1963b. Contact chemoreceptors, mating behavior and reproductive isolation in the rotifer genus *Brachionus*. Journal of Experimental Biology 40:625–641.

Gilbert, J. J. 1966. Rotifer ecology and embryological induction. Science 151:1234–1237.

Gilbert, J. J. 1967. *Asplanchna* and postero-lateral spine production in *Brachionus calyciflorus*. Archiv für Hydrobiologie 64:1–62.

Gilbert, J. J. 1970. Monoxenic cultivation of the rotifer *Brachionus calyciflorus* in a defined medium. Oecologia 4:89–101.

Gilbert, J. J. 1974. Dormancy in rotifers. Transactions of the American Microscopical Society 93:490–513.

Gilbert, J. J. 1977. Effect of the non-tocopherol component of the diet on polymorphism, sexuality and reproductive rate of the rotifer *Asplanchna sieboldi*. Archiv für Hydrobiologie 80:375–397.

Gilbert, J. J. 1980a. Female polymorphisms and sexual reproduction in the rotifer *Asplanchna*: evolution of their relationship and control by dietary tocopherol. American Naturalist 116:409–431.

Gilbert, J.J. 1980b. Feeding in the rotifer *Asplanchna*: behavior, cannibalism, selectivity, prey defenses, and impact on rotifer communities. Pages 158–172 *in*: W.C. Kerfoot, editor. Evolution and ecology of zooplankton communities. University Press of New England, Hanover, New Hampshire.

Gilbert, J. J. 1983. Rotifera. Pages 181-193 *in*: K.G. Adiyodi and R.G. Adiyodi, editors. Reproductive biology of invertebrates. Vol. 2: Spermatogenesis and sperm function. Wiley, New York.

Gilbert, J. J. 1985a. Escape response of the rotifer *Polyarthra*: a high-speed cinematographic analysis. Oecologia (Berlin) 66:322–331.

Gilbert, J. J. 1985b. Competition between rotifers and *Daphnia*. Ecology 66:1943–1950.

Gilbert, J. J. 1987. The *Polyarthra* escape response: defense against interference from *Daphnia*. Hydrobiologia 147:235–238.

Gilbert, J. J. 1988. Suppression of rotifer populations by *Daphnia*: a review of the evidence, the mechanisms, and the effects on zooplankton community structure. Limnology and Oceanography 33(6, Part 1):1286–1303.

Gilbert, J. J. 1989a. The effect of *Daphnia* interference on a natural rotifer and ciliate community: short-term bottle experiments. Limnology and Oceanography 34(3):606–617.

Gilbert, J. J. 1989b. Competitive interactions between the rotifer *Synchaeta oblonga* and the cladoceran *Scapholeberis kingi* Sars. Hydrobiologia 186/187:75–80.

Gilbert, J. J., and K. G. Bogdan. 1981. Selectivity of *Polyarthra* and *Keratella* for flagellate and aflagellate cells. Internationale Vereinigung für theoretische und angewandte Limnologie, Verhandlungen 21:1515–1521.

Gilbert, J. J., and K. G. Bogdan. 1984. Rotifer grazing: in situ studies on selectivity and rates. Pages 97–133 *in* D. G. Meyers and J. R. Strickler, editors, Trophic Interactions within Aquatic Ecosystems. American Association for the Advancement of Science Selected Symposium 85. Westview, Boulder, Colorado, USA.

Gilbert, J. J., and K. L. Kirk. 1988. Escape response of the rotifer *Keratella*: description, stimulation, fluid dynamics, and ecological significance. Limnology and Oceanography 33(6, Part 2):1440–1450.

Gilbert, J. J., and P. L. Starkweather. 1977. Feeding in the rotifer *Brachionus calyciflorus* I. Regulatory mechanisms. Oecologia (Berlin) 28:125–131.

Gilbert, J. J., and P. L. Starkweather. 1978. Feeding in the rotifer *Brachionus calyciflorus* III. Direct observations on the effects of food type, food density, change in food type, and starvation on the incidence of pseudotrochal screening. Internationale Vereinigung für theoretische und angewandte Limnologie, Verhandlungen 20:2382–2388.

Gilbert, J. J., and R. S. Stemberger. 1985a. Control of *Keratella* populations by interference competition from *Daphnia*. Limnology and Oceanography 30:180–188.

Gilbert, J. J., and R. S. Stemberger. 1985b. Prey capture in the rotifer *Asplanchna girodi*. Internationale Vereinigung für theoretische und angewandte Limnologie, Verhandlungen 22:2997–3000.

Gilbert, J. J., and R. S. Stemberger. 1985c. The costs and benefits of gigantism in polymorphic species of the rotifer *Asplanchna*. Archiv für Hydrobiologie, Beiheft 21:185–192.

Gilbert, J. J., and C. E. Williamson. 1978. Predator-prey behavior and its effect on rotifer survival in associations of *Mesocyclops edax, Asplanchna girodi, Polyarthra vulgaris,* and *Keratella cochlearis*. Oecologia (Berlin) 37:13–22.

Green, J. 1972. Latitudinal variations in associations

of planktonic rotifers. Journal of Zoology, London 167:31–39.

Green, J. 1977. Dwarfing of rotifers in tropical crater lakes. Archiv für Hydrobiologie Beiheft 8:232–236.

Green, J. 1985. Horizontal variations in associations of zooplankton in Lake Kariba. Journal of Zoology, London 206:225–239.

Gvozdov, A.O. 1986. Phototaxis as a test function in bioassay. Gidrobiologichesky Zhurnal 22:65–68.

Halbach, U. 1970. Die Ursachen der Temporal Variation von *Brachionus calyciflorus* Pallas (Rotatoria). Oecologia 4:262–318.

Halbach, U. 1984. Population dynamics of rotifers and its consequences for ecotoxicology. *Hydrobiologia* 109:79–96.

Halbach, U., and G. Halbach-Keup. 1974. Quantitative Beiehungen zwischen Phytoplankton und der Populationdynamik des Rotators *Brachionus calyciflorus* Pallas. Befunde aus Laboratoriums-experimenten und Freilanduntersuchungen. Archiv für Hydrobiologie 73:273–309.

Halbach, U., M. Wiebert, C. Wissel, J. Kalus, K. Beuter, and M. Delion. 1981. Population dynamics of rotifers as bioassay tools for toxic effects of organic pollutants. Internationale Vereinigung für theoretische und angewandte Limnologie, Verhandlungen 21:1141–1146.

Halbach, U., M. Wiebert, M. Westermayer, and C. Wissel. 1983. Population ecology of rotifers as a bioassay tool for ecotoxicological tests in aquatic environments. Ecotoxicology and Environmental Safety 7:484–513.

Hall, D. J., S. T. Threlkeld, C. W. Burns, and P. H. Crowley. 1976. The size-efficiency hypothesis and the size structure of zooplankton communities. Annual Review of Ecology and Systematics 7:177–208.

Harring, H. K., and F. J. Myers. 1922. The rotifers of Wisconsin. Transactions of the Wisconsin Academy of Sciences, Arts, and Letters 20:553–662.

Hertel, E. W. 1942. Studies on vigor in the rotifer *Hydatina senta*. Physiological Zoology 15:304–324.

Herzig, A. 1983. Comparative studies on the relationship between temperature and the duration of embryonic development of rotifers. Hydrobiologia 104:237–246.

Herzig, A. 1987. The analysis of planktonic rotifer populations: a plea for long-term investigations. Hydrobiologia 147:163-180.

Hewitt, D. P., and D. G. George. 1987. The population dynamics of *Keratella cochlearis* in a hypereutrophic tarn and the possible impact of predation by young roach. Hydrobiologia 147:221–227.

Hillbricht-Ilkowska, A. 1983. Response of planktonic rotifers to the eutrophication process and to the autumnal shift of blooms in Lake Biwa, Japan. I. Changes in abundances and composition of rotifers. The Japanese Journal of Limnology 44:93–106.

Hirata, H., and W. Nagata. 1982. Excretion rates and excreted compounds of the rotifer *Brachionus plicatilis* O. F. Müller in culture. Memoirs of the Faculty of Fisheries Kagoshima University 31:161–174.

Hirayama, K., and S. Ogawa. 1972. Fundamental studies on the physiology of the rotifer for its mass culture I. Filter feeding of the rotifer. Bulletin of the Japanese Society of Scientific Fisheries 38:1207–1214.

Hofmann, W. 1977. The influence of abiotic factors on population dynamics in planktonic rotifers. Archiv für Hydrobiologie, Beiheft 8:77–83.

Hofmann, W. 1982. On the coexistence of two pelagic *Filinia* species (Rotatoria) in Lake Plußee I. Dynamics of abundance and dispersion. Archiv für Hydrobiologie 95:125–137.

Hurlbert, S.H., M.S. Mulla, and H.R. Wilson. 1972. Effects of an organophosphorus insecticides on the phytoplankton, zooplankton and insect populations of fresh-water ponds. Ecological Monographs 42:269–299.

Hutchinson, G.E. 1967. A treatise on limnology. Vol. 2: Introduction to lake biology and the limnoplankton. Wiley, New York.

Hyman, L. H. 1951. The invertebrates: Acanthocephala, Aschelminthes, and Entoprocta. The pseudocoelomate Bilateria. Vol. 3. McGraw-Hill, New York.

Iltis, A., and S. Riou-Duvat. 1971. Variations saisonniéres du peuplement en rotifères des eaux natronées du Kanem (Tchad). Cah. O.S.T.R.O.M. ser. Hydrobiol. 5(2):101–112.

Ito, T. 1960. On the culture of the mixohaline rotifer *Brachionus plicatilis* O. F. Müller, in seawater. Report of the Faculty of Fisheries, Prefectural University of Mie 3:708–740.

Jenkins, D. G., and A. L. Buikema. 1985. Plankton rotifer responses to herbicide stress in in situ microcosms. Virginia Journal of Science 36:144.

Jennings, H. S., and R. S. Lynch. 1928. Age, mortality, fertility and individual diversities in the rotifer *Proales sordida* Gosse. I. Effects of the age of the parent on characteristics of the offspring. Journal of Experimental Zoology. 50:345–407.

Johansson, S. 1983. Annual dynamics and production of rotifers in an eutrophication gradient in the Baltic Sea. Hydrobiologia 104:335–340.

Kaushik, N.K., G.L. Stephenson, K.R. Solomon, and K.E. Day. 1985. Impact of permethrin on zooplankton communities in limnocorrals. Canadian Journal of Fisheries and Aquatic Sciences 42:77–85.

Keshmirian, J., and T. Nogrady. 1988. Histofluorescent labelling of catecholaminergic structures in rotifers (Aschelminthes) II. Males of *Brachionus plicatilis* and structures from sectioned females. Histochemistry 89:189–192.

King, C. E. 1967. Food, age and the dynamics of a laboratory population of rotifers. *Ecology* 48:111–128.

King, C. E. 1969. Experimental studies of aging in rotifers. Experimental Gerontology 4:63–79.

King, C. E. 1972. Adaptation of rotifers to seasonal variation. Ecology 53:408–418.

King, C. E. 1977. Genetics of reproduction, variation, and adaptation in rotifers. Archiv für Hydrobiologie Beiheft 8:187–201.

King, C. E. 1980. The genetic structure of zooplankton populations. Pages 315-328 in: W.C. Kerfoot, editor. Evolution and Ecology of Zooplankton Communities.

University Press of New England, Hanover, New Hampshire.

King, C. E. 1983. A re-examination of the Lansing effect. Hydrobiologia 104:135–139.

King, C. E., and M.R. Miracle. 1980. A perspective on aging in rotifers. Hydrobiologia 73:13–19.

King, C. E., and T. W. Snell. 1977a. Sexual recombination in rotifers. Heredity 39:357–360.

King, C. E., and T.W. Snell. 1977b. Culture media (natural and synthetic): Rotifera. Pages 71–75 *in:* M. Rechcigal, Jr., editor. CRC Handbook Series in Nutrition and Food. Sect. G: Diets, Culture Media, Food Supplements. Vol II: Food Habits of, and Diets for Invertebrates and Vertebrates—Zoo Diets. CRC Press, Cleveland, Ohio.

King, C. E., and Y. Zhao. 1987. Coexistence of rotifer (*Brachionus plicatilis*) clones in Soda Lake, Nevada. Hydrobiologia 147:57–64.

Koehler, J. K. 1966. Some comparative fine structure relationships of the rotifer integument. Journal of Experimental Zoology 162:231–244.

Koehler, J. K., and T. L. Hayes. 1969. The rotifer jaw: a scanning and transmission electron microscope study. Journal of Ultrastructural Research 27:402–418.

Koste, W. 1976. Über die Rädertierbestände (Rotatoria) der oberen und mittleren Hase in de Jahren 1966–1969. Osnabrücker Naturwissenschaftliche. Mitteilungen 4:191–263.

Koste, W. 1978. Rotatoria. Die Rädertiere Mitteleuropas. 2 vols. Borntraeger, Berlin.

Koste, W. 1989. *Octotrocha speciosa*, eine seltene, sessile Art mit einem merkwürdigen Räderorgan. Mikrokosmos 78:115–121.

Koste, W., and S. J. De Paggi. 1982. Rotifera of the superorder Monogononta recorded from the Neotropics. Gewasser Abwasser 68/69:71–102.

Koste, W., and R.J. Shiel. 1986. Rotifera from Australian inland waters. I. Bdelloidea (Rotifera: Digononta). Australian Journal of Marine and Freshwater Research 37:765–792.

Koste, W., and R.J. Shiel. 1987. Rotifera from Australian inland waters. II. Epiphanidae and Brachionidae (Rotifera: Monogononta). Invertebrate Taxonomy 7:949–1021.

Lorenzen, S. 1985. Phylogenetic aspects of pseudocoelomate evolution. Pages 210–223 *in:* S. Conway Morris, J. D. George, and H. M. Platt, editors. The origins and relationships of lower invertebrates. Systematics Association, Vol. 28. Clarendon Press, Oxford.

Lubzens, E. 1987. Raising rotifers for use in aquaculture. Hydrobiologia 147:245–255.

Lubzens, E., G. Minkoff, and S. Maron. 1985. Salinity dependence of sexual and asexual reproduction in the rotifer *Brachionus plicatilis*. Marine Biology 85:123–126.

Lubzens, E., A. Tandler, and G. Minkoff. 1989. Rotifers as food in aquaculture. *Hydrobiologia* 186/187:387–400.

Luciani, A., J.-L. Chasse, and P. Clément. 1983. Aging in *Brachionus plicatilis:* the evolution of swimming as a function of age at two different calcium concentrations. Hydrobiologia 104:141–146.

MacIsaac, H. J., T. C. Hutchinson, and W. Keller. 1987. Analysis of planktonic rotifer assemblages from Sudbury, Ontario, area lakes of varying chemical composition. Canadian Journal of Fisheries and Aquatic Sciences 44:1692–1701.

Mäemets, A. 1987. Rotifers as indicators of lake types in Estonia. Hydrobiologia 104:357–361.

Magnien, R. E., and J. J. Gilbert. 1983. Diel cycles of reproduction and vertical migration in the rotifer *Keratella crassa* and their influence on the estimation of population dynamics. Limnology and Oceanography 28:957–969.

Makarewicz, J. C., and G. E. Likens. 1975. Niche analysis of a zooplankton community. *Science* 190:1000–1003.

Makarewicz, J. C., and G. E. Likens. 1979. Structure and function of the zooplankton community of Mirror Lake, New Hampshire. Ecological Monographs 49:109–127.

May, L. 1980. On the ecology of *Notholca squamula* Müller in Loch Leven, Kinross, Scotland. *Hydrobiologia* 73:177–180.

May, L. 1985. The use of procaine hydrochloride in the preparation of rotifer samples for counting. Internationale Vereinigung für theoretische und angewandte Limnologie, Verhandlungen 22:2987–2990.

May, L. 1986. Rotifer sampling—a complete species list from one visit? Hydrobiologia 134:117–120.

May, L. 1987. Effect of incubation temperature on the hatching of rotifer resting eggs collected from sediments. Hydrobiologia 147:335–338.

May, L. 1989. Epizoic and parasitic rotifers. Hydrobiologia 186/187:59–67.

May, L., and Jones, D. H. 1989. Does interference competition from *Daphnia* affect populations of *Keratella cochlearis* in Loch Leven, Scotland? Journal of Plankton Research 11:445–461.

Meadow, N. D., and C. H. Barrows, Jr. 1971. Studies on aging in a bdelloid rotifer. Journal of Experimental Zoology 176:303–314.

Miracle, M. R. 1974. Niche structure in freshwater zooplankton: a principal components approach. Ecology 55:1306–1316.

Miracle, M. R., M. Serra, E. Vincente, and C. Blanco. 1987. Distribution of *Brachionus* species in Spanish mediterranean wetlands. Hydrobiologia 147:75–81.

Moore, M.V. 1988. Density-dependent predation of early instar *Chaoborus* feeding on multispecies prey assemblages. Limnology and Oceanography 33:256–269.

Moore, M. V., and J. J. Gilbert. 1987. Age-specific *Chaoborus* predation on rotifer prey. Freshwater Biology 17:223–236.

Myers, F. J. 1936. Psammolittoral rotifers of Lenape and Union Lakes, New Jersey. American Museum Novitates 830:1–22.

Myers, F. J. 1937. A method of mounting rotifer jaws for study. Transactions of the American Microscopical Society 56:256–257.

Neel, J. K. 1948. A limnological investigation of the psammon in Douglas Lake, Michigan, with especial reference to shoal and shoreline dynamics. Transactions of the American Microscopical Society 67:1–53.

Neill, W. E. 1984. Regulation of rotifer densities by crustacean zooplankton in an oligotrophic montane lake in British Columbia. Oecologia 61:175–181.

Nipkow, F. 1952. Die Gattung *Polyarthra* Ehernberg im Plankton des Zürichsees und einiger anderer Schweizer Seen. Schweizer Journal Hydrobiologie 14:135–181.

Nipkow, F. 1961. Die Radertiere im Plankton des Zürchsee und ihre Entwisklungsphasen. Schweizer Journal Hydrobiologie 23:398–461.

Nogrady, T. 1982. Rotifera. Pages 865-872 *in:* S. P. Parker, editor. Synopsis and classification of living organisms. McGraw-Hill, New York.

Nogrady, T. 1988. The littoral rotifera plankton of the Bay of Quinte (Lake Ontario) and its horizontal distribution as indicators of trophy I. A full season study. Archiv für Hydrobiologie (Supplement) 79:145–165.

Nogrady, T., and M. Alai. 1983. Cholinergic neurotransmission in rotifers. Hydrobiologia 104:149–153.

Nogrady, T., and J. Keshmirian. 1986a. Rotifer neuropharmacology I. Cholinergic drug effects on oviposition of *Philodina acuticornis* (Rotifera, Aschelminthes). Comparative Biochemistry and Physiology 83C:335–338.

Nogrady, T., and J. Keshmirian. 1986b. Rotifer neuropharmacology II. Synergistic effect of acetyl-choline on local anesthetic activity in *Brachionus calyciflorus* (Rotifera, Aschelminthes). Comparative Biochemistry and Physiology 83C:339–344.

O'Brien, W.J. 1979. The predator–prey interaction of planktivorous fish and zooplankton. American Scientist 67:572–581.

O'Brien, W.J. 1987. Planktivory by freshwater fish: thrust and parry in pelagia. Pages 3–16 *in:* C. W. Kerfoot and A. Sih, editors. Predation: Direct and Indirect Impacts on Aquatic Communities. University Press of New England, Hanover, New Hampshire.

Paloheimo, J.E. 1974. Calculation of instantaneous birth rates. Limnology and Oceanography 19:692–694.

Pejler, B. 1956. Introgression in planktonic Rotatoria with some points of view on its causes and results. Evolution 10:246–261.

Pejler, B. 1977a. General problems on rotifers taxonomy and global distribution. Archiv für Hydrobiologie Beiheft 8:212–220.

Pejler, B. 1977b. On the global distribution of the family Brachionidae (Rotatoria). Archiv für Hydrobiologie (Supplement) 53:255–306.

Pennak, R. W. 1940. Ecology of the microscopic metazoa inhabiting the sandy beaches of some Wisconsin lakes. Ecological Monographs 10:537–615.

Pennak, R. W. 1962. Quantitative zooplankton sampling in littoral vegetation areas. Limnology and Oceanography 7:487–489.

Pennak, R. W. 1989. Fresh-water invertebrates of the United States. 3rd Edition. Wiley, New York.

Persoone, G., A. Van De Vel, M. Van Steertegem, and B. De Nayer. 1989. Predictive value of laboratory tests with aquatic invertebrates: influence of experimental conditions. Aquatic Toxicology 14:149–166.

Pilarska, J. 1972. The dynamics of growth of experimental populations of the rotifer *Brachionus rubens* Ehrbg. Polish Archives for Hydrobiology 19:265–277.

Pontin, R. S. 1978. A Key to the British Freshwater Planktonic Rotifera. Scientific Publications No. 38. Freshwater Biological Association, Cumbria, England.

Pourriot, R. 1958. Sur l'élevage des Rotifères au laboratoire. Hydrobiologia 11:189–197.

Pourriot, R. 1965. Recherches sur l'ecologie des Rotifères. Vie et Milieu (Supplement) 21:1–224.

Pourriot, R. 1977. Food and feeding habits of Rotifera. Archiv für Hydrobiologie Beiheft 8:243–260.

Pourriot, R. 1979. Rotifères du sol. Revue d'Ecologie et de Biologie du Sol. 16:279–312.

Pourriot, R. 1986. Les rotifères—Biologie. *in:* G. Barnabé, editor. Aquaculture Volume 1. Technique et documentation. Lavorsier, Paris.

Pourriot, R., and P. Clément. 1981. Action de facteurs externes sur la reproduction et le cycle reproducteur des Rotifers. Acta Oecologia Generale 2:135–151.

Pourriot, R., and C. Rougier. 1975. Dynamique d'une population experimentale de *Brachionus dimidatus* (Bryce) (Rotifère) en fonction de la nourriture et de la temperature. Annals Limnology 11:125–143.

Pourriot, R., and T. W. Snell. 1983. Resting eggs in rotifers. Hydrobiologia 104:213–214.

Pourriot, R., C. Rougier, and D. Benest. 1986. Food quality and mictic female control in the rotifer *Brachionus rubens* Ehr. Bulletin Society Zoologie Frances 111:105–112.

Preissler, K. 1980. Field experiments on the optical orientation of pelagic rotifers. Hydrobiologia 73:199–203.

Rao, T. R., and S. S. S. Sarma. 1986. Demographic parameters of *Brachionus patulus* Muller (Rotifera) exposed to sublethal DDT concentrations at low and high food levels. Hydrobiologia 139:193–200.

Rees, B. 1960. *Alberta vermicularis* (Rotifera) parasitic in the earthworm *Allolobophora caliginosa*. Parasitology 50:61–65.

Remane, A. 1963. The systematic position and phylogeny of the pseudocoelomates. Pages 247–255 *in:* E. C. Dougherty, editor. The Lower Metazoa. University of California Press, Berkeley.

Ricci, C. 1983. Life histories of some species of Rotifera Bdelloida. Hydrobiologia 104:175–180.

Ricci, C. 1984. Culturing of some bdelloid rotifers. Hydrobiologia 112:45–51.

Ricci, C. 1987. Ecology of bdelloids: how to be successful. Hydrobiologia 147:117–127.

Ricci, C., L. Vaghi, and M.L. Manzini. 1987. Desiccation of rotifers (*Macrotrachela quadricornifera*): survival and reproduction. Ecology 68:1488–1494.

Robb, E. J., and G. L. Barron. 1982. Nature's ballistic missile. Science 218:1221–1222.

Rogerson, A., J. Berger, and C.M. Grosso. 1982. Acute toxicity of ten crude oils on the survival of the rotifer *Asplanchna sieboldi* and sublethal effects on rates of prey consumption and neonate production. Environmental Pollution (Series A) 29:179–187.

Rothhaupt, K. O. 1985. A model approach to the population dynamics of the rotifer *Brachionus rubens* in two-stage chemostat culture. Oecologia 65:252–259.

Rothhaupt, K. O. 1990a. Differences in particle size-dependent feeding efficiencies of closely related rotifer species. Limnology and Oceanography 35:16–23.

Rothhaupt, K. O. 1990b. Changes of the functional responses of the rotifers *Brachionus rubens* and *Brachionus calyciflorus* with particle size. Limnology and Oceanography 35:24–32.

Rougier, C., and R. Pourriot. 1977. Aging and control of the reproduction in *Brachionus calyciflorus* (Pallas) (Rotatoria). Experimental Gerontology 12:137–151.

Ruttner-Kolisko, A. 1963. The interrelationships of the Rotatoria. Pages 263–272 *in:* E.C. Dougherty, editor. The Lower Metazoa. University of California Press, Berkeley.

Ruttner-Kolisko, A. 1969. Kreuzungexperimente zwischen *Brachionus urceolaris* and *Brachionus quadridentatus*, ein Beitrag zur Fortpflanzungbiologie der heterogonen Rotatoria. Archiv für Hydrobiologie 65:397–412.

Ruttner-Kolisko, A. 1974. Planktonic rotifers biology and taxonomy. Die Binnengewässer (Supplement) 26:1–146.

Ruttner-Kolisko, A. 1977a. The effect of the microsporid *Plistophora asperospora* on *Conochilus unicornis* in Lunzer Untersee (LUS). Archiv für Hydrobiologie Beiheft 8:135–137.

Ruttner-Koliski, A. 1977b. Amphoteric reproduction in a population of *Asplanchna priodonta*. Archiv fur Hydrobiologie Beifheft 8:178–181.

Salt, G. W. 1987. The components of feeding behavior in rotifers. Hydrobiologia 147:271–281.

Salt, G. W., G. F. Sabbadini, and M. L. Commins. 1978. Trophi morphology relative to food habits in six species of rotifers (Asplanchnidae). Transactions of the American Microscopical Society 97:469–485.

Sawada, M., and H. E. Enesco. 1984. A study of dietary restriction and lifespan in the rotifer *Asplanchna brightwelli* monitored by chronic neutral red exposure. Experimental Gerontology 19:329–334.

Schlüter, M. 1980. Mass culture experiments with *Brachionus rubens*. Hydrobiologia 73:45–50.

Schlüter, M., and J. Groeneweg. 1981. Mass production of freshwater rotifers on liquid wastes. I. The influence of some environmental factors on population growth on *Brachionus rubens* Ehrenberg 1838. Aquaculture 25:17–24.

Schlüter, M., and J. Groeneweg. 1985. The inhibition by ammonia of population growth of the rotifer, *Brachionus rubens*, in continuous culture. Aquaculture 46:215–220.

Schlüter, M., J. Groeneweg, and C. J. Soeder. 1987. Impact of rotifer grazing on population dynamics of green microalgae in high-rate ponds. Water Research 10:1293–1297.

Schnese, W. 1973. Relations between phytoplankton and zooplankton in brackish coastal water. Oikos (Supplement) 15:28–33.

Schramm, U., and W. Becker. 1987. Anhydrobiosis of the bdelloid rotifer *Habrotrocha rosa* (Aschelminthes). Zeitschrift fuer Mikroskopisch–Anatomische Forschung 101:1-17.

Scott, J. M. 1983. Rotifer nutrition using supplemented monoxenic cultures. Hydrobiologia 104:155–166.

Scott, J. M. 1988. Effect of growth rate on the physiological rates of a chemostat-grown rotifer *Encentrum linnhei*. Journal of the Marine Biological Association of the United Kingdom 68:165–177.

Seaman, M. T., M. Gophen, B. Z. Cavari, and B. Azoulay. 1986. *Brachionus calyciflorus* Pallas as agent for removal of *E. coli* in sewage ponds. Hydrobiologia 135:55–60.

Serra, M., and M. Miracle. 1985. Enzyme polymorphisms in *Brachionus plicatilis* populations from several Spanish lagoons. Internationale Vereinigung für theoretische und angewandte Limnologie, Verhandlungen 22:2991–2996.

Serrano, L., M. R. Miracle, and M. Serra. 1986. Differential response of *Brachionus plicatilis* (Rotifera) ecotypes to various insecticides. Journal of Environmental Biology 7:259–275.

Sladecek, V. 1983. Rotifers as indicators of water quality. Hydrobiologia 100:169–201.

Snell, T. W. 1977. Lifespan of male rotifers. Archiv für Hydrobiologie Beiheft 8:65–66.

Snell, T. W. 1979. Intraspecific competition and population structure in rotifers. Ecology 60:494–502.

Snell, T. W. 1980. Blue-green algae and selection in rotifer populations. Oecologia 46:343–346.

Snell, T. W. 1986. Effects of temperature, salinity and food level on sexual and asexual reproduction in *Brachionus plicatilis* (Rotifera). Marine Biology 92:157–162.

Snell, T. W. 1989. Systematics, reproductive isolation and species boundaries in monogonont rotifers. Hydrobiologia 186/187:299–310.

Snell, T. W., and M. J. Childress. 1987. Aging and loss of fertility in male and female *Brachionus plicatilis* (Rotifera). International Journal of Invertebrate Reproduction and Development 12:103–110.

Snell, T. W., and B. L. Garman. 1986. Encounter probabilities between male and female rotifers. Journal of Experimental Marine Biology and Ecology 97:221–230.

Snell, T. W., and C. A. Hawkinson. 1983. Behavioral reproductive isolation among populations of the rotifer *Brachionus plicatilis*. Evolution 37:1294–1305.

Snell, T. W., and F. H. Hoff. 1985. The effect of environmental factors on resting egg production in the rotifer *Brachionus plicatilis*. Journal of the World Mariculture 16:484–497.

Snell, T. W., and F. H. Hoff. 1987. Fertilization and male fertility in the rotifer *Brachionus plicatilis*. Hydrobiologia 147:329–334.

Snell, T. W., and C. E. King. 1977. Lifespan and fecundity patterns in rotifers: the cost of reproduction. Evolution 31:882–890.

Snell, T. W., and G. Persoone. 1988a. Acute toxicity bioassays using rotifers. I. A test for marine and brackish environments with *Brachionus plicatilis*. Aquatic Toxicology 14:65–80.

Snell, T. W., and G. Persoone. 1988b. Acute toxicity bioassays using rotifers. II. A freshwater test with *Brachionus rubens*. Aquatic Toxicology 14:81–92.

Snell, T. W., and B. C. Winkler. 1984. Isozyme analysis of rotifer protein. Biochemical Systematics and Ecology 12:199–202.

Snell, T. W., B. E. Burke, and S. D. Messur. 1983. Size and distribution of resting eggs in a natural population of the rotifer *Brachionus plicatilis*. Gulf Research Reports 7:285–288.

Snell, T. W., M. J. Childress, E. M. Boyer, and F. H. Hoff. 1987. Assessing the status of rotifer mass cultures. Journal of the World Aquaculture Society 18:270–277.

Snell, T. W., M. J. Childress, and B.C. Winkler. 1988. Characteristics of the mate recognition factor in the rotifer *Brachionus plicatilis*. Comparative Biochemistry and Physiology 89A:481–485.

Sohlenius, B. 1982. Short-term influence of clear-cutting on abundance of soil-microfauna (Nematoda, Rotatoria and Tardigrada) in a Swedish pine forest soil. Journal of Applied Ecology 19:349–359.

Spoon, D. M. 1978. A new rotary microcompressor. Transactions of the American Microscopical Society 97:412–416.

Starkweather, P. L. 1980. Aspects of the feeding behavior and trophic ecology of suspension feeding rotifers. Hydrobiologia 73:63–72.

Starkweather, P. L. 1987. Rotifera. Pages 159–183 in: T. J. Pandian and F.J. Vernberg, editors. Animal Energetics. Vol. 1: Protozoa through Insecta. Academic Press, Orlando, Florida.

Starkweather, P. L., and P. E. Kellar. 1987. Combined influences of particulate and dissolved factors in the toxicity of *Microcyctis aeruginosa* (NRC-SS-17) to the rotifer *Brachionus calyciflorus*. Hydrobiologia 147:375–378.

Stemberger, R. S. 1979. A Guide to Rotifers of the Laurentian Great Lakes. U.S. Environmental Protection Agency, Cincinnati, Ohio. (Available from National Technical Information Service, Springfield, Virginia. PB80-101280)

Stemberger, R. S. 1981. A general approach to the culture of planktonic rotifers. Canadian Journal of Fisheries and Aquatic Sciences 38:721–724.

Stemberger, R. S. 1984. Spine development in the rotifer *Keratella cochlearis*: induction by cyclopoid copepods and *Asplanchna*. Freshwater Biology 14:639–647.

Stemberger, R. S. 1987. The potential for population growth of *Ascomorpha ecaudis*. Hydrobiologia 147:297–301.

Stemberger, R. S. 1988. Reproductive costs and hydrodynamic benefits of chemically induced defenses in *Keratella testudo*. Limnology and Oceanography 33:593–606.

Stemberger, R. S. 1990. Food limitation, spination, and reproduction in *Brachionus calyciflorus*. Limnology and Oceanography 35:33–44.

Stemberger, R. S., and J. J. Gilbert. 1984a. Body size, ration level, and population growth in *Asplanchna*. Oecologia (Berlin) 64:355–359.

Stemberger, R. S., and J.J. Gilbert. 1984b. Spine development in the rotifer *Keratella cochlearis*: induction by cyclopoid copepods and *Asplanchna*. Freshwater Biology 14:639–647.

Stemberger, R. S., and J. J. Gilbert. 1985. Body size, food concentration and population growth in planktonic rotifers. *Ecology* 66:1151–1159.

Stemberger, R. S., and J. J. Gilbert. 1987a. Defenses of planktonic rotifers against predators. Pages 227–239 in: W. C. Kerfoot and A. Sih, editors. Predation: Direct and Indirect Impacts on Aquatic Communities. University Press of New England, Hanover, New Hampshire.

Stemberger, R. S., and J. J. Gilbert. 1987b. Multiple species induction of morphological defenses in the rotifer *Keratella testudo*. Ecology 68:370–378.

Threlkeld, S. T., J. T. Rybock, M. D. Morgan, C. L. Folt, and C. R. Goldman. 1980. The effects of an introduced invertebrate predator and food resource variation on zooplankton dynamics in an ultraoligotrophic lake. Pages 555–568 in: W. C. Kerfoot, editor. Evolution and Ecology of Zooplankton Communities. University Press of New England, Hanover, New Hampshire.

Torrans, E. L. 1986. Fish/plankton interactions. Pages 67–81 in: J. E. Lannan, R. O. Smitherman, and G. Tchobanoglous, editors. Principles and Practices of Pond Aquaculture. Oregon State University Press, Corvallis, Oregon.

Tzean, S. S., and G. L. Barron. 1983. A new predatory hypomycete capturing bdelloid rotifers in soil. Canadian Journal of Botany 61:1345–1348.

Tzschaschel, G. 1980. Verteilung, Abundanzdynamik und Biologie mariner interstitieller Rotatoria. Mikrofauna Meeresboden 81:1–56.

Vidrine, M. F., R. E. McLaughlin, and O.R. Willis. 1985. Free-swimming colonial rotifers (Monogononta: Floscularidea: Flosculariidae) in Southwestern Louisiana rice fields. Freshwater Invertebrate Biology 4:187–193.

Wagstaffe, R., and J. H. Fidler. 1961. The Preservation of Natural History Specimens. Philosophical Library, New York.

Wallace, R. L. 1975. Larval behavior of the sessile rotifer *Ptygura beauchampi* (Edmondson). Internationale Vereinigung für theoretische und angewandte Limnologie, Verhandlungen 19:2811–2815.

Wallace, R. L. 1977a. Adaptive advantages of substrate selection by sessile rotifers. Archiv für Hydrobiologie Beiheft 8:53–55.

Wallace, R. L. 1977b. Distribution of sessile rotifers in an acid bog pond. Archiv für Hydrobiologie 79:478–505.

Wallace, R. L. 1978. Substrate selection by larvae of the sessile rotifer *Ptygura beauchampi*. Ecology 59:221–227.

Wallace, R. L. 1980. Ecology of sessile rotifers. Hydrobiologia 73:181–193.

Wallace, R. L. 1987. Coloniality in the phylum Rotifera. Hydrobiologia 147:141–155.

Wallace, R. L., and R. A. Colburn. 1989. Phylogenetic relationships within phylum Rotifera: orders and genus *Notholca*. Hydrobiologia 186/187:311–318.

Wallace, R. L., and W. T. Edmondson. 1986. Mechanism and adaptive significance of substrate selection by a sessile rotifer. Ecology 67:314–323.

Wallace, R. L., and P. L. Starkweather. 1985. Clearance rates of sessile rotifers: in vitro determinations. Hydrobiologia 121:139–144.

Wallace, R. L., W. K. Taylor, and J. R. Litton. 1989. Invertebrate zoology. Macmillan, New York. 337 pp.

Walz, N. 1983a. Continuous culture of the pelagic rotifers *Keratella cochlearis* and *Brachionus angularis*. Archiv für Hydrobiologie 98:70–92.

Walz, N. 1983b. Individual culture and experimental population dynamics of *Keratella cochlearis* (Rotatoria). Hydrobiologia 107:35–45.

Walz, N. 1987. Comparative population dynamics of the rotifers *Brachionus angularis* and *Keratella cochlearis*. Hydrobiologie 147:209–213.

Walz, N., H.-J. Elster, and M. Mezger. 1987. The development of the rotifer community structure in Lake Constance during its eutrophication. Archiv für Hydrobiologie (Supplement) 74:452–487.

Watanabe, T., C. Kitajima, and S. Fujita. 1983. Nutritional values of live food organisms used in Japan for mass propogation of fish: a review. Aquaculture 34:115–143.

Whitfield, P. J. 1971. Phylogenetic affinities of Acanthocephala: an assessment of ultrastructural evidence. Parasitology 63:49–58.

Williamson, C. E. 1983. Invertebrate predation on planktonic rotifers. Hydrobiologia 104:385–396.

Williamson, C. E. 1987. Predator–prey interactions between omnivorous diaptomid copepods and rotifers: the role of prey morphology and behavior. Limnology and Oceanography 32:167–177.

Wurdak, E. 1987. Ultrastructure and histochemistry of the stomach of *Asplanchna sieboldi*. Hydrobiologia 147:361–371.

Wurdak, E., and J.J. Gilbert. 1976. Polymorphism in the rotifer *Asplanchna sieboldi:* fine structure of saccate, cruciform and campanulate females. Cell Tissue Research 169:435–448.

Wurdak, E., J. J. Gilbert, and R. Jagels. 1977. Resting egg ultrastructure and formation of the shell in *Asplanchna sieboldi* and *Brachionus calyciflorus*. Archiv für Hydrobiologie Beiheft 8:298–302.

Wurdak, E., P. Clément, and J. Amselem. 1983. Sensory receptors involved in the feeding behavior of the rotifer *Asplanchna brightwelli*. Hydrobiologia 104:203–212.

Nematoda and Nematomorpha

9

George O. Poinar, Jr.

Department of Entomology
University of California
Berkeley, California 94720

Chapter Outline

NEMATODA

I. INTRODUCTION

II. MORPHOLOGY AND PHYSIOLOGY
   A. Cuticle
   B. Hypodermis
   C. Muscular System
   D. Digestive System
   E. Excretory System
   F. Respiration
   G. Nervous System
   H. Sense Organs
   I. Reproductive System

III. DEVELOPMENT AND LIFE HISTORY

IV. ECOLOGY
   A. Fossil Record
   B. Habitats of Freshwater Nematodes
   C. Enemies

V. COLLECTING AND REARING TECHNIQUES
   A. Sampling Methods
   B. Extraction
   C. Fixing and Mounting
   D. Culturing

VI. IDENTIFICATION
   A. Taxonomic Key to Families and Selected Genera of Freshwater Nematoda
   B. Taxonomic Key to Genera of Freshwater Mermithidae

NEMATOMORPHA

VII. INTRODUCTION

VIII. MORPHOLOGY AND PHYSIOLOGY

IX. DEVELOPMENT AND LIFE HISTORY

X. SAMPLING

XI. IDENTIFICATION
   A. Taxonomic Key to Genera of Nematomorpha
   Literature Cited
   Appendix 9.1 Systematic Arrangement of the Nematode Genera

NEMATODA

I. INTRODUCTION

The phylum Nematoda comprises a wide range of roundworms that can be categorized either in nutritional categories (microbotrophic, predaceous, parasitic in plants, invertebrates, or vertebrates) or ecological groupings (soil, marine, freshwater, ectoparasitic, endoparasitic, or free-living) (Poinar 1983). Relatively little attention has been devoted to this important group of invertebrates by freshwater biologists. Difficulties associated with sampling, extraction, and identification are the main reasons for this lack of attention.

The nematodes treated here are those which during all or a large part of their life cycles occur in freshwater habitats (these habitats are discussed elsewhere). Nematode representatives of essentially all of the nutritional categories just mentioned occur in freshwater during one or more life stages. Nematode parasites of vertebrates that live in or frequent freshwater usually occur in the freshwater habitat only as eggs or within intermediate hosts and, therefore, are not discussed in this chapter. Mermithid parasites of aquatic insects are included since the eggs, infective stages, postparasitic juveniles, and adults are frequently encountered in freshwater habitats.

Only nematode genera containing obligate aquatic species (those which live nowhere else) that have been reported from North America are included in the keys in the present work. The systematic arrangement of these genera is listed in Appendix 9.1. As more studies are undertaken, it is likely that genera described from other continents will also be represented in North America. In addition, the reader should note that undescribed genera may be encountered in routine sampling.

The related group, Nematomorpha, or hairworms, are covered here since they are frequently encountered by freshwater biologists. Superficially, they resemble mermithid nematodes and share with the latter group similar life histories.

II. MORPHOLOGY AND PHYSIOLOGY

Nematodes are nonsegmented, wormlike invertebrates lacking jointed appendages but possessing a body cavity and a complete alimentary tract (Fig. 9.1). While they lack specialized respiratory and circulatory systems, nematodes possess a well-developed nervous system, an excretory system, and a set of longitudinal muscles.

With the exception of the family Mermithidae, most freshwater nematodes are under 1 cm in length. Although the basic body shape and anatomic plan of all nematodes are similar, there are some characteristics of most freshwater nematodes that are lacking in many terrestrial forms. One of these is the presence of three unicellular hypodermal glands commonly located in the upper part of the tail, immediately behind the anus (Figs. 9.1 and 9.2d). These caudal glands produce a secretion that is carried by a canal to the tip of the tail. At this point, the canal is attached to a valve (spinneret), which passes through a pore in the tail terminus (Figs. 9.1 and 9.6d). The secretion produces an adhesive deposit by which the nematodes can attach themselves to the surfaces of submerged rocks, plants, invertebrates, and debris.

Mucous may also be produced by pharyngeal glands of aquatic nematodes. The deposits from the caudal and pharyngeal glands can be used to form slimy traces over the substratum. Microorganisms that become entrapped in these deposits serve as food when the nematodes ingest the mucous together with the attached microorganisms (Riemann and Schrage 1978).

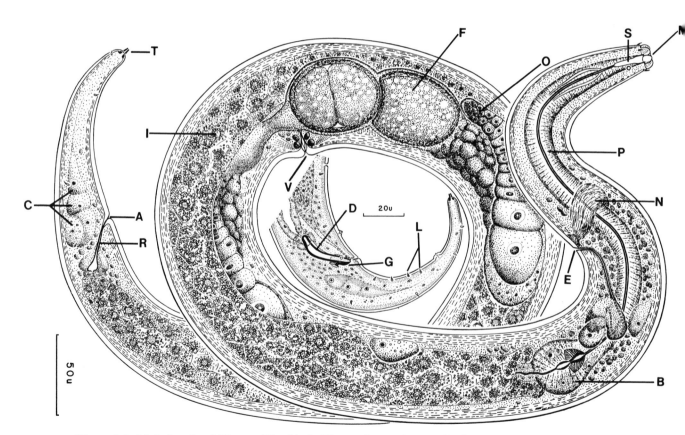

Figure 9.1 Adult female of *Plectus* (Plectidae). [Center Insert shows tail of male.] M, mouth; S, stoma; P, pharynx; N, nerve ring; E, excretory pore; B, valvated basal bulb of pharynx; I, intestine; R, rectum; A, anus; C, caudal glands; T, spinneret; V, vulva; O, ovary; F, egg; D, spicule; G, gubernaculum; L, genital papillae. (Drawing by A. Maggenti, University of California, Davis; labeling by G. Poinar.)

A. Cuticle

The exterior body covering on all nematodes is a noncellular, flexible, multilayered structure called the cuticle. It is secreted by a layer of underlying hypodermal cells and covers the entire external surface of the nematode, with portions entering various external openings such as the stoma, rectum, vagina, amphids, and sometimes the excretory pore. The cuticle is composed of an outer cortex, a middle matrix, and an inner fiber or basal zone. All of these regions are constructed of sublayers, which vary in thickness according to the species and stage. The outermost lipoidal layer of the cortex is important because it serves as a semipermeable membrane, allowing water and solutes to pass in and out of the body cavity.

The cuticle of freshwater nematodes may bear longitudinal striations, punctations, bristles, setae, or somatic alae (Figs. 9.6b and 9.7a). When present, the bursa is attached to the male tail and is composed of flattened flaps of cuticle which guide and hold the male in place during mating (Fig. 9.6c). All of these cuticular structures are useful characters for identifying nematode groups or species. The cuticle is normally shed four times during the development of all nematodes. At each molt (ecdysis), the old cuticle splits and is discarded. In a few forms, the old cuticle is retained, producing a double cuticle or sheath around the nematode (Fig. 9.8b).

B. Hypodermis

The hypodermis is a layer of tissue that is located beneath the cuticle and is responsible for the formation of the cuticle. It is generally a thin layer of tissue that expands into the coelom to form longitudinal cords between the muscle fields. Most nematodes possess lateral, dorsal, and ventral cords that contain the nuclei and other cytoplasmic inclusions of the hypodermis. The hypodermal cords originate as distinct cells but later lose their identity and form a syncytium.

C. Muscular System

Nematode movement is controlled by longitudinal muscles because these organisms do not possess circular muscles. These somatic muscles are composed of many adjacent cells which are innervated by nerves lodged in the ventral and dorsal hypodermal cords. Most nematode movement superficially resembles that of many other elongate, stiff-bodied, appendageless animals, whether they be vertebrates (snakes) or invertebrates (biting midge larvae). In locomotion, the side-to-side bending of the body is produced by alternating contractions and relaxations of opposite muscle sectors that are controlled by the central nervous system. When placed on flat surfaces in the laboratory, nematodes normally crawl on their sides. Some aquatic nematodes can swim by rapid vibratory or thrashing side-to-side movements of the body.

D. Digestive System

All freshwater nematodes except the Mermithidae contain a continuous alimentary tract composed of a stoma, pharynx, and intestine (Fig. 9.1).

The stoma (mouth) is the most anterior part of the digestive tract and is usually lined with cuticle. The stomal walls are formed by a fusion of various smaller segments or rings (rhabdions), which may be separated as is found in the Cephalobidae, or fused, as is found in the Rhabditidae (Fig. 9.8a). The structure of the stoma generally reflects the food selection of the bearer. For example, predaceous nematodes tend to have a wide stoma armed with teeth (Fig. 9.7b), a stylet, or a spear (Fig. 9.9b), whereas microbotrophic forms generally have a small tubular stoma (Fig. 9.8a). Plant parasites have protrusible stylets which can be inserted into plant tissues (Fig. 9.8b); and the infective stages of invertebrate parasites often have a small stylet used in penetrating the cuticle or intestinal wall of the host.

The pharynx (or esophagus) lies between the stoma and intestine and passes food from the mouth into the intestine. It is usually composed of radial muscles, which upon dilation allow the pharynx to function as a pump or vacuum cleaner. The pharynx also contains glands that produce secretions used in digestion, escape from the egg shell, molting, and penetration of host tissue. There are normally three portions to the pharynx: (1) the corpus (sometimes enlarged to form a distinct metacorpus or median bulb); (2) the isthmus (usually surrounded by the nerve ring); and (3) the basal bulb (often containing valves to keep food particles from reversing their direction of flow) (Fig. 9.8a). When nematodes feed, material is sucked up into the mouth by reduced pressure resulting from contraction of the radial muscles of the pharynx. Food is passed along the lumen by waves of alternating contractions and relaxations of the pharyngeal muscles. The pharyngeal lumen opens behind the food, thus creating suction pressure, and closes in front of the food, forcing it through.

The intestine is a single-cell-thick, cylindrical tube that runs from the pharynx to the anal opening (cloacal opening in males). It is composed of an anterior ventricular region, a midintestine or intestine proper, and a posterior rectal region (not always

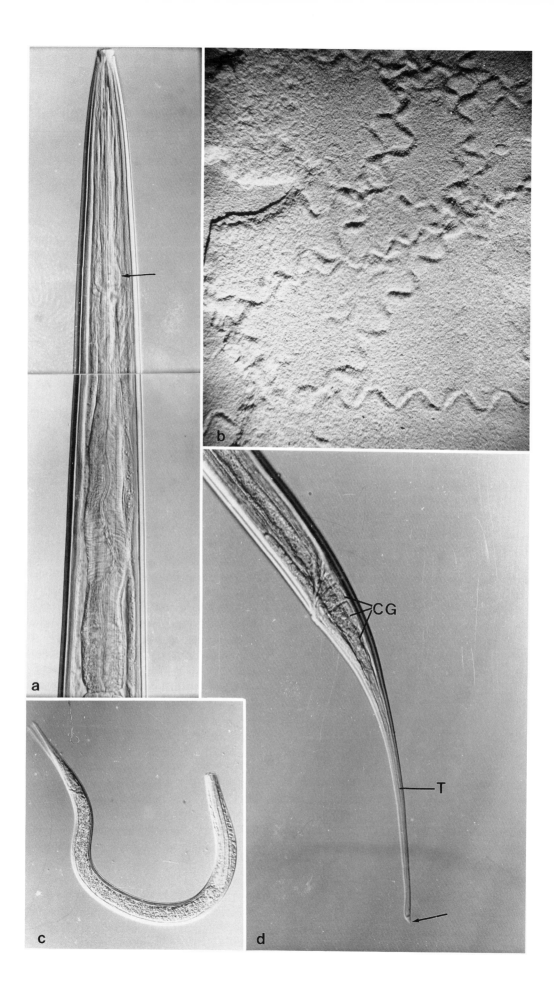

obvious). The inner surfaces of the cells facing the lumen are lined with microvillae which seem to be responsible for the major uptake of nutrients. A pharyngeal–intestinal valve is located in the anterior portion, behind the basal bulb of the pharynx. In some freshwater forms, this valve is quite distinct and elongated. An intestinal–rectal valve is located at the junction of the intestine and rectum. This unicellular sphincter muscle controls the amount of water passing out of the body. Many rhabditid-type nematodes have three rectal glands, which empty into the rectum. These glands are not to be confused with the caudal glands, which produce secretions from the tail terminus.

The Mermithidae are unusual in having the intestine detached from the remainder of the alimentary tract and modified into a trophosome or food storage organ. During its development, the trophosome separates from the pharynx and rectum, thus remaining as a separate, isolated organ.

E. Excretory System

Although variable in structure, the excretory system of nematodes falls into two basic types. The first is frequently found in freshwater forms and consists of a ventral excretory gland (renette cell) connected by an excretory duct to the excretory pore on the surface of the cuticle. The ventral pore usually opens in the pharyngeal or anterior intestinal region of the body. The second type is a tubular system consisting of a series of longitudinal excretory canals; pooled contents of these canals pass contents into a joining canal which connects with the excretory duct and pore. Most freshwater nematodes excrete nitrogen in the form of ammonia or urea. The excretory system also has an osmoregulatory function of removing water that diffuses into the pseudocoelom and so regulates the amount of turgor pressure in the body cavity; some scientists believe that this is the primary function of the "excretory" system.

F. Respiration

Most nematodes are aerobic organisms, at least during their developmental period. Most nematodes can survive short periods of anaerobic conditions but only a few forms can survive anoxia indefinitely. Strayer (1985) listed representatives of three nematode genera that he considered as benthic species found under anoxic conditions in Mirror Lake. A specialized respiratory system is lacking in nematodes. Instead, they obtain and lose gases by simple diffusion. Body cavity cells, when present, appear to play no role in oxygen transport.

G. Nervous System

The nervous system of nematodes centers around a central anterior mass of ganglia or "brain," which connects with nerve cords extending anteriorly and posteriorly through the body. The central ganglionic mass is closely associated with the circumpharyngeal commissure or nerve ring (Fig. 9.2a). Nerves extending anteriorly from the nerve ring innervate the amphids and cephalic sense organs, while those running posteriorly control the male tail papillae.

H. Sense Organs

The size, shape, position, and spatial arrangement of sense organs are all characteristic of nematode species or groups and are therefore frequently used as taxonomic characters.

The anterior sensory papillae are found on the nematode head. The basic number is 16 but usually some have become vestigial. There is usually an inner circle of six labial papillae around the mouth, a second ring of six labial papillae slightly behind the first set, and a third circle of four cephalic papillae still further back on the head. In mermithids, the cephalic papillae may also include adjacent enlargements of the hypodermal tissue.

The amphids are paired, lateral sense organs that open to the exterior on the nematode cuticle. They may be located on the same circle as the labial or cephalic papillae or further back in the neck region. The term amphid generally refers to the exterior configuration of the amphidial opening, although the amphid also consists of an amphidial pouch, amphidial gland, and amphidial nerves. The structure of the amphidial openings and their location are very important taxonomic characters. They can vary from small pores to large circles, packets, or spirals (Figs. 9.7c, d). Sexual dimorphism may be evident with either the male or female possessing a considerably larger (usually the same general shape) amphid.

Figure 9.2 (a) Gradually tapering pharynx of a freshwater *Dorylaimus* sp. Note absence of distinct valve in basal portion of pharynx. Arrow shows the nerve ring. (b) Fossil tracks thought to be made by freshwater nematodes. From the 50 million year old (Middle Eocene) Green River sediments of Lake Uinta in the Soldier summit area near Provo, Utah. (c) A juvenile of a freshwater nematode capable of swimming by rapid, vibratory thrashing movements. (d) Tail end of a freshwater nematode showing the three caudal glands (CG), the caudal gland tube (T) and spinneret (arrow).

Amphids may be multifunctional and evidence suggests that they serve as photoreceptors, olfactors, or sensory glands. Some are sensitive to pH and ions.

Other innervated papillae on nematodes are deirids (paired papillae located laterally near the nerve ring), postdeirids (similar structures located at the midbody), and genital papillae (found on the ventral surface around the cloacal opening of males). When elongate, the genital papillae are usually attached to thin membranous outgrowths of cuticle (bursa) (Fig. 9.6c). The presence or absence of a bursa and its structure are important taxonomic characters.

A few freshwater nematodes possess what are thought to be light receptor organs. These receptors normally consist of pigmented areas in the neck region, which may or may not be associated with a cuticular-structured lens. If a lensatic body is present, then the organ is sometimes referred to as an ocellus or pseudocellus.

I. *Reproductive System*

The majority of nematodes, including the freshwater forms, reproduce by amphimixis (sperm and eggs come from separate individuals). However, some forms have developed uniparental reproduction (i.e., autotoky) and this condition also occurs in the aquatic genera. In the latter, autotoky expresses itself as parthenogenesis, where progeny arise from unfertilized eggs. There is no record of hermaphroditism in any of the truly aquatic nematode genera (Poinar and Hansen 1983) and asexual reproduction does not occur in this phylum.

In amphimictic reproduction, the male nematode always places sperm inside the female; external fertilization has never been demonstrated. Some aquatic species exhibit "traumatic insemination" when the male penetrates the female cuticle with his spicules and releases sperm into her body cavity.

Female nematodes contain one or two gonads, which open to the exterior on the ventral side of the body at the vulva (Figs. 9.1 and 9.6a). The vulva is connected to a muscular tube, the vagina, which in turn leads into the uterus followed by the oviduct and ovary. A spermatheca (sperm-collecting area) is usually present between the uterus and oviduct.

With double-ovary species (didelphic), the gonads are usually opposite and join at a common vagina (Fig. 9.6a). Single-ovary forms (monodelphic) have retained the anterior ovary, which is often reflexed in the posterior part of the body (Fig. 9.8d).

The male may have a single testis (monorchic) or two testes (diorchic), which lead to a common seminal vesicle and vas deferens before entering the cloacal chamber, a common opening for the reproductive and digestive systems. Males of freshwater nematodes all possess one or two sclerotized structures termed spicules (Figs. 9.1 and 9.6c). These are inserted into the vagina of the female at insemination (except in traumatic insemination when they penetrate the cuticle of the female). Some males also have another sclerotized structure, the gubernaculum. The spicules are contained within the walls of the cloacal chamber while the gubernaculum is attached to the floor of the cloacal chamber. The latter structure supports the spicules during their movement in and out of the cloacal opening (Fig. 9.6c). On the ventral surface of the male tail is usually some type of genital papillae, which may or may not be supported by a bursa (Fig. 9.6c).

Through the release and detection of sex attractants, male and female nematodes are able to locate each other. Once together, the nematodes intertwine and the posterior region of the male contacts the female vulva. Sometimes an adhesive compound is produced to seal the union. Spicules are then inserted and sperm is transferred from the vas deferens of the male into the vagina and uterus of the female. The pair may remain joined for only several minutes or may be united for days. It is common to find "balls" of mermithids in stream bottoms all coiled together in continuous copulation. Individual sperm as well as secondary spermatocytes may be transferred during mating. In many nematodes, sperm maturation is completed in the female. Nematode sperm is variable in size and shape but is always nonflagellated.

Parthenogenetic forms can reproduce by mitotic or meiotic parthenogenesis. In the former, the diploid somatic number of chromosomes is retained, whereas in the latter, two maturation divisions occur which are similar to oogenesis in amphimictic species. A diploid chromosome number is established by fusion of the nonextruded polar nucleus with the egg pronucleus or by doubling of the chromosome before the first cleavage division occurs.

III. DEVELOPMENT AND LIFE HISTORY

The eggs of nematodes differ from all other stages in containing chitin in their shells. During embryonic development, nematodes show predeterminate cleavage, which means that cells destined to form specific tissues appear early in the embryo. Freshwater forms like the mermithids may deposit eggs with fully formed juveniles ready to hatch upon receiving the right stimuli. Other species deposit eggs in the single-cell stage, with embryonic development occurring in the environment.

Postembryonic development in nematodes (development after hatching) is similar to the gradual type of metamorphosis in insects. Aside from an increase in size, proportional changes of various organs, and the development of the gonads, there are few differences between juvenile and adult nematodes (Fig. 9.10c). For this reason, the immature stages of nematodes should be called juveniles and not larvae, because the latter usually implies complete metamorphosis where the immature stages differ radically from the adults. [Despite these differences, however, the term larva is widely used today in referring to the immature, postembryonic stages of nematodes.]

Hatching is influenced by temperature and external stimuli. The eggs of the mermithid *Pheromermis pachysoma* hatch when ingested by aquatic insects. Eggs of other freshwater nematodes hatch when the water reaches a certain temperature. Eggs that undergo anabiosis as a result of desiccation will not hatch until the area is flooded. This probably enhances survival for nematodes in temporary water sources, although this point still requires elucidation. In most cases, it is the first-stage juvenile that emerges from the egg; however in mermithids, juveniles molt once in the egg and then emerge as second-stage juveniles. If equipped with a spear or tooth, the young nematode will use these to break out of the shell, otherwise it simply employs pressure and pharyngeal secretions to soften and penetrate the shell wall.

All nematodes undergo four molts and have six stages during their development (i.e., egg, first-stage juvenile, second-stage juvenile, third-stage juvenile, fourth-stage juvenile, and adult). During each molt, the cuticle is shed and replaced by another secreted by the hypodermis. Sometimes two molts may occur almost simultaneously, as in postparasitic juvenile Mermithidae (Fig. 9.9a), and occasionally the shed skin is retained around the body of the next stage. Free-living nematodes generally complete their development rapidly in comparison to other metazoans. This can mean in 3–5 days for rhabditids (under optimum conditions) and generally from 1–6 weeks for most other freshwater forms. Therefore, nematode populations can turn over very quickly (see Schiemer 1983, 1987 for energetic studies on nematodes).

Aside from premolt and quiescent stages, most nematodes feed continuously throughout the growth period. Most freshwater nematodes take food in through the mouth and absorb nutrients into the intestinal cells. During their parasitic development in the hemolymph of invertebrates, mermithids absorb nutrients directly through their body walls. Nematode growth entails both enlargement and mul-

tiplication of cells. This growth is frequently nonproportional from stage to stage. For example, the intestine often will grow faster than the pharynx, so the proportion of the two organs differs in each stage. This information can be used to determine nematode stages, along with gonad development.

The types of food vary among the freshwater nematode groups. Microbotrophic forms ingest bacteria, algae, single-celled fungi, and protozoa. Predaceous species attack small metazoans such as nematodes, annelids, early insect stages, and molluscs. Many are omnivorous and will consume microbial life as well as metazoans. Plant parasites feed on the cytoplasm of higher plants and filamentous fungi. Parasites of invertebrates absorb nutrients from the hemolymph of various taxa, especially insects in the case of the Mermithidae. Vertebrate parasites may occur in the guts or body cavities of fish, aquatic birds, or amphibians, where they feed on host fluids and tissues.

The dependency of nematodes on moisture has brought a tremendous selection pressure to withstand periods of desiccation. Some nematodes form resistant stages, while in others, all life stages tolerate drought conditions for various periods. Resistant stages include the egg and the second- through fourth-stage juveniles. Some nematodes have been maintained for years (up to 25 years) in anabiosis and then revived in water.

Very little is known about the resistant stages, dispersal, and survival of freshwater nematodes. Jacobs (1984) mentions the "ability for passive dispersal" of resistant forms, but does not discuss how this dispersal occurs. It is very likely that the eggs and possibly some juvenile stages of forms adapted to semitemporary water sources can enter a resistant, quiescent stage capable of surviving desiccation and possibly temperature extremes; these have been demonstrated in some soil and plant nematodes. Passive dispersal may occur when these resistant stages are blown by heavy winds or washed to new areas by flash floods. The possibility of freshwater nematodes being carried in mud attached to the body parts of various water-frequenting animals is also feasible.

IV. ECOLOGY

A. Fossil Record

The oldest known possible fossil record of aquatic nematodes consists of tracks dating from the Upper Precambrian of Australia and Europe (Hantzschel 1975). Tracks similar to these in the Green River sediments of Lake Uinta (Middle Eocene) (Fig. 9.2b) were considered to be made by nematodes

(Moussa 1969). These more recent tracks occur in the Soldier summit area near Provo, Utah and were made during the second lacustrine phase of the Green River sediments. Lake Uinta was a large freshwater lake that occupied the Uinta and Piceance Creek Basins in Utah and Colorado. These tracks could have marked the passage of fairly large freshwater nematodes moving near the shore. The early Precambrian tracks could well be nematodes since it is highly probable that the phylum dates back to that period when representatives of the now predominantly aquatic order Araeolaimida probably existed (Poinar 1983).

B. Habitats of Freshwater Nematodes

Nematodes constitute an important and significant portion of the zoobenthic community of freshwater habitats. This community has been separated by benthic ecologists into three size categories (Strayer 1985). Animals retained on sieves with a mesh ranging from 200–2000 μm compose the macrofauna, those that are retained on sieves with a mesh ranging from 40–200 μm construe the meiofauna, and those that pass through a 40 μm mesh sieve make up the microfauna. Freshwater nematodes covered in the present chapter could be considered as belonging to all groups; the large free-living stages of the Mermithidae would be considered macrofauna, the majority of the adults and juveniles of the other groups would fall under meiofauna, and the young juveniles and eggs of many forms would fall into the microfaunal category. Thus, for nematodes in general, these size categories are little used, although they are helpful if only certain stages are being studied.

Nematodes have constituted 60% of all the benthic metazoans in Mirror Lake, New Hampshire, and are probably the most abundant benthic animals in freshwater (Strayer 1985). In Mirror Lake, nematode abundance showed a distinct depth distribution, with a strong minimum at 7.5 m. More than 90% of the nematode biomass at 7.5 m was composed of species of the large predatory nematodes belonging to the genus *Mononchus*. Strayer (1985) also reported that 63% of the nematodes in Mirror Lake lived in the top 2 cm of sediment while only 18% penetrated to a depth greater than 4 cm. Species of the genera *Monhystera* and *Ethmolaimus,* which constituted 55% of the nematodes in Mirror Lake, were found in the benthic zone. Although the nematodes outranked all other animals in abundance (constituting 59% of all metazoan individuals in the benthos), they still represented only 1% of the zoobenthic biomass. Nematodes normally contribute between 1 and 15% of the zoobenthic biomass in lakes, and other studies show that they consti-

tute from 40–80% of all meiobenthic animals (Holopainen and Paasivirta 1977, Oden 1979, Nalepa and Quigley 1983).

Nematodes that occur in freshwater have traditionally been called free-living nematodes. The term free-living is an ecological ranking, which means that these nematodes have no symbiotic association with multicellular plants or animals. Such nematodes obtain nourishment from bacteria, algae, protozoa, and other unicellular organisms. Nematodes with this type of nutrition are best referred to as microbotrophic, although they have been called saprophytic, saprophagous, bacteriophagous, microbivorous, and microphagous.

The term free-living, however, is also used to refer to a particular stage in the life cycle of a nematode—often a stage in the life cycle of a plant-parasitic or animal-parasitic nematode—when it has either ended or not yet begun its parasitic relationship. Examples in the freshwater habitat are the mermithids. The egg, infective second-stage juvenile, postparasitic juvenile, and adult mermithids are free-living and nonfeeding. Nourishment is obtained from the hemocoel of insects only during a portion of their juvenile development. Thus, free-living in the present work refers to those nematodes that have some developmental stages occurring in freshwater, irrespective of their nutritional requirements or their status at other developmental stages.

The habitats of freshwater nematodes vary as do the categories of freshwater resources. Two important requirements for freshwater nematodes to complete their development are continuation of the water source and availability of food and oxygen. Many freshwater nematodes feed on microorganisms that flourish in the gyttja, an organic deposit composed of excretory material from benthic animals. It is most obvious in lakes and ponds, where it is often combined with decomposing plant debris. This sediment increases in density with increasing depth. Core samples are used for quantitative measurements of zoobenthos in the gyttja (Strayer 1985).

Annual productivity of freshwater nematodes in an Austrian alpine lake was studied by Bretschko (1973). In the deeper benthos below 20 m, there were 2–4 annual generations with the total yield of nematode biomass estimated at 66 kg per year. It is interesting that the production was higher in shallow water and during the winter when the lake was covered with ice. For example, *Tobrilus grandipapellatus* was collected at populations of 235,000/m^2 in the winter, but only 60,000/m^2 in the summer.

Nematodes of all environments, including freshwater, are among the most numerous animals feeding on both primary decomposers such as bacteria and fungi, as well as primary producers such as algae and higher plants (Nicholas 1984). The significance of nematodes in the general economy of the ecosystem is not known, and attempts to estimate their contributions to energy flow have been made only with terrestrial and marine forms. Such estimates have used the equation:

$$C = P + R + E + U$$

where C is consumption, P is production, R is respiration, and E is ejecta (feces) and U is excretion.

Although freshwater nematodes have not been examined in detail, estuarine forms have been examined (Warwick and Price 1979). Some 40 species of nematodes occur on a mud flat in the Lynher Estuary in southern Britain. The population varies from $8-9 \times 10^6$ individuals/m^2 in winter to nearly 23×10^6 individuals/m^2 in late spring (15% of the macrofauna). Warwick and Price calculated a total annual oxygen (O_2) consumption of 28 liter $O_2/m^2/$ yr, which is equivalant to some 11 g of metabolized carbon (C). The annual production was calculated at about 6.6 g C/m^2/yr. Previous calculations with the soil nematode *Caenorhabditis briggsae* showed that the conversion of bacteria into nematode tissue varied from 13–20% (Nicholas 1984).

The question of nematodes living under anoxic conditions has been addressed by several workers. In deep lakes, the hypolimnion may become anaerobic for certain periods, resulting in a reduction of nematode fauna, but not eradication. In Lake Tiberias in Israel, Por and Moary (1968) found large numbers of *Eudorylaimus andrassyi* at a depth of 43 m during the winter oxygen-free period.

In the Neusiedlersee in Austria, *Tobrilus gracilus* occurs in muddy anaerobic zones beyond the limits of emergent vegetation (Schiemer 1978). In his study of the nematodes in Mirror Lake, Strayer (1985) found most species in the littoral zone, but species of *Ethmolaimus, Monhystera,* and unidentified Tylenchidae were very abundant in the anaerobic sediments at a depth of 10.5 m.

Extreme habitats for freshwater nematodes include low-temperature snow pools and high-temperature hot springs. Table 9.1 lists some freshwater nematodes collected at unusually high temperatures. The listing of 61.3°C for *Aphelenchoides* sp. is not only a record for nematodes but for all metazoan life forms!

Other specialized habitats encompass brackish and estuarine waters, inland saline lakes, cave streams, and associations with specific aquatic plants. ''Artificial water sources'' include canals, waste water, tap water, wells, filter beds of sewage treatment plants, and temporary water sources. Nematodes found in these habitats often do not have any relation to true freshwater forms or a freshwater habitat, and the majority are not treated in this work.

Previous workers have attempted to separate the freshwater-dependent nematodes from the facultative forms. Micoletzky (1922) distinguished between completely aquatic, principally aquatic, amphibious, and principally terrestrial species; while more recently, Jacobs (1984) presented an ecological classification with stenohygrophilic nematodes representing the strictly aquatic forms and euhygrophilic nematodes representing the semi-aquatic forms. The strictly aquatic or stenohygrophilic taxa covered here occur in many microhabitats (Fig. 9.3). As examples, periphytic forms live among plant roots, bryophilic taxa associate with moss and liverworts, endobenthic species thrive within the sediment and bottom debris, epibenthic groups live on

Table 9.1 Survival of Freshwater Nematodes in Hot Water Springs at High Temperatures

Nematode Species	Temp (°C)	Location	Reference
Aphelenchoides sp.	61.3	New Zealand	Rahm (1937)
Aphelenchoides sp.	35.0	New Zealand	Winterbourn and Brown (1967)
Aphelenchus sp.	57.6	Chile	Rahm (1937)
Dorylaimus atratus Linstow	47.0	Italy	Issel (1906)
Dorylaimus atratus Linstow [*D. thermae* Cobb]	40.0	Wyoming, United States	Hoeppli (1926)
Dorylaimus atratus Linstow (*D. thermae* Cobb)	53.0	Wyoming, United States	Cobb, in Hoeppli (1926)
Euchromadora striata (Eberth) (normally a marine genus)	52.0	Italy	Meyl (1954)
Monhystera ocellata Bütschli	52.0	Italy	Meyl (1954)
Monhystera gerlachii Meyl	52.0	Italy	Meyl (1954)
Plectus sp.	57.6	Chile	Rahm (1937)
Rhabdolaimus brachyuris Meyl	52.0	Italy	Meyl (1954)
Theristus pertenuis Bresslau and Schuurmans Stekhoven	52.0	Italy	Meyl (1954)
Tylocephalus sp.	45.3	New Zealand	Winterbourn and Brown (1967)

Figure 9.3 Scene of the bottom of a fast-flowing mountain stream in California showing several nematode microhabitats. A, stream bed containing endobenthic forms; B, bed surface containing epibenthic groups; C, rock surfaces containing haptobenthic groups; D, *Nostac* algal pads containing mermithids that parasitize *Cricotopus* midges (Chironomidae) living inside the pads.

the surface of the bottom, haptobenthic forms exist on the surface of submerged rocks, debris, and aquatic plants, and planktonic taxa live continuously in the water column. The latter group is poorly known and mainly found in turbid waters where their occurrence may represent more of a dispersal than a habitat mode.

Many genera of freshwater stenohygrophilic nematodes are widely dispersed, occurring on more than one continent. Some species are eurytopic, occurring in diverse habitats, while others are stenotopic and restricted to a few specialized habitats.

Freshwater nematodes have been considered as indicators of water pollution (Zullini 1976). Heavy metals and other pollutants settle and are taken up by organic matter in the sediments. Nematodes ingest this material and those that are sensitive to the pollutants may die. Representatives of the Chromodorida are apparently more sensitive to pollution that those of the Rhabditida (Zullini 1976). In extremely polluted waters, the Rhabditoidea may be quite abundant. A two-year study sampling nematodes along two streams in Indiana indicated that an analysis of the benthic nematode community struc-

ture could be useful in evaluating disturbance to aquatic habitats (Ferris and Ferris 1972).

C. Enemies

Perhaps the greatest enemies of freshwater nematodes are other predaceous nematodes. Representatives of the Mononchida, Dorylaimida, and Enoplida are known to attack and devour a range of small invertebrates in their surroundings.

The crayfish *Pacifastacus leniusculus* was observed to feed occasionally on nematodes (Flint 1976). The freshwater, rhabdocoel turbellarian, *Microstomum* feeds on nematodes (Stirewalt 1937), as will the freshwater nemertean worm, *Prostoma* (Coe 1937). The role of disease organisms in regulating populations of freshwater nematodes is not known, but the author has frequently collected specimens of *Tobrilus* infested with what appeared to be microsporidian spores (Fig. 9.9c). Other records of probable protozoan diseases in freshwater nematodes have been reported in *Chromadora, Dorylaimus, Ironus, Monhystera, Plectus, Theristus, Trilo bus, Tripyla, Prodesmodora, Achromadora,*

Paraphanolaimus, and *Desmolaimus* (Poinar and Hess 1988).

V. COLLECTING AND REARING TECHNIQUES

A. Sampling Methods

There are two basic types of sampling: qualitative and quantitative. The former method is used in preliminary studies to determine which kinds of nematodes are present in the ecosystem. The latter method is used to evaluate the components of a population, the proportion of each in the environment, and the dynamics of the population over time.

Qualitative methods for sampling freshwater nematodes depend on the water flow and the zone to be sampled. In sampling nematodes from beds of fast-flowing streams, a net, an instrument to turn over rocks and debris on the stream bed, and a set of sieves are necessary (Fig. 9.4a). The net is first anchored with the opening facing upstream. A pick or geologist's hammer is used to disturb the stream bed slightly upstream from the net (Fig. 9.4b). The stream washes nematodes and other debris from the disturbed area into the net. The contents of the net are then placed in enamel collecting pans with water

Figure 9.4 (a) Equipment needed to sample freshwater nematodes. A, pail; B, set of gradated sieves; C, net; D, pick; E, squeeze bottle; F, shovel; G, geologist's hammer. (Hip boots are not shown.) (b) With a geologist's hammer, the stream bed is disturbed upstream from the positioned net. (c) With a shovel, samples of the stream bed are removed together with roots of aquatic plants. (d) Nematodes and debris are removed from the net into an enamel pan.

and set aside for extraction. Nets can be pulled over aquatic plants to obtain nematodes living on their surfaces. The type of nematode being collected will determine the mesh size of the net selected. It is recommended that in order to obtain a fair sample of all nematodes, a mesh size no larger than 40 μm be used.

A number of devices have been devised to collect freshwater nematodes (Filipjev and Schuurmans Stekhoven 1959). These include dredges, wormnets, bottom catchers, mudsuckers, swab or sledge trawls, and plankton nets. Their use depends on the type of material desired. For sluggish or stationary water sources, samples of the bottom (mud, sand) can be placed in a water-filled container (Fig. 9.4c) which is then shaken, the suspension allowed to settle for 3–4 sec, and the supernatant then poured into a second container for extraction. Rocks and other submerged material can be placed in buckets and the nematodes washed off with a water spray or by vigorous shaking or brushing.

Quantitative sampling methods involve the use of augers or probes to remove core samples of measurable sizes from the beds (Strayer 1985). Other quantitative methods can be devised on the basis of need and the desired biotype to be investigated. Techniques applied to marine nematodes may also be suitable for freshwater forms. For further information, see Hulings and Gray (1971), Holme and McIntyre (1971), and Downing and Rigler (1984). It should be remembered that any estimate of the nematode population depends on the type of sample taken. Thus, a biased sample will produce a biased estimate of the nematode population. Also, it should be noted that nematodes are sensitive animals and all sampling and extraction techniques should be as gentle as possible.

B. Extraction

The process of extraction involves removing nematodes from samples collected in various freshwater habitats (Figs. 9.4d, 9.5a–d). The basic principles of nematode extraction are based on the fact that most nematodes have a specific gravity between 1.10 and 1.14, thus they will slowly sink in freshwater. They are denser than fine clay particles but lighter than sand. The extraction methods used for soil and plant-parasitic nematodes are wholly adequate for freshwater forms.

Although the majority of freshwater nematodes are too small to view with the naked eye, the free-living stages of many aquatic mermithids are large enough to be handpicked from the samples (Fig. 9.5c). They can be carefully lifted with a fine needle containing an L-shaped bend at the tip or with a pair

of fine forceps (Fig. 9.5d). Other mermithids that are small enough to escape visual spotting can be collected with the extraction processes.

One of the oldest and simplest methods of extracting nematodes from a wide range of samples is the Cobb sieving and gravity method, or a modification of the latter. This consists of making a water suspension of the nematodes and pouring it through a series of screens of different mesh to collect the nematodes. The suspension is made by placing the samples in a pail or other suitably large container (Fig. 9.4c), adding 2–3 liters of water (preferably from the original source), allowing the mixture to set for 10–20 sec to allow the heavier particles to sink, and then pouring the suspension containing the nematodes through a series of screens (Fig. 9.5a). If extraneous large debris (wood particles, rocks) is present, the samples should first be poured through a 20–40 mesh sieve (openings are 0.840–0.350 mm). The sizes of sieves selected will, of course, influence the size range of nematodes extracted. If a sample of all available nematodes is desired, then a final fine screen of 400–500 mesh (openings are 0.035–0.026 mm) can be used. The problem with the fine screens is that they become easily clogged with debris. Such screens should be tapped rapidly and gently on the under surface with the fingers to assist passage of the water. The nematodes, together with debris, will be trapped on the surface of the 400 or 500 mesh sieve. Both can be removed by directing a small jet of water on the back of the sieve and washing the contents into a separate container (Fig. 9.5b). Certain biases are inherent in sieving extraction methods (Hummon 1981); since they may be unsuitable for quantitative studies (Viglierchio and Schmidt 1983), the more tedious method of sorting through unsieved samples under a dissecting microscope may be desirable (Strayer 1985).

The final step consists of separating the nematodes from the fine debris. For this, the Baermann funnel method, or a modification thereof, can be used. All that is needed is a funnel with a piece of rubber tubing attached to the stem. The tube should be closed with a screw clamp. The nematode–debris mixture is placed on a fine cloth or facial tissue, which is supported by a piece of screen held in the top, wide portion of the funnel. Water is slowly added to the funnel until the nematode–debris mixture is just covered. The clamp should be opened to eliminate the air trapped in the funnel stem and to allow a continuous column of water to flow into the funnel. The nematodes crawl through the debris and facial tissue, drop through the screen, and settle at the base of the funnel stem, where they can be drawn off by releasing the clamp. The nematodes should be drawn off every 6 hr; the apparatus can be operated

Figure 9.5 (a) Nematodes and debris collected from the stream bed are passed through a series of three sieves with openings ranging from 0.4–0.04 mm². (b) Nematodes are washed off of the surface of the final screen with a spray of water from the plastic squeeze bottle. (c) Free-living stages of mermithids are large enough to be seen with the naked eye and can be picked up with forceps. (d) Larger nematodes can be transferred directly to a small collecting vial.

in this fashion for several days. Then the nematodes can be counted or handpicked under a dissecting microscope.

A more recently devised method of isolating nematodes from samples involves the principle of increasing the specific gravity of the solution to make the nematodes float to the surface. Substances such as sugar, salt, or Ludox can be added to the nematode–water mixture in the centrifugal-flotation and sugar-flotation techniques (Ayoub 1977). If sugar is used, then 673 g added to 1 liter of water will result in a specific gravity of 1.18, which is greater than that of nematodes (causing them to float on the surface). Detailed instructions for these and other methods of extracting nematodes from soil and plant samples are presented by Ayoub (1977) and Nichols (1979).

Quantitative methods of extraction for monitoring nematode populations over a period of time often involve the use of elutriators, which use an upcurrent of water to separate nematodes from the medium. Such methods are rather elaborate and are discussed by Southey (1978).

C. *Fixing and Mounting*

After the nematodes have been extracted, they should be held in water (preferably water from their original collecting source) prior to fixation. The nematodes should be killed before being placed in fixative; otherwise they become distorted and difficult to examine. They are most easily killed with heat, which also tends to relax and extend them. Too much heat will destroy the internal tissues, but good results can be obtained by pouring hot water (60°–70°C) over the nematodes. They should then be transferred to the fixative as soon as possible. The best fixative for nematodes is TAF (7 ml 40% formalin, 2 ml triethanolamine, and 91 ml distilled water). However, 3–5% formalin or 70% ethanol can also be used if TAF is not available. Some distortion may occur with ethanol, however.

The nematodes should be fixed for at least 2–3 days before transfer to glycerin for mounting on microscope slides. Nematodes are most easily conveyed to glycerin by the evaporation method, which requires little handling. Fixed specimens are transferred to a dish containing a solution of 70 ml ethanol (95%), 5 ml glycerol, and 25 ml water. The dish is partly covered for the first 3 days to allow the alcohol and water to evaporate at a slow rate; then the cover is removed for the next 14 days. Finally, the containers (now with mostly glycerin) are placed in a desiccator or an oven (35°C) for another 2 weeks to drive out the remaining water. The nematodes will then be in a relatively pure solution of glycerin and can be mounted directly on microscope slides. Such slides are termed permanent since they can be kept for years for continuous study. Eventually, the nematode tissues tend to become transparent. Temporary slides are made by mounting the nematodes directly after fixation in the fixing solution (formalin or alcohol). They may last several weeks if the ringing seal is tight.

The following steps can be taken to make permanent slides with specimens transferred to glycerin (modified from Poinar 1983).

1. With a small pointed instrument (dental pulp canal file, small insect pin mounted on a wooden splint, needle), transfer the nematode(s) to a small drop of glycerin placed in the center of a microscope slide.
2. Push the nematode to the bottom of the drop.

Then add at three equidistant points around the nematode, small supports (coverslip pieces, wire) with a width equal to or slightly wider than that of the nematode. Push these also to the bottom of the drop.
3. Place a cover slip over the drop and lower it slowly at a 30° angle so that air bubbles will not become entrapped in the glycerin as the slide touches the mounting medium.
4. Add more glycerin, if needed, by placing a small drop at the edge of the coverslip and allowing it to move under the glass and spread throughout. If the coverslip is floating on glycerin, then remove the excess by blotting it with a moistened (water) piece of tissue.
5. Carefully seal the edges of the coverslip with a ringing compound such as nail polish or Turtox slide ringing cement (General Biological Supply House, Chicago, Illinois).
6. Label the slide with information regarding the date, locality, and collector.

D. *Culturing*

Many free-living, microbotrophic freshwater nematodes can be cultured in the laboratory if they are presented with a growing colony of their preferred food (i.e., bacterial or algal species) and maintained under environmental conditions (temperature, oxygen) comparable to those present at the collecting site. Nuttycombe (1937) successfully cultured several species of freshwater nematodes using a wheat grain infusion method. Between 200–300 grains of wheat seeds were added to 250 ml of spring water in a flask. This mixture and a separate flask of pure spring water were heated to a boil and allowed to cool. The spring water was poured into 200 ml petri dishes with 3–4 grains of the boiled wheat. The dishes with the wheat seeds were allowed to stand for 2–3 days before the nematodes were added. Microorganisms brought in with the nematodes grew on the wheat seeds, and the microbotrophic nematodes fed on the microbes.

Postparasitic juveniles of freshwater mermithids can be held in the laboratory until they mature for identification. Care should be taken so that both the temperature and the amount of dissolved oxygen parallel those of the collection site. Periodic transfers to freshwater may be necessary to prevent the buildup of fungi which will destroy the nematodes. Petersen (1972) devised a method of culturing a mosquito mermithid parasite (*Romanomermis culicivorax*) through its entire life cycle. The mosquito chosen as a laboratory host was *Culex pipiens quinquefasciatus* (Say), because it could be reared continuously in crowded conditions, was easily maintained in colony, and was highly susceptible to the nematode. Postparasitic juvenile mermithids were placed in wet sand within plastic trays (36 × 25 × 100 cm). The postparasites molted to the adult stage, mated, and oviposited in the sand. The eggs, which embryonated in the moist sand, could be maintained in their resting state for several months. When the trays were flooded with water, the infective stages emerged from the sand and were capable of infecting newly hatched mosquito larvae. The infected insect larvae were kept in a separate container and fed. By the time the mosquitoes were ready to pupate, the nematodes had completed their development and had started to emerge. The emerging postparasitic juveniles were then transferred to wet sand to continue the cycle. Adequate numbers of mermithids were available for both scientific investigations and biologic control studies with this rearing method.

VI. IDENTIFICATION

Once the specimens are mounted, they can be examined and identified with a compound microscope. Because the identification of nematodes involves the use of many fine details, a microscope equipped with oil immersion should be employed (1000X). The use of dioscopic illumination after Nomarski or differential interference contrast better reveals minute features of the cuticle (such as the amphid structure). If possible, living nematodes should also be examined under the microscope. Their movements can be reduced by removing water from under the coverslip and compressing the nematode between the microscope slide and coverslip. Certain charac-

ters are much clearer on living specimens and there is a certain thrill in watching these creatures move under a microscope that is lacking with preserved material. However, fine details can best be observed on fixed nematodes.

In preparing the taxonomic key presented here, the following works dealing with freshwater nematodes were consulted: Cobb (1914); Chitwood and Allen (1959); Ferris *et al.* (1973); Tarjan *et al.* (1977); Pennak (1953, 1978); Esser *et al.* (1985); Esser and Buckingham (1987); Jacobs (1984); Tsalolikhin (1983); Gerber and Smart (1987). The first five references also have keys which can be consulted by the interested reader for cross-reference. For descriptions of various genera and a listing of species, Goodey (1963) is still very helpful, although now out of date.

Some 66 genera of freshwater nematodes are included in the present key. Between 300 and 500 species of nematodes are probably included within these 66 genera. Only those genera that contain species that are obligately aquatic (live nowhere else) and have been reported from North America are included. Over 95% of the aquatic nematodes encountered should be included in the following genera.

A. Taxonomic Key to Families and Selected Genera of Freshwater Nematoda

The following key pertains to the families and selected genera of freshwater nematodes and free-living stages of parasitic nematodes found in freshwater habitats of North America. Following this is a separate key to the genera of aquatic Mermithidae.

1a.	Elongated and thread-like, generally over 1 cm in length and usually observable with the naked eye; found on the bottom surface or several centimeters within the bed (Figs. 9.5c and 9.10a, b)	2
1b.	Not elongate, generally under 1 cm in length and observable only with a lens (Figs. 9.2c and 9.10c); found on all exposed surfaces (rocks, on or in plants) as well as on the bottom surface and within the bed	3
2a(1a).	Forms usually long (6 cm or more in length), leathery body wall, range in color from light brown to black (Fig. 9.11a)	Hairworms (phylum Nematomorpha)
2b.	Forms usually smaller (under 6 cm in length), body wall fragile, range in color from white to rose, green and yellow (Fig. 9.5c)	Mermithidae (see Section VI.B.)
3a(1b).	Head of nematode bearing a stylet (Figs. 9.8b and 9.9b)	4
3b.	Teeth and other armature may be present but a stylet is absent (Figs. 9.7a, b and 9.8a)	16
4a(3a).	Stylet knobs usually absent (Fig. 9.9b); if present, then pharynx lacking a valvated metacorpus; pharynx narrow anterior and thickened posterior at junction with intestine (Fig. 9.2a); valvated metacorpus absent	Dorylaimida 11

4b.	Stylet knobs usually present (Fig. 9.8b); pharynx composed of a valvated metacorpus followed by a slender isthmus and basal glandular bulb leading into the intestine (Figs. 9.8b, c)	5
5a(4b).	Dorsal gland outlet in precorpus; metacorpus moderate in size (Fig. 9.8b) (less than 3/4 body width) .... Tylenchida	8
5b.	Dorsal gland outlet in metacorpus; metacorpus large (3/4 of body width or more) (Fig. 9.8c) .... Aphelenchida	6
6a(5b).	Stylet usually with faint or inconspicuous knobs (Fig. 9.8c); tail tip usually pointed; males lacking bursa and gubernaculum .... Aphelenchoididae	7
6b.	Stylet without knobs; tail tip usually rounded; males with a bursa (Fig. 9.6c) and gubernaculum .... Aphelenchidae *Aphelenchus* Bastian, 1865	
7a(6a).	Tail shape elongate, filiform .... *Seinura* Fuchs, 1931	
7b.	Tail shape short, conical .... *Aphelenchoides* Fischer, 1894	
8a(5b).	Head bearing distinct setae (Fig. 9.7a) .... Atylenchidae *Atylenchus* Cobb, 1913	
8b.	Head without setae	9
9a(8b).	Procorpus fused with large, oval metacorpus; cuticle strongly annulated; adult female with cuticular sheath and well-developed stylet (Fig. 9.8b) .... Criconematidae *Hemicycliophora* de Man, 1921	
9b.	Procorpus distinct and narrow before reaching the expanded metacorpus, cuticle not strongly annulated; adult female without cuticular sheath; stylet may or may not be well developed .... Tylenchidae	10
10a(9b).	Large distinct stylet, ovaries paired, amphidelphic (Fig. 9.6a); tail tapering but not filiform .... *Hirschmanniella* Luc and Goodey, 1963	
10b.	Stylet small and inconspicuous, ovary single, prodelphic; tail filiform .... *Tylenchus* Bastian, 1865	
11a(4a).	Pharynx gradually widens toward basal bulb (Fig. 9.2a)	12
11b.	Pharynx abruptly widens to form the basal bulb which contains a valvular chamber (Fig. 9.8a) .... Campydoridae *Aulolaimoides* Micoletzky, 1915	
12a(11a).	Head bearing a narrow protrusible mural spear that is symmetrically pointed at tip .... Nygolaimidae *Nygolaimus* Cobb, 1913	
12b.	Head bearing a thick protrusible axial spear that is asymmetrically pointed at tip (sloped on one side) (Fig. 9.9b) .... Dorylaimidae	13
13a(12b).	Stomal area bearing distinct teeth or denticles around the tip of the stylet	14
13b.	Stomal area lacking distinct teeth or denticles in the stylet tip region	15
14a(13a)	Four large teeth together with mural denticles present .... *Paractinolaimus* Meyl, 1957	
14b.	Four large teeth only present .... *Actinolaimus* Cobb, 1913	
15a(13b).	Cuticle thick and longitudinally ridged (Fig. 9.9b) .... *Dorylaimus* Dujardin, 1845	
15b.	Cuticle smooth .... *Mesodorylaimus* Andrássy, 1959	
16a(3b).	Amphids minute and borne on the lateral lips, excretory duct cuticularized (canal lined with a fine layer of cuticle); caudal glands absent, conspicuous cephalic setae absent, bursa present or absent	17
16b.	Amphids usually enlarged and located on the neck region, excretory duct rarely cuticularized (except in Plectidae); head often with conspicuous setae, bursa usually absent	21
17a(16a).	Pharynx with an expanded median bulb (metacorpus) containing a longitudinal valve and a glandular nonvalvated basal bulb ...... Diplogasteridae *Rhabditolaimus* Fuchs, 191	
17b.	Pharynx may or may not contain a metacorpus, but if it does, it does not contain a valve; basal pharyngeal bulb with or without a basal valve	18
18a(17b).	Pharynx elongate and lacking a valve in the basal bulb or bulb area, slender; slow moving forms (Fig. 9.9d) .... Daubayliidae *Daubaylia* Chitwood and Chitwood, 1934	
18b.	Pharynx normal in length, with a basal bulb containing a valve (Fig. 9.8a); stouter, quick-moving forms	19
19a(18b).	Stoma with rhabdions separate, female with a single ovary, bursa absent .... Cephalobidae	20

19b.	Stoma with rhabdions fused, cylindrical (Fig. 9.8a), female with a single or paired ovaries, bursa usually present (Fig. 9.6c)	Rhabditidae

[Most of the genera in this large family are soil forms, some of which appear from time to time in aquatic habitats. *Mesorhabditis* (Osche 1952), which has a single ovary and a bursa, occurs i.n sewage beds and run-off water. *Pellioditis* (Doughtery 1953), which has paired ovaries also occurs in aquatic habitats.]

20a(19a).	Lateral fields reach to tail tip; female tail rounded	*Cephalobus* Bastian, 1865
20b.	Lateral fields end at the phasmids; female tail usually pointed	*Eucephalobus* Steiner, 1936
21a(16b).	Amphids spiral (Fig. 9.7d), loop-like, or circular (Fig. 9.7c); pharynx usually with a basal bulb (Fig. 9.8a); caudal glands and spinneret usually present (Figs. 9.1, 9.2d, 9.6d)	22
21b.	Amphids pore-like to pocket-like, pharynx usually without a distinct basal bulb	41
22a(21a).	Ovaries outstretched (Fig. 9.8d), amphids usually circular (Fig. 9.7c) . . . . Monhysterida	23
22b.	Ovaries reflexed (Fig. 9.8e) (except in *Odontolaimus* which also has circular amphids), amphids variable (including circular)	25
23a(22a).	Amphids slit-like, faint; stoma with denticulated cushions (padded areas covered with minute projections)	*Prismatolaimus* de Man, 1880
23b.	Amphids circular, distinct, stoma lacking denticulated cushions	24
24a(23b).	Stoma shallow; pharynx without distinct basal bulb (as in Fig. 9.2a)	*Monhystera* Bastian, 1865
24b.	Stoma elongate; pharynx with distinct basal bulb (as in Fig. 9.8a)	*Monhystrella* Cobb, 1918
25a(22b).	Caudal glands and spinneret present (Fig. 9.2c); stoma usually armed with teeth (Fig. 9.7b); knobs, bristles (Fig. 9.7a), or punctations usually present on cuticle	Chromadorida	26
25b.	Caudal glands and spinneret present or absent; stoma may or may not be armed with teeth; bristles sometimes present on cuticle	Araeolaimida	33
26a(25a).	Amphids circular (Fig. 9.7c); cuticular punctations minute . . . . Microlaimidae	*Microlaimus* de Man, 1880
26b.	Amphids spiral (Fig. 9.7d), kidney shaped, or circular; cuticular punctations coarse	27
27a(26b).	Amphids circular (as in Fig. 9.7c) or spiral (as in Fig. 9.7d), pharyngeal–intestinal junction large, tri-radiate	Cyatholaimidae	29
27b.	Amphids spiral or kidney shaped, pharyngeal–intestinal junction small, not tri-radiate	Chromadoridae	28
28a(27b).	Amphids represented as a broad transverse slit; five or more preanal genital papillae in male	*Chromadorita* Filipjev, 1922
28b.	Amphids represented as a flattened, broken ring; no more than three preanal genital papillae in male	*Punctodora* Filipjev, 1930
29a(27a).	Amphids circular (Fig. 9.7c)	30
29b.	Amphids spiral (Fig. 9.7d)	31
30a(29a).	Stoma cup-like; length shorter than neck width	*Prodesmodora* Micoletzky, 1923
30b.	Stoma tubular (Fig. 9.7a); length two or more times neck width	*Odontolaimus* de Man, 1880
31a(29b).	Stoma tubular (Fig. 9.7a), there may be three anterior equal teeth in the stoma, but single large dorsal tooth is lacking	*Ethmolaimus* de Man, 1880
31b.	Stoma cup- or funnel-shaped, armed with a prominent dorsal tooth (as in Fig. 9.7b) much larger than any other teeth	32
32a(31b).	Amphids located at level of stoma, small subventral teeth absent	*Paracyatholaimus* Micoletzky, 1924
32b.	Amphids located posterior to stoma, one or two small subventral teeth present	*Achromadora* Cobb, 1913
33a(25b).	Caudal glands absent; stoma cylindrical with separate rhabdions	Teratocephalidae	34
33b.	Caudal glands present (Fig. 9.2a); stoma tubular with rhabdions fused	35

34a(33a).	Amphids pore-like; cuticular annulation strong (Fig. 9.8b)	*Teratocephalus* de Man, 1876
34b.	Amphids spiral (Fig. 9.7d); cuticular annulation weak	*Euteratocephalus* Andrássy, 1958
35a(33b).	Ovaries outstretched (Fig. 9.8d); amphids circular .... Axonolaimidae	*Cylindrolaimus* de Man, 1880
35b.	Ovaries reflexed (Fig. 9.8e)	36
36a(35b).	Pharynx usually with a basal valvated bulb (Figs. 9.1 and 9.8a) .... Plectidae	37
36b.	Pharynx with or without a basal bulb, if bulb present, then valve absent	39
37a(36a).	Basal bulb of pharynx lacking	*Anonchus* Cobb, 1913
37b.	Basal bulb of pharynx with valves (Fig. 9.8a)	38
38a(37b).	Anterior portion of stoma funnel-shaped, posterior portion tubular; long pharyngeal–intestinal valve	*Chronogaster* Cobb, 1913
38b.	Stoma tubular and narrowing at the base; short pharyngeal–intestinal valve (Fig. 9.1)	*Plectus* Bastian, 1865
39a(36b).	Basal bulb lacking .... Bastianiidae	*Bastiania* de Man, 1876
39b.	Basal bulb present .... Camacolaimidae	40
40a(39b).	Amphids circular (Fig. 9.7c), basal pharyngeal bulb absent	*Aphanolaimus* de Man, 1880
40b.	Amphids spiral (Fig. 9.7d), a slight basal pharyngeal bulb present	*Paraphanolaimus* Micoletzky, 1923
41a(21b).	Stoma usually oval in outline (Fig. 9.7b), heavily sclerotized and armed with one or more teeth, setae and bristles absent .... Mononchidae	50
41b.	Stoma not in the form of a heavily sclerotized oval cavity, teeth or denticles may be present, setae and bristles may be present (Fig. 9.7a)	42
42a(41b).	Cuticle of head double .... Oncholaimidae	43
42b.	Cuticle of head normal	44
43a(42a).	Setae or bristles present on tail	*Oncholaimus* Dujardin, 1845
43b.	Setae and bristles usually lacking on tail	*Mononchulus* Cobb, 1918
44a(42b).	Stoma heavily sclerotized, cylindrical .... Ironidae	49
44b.	Stoma not heavily sclerotized, funnel-shaped or tubular	45
45a(44b).	Stoma vestigial and unarmed; pharynx base expanded and set off to form a slight, elongate bulb .... Alaimidae	46
45b.	Stoma distinct, or if vestigial then armed with an inconspicuous median tooth; pharynx may be expanded at base, but rarely set off from the remainder in the form of a bulb	47
46a(45a).	Amphids pore-like, minute	*Alaimus* de Man, 1880
46b.	Amphids cup-shaped, distinct	*Amphidelus* Thorne, 1939
47a(45b).	Base of pharynx expanded to form a valvated bulb; three rod-like thickenings compose posterior part of stoma .... Leptolaimidae	*Rhabdolaimus* de Man, 1880
47b.	Pharynx cylindrical, not with basal valvated bulb; stoma a simple tube without rod-like thickenings .... Tripylidae	48
48a(47b).	Three lips; amphids pore-like, minute; stoma cylindrical	*Tripyla* Bastian, 1865
48b.	Six lips; amphids cup-shaped, distinct; stoma funnel-shaped	*Tobrilus* Andrássy, 1959
49a(44a).	Stoma tubular, with two minute teeth at the base; excretory pore opening posterior to head	*Cryptonchus* Cobb, 1913
49b.	Stoma tubular, with three anterior hook-like teeth; excretory pore opening in head area	*Ironus* Bastian, 1865
50a(41a).	Pharyngeal–intestinal junction tuberculate; dorsal tooth on stoma pointing posteriorally	*Anatonchus* Cobb, 1916
50b.	Pharyngeal–intestinal junction not tuberculate; dorsal tooth on stoma pointing anteriorally	51
51a(50b).	Ventral stomatal ridge with longitudinal row of denticles	*Prionchulus* Cobb, 1916
51b.	Ventral stomatal ridge absent, or if present then unarmed	*Mononchus* Bastian, 1865

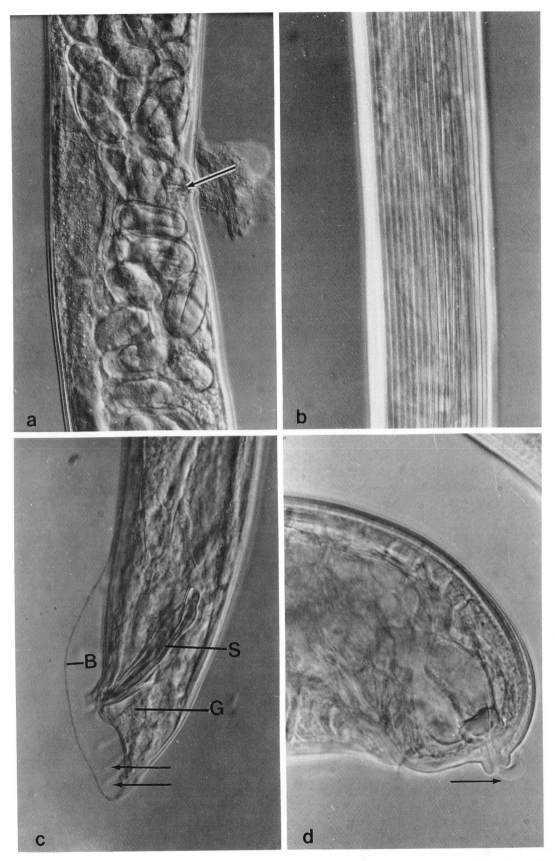

Figure 9.6 (a) Midbody region of a female *Pellioditis* sp. (Rhabditidae) showing the vagina (arrow) and paired, opposite uteri filled with developing eggs. Note copulation deposit surrounding the vulva opening. (b) Longitudinal striations lining the cuticle of *Dorylaimus* sp. (Dorylaimidae). (c) Tail of male *Pellioditis* (Rhabditidae) showing spicule (S), gubernaculum (G), bursa (B), and bursal papillae or rays (arrows). (d) Tail of an aquatic nematode showing caudal gland mucous being emitted from the tail terminus (arrow).

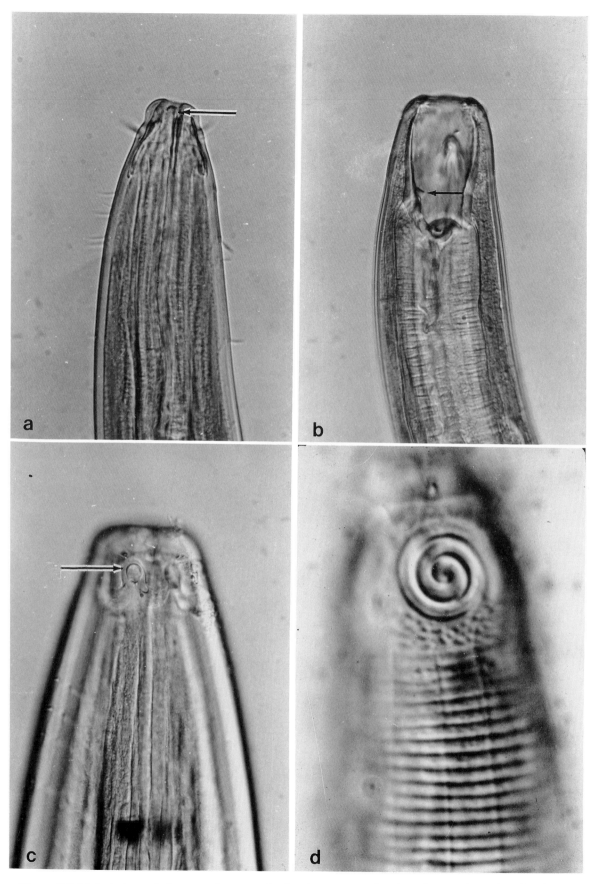

Figure 9.7 (a) Anterior end of an aquatic nematode showing narrow tube-shaped stoma (arrow) and cuticular bristles or setae. (b) Large oval, cup-shaped stoma of a predatory aquatic nematode. Note dorsal tooth (arrow) on stomal wall. (c) Nematode bearing a medium-sized, circular amphid. (d) Nematode bearing a large, spiral-shaped amphid.

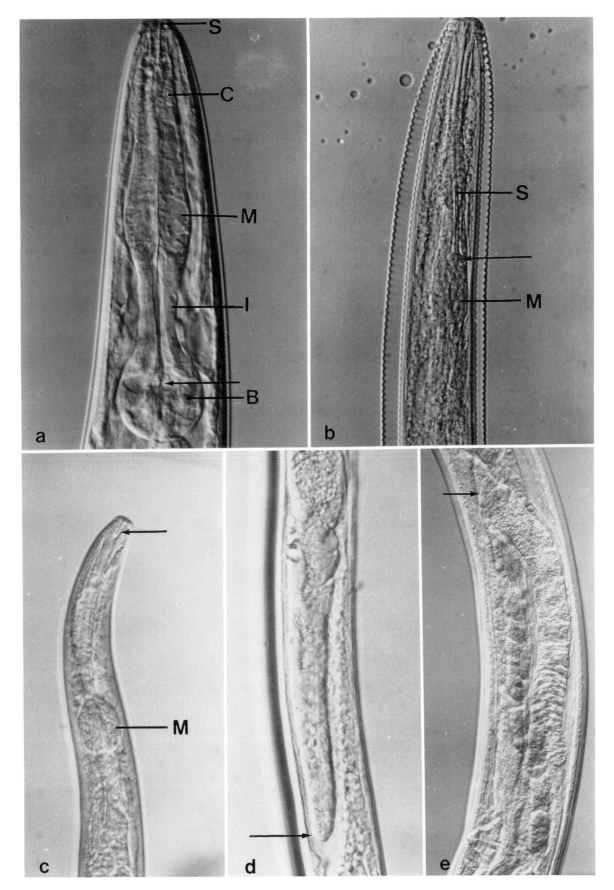

Figure 9.8 (a) Pharyngeal region of *Pellioditis* (Rhabditidae) showing tubular stoma (S), corpus (C), metacorpus (M), isthmus (I), and basal bulb (B). Arrow shows valve in basal bulb. (b) Outer ensheathing cuticle of *Hemicycliophora* (Hemicycliophoridae). Note long stylet (S) with prominent basal bulbs (arrow) and valvated metacorpus (M). (c) Anterior of *Aphelenchoides* (Aphelenchoididae) showing small stylet (arrow) and large metacorpus (M) with valve. (d) Outstretched ovary (arrow denotes tip). (e) Reflexed ovary (arrow shows point of reflexion).

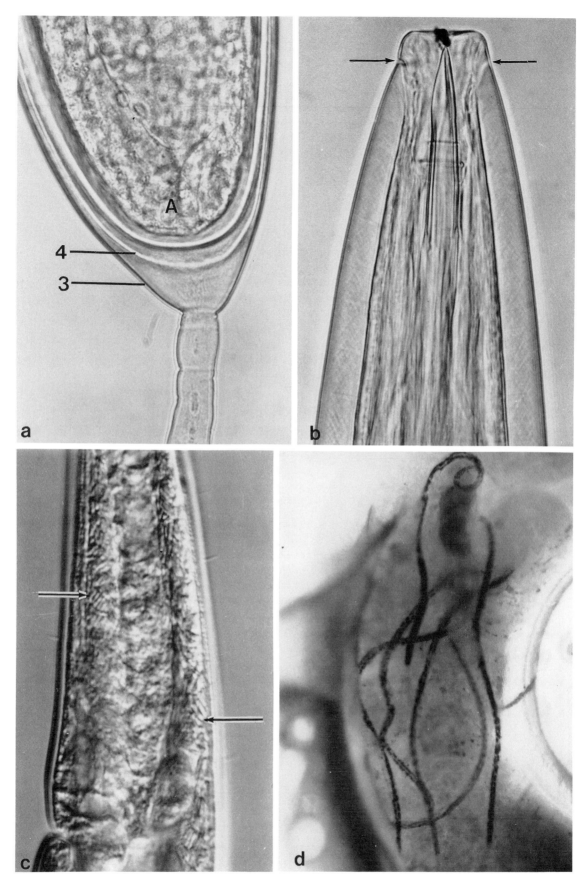

Figure 9.9 (a) Posterior portion of a postparasitic juvenile mermithid (Mermithidae) in the process of molting. Note shedding of third- (3) and fourth-stage (4) cuticle. A, adult. (b) Head of *Dorylaimus* sp. (Dorylaimidae) showing small pore-like amphids (arrows) and stylet lacking basal bulbs. (c) Tail region of a *Tobrilus* sp. (Tobrilidae) showing spores of a microsporidian parasite filling the hypodermal tissue (arrows). (d) Adults of *Daubaylia* inside an aquatic snail.

B. Taxonomic Key to Genera of Freshwater Mermithidae

The mermithids represent a family of nematodes that have become parasitic on other invertebrates. Over half of the described genera parasitize only aquatic insects. Two genera, *Aranimermis* and *Pheromermis,* enter the immature stages of aquatic insects which are later consumed by terrestrial arthropods. However, the free-living stages of these and all other aquatic mermithids (postparasitic juvenile, adults, eggs, and infective juveniles) are found in the aquatic habitat. Only the developing second- and third-stage juveniles parasitize the aquatic stages of insects and other invertebrates; the other stages do not take nourishment. Eggs and infective-stage juveniles are microscopic and normally not taken in samples but the postparasitic juveniles and adults are easily seen and collected. Since the keys are based on adult characters, postparasitic juveniles should be held in water until they molt to the adult stage. One important character in identifying mermithids is the number of hypodermal cords present. These can be determined by cutting thin sections with a razor blade by hand, mounting them in cross-sections, and examining them under the microscope. The hypodermal cords protrude through the muscle fields in dorsal, ventral, lateral, and often the subdorsal and subventral regions.

1a.	Adult cuticle with cross-fibers; thick, robust white nematodes found along the edges of bogs, springs, and streams (parasites of wasps, ants, and horseflies) (Fig. 9.10b) .... *Pheromermis* Poinar, Lane, and Thomas, 1976	
1b.	Adult cuticle without cross-fibers, robust or slender, white, green, pink, brown, or yellow nematodes found in lake, stream, and pond beds	2
2a(1b).	With four cephalic papillae .... *Pseudomermis* de Man, 1903 (syn. *Tetramermis* Steiner, 1925)	
2b.	With six cephalic papillae	3
3a(2b).	With single or fused (rare) spicules	4
3b.	With paired, separate spicules	8
4a(3a).	Spicule medium to long, more than twice anal body width	5
4b.	Spicule short, less than twice anal body width	6
5a(4a).	Mouth terminal; spicules J-shaped; vulval flap present; [parasites of midges (Chironomidae)] .... *Lanceimermis* Artyukhovsky, 1969	
5b.	Mouth normally shifted ventrally; spicule curved but not J-shaped; vulva flap absent; [parasites of blackflies (Simuliidae), midges (Chironomidae), and mayflies (Ephemeroptera)] .... *Gastromermis* Micoletzky, 1923	
6a(4b).	Spicule shorter than anal body width; amphids small; [parasites of mosquitoes (Culicidae)] .... *Perutilimermis* Nickle, 1972	
6b.	Spicule longer than anal body width; amphids medium to large	7
7a(6b).	Tail pointed; eight hypodermal cords; [parasites of midges (Chironomidae) and mosquitoes (Culicidae)] .... *Hydromermis* Corti, 1902	
7b.	Tail rounded; six hypodermal cords; [parasites of midges (Chironomidae) and blackflies (Simuliidae)] .... *Limnomermis* Daday, 1911	
8a(3b).	Vagina straight or nearly so, barrel- or pear-shaped	9
8b.	Vagina S-shaped, U-shaped, or elongate with both ends curved	13
9a(8a).	Spicules shorter than cloacal body width	10
9b.	Spicules equal to or longer than cloacal body width	11
10a(9a).	Head expanded into a bulb-like shape with very thick cuticle; male lacking lateral-dorsal genital papillae; [parasites of midges (Chironomidae)] .... *Capitomermis* Rubtsov, 1968	
10b.	Head not expanded into a bulb-like shape with thick cuticle; male with genital papillae on lateral-dorsal surface; [parasites of biting midges (Ceratopogonidae)] .... *Heleidomermis* Rubtsov, 1970	
11a(9b).	Six hypodermal cords; [parasites of blackflies (Simuliidae)] .... *Mesomermis* Daday, 1911	
11b.	Eight hypodermal cords	12
12a(11b).	Spicules 2–4 times anal body width; [parasites of mosquitoes (Culicidae)] .... *Romanomermis* Coman, 1961	

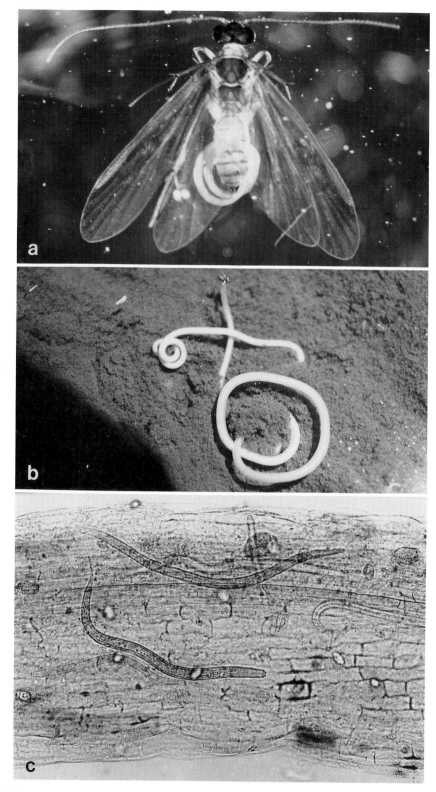

Figure 9.10 (a) A postparasitic juvenile aquatic mermithid emerging from an adult caddisfly (Trichoptera). (Collected by Henk Wolda and submitted by Robert Schuster.) (b) Adults of *Pheromermis pachysoma* (Mermithidae) in a California spring bed. (c) Various juvenile stages and eggs of a freshwater nematode that develop in decaying aquatic vegetation.

12b.	Spicules 1–2 times anal body width; [parasites of mosquitoes (Culicidae) and midges (Chironomidae)] .... *Octomyomermis* Johnson, 1963	
13a(8b).	Eight hypodermal cords	16
13b.	Six hypodermal cords	14
14a(13b).	Spicule length less than two times anal diameter; [parasites of midges (Chironomidae) and mosquitoes (Culicidae)] .... *Strelkovimermis* Rubtsov, 1969	
14b.	Spicule length three or more times cloacal diameter	15
15a(14b).	Spicules ten or more times body width at cloaca; vagina elongate with bends at both ends; postparasitic juveniles with distinct tail appendage; [parasites of diving beetles (Dytiscidae)] .... *Drilomermis* Poinar and Petersen, 1978	
15b.	Spicules 3–10 times body width at cloaca; vagina elongate with 3–6 irregular bends; postparasitic juveniles with indistinct or no tail appendage; [parasites of spiders] .... *Aranimermis* Poinar and Benton, 1986	
16a(13a).	Vagina elongate, slightly curved; spicules shorter than anal body diameter; [parasites of mosquitoes (Culicidae)] .... *Culicimermis* Rubtsov and Isaeva, 1975	
16b.	Vagina S-shaped, ends distinctly curved; spicules range from shorter than anal body diameter to about 10 times tail diameter	17
17a(16b).	Spicule length equal to or shorter than tail diameter; amphids small; cephalic crown well developed; [parasites of mosquitoes (Culicidae)] .... *Empidomermis* Poinar, 1977	
17b.	Spicule length between one and ten times anal diameter; amphids medium-large, cephalic crown absent or only slightly developed; [parasites of blackflies (Simuliidae)] .... *Isomermis* Coman, 1953	

NEMATOMORPHA

VII. INTRODUCTION

Members of the poorly known phylum Nematomorpha (Gordiacea) constitute a relict group that has no clear or close relationship with any other living forms. Even though the first known fossil representative, *Gordius tenuifibrosus* (Voigt 1938) (identified from a 15 mm long fragment of subcuticular tissue in the brown coals of the Geisel Valley near Halle, Germany) dates from the Eocene, the group probably originated considerably earlier (lower Paleozoic) as an offshoot of now extinct lines. Chitwood (1950) considered that Nematomorpha had the closest ties (albeit distant) with nematodes and rotifers (the Acanthocephala and Echinodermata were tied for the next closest kin). However, it is difficult to establish lineages when such basic characters as the presence or absence of molts during the growth phase has not been established.

Some myths surround this group, perhaps associated with the proverbial "gordian knot" (representing a problem solvable only by drastic action and reflected by Alexander the Great cutting the knot that could not be untied, which bound a chariot to a pole at Gordium, the capital of Phrygia, in 333 BC). Another myth, still believed by some today, is indicated by the common name given to these forms, "hairworms" or "horsehair worms." They were supposed to have arisen from horse hairs that fell into water. This belief was scientifically disproven by Leidy in 1870, when he observed horse hairs placed in water over a period of many months without "...having had the opportunity of seeing their vivification."

Adult hairworms have been associated with the digestive and urogenital tract of humans (Watson 1960) and larval hairworms will burrow into a wide range of invertebrate and vertebrate tissue, including human facial tissue sometimes resulting in orbital tumors (Poinar and Doelman 1974).

VIII. MORPHOLOGY AND PHYSIOLOGY

When referring to hairworms, the free-living adult stages are normally described because they are most frequently encountered in sampling. These are dark, slender, worm-like forms ranging in length from several cm to 1 m and in width from 0.25–3 mm (Fig. 9.11a). The color of most adults varies from yellowish to black; and although the anterior end is generally attenuated, both tips tend to be obtusely rounded or blunt. Because the cuticle is normally opaque, it is impossible to examine the internal organs from the outside. Ultrastructural studies of a North American *Gordius* sp. revealed some interesting features of the morphology of adult nematomorphs (Eakin and Brandenburger 1974). The cuticular structure that is such an important taxo-

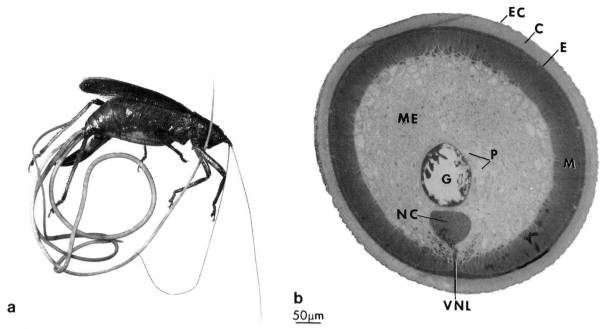

Figure 9.11 (a) A hairworm (Nematomorpha) emerging from its orthopteran host. (b) Transverse section of *Gordius* sp. C, cuticle; E, epidermis; EC, epicuticle; G, gut; M, muscles; ME, mesenchyme; NC, nerve cord; P, pseudocoel; S, debris on cuticle; VNL, ventral neural lamella. (Courtesy of R. M. Eakin and J. L. Brandenburger.)

nomic character is actually a sculpturing on the surface of the thin, superficial epicuticle (Figs. 9.12a, b). This epicuticle is normally crisscrossed by grooves or furrows, leaving small elevations of irregular areas (areoles) between them. The surfaces of the areoles may be smooth or may bear setae (bristles) or cuticular projections (tubercles), arranged singly or in clusters. These bristles and tubercles may also occur in the interareolar furrows.

The cuticle, beneath the epicuticle, is composed of many layers of cylindrical, nonperiodic collagenous fibers, spirally coiled along the length of the adult (Fig. 9.12c). Beneath this is the epidermal layer, constructed of a single layer of interdigitated cells. The musculature consists only of overlapping, longitudinal, flat cells; circular muscles are absent. The pseudocoel is nearly filled with mesenchymal cells containing large clear vacuoles, which give the tissue a foamy appearance. These cells are embedded in a supporting collagenous matrix (Fig. 9.11b).

The mouth is located at the calotte or anterior tip of the body (Fig. 9.13a). This region is often lighter in color than the rest of the body and contains the nonfunctional pharynx, which, in turn, leads into a degenerate intestine lined with epithelium that is one cell in thickness. The epithelial cells of *Gordius* sp. contain numerous microvilli that project into the lumen (Eakin and Brandenburger 1974).

In both sexes, the genital ducts empty into the intestine, forming a cloaca lined with cuticle (Fig. 9.13b). Because adults do not feed, the cloaca is probably used solely for reproductive purposes. All nematomorphs are amphimictic. Males have paired cylindrical testes, each of which connects with the cloaca via a separate sperm duct. Females possess paired ovaries, which after passing through oviducts, enter the cloaca independently.

Adults become sexually mature soon after emerging from their hosts and copulation occurs after the male coils its posterior end around the terminus of the female. The spermatozoa are either deposited as a spermatophore on the female terminus, from where they migrate into the cloaca, or are deposited directly into the female cloaca. The eggs are deposited singly or in clusters, in large strings held together by secretions produced by the antrum (anterior portion of the female cloaca).

The preparasitic stage that hatches from the egg is morphologically quite different from the adult worm and can, therefore, properly be called a larva. [Recall that the term larva implies that some type of metamorphosis occurs before the adult stage.] The larva consists of a presoma (featuring an evaginable proboscis armed with cuticular spines) and a body containing adult tissue primordia (Figs. 9.13c, d). It is unfortunate that so few larvae of described nematomorph species are known, since they probably possess their own specific characters which could be of taxonomic value.

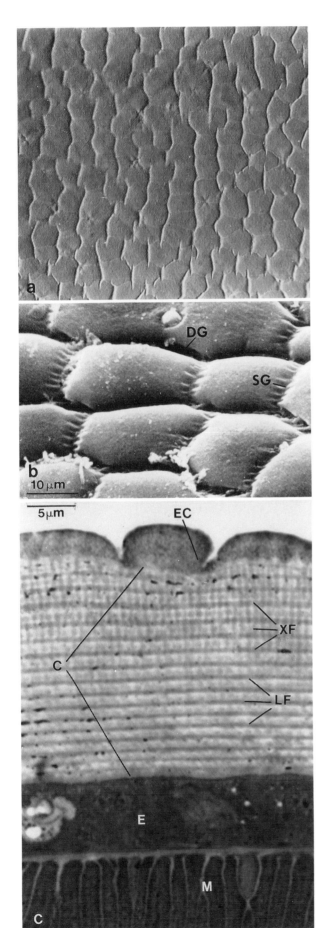

IX. DEVELOPMENT AND LIFE HISTORY

The four stages of the life history of nematomorphs are the egg, the preparasitic larva that hatches from the egg, the parasitic larva that develops within an invertebrate, and the free-living, aquatic adult. Parasitic larvae of all Nematomorpha develop within invertebrates; such invertebrates can be called definitive hosts since the adults are formed in them. A second type of host involved in the life cycle is a transport or paratenic host. The preparasitic larva enters the paratenic host, but does not develop further until eaten by a scavenger or predator. The paratenic host is usually an invertebrate but can also be a vertebrate (e.g., tadpoles and fish).

Three types of life cycles are known for nematomorphs. The first is the direct type where the egg hatches in water and the preinfective larva is ingested by, and develops in, the definitive invertebrate host. Such a cycle is illustrated with Trichoptera larvae (*Stenophylax* sp.) infected by *Gordius* (Dorier 1930). The second type of cycle could be called indirect-free-living, where the preparasitic larva hatches from the egg, then encysts on leaves or other detritus as the water dries up (found in forms inhabiting temporary ponds). The definitive host is infected by ingesting the cysts on the vegetation. Dorier (1930) showed this type of cycle in *Gordius* sp. that had infected millipedes. The third type of cycle can be called indirect-paratenic, where after hatching, the preparasitic larva is ingested by a small-bodied invertebrate host or a vertebrate. The parasite burrows into the tissue of this paratenic host but then encysts and does not develop further. Only when the paratenic host is eaten by a predator (carabid beetle, praying mantis, dragonfly) or omnivore (various Orthoptera) does the parasite initiate development. This type of cycle occurs in *Chordodes* developing in praying mantids that ingest infected mayflies (Inoue 1962), *Gordius* developing in *Dytiscus* that feed on infected tadpoles of *Rana temporaria* (Blunck 1922), and probably *Neochordodes*, which readily infects and encysts in mosquitoes and other aquatic insects (Poinar and Doelman 1974).

Figure 9.12 (a) Surface view of the areolar pattern on the epicuticle of *Neochordodes* sp. (b) Raised areoles and intra-areolar furrows on the epicuticle of a *Gordius* sp. (Courtesy of R. M. Eaken and J. L. Brandenburger.) (c) Cross-section through the body wall of a *Gordius* sp. C, cuticle; E, epidermis; EC, epicuticle; LF, cuticular fibers cut longitudinally; M, muscle plates; XF, cuticular fibers cut crosswise. (Courtesy of R. M. Eaken and J. L. Brandenburger.)

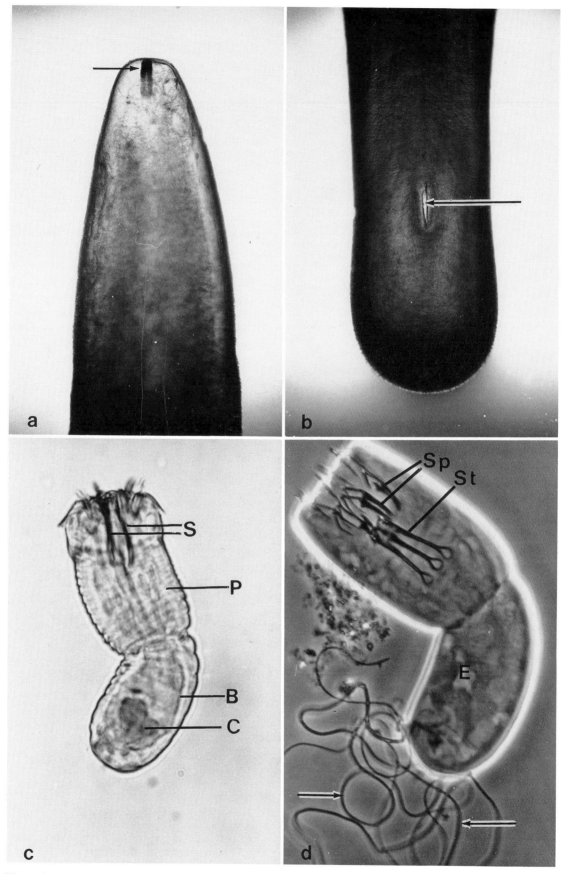

Figure 9.13 (a) Anterior end of an adult *Neochordodes*. Note the degenerate pharynx (arrow). (b) Ventral view of male tail of *Neochordodes* showing the elongate cloacal aperture (arrow). (c) Preparasitic larval of *Neochordodes* showing presoma (P), body (B), intestinal gland (C), and stylets (S). (d) Preparasitic larva of *Neochordodes* showing extruded secretions from the intestinal gland (arrows), stylet (ST), and spines (SP).

There is no evidence that the preparasitic larvae can burrow directly through the outer body wall of either the paratenic or definitive host. In all cases, they enter by way of mouth and encyst in the midgut or burrow through the midgut and encyst in the tissues of the body cavity.

When parasitic development has been completed and the hairworm is ready to emerge, the host must come into contact with water. This poses no problem for aquatic hosts but can present obstacles when the host is a terrestrial arthropod. Although entry into water may occur accidentally, it is more likely that the hosts are possessed with a desire to reach water. This may result from partial desiccation due to the actions of the parasite or be a consequence of some abnormal stimuli from parasite products that affect physiological centers of the host. When the host enters water, the hairworms emerge and either gradually sink or initiate undulating body movements. Males are claimed to be more active than females. As the parasites mature in the host, they change from a light, cream color to a yellowish brown or dark brown color. At the time of emergence, most have already turned their natural color, but in some cases, further darkening occurs after emergence.

Since there are so few records of identified gordiids from identified hosts, very little is known about host selection and specificity. Some hairworms are probably restricted to certain invertebrate genera, while others may be able to develop in a wide range of hosts. Most definitive hosts are medium to large-bodied predaceous or omnivorous arthropods. The great majority of freshwater nematomorphs have been collected from representatives of the insect orders Coleoptera and Orthoptera. Other host groups include spiders (Araneae), Myriopods (Diplopoda and Chilopoda), crustaceans, and leeches. A list of invertebrate families known to contain developing stages of hairworms is presented in Table 9.2. The diversity of paratenic hosts is quite great, extending from trematodes (Cort 1915) to vertebrates. Preparasites will attempt to burrow and encyst into any organism that will ingest them.

X. SAMPLING

Except for the marine genus *Nectonema,* all known representatives of the Nematomorpha occur in freshwater. Their habitats range from watering

Table 9.2 List of Host Families Reported to Contain the Developing Stages of Nematomorpha[a]

Taxon	Reference
Phylum Annelida	
Class Hirudinea	
Order Rhynchobdellida	
Family Glossiphoniidae	Leidy (1878)
Order Pharyngobdellida	
Family Erpobdellidae	Sawyer (1971)
Phylum Arthropoda	
Class Arachnida	
Order Scorpionida	
Family Vaejovidae	S. Williams (personal communication)
Order Araneae	
Family Ctenidae	Acholonu (1968)
Family Gnaphosidae	Blunck (1922)
Order Ambylpygi	
Family Phrynidae	Sciacchitano (1958)
Class Crustacea	
Order Notostraca	
Family Apodidae	Linstow (1878)
Order Decapoda	
Family Caridae	Linstow (1878)
Class Chilopoda	
Order Lithobiomorpha	
Family Lithobiidae	Dorier (1930)
Order Scolopendromorpha	
Family Scolopendridae	Dorier (1930)
Class Diplopoda	
Order Glomerida	
Family Glomeridae	Dorier (1930)

(continues)

Table 9.2 (Continued)

Taxon	Reference
Order Spirobolida	
Family Spirobolidae	Cooper and Storck (1973)
Order Julida	
Family Julidae	Sahli (1972)
Class Insecta	
Order Odonata	
Family Libellulidae	Heinze (1937)
Order Ephemerida	
Family Baetidae	White (1966)
Family Heptageniidae	Dorier (1965)
Order Orthoptera	
Family Acrididae	Blunck (1922)
Family Blattidae	Sciacchitano (1958)
Family Gryllacrididae	Poinar (1989)
Family Gryllidae	Montgomery (1907)
Family Stenopelmatidae	Poinar (1991)
Family Tettigonidae	Thorne (1940)
Order Mantodea	
Family Mantidae	Sciacchitano (1958)
Order Phasmatida	
Family Phasmatidae	Römer (1895)
Order Dermaptera	
Family Forficulidae	Sciacchitano (1958)
Order Hemiptera	
Family Corixidae	White (1966)
Order Neuroptera	
Family Sialidae	Mellanby (1951)
Order Trichoptera	
Family Brachycentridae	White (1969)
Family Phryganeidae	Dorier (1930)
Order Lepidoptera	
Family Saturnidae	Camerano (1897)
Order Coleoptera	
Family Carabidae	Blunck (1922)
Family Chrysomelidae	Dorier (1930)
Family Dytiscidae	Blunck (1922)
Family Silphidae	Blunck (1922)
Family Tenebrionidae	Baylis (1944)
Order Hymenoptera	
Family Apidae	Cury (1946)
Family Formicidae	Donisthorpe (1927)

[a]A single reference is provided to document one record for each family.

troughs and puddles to rivers and subterranean streams. No special sampling technique has been developed for hairworms. They are large enough to be seen with the naked eye and can be netted or lifted out of the water by hand. Stream forms often accumulate in slower moving water off to the side of the main current. Holes dug into small streamlets will often catch and hold adults that are being carried downstream. Late summer and spring are the best times to look for the free-living adults. Although the long, whitish egg strands of some species can sometimes be spotted in water, the preinfectives are too small to be found and are usually noted only by chance when examining the paratenic hosts. Devel-

opmental larvae can be found within respective hosts throughout the year but are nearly impossible to keep alive if they are removed from their host prior to completion of their development. No methods are available for *in vitro* development or culture of nematomorphs. May (1919) was able to monitor development by injecting preparasitic larvae of *Gordius* into the body cavity of long-horned grasshopper hosts (Tettigonidae). However, because of the fragile character of the preparasitic forms and the relatively long period of parasitic development (several months) coupled with the problem of maintaining the hosts, few workers have even bothered to culture hairworms *in vivo*.

XI. IDENTIFICATION

Adult hairworms can be preserved in 5% formalin or 70% alcohol for identification purposes. All diagnostic characters are external features associated with the head, tail, and epicuticular surfaces. The extremities can be removed and mounted in lactophenol or glycerin after dehydration in an alcohol series. For cuticular examination, small slivers of epicuticle can be removed from the midbody region with a razor blade and placed in lactophenol or dehydrated in glycerin. The underlying epidermis and muscle tissue can then be scraped away and the cuticular slice mounted (outer surface up) in lactophenol or glycerin. This will expose the areoles and their ornamentation. Both males and females should be available for accurate identification; however, some males can be identified alone with the following key.

The classification used here is that presented by Dorier (1965) in the Traité de Zoologie. It incorporates the contributions of May (1919), Heinze (1952), and others. Studies on North American forms include works by May (1919), Carvalho (1942), Montgomery (1898a, 1898b, 1899), Poinar and Doelman (1974), Eakin and Brandenburger (1974), and Redlich (1980).

A. Taxonomic Key to Genera of Nematomorpha

Although some of the genera included have not been reported from North America, it is highly probable that they and other still undescribed genera occur in this area. Those genera with North American representatives are marked with an asterisk.

1a.	Pseudocoel hollow; gonad single; marine forms	Nectonematoidea (Rauther 1930)
		Nectonema Verrill, 1879
1b.	Pseudocoel filled with mesenchymatous cells; gonads paired; freshwater forms	Gordioidea Rauther, 1930 2
2a(1b).	Cuticle covered with ridges; male cloaca terminal	*Chordodiolus* Heinze, 1935
2b.	Cuticle not covered with ridges, male cloaca subterminal	3
3a(2b).	Cuticle smooth or with flat, smooth areoles; male tail bilobed, with a postcloacal cresent; female tail entire	Gordiidae May, 1919 4
3b.	Cuticle with distinct areoles usually ornamented with bristles or tubercles; male tail entire or if bilobed, then without a postcloacal cresent; female tail entire or trilobed	Chordodidae May, 1919 5
4a(3a).	Lobes of male tail rounded; head region not attenuated	* *Gordius* Linneaus, 1766
4b.	Lobes of male tail pointed; head region attenuated	*Acutogordius* Heinze, 1952.
5a(3b).	Male tail entire; female tail entire	6
5b.	Male tail bilobed; female tail entire or trilobed	9
6a(5a).	Epicuticle with one type of areole, containing short bristles and tubercles	* *Neochordodes* Carvalho, 1942
6b.	Epicuticle with two or more types of areoles	7
7a(6b).	Areoles containing prominent tubercles or papillae; female cloaca terminal	*Chordodes* Creplin, 1847
7b.	Areoles without prominent tubercles or papillae; female cloaca subterminal	8
8a(7b).	Spines present on cuticle and in furrows between areoles; slender, small forms	*Euchordodes* Heinze, 1937
8b.	Interareolar furrows without spines; normal-sized forms	* *Pseudochordodes* Carvalho, 1942
9a(5b).	Female tail trilobed	10
9b.	Female tail entire	12
10a(9a).	One type of areole present	* *Paragordius* Camerano, 1897
10b.	More than one type of areole present	11
11a(10b).	Two types of areoles present	*Progordius* Kirjanova, 1950
11b.	Three types of areoles present	*Digordius* Kirjanova, 1950
12a(9b).	Areoles arranged in rows separated by ridges	*Beatogordius* Heinze, 1934
12b.	Areoles randomly arranged	13
13a(12b).	Interareolar furrows large, forming a network around the areoles	*Semigordionus* Heinze, 1952

13b.	Interareolar furrows normal	14
14a(13b).	A single type of areole present	*Gordionus* G. W. Müller, 1927
14b.	Two types of areoles present	15
15a(14b).	Lobes on male tail longer than wide; adult heads clearly flattened; pore canals present on larger areoles in males	*Parachordodes* Camerano, 1897
15b.	Lobes on male tail approximately as long as wide; adult heads slightly flattened; pore canals absent or if present, then only in interareolar areas in males	*Paragordionus* Heinze, 1935

LITERATURE CITED

Acholonu, A. D. 1968. *Neochordodes* sp. (Nematomorphs) as a parasite of the spider (*Ctenus bryrrbus*) in Costa Rica. Journal of Parasitology 54:1233–1234.

Ayoub, S. M. 1977. Plant Nematology. An agricultural training aid. California Department of Food and Agriculture, Sacramento. 157 pp.

Baylis, H. A. 1944. Notes on the distribution of hairworms (Nematomorpha: Gordiidae) in the British Isles. Proceedings of the Zoological Society of London 113:193–197.

Blunck, H. 1922. Die Lebenageschichte der im Gelbrand schmarotzender Saitenewürner. Zoologischer Aneizer 54:111–132, 145–162.

Bretschko, G. 1973. Benthos production of a high-mountain lake: Nematoda. Verhandlungen des Instituts für Vereinforschung in Limnologica 18:1421–1428.

Camerano, L. 1897. Monografia dei Gordii. Memoires Royal Academia delle Scienze de Torino 47:339–419.

Carvalho, J. M. C. 1942. Studies on some Gordiaceae of North and South America. Journal of Parasitology 28:231–222.

Chitwood, B. G. 1950. Nemic relationships. Pages 227–241 *in:* B. G. Chitwood and M. B. Chitwood, editors. Introduction to Nematology. University Park Press, Baltimore, Maryland.

Chitwood, B. G., and M. W. Allen. 1959. Nemata. Pages 368–401 *in:* W. T. Edmondson, editor. Fresh-Water Biology. 2nd Edition. Wiley, New York.

Cobb, N. A. 1914. North American fresh-water nematodes. Transactions of the American Microscopical Society 33:35–100.

Coe, W. R. 1937. Methods for the laboratory culture of nemerteans. Pages 162–165 *in:* J. G. Needham, editor. Culture methods for invertebrate animals. Dover, New York.

Cooper, C. L., and T. W. Storck. 1973. *Gordius* sp., a new host record. Ohio Journal of Science 73:228.

Corbel, J.-C. 1967. Les parasites des Orthopteres. Annales de Biologie 6:391–426.

Cort, W. W. 1915. *Gordius* larvae parasitic in a trematode. Journal of Parasitology 1:198–199.

Cury, R. 1946. Moléstias das abelhas. Biológico, 12:241–254.

Donisthorpe, H. J. K. 1927. The Guests of British Ants, their habits and life-histories. Routledge, London.

Dorier, A. 1930. Recherches biologiques et systematiques sur les Gordiacés. Annales de la Universite de Grenoble, nouve série 7:1–183.

Dorier, A. 1965. Classe Gordiacés. Pages 201–1222 *in:* P.-P. Grassé, editor. Traité de Zoologie. Vol. 4. Masson, Paris.

Downing, J. A., and F. H. Rigler, editors. 1984. A manual on methods for the assessment of secondary production in freshwaters. Blackwell, Oxford.

Eakin, R. M., and J. L. Brandenburger. 1974. Ultrastructural features of a gordian worm (Nematomorpha). Journal of Ultrastructure Research 46:351–374.

Esser, R. P., and G. R. Buckingham. 1987. Genera and species of free-living nematodes occupying fresh-water habitats in North America. Pages 477–487 *in:* J. A. Veech and D. W. Dickson, editors. Vistas on Nematology. Society of Nematologists, Hyaltsville, Maryland.

Esser, R. P., G. R. Buckingham, C. A. Bennett, and K. J. Harkcom. 1985. A survey of phytoparasitic and free living nematodes associated with aquatic macrophytes in Florida. Proceedings of the Florida Soil and Crop Science Society 44:150–155.

Ferris, V. R., and J. M. Ferris. 1972. Nematode community structure: a tool for evaluating water resource environments. Technical report No. 30. Water Resources Research Center, Purdue University, Lafayette, Indiana.

Ferris, V. R., J. M. Ferris, and J. P. Tjepkema. 1973. Genera of freshwater nematodes (Nematoda) of Eastern North America. U.S. Environmental Protection Agency Identification Manual No. 10. 38 pp.

Filipjev, I. N., and J. H. Schuurmans Stekhoven, Jr. 1959. Agricultural Helminthology. Brill, Leiden, Netherlands.

Flint, R. W. 1976. The natural history, ecology and production of the crayfish, *Pacifastacus leniusculus,* in a subalpine lacustrine environment. Ph.D. Thesis, University of California, Davis.

Gerber, K., and G. C. Smart, Jr. 1987. Plant-parasitic nematodes associated with aquatic vascular plants. Pages 488–501 *in:* J. A. Veech and D. W. Dickson, editors. Vistas on Nematology. Society of Nematologists, Hyaltsville, Maryland.

Goodey, J. B. 1963. Soil and Freshwater Nematodes. Methuen, London.

Hantzschel, W. 1975. Trace fossils and problematics. Pages 1–269 *in:* R. C. Moore, editor. Treatise on Invertebrate Paleontology. Part W, Supplement 1. The Geology Society of America. Boulder, Colorado.

Heinze, K. 1937. Die Saitenwürmer (Gordioidea) Deutschlands. Eine systematisch-faunistische Studie über insektenparasiten aus der Gruppe der Nematomorpha.

Zeitschrift für Parasitenkunde 9:263–344.

Heinze, K. 1952. Über Gordioidea, eine systematische Studie über Insektenparasiten aus der gruppe der Nematomorpha. Zeitschrift für Parasitenkunde 7:657–678.

Hoeppli, R. J. C. 1926. Studies of free-living nematodes from the thermal waters of Yellowstone Park. Transactions of the American Microscopical Society 45:234–255.

Holme, N. A., and A. O. McIntyre, editors. 1971. Methods for the study of marine benthos. IBP Handbook No. 16. Blackwell, Oxford.

Holopainen, I. J., and L. Paasivirta. 1977. Abundance and biomass of the meiozoobenthos in the oligotrophic and mesohumic lake Paajarvi, southern Finland. Annales Zoologica Fennland 14:124–134.

Hulings, N. C., and J. S. Gray, editors. 1971. A manual for the study of meiofauna. Smithsonian Contributions to Zoology 78:84 pp.

Hummon, W. D. 1981. Extraction by sieving: a biased procedure in studies of stream meiofauna. Transactions of the American Microscopical Society 100:278–284.

Inoue, I. 1962. Studies on the life history of *Chordodes japonensis*, a species of Gordiaceae. III. The mode of infection. Annotationes Zoological Japonenses 35:12–19.

Issel, R. 1906. Sulla termobiosi negli animali acquatici. Ricerche faunistiche e biologiche. Attidella Societa Ligustica di Scienze naturali e geografiche 17:3–72.

Jacobs, L. J. 1984. The free-living inland aquatic nematodes of Africa—a review. Hydrobiologia 113:259–291.

Kirjanova, E. S. 1958. On the structure of the copulative organs of males of the freshwater hairworms (Nematomorpha, Gordioidea). Zoologischer Zhurnal, 37:359–372. (In Russ.)

Leidy, J. 1870. The gordius, or hairworm. The American Entomologist and Botanist 2:193–197.

Leidy, J. 1878. On *Gordius* infesting the cockroach and leech. Proceedings of the Academy of Natural Sciences of Philadelphia 30:383.

Linstow, O. F. B. von. 1878. Compendium der Helminthologie. Hannover.

May, H. G. 1919. Contributions to the life histories of *Gordius robustus* Leidy and *Paragordius varius* (Leidy). Illinois Biological Monographs 5:1–119.

Mellanby, H. 1951. Animal life in fresh water. A guide to fresh-water invertebrates. 4th edition. Methuen, London.

Meyl, A. H. 1954. Beiträge zur Kenntnis der Nematodenfauna vulkanisch erhitzter Biotope. Zeitschrift für Morphologie und Ökologie der Tiere 42:421–448.

Micoletzky, H. 1922. Die freilebenden Erd-Nematoden. Archiv für Naturgeschichte A87:1–650.

Montgomery, T. H., Jr. 1898a. The Gordiacea of certain American collections with particular reference to the North American fauna. Bulletin of the Museum of Comparative Zoology 32:1–59.

Montgomery, T. H., Jr. 1898b. The Gordiacea of certain American collections, with particular reference to the North American fauna. II. Proceedings of the Califor-

nia Academy of Science 1:333–344.

Montgomery, T. H., Jr. 1899. Synopses of North American invertebrates. II. Gordiaces (hair worms). American Naturalist 33:647–652.

Montgomery, T. H., Jr. 1907. The distribution of the North American Gordiacea, with descriptions of a new species. Proceedings of the Academy of Natural Sciences of Philadelphia, 59:270–272.

Moussa, M. T. 1969. Nematode fossil tracks of Eocene age from Utah. Nematologica 15:376–380.

Nalepa, T. F., and M. A. Quigley. 1983. Abundance and biomass of the meiobenthos in nearshore lake Michigan with comparisons to the macrobenthos. Journal of Great Lakes Research 9:530–547.

Nicholas, W. L. 1984. The biology of free-living nematodes. Clarendon Press, Oxford.

Nichols, J. A. 1979. A simple flotation technique for separating meiobenthic nematodes from fine-grained sediments. Transactions of the American Microscopical Society 98:127–130.

Nuttycombe, J.W. 1937. Wheat-grain infusion. Pages 135–136 *in:* J. G. Needham, editor. Culture methods for invertebrate animals. Dover, New York.

Oden, B. J. 1979. The freshwater littoral meiofauna in a South Carolina reservoir thermal effluents. Freshwater Biology 9:291–304.

Pennak, R. W. 1953. Fresh-water invertebrates of the United States. Ronald Press, New York.

Pennak, R. W. 1978. Fresh-water invertebrates of the United States. 2nd Edition. Wiley, New York.

Petersen, J. J. 1972. Procedures for the mass rearing of a mermithid parasite of mosquitoes. Mosquito News 32:226–230.

Poinar, G. O., Jr. 1991. Hairworm (Nematomorpha: Gordioidea) parasites of New Zealand wetos (Orthoptera: Stenopelmatidae) Canadian Journal of Zoology (In Press).

Poinar, G. O., Jr. 1983. The Natural History of Nematodes. Prentice-Hall. Englewood Cliffs, New Jersey.

Poinar, G. O., Jr. 1989. Unpublished observations in Australia.

Poinar, G. O., Jr., and J. J. Doelman. 1974. A reexamination of *Neochordodes occidentalis* (Montg.) comb. n. (Chordodidae: Gordioidea): Larval penetration and defense reaction in *Culex pipiens* L. Journal of Parasitology 60:327–335.

Poinar, G. O., Jr., and E. Hansen. 1983. Sex and reproductive modifications in nematodes. Helminthological Abstract - Series B 52:145–163.

Poinar, G. O., Jr., and R. Hess. 1988. Protozoan diseases of nematodes. Pages 103–131 *in:* G. O. Poinar, Jr. and H.-B. Jansson, editors. Diseases of Nematodes. Vol. 1. CRC Press, Boca Raton, Florida.

Por, F. D., and D. Moary. 1968. Survival of a nematode and an oligochaete species in the anaeorbic benthal of Lake Tiberios. Oikos 19:388–391.

Rahm, G. 1937. Grenzen des Lebens? Studien in heissen Quellen. Forschungen und Fortschritte 13:381–387.

Redlich, A. 1980. Description of *Gordius attoni* sp.n. (Nematomorpha, gordiidae) from Northern Canada. Canadian Journal of Zoology 58:382–385.

Riemann, F., and M. Schrage. 1978. The mucus-trap hypothesis on feeding of aquatic nematodes and implications of biodegradation and sediment texture. Oecologia 34:75–88.

Römer, F. 1895. Die Gordiiden des Naturhistorischen Museums in Hamburg. Zoologische Jahrbücher, Abtheilung für Systematik, Geographie und Biologie der Thiere 8:790–803.

Sahli, F. 1972. Modifications des caracteres sexuel secondaires mâles chez les Julidae (Myriapods, Diplopoda) sous lbinfluence de Gordiacés parasites. Comptes Rendus de l'Academie des Sciences 274:900–903.

Sawyer, R. T. 1971. Erpobdellid leeches as new hosts for the Nematomorpha, *Gordius*. Journal of Parasitology 57:285.

Schiemer, F. 1978. Verteilung und Systematik der freilebenden Nematoden des Neusiedlersee. Hydrobiologie 58:167–194.

Schiemer, F. 1983. Comparative aspects of food dependence and energetics of freeliving nematodes. Oikos 41:32–42.

Schiemer, F. 1987. Nematoda. Pages 185–215 *in:* T. J. Pandian and F. J. Vernberg, editors. Animal energetics. Vol. 1: Protozoa through Insecta. Academic Press, New York.

Sciacchitano, I. 1958. Gordioidea del Congo Belga. Annales Museum de la Republic de Congo Belge 67:1–111.

Southey, J. F. 1978. Laboratory methods for work with plant and soil nematodes. Tech. Bull. No. 2. Ministry of Agriculture, Fisheries and Food, London.

Stirewalt, M. A. 1937. The culture of *Microstomum*. Pages 149–150 *in:* J. G. Needham, editor. Culture methods for invertebrate animals. Dover, New York.

Strayer, D. 1985. The benthic micrometazoans of Mirrow Lake, New Hamphire. Archives für Hydrobiologie, Supplement 72:287–426.

Thorne, G. 1940. The hairworm, *Gordius robustus* Leidy, as a parasite of the Mormon cricket, *Anabrus simplex* Haldeman. Journal of the Washington Academy of Science 30:219–231.

Torjan, A. C., R. P. Esser, and S. L. Chang, 1977. An illustrated key to nematodes found in freshwater. Journal of Water Pollution Control Federation 49:2318–2337.

Tsalolikhin, S. Y. 1983. The nematode families Tobrilidae and Tripylidae: World fauna. Nauka, Leningrad. (In Russ.)

Viglierchio, D. R., and R. V. Schmidt. 1983. On the methodology of nematode extraction from field samples: comparison of the methods for soil extraction. Journal of Nematology 15:450–454.

Voigt, E. 1935. Ein fossiler Saitenwrm (*Gordius tenuifibrosus* n.sp.) aus der Eozänen Braunkohle des Geiseltales. Nova Acta Leopoldina, new series 5:351–360.

Warwick, R. M., and R. Price. 1979. Ecological and metabolic studies on free-living nematodes from an estuarine mudflat. Estuary Coastal Marine Science 9:257–271.

Watson, J. M. 1960. Medical Helminthology. Baillière, London.

White, D. A. 1966. A new host record for *Paragordius varius* (Nematomorpha). Transactions of the American Microscopical Society 85:579.

White, D. A. 1969. The infection of immature aquatic insects by larval *Paragordius* (Nematomorpha). The Great Basin Naturalist 29:44.

Winterbourn, M. J., and T. J. Brown. 1967. Observations on the faunas of two warm streams in the Taupo Thermal Region. New Zealand Journal of marine and freshwater Resources 1:38–50.

Zullini, A. 1976. Nematodes as indicators of river pollution. Nematology Mediterranean 4:13–22.

Appendix 9.1 Systematic Arrangement of the Nematode Genera

Phylum Nematoda
  Class Adenophorea
    Subclass Chromadorida
      Order Araeolaimida
        Suborder Araeolaimina
          Superfamily Axonolaimoidea
            Family Axonolaimidae (*Cylindrolaimus*)
          Superfamily Leptolaimoidea
            Family Bastianiidae (*Bastiania*)
            Family Leptolaimidae (*Rhabdolaimus*)
          Superfamily Camacolaimidae
            Family Camacolaimidae (*Aphanolaimus, Paraphanolaimus*)
          Superfamily Plectoidea
            Family Plectidae (*Anonchus, Chronogaster, Plectus*)
            Family Teratocephalidae (*Euteratocephalus, Teratocephalus*)
      Order Monhysterida
        Superfamily Monhysteroidea
          Family Monhysteridae (*Monhystera, Monhystrella, Prismatolaimus*)
      Order Chromadorida
        Suborder Chromadorina
          Superfamily Chromadoridea
            Family Chromadoridae (*Chromadorita, Punctodora*)
        Suborder Cyatholaimina
          Superfamily Cyatholaimoidea
            Family Cyatholaimidae (*Achromadora, Ethmolaimus, Odontolaimus, Paracyatholaimus, Prodesmodora*)
            Family Microlaimidae (*Microlaimus*)
      Order Enoplida
        Suborder Enoplina
          Superfamily Tripyloidea
            Family Tripylidae (*Tobrilus, Tripyla*)
            Family Alaimidae (*Alaimus, Amphidelus*)
            Family Ironidae (*Cryptonchus, Ironus*)
        Suborder Oncholaimina
          Superfamily Oncholaimoidea
            Family Oncholaimidae (*Mononchulus, Oncholaimus*)
      Order Dorylaimida
        Suborder Dorylaimina
          Superfamily Dorylaimoidea
            Family Dorylaimidae (*Actinolaimus, Dorylaimus, Mesodorylaimus, Paractinolaimus*)
            Family Nygolaimidae (*Nygolaimus*)
          Superfamily Leptonchoidea
            Family Campydoridae (*Aulolaimoides*)

Order Mononchida
  Suborder Mononchina
    Superfamily Mononchoidea
      Family Mononchidae (*Anatonchus, Mononchus, Prionchulus*)
Order Mermithida
  Suborder Mermithina
    Superfamily Mermithoidea
      Family Mermithidae (*Aranimermis, Capitomermis, Culicimermis, Drilomermis, Empidomermis, Gastromermis, Heleidomermis, Hydromermis, Isomermis, Lanceimermis, Limnomermis, Mesomermis, Octomyomermis, Perutilimermis, Pheromermis, Pseudomermis, Romanomermis, Strelkovimermis*)
Class Secernentea
  Order Tylenchida
    Suborder Tylenchina
      Superfamily Hoplolaimoidea
        Family Hoplolaimidae (*Hirschmanniella*)
      Superfamily Tylenchoidea
        Family Tylenchidae (*Tylenchus*)
      Superfamily Atylenchoidea
        Family Atylenchidae (*Atylenchus*)
    Suborder Criconematina
      Superfamily Hemicyclcophoroidea
        Family Hemicycliophoridae (*Hemicycliophora*)
  Order Aphelenchida
    Suborder Aphelenchina
      Superfamily Aphelenchoidoidea
        Family Aphelenchoididae (*Aphelenchoides, Seinura*)
      Superfamily Aphelenchidae
        Family Aphelenchidae (*Aphelenchus*)
  Order Rhabditida
    Suborder Diplogasteroidea
      Superfamily Diplogasteroidea
        Family Diplogasteridae (*Rhabditolaimus*)
      Superfamily Rhabditoidea
        Family Cephalobidae (*Cephalobus, Eucephalobus*)
        Family Rhabditidae (*Mesorhabditis, Pellioditis*)
        Family Daubayliidae (*Daubaylia*)

---

[a] Classification includes genera cited in this chapter and is based on a synthesis of data from several workers.

Mollusca: Gastropoda

Kenneth M. Brown

Department of Zoology and Physiology
Louisiana State University
Baton Rouge, Louisiana 70803

Chapter Outline

I. INTRODUCTION
II. ANATOMY AND PHYSIOLOGY
A. External and Internal Morphology
1. Shell
2. Soft Parts
B. Organ System Function
1. Reproductive System
2. Digestive System
3. Respiratory and Circulatory Systems
4. Excretory System
C. Environmental Physiology
III. ECOLOGY AND EVOLUTION
A. Diversity and Distribution
B. Reproduction and Life History
C. Ecological Interactions
1. Habitat and Food Selection: Effects on Producers
2. Factors Regulating Population Size
3. Production Ecology
4. Ecological Determinants of Distribution
5. Suggestions for Further Work
D. Evolutionary Relationships
IV. COLLECTING AND CULTURING FRESHWATER GASTROPODS
V. IDENTIFICATION OF THE FRESHWATER GASTROPODS OF NORTH AMERICA
A. Taxonomic Key to Families and Selected Genera of Gastropoda
Literature Cited

I. INTRODUCTION

Snails are among the most ubiquitous organisms of shallow littoral zones in lakes and streams. They feed on detritus, graze on the periphyton covering of macrophytes or cobble, or even float upside down at the water surface (supported by the surface tension) and feed on algae trapped at the surface (Fig. 10.1). They are also found at considerable depths in many lakes and form the basis of food chains dominated by sport fish. One of the most intriguing aspects of the biology of freshwater snails is their adaptation to the relative ephemerality (in both space and time) of their habitats. Extensive intraspecific variation in life histories, productivity, morphology, and feeding habits adapt freshwater gastropods to these uncertain habitats. This flexibility has been termed adaptive plasticity (Russell-Hunter 1970, 1978).

Two other aspects of the biology of freshwater gastropods are currently garnering tremendous interest: their role in predator–prey interactions (Vermeij and Covich 1978, Lodge *et al.*1987), and their role as grazers in freshwater systems. We have recently learned for example that predators, by controlling snail populations, may indirectly facilitate algal producers.

Gastropods are the most diverse class of the phylum Mollusca, comprising almost three quarters of the 110,000 or so described species of molluscs. Over 50,000 of these species belong to the mostly marine and freshwater subclass Prosobranchia, while another 20,000 species belong to the subclass Pulmonata, of which most are terrestrial.

Molluscs are soft-bodied, unsegmented animals, with a body organized into a muscular foot, a head region, a visceral mass, and a fleshy mantle that secretes the calcareous shell. Gastropods have a univalve shell, and possess a file-like radula used in feeding on the periphyton coverings of rocks or plants.

In North America, there are 49 genera and 349

Figure 10.1 An adult *Lymnaea stagnalis* (length about 5 cm) foraging upside down at the surface of the water, supported by the surface tension.

species of prosobranch snails and 29 genera and 150 species of pulmonate snails (Burch 1982). Prosobranch snails possess a gill (ctenidium) and a horny (flexible) or calcareous operculum, or "trap door," which is pulled in after the foot to protect the animal. Pulmonates have secondarily re-invaded freshwaters from terrestrial habitats; they use a modified portion of the mantle cavity as a lung and lack an operculum (Pechenik 1985). Students seeking a detailed introduction to gastropod biology are referred to Hyman (1967), Fretter and Graham (1962), and the volumes on the general biology of molluscs edited by Wilbur (1983) or the biology of pulmonates edited by Fretter and Peake (1975, 1978).

II. ANATOMY AND PHYSIOLOGY

A. External and Internal Morphology

1. Shell

The most striking aspect of gastropod anatomy is the shell; its structure plays an important role in systematics. Shell geometry falls into three broad catego-

ries (Fig. 10.2A–C). Shells can have a simple conical shape, as in the limpets (family Ancylidae), with new shell material secreted at the margin. Second, the shell can be secreted in a spiral, but with the whorls all in one plane (usually termed planospiral) as illustrated by the pulmonate family Planorbidae. Finally, whorls may be elevated into a "spire," as in the pulmonate families Physidae and Lymnaeidae and in various prosobranch families. An excellent discussion of how new shell material is secreted by the mantle for each of these three basic shapes is given by Russell-Hunter (1983). If shells are placed with the spire facing away from the observer and the aperture (opening from which the foot extends) upwards, shells with the aperture on the left are sinistral (e.g., the physids), whereas those on the right are dextral (e.g., the lymnaeids). Although not externally obvious, spiral shells have a central supporting member, the columella, similar to the center support of a spiral staircase. The columella adds to the strength of the shell and provides an attachment point for the soft parts via the columellar muscle.

A spiral shell (Fig. 10.2D) is convenient for illustrating some of the terminology used in systematics. The pointed end of the shell, opposite the aperture, is the apex. Shell length in spiral shells is measured from the apex to the lower tip of the aperture, while the greatest diameter is used in planospiral shells. The spire is separated into a number of whorls by sutures (Fig. 10.2). The most apical whorl is the nuclear whorl or protoconch, the initial shell of the newly hatched snail or "spat." The final and biggest whorl (representing the most recent shell accretion) is the body whorl, ending in the aperture. Whorls

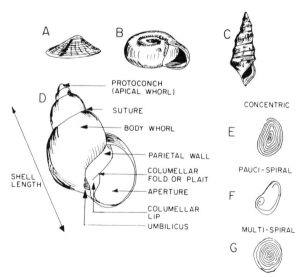

Figure 10.2 Basic anatomy of the shell, including shell architecture (conical, A; planospiral, B; spiral C,D), major features of the shell (D), and three types of opercula (concentric, E; paucispiral, F; multispiral, G).

may be rounded and sutures deep and well defined (as in a typical lymnaeid shell, Fig. 10.2D), or whorls may be flattened and sutures shallow, as in the prosobranch family Pleuroceridae (Fig. 10.2C).

Part of the aperture is often reflected over the body whorl (Fig. 10.2D) to form an inner lip. If there is a channel between the inner lip and the body whorl, the shell is umbilicate or perforate (the opening of the channel is called the umbilicus and leads up and inward into the columella). Imperforate shells lack an umbilicus. While freshwater shells are not as ornate as their marine relatives (Vermeij and Covich 1978), shells may have spines, ridges, colored bands, or small malleations (hammerings) on the surface. The operculum is often useful in classifying prosobranchs. If the growth lines lie completely within each other, the operculum is concentric; whereas, lines arranged in a spiral are termed multispiral or paucispiral (see examples in Fig. 10.2E–G). Further discussion on shell sculpture is given in Fretter and Graham (1962).

The shell is composed of an outer periostracum of organic (mostly protein) composition which may limit shell abrasion or dissolution of shell calcium carbonate by acid waters. Beneath the periostracum is a thick layer of crystalline calcium carbonate with some protein material interlaced (Russell-Hunter 1978). Calcium carbonate is either absorbed directly from water or is sequestered from food (McMahon 1983). While the accepted relationship has been that harder water results in thicker and more sculptured shells (Pennak 1978), this may be an oversimplification (see discussions in McMahon 1983, Lodge *et al.* 1987).

2. Soft Parts

As in all gastropods, the soft parts of freshwater snails are separated into four basic areas: head, foot, visceral mass, and mantle (Fig. 10.3). Aquatic pulmonates and prosobranchs possess eyes at the bases of their tentacles, unlike terrestrial pulmonates whose eyes are at the tips of the tentacles. The muscular foot is provided with both cilia and secretory epithelium to secrete mucus for locomotion, as well as pedal muscles which produce waves of contraction to push the animal forward. The visceral hump includes most of the organs of digestion and reproduction. The mantle covers the visceral mass and underlays the shell, which it secretes. The anterior mantle, over the head, possesses a mantle cavity where the gill or ctenidium is located in prosobranchs. Gastropods have ganglia innervating each of these areas. Further information on internal anatomy can be gleaned from texts of invertebrate zoology such as Fretter and Graham (1962), Barnes (1987), Pechenik (1985), or Hyman (1967).

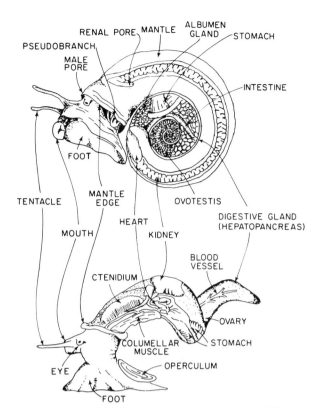

Figure 10.3 Basic internal anatomy of a planorbid pulmonate (above, after Burch 1982) and a pleurocerid prosobranch (below, after Pechenik 1985).

B. Organ System Function

1. Reproductive System

Prosobranch snails are usually dioecious, and males use the enlarged right tentacle as a copulatory organ (in the viviparids), or possess a specialized penis or verge (in the hydrobiids, pomatiopsids, and valvatids) or have no copulatory organ (thiarids and pleurocerids). Many pleurocerids lay clutches of a few eggs, whereas viviparids lay eggs that hatch and develop in a fold of the anterior mantle (the pallial oviduct), and are born free living. Some viviparids in the genus *Campeloma* are parthenogenetic (Van Cleave and Altringer 1937, Vail 1978). Parthenogenesis is considered an adaptation for colonizing low-order streams or other unpredictable habitats where densities are low and chances of finding mates rare (Vail 1978). One group of freshwater prosobranchs, the family Valvatidae, is hermaphroditic.

Pulmonates, in contrast, are all monoecious. The basic components of the plumonate reproductive system are shown in Figure 10.4. Sperm and eggs are produced in the ovotestis and exit via a common hermaphroditic duct in all pulmonates except ancylids, which have two openings and thus obligate cross-fertilization. The albumen gland adds protein and nutrients to the egg. Eggs are either fertilized in

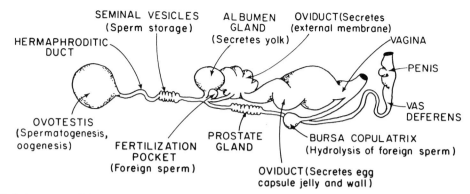

Figure 10.4 Anatomy of the reproductive system of the monoecious pulmonate *Physa*. (After Duncan 1975.) Sperm are produced in the ovotestis, stored in the seminal vesicle, exit via vas deferens, and are inserted into the vagina of a second individual via the introvertible penis, or may fertilize the same individual's eggs in the hermaphroditic duct. Most foreign sperm are hydrolyzed in *Physa* in the bursa copulatrix, but some do fertilize eggs in the fertilization pocket. The albumen gland secretes yolk around the egg, and the oviduct wall secretes the egg membrane and the jelly and external wall of the egg case.

the hermaphroditic duct by the same individual's sperm or by sperm from another individual near the junction of the hermaphroditic duct and the oviduct. The external egg membranes are then secreted in the oviduct. Eggs are laid in gelatinous egg cases and attached to plants or rocks. A more detailed discussion of egg and sperm formation, copulation and fertilization, and egg capsule deposition can be found in Duncan (1975). Although pulmonates are usually simultaneous hermaphrodites, most species outcross when possible. Pulmonates that "self" usually mature at later ages and have lower fecundity (Duncan 1975, Brown 1979). Males have an introvertible penis and fertilization is internal.

The adaptive value of hermaphroditism in pulmonates is usually explained in the following way (Calow 1983, Brown 1983). Pulmonates are slow-moving organisms and populations often go through seasonal "bottlenecks" (precipitous declines in density). The chances of finding a mate in such situations are small, providing a selective advantage for monoecy (see Maynard-Smith 1976 for a general discussion). Because pulmonates are dispersed passively as spat trapped in mud on the feet of birds (see Boag 1986 and references therein), being monoecious (if one is the "only game in town") is obviously beneficial. The costs of producing two reproductive systems (even when some parts are shared) as well as any inbreeding depression are evidently less than not being able to reproduce at all in such situations.

2. Digestive System

While food preferences of different gastropod groups are discussed later, a general description of the digestive system here provides an introduction.

Food is brought into the mouth by rasping movements of the radula. The radula is a file-like structure (Fig. 10.5) resting on a cartilage (the odontophore) to which radular protractors (muscles that extend the radula) and retractors are attached. When the radula is protracted, it contacts the substratum, and algal particles are scraped off when retractors pull the radula back into the mouth. The radula may also pulverize food particles by grinding them against the roof of the mouth. Food then travels via a long esophagus to the stomach, located in the visceral mass. Some gastropods possess a specialized crop where sand grains further abrade food particles. Emptying into the stomach are digestive enzymes produced by the molluscan equivalent of a liver, the digestive gland or hepatopancreas. Considerable digestion also occurs intracellularly in the hepatopancreas. Snails are one of the few animal groups to possess cellulases (see Kesler 1983, Kesler *et al.* 1986, and Chapter 18), which can degrade the cell walls of detritus or algae. The feeding behavior of gastropods is reviewed by Kohn (1983).

3. Respiratory and Circulatory Systems

The respiratory system differs radically between prosobranchs and pulmonates. Freshwater prosobranchs have a single ctenidium or gill (Aldridge 1983). The ctenidium, usually located in the mantle cavity, has leaf-like triangular plates richly supplied with blood vessels where oxygen transfer occurs. Oxygen-poor blood is brought to the ctenidium by a vessel, where it passes across the plates in the opposite direction from oxygen-rich water currents (generated by ctenidial cilia). This counter-current mechanism assures a net positive diffusion of oxygen from water into the blood. Pulmonates, on the

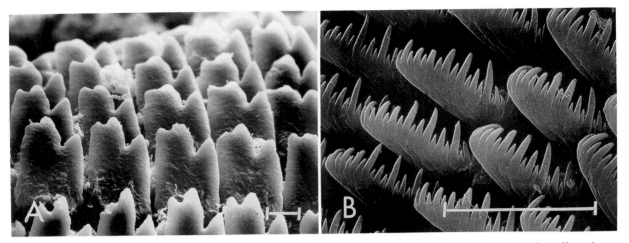

Figure 10.5 Scanning electron microscopic photographs of radular teeth of (A) *Pseudosuccinea columella* and (B) *Physa vernalis*. (From Kesler *et al.* 1986.) Note the lymnaeid has fairly large teeth with few cusps. These teeth, along with large, cropping jaws in the buccal mass, a grinding gizzard equipped with sand, and high cellulase levels allow it to specialize on filamentous green algae (Kesler *et al.* 1986). *Physa* has many small teeth with long cusps, arranged in chevron-shaped rows, perhaps useful for piercing detritus and attached bacteria. Bar = 10 μm.

other hand, lost their gill during their intermediate terrestrial phase. Instead, they have a richly vascularized pocket in the mantle, which is used as a lung (hence their name). The opening of the lung is the pneumostome. Pulmonates either rely on surface breathing or have a limited capacity for oxygen transfer across their epithelial tissues (McMahon 1983). Certain species of physids and lymnaeids have adapted to benthic conditions by filling the mantle pocket with water and using it as a derived gill (Russell-Hunter 1978). Ancylids and planorbids have re-adapted even further to aquatic conditions by evolving a conical extension of epithelium that is used as a gill, and planorbids also have a respiratory pigment, hemoglobin, which increases the efficiency of oxygen transport (McMahon 1983).

4. Excretory System

Gastropods have a permeable epidermis and are subject to osmotic inflow of water from their hypoosmotic surroundings. Thus they must pump out excess water in their urine. The gastropod coelom is little more than a small cavity (pericardium) surrounding the heart. The coelomic fluid is largely a filtrate of the blood containing waste molecules such as ammonia, which are filtered across the wall of the heart. Additional wastes are actively secreted into the coelom by the walls of the pericardium. The coelomic fluid then enters a metanephridial tubule (the coelomoduct) where selective resorption of salts and further secretion of wastes occur. The urine is then discharged into the mantle cavity. Fur-

ther discussion of excretion can be found in Martin (1983) for gastropods in general and in Machin (1975) for pulmonates.

C. Environmental Physiology

To describe physiological mechanisms available for coping with environmental changes, two terms need to be defined. Tolerance adaptations refer to the range of variation in a certain parameter that a species can withstand, while capacity adaptations refer to the changes in physiological processes that occur "within" that range. Pulmonates usually face wider extremes of temperature variation than do most prosobranchs (McMahon 1983) and therefore have wider tolerance adaptations to temperature. Most temperate pulmonates, for example, can withstand temperatures near 0°C for several months, whereas tropical pulmonates can withstand temperatures near 40°C for extended periods. This is undoubtedly adaptive, not only because of greater seasonal variation in the ponds where pulmonates are more common, but also because of the dramatic diurnal variation in temperature in many pond habitats.

Pulmonates also have considerable capacity adaptations to changing temperature. For example, pulmonate snails generally have smaller changes in metabolic rate with changing temperature than do prosobranchs. For instance, Q_{10} values for 18 pulmonates (averaging 2.2) were less than in 13 prosobranchs (averaging 2.8), and pulmonates apparently accomplish this greater degree of regulation through acclimation (McMahon 1983). Differences in adaptation to fluctuating temperatures may also occur within each of these two taxonomic groups. For example, stream ancylids are adapted to lower and more rapidly fluctuating levels of temperatures than are planorbids, because planorbids are found in ponds with relatively warmer water (Calow 1975).

Finally, temperature causes important capacity adaptations in the life histories of freshwater gastropods (McMahon 1983). As average temperatures increase, snails grow faster and reproduce at an earlier age, often with more generations resulting per year. Increasing water temperature is also considered to be the cue for onset of reproduction in many temperate pulmonates. The ability of pulmonates to reproduce in cold water temperatures may be adaptive in allowing them to breed early in the spring. The resulting juveniles grow rapidly to adult size during the summer, assuring the availability of adults that can then breed in the next spring.

Pulmonates also show greater degrees of resistance adaptations to desiccation than do prosobranchs (McMahon 1983; but for an exception see Davis 1981). Pulmonates can secrete a mucous covering over the aperture, called an epiphragm, to retard moisture loss (Boss 1974, Jokinen 1978). A more subtle adaptation concerns differences in excretion products. Freshwater gastropods excrete nitrogen both as ammonia and as urea. Ammonia is adaptive in aquatic environments because, although it is toxic, it is extremely soluble and readily diffuses away. Pulmonates often produce urea, which is better in terrestrial situations or during hibernation or estivation because it is relatively nontoxic and can be stored in the blood until it is able to be excreted (McMahon 1983).

In terms of adaptation to hypoxia, pulmonates may have greater levels of resistance adaptations than do prosobranchs (possibly because prosobranchs rarely experience hypoxia in the fast- flowing rivers where they are common) and also regulate oxygen consumption at varying levels of dissolved oxygen better than prosobranchs, which are "oxyconformers." Pulmonates apparently withstand lower oxygen tensions by surface breathing and reliance on anaerobic metabolism (McMahon 1983).

Approximately 45% of all freshwater gastropods are restricted to waters with calcium concentrations greater than 25 mg/liter, and 95% to levels greater than 3 mg/liter. Although it may not take much energy to absorb calcium from water, assuming adequate water hardness, it may be energetically costly to secrete calcium into the shell against an electrochemical gradient. McMahon (1983) and Lodge *et al.* (1987) discuss the degree of calcium regulation

and the relationship of external calcium level to shell thickness and growth, a subject complex enough to be beyond the scope of this chapter. Interestingly, shell accretion may lag behind tissue growth in eutrophic habitats, producing thinner-shelled individuals (McMahon 1983).

III. ECOLOGY AND EVOLUTION

A. Diversity and Distribution

Pleurocerids are widespread in Africa and Asia, but have reached their greatest abundance in the rivers and streams of the southeastern United States (Table 10.1). Females possess an egg-laying sinus on the right side of the foot (Dazo 1965). Shell anatomy is used to classify genera, although many species show considerable variation in shell characters. The shells of pleurocerids are solid and the aperture may bear a canal anteriorly. The operculum is corneous and paucispiral.

Viviparids are worldwide in distribution and are fairly diverse throughout the eastern states and Canadian provinces. *Campeloma*, *Lioplax*, and *Tulotoma* are endemic to North America. Although the first two genera are widespread, *Tulotoma* is found only in Alabama rivers. *Viviparus* is quite common in rivers and lakes throughout eastern North America. Ampullarids are a tropical, mostly amphibious family with a mantle cavity provided with both gill and lung. The two genera present in Florida, *Pomacea* and *Marisa*, are quite large snails (50–60 mm).

The Neritinidae are a marine, tropical group. A few species have invaded estuarine and freshwater habitats, for example *Neritina reclivata* in Florida, Georgia, Alabama, and Mississippi. The paucispiral, calcareous operculum has a pair of projections, which lock the operculum against the teeth on the aperture, providing a stronger defense.

Of the 20 species of valvatids in the northern hemisphere, 11 are found in North America. Valvatids are egg-laying hermaphrodites, with a single, featherlike gill carried on the left side, and a pallial tentacle carried on the right side of the shell as the animal crawls. Valvatids have small (ca. 5 mm diameter) dextral shells, with a corneous, slightly concave and thin, multispiral operculum. The shells are sometimes carinated.

Hydrobiids are extremely diverse and widespread in freshwater, with brackish water and marine representatives as well. There are 103 genera worldwide (Burch 1982) and they are diverse in North America as well. The small, dextral shells generally have a paucispiral operculum. Because of

Table 10.1 The Diversity of Freshwater Snail Families in North America[a]

Subclass	Family	Number of Genera	Number of Species	Area of Greatest Diversity[b]
Prosobranchia	Ampullaridae	2	3	SE
	Bithyniidae	1	1	NE
	Hydrobiidae	28	152	U
	Micromelaniidae	1	1	SE
	Neritinidae	1	1	SE
	Pleuroceridae	7	153	U
	Pomatiopsidae	1	6	SE,W
	Thiaridae	2	2	SW,SE
	Valvatidae	1	11	U
	Viviparidae	5	19	SE,C
Pulmonata	Acroloxidae	1	1	NE
	Ancylidae	4	11	U
	Lymnaeidae	9	57	N
	Physidae	4	37	U
	Planorbidae	11	44	U

[a]Data compiled from Burch (1982).
[b]N, North; E, East; W, West; S, South; U, Ubiquitous; C, Central.

the similarity of their shells, the internal anatomy, principally the structure of the verge (penis), must be used in classification.

The six North American pomatiopsids are similar in general anatomy to the hydrobiids. Pomatiopsids are, however, amphibious, inhabiting stream banks, while hydrobiids are truly aquatic (Burch 1982). Their systematics are discussed by Davis (1979).

Thiarid females are parthenogenetic, brooding eggs in a pouch in the neck region, which opens on the right side. In contrast, the similarly shelled pleurocerids, discussed above, are dioecious and oviparous.

Ancylids have a worldwide distribution and all possess a simple cone-shaped shell. In North America, they have reached moderate diversity. Ancylids have sinistral shells, with the apex inclined slightly to the right, and the gill (pseudobranch) and many of the internal organs opening on the left side of the body.

The family Acroloxidae occurs mainly in Eurasian lakes and ponds. Only one species of *Acroloxus* occurs in the United States (in Colorado and southeastern Canada). Since the apex in *Acroloxus* is tipped to the left, the aperture is considered to be dextral, unlike the ancylid limpets.

Lymnaeids are worldwide in distribution, and are the most diverse pulmonate group in the northern United States and Canada. Lymnaeids have broad triangular tentacles and lay long, sausage-shaped egg masses. One group of Lymnaeids, found along the Pacific coast of North America, has limpet-shaped shells, but they are larger than ancylids.

Physids also have a worldwide distribution and are ubiquitous in North America. Their shells are small, sinistral, with raised spires. Their tentacles and foot are slender and they have fingerlike mantle extensions and lay soft, crescent-shaped egg masses.

Planorbids are widespread and fairly diverse snails and range in size from minute (1 mm) to quite large (30 mm) in North America. They possess hemoglobin as a respiratory pigment, sometimes giving the tissue a red hue. Planorbids are considered closely related to the ancylids.

B. Reproduction and Life History

Studies of the reproductive behavior of freshwater snails are common because they are fairly easy to sample, have relatively short life cycles, and are easy to rear in the laboratory. However, they are also extremely interesting because of the variety of observed life-history patterns. For example, freshwater pulmonates are oviparous hermaphrodites, and are usually annual and semelparous. On the other hand, almost all prosobranchs are dioecious, and are often iteroparous, with perennial life cycles. Prosobranchs can be oviparous or ovoviviparous (Russell-Hunter 1978, Calow 1978, 1983, Brown 1983). Annual species have essentially a one year life cycle, whereas perennials often live and reproduce for 4–5 years. A more thorough description of these life-history traits, as well as their general adaptive basis, can be found in Partridge and Harvey (1988).

Russell-Hunter (1978) and Calow (1978) have developed a classification of life-history types for freshwater gastropods (Fig. 10.6). At one end of the spectrum are annual adults that reproduce in the spring and die (i.e., there is complete replacement of generations). Most pulmonates belong to this group, including species from the genera *Lymnaea, Physa,* and *Aplexa* (the original studies are listed in Calow 1978). In the second category (Fig. 10.6B), reproduction occurs in both spring and late summer with both cohorts surviving the winter, or (Fig. 10.6C) where there again is complete replacement of generations. In Fig. 10.6D–F, there are three reproductive intervals, with varying degrees of replacement of generations. These would predominantly be populations in subtropical or tropical environments (see, e.g., McMahon 1975a). Finally, there are populations that can truly be considered perennial and iteroparous (Fig. 10.6G); most are prosobranchs (Calow 1978, 1983).

Clear differences in life-cycle patterns occur between marine and freshwater snails (Calow 1978). Marine snails have enormous fecundity, but individual eggs are extremely small. Most marine snails are prosobranchs, and freshwater prosobranchs (their descendants) probably have relatively small eggs as a result. In fact, viviparids and thiarids may have evolved ovoviviparity (and in some cases the production of larger embryos) to cope with the much more unpredictable conditions in freshwaters (Calow 1978, 1983). Pulmonates also produce fewer but larger eggs than marine prosobranchs. The loss of the planktonic veliger and the shortening of the developmental period in freshwater snails have been attributed to the more variable physicochemical conditions (Calow 1978).

Pulmonate families, on the average (1) reproduce at smaller sizes and earlier ages; (2) produce more eggs; (3) have larger clutch sizes; (4) have greater shell growth rates; and (5) have shorter life cycles and smaller final shell sizes than viviparid prosobranchs. Pulmonate families are also semelparous, with relatively high reproductive output (Brown 1983). Viviparid prosobranchs, on the other hand, have a long reproductive interval and are iteroparous with relatively small clutch sizes in most cases.

Reproductive effort (percentage of energy devoted to reproduction) is therefore lower in iteroparous freshwater snails than in semelparous ones (Browne and Russell-Hunter 1978), which has been predicted on theoretical grounds (Partridge and Harvey 1988). Presumably, the reduced fecundity and the increased parental care found in viviparids (versus semelparous pulmonates) have evolved to increase offspring survival. Life tables for *Viviparus georgianus* (Jokinen *et al.* 1982) and for the pulmonate *Lymnaea elodes* (Brown *et al.* 1988) indicate that survival to maturity is indeed much less than 1% in all of the pulmonate populations, but over 40% in the ovoviviparous prosobranch.

Prosobranchs are often sexually dimorphic in life-history patterns. Females live longer (males usually survive for only one reproductive season) and reach larger sizes (Van Cleave and Lederer 1932, Browne 1978, Jokinen *et al.* 1982, Pace and Szuch 1985, Brown *et al.* 1989). Browne (1978) suggested that males expend more energy than females to locate mates in their first reproductive season, resulting in less time spent feeding and thus lower survivorship. In support of this hypothesis, Ribi and Ater (1986) found that marked males of *Viviparus ater* moved greater distances than females.

The relative importance of environmental and genetic factors in explaining life-history variation in freshwater snails is still relatively unknown. A large number of studies have implicated environmental factors like periphyton productivity (Eisenberg 1966, 1970, Burky 1971, Hunter 1975, Browne 1978, Aldridge 1982, Brown 1985) in determining voltinism patterns, growth rates, fecundity, and gastropod secondary production (see review in Russell-Hunter 1983). Other important factors include physicochemical variables such as water hardness and water temperature (see discussion in McMahon 1983). For example, populations of *Lymnaea stagnalis* in Canada often take several seasons to

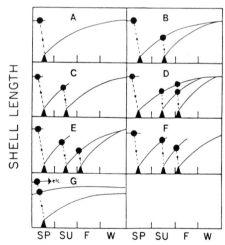

Figure 10.6 The variety of life cycles in freshwater snails. Curves represent the growth of individual cohorts (in mm shell length) versus season of the year (SP, spring; SU, summer; F, fall; W, winter). Circles represent size at maturity while triangles represent appearance of egg cases in samples. Panel A represents the simplest life cycle, annual semel parity; Panel B–F, increasing numbers of cohorts per year, and Panel G, perennial iteroparity. See discussion in text. (From Calow 1978.)

complete their life cycle (Boag and Pearlstone 1979), while populations are annual in the warmer waters of Iowa (Brown 1979).

Populations of *Lymnaea peregra* in England in wave-swept habitats have life-history traits characteristic of *r*-selected populations (early reproduction and high reproductive output) in comparison to populations in less harsh habitats (Calow 1981, Lam and Calow 1989a). However, other studies of life history variation in molluscs do not agree as well with the predictions of *r* and *K* theory (see discussion in Burky 1983). Transplant studies, where individuals from separate populations are reared in a common environment, have usually indicated that environmental effects on life histories are much more important than genetic differences between populations. For example, populations of *Lymnaea elodes* in more productive ponds lay nine times as many eggs, have an annual versus a biennial reproductive cycle, and reach larger individual sizes (Brown 1985, but also see Lam and Calow 1989b).

Genetic polymorphism has been studied extensively (using gel electrophoresis) in terrestrial pulmonates and freshwater prosobranchs, but rarely in aquatic pulmonates (see references in Brown and Richardson 1988). Freshwater snails have levels of genetic polymorphism that are intermediate between terrestrial and marine species (Brown and Richardson 1988). A possible reason is that terrestrial snails inhabit microclimates, which, because they are patchily distributed, increase chances for low population densities and self-fertilization, resulting in low levels of genetic polymorphism within populations. Freshwater snails may still experience low densities because of periodic seasonal bottlenecks and thus may self-fertilize. Marine environments are less seasonal, and many marine snails have planktonic larvae, facilitating gene flow even further and increasing levels of polymorphism within populations. However, more electrophoretic data are needed for freshwater pulmonates before stronger comparisons can be made. These types of comparisons may also be confounded by systematic differences, because most terrestrial snails are pulmonates, while freshwater snails contain both pulmonates and prosobranchs, and marine snails are almost all prosobranchs (Brown and Richardson 1988).

C. Ecological Interactions

1. Habitat and Food Selection: Effects on Producers

Substratum selection is well documented in snails. *Campeloma decisum* is positively rheotactic (moves upstream) and forms aggregations at any barrier to movement (e.g., logs, riffle zones, Bovbjerg 1952). Slow-moving, silty habitats are occupied predominantly by pulmonates or detritivorous prosobranchs, whereas fast-current areas are dominated by limpets or prosobranch grazers (Harman 1972). *Physa (Physella) integra* prefers cobble substrata with attached periphyton, while *Helisoma anceps* prefers sand (Clampitt 1973). Most gastropods in northern Wisconsin lakes, with the exception of *Campeloma decisum*, prefer periphyton-covered cobble over sand or macrophytes (Weber and Lodge 1990). In fact, substratum selection may even occur on a finer level, as gastropods from an English pond preferred periphyton isolated from the macrophytes on which the snails were most common in the field (Lodge 1986).

Seasonal migrations are common in lakes, with snails moving to deeper water in the fall and back to shallow littoral areas in the spring (Cheatum 1934). For example, *Physa (Physella) integra* move up-current and up-slope in the spring, halting when they reach periphyton-rich cobble in the shallow littoral zone (Clampitt 1974). *Lymnaea stagnalis, Physa (Physella) gyrina,* and *Helisoma (Planorbella) trivolvis* move into deeper water with declining water temperature, but *Lymnaea (Stagnicola) elodes* move into shallower water (Boag 1981). Extensive vertical migrations cannot occur in shallow ponds, but pulmonates may burrow into the substratum with declining temperatures (Boerger 1975).

Freshwater gastropods are either herbivores or detritivores, although they occasionally ingest carrion (Bovbjerg 1968) or passively consume small invertebrates associated with periphyton (Cuker 1983a). They evidently prefer periphyton because it is easier to scrape than macrophyte tissue and contains higher concentrations of nitrogen and other limiting nutrients (Russell-Hunter 1978, Aldridge 1983). Carbon to nitrogen (e.g., carbohydrate to protein) ratios in periphyton range from $3.7–10.1:1$, while macrophytes have ratios of $24.1:1$ (McMahon *et al.* 1974). Algal and diatom remains also predominate in the guts of snails (Calow 1970, 1973a, 1975, Calow and Calow 1975, Reavell 1980, Kesler *et al.* 1986, Lodge 1986).

Some gastropods consume macrophytes, however, and may indeed suppress macrophyte species richness if the snails reach high enough densities (Sheldon 1987). Macrophytes preferred in laboratory grazing experiments were usually the ones that were lost under increased grazing pressure in these field experiments (Sheldon 1987). Fish predators were excluded from small cages, producing roughly a tenfold increase in the snail density. However, at the increased densities, the snails may still have consumed available periphyton first and only then consumed macrophyte tissue.

Feeding preferences for the freshwater gastropod families are summarized in Table 10.2. Analyses of gut contents of field-collected individuals indicate that lymnaeids are "micro-herbivores," scraping algae and diatoms from rocks or macrophytes. Lymnaeids do, however, grow more rapidly when both plant and animal tissues are included in their diet (Bovbjerg 1968). *Pseudosuccinea columella* (a lymnaeid) is an omnivore but takes more algae than the sympatric *Physa (Physella) vernalis* (Kesler *et al.* 1986). The lymnaeid also possesses higher levels of cellulases, a radula and jaws well adapted for cropping algae (Fig. 10.5), and a gizzard filled with sand that can macerate food. The physid, a detritivore, lacks these adaptations.

Both the limpet family Ancylidae and the prosobranch family Pleuroceridae are also considered to be grazers on periphyton and perilithon. *Ancylus fluviatilis* selectively grazes diatoms, but the limpet apparently has little effect on periphyton communities due either to adaptations of algal and diatom species to grazing or to relatively low limpet abundances (Calow 1973a, 1973b). Aldridge (1983) also suggested that pleurocerid grazers feed on periphyton rather than macrophyte tissue because of higher levels of nitrogen.

The prosobranch *Bithynia tentaculata* can both graze on periphyton and use its ctenidium to filter phytoplankton. Indeed, filter feeding may be more efficient than scraping, explaining why this species has become so abundant in nutrient-rich, eutrophic lakes in New York (Tashiro 1982, Tashiro and Colman 1982).

Viviparus georgianus is a micro-algivore in some cases (Duch 1976, Jokinen *et al.* 1982), but a detritivore on fine particulate organic matter in others (Pace and Szuch 1985). Most viviparids, however, are probably detritivores or may utilize bacteria associated with detritus. As macrophytes decompose,

nitrogen levels increase, again increasing their value as food resources. For example, *Viviparus georgianus* reaches extremely high densities (151–608 per m^2) in a detritus-laden stream in Michigan (Pace and Szuch 1985). In detritus-rich bayous in Louisiana, *Viviparus subpurpureus* reaches densities as high as 1700 per m^2, and densities of *Campeloma decisum* are as high as 900 per m^2 (Brown *et al.* 1989).

Available data also suggest that physids and planorbids are detritivores and/or bacterial feeders. As discussed earlier, both physids (Kesler *et al.* 1986) and planorbids (Calow 1973b, 1974a, 1974b) are thought to prefer detritus. Although in laboratory experiments *Physa (Physella) gyrina* and *Helisoma (Planorbella) trivolvis* did not select detritus over periphyton, another physid, *Aplexa hypnorum*, did. *Aplexa* was also much more common in wooded ponds with a rich detritus food base (Brown 1982).

Originally, snail algivores were considered indiscriminate grazers, taking all components of the periphyton (see discussions in Hunter 1980, Hunter and Russell-Hunter 1983). However, limpets and planorbids are selective (Calow 1973 a, 1973b) and *Lymnaea peregra* grazes selectively on filamentous green algae (Lodge 1985), supporting Bovbjerg's (1965) field observations that *Lymnaea (Stagnicola) reflexa* aggregates on patches of the filamentous green alga *Spirogyra*. *Planorbis vortex* ingests diatoms in greater quantities than are found in the periphyton, but is still predominantly a detritivore. In laboratory choice experiments, snails also prefer periphyton washings of plants on which they are most abundant in the field (Lodge 1986).

Bovbjerg (1968) found a lymnaeid could detect meat juices at relatively long distances, but concluded the snails could not orient to plant extracts and instead simply ceased moving when algae were encountered. However, more recent work (reviewed in Croll 1983) argues that gastropods do have the ability to locate macrophytes through distant chemoreception. For example, *Lymnaea peregra* is

Table 10.2 Feeding Preferences of Freshwater Snails

Feeding Type	Family	Reference
Algivores	Ancylidae	Calow (1973a, 1973b, 1975)
	Lymnaeidae	Bovbjerg (1968, 1975), Brown (1982), Calow (1970), Cuker (1983a), Hunter (1980), Kairesalo and Koskimies (1987), Kesler *et al.* (1986), Lodge (1986)
	Neritinidae	Jacoby (1985)
	Pleuroceridae	Aldridge (1982, 1983), Dazo (1965), Goodrich (1945)
	Viviparidae	Duch (1976), Jokinen *et al.* (1982)
Detritivores or bacterial feeders	Physidae	Brown (1982), Kesler *et al.* (1986), Townsend (1975)
	Planorbidae	Calow (1973b, 1974a, 1974b)
	Viviparidae	Chamberlain (1958), Pace and Szuch (1985), Reavell (1980)

positively attracted to *Ceratophylum demersum*, indicating that dissolved organic materials excreted by macrophytes attract grazers (Bronmark 1985b). Similarly, *Potamopyrgus jenkinsi* detects both plant and animal extracts and orients toward the source (Haynes and Taylor 1984), while *Biomphalaria glabrata* may either orient toward or away from a plant source, the strength of the movement dependent on the specific macrophyte (Bousefield 1979).

Gastropods can have important effects on algal producers, such as determining periphyton composition. Snail grazers selectively remove larger filamentous green algae and leave smaller, adnate species behind (Patrick 1970, Sumner and McIntire 1982, Cuker 1983b, Lowe and Hunter 1988, McCormick and Stevenson 1989, Barnese *et al.* 1990). Under slight gastropod grazing pressure, periphyton assemblages are dominated by filamentous green algae, but more intensely grazed assemblages are dominated by more tightly adhering species, or toxic species such as blue-green algae (Cuker 1983b, Cattaneo 1983, Steinman *et al.* 1987; see also discussion in Bronmark 1989).

Snail grazers can also control periphyton abundance, decreasing standing crops but in some cases increasing chlorophyll *a* concentration (Kehde and Wilhm 1972, Doremus and Harman 1977, Barnese *et al.* 1990). Pulmonate gastropods can significantly alter the quality (e.g., nitrogen to carbon ratios and chlorophyll *a* levels) of periphyton (Hunter 1980), as can prosobranchs in streams (Kesler 1981, Sumner and McIntire 1982, Jacoby 1985, Hawkins and Furnish 1987). Studies with radioactive tracers have provided independent evidence that gastropods are significant grazers (Kairesalo and Koskimies 1987).

Snail grazers may in fact indirectly facilitate macrophytes. If periphyton coverings shade and limit macrophyte growth and snails prefer periphyton over macrophytes, then grazing might actually increase the growth rates of macrophytes. For example, the growth rate of *Ceratophylum demersum* is increased when gastropod grazers are present (Bronmark 1985b). Pumpkinseed sunfish depress snail abundances, resulting in increased periphyton abundance in fish enclosures versus exclosures (Bronmark 1989). Thus, interactions between gastropod predators, snails, and periphyton and macrophyte abundances may be very complex in natural systems.

2. Factors Regulating Population Size

The first experimental demonstration of population regulation using field manipulations was with *Lymnaea (Stagnicola) elodes* (Eisenberg 1966, 1970).

When the densities of adult snails in pens in a small pond were increased, adult fecundity declined, as did juvenile survival. With addition of a high quality resource, spinach, an increase in the number of eggs per mass occurred. Evidently, the availability of micronutrients in periphyton was the crucial variable (Eisenberg 1970).

Brown (1985) provided additional evidence on the role of habitat productivity by transferring juvenile *Lymnaea (Stagnicola) elodes* to a series of ponds differing in periphyton productivity. There was an exponential increase in growth with increasing pond productivity, and snails in the most productive pond laid nine times as many eggs as snails in the two less productive ponds. Thus, both food level and population density affect growth and fecundity in *Lymnaea (Stagnicola) elodes*.

A number of field studies of life cycles have also provided indirect evidence on the importance of resource abundance to snail population biology. Limpets in more eutrophic habitats have more generations per year, more rapid shell growth, and lay more eggs (Burky 1971, McMahon 1975b). Similarly, *Lymnaea (Stagnicola) palustris (elodes)* in more eutrophic habitats have more generations per year, more rapid shell growth, and greater individual fecundities (Hunter 1975). *Helisoma (Planorbella) trivolvis* at eutrophic sites also have shorter life cycles and greater growth rates (Eversole 1978). Highly eutrophic sites may, however, be detrimental to gastropods, as gastropod diversity declined over a fifty-year period as Lake Oneida in New York became highly productive (Harman and Forney 1970). Finally, viviparid detritivore abundances increase in southern Louisiana habitats with greater detritus levels (K. M. Brown, unpublished).

Other potential regulators of snail populations, unfortunately overlooked by most malacologists, are the parasitic larvae of digenetic trematodes (Holmes 1983). Infections have dramatic effects on individual snails; infection either accelerates or decelerates growth, depending on the host–parasite system (Brown 1978, Anderson and May 1979, Holmes 1983; see also discussion in Minchella *et al.* 1985). Immediately after infection, or even after only exposure, snail egg production rates may increase dramatically (Minchella and Loverde 1981). Eventually, however, both growth and egg production of "patent" snails (those infections in the final stage where cercaria are emerging from snails) drop below that of uninfected snails (Minchella *et al.* 1985).

Do these striking effects on individuals translate into regulation of numbers at the population level? First of all, to control gastropod populations, trematode prevalence (percentage of infection levels) should be fairly high. However, prevalence varies from as low as 1% in some populations to over 50% in other host–parasite systems (Brown 1978, Holmes 1983). For example, in *Lymnaea (Stagnicola) elodes*, prevalence in Indiana ponds varied from as low as 4% to as high as 49% (Brown *et al.* 1988). Prevalence was higher in less productive ponds, evidently because the chances that a snail would be found by miracidia (the fluke stage that infects snails) rose with the longer snail life cycles observed in these ponds due to lower food levels. Life table models predicted that the number of offspring produced per adult in the next generation declined by 14–21% in parasitized populations of *L. elodes*. Trematode parasites can thus potentially have important effects on snail populations, and future studies of gastropods need explicitly to take prevalence levels into account.

Trematode parasites and their snail hosts are also extremely interesting from a coevolutionary viewpoint (Holmes 1983, Minchella *et al.* 1985). Because invertebrates cannot easily acquire resistance to parasites, frequency-dependent selection may operate to ensure the fitness of any genotype less vulnerable to a particular trematode (Holmes 1983). In addition, parasites may also cause a shift in investment of resources from costly reproduction to growth and maintenance and even result in increased survivorship (see also Baudoin 1975, Minchella *et al.* 1985).

3. Production Ecology

The production ecology and bioenergetics of freshwater gastropods have been extensively studied (see review in Russell-Hunter and Buckley 1983). Average standing crop biomass, productivity, and turnover times (these terms are defined in the glossary) are compared for prosobranchs and pulmonates in Table 10.3 (summarized from Russell-Hunter and Buckley 1983). Standing crops are greater on the average for prosobranchs than for pulmonates, probably because the prosobranch populations studied have, like *Viviparus* and *Gonio-*

basis (Elimia), had relatively large individuals, rather than genera with smaller individuals such as *Amnicola* and *Valvata*.

Although pulmonates possess lower standing stocks, their rapid growth and short life cycles still result in higher average production rates and shorter turnover times. For example, two pleurocerid snails in an Alabama stream, *Elimia cahawbensis* and *E. clara*, have considerable biomasses of 2–5 g ash-free dry mass (AFDM) per m^2, but slow growth rates and long life cycles result in relatively low rates of secondary production (0.5–1.5 g AFDM/m^2). The combination of high biomass and low production results in a low production to biomass ratio of 0.3 (Richardson *et al.* 1988).

Some prosobranch detritivores may, however, have exceptionally high production rates. *Viviparus subpurpureus* and *Campeloma decisum*, due to high densities and short life cycles in subtropical Louisiana bayous, have high standing crop biomasses (10–20 g AFDM/m^2) and production rates (20–40 g AFDM/m^2/yr). These production estimates are among the highest known for freshwater molluscs (Richardson and Brown, 1989).

4. Ecological Determinants of Distribution

Water hardness and pH are often considered the major factors determining the distributions of freshwater snails (see, e.g., Boycott 1936, Macan 1950, Russell-Hunter 1978, Okland 1983, Pip 1986). However, these studies often dealt with lake districts with fairly soft water, and in lake districts with adequate calcium (above about 5 mg/liter $CaCO_3$), little relationship exists between physicochemical parameters and gastropod diversity (Lodge *et al.* 1987). For example, individual species in New York lakes overlap broadly in their tolerance adaptations to every physicochemical variable measured (Harman and Berg 1971).

Lodge *et al.* (1987) proposed a hierarchial framework for the biotic and abiotic factors determining the geographic and local distributions of snails. They suggested that calcium and other physicochemical

Table 10.3 Comparison of Average Standing Stocks, Productivity, and Turnover Times for Populations of Pulmonate and Prosobranch Snails[a]

Subclass	Mean Biomass ± SE (g C m^{-2}, N)[b]	Mean Production ± SE (mg C m^{-2} day^{-1})	Mean Turnover Time (days) ± SE (N)
Pulmonata	0.98 ± 0.50 (6)	5.71 ± 2.44 (10)	98.0 ± 9.46 (10)
Prosobranchia	4.64 ± 1.80 (4)	3.56 ± 0.51 (5)	385.3 ± 33.5 (4)

[a]Reported in Russell-Hunter and Buckley (1983).
[b]N, Number of species averaged.

variables may act only to limit species from successfully invading habitats with extreme levels of these factors (for example, lakes with extremely soft water or low pH, or lakes in arid regions that have extremely high salinities).

Another factor determining whether a gastropod will be present is its dispersal ability. Studies of lakes and ponds as "islands" within a terrestrial "sea" have revealed much about the dispersal powers of aquatic gastropods (Lassen 1975, Browne 1981, Bronmark 1985a). Island biogeographic theory (MacArthur and Wilson 1967) assumes that diversity on an island is a balance between immigration and extinction of species already present. Since immigration rates generally increase and extinction rates decrease with island size, larger "islands," all else being equal, usually support more species. This rule appears to apply to gastropod diversity in American and Danish ponds and lakes (Lassen 1975, Browne 1981). A significant relationship between snail diversity and habitat area also occurs in Scandinavian gastropods (Bronmark 1985a), although macrophyte diversity and the diversity of gastropods in surrounding habitats are also important.

Given the fact that at least pulmonates evidently have strong passive dispersal powers, the next potential factor determining successful colonization is availability of adequate substrata (Lodge *et al.* 1987). For example, a significant relationship exists between the number of coexisting gastropod species and the number of substrata available in lakes and streams in New York state (Harman 1972). Differences in substratum preferences for several species of snails common in ponds in the midwestern United States also occur, and these preferences are good predictors of the types of habitats in which the snail species are common (Brown 1982).

Given successful colonization and adequate substrata, disturbance may be the next factor important in determining the assemblage of snails present (Lodge *et al.* 1987). In temporary ponds, diversity is lowered by frequent drying, and in habitats that become anoxic, diebacks of macrophyte and gastropod populations will also occur (Lodge and Kelly 1985, Lodge *et al.* 1987). In larger lakes, disturbance may limit some species from disturbance prone areas (wave-swept shores, littoral zones of reservoirs, etc.).

In large lakes, biotic interactions such as interspecific competition or predation will be more important in determining gastropod diversity and abundance (Lodge *et al.* 1987). The role of interspecific competition in natural communities has probably been overstated in many cases (see discussion in Strong *et al.* 1984). In freshwater snails, no field-based, experimental evidence as yet suggests an im-

portant role for interspecific competition. In pulmonate pond snails in the midwestern United States, substantial overlap on food and habitat dimensions of the niche occurred in only one out of six possible pairs of coexisting species (Brown 1982). Both of these species were common only in temporary ponds, where drying probably precluded long term effects of competition.

Other studies have also indicated little overlap in resource utilization. For example, coexisting populations of *Ancylus fluviatilis* and *Planorbis contortus* differ both in feeding mechanisms and habitat use. The ancylid prefers diatoms and is found on the top of cobbles where periphyton is abundant, whereas the planorbid is a detritivore and occurs more often under stones, where detritus accumulates (Calow 1973a,b, 1974a,b). The limpet *Ferrissia fragilis* prefers understory diatoms and avoids overstory species (perhaps due to its closefitting shell and the small teeth on its radula) perhaps allowing it to coexist with other gastropods that graze preferentially on overstory species (Blinn *et al.* 1989). Similarly, *Pseudosuccinea columella* and *Physa (Physella) vernalis* possess differences in their gut and radulae, allowing coexistence in New England ponds (Kesler *et al.* 1986). Niche partitioning may also be facilitated by differences in cellulase activity (Calow and Calow 1975). Kesler (1983), however, could find no differences in cellulase activity among species populations reported by Brown (1982) to differ in feeding preferences.

Some indirect evidence exists, however, for competition between freshwater snails. For instance, there are often fewer coexisting congeners in field samples than would be expected, based on model simulations (Dillon 1981, 1987). Second, competition has been inferred from changes in relative abundance of gastropod species through time, that is, apparent competitive exclusion (Harman 1968). Third, pulmonates are often common in ponds or in vegetated areas of lakes, while prosobranchs are rare in ponds and common in lakes. One explanation could be competitive exclusion of pulmonates from lakes by prosobranchs, or exclusion of prosobranchs from ponds by the same mechanism (Lodge *et al.* 1987). In fact, the prosobranch *Elimia livescens* was less effective than five pulmonate species in grazing algae in an experimental study (Barnese *et al.* 1990). However, prosobranchs are, again, probably more vulnerable to hypoxic conditions common in ponds and the relatively thin-shelled pulmonates are more vulnerable to shell-crushing fish common in lakes.

Given the rarity of coexisting congeners (Dillon 1981, 1987, Brown 1982), the differences in food utilization among coexisting species (Calow

1973a,b, Brown 1982, Kessler *et al.* 1986), and the absence of clear experimental evidence from manipulative field experiments, the role of interspecific competition in structuring freshwater snail assemblages is still unclear (Lodge *et al.* 1987).

What about predation? Lodge *et al.* (1987) argued that predators may determine the composition of gastropod assemblages in lakes. First, there is evidence that gastropods have a number of antipredator adaptations. These include thick shells to protect against shell-crushing predators (Vermeij and Covich 1978, Stein *et al.* 1984, Brown and DeVries 1985), as well as escape behaviors such as shaking the shell or crawling above the water to protect against shell-invading invertebrate predators (Townsend and McCarthy 1980, Bronmark and Malmqvist 1986, Brown and Strouse 1988). Second, experimental studies are beginning to clearly show the impact of predators on snail diversity and abundance, particularly in permanent ponds and lakes. Predators of snails can be separated into those that crush the shell and those that invade it (Table 10.4). Centrarchids, such as the pumpkinseed sunfish *Lepomis gibbosus,* and the red-ear or shell-cracker sunfish *Lepomis microlophus,* specialize on gastropod prey and have pharyngeal teeth adapted to crush shells. The thick-shelled prosobranch *Oxytrema (Goniobasis* or *Elimia)* is more resistant to crushing than *Helisoma,* while thin-shelled *Physa* are least resistant, and red-ear sunfish prefer the weaker-shelled species (Stein *et al.* 1984). Crayfish will also select snails instead of grazing on macrophytes if given a choice (Covich 1977), using their mandibles to chip shells back from the aperture.

In addition, large-scale experimental manipulations point to the importance of shell-crushing predators. For example, the central mud minnow can significantly lower the density of relatively thin-shelled snails in permanent ponds (Brown and De-Vries 1985). When fish densities were manipulated in pens, the numbers of eggs and juveniles of *Lymnaea (Stagnicola) elodes* were significantly less in the presence of fish. The small, gape-limited fish feed on eggs and juveniles and may restrict *Lymnaea (Stagnicola) elodes* from permanent habitats such as marshes or lakes.

Snail assemblages shift toward species with small size but thick shells in the presence of larger, molluscivorous fish, whereas resource limitation favors species with slower growth rates (Osenberg 1989). Pumpkinseed sunfish strongly prefer large, weak-shelled gastropod species in laboratory experiments, and thin-shelled species also declined dramatically in pumpkinseed enclosures in lakes in Wisconsin (S. Klosiewski and R. Stein, unpublished). Another shell-crushing predator, the crayfish *Orconectes rusticus,* significantly reduced snail abundance in enclosure experiments in Wisconsin lakes, and an interlake survey indicated that snail abundance is negatively correlated with crayfish catches (D. Lodge, unpublished). Crayfish also shift size distributions of *Physa virgata* upward in Oklahoma streams, due, possibly, both to size-selective predation and to the snail's diversion of energy from reproduction to growth (Crowl and Covich 1990).

Invertebrate, shell-invading predators may also limit snail abundance, or cause shifts in relative abundance. Although some leeches, such as *Nephelopsis obscura,* have fairly low feeding rates of one snail per day (Brown and Strouse 1988), crayfish can consume as many as 100 or more snails per night (A. Covich, unpublished). Belostomatid bugs have been shown to eat up to 5 snails per day in the laboratory (Crowl and Alexander 1989) and 0.5 snails per day under field conditions (Kesler and Munns 1989).

Table 10.4 Two Major Categories of Molluscivores and Examples of Studies Dealing with Each Type from the Literature

Predator Type	Predator Species	Reference
Shell-crushing	Pumpkinseed sunfish	Etnier (1971), Keast (1978), Laughlin and Werner (1980), Mittelbach (1984), Seaburg and Moyle (1964)
	Red-ear sunfish	Carrothers and Allison (1968), Chable (1947), Cross and Collins (1975), Huish (1957), Pflieger (1975), Stein *et al.* (1984), Trautman (1957)
	Central mud minnow	Brown and Devries (1985)
	Crayfish (spp.)	Covich (1977, 1981), Vermeij and Covich (1978), Crowl and Covich (1990)
Shell-invading	Aquatic Coleopterans, Diptera, Hemiptera	Crowl and Alexander (1989), Eckblad (1973), Eisenberg (1966), Harman and Berg (1971), Kesler and Munns (1989)
	Leeches, Flatworms	Bronmark and Malmqvist (1986), Brown and Strouse (1988), Davies *et al.* (1978, 1979, 1981), Townsend and McCarthy (1980), Young and Ironmonger (1980), Young (1981)

To summarize the conceptual model of Lodge *et al.* (1987), the importance of calcium in determining snail distributions has probably been overstated. If calcium is above 5 mg/liter, and the habitat is above 0.1 ha in size, most species can successfully colonize the site. If disturbances like habitat drying and hypoxia are rare and suitable substrata are available, then predation will probably be the major force structuring gastropod assemblages. Thus, most pulmonates should occur in lakes only in macrophyte beds, where they have a refuge from visual fish predators, and sandy areas should be dominated either by thicker-shelled pulmonates like *Helisoma* or by prosobranchs. Indeed, gastropod distributions among habitats within Indiana and Wisconsin lakes do follow these patterns (Lodge *et al.* 1987).

5. Suggestions for Further Work

Additional studies of the production ecology of smaller prosobranchs and investigations of populations in areas outside the northeastern United States are required. Little is known of how pulmonates are physiologically adapted to relatively ephemeral or anoxic aquatic habitats, and studies of differences in metabolic pathways, nitrogen excretion, and how these differences are adaptive are still needed (McMahon 1983). Another area of interest is detailed studies of the physiological and ecological constraints to shell formation versus tissue growth, given the remarkable intraspecific variation in shell structure and composition in freshwater gastropods. Studies of possible physiological and anatomical bases of feeding preferences (such as relative cellulase activities) should be expanded. We know little of micropreferences (e.g., preferences for specific algae or diatom species), or whether detritivores are consuming leaf material or more nutritious bacteria or fungi that have colonized the leaves.

We need to study food limitation in more species, especially detritivores. In most cases, the relative roles of food, physicochemical variables, genetic divergence, or drift in explaining snail life-history variation are poorly known, particularly in prosobranchs. To determine the genetic basis for phenotypic variation in life histories, these studies should use both quantitative genetics and techniques of electrophoretic analysis of enzyme polymorphisms (Brown and Richardson 1988).

Further research on abiotic and biotic factors influencing snail assemblages is necessary, testing the ideas presented in Lodge *et al.* (1987). Physical disturbances such as pond drying, anoxia, or macrophyte die-offs could explain as much about gastropod population dynamics or species composition as

the often mentioned role of calcium (see discussions in Lodge *et al.* 1987, Lodge and Kelly 1985).

Further work is necessary to determine whether the nonoverlapping distributions of pulmonates and prosobranchs are due to competition. Competition experiments with an annual pulmonate algivore, say a lymnaeid, and an annual prosobranch algivore, a valvatid for example, might give us some interesting comparisons. Studies of food and habitat niche differences in freshwater gastropods may also present more indirect evidence for the role of competition.

Further work is also needed on the relative role of invertebrate predators like leeches, belostomatid bugs, and scyomyzid fly larvae (Eckblad 1973) in structuring snail assemblages. A comparative study encompassing temporary and permanent habitats and involving several different predators would be extremely interesting. The role of vegetated areas in lakes as refuges for snails should also be studied, as should the question of whether predators are responsible for the frequently observed patchy distributions of snails.

Finally, we need more information on whether gastropods prefer periphyton over macrophytes, possibly facilitating the growth and spread of macrophytes. Laboratory experiments, where snails are offered choices between macrophytes with and without periphyton coverings and detailed field experiments, documenting whether snails first remove periphyton before negatively affecting macrophyte abundance or composition, are both necessary. Experiments evaluating the roles of nutritional quality or toxicity of different periphyton or macrophyte species in determining preferences by gastropods would also be extremely interesting.

D. Evolutionary Relationships

Prosobranchs are the most primitive gastropods, having given rise both to marine opistobranchs and terrestrial and freshwater pulmonates. Freshwater prosobranchs evolved the ability to withstand the dilute osmotic conditions of estuaries and then rivers, and most modern families are widespread because their adaptive radiations predated continental breakups and drift.

Davis (1979, 1982) and Clarke (1981) pointed out that prosobranch dispersal is limited to slow movement along streams and rivers. Such populations are more likely to become isolated, promoting chances of speciation and adaptive radiation. Examples include the dramatic adaptive radiation of hydrobiids in lentic habitats (Davis 1982) and pleurocerids in lotic habitats in the southeastern United States (Burch 1982). Bithyniids, viviparids, and ampullarids have not radiated as much as hydrobiids or pleurocerids, perhaps because their filter feeding limits the range of habitats that they can invade (Davis 1982).

In contrast, pulmonates, due to their passive dispersal on birds and insects, often have broader distributions. For example, they are particularly widespread in the northern and northeastern United States as well as in Canada (Harman and Berg 1971, Clarke 1981). Immigration rates to ponds can be as high as nine species per year (see discussion in Davis 1982). A study of average immigration and extinction rates of pulmonates in ponds indicated both were similar, at 0.8 species per year (Lassen 1975). Pulmonates also predominate in shallow, more ephemeral habitats, usually in areas of less than 100 km^2 and with durations of less than 10^3 years. Therefore, populations probably do not exist in any habitat long enough for speciation to occur, explaining why pulmonates are relatively less speciose than prosobranchs (Russell-Hunter 1983, Clarke 1981, Davis 1982).

Pulmonates apparently evolved from intertidal prosobranchs that came to rely less and less on aquatic respiration (Morton 1955; see also discussion in McMahon 1983). Modern estuarine pulmonates such as *Melampus* may resemble these ancestral species. The intermediate, terrestrial pulmonates lost the ctenidium and gave rise both to modern terrestrial pulmonates (order Stylommatophora) and the aquatic pulmonates (order Basommatophora). Within the aquatic pulmonates, there has been a progressive readaptation to aquatic habitats (see discussion in McMahon 1983). Lymnaeids are in many cases still amphibious, while physids are intermediate in their adaptation to aquatic habitats. Ancylids and small planorbids have readapted most successfully, with the evolution of secondary gills, and in the planorbids, hemoglobin.

Finally, several caveats about the evolution of freshwater snails should be kept in mind (Davis 1982). First, phylogenetic analyses, and therefore biogeographic studies, are often limited by the poor systematic information available in many groups as well as the poor fossil record (only shells are preserved, which often show convergence). Second, a number of factors can explain current distributions, as illustrated for hydrobiids and pomatiopsids by Davis (1982): (1) phylogenetic events such as centers of origin and adaptive radiation; (2) past historic events such as continental drift and geological alteration of stream and river flow; (3) dispersal powers; and (4) ecological factors. Based on these factors (Davis 1982), one can rank gastropod groups from those with broad distributions but low diversity to the reverse: monoecious pulmonates > parthenogenetic prosobranchs > viviparous prosobranchs > oviparous prosobranchs.

IV. COLLECTING AND CULTURING FRESHWATER GASTROPODS

A number of gastropod sampling techniques exist (reviewed in Russell-Hunter and Buckley 1983), although many are not quantitative. The least quantitative technique, but one that often gives large numbers of individuals and a good idea of species composition is sweep-netting with a net of 1 mm mesh (Brown 1979).

In soft sediments or sand, quantitative samples can be collected with Ekman or Peterson grabs, although perhaps the least biased estimates come from core samples (Lodge *et al.* 1987). When sampling macrophytes, the sampler must collect both plants with attached snails and the substratum with any bottom-dwelling species. Examples of such samplers are described in Gerking (1957) and Savino and Stein (1982). In coarse sand or when silt decreases visibility, samplers like Ekman and Ponar grabs or some type of automated corer are the best alternatives. In cobble, little recourse is available, other than direct counts of given areas by visual search. By estimating the surface area of the cobble (for example by covering the rocks with aluminum foil and then estimating the area from weight to area regressions for the foil), one can estimate densities.

In most cases, the best sorting technique for gastropods is simply hand sorting. Large adults can be removed visually and samples are then washed through a graded series of sieves (the smallest having a mesh of about 0.5 mm) to remove mud. Samples should then be placed on flat white enamel trays, the vegetation teased apart, and the whole tray examined in a systematic fashion to remove small gastropods and egg cases.

Temperate gastropods will grow well at 15–20°C, while subtropical species grow better at 20–25°C. Any hard substrate with a dense periphyton covering can be added for food, although artificial foods such as lettuce, cereal, or spinach are sometimes used. Provide food as needed to avoid fouling containers. Detritivores should be fed leaf litter colonized with bacteria and fungi (i.e., held for at least two weeks in a pond or stream).

Avoid crowding of snails, as growth and reproduction are sensitive to density. An approximate rule of thumb is one snail per liter. Water should be recirculated through a gravel or charcoal filter or at least be changed weekly. Prosobranchs and pulmonates should be paired so that mating can occur. Culturing of "weedy" species like physids is best done at low temperatures to retard egg production and constant removal of egg cases is necessary to prevent population explosions. Adequate lighting (with a 12 hr light : 12 hr dark cycle) is necessary to promote periphyton growth in aquaria (gro-lites work well), with the lighting regime as even as possible.

V. IDENTIFICATION OF THE FRESHWATER GASTROPODS OF NORTH AMERICA

The following taxonomic key for the most part is based on Burch's (1982) treatment of North American freshwater snails, the most exhaustive key and species list compiled to date. Readers interested in identifying specimens to the species level should consult Burch (1982) or several excellent keys addressing regional snail fauna (Harman and Berg 1971, Jokinen 1983, Thompson 1984). Where Burch's (1982) taxonomy differed from that used by authors whose work was reviewed earlier in the chapter, I have placed Burch's designation in parentheses.

Taxa are keyed to genera in all families except the diverse hydrobiids. While collecting specimens, try to collect and preserve (in 70% ethyl alcohol) living specimens for identification, as possession of an operculum will be the first determination necessary. Also, collect a good number of specimens because some closely related genera unfortunately differ only in adult shell size. Following Burch (1982), small shells are considered here to be less than 10 mm in length, medium are between 11 and 29 mm, and large are greater than 30 mm. Refer to the glossary for shell sculpture terminology such as costae, lirae, etc. In some cases, examination of soft tissues is necessary for identification. For example, shell morphology is very conservative in the hydrobiids, and the structure of the verge is important. Either gently remove the shells of preserved individuals, or relax living snails by sprinkling granulated menthol on the water surface. Examine the verge under a dissecting microscope. If examination of the radula is necessary, dissect out the buccal mass (a muscular mass behind the mouth that surrounds the radula) from a relaxed specimen and place it in a 10% potassium hydroxide solution. After several hours, tease the radula from the remaining tissue, rinse it in 70% alcohol, stain it, and mount it on a microscope slide for examination under a compound microscope.

A. Taxonomic Key to Families and Selected Genera of Gastropoda

1a.	Operculum present in the shell aperture (Fig. 10.2); gills in dorsal mantle cavity	subclass Prosobranchia	2
1b.	Without operculum, mantle cavity modified into lung, or pseudobranch (false gill) present outside the mantle cavity	subclass Pulmonata	27
2a(1a).	Aperture with small teeth or projections on the parietal inner margin; operculum calcareous, paucispiral (Fig. 10.2), with two projections on inner surface, adult shell length about 20 mm; gill featherlike; Florida and southern Georgia (Fig. 10.7A)	family Neritinidae	*Neritina*
2b.	Shell without teeth on parietal wall, corneous operculum without projections		3
3a(2b).	Shell 8 mm or less in diameter, spire depressed, some species with carina; operculum multispiral (Fig. 10.2); gill featherlike, visible outside shell when snail is active (Fig. 10.7B, C)	family Valvatidae	*Valvata*
3b.	Shell and spire length variable, operculum multispiral, paucispiral, or concentric (Fig. 10.2)		4
4a(3b).	Operculum multispiral or paucispiral (Fig. 10.2), outer portion not concentric		5
4b.	Operculum concentric (but nucleus may be paucispiral)		20
5a(4a).	Adult shell length usually less than 7 mm, males with penis (verge)		6
5b.	Adult shell length usually more than 15 mm, males without verge		12
6a(5a).	Shell globe- to conic-shaped, with numerous, spiral epidermal ridges; in cave streams in Indiana and Kentucky	family Micromelaniidae	*Antroselates*
6b.	Shell shape variable, surface smooth, never with spiral epidermal ridges		7
7a(6b).	High-spired; body subdivided on each side by longitudinal groove in head–foot region; eyes in prominent swellings on outer bases of tentacles, amphibious, crawls with step-like movement (Fig. 10.7D)	family Pomatiopsidae	*Pomatiopsis*

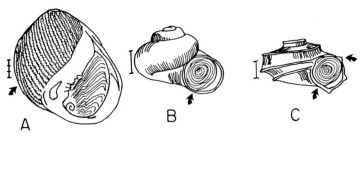

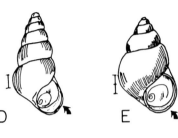

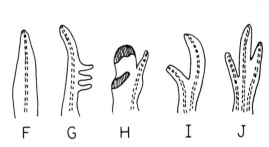

Figure 10.7 Representative neritinids, valvatids, and hybrobiids. (A) *Neritina reclivata* (note teeth on parietal wall); (B) *Valvata sincera* (note multispiral operculum); (C) *Valvata tricarinata* (note carina); (D) *Pomatiopsis lapidaria* (note paucispiral operculum); (E) *Bithynia tentaculata*; (F) simple verge of hydrobiid subfamily Lithoglyphinae; (G) verge with accessory lobes in hydrobiid subfamily Hydrobiinae; (H) glandular crests of verge of hydrobiid subfamily Nymphophilinae; (I) two-ducted verge of hydrobiid subfamily Amnicolinae; (J) three-ducted verge of hydrobiid subfamily Fontigentinae. (A–J after Burch 1982.) All bars in this figure and subsequent figures are marked in mm.)

7b.   Spire length variable; body without longitudinal groove; eye location same but without prominent swellings, aquatic, moves with gliding motion .... family Hydrobiidae   8

8a(7b).   Males with single-ducted verges (Fig. 10.7F, G, H) 9
8b.   Males with two- or three-ducted verges (Fig. 10.7I, J) 11

9a(8a).   Males with flat, blade-like verge (Fig. 10.7F) .... subfamily Lithoglyphinae
             Lepyrium, Cochliopina, Fluminicola, Antrobia, Clappia, Somatogyrus

9b.   Verge has accessory lobes or glandular apical and subapical crests (Fig. 10.7G, H) 10

10a(9b).   Verge has accessory lobes (Fig. 10.7G) .... subfamily Hydrobiinae (sometimes referred to as Littoridininae)   *Probythinella, Hoyia, Tryonia, Pyrogophorus, Littoridinops, Aphaostracon, Hyalopyrgus*
10b.   Verge has glandular, terminal lobe (Fig. 10.7H) .... subfamily Nymphophilinae   *Orygocerus, Birgella, Striobia, Marstonia, Rhapinema, Notogillia, Spilochlamys, Cincinnatia, Pyrgulopsis, Fontelicella, Natricola*

11a(8b).   Two-ducted verge (Fig. 10.7I) .... subfamily Amnicolinae   *Amnicola, Lyogyrus, Hauffenia, Horatia*

| 11b. | Three-ducted verge; common in eastern North America (Fig. 10.7J) .... subfamily Fontigentinae |
| | *Fontigens* |

| 12a(5b). | Mantle edge smooth; dioecious (males present), females oviparous, with egg-laying sinus on the right side of the foot .... family Pleuroceridae   14 |
| 12b. | Mantle edge papillate; parthenogenetic; females brood young in a brood pouch posterior to the head; introduced from Florida to Texas .... family Thiaridae   13 |

| 13a(12b). | Rounded whorls have spiral grooves and transverse raised lines (costae); Florida to Arizona (Fig 10.8B) .... *Melanoides* |
| 13b. | Whorls flattened near spire, spiral rows of nodules on shell; Florida and Texas (Fig. 10.8A) .... *Thiara* |

| 14a(12a). | Large shell often with elongated spines, and long, anterior canal on aperture; rivers in eastern Tennessee and western Virginia (Fig. 10.8C) .... *Io* |
| 14b. | Shell size and sculpture extremely variable, shape conical to subglobose; aperture sometimes ends in a short canal .... 15 |

| 15a(14b). | Body whorl with a slit along the suture; Alabama (Fig. 10.8D) .... *Gyrotoma* |
| 15b. | Body whorl without a slit along the suture .... 16 |

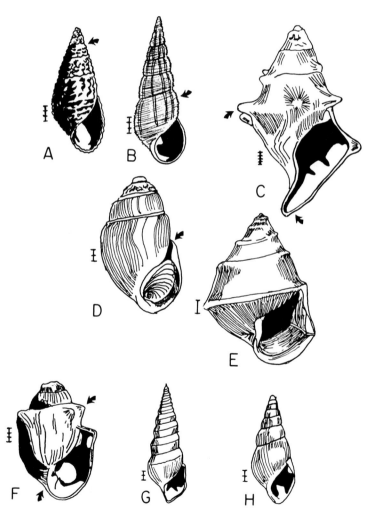

Figure 10.8 Representative thiarids and pleurocerids. (A) *Thiara granifera* (note tubercules and flattened whorls near apex); (B) *Melanoides tuberculata* (note costae and lirae); (C) *Io fluvialis* (length of spines variable); (D) *Gyrotoma excisum* (note slit in suture above aperture); (E) *Leptoxis carinata;* (F) *Lithasia geniculata* (note thickened anterior aperture lip); (G) *Pleurocera acuta* (note acute angle on anterior aperture); (H) *Elimia* (*Goniobasis, Oxytrema*) *livescens* (note: there is tremendous variation in shell sculpture in this genus). (A–H after Burch 1982.)

16a(15b).   Lateral radular teeth with broad, bluntly rounded median cusps; shell
            medium to small, subglobose to broadly conic or ovate (Fig. 10.8E) .... *Leptoxis*

16b.   Lateral radular teeth with narrow, triangular median cusps; shell usually
            a narrow cone 17

17a(16b).   Shell length medium, either cone-shaped or subglobose, often with
            spines or prominent nodules; inside margin of the aperture thickened,
            basal lip of aperture with a channel or strong angle (Fig. 10.8F)
            .... *Lithasia*

17b.   Shell length variable, usually an elongated cone; sculpture variable;
            base of aperture either rounded or with a short canal, inner edge of
            aperture not thickened 18

18a(17b)   Anterior (basal) end of aperture with short canal, giving shell "auger"
            shape; Mississippi, Great Lakes, and Hudson River drainages
            (Fig. 10.8G) .... *Pleurocera*

18b.   Anterior aperture not auger-shaped 19

19a(18b)   In southern rivers east of the Mississippi, and in the Great Lakes,
            St. Lawrence River, or Hudson Bay (Fig 10.8H) .... *Elimia*

19b.   West of the Mississippi, in river drainages in the Great Basin or Pacific
            slope .... *Juga*

20(4b).   Adults more than 20 mm in length (in some up to 50 or 60 mm),
            operculum corneous 21

20b.   Adults less than 15 mm in length; operculum calcareous; from
            Wisconsin to Pennsylvania and New York (Fig. 10.7E) .... family Bithyniidae   *Bithynia*

21a(20a).   Shell length up to 60 mm, globose or planospiral; penis to right of
            mantle, calcareous (*Pomacea*) or gelatinous (*Marisa*) eggs; Florida. .... family Pilidae
                                                                                    (Ampullariidae)   26

21b.   Shell subglobose, right tentacle in males thicker and used as a
            penis, females ovoviviparous; widespread in the United States
            and Canada .... family Viviparidae   22

22a(21b).   Adult shell over 35 mm in length, relatively thin; introduced to
            northeastern United States (Fig. 10.9A) .... *Cipangopaludina*

22b.   Adult shell thick and less than 35 mm 23

23a(22b).   Shell sometimes with one or two spiral rows of nodules, outer edge of
            aperture concave (when observed from the side), inside (columellar)
            edge of operculum folded inward; Alabama rivers (Fig. 10.9B) .... *Tulotoma*

23b.   Shell without nodules, aperture not as above 24

24a(23b).   Operculum with spiral nucleus and concentric periphery; whorls with a
            median, low ridge (Fig. 10.9C) .... *Lioplax*

24b.   Operculum entirely concentric; whorls without ridge 25

25a(24b).   Shell sometimes with spiral color bands (especially in small specimens);
            aperture nearly circular (Fig. 10.9D) .... *Viviparus*

25b.   Shell without spiral bands, whorls and aperture more shouldered;
            aperture and operculum longer than wide (Fig. 10.9E) .... *Campeloma*

26a(21a).   Shell globose; Florida (Fig. 10.9G) .... *Pomacea*

26b.   Shell planospiral; southern Florida, and introduced to rivers in central
            Texas (Fig. 10.9F) .... *Marisa*

27a(1b).   Shell coiled 28

27b.   Shell cone-, limpet-, or cap-shaped (Fig. 10.2) 49

28a(27a).   Shell dextral (aperture to right when viewed with spire
            pointing away) .... family Lymnaeidae   29

28b.   shell sinistral (aperture to left) 35

29a(28a).   Adult with large, globose body whorl and flared aperture (Fig 10.10A) .... *Radix*

29b.   Adult with narrower body whorl, and minute spiral striations 30

30a(29b).   Shell extremely laterally compressed; southern Canada and northcentral
            United States east to New England (Fig. 10.10D) .... *Acella*

30b.   Shell not especially narrow 31

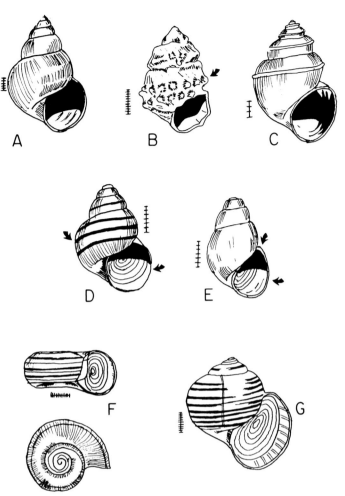

Figure 10.9 Representative viviparids and ampullarids. (A) *Cipangopaludina japonica* (an introduced but now widespread species in North America); (B) *Tulotoma magnifica* (note tubercules which may be absent in some morphs or species); (C) *Lioplax subcarinata;* (D) *Viviparus georgianus* (note circular operculum, and bands which may disappear in adults); (E) *Campeloma decisum* (sometimes referred to as *decisa;* note operculum is longer than it is wide and shouldered junction of aperture and body whorl); (F) *Marisa cornuarietis* (note large, planospiral shell); (G) *Pomacea paludosa* (A–G after Burch 1982.)

31a(30b).	Shell transparent and extremely fragile, with large, oval aperture and body whorl, small spire, amphibious; eastern United States (Fig. 10.10c)	*Pseudosuccinea*
31b.	Shell not as thin or fragile	32
32a(31b).	Adults more than 40 mm in length	33
32b.	Adults smaller	34
33a(32a).	Shell with a narrow, pointed spire, moderately fragile, in marshes or streams in Canada, northcentral and eastern United States (Fig. 10.10G)	*Lymnaea*
33b.	Shell globose, thick, with a relatively wide spire; Great Lakes and St. Lawrence River system, Canadian interior (Fig. 10.10B)	*Bulimnea*
34a(32b).	Adult more than 13 mm in length, surface sculptured with microscopic spiral striations or with malleations, columellar edge of aperture with a well-developed twist or plait; widely distributed in ponds and marshes in northern United States or in alpine regions of western United States (Fig. 10.10F)	*Stagnicola*
34b.	Shell less than 13 mm in length, spiral sculpture usually absent; columella generally without a twist or plait (Fig. 10.10E)	*Fossaria*
35a(28b).	Adult shell with raised spire, pseudobranch or false gill absent; mantle margin often digitate or lobed . . . . family Physidae	36

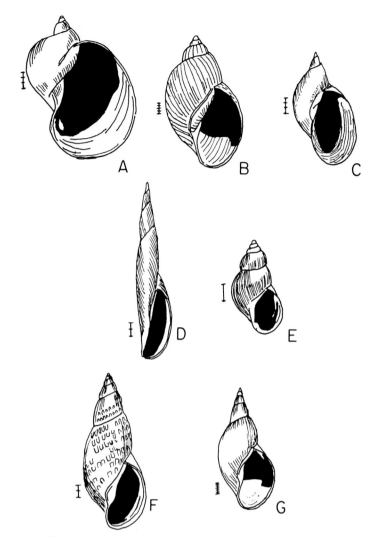

Figure 10.10 Representative lymnaeids. (A) *Radix* (*Lymnaea*) *auricularia* (note expanded body whorl); (B) *Bulimnea megasoma* (note thick, large shell); (C) *Pseudosuccinea columella* (note thin, transparent shell and amphibious habit); (D) *Acella haldemani* (note extremely narrow shell); (E) *Fossaria* (*Lymnaea*) *humilis* (note small size and amphibious habit); (F) *Stagnicola* (*Lymnaea*) *elodes* (malleations sometimes present, this species is common only in ponds and marshes, rarely lakes); (G) *Lymnaea stagnalis* (note large and fragile shell). (A–G after Burch 1982.)

35b.	Shell planospiral (Fig. 10.2), false gill near pneumostome or anus; mantle margin simple .... family Planorbidae	39
36a(35a).	Mantle edge with fingerlike projections	37
36b.	Mantle edge without projections, but may be serrated	38
37a(36a).	Projections on both sides of mantle .... *Physa*	
37b.	Projections on parietal side of mantle only (Fig. 10.11 A–C) .... *Physella*	
38a(36b).	Serrated mantle edge extends beyond apertural lip, partly overlapping shell; Texas .... *Stenophysa*	
38b.	Shell elongate, surface black and glossy, sutures smooth, in ponds in wood lots in Canada and northern United States (Fig. 10.11D) .... *Aplexa*	
39a(35b).	Adult shells less than 8 mm in diameter	40
39b.	Adult shell more than 30 mm in diameter	46
40a(39a).	Shell costate (with transverse raised ridges); Canada and northern United States (Fig. 10.11E) .... *Armiger*	
40b.	Shell not costate	41

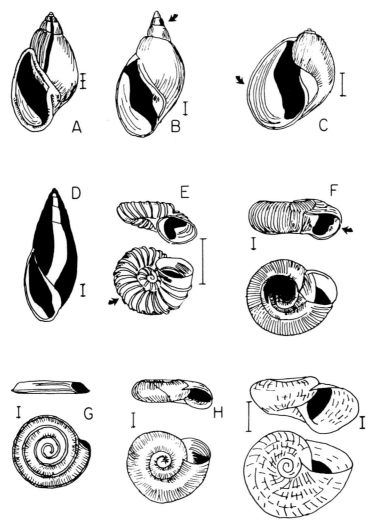

Figure 10.11 Representative physids and planorbids. (A) *Physella (Physa)gyrina* (very similar and possible subspecies are *P. anatina* and *P. virgata* of the midwest and south, the only physid to reach 20 mm shell length); (B) *Physella(Physa) integra* (note the more elevated spire than *P. gyrina*, rarely reaches 10 mm shell length); (C) *Physella (Petrophysa)zionis* (note the large aperture); (D) *Aplexa hypnorum* (note the bullet shape and lustrous black shell); (E) *Armiger crista* (note costae); (F) *Planorbula armigera (jenksii)* (note teeth in aperture); (G) *Drepanotrema kermatoides;* (H) *Gyraulus deflectus;* (I) *Menetus dilatatus* (A–I after Burch 1982.)

41a(40b).	Adult shell 2 mm or less in diameter; Alabama	*Neoplanorbis*
41b.	Adult shell more than 2 mm in diameter	42
42a(41b).	Shell laterally compressed, aperture or body whorl without ''teeth'' or lamellae	43
42b.	Shell higher, teeth inside aperture or body whorl (Fig. 10.11F)	*Planorbula*
43a(42a).	Shell either extremely flattened and multiwhorled or with many small, close-set spiral ridges (lirae); Florida, Texas, and southern Arizona (Fig. 10.11G)	*Drepanotrema*
43b.	Shell not excessively flattened, with fewer whorls and without lirae	44
44a(43b).	Body whorl increases in height toward the aperture	45
44b.	Body whorl nearly equal on each side (Fig. 10.11H)	*Gyraulus*
45a(44a).	Shell diameter less than 3 mm, periphery round or carinate (angular) (Fig. 10.11I)	*Menetus*
45b.	Larger, carinate shell (Fig. 10.12a)	*Promenetus*

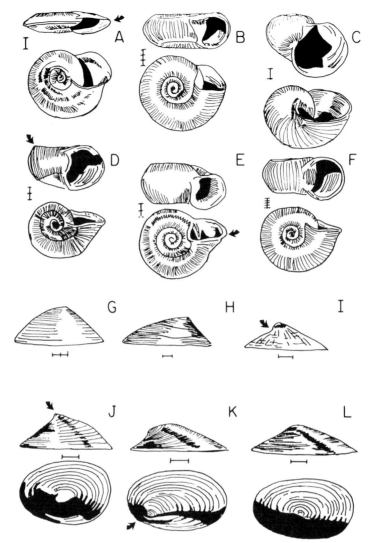

Figure 10.12 Representative planorbids and limpets. (A) *Promenetus exacuous* (note flared body whorl and carina); (B) *Biomphalaria glabrata;* (C) *Vorticifex (Parapholyx) effusa;* (D) *Helisoma anceps* (note strong growth lines and carina); (E) *Planorbella (Helisoma) companulata* (sometimes called *companulatum,* note flared lip of aperture); (F) *Planorbella (Helisoma) trivolvis* (this species reaches 20 mm in diameter and is extremely common); (G) the lymnaeid limpet *Lanx patelloides* (note large size, west coast distribution); (H) lymnaeid limpet *Fisherola nutalli;* (I) Ancylid limpet *Rhodacmea rhodacme (hinkezi)* (note notched depression and southeastern distribution); (J) ancylid limpet *Ferrissia rivularis* (note elevated shell, may possess posterior "shelf" in shell, wide distribution); (K) ancylid limpet *Hebetancylus excentricus* (note depressed apex to the right of midline, colorless tentacles, and southern distribution); (L) *Laevapex fuscus* (note obtuse apex near midline of shell, black-pigmented tentacles, and widespread distribution in eastern backwaters and southern, slow-flowing streams). (A–L after Burch 1982.)

46a(39b).	Fragile, thin shell with relatively depressed body whorl; Florida, Texas, Arizona (Fig. 10.12B)	*Biomphalaria*
46b.	Shell thicker, solid, body whorl often high	47
47a(46b).	Shell with exceptionally large body whorl; western in distribution (Fig. 10.12C)	*Vorticifex*
47b.	Shell with many whorls, body whorl relatively similar in size	48
48a(47b).	Shell spire (left side) with a conical depression; body whorl often with strong carina (Fig. 10.12D)	*Helisoma*

48b.   Shell spire (left side) with a shallow depression, or even raised above body whorl, body whorl more rounded, aperture lip may be flared (Fig. 10.12 E, F) .... *Planorbella*

49a(27b).   Adult shell up to 12 mm in length, apex not distinctly to the right or left of the median line; animal dextral. Pacific drainage .... family Lymnaeidae   50

49b.   Adult shell 7 mm or less in length, apex often to the right or left of the median line, animal dextral or sinistral 51

50a(49a).   Apex subcentral (Fig. 10.12G) .... *Lanx*
50b.   Apex closer to anterior; Columbia river drainage (Fig. 10.12H) .... *Fisherola*

51a(49b).   Shell dextral; Rocky Mountain lakes, northeastern Ontario and northcentral Quebec .... family Acroloxidae   *Acroloxus*
51b.   Shell sinistral; widespread .... family Ancylidae   52

52a(51b).   Shell elevated, apex in midline and tinged with pink or red inside and out; shell radially striate, with a notch-shaped depression in unworn specimens; aperture lip broad and flat; in southeastern rivers (Fig. 10.12I) .... *Rhodacmea*
52b.   Shell height variable, apex in midline or to the right, but same color as the rest of shell, which has fine radial striations or is smooth; widely distributed in running or standing water 53

53a(52b).   Apex with fine radial striae, but eroded in older specimens; aperture narrow to broadly ovate, rarely with a horizontal shelf on the posterior; one lobed, flat pseudobranch. Widely distributed in lentic and lotic habitats (Fig. 10.12J) .... *Ferrissia*
53b.   Shell more depressed; apex without radial striae; aperture ovate to subcircular, always open; second, lower lobe of pseudobranch elaborately folded; common in lentic habitats in eastern United States and south 54

54a(53b).   Apex tipped to the right, tentacles colorless; in southern Florida, or Texas, in canals (Fig. 10.12K) .... *Hebetancylus*
54b.   Apex in midline of the shell; tentacles with black core; east of the Mississippi in lentic habitats, occasionally in southcentral streams (Fig. 10.12L) .... *Laevapex*

LITERATURE CITED

Aldridge, D. W. 1982. Reproductive tactics in relation to life-cycle bioenergetics in three natural populations of the freshwater snail, *Leptoxis carinata*. Ecology 63:196–208.

Aldridge, D. W. 1983. Physiological ecology of freshwater prosobranchs. Pages 329–358 *in*: W. D. Russell-Hunter, editor. The Mollusca. Vol. 6: Ecology. Academic Press, Orlando, Florida.

Anderson, R. M., and R. M. May. 1979. Prevalence of schistosome infections within molluscan populations: observed patterns and theoretical predictions. Parasitology 79:63–94.

Barnes, R. D. 1987. Invertebrate Zoology. 5th edition. Saunders, Philadelphia, Pennsylvania.

Barnese, L. E., and R. L. Lowe. 1990. Comparative grazing efficiency of pulmonate and prosobranch snails. Journal of the North American Benthological Society 9:35–44.

Baudoin, M. 1975. Host castration as a parasite strategy. Evolution 29:335–352.

Blinn, D. W., R. E. Truitt, and A. Pickart. 1989. Feeding ecology and radular morphology of the freshwater limpet *Ferrissia fragilis*. Journal of the North American Benthological Society 8:237–242.

Boag, D. A. 1981. Differential depth distribution among freshwater pulmonate snails subjected to cold temperatures. Canadian Journal of Zoology 9:733–737.

Boag, D. A. 1986. Dispersal in pond snails: potential role of waterfowl. Canadian Journal of Zoology 64:904–909.

Boag, D. A., and P. S. M. Pearlstone. 1979. On the life cycle of *Lymnaea stagnalis* (Pulmonate:Gastropoda) in Southwestern Alberta. Canadian Journal of Zoology 52:353–362.

Boerger, H. 1975. Movement and burrowing of *Helisoma trivolvis* (Say) (Gastropoda: Planorbidae) in a small pond. Canadian Journal of Zoology 53:456–464.

Boss, K. J. 1974. Oblomovism in the Mollusca. Transactions of the American Microscopical Society 93:460–481.

Bousefield, J. D. 1979. Plant extracts and chemically triggered positive rheotaxis in *Biomphalaria glabrata* (Say), snail intermediate host of *Schistosoma mansoni*. Journal of Applied Ecology 16:681–690.

Bovbjerg, R. V. 1952. Ecological aspects of dispersal of the snail *Campeloma decisum*. Ecology 33:169–176.

Bovbjerg, R. V. 1965. Feeding and dispersal in the snail *Stagnicola reflexa*. Malacologia 2:199–207.

Bovbjerg, R. V. 1968. Responses to food in lymnaeid snails. Physiological Zoology 41:412–423.

Bovbjerg, R. V. 1975. Dispersal and dispersion of pond snails in an experimental environment varying in three factors, singly and in combination. Physiological Zoology 48:203–215.

Boycott, A. E. 1936. The habitats of freshwater mollusca in Britain. Journal of Animal Ecology 5:116–186.

Bronmark, C. 1985a. Freshwater snail diversity: effects of pond area, habitat heterogeneity and isolation. Oecologia 67:127–131.

Bronmark, C. 1985b. Interactions between macrophytes, epiphytes, and herbivores: an experimental approach. Oikos 45:26–30.

Bronmark, C. 1989. Interactions between epiphytes, macrophytes and freshwater snails: a review. Journal of Molluscan Studies 55:299–311.

Bronmark, C., and B. Malmqvist. 1986. Interactions between the leech *Glossiphonia complanata* and its gastropod prey. Oecologia 69:268–276.

Brown, D. S. 1978. Pulmonate molluscs as intermediate hosts for digenetic trematodes. Pages 287–333 *in:* V. Fretter and J. Peake, editors. Pulmonates. Vol. 2A: Systematics, Evolution and Ecology. Academic Press, New York.

Brown, K. M. 1979. The adaptive demography of four freshwater pulmonate snails. Evolution 33:417–432.

Brown, K. M. 1982. Resource overlap and competition in pond snails: an experimental analysis. Ecology 63:412–422.

Brown, K. M. 1983. Do life history tactics exist at the intra-specific level? Data from freshwater snails. American Naturalist 121:871–879.

Brown, K. M. 1985. Intraspecific life history variation in a pond snail: The roles of population divergence and phenotypic plasticity. Evolution 39:387–395.

Brown, K. M., and D. R. DeVries. 1985. Predation and the distribution and abundance of a pulmonate pond snail. Oecologia 66:93–99.

Brown, K. M., and T. D. Richardson. 1988. Genetic polymorphism in gastropods: a comparison of methods and habitat scales. American Malacological Bulletin 6:9–17.

Brown, K. M., and B. H. Strouse. 1988. Relative vulnerability of six freshwater gastropods to the leech *Nephelopsis obscura* (Verrill). Freshwater Biology 19:157–166.

Brown, K. M., B. K. Leathers, and D. J. Minchella. 1988. Trematode prevalence and the population dynamics of freshwater pond snails. American Midland Naturalist 120:289–301.

Brown, K. M., D. E. Varza, and T. D. Richardson. 1989. Life histories and population dynamics of two subtropical snails (Prosobranchia: Viviparidae). Journal of the North American Benthological Society, 8:222–228.

Browne, R. A. 1978. Growth, mortality, fecundity, and productivity of four lake populations of the prosobranch snail, *Viviparus georgianus*. Ecology 59:742–750.

Browne, R. A. 1981. Lakes as islands: biogeographic distribution, turnover rates, and species composition in the lakes of central New York. Journal of Biogeography 8:75–83.

Browne, R. A., and W. D. Russell-Hunter. 1978. Reproductive effort in molluscs. Oecologia 27:23–27.

Burch, J. B. 1982. Freshwater Snails (Mollusca: Gastropoda) of North America. United States Environmental Protection Agency Publication 600/3-82-026.

Burky, A. J. 1971. Biomass turnover, respiration, and interpopulation variation in the stream limpet *Ferrissia rivularis*. Ecological Monographs 41:235–251.

Burky, A. J. 1983. Physiological ecology of freshwater bivalves. Pages 281–327 *in:* W. D. Russell-Hunter, editor. The Mollusca. Vol. 6: Ecology. Academic Press, Orlando, Florida.

Calow, P. 1970. Studies on the natural diet of *Lymnaea peregra obtusa* (Kobelt) and its possible ecological implications. Proceedings of the Malacological Society of London 39:203–215.

Calow, P. 1973a. Field observations and laboratory experiments on the general food requirements of two species of freshwater snail, *Planorbis contortus* (Linn.) and *Ancylus fluviatilis* Bull. Proceedings of the Malacological Society of London 40:483–489.

Calow, P. 1973b. The food of *Ancylus fluviatilis* (Mull.), a littoral, stone-dwelling herbivore. Oecologia 13:113–133.

Calow, P. 1974a. Evidence for bacterial feeding in *Planorbis contortus* Linn. (Gastropoda: Pulmonata). Proceedings Malacological Society of London 41:145–156.

Calow, P. 1974b. Some observations on the dispersion patterns of two species populations of littoral, stone-dwelling gastropods (Pulmonata). Freshwater Biology 4:557–576.

Calow, P. 1975. The respiratory strategies of two species of freshwater gastropods (*Ancylus fluviatilis* Mull. and *Planorbis contortus* Linn) in relation to temperature, oxygen concentration, body size, and season. Physiological Zoology 48:114–129.

Calow, P. 1978. The evolution of life-cycle strategies in fresh-water gastropods. Malacologia 17:351–364.

Calow, P. 1981. Adaptational aspects of growth and reproduction in *Lymnaea peregra* from exposed and sheltered aquatic habitats. Malacologia 21:5–13.

Calow, P. 1983. Life-cycle patterns and evolution. Pages 649–680 *in:* W. D. Russell-Hunter, editor. The Mollusca. Vol. 6: Ecology. Academic Press, Orlando, Florida.

Calow, P., and L. J. Calow. 1975. Cellulase activity and niche separation in freshwater gastropods. Nature (London) 255:478–480.

Carrothers, J. L., and R. Allison. 1968. Control of snails by the redear (shellcracker) sunfish, *In:* T. V. R. Pillay, editor. Proceedings of the Food and Agriculture Symposium on Warm-Water Pond Fish Culture. Fisheries Report 44, No. 5. Food and Agriculture Organization, Rome.

Cattaneo, A. 1983. Grazing on epiphytes. Limnology and Oceanography 28:124–132.

Chable, A. C. 1947. A study of the food habitats and ecological relationships of the sunfishes of northern Florida. Ph.D. Thesis, Univ. of Florida, Gainesville, Florida.

Chamberlain, N. A. 1958. Life history studies of *Campeloma decisum*. Nautilus 72:22–29.

Cheatum, E. P. 1934. Limnological investigations on respiration, annual migratory cycle, and other related phenomena in freshwater pulmonate snails. Transactions of the American Microscopical Society 53:348–407.

Clampitt, P. T. 1973. Substratum as a factor in the distribution of pulmonate snails in Douglas Lake, Michigan. Malacologia 12:379–399.

Clampitt, P. T. 1974. Seasonal migratory cycle and related movements of the fresh-water pulmonate snail, *Physa integra*. American Midland Naturalist 92:275–300.

Clarke, A. H. 1981. The Freshwater Molluscs of Canada. National Museum of Natural Sciences, Ottawa.

Covich, A. P. 1977. How do crayfish respond to plants and mollusca as alternate food resources? Freshwater Crayfish 3:165–169.

Covich, A. P. 1981. Chemical refugia for thin-shelled gastropods in a sulfide-enriched stream. Internationale Vereinigung fur Theoretische und Angewandte Limnologie 21:1632–1636.

Croll, R. P. 1983. Chemoreception. Biological Reviews 58:293–319.

Cross, F. B., and J. T. Collins. 1975. Fishes in Kansas. Univ. of Kansas Printing Service, Lawrence.

Crowl, T. A., and J. E. Alexander. 1989. Parental care and foraging ability in *Belastoma flumineum*. Canadian Journal of Zoology 67:513–515.

Crowl, T. A., and A. P. Covich. 1990. Predator-induced life history shifts in a freshwater snail. Science 247:949–951.

Cuker, B. E. 1983a. Competition and coexistence among the grazing snail *Lymnaea*, Chironomidae, and microcrustacea in an arctic epilithic lacustrine community. Ecology 64:10–15.

Cuker, B. E. 1983b. Grazing and nutrient interactions in controlling the activity and composition of the epilithic algal community of an arctic lake. Limnology and Oceanography 28:133–141.

Davies, R. W., F. J. Wrona, and R. P. Everett. 1978. A serological study of prey selection by *Nephelopsis obscura* Verill (Hirudinoidea). Canadian Journal of Zoology 56:587–591.

Davies, R. W., F. J. Wrona, and L. Linton. 1979. A serological study of prey selection by *Helobdella stagnalis*. Journal of Animal Ecology 48:181–194.

Davies, R. W., F. J. Wrona, L. Linton, and J. Wilkias. 1981. Inter- and intra-specific analyses of the food niches of 2 sympatric species of Erpobdellidae (Hirudinoidea) in Alberta, Canada. Oikos 37:105–111.

Davis, G. M. 1979. The Origin and Evolution of the Pomatiopsidae, with Emphasis on the Mekong River Hydrobioid Gastropods. Monographs of the Academy of Natural Sciences, Philadelphia. Number 20.

Davis, G. M. 1981. Introduction to the second international symposium on evolution and adaptive radiation of Mollusca. Malacologia 21:1–4.

Davis, G. M. 1982. Historical and ecological factors in the evolution, adaptive radiation, and biogeography of freshwater molluscs. American Zoologist 22:375–395.

Dazo, B. Z. 1965. The morphology and natural history of *Pleurocera acuta* and *Goniobasis livescens*. Malacologia 2:1–80.

Dillon, R. T. 1981. Patterns in the morphology and distribution of gastropods in Oneida Lake, New York, detected using computer-generated null hypotheses. American Naturalist 118:83–101.

Dillon, R. T. 1987. A new monte carlo method for assessing taxonomic similarity within faunal samples: re-analysis of the gastropod community of Oneida Lake, New York. American Malacological Bulletin 5:101–104.

Doremus, C. M., and W. N. Harman. 1977. The effects of grazing by physid and planorbid freshwater snails on periphyton. Nautilus 91:92–96.

Duch, T. M. 1976. Aspects of the feeding habits of *Viviparus georgianus*. Nautilus 90:7–10.

Duncan, C. J. 1975. Reproduction. Pages 309–366 *in:* V. Fretter and J. Peake, editors. Pulmonates. Vol. 1: Functional Anatomy and Physiology. Academic Press, Orlando, Florida.

Eckblad, J. W. 1973. Experimental predation studies of malacophagous larvae of *Sepedon fuscipennis* (Diptera: Sciomyzidae) and aquatic snails. Experimental Parasitology 33:331–342.

Eisenberg, R. M. 1966. The regulation of density in a natural population of the pond snail, *Lymnaea elodes*. Ecology 47:889–906.

Eisenberg, R. M. 1970. The role of food in the regulation of the pond snail, *Lymnaea elodes*. Ecology 51:680–684.

Etnier, D. A. 1971. Food of three species of sunfishes (*Lepomis*, Centrarchidae) and their hybrids in three Minnesota lakes. Transactions of the American Fisheries Society 100:124–128.

Eversole, A. G. 1978. Life cycles, growth and population bioenergetics in the snail *Helisoma trivolvis* (Say). Journal of Molluscan Studies 44:209–222.

Fretter, V., and A. Graham. 1962. British Prosobranch Molluscs. Ray Society, London.

Fretter, V., and J. Peake, editors. 1975. Pulmonates. Vol. 1: Functional Anatomy and Physiology. Academic Press, Orlando, Florida.

Fretter, V., and J. Peake, editors. 1978. Pulmonates. Vol. 2A: Systematics, Evolution and Ecology. Academic Press, Orlando, Florida.

Gerking, S. D. 1957. A method of sampling the littoral macrofauna and its application. Ecology 38:219–225.

Goodrich, B. 1945. *Goniobasis livescens* of Michigan. Miscellaneous Publications of the Museum of Zoology of the University of Michigan 64:1–26.

Harman, W. N. 1968. Replacement of pleurocerids by *Bithynia* in polluted waters of central New York. Nautilus 81:77–83.

Harman, W. N. 1972. Benthic substrates: their effect on fresh-water mollusca. Ecology 53:271–277.

Harman, W. N., and C. O. Berg. 1971. The freshwater snails of central New York. Search (Agriculture) 1:1–68.

Harman, W. N., and J. L. Forney. 1970. Fifty years of change in the molluscan fauna of Oneida Lake, New York. Limnology and Oceanography 15:454–460.

Hawkins, C. P., and J. K. Furnish. 1987. Are snails important competitors in stream ecosystems? Oikos 49:209–270.

Haynes, A., and B. J. R. Taylor. 1984. Food finding and food preference in *Potamopyrgus jenkensi* (E. A. Smith) (Gastropoda: Prosobranchia) Archives für Hydrobiologie 100:479–491.

Holmes, J. C. 1983. Evolutionary relationships between parasitic helminths and their hosts. Pages 161–185 *in:* D. J. Futuyma, and M. Slatkin, editors. Coevolution. Sinauer, Sunderland, Massachusetts.

Huish, M. T. 1957. Food habits of three Centrarchidae in Lake George, Florida. Proceedings of the Annual Conference of the Southeastern Association of Fisheries Game Commission. 11:293–302.

Hunter, R. D. 1975. Growth, fecundity, and bioenergetics in three populations of *Lymnaea palustris* in upstate New York. Ecology 56:50–63.

Hunter, R. D. 1980. Effects of grazing on the quantity and quality of freshwater aufwuchs. Hydrobiologia 69:251–259.

Hunter, R. D., and W. D. Russell-Hunter. 1983. Bioenergetic and community changes in intertidal aufwuchs grazed by *Littorina littorea*. Ecology 64:761–769.

Hyman, L. H. 1967. The Invertebrates. Vol. 6: Mollusca I. McGraw-Hill, New York.

Jacoby, J. M. 1985. Grazing effects on periphyton by *Theodoxus fluviatilis* (Gastropoda) in a lowland stream. Journal of Freshwater Ecology 3:265–274.

Jokinen, E. H. 1978. The aestivation pattern of a population of *Lymnaea elodes*. American Midland Naturalist 100:43–53.

Jokinen, E. H. 1983. The Freshwater Snails of Connecticut. Publ. No. 109. Department of Environmental Protection, State Geological and Natural History Survey of Connecticut.

Jokinen, E. H., J. Guerette, and R. W. Kortmann. 1982. The natural history of an ovoviviparous snail, *Viviparus georgianus* (Lea), in a soft-water eutrophic lake. Freshwater Invertebrate Biology 1:2–17.

Kairesalo, T., and I. Koskimies. 1987. Grazing by oligochaetes and snails on epiphytes. Freshwater Biology 17:317–324.

Keast, A. 1978. Feeding interrelations between age-groups of pumpkinseed (*Lepomis gibbosus*) and comparisons with bluegill (*Lepomis macrochirus*). Journal of the Fisheries Research Board of Canada. 35:12–27.

Kehde, P. M., and J. L. Wilhm. 1972. The effects of grazing by snails on community structure of periphyton in laboratory streams. American Midland Naturalist 87:8–24.

Kesler, D. H. 1981. Periphyton grazing by *Amnicola limosa:* an enclosure-exclosure experiment. Journal of Freshwater Ecology 1:51–59.

Kesler, D. H. 1983. Cellulase activity in gastropods: should it be used in niche separation? Freshwater Invertebrate Biology 2:173–179.

Kesler, D. H., and W. R. Munns, Jr. 1989. Predation by *Belastoma flumineum* (Hemiptera): an important cause of mortality in freshwater snails. Journal of the North American Benthological Society 8:342–350.

Kesler, D. H., E. H. Jokinen, and W. R. Munns, Jr. 1986. Trophic preferences and feeding morphology of two pulmonate snail species from a small New England pond, U.S.A. Canadian Journal of Zoology 64:2570–2575.

Kohn, A. J. 1983. Feeding biology of gastropods. Pages 2–64 *in:* K. M. Wilbur, editor. The Mollusca. Vol. 5. Academic Press, Orlando, Florida.

Lam, P. K. S., and P. Calow. 1989a. Intraspecific life history variation in *Lymnaea peregra* (Gastropoda: Pulmonata) I. Field Study. Journal of Animal Ecology 58:571–588.

Lam, P. K. S., and P. Calow. 1989b. Intraspecific life history variation in *Lymnaea peregra* (Gastropoda): Pulmonata). II. Environmental or genetic variance? Journal of Animal Ecology 58:589–602.

Lassen, H. H. 1975. The diversity of freshwater snails in view of the equilibrium theory of island biogeography. Oecologia 19:1–8.

Lauglin, D. R., and E. E. Werner. 1980. Resource partitioning in two coexisting sunfish: pumpkinseed (*Lepomis gibbosus*) and northern longear sunfish (*Lepomis megalotis peltastes*). Canadian Journal of Fisheries and Aquatic Science 37:1411–1420.

Lodge, D. M. 1985. Macrophyte-gastropod associations: observations and experiments on macrophyte choice by gastropods. Freshwater Biology 15:695–708.

Lodge, D. M. 1986. Selective grazing on periphyton: a determinant of fresh-water gastropod microdistributions. Freshwater Biology 16:831–841.

Lodge, D. M., and P. Kelly. 1985. Habitat disturbance and the stability of freshwater gastropod populations. Oecologia 68:111–117.

Lodge, D. M., K. M. Brown, S. P. Klosiewski, R. A. Stein, A. P. Covich, B. K. Leathers, and C. Bronmark. 1987. Distribution of freshwater snails: spatial scale and the relative importance of physicochemical and biotic factors. American Malacological Bulletin 5:73–84.

Lowe, R. L., and R. D. Hunter. 1988. Effect of grazing by *Physa integra* on periphyton community structure. Journal of the North American Benthological Society 7:29–36.

Macan, T. T. 1950. Ecology of freshwater Mollusca in the English Lake District. Journal of Animal Ecology 19:124–146.

MacArthur, R. H., and E. O. Wilson. 1967. The Theory of Island Biogeography. Princeton University Press, Princeton, New Jersey.

Machin, J. 1975. Water relationships. Pages 105–164 *in:* V. Fretter, and J. Peake, editors. Pulmonates. Vol. 1: Functional Anatomy and Physiology. Academic Press, Orlando, Florida.

Martin, A. W. 1983. Excretion. Pages 353–407 *in:* K. M. Wilbur, editor. The Mollusca. Vol. 5: Physiology. Academic Press, Orlando, Florida.

Maynard-Smith, J. 1976. The Evolution of Sex. Cambridge University Press, London.

McCormick, P. U., and R. J. Stevenson. 1989. Effect of snail grazing on benthic algal community structure in different nutrient environments. Journal of the North American Benthological Society 8:162–172.

McMahon, R. F. 1975a. Effects of artificially elevated water temperatures on the growth, reproduction and life cycle of a natural population of *Physa virgata* Gould. Ecology 56:1167–1175.

McMahon, R. F. 1975b. Growth, reproduction, and bioenergetic variation in three natural populations of a freshwater limpet *Laevapex fuscus*. Proceedings of the Malacological Society of London 41:331–352.

McMahon, R. F. 1983. Physiological ecology of freshwater pulmonates. Pages 359–430 *in:* W. D. Russell-Hunter, editor. The Mollusca. Vol. 6: Ecology. Academic Press, Orlando, Florida.

McMahon, R. F., R. D. Hunter, and W. D. Russell-Hunter. 1974. Variation in aufwuchs at six freshwater habitats in terms of carbon biomass and of carbon : nitrogen ratio. Hydrobiologia 45:391–404.

Minchella, D. J., and P. T. Loverde. 1981. A cost of increased early reproductive effort in the snail *Biomphalaria glabrata*. American Naturalist 118:876–881.

Minchella, D. J., B. K. Leathers, K. M. Brown, and J. K. McNair. 1985. Host and parasite counter-adaptations: An example from a freshwater snail. American Naturalist 126:843–854.

Mittelbach, G. G. 1984. Predation and resource partitioning in two sunfishes (Centrarchidae). Ecology 65:499–513.

Morton, J. E. 1955. The evolution of the Ellobiidae with a discussion on the origin of the pulmonata. Proceedings of the Zoological Society of London 125:127–168.

Okland, J. 1983. Factors regulating the distribution of freshwater snails (Gastropoda) in Norway. Malacologia 24:277–288.

Osenberg, C. W. 1989. Resource limitation, competition and the influence of life history in a freshwater snail community. Oecologia 79:512–519.

Pace, G. L., and E. J. Szuch. 1985. An exceptional stream population of the banded applesnail, *Viviparus georgianus*, in Michigan. Nautilus 99:48–53.

Partridge, L., and P. H. Harvey. 1988. The ecological context of life history evolution. Science 241:1449–1455.

Patrick, R. 1970. Benthic stream communities. American Scientist 59:546–549.

Pechenik, J. A. 1985. Biology of the Invertebrates. Prindle, Weber, & Schmidt, Boston, Massachusetts.

Pennak, R. W. 1978. Fresh-water Invertebrates of the United States. 2nd edition. Wiley, New York.

Pflieger, W. L. 1975. The Fishes of Missouri. Missouri Department of Conservation, Western Publications, Columbia.

Pip, E. 1986. The ecology of freshwater gastropods in the central Canadian region. Nautilus 100:56–66.

Reavell, P. E. 1980. A study of the diets of some British freshwater gastropods. Journal of Conchology 30:253–271.

Ribi, G., and H. Ater. 1986. Sex related differences of movement speed in the freshwater snail *Viviparus ater*. Journal of Molluscan Studies 52:91–96.

Richardson, T. D., and K. M. Brown. 1989. Secondary production of two subtropical viviparid prosobranchs.

Journal of the North American Benthological Society 8:229–236.

Richardson, T. D., J. F. Scheiring, and K. M. Brown. 1988. Secondary production of two lotic snails (Pleuroceridae: *Elimia*). Journal of the North American Benthological Society 7:234–245.

Russell-Hunter, W. D. 1970. Aquatic Productivity. Macmillan, New York.

Russell-Hunter, W. D. 1978. Ecology of freshwater pulmonates. Pages 335–383 *in:* V. Fretter, and J. Peake, editors. The Pulmonates. Vol. 2A: Systematics, evolution and ecology. Academic Press, Orlando, Florida.

Russell-Hunter, W. D. 1983. Overview: Planetary distribution of and ecological constraints upon the mollusca. Pages 1–28 *in:* W. D. Russell-Hunter, editor. The Mollusca. Vol. 6: Ecology. Academic Press, Orlando, Florida.

Russell-Hunter, W. D., and D. E. Buckley. 1983. Actuarial bioenergetics of nonmarine molluscan productivity. Pages 463–503 *in:* W. D. Russell-Hunter, editor. The Mollusca. Vol. 6: Ecology. Academic Press, Orlando, Florida.

Savino, J. F., and R. A. Stein. 1982. Predator-prey interaction between large-mouth bass and bluegills as influenced by simulated submerged vegetation. Transactions of the American Fisheries Society 111:255–266.

Seaburg, K. G., and J. B. Moyle. 1964. Feeding habits, digestive rates, and growth of some Minnesota warmwater fishes. Transactions of the American Fisheries Society 93:269–285.

Sheldon, S. P. 1987. The effects of herbivorous snails on submerged macrophyte communities in Minnesota lakes. Ecology 68:1920–1931.

Stein, R. A., C. G. Goodman, and E. A. Marschall. 1984. Using time and energetic measures of cost in estimating prey value for fish predators. Ecology 65:702–715.

Steinman, A. D., C. D. McIntire, S. V. Gregory, G. A. Lamberti, and L. R. Ashkenas. 1987. Effects of herbivore type and density on taxonomic structure and physiognomy of algal assemblages in laboratory streams. Journal of the North American Benthological Society 6:125–188.

Strong, D. R., Jr., D. Simberloff, L. G. Abele, and A. B. Thistle. 1984. Ecological communities: Conceptual issues and the evidence. Princeton Univ. Press, Princeton, New Jersey.

Sumner, W. T., and C. D. McIntire. 1982. Grazer–periphyton interactions in laboratory streams. Archives fur Hydrobiologie 93:135–157.

Tashiro, J. S. 1982. Grazing in *Bithynia tentaculata:* age specific bioenergetic patterns in reproductive partitioning of ingested carbon and nitrogen. American Midland Naturalist 107:133–150.

Tashiro, J. S., and S. D. Colman. 1982. Filter feeding in the freshwater prosobranch snail *Bithynia tentaculata:* bioenergetic partitioning of ingested carbon and nitrogen. American Midland Naturalist 107:114–132.

Thompson, F. G. 1984. The freshwater snails of Florida: A manual for identification. University of Florida Press, Gainesville.

Townsend, C. R. 1975. Strategic aspects of time allocation in the ecology of a freshwater pulmonate snail. Oecologia 19:105–115.

Townsend, C. R., and T. K. McCarthy. 1980. On the defense strategy of *Physa fontinalis* (L.), a freshwater pulmonate snail. Oecologia 46:75–79.

Trautman, B. B. 1957. The Fishes of Ohio. Ohio State University Press, Columbus.

Vail, V. A. 1978. Seasonal reproductive patterns in three viviparid gastropods. Malacologia 17:73–97.

Van Cleave, H. J., and D. A. Altringer. 1937. Studies on the life cycle of *Campeloma rufrum,* a freshwater snail. American Naturalist 71:167–184.

Van Cleave, H. J., and L. G. Lederer. 1932. Studies on the life cycle of the snail, *Viviparus contectoides.* Journal of Morphology 53:499–522.

Vermeij, G. J., and A. P. Covich. 1978. Coevolution of freshwater gastropods and their predators. American Naturalist 112:833–843.

Weber, L. M., and D. M. Lodge. 1990. Periphyton food and crayfish predators: relative roles in determining snail distributions. Oecologia 82:33–39.

Wilbur, K. M., series editor. 1983. The Mollusca. Vols. 1-6. Academic Press, Orlando, Florida.

Young, J. O. 1981. A comparative study of the food niches of lake-dwelling triclads and leeches. Hydrobiologia 84:91–102.

Young, J. O., and J. W. Ironmonger. 1980. A laboratory study of the food of three species of leeches occurring in British lakes. Hydrobiologia 68:209–215.

Mollusca: Bivalvia

11

Robert F. McMahon
Department of Biology
The University of Texas at Arlington
Arlington, Texas 76019

Chapter Outline

I. INTRODUCTION

II. ANATOMY AND PHYSIOLOGY
  A. External Morphology
    1. Shell
    2. Locomotory Structures
  B. Organ System Function
    1. Circulation
    2. Gills and Gas Exchange
    3. Excretion and Osmoregulation
    4. Digestion and Assimilation
    5. Reproductive Structures
    6. Nervous System and Sense Organs
  C. Environmental Physiology
    1. Seasonal Cycles
    2. Diurnal Cycles
    3. Other Factors Affecting Metabolic Rate
    4. Dessication Resistance
    5. Gill Calcium Phosphate Concretions in Unionaceans
    6. Water and Salt Balance

III. ECOLOGY AND EVOLUTION
  A. Diversity and Distribution
  B. Reproduction and Life History
    1. Unionacea
    2. Sphaeriidae
    3. *Corbicula fluminea*
    4. *Dreissena polymorpha*
  C. Ecological Interactions
    1. Behavioral Ecology
    2. Feeding
    3. Population Regulation
    4. Functional Role in the Ecosystem
    5. Bivalves as Biomonitors
  D. Evolutionary Relationships

IV. COLLECTING, PREPARATION FOR IDENTIFICATION, AND REARING
  A. Collecting
  B. Preparation for Identification
  C. Rearing Freshwater Bivalves

V. IDENTIFICATION OF THE FRESHWATER BIVALVES OF NORTH AMERICA
  A. Taxonomic Key to Superfamilies of Freshwater Bivalvia
  B. Taxonomic Key to Genera of Freshwater Corbiculacea
  C. Taxonomic Key to Genera of Freshwater Unionacea
  Literature Cited

I. INTRODUCTION

Freshwater bivalve molluscs (class Bivalvia) fall within the subclass Lamellibranchia and are characterized by greatly enlarged gills with elongated, ciliated filaments for filter feeding. As the name implies, the mantle tissue underlying the shell is separated into left and right shell-secreting centers or lobes. However, the bivalved shell is a single structural entity. In early development, the bivalve mantle (tissue outgrowth from the dorsal side of the visceral mass enfolding the body and secreting the shell) is dorsally divided by anterior and posterior bifurcations into right and left portions, which always remain connected by a mid-dorsal isthmus (Allen 1985). Bivalves, as do all molluscs, secrete a shell made up of proteinaceous and crystalline calcium carbonate elements (Wilbur and Saleuddin 1983). The right and left mantle halves secrete shell material with a high proportion of crystalline calcium carbonate to form the left and right valves. The adjoining isthmus primarily secretes proteinaceous material to form an elastic hinge ligament dorsally connecting the two valves (Fig. 11.1). The hinge ligament is external in all freshwater bivalves. Its elastic properties force the valves apart when shell adductor muscles are relaxed. Anterior and posterior shell adductor muscles (Fig. 11.2) run between the valves and function in opposition to the hinge ligament to close the valves on contraction.

Ecology and Classification of North American Freshwater Invertebrates
Copyright © 1991 by Academic Press, Inc.

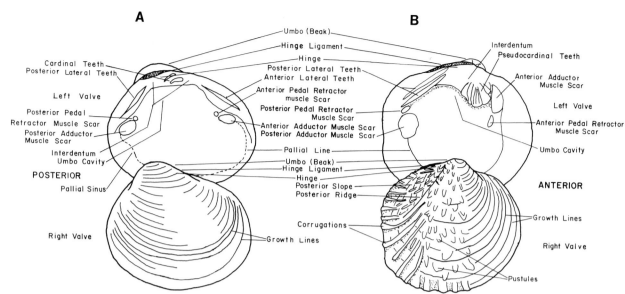

Figure 11.1 General morphologic features of the shells of (A) corbiculacean and (B) unionacean freshwater bivalves.

The left and right mantle lobes and overlying shell valves extend anteriorly, posteriorly, and ventrally to enclose the entire body of the bivalve. Surrounded by the valves, cephalic sensory structures have become vestigial or lost and external sensory functions relegated to the mantle edge, which is exposed directly to the external environment. Bivalves, compared to other molluscs, are laterally compressed and expanded dorso-ventrally. Being entirely enclosed within the shell valves and mantle protects soft tissues from sediment abrasion and prevents invasion of the mantle cavity by fine sediment, which could interfere with ciliated gill-filtering mechanisms. This adaptation, in conjunction with both lateral body compression and evolution of a highly extendable, spadelike foot adapted for burrowing, has allowed bivalves to become perhaps the most successful infaunal filter feeders of marine and freshwater habitats.

The bivalve fauna of North American freshwaters is the most diverse in the world, consisting of 260 native and 6 introduced species (Burch 1975a,b). This diversity is greatest among unionacean mussels (superfamily Unionacea, 227 native species in 44 genera), many species of which have unique morphological adaptations and highly endemic, often endangered populations. In contrast, the sphaeriids (family Sphaeriidae, 33 native and 4 introduced species in four genera) have far fewer species and genera but are more widely distributed and cosmopolitan than most unionaceans (several sphaeriid genera and species have pandemic distributions). North American freshwaters have also been invaded by an exotic southeast Asian species, *Corbicula fluminea*

(family Corbiculidae), which has spread throughout the freshwater drainage systems of the coastal and southern United States and Mexico, becoming the dominant benthic species in many habitats (Counts 1986, McMahon 1983a). Another exotic species, *Dreissena polymorpha,* the zebra mussel, was discovered in Lake St. Clair and Lake Erie in 1988. Based on the size–age structure of present populations, it appears to have been introduced to the Great Lakes from Europe/Asia in 1985 or 1986 (Hebert *et al.* 1989, Mackie *et al.* 1989).

II. ANATOMY AND PHYSIOLOGY

The majority of North American freshwater bivalve species fall into two superfamilies: the Corbiculacea and the Unionacea. The general external and internal features of these two groups are relatively similar (Fig. 11.2). Indeed, the anatomical features of lamellibranch bivalves are highly sterotypic and, therefore, will be discussed in general terms here.

A. External Morphology

1. Shell

The bivalve shell is composed of nonliving calcium carbonate ($CaCO_3$) crystals embedded in a proteinaceous matrix, both secreted by the underlying living mantle. The shell of most bivalves consists of three distinct portions: an outer proteinaceous periostracum secreted from the periostracal groove in the mantle edge, an underlying prismatic layer, and

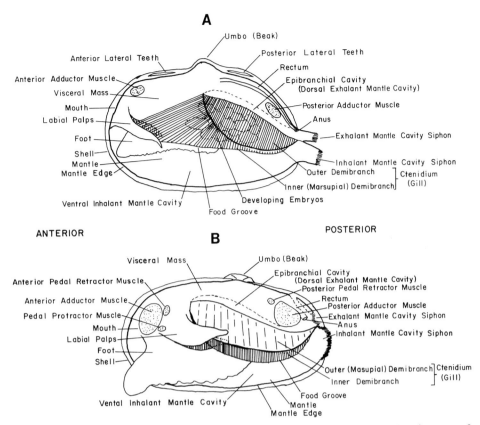

Figure 11.2 General external anatomy of the soft tissues of (A) corbiculacean and (B) unionacean freshwater bivalves.

inner layer of nacre. The last two layers are formed from $CaCO_3$ crystals in an organic matrix. Sphaeriids uniquely lack the nacreous layer. The periostracum is initially secreted free from other shell material, but soon fuses with the underlying, primarily calcareous prismatic layer secreted by a portion of the mantle edge just external to the periostracal groove (Fig. 11.3). The prismatic shell layer is a single layer of elongated calcium carbonate crystals oriented at 90° to the horizontal plane of the shell (Fig. 11.3). The free edge of the periostracum seals the extrapallial space between the mantle and shell from contact with the external medium, allowing $CaCO_3$ concentrations in extrapallial fluid to reach the saturation level required for crystal deposition (Saleuddin and Petit 1983). In unionaceans, the tripartite periostracum is secreted as an outer layer forming the external proteinaceous surface of the shell, and middle layer apparently involved with formation and organization of the prismatic layer, and an inner layer association with the initial formation of the nacreous layer at the growing shell edge (Saleuddin and Petit 1983). The periostracum is relatively impermeable to water, preventing dissolution of $CaCO_3$ from the shell surface. The nacreous (pearly) shell layer is continuously secreted by the

underlying mantle epithelium and consists of consecutive layers of small $CaCO_3$ crystals parallel to the plane of the shell (Fig. 11.3). Accumulation of nacreous layers through time thickens the shell, accounting for its strength and rigidity.

Calcium (Ca^{2+}) and bicarbonate (HCO_3^-) ions necessary for deposition of shell $CaCO_3$ crystals are transported from the external medium across the external epithelium into the hemolymph (blood). Bicarbonate ions are also generated from metabolically released CO_2 ($CO_2 + H_2O \leftrightarrow H^+ + HCO_3^-$). The mantle transports these ions from the hemolymph into the extrapallial fluid where they are deposited as shell crystals (Wilbur and Saleuddin 1983).

Formation of $CaCO_3$ crystals requires release of protons (H^+) ($Ca^{2+} + HCO_3^- \leftrightarrow CaCO_3 + H^+$), which must be removed from the extrapallial fluid to maintain the high pH required for $CaCO_3$ deposition (pH range 7.4–8.3). A proposed mechanism involves the combination of H^+ with HCO_3^- to form H_2CO_3, then its dissociation into CO_2 and H_2O, which diffuse into the hemolymph. The presence in mantle tissue of the enzyme carbonic anhydrase, which catalyzes this latter reaction, is evidence for this mechanism (Wilbur and Saleuddin 1983). Extra-

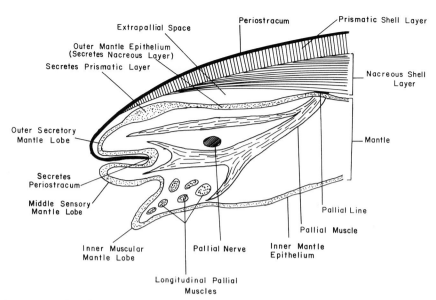

Figure 11.3 A cross-section through the mantle and shell edges of a typical freshwater unionacean bivalve displaying the anatomic features of the shell, mantle, and mantle edge. Sphaeriids have a complexed cross-lamellar shell structure and lack nacre but their mantle edge has a similar structure.

pallial fluid pH is higher is freshwater bivalves than marine bivalves, favoring $CaCO_3$ deposition at the lower Ca^{2+} concentrations characteristic of the dilute hemolymph concentration of freshwater species (Wilbur and Saleuddin 1983).

Shell $CaCO_3$ and organic matrix material precipate from the extrapallial fluid. In freshwater bivalves, Ca^{2+} and HCO_3^- are actively concentrated in the extrapallial fluid, favoring deposition as $CaCO_3$ crystal (Wilbur and Saleuddin 1983). The organic shell matrix, which separates individual crystals and binds them and crystal layers into a unified structure, is also involved with crystal formation. It has crystal-nucleating sites (possibly calcium-binding polypeptides) on which $CaCO_3$ crystals initially develop, eventually growing to form a new, nacreous shell layer (Wilbur and Salueddin 1983).

It has been proposed that a minimum of one ATP molecule is required for every two Ca^{2+} ions deposited, with additional ATP required for active HCO_3^- transport (Wilbur and Saleuddin 1983). Thus, fast-growing species or those with massive shells must devote a relatively high proportion of maintenance energy to shell mineral deposition. Deposition of the proteinaceous shell matrix (including the periostracum) also demands energy, requiring four ATP for each peptide bond formed. Although rarely more than 10% of shell dry mass, the highly condensed shell matrix can account for one-third to one-half of the total body dry organic matter (shell + tissue organic matter) or up to one-third of the total energy devoted to growth (Wilbur and Saleuddin 1983).

Thus, fast-growing, thin-shelled species devote proportionately far less energy to shell production than do slower-growing, thick-shelled species, allowing allocation of a greater amount of energy to growth and reproduction (thereby increasing fitness). However, a thinner, more fragile shell increases the probability of predation or lethal desiccation during emergence, reducing fitness. Therefore, the balance struck between energy allocation to shell formation and tissue growth in a species represents an adaptive strategy, evolved under species-specific niche selection pressures (see Section III.B).

The shell surface may be marked with concentric or radial corrugations, ridges, or pigmented rays or blotches in the periostracum. These, along with the shape of the posterior ridge and outline shape of the shell are diagnostic taxonomic characters. Another major external feature is the umbos or beaks, anteriorly curving, dorsally expanded structures, representing the oldest portions of the shell valves (Fig. 11.1).

Internally, major shell features include the hinge and projecting hinge teeth, which interlock to hold the valves in exact juxtaposition and form the fulcrum on which they open and close, and various muscle insertion scars. These are also major diagnostic characters (see Section V). In corbiculaceans, massive conical cardinal teeth form just below the umbos (one in the right valve and two in the left). Anterior and posterior to these lie the lateral teeth, usually elongated lamellae (Fig. 11.1A). Unionaceans have no true cardinal teeth. Instead, massive, raised, pseudocardinal teeth develop from anterior lateral teeth just anterior to the umbos and serve a similar function. Elongated, lamellar, posterior lateral teeth extend posterior to the umbos (Fig. 11.1B). In the Anodontinae, hinge teeth are vestigial or lost. The internal nacreaus layer may have species-specific colors. Muscle scars mark the insertion points of the anterior and posterior shell adductor muscles, the anterior and posterior pedal retractor muscles, and anterior pedal protractor muscles, and the pallial line muscles, which attach the margin of the mantle to the shell (Figs. 11.2 and 11.3). Posteriorly, the pallial line may be indented, marking the pallial sinus into which the siphons are withdrawn (Fig. 11.1A).

2. Locomotory Structures

With the exception of the epibenthic *Dreissena polymorpha*, the vast majority of North American freshwater bivalves are burrowers in benthic sediments. Some species can be found above the substratum on exposed rocky bottoms, but they still use the foot to wedge into crevices or under rocks. In all cases, locomotion is achieved with a highly muscular, flexible, protrusible anterioventrally directed foot (Figs. 11.1 and 11.2).

Bivalve burrowing mechanics have been described in detail by Trueman (1983). The burrowing cycle is initiated with relaxation of the adductor muscles, allowing shell valves to be forcibly opened against the surrounding substratum by expansion of the hinge ligament, anchoring the bivalve in place. Contraction of the transverse and circular muscles around the foot hemolymph sinuses, acting as hydraulic skeletons, then causes the foot to decrease in diameter and to elongate, forcing it forward into the substratum as the open valves are wedged in place against the burrow walls. Once extended into the substratum, the distal end of the foot is expanded by a flow of blood into its hemocoels, anchoring it in its new position in the substratum. Adductor muscle contraction then rapidly closes the valves, releasing their hold on the burrow walls and forcibly expelling water from the pedal gape downward into the sediments. This blast of water loosens compacted sediments at the anterior edge of the shell. Thereafter, anterior and posterior pedal retractor muscle contractions pull the shell forward into the loosened sediments against the anchored foot tip. Once a new position is achieved, the adductor muscles again relax to reanchor the shell valves against the substratum and the burrowing cycle is repeated. As North American bivalves have relatively short siphons (highly reduced or absent in some *Pisidium*), they generally burrow to depths where the posterior shell

margin is either buried just beneath the sediment surface or extended slightly above it.

Many juvenile freshwater bivalves can crawl over the substratum surface for considerable distances before settlement. Indeed, juveniles of some species can traverse even relatively smooth vertical surfaces. Crawling is achieved by extension of the foot, anchoring its tip with mucus and/or a muscular attachment sucker, followed by contraction of pedal muscles to pull the body forward. The capacity for surface locomotion is greatly reduced in most adult unionaceans, but retained to varying degrees in adult sphaeriids and *Corbicula fluminea*. The zebra mussel, *Dreissena polymorpha*, though normally attached to hard substrata by proteinaceous threads (the byssus), is capable of extensive surface locomotion. Such locomotion is particularly common in juvenile and immature specimens. After shedding its byssus, this mussel employs pedal locomotion to move to a new location where the byssal attachment is reformed (Mackie *et al.* 1989).

B. Organ System Function

1. Circulation

Circulation in bivalves has been reviewed by Jones (1983). Bivalves have an open circulatory system, in which the circulatory fluid is not continually enclosed in vessels (i.e., a closed circulatory system), but rather passes through open, spongy hemocoels (blood sinuses) in which the hemolymph bathes the tissues directly before returning to the heart. The open circulatory system of bivalves is associated with large hemolymph volumes, which account for 49–55% of total corporal water (Burton 1983, Jones 1983). The ventricle of the heart uniquely surrounds the rectum and pumps oxygenated hemolymph from the gills and mantle via the kidney through anterior and posterior blood vessels (Fig. 11.4), which subdivide into many smaller vessels including pallial arteries to the mantle and visceral arteries to the body organs and foot. The arteries further subdivide into many tiny vessels that eventually open into the hemocoels where cellular exchange of nutrients, gases, and wastes occurs. Blood returns via hemocoels to the gills and mantle and thence to the heart. Evolution of an open circulatory system in molluscs has been associated with reduction of the coelom. The ventricle is surrounded by one of the few coelomic remnants, the pericardial cavity, surrounded by the pericardial epithelium (pericardium). The only other coelomic remnants are spaces comprising the kidneys and gonads.

The hemolymph of freshwater bivalves, as in most lamellibranch species, has no specialized re-

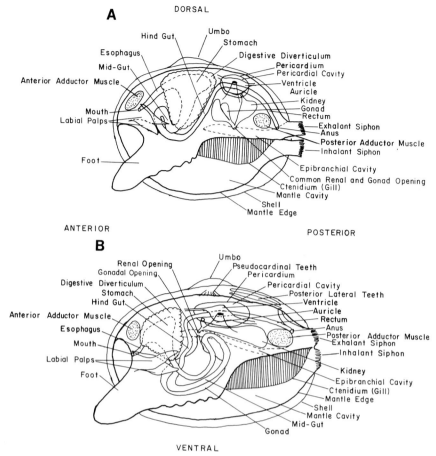

Figure 11.4 General internal anatomy, organs, and organ systems of the soft tissues of (A) corbiculacean and (B) unionacean freshwater bivalves.

spiratory pigments for O_2 transport (Bonaventura and Bonaventura 1983). Instead, oxygen is dissolved directly in hemolymph fluid, making its O_2 carrying capacity essentially that of water. However, the very low metabolic rates, reduced oxygen demands, and extensive gas-exchange surfaces of bivalves allow maintenance of a primarily aerobic metabolism, in spite of reduced hemolymph O_2 carrying capacity.

2. Gills and Gas Exchange

In lamellibranch bivalves, including all North American species, the gills (Fig. 11.2) are greatly expanded beyond requirements for respiratory gas exchange, as they are also utilized for filter feeding (i.e., suspension feeding), the main method of food acquisiton for the majority of species (see Section III.C.2). The gill axes extend anteriolaterally along either side of the dorsal portion of the visceral mass. Many long, thin, inner and outer gill filaments extend from either side of the gill axes. The gill filaments are fused together (an evolutionarily ad-

vanced condition) but penetrated by a series of pores or ostia (i.e., the eulamellibranchiate condition). From the axis, the filaments first extend ventrally (descending filament limbs) and then reflect dorsally (ascending filament limbs), attaching distally to either the dorsal mantle wall (outer gill filaments) or the dorsal side of the visceral mass (inner gill filaments) to form two v-shaped curtains known as the inner and outer demibranchs. The demibranchs completely separate the mantle cavity into ventral inhalent and dorsal exhalant portions (Fig. 11.5). The descending and ascending portions of the filaments are periodically cross-connected by tissue bridges called interlamellar junctions. The area between descending and ascending filaments is called a water tube or interlamellar space (Fig. 11.5). Gill feeding and respiratory currents are maintained by lateral cilia on the sides of the external surfaces of the filaments (Fig. 11.5A), which force water through the ostia into the water tubes. Inhalent water passes through the gill ostia to the water tubes to be carried to the dorsal exhalant mantle cavity or epibranchial cavity (formed by connection of the

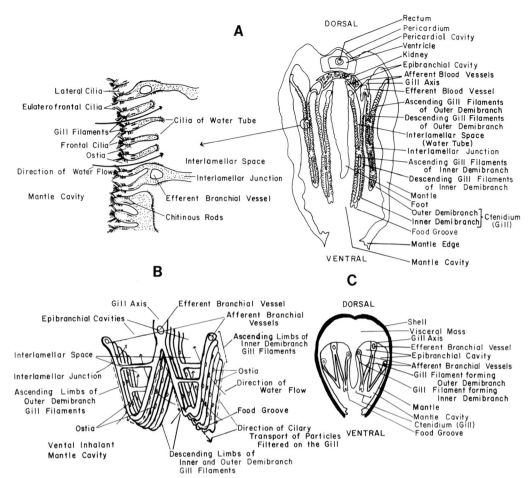

Figure 11.5 The structural features of the gills (ctenidia) of freshwater bivalves. (A) Cross-section through the central visceral mass, ctenidia, and mantle of a typical freshwater unionid, with high-magnification view showing details of filaments, ostia, and ciliation. (B) Diagrammatic representation of the respiratory and feeding water currents across the ctenidium. (C) Diagrammatic cross-sectional representation of lamellibranch bivalve ctenidia.

distal ends of the inner and outer demibranchs to the mantle wall or visceral mass, respectively) where it flows posteriorly to exit via the exhalant siphon (Figs. 11.2 and 11.5).

3. Excretion and Osmoregulation

Freshwater bivalves, as all freshwater animals, have hemolymph and tissue osmotic concentrations greater than their freshwater medium, resulting in a constant ion loss and water gain. This osmotic problem is compounded in bivalves by extensive mantle and gill surface areas over which such water and ion flux can occur (Burton 1983, Dietz 1985). To reduce ion and water fluxes, freshwater bivalves have the lowest hemolymph and cell osmotic concentrations of any metazoan, being 25–50% of that found in most other freshwater species (Burton 1983, Dietz 1985). In spite of these characteristics, the extensive

epithelial surface areas of bivalves cause water and ion fluxes to be greater than in other freshwater species (urine clearance is 20–50 ml/kg/hr) (Dietz 1985).

In unionaceans, sodium is taken up in exchange for outward transport of hemolymph cations such as H^+ or NH_4^+ and, perhaps, Ca^{2+}. Chloride ion uptake is in exchange for HCO_3^- or OH^-. Active Ca^{2+} uptake has also been reported (Burton 1983). In freshwater snails, the major source of calcium ions is ingested food (McMahon 1983b), however, the relative roles of food and external medium as sources for shell Ca^{2+} are unknown in freshwater bivalves. Sodium uptake does not require the presence of Cl^-, indicating that the transport systems of these two ions operate independently (Burton 1983). Active transport appears to be the major route by which unionaceans gain ions from the medium, but in *Corbicula fluminea* exchange diffusion (transport of an ion in one direction in exchange for diffusion of a second ion species down its concentration gradient in the opposite direction) accounts for 67% of Na$^+$ uptake. *Corbicula fluminea* has much higher ion transport rates and hemolymph ionic concentrations than do unionaceans, reflecting its geologically recent penetration of freshwaters (Dietz 1985). Interestingly, exchange diffusion may account for up to 90% of Cl$^-$ turnover in unionaceans in pond water, but when individuals are salt-depleted, active transport dominates Cl$^-$ uptake (Dietz 1985). Na$^+$ and Cl$^-$ uptake can occur over the general epithelial surface of unionaceans, but the majority occurs over the gills (Dietz 1985).

Excess water is eliminated through the renal organs or kidneys. The walls of the auricles initially ultrafilter the blood. Under the hydrostatic pressure generated by auricular contraction, blood fluid containing ions and small organic molecules passes as a fluid essentially isosmotic and isoionic to the hemolymph through the auricle walls into the pericardial cavity. Filtration appears to occur through podocyte cells of the pericardial gland lining the inner auricular surface and perhaps through the efferent branchial vein, carrying hemolymph from the longitudinal vein of the kidney to the auricles (Martin 1983) (Fig. 11.6). Only larger blood proteins, lipids, and carbohydrate molecules cannot pass the pericardial gland filter. The filtrate passes from the pericardial cavity through left and right renopericardial openings in the pericardial wall into the renopericardial canals and thence into the left and right renal organs or kidneys. Larger organic waste molecules enter the filtrate by active transport across the kidney walls. The excretory fluid is released through nephropores (Fig. 11.6) opening into the epibranchial cavities to be carried out the exhalant siphon (Figs. 11.2 and 11.4A) (Martin 1983).

While not studied in freshwater bivalves, it is presumed that the bivalve kidney is the site of major active ion resorption from the filtrate back into the hemolymph as in freshwater snails (Little 1985). As kidney walls appear relatively impermeable to water, active ion resorption from the filtrate allows formation of a dilute excretory fluid (filtrate osmo-

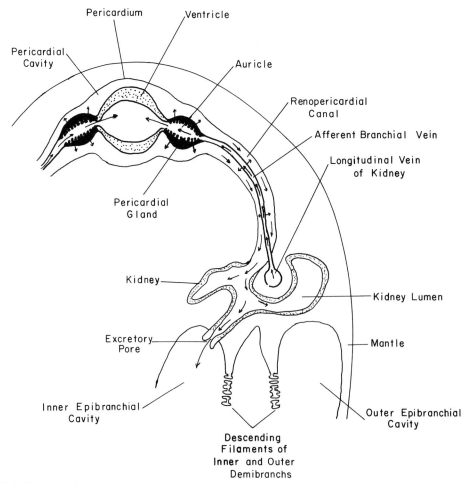

Figure 11.6 Cross-sectional representation of the anatomic features of the excretory system of a typical freshwater bivalve. Arrows indicate pathways for the excretion of excess water in the hemolymph through the excretory system to be eliminated at the excretory pore. (Redrawn from Martin 1983.)

larity is 50% that of the hemolymph in the unionid *Anodonta cygnea*, Martin 1983) facilitating removal of excess water. Filtrate ion resorption is energetically less expensive than recovery from freshwater because concentration gradients between the kidney filtrate and hemolymph are reduced relative to those between hemolymph and freshwater. Excretory fluid production is high, approximately 0.03 ml/g wet tissue/day in *A. cygnea*. In spite of renal absorption of major ions, the excreted filtrate has a higher ionic concentration than freshwater, thus ions are lost with urine excretion. These ions and those lost by diffusion over body and gill surfaces must be recovered by active transport from the medium across epithelial surfaces (particularly those of the gills) to the hemolymph (for details see Section II.C.6).

4. Digestion and Assimilation

As lamellibranchs, the vast majority of freshwater bivalves are suspension feeders, filtering unicellular algae, bacteria, and suspended detrital particles from the pallial water flow across the gill. Material filtered on the gill is passed to the labial palps for cilia-mediated sorting of food from nonfood before being carried on ciliary tracts to the mouth. Some species also have mechanisms for gathering organic detrital particles from the substratum. Filter and detritus feeding are described in Section III.C.2. The present discussion is devoted to the processes of digestion and assimilation.

The bivalve mouth is a simple opening flanked laterally by left and right pairs of labial palps, whose ciliary tracts deliver food to the mouth as a constant stream of fine particles. Food entering the mouth passes a short ciliated esophagus where it is bound into a mucus rope before entering the stomach. The stomach, which lies in the anterior-dorsal portion of the visceral mass (Fig. 11.4), is a complex structure, containing ciliated sorting surfaces and openings to a number of digestive structures and organs. On its ventral floor, posterior to the midgut opening, is an elongated evagination of the stomach wall called the style sac. Cells at the base of the style sac secrete the crystalline style, a long mucopolysaccharide rod projecting dorsally from the style sac into the stomach. The cells lining the style sac secrete digestive enzymes into the style matrix. Their cilia function to rotate the crystalline style slowly. The stomach and style have pH levels ranging from 6.0–6.9, style acidity varying with phase of digestion (Morton, 1983).

The free end of the style projects against the roof of the stomach where it rotates against a chitinous plate, the gastric shield. The gastric shield is pene-

trated by microvilli from epithelial cells underlying it. These cells are considered to secrete digestive enzymes through these microvilli onto the shield surface (Morton 1983). Rotation of the style mixes stomach contents and winds the esophageal mucus thread containing freshly ingested food onto the free end of the style where slow release of its embedded enzymes begins the process of extracellular digestion. Wear from abrasion of the style tip against the gastric shield causes attached food particles to be carried dorsally to the gastric shield. Here, style tip rotation causes food to be triturated in direct contact with digestive enzymes concentrated by release from the eroding style matrix and epithelia underlying the gastric shield.

After this initial trituration, food particles released into the stomach may have several fates. Particles entering the midgut may either be passed to the hindgut and thence to the rectum to be egested, or returned back to the stomach from the midgut for further extracellular breakdown. This size-based particle sorting is effected by a ciliated ridge, the typhlosole, running the length of the midgut. Sufficiently small particles are eventually carried on ciliary tracts in the stomach to the digestive diverticulum for the final intracellular phase of digestion (Fig. 11.4); these diverticula have the lowest fluid pH of any portion of the gut (Morton 1983). Larger particles are recycled to the stomach for further trituration and enzymatic digestion. Thus, particles may pass over gut ciliated sorting surfaces and be exposed to digestive processing several times before acceptance for assimilation or rejection as feces.

Digestive cells lining the lumina of terminal tubules in the digestive diverticulum take up fine food particles by endocytosis into food vacuoles, where the final stages of digestion and absorption take place. After completion of this intracellular phase of digestion, the apical portions of digestive cells, which contain vacuoles with undigested wastes, are shed into the tubule lumina to be carried as fragmentation spherules into the stomach by ciliated rejection pathways. The breakdown of these spherules in the stomach is hypothesized to be a major source of stomach acidity and extracellular digestive enzymes (Morton 1983).

Undigestible matter passes through the relatively short hindgut into the rectum and out the anus, which opens into the epibranchial cavity on the posterior face of the posterior adductor muscle near the exhalant siphon. Feces are expelled on exhalant pallial currents (Fig. 11.4). Mucous secreted by hindgut and rectal cells binds undigested particles into discrete feces before egestion, preventing recirculation of fecal material with inhalant currents.

The cerebropleural and visceral ganglia of

freshwater bivalves release neurohormones influencing glycogenesis (Joose and Geraerts 1983). The
vertebrate glycogenetic hormones, insulin and
adrenalin, have effects similar to those in vertebrates when administered to unionaceans. Injection
of vertebrate insulin into *Lamellidens corrianus*
caused decreased blood glucose levels and an increase in foot and digestive diverticulum glycogen
stores, while injection of adrenaline induced
breakdown of glycogen stores and increased blood
sugar levels (Jadhav and Lomte 1982b). Gut epithelial cells produce an insulin-like substance (ILS) in
Unio pictorum and *Anodonta cygnea,* the release of
which is stimulated by elevated hemolymph glucose
levels. ILS, in turn, stimulates activity of glycogen
synthetase, an enzyme involved in the uptake of
glucose into cellular glycogen stores, leading to return of hemolymph glucose to normal levels (Joose
and Geraerts 1983). Cerebropleural ganglionic neurosecretory hormones regulate accumulation and release of proteinaceous and nonproteinaceous energy
stores from the digestive diverticulum and foot of
Lamellidens corrianus (Jadhav and Lmote 1983).

5. *Reproductive Structures*

The gonads of freshwater bivalves, as in almost all
bivalves, are paired and lie close to the digestive
diverticulum. In unionaceans, they lie so closely
together that the paired condition is often hard to
discern. Unionacean gonads envelop the lower portions of the intestinal tract and sometimes extend
into proximal portions of the foot (Fig. 11.4B), while
in sphaeriids and *Corbicula fluminea,* gonads lie
more dorsally in the visceral mass, extending along
the stomach, intestine, and digestive diverticulum
(Fig. 11.4A) (Mackie 1984). The gonoducts from
each gonad are short, and open into paired gonopores in the epibranchial mantle cavity allowing released gametes to be discharged from the exhalant
siphon (Fig. 11.4). In freshwater unionaceans,
which are generally gonochoristic (except for a few
hermaphorditic species of *Anodonta*), the tracts and
openings of the renal and reproductive systems are
entirely separate. This is considered to be an evolutionarily advanced condition (Fig. 11.7C). In sphaeriids and *C. fluminea,* which are hermaphroditic, the
gonads are comprised of distinct regions or zones in
which either male or female acini produce eggs or
sperm. In these groups, ducts carrying eggs or
sperm unite into a single gonoduct from each gonad.
These open either into the distal end of the kidney—
allowing discharge of gametes through the renal canal and nephridiopore (Fig. 11.7A)—or into the
gonoduct and nephridial canal, which discharge
through a common pore on a papilla extending into

the epibranchial cavity (Fig. 11.7B) (Kraemer *et al.*
1986, Mackie 1984). The zebra mussel, *Dreissena
polymorpha,* is gonochoristic.

Almost all North American freshwater bivalve
species are ovoviviparous, brooding embryos
through early development stages in the gill. The
single exception is *D. polymorpha,* which releases
both sperm and eggs externally, leading to external
fertilization and development of a free-swimming
veliger larval stage (Mackie *et al.* 1989). In brooding
species, the interlamellar spaces of the demibranchs
are modified to form marsupia (brood chambers).
Fully formed juveniles are released from the exhalant siphon in sphaeriids and *C. fluminea.* In unionaceans, bivalved glochidium larvae are released
from the exhalant siphon, from specialized gill
pores, or by rupture of the ventral portion of the gill.
Glochidia parasitize a fish host before metamorphosing into a juvenile. The outer demibranchs form
the marsupium in most unionacean species except
for members of the subfamily Ambleminae, in which
marsupia are formed in both the inner and outer
demibranches (Burch 1975b). The inner demibranchs serve this purpose in sphaeriids and *C.
fluminea.* Members of the unionid genus *Lampsilis*
develop marsupia only in the posterior portion of the
outer demibranchs and release glochidia from small
pores in the marsupial demibranch directly into the
inhalant mantle cavity. In *C. fluminea,* developing
embryonic stages are brooded directly within inter-

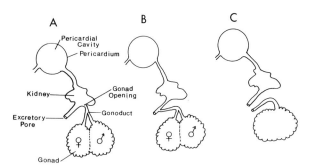

Figure 11.7 Schematic representations of typical reproductive systems of bivalves. (A) The primitive condition in some marine bivalve species: male or female gametes are released from the gonoduct opening into the
kidney and passed externally through the excretory pore
(shown here is a primitive marine hermaphroditic bivalve;
the anatomy is essentially similar in gonochoristic species). (B) Hermaphroditic freshwater corbiculacean bivalves (*Corbicula fluminea* and sphaeriids): male and female gametes are passed through the gonoduct into the
kidney duct close to the excretory pore from which gametes are shed. (C) Gonochoristic freshwater unionids:
gametes pass to the outside through a gonoduct and
gonopore totally separate from the kidney duct and excretory pore. (After Mackie 1984).

lamellar spaces without other specialized brooding structures. In sphaeriids, fertilized eggs are enclosed in specialized brood chambers formed from evaginations of the gill filaments into the interlamellar space. In anodontid unionaceans, the interlamellar space of the marsupial demibranchs is divided by septa into a separate interlammelar cavity associated with each filament. Each interlamellar cavity is further divided into inner and outer water tubes carrying water to the epibranchial cavity and a central marsupium containing developing glochidia, an advanced tripartite structure that does not occur in other unionaceans. In the primitive Margartiferidae and the Ambleminae, the entire outer demibranch forms the marsupium, while in the Pleurobemini and Lampsilini, marsupia are limited to just a portion of outer demibranch (Burch 1975b, Mackie 1984).

Some unionaceans display sexual dimorphism, a characteristic generally rare in other freshwater bivalves. In such species, incubation of glochidia results in massive distension of the outer demibranch marsupium (Mackie 1984). In female lampsilids and dysonomids, the posterior portions of the valves are greatly inflated (relative to the condition in males) to afford space for the expanded posterior marsupia of the outer demibranchs (Fig. 11.28A,H). Similar general inflation of female shells, but much less obvious, occurs in some anodontids (Burch 1975b, Mackie 1984).

Hermaphroditic unionaceans and all sphaeriids are generally simultaneous hermaphrodites, producing mature eggs and sperm concurrently in the gonads. *C. fluminea* has an unusual pattern of producing only eggs at earliest maturity (shell length ≈ 6 mm) followed somewhat later by sperm production. It then remains simultaneously hemaphroditic throughout the rest of life (Kraemer and Galloway, 1986).

As bivalves are relatively sessile and individuals can be widely separated from each other, copulatory organs would be useless and are therefore lacking. In freshwater species other than *D. polymorpha*, in which eggs are fertilized externally (Mackie *et al.* 1989), sperm is released to surrounding water to be taken up on the inhalant currents of other individuals and carried to unfertilized eggs retained in the gill marsupia. Self-fertilization appears to be relatively common in sphaeriids, occurring near the conjunction of male and female gonoducts (Mackie 1984). In *C. fluminea,* developing embryos occur in the lumina of gametogenic follicles and gonoducts, suggesting that self-fertilization can take place within the gonad itself (Kraemer *et al.* 1986). The ability to self-fertilize makes hermaphroditic species highly invasive, as described in Section III.B.

The main stimulus for reproduction appears to be temperature. Gametogenesis and fertilization begin when ambient temperature rises above a critical level or falls within critical limits. Other environmental factors that may affect reproduction but require further study are neurosecretory controls, density-dependent factors, diurnal rhythms, and parasites. While clear evidence for neurosecretory and density controls of reproductive cycles has been demonstrated in sphaeriids (Mackie 1984), such observations have not been made for other freshwater bivalves. Certainly, evidence of increasing activity and metabolic rate during dark hours in unionaceans (McCorkle *et al.* 1979) and *C. fluminea* (McCorkle-Shirley 1982) suggests that spawning activity and glochidial or juvenile release rates may also display diurnal rhythmicity.

The sperm of freshwater bivalves may have ellipsoid or conical nuclei and an acrosome of variable complexity depending on species (Mackie 1984). The sperm of corbiculacean species have elongate heads and that of *C. fluminea* is unusually biflagellate (Kraemer *et al.* 1986). Sperm with elongate heads appear adapted for swimming in gonadal and oviductal fluids more viscous than water and are associated with internal fertilization in gonadal ducts rather than externally in marsupia (Mackie 1984).

The eggs of freshwater bivalves are round and generally larger with greater yolk stores than marine species with planktonic larval stages. Again, the exception is *D. polymorpha* with relatively small eggs (40–70 μm diameter) associated with retention of a free-swimming veliger, which feeds and grows considerably in the water column before settlement and juvenile metamorphosis (Mackie *et al.* 1989). The larger, yolky eggs of all other species contain nourishment required for development to the juvenile or glochidium stage. The Sphaeriacea produce the largest eggs and have, correspondingly, the smallest brood sizes ranging from 6–24/adult in *Sphaerium*, 1–135/adult for *Musculium* and 3.3–6.7/adult for *Pisidium* (Burkey 1983). Adult *Musculium partumeium,* which are only 4 mm in length release juveniles 1.4 mm long (Hornbach *et al.* 1980, Way *et al.* 1980). In contrast, unionaceans and *C. fluminea* have smaller eggs and release smaller glochidia or juveniles than sphaeriids (generally < 0.2 mm diameter) and have much larger brood sizes (10^3–10^6/adult) (Burky 1983, McMahon 1983a). The evolutionary implications of these major differences in fecundities of freshwater bivalves are discussed in Section III.B.

The egg is surrounded by a vitelline membrane that is relatively thin in sphaeriids and thicker in unionaceans and *C. fluminea* (Mackie 1984). In the latter two groups, it remains intact throughout most of development, but disintegrates during early development in sphaeriids (Heard 1977, Mackie 1984), an adaptation that may allow developing embryos to absorb maternally supplied nutrients from brood sacs without embryos and/or from nutrient cells lining the interlamellar space (Mackie 1978).

6. Nervous System and Sense Organs

In bivalves, the head is entirely enclosed within the valves and is, therefore, no longer in direct contact with the external environment. This condition has led to loss of cephalic structure, the head consisting of only the mouth opening and attachment points of the labial palps (Fig. 11.2). The cephalic sense organs found in other molluscan classes have been lost in bivalves, along with the associated cephalic concentration of central nervous system ganglia. Thus, the bivalve nervous system is far less centralized than in the majority of molluscs. A pair of cerebropleural ganglia lie on each side of the esophagus near the mouth, interconnected by commissures dorsal to the esophagus (Fig. 11.8). From these extend two pairs of nerve chords. A pair of dorsal nerve chords extend posteriorly through the visceral mass to a pair of visceral ganglia on the anterior-ventral surface of the posterior adductor muscle (Barnes 1986). The second pair of nerve chords extend ventrally to innervate a pair of pedal ganglia in the foot.

The pedal and cerebropleural ganglia exert motor control over the pedal and anterior adductor muscles, while control of the posterior adductor muscles and siphons is mediated by the visceral ganglia. Coordination of foot and valve movements, as occurs in burrowing behavior (see Section II.A.2), resides in the cerebrospinal ganglia.

Associated with enclosure and reduction of head sense organs is the development of sense organs in those tissues most directly exposed to the external environment, the mantle edge and siphons. On the mantle edge, sense organs reach their highest concentrations in the middle sensory mantle lobe (Fig. 11.3). Photoreceptor cells (but not distinct eyes as in some marine genera) detect changes in light intensity associated with shadow reflexes, phototaxis, and diurnal rhythms; while tentacles and stiffened immobile cilia associated with tactile mechanoreceptor sense organs perceive direct contact displacement (touch) or vibrations passing through the water. On the siphon margins, such tactile receptors prevent drawing of large particles on inhalant currents into the mantle cavity. When large particles contact these tentacles, the siphons are closed by sphincter muscles and then retract. Stronger stimuli cause the valves to close rapidly forcing water under high pressure from the siphons, thereby ejecting any impinging material. Under intense stimulation of the mantle or siphons, the siphons are withdrawn and valves closed tightly, a common predator defense in all freshwater bivalve species.

A pair of statocysts lying near or within the pedal ganglion of the foot are innervated by commissures from the cerebropleural ganglion (Fig. 11.8). Such statocysts are greatly reduced in sessile marine species (e.g., oysters), suggesting their importance to locomotion and burrowing in free-living freshwater species. Statocysts in freshwater bivalves have been described by Kraemer (1978). They are lined with ciliary mechanoreceptors responding to pressure exerted by a calcareous statolith or series of smaller

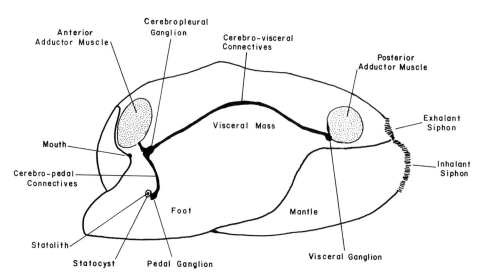

Figure 11.8 The anatomic features of the central nervous system of a typical unionid freshwater bivalve (central nervous system anatomy is essentially similar in freshwater corbiculacean bivalves).

granular statoconia within the statocyst vesicle. As gravity sensing organs, statocysts detect body orientation and thus function in bivalve geotactic and positioning responses particularly during burrowing and pedal locomotion.

C. Environmental Physiology

1. Seasonal Cycles

Freshwater bivalves display seasonal variation in physiological response associated with both temperature and reproductive cycles. While such cycles have been studied more thoroughly in marine species (Gabbott 1983), limited research on freshwater taxa has revealed interesting observations. Metabolic rates show major seasonal variation in freshwater bivalves (for reviews see Burky 1983, Hornbach 1985). As in almost all temperate zone animals, the metabolic rate in freshwater bivalves is generally greatest in summer and least in winter due to temperature effects (Burky 1983, Hornbach 1985). Annual variation in metabolic rate can be extensive (Table 11.1). Maximal summer oxygen consumption rates, or $\dot{V}_{O_2}$ may be 20–33 times the minimal winter rates over a seasonal range of

2–22°C in *Sphaerium striatinum* (Hornbach *et al.* 1983) or as little as 2.3 times the minimal winter rates in *Pisidium walkeri* (Burky and Burky 1976) (Table 11.1).

An immediate temperature increase causes a corresponding increase in the metabolic rate of ectothermic animals such as bivalves. Acute, temperature-induced changes in the metabolic rate or $\dot{V}_{O_2}$ (or in any rate function) are described by Q_{10} values (i.e., the factor by which a rate function changes with a 10°C increase in temperature) as follows:

$$Q_{10} = \frac{RATE_2{}^{(10/Temp_2 - Temp_1)}}{RATE_1}$$

where Rate$_1$ is the rate at the higher temperature, Rate$_2$, the rate at the lower temperature, Temp$_2$, the higher temperature and Temp$_1$, the lower temperature (°C). The Q_{10} for metabolic rate in the majority of ectothermic animals is 2–2.5, essentially that for chemical reaction rates. Thus, Q_{10} values outside this range indicate active metabolic regulation, values less than 1.5 suggest metabolic suppression with increasing temperature, and above 3.0, metabolic stimulation. Freshwater bivalve Q_{10} values are

Table 11.1 Seasonal Variation in the Oxygen Consumption Rates ($\dot{V}_{O_2}$) of Selected Species of Freshwater Bivalves

Species and Source	mg Dry Tissue Weight	Max. $\dot{V}_{O_2}$ μl O$_2$/hr	Min. $\dot{V}_{O_2}$ μl O$_2$/hr	Ratio Min:Max $\dot{V}_{O_2}$	$Q_{10(Acc.)}$[a]	Seasonal Temperature Range (°C)
Sphaerium striatinum	25 mg	31.5	1.40	22.5:1	4.7	2–22
Hornbach *et al.*	7 mg	15.8	0.78	20.3:1	4.5	2–22
(1983)	2 mg	8.7	0.24	33.6:1	5.8	2–22
Pisidium compressum	3 mg	0.71	0.04	17.5:1	—	—
Way and Wissing	0.9 mg	0.27	0.03	9.0:1	—	—
(1984)	0.05 mg	0.13	0.02	6.5:1	—	—
Pisidum variabile	3 mg	1.13	0.22	5.14:1	—	—
Way and Wissing	0.9 mg	0.42	0.10	4.20:1	—	—
(1984)	0.05mg	0.02	0.15	7.50:1	—	—
Pisidium walkeri	1.3 mg	0.85	1.11	2.3:1	1.4	1–26
Burky and Burky (1976)	0.02 mg	0.017	0.0074	2.3:1	1.4	1–26
Corbicula fluminea	348 mg	430.6	34.4	12.5:1	3.2	7–29
Williams (1985)	204 mg	308.2	31.5	9.8:1	2.8	7–29
	60 mg	143.3	25.8	5.6:1	2.2	7–29
Anodonta grandis	10 g	4690.0	400.0	11.7:1	2.7	6–31
Huebner (1982)	5 g	2690.0	250.0	10.8:1	2.6	6–31
Lampsilis radiata	5 g	2090.0	170.0	12.3:1	2.8	6–31
Huebner (1982)	2 g	810.0	80.0	10.1:1	2.6	6–31

[a] $Q_{10(Acc.)}$is the respiratory Q_{10} value computed from a change in the $\dot{V}_{O_2}$ value recorded for acclimated individuals at respective acclimation temperatures.

highly variable between and within species, ranging from 0.2–14.8 for 20 species of sphaeriid (Hornbach 1985). Far less information is available for union-aceans. In *Lampsilis radiata*, Q_{10} values range from 1.88–4.98 and for *Anadonta grandis,* from 1.27–10.35 (Huebner 1982). Temperature range of determination, body mass, and season may affect Q_{10} values in some species but not in others such that no general patterns emerge. Rather, metabolic response to temperature appears to have evolved under species-specific microhabitat selection pressures in freshwater bivalves (Hornbach 1985).

Without a capacity for regulation of metabolic rate, massive seasonal metabolic fluctuations could cause energetic inefficiency; rates being suboptimal during colder months and supraoptimal during warmer months. Thus, many ectothermic species display a capacity for metabolic temperature acclimation or compensation involving adjustment of the metabolic rate over a period of a few days to several weeks in a new temperature regime. Typically, metabolic rates are adjusted upward upon acclimation to colder temperatures and downward upon acclimation to warmer temperatures, which dampens metabolic fluctuation with seasonal temperature change, allowing maintenance of a more optimal metabolic rate throughout the year.

For most species, such typical seasonal acclimation is only partial, with metabolic rates not returning to absolutely optimal levels. Such partial metabolic temperature compensation can be detected by comparing Q_{10} values of $\dot{V}_{O_2}$ in instantaneous response to acute temperature change with those measured at the temperature of acclimation over a wide range of ambient temperatures (Acclimation Q_{10} or $Q_{10(Acc.)}$). If the $Q_{10(Acc.)}$ approximates 1.0, metabolic temperature compensation is nearly perfect with $\dot{V}_{O_2}$ regulated near the optimal level throughout the year. If $Q_{10(Acc.)}$ is less than 2.0 or considerably less than acute Q_{10}, acclimation is partial, with the metabolic rate approaching, but not reaching, the optimal level. If $Q_{10(Acc.)}$ is approximately equal to 2.0–2.5 or the acute Q_{10}, the species is incapable of temperature acclimation. If the $Q_{10(Acc.)}$ is greater than 3.0 or acute Q_{10}, inverse or reverse acclimation is displayed, in which acclimation to colder temperatures further depresses metabolic rate and acclimation to warmer temperatures further stimulates metabolic rate.

Three general patterns of metabolic temperature acclimation occur in freshwater bivalves; (1) no capacity for acclimation ($Q_1^0{}_{(Acc.)}$ equivalent to acute Q_{10}) displayed by the unionids *A. grandis* and *L. radiata* (Huebner 1982), (2) partial acclimation by Pisidium walkeri ($Q_{10(Acc.)}$ is considerably less

than maximal acute Q_{10}) (Burky and Burky 1976), and (3) reverse acclimation ($Q_{10(Acc.)}$ greater than acute Q_{10}) by *Sphaerium striatinum* (Hornbach *et al.* 1983) and *C. fluminea* (Williams 1985) (Table 11.1). The adaptive significance of reverse acclimation in bivalves is unclear (for hypotheses regarding its selective advantage, see McMahon 1983b), but for freshwater bivalves, that of reduction of energy store catabolism while inactive over winter appears most parsimonious (Burky 1983).

The $\dot{V}_{O_2}$ is also related to individual size or biomass in all animals as follows:

$$\dot{V}_{O_2} = aM^b$$

were $\dot{V}_{O_2}$ is oxygen consumption rate or metabolic rate, $M,$ individual biomass, and a and b are constants. The equation may be rewritten as a linear regression with $\dot{V}_{O_2}$ and M transformed into logarithmic values:

$$\text{Log}_{10}\,\dot{V}_{O_2} = a + b\,(\text{Log}_{10}\,M)$$

in which a and b are the Y-intercept ($\dot{V}_{O_2}$ at $\text{Log}_{10}\,M = 0$ or $M = 1$) and the slope (increase in $\text{Log}_{10}\,\dot{V}_{O_2}$ for each unit increase in $\text{Log}_{10}\,M$), respectively. Thus, a is a measure of the relative magnitude of $\dot{V}_{O_2}$ and b, the rate of increase with increasing biomass. If $b = 1$, $\dot{V}_{O_2}$ increases in direct proportion with M. If $b > 1$, $\dot{V}_{O_2}$ increases at a proportionately greater rate than M and if $b < 1$, $\dot{V}_{O_2}$ increases at a proportionately lesser rate than M. Thus, b values of less than one indicate that weight-specific $\dot{V}_{O_2}$ (O_2 uptake per unit body mass) decreases with increasing size and values of greater than one indicate that it increases with increasing size. Conventional wisdom states that animal b values range between 0.5–0.8. While generally true for vertebrates, it is less characteristic of invertebrates, including freshwater bivalves. Among 14 species of sphaeriids, b values ranged from 0.12–1.45 (Hornbach 1985). Limited data suggest that unionaceans have more typical b values of 0.90 for *L. radiata* and 0.77 for *Anodonta grandis* (Huebner 1982). The b value can also change with season in some species (Hornbach *et al.* 1983, Way and Wissing 1984), but remains constant in others (Burky and Burky 1976, Huebner 1982). It can vary with reproductive condition, increasing when adults brood juveniles in *P. compressum* (Way and Wissing 1982) and *Musculium lacustre* (Alexander 1982), but is not correlated with reproductive cycles in other species (Burky 1983, Hornbach 1985). The metabolic rate may also vary with physiologic state, declining in *Musculium partumeium* during midsummer when the habitats are dry and clams estivate (Burky 1983, Burky *et al.* 1985b, Way *et al.* 1981).

Comparison of *a* values in sphaeriids indicates that the metabolic rate of species of *Pisidium* (mean *a* = 0.399) is about 1/3 that of species of *Musculium* (means *a* = 1.605) or *Sphaerium* (mean *a* = 1.439) (Hornbach 1985). Reduced metabolic rate in *Pisidium* species may be related to their reduced gill surface area (Hornbach 1985), hypoxic mud burrowing, and interstitial suspension-feeding habits (Lopez and Holopainen 1987). The reduced metabolic demand of profundal pisidiids could account for their generally high tolerance of hypoxia (Burky 1983, Holopainen 1987, Holopainen and Jonasson 1983, Jonasson 1984a, 1984b).

Annually, values of *a* in *C. fluminea* varied from −0.12–1.43 (mean = 0.72) (Williams 1985). The annual range in values of *a* for *A. grandis* was (−0.13)–(−1.098) (mean *a* = −0.563) and for *L. radiata*, (−0.403)–(−1.331) (mean *a* = −0.800) (Huebner 1982). The high *a* values of *C. fluminea* relative to unionids and sphaeriids (range = 0.399–1.605) indicate that it has a higher $\dot{V}_{O_2}$ relative to other species, while the low values for unionaceans indicate a relatively reduced $\dot{V}_{O_2}$. Shell and tissue growth account for a major portion of metabolic energy utilization, being over 20% of total metabolic rate in young *Mytilus edulis* (Hawkins *et al.* 1989). Thus, in slow-growing unionacean species (see Section III.B.2), metabolic costs are depressed relative to the rapidly growing *C. fluminea*.

In some sphaeriid species, $\dot{V}_{O_2}$ is influenced by growth and reproductive cycles (Burky 1983, Hornbach 1985); maximal metabolic rates occur during periods of peak adult and brooded juvenile growth (Burky and Burky 1976, Hornbach *et al.* 1983, Way *et al.* 1981, Way and Wissing 1984). Peaking of metabolic rates during maximal growth and development of brooded juveniles may result from the elevated metabolic demands associated with accelerated tissue growth (Hawkins *et al.* 1989), the higher respiratory rates of brooded juveniles, and the energetic costs of providing maternal metabolites to brooded juveniles (Burky 1983, Mackie 1984). In contrast, metabolic rates in *C. fluminea* (Williams 1985) and unionids (Huebner 1982) are unaffected by embryo brooding. These species do not provide embryos with maternal nourishment (Mackie 1984), suggesting that it must have a high metabolic cost in sphaeriids.

Rates of filtration also vary seasonally in sphaeriids. In both *S. striatinum* (Hornbach *et al.* 1984b) and *M. partumeium* (Burky *et al.* 1985a), filtration rates decreased with increased particle concentration and decreased temperature. Elevated ambient temperatures induce maximal filtration during summer months, but there was also a tendency for filtra-tion to peak during periods of peak reproduction. In *M. partumeium,* the filtration rate declined in midsummer coincidently with the metabolic rate (Way *et al.* 1981) as individuals estivated during habitat drying (Burky 1983).

Freshwater bivalves also display distinct seasonal cycles in tissue biochemical composition. While data are sparse, variation in biochemical composition appears to be related primarily to the reproductive cycle, as occurs in marine bivalves (Gabbot 1983). In the freshwater unionid *Lamellidens corrianus,* whole body protein, glycogen, and lipid contents reach maximal levels during gametogenesis and gonad development and minimum levels during glochidial release (Fig. 11.9A); a pattern repeated particularly for protein and lipids in the majority of individual tissues (Fig. 11.9B–D) (Jadhav and Lomte 1982a). Similarly, overwintering, nonreproductive individuals of *C. fluminea* have twice the biomass and higher levels of nonproteinaceous energy stores than reproductive individuals in summer (Williams 1985, Williams and McMahon 1989). Thus, reproductive effort appears to require massive mobilization of organic energy stores from somatic as well as gonadal tissues to support gamete production in this species.

In the sphaeriids *Sphaerium corneum* and *Psidium amnicum,* tissue glycogen content increases after reproduction in the fall and was associated with a 2–3 fold increase in survival of winter-conditioned individuals under anoxic conditions. As glycogen is a major substrate for anaerobic metabolism (de Zwaan 1983), glycogen-poor, summer-conditioned individuals were relatively intolerant of anoxia (Holopainen 1987). Thus, fall accumulation of glycogen stores not only supports gametogenesis the following spring but also provides the anaerobic substrate for survival of prolonged anoxia when ice cover prevents surface gas exchange (Holopainen 1987).

2. Diurnal Cycles

Freshwater bivalves also display diurnal cycles of metabolic activity. Active uptake of Na^+ is greatest during dark hours in both *C. fluminea* (McCorkle-Shirley 1982) and the unionid, *Carunculina parva texasensis* (Graves and Dietz 1980). Diurnal Na^+ transport rhythm was lost in constant light, suggesting that rhythmicity of ion uptake is driven exogenously by changes in light intensity. Such diurnal ion transport rhythms appear to be closely linked to activity rhythms. In the unionid *Ligumia subrostrata,* valve gaping activity peaked 1–2 hr after onset of darkness and remained elevated during dark pe-

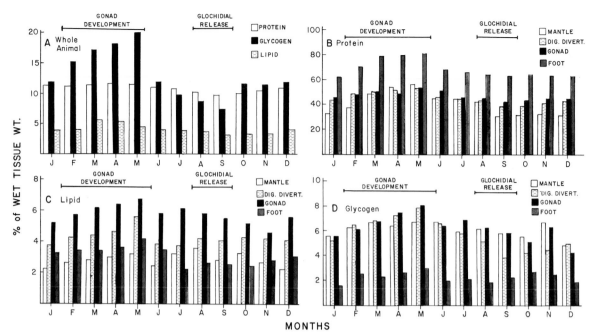

Figure 11.9 Seasonal variation in the protein, lipid, and glycogen contents of the whole body and various tissues of the freshwater unionid mussel, *Lamellidens corrianus,* relative to the reproductive cycle. All organic contents are expressed as percentage of wet tissue weight. (A) Annual variation in whole body contents of protein (open histograms), glycogen (solid histograms), and lipid (cross-hatched histograms). Remaining figures represent levels of protein (B), lipid (C), or glycogen (D) in the mantle (open histograms), digestive diverticulum (cross-hatched histograms), gonad (solid histograms), and foot (stippled histograms). Horizontal bars at the top of each figure represent reproductive cycles, indicating periods during which either gonads develop or glochidia are released. Gonad development is associated with increases in organic content, and glochidial release, with decreases in organic content of the whole body and various tissues. (From data of Jadhav and Lomte 1982a.)

riods. Rhythmic valve gaping behavior was lost in constant light, again suggesting primary responsiveness to exogenous changes in light intensity (McCorkle *et al.* 1979). Rhythmic patterns of oxygen consumption in *L. subrostrata* appeared driven by changes in light intensity, declining immediately after an increase in intensity and increasing after a decrease in intensity. Rhythmic respiratory behavior, however, had an endogenous component as it persisted in constant light for 14 days (McCorkle *et al.* 1979). This evidence suggests that at least some freshwater bivalves may be more active during dark hours. Such activity rhythms may be correlated with diurnal feeding and vertical migration cycles, individuals coming to the surface to feed at night and retreating below it during the day, thus avoiding visual predators; however this hypothesis requires future research.

3. Other Factors Affecting Metabolic Rate

Other than seasonal and temperature effects (described earlier), $\dot{V}_{O_2}$ can be suppressed by chemical pollutants such as heavy metals, ammonia, and cy-

anide, which degrade metabolic processes (Lomte and Jadhav 1982a). Increased levels of suspended solids impaired $\dot{V}_{O_2}$ and induced apparent starvation in three unionid species, indicating that suspended solids interfered with maintenance of gill respiratory and filter-feeding currents (Aldridge *et al.* 1987). Individual metabolic rates may also be dependent on population density. The metabolic rates of specimens of *Elliptio complanata* declined with density, the $\dot{V}_{O_2}$ of a single individual being three times that of individuals held in groups of seven or more. This species may release a pheromone that induces the reduction of the metabolic rate in nearby individuals (Paterson 1983).

A number of unionaceans and sphaeriids display varying degrees of respiratory oxygen dependency such that, when subjected to declining oxygen concentrations, $\dot{V}_{O_2}$ declines proportionately with a decline in ambient partial pressure of oxygen (P_{O_2}) (Burky 1983). Such species are generally intolerant of prolonged hypoxia and, thus, are restricted to well-oxygenated habitats. In contrast, some species are oxygen independent and regulate $\dot{V}_{O_2}$ at relatively constant levels with declining oxygen tension until a critical P_{O_2} is reached, below which $\dot{V}_{O_2}$ declines proportionately with further decline in O_2 concentration. Such species survive in aquatic habitats periodically subjected to prolonged hypoxia. The profundal sphaeriids *Sphraeium simile* and *Pisidium casertanum* are periodically exposed to hypoxia and are relatively oxygen independent (Burky 1983). Two unionids, *Elliptio companata* and *Anodonta grandis,* inhabiting mud and sand in a small eutrophic Canadian lake, were extreme oxygen regulators maintaining a nearly constant $\dot{V}_{O_2}$ down to 1 mg O_2/liter ($P_{O_2} \approx 18$ torr) (Fig. 11.10) (Lewis 1984). In these northern temperate species, the ability to regulate $\dot{V}_{O_2}$ is highly adaptive, as ice cover causes lentic habitats to become severely hypoxic during winter months and individuals overwinter burrowed deeply into hypoxic sediments (Lewis 1984). In contrast, more tropical species not experiencing winter hypoxia are more oxygen dependent (Das and Venkatachari 1984, McMahon 1979a). The $\dot{V}_{O_2}$ of the subtropical species, *C. fluminea,* approaches near zero levels after a decline in O_2 tension of just 30% of full air saturation, indicative of extreme oxygen dependence (McMahon 1979a). This response may account for the restriction of *C. fluminea* to well-oxygenated habitats and its intolerance of hypoxic waters receiving organic wastes (McMahon 1983a).

Many species of freshwater bivalves are very tolerant of acute hypoxia or even anoxia. Such tolerance is highly adaptive, as hypoxic conditions may occur below the thermocline of stratified lakes or above reducing substrata with heavy organic loads and/or dense animal populations (Butts and Sparks 1982). Profundal sphaeriids tolerate acute hypoxia throughout summer months after thermocline formation (Holopainen and Jonasson 1983, Jonasson 1984a, 1984b), surviving 4.5 to greater than 200 days of complete anoxia, depending on season and temperature (Holopainen 1987). The unionid, *Anodonta cygnea* can survive 22 days of anoxia (Zs.-Nagy *et al.* 1982).

When anoxic, bivalves rely on anaerobic metabolic pathways. These pathways are not those of glycolysis; instead, they involve alternative pathways, simultaneously degrading glycogen and

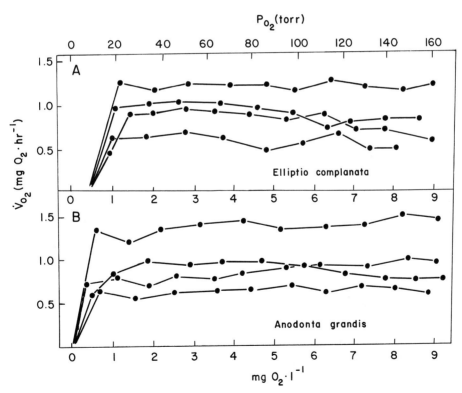

Figure 11.10 Respiratory responses of the freshwater unionid mussels, *Elliptio complanata* and *Anodonta grandis* to declining ambient oxygen tensions from near full air saturation (8–9 mg O_2/liter, lower horizontal axis, $P_{O_2} = 140$–160 torr or mg Hg, upper horizontal axis) to the concentration at which O_2 uptake ceases. Respiratory responses of four individuals of *E. complanta* (A) and four individuals of *A. grandis* (B). Both species maintained normal oxygen uptake rates at a P_{O_2} as low as 15–20 torr (9–13% of full air saturation) suggesting high capacity for oxygen regulation. (Redrawn from Lewis 1984.)

aspartate or other amino acids to yield the end-products alanine and succinate. During prolonged anaerobiosis, succinate can be further degraded into volatile fatty acids such as propionate or acetate (van den Thillart and de Vries 1985, de Zwaan 1983, Zs.-Nagy *et al.* 1982). Anoxic for 6 days, *Anodonta cygnea* maintained 52–94% of aerobic ATP levels, higher than could occur by typical glycolytic pathways, and associated with its ability to anaerobically oxidize succinate while producing ATP molecules (Zs.-Nagy *et al.* 1982). These alternative pathways are more efficient than glycolysis (which yields only 2 moles of ATP per mole of glucose catabolized), producing 4.71–6.43 moles of ATP per mole of glucose (de Zwaan 1983). Thus, they allow tolerance of more extended anaerobiosis than does the less efficient glycolysis and the higher energy yields of these alternate pathways allow excretion of anaerobic endproducts, preventing retention of lactate to lethal levels as occurs in glycolytic species during anaerobiosis (de Zwaan 1983).

During anaerobiosis, the buildup of acidic, anaerobic endproducts in the tissues and hemolymph can cause considerable acidosis (decline in pH). As freshwater bivalves have no respiratory pigments, their blood has little inherent buffering capacity (Byrne 1988, Heming *et al.* 1988). Instead, bivalves mobilize calcium carbonate from the shell to buffer respiratory acidosis. Thus, when anaerobic, the pH of the pallial fluid of *Margaritifera margaritifera* remains highly constant, but its Ca^{2+} concentration increases (Heming *et al.* 1988). Similarly, blood Ca^{2+} levels rise eight-fold in *Ligumia subrostata* exposed to an atmosphere of nitrogen (Dietz 1974). Emersed in nitrogen for 72 hr, the blood Ca^{2+} concentrations of *C. fluminea* rose nearly five-fold, partially buffering accumulated respiratory acidosis (Byrne 1988).

The gills of unionaceans (Silverman *et al.* 1983; Steffens *et al.* 1985) harbor extensive extracellular calcium phosphate concretions that could also buffer hemolymph pH. However, rather than releasing calcium, the mass of these concretions increases during prolonged anoxia. Indeed, their mass is related inversely to blood pH and directly to blood calcium concentration. This suggests that calcium released from the shell during anoxia is sequestered in the gill concretions, preventing loss of diffusion to the external medium. This response is adaptive as it reduces the necessity for Ca^{2+} uptake to replace lost shell calcium on return to aerobic conditions (Silverman *et al.* 1983).

4. Desiccation Resistance

Freshwater bivalves may be exposed to air for weeks or months during seasonal dry periods or unpredictable periods of extreme drought (McMahon 1979b, 1983a, McMahon and Williams 1984). Lack of mobility leaves some species stranded in air as water levels recede, while other species occur in habitats that dry completely. Unlike other freshwater species, bivalves have no obvious adaptations or structures for maintenance of aerial gas exchange when out of water.

Many species of sphaeriids inhabit temporary ponds and survive periods of several months in air during drying (for a review see Burky 1983). In some cases, both adults and juveniles survive emersion (Collins 1967, McKee and Mackie 1980; in others, only recently hatched juveniles survive (McKee and Mackie 1980, Way *et al.* 1981). Sphaeriids burrow into sediments prior to air exposure. Oxygen consumption and filtration rates decline in *M. partumeium* prior to pond drying as individuals begin to estivate prior to prolonged emersion (Way *et al.* 1981, Burky *et al.* 1985a,b). Reduction in metabolic demand while emerged allows long-term maintenance in air on limited energy reserves.

Sphraeium occidentale (Collins 1967) is exposed in air for several months in its ephemeral pond habitats. In air, the $\dot{V}_{O_2}$ of *S. occidentale* is 20% of aquatic rates, gas exchange apparently taking place across specialized pyramidal cells extending through punctae in the shell. Gas exchange through shell punctae allows continual valve closure, minimizing water loss. In *C. fluminea*, aerial $\dot{V}_{O_2}$ is 21% of the aquatic rate (McMahon and Williams 1984). In air, *C. fluminea* periodically gapes the valves and exposes mantle edge tissues fused together with mucous (McMahon 1979b, Byrne 1988, Byrne *et al.* 1988). Mantle edge exposure is associated with high rates of aerial O_2 uptake while no O_2 consumption occurs during valve closure (Byrne 1988, McMahon and Williams 1984). During mantle edge exposure, bursts of metabolic heat production occur, suggesting that mantle edge exposure allows periodic aerobic recharging of spent ATP and phosphagen stores depleted during longer periods of valve closure (Byrne 1988). Periodic exposure of mucus-sealed mantle edges greatly reduces the tissue surface area exposed to the atmosphere and the duration of such exposure. In contrast, intertidal species gape more continually in air exposing moist mantle tissues directly to the atmosphere through parted mantle edges or open inhalant siphons. Consequently, they generally have higher levels of evaporative water loss than freshwater species when emersed (McMahon 1988). The degree of mantle edge exposure activity in *C. fluminea* is reduced with increased temperature, decreased relative humidity, and increasing duration of emersion. This suggests that individuals respond to increasing dessication pressure by greater reliance on anaerobic metabolism, reducing the degree of mantle edge exposure and associated evaporative water loss (Byrne 1988, Byrne *et al.* 1988, McMahon 1979b). Emersed unionaceans also display periodic mantle edge exposure (Heming *et al.* 1988), including *Ligumia subrostrata* whose aerial oxygen consumption is 21–23% of aquatic rates (Dietz 1974). In air, unionaceans and *C. fluminea* utilize shell Ca^{2+} to buffer accumulating HCO_3^- (Byrne, 1988, Dietz 1974, Heming *et al.* 1988). Both Ca^{2+} and HCO_3^- accumulate in the mantle cavity fluids of emersed pearl mussels, *Margaritifera margaritifera* (Heming *et al.* 1988). Mantle edge exposure is also associated with release of CO_2 generated by metabolic and shell-buffering processes (Byrne 1988, Heming *et al.* 1988).

Oxygen uptake rates are elevated on resubmersion after prolonged emersion in both *C. fluminea* (Byrne 1988) and the unionid *Lamellidens corrianus* (Lomte and Jadhav 1982b). In marine intertidal bivalves, elevated $\dot{V}_{O_2}$ 1–2 hr after reimmersion has been considered payment of an oxygen debt resulting from oxidation of anaerobic endproducts accumulated during air exposure (de Zwann, 1983). However, in *C. fluminea*, $\dot{V}_{O_2}$ remained elevated for at least 12 hr after reimmersion, suggesting that it was the result of a process other than typical oxygen debt payment, such as the increased metabolic demands associated with tissue damage repair or excretion of accumulated metabolic wastes. Lack of typical oxygen debt payment suggests that *C. fluminea* remains primarily aerobic when in air (Byrne 1988).

Among freshwater species, some unionaceans appear most tolerant of emersion and can survive for months or even years in air (Dance 1958, Dietz 1974, Hiscock 1953, White 1979). Certainly, the ability of some species of unionaceans to tolerate extraordinary periods of emersion and/or to migrate up and down the shore with changing water levels (White 1979) may partially account for the dominance of unionaceans in larger North American river drainages characterized by major seasonal water level fluctuations. Patterns of bivalve growth, reproduction and other important life-history phenomena may be driven in part by seasonal water level variation. Thus, human regulation of river flow and level may have contributed to the extirpation of many unionacean species from North American drainage systems.

The mode of nitrogen excretion or detoxification during emersion is an unresolved question in freshwater bivalves. Ammonium ion (NH_4^+) is the major nitrogenous excretory product of aquatic molluscs (Bishop *et al.* 1983). Due to its toxic effects on

oxidative phosphorylation even at low concentrations, ammonium ion is generally not accumulated in emersed molluscs and its high solubility in water precludes its release to the atmosphere as ammonia gas. Instead, when emersed, many aquatic species detoxify ammonium ion by conversion to less toxic compounds such as urea or uric acid, which are accumulated in the hemolymph and kidneys to be excreted on resubmersion. However, freshwater bivalves do not appear to have the capacity to produce urea or uric acid (Bishop *et al.* 1983, Vitale and Friedl 1984).

Without the capacity to detoxify accumulating ammonia, how do freshwater bivalves tolerate prolonged emersion? Recent studies of ammonia excretion in emersed *C. fluminea* indicate that, unlike intertidal bivalves (Bishop *et al.* 1983), this species does not catabolize amino acids during emersion, preventing ammonium ion accumulation (Byrne 1988). Further evidence for protein catabolism suppression in emersed freshwater bivalves is the near total dependence of the unionids *Lamellidens corrianus* and *L. marginalis* on "carbohydrate catabolism while emersed (Lomte and Jahav 1982c, Sahib *et al.* 1983).

Of interest is the report that individuals of *Sphraeium occidentale* and *Musculium securis* were more tolerant of emersion when drawn from populations estivating in a dry pond than when drawn from a submerged population in a permanent pond (McKee and Mackie 1980). Thus, tolerance of prolonged air exposure may depend on physiological and biochemical alterations in estivating individuals including, perhaps, increased dependence on carbohydrate catabolism previously described for *C. fluminea* and unionaceans. Entrance into estivation in response to emersion may be controlled by neurosecretory hormones (Lomte and Jadhav 1981a).

5. Gill Calcium Phosphate Concretions in Unionaceans

Dense calcium phosphate concretions occur in the tissues of unionid bivalves. Intracellular calcium phosphate concretions in unionacean mantle tissue may be a source of calcium for shell deposition (Davis *et al.* 1982, Jones and Davis 1982). Dense deposits of extracellular calcium phosphate concretions (1–3 μm in diameter) occur in the gills of unionaceans (Silverman *et al.* 1983, 1988, Steffens *et al.* 1985). Gill calcium phosphate concretions account for up to 60% of gill dry weight in some species (Silverman *et al.* 1985, Steffens *et al.* 1985). They are most dense along a series of parallel nerve tracts running 90° to the long axis of the gill filaments, density declining in the ventral portions of the demibranchs (Silverman *et al.* 1983, 1985, Steffens *et al.* 1985).

Besides storing shell Ca^{2+} that is released to buffer respiratory acidosis during hypoxia (see Section II.C.3), gill concretions provide a ready source of maternal calcium available for rapid shell development in brooded glochidia (Silverman *et al.* 1985, 1987). Thus, during glochidial incubation, the mass and density of gill concretions decline, particularly in the outer marsupial gill. The gill concretion masses of individuals of *L. subrostrata* and *A. grandis grandis* brooding glochidia were only 47% and 70% of that during nonreproductive periods, respectively (Silverman *et al.* 1985). Indeed, ^{45}Ca tracer studies indicated that 90% of glochidial shell calcium was of maternal origin in *A. grandis grandis*, the most likely source being gill mineral concretions, with nonmaternal calcium accounting for only 8% of glochidial shell calcium in *L. subrostrata* (Silverman *et al.* 1987).

6. Water and Salt Balance

Living in an extremely dilute medium, freshwater bivalves constantly gain water and lose ions. Excess water is eliminated as a fluid hypo-osmotic to the tissues via the kidneys and lost ions are recovered via active transport over the gills and other external epithelial surfaces as described in Section II.B.3. *C. fluminea* is relatively tolerant of salinity compared to other freshwater species, surviving long-term exposures to 10–14 ppt sodium (McMahon 1983a, Morton and Tong 1985). Above this salinity, it is incapable of maintaining hemolymph ion or osmotic concentrations and becomes isosmotic with the medium (Gainey and Greenberg 1977). Freshwater unionids, with much lower hemolymph osmotic concentrations, generally lose osmoregulatory capacity and cannot regulate volume above 3 ppt sodium (Hiscock 1953). In contrast, hyperosmotically stressed specimens of *C. fluminea* increase the blood osmotic concentration by actively increasing the free amino acid pools in the blood. This phenomenon also occurs in a number of estuarine species (Gainey 1978a, 1978b, Gainey and Greenberg 1977, Matsushima *et al.* 1982), preventing water loss by remaining isosmotic with the medium. This capacity is unexpected in freshwater species, which are never exposed to hyperosmotic medium and probably reflects the recent penetration of freshwater by *C. fluminea* from an estuarine ancestor.

Both unionaceans and *C. fluminea* respond to hemolymph ion depletion (via maintenance in an extremely dilute medium) by increasing the rate of active Na^+ uptake, allowing maintenance of hemo-

lymph ion concentration (Dietz 1985). The activity of $(Na^+ + K^+)$-activated ATPase, an enzyme associated with active sodium transport, increased in the mantle and kidney tissues of salt-depleted individuals of *C. fluminea*, suggesting activation of sodium transport. This response did not occur in the unionid *Lampsilis claibornesis*, indicating that regulation of active ion uptake has been lost in unionids with a much longer evolutionary history in freshwaters (Deaton 1982). In *Anodonta woodiana*, mantle cavity water had an osmotic concentration of 34 mosmol/liter, 76% that of hemolymph (45 mosmol/liter with pallial concentrations of Na^+, K^+, and Cl^- maintained at 71%, 76%, and 72% of blood levels, respectively, even in an extremely dilute medium (less than 1 mosmol/liter). Maintenance of elevated mantle water ion concentrations suggests that it acts as a buffer, reducing the gradient for and thus the rate of diffusive ion loss from tissues to the dilute freshwater medium (Matsushima and Kado 1982).

In both unionaceans and *C. fluminea*, the enzyme carbonic anhydrase (CA), which catalyzes formation of carbonic acid (H_2CO_3) from water and carbon dioxide, occurs in gill and mantle tissues. H_2CO_3 degrades into H^+ and HCO_3^- (bicarbonate ion), which appear to function as counter ions exchanged for Na^+ and Cl^- actively taken up from the medium. Bivalves in an extremely dilute medium increase CA activity specifically in the gill and mantle. Inhibition of CA activity by acetazolamide causes reduction in both hemolymph Na^+ and Cl^- concentrations and net Na^+ and Cl^- uptake rates, which is strong evidence for its ion regulatory role (Henery and Saintsing 1983).

The osmotic concentration of freshwater bivalves, particularly unionaceans, is the lowest recorded for multicellular freshwater invertebrates, that for *Anodonta cygnea* being 40–50 mosmol/liter or 4–5% of seawater, while that for other freshwater species is 100–400 mosmole/liter or 10–40% of seawater (Burky 1983). The apparent adaptive advantage of such low hemolymph osmotic concentrations is a reduction of the gradient for water gain from, and ion loss to, the dilute freshwater medium. This reduces transepithelial osmosis and ion diffusion rates across extensive mantle and gill surfaces to those which can be balanced by water excretion and active ion recovery mechanisms at energetically feasible levels (Burton 1983). Studies have shown that the hydrostatic pressure generated by the ventricle and auricles in *A. cygnea* is great enough to allow sufficient filtration of hemolymph plasma into the pericardial space to account for urine formation. As ventricular hydrostatic pressures in *A. cygnea* are approximately twice those of the marine clam *Mya arenaria*, capacity for excretion of excess water is greater in freshwater species compensating for the increased water gain associated with hyperosmotic regulation (Jones and Peggs 1983). Exposure of *Anodonta* sp. to a very dilute medium caused the appearance of extensive extracellular membrane spaces in the deep infoldings of kidney epithelial cells, perhaps allowing active ion uptake from the excretory fluid, producing a dilute urine or increased transport of excess water into the kidney (Khan *et al.* 1986).

The hormonal control of osmoregulation in freshwater bivalves has been reviewed by Dietz (1985). Cyclic AMP (cAMP) stimulates active uptake of Na^+ by unionaceans, while prostaglandin inhibits it and prostaglandin inhibitors stimulate it (Dietz *et al.* 1982, Graves and Dietz 1982, Saintsing and Dietz 1983). Serotonin stimulates tissue accumulation of cAMP, inducing increased Na^+ uptake (Saintsing and Dietz 1983). Thus, an antagonistic relationship exists between serotonin and prostaglandins in modulating adenylate cyclase-catalyzed cAMP stimulation of Na^+ active uptake. Not surprisingly, high concentrations of serotonin are found in the gill nerve tracts of unionaceans (Dietz 1985). The circadian rhythms of Na^+ uptake in freshwater clams (Graves and Dietz 1980, McCorkle-Shirley 1982) may be mediated by this antagonistic hormonal system (Dietz 1985).

When either their cerebropleural or visceral ganglia were ablated, individuals of the unionid *Lamellidens corrianus* rapidly gained water, indicating loss of osmoregulatory ability. Injection of ganglionic extracts restored normal osmoregulatory capacity, suggesting that ganglionic neurosecretory hormones are involved in regulation of water balance (Lmote and Jadhav 1981b). The pedal ganglion of *A. cygnea* has a higher affinity for serotonin and other monoamines controlling ion and water balance at low temperatures, indicating a seasonal component to hormonal control of osmoregulation (Hiripi *et al.* 1982).

III. ECOLOGY AND EVOLUTION

A. Diversity and Distribution

The distributions of freshwater bivalves, particularly unionaceans, in North America have been well described. Species distribution maps for unionaceans and sphaeriids have been published for Canada (Clarke 1973) and the United States (LaRocque 1967a). North American distribution records for the vast majority of species are provided in Burch (1975a, 1975b). LaRocque (1967b) also describes living and Pleistocene fossil assemblages at specific North American localities. In addition, there is a

massive literature, too numerous to cite here, describing species occurrences at specific sites or species assemblages in various drainage systems in North America.

Distribution data for North American sphaeriids indicate that all native (non-introduced) species have broad distributions, often extending from the Atlantic Coast to the Pacific coast. Introduced to North America from southeast Asia in early 1900s (McMahon 1982), *Corbicula fluminea* (i.e., the light-colored shell morph of *Corbicula*) has a similarity widespread North American distribution. It extends into the drainages of the west coast of the United States and the southern tier of states, and throughout drainages east of the Mississippi River, with the exception of the most northernly states (Counts 1986, McMahon 1982), and into northern Mexico (Hillis and Mayden 1985) (Fig. 11.11). A second, unidentified species of *Corbicula* (i.e., the dark-colored shell morph) is restricted to isolated, spring-fed drainages in southcentral Texas and southern California and Arizona (Fig. 11.11). (Hillis and Patton 1982, Britton and Morton 1986, McLeod 1986). In contrast, North American unionacean species generally have more restricted distributions. Few species range on both sides of the continental divide and a surprisingly large number are limited to single drainage systems (Burch 1975b, LaRocque 1967a).

The widespread distributions of sphaeriids and *Corbicula* relative to unionaceans in North America may reflect fundamental differences in their capacities for dispersal. Unionaceans depend primarily on host fish transport of the glochidium for dispersal (Kat 1984), thus their ranges reflect those of their specific glochidial host fish species. While host fish transport of glochidia increases the probability of their dispersal into favorable habitats, as host fish and adult unionacean habitat preferences generally closely coincide (Kat 1984), it greatly limits the extent to which such dispersal can occur, leading to development of highly endemic species. For example, electrophoretic studies of peripheral populations of Nova Scotian unionid species suggest that invasion of new habitats is primarily by host fish dispersal (Kat and Davis 1984); therefore, barriers to fish dispersal are also barriers to unionid dispersal. Thus, the distribution of modern and fossil North American interior basin unionacean assemblages are limited to areas below major waterfalls in the drainage systems of Lake Champlain (New York, Vermont, Quebec), because these falls act as upstream migration barriers to host fish dispersal (Smith 1985a). Further, the recent re-establishment of *Anodonta impicata* populations in the upper portions of the Connecticut River Drainage closely fol-

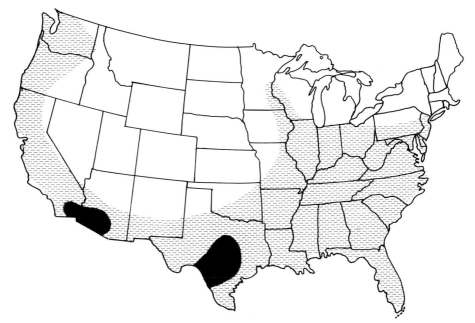

Figure 11.11 Distribution of *Corbicula* in the United States. Hatched area is the distribution of the light-colored shell morph of *Corbicula, Corbicula fluminea.* The darkly stippled areas are distribution of the dark-colored shell morph of *Corbicula,* yet to be assigned a species designation.

lowed the restoration of its anadromous glochidial clupeid fish host populations, by the building of fishways past numerous manmade impoundments that previously prevented upstream fish host dispersal (Smith 1985b).

Sphaeriids and *C. fluminea* have evolved mechanisms allowing dispersal between drainage systems, making them more invasive than unionaceans and accounting for their more cosmopolitan distributions. Juvenile sphaeriids disperse between drainage systems by clamping their shell valves onto limbs of aquatic insects, the feathers of water fowl (Burky 1983), or even the limbs of salamanders (Davis and Gihen 1982). Some sphaeriid species survive ingestion and regurgitation by ducks, which commonly feed on them, allowing long-distance dispersal (Burky 1983). The rapid spread of *C. fluminea* through North American drainage systems (McMahon 1982), while in part mediated by human vectors, also has resulted from the natural dispersal capacities of this species. The long mucilaginous byssal thread produced by juveniles or the filamentous algae on which they can settle becomes entangled in the feet or feathers of shore birds or water fowl, making them juvenile transport vectors between drainages (McMahon 1982, 1983a). Its natural capacity for dispersal is highlighted by its spread into northern Mexico drainage systems where human-mediated transport is highly unlikely (Hillis and Mayden 1985), and into southern Britain during interglacial periods (McMahon 1982). Adult zebra mussels, *Dreissena polymorpha,* attach to

floating wood or boat hulls with byssal threads, facilitating transport over long distances (Mackie *et al.* 1989), and can also be transported between drainage systems attached to macrophytic vegetation utilized by nesting shore birds and water fowl.

Juveniles of *C. fluminea* can be transported long distances downstream passively suspended in water currents (McMahon and Williams 1986b, Williams and McMahon 1986). Water currents also disperse the actively swimming veliger stage of *D. polymorpha* (Mackie *et al.* 1989). Adult *C. fluminea* can also leave sediments to be carried downstream over the substratum by water currents (Williams and McMahon 1986). This process is assisted by production of a mucus dragline from the exhalant siphon, which increases the drag exerted on individuals by water currents (Prezant and Chalermwat 1984). Such passive dispersal of juvenile and adult *C. fluminea* on water currents not only accounts for the extraordinary ability of this species to invade the downstream portions of drainage systems after introduction (McMahon 1982), but also is the basis for its impingement and fouling of small-diameter piping and other components of industrial, agricultural, and municipal raw-water systems (McMahon 1983a). Similarly, current-mediated transport of free-swimming veligers and adults of *D. polymorpha* (adults attached to floating substrata or carried over the bottom as byssally attached clumps of individuals) on water currents accounts for its rapid dispersal through European drainage systems after it escaped from the Caspian Sea in the late eighteenth century (Morton 1969). This species has rapidly spread downstream throughout Lake Erie from its original upstream introduction in Lake St. Clair in 1985–1986 (where it was apparently transported from Europe in ship ballast water) (Mackie *et al.* 1989) (Fig. 11.12). Zebra mussels had invaded Lake Ontario, the St. Lawrence River, portions of lakes Superior, Huron, and Michigan and the western portion of the Erie–Borge Canal by the fall of 1990 when the final revision of this chapter was completed. Capacity for downstream transport and byssal attachment to hard surfaces make *D. polymorpha* destined to be a major North American biofouling pest species, recapitulating its recent history in Europe. Major incidents of zebra mussel fouling are already being reported in raw-water facilities on Lakes St. Clair, Erie, and Ontario (Mackie *et al.* 1989). Passive, current-mediated downstream transport is also reported for juvenile sphaeriids (*Pisidium punctiferum*) (McKillop and Harrison 1982) and may be an important but uninvestigated means of dispersal for many species in this family. In contrast, passive downstream transport is extremely rare in unionaceans (Imlay 1982), making them reproductively isolated.

As both sphaeriids and *C. fluminea* are self-fertilizing hermaphrodites (see Section II.B.5, introduction of only a single individual can found a new population. In contrast, *D. polymorpha* and the ma-jority of unionids are gonachoristic, requiring simultaneous introduction of males and females to found a new population, thus reducing the probability of successful invasion of isolated drainage systems.

There have been major declines of unionacean populations and species diversity in North America over the last century. Over 25 unionacean species are on the Federal Register's Endangered Species list, with additional species on state lists and new species being added yearly (Laycock 1983, Taylor and Horn 1983, White 1982). Massive historical losses of unionacean species from river drainage systems are revealed by comparison of present day species assemblages with those of earlier surveys of living species or with recent fossil assemblages (Ahlstedt 1983, Hartfield and Rummel 1985, Havlik 1983, Hoeh and Trdan 1984, Miller *et al.* 1984, Neves and Zale 1982, Parmalee and Klippel 1982, 1984, Starnes and Bogan 1988, Stern 1983, Taylor 1985). Extirpations of sphaeriid faunae are far less common, but have occurred (Mills *et al.* 1966, Paloumpis and Starrett 1960). Unionacean assemblages in Indian middens near the upper Ohio River yielded at least 32 species, while a 1921 survey of its bivalve fauna yielded only 25 midden species and a 1979 survey, only 13 midden species in the same area—indicating massive species extirpation (Taylor and Spurlock 1982). Similar historical loss of Indian midden species has occurred in the Tennes-

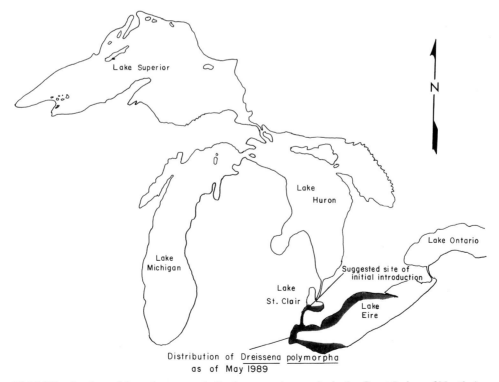

Figure 11.12 Distribution of the zebra mussel, *Dreissena polymorpha* in the Great Lakes of North America as of May 1989.

ses River Drainage (Parmalee 1988, Parmalee *et al.* 1982). Further evidence of the changing environmental conditions in the upper Ohio River is demonstrated by the establishment of 15 unionid species previously unreported from Indian middens or earlier surveys (Taylor and Spurlock 1982).

Postulated causes for the massive decline in North American unionacean populations are numerous. The freshwater pearling industry can extirpate entire populations (Laycock 1983), overfishing for pearls being a major factor in the recent decline of the pearl mussel, *Margaritifera magaritifera* in Great Britain (Young and Williams 1983a). Extensive artificial impoundments of drainage systems slow flow velocity and subsequent accumulation of silt causes reductions in mussel faunae (Duncan and Thiel 1983, Parmalee and Klippel 1984, Starnes and Bogan 1988, Stern 1983). Impoundments may also eliminate fish glochidial hosts (Mathiak 1979) or prevent dispersal of glochidia by fish hosts (see earlier discussion). They may also damage downstream unionacean populations by releasing cold hypolimnic water (Ahlstedt 1983, Clarke 1983) or by inducing major short-term oscillations in flow rate, either scouring the bottom of suitable substrata for mussels during periods of high flow or causing prolonged aerial exposure of mussels during periods of low flow (Miller *et al.* 1984). Channelization of drainage systems for navigation or flood control is detrimental to unionaceans. Increased flow velocity and propeller wash elevate suspended solids, which interfere with mussel filter feeding and oxygen consumption (Aldridge *et al.* 1987, Payne and Miller 1987). It also reduces availability of stabilized sediments, sand bars, and low flow areas, all preferred unionacean habitats (Hartfield and Ebert 1986, Payne and Miller 1989, Stern 1983, Way *et al.* 1990a).

Pollution adversely affects bivalves. Mussel fauna receiving industrial pollution (Zeto *et al.* 1987), urban waste water effluents (sewage, silt, pesticides) (Gunning and Suttkus 1985, St. John 1982, Neves and Zale 1982), or silt and acid discharges from mines (Taylor 1985, Warren *et al.* 1984) become severely depauperate or totally extirpated. The advent of modern sewage treatment on the Pearl River, Louisiana, allowed re-establishment of five unionid species previously absent for at least 20 years (Gunning and Suttkus 1985).

Physical factors also influence bivalve distributions. While environmental requirements are species specific, a number of generalities appear warranted. Sediment type clearly affects distribution patterns. Unionaceans are generally most successful in stable, coarse sand or sand–gravel mixtures and are generally absent from substrata with heavy silt loads (Cooper 1984, Salmon and Green 1983, Stern 1983, Way *et al.* 1990a). In the Wisconsin and St. Croix Rivers only 7 of 28 unionid species occurred in sand–mud sediments, the majority preferring sand–gravel mixtures. Only three species, *Anodonta grandis, Lampsilis anodontes,* and *L. radiata,* typically inhabited sand–mud substrata (Stern 1983). In contrast, *C. fluminea* has much broader sediment preferences, successfully colonizing habitats ranging from bare rock through gravel and sand to sediments with relatively high silt loads (McMahon 1983a). Broad sediment preference has allowed this species to invade a wide variety of North American drainage systems; however, its optimal habitat is well-oxygenated fine sands or gravel–sand mixtures (Belanger *et al.* 1985).

In contrast to unionaceans, species diversity in the genus *Pisidium* increases with decreasing particle size (Fig. 11.13A), becoming maximal at a mean particle diameter of 0.18 mm (Kilgour and Mackie 1988). In southeastern Lake Michigan, *Pisidium* density and diversity were maximal in very fine sand–clay and silt–clay sediments, while peak *Sphaerium* diversity occurred at somewhat larger particle sizes (Zdeba and White 1985). These observations suggest differences in substratum preference among sphaeriid genera, perhaps associated with sediment oranic detritus feeding mechanisms in *Pisidium* (see Section III.C.2).

Apparent differences in substratum preferences may be associated with species-specific differences in optimal water velocities. Unionaceans are most successful where water velocities are low enough to allow sediment stability, but high enough to prevent excessive siltation (Salmon and Green 1983, Stern 1983, Way *et al.* 1990a), making well-oxygenated, coarse sand and sand–gravel beds optimal habitats for riverine species. Low or variable velocities allow silt accumulations that either make sediments too soft for maintenance of proper position (Lewis and Riebel 1984, Salmon and Green 1983) or interfere with filter feeding and gas exchange in unionaceans (Aldridge *et al.* 1987). In contrast, periodic scouring of substrata exposed to high flow velocities can both remove substrate and mussels and prevent their successful resettlement (Young and Williams 1983b). Sediment type did not affect the burrowing ability of three lotic unionid species (*A. grandis, Elliptio complanata,* and *Lampsilis radiata*) (Lewis and Riebel 1984), suggesting that it is not involved in substratum preferences. *C. fluminea*, with its relatively heavy, ridged shell and rapid burrowing ability, is better adapted for life in high current velocities and unstable substrata than are most unionaceans (McMahon 1983a). Indeed, in the Tangipahoa River, Mississippi, it successfully colonizes

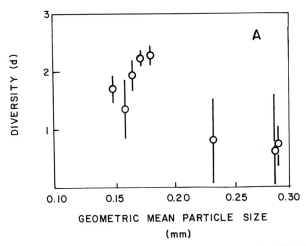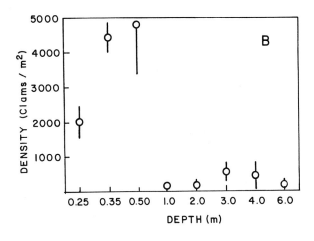

Figure 11.13 Sediment relationships in sphaeriid clam communities from sites along a depth transect in Britannia Bay, Ottawa River, Canada. (A) Mean sphaeriid diversity (Shannon-Weaver d) values for various sphaeriid communities in relation to mean sediment geometric particle size. (B) Mean sphaeriid density values at various depths. Vertical bars about points are 95% confidence limits. Note increase in diversity with decrease in mean particle size of sediments and maximization of density at depths of less than one meter. (Redrawn from data of Kilgour and Mackie 1988.)

unstable substrata from which unionaceans are excluded (Miller *et al.* 1986). In contrast to the majority of unionaceans and *C. fluminea*, many sphaeriid species occur in small ponds and the deeper portions of large lakes, where water flow is negligible and the substratum has both a high silt content and a heavy organic load. The preference of a number of sphaeriids for low flow habitats and silty sediments may be associated with their interstitial sediment feeding mechanisms, particularly in the genus *Pisidium* (Lopez and Holopainen 1987) (see Section III.C.2). Adult byssal thread attachment to hard substrata not only allows *D. polymorpha* to inhabit relatively high flow areas compared to other bivalves (the postveligor successfully settles at flow rates up to 1.5m/sec.), but also makes it a highly successful epibenthic species in lentic habitats characterized by a preponderance of hard substrata from which native North American species are generally eliminated (Mackie *et al.* 1989).

Water depth also affects freshwater bivalve distributions. Most species of unionaceans prefer shallow water habitats generally less than 4–10 m in depth (Machena and Kautsky 1988, Salmon and Green 1983, Stone *et al.* 1982, Way *et al.* 1990a), although some species can be found in the deeper regions of lotic habitats if they are well oxygenated. *C. fluminea* is also restricted to shallow, near-shore habitats in lentic waters (although they can be found in deeper waters if they are well oxygenated) (McMahon 1983a), as are the majority of *Sphaerium* and *Musculium* species (Fig. 11.13B) (Kilgour and Mackie 1988, Zdeba and White 1985). In contrast, some species of *Pisidium* inhabit the profundal re-

gions of lakes (Holopainen and Jonasson 1983, Kilgour and Mackie 1988, Zdeba and White 1985; for a review see Burky 1983). The depth distributions of *D. polymorpha* vary between habitats; however, adults are rarely found in great numbers above 2 m and dense populations can extend to depths of 4–60 m, but always occur in well-oxygenated waters above the epilimnion. Younger, recently settled individuals tend to migrate toward deeper water after settlement (Mackie *et al.* 1989).

The limitation of most bivalves to shallow habitats when they occur in lentic waters may be associated with their relatively poor tolerance of hypoxia. In lentic habitats, waters below the epilimnion are often depleted of dissolved oxygen. As with the vast majority of unionaceans *Sphaerium* and *Musculium* species (Burky 1983) and *C. fluminea* (McMahon 1983a) cannot maintain normal rates of O_2 uptake under severely hypoxic conditions. Thus, they are mostly restricted to shallow, well-oxygenated habitats. In contrast, many species of *Pisidium* are extreme regulators of $\dot{V}_{O_2}$ when hypoxic (see Burky 1983 and references therein), allowing them to inhabit the deeper hypolimnetic regions of lakes where summer ambient O_2 tensions fall to near zero levels (Holopainen 1987, Holopainen and Jonasson 1983, Jonasson 1984a, 1984b). However, during summer hypoxic periods, growth and reproduction are retarded in profundal *Pisidium* populations, indicating that low oxygen tensions can have deleterious effects on even hypoxia-tolerant species (Halopainen and Jonasson 1983). Hypoxia-intolerant *C. fluminea* invaded the profundal regions of a small lake only after artificial aeration eliminated hypoxic hypolimnetic waters (McMahon 1983a). Sewage-induced hypoxia in the Pearl River, Louisiana, eliminated its unionid fauna (Gunning and Suttkus 1985). Even highly hypoxia-tolerant profundal *Pisidiid* communities have been exterminated by extreme hypoxia induced by sewage effluents (Jonasson 1984a). The low tolerance of most unionaceans to even moderate hypoxia (Burky 1983) renders them highly suceptible to wastewater release, a factor implicated in the decline of North American faunae.

Ambient pH does not greatly limit the distribution of freshwater bivalves. The majority of species prefer alkaline waters with the pH above 7.0; species diversity declines in more acidic habitats (Okland and Kuiper 1982). However, unionaceans can grow and reproduce over a pH range of 5.6–8.3, a pH of less than 4.7–5.0 being the absolute lower limit (Fuller 1974, Hornbach and Childers 1987, Kat 1982, Okland and Kuiper 1982). Some sphaeriid species are relatively insensitive to pH or alkalinity. No differences in species richness or growth and reproduction occurred in sphaeriid fauna from six lakes of extremely low alkalinity relative to those with higher alkalinity levels (Rooke and Mackie 1984a, 1984b, Servos *et al.* 1985). Maximal laboratory growth and reproduction in *Musculium partumeium* occurred at pH 5.0, suggesting adaptation of this species to moderately acidic habitats (Hornbach and Childers 1987).

Habitats of low pH generally also have low calcium concentrations. Low pH leads to shell dissolution and eventual mortality in older individuals if shell penetration occurs (Kat 1982). Sphaeriids have been reported from waters with calcium concentrations as low as 2 mg Ca/liter, while the unionid *Elliptio companata* occurs in lakes with 2.5 mg Ca/liter (Rooke and Mackie 1984a). Freshwater bivalves can actively take up Ca^{2+} from the medium at concentrations as low as 0.5 mM Ca/liter (0.02 mg Ca/liter, see Section II.A.1), a level far below the minimal tolerated ambient Ca^{2+} concentrations of 2–2.5 mg/liter. Thus, minimal environmental calcium limits are much greater than those allowing active Ca^{2+} uptake. As such, the minimum ambient calcium concentration tolerated by freshwater bivalves appears to be the concentration at which the rates of calcium uptake and deposition to the shell exceed the calcium loss rate from shell dissolution and diffusion, allowing maintenance of shell integrity and growth. As many factors affect shell deposition and dissolution rates (e.g., temperature, pH, and calcium concentration), the minimal calcium concentration and/or pH tolerated by a species may vary greatly between habitats dependent on interacting biotic and abiotic parameters and are often species specific. Waters with low calcium con-

centrations usually have low concentrations of other biologically important ions, making them inhospitable to bivalves even if calcium concentrations are suitable for maintaining shell growth.

Temperature influences bivalve species distributions; species have specific upper and lower limits for survival and reproduction (Burky 1983). For example, intolerance of temperatures below 2°C prevents *C fluminea* from expanding into drainages in the north central United States, which reach 0°C in winter (Fig. 11.11) (Counts 1986, McMahon 1983a). This results in massive low temperature winter kills in populations on the northern edge of its range (Cherry *et al.* 1980, Sickel 1986). Thus, *C. fluminea* populations north of the 2°C winter water temperature isotherm are restricted to areas receiving heated effluents (Counts 1986, McMahon 1982, 1983a). In contrast, the maximal temperature for the development of *D. polymorpha* eggs is 24°C and for larval development, 25°–27°C (Mackie *et al.* 1989). Such temperature maxima make this species unlikely to colonize drainage systems in the extreme southern and southwestern United States where ambient summer water temperatures routinely reach 30°–32°C.

Water level variation can affect bivalve distributions. Declining water levels during droughts or dry periods expose relatively immotile bivalves for weeks or months to air. Some species are adapted to withstand prolonged emersion, while others are emersion-intolerant (McMahon 1983a) (see Section II.C.4). The tendency for restriction of many bivalve populations to shallow near-shore waters makes them highly susceptible to emersion when water levels decline. Many sphaeriid species and a few unionacean taxa are highly tolerant of air exposure and thus able to survive prolonged seasonal emersion in ephemeral or variable level habitats (Burky 1983, White 1979). These species display unique adaptations to emergence described in Section II.C.4.

Freshwater bivalve distribution is also related to stream size or order. The number of unionacean species in drainage systems in southeastern Michigan increased proportionately with the size of the drainage area; this was mainly in response to species additions (Strayer 1983) (Fig. 11.14). However, variation in species richness could not be completely accounted for by drainage area size (note the high degree of variation in species richness values at specific drainage area values in Fig. 11.14). This suggests that other environmental variables may affect mussel distribution patterns. Among these are surface geology and soil porosity. Porous soils retain water, buffering runoff so streams draining them have relatively constant flow and are rarely dry. Therefore, they support greater numbers of mussel

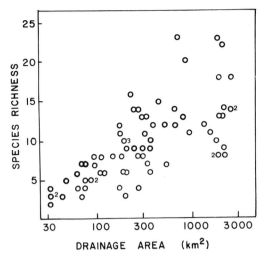

Figure 11.14 Unionid mussel species richness (total number of species present at a particular site) as a function of stream size measured by its total drainage area in km² in southeastern Michigan, United States. Numbers next to points indicate the number of observations falling on that point. The relationship between drainage area and mussel species richness was statistically significant ($r = 0.68$, $P < 0.001$). (Redrawn from Strayer 1983.)

species than streams draining soils of poor water infiltration capacity, whose drainages are prone to flooding–drying cycles. In such streams, exposure of mussels to air or low oxygen in stagnant pools during dry periods, and bottom scouring and high silt loads during floods all induce mortality, reducing species richness (Strayer 1983). Indeed, the stability of water flow, high O_2 concentrations, reduced risk of flooding, and reduction of silt loading associated with larger stream size appear to account for the increased bivalve species richness associated with them (Fig. 11.14) (Strayer 1983). However, some unionacean species, such as *Amblema plicata,* are adapted to small, variable flow streams and, in southeastern Michigan, occur almost exclusively in such habitats from which other unionaceans are virtually excluded (Strayer 1983).

B. Reproduction and Life History

North American freshwater bivalves display extraordinary variation in life history and reproductive adaptations. Life-history traits (e.g., those affecting reproduction and survival, including growth, fecundity, life span, age to maturity, and population energetics) have been reviewed for freshwater molluscs (Calow 1983, Russell-Hunter and Buckley 1983) and specifically for freshwater bivalves (Burkey 1983, Mackie 1984, Mackie *et al.* 1989, McMahon 1983a). Most research on life-history traits involves the Sphaeriidae, with a paucity of information for union-

aceans. Sphaeriids are good subjects for studies of life histories because of their greater abundances, ease of collection and laboratory maintenance, relatively simple hermaphroditic life cycles, ovoviviparity, semelparity, release of completely formed immature adults, and relatively short life spans. In contrast, unionaceans are more difficult subjects for life-history studies because they are gonochoristic, long-lived, interoparous, often rare and difficult to collect, and have life cycles complicated by the parasitic glochidial stage. The life-history traits of *C. fluminea* and *D. polymorpha* have been intensely studied due to their invasive nature and economic importance as fouling organisms (Mackie *et al.* 1989, McMahon 1983a).

1. Unionacea

The life-history characteristics of freshwater unionaceans are clearly different from those of the sphaeriids, *C. fluminea* or *D. polymorpha* (Table 11.2). The majority of unionid species live in large, stable aquatic habitats where they are generally buffered from periodic catastrophic population reductions that are typical of smaller, unstable aquatic environments (see Section III.A). In such stable habitats, long-lived adults accumulate in large numbers (Payne and Miller 1989), which can lead to competition for space and food. A few species preferentially inhabit ponds (Burch 1975b), but their life history traits have not been studied.

An important aspect of the unionacean life-history traits is their unique parasitic larval stage, the glochidium (Fig. 11.15B). With the single exception of *Simpsonichoncha ambigua,* whose glochidial host is the aquatic salamander, *Necturus maculosus,* all other North American unionaceans have glochidia that parasitize fish. The significance of the glochidial stage to unionacean reproduction has been reviewed by Kat (1984). Details of glochidial incubation and development in outer demibranch marsupial brood pouches were described previously in Section II.B.5. The glochidium has a bivalved shell adducted by a single muscle. Its mantle contains sensory hairs. In the genera *Unio, Anodonta, Megalonaias,* and *Quadrula,* a long threadlike structure projects from the center of the mantle tissue beyond the ventral edge of the valves. Its function is unknown, but it may be involved with the detection of, and/or attachment to, fish hosts.

There are three general forms of glochidia. In the subfamily Anodontina, "hooked glochidia" occur, with triangular valves from whose ventral edges project an inward curving hinged hook covered with smaller spines (Fig. 11.15B). On valve closure, the hooks penetrate the skin, scales, or fins of fish hosts

Table 11.2 Summary of the Life History Characteristics of North American Freshwater Bivalves, Unionacea, Sphaeriidae, *Corbicula fluminea*, and *Dreissena polymorpha*

Life History Trait[a]	Unionacea	Sphaeriidae	Corbicula fluminea	Dreissena polymorpha
Life span	< 6–> 100 yr (species dependent)	< 1–> 5 yr (species dependent)	1–5 yr	4–7 yr
Age at Maturity (yr)	6–12 yr	> 0.17–< 1.0 yr (1 yr in some species)	0.25–0.75 yr	1–2 yr
Reproductive Mode	Gonochoristic (a few hermaphroditic species)	Hermaphroditic	Hermaphroditic	Gonochoristic
Growth Rate	Rapid prior to maturity, slower thereafter	Slow relative to Unionids or *C. fluminea*	Rapid throughout life	Rapid throughout life
Fecundity (young/ average adult/ breeding season)	200,000–17,000,000	3–24 (*Sphaerium*) 2–136 (*Musculium*) 3–7 (*Pisidium*)	35,000	30,000–40,000/ female
Juvenile size at Release	Very small 50–400μm	Large 600–4150 μm	Very small 250 μm	Extremely small 40–70 μm
Relative Juvenile Survivorship	Extremely low	High	Extremely low	Extremely low
Relative Adult Survivorship	High	Intermediate	Low 2–41%/yr	Intermediate 26–88%/yr
Semelparous/ Iteroparous	Iteroparous	Semelparous or iteroparous	Generally iteroparous	Iteroparus
No. of Reproductive Efforts/Year	1	1–3 (continuous in some species)	2	1 (2–8 months long)
Assimilated Energy Respired (%)	–	21–91% (avg. = 45%)	11–42%	–
Nonrespired Energy in Growth (%)	85.2–97.5	65–96% (avg. = 81%)	58–71%	96.1%
Nonrespired Energy in Reproduction (%)	2.8–14.8	4–35% (avg. = 19%)	15%	4.9%
Turnover Time in Days (Mean standing crop biomass : biomass production/day ratio	1790–2849	27–1972 (generally < 80)	73–91	53–869 (dependent on habitat)

[a]See text for literature citations for data on which this table was based.

allowing glochidial attachment and encystment on host external surfaces. The majority of North American unionaceans produce "hookless glochidia," characterized by more rounded valves bearing reinforcing structures and/or a series of small spines or stylets on their ventral margins. These generally attach and encyst on the gills of fish hosts. "Axe-head glochidia" of the genus *Proptera* have a distinctly flared ventral valve margin, and near-rectangular valves which may have hooklike structures on each corner. Their host attachment sites are unknown (Kat 1984).

Glochidia initially attach to fish hosts by clamping (or snapping) the valves onto fins, scales, and/or gill filaments. Glochidia do not appear to be host-specific in attachment; rather, they attach to any fish they contact (Kat 1984). Released glochidia display snapping behavior (i.e., host attachment behavior), whereby valves are periodically rapidly and repeatedly opened and shut. Valve snapping by *M. margaritifera* glochidia is greatly stimulated by the presence of mucus, blood, gill tissue, or fins of their brown trout host, but not by water currents or direct tactile stimulation (the latter causes prolonged valve closure) (Young and Williams 1984a). This suggests that glochidia use chemical cues to detect and attach

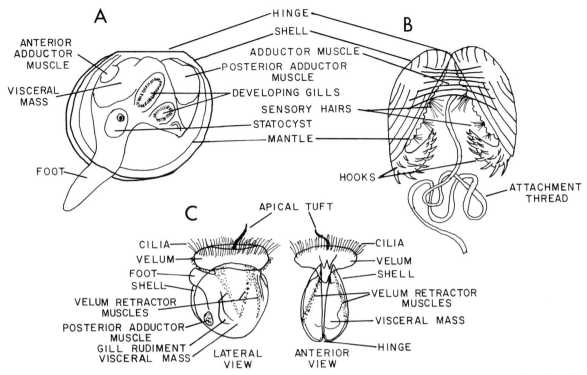

Figure 11.15 Anatomic features of freshwater bivalve larval stages. (A) The D-shaped juvenile of *Corbicula fluminea*, the freshwater Asian clam (shell length = 0.2 mm). (B) The glochidium larva of unionids, which is parasitic on fish. Depicted is a glochidium of *Anodonta* characterized by the presence of paired spined hooks projecting medially from the ventral edges of the shell; not all unionid species have glochidia with such hooks (50–400 μm in diameter depending on species. (C) Lateral and anterior views of the free-swimming, planktonic, veliger larva of *Dreissena polymorpha*, the zebra mussel; the veliger is 40–290 μm in diameter and uses the ciliated velum to swim and feed on phytoplankton. The juveniles of freshwater sphaeriid species are large and highly developed, having essentially adult features at birth (Figs. 11.2A and 11.4A).

to host fish. Fuller (1974) provided an extensive list of suitable fish hosts for the glochidia of many unionacean species.

Glochidia encyst in host tissues within 2–36 hr of attachment and may or may not grow during encystment, depending on the species. The time to juvenile metamorphosis and excystment is also species-dependent, ranging from 6–160 days; however, it is reduced at higher temperatures (Kat 1984, Zale and Neves 1982). Unsuitable host fish reject glochidia, sloughing them off after encystment (Kat 1984), with the fish blood serum components apparently dictating host suitability (Neves *et al.* 1985). As glochidia attach readily to unsuitable hosts (Kat 1984, Neves *et al.* 1985, Trdan and Hoeh 1982), host suitability appears to be more dependent on fish immunity mechanisms than on glochidial host recognition. Indeed, even suitable host fish can reject glochidia. Prior to metamorphosis, the number of glochidia of *M. margaritifera* encysted on a natural population of brown trout declined, suggestive of host rejection (Young and Williams 1984b). In the laboratory, only 5–12% of *M. margaritifera* glochi-

dia that successfully encysted in appropriate host fish actually completed development to the point of excystment as free-living juveniles, indicating host rejection of most encysted individuals (Young and Williams 1984a).

Unionaceans display a number of adaptations that increase the likelihood that glochidia will come into contact with fish hosts (Kat 1984). Glochidial release occurs once per year, but the duration of release is species dependent. Cycles of gametogenesis and glochidial release may be controlled by neurosecretory hormones (Nagabhushanam and Lomte 1981). Tachytictic mussels are short-term breeders, whose glochidial development and release take place between April and August; in many of these species, shedding of glochidia corresponds with either migratory periods of anadromous fish hosts, or the reproductive and nesting periods of host fish species. The fish hosts of these species often construct nests in areas where unionacean populations are the most dense. Residence of adult unionaceans on host nesting sites, host nest construction by fanning away of substrata, and the fanning of developing embryos in nests all provide optimal conditions for glochidial–host contact and encystment. Thus, a high proportion of nest-building fish species, such as centrarchids, are common hosts for North American unionaceans (Fuller 1974, Kat 1984). In contrast, bradytictic unionacean species retain developing glochidia in gill marsupia throughout the year, releasing them in summer (Kat 1984).

When shed from adult mussels, glochidia are generally bound together by mucous into discrete packets, which either dissolve (releasing glochidia) or are maintained intact as discrete glochidial "conglutinates" of various species-specific forms and colors. Glochidia with attachment threads (Fig. 11.15B) are released in tangled mucus threads, forming loosely organized webs that dissolve relatively rapidly. Many of these glochidia possess hooks and attach to the external surfaces of fish hosts. In some unionaceans, such mucilaginous networks of glochidia persist, suspending glochidia above the substratum, thus enhancing the possibility of host contact.

Unionids with hookless glochidia that attach to fish gills may release conglutinates that mimic the vermiform food items of their fish hosts, and thus resemble brightly colored oligochaetes, flatworms, or leeches. Some species hold their vibrant, worm-like conglutinates partially extruded from the exhalant siphon, making them more obvious to fish hosts. Consumption of such conglutinates releases glochidia within the buccal cavity of the fish, where they can be carried directly onto gill filament attachment sites by respiratory currents.

The most unusual form of unionid host food mimicry involves use of pigmented muscular extensions of the mantle edges in female lampsilids. These mantle flaps (Fig. 11.16) resemble the small fish prey of their piscivorous fish hosts. Gravid females extend the posterior shell margins well above the substratum and periodically pulsate the mantle flaps to mimic a small, actively swimming fish. When a fish attacks these mantle flap lures, glochidia are forcibly released through pores in the posterior portion of the marsupial gill (often projected between the mantle flaps, Fig. 11.16) assuring glochidial contact with the fish host (Kat 1984).

As the glochidia of some species do not grow while encysted, the degree to which they are parasitic on fish hosts has been questioned. However, recent *in vitro* glochidial culture experiments suggest that glochidia both absorb organic molecules from fish tissues and require fish plasma for development and metamorphosis (Isom and Hudson 1982) in a true host–parasite relationship.

As in other parasitic species, glochidia are shed in huge numbers to ensure the maximum potential for host contact and attachment. Fecundity in unionaceans is reported to range from 200,000–17,000,000 glochidia/female/breeding season (Parker *et al.* 1984, Paterson 1985, Paterson and Cameron 1985, Young and Williams 1984b). Nonetheless, chances for glochidial survival to metamorphosis are extremely small. In a natural population of *M. margaritifera* that does not produce glochidial conglutinates, only 0.0004% of released glochidia successfully encysted in fish hosts. Of these, only 5% were not rejected before full development and excystment; and, of those successfully metamorphosing, only 5% successfully became established as juveniles in the substratum (Young and Williams 1984b). Thus, overall, only 1 in every 100,000,000 shed glochidia became settled juveniles. High glochidial mortality makes the effective fecundity of unionaceans extremely low, which is not unusual for a species in a stable habitat. Based on these data and the fecundity ranges listed in Table 11.2, only 0.002–0.17 juveniles from each female unionacean's annual reproductive effort would successfully settle in the sediment. Therefore, the main advantages of the parasitic glochidial stage appear to be directed dispersal by fish hosts into favorable habitats (Kat 1984) and utilization of fish host energy resources by glochidia to complete development to a juvenile size that is large enough to complete effectively for limited resources after settlement in adult habitats. Utilization of fish host energy stores by glochidia prevents their direct competition with adults for limited food and space resources during early development, as occurs in juvenile corbiculaceans and *D. polymorpha*. Glochidial parasitism of fish hosts also allows female unionaceans to devote relatively small amounts of nonrespired, assimilated energy to reproduction (2.8–14.5% of total nonrespired, assimilated energy), leaving the majority for somatic tissue growth (Table 11.2) (James 1985, Negus 1966, Paterson 1985). This allocation of a high proportion of energy to tissue growth is characteristic of species adapted to stable habitats. As unionaceans are long-lived and highly iteroparous, with greater than 6–10 reproductive periods throughout life, allocation of the majority of nonrespired energy to growth increases the probability of adult survival to the next reproductive period. Increased growth rate and reduction of reproductive effort increases the ability to compete and reduces the probability of predation and/or mortality that is associated with reproductive effort or removal from the substratum during periods of high water flow. All of these characteristics increase fitness in stable habitats (Sibly and Calow 1986).

In the majority of unionaceans, the greatest shell growth occurs in immature individuals in the first

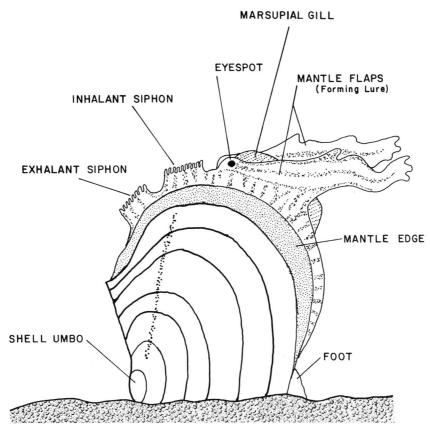

Figure 11.16 The modified mantle flaps of a female specimen of *Lampsilis ventricosa*, which mimic the small fish prey (note eye spot and lateral line-like pigmentation) of the predatory fish species that are hosts to the glochidia of this species. The posterior portions of the marsupial outer demibranchs are projected from the mantle cavity to lie between the mantle flap lures. When the mantle flap lures are disturbed by an attacking fish, glochidia are released through pores in the marsupial gill ensuring maximal contact with the fish host. Mantle flap lures are characteristic of the unionid genus *Lampsilis*.

four years of life (Fig. 11.17A). Indeed, relative shell growth rate in young unionids is greater than in sphaeriids or even the fast-growing species, *C. fluminea* and *D. polymorpha* (Table 11.2). In unionaceans, the shell growth rate declines exponentially with age, but the rate of tissue biomass accumulation remains constant or actually increases with age (Figs. 11.17A, B; see also Houkioja and Hakala 1978). Thus, early in life, increases in shell size and biomass occur preferentially over tissue accumulation; whereas after maturity (> 6 yr), shell growth slows and tissue is accumulated at a proportionately higher rate. The delayed maturity of unionids (6–12 yr, Table 11.2) allows all available nonrespired assimilation to be devoted to growth early in life.

Once mature, large adult unionaceans display high age-specific survivorship between annual reproductive efforts, being 81–86% in 5–7 year old *Anodonta anatina* (Negus 1966) and generally greater than 80% in mature *M. margaritifera* (12–90 yr) (Bauer 1983). High adult survivorship, long life spans, and low juvenile survivorship of union-

aceans accounts for the preponderance of large adult individuals in natural populations (Bauer 1983, James 1985, Negus 1966, Paterson 1985, Paterson and Cameron 1985, Tevesz *et al.* 1985). Populations dominated by adults are characteristic of stable, highly competitive habitats (Sibly and Calow 1986).

The preponderance of large, long-lived adults in unionacean populations causes them to be characterized by high proportions of standing crop biomass relative to biomass production rates. This relationship between standing crop biomass and biomass production rate can be expressed as turnover times, the time in days required for the population production of biomass to be equivalent to the average population standing crop biomass. Such turnover times can be computed in days as the average standing crop of a population divided by its mean daily productivity rate (Russell-Hunter and Buckley 1983). Turnover times have been based on dry weight, organic carbon, or nitrogen biomass units or energetic units in studies of freshwater molluscs (Russell-Hunter and Buckley 1983). Long-lived unionids,

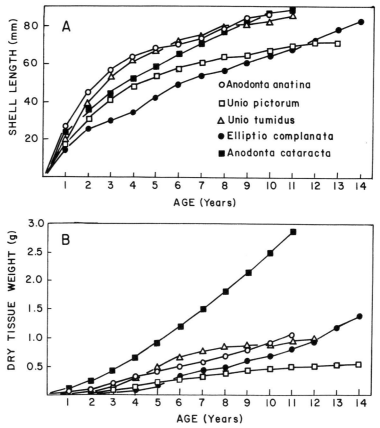

Figure 11.17 The shell and tissue growth of selected species of unionids (*Anodonta anatina,* open circles, *Unio pictorum,* open squares; *Unio tumidus,* open triangles; *Elliptio complanata,* solid circles; and *Anodonta cataracta,* solid squares). (A) Mean shell length increase with increasing age over the entire life span of each species. (B) Mean dry tissue weight increase with increasing age over the entire life span of each species. Note that increase in shell length declines with age in most species, while dry tissue weight increases either linearly or exponentially with age. [From data of Negus (1966), Paterson (1985), and Paterson and Cameron (1985)].

with populations dominated by large adults, have extremely long turnover times ranging from 1790–2849 days (computed from data in James 1985, Negus 1966, Paterson 1985) compared with sphaeriids (27–1972 days), *C. fluminea* (73–91 days), or *D. polymorpha* (53–869 days) (Table 11.2). Such extended turnover times are characteristic of long-lived, iteroparous species from stable habitats (Burky 1983, Russell-Hunter and Buckley 1983).

In only one aspect do unionaceans deviate from the life-history traits expected of species inhabiting stable habitats and experiencing extensive competition. That is in the production of very large numbers of very small young (glochidia). However, as described earlier, this is essentially an adaptation that ensures a sufficiently high probability of glochidial contact with appropriate fish hosts to maintain adequate juvenile recruitment rates. Accordingly, those species producing conglutinates that resemble fish host prey items have a higher probability of glochidial–host contact and, thus, produce fewer and larger glochidia. *M. margaritifera* releases very

small, free-living glochidia and has extraordinarily high fecundities (up to 17,000,000 glochidia/female) (Young and Williams 1984b), while species with conglutinates mimicking host prey items have much lower fecundities (200,000–400,000 glochidia/female) and larger glochidia (Kat 1984).

Extended life spans, delayed maturity, low effective fecundities, reduced powers of dispersal, high habitat selectivity, poor juvenile survival and extraordinarily long turnover times make unionaceans highly susceptible to human perturbations. Because of these life-history traits (particularly long life spans and low effective fecundities), unionacean populations do not recover rapidly once decimated by pollution or other human-mediated habitat disturbances (see Section III.A). Successful settlement of juveniles appears to be particularly affected by such disturbance, with population age–size structures marked by periods when entire annual generations are not recruited (Bauer 1983, Negus 1966, Payne and Miller 1989). Disturbance-induced lack of juvenile recruitment raises the specter of many North

American unionacean populations being composed of slowly dwindling numbers of long-lived adults destined for extirpation as pollution or other disturbance prevents juvenile recruitment to aging populations.

2. *Sphaeriidae*

The Sphaeriidae display great intra- and interspecific variation in life-history characteristics (for reviews see Burky 1983, Holopainen and Hanski 1986, Mackie 1984, Way 1988). Like unionaceans, their life-history traits do not fall neatly into suits associated with life in either stable or unstable habitats. Instead, they are a mixed bag, including the short life spans, early maturity, small adult size, and increased energetic input to reproduction associated with adaptation to unstable habitats, and the slow growth, low fecundity, and release of extremely large, fully developed young associated with adaptation to highly stable habitats (Sibly and Calow 1986) (Table 11.2). Sphaeriids are very euryoecic, with some members inhabiting stressful habitats such as periodically drying ephemeral ponds, and small streams prone to flash flooding and drying while others live in highly stable, profundal lake habitats (Burky 1983). [Here I attempt to generalize the life-history traits of sphaeriids within adaptive and evolutionary frameworks. However, the degree of inter- and intraspecific life-history variation within this group is such that for every generality drawn, specific exceptions can be cited.]

Of prime importance in understanding sphaeriid life-history traits is their ovoviviparous mode of reproduction. All species brood developing embryos in specialized brood chambers formed from evaginations of the exhalant side of the inner demibranch gill filaments. Maternal nutrient material is supplied to embryos developing in marsupia allowing considerable growth during development and release as fully formed miniature adults (see Section II.B.5). Thus, even though the sphaeriids have the smallest adult sizes of all North American freshwater bivalves, they release, by far, the largest young (Mackie 1984, Burky 1983). Based on data for 13 species, average birth shell length in sphaeriids ranges from 0.6–4.15 mm (Burky 1983, Holopainen and Hanski 1986, Hornbach and Childers 1986, Hornbach *et al.* 1982, Mackie and Flippance 1983a) making them much larger than unionacean glochidia (0.05–0.4 mm), juveniles of *C. fluminea* (0.25 mm), or veliger larvae of *D. polymorpha* (0.04–0.07 mm) (Table 11.2). Based on these shell length values, newborn sphaeriids have 3.4 to $2.1(10^5)$ times greater biomass than recently hatched individuals of other groups. Ratios of maximum adult shell

length : birth shell length (adult SL : birth SL) in sphaeriids range from 2.8 : 1 to 5.4 : 1, suggesting that newly released juveniles have biomasses that are 0.6–4.6% of the maximum adult biomass. There is a significant direct relationship between these two parameters, shown in Fig. 11.18.

The extremely large size of their offspring greatly reduces the fecundity of sphaeriids. Published values for average clutch sizes range from 3–24 young/adult for *Sphaerium*, 2–136 young/adult for *Musculium*, and 1.3–16 young/adult for *Pisidium* (Burky 1983, Holopainen and Hanski 1986). Even with reduced fecundity, the biomass of the large juveniles produced requires allocation of relatively larger amounts of nonrespired energy for reproduction ($\bar{x} = 19\%$, for a review see Burky 1983) compared to either unionaceans (< 14.8%), *C. fluminea* (15%), or *D. polymorpha* (4.9%) (Table 11.2). As developing juveniles have relatively high metabolic rates (Burky 1983, Hornbach 1985, Hornbach *et al.* 1982) and are supported by energetic transfer from adults (Mackie 1984), estimates of reproductive costs based solely on biomass of released juveniles

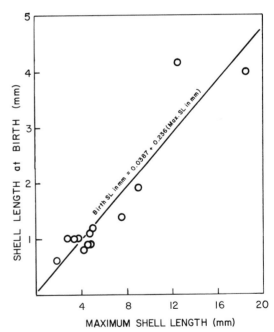

Figure 11.18 The relationship between shell length (SL) of juveniles at birth and maximal SL of adults for 13 species of sphaeriid freshwater clams. Note that juvenile birth length increases linearly with maximal adult size, suggesting that adult size may limit juvenile birth size in sphaeriids. The solid line represents the best fit of a linear regression relating birth size to maximal adult size as follows: Birth SL in mm = 0.0387 + 0.236 (maximal adult SL in mm) ($n = 13$, $r = 0.935$, $F = 76.3$, $P < 0.0001$). [Data from Holopainen and Hanski (1986), Hornbach and Childers (1986), Hornbach *et al.* (1982), and Mackie and Flippance (1983a).]

may be gross underestimates of actual costs in this group.

Life history hypotheses would predict the low fecundity and large birth size characteristic of sphaeriids to be adaptations maximizing fitness in stable habitats. However, the majority of sphaeriid species inhabit highly variable small ponds and streams or the profundal portions of large lakes subject to summer episodes of prolonged hypoxia (Burky 1983, Holopainen and Hanski 1986). Burky (1983) and Way (1988) have argued that while such habitats appear unstable, the harsh conditions associated with them are seasonally predictable, making them, in reality, stable. Thus, sphaeriid species inhabiting them have evolved adaptations preventing catastrophic population reductions during seasonally predictable epidoses of environmental stress (see Section II.C). This allows populations to reach carrying capacity and has lead to selection of life-history adaptations generally associated with the intense intraspecific competition that is characteristic of life in stable habitats.

Seasonal episodes of water level fluctuation and hypoxia can more severely impact smaller individuals. Thus, production of large, well-developed juveniles by sphaeriids may increase their probability of surviving predictable epidsodes of environmental stress. As such stresses can cause high adult mortality (Burky *et al.* 1985b, Holopainen and Jonasson 1983, Hornbach *et al.* 1982, Jonasson 1984b), fitness would also be increased by devoting greater proportions of nonrespired assimilation to production of greater numbers of young in any one reproductive effort, as chances of adult survival to the next reproduction are low. Thus, sphaeriids as a group devote relatively higher proportions of energy to reproduction than do other freshwater bivalves (Burky 1983) (Table 11.2). This combination of *r*-selected and K-selected traits, allowing optimization of fitness in habitats subject to periodic, predictable stress as displayed by sphaeriids, is a life-history strategy referred to as bet hedging (Stearns 1977, 1980).

The early maturation that is characteristic of many sphaeriid species (Burky 1983, Holopainen and Hanski 1986) (Table 11.2) is also adaptive, as it allows reproduction to occur before onset of seasonal episodes of environmental stress. This trait is displayed by sphaeriid species inhabiting ephemeral ponds and streams. These species are characterized by rapid growth, early maturity, and reduced numbers of reproductive efforts. Thus, *Musculium lacustre*, *Pisidium clarkeanum*, and *P. annandalie* from temporary drainage furrows in Hong Kong live less than one year and have only one or two re-

productive efforts (Morton 1985, 1986). An ephemeral pond population of *Musculium partumeium* had a life span of one year, was semelparous, reproduced just prior to pond-drying, and devoted a relatively large proportion of nonrespired assimilation (18%) to reproduction (Burky *et al.* 1985b). Ephemeral pond populations of *M. lacustre* are similarly univoltine and semelparous, devoting 19% of nonrespired energy to reproduction (Burky 1983). *Sphaerium striatinum*, in a population subject to periodic stream flooding, had life spans of one year or less and reproduced biannually (Hornbach *et al.* 1982), devoting 16.1% of nonrespired assimilation to reproduction (Hornbach *et al.* 1984a). In contrast, when populations of these species occur in more permanent habitats, they become bivoltine, reproducing in both spring and fall. Spring generations are iteroparous and reproduce in the fall and following spring; whereas fall generations are semelparous and reproduce only in the spring (Burky 1983, Burky *et al.* 1985b).

Some species of *Pisidium* live in the deeper, profundal portions of larger, more permanent lentic habitats where they are subjected to hypoxia after formation of a hypolimnion. A number of *Pisidium* species tolerate these hypoxic periods (see Section II.C.3, Burky 1983, Holopainen 1987, Holopainen and Hanski 1986, Holopainen and Jonasson 1983). As profundal species are tolerant of prolonged hypoxia, such environments are more stable than small ponds and streams, although they are far less productive, reducing food availability (Holopainen and Hanski 1986). Thus, sphaeriid populations in profundal habitats display much different life-history strategies than those in small ponds and streams (Burky 1983, Holopainen and Hanski 1986). Profundal populations display reduced growth rates, longer life spans of 4–5 yr, delayed maturation often exceeding one year, high levels of interoparity, and univoltine reproductive patterns (Holopainen and Hanski 1986), all life-history traits characteristic of more stable habitats (Sibly and Calow 1986). However, shallow-water and profundal populations of the same species of *Pisidium* may display quite different life-history tactics. Compared to profundal populations, shallow-water populations of the same species grow more rapidly, mature earlier, have shorter life spans, and tend toward semelparity (Holopainen and Hanski 1986), all life-history traits associated with unstable habitats (Sibly and Calow 1986). As pisidiids have well-developed dispersal capacities, it is unlikely that the broad variation in the life-history tactics of populations of the same species occupying different habitats primarily results from genetic adaptation to specific microhabitats. Rather, the majority of such variation is likely to represent environmentally induced, nongenetic, ecophenotypic plasticity. This wide inter- and intra-population plasticity in life-history traits is reflected in the highly variable turnover times reported for sphaeriids (27–1972 days, Table 11.2).

The growth rates of freshwater bivalves are highly dependent on ecosystem productivity. Populations from productive habitats have greater levels of assimilation and thus allocate greater absolute amounts of nonassimilated energy to growth (Burky 1983). Shallow, freshwater habitats are usually highly productive, warm, and rarely oxygen-limited; therefore, they support higher growth rates. Conversely, profundal environments are less productive, cooler, and often oxygen-limited, leading to lower bivalve growth rates. Since sphaeriids mature to a species-specific size irrespective of growth rate (Burky 1983, Holopainen and Hanski 1986), rapid growth leads to early maturity in shallow-water habitats and slow growth to delayed maturity in profundal habitats. Sphaeriids also have species-specific terminal sizes at which individuals die whether that size is attained rapidly or slowly. As growth rate determines the time required to reach terminal size, fast-growing individuals from shallow water habitats reach terminal sizes more rapidly (often within less than one year) allowing participation in only one or two reproductive efforts, while slow-growing individuals from profundal habitats (Holopainen and Hanski 1986) reach terminal size more slowly, allowing participation in a greater number of reproductive efforts.

This fundamentally ecophenotypic nature of intraspecific life-history variation in sphaeriids has been demonstrated for *Pisidium casertanum*. When individuals of this species were reciprocally transferred between two populations with different life-history traits or co-reared under similar laboratory conditions, the majority of life-history trait differences proved to be environmentally induced. However, electrophoresis indicated genetic differences between the populations and transfer and laboratory co-rearing experiments indicated that a small portion of the observed life-history variation could be genetically based (Hornbach and Cox 1987).

Such extensive capacity for ecophenotypic plasticity may account for the euryoecic nature and cosmopolitan distributions of sphaeriids (see Section III.A). Certainly, the capacity to adjust growth rates, maturity, reproductive cycles, life cycles, and energetic allocation patterns to compensate for broad habitat variation in biotic and abiotic factors allows species in this group to have relatively wide niches and thus broad distributions.

3. Corbicula fluminea

The introduced Asian freshwater clam, *Corbicula fluminea,* unlike unionaceans and sphaeriids, displays life-history traits clearly adapted for life in unstable, unpredictable habitats (McMahon 1983a). As such, it has been the most invasive of all North American freshwater bivalve species. *C. fluminea* grows very rapidly, in part because it has higher filtration and assimilation rates than other bivalve species (Foe and Knight 1986a, Lauritsen 1986a, Mattice 1979). In a natural population, only a relatively small proportion of assimilation (29%) was devoted to respiration (Table 11.2), the majority (71%) being allocated to growth and reproduction (Aldridge and McMahon 1978). These data were confirmed by laboratory studies showing that 59–78% (Lauritsen 1986a) or 58–89% of assimilation (Foe and Knight 1986a) went to tissue production. Thus, *C. fluminea* has the highest net production efficiencies recorded for any freshwater bivalve species, which is reflected by its very low turnover times, ranging from 73–91 days (Table 11.2).

The very high proportion of nonrespired assimilation (85–95%) devoted to growth in *C. fluminea* (Aldridge and McMahon 1978, Long 1989), sustains high rates of growth (shell length = 15–30 mm in the first year of life, 35–50 mm in the terminal third to fourth year) (McMahon 1983a). High growth rates decrease the probability of predation, as many fish and bird predators feed only on small individuals (McMahon 1983a; see also Section III.C.3) and, thus, increase the probability of survival to the next reproduction in this iteroparous species. Indeed, the increase in shell size occurs at the expense of tissue production during summer maximal growth periods (Long 1989), suggesting that larger shells optimize fitness. The high growth rates of *C. fluminea* allow it to sustain the highest population production rates (10.4–14.6 g organic carbon/m² yr, Aldridge and McMahon 1978) reported for any species of freshwater bivalve (Burky 1983).

Newly released juveniles are small (shell length = 250 μm) but completely formed, having a well-developed and characteristically D-shaped bivalved shell, adductor muscles, foot, statocysts, gills, and digestive system (Kraemer and Galloway 1986) (Fig. 11.15A). As juveniles are denser than water, they settle and anchor to sediments or hard surfaces with a mucilaginous byssal thread. However, they are small enough (0.25 mm) to be suspended on turbulent water currents and dispersed great distances downstream (McMahon 1983a). A relatively low amount of nonrespired assimilation is allocated to reproduction (5–15%, Aldridge and McMahon 1978, Long 1989), equivalent to that expended by unionids but less than the average expended by sphaeriids (19%, Table 11.2). However, the elevated assimilation rates of this species allow allocation of higher actual values of energy to reproduction than occur in other freshwater bivalves.

Because the juvenile of *C. fluminea* is small (organic carbon biomass = 0.136 μg, Long 1989), fecundity is large, ranging from 97–570 juveniles/adult/day during reproductive seasons, for an average annual fecundity estimate of 68,678 juveniles/adult/yr (McMahon 1983a). Juvenile surivorship to successful settlement is extremely low and mortality rates remain high throughout adult life (74–98% in the first year, 59–69% in the second year, and 93–97% in the third year of life, computed from the data of McMahon and Williams 1986a, Williams and McMahon 1986), making the vast majority of individuals in populations juveniles and immatures. High adult mortality and population dominance by immature individuals is characteristic of species adapted to unstable habitats (Charlesworth 1980, Sibly and Calow 1986, Stearns 1980).

The majority of North American *C. fluminea* populations display two annual reproductive periods, one in spring and early summer and the second in late summer (McMahon 1983a). *C. fluminea* is hermaphroditic and capable of self-fertilization (Kraemer and Galloway 1986, Kraemer *et al.* 1986) such that single individuals can found a new population. Spermiogenesis occurs only during reproductive periods, but gonads contain mature eggs throughout the year (Kraemer and Galloway 1986, Long and McMahon 1987).

C. fluminea matures within 3–6 months at a small shell length of 6–10 mm (Kraemer and Galloway 1986). Thus, juveniles born in the spring may grow to maturity and participate in the reproductive effort the following fall (Aldridge and McMahon 1978, McMahon 1983a) (Table 11.2). The maximum life span is highly variable between populations and temporally within populations, ranging from 1–4 yr (McMahon 1983a, McMahon and Williams 1986a). Early maturity allows individuals of this iteroparous species to participate in 2–7 reproductive efforts, depending on life span.

The relatively short life span, early maturity, high fecundity, bivoltine juvenile release patterns, high growth rates, small juvenile size, and capacity for downstream dispersal of *C. fluminea* makes it both highly invasive and well adapted for life in truly unstable, disturbed lotic habitats that are subject to unpredictable catastrophic faunal reductions. Its ex-

tremely high reproductive potential and growth rates allow it to reach or re-establish high densities after invading a new habitat or after catastrophic population declines. Thus, it is highly successful in North American drainage systems that are subjected to periodic human interference such as channelization, navigational dredging, pearling, sand and gravel dredging, commercial and/or recreational boating, and organic and/or chemical pollution, compared to far less resilient unionaceans or sphaeriids (McMahon 1983a).

Surprisingly, *C. fluminea* is more susceptible to environmental stresses such as temperature extremes, hypoxia, drying, and low pH than are most sphaeriids and unionaceans (Byrne 1988, Byrne *et al.* 1988, Kat 1982, McMahon 1983a), making its populations more susceptible to declines from human disturbance. With only a limited capacity to tolerate unpredictable environmental stress, why is *C. fluminea* so successful in disturbed habitats? The answer lies in its ability to recover from disturbance-induced catastrophic population crashes much more rapidly than either sphaeriids or unionids. *C. fluminea* rapidly re-establishes populations even if disturbance has reduced them to a few widely separated individuals, as all individuals are hermaphrodites capable of self-fertilization and have high fecundities. Downstream dispersal of juveniles from viable upstream populations also allows rapid reinvasion of decimated populations. After juvenile reinvasion of depauperated populations, the accelerated growth, high fecundity, and relatively short life spans of this species allows rapid population re-establishment, including normal age–size distributions and densities within 2–4 years (for examples see McMahon 1983a). Biannual reproduction in *C. fluminea* also increases the probability of surviving catastrophic density reductions, as it prevents loss of an entire generation to a chance environmental disturbance (bet hedging, Stearns 1980). The capacity for rapid recolonization of habitats from which populations have been extirpated allows *C. fluminea* to sustain populations in substrata that are subject to periodic scouring during floods; the slower growing and maturing unionids are eliminated from such habitats (Way *et al.* 1990a).

Like sphaeriids, North American *C. fluminea* populations display an extraordinary degree of interpopulation variation in life-history traits. As there is little or no genetic variation among North American populations (McLeod 1986), this interpopulation life-history variation must be ecophenotypic. Growth rates increase and time to maturity and life spans decrease in populations from more productive habitats (McMahon 1983a). On the northern edge of its North American range, low temperatures reduce growth and reproductive periods, making populations univoltine rather than bivoltine in reproduction. I have observed univoltine reproduction and semelparity in a slow-growing population of *C. fluminea* within an oligotrophic Texas lake. Even within populations, life-history tactics vary greatly, dependent on year-to-year variations in temperature and primary productivity (McMahon and Williams 1986a, Williams and McMahon 1986). As in sphaeriids, the capacity for substantial ecophenotypic life-history trait variation is highly adaptive in *C. fluminea*, as it allows colonization of a broad range of habitats. Being at once highly euryoecic and highly invasive has made it the single most successful and economically costly aquatic animal species introduced to North America (Isom 1986).

4. Dreissena polymorpha

The zebra mussel, *Dreissena polymorpha,* is the most recently introduced bivalve species to North American freshwaters (Hebert *et al.* 1989). Like *C. fluminea,* many of its life-history characteristics (reviewed in Mackie *et al.* 1989) make it highly invasive. Unlike all other North American bivalve species, it releases sperm and eggs to the surrounding medium such that fertilization is completely external. A free-swimming planktonic veliger larva (Fig. 11.15C) hatches from the egg and remains suspended in the water column where it feeds and grows for 8–10 days before settling to the substratum. This behavior enhances the dispersal ability of the zebra mussel. Adults attached to floating objects by the byssus can also be transported long distances downstream. Adults become sexually mature in the second year of life (first year in some North American populations) and typically have life spans ranging from 5–6 yr and, like *C. fluminea,* sustain high growth rates throughout life (Fig. 11.19). *Dreissena polymorpha* is iteroparous and univoltine; an individual participates in 3–4 annual reproductive periods over the course of its life. The egg and freshly hatched veliger are small (diameter = 40–70 μm), but the post-veliger grows to 180–290 μm just prior to settlement and juvenile metamorphosis (Mackie *et al.* 1989), indicative of a 100–400 fold increase in biomass during planktonic growth.

Maximal *D. polymorpha* adult size ranges from 3.5–5 cm depending on growth rate, which, like terminal size, is dependent on the primary productivity and temperature of the habitat. Like *C. fluminea, D. polymorpha* allocates an extremely high percentage (96.1%) of nonrespired assimilation to somatic growth, leaving only 3.9% for reproduction (Mackie *et al.* 1989) (Table 11.2). Allocation of a large proportion of nonrespired assimilation to growth allows

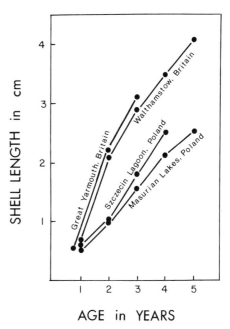

Figure 11.19 Shell growth rates in European populations of the zebra mussel, *Dreissena polymorpha.* (Redrawn from Morton 1969.) Growth rates in North American populations in Lake Erie are similar to or faster than those depicted here for fast-growing British populations.

individuals to rapidly increase in size, making them more competitive and less subject to predation (Sibly and Calow 1986). Zebra mussel veligers settle on the shells of established individuals, forming thick mats or clusters of individuals many shells deep (Mackie *et al.* 1989). In such mats, competition for space and food are very intense. Thus, rapid growth of an individual to a large size is highly adaptive as it increases the probability of development of a stable byssal holdfast to the substratum and the positioning of siphons at the mat surface, where food and oxygenated water are most available. In spite of the low levels of energy devoted to reproductive effort by *D. polymorpha,* its very small eggs make individual fecundity large, ranging from 30,000–40,000 eggs/female.

Dreissena polymorpha population densities range from 7000–114,000/m² and standing crop biomasses from 0.05–15 kg/m² (Mackie *et al.* 1989). The high population densities and biomass result from the tendency of juveniles to settle on substrata already inhabited by adults. Also, adults attach to the shells of other adults by the byssus to form dense mats or clumps that are many layers of individuals thick. High individual growth rates and population densities lead to very high population productivities, estimated to be 0.05–14.7 g C/m²/yr (computed from dry tissue values in Mackie *et al.* 1989), values comparable to those of highly productive *C. fluminea* populations. However, as population growth and

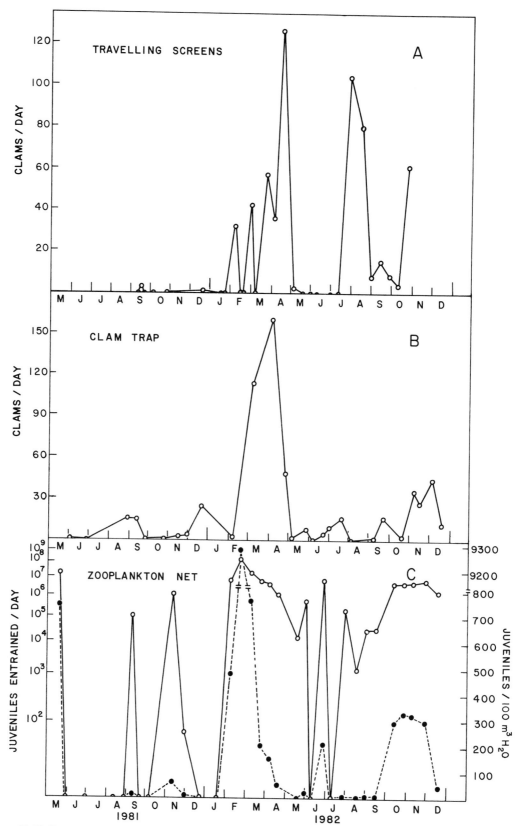

Figure 11.20 Seasonal variation in downstream dispersal behavior by juvenile, subadult, and adult *Corbicula fluminea* in the intake canal of a power station. (A) Rate of impingement of adults dispersing downstream onto traveling screens in front of water intake embayments (shell length > 10 mm). (B) Rate of retention of subadults (shell length = 1–7 mm) in a clam trap held on the substratum surface of the intake canal. (C) Juveniles suspended in the water column (shell length < 2 mm). Right ventrical axis is density of juveniles in the intake

productivity are highly habitat-dependent in this species, turnover times are variable, ranging from extremely low values of 53 days to relatively high values of 869 days (Table 11.2).

The life-history traits of a high growth rate throughout life, high fecundity, short life spans, and the capacity for both adult and larval stage downstream dispersal make *Dreissena* (like *Corbicula*) a highly invasive species. However, unlike *C. fluminea*, *D. polymorpha* populations tend to be restricted to much more stable habitats such as larger, permanent lakes and rivers. Its apparent preference for more stable habitats is reflected by its original distribution in the Caspian Sea and Ural River (large stable habitats), avoidance of shallow, near-shore habitats, relatively long age to maturity (generally in the second year of life), iteroparity, gonochorism, and relatively high adult survivorship (26–88% per year) (Mackie *et al.* 1989).

Restriction of its original range to the Caspian Sea and Ural River also suggests a limited natural capacity for dispersal between isolated drainage systems. Dispersal of the species through Europe occurred only in the nineteenth century (and continues in western Asia today). This recent dispersal was accomplished primarily by human vectors, including transport of adults attached to boat hulls, ballast water dumping, and transport of veligers through canal systems interconnecting catchments. Adults have a relatively low tolerance to prolonged air exposure (Mackie *et al.* 1989), precluding extensive natural overland dispersal unless it is human-mediated. Thus, the future dispersal of this species between catchments in North America will be primarily by human vectors, with larger, permanent bodies of water most susceptible. Without natural dispersal vectors, dispersal of *D. polymorpha* through North American drainages may proceed at a somewhat slower pace than that recorded for *C. fluminea,* but human activities will ensure that this species will eventually be widely distributed in North American freshwaters.

C. Ecological Interactions

1. Behavioral Ecology

Other than detailed studies of burrowing (see Section II.A.2), information on bivalve behavior is sparse. Indeed, a recent major review of molluscan neurobiology and behavior (Willows 1985, 1986) was entirely devoid of bivalve references. Lack of information about bivalves reflects the difficulties associated with making behavioral observations on predominantly sessile, infaunal species completely surrounded by a shell, rather than a lack of complex and intriguing behaviors.

A number of interesting behaviors are associated with reproduction in freshwater bivalves, including those involved with ensuring glochidial contact with fish hosts in unionaceans (see Section II.B). Adult *C. fluminea* display unique downstream dispersal behavior associated with reproductive periods. While juvenile clams (SL < 2 mm) are found suspended in the water column throughout the year, immatures (SL = 2–7 mm) and adults (SL > 7 mm) leave the substratum to be carried passively downstream over the sediment surface (''rolling'') on water currents only prior to reproductive periods (Fig. 11.20). Dispersing adults have lower dry tissue weights, lower tissue organic carbon to nitrogen ratios (Williams and McMahon 1989), higher levels of ammonia excretion, and reduced molar oxygen consumption to nitrogen excretion ratios than those remaining in the substratum (Williams 1985, Williams and McMahon 1985), which are indicative of poor nutritional condition. Thus, downstream dispersal allows starving individuals to move away from areas of low food availability and high intraspecific competition into areas more nutritionally favorable for reproductive efforts (Williams 1985, Williams and McMahon 1986, 1989).

Some adult unionacean bivalves and *C. fluminea* display surface locomotory behavior. Surface locomotion involves the same movements of the foot and valves described in Section II.B.2 for burrowing but is horizontal rather than vertical. Indeed, surface locomotion by unionaceans is fairly common (Imlay 1982). Tracts left in sediments by surface locomoting uninaceans are 3–10 m long (Golightly 1982), indicative of major short-term horizontal displacement. The adaptive significance of surface locomoation through sediments in freshwater bivalves is not well understood. Such behavior could attract potential predators. However, it may be involved with pedal feeding on organic sediment deposits (for details see Section III.C.2).

canal water column (solid circles connected by dashed lines). Left vertical axis is number of juveniles entrained daily with intake water (open circles connected by solid lines). Note that adult and subadult clams display significant downstream dispersal behavior during only two periods, March–May and July–August just prior to the spring and fall reproductive periods of the population (April–July and September–November). Juveniles occurred in the water column throughout the year; peak juvenile water column densities occurred during reproductive periods and midwinter periods of low ambient water temperature. (From Williams and McMahon 1986.)

Some species, such as *Anodonta grandis,* migrate vertically on the shore with seasonal changes in water level (White 1979), thereby avoiding prolonged emersion. Other species such as *Uniomerus tetralasmus, C. fluminea,* and some sphaeriids remain in position and suffer prolonged emersion during periods of receding water (see Section II.C.4). In Texas, I have observed fire ants, *Solenopsis invincta,* killing bivalves exposed to air by receding water levels. [Hence, this introduced insect species may represent a new threat to unionaceans and sphaeriids throughout its expanding range in the southeastern United States.]

Some sphaeriids, *C. fluminea,* and *D. polymorpha* are also capable of crawling over hard substrata or macrophytes. *Sphaerium corneum* holds the shell valves erect and moves over hard substrata or macrophytes by extending the foot tip, anchoring it with mucus, and then contracting pedal muscles to draw the body forward (Wu and Trueman 1984). Juvenile and young specimens of *C. fluminea* move over hard substrata in the same manner, while adults similarly crawl over hard surfaces lying on one of the valves (Cleland 1988). Small specimens of *D. polymorpha* are highly active crawlers and can climb smooth vertical surfaces. Young individuals of this species routinely discard their byssus and migrate to new positions where they resecrete attachment threads. As an example, juveniles settling in shallow water during the summer may migrate to deeper waters in the winter (Mackie *et al.* 1989).

Freshwater bivalves also detect and respond to a number of external environmental cues. Chief among these responses is valve closure in response to irritating external stimuli. Stimuli for valve closure are likely detected by sense organs concentrated on the mantle edge and the siphons (see Section II.B.6). Valve closure effectively seals internal tissues from the damaging effects of external irritants. Almost all freshwater bivalves tolerate some degree of facultative anaerobiosis (see Section II.C.3). Thus, individuals exposed to irritants can remain anaerobic with the valves clamped shut for relatively long periods until external conditions become more favorable. Freshwater bivalves close the valves on exposure to heavy metals (Doherty *et al.* 1987), chlorine, and other biocides (Mattice 1979, Mattice *et al.* 1982, McMahon and Lutey 1988), and high levels of suspended solids (Aldridge *et al.* 1987). This ability allows *C. fluminea* to avoid intermittent exposure to chlorination or other biocides, making chemical macrofouling control of this species extremely difficult (Goss *et al.* 1979, Mattice *et al.* 1982). Valve closure in immediate response to tactile stimulation of the mantle edge or siphons is

also a predator defense mechanism common to all freshwater species.

Also of interest are the reactions of freshwater bivalves to prolonged emersion. The physiological adaptations to emersion are discussed in Section II.C.4. Here, behavioral responses are described in greater detail. *C. fluminea,* when exposed in air, displays four major responses: (1) escape behavior involving extending the foot in an attempt to burrow; (2) valves gaped widely with mantle edges parted, opening the mantle cavity directly to the atmosphere; (3) valves narrowly gaped with mantle edges exposed, but cemented together with mucus; and (4) valves clamped shut. Behaviors (1) and (2) are never displayed more than 6% of the time in air. Exposure of sealed mantle edges is associated with aerial gas exchange (Byrne 1988, McMahon and Williams 1984), but results in evaporative water loss. When valves are closed, water loss is minimized; but oxygen uptake ceases (Byrne *et al.* 1988, McMahon and Williams 1984). As temperature increases (Table 11.3) or relative humidity decreases, duration of mantle edge exposure decreases relative to that with the valves clamped shut (Byrne 1988, Byrne *et al.* 1988, McMahon, 1979b). Hence, behaviors associated with water loss are reduced in response to increased desiccation pressure. Indeed, relative humidities near zero or temperatures above 30°–35°C cause the valves to remain continually closed (Byrne 1988, Byrne *et al.* 1988, McMahon 1979b) (Table 11.3), preventing excessive water loss but making individuals continually anaerobic. I have observed similar behaviors in emerged unionaceans. Some species plug the siphons with mucus to further reduce water loss and spend less time with the mantle edges exposed than does *C. fluminea.* Ability of emersed individuals to adjust behaviors associated with aerobic gas exchange to external desiccation

Table 11.3 The Effects of Temperature on the Percentage of Time Spent in Various Valve Movement Behaviors in Emerged Specimens of *Corbicula fluminea*[a]

Temperature (°C)	Time with Valves Closed (%)	Time with Mantle Edge Exposed (%)	Time with Mantle Edge Parted or Attempting to Burrow (%)
15	29.5	65.8	4.7
25	51.2	43.5	5.3
35	90.5	9.1	0.4

[a]From Byrne (1988).

pressures is a complex behavior, requiring the capacity to sense, integrate, and respond to external temperature and relative humidity levels and internal osmotic concentration. Certainly, the capacity for such behaviors is worthy of further experimental investigation.

Freshwater bivalves may also have as yet undiscovered circadian patterns of behavior. Both ion uptake and oxygen consumption rates are greater during dark than light hours in unionaceans and *C. fluminea* (Graves and Dietz 1980, McCorkle *et al.* 1979, McCorkle-Shirley 1982; see also Section II.C.2), strongly suggesting the presence of circadian-activity patterns in feeding, reproduction, and burrowing. I have recorded increases in the density of juvenile *C. fluminea* in the water column during dark hours, suggesting that adults preferentially release juveniles at night. Freshwater bivalves may have circadian burrowing cycles, retreating more deeply into sediments during light hours to avoid visually oriented predators like fish and birds. These speculations require further investigation.

2. Feeding

Most freshwater bivalves primarily feed by filtering suspended material from water as it passes over the gills. The anatomy of bivalve gills is described in Section II.B.2. Water flow across the gills is maintained by the beating of powerful lateral cilia located along each side of the filaments. Projecting laterally from each side of the leading edge of the filament are tufts or cirri of partially fused eulaterofrontal cilia at intervals of 2–3.5 μm (Fig. 11.5A and 11.21). The primary filtering mechanism is composed of eulaterofrontal cirri on adjacent filaments, which project toward each other forming a stiffened grid. Water is driven through this grid by the lateral cilia, and suspended seston (phytoplankton, bacteria, and fine detritus) is retained on the eulaterofrontal cilia. The filtering mechanism probably functions both by forming a filtering mesh and creating eddies in which seston particles settle. The mesh size of the eulaterofrontal cirri (2–3.5 μm) is highly correlated with the size of particles filtered. Particles removed range in size from 1.5–10 μm, with *D. polymorpha* and *C. fluminea* both able to efficiently filter much smaller particles ($\leq$ 1.0 μm) than the majority of freshwater species (Fig. 11.22) (Jorgensen *et al.* 1984, Paterson 1984, Way 1989, Way *et al.* 1989).

Some investigators claim that the gill is covered with a fine mucus net acting as the primary filter (for a review see Morton 1983). While the degree to which mucus is involved in the filtering mechanism continues to be debated, there is little doubt that the dense meshwork formed by the eulaterofrontal cilia could act as an effective filter (Morton 1983, Way 1989).

The basal portions of the eulaterofrontal cilia contain stiffening elements, but the free distal ends beat at right angles to the filament axis, apparently driving particles trapped by the eulaterofrontal cirri onto the leading edge of the gill filament. There, particles are carried along bands of frontal cilia running the length of the filament (Figs. 11.5A and 21), either dorsally or ventrally. Particles are deposited in specialized food grooves on the ventral edge of the demibranchs and/or at the base of the gill axis (depending on the species) (Figs. 11.2 and 11.5A–C). In sphaeriids, *C. fluminea*, and presumably unionaceans, the cilia of food grooves are differentiated from other forms of gill ciliation (Way *et al.* 1989). Unique, long frontal cirri of fused cilia beating at a right angle to the long axis of the filaments are also associated with the frontal cilia of sphaeriid and corbiculacean bivalves (Fig. 11.21A,B). Their function has yet to be elucidated (Way *et al.* 1989).

Cilia in the food groove carry filtered particles anteriorly to the labial palps, which are paired triangular flaps located on each side of the mouth (Figs. 11.2 and 11.4). The outer labial palp lies against the outer ventral side of the outer demibranch, and the inner palp, against the inner ventral side of the inner demibranch (Fig. 11.2). The epithelial surfaces of the palps facing away from the demibranchs are smooth. In contrast, the palp surfaces adjacent to the demibranchs are highly corrugated into a series of parallel ridges and channels lying obliquely to the ciliated oral groove that leads to the mouth. The oral groove is formed between the fused dorsal junction of the inner and outer labial palps (Morton 1983). The corrugated palp surface sorts filtered particles into those accepted for ingestion and those rejected. Generally, smaller particles are carried over the upper edges of palp corrugations by cilia beating at right angles to the long axis of the corrugations. Particles carried over palp corrugations are deposited in the oral groove ciliary tract, which carries them to the mouth for ingestion. Denser, larger particles fall between the corrugations where cilia carry them to the ventral edge of the palp. Here, they are bound in mucus and carried by a ciliated ventral groove to the tip of the palp for release onto the mantle as pseudofeces (Morton 1983). Pseudofeces are carried by mantle cilia to accumulate at the base of the inhalant siphon (Fig. 11.2), where they are periodically expelled with water forced violently from the inhalant siphon by rapid valve adduction (i.e., valve clapping) (Morton 1983).

The palps also sort medium-sized particles.

A

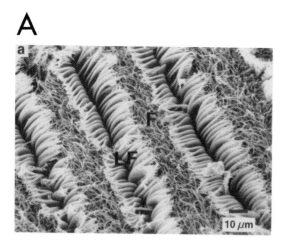

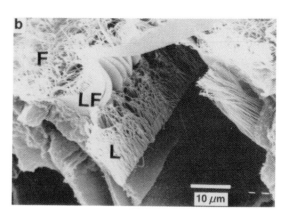

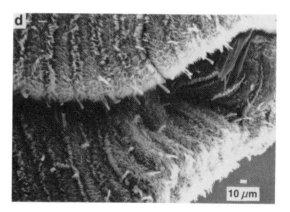

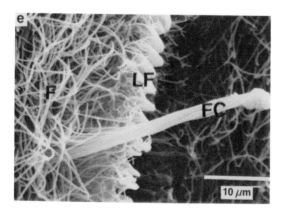

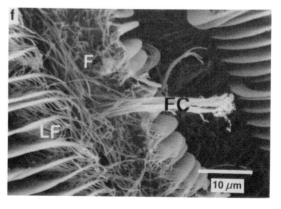

Figure 11.21 Scanning electron micrographs of the gill ciliation of representative freshwater bivalves. (A) *Musculium transversum* (Sphaeriidae): (a) arrangement of frontal and eulaterofrontal ciliation on the leading edges of the filaments in the mid-gill region; (b) oblique view of a cross-sectional fracture of the gill showing frontal and eulaterofrontal cilia on the leading edge of the filament and lateral cilia on the side of the filament; (c) ciliation of the food groove on the ventral edge of the gill; (d) posterior portion of the outer (foreground) and inner (background) demibranchs of the gill, (e) a frontal cirrus formed from fused cilia emerging between the eulaterofrontal and frontal cilia on the leading edge of a filament; (f) a frontal cirrus emerging from a band of

B

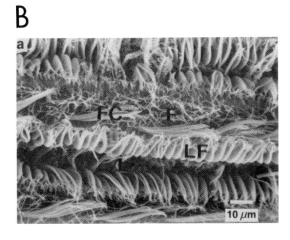

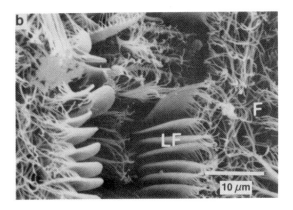

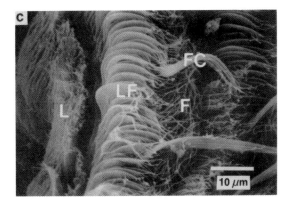

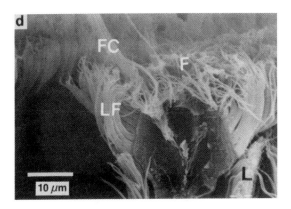

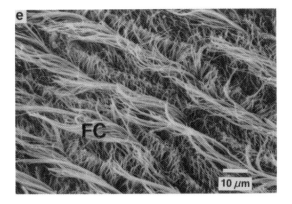

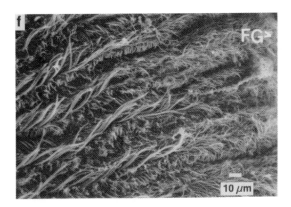

frontal cilia. (B) *corbicula fluminea* (Corbiculidae): (a) leading edge of a gill filament showing frontal, eulatero-frontal, and lateral cilia and frontal cirri; (b) high-magnification view of frontal and eulaterofrontal cilia; (c) oblique view of a longitudinal fracture through the mid-gill region showing all four types of ciliation, including lateral cilia lining the sides of the gill filament; (d) cross-section of the mid-gill region showing the origin of the frontal cirrus; (e) lower region of the gill showing the presence of less well organized frontal cilia; (f) food

(Figure continues)

C

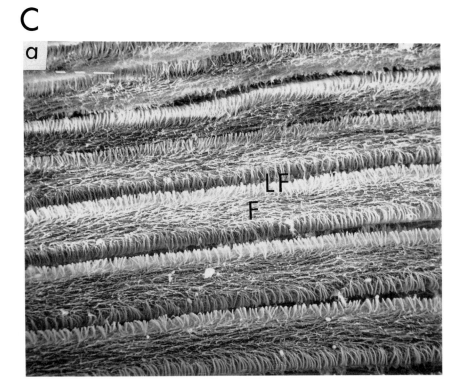

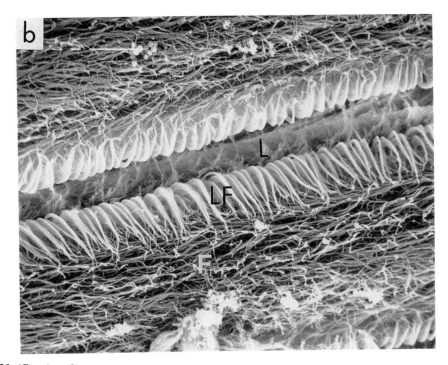

Figure 11.21 (Continued)

groove region at the ventral edge of the inner demibranch; note loss of ciliary organization in food groove, including lack of frontal cirri. (C) *Oliquaria reflexa* (Unionidae): (a) frontal and eulaterofrontal ciliation on the leading edges of several gill filaments (500X); (b) high-power view showing lateral cilia on sides of the gill filaments between tracts of frontal and eulaterofrontal cilia on adjacent filaments. Note lack of frontal cirri characteristic of the gill ciliation of Sphaeriidae (A) and *Corbicula* (B). Label key for all parts is: F, frontal cilia; FC, frontal cirrus; L, lateral cilia; LF, eulaterofrontal cilia; and FG, food groove. (Photomicrographs supplied by Tony Deneka and Daniel J. Hornbach (Macalester College) and Carl M. Way (U.S. Army Corps of Engineers, Waterways Experiment Station).

Smaller medium-sized particles tend to remain longer on the corrugation crests than in the intervening grooves, and eventually reach the oral groove for ingestion. Larger medium-sized particles tend to remain longer in the grooves between corrugations and are eventually rejected as pseudofeces. Thus, particle sorting on the labial palps appears to be strictly a function of particle size and, perhaps, density.

The width of palp corrugations and gill filaments can be adjusted by both muscular activity and degree of hemolymph distension of gill and palp blood sinuses, thus allowing active size selection of particles filtered and accepted for ingestion (Morton 1983). When individuals of the mussel *Elliptio companata* were fed on natural seston, the percentage of particles that were retained declined linearly over a particle size range of 5 μm (nearly 100% retention) to > 10 μm (less than 10% retention), suggesting size selection (Paterson 1984). Similar selection for particles most common in the ambient water column has been reported for *Musculium transversum* (Way 1989) and *C. fluminea* (Way *et al.* 1990b). In addition, different mussel species are reported to select quite different algal types. In the same river pool, a population of *Amblema plicata* ingested a greater proportion of green algae, a lower diversity of algal species, and different algal species compared to a sympatric population of *Ligumia recta* (Bisbee 1984). Such interspecific differences in particle selection may allow sympatric species to avoid intense competition for food resources.

The filtering rate of bivalves can be computed as follows:

$$C = M/t(\ln C_o/C_t)$$

where M = volume of the suspension filtered, t = time of filtration, and C_o and C_t = concentrations of filtered particles at times O and t, respectively (Jorgensen *et al.* 1984). Filtration rates appear to be relatively elevated in *C. fluminea* compared to other freshwater bivalves. For a 25 mm SL specimen, the rate ranged from 300–2500 liter/hr depending on temperature and seston concentration (Foe and Knight 1986b, Lauritsen 1986b, Long 1989, Way *et al.* 1990b). Increased filtration rate may be associated with specialized frontal cirri on the gill filaments of this species (Fig. 11.21B) (Way *et al.* 1989). The filtering rate of specimens of *C. fluminea* fed on nautral seston concentrations was not significantly correlated with field ambient water temperature (Long 1989) and was independent of temperature between 20° and 30°C in the laboratory (Lauritsen 1986b), suggesting an unusual capacity for temperature compensation of gill ciliary filtering activity. The filtration rate is also somewhat independent of

temperature in *Sphaerium striatinum* and *Musculium partumeium,* reaching maximal values during spring and fall reproductive periods when ambient water temperatures were well below midsummer highs (Burky *et al.* 1981, 1985a, Way *et al.* 1981).

Above an upper critical limit, high concentrations of suspended particles reduce filtering rates (Burky 1983, Burky *et al.* 1985a, Morton 1983, Way 1989, Way *et al.* 1990b). Increased particle concentration depressed the filtering rate of *S. striatinum* (Hornbach *et al.* 1984b) and *M. partumeium* (Burky *et al.* 1985a) at concentrations well below those of natural seston. In both species, however, rate of particle ingestion becomes nearly constant above the critical particle concentration, suggesting that the observed reduction in filtering rate is associated with control of the ingestion rate. Particle concentrations within ambient seston concentrations do not affect the filtering rate in *C. fluminea* (Mattice 1979, Long 1989); ingestion rates increase directly with concentration (Long 1989). However, the filtration rate in *C. fluminea* declined 3-fold with increasing particle concentration between 0.33 and 2.67 μl algal volume/liter suggesting inhibition at very high concentrations. Increasing algal concentration over this range still resulted in a 3.5-fold rise in ingestion rate (Lauritsen 1986b). Similar reduction in filtration rate at particle concentrations above natural levels has been reported for both *C. fluminea* (Way *et al.* 1990b) and *M. transversum* (Way 1989). In contrast to *C. fluminea,* the filtration rate of the mussel *E. complanta* was depressed by increasing natural seston concentrations (Paterson 1984). The filtration rate varied between different populations of *C. fluminea,* but the amount ingested was highly similar, suggesting regulation of the filtration rate to control ingestion rates (Way *et al.* 1990b). Such data indicate both that the seasonal and environmental responses in filter feeding of freshwater bivalves are species-specific and laboratory measurements of filtration and consumption rates are likely to be invalid unless carried out at natural seston concentrations and sizes.

Suspended silt in the water column can inhibit filtering and consumption rates in freshwater bivalves, perhaps by overwhelming ciliary filtering and sorting mechanisms. Very high levels of suspended silt caused significant reduction in the filtering and metabolic rates of three species of unionaceans whether exposure was infrequent (once every 3 hr or frequent (once every 0.5 hr), suggesting interference with ciliary maintenance of water flow over the gills. Furthermore, those individuals experiencing frequent exposure greatly increased reliance on carbohydrate catabolism (Aldridge *et al.* 1987), a phenomenon symptomatic of short-term starvation (Cleland 1988). For these reasons, union-aceans can be eliminated from habitats with high silt loads (Adam 1986). In contrast, suspended sediments within a naturally occurring concentration range did not affect shell or tissue growth in *C. fluminea* (Foe and Knight 1985). This may account in part for its ability to colonize lotic environments with greater flow rates and levels of suspended solids than are tolerated by most unionacean species (Payne and Miller 1987; see also Section III.A). Surprisingly, growth in juvenile unionaceans was stimulated rather than inhibited in the laboratory by small amounts of suspended silt in their algal food (Hudson and Isom 1984).

In smaller streams, where phytoplankton productivity may be extremely low, most suspended organic matter available to filter feeders is either particulate organic detritus or heterotrophic bacteria and fungi (Nelson and Scott 1962). While experiments are few, at least some freshwater bivalves can efficiently filter suspended bacteria. Certainly, *D. polymorpha* appears capable of efficiently extracting particles of bacterial size (< 1.0 μm in diameter) (Fig. 11.22) as can *C. fluminea* (Way *et al.* 1990b). In contrast, *M. transversum* (Way 1989) and unionaceans (Fig. 11.22) cannot efficiently filter bacterial-sized particles. As bacteria and detrital particles may comprise the vast majority of organic matter in some aquatic habitats, this mode of feeding is deserving of greater experimental attention.

At least some freshwater bivalve species may also supplement filter feeding by consuming organic detritus or interstitial bacteria from sediments. The infaunal habit of many members of *Pisidium* suggests dependence on food sources other than plytoplankton. Many species of *Pisidium* can efficiently filter small interstitial bacteria. They burrow continually through sediments with the shell hinge downward. In this position, dense bacterial populations are drawn with interstitial water through the parted ventral mantle edges into the mantle cavity and filtered on the gills, thus making up the majority of ingested organic matter (Lopez and Halopainen 1987). Both *Pisidium casertanum* and *P. conventus* feed in this manner and can filter, ingest, and assimilate interstitial bacteria of much smaller diameter (< 1 μm) than considered filterable by most bivalves (Lopez and Holopainen 1987) (Fig. 11.22). The capacity to filter rich interstitial bacterial flora may be associated with the characteristically reduced ctenidial surface areas as well as low filtration and metabolic rates of *Pisidium* species compared to those of *Sphaerium* or *Musculium* (Hornbach 1985, Lopez and Holopainen 1987). In another mode of deposit feeding, *M. transversum* reportedly uses its

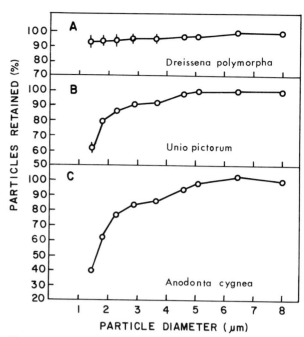

Figure 11.22 Relationship between the percentage of suspended particles regained by the gill filtering mechanism of freshwater bivalves and particle size, as particle diameter in micrometers. Horizontal axis for all figures is the percentage of total particles retained from suspensions passing over clam gill filtering systems in *Dreissena polymorpha* (A), and the unionid mussels, *Unio pictorum* (B), and *Anodonta cygnea* (C). Vertical lines about points are standard deviations. Note that all three species efficiently retained particles of 2.5–8 μm, equivalent to most unicellular algae, but only *D. polymorpha* efficiently retained particles of 1 μm equivalent to bacteria. (Redrawn from Jorgensen *et al.* 1984.)

long inhalant siphon to vacuum detrital particles from the sediment surface on inhalant currents for filtration (Way 1989).

Pedal feeding on sediment organic detritus may be another important feeding mechanism in some freshwater species. Such pedal feeding is known to occur in a few marine and estuarine bivalves (Morton 1983), but has been little studied in freshwater species. *Corbicula fluminea* draws sediment detrital particles dorsally over the ciliated foot epithelium into the mantle cavity, where they accumulate in the food groove on the ventral edge of the inner demibranch. Eventually, they are carried with particles filtered on the gill to the palps for sorting and ingestion (Cleland 1988, Way *et al.* 1990b). Cellulose particles (40–120 μm in diameter) soaked in various amino acid solutions and mixed with sediments will accumulate in the stomachs of individuals of *C. fluminea* exhibiting pedal-feeding behavior (R. F. McMahon personal observation). Similar pedal feeding has been reported in *M. transversum*

(Way 1989). Such observations suggest that pedal deposit feeding is an important auxiliary feeding mode, at least in corbiculacean species.

Pedal feeding may be much more common in freshwater species than previously suspected. In a stream *S. striatinum* population, only 35% of the total organic carbon assimilated could be accounted for by filter feeding, leaving the remaining 65% to come from sediment detrital sources (Hornbach *et al.* 1984a), presumably from pedal feeding. Other sphaeriid species reach highest densities in silt-laden sediments of high organic content (Zdeba and White 1985) or in habitats receiving organic sewage effluents, both strongly suggestive of dependence on pedal deposit-feeding mechanisms (Burky 1983). Pedal deposit-feeding mechanisms may explain the extensive horizontal locomotion through sediments observed in a number of freshwater bivalves (Imlay 1982, Way 1989; see also Section II.D), as it would allow sediments laden with organic detritus to be renewed continually at the foot surface. Certainly, this unique feeding mode deserves further investigation.

3. Population Regulation

There has been little detailed research regarding regulation of freshwater bivalve populations. Most of the information is anecdotal. Fuller (1974) reviews the known abiotic and biotic factors affecting population density and reproductive success of freshwater bivalves. Catastrophic abiotic factors recognized as causing periodic reductions in bivalve populations are reviewed in Section III.B. Among these are accumulations of silt in the sediments of impounded rivers and silt suspended in the water column during periods of flooding. Silt interferes with gill filter feeding and gas exchange mechanisms (Aldridge *et al.* 1987, Payne and Miller 1987) and can cause massive mortality in unionacean populations (Adam 1986). In contrast, many sphaeriids thrive in sediments with high silt loads (Burky 1983, Kilgour and Mackie 1988), associated with their partial or complete dependence on detrital food sources (see Section III.C.2). Even the shell growth of *C. fluminea,* a species tolerant of high turbulence and suspended silt, can be temporarily inhibited by periods of high flow and turbidity (Fritz and Lutz 1986).

Temperature extremes also affect freshwater bivalve populations. The tolerated temperature range of *C. fluminea* is 2°C (Mattice 1979) to 36°C (McMahon and Williams 1986b). Both cold-induced winter kills (Blye *et al.* 1985, Sickel 1986) and heat-induced summer kills (McMahon and Williams

1986b) are reported for this species. Temperature limitations for other North American species are not as well studied; but on average they appear to have broader upper and lower temperature limits than *C. fluminea* (Burky 1983). Even within the tolerated range, temperature may have detrimental effects on reproductive success. Sudden temperature decreases cause abortive release of developing glochidia in unionaceans (Fuller 1974), and temperatures above 30–33°C inhibit reproduction in *C. fluminea* (Long and McMahon 1987). Sphaeriid densities were reduced in midsummer in an area receiving heated effluents from an electric generating station (Winnell and Jude 1982). Elevated temperatures in areas receiving heated effluents may also stimulate bivalve growth, inducing early maturity and increasing reproductive effort. They may provide heated refugia in which populations may overwinter as occurs in the northernmost populations in *C. fluminea* in North America (Counts 1986, Cairns and Cherry 1983, McMahon 1982, 1983a).

Low ambient pH can also depauperate or completely extirpate bivalve populations. Bivalve populations have been eliminated from habitats receiving acid mine drainage (Taylor 1985, Warren *et al.* 1984). Highly acidic waters cause shell erosion and eventual death (Kat 1982). Perhaps of more importance than pH in regulating freshwater bivalve populations are water hardness and alkalinity. Unionaceans do not occur in New York drainage systems with calcium concentrations of less than 8.4 mg Ca/liter (Fuller 1974). In waters of extremely low alkalinity, calcium concentrations may be too low for shell calcium deposition; the lower limit for most bivalves is 2–2.5 mg Ca/liter (Okland and Kuiper 1982, Rooke and Mackie 1984a). Growth and fecundity rates were suppressed in a *Pisidium casertanum* population in a pond with a low calcium concentration relative to a population in a pond with a higher calcium concentration (Hornbach and Cox 1987). Waters of low alkalinity have little pH-buffering capacity and are, therefore, subject to major seasonal pH variation. This makes them particularly sensitive to acid rain, which detrimentally affects bivalve faunae (Rooke and Mackie 1984a). Shell calcium content can be related to water calcium concentration. Of ten species of Canadian freshwater bivalves, two sphaeriid and one unionacean species displayed no relationship between shell calcium content and water calcium concentration. In contrast, shell calcium content decreased with increased water calcium concentration in two sphaeriid species and increased with water calcium concentration in four sphaeriid and two unionacean species (Mackie and Flippance 1983b).

Lowering of habitat water levels during extreme droughts or dry periods can produce massive mortalities by exposing bivalves to air. Lowering of lake levels caused near 100% mortality in emerged *C. fluminea* populations (White 1979), a result that probably reflects its poor dessication tolerance (Byrne *et al.* 1988). Many unionacean species are relatively tolerant of prolonged emersion. When sympatric populations of nine unionacean species, the sphaeriid *Musculium transversum*, and *C. fluminea* were exposed on land for several months, both the *M. transversum* and *C. fluminea* populations suffered 100% mortality. The unionaceans, however, suffered only 50% mortality because they either migrated downshore or resisted dessication (White 1979).

Low environmental O_2 concentrations can be detrimental to freshwater bivalves. Continual eutrophication of Lake Estrom, Denmark, so reduced ambient profundal O_2 concentrations (0–0.2 mg O_2/liter for three months) that massive reductions of sphaeriid populations ensued (Holopainen and Jonasson 1983, Jonasson 1984a, 1984b), even in species generally highly tolerant of hypoxia such as *Pisidium casertanum* and *P. subtruncatum*. The lower critical O_2 limit for maintenance of aerobic respiration in both species was 1.7 mg O_2/liter (Jonasson 1984b).

Various forms of pollution are also highly detrimental to freshwater bivalves (for a review see Fuller 1974) including chemical wastes (Zeto *et al.* 1987), asbestos (Belanger *et al.* 1986a), organic sewage effluents (Gunning and Sutkkus 1985, Neves and Zale 1982, St. John 1982), heavy metals (Belanger *et al.* 1986b, Fuller 1974, Lomte and Jadhav 1982a), chlorine and paper mill effluents (Fuller 1974), and acid mine drainage (Taylor 1985, Warren *et al.* 1984). Potassium ion, even in low concentration (> 4–7 mg K^+/liter), can be lethal to freshwater bivalves. Populations can be seriously affected or even extirpated in watersheds where potassium is naturally abundant (as occurs in the western Mississippi basin) (Fuller 1974).

Little hard experimental information is available on the biotic factors controlling freshwater bivalve populations. In some *C. fluminea* populations, massive die-offs (particularly of older individuals) have been observed after reproductive efforts (Aldridge and McMahon 1978, McMahon and Williams 1986a, Williams and McMahon 1986), probably as a result of major reductions in the tissue energy reserves of postreproductive individuals (Williams and McMahon 1989). Such postreproductive mortality in adult individuals also occurs in sphaeriids (Burky 1983).

Freshwater bivalves are hosts for a number of parasites. They are intermediate hosts for digenetic trematodes (Fuller 1974). While such infections cause sterility in gastropods, their effects on freshwater bivalves are unknown. Parasitic nematode worms inhabit the guts of unionaceans (Fuller 1974). The external oligochaete parasite *Chaetogaster limnaei* resides in the mantle cavities of unionaceans (Fuller 1974) and *C. fluminea* (Sickel 1986). All of these parasites probably contribute to the regulation of freshwater bivalve population densities, but the degree to which they do has received little experimental attention.

Water mites of the family Unionicolidae, including the genera *Unionicola* and *Najadicola*, are important external parasites of unionaceans. Both mature and pre-adult mites are parasitic, attaching to gills, mantle, palps, and the visceral epithelium (depending on species) (for life histories of the Unionicolidae see Mitchell 1955; see also Chapter 16). Heavy mite infestations of unionacean gills can cause portions of the gills to be shed, abortion of developing glochidia, or even death of infected hosts (Fuller 1974), perhaps making unionicolid mites a major regulator of unionacean populations.

Disease has been little studied. Available evidence indicates little viral or bacterial involvement in massive die-offs of *C. fluminea* (Sickel 1986) or unionacean (Fuller 1974) populations.

Predation may be the most important regulator of fresh water bivalve populations. Shore birds and ducks are major feeders on sphaeriids and small specimens of *C. fluminea* (Dreier 1977, Fuller 1974, Paloumpis and Starrett 1960, Smith *et al.* 1986, Thompson and Sparks 1977, 1978). Indeed, *C. fluminea* densities were 3–5 times greater in enclosures excluding diving ducks (Smith *et al.* 1986). Water fowl feeding also significantly reduces *D. polymorpha* populations (Mackie *et al.* 1989). Crayfish also feed on small bivalves including *C. fluminea* (Covich *et al.* 1981) and *D. polymorpha* (Mackie *et al.* 1989), and in some habitats may have significant influence on bivalve population densities. Fire ants (*Solenopis invincta*) prey heavily on clams periodically emersed by receding water levels. In addition, turtles, frogs, and the mudpuppy salamander *Necturus maculosus* have all been reported to feed to a limited extent on small or juvenile bivalves (Fuller 1974). Free-living oligochaetes prey on freshly released glochidia (Fuller 1974).

Perhaps the major predators involved in regulation of freshwater bivalve populations are fish. A number of North American fish species have been identified as molluscivores (Table 11.4). While the majority of molluscivorous fish feed on small bivalve species or immature specimens (SL < 7 mm) of larger bivalve species, several routinely take larger

Table 11.4 List of Major Fish Predators of Freshwater Bivalves[a]

Family	Genus and Species	Common Name
Clupeidae	*Alosa sapidissima*	American Shad
Cyprinidae	*Cyrinus carpio*	Common Carp
Catostomidae	*Ictiobus bubalus*	Smallmouth Buffalo
	Ictiobus niger	Black Buffalo
	Minytrema melanopus	Spotted Sucker
	Moxostoma carinatum	River Redhorse
Percichthyidae	*Roccus saxatilis*	Striped Bass
Ictaluridae	*Ictalurus furcatus*	Blue Catfish
	Ictalurus punctatus	Channel Catfish
Centrarchidae	*Lepomis gulosus*	Warmouth
	Lepomis macrochirus	Bluegill
	Lepomis microlophus	Red-ear Sunfish
Sciaenidae	*Aplodinotus grunniens*	Freshwater Drum
Acipenseridae	*Acipenser fulvesens*	Lake Sturgeon

[a]Data from Fuller (1974), McMahon (1983a), Robinson and Wellborn (1988), and Sickel (1986).

adult bivalves, including carp (*Cyprinus carpio*), channel catfish (*Ictalurus punctatus*) and freshwater drum (*Aplodinotus grunniens*). This latter species crushes large bivalve shells with three massive, highly muscularized pharyngeal plates, as do carp to a lesser extent. In contrast, channel catfish swallow bivalves intact (J. C. Britton, personal communication). The majority of molluscivorous fish limit predation to smaller bivalves because the shells of larger individuals are too strong to crack or crush. The shell strength of *C. fluminea* increases exponentially with size [$\log_{10}$ force in newtons to crack the shell = $-0.76 + 2.31$ ($\log_{10}$ SL in mm)]; it is 6.5 times stronger than that of the very thick-shelled estuarine bivalve *Rangia cuneata* (Fig. 11.23) (Kennedy and Blundon 1983).

In some instances, fish predation significantly depletes freshwater bivalve populations. The diversity and density of sphaeriids increased in habitats from which molluscivorous fish were excluded (Dyduch-Falniowska 1982). Robinson and Wellborn (1988) reported that 11 months after settlement, the densities of a spring cohort of *C. fluminea* were 29 times greater in enclosures excluding fish.

As fish are intermediate hosts for the glochidia of unionaceans, the size of fish host populations can significantly influence mussel reproductive success. The absence of appropriate fish hosts has caused the extirpation of unionaceans in a number of North American habitats (Fuller 1974, Kat and Davis 1984, Smith 1985a, 1985b; see also Section III.A). Indeed, restoration of fish host populations has produced remarkable recoveries in some endangered unionacean populations (Smith 1985b). Thus, human activities reducing population densities of fish glochidial hosts can result in the destruction of unionacean populations. Therefore, relationships between unionaceans and their glochidial host fish should be

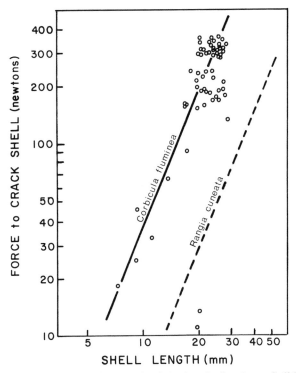

Figure 11.23 Shell strength of *Corbicula fluminea*. Solid line is the best fit of a geometric mean estimate of a least squares log–log linear regression, relating shell length (SL) to force required to crack the shell (open circles) as follows: $\log_{10}$ force to crack the shell in newtons = $-0.76 + 2.31$ ($\log_{10}$ mm SL). Dashed line is the best fit of a similar log–log linear regression for the estuarine wedge clam, *Rangia cuneata,* as follows: $\log_{10}$ force to crack the shell in newtons = $-1.73 + 2.42$ ($\log_{10}$ SL in mm). *R. cuneata,* with the strongest shell of eight tested estuarine bivalve species, has a thicker shell than does *C. fluminea.* However, as the regression slope values for the two species are nearly equal, the elevated y-intercept for *C. fluinea* (-0.76 compared to -1.73 for *R. cuneata*) indicates that its shell is approximately an order of magnitude stronger. (Redrawn from Kennedy and Blundon 1983.)

an important consideration in the future management of drainage systems and their fisheries to ensure continued health of remaining unionacean faunae.

Mammals also prey on freshwater bivalves. Otters, minks, muskrats, and raccoons all routinely include bivalves in their diets and may limit population growth in some species (Fuller 1974). Raccoons and muskrats feed extensively on *C. fluminea* and may account for the large reductions of adult clam densities in the shallow, near-shore waters of some Texan rivers, where favored feeding sites are marked by accumulations of thousands of open shells (personal observation). Even wild hogs have been reported to root in and destroy shallow water beds of unionacean mussels (Fuller 1974). Muskrats annually ate 3% of individuals in a population of *Anodonta grandis simpsoniana* in a small Canadian lake or 31% of its annual tissue production. The muskrats preferentially consumed larger mussels (shell length > 55 mm), strongly affecting the size–age structure of the population; this resulted in a decline in the biomass of large individuals in the mussel population and a significant reduction in the reproductive effort of the mussel population (Hanson *et al.* 1989).

There have been no experimental evaluations of interspecific or intraspecific competition among freshwater bivalves; only highly anecdotal observations have been made. Unionacean or sphaeriid population declines coincident with the establishment of *C. fluminea* populations have occurred in a number of North American habitats (McMahon 1983a, Sickel 1986). However, invasion by *C. fluminea* does not appear to be a primary cause of reductions in native bivalve populations. Rather, habitats are first made inhospitable to indigenous species by channelization, dredging, pollution, over-fishing or other water management practices. *C. fluminea*, which is highly invasive and more tolerant of the higher current velocities, increased levels of suspended solids, and silting caused by such practices, rapidly colonize such disturbed habitats from which other species have been extirpated (McMahon 1983a). In contrast, establishment of *C. fluminea* populations in drainage systems not subject to human interference has had little effect on native bivalve populations (McMahon 1983a, Sickel 1986). This may suggest an inability by *Corbicula* to outcompete native species in most natural North American habitats.

Published experimental investigations of intraspecific, density-dependent, population regulation mechanisms in freshwater bivalves are also lacking. However, some evidence is available from descriptive field studies. In *C. fluminea*, extremely successful recruitment of newly settled juveniles is known to occur after adult populations have been decimated (Cherry *et al.* 1980, McMahon and Williams 1986a, 1986b); whereas, juvenile settlement is generally far less successful in habitats harboring dense adult populations. High adult density may prevent successful recruitment of juveniles, thus preventing extensive juvenile–adult competition for limited food and space resources. This hypothesis was tested by placing sand-filled trays with different densities of adult *C. fluminea* in a lake harboring a relatively dense *Corbicula* population. The results indicated that while adult density had no effects on either adult or juvenile growth rates, it clearly affected success of juvenile settlement. Settlement in the tray without adults was 8702 juveniles/m². Maximal settlement (26,996 juveniles/m²) occurred in the tray with 329 adults/m²; while juvenile settlement was least at the highest adult densities of 659 and 1976 clams/m² being only 8642 and 3827 juveniles/m², respectively (R. F. McMahon unpublished). This clearly indicated a significant inhibitory effect of adult density on successful juvenile settlement. The mechanism of density-dependent juvenile settlement has not yet been clarified for this species. Mackie *et al.* (1978) found that high adult densities in *Musculium securis* caused significant reductions in the number of juveniles incubated, implying that intraspecific adult competition can regulate reproductive effort. In addition, interspecific competition between *M. securis* and *M. transversum* caused reduction in reproductive capacity in the subdominant species, with species dominance dependent on habitat. Such experiments clearly show that intraspecific density effects and interspecific competition both contribute to regulation of freshwater bivalve populations and, therefore, are deserving of greater experimental attention.

4. Functional Role in the Ecosystem

Bivalves have a number of important roles in freshwater ecosystems. Because they can achieve very high densities and are filter feeders, they are important second trophic level consumers of phytoplanktonic primary productivity. Unionaceans, with filtering rates on the order of 300 ml/g dry tissue/hr (Paterson 1984), can account for a large proportion of total animal consumption of phytoplankton productivity. As unionacean densities are reported to range from 15 individuals/m² (Paterson 1985) to 28 individuals/m² (Negus 1966) and dry tissue standing crop biomasses from 2.14 g/m² (Paterson 1985) to 17.07 g/m² (Negus 1966), filtering by unionacean communities could be as high as 15–122.9 liter/m²/day. Thus, on an annual basis, a unionacean community in a eutrophic European lake was estimated to filter 79% of total lake volume, removing 92–100% of all suspended seston from the filtered water. However, even with such high cropping rates, the high reproductive rates of phytoplankton resulted in unionaceans reducing seston concentrations by only 0.44% (Kasprzak 1986). While the unionacean community accounted for only 0.46% of seston removed by all phytoplanktivorous animals in the lake, they accounted for 85% of the standing crop biomass of the phytoplanktivorous community. They also contained in their biomass a high proportion of the total phosphorus load of the lake, greatly limiting the phosphorus available for phytoplankton production. Similarly, populations of *D. polymorpha* in five European lakes had an average dry weight biomass 38 times that of submerged macrophytes and average standing crop phosphorus and nitrogen contents 2.94 (range = 0.39–7) and 4.21 (range = 0.59–9.4) times that of submerged macrophytes, respectively, greatly reducing availability of these growth-limiting inorganic nutrients to the aquatic plant community (Stanczykowska 1984).

Sphaeriids also have important impacts on phytoplankton communities in lentic habitats. The sphaeriid community of an oligotrophic lake comprised only 0.12% of the total biomass of the phytoplanktivores and 0.2% of the bivalve biomass, but accounted for 51% of total seston consumption (Kasprzak 1986). This probably resulted from their relatively high rates of filtration and population productivity (Burky *et al.* 1985a, Hornbach *et al.* 1984a,b). In smaller lotic and lentic habitats, sphaeriids can consume a major portion of primary productivity. In a canal community dominated by sphaeriids (98% of the biomass of phytoplanktivores), they accounted for 96% of total seston consumption (Krasprzak 1984). A stream population of *S. straitinum* was estimated to filter 3.67 g organic carbon/m²/yr or 0.0004% of organic carbon in the seston flowing over them (Hornbach *et al.* 1984a). In contrast, filtration by a pond population of *M. partumeium* removed 13.8 g C/m²/yr as seston (Burky *et al.* 1985a). Using the data of Burky *et al.* (1985b), standing crop seston levels for the entire pond were estimated to range annually between 46 and 550 g carbon. Average annual consumption of seston organic carbon by the *M. partumeium* community was estimated to be roughly 3.9 g C/day, yielding a population feeding rate of 0.7–8.5% of total standing crop seston carbon per day. These data indicate that sphaeriid communities crop a significant portion of the primary productivity of their small lentic habitats on a daily basis.

With its extremely high filtering rates and capacity to develop very dense populations (McMahon 1983a), *C. fluminea* is potentially a major consumer of phytoplankton productivity. In the Potomac River, phytoplankton densities and chlorophyll *a* concentrations declined in the water column as it passed over dense beds of *C. fluminea,* both measures falling to levels 20–75% lower than upstream values; current phytoplankton levels in this river section are considerably lower than those recorded prior to *C. fluminea* invasion. Based on filtering rates, population size distributions, and adult densities, it was estimated that the *C. fluminea* population filtered the entire water column in the 3–4 day time period required for it to pass over the river reach where it was most dense (Cohen *et al.* 1984).

The great reduction in phytoplankton density and chlorophyll *a* concentration must have been due to clam filter feeding, particularly as discharge volume variation, zooplankton feeding, toxic substances, and nutrient limitations were not different from other river sections (Cohen *et al.* 1984). Similarly, Lauritsen (1986b) estimated that a *C. fluminea* population of 350 clams/m² in the Chowan River, North Carolina, at an average summer depth of 5.25 m and clam filtering rate of 564–1010 ml/hr filtered the equivalent of the entire overlying water column every 1–1.6 days. At an average depth of 0.25 m, an average current flow of 18.5 m/min, and an average clam filtering rate of 750 ml/hr (Lauritsen 1986b), the equivalent of the entire water column of the Clear Fork of the Trinity River flowing over a *C. fluminea* population with an average adult density of 3750 clams/m² would be filtered every 304 m of river reach or every 16 min (R. F. McMahon personal observation). Such massive turnover of the water column by dense bivalve populations could keep seston concentrations at minimal levels and greatly limit the energy available to other seston-feeding species. Consumption of the majority of primary productivity by dense bivalve populations and the accumulation of that production in large, relatively long-lived, predator-resistant adult clams may not only limit the inorganic nutrients available for primary productivity and the energy available to other primary consumers but may also divert energy flow away from higher trophic levels. Such diversion of energy flow by massive freshwater bivalve populations away from higher trophic levels could lead to reductions in predatory game fish stocks in habitats with dense bivalve populations. This hypothesis has interesting implications for management of game fish stocks, but it is one that awaits field and experimental confirmation. In this regard, the long-term ecological impacts of the apparent reduction of phytoplankton biomass in Lake Erie by zebra mussels should prove most interesting.

Bivalves, by filtering suspended seston, act as water clarifiers and organic nutrient sinks. When co-cultured with channel catfish, *C. fluminea* significantly increased ambient water O_2 concentrations by reducing seston and turbidity levels (Buttner 1986). Dense bivalve populations, by removing phytoplankton and other suspended material from natural water columns, significantly increase clarity; and, by binding suspended sediments into pseudofeces, they accelerate sediment deposition rates (Prokoprovich 1969). Dense populations of *D. polymorpha* clarify water and increase sedimentation rates in European canals (see Mackie *et al.* 1989 and references therein) and may similarly improve the water quality in Lake Erie as populations eventually reach maximal densities. By increasing water clarity and therefore depth of light penetration, bivalves may stimulate the growth of rooted aquatic macrophytes.

Because freshwater clams filter suspended organic detritus and bacteria and consume interstitial bacteria and organic detritus in the sediments (see Section III.D.2), they may also be significant members of the aquatic decomposer assemblage. Lopez and Holopainen (1987) presented evidence that *Pisidium casertanum* and *P. corneum* feed primarily by filtering interstitial bacteria from the sediments. Hornbach *et al.* (1982, 1984a) estimated that only 24–35% of the total energy needs of the population of *Sphaerium striatinum* are met by filter feeding, the remaining coming from assimilation of sediment organic detritus. *C. fluminea* pedally feeds on sediment detritus and can reduce sediment organic contents by 50% within 25 days (Cleland 1988), thus clearly exerting a major impact on the decomposer community. *Musculium transversum* utilizes its long inhalant siphon to vacuum organic detritus from the sediment surface (Way 1989). The detritivorous habit of many freshwater bivalve species may divert primary productivity ordinarily lost to respiration of the detritivorous community back into bivalve tissue where it can become reavailable to higher trophic levels.

The activity of freshwater bivalves may also directly affect the physical characteristics of their habitats. Deposition of calcium in the shells of growing clams may reduce ambient water calcium concentrations. In a population of *C. fluminea* with an average density of 32 clams/m^2, annual fixation of shell $CaCO_3$ was 0.32 kg $CaCO_3/m^2/yr$ (Aldridge and McMahon 1978). At this rate, annual shell $CaCO_3$ fixation in populations of 100,000 clams/m^2 (Eng 1979) could be as great as 50–60 kg $CaCO_3/m^2/yr$. Such rapid removal of calcium to shells accumulating in the sediments could lead to a considerable reduction in water hardness, particularly in lentic

habitats. Seasonal cycles of shell growth could induce seasonal cycles in ambient water calcium concentration, with calcium concentrations being greatest in winter when shell growth is minimal and least in summer when shell growth is maximal (Rooke and Mackie 1984c). Bivalve activity can also affect the flux rates of solutes between sediments and the overlying water column. Unionacean mussels enhanced the release of nitrate and chloride and inhibited $CaCO_3$ release from surrounding sediments (Matisoff *et al.* 1985). Dense populations of sphaeriids can be the principal effectors of sediment dissolved oxygen demand, in some habitats reported to reach levels characteristic of semipolluted and polluted streams (Butts and Sparks 1982).

Small bivalves (sphaeriids, juvenile unionaceans, *D. polymorpha,* and *C. fluminea*) can be a major food source for a number of second trophic level carnivores, including fish and crayfish (see Section III.C.3). Bivalve flesh has a relatively low caloric content, 3.53–5.76 kcal/g ash-free dry weight (Wissing *et al.* 1982), reflecting low lipid and high protein content. *C. fluminea* has tissue organic C : N ratios ranging from 4.9–6.1 : 1, indicative of flesh protein contents of 51–63% of dry weight (Williams and McMahon 1989). High protein content makes bivalve flesh an excellent food source to sustain predator tissue growth.

Many freshwater bivalve populations are highly productive (particularly sphaeriids, *C. fluminea,* and *D. polymorpha*), thus they rapidly convert primary productivity into tissue energy available to third trophic level predators. Productivity values range from a low of 0.019 g C/m^2/yr for *Pisidium crassum* population to a high of 10.3 g C/m^2/yr for a *C. fluminea* population; the average for 11 sphaeriid species and 2 species of *Corbicula* was 2.5 g C/m^2/yr (Burky 1983). Other more recent values include 12.8–14.6 g C/m^2/yr for an Arizona canal population of *C. fluminea* (computed from the data of Marsh 1985) and 1.07 g C/m^2/yr for a New Brunswick lake population of the unionacean *Elliptio complanata* (computed from the data of Paterson 1985).

Furthermore, bivalves are generally more highly efficient at converting consumed and assimilated food energy into new tissue than most second trophic level aquatic animals because their sessile, filter-feeding habits minimize energy expended in food acquisition, allowing more efficient transfer of energy from primary production to bivalve predators. Thus, bivalves can act as important conduits of energy fixed by photosynthesis in phytoplankton to higher trophic levels in the ecosystem. However, such trophic energy transfer is mainly through smaller species and juvenile specimens, as large adult bivalves are relatively immune to predation.

The measure of efficiency of conversion of assimilated energy (energy absorbed across the gut wall) into energy fixed in new tissue growth is net growth efficiency:

$$\text{\% Net Growth Efficiency} = P/A(100)$$

where P = productivity rate or rate of energy or organic carbon fixation into new tissue growth by an individual or population and A = assimilation rate or rate of energy or organic carbon assimilated by an individual or population. The greater the net growth efficiency of a second trophic level species, the more efficient is its conversion of assimilated energy into flesh, and, therefore, the greater the potential for energy flow through it to third trophic level predators (for detailed discussions of bivalve energetics see Burky 1983, Holopainen and Hanski 1986, Hornbach 1985, Russell-Hunter and Buckley 1983). Net growth efficiencies, computed as 100% − the percentage of assimilated energy respired values (Table 11.2), are 9–79% in sphaeriids (average = 55%) and 58–89% in *C. fluminea*, indicating relatively efficient conversion of assimilated energy into flesh compared to other aquatic second trophic level animals.

5. Bivalves as Biomonitors

Freshwater bivalves, particularly larger unionaceans and *C. fluminea*, may also be excellent biomonitors of water pollution and other environmental perturbations. Characteristics making them excellent biomonitors (for a review see Imlay 1982) include long life span, and growth and reproductive rates sensitive to environmental perturbation (Burky 1983). They are readily held in field enclosures without excessive maintenance. Shell growth is sensitive to environmental variation and/or disturbance (Belanger *et al.* 1986a, 1986b, Burky 1983, Fritz and Lutz 1986, Way and Wissing 1982, Way 1988), and the valves remain as evidence of death, allowing mortality rates to be estimated. Shells are easily marked by tags (Young and Williams 1983c), paint, or shell-etching (McMahon and Williams 1986a).

Bivalves are collectible throughout the year and are easily shipped alive long distances from field sites to the laboratory. Because adults of large species tend to remain in place (exception is *C. fluminea*), they are subjected to conditions representative of the monitored environment throughout life (Imlay 1982). Shell growth in annual increments allows determination of annual variation in heavy metal pollutant levels over long periods by analysis of metal levels in successively secreted shell layers (Imlay 1982). Large size allows analysis of pollutant

levels in single individuals, and the wide geographical ranges of some species (Imlay 1982) allow direct comparisons across drainage systems. Table 11.5 indicates the types of pollutants and environmental perturbations for which freshwater bivalves have been used as biomonitors.

D. Evolutionary Relationships

Sphaeriids, dreissenids, corbiculids, and unionaceans represent separate evolutionary invasions of freshwaters. Sphaeriids, dreissenids, and corbiculids, all in the order Veneroida, are quite distinct from unionaceans in the order Unionoida (Allen 1985). The Veneroida became successful infaunal filter feeders through evolution of inhalant and exhalant siphons, allowing sediment burrowing while maintaining respiratory and feeding currents from the overlying water column. Their eulamellibranch gills with fused filaments (Fig. 11.5) were a clear advancement over the primitive (i.e., filibranch) condition with separate gill filaments attached only by interlocking cilia (Allen 1985).

The family Sphaeriidae in the superfamily Corbiculacea has a long fossil history in freshwater extending from the Cretaceous (Keen and Dance 1969) and has evolved along two major lines. The first, represented by *Pisidium*, involves adaptation to life in organically rich sediments with the capacity to filter interstitial bacteria (Lopez and Holopainen 1987); this makes them dominant in the profundal regions of lakes. The second, represented by *Sphaerium* and *Musculium*, involves adaptation to life in small, shallow lentic or lotic habitats subject to predictable seasonal perturbation such as habitat-drying. Their adaptations include estivation during prolonged emergence (see Section III.A). Like many of the Veneroida, the byssus is absent in the adults of most species. They have also evolved an unusual shell made fragile by lack of a nacreous layer.

The superfamily Unionacea, like the Sphaeriidae, has a long fossil history in freshwater, extending at least from the Triassic (Hass 1969). A long fossil history and tendency for reproductive isolation due to gonochorism and the parasitic glochidial stage (see Section III.A) has led to extensive radiation in the Unionacea, represented worldwide by 150 genera and a great number of species (Allen 1985). Their origin remains unclear. Similarity of shell structure suggests a possible relationship to shallow-burrowing marine fossil species in the order Trigonida (Allen 1985). Like the majority of sphaeriid species, adults lack a byssus, which would be of no use in their soft sediment, infaunal habitats. They are the dominant bivalve fauna in shallow regions of

Table 11.5 List of Recent Investigations Involving Utilization of Bivalves to Monitor Effects or Levels of Pollutants in Freshwater Habitats

Pollutant Monitored	Species Utilized	Literature Citation
Arsenic	Corbicula fluminea	Elder and Mattraw (1984), Price and Knight (1978), Tatem (1986)
Cadmium	8 Unionid species	Price and Knight (1978)
	Anodonta anatina	Hemelraad et al. (1985)
	Anodonta cygnea	Hemelraad et al. (1985).
	Corbicula fluminea	Elder and Mattraw (1984), Graney et al. (1984), Price and Knight (1978), Tatem (1986)
Chromium	8 Unionid species	Price and Knight (1978)
	Corbicula fluminea	Elder and Mattraw (1984), Tatem (1986)
Copper	Corbicula fluminea	Annis and Belanger (1986), Elder and Mattraw (1984), Foe and Knight (1986c), Tatem (1986)
Iron	Corbicula fluminea	Tatem (1986)
Lead	Corbicula fluminea	Annis and Belanger (1986), Elder and Mattraw (1984), Price and Knight (1978), Tatem (1986)
Manganese	8 Unionid species	Price and Knight (1978)
	Corbicula fluminea	Elder and Mattraw (1984), Tatem (1986)
Mercury	Corbicula fluminea	Elder and Mattraw (1984), Price and Knight (1978)
Tin	8 Uniondid species	Price and Knight (1978)
	Anodonta sp.	Herwig et al. (1985)
Zinc	Corbicula fluminea	Belanger et al. (1986b), Elder and Mattraw (1984), Foe and Knight (1986c)
Asbestos	Corbicula fluminea	Belanger et al. (1986a, 1987)
Octachlorostyrene	Lampsilis radiata siliquiodea	Pugsley et al. (1985)
Polychlorinated Biphenols (PCBs)	Corbicula fluminea	Elder and Mattraw (1984), Tatem (1986)
Pesticides	Lampsilis radiata siliquiodea	Pugsley et al. (1985)
	Corbicula fluminea	Elder and Mattraw (1984), Hartley and Johnston (1983), Tatem (1986)
Sewage Effluents	Corbicula fluminea	Foe and Knight (1986c), Horne and MacIntosh (1979), Weber (1973)
Power Station Thermal Effluents	Corbicula fluminea	Dreier and Tranquilli (1981), Farris et al. (1988), Foe and Knight (1987), McMahon and Williams (1986b)

larger, relatively stable lentic and lotic habitats, particularly in sand–gravel substrata not subject to disturbance by currents (see Section III.A).

The Sphaeriidae and Unionacea, with long fossil histories in freshwater, have evolved quite distinct niches. While both occur in relatively stable habitats, the majority of sphaeriids live in smaller ponds and streams (including those with predictable seasonal perturbation) or in the profundal regions of large lakes; whereas, the majority of unionaceans colonize stable sediments of shallow portions of larger rivers and lakes (see Sections III.A and III.B for descriptions of the adaptations of two of these groups to their respective habitats).

C. fluminea has entered freshwater only in recent (Pleistocene) times (Keen and Casey 1969) and, therefore, is unlikely to complete effectively with sphaeriids or unionaceans, as the long fossil histories of these latter groups have made them highly adapted to their preferred stable freshwater habitats. Indeed, C. fluminea does not appear to have a serious impact on native North American bivalve fauna in undisturbed freshwaters (see McMahon 1983a and references therein). Rather, this species is adapted for life in highly unstable habitats that are subject to periodic catastrophic perturbation, particularly flood-induced sediment disturbance; such environments are generally unsuitable for sphaeriids or unionaceans (see Section III.A). Among its adaptations to high-flow lotic habitats with unstable substrata are: (1) capacity for rapid burrowing; (2) a strong, heavy, concentrically sculptured, inflated shell; and (3) juvenile retention of a byssal thread— all of which allow maintenance of position in sediments (Vermeij and Dudley 1985). Additionally, its high fecundity, elevated growth rate, early maturity, and extensive capacity for dispersal (see Section III.B) allow it to both invade habitats from which other bivalves have been extirpated and rapidly recover from catastrophic population reductions. Thus, *C. fluminea* is a successful recent invader of freshwaters because it has evolved a niche in unstable, disturbed habitats not greatly utilized by either sphaeriids or unionaceans. Freshwater *Corbicula* are considered to have evolved from an estuarine *Corbicula* ancestor inhabiting the upper, near freshwater portions of estuaries (Morton 1982).

The superfamily Dreissenacea containing *D. polymorpha,* while superficially resembling mytilacean mussels (superorder Pterioida), is placed in the Veneroida because of its advanced eulamellibranch ctenidium. Adult dreissenaceans retain an attachment byssus, considered a primitive condition. In contrast, the mytiliform shell with its reduction of the anterior ends of the valves and the anterior adductor muscle, is considered to be a derived characteristic, resulting from adaptation to an epibenthic niche characterized by byssal attachment to hard surfaces (Allen 1985, Mackie *et al.* 1989). *D. polymorpha* probably evolved from an estuarine dreissenacean ancestor of the genus *Mytilopsis.* As with *C. fluminea,* its fossil record indicates a recent Pleistocene introduction to freshwaters (Mackie *et al.* 1989). Also like *C. fluminea, D. polymorpha* appears to have successfully colonized freshwaters because its niche (an epibenthic species attached to hard substrata) is not occupied by the infaunal, burrowing sphaeriids and unionaceans, minimizing competition with these more advanced groups. However, *D. polymorpha* may have an indirect detrimental competitive impact on native North American unionacean faunae, because of its tendency to settle and grow on portions of unionid shells exposed above the substratum. Dense mats of *D. polymorpha* byssally attached to the exposed siphonal area of unionacean shells could interfere with their respiratory and feeding currents, leading to eventual death. This aspect of the interaction of *D. polymorpha* with unionaceans merits further study.

Freshwater bivalves display a number of characteristics that are uncommon in comparable marine species. Most shallow-burrowing marine species are gonochoristic with external fertilization and free-swimming planktonic veligers. A planktonic veliger is nonadaptive in lotic freshwater habitats, as it would be carried downstream before settlement, eventually leading to elimination of upstream adult populations. Thus, the veliger stage has been completely suppressed in sphaeriids, which brood developing embryos in gill marsupia and release large, fully formed juveniles immediately ready to take up life in the sediment. In unionids, eggs are retained in gill marsupia and hatch into the glochicia, whose parasitism of fish hosts both allows upstream as well as downstream dispersal and permits release of well-developed juveniles into favorable habitats (see Section III.B). Even the recently evolved *C. fluminea* has suppressed the veliger stage of its estuarine ancestors (Morton 1982), producing a small, but fully formed juvenile whose byssal thread allows immediate settlement and attachment to the substratum. Retention of external fertilization and a free-swimming veliger in *D. polymorpha* are primitive characteristics reflecting its relatively recent evolution from an estuarine ancestor.

Downstream veliger dispersal would appear to preclude zebra mussels from high-flow lotic habitats, unless upstream impoundments provide a source of replacement stock. In fact, in Europe and Asia, this species is most successful in large lentic or low-flow lotic habitats such as canals or large rivers (Mackie *et al.* 1989). Serial impoundment of rivers will provide lentic refugia for this species in otherwise lotic drainages, facilitating its invasion of North American waters.

Another common adaptation in freshwater species is evolution of hermapharoditism with a capacity for self-fertilization. Hermaphroditism makes all individuals in a population reproductive, allowing rapid population expansion during favorable conditions. Hermaphroditism allows sphaeriids to reestablish populations after predictable seasonal perturbation and in *C. fluminea,* along with high fecundity, allows rapid re-establishment of populations after catastrophic habitat disturbance. It makes both sphaeriids and *C. fluminea* invasive, as the introduction of a single individual can lead to the founding of a new population. In contrast, the gonochorism of the majority of unionaceans and *D. polymorpha* limits their invasive capacity, reflected by their tendency to inhabit stable, relatively undisturbed waters.

Finally, unionaceans and sphaeriids have different shell morphologies relative to comparable shallow-burrowing marine bivalves. Freshwater species generally do not have denticulated or crenulated inner valve margins, extensive radial or concentric shell ridges, tightly sealing valve margins, well-developed hinge teeth, overlapping shell margins, or uniformly thick, inflated, strong shells to the degree displayed by shallow-burrowing marine species (Vermeij and Dudley 1985). These shell characteristics allow shallow-burrowing marine species to maintain position in unstable substrata and/or resist to shell cracking or boring by large predators common in marine habitats. Lack of such structures in unionaceans and sphaeriids reflects their preference for stable sediments and the general absence of effective predators on adult bivalves; certainly, there are no shell-boring predators of North American freshwater species. Indeed, the shells of freshwater unionaceans display considerably less nonlethal, predator-induced damage than shallow-burrowing marine species (Vermeij and Dudley 1985). While retention by *C. fluminea* of a thick, extremely strong shell (Kennedy and Blundon 1983) lacking pedal and siphonal gapes, well-developed hinge teeth, shell inflation, and concentric ornamentation may represent a primitive condition, it appears to allow this species to be successful in sediments too unstable to support unionacean or sphaeriid faunae. Such primitive characteristics may reflect similar adaptations in the immediate estuarine ancestor of this recently evolved freshwater species.

IV. COLLECTING, PREPARATION FOR IDENTIFICATION, AND REARING

A. Collecting

Small sphaeriid clams are best collected by removing sediments containing clams with a: (1) trowel or shovel (shallow water); (2) long handled dip net or shell scoop with a mesh of less than 0.35 mm (moderate depths) or by Ekman, Ponar, or Peterson dredges (deeper water), the heavier Ponar and Peterson dredges being better in lotic habitats. Drag dredges may also be used but must have sediment catch bags of small mesh (1 mm or less). Williams (1985) described an easily and inexpensively made drag dredge for use from a small boat. Use of scuba is also an effective way of collecting sphaeriids. Larger sphaeriid species can be separated from fine sediments with a 1 mm mesh sieve, but a 0.35 mesh is required for smaller specimens. Fragile sphaeriids should be separated from sediments by gentle vertical agitation of the sieve at the surface of the water to prevent shell damage. All coarse material should be removed from the sample prior to sieving to avoid

shell breakage. Exceptionally small, fragile species can be collected by washing small quantities of sediment into settlement pans, specimens being revealed after sediments settle and the water clears.

Ekman, Ponar, or Peterson dredges are best for quantitative samples of sphaeriids because they remove a specific area of surface sediments. Quadrat frames can also be used (Miller and Payne 1988). These are driven into or placed on the substratum and all surface sediments within the frame collected for sorting. Such frames can be utilized in shallow or deeper waters (the latter in conjunction with scuba equipment). Drag dredging at specified speeds for known intervals allows partial sample quantification. Core sampling devices consisting of tubes driven into the substratum to remove sediment segments of specific depth can yield both accurate estimates of density and sediment depth distributions of sphaeriid populations. Core samplers are limited by sampling a small surface area and are, therefore, best utilized with dense populations and/or repetitive sampling.

The low densities of some unionacean populations can make their collection difficult. Where populations are dense, shoveling sediments through a sieve allows collection of a broad range of sizes and age classes (Miller and Payne 1988). As unionaceans are larger than sphaeriids, sieve mesh sizes of 0.5–1 cm may be appropriate unless recently settled juveniles must be collected. In less dense, shallow water populations, hand-picking using a glass-bottomed bucket to locate specimens can often suffice. In turbid waters and/or soft sediments (sand or mud), shell rakes may be utilized. Hand-picking and shell rakes select for larger, older individuals. In shallow, turbid water, raking of hands and fingers systematically through soft sediment while feeling for burrowed mussels can be effective, but also selects for larger individuals and can lead to cut fingers. On rocky or boulder bottoms, unionaceans generally accumulate in crevices or near downstream bases of large rocks. Here, only hand-picking (in conjunction with scuba techniques in deeper water) is effective. Unionaceans may also be efficiently collected by various forms of drag or brail dredges. Mussel brails (often used by commercial shellers) consist of a bar with attached lines terminating in blunt-tipped gang hooks. When dragged over unionid beds, mussels clamp the valves onto the hooks allowing them to be brought to the surface.

Quantitative sampling of unionaceans can be accomplished with quadrats, along with scuba in deeper waters (Miller and Payne 1988). Only heavy Ponar and Peterson dredges bite deeply enough into the substratum to take large unionacean species.

In sparse populations, errors in density estimates can result from the small surface areas that these dredges sample, requiring repetitive samples to improve accuracy. Drag dredges sample larger areas and may be partially quantitative, but generally do not provide accurate density estimates. Unionaceans may also be collected during natural or planned water level draw-downs (White 1979) by collecting either emerged individuals or those migrating downshore and accumulating at the edge of the water. As a great many North American unionacean species are presently endangered, one should ascertain the endangered status of any species before permanently removing specimens from their natural habitats. Take only a few living specimens for identification, making the remaining collection from dead shells.

C. fluminea is easily collected because it occurs in high densities, prefers shallow waters, and is easily recognized in the field. Individuals are readily separated from sediments with a 1 mm mesh sieve. Qualitative samples are best obtained by shovel or drag dredge (Williams 1985, Williams and McMahon 1986) and quantitative samples by quadrat frame (Miller and Payne 1988) or Peterson (Aldridge and McMahon 1978), Ponar, or Ekman dredges (Williams and McMahon 1986). In rock or gravel substrata, *C. fluminea* are best taken by hand, hand trowel, or shovel; collected sediments should be passed through a coarse mesh sieve to separate specimens from sediments. In fast-flowing streams, specimens of *C. fluminea* accumulate in crevices or behind the downstream sides of large rocks.

Because if its byssal attachment to rocks and other hard surfaces and its preference for deeper waters (< 1–2 m), *D. polymorpha* is best collected by scuba with manual removal, using quadrats for quantification. Ripping of individuals from the byssus attachment damages tissue, thus the byssus should be cut with a knife or sharp trowel before removal. Juveniles can be collected after they have settled on bare settlement blocks or submerged buoys set out during the reproductive season. Ekman, Ponar, Peterson, and drag dredges are generally not suitable for collection of attached epibenthic species such as *D. polymorpha*.

Juveniles of sphaeriids and *C. fluminea* can either be surgically removed from adult brood sacs or collected from sediments just after release with a fine mesh sieve (mesh size < 0.35 mm). Juvenile *C. fluminea* can also be obtained by release from freshly collected, gravid adults left in water for 12–24 hr (Aldridge and McMahon 1978). *C. fluminea* juveniles and *D. polymorpha* veligers may be collected from the water column with a zooplankton net towed behind a boat or held in a current. Their density in the water column may be quantitatively sampled by passing known volumes of water through a zooplankton net. Recently settled juveniles of unionaceans may be sieved from sediments. Glochidia can be surgically removed from demibranchs of gravid females or encysted glochicia from the fins, pharyngeal cavity, or gills of their fish hosts (for a list of unionacean fish hosts see Fuller 1974).

B. *Preparation for Identification*

To preserve bivalves, larger individuals should first be narcotized or relaxed, allowing tissues to be preserved in a life-like state and preservatives to penetrate tissues through gaped shell valves. Live bivalves placed directly in fixatives clamp the valves, which prevents preservative penetration, causing incomplete tissue fixation. There are a number of bivalve relaxing agents (Lincoln and Sheals 1979, Russel 1963), including alcoholized water (either 3% ethyl alcohol by volume with water or 70% ethyl alcohol added slowly drop by drop to the medium until bivalves gape), chloroform added slowly to the medium, methol crystals (one level teaspoon per liter, scattered on the water surface), propylene phenoxetol and phenoxetol BPC (5 ml of product emulsed with 15–20 ml of water added to water containing bivalves or introduction of a droplet equal to 1% of holding water volume), phenobarbitol added in small amounts to the holding medium, magnesium sulfate (introduced into holding medium over a period of several hours to form a 20–30% solution by weight), magnesium chloride (7.5% solution by weight), and urethane.

None of these relaxing agents works equally well with all species. My laboratory tests of narcotizing agents against *C. fluminea* revealed propylene phenoxetol to be the only agent capable of relaxing this species for experimental surgery, and allowing recovery on return to fresh medium. Heating bivalves to 50°C for 30–60 min causes most species to relax and gape widely, but is usually lethal. In larger specimens, wooden pegs or portions of matchsticks forced between the valves prior to fixation allows good preservative penetration of tissues.

The best long-term preservative for freshwater bivalves is 70% ethyl alcohol (by volume with water). Specimens may be initially fixed in 5–10% formaldehyde solutions (by volume with 40% formaldehyde solutions) for 3–7 days. But formaldehyde is acidic and will dissolve calcareous portions of shells unless pH-neutralized by addition of either powdered calcium carbonate ($CaCO_3$) to make a saturated solution, 5 g powdered sodium bicarbonate

(NaHCO$_3$) per liter, or 1.65 g potassium dihydrogen orthophosphate and 7.75 g disodium hydrogen orthophosphate per liter (Smith and Kershaw 1979). After 3–7 days in formaldehyde, specimens should be transferred to 70% ethyl alcohol for permanent preservation. Addition of 1–3% glycerin (by volume) to alcohol preservatives keeps tissues soft and pliable (Smith and Kershaw 1979). Smaller bivalve species (shell length < 15 mm) generally do not require relaxation before fixation. Tissues to be utilized in microscopy should be preserved in gluteraldehyde or Bouin's solution.

Dead shells or those from which tissues have been removed can be cleaned with a mild soap solution and soft brush. Organic material can be digested from shell surfaces by immersion in a dilute (3% by weight with water) solution of sodium or potassium hydroxide at a temperature of 70°–80°C, thereafter any remaining organic matter is readily removed with a soft brush. For dry-keeping, the shell periostracal surface should be varnished or covered with petroleum jelly to prevent drying, cracking, and peeling. Numbers identifying collection and specimen can be marked on the inner shell surface with India ink.

To remove soft parts from living bivalves, immerse them in boiling water and remove tissues after valves fully gape. Separated flesh can be fixed in 70% alcohol. For fragile sphaeriids, flesh is best removed with the tip of a fine needle, manipulating specimens with a fine brush. Shells should be dried in air at room temperature, not in an oven; heat causes the shell to crack and the periostracum to crack and peel.

Both soft tissue and shell characteristics are utilized in the identification of freshwater bivalves (see Section V), so both must be preserved for species identification. For unionaceans, C. fluminea, and D. polymorpha, most diagnostic taxonomic characteristics can be seen by eye or with a 10X hand lens. For sphaeriids, a dissecting microscope with at least 10–30X power or a compound microscope is required (Ellis 1978). Anatomical details are most clearly observed when shells are dry and when soft tissues are held under water.

Identification of recently released juvenile specimens of freshwater bivalves is difficult and may require preparation of stained slide whole mounts. Glochidia are best identified by removal from gravid adults as are juvenile sphaeriids. The D-shaped juvenile of C. fluminea (Fig. 11.15A) is highly recognizable and the only juvenile freshwater bivalve to occur in large numbers in the plankton (McMahon 1983a). The planktonic veliger of D. polymorpha is clearly distinguishable by its ciliated velum (Fig. 11.15C). The glochidia of many unionacean species have specific fish host species (Fuller 1974), whose identification will assist glochidial identification.

C. Rearing Freshwater Bivalves

For artificial rearing of freshwater bivalves, water in holding tanks should be temperature regulated. Adequate aeration, filtration, and ammonia removal systems are required, as some species have low tolerances of hypoxia and ammonia (Byrne 1988).

As unionaceans and C. fluminea are filter feeders of phytoplankton, they require a constant supply of filterable food to remain healthy and growing for long periods. The best food under artificial conditions appears to be algal cultures. When fed monoalgal cultures of the green algae, Ankistrodesmus and Chlorella vulgaris or the cyanobacterium, Anabaena oscillarioides, assimilation efficiencies in C. fluminea were 47.4–57.7% and net growth efficiencies, 59.4–78.2%, making them excellent food sources (Lauritsen 1986a).

Artificial diets do not appear to be as successful as algal cultures in maintaining clam growth. When fed either ground nine-grain cereal, rice flour, rye bran, brewers' yeast, or artificial trout food, small C. fluminea (5–8 mm SL) lost tissue mass on all but nine-grain cereal, the latter supporting little tissue growth. Supplementing these grain diets with live green algae (Ankistrodesmus sp.) greatly enhanced tissue growth, but greatest growth occurred in individuals fed pure Ankistrodesmus cultures (Foe and Knight 1986b). C. fluminea that were fed mixed cultures of the green algae Pedinomonas sp., Ankistrodesmus sp., Chlamydomonas sp., Chlorella sp., Scenedesmus sp., and Selenastrum sp. had maximal growth when the mixtures did not include Selenastrum, which proved toxic to this species. Tissue growth rates were greatest in clams fed mixed cultures containing all five of the remaining species and generally declined with the number of algal species in feeding cultures; feeding with cultures of only two algal species generally resulted in starvation (Foe and Knight 1986b). Thus, artificial bivalve cultures appear to be best supported on mixed algal diets, but certain toxic algal species must be avoided.

Temperature also affects growth rate. When 5–8 mm SL specimens of C. fluminea were fed mixed algal cultures of Chlamydomonas, Chlorella, and Ankistrodesmus at 10^5 cells/ml, assimilation efficiencies were greatest at 16 and 20°C (48–51%), declining above and below this range. Tissue growth was maximal at 18–20°C and became negative (tissue loss) at 30°C and above (Foe and Knight 1986a), suggesting 18°–20°C to be the ideal culture temperature for this species and, perhaps, most other North American bivalves. However, tissue growth in natural populations of *C. fluminea* increases with rising temperatures, up to 30°C (McMahon and Williams 1986a), indicating that artificial culture systems are not equivalent to field conditions in supporting bivalve growth. In this regard, excellent tissue growth in *C. fluminea* was supported by algal cultures produced by exposure to sunlight of water taken from the natural habitat of the clam for several days to increase its algal concentration (Foe and Knight 1985). Thus, ideal culture conditions for unionaceans and *C. fluminea* would appear to be a 20°C holding temperature and a feeding medium of natural algal assemblages whose growth has been promoted with inorganic nutrients and exposure to sunlight.

Specimens of both *C. fluminea* and unionaceans may be held for long periods in the laboratory without feeding. *C. fluminea* (light and dark morphs) survived 154 days of starvation at room temperature (22°–24°C) while sustaining tissue weight losses ranging from 41–71% (Cleland *et al.* 1986). Similarly, I have held unionaceans in the laboratory for many months without feeding them. Maintenance at low temperatures (<10°C) greatly prolongs the time bivalves may be held in the laboratory without feeding.

C. fluminea has never been reared successfully to maturity or carried through a reproductive cycle in artificial culture, although field-collected, nongravid adults released juveniles after four months in laboratory culture (King *et al.* 1986). The glochidium stage makes unionaceans difficult to rear in the laboratory as it requires encystment in an appropriate fish host for successful juvenile metamorphosis, but it can be accomplished (Young and Williams 1984a). Glochidia of several unionacean species have been transformed into juveniles *in vitro* in a culture medium containing physiologic salts, amino acids, glucose, vitamins, antibiotics, and host fish plasma (Isom and Hudson 1982). Juvenile *Anodonta imbecilis* and *Dysnomia triquetra* have been successfully cultured in a medium of river water exposed to sunlight for 1–4 days to enhance algal concentration. Addition of silt to this medium enhanced juvenile growth rates in both species, while feeding artificial, mixed cultures of three algal species resulted in starvation (Hudson and Isom 1984).

Many sphaeriid species can be easily maintained in simple artificial culture systems. Ease of artificial culture in this group may relate to their sediment organic detritus (Burkey *et al.* 1985b, Hornbach *et al.* 1984a, 1984b) and sediment interstitial bacteria feeding mechanisms (Lopez and Holopainen 1987), making maintenance of algal cultures as food sources unnecessary. Hornbach and Childers (1987) maintained *Musculium partumeium* through successful reproduction in beakers with 325 ml of filtered river water, without sediments, on a diet of 0.1 mg of finely ground Tetra Min fish food/clam/day. The first generation in this simple culture system survived 380–500 days, a life span essentially equivalent to that in natural populations (Hornbach *et al.* 1980). Similarly, Rooke and Mackie (1984c) maintained 20 adult *Pisidium casertanum* in a 6 liter aquaria with sediments for 35 weeks without feeding, suggesting that individuals fed on sediment bacteria or organic deposits. Live molluscs rapidly remove dissolved calcium from culture media (Rooke and Mackie 1984c), thus calcium levels in holding media should be augmented by addition of a source of $CaCO_3$. The ease with which sphaeriids can be artificially cultured through many generations makes them ideal for laboratory microcosm experiments. Ideal culture conditions appear to include provision of natural sediments, a source of calcium (i.e., ground $CaCO_3$), and finely ground food of reasonable protein content, such as ground aquarium fish food or brewers' yeast, which may be directly assimilated by clams or support the growth of the interstitial bacteria upon which they feed (Lopez and Holopainen 1987).

V. IDENTIFICATION OF THE FRESHWATER BIVALVES OF NORTH AMERICA

A. Taxonomic Key to Superfamilies of Freshwater Bivalvia

There are five bivalve superfamilies with freshwater representatives in North America. Of these, the Unionacea, Corbiculacea, and Dreissenacea contain the true freshwater species and comprise the vast majority of freshwater bivalve fauna. The remaining two superfamilies, Cyrenoidacea and Mactracea contain but one brackish water species, however, each can extend into coastal freshwater drainages and so are included here. In North America, the Dreissenacea is represented by a recently introduced freshwater species and an estuarine species; the Corbiculacea includes 33 native and 5 introduced species in 6 genera; the Unionacea are composed of approximately 227 native species in 44 genera (Burch 1975a, 1975b). Separate taxonomic keys to the genera are provided here for the latter two superfamilies.

1a.	Shell hinge ligament is external	2
1b.	Shell hinge ligament is internal	4

2a(1a). Shell with lateral teeth extending anterior and posterior of true cardinal teeth (Fig. 11.1), shells of adults generally small (<25 mm in shell length, shell thin and fragile; exceptions are the genera *Polymesoda* and *Corbicula*) .... superfamily Corbiculacea
     [See Section V.B.]

2b. Shell without lateral teeth extending anterior and posterior of cardinal teeth 3

3a(2b). Shell hinge with two cardinal teeth and without lateral teeth; shell thin and fragile, 12–15 mm long with small umbos; *Cyrenoida floridana* (Dall) (extends from brackish into coastal freshwater drainages in Florida .... superfamily Cyrenoidacea

3b. Shell without true cardinal teeth; when present, lateral teeth only occur posterior to usually well-developed pseudocardinal teeth (Fig. 11.1), pseudocardial teeth absent or vestigial in some species; Shells of adults are generally large (> 25 mm in shell length) .... superfamily Unionacea
     [See Section V.C.]

4a(1b). Hinge with anterior and posterior teeth on either side of cardinals; Shell massive, adults 25–60 mm shell length, obliquely ovate; *Rangia cuneata* (Gray) (extends from brackish into coastal freshwater drainages from Delaware to Florida to Veracruz, Mexico) .... superfamily Mactracea

4b. Hinge without teeth; shell mytiloid in shape, anterior end reduced and pointed, hinge at anterior end, posterior portion of shell expanded, anterior adductor muscle attached to internal apical shell septum, attached to hard substrata by byssal threads .... superfamily Dreissenacea   5

5a(4b). Periostracum bluish brown to tan without a series of dorsoventrally oriented black zigzag markings, anterior end hooked sharply ventrally, ventral shell margins not distinctly flattened over entire ventral side of valves; restricted to brackish water habitats. *Mytilopsis leucophaeata* (Conrad) (extends into coastal freshwater drainages from New York to Florida to Texas and Mexico) .... *Mytilopsis*

5b. Periostracum light tan and marked with a distinct series of black vertical zigzag markings; anterior portion of shell not ventrally hooked, ventral shell margins extremely flattened, restricted to freshwaters, *Dreissena polymorpha* (Pallas) (Fig. 11.24); (a European species introduced into the Great Lakes in Lake St. Clair and present by August 1990 throughout Lake Erie, extending into western Lake Ontario and the St. Lawrence river. Isolated populations also occur in lakes Huron, Superior, and Michigan; it is likely to spread rapidly through North American freshwaters) .... *Dreissena*

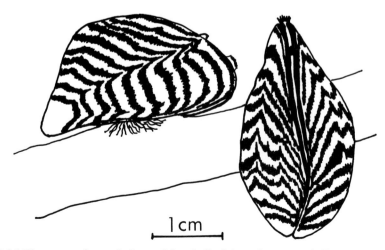

Figure 11.24 The external morphology of the shell of the zebra mussel, *Dreissena polymorpha*.

B. Taxonomic Key to Genera of Freshwater Corbiculacea

This key is based on the excellent species key for North American freshwater Corbiculacea by Burch (1975a) with additional material from Clarke (1973) for the Canadian Interior Basin, and Mackie *et al.* (1980) for the Great Lakes. The Corbiculacea have ovate, subovate, or trigonal shells with lateral hinge teeth anterior and posterior to the cardinal teeth. All North American species except *Corbicula fluminea* are in the family Sphaeriidae. The family designation

Pisidiidae has also been commonly applied to this group, but the International Commission of Zoological Nomenclature (ICZN) placed the Sphaeriidae (Name number 573) on the Official List of Family Names (Opinion 1331) in 1985; hence, Sphaeriidae is used as the family designation in this key and the rest of the chapter. In North America, the Sphaeriidae comprise the dominant bivalve fauna in small, often ephemeral ponds, lakes, and streams, the profundal portions of lakes and in silty substrata. Identification is generally based upon shell morphology but requires, in some cases, soft tissue morphology.

1a.	Shells large (maximum adult shell length > 25 mm), thick and massive	family Corbiculidae	2
1b.	Shells generally small (maximum adult shell length < 25 mm), thin and generally fragile	family Sphaeriidae	4
2a(1a).	Maximum adult shell length generally < 50 mm, shell ornamented by distinct, concentric sulcations, anterior and posterior lateral teeth with many fine serrations, simultaneous hermaphrodites, massive numbers of small (length < 0.3 mm) developmental stages (> 1000) incubated directly in inner demibranchs, released juveniles (< 5 mm SL) anchor to substratum with a single mucilaginous byssal thread (Fig. 11.25)	*Corbicula*	3
2b.	Maximum adult shell length generally > 50 mm, shell ornamentation of many fine, closely spaced concentric striations, embryos not incubated in demibranchs, dioecious, periostracum deep brown in color, three cardinal teeth, estuarine, restricted to brackish waters in the tidal portions of rivers. *Polymesoda caroliniana* (Bosc) (Virginia to northern Florida to Texas)	*Polymesoda*	
3a(2a).	Shell nacre white with light blue, rose, or purple highlights, particularly at shell margin, muscle scars of same color intensity as rest of nacre, periostracum yellow to yellow-green or brown with outer margins always yellow or yellow-green in healthy, growing specimens, shell trigonal to ovate, umbos inflated and distinctly raised above dorsal shell margin, shell length : shell height ratio ≈ 1.06, shell length : shell width ratio ≈ 1.47, shell height : shell width ratio ≈ 1.38, concentric shell sulcations widely spaced, 1.5 sulcations/mm shell height (Hillis and Patton, 1982); introduced in the early 1900's, it has spread throughout drainage systems of the United States and coastal northern Mexico (Fig. 11.11), the "light colored shell morph" of *Corbicula* or Asian clam (Fig. 11.25B,C)	*Corbicula fluminea* (Müller)	
3b.	Shell nacre uniformly royal blue to deep purple over entire internal surface, muscle scars more darkly pigmented than rest of nacre, periostracum dark olive green to black, edges of valves in healthy, growing specimens not yellow or yellow-green, shell more ovate and laterally compressed with umbos less inflated and less distinctly raised above the dorsal shell margin than in *C. fluminea,* shell length : height ratio ≈ 1.15, shell length to width ratio ≈ 1.65, shell height : width ratio ≈ 1.43, concentric shell sulcations narrowly spaced, particularly at umbos, 2.1 sulcations/mm shell height (Hillis and Patton 1982); introduced, distribution limited to highly oligotrophic, permanent, spring-fed, calcium carbonate-rich streams in the southwestern United States (Britton and Morton 1986); (Fig. 11.11). Called the dark-colored shell morph of *Corbicula,* its taxonomic status is uncertain, electrophoretic (Hillis and Patton 1982, McLeod 1986) and physiological evidence (Cleland *et al.* 1986) suggest it to be distinct from *C. fluminea* (Fig. 11.25A,D)	*Corbicula* sp.	

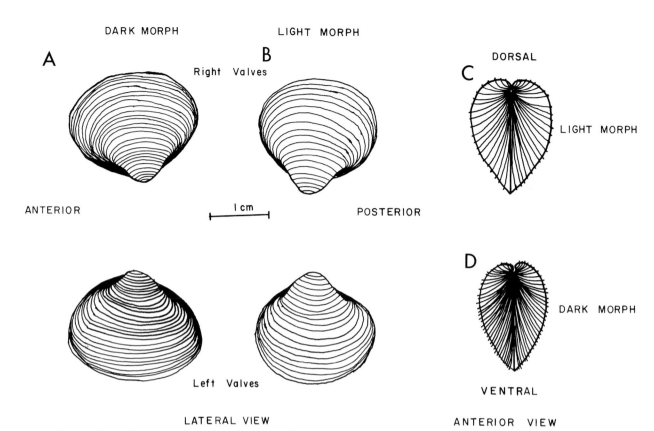

Figure 11.25 External morphology of the shell valves of the light-colored shell morph (*Corbicula fluminea*) and dark-colored shell morph (*Corbicula* sp.) of the North American *Corbicula* species complex. (A) Right and left valves of *Corbicula* sp. (dark-colored morph). (B) Right and left valves of *C. fluminea* (light-colored morph) (C) Anterior view of the shell valves of *C. fluminea* (light-colored morph). (D) Anterior view of the shell valves of *Corbicula* sp. (dark-colored morph). Note the distinguishing shell characteristics of these two species. *C. fluminea,* the white morph, which is widely distributed in North America (Fig. 11.11), has a more nearly trigonal shell, taller umbos, a greater relative shell width and more widely spaced concentric sulcations than does *Corbicula* sp., the dark morph that is limited to spring-fed, alkaline, lotic habitats in the southwestern United States (Fig. 11.11). The dark-colored shell morph also has a dark olive green to black periostracum and deep royal blue nacre, while the light-colored shell morph has a yellow-green to light brown periostracum and white to light blue or light purple nacre.

4a(1b).	Both inhalant and exhalant mantle cavity siphons present and well developed, umbos lie anterior of center		5
4b.	Only exhalant mantle cavity siphon present, inhalant siphon either absent or formed as a slit in the posterior-ventral mantle edges, umbos posterior of center, generally small, shell length 0.5–12 mm, embryos in inner demibranch held in thick-walled sacs, each with individual chambers for embryos, no byssal gland, 24 species widely distributed in North America; for species identifications and distributions see Burch (1975a) (Fig. 11.26A) .... subfamily Pisidiinae	*Pisidium*	
5a(4a).	Inhalant and exhalant mantle cavity siphons partially fused, embryos incubated in inner demibranchs in thin-walled longitudinal pouches, no byssal gland, shell with two cardinal teeth in each valve, without external mottling .... subfamily Sphaeriinae		6
5b.	Inhalant and exhalant siphons not fused, embryos develop in individual chambers formed between inner and outer lamellae of inner demibranchs, functional byssal gland present, only one cardinal tooth in each shell valve, with external, mottled pigmentation, *Eupera cubensis* (Prime) (Atlantic coastal plain drainages from southern Texas to North Carolina, Caribbean Islands) (Fig. 11.26B) .... subfamily Euperinae	*Eupera*	

6a(5a). Shell sculptured with relatively coarse or widely spaced striae ($\leq$ 8 striae/mm in middle of shell), shell relatively massive and strong, *Sphaerium simile* (Say) (Southern Canada from New Brunswick to British Columbia, south from Virginia to Wyoming, *S. striatinum* (Lamarck) (Canada from New Brunswick to the upper Yukon River, throughout the United States, Mexico, and Central America), *S. fabale* (Prime) (Southern Ontario to Georgia and Alabama) (Fig. 11.26C) .... *Sphaerium*

6b. Shell relatively thin, often fragile, with many fine, narrowly spaced striae ($\geq$ 12 striae/mm in middle of shell) 7

7a(6b). Shell of adults < 8 mm in length 8
7b. Shell of adults > 8 mm in length 9

8a(7a). Posterior valve margin at near right angle to dorsal margin, shells roughly rhomboidal, umbos large and distinctly elevated above dorsal shell margin, *Musculium partumeium* (Say) (United States and southern Canada), *M. transversum* (Say) (Canada and the United States east of the continental divide, extending into Mexico), *M. securis* (Prime) (Nova Scotia to British Columbia southwestern Northwest Territories in Canada, United States except for southwest) (Fig. 11.26D) .... *Musculium*

8b. Posterior and dorsal margins rounded or forming an obtuse angle, shells ovate, *Sphaerium corneum* (Linnaeus) (introduced from Europe, localities in southern Ontario and Lakes Champlain and Erie), *S. nitidum* Westerlund (distribution is holarctic, northern Canada to northern United States), *S. occidentale* (Prime) (Canada from New Brunswick to southeastern Manitoba, northern United States south to Florida, west to Utah and Colorado) (Fig. 11.26E, F) .... *Sphaerium*

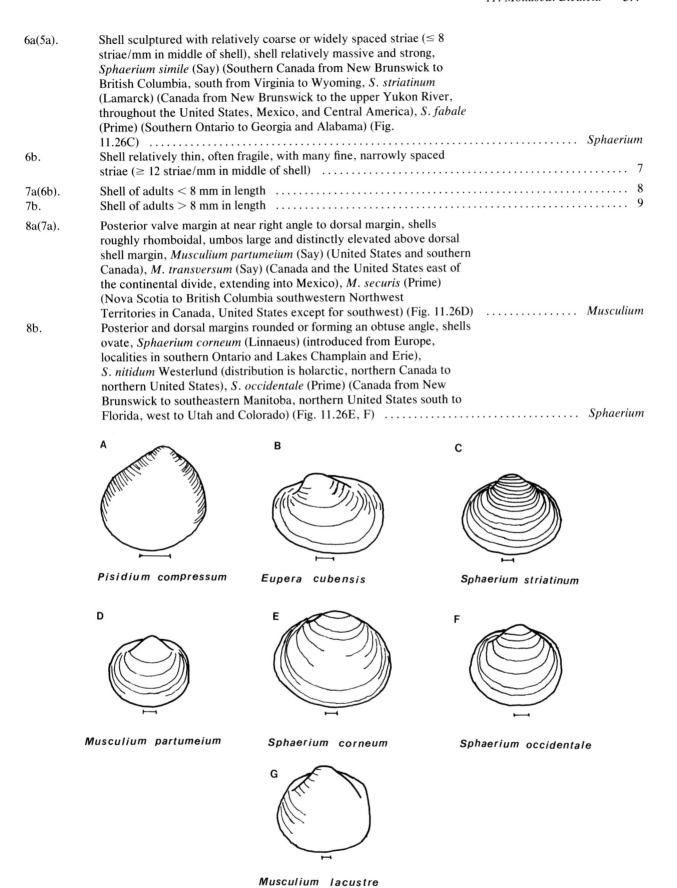

Figure 11.26 Diagrams of the external morphology of the left shell valve of species representative of the North American genera of the freshwater bivalve family, Sphaeriidae. Size scaling bar is 1 mm long.

9a(7b).   Umbos large, distinctly elevated above the dorsal shell margin 10
9b.   Umbos small, indistinctly elevated above the dorsal shell margin,
           Sphaerium corneum (Linnaeus), (introduced, localities in Ontario,
           Lakes Champlain and Erie) (Fig. 11.26E) .... *Sphaerium*

10a(9a).   Shell rounded, umbos not prominent, *Sphaerium occidentale* (Prime)
           (New Brunswick to southeastern Manitoba, northern United States
           south to Florida in the east and Utah and Colorado in the west) (Fig.
           11.26F) .... *Sphaerium*
10b.   Posterior end of shell truncate, shell rhomboidal, umbos prominent,
           Musculium lacustre (Müller) (From treeline in Canada south throughout
           all but southwestern United States into central America) (Fig.
           11.26G) .... *Musculium*

C. Taxonomic Key to Genera of Freshwater Unionacea

This key is based primarily on the excellent key to the species of North American Unionacea by Burch (1975b) with additional material from Clarke (1973) and Mackie *et al.* (1980). North America has the richest and most diverse unionacean fauna in the world including, conservatively, 227 species in 46 genera. Unionacean taxonomy remains very uncertain because intraspecific, interpopulation ecophenotypic variability often makes identification and systematics difficult. Unionacean shells lack true cardinal teeth and, when present, lateral teeth occur only posterior to pseudocardinal teeth. The Unionacea make up the large bivalve fauna (shell length > 25 mm) of permanent freshwater lakes, rivers, and ponds. Figure 11.27 displays shell-shape outlines and external shell ornamentations referred to in these taxonomic keys.

1a.   Posterior mantle margins not fused to form separate anal opening,
           posterior mantle margins display no thickenings or other structures
           associated with development of inhalant or exhalant siphons, shell
           laterally compressed and elongated anterio-posteriorly, adults 80–
           175 mm in shell length .... family Margaritiferidae   2
1b.   Posterior mantle margins forming separate anal opening, posterior
           mantle margins display thickenings and other structures associated with
           development of distinct inhalent and exhalant siphons .... family Unionidae   3
2a(1a).   Gill interlamellar junctions scattered and in interrupted rows, but
           developed as continuous septa, which run obliquely forward (Burch
           1975b); shell thin, fragile and highly elongated, pseudocardinal teeth
           reduced, umbos not distinctly elevated above dorsal shell margin, shell
           surface usually with heavy growth lines, *Cumberlandia monodonta*
           (Say) (Ohio, Tennessee and Mississippi River drainages, central United
           States); (Fig. 11.28A) .... subfamily Cumberlandiidae   *Cumberlandia*
2b.   Gill interlamellar junctions discontinuous, irregularly scattered or falling
           into oblique rows (Burch 1975b), shell generally more massive, and
           relatively deeper dorsoventrally than that of *Cumberlandia,* well-
           developed pseudocardinal teeth. External posterior surface of valves
           corrugated, *Margaritifera hembeli* (Conrad) (Alabama and Louisiana),
           external posterior surface of valves smooth, *M. margaritifera*
           (Linnaeus) (northern United States, east of continental divide,
           M. falcata (Gould) (drainages west of continental divide, southern
           Arkansas to southern California); (Fig. 11.28B) .. subfamily Margaritiferinae   *Margaritifera*
3a(1b).   Embryos incubated in all four demibranchs so all four demibranchs are
           swollen in gravid females .... subfamily Ambleminae   4
3b.   Embryos incubated only in marsupia formed in the outer demibranchs,
           so these are swollen in gravid females .... subfamily Unioninae   16
4a(3a).   Hinge teeth well developed 5
4b.   Hinge teeth rudimentary or absent, *Gonidea angulata* (Lea) (west coast
           drainages from British Columbia to central California, east to Nevada
           and Idaho); (Fig. 11.28C) .... tribe Gonideini   *Gonidea*

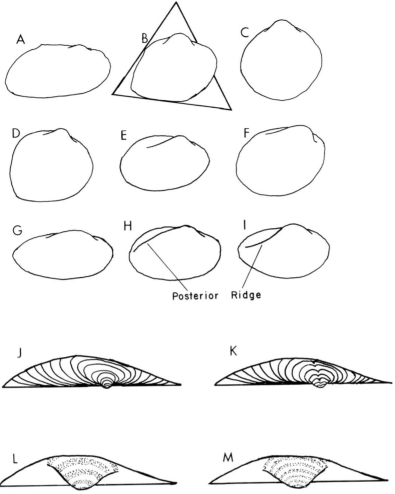

Figure 11.27 Illustrations of the diagnostic shell features or characters used for taxonomic identification in Section V.C. Shell shape descriptions: rhomboidal (A); triangular or trigonal (B); round (C); quadrate (D); oval or ovoid (E and F); and elliptical (G). Posterior shell ridge morphology: posterior ridge convex (H); and posterior ridge concave (I). Concentric ridge structures of umbos: single-looped concentric ridges (J): double-looped concentric ridges (K); coarse concentric ridges (L); and fine concentric ridges (M). (Redrawn from Burch 1975b.)

5a(4a).	Posterior shell slope with distinct pustules or corrugations 6	
5b.	Posterior shell slope without distinct pustules or corrugations 14	
6a(5a).	Posterior slope of shell abbreviated and steep, shell height nearly equal to shell length, *Quadrula stapes* (Alabama and Tombigbee Rivers) .... tribe Amblemini	*Quadrula*
6b.	Posterior slope of shell not steep, shell generally longer than high (if not, posterior slope is not steep) 7	
7a(6b).	Posterior external shell surface with distinct pustules 8	
7b.	Posterior external shell surface without distinct pustules 10	
8a(7a).	Shell roundly oval (shell length : height ratio slightly > 1), angular at intersection of posterior and ventral margins, *Quadrula intermedia* (Tennessee River Drainage) .... tribe Amblemini	*Quadrula*
8b.	Shell length distinctly greater than height, rhomboidally shaped 9	
9a(8b).	Shell with well-developed posterior ridge, small pustules covering middle and anterior portions of shell, purple to purple-pink nacre, *Tritogonia verrucosa* (Rafinesque) (drainages of the Mississippi River and coastal Gulf of Mexico slope from the Alabama River west to Texas); (Fig. 11.28D) .... tribe Amblemini	*Tritogonia*

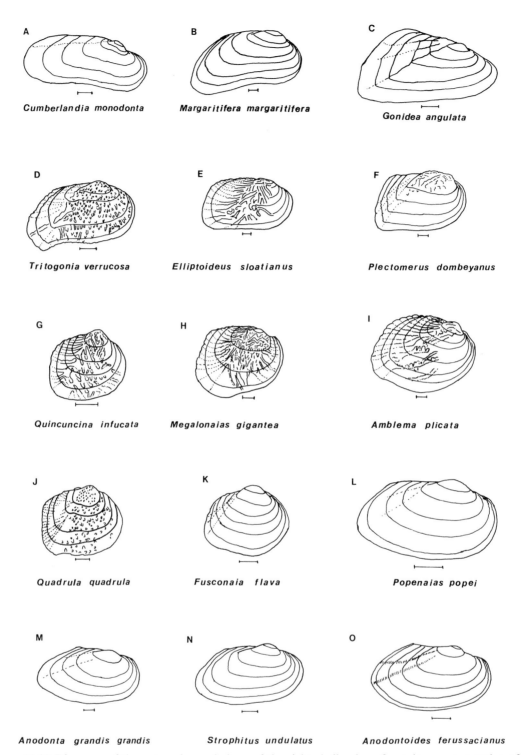

Figure 11.28 Diagrams of the external morphology of the right shell valve of species representative of the North American genera of the freshwater divalve superfamily Unionacea (Fig. 11.28A–AS). Size scaling bar is 1 cm long.

9b.	Posterior ridge low, parallel row of large pustules just anterior and ventral to it, pustules less developed on anterior end, white nacre, *Quadrula cylindrica* (Say) (Ohio, Cumberland, and Tennessee River systems, south to Arkansas and Oklahoma, west to Nevada ... tribe Amblemini *Quadrula*
10a(7b).	Shell rhomboidally shaped, posterior margin relatively straight, pale violet to bronze nacre 11

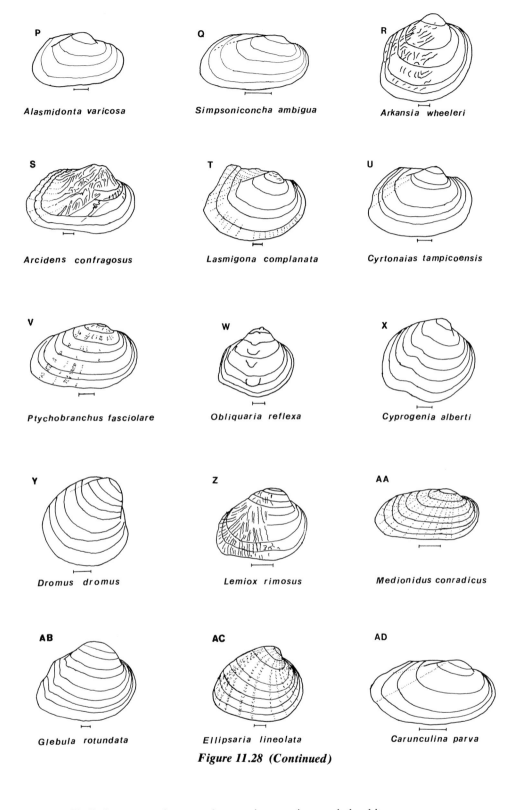

P **Alasmidonta varicosa**

Q **Simpsoniconcha ambigua**

R **Arkansia wheeleri**

S **Arcidens confragosus**

T **Lasmigona complanata**

U **Cyrtonaias tampicoensis**

V **Ptychobranchus fasciolare**

W **Obliquaria reflexa**

X **Cyprogenia alberti**

Y **Dromus dromus**

Z **Lemiox rimosus**

AA **Medionidus conradicus**

AB **Glebula rotundata**

AC **Ellipsaria lineolata**

AD **Carunculina parva**

Figure 11.28 (Continued)

10b.   Shell elongate oval or round, posterior margin rounded, white nacre 12

11a(10a or 14a).   Umbos indistinctly demarked from shell, anterior adductor muscle scar smooth surfaced, *Elliptoideus sloatianus* (Lea) (Apalachicola and Ochlockonee River drainages of Georgia and Florida); (Fig. 11.28E) .... tribe Amblemini   *Elliptoideus*

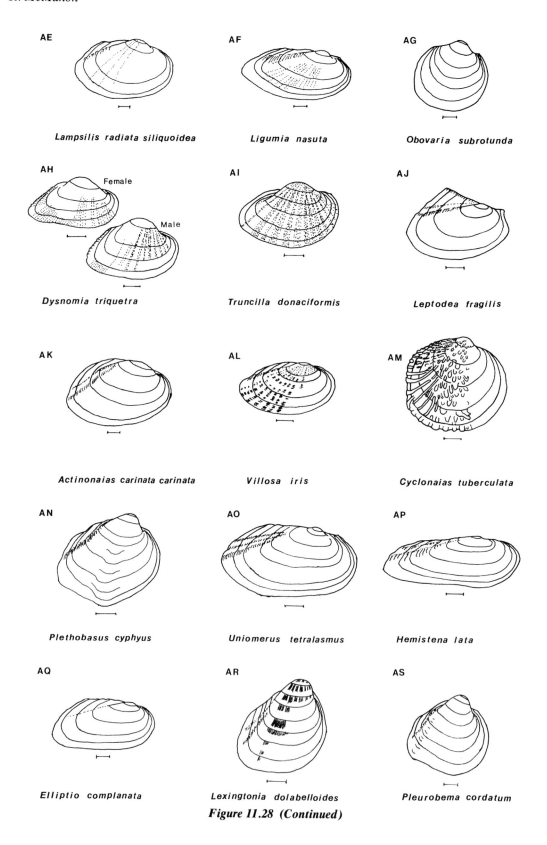

AE

Lampsilis radiata siliquoidea

AF

Ligumia nasuta

AG

Obovaria subrotunda

AH

Female

Male

Dysnomia triquetra

AI

Truncilla donaciformis

AJ

Leptodea fragilis

AK

Actinonaias carinata carinata

AL

Villosa iris

AM

Cyclonaias tuberculata

AN

Plethobasus cyphyus

AO

Uniomerus tetralasmus

AP

Hemistena lata

AQ

Elliptio complanata

AR

Lexingtonia dolabelloides

AS

Pleurobema cordatum

Figure 11.28 (Continued)

11b.   Umbos distinctly demarked from shell, anterior adductor muscle scar
            rough surfaced, *Plectomerus dombeyanus* (Valenciennes) (coastal Gulf
            of Mexico slope drainages from Alabama River to eastern Texas,
            Mississippi drainage north to Tennessee); (Fig. 11.28F) . . . . . tribe Amblemini   *Plectomerus*

12a(10b).	Shell small (< 6 cm shell length) with either relatively indistinct or no corrugations, *Quincuncina burkei* (Walker) (Choctawhatcee River drainage, Florida), *Q. infucata* (Conrad) (Suwannee River, west to Apalachicola River, Florida), *Q. guadalupensis* Wurtz (Guadalupe and Leon Rivers, Texas); (Fig. 11.28G) .... tribe Amblemini **Quincuncina**
12b.	Shell very large and massive, 13–18 cm in length with deep corrugations **13**
13a(12b).	Corrugations extend from umbos towards anterior margin, umbos low with large ridges extending to posterior shell margin, white to pink nacre, *Megalonaias gigantea* (Barnes), (throughout Mississippi River drainage, coastal Gulf of Mexico slope drainages from Tombigbee River, Alabama to Nuevo Leon, Mexico, and Ochlockonee River west to Escambia River in Florida Panhandle if *M. boykiniana* (Lea) is a synonym of *M. gigantea*); (Fig. 11.28H) .... tribe Megalonaiadini **Megalonaias**
13b.	Corrugations not extending anterior of umbos, shell sculpture consisting of large ridges running from below prominent umbos towards the posterior shell margin, periostracum usually dark brown or black, but may be lighter in color, *Amblema neisleri* Lea (Apalachicola, Chipola, and Flint Rivers, Florida and Georgia), *A. plicata* Say (Mississippi drainage, western New York to Minnesota and eastern Kansas, Gulf of Mexico slope drainages from Texas to western Florida Panhandle, Saint Lawrence River and Great Lakes drainages with exception of those of Lake Superior, Red River of the North, and some other tributaries of Lake Winnipeg, central Canada; (Fig. 11.28I) .... tribe Amblemini **Amblema**
14a(5b).	Shell rhomboidal, posterior margin relatively straight **11**
14b.	Shell round, oval or trigonal, posterior margin curved **15**
15a(14b).	Shell surface pustulose, often with corrugations, umbos well developed, posterior ridge usually well developed, nine species, occurring in the St. Lawrence and Mississippi Rivers and Gulf of Mexico slope drainages of the United States; a single species, *Quadrula quadrula* (Rafinesque) extends into Canada in the Red River of the North drainage (Clarke 1973); (Fig. 11.28J) .... tribe Amblemini **Quadrula**
15b.	Shell surface smooth without pustules, with moderate posterior ridge, umbos well developed, curving anteriorly and medially with few coarse concentric ridges, nacre white, salmon, or pink, 13 species restricted to Mississippi and Gulf of Mexico slope drainages (exception is *Fusconaia flava* (Rafinesque) in the St. Lawrence River drainage, and Red River of the North and Nelson River drainages of central Canada); (Fig. 11.28K) .... tribe Amblemini **Fusconaia**
16a(3b).	Exhalant water channels of demibranchs in gravid females undivided by secondary septa, glochidia without hooks (exception is genus *Proptera*, with axehead-shaped glochidia) **17**
16b.	Exhalant water channels of demibranchs of gravid females divided by secondary cross septa between adjacent primary septa into three channels, glochidia brooded only in middle channels (exception is genus *Strophitus*), glochidia with hooks .... tribe Anodontini **19**
17a(16a).	Glochidia brooded throughout smoothly swollen outer demibranchs, shells of males and females morphologically similar **18**
17b.	Glochidia brooded only in specific portions of outer demibranch, usually ventral portions or alternating gill lamellae, gill marsupia not smoothly swollen, instead marked by external annuli, shells of males and females usually morphologically dissimilar .... tribe Lampsilini **36**
18a(17a).	Bradytictic breeders retaining developing glochidia in gill marsupia throughout the year except during summer, *Popenaias buckleyi* (Lea) (Florida peninsula), *P. popei* (Lea) (southern Texas and northeast Mexico); (Fig. 11.28L) .... tribe Popenaiadini **Popenaias**
18b.	Tachytictic breeders retaining glochidia in gill marsupia only during the spring and summer .... tribe Pleurobemini **58**

19a(16b).	Hinge teeth absent or, if present, very reduced with only rudimentary pseudocardinal teeth, shell thin and fragile	20
19b.	Hinge teeth distinct although poorly developed in some species, shell thin to moderately thick but not fragile	33
20a(19a).	Pseudocardinal teeth absent	21
20b.	Rudimentary pseudocardinal teeth present	29
21a(20a).	Found in drainages east of the continental divide	22
21b.	Found in drainages west of the continental divide, eight species, for identifications and distributions see Burch 1975b) .... *Anodonta*
22a(21a).	Umbos distinctly elevated above dorsal margin	23
22b.	Umbos not distinctly elevated above dorsal margin, four species, *Anodonta suborbiculata* Say (Mississippi and Escambia River drainages), *A. imbecilus* Say (United States east of continental divide), *A. peggyae* Johnson (eastern Alabama to Florida), *A. couperiana* Lea (Florida to North Carolina) .... *Anodonta*
23a(22a).	Shell evenly inflated, not noticeably inflated posteriorly	24
23b.	Shell highly inflated posteriorly down to ventral shell margin, *Anodonta gibbosa* Say (Altamaha River drainage, Georgia) .... *Anodonta*
24a(23a).	Umbo sculptured with concentric ridges, each of uniform height along its length	25
24b.	Umbo sculptured with concentric ridges with distinct nodules (nodules formed because ridge height not uniform along ridges), *Anodonta grandis grandis* Say (Canadian Interior Basin, St. Lawrence River drainage, and Gulf of Mexico drainages in Louisiana and Texas); (Fig. 11.28M) .... *Anodonta*
25a(24a).	Umbos with 3–6 concentric ridges	26
25b.	Umbos with 7–10 concentric ridges, *Anodonta kennerlyi* Lea (found east of the continental divide only in western Alberta .... *Anodonta*
26a(25a).	Umbos with single-looped or faintly double-looped concentric ridges; (Fig. 11.27)	27
26b.	Umbos with distinctly double-looped concentric ridges (Fig. 11.27), *Anodonta implicata* Say (coastal drainages from New Brunswick and Nova Scotia to Maryland), *Anodonta cataracta* Say (Gulf of Mexico drainages of Alabama and western Florida, Georgia to lower St. Lawrence River drainage, west to Michigan .... *Anodonta*
27a(26a).	Concentric ridges of umbos relatively fine	28
27b.	Concentric ridges of umbos relatively coarse, *Strophitus undulatus* (Say) (Interior Basin from Texas to western Ontario to Saskatchewan, Atlantic coast from Nova Scotia to South Carolina); (Fig. 11.28N) .... *Strophitus*
28a(27a).	Concentric ridges on umbos are parallel to growth lines, *Anodonta grandis simpsoniana* Lea (Hudson Bay and arctic Canada) .... *Anodonta*
28b.	Concentric ridges on umbos cross growth lines obliquely, *Anodontoides ferussacianus* (Lea) (widely distributed in the North American Interior Basin); (Fig. 11.28O) .... *Anodontoides*
29a(20b).	Pseudocardinal teeth very thin, bladelike	30
29b.	Pseudocardinal teeth more massive, tubercular	31
30a(29a).	Shell rhomboidal with distinct posterior ridge, fine, concentric corrugations on posterior slope, *Alasmidonta varicosa* (Lamarck) (lower St. Lawrence River drainage to South Carolina; (Fig. 11.28P) .... *Alasmidonta*
30b.	Shell smooth, distinctly ovate without posterior ridge, *Anodontoides radiatus* (Conrad) (Gulf of Mexico drainages of Alabama, Georgia and Florida) .... *Anodontoides*
31a(29b).	Posterior external slope of shell without corrugations	32

31b.	Posterior external slope of shell with distinct concentric corrugations, *Alasmidonta marginata* Say (upper Mississippi drainage in Ohio, Cumberland and Tennessee Rivers, St. Lawrence drainage from Lake Huron to the Ottawa River), *A. raveneliana* (Lea) (Tennessee and Cumberland River drainages), *A. varicosa* (Lamarck) (lower St. Lawrence River drainage south to upper Savannah River drainage of South Carolina .... *Alasmidonta*

32a(31a). Shell elongately ovate, shell length : height ratio > 2, posterior ridge reduced, *Simpsoniconcha ambigua* (Say) (Ohio River drainage); (Fig. 11.28Q) .... *Simpsoniconcha*

32b. Shell roundly ovate, shell length : height ratio < 1.6 with distinct posterior ridge, *Strophitus subvexus* (Conrad) (Mississippi, Georgia and western Florida) .... *Strophitus*

33a(19b). Shell with large corrugations on disc or posterior slope 34
33b. Shell without corrugations on disc or posterior slope 35

34a(33a). Tubercules occur on first 3–4 mm of umbos, pseudocardinal teeth massive and triangular, *Arkansia wheeleri* Ortmann and Walker (Ouachita Mountains, Kiamichi and Old Rivers, Arkansas and Oklahoma); (Fig. 11.28R) .... *Arkansia*

34b. Tubercules on beak extend down onto shell, pseudocardinal teeth compressed and closely adjacent, *Arcidens confragosus* (Say) (Mississippi drainage southward from Wisconsin and Ohio, Colorado River, Texas and Bayou Teche, Louisiana; (Fig. 11.29S) .... *Arcidens*

35a(33b). Ridges of umbos smoothly concentric without a dorsally directed indentation (single-looped, Fig. 11.27), 11 species, identifications and distributions in Burch (1975b) .... *Alasmidonta*

35b. Concentric ridges of umbos not smoothly concentric but interrupted by a dorsally directed identation (double-looped, Fig. 11.27), five species, identifications and distributions in Burch (1975b); (Fig. 11.28T) .... *Lasmigona*

36a(17b). Glochidia incubated only in ventral portion of outer demibranchs 37
36b. Glochidia incubated only in central or posterior portion of outer demibranchs 38

37a(36a). Ventral, outer marsupial demibranch edge smooth, without folds, shell ovate, inflated, *Cyrtonaias tampicoensis* (Lea) (southeastern Texas to northeastern Mexico and Honduras); (Fig. 11.28U) .... subtribe Longenae *Crytonaias*

37b. Ventral edge of outer marsupial demibranch with 6–20 distinct folds, shell laterally compressed, elongated, subelliptical or rhomboidal, five species, identifications and distributions in Burch (1975a); (Fig. 11.28V) .... subtribe Ptychogenae, *Ptychobranchus*

38a(36b). Marsupium restricted to central portion of outer demibranch ...... subtribe Mesogenae 39
38b. Marsupium restricted to posterior portion of outer demibranch .... subtribe Heterogenae 40

39a(38a). Shell with 3–5 very large tubercules in single medial row extending dorsoventrally, shell roundly ovate, massive hinge teeth, *Obliquaria reflexa* Rafinesque (Mississippi River drainage); (Fig. 11.28W) .... *Obliquaria*

39b. Shell with numerous, various-sized tubercules, pustules or corrugations, without single medial row of large tubercules, *Cyprogenia aberti* (Conrad) (Kansas, Missouri, Oklahoma and Arkansas), *C. irrorata* (Lea) (Ohio, Cumberland, and Tennessee River drainages); (Fig. 11.28X) .... *Cyprogenia*

40a(38b). Marsupium occupies entire posterior portion of the outer demibranch .... subtribe Heterogenae 41

40b.	Marsupium restricted to ventral posterior portion of outer demibranch, *Dromus dromus* (Lea) (Tennessee and Cumberland River drainages); (Fig. 11.28Y) .... subtribe Eschatigenae *Dromus*
41a(40a).	Posterior slope of shell transversely corrugated 42
41b.	Posterior slope of shell either smooth or radially corrugated 43
42a(41a).	Shell roundly ovate, corrugations cover posterior half of shell extending to ventral margin, *Lemiox rimosus* Rafinesque (Tennessee River drainage); (Fig. 11.28Z) .... *Lemiox*
42b.	Shell elongately ovate without radiating corrugations extending to ventral margin, five species (identifications and distributions in Burth 1975b); (Fig. 11.28AA) .... *Medionidus*
43a(41b).	Posterior pseudocardinal teeth are deeply divided into parallel, vertical, plicate lamellae (Fig. 11.29), umbos large, inflated, *Glebula rotundata* (Lamarck) (Gulf of Mexico drainages from eastern Texas to Florida panhandle); (Fig. 11.28AB) .... *Glebula*
43b.	Posterior pseudocardinal teeth not deeply divided into parallel, vertical lamellae, umbos relatively small 44
44a(43b).	Shell trigonal, dorsal margin arched, laterally compressed, posterior transverse slope reduced and at 90° angle to disc, no wing (Fig. 11.30) extending from dorsal shell margin posterior to hinge, hinge teeth large and massive, *Ellipsaria lineolata* (Rafinesque) (Mississippi, Tombigbee, and Alabama River drainages): (Fig. 11.28AC) .... *Ellipsaria*

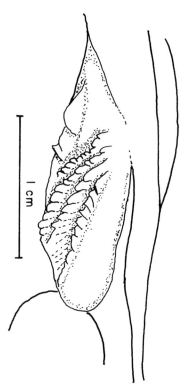

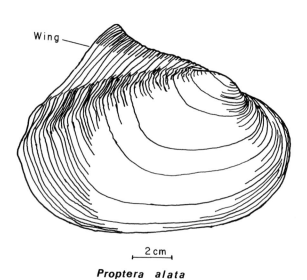

Figure 11.29 Structure of the posterior pseudocardinal teeth of the right valve of *Glebula rotundata*. Note that the posterior pseudocardinal teeth are deeply divided into parallel, vertical, plicate lamellae, a tooth arrangement uniquely characteristic of this species. (Redrawn from Burch 1975b.)

Figure 11.30 Right shell valve of the unionid, *Proptera alata,* showing the thin, extensive dorsal projection of the shell posterior to the umbos forming a "wing" whose presence or absence is a diagnostic characteristic valuable for identification of a number of unionid species. (Redrawn from Burch 1975b.)

44b.	Shell does not display above suite of characteristics, if high, dorsal margin is not generally greatly arched; if arched, shell is inflated, not laterally compressed with a more obtuse posterior slope 45
45a(44b).	In females, inner mantle edges have a well-developed caruncle (Fig. 11.31A) formed from a group of short, crowded papillae just ventral and anterior to the inhalant siphon, adults are small, usually < 4 cm in length, *Carunculina parva* (Barnes) (Mississippi drainage and Florida), *C. pulla* (Conrad) (Georgia to North Carolina); (Fig. 11.28AD) .... *Carunculina*
45b.	Females without caruncles on inner mantle edges just ventral and anterior to inhalant siphon, adult shell length generally > 4 cm 46
46a(45b).	Shell elongately ovate, shell length : height ratio > 2 47
46b.	Shell roundly ovate, shell length : height ratio < 2 48
47a(46a, 51b, or 57b).	In females, there is a flap of tissue projecting from the inner mantle edge just ventral and posterior to the inhalant mantle cavity siphon; on medial side, flaps are often colored, and have a black streak and a darkly pigmented spot (Figs. 11.16 and 11.31B); in males, the flaps are rudimentary in structure, 22 species distributed throughout North America, east of the continental divide, species identifications and distributions in Burch (1975b); (Figs. 11.16 and 11.28AE) .... *Lampsilis*
47b.	In females, there are a distinct series of papillae projecting from the inner mantle edges just ventral and anterior to the inhalant mantle cavity siphon (Fig. 11.31C); papillae are rudimentary in males, *Ligumia nasuta* (Say) (eastern United States north to the St. Lawrence River drainage of Canada, if *L. Subrostrata* (Say) is a synonym of *L. nasuta*), *L. recta* (Lamarck) (Mississippi, Alabama, and St. Lawrence River drainages, Winnipeg and Red River of the North drainages of Canadian Interior Basin); (Fig. 11.28AF) .... *Ligumia*
48a(46b).	Shell round or high oval dorsoventrally (exception is *Obovaria jacksoniana* Frierson, which has a somewhat elongated posterior slope), pseudocardinal teeth massive, five species, identifications and distributions in Burch (1975b); (Fig. 11.28AG) .... *Obovaria*

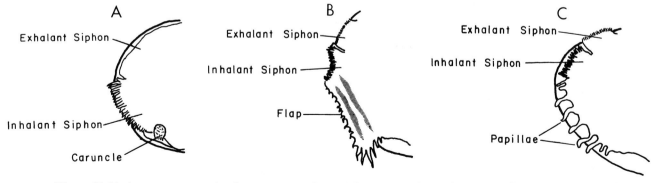

Figure 11.31 Accessory reproductive structures formed from extensions of the posterior mantle edge of female unionids: (A) a caruncle characteristic of the genus *Carunculina;* (B) a mantle flap characteristic of the genus *Lampsilis;* (C) fine papillae or projections characteristic of the genus *Villosa.* In some species, these projections mimic prey of the glochidial fish hosts of a particular unionid species, luring them close to the point of glochidial release. (Redrawn from Burch 1975b.)

48b.   Shell elongate, subelliptical, subrhomboidal, trigonal or
                       oval, pseudocardinal teeth may or may not be massive 49

49a(48b).   Adult shell small, usually < 6 cm in length and massive,
                       sexual dimorphism, females with shell inflated over
                       marsupial portion of outer demibranchs, in some species
                       radiating ridges on posterior slope, 19 species, (southern
                       Canada and eastern United States, Mississippi drainages
                       and Gulf of Mexico slope drainages east of Mississippi
                       River), identifications and distributions in Burch (1975b);
                       (Fig. 11.28AH) . . . . *Dysnomia*
49b.   Shell without above described suite of characteristics 50

50a(49b).   Posterior ridge well developed and angular 51
50b.   Posterior ridge rounded or absent 52

51a(50a).   Shell laterally compressed, umbo cavities shallow, with
                       radiating color bands with or without v-shaped markings,
                       Truncilla donaciformis (Lea) (Mississippi, Lake Erie, and
                       Lake St. Clair drainages), *T. macrodon* (Lea) (Texas and
                       Oklahoma), *T. truncilla* Rafinesque (Mississippi, Lake
                       Erie, and Lake St. Clair drainages); (Fig. 11.28AI) . . . . *Truncilla*
51b.   Shell inflated with deep umbo cavities, some species with
                       radiating color bands (without v-shaped markings),
                       others without color bands 47

52a(50b).   Pseudocardinal teeth well developed 53
52b.   Pseudocardinal teeth small or vestigial, four species
                       (St. Lawrence drainage of Canada and United States east of
                       the continental divide), for species identifications and
                       distributions see Burch (1975b); (Fig. 11.28AJ) . . . . *Leptodea*

53a(52a).   Shell with an extensive wing (dorsal extension of the shell
                       posterior to umbos, Fig. 11.30), *Proptera alata* (Say)
                       (Mississippi drainages, St. Lawrence drainage from Lake
                       Huron to Lake Champlain, Red River of the North and
                       Winnipeg River drainages of Canadian Interior Basin);
                       (Fig. 11.30) . . . . *Proptera*
53b.   Shell wing absent or poorly developed 54

54a(53b).   Shell extremely inflated with fine sculpture on umbos,
                       Proptera capax (Green) (lower Ohio River drainage) . . . . *Proptera*
54b.   Shell not extremely inflated, or if inflated, has coarse
                       sculpture on umbos 55

55a(54b).   Adult shell large (up to 11.5 cm), nacre purple, *Proptera
                       purpurata* (Lamarck) (eastern Texas, north to Kansas and
                       southern Missouri, western Tennessee to Alabama River
                       drainage) . . . . *Proptera*
55b.   Shell nacre usually white but not purple, may be pinkish-
                       purple in smaller specimens (< 6 cm in SL) 56

56a(55b).   May have crenulations on inner mantle edge just ventral
                       and anterior to inhalant mantle cavity siphon, but without
                       projecting papillae or flaps in this area (see Figs. 11.31B
                       and C for descriptions of mantle papillae or flaps), four
                       species, identifications in Burch (1975b) (Mississippi, Ohio,
                       Tennessee, Cumberland, and St. Lawrence River
                       drainages); (Fig. 11.28AK) . . . . *Actinonaias*
56b.   Distinct papillae or flaps project from inner mantle margin
                       just ventral and anterior to the inhalant mantle cavity
                       siphon (Fig. 11.31B, C) 57

57a(56b).   Long papillae project from posterior inner mantle margins
                       (Fig. 11.31C), 16 species, identifications and distributions
                       in Burch (1975b); (Fig. 11.28AL) . . . . *Villosa*

57b. Distinct flaps formed on inner mantle edges (Fig. 11.25B) 47

58a(18b). Pustules on shell surface 59
58b. Shell surface without pustules 60

59a(58a). Shell rounded with purple nacre, *Cyclonaias tuberculata*
(Rafinesque) (Mississippi drainage, Lake St. Clair drainage,
Detroit River and Lake Erie; (Fig. 11.28AM) .... *Cyclonaias*
59b. Shell more ovate, junction of posterior and ventral
margins more angular, pustules either cover posterior
slope in *Plethobasus cooperianus* (Lea) (Cumberland and
Tennessee River drainages) or are limited to a central
medial row in *P. cyphyus* (Rafinesque) (Ohio, Cumberland,
and Tennessee River drainages, Mississippi drainage north
to Missouri and Minnesota; (Fig. 11.28AN) .... *Plethobasus*

60a(58b). Pseudocardinal teeth reduced in size, poorly developed or
vestigial 61
60b. Pseudocardinal teeth robust and well developed 62

61(60a). Pseudocardinal teeth present, although reduced in size,
Uniomerus tetralasmus (Say) (Mississippi drainage north
to Ohio River, and west to Colorado, Rio Grande drainage,
Gulf of Mexico drainages from Texas east to Florida, north
to North Carolina; (Fig. 11.28AO) .... *Uniomerus*
61b. Pseudocardinal teeth poorly developed or vestigial,
Hemistena lata (Rafinesque) (Ohio, Cumberland, and
Tennessee River drainages); (Fig. 11.28AP) .... *Hemistena*

62a(60b). Shells generally elevated dorsoventrally, dorsoventrally
ovate or roundly oval with some species elliptical, umbos
prominent and curving anteriorly well above dorsal
margin, white to occasionally pinkish nacre 63
62b. Shells elongately ovate or rhomboidal (if low triangular,
are broadly elliptical or oval with purple nacre), umbos
not distinctly elevated above dorsal margin or curving
anteriorly, nacre usually purple but may be pink or
irridescent, 19 species broadly distributed in North
America east of the Continental Divide from Hudson Bay
drainages on the north, south to drainages on the Gulf of
Mexico coastal slope of The United States, identifications
and distributions in Burch (1975b); (Fig. 11.28AQ) .... *Elliptio*

63a(62a). Outer (marsupial) demibranchs of females are deep orange
or red when incubating developmental stages and
glochidia, *Lexingtonia dolabelloides* (Lea) (Tennessee
River drainage), *L. subplana* (Conrad) (North
River, Virginia); (Fig. 11.28AR) .... *Lexingtonia*
63b. Outer (marsupial) demibranchs of females are white,
grayish, yellowish or, rarely, pale orange, not deep orange
or red when incubating developmental stages and
glochidia; the systematic status of this genus has not
been thoroughly addressed, making species identification
difficult, Burch (1975b) lists 32 species, widely distributed
in the eastern United States, species distributions in
Burch (1975b); (Fig. 11.28AS) .... *Pleuroblema*

ACKNOWLEDGMENTS

This chapter is dedicated to David W. Aldridge, Roger A. Byrne, John D. Cleland, David F. Holland, Diana M. Kropf, David P. Long, and Carol J. Williams, all graduate student associates involved with my studies of freshwater bivalves. Brad Shipman and Joseph C. Britton, Gerald Elick, Thomas H. Dietz, Barry S. Payne, and Neal J. Smatresk continue to collaborate in these studies. W. D. Russell-Hunter first introduced me to the study of freshwater molluscs and has been an unfailing colleague through the years. Carl M. Way and Daniel J. Hornbach kindly provided the scanning electron micrographs of freshwater bivalve gill ciliation presented in this chapter. Sussana Lamers assisted with the literature search and organization of library resources. Paula Smallwood provided secretarial assistance and technical support. Joseph C. Britton (Texas Christian University) Gerry L. Mackie (University of Guelph) and Carl M. Way (U.S. Army Corps of Engineers, Waterways Experiment Station, Environmental Laboratory), critically reviewed and made important contributions to an initial draft of the manuscript. An invaluable critical review of the penultimate draft of the manuscript was provided by James H. Thorp, University of Louisville. David F. Holland, Michael L. Moeller, and Randal T. Melton assisted with editing and preparation of the final draft of the manuscript.

LITERATURE CITED

Adam, M. E. 1986. The Nile bivalves: How do they avoid silting during the flood? Journal of Molluscan Studies 52:248–252.

Ahlstedt, S. A. 1983. The Mollusca of the Elk River in Tennessee and Alabama. American Malacological Bulletin 1:43–50.

Aldridge, D. W., and R. F. McMahon. 1978. Growth, fecundity, and bioenergetics in a natural population of the Asiatic freshwater clam, *Corbicula manilensis* Philippi, from north central Texas. Journal of Molluscan Studies 44:49–70.

Aldridge, D. W., B. S. Payne, and A. C. Miller. 1987. The effects of intermittent exposure to suspended solids and turbulence on three species of freshwater mussels. Environmental Pollution 45:17–28.

Alexander, J. P. 1982. Energetics of a lake population of the freshwater clam, *Muculium lacustre* Müller (Bivalvia: Pisidiidae), with special reference to seasonal patterns of growth, reproduction and metabolism. Master's Thesis, University of Dayton, Dayton, Ohio.

Allen, A. J. 1985. Recent Bivalva: their form and evolution. Pages 337–403 *in:* E. R. Trueman and M. R. Clarke, editors. The Mollusca. Vol. 10: Evolution. Academic Press, New York.

Annis, C. G., and T. V. Belanger. 1986. *Corbicula manilensis,* potential bio-indicator of lead and copper pollution. Florida Scientist 49:30.

Barnes, R. D. 1986. Invertebrate zoology. 5th Edition. Saunders, Philadelphia, Pennsylvania.

Bauer, G. 1983. Age structure, age specific mortality rates and population trend of the freshwater pearl mussel (*Margaritifera margaritifera*) in north Bavaria. Archiv fuer Hydrobiologie 98:523–532.

Belanger, S. E., J. L. Farris, D. S. Cherry, and J. Cairns, Jr. 1985. Sediment preference of the freshwater Asiatic clam, *Corbicula fluminea.* Nautilus 99:66–73.

Belanger, S. E., D. S. Cherry, and J. Cairns, Jr. 1986a. Uptake of chrysotile asbestos fibers alters growth and reproduction of Asiatic clams. Canadian Journal of Fisheries and Aquatic Sciences 43:43–52.

Belanger, S. E., J. L. Farris, D. S. Cherry, and J. Cairns, Jr. 1986b. Growth of Asiatic clams (*Corbicula* sp.) during and after long-term zinc exposure in field-located and laboratory artificial streams. Archives of Environmental Contamination and Toxicology 15:427–434.

Balanger, S. E., D. S. Cherry, J. Cairns, Jr., and M. J. McGuire. 1987. Using Asiatic clams as a biomonitor for chrysotile asbestos in public water supplies. Journal of the American Water Works Association 79:69–74.

Bisbee, G. D. 1984. Ingestion of phytoplankton by two species of freshwater mussels, the black sandshell, *Ligumia recta,* and the three ridger, *Amblema plicata,* from the Wisconsin River in Oneida County, Wisconsin. Bios 58:219–225.

Bishop, S. H., L. L. Ellis, and J. M. Burcham. 1983. Amino acid metabolism in molluscs. Pages 243–327 *in:* P. W. Hochachka, editor. The Mollusca. Vol. 1: Metabolic biochemistry and molecular biomechanics. Academic Press, New York.

Blye, R. W., W. S. Ettinger, and W. N. Nebane. 1985. Status of Asiatic clam (*Corbicula fluminea*) in south eastern Pennsylvania—1984. Proceedings of the Pennsylvania Academy of Science 59:74.

Bonaventura, C., and J. Bonaventura. 1983. Respiratory pigments: Structure and function. Pages 1–50 *in:* P. W. Hochachka, editor. The Mollusca. Vol. 2: Environmental chemistry and physiology. Academic Press, New York.

Britton, J. C., and B. Morton. 1986. Polymorphism in *Corbicula fluminea* (Bivalvia: Corbiculidae) from North America. Malacological Review 19:1–43.

Burch, J. B. 1975a. Freshwater sphaeriacean clams (Mollusca: Pelecypoda) of North America. Malacological Publications, Hamburg, Michigan.

Burch, J. B. 1975b. Freshwater unionacean clams (Mollusca: Pelecypoda) of North America. Malacological Publications, Hamburg, Michigan.

Burky, A. J. 1983. Physiological ecology of freshwater bivalves. Pages 281–327 *in:* W. D. Russell-Hunter, editor. The Mollusca. Vol. 6: Ecology. Academic Press, New York.

Burky, A. J., and K. A. Burky. 1976. Seasonal respiratory variation and acclimation in the pea clam, *Pisidium walkeri* Sterki. Comparative Biochemistry and Physiology. 55A:109–114.

Burky, A. J., D. J. Hornbach, and C. M. Way. 1981. Growth of *Pisidium casertanum* (Poli) in west central Ohio. Ohio Journal of Science 81:41–44.

Burky, A. J., R. B. Benjamin, D. G. Conover, and J. R. Detrick. 1985a. Seasonal responses of filtration rates to temperature, oxygen availability, and particle concentration of the freshwater clam *Musculium partumeium* (Say). American Malacological Bulletin 3:201–212.

Burky, A. J., D. J. Hornbach, and C. M. Way. 1985b. Comparative bioenergetics of permanent and temporary pond populations of the freshwater clam, *Musculium partumeium* (Say). Hydrobiologia 126:35–48.

Burton, R. F. 1983. Ionic regulation and water balance. Pages 292–352 *in:* A. S. M. Saleuddin and K. M. Wilbur, editors. The Mollusca. Vol. 5: Physiology, Part 2. Academic Press, Orlando, Florida.

Buttner, J. K. 1986. *Corbicula* as a biological filter and polyculture organism in catfish rearing ponds. Progressive Fish-Culturist 48:136–139.

Butts, T. A., and R. E. Sparks. 1982. Sediment oxygen demand—fingernail clam relationship in the Mississippi River Keokuk Pool. Transactions of the Illinois Academy of Sciences 75:29–39.

Byrne, R. A. 1988. Physiological and behavioral responses to aerial exposure in the Asian clam, *Corbicula fluminea* (Müller). Ph.D. Thesis, Louisiana State University, Baton Rouge.

Byrne, R. A., R. F. McMahon, and T. H. Dietz. 1988. Temperature and relative humidity effects on aerial exposure tolerance in the freshwater bivalve *Corbicula fluminea*. Biological Bulletin (Woods Hole, Mass.) 175:253–260.

Cairns, J., Jr., and D. S. Cherry. 1983. A site-specific field and laboratory evaluation of fish and Asiatic clam population responses to coal fired power plant discharges. Water Science and Technology 15:31–58.

Calow, P. 1983. Life-cycle patterns and evolution. Pages 649–678 *in:* W. D. Russell-Hunter, editor. The Mollusca. Vol. 6: Ecology. Academic Press, New York.

Charlesworth, B. 1980. Evolution in age structured-populations. Cambridge University Press, London.

Cherry, D. S., J. H. Rodgers, Jr., R. L. Graney, and J. Cairns, Jr. 1980. Dynamics and control of the Asiatic clam in the New River, Virginia. Bulletin of the Virginia Water Resources Center 123:1–72.

Clarke, A. H. 1973. The freshwater molluscs of the Canadian interior basin. Malacologia 13:1–509.

Clarke, A. H. 1983. The distribution and relative abundance of *Lithasia pinguis* (Lea), *Pleurobema plenum* (Lea), *Villosa trabalis* (Conrad), and *Epioblasma sampsoi* (Lea). American Malacological Bulletin 1: 27–30.

Cleland, J. D. 1988. Ecological and physiological considerations of deposit-feeding in a freshwater bivalve, *Corbicula fluminea*. Master's Thesis, The University of Texas at Arlington.

Cleland, J. D., R. F. McMahon, and G. Elick. 1986. Physiological differences between two morphotypes of the Asian clam. *Corbicula*. American Zoologist 26: 103A.

Cohen, R. R. H., P. V. Dresler, E. P. J. Phillips, and R. L. Cory. 1984. The effect of the Asiatic clam, *Corbicula fluminea*, on phytoplankton of the Potomac River, Maryland. Limnology and Oceanography 29:170–180.

Collins, T. W. 1967. Oxygen-uptake, shell morphology and dessication of the fingernail clam, *Sphaerium occidentale* Prime. Ph.D. Thesis, University of Minnesota, Minneapolis.

Cooper, C. M. 1984. The freshwater bivalves of Lake Chicot, an oxbow of the Mississippi in Arkansas. Nautilus 98:142–145.

Counts, C. L., III. 1986. The zoogeography and history of the invasion of the United States by *Corbicula fluminea* (Bivalvia: Corbiculidae). American Malacological Bulletin, Special Edition No. 2:7–39.

Covich, A. P., L. L. Dye, and J. S. Mattice. 1981. Crayfish predation on *Corbicula* under laboratory conditions. American Midland Naturalist 105:181–188.

Dance, S. P. 1958. Drought resistance in an African freshwater bivalve. Journal of Conchology 24:281–282.

Das, V. M. M., and S. A. T. Venkatachari. 1984. Influence of varying oxygen tension on the oxygen consumption of the freshwater mussel *Lamellidens marginalis* (Lamarck) and its relation to body size. Veliger 26:305–310.

Davis, D. S., and J. Gilhen. 1982. An observation of the transportation of pea clams, *Pisidium adamsi,* by blue-spotted salamanders, *Ambystoma laterale.* The Canadian Naturalist 96:213–215.

Davis, W. L., R. G. Jones, J. P. Knight, and H. K. Hagler. 1982. An electron microscopic histochemical and x-ray microprobe study of spherites in a mussel. Tissue and Cell 14:61–67.

Deaton, L. E. 1982. Tissue (NA + K)-activated adenosinetriphosphatase activities in freshwater and brackish water bivalve molluscs. Marine Biology Letters 3:107–112.

de Zwaan, A. 1983. Carbohydrate catabolism in bivalves. Pages 137–175 *in:* P. W. Hochachka, editor. The Mollusca. Vol. 1: Metabolic biochemistry and molecular biomechanics. Academic Press, New York.

Dietz, T. H. 1974. Body fluid composition and aerial oxygen consumption in the freshwater mussel, *Ligumia subrostrata* (Say): effects of dehydration and anoxic stress. Biological Bulletin (Woods Hole, Mass.) 147:560–572.

Dietz, T. H. 1985. Ionic regulation in freshwater mussels: a brief review. American Malacological Bulletin 3:233–242.

Dietz, T. H., J. I. Scheide, and D. G. Saintsing. 1982. Monoamine transmitters and cAMP stimulation of Na transport in freshwater mussels. Canadian Journal of Zoology 60:1408–1411.

Doherty, F. G., D. S. Cherry, and J. Cairns, Jr. 1987. Valve closure responses of the Asiatic clam *Corbicula fluminea* exposed to cadmium and zinc. Hydrobiologia 153:159–167.

Dreier, H. 1977. Study of *Corbicula* in Lake Sangchris. Pages 7.1–7.52 *in:* The annual report for fiscal year 1976, Lake Sangchris project, Section 7. Illinois Natural History Survey, Urbana.

Dreier, H., and J. A. Tranquilli. 1981. Reproduction, growth, distribution, and abundance of *Corbicula* in an Illinois cooling lake. Illinois Natural History Survey Bulletin 32:378–393.

Duncan, R. E., and P. A. Thiel. 1983. A survey of the mussel densities in Pool 10 of the upper Mississippi River. Technical Bulletin No. 139. Wisconsin Department of Natural Resources, Madison. 14 pp.

Dyduch-Falniowska, A. 1982. Oscillations in density and diversity of *Pisidium* communities in two biotopes in southern Poland. Hydrobiological Bulletin 16:123–132.

Elder, J. F., and H. C. Mattraw, Jr. 1984. Accumulation of trace elements, pesticides, and polychlorinated biphenyls in sediments and the clam *Corbicula manilensis* of the Apalachicola River, Florida. Archives of Environmental Contamination and Toxicology 13:453–469.

Ellis, A. E. 1978. British freshwater bivalve Mollusca. Keys and notes for identification of the species. Academic Press, New York.

Eng, L. L. 1979. Population dynamics of the Asiatic clam, *Corbicula fluminea* (Müller), in the concrete-lined Delta-Mendota Canal of central California. Pages 39–68 *in:* J. C. Britton, editor. Proceedings, first international *Corbicula* symposium. Texas Christian University Research Foundation, Fort Worth.

Farris, J. L., J. H. van Hassel, S. E. Belanger, D. S. Cherry, and J. Cairns, Jr. 1988. Application of cellulolytic activity of Asiatic clams (*Corbicula* sp.) to instream monitoring of power plant effluents. Environmental Toxicology and Chemistry 7:701–713.

Foe, C., and A. Knight. 1985. The effect of phytoplankton and suspended sediment on the growth of *Corbicula fluminea* (Bivalvia). Hydrobiologia 127:105–115.

Foe, C., and A. Knight. 1986a. A thermal energy budget for juvenile *Corbicula fluminea*. American Malacological Bulletin, Special Edition. No. 2:143–150.

Foe, C., and A. Knight. 1986b. Growth of *Corbicula fluminea* (Bivalvia) fed on artificial and algal diets. Hydrobiologia 133:155–164.

Foe, C., and A. Knight. 1986c. A method for evaluating the sublethal impact of stress employing *Corbicula fluminea*. American Malacological Bulletin, Special Edition No. 2:133–142.

Foe, C., and A. Knight. 1987. Assessment of the biological impact of point source discharges employing Asiatic clams. Archives of Environmental Contamination and Toxicology 16:39–51.

Fritz, L. W., and R. A. Lutz. 1986. Environmental perturbations reflected in internal shell growth patterns of *Corbicula fluminea* (Mollusca: Bivalvia). Veliger 28:401–417.

Fuller, S. L. H. 1974. Clams and mussels (Mollusca: Bivalvia). Pages 215–273 *in:* C. W. Hart, Jr. and S. L. H. Fuller, editors. Pollution ecology of freshwater invertebrates. Academic Press, New York.

Gabbott, P. A. 1983. Developmental and seasonal metabolic activities in marine molluscs. Pages 165–217 *in:* P. W. Hochachka, editor. The Mollusca. Vol. 2: Environmental biochemistry and physiology. Academic Press, New York.

Gainey, L. F., Jr. 1978a. The response of the Corbiculidae (Mollusca: Bivalvia) to osmotic stress: the organismal response. Physiological Zoology 51:68–78.

Gainey, L. F., Jr. 1978b. The response of the Corbiculidae (Mollusca: Bivalvia) to osmotic stress: the cellular response. Physiological Zoology 51:79–91.

Gainey, L. F., Jr., and M. J. Greenberg. 1977. Physiological basis of the species abundance-salinity relationship in molluscs: a speculation. Marine Biology (Berlin) 40:41–49.

Golightly, C. G., Jr. 1982. Movement and growth of Unionidae (Mollusca: Bivalvia) in the Little Brazos River, Texas. Ph.D. Thesis, Texas A&M University, College Station.

Goss, L. B., J. M. Jackson, H. B. Flora, B. G. Isom, C. Gooch, S. A. Murray, C. G. Burton, and W. S. Bain. 1979. Control studies on *Corbicula* for steam-electric generating plants. Pages 139–151 *in:* J. C. Britton, editor. Proceedings, first international *Corbicula* symposium. Texas Christian University Research Foundation, Fort Worth.

Graney, R. L., Jr., D. S. Cherry, and J. Cairns, Jr. 1984. The influence of substrate, pH, diet and temperature upon cadmium accumulation in the Asiatic clam (*Corbicula fluminea*) in laboratory streams. Water Research 18:833–842.

Graves, S. Y., and T. H. Dietz. 1980. Diurnal rhythms of sodium transport in the freshwater mussel. Canadian Journal of Zoology 58:1626–1630.

Graves, S. Y., and T. H. Dietz. 1982. Cyclic AMP stimulation and prostaglandin inhibition of Na transport in freshwater mussels. Comparative Biochemistry and Physiology A 71:65–70.

Gunning, G. E., and R. D. Suttkus. 1985. Reclamation of the Pearl River. A perspective of unpolluted versus polluted waters. Fisheries 10:14–16.

Haas, F. 1969. Superfamily Unionacea Fleming, 1828. Pages 411–467 *in:* R. C. Moore, editor. Treatise on invertebrate paleontology. Part N: Mollusca. The Geological Society of America, Boulder, Colorado.

Hanson, J. M., W. C. MacKay, and E. E. Prepas. 1989. Effect of size-selective predation by muskrats (*Ondatra zebithicus*) on a population of unionid clams (*Anodonta grandis simpsoniana*). Journal of Animal Ecology 58:15–28.

Hartfield, P., and D. Ebert. 1986. The mussels of southwest Mississippi streams. American Malacological Bulletin 4:21–23.

Hartfield, P., and R. G. Rummel. 1985. Freshwater mussels (Unionidae) of the Big Black River, Mississippi. Nautilus 99:116–119.

Hartley, D. M., and J. B. Johnston. 1983. Use of the freshwater clam *Corbicula manilensis* as a monitor for organochlorine pesticides. Bulletin of Environmental Contamination and Toxicology 31:33–40.

Haukioja, E., and T. Hakala. 1978. Life-history evolution in *Anodonta piscinalis* (Mollusca, Pelecypoda). Oecologia 35:253–266.

Havlik, M. E. 1983. Naiad mollusk populations (Bivalvia: Unionidae) in Pools 7 and 8 of the Mississippi River near La Crosse, Wisconsin. American Malacological Bulletin 1:51–60.

Hawkins, A. J. S., J. Widdows, and B. L. Bayne. 1989. The relevance of whole-body protein metabolism to measured costs of maintenance and growth in *Mytilus edulis*. Physiological Zoology 62:745–763.

Heard, W. H. 1977. Reproduction of fingernail clams (Sphaeriidae: *Sphaerium* and *Musculium*). Malacologia 16:421–455.

Hebert, P. D. N., B. W. Muncaster, and G. L. Mackie. 1989. Ecological and genetic studies on *Dreissena polymorpha* (Pallas): a new mollusc in the Great Lakes. Canadian Journal of Fisheries and Aquatic Sciences 46:1587–1591.

Hemelraad, J., H. J. Herwig, D. A. Holwerda, and D. I. Zandee. 1985. Accumulation, distribution and localization of cadmium in the freshwater clam *Anodonta* sp. Marine Environmental Research 17:196.

Heming, T. A., G. A. Vinogradov, A. K. Klerman, and V. T. Komov. 1988. Acid-base regulation in the freshwater pearl mussel *Margaritifera margaritifera:* effects of emersion and low water pH. Journal of Experimental Biology 137:501–511.

Henery, R. P., and D. G. Saintsing. 1983. Carbonic anhydrase activity and ion regulation in three species of osmoregulating bivalve molluscs. Physiological Zoology 56:274–280.

Herwig, H. J., J. Hemelraad, D. A. Howerda, and D. I. Zandee. 1985. Cytochemical localization of cadmium and tin in bivalves. Marine Environmental Research 17:196–197.

Hillis, D. M., and R. L. Mayden. 1985. Spread of the Asiatic clam. *Corbicula* (Bivalvia: Corbiculacea), into the new world tropics. Southwestern Naturalist 30:454–456.

Hillis, D. M., and J. C. Patton. 1982. Morphological and electrophoretic evidence for two species of *Corbicula* (Bivalvia: Corbiculacea) in north Central America. American Midland Naturalist 108:74–80.

Hiripi, L., D. E. Burrell, M. Brown, P. Assanah, A. Stanec, and G. B. Stefano. 1982. Analysis of monoamine accumulations in the neuronal tissues of *Mytilus edulis* and *Anodonta cygnea* (Bivalvia)—III. Temperature and seasonal influences. Comparative Biochemistry and Physiology, *C* 71:209–213.

Hiscock, I. D. 1953. Osmoregulation in Australian freshwater mussels (Lamellibranchiata). I. Water and chloride exchange in *Hyridella australis* (Lam.). Australian Journal of Marine and Freshwater Research 4:317–329.

Hoeh, W. R., and R. J. Trdan. 1984. The freshwater mussels (Pelecypoda: Unionidae) of the upper Tittabawassee River drainage, Michigan. Malacological Review 17:97–98.

Holopainen, I. J. 1987. Seasonal variation of survival time in anoxic water and the glycogen content of *Sphaerium corneum* and *Pisidium amnicum* (Bivalvia: Pisidiidae). American Malacological Bulletin 5:41–48.

Holopainen, I. J., and I. Hanski. 1986. Life history variation in *Pisidium*. Holarctic Ecology 9:85–98.

Holopainen, I. J., and P. M. Jonasson. 1983. Long-term population dynamics and production of *Pisidium* (Bivalvia) in the profundal of Lake Esrom, Denmark. Oikos 41:99–117.

Hornbach, D. J. 1985. A review of metabolism in the Pisidiidae with new data on its relationship with life history traits in *Pisidium casertanum*. American Malacological Bulletin 3:187–200.

Hornbach, D. J., and D. L. Childers. 1986. Life-history variation in a stream population of *Musculium partumeium* (Bivalvia: Pisidiidae). Journal of the North American Benthological Society 5:263–271.

Hornbach, D. J., and D. L. Childers. 1987. The effects of acidification on life-history traits of the freshwater clam *Musculium partumeium* (Say, 1822) (Bivalvia: Pisidiidae). Canadian Journal of Zoology 65:113–121.

Hornbach, D. J., and C. Cox. 1987. Environmental influences on life history traits in *Pisidium casertanum* (Bivalvia: Psidiidae): field and laboratory exper-

imentation. American Malacological Bulletin 5: 49–64.

Hornbach, D. J., C. M. Way, and A. J. Burky. 1980. Reproductive strategies in the freshwater sphaeriid clam, *Musculium partumeium* (Say), from a permanent and a temporary pond. Oecologia 44:164–170.

Hornbach, D. J., T. E. Wissing, and A. J. Burky. 1982. Life-history characteristics of a stream population of the freshwater clam *Sphaerium striatinum* Lamarck (Bivalvia: Pisidiidae). Canadian Journal of Zoology 60:249–260.

Hornbach, D. J., T. E. Wissing, and A. J. Burky. 1983. Seasonal variation in the metabolic rates and Q_{10}-values of the fingernail clam, *Sphaerium striatinum* Lamark. Comparative Biochemistry and Physiology. 76A:783–790.

Hornbach, D. J., T. E. Wissing, and A. J. Burky. 1984a. Energy budget for a stream population of the freshwater clam, *Sphaerium striatinum* Lamarck (Bivalvia: Pisidiidae). Canadian Journal of Zoology 62:2410–2417.

Hornbach, D. J., C. M. Way, T. E. Wissing, and A. J. Burky. 1984b. Effects of particle concentration and season on the filtration rates of the freshwater clam, *Sphaerium striatinum* Lamarck (Bivalvia: Pisidiidae). Hydrobiologia 108:83–96.

Horne, F. R., and S. MacIntosh. 1979. Factors influencing distribution of mussels in the Blanco River of central Texas. Nautilus 94:119–133.

Hudson, R. G., and B. G. Isom. 1984. Rearing of juveniles of the freshwater mussels (Unionidae) in a laboratory setting. Nautilus 98:129–135.

Huebner, J. D. 1982. Seasonal variation in two species of unionid clams from Manitoba, Canada: Respiration. Canadian Journal of Zoology 60:560–564.

Imlay, M. J. 1982. Use of shells of freshwater mussels in monitoring heavy metals and environmental stresses: a review. Malacological Review 15:1–14.

Isom, B. G. 1986. Historical review of Asiatic clam (*Corbicula*) invasion and biofouling of waters and industries in the Americas. American Malacological Bulletin, Special Edition No. 2:1–5.

Isom, B. G., and R. G. Hudson. 1982. *In vitro* culture of parasitic freshwater mussel glochidia. Nautilus 96:147–151.

Jadhav, M. L., and V. S. Lomte. 1982a. Seasonal variation in biochemical composition of the freshwater bivalve, *Lamellidens corrianus*. Rivista di Idrobiologia 21:1–17.

Jadhav, M. L., and V. S. Lomte. 1982b. Hormonal control of carbohydrate metabolism in the freshwater bivalve, *Lamellidens corrianus*. Rivista di Idrobiologia. 21:27–36.

Jadhav, M. L., and V. S. Lomte. 1983. Neuroendocrine control of midgut gland in the bivalve, *Lamellidens corrianus* (Prasad) (Mollusca: Lamellibranchiata). Journal of Advanced Zoology 4:97–104.

James, M. R. 1985. Distribution, biomass and production of the freshwater mussel, *Hyridella menziesi* (Gray), in Lake Taupo, New Zealand. Freshwater Biology 15:307–314.

Jonasson, P. M. 1984a. Decline of zoobenthos through five decades of euthrophication in Lake Esron. Verhandlungen Internationale Vereinigung fur Theoretische und Angewandte Limnologie 22:800–804.

Jonasson, P. M. 1984b. Oxygen demand and long term changes in zoobenthos. Hydrobiologia 115:121–126.

Jones, H. D. 1983. Circulatory systems of gastropods and bivalves. Pages 189–238 *in:* A. S. M. Saleuddin and K. M. Wilbur, editors. The Mollusca. Vol. 5: Physiology, Part 2. Academic Press, New York.

Jones, H. D., and D. Peggs. 1983. Hydrostatic and osmotic pressures in the heart and pericardium of *Mya arenaria* and *Anodonta cygnea*. Comparative Biochemistry and Physiology 76A:381–385.

Jones, R. G., and W. L. Davis. 1982. Calcium containing lysosomes in the outer mantle epithelial cells of *Amblema*, a fresh-water mollusc. Anatomical Record 203:337–343.

Joosse, J. and W. P. M. Geraerts. 1983. Endocrinology. Pages 318–406 *in:* A. S. M. Saleuddin and K. M. Wilbur, editors. The Mollusca, Vol. 4: Physiology, Part 1. Academic Press, New York.

Jorgensen, C. B., T. Kiorboe, F. Mohlenberg, and H. U. Riisgard. 1984. Ciliary and mucus-net filter feeding, with special reference to fluid mechanical characteristics. Marine Ecology Progress Series 15:283–292.

Kasprzak, K. 1986. Role of the Unionidae and Sphaeriidae (Mollusca, Bivalvia) in the eutrophic Lake Zbechy and its outflow. Internationale Revue Der Gesamten Hydrobiologie 71:315–334.

Kat, P. W. 1982. Shell dissolution as a significant cause of mortality for *Corbicula fluminea* (Bivalvia: Corbiculidae) inhabiting acidic waters. Malacological Review 15:129–134.

Kat, P. W. 1984. Parasitism and the Unionacea (Bivalvia). Biological Review 59:189–207.

Kat, P. W., and G. M. Davis. 1984. Molecular genetics of peripheral populations of Nova Scotian Unionidae (Mollusca: Bivalva). Biological Journal of the Linnean Society 22:157–185.

Keen, M., and R. Casey. 1969. Family Corbiculidae Gray, 1847. Pages 664–669 *in:* R. C. Moore, editor. Treatise on invertebrate paleontology. Part N: Mollusca. The Geological Society of America., Boulder, Colorado.

Keen, M., and P. Dance. 1969. Family Pisidiidae, Gray, 1857. Pages 669–670 *in:* R. C. Moore, editor. Treatise on invertebrate paleontology. Part N: Mollusca. The Geological Society of America, Boulder, Colorado.

Kennedy, V. S., and J. A. Blundon. 1983. Shell strength in *Corbicula* sp. (Bivalvia: Corbiculacea) from the Potomac River, Maryland. Veliger 26:22–25.

Khan, H. R., M. L. Aston, and A. S. M. Saleuddin. 1986. Fine structure of the kidneys of osmotically stressed *Mytilus, Mercenaria*, and *Anodonta*. Canadian Journal of Zoology 64:2779–2787.

Kilgour, B. W., and G. L. Mackie. 1988. Factors affecting the distribution of sphaeriid bivalves in Britannia Bay of the Ottawa River. Nautilus 102:73–77.

King, C. A., C. J. Langdon, and C. L. Counts, III. 1986. Spawning and early development of *Corbicula flu-minea* (Bivalvia: Corbiculidae) in laboratory culture. American Malacological Bulletin 4:81–88.

Kraemer, L. R. 1978. Discovery of two distinct kinds of statocysts in freshwater bivalved mollusks: some behavioral implications. Bulletin of the American Malacological Union pp. 24–28.

Kraemer, L. R., and M. L. Galloway. 1986. Larval development of *Corbicula fluminea* (Müller) (Bivalvia: Corbiculacea): an appraisal of its heterochrony. American Malacological Bulletin 4:61–79.

Kraemer, L. R., C. Swanson, M. Galloway, and R. Kraemer. 1986. Biological basis of behavior in *Corbicula fluminea*, II. Functional morphology of reproduction and development and review of evidence for self-fertilization. American Malacological Bulletin, Special Edition. No. 2:193–201.

LaRocque, A. L. 1967a. Pleistocene Mollusca of Ohio. Part 2. State of Ohio Division of Geological Survey, Department of Natural Resources, Columbus. Pages 113–356.

LaRocque, A. L. 1967b. Pleistocene Mollusca of Ohio. Part 1. State of Ohio Department of Natural Resources, Division of Geological Survey, Columbus. Pages 1–111.

Lauritsen, D. D. 1986a. Assimilation of radiolabeled algae by *Corbicula*. American Malacological Bulletin, Special Edition No. 2:219–222.

Lauritsen, D. D. 1986b. Filter-feeding in *Corbicula fluminea* and its effect on seston removal. Journal of the North American Benthological Society 5:165–172.

Laycock, G. 1983. Vanishing naiads. Audubon 85:26–28.

Lewis, J. B. 1984. Comparative respiration in two species of freshwater unionid mussels (Bivalvia). Malacological Review 17:101–102.

Lewis, J. B., and P. N. Riebel. 1984. The effect of substrate on burrowing in freshwater mussels (Unionidae). Canadian Journal of Zoology 62:2023–2025.

Lincoln, R. J., and J. G. Sheals. 1979. Invertebrate animals. Collection and Preservation. Cambridge University Press, London.

Little, C. 1985. Renal adaptations of prosobranchs to the freshwater environment. American Malacological Bulletin 3:223–231.

Lomte, V. S., and M. L. Jadhav. 1981a. Effect of desiccation on the neurosecretory activity of the freshwater bivalve, *Parreysia corrugata*. Science and Culture 47:437–438.

Lomte, V. S., and M. L. Jadhav. 1981b. Neuroendocrine control of osmoregulation in the freshwater bivalve, *Lamellidens corrianus*. Journal of Advanced Zoology 2:102–108.

Lomte, V. S., and M. L. Jadhav. 1982a. Effects of toxic compounds on oxygen consumption in the fresh water bivalve, *Corbicula regularis* (Prime, 1860). Comparative Physiology and Ecology 7:31–33.

Lomte, V. S., and M. L. Jadhav. 1982b. Respiratory metabolism in the freshwater mussel *Lamellidens corrianus*. Marathwada University Journal of Science 21:87–90.

Lomte, V. S., and M. L. Jadhav. 1982c. Bichemical composition of the freshwater bivalve, *Lamellidens corrianus* (Prasad, 1922). Rivista di Idrobiologia 21: 19–25.

Long, D. P. 1989. Seasonal variation and the influence of thermal effluents on the bioenergetics of the introduced Asian clam, *Corbicula fluminea* (Müller). Masters Thesis, The University of Texas at Arlington.

Long, D. P., and R. F. McMahon. 1987. High temperature inhibition of growth and reproduction in a natural field population of *Corbicula fluminea* (Müller). American Zoologist 27:39A.

Lopez, G. R., and I. J. Holopainen. 1987. Interstitial suspension-feeding by *Pisidium* sp. (Pisididae: Bivalvia): a new guild in the lentic benthos? American Malacological Bulletin 5:21–30.

Machena, C., and N. Kautsky. 1988. A quantitative diving survey of benthic vegetation and fauna in Lake Kariba, a tropical man-made lake. Freshwater Biology 19:1–14.

Mackie, G. L. 1978. Are sphaeriid clams ovoviparous or viviparous? Nautilus 92:145–147.

Mackie, G. L. 1984. Bivalves. Pages 351–418 *in:* A. S. Tompa, N. H. Verdonk, and J. A. M. van der Biggelaar, editors. The Mollusca. Vol. 7: Reproduction. Academic Press, New York.

Mackie, G. L., and L. A. Flippance. 1983a. Life history variations in two populations of *Sphaerium rhombiodeum* (Bivalvia: Pisidiidae). Canadian Journal of Zoology 61:860–867.

Mackie, G. L., and L. A. Flippance. 1983b. Intra- and interspecific variations in calcium content of freshwater Mollusca in relation to calcium content of the water. Journal of Molluscan Studies 49:204–212.

Mackie, G. L., S. U. Quadri, and R. M. Reed. 1978. Significance of litter size in *Musculium securis* (Bivalvia: Sphaeriidae). Ecology 59:1069–1074.

Mackie, G. L., D. S. White, and T. W. Zdeba. 1980. A guide to the freshwater mollusks of the Laurentian Great Lakes with special emphasis on the genus, *Pisidium*. EPA-600/3-80-068. United States Environmental Protection Agency, Duluth, Minnesota.

Mackie, G. L., W. N. Gibbons, B. W. Muncaster, and I. M. Gray. 1989. The zebra mussel, *Dreissena polymorpha*: a synthesis of European experiences and a preview for North America. 0-7729-5647-2. Water Resources Branch, Ontario Ministry of the Environment, Ontario, Canada.

Marsh, P. C. 1985. Secondary production of introduced Asiatic clam, *Corbicula fluminea*, in a central Arizona canal. Hydrobiologia 124:103–110.

Martin, W. A. 1983. Execretion. Pages 353–405 *in:* A. S. M. Saleuddin and K. M. Wilbur, editors. The Mollusca. Vol. 5: Physiology, Part 2. Academic Press, New York.

Mathiak, H. A. 1979. A river survey of the unionid mussels of Wisconsin 1973–1977. Sand Shell Press, Horicon, Wisconsin.

Matisoff, G., J. B. Fisher, and S. Matis. 1985. Effects of macroinvertebrates on the exchange of solutes between sediments and freshwater. Hydrobiologia 122:19–33.

Matsushima, O., and Y. Kado. 1982. Hyperosmoticity of the mantle fluid in the freshwater bivalve, *Anodonta woodiana*. Journal of Experimental Biology 221:379–381.

Matsushima, O., F. Sakka, and Y. Kado. 1982. Free amino acid involved in intracellular osmoregulation in the clam, *Corbicula*. Journal of Science of Hiroshima University, Series B, Division *1* 30:213–219.

Mattice, J. S. 1979. Interactions of *Corbicula* sp. with power plants. Pages 119–138 *in:* J. C. Britton, editor. Proceedings, first international *Corbicula* symposium. Texas Christian University Research Foundation, Forth Worth.

Mattice, J. S., R. B. McLean, and M. B. Burch. 1982. Evaluation of short-term exposure to heated water and chlorine for control of the Asiatic clam (*Corbicula fluminea*). Oak Ridge National Laboratory, Environmental Sciences Division, Publication No. 1748. U.S. National Technical Information Service, Department of Commerce, Springfield, Virginia. 33 pp.

McCorkle, S., T. C. Shirley, and T. H. Dietz. 1979. Rhythms of activity and oxygen consumption in the common pond clam, *Ligumia subrostrata* (Say). Canadian Journal of Zoology 57:1960–1964.

McCorkle-Shirley, S. 1982. Effects of photoperiod on sodium flux in *Corbicula fluminea* (Mollusca: Bivalvia). Comparative Biochemistry and Physiology. 71A: 325–327.

McKee, P. M., and G. L. Mackie. 1980. Dessication resistance in *Sphaerium occidentale* and *Musculium securis* (Bivalvia: Sphaeriidae) from a temporary pond. Canadian Journal of Zoology 58:1693–1696.

McKillop, W. B., and A. D. Harrison. 1982. Hydrobiological studies of the eastern Lesser Antillean Islands. VII. St. Lucia: behavioural drift and other movements of freshwater marsh molluscs. Archiv fuer Hydrobiologie 94:53–69.

McLeod, M. J. 1986. Electrophoretic variation in North American *Corbicula*. American Malacological Bulletin, Special Edition No. 2:125–132.

McMahon, R. F. 1979a. Response to temperature and hypoxia in the oxygen consumption of the introduced Asiatic freshwater clam *Corbicula fluminea* (Müller). Comparative Biochemistry and Physiology 63A:383–388.

McMahon, R. F. 1979b. Tolerance of aerial exposure in the Asiatic freshwater clam, *Corbicula fluminea* (Müller). Pages 227–241 *in:* J. C. Britton, editor. Proceedings, first international *Corbicula* symposium. Texas Christian University Research Foundation, Fort Worth.

McMahon, R. F. 1982. The occurrence and spread of the introduced Asiatic freshwater clam, *Corbicula fluminea* (Müller), in North America: 1924–1982. Nautilus 96:134–141.

McMahon, R. F. 1983a. Ecology of an invasive pest bivalve, *Corbicula*. Pages 505–561 *in:* W. D. Russell-Hunter, editor. The Mollusca. Vol. 6: Ecology. Academic Press, New York.

McMahon, R. F. 1983b. Physiological ecology of freshwater pulmonates. Pages 360–430 *in:* W. D.

Russell-Hunter, editor. The Mollusca. Vol. 6: Ecology. Academic Press, New York.

McMahon, R. F. 1988. Respiratory response to periodic emergence in intertidal molluscs. American Zoologist 28:97–114.

McMahon, R. F., and R. W. Lutey. 1988. Field and laboratory studies of the efficacy of poly[oxyethylene(dimethyliminio)ethylene(dimethyliminio)ethylene dichloride] as a biocide against the Asian clam, *Corbicula fluminea*. Pages 61–72 *in:* Proceedings: service water reliability improvement seminar. Electric Power Research Institute, Palo Alto, California.

McMahon, R. F., and C. J. Williams. 1984. A unique respiratory adaptation to emersion in the introduced Asian freshwater clam *Corbicula fluminea* (Müller) (Lamellibranchia: Corbiculacea). Physiological Zoology 57:274–279.

McMahon, R. F., and C. J. Williams. 1986a. A reassessment of growth rate, life span, life cycles and population dynamics in a natural population and field caged individuals of *Corbicula fluminea* (Müller). American Malacological Bulletin, Special Edition No. 2:151–166.

McMahon, R. F., and C. J. Williams. 1986b. Growth, life cycle, upper thermal limit and downstream colonization rates in a natural population of the freshwater bivalve mollusc, *Corbicula fluminea* (Müller) receiving thermal effluents. American Malacological Bulletin, Special Edition No. 2:231–239.

Miller, A. C., and B. S. Payne. 1988. The need for quantitative sampling to characterize demography and density of freshwater mussel communities. American Malacological Bulletin 6:49–54.

Miller, A. C., L. Rhodes, and R. Tipit. 1984. Changes in the naiad fauna of the Cumberland River below lake Cumberland in central Kentucky. Nautilus 98:107–110.

Miller, A. C., B. S. Payne, and D. W. Aldridge 1986. Characterization of a bivalve community in the Tangipahoa River, Mississippi. Nautilus 100:18–23.

Mills, H. B., W. C. Starrett, and F. C. Belrose. 1966. Man's effect on the fish and wildlife of the Illinois River. Biological Notes (Illinois Natural History Survey) No. 57:24 pp.

Mitchell, R. D. 1955. Anatomy, life history, and evolution of the mites parasitizing fresh-water mussels. Miscellaneous Publications Museum of Zoology University of Michigan 89:1–28.

Morton, B. 1969. Studies of the biology of *Dreissena polymorpha* Pall. III. Population dynamics. Proceedings of the Malacological Society of London 38:471–482.

Morton, B. 1982. Some aspects of the population structure and sexual strategy of *Corbicula* cf. *fluminalis* (Bivalvia: Corbiculacea) from the Pearl River, Peoples Republic of China. Journal of Molluscan Studies 48:1–23.

Morton, B. 1983. Feeding and digestion in bivalvia. Pages 65–147 *in:* A. S. M. Saleuddin and K. M. Wilbur, editors. The Mollusca. Vol. 5: Physiology, Part 2. Academic Press, New York.

Morton, B. 1985. The population dynamics, reproductive strategy and life history tactics of *Musculium lacustre*

(Bivalvia: Pisidiidae) in Hong Kong. Jounral of Zoology (London) (A) 207:581–603.

Morton B. 1986. The population dynamics and life history tactics of *Pisidium clarkeanum* and *P. annandalei* (Bivalvia: Pisidiidae) sympatric in Hong Kong. Journal of Zoology (London) (A) 210:427–449.

Morton, B., and K, Y. Tong. 1985. The salinity tolerance of *Corbicula fluminea* (Bivalvia: Corbiculoidea) from Hong Kong. Malacological Review 18:91–95.

Nagabhushanam, R., and V. S. Lomte. 1981. Observations on the relationship between neurosecretion and periodic activity in the mussel, *Parreysia corrugata*. Indian Journal of Fisheries 28:261–265.

Negus, C. 1966. A quantitative study of growth and production of unionid mussels in the River Thames at Reading. Journal of Animal Ecology 35:513–532.

Nelson, D. J., and D. C. Scott. 1962. Role of detritus in the productivity of a rock-outcrop community in a Piedmont stream. Limnology and Oceanography 7:396–413.

Neves, R. J., and A. V. Zale. 1982. Freshwater mussels (Unionidae) of Big Moccasin Creek, Southwestern Virginia. Nautilus 96:52–54.

Neves, R. J., L. R. Weaver, and A. V. Zale. 1985. An evaluation of host fish suitability for glochidia of *Villosa vanuxemi* and *V. nebulosa* (Pelecypoda: Unionidae). American Midland Naturalist 113:13–18.

Okland, K. A., and J. G. J. Kuiper. 1982. Distribution of small mussels (Sphaeriidae) in Norway, with notes on their ecology. Malacologia 22:469–477.

Palomphis, A. A., and W. C. Starrett. 1960. An ecological study of benthic organisms in three Illinois River flood plain lakes. American Midland Naturalist 64:406–435.

Parker, R. S., C. T. Hackney, and M. F. Vidrine. 1984. Ecology and reproductive strategy of a south Louisiana freshwater mussel, *Glebula rotundata* (Lamarck) (Unionidae: Lampsilini). Freshwater Invertebrate Biology 3:53–58.

Parmalee, P. W. 1988. A comparative study of late prehistoric and modern molluscan faunas of the Little Pigeon River System, Tennessee. American Malacological Bulletin 6:165–178.

Parmalee, P. W., and W. E. Klippel. 1982. A relic population of *Obovaria retusa* in the middle Cumberland River, Tennessee. Nautilus 96:30–32.

Parmalee, P. W., and W. E. Klippel. 1984. The naiad fauna of the Tellico River, Monroe County, Tennessee. American Malacological Bulletin 3:41–44.

Parmalee, P. W., W. E. Klippel, and A. E. Bogan. 1982. Aboriginal and modern freshwater mussel assemblages (Pelecypoda: Unionidae) from the Chickamauga Reservoir, Tennessee. Brimleyana 8:75–90.

Paterson, C. G. 1983. Effect of aggregation on the respiration rate of the freshwater unionid bivalve, *Elliptio complanata* (Solander). Freshwater Invertebrate Biology 2:139–146.

Paterson, C. G. 1984. A technique for determining apparent selective filtration in the fresh-water bivalve *Elliptio companata* (Lightfoot). Veliger 27:238–241.

Paterson, C. G. 1985. Biomass and production of the unionid, *Elliptio complanata* (Lightfoot) in an old reservoir in New Brunswick, Canada. Freshwater Invertebrate Biology 4:201–207.

Paterson, C. G., and I. F. Cameron. 1985. Comparative energetics of two populations of the unionid, *Anodonta cataracta* (Say). Freshwater Invertebrate Biology 4:79–90.

Payne, B. S., and A. C. Miller. 1987. Effects of current velocity on the freshwater bivalve *Fusconaia ebena*. American Malacological Bulletin 5:177–179.

Payne, B. S., and A. C. Miller. 1989. Growth and survival of recent recruits to a population of *Fusconaia ebena* (Bivalvia: Unionidae) in the lower Ohio River. 121:99–104.

Prezant, R. S., and K. Chalermwat. 1984. Flotation of the bivalve *Corbicula fluminea* as means of dispersal. Science 225:1491–1493.

Price, R. E., and L. A. Knight, Jr. 1978. Mercury, cadmium, lead and arsenic in sediments, plankton, and clams from Lake Washington and Sardis Reservoir, Mississippi, October 1975–May 1976. Pesticides Monitoring Journal 11:182–189.

Prokopovich, N. P. 1969. Deposition of clastic sediments by clams. Sedimentary Petrology 39:891–901.

Pugsley, C. W., P. D. N. Hebert, G. W. Wood, G. Brotea, and T. W. Obal. 1985. Distribution of contaminants in clams and sediments from the Huron-Erie corridor. I—PCBs and octachlorostyrene. Journal of Great Lakes Research 11:275–289.

Robinson, J. V., and G. A. Wellborn. 1988. Ecological resistance to the invasion of a freshwater clam, *Corbicula fluminea:* fish predation effects. Oecologia 77:445–452.

Rooke, J. B., and G. L. Mackie. 1984a. Mollusca of six low-alkalinity lakes in Ontario. Canadian Journal of Fisheries and Aquatic Science 41:777–782.

Rooke, J. B., and G. L. Mackie. 1984b. Growth and production of three species of molluscs in six low alkalinity lakes in Ontario, Canada. Canadian Journal of Zoology 62:1474–1478.

Rooke, J. B., and G. L. Mackie. 1984c. Laboratory studies of the effects of Mollusca on alkalinity of their freshwater environment. Canadian Journal of Zoology 62:793–797.

Russell, H. D. 1963. Notes on methods for the narcotization, killing, fixation, and preservation of marine organisms. Systematics-Ecology Program, Marine Biological Laboratory, Woods Hole, Massachusetts.

Russell-Hunter, W. D., and D. F. Buckley. 1983. Actuarial bioenergetics of nomarine molluscan productivity. Pages 464–503 *in:* W. D. Russell-Hunter, editor. The Mollusca. Vol. 6: Ecology. Academic Press, New York.

Sahib, I. K. A., B. Narasimhamurthy, D. Sailatha, M. R. Begum, K. S. Prasad, and K. V. R. Roa. 1983. Orientation in the glycolytic potentials of carbohydrate metabolism in the selected tissues of the aestivating freshwater pelecypod, *Lamellidens marginalis* (Lamarck). Comparative Physiology and Ecology 8:180–184.

Saintsing, D. G., and T. H. Dietz. 1983. Modification of sodium transport in freshwater mussels by prostaglandins, cyclic AMP and 5-hydroxytryptamine: Effects of inhibitors of prostaglandin synthesis. Comparative Biochemistry and Physiology 76C:285–290.

Saleuddin, A. S. M., and H. P. Petit. 1983. The mode of formation and the structure of the periostracum. Pages 199–234 *in:* A. S. M. Saleuddin and K. M. Wilbur, editors. The Mollusca. Vol. 4: Physiology, Part 1. Academic Press, New York.

Salmon, A., and R. H. Green. 1983. Environmental determinants of unionid clam distribution in the Middle Thames River, Ontario. Canadian Journal of Zoology 61:832–838.

Servos, M. R., J. B. Rooke, and G. L. Mackie. 1985. Reproduction of selected Mollusca in some low alkalinity lakes in south-central Ontario. Canadian Journal of Zoology 63:511–515.

Sibly, R. M., and P. Calow. 1986. Physiological ecology of animals: an evolutionary approach. Blackwell, London.

Sickel, J. B. 1986. *Corbicula* population mortalities: factors influencing population control. American Malacological Bulletin, Special Edition. No. 2:89–94.

Silverman, H., W. L. Steffens, and T. H. Dietz. 1983. Calcium concretions in the gills of a freshwater mussel serve as a calcium reservoir during periods of hypoxia. Journal of Experimental Zoology 227:177–189.

Silverman, H., W. L. Steffens, and T. H. Dietz. 1985. Calcium from extracellular concretions in the gills of freshwater unionid mussels is mobilized during reproduction. Journal of Experimental Zoology 236:137–147.

Silverman, H., W. T. Kays, and T. H. Dietz. 1987. Maternal calcium contribution to glochidial shells in freshwater mussels (Eulamellibranchia: Unionidae). Journal of Experimental Biology 242:137–146.

Silverman, H., L. D. Silby, and W. L. Steffens. 1988. Calmodulin-like calcium-binding protein identified in the calcium-rich mineral deposits from freshwater mussel gills. Journal of Experimental Zoology 247:224–231.

Smith, B. J., and R. C. Kershaw. 1979. Field guide to the non-marine molluscs of south eastern Australia. Australian National University Press, Canaberra.

Smith, D. G. 1985a. A study of the distribution of freshwater mussels (Mollusca: Pelecypoda: Unionidae) of the Lake Champlain drainage in northwestern New England. American Midland Naturalist 114:19–29.

Smith, D. G. 1985b. Recent range expansion of the freshwater mussel, *Anadonta implicata* and its relationship to clupeid fish restoration in the Connecticut River system. Freshwater Invertebrate Biology 4:105–108.

Smith, L. M., L. D. Vangilder, R. T. Hoppe, S. J. Morreale, and I. L. Brisbin, Jr. 1986. Effect of diving ducks on benthic food resources during winter in South Carolina, U.S.A. Wildfowl 37:136–141.

Stanczykowska, A. 1984. Role of bivalves in the phosphorus and nitrogen budget in lakes. Verhandlungen—Internationale Vereinigung fur Theoretische und Angewandte Limnologie 22:982–985.

Starnes, L. B., and A. E. Bogan. 1988. The mussels (Mollusca: Unionidae) of Tennessee. American Malacological Bulletin 6:9–37.

Stearns, S. C. 1977. The evolution of life history traits: a critique of the theory and a review of the data. Annual Reviews of Ecology and Systematics 8:145–171.

Stearns, S. C. 1980. A new view of life-history evolution. Oikos 35:266–281.

Steffens, W. L., H. Silverman, and T. H. Dietz. 1985. Localization and distribution of antigens related to calcium-rich deposits in the gills of several freshwater bivalves. Canadian Journal of Zoology 63:348–354.

Stern, E. M. 1983. Depth distribution and density of freshwater mussels (Unionidae) collected with scuba from the lower Wisconsin and St. Croix Rivers. Nautilus 97:36–42.

St. John, F. L. 1982. Crayfish and bivalve distribution in a valley in southwestern Ohio. Ohio Journal of Science 82:242–246.

Stone, N. M., R. Earl, A. Hodgson, J. G. Mather, J. Parker, and F. R. Woodward. 1982. The distributions of three sympatric mussel species (Bivalvia: Unionidae) in Budworth Mere, Cheshire. Journal of Molluscan Studies 48:266–274.

Strayer, D. 1983. The effects of surface geology and stream size on freshwater mussel (Bivalvia: Unionidae) distribution in southeastern Michigan, U.S.A. Freshwater Biology 13:253–264.

Tatem, H. E. 1986. Bioaccumulation of polychlorinated biphenyls and metals from contaminated sediment by freshwater prawns, *Macrobrachium rosenbergii* and clams, *Corbicula fluminea*. Archives of Environmental Contamination and Toxicology 15:171–183.

Taylor, R. W. 1985. Comments on the distribution of freshwater mussels (Unionacea) of the Potomac River headwaters in West Virginia. Nautilus 99:84–87.

Taylor, R. W., and K. J. Horn. 1983. A list of freshwater mussels suggested for designation as rare, endangered or threatened in West Virginia. Proceedings of the West Virginia Academy of Science (Biology Section) 54:31–34.

Taylor, R. W., and B. D. Spurlock. 1982. The changing Ohio River naiad fauna: a comparison of early Indian middens with today. Nautilus 96:49–51.

Tevesz, M. J. S., D. W. Cornelius, and J. B. Fisher. 1985. Life habits and distribution of riverine *Lampsilis radiata luteola* (Mollusca: Bivalvia). Kirtlandia 41:27–34.

Thompson, C. M., and R. E. Sparks. 1977. Improbability of dispersal of adult Asiatic clams, *Corbicula manilensis,* via the intestinal tract of migratory water fowl. American Midland Naturalist 98:219–223.

Thompson, C. M., and R. E. Sparks. 1978. Comparative nutritional value of a native fingernail clam and the introduced Asiatic clam. Journal of Wildlife Management 42:391–396.

Trdan, R. J., and W. R. Hoeh. 1982. Eurytopic host use by two congeneric species of freshwater mussel (Pelecypoda: Unionidae: *Anodonta*). American Midland Naturalist 108:381–388.

Trueman, E. R. 1983. Locomotion in molluscs. Pages 155–198. *in:* A. S. M. Saleuddin and K. M. Wilbur,

editors. The Mollusca. Vol. 4: Physiology. Part 1. Academic Press, New York.

van den Thillart, G., and I. de Vries. 1985. Excretion of volatile fatty acids by anoxic *Mytilus edulis* and *Anodonta cygnea.* Comparative Biochemistry and Physiology 80B:299–301.

Vermeij, G. J., and E. C. Dudley. 1985. Distribution of adaptations: a comparison between the functional shell morphology of freshwater and marine pelecypods. Pages 461–478 *in:* E. R. Trueman and M. R. Clarke, editors. The Mollusca. Vol. 10: Evolution. Academic Press, New York.

Vitale, M. A., and F. E. Friedl. 1984. Ammonia production by the freshwater bivalve *Elliptio buckleyi* (Lea): intact and monovalve preparations. Comparative Biochemistry and Physiology 77A:113–116.

Warren, M. L., Jr., D. R. Cicerello, K. E. Camburn, and G. J. Fallo. 1984. The longitudinal distribution of the freshwater mussels (Unionidae) of Kinniconick Creek, northeastern Kentucky. American Malacological Bulletin 3:47–53.

Way, C. M. 1988. An analysis of life histories in freshwater bivalves (Mollusca: Pisidiidae). Canadian Journal of Zoology 66:1179–1183.

Way, C. M. 1989. Dynamics of filter-feeding in *Musculium transversum* (Bivalvia: Sphaeriidae). Journal of the North American Benthological Society 8:243–249.

Way, C. M., and T. E. Wissing. 1982. Environmental heterogeneity and life history variability in the freshwater clams, *Pisidium variabile* (Prime) and *Pisidium compressum* (Prime) (Bivalvia: Pisidiidae). Canadian Journal of Zoology 60:2841–2851.

Way, C. M., and T. E. Wissing. 1984. Seasonal variability in the respiration of the freshwater clams, *Pisidium variabile* (Prime) and *P. compressum* (Prime) (Bivalvia: Pisidiidae). Comparative Biochemistry and Physiology 78A:453–457.

Way, C. M., D. J. Hornbach, and A. J. Burky. 1980. Comparative life history tactics of the sphaeriid clam, *Musculium partumeium* (Say), from a permanent and temporary pond. American Midland Naturalist 104:319–327.

Way, C. M., D. J. Hornbach, and A. J. Burky. 1981. Seasonal metabolism of the sphaeriid clam, *Musculium partumeium,* from a permanent and temporary pond. Nautilus 95:55–58.

Way, C. M., D. J. Hornbach, T. Deneka, and R. A. Whitehead. 1989. A description of the ultrastructure of the gills of freshwater bivalves, including a new structure, the frontal cirrus. Canadian Journal of Zoology 67:357–362.

Way, C. M., A. C. Miller, and B. S. Payne. 1990a. The influence of physical factors on the distribution and abundance of freshwater mussels (Bivalvia: Unionacea) in the lower Tennessee River. Nautilus 103:96–98.

Way, C. M., D. J. Hornbach, C. A. Miller-Way, B. S. Payne, and A. C. Miller. 1990b. Dynamics of filter-feeding in *Corbicula fluminea* (Bivalvia: Corbiculidae). Canadian Journal of Zoology. 68:115–120.

Weber, C. I. 1973. Biological field and laboratory methods

for measuring the quality of surface waters and effluents (macroinvertebrate section). EPA-670/4-73-001. National Environmental Research Center, Office of Research and Development, United States Environmental Protection Agency, Cincinnati, Ohio.

White, C. P. 1982. Endangered and threatened wildlife of the Chesapeake Bay region. Tidewater Publishers, Centerville, Maryland.

White, D. S. 1979. The effect of lake-level fluctuations on *Corbicula* and other pelecypods in Lake Texoma, Texas and Oklahoma. Pages 82–88 *in:* J. C. Britton, editor. Proceedings, first international *Corbicula* symposium. Texas Christian University Research Foundation, Fort Worth.

Wilbur, K. M., and A. S. M. Saleuddin. 1983. Shell formation. Pages 236–287 *in:* A. S. M. Saleuddin and K. M. Wilbur, editors. The Mollusca. Vol. 4: Physiology, Part 1. Academic Press, New York.

Williams, C. J. 1985. The population biology and physiological ecology of *Corbicula fluminea* (Müller) in relation to downstream dispersal and clam impingement on power station raw water systems. Master's Thesis, The University of Texas at Arlington.

Williams, C. J., and R. F. McMahon. 1985. Seasonal variation in oxygen consumption rates, nitrogen excretion rates and tissue organic carbon: nitrogen ratios in the introduced Asian freshwater bivalve, *Corbicula fluminea* (Müller) (Lamellibranchia: Corbiculacea). American Malacological Bulletin 3:267–268.

Williams, C. J., and R. F. McMahon. 1986. Power station entrainment of *Corbicula fluminea* (Müller) in relation to population dynamics, reproductive cycle and biotic and abiotic variables. American Malacological Bulletin Special Edition No. 2:99–111.

Williams, C. J., and R. F. McMahon. 1989. Annual variation of tissue biomass and carbon and nitrogen content in the freshwater bivalve, *Corbicula fluminea*, relative to downstream dispersal. Canadian Journal of Zoology 67:82–90.

Willows, A. O., editor. 1985. The Mollusca. Vol. 8: Neurobiology and behavior, Part 1. Academic Press, New York.

Willows, A. O., editor. 1986. The Mollusca. Vol. 9: Neurobiology and behavior, Part 2. Academic Press, New York.

Winnell, M. H., and D. J. Jude. 1982. Effects of heated discharge and entrainment on benthos in the vicinity of the J. H. Campbell Plant, Eastern Lake Michigan, 1978–1981. Special Report No. 94. Great Lakes Research Division, University of Michigan, Ann Arbor. 202 pp.

Wissing, T. E., D. J. Hornbach, M. S. Smith, C. M. Way, and J. P. Alexander. 1982. Caloric contents of corbiculacean clams (Bivalvia: Heterodonta) from freshwater habitats in the United States and Canada. Journal of Molluscan Studies 48:80–83.

Wu, W.-L., and E. R. Trueman. 1984. Observation on surface locomotion of *Sphaerium corneum* (Linneaus) (Bivalvia: Sphaeriidae). Journal of Molluscan Studies 50:125–128.

Young, M. R., and J. Williams. 1983a. The status and conservation of the freshwater pearl mussel *Margaritifera margaritifera* Linn. in Great Britain. Biological Conservation 25:35–52.

Young, M. R., and J. Williams. 1983b. Redistribution and local recolonization by the freshwater pearl mussel *Margaritifera margaritifera* (L.). Journal of Conchology 31:225–234.

Young, M. R., and J. C. Williams. 1983c. A quick secure way of marking freshwater pearl mussels. Journal of Conchology 31:190.

Young, M. R., and J. Williams. 1984a. The reproductive biology of the freshwater pearl mussel *Margaritifera margaritifera* (Linn.) in Scotland II. Laboratory studies. Archiv fuer Hydrobiologie 100:29–43.

Young, M. R., and J. Williams. 1984b. The reproductive biology of the freshwater pearl mussel *Margaritifera margaritifera* (Linn.) in Scotland I. Field studies. Archiv fuer Hydrobiologie 99:405–422.

Zale, A. V., and R. J. Neves. 1982. Fish hosts of four species of lampsiline mussels (*Mollusca unionidae*) in Big Moccasin Creek, Virginia. Canadian Journal of Zoology 60:2535–2542.

Zbeda, T. W., and D. S. White. 1985. Ecology of the zoobenthos of southeastern Lake Michigan near the D. C. Cook Nuclear Plant. Part 4: Pisidiidae. Special Report No. 13. Great Lakes Research Division, Institute of Science and Technology, University of Michigan, Ann Arbor.

Zeto, M. A., W. A. Tolin, and J. E. Smith. 1987. The freshwater mussels of the upper Ohio River, Greenup and Belleville Pools, West Virginia. Nautilus 101:182–185.

Zs.-Nagy, I., D. A. Holwerda, and D. I. Zandee. 1982. The cytosomal energy production of molluscan tissues during anaerobiosis. Basic and Applied Histochemistry 26 (Suppl.):8–10.

Annelida: Oligochaeta and Branchiobdellida

12

Ralph O. Brinkhurst
Ocean Ecology Laboratory
Institute of Ocean Sciences
Fisheries and Oceans
Sidney, British Columbia
V8L 3B5 Canada

Stuart R. Gelder
Department of Biology
University of Maine at Presque Isle
Presque Isle, Maine 04769

Chapter Outline

OLIGOCHAETA

I. **INTRODUCTION TO OLIGOCHAETA**

II. **OLIGOCHAETE ANATOMY AND PHYSIOLOGY**
   A. Anatomy
   B. Physiology

III. **ECOLOGY AND EVOLUTION OF OLIGOCHAETA**
   A. Diversity and Distribution
   B. Reproduction and Life History
   C. Ecological Interactions
   D. Evolutionary Relationships

IV. **COLLECTING, REARING, AND PREPARATION OF OLIGOCHAETES FOR IDENTIFICATION**
   A. Collection
   B. Rearing Oligochaetes
   C. Preparation for Identification

V. **TAXONOMIC KEYS FOR OLIGOCHAETA**
   A. Taxonomic Key to Families of Freshwater Oligochaeta and Aphanoneura
   B. Taxonomic Key to Genera and Selected Species of Freshwater Lumbriculidae
   C. Taxonomic Key to Genera and Selected Species of Freshwater Naididae
   D. Taxonomic Key to Genera and Selected Species of Freshwater Tubificidae

BRANCHIOBDELLIDA

VI. **INTRODUCTION TO BRANCHIOBDELLIDA**

VII. **ANATOMY AND PHYSIOLOGY OF BRANCHIOBDELLIDANS**
   A. External Morphology
   B. Organ System Function
   C. Environmental Physiology

VIII. **ECOLOGY AND DISTRIBUTION OF BRANCHIOBDELLIDA**
   A. Diversity and Distribution
   B. Reproduction and Life History
   C. Ecological Interactions
   D. Evolutionary Relationships

IX. **IDENTIFICATION OF BRANCHIOBDELLIDANS**
   A. Collecting, Rearing, and Preparation for Identification
   B. Taxonomic Key to Genera of Freshwater Branchiobdellida
   Literature Cited

OLIGOCHAETA

I. INTRODUCTION TO OLIGOCHAETA

One family of aquatic Oligochaeta, the Tubificidae (sludge worms) has been recognized since the time of Aristotle for its ability to develop dense colonies in organically polluted waters (Fig. 12.1). Beginning

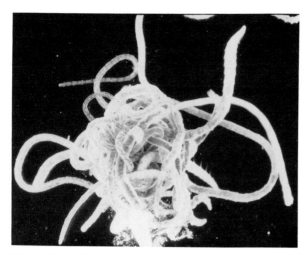

Figure 12.1 Clump of living tubificids (Photograph by P. M. Chapman).

with the concept that all sludge worms are pollution-tolerant indicators, recent work has made biologists aware of the whole range of ecological niches occupied by aquatic annelids in general, and oligochaetes and branchiobdellidans in particular. They are found in every kind of freshwater and estuarine habitat, not only in organic mud. Specialized forms can be found in groundwater, in oligotrophic lakes and streams, and as commensals or symbionts on crayfish, molluscs, and even tree frogs. Several North American genera prey on other worms or a variety of small invertebrates. The most surprising recent discovery has been the description of a rich diversity of tubificids at all depths in the oceans of the world.

While most students are familiar with the terms Oligochaeta and Clitellata, there is no general agreement about the equivalent rankings of the various annelid groups. Some authors place the arthropods and annelids in a single phylum, but most texts separate them. Within the Annelida, it is possible to classify the Aclitellata (including Polychaeta) and the Clitellata as superclasses or classes. The former arrangement allows us to refer to the Polychaeta, Oligochaeta, Hirudinea, and Branchiobdellida as classes, whereas the latter would demote these to orders. Relationships among the last three taxa, as well as the monotypic Acanthobdellida, are presently under investigation using cladistic methods, which greatly assist the process of determining sister-group relationships. Biologists who reject Hennigian methods may well classify the various groups in a different way from those who do not. The various terms will be used here based on the concept of a superclass Clitellata with a series of classes, although we strongly suspect that the rankings of the Branchiobdellida and Acanthobdellida will shortly be revised.

II. OLIGOCHAETE ANATOMY AND PHYSIOLOGY

A. Anatomy

Oligochaete worms are bilaterally symmetrical, segmented coelomates with four bundles of chaetae on every segment except the first. In earthworms, the chaetal "bundles" usually consist of two chaetae each, but in aquatic species there are often several, and there may even be more than a dozen per bundle. In a few exceptional forms, the chaetae are absent in some or even all segments. Aquatic species (superorder Microdrili) are smaller than earthworms (Megadrili) as the name suggests, and they have simple body walls and none of the specialized regions of the digestive tract found in the earthworms. The clitellum is a single-layer thick and is limited to the genital region in microdriles, and the eggs are large and yolky. In earthworms, secretions from the multilayered clitellum provide an alternative source of food for the eggs, which are small and less yolky. The clitellum is located well behind the gonadal segments. In aquatic species, the male and female ducts are connected to external genital pores located either in the same segment as the gonads they serve, or in the segment immediately behind them. In earthworms, the male ducts are extended back to open well behind the gonadal segments. It is generally thought that there were originally four gonadal segments, two with paired testes, and two with paired ovaries, in X–XIII (Roman numerals are used for segment number by convention). These have been reduced in number in various ways as gamete storage and sperm transmission through copulation have evolved (see Section III.D). In some entire families and in a few individual species in other families, the gonadal segments may be found much nearer to the anterior end than usual. This may be due to a tendency to regenerate fewer than the normal number of anterior segments in the predominantly asexual-reproducing forms in which this happens (family Naididae). This explanation will not suffice for sexually reproducing forms in which the forward shift involves just one segment (family Lumbriculidae). In the family Opistocystidae, the gonads are located behind XX in all but the one species seen here.

The arrangement of the various reproductive organs of some typical aquatic oligochaetes is illustrated in Fig. 12.2. The reproductive system and chaetae of one side are illustrated, but in life these animals are bilaterally symmetrical. The anatomy is easy to understand if the stages in reproduction from sperm development to egg hatching are described. The sperm mother cells leave the testes to float free

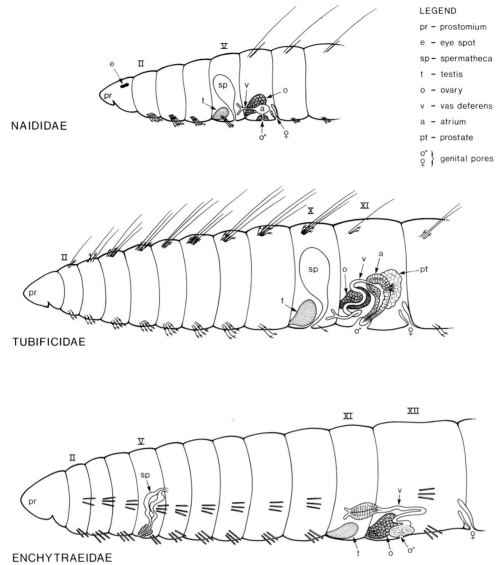

LEGEND

pr – prostomium
e – eye spot
sp – spermatheca
t – testis
o – ovary
v – vas deferens
a – atrium
pt – prostate
$\left. \begin{matrix} \text{♂} \\ \text{♀} \end{matrix} \right\}$ genital pores

NAIDIDAE

TUBIFICIDAE

ENCHYTRAEIDAE

Figure 12.2 Reproductive systems and external characteristics of three major families of Oligochaeta. The worms are shown with the anterior to the left, and only the structures of the left side are visible. The dorsal and ventral chaetae are illustrated. In the Naididae, an example with dorsal chaetae beginning in VI is shown. Segments are identified with Roman numerals by convention (septa would be VII/VIII, VIII/IX, etc.). a, Atrium; e, eye spot; o, ovary; pr, prostomium; pt, prostate; sp, spermatheca; t, testis; v, vas deferens; ♀ ♂, genital pores.

in the coelom. The amount of space available to the developing sperm masses, or morulae, is increased by the development of outpushings of the testicular segments, slightly anteriad but more extensively posteriad through several segments. These are called sperm sacs. The sperm eventually cluster around large funnels suspended on the posterior walls of the testicular segment, sometimes well within the sperm sacs. These sperm funnels lead into vasa deferentia that originally drained via the male pores, located near the ventral chaetae of the post-testicular segment. Various structures have evolved, presumably at the point where the vasa

deferentia contact the original body wall. Inversion of the body wall has led to the evolution of atria. These muscular bodies were originally lined with ectodermal cells, many of which had a glandular function, which presumably fed and lubricated the sperm now stored in the atria ready for copulation. The glandular cells are thought to have migrated inward, through the muscle layers, so that their cell bodies now lie in the coelom. This achieves both an increase in the space available to the glandular cell bodies and an increase in the space available for sperm within the atria. The ways in which the gland cells have migrated through the muscles of the atria to form the prostate glands vary, and this provides a very useful taxonomic feature (though not one that can be seen without some effort). The term prostate gland is used for very different structures in earthworms, and a third set of terms is used in the Branchiobdellida. In a further evolutionary development, another section of the body wall becomes part of the male reproductive structure between the atria and the male pores. Various types of penes are found in what are considered to be the more advanced groups (family Tubificidae, family Lumbriculidae), and have evolved quite independently several times. In some penes, the normal cuticular layer over the epidermis has become hypertrophied to form cuticular penis sheaths. Sometimes this cuticular layer is hardly noticeable, but the cuticular tubes may become thick and rigid with soft penes free within them. If the cuticle over the penes can be seen in cleared, whole-mounted worms and has a consistent shape, cuticular penis sheaths are said to be present. In some species, basement membranes or cuticular linings of eversible penes appear on whole mounts, but these are usually irregular in form. These are called apparent penis sheaths.

When worms copulate, sperm is passed to the partner and is usually deposited in sac-like spermathecae. There may be one or more pairs of these structures located in or near the gonadal segments, commonly in front of the testicular segments or in them. The number and position of the spermathecae may vary within the Lumbriculidae, but there is usually one pair in the single testicular segment (family Tubificidae, family Naididae), or separated from the other genitalia in V (family Enchytraeidae). In a few species, spermathecae may be present in some specimens but not others, or they may be missing altogether. In *Bothrioneurum* (Tubificidae), the sperm are placed in spermatophores that are attached externally to the body wall of the partner, usually near the gonopores, during copulation. In members of the tubificid subfamily Tubificinae, there are two types of sperm, which come together in organized masses called spermatozeugmata. These can be recognized within the spermathecae of mated worms.

After a worm has copulated and has sperm in its spermathecae, the clitellum secretes a cocoon. The eggs have already developed in the coelomic space of the ovarian segment. This segment is extended posteriad into egg sacs. As the sperm sacs already displace the septa of several segments behind the testes, the egg sacs and their contained eggs are only detectable in segments beyond the sperm sacs. The egg and sperm sacs have been omitted from Fig. 12.2 because it becomes confusing to follow a whole series of fingerlike protrusions running through several segments, all contained within each other. [It is worth noting here that the egg and sperm sacs of the Acanthobdellida and Hirudinea are not contained within each other, but lie parallel to each other. They are paired, bilateral structures just like the rest of the reproductive system.] Eggs are deposited into the cocoon via the female ducts. These consist of nothing more than ciliated funnels on the posterior walls of the ovarian segment that open into the intersegmental furrow. In a few primitive forms, the female ducts actually penetrate the postovarian segment and open on its anterior surface. It is assumed that fertilization occurs in the cocoon. The basic position for the spermathecal, male, and female pores is in the longitudinal line of the ventral (ventrolateral) chaetal bundles, but there is some variation in this due to midventral fusion, the development of median copulatory bursae, or the migration of the spermathecal pores to a more dorsal position. While the spermathecal pores may open near the anterior furrow of the segment they occupy, all the genital pores commonly lie close to the ventral chaetae.

The worm finally releases the cocoon, which formed as a ring of material around it, secreted (presumably) by the clitellum. The ends of the cocoon close, and the fertilized eggs develop inside. Small worms eventually escape through the openings at the ends of the cocoons, and are recognizable as immature specimens, which can be identified if they happen to belong to one of the species with characteristic chaetae.

The various reproductive organs and the modified chaetae that may develop close to the genital pores can best be found by looking carefully in the correct segment. When the keys mention genital structures, these will only be found on mature specimens. An immature aquatic oligochaete appears to consist of very little more than the body wall with its chaetae and a simple tubular gut running through the coelom. Very few specialized anatomical features have been added to the basic vermiform shape, unlike the situation in the Polychaeta. That is why it is generally useless to illustrate the general habit of oligochaetes. Useful illustrations are those of the chaetae and reproductive systems. In contrast to immature specimens, a mature worm will have an external clitellum, though this is not as prominent as that of an earthworm, and the egg and sperm sacs are full and therefore visible. The worm is often somewhat distended in the genital region, and the reproductive organs obscure the normal view of the gut.

As most worms burrow through soft substrates and extract their food by ingesting them, the range of externally visible anatomical adaptations are few because this habit imposes severe limitations on body shape. The prostomium lies in front of the mouth, and while its body cavity is clearly coelomic, it is considered to be presegmental. The prostomium may bear a median anterior prolongation, the proboscis. In *Bothrioneurum,* there is a median dorsal pocket containing sensory cells, the prostomial pit (Fig. 12.3). Segment numbering begins with the peristomium, the small segment around the mouth. In aquatic species, the peristomium is completely separate from the prostomium. The first segment bears no chaetae. This is important to remember when using cleared, whole-mounted specimens as it is often easier to locate specific segments by counting chaetal bundles, which begin in segment II.

The chaetae vary in form as well as number. This variation provides very useful key characters, but there is increasing evidence that the fine details of chaetal form, or even the presence or absence of a particular type of chaetae such as hairs, can be affected by quite simple environmental variables such as pH or conductivity (Chapman and Brinkhurst 1987). The Enchytraeidae have simple-pointed chaetae, which may be straight or sigmoid. The chaetae in a bundle may vary in size (Fig. 12.4). Hair chaetae are elongate, simple shafts. They are found only in the dorsal bundles of some species of Naididae, Tubificidae, and in *Crustipellis* (Opistocystidae). They are present in both dorsal and ventral bundles in the Aeolosomatidae (Aphanoneura, Aclitellata), this being one of a series of fundamental differences between these minute freshwater worms and true oligochaetes. Hair chaetae may bear a series of fine lateral hairs; then, they are termed hispid or serrate. This condition may vary within a species depending on seasonal or environmental variables, but it is still used as a taxonomic character for many naidids for which the limits of intraspecific variability have yet to be established. In a few naidids, the lateral hairs may be especially prominent along one side of the hair chaeta, and these are called plumose hairs from their featherlike appearance. In one or two instances, minute bifid tips have been observed at the ectal end of hair chaetae, indicating their probable origin from the much more ubiquitous bifid chaetae. This bifid form, with the ectal end divided into two teeth, is characteristic of aquatic as opposed to terrestrial oligochaetes, although the functional reason for this is unknown. Bifid chaetae may be the only type present in all bundles, and they usually fill all the ventral bundles. Some simple-pointed chaetae may be found in a few anterior ventral bundles (in the tubificid genus *Spirosperma,* for instance), or they may accompany the hair chaetae in the dorsal bundles in some Naididae. A common combination in the dorsal bundles is hair chaetae accompanied by pectinate chaetae in at least the preclitellar bundles. Pectinate chaetae are usually of the same general form as the bifids for the particular taxon, but the space between the two teeth is filled with a comb of thin teeth or a ridged web. In a few naidids and tubificids, these webbed chaetae may be expanded as palmate chaetae. In general, the dorsal chaetae accompanying the hair chaetae in the majority of naidid species differ markedly from the ventral chaetae. For this reason, they have come to be termed needles, whether they are palmate, pectinate, bifid, or simple-pointed. Some authors like to substitute the terms fascicle, capilliform, bifurcate, and crotchet for bundle, hair, bifid, and chaeta respectively. They also use the terms distal and proximal for upper and lower tooth, the designations used in describing the variations in the length and thickness of the teeth, often a useful characteristic. [Anglo-Saxon terms are short and to the point, and are used internationally, but the longer Latin terms are more readily translated into French and Spanish.] The normal ventral chaetae may be replaced by specialized genital chaetae, usually beside the spermathecal or penial pores. This will be illustrated where appropriate in the keys. These do not vary much in form in the Naididae, where there are often two or three blunt penial chaetae in each bundle beside the male pore. These chaetae are quite long proximally, but the distal end beyond the nodulus (the swelling on the chaeta at the point where it emerges from the chaetal sac) is short. Penial chaetae are usually somewhat longer and

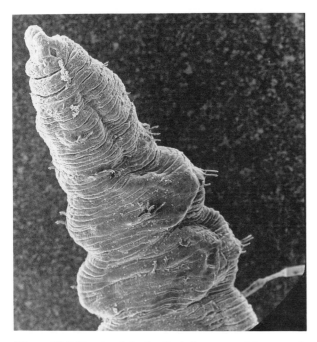

Figure 12.3 Prostomial pit, *Bothrioneurum* (photograph by P. M. Chapman).

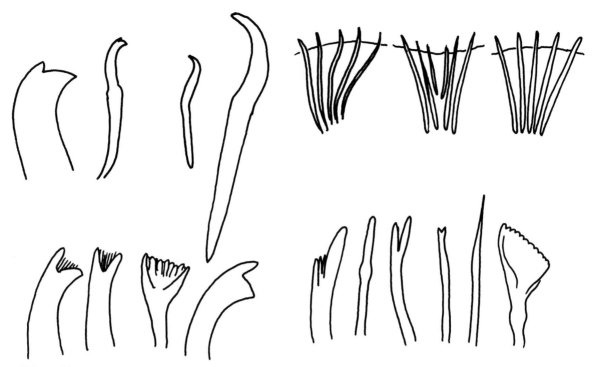

Figure 12.4 Representative chaetae. From the left, top row: lumbriculid bifid with reduced upper tooth, head end, and entire; small dorsal and large ventral sickle-shaped chaetae of *Haplotaxis;* three types of bundles from the Enchytraeidae. Bottom row: two pectinates, a palmate, and a bifid dorsal chaeta, Tubificidae; six types of dorsal needles, Naididae.

thicker than the normal ventral chaetae. Penial chaetae, like those of the Naididae may be found in the Rhyacodrilinae, a subfamily of Tubificidae, though there are usually several to a bundle. They are usually arranged with the outer ends close together, the inner ends spread out. In Tubificinae, the ventral chaetae in the penial segment are usually lost but the spermathecal chaetae are often modified. These have elaborate distal ends shaped like trowels. They may be used to introduce the modified spermatozeugmata into the spermathecae during copulation. The functions of both these and the penial chaetae as claspers are based on speculation because aquatic oligochaetes have rarely been observed mating. Spermathecal chaetae are often associated with special glands. Where these are large, and there is a great deal of competition for space in the genital segments, the spermathecal chaetae and glands lie immediately in front of the spermathecal segment (*Rhizodrilus*, Tubificidae). In general, the number and diversity of chaetae is reduced toward the posterior end of the worm, and modified teeth on the bifids commonly become "normal." There are a few exceptions. The upper teeth of the chaetae of *Amphichaeta* (Naididae) get longer posteriorly, and the posterior segments of *Telmatodrilus vejdovskyi* (Tubificidae) bear chaetae with almost brushlike tips (see illustrations in the key).

Few other anatomical features are used in aquatic oligochaete taxonomy. External gills are found in a few taxa. The posterior segments of the tubificid *Branchiura* bear single dorsal and ventral gill filaments. These are longest in the median segments of the gill-bearing region, very short at each end of the row. *Dero* (Naididae) has gills at the posterior end of the body, usually enclosed in a gill chamber (branchial fossa). These are best observed in living specimens. All aquatic oligochaetes have red blood pigments which aid oxygen uptake and transport. In most species, respiration takes place through the posterior body wall or even within the pre-anal part of the intestine. Water may be pumped in and out of the anus for this purpose. Most aquatic oligochaetes protrude the tail end out of the sediment into the water, and may wave the tail to create enough disturbance to maintain high concentrations of oxygen around the tail. The body wall of posterior segments in many lumbriculids is penetrated by special branching blood vessels, presumably to improve respiration.

External sense organs are mostly limited to ultrastructural features of the body wall (Jamieson 1981). These may become prominent in those species that protect the body wall with secretions that trap bacteria and foreign matter in an external crust. Sensory papillae can be seen protruding through such layers in living specimens, but they are usually retracted in fixed material. This has led to arguments about their presence or absence in certain species, but we may assume that any worm with its normal body-wall sensory system denied access to information has developed localized retractile sensory papillae. It is also easy to understand why the necessarily sensitive prostomium and first two or three segments may be naked in taxa that have protected body walls, and why the anterior end is often retractile within the rest of the protected body. Simple pigmented eye spots are located on the lateral surface of segment I in a number of naidids, but may be present or absent in the same species.

The majority of aquatic oligochaetes feed by ingesting sediment, extracting their nutrition from the organic matter and especially the bacteria, which comprise less than 10% of it. They feed continuously in a conveyor-belt fashion, often ingesting recolonized feces. The food is obtained by everting the roof of the pharynx through the mouth. The glandular cells associated with the pharynx have evolved into pharyngeal or septal glands, in which the cell bodies lie in the coelomic cavities of several anterior segments, attached to the posterior septa. Intracellular canals run forward from these cell bodies along the lateral sides of the gut to open dorsally on the pharyngeal pad, which is one of the characteristics of worms of this group. Again, this feature is absent in the Aeolosomatidae. The pharyngeal glands are used in the taxonomy of the Enchytraeidae, but not in other families. The pharyngeal glands are lost in predatory oligochaetes, but traces of the dorsal pad can still be found in some (*Chaetogaster*, Naididae) although it has been lost in others (*Phagodrilus*, Lumbriculidae, and *Haplotaxis*, Haplotaxidae).

B. Physiology

The primary emphasis of studies on worm physiology has been on respiration as it provides a simple model of the value of red blood pigment function and anaerobic metabolism. A recent review of respiration studies with reference to their use in determining production in worms is provided by Bagheri and McLusky (1984), who studied the estuarine tubificid *Tubificoides benedii* (*benedeni*). Oxygen consumption varied from > 0.01 to < 0.5 μl mg^{-1} h^{-1}, but while there were small, apparent linear increases in rate with temperature, these were not statistically significant. These results are among the lowest reported for worms, which is usually taken to indicate that the experimental technique is not creating stress in the test animals. The response of respiration levels to stress has been suggested as a tool in pollution biology (Brinkhurst *et al.* 1983). Most tubificids respire at a more or less constant rate as the external

oxygen concentration is lowered. Once a critical value is reached, respiration rate falls dramatically. Stress can elevate or depress the constant respiration rate, or it may shift the critical value. It may also cause a reduction in or a total loss of respiratory control above the critical value. In some instances, respiratory control actually increased in the face of supposed stress factors. A unique phenomenon is that mixing certain tubificid species together will actually reduce their respiration rates by a significant amount (Chapman *et al.* 1982a). This result was anticipated by earlier studies of worm production using monocultures and mixed cultures of three tubificid species. In those studies (summarized in Brinkhurst and Austin 1978), when respiration went down, growth rates went up, as did assimilation efficiencies, in mixed populations. These experiments seemed to provide an explanation for the way in which tubificids live in small clumps of mixed species. Some worms prefer the recolonized feces of other species as food, and seek the company of that species. Even when they are only in water from monocultures of the appropriate second species, worms move less, eat more, and (in the presence of the real thing) ingest more of their preferred food. This causes a shift in all of the variables measured in production work based on estimating calorific or carbon budgets. The work provides an unusual example of the value of multispecies flocks.

Freshwater oligochaetes, such as the pollution-tolerant tubificids *Tubifex tubifex* and *Limnodrilus hoffmeisteri*, can withstand exposure to 10 ppt salinity, but others can only survive at 5 ppt (Chapman *et al.* 1982b, 1982c). A small amount of salinity actually seems to improve the ability of freshwater worms to withstand stress factors. It is less surprising that the provision of sediment improves stress resistance. Estuarine species may withstand a very wide range of environmental stress, including extreme salinity values (*Paranais*, Naididae; *Monopylephorus*, Tubificidae). It would be a mistake to believe that these animals are exposed to a wide range of salinities due to daily tidal excursions, however. In muddy situations in rivers such as the Fraser in British Columbia, the "interstitial" salinity goes through a seasonal cycle related to the capture and release of snow in the headwaters. As temperatures increase, the flow of freshwater increases and tidal sea water incursions into the estuary are limited to the outer estuary. All of the benthic communities move slowly up and down the estuary in response, maintaining themselves at what are presumably preferred salinity ranges for each species. It is selection for these optimal salinities (that remain quite constant within the mud on a short-term basis) that causes the sequence of species observed along the salinity gradient in estuaries (Chapman 1981, Chapman and Brinkhurst 1981).

While worms in cocoons may be able to survive desiccation, there is no experimental evidence of this. Adult *Tubifex tubifex* (Tubificidae) are able to form cysts for adult worms that can survive unmoistened for 14 days, or for as long as 70 days with occasional dampening. Once the worms have been moistened they may emerge in 20 hr (Kaster and Bushnell 1981). The lumbriculid *Lumbriculus variegatus* can also encyst, but it only survived for 4 days under experimental conditions. In both instances, the covering of foreign matter, which helps to disguise the cyst, is also essential to the survival of the worms. There are records of a handful of other lumbriculids undergoing fission in cysts, but no physiological work has been done. *Lumbriculus* also forms a different type of cyst to withstand freezing (Olsson 1981). Specimens of the tubificid *Limnodrilus profundicola* were collected by R. Brinkhurst at 4000 m on Mount Evans, Colorado. After being accidentally frozen into a solid block of ice, the worms emerged quite unscathed when thawed the next day. As so many aquatic Oligochaeta, especially the Naididae, are cosmopolitan or at least very widely distributed, one would expect more evidence of adaptations to counter desiccation. Dispersal mechanisms of oligochaetes are still a matter for speculation.

The development of eye spots is unique to the Naididae, a family of small, delicate worms, some of which can make rudimentary swimming movements. Other naidid species, while not freeliving, occupy tubes attached to aquatic plants and may protrude the anterior end, unlike the majority of aquatic oligochaetes, which avoid the light by keeping the head end buried in their food supply, the sediment. Most aquatic worms have quite sensitive tails and have a rapid withdrawal reflex based on a mechanosensory detection system that is responsive to the approach of potential predators (Zoran and Drewes 1987). *Branchiura* has the fastest escape reaction reported for any invertebrate. This speed is associated with the development of the lateral giant nerve fibers in the ventral nerve cord. These are responsible for rapid forward transmission of sensory information. The median giant nerve fiber transmits posteriad from the head end. In *Lumbriculus,* which lives in shallow littoral habitats, the body is extended up to the air–water interface, with the tail lying horizontally in it. Not surprisingly, these worms have what Drewes and Fourtner (1988) term hindsight or light-sensitive tails, and they respond to passing shadows. Light can be damaging to worms.

A great deal of work has been done in the labora-tory on the effect of toxic substances on worms, summarized by Chapman and Brinkhurst (1984) and Wiederholm *et al.* (1987).

III. ECOLOGY AND EVOLUTION OF OLIGOCHAETA

A. Diversity and Distribution

Many aquatic oligochaetes are cosmopolitan, or at least widely distributed. This is especially true of the Naididae, but even here significant patterns are beginning to emerge. The supposedly primitive naidid subfamily Stylarinae are scarce in the southern United States, such genera as *Stylaria, Arcteonais, Ripistes,* and *Vejdovskyella* being common in northern locations. *Dero* and *Allonais* species (members of the Naidinae) are well represented in subtropical and tropical regions. There are other north–south distribution patterns. The tubificid *Arctodrilus wulikensis* is found in Alaska, as are several lumbriculid species that are part of the Pacific Rim fauna mentioned later. *Phallodrilus hallae* is a tubificid limited to Lake Superior that is related to a large number of marine taxa in the subfamily Phallodrilinae. Unlike the northern forms, the tubificid species that are restricted to southern or southeastern localities belong to genera that are otherwise more widely distributed.

Among the tubificids and lumbriculids, some taxa are limited to the western United States and Canada. Some of these are related to or are the same as Asian Pacific Rim counterparts, suggesting an evolutionary distribution pattern. Several tubificids are known only from Lake Tahoe, which may prove to be home to a number of endemic species, though some forms first described there have since been found elsewhere, mostly still in California.

The apparent limitation of some species to the eastern United States may simply reflect a greater frequency of caves or more interest in cave faunae in those areas. At least six lumbriculid species fall into this category, but there are also some eastern surface water tubificid species. The groundwater tubificid *Rhyacodrilus falciformis* is a European species now known to be widespread in such habitats on Vancouver Island, British Columbia, and there is an undescribed lumbriculid known from a well in California; so there may be as rich a western groundwater fauna (awaiting description) as there is in the east. Many groundwater species have very limited distributions, although the distinction between species in genera such as *Trichodrilus* is often based on minute differences. This may mean that biologists look more carefully for differences where they anticipate that they may exist.

The reasons for these geographic patterns have not been studied, partly because they have only been recognized for a few years. There may be ecological reasons for the distribution of some taxa. Species of *Aulodrilus* (Tubificidae) may be widely distributed, but in the more temperate zones they appear to reproduce asexually. This may suggest a recent spread from warmer areas where they do breed sexually. Some species can be shown to be recent introductions by man. The North American *Limnodrilus cervix* has been introduced to Britain and Sweden via ports and canals (Milbrink 1983). It is possible that the diverse collection of *Potamothrix* (Tubificidae) species in the St. Lawrence Great Lakes represents a series of introductions of common European species into North America. This is more certain for *Psammoryctides barbatus* (Tubificidae), a very recent discovery in the St. Lawrence River in Quebec (Vincent *et al.* 1978), as well as the estuarine and coastal species *Tubificoides benedii* found recently by the author in Vancouver Harbor, British Columbia. One of the most common tropical tubificids, *Branchiura sowerbyi,* was first described from a Victorian greenhouse in Britain, where it had been introduced along with the tropical plants on display. It has now become adapted to temperate waters, though it still thrives best in manmade lakes and streams, especially where the temperature is elevated. It has apparently spread all across North America since it was first recorded in Ohio in 1931; but, this could simply reflect a gradual spread of awareness of this easily recognized species (Brinkhurst 1965). Some of these regional distributions are mentioned in the keys, but the reader should always be prepared for the unusual.

Niche discrimination in aquatic oligochaetes is less obvious than zoogeographic patterns. The majority of these worms are adapted to live in sediments ranging from sand to mud. They can be found in pockets of such sediments in stony habitats as well as in lowland rivers, lakes, and ponds where soft substrates are the norm. There is none of the usual distinction between lacustrine and riverine species. The best guess that can be made about niche discrimination is that the precise nature of the organic matter in sediments and the microflora it supports determines the outcome of interspecific competition. The ecophysiological work on multispecific flocks described earlier is probably applicable to most situations and not limited to the *Tubifex–Limnodrilus–Quistadrilus* association in which it was described. In most habitats, between 5 and 15 species can be found. There are some special adaptations, such as the groundwater fauna already mentioned. Several of the genera of the Naididae are no longer restricted to life in the sediment, but may be found among aquatic weeds in quiet rivers and ponds. *Lumbriculus* lives among vegetation in pond and stream margins. *Dero (Allodero)* species are symbionts of frogs, which, once released from their host come to resemble specimens of *Dero (Aulophorus)*, rendering the taxonomic distinction suspect. The predators *Haplotaxis* and *Phagodrilus* are found in springs, seepages, and groundwater, but *Chaetogaster* is more ubiquitous. Estuarine species form another group, but a number of species are restricted to tidal freshwater where there is little, if any, detectable saltwater. None of these exhibit any marked anatomical adaptations to their way of life, with the exception of the pharyngeal structures of the predators and loss of chaetae and gills in *Allodero*.

The significance of food in the ecological specialization of the aquatic oligochaetes, especially the tubificids, is demonstrated by the sequence of species groups that inhabit progressively more organically polluted stretches of rivers or more eutrophic lakes. The Lumbriculidae and some naidid genera such as *Pristina* and *Pristinella* are found in streams with low organic matter in the sediments, and they may be accompanied by tubificids belonging to the Rhyacodrilinae in particular (*Rhyacodrilus, Bothrioneurum, and Rhizodrilus*). Tubificinae, such as *Potamothrix* and *Aulodrilus* can be found in situations with a reasonable supply of organic matter but also a good oxygen supply; but, *Tubifex tubifex,* several *Limnodrilus* species, and *Quistadrilus* are found in areas where the oxygen supply can become reduced. This basic pattern has been expanded, quantified, and recognized in Europe as well as North America (see, e.g., Lang 1984). At first, this type of work followed the much earlier pattern of establishing lists of chironomid midge species characteristic of lake types (reviewed in Brinkhurst 1974). The maximum value in identifying aquatic oligochaetes in pollution studies today is to add information for use in data matrices to be analyzed by multivariate statistics like cluster analyses. There are relatively few invertebrate species in freshwater muddy habitats, in contrast to stony streams (where there are far more insects) or marine habitats (where there are many more mud-dwelling taxa). The taxonomic identity of the communities from each sample site are retained in such analyses, whereas they are lost in simpler diversity models.

B. Reproduction and Life History

Field studies on reproduction in the Naididae were reviewed by Loden (1981). Asexual reproduction, usually by paratomy, predominates but sexual reproduction apparently enables the worms to exist through periods of unsuitable conditions. Abundance varies seasonally depending on the geographic region, with spring, summer, or winter maxima being produced asexually. Most species apparently cease feeding as they mature because the gut degenerates and the mouth may even become sealed. According to Loden, some species incorporate the anterior part of the worm in the cocoon, but Lochhead and Learner (1984) dispute this in describing a life history for *Nais*. They report that worms from this and other naidid genera are able to shed their cocoons. According to these investigators, the adult worms are more resistant to environmental stress than the cocoons, and so in the temperate latitudes in which they work, worms may not use sexual reproduction to avoid stressful periods. All cocoons of aquatic oligochaetes may be disguised with mud particles, or may be placed out of reach of predators (Newrlka and Mutayoba 1987).

Most tubificids reproduce sexually, but there are parthenogenetic species. These often lack spermathecae and the testes may be reduced or absent. The vasa deferentia may be reduced to solid strands in these species, but the atria are usually retained. A few species seem able to reproduce asexually by fission. The results of various field studies produce conflicting interpretations of the life histories of such common genera as *Limnodrilus*. There is less disagreement about events in less adaptable species found in more restricted habitats. The worms may mature a few weeks after hatching if temperature and food supplies are adequate, and they may breed more than once in a season. Alternative studies suggest that many species mature in their second year. After breeding, these forms resorb the gonads and gonoducts and would be recognized as immature specimens. The next year they develop a new set of reproductive organs and breed again, after which most species die. Some published examples are the studies of Block *et al.* (1982) and Lazim and Learner (1986a, 1986b) for *Limnodrilus* species, Christensen (1984) on asexual reproduction, and Poddubnaya (1984) on parthenogenesis. Because of the problems encountered in interpreting the results of field studies on life histories, considerable effort has been put into the development of laboratory cohort studies (Bonacina *et al.* 1987). In general, oligochaete life histories are probably not as precisely tuned to environmental variables like day length as they are in more highly evolved forms such as insects. This may be attributable to simpler neurosecretory systems in worms, but there is no evidence to support this idea. Food and temperature may cause considerable local variation in life cycles, especially in the opportunistic species like *Limnodrilus hoffmeisteri*. Unfortunately, these abundant, cosmopolitan species are the most convenient laboratory animals, but they are so adaptable and perhaps atypical that comparable work should be done with species found in more restricted habitats.

C. Ecological Interactions

There is some published work on the food and feeding behavior of the Naididae, but a much larger literature exists on the relationship between tubificids (and one lumbriculid species *Stylodrilus heringianus*) and the sediments that they ingest (and therefore defecate) continuously.

Algae and other epiphytic material are the main food of many naidids, but *Nais* feeds on heterotrophic, aerobic bacteria like the tubificids. There is no specialized gut flora in either group (Brinkhurst and Chua 1969, Harper *et al.* 1981). There is no evidence of selective ingestion or digestion of specific bacteria in *Nais* but selective digestion is important in tubificids. *Nais elinguis* will select algal filaments in preference to monofilament nylon thread and prefers some algae to others or to glass beads (Bowker *et al.* 1985). Algae also form part of the food supply of the tubificids *Limnodrilus claparedeianus* and *Rhyacodrilus sodalis* and the lumbriculid *Lumbriculus variegatus*.

Studies of defecation rate and the chemical and bacterial content of feces in tubificids have usually exploited the fact that these worms will continue to feed with their heads in the sediment and their tails in the water column, even when such colonies in small containers are inverted. If a layer of cotton or fine netting is used to prevent the sediment drifting down past the worms into the collecting funnel placed beneath the culture, it is easy to determine the amount and the chemical content of the feces. If the feces are collected over very short time periods, the bacterial flora of the feces can be studied. Similar analyses of the sediment provided as food make it possible to study the effect of passage through the gut on the organic matter content and bacterial flora (Brinkhurst and Austin 1978). Fecal samples can also be collected from cultures oriented normally, but the feces then have to be collected from the artificial sediment surface instead of being allowed to drop into a container. Both methods are potentially liable to error and cultures must be handled very carefully. The behavior of the worms is affected by changes in temperature and light and by physical disturbance. The results derived from cultures maintained in both orientations have usually shown that the worms are not affected by being inverted, and that, if either method is subject to error in determining the amount of fecal material produced, these errors cancel out. Only Kaster *et al.*

(1984) suggested that worms in the normal position produce more feces (45–110%) than those in inverted cultures. Such obvious differences would have been detectable in earlier studies if they had existed. Defecation rate seems to be very difficult to determine. The nutritional value of the food provided may decline as the experiment proceeds, and this may cause an increase in the amount of sediment processed in order to feed the colony. Experimental results are affected by periods of acclimation to the artificial sediments, which may be as long as a month. Two supposedly identical cultures maintained side by side may behave quite differently (Appleby and Brinkhurst 1970). Defecation rates may be used in studies of worm production when associated with estimates of organic matter content and calorific value of the food and feces. They may also be used in studies of sedimentation and bioperturbation of sedimentary layers in freshwater ecosystems. The lumbriculid *Stylodrilus heringianus* builds up large populations in many regions of the St. Lawrence Great Lakes, where it behaves in the same way as a tubificid. Krezoski *et al.* (1984) used gamma spectroscopy methods to measure the rates at which a layer of sediment, labeled with radioactive cesium, and overlying water, labeled with radioactive sodium, were transported by the burrowing activity of worms and other benthos. The results of these and similar studies all support the idea that the superficial sediment layers in lakes are thoroughly mixed by benthic organisms, and that solute transport across the mud–water interface is greatly enhanced by them. Worms are probably the most important animals involved in this because of the depths to which they penetrate the sediment, the large populations that they build up, and their ceaseless activity. Gardner *et al.* (1981) showed that invertebrates could be responsible for most of the phosphorus released from aerobic sediments in Lake Michigan. Chatarpaul *et al.* (1979) showed that worms accelerated nitrogen loss from coarse stream sediments, indicating their importance in rivers as well as lakes. There is now a growing realization that the sediments are not just the final resting place for organic matter and contaminants, but that they may become a source for them (Karlckhoff and Morris 1985).

D. Evolutionary Relationships

There are contradictory ideas about the evolution of any biological group, many of which are due to fundamental differences in the methods used. Even among those who employ cladistic methods and those who agree with at least the basic concepts of Hennig, there are strongly held opinions about the way in which relationships should be determined. The following account is a personal one, based on experience with most of the families of aquatic oligochaetes, an exposure to marine and freshwater worms worldwide, and some familiarity with cladistics. This is not said in order to validate this particular view, but to help the reader identify the biases it inevitably includes. At least it should not suffer from the geographic or taxonomic parochialism that tends to cause taxonomic inflation—the tendency to elevate the taxa we are most interested in to the highest level possible.

We should perhaps consider what the original worm looked like, rather than asking whether polychaetes or oligochaetes came first. It seems quite reasonable to suppose that an animal looking a lot like a very simple earthworm, but with a reproductive system that released eggs and sperm freely into the environment provides a common ancestor to both groups (Fig. 12.5). There has been some support for this idea among polychaete biologists, but in general, they seem to have difficulty deciding on relationships above the family level and on what constitutes a primitive worm. The "simplified earthworm," with two chaetae per bundle and no parapodia is adopted following the ideas of Clark (1964) regarding the evolution of segments and coelom. There is, of course, no reason to suppose that every group evolved in the ocean. The teleost fish provide a good example of a group that may well have undergone a very large diversification in freshwater before re-invading the ocean, so that the existence of a large number of marine Tubificidae is not prima facie evidence for their antiquity, even within that family, though the possibility should not be excluded. All possible combinations of the reproductive system segmental arrangement within the oligochaetes (and other Clitellata) can be derived from a worm with four genital segments, two with paired testes in each, followed by two with paired ovaries. The original position of this series, termed GI–GIV (Brinkhurst 1982), seems to have been in X–XIII. The organisms with this arrangement that exist today are found in the Haplotaxidae, with only one or two exceptions that are thought to be primitive members of groups descended from that family directly (Fig. 12.6). There are some exceptions that have been explained elsewhere, but only the barest outline will be presented here. [Challenges to this largely Austral group as being ancestral come from the fact that *Haplotaxis* itself is a highly modified genus of predatory worms (see the taxonomic key).] It is then thought that a number of families have quite independently lost the anterior testes (GI) and the posterior ovaries (GIV), making copulation easier (there are fewer pores to bring into contact with

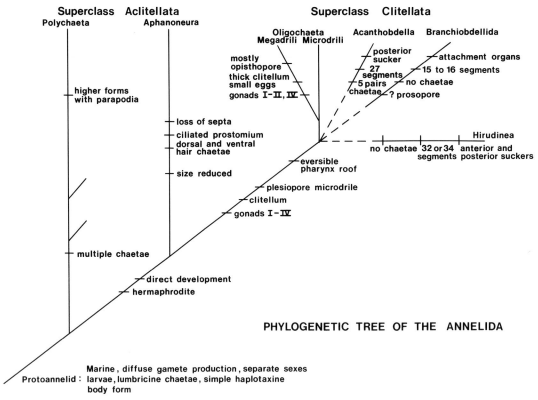

Figure 12.5 Phylogenetic tree of the Annelida. This diagram represents an attempt to outline major (potential) monophylies and the apomorphic character states that define them (solid bars across lines). Note the tentative positions for the leeches, branchiobdellidans and *Acanthobdella,* shown rotated (for convenience) in Fig. 12.6, following. The taxa below the superclass headings are probably best considered to be classes at present (hence Acanthobdellida would apply, although this monotypic genus is probably an aberrant leech) but relative rankings are currently under investigation. Only the most general concepts are included.

each other). This happens because the development of egg and sperm sacs enables more gametes to be stored in the coelom from a single pair of gonads than could be achieved in a single segment when the sacs were small or undeveloped. Similarly, the evolution of multiple chaetae per bundle seems to have arisen independently a number of times, most notably in the Enchytraeidae, Tubificidae, and Naididae. In the former, most species have two chaetae per bundle, a few specialists have no chaetae in some or all bundles (another common but independent evolutionary step found in several families), but some have several chaetae. These are all simple-pointed, however, which seems to be correlated with the fact that most members of this family are at best only semiaquatic if not totally terrestrial. Chaetae with bifid tips are a hallmark of aquatic oligochaetes, as are more complex pectinate and palmate chaetae and their derivatives, hair chaetae. The hair chaetae of the Phreodrilidae (a small, southern hemisphere family) seem to have evolved independently of those of the Naididae and Tubificidae. These are the two most closely related families of aquatic oligochaetes, and they are apparently the most highly

evolved in many respects. The Naididae exhibit several adaptations to an aquatic life as opposed to one restricted to burrowing in moist sediments. They have simple eyes, a few have specialized chaetae in a few anterior segments, and some can swim in a very simple way. The predominance of asexual reproduction enables them to exploit a more changeable environment than the sediments, and in some as yet unidentified way has enabled most of them to become very widely distributed. Their diet now includes algae as well as bacteria, and we can assume that they can take up macromolecules through the body wall as food as well as the freshwater and marine tubificids. The male reproductive system has become progressively more complex through these groups. In the Enchytraeidae, the male ducts may end in what appear to be more glandular than muscular structures, which are not yet identifiable as atria. The tubificids have muscular atria and the higher forms have stalked prostate glands, elaborate penes, and either penis sheaths, penial chaetae, or spermathecal chaetae.

The Lumbriculidae are more of a puzzle. They have usually retained both testicular segments, but

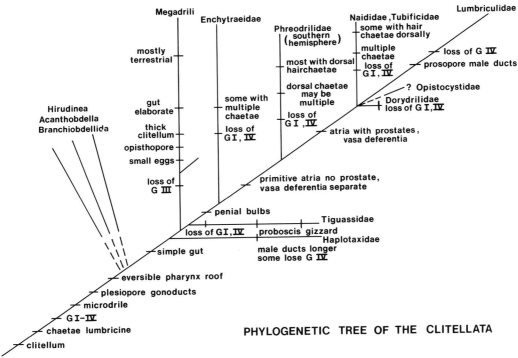

Figure 12.6 Evolution of the Clitellata. As in Fig. 12.5, general concepts are included. The three basal taxa, including Hirudinea, are not supposed to be more primitive than the oligochaetes. Their position is conjectural, and represents a simple rotation of that shown in Fig. 12.5 (for convenience). Tiguassidae is a monotypic family for a single species of *Tiguassu,* a South American genus.

have lost the second ovaries (GIV). Many advanced forms secondarily loose GI, and a few parthenogenetic forms have an irregular, asymmetrical, or even exaggerated number of gonads. The chaetae are restricted to two per bundle, but these are usually bifid. Some biologists see the paired chaetae as evidence of a close relationship between the lumbriculids and the earthworms, suggesting that the latter are the derivatives of these aquatic forms. This puts an enormous weight on just one character, but these people also regard hair chaetae as being retained from a polychaete ancestor. The difficulty with relating the lumbriculids to the earthworms is that they have a unique way of reducing the number of male pores. They have their atria in GII and both pairs of male ducts drain into them. The prostate glands consist of a diffuse covering of gland cells and none of them have developed stalked prostates. Cuticular penis sheaths are very rare; genital chaetae are limited to a single species. Too much has been made of the apparent diversity of reproductive segments in this family. While the variation is greater than that of other families, the most common form does seem to be the ancestral one (Brinkhurst 1989). While the aquatic oligochaetes closest to their presumed ancestors seem to be found in Australia and New Zealand, not unlike many other freshwater invertebrate groups, other families are widely distributed. Ac-

cording to a cladistic analysis of some Enchytraeidae (Coates in 1989), South America seemed to contain the basal forms. In the Naididae, it is the northern Stylarinae that appear to be the oldest (Nemec and Brinkhurst 1987). The Lumbriculidae inhabit the northern hemisphere, with a very high proportion of them being limited to, or extending eastward from Lake Baikal in Siberia. The European *Cookidrilus* seems to most resemble the presumed ancestor, with the widespread genera *Stylodrilus* and *Trichodrilus* being quite generalized (Brinkhurst 1989). There has yet to be a detailed analysis of the more cosmopolitan Tubificidae, but there is a greater diversity of genera in the northern hemisphere, with very few restricted to the south (as *Antipodrilus*) and very few adapted to the subtropical and tropical regions, apart from *Branchiura,* and perhaps *Aulodrilus.*

The authors of this chapter are currently trying to assess the relationships between the clitellate taxa Oligochaeta, Branchiobdellida, Hirudinea, and Acanthobdellida. In a preliminary revision (Brinkhurst and Gelder 1989), it was shown that the supposed link between the predatory lumbriculid worm *Agriodrilus* from Lake Baikal (USSR) and the nonoligochaete taxa in this group is purely circumstantial, made even clearer by the recent discovery of a parallel evolution in the North American *Phagodri-*

lus. Without presenting all of the data here, it now looks as if the Branchiobdellida and Lumbriculidae may share a common ancestor, both having two pairs of male ducts draining through atria in GII, unless this character proves to be a convergence. It remains to be seen if the male ducts of the Hirudinea and the Acanthobdellida are at all related to what we might call the semiprosopore state of the male ducts in the lumbriculids (only one pair of male ducts are now thought to be in their testicular segment in the ancestor of the family according to a cladistic analysis). The extreme modification of the male ducts in the two leech groups, in which the testes lie in sacs behind the two successive segments bearing male and female pores is considered to be derived from a system with the testes in front of the ovaries, as in the oligochaetes. With the loss of the coelom, the testes are supposed to be transferred into the surviving sperm sacs as a series of small gonads replacing a single original pair. Complex male ducts have been developed but these are in the form of a continuous tube in which the sperm are shed internally. These ducts need to be re-examined for any trace of relict sperm funnels, to see if there is any evidence that they were ever prosoporous like the lumbriculids. The acanthobdellids, represented by one species with one probable synonym, have four pairs of chaetae in a few anterior segments including the oral segment. There is no trace of a prostomium, but all this is being investigated from fresh material. *Acanthobdella* also retains or possibly recovers the coelom, which is lost in leeches, but the testes still lie within the sperm ducts. Two intriguing aspects are the pattern of eyes and the unique sperm-receiving device shared by this genus and the piscicolid leeches. Other revised information about the branchiobdellidans, discussed later, also weakens the argument for a relationship between the leech groups and them, which was based on a very superficial examination of the detailed facts available. Textbook accounts in this and many other areas of supposed evolutionary relationships should be read with caution, especially if they give no clear indication of the theoretical basis on which the opinions are based.

IV. COLLECTING, REARING, AND PREPARATION OF OLIGOCHAETES FOR IDENTIFICATION

A. Collection

Oligochaetes may be collected using any of the standard array of samplers from dip nets to grabs and core tubes. Strictly quantitative samples from soft sediments are best obtained with narrow core tubes without core retainers. Using core tube lengths appropriate to the type of substrate can save a great deal of frustration, but most people seem to try to work with a single set of tubes. Soft sediments can be washed through screens and sorted in the laboratory, preferably after being preserved in their entirety in the field. If cores are used, the volume to be carried back from the field may be quite small. Field sorting is relatively inaccurate, and small specimens may well be missed. The best preservative is 10% buffered formalin (4% formaldehyde solution), but it is easier to carry 40% in the field and add this to an aqueous sample to approximate 10% in the whole sample. Addition of rose bengal or phloxine B in very small amounts aids sorting, and the stain can be washed out of the worms when they are transferred to 70% alcohol for storage. Sorting should be done with stereomicroscopes under good lighting. Care should be taken to avoid diluting the alcohol when transferring worms into the storage vial. Many North American collections are sent for identification with the worms half decomposed because the original alcohol has not been drained and replaced with 70%. Storage can be quite permanent in patent lip vials with neoprene stoppers. The stoppers tend to swell and seal the vial against alcohol loss. We have some in our laboratories that have not needed attention for 20 years. If these vials and stoppers are not available, vials must be sealed with tape, wax, or parafilm.

If the collector is not taking quantitative samples for ecological or environmental studies and is working in a remote or unusual location, then it may be worth the effort of sorting and preserving some specimens while in the field if they are readily visible. Putting the sample in an enamel dish and allowing everything to settle will reveal even small worms as they begin to move. Carrying whole samples back to the laboratory to sort them while the worms are alive can cause selective mortalities, some species surviving elevated temperatures and lowered oxygen concentration better than others (and the latter always seem to be the more interesting species from a taxonomic viewpoint). Live worms can be sorted from sediment and detritus by spreading clean sand over the sample or putting the sample on a screen set over clean water. The worms will then actively migrate into the sand or water. This can be useful in samples full of plant fragments.

Samples taken with dip nets or other methods that include coarse substrates can be processed in the field so as to reduce the volume and weight of the sample. Repeated, careful elutriation into a screen, screen bucket, or back through the original dip net can be used, imitating a gold-panning technique. The amount of vigor used can be selected in order to separate enough of the fine fraction including all the specimens from the heavy material. The coarse residues should, of course, be carefully checked for large, heavy, or attached organisms, or specimens trapped in empty shells and other unusual situations.

Screen sizes used to remove and wash sediments should be consistent. There seems to be little value in using a smaller screen size in the laboratory than the screens used in the field (if any) unless shrinkage of the specimens after fixation is a possibility.

B. Rearing Oligochaetes

Culture of worms can be as simple as placing the substrate and the worms in an aerated aquarium, preferably where temperatures can be kept at appropriate levels, usually between 4° and 10°C depending on the original habitat. Small cultures used for laboratory experiments can be maintained in shallow vessels half filled with sterilized lake sand. Filtered lake water is added every two weeks and frozen lettuce strips are buried in the sand to provide bacteria as food. Mass culture to provide fish food can utilize manure to feed the worms. Marian and Pandian (1984) describe raising 7.5 mg *Tubifex tubifex* every 42 days on a mixture of 75% cow dung and 25% fine sand, maintaining the oxygen level at 3 mg liter^{-1}. Fresh cow dung is added every four days, and harvesting is supposedly best done at night. Aston (1984) cultured *Branchiura sowerbyi* at 25°C. Leitz (1987) reviewed attempts to mass culture *Lumbriculus variegatus*. Production rates of 15 kg m^2 and population doubling times of 11–42 days are reported. The use of live worms as fish food has many advantages over other diets. Most of the commercially bought live worms are now *Lumbriculus* (called the black worm in commerce) rather than *Limnodrilus*.

C. Preparation for Identification

Many biologists believe that these animals are hard to identify. It is worth indicating that, as they are hermaphrodites, there is no difficulty with keys for males and females, and there are no larval forms. About 60% of the species can be identified from superficial somatic characters, but the rest require mature specimens. It is, of course, always preferable to confirm identities from mature specimens wherever possible, but it is not essential. Those used to identifying organisms with stereomicroscopes may be bothered by the need to make slide mounts, but simple methods can be used in most instances. The serious taxonomist will want to make more elaborate preparations and may want to section

specimens to check details, but this is not necessary for routine identification. There are only about 150 freshwater oligochaetes and a small number of branchiobdellidans in North America, and many of these are rare and local. Recent keys have been written with the beginner in mind, and so they emphasize external or readily visible internal characters. Such keys work well if they proceed directly to the specific level, but it is a little more difficult to write keys to the genera, as preferred in this volume. Generic characteristics are mostly detailed aspects of the soft parts of the male reproductive structures. These may be difficult to see on rapidly made preparations in which the soft parts are dissolved. The keys presented here do not always proceed directly to genera for this reason, but we have managed to write keys involving externally visible characteristics. Some reference to geographic distribution is made in the keys but this has been done only where the information available seems reliable. Fully illustrated keys with supporting descriptions of all the species are available elsewhere (Brinkhurst 1986).

Once the worms are preserved in alcohol, they must be prepared for study with a compound microscope. Examination of specimens from remote or unusual locations should begin with a study of the external characteristics, such as position of the gonopores and other obvious features. For routine identifications of specimens from ordinary localities, very little benefit can be derived from this step in the process. In this case, the worms can either be immersed in a temporary mounting medium by replacing the alcohol with Amman's lactophenol, or they may be washed with water and then placed in separate drops of lactophenol on slides. It is usually possible to mount five worms under each coverslip, two coverslips per slide, unless the worms are very large. These wet mounts should be set aside for 1–2 days on stacking trays. Amman's medium is quite corrosive, so it should be handled with care and any spillage should be wiped off microscope stages immediately. It is made up with 20% phenol, 20% lactic acid, 20% water, and 40% glycerine. It is best stored in dark bottles; however, it is hygroscopic and it does not have a good shelf-life. Somewhat more permanent mounts can be made with media such as CMCP and polyvinyl lactophenol. Such slides should usually be sealed with a ring of material (Glyceel, nail varnish, etc.) around the edges of the coverslips to prevent the media from drying up.

None of these methods produce satisfactory permanent slides, nor are they adequate for working with the Enchytraeidae or the marine Tubificidae, in which internal anatomy must be seen. For this type of material, permanent whole mounts using stained and dehydrated worms mounted in Canada Balsam or one of its more recent substitutes are used. To prevent the worms from becoming brittle, they should be cleared in methyl salicylate rather than xylol. This will enable advanced students to try cutting the separated genital region of each worm in two (sagitally) or even dissecting out the male ducts when necessary. This should be limited to attempts to identify new taxa to the generic level, when details of the male reproductive system need to be established. Serial sections (usually sagital longitudinal) are also employed in taxonomic work, but not in routine identification of well-known species. It is important to be knowledgable about the expected locations of the various organs before attempting dissection or interpretation of serial sections.

The procedure for examining a worm on a slide is as follows. Check the prostomium for a proboscis or sense organ. Next, examine the ventral chaetae of the first two or three bundles and determine the form of the dorsal bundles and where they begin (usually in II or between IV and VI). Establish the number and form of these chaetae (bifid, pectinate, hair chaetae, etc.). Determine the relative lengths of the teeth. As these are illustrated by taxonomists with the ectal or outer end of the whole chaeta toward the top edge or corner of the figure, the upper tooth is always drawn lying above the lower tooth. When viewing a whole-mounted specimen, the chaetae are often lying flat on the body wall and so no distinction of position can be made; therefore, always visualize a chaeta in the upright position. Care should be taken to examine several chaetae from an exactly lateral aspect, because slight deviations can produce apparent distortion of the relative lengths of the teeth. Measurements often betray the bias of the eye

of the observer and are recommended where an eyepiece micrometer scale is available. The worm should then be searched for genital characters. Carefully check the appropriate segments (usually X–XI in a tubificid, for example) to see if the ventral chaetae are modified. If the dorsal and ventral chaetae are alike, make sure that you have examined at least three bundles in those segments to ensure a ventral bundle has been examined. Check the penial segment to see if there is a penis sheath (they should be paired of course) even if it is thin and inconspicuous. Check the posterior chaetae and the body itself for gills or any other special features. Then proceed to check the key. If the worm has distinct features, like single dorsal hairs with single simple needles (these beginning in V) and the anterior ventral (II–V) and posterior ventral (VI onward) bundles have chaetae that differ in length, width, and shape, it will rapidly become possible to skip the key altogether and proceed straight to *Nais* or *Dero* depending on the presence of gills. Many such snap identifications to major groups or to genera can be made with a little practice.

When instructing classes on how to identify worms, the most important thing to teach is good microscopy. Lenses must be clean, not covered with dried mounting medium. Illumination should be parallel, and the condenser should be correctly focussed and then left alone. Light levels should be adjusted with the iris diaphragm at every lens change, and this should never be done by raising and lowering the condenser. Nothing will compensate for very bad, misaligned microscopes; however, miracles were achieved in the past by people with simple systems and a thorough knowledge of their use. Chaetae and penis sheaths have a refractive index

Table 12.1 Some Characteristics of Microdrile Families

Character	Haplotaxidae	Lambriculidae	Naididae	Tubificidae	Enchytraeidae	Opistocystidae
Chaetal number	Single	Paired	Multiple	Multiple	Paired/several	Multiple
Bifid chaetae	Absent	Present/absent	Present	Present	Absent[a]	Present
Dorsal hair chaetae	Absent	Absent	Absent/present	Absent/present	Absent	Present
Proboscis	Absent	Present/absent	Present/absent	Absent	Absent	Present
Eye spots	Absent	Absent	Present/absent	Absent	Absent	Absent
Gills	Absent	Absent	Present in *Dero*	Present in *Branchiara*	Absent	Absent
Testes (pairs)	2	1–2	1	1	1	1
Ovaries	1–2	1	1	1	1	1
Male ducts	Plesiopore	Prosopore	Plesiopore	Plesiopore	Plesiopore	Plesiopore
Atria	Absent	Present	Present	Present	Absent	Absent
Gonadal segments (majority)	X–XII/XIII	IX–XI	IV/V, V/VI, or VI/VII	X/XI	XI/XII	XI–XII

[a] Apart from anterior chaetae of *Barbidrilus*.

that does not differ much from the background, and too much light makes them impossible to see. Oil immersion lenses must be used to see pectinations on chaetae in tubificids (but with practice they can be identified without), and they must also be used to determine the form of the needles (dorsal chaetae) in naidids.

V. TAXONOMIC KEYS FOR OLIGOCHAETA

The first key is written for the identification of families. Where there is only a single taxon in the North American faunae in a given family, that taxon is identified at that point and not dealt with subsequently. Three families are treated in subsequent keys. These keys proceed to suprageneric group, genus, or infrageneric group, but individual species are named where the group is monotypic. To do otherwise would involve using internal organ systems usually difficult to see without dissecting or sectioning. Most of the information in this key is summarized in Table 12.1.

A. Taxonomic Key to Families of Freshwater Oligochaeta and Aphanoneura

1a.	Minute worms, 1–2 mm or chains of animals up to 10 mm; hair chaetae in both dorsal and ventral bundles; worms move in a gliding motion using cilia; body wall with colored or refractile epidermal glands; no eversible pharyngeal pad; no clitellum, only ventral copulatory glands in rare mature specimens; nervous system ladder-like (Fig. 12.7) .... Aphanoneura	Aeolosomatidae	
	[Key to world species in Brinkhurst and Jamieson (1971).]		
1b.	Larger worms; hair chaetae restricted to dorsal bundles or absent; worms move by alternate contraction of longitudinal and circular muscles; no colored epidermal glands; eversible pharyngeal roof present (except in aquatic earthworms); clitellum present in mature worms; nervous system fused ventrally .... Oligochaeta	2	
2a(1b).	Ventral chaetae large, sickle-shaped, single (i.e., two per segment); dorsal chaetae small, straight, often missing from some or all segments; prostomium very long, furrowed, no proboscis; mouth large, muscular pharynx present; body elongate, slender, resembling a gordian worm (Fig. 12.8) .... Haplotaxidae	*Haplotaxis*	
	[Worms in this genus resemble the European *H. gordioides* (Hartman, 1821), but mature specimens have yet to be described from North America. There may be several species (now regarded as synonyms) differing by the number and distribution of dorsal chaetae.]		
2b.	All chaetae paired, or multiple in a bundle, unless partially or totally absent; prostomium short, often conical, without a transverse furrow but with or without a proboscis; body form very rarely elongate and slender 3		
3a(2b).	Chaetae all paired from II or, if more numerous, all simple-pointed 4		
3b.	Chaetae more than two per bundle, usually bifid, sometimes with pectinate and hair chaetae dorsally, simple-pointed chaetae rare (limited to a few anterior ventral bundles, or single needles with the hairs dorsally) 7		
4a(3a).	Chaetae paired, bifid, with small to rudimentary upper teeth (Fig. 12.9) .... Lumbriculidae (in part)	(Section IV.B)	
4b.	Chaetae simple-pointed 5;		
5a(4b).	Thick-bodied worms with clitellum several segments behind the gonopores (XII–XIV) .... *Megadrili*		
	[Mostly terrestrial but some aquatic species. Most aquatic species are members of the Sparganophilidae (*Sparganophilus*) or the Lumbricidae (*Eiseniella*).]		
5b.	Thin-bodied worms with clitellum one cell layer thick and in region of gonopores (X–XII or further forward) 6		

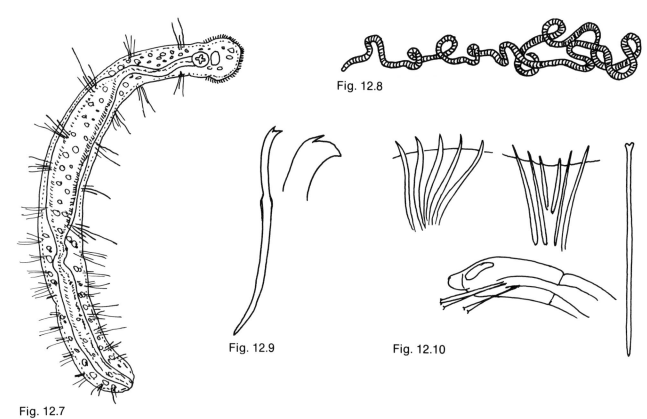

Fig. 12.7

Fig. 12.8

Fig. 12.9

Fig. 12.10

Figure 12.7. *Aeolosoma* sp. Note ciliated prostomium, lack of septa, simple pharynx, dorsal and ventral hair chaetae, and colored body wall inclusions (irregular circles). *Figure 12.8.* *Haplotaxis* sp. Whole specimens can be even longer in relation to the breadth. *Figure 12.9* Bifid chaetae, Lumbriculidae. *Figure 12.10* Enchytraeid chaetae. These may be straight, curved, with or without nodulae, and may differ in length within a bundle. All chaetae are usually identical. The anterior end of *Barbidrilus* (below) has ventral bundles of strange rod-like chaetae with bifid tips only in II and III, shown on the right.

6a(5b).		Larger worms (difficult to get under a coverslip) with sigmoid, nodulate chaetae; proboscis present or absent; spermathecae in or adjacent to the gonadal segments, male pores between VIII and X .... Lumbriculidae (in part)	(Section V.B)
6b.		Smaller worms with chaetae often straight, commonly not nodulate, sometimes missing from some if not all segments, (in *Barbidrilus* Loden and Locy, 1981 with uniquely forked chaetae in II–III ventrally, others missing). No proboscis. Spermathecae open in V, male pores in XII (Figs. 12.2 and 12.10) .... Enchytraeidae	
7a(3b).		Worms (1.7–3.0 mm) with posterior end bearing one median and two lateral processes. Genital region with testes in XI, ovaries and atria in XII, spermathecae in XIII (Fig. 12.11) .... Opistocystidae	
		[The only definite North American record is for *Crustipellis tribranchiata* (Harman, 1970), restricted to Louisiana, Mississippi, Florida, and North Carolina. The tail processes make this obvious (Harman and Loden 1978).]	
7b.		Posterior end naked, or with gills, or with two lateral processes plus gills, no median process (Fig. 12.26); spermathecae in the testicular segment, atria in the ovarian segment, these being in X–XI or further forward (Fig. 12.2) 8	
8a(7b).		Length usually < 1 cm; hair chaetae usually present in dorsal bundles but absent in some genera; hair chaetae commonly associated with simple-pointed or bifid chaetae rather than the rarer palmate or pectinate chaetae, these dorsal chaetae (needles) very often differ from the	

ventrals in form (Fig. 12.12); dorsal chaetae may begin behind II, often in V or VI, sometimes elsewhere; dorsal chaetae often limited to one or two needles, one or two hairs per bundle; ventral bundles with numerous bifid chaetae, those of II or even II–V often differ in form and thickness from the rest; asexual reproduction by budding forms chains of individuals; mature specimens have spermathecae in IV, V, or VI with the male pores one segment behind them (Fig. 12.2); eye spots may be present; gills may surround the anus (Fig. 12.26) and the prostomium may bear a proboscis (Fig. 12.23) .... Naididae (Section V.C)

8b. Length usually > 1 cm, width usually 0.5–1.0 mm; when hair chaetae are present dorsally, they are usually accompanied by pectinate chaetae that closely resemble the ventral chaetae, apart from having intermediate teeth (Fig. 12.13); when hair chaetae are absent dorsally, all bundles usually contain similar bifid chaetae; dorsal chaetae begin in II, normally several per bundle; reproduction normally sexual, rarely by fragmentation; spermathecal pores normally on X, male pores on XI (Fig. 12.2); no eyes or proboscis; no gills around anus, single dorsal and ventral gill filaments on some posterior segments of one species (Fig. 12.30) .... Tubificidae (Section V.D)

B. Taxonomic Key to Genera and Selected Species of Freshwater Lumbriculidae

1a.   Prostomium with a proboscis (Figs. 12.14 and 12.15) 2
1b.   Prostomium without a proboscis 6

2a(1a). Chaetae bifid, at least in anterior segments 3
2b.   Chaetae all simple-pointed 4

3a(2a). Proboscis elongate; chaetae bifid in anterior segments (Fig. 12.14) .... *Kincaidiana hexatheca* Altman, 1936

[This species is limited to the Pacific northwest.]

3b.   Proboscis short; all chaetae bifid, but with colateral teeth (set side by side) (Fig. 12.15) .... *Rhynchelmis brooksi* Holmquist, 1976

[Known only from northern Alaska.]

Fig. 12.11

Fig. 12.12

Fig. 12.13

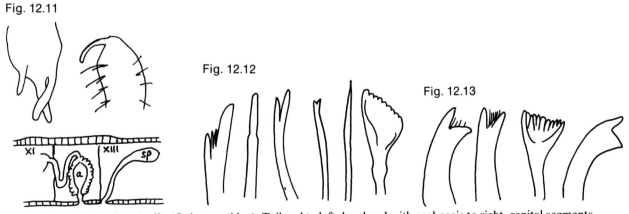

Figure 12.11 *Crustipellis* (Opistocystidae). Tail end to left, head end with proboscis to right, genital segments below (a, atrium, sp, spermatheca, both of left side). ***Figure 12.12*** Various forms of dorsal chaetae, needles, of Naididae. These usually differ from the more normally bifid ventrals. ***Figure 12.13*** Dorsal chaetae of Tubificidae. Pectinates, on the left, differ relatively little from a bifid, shown on the right. Ventral bundles usually consist of bifids, sometimes with minute pectinations in those species with pectinate dorsals. Pectinates are usually accompanied by hairs in the dorsal bundles, rarely are bifid dorsals accompanied by hairs. All chaetae may be bifid.

4a(2b).   Cave-dwelling species with short proboscis .... *Trichodrilus allegheniensis* Cook, 1971
            [Known only from Tennessee.]

4b.   Surface water species with long proboscis 5

5a(4b).   Atria of male reproductive system with spiral muscles (Fig. 12.16) .... *Eclipidrilus* (in part)
            [This genus was reviewed by Wassell (1984). The species with
            probosces are known mostly from the southeastern United States,
            with one questionable record from Montana.]

5b.   Atria without spiral muscles .... *Rhynchelmis* (in part)
            [Four, possibly fewer, species distributed from California–Nevada to
            Alaska and Northwest Territories of Canada.]

6a(1b).   Chaetae bifid 7
6b.   Chaetae simple-pointed 8

7a(6a).   Elongate (to 100 mm or more) slender worms, the front end often green
            in life, the rest dark red to black; elaborately branched blood vessels
            laterally in the body wall of posterior segments (Fig. 12.17); reproduces
            sexually and asexually (fragmentation with or without encystment); no
            permanent everted penes .... *Lumbriculus*
            [One widely distributed species, *Lumbriculus variegatus* (Muller
            1774), inhabits shallow water, where it floats the tail in the air–water

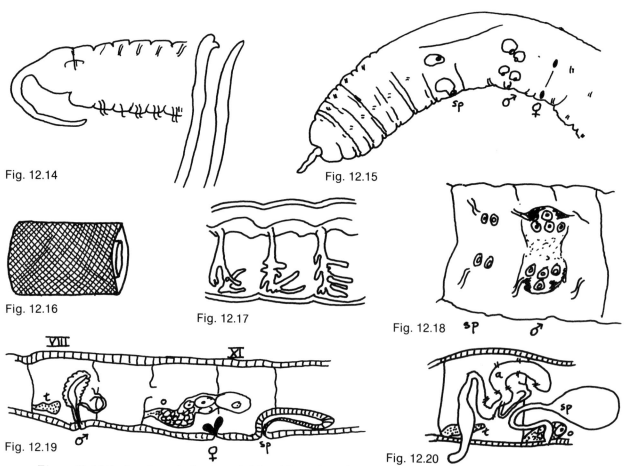

Fig. 12.14

Fig. 12.15

Fig. 12.16

Fig. 12.17

Fig. 12.18   sp   ♂

Fig. 12.19   ♂   ♀   sp

Fig. 12.20

Figure 12.14 Proboscis and chaetae of *Kincaidiana hexatheca*. **Figure 12.15** Proboscis of *Rhynchelmis brooksi*. **Figure 12.16** Spiral atrial muscles, *Eclipidrilus*. **Figure 12.17** Blood vessels of the lateral body wall of posterior segments, *Lumbriculus*. **Figure 12.18** Ventral view of segments IX (to left) and X showing unique gonopores of *Spelaedrilus* (sp, spermathecal pores; male symbol, male pores, 4 and 8 of each, respectively). **Figure 12.19** Reproductive system of *Styloscolex*, anterior to left, left side shown. Note testis (t) and atrium in VIII, ovary (o) and female funnel in X, and spermatheca (sp) in XI. **Figure 12.20** Reproductive system, *Kincaidiana freidris*, anterior to left, organs of left side shown. Note testis (t) and atrium (a) in VIII, prostate gland stalks shown as well as male duct, ovary (o), and spermatheca (sp) in IX.

interface. Several other possible taxa, mostly from Alaska, are almost
certainly variants of this (see *Lumbriculus* and *Thinodrilus,* in
Holmquist 1976).]

7b.   Short (25–40 mm), tapering worms, pale to white color; blood vessels in
the posterior lateral body wall with short, unbranched lateral
diverticulae; reproduction sexual; permanently everted soft penes on X
close together near the midline on mature specimens ...... *Stylodrilus heringianus* Claparede, 1862
[Widespread in the St. Lawrence Great Lakes and Canada, possibly a
European introduction.]

8a(6b).   Groundwater species 9

8b.   Surface water species 11

9a(8a).   Four pairs of male pores on X, two pairs of spermathecal pores on IX (Fig. 12.18) ..... *Spelaedrilus multiporus* Cook, 1975
[Known only from a cave in Virginia.]

9b.   Male pores paired on X, spermathecal pores paired in either IX, XI, or XI and XII 10

10a(9b).   Spermathecal pores paired on IX .... *Stylodrilus beatiei* Cook, 1975

10b.   Spermathecal pores paired in XI or XI and XII .... *Trichodrilus* (in part)
[From caves in West Virginia, Illinois, Washington.]

11a(8b).   Male pores paired in VIII 12

11b.   Male pores paired in X 13

12a(11a).   Spermathecal pores in XI, ovaries in X (Fig. 12.19) .... *Styloscolex opisthothecus* Sokol'skaya, 1969
[This Asian species is restricted to Northern Alaska.]

12b.   Spermathecal pores paired in IX, ovaries in IX (Fig. 12.20) .... *Kincaidiana freidris* Cook, 1966
[This species may not be congeneric with *K. hexatheca,* but cladistic
analyses suggest they are quite closely related. Both are Pacific
Northwest species, this one from California.]

13a(11b).   Spermathecal pores paired in IX .... *Stylodrilus sovaliki* (Holmquist, 1976)
[This is another species restricted to Alaska, but it closely resembles
its European congener *S. absoloni* (Hrabe, 1970).]

13b.   Spermathecal pores either single median in VIII–IX or in IX, or paired in IX ... *Eclipidrilus* (in part)
[One species, (*E. frigidus* Eisen, 1881) is from Idaho and California,
the other two, (*E. lacustris* Verrill, 1871 and *E. fontanus* Wassell,
1984) are from Ontario, Quebec, New York, and Pennsylvania.]

C. Taxonomic Key to Genera and Selected Species of Freshwater Naididae

1a.   No dorsal chaetae present (in some rare specimens the dorsal chaetae
are recovered, but these can be recognized because of the absence of
ventral chaetae in III–V and the presence of an enlarged pharynx and
reduced prostomium associated with a predatory habit) .... *Chaetogaster*

1b.   Dorsal chaetae present (rare in *Ophidonais*) 2

2a(1b).   Dorsal chaetal bundles without hair chaetae 3

2b.   Dorsal chaetal bundles with hair chaetae (Fig. 12.2) 7

3a(2a).   Dorsal chaetae begin in II or III 4

3b.   Dorsal chaetae begin in V or VI 5

4a(3a).   Dorsal chaetae begin in II; no gap between chaetal bundles of III and
IV .... *Homochaeta naidina* Bretscher, 1896

4b.   Dorsal chaetae begin in III; gap between chaetal bundles of III and IV
(Fig. 12.21) .... *Amphichaeta*

5a(3b).   Dorsal chaetae begin in V; estuarine species .... *Paranais*

5b.   Dorsal chaetae begin in VI; freshwater species 6

6a(5b).   Dorsal chaetae stout, solitary, bluntly simple-pointed or notched at the
outer end (Fig. 12.22) .... *Ophidonais serpentina* (Muller, 1773)

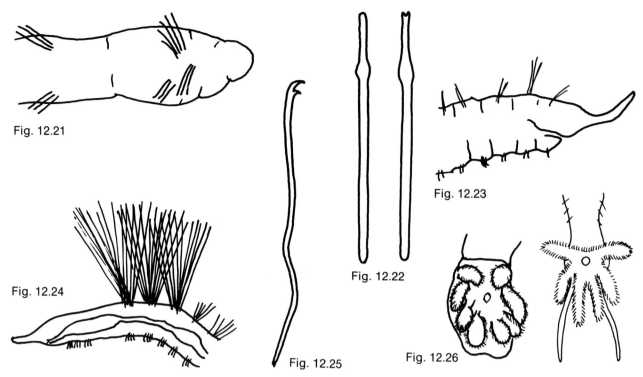

Fig. 12.21

Fig. 12.24

Fig. 12.22

Fig. 12.23

Fig. 12.25

Fig. 12.26

Figure 12.21 Anterior end (to right), *Amphichaeta,* showing dorsal chaetae beginning in III, and gap between chaetae of III and IV, particularly pronounced in this species, but noticeable throughout the genus. *Figure 12.22* Dorsal chaetae, *Ophidonais.* These are single, or even absent in many segments. *Figure 12.23* Proboscis on the prostomium of a naidid, this one with dorsal chaetae from II, but this varies (see key). *Figure 12.24* Dorsal chaetae of *Ripistes* begin in VI, with giant hairs in three segments. *Figure 12.25* Ventral chaeta of *Stylaria lacustris.* *Figure 12.26* Perianal gills of *Dero. Dero (Aulophorus)* with long palps (right) and *Dero (Dero)* without palps (left).

6b.	Dorsal chaetae curved, with bifid tips, 2–4 per bundle. [See also *Piguetiella,* in which hair chaetae may be missing in most, if not all, dorsal bundles; some *Uncinais* specimens are said to have dorsal chaetae in V.] .... *Uncinais uncinata* (Orsted, 1842)
7a(2b).	With a proboscis on the prostomium (visible as a stump if broken off) (Fig. 12.23) 8
7b.	Without a proboscis on the prostomium 11
8a(7a).	Dorsal chaetae begin in II .... *Pristina*
8b.	Dorsal chaetae begin in VI (observe with care as ventral chaetae of IV and/or V may be missing) 9
9a(8b).	Dorsal bundles of VI–VIII with 2–16 giant hair chaetae (Fig. 12.24) .... *Ripistes parasita* (Schmidt, 1874)
9b.	No giant hair chaetae 10
10a(9b).	Ventral chaetae with a characteristic double bend (Fig. 12.25); dorsal bundles with 1–3 hairs and 3–4 shorter, hairlike needles .... *Stylaria lacustris* (Linnaeus, 1767)
10b.	Ventral chaetae slightly curved; dorsal bundles with 8–18 hairs and 9–12 hairlike needles .... *Arcteonais lomondi* (Martin, 1907)
11a(7b).	Dorsal chaetae begin in II or III 12
11b.	Dorsal chaetae begin in IV or further back 14
12a(11a).	Body wall covered with foreign matter adhering to glandular secretions .... *Stephensoniana*
12b.	Body wall naked 13
13a(12b).	Dorsal chaetae begin in II; hair chaetae in all bundles, none especially thin, either none especially elongate "or" elongate hair chaetae in III .... *Pristinella*

13b.   Dorsal chaetae either begin in III, hairs of III elongate "or" begin in II, the hairs being longest in midbody, and often missing from a number of segments, and exceptionally thin when present .... *Bratislavia*

14a(11b). Posterior end of the body with a branchial fossa, normally with gills (Fig. 12.26); some species symbionts in tree frogs, these develop gills once outside their host. [There are three subgenera, *Allodero* containing the symbiotic species, *Aulophorus* with long palps on each side of the branchial fossa, and *Dero* that lacks palps. In a strict cladistic sense, *Allodero* would not be recognized.] .... *Dero*

14b.   No gills; free-living species 15

15a(14b). Dorsal chaetae begin in XVIII–XX, each bundle with a single short hair and a robust needle .... *Haemonais waldvogeli* Bretscher, 1900

15b.   Dorsal chaetae begin in V, VI, or VII (note that ventral chaetae of IV and/or V may be missing) 16

16a(15b). Hair chaetae up to 9 per bundle, thick, with long lateral hairs (Fig. 12.27) .... *Vejdovskyella*

16b.   Hair chaetae 1–3 per bundle at most, thin with or without fine lateral hairs 17

17a(16b). One to three especially long hair chaetae in each dorsal bundle of VI, 1–2 ordinary hair chaetae in the remainder (Fig. 12.28) ..... *Slavina appendiculata* (d'Udekem, 1855)

17b.   No elongate hair chaetae 18

18a(17b). Hair chaetae very short (84–120 μm), present in only one or two bundles, some with no hair chaetae at all (see 6) .... *Piguetiella*

18b.   Hair chaetae mostly 1–2 per bundle (two species with up to 5), long (180–200 μm or more), present in all bundles from V or VI 19

19a(18b). Characteristic needle chaetae with thick, unequal teeth, sometimes clearly pectinate (Fig. 12.29); ventral chaetae change form and length slightly between V and VI .... *Allonais*

19b.   Needle chaetae simple-pointed, bifid, or faintly pectinate; either ventral chaetae of II differ from the rest "or" those of II–V differ markedly in length, width, and form from the rest 20

20a(19b). Needle chaetae bifid or faintly pectinate; ventral chaetae of II may differ from the rest .... *Specaria*

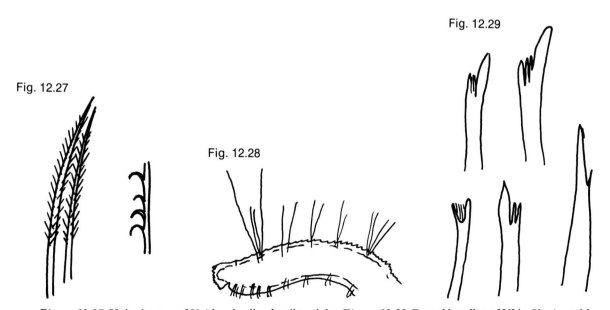

Fig. 12.29

Fig. 12.27

Fig. 12.28

Figure 12.27 Hair chaetae of *Vejdovskyella,* detail to right. **Figure 12.28** Dorsal bundles of VI in *Slavina* with elongate hairs. **Figure 12.29** Needle chaetae of various *Allonais* species.

20b.   Needle chaetae simple-pointed, bifid, or pectinate; ventral chaetae of
              II–V differ strongly in length, thickness, and form from the rest in the
              majority of species. [Species of *Dero* with the posterior end of the body
              missing may key out here because there will be no gills. Compare
              specimens to those with gills; the anterior ventral chaetae in *Dero*
              species have very long upper teeth.] .... *Nais*

D. Taxonomic Key to Genera and Selected Species of Freshwater Tubificidae

This key is designed to maximize the use of external characters. As it is primarily a key to genera, it is necessary to refer mostly to characters found on mature specimens. Immature specimens can often be identified on the basis of chaetal form alone, but then the key must proceed directly to the specific level (Brinkhurst, 1986). In this key, there are two genus groups that are difficult to separate on external characters; and where this is true, evidence from distribution patterns is introduced as an additional aid. Keys for worms of this family are traditionally written at the species level.

1a.   Single dorsal and ventral gill filaments on posterior segments (Fig.
              12.30). [Broken anterior fragments may be mistaken for *Aulodrilus
              pluriseta* (see 18b).] .... *Branchiura sowerbyi* Beddard, 1892
1b.   No gill filaments 2

2a(1b).   Body wall papillate (with projections usually covered with foreign
              matter attached by secretions) (Fig. 12.31) .... *Telmatodrilus* (in part), *Spirosperma, Quistadrilus*
              [*Telmatodrilus (Alexandrovia) onegensis* Hrabe, 1962 is recorded from
                 Alaska; the subgenus is often elevated to generic status, which may
                 prove correct.]
2b.   Body wall naked 3

3a(2b).   Spermathecal chaetae replace normal ventral chaetae in the
              spermathecal segment, very rarely in adjacent segments also (Fig.
              12.34); with or without modified penial chaetae (Figs. 12.39, 12.41,
              12.42) and cuticular penis sheaths (Fig. 12.35) 4
3b.   No modified spermathecal chaetae 8

4a(3a).   Thin, hollow-tipped spermathecal chaetae in X, similar chaetae in XI
              together with apparent cuticular penis sheaths (Fig. 12.32) .... *Haber speciosus* (Hrabe, 1931)
4b.   Spermathecal chaetae not duplicated (or, if so, on VII–VIII and very
              rarely scattered from VI–XII); with or without cuticular penis
              sheaths 5

5a(4b).   Spermathecal chaetae broad, spatulate, one each side of IX associated
              with long tubular glands; ventral chaetae of X similar or absent; several
              penial chaetae in each ventral bundle of XI with short, knobbed distal
              ends; male ducts open into large median inversion of the body wall of
              XI (Fig. 12.33) .... *Rhizodrilus lacteus* Smith, 1900
                 [Known from Havana, Illinois, and Aiken County, South Carolina.]
5b.   Without this unique set of characters combined 6

6a(5b).   Spermathecal chaetae much broader and longer than the normal ventral
              chaetae, distal ends hollow, located on X (or in one instance on VII–
              VIII or even irregularly between VI and XII); no cuticular penis
              sheaths .... *Potamothrix*
                 [Most species are restricted to the St. Lawrence Great Lakes basin or
                 extend into adjacent states; only one is widespread. These may be
                 introduced from Europe, where they are characteristic of productive
                 waters.]
6b.   Spermathecal chaetae longer than the normal ventral chaetae, slender
              distally, somewhat like a hypodermic needle (Fig. 12.34); true or
              apparent cuticular penis sheaths present 7

7a(6b).   True cuticular penis sheaths present, but not much thicker than the
              normal cuticular layer of the body wall (Fig. 12.35) .... *Isochaetides*

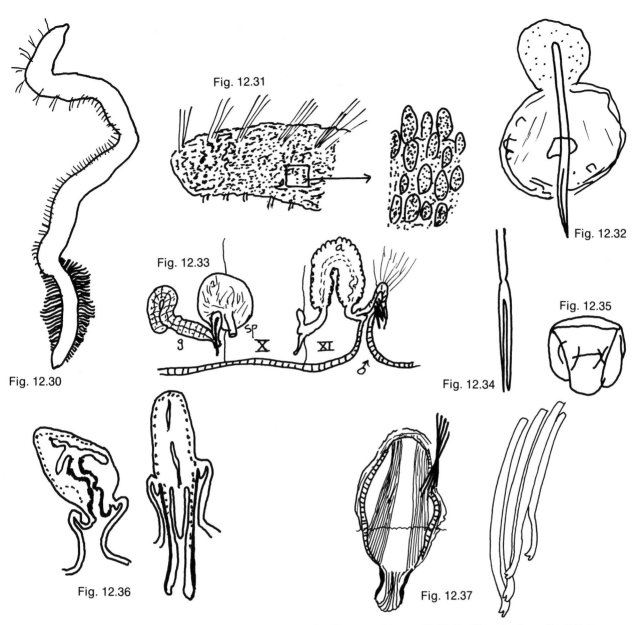

Figure 12.30 *Branchiura,* showing posterior fan of gill filaments. **Figure 12.31** Papillate body wall of *Spirosperma.* **Figure 12.32** Spermathecal chaeta of *Haber* (in which penial chaetae similar). **Figure 12.33** Reproductive organs, left side (anterior to left) of *Rhizodrilus lacteus.* Note spermathecal chaeta and gland (g) in IX, spermatheca (sp) in X, atrium (a) in XI discharging into median chamber that also receives penial chaetae (on right, opposite atrial pore). **Figure 12.34** Spermathecal chaeta of *Isochaetides.* **Figure 12.35** Penis sheath of *Isochaetides.* **Figure 12.36** Inverted (left) and everted penis of *Psammoryctides.* **Figure 12.37** *Varichaetadrilus,* penis (left) with penial sheath on tip (shown pointing down) and penial chaetae, top right and (enlarged) to right.

7b.	Apparent penis sheaths present (probably the cuticular linings of eversible penes), look like crumpled cylinders in whole-mounts of most specimens (Fig. 12.36) .... *Psammoryctides*
	[One species is a recent introduction to the St. Lawrence River (Quebec) from Europe, another is known only from coastal swamps of the Gulf of Mexico, and a third is known from California and the St. Lawrence Great Lakes.]
8a(3b).	Penial chaetae present replacing the normal ventral chaetae of XI (Figs. 12.37, 12.39, 12.41, 12.42) 9

8b.   Penial chaetae absent, ventral chaetae of XI usually missing in mature specimens 12

9a(8a).   Penial chaetae bifid but with shortened distal ends and elongate proximal ends, somewhat thicker than the normal ventral chaetae; with short penis sheaths on the ends of erectile penes (Fig. 12.37) .... *Varichaetadrilus* (in part)
           [Distributed from Alaska to California and Alberta (but see also 14).]

9b.   Penial chaetae strongly modified, with knobbed or hooked tips to short distal ends, elongate proximal ends, arranged fanwise with heads close together (as if they function as claspers) "or" large, single and sickle-shaped (Figs. 12.39, 12.41, 12.42) 10

10a(9b).   Male pores open into an eversible chamber in XI that contains so-called paratria (Fig. 12.38) and penial chaetae when present (Fig. 12.39); sensory pit on the prostomium (Fig. 12.3); sperm attached to the body wall close to the male pore in external spermatophores (Fig. 12.40) .... *Bothrioneurum vejdovskyanum* Stolc, 1888
           [Specimens lacking penial chaetae may be classified as *B. americanum* Beddard, 1894, a South American species recorded from Georgia and Louisiana. This may be a synonym of the preceeding species]

10b.   No median copulatory chamber or sensory pit; sperm loose in spermathecae after copulation 11

11a(10b). Penial chaetae in XI, 3–6 simple-pointed chaetae with hooked tips (Fig. 12.41); somatic chaetae 3–5 bifids from II–XX, simple-pointed chaetae from XXV–XXX posteriad, no dorsal hair chaetae ..... *Phallodrilus hallae* Cook and Hiltunen, 1975
           [This member of a subfamily of marine species is recorded from Lake Huron and Lake Superior.]

11b.   Penial chaetae with knobbed tips, or single, sickle-shaped (Fig. 12.42); no simple-pointed somatic chaetae (except in the ventral bundles of one species, when accompanied by bifid chaetae, and the dorsal bundles of that species possess hair chaetae) .... *Rhyacodrilus*

12a(8b). Cuticular penis sheaths present in XI (Figs. 12.43–12.45) 13
12b.   Cuticular penis sheaths absent 15

13a(12a). Penis sheaths more or less elongate, cylindrical, with penes free within them (Fig. 12.43); all chaetae bifid .... *Limnodrilus*
13b.   Penis sheaths annular to conical, attached to the surface of the penis (Figs. 12.44 and 12.45) 14

14a(13b). Short cuticular penis sheaths, annular to tub-shaped (Fig. 12.44) .... *Varichaetadrilus* (in part), *Tubifex*, *Ilyodrilus* (in part)
           [The *Varichaetadrilus* species included here are reported from the southeastern United States; *I. frantzi* Brinkhurst, 1965 is restricted to freshwater at the head of estuaries from British Columbia (Canada) to California.]
14b.   Penis sheaths conical (Fig. 12.45) .... *Tasserkidrilus*, *Ilyodrilus templetoni* (Southern, 1904)

15a(12b). Chaetae simple-pointed anteriorly, behind the clitellum the chaetae have brushlike tips (Fig. 12.46) .... *Telmatodrilus vejdovskyi* Eisen, 1879
           [This species is known from British Columbia (Canada) to California.]
15b.   Chaetae bifid, or with upper tooth divided, or pectinate to palmate, all with the teeth in a single plane 16

16a(15b). Pharynx and mouth enlarged, prostomium reduced (Fig. 12.47); chaetae of III broad, curved, with large lower teeth, the rest thinner, straighter, and with teeth more nearly equal; no spermathecae, presumably parthenogenetic .... *Teneridrilus mastix* (Brinkhurst, 1978)
           [From British Columbia (Canada) to California, in low-salinity regions of estuaries.]
16b.   Pharynx and mouth not enlarged, prostomium well developed; chaetae of III not modified; with spermathecae 17

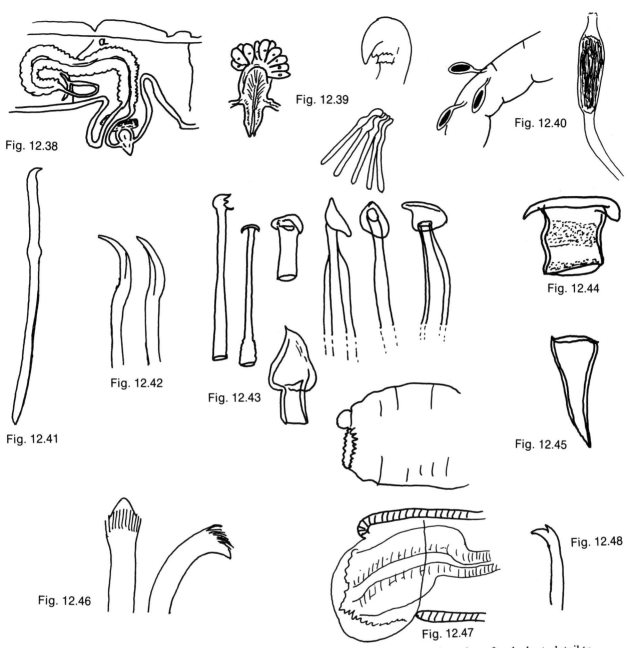

Figure 12.38 Paratrium of *Bothrioneurum*, shown left *in situ* in everted terminal portion of male duct, detail to right. **Figure 12.39** Penial chaetae, *Bothrioneurum*, detail of tip above, bundle below. **Figure 12.40** Spermatophores of *Bothrioneurum*, attached to body wall around gonopores (left) and detail (right). Note sperm enclosed, with no duct through attachment stalk. **Figure 12.41** Penial chaeta, *Phallodrilus hallae*. **Figure 12.42** Penial chaetae, *Rhyacodrilus falciformis*. **Figure 12.43** Penes of various *Limnodrilus* species, three to right much longer than shown (40 times the length when fully developed). **Figure 12.44** Penis sheath of *Tubifex tubifex*, often thin and difficult to see, but granular surface often a clue. **Figure 12.45** Penis sheath of *Ilyodrilus*, short conical form. The terminal opening often looks torn, and individuals with a second portion as long as that illustrated here may be seen, this second part presumably shed before mating. **Figure 12.46** Posterior brushlike chaetae, *Telmatodrilus*. **Figure 12.47** Inverted (above) and everted pharynx of *Teneridrilus mastix*. **Figure 12.48** Ventral chaeta, *Aulodrilus*.

17a(16b).	Ventral chaetal bundles with up to 4 simple-pointed or bifid chaetae with reduced upper teeth. [In this species from Lake Tahoe there are apparently no penial chaetae, although these are usually present in this genus (see 11b).] .... *Rhyacodrilus brevidentatus* Brinkhurst, 1965
17b.	No simple-pointed ventral chaetae 18

18a(17b).   Dorsal chaetal bundles with 2–4 pectinate to palmate chaetae from II–XIV, from there posteriad all dorsal chaetae bifid; anterior ventral chaetae 3–5 per bundle, bifid with the upper teeth thinner than but only a little longer than the lower, but beyond XIV the upper teeth become much longer than the lower .... *Arctodrilus wulikensis* Brinkhurst and Kathman, 1983
           [This species is recorded from the Brooks Range, Alaska.]
18b.   Chaetae progressively enlarged from III posteriad, becoming very broad with large lower teeth and recurved distal ends, "or" no hair chaetae dorsally and large numbers of bifid chaetae with thin, short upper teeth, becoming palmate beyond VI in one species, "or" hair chaetae present but all ventral chaetae again characteristically numerous and with short upper teeth (Fig. 12.48) .... *Aulodrilus*

BRANCHIOBDELLIDA

VI. INTRODUCTION TO BRANCHIOBDELLIDA

Branchiobdellidans are leech-like, ectosymbionts living primarily on astacid crayfish and a few other crustaceans. A single host can be infested with over 350 individuals and a taxonomic diversity of up to 7 species representing 4 genera. Branchiobdellidans have an Holarctic distribution with currently 15 genera containing approximately 100 species reported from North America (Holt 1986, 1988, Gelder and Hall 1990).

The Branchiobdellida is a monophyletic group and is considered to be an independent taxon of equivalent rank to the oligochaetes and leeches (Holt 1986, Gelder and Brinkhurst 1990). Despite the phylogenetic importance of the branchiobdellidans and their ecophysiologically distinct niche, the taxon remains largely unstudied. A detailed review and bibliography of the taxon has been made (Sawyer 1986); however, the interpretations expressed are not accepted generally (Holt 1989). As there is no single publication dealing with all known species of branchiobdellidans, the present work will deal only with genera (but citations are included through which species descriptions can be found).

VII. ANATOMY AND PHYSIOLOGY OF BRANCHIOBDELLIDANS

A. External Morphology

Adult branchiobdellidans range from 0.8–10 mm in length. In most species the body is rod-shaped (terete), although some have a characteristic pyriform or flask-shape with either ventral or dorsoventral flattening. The segment number is constant and, based on the number of paired ganglia, is accepted to be 17 (as shown by Roman numerals in Fig. 12.49). The external morphology shows a peristomium and three segments forming a head, and the remaining body segments are numbered in Arabic numerals, 1–11. Segment 11 is modified into the posterior, disc-shaped attachment organ (sucker) and contains ganglia XV–XVII. Peristomial tentacles or lobes may be present on the dorsal lip, and the mouth is usually surrounded by 16 oral papillae. Each body segment is divided into a major anterior and a minor posterior annulus, although a few species have the major annulus subdivided into two. Dorsal transverse ridges across the major annuli are produced by supernumerary longitudinal muscles (sl in Fig. 12.49) and sometimes support dorsal digitiform or multibranched appendages (d).

B. Organ System Function

The alimentary canal consists of a pair of sclerotized jaws (j in Fig. 12.49) situated in the anterior pharynx followed by one to three pairs of sulci (su), a short esophagus (e), stomach (s), intestine (i), and an anus (a) opening dorso-medially onto segment 10. Locomotory movements are leech-like and involve an anterior and a posterior attachment organ (Gelder and Rowe 1988). The anterior attachment site is on the ventral surface of the ventral lip and is not the mouth, as is generally assumed. The muscle cells have the nucleus and cytoplasm in the medulla surrounded by a contractile cortex. This cylindrical, circomyarian arrangement, or variations of it, are found in branchiobdelliians, along with a T-system (de Eguileor and Ferraguti, 1980). The pore(s) of the asymmetrically arranged anterior pair of nephridia open on segment 3 (np) and the posterior pair are located in segment 8 but open on segment 9. The spermatheca (sp), when present, is situated in segment 5 and consists of a glandular ental and ductal ectal regions. A pair of testes are located in segments 5 and 6, and the sperm develop (ds) in the coelom of those segments. Two vasa efferentia, both with a ciliated funnel, merge into a vas deferens, and the two vasa deferentia, one from each segment, enter the spermiducal gland (glandular atrium)

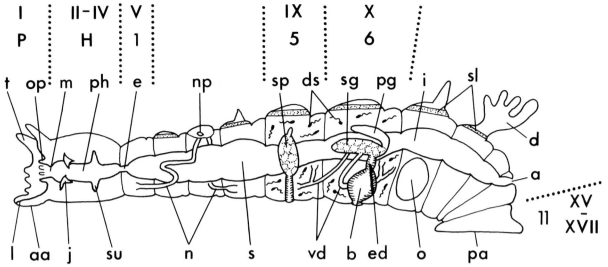

Figure 12.49 Diagram of the lateral aspect of a hypothetical branchiobdellidan showing anatomic characters used in the taxonomy of this group: a, anus; aa, anterior attachment surface; b, bursa; d, digitiform dorsal appendages; ds, developing sperm; e, esophagus; ed, ejaculatory duct; i, intestine; j, jaw; l, lip; m, mouth; n, nephridia; np, nephridial pore; o, ovum; op, oral papillae; pa, posterior attachment surface; pg, prostate gland; ph, pharynx; s, stomach; sg, spermiducal gland; sl, supernumerary longitudinal muscles; sp, spermatheca; su, sulcus; t, tentacle; vd, vasa deferentia; H, head; P, peristomium; 1–11, body segments; I–XVII, segments based on paired ganglia.

separately in segment 6. The prostate gland forms a protuberance or a diverticulum from the spermiducal gland. The ejaculatory duct (muscular atrium) passes into the penis which is surrounded by a muscular bursa (b). The penis may be either eversible or protrusible, although the mechanism employed for sperm transfer is unknown where the penis is reduced. The spermatheca and bursa open mesad on the ventral surfaces of segents 5 and 6, respectively.

The vascular system consists of single dorsal and ventral longitudinal vessels that are connected by four paired, lateral branches in the head, and a single pair of branches in segments 1, 7, and 11, respectively. The dorsal vessel becomes a sinus surrounding the gut wall from segments 3–7.

The central nervous system is composed of a dorsal brain, circumesophageal connectives, and a paired ventral cord extending into the last segment. The head contains four pairs of ganglia; each body segment contains one pair of ganglia; and three pairs appear to be located in segment 11. The nerve cord is displaced laterally in segments 5 and 6, where the spermatheca and male genitalia pass mesad through the ventral body wall.

C. Environmental Physiology

Most branchiobdellidans browse on the epibionts attached to the host and are opportunistic commensals when fragments of the food of the host are available (Jennings and Gelder 1979). Contrary to the reputation of these worms, a parasitic regime has been demonstrated only in *Branchiobdella hexodonta,* and inferred in a few other species. Food is sucked into the mouth and then the jaws crop small microorganisms from the substratum or fragment large particles. Various mucoid secretions assist in ingestion, but the term "salivary gland" quoted by Sawyer (1986) is inappropriate and has been corrected by Gelder and Rowe (1988). Digestion is extracellular in the stomach and intestine with uptake occurring in the cells of the latter region.

Low oxygen tensions do not appear to exert an adverse effect on branchiobdellidans, although some species can tolerate higher water temperatures better than others.

VIII. ECOLOGY AND EVOLUTION

A. Diversity and Distribution

The branchiobdellidans are characterized by a lack of anatomic and ecologic diversity in comparison with equivalent taxa. However, recent electron microscope studies on spermatozoa have shown significant variations in ultrastructure between different genera. Branchiobdellidans occur in North America from southern Canada to Costa Rica. A few species are found distributed throughout the region, but the remainder have a more limited range (see Section IX.B). The zoogeography of species is becoming more confused and obscure as crayfish and

Fig. 12.50   Fig. 12.51   Fig. 12.52   Fig. 12.53

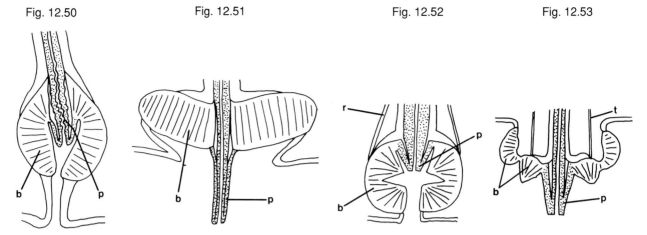

Figure 12.50 Diagrammatic section of a withdrawn, eversible penis: b. bursa, p. penis. **Figure 12.51** Diagrammatic section of an extended, eversible penis; b, bursa; p, penis. **Figure 12.52** Diagrammatic section of a withdrawn, protrusible penis: b, bursa; p, penis; r, retractor muscle. **Figure 12.53** Diagrammatic section of an extended, protrusible penis; b, bursa, p, penis; r, retractor muscle.

their symbionts are introduced into new areas for aquaculture, live food, research, and teaching purposes.

B. *Reproduction and Life History*

Details on the reproductive biology of branchiobdellidans are scarce and limited to just a few species. Information on chromosome numbers is restricted to three species: *Branchiobdella astaci* (2n = 14), *B. kozarovi* (2n = 12), and *B. parasita* (2n = 10) (Mihailova and Subchev 1981); this research refutes previous reports of n = 4 and 8 for the first species. The ultrastructure of the developing sperm in *B. pentodonta* and *Cambarincola fallax* show significant differences in morphology between the two species (Ferraguti *et al.* 1986). Spermatogenesis occurs in the coelom and the mature spermatozoa congregate at the mouth of the sperm funnels prior to passage through the genitalia. Copulation occurs between two individuals with the penis of one, being either everted or protruded (Figs. 12.50–12.53), depositing sperm in the spermatheca of the other. Whether this is a reciprocal operation is not known. As members of two genera do not have a spermatheca, their method for sperm transfer is open to speculation, but the use of spermatophores is suspected.

The secretions from the clitellar gland cells have been characterized histochemically and compared with those in other clitellate annelids (Gelder and Rowe 1988). These secretions produce the pedunculate cocoon and nourishment to sustain the embryo while it develops. The embryo is lecithotrophic and then becomes an albuminotrophic cryptolarva. In this latter phase, precocious feeding of the surrounding nutrient fluid starts with ingestion by a transitory pharynx and digestion by the yolk-sac epithelium.

Branchiobdellidans will only deposit their cocoons on live, hosts, and this has so far prevented an *in vitro* study of the life cycle. Hatching is temperature dependent and takes 10–12 days at 20°–22°C. There are no reports of growth rates or the time required to reach sexual maturity.

C. *Ecological Interactions*

Branchiobdellidans have been reported only on freshwater crustaceans, primarily astacid crayfish. Although all crayfish appear to be potential hosts, some species actively remove any branchiobdellidans from their exoskeleton (Gelder and Smith 1987). Non-crayfish hosts are certainly acceptable in troglobitic habitats (Holt 1973b) and at the southern limit of the geographical distribution of the symbiont—which is beyond that of astacid crayfish.

Branchiobdellidans also harbor ectosymbionts and parasites. Externally, these include rotifers and stalked, solitary, and colonial ciliate protozoans; internally, these encompass the third juvenile stage of the nematode *Dioctophyme renale* (giant kidney worm) (Schmidt and Roberts, 1989) and the trematode mesocercaria of *Pharyngostomoides procyonis* [reviewed in Schell (1985)].

D. *Evolutionary Relationships*

The taxonomic rank of the branchiobdellidans is currently accepted to be equivalent to that of the oligochaetes and leeches. The phylogenetic position of the Branchiobdellida in the Clitellata is enigmatic and will remain so until the latter taxon is demonstrated to be monophyletic. Until then, it appears prudent for the monophyletic branchiobdellidans (Gelder and Brinkhurst 1990) to continue to be considered as an independent and intermediately located taxon between the oligochaetes and the leeches.

IX. IDENTIFICATION OF BRANCHIOBDELLIDANS

A. Collecting, Rearing, and Preparation for Identification

Crayfish and other crustacean hosts are collected by netting or baiting a trap (see Chapter 22) and then placed directly into a container with preservative. The recommended one is formalin–alcohol–acetic acid (FAA) which is made by adding together 70 ml ethanol + 10 ml formaldehyde (full strength) + 15 ml distilled water + 5 ml glacial acetic acid. The volume of preservative should be at least twice that of the crustaceans to ensure thorough preservation of the symbionts and host. Branchiobdellidans, which can be removed from live hosts with the aid of a dissecting microscope, have been maintained *in vitro* for several months.

Microscopic examination of a living branchiobdellidan from a dorsal or ventral aspect enables one to observe the peristomium, jaws, and the location of the anterior nephridial pore(s). The specimen can then be rolled to show a lateral aspect, allowing the spermatheca and genitalia to be studied.

Live specimens can be relaxed using carbonated drinks (i.e., Club Soda) or a 1–2% magnesium chloride solution (Delly 1985), straightened or flattened, and then immediately preserved. Unstained specimens are dehydrated in graded ethanol solutions, cleared in clove oil or oil of wintergreen (methyl salicylate), and then mounted in Canada balsam for permanent preparations. Specimens usually curl when preserved on the host, and so these worms can only be examined from the lateral aspect. Large branchiobdellidans (4–10 mm long) will require microdissection to expose their genitalia. The necessary instruments can be made from fragments of a razor blade and insect pins glued to the end of cocktail sticks. For techniques to prepare serial sections and to stain specimens, refer to a book on animal microtechniques, such as Humason (1979).

Consult Holt (1986, 1988) and Gelder and Hall (1990) for citations of the diagnoses and descriptions of genera and species.

B. Taxonomic Key to Genera of Freshwater Branchiobdellida

1a.	Two anterior nephridial pores	2
1b.	One anterior nephridial pore	4
2a(1a).	Ventral flattening of body	3
2b.	Dorso-ventral flattening of body (Fig. 12.54) .... *Xironogiton* [British Columbia, Oregon, Washington, California, Idaho, Wyoming, and the Appalachian region from Maine to Georgia]	
3a(2a).	Vasa deferentia entering spermiducal gland mesad .... *Ankyrodrilus* [Tennessee, Virginia]	
3b.	Vasa deferentia entering spermiducal gland entad .... *Xironodrilus* [Central and Eastern United States]	
4a(1b).	Vasa deferentia entering spermiducal gland mesad	5
4b.	Vasa deferentia entering spermiducal gland entad	7
5a(4a).	Body surface with indistinct segments	6
5b.	Body surface with distinct segments (Fig. 12.59) .... *Cronodrilus* [Georgia]	
6a(5a).	Lateral segmental glands, segments 1–9 (Fig. 12.57) .... *Bdellodrilus* [Southern Canada to southern Mexico]	
6b.	Lateral segmental glands not in segments 1–9 .... *Uglukodrilus* [Oregon]	
7a(4b).	Spermatheca present	8
7b.	Spermatheca absent .... *Ellisodrilus* [Indiana, Kentucky, Michigan, and Tennessee]	
8a(7a).	Penis protrusible (Figs. 12.52 and 12.53)	9
8b.	Penis eversible (Figs. 12.50 and 12.51)	10
9a(8a).	Prostate body present	11

Fig. 12.54   Fig. 12.55   Fig. 12.56

Fig. 12.57   Fig. 12.58

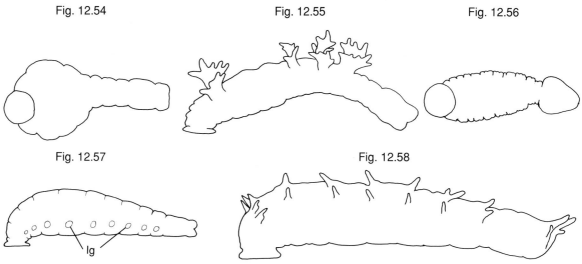

Figure 12.54 Ventral view of *Xironogiton instabilis*, length about 2 mm. **Figure 12.55** Lateral view of *Ptero-drilus alcicornis*, length about 1.5 mm. (Redrawn from Holt, 1986b.) **Figure 12.56** Ventral view of *Triannulata magna*, length about 4.5 mm. (Redrawn from Holt 1974.) **Figure 12.57** Lateral view of *Bdellodrilus illumi-natus*, length about 3 mm, lg, lateral glands. **Figure 12.58** Lateral view of *Ceratodrilus thysanosomus*, length about 3 mm. (Redrawn and modified from Holt 1960.)

Fig. 12.59   Fig. 12.60   Fig.12.61

Fig. 12.62   Fig. 12.63   Fig. 12.64

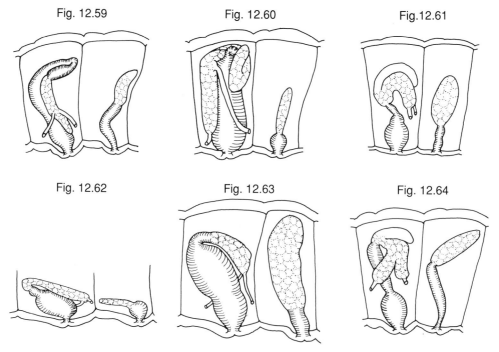

Figure 12.59 Lateral view of the male genitalia and spermatheca in *Cronodrilus ogyguis*. (Redrawn and modified from Holt 1968a.) **Figure 12.60** Lateral view of the male genitalia and spermatheca in *Magmatodrilus obscurus*. (Redrawn and modified from Holt 1967.) **Figure 12.61** Lateral view of the male genitalia and spermatheca in *Cambarincola fallax*. **Figure 12.62** Lateral view of the male genitalia and spermatheca in *Sathodrilus hortoni*. (Redrawn from Holt 1973a.) **Figure 12.63** Lateral view of the male genitalia and sper-matheca in *Sathodrilus carolinensis*. (Redrawn from Holt 1968a.) **Figure 12.64** Lateral view of the male genitalia and spermatheca in *Sathodrilus attenuatus*. (Redrawn from Holt 1981.)

9b.	Prostate body absent (Fig. 12.60)	*Magmatodrilus*
	[California]	
10a(8b).	Dorsal appendages present, range from paired digitiform unit to single transverse ridge on segment 8 (Fig. 12.55)	*Pterodrilus*
	[Eastern United States]	
10b.	Dorsal appendages absent (Fig. 12.61)	*Cambarincola*
	[Southern Canada to Costa Rica]	
11a(9a).	Prostate body absent	12
11b.	Prostate body present	13
12a(11a).	Two annuli per segment (Fig. 12.62)	*Sathodrilus* (in part)
	[California, Florida, and Mexico]	
12b.	Three annuli per segment (Fig. 12.56)	*Triannulata*
	[Oregon and Washington]	
13a(11b).	Dorsal appendages present (Fig. 12.58)	*Ceratodrilus*
	[Idaho, Oregon, Utah, and Wyoming]	
13b.	Dorsal appendages absent. [This grouping cannot be resolved further without detailed histologic preparation.]	
	(see Fig. 12.63)	*Sathodrilus* (in part)
	[South Carolina, Georgia, and Mexico]	
	(see Fig. 12.64)	*Sathodrilus* (in part)
	[Southern Canada to Mexico]	
	....	*Oedipodrilus*
	[Illinois, Indiana, Kentucky, Ohio, Pennsylvania, and Tennessee]	
	....	*Tettodrilus*
	[Tennessee]	

LITERATURE CITED

Appleby, A. G., and R. O. Brinkhurst. 1970. Defecation rate of three tubificid oligochaetes found in the sediment of Toronto Harbour, Ontario. Journal of the Fisheries Research Board of Canada 27:1971–1982.

Aston, R. J. 1984. The culture of *Branchiura sowerbyi* (Tubificidae, Oligochaeta) using cellulose substrate. Aquaculture 40:89–94.

Bagheri, E. A., and D. S. McLusky. 1984. The oxygen consumption of *Tubificoides benedeni* (Udekem) in relation to temperature, and its application to production biology. Journal of Experimental Marine Biology 78:187–197.

Block, E. M., G. Moreno, and C. J. Goodnight. 1982. Observations on the life history of *Limnodrilus hoffmeisteri* (Annelida, Tubificidae) from the Little Calumet River in temperate North America. International Journal of Invertebrate Reproduction 4:239–247.

Bonacina, C., G. Bonomi, and C. Monti. 1987. Progress in cohort cultures of aquatic Oligochaeta. Hydrobiologia 155:163–169.

Bowker, D. W., M. T. Wareham, and M. A. Learner. 1985. A choice chamber experiment on the selection of algae as food and substrata by *Nais elinguis* (Oligochaeta, Naididae). Freshwater Biology 15:547–557.

Brinkhurst, R. O. 1965. Studies on the North American aquatic Oligochaeta II. Tubificidae. Proceedings of the Academy of Natural Sciences, Philadelphia 117:117–172.

Brinkhurst, R. O. 1974. The benthos of lakes. Macmillan, London.

Brinkhurst, R. O. 1982. Evolution in the Annelida. Canadian Journal of Zoology 60:1043–1059.

Brinkhurst, R. O. 1986. Guide to the freshwater aquatic microdrile oligochaetes of North America. Canadian Special Publication of Fisheries and Aquatic Sciences 84:1–259.

Brinkhurst, R. O. 1990. A phylogenetic analysis of the Lumbriculidae (Annelida, Oligochaeta). Canadian Journal of Zoology. submitted.

Brinkhurst, R. O., and M. J. Austin. 1978. Assimilation by aquatic oligochaetes. Internationale Revue der gesamten Hydrobiologie 63:863–868.

Brinkhurst, R. O., and K. E. Chua. 1969. Preliminary investigation of the exploitation of some potential nutritional resources by three sympatric tubificid oligochaetes. Journal of the Fisheries Research Board of Canada 26:2659–2668.

Brinkhurst, R. O., and S. R. Gelder. 1989. Did the lumbriculids provide the ancestors of the branchiobdellids, acanthobdellids and leeches? Hydrobiologia 180:7–15.

Brinkhurst, R. O., and B. G. M. Jamieson. 1971. Aquatic Oligochaeta of the world. Oliver and Boyd, Edinburgh.

Brinkhurst, R. O., P. M. Chapman, and M. J. Farrell. 1983. A comparative study of respiration rates of some aquatic oligochaetes in relation to sublethal stress. Internationale Revue der gesamten Hydrobiologie 68:683–699.

Chapman, P. M. 1981. Measurement of the short-term stability of interstitial salinities in subtidal estuarine sediments. Estuarine Coastal and Shelf Science 12:67–81.

Chapman, P. M., and R. O. Brinkhurst. 1981. Seasonal changes in interstitial salinities and seasonal movements of subtidal benthic invertebrates in the Fraser River Estuary. Estuarine and Coastal Shelf Science 12:49–66.

Chapman, P. M., and R. O. Brinkhurst. 1984. Lethal and sublethal tolerances of aquatic oligochaetes with reference to their use as a biotic index of pollution. Hydrobiologia 115:139–144.

Chapman, P. M., and R. O. Brinkhurst. 1987. Hair today and gone tomorrow. Induced chaetal changes in tubificid oligochaetes. Hydrobiologia 155:45–55.

Chapman, P. M., M. A. Farrell, and R. O. Brinkhurst. 1982a. Effects of species interactions on the survival and respiration of *Limnodrilus hoffmeisteri* and *Tubifex tubifex* (Oligochaeta, Tubificidae) exposed to various pollutants and environmental factors. Water Research 16:1405–1408.

Chapman, P. M., M. A. Farrell, and R. O. Brinkhurst. 1982b. Relative tolerance of selected aquatic oligochaetes to individual pollutants and environmental factors. Aquatic Toxicology 2:47–67.

Chapman, P. M., M. A. Farrell, and R. O. Brinkhurst. 1982c. Relative tolerances of selected aquatic oligochaetes to combinations of pollutants and environmental factors. Aquatic Toxicology 2:69–78.

Chatarpaul, L., J. B. Robinson, and N. K. Kaushik. 1979. Role of tubificid worms on nitrogen transformations in stream sediment. Journal of the Fisheries Research Board of Canada 36:673–678.

Christensen, B. 1984. Asexual propagation and reproductive strategies in aquatic oligochaetes. Hydrobiologia 115:91–95.

Clark, R. B. 1964. Dynamics in metazoan evolution. Clarendon Press, Oxford.

Coates, K. A. 1990. Phylogeny and origins of the Enchytraeidae. Hydrobiologia in press.

de Eguileor, M. and M. Ferraguti. 1980. Architecture of the T-system in helical paramyosin muscles of an annelid (*Branchiobdella*). Tissues and Cell 12:739–747.

Delly, J.G. 1985. Narcosis and preservation of freshwater animals. American Laboratory pp. 31–40.

Drewes, C. D., and C. R. Fourtner. 1988. Hindsight and rapid escape in an aquatic oligochaete. Page 278.5 *in:* Society for Neuroscience, Abstracts of the 18th Annual Meeting, Toronto, Ontario, Nov. 13–18 1988. 14(1).

Ferraguti, M., S. R. Gelder, and G. Bernardini. 1986. On helices, sperm homologies and phylogeny: the case of branchiobdellids. Pages 199–204 *in:* M. Cresti and R. Dallai, editors. Biology of Reproduction and Cell Motility in Plants and Animals. University Press of Siena, Siena, Italy.

Gardner, W. S., T. F. Nalepa, M. A. Quigley, and J. M. Malczyk. 1981. Release of phosphorus by certain benthic invertebrates. Canadian Journal of Fisheries and Aquatic Science 38:978–981.

Gelder, S. R., and R. O. Brinkhurst. 1990. An assessment of the phylogeny of the Branchiobdellida (Annelida: Clitellata) using PAUP. Canadian Journal of Zoology 68:1318–1326.

Gelder, S. R. and L. A. Hall. 1990. Description of *Xironogiton victoriensis* n.sp. from British Columbia, Canada with some remarks on other species and a Wagner analysis of *Xironogiton* (Clitellata: Branchiobdellia). Canadian Journal of Zoology 68:2352–2359.

Gelder, S. R., and J. P. Rowe. 1988. Light microscopical and cytochemical study on the adhesive and epidermal gland cell secretions of the branchiobdellid *Cambarincola fallax* (Annelida: Clitellata). Canadian Journal of Zoology 66:2057–2064.

Gelder, S. R., and R. C. Smith. 1987. Distribution of branchiobdellids (Annelida: Clitellata) in northern Maine, U.S.A. Transactions of the American Microscopical Society 106:85–88.

Harman, W. J., and M. S. Loden. 1978. A re-evaluation of the Opitocystidae (Oligochaeta) with descriptions of two new genera. Proceedings of the Biological Society of Washington 91:453–462.

Harper, R. M., J. C. Fray, and M. A. Learner. 1981. A bacteriological investigation to elucidate the feeding biology of *Nais variabilis* (Oligochaeta, Naididae). Freshwater Biology 11:227–236.

Holmquist, C. 1976. Lumbriculids (Oligochaeta) of Northern Alaska and Northwestern Canada. Zoologische Jahrbucher (Systematik) 103:377–431.

Holt, P. C. 1960. The genus *Ceratodrilus* Hall (Branchiobdellidae, Oligochaeta) with the description of a new species. Virginia Journal of Science 11:53–77.

Holt, P. C. 1967. Status of the genera *Branchiobdella* and *Stephanodrilus* in North America with description of a new genus (Clitellata: Branchiobdellida). Proceedings of the United States National Museum 124:1–10.

Holt, P. C. 1968a. New genera and species of branchiobdellid worms (Annelida: Clitellata). Proceedings of the Biological Society of Washington 81:291–318.

Holt, P. C. 1968b. The genus *Pterodrilus* (Annelida: Branchiobdellida). Proceedings of the United States National Museum 125:1–44.

Holt, P. C. 1973a. Epigean branchiobdellids (Annelida: Clitellata) from Florida. Proceedings of the Biological Society of Washington 86:79–104.

Holt, P.C. 1973b. Branchiobdellids (Annelida: Clitellata) from some eastern North American caves, with descriptions of new species of the genus *Cambarincola*. International Journal of Speleology 5:219–256.

Holt, P. C. 1974. An emendation of the genus *Triannulata* Goodnight, 1940, with the assignment of Triannulata montana to *Cambrincola* Ellis 1912 (Clitellata: Branchiobdellida). Proceedings of the Biological Society of Washington 87:57–72.

Holt, P. C. 1981. New species of *Sathodrilus* Holt, 1968, (Clitellata: Branchiobdellida) from the Pacific drainage of the United States, with the synonymy of *Sathodrilus virgiliae* Holt, 1977. Proceedings of the Biological Society of Washington 94:848–862.

Holt, P. C. 1986. Newly established families of the order Branchiobdellida (Annelida: Cllitellata) with a synopsis of the genera. Proceedings of the Biological Society of Washington 99(4):676–702.

Holt, P. C. 1988. Four new species of cambarincolids (Clitellata: Branchiobdellida) from the souhteastern United States with a redescription of *Oedipodrilus macbaini* (Holt, 1955). Proceedings of the Biological Society of Washington 101:794–808.

Holt, P. C. 1989. Comments on the classification of the Clitellata. Hydrobiologia 180:1–5.

Humason, G. L. 1979. Animal Tissue Techniques. 4th Edition Freeman, San Francisco, California.

Jamieson, B. G. M. 1981. The Ultrastructure of the Oligochaeta. Academic Press, London.

Jennings, J. B., and Gelder, S. R. 1979. Gut structure, feeding and digestion in the branchiobdellid oligochaeta *Cambarincola macrodonta* Ellis 1912, an ectosymbiote of the freshwater crayfish *Procambarus clarkii*. Biological Bulletin 156:300–314.

Karlckhoff, S. W., and K. R. Morris. 1985. Impact of tubificid oligochaetes on pollution transport in bottom sediments. Environmental Science and Technology 19:51–56.

Kaster, J. L., and J. H. Bushnell. 1981. Cyst formation by *Tubifex tubifex* (Tubificidae). Transactions of the American Microscopical Society 100:34–41.

Kaster, J. L., J. Val Klump, J. Meyer, J. Krezoski, and M. E. Smith. 1984. Comparison of defecation rates of *Limnodrilus hoffmeisteri* Claparede (Tubificidae) using two different methods. Hydrobiologia 111:181–184.

Krezoski, J. R., J. A. Robins, and D. S. White. 1984. Dual radiotracer measurement of zoobenthos-mediated solute and particle transport in freshwater sediments. Journal of Geophysical Research 89:7937–7947.

Lang, C. 1984. Eutrophication of Lakes Leman and Neuchatel (Switzerland) indicated by oligochaete communities. Hydrobiologia 115:131–138.

Lazim, M. N., and M. A. Learner. 1986a. The life-cycle and production of *Limnodrilus hoffmeisteri* and *L. udekemianus* (Oligochaeta, Tubificidae) in the organically enriched Moat-Feeder stream, Cardiff, South Wales. Archiv fur Hydrobiologie (Supplement) 4:200–225.

Lazim, M. N., and M. A. Learner. 1986b. The life-cycle and productivity of *Tubifex tubifex* (Oligochaeta, Tubificidae) in the Moat-Feeder stream, Cardiff, South Wales. Holarctic Ecology 9:185–192.

Leitz, D. M. 1987. Potential for aquatic oligochaetes as live food in commercial aquaculture. Hydrobiologia 155:309–310.

Lochhead, G., and M. A. Learner. 1984. The cocoon and hatchling of *Nais variabilis* (Naididae, Oligochaeta). Freshwater Biology 14:189–193.

Loden, M. S. 1981. Reproductive ecology of Naididae. Hydrobiologia 83:115–123.

Marian, M. P., and T. J. Pandian. 1984. Culture and harvesting techniques for *Tubifex tubifex*. Aquaculture 42:303–315.

Mihailova, P. V., and M. A. Subchev. 1981. On the karyotype of three species of the family Branchiobdellidae (Annelida: Oligochaeta). Comptes Rendus l'Academie Bulgare des Sciences 34:265–267.

Milbrink, G. 1983. An improved environmental index based on the relative abundance of oligochaete species. Hydrobiologia 102:89–97.

Nemec, A. F. L., and R. O. Brinkhurst. 1987. A comparison of methodological approaches to the subfamilial classification of the Naididae (Oligochaeta). Canadian Journal of Zoology 65:691–707.

Newrlka, P., and S. Mutayoba. 1987. Why and where do oligochaetes hide their cocoons? Hydrobiologia 155:171–178.

Olsson, T. I. 1981. Overwintering of benthic macroinvertebrates in ice and frozen sediment in a North Swedish River. Holarctic Ecology 4:161–166.

Poddubnaya, T. L. 1984. Parthenogenesis in Tubificidae. Hydrobiologia 115:97–99.

Sawyer, R. T. 1986. Leech Biology and Behaviour. Clarendon, Oxford.

Schell, S. C. 1985. Handbook of Trematodes of North America, North of Mexico. University of Idaho Press, Moscow.

Schmidt, G. D., and L. S. Roberts. 1989. Foundations of Parasitology. Times Mirror/Morsby, Boston, Massachusetts.

Vincent, V., G. Vaillancourt, and S. McMurray. 1978. Premiere mention de *Psammoryctides barbatus* (Grube) (Annelida, Oligochaeta) en Amerique du Nord et note sur sa distribution dans le haut estuaire du Saint-Laurent. Le Naturaliste Canadien 105:77–80.

Wassell, J. T. 1984. Revision of the lumbriculid oligochaete *Eclipidrilus* Eisen, 1881 with description of three subgenera and *Eclipidrilus (Leptodrilus) fontanus* n. subgen. n. sp. from Pennsylvania. Proceedings of the Biological Society of Washington 97:78–85.

Wiederholm, T., A.-M. Wiederholm, and G. Milbrink. 1987. Bulk sediment bioassays with five species of fresh-water oligochaetes. Water, Air and Soil Pollution 36:131–154.

Zoran, M. J., and C. D. Drewes. 1987. Rapid escape reflexes in aquatic oligochaetes: variations in design and function of evolutionarily conserved giant fiber systems. Journal of Comparative Physiology (*A*) 161:729–738.

ADDITIONAL SUGGESTED READINGS

In addition to the literature actually cited in the text, the following publications will provide more specific information on taxonomy (Brinkhurst and Wetzel 1984) and data derived from a series of international conferences on aquatic oligochaete biology.

Brinkhurst, R. O., and D. G. Cook, editors. 1980. Aquatic Oligochaete Biology. Plenum, New York.

Brinkhurst, R. O., and M. J. Wetzel. 1984. Aquatic Oligochaeta of the World: Supplement. A Catalogue of New Freshwater Species, Descriptions, and Revisions. Canadian Technical Report of Hydrography and Ocean Sciences 44:i–v + 101 pp.

Bonomi, G., and C. Erseus, editors. 1984. Aquatic Oligochaeta. Developments in Hydrobiology, No. 180. Junk, Dordrecht, Netherlands.

Brinkhurst, R. O., and R. J. Diaz, editors. 1987. Aquatic Oligochaeta. Developments in Hydrobiology, No. 40. Junk, Dordrecht, Netherlands.

Kaster, J. L., editor. 1989. Aquatic Oligochaeta. Developments in Hydrobiology, No. 51. Junk, Dordrecht, Netherlands.

Annelida: Leeches, Polychaetes, and Acanthobdellids

13

Ronald W. Davies
Department of Biological Sciences
University of Calgary
Calgary, Alberta T2N1N4 Canada

Chapter Outline

I. INTRODUCTION
II. TAXONOMIC STATUS AND CHARACTERISTICS
III. LEECHES AND ACANTHOBDELLIDAE
   A. General Anatomy
   B. Physiology
IV. ECOLOGY
   A. Diversity
   B. Foraging Relationships
   C. Reproduction and Life History
   D. Dispersal
   E. Predation
   F. Behavior
   G. Population Regulation
V. COLLECTION AND REARING
   A. Collection
   B. Rearing
VI. IDENTIFICATION
   A. Preparation of Specimens
   B. Important Features for Identification
VII. POLYCHAETA
   A. General Anatomy
   B. Ecology
   C. Collection and Identification
VIII. TAXONOMIC KEYS
   A. Taxonomic Key to Higher Taxa of Freshwater Annelida
   B. Taxonomic Key to Species of Freshwater Hirudinoidea
   C. Taxonomic Key to Species of Freshwater Polychaeta
   Literature Cited
   Appendix 13.1. Distribution of Leech and Acanthobdellid Taxa in North America
   Appendix 13.2. Distribution of Polychaete Taxa in North America

I. INTRODUCTION

Leeches form an important component of the benthos of most lakes and ponds, and some species commonly occur in the quieter flowing sections of streams and rivers. There are a total of 69 species presently recorded from North America, of which the majority are predators feeding on chironomids, oligochaetes, amphipods, and molluscs. The young of several species feed on zooplankton, but *Erpobdella montezuma* is unique in specializing on planktonic amphipods using mechanoreception to detect prey. Other species of leeches are temporary sanguivorous (i.e., blood feeding) ectoparasites of fish, turtles, amphibians, water birds, or occasionally humans. The majority of predaceous leeches have an annual or biannual life cycle, breeding once and then dying (semelparity). However, at least one species has been shown to be genetically capable of breeding several times (iteroparity) although it exhibits phenotypic semelparity in the field. The majority of sanguivorous leeches are iteroparous showing saltatory growth after the three or more blood meals required to reach mature size. Because of the diversity of physicochemical conditions encountered in different aquatic ecosystems and temporally within a single habitat, leeches show considerable physiological plasticity. They are able to live in waters with very low salt concentrations as well as waters with salinities that exceed sea water. Similarly, some species are capable of surviving in the absence of oxygen (anoxia) for more than 60 days and in the presence of supersaturated water (hyperoxia) for similar periods. In many small ponds and lakes, leeches are the top predators and are ideal for studying intra- and interspecific resource competition, niche overlap, and predator–prey interactions. In larger lakes and rivers, leeches may form an important component of the diet of fish; leeches are also grown commercially for fish bait. Sanguivorous leeches are being investigated extensively in relation to the pharmacological properties of their salivary secretions, especially their anticoagulants.

The leech-like *Acanthobdella peledina* is a temporary ectoparasite of salmonids. Its ecology and feeding have not been extensively studied but are likely to be similar to the Piscicolidae—a family of leeches ectoparasitic on fish.

The freshwater polychaetes are also poorly known. So far, 12 species have been identified, but this paucity probably reflects the intensity of sampling and collection difficulties. The Nereidae are predators or omnivores, while the remaining families are either filter feeders or deposit feeders. The physiological and ecological adaptations to freshwater of this predominantly marine group will make interesting investigations.

II. TAXONOMIC STATUS AND CHARACTERISTICS

There is an undisputed close taxonomic affinity between leeches and oligochaetes, although there are differences of opinion about the exact nature of the relationship. This has resulted in leeches being variously classified as Hirudinea, Hirudinoidea, or Euhirudinea.

The diagnostic features of leeches are (1) 32 postoral metameres (Fernandez 1980, Weisblat *et al.* 1980) plus a nonsegmental prostomium; (2) a reduced or obliterated coelom; (3) absence in the adult of chaetae and septa; (4) the presence of an anterior (oral) sucker and a usually larger posterior (anal) sucker (Fig. 13.1); (5) superficial subdivision of the metameres into annuli (Figs. 13.2a and 13.34a, b); and (6) a median, unpaired, male genital pore placed anterior to the single median, ventral, female genital pore (Fig. 13.3). The phylogenetic position of the Acanthobdellida is uncertain, because they share some morphological, behavioral, and ecological attributes with leeches, while also showing significant differences. Several workers (Mann 1961, Elliott and Mann 1979, Klemm 1982, 1985, Sawyer 1972, 1986) include the Acanthobdellida within the leeches. Here, they are considered as a separate class with an independent origin from the Oligochaeta or early Annelida, but showing parallel or convergent evolution with the leeches.

A similar argument has been forwarded by Holt (1969) for the systematic position of the leech-like Branchiobdellidae (Chapter 12). Acanthobdellida are temporary ectoparasites of salmonid fishes. Unlike leeches, the Acanthobdellida have five anterior coelomic compartments separated by distinct septa.

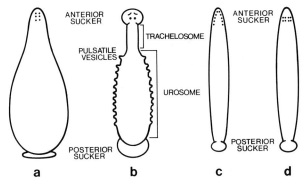

Figure 13.1 Dorsal view of the general body shape of a member of each of the families: (a) Glossiphoniidae; (b) Piscicolidae; (c) Hirudinidae; and (d) Erpobdellidae.

These anterior segments bear two pairs of chaetae (setae) (Fig. 13.4). They do not have an anterior sucker, but a small posterior sucker is formed from 4 of the 29 postoral metameres forming the body. Each segment is superficially subdivided into four annuli (Fig. 13.2b).

Classically, oligochaetes and leeches are placed within the phylum Annelida either in the order Hirudinea, class Clitellata, or in the class Hirudinea. Soos (1965, 1966a, 1966b, 1966c, 1967, 1968, 1969a, 1969b) placed leeches in the taxonomically undefined Hirudinoidea, but subsequently both Soos (1969b) and Richardson (1969) referred to the class Hirudinoidea. Correctly, Klemm (1985) and Sawyer (1986) indicated that in accordance with the Zoological Code the suffix "-oidea" should only be added to the stem name of a superfamily and recommended the suppression of the class Hirudinoidea. Davies (1971 and subsequent references) classified leeches as Hirudinoidea implicitly as a superfamily of the class Oligochaeta.

Until recently, these differences were primarily academic because everyone agreed that as either Hirudinea or Hirudinoidea, leeches were in the phylum Annelida and closely related to oligochaetes. However, Sawyer (1986) defined the class Hirudinea in a significantly different way, with Euhirudinea (leeches), Acanthobdellidae, Branchiobdellidae, and Agriodrilidae as subclasses within the phylum Uniramia. Thus, the class Hirudinea (*sensu* Sawyer

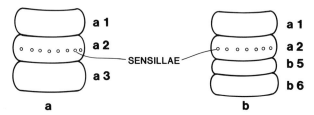

Figure 13.2 Annulation showing (a) leech 3-annulate condition and (b) *Acanthobdella peledina* 4-annulate condition.

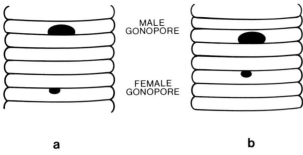

Figure 13.3 Ventral view of the anterior male and posterior female gonopores separated by (a) four annuli and (b) two annuli.

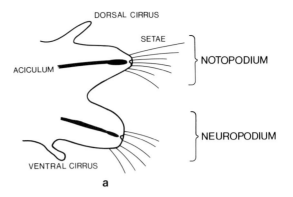

a

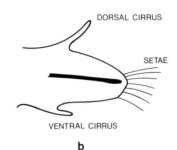

b

Figure 13.5 Parapodia of Polychaeta: (a) biramous, (b) uniramous.

1986) requires acceptance of a monophyletic relationship between leeches, acanthobdellids, and branchiobdellids, removes it from the Annelida, and places it in a phylum with the Onychophora, Myriapoda, and Hexapoda. As the evidence to support this is scant (Sawyer 1984, 1986) and contentious (Davies 1987), the term Hirudinea will not be used in this chapter. While the evidence that leeches are a superfamily (i.e., Hirudinoidea) is not conclusive, this term will be used because it has at least the advantage of clearly maintaining the leeches within the Annelida, separating it from the Uniramia, and does not suggest a monophyletic relationship with other groups.

Polychaetes are metamerically segmented with each metamere bearing a pair of lateral, paddle-like parapodia (Figs. 13.5a, b, and 13.36). At the anterior is a well-developed head consisting of the prostomium (which bears antennae), a pair of palps, zero to several pairs of eyes, and the peristomium with its ventrally located mouth (Fig. 13.6). The prostomium projects forward over the mouth, and the peristomium is frequently modified with sensory, peristomial cirri. The body segments are basically similar

but in some species the body is differentiated into distinct regions (thorax and abdomen) (Fig. 13.7a).

III. LEECHES AND ACANTHOBDELLIDAE

A. General Anatomy

The body of a leech consists of 2 preoral, nonmetameric segments and 32 postoral metameres designated I through XXXIV, while the Acanthobdellidae

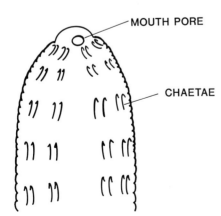

Figure 13.4 Ventral view of the anterior of *Acanthobdella peledina* showing the absence of an anterior sucker, the presence of a mouth pore, and five pairs of hooked chaetae.

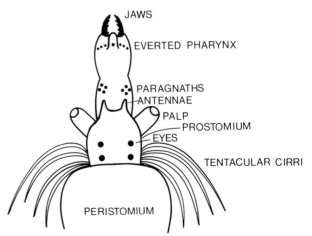

Figure 13.6 Anterior of *Nereis succinea* showing prostomium with antennae, palps and eyes, and everted pharynx.

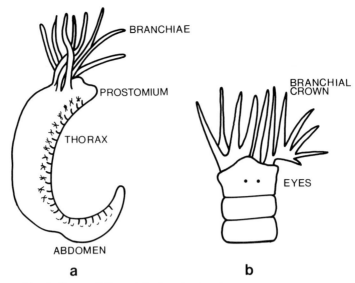

Figure 13.7 (a) General body form of *Hypaniola florida* showing body divided into thorax and abdomen and four pairs of branchiae; (b) anterior of *Manayunkia speciosa* showing the branchial crown.

have only 29 postoral metameres (designated I through XXIX). Each segment is usually divided externally by superficial furrows into 2–16 annuli (Figs. 13.2a, 13.34a,b). Segments with the full number of annuli (complete segments) are found in the middle of the body; because this number is generally characteristic of the genus or species, it is referred to in the keys. Annuli features can be most easily seen in the lateral margins of the ventral surface.

Moore (1898) numbered the three primary annuli a_1, a_2, and a_3 from the anterior, with annulus a_2 (neural annulus) containing the nerve cord ganglion. The neural annulus (a_2) is marked externally by transverse rows of cutaneous sensillae (Fig. 13.2a). Repeated bisection of the primary annuli gives more complex annulation (Table 13.1; Figs. 13.2b and 13.34a, b).

The number and position of the eyes are also important diagnostic features. In some genera, coalescence of eyes sometimes occurs, but the lobed nature of the eyes usually indicates the original condition (Fig. 13.8b). Eye spots (oculiform spots) can also occur on the lateral margins of the body and on the posterior sucker (Figs. 13.8c, 13.9b, 13.23).

Papillae, small protrusible sense organs, are often widely scattered over the dorsal surface. Tubercles, which are larger protrusions that include some of the dermal tissues and muscles, may also be present and can bear papillae.

The body of most leeches and acanthobdellids is not divisible into regions. With the genera *Illinobdella* and *Piscicolaria* as exceptions, the body of the family Piscicolidae is divided into a narrow anterior trachelosome and a longer and wider posterior urosome (Figs. 13.1b and 13.9a, b). Lateral gills are

absent, but some genera of Piscicolidae have paired pulsatile vesicles on the neural annuli of the urosome (Figs. 13.1b and 13.9a, b).

The anterior sucker of leeches may be very prominent (Figs. 13.1b, 13.9a, b, 13.10b) or simple (Figs. 13.1a, c, d, 13.10a, c), consisting only of the expanded lips of the mouth. In the acanthobdellids, anterior suckers are entirely absent (Fig. 13.4). The

Table 13.1 Nomenclature for the Annuli Produced by Repeated Bisection of the Three Primary Annuli[a]

Primary	Secondary	Tertiary	Quaternary
		c_1	d_1, d_2
	b_1		
		c_2	d_3, d_4
a_1			
		c_3	d_5, d_6
	b_2		
		c_4	d_7, d_8
		c_5	d_9, d_{10}
	b_3		
		c_6	d_{11}, d_{12}
a_2			
		c_7	d_{13}, d_{14}
	b_4		
		c_8	d_{15}, d_{16}
		c_9	d_{17}, d_{18}
	b_5		
		c_{10}	d_{19}, d_{20}
a_3			
		c_{11}	d_{21}, d_{22}
	b_6		
		c_{12}	d_{23}, d_{24}

[a]After Moore (1898).

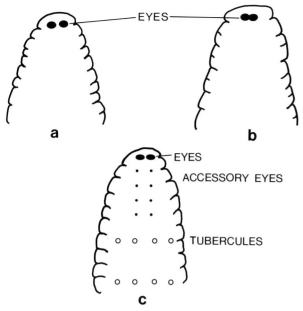

Figure 13.8 Dorsal view of the head of *Batracobdella* with a single pair of eyes (a) close together; (b) lobed; and (c) *Placobdella hollensis* with one pair of eyes followed by pairs of accessory eyes.

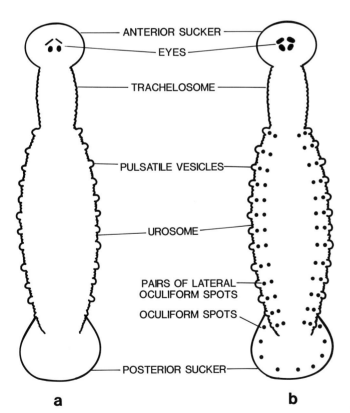

a   **b**

Figure 13.9 (a) *Cystobranchus verrilli* and (b) *Cystobranchus meyeri* showing the pulsatile vesicles on the urosome and the presence on *C. meyeri* of oculiform spots on the posterior sucker and the paired lateral oculiform spots on the urosome.

posterior sucker of a leech is generally directed ventrally and is wider than the body at the point of attachment to the substrate (Figs. 13.1, 13.9, 13.23, 13.26–13.29). The mouth is either a small pore on the edge (Fig. 13.10a) or center (Fig. 13.10b) of the ventral surface of the anterior sucker, or is large and occupies the entire cavity (Fig. 13.10c) of the anterior sucker. In the family Hirudinidae, the buccal cavity (which may or may not contain jaws) is separated from the cavity of the anterior sucker by a flap of skin (the velum) (Fig. 13.11). When present, there are three muscular jaws (two ventrolateral and one dorsomedial) (Fig. 13.11b, c) bearing teeth arranged in one (monostichodont) (Fig. 13.12a) or two (distichodont) rows (Fig. 13.12b).

In the Erpobdellidae, the buccal cavity contains three muscular ridges; and in the Glossiphoniidae and Piscicolidae, the pharynx is modified to form an eversible muscular proboscis. Opening into the pharynx are the salivary glands which secrete some digestive enzymes; but, in bloodsucking leeches these glands are primarily related to the process of bloodsucking rather than to digestion. The components of the salivary secretions and their functions are presented in Sawyer (1986) but can be summarized as lubrication, proteolytic inhibition, spreading (hyaluronidase), vasodilation, and anticoagulation (e.g., hirudin or hementin). The anticoagulants produced by North American sanguivores have not yet been investigated with work so far

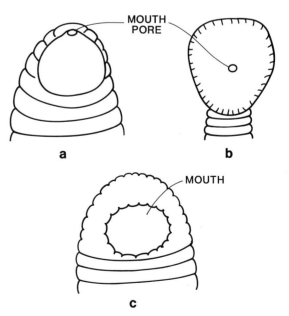

Figure 13.10 Ventral view of the anterior sucker of (a) the mouth pore on the anterior rim of the sucker (e.g., *Marvinmeyeria, Placobdella, Oligobdella*); (b) near the center of the sucker (Piscicolidae, *Batracobdella, Helobdella*); and (c) the mouth occupying the entire cavity (e.g., Hirudinidae, Erpobdellidae).

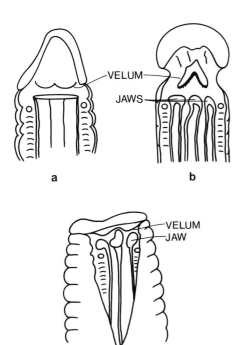

Figure 13.11 Ventral view of the dissection of the mouth and buccal cavity of (a) *Mollibdella grandis;* (b) *Percymoorensis marmorata;* and (c) *Macrobdella decora* showing the velum, and the relative size of the jaws in *Macrobdella decora* and *P. marmorata* and the absence of jaws in *Mollibdella grandis.*

concentrating on the European *Hirudo medicinalis* and the South American *Haementaria ghilianii.* The pharynx leads to the tube-like crop, which is adapted for storage of fluids with 6–11 pairs of lateral caecae. Between the last pair of crop caecae, the intestine leads to the anus (Fig. 13.13a). Only in the Glossiphoniidae does the anterior portion of the intestine give off four pairs of caecae (Fig. 13.13b). The anus usually opens on the dorsal surface on or near segment XXVII, just anterior to the posterior

sucker. In a few species, the anus is displaced anteriorly.

Digestion and absorption mainly occur in the intestine although the crop is also used in predatory species. In sanguivores, digestion is primarily by enzymes produced by endosymbiotic bacteria. The bacteria involved are species specific for each species of sanguivorous leech (Jennings and Vanderlande 1967). For example, *Hirudo medicinalis* utilizes *Aeromonas hydrophila* (*Pseudomonas hirudinis*) (Büsing 1951, Büsing *et al.* 1953, Jennings and Vanderlande 1967). Endopeptidases are absent or rare in leeches, but exopeptidases occur abundantly in predatory species, which may also harbor several species of endosymbiotic bacteria (Jennings and Vanderlande 1967). Because of their dependence on endosymbiotic bacteria, digestion in sanguivorous leeches takes weeks or months; but for predatory species, digestion requires just a few days (Davies *et al.* 1977, 1978, 1981, 1982a, Wrona *et al.* 1979, 1981).

The alimentary canal of the acanthobdellids is similar to that of the glossiphoniid leeches. A small eversible proboscis leads into the muscular phar-

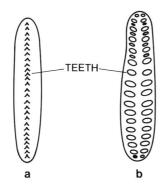

Figure 13.12 Surface view of the jaws of Hirudinidae showing the teeth arranged in (a) one (monostichodont) or (b) two (distichodont) rows.

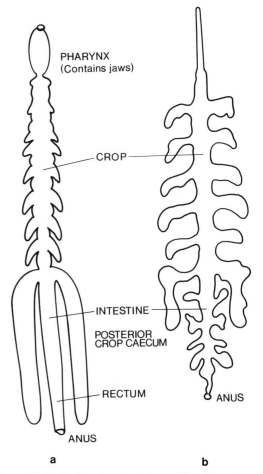

Figure 13.13 Stylized ventral view of the alimentary canal of (a) Hirudinidae and (b) Glossiphoniidae.

ynx, which in turn leads into a large tubular crop lacking caecae. The crop leads into the intestine, which bears six pairs of lateral caecae; the anus is located dorsally between segments XXV and XXVI. Digestion is probably similar to sanguivorous leeches with the use of endosymbiotic bacteria.

The male and female gonopores are visible in mature leeches on the midline of the ventral surface of segments XI and XII, respectively, and are separated by a species-specific number of annuli (Figs. 13.3a, b, 13.32). The anterior male gonopore is larger and more easily seen than the female gonopore. Leeches do not have true testes or ovaries but instead possess testisacs and ovisacs. The testisacs are located posterior to segment XI and are either discrete, paired spherical structures (Fig. 13.14b) (e.g., Hirudinidae, Glossiphoniidae, and Piscicolidae) located intersegmentally or multifollicular columns resembling bunches of grapes lying on either side of the ventral nerve cord (Fig. 13.14a) (e.g., Erpobdellidae, Singhal and Davies 1985, Davies *et al*. 1985). Short vasa efferentia connect the testisacs to the vasa deferentia on each side of the body, which run anteriorly to form large coiled epidymes (sperm vesicles). Paired ejaculatory ducts run from the epidymes through the atrial cornua and unite to form a medium atrium consisting of a bulb and ever-

sible penis. There is a single pair of ovisacs, which are either small, spherical organs confined to segment XII or elongate, coiled tubes sometimes bent back on themselves such that their blind ends lie close to the female gonopore (Singhal and Davies 1985, Davies *et al*. 1985). Two oviducts run from the ovisacs to converge and form the common oviduct leading to the female gonopore (Fig. 13.14).

Leeches are frequently recorded as sequential protandrous hermaphrodites (Van Damme 1974, Lasserre 1975). *Nephelopsis obscura* shows sequential protandry during its first bout of gametogenesis, although there is a short transition period when gametogenic stages from both sexes co-occur. In the second cycle of gametogenesis, mature sperm and ova occur simultaneously (Davies and Singhal 1988) and thus this cycle demonstrates simultaneous hermaphroditism.

At the time of egg laying, a cocoon is secreted by the clitellum, a specialized region of epidermis usually posterior to the gonopores. As the cocoon moves toward and passes over the head, fertilized eggs pass into the cocoon from the female gonopore.

The reproductive system of *Acanthobdellida peledina* is basically similar with a pair of elongate testisacs and ovisacs and median unpaired male and female gonopores.

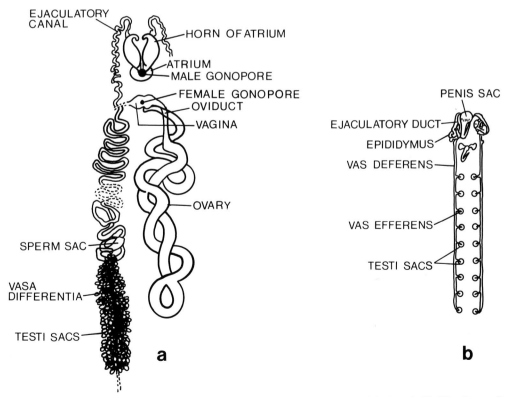

Figure 13.14 Reproductive systems of (a) *Nephelopsis obscura* (Erpobdellidae) and (b) *Hirudo medicinalis* (Hirudinidae).

The excretory system of leeches and acanthobdellids consists of a maximum of 17 pairs of highly modified metanephridia opening along the body between segments VII and XXII. Many species do not exhibit the full number of metanephridia, showing reductions in the more anterior metameres and/or the segments containing gonads. In addition to excretory nitrogenous wastes (mainly ammonia), the nephridia help maintain water and salt balance in the body. With the osmotic pressure of the body higher than the surrounding freshwater, there is a tendency for an inward flow of water through the epidermis. The nephridia pass out excess water and help retain inorganic ions. Only about 6% of the salts found in the primary urine are excreted, the rest being reabsorbed (Zerbst-Boroffka 1975).

In the leeches and acanthobdellids, the large coelom, which in Annelida is typically divided by intersegmental septa, has to varying degrees been obliterated (Fig. 13.15). In the Glossiphoniidae (Fig. 13.15a), Piscicolidae (Fig. 13.15d), and acanthobdellids, the blood vascular system consists of: (1) a ventral vessel enclosed with the nerve cord in the ventral lacuna; (2) a dorsal blood vessel enclosed in the dorsal lacuna; and (3) transverse blood vessels connecting the dorsal and ventral vessels in the anterior and posterior parts of the body. In the Hirudinidae (Fig. 13.15b) and Erpobdellidae (Fig. 13.15c), the true blood vascular system is completely lost; instead the blood circulates in the coelomic lacuna system, forming a haemocoel.

Oxygen uptake in freshwater leeches and acanthobdellids is through the general body surface, although some species of Piscicolidae have small pulsatile vesicles filled with coelomic fluid that function as accessory respiratory organs. Most leeches can also ventilate their body surfaces by dorsoventral undulations along the body reminiscent of swimming but with the posterior sucker firmly attached. However, *Acanthobdella peledina* does not ventilate (Dahm 1962).

All leeches and acanthobdellids show typical looping; locomotory movements consisting of body elongation and shortening, with the anterior sucker (leeches only) and posterior sucker serving alternately as points of attachment. Many leech species are also able to swim using dorsoventral undulations. All Erpobdellidae and Hirudinidae are good swimmers, but only a few species of Glossiphoniidae and Piscicolidae are efficient swimmers. Some species (*Placobdella hollensis*) swim readily as juveniles and adults, but others (*Placobdella ornata* and *Placobdella parasitica*) swim only as juveniles (Sawyer 1981).

B. Physiology

Perhaps the greatest variability in the abiotic environment experienced by leeches occurs in reference to temperature, dissolved oxygen concentrations, ionic content, and total dissolved salts.

In stratified lakes, the epilimnion is generally at or above saturation during the summer, while in the hypolimnion oxygen depletion occurs. Under ice,

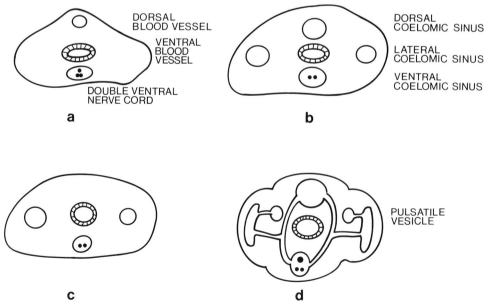

Figure 13.15 Stylized transverse sections of (a) Glossiphoniidae; (b) Hirudinidae; (c) Erpobdellidae; and (d) Piscicolidae showing the coelomic sinuses, dorsal and ventral blood vessels, and the double ventral nerve cord.

especially at the substrate–water interface, hypoxia and possibly anoxia occur (Babin and Prepas 1985, Baird *et al.* 1987a). Thus, hypoxia or anoxia can occur on a diel basis during the open-water season and on a long-term basis during ice cover. As spring approaches, the water temperature rises above 5°C, snow cover melts and under-ice primary productivity can result in short-term (1–2 days) hyperoxia. Similarly during the summer, long-term hyperoxia can persist for several days. Both anoxia and hyperoxia can cause mortality of vertebrates and invertebrates (Casselman and Harvey 1975). In Europe, Mann (1956) showed that oxygen uptake by some leech species was proportional to the oxygen concentration of the water (i.e., they were conformers), while other species (*Glossiphonia complanata*, *Helobdella stagnalis*) showed a comparatively constant oxygen uptake over a range of dissolved oxygen concentrations of the water (i.e., they were regulators). *Nephelopsis obscura* and *Erpobdella punctata* exposed to short-term hypoxia both showed a decline in oxygen uptake with reduced oxygen tension in the water (Wrona and Davies 1984). When exposed to hypoxic conditions, most leeches, but not *Acanthobdella peledina* (Dahm 1962), show ventilation movements with the posterior sucker attached to the substrate and the body thrown into dorsoventral flexions. It is assumed that the primary function of ventilation is respiratory since ventilation increases under conditions of low-oxygen availability. While *N. obscura* and *E. punctata* ventilate in hypoxic conditions, ventilations also occur under saturated (normoxic) conditions and always cease when oxygen decreases below 20% saturation. This suggests ventilation may have additional roles.

Many leeches can withstand anaerobic conditions (von Brand 1946); but little research has been done on the ecological effect of anoxia on the leeches, with the majority of work related to use of leeches as indicator species in the European saprobic system (reviewed in Sladacek and Kosel 1984).

In flow-through experiments, Davies *et al.* (1987) showed inter- and intraspecific differences in the survivorship of *N. obscura* and *E. punctata* to anoxic conditions and demonstrated the errors of using static experiments for this type of study (Fig. 13.16). For both species, survivorship decreased with increased water temperature. Large *N. obscura* had longer survival times (> 50 days) than small individuals (12 days) at 5°C and also at 20°C (8 and 4 days, respectively). Conversely, at 5°C large *E. punctata* had shorter survival times (25 days) than small individuals (> 50 days), but at 20°C, large individuals survived longer (8 days) than small individuals (2 days). As young (small) *N. obscura* have a

low probability of surviving anoxia at winter temperatures (5°C), it is evident that large *N. obscura* would be taking less risk if they overwintered rather than hazarding cocoon production and hatching of young up to or close to ice formation. In contrast, large *E. punctata,* which have a low probability of surviving winter anoxia, would be taking less risk to continue cocoon production later in the season since their offspring can survive winter anoxic conditions. These results correlate well with the patterns of reproduction observed in the field (Davies 1978, Davies *et al.* 1977).

Even less work has been done on the effect of hyperoxia on leeches. Survival of *N. obscura* and *E. punctata* exposed to hyperoxia (200%, 300%) increased with leech size and decreased with increased temperature. The survivorship time for all size classes of *N. obscura* and *E. punctata* greatly exceeded the maximum recorded duration (1–2 days) of hyperoxia in the field during the spring. However, during the summer (20°C), if hyperoxia in the 200–300% range persisted for more than 20 days, medium and large *N. obscura* would show higher mortality than comparable sizes of *E. punctata*.

With very low or undetectable levels of superoxide dismutase in *N. obscura*, Singhal and Davies (1987) concluded that *N. obscura* had two possible defenses against hyperoxia. These were to move away from areas of hyperoxia or, as recorded by

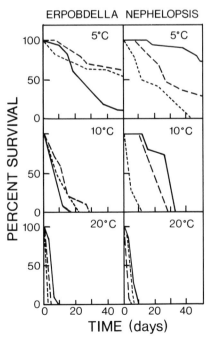

Figure 13.16 Percentage of survival of three size ranges of *Nephelopsis obscura* and *Erpobdella punctata* in flow-through anoxic conditions at three temperatures. Large (150–250 mg) (___); medium (50–140 mg) (-----); small (10–15 mg) (. . . . .).

Singhal and Davies (1987), to secrete large amounts of mucus over the epidermis, which reduces oxygen diffusion into the body and thereby the effects of oxygen toxicity. Davies and Everett (1977) and Davies *et al.* (1977) recorded intra- and interspecific differences in the timing of seasonal movements between lentic microhabitats of *N. obscura* and *E. punctata*. The results from these investigations on hyperoxia show that small leeches are highly intolerant of hyperoxia, especially at higher temperatures, and suggest that their seasonally earlier movement to deeper water is related to avoidance of potentially lethal oxygen conditions. Similarly, the presence of only large leeches in the macrophyte zone, where hyperoxia is more common and more persistent, is compatible with the greater tolerance of the large size classes to hyperoxic conditions.

Ultrastructural investigations of the neurons of *N. obscura* exposed to anoxia and hyperoxia (Singhal *et al.* 1988) showed a marked reduction in numbers of mitochondria, ribosomes, and neurotransmitter vesicles with swollen and fragmented cristae. These changes are more severe in anoxia than hyperoxia (Fig. 13.17).

In most lakes and rivers inhabited by leeches, ionic content and total dissolved salts (TDS) in the water show little temporal variability. However, significant geographic variability occurs between lakes and rivers primarily related to the edaphic features of the underlying geological formations. It has been shown that the distribution of leeches is, at least in part, related to TDS of the habitat waters (Herrmann 1970a, 1970b, Scudder and Mann 1968, Reynoldson and Davies 1976). Similarly, in Europe, Mann (1955) showed that the composition of the leech fauna in lentic habitats was a function of water hardness. It has been demonstrated that the relative concentrations of ions in the medium modify the mortality of aquatic animals (Croghan 1958, Beadle 1939, Linton *et al.* 1983a, 1983b).

The blood of leeches is relatively concentrated compared to oligochaetes and is hyperosmotic to the water in which they live. *Nephelopsis obscura* and *Helobdella stagnalis* were able to maintain themselves hyperosmotically in medium between 15.6 and 59.5 mosmol/L^{-1} and 47 and 112.7 mosmol/L^{-1} respectively but were conformers at higher medium concentrations. *Theromyzon rude* was a regulator in both hypo- and hyperosmotic media. Leeches are also capable of regulating body volume when exposed to varying salinities (Rosca 1950, Madanmohanrao 1960, Smiley and Sawyer 1976, Reynoldson and Davies 1980, Linton *et al.* 1982). *Nephelopsis obscura* was able to maintain its weight in medium from 100–207 mosmol/L^{-1} while *E. punctata* could only regulate between 58 and 200

mosmol/L^{-1}. Weight regulation and volume regulation are probably not regulated per se, although these variables change in a predictable fashion with change in the osmotic concentration of the environment. It is more likely that leeches regulate the osmotic pressure of the body fluids; hence with influxes and effluxes of water and salts, weight, and volume change.

Using a three-way factorial experimental design, the effect of water temperature, ionic content, and TDS on the mortality and reproduction of *N. obscura* and *E. punctata* were examined (Linton *et al.* 1983a, 1983b). A strong interaction among temperature, ionic content, and TDS was observed on the mortality of *E. punctata*, but *N. obscura* showed uniformly low mortality unaffected by temperature, ionic content, or TDS. For both species, temperature showed the greatest influence on cocoon production and ionic content showed the least. Cocoon production by *E. punctata* was reduced in waters with low TDS and the proportion of nonviable cocoons was higher in low TDS than in high TDS.

IV. ECOLOGY

A. Diversity

Leeches are represented in North America by four families. Glossiphoniidae are either predators of macroinvertebrates or temporary ectoparasites of freshwater fish, turtles, amphibians, or water birds; Piscicolidae are parasites of fishes as well as Crustacea; Erpobdellidae are primarily predators of macroinvertebrates and zooplankton; and, Hirudinidae are either predators of macroinvertebrates (e.g., oligochaetes, snails) or bloodsucking ectoparasites of freshwater and terrestrial Amphibia and mammals. Sanguivorous (bloodsucking) leeches spend a relatively short period of time taking a blood meal on the host and are more frequently found free living in the benthos. For this reason, host specificity is not a very useful feature for identification of leeches.

While the majority of leeches parasitic on fish belong to the family Piscicolidae, two species of Glossiphoniidae (*Actinobdella inequiannulata* and *Placobdella pediculata*) show a strong preference for fish hosts, and two other species of glossiphoniids (*P. montifera* and *P. phalera*) are also occasionally found on fishes. Members of the Erpobdellidae and Hirudinidae have been recorded on fishes, but it is doubtful that any species in these families regularly feeds on these hosts. All leeches require a firm substrate for attachment and are capable of feeding on body fluids. If a healthy or injured fish can provide either of these requirements, all

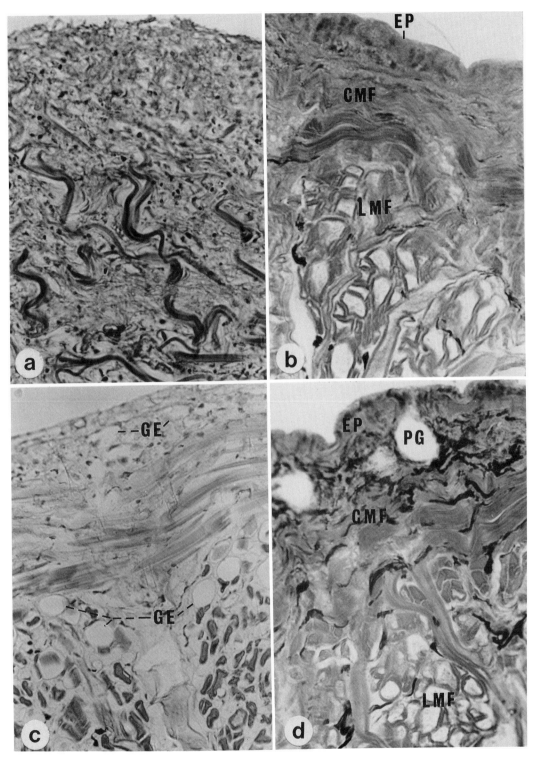

Figure 13.17 Cross-sections of the body wall of *Nephelopsis obscura* exposed to supersaturated (300%) (a, c) and saturated (100%) (b, d) oxygen conditions for 14 days showing (a) dissolution of muscle fibers; (b) the presence of intact muscle fibers; (c) absence of pear-shaped glands in the epithelium (EP) and the presence of gas emboli (GE) in animals exposed to supersaturation; and (d) presence of pear-shaped glands (PG) and absence of gas emboli in animals exposed to normoxic saturated conditions. CMF, circular muscle fibers; LMF, longitudinal muscle fibers.

freshwater leech species will attach themselves. Parasitic leeches generally attach periodically to the fishes, take a blood meal, and then move back into the benthos. Acanthobdellids feed similarly on cold-water fishes.

B. Foraging Relationships

Leeches are either predators, consuming macroinvertebrates in the benthos and plankton, or ectoparasitic sanguivores, feeding mostly on vertebrates. Acanthobdellids are strictly ectoparasites. In all cases, leeches and acanthobdellids are primarily fluid feeders; because, even though some predators ingest whole prey, they quickly evacuate the hard skeletal parts through the mouth or anus after extracting body fluids. Sanguivory occurs in the Hirudinidae, which have three-toothed jaws used to bite and penetrate the host's skin; it is also present in the Piscicolidae and some Glossiphoniidae, which in contrast, lack jaws and penetrate the

host with a proboscis. The acanthobdellid similarly has a short proboscis-like specialization of the anterior foregut.

Theromyzon rude and *T. tessulatum*, sanguivorous glossiphoniids feeding on waterfowl, require a minimum of three blood meals (Davies 1984, Wilkialis and Davies 1980). In comparison, *Placobdella papillifera* and *Haementaria ghilianii* require a minimum of four (Davies and Wilkialis 1982, Sawyer *et al.* 1981). The glossiphoniids generally appear to require significantly fewer blood meals than *Hirudo medicinalis,* which takes ten or more meals (Blair 1927, Pütter 1907, 1908)—although in the laboratory at 26°C, only 4–5 blood meals are required to reach maturity (R. W. Davies unpublished). For *T. rude,* over 80% of the population takes three meals in the first six months after hatching (Fig. 13.18). The remainder of the population overwinters after two meals and take the third in the spring so that all the population reproduces approximately 12 months after hatching. Some individuals that overwinter after three meals decline in weight before spring and

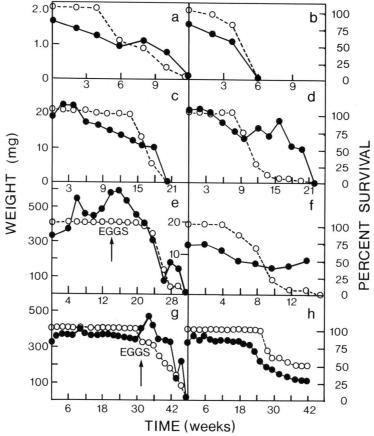

Figure 13.18 Changes in mean weight (●—●) and the percentage of survival (o----o) for populations of *Theromyzon rude* exposed to different water temperatures and feeding regimes: (a) fed once, maintained at 20°C; (b) fed once, maintained at 5°C; (c) fed twice, maintained at 20°C; (d) fed twice, maintained at 5°C; (e) fed three times, maintained at 20°C; (f) immature (14.7 mg) collected from the field, fed once, maintained at 20°C; (g) immature (14.7 mg) collected from the field, fed once, maintained at 5°C.

require a fourth meal before reproduction commences.

Predatory leeches either suck fluids with a proboscis (Glossiphoniidae) or have a suctorial mouth (Erpobdellidae and Hirudinidae). The range of prey species is normally quite diverse and changes seasonally with prey availability. In some instances, prey selection varies intraspecifically with body size and/or age class (Davies *et al.* 1978, 1981, 1988, Wrona *et al.* 1979, 1981). Because leeches either do not ingest identifiable hard parts or quickly evacuate them, however, the only accurate and efficient method of examining feeding is with serologic techniques. Antisera against prey are produced (Davies 1969) and then used for a serologic test of the gut contents.

The feeding of *Nephelopsis obscura* and *Erpobdella punctata* was investigated using specific rabbit antisera against Cladocera/Copepoda, Chironomidae, Oligochaeta, Amphipoda, and Gastropoda. Apart from the absence of Gastropoda in the diet of *E. punctata,* there were no significant differences at the species level in prey utilization, niche breadth, and evenness (Fig. 13.19). However, temporal differences in prey utilization were evident, and both intra- and interspecific resource partitioning oc-

curred as a function of different weight class utilization of prey (Davies *et al.* 1981). Similarly, temporal differences in feeding and intraspecific weight class differences in prey utilization were found in sympatric *Glossiphonia complanata* and *Helobdella stagnalis* (Wrona *et al.* 1981) (Fig. 13.20).

While most predatory leech species utilize a variety of prey, *Erpobdella montezuma* is an exception. The pelagic amphipod *Hyalella montezuma* comprises nearly 90% of the diet of the endemic *E. montezuma* in the thermally constant environment of Montezuma Well (Arizona), even though numerous other potential prey are abundant throughout the year (Blinn *et al.* 1987, Davies *et al.* 1988). This restricted diet was confirmed by both gut content and serological analyses.

C. Reproduction and Life History

All leeches are hermaphrodites showing protandry or cosexuality (Davies and Singhal 1988), with reciprocal cross-fertilization as the general rule. Fertilization, which is internal, is accomplished in the majority of the Glossiphoniidae and all of the Piscicolidae and Erpobdellidae by attaching a spermatophore to the body of the partner. The spermatozoa

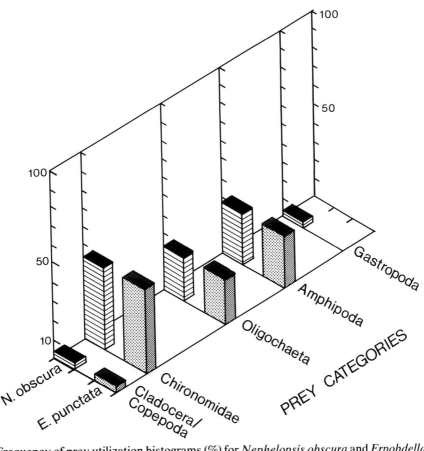

Figure 13.19 Frequency of prey utilization histograms (%) for *Nephelopsis obscura* and *Erpobdella punctata*.

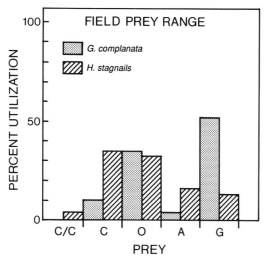

Figure 13.20 Frequency of prey utilization histograms (%) for *Glossiphonia complanata* and *Helobdella stagnalis* (C/C, Cladocera/Copepoda; C, Chironomidae; O, Oligochaeta; A, Amphipoda; G, Gastropoda).

penetrate the body wall and make their way to the ovisacs via the coelomic sinuses. The clitellar region is the most frequent site for the deposition of spermatophores, but they can be attached to any part of the body of the partner. In some species of Piscicolidae, there is a specialized region for the reception of spermatophores; fertilization is affected only by spermatophores deposited there. In the Hirudinidae, reciprocal internal fertilization is brought about by the insertion of an eversible penis into the vagina of the partner.

Once fertilization occurs, the eggs are deposited into a cocoon secreted by the clitellum (which is not always easily visible). There can be a considerable delay between copulation and cocoon deposition. For example, in field populations of *Helobdella stagnalis,* copulation occurs in the fall and cocoon deposition takes place in the spring (Davies and Reynoldson 1976). Erpobdellid cocoons are thick-walled, oval in shape, and attached to a firm substrate (stone, leaf, wood). The cocoons of Piscicolidae and Hirudidae are only loosely attached to the substrate and are usually spherical. In species of hirudinids depositing their cocoons in moist habitats out of water, the outer wall of the cocoon is spongy, which is thought to reduce water loss. The cocoons of Glossiphoniidae are very thin-walled (sometimes called egg cases) and are either deposited on the substrate and immediately covered by the body of the parent or are attached to the ventral surface of the parent. In both cases, the hatchlings are attached to the ventral body wall and are carried around by the parent for a long period, which for *Theromyzon*

tessulatum can be five months and for *T. rude* can be one month (Wilkialis and Davies 1980a).

The life cycles of all leeches and acanthobdellids consist of egg, juvenile, and mature hermaphrodite adult, which reproduces and produces more eggs. Adults of predatory species usually reproduce only once before death (i.e., semelparity), although some sanguivorous species reproduce several times over a few years (i.e., iteroparity). Semelparity and iteroparity are generally considered genetically different reproductive strategies determined through evolutionary processes. Studies on the reproductive biology of leeches have shown considerable variation in life history with annual or biennial life cycles in predatory species. Some sanguivorous species (e.g., *T. rude*) live for several years with the duration of the life cycle dependent on the interval between blood meals (Davies 1984). Baird *et al.* (1986, 1987) showed that changes in water temperature, size at reproduction, energy loss during reproduction, and postreproductive feeding of the erpobdellid *Nephelopsis obscura* significantly affected postreproductive mortality. Although genetically iteroparous, *N. obscura* is like most predatory leeches and some sanguivorous species in being almost always semelparous in the field because of high postreproductive mortality. Thus, it is quite possible that the range of life histories reported for various species of leeches falls within the range of environmentally induced variability. Indeed, it is conceivable that all leeches are genetically iteroparous but phenotypically exhibit semelparity under most conditions. This flexibility would ensure the long-term persistence of genotypes in a variable environment, where a more rigid strategy might have a low or zero fitness in some years. Freshwater ecosystems are highly variable in terms of both abiotic and biotic environmental parameters, and Bulmer (1985) suggested that iteroparity is found in variable environments as a bet-hedging or risk-avoidance strategy.

The life cycle and life history of comparatively few freshwater leech species in North America have been examined. In Alberta, two generations of *N. obscura* are produced annually, one in the spring and one in the late summer. The spring generation is the progeny of the heavier individuals from the spring generation of the previous year and from the late-summer generation produced two years previously. The late-summer generation is produced by portions of the previous year's spring and late-summer generations. Thus each generation produces young after either 12 and 15 months or 12 and 19 months, although each individual reproduced only once (Davies and Everett 1977) (Fig. 13.21). In Minnesota, *N. obscura* delays its age of reproduction to two years and in so doing attains a larger body size (Peterson 1982, 1983). *Erpobdella punctata* is usually subdominant to *N. obscura* and reproduces after one year. In the rare instances when *E. punctata* is dominant, however, it has a more complex cycle with reproduction after one or two years, at a larger size, over a shorter season (Davies *et al.* 1977). *Helobdella stagnalis* can produce either one or two generations per year depending on the water temperature regime (Davies and Reynoldson 1976). In warmer conditions, the overwintering population reproduces in the spring and dies on completion of brooding. The spring generation grows rapidly and a second generation of young is produced in the summer, which forms the overwintering population. In colder temperature regimes, members of the overwintering population reproduce in the late spring/early summer and die after brooding. After summer growth, this generation forms the next overwintering population. Davies (1978) showed that populations of *H. stagnalis* from cold water regimes are able to produce two generations per year in appropriate conditions.

Several environmental factors have been implicated in affecting the reproduction of leeches, including food availability and water temperature. Water temperature regime can affect the relative reproductive success and growth of species (Davies and Reynoldson 1976, Davies 1978, Wrona *et al.* 1987), as well as depth distribution and seasonal movement (Gates and Davies 1987). A bioenergetics

simulation model of the growth and life history of *N. obscura* (Linton and Davies 1987) showed that its growth is more sensitive to prey variation between years than to temperature variation.

The life cycle of *Acanthobdella peledina* is basically annual with young immature individuals first occurring on its host in June. These individuals leave the fish in October on reaching sexual maturity. Reproduction with egg and cocoon production occur off the fish and have not been observed (Holmquist 1974).

D. Dispersal

The dispersal of freshwater leeches and acanthobdellids is generally assumed to be the result of passive transfer from water body to water body by other animals. This seems probable for parasitic sanguivorous species but improbable for the majority of freshwater leeches which are predators of benthic invertebrates, although one case has been recorded by Daborn (1976). Davies *et al.* (1982b) examined passive transfer by ducks of two parasitic species (*Theromyzon rude* and *Placobdella papillifera*) and two predatory species (*Helobdella stagnalis* and *Nephelopsis obscura*). *Theromyzon rude* adults were transferred both in the nares, while taking a blood meal, and on the duck's body under the feathers. *Placobdella papillifera* were also conveyed on the body of the duck, but adult *H. stagnalis* and *N. obscura* were not transported. Adult leeches ingested by ducks were not recovered from the duck feces, but viable *N. obscura* cocoons were recovered from the feces of fed ducks.

Passive dispersal by wind has never been recorded, but the transport of cocoons attached to macrophytes is possible. Davies (1979) showed that *Helobdella stagnalis*, *H. triserialis*, *Glossiphonia complanata*, and *Mooreobdella fervida* were dispersed to Anticosti Island by sea currents from the Quebec north shore of the Gulf of St. Lawrence. Dispersal by currents in large lakes or through rivers is also highly probable, and leeches have been recorded in the drift (Elliott 1973). Accidental transport of leeches by humans has not yet been shown in North America but has been demonstrated in other areas of the world (Sawyer 1986). Active directional dispersion up rivers was recorded for *Percymoorensis marmorata* by Richardson (1942) and for *E. punctata* by Sawyer (1970).

E. Predation

In North America, predators of leeches include fish, birds, garter snakes, newts, and salamanders (Able

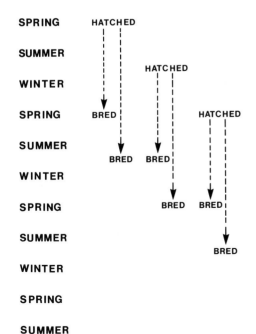

Figure 13.21 Life cycles of spring and summer hatchling *Nephelopsis obscura* in Alberta, Canada.

1976, Arnold 1981, Bartonek 1972, Bartonek and Hickey 1969a, 1969b, Bartonek and Murdy 1970, Bartonek and Trauger 1975, Davies *et al.* 1982b, Kephart 1982, Kephart and Arnold 1982, Pearse 1932), as well as insects, gastropods, and amphipods (Pritchard 1964, Hobbs and Figueroa 1958, Sawyer 1970, Anderson and Raasveldt 1974). Young and Spelling (1986), Spelling and Young (1987) and Young (1987), reviewing predation on three lentic leech species in Great Britain, recorded a similar range of predators as well as interspecific leech predation and cannibalism.

Leeches lack hard body parts and are thus difficult, if not impossible, to identify from the gut contents of predators. Serology is the most appropriate method for analyzing predation on soft-bodied animals (Davies 1969), but this approach requires prior knowledge of the range and size of potential predators to be collected from the field and tested.

In the laboratory, Cywinska and Davies (1989) found that four species of dytiscids, one species of Odonata, two species of hemipterans, and two species of amphipods fed on one or more size classes of *N. obscura*. In Great Britain, Young and Spelling (1986) also noted that species within these taxa fed on *E. octoculata*, *Glossiphonia complanata*, and *Helobdella stagnalis* in addition to triclads, trichopterans, and megalopterans. It would thus appear that larval and adult coleopterans and nymphal odonates and hemipterans are potentially voracious predators of leeches throughout the Holarctic. Young and Spelling (1986) did not find amphipods feeding on leeches, but they did not test for predation on leech cocoons.

In experiments with predation on cocoons, Cywinska and Davies (1989) recorded two methods of cocoon consumption: either the embryos and eggs within the cocoons were fed upon, by piercing holes through the cocoon wall (e.g., Coleoptera larvae and adult Hemiptera), or the whole cocoon (sometimes plus attached plant leaf) was consumed (e.g., amphipods and adult Coleoptera). Predation has also been recorded by gastropods on *E. punctata* cocoons (Sawyer 1970) and by a mite on *Haemopsis sanguisuga* cocoons (Bennike 1943). The only known vertebrate predators of leech cocoons are salamanders (Pearse 1932) and waterfowl (Davies *et al.* 1982b).

Predation rates on *N. obscura* in the laboratory were inversely related to size, with individuals > 30 mg being consumed only at very low rates by certain coleopterans. Young and Spelling (1986) found a similar trend of increased consumption of leeches by most predators with declining leech weight. This can be explained by the inability of some potential predators to capture and overcome the more powerful larger leeches, rapid satiation of predators with larger prey, or the ability of larger leeches to produce more mucus and so impede the movements of the potential predators.

A caveat of these feeding experiments was that predators were starved, had no alternative food available, and had no choice in size of *N. obscura* prey. In such conditions, the results of the experiments probably showed the maximum potential predation rates and suggested that predation on cocoons and size classes < 10 mg might be substantial. Considering the very high densities of potential predators in many lakes and ponds and the ability of some predators to consume large numbers of leeches and/or their cocoons, the effects of predation in the field are potentially significant in macrophyte zones.

The rapid decline in density of hatchling and small (< 10 mg) *N. obscura* observed in the field by Davies and Everett (1977) could be explained by predation in the shallow macrophyte zones where cocoons are deposited and hatch. Placement of cocoons by *E. montezuma* on deeply submerged *Potamogeton* stems at least 3.5 m below the water surface at the interface of the littoral–pelagic zone in Montezuma Well may be a strategy to reduce predation (Davies *et al.* 1988).

Predation in the water column is almost entirely restricted to fish, amphibia, and birds. The presence of large numbers of leeches in the gut contents of fish (particularly from rivers) and the development of bait-leeches in Minnesota (Peterson 1982) suggest predation can be at least occasionally heavy.

F. Behavior

Sawyer (1981) suggested that the behavior of leeches and acanthobdellids could be divided into a number of elementary responses: crawling (vermiform, inchworm), swimming, shortening (fast, slow), searching (head movement, body waving), and alert posture when the body is extended and held motionless for some time. However, species differ markedly in the frequency of expression of these different responses. Behavior of leeches can be modified by size (age) and physiological state (starved or fed, and mature or immature) as well as by extrinsic environmental parameters monitored by eyes, sensillae, chemoreceptors, and mechanoreceptors (T-cells).

The foraging behaviors of sanguivores and predators are usually different. Sanguivorous leeches in the proximity of a potential host respond to changes in illumination (shadows), vibrations in the water, chemical stimuli, and possibly differences in temperature between the host and the surrounding water. Compared to predatory leeches, the sanguivores are less negatively phototactic and nocturnal, especially when hungry. *Theromyzon tessulatum, T. rude, Piscicola geometra,* and *Calliobdella vivida,* when hungry, have all been found in the open water ready to attach to a host (Herter 1929a, 1929b, Meyer and Moore 1954, Sawyer and Hammond 1973). *Hirudo medicinalis* and *Theromyzon tessulatum* are stimulated to attach to and bite hosts with temperatures of 37°–40°C (Dickinson and Lent 1984) and *Placobdella costata* is attracted from as far as 15 cm to hosts with temperatures of 33°–35°C (Mannsfield 1934). *Haementaria ghilianii* and *Hirudo medicinalis* are both sensitive to water disturbances (traveling surface waves) (Sawyer 1981, Young *et al.* 1981), and it is likely that most sanguivores respond in a similar manner.

Water disturbance is frequently the first indication of the presence of a potential host; the leech moves toward the disturbance using cues detected by their sensilla (Friesen and Dedwylder 1978, Friesen 1981, Derosa and Friesen 1981). *Theromyzon tessulatum* and *T. rude* respond to water disturbance in a similar manner but adult brooding *T. tessulatum* are unique in that they move toward a potential host even though they never take a blood meal. If the parent *T. tessulatum* locates a host, it attaches to it and the brooding young leave to seek their first blood meal (Wilkialis and Davies 1980a).

Although movements toward a potential host by sanguivorous leeches are not specific, the next step of determining whether or not the potential host is an appropriate source of blood involves some selectivity. After attachment to the potential host with the posterior sucker, the leech makes a number of searching movements with the anterior portion of the body. If the host is not appropriate, the leech detaches. Attachment to a host does not always occur at an appropriate location. For example, *T. rude* primarily feed in the nares; but if one attaches to the feathers rather than the beak it will remain on the bird for up to 30 min, moving about in search of a suitable location to feed. Host recognition by sanguivores presumably results from chemoreception and/or mechanoreception.

Predatory leeches are generally more nocturnal and negatively phototactic than sanguivores (Mann 1961); and while some predators also show increased activity in response to disturbances in the water, they are generally opportunistic, feeding on prey encountered through either their own locomotion or the movements of the prey. Prey detection by predator leeches has received little attention since Gee (1913) noted that *Mooreobdella microstoma* responded to vibrations in the water and chemical extracts of their prey by unidirectional swimming and random movements of the anterior. Davies *et al.* (1982a) examined the chemosensory detection of prey by *N. obscura* and showed that without direct tactile contact, *N. obscura* were unable to detect and react to any of the common prey types tested (i.e., they did not show clinotaxis). Whether or not a particular prey appears in the diet of a predatory leech depends on the probability that it will be encountered, the probability that it will be attacked once encountered, and the probability that an attack will result in capture.

An exception to the general rule that the diet of predatory leeches is usually nonspecific and contains a wide variety of prey is demonstrated by *Erpobdella montezuma.* Blinn *et al.* (1988) showed that *E. montezuma* forages selectively on the pelagic amphipod *Hyalella montezuma* despite the presence of a wide variety of alternate prey. *Erpobdella montezuma* discriminates between two congeneric amphipod prey by mechanoreception, exhibiting a high response to the acoustic vibration signals of the endemic *H. montezuma* but not to *H. azteca.* Not only does *E. montezuma* discriminate between species by mechanoreception, but it also selectively discriminates between size classes, feeding principally on juveniles. Although amphipods form part of the diet of *N. obscura* and *E. punctata*, neither species shows a significant response to their acoustic vibration signals (Blinn and Davies 1989).

G. Population Regulation

On a geological time scale, only a very few freshwater ecosystems are long-lived and even on a shorter time scale of 1–20 years, many lentic ecosystems are ephemeral. In ephemeral ecosystems, leeches may not have sufficient time to reach carrying capacity. Evidence for competition between leech species has been presented by Davies *et al.* (1982c) for *Nephelopsis obscura* and *Erpobdella punctata* and by Wrona *et al.* (1981) for *Glossiphonia complanata* and *Helobdella stagnalis.* However, only the studies on competition between *N. obscura* and *E. punctata* are comprehensive in relation to the established criteria (Reynoldson and Bellamy 1970, Williamson 1972).

Reynoldson and Davies (1976, 1980) showed a considerable overlap in the range of water conductances in which *N. obscura* and *E. punctata* can regulate their weights. The sympatric distributions of the species therefore appear to be related to common salinity tolerances. However, among sympatric populations, there is a strong inverse relationship between the numerical abundances of *N. obscura* and *E. punctata* in lentic ecosystems (Davies *et al.* 1977). Dominance of *E. punctata* over *N. obscura* is a rare temporary event and the switch to dominance by *N. obscura* is rapid. The reproductive strategies of *E. punctata* vary depending on whether it is dominant or subordinate (Davies *et al.* 1977). Thus, the biological interactions between *N. obscura* and *E. punctata* are amenable to explanation based on competition.

The utilization of common food resources by *N. obscura* and *E. punctata* has been demonstrated by numerous laboratory and field studies (Davies and Everett 1977, Davies *et al.* 1978, 1981, 1982c). At the species level, there were no significant differences in the diets of either species between years and the trends in prey utilization were similar each year (Fig. 13.19). Intraspecific food resource partitioning, as a function of differential weight class utilization of prey was shown by both species. Such resource partitioning minimizing intraspecific competition, increases population fitness (Giesel 1974). The change in feeding ecology of *E. punctata* with changes in numerical dominance provided further evidence for interspecific competition. Furthermore, a significant decrease in mean niche overlap and niche breadth was found for both species when *N. obscura* became dominant. The majority of criteria established to determine interspecific competition have been supported by the data collected on *N. obscura* and *E. punctata;* however, the manipulation of food resources has not yet been completed.

V. COLLECTION AND REARING

A. Collection

Leech and acanthobdellid ectoparasites are sometimes found attached to hosts and are usually easily removed. Most leech species (but not *Acanthobdella peledina*) are free living for the majority of their life cycle and are normally attached to a firm substrate or free swimming, as are the predatory leech species. They can be collected with bottom samplers and dip or plankton nets.

Free-living leeches can be sampled quantitatively using colonization chambers, grab samplers, or timed collections from rocks and stones. As a general rule, the plan area of the largest leech collected, or its area of escape response, should not exceed 5% of the sampler area (Green 1979). The only benthic samplers that even approximately meet these criteria are grab samplers such as the Ekman grab (Ekman 1911). Another problem faced when quantitatively sampling free-living leeches is that they occur in both the sediment and the vegetation. The Gerking box sampler (Gerking 1957) is one of the few methods of quantitatively estimating population density, separable into these two components. This technique has, however, a number of technical problems including: the difficulty of clipping macrophytes close to the substrate, the entrainment of sediments and associated fauna in the macrophyte sample, the invasive influence of the operator, and the difficulty of integrating the macrophyte and sediment samples because the benthic samples are taken from inside the area from which the macrophyte samples are collected. These problems are overcome by using a sampler consisting of a bottom-mounted detachable grab connected to a box sampler with levered, spring-loaded jaws for cutting macrophytes (Gates *et al.* 1987). The sampler is triggered from above the water and simultaneously compartmentalizes the soft sediments in the grab and the phytomacrofauna in the water column above.

B. Rearing

Leeches are generally easy to maintain in the laboratory provided they are maintained under approximately similar conditions encountered in the field. Appropriate dissolved oxygen and temperature regimes are particularly important. Light regime is less critical but if an ambient light regime cannot be provided, darkened conditions are preferable. Leeches are very susceptible to rapid changes in ionic content, temperature, and dissolved oxygen concentrations. If field conditions cannot be provided, they should be slowly acclimated to the new conditions (e.g., temperature changes exceeding 5°C per day will usually cause very high mortality, but a temperature change of 1°C per day within the normal field range will usually result in little mortality).

Predatory leech species are easily fed in the laboratory. If suitable prey at high densities are provided, most leeches will feed avidly when hungry. For maintenance, *ad libitum* feeding one day per week is sufficient.

Most predatory species brought into the laboratory and provided with ambient field conditions will feed, grow, and reproduce. The viability of the cocoons produced is sometimes decreased if the parent has been maintained in the laboratory for a long period. It is much more difficult to raise hatchlings produced in the laboratory through to mature size capable of producing viable cocoons. Each species requires special conditions and there is much research to be done in this area.

Conversely, while some sanguivorous leeches are more difficult to feed in the laboratory, reproduction, cocoon production, and hatching are generally easier. Some sanguivores (e.g., *Placobdella papillifera*) will eat if immersed directly into blood. Many others (e.g., *Hirudo medicinalis*) will feed through a membrane if blood at an appropriate temperature (30°–40°C) is on the other side. For these species, the simplest technique is to provide the hungry sanguivore with a sausage made from intestinal epithelium filled with warmed blood. Other species (e.g., *Theromyzon rude*) have not yet been induced to feed artificially and must be provided with a live host and given the opportunity to reach an appropriate feeding site (the nares of waterbirds for *T. rude*). However, when suitable sites are available, cocoon production occurs readily.

VI. IDENTIFICATION

A. Preparation of Specimens

Some species can be studied live, but permanently stained slides and serial sections are required for identification of a few species. Clearing, without permanent staining, is often required for determining the ocular number and arrangement.

Most leeches contract strongly when placed in cold fixative and should be narcotized first. Before narcotizing the specimens, the colors of the dorsal and ventral surfaces should be recorded, as chromatophores are dissolved or altered by most narcotizing agents. Leeches are generally difficult to relax properly, and any of the following three methods are recommended:

1. Add drops of 95% methanol slowly to the water containing the leech, gradually increasing the concentration for about 30 min until movement ceases. When the leech is limp and no longer responds to touch, pass it between the fingers to straighten it and remove excess mucous.
2. Add carbonated water or bubble in CO_2 until movement of the leech stops. Straighten the leech out on a slide and slowly add warm 70% ethanol and a few drops of glacial acetic acid until it is covered.
3. Add a drop or two of 6% nembutal until movements stop and then straighten the leech on a slide.

For some specimens, slight flattening is occasionally desirable. This is best done between two glass slides with weights added if necessary.

After relaxation, the leeches should be fixed for 24 hr (depending on size) in 5 or 10% formalin. For histological preparations, Fleming's or Bouin's fixatives should be used. Large leeches should be injected with 70% ethanol to ensure preservation of the internal organs.

To preserve a leech, wash the fixed specimen with deionized water to remove the formalin and then add either 70% ethanol or 5% buffered formalin. Although formalin preserves the colors longer, ethanol is recommended since it is safer to use and less destructive to soft tissues.

Leeches fixed in formalin or Bouin's or Fleming's fixative can be stained in Mayer's paracarmine, borax carmine, or Harris' hematoxylin for 12–24 hr, destained in a 1% HCl–70% ethanol solution until the leech epidermis is free of stain, and then neutralized in a 1% NH_4OH–70% ethanol solution. Fast green or eosin are routinely used as counterstains. Stained specimens should be dehydrated in progressively higher concentrations of ethanol, cleared in methyl salicylate, and mounted in a neutral pH mounting medium.

B. Important Features for Identification

Identification of leeches can be difficult, if not impossible, with poorly preserved specimens. External features such as the following are used whenever possible: annulation; number of eyes and their arrangement; presence or absence of papillae, tubercles, and pulsatile vesicles; relative size of anterior (oral) and posterior (anal) suckers; size and position of the mouth; and the relative positions of the male and female gonopores. The only notes on coloration included in the following keys are those known to persist after preservation. In a few instances, reference is made to internal anatomical features. These have been kept to a minimum and are used only when positive identification is impossible without them. To distinguish between *Nephelopsis obscura* and the *Dina–Mooreobdella* complex, the atrial cornua must be examined (Fig. 13.22). The preserved specimen should be pinned out with the ventral surface up and a transverse incision made across the body three or four annuli posterior to the male gonopore. Cuts should be made anteriorly up the lateral margins of the body for about 20 annuli and the posterior edge of the flap lifted forward exposing the inner tissues, which must be cleared to expose the atrium.

To examine the velum and jaws, the specimen should be pinned out and a median ventral incision made from the lower lip of the anterior sucker back far enough for the margins to be pinned out to expose the inner surface of the pharynx (Fig. 13.11). Details of the teeth (Fig. 13.12) can only be seen by removal of a jaw and making a temporary mount on a microscope slide.

The length of leeches and acanthobdellids is dependent on species, age, physiological state, and the degree of relaxation. The measurements presented in the keys are maximum total lengths of relaxed individuals.

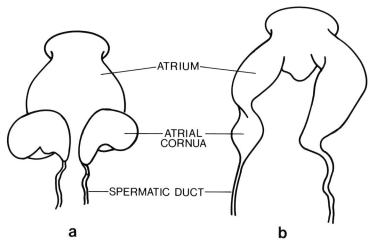

Figure 13.22 (a) Spirally curved atrial cornua of *Nephelopsis obscura* and (b) the simply curved atrial cornua of *Dina/Mooreobdella.*

1. Family Glossiphoniidae

Confusion exists in the systematics of these genera, and many species have undergone considerably synonymy. Although some taxonomists consider that the genus *Haementeria* (de Filippi 1849) has priority over *Placobdella* (Blanchard 1893), the latter name is widely recognized and accepted as a genus in North America; it is well represented with seven species. Lukin (1976) recognized important differences between *Glossiphonia complanata* and *G. heteroclita* and suggested that the latter species be placed in the subgenus *Alboglossiphonia;* Klemm (1982) elevated *Alboglossiphonia* to generic level. Although the identification of *Batracobdella paludosa* is based on a single identification by Pawlowski (1948) from Nova Scotia, Canada and additional specimens have not yet been recorded, it is included for completeness. *Batracobdella picta* is sometimes placed in the genus *Placobdella,* and conversely *P. phalera* is sometimes classified in the genus *Batracobdella. Theromyzon tessulatum* is characteristically a European species, but its presence in North America has been well documented. Another primarily European species, *T. maculosum,* has also been recorded in North America, but Klemm (1977) questioned the validity of this species and suggested a new species (*T. biannulatum*) for specimens with two annuli between the gonopores.

The genus *Actinobdella* was at first erroneously placed in the family Piscicolidae by Moore (1901), who was misled by the unusual annulation of *Actinobdella inequiannulata.* Complete segments of *Actinobdella* are 3-annulate (a_1, a_2, a_3), which are further divided into six unequal annuli (b_1, b_2, b_3, b_4, b_5, b_6).

2. Family Piscicolidae

Some workers have suggested that *Piscicola geometra* and *Piscicola milneri* are synonymous, although *Piscicola milneri* has 10–12 punctiform eyespots, whereas *Piscicola geometra* has 12–14 punctiform eyespots and dark pigmented rays on the posterior sucker (Fig. 13.23c, d). Klemm (1977) indicated that the number of punctiform eyespots on the caudal sucker and the presence or absence of the pigmented rays varies with the age of the specimens. As the gonopores of *Piscicola milneri* are separated by two annuli and by three annuli in *Piscicola geometra,* they are here considered distinct species.

The genera *Myzobdella* and *Illinobdella* are very similar (Meyer 1940, Daniels and Sawyer 1973), but their relationship needs to be thoroughly investigated before the synonymy suggested by Daniels and Sawyer (1973) and Sawyer (1986) can be fully accepted. In the present chapter, a conservative approach has been used. If later evidence supports the propositions of Daniels and Sawyer (1973) and Sawyer (1986), then it will be relatively simple to combine the records for *Illinobdella* with those for *Myzobdella.* However, if supporting data are not forthcoming, subsuming *Illinobdella* with *Myzobdella* will result in an irretrievable loss of information.

3. Family Hirudinidae

Richardson (1969, 1971) divided the North American haemopisoid leeches into the genera *Mollibdella, Bdellarogatis,* and *Percymoorensis,* rather than placing them all in the genus *Haemopis.* This was accepted by Soos (1969b), Davies (1971, 1972, 1973), and Klemm (1972a), and is accepted here.

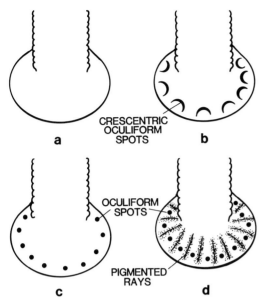

Figure 13.23 Posterior suckers of (a) *Piscicola punctata* without oculiform spots; (b) *Piscicola salmositica* with 8–10 crescentiform oculiform spots; (c) *Piscicola geometra* with 12–14 oculiform spots separated by pigmented rays.

Sawyer and Shelley (1976) and Sawyer (1986), for unspecified reasons, recombined them all back into the genus *Haemopis,* a view also adopted by Klemm (1977, 1982, 1985).

4. Family Erpobdellidae

There is some disagreement as to whether *Dina* and *Mooreobdella* are distinct genera (Moore 1959) or subgenera of the genus *Dina* (Soos 1968). The diagnostic feature is whether the vasa deferentia form preatrial loops extending to ganglion XI (*Dina*) or do not have preatrial loops (*Mooreobdella*). This is difficult to determine, and thus in this key, the genera (or subgenera) are lumped together in a *Dina–Mooreobdella* complex, separated on external features.

VII. POLYCHAETA

A. General Anatomy

Polychaetes, like leeches and acanthobdellids, belong to the phylum Annelida, but have a variable number of body metameres. Typically, each polychaete metamere bears a pair of lateral lobes (parapodia) with tufts of chaetae at the distal ends (Figs. 13.5 and 13.36). Basically each metamere is identical but some segments become fused for specialized functions. The head consists of the preoral prostomium and several fused segments (Fig. 13.6). The prostomium bears several different types of sensory appendages: antennae with a tactile function; palps (tentacles) with a chemosensory and/or feeding function; and pairs of eyes each consisting of a lens and retinal cup. The rest of the body is composed of bilaterally symmetrical metameres, each containing a pair of gonads, excretory system, a single unpaired ventral segmental nerve ganglion and a pair of parapodia. In some tube-dwelling species, the body metameres are subdivided into an anterior thorax with gills and a posterior abdomen (Fig. 13.7a).

Each parapodium is divided into a dorsal notopodium and a ventral neuropodium, each supported internally by a chitinous rod (aciculum). A dorsal and ventral cirrus project from the base of the notopodium and neuropodium, respectively (Fig. 13.5). In some species, the dorsal cirrus may be modified to form a gill (branchia). The distal ends of the parapodia bear chaetae which may be jointed (compound) (Fig. 13.24a) or unjointed (simple). Simple, long tapering chaetae are termed capillaries (Fig. 13.24b, c); these may also be hooked (Fig. 13.24d). Burrowing and tubiculous species have uncini (Fig. 13.24d), which are simple chaetae with expanded multidentate heads protruding through the body wall.

In some polychaetes, the anterior part of the gut can be everted through the mouth and used to capture food. The proboscis has a pair of cuticular jaws, and may also carry rows of soft papillae and hard cuticular thickenings (paragnaths) (Fig. 13.6). Species that live in permanent burrows or tubes capture food with grooved tentacular palps; suspension feeders utilize a semirigid crown of grooved tentacles (Fig. 13.7).

The tubes are simple flexible sheaths of mucous (secreted by glands located along the body and on the parapodia), sometimes made more rigid by polysaccharides and proteins that harden in water, and strengthened by incorporation of sediment particles, or secretion of calcium carbonate.

Polychaetes are typically marine and of the over 80 families (George and Hartmann-Schröder 1985), only four—Nereididae, Amphoretidae, Sabellidae, and Serpulidae—have been recorded in North American freshwaters.

1. Nereididae

Elongate with numerous metameres; prostomium bearing antennae, biarticulated palps, and eyes; eversible proboscis with a pair of jaws, sometimes with chitinous paragnaths or soft papillae; parapodia typically biramous but some species show reduction; chaetae compound or simple.

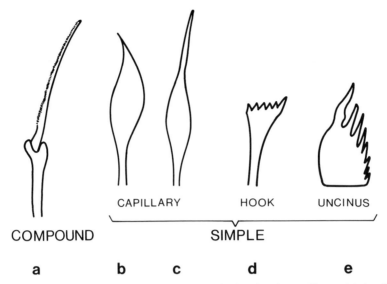

Figure 13.24 Types of polychaete chaetae: (a) compound; (b) simple capillary; (c) hooked capillary; (d) uncinus.

2. Amphoretidae

Body divided into head, thorax, and abdomen; biramous parapodia on thorax but notopodia absent or very reduced on abdomen; prostomium small and trilobed; numerous grooved buccal feeding tentacles which are retractable into the mouth; four pairs of branchiae; notopodium with capillary chaetae; neuropodia with uncini in both thorax and abdomen; tubiculous.

3. Sabellidae

Prostomium small and fused to the peristomium which forms a large tentacular crown; body cylindrical, divided into a thorax and abdomen; parapodia greatly reduced; notopodia with capillary chaetae and neuropodia with uncini on the thorax; notopodia with uncini and neuropodium with capillaries on abdomen; tubiculous.

4. Serpulidae

Prostomium reduced and fused to the peristomium which forms a large tentacular crown, one radiale of which is modified to form an operculum; body divided into thorax and abdomen; thorax with capillary chaetae in notopodium and uncini in neuropodia; chaetae positions reversed on abdomen with capillaries in the neuropodium and uncini in the notopodium; tube calcareous.

B. Ecology

The relatively few species of polychaetes recorded in freshwaters of North America either inhabit streams linked to the ocean or live in lakes that were connected to the sea in the relatively recent past. Polychaetes can be divided into two ecological groups: Errantia (e.g., Nereididae), which are active crawlers and swimmers and Sedentaria (e.g., Amphoretidae, Sabellidae, and Serpullidae), which are tube-dwellers. This separation is based on differences both in the structure of the anterior portion of the body and in their life habits. Unfortunately, it is artificial, and several phylogenetic classifications (Dales 1962, Clark 1969, Mileikovsky 1977, Fauchold 1977, Pettibone 1982) have been proposed.

A number of methods of feeding have been adopted by the polychaetes. The errant Nereididae have an eversible, jawed proboscis (Fig. 13.6) but are not necessarily predators; indeed, the majority are omnivores. The sedentary tubiculous Sabellidae and Serpulidae are filter feeders, collecting fine particles suspended in the water column. The Ampharetidae are deposit feeders using their grooved buccal feeding tentacles to collect sediment. The Errantia are found in lakes and rivers in or on soft substrates, while the Sedentaria are most frequently found in lakes. Within the latter group, the Ampharetidae and Sabellidae build their tubes within the soft substrates, and the Serpulidae attach their calcareous tubes to rocks, shells, or macrophytes. The sabellid *Manayunkia speciosa* has also been recorded at the mouth of rivers (Mackie and Qadri 1971, Hazel 1966, Hiltunen 1965) where the water velocity does not exceed 0.5 m sec^{-1} and the dissolved oxygen concentration exceeds 5.0 mg/liter.

Many polychaetes show both asexual and sexual

reproduction and are able to regenerate parts of their bodies after damage or predation; asexual reproduction has, however, never been recorded in a freshwater species (Schroeder and Hermans 1975). The sexes are usually separate, but some species of freshwater nereids (e.g., *Nereis limnicola*) are hermaphrodites. Just prior to spawning, some nereids (e.g., *Nereis succinea*) undergo considerable morphological changes including enlargement of the eyes, parapodia, and cirri. On the correct spawning cue, these modified individuals (epitokes) (Kinne 1954) leave the substrate and swim toward the water surface. It is not known whether the lunar rhythm of swarming observed for *N. succinea* in marine habitats (Lillie and Just 1913) also occurs in freshwater ecosystems. Many freshwater species have direct development with suppression of the normal free-living trochophore larvae, and *N. limnicola* develops its larvae within the coelom (Smith 1950). The eggs of *Manayunkia speciosa* are produced by the females within the parent's tube; the eggs pass through their early developmental stages and are retained within the tube until the young have at least 8–9 metameres (Pettibone 1953).

The only polychaetes able to live in low salinity and freshwaters have the ability to regulate their body salt concentration. Nereids have been particularly successful in colonizing freshwaters (Wesenberg-Lund 1958), and their low salinity adaptations have been the subject of numerous investigations (reviewed in Oglesby 1969). In freshwaters, polychaetes maintain hyperosmotic coelomic fluid concentrations by volume regulation (Oglesby 1969) with the nephridia involved in salt and water balance (Koechlin 1972, 1975, 1977, Kamemoto and Larson 1964). While Krishnan (1952) noted that nereids from freshwater had larger and more vascularized nephridia compared to marine species, Jones (1967), working with two populations of *N. limnicola*, found that those from the more saline environment had larger excretory surfaces.

C. Collection and Identification

Specimens from soft sediments are best sampled with a grab or corer or by digging with a spade and sieving the substrate collected through a fine mesh (0.6–2.0 mm). As many freshwater species are only a few mm long and easily overlooked, Light (1969) suggested that samples of surface sediments be allowed to settle in shallow dishes overnight, permitting the animals to reconstruct their tubes or burrows, making detection much easier. Serpulids in their calcareous tubes attached to rocks, stones, or macrophytes are best removed by hand. Free–living larvae can be collected in plankton nets.

Collected specimens should be narcotized in 7% $MgCl_2$ and, if possible, some examined live to determine the character and number of anterior appendages, the types of parapodia and chaetae, whether chaetae type changes along the body, and the presence or absence of gills (branchiae). These features are more easily determined in live specimens.

For preservation, specimens should be fixed in 7–10% neutral formalin for 24 hr before being washed in freshwater and preserved in 70% ethanol.

It is always necessary to examine the chaetae and parapodia for identification. Small specimens can be mounted laterally on a slide but for larger specimens, the parapodia can be removed and mounted separately.

For the Nereididae, an examination of the proboscis and jaws is required. If the specimens do not spontaneously evert their proboscis when narcotized, this can be induced by applying gentle pressure behind the head. If after narcotization the proboscis is not everted, make a longitudinal slit along the dorsum just lateral to the midline and peel open the skin to show the inverted proboscis. Make a similar insertion through the proboscis and fold back the walls to show the jaws, paragnaths, and papillae.

VIII. Taxonomic Keys

A. Taxonomic Key to Higher Taxa of Freshwater Annelida

1a.	Body divided into metameric segments; paired lateral, muscular projections (parapodia) on all or most segments (Fig. 13.5); parapodia bearing groups of chaetae; prostomium with or without appendages (antennae or palps) (Fig. 13.6); usually dioecious; gonads not distinct, occurring in numerous segments; eggs shed into water or attached to substrate; no cocoon; no clitellum . . . . Polychaeta
1b.	Body divided into metameric segments; no paired parapodia; chaetae present or absent; prostomium without appendages; hermaphrodite; gonads distinct and restricted to specialized segments; eggs laid in cocoons secreted by a clitellum 2

2a(1b).   Body divided into variable number of postoral metameric segments; segments not superficially divided into annuli; setae in four bundles (two dorsolateral and two ventrolateral) on majority of segments; no anterior (oral) or posterior (anal) suckers; great majority have paired male and female gonopores and spermathecal pores; testes anterior to ovaries; large coelom; mostly detritivores but a few predators .... Oligochaeta (Chapter 12)

2b.   Body divided into a fixed number of postoral segments (either 29 or 32); segments subdivided superficially into annuli; chaetae absent or restricted to two pairs on each of five anterior segments (Fig. 13.4); anterior sucker present or absent; posterior sucker present; single median ventral female gonopore; paired or single ventral male gonopore; testisacs posterior to ovisacs; reduced coelom; carnivores or sanguivores 3

3a(2b).   Body divided into 29 postoral segments; segments subdivided superficially into four annuli (a_1, a_2, b_5, b_6) (Fig. 13.2b); no anterior sucker; mouth on ventral surface of segment III; no jaws; posterior sucker present consisting of four segments; two pairs of chaetae on five consecutive anterior segments; distal ends of chaetae bent to form hooks; (Fig. 13.4); chaetae absent from remainder of the body; paired ventral male gonopores; single median ventral male gonopore; length 22 mm; ectoparasite of salmonids .... *Acanthobdella peledina* Grube 1851

3b.   Body divided into 32 postoral segments; segments subdivided superficially into 3–16 annuli (Figs. 13.2a and 13.34), anterior sucker present consists of four segments; mouth on ventral surface of anterior sucker (Fig. 13.10); jaws present or absent; posterior sucker consists of seven segments; chaetae absent from entire body; median ventral unpaired male and female gonopores (Figs. 13.3 and 13.32) .... Hirudinoidea

B. Taxonomic Key to Species of Freshwater Hirudinoidea

1a.   Mouth a small pore on ventral surface of anterior sucker through which a muscular pharyngeal proboscis can be protruded (Fig. 13.10a, b); no jaws or teeth .... Rhynchobdellida   2

1b.   Mouth large, occupying the entire cavity of anterior sucker (Fig. 13.10c); no protrusible proboscis; jaws with teeth either present or absent (Fig. 13.11) .... Arhynchobdellida   3

2a(1a).   Body flattened dorsoventrally and much wider than head (Fig. 13.1a) (except *Placobdella montifera* and *Placobdella nuchalis*); body not cylindrical (except for *Helobdella elongata* which is subcylindrical); not differentiated into two body regions; anterior sucker ventral, more or less fused to body and narrower than body; body never divided into anterior trachelosome and posterior urosome; eggs in membranous cocoons and young brooded on ventral surface of parent; one, two, three or four pairs of eyes; no oculiform eye spots on posterior sucker; segments 3-annulate (a_1, a_2, a_3) (Fig. 13.2a) (except *Oligobdella biannulata* which is 2-annulate .... family Glossiphoniidae   4

2b.   Body cylindrical and usually long and narrow; body sometimes divided into a narrow anterior trachelosome and a wider posterior urosome (*Illinobdella* and *Piscicolaria*) (Fig. 13.1b); anterior sucker expanded and distinct from body; zero, one, or two pairs of eyes; pulsatile vesicles along the lateral margins present (*Piscicola* and *Cystobranchus*) (Fig. 13.1b) or absent; seven or more annuli per segment (except *Piscicolaria reducta* which is 3-annulate); oculiform eye spots sometimes present on posterior sucker (Figs. 13.9b, 13.23a, b, c); no brooding of cocoons or young .... family Piscicolidae   34

3a(1b).   Five pairs of eyes arranged in an arch on segments II–VI with the third
           and fourth pairs of eyes separated by one annulus (Fig. 13.25a); body
           elongate (Fig. 13.1c); jaws with teeth either present or absent; nine or
           ten pairs of testes arranged metamerically (Fig. 13.14b); pharynx
           short .... family Hirudinidae   45
3b.   Zero, three, or four pairs of eyes in separate labial and buccal groups
           (Fig. 13.25b); body elongate (Fig. 13.1d); no jaws; testes small and very
           numerous (Fig. 13.14a); pharynx about one-third of body
           length .... family Erpobdellidae   58

4a(2a).   Family Glossiphoniidae—Posterior sucker conspicuous with a marginal
           circle of 30–60 glands and retractile papillae, their positions being
           indicated dorsally by faint radiating ridges (Fig. 13.26) .... *Actinobdella*   5
4b.   Posterior sucker without a marginal circle of glands or retractile
           papillae   6

5a(4a).   Posterior sucker on short, distinct pedicel with 29–31 digitate processes
           on rim (Fig. 13.26); somites 3- or 6-annulate; dorsal papillae in 1–5
           longitudinal rows; length 22 mm .... *Actinobdella inequiannulata* Moore 1901
5b.   Posterior sucker on short distinct pedicel with about 60 digitate
           processes on rim; somites six-annulate with b₃ and b₅ the largest and
           most conspicuous; length 11 mm; dorsal papillae in five longitudinal
           rows .... *Actinobdella annectens* Moore 1906

6a(4b).   Zero, one, or two pairs of eyes; a series of paired accessory eyes
           sometimes present along body (Fig. 13.8c)   7
6b.   Three or four pairs of eyes   29

7a(6a).   Mouth apical or subapical on rim of anterior sucker (Fig. 13.10a); zero
           or one pair of eyes   8
7b.   Mouth within anterior sucker and clearly not on rim (Fig. 13.10b); one
           or two pairs of eyes; gonopores separated by one or two annuli (Fig.
           13.3b)   19

8a(7a).   Male and female gonopores united in common bursal pore; one pair of
           eyes well separated; body smooth without papillae; six pairs of crop
           caecae; length 22 mm .... *Marvinmeyeria lucida* (Moore 1954)
8b.   Male and female gonopores separated by two annuli (Fig. 13.3b); zero
           or one pair of eyes (Fig. 13.8a) [*Placobdella hollensis* has several pairs
           of accessory eyes on the neck (Fig. 13.8c)]; eyes close together or
           confluent (Fig. 13.8b) (except *Placobdella montifera* and *Placobdella
           nuchalis* which have eyes well separated); body usually papillated;
           seven pairs of crop caecae .... *Placobdella/Oligobdella*   9

9a(8b).   One pair of eyes on segment III; no supplementary eyes   10
9b.   One pair of eyes on segment III followed by an indefinite number of
           pairs of accessory eyes (Fig. 13.8c); tubercles large and rough; length
           30 mm .... *Placobdella hollensis* (Whitman 1892)

Fig. 13.25

Fig. 13.26

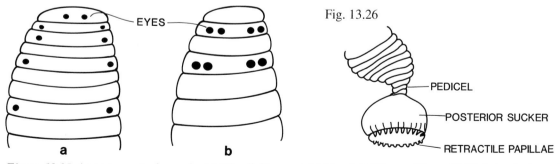

Figure 13.25 Arrangement of eyes in (a) Hirudinidae and (b) Erpobdellidae. ***Figure 13.26*** Lateral view of
the posterior sucker of *Actinobdella inequiannulata* showing the pedicel and the retractile papillae around the
margin.

10a(9a).	Anus between segments XXIII and XXIV with the 16 postanal annuli forming a slender stalk (pedicel) which bears the posterior sucker; no papillae; length 35 mm .... *Placobdella pediculata* (Hemingway 1908)	
10b.	Anus close to the posterior sucker; no pedicel	11
11a(10b).	Margins of posterior sucker denticulate (Fig. 13.27)	12
11b.	Margins of posterior sucker not denticulate	14
12a(11a).	Head expanded and discoid and set off from body by a narrow neck (Fig. 13.28)	13
12b.	Head not discoid; prominent yellow or white band on somite VI; length 10 mm .... *Placobdella phalera* (Graf 1899)	
13a(12a).	Dorsum with three prominent tuberculate keels or ridges (Fig. 13.28); length 16 mm .... *Placobdella montifera* Moore 1906	
13b.	Dorsum smooth; no keels or ridges; length 25 mm .... *Placobdella nuchalis* Sawyer and Shelley 1976	
14a(11b).	Medium longitudinal row of tubercles on dorsal surface are large and conspicuous; tubercles bear several papillae giving a rough, warty appearance; ventrum unstriped; length 40 mm .... *Placobdella ornata* (Verrill 1872)	
14b.	Dorsal papillae, if present, simple cones with smooth domes	15
15a(14b).	Dorsum smooth; dorsal papillae inconspicuous or absent	16
15b.	Dorsal papillae small and conical but distinct	18
16a(15a).	Dorsum smooth; no papillae or tubercles; dorsum with conspicuous white genital and anal patches; one or more medial white patches; white bar on neck	17
16b.	Dorsal papillae inconspicuous or absent; when present, papillae arranged in five longitudinal rows; ventral surface with 11–12 blue, brown, or green stripes; annulus a$_3$ in middle of body without distinct cross furrow; length 65 mm .... *Placobdella parasitica* (Say 1824)	
17a(16a).	3-annulate (Fig. 13.2a); posterior sucker not conspicuous; length 11 mm .... *Placobdella translucens* Sawyer and Shelley 1976	
17b.	2-annulate; posterior sucker large; length 7 mm .... *Oligobdella biannulata* (Moore 1900)	
18a(15b).	Dorsum with 5–7 longitudinal rows of small papillae; papillae on posterior sucker; annulus a$_3$ in middle of body with distinct cross furrow; length 45 mm .... *Placobdella papillifera* (Verrill 1872)	
18b.	Dorsum with few or numerous small papillae in five longitudinal rows; length 50 mm .... *Placobdella multilineata* Moore 1953	
19a(7b).	One or two pairs of eyes; if only one pair of eyes present, these are close together or lobed indicating coalescence (Fig. 13.8a, b); gonopores separated by two annuli (Fig. 13.3b) .... *Batracobdella*	20

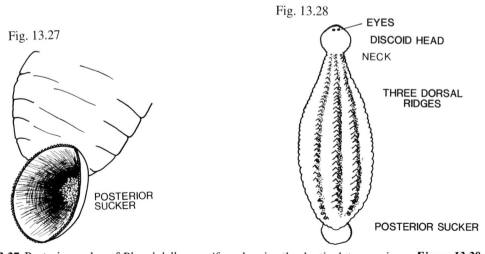

Figure 13.27 Posterior sucker of *Placobdella montifera* showing the denticulate margins. **Figure 13.28** *Placobdella montifera* showing the expanded discoid head and dorsum with three prominent tuberculate dorsal ridges (keels).

19b.   One pair of eyes which are well separated (Fig. 13.14); gonopores
         separated by one annulus .... *Helobdella*   23

20a(19a). Two pairs of eyes; one small pair of eyes on segment III and a larger
         pair of eyes on segment IV; eyes sometimes coalesced (Fig. 13.8a, b);
         seven pairs of crop caecae; papillae absent; length 20 mm .... *Batracobdella paludosa* (Carena 1824)

20b.   One pair of eyes on somite III only; eyes sometimes confluent (Fig.
         13.8b) 21

21a(20b). Dorsum with white genital and anal patches, one or more medial white
         patches and a white bar on neck; no papillae on dorsum; five
         longitudinal rows of white prominences surrounded by yellowish dots,
         equidistant longitudinally and transversely; body very flattened; length
         10 mm .... *Batracobdella michiganensis* Sawyer 1972

21b.   Dorsum without white patches and bar on neck 22

22a(21b). Posterior sucker separated from body on a short pedicel; dorsum
         smooth; eight rows of inconspicuous sensillae on dorsal annuli; only
         recorded parasitism on *Cryptobranchus alleganiensis;* length
         17 mm .... *Batracobdella cryptobranchii* Johnson and Klemm 1977

22b.   Posterior sucker not on pedicel; posterior sucker small; low small
         papillae present; 3-annulate (Fig. 13.3a); length 25 mm .... *Batracobdella picta* (Verrill 1872)

23a(19b). Brown, horny, chitinoid scute (nuchal plate) on the dorsal surface of
         segment VIII (Fig. 13.29) 24

23b.   Without a nuchal plate 25

24a(23a). Pigment pattern arranged in longitudinal stripes on dorsum; six branched
         pairs of crop caecae; posterior sucker pigmented on dorsum; length
         18 mm .... *Helobdella california* Kutschera 1988

24b.   No longitudinal stripes on dorsum; six unbranched pairs of crop caecae,
         the last pair direct posteriorly; posterior sucker unpigmented; length
         14 mm .... *Helobdella stagnalis* (Linnaeus 1758)

25a(23b). Dorsal surface smooth 26

25b.   Dorsal surface with 3–7 longitudinal series of papillae, or with scattered
         papillae 28

26a(25a). Body unpigmented and translucent; body rounded and subcylindrical;
         lateral margins almost parallel; posterior sucker small and terminal; one
         pair of crop caecae; length 25 mm .... *Helobdella elongata* (Castle 1900)

26b.   Body pigmented, with or without longitudinal or transverse bands; body
         flat with posterior wider than tapering anterior; six pairs of crop
         caecae 27

27a(26b). Dorsum without transverse pigmentation; dorsum with six prominent
         longitudinal white stripes alternating with brown stripes; length
         14 mm .... *Helobdella fusca* (Castle 1900)

27b.   Dorsum with transverse brown interrupted stripes alternating with
         irregular white bands; length 10 mm .... *Helobdella transversa* Sawyer 1972

28a(25b). Dorsal surface with 5–9 longitudinal rows of papillae; papillae large,
         conspicuous, and rounded; lightly or unpigmented; length 14 mm .... *Helobdella papillata*
                                                                                         (Moore 1906)

28b.   Dorsal surface with three or fewer incomplete series of small papillae or
         with scattered papillae; body dorsoventrally flattened; pigmentation
         variable; length 30 mm .... *Helobdella triserialis* (Blanchard 1849)

29a(7b). Four pairs of eyes on paramedian lines of segments II–V (Fig. 13.30a);
         body very soft .... *Theromyzon*   30

29b.   Three pairs of eyes (Fig. 13.30b, c) (coalesced eyes sometimes occur but
         the lobed nature indicates the original condition); body firm 32

30a(29a). Gonopores separated by two annuli (Fig. 13.3b); length
         26 mm .... *Theromyzon biannulatum* Klemm 1977

30b.   Gonopores separated by three or four annuli (Fig. 13.3a) 31

31a(30b). Gonopores separated by three annuli; length 30 mm .... *Theromyzon rude* (Baird 1869)

Fig. 13.29

Fig. 13.30

Fig. 13.31

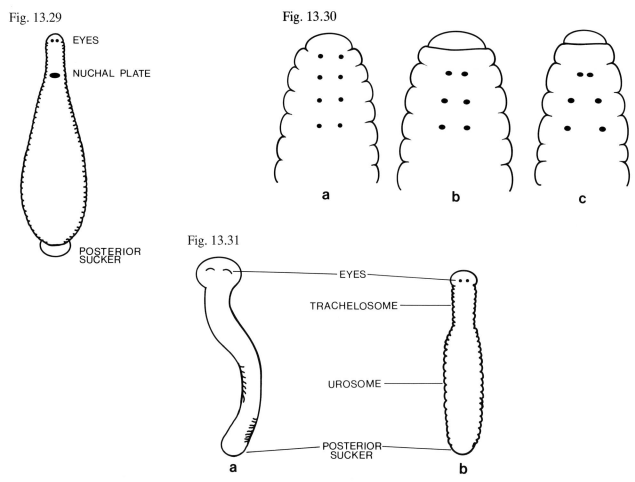

Figure 13.29 *Helobdella stagnalis* (or *H. california*) showing the chitinous scute (nuchal plate) on the dorsal surface and the single pair of eyes. **Figure 13.30** Dorsal view of (a) *Theromyzon rude*; (b) *Glossiphonia complanata*; and (c) *Alboglossiphonia heteroclita* showing the arrangement of the eyes. **Figure 13.31** (a) *Myzobdella lugubris* and (b) *Piscicolaria reducta* showing the trachelosome and urosome without pulsatile vesicles, the relative sizes of the eyes, and the weakly developed posterior sucker.

31b.	Gonopores separated by four annuli (Fig. 13.3a); length 30 mm .... *Theromyzon tessulatum* (Muller 1774)
32a(29b).	First pair of eyes closer together than succeeding two pairs, i.e., eyes arranged in triangular pattern (Fig. 13.30c); no papillae; male and female ducts open into a common gonopore; little pigmentation; generally amber colored; length 10 mm .... *Alboglossiphonia heteroclita* (Linnaeus 1761)
32b.	Eyes equidistant in two paramedian rows (Fig. 13.30b) 33
33a(32b).	Dorsum sometimes with papillae on annulus a₂ in six longitudinal rows; pair of paramedial stripes on dorsum and ventrum; six pairs of crop caecae; gonopores separated by two annuli (Fig. 13.3b); length 25 mm .... *Glossiphonia complanata* (Linnaeus 1758)
33b.	Dorsum with large, distinct papillae on annuli a₂ and a₃; dorsum with numerous, irregularly shaped whitish spots; seven pairs of crop caecae; length 25 mm .... *Boreobdella verrucata* (Muller 1844)
34a(2b).	Posterior sucker flattened, as wide or wider than the widest part of body (Fig. 13.1b); pulsatile vesicles on lateral margins of neural annuli of urosome (Fig. 13.9a, b); zero or two pairs of eyes 35
34b.	Posterior sucker concave, weakly developed, and narrower than widest part of the body; no pulsatile vesicles (Fig. 13.31); zero or one pair of eyes 42

35a(34a). Body divided into anterior trachelosome and posterior urosome; 11 pairs of pulsatile vesicles not very apparent in preserved specimens; each pulsatile vesicle covers two annuli; body cylindrical or sometimes slightly flattened; two pairs of eyes; both anterior and posterior suckers wider than body; oculiform spots present on posterior sucker (Fig. 13.23b, c, d) (except *Piscicola punctata*) (Fig. 13.23a); 14-annulate (except *Piscicola punctata* which is 3-annulate) .... *Piscicola* 36

35b. Body divided into distinct anterior trachelosome and posterior urosome; 11 pairs of large and distinct pulsatile vesicles easily seen in preserved and live specimens (Figs. 13.1b, 13.9a, b); each pulsatile vesicle covers four annuli; well-developed anterior and posterior suckers; no papillae; zero or two pairs of eyes (Fig. 13.9); oculiform spots on posterior sucker present or absent (Fig. 13.9a, b); 7-annulate; length 80 mm .... *Cystobranchus* 39

36a(35a). Posterior sucker with 8–14 oculiform spots (Fig. 13.23b, c, d) 37
36b. Posterior sucker without oculiform spots (Fig. 13.23a); 3-annulate (Fig. 13.2a); two pairs of crescent-shaped eyes; gonopores separated by three or four annuli; length 16 mm .... *Piscicola punctata* (Verrill 1872)

37a(36a). Posterior sucker with 10–14 punctiform oculiform spots (Fig. 13.23c, d) 38
37b. Eight to ten crescent-shaped oculiform spots on the posterior sucker (Fig. 13.23b); gonopores separated by two annuli; 14-annulate; length 31 mm .... *Piscicola salmositica* Meyer 1946

38a(37a). With 10–12 (usually 10) punctiform oculiform spots on posterior sucker; dark rays absent from posterior sucker (Fig. 13.23c); gonopores separated by two annuli (Fig. 13.3a); anterior pair of eyes like heavy dashes twice as long as wide; length 24 mm .... *Piscicola milneri* (Verrill 1874)
38b. With 12–14 punctiform oculiform spots on posterior sucker, separated by an equal number of dark pigmented rays (Fig. 13.23d); gonopores separated by three annuli; anterior pair of eyes like fine dashes five times as long as wide; length 30 mm .... *Piscicola geometra* (Linnaeus 1758)

39a(35b). With oculiform spots on posterior sucker 40
39b. Without oculiform spots on posterior sucker 41

40a(39a). Eight oculiform spots on posterior sucker; 2 rows of 12 lateral oculiform spots on each side of body (Fig. 13.9b); two pairs of eyes; length 7 mm .... *Cystobranchus meyeri* Hayunga and Grey 1976
40b. Ten oculiform spots on posterior sucker; no lateral ocelli; two pairs of eyes; length 15 mm .... *Cystobranchus virginicus* Hoffman 1964

41a(39b). Two pairs of eyes; the first pair forming conspicuous dashes at 45° to the longitudinal axis; the second pair of eyes ovoid (Fig. 13.9a); gonopores separated by two annuli (Fig. 13.3b); length 30 mm .... *Cystobranchus verrilli* Meyer 1940
41b. No eyes; length 30 mm .... *Cystobranchus mammillatus* (Malm 1863)

42a(34b). Body divided into small trachelosome and larger urosome (Fig. 13.31b); length-to-width ratio 3–6:1; one pair of eyes; 12- or 14-annulate .... *Myzobdella lugubris* Leidy 1851
42b. Body not divided into trachelosome and urosome (Fig. 13.31a); zero or one pair of eyes 43

43a(42b). Length-to-width ratio ≥ 10:1; zero or one pair of eyes in posterior half of the head; 12- or 14-annulate .... *Illinobdella* 44
43b. Length-to-width ratio 4–5:1 (Fig. 13.31b); one pair of eyes; 3-annulate (Fig. 13.2a); dorsum with six black/brown longitudinal stripes; length 8 mm .... *Piscicolaria reducta* Meyer 1940

44a(43a). Length-to-width ratio about 10:1; 14-annulate .... *Illinobdella alba* Meyer 1940
44b. Length-to-width ratio ≥ 15:1 45

45a(44b). Length-to-width ratio 15:1; one pair of eyes (sometimes absent); anus 15 annuli anterior to posterior sucker; 14-annulate; posterior sucker little more than concavity of the posterior body .... *Illinobdella richardsoni* Meyer 1940

Fig. 13.32   Fig. 13.33

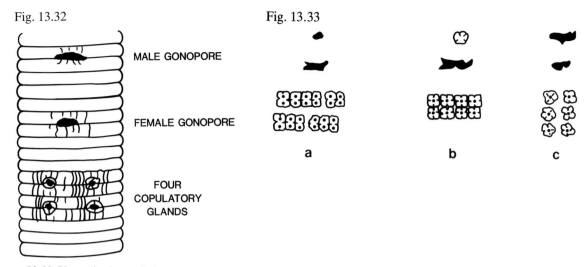

Figure 13.32 Ventral view of the male and female gonopores of *Macrobdella decora* with copulatory glands.   **Figure 13.33** Diagrammatic representation of the relative slopes and positions of the male and female gonopores and copulatory glands of (a) *Macrobdella sestertia;* (b) *Macrobdella ditetra;* and (c) *Macrobdella diplotertia.*

45b.	Length-to-width-ratio 18:1; one pair of eyes; anus ten annuli or less anterior to posterior sucker; 12-annulate; posterior sucker half as wide as posterior body .... *Illinobdella elongata* Meyer 1940
46a(32a).	With copulatory gland pores on the ventral surface, ten or eleven annuli posterior to the male gonopores (Figs. 13.32 and 13.33) .... *Macrobdella* 47
46b.	Ventral copulatory gland pairs absent 50
47a(46a).	With 24 copulatory gland pores (four rows with six gland pores each) (Fig. 13.33a) on raised pads; dorsum with red/orange spots; 2–2 1/2 annuli between gonopores; length 150 mm .... *Macrobdella sestertia* Whitman 1886
47b.	With four, six, or eight copulatory gland pores (Fig. 13.33); with or without red/orange spots on dorsum 48
48a(47b).	Eight copulatory gland pores (Fig. 13.33b) (two rows of four); two annuli between gonopores; dorsum without red/orange spots; length 150 mm .... *Macrobdella ditetra* Moore 1953
48b.	Four or six copulatory gland pores; dorsum with red/orange spots 49
49a(48b).	Four copulatory gland pores (Fig. 13.32) (two rows of two); 5–5 1/2 annuli between gonopores; jaws well developed with about 65 monostichodont acute teeth (Fig. 13.12a); ten pairs of testes; ventrum red/orange; length 150 mm .... *Macrobdella decora* (Say 1824)
49b.	Six copulatory gland pores (Fig. 13.33c) (three rows of two); 4–5 annuli between gonopores; ventrum yellow/grey; length 150 mm ...... *Macrobdella diplotertia* Meyer 1975
50a(46b).	Glandular area around gonopores; gonopores separated by three or four annuli .... *Philobdella* 51
50b.	No glandular area around gonopores; gonopores separated by 5–7 annuli 52
51a(50a).	20–26 distichodont teeth per jaw (Fig. 13.12b); lateral margins of dorsum without discrete spots; length 85 mm .... *Philobdella floridana* (Verrill 1874)
51b.	35–48 distichodont teeth per jaw; seven pairs of testes; irregular black spots on margins of dorsum; length 85 mm .... *Philobdella gracilis* Moore 1901
52a(50b).	Jaws absent (Fig. 13.11a) 53
52b.	Jaws present (Fig. 13.11b, c) and denticulate (Fig. 13.23b, c) 54
53a(52a).	Lower surface of velum smooth (Fig. 13.11a); gonopores in the furrows between the annuli and separated by five annuli; pharnyx with 12 internal ridges; length 300 mm .... *Mollibdella grandis* (Verrill 1874)

53b.	Lower surface of velum papillate; gonopores in the middle of annuli and separated by 5 annuli; pharnyx with 15 internal ridges .... *Bdellarogatis plumbeus* (Moore 1912)
54a(52b).	Jaws small and retractable into narrow-mouthed tubular pits; 9–25 distichodont (Fig. 13.12b) teeth per jaw .... *Percymoorensis* 55
54b.	Jaws large; 35–100 acute monostichodont (Fig. 13.12a) teeth; length 100 mm .... *Hirudo medicinalis* Linnaeus 1758
55a(54a).	Gonopores separated by 5–5 1/2 annuli; female gonopore small 56
55b.	Gonopores separated by 6 1/2–7 annuli; female gonopore large and conical; length 200 mm .... *Percymoorensis septagon* Sawyer and Shelley 1976
56a(55a).	Dorsum with median black stripe 57
56b.	Dorsum with irregular, scattered black spots 58
57a(56a).	Dorsum uniformly black or slate grey; posterior sucker narrower than body; jaws with 20–25 pairs of teeth; length 250 mm .... *Percymoorensis lateralis* (Say 1824)
57b.	Dorsum brown/green; posterior sucker as wide as body; 9–14 pairs of teeth .... *Percymoorensis kingi* (Mather 1954)
58a(56b).	Jaws with 10–12 teeth; posterior sucker about 75% of maximal body width and attached by very short pedicel; length 75 mm .... *Percymoorensis lateromaculata* (Mather 1963)
58b.	Jaws with 12–16 pairs of teeth; posterior sucker about half maximum body width; length 150 mm .... *Percymoorensis marmorata* (Say 1824)
59a(32b).	Somites 5-annulate (b_1, b_2, a_2, b_5, b_6) (Fig. 13.34a) with all annuli equal in length; three pairs of eyes; gonopores separated by two annuli (Fig. 13.3b); length 100 mm .... *Erpobdella* 60
59b.	Somites 6- or 7-annulate (Fig. 13.34b); annuli of unequal length units will be either subdivided or longer than the others; in any group of six consecutive annuli at least one annulus narrower or wider than the others; three or four pairs of eyes 61
60a(59a).	Eyes all similar in size; annuli not raised on dorsum and without papillae; anus located three or four segments anterior to posterior sucker; mouth small; widespread .... *Erpobdella punctata* (Leidy 1870)
60b.	Eyes differ in size with the second and third pairs smaller than the anterior pair; each annulus raised on dorsum with 14–18 small white-tipped papillae; anus located at base of posterior sucker; mouth large; endemic to Montezuma's Well, Arizona .... *Erpobdella montezuma* Davies et al. 1987
61a(59b).	Four pairs of eyes, two labial and two buccal 62
61b.	Zero or three pairs of eyes 64

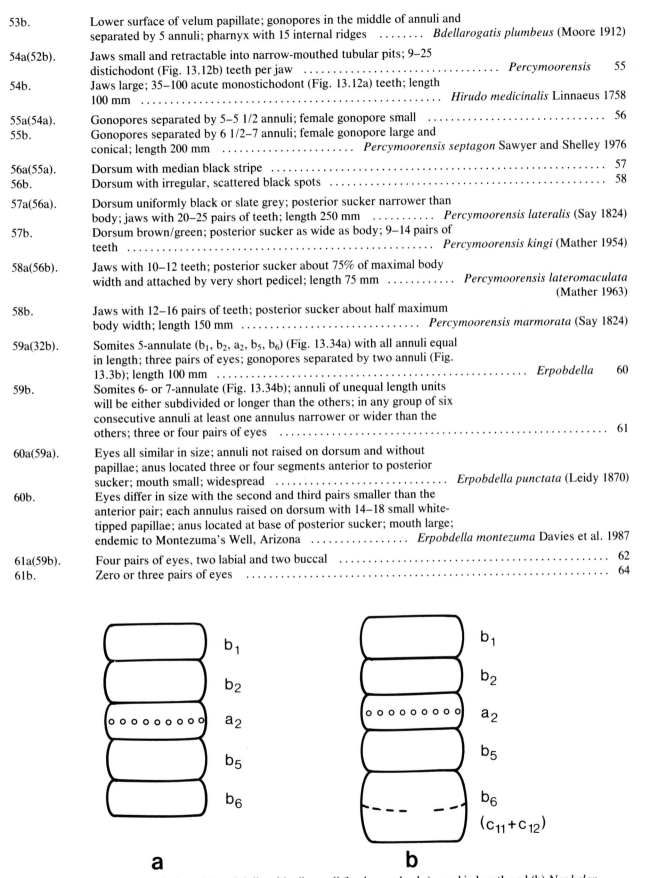

Figure 13.34 (a) Annulation of *Erpobdella* with all annuli (b_1, b_2, a_2, b_5, b_6) equal in length and (b) *Nephelopsis, Dina,* or *Mooreobdella* with the annuli (b_1, b_2, a_2, b_5, c_{11}, c_{12}) of unequal lengths.

62a(61a).   Two annuli between gonopores (Fig. 13.3a); atrial cornua spirally coiled like the horn of a ram (Fig. 13.22a); length 100 mm .... *Nephelopsis obscura* Verrill 1872

62b.   Three or more annuli between gonopores; atrial cornua simply curved (Fig. 13.22b) .... *Dina/Mooreobdella*   63

63a(62b).   3 1/2–4 annuli between gonopores; body heavily blotched with a median stripe; anus large, opening on a conical tubercle; length 60 mm .... *Dina dubia* (Moore and Meyer 1951)

63b.   3–3 1/2 annuli between gonopores; nearly pigmentless or with a few dark spots; anus small, not on tubercle; length 30 mm .... *Dina parva* Moore 1912

64a(61b).   Dorsum lacking scattered black pigment 65

64b.   Dorsum with scattered black pigment; two annuli between gonopores (Fig. 13.3a); length 55 mm .... *Mooreobdella melanostoma* Sawyer and Shelley 1976

65a(64a).   2–2 1/2 annuli between gonopores 66

65b.   3–4 1/2 annuli between gonopores 68

66a(65a).   Male gonopore surrounded by circle of papillae; zero or three pairs of eyes; two annuli between gonopores (Fig. 13.3a); length 15 mm .... *Dina anoculata* Moore 1898

66b.   No papillae around male gonopore; three (rarely four) pairs of eyes 67

67a(66b).   Two annuli between gonopores (Fig. 13.3a); three (rarely four) pairs of eyes; length 50 mm .... *Mooreobdella fervida* Verrill 1872

67b.   Either 2 or 2 1/2 annuli between gonopores; three pairs of eyes; length 30 mm .... *Mooreobdella bucera* Moore 1949

68a(65b).   Three pairs of eyes; gonopores separated by three - 3 annuli; length 50 mm .... *Mooreobdella microstoma* (Moore 1901)

68b.   Gonopores separated by 4–4 1/2 annuli; three pairs of eyes; length 40 mm .... *Mooreobdella tetragon* Sawyer and Shelley 1976

C. Taxonomic Key to Freshwater Polychaeta

1a.   Body divided into distinct thorax and abdomen (Fig. 13.7a); prostomium reduced and often fused to the peristomium which may bear palps, tentacular cirri, or a branchial crown; parapodia reduced; simple capillary setae with hooks or uncini (Fig. 13.24); pharynx without jaws .... Sedentaria   2

1b.   Body segments similar; body not divided into regions; prostomium well developed usually with antennae, palps, and sometimes eyes (Fig. 13.6); palps divided into thick basal joint and small distal joint; parapodia well developed (Fig. 13.5a); usually with compound as well as simple setae (Fig. 13.24); pharynx eversible with toothed jaws (Fig. 13.6) .... Errantia, family Nereididae   5

2a(1a).   Family Ampharetidae—Prostomium small, trilobed with retractile oral tentacles; four pairs of branchiae on anterior three segments (Fig. 13.7a); notopodium with capillary chaetae and neuropodium with uncini (Fig. 13.24) in both thorax and abdomen; tube composed of sandy debris .... *Hypaniola florida* (Hartman 1952)

2b.   Anterior end of body modified to form branchial crown of filaments surrounding mouth (Fig. 13.7b); notopodium with capillary chaetae and neuropodium with uncini in the thorax, but with uncini in the notopodium and capillaries in the neuropodium in the abdomen 3

3a(2b).   Family Sabellidae—Branchial crown without stalked operculum; branchiae absent behind head; thorax without collar; tube membranous or sandy; sandy–silty tube; 3–4 mm .... *Manayunkia speciosa* Leidy 1858

3b.   With stalked operculum closing mouth of tube; well-developed thoracic collar and thoracic membrane; tube calcareous .... Serpulidae   4

4a(3b).   Operculum with curved spines; calcareous (Fig. 13.35); 25 mm .... *Ficopomatus enigmaticus* (Fauvel 1922)

4b.   Operculum smooth .... *Ficopomatus miamiensis* (Treadwell 1914)

Fig. 13.35

Fig. 13.36

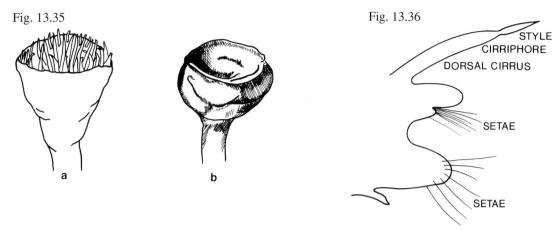

Figure 13.35 Dorsal view of (a) *Ficopomatus enigmaticus* showing spiny operculum; (b) *Ficopomatus miamiensis* showing the mouth operculum. **Figure 13.36** Parapodium of *Stenoninereis martini* showing modification of the dorsal curves with cirriphore and styli.

5a(1b).	Family Nereidae—Parapodia sub-biramous or uniramous (Fig. 13.1b); notopodia represented by internal noto aciculum; no noto chaetae; pharynx without paragnaths or papillae 6
5b.	Parapodia distinctly biramous (Fig. 13.35a); well-developed notopodium; both noto- and neuropodium with setae; pharynx with or without paragnaths or papillae 9
6a(5a).	Prostomium with four eyes; tentacular cirri short and smooth; dorsal cirri increasing in size posteriorly becoming large, flattened, and leaflike .... *Namalycastis abiuma* (Müller 1871)
6b.	Prostomium without eyes; dorsal cirri not increasing in size posteriorly 7
7a(6b).	Prostomium not bilobed; without eyes; with three pairs of tentacular cirri; proboscis without papillae; with elevated pads on dorsal side of basal ring .... *Lycastopsis hummelincki* Augener 1933
7b.	Prostomium bilobed; with four pairs of tentacular cirri 8
8a(7b).	With tentacular cirri generally uniform in length; short, tapering; with eyes; compound chaetae with long, smooth tips (Fig. 13.24c) .... *Namanereis hawaiiensis* (Johnson 1903)
8b.	With tentacular cirri that appear articulated; not uniform in length; without eyes .... *Lycastoides alticola* Johnson 1903
9a(5b).	Dorsal cirri with elongated cirrophores and distal styles (Fig. 13.36); pharynx without papillae or paragnaths .... *Stenoninereis martini* Wesenberg-Lund 1958
9b.	Dorsal cirri attached on basal parts of notopodium 10
10a(9b).	Pharynx with groups of soft papillae; no paragnaths .... *Laenonereis culveri* (Webster 1879)
10b.	Pharynx with groups of paragnaths 11
11a(10b).	Parapodia biramous; well developed; posterior dorsal notopodia much enlarged and often foliaceous; proboscis with large paragnaths; compound setae with straight blades; 50 mm .... *Nereis succinea* Frey and Leuckart 1847
11b.	Parapodia biramous; well developed but without enlarged posterior notopodia; proboscis with small paragnaths; compound chaetae .... *Nereis limnicola* Johnson 1903

ACKNOWLEDGMENTS

It is a pleasure to acknowledge the dedicated assistance of Erin Moloney, who typed and proofread this chapter, and the Department of Zoology, James Cook University of Northern Queensland, Townsville, Australia for providing office space during the majority of the writing.

LITERATURE CITED

Able, K. W. 1976. Cleaning behaviour in the cyprinodont fishes: *Fundus majalis, Cyprinodon variegatus* and *Lucania parva*. Chesapeake Science 17:35–39.

Anderson, R. S., and L. G. Raasveldt. 1974. *Gammarus* predation and the possible effects of *Gammarus* and

Chaoborus feeding on the zooplankton composition in some small lakes and ponds in western Canada. Canadian Wildlife Service Occasional Paper No. 18.

Arnold, S. J. 1981. Behavioural variation in natural populations. I. Phenotypic, genetic and environmental correlations between chemoreceptive responses to prey in the garter snake *Thamnophis elegans*. Evolution 35:489–509.

Augener, H. 1933. Susswasser Polychaeten von Bonaire. Zoologische Jahrücher. Systematik, Okologie und Geographie der Tierre 64:351–356.

Babin, J., and E. E. Prepas. 1985. Modelling winter oxygen depletion rates in ice-covered temperate zone lakes in Canada. Canadian Journal of Fisheries and Aquatic Sciences 42:239–249.

Baird, D. J., L. R. Linton, and R. W. Davies. 1986. Life-history evolution and post-reproductive mortality risk. Journal of Animal Ecology 55:295–302.

Baird, D. J., T. E. Gates, and R. W. Davies. 1987a. Oxygen conditions in two prairie pothole lakes during winter ice cover. Canadian Journal of Fisheries and Aquatic Sciences 44:1092–1095.

Baird, D. J., L. R. Linton, and R. W. Davies. 1987b. Life-history flexibility as a strategy for survival in a variable environment. Functional Ecology 1:45–48.

Baird, W. 1869. Descriptions of some new suctorial Annelids in the collection of the British Museum. Proceedings of the Zoological Society of London. pp. 310–318.

Bartonek, J. C. 1972. Summer foods of American Widgeon, Mallards and Green-winged Teal near Great Slave Lake, Northwest Territories. Canadian Field-Naturalist 86:373–376.

Bartonek, J. C., and J. J. Hickey. 1969a. Selective feeding by juvenile diving ducks in summer. The Auk 86:457–493.

Bartonek, J. C., and J. J. Hickey. 1969b. Food habits of canvasbacks, redheads and lesser scaup in Montana. Condor 71:280–290.

Bartonek, J. C., and H. W. Murdy. 1970. Summer foods of lesser scaup in subarctic taiga. Arctic 23:35–44.

Bartonek, J. C., and D. L. Trauger. 1975. Leeches (Hirudinea) infestations among waterfowl near Yellowknife, Northwest Territories. Canadian Field-Naturalist 89:234–243.

Beadle, L. D. 1939. Regulation of the haemolymph in the saline water mosquito larva *Aedes detritus* (Edw.). Journal of Experimental Biology 16:346–362.

Bennike, S. A. B. 1943. Contributions to the ecology and biology of the Danish freshwater leeches. Folia Limnologica Scandinavica 2:1–109.

Blair, W. N. 1927. Notes on *Hirudo medicinalis*, the medicinal leech, as a British species. Proceedings of the Zoological Society of London 11:999–1002.

Blanchard, E. 1849. Annelides. *In:* Gay's historia fisca y politico de Chile. Zoologia, Paris 3:43–50.

Blanchard, R. 1893. Courtes notices sur les Hirudinees: X. Hirudinees de l'Europe boreale. Bulletin Société Zoologique de France 18:93–94.

Blinn, D. W., and R. W. Davies. 1989. The evolutionary importance of mechanoreception in three erpobdellid leech species. Oecologia 79:6–9.

Blinn, D. W., R. W. Davies, and B. Dehdashti. 1987. Specialized pelagic feeding by *Erpobdella montezuma* (Hirudinea). Holarctic Ecology 10:235–240.

Blinn, D. W., C. Pinney, and V. T. Wagner. 1988. Intraspecific discrimination of amphipod prey by a freshwater leech through mechanoreception. Canadian Journal of Zoology 66:427–430.

Büsing, K. H. 1951. *Pseudomonas hirudinis,* ein bakterieller Darmsybiont des Blutegels (*Hirudo officinalis*). Zentralblatt fuer Bakteriologie Parasitenkunde 157:478–484.

Büsing, K. H., W. Döll, and K. Freytag. 1953. Die Bakterienflora der medizinischen Blutegel. Archiv für Mikrobiologie 19:52–86.

Bulmer, M. G. 1985. Selection for iteroparity in a variable environment. American Naturalist 136:63–71.

Carena, H. 1824. Monographic due genre *Hirudo*. Supplement. Memoire Accademia Science, Torino 28:331–337.

Casselman, J. M., and H. H. Harvey. 1975. Selective fish mortality resulting from low winter oxygen. Internationale Vereinigung für Theoretische und Angewandte Limnologie 19:2418–2429.

Castle, W. E. 1900. Some North American fresh-water Rhynchobdellidae, and their parasites. Bulletin of the Museum of Comparative Zoology, Harvard 36:17–64.

Clark, R. B. 1969. Systematics and phylogeny: Annelida, Echiura, Sipuncula. Pages 1–68 *in:* M. Florkin and B. Scheer, editors. Chemical zoology. Vol. 4. Academic Press, New York.

Croghan, P. C. 1958. The survival of *Artemia salina* (Linn.) in various media. Journal of Experimental Biology 35:213–216.

Cywinska, A., and R. W. Davies. 1989. Predation on the erpobdellid leech *Nephelopsis obscura* in the laboratory. Canadian Journal of Zoology 67:2689–2693.

Daborn, G. R. 1976. Colonization of isolated aquatic habitats. Canadian Field-Naturalist 90:56–57.

Dahm, A. G. 1962. Distribution and biological patterns of *Acanthobdella peledina* Grube from Sweden (Hirudinea, Acanthobdellidae). Acta Universitatis Lundensis Arrskrift 58:1–35.

Dales, R. P. 1962. The polychaete stomadium and the interrelationships of the families of Polychaeta. Proceedings of the Zoological Society (London) 139:398–428.

Daniels, B. A., and R. T. Sawyer. 1973. Host-parasite relationship of the fish leech *Illinobdella moorei* Meyer and the white catfish *Ictalurus catus* (Linnaeus). Bulletin of the Association of Southeastern Biologists 20:48.

Davies, R. W. 1969. The production of antisera for detecting specific triclad antigens in the gut contents of predators. Oikos 20:248–260.

Davies, R. W. 1971. A key to the freshwater Hirudinoidea of Canada. Journal of the Fisheries Research Board of Canada 28:543–552.

Davies, R. W. 1972. Annotated bibliography to the freshwater leeches (Hirudinoidea) of Canada. Fisheries Research Board of Canada., Technical Report No. 306:15 pp.

Davies, R. W. 1973. The geographic distribution of

freshwater Hirudinoidea in Canada. Canadian Journal of Zoology 51:531–545.

Davies, R. W. 1978. Reproductive strategies shown by freshwater Hirudinoidea. Internationale Vereinigung für Theoretische und Angewandte Limnologie 20: 2378–2381.

Davies, R. W. 1979. Dispersion of freshwater leeches (Hirudinoidea) to Anticosti Island, Quebec. Canadian Field-Naturalist 93:310–313.

Davies, R. W. 1984. Sanguivory in leeches and its effects on growth, survivorship and reproduction of *Theromyzon rude*. Canadian Journal of Zoology 62:589–593.

Davies, R. W. 1987. All about leeches. *Nature (London)* 325:585.

Davies, R. W., and R. P. Everett. 1977. The life history, growth and age structure of *Nephelopsis obscura* Verrill, 1872 (Hirudinoidea) in Alberta. Canadian Journal of Zoology 55:620–627.

Davies, R. W., and T. B. Reynoldson. 1976. A comparison of the life-cycle of *Helobdella stagnalis* (Linn. 1758) (Hirudinoidea) in two different geographic areas in Canada. Journal of Animal Ecology 45:457–470.

Davies, R. W., and R. N. Singhal. 1988. Cosexuality in the leech, *Nephelopsis obscura* (Erpobdellidae). International Journal of Invertebrate Reproduction and Development 13:55–64.

Davies, R. W., and J. Wilkialis. 1982. Observations on the ecology and morphology of *Placobdella papillifera* (Verrill) (Hirudinoidea: Glossiphoniidae) in Alberta, Canada. American Midland Naturalist 107:316–324.

Davies, R. W., T. B. Reynoldson, and R. P. Everett. 1977. Reproductive strategies of *Erpobdella punctata* (Hirudinoidea) in two temporary ponds. Oikos 29:313–319.

Davies, R. W., F. J. Wrona, and R. P. Everett. 1978. A serological study of prey selection by *Helobdella stagnalis* (Hirudinoidea). Journal of Animal Ecology 48:181–194.

Davies, R. W., F. J. Wrona, L. Linton, and J. Wilkialis. 1981. Inter- and intra-specific analyses of the food niches of two sympatric species of Erpobdellidae (Hirudinoidea) in Alberta, Canada. Oikos 37:105–111.

Davies, R. W., L. R. Linton, W. Parsons, and E. S. Edgington. 1982a. Chemosensory detection of prey by *Nephelopsis obscura* (Hirudinoidea:Erpobdellidae). Hydrobiologia 97:157–161.

Davies, R. W., L. R. Linton, and F. J. Wrona. 1982b. Passive dispersal of four species of freshwater leeches (Hirudinoidea) by ducks. Freshwater Invertebrate Biology 1:40–44.

Davies, R. W., F. J. Wrona, and L. Linton. 1982c. Changes in numerical dominance and its effects on prey utilization and inter-specific competition between *Erpobdella punctata* and *Nephelopsis obscura* (Hirudinoidea): an assessment. Oikos 39:92–99.

Davies, R. W., R. N. Singhal, and D. W. Blinn. 1985. *Erpobdella montezuma* (Hirudinoidea: Erpobdellidae), a new species of freshwater leech from North America. Canadian Journal of Zoology 63:965–969.

Davies, R. W., T. Yang, and F. J. Wrona. 1987. Inter- and intra-specific differences in the effects of anoxia on

erpobdellid leeches using static and flow-through systems. Holarctic Ecology 10:149–153.

Davies, R. W., D. W. Blinn, B. Dehdashti, and R. N. Singhal. 1988. The comparative ecology of three species of Erpobdellidae (Annelida-Hirudinoidea). Archiv für Hydrobiologie 111:601–614.

De. Filippi, F. 1849. Sopra un nuovo genere (*Haementeria*) di Annelidi della famiglia delle Sanguisughe. Memoires de Academie de Science., Torino 10:1–14.

Derosa, X. S., and W. O. Friesen. 1981. Morphology of leech sensilla: observations with the scanning electron microscope. Biological Bulletin 160:383–393.

Dickinson, M. H., and C. M. Lent. 1984. Feeding behaviour of the medicinal leech, *Hirudo medicinalis* L. Journal of Comparative Physiology 154:449–455.

Ekman, S. 1911. Die Bodenfauna des Vattern, Qualitative und Quantitativ Untersucht. Internationale Revue der gesamten Hydrobiologie 7:146–204.

Elliott, J. M. 1973. The diel activity pattern, drifting and food of the leech *Erpobdella octoculata* (L.) (Hirudinea: Erpobdellidae) in a Lake District stream. Journal of Animal Ecology 42:449–459.

Elliott, J. M., and K. H. Mann. 1979. A key to the British freshwater leeches. Freshwater Biological Association Scientific Publication No. 40:72 pp.

Fauchold, K. 1977. The polychaete worms. Definitions and keys to the orders, families and genera. Natural History Museum of Los Angeles County, Science Series 28:1–190.

Fauvel, P. 1922. Un nouveau serpulien d'eau Saumatre *Mercierella enigmatica*. Bulletin de la Société Zoologique de France 47:425–431.

Fernandez, J. 1980. Embryonic development of the Glossiphoniid leech *Theromyzon rude*: characterization of developmental stages. Developmental Biology 76:245–262.

Foster, N. 1972. Freshwater Polychaetes (Annelida) of North America. United States Environmental Protection Agency Identification Manual No. 4: 15 pp.

Frey, J., and K. Leuckart. 1847. Cited in Hartman (1951).

Friesen, W. O. 1981. Physiology of water motion detection in the medicinal leech. Journal of Experimental Biology 92:255–275.

Friesen, W. O., and R. D. Dedwylder. 1978. Detection of low-amplitude water movements: a new sensory modality in the medicinal leech. Neuroscience Abstracts 4:380.

Gates, T. E., and R. W. Davies. 1987. The influence of temperature on the depth distribution of sympatric Erpobdellidae (Hirudinoidea). Canadian Journal of Zoology 65:1243–1246.

Gates, T. E., D. J. Baird, F. J. Wrona, and R. W. Davies. 1987. A device for sampling macroinvertebrates in weedy ponds. Journal of the North American Benthological Society 6:133–139.

Gee, W. 1913. The behavior of leeches with special reference to its modifiability. University of California Publications in Zoology 11:197–305.

George, J. D., and G. Hartmann-Schröeder. 1985. Polychaetes: British Amphinomida, Spintherida and

Eunicida. *In:* D. M. Kermack and R. S. K. Barnes, editors. Synopsis of the British Fauna. The Linnean Society of London and The Estuarine and Brackish-Water Science Association.

Gerking, S. D. 1957. A method of sampling the littoral macrofauna and its application. Ecology 38:219–266.

Giesel, J. T. 1974. The biology and adaptability of natural populations. Mosley, St. Louis, Missouri.

Graf, A. 1899. Hirudineenstudien. Acta Acadamemiae Leopoldensia, Halle 72:215–404.

Green, R. H. 1979. Sampling design and statistical methods for environmental biologists. Wiley, New York.

Grube, A. E. 1851. Annulaten reise in den Aussersten Norden und Ostens Sibiriens. Herausgegeben von Dr. A. Th. Middendorff, St. Petersburg 1:1–254.

Hartman, O. 1951. The Literature of the Polychaetous Annelids. Vol 1. Allan Hancock Foundation, University of Southern California, Los Angeles. 290 pp.

Hartman, O. 1952. Iphitime and Ceratocephala (Polychaetous Annelids) from California. Bulletin of the Southern California Academy of Science 51:9–20.

Hayunga, E. G., and A. J. Grey. 1976. *Cystobranchus meyeri* sp. n. (Hirudinea: Piscicolidae) from *Catostomus commersoni* Lacépède in North America. Journal of Parasitology 62:621–627.

Hazel, C. R. 1966. A note on the freshwater polychaete *Manayunkia speciosa* Leidy, from California and Oregon. Ohio Journal of Science 66:533–535.

Hemingway, E. E. 1908. The anatomy of *Placobdella pediculata* Pages 29–63 *in:* The Leeches of Minnesota. Geological and Natural History of Minnesota., Zoology Series No. 5.29–63.

Herrmann, S. J. 1970a. Systematics, distribution and ecology of Colorado Hirudinea. American Midland Naturalist 83:1–37.

Herrmann, S. J. 1970b. Total residual tolerances of Colorado Hirudinea. Southwestern Naturalist 15:261–273.

Herter, K. 1929a. Temperaturversuche mit Egeln. Zeitschrift fuer Vergleichende Physiologie 10:248–271.

Herter, K. 1929b. Reizphysiologisches Verhalten und Parasitismus des Entenegels *Protoclepsis tesselata* O.F. Müller. Zeitschrift fuer Vergleichende Physiologie 10:272–308.

Hiltunen, J. H. 1965. Distribution and abundance of the polychaete, *Manayunkia speciosa* Leidy, in Western Lake Erie. Ohio Journal of Science 65:183–185.

Hobbs, H. H., and H. V. Figueroa. 1958. The exoskeleton of a freshwater crab as a microhabitat for several invertebrates. Virginia Journal of Science 9:395–396.

Hoffman, R. L. 1964. A new species of Cystobranchus from southwestern Virginia (Hirudinea: Piscicolidae). American Midland Naturalist 72:390–395.

Holmquist, C. 1974. A fish leech of the genus *Acanthobdella* found in North America. Hydrobiologia 44:241–245.

Holt, P. C. 1969. The systematic position of the Branchiobdellidae (Annelida: Clitellata). Systematic Zoology 14:25–32.

Jennings, J. B., and V. M. Vanderlande. 1967. Histochem-

ical and bacteriological studies on digestion in nine species of leeches (Annelida: Hirudinea). Biological Bulletin 133:166–183.

Johnson, G. M., and D. J. Klemm. 1977. A new species of leech, *Batracobdella cryptobranchii* n. sp. (Annelida: Hirudinea) parasitic on the Ozark hillbender. Transactions of the American Microscopical Society 90:327–331.

Johnson, H. P. 1903. Freshwater nereids from the Pacific coast and Hawaii, with remarks on freshwater Polychaeta in general. Mark Anniversary Volume. Holt, New York. Pages 205–223.

Jones, M. L. 1967. On the morphology of the nephridia of *Nereis limnicola* Johnson. Biological Bulletin 132:362–380.

Kamemoto, F. I., and E. J. Larson. 1964. Chloride concentrations in the coelomic and nephridia of fluids of the Sipunculia, *Dendrostomum signifer*. Comparative Biochemistry and Physiology A 13A:477–480.

Kephart, D. G. 1982. Microgeographic variation in the diets of garter snakes. Oecologia (Berlin) 52:287–291.

Kephart, D. G., and S. J. Arnold. 1982. Garter snake diets in a fluctuating environment: a seven-year study. Ecology 63:1232–1236.

Kinne, O. 1954. Uber da Schwärmen und die Larvalentwicklung von *Nereis succinea* Leuckart. Zoologischer Anzeiger 153:114–126.

Klemm, D. J. 1972a. Freshwater Leeches (Annelida:Hirudinea) of North America. United States Environmental Protection Agency Identification Manual No. 8:53 pp.

Klemm, D. J. 1972b. Freshwater Polychaetes (Annelida) of North America. United States Environmental Protection Agency Identification Manual No. 4: 15 pp.

Klemm, D. J. 1977. A review of the leeches (Annelida:Hirudinea) in the Great Lakes region. Michigan Academician 9:397–418.

Klemm, D. J. 1982. Leeches (Annelida:Hirudinea) of North America. United States Environmental Protection Agency 600/3–82–025:177 pp.

Klemm, D. J. 1985. A Guide to the Freshwater Annelida (Polychaeta, Naidid and Tubificid Oligochaeta, and Hirudinea) of North America. Kendall/Hunt, Dubuque, Iowa. 198 pp.

Koechlin, N. 1972. Etude histochemique et ultrastructurales des pigments néphridiens et coelomiques d'une Annélide Polychète (*Sabella*). Annales d'Histochemie 17:27–54.

Koechlin, N. 1975. Micropuncture studies of urine formation in a marine invertebrate, *Sabella pavonina* Savigny (Polychaeta, Annelida). Comparative Biochemistry and Physiology A 52A:459–465.

Koechlin, N. 1977. Micropuncture study of the sodium and potassium content in the nephridial and coelomic fluids in a marine invertebrate *Sabella pavonina* (Polychaeta, Annelida). Comparative Biochemistry and Physiology A 57A: 353–359.

Krishnan, G. 1952. On the nephridia of *Nereidae* in relation to habitat. National Institute of Science, India 18:241–255.

Kutschera, U. 1988. A new leech species from North

America, *Helobdella californica* nov. sp. (Hirudinea: Glossiphoniidae). Zoologischer Anzeiger 220:173–178.

Lasserre, P. 1975. Clitellata. Pages 215–275 *in:* A. C. Giese and J. S. Pearse, editors. Reproduction of marine invertebrates. Vol. 3: Annelids: Echiurans. Academic Press, New York.

Leidy, J. 1851. Description of *Myzobdella.* Proceedings of the Academy of Natural Sciences, Philadelphia 1851–1853:243.

Leidy, J. 1858. *Manayunkia speciosa.* Proceedings of the Academy of Natural Science, Philadelphia 10:90.

Leidy, J. 1870. Description of *Nephelis punctata.* Proceedings of the Academy of Natural Sciences, Philadelphia 22:89.

Light, W. J. 1969. Extension of range of *Manayunkia aestuarina* (Polychaeta: Sabellida) to British Columbia. Journal of the Fisheries Research Board of Canada 26:3088–3091.

Lillie, F. R., and E. Just. 1913. Breeding habits of the heteranereis form of *Nereis limbata.* Biological Bulletin 24:147–168.

Linnaeus, C. 1758. Systema naturae. Lipsiae. 10th Edition.

Linnaeus, C. 1761. Fauna suecia. Stockholm. 578 pp.

Linton, L. R., and R.W. Davies. 1987. An energetics model of an aquatic predator and its application to life-history optima. Oecologia 71:552–559.

Linton, L. R., R. W. Davies, and F. J. Wrona. 1982. Osmotic and respirometric responses of two species of Hirudinoidea to changes in water chemistry. Comparative Biochemistry and Physiology 71A:243–247.

Linton, L. R., R. W. Davies, and F. J. Wrona. 1983a. The effects of water temperature, ionic content and total dissolved solids on *Nephelopsis obscura* and *Erpobdella punctata* (Hirudinoidea, Erpobdellidae). I. Mortality. Holarctic Ecology 6:59–63.

Linton, L. R., R. W. Davies, and F. J. Wrona. 1983b. The effect of water temperature, ionic content and total dissolved solids on *Nephelopsis obscura* and *Erpobdella punctata* (Hirudinoidea, Erpobdellidae). II. Reproduction. Holarctic Ecology 6:64–68.

Lukin, E. I. 1976. Leeches of fresh and brackish water bodies. Fauna of the Union of Soviet Socialist Republics. Leeches. Vol. 1. Academy of Science, USSR, Moscow. 484 pp.

Mackie, G. L., and S. U. Qadri. 1971. A polychaete *Manayunkia speciosa*, from the Ottawa River, and its North American distribution. Canadian Journal of Zoology 49:780–782.

Madanmohanrao, G. 1960. Salinity tolerances and oxygen consumption of the cattle leech, *Hirudinaria granulosa.* Proceedings of the Indian Academy of Science 11:211–218.

Malm, C. W. 1863. Svenska Iglar, Disciferae. Götheborg. Vetenskapsakademien Arhandlingar i Naturskosarendon 8:153–263.

Mann, K. H. 1955. Some factors influencing the distribution of freshwater leeches in Britain. Internationale Vereinigung für Theoretische und Angewandte Limnologie 12:582–587.

Mann, K. H. 1956. A study of the oxygen consumption of

five species of leech. Journal of Experimental Biology 33:615–626.

Mann, K. H. 1961. Leeches (Hirudinea) their structure, physiology, ecology and embryology. Pergamon, Oxford. 201 pp.

Mannsfeld, W. 1934. Pages 539–642 *in:* Literature uber Hirudineen bis zum Jahre 1938. Bronn's Klassen und Ordnungen des Tierreichs. Bd. 4, Abt. III: Buch 4, T.2.

Mather, C. K. 1954. *Haemopis kingi,* new species (Annelida: Hirudinea). American Midland Naturalist 52:460–468.

Mather, C. K. 1963. *Haemopis latero-maculatum,* new species (Annelida: Hirudinea). American Midland Naturalist 70:168–174.

Meyer, M. C. 1940. A revision of the leeches (Piscicolidae) living on freshwater fishes of North America. Transactions of the American Microscopical Society 59:354–376.

Meyer, M. C. 1946. A new leech *Piscicola salmositica,* n. sp. (Piscicolidae), from steelhead trout (*Salmo gairdneri gairdneri* Richardson, 1838). Journal of Parasitology 32:467–476.

Meyer, M. C. 1975. A new leech, *Macrobdella diplotertia* sp. n. (Hirudinea: Hirudinidae), from Missouri. Proceedings of the Helminthological Society of Washington 42:83–85.

Meyer, M. C., and J. P. Moore. 1954. Notes on Canadian leeches (Hirudinea), with the description of a new species. Wasmann Journal of Biology 12:63–96.

Mileikovsky, S. A. 1977. On the systematic interrelationships within the Polychaeta and Annelida—an attempt to create an integrated system based on their larval morphology. Pages 503–524 *in:* D. J. Reish and K. Fauchald, editors. Essays on Polychaetae Annelids in memory of Dr. Olga Hartman Allan Hancock Foundation, Los Angeles, California.

Moore, J. P. 1898. The leeches of the U.S. National Museum. Proceedings of the United States National Museum 21:543–563.

Moore, J. P. 1900. A description of *Microbdella biannulata* with especial regard to the constitution of the leech somite. Proceedings of the Academy of Natural Science, Philadelphia 52:50–73.

Moore, J. P. 1901. The Hirudinea of Illinois. Illinois State Laboratory of Natural History Bulletin 5:479–547.

Moore, J. P. 1906. Hirudinea and Oligochaeta collected in the Great Lakes region. Bulletin of the Bureau of Fisheries 25:153–171.

Moore, J. P. 1912. Classification of the leeches of Minnesota. *in:* The Leeches of Minnesota. Geology and Natural History of Minnesota, Zoology Series No. 5:63–150.

Moore, J. P. 1949. Hirudinea. Pages 38–39 *in:* R. Kenk, editor. The Animal Life of Temporary and Permanent Ponds in Southern Michigan. Miscellaneous Publications of the Museum of Zoology, University of Michigan No. 71:38–39.

Moore, J. P. 1953. Three undescribed North American leeches (Hirudinea). Notulae National Academy of Natural Science, Philadelphia No. 250:1–13.

Moore, J. P. 1954. Cited in Meyer and Moore (1954).

Moore, J. P. 1959. Hirudinea. Pages 542–557 *in:* W. T. Edmonson, editor. Freshwater Biology. 2nd Edition. Wiley, New York.

Moore, J. P., and M. C. Meyer. 1951. Leeches (Hirudinea) from Alaskan and adjacent waters. Wasmann Journal of Biology 9:11–77.

Muller, F. 1844. De Hirudinea circa Berolinum hueusque observatis. Dissitation Inaug., Berlin. Pages 23–25.

Müller, F. 1871. Cited in Grube (1871).

Muller, O. F. 1774. Vermium terrestrium et fluviatilium, seu animalium infusoriorum helminthicorum, et testaceorum, non marinorum succincta historia. *Havniae et Lipsiae (Helminthica)* 1:37–51.

Oglesby, L. C. 1969. Inorganic components and metabolism, ionic and osmotic regulation. Annelida, *Sipuncula* and Echiura. Pages 211–310 *in:* M. Florkin and B. T. Scheer, editors. Chemical Zoology. Vol. 4. Academic Press, New York.

Pawlowski, L. K. 1948. Contribution à la connaissance des sangsus (Hirudinea) de la Nouvelle-Ecosse, de Terre-Neuve et des iles francais Saint-Pierre et Miquelon. Fragmenta faunistica Musei Zoologici Polonici 5:317–353.

Pearse, A. S. 1932. Parasites of Japanese salamanders. Ecology 13:135–152.

Peterson, D. L. 1982. Management of ponds for bait-leeches in Minnesota. Minnesota Department of Natural Resources, Section of Fisheries Investigation. Report No. 357:43.

Peterson, D. L. 1983. Life cycle and reproduction of *Nephelopsis obscura* Verrill (Hirudinea; Erpobdellidae) in permanent ponds of northwestern Minnesota. Freshwater Invertebrate Biology 2:165–172.

Pettibone, M. H. 1953. Fresh-water polychaetous annelid, *Manayunkia speciosa* Leidy, from Lake Erie. Biological Bulletin (*Woods Hole, Mass*) 105:149–153.

Pettibone, M. H. 1982. Annelida. Pages 1–61 *in:* S. P. Parker, editor. Synopsis and classification of living organisms. Vol. 2. McGraw-Hill, New York.

Pritchard, G. 1964. The prey of dragonfly larvae (Odonata: Anisoptera) in ponds in northern Alberta. Canadian Journal of Zoology 42:785–800.

Pütter, A. 1907. Der Stoffwechsel der Blutegels (*Hirudo medicinalis* L.). I. Zeitschrift fuer Allgemeine Physiologie 6:217–286.

Pütter, A. 1908. Der Stoffwechsel der Blutegels (*Hirudo medicinalis*) II. Zeitschrift fuer Allgemeine Physiologie 7:16–61.

Reynoldson, T. B., and L. S. Bellamy. 1970. The establishment of interspecific competition in action between *Polycelis nigra* (Müll.) and *P. tenuis* (Ijima) (Turbellaria, Tricladida). Pages 282–297 *in:* Proceedings of the Advanced Study Institute, Dynamics Number Population, Oosterbruck.

Reynoldson, T. B., and R. W. Davies. 1976. A comparative study of the osmoregulatory ability of three species of leech (Hirudinoidea) and its relationship to their distribution in Alberta. Canadian Journal of Zoology 54:1908–1911.

Reynoldson, T. B., and R. W. Davies. 1980. A comparative study of weight regulation in *Nephelopsis obscura*

and *Erpobdella punctata* (Hirudinoidea). Comparative Biochemistry and Physiology 66:711–714.

Richardson, L. R. 1942. Observations on migratory behaviour of leeches. Canadian Field-Naturalist 56:67–70.

Richardson, L. R. 1969. A contribution to the systematics of the hirudinid leeches with descriptions of new families genera and species. Acta Zoologica Academiae Scientiarum Hungaricae 15:97–149.

Richardson, L. R. 1971. A new species from Mexico of the Nearctic genus *Percymoorensis* and remarks on the family Haemopidae (Hirudinoidea). Canadian Journal of Zoology 49:1095–1103.

Rosca, D. I. 1950. Duration of survival and variations in body weight in *Hirudo medicinalis* placed in solutions of increasing salinity. Studii si Cercetari de Biologie Cluj. 1:211–222.

Sawyer, R. T. 1970. Observations on the natural history and behaviour of *Erpobdella punctata* (Leidy) (Annelida: Hirudinea). American Midland Naturalist 83:65–80.

Sawyer, R. T. 1972. North America freshwater leeches, exclusive of the Piscicolidae, with a key to all species. Illinois Biological Monograph No. 46:154 pp.

Sawyer, R. T. 1981. Leech biology and behavior. Pages 7–26 K. J. Muller, J. G. Nicholls, and G. S. Stent, editors. *in:* The neurobiology of the leech. Cold Spring Harbor Lab., Cold Spring Harbor, New York.

Sawyer, R. T. 1984. Arthropodization in the Hirudinea: evidence for a phylogenetic link with insects and other Uniramia. Zoological Journal of the Linnean Society, London 80:303–322.

Sawyer, R. T. 1986. Leech biology and behaviour. Clarendon, Oxford. 1065 pp.

Sawyer, R. T., and D. L. Hammond. 1973. Distribution, ecology and behavior of the marine leech *Calliobdella carolinensis* (Annelida: Hirudinea), parasitic on the Atlantic menhaden in epizootic proportions. Biological Bulletin (*Woods Hole, Mass.*) 145:373–388.

Sawyer, R. T., and R. H. Shelley. 1976. New records and species of leeches (Annelida: Hirudinea) from North and South Carolina. Journal of Natural History 10:65–97.

Sawyer, R. T., F. LePont, D. K. Stuart, and A. P. Kramer. 1981. Growth and reproduction of the giant glossiphoniid leech *Haementaria ghilianii*. Biological Bulletin (*Woods Hole, Mass.*) 160:322–331.

Say, T. 1824. Sur les *Hirudo parasitica, laterialis, marmorata* et *decora* dans le voyage du Major Long. Natural History, Zoology Philadelphia 2:253–387.

Schroeder, P. C., and C. O. Hermans. 1975. Annelida: Polychaeta. Pages 1–213 *in:* A. C. Giese and J. S. Pearse, editors. Reproduction of marine invertebrates. Vol. 3. Annelids: Echiurans. Academic Press, New York.

Scudder, G. G. E., and K. H. Mann. 1968. The leeches of some lakes in the southern interior plateau region of British Columbia. Syesis 1:203–209.

Singhal, R. N., and R. W. Davies. 1985. Descriptions of the reproductive organs of *Nephelopsis obscura* and *Erpobdella punctata* (Hirudinoidea: Erpobdellidae). Freshwater Invertebrate Biology 5:91–97.

Singhal, R. N., and R. W. Davies. 1987. Histopathology of hyperoxia in *Nephelopsis obscura* (Hirudinodiea: Erpobdellidae). Journal of Invertebrate Pathology 50:33–39.

Singhal, R. N., H. B. Sarnat, and R. W. Davies. 1988. Effect of anoxia and hyperoxia on the neurons in the leech *Nephelopsis obscura* (Erpobdellidae): ultrastructural studies. Journal of Invertebrate Pathology 52:409–418.

Sladacek, V., and V. Kosel. 1984. Indicator value of freshwater leeches (Hirudinea) with a key to the determination of European species. Acta Hydrochimie und Hydrobiologie 12:451–461.

Smiley, J. W., and R. T. Sawyer. 1976. Osmoregulation in the leeches *Macrobdella dietra* and *Haemopis marmorata*. Bulletin of the Association of Southeastern Biologists 23:96.

Smith, R. I. 1950. Embryonic development in the viviparous nereid polychaete *Neanthes lighti*, Hartman. Journal of Morphology 87:417–466.

Soos, A. 1965. Identification key to the leech (Hirudinoidea) genera of the world, with a catalogue of the species. I. Family: Piscicolidae. Acta Zoologica Academiae Scientarium Hungaricae 2:417–463.

Soos, A. 1966a. Identification key to the leech (Hirudinoidea) genera of the world, with a catalogue of the species. II. Families: Semiscolecidae, Trematobdellidae, Americobdellidae, Diestecostomatidae. Acta Zoologica Academiae Scientarium Hungaricae 12:145–160.

Soos, A. 1966b. Identification key to the leech (Hirudinoidea) genera of the world, with a catalogue of the species. III. Family: Erpobdellidae. Acta Zoologica Academiae Scientarium Hungaricae 12:371–407.

Soos, A. 1966c. On the genus *Glossiphonia* Johnson, 1816, with a key and catalogue to the species Hirudinoidea: Glossiphoniidae. Annales Historico-Naturales Musei Nationalis Hungarici 58:271–279.

Soos, A. 1967. Identification key to the leech (Hirudinoidea) genera of the world, with a catalogue of the species. IV. Family: Haemodipsidae. Acta Zoologica Academiae Scientiarum Hungaricae 13:417–432.

Soos, A. 1968. Identification key to the species of the genus *Erpobdella* de Blainville, 1818 (Hirudinoidea:Erpobdellidae). Annales Historico-Naturales Musei Nationalis Hungarici 60:141–145.

Soos, A. 1969a. Identification key to the leech (Hirudinoidea) genera of the world, with a catalogue of the species. V. Family: Hirudinidae. Acta Zoologica Academiae Scientiarum Hungaricae 15:151–201.

Soos, A. 1969b. Identification key to the leech (Hirudinoidea) genera of the world, with a catalogue of the species. VI. Family: Glossiphoniidae. Acta Zoologica Academiae Scientiarum Hungaricae 15:397–454.

Spelling, S. M., and J. O. Young. 1987. Predation on lake-dwelling leeches (Annelida: Hirudinea): An evaluation by field experiment. Journal of Animal Ecology 56:131–146.

Treadwell, A. L. 1914. Polychaetous annelids of the Pacific coast in the collections of the zoological museum of the University of California. University of California Publications in Zoology 13:175–234.

Van Damme, N. 1974. Organogénèse de l'appareil genital chez la sangsue *Erpobdella octoculata* L. (Hiridinée-Pharyngobdelle). Archives de Biologie 18:373–377.

Verrill, A. E. 1872. Descriptions of North American freshwater leeches. American Journal of Science 3:126–139.

Verrill, A. E. 1874. Synopsis of the North American freshwater leeches. Report to the United States Fisheries Commission 1872/73:666–689.

von Brand, T. 1946. Anaerobiosis in invertebrates. Missouri.

Webster, H. 1879. Annelida Chaetopoda of New Jersey. New York State Museum of Natural History Report 32:101–125.

Weisblat, D. A., G. Harper, G. S. Stent, and R. T. Sawyer. 1980. Embryonic cell lineages in the nervous system of the glossiphoniid leech *Helobdella triserialis*. Developmental Biology 76:58–78.

Wesenberg-Lund, E. 1958. Lesser Antillean polychaetes, chiefly from brackish water, with a survey and a bibliography of fresh and brackish-water polychaetes. Studies on the Fauna of Curacao and other Caribbean Islands 8:1–41.

Whitman, C. O. 1886. The leeches of Japan. Quarterly Journal of Microscopical Science 26:317–416.

Whitman, C. O. 1892. The metamerism of Clepsine. Pages 285–295 *in*: Festschrift 70 sten Geburtstag R. Leuckarts. 285–295.

Wilkialis, J., and R. W. Davies. 1980a. The reproductive biology of *Theromyzon tessulatum* (Glossiphoniidae: Hirudinoidea), with comments on *Theromyzon rude*. Journal of Zoology (London) 192:421–429.

Wilkialis, J., and R. W. Davies. 1980b. The population ecology of the leech (Hirudinoidea: Glossiphoniidae) *Theromyzon tessulatum*. Canadian Journal of Zoology 58:906–912.

Williamson, M. 1972. The analysis of biological populations. Arnold, London.

Wrona, F. J., and R. W. Davies. 1984. An improved flow-through respirometer for aquatic macroinvertebrate bioenergetic research. Canadian Journal of Fisheries and Aquatic Sciences 41:380–385.

Wrona, F. J., R. W. Davies, and L. Linton. 1979. Analysis of the food niche of *Glossiphonia complanata* (Hirudinoidea: Glossiphoniidae). Canadian Journal of Zoology 57:2136–2142.

Wrona, F. J., R. W. Davies, L. Linton, and J. Wilkialis. 1981. Competition and co-existence between *Glossiphonia complanata* and *Helobdella stagnalis* (Glossiphoniidae: Hirudinoidea). Oecologia 48:133–137.

Wrona, F. J., L. R. Linton, and R. W. Davies. 1987. Reproductive success and growth of two species of Erpobdellidae: the effects of water temperature. Canadian Journal of Zoology 65:1253–1256.

Young, J. O. 1987. Predation on leeches in a weedy pond. Freshwater Biology 17:161–167.

Young, J. O., and S. M. Spelling. 1986. The incidence of predation on lake-dwelling leeches. Freshwater Biology 16:465–477.

Young, S. R., R. Dedwylder, and W. O. Friesen. 1981. Response of the medicinal leech to water waves. Journal of Comparative Physiology 144:111–116.

Zerbst-Boroffka, I. 1975. Function and ultrastructure of the nephridium in *Hirudo medicinalis* L. III. Mechanisms of the formation of primary and final urine. Journal of Comparative Physiology B 100:307–316.

APPENDIX 13.1: DISTRIBUTION OF LEECH AND ACANTHOBDELLID TAXA IN NORTH AMERICA

The following information is a list of freshwater species of Hirudinoidea, Acanthobdellidae, and Polychaeta recorded from North America and their distributions on a state (United States), province, and territory (Canada) basis. Information is based on published literature, primarily Davies (1971, 1972, 1973), Sawyer (1972, 1986), Klemm (1972a, 1972b, 1985), and personal records.

Family Glossiphoniidae

Genus: *Actinobdella*

Actinobdella annectens: Ontario.

Actinobdella inequiannulata: Alabama; British Columbia; Georgia; Idaho; Indiana; Illinois; Louisiana; Maine; Manitoba; Massachusetts; Michigan; Minnesota; Mississippi; New Hampshire; New York; North Dakota; Ohio; Ontario; Pennsylvania; Virginia; Washington; Wyoming.

Genus: *Alboglossiphonia*

Alboglossiphonia heteroclita: Alberta; British Columbia; Connecticut; Illinois; Indiana; Iowa; Maine; Manitoba; Maryland; Massachusetts; Michigan; Newfoundland; New York; North Carolina; Ohio; Ontario; Pennsylvania; Quebec; Saskatchewan; South Carolina; Wisconsin.

Genus: *Batracobdella*

Batracobdella cryptobranchii: Missouri.

Batracobdella michiganensis: Michigan.

Batracobdella paludosa: Nova Scotia.

Batracobdella picta: Arkansas; British Columbia; Colorado; Connecticut; Idaho; Illinois; Kansas; Louisiana; Manitoba; Massachusetts; Michigan; Missouri; New York; North Carolina; North Dakota; Ohio; Ontario; Quebec; Saskatchewan; South Dakota; Texas; Utah; Virginia; Wisconsin; Wyoming.

Genus: *Boreobdella*

Boreobdella verrucata: Alaska; British Columbia.

Genus: *Glossiphonia*

Glossiphonia complanata: Alaska; Alberta; British Columbia; California; Colorado; Connecticut; Idaho; Illinois; Indiana; Iowa; Kansas; Maine; Manitoba; Maryland; Massachusetts; Michigan; Minnesota; Missouri; Montana; Nebraska; Nevada; Newfoundland; New Brunswick; New Hampshire; New Jersey; New York; North Dakota; Northwest Territories; Nova Scotia; Ohio; Ontario; Oregon; Pennsylvania; Prince Edward Island; Quebec; Rhode Island; Saskatchewan; South Dakota; Utah; Vermont; Washington; Wisconsin; Wyoming; Yukon.

Genus: *Helobdella*

Helobdella californica: California.

Helobdella elongata: Alabama; Alberta; Colorado; Connecticut; Florida; Georgia; Idaho; Illinois; Indiana; Iowa; Kansas; North Dakota; Nova Scotia; Ohio; Ontario; Pennsylvania; Quebec; Louisiana; Maine; Massachusetts; Michigan; Minnesota; Mississippi; Missouri; New York; North Carolina; North Dakota; Ohio; Ontario; Oregon; South Carolina; Tennessee; Texas; Vermont; Virginia; Washington; Wisconsin.

Helobdella fusca: Alaska; Alberta; California; Colorado; Connecticut; Florida; Illinois; Indiana; Iowa; Louisiana; Maine; Manitoba; Maryland; Massachusetts; Michigan; Minnesota; Mississippi; Missouri; Nebraska; New Jersey; New York; Northwest Territories; Ohio; Oklahoma; Ontario; Pennsylvania; Quebec; Rhode Island; Saskatchewan; Texas; Vermont; Virginia; West Virginia; Wisconsin.

Helobdella papillata: Connecticut; Illinois; Ohio; Pennsylvania; Maryland; Michigan; Minnesota; Ohio; Wisconsin.

Helobdella stagnalis: Alaska; Alberta; Arizona; Arkansas; British Columbia; California; Colorado; Connecticut; Delaware; Florida; Georgia; Idaho; Illinois; Indiana; Iowa; Kansas; Kentucky; Louisiana; Maine; Manitoba; Maryland; Massachusetts; Michigan; Minnesota; Mississippi; Missouri; Montana; Nebraska; Nevada; New Brunswick; New Hampshire; New Jersey; New York; Newfoundland; North Carolina; North Dakota; Northwest Territories; Nova Scotia; Ohio; Ontario; Oregon; Pennsylvania; Prince Edward Island; Quebec; Rhode Island; Saskatchewan; South Carolina; South Dakota; Tennessee; Texas; Utah; Vermont; Virginia; Washington; West Virginia; Wisconsin; Wyoming; Yukon.

Helobdella transversa: Michigan.

Helobdella triserialis: Alabama; Alberta; Arkansas; California; Florida; Georgia; Illinois; Indiana; Iowa; Kansas; Louisiana; Manitoba; Massachusetts; Michigan; Minnesota; Mississippi; Missouri; Nebraska; New Hampshire; New York; North Carolina; Northwest Territories; Nova Scotia; Ohio; Oklahoma; Ontario; Pennsylvania; Quebec; Rhode Island; South Carolina; Tennessee; Texas; Vermont; Virginia; West Virginia; Wisconsin.

Genus: *Marvinmeyeria*

Marvinmeyeria lucida: Alberta; British Columbia; Manitoba; Michigan; Ontario; Quebec; Saskatchewan.

Genus: *Oligobdella*

Oligobdella biannulata: North Carolina; South Carolina.

Genus: *Placobdella*

Placobdella hollensis: Florida; Georgia; Iowa; Massachusetts; Michigan; Minnesota; Ontario; Wisconsin.

Placobdella montifera: Alabama; Alberta; British Columbia; Connecticut; Georgia; Illinois; Indiana; Iowa; Kansas; Kentucky; Louisiana; Maryland; Michigan; Minnesota; Missouri; New York; North Carolina; North Dakota; Oklahoma; Ohio; Ontario; Pennsylvania; Quebec; Saskatchewan; Tennessee; Texas; Vermont; Virginia; Washington; Wisconsin.

Placobdella multilineata: Alabama; Florida; Georgia; Illinois; Iowa; Kansas; Louisiana; Michigan; Nebraska; North Carolina; Oklahoma; South Carolina; Tennessee; Texas; Utah.

Placobdella nuchalis: Florida; Georgia; Kentucky; North Carolina; Pennsylvania; South Carolina; Tennessee.

Placobdella ornata: Alberta; British Columbia; California; Colorado; Connecticut; Florida; Idaho; Illinois; Indiana; Iowa; Kansas; Louisiana; Maine; Manitoba; Maryland; Massachusetts; Michigan; Minnesota; Missouri; Montana; Nebraska; New Jersey; New Mexico; New York; North Carolina; North Dakota; Nova Scotia; Ohio; Oklahoma; Ontario; Oregon; Pennsylvania; Quebec; Saskatchewan; Tennessee; Texas; Utah; Virginia; Washington; Wisconsin; Yukon.

Placobdella papillifera: Alabama; Alberta; Arkansas; Connecticut; Florida; Georgia; Illinois; Kentucky; Louisiana; Manitoba; Massachusetts; Michigan; Mississippi; North Carolina; Northwest Territories; Ontario; Pennsylvania; South Carolina; Texas; Virginia; West Virginia.

Placobdella parasitica: Alabama; Alberta; Arizona; Arkansas; Colorado; Connecticut; Delaware; Florida; Georgia; Illinois; Indiana; Iowa; Kansas; Kentucky; Louisiana; Maine; Manitoba; Maryland; Massachusetts; Michigan; Minnesota; Mississippi; Missouri; Nebraska; Nevada; New Jersey; New York; North Carolina; North Dakota; Nova Scotia; Ohio; Ontario; Pennsylvania; Saskatchewan; South Carolina; South Dakota; Tennessee; Texas; Virginia; West Virginia; Wisconsin.

Placobdella pediculata: Illinois; Iowa; Ohio; Iowa; Kentucky; Maine; Michigan; Minnesota; Missouri; Oklahoma; Tennessee; Wisconsin.

Placobdella phalera: Colorado; Connecticut; Florida; Georgia; Idaho; Illinois; Iowa; Louisiana; Maine; Manitoba; Massachusetts; Michigan; Minnesota; New York; North Carolina; Nova Scotia; Ohio; Ontario; Pennsylvania; Quebec; Rhode Island; South Carolina; Texas; Virginia; Wisconsin.

Placobdella translucens: Alabama; Florida; Georgia; Louisiana; North Carolina; South Carolina.

Genus: *Theromyzon*

Theromyzon biannulatum: Alberta; Illinois; Iowa; Michigan; Minnesota; North Dakota; Pennsylvania; Prince Edward Island; Saskatchewan; South Dakota; Yukon.

Theromyzon rude: Alaska; Alberta; British Columbia; California; Colorado; Idaho; Iowa; Manitoba; Michigan; Montana; Nevada; Northwest Territories; Oregon; Saskatchewan; South Dakota; Utah; Yukon.

Theromyzon tessulatum: Alaska; British Columbia; Colorado; Manitoba; Nova Scotia; Quebec; Saskatchewan.

Family Piscicolidae

Genus: *Cystobranchus*

Cystobranchus mammillatus: Northwest Territories.

Cystobranchus meyeri: Maryland; New York.

Cystobranchus verrilli: Arkansas; Illinois; Iowa; Manitoba; Michigan; New York; Ontario; Saskatchewan; West Virginia.

Cystobranchus virginicus: Virginia; West Virginia.

Genus: *Myzobdella*

Myzobdella lugubris: Alabama; Alberta; Arkansas; Arizona; California; Colorado; Connecticut; Florida; Idaho; Illinois; Iowa; Kentucky; Louisiana; Maine; Maryland; Massachusetts; Michigan; Minnesota; Missouri; New Jersey; New Mexico; New York; North Carolina; North Dakota; Ohio; Ontario; Pennsylvania; Saskatchewan; South Carolina; Tennessee; Texas; Virginia; Washington; Wisconsin.

Genus: *Illinobdella*

Illinobdella alba: Connecticut; Illinois; Michigan; New York; Ontario; Tennessee.

Illinobdella elongata: Illinois; Michigan; Minnesota; Ontario.

Illinobdella richardsoni: California; Illinois; Kansas; Michigan; Minnesota; New York; Ohio; Tennessee.

Genus *Piscicola*

Piscicola geometra: Alaska; Manitoba; Michigan; Minnesota; New York; Northwest Territories; Ontario; Quebec; Saskatchewan; Wisconsin; Yukon.

Piscicola milneri: Alaska; British Columbia; California; Illinois; Maine; Manitoba; Michigan; Minnesota; New York; Northwest Territories; Ontario; Pennsylvania; Quebec; Saskatchewan; Wisconsin; Yukon.

Piscicola punctata: Alberta; British Columbia; Connecticut; Georgia; Illinois; Kentucky; Louisiana; Maryland; Michigan; Minnesota; Mississippi; Missouri; Montana; New Jersey; New York; North Carolina; Nova Scotia; Ohio; Ontario; Pennsylvania; Quebec; Rhode Island; Saskatchewan; South Dakota; Tennessee; Virginia; Wisconsin.

Piscicola salmositica: British Columbia; California; Oregon; Washington; Wyoming.

Genus: *Piscicolaria*

Piscicolaria reducta: Arkansas; Connecticut; Florida; Illinois; Indiana; Kansas; Kentucky; Louisiana; Maine; Massachusetts; Michigan; Minnesota; New Jersey; New York; Oklahoma; Ontario; Pennsylvania; Rhode Island; Tennessee; West Virginia; Wisconsin.

Family Hirudinidae

Genus: *Bdellarogatis*

Bdellarogatis plumbea: Iowa; Michigan; Minnesota; Ohio; Ontario; Quebec; Wisconsin.

Genus: *Hirudo*

Hirudo medicinalis: Alberta; New Jersey; Newfoundland; Pennsylvania.

Genus: *Mollibdella*

Mollibdella grandis: Alberta; Connecticut; Idaho; Kansas; Maine; Manitoba; Massachusetts; Michigan; Minnesota; Nebraska; New Brunswick; New Jersey; New York; Northwest Territories; Ohio; Ontario; Prince Edward Island; Quebec; Saskatchewan; Virginia; West Virginia; Wisconsin.

Genus: *Percymoorensis*

Percymoorensis kingi: Arizona; Colorado; Iowa; New Mexico.

Percymoorensis lateralis: Alabama; Arkansas; Colorado; Florida; Illinois; Indiana; Iowa; Kansas; Kentucky; Louisiana; Maine; Michigan; Minnesota; Mississippi; Missouri; North Carolina; Ohio; Ontario; Tennessee.

Percymoorensis lateromaculata: Iowa; Minnesota.

Percymoorensis marmorata: Alaska; Alberta; Arizona; British Columbia; California; Colorado; Connecticut; Delaware; Idaho; Illinois; Indiana; Iowa; Kansas; Kentucky; Manitoba; Maryland; Massachusetts; Michigan; Minnesota; Missouri; Montana; Nebraska; Nevada; Newfoundland; New Brunswick; New Jersey; New Mexico; New York; North Carolina; North Dakota; Northwest Territories; Nova Scotia; Ohio; Ontario; Pennsylvania; Prince Edward Island; Quebec; Rhode Island; Saskatchewan; Tennessee; Utah; Virginia; Wisconsin; Wyoming; Yukon.

Percymoorensis septagon: North Carolina; South Carolina; Virginia.

Genus: *Macrobdella*

Macrobdella decora: Alberta; Colorado; Connecticut; Georgia; Illinois; Iowa; Kansas; Maine; Manitoba; Maryland; Massachusetts; Michigan; Minnesota; Nebraska; New Brunswick; New Hampshire; New Jersey; New Mexico; New York; Newfoundland; North Carolina; North Dakota; Nova Scotia; Ohio; Ontario; Pennsylvania; Prince Edward Island; Quebec; Saskatchewan; South Carolina; Vermont; Virginia; Wisconsin.

Macrobdella diplotertia: Kansas; Missouri.

Macrobdella ditetra: Alabama; Arkansas; Florida; Georgia; Louisiana; Maryland; Mississippi; Missouri; North Carolina; South Carolina; Texas; Virginia.

Macrobdella sestertia: Louisiana; Massachusetts.

Genus: *Philobdella*

Philobdella floridana: Florida; Louisiana; North Carolina.

Philobdella gracilis: Florida; Georgia; Illinois; Louisiana; Michigan; Mississippi; Missouri; North Carolina; South Carolina; Texas.

Family Erpobdellidae

Genus: *Erpobdella*

Erpobdella montezuma: Arizona.

Erpobdella punctata: Alaska; Alberta; Arizona; British Columbia; California; Colorado; Connecticut; Delaware; Georgia; Idaho; Illinois; Indiana; Iowa; Kansas; Kentucky; Louisiana; Maine; Manitoba; Maryland; Massachusetts; Michigan; Minnesota; Missouri; Montana; Nebraska; New Brunswick; New Hampshire; New Jersey; New York; New-

foundland; North Carolina; North Dakota; Northwest Territories; Nova Scotia; Ohio; Ontario; Oregon; Pennsylvania; Quebec; Rhode Island; Saskatchewan; South Carolina; Tennessee; Utah; Vermont; Virginia; Washington; West Virginia; Wisconsin; Wyoming; Yukon.

Genus: *Dina*

Dina anoculata: British Columbia; California; Oregon.

Dina dubia: Alaska; Alberta; Colorado; Idaho; Illinois; Indiana; Iowa; Maine; Michigan; Missouri; Northwest Territories; Ontario; Quebec; Saskatchewan; Utah; Wyoming.

Dina parva: Alberta; California; Colorado; Idaho; Illinois; Indiana; Iowa; Maine; Manitoba; Michigan; Minnesota; Missouri; New York; Ontario; Oregon; Quebec; Saskatchewan; Virginia; Wisconsin; Wyoming.

Genus: *Mooreobdella*

Mooreobdella bucera: Michigan.

Mooreobdella fervida: Alberta; British Columbia; Colorado; Idaho; Illinois; Indiana; Iowa; Kansas; Maine; Massachusetts; Michigan; Minnesota; Missouri; Montana; New Brunswick; New York; Newfoundland; North Dakota; Nova Scotia; Ohio; Ontario; Pennsylvania; Prince Edward Island; Quebec; Saskatchewan; Wisconsin.

Mooreobdella melanostoma: Louisiana; Massachusetts; North Carolina; Quebec; South Carolina.

Mooreobdella microstoma: Alabama; Arkansas; California; Colorado; Florida; Idaho; Illinois; Indiana; Iowa; Kansas; Louisiana; Massachusetts; Michigan; Missouri; Nebraska; New York; North Carolina; Ohio; Oklahoma; Ontario; Pennsylvania; Quebec; South Carolina; Texas; Wyoming.

Mooreobdella tetragon: Alabama; Florida; Georgia; Massachusetts; North Carolina; Rhode Island; South Carolina.

Genus: *Nephelopsis*

Nephelopsis obscura: Alaska; Alberta; British Columbia; Colorado; Idaho; Illinois; Iowa; Maine; Manitoba; Maryland; Massachusetts; Michigan; Minnesota; Mississippi; Montana; New Brunswick; New York; Newfoundland; North Dakota; Northwest Territories; Nova Scotia; Ontario; Prince Edward Island; Quebec; Saskatchewan; Utah; Washington; Wisconsin; Wyoming.

Class Acanthobdellida

Family Acanthobdellidae

Genus: *Acanthobdella*

Acanthobdella peledina: Alaska.

APPENDIX 13.2: DISTRIBUTION OF POLYCHAETE TAXA IN NORTH AMERICA

Freshwater species of Polychaeta recorded from North America and their distribution on a state (United States) and province (Canada) basis. Information is based on published literature (Klemm 1985, Foster 1972) and personal records.

Family Ampharetidae

Genus: *Hypaniola*

Hypaniola florida: Connecticut; Delaware; Florida; Georgia; Maine; Maryland; Massachusetts; New Hampshire; New Jersey; New York; North Carolina; Rhode Island; South Carolina; Virginia.

Family Nereididae

Genus: *Laeonereis* Hartman

Laeonereis culveri: Alabama; Connecticut; Delaware; Florida; Georgia; Louisiana; Maryland; Mississippi; New Jersey; New York; North Carolina; South Carolina; Texas; Virginia.

Genus: *Lycastoides*

Lycastoides alticola: California.

Genus: *Lycastopsis*

Lycastopsis hummelincki: Florida.

Genus: *Namalycastis*

Namalycastis abiuma: California; Florida; Georgia; North Carolina; South Carolina.

Genus: *Namanereis*

Namanereis hawaiiensis: California.

Genus: *Nereis*

Nereis limnicola: British Columbia; California; Oregon; Washington.

Nereis succinea: British Columbia; California; Connecticut; Delaware; Florida; Georgia; Maine; Maryland; Massachusetts; New Brunswick; New Hampshire; New Jersey; New York; Newfoundland; North Carolina; Nova Scotia; Oregon; Prince Edward Island; Quebec; Rhode Island; South Carolina; Virginia; Washington.

Genus: *Stenoninereis*

Stenoninereis martini: California; Florida.

Family Sabellidae

Genus: *Manayunkia*

Manayunkia speciosa: Alaska; British Columbia; California; Connecticut; Delaware; Georgia; Illinois; Indiana; Manitoba; Maryland; Massachusetts; Michigan; Minnesota; New Hampshire; New Jersey; New York; North Carolina; Ohio; Ontario; Oregon; Pennsylvania; Quebec; South Carolina; Vermont; Virginia; Washington; Wisconsin.

Family Serpulidae

Genus: *Ficopomatus*

Ficopomatus enigmaticus: California; Florida.

Ficopomatus miamiensis: Alberta; Florida.

Bryozoans

14

Timothy S. Wood
Department of Biological Sciences
Wright State University
Dayton, Ohio 45435

Chapter Outline

I. INTRODUCTION
II. ANATOMY AND PHYSIOLOGY
  A. External Morphology
    1. Zooids and Lophophores
    2. Statoblasts
  B. Organ System Function
    1. Coelom, Neural System, and Body Wall
    2. Feeding Mechanics
    3. Reproductive Systems and Larvae
  C. Environmental Physiology
III. ECOLOGY AND EVOLUTION
  A. Diversity and Distribution
  B. Reproduction and Life History
  C. Ecological Interactions
  D. Evolutionary Relationships
IV. ENTOPROCTA
V. STUDY METHODS
  A. Taxonomic Key to Species of Freshwater Bryozoans
  Literature Cited

I. INTRODUCTION

Bryozoans are among the most commonly encountered animals that attach to submerged surfaces in freshwater. During warm months of the year, they are found in almost any lake or stream where there are suitable attachment sites. The variety of forms ranges from wisps of stringy material to massive growths weighing several kilograms. In the days before sand filtration, bryozoans were notorious for clogging the distribution pipelines of public water systems, and today they can still foul water intake lines. But for many biologists, bryozoans remain a biological oddity, with much of their ecology and physiology virtually unknown.

In general, bryozoans are sessile, colonial invertebrates with ciliated tentacles for capturing suspended food particles. Three distinct groups are recognized: the two major classes of the phylum Ectoprocta (Gymnolaemata and Phylactolaemata), and the phylum Entoprocta (also called Kamptozoa).[1] Their interrelationships are far from clear. Among the Gymnolaemata, only a few species in the subclass Ctenostomata are found in fresh or brackish water, while Phylactolaemata is an exclusively freshwater group. This chapter is devoted primarily to phylactolaemate bryozoans, with gymnolaemates included in the general discussion. Entoprocts are dealt with under a separate heading.

II. ANATOMY AND PHYSIOLOGY

A. External Morphology

1. Zooids and Lophophores

The phylactolaemate bryozoan colony is composed of identical zooids fused seamlessly together into a single structure (Fig. 14.1). Major anatomic features

---

[1] Some authors prefer the term "Bryozoa" to designate the phylum, which may or may not include entoprocts.

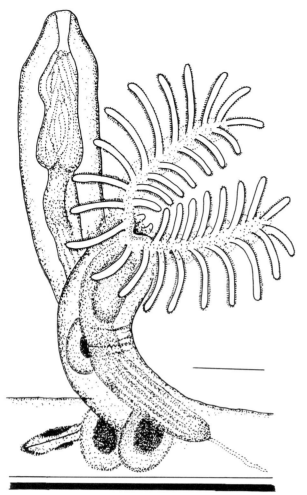

Figure 14.1 Two zooids of a tubular bryozoan colony, one retracted and the other extended in a feeding position. Scale bar = 0.5 mm.

are shown in Figure 14.2. Each zooid in the colony has two basic parts: an organ system, or polypide, which can be partially protruded as a unit into the surrounding water; and a body wall, which can enclose the entire polypide and separates the interior of the colony from the surrounding water. These parts are joined in different places by three structures: a tentacular sheath, prominent retractor muscles, and a slender, tube-like funiculus loosely running from the base of the gut caecum to a point on the interior of the body wall.

The polypide bears a prominent lophophore composed of ciliated tentacles arranged around a central mouth. In *Fredericella* species, the tentacles simply encircle the mouth (Fig. 14.3). In all other phylactolaemates, the lophophore extends dorsally in a pair of bilateral arms, forming a U-shaped structure with an outer row of long tentacles and an inner row of shorter ones (Fig. 14.4). In most species, each tentacle bears one medial and two lateral tracts

of cilia, which beat in metachronal waves. Tentacles are loosely joined near their bases by an intertentacular membrane, which thus forms a groove between rows of tentacles. Two such grooves converge along the arms towards the mouth. Overhanging the mouth is the epistome, a small, heavily ciliated lobe believed to function in food selection.

In gymnolaemate (ctenostome) bryozoans, the zooids are tubular and look rather delicate, with small, conical lophophores (Fig. 14.8a–c). The body wall is thin and transparent. Colonies are diffuse, consisting of creeping chains of zooids sometimes joined by narrow pseudostolons.

2. Statoblasts

A conspicuous feature of phylactolaemate bryozoans is the presence of encapsulated dormant buds, called statoblasts, formed within the colony. Three distinct types are recognized: the buoyant floatoblast, bearing a wide peripheral band of gas-filled chambers; the sessoblast, cemented to the firm objects on which the colony grows; and the piptoblast, with no intrinsic structures for either attachment or buoyancy. All statoblasts have a thin outer chitinous shell composed of paired valves joined at an equatorial suture. Inside is a mass of yolky material and germinal tissue.

Most floatoblast valves have two structural layers (Fig. 14.5). An inner capsule contains germinal tissue and food reserves. The outer periblast covers the capsule completely. In floatoblasts (Fig. 14.5a), the periblast includes buoyant chambers which form the annulus. A transparent central area is called the fenestra. In most species, the so-called dorsal (or cystigenic) valve generally has a wider annulus and smaller fenestra than the ventral (deutoplasmic) valve.

Sessoblast valves also have two layers (Fig. 14.5b), but they are not so easily separated. The periblast of the frontal valve often bears rough markings. A thin peripheral lamella is believed to be homologous with the annulus in floatoblasts. The basal sessoblast valve provides a large, irregular ring which adheres to the substrate.

Gymnolaemate bryozoans do not produce statoblasts. Instead, those species restricted to fresh or brackish water form special thick-walled hibernacula. The irregularly shaped bodies are integrated into the colony structure. Their appearance in the colony is usually associated with suboptimal environmental conditions, such as low temperature, organic pollution, or changes in salinity.

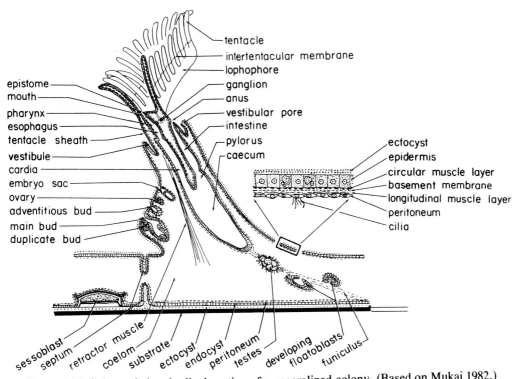

epistome
mouth
pharynx
esophagus
tentacle sheath
vestibule
cardia
embryo sac
ovary
adventitious bud
main bud
duplicate bud

tentacle
intertentacular membrane
lophophore
ganglion
anus
vestibular pore
intestine
pylorus
caecum

ectocyst
epidermis
circular muscle layer
basement membrane
longitudinal muscle layer
peritoneum
cilia

sessoblast
septum
retractor muscle
coelom
substrate
ectocyst
endocyst
peritoneum
testes
developing floatoblasts
funiculus

Figure 14.2 Schematic longitudinal section of a generalized colony. (Based on Mukai 1982.)

B. Organ System Function

1. Coelom, Neural System, and Body Wall

Most ectoproct colonies have a spacious coelom shared by all zooids. A clear coelomic fluid is circulated by cilia on the peritoneum. The main gastric coelom communicates through an incomplete diaphragm with the coelom of the lophophore, which extends fully into every tentacle.

A small nerve ganglion is located on the diaphragm at the base of the epistome between the mouth and the anus. A single nerve tract extends to

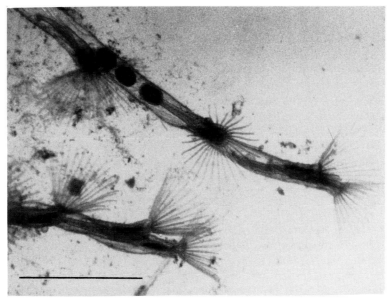

Figure 14.3 Portion of a *Fredericella* colony showing characteristic circular outline of the lophophore. Scale bar = 2 mm.

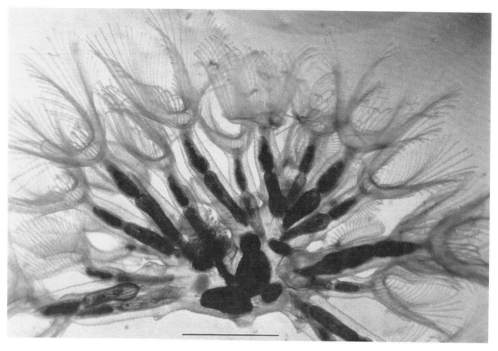

Figure 14.4 Portion of a *Lophopodella* colony showing the typical horseshoe-shaped lophophore. Scale bar = 2 mm.

each of the paired arms of the lophophore, with nerves branching off to each tentacle. Other nerves from the ganglion innervate the epistome, the tentacle sheath, and the digestive tract. There is no nervous communication between zooids.

The body wall is composed of living tissues collectively called the endocyst, together with a nonliving outer ectocyst (Fig. 14.2). Lining the main coelom is a peritoneum bearing scattered tracts of cilia. Behind the peritoneum lies a thin longitudinal muscle layer followed by a basement membrane. Overlying this is a thin layer of circular muscles and a single layer of epidermis. Near the apex of the zooid, the endocyst folds inward to form a flat pocket, the vestibule, then joins the lophophore as an eversible tentacle sheath. When the lophophore is completely retracted, the tip of the living zooid contains a round opening, or orifice. A pore in the vestibular body wall of *Plumatella emarginata* (and presumably in certain other species as well) permits the release of floatoblasts from living colonies.

In some tubular species, a portion of the body wall grows inward to form incomplete septa. Those branches of tubular colonies that are attached to a substrate may have a raised line, or keel, running along the outer surface. The peak of the keel, usually free of encrusted particles, is known as a furrow.

The composition of the ectocyst is highly variable

among species. In tubular colonies, it is sclerotized and remains intact for a time after living parts have died. In at least two species, the ectocyst is composed largely of chitinous microfibrils with random orientation. A thin, proximal electron-dense layer is overlaid by a thicker stratum of looser fibrils (Goethals *et al.* 1984). Many organic and inorganic particles adhere to this outer coat including bacteria, which also penetrate the interior of the loose layer. The ectocyst in these species can vary from being thin and flexible to leathery, opaque, and brittle. With age, it may also become darker as new chitin is laid down. Thus, the growing tips of the zooids are often distinctly lighter in color than the rest of the colony. Rao *et al.* (1978) found that with increasing age a second, separate, sclerotized layer can form in the inner body wall of *Plumatella casmiana*.

Mukai *et al.* (1984) have shown that the body wall breaks down when contact is made between adhesive pads from germinating statoblasts of the same species. The resulting colonies are confluent, possibly allowing cross-fertilization from an eventual exchange of sperm.

In higher phylactolaemates, the ectocyst is gelatinous and often restricted to certain parts of the colony. Mukai and Oda (1980) reported many vacuolar cells, apparently with a secretory function, in the epidermis of these species. The massive jellylike substance produced by *Pectinatella magnifica* is

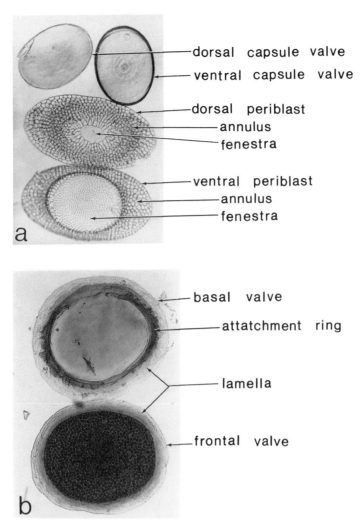

Figure 14.5 Sclerotized components of statoblasts separated after a brief exposure to hot saturated potassium hydroxide. (a) Floatoblast of *Plumatella emarginata;* (b) sessoblast of *Plumatella repens.*

more than 99% water and contains a protein similar to egg albumen, along with some chitin, calcium, and sodium chloride (Morse 1930).

2. Feeding Mechanics

Bryozoan feeding mechanics are complex. Although the ciliated lophophore brings in many suspended particles, those directly in line with the intertentacular groove have the best chance of being ingested. Flicking movements of individual tentacles sometimes throw particles toward the mouth; at other times, all tentacles are brought together, as if to prevent the escape of an active organism. Bullivant (1968) explained lophophore operation on the principle of "impingement," in which particles accelerating into the lophophore are impelled the final short distance to the intertentacular groove by their own momentum while the water is deflected abruptly to the sides. Gilmour (1978) showed that differential beating of the lateral and frontal cilia sort particles by density, rejecting heavier, inedible materials before they reach the epistome.

Food particles entering the mouth collect momentarily in a short, ciliated pharynx, then pass down a narrow esophagus to the Y-shaped stomach. A short unciliated cardia leads to the long, cylindrical caecum, where particles are churned by rhythmic peristaltic contractions. In well-fed colonies, waste products collect in brownish, longitudinal bands, giving the caecum a striped appearance. A short proximal part of the caecum, the pylorus, leads through a narrow sphincter to a straight intestine. Here, processed particles are consolidated and surrounded by mucous, then ejected as a fecal pellet through the anus, which lies just outside the whorl of tentacles. Encased in its mucous capsule, the eliminated material falls away from the colony and cannot be reingested by neighboring zooids.

3. Reproductive Systems and Larvae

Phylactolaemate bryozoans are sexually active during a single brief period of the year. Sperm develop in conspicuous masses on the funiculus of certain zooids, and later they circulate passively in the colony coelom. Egg clusters develop on the peritoneum ventrally within the zooid. There is no evidence that sperm leave the colony, so self-fertilization is presumed to occur except in instances where two colonies have become confluent. An ingrowth of the body wall forms a special sac into which a single zygote migrates. The embryo develops into a specialized free-swimming structure usually called a "larva." Embryology of the planktonic larva and the steps leading to its metamorphosis are summarized by Hyman (1959, pp. 451–455). Fine structure is described by Franzén and Sensenbaugh (1983). The larva is composed of two parts: a heavily ciliated outer mantle and an inner pear-shaped mass (Fig. 14.6a). The inner parts include one to four fully formed polypides together with their funiculi, body walls, and associated musculature. The larva swims with its aboral pole forward, which contains a neural center and special gland cells. Settling usually occurs within an hour (Fig. 14.6b–d).

Among Gymnolaemata, sexual reproduction leads to the development of trochophore-like larvae totally unlike those of Phylactolaemata. Settling and metamorphosis occur shortly after release, since the larvae carry no feeding or digestive organs. Virtually nothing is known of larval ecology or behavior in ctenostome bryozoans.

All bryozoan zooids are capable of budding new individuals. In the Phylactolaemata, buds arise from a specific site on the midventral wall of the mother zooid (Figs. 14.2, 14.6c). The development of a new zooid is accompanied by the appearance of new bud primordia, which may or may not develop further. Brien (1953) showed that every zooid bears two bud primordia: a main bud, which forms the first daughter zooid, and an adventitious bud between the main bud and mother zooid, which becomes the second daughter zooid. In addition, each main bud itself has a small duplicate bud primordium on its ventral side. In the process of budding, duplicate and adventitious buds become main buds, and new duplicate and adventitious buds are formed. These relations have been nicely clarified by Mukai *et al.* (1987).

In Gymnolaemata, the new zooidal bud originates from an outswelling of the body wall of a parental zooid. An interior wall then grows across the base of the bud to separate it from the body cavity of the parent. Most often, a new bud develops at the distal end of a lineal series of zooids, and as it grows a wall forms at the tip to separate it from the body cavity of the next distal bud in the series.

C. Environmental Physiology

Little is known about the digestive physiology of phylactolaemate bryozoans, except that the stomach environment is acidic, the esophagus is alkaline, and the intestinal pH is neutral. Passage time through the gut varies from 1–24 hr, depending on the ingestion rate. Zones of basophilic and acidophilic cells line the caecum, but there is no agreement among workers as to their function. The ability of some species to flourish in turbid water, ingesting little except inorganic silt, suggests nourishment by bacteria that colonize the surface of such inert particles. If these are mainly bacteriophagous animals, intracellular digestion would be a distinct possibility.

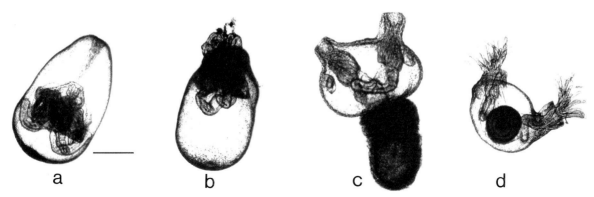

Figure 14.6 Settling and transformation of a larva. Scale bar for all figures = 500 μm. (a) Swimming larva containing two fully developed zooids; the more tapered end is the oral pole; (b) the mantle peels back immediately after the larva attaches to a solid substrate; (c) constricted mantle appears as a dark mass; the two zooids are now clearly visible, each with its own main bud; (d) the mantle is drawn into the body cavity, lophophores are extended, and the two zooids begin feeding. (From Zimmerman 1979.)

Statoblasts have been the subject of considerable study. The period of obligate dormancy imposed on most statoblasts serves to maintain populations through periods of unfavorable environmental conditions. Those statoblasts not fixed to immobile substrates may also function in passive dispersal, carried either by water currents or by migrating animals. Brown (1933) demonstrated survival of *Plumatella* statoblasts after they had passed through the separate digestive tracts of a salamander, frog, turtle, and duck.

Like statoblasts, the hibernacula of gymnolaemates maintain the populations during periods of unfavorable conditions. Unlike most statoblasts, however, they are permanently attached to the colony and thus cannot normally function as independent disseminules.

The resistance of statoblasts to environmental stress has been reviewed by Bushnell and Rao (1974). Most *Plumatella* and *Fredericella* statoblasts survive desiccation or freezing for periods of 1–2 years. *Lophopodella carteri* statoblasts have remained viable during dry storage for over 6 years. Other less hardy statoblasts include those of *Pectinatella magnifica*, which do not normally withstand desiccation and survive only brief periods of freezing temperatures.

Statoblasts stored in water at a favorable temperature eventually germinate after a variable dormant period. However, when that period is prolonged by continued exposure to cold, dry, anaerobic, or other unfavorable conditions, a nearly simultaneous germination can be achieved when a suitable environment is finally restored. In some species, light is an important factor to promote germination (Oda 1980). Ultrastructure of the statoblast suture and its relation to germination has been studied by Bushnell and Rao (1974) and Rao and Bushnell (1979).

The budgeting of energy resources for colony growth and reproduction is poorly understood. In many cases, gametogenesis precedes statoblast production, but the two processes may also operate concurrently. The first zooid to germinate from a statoblast never forms statoblasts itself, but it may participate in gametogenesis. From laboratory colonies of *Plumatella emarginata*, Mukai and Kobayashi (1988) found sexual activity only in certain colonies, which varied in size down to ten zooids. Only those colonies with the most vigorous development of testes eventually retained and developed mature larvae.

The control of sessoblast development in Plumatellidae has not yet been clarified. Apparently, statoblast primordia can differentiate into either floatoblasts or sessoblasts, and this differentiation becomes evident during the late epidermal-disk stage (Mukai and Kobayashi 1988). In natural populations, Wood (1973) noted that most sessoblasts were formed either early or late in the season, leading him to suppose that this was a response to suboptimal conditions. However, laboratory-reared *Plumatella* colonies frequently form both floatoblasts and sessoblasts under apparently favorable growing conditions.

III. ECOLOGY AND EVOLUTION

A. Diversity and Distribution

Freshwater bryozoans are generally restricted to relatively warm water, flourishing at 15°–28°C. However, *Cristatella mucedo* has been found at 6°C (Bushnell 1966), and *Lophopus crystallinus* survives at 0°C for brief periods (Marcus 1934). *Fredericella indica* lives through the winter in most of North America, budding new zooids at 3°C and producing piptoblasts above 8°C. All of these species also grow well at higher temperatures. Only *Stephanella hina* has never been found above 17°C and is probably restricted to cool water. Temperatures as high as 37°C were recorded for living *Plumatella repens* and *P. fruticosa* (Bushnell 1966). Shrivastava and Rao (1985) noted that *Plumatella emarginata* in India shows only meager growth at 34°C.

Bryozoans tolerate a wide range of pH, but they seem to favor slightly alkaline water. This was the conclusion of Tenney and Woolcott (1966) in a limited North Carolina study where pH values ranged from 6.3–8.0. A *Fredericella* species (probably *F. indica*) has been collected in acidic conditions as low as pH 4.9 (Everitt 1975), *Plumatella fruticosa* at pH 5.7 (Bushnell 1966), and both *P. repens* and *Hyalinella punctata* at pH 6.3 (Tenney and Woolcott 1966).

Most species occur in both still and running water, but *Plumatella emarginata* and *Fredericella indica* grow especially well in lotic habitats. *Plumatella repens*, *Lophopus crystallinus*, *Hyalinella punctata*, and *H. orbisperma* are among those species associated with very still waters. Many common species tolerate turbid water, but *Pectinatella magnifica* does not, possibly because its large colonies are necessarily more exposed to settling particles (Cooper and Burris 1984).

The pollution ecology of freshwater bryozoans has been thoroughly reviewed by Bushnell (1974). *Fredericella sultana*, *Plumatella emarginata*, and *P. repens* are noted as particularly tolerant of extreme contamination from sewage and industrial wastes. According to Sládeček (1980), both *Plumatella repens* and *P. fungosa* survive at only 30% saturation of dissolved oxygen saturation. *Lophopodella carteri* flourishes both in the laboratory and the field when supplied with large quantities of suspended organic particles. In Europe, the distribution of *Plumatella fungosa* is correlated with nutrient-enriched water (Job 1976). *Pectinatella magnifica* in the United States and Japan becomes luxuriant in areas that are visibly eutrophic.

One clearly limiting factor for all species is the availability of suitable surface on which to grow. Almost any solid, biologically inactive material is acceptable, including rocks, glass, plastics, automobile tires, and aged wood. Surfaces that are never successfully colonized include newly dead wood, corroded metal, and oily or tarred materials. Aquatic plants are common substrates, although in a detailed survey of bryozoan–plant relations, Bushnell (1966) found little evidence for species specificity.

Colonies derived from free floatoblasts are carried passively to new substrates. Swimming larvae, on the other hand, actively select attachment sites. Hubschman (1970) demonstrated particle size discrimination in larvae of *Pectinatella magnifica*, concluding that rock particles smaller than 1 mm diameter were avoided. Additional selection criteria appear to be likely. Curry *et al.* (1981) found evidence for selective settling of *Paludicella articulata* larvae on mussel shells.

In general, the distribution of freshwater bryozoans in North America is consistent with global patterns. Some of our most common species (*Plumatella repens, P. emarginata, Hyalinella punctata*) are also found on every continent but Antarctica. *Plumatella casmiana* and *Paludicella articulata* also are abundant both in North America and worldwide. Other common American species include *Fredericella indica* and *Pectinatella magnifica*. Two species, *Hyalinella orbisperma* and *Plumatella reticulata,* are unknown outside North America. Two other species have been collected only from single sites in North America: these are *Stolella evelinae* in Kalamazoo County, Michigan (Bushnell 1965c) and *Stephanella hina* in Hampshire County, Massachusetts (Smith 1989a). Colonies of *Lophopus crystallinus* have not been reported from North America since 1898 despite systematic searches, although statoblast fragments were recently seen in Mississippi. This species is also very scarce in Europe, and is probably seriously endangered throughout its range.

All known North American species can be found east of the Mississippi River and north of the 39th parallel. Sixteen of the twenty-two species occur in Michigan and Ohio, and ten of these can be found in the western basin of Lake Erie. An additional two species are described only from New England. Only the brackish species, *Victorella pavida,* has not been found further north than Chesapeake Bay.

The three most recent statewide surveys are those of Bushnell (1965a, 1965b, 1965c) in Michigan, Wood (1989) in Ohio, and Smith (1989b) in Massachusetts. Each of these has revealed significant new species records for North America. What appear to be disjunct distributional patterns in a few cases may be due to the lack of such systematic surveys.

Relatively little study has been made of bryozoans west of Ontario and the Mississippi River. Most states and provinces have no published records of bryozoans. However, it is known that parts of Colorado and Utah have natural populations of *Plumatella repens, P. casmiana, P. emarginata, P. fruticosa, P. fungosa, Fredericella indica,* and *Cristatella mucedo.* Other western reports include *Cristatella mucedo* in British Columbia (Reynolds 1976), and *Plumatella repens* in Arizona and Nevada (Wilde *et al.* 1981, Dehdashti and Blinn 1986). Large river impoundments in eastern Texas have recently begun to support populations of *Pectinatella magnifica.*

Patterns of distribution should become clearer as taxonomic questions are resolved. For example, it seems increasingly likely that both *Plumatella repens* and *P. emarginata* include several subspecies. The recent distinction of *Fredericella indica* from the European *F. sultana* has altered distributional schemes for this genus (Wood and Backus submitted).

B. Reproduction and Life History

In most parts of the United States, bryozoan populations have two or more generations during the growing season. Only *Fredericella indica* typically is found throughout the year, living even under a cover of ice. Populations of other species are normally maintained during unfavorable seasons by dormant statoblasts. An unusual exception to this pattern was reported by Dehdashti and Blinn (1986) in a population of *Plumatella repens* inhabiting an Arizona cave, where nearly constant conditions permitted uninterrupted, active growth.

Winter dormancy in statoblasts is broken by conditions favorable to colony growth. In temperate climates, the principal triggering factor appears to be temperature. Viable statoblasts often germinate within a few days of each other when the water temperature rises above 8°C. In those regions where winter water temperature never approaches freezing, the time of germination may be more variable.

The zooid that emerges from a statoblast is termed the ancestrula. It eventually buds from 1–5 new zooids, each similarly capable of multiple zooid budding. Colonies grow rapidly as the water temperature rises. Bushnell (1966) described naturally occurring colonies of *Plumatella repens* that tripled and quadrupled in size within one week. Doubling times of five days have been reported for both *Plumatella casmiana* and *Fredericella indica* (Wood 1973).

In *Plumatella* and *Fredericella* species, statoblasts may appear in colonies having fewer than five zooids; in most other species, they are not formed until colonies are much larger. Gametogenesis, when it occurs, is normally encountered in the first spring generation of colonies, and larvae are released 2–4 weeks after fertilization.

In natural populations, a second generation usually arises either from the statoblasts or larvae of early spring colonies, or even from fragments of colonies surviving from the first period of growth. Third generation colonies also are known, but any statoblasts they produce normally remain in diapause until the following spring.

Sexual activity occurs once a year during a relatively brief period. Sperm masses develop on the funiculus as ova appear on the peritoneum. The sperm later break free to circulate in the coelom for 1–2 weeks. This is followed by larval development and eventual release. There appears to be little overlap in larval release dates among closely related species living in the same area. In two Ohio lakes, free-swimming larvae of *Plumatella repens* appeared throughout June; *P. casmiana* larvae were released during July; then came *Hyalinella punctata* in late July-early August, followed by *Plumatella emarginata* in late August-early September (Zimmerman 1979).

In the Plumatellidae, with their prolific production of statoblasts, sexual reproduction seems a relatively unimportant means to increase colony numbers. However, for *Pectinatella magnifica,* which has only one statoblast generation per year, sexually produced larvae are released throughout much of the season to play a significant role in establishing new colonies. Likewise, in *Fredericella* species, because of their fixed statoblasts, larvae must play an important role in both population growth and dispersal.

C. Ecological Interactions

The constant flow of water through bryozoan lophophores creates a favorable environment for the microscopic aufwuchs community. Protozoans, rotifers, gastrotrichs, microcrustaceans, and other small animals congregate especially on and around branching tubular colonies of bryozoans. Flatworms, oligochaetes, snails, orbatid mites, and such insect larvae as caddisflies and midges occasionally graze on the living zooids.

Chironomid larvae enter old tubes of *Plumatella repens,* hastening their disintegration and damaging living portions of the colony. Some workers consider such damage to be accidental rather than the result of direct feeding, while others regard chironomid larvae as important predators. Colonies of *P. casmiana* escape such harm, possibly because larvae cannot fit themselves inside the narrow, branching tubules but instead, build detritus tubes alongside the exterior colony wall where they cause little damage. *Pectinatella magnifica* is often a host to chironomid larvae, which find shelter by burrowing into the gelatinous base close to the substrate and causing no apparent damage. One of these, *Tendipes pectinatellae,* is specifically commensal on *P. magnifica* (Dendy and Sublette 1959).

Predation by fish is never extensive. Dendy (1963) observed bluegills biting off pieces of *Plumatella* sp., but believed the fish were doing so only to obtain insect larvae associated with the colonies. Coelomic fluid of *Lophopodella carteri* is selectively toxic to fish (Tenney and Woolcott 1964), and attempted feeding by fish on this or any other gelatinous species has never been observed. Freshwater prawns will graze on colonies of *Hyalinella vaihiriae* but only when no other food is available (Bailey-Brock and Hayward 1984), suggesting a possible lack of nutritional value or a repellant chemistry of these colonies.

As sessile suspension feeders, bryozoans handle a wide variety of food in terms of both quality and quantity. Unfortunately, knowledge of bryozoan feeding and nutrition is still largely circumstantial. Ingested particles include diatoms, desmids, green algae, cyanobacteria, nonphotosynthetic bacteria, dinoflagellates, rotifers, small nematodes, protozoa, and even microcrustaceans, along with bits of detritus and inorganic materials. In a comparative study of three species, Kaminski (1984) found that 95% of ingested particles were under 5 μm in diameter. *Plumatella repens,* with its wide mouth and strong gut musculature, ingested larger organisms than did either *Cristatella mucedo* or *Plumatella fruticosa.* These included rotifers (*Keratella* sp.) and colonial green algae and cyanobacteria as large as 75 μm. In general, organisms with long body extensions were able to avoid ingestion by bryozoans, while small, rounded shapes were taken most easily. Analysis of stomach contents alone, however, does not reveal the important sources of bryozoan nutrition. Rotifers and green algae have been known to pass through the gut completely unharmed (Hyman 1959, p. 489), although at other times these organisms are apparently digested (Rüsche 1938).

D. Evolutionary Relationships

Ectoprocta is one of three animal phyla collectively known as lophophorates. The other two groups,

Brachiopoda and Phoronida, are represented by a small number of solitary, mostly sessile, marine animals. The unifying feature is a lophophore with hollow tentacles containing an extension of the coelom, and with cilia beating in metachronal waves to bring water and suspended food toward the mouth.

As a group, the lophophorates show no clear affinity with any other invertebrates. The trochophore larvae of phoronids and marine ectoprocts suggest a protostome lineage, but the radial cleavage and mesoderm formation in brachiopods is distinctly deuterostome.

Structural similarities have been noted between ectoprocts and the worm-like sipunculids. The anterior introvert of a sipunculid is protruded and withdrawn in a manner very similar to that of the ectoproct polypide, using coelomic pressure and retractor muscles. The sipunculid introvert ends in a mouth surrounded by hollow, ciliated, tentacular outgrowths. The gut is U-shaped, and the anus is situated near the base of the introvert.

From whatever ancestral lophophorate, one possible line may have led directly to phoronids and phylactolaemate ectoprocts. Only these two groups have a crescentic lophophore and an epistome. They also both produce new buds from a region on the oral (ventral) side of the adult, while in other ectoprocts budding is in an anal direction (Jebram 1973). Finally, many phoronids produce special fat bodies for energy storage, which are strikingly similar to early developmental stages of phylactolaemate statoblasts (B. T. Backus personal communication).

Phoronids and gymnolaemate ectoprocts are probably more distantly related, as evidenced especially by their different embryologies and larval metamorphoses. For example, the phoronid actinotroch larva has a distinct coelom, the larval gut is retained in the adult, and the lophophore develops from the metatroch ring of cilia. By contrast, the gymnolaemate cyphonautes larva has no larval coelom, the larval gut is not retained, and tentacles develop from the episphere region.

Thus, among the ectoprocts themselves, it is possible that phylactolaemates and gymnolaemates evolved independently from a phoronid-like ancestor. The phylactolaemate line would have invaded freshwater habitats, losing completely the planktotrophic larval stage and deriving statoblasts from its fat bodies. This is consistent with the adaptations characteristic of many other freshwater metazoan groups. The gymnolaemate line would have retained a trochophore type of larva and the basic adult body plan.

Within the Phylactolaemata, similarities in the mode of colony growth and statoblast morphology have long been regarded as the basis for evo-

lutionary relationships. *Fredericella* species are thought to exhibit primitive features with their simple statoblasts and an open, dendritic pattern of colony branching. The circular outline of the lophophore, associated with a relatively small number of tentacles, may also be a primitive character.

In general, evolutionary trends in phylactolaemates include greater compactness of the colonies. The lophophore accommodates an increasing number of tentacles by an inward dorsal deflection, giving the U-shape usually associated with freshwater bryozoans. Statoblasts serve the two seemingly conflicting roles of dispersal to new habitats versus retention of a position on proven favorable substrates.

Among the Plumatellidae, there is a wide range in colony morphology, but all species retain a basically tubular design. The two statoblast functions are served by distinctly different statoblast types: floatoblasts and sessoblasts.

In Lophopodidae and Cristatellidae, the pattern of budding is modified from that of the lower species, but the colonies are gelatinous and globular, with no trace of branching. This development is coupled with a larger lophophore and a single remarkable statoblast design that incorporates a buoyant ring with marginal hooks. These provide both an anchor and a means for dispersal. Radiating spines and hooks were apparently acquired independently by the two families in an interesting case of parallel evolution.

Hyalinella species, normally classified with the Plumatellidae, share many features with the Lophopodidae and may actually link the two groups. Such features include a gelatinous colony wall, the absence of a sessoblast, and large floatoblasts which, like those of *Lophopodella*, are initially nonbuoyant upon release.

The freshwater Gymnolaemata, all ctenostomes, are members of an ancient marine group. They share with the Cheilostome bryozoans many features expressed in the development of both larvae and colonies. At this point, however, phylogenetic relationships are still speculative.

IV. ENTOPROCTA

Entoprocta is a small group of about 60 species distinct from the Ectoprocta but often included with them under the name Bryozoa. The only freshwater genus is *Urnatella*, with two or three species. Of these, *Urnatella gracilis* is by far the most common and widely distributed; it is also the only species in North America (Fig. 14.7). Although normally classified as Urnatellidae, the genus is similar to the marine species *Barentsia* and may be united with it in the family Pedicellinidae.

The zooid is a bulbous calyx borne on a flexible, segmented stalk measuring up to 5 mm long. Several such stalks, either solitary or sparingly branched, may arise from a basal plate. The calyx bears a single whorl of 8–16 uniformly short, ciliated tentacles. The area of the calyx enclosed by the tentacles, called an atrium or vestibule, includes both the mouth and the anus. When the zooid is disturbed, the tentacles fold over the vestibule and are covered by a tentacular membrane.

Internal organs include a fully ciliated digestive tract and a large medial ganglion. A number of excretory flame bulbs communicate through short ducts to a common nephridiopore; additional flame bulbs occur in the stalk. The body cavity is a pseudocoel filled with loose mesenchyme and extending into each tentacle.

The calyx is deciduous, dropping off at the onset of cold temperatures or other unfavorable conditions. The basal segments of stalks, containing food and germinal tissue, can survive the winter much like ectoproct statoblasts. They form new calyces when the water temperature rises to around 15°C. New zooids develop by budding from the stalk. It is reported that young colonies can disperse locally by crawling over the substrate (Protosov 1980). Sexual reproduction leads to the development of swimming larvae, presumably similar to those of other Pedicellinidae; details are unknown.

Like other bryozoans, *Urnatella gracilis* is a suspension feeder, consuming organic particles, unicellular algae, and protozoans. Two especially important foods are the diatom *Melosira* and two green algae species, *Pediastrum duplex* and *P. simplex* (Weise 1961).

Urnatella gracilis is found on every continent but Antarctica and Australia. In North America, its known distribution ranges from the east coast to the west coast and from Florida, Louisiana, and Texas to as far north as Michigan. Zooids attach to almost any substrate, including rocks, sticks, aquatic plants, bivalve shells, ectoprocts, and such debris as nails, beverage cans, and lead fishing weights (Eng 1977). Individuals occur most frequently in flowing water or in shallow areas of large lakes where there is extensive water movement. Tolerance to a wide range of chemical and physical conditions has been noted most recently by Hull *et al.* (1980) and Cusak and McCullough (1985).

In some areas, especially on new substrate, entoprocts may comprise a sizeable fraction of the macroinvertebrate community. Stalk densities of over 225,000/m^2 and an average biomass of 868 mg/m^2 (dry weight) were reported from a seven month old artificial stream in Mississippi (King *et al.* 1988).

Entoproct phylogeny is even less clear than that of the Ectoprocta. Traditionally, they have been grouped with the Ectoprocta under the phylum Bryozoa because of their sessile habits, colonial structure, and ciliated tentacles. Nielsen (1977) supported this idea by pointing out many additional similarities between entoprocts and gymnolaemates, mostly involving larval structure, metamorphosis, and zooid budding patterns. However, others view these as only superficial links, more likely the result of convergent evolution than common ancestry. A close comparison of entoproct tentacles and the ectoproct lophophore reveals little in common, despite their outwardly similar structure and function. Body cavities of the two groups are likewise difficult to reconcile: a pseudocoel on one hand and a true coelom on the other. Nevertheless, puzzles such as this are expected in animals so highly modified by sessile life style and colonial architecture, and for now the question of entoproct origins remains open.

V. STUDY METHODS

Bryozoans are normally found in areas of shallow water where there is suitable firm substrate. Although most species are plainly visible to the unaided eye, a Coddington-type magnifier is useful for field identification. The presence of many species can be detected by examining the edges of floating objects for free statoblasts. Most gelatinous colonies can be scraped gently from the substrate with little damage, but with other species it is better to collect pieces of substrate with colonies attached.

To obtain swimming larvae, place large colonies in a shallow tray of water. If larvae are present, they will generally be released during the night. However, larvae can sometimes be expelled at any time when the colony is suddenly disturbed. They are best seen against a dark background.

Bryozoans are not difficult to rear in the laboratory, but the substrate on which they attach must always be inverted so that fecal material and other settling debris will not accumulate around the zooids. Most species can be kept at room temperature. At lower temperatures, colony growth is slower and there are fewer problems with fouling organisms. Food appears to be the most important variable for success with laboratory-reared bryozoans. Oda (1980) maintained colonies *Lophopodella carteri* on pure cultures of *Chlamydomonas reinhardtii*. Wayss (1968) kept *Plumatella repens* for three years using a variety of unicellular green algae in 1:1

Knop's solution and soil extract. Others have achieved limited short-term success using mixed protozoan cultures, pet fish food, and fine detritus. One very trouble-free source of food is simply the suspended organic particles circulated through a rearing tank from a large aquarium in which active fish are maintained (Wood 1971).

If specimens are to be preserved, they should be anaesthetized before fixing so the lophophores will remain extended. The most convenient method is to confine colonies to a small covered dish of water with thin wafers of menthol floating on the surface. The menthol diffuses slowly into the water and relaxes the zooids within 1–2 hr. For most purposes, 10% formalin is a suitable fixative, diluted to 5% for long-term preservation.

Species identification usually requires the presence of statoblasts inside the colony. In specimens stored for an appreciable time, any gas in the floatoblast annulus may be replaced by liquid, making the entire capsule clearly visible through the periblast. In this case, it is important not to mistake capsule dimensions for those of the fenestra.

For detailed examination of a statoblast, you can separate the component parts by placing it briefly in a hot solution of potassium hydroxide. Then transfer it to distilled water and use fine needles to open the valves and tease the capsule away from the periblast. I find it convenient to arrange these parts under the microscope and photograph them with color slide film for my working records.

A. *Taxonomic Key to Species of Freshwater Bryozoans*

1a.	Zooid composed of bulbous body on externally segmented stalk; tentacles folding individually toward center when zooid is disturbed; statoblasts absent (Fig. 14.7) .... Entoprocta	*Urnatella gracilis* Leidy
1b.	Zooid without externally segmented stalk; tentacles withdrawn together when zooid is disturbed; statoblasts may be present 2	
2a(1b).	Colony composed of branching tubules (Fig. 14-10)a–d; body wall transparent to opaque; tentacles fewer than 65 3	
2b.	Colony smooth or lobed in outline, but never branching; body wall transparent; floatoblasts with peripheral spines, hooks, or pointed polar extensions; sessoblasts absent; tentacles more than 65 21	
3a(2a).	Extended lophophore circular in outline; statoblasts, if present, piptoblasts only (Fig. 9a–b); tentacles fewer than 30 4	
3b.	Extended lophophore U-shaped in outline; statoblasts either floatoblasts or sessoblasts (or both); tentacles more than 24 .... Plumatellidae 9	
4a(3a).	Statoblasts never formed; ectocyst stiff, shiny, and transparent; tentacles fewer than 20; individual zooids clearly demarcated by internal septa; colony retains its shape when withdrawn from the water; orifice appears quadrangular when lophophore is withdrawn .... Gymnolaemata 5	
4b.	Statoblasts formed, but possibly missing in some specimens; ectocyst not stiff and shiny; tentacles more than 19 .... *Fredericella* 7	
5a(4a).	Zooids arising from narrow stolons; orifice terminal; uncommon in North America 6	
5b.	Zooids branching from each other at nearly right angles; stolons absent; orifice lateral; widely distributed in North America (Fig.14.8c) .... *Palucidella articulata* (Ehrenberg 1831)	
6a(5a).	Fewer than ten tentacles; occurring mainly in brackish water; sometimes developing into fuzzy, branching masses; reported in North America south from Chesapeake Bay and in southeastern Louisiana (Fig. 14.8b) .... *Victorella pavida* (Kent, 1870)	
6b.	More than 15 tentacles; never found in brackish water (Fig. 14.8a) .... *Pottsiella erecta* (Potts, 1884)	
7a(4b).	Statoblast surface appearing smooth and shiny when dry 8	
7b.	Statoblast surface appearing dull and granular when dry (Fig. 14.9a) .... *Fredericella indica* (Annandale, 1909)	

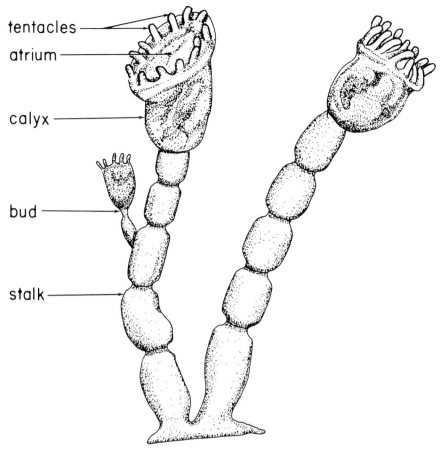

tentacles

atrium

calyx

bud

stalk

Figure 14.7 Small entoproct colony (*Urnatella gracilis*) showing external structures. × 25.

8a(7a).	Statoblast nearly round, thick-rimmed; typically more than one statoblast per zooid (Fig. 14.9b); rare, but locally abundant; known from Maryland, Ohio, Illinois, and Utah .... *Fredericella australiensis* Goddard 1909
8b.	Statoblast oval to elongate, thin-rimmed; seldom more than one statoblast per zooid; common in Europe, suspected but not confirmed in North America .... *Fredericella sultana* (Blumenbach, 1779)
9a(3b).	Zooids clustered in groups of 2–5, cluster joined by stolon-like parts; colony wall delicate and transparent; rare .... *Stolella* 10
9b.	Zooids not arranged in clusters; colony wall transparent to opaque; common 11
10a(9a).	Polypides in groups of two; ectocyst often wrinkled; zooids bent sharply in various directions; reported only from single sites in Maryland, Pennsylvania, and Michigan .... *Stolella indica* Annandale, 1909
10b.	Polypides in groups of 2–5; clusters oriented vertically to substrate, zooids not bent sharply; reported in North America from only one site in Michigan (Fig. 14.9l) .... *Stolella evelinae* Marcus, 1941
11a(9b).	Body wall hyaline, thick, appearing swollen or gelatinous, at least in distal branches 12
11b.	Body wall thin, not gelatinous 15
12a(11a).	Lophophore arms so short that U-shape is indistinct; floatoblasts nearly circular (Fig. 14.9m), sessoblast surface with distinctive ridges and tubercular projections; cold-water species known in North America only from Hampshire County, Massachusetts .... *Stephanella hina* Oka 1908
12b.	U-shape of lophophore is prominent; floatoblasts circular to oval, sessoblasts normally absent .... *Hyalinella* 13

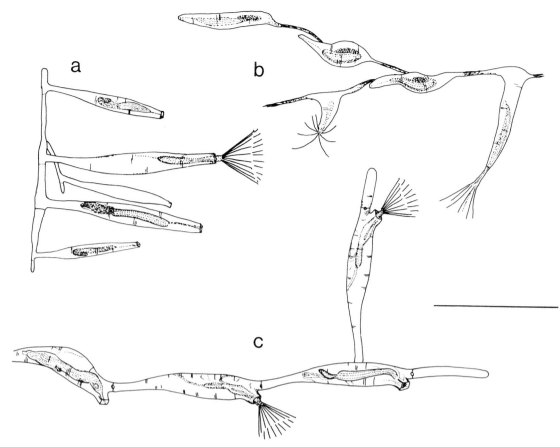

Figure 14.8 Gymnolaemate colonies occurring in freshwater. Scale bar = 1 mm. (a) *Pottsiella erecta;* (b) *Victorella pavida;* (c) *Paludicella articulata.*

13a(12b).	Floatoblasts circular or nearly so (Fig. 14.9k); known only from scattered sites in Michigan and Ontario .... *Hyalinella orbisperma* (Kellicott, 1882)
13b.	Floatoblasts distinctly oval 14
14a(13b).	Floatoblasts rounded at the ends, buoyant only after being dried, measuring over 440 μm long $\times$ 375 μm wide (Fig. 14.9i); common and widely distributed (Fig. 14.10d) .... *Hyalinella punctata* (Hancock, 1850)
14b.	Floatoblasts tapered at ends, buoyant upon release from the colony, measuring less than 470 μm long $\times$ 400μm wide (Fig. 14.9j); known in North America only from Box Elder County, Utah, also collected in Hawaii, eastern Australia, and Tahiti .... *Hyalinella vaihiriae* Hastings, 1929
15a(11b).	Width of floatoblast annulus at poles of dorsal valve greater than or equal to length of dorsal fenestra (Figs. 14.9g–h) 16
15b.	Width of floatoblast annulus at poles of dorsal valve less than length of dorsal fenestra (Fig. 14.9e–f, i–m) 17
16a(15a).	Floatoblast dorsal valve nearly flat; suture between valves visible in dorsal view; sessoblast surface uniformly granular, tentacles about 39; common and widely distributed (Fig. 14.5a, 14.9h) .... *Plumatella emarginata* Allman, 1844
16b.	Floatoblast valves almost equally convex, sessoblast surface with network of dark lines; tentacles about 32; known only from Ohio, Indiana, and Illinois (Fig. 14.9g) .... *Plumatella reticulata* Wood, 1988
17a(15b).	Colonies with thin branches largely free of substrate; floatoblast and sessoblast length more than twice width (Fig. 14.9f); widely distributed but generally uncommon .... *Plumatella fruticosa* Allman, 1884
17b.	Colonies seldom with branches free from substrate, statoblasts otherwise 18

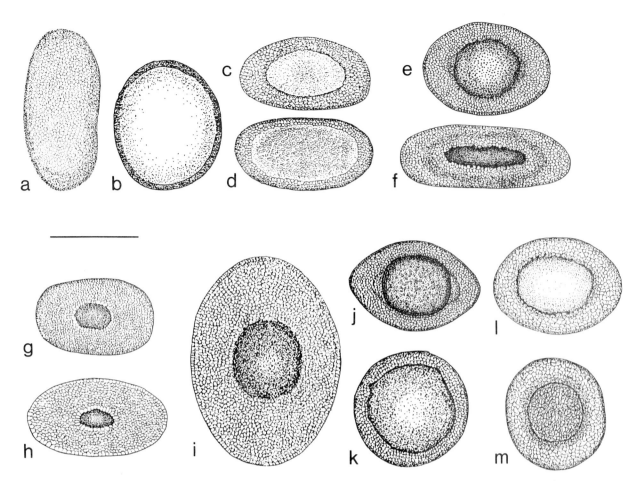

Figure 14.9 Free statoblasts from the families Plumatellidae and Fredericellidae. Scale bar for all figures = 250 μm. (a) *Fredericella indica;* (b) *Fredericella australiensis;* (c) *Plumatella casmiana* (intermediate type); (d) *Plumatella casmiana* (leptoblast); (e) *Plumatella repens;* (f) *Plumatella fruticosa;* (g) *Plumatella reticulata;* (h) *Plumatella emarginata;* (i) *Hyalinella punctata;* (j) *Hyalinella varhiriae;* (k) *Hyalinella orbisperma;* (l) *Stolella evelinae;* (m) *Stephanella hina.* [a–e and g–i modified from Wood (1989); j based on Rogick and Brown (1942); k and m based on Bushnell (1965b).]

18a(17b).	Tentacles 23-41; floatoblast length to width greater than 1.5 (Fig. 14.9c); delicate thin-walled floatoblast sometimes present (Fig. 14.9d); colony composed of short, profusely branched tubes closely applied to substrate (Fig. 14.10b), tubes becoming erect and fused when crowded .... *Plumatella casmiana* Oka, 1907
18b.	Tentacles 40-65; floatoblast length to width less than 1.5; each cell in dorsal fenestra of floatoblast with small dark tubercle (Fig. 14.9e); zooids tending to be erect and often elongate 19
19a(18b).	Erect zooids in crowded colonies may be in contact, but are never fused; colonies mostly with linear series of zooids ranging over the substrate (Fig. 14.10c); colony wall hyaline; floatoblast valves almost equally convex; widely distributed and very common .... *Plumatella repens* (Linnaeus, 1758)
19b.	Erect zooids in crowded colonies up to 2.0 mm long and fused along part or all of their length 20
20a(19b).	Zooids massed together and fused along entire length; colony forming thick, pillow-like growths; dorsal valve of floatoblast nearly flat, ventral valve strongly convex; conspicuous dark septa .... *Plumatella fungosa* (Pallas 1768)
20b.	Erect zooids in crowded colonies fused at their bases but free at the tips; ectocyst usually described as dark brown; reported in North America only from New England .... *Plumatella corralloides* Allman, 1850

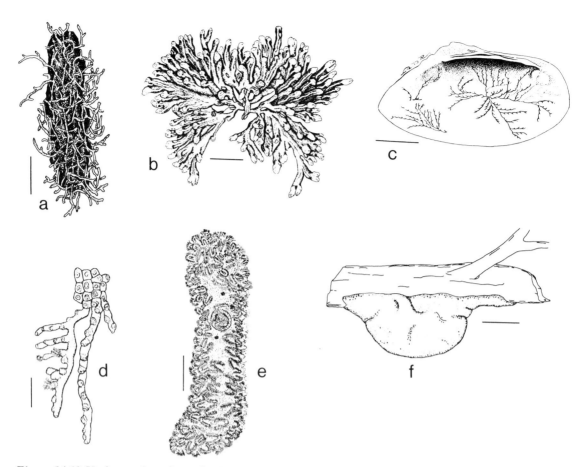

Figure 14.10 Various colony forms in Phylactolaemata. (a) *Fredericella indica* growing on a piece of wood, showing many free branches, scale bar = 5 cm; (b) *Plumatella casmiana,* with zooids attached throughout their length to the substrate or to each other, scale bar = 2 mm; c) *Plumatella repens* growing on the inner surface of a mussel valve, scale bar = 2 cm; (d) *Hyalinella punctata,* showing both densely packed and free-ranging growth, all zooids firmly attached to the substrate, scale bar = 5 mm; (e) *Cristatella mucedo,* showing a single statoblast inside, scale bar = 5 mm; (f) *Pectinatella magnifica* growing on the underside of a small log, scale bar = 2 cm. [b from Rogick (1941); c from Rogick and Brown (1942), d–f from Wood (1989).]

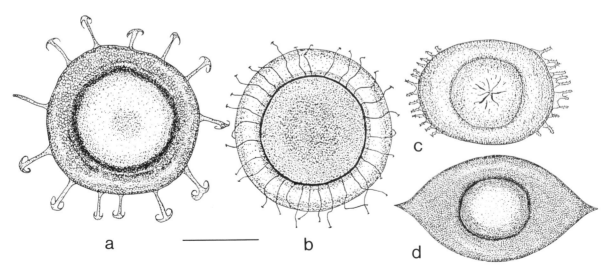

Figure 14.11 Statoblasts in the families Lophopodidae and Cristatellidae. Scale bar for all figures = 500 μm. (a) *Pectinatella magnifica;* (b) *Cristatella magnifica;* (c) *Lophopodella carteri;* (d) *Lophopus crystallinus.* (a–c modified from Wood 1989.)

21a(2b).   Colony large, becoming football-sized, gelatinous, and slimy (Fig. 14.10f); mouth region with red pigmentation; prominent pair of white spots at end of each arm of lophophore; statoblasts with hooked spines radiating from outer margin of annulus (Fig. 14.11a) . . . . *Pectinatella magnifica* (Leidy, 1851)

21b.   Colony small; zooids without red pigmentation; statoblasts not as above 22

22a(21b).   Colony linear, often longer than 2 cm (Fig. 14.10e); statoblasts with wiry, hooked spines radiating beyond periphery from margin of fenestrae of both valves (Fig. 14.11b) . . . . *Cristatella mucedo* Cuvier, 1798

22b.   Colony globular, never more than 1 cm diameter; statoblasts without radiating spines 23

23a(22b).   Statoblast with series of small hooks localized along the margin at the poles (Fig. 14.11c); uncommon, but can be locally abundant (Fig. 14.4) . . . . *Lophopodella carteri* (Hyatt, 1866)

23b.   Statoblast tapering to a single point at each pole (Fig. 14.11d); extremely rare worldwide and possibly endangered . . . . *Lophopus crystallinus* (Pallas, 1768)

LITERATURE CITED

Bailey-Brock, J. H., and P. J. Hayward. 1984. A freshwater bryozoan, *Hyalinella vaihiriae* Hastings (1929), from Hawaiian prawn ponds. Pacific Science 38:199–204.

Brien, P. 1953. Etude sur les Phylactolémates. Annales de la Société Royale Zoologique de Belgique 84:301–444.

Brown, C. J. D. 1933. A limnological study of certain fresh-water Polyzoa with special reference to their statoblasts. Transactions of the American Microscopical Society 52:271–313.

Bullivant, J. S. 1968. The rate of feeding of the bryozoan, *Zoobotryon verticillatum*. New Zealand Journal of Marine and Freshwater Research. 2:111–134.

Bushnell, J. H. 1965a. On the taxonomy and distribution of freshwater Ectoprocta in Michigan. Part I. Transactions of the American Microscopical Society 84:231–244.

Bushnell, J. H. 1965b. On the taxonomy and distribution of freshwater Ectoprocta in Michigan. Part II. Transactions of the American Microscopical Society 84:339–358.

Bushnell, J. H. 1965c. On the taxonomy and distribution of freshwater Ectoprocta in Michigan. Part III. Transactions of the American Microscopical Society 84:529–548.

Bushnell, J. H. 1966. Environmental relations of Michigan Ectoprocta, and dynamics of natural populations of *Plumatella repens*. Ecological Monographs 36:95–123.

Bushnell, J. H. 1974. Bryozoa (Ectoprocta). Pages 157–194 *in*: C. W. Hart and S. L. H. Fuller, editors. Pollution ecology of freshwater invertebrates. Academic Press, New York.

Bushnell, J. H., and K. S. Rao. 1974. Dormant or quiescent stages and structures among the Ectoprocta: physical and chemical factors affecting viability and germination of statoblasts. Transactions of the American Microscopical Society 93:524–543.

Cooper, C. M. and J. W. Burris. 1984. Bryozoans—possible indicators of environmental quality in Bear Creek, Mississippi. Journal of Environmental Quality 13(1):127–130.

Curry, M. G., B. Everitt, and M. F. Vidrine. 1981. Haptobenthos on shells of living freshwater clams in Lousiana. The Wassman Journal of Biology 39:56–62.

Cusak, T. M., and J. D. McCullough. 1985. *Urnatella gracilis* (Entoprocta) from Caddo Lake, Texas and Louisiana. The Texas Journal of Science 37:141–142.

Dehdashti, B., and D. W. Blinn. 1986. A bryozoan from an unexplored cave at Montezuma Well, Arizona. The Southwestern Naturalist 31:557–558.

Dendy, J. S. 1963. Observations on bryozoan ecology in farm ponds. Limnology and Oceanography 8:478–482.

Dendy, J. S., and J. E. Sublette. 1959. The Chironomidae (=Tendipedidae: Diptera) of Alabama with descriptions of six new species. Annals of the Entomological Society of America 52:506–519.

Eng, L. L. 1977. The freshwater entoproct, *Urnatella gracilis* Leidy, in the Delta-Mendota Canal, California. The Wassman Journal of Biology 35:196–202.

Everitt, B. C. 1975. Fresh-water Ectoprocta: distribution and ecology of five species in southeastern Louisiana. Transactions of the American Microscopical Society 94:130–134.

Franzén, A., and T. Sensenbaugh. 1983. Fine structure of the apical plate in the larva of the freshwater bryozoan *Plumatella fungosa* (Pallas) (Bryozoa: Phylactolaemata). Zoomorphology 102:87–98.

Gilmour, T. H. J. 1978. Ciliation and function of the food-collecting and waste-rejecting organs of lophophorates. Canadian Journal of Zoology 56:2142–2155.

Goethals, B., M. F. Voss-Foucart, and G. Goffinet. 1984. Composition chimique et ultrastructure de l'ectocyste et de la coque des statoblastes de *Plumatella repens* et *Plumatella fungosa* (bryozoaires phylactolèmes). Annales des Sciences Naturelles, Zoologie, Paris 13ᵉ Serie 6:197–206.

Hubschman, J. H. 1970. Substrate discrimination in *Pectinatella magnifica* Leidy (Bryozoa). Journal of Experimental Biology 52:603–607.

Hull, H. C., L. F. Bartos, and R. A. Martz. 1980. Occurrence of *Urnatella gracilis* Leidy in the Tampa Bypass Canal, Florida. Florida Scientist 43:2–13.

Hyman, L. H. 1959. The lophophorate coelomates: Phylum Electoprocta. The Invertebrates. McGraw-Hill, New York.

Jebram, D. 1973. The importance of different growth directions in the Phylactolaemata and Gymnolaemata for reconstructing the phylogeny of the Bryozoa. Pages 565–576 *in:* G. P. Larwood, editor. Living and Fossil Bryozoa. Academic Press, London.

Job, P. 1976. Intervention des populations de *Plumatella fungosa* (Pallas) (Bryozoaire phylactolème) dans l'autoépuration des eaux d'unétang et d'un ruisseau. Hydrobiologica 48:257–261.

Kaminski, M. 1984. Food composition of three bryozoan species (Bryozoan Phylactolaemata) in a mesotrophic lake. Polskie Archiwum Hydrobiologii 31(1):45–53.

King, D. K., R. H. King, and A. C. Miller. 1988. Morphology and ecology of *Urnatella gracilis* Leidy, (Entoprocta), a freshwater macroinvertebrate from artificial riffles of the Tombigbee River, Mississippi. Journal of Freshwater Ecology 4(3):351–359.

Marcus, E. 1934. Uber *Lophopus crystallinus* (Pall.). Zoologische Jahrbücher. Abteilung für Anatomie und Ontogenie der Tiere 58:501–606.

Morse, W. 1930. The chemical constitution of *Pectinatella*. Science 71:265.

Mukai, H. 1982. Development of freshwater bryozoans (Phylactolaemata). Pages 535–576 *in:* F. W. Harrison and R. R. Cowden, editors. Developmental biology of freshwater invertebrates. Alan R. Liss, New York.

Mukai, H., and K. Kobayashi. 1988. External observations on the formation of statoblasts in *Plumatella emarginata* (Bryozoa, Phylactolaemata). Journal of Morphology 196:205–216.

Mukai, H., and S. Oda. 1980. Histological and histochemical studies on the epidermal system of higher phylactolaemate bryozoans. Annotationes Zoologicae Japonenses 53:1–17.

Mukai, H., M. Tsuchiya, and K. Kimoto. 1984. Fusion of ancestrulae germinated from statoblasts in plumatellid freshwater bryozoans. Journal of Morphology 179:197–202.

Mukai, H., M. Fukushima, and Y. Jinbo. 1987. Characterization of the form and growth pattern of colonies in several freshwater bryozoans. Journal of Morphology 192:161–179.

Nielsen, C. 1977. Phylogenetic considerations: the protostomian relationships. Pages 519–534 *in:* R. Woollacott and R. Zimmer, editors. Biology of Bryozoans. Academic Press, New York.

Oda, A. 1980. Effects of light on the germination of statoblasts in freshwater Bryozoa. Annotationes Zoologicae Japonenses 53:238–253.

Protosov, A. A. 1980. On distribution of *Urnatella gracilis* (Kamptozoa) with reference to discharges of heated waters by thermal electrical power stations. Zoologicheskii Zhurnal 59:1569–1571. (In Russ.)

Rao, K. S., and J. H. Bushnell. 1979. New structures in binding designs of freshwater Ectoprocta dormant bodies (statoblasts). Acta Zoologica (Stockholm) 60:123–127.

Rao, K. S., A. P. Diwan, and P. Shrivastava. 1978. Structure and environmental relations of sclerotized structures in fresh water Bryozoa. III. Observations on *Plumatella casmiana* (Ectoprocta: Phylactolaemata). Journal of Animal Morphology and Physiology 25:8–15.

Reynolds, J. D. 1976. Occurrence of the fresh-water Bryozoan, *Cristatella mucedo* Cuvier, in British Columbia. Syesis 9:365–366.

Rogick, M. D. 1941. Studies on freshwater Bryozoa. X. The occurrence of *Plumatella casmiana* in North America. Transactions of the American Microscopical Society 60(2):211–220.

Rogick, M. D., and C. J. D. Brown. 1942. Studies on fresh-water Bryozoa XII. A collection from various sources. Annals of the New York Academy of Sciences 43:123–144.

Rüsche, E. 1938. Nahrungsaufname und Nahrungsauswertung bei *Plumatella fungosa* (Pallas). (Ein Beitrag zur Frage des Stoffkreislaufs im Weiher). Archiv für Hydrobiolgie 33:271–293.

Shrivastava, P., and K. S. Rao. 1985. Ecology of *Plumatella emarginata* (Ectoprocta: Phylactolaemata) in the surface waters of Madhya Pradesh with a note on its occurrence in the protected waterworks of Bhopal (India). Environmental Pollution (Series A) 39:123–130.

Sládeček, V. 1980. Indicator value of freshwater Bryozoa. Acta Hydrochimica and Hydrobiologica 8:273–276.

Smith, D. G. 1989a. On *Stephanella hina* Oka (Ectoprocta: Phylactolaemata) in North America, with notes on its morphology and systematics. Journal of the American Benthological Society 7:253–259.

Smith, D. G. 1989b. Keys to the freshwater macroinvertebrates of Massachusetts. No. 4: Benthic colonial phyla, including the Cnidaria, Entoprocta, and Ectoprocta (colonial hydrozoans, moss animals). Massachusetts Div. Water Pollut. Control, Westborough.

Tenney, W. R., and W. S. Woolcott. 1964. A comparison of the responses of some species of fishes to the toxic effect of the bryozoan, *Lophopodella carteri* (Hyatt) in Virginia. American Midland Naturalist 68:247–248.

Tenney, W. R., and W. S. Woolcott. 1966. The occurrence and ecology of freshwater bryozoans in the headwaters of the Tennessee, Savannah, and Saluda River systems. Transactions of the American Microscopical Association 85:241–245.

Wayss, K. 1968. Quantitative Untersuchungen über Wachstum und Regeneration bei *Plumatella repens* (L.). Zoologische Jahrbucher. Abteilung für Anatomie und Ontogenie der Tiere 85:1–50.

Weise, J. G. 1961. The ecology of *Urnatella gracilis* Leidy: Phylum Entoprocta. Limnology and Oceanography 6:228–230.

Wilde, G. A., T. A. Burke, and C. Keenan. 1981. Bryozoa from Lake Mead and Lake Mohave, Nevada–Arizona. The Southwestern Naturalist 26:201.

Wood, T. S. 1971. Laboratory culture of freshwater Ectoprocta. Transactions of the American Microscopical Society 90:229–231.

Wood, T. S. 1973. Colony development in species of *Plumatella* and *Fredericella*(Ectoprocta: Phylactolaemata). Pages 395–432 *in:* R. S. Boardman, A. Cheetham, and J. Oliver, editors. Development and function of animal colonies through time. Dowden, Hutchinson & Ross, Stroudsburg, Pennsylvania.

Wood, T. S. 1989. Ectoproct bryozoans of Ohio. Bulletin of the Ohio Academy of Science, New Series 8: 1–66.

Wood, T. S. and B. T. Backus. The North American and European forms of *Fredericella sultana* are different species. (Submitted).

Zimmerman, M. 1979. Larval release and settlement behavior in three species of plumatellid Bryozoa (Ectoprocta). Master's Thesis, Wright State University, Dayton, Ohio.

Tardigrada

15

Diane R. Nelson
Department of Biological Sciences
East Tennessee State University
Johnson City, Tennessee 37614

Chapter Outline

I. INTRODUCTION

II. ANATOMY
    A. Cuticle
    B. Claws
    C. Digestive System
        1. General Anatomy
        2. Buccal Apparatus
        3. Esophagus, Midgut, and Hindgut
    D. Other Anatomical Systems

III. PHYSIOLOGY (LATENT STATES)

IV. REPRODUCTION AND DEVELOPMENT
    A. Reproductive Modes
    B. Reproductive Apparatus
    C. Mating and Fertilization
    D. Eggs
    E. Parthenogenesis
    F. Hermaphroditism
    G. Development

V. ECOLOGY
    A. Molting and Life History
    B. Habitats
    C. Population Density
    D. Distribution
    E. Dissemination

VI. TECHNIQUES FOR COLLECTION, EXTRACTION, AND MICROSCOPY

VII. IDENTIFICATION
    A. Taxonomic Key to Genera of Freshwater and Terrestrial Tardigrada
    Literature Cited

I. INTRODUCTION

Since 1773, when J. A. E. Goeze reported the first observation of a "Kleiner Wasser Bär," tardigrades have commonly been called "water bears" (Goeze, 1773). Their bear-like appearance, legs with claws, and slow lumbering gait formed the basis for the descriptive name. In 1776, the naturalist Lazzaro Spallanzani introduced the name "il Tardigrado" (slow-stepper) to describe the slow, tortoise-like movement of the animal (Spallanzani, 1776). The systematic position of the Tardigrada, the taxon name assigned to the group by Doyère (1840), has ranged from inclusion within the Infusoria to the Annelida to the Arthropoda (Acarina, Crustacea, Insecta), as similarities to both the onychophoran–arthropod complex and the aschelminth complex were recognized. The phylum Tardigrada was recognized by Ramazzotti (1962) in his first monograph, but the phylogenetic affinities of the phylum remain debatable. Approximately 600 species have been described from marine, freshwater, and terrestrial habitats.

Tardigrades are hydrophilous micrometazoans with a bilaterally symmetrical body and four pairs of legs usually terminating in claws (Figs. 15.1 and 15.2). Generally convex on the dorsal side and flattened on the ventral side, the body is indistinctly divided into a head (cephalic) segment, three trunk segments each bearing a pair of legs, and a caudal segment with the fourth pair of legs directed posteriorly. Although tardigrades range in body length from 50 μm in juveniles to 1200 μm in adults (both excluding the last pair of legs), mature adults average 250–500 μm with a few species exceeding 500 μm.

Despite their overall abundance and cosmopolitan distribution, the Tardigrada have been relatively neglected by invertebrate zoologists. Frequently categorized as one of the "minor phyla," they are more appropriately termed one of the "lesser-

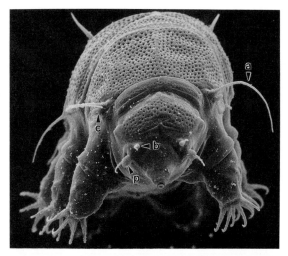

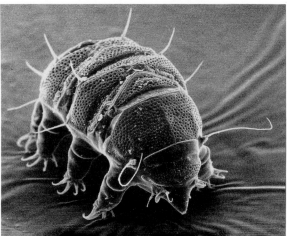

Figure 15.1 Scanning electron micrographs of a hetero-tardigrade, *Echiniscus spiniger*. a, Cirrus A; b, buccal cirrus; c, clava; p, cephalic papilla. (From Nelson 1982a.)

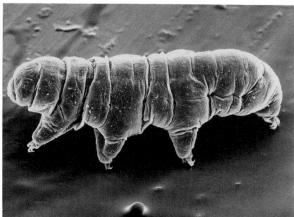

Figure 15.2 Scanning electron micrographs of a eutard-igrade, *Macrobiotus tonollii*. (From Nelson 1982a.)

known phyla." Because of difficulties in collecting and culturing the organisms and their apparent lack of economic importance to humans, our knowledge of tardigrades has remained in the nascent state since their discovery over 200 years ago. The lack of basic biological information has hindered the growth of interdisciplinary studies, which are just beginning to develop.

Today, new insights into the biology of the Tardigrada are evolving from more comprehensive, contemporary investigations utilizing transmission and scanning electron microscopy as well as histochemistry, cytological techniques, and phase- and interference-contrast microscopy. The "state of the art," although still primitive, provides a vast array of opportunities for classical and contemporary biological investigations. Within the last decade, an upsurge in interest and activity has led to a boom in tardigrade research.

II. ANATOMY

A. Cuticle

The cuticle, which is highly permeable to water, covers the body surface and lines the foregut and hindgut. Secreted by the underlying hypodermis, the cuticle consists of several layers which may be further subdivided into: (1) an outer complex epicuticle; (2) an intracuticle (not present in all species); and (3) an inner procuticle, which contains chitin (Greven and Greven 1987, Greven and Peters 1986). Although most freshwater tardigrades are white or colorless, some terrestrial species exhibit brown, yellow, orange, pink, red, or green coloration due to intestinal contents, body cavity cells and granules, or pigmentation in the hypodermis and cuticle (Ramazzotti and Maucci 1983). Based on cuticular structures, three classes, formerly considered orders, can be distinguished (Table 15.1). The class Heterotardigrada (Figs. 15.1 and 15.3) includes mainly the marine and armored terrestrial tardigrades, while the class Eutardigrada (Figs. 15.2 and

Table 15.1 Subdivision of the Phylum Tardigrada with Habitat Classifications

I. Class Heterotardigrada	
A. Order Arthrotardigrada	Marine, except *Styraconyx hallasi* (freshwater)
B. Order Echiniscoidea	Marine, freshwater, terrestrial
1. Family Echiniscoididae	Marine; genera *Echiniscoides* and *Anisonyches*
2. Family Oreellidae	Terrestrial; genus *Oreella*
3. Family Carphanidae	Freshwater; genus *Carphania*
4. Family Echiniscidae	Terrestrial, except for a few species of *Echiniscus, Hypechiniscus,* and *Pseudechiniscus* occasionally in freshwater; 12 genera
II. Class Mesotardigrada	
A. Order Thermozodia	
1. Family Thermozodiidae	Hot springs; genus *Thermozodium*
III. Class Eutardigrada	
A. Order Parachela	
1. Family Macrobiotidae	Terrestrial and freshwater; *Macrobiotus, Minibiotus, Dactylobiotus, Pseudodiphascon, Adorybiotus*
2. Family Calohypsibiidae	Terrestrial; genera *Calohypsibius, Hexapodibius, Haplomacrobiotus, Microhypsibius*
3. Family Eohypsibiidae	Terrestrial and freshwater; *Eohypsibius, Amphibolus*
4. Family Hypsibiidae	Marine; genus *Halobiotus* and some *Isohypsibius*. Terrestrial and freshwater; genera *Hypsibius, Isohypsibius, Ramazzottius, Doryphoribius, Apodibius, Pseudobiotus, Thulinia, Diphascon, Platycrista, Mesocrista, Hebesuncus,* and *Itaquascon*
5. Family Necopinatidae	Terrestrial; genus *Necopinatum*
B. Order Apochela	
1. Family Milnesiidae	Terrestrial; *Milnesium* and *Limmenius*

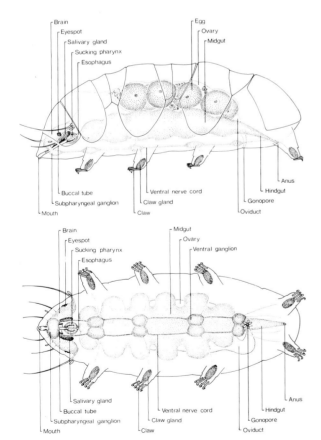

Figure 15.3 Heterotardigrade internal anatomy (*Bryodelphax parvulus*, female); lateral view (*above*) and ventral view (*below*). (From Nelson 1982a; drawing by R. P. Higgins.)

15.4) primarily encompasses freshwater and other terrestrial species. A third class, Mesotardigrada (Rahm 1937) is based on a single species, *Thermozodium esakii,* from a hot spring in Japan; it is difficult to evaluate the validity of this class due to the lack of type specimens and to the destruction of the type locality by an earthquake (Ramazzotti and Maucci 1983).

Heterotardigrades (Fig. 15.1) are characterized by paired cephalic sensory appendages (Fig.15.1): internal buccal cirri, cephalic papillae, external buccal cirri, clavae, and lateral cirri (cirri A). Lateral cirri A mark the junction of the cephalic and scapular plates. Some marine species also have a single median cirrus, particularly in the most primitive order, Arthrotardigrada. In the class Eutardigrada (Fig. 15.2), cephalic sensory structures, which are not homologous to those in heterotardigrades, are present in the order Apochela; *Milnesium* has two lateral papillae and six peribuccal (oral) papillae, and *Lemmenius* has two lateral papillae. Other eutardigrades (order Parachela) lack external sensory appendages but have sensory regions on the head (Walz 1978).

The cuticle and its processes are taxonomically important in identifying genera and species and also form the basis for separating tardigrades into two large groups. The "armored" tardigrades are heterotardigrades that have a thickened dorsal cuticle divided into individual plates, which have a char-

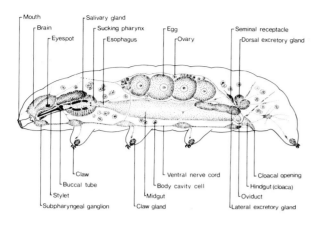

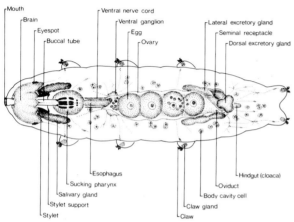

Figure 15.4 Eutardigrade internal anatomy (*Macrobiotus hufelandi*, female); lateral view (*above*) and dorsal view (*below*). (From Nelson 1982a; drawing by R. P. Higgins.)

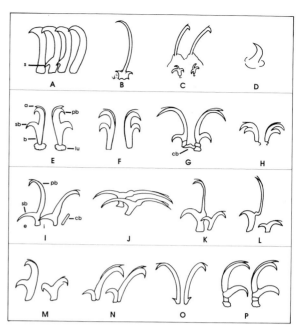

Figure 15.5 Claw types. All claws are shown in ventral view of the fourth right leg. (A) *Echiniscus*; (B) *Carphania*; (C) *Milnesium, Limmenius*; (D) *Necopinatum*; (E) *Macrobiotus, Adorybiotus*; (F) *Pseudodiphascon*; (G) *Dactylobiotus*; (H) *Minibiotus*; (I) *Isohypsibius, Thulinia, Doryphoribius*; (J) *Pseudobiotus*; (K) *Hypsibius, Platycrista, Mesocrista, Hebesuncus, Diphascon, Itaquascon*; (L) *Ramazzottius*; (M) *Calohypsibius, Microhypsibius*; (N) *Hexapodibius*; (O) *Haplomacrobiotus*; (P) *Eohypsibius, Amphibolus*. a, Accessory point; b, base; cb, cuticular bar; e, external claw; i, internal claw; lu, lunule; pb, primary branch; s, spur; sb, secondary branch. (Redrawn from Nelson and Higgins 1990.)

acteristic pattern of sculpture within a species. Armored tardigrades include both marine species (Renaudarctidae and Stygarctidae, order Arthrotardigrada) and terrestrial species (Echiniscidae, order Echiniscoidea). Kristensen (1987) revised the genera of the Echiniscidae, adding four new genera to the existing eight. Of these, some species in the genus *Pseudechiniscus* are found in freshwater, but the dorsal plates are nonsclerotized. The "naked" or unarmored tardigrades include the class Eutardigrada (freshwater and terrestrial) as well as the heterotardigrade families Oreellidae (terrestrial) and Carphanidae (freshwater) (Binda and Kristensen 1986, Kristensen 1987). They have a thin, smooth cuticle or a sculptured one with pores, granulations, reticulations, tubercles, papillae, or spines, but lack plates.

B. Claws

In addition to cuticular structures, the size, shape, and number of claws (Figs. 15.5 and 15.6) are important in tardigrade systematics and have been used for revision of the eutardigrades (Pilato 1969, 1982,

Bertolani and Kristensen 1987, Kristensen and Higgins 1984a, 1984b, Ramazzotti and Maucci 1983). In the heterotardigrade family Echiniscidae, each leg terminates in four single claws in the adult and juvenile or two claws in the larva. Frequently, the two internal claws have a ventral spur, and occasionally one or more spurs may be present on the external claw near the base (Kristensen 1987, Ramazzotti and Maucci 1983). In the heterotardigrade monotypic family Carphanidae (formerly in Oreellidae), the freshwater genus *Carphania* has two claws on legs I–III and only one claw on reduced leg IV; all claws are flexible and have two basal spurs (Binda and Kristensen 1986).

Eutardigrades usually possess two double claws on each leg, arranged as external and internal claws (Schuster *et al.* 1980a, Bertolani 1982a). Each double claw consists of a base with a long primary branch with two accessory points and a secondary branch without accessory points. In the order Apochela (family Milnesiidae), the primary branch is long, thin, and well separated from the stout, short secondary branch which bears hooks

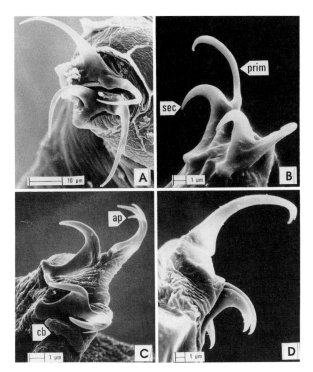

Figure 15.6 Scanning electron micrographs of eutardigrade claws. (A) *Pseudobiotus*; (B) *Hypsibius*, prim, primary branch; sec, secondary branch; (C) *Itaquascon*. ap, Accessory point; cb, cuticular bar; D, *Milnesium*. (From Schuster *et al.* 1980a.)

or spurs. In the order Parachela, the secondary branch is attached to the leg and the primary branch arises from the secondary branch. The sequence of claw branches, with respect to the midline of extended legs, in some genera is alternate (2-1-2-1) (i.e., secondary-primary-secondary-primary). In other genera, the two primary branches are next to one another (2-1-1-2) (i.e., secondary-primary-primary-secondary). In some eutardigrades (e.g., *Macrobiotus*), a cuticular thickening called a lunule surrounds the base of each double claw. Lunules vary in size and shape; they may be smooth or dentate. In several genera, one or two cuticular bars may be present on the legs, separate from the claws but located between the two double claws, just below the base of the claws, or along the sides of the internal claws.

Six basic claw types have been identified. In the *Dactylobiotus*-type, the two double claws on each leg are similar in size and shape, with a 2-1-1-2 sequence, lack lunules, and are connected at the base by a cuticular band. The secondary branch is much shorter than the primary and is inserted near the base. The *Macrobiotus*-type claws also are similar in size and shape and have a 2-1-1-2 sequence, but usually possess lunules and are not connected at the base. They may be either V-shaped or Y-shaped

claws. The *Calohypsibius*-type claws are similar in size and shape but exhibit the alternate (2-1-2-1) sequence; the primary branch is rigidly attached to the secondary branch. The *Hypsibius*-type claws are usually dissimilar in size and shape, with the external double claw longer and more slender than the internal claw. The primary branch of the external double claw is inserted on the basal branch, which is continuous with the secondary branch, forming a flexible junction. The *Isohypsibius*-type claws resemble those of *Hypsibius;* however, they are usually similar in size and shape. To distinguish the two types, observe the profile of the secondary branch of the external claw from one of the first three pairs of legs. In the *Hypsibius*-type, the secondary branch and basal branch run in a continuous arc or curve; whereas, in the *Isohypsibius*-type, the secondary branch forms a right angle with the basal branch. The *Eohypsibius*-type claws are divided into three distinct parts: the base, the secondary branch, and the primary branch; they are usually arranged in a 2-1-2-1 sequence, however, the internal claw can rotate 180°, resulting in a 2-1-1-2 sequence. The external and internal claws are similar in size, but the angle between the primary and secondary branches is about 45° on the external claw and about 80° on the internal claw.

C. Digestive System

1. General Anatomy

The digestive system consists of a foregut, a midgut, and a hindgut (Figs. 15.3 and 15.4). Both the foregut (buccal apparatus and esophagus) and the hindgut have a cuticular lining. The hindgut is a true cloaca in eutardigrades and the opening is a ventral transverse slit just anterior to the fourth pair of legs. In heterotardigrades, however, reproductive and digestive systems are separate, and the gonopore opens anterior to the anus, which is a longitudinal slit between the fourth pair of legs. In the family Echiniscidae (heterotardigrades), first instar individuals have a mouth but no gonopore or anus.

2. Buccal Apparatus

The buccal apparatus is a complex structure with considerable taxonomic significance in the eutardigrades (Pilato 1972, 1982, Schuster *et al.* 1980a). Basically, it consists of a buccal tube, a muscular sucking pharynx, and a pair of piercing stylets that can be extended through the mouth opening (Figs. 15.3, 15.4, 15.7, and 15.8). The position of the mouth may be terminal (as in *Macrobiotus* and *Milnesium*) or subterminal (as in *Hypsibius*). In some eutardigrades, cuticular structures surround the mouth

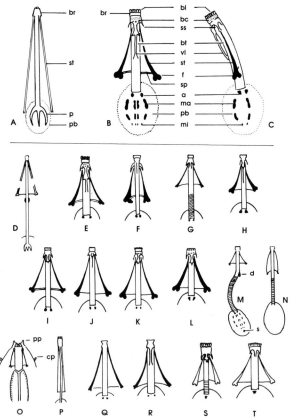

Figure 15.7 Buccal apparatus. (A) Heterotardigrada, *Echiniscus* (ventral); (B) Eutardigrada, *Macrobiotus* (ventral); (C) Eutardigrada, *Macrobiotus* (lateral); (D) *Carphania*; (E) *Macrobiotus, Dactylobiotus*; (F) *Adorybiotus*; (G) *Pseudodiphascon*; (H) *Minibiotus*; (I) *Hypsibius, Isohypsibius, Ramazzottius, Necopinatum*; (J) *Doryphoribius, Apodibius*; (K) *Pseudobiotus*; (L) *Thulinia*; (M) *Diphascon*; (N) *Itaquascon*; (O) *Milnesium*; (P) *Limmenius*; (Q) *Calohypsibius, Microhypsibius*; (R) *Hexapodibius, Haplomacrobiotus*; (S) *Eohypsibius*; (T) *Amphibolus*. a, Apophysis; bc, buccal cavity; bl, buccal lamellae; br, buccal ring, bt, buccal tube; cp, cephalic papilla; d, drop-like thickening; f, furca; ma, macroplacoid; mi, microplacoid; p, placoid; pb, pharyngeal bulb; pp, peribuccal papilla; s, septulum; sp, stylet support; ss, stylet sheath; st, stylet; vl, ventral lamina. (Redrawn from Nelson and Higgins 1990.)

such as elongate peribuccal papillae (six in *Milnesium*), flattened peribuccal papulae (ten in *Minibiotus* and *Haplomacrobiotus* and six in *Calohypsibius*), or lobes (six in *Isohypsibius* and others) (Schuster *et al.* 1980a).

The buccal tube begins with a buccal ring, which bears lamellae in some genera. The number of lamellae varies with the genus; e.g., there are 10 in *Macrobiotus*, 12 in *Thulinia*, 14 in *Amphibolus*, and about 30 in *Pseudobiotus*. In *Milnesium*, the mouth is surrounded by six triangular lamellae, which close the opening. The interior of the buccal cavity may

have anterior and posterior rows of small cuticular teeth (mucrones) followed by dorsal and ventral transverse ridges (baffles), which have significant taxonomic value as species-specific characteristics (Pilato 1972, Schuster *et al.* 1980a). Most eutardigrades have a completely rigid buccal tube. However, in some genera (the *Diphascon* assemblage, *Pseudodiphascon, Itaquascon*), there is a rigid anterior buccal tube between the buccal opening and the stylet supports, and a flexible, spiral-walled posterior pharyngeal tube from the stylet supports to the pharynx. In *Macrobiotus* and other genera, there is a median ventral lamina (sometimes called a buccal tube support or reinforcement rod) that extends from the buccal ring to the midregion of the tube.

Four basic types of buccal apparatus have been described based on the genera *Macrobiotus, Hypsibius, Diphascon*, and *Pseudodiphascon*. The *Macrobiotus*-type has a rigid buccal tube with a ventral lamina (reinforcement rod), whereas the *Hypsibius*-type has a rigid buccal tube but lacks a ventral lamina. The *Diphascon*-type has a rigid anterior portion and a flexible, annulated posterior portion; it also lacks a ventral lamina. The *Pseudodiphascon*-type has a buccal tube like *Diphascon* but has a ventral lamina. More types may be identified in the future as additional genera are described. The buccal apparatus was used by Schuster *et al.* (1980a) to revise the families of eutardigrades. Pilato (1982) stated his objections to the revision, and Ramazzotti and Maucci (1983) basically followed Pilato's classification scheme based on claw structure.

The stylets are paired, piercing structures lateral to the buccal tube. Composed of cuticular and calcified substances that dissolve in some mounting media, the stylets lie in the lumen of the salivary glands, which are dorsolateral to the pharynx. The salivary epithelium secretes the new buccal tube, stylets, and stylet supports after the tardigrade molts. The stylets and salivary secretions enter the anterior end of the buccal tube through openings ventral to the lateral flanges (stylet sheaths). Protractor and retractor muscles operate the stylets and are attached to the furca (condyle), the thickened base at the posterior end of each stylet. Protractor muscles extend from the furca to the buccal tube; retractor muscles insert on the pharynx. Some genera have apophyses for the insertion of the stylet protractor muscles on the buccal tube. In *Hypsibius*, the dorsal and ventral apophyses are hook-shaped; in *Isohypsibius* and other genera, they form dorsal and ventral crests. Pilato (1987) divided the genus *Diphascon* into four genera based on differences in the shape of the apophyses and other characteristics of the buccal–pharyngeal apparatus. The stylet support is a flexible lateral extension that attaches the

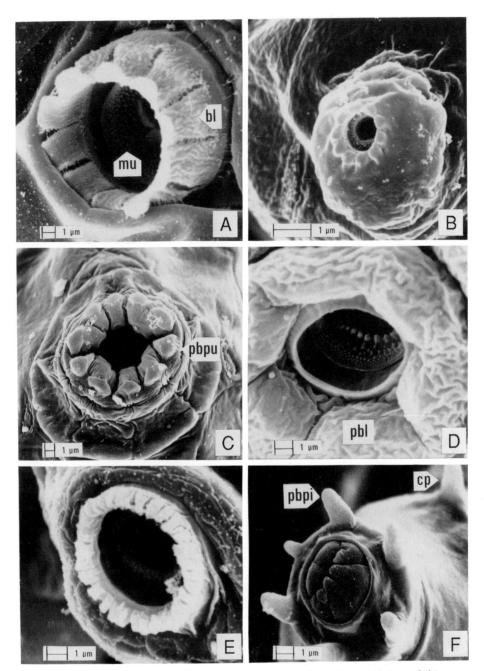

Figure 15.8 Scanning electron micrographs of eutardigrade buccal openings. (A) *Dactylobiotus*. mu, Buccal mucrones; bl, buccal lamella. (B) *Minibiotus*, buccal opening and papulae. (C) *Haplomacrobiotus*. pbpu, Peribuccal papulae. (D) *Isohypsibius*. pbl, Peribuccal lobe. (E) *Pseudobiotus*, buccal opening and buccal lamellae. (F) *Milnesium*. cp, Cephalic (lateral) papilla; pbpi, peribuccal papilla. (From Schuster *et al*. 1980a.)

furca of the stylet to the buccal tube, but it is poorly developed in the genus *Itaquascon*. In the heterotardigrades, the long, thin stylets may have delicate stylet supports or may be attached to the pharynx by a muscular band. In general, the buccal apparatus in heterotardigrades is of less taxonomic importance than in eutardigrades.

The buccal tube enters the muscular, tripartite pharyngeal bulb (sucking pharynx), which, in most eutardigrades, is lined with three double rows of cuticular thickenings called placoids. The placoids alternate in position with three equal cuticular apophyses, which are posterior to the end of the buccal tube. The larger, anterior placoids, called macroplacoids, are present in either two or three transverse rows. Smaller, posterior placoids are termed microplacoids; when present, they are arranged in a single row. In some species of the genus

Diphascon, a median septulum is present posterior to, and alternating in position with, the placoids. The number, size, and shape of the placoids are important species characters in the eutardigrades; however, placoids are reduced or absent in *Itaquascon, Milnesium,* and *Limmenius.*

3. Esophagus, Midgut, and Hindgut

The pharyngeal bulb is followed by a short esophagus, which is tripartite and lined with cuticle. A mucoid substance is produced by esophageal cells, but its function is unknown.

The midgut is derived from endoderm and is not lined with cuticle. In the Heterotardigrada, the midgut has five or six lateral diverticula; no excretory glands occur at the junction of the midgut and hindgut. Eutardigrades have a straight midgut, and three or four excretory glands empty into the midgut–hindgut junction. Food is digested in the midgut, which has a convoluted single layer of epithelial cells. The intestinal content varies in color depending on the type of food consumed, the state of the digestive process, and the nutritional status of the animal.

The hindgut, which is lined with cuticle, receives the contents of the midgut and is specialized for osmoregulation. In eutardigrades, the genital ducts also empty into the hindgut, which is therefore a true cloaca. In heterotardigrades, the anus and gonopore are separate openings.

D. Other Anatomical Systems

Reviews of the morphology of other tardigrade systems are found in Ramazzotti and Maucci (1983), Greven (1980), Bertolani (1982a), Nelson (1982a), and Nelson and Higgins (1990).

The nervous system consists of a dorsal lobed brain (cerebral ganglion), a circumesophageal connective around the buccal tube, a subesophageal ganglion, a circumpharyngeal connective around the pharynx, and four ventral ganglia joined by paired nerve tracts (Figs. 15.3 and 15.4). Many eutardigrades and some heterotardigrades have a pair of eye spots, cup-shaped structures with pigment granules, in the lateral lobes of the brain. Walz (1978) described four sensory areas that function as chemoreceptors or mechanoreceptors in the anterior head region of the eutardigrade *Macrobiotus:* the anteriolateral sensory field, the circumoral sensory field, the suboral sensory region, and the pharyngeal organ. The cephalic appendages in heterotardigrades also function as sense organs.

Tardigrade musculature has been studied in detail in only a few species. The muscular system consists of dorsal and ventral longitudinal fibers, transverse muscles for movement of the legs, stylet muscles, pharyngeal and intestinal muscles, and muscles for defecation and deposition of eggs. Muscles are attached to the cuticle by protrusions (apodemes) that are reformed at each molt. The somatic muscles are single elongate cells, a smooth-muscle type with some characteristics of obliquely striated muscle (Walz 1974, 1975).

According to Ramazzotti and Maucci (1983), excretion in tardigrades probably occurs in four ways: (1) through the salivary glands at molting; (2) through shedding the cuticle containing accumulated excretory granules; (3) through the wall of the midgut; and (4) through excretory glands (in eutardigrades).

Specialized respiratory and circulatory systems are absent. Respiration occurs through the cuticle. Circulation is accomplished by the movement of fluid and coelomocytes in the body cavity when the animal moves.

III. PHYSIOLOGY (LATENT STATES)

All tardigrades are aquatic, regardless of their specific habitat, since they require a film of water surrounding the body to be active. They are capable, however, of entering a latent state (cryptobiosis) when environmental conditions are unfavorable. Crowe (1975) identified five types of latency: encystment, anoxybiosis, cryobiosis, osmobiosis, and anhydrobiosis. When a tardigrade is in a latent state, metabolism, growth, reproduction, and senescence are reduced or cease temporarily, and resistance to environmental extremes such as cold, heat, drought, chemicals, and ionizing radiation is increased. Latent states thus have a significant impact on the ecological role of the organism.

Encystment commonly occurs in freshwater tardigrades living in permanent pools, although cysts can also be formed in tardigrades inhabiting soil and moss. The conditions for encystment are not fully known; however, Weglarska (1957) determined that a gradual worsening of the environment is one inducing factor for encystment in the freshwater tardigrade *Dactylobiotus dispar.* Specimens encysted regularly within 12 hr if placed in water containing cyanophytes or decomposing leaves, but a "swift worsening" of the environment caused the animals to go into an asphyxial state (anoxybiosis).

Before encystment, the tardigrade ingests large amounts of food for storage in body cavity cells. Proceeding through stages similar to molting, the animal first ejects the buccal apparatus and enters the simplex stage. After this process, the gut is emptied by defecation, the hindgut lining is shed, and the organism contracts within the cuticle, becoming immobile and barrel-shaped. A cyst membrane is formed and gradually becomes smooth and dark, replacing the old cuticle which sloughs off. The cyst is encased in strong walls, often opaque and encrusted with debris. Metabolism is very reduced in the cysts. Before emergence from the cyst when environmental conditions improve, a new cuticle is produced, including buccal apparatus and claws. Encystment is not obligatory in the life cycle of a tardigrade, and reasons for emergence are not fully known. The cysts are not able to withstand high temperatures, as in anhydrobiosis, because of the high water content; however, the cysts can survive for over a year in nature without depleting food reserves.

Anoxybiosis, or asphyxia, is a cryptobiotic state induced by low oxygen tensions in the environmental water. The tardigrade becomes immobile, transparent, rigid, and extended due to water absorption resulting from loss of osmotic control. The asphyxial state is temperature dependent; the time required for inducement varies from a few hours to several days. Some species can survive up to five days in anoxybiosis; however, strictly aquatic species live only from a few hours to three days in the asphyxial state. Prior to this time limit, specimens can be revived by the addition of oxygen to the water; return to the active state requires a few minutes to several hours. The metabolic rate of tardigrades in anoxybiosis is unknown.

Induced by low temperatures, cryobiosis is a type of cryptobiosis that enables tardigrades to survive freezing and thawing. Slow cooling rates and the formation of barrel-shaped tuns enhance survival, but little is known about the process of cryoprotection.

Osmobiosis is cryptobiosis that results from elevated osmotic pressures and decreased water activity at the surface of the animal. Placed in salt solutions of various ionic strengths, freshwater and terrestrial tardigrades form contracted tuns. The marine tardigrade *Echiniscoides sigismundi* becomes turgid in freshwater but can survive up to three days. Activity resumes when the animals are returned to water with the former osmotic concentrations.

Anhydrobiosis (formerly called anabiosis) is a type of cryptobiosis induced by evaporative water loss from terrestrial eutardigrades and echiniscids. [Nearly all marine and true freshwater tardigrades lack the ability to survive dehydration and undergo cryptobiosis, but experimental evidence is sparse.] As the environmental water surrounding the animal evaporates, the tardigrade contracts, retracting the head and legs and becoming the characteristic barrel-shaped tun. The rate of desiccation must be slow to ensure survival and return to active life with the addition of water. Survival of dehydration is correlated with synthesis of trehalose, a disaccharide of glucose, during dehydration or its degradation following rehydration. Trehalose alters the physical properties of membrane phospholipids and stabilizes dry membranes in anhydrobiotic organisms (Crowe *et al.* 1984). In the anhydrobiotic state, the tardigrade is highly resistant to extreme environmental factors (temperature, radiation, chemicals). Anhydrobiotic animals have remained viable on herbarium specimens for as long as 120 years. The eggs of terrestrial tardigrades can also withstand desiccation, but little is known about the eggs of the distinctly aquatic species. The lifespan of anhydrobiotic tardigrades can be greatly extended by going in and out of this cryptobiotic state.

IV. REPRODUCTION AND DEVELOPMENT

A. Reproductive Modes

Sexual reproduction, parthenogenesis, and hermaphroditism are reproductive modes in the Tardigrada (Ramazzotti and Maucci 1983, Nelson 1982a, Bertolani 1982a, 1982b, 1983a, 1983b, 1987a). Marine tardigrades are bisexual with sexual dimorphism evident in the gonopore structure. Nonmarine species, especially the eutardigrades, are gonochoristic (dioecious) with unisexual or bisexual populations, but external sexual dimorphism is rarely observed. Males are generally smaller than females; however, because sexual maturity may be reached before the maximum size is attained, mature males may be larger than immature females in the same population. Males may also exhibit modifications of the claws, especially on the first pair of legs, such as in *Milnesium* and the freshwater *Pseudobiotus augusti* and *P. megalonyx*. Meiotic development of the gametes occurs in tardigrades with separate sexes. In nonmarine tardigrades, unisexual populations are the most common type, and parthenogenesis is the mode of reproduction. Simultaneous hermaphroditism is sporadic but occurs in freshwater and terrestrial eutardigrades.

B. Reproductive Apparatus

A single (unpaired) gonad is present in both males and females (Figs. 15.3 and 15.4). This sac-like structure is dorsal to the intestine and attached to the body wall by two anterior ligaments in adult eutardigrades or by a single median ligament in heterotardigrades (Ramazzotti and Maucci 1983, Nelson 1982a). The morphology of the gonad varies with age, sex, species, and reproductive activity of the tardigrade. The male eutardigrade has two sperm ducts which open into the hindgut (cloaca); the female has only one oviduct on either the right or the left side of the intestine. A small seminal receptacle, which opens ventrally into the hindgut next to the oviduct opening, is present rarely in some females of *Macrobiotus* and *Hypsibius* (Marcus 1929). In eutardigrades, the cloacal opening is a ventral transverse slit anterior to the fourth pair of legs. In heterotardigrades, the preanal gonopore is rosette-shaped in females but is a protruding, round or oval-shaped aperture in males.

C. Mating and Fertilization

Tardigrade spermatozoa have been described in only a few species (see reviews in Baccetti 1987, Bertolani 1983b, Ramazzotti and Maucci 1983). Henneke (1911) first observed the presence of two kinds of sperm in the freshwater tardigrade *Dactylobiotus macronyx*. The ultrastructure of the spermatozoa of *Macrobiotus hufelandi* was first investigated by Baccetti *et al.* (1971). Other workers reported that the morphology of spermatozoa may be correlated with a specialized mode of sperm transfer. For example, in *Macrobiotus hufelandi*, thread-shaped spermatozoa with modified mitochondria are associated with internal fertilization; the female has a seminal receptacle that contains agglutinated sperm (Kristensen 1979, Ramazzotti and Maucci 1983). In the freshwater eutardigrade *Isohypsibius granulifer*, transfer of modified sperm occurs by copulation, and spermatozoa are found in the ovary of the female (Wolburg-Buchholz and Greven 1979). In some heterotardigrades, the spermatozoa are less specialized, the female has no seminal receptacle, and fertilization is external (Kristensen 1979).

Mating and fertilization have been observed in only a few species of tardigrades (Ramazzotti 1972, Nelson 1982a). In some freshwater species, one or more males deposit sperm into the cloacal opening of the old cuticle of the female prior to, or during, a molt. External fertilization occurs as the eggs are deposited in the old cuticle. In other species, internal fertilization occurs when sperm deposited in the cloaca fertilize eggs in the ovary.

D. Eggs

Eggs are often essential for the identification of freshwater and terrestrial species, especially in some genera of eutardigrades (Fig. 15.9). Orna-

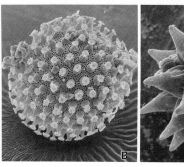

Figure 15.9 Scanning electron micrographs of *Macrobiotus* eggs. (A) *Macrobiotus areolatus*; (B) *Macrobiotus hufelandi*; (C) *Macrobiotus tonollii*.

mented eggs, characteristic of *Macrobiotus, Dactylobiotus, Amphibolus,* and some *Hypsibius* species, are usually deposited freely and singly or in small groups. The ornamentation may include pores, reticulations, and processes. The different types of egg ornamentations and their taxonomic significance have been discussed by various authors (Grigarick *et al.* 1973, Toftner *et al.* 1975, Greven 1980, Ramazzotti and Maucci 1983). In some genera, smooth oval eggs with a thin shell are usually deposited in the molted cuticle. These smooth eggs are typical of the armored heterotardigrades, most eutardigrades in the genera *Pseudobiotus, Hypsibius, Isohypsibius,* and *Milnesium,* and only a few of the *Macrobiotus* species. Some aquatic tardigrades deposit eggs inside the empty exoskeletons of cladocerans, ostracods, or insects. The aquatic eutardigrades *Isohypsibius annulatus* and two species of *Pseudobiotus* carry the exuvium containing eggs attached to the body for some time. The method of egg deposition is unknown for many species, especially in marine tardigrades.

E. Parthenogenesis

Parthenogenesis, a common mode of reproduction in nonmarine tardigrades, is correlated with the ability to undergo cryptobiosis (Bertolani 1982b). According to Kristensen (1987), in echiniscid heterotardigrades, males have not been recorded in *Echiniscus, Bryochoerus, Parechiniscus,* and many *Pseudechiniscus* species; males are rare in *Bryodelphax, Testechiniscus, Cornechiniscus,* and *Mopsechiniscus* but common (50% of the population) in *Oreella, Hypechiniscus, Proechiniscus,* and *Antechiniscus.* In the absence of males in a tardigrade population, parthenogenesis is assumed to be the mode of reproduction. Dastych (1987a), however, reported the presence of males in four species of the genus *Echiniscus,* thus negating the hypothesis that the genus is entirely parthenogenetic. R. O. Schuster (personal communication) has observed males in many species of *Echiniscus.* Grigarick *et al.* (1975) observed two types of gonopores and published the first scanning electron micrograph of a male gonopore of a heterotardigrade originally determined as *Echiniscus oihonnae* but later described as a new species, *Echiniscus laterculus* (Schuster *et al.* 1980b). Nelson (1982a) stated, "Whether the difference is due to an alteration following oviposition or to the presence of males is unknown. If males were indeed present, this would be the first observation of males in any *Echiniscus* species..." Based on the presence of ventral plates, Dastych (1987a) placed the species studied by Grigarick *et al.* (1975) in Kristensen's (1987) new genus *Testechiniscus,* in which males are rare.

Polyploidy is often associated with parthenogenesis and is frequently found in eutardigrades in various genera that inhabit different microhabitats (Nelson 1982a). Meiotic (automictic) parthenogenesis occurs in diploid biotypes of the freshwater eutardigrades *Dactylobiotus dispar* and *Hypsibius dujardini.* Ameiotic (apomictic) parthenogenesis occurs in triploid and tetraploid biotypes of some freshwater and terrestrial eutardigrades (*Macrobiotus, Hypsibius, Pseudobiotus*). Some species undergo sexual reproduction when males are present in the spring and parthenogenesis when males are absent in other seasons. According to Pilato (1979), ". . . many species in which males do occur are actually an association of amphigonic with one or more parthenogenetic, generally polyploid and reproductively isolated strains, often distinguishable by their mechanism of egg deposition."

F. Hermaphroditism

Most eutardigrade populations are unisexual (thelytokus) although some are bisexual. Hermaphroditism is rare in eutardigrades and unknown in heterotardigrades (Bertolani 1987a). In hermaphrodites, a single ovotestis with only one gonoduct is present.

Both mature and almost mature spermatozoa and oocytes may be found in a single individual, indicating the possibility of self-fertilization. Cyclic maturation probably occurs, with the initial prevalence of one type of gamete and then another. Hermaphroditic eutardigrades include *Parhexapodibius pilatoi* from grasslands, *Macrobiotus joannae, Amphibolus weglarskae* from leaf litter, and four species of *Isohypsibius* from freshwater (Bertolani and Manicardi 1986).

G. Development

Developmental time, from deposition of the egg to hatching, varies considerably within and between species and with environmental conditions. A minimum of five to a maximum of forty days have been required for egg development under experimental conditions (Marcus 1929). Eggs can be found at any time of year in mosses, lichens, and soil; however, few observations of eggs of freshwater tardigrades have been reported (Ramazzotti and Maucci 1983).

When embryonic development is complete, the immature tardigrade emerges from the egg by the action of the stylets or hindlegs or by an increase in hydrostatic pressure (Ramazzotti and Maucci 1983). Postembryonic development occurs through several successive molts, primarily by the growth of the individual cells rather than by cell division. Although tardigrades have been considered cell constant, Bertolani (1970, 1982b) observed mitosis in eutardigrades, but not in heterotardigrades.

In eutardigrades, development is direct and the first-instar animals (juveniles) are similar to the adults but have immature gonads and slight differences in the claws and buccal apparatus. Bertolani *et al.* (1984) concluded that heterotardigrades undergo indirect development characterized by two larval stages. The first larval stage lacks an anus and a gonopore and has two claws less on each leg than the adult. The second larval stage (sometimes called a juvenile) has an anus and the same number of claws as the adult but the gonopore is rudimentary or lacking. The number of filamentous and spinous cuticular appendages generally increases with age (but decreases in *Mopsechiniscus*), although variations may also occur (Grigarick *et al.* 1975, Ramazzotti and Maucci 1983, Nelson 1982a).

V. ECOLOGY

A. Molting and Life History

Molting, which usually requires 5–10 days, occurs periodically throughout the life of the tardigrade (Walz 1982). The entire cuticular lining of the foregut, including the stylets and stylet supports, is ejected through the expanded buccal opening. The mouth opening closes and the animal cannot feed. This is the simplex stage, characterized by the absence of the buccal apparatus. The salivary glands reform the cuticular structures of the buccal apparatus; the esophagus regenerates its own cuticular lining. Concomitantly, new body cuticle, including the hindgut lining, is synthesized by the underlying epidermis and new claws are produced by claw glands in the legs (Walz 1982). During the molt, the old body cuticle is shed, including the claws and the lining of the hindgut (Fig. 15.10). Generally, body length increases with each molt until maximum size is attained, although lack of food can result in a decrease in size. Although growth is more rapid during the earlier molts, molting and growth continue even after sexual maturity (Nelson 1982a).

Life histories of certain tardigrade species have been reviewed by Walz (1982), Ramazzotti and Maucci (1983), and Nelson (1982a). Based on frequency distributions of body length and buccal tube length, the number of molts ranges from four to twelve. Problems inherent in the method have been discussed by Baumann (1961), Morgan (1977), and Ramazzotti and Maucci (1983). Considerable variation and overlapping of the stages may occur within a species or due to environmental conditions. Sexual maturity is reached with the second or third molt, and egg production continues throughout life, although molting can occur without egg production.

The lifespan of active tardigrades (excluding cryptobiotic periods) has been estimated to be from 3–30 months (Higgins 1959, Franceschi *et al.* 1962–1963, Pollock 1970, Ramazzotti and Maucci 1983). The total lifespan, from hatching until death, may be

greatly extended by periods of latent life (encystment and anhydrobiosis). Since aquatic tardigrades have little or no ability to undergo anhydrobiosis, the period of active life generally corresponds to that of total life. Ramazzotti and Maucci (1983) concluded that freshwater species of *Macrobiotus* and *Hypsibius* live about 1–2 yr; whereas moss-inhabiting species of the same genera average 4–12 yr. Shorter lifespans of 3–4 months were proposed by Pollock (1970) for marine species and Franceschi *et al.* (1962–1963) for moss-inhabiting species. Morgan (1977) estimated a lifespan of 3–7 months for *Macrobiotus hufelandi* and up to 3 months for *Echiniscus testudo* in roof mosses.

B. Habitats

Although active individuals require water, the environments in which tardigrades live are generally divided into: (1) marine and brackish water; (2) freshwater; and (3) terrestrial habitats. The distinction between freshwater and terrestrial species is sometimes unclear because certain tardigrades can live in a wide range of habitats.

With few exceptions, marine tardigrades belong to the class Heterotardigrada, either in the order Arthrotardigrada or the order Echiniscoidea (Table 15.1). In the class Eutardigrada, only five species of the genus *Halobiotus* (formerly *Isohypsibius*) and two species of *Isohypsibius* are marine (Crisp and Kristensen 1983). Studies on marine tardigrades are relatively few but indicate high morphological diversity, with most of the genera having one or two species (Renaud-Mornant 1982, 1987). Three major ecological groups are recognized:

1. a small number of species on intertidal algae or other substrates including facultative and obligatory invertebrate ectoparasites and commensals;
2. a larger group of species inhabiting the interstitial biotype;
3. a highly diversified group of abyssal and deep-sea ooze dwellers.

Hydrophilous tardigrades are "distinctly aquatic" species that live only in permanent freshwater habitats. Although they are occasionally found in plankton samples, they are benthic organisms that crawl on the surfaces of aquatic plants or in the interstitial spaces of sediments in ponds, lakes, rivers, and streams. Most species live in the littoral zone; however, some individuals have been collected from lakes up to 150 m deep (Ramazzotti and Maucci 1983). Hygrophilous tardigrades, which inhabit moist mosses, and eurytopic species, which

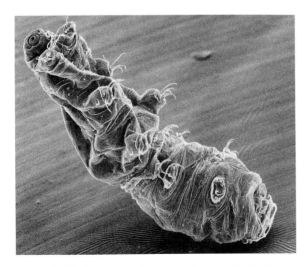

Figure 15.10 Scanning electron micrograph of *Pseudobiotus augusti* undergoing a molt.

live in a wide range of moisture conditions, are often found in aquatic habitats.

In the Heterotardigrada, *Styraconyx hallasi* is the only freshwater arthrotardigrade. In the order Echiniscoidea, *Carphania fluviatilis* in the family Carphanidae (Binda and Kristensen 1986), and a few species of *Echiniscus, Hypechiniscus,* and *Pseudechiniscus* in the family Echiniscidae are occasionally limnic (Dastych 1987a, Kristensen 1987). The single specimen of *Echinursellus longiunguis,* originally described as an arthrotardigrade from the shores of Lago Chungara on the Altiplano of Chile, is a cyst or cyclomorphic stage of a eutardigrade (Kristensen 1987). The genus *Echinursellus,* formerly considered the missing link between Arthrotardigrada and Echiniscoidea, is designated as a junior synonym of the eutardigrade genus *Pseudobiotus,* establishing the new combination *Pseudobiotus longiunguis.* The status of the mesotardigrade *Thermozodium esakii* from a Japanese hot spring is dubious.

The class Eutardigrada contains both terrestrial and freshwater species (Table 15.1). Three genera (*Dactylobiotus, Pseudobiotus,* and *Thulinia*) are exclusively freshwater (Fig. 15.11). Other genera have one or more species that are typically aquatic, i.e., *Macrobiotus, Amphibolus, Doryphoribius, Isohypsibius, Hypsibius, Itaquascon,* and the *Diphascon* group (Bertolani 1982a). Several species of the genus *Hypsibius,* e.g., *H. dujardini* and *H. convergens,* are not obligate freshwater organisms but are often found in this habitat. *Milnesium tardigradum* and *Macrobiotus hufelandi* are cosmopolitan eurytopic terrestrial species that also inhabit freshwater. The tardigrade component of the psammon community in lakes and river banks is poorly known, but it consists primarily of hydrophilous and hygrophilous species that show a distinct seasonality in population densities, with peak periods mainly in the winter and spring (Ramazzotti and Maucci 1983).

The largest number of tardigrade species are terrestrial and live in moist habitats, such as in soil and leaf litter or among mosses, lichens, liverworts, and cushion-shaped flowering plants. Most terrestrial tardigrades inhabit the moss environment, which can be divided into three groups: wet, moist, and dry mosses (Ramazzotti and Maucci 1983). Wet mosses such as submerged and emergent species in lakes, ponds, and streams never dehydrate completely. In contrast, moist and dry mosses living on tree trunks, rocks, walls, roofs, and soil are subject to total desiccation for various periods of time and are dependent on atmospheric precipitation for moisture. The typical terrestrial habitat of a tardigrade has: (1) sufficient aeration (tardigrades are hypersensitive to low oxygen levels); (2) alternate periods of wet and dry conditions; and (3) suitable and sufficient food. Tardigrades are less likely to inhabit impermeable clay soils or dense cushion mosses with thick cell walls. A distinct insular distribution of tardigrades within patches of mosses and soil has been frequently noted (Ramazzotti and Maucci 1983).

C. Population Density

Population densities of tardigrades are highly variable; however, neither minimal nor optimal conditions for population growth are known. Changes in tardigrade population densities have been correlated with a variety of environmental conditions, including temperature and moisture (Franceschi *et al.* 1962–1963, Morgan 1977) and food availability (Hallas and Yeates 1972). Other factors such as competition, predation, and parasitism may also play a role. Predators include nematodes, other tardigrades, mites, spiders, and insect larvae; parasitic protozoa and fungi often infect tardigrade populations (Ramazzotti and Maucci 1983). Considerable variation in adjacent, seemingly identical, microhabitats occurs in both population density and species diversity. Usually 2–6 species are present, but 10 or more species have been found in a single sample.

Populations of tardigrades from soil and leaf litter habitats range from 5–40/cm^2 (see Nelson and Higgins 1990 for review). Similarly, population densities in mosses and other terrestrial habitats vary widely, from 0 to 50–200/cm^2 (for review see Ramazzotti and Maucci 1983). Large populations of both heterotardigrades and eutardigrades are often found in thin, spreading mosses growing in open, exposed places; in contrast, very few heterotardigrades but many eutardigrades can be collected in

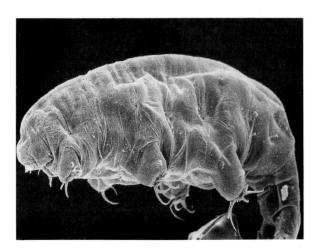

Figure 15.11 Scanning electron micrograph of an aquatic tardigrade, *Pseudobiotus augusti.*

mosses from moist shady areas. Studies of the population dynamics of freshwater tardigrades are less common (Schuster *et al.* 1977, Wainberg and Hummon 1981, Kathman and Nelson 1987, Nelson *et al.* 1987). Ramazzotti and Maucci (1983) and Nelson *et al.* (1987) reported that freshwater eutardigrades often reach a very high population density in the spring, whereas Schuster *et al.* (1977) and Wainberg and Hummon (1981) noted population peaks in the fall. Kathman and Nelson (1987) stated: "Several previous life-history studies indicated wide variation in total numbers of animals obtained from month to month, but few studies extended beyond 12 months or presented data on replicates, means, or standard errors for determining the validity of these variations. Unless there are recurring patterns over several years, abundance estimates may be invalid if adequate numbers of quantitative replicates are not obtained, as our data show that within-period variation was often greater than between-period variation."

D. Distribution

Due to the paucity of data, the biogeography of the Tardigrada remains unknown. As a phylum, they have a worldwide distribution. Many species with broad ecological requirements are considered to be cosmopolitan, whereas others with more narrow tolerances are rare or endemic. Moss-inhabiting tardigrades are more common in polar and temperate zones than in the tropics. Arctic and Antarctic species live in algal mats in freshwater lakes and pools as well as in terrestrial moss turfs (McInnes and Ellis-Evans 1987, Dastych 1984, Usher and Dastych 1987, Bertolani and Kristensen 1987, Crisp and Kristensen 1983). Deep-sea tardigrades have been reported from abyssal and bathyal depths as well as shallower waters off the continental shelf and the intertidal zone (Renaud-Mornant 1982, 1987). Significant differences in altitudinal distributions of tardigrades have been reported by Nelson (1975), Dastych (1987b), Manicardi and Bertolani (1987), and Ramazzotti and Maucci (1983). More data are required, however, before generalizations can be made about the biogeographical distributions of the phylum.

E. Dissemination

Eggs, cysts, and barrel-shaped anhydrobiotic specimens of terrestrial tardigrades are dispersed predominantly by wind, although rain, floodwaters, and melting snow can transport active as well as cryptobiotic forms (Ramazzotti and Maucci 1983). Other animals, associated with terrestrial communities, e.g., birds, snails, isopods, mites, and insects, can also assist in the dispersal of tardigrades. Active dissemination by crawling is limited by the necessity for a constant film of water covering the body of the tardigrade and by its size and speed of movement.

Little information is available on dispersal of the distinctly aquatic species, which probably do not undergo cryptobiosis. Transport of eggs or encysted forms from dry temporary ponds is possible. Active tardigrades may be distributed through river systems, particularly during periods of heavy rainfall and flooding. The mechanism for repopulation of aquatic habitats after the summer disappearance of tardigrades or after habitat disturbances has not been investigated (Nelson *et al.* 1987).

VI. TECHNIQUES FOR COLLECTION, EXTRACTION, AND MICROSCOPY

Qualitative collections of tardigrades from aquatic habitats with sandy bottoms can be made by stirring the sand in a container of water and decanting immediately after the sand settles (Schuster *et al.* 1977). The decanted water is passed through a sieve (U.S. Standard No. 325) with pores of about 44 μm, and the specimens are rinsed from the screen to a sample jar. Tardigrades in large volumes of water can be preserved by the addition of formalin (5%) or gluteraldehyde.

Collections of tardigrades from vegetation can be made by removing the plant (moss, lichen, liverwort, etc.) from the substrate (tree, rock, soil) and placing the sample in a small paper bag for dry storage until processing. The sample is then placed in a container of water for several hours to reverse anhydrobiosis. The sample is agitated and squeezed to remove specimens, eggs, and cysts. After the excess water is decanted, the remaining sediment containing the live tardigrades can be examined with a dissecting microscope at a magnification of 30X or greater. Boiling water or boiling alcohol (85% or greater) added to specimens in small amounts of water is adequate for tardigrade fixation. Alcohol is essential to preserve specimens to be mounted later. The sample may also be washed through sieves to remove particles smaller than 50 μm or larger than 1200 μm. R. D. Kathman (personal communication) found that mounting the live tardigrade directly in Hoyer's medium was more efficient and produced a better slide preparation.

Core samples of soil and leaf litter can also be placed in paper bags for dry storage and processed according to the traditional procedure for vegeta-

Table 15.2 Comparison of Systematic Arrangements of the Eutardigrada[a]

Pilato[b]		Schuster *et al.* [c]	
Family	**Genera**	**Family**	**Genera**
Macrobiotidae	*Macrobiotus*	Macrobiotidae	*Macrobiotus*
	Minibiotus		*Minibiotus*
	Dactylobiotus		*Dactylobiotus*
	Pseudodiphascon		*Pseudodiphascon*
			Haplomacrobiotus
			Hexapodibius
			Parhexapodibius
			Doryphoribius
Calohypsibiidae	*Haplomacrobiotus*		
	Hexapodibius		
	Parhexapodibius		
	Calohypsibius		
Hypsibiidae	*Doryphoribius*	Hypsibiidae	*Isohypsibius*
	Isohypsibius		*Hypsibius*
	Pseudobiotus		*Pseudobiotus*
	Hypsibius		*Calohypsibius*
	Diphascon		*Necopinatum*
	Itaquascon		*Diphascon*
			Itaquascon
Milnesiidae	*Milnesium*	Milnesiidae	*Milnesium*
	Limmenius		*Limmenius*
Incertae sedis	*Necopinatum*		

[a]Based on Pilato 1982.
[b]Pilato 1969, 1982.
[c]Schuster *et al.* 1980a.

tion. The sample is thoroughly soaked in water for several hours and examined in small quantities of thin aqueous suspensions since eggs and cysts are easily overlooked. The technique is extremely laborious and time-consuming.

Modifications of techniques to extract soil nematodes by centrifugation in water have been used to remove tardigrades from aquatic and terrestrial habitats (Hallas 1975, Hallas and Yeates 1972, Bertolani 1982a). A watery suspension of the sample is prepared and filtered. The portion containing particles of the appropriate size range is centrifuged with water or a sucrose solution. The supernatant is filtered and the residue washed into a petri dish for examination with a dissecting microscope.

After fixation, individual specimens are mounted on microscope slides and observed with phase microscopy. Hoyer's mounting medium produces cleared, distended specimens, but the coverslip must be sealed to ensure permanence. The amount of chloral hydrate may be reduced to decrease the extent of clearing of specimens. CMC and CMCP media are useful for temporary mounts; however, excessive stain may obscure important structures, and the specimens may clear completely after a period of time. Other media used include Faure's medium and polyvinyl lactophenol. More perma-

nent mounts with balsam or synthetic resins are not as useful for phase microscopy. Kristensen and Higgins (1984a) recommend transferring animals with a drop of 2% formalin to slides and covering with a coverslip. The formalin preparation is infused with a 10% solution of glycerin and 90% ethyl alcohol and allowed to evaporate to glycerin over several days. Then the coverslip is sealed with Murrayite.

For scanning electron microscopy (SEM), fixed tardigrades are dehydrated to absolute alcohol, transferred to an intermediary liquid such as amyl acetate, and dried by the critical point method (Grigarick *et al.* 1975). Individuals are mounted on SEM stubs and coated with 200–300 nm of gold or gold–palladium.

VII. IDENTIFICATION

Identification of species in the phylum Tardigrada is based primarily on the morphology of the cuticle, claws, buccal apparatus, and eggs, which have been discussed in previous sections. Early European workers described the majority of tardigrade species from temporary wet mounts of live specimens

Table 15.3 Systematic Arrangement of the Eutardigrada[a]

Order	Family	Genera
Parachela	Macrobiotidae	*Macrobiotus*
		Pseudodiphascon
		Dactylobiotus
		Adorybiotus
		Haplomacrobiotus
	Calohypsibiidae	*Calohypsibius*
		Hexapodibius
		Microhypsibius
		Eohypsibius
	Hypsibiidae	*Hypsibius*
		Doryphoribius
		Isohypsibius
		Pseudobiotus
		Itaquascon
		Diphascon
	Amphibolidae	*Amphibolus*
	Necopinatidae	*Necopinatum*
Apochela	Milnesiidae	*Milnesium*
		Limmenius

[a]Based on Ramazzotti and Maucci 1983.

observed with the light microscope, often resulting in inadequate, incorrect, or incomplete descriptions and illustrations. Morphology, systematics, and natural history were focal points of their studies. Ramazzotti (1962, 1972) published the definitive monograph, in Italian, which provides keys and illustrated diagnoses of the species, thorough sections on morphology and biology, and an extensive bibliography. A supplement was produced in 1974, and a revision of the monograph was published in 1983 by Ramazzotti and Maucci. Earlier monographs by Marcus (1936) in German and Cuénot (1932) in French formed the basis of tardigrade systematics for many years. Useful references in English include those by Pennak (1979), Morgan and King (1976), Nelson (1982a), and the four international symposia volumes (Higgins 1975, Weglarska 1979, Nelson 1982b, Bertolani 1987b).

The system of classification presented in Ramazzotti's monograph (1972) reflected previous changes in tardigrade systematics and was generally accepted, although the need for a thorough revision was recognized. Pilato (1969) revised the systematics of the Eutardigrada, emphasizing claw structure for his classification of the genera and families. Subsequently, Schuster *et al.* (1980a) presented another revision of Eutardigrada systematics based on a multicharacter analysis of the complex buccal apparatus, and they proposed two orders (Parachela and Apochela), three families (Macrobiotidae, Hypsibiidae, and Milnesiidae), and three new genera (*Dactylobiotus*, *Minibiotus*, and *Pseudobiotus*). Pilato (1982) compared his proposal with that of Schuster *et al.* (1980a), stating his reasons for disagreement (Table 15.2). Ramazzotti and Maucci (1983) incorporated parts of both schemes, retaining the orders Parachela and Apochela and designating 6 families and 19 genera based primarily on claw structure rather than buccal apparatus (Table 15.3). Figures 15.5–15.8 illustrate the claw types and buccal apparatus of genera of eutardigrades, basically according to the classification system of Ramazzotti and Maucci but with the inclusion of *Minibiotus* and other new genera. The figures illustrate the range of variation in morphological structures that are considered of taxonomic importance in defining genera and also the need for further analysis.

The following key includes genera previously found and likely to be found in freshwater and terrestrial habitats. Positive identification with the key may be difficult because of inadequate generic descriptions and the need for further revision of tardigrade systematics.

A. Taxonomic Key to Genera of Freshwater and Terrestrial Tardigrada

1a.	Lateral cirri A present	2
1b.	Lateral cirri A absent	4
2a(1a).	Cuticle without dorsal plates	3
2b.	Cuticle with complete or partial dorsal plates; terrestrial .... family Echiniscidae
	[*Bryochoerus, Bryodelphax, Cornechiniscus, Echiniscus,*	
	Hypechiniscus, Mopsechiniscus, Parechiniscus, Pseudechiniscus (see	
	Kristensen 1987 for revision).]	
3a(2a).	Body indistinctly divided into eight segments; clavae present; four claws	
	on all legs in adults; terrestrial .... *Oreella*
3b.	Body eutardigrade-like in shape; clavae absent; two claws on legs I–III,	
	one claw on legs IV; freshwater .... *Carphania*

4a(1b).	Head with cephalic papillae	5
4b.	Head without cephalic papillae	6
5a(4a).	Head with six peribuccal papillae and two lateral papillae; terrestrial, occasionally in freshwater .... *Milnesium*	
5b.	Head without peribuccal papillae; two lateral papillae present; terrestrial .... *Limmenius*	
6a(4b).	Claws absent on all legs; legs very short; terrestrial (Binda, 1984) .... *Apodibius*	
6b.	Claws present on at least one pair of legs	7
7a(6b).	Single unbranched claws without secondary branches	8
7b.	Double claws, with primary and secondary branches	9
8a(7a).	One unbranched conical claw on leg I, no claws on legs II, III, IV; terrestrial .... *Necopinatum*	
8b.	Two unbranched claws on legs I–III, small basal spur-like claws on leg IV; terrestrial .... *Haplomacrobiotus*	
9a(7b).	Two double claws on each leg similar in size and shape and symmetrical with respect to the median plane of the leg; claw branch sequence is 2-1-1-2 (secondary, primary, primary, secondary)	10
9b.	Two double claws on each leg usually dissimilar in size and shape and asymmetrical with respect to the median plane of the leg; claw branch sequence is 2-1-2-1 (secondary, primary, secondary, primary)	14
10a(9a).	Buccal tube of distinctly greater length between stylet supports and placoids than between stylet supports and buccal opening; posterior part of tube of spiral composition; terrestrial .... *Pseudodiphascon*	
10b.	Buccal tube between stylet supports and placoids not longer than distance between stylet supports and buccal opening; tube without spiral composition	11
11a(10b).	Stylet supports attach near middle of buccal tube; with ten peribuccal papulae (not lamellae); terrestrial .... *Minibiotus*	
11b.	Stylet supports attach in posterior half of buccal tube	12
12a(11b).	Branches of claws at right angles so that tips are remote; two claws structurally connected at base; claws without lunules; cuticle without pores; freshwater .... *Dactylobiotus*	
12b.	Both branches of claw follow similar axis so that tips are close together; two claws not structurally connected at base; claws with lunules; cuticle with pores	13
13a(12b).	Ventral lamina (reinforcement rod, buccal tube support) on buccal tube well developed; no anterior crests on buccal tube; ten peribuccal lamellae; terrestrial or freshwater .... *Macrobiotus*	
13b.	Ventral lamina narrow or absent; unequal dorsal and ventral crests on buccal tube; lamellae unknown; terrestrial .... *Adorybiotus*	
14a(9b).	Buccal tube with median longitudinal anteroventral lamina (ventral lamina or buccal tube support)	15
14b.	Buccal tube without median longitudinal anteroventral lamina (ventral lamina or buccal tube support)	16
15a(14a).	Legs IV reduced or absent, with no claws or one or two minute double claws; terrestrial .... *Hexapodibius*	
15b.	Legs IV normally developed; terrestrial and freshwater .... *Doryphoribius*	
16a(14b).	Claw of *Calohypsibius*-type, primary and secondary branches rigidly connected, claws equal in size	17
16b.	Claw with two or three distinct parts	18
17a(16a).	Cuticle sculptured, two transverse rows of macroplacoids; terrestrial .... *Calohypsibius*	
17b.	Cuticle smooth; three transverse rows of macroplacoids; terrestrial .... *Microhypsibius*	

18a(16b).   Each double claw with three distinct parts 19
18b.   Each double claw with two distinct parts 20

19a(18a).   Posterior part of buccal tube of spiral composition; 14 (16?) lamellae
             present; terrestrial .... *Eohypsibius*
19b.   Buccal tube rigid, without spiral composition; 14 lamellae present;
             terrestrial or freshwater .... *Amphibolus*

20a(18b).   Buccal tube rigid, without posterior part of spiral composition 21
20b.   Buccal tube with posterior part of spiral composition 25

21a(20a).   Internal and external claws of dissimilar size and shape, with secondary
             branch continuous with the base; no lunules; no peribuccal
             lamellae 22
21b.   Internal and external claws of similar size and shape, with secondary
             branch forming right angle with base; lunules and/or peribuccal lamellae
             present or absent 23

22a(21a).   Claws of *oberhaeuseri* type: *Hypsibius*-type with internal and external
             claws of very dissimilar size and shape, with primary branch of external
             claw very long and narrow, with very thin flexible connection to upper
             basal portion; elliptical dorsolateral sense organs on head; terrestrial
             (Binda and Pilato 1986) .... *Ramazzottius*
22b.   Claws of *dujardini* type: *Hypsibius*-type with external claw slightly
             larger than internal claw, with primary branch attached to median basal
             portion of claw; no sense organs on head; terrestrial or
             freshwater .... *Hypsibius*

23a(21b).   Dorsal and ventral crests on buccal tube absent; peribuccal lamellae
             and/or lunules rarely present; cuticle smooth or with granulations and/or
             gibbosities; terrestrial or freshwater .... *Isohypsibius*
23b.   Dorsal and ventral crests on buccal tube present 24

24a(23b).   With 12 peribuccal lamellae; freshwater .... *Thulinia*
24b.   With 30 peribuccal lamellae; freshwater .... *Pseudobiotus*

25a(20b).   Pharynx with placoids and stylet supports reduced or absent; terrestrial
             or freshwater .... *Itaquascon*
25b.   Pharynx with normal placoids; buccal tube with obvious stylet supports;
             terrestrial or freshwater; (formerly *Diphascon;* revised in Pilato
             1987) 26

26a(25b).   Apophyses for insertion of stylet muscles in the shape of a
             "hook" 27
26b.   Apophyses for insertion of stylet muscles in the shape of wide, flat
             "ridges" 28

27a(26a).   Stylet apophyses in the shape of a semilunar hook; pharyngeal tube
             longer than the buccal tube; with (subgenus *Diphascon*) or without
             (subgenus *Adropion*) drop-like thickening on buccal tube below insertion
             of stylet supports .... *Diphascon*
27b.   Stylet apophyses in the shape of a blunt hook; pharyngeal tube shorter
             than the buccal tube .... *Hebesuncus*

28a(26b).   Stylet furcae with posteriolateral processes thickened at
             apices .... *Mesocrista*
28b.   Stylet furcae with posteriolateral processes spoon-like and tapering at
             their apices .... *Platicrista*

LITERATURE CITED

Baccetti, B. 1987. The evolution of the sperm cell in the phylum Tardigrada (Electron microscopy of Tardigrades. 5). Pages 87–91 *in:* R. Bertolani, editor. Biol-

ogy of Tardigrades. Selected Symposia and Monographs U.Z.I., 1. Mucchi, Modena, Italy.

Baccetti, B., F. Rosati, and G. Selmi. 1971. Electron microscopy of Tardigrades. 4. The spermatozoon. Monitore Zoologico Italiano 5:231–240.

Baumann, H. 1961. Der Lebensablauf von *Hypsibius (H.)*

convergens Urbanowicz (Tardigrada). Zoologischer Anzeiger 167:362–381.

Bertolani, R. 1970. Mitosi somatiche e costanza cellulare numerica nei Tardigradi. Rendiconti Accademia Nazionale Lincei, Ser. VIII, 48:739–742.

Bertolani, R. 1982a. 15. Tardigradi. Guide per il riconoscimento delle specie animali delle acque interne Italiane. Consiglio Nazionale Delle Ricerche, Verona, Italy. 104 pp.

Bertolani, R. 1982b. Cytology and reproductive mechanisms in tardigrades. Pages 93–114 *in:* D. Nelson, editor. Proceedings of the Third International Symposium on the Tardigrada. East Tennessee State University Press, Johnson City.

Bertolani, R. 1983a. 17. Tardigrada. Pages 431–441 *in:* K. G. and R. G. Adiyodi, editors. Reproductive biology of invertebrates. Vol. I: Oogenesis, oviposition, and oosorption. Wiley, New York.

Bertolani, R. 1983b. 19. Tardigrada. Pages 387–396 *in:* K. G. and R. G. Adiyodi, editors. Reproductive biology of invertebrates. Vol. II: Spermatogenesis and sperm function. Wiley, New York.

Bertolani, R. 1987a. Sexuality, reproduction, and propagation in tardigrades. Pages 93–101 *in:* R. Bertolani, editor. Biology of Tardigrades. Selected Symposia and Monographs U.Z.I., 1. Mucchi, Modena, Italy.

Bertolani, R., editor. 1987b. Biology of Tardigrades (Proceedings of the Fourth International Symposium on the Tardigrada). Selected Symposia and Monographs U.Z.I., 1. Mucchi, Modena, Italy.

Bertolani, R., and R. Kristensen. 1987. New records of *Eohypsibius nadjae* Kristensen, 1982, and revision of the taxonomic position of two genera of Eutardigrada (Tardigrada). Pages 359–372 *in:* R. Bertolani, editor. Biology of Tardigrades. Selected Symposia and Monographs U.Z.I., 1. Mucchi, Modena, Italy.

Bertolani, R., and G. C. Manicardi. 1986. New cases of hermaphroditism in tardigrades. International Journal of Invertebrate Reproduction 9:363–366.

Bertolani, R., S. Grimaldi de Zio, M. D'Addabbo Gallo, and M. R. Morone De Lucia. 1984. Postembryonic development in heterotardigrades. Monitore Zoologico Italiano 18:307–320.

Binda, M. G. 1984. Notizie sui Tardigradi dell'Africa Meridonale con descrizione di una nuova specie di *Apodibius* (Eutardigrada). Animalia 11:5–15.

Binda, M. G., and R. Kristensen. 1986. Notes on the genus *Oreella* (Oreellidae) and the systematic position of *Carphania fluviatilis* Binda, 1978 (Carphanidae fam. nov., Heterotardigrada). Animalia 13(1/3):9–20.

Binda, M. G., and G. Pilato. 1986. *Ramazzottius,* nuovo genere di Eutardigrado (Hypsibiidae). Animalia 13:159–166.

Crisp, M., and R. Kristensen. 1983. A new marine interstitial eutardigrade from East Greenland, with comments on habitat and ecology. Videnskabelige Meddeleser fra Dansk Naturhistorisk Forening 144:99–114.

Crowe, J. 1975. The physiology of cryptobiosis in tardigrades. Memorie dell'Istituto Italiano di Idrobiologia 32(Suppl.):37–59.

Crowe, J., L. Crowe, and D. Chapman. 1984. Preservation of membranes in anhydrobiotic organisms: the role of trehalose. Science 223:701–703.

Cuénot, L. 1932. Tardigrades. In: P. Lechevalier, editor. Faune de France 24:1–96.

Dastych, H. 1984. The Tardigrada from Antarctic with description of several new species. Acta Zoologica Cracoviensia 27:377–436.

Dastych, H. 1987a. Two new species of Tardigrada from the Canadian Subarctic with some notes on sexual dimorphism in the family Echiniscidae. Entomologische Mitteilungen aus dem Zoologischen Museum Hamburg 8(129):319–334.

Dastych, H. 1987b. Altitudinal distribution of Tardigrada in Poland. Pages 169–176 *in:* R. Bertolani, editor. Biology of Tardigrades. Selected Symposia and Monographs U.Z.I., 1. Mucchi, Modena, Italy.

Doyère, L. 1840. Memorie sur les Tardigrades. Annales des Sciences Naturelles, Zool., Paris, Series 2, 14:269–362.

Franceschi, T., M. L. Loi, and R. Pierantoni. 1962–1963. Risultati di una prima indagine ecologica condotta su popolazioni di Tardigradi. Bollettino dei Musei e degli Istituti Biologici dell'Universita di Genova 32:69–93.

Goeze, J. A. E. 1773. Uber den Kleinen Wasserbär. Page 67 *in:* H. K. Bonnet, editor. Abhandlungen aus der Insectologie, Ubers. Usw, 2. Beobachtg.

Greven, H. 1980. Die Bärtierchen. Die Neue Brehm-Bucherei, Vol. 537. Ziemsen Verlag, Wittenberg Lutherstradt, G. D. R.

Greven, H., and W. Greven. 1987. Observations on the permeability of the tardigrade cuticle using lead as an ionic tracer. Pages 35–43 *in:* R. Bertolani, editor. Biology of Tardigrades. Selected Symposia and Monographs U.Z.I., 1. Mucchi, Modena, Italy.

Greven, H., and W. Peters. 1986. Localization of chitin in the cuticle of Tardigrada using wheat germ agglutinin gold conjugate as a specific electron dense marker. Tissue Cell 18:297–304.

Grigarick, A. A., R. O. Schuster, and E. C. Toftner. 1973. Descriptive morphology of eggs of some species in the *Macrobiotus hufelandii* group. The Pan-Pacific Entomologist 49(3):258–263.

Grigarick, A. A., R. O. Schuster, and E. C. Toftner. 1975. Morphogenesis of two species of *Echiniscus.* Memorie dell'Istituto Italiano di Idrobiologia 32(Suppl.):133–151.

Hallas, T. E. 1975. A mechanical method for the extraction of Tardigrada. Memorie dell'Istituto Italiano di Idrobiologia 32(Suppl.):153–158.

Hallas, T. E., and G. W. Yeates. 1972. Tardigrada of the soil and litter of a Danish beech forest. Pedobiologia 12:287–304.

Henneke, J. 1911. Beiträge zur Kenntnis der Biologie und Anatomie der Tardigraden (*Macrobiotus macronyx* Duj.). Zeitschrift fuer Wissenschaftliche Zoologie 97:721–752.

Higgins, R. P. 1959. Life history of *Macrobiotus islandicus* Richters with notes on others tardigrades from Colorado. Transactions of the American Microscopical Society 78:137–154.

Higgins, R. P., editor. 1975. International Symposium on

Tardigrades, Pallanza, Italy, June 17–19, 1974. Memorie dell'Istituto Italiano di Idrobiologia 32(Suppl.):469 pp.

Kathman, R. D., and D. R. Nelson. 1987. Population trends in the aquatic tardigrade *Pseudobiotus augusti* (Murray). Pages 155–168 *in:* R. Bertolani, editor. Biology of Tardigrades. Selected Symposia and Monographs U.Z.I., 1. Mucchi, Modena, Italy.

Kristensen, R. 1979. On the fine structure of *Batillipes noerrevangi* Kristensen 1976. (Heterotardigrada). 3. Spermiogenesis. *In:* B. Weglarska, editor. Second international symposium on tardigrades, Kraków, Poland, July 28–30, 1977. Zeszyty Naukowe Uniwersytetu Jagiellonskiego, Prace Zoologiczne 25:97–105.

Kristensen, R. 1987. Generic revision of the Echiniscidae (Heterotardigrada), with a discussion of the origin of the family. Pages 261–335 *in:* R. Bertolani, editor. Biology of Tardigrades. Selected Symposia and Monographs U.Z.I., 1. Mucchi, Modena, Italy.

Kristensen, R., and R. P. Higgins. 1984a. Revision of *Styraconyx* (Tardigrada: Halechiniscidae) with descriptions of two new species from Disko Bay, West Greenland. Smithsonian Contributions to Zoology 391:1–40.

Kristensen, R., and R. P. Higgins. 1984b. A new family of Arthrotardigrada (Tardigrada: Heterotardigrada) from the Atlantic coast of Florida, U.S.A. Transactions of the American Microscopical Society 103:295–311.

Manicardi, C., and R. Bertolani. 1987. First contribution to the knowledge of Alpine grassland tardigrades. Pages 177–185 *in:* R. Bertolani, editor. Biology of Tardigrades. Selected Symposia and Monographs U.Z.I., 1. Mucchi, Modena, Italy.

Marcus, E. 1929. Tardigrada. *In:* H. G. Bronn, editor. Klassen und Ordnungen des Tierreichs. Vol. 4. Akademische Verlagsgesellschaft, Leipzig. 608 pp.

Marcus, E. 1936. Tardigrada. *In:* F. Schultze, editor. Das Tierreich. W. de Gruyter, Berlin. 340 pp.

McInnes, S., and J. C. Ellis-Evans. 1987. Tardigrades from maritime Antarctic freshwater lakes. Pages 111–123 *in:* R. Bertolani, editor. Biology of Tardigrades. Selected Symposia and Monographs U.Z.I., 1. Mucchi, Modena, Italy.

Morgan, C. 1977. Population dynamics of two species of Tardigrada, *Macrobiotus hufelandii* (Schultze) and *Echiniscus Echiniscus testudo* (Doyère), in roof moss from Swansea. Journal of Animal Ecology 46:263–279.

Morgan, C., and P. King. 1976. British Tardigrada. Tardigrada keys and notes for the identification of the species. Synopses of the British Fauna, No. 9. Academic Press, London. 132 pp.

Nelson, D. R. 1975. Ecological distribution of tardigrades on Roan Mountain, Tennessee–North Carolina. Memorie dell'Istituto Italiano di Idrobiologia 32(Suppl.):225–276.

Nelson, D. R. 1982a. Developmental biology of the Tardigrada. Pages 363–368 *in:* F. Harrison and R. Cowden, editors. Developmental biology of freshwater invertebrates. Alan R. Liss, New York.

Nelson, D. R., editor. 1982b. Proceedings of the Third International Symposium on the Tardigrada, August 3–6, 1980, Johnson City, Tennessee. East Tennessee State University Press, Johnson City. 236 pp.

Nelson, D. R., and R. P. Higgins. 1990. Tardigrada. Pages 393–419 *In:* D. Dindal (editor), Soil biology guide. Wiley, New York.

Nelson, D. R., C. J. Kincer, and T. C. Williams. 1987. Effects of habitat disturbances on aquatic tardigrade populations. Pages 141–153 *in:* R. Bertolani, editor. Biology of Tardigrades. Selected Symposia and Monographs U.Z.I., 1. Mucchi, Modena, Italy.

Pennak, R. 1979. Tardigrada. Fresh-water invertebrates of the United States, 2nd Edition. Wiley, New York. pp. 239–253.

Pilato, G. 1969. Schema per una nuova sistemazione delle famiglie e dei generi degli Eutardigrada. Bolletino delle Sedute delle Accademia Gioenia di Scienze Naturali in Catania, Serie 4, 10:181–193.

Pilato, G. 1972. Structure, intraspecific variability and systematic value of the buccal armature of eutardigrades. Zeitschrift fuer Zoologische Systematik und Evolutionsforschung 10:65–78.

Pilato, G. 1979. Correlations between cryptobiosis and other biological characteristics in some soil animals. Bollettino di Zoologia 46:319–332.

Pilato, G. 1982. The systematics of eutardigrades. A comment. Zeitschrift fuer Zoologische Systematik und Evolutionsforschung 20:271–284.

Pilato, G. 1987. Revision of the genus *Diphascon* Plate, 1889, with remarks on the subfamily Itaquasconinae (Eutardigrada, Hypsibiidae). Pages 337–357 *in:* R. Bertolani, editor. Biology of Tardigrades. Selected Symposia and Monographs U.Z.I., 1. Mucchi, Modena, Italy.

Pollock, L. W. 1970. Distribution and dynamics of interstitial Tardigrada at Woods Hole, Massachusetts, USA. Ophelia 7:145–165.

Rahm, G. 1937. A new ordo of tardigrades from the hot springs of Japan (Furu-Section, Unzen). Annotationes Zoologicae Japonenses 16:345–352.

Ramazzotti, G. 1962. Il Phylum Tardigrada. Memorie dell'Istituto Italiano di Idrobiologia 16:595 pp.

Ramazzotti, G. 1972. Il Phylum Tardigrada, 2nd Ed. Memorie dell'Istituto Italiano di Idrobiologia 28:732 pp.

Ramazzotti, G. 1974. Supplement A, Il Phylum Tardigrada, Seconda Edizione, 1972. Memorie dell'Istituto Italiano di Idrobiologia 31:69–179.

Ramazzotti, G., and W. Maucci. 1983. Il Phylum Tardigrada, III edizione riveduta e aggiornata. Memorie dell'Istituto Italiano di Idrobiologia 41:1–1012.

Renaud-Mornant, J. 1982. Species diversity in marine Tardigrada. Pages 149–178 *in:* D. Nelson, editor. Proceedings of the Third International Symposium on the Tardigrada. East Tennessee State University Press, Johnson City.

Renaud-Mornant, J. 1987. Bathyl and abyssal Coronarctidae (Tardigrada), description of new species and phylogenetical significance. Pages 229–252 *in:* R. Bertolani, editor. Biology of Tardigrades. Selected Symposia and Monographs U.Z.I., 1. Mucchi, Modena, Italy.

Schuster, R. O., E. C. Toftner, and A. A. Grigarick. 1977.

Tardigrada of Pope Beach, Lake Tahoe, California. The Wasmann Journal of Biology 35:115–136.

Schuster, R. O., D. R. Nelson, A. A. Grigarick, and D. Christenberry. 1980a. Systematic criteria of the Eutardigrada. Transactions of the American Microscopical Society 99:284–303.

Schuster, R. O., A. A. Grigarick, and E. C. Toftner. 1980b. A new species of *Echiniscus* from California (Tardigrada: Echiniscidae). The Pan-Pacific Entomologist 56:265–267.

Spallanzani, L. 1776. Opuscoli di fisica animale e vegetabile, Vol. 2, il Tardigrado etc. Opuscolo 4:222.

Toftner, E. C., A. A. Grigarick, and R. O. Schuster. 1975. Analysis of scanning electron microscope images of *Macrobiotus* eggs. Memorie dell'Istituto Italiano di Idrobiologia 32(Suppl.):393–411.

Usher, M. B., and H. Dastych. 1987. Tardigrada from the maritime Antarctic. British Antarctic Survey Bulletin 77:163–166.

Wainberg, R., and W. Hummon. 1981. Morphological variability of the tardigrade *Isohypsibius saltursus*. Transactions of the American Microscopical Society 100:21–33.

Walz, B. 1974. The fine structure of somatic muscles of Tardigrada. Cell and Tissue Research 149:81–89.

Walz, B. 1975. Ultrastructure of muscle cells in *Macrobiotus hufelandi*. Memorie dell'Istituto Italiano di Idrobiologia 32(Suppl.):425–443.

Walz, B. 1978. Electron microscopic investigation of cephalic sense organs of the tardigrade *Macrobiotus hufelandi* Schultze. Zoomorphologie 89:1–19.

Walz, B. 1982. Molting in Tardigrada. A review including new results on cuticle formation in *Macrobiotus hufelandi*. Pages 129–147 *in:* D. Nelson, editor. Proceedings of the Third International Symposium on the Tardigrada. East Tennessee State University Press, Johnson City.

Weglarska, B. 1957. On the encystation in Tardigrada. Zoologica Poloniae 8:315–325.

Weglarska, B., editor. 1979. Second international symposium on tardigrades, Kraków, Poland, July 28–30, 1977. Zeszyty Naukowe Uniwersytetu Jagiellonskiego, Prace Zoologiczne 25:197 pp.

Wolburg-Buchholz, K., and H. Greven. 1979. On the fine structure of the spermatozoon of *Isohypsibius granulifer* Thulin 1928 (Eutardigrada) with reference to its differentiation. *In:* B. Weglarska, editor. Second international symposium on tardigrades, Kraków, Poland, July 28–30, 1977. Zeszyty Naukowe Uniwersytetu Jagiellonskiego, Prace Zoologiczne 25:191–197.

Water Mites

16

Ian M. Smith
Biosystematics Research Centre
Central Experimental Farm
Agriculture Canada
Ottawa, Ontario K1A OC6
Canada

David R. Cook
Department of Biological Sciences
Wayne State University
Detroit, Michigan 48202

Chapter Outline

I. INTRODUCTION
  A. General Relationships
  B. Origin
  C. Diversity and Classification
  D. Zoogeography
    1. Global Patterns
    2. Nearctic Patterns

II. EXTERNAL MORPHOLOGY AND INTERNAL ANATOMY
  A. Morphology of Larvae
  B. Morphology of Deutonymphs and Adults
  C. Evolution of Adult Exoskeleton
  D. Internal Anatomy
    1. Digestive System
    2. Excretory System
    3. Respiratory System
    4. Neural System
    5. Reproductive System

III. ECOLOGY
  A. Life History
    1. Egg
    2. Larva
    3. Nymphochrysalis
    4. Deutonymph
    5. Imagochrysalis
    6. Adult
  B. Habitats and Communities
    1. Springs (Including Seepage Areas)
    2. Riffle Habitats
    3. Interstitial Habitats
    4. Stenothermic Pools

    5. Lakes
    6. Temporary Pools
  C. Ecological Importance
    1. Impact as Parasites
    2. Impact as Predators
    3. Importance as Prey
    4. Potential as Indicators of Environmental Quality

IV. COLLECTING, REARING, AND PREPARATION FOR STUDY
  A. Collecting and Extracting Techniques
    1. Deutonymphs and Adults
    2. Larvae
  B. Rearing
  C. Preservation and Preparation for Study

V. TAXONOMIC KEYS TO SUBFAMILIES OF WATER MITES IN NORTH AMERICA
  A. Taxonomic Key to Subfamilies for Known Larvae of Water Mites
  B. Taxonomic Key to Subfamilies for Adults of Water Mites
  Literature Cited

I. INTRODUCTION

Mites belonging to five unrelated groups are commonly found in freshwater habitats. The Hydrachnida (Hydrachnellae, Hydracarina, or Hydrachnidia of other authors) and some Halacaridae, Oribatida, Acaridida, and Mesostigmata have independently invaded freshwater and become fully adapted for living there. Hydrachnida have flour-

Ecology and Classification of North American Freshwater Invertebrates

ished to become by far the most numerous, diverse, and ecologically important group of freshwater Acari, and in this chapter we refer to them simply as water mites. The others, though of considerable biological interest, usually are less abundant and are relatively conservative in morphology and habits.

Water mites are among the most abundant and diverse benthic arthropods in many habitats. One square meter of substratum from littoral weed beds in eutrophic lakes may contain as many as 2000 deutonymphs and adults representing up to 75 species in 25 or more genera (Fig. 16.285). Comparable samples from an equivalent area of substratum in rocky riffles of streams often yield over 5000 individuals of more than 50 species in over 30 genera (including both benthnic and hyporheic forms). Mites have coevolved with some of the dominant insect groups in freshwater ecosystems, especially nematocerous Diptera, and interact intimately with these insects at all stages of their life histories.

We have tried to avoid unnecessary use of specialized acarological terms in this chapter. We provide a glossary at the end of this book of the essential terminology needed for describing morphological and anatomical features. Those wishing to consult a more comprehensive account of mite structure and function are referred to the appropriate sections of books by Cook (1974) and Krantz (1978).

A. General Relationships

Water mites, along with the terrestrial Calyptostomatoidea, Trombidioidea, and Erythraeoidea, belong to a remarkably diverse natural group of actinedid mites, the Parasitengona. The complex and essentially holometabolous type of development (Fig 16.1) of this group is unique among Acari. After emerging from the egg membrane, the hexapod larva seeks out an appropriate host and becomes an ectoparasite, which is passively transported while feeding on host fluids. When fully engorged, the larva transforms to the quiescent nymphochrysalis (corresponding to the protonymphal instar). Radical structural reorganization occurs during this stage, giving rise to the active deutonymph, which resembles the adult in being octopod and typically predaceous, but is sexually immature and exhibits incomplete sclerotization and chaetotaxy. The deutonymph feeds and grows in size before entering another quiescent stage, the imagochrysalis (corresponding to the tritonymphal instar). After completion of metamorphosis during this stage, the mature adult emerges.

B. Origin

Water mites evolved from terrestrial stock, and hypotheses on their origin usually presume an ancestral terrestrial parasitengone (Mitchell 1957a, Davids and Belier 1979). A recently proposed alternative hypothesis (Wiggins et al. 1980, Smith and Oliver 1986) suggests taht the primitive parasitengone stock may have been water mites resembling certain extant Hydryphantoidea. According to this postulate, water mites diverged from terrestrial ancestors with direct development, perhaps resembling extant Anystoidea, while evolving the basic parasitengone life-history pattern as a set of adaptations for exploiting spatially and temporally intermittent aquatic habitats.

The fossil record provides little insight into the evolutionary history of water mites. The only reported fossils that undoubtedly are water mites (Cook 1957, Poinar 1985) are larval specimens from Tertiary deposits representing highly derived taxa. However, distributional data and host associations indicate to us that the group originated no later than the Triassic or Jurassic period.

Available morphological and behavioral data suggest that extant water mites, possibly excluding the enigmatic Stygothrombidioidea and Hydrovolzioidea, are monophyletic (Barr 1972, Cook 1974, Smith and Oliver 1976, 1986), and that the major phyletic lineages of water mites, and possibly all Parasitengona, were derived from hydryphantoid-like ancestors (Mitchell 1957a, Smith and Oliver 1986).

C. Diversity and Classification

Well over 5000 species of water mites are currently recognized worldwide, representing more than 300 genera and subgenera in over 100 families and subfamilies (Viets 1987). Acarologists usually consider water mites to be a taxon of intermediate rank between superfamily and suborder. Interestingly, the group appears to rival several orders of aquatic insects in diversity, and is comparable to some of them in age. Most of the genera and families are reasonably stable and appear to approximate closely holophyletic lineages. The families are conservatively grouped into eight superfamilies. Four of these—Stygothrombidioidea, Hydrovolzioidea, Hydrachnoidea, and Eylaoidea—probably represent natural groupings. The others—Hydryphantoidea, Lebertioidea, Hygrobatoidea, and Arrenuroidea—are all either paraphyletic or polyphyletic assemblages subject to future revision. A comprehensive, detailed cladistic analysis of extant

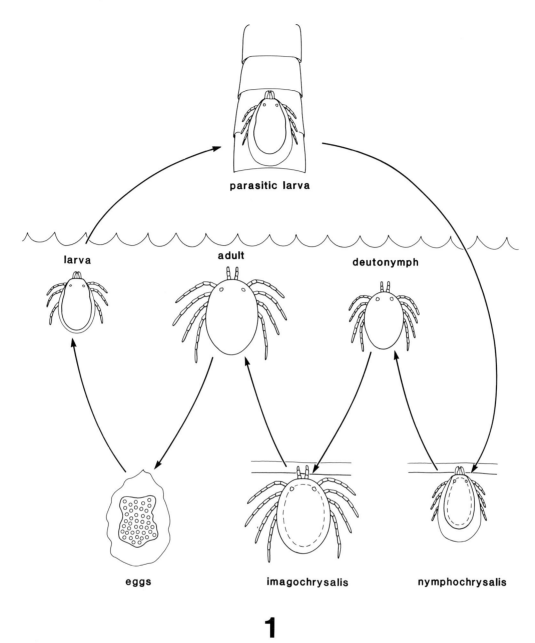

parasitic larva

larva

adult

deutonymph

eggs

imagochrysalis

nymphochrysalis

1

Figure 16.1 Diagrammatic illustration of a generalized water mite life history. (Redrawn and modified from Smith 1976.)

fauna is required to permit construction of more stable and informative superfamily groupings. The phylogenetic studies that prove most useful for understanding water mite evolution and developing a more natural classification are ones that provide syntheses of information based upon the morphology of all instars, behavioral attributes, and life-history data (Mitchell 1957a, Cook 1974, Smith 1976, Smith and Oliver 1976, 1986).

The over 1500 species currently estimated to occur in North America north of Mexico represent 124 genera (one of which has yet to be named) in 66 subfamilies, 38 families, and all 8 superfamilies. The fauna remains incompletely known even at the generic level. Species of over 15 new or unreported genera have been discovered in North America since the publication of Cook's (1974) review. About half of the species expected to occur in North America are not yet named, and many described species are known from only a few specimens in collections. The most diverse genera, such as *Sperchon* and *Aturus* in lotic habitats and *Piona* and *Arrenurus* in lentic habitats, probably contain well over 100 nearctic species each. Additional taxa await discovery throughout the continent, especially in springs and interstitial habitats. The least explored areas, such as southern Appalachia and the boreal and arctic regions of northern Canada and Alaska, undoubtedly will yield some unexpected taxa when thoroughly studied.

D. Zoogeography

1. Global Patterns

Water mites occur throughout the world, except Antarctica (Cook 1974). All superfamilies, except Stygothrombidioidea and Hydrovolzioidea, as well as many early derivative families and subfamilies, are richly represented in all zoogeographic regions. Knowledge of global distribution patterns of water mite taxa has been substantially improved by recent studies of the fauna of austral regions, especially those by K. O. Viets and Cook. The basic patterns can be explained as the result of vicariance due to plate tectonics (Cook 1986, 1988). Dispersal (along with hosts) between adjacent land masses (e.g., southeast Asia and Australia) or across continents (e.g., Africa) significantly altered these patterns as more recent groups progressively displaced ancient ones.

2. Nearctic Patterns

Four major groups of family-group taxa can be recognized among extant North American water mites, reflecting successive historical stages in the development of the modern fauna (see Table 16.1).

a. Pangean Origin

Twenty-five early-derivative, cosmopolitan taxa exhibiting substantial endemism in both the northern and southern hemispheres apparently originated in Pangea, and were represented in both Laurasia and Gondwanaland at the time of their separation during the Jurassic.

b. Laurasian Origin

Twenty-eight taxa with predominantly northern hemispheric distributions, exhibiting comparable diversity and endemism in the Nearctic and Palearctic, probably originated in Laurasia.

The taxa of Pangean and Laurasian origin, including representatives of many extant genera, probably were distributed throughout Laurasia during the late Cretaceous. The vast majority of species currently inhabiting North America belong to these taxa and descended from ancestors that either were present on this continent when it separated from Eurasia during the early Tertiary or immigrated from Eurasia via land bridges later in that Period.

c. Gondwanan or South American Origin

Eight taxa with predominantly southern hemispheric distributions, exhibiting substantial diversity and endemism in the Neotropics, are probably of Gondwanan or South American origin. Members of the Gondwanan taxa invaded southern North America from Central and South America after the Panamanian isthmus appeared during the Pliocene.

d. North American Origin

Five taxa are currently known only from North America. Laversiidae may be an endemic group of recent origin. The other four of these taxa—Cowichaniinae, Cyclomomoniinae, Horreolaninae, Stygameracarinae—appear to have Tertiary relict distributions, suggesting that they originated in Laurasia, and may have undiscovered sister taxa inhabiting poorly known areas of the Palearctic or Oriental regions.

Distributions of water mite taxa within North America were dramatically influenced by climatic cooling and glaciation during the late Pliocene and Pleistocene. Taxa that show evidence of recent adaptive radiation have reestablished pan-continental distributions, while those with relatively few species have remained within more limited areas. Many species groups came to be restricted to either temperate or boreal areas of the continent, as reflected to some extent in the distributions of families and subfamilies (Table 16.1). As a result, typical "Tertiary-relict" distributions in, or adjacent to, unglaciated refugia such as coastal California and Oregon, the Ozark Plateau, and parts of the Appalachian and Rocky Mountains are exhibited by some taxa (e.g., Cyclothyadinae, Stygomomoniinae, Uchidastygacarinae, Morimotacarinae, Neoacaridae, Chappuisididae). Other groups (e.g., Teutoniidae, Mideidae, Acalyptonotidae) have characteristic "boreal" or "arctic–alpine" distributions at higher latitudes and elevations. Distributional limitations are frequently correlated with habitat specializations. For example, many interstitial taxa have Tertiary-relict distributions because their habitats were destroyed in most northern regions of the continent by Pleistocene glaciation and will require a long period of climatic stability to become reestablished there. The sister species of many nearctic Tertiary relicts now inhabit similar areas in temperate Asia, resulting in strikingly discontinuous distributions for certain taxa. In contrast, species adapted to cold stenothermic pools inhabit subarctic areas, high mountains near permanent ice-fields, and temperate lowlands where groundwater springs create similar conditions. The sister species of many stenophilic nearctic water mites live in northern

Eurasia, yielding essentially circumpolar or boreal holarctic distribution patterns for these groups.

II. EXTERNAL MORPHOLOGY AND INTERNAL ANATOMY

Water mites exhibit the characteristic acarine body plan comprising the gnathosoma, or mouth region, and the idiosoma, or body proper, representing the fused cephalothorax and abdomen.

A. Morphology of Larvae

The morphological structure of water mite larvae are illustrated in Figs. 16.2–16.8. The short gnathosoma (Fig. 16.2) bears the stocky pedipalps, which have five free segments (trochanter, femur, genu, tibia, tarsus) that flex ventrally. The tarsus of the pedipalp is relatively long and cylindrical in some ancient taxa (e.g., Hydryphantinae, Thyadinae) (Figs. 16.15 and 16.19), but is typically reduced to a dome- or button-shaped pad in most derivative groups (Figs. 16.72 and 16.92). A highly modified thick, curved seta is present dorsally at the end of the tibia (Fig. 16.2), the homologue of the tibial "claw," which characterizes the pedipalp of most terrestrial Anystoidea, Parasitengona, and related groups. The paired chelicerae (Fig. 16.3), each consisting of a cylindrical basal segment and a movable terminal claw, lie between the pedipalps.

The idiosoma is plesiotypically mainly unsclerotized, as in extant Hydryphantoidea. The dorsal integument (Fig. 16.23) bears a medial eye-spot, two pairs of lens-like lateral eyes, four pairs of propodosomal setae (AM, AL, SS, and PL), four pairs of mediohysterosomal setae (Mh1–Mh4), one pair of humeral setae (Hu), three pairs of laterohysterosomal setae (Lh1–Lh3), and four pairs of lyrifissures. Ventrally (Fig. 16.24), the integument bears the paired coxal plates, four pairs of coxal plate setae (C1–C4), paired urstigmata (Ur) between coxal plates I and II, the excretory pore (EP), two pairs of excretory pore plate setae (E1 and E2), four pairs of ventral setae (V1–V4), and one pair of lyrifissures. The size, shape, and position of these dorsal and ventral structures and the degree of fusion of the sclerites associated with them provide useful taxonomic characters. The legs (Fig. 16.18) are inserted laterally on the coxal plates, and in the plesiotypical condition have six movable segments, namely trochanter (Tr), basifemur (BFe), telofemur (TFe), genu (Ge), tibia (Ti), and tarsus (Ta), which articulate to permit ventral flexion. The segments have characteristic complements of setae and

solenidia (as in Figs. 16.6–16.8). The tarsi bear paired claws and claw-like empodia terminally.

The soft-bodied condition, retained in all known larval Hydryphantoidea, apparently provides adequate support for the muscles used in locomotion on the surface film. In all other major lineages, the dorsum (Figs. 16.2, 16.34, 16.49, 16.66, 16.101, 16.154) bears an extensive plate, which incorporates the bases of the propodosomal setae and, in some cases, one or more hysterosomal pairs. Similarly, on the venter, the coxal plates are enlarged and variously fused, and the excretory pore plate is expanded to incorporate the bases of both pairs of excretory pore plate setae. The coxal plates and excretory pore plate may also bear various combinations of ventral setae.

Dorsal plates and expanded coxal and excretory pore plates developed independently in Hydrovolzioidea and most Eylaoidea to provide support for muscles used in running and crawling on the surface film, and again in both Hydrachnoidea and the ancestral stock of the three "higher" superfamilies to anchor muscles used in swimming or crawling under water. In Hygrobatoidea, fused coxal plates II + III often have conspicuous lateral coxal apodemes (LCA), medial coxal apodemes (MCA), and transverse muscle attachment scars (TMAS) (Fig. 16.177).

Larvae of Hydryphantoidea and Eylaoidea retain six movable leg segments (Figs. 16.18 and 16.38), but those of all other groups have the basifemoral and telofemoral segments fused (Figs. 16.6–16.8, 16.54). Larval Hydryphantoidea also retain the most plesiotypical complement of setae on the segments of the legs. Larvae of each of the more derivative groups exhibit characteristic leg chaetotaxies that reflect reductions from the plesiotypical complement. Certain ventral setae on the genua, tibiae, and tarsi are elongate and plumose in larvae adapted for swimming.

B. Morphology of Deutonymphs and Adults

The gnathosoma (Fig. 16.186) is plesiotypically a simple, short channel, derived from extensions of the pedipalpal coxae and leading to the esophagus (Mitchell 1962). A protrusible tube of integument connecting the gnathosoma to the idiosoma has developed independently in several distantly related groups (e.g., Rhyncholimnocharinae, Clathrosperchoninae, certain Krendowskiidae). The paired pedipalps (Figs. 16.186 and 16.199), inserted on the gnathosoma, have both tactile and raptorial functions. In the plesiomorphic condition, the pedipalps

Table 16.1 Summary of Some Important Zoogeographic and Ecological Characteristics of North American Water Mite Subfamilies

Taxa (Number of North American Genera in Parentheses)	Zoogeography		Larvae			Ecology (Deutonymphs and Adults)			Adults
	Origin	Nearctic Distribution	Type	Order of Hosts	Attachment Site	Habitat Type	Mode of Locomotion	Prey	Spermatophore Transfer
STYGOTHROMBIDIOIDEA									
Stygothrombididae (1)	Laurasian	Temperate, Boreal	Aquatic	Plecoptera	Thorax, Abdomen	Interstitial	Crawling	Unknown	Unknown
HYDROVOLZIOIDEA									
Hydrovolziidae									
Hydrovolziinae (1)	Laurasian	Temperate, Boreal	Terrestrial	Hemiptera, Diptera	Thorax, Abdomen	Springs, Riffles, Interstitial	Crawling	Unknown	Unknown
HYDRACHNOIDEA									
Hydrachnidae (1)	Pangean	Throughout	Aquatic	Coleoptera, Hemiptera	Thorax, Abdomen	Lakes, Temporary pools	Swimming	Hemiptera eggs	Indirect
EYLAOIDEA									
Limnocharidae									
Limnocharinae (2)	Pangean	Temperate, Boreal	Terrestrial	Hemiptera, Odonata, Coleoptera	Thorax, Abdomen	Riffles, Pools, Lakes Interstitial	Crawling, Swimming	Diptera larvae	Unknown
Rhyncholimnocharinae (1)	Gondwanan	Temperate	Aquatic	Coleoptera	Abdomen	Pools, Lakes,	Crawling	Unknown	Unknown
Eylaidae (1)	Pangean	Throughout	Terrestrial	Hemiptera	Thorax, Abdomen	Temporary pools	Swimming	Ostracoda	Indirect, Direct
Piersigiidae									
Piersigiinae (1)	Laurasian	Temperate, Boreal	Terrestrial	Coleoptera	Abdomen	Springs, Interstitial, Temporary pools	Crawling	Cladocera	Unknown
HYDRYPHANTOIDEA									
Hydryphantidae									
Hydryphantinae (1)	Pangean	Throughout	Terrestrial	Hemiptera, Odonata, Diptera	Thorax, Abdomen	Lakes, Temporary pools	Swimming	Diptera eggs	Indirect
Thyadinae (13)	Laurasian	Throughout	Terrestrial	Collembola, Thysanoptera, Hemiptera, Diptera, Trichoptera	Thorax	Springs, Riffles, Temporary pools	Walking, Crawling	Diptera eggs	Indirect
Protziinae (2)	Laurasian	Temperate, Boreal	Terrestrial	Plecoptera, Diptera, Trichoptera	Thorax, Abdomen	Riffles	Walking	Unknown	Unknown
Wandesiinae (1)	Pangean	Temperate	Aquatic	Plecoptera	Thorax	Springs, Interstitial	Crawling	Unknown	Unknown
Tartarothyadinae (1)	Pangean	Temperate	Terrestrial	Unknown	Unknown	Springs, Riffles	Walking, Crawling	Unknown	Unknown
Pseudohydryphantinae(1)	Pangean	Boreal	Unknown	Unknown	Unknown	Pools, Lakes	Swimming	Unknown	Unknown
Cyclothyadinae (1)	Laurasian	Temperate (West)	Unknown	Unknown	Unknown	Springs, Interstitial	Crawling	Unknown	Unknown
Cowichaniinae (1)	Laurasian or North American or Pangean	Temperate (West)	Unknown	Unknown	Unknown	Interstitial	Crawling	Unknown	Unknown
Hydrodromidae (1)	Pangean	Throughout	Terrestrial	Diptera	Thorax	Riffles, Pools, Lakes	Swimming	Diptera eggs	Indirect
Rhynchohydracaridae									
Clathrosperchoninae (1)	Gondwanan	Temperate	Unknown	Unknown	Unknown	Riffles, Interstitial	Crawling	Unknown	Unknown
Thermacaridae (1)	Pangean or Laurasian	Temperate	Terrestrial	Anura (Amphibia)	Body dorsum	Hot springs	Crawling	Unknown	Unknown

Taxon									
LEBERTIOIDEA									
Sperchontidae									
Sperchontinae (2)	Laurasian	Throughout	Aquatic	Diptera, Trichoptera	Thorax	Springs, Riffles, Pools, Lakes	Crawling	Diptera larvae	Unknown
Teutoniidae (1)	Laurasian	Boreal	Aquatic	Diptera	Abdomen	Pools	Crawling, Swimming	Unknown	Unknown
Rutripalpidae (1)	Laurasian	Boreal (East)	Unknown	Unknown	Unknown	Springs, Pools	Crawling	Unknown	Unknown
Anisitsiellidae									
Anisitsiellinae (6)	Pangean	Temperate, Boreal	Aquatic	Diptera	Thorax, Abdomen	Springs, Riffles, Interstitial, Pools	Crawling	Unknown	Unknown
Lebertiidae (2)	Laurasian	Throughout	Aquatic	Diptera	Thorax	Springs, Riffles, Pools, Lakes	Crawling, Running, Swimming	Diptera larvae	Indirect
Oxidae (2)	Pangean	Temperate, Boreal	Aquatic	Diptera	Thorax	Pools, Lakes	Swimming	Diptera larvae	Unknown
Torrenticolidae									
Testudacarinae (1)	Laurasian	Temperate, Boreal	Aquatic	Diptera	Thorax	Riffles, Interstitial	Crawling	Diptera larvae	Unknown
Neoatractidinae (1)	Gondwanan	Temperate (Southwest)	Unknown	Unknown	Unknown	Riffles	Crawling	Unknown	Unknown
Torrenticolinae (2)	Laurasian	Temperate, Boreal	Aquatic	Diptera	Thorax	Riffles, Interstitial, Pools, Lakes	Crawling	Diptera larvae	Unknown
HYGROBATOIDEA									
Limnesiidae									
Neomamersinae (2)	Gondwanan	Temperate	Unknown	Unknown	Unknown	Interstitial	Crawling, Running	Unknown	Unknown
Kawamuracarinae (1)	Laurasian	Temperate (Southwest)	Unknown	Unknown	Unknown	Interstitial	Crawling, Running	Unknown	Unknown
Tyrrelliinae (2)	Gondwanan	Temperate, Boreal	Aquatic	Diptera	Abdomen	Springs, Pools, Lakes	Crawling	Unknown	Unknown
Protolimnesiinae (1)	Gondwanan	Temperate (Southwest)	Unknown	Unknown	Unknown	Interstitial	Crawling, Running	Unknown	Unknown
Limnesiinae (2)	Pangean	Temperate, Boreal	Aquatic	Diptera	Thorax	Riffles, Interstitial, Pools, Lakes	Crawling, Swimming	Copepoda, Cladocera, Insect eggs and larvae	Unknown
Omartacaridae									
Omartacarinae (1)	Gondwanan	Temperate (Southwest)	Unknown	Unknown	Unknown	Interstitial	Crawling	Unknown	Unknown
Hygrobatidae (5)	Pangean	Temperate, Boreal	Aquatic	Diptera, Trichoptera	Abdomen	Springs, Riffles, Interstitial, Pools, Lakes	Crawling, Swimming	Cladocera, Diptera larvae	Unknown
Unionicolidae									
Unionicolinae(1)	Pangean	Temperate, Boreal	Aquatic	Diptera, Trichoptera	Abdomen, Legs	Pools, Lakes, Mollusc parasites	Crawling, Swimming	Copepoda, Cladocera, Diptera larvae, Mollusc tissue	Indirect, Direct
Pionatacinae (2)	Pangean	Temperate, Boreal	Aquatic	Diptera, Trichoptera	Abdomen	Riffles, Pools, Lakes	Swimming	Unknown	Unknown
Feltriidae (1)	Laurasian	Temperate, Boreal	Aquatic	Diptera	Abdomen	Springs, Riffles, Interstitial	Crawling	Unknown	Direct
Pionidae									
Wettininae (1)	Pangean	Temperate, Boreal	Aquatic	Diptera	Abdomen	Pools, Lakes	Swimming	Unknown	Indirect
Hydrochoreutinae (1)	Laurasian	Temperate, Boreal	Aquatic	Diptera	Abdomen	Pools, Lakes	Swimming	Unknown	Direct

(continued)

Table 16.1 (Continued)

Taxa (Number of North American Genera in Parentheses)	Zoogeography			Ecology						
			Larvae			Deutonymphs and Adults			Adults	
	Origin	Nearctic Distribution	Type	Order of Hosts	Attachment Site	Habitat Type	Mode of Locomotion	Prey	Spermatophore Transfer	
Foreliinae (3)	Laurasian	Temperate, Boreal	Aquatic	Diptera	Abdomen	Springs, Interstitial, Pools, Lakes	Crawling, Swimming	Diptera larvae	Direct	
Huitfeldtiinae (1)	Pangean	Boreal	Aquatic	Diptera	Abdomen	Lakes (profundal)	Swimming	Unknown	Unknown	
Tiphyinae (3)	Pangean	Throughout	Aquatic	Diptera	Abdomen	Pools, Lakes, Temporary pools	Swimming	Diptera larvae	Direct	
Pioninae (2)	Laurasian	Throughout	Aquatic	Diptera	Abdomen	Springs, Riffles, Pools, Lakes Temporary pools	Crawling, Swimming	Copepoda, Cladocera Diptera larvae	Direct	
Najadicolinae (1)	Laurasian	Temperate	Aquatic	Diptera	Abdomen	Lakes (Mollusc parasites)	Crawling	Mollusc tissue	Unknown	
Aturidae										
Frontipodopsinae (1)	Pangean	Temperate (West)	Unknown	Unknown	Unknown	Interstitial	Crawling, Running	Unknown	Unknown	
Axonopsinae (12)	Pangean	Throughout	Aquatic	Diptera	Abdomen	Riffles, Interstitial, Pools, Lakes	Crawling, Running, Swimming	Diptera larvae	Indirect, Direct	
Albiinae (1)	Laurasian	Temperate,	Aquatic	Trichoptera	Abdomen	Springs, Riffles, Interstitial, Pools, Lakes	Swimming	Unknown	Unknown	
Aturinae (4)	Laurasian	Temperate, Boreal	Aquatic	Diptera, Trichoptera	Abdomen		Crawling	Diptera larvae	Direct	
ARRENUROIDEA										
Mideidae (1)	Laurasian	Throughout	Aquatic	Diptera	Abdomen	Pools, Lakes	Swimming	Diptera larvae	Direct	
Momoniidae										
Cyclomomoniinae (1)	Laurasian or North American	Temperate (West)	Unknown	Unknown	Unknown	Interstitial	Crawling	Unknown	Unknown	
Stygomomoniinae (1)	Laurasian	Temperate	Aquatic	Trichoptera	Abdomen	Interstitial	Crawling	Unknown	Unknown	
Momoniinae (1)	Pangean	Temperate	Aquatic	Trichoptera	Thorax, Abdomen	Pools, Lakes	Swimming	Diptera larvae	Direct	

Taxon	Distribution	Climate/Zone			Body region	Habitat	Locomotion		Development
Nudomideopsidae (3)	Pangean	Temperate, Boreal	Aquatic	Diptera	Thorax	Springs, Interstitial	Crawling	Unknown	Unknown
Mideopsidae Mideopsinae (1)	Laurasian	Temperate, Boreal	Aquatic	Diptera	Thorax	Springs, Riffles, Interstitial, Pools, Lakes	Crawling, Swimming	Diptera larvae	Unknown
Uchidastygacaridae Uchidastygacarinae (1)	Laurasian	Temperate	Unknown	Unknown	Unknown	Interstitial	Crawling, Running	Unknown	Unknown
Morimotacarinae (2)	Laurasian	Temperate (West)	Unknown	Unknown	Unknown	Interstitial	Crawling, Running	Unknown	Unknown
Neoacaridae (2)	Laurasian	Temperate, Boreal	Aquatic	Diptera	Thorax	Riffles, Interstitial, Lakes	Crawling	Unknown	Direct
Bogatiidae Horreolaninae (1)	Laurasian or North American	Temperate	Unknown	Unknown	Unknown	Interstitial	Crawling	Unknown	Unknown
Chappuisididae Chappuisidinae (1)	Laurasian	Temperate	Unknown	Unknown	Unknown	Interstitial	Crawling, Running	Unknown	Unknown
Acalyptonotidae (2)	Laurasian	Boreal, Arctic	Aquatic	Diptera	Thorax	Springs, Pools, Lakes	Crawling	Unknown	Unknown
Athienemanniidae Athienemanniinae (2)	Pangean	Temperate, Boreal	Aquatic	Diptera	Thorax	Springs, Riffles, Interstitial, Pools	Crawling	Diptera larvae	Unknown
Stygameracarinae (1)	Laurasian or North American	Temperate	Unknown	Unknown	Unknown	Interstitial	Crawling, Running	Unknown	Unknown
Arenohydracaridae (1)	Gondwanan	Temperate (Southwest)	Unknown	Unknown	Unknown	Interstitial	Crawling	Unknown	Unknown
Laversiidae (1)	North American	Temperate, Boreal	Aquatic	Diptera	Thorax	Springs, Riffles, Lakes (profundal)	Crawling	Unknown	Unknown
Krendowskiidae (2)	Pangean	Temperate	Aquatic	Diptera	Abdomen	Pools, Lakes	Swimming	Unknown	Unknown
Arrenuridae Arrenurinae (1)	Pangean	Throughout	Aquatic	Diptera, Odonata	Thorax, Abdomen	Springs, Riffles, Interstitial, Pools, Lakes, Temporary pools	Running, Swimming	Ostracoda, Cladocera, Diptera larvae	Indirect

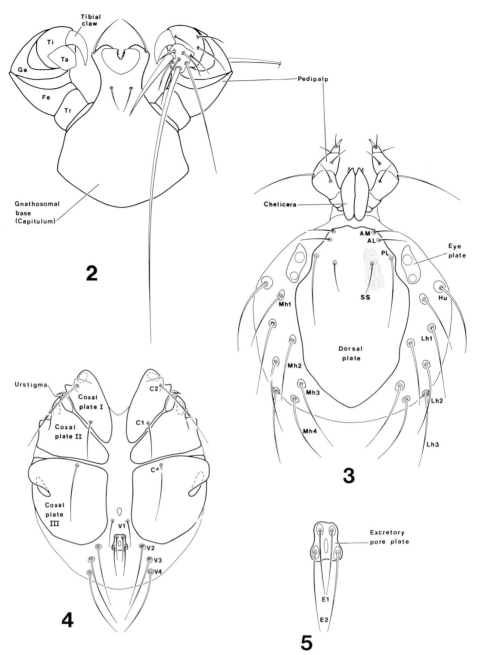

Figures 16.2–16.5 *Sperchonopsis ecphyma* Prasad and Cook (Sperchontidae), larva. Fig. 16.2, venter of gnathosoma. Fig. 16.3, dorsum of idiosoma and gnathosoma. Fig. 16.4, venter of idiosoma. Fig. 16.5, excretory pore plate. AL, AM, PL, anterolateral, anteromedial, posterolateral propodosomal setae respectively; C1, C2, C4, coxal plate setae; E1, E2, excretory pore plate setae; Fe, femur; Ge, genu; Hu, humeral seta; Lh1–Lh3, laterohysterosomal setae; Mh1–Mh4, mediohysterosomal setae; SS, sensilla; Ta, tarsus; Ti, tibia; Tr, trochanter; V1–V4, ventral setae. (Redrawn and modified from Smith 1982.)

have five movable segments, namely trochanter, femur, genu, tibia, and tarsus, which are essentially cylindrical and articulate to allow ventral flexion. The tibia bears a thick, blade-like, dorsal seta distally in many ancient groups (e.g., Hydryphantinae, Tartarothyadinae, Pseudohydryphantinae). As in larvae, this seta is the homologue of the tibial "claw" of terrestrial relatives, and it often makes

the pedipalps appear chelate. In derivative groups, other setae along with various denticles and tubercles may be elaborated to enhance the raptorial function of the pedipalps. Segmentation of the pedipalps is reduced by fusion in a few groups. A modification that has developed independently in various taxa of Arrenuroidea is the so-called uncate condition. In these groups, the tibia is expanded and pro-

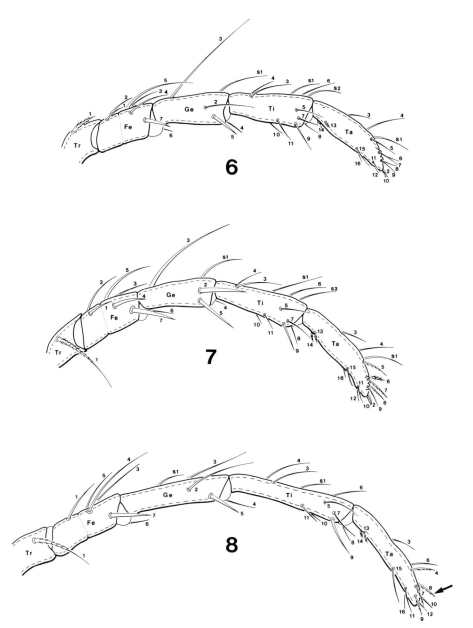

Figures 16.6–16.8 *Sperchonopsis ecphyma* Prasad and Cook (Sperchontidae), larva, anterolateral views of legs. Fig. 16.6, leg I; Fig. 16.7, leg II; Fig. 16.8, leg III. Fe, femur; Ge, genu; Ta, tarsus; Ti, tibia; Tr, trochanter. (Redrawn and modified from Smith 1982.)

duced ventrally to oppose the tarsus, permitting the mites to grasp and hold slender appendages of prey organisms securely. The paired chelicerae lie in longitudinal grooves between the pedipalps on the dorsal surface of the gnathosoma. Plesiotypically, they consist of a cylindrical basal segment bearing a movable terminal claw. This cheliceral structure is designed for tearing the integument of prey organisms and is retained in nearly all derivative groups. Hydrachnidae are unique in having unsegmented and stilettoform chelicerae, an obvious adaptation for piercing the insect eggs upon which they feed. The chelicerae are separate in all groups except

Limnocharidae and Eylaidae, where they are fused medially.

The most comprehensive comparative studies of water mite mouthparts were published by Motas (1928) and Mitchell (1962).

The idiosoma, or body proper, is plesiotypically round or ovoid in outline, slightly flattened dorsoventrally, and mostly unsclerotized, as in certain extant members of ancient taxa such as Hydryphantinae, Thyadinae, and Pseudohydryphantinae. The dorsal integument (Fig. 16.209) bears an unpaired medial eye, paired lateral eyes that are usually enclosed in capsules, paired preocular and postocular setae, and longitudinal series of paired glandularia (six dorsoglandularia, five lateroglandularia), muscle attachment sites (five dorsocentralia, four dorsolateralia), and lyrifissures (five). Ventrally (Fig. 16.207), the integument bears the paired coxal plates (fused into anterior and posterior groups on each side), the genital field (comprising the gonopore, three pairs of acetabula, and paired genital valves), five pairs of ventroglandularia (including coxoglandularia I between the anterior and posterior coxal groups and coxoglandularia II behind the posterior groups), and the excretory pore. As in larvae, these idiosomal structures provide a wealth of useful taxonomic characters.

The legs (Figs. 16.186 and 16.276) are inserted laterally on the coxae, and plesiotypically articulate on a vertical major axis and have six movable segments that articulate to permit ventral flexion. The segments are essentially cylindrical and have variable complements of setae. Though chaetotaxy of the legs provides a variety of taxonomic characters in deutonymphs and adults, the expression and position of individual setae are highly variable within taxa. Consequently, the rigorous analysis of chaetotactic patterns that proves so useful in the case of larvae is not practicable for later instars. The leg tarsi plesiotypically bear paired claws terminally.

The plesiotypical soft-bodied condition is retained in early derivative groups adapted for walking on the substratum (e.g., Thyadinae) or swimming in shallow habitats such as seepage pools or temporary pools (e.g., Hydryphantinae, Pseudohydryphantinae). Walking forms have relatively short, stocky leg segments and setae, whereas swimmers have longer segments bearing fringes of slender swimming setae. As pointed out by Mitchell (1957a, 1957b, 1958, 1964b), the soft integument of these mites permits the highly precise local control of body shape and internal pressure that is needed to produce the walking and swimming movements of the legs.

C. Evolution of Adult Exoskeleton

The major lineages of water mites apparently differentiated from ancestral stock that resembled extant soft-bodied Hydryphantoidea (Mitchell 1957a, 1957b; Smith and Oliver 1986) primarily through development of apotypical larval behavioral patterns and life histories. Hydrachnidae and Eylaidae became specialized for exploiting standing water habitats, especially temporary pools, whereas Hydryphantoidea, Lebertioidea, Hygrobatoidea, and Arrenuroidea ultimately diversified and radiated in a wide range of habitats. In each of these four groups, and presumably in Hydrovolzioidea, sclerotization

of the integument came about as soft-bodied ancestral stock began to invade new habitats and diverge.

Adaptation to habitats such as seepage areas and springs or substrata in streams required a change in locomotor habits from walking or swimming. Mites adapting to moss mats and wet litter habitats developed a hydrophilic integument, which draws a film of water over the dorsal surface, creating sufficient downward force to press the body to the substratum and prevent efficient walking (Mitchell 1960). Groups invading streams evolved a wedge-shaped body designed to produce similar downward pressure in response to water flowing over it. Locomotion in both of these types of habitats necessitated evolution of a crawling gait. This change involved a shift in orientation of the major axis of the legs, shortening and thickening of leg segments and setae, enlargement of tarsal claws, and development of stronger, more massive muscles to control leg movements. Expansion of coxal plates and sclerites associated with glandularia, setal bases, and the genital field, along with sclerotization of the dorso- and laterocentralia, occurred to provide rigid exoskeletal support for these muscles. Fusion of these sclerotized areas led to development of complete dorsal and ventral shields in adults of certain groups. Multiple invasions of helocrene, rheocrene, and lotic habitats gave rise to a diverse array of dorsoventrally flattened, sclerotized, crawling groups of Hydryphantoidea, Lebertioidea, Hygrobatoidea and Arrenuroidea (Fig. 16.287).

During the early evolution of each of these lineages, certain groups became adapted for exploiting interstitial habitats in subterranean waters and the hyporheic zones of rheocrenes and streams. These mites tended to lose eyes and integumental pigmentation and required further streamlining of the body. Consequently, soft-bodied forms adopted a vermiform shape, while sclerotized mites became extremely compressed laterally or dorsoventrally (Fig. 16.288). The crawling mode of locomotion became secondarily modified in several groups of well-sclerotized interstitial species (e.g., Neomamersinae, Frontipodopsinae, Chappuisidinae) to permit rapid and agile running in hyporheic and groundwater habitats.

By late Pangean times, evolution within water mites apparently had produced basic communities of essentially soft-bodied species living in temporary and permanent standing water, and partly to fully sclerotized species living in emergent groundwater, lotic, and interstitial habitats. Subsequent phylogeny of Hydryphantoidea, Lebertioidea, Hygrobatoidea, and Arrenuroidea during late Mesozoic, Tertiary, and Quaternary times appears to have followed a recurring pattern of extended periods of gradual evolution leading to specialization within particular habitats, punctuated by episodes of relatively rapid and dramatic diversification. Divergence of many of the clades that originated during this time appears to have been correlated with successful invasion of new habitats. Repeated habitat diversification within the major clades of these four superfamilies, in response to opportunities provided by the explosive evolution of their primary host groups and by prolonged geologic and climatic instability, resulted in several parallel and convergent trends in body sclerotization. Members of different clades developed superficially similar sclerite arrangements in adapting to lotic or interstitial habitats (e.g., certain Hygrobatidae and Axonopsinae); others underwent homoplastic reduction or loss of sclerites in secondarily invading lentic habitats (e.g., various Foreliinae and Pioninae). Finally, some groups retained extensive sclerotization while becoming adapted for swimming in standing water (e.g., certain Axonopsinae and Mideopsinae). Consequently, modern communities are highly heterogeneous phylogenetically and morphologically, with each monophyletic component exhibiting unique exoskeletal adaptations for living in the particular habitat.

Evolutionary trends involving other readily observed and taxonomically useful characters—such as position and number of genital acetabula, positions of glandularia and lyrifissures, number and arrangement of setae on the body plates, and modification of claws on the tarsi of the legs—are still not adequately explained. For example, although the genital acetabula are plesiotypically borne in the gonopore in most Hydryphantoidea and Lebertioidea, they are either fused with the genital flaps or incorporated into acetabular plates flanking the gonopore in most taxa of other superfamilies. In addition, the number of acetabula has independently proliferated several times from the plesiotypical complement of three pairs in each of the six largest superfamilies. Alberti (1977, 1979) has recently proposed that acetabula in water mites may function as an osmoregulatory chloride epithelium, and Barr (1982) concluded that analysis of morphological data for early derivative taxa supports this hypothesis. However, until the function of acetabula is clearly understood on the basis of well-designed experiments, the significance of the trend to increasing numbers of acetabula will remain speculative.

The striking and elaborate color patterns of adult water mites present an intriguing enigma. Most ancient taxa (e.g., Hydrachnoidea, Eylaoidea, Hydryphantoidea) have colorless integument but are red, similar to terrestrial parasitengones, due to the presence of pigment granules that appear to be distributed throughout the body. Members of other superfamilies, except for taxa in interstitial habitats, exhibit highly distinctive patterns resulting from symmetrically arranged concentrations of pigment granules of various colors. In soft-bodied taxa, the pigments are located beneath the integument, but in sclerotized groups, they are incorporated into the plates. Well-sclerotized taxa of Lebertioidea, Hygrobatoidea, and Arrenuroidea may be various shades of red, orange, yellow, green, or blue and the dorsal shield frequently exhibits an intricate pattern combining several contrasting colors. The dorsal patterns of these mites can readily be interpreted as disruptive camouflage, but the adaptive value of their bright colors is more difficult to understand. The evolution of a wide range of highly colored pigments in these taxa certainly suggests that distinctive coloration confers some selective advantage. Though protection from predation may be part of the explanation, it is tempting to speculate that these mites, despite their apparently simple eye structure, may use color as a cue for recognizing potential mates. Research is needed to determine the extent of their ability to detect and respond to the subtle differences in color that distinguish closely related taxa.

D. Internal Anatomy

The organ systems of water mites occupy the hemocoel, bathed by hemolymph, which is circulated by movements of the body musculature (Schmidt 1935, Bader 1938, Mitchell 1964c). There is no closed circulatory system.

1. Digestive System

Digestion of food materials begins preorally, and only fluids are ingested. Food is drawn into the mouth (or buccal cavity) by the muscular pharynx, then passed through the tubular esophagus to the lobed midgut (Bader 1938) where digestion and absorption occur. Undigested material accumulates as insoluble particles in a posterodorsal lobe of the midgut. All lobes of the midgut end blindly, and there is no connection between the gut and the excretory pore.

2. Excretory System

This system consists of a large, thin-walled excretory tubule apparently derived from the primitive hindgut that lies dorsal to and in close contact with the midgut. Waste products are absorbed from hemolymph, and stored in the excretory tubule as insoluble, whitish or yellowish crystals of unknown chemical composition. When filled with crystalline material, the excretory tubule may be visible through the dorsal integument as a T- or Y-shaped structure. The excretory tubule connects ventrally with the excretory pore, and is evacuated periodically by pressure generated through movements of the body muscles.

Osmoregulation is apparently accomplished by the urstigmata in larvae, and by the genital acetabula in deutonymphs and adults (Barr 1982). Both of these structures have porous caps which apparently provide an ample surface area of chloride epithelium for maintaining water balance.

3. Respiratory System

Many groups of water mites retain paired stigmata located between the bases of the chelicerae in all active instars. Plesiotypically, the stigmata lead to tracheal trunks that anastomose repeatedly into tracheolar tubules extending to all part of the body. However, they appear to be nonfunctional in deutonymphs and adults, and respiration occurs by diffusion through the integument. In adults of many large species inhabiting standing water, a network of closed tubes of tracheolar dimensions lying beneath the integument transports gases to and from the tracheae that lead to internal organs (Mitchell 1972). Heavily sclerotized mites have the body plates well supplied with regularly arranged "pores" that permit diffusion of gases between the tracheolar loops and surrounding water through areas of thin integument. In some species, there are regions of tracheal anastomosis lateral to the brain (Wiles 1984).

4. Neural System

The so-called brain, a fused, undifferentiated central ganglionic mass, surrounds the esophagus. Nerve trunks dorsal to the esophagus lead to the anterior sense organs and mouthparts, while ventral trunks lead to the legs and genital region.

The lateral eyes appear to be the primary light-sensing structures in water mites. There are typically two pairs, with the eyes of each side located close together and often enclosed in lens-like capsules that may lie above or beneath the integument. In post-larval instars of Eylaoidea, the eyes are borne on a medial sclerite. Many Hydryphantoidea also have a medial eye spot between the lateral eyes, but this structure is absent in most members of the other superfamilies. The eyes seem to function as ocelli, permitting the mites to detect the intensity and direction and, at least in some taxa, the wavelength of incident light. Experiments with Unioni-

colinae have demonstrated that some species respond differentially to various wavelengths (Dimock and Davids 1985) and the response of the mites can be influenced by chemicals produced and released by their mollusc hosts (Roberts *et al.* 1978). Retinal development, though rudimentary, has been reported for a variety of taxa (Lang 1905).

Members of the various taxa have highly characteristic complements of setae on the idiosoma and appendages. Setae are the main tactile receptors, and many of them have also assumed secondary functions in locomotion, feeding, or mating. All water mites have specialized tactile organs, called glandularia, distributed in pairs on both the dorsum and venter of the body. Each glandularium consists of a small, goblet-shaped sac, or "gland," with a tiny external opening and an associated seta. The gland contains a milky, viscous fluid that is ejected abruptly when the seta is stimulated. This substance quickly stiffens to a sticky gel after it comes in contact with water and apparently evolved as a deterrent to attackers. In males of Arrenurinae, however, material secreted by the glandularia seems to function also in cementing the body of the female in place during copulation.

The five pairs of lyrifissures, which are arranged in series on the dorsum and venter of the body, are thought to be proprioceptors (Krantz 1978). In addition, the distal segments of the pedipalps and legs are well supplied with a variety of specialized setiform structures, apparently derived from outgrowths of the integument, which function as chemoreceptors. The most conspicuous of these organs are the solenidia that adorn the dorsal surfaces of the genua, tibiae, and tarsi of the legs and the tarsi of the pedipalps.

5. Reproductive System

Males have paired testes and vasa deferentia, which lead to an elaborate ejaculatory complex consisting of a series of membranous chambers attached to a sclerotized framework (Barr 1972). Positioned immediately above the genital field, the ejaculatory complex functions as a syringe-like organ for compacting masses of spermatozoa, assembling spermatophores, and expelling them from the genital tract through the gonopore.

Females have paired ovaries that are more or less fused and paired oviducts that also fuse to form a single duct leading to the genital chamber within the gonopore. Paired spermathecae, for storing spermatozoa picked up in spermatophores, flank the genital chamber and are connected to it by short ducts.

III. ECOLOGY

A. Life History

The basic life-history pattern of water mites (Fig. 16.1) was first correctly inferred by Wesenburg-Lund (1918) and has been confirmed by subsequent studies on a wide range of taxa (see Smith and Oliver 1986 for a summary of relevant literature). The first well-documented life-history studies were of palearctic species, especially Arrenurinae, investigated by Münchberg (1935a, 1935b). Subsequently, Mitchell (1959, 1964a) dealt intensively with a number of nearctic species of Arrenurinae, and Böttger (1962, 1972a, 1972b) treated palearctic species of Hydrachnidae, Limnocharinae, Eylaidae, Limnesiinae, Unionicolinae, Pioninae, and Arrenurinae in considerable detail. A meticulously comprehensive account of the life history of *Hydrodroma despiciens* (Müller) (family Hydrodromidae) was published by Meyer (1985).

1. Egg

Eggs are typically laid in masses in a gelatinous matrix and attached to plants, wood particles, or stones. Females of Hydrachnidae use a uniquely developed, elongate ovipositor to lay eggs individually in stems of aquatic plants, and those of certain Unionicolinae use short stylets to oviposit in the tissues of sponges or mussels. The most comprehensive morphological study of water mite eggs was published by Sokolow (1977).

Within the egg membrane, water mites pass through a transitory prelarval instar (Meyer 1985) and then usually develop rapidly and directly into larvae. Fully formed individuals can be observed moving within the egg membrane just prior to emerging 1–3 weeks after oviposition. Arrested larval development has been reported in certain species of Unionicolinae (Mitchell 1955), Teutoniidae, and Anisitsiellinae (Smith 1982), whose larvae do not become active and emerge until at least six months after eggs are laid.

Female mites presumably possess adaptations for selecting appropriate oviposition sites, as the larvae that emerge in closest proximity to an abundant supply of potential hosts must have the highest probability of survival.

2. Larva

a. *Host Selection and Parasitic Associations*

After emerging, larvae immediately begin to exhibit host-seeking behavior. To develop further, they must quickly locate hosts that both provide ade-quate sites for attachment and feeding and regularly visit habitats that are suitable for postlarval development. Insect hosts provide water mites with both the source of nutrition necessary for larval growth and their primary dispersal mechanism. Host associations of larval water mites, reviewed by Smith and Oliver (1976, 1986), are summarized in Table 16.1.

Larvae of Hydrovolzioidea and most Eylaoidea and Hydryphantoidea rise to search for potential hosts on the surface film, exhibiting plesiotypical terrestrial behavior. In contrast, larvae of Stygothrombidiidae, Hydrachnoidea, Lebertioidea, Hygrobatoidea, and Arrenuroidea begin to swim in the water column or crawl on the substratum as fully adapted aquatic organisms. This apotypical behavior developed in ancestral Hydrachnidae, and again in ancestors of the other three superfamilies. The development of fully aquatic larvae can best be understood as an important adaptation for locating potential hosts during their final preadult instar.

Surface-dwelling larvae are essentially terrestrial, and can locate hosts only when they visit or pass through the surface film. These opportunities are transitory events occurring irregularly in space and time. Some water mites with terrestrial larvae overcome the risks of failure by utilizing a relatively broad range of host organisms and producing large numbers of eggs. For example, larvae of Hydrovolziinae and certain subfamilies of Hydryphantidae apparently exploit a wide range of hosts, including insects that are only casual visitors to the habitat of the mites. This strategy presumably results in considerable wastage, as larvae attaching to hosts that do not return to suitable mite habitats will die.

Larvae of other hydryphantids and the various subfamilies of Eylaoidea have developed strong specificities for particular groups of insect hosts that live on the surface film (e.g., Hydrometridae, Gerridae), regularly visit it to replenish air supplies (e.g., aquatic Hemiptera and Coleoptera), or pass through it just before ecdysis and regularly revisit it after emergence (e.g., Odonata, Trichoptera, Diptera). This strategy improves the probability that a larva finding a host will be returned to an appropriate habitat, but limits the availability of potential hosts. These mites apparently compensate for the high risk of failure in finding hosts by producing many hundred eggs per female.

Terrestrial larvae seem to use both tactile and visual cues to locate hosts at the surface film. Those of several Hydryphantidae have the hind legs modified for jumping toward objects positioned above them, an obvious adaptation for locating their hosts on the surface film. After moving onto a potential host, terrestrial larvae seek out an appropriate attachment site, embed their chelicerae, and begin to feed.

In contrast, aquatic larvae have access to potential hosts throughout the penultimate life-history stage and can actively seek and select hosts in the water column or substratum prior to ecdysis. As a result, species with aquatic larvae can develop highly specialized associations with particular hosts that are intimately dependent upon the same habitat conditions as the mites for subsequent development. Larvae of Hydrachnidae exploit virtually the same range of hosts as Eylaidae, but locate them just beneath the surface film. Members of certain Anisitsiellinae and Teutoniidae (Lebertioidea), as well as Krendowskiidae and Arrenurinae exhibit a similar approach to seeking hosts. Their larvae locate swimming pupae of tanypodine Chironomidae, Culicidae, Ceratopogonidae, or Chaoboridae, or final instar larvae of Odonata in the water column below the surface film. Mites with this type of behavior apparently exploit appropriate hosts more efficiently than do those with terrestrial larvae, as their females tend to lay smaller though still substantial numbers of eggs.

Larvae of Stygothrombidiidae, Wandesiinae, and Rhyncholimnocharinae apparently locate their hosts, various Plecoptera and elmid Coleoptera, respectively, in the substratum. Larval stygothrombidiids and wandesiines are phoretic on nymphal stoneflies before transferring to the emerging adults and becoming parasites, while larval rhyncolimnocharines are subelytral parasites of adult elmids that remain in the water. Larvae of the remaining families and subfamilies of Lebertioidea, Hygrobatoidea, and Arrenuroidea locate hosts as final instar larvae or pupae on or in the substratum. The hosts, various Chironomidae or Trichoptera, construct cases or retreats where they remain during prepupal and pupal stages. These mites have developed precise larval adaptations to exploit this sedentary behavior, greatly increasing their probability of successfully locating and attaching to appropriate hosts. Details of the sensory mechanisms used by these mites to locate hosts remain unclear, but there is evidence that both olfactory and tactile cues are important (Böttger 1972b). After locating a host, these larvae cling to the integument of the pupa until it rises to the surface, then transfer to the imago as it emerges, embed their chelicerae, and enter the parasitic phase. Individual females in these groups lay relatively small numbers of eggs (often less than ten), indicating that the larvae efficiently find hosts with a high probability of returning to a suitable habitat.

The development of fully aquatic larvae had important consequences each time it occurred during water mite evolution. In particular, it was a crucial preadaptation for ancestors of the various groups of Lebertioidea, Hygrobatoidea, and Arrenuroidea that became specialized to exploit nematocerous Diptera as hosts. The origin and evolution of modern genera and families of these mites apparently followed the explosive radiation of the Nematocera, and especially Chironomidae, during the Jurassic and early Cretaceous. Refinement of their elegant adaptations for parasitizing these flies permitted the so-called "higher" water mites to coevolve with them and to exploit their potential for dispersal, invasion into new habitats, and rapid evolution.

b. *Site Selection, Engorgement, and Dispersal*

Larvae of many subfamilies of water mites exhibit strong selectivity for attaching to particular sites on the body of the host (Smith and Oliver 1976, 1986), as shown in Table 16.1. These preferences illustrate another adaptive process that has permitted mites to coevolve successfully with their hosts.

Larvae of Hydrovolziinae, and certain Hydryphantinae that parasitize Hemiptera, Plecoptera, and Trichoptera, seem to attach rather indiscriminately to various parts of the body of the host. Those of all other groups are more selective. Larval Eylaidae are essentially terrestrial organisms and select attachment sites bathed by the air supplies of the host, under the elytra of aquatic bugs and beetles (Fig. 16.182). Significantly, the fully aquatic larvae of Hydrachnidae are able to utilize a much wider range of sites on the same hosts (Fig. 16.183).

Larval Hydryphantoidea that parasitize Diptera tend to attach to thoracic sites, exhibiting apparently plesiotypical behavior. Larvae of many early derivative groups of Lebertioidea and most Arrenuroidea parasitizing Diptera also prefer thoracic sites. These larvae tend to be relatively large, and only a few individuals can share a single host. When more than one of these larvae attach to the same part of the thorax of the host, they are usually symmetrically arranged on either side of the body. Larvae of some species of Sperchontinae, Lebertiidae, Oxidae, and Torrenticolidae can use the anterior abdominal segments of hosts when thoracic sites are already occupied.

Larvae of virtually all Hygrobatoidea (except Limnesiidae, which may not be closely related to other taxa in the superfamily) and a few families of Arrenuroidea attach exclusively to abdominal sites on their hosts, usually nematocerous flies but in some cases caddisflies. Larvae of Arrenurinae that exploit Odonata are also predominantly abdominal parasites (Fig. 16.184). This apotypical behavior correlates with a marked trend to smaller size, allowing several larvae to share each attachment site on a host. These small larvae occupying abdominal attachment sites can exploit even the smallest species of Chironomidae as hosts without seriously impairing their ability to fly and disperse (Fig. 16.185).

The duration of the parasitic phase of larval development varies considerably. Larvae of several early derivative groups such as certain Limnocharinae and Hydryphantinae retain their mobility and can transfer from one instar to the next of the same host individual, completing engorgement only after the host reaches adulthood. This behavior may be the plesiotypical condition in water mites. Larval Hydrachnidae and Eylaidae remain attached to the same adult bug or beetle throughout the parasitic phase, for two weeks or more. Larvae of species in these families that are adapted to temporary pools often remain on hosts throughout the dry phase of the habitat, and only engorge after a period of arrested development that may last as long as ten months. Larvae of species parasitizing nematocerous Diptera or other aerial insects typically engorge rapidly and mature within a few hours. Fully fed larvae enter the quiescent nymphochrysalis stage.

Interestingly, the dispersal function of the larva appears to have assumed dominance over feeding and growth during water mite evolution. Larvae of early derivative groups such as Hydrachnidae, Eylaoidea, and some Hydryphantidae grow substantially, increasing several hundred-fold in size in some cases (Fig 16.182), while feeding on their relatively long-lived hosts. Larvae of more recently evolved groups associated with short-lived insects, especially nematocerous Diptera, still engorge on fluids from the host but increase only slightly in size in the process. The primary strategic role of the larval instar in the life history of most Hydryphantoidea, Lebertioidea, Hygrobatoidea, and Arrenuroidea is to ensure dispersal. Most feeding and growth occur during the deutonymph and adult stages in these mites.

The larva may be facultatively suppressed as an active instar in a few unrelated species in widely divergent groups (Thyadinae, Lebertiidae, Limnesiinae, Hygrobatidae, Pionidae, Axonopsinae, Arrenuridae), permitting the mites to accelerate development to exploit ephemeral optimal conditions (Smith and Oliver 1986). In these cases, larvae either remain within egg membranes or emerge for a brief period of activity before entering the nymphochrysalis stage.

3. Nymphochrysalis

Engorged larvae of all groups of water mites enter a quiescent stage, representing the suppressed protonymph, during which larval tissues are resorbed and reorganized and the deutonymph develops. Mature larvae of most groups drop off their aerial hosts into water, in response to environmental stimuli that are poorly understood. Larvae of various taxa may use host behavior, along with mechanical, visual, and chemical stimuli, as cues to the proximity of water (Böttger 1972a, 1972b). Engorged larvae are capable of only limited movement and those successfully reentering water typically seek out plant material, immediately attach themselves by their chelicerae, become quiescent, and transform to nymphochrysales. After a few days, fully formed deutonymphs emerge from the larval skins and become active.

Mites that parasitize large, long-lived insects, including Hydrachnidae, most Eylaoidea, and certain Hydryphantinae, apotypically pass through the nymphochrysalis stage while still attached to the host, within the larval integument. This trait, along with other adaptations, allows many of these mites to exploit periodically temporary habitats efficiently.

Normally, only larvae that are returned by their hosts to the parental habitat, or a very similar one, are able to transform to nymphochrysales and continue development. Larvae attached to hosts that die without returning to water or detaching from hosts in unfavorable habitats perish.

4. Deutonymph

In all known species, the deutonymphal instar is active and predaceous and resembles the adult, but is sexually immature. Deutonymphs typically exhibit relatively undeveloped idiosomal sclerites and chaetotaxy compared to adults. They have only a rudimentary (or provisional) genital field bearing an incomplete complement of acetabula. They feed voraciously, often preying upon immature instars of the same taxa of insects that they parasitized as larvae (Table 16.1). The deutonymphal stage varies in duration from a few days or weeks in early derivative groups such as Hydrachnidae, Eylaoidea, and certain Hydryphantidae, to several months in many groups of Lebertioidea, Hygrobatoidea, and Arrenuroidea. Deutonymphs of some hygrobatoid families are especially long-lived. Those of Pionidae typically live for many months through summer and the following winter, and this trait has preadapted certain species to exploit annually temporary pools. In these pionids, the deutonymphs are able to endure the dry phase of the cycle in damp retreats in the substrate (Smith 1976, Wiggins *et al.* 1980).

The deutonymph is the primary growth instar in most groups of water mites, and body size increases dramatically during this stage. On attaining adult size, deutonymphs prepare to transform to the imagochrysalis instar by embedding their chelicerae in plant tissue or soft detritus and becoming inactive.

5. *Imagochrysalis*

During this stage, further structural reorganization occurs to produce the adult. This final metamorphosis is rapid, and adults usually are ready to emerge within a few days.

6. *Adult*

Adults emerge from the deutonymphal integument in teneral condition with soft, pliable, and colorless sclerites. They immediately become active, crawling or swimming about while sclerotization of the body is completed and distinctive color patterns become evident. Adults of the various groups display a wide range of body shapes and arrangements of idiosomal sclerites, primarily as adaptations for living in different aquatic habitats. Shortly after emergence, both males and females mature sexually and begin to mate. In many taxa, males tend to emerge and become mature a few days earlier than conspecific females, and are ready to mate as soon as females become active in the habitat. Adults of certain Limnocharinae (Böttger 1972a) and possibly some Arrenurinae (Imamura 1952) undergo supernumerary moults.

The adult is a highly specialized reproductive instar in water mites, as clearly evidenced by the remarkable behavioral modifications and associated morphological adaptations that have developed in various groups to facilitate successful spermatophore transfer (reviewed in Smith and Oliver 1986).

The plesiotypical behavior pattern in water mites—indirect transfer with no contact between the sexes—has been observed in certain Hydrachnidae (Davids and Belier 1979), Eylaidae (Lanciani 1972), Hydryphantinae (Mitchell 1958), Hydrodromidae (Meyer 1985), and free-living Unionicolinae (Hevers 1978). Males deposit stalked spermatophores on the substratum for subsequent retrieval by conspecific females. Isolated males of some species will deposit large numbers of spermatophores. In other cases, the presence of females appears to stimulate males to deposit them. After finding the spermatophores, perhaps using pheromonal cues, females pick them up in the gonopore. This method of spermatophore transfer is widespread in actinedid mites and probably occurs in most groups of water mites with unmodified males.

In more interactive types of indirect transfer, males deposit spermatophores on the substratum, then actively assist females to find them and pick them up. Males of some Eylaidae lead receptive females in a circular dance, stopping repeatedly to deposit spermatophores at the same location on the circumference of the route. Halfway through each revolution of the dance, the female pauses with the male over his mass of spermatophores and picks up one or more of them in her gonopore (Lanciani 1972). In certain species of Unionicolinae that inhabit the mantle cavity of freshwater mussels, males retrieve their own spermatophores and carry them between the genua or tibiae of their hind legs while seeking receptive females. On finding a female, the male crawls beneath her and places his spermatophores in her gonopore (Hevers 1978).

The most complex modes of indirect spermatophore transfer occur in Arrenurinae. Males of the various species have the idiosoma modified posteriorly to form a characteristically shaped cauda (Fig. 16.286). A receptive female mounts a male, positioning herself so that her gonopore rests over the back edge of his cauda. Secretions from glandularia of the male temporarily cement the pair together, and the male then deposits one or more stalked spermatophores on the substratum. In some species, the male then rocks his body down and forward to bring the gonopore of the female in contact with the spermatophore (Lundblad 1929, Böttger and Schaller 1961, Böttger 1962). In others, the male uses his petiole, a cup-shaped posterior appendage of the cauda, to scoop up spermatophores into masses which he then inserts into the gonopore of the female (Böttger 1965). Variations of this behavioral pattern occur throughout the many species of Arrenurinae, explaining the multitude of different modifications in caudal morphology that characterize the males. This type of mating probably can be inferred in all groups of water mites with elaborate caudal development in males.

Males of many groups of Hygrobatoidea produce unstalked spermatophores, which they carry between modified claws on the tarsi of their flexed third pair of legs before transferring them directly into the gonopore of the female. This type of mating has been observed in various Feltriidae (Motas 1928), Foreliinae (Uchida 1940), Tiphyinae (Viets 1914, Mitchell 1957c), Pioninae (Koenike 1891, Mitchell 1957c, Böttger 1962, Smith 1976), and Axonopsinae (Motas 1928, Halik 1955). In each case, the male uses specially modified appendages (usually the fourth pair of legs) to hold the female in a characteristic posture so that he can transfer spermatophores to her gonopore by simply extending his third pair of legs. This mating behavior probably occurs in all water mites whose males have highly modified leg segments but no elaborate caudal development. Finally, direct transfer of spermatophores from gonopore to gonopore during copulation occurs in certain Eylaidae (Böttger 1962, Lanciani 1972) and Mideidae (Lundblad 1929). In these cases, males either have the gonopore protruding from the venter on a cone-shaped projection (Eylaidae) or are equipped with wing-like appendages flanking the gonopore (Mideidae) to facilitate transfer.

Clearly, each type of direct spermatophore transfer has evolved more than once from an indirect mode involving complex sexual interaction. Direct transfer by specialized appendages of the male during copulation has developed independently several times in the superfamily Hygrobatoidea, including at least three times within the family Pionidae (Mitchell 1957c, Smith 1976). Direct transfer from gonopore to gonopore certainly developed independently in Eylaidae and Mideidae.

The often profound modifications of the male idiosoma and appendages that are associated with diverse modes of spermatophore transfer emphasize the highly specialized mating function of the adult instar in water mites. Males in many groups mate and die within a few days of emerging. Those of some groups, such as Arrenurinae, may live for several weeks. Males of some Foreliinae and Tiphyinae are so modified morphologically that they are unable to move about easily, and are restricted to crawling slowly over the substratum using the anterior pairs of legs.

Mated females may live for many months, continuing to feed while they produce several clutches of eggs. In many groups, mating occurs in late summer and fertilization is delayed while females overwinter. These females lay their eggs after they have been fertilized early in the following spring or early summer. In contrast, females of species inhabiting temporary pools must lay their eggs within a few days of mating to ensure that their offspring reach a life-history stage capable of avoiding or withstanding the dry period before water disappears from the habitat (Smith 1976, Wiggins *et al.* 1980).

B. Habitats and Communities

Throughout water mite evolution, successful invasion and exploitation of new habitats has depended on the development of compatible adaptive strategies for the various instars. Larval traits tend to promote parasitism and dispersal on hosts, while those of deutonymphs and adults favor feeding, growth, and reproduction in water. Water mite species are regularly provided with opportunities to colonize new habitats by the passive transport of their larvae on hosts, and this has been the primary mech-

anism promoting speciation and divergence over time. Nevertheless, in modern communities, most species and many monophyletic groups representing genus or family level taxa are restricted to one or a few similar types of habitats. The strong correlation between certain clades and habitats suggests that conservative factors such as adaptive requirements for locating hosts, prey, mates, and oviposition sites tend to constrain adaptive radiation. It also indicates that successful invasion of new habitats usually resulted in the rapid evolution of significant new life-history, behavioral, and morphological traits.

The communities living in the major types of freshwater habitats in North America are outlined next and in Table 16.1.

1. Springs (Including Seepage Areas)

Species of 23 family-group taxa live in moss and wet detritus associated with rheocrenes and helocrenes. Members of ancient taxa that may well have originated in this type of habitat are soft-bodied (e.g., Tartarothyadinae), or partly to fully sclerotized (e.g., Thyadinae). Adults of derivative groups that apparently invaded spring and seepage habitats from flowing water more recently have entire dorsal and ventral shields (e.g., Nudomideopsidae, Mideopsinae, Laversiidae). Most spring-inhabiting species are cold-adapted stenophiles, but some Wandesiinae (subgenus *Partnuniella*) and all Thermacaridae live only in hot springs.

2. Riffle Habitats

Species of 25 taxa live on the substratum in rapidly flowing areas of streams and rivers. Members of a few early derivative taxa are soft-bodied (e.g., Protziinae), but adults of most groups are both strongly flattened and well sclerotized. Most rheophilic species are cold- or cool-adapted stenophiles.

3. Interstitial Habitats

Species of 34 taxa live in sand and gravel deposits to depths of 50 cm or more, mainly in the hyporheic zone of streams and rheocrenes. Deutonymphs and adults crawl through the spaces between particles, often with considerable speed and agility. Members of most early derivative taxa are soft-bodied (e.g., Rhyncholimnocharinae) and some are also vermiform (e.g., Stygothrombidiidae, Wandesiinae). In adults of more recently derived groups, the dorsum and venter are partly covered by various plates or small shields (e.g., Clathrosperchoninae, Kawamuracarinae, Hygrobatidae), or are well sclerotized with either entire dorsal and ventral shields or greatly expanded coxal plates (e.g., Aturidae, Ar-

renuroidea). Mites adapted to interstitial habitats
have reduced eyes and lack pigmentation of the integument.

4. Stenothermic Pools

Species of 31 taxa live in silty substrata in spring-fed
pools, fen pools and depositional areas of streams.
Members of ancient taxa are soft-bodied (e.g.,
Pseudohydryphantinae), while adults of early derivative clades exhibit various degrees of sclerotization ranging from slight enlargement of dorsal and
ventral plates to development of entire dorsal and
ventral shields (e.g., Teutoniidae, Wettininae,
Foreliinae, Pioninae, Momonniinae, Acalyptonotidae). Adults of certain taxa that invaded pools relatively recently from lotic and hyporheic habitats
have retained dorsal and ventral shields, though
they have become secondarily adapted for swimming (e.g., Axonopsinae, Mideopsinae). Certain
groups inhabiting depositional areas of streams
(e.g., Teutoniidae, Oxidae, Wettininae) include the
fastest and most agile swimmers among the water
mites. Most inhabitants of pools are cool-adapted
stenophiles.

5. Lakes

Species of 33 taxa inhabit large bodies of standing
water (including permanent ponds, marshes,
swamps, and bogs). Members of ancient and early
derivative groups are generally soft-bodied (e.g.,
Hydrodromidae, Limnesiinae, Hygrobatidae, certain Pionidae), while adults of several taxa that have
secondarily invaded standing water from lotic or
hyporheic habitats retain extensive dorsal and ventral shields (e.g., Aturidae and most Arrenuroidea).
Most species are good swimmers and have fringes
of long, slender setae on the genua and tibiae of
the second, third, and fourth pairs of legs. These
swimming setae help to propel the mites as they
move about at the surface of the substratum or
among plants. Some species of Unionicolinae and
Pioninae have especially long swimming setae, and
are essentially pelagic or planktonic (Riessen 1982).
A few species that lack swimming setae either walk
or crawl actively on plants and detritus, reflecting
recent rheobiontic ancestry (e.g., Sperchontinae,
Torrenticolinae, Mideopsinae, Neoacaridae). Members of Tyrrelliinae are unusual in that they inhabit
wet litter at the edges of pools, ponds, and lakes,
where they crawl about at the surface film. Many
species of Unionicolinae and the only known species
of Najadicolinae are obligate parasites or commensals in the mantle cavity of molluscs in both lentic
and lotic habitats (Mitchell 1955, Simmons and
Smith 1984, Gledhill 1985, Vidrine 1986).

Arrenurinae and Pioninae are remarkable in that
25 or more species of each group are often sympatric
in a single small lake. Mitchell (1964a, 1969) concluded that behavioral flexibility exhibited by larval
Arrenurinae in selecting and exploiting their odonate hosts is the main factor permitting such high
levels of sympatry, and that competition between
deutonymphs and adults of the various species is
probably not significant in natural communities.

Most limnophilic taxa tolerate a broad range of
temperatures which they encounter in weed beds
and silty substrata in the littoral and sublittoral
zones. However, a few are cold-adapted stenophiles
restricted to either arctic–alpine glacial lakes (e.g.,
Acalyptonotidae) or the profundal zone of oligotrophic lakes (e.g., Huitfeldtiinae, Laversiidae). Certain species of various taxa are highly tolerant of
extreme thermal and chemical regimes, and are able
to inhabit desert sloughs, alkaline lakes, or acid bogs
(e.g., Limnocharinae, Hydrodromidae). The water
mite fauna of wetland habitats in Canada was treated
by Smith (1987).

6. Temporary Pools

Species of eight taxa live in annually temporary
pools. Members of ancient taxa that may have originated in this type of habitat range from soft-bodied
walkers (e.g., Thyadinae) and swimmers (e.g., Eylaidae) to mites with unique arrangements of integumental plates that either crawl (e.g., Piersigiinae) or
swim awkwardly (e.g., Hydrachnidae, Hydryphantinae). Species of derivative groups that apparently
invaded temporary pools secondarily tend to be relatively good swimmers (e.g., Tiphyinae, Pioninae,
Arrenurinae). Water mites inhabiting temporary
pools are adapted either to avoid (e.g., Hydrachnidae, Eylaidae) or to endure the dry phase of the
annual cycle (Wiggins *et al.* 1980).

C. Ecological Importance

The parasitic larvae and predaceous deutonymphs
and adults of water mites have direct and almost
certainly significant effects on the size and structure
of insect populations in many habitats (B. P. Smith
1983, 1988, Lanciani 1983). Unfortunately, their impact has rarely been measured accurately because of
the routine neglect of mites in ecological studies of
freshwater communities. Due to their small size and
often cryptic habits, mites are usually absent or seriously under-represented in samples that are collected using standard techniques for capturing insects and crustaceans. Most entomologists are
unfamiliar with mites and tend either to disregard
them as too poorly known ecologically and difficult

taxonomically or to lump them together in a meaningless way, when conducting community studies. Failure to include a realistic assessment of the roles played by water mites often results in seriously flawed analyses of freshwater communities.

1. Impact as Parasites

The impact of larval water mites on host insects was reviewed recently by Smith (1988). Larval water mites regularly parasitize 20–50% of adults in natural populations of aquatic insects in such diverse families as Corixidae (Hemiptera), Dytiscidae (Coleoptera), Libellulidae (Odonata), Culicidae, and Chironomidae (Diptera). In some host populations, almost all individuals are parasitized.

Experimental studies have demonstrated that parasitism by larval water mites impairs the vitality, growth, mobility, and fecundity of host insects and these effects are proportional to the parasite load (B. P. Smith 1983, 1988, Lanciani 1983).

The dynamics of natural water mite populations in North American habitats are poorly understood, and well-documented, long-term studies are needed. Casual observations suggest that the species composition of communities in undisturbed habitats remains quite constant for periods of 25 years or more. Species that become well established in a habitat often maintain apparently stable populations for many generations. In these cases, the life cycles of the mites must be synchronized with those of their hosts so that the periods of emergence of mite larvae and adult insects coincide. In general, host loading by parasitic larvae must be limited to levels that permit some hosts to reproduce, though there is evidence of occasional mite-induced crashes of host populations (Smith 1988). Many species of mites parasitize several related species of host insects. This often results in species of insects within a habitat, especially among Chironomidae, being parasitized simultaneously by larvae of several species of mites. In extreme cases, up to six or seven species of mites representing as many different genera may attack the same chironomid host. Maintenance of viable populations of mites and their hosts must then depend upon a dynamic equilibrium involving all interacting species of the community.

Research is needed to identify the factors that influence the stability of natural populations, and to investigate the mechanisms that may promote balanced interaction between populations of mites and hosts.

2. Impact as Predators

The feeding habits of deutonymphal and adult water mites were reviewed by Böttger (1970) and summa-

rized more recently by Gledhill (1985). Data for species inhabiting Canadian wetlands were reviewed by Smith (1987). Deutonymphs and adults of free-living species are typically voracious predators of small aquatic organisms, including eggs of insects and fish, larvae of nematocerous Diptera and other insects, and a wide range of ostracod, cladoceran, and copepod crustaceans. Some species at least occasionally scavenge on dead organisms. Most species exhibit marked preferences for prey of a particular size with particular morphological and behavioral attributes, although some Limnesiinae and Pioninae become more opportunistic under laboratory conditions. Walking and crawling mites tend to feed on sedentary organisms such as insect eggs or tube-dwelling larval chironomids, while agile swimmers can hunt active prey such as mosquito larvae or cladoceran crustaceans. Adults of some groups, especially Limnesiinae, attack other water mites when confined in crowded containers, but there is no evidence to suggest that they do so regularly in natural habitats.

In experimental studies, some deutonymphs and adults exhibit feeding rates that are high enough to be considered as potentially important influences on the size and structure of prey populations in nature (Paterson 1970, Mullen 1975, Lanciani 1979, Wiles 1982, Winkel and Davids 1985, 1987).

Interestingly, deutonymphs and adults of mite species commonly prey upon the immature instars of the same species of insects that they parasitize as larvae (Davids 1973, Wiles 1982). This phenomenon requires further investigation, but preliminary indications suggest that many species of mites are specialized to exploit one group of insects throughout their life history.

3. Importance as Prey

Water mites are not a major prey group for freshwater vertebrates, though deutonymphs and adults occasionally form a significant part of the diets of fish and turtles (Marshall 1933, 1940). We have seen examples of stomach contents from individual brook trout and coregonids that consisted exclusively of several hundred adults of one species of Pioninae and Hygrobatidae, respectively.

Experiments have shown that red-colored water mites are distasteful to fish, which quickly learn to reject them as potential prey (Elton 1922, Kerfoot 1982). Interestingly, most red mites live in marginal habitats or temporary pools where fish are absent. This suggests that red pigments evolved in these mites mainly in response to other environmental factors such as the need to absorb heat from incident radiation in cold-water habitats. Seemingly, aposematic coloration is a secondary benefit to the mites.

4. Potential as Indicators of Environmental Quality

Species of water mites are specialized to exploit narrow ranges of physical and chemical regimes, as well as the particular biologic attributes of the organisms they parasitize and prey upon. Consequently, water mites should be exceptionally sensitive indicators of the impact of environmental changes on freshwater communities. Preliminary studies of physicochemical and pollution ecology of the relatively well-known fauna of Europe (Schwoerbel 1959, Pieczynski 1976, Biesiadka 1979, Kowalik and Biesiadka 1981, Kowalik 1981, Bagge and Merilainen 1985) have demonstrated that water mites are excellent indicators of habitat quality. The results of these studies, along with our own observations in sampling a wide variety of habitats in North America and elsewhere, lead us to conclude that water mite diversity is dramatically reduced in habitats that have been degraded by chemical pollution or physical disturbance.

Progress toward the establishment of essential baseline documentation on water mite communities in North America, however, continues to be hampered by lack of sufficient systematic and ecological information at the species level and by inadequate sampling and extraction procedures. Application of potentially useful information on water mites in assessing and monitoring the well-being of freshwater habitats and communities demands clearer understanding of the individual niche requirements and trophic relationships of species. This can be achieved by improving systematic knowledge of the group and by developing more appropriate collecting, observing, and rearing techniques.

IV. COLLECTING, REARING, AND PREPARATION FOR STUDY

These subjects were treated comprehensively by Barr (1973), and we discuss here only those aspects needing further emphasis or refinement.

A. Collecting and Extracting Techniques

Thus far, field work on water mites in North America has focused largely on qualitative sampling to obtain specimens required for systematic or life-history studies.

1. Deutonymphs and Adults

In most habitats and for virtually all types of substrata, the best results are achieved using a proce-

dure that requires a strong net with a wide opening (25–40 cm diameter) and fine-mesh size (250 mm), a supply of strong, clear polyethylene bags (4–5 kg capacity), a set of standard sieves of various mesh size (ranging from 250–2 mm), six large, leakproof polyethylene containers (1 liter), an ice chest, six rectangular, white photographic trays (30 × 40 cm), a flashlight, a set of eye droppers with a range of differently sized tip openings, and a supply of small, leakproof polyethylene containers (50 ml).

Once the desired habitat or microhabitat has been selected, the substratum must be thoroughly stirred. In flowing water, the net should be positioned and secured so that the current will carry dislodged organisms and particles of substratum into it. In lentic habitats, it must be swept back and forth through the column of mixed water and substratum to pick up the organisms.

In very shallow habitats, such as seepage areas, small rheocrenes, fens, and the edges of pools and ponds, the substratum must be gathered and stirred by hand. The net is worked beneath the edge of the wet moss, leaf litter, or detritus that is to be sampled so that mats of wet substratum can be separated and rinsed over the opening. Where there is sufficient surface water, successive handfuls of loose material are picked or scooped up with the aid of a small trowel, thoroughly picked apart and washed in the net. Larger pieces of detritus are discarded after careful washing. In seepage areas, spring edges, fens, and bog margin habitats where there is often an extensive carpet of wet moss and associated plants, mites can also be collected by methodically and vigorously treading down the plants while dragging the net along to scoop up the detritus, silt, and organisms that become temporarily dislodged as water is forced up through the disturbed layer of vegetation. In either case, when a mixture of fine silt, plant fragments, and organisms begins to fill the bottom of the net, it is transferred to a plastic bag containing a small amount of water. This process is repeated until the bag is approximately half full of material. The bag is then nearly filled with clean water and the contents stirred gently but thoroughly to allow heavy inorganic material such as gravel and sand to settle to the bottom. The contents of the bag are then poured carefully through a set of sieves so that gravel and sand remain in the bag. This operation is repeated until all loose organic material is in the sieves. We find that a pair of upper and lower sieves with mesh sizes of 1.4 mm and 250 μm, respectively, yields good results. The coarse material in the upper sieve is thoroughly stirred and rinsed with clean water. The fine silt and small organisms accumulating in the lower sieve are regularly transferred to a large polyethylene container with enough water to just cover the surface of the silt. When the container is about half full of fine silt, it is closed and placed in an ice chest to be transported back to the lab or sorting area. In situations where there is insufficient surface water to permit washing and rinsing of samples on site, accumulations of unsorted substratum material must be gathered, stored in plastic bags, and transported on ice to a source of clean water for subsequent processing.

With few modifications, this procedure can be used effectively in other types of habitats. Where the water column is between 25 cm and 1 m in depth and the bottom is firm enough to permit wading, the feet, hands, and a net can be used to agitate the substratum (with the aid of a spade when necessary). In stream riffles with a heterogeneous substratum of rocks, gravel, sand, and detritus, vigorous and persistent agitation is usually necessary to dislodge the large number and variety of mites that cling tenaciously to rocks or burrow in the silt under larger stones. We get good results by securing the net approximately 1–2 m downstream, and then systematically working through the substratum to loosen and thoroughly rinse all particles so that organisms are carried into the net. Where well-developed interstitial habitats occur in a stream bed, this approach can be extended to yield rich collections of mites including both surface and hyporheic species. This is accomplished by simply digging down through the gravel and sand beneath the riffle zone and removing all large stones to allow the current to flush out the pockets of interstitial silty detritus and the organisms that live there. By digging out an area of substratum of approximately 1 m² to a depth of 0.5 m, we are often able to recover good numbers of many hyporheic species. This technique can often be used to sample the hyporheic zone in habitats where the standard Karaman-Chappuis method (Chappuis 1942, Gledhill 1982) is not practical.

In littoral lentic habitats such as depositional pools in streams, ponds, and sheltered bays in lakes, the feet can often be used to thoroughly stir up the top few centimeters of sediment and any rooted aquatic plants that may be present. The net is then repeatedly swept back and forth through the entire volume of disturbed water column to ensure that both swimming and crawling species are captured. As some mites in lentic habitats are too large to pass readily through 1.4 mm mesh, we recommend use of an upper sieve with 2.0 mm mesh when sampling ponds and lakes.

The procedure outlined here can also be used by scuba divers and a boat operator to collect mites and insects from the sublittoral and profundal zones of lakes down to depths of 30 m or more, at a much higher level of efficiency than is possible with conventional grabs or dredges.

In the sorting area, a number of white photographic trays are set up on tables and filled to a depth of 2–4 cm with clean water. Once the water has settled, the contents of one polyethylene container are carefully decanted into each tray, so that the silt spreads out in the center of the tray but does not reach the edges. As the mites become active and begin to move around in the trays, they are picked up individually using an eye dropper and transferred to small polyethylene containers. Many swimming species are positively phototropic and can be recovered efficiently from trays kept in bright light. However, most crawling mites and a considerable number of swimming species are negatively phototropic and become active only in darkness. The effectiveness and rate of recovering mites from all types of habitats can be substantially increased by reducing ambient lighting to low levels indoors or waiting until after sunset outdoors, and then scanning the trays systematically and periodically with a flashlight beam. An experienced collector can then spot the mites as they move across a contrasting background provided by either the darkly colored silt or the white bottom of the tray. Mites remain active and continue to emerge from silt in the trays for up to 72 hr, as deoxygenation of the water proceeds at room temperature. For this reason, we suggest that, whenever possible, mites should be extracted from samples in situations that permit periodic observation of undisturbed trays over a period of at least 4–6 hr with controlled lighting. Under these conditions, mites tend to congregate around the edges of the trays, especially in the corners, where they can be easily spotted and picked out.

Although water mites frequently are found in samples collected for quantitative analysis in ecological studies, they are almost always underrepresented in terms of both numbers and diversity. This is often because collecting devices are designed or selected to capture organisms that are larger or more sedentary than mites. Due to their small size and ability to cling tenaciously to substrata, mites in lotic habitats can often resist casual efforts to dislodge them or can simply pass through nets with coarse mesh. By running or swimming, mites in hyporheic or lentic habitats can avoid capture by traps that are designed to sample instantaneously small areas of superficial substrate. Mites specimens are frequently overlooked when preserved samples are sorted because extraction techniques concentrate on larger organisms or depend upon flotation, a method that is ill-suited for separating water mites from detritus.

Recently, a promising technique has been developed for extracting water mites and other aquatic arthropods quantitatively from samples of substratum by inducing them to respond actively to a temperature gradient (Fairchild *et al.* 1987). This method has yielded very good results for samples of baled sphagnum from bog ponds, and could probably be adapted for use with any type of substratum inhabited by mites.

2. Larvae

Free-living larvae are often collected from aquatic habitats along with deutonymphs and adults, but these specimens must be preserved in order to be identified and often cannot be reliably associated with other instars of the same species. Parasitic larvae can be recovered from insect hosts collected at or near the parental habitats. Aquatic hosts are usually captured using a dip net, whereas aerial adults can be trapped as they emerge from water, netted while swarming (especially over water at dusk), swept from vegetation, or attracted to lights at night.

B. Rearing

Larvae needed for taxonomic or life-history studies can be reared from females using the techniques described by Prasad and Cook (1972), Barr (1973), and Smith (1976). Fertilized females of many species can be collected in spring and early summer. Females needed for rearing should be placed individually in small dishes or vials containing water and a few fragments of moss or wood as a substratum. The containers should then be covered and placed in a shaded, well-ventilated location where the temperature can be maintained near levels experienced in the natural habitat of the mites. Mites from cold, stenothermic habitats must be refrigerated. The rearing containers should be checked periodically for the presence of egg masses. Females should be preserved immediately after they have laid their eggs to ensure that good quality specimens are available for confirming species identifications. Larvae should be either preserved or slide-mounted directly as they emerge.

C. Preservation and Preparation for Study

Deutonymphal and adult mites must be preserved in Modified Koenike's Solution (or GAW), consisting of 5 parts glycerin, 4 parts water, and 1 part glacial acetic acid, by volume (see Barr 1973), so that they can be properly cleared, dissected, and slide-mounted for identification and study. Specimens preserved in alcohol or other hygroscopic agents become so distorted and brittle that subsequent preparation is difficult or impossible. Deutonymphs and adults must be cleared in either acetic corrosive or 10% KOH, dissected, and mounted in glycerin jelly, following the techniques described by Barr (1973) and Cook (1974).

Reared larvae are often slide-mounted directly, but can be preserved in 70% alcohol for subsequent mounting without being seriously damaged. Parasitic larvae should be preserved with the host in 70% alcohol, although larvae removed from hosts that have been pinned and dried often can be slide-mounted successfully. Before larvae are removed from hosts, the attachment sites should be noted and recorded. Larvae should be mounted whole in Hoyer's medium (see Barr 1973).

V. TAXONOMIC KEYS TO SUBFAMILIES OF WATER MITES IN NORTH AMERICA

The following keys for larvae and adults permit identification of specimens to superfamily, family, and subfamily level. Subfamily names are followed by a list of known nearctic genera that are diagnosed by the character states in the respective couplets. The family name alone is used where families include only a single subfamily. The larval key does not include all taxa that appear in the adult key, as larvae of several subfamilies remain unknown (Table 16.1). Knowledge of larvae in a number of other subfamilies is based on only a few species. Consequently, the key for larvae will be subject to considerable revision and refinement as larvae of more species are reared and described.

Those wishing to identify specimens to genus are referred to Prasad and Cook (1972) for larvae, and Cook (1974) for adults, and to the series of more recent papers by Smith, listed by Viets (1982, 1987).

The keys presented here are designed for use with a minimum of specimen preparation. Larvae should be whole-mounted on slides. The legs should be spread out from the body as much as possible by gently pressing down on the coverslip, as the chaetotaxy of leg segments is frequently used as a key character. Adults require clearing and some dissection prior to slide-mounting. The mouth parts (capitulum and appendages) should be removed to permit examination of the pedipalps in medial and lateral view. Dorsal and ventral shields, when present, should be separated to allow clear viewing of certain key characters.

We follow the convention for numbering the setiform structures on the leg segments of larvae that was established by Prasad and Cook (1972) and modified slightly by Smith (1976) (Figs. 16.6–16.8). The setae are numbered starting proximally on the posterolateral surface, proceeding distally along the dorsal half of the segment to the end, then returning proximally along the ventral half of the segment. Exceptions are made in the cases of solenidia (setiform chemoreceptors, see Fig. 16.30, arrow) which are numbered first, that is as "s1" on the genua, tibia III, and the tarsi, and as "s1" and "s2" on tibiae I and II, and tarsal eupathids (specialized setae, see Fig. 16.96, arrow) which are numbered second, that is as "2" on the tarsi. Where it has been necessary to use leg chaetotaxy in the key, the number of solenidia present on a segment is indicated in brackets (e.g., "+2s") immediately following the information on the number of setae, so that we can account for all setiform structures. We usually have omitted the empodia and claws from the illustrations of the legs of larvae so that the number and positions of distal setae on the tarsi can be seen clearly. The names of the segments of the appendages are consistently abbreviated as follows: trochanter, Tr; femur, Fe (basifemur, BFe; telofemur, TFe); genu, Ge; tibia, Ti; and tarsus, Ta.

A. Taxonomic Key to Subfamilies for Known Larvae of Water Mites

1a.	Legs with six movable segments [basifemur (BFe) and telofemur (TFe) separated] (Figs. 16.9, 16.18, 16.38, 16.56)	2
1b.	Legs with five movable segments (basifemur and telofemur fused) (Figs. 16.6–16.8, 16.54, 16.55, 16.60)	13
2a(1a).	Dorsal plate absent (Figs. 16.27 and 16.33), or small, covering less than one-third of length of idiosoma and usually with setae SS at posterior edge (Figs. 16.11, 16.23, 16.25, 16.26); lateral eyes borne on separate platelets (Figs. 16.23, 16.25–16.27, 16.33); leg tarsi with unmodified claws and empodium (Emp) (Fig. 16.22)	3
2b.	Dorsal plate large, covering well over one-third of length of idiosoma and with setae SS near midlength (Figs. 16.34, 16.36, 16.37); lateral eyes borne on single eye plates (Figs. 16.34, 16.36, 16.37); leg tarsi with claws modified or reduced (Figs. 16.40 and 16.45) .... Eylaoidea 10
3a(2a).	Dorsal plate present, bearing only setae AM and AL near anterior edge (Fig. 16.11); bearing irregular rows of urstigmata at anterior edges of coxal plates II (Fig 16.12, arrow); pedipalp tibia (Ti) reduced, bearing four thick setae that overlap tarsus (Fig. 16.10); humeral setae (Hu) foliate (Fig. 16.11) .... Hydrovolzioidea .... Hydrovolziidae Hydrovolziinae *Hydrovolzia*
3b.	Dorsal plate absent (Figs. 16.27 and 16.33), or present and bearing at least one pair of setae (SS) in posterior half near edge (Figs. 16.23, 16.25, 16.26); bearing one or two pairs of urstigmata (Figs. 16.16 and 16.24; arrows); pedipalp tibia well developed, bearing thick, claw-like seta distidorsally that extends above cylindrical, thumb-like tarsus (Figs. 16.15 and 16.19; arrows); humeral setae (Hu) not foliate (Figs. 16.23, 16.25, 16.27) .... Hydryphantoidea 4
4a(3b).	Dorsal plate roughly quadrangular, bearing all four pairs of propodosomal setae (Figs. 16.23 and 16.26), or three pairs with setae PL borne on separate platelets near posterolateral angles of plate	5
4b.	Dorsal plate triangular, fragmented, or absent, bearing fewer than three pairs of propodosomal setae (Figs. 16.25, 16.27, 16.33)	8
5a(4a).	Claw (movable digit) of chelicerae over one-half of length of basal segment (Fig. 16.14); excretory pore plate setae and their alveoli absent (Fig. 16.13) .... Thermacaridae *Thermacarus*
5b.	Claw (movable digit) of chelicerae less than one-third of length of basal segment (Fig. 16.23); excretory pore plate setae, (E1 and E2) or at least their alveoli, present (Figs. 16.16 and 16.24) ...... Hydryphantidae (in part) 6
6a(5b).	Coxal plates II lacking setae (C3 absent) (Fig. 16.16); pedipalp tibia with terminal claw-like seta deeply bifurcate (Fig. 16.15, arrow) and tarsus with only one thickened seta; excretory pore plate absent but setae E1 and E2 present (Fig. 16.16) ...... Wandesiinae *Wandesia*

6b.   Coxal plates II bearing setae C3 (Fig. 16.24); pedipalp tibia with terminal claw-like seta only slightly bifurcate or undivided, and tarsus with two thickened setae (Fig. 16.17, arrows); excretory pore plate present; setae E1 and E2 present or absent, but their alveoli always present (Fig. 16.24) 7

7a(6b).   Medial eye present (Fig. 16.23, arrow) .... Thyadinae   *Panisus, Panisopsis, Thyas, Thyasides, Euthyas, Zschokkea*

7b.   Medial eye reduced (Fig. 16.26) .... Tartarothyadinae   *Tartarothyas*

8a(4b).   Dorsal plate triangular, bearing setae PL and SS, with setae AM and AL borne on small platelets flanking apex of plate (Fig. 16.25); excretory pore plate setae present, with E1 borne on plate (Fig. 16.20) and E2 borne on small separate platelets; basal segment of chelicerae massive and longitudinally striate (Fig. 16.28) .... Hydryphantidae (in part)
Hydryphantinae   *Hydryphantes*

8b.   Dorsal plate fragmented or absent, with propodosomal setae borne on paired platelets (Figs. 16.27, 16.33); excretory pore plate setae reduced to vestiges or absent but their alveoli present (Figs. 16.21 and 16.31); basal segment of chelicerae relatively slender and smooth (Fig. 16.27) 9

9a(8b).   Medial eye present (Fig. 16.33, arrow); excretory pore plate well sclerotized and nearly triangular (Fig. 16.21); pedipalp tarsus with all setae slender; solenidia on leg tarsi slender (Fig. 16.32, arrow) .... Hydryphantidae (in part)
Protziinae   *Protzia, Partnunia*

9b.   Medial eye absent (Fig. 16.27); excretory pore plate poorly sclerotized and oblong (Fig. 16.31); pedipalp tarsus with two thickened, blade-like setae (Fig. 16.29, arrows); solenidia on tarsi of legs I and II very thick (Fig. 16.30, arrow) .... Hydrodromidae   *Hydrodroma*

10a(2b).   Coxal plates I, II, and III on each side bearing two, two, and two setae, respectively (Figs. 16.35 and 16.43); tarsi of legs bearing paired claws and claw-like empodium (Emp) (Fig. 16.40) 11

10b.   Coxal plates I, II, and III on each side bearing two or one, one, and one setae, respectively (Figs. 16.39 and 16.53); tarsi of legs bearing two dissimilar claw-like structures terminally (Fig. 16.45) .... Limnocharidae   12

11a(10a).   Dorsum of idiosoma nearly covered by single, elongate dorsal plate in unengorged larvae, bearing seven pairs of setae (or their alveoli) (Fig. 16.34); coxal plate setae simple (Fig. 16.35); excretory pore plate elongate (Fig. 16.44) .... Eylaidae   *Eylais*

11b.   Dorsum of idiosoma with three separate plates bearing five, one, and one pairs of setae, respectively (Fig. 16.36); coxal plate setae C1, C2, C5, and C6 blunt and conical in shape (Fig. 16.43); excretory pore plate obcordate (Fig. 16.43) .... Piersigiidae .. Piersigiinae   *Piersigia*

12a(10b).   Dorsal plate covering no more than anterior half of idiosomal dorsum, bearing four pairs of propodosomal setae (Fig. 16.37); coxal plates I bearing two pairs of setae, C1 and C2 (Fig. 16.39); excretory pore plate diamond-shaped (Fig. 16.41) .... Limnocharinae   *Limnochares, Neolimnochares*

12b.   Dorsal plate covering entire idiosomal dorsum in unengorged larvae, bearing four pairs of propodosomal setae and three pairs of hysterosomal setae (Fig. 16.47); coxal plates I bearing one pair of setae (Fig. 16.53); excretory pore plate pyriform (Fig. 16.53) .... Rhyncholimnocharinae   *Rhyncholimnochares*

13a(1b).   Dorsal plate bearing three pairs of propodosomal setae and an unpaired anteromedial seta (Fig. 16.52, arrow); urstigmata (Ur) stalked (Figs. 16.51 and 16.52); pedipalp tibiae with terminal claw-like seta four-pronged (Fig. 16.48, arrow), leg tarsi bearing paired pectinate claws and a simple empodium (Fig. 16.54) .... Stygothrombidioidea
Stygothrombidiidae   *Stygothrombium*

13b. Dorsal plate bearing four pairs of propodosomal setae, and in some cases one or more pairs of hysterosomal setae, but no unpaired setae (Figs. 16.49, 16.57, 16.59); pedipalp tibiae with terminal claw-like seta undivided (Fig. 16.58, arrow A), or deeply bisected (Figs. 16.46 and 16.64, arrows); leg tarsi bearing paired simple claws (Fig. 16.78) or lacking claws (Fig. 16.42), but always bearing an empodium 14

14a(13b). Dorsal plate bearing four pairs of propodosomal setae, the humeral setae, and three pairs of mediohysterosomal setae (Fig. 16.49); leg tarsi lacking paired claws (Fig. 16.42); excretory pore plate tiny, and bearing neither setae nor their alveoli (Fig. 16.50, arrow) .... Hydrachnoidea
                             Hydrachnidae *Hydrachna*

14b. Dorsal plate bearing four pairs of propodosomal setae (Fig. 16.59), and in some cases one or at most two pairs of mediohysterosomal setae (Fig. 16.57); leg tarsi bearing paired claws (Fig. 16.78); excretory pore plate small to large, always bearing setae E1 and E2 or at least their alveoli (Figs. 16.62, 16.68–16.69, 16.75, 16.87, 16.100, 16.119, 16.130–16.132) 15

15a(14b). Coxal plates I, II, and III on each side all separate (Figs. 16.61, 16.63, 16.65, 16.68, 16.71, 16.81, 16.88), or all fused (Fig. 16.69) ''and'' plates of two sides fused medially ''and'' dorsal plate round and bearing setae Mhl laterally (Fig. 16.67) 16

15b. Coxal plates I and II on each side separate and plates II and III fused at least medially (Figs. 16.102, 16.126, 16.143, 16.153), or all plates on each side fused (Figs. 16.113 and 16.120) ''but'' plates of two sides separate medially ''and'' dorsal plate elliptical and not bearing setae Mhl (Figs. 16.112, 16.121, 16.133, 16.139, 16.149, 16.154) 29

16a(15a). Tarsi of legs III bearing 12 or more setae, including Ta8 (Fig. 16.8, arrow); dorsal plate usually elongate and elliptical and bearing only propodosomal setae (Figs. 16.3, 16.59, 16.70); ''when'' dorsal plate round and bearing Mh1 laterally (Fig. 16.57) ''then'' coxal plates IV bearing both setae V1 and V2 in addition to C4 (Fig. 16.63) ''and'' all setae on pedipalp tarsi short and blade-like (Fig. 16.64) .... Lebertioidea (in part) 17

16b. Tarsi of legs III usually bearing 11 or fewer setae, lacking at least Ta8 (Figs. 16.77, 16.79, 16.89, 16.90); dorsal plate usually nearly round (Figs. 16.67, 16.73, 16.74, 16.83, 16.86), rarely elongate and elliptical (Fig. 16.66); ''when'' tarsi of legs III bearing 12 setae ''then'' dorsal plate round and bearing Mh1 laterally (Fig. 16.74), ''and'' coxal plates IV bearing only setae C4 (Fig. 16.75), ''and'' at least one seta on pedipalp tarsi long and whip-like (similar to Fig. 16.82) .... Arrenuroidea 19

17a(16a). Pedipalp tarsi with no setae as long as pedipalp (Fig. 16.64); tarsi of legs I, II, and III bearing 13 or 14 (+1s), 13 or 14 (+1s), and 12 setae respectively, lacking Ta15 .... Anisitsiellidae
                   Anisitsiellinae (in part) *Bandakiopsis, Utaxatax*

17b. Pedipalp tarsi with at least one seta as long as pedipalp (Figs. 16.2 and 16.58); tarsi of legs I, II, and III bearing 15 (+1s), 15 (+1s), and 13 or 14 setae, respectively, including Ta15 (Figs. 16.6–16.8, 16.60) 18

18a(17b). Dorsal plate longitudinally striate (Figs. 16.3 and 16.59); coxal plates III bearing only setae C4 (Figs. 16.4 and 16.65); excretory pore plate small and usually quadrangular (Figs. 16.5 and 16.62) .... Sperchontidae *Sperchon, Sperchonopsis*

18b. Dorsal plate reticulate (Fig. 16.70); coxal plates III bearing setae C4 and supernumerary setae C5 (Fig. 16.61); excretory pore plate relatively large and diamond-shaped (Fig. 16.61) .... Teutoniidae *Teutonia*

19a(16b). Coxal plates I, II, and III on each side all fused and coxal plates of two sides fused medially, coxal plate setae C1, C3, and C4 reduced to vestiges (Fig. 16.69); pedipalp tarsi with no setae as long as pedipalp (Fig. 16.72); dorsal plate with setae SS reduced to vestiges Fig. 16.67) .... Acalyptonotidae *Acalyptonotus*

19b. Coxal plates I, II, and III on each side all separate and coxal plates of two sides separate (Figs. 16.68, 16.71, 16.75, 16.93, 16.97); pedipalp tarsi with at least one seta as long as pedipalp (Fig. 16.82) 20

20a(19b).   Idiosoma moderately flattened dorsoventrally; gnathosoma projecting beyond anterior edge of dorsal plate, entirely exposed in dorsal view (Figs. 16.66 and 16.74); excretory pore plate variously shaped (Figs. 16.68, 16.75, 16.76) but not attenuate anteriorly; pedipalp tarsi with long, thick, distal seta straight or only slightly bowed basally; leg genua with setae Ge5 borne on tubercles that usually are prominent (Fig. 16.80, arrow) 21

20b.   Idiosoma extremely flattened dorsoventrally; gnathosoma recessed beneath anterior edge of dorsal plate, partially or entirely concealed in dorsal view (in unmounted specimens) (Fig. 16.73); excretory pore plate triangular with anterior apex attenuate (Figs. 16.71, 16.81, 16.87–16.88, 16.93, 16.97); pedipalp tarsi with long, thick, distal seta strongly bowed, and usually lobed, basally (Figs. 16.82 and 16.85; arrows); leg genua with setae Ge5 not borne on tubercles 24

21a(20a).   Dorsal plate bearing four pairs of propodosomal setae only, with setae Mh1 on lateral membranous integument (Fig. 16.66); setae Lh1 and Lh2 thick and long relative to other hysterosomal setae (Fig. 16.66); excretory pore plate setae E1 and E2 absent, represented by their alveoli on excretory pore plate (Fig. 16.68); tibiae of legs I bearing eight setae (+2s), including Ti11; leg tarsi lacking setae Ta14 (Figs. 16.78 and 16.80) .... Momoniidae   22

21b.   Dorsal plate bearing five pairs of setae, including the propodosomals and setae Mh1 laterally (Fig. 16.74); setae Lh1 and Lh2 similar to other hysterosomal setae (Fig. 16.74); excretory pore plate setae E1 and E2 present on plate (Fig. 16.75 and 16.76), tibiae of legs I bearing seven setae (+2s), lacking Ti11; leg tarsi bearing setae Ta14 (Figs. 16.77 and 16.79; arrows) 23

22a(21a).   Legs with setae Tr1, Ge5, Ti9, and Ti11 much longer than respective segments; setae IITi9, IITi11, IIIGe5, IIITi9, and IIITi11 plumose (Fig. 16.80) .... Momoniinae   *Momonia*

22b.   Legs with setae Tr1, Ge5, Ti9, and Ti11 shorter than respective segments; setae IITi9, IITi11, IIIGe5, IIITi9, and IIITi11 blade-like (Fig. 16.78) .... Stygomomoniinae   *Stygomomonia*

23a(21b).   Tarsi of legs III bearing nine setae, lacking Ta12 (Fig. 16.77); tarsi of legs II and III with setae Ta4 and Ta6 long (usually as long as respective segments) (Fig. 16.77); coxal plates I and II with conspicuous denticulate projections on posterior edges (Fig. 16.76, arrows), coxal plates III with transverse muscle attachment scars (TMAS), excretory pore plate subtriangular (Fig. 16.76) .... Krendowskiidae   *Krendowskia, Geayia*

23b.   Tarsi of legs III bearing at least 11 setae, including Ta12 (Fig. 16.79); tarsi of legs II and III with setae Ta4 and Ta6 usually short (conspicuously shorter than respective segments) (Fig. 16.79); coxal plates I and II lacking denticulate projections on posterior edges, coxal plates III lacking transverse muscle attachment scars (TMAS), excretory pore plate variously shaped, but rarely triangular (Fig. 16.75) .... Arrenuridae

Arrenurinae   *Arrenurus*

24a(20b).   Dorsal plate bearing five pairs of setae, including the propodosomal setae and setae Mh1 laterally (Fig. 16.73) .... Mideopsidae

Mideopsinae   *Mideopsis*

24b.   Dorsal plate bearing four pairs of propodosomal setae only, with setae Mh1 on lateral membranous integument (Figs. 16.83 and 16.86) 25

25a(24b).   Tibiae of legs II and III bearing nine setae (+2s and +1s, respectively), including both Ti10 and Ti11 (Figs. 16.89 and 16.90) 26

25b.   Tibiae of legs II and III bearing fewer than nine setae (+2s and +1s, respectively), lacking either Ti10 or Ti11, or both (Figs. 16.84 and 16.95) 27

26a(25a). Pedipalp tarsi with long, thick, distal seta bowed, but not deeply lobed or fringed, basally, and with most medial seta moderately thick but not fringed (Fig. 16.85, arrow); legs II and III with setae Ge5, Ti9, and Ti11 much longer than respective segments and plumose, and setae Ta4 and Ta6 much longer than respective segments (Fig. 16.89) . . . . Mideidae   *Midea*

26b. Pedipalp tarsi with long, thick, distal seta bowed, deeply lobed and fringed basally, and most medial seta thick and fringed (Fig. 16.82, arrow); legs II and III with setae Ge5, Ti9, and Ti11 shorter than respective segments and simple, and setae Ta4 and Ta6 shorter than respective segments (Fig. 16.90) . . . . Nudomideopsidae   *Nudomideopsis, Paramideopsis*

27a(25b). Tibiae of legs II bearing seven setae (+2s), lacking Ti10 and Ti11; tibiae of legs III bearing eight setae (+1s), lacking Ti10 (Fig. 16.84) . . . . Laversiidae   *Laversia*

27b. Tibiae of legs II and III bearing eight setae (+2s and +1s, respectively), including Ti10 but lacking Ti11 (Fig. 16.95) 28

28a(27b). Dorsal plate with setae SS well in anterior half, setae Mh1 located in lateral integument near midlength of plate, posterior to level of setae SS (Fig. 16.83) . . . . Neoacaridae   *Neoacarus*

28b. Dorsal plate with setae SS near midlength, setae Mh1 located in lateral integument near humeral setae, anterior to level of setae SS (Fig. 16.91) . . . . Athienemanniidae
Athienemanniinae   *Chelomideopsis, Platyhydracarus*

29a(15b). Coxal plates III bearing setae C4 and at least one other pair of setae (V1 or C5) "and" excretory pore plate bearing only setae E1 and E2 (Figs. 16.102, 16.107, 16.109, 16.113) . . . . Lebertioidea (in part)   30

29b. Coxal plates III usually bearing only setae C4 (Figs. 16.100, 16.114, 16.120, 16.142, 16.153), "when" coxal plates III also bearing setae V1 "then" excretory pore plate bearing setae V2 in addition to E1 and E2 (Figs. 16.119, 16.143, 16.148) 33

30a(29a). Coxal plates III truncate posteriorly, bearing three pairs of setae, including both V1 and V2 on posterior edges (Fig. 16.102); tibiae of legs I, II, and III bearing nine setae (+2s, +2s, and +1s, respectively), including both Ti9 and Ti10 (Fig. 16.103); tarsi of legs I and II bearing 14 setae (+1s), including Ta14 and Ta15 (Fig. 16.103); cheliceral bases separate (Fig. 16.101, arrow) . . . . Lebertiidae   *Lebertia*

30b. Coxal plates III rounded or pointed posteriorly, bearing two pairs of setae, including either C5 laterally or V1 posteromedially (Figs. 16.107, 16.109, 16.113); tibiae of legs I, II, and III bearing eight setae (+2s, +2s, and +1s, respectively), lacking Ti9 or Ti10 (Figs. 16.104 and 16.106); tarsi of legs I and II bearing 13 setae (+1s), lacking Ta15 (Fig. 16.106), or 12 setae (+1s), lacking both Ta15 and Ta14 (Fig. 16.104); cheliceral bases fused (Figs. 16.105, 16.108, 16.112) 31

31a(30b). Dorsal plate bearing five pairs of setae including the propodosomal setae and setae Mh1 laterally, integument beneath posterior edge of shield intricately folded (Fig. 16.105, arrow); coxal plates I elongate, extending posteriorly nearly to level of excretory pore plate, and plates III bearing setae C5 laterally (Fig. 16.107); tarsi of legs I and II bearing 12 setae (+1s), lacking Ta14 (Fig. 16.104) . . . . Oxidae   *Oxus, Frontipoda*

31b. Dorsal plate bearing four pairs of setae, or five pairs including the propodosomals and setae Mh2 laterally, integument beneath posterior edge of shield not intricately folded (Figs. 16.108 and 16.112); coxal plates I not elongate, extending posteriorly only to level of insertion of legs III (Fig. 16.109), or fused with plates II (Fig. 16.113); plates III bearing setae V1 posteromedially (Figs. 16.109 and 16.113); tarsi of legs I and II bearing 13 setae (+1s), including Ta14 (Fig. 16.106) . . . . Torrenticolidae   32

32a(31b).   Dorsal plate bearing five pairs of setae including the propodosomals and setae Mh2 laterally, and with setae PL long and thick (Fig. 16.108); coxal plates I clearly delineated by complete suture lines, and plates II and III fused with suture line incomplete medially (Fig. 16.109); excretory pore plate with convex projection posteromedially (Fig. 16.109); tibiae of all legs lacking setae Ti10, genua of legs III bearing four setae (+1s) including Ge3, and tarsi of legs III bearing 11 setae, lacking Ta10 (Fig. 16.110) .... Testudacarinae   *Testudacarus*

32b.   Dorsal plate bearing only four pairs of propodosomal setae, and with setae PL relatively short and slender (Fig. 16.112); coxal plates of each side all fused, with suture lines obliterated except laterally (Fig. 16.113); excretory pore plate without projection posteromedially (Fig. 16.113); tibiae of all legs lacking setae Ti9, genua of legs III bearing three setae (+1s), lacking Ge3, tarsi of legs III bearing 12 setae, including Ta10 (Fig. 16.111) .... Torrenticolinae   *Torrenticola*

33a(29b).   Cheliceral bases separate (Fig. 16.98, arrow); tarsi of legs III bearing 12 setae, including Ta9 (Fig. 16.99, arrow); excretory pore plate very large (equal in width to coxal plates of one side), bearing excretory pore and setae E1 and E2 anteromedially and occasionally also setae V2 at posterolateral angles (Fig. 16.94), ''and'' coxal plates I separate from posterior coxal groups, ''and'' coxal plates III bearing only setae C4 (Fig. 16.100) .... Lebertioidea (in part) .... Anisitsiellidae (in part)
                                                                          Anisitsiellinae (in part)   *Bandakia*

33b.   Cheliceral bases fused (Figs. 16.121, 16.125, 16.127, 16.139, 16.154); tarsi of legs III bearing 11 or fewer setae, lacking Ta9 (Fig. 16.168); excretory pore plate small to very large, ''when'' equal in width to coxal plates of one side (measured from midline to insertion of leg III) (Figs. 16.117, 16.119, 16.120, 16.142, 16.148) ''then'' either excretory pore located near posterior edge of plate (Fig. 16.142, arrow), ''or'' coxal plates I fused to posterior coxal groups (Fig. 16.120), ''or'' coxal plates III bearing setae V1 in addition to setae C4 (Fig. 16.148); .... Hygrobatoidea   34

34a(33b).   Two pairs of urstigmata borne distally between coxal plates I and II (Figs. 16.114, arrows A; 16.116, arrows); tibiae of legs I and II bearing seven or eight setae (+2s), including only three ventral setae, lacking either Ti8 and Ti9 or Ti10 and Ti11 (Figs. 16.118 and 16.124) .... Limnesiidae   35

34b.   One pair of urstigmata borne distally between coxal plates I and II (Figs. 16.120, arrow, 16.126, 16.128, 16.143, 16.174); tibiae of legs I and II bearing nine or more setae (+2s), including five ventral setae (Ti7, Ti8, Ti9, Ti10, and Ti11) (Figs. 16.123, 16.158, 16.159) 36

35a(34a).   Coxal plates I and II completely separate (Fig. 16.116); tarsi of legs III bearing nine setae, lacking Ta9, Ta13, and Ta14; dorsal plate extending laterally well beyond level of eye plates (contrast with Fig. 16.115) .... Tyrrelliinae   *Tyrrellia*

35b.   Coxal plates I and II on each side fused, with suture lines obliterated medially (Fig. 16.114, arrow B); tarsi of legs III bearing 12 setae, including Ta9, Ta13, and Ta14 (Fig. 16.122); dorsal plate narrow, usually confined to region between eye plates (Fig. 16.115) .... Limnesiinae   *Limnesia*

36a(34b).   Coxal plates I, II, and III on each side all fused (Fig. 16.120) .... Hygrobatidae   *Hygrobates, Atractides*

36b.   Coxal plates I separate from posterior coxal group on each side (Figs. 16.126, 16.128, 16.136, 16.142, 16.153) 37

37a(36b).   Tarsi of legs I and II bearing 10 setae (+1s), lacking Ta14, and either Ta12 (Fig. 16.129) or Ta9 (Figs. 16.137 and 16.138) 38

37b.   Tarsi of legs I and II bearing 11 or more setae (+1s), including Ta9 (Figs. 16.140; 16.179, arrows; 16.141, 16.146, arrows A) 40

38a(37a).	Dorsal and coxal plates, and leg sclerites, conspicuously longitudinally striate (as in Fig. 16.125); coxal plates III rounded posteriorly (Fig. 16.126); tibiae of legs I bearing seven setae (+2s), lacking Ti10 and Ti11 (Fig. 16.129) .... Feltriidae *Feltria*
38b.	Dorsal and coxal plates reticulate (Fig. 16.127); coxal plates III bearing pointed projections posteriorly (Fig. 16.128, arrow); tibiae of legs I bearing nine setae (+2s), including Ti10 and Ti11 (Figs. 16.137 and 16.138) .... Unionicolidae 39
39a(38b).	Empodium undivided (Fig. 16.134) .... Pionatacinae *Neumania, Koenikea*
39b.	Empodium bifurcate (Fig. 16.135, arrow) .... Unionicolinae *Unionicola*
40a(37b).	Tibiae of all legs bearing seven setae (+2s, +2s, and +1s, respectively), lacking Ti9 and Ti11 (Fig. 16.140) .... Aturidae (in part) ...... Aturinae *Aturus*
40b.	Tibiae of all legs bearing eight or more setae (+2s, +2s, and +1s respectively), including Ti9 and Ti11 (Fig. 16.141) 41
41a(40b).	Coxal plates III with pointed or lobed projections posteriorly, bearing two pairs of setae, C4 anteromedially and V1 posteromedially (Figs. 16.143 and 16.148); excretory pore plate large, bearing three pairs of setae, E1 and E2, and V2 posterolaterally (Figs. 16.143 and 16.148) .... Aturidae (in part) ..... Axonopsinae (in part) *Axonopsis, Estellacarus, Woolastookia, Brachypoda*
41b.	Coxal plates III without or with small projections posteriorly, bearing only setae C4 (Figs. 16.142, 16.145, 16.151, 16.153); excretory pore plate small or large, bearing only setae E1 and E2 (Figs. 16.142, 16.144, 16.152, 16.157, 16.160–16.165) 42
42a(41b).	Coxal plates III with small projections posteriorly (Fig. 16.145, arrow); excretory pore plate small, little larger than combined area of excretory pore and bases of setae E1 and E2 (Fig. 16.144) .... Pionidae (in part) Wettininae *Wettina*
42b.	Coxal plates III without projections posteriorly (Figs. 16.142, 16.151, 16.153, 16.156, 16.166, 16.178); excretory pore plate considerably larger than combined area of excretory pore and bases of setae E1 and E2 (Figs. 16.142, 16.152, 16.161, 16.173, 16.175, 16.181) 43
43a(42b).	Tarsi of legs I bearing 13 setae (+1s), including Ta8 (Figs. 16.141 and 16.146; arrows B); excretory pore plate large, and usually triangular or obcordate (Figs. 16.142, 16.152, 16.157, 16.161–16.163) 44
43b.	Tarsi of legs I bearing 12 setae (+1s), lacking Ta8 (Fig. 16.179); excretory pore plate small to large, and variously shaped (Figs. 16.164–16.165, 16.169–16.170, 16.175, 16.181) 47
44a(43b).	Excretory pore plate with setae E1 close together adjacent to excretory pore in posterior half of plate (Figs. 16.152, 16.155, 16.157, 16.160–16.163) .... Pionidae (in part) 45
44b.	Excretory pore plate with setae E1 widely separated and removed from excretory pore in anterior half of plate (Figs. 16.142 and 16.151, arrows) .... Aturidae (in part) 46
45a(44a).	Tarsi of legs II bearing 13 setae (+1s), including Ta8 (Fig. 16.158); tibiae of legs with setae Ti9 and Ti11 simple (Fig. 16.158) .... Najadicolinae *Najadicola*
45b.	Tarsi of legs II usually bearing 12 setae (+1s), lacking Ta8 (Fig. 16.159), "when" bearing 13 setae "then" tibiae of legs with Ti9 and Ti11 long and plumose (as in Fig. 16.159) .... Pioninae *Nautarachna, Piona*
46a(44b).	Excretory pore plate broadly obcordate (Fig. 16.142) .... Axonopsinae (in part) *Ljania*
46b.	Excretory pore plate triangular (Fig. 16.151) .... Albiinae *Albia*
47a(43b).	Excretory pore plate small and subtriangular, with setae E2 borne near posterolateral angles of plate (Fig. 16.150); suture lines between coxal plates II and III parallel to anterior edge of plate II, coxal plates without distinct lateral coxal apodemes and coxal plates III lacking medial coxal apodemes and transverse muscle attachment scars (Fig. 16.150) .... Aturidae (in part) ..... Axonopsinae (in part) *Neobrachypoda*

47b.   Excretory pore plate variously shaped, "when" subtriangular with setae
            E2 borne near posterolateral angles of plate, "then" suture lines
            between coxal plates II and III terminating in distinct lateral coxal
            apodemes (LCA) that are usually nearly transverse "and" coxal plates
            III bearing at least medial coxal apodemes (MCA)(Figs. 16.166, 16.171–
            16.172, 16.174, 16.177–16.178, 16.180) .... Pionidae (in part)   48

48a(47b).   Suture line between coxal plates II and III on each side extending
            medially beyond lateral coxal apodeme nearly to midline (Fig. 16.166,
            arrow); excretory pore borne on elevated projection which
            extends to or beyond posterior edge of plate (Figs. 16.164–
            16.165) .... Hydrochoreutinae   *Hydrochoreutes*

48b.   Suture line between coxal plates II and III on each side extending
            medially only as far as lateral coxal apodeme (Figs. 16.171, 16.172,
            16.174, 16.177–16.178, 16.180); excretory pore usually not borne on
            elevated projection, "when" on small projection, this does not extend
            to posterior edge of plate   49

49a(48b).   Trochanters of legs III bearing two setae, including a supernumerary
            dorsal seta (Fig. 16.167, arrow), "or" tibiae of legs II and III
            bearing 11 or 12 setae (+2s and +1s, respectively), including two
            supernumerary posteroventral setae (Fig. 16.168, arrows), "or" both
            coxal plates III lacking transverse muscle attachment scars "and"
            excretory pore plate with setae E2 near posterior edge of plate
            (Fig. 16.170) .... Foreliinae   *Pseudofeltria, Forelia*

49b.   Trochanters of legs III bearing only one seta; tibiae of legs II and III
            bearing only nine setae (+2s and +1s, respectively) (as in Figs. 16.158
            and 16.159); coxal plates III usually with transverse muscle attachment
            scars (Figs. 16.172, 16.177–16.178), "when" coxal plates lacking these
            scars (as in Fig. 16.180) "then" excretory pore plate with setae E2
            removed from posterior edge of plate (Fig. 16.181)   50

50a(49b).   Excretory pore plate transversely elliptical or semicircular
            (Fig. 16.181) .... Huitfeldtiinae   *Huitfeldtia*

50b.   Excretory pore plate broadly elliptical (Figs. 16.175 and 16.178), circular
            (Figs. 16.172 and 16.176), or subtriangular .... Tiphyinae   *Neotiphys, Pionopsis, Tiphys*

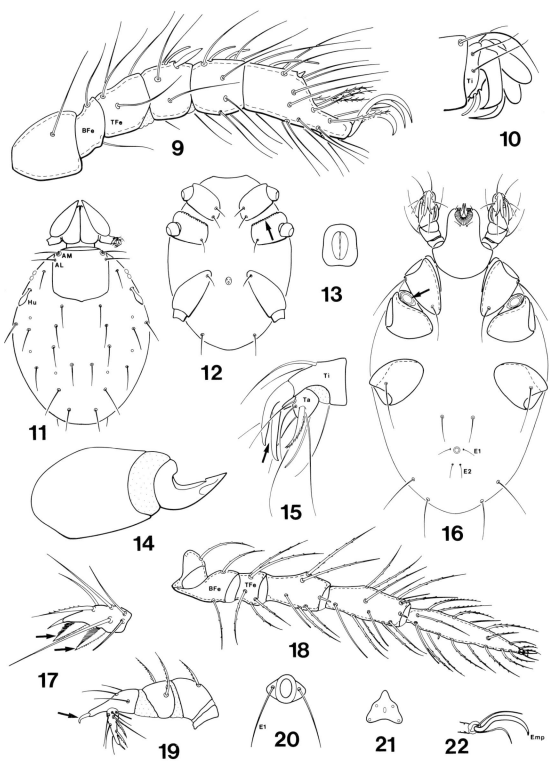

Figures 16.9–16.12 *Hydrovolzia gerhardi* Mitchell (Hydrovolziinae), larva. Fig. 16.9, leg I, anterolateral view; Fig. 16.10, distal segments of pedipalp; Fig. 16.11, dorsum of idiosoma and gnathosoma; Fig. 16.12, venter of idiosoma. (Redrawn and modified from Wainstein 1980.) **Figures 16.13–16.14** *Thermacarus nevadensis* Marshall (Thermacaridae), larva. Fig. 16.13, excretory pore plate; Fig. 16.14, chelicera. **Figures 16.15–16.16** *Wandesia* sp. (Wandesiinae), larva. Fig. 16.15, tibia and tarsus of pedipalp; Fig. 16.16, venter of idiosoma and gnathosoma. **Figures 16.17–16.19, 16.22** *Thyas stolli* Koenike (Thyadinae), larva. Fig. 16.17, tarsus of pedipalp; Fig. 16.18, leg I; Fig. 16.19, pedipalp; Fig. 16.22, claws and empodium of leg I. (Redrawn and modified from Prasad and Cook 1972.) **Figure 16.20** *Hydryphantes ruber* (de Geer) (Hydryphantinae), larva, excretory pore plate. (Redrawn from Prasad and Cook 1972.) **Figure 16.21** *Protzia* sp. (Protziinae), larva, excretory pore plate. AL, AM, anterolateral and anteromedial propodosomal setae; BFe, basifemur; E1 and E2, exrectory pore plate setae; Emp, empodium; Hu, humeral seta; Ta, tarsus; TFe, telofemur; Ti, tibia.

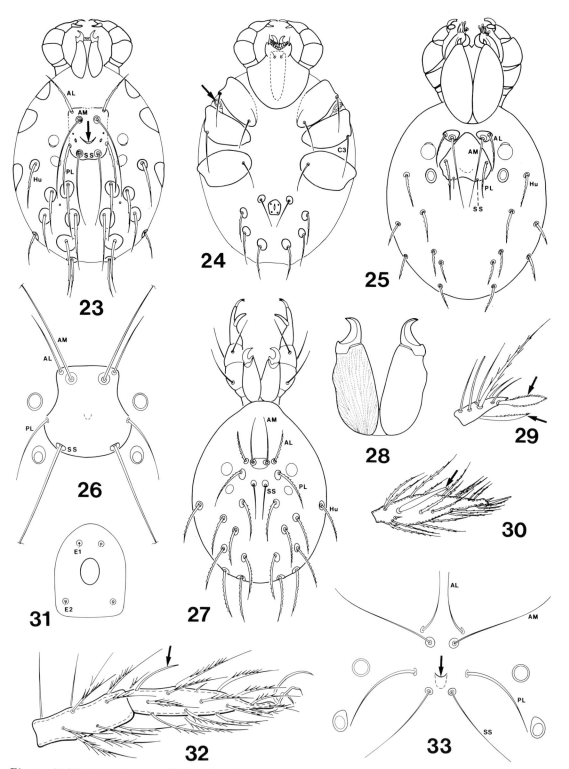

Figures 16.23–16.24 Thyas stolli Koenike (Thyadinae), larva. Fig. 16.23, dorsum of idiosoma and gnathosoma; Fig. 16.24, venter of idiosoma and gnathosoma. (Redrawn and modified from Prasad and Cook 1972.) *Figures 16.25, 16.28* Hydryphantes ruber (de Geer) (Hydryphantinae), larva. Fig. 16.25, dorsum of idiosoma and gnathosoma; Fig. 16.28, chelicerae. (Redrawn and modified from Prasad and Cook 1972.) *Figure 16.26* Tartarothyas sp. (Tartarothyadinae), larva, dorsal plate. *Figures 16.27, 16.29–16.31* Hydrodroma despiciens (Müller) (Hydrodromidae), larva. Fig. 16.27, dorsum of idiosoma and gnathosoma; Fig. 16.29, tarsus of pedipalp; Fig. 16.30, tarsus of leg I; Fig. 16.31, excretory pore plate. (Redrawn and modified from Prasad and Cook 1972.) *Figures 16.32–16.33* Protzia sp. (Protziinae), larva. Fig. 16.32, tibia and tarsus of leg I; Fig. 16.33, prodorsal region of idiosoma. AL, AM, PL, anterolateral, anteromedial, posterolateral propodosomal setae respectively; E1, E2, excretory pore plate setae, Hu, humeral seta; SS, sensilla.

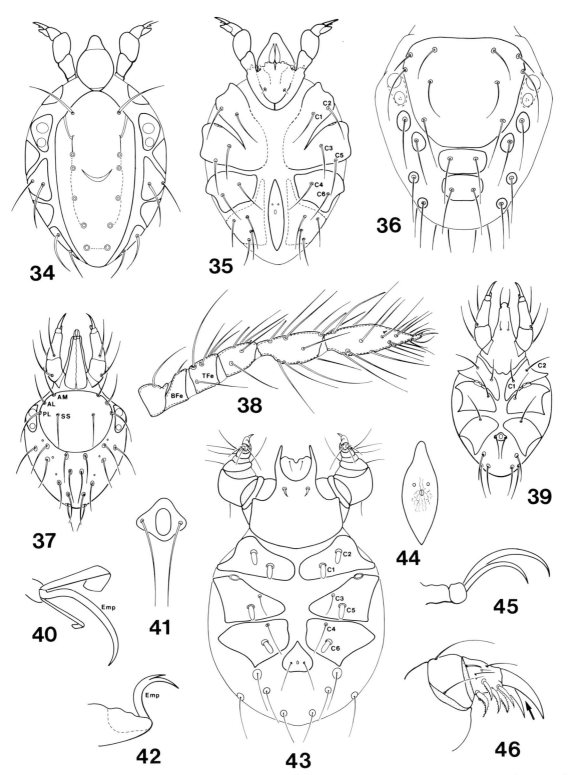

Figures 16.34–16.35, 16.38, 16.40, 16.44 *Eylais major* Lanciani (Eylaidae), larva. Fig. 16.34, dorsum of idiosoma and gnathosoma; Fig. 16.35, venter of idiosoma and gnathosoma; Fig. 16.38, leg I; Fig. 16.40, claws and empodium of leg I; Fig. 16.44, excretory pore plate. (Redrawn and modified from Prasad and Cook 1972.) **Figures 16.36, 16.43** *Piersigia* sp. (Piersigiidae), larva. Fig. 16.36, dorsum of idiosoma; Fig. 16.43, venter of idiosoma and gnathosoma. **Figures 16.37, 16.39, 16.41, 16.45** *Limnochares americana* Lundblad (Limnocharinae), larva. Fig. 16.37, dorsum of idiosoma and gnathosoma; Fig. 16.39, venter of idiosoma and gnathosoma; Fig. 16.41, excretory pore plate; Fig. 16.45, claws and empodium of leg I. (Redrawn and modified from Prasad and Cook 1972.) **Figures 16.42, 16.46** *Hydrachna magniscutata* Marshall (Hydrachnidae), larva. Fig. 16.42, empodium of leg I; Fig. 16.46, distal segments of pedipalp. (Redrawn from Prasad and Cook 1972.) AL, AM, PL, anterolateral, anteromedial, posterolateral propodosomal setae respectively; BFe, basifemur, C1-C6, coxal plate setae; Emp, empodium; ss, sensilla; TFe, telofemur.

557

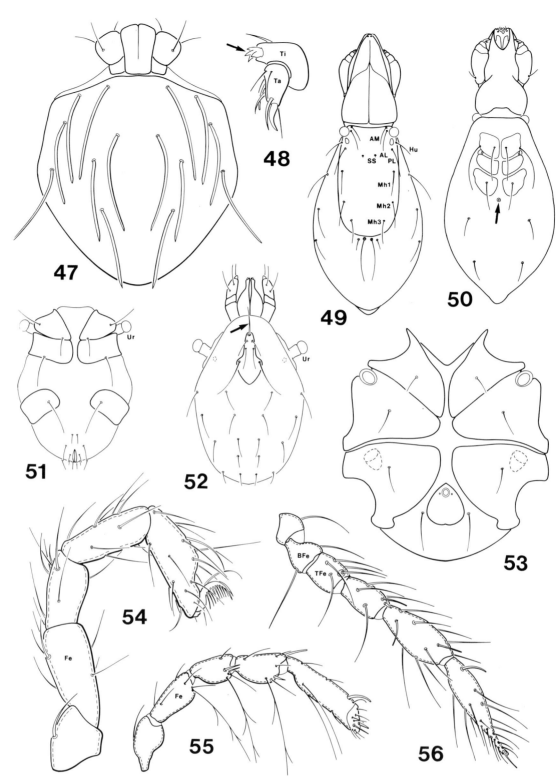

Figures 16.47, 16.53 *Rhyncholimnochares kittatinniana* Habeeb (Rhyncholimnocharinae), larva. Fig. 16.47, dorsum of idiosoma and gnathosoma; Fig. 16.53, venter of idiosoma. **Figures 16.48, 16.51–16.52, 16.54** *Stygothrombium* sp. (Stygothrombidiidae), larva. Fig. 16.48, tibia and tarsus of pedipalp; Fig. 16.51, venter of idiosoma; Fig. 16.52, dorsum of idiosoma and gnathosoma; Fig. 16.54, leg I, anterolateral view. **Figures 16.49–16.50, 16.55** *Hydrachna magniscutata* Marshall (Hydrachnidae), larva. Fig. 16.49, dorsum of idiosoma and gnathosoma; Fig. 16.50, venter of idiosoma and gnathosoma; Fig. 16.55, leg I. (Redrawn and modified from Prasad and Cook 1972.) **Figure 16.56** *Limnochares americana* Lundblad (Limnocharinae), larva, leg I. (Redrawn and modified from Prasad and Cook 1972.) AL, AM, PL, anterolateral, anteromedial, posterolateral propodosomal setae respectively; BFe, basifemur; Fe, femur; Hu, humeral seta; Mh1-Mh3, mediohysterosomal setae; SS, sensilla; Ta, tarsus; TFe, telofemur; Ti, tibia; Ur, urstigma.

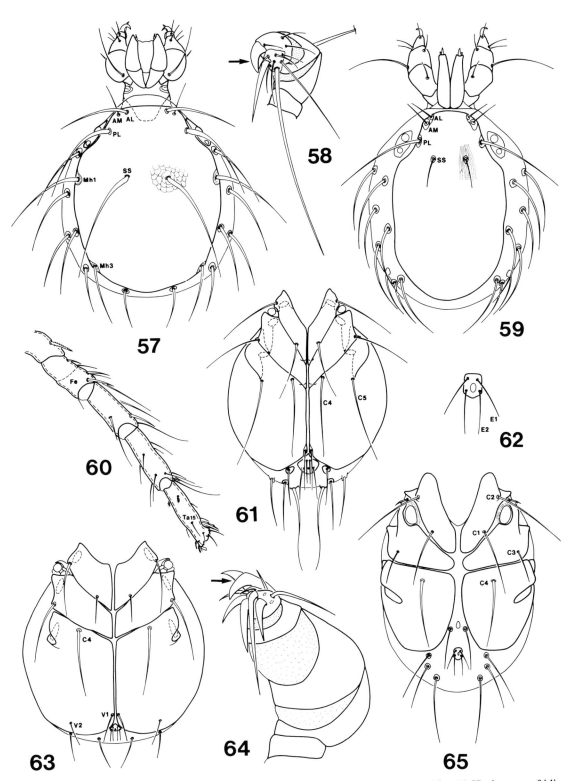

Figures 16.57, 16.63–16.64 *Bandakiopsis fonticola* Smith (Anisitsiellinae), larva. Fig. 16.57, dorsum of idiosoma and gnathosoma; Fig. 16.63, venter of idiosoma; Fig. 16.64, pedipalp. (Redrawn from Smith 1982.) **Figures 16.58–16.60, 16.62, 16.65** *Sperchon glandulosus* Koenike (Sperchontidae), larva. Fig. 16.58, pedipalp; Fig. 16.59, dorsum of idiosoma and gnathosoma; Fig. 16.60, leg I, anterolateral view; Fig. 16.62, excretory pore plate; Fig. 16.65, venter of idiosoma. (Redrawn from Smith 1982.) **Figure 16.61** *Teutonia lunata* Marshall (Teutoniidae), larva, venter of idiosoma. (Redrawn from Smith 1982.) AL, AM, PL, anterolateral, anteromedial, posterolateral propodosomal setae respectively; C1-C4, coxal plate setae; E1 and E2, excretory pore plate setae; Fe, femur; Mh1 and Mh3, mediohysterosomal setae; SS, sensilla; Ta, tarsus; V1 and V2, ventral setae.

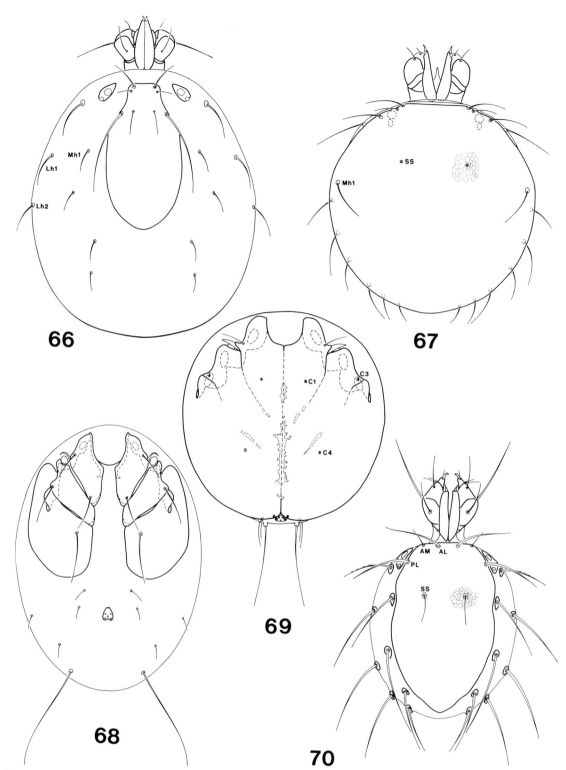

Figures 16.66, 16.68 *Stygomomonia (?) mitchelli* Smith (Stygomomoniinae), larva. Fig. 16.66, dorsum of idiosoma and gnathosoma; Fig. 16.68, venter of idiosoma. (Redrawn from Smith 1989c.) **Figures 16.67, 16.69** *Acalyptonotus neoviolaceus* Smith (Acalyptonotidae), larva. Fig. 16.67, dorsum of idiosoma and gnathosoma; Fig. 16.69, venter of idiosoma. (Redrawn from I. M. Smith 1983e.) **Figure 16.70** *Teutonia lunata* Marshall (Teutoniidae), larva, dorsum of idiosoma and gnathosoma. (Redrawn from Smith 1982.) AL, AM, PL, anterolateral, anteromedial, posterolateral propodosomal setae respectively; C1, C3, C4, coxal plate setae; Lh1 and Lh2, laterohystersomal setae; Mh1, mediohysterosomal seta; SS, sensilla.

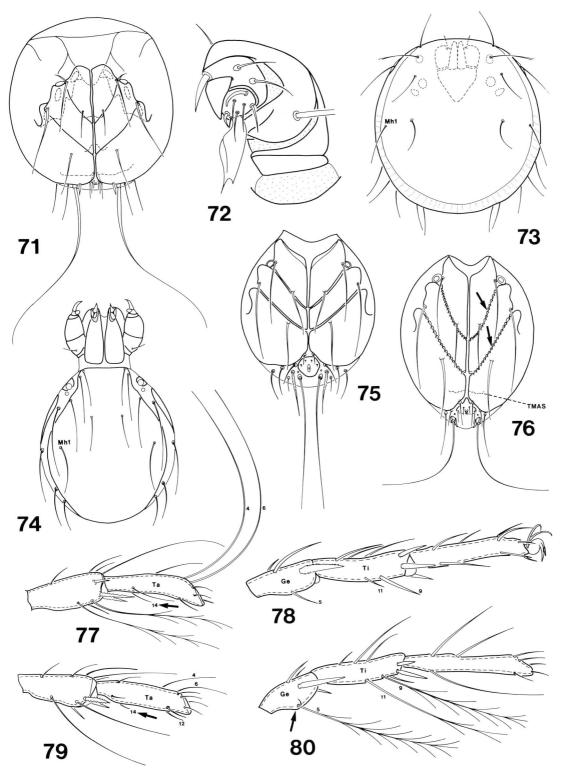

Figures 16.71, 16.73 *Mideopsis borealis* Habeeb (Mideopsinae), larva. Fig. 16.71, venter of idiosoma; Fig. 16.73, dorsum of idiosoma and gnathosoma. (Redrawn from Smith 1978.) **Figure 16.72.** *Acalyptonotus neoviolaceus* Smith (Acalyptonotidae), larva, pedipalp. (Redrawn from I. M. Smith 1983e.) **Figures 16.74–16.75, 16.79** *Arrenurus planus* Marshall (Arrenurinae), larva. Fig. 16.74, dorsum of idiosoma and gnathosoma; Fig. 16.75, venter of idiosoma; Fig. 16.79, tibia and tarsus of leg III, anterolateral view. (Redrawn and modified from Prasad and Cook 1972.) **Figures 16.76–16.77** *Krendowskia similis* Viets (Krendowskiidae), larva. Fig. 16.76, venter of idiosoma; Fig. 16.77, tibia and tarsus of leg III, anterolateral view. (Redrawn and modified from Prasad and Cook 1972.) **Figure 16.78** *Stygomomonia (?) mitchelli* Smith (Stygomomoniinae), larva, distal segments of leg III, anterolateral view. (Redrawn from Smith 1989c.) **Figure 16.80** *Momonia marciae* Habeeb (Momoniinae), larva, distal segments of leg III, anterolateral view. (Redrawn from Smith 1978.) Ge, genu; Mh1, mediohysterosomal seta; Ta, tarsus; Ti, tibia; TMAS, transverse muscle attachment scar.

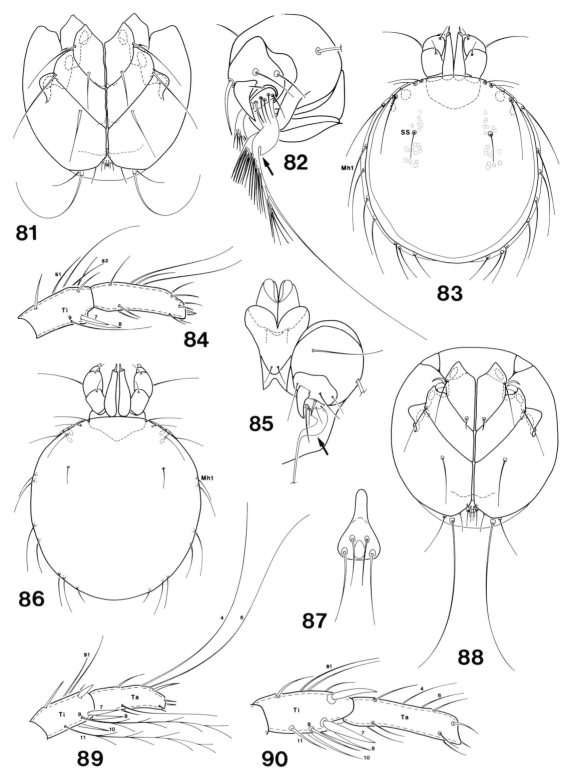

Figures 16.81, 16.85, 16.89 *Midea expansa* Marshall (Mideidae), larva. Fig. 16.81, venter of idiosoma; Fig. 16.85, venter of gnathosoma; Fig. 16.89, tibia and tarsus of leg III, anterolateral view. (Redrawn and modified from Smith 1978.) **Figures 16.82, 16.90** *Paramideopsis susanae* Smith (Nudomideopsidae), larva. Fig. 16.82, pedipalp; Fig. 16.90, tibia and tarsus of leg III, anterolateral view. (Redrawn from I. M. Smith 1983c.) **Figure 16.83** *Neoacarus occidentalis* Cook (Neoacaridae), larva, dorsum of idiosoma and gnathosoma. (Redrawn from I. M. Smith 1983a.) **Figure 16.84** *Laversia berulophila* Cook (Laversiidae), larva, tibia and tarsus of leg II, anterolateral view. (Redrawn from Smith 1978.) **Figures 16.86–16.88** *Nudomideopsis magnacetabula* (Smith) (Nudomideopsidae), larva. Fig. 16.86, dorsum of idiosoma and gnathosoma; Fig. 16.87, excretory pore plate; Fig. 16.88, venter of idiosoma. (Redrawn from I. M. Smith 1983d.) Mh1, mediohysterosomal seta; SS, sensilla; Ta, tarsus; Ti, tibia.

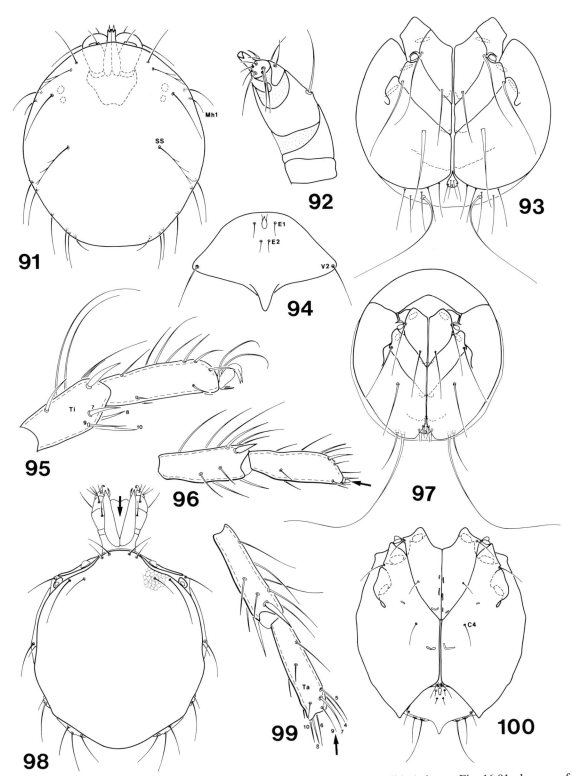

Figures 16.91, 16.95, 16.97 *Platyhydracarus juliani* Smith (Athienemanniidae), larva. Fig. 16.91, dorsum of idiosoma and gnathosoma; Fig. 16.95, tibia and tarsus of leg II, anterolateral view; Fig. 16.97, venter of idiosoma. (Redrawn from Smith 1989b.) ***Figures 16.92, 16.94, 16.96, 16.98–16.100*** *Bandakia phreatica* Cook (Anisitsiellinae), larva. Fig. 16.92, pedipalp; Fig. 16.94, excretory pore plate; Fig. 16.96, tibia and tarsus of leg I, anterolateral view; Fig. 16.98, dorsum of idiosoma and gnathosoma; Fig. 16.99, tibia and tarsus of leg III, anterolateral view; Fig. 16.100, venter of idiosoma. (Redrawn from Smith 1982.) ***Figure 16.93*** *Laversia berulophila* Cook (Laversiidae), larva, venter of idiosoma. (Redrawn from Smith 1978.) C4, coxal plate seta; Mh1, mediohysterosomal seta; SS, sensilla; Ta, tarsus; Ti, tibia.

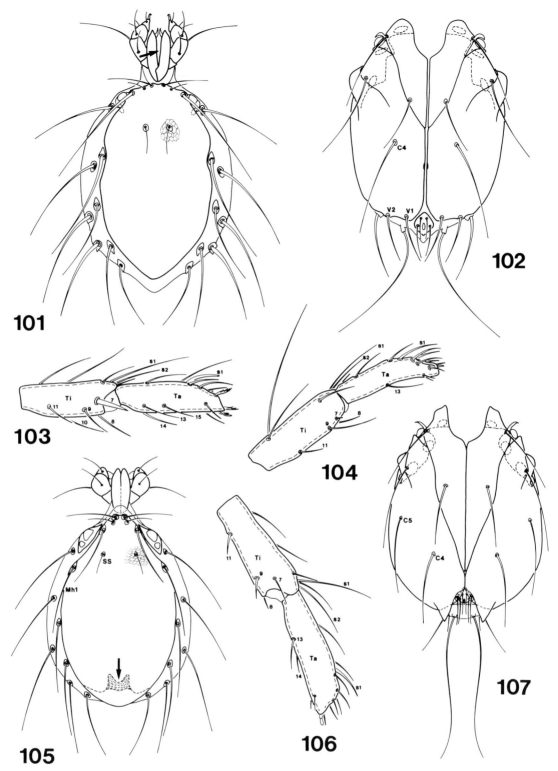

Figures 16.101–16.103 *Lebertia* sp. (Lebertiidae), larva. Fig. 16.101, dorsum of idiosoma and gnathosoma; Fig. 16.102, venter of idiosoma; Fig. 16.103, tibia and tarsus of leg I, anterolateral view. (Redrawn from Smith 1982.) **Figures 16.104–16.105, 16.107** *Oxus elongatus* Marshall (Oxidae), larva. Fig. 16.104, tibia and tarsus of leg I, anterolateral view; Fig. 16.105, dorsum of idiosoma and gnathosoma; Fig. 16.107, venter of idiosoma. (Redrawn from Smith 1982.) **Figure 16.106** *Testudacarus* sp. (Testudacarinae), larva, tibia and tarsus of leg I, anterolateral view. (Redrawn from Smith 1982.) C4 and C5, coxal plate setae; Mh1, mediohysterosomal seta; SS, sensilla; Ta, tarsus; Ti, tibia; V1 and V2, ventral setae.

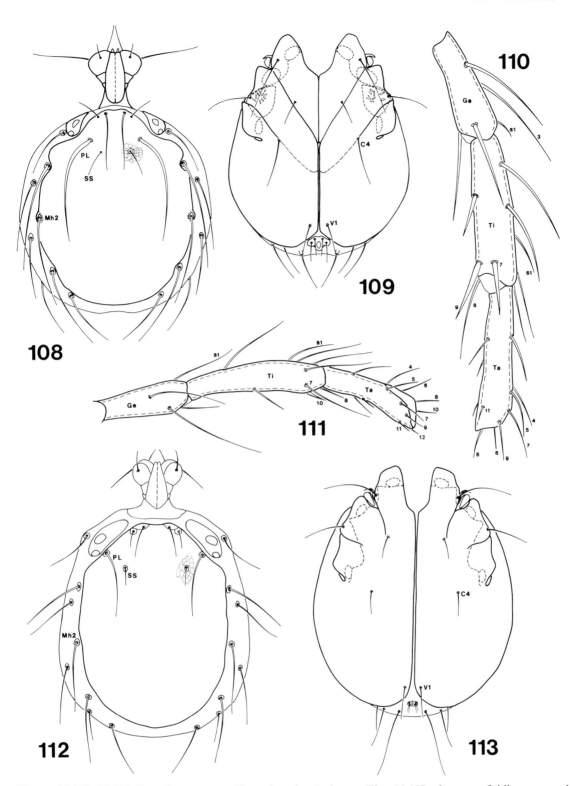

Figures 16.108–16.110 *Testudacarus* sp. (Testudacarinae), larva. Fig. 16.108, dorsum of idiosoma and gnathosoma; Fig. 16.109, venter of idiosoma; Fig. 16.110, distal segments of leg III, anterolateral view. (Redrawn from Smith 1982.) ***Figures 16.111–16.113*** *Torrenticola* sp. (Torrenticolinae), larva. Fig. 16.111, distal segments of leg III, anterolateral view; Fig. 16.112, dorsum of idiosoma and gnathosoma; Fig. 16.113, venter of idiosoma. (Redrawn from Smith 1982.) C4, coxal plate setae; Ge, genu; Mh2, mediohysterosomal seta; PL, posterolateralpropodosomal seta; SS, sensilla; Ta, tarsus; Ti, tibia; V1, ventral setae.

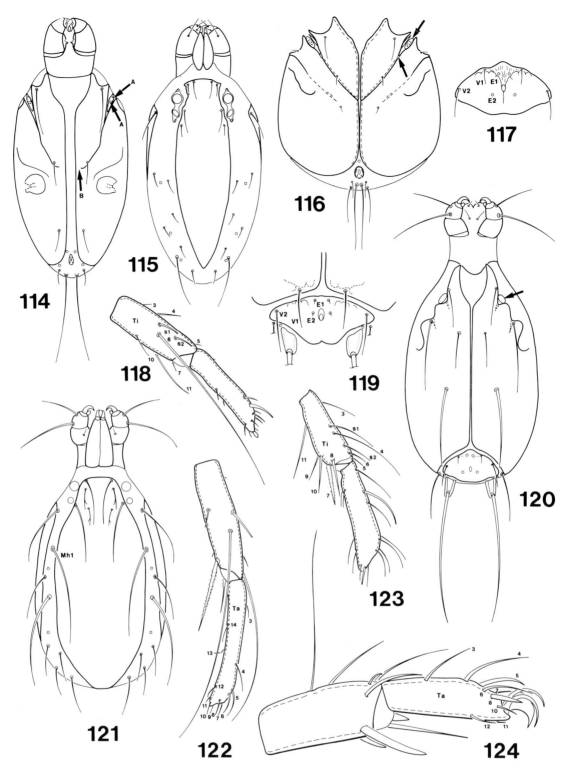

Figures 16.114–16.115, 16.118, 16.122 *Limnesia marshalliana* Lundblad (Limnesiinae), larva. Fig. 16.114, venter of idiosoma and gnathosoma; Fig. 16.115, dorsum of idiosoma and gnathosoma; Fig. 16.118, tibia and tarsus of leg II, posterolateral view; Fig. 16.122, tibia and tarsus of leg III, posterolateral view. (Redrawn and modified from Prasad and Cook 1972.) **Figures 16.116, 16.124** *Tyrrellia* sp. (Tyrrelliinae), larva. Fig. 16.116, venter of idiosoma; Fig. 16.124, tibia and tarsus of leg III, posterolateral view. **Figures 16.117, 16.120–16.121** *Atractides grouti* Habeeb (Hygrobatidae), larva. Fig. 16.117, excretory pore plate; Fig. 16.120, venter of idiosoma; Fig. 16.121, dorsum of idiosoma and gnathosoma. (Redrawn and modified from Prasad and Cook 1972.) **Figure 16.119** *Hygrobates neocalliger* Habeeb (Hygrobatidae), larva, excretory pore plate region. (Redrawn and modified from Prasad and Cook 1972.) **Figure 16.123** *Hygrobates* sp. (Hygrobatidae), larva, tibia and tarsus of leg II. (Redrawn and modified from Prasad and Cook 1972.) E1 and E2, excretory pore plate setae; Mh1, mediohysterosomal setae; Ta, tarsus; Ti, tibia; V1 and V2, ventral setae.

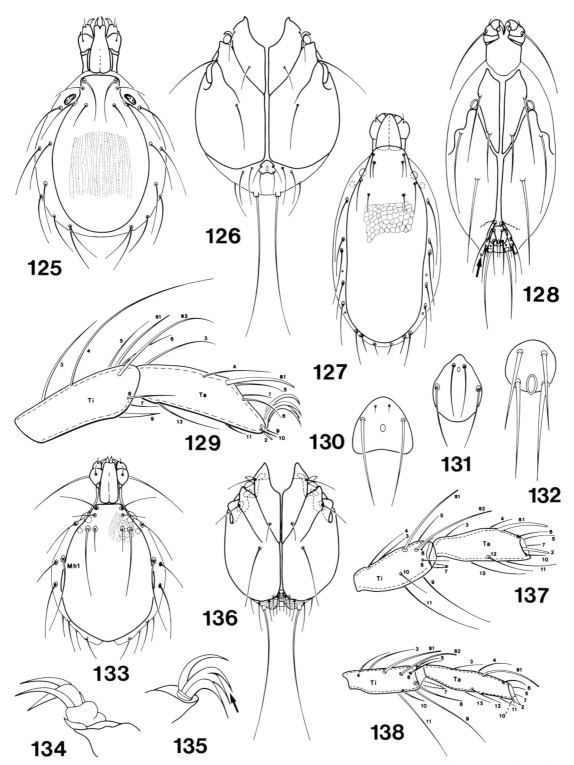

Figures 16.125–16.126, 16.129 *Feltria* sp. (Feltriidae), larva. Fig. 16.125, dorsum of idiosoma and gnathosoma; Fig. 16.126, venter of idiosoma; Fig. 16.129, tibia and tarsus of leg I, posterolateral view. **Figures 16.127–16.128, 16.132, 16.135, 16.137** *Unionicola* sp. (Unionicolinae), larva. Fig. 16.127, dorsum of idiosoma and gnathosoma; Fig. 16.128, venter of idiosoma and gnathosoma; Fig. 16.132, excretory pore plate; Fig. 16.135, claws and empodium of leg I; Fig. 16.137, tibia and tarsus of leg I, posterolateral view. (Redrawn and modified from Prasad and Cook 1972.) **Figure 16.130** *Koenikea marshalli* Viets (Pionatacinae), larva, excretory pore plate. (Redrawn and modified from Prasad and Cook 1972.) **Figures 16.131, 16.134, 16.138** *Neumania punctata* Marshall (Pionatacinae), larva. Fig. 16.131, excretory pore plate; Fig. 16.134, claws and empodium of leg I; Fig. 16.138, tibia and tarsus of leg I, posterolateral view. (Redrawn and modified from Prasad and Cook 1972.) **Figures 16.133, 16.136** *Aturus* sp. (Aturinae), larva. Fig. 16.133, dorsum of idiosoma and gnathosoma; Fig. 16.136, venter of idiosoma. (Redrawn from Smith 1984.) Mh1, mediohysterosomal seta; Ta, tarsus; Ti, tibia.

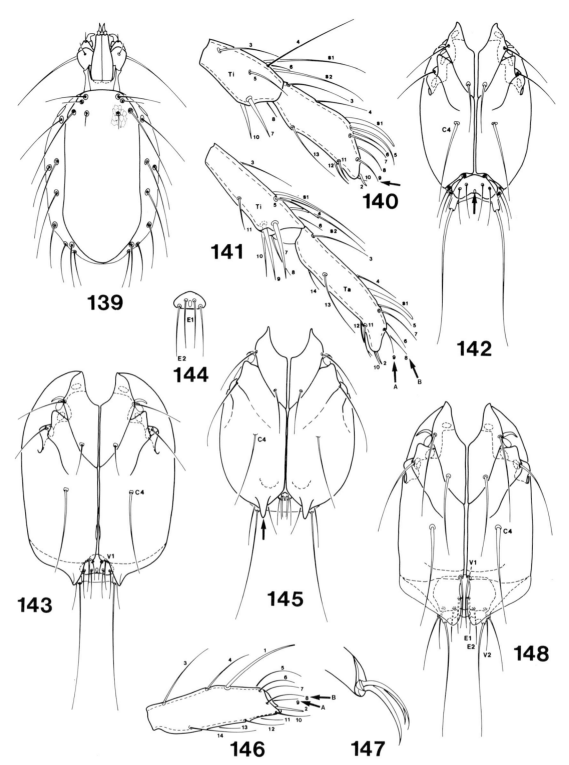

Figures 16.139, 16.141–16.142 *Ljania bipapillata* Thor (Axonopsinae), larva. Fig. 16.139, dorsum of idiosoma and gnathosoma; Fig. 16.141, tibia and tarsus of leg I, anterolateral view; Fig. 16.142, venter of idiosoma. (Redrawn from Smith 1984.)   **Figure 16.140** *Aturus deceptor* Habeeb (Aturinae), larva, tibia and tarsus of leg I, anterolateral view. (Redrawn from Smith 1984.)   **Figure 16.143** *Woolastookia setosipes* Habeeb (Axonopsinae), larva, venter of idiosoma. (Redrawn from Smith 1984.)   **Figures 16.144–16.145** *Wettina ontario* Smith (Wettininae), larva. Fig. 16.144, excretory pore plate; Fig. 16.145, venter of idiosoma. **Figure 16.146** *Piona mitchelli* Cook (Pioninae), larva, tarsus of leg I, posterolateral view. (Redrawn and modified from Prasad and Cook 1972.)   **Figure 16.147** *Piona carnea* (Koch) (Pioninae), larva, claws and empodium of leg III. (Redrawn and modified from Prasad and Cook 1972.)   **Figure 16.148** *Brachypoda setosicauda* Habeeb (Axonopsinae), larva, venter of idiosoma. (Redrawn from Smith 1984.) C4, coxal plate seta; E1 and E2, excretory pore plate setae; Ta, tarsus, Ti, tibia; V1 and V2, ventral setae.

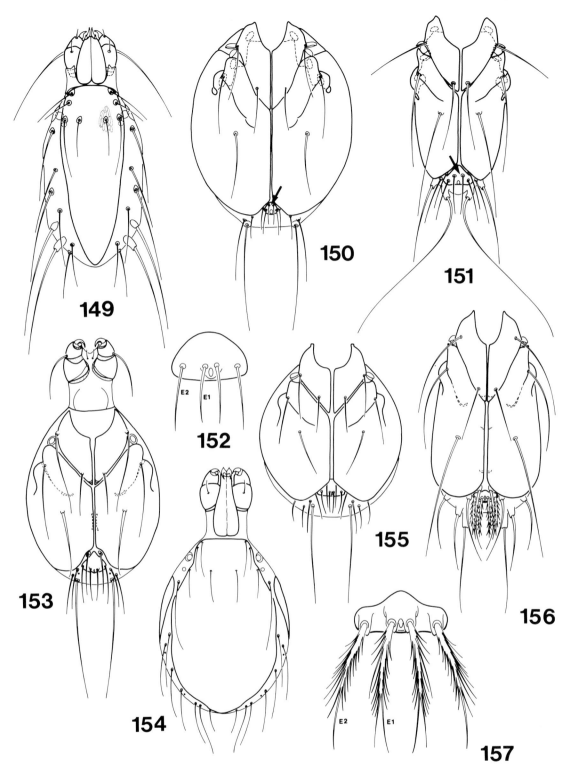

Figures 16.149, 16.151 *Albia neogaea* Habeeb (Albiinae), larva. Fig. 16.149, dorsum of idiosoma and gnathosoma; Fig. 16.151, venter of idiosoma. (Redrawn from Smith 1984.)   **Figure 16.150** *Neobrachypoda ekmani* (Walter) (Axonopsinae), larva, venter of idiosoma. (Redrawn from Smith 1984.)   **Figures 16.152, 16.155** *Najadicola ingens* Koenike (Najadicolinae), larva. Fig. 16.152, excretory pore plate; Fig. 16.155, venter of idiosoma. (Redrawn from Simmons and Smith 1984.)   **Figures 16.153–16.154** *Piona interrupata* Marshall (Pioninae), larva. Fig. 16.153, venter of idiosoma and gnathosoma; Fig. 16.154, dorsum of idiosoma and gnathosoma. (Redrawn and modified from Prasad and Cook 1972.)   **Figures 16.156–16.157** *Nautarachna muskoka* Smith (Pioninae), larva. Fig. 16.156, venter of idiosoma; Fig. 16.157, excretory pore plate. E1 and E2, excretory pore plate setae.

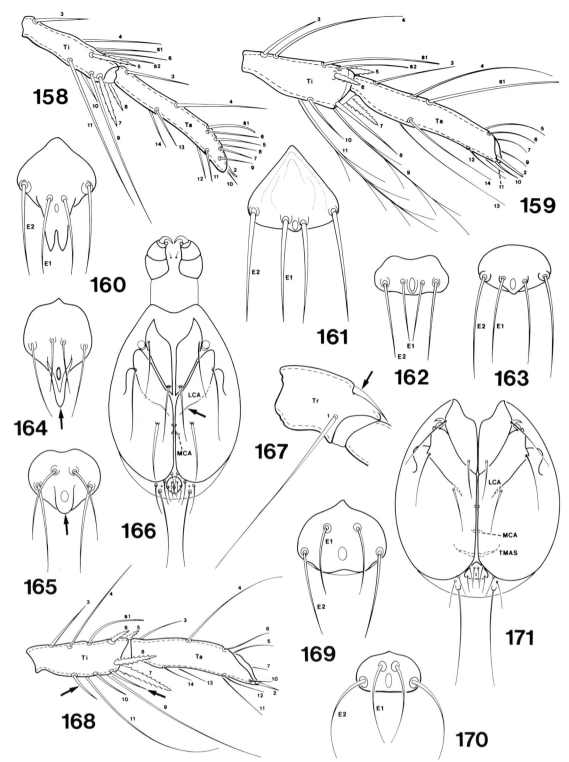

Figure 16.158 *Najadicola ingens* Koenike (Najadicolinae), larva, tibia and tarsus of leg II, anterolateral view. (Redrawn from Simmons and Smith 1984.) ***Figure 16.159*** *Piona mitchelli* Cook (Pioninae), larva, tibia and tarsus of leg II, posterolateral view. (Redrawn and modified from Prasad and Cook 1972.) ***Figure 16.160*** *Piona carnea* (Koch) (Pioninae), larva, excretory pore plate. (Redrawn from Prasad and Cook 1972.) ***Figure 16.161*** *Piona interrupta* Marshall (Pioninae), larva, excretory pore plate. (Redrawn from Prasad and Cook 1972.) ***Figures 16.162–16.163*** *Piona constricta* (Wolcott) (Pioninae), larvae, excretory pore plates. (Redrawn from Prasad and Cook 1972.) ***Figure 16.164*** *Hydrochoreutes microporus* Cook (Hydrochoreutinae), larva, excretory pore plate. (Redrawn and modified from Prasad and Cook 1972.) ***Figures 16.165–16.166*** *Hydrochoreutes minor* Cook (Hydrochoreutinae), larva. Fig. 16.165, excretory pore plate; Fig. 16.166, venter of idiosoma and gnathosoma. (Redrawn and modified from Prasad and Cook 1972.) ***Figures 16.167, 16.171*** *Pseudofeltria multipora* Cook (Foreliinae), larva. Fig. 16.167, trochanter of leg III; Fig. 16.171, venter of idiosoma. ***Figures 16.168–16.169*** *Forelia ovalis* Marshall (Foreliinae), larva. Fig. 16.168, tibia and tarsus of leg III, posterolateral view; Fig. 16.169, excretory pore plate. (Redrawn and modified from Prasad and Cook 1972.) ***Figure 16.170*** *Forelia onondaga* Habeeb (Foreliinae), larva, excretory pore plate. E1 and E2, excretory pore plate setae; LCA and MCA, lateral and medial coxal apodemes; Ta, tarsus; Ti, tibia; TMAS, transverse muscle attachment scar; Tr, trochanter.

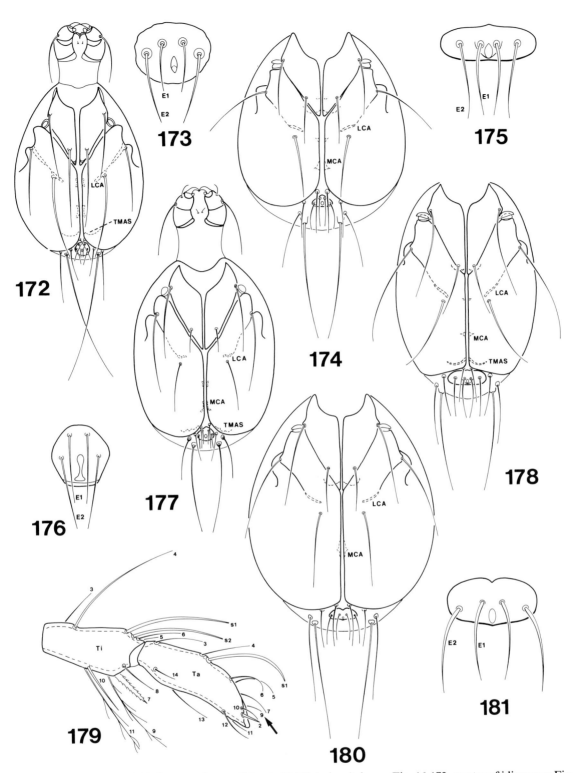

Figures 16.172, 16.176 *Tiphys americanus* (Marshall) (Tiphyinae), larva. Fig. 16.172, venter of idiosoma; Fig. 16.176, excretory pore plate. (Redrawn and modified from Prasad and Cook 1972.) ***Figures 16.173, 16.179*** *Tiphys ornatus* Koch (Tiphyinae), larva. Fig. 16.173, excretory pore plate; Fig. 16.179, tibia and tarsus of leg I, posterolateral view. (Redrawn and modified from Prasad and Cook 1972.) ***Figure 16.174*** *Forelia onondaga* Habeeb (Foreliinae), larva, venter of idiosoma. ***Figures 16.175, 16.178*** *Neotiphys pionoidellus* (Habeeb) (Tiphyinae), larva. Fig. 16.175, excretory pore plate; Fig. 16.178, venter of idiosoma. ***Figure 16.177*** *Forelia ovalis* Marshall (Foreliinae), larva, venter of idiosoma. (Redrawn and modified from Prasad and Cook 1972.) ***Figures 16.180–16.181*** *Huitfeldtia rectipes* Thor (Huitfeldtiinae), larva. Fig. 16.180, venter of idiosoma; Fig. 16.181, excretory pore plate. E1 and E2, excretory pore plate setae; LCA and MCA, lateral and medial coxal apodemes; Ta, tarsus; Ti, tibia; TMAS, transverse muscle attachment scar.

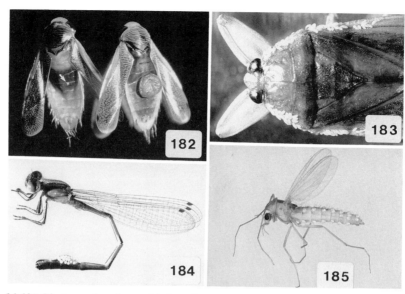

Figure 16.182–16.185 Photographs of parasitic larval water mites attached to host insects. Fig. 16.182, larvae of *Eylais* (Eylaidae) attached to the subelytral abdominal integument of adult waterboatmen (Hemiptera: Corixidae), illustrating larval size before and after engorgement (Courtesy B. P. Smith); Fig. 16.183, larvae of *Hydrachna* (Hydrachnidae) attached to the thorax and elytra of a giant water bug (Hemiptera: Belostomatidae); Fig. 16.184, larvae of *Arrenurus* (Arrenurinae) attached to the abdominal segments of a damselfly (Odonata: Zygoptera); Fig. 16.185, larvae of *Feltria* (Feltriidae) and *Aturus* (Aturinae) attached to the abdominal segments of a chironomid midge (Diptera: Chironomidae).

B. Taxonomic Key to Subfamilies for Adults of Water Mites

1a.	Tarsi of legs each with empodium and paired claws (Fig. 16.192); tarsus of pedipalp with a single, long terminal seta (Fig. 16.189, arrow); body soft and elongate; interstitial .... Stygothrombidioidea	
	Stygothrombidiidae *Stygothrombium*	
1b.	Tarsi of legs with paired claws but lacking empodium; tarsus of pedipalp without a single long terminal seta as above; body various; occurring in various types of habitats 2	
2a(1b).	Glandularia divided into gland- and seta-bearing portions (Fig. 16.190); genital acetabula absent; dorsum as shown in Figure 16.188 .... Hydrovolzioidea .... Hydrovolziidae	
	Hydrovolziinae *Hydrovolzia*	
2b.	Glandularia not divided; genital acetabula present; dorsum not as illustrated in Figure 16.188 3	
3a(2b).	Tibia of pedipalp much shorter than genu and bearing a dorsodistal projection extending well beyond insertion of tarsus (Fig. 16.187); chelicera one-segmented (Fig. 16.191) .... Hydrachnoidea	
	Hydrachnidae *Hydrachna*	
3b.	Tibia of pedipalp usually longer than genu, "when" shorter "then" lacking a dorsodistal projection; chelicera two-segmented (Fig. 16.196) 4	
4a(3b).	Lateral eye capsules fused medially into a common eye plate (Figs. 16.193, 16.194, 16.195) .... Eylaoidea 5	
4b.	Lateral eye capsules, when present, well separated from each other and not on a common plate (Figs. 16.205 and 16.206) 8	
5a(4a).	Lenses of eye plate well separated from each other on their respective sides; eye plate bearing one pair of setae (Fig. 16.193) .... Eylaidae *Eylais*	
	16.193) .... Eylaidae *Eylais*	
5b.	Lenses of eye plate relatively close together on their respective sides; eye plate bearing four pairs of setae (Figs. 16.194 and 16.195) 6	
6a(5b).	Eye plate much longer than wide (Fig. 16.194); genital acetabula borne scattered in ventral integument .... Limnocharidae 7	

6b.   Eye plate approximately as wide as long (Fig. 16.195); genital acetabula borne on, or surrounded by, acetabular plates .... Piersigiidae
Piersigiinae   *Piersigia*

7a(6a).   Pedipalp four- or five-segmented, with tarsus inserted at distal end of tibia (Fig. 16.197) .... Limnocharinae   *Limnochares, Neolimnochares*

7b.   Pedipalp three-segmented, with tarsus (Fig. 16.198, arrow) reduced and inserted dorsally on tibia .... Rhyncholimnocharinae   *Rhyncholimnochares*

8a(4b).   Dorsum with a series of reticulate platelets as shown in Figure 16.201 .... Rhynchohydracaridae
Clathrosperchontinae   *Clathrosperchon*

8b.   Dorsum various, but not with an arrangement of platelets as shown in Figure 16.201 9

9a(8b).   Pedipalp chelate, with dorsodistal portion of tibia extending well beyond insertion of tarsus (Figs. 16.199 and 16.203, arrows); capitulum without an anchoral process (Fig. 16.283, arrow) 10

9b.   Pedipalp not chelate, "when" appearing chelate (as in some Tiphyinae) (Fig. 16.200) "then" capitulum with a well-developed anchoral process (as in Fig. 16.280, arrow A) 18

10a(9a).   Dorsodistal extension of pedipalp tibia relatively long (Fig. 16.203, arrow) .... Hydrodromidae   *Hydrodroma*

10b.   Dorsodistal extension of pedipalp tibia relatively short (Fig. 16.199, arrow) .... Hydryphantidae   11

11a(10b).   Swimming setae present 12
11b.   Swimming setae absent 13

12a(11a).   Dorsum with a medial eye plate bearing a pigmented medial eye and two pairs of setae, between lateral eyes (Figs. 16.202 and 16.205) ..... Hydryphantinae   *Hydryphantes*

12b.   Dorsum without a medial eye plate between lateral eyes .... Pseudohydryphantinae   *Pseudohydryphantes*

13a(11b).   Third and fourth coxal plates with medial margins extensive and close together, not separated by genital field (Fig. 16.207); lateral eyes in capsules; body not elongated .... Cowichaniinae   *Cowichania*

13b.   Third and fourth coxal plates usually well separated, with genital field extending between them, "when" these coxal plates close together "then" lateral eyes not in capsules "and" body greatly elongated (Fig. 16.204) 14

14a(13b).   Lateral eyes in capsules (Fig. 16.206, arrow) 15
14b.   Lateral eyes not in capsules 17

15a(14a).   Dorsum without dorsalia (Fig. 16.206); claws with palmately arranged clawlets (in most commonly collected species) (Fig. 16.214), or simple .... Protziinae   *Protzia, Partnunia*

15b.   Dorsum with dorsalia (muscle attachment platelets) of various sizes and arrangements (Figs. 16.208 and 16.209); claws simple, without palmately arranged clawlets 16

16a(15b).   Dorsum with all dorsalia peripheral in position (Fig. 16.212, arrow) .... Cyclothyadinae   *Cyclothyas*

16b.   Dorsum with some dorsalia more medial in position (Figs. 16.208 and 16.209) .... Thyadinae   *Trichothyas, Thyopsis, Thyopsella, Panisus, Panisopsis, Notopanisus, Thyasella, Thyas, Thyasides, Zschokkea, Euthyas, Tjadikothyas*

17a(14b).   Body not greatly elongated; genital flaps well developed (Fig. 16.211, arrow) .... Tartarothyadinae   *Tartarothyas*

17b.   Body greatly elongated (Fig. 16.204); genital flaps poorly developed or absent .... Wandesiinae   *Wandesia*

18a(9b).   Femur of pedipalp with two long medial setae (Fig. 16.213, arrow); occurring only in hot springs .... Thermacaridae   *Thermacarus*

18b.   Femur of pedipalp usually without two medial setae and never as illustrated in Figure 16.213; not occurring in hot springs 19

19a(18b).   Movable genital flaps flanking gonopore and when closed either partially or completely covering gonopore; genital acetabula lying free in gonopore (not on flaps or acetabular plates) (Figs. 16.210, 16.219, 16.226) .... Lebertioidea   20

19b.   Movable genital flaps usually absent, "when" flaps present "then" acetabula lie on flaps rather than in gonopore (Fig. 16.186) 29

20a(19a).   Medial margins of fourth coxal plates reduced to narrow angles and bearing a pair of glandularia (Fig. 16.218, arrow); tarsus of pedipalp bearing pad-shaped setae and appearing spatulate distally (Fig. 16.216, arrow) .... Rutripalpidae   *Rutripalpus*

20b.   Medial margins of fourth coxal plates usually extensive, "when" reduced to narrow angles "then" lacking glandularia as above 21

21a(20b).   Dorsal and ventral shields present; dorsal shield usually comprising a large plate and a series of smaller platelets (Figs. 16.215, 16.217, 16.223); gonopore usually with six pairs of genital acetabula (Fig. 16.221), "when" with only three pairs "then" dorsal shield has a complete ring of marginal platelets (Fig. 16.217); one known North American species has an entire dorsal shield and six pairs of genital acetabula .... Torrenticolidae   22

21b.   Dorsal and ventral shields present or absent; when present, dorsal shield entire, without smaller platelets; gonopore with three pairs of genital acetabula 24

22a(21a).   Dorsal shield with a single anteromedial platelet and a series of small marginal platelets (Fig. 16.217); gonopore with three pairs of genital acetabula .... Testudacarinae   *Testudacarus*

22b.   Dorsal shield with smaller platelets all paired and anterior in position (Figs. 16.215 and 16.223), or in one species fused with the large central plate; gonopore with six pairs of genital acetabula 23

23a(22b).   Third coxal plates bearing a pair of glandularia (Fig. 16.219, arrow); pedipalp four-segmented (Fig. 16.220) .... Neoatractidinae   *Neoatractides*

23b.   Third coxal plates lacking glandularia (Fig. 16.221); pedipalp five-segmented (as in Fig. 16.222) .... Torrenticolinae   *Torrenticola, Pseudotorrenticola*

24a(21b).   Fourth coxal plates bearing glandularia (Fig. 16.224, arrow) .... Teutoniidae .. Teutoniinae   *Teutonia*

24b.   Fourth coxal plates lacking glandularia (Figs. 16.226, and 16.230) 25

25a(24b).   Tarsi of fourth pair of legs bearing claws (Fig. 16.229) 26

25b.   Tarsi of fourth pair of legs lacking claws (Fig. 16.231) 28

26a(25a).   Suture lines between second and third coxal plates incomplete (Fig. 16.226, arrow), but touching the genital field area posteriorly; genu of pedipalp usually with at least five long setae on medial surface (Fig. 16.225), but, in one known species, with four long setae on the dorsal surface .... Lebertiidae   *Lebertia, Scutolebertia*

26b.   Suture lines between second and third coxal plates complete (Fig. 16.230), these coxal plates often completely separated on their respective sides (Fig. 16.210); genu of pedipalp without long setae as illustrated above 27

27a(26b).   Complete dorsal "and" ventral shields present .... Anisitsiellidae (in part)
                                  Anisitsiellinae (in part)   *Bandakia, Bandakiopsis,*   *Cookacarus, Oregonacarus, Utaxatax*

27b.   Dorsal and ventral sclerites variously developed, but complete dorsal "and" ventral shields never both present .... Sperchontidae   Sperchontinae   *Sperchon, Sperchonopsis*

28a(25b).   Without a medial ventral suture line (Fig. 16.228); all legs inserted near anterior end of idiosoma; idiosoma circular in cross-section or higher than wide .... Oxidae .... Oxinae   *Oxus, Frontipoda*

28b.   With a medial ventral suture line; fourth legs inserted laterally near middle of idiosoma (similar to Fig. 16.230, arrow); idiosoma dorsoventrally flattened .... Anisitsiellidae (in part) .... Anisitsiellinae (in part)   *(Mamersellides)*

29a(19b).   All coxal plates grouped together, with fourth coxal plates rounded posteriorly and suture lines between third and fourth plates extending posterolaterally (Fig. 16.235); interstitial .... Omartacaridae   Omartacarinae   *Omartacarus*

29b.	Coxal plates not as described and illustrated above; mostly inhabiting superficial waters		30
30a(29b).	Femur of pedipalp bearing a single ventral seta (Figs. 16.227 and 16.244; arrows); movable genital flaps that cover gonopore when closed present in females (Fig. 16.186) and in some males	Limnesiidae	31
30b.	Femur of pedipalp lacking a ventral seta or, rarely, bearing three setae (Fig. 16.234, arrow); lacking movable genital flaps that cover gonopore when closed		35
31a(30a).	Tibia of pedipalp with five or more long, pointed ventral tubercles and a long medial seta (Fig. 16.227) .... Kawamuracarinae		*Kawamuracarus*
31b.	Tibia of pedipalp either with four long, pointed tubercles and no long medial seta, or without long ventral tubercles		32
32a(31b).	Tarsi of fourth legs lacking claws		33
32b.	Tarsi of fourth legs bearing claws		34
33a(32a).	Genital flaps on each side divided into anterior and posterior sclerites (Fig. 16.233); dorsum with large, transversely divided dorsal shield (Fig. 16.232) .... Neomamersiinae		*Neomamersa, Meramecia*
33b.	Genital flaps of females (and acetabular plates of males) entire on each side (Fig. 16.186); dorsum usually without dorsal shield, but rarely with entire (undivided) shield .... Limnesiinae		*Limnesia, Centrolimnesia*
34a(32b).	Femur of pedipalp with a ventral seta inserted directly on segment (as in Fig. 16.227); dorsum nearly covered by two large plates (Fig. 16.237) .... Protolimnesiinae		*Protolimnesia*
34b.	Femur of pedipalp with a short, thickened ventral seta inserted on a tubercle (Fig. 16.244, arrow); dorsum bearing an entire dorsal shield, or smaller platelets covering less than one-half its area .... Tyrrelliinae		*Tyrrellia, Neotyrrellia*
35a(30b).	Integument between fourth coxal plates and genital field with two pairs of glandularia arranged more or less in a row, and either free (Fig. 16.238, arrows) or variously fused with other ventral sclerites; dorsal shield absent, or present and bearing fewer than five pairs of glandularia .... Feltriidae		*Feltria*
35b.	Integument between fourth coxal plates and genital field usually without two pairs of glandularia arranged in a row, "when" glandularia in this position (some Pionatacinae, *Koenikea*) "then" dorsal shield bearing six pairs of glandularia (Fig. 16.236)		36
36a(35b).	Idiosoma laterally compressed and much higher than wide (Fig. 16.242); fourth legs laterally compressed (Fig. 16.239) .... Aturidae (in part) Frontipodopsinae		*Frontipodopsis*
36b.	Idiosoma not laterally compressed, either spherical or dorsoventrally flattened		37
37a(36b).	Both dorsal and ventral shields present; tibia of first leg without two thickened setae and a downturned seta distoventrally (i.e., not as shown in Fig. 16.241)		38
37b.	Either dorsal or ventral shield, or both, usually absent, "when" both shields present [males of some Hygrobatidae,(*Atractides*)] "then" first leg with tarsus bowed and tibia bearing two thickened setae and a downturned seta distoventrally (Fig. 16.241)		63
38a(37a).	First leg with tarsus much shorter than tibia, and claws dorsal in position and flexing proximally (Fig. 16.240, arrow) .... Momoniidae		39
38b.	First leg with distal segments not modified as described and illustrated above; claws terminal in position and flexing ventrally		41
39a(38a).	Dorsal shield composed of a large plate flanked by numerous smaller platelets (Fig. 16.245) .... Cyclomomoniinae		*Cyclomomonia*
39b.	Dorsal shield entire or comprising large anterior and posterior plates, but not flanked by smaller platelets (i.e., not as in Fig. 16.245)		40
40a(39b).	A pair of glandularia located close to anterior end of genital field (Fig. 16.243, arrow) .... Momoniinae		*Momonia*
40b.	No glandularia located near anterior end of genital field; interstitial .... Stygomomoniinae		*Stygomomonia*

41a(38b).   Dorsal shield large and transversely divided (somewhat as in Fig. 16.232);
suture lines between third and fourth coxal plates noticeably looped
around a pair of glandularia (Fig. 16.248, arrow) .... Hygrobatidae (in part)   *Diamphidaxona*

41b.   Dorsal shield usually entire, never transversely divided as above; suture
lines between third and fourth coxal plates not looped as above 42

42a(41b).   Outer rim of coxal plates forming a smooth semicircle that does not
extend beyond edge of idiosoma (Fig. 16.247, arrow);
interstitial .... Uchidastygacaridae (in part)
Uchidastygacarinae   *Uchidastygacarus*

42b.   Coxal plates not as described and illustrated above 43

43a(42b).   A pair of glandularia located near midline at junction of third and fourth
coxal plates (Fig. 16.246, arrow) .... Chappuisididae .... Chappuisidinae
Chappuisides

43b.   No glandularia in position described and illustrated above 44

44a(43b).   Suture lines between third and fourth coxal plates extending distinctly
posteromedially to, and touching, genital field region, and well separated
from each other medially (Fig. 16.252, arrow); genital acetabula borne in
a single row on each edge of gonopore .... Neoacaridae   *Neoacarus, Volsellacarus*

44b.   Not with above combination of characters 45

45a(44b).   Coxoglandularia I (Fig. 16.256, arrow) shifted far forward, near suture
lines between first and second coxal plates; genital field projecting
between, and widely separating, fourth coxal plates .... Mideidae   *Midea*

45b.   Not with above combination of characters 46

46a(45b).   Pedipalp distinctly uncate (Figs. 16.249, arrow A; 16.250, arrow) 47

46b.   Tibia of pedipalp with either a slight distoventral projection (Fig. 16.261,
arrow), a distinct ventral projection near middle of segment (Fig.
16.253), or no ventral projection 49

47a(46a).   Genital acetabula numerous; fourth coxal plates lacking glandularia;
genital field extending at most only slightly between fourth coxal plates,
and not separating them medially (Fig. 16.254) 48

47b.   Usually three or four pairs of genital acetabula present (but up to six
pairs in one species); fourth coxal plates bearing a pair of glandularia
(Fig. 16.251, arrow); genital field extending between fourth coxal plates
and widely separating them medially .... Krendowskiidae   *Krendowskia, Geayia*

48a(47a).   Genital field with at least some and often all acetabula lying free in
gonopore (Fig. 16.254), or with acetabula lying on plates that flank
gonopore but are not fused with ventral shield (similar to Fig. 16.256);
capitulum with a pair of long ventral setae (Fig. 16.255, arrow); tibia of
pedipalp rotated (up to 90°) relative to genu .... Athienemanniidae   49

48b.   Genital acetabula borne on acetabular plates incorporated into ventral
shield, never in gonopore (Fig. 16.257); capitulum with all setae short;
tibia of pedipalp not greatly rotated relative to genu (Fig. 16.250);
posterior end of male body variously (and often highly) modified
(Fig. 16.286) .... Arrenuridae .... Arrenurinae
Arrenurus

49a(48a).   Genital acetabula in one or two rows on each side of gonopore (Fig.
16.263); genu of pedipalp with a pronounced ventral projection (Fig.
16.249, arrow B) .... Stygameracarinae   *Stygameracarus*

49b.   Genital acetabula in three or more rows on each side of gonopore (Fig.
16.254); genu of pedipalp without pronounced ventral projection
(Fig. 16.255) .... Athienemanniinae   *Chelomideopsis, Platyhydracarus*

50a(46b).   Fourth leg with tarsus indented or curved and bearing short peg-like
setae (Fig. 16.260) .... Pionidae (in part) .... Foreliinae (in part)
Pseudofeltria (males), *Forelia* (males of some),
Pionacercus (males of some)

50b.   Fourth leg not as described and illustrated above 51

51a(50b).   Fourth leg with genu noticeably notched and bearing short peg-like setae
associated with this notch (Fig. 16.258) .... Pionidae (in part) .... Pioninae (in part)
Nautarachna (males), *Piona* (males of some)

51b. Fourth leg not as described and illustrated above 52

52a(51b). Segments of pedipalp short and partially fused, femur with three ventral
setae (Fig. 16.234, arrow) .... Bogatiidae .... Horreolaninae
Horreolanus

52b. Pedipalp not as described and illustrated above 53

53a(52b). Coxoglandularia I shifted far forward, near suture lines between first
and second coxal plates (Fig. 16.259, arrow) .... Nudomideopsidae *Nudomideopsis,*
Neomideopsis, Paramideopsis

53b. Coxoglandularia I between second and third coxal plates (Fig. 16.265,
arrow) or shifted only slightly onto second coxal plates (Fig. 16.262, arrow A) 54

54a(53b). Three or four pairs of genital acetabula present 55
54b. More than four pairs of genital acetabula present 58

55a(54b). Genital acetabula borne either in gonopore completely surrounded by
ventral shield (Fig. 16.265) or on ventral shield in a row on each side of
gonopore, well anterior to posterior edge of idiosoma (Fig. 16.264) 56

55b. Genital acetabula not borne in gonopore completely surrounded by
ventral shield, but arranged in either an arc or a triangle on each side of
gonopore (Fig. 16.271); at least some acetabula located near posterior
end of idiosoma .... Aturidae (in part) ...... Axonopsinae (most
North American genera) *Albaxona, Stygalbiella,*
Axonopsis, Erebaxonopsis, Brachypoda, Neobrachypoda,
Estellacarus, Woolastookia, Ljania, Lethaxona

56a(55a). Suture lines between third and fourth coxal plate oriented at right angles
to long axis of body where they join midline (Fig. 16.264,
arrow) .... Uchidastygacaridae (in part)
Morimotacarinae *Morimotacarus, Yachatsia*

56b. Suture lines between third and fourth coxal plates oriented at an oblique
angle to long axis of body where they join midline 57

57a(56b). Insertion of tarsus of pedipalp on distal end of tibia decidedly dorsal in
position (Fig. 16.261); openings for insertion of fourth legs covered by
projections of fourth coxal plates when viewed ventrally (Fig. 16.266,
arrow) .... Acalyptonotidae (in part) *Paenecalyptonotus*

57b. Insertion of tarsus of pedipalp on tibia not decidedly dorsal in position
(Fig. 16.253); openings for insertion of fourth legs slightly or not at all
covered by projections of fourth coxal plates (Fig.
16.265) .... Mideopsidae .... Mideopsinae
Mideopsis

58a(54b). Dorsal shield bearing six pairs of glandularia, with three pairs grouped
close together in a line or triangle on each side (Fig. 16.236, circled
area) .... Unionicolidae (in part) .... Pionatacinae (in part)
Koenikea

58b. Dorsal shield not as described and illustrated above 59

59a(58b). Gonopore located very near center of ventral shield (Fig. 16.262) 60
59b. Gonopore located close to posterior end of ventral shield 61

60a(59a). Ventral shield incorporating excretory pore (Fig. 16.262, arrow B) .... Laversiidae *Laversia*
60b. Ventral shield not incorporating excretory pore .... Pionidae (in part)
Pioninae (in part) *Nautarachna* (females of some)

61a(59b). First coxal plates projecting beyond anterior edge of ventral shield
(Figs. 16.268 and 16.270) 62

61b. First coxal plates not projecting beyond anterior edge of ventral shield
(Fig. 16.269) 63

62a(61a). Three pairs of glandularia on dorsal shield, all located near edge
(Fig. 16.267) .... Arenohydracaridae *Arenohydracarus*

62b. One to several pairs of glandularia on dorsal shield, with at least one
pair located well medial to edge .... Aturidae (in part)
Aturinae *Aturus, Kongsbergia, Bharatalbia,*
Phreatobrachypoda

63a(61b).   Ventral shield usually with projections partially covering openings for
insertion of fourth legs when viewed ventrally, "when" lacking these
projections "then" genital field with acetabula in a single row on each
side (much as shown in Fig. 16.268) .... Aturidae (in part)
Axonopsinae (in part)   *Submiraxona, Lethaxonella*

63b.   Ventral shield lacking projections associated with openings for insertion
of fourth legs "and" genital field with acetabula arranged in several
rows (Fig. 16.269) .... Aturidae (in part) .... Albiinae
Albia

64a(37b).   Fourth coxal plates bearing glandularia (Fig. 16.273,
arrow) .... Hygrobatidae (in part)   *Hygrobates, Mesobates,*
Atractides, Corticacarus

64b.   Fourth coxal plates not bearing glandularia 65

65a(64b).   Apodemes of anterior coxal groups extending to approximately middle
of fourth coxal plates (Fig. 16.277, arrow) .... Unionicolidae (in part)
Pionatacinae (in part)   *Neumania*

65b.   Apodemes of anterior coxal groups not extending to middle of fourth
coxal plates 66

66a(65b).   Genital field surrounded by a well developed ventral
shield .... Acalyptonotidae (in part)   *Acalyptonotus*

66b.   Genital field not surrounded by a ventral shield 67

67a(66b).   Medial margins of fourth coxal plates more or less reduced to angles
(Fig. 16.274, arrow) 68

67b.   Medial margins of fourth coxal plates not reduced to angles (Fig. 16.275,
arrow) 69

68a(67a).   Three or four pairs of genital acetabula present; tarsus of first leg with
claw socket occupying much more than one-half of dorsal surface of
segment (Fig. 16.272); males with tarsus of fourth leg
unmodified .... Pionidae (in part) .... Wettininae
Wettina

68b.   Three or more pairs of genital acetabula present; tarsus of first leg with
claw socket occupying one-half or less of dorsal surface of segment;
males with tarsus of fourth leg indented or bowed and bearing short peg-
like setae (Fig. 16.260) .... Pionidae (in part) ...... Foreliinae (in part)
Pionacercus, Forelia, Pseudofeltria
(all females and males of some)

69a(67b).   Posterior end of capitulum without an anchoral process (Fig. 16.283,
arrow); symbionts of clams .... Pionidae (in part) .... Najadicolinae
Najadicola

69b.   Posterior end of capitulum with a distinct anchoral process (Fig. 16.280,
arrow A); symbionts of clams or free living 70

70a(69b).   Segments of pedipalps comparatively very long and slender (Fig.
16.278); three pairs of genital acetabula present; males with genu of
third leg modified (Fig. 16.279) and a petiole posterior to genital field
(Fig. 16.282) .... Pionidae (in part)
Hydrochoreutinae   *Hydrochoreutes*

70b.   Segments of pedipalps comparatively short and stocky; three to many
pairs of genital acetabula present; males with genu of third leg not
modified 71

71a(70b).   Suture lines between third and fourth coxal plates extending only a
short distance towards midline (Fig. 16.284, arrow A); with a pair of
glandularia present at, and usually fused with, posteromedial angles of
fourth coxal plates (Fig. 16.284, arrow B); first leg with long movable
setae in free-living species (Fig. 16.276); symbionts of clams or sponges,
or free living .... Unionicolidae (in part) .... Unionicolinae
Unionicola

71b.   Suture lines between third and fourth coxal plates extending to (or
nearly to) medial margins of coxal plates (usually longer than shown in
Fig. 16.275); without glandularia at posteromedial angles of fourth coxal
plates; first leg not as above; free living 72

72a(71b). Usually three, but rarely up to six pairs of genital acetabula present .... Pionidae (in part)
Tiphyinae *Tiphys, Pionopsis, Neotiphys*

72b. Seven or more pairs of genital acetabula present 73

73a(72b). Genu of pedipalp medially with a very long seta (Fig. 16.281, arrow); second and third coxal plates each with a thickened seta (Fig. 16.280, arrows B); legs of males unmodified .... Pionidae (in part)
Huitfeldtiinae *Huitfeldtia*

73b. Genu of pedipalp lacking a long seta; second and third coxal plates lacking heavy setae (Fig. 16.275); males with a notch on genu of fourth leg bearing associated peg-like setae (Fig. 16.258) .... Pionidae (in part)
Pioninae (in part) *Piona* (females, males of most), *Nautarachna* (females of some)

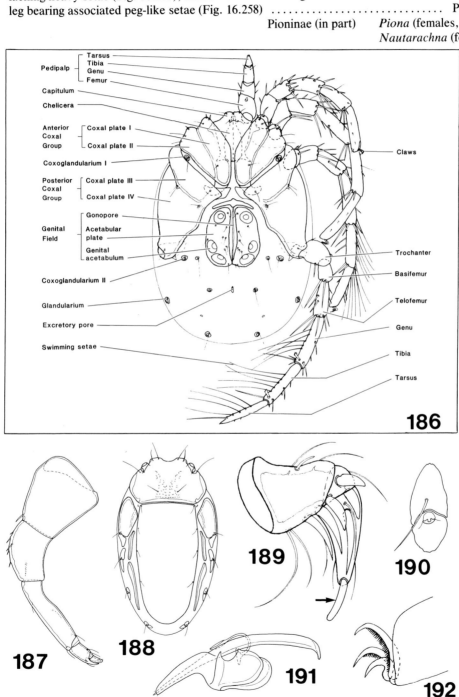

Figure 16.186 *Limnesia* sp. (Limnesiinae), adult female, ventral view. **Figures 16.187, 16.191** *Hydrachna* sp. (Hydrachnidae), adult. Fig. 16.187, pedipalp; Fig. 16.191, capitulum and chelicera, lateral view. (From Cook 1980.) **Figures 16.188, 16.190** *Hydrovolzia marshallae* Cook (Hydrovolziinae), adult. Fig. 16.188, dorsum of idiosoma; Fig. 16.190, glandularium. (From Cook 1974.) **Figures 16.189, 16.192** *Stygothrombium* sp. (Stygothrombidiidae), adult. Fig. 16.189, pedipalp; Fig. 16.192, claws and empodium of leg I.

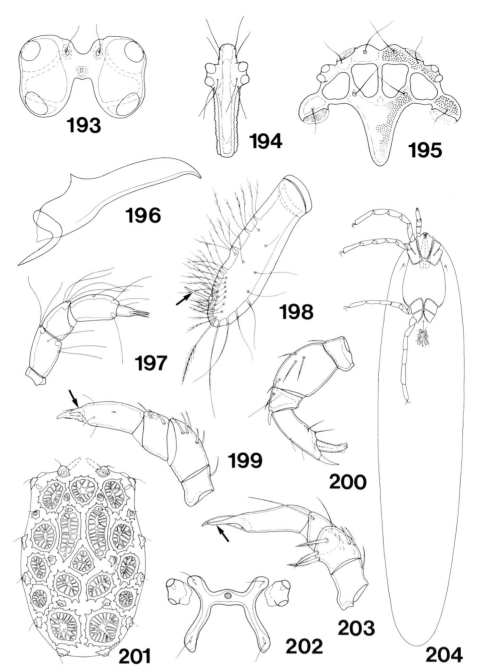

Figure 16.193 *Eylais* sp. (Eylaidae), adult, eye plate. (From Cook 1980.) *Figures 16.194, 16.197* *Limnocharres* sp. (Limnocharinae), adult. Fig. 16.194, eye plate; Fig. 16.197, pedipalp. (From Cook 1974.) *Figure 16.195* *Piersigia limnophila* Protz (Piersigiidae), adult, eye plate. (From Cook 1974.) *Figure 16.196* *Hygrobates* sp. (Hygrobatidae), adult, chelicera. *Figure 16.198* *Rhyncholimnochares* sp. (Rhyncholimnocharinae), adult, pedipalp. (From Cook 1980.) *Figures 16.199, 16.204* *Wandesia* sp. (Wandesiinae), adult. Fig. 16.199, pedipalp; Fig. 16.204, ventral view of female. (From Cook 1988 and Cook 1974, respectively.) *Figure 16.200* *Tiphys weaveri* Cook (Tiphyinae), adult, pedipalp. (From Cook 1974.) *Figure 16.201* *Clathrosperchon ornatus* Cook (Clathrosperchontinae), adult, dorsum of idiosoma. (From Cook 1974.) *Figure 16.202* *Hydryphantes* sp. (Hydryphantinae), adult, eye plate. (From Cook 1974.) *Figure 16.203* *Hydrodroma* sp. (Hydrodromidae), adult, pedipalp. (From Cook 1980.)

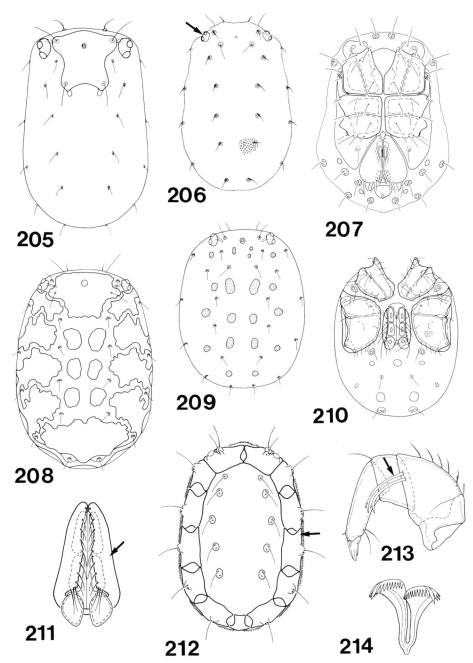

Figure 16.205 *Hydryphantes ruber* (de Geer) (Hydryphantinae), adult, dorsum of idiosoma. (From Cook 1974.)   **Figure 16.206** *Partnunia steinmanni* Walter (Protziinae), adult, dorsum of idiosoma. (From Cook 1974.)   **Figure 16.207** *Cowichania interstitialis* Smith (Cowichaniinae), adult female, venter of idiosoma. (Redrawn from I. M. Smith 1983b.)   **Figure 16.208** *Panisus condensatus* Habeeb (Thyadinae), adult, dorsum of idiosoma. (From Cook 1974.)   **Figure 16.209** *Thyas stolli* Koenike (Thyadinae), adult, dorsum of idiosoma. (From Cook 1974.)   **Figure 16.210** *Sperchon* sp. (Sperchontinae), adult male, venter of idiosoma. (From Cook 1974.)   **Figure 16.211** *Tartarothyas* sp. (Tartarothyadinae), adult female, genital field.   **Figure 16.212** *Cyclothyas siskiyouensis* Smith (Cyclothyadinae), adult, dorsum of idiosoma. (Redrawn from I. M. Smith 1983f.)   **Figure 16.213** *Thermacarus nevadensis* Marshall (Thermacaridae), adult, pedipalp. (From Cook 1974.)   **Figure 16.214** *Protzia* sp. (Protziinae), adult, claws of leg I. (From Cook 1974.)

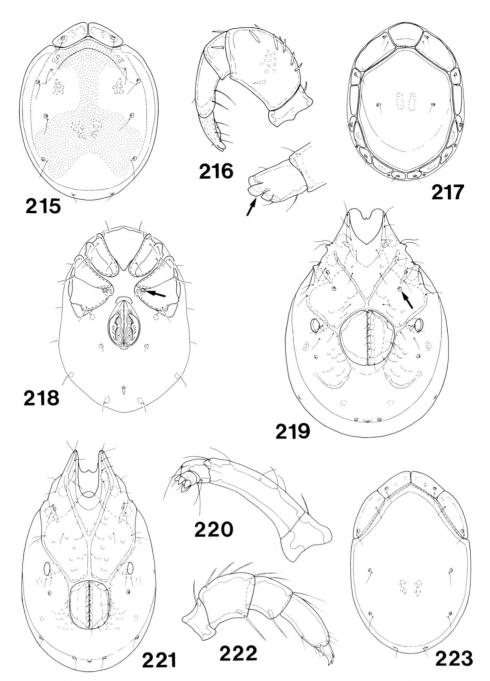

Figures 16.215, 16.219, 16.220 *Neoatractides* sp. (Neoatractidinae), adult. Fig. 16.215, dorsum of idiosoma; Fig. 16.219, venter of idiosoma, female; Fig. 16.220, pedipalp. (From Cook 1980.) ***Figures 16.216, 16.218*** *Rutripalpus* sp. (Rutripalpidae), adult. Fig. 16.216, pedipalp; Fig. 16.218, venter of idiosoma, female. (From Cook 1974 and Smith 1991b, repectively.) ***Figures 16.217, 16.222*** *Testudacarus americanus* Marshall (Testudacarinae), adult. Fig. 16.217, dorsum of idiosoma; Fig. 16.222, pedipalp. (From Cook 1974.) ***Figures 16.221, 16.223*** *Torrenticola* sp. (Torrenticolinae), adult. Fig. 16.221, venter of idiosoma, male; Fig. 16.223, dorsum of idiosoma. (From Cook 1980.)

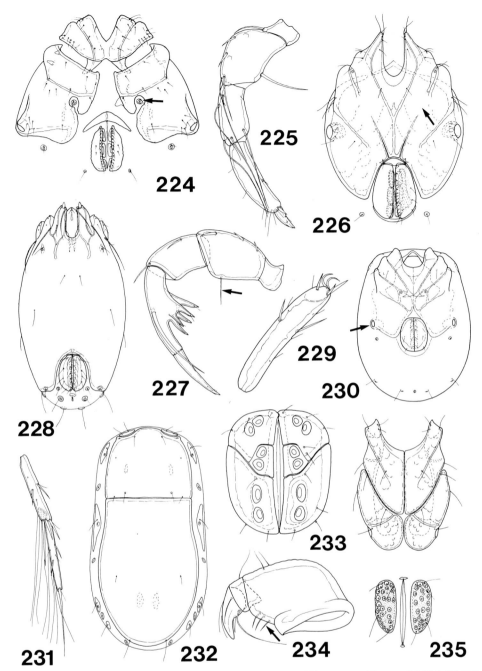

Figure 16.224 Teutonia sp. (Teutoniidae), adult female, venter of idiosoma. *Figures 16.225–16.226 Lebertia* sp. (Lebertiidae), adult. Fig. 16.225, pedipalp; Fig. 16.226, ventral plates, female. (From Cook 1980.) *Figure 16.227 Kawamuracarus* sp. (Kawamuracarinae), adult, pedipalp. (From Cook 1980.) *Figures 16.228, 16.231 Oxus* sp. (Oxidae), adult. Fig. 16.228, venter of idiosoma, female; Fig. 16.231, distal segments of leg IV. (From Cook 1980.) *Figure 16.229 Utaxatax ovalis* Cook (Anisitsiellinae), adult, tarsus of leg IV. (From Cook 1974.) *Figure 16.230 Bandakia phreatica* Cook (Anisitsiellinae), adult male, venter of idiosoma. (From Cook 1974.) *Figures 16.232–16.233 Neomamersa* sp. (Neomamersinae), adult. Fig. 16.232, dorsum of idiosoma; Fig. 16.233, genital field, male. (From Cook 1980.) *Figure 16.234 Horreolanus orphanus* Mitchell (Horreolaninae), adult, pedipalp. (From Cook 1974.) *Figure 16.235 Omartacarus* sp. (Omartacarinae), adult female, ventral plates. (From Cook 1980.)

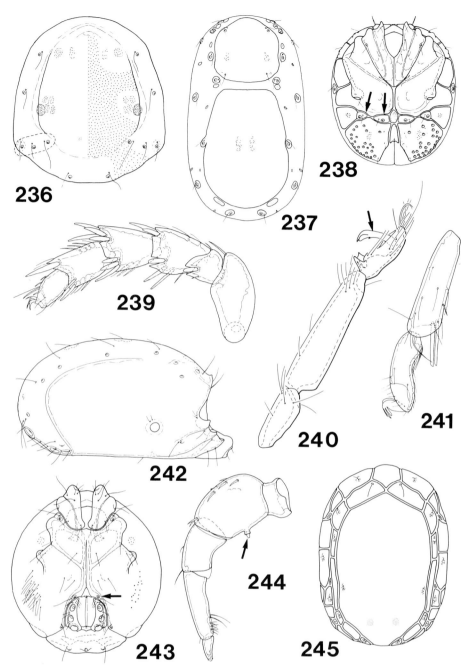

236

237

238

239

240

241

242

243

244

245

Figure 16.236 *Koenikea* sp. (Pionatacinae), adult, dorsal shield. (From Cook 1988.)   *Figure 16.237* *Protolimnesia* sp. (Protolimnesiinae), adult, dorsum of idiosoma. (From Cook 1980.)   *Figure 16.238* *Feltria exilis* (Cook) (Feltriidae), adult female, venter of idiosoma. (From Cook 1974.)   *Figures 16.239, 16.242* *Frontipodopsis* sp. (Frontipodopsinae), adult. Fig. 16.239, leg IV; Fig. 16.242, lateral view of idiosoma, female. (From Cook 1980.)   *Figure 16.240* *Stygomomonia riparia* Habeeb (Stygomomoniinae), adult, distal segments of leg I. (Redrawn from Smith 1991a.)   *Figure 16.241* *Atractides* sp. (Hygrobatidae), adult, distal segments of leg I. (From Cook 1980.)   *Figure 16.243* *Momonia projecta* Cook (Momoniinae), adult female, venter of idiosoma. (From Cook 1974.)   *Figure 16.244* *Tyrrellia ovalis* Marshall (Tyrrelliinae), adult, pedipalp. (From Cook 1980.)   *Figure 16.245* *Cyclomomonia andrewi* Smith (Cyclomomoniinae), adult, dorsum of idiosoma. (Redrawn from Smith 1989a.)

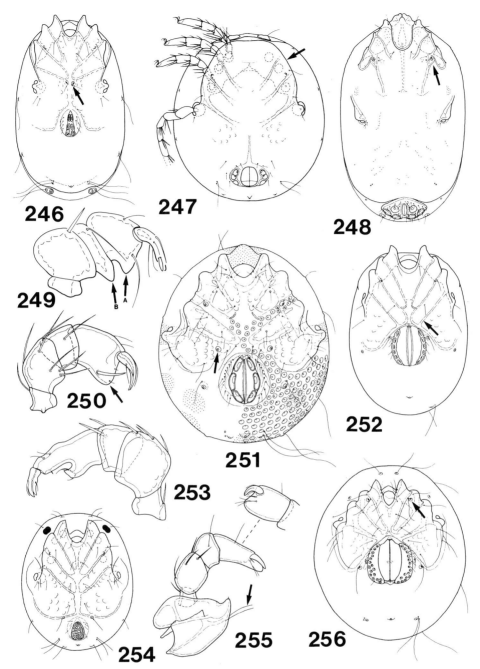

Figure 16.246 *Chappuisides eremitus* Cook (Chappuisididae), adult male, venter of idiosoma. (From Cook 1974.) *Figure 16.247* *Uchidastygacarus ovalis* Cook (Uchidastygacarinae), adult female, venter of idiosoma. (From Cook 1974.) *Figure 16.248* *Diamphidaxona* sp. (Hygrobatidae), adult male, ventral view. (From Cook 1980.) *Figure 16.249* *Stygameracarus cooki* Smith (Stygameracarinae), adult, pedipalp. (Redrawn from Smith 1990.) *Figure 16.250* *Arrenurus* sp. (Arrenurinae), adult, pedipalp. (From Cook 1974.) *Figure 16.251* *Geayia* sp. (Krendowskiidae), adult female, venter of idiosoma. (From Cook 1980.) *Figure 16.252* *Volsellacarus sabulonus* Cook (Neoacaridae), adult female, venter of idiosoma. (From Cook 1974.) *Figure 16.253* *Mideopsis* sp. (Mideopsinae), adult, pedipalp. (From Cook 1980.) *Figures 16.254–16.255* *Chelomideopsis besselingi* (Cook) (Athienemanniinae), adult. Fig. 16.254, venter of idiosoma, male; Fig. 16.255, capitulum and pedipalp. (From Cook 1974.) *Figure 16.256* *Midea expansa* Marshall (Mideidae), adult female, venter of idiosoma. (From Cook 1974.)

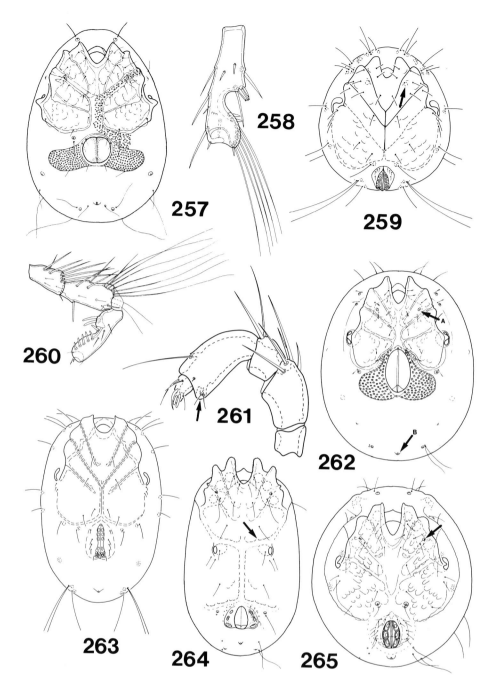

Figure 16.257 *Arrenurus* sp. (Arrenurinae), adult female, venter of idiosoma. (From Cook 1986.)   *Figure 16.258* *Piona* sp. (Pioninae), adult male, genu of leg IV. (From Cook 1980.)   *Figure 16.259* *Paramideopsis susanae* Smith (Nudomideopsidae), adult male, venter of idiosoma. (Redrawn from I. M. Smith 1983c.)   *Figure 16.260* *Forelia floridensis* Cook (Foreliinae), adult male, distal segments of leg IV. (From Cook 1974.)   *Figure 16.261* *Acalyptonotus neoviolaceus* Smith (Acalyptonotidae), adult, pedipalp. (Redrawn from I. M. Smith 1983e.)   *Figure 16.262* *Laversia berulophila* Cook (Laversiidae), adult female, venter of idiosoma. (From Cook 1974.)   *Figure 16.263* *Stygameracarus cooki* Smith (Stygameracarinae), adult male, venter of idiosoma. (Redrawn from Smith 1990.)   *Figure 16.264* *Yachatsia mideopsoides* Cook (Morimotacarinae), adult female, venter of idiosoma. (From Cook 1974.)   *Figure 16.265* *Mideopsis* sp. (Mideopsinae), adult female, venter of idiosoma. (From Cook 1980.)

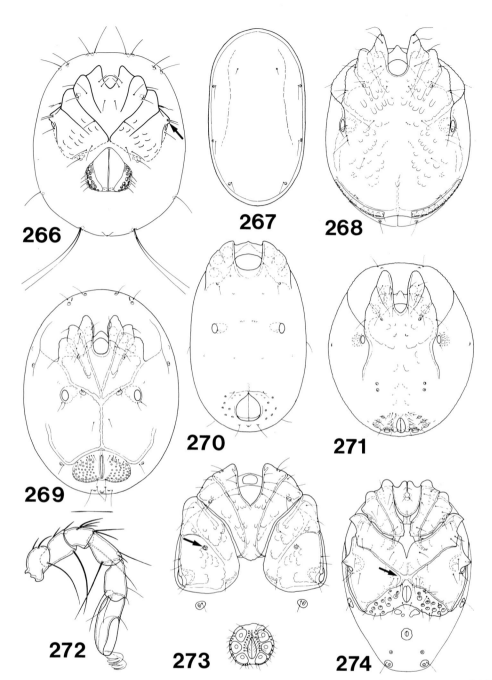

Figure 16.266 *Acalyptonotus neoviolaceus* Smith (Acalyptonotidae), adult female, venter of idiosoma. (Redrawn from I. M. Smith 1983e.) **Figures 16.267, 16.270** *Arenohydracarus* sp. (Arenohydracaridae), adult. Fig. 16.267, dorsum of idiosoma; Fig. 16.270, venter of idiosoma, female. (From Cook 1980.) **Figure 16.268** *Aturus* sp. (Aturinae), adult female, venter of idiosoma. (From Cook 1980.) **Figure 16.269** *Albia* sp. (Albiinae), adult male, venter of idiosoma. (From Cook 1980.) **Figure 16.271** *Axonopsis* sp. (Axonopsinae), adult male, venter of idiosoma. (From Cook 1980.) **Figure 16.272** *Wettina octopora* Cook (Wettininae), adult, leg I. (From Cook 1974.) **Figure 16.273** *Atractides* sp. (Hygrobatidae), adult male, venter of idiosoma. (From Cook 1980.) **Figure 16.274** *Forelia floridensis* Cook (Foreliinae), adult male, venter of idiosoma. (From Cook 1974.)

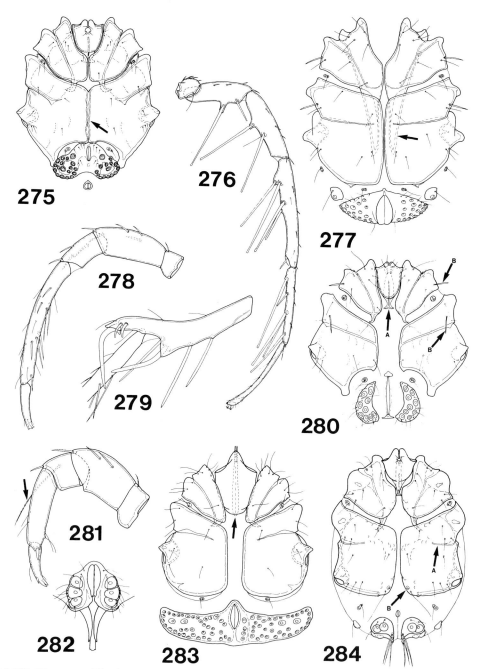

Figure 16.275 *Piona* sp. (Pioninae), adult male, venter of idiosoma. (From Cook 1980.)   **Figures 16.276, 16.284** *Unionicola* sp. (Unionicolinae), adult. Fig. 16.276, leg I; Fig. 16.284, venter of idiosoma, female. (From Cook 1980 and Cook 1986, respectively.)   **Figure 16.277** *Neumania* sp. (Pionatacinae), adult male, venter of idiosoma. (From Cook 1980.)   **Figures 16.278–16.279, 16.282** *Hydrochoreutes intermedius* Cook (Hydrochoreutinae), adult. Fig. 16.278, pedipalp; Fig. 16.279, genu of leg III, male; Fig. 16.282, genital field region, male. (From Cook 1974.)   **Figures 16.280–16.281** *Huitfeldtia rectipes* Thor (Huitfeldtiinae), adult. Fig. 16.280, venter of idiosoma, female; Fig. 16.281, pedipalp. (From Cook 1974.)   **Figure 16.283** *Najadicola ingens* Koenike (Najadicolinae), adult male, venter of idiosoma. (From Cook 1974.)

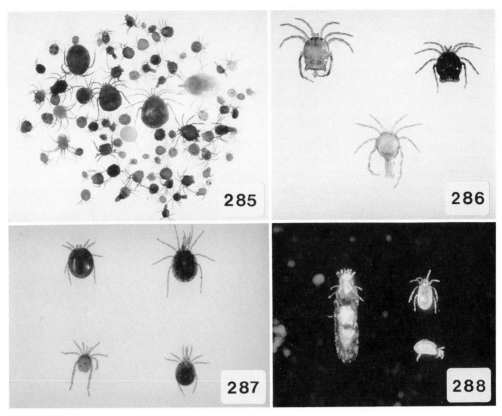

Figures 16.285–16.288 Photographs of adult water mites. Fig. 16.285, typical collection of water mites from a weedy littoral habitat in a shallow lake; Fig. 16.286, males of several different species of the genus *Arrenurus* (Arrenurinae), illustrating a variety of characteristic caudal modifications associated with copulation during spermatophore transfer; Fig. 16.287, species representing various taxa that are adapted for living in spring or small stream habitats; Fig. 16.288, species representing various taxa that are adapted for living in interstitial habitats.

LITERATURE CITED

Alberti, G. 1977. Zur Feinstruktur und Funktion der Genitalnäpfe von *Hydrodroma despiciens* (Hydrachellae, Acari). Zoomorphologie 87:155–164.

Alberti, G. 1979. Fine structure and probable function of genital papillae and Claparede organs of Actinotrichida. Pages 501–507 *in:* J. G. Rodriguez, editor. Recent Advances in Acarology 2:569 pp.

Bader, C. 1938. Beitrag zur Kenntnis der Verdauungsvorgänge bei Hydracarinen. Revue Suisse de Zoologie 45:721–806.

Bagge, P., and J. J. Merilainen. 1985. The occurrence of water mites (Acari: Hydrachnellae) in the estuary of the River Kyronjoki (Bothnian Bay). Annales Zoologici Fennici 22:123–127.

Barr, D. W. 1972. The ejaculatory complex in water mites (Acari: Parasitengona): Morphology and potential value for systematics. Life Science Contributions R. Ontario Museum 81:87 pp.

Barr, D. W. 1973. Methods for the collection, preservation, and study of water mites (Acari: Parasitengona). Life Science Miscellaneous Publications R. Ontario Museum 28 pp.

Barr, D. W. 1982. Comparative morphology of the genital acetabula of aquatic mites (Acari, Prostigmata): Hydrachnoidea, Eylaoidea, Hydryphantoidea, Lebertioidea. Journal of Natural History 16:147–160.

Biesiadka, E. 1979. Wodopojki (Hydracarina) Pienin. Fragmenta Faunistica 24:97–173.

Böttger, K. 1962. Zur Biologie und Ethologie der einheimischen Wassermilben *Arrenurus* (*Megaluracarus*) *globator* (Müll.), 1776, *Piona nodata nodata* (Müll.), 1776 und *Eylais infundibulifera meriodionalis* (Thon), 1989. (Hydrachnellae, Acari). Zoologische Jahrbuecher Syst. 89:501–584.

Böttger, K. 1965. Zur Ökologie und Fortpflanzungsbiologie von *Arrenurus valdiviensis* K. O. Viets 1964 (Hydrachnellae, Acari). Zeitschrift fuer Morphologie und Oekologie der Tiere 55:115–141.

Böttger, K. 1970. Die Ernährungsweise der Wassermilben

(Hydrachnellae, Acari). Internationale Revue der Gesamten Hydrobiologie 55:895–912.

Böttger, K. 1972a. Vergleichend biologisch-ökologische Studien zum Entwicklungszyklus der Süsswassermilben (Hydrachnellae, Acari). I. Der Entwicklungszyklus von *Hydrachna globosa* und *Limnochares aquatica*. Internationale Revue der Gesamten Hydrobiologie 57:109–152.

Böttger, K. 1972b. Vergleichend biologisch-ökologische Studien zum Entwicklungszyklus der Süsswassermilben (Hydrachnellae, Acari). II. Der Entwicklungszyklus von *Limnesia maculata* und *Unionicola crassipes*. Internationale Revue der Gesamten Hydrobiologie 57:263–319.

Böttger, K., and F. Schaller. 1961. Biologische und ethologische Beobachtungen an einheimischen Wassermilben. Zoologischer Anzeiger 167:46–50.

Chappuis, P. A. 1942. Eine neue Methode zur Untersuchung der Grundwasser fauna. Acta Sci. Math. Natur. Kolozsvar 6:3–7.

Cook, D. R. 1957. Order Acarina. Suborder Hydracarina. Genus *Protoarrenurus* Cook, n. gen. Pages 248–249 *In:* A. R. Palmer, editor. Miocene Arthropods from the Mojave Desert, California. Geological Survey Professional Paper (U.S.) No. 294-G.

Cook, D. R. 1974. Water mite genera and subgenera. Memoirs of the American Entomological Institute 21:860 pp.

Cook, D. R. 1980. Studies on Neotropical water mites. Memoirs of the American Entomological Institute 31:645 pp.

Cook, D. R. 1986. Water mites from Australia. Memoirs of the American Entomological Institute 40:568 pp.

Cook, D. R. 1988. Water mites from Chile. Memoirs of the American Entomological Institute 42:356 pp.

Davids, C. 1973. The water mite *Hydrachna conjecta* Koenike, 1895 (Acari, Hydrachnellae), bionomics and relation to species of Corixidae (Hemiptera). Netherlands Journal of Zoology 23:363–429.

Davids, C., and R. Belier. 1979. Spermatophores and sperm transfer in the water mite *Hydrachna conjecta* Koen. Reflections of the descent of water mites from terrestrial forms. Acarologia 21:84–90.

Dimock, R. V., Jr., and C. Davids. 1985. Spectral sensitivity and photo-behaviour of the water mite genus *Unionicola*. Journal of Experimental Biology 119:349–363.

Elton, C. S. 1922. On the colours of water-mites. Proceeding of the Zoological Society of London 82:1231–1239.

Fairchild, W. L., M. C. A. O'Neill, and D. M. Rosenberg. 1987. Quantitative evaluation of the behavioral extraction of aquatic invertebrates from samples of sphagnum moss. Journal of the North American Benthological Society 6:281–287.

Gledhill, T. 1982. Water-mites (Hydrachnellae, Limnohalacaridae, Acari) from the interstitial habitat of riverine deposits in Scotland. Polish Archives of Hydrobiology 29:439–451.

Gledhill, T. 1985. Water mites—Predators and parasites. Freshwater Biology Association Annual Report 53: 45–59.

Halik, L. 1955. O kopulach vodule *Brachypoda versicolor* (Müll.). Biologia (Bratislava) 10:464–474.

Hevers, J. 1978. Zur Sexualbiologie der Gattung *Unionicola* (Hydrachnellae, Acari). Zoologische Jahrbuecher Syst. 105:33–64.

Imamura, T. 1952. Notes on the moulting of the adult of the water mite, *Arrenurus uchidai* n. sp. Annotationes Zoologicae Japonenses 25:447–451.

Kerfoot, W. C. 1982. A question of taste: crypsis and warning coloration in freshwater zooplankton communities. Ecology 63:538–554.

Koenike, F. 1891. Seltsame Begattung unter der Hydrachniden. Zoologischer Anzeiger 14:253–256.

Kowalik, W. 1981. Wodopojki (Hydracarina) rzek dorzecza Wieprza. Annales Universitatis Mariae Curie-Sklodowska, Section C 36:327–352.

Kowalik, W., and E. Biesiadka. 1981. Occurrence of water mites (Hydracarina) in the River Wieprz polluted with domestic-industry sewage. Acta Hydrobiologica 23:331–347.

Krantz, G. W. 1978. A manual of Acarology. 2nd Edition. Oregon State University Press, Corvallis. 335 pp.

Lanciani, C. 1972. Mating behaviour of water mites of the genus *Eylais*. Acarologia 14:631–637.

Lanciani, C. 1979. The food of nymphal and adult water mites of the species *Hydryphantes tenuabilis*. Acarologia 20:563–565.

Lanciani, C. 1983. Overview of the effects of water mite parasitism on aquatic insects. Pages 86–90 *in:* M. A. Hoy, G. L. Cunningham, and L. Knutson, editors. Biological Control of Pests by Mites. Special Publications of the University of California Agricultural Experimental Station 3304.

Lang, p. 1905. Über den Bau der Hydrachnidenaugen. Zoologische Jahrbuecher Anatomie 21:453–494.

Lundblad, O. 1929. Über den Begattungsvorgang bei einigen *Arrhenurus*-Arten. Zietschrift fuer Morphologie und Oekologie der Tiere 15:705–722.

Marshall, R. 1933. Water mites from Wyoming as fish food. Transactions of the American Microscopal Society 52:34–41.

Marshall, R. 1940. On the occurrence of water mites in the food of turtles. American Midland Naturalist 24:361–364.

Meyer, E. 1985. Der Entwicklungszyklus von *Hydrodroma despiciens* (O. F. Müller, 1776) (Acari: Hydrodromidae). Archiv fuer Hydrobiologie, Supplement 66:321–453.

Mitchell, R. 1955. Anatomy, life history, and evolution of the mites parasitizing fresh-water mussels. Miscellaneous Publications of the Museum of Zoology, University of Michigan 89:1–41.

Mitchell, R. 1957a. Major evolutionary lines in water mites. Systematic Zoology 6:137–148.

Mitchell, R. 1957b. Locomotor adaptations of the family Hydryphantidae (Hydrachnellae, Acari). Abhandlungen Herausgegben vom Naturwissenschaftlichen Verein zu Bremen 35:75–100.

Mitchell, R. 1957c. The mating behaviour of pionid water mites. American Midland Naturalist 58:360–366.

Mitchell, R. 1958. Sperm transfer in the water mite *Hydry-*

phantes ruber Geer. American Midland Naturalist 60:156–158.

Mitchell, R. 1959. Life histories and larval behavior of arrenurid water-mites parasitizing Odonata. Journal of the New York Entomological Society 67:1–12.

Mitchell, R. 1960. The evolution of thermophilous water mites. Evolution 14:361–377.

Mitchell, R. 1962. The structure and evolution of water mite mouthparts. Journal of Morphology 110:41–59.

Mitchell, R. 1964a. A study of sympatry in the water mite genus *Arrenurus* (family Arrenuridae). Ecology 45:546–558.

Mitchell, R. 1964b. An approach to the classification of water mites. Acarologia 6:75–79.

Mitchell, R. 1964c. The anatomy of an adult chigger mite *Blankaartia* acuscutellaris (Walch). Journal of Morphology 114:373–392.

Mitchell, R. 1969. A model accounting for sympatry in water mites. American Naturalist 103:331–346.

Mitchell, R. 1972. The tracheae of water mites. Journal of Morphology 136:327–335.

Motas, C. 1928. Contribution à la connaissance des Hydracariens français particilièrement du Sud-Est de la France. Travaux du Laboratoire d'Hydrobiologie Piscic. University of Grenoble 20:373 pp.

Münchberg, P. 1935a. Über die bisher bei einigen Nematocerenfamilien (Culicidae, Chironomidae, Tipulidae) beobachteten ektoparasitären Hydracarinenlarven. Zeitschrift fuer Morphologie und Oekologie der Tiere 29:720–749.

Münchberg, P. 1935b. Zur Kenntnis der Odonatenparasiten, mit ganz besonderer Berucksichtigung der Ökologie der in Europa an Libellen schmarotzenden Wassermilbenlarven. Archiv fuer Hydrobiologie 29:1–122.

Mullen, G. R. 1975. Predation by water mites (Acarina: Hydrachnellae) on immature stages of mosquitoes. Mosquito News 35:168–171.

Paterson, C. G. 1970. Water mites (Hydracarina) as predators of chironomid larvae (Insecta: Diptera). Canadian Journal of Zoology 48:610–614.

Pieczynski, E. 1976. Ecology of water mites (Hydracarina) in lakes. Polish Ecological Studies 2:5–54.

Poinar, G. O. 1985. Fossil evidence of insect parasitism by mites. International Journal of Acarology 11:37–38.

Prasad, V., and D. R. Cook. 1972. The taxonomy of water mite larvae. Memoirs of the American Entomological Institute 18:326 pp.

Riessen, H. P. 1982. Pelagic water mites: Their life history and seasonal distribution in the zooplankton community of a Canadian lake. Archiv fuer Hydrobiologie, Supplement 62:410–439.

Roberts, E. A., R. V. Dimock, Jr., and R. B. Forward, Jr. 1978. Positive and host-induced negative phototaxis of the symbiotic water mite *Unionicola formosa*. Biological Bulletin (*Woods Hole, Mass.*) 155:599–607.

Schmidt, U. 1935. Beiträge zur Anatomie und Histologie der Hydracarinen, besonders von *Diplodontus despiciens* O. F. Müller. Zeitschrift fuer Morphologie und Oekologie der Tiere 30:99–175.

Schwoerbel, J. 1959. Ökologische und tiergeographische Untersuchungen über die Milben (Acari, Hydrachnellae) der Quellen und Bäche des südlichen Schwarzwaldes und seiner Randgebiete. Archiv fuer Hydrobiologie, Supplement 24:385–546.

Simmons, T. W., and I. M. Smith. 1984. Morphology of larvae, deutonymphs, and adults of the water mite *Najadicola ingens* (Prostigmata: Parasitengona: Hygrobatoidea) with remarks on phylogenetic relationships and revision of taxonomic placement of Najadicolinae. Canadian Entomologist 116:691–701.

Smith, B. P. 1983. The potential of mites as biological control agents of mosquitoes. Pages 79–85 *in:* M. A. Hoy, G. L. Cunningham, and L. Knutson, editors. Biological Control of Pests by Mites. Special Publication of the University of California Agricultural Experimental Station. No. 3304.

Smith, B. P. 1988. Host–parasite interaction and impact of larval water mites on insects. Annual Review of Entomology 33:487–507.

Smith, I. M. 1976. A study of the systematics of the water mite family Pionidae (Prostigmata: Parasitengona). Memoirs of the Entomological Society of Canada 98:249 pp.

Smith, I. M. 1978. Descriptions and observations on host associations of some larval Arrenuroidea (Prostigmata: Parasitengona), with comments on phylogeny in the superfamily. Canadian Entomologist 110:957–1001.

Smith, I. M. 1982. Larvae of water mites of the superfamily Lebertioidea (Prostigmata: Parasitengona) in North America with comments on phylogeny and higher classification of the superfamily. Canadian Entomologist 114:901–990.

Smith, I. M. 1983a. Description of larvae of *Neoacarus occicentalis* (Acari: Arrenuroidea: Neoacaridae). Canadian Entomologist 115:221–226.

Smith, I. M. 1983b. Description of *Cowichania interstitialis* n. gen., n. sp., and proposal of Cowichaniinae n. subfam., with remarks on phylogeny and classification of Hydryphantidae (Acari: Parasitengona: Hydryphantoidea). Canadian Entomologist 115:523–527.

Smith, I. M. 1983c. Description of larvae and adults of *Paramideopsis susanae* n. gen., n. sp., with remarks on phylogeny and classification of Mideeopsidae (Acari: Parasitengona: Arrenuroidea). Canadian Entomologist 115:529–538.

Smith, I. M. 1983d. Description of larvae of *Nudomideopsis magnacetabula* (Acari: Arrenuroidea: Mideopsidae) with new distributional records for the species and remarks on the classification of *Nudomideopsis*. Canadian Entomologist 115:913–919.

Smith, I. M. 1983e. Descriptions of two new species of *Acalyptonotus* from western North America, with a new diagnosis of the genus based on larvae and adults, and comments on phylogeny and taxonomy of Acalyptonotidae (Acari: Parasitengona: Arrenuroidea). Canadian Entomologist 115:1395–1408.

Smith, I. M. 1983f. Descriptions of adults of a new species of *Cyclothyas* (Acari: Parasitengona: Hydryphantidae) from western North America, with comments on phylogeny and distribution of mites of the genus. Canadian Entomologist 115:1433–1436.

Smith, I. M. 1984. Larvae of water mites of some genera of Aturidae (Prostigmata: Hygrobatoidea) in North America with comments on phylogeny and classification of the family. Canadian Entomologist 116:307–374.

Smith, I. M. 1987. Water mites of peatlands and marshes in Canada. Pages 3l–46 *in:* D. M. Rosenberg, and H. V. Danks, editors. Aquatic Insects of Peatlands and Marshes in Canada. Memoirs of the Entomological Society of Canada l40:l74 pp.

Smith, I. M. 1989a. North American water mites of the family Momoniidae Viets (Acari: Arrenuroidea). I. Description of adults of *Cyclomomonia andrewi* gen. nov., sp. nov., and key to world genera and subgenera. Canadian Entomologist 121:543–549.

Smith, I. M. 1989b. Description of two new species of *Platyhydracarus* gen. nov. from western North America, with remarks on classification of Athienemanniidae (Acari: Parasitengona: Arrenuroidea). Canadian Entomologist 121:709–726.

Smith, I. M. 1989c. North American water mites of the family Momoniidae Viets (Acari: Arrenuroidea). III. Revision of species of *Stygomomonia* Szalay, 1943, subgenus *Allomomonia* Cook, 1968. Canadian Entomologist 121:989–1025.

Smith, I. M. 1990. Description of two new species of *Stygameracarus* gen. nov. from North America, and proposal of Stygameracarinae subfam. nov. (Acari: Arrenuroidea: Athienemanniidae). Canadian Entomologist 122:181–190.

Smith, I. M. 1991a. North American water mites of the family Momoniidae Viets (Acari: Arrenuroidea). IV. Review of species of *Stygomomonia* Szalay, 1943, *sensu stricto*. Canadian Entomologist 123. In press.

Smith, I. M. 1991b. Descriptions of new species representing new or unreported genera and subgenera of Lebertioidea from North America. Canadian Entomologist 123. In press.

Smith, I. M., and D. R. Oliver. 1976. The parasitic associations of larval water mites with imaginal aquatic insects, especially Chironomidae. Canadian Entomologist 108:1427–1442.

Smith, I. M., and D. R. Oliver. 1986. Review of parasitic associations of larval water mites (Acari: Para-sitengona: Hydrachnida) with insect hosts. Canadian Entomologist 118:407–472.

Sokolow, I. I. 1977. The protective envelopes in the eggs of Hydrachnellae. Zoologischer Anzeiger 198:36–42.

Uchida, T. 1940. Über die Begattungsakt bei *Forelia variegator*. Annotationes Zoologicae Japoneses 19:280–282.

Vidrine, M. F. 1986. Revision of the Unionicolinae (Acari: Unionicolidae). International Journal of Acarology 12:233–243.

Viets, K. H. 1914. Über die Begattungsvorgänge bei *Acercus*-Arten. Internationale Revue Gesamten Hydrobiologie Hydrogr., Biological Supplement 6:1–10.

Viets, K. O. 1982. Die Milben des Süsswassers (Hydrachnellae und Halacaridae [part.], Acari). 1: Bibliographie. Sonderbande Naturwiss Vereins Hamburg 6:116 pp.

Viets, K. O. 1987. Die Milben des Süsswassers (Hydrachnellae und Halacaridae [part.], Acari). 2. Katalog. Sonderbände Naturwiss. Vereins Hamburg 8: 1012 pp.

Wainstein, B. A. 1980. Opredelitel lichinok vodjanych kleshchei (Key to larval water mites). Institut Biologii Vnutrennikh Vod. (Akademiia Nauk SSSR) 238 pp.

Wesenburg-Lund, C. J. 1918. Contributions to the knowledge of the postembryonal development of the Hydracarina. Videnskabelige Meddelelser fra Dansk Naturhistorisk Forening I Kjøbenhaun. 70:5–57.

Wiggins, G. B., R. J. Mackay, and I. M. Smith. 1980. Evolutionary and ecological strategies of animals in annual temporary pools. Archiv fuer Hydrobiologie, Supplement 58:97–206.

Wiles, P. R. 1982. A note on the watermite *Hydrodroma despiciens* feeding on chironomid egg masses. Freshwater Biology 12:83–87.

Wiles, P. R. 1984. Watermite respiratory systems. Acarologia 25:27–31.

Winkel, E. H. ten, and C. Davids. 1985. Bioturbation by cyprinid fish affecting the food availability for predatory water mites. Oecologia 67:218–219.

Winkel, E. H. ten, and C. Davids. 1987. Chironomid larvae and their food web relations in the littoral zone of Lake Maarsseveen. Kaal Boek, Amsterdam. 145 pp.

17

Diversity and Classification of Insects and Collembola[1]

William L. Hilsenhoff
Department of Entomology
University of Wisconsin
Madison, Wisconsin 53706

Chapter Outline

I. INTRODUCTION
  A. General Morphology of Aquatic Insects
  B. Common Techniques for Studying Freshwater Insects
  1. Qualitative Sampling and Rearing of Aquatic Insects
  2. Preserving and Storing Aquatic Insects
  3. An Introduction to the Biological Literature

II. AQUATIC ORDERS OF INSECTS
  A. Ephemeroptera—Mayflies
  1. Families of Ephemeroptera—Suborder Pannota
  2. Families of Ephemeroptera—Suborder Schistonota
  B. Odonata—Dragonflies and Damselflies
  1. Families of Odonata—Suborder Anisoptera (Dragonflies)
  2. Families of Odonata—Suborder Zygoptera (Damselflies)
  C. Plecoptera—Stoneflies
  1. Families of Plecoptera—Suborder Euholognatha
  2. Families of Plecoptera—Suborder Systellognatha
  D. Trichoptera—Caddisflies
  1. Families of Trichoptera—Suborder Annulipalpia
  2. Families of Trichoptera—Suborder Spicipalpia
  3. Families of Trichoptera—Suborder Integripalpia
  E. Megaloptera—Fishflies and Alderflies
  1. Families of Megaloptera

III. PARTIALLY AQUATIC ORDERS OF INSECTS
  A. Aquatic Heteroptera—Aquatic and Semiaquatic Bugs
  1. Families of Aquatic Heteroptera—Suborder Nepomorpha
  2. Families of Semiaquatic Heteroptera—Suborder Gerromorpha
  B. Aquatic Neuroptera—Spongillaflies
  C. Aquatic Lepidoptera—Aquatic Caterpillars
  D. Aquatic Coleoptera—Water Beetles
  1. Families of Aquatic Coleoptera—Suborder Adephaga
  2. Families of Aquatic Coleoptera—Suborder Myxophaga
  3. Families of Aquatic Coleoptera—Suborder Polyphaga
  E. Aquatic Diptera—Flies and Midges
  1. Families of Aquatic Diptera—Suborder Nematocera
  2. Families of Aquatic Diptera—Suborder Brachycera

IV. SEMIAQUATIC COLLEMBOLA—SPRINGTAILS

V. IDENTIFICATION OF THE FRESHWATER INSECTS AND COLLEMBOLA
  A. Taxonomic Key to Orders of Freshwater Insects
  B. Taxonomic Key to Families of Freshwater Ephemeroptera Larvae
  C. Taxonomic Key to Families of Freshwater Odonata Larvae

[1] Much of the material on Collembola was contributed by Barbara L. Peckarsky, Department of Entomology, Cornell University. The author and editors greatly appreciate her generous contribution of this material.

D. Taxonomic Key to Families of Freshwater Plecoptera Larvae

E. Taxonomic Key to Families of Freshwater Trichoptera Larvae

F. Taxonomic Key to Families of Freshwater Megaloptera Larvae

G. Taxonomic Key to Families of Aquatic, Semiaquatic, and Riparian Adults Heteroptera

H. Taxonomic Key to Families of Freshwater Coleoptera
    1. Adult Water Beetles
    2. Larval Water Beetles

I. Taxonomic Key to Families of Freshwater Diptera Larvae

J. Taxonomic Key to Families of Semiaquatic Collembola

Literature Cited

I. INTRODUCTION

This chapter includes orders and families of insects in which one or more life stages are truly aquatic and are adapted for living under or on the surface of the water.[2] Families that live on or burrow into emergent aquatic vegetation, internal parasites of aquatic animals, and riparian families that are closely associated with water but do not inhabit it are not included. Terrestrial stages of aquatic families are discussed briefly, but only as they relate to the general life history and ecology of the family. In addition, brief mention is made of another arthropod group, the semiaquatic springtails (class Entognatha, order Collembola) (see Sections IV and V.J).

There are ten orders of insects that contain aquatic species. Five of them (Ephemeroptera, Odonata, Plecoptera, Trichoptera, and Megaloptera) are aquatic orders, in which almost all species have aquatic larvae. The remaining five orders (Heteroptera, Coleoptera, Diptera, Lepidoptera, and Neuroptera) are partially aquatic orders, in which most species are terrestrial, but there are species or entire families that have one or more life stages adapted for living in the aquatic environment.

Three aquatic orders (Ephemeroptera, Odonata, and Plecoptera) have a hemimetabolous life cycle, which includes three developmental stages: egg, larva, and adult. The term larva is being used instead of nymph or naiad because in these orders, the immature developmental stage is adapted for survival

in aquatic environments and does not resemble the terrestrial adult, except in Plecoptera. The other two aquatic orders (Trichoptera and Megaloptera) have a holometabolous life cycle, which includes four developmental stages: egg, larva, pupa, and adult. In these orders, the larva bears no resemblance to the adult.

Four of the five partially aquatic orders (Coleoptera, Diptera, Lepidoptera, and Neuroptera) also have holometabolous life cycles and larvae that do not resemble the adult. The fifth order, Heteroptera (Hemiptera), has a paurometabolous life cycle, which includes three developmental stages: egg, nymph, and adult. In this order, nymphs live in the same habitat as adults and resemble them in many respects, differing mostly by being smaller and less sclerotized, having wing-pads instead of wings, and lacking genital structures.

A. General Morphology of Aquatic Insects

Morphological terms used for identification of aquatic insects vary somewhat from order to order, but several terms are commonly used for all orders. A Plecoptera larva (Figs. 17.1 and 17.2) is used to illustrate these terms. Insects have three body regions: head, thorax, and abdomen. The anterior region is the head, which dorsally (Fig. 17.1) has a pair of antennae, a pair of compound eyes, and often three ocelli or simple eyes. On the underside of the head are several mouthparts (Fig. 17.2), including a labrum or upper lip, a pair of mandibles, a pair of maxillae, and a labium or lower lip. The mandibles usually are heavily sclerotized, tooth-like structures. The maxillae and labium are usually multisegmented and frequently have a pair of elongate, segmented structures called palpi.

The thorax is immediately posterior to the head and consists of three large segments (Fig. 17.1); a prothorax (anterior), a mesothorax, and a metathorax (posterior). The dorsal sclerite of each segment is referred to as a notum, and the ventral sclerite as a sternum. Each segment has a pair of legs that are divided into five parts (Fig. 17.2); a coxa, trochanter, femur, tibia, and tarsus. The femur and tibia are usually elongate and the tarsus has one to five segments, depending on the order. Each tarsus has one or two tarsal claws. Prefixes pro-, meso-, and meta- refer to the respective thoracic segments. Thus, a pronotum would be on the prothorax and a mesofemur would be on the mesothorax. In hemimetabolous and paurometabolous insect larvae or nymphs, wing-pads (developing wings) are usually present dorsally on the meso- and metathorax (Fig. 17.2), especially in later instars.

---

[2] Because aquatic ecologists have traditionally studied insects by orders, this chapter is organized taxonomically, rather than by structure or function as in other chapters—the Editors.

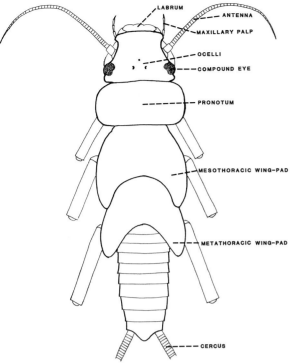

Figure 17.1 Plecoptera larvae (Perlidae); *Acroneuria* larva (dorsal) showing various structures (thoracic gills and distal segments of cerci not shown).

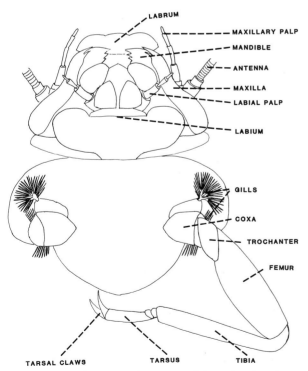

Figure 17.2 Plecoptera larvae (Perlidae); head and prothorax (ventral) of *Acroneuria* larva showing mouthparts and structures of a leg (distal segments of right leg and antennae not shown).

The abdomen is the posterior body region, and may have as many as ten visible segments, which are numbered posteriorly from the thorax. The dorsal sclerite of each segment is called a tergum, the ventral sclerite a sternum, and the lateral sclerite, when present, a pleuron. Terminal abdominal appendages vary widely or may be absent; cerci are present in Plecoptera and several other orders. In some orders, gills of various types may be present on the abdomen, thorax, or even on the head.

A taxonomic key to the orders of aquatic insects is provided in Section V.A. Family level keys are also provided for aquatic larvae and aquatic adults; keys to eggs, Heteroptera nymphs, pupae, and terrestrial stages of aquatic species are not included. The most recent keys to genera of aquatic insects in North America appear in Merritt and Cummins (1984). Regional keys to genera or species of aquatic insects include those in Usinger (1956), Brigham *et al.* (1981), Hilsenhoff (1981), and Peckarsky *et al.* (1990). Recent keys to aquatic stages of North American species are listed at the end of the discussion of each order. Reliable species keys do not exist for many families and genera, because larvae of these taxa have not been reared for association with adults. Subsequent descriptions of new species are not listed, but should be consulted when species are being identified. References to recent revisions and descriptions of new species may be found in bibliographies such as those published annually by the North American Benthological Society and through computer searches of data bases such as Zoological Record and Dissertation Abstracts Online.

B. Common Techniques for Studying Freshwater Insects

1. Qualitative Sampling and Rearing of Aquatic Insects

A D-frame or rectangular aquatic net with a mesh size of 1 mm or less is recommended for qualitative sampling of lotic habitats. Shallow, rock or gravel riffles can be sampled by placing the net firmly against the bottom and disturbing the substrata immediately upstream from the net with hands or feet, allowing insects and debris washed into the water to be carried into the net. This is commonly called a kick sample. The contents of the net are emptied into a shallow white pan containing 2–3 cm of water, taking care not to collect so much debris that it completely obscures the bottom of the pan. Insects that swim or crawl from the debris are most easily collected with a curved forceps and placed into a collecting jar containing 70% ethanol or Kahle's fluid (15 parts 95% ethanol, 30 parts distilled water, 6 parts formalin, and 1 part glacial acetic acid). The empty net should be examined for insects that remain clinging to it. Alternatively, the entire sample with associated debris can be preserved for later, more quantitative sorting in the laboratory. Samples can also be collected with a net by pushing it under the stream bank or through silt and debris in pools and rinsing excess silt from the net before placing the sample in the pan. It is also important to remove pieces of decaying wood from the stream and to collect insects that crawl from the wood as it dries. Additional insects may be collected by examining stones that are temporarily removed from the stream bed. Deep areas of larger streams can be sampled by using grabs or dredges. A Ponar grab or Ekman grab is most commonly used, but the latter can be used only in soft sediments.

Shallow lentic habitats such as ponds, marshes, swamps, and vegetated lake margins are also best sampled qualitatively with a flat-bottomed aquatic net. This is accomplished by pushing, pulling, or sweeping the net through the vegetation or debris, rinsing excess silt from the debris, and processing as just described. Additional larger collections can be processed by placing them on a half-inch (13 mm) mesh screen placed over a large white pan and allowing at least 15 min for insects to crawl from the debris and drop into the pan. This method is especially effective for collecting Coleoptera and Heteroptera. Wave-swept shorelines of lakes can be sampled by disturbing the sand or gravel bottom and immediately collecting suspended material with a net. As in streams, removal and examination of decaying wood or stones from lake shores should not be overlooked, because these substrata often harbor species that normally will not be collected with a net. Deeper waters of lakes must be sampled with grabs, dredges, or corers. Special habitats such as tree holes or pitcher plants are best sampled with a large syringe or poultry baster.

Since immature stages in many families of aquatic insects cannot be identified to species, it is often necessary to rear them to the adult stage for identification. This can be readily accomplished in most orders, but the presence of a terrestrial pupal stage in Coleoptera, Megaloptera, Neuroptera, and some Diptera makes rearing of larvae in these orders more difficult. Chapters on qualitative and quantitative sampling techniques and rearing procedures appear in Downing and Rigler (1984), Merritt and Cummins (1984), McCafferty (1981), and Usinger (1956).

2. Preserving and Storing Aquatic Insects

Aquatic insects can most conveniently be killed and preserved by placing them in a jar containing a liquid preservative. Many preservatives have been recommended and used; 70–80% ethanol or isopropanol, or Kahle's fluid are the most popular. The latter penetrates insect larvae more rapidly and preserves some colors better, but should be replaced with 70% ethanol after one or two days.

For permanent storage, 70% ethanol containing 3% glycerine is recommended, especially for larvae. The glycerine serves to preserve larvae if the alcohol evaporates due to a poorly fitting stopper. Most aquatic entomologists store specimens in patent-lip vials with neoprene stoppers. Screw-cap vials and cork-stoppered shell vials should be avoided because of gradual evaporation of the alcohol, although screw-cap vials with beveled polyethylene inserts are satisfactory. One-dram shell vials with polyethylene stoppers, which are commonly used for liquid scintillation studies, are an inexpensive substitute for neoprene-stoppered vials. Quality control of these polyethylene-stoppered vials was initially poor and many leaked, but those purchased in recent years do not leak and retain fluid as well as neoprene-stoppered vials. They have a wider opening than patent-lip or screw-cap vials, which facilitates removal of insects. Because they are optically clear, insects that are stored in these vials can be examined and often identified under a microscope while still in the vial if the vial is only half-filled with alcohol, so that a meniscus or bubble does not interfere with viewing. Adult aquatic Coleoptera and Heteroptera are normally preserved on pins, but study collections can be preserved in liquid, with several of the same species kept in a single vial to reduce storage space and curating time. Adult Odonata are preserved on pins or in envelopes. Adults of other orders also may be pinned, but are most frequently preserved in alcohol. Storage in liquid also facilitates removal of genitalia for study, which is important for identification of many species.

For examination under a dissecting microscope, specimens stored in liquid should be placed in a Syracuse watch glass or similar container and completely covered with 95% alcohol. If 70% alcohol is used, insects may float and the alcohol will become cloudy as it evaporates. If insects are not completely covered with alcohol, a meniscus will cause distortions. A layer of very small glass beads or silica sand in the bottom of the watch glass permits insects to be embedded for viewing at any angle.

3. An Introduction to the Biological Literature

A discussion of the biology and ecology of insects in each family is followed by a key to families in each order, but it is often difficult to make generalizations at the family level. Literature on biology and ecology of the numerous genera and species is voluminous. References to ecological literature pertaining to each genus are listed in Merritt and Cummins (1984), and the ecology and behavior of aquatic insects is comprehensively treated in a recent book edited by Resh and Rosenberg (1984). Several review articles treat specific aspects of biology and ecology such as drift of stream insects (Waters 1972), trophic relations of aquatic insects (Cummins 1973), predator–prey relationships (Bay 1974), detritus processing (Anderson and Sedell 1979), and filter feeding (Wallace and Merritt 1980). Additionally, references to studies of general aquatic ecology appear in Current and Selected Bibliographies on Benthic Biology that is published each year by the North American Benthological Society.

II. AQUATIC ORDERS OF INSECTS

A. Ephemeroptera—Mayflies

With only about 575 species in 17 families known from North America, this is a relatively small hemimetabolous order, but it is widespread and abundant in most regions. Larvae of the various families often differ so greatly from one another that they can be easily recognized in the field. All species have aquatic larvae; and while the great majority live in streams, several inhabit a variety of permanent and temporary lentic habitats. Ephemeroptera larvae can be found in streams of all sizes and water temperatures, and may even be present in temporary streams. In lakes, they occur among vegetation in littoral areas and sometimes also may be found on the bottom in deep water or along wave-swept shores. Frequently, they occur in vernal or permanent ponds and marshes. They are often an important source of food for fish in streams. Because of the wide range of dissolved oxygen requirements among the species, they are very important in biological monitoring of streams. Some species that develop in lakes or large rivers may be sufficiently abundant that emerging subimagos (preadults) and adults create severe nuisance problems.

Except in warm waters of the southern United States, most species are univoltine, with eggs either hatching shortly after being laid and larvae developing slowly over a one-year period (slow seasonal life cycle), or with eggs diapausing followed by rapid development of the larvae after hatching several months later (fast seasonal life cycle). Other species are bivoltine, with rapid larval development and with eggs of the second generation diapausing over the winter. A few species are multivoltine or semivoltine, with multivoltinism being especially prevalent in the South. The number of larval instars

varies widely among species and even within a species, with 12 or more having been recorded for species that have been studied. After completion of larval development, the larva either crawls from the water or swims or floats to the surface where the sexually mature subimago emerges. This winged subadult or "dun" stage is unique among insects. Subimagos of most species fly to nearby vegetation or supports where they spend about a day before molting into the adult (imago) stage. A few species molt almost immediately to the adult, sometimes while in flight. Others never molt to the adult.

Adult mayflies do not feed, and generally live only a few days. A few may live two weeks or more, and some live only an hour or two. Mating normally takes place when adult males form an aerial swarm to attract females and seize them when they enter the swarm. In some species, adult males mate with subimago females as they emerge from the water. Eggs are laid mostly on the surface of the water by females that repeatedly touch the water with their abdomen while in flight or by females alighting briefly on the surface to oviposit. Females of some species enter the water on substrata to oviposit on the substrata under water. Most females die on the surface of the water upon completion of oviposition.

Mayfly larvae can be readily distinguished from other aquatic insects by their long, filamentous caudal filaments and the presence of lateral or ventrolateral gills on most of the first seven abdominal segments (Figs. 17.3–17.26). Usually three caudal filaments are present (a pair of cerci and a median filament); in a few species the median filament is so reduced that only the cerci are visible. Larvae most closely resemble Plecoptera (Figs. 17.1, 17.44, and 17.45), which always have only cerci, lack gills on the middle abdominal segments, and have two tarsal claws. Ephemeroptera larvae have compound eyes, ocelli, filamentous antennae, and one-segmented tarsi with a single claw (Fig. 17.9). Subimagos and adults differ greatly from the larvae, having one or two pairs of wings that they hold vertically above the body when at rest, very large compound eyes, and no gills or functional mouthparts. In the subimago, unmarked areas of the wings are opaque; in adults, these areas are transparent.

While mayfly larvae mostly crawl on the substratum, many are able and rapid swimmers. They swim by moving their abdomen and caudal filaments up and down rather than horizontally as in stoneflies. Their gills aid significantly in the uptake of oxygen, and most species are capable of moving their gills to increase water circulation. This enables some species to inhabit lentic habitats or lotic habitats that are less rich in oxygen. Almost all mayfly larvae are herbivores or detritivores; only a few species are known to be predators on other invertebrates. Subimagos and adults do not feed.

Larvae may be readily collected with a net by techniques already described. Those that inhabit deep water must be collected with grabs, and those in shallow sand or silt areas can only be obtained by sieving these substrata. Subimagos and adults may be collected by sweeping stream-side vegetation, and large numbers are often collected at lights.

In recent years, there have been several changes in the higher classification of Ephemeroptera, with the addition of several new families. Although new species and genera are still being discovered, recent revisions have often synonymized species, sometimes resulting in a reduction of the number of known species. In general, larvae are poorly known and cannot be reliably identified to species in most families. Early books by Needham *et al.* (1935) and Burks (1953) provided an essential background for more recent studies. The book by Edmunds *et al.* (1976) provides recent generic keys to adults and larvae as well as information on collecting, rearing, and mayfly biology. The biogeography and evolution of Ephemeroptera was reviewed by Edmunds (1972) and Brittain (1982) reviewed the biology of this order. However, much additional study is needed.

The order is divided into two suborders. Larvae in the suborder Pannota are relatively stout, their gills are usually protected by an operculum or carapace, and most move slowly. Larvae in the more diverse Schistonota are more streamlined and are excellent swimmers. In several families of Schistonota, larvae are modified for burrowing into soft substrata. Aspects of the biology of the 17 families in these two suborders are given as follows. See Section V.B. for a key to the families of Ephemeroptera and Table 17.1 for a list of references for the identification of mayfly species.

1. Families of Ephemeroptera—Suborder Pannota

a. *Baetiscidae (1 Genus)*

Larvae are distinctive, having a mesonotal carapace that covers most of the abdomen and all abdominal gills (Fig. 17.9). They occur throughout eastern and central North America, and are rare or absent in the West. Larvae partially burrow into silty margins or eddies of streams and clean lakes, especially when the primary substratum is sand. All species are probably univoltine and overwinter as larvae. Larvae crawl from the water onto a substratum to emerge, with emergence occurring in spring or early summer.

Table 17.1 Literature References for Identification of Mayfly Species

Regional keys to larvae that include more than one family
  California—Day (1956)
  Illinois—Burks (1953)
  North and South Carolina—Unzicker and Carlson (1981)
  United States—(Ephemeroidea) McCafferty (1975)

Keys to larvae listed by family
  Ametropodidae—Allen and Edmunds (1976)
  Baetidae
    Baetis—(Wisconsin) Bergman and Hilsenhoff (1978), Morihara and McCafferty (1979)
    Callibaetis—Check (1982)
    Cloeodes—Waltz and McCafferty (1987a)
  Baetiscidae—Pescador and Berner (1981)
  Behningiidae—Edmunds and Traver (1959)
  Caenidae
    Brachycercus and *Cercobrachys*—Soldan (1984)
  Ephemerellidae
    Attenella—Allen and Edmunds (1961b)
    Caudatella—Allen and Edmunds (1961a)
    Dannella—Allen and Edmunds (1962a), McCafferty (1977)
    Drunella—Allen and Edmunds (1962b)
    Ephemerella—Allen and Edmunds (1965)
    Eurylophella—Allen and Edmunds (1963b)
    Serratella—Allen and Edmunds (1963a)
    Timpanoga—Allen and Edmunds (1959)
  Heptageniidae—(Wisconsin) Flowers and Hilsenhoff (1975)
    Heptagenia—(Rocky Mountains) Bednarik and Edmunds (1980)
    Macdunnoa—Flowers (1982)
    Stenonema—Lewis (1974), Bednarik and McCafferty (1979)
  Leptophlebiidae
    Paraleptophlebia—(Oregon) Lehmkuhl and Anderson (1971)
    Traverella—Allen (1973)
  Neoephemeridae—Berner (1956)
  Oligoneuriidae
    Isonychia—Kondratieff and Voshell (1984)
  Tricorythidae
    Leptohyphes—Allen (1978); Allen and Murvosh (1987)

b. *Caenidae (4 Genera)*

The small larvae are easily recognized by their nearly square operculate gills that are not mesally fused (Fig. 17.14). They are widespread and common in a wide variety of lotic and lentic habitats, including streams of all sizes, spring seeps, marshes, swamps, ponds, and lakes. Here, they frequent the sediments and often are partially covered with silt. They are generally more tolerant of low levels of dissolved oxygen than mayflies in any other family. Life cycles are poorly known. In the North, most species are probably univoltine with overlapping cohorts; in the South, two or more generations per year are likely. Subimagos emerge at the surface, fly to a support, and molt immediately to adult. Adults, which live only a few hours, swarm and mate shortly after emerging.

c. *Ephemerellidae (8 Genera)*

Larvae are recognized by the absence of gills on abdominal segment 2 and lamellate or operculate gills on segments 3- or 4–7 (Fig. 17.15). Most species inhabit clean streams, where they are often abundant in leaf litter, eddies, or near the banks. A few species may persist in organically enriched streams, and other species may be found on wave-swept lake shores. While most larvae are herbivore–detritivores, a few species are omnivorous and also feed on small invertebrates. Most species are univoltine, with a spring emergence period and a summer egg diapause or with emergence in summer and no egg diapause. Subimagos emerge from the surface or from protruding substrates.

d. *Neoephemeridae (1 Genus)*

This relatively uncommon family is confined to the East. Larvae have square operculate gills and resemble Caenidae, but the gills are fused mesally (Fig. 17.13), larvae grow to a larger size than Caenidae, and they have metathoracic wing-pads, which Caenidae lack. Larvae cling to vegetation, debris, or undersides of rocks in slow to fairly rapid streams. They apparently are univoltine with a slow seasonal development and emergence in spring or early summer, depending on latitude.

e. *Tricorythidae (2 Genera)*

Larvae are recognized by their triangular or oval operculate gills on abdominal segment 2 (Figs. 17.11 and 17.12), which are well separated distally, and by the absence of fringed margins on other gills as found in Caenidae and Neoephemeridae. Larvae are widespread, inhabiting detritus, silt, and gravel in streams of all sizes; some species are quite resistant to lowered levels of dissolved oxygen. They have univoltine to multivoltine life cycles, with eggs often diapausing over winter and emergence occurring in spring and summer after rapid development of each generation.

2. Families of Ephemeroptera—Suborder Schistonota

a. *Ametropodidae (1 Genus)*

Larvae are relatively large and flat, with very long meso- and metathoracic claws and shorter prothoracic claws. They partially burrow into sand in eddies of medium to large streams in western and northcentral North America and are generally rare. Larvae have a unique method for feeding on suspended algae and detritus, which they gather with their prothoracic legs from a vortex created in a pit that they construct in the sand. Their life cycle is one year, with a relatively long emergence period in late spring or early summer.

b. *Baetidae (10 Genera)*

Larvae are relatively small, active swimmers, with lamellate abdominal gills (Fig. 17.26). They differ from other species with lamellate gills on most of the first seven abdominal segments in having relatively long antennae (most twice width of head), usually no pronounced posterolateral spines on abdominal segment 9, and a median caudal filament that is often shorter than the cerci or vestigial. Baetidae is a widespread and abundant family that occurs in a variety of streams and also in permanent and temporary ponds or littoral zones of lakes. Lotic species are sometimes univoltine, but more frequently are bivoltine; most species overwinter as diapausing eggs. Emergence of various cohorts and species occurs from early spring to late autumn. Lentic species are often multivoltine. Recent studies of larvae (Waltz and McCafferty 1987 b,c) have led to changes in the placement of species within genera, with an indication that additional changes will follow. Northern populations of some species are known to be parthenogenetic.

c. *Behningiidae (1 Genus)*

Found mostly in the extreme southeastern United States, with a single record from northwest Wisconsin, the unusual larvae burrow in the clean shifting sands of larger streams with a fairly rapid current. The relatively large larvae are unique, possessing crowns of bristles on the head and pronotum (Fig. 17.10), having ventrolateral filamentous gills, and lacking claws on their legs, which are highly modified for digging. The life cycle is two years, with a very long diapausing egg stage followed by slow larval growth and emergence in late April and early May. Unlike most other mayflies, larvae are predators, feeding mostly on larvae of Chironomidae.

d. *Ephemeridae (3 Genera)*

Larvae are recognized by their relatively large size, filamentous gills (Fig. 17.4) that extend upward over their back, and mandibular tusks that curve upward and outward and lack rows of spines. Larvae burrow into sand or eddies in riffle areas of small- to medium-sized streams or inhabit silt bottoms of medium to large streams; they also inhabit sandy or silt substrata in relatively clean lakes. In lakes and large rivers, they may occur in such tremendous numbers that their synchronized emergences create severe nuisance problems. Most species are apparently univoltine with emergence mostly in the summer, but two-year life cycles are known to occur in Canada. Larvae are primarily filter feeders that circulate water through their burrows; they also graze algae and detritus from the substratum.

e. *Heptageniidae (15 Genera)*

The distinctive larvae are readily identified by their flattened appearance and the dorsal location of their eyes and antennae (Fig. 17.16). Some genera lack the median caudal filament. They are widespread and abundant in streams; some also occur on wave-swept shorelines of lakes or in vernal ponds adjacent to streams. Larvae inhabit rocks, wood, debris, and other substrata to which they cling; a few rare genera burrow into sandy substrata. Most species are probably univoltine, but a few may be bivoltine, especially in the South. Emergence occurs from spring to autumn, depending on the species, with most emerging in the summer. Larvae of some of the rarer genera are primarily carnivores, a substantial departure from normal ephemeropteran feeding habits. McCafferty and Provonsha (1988) recently changed the generic placement of a few rare species.

f. *Leptophlebiidae (8 Genera)*

The medium-sized larvae characteristically have gills on segments 2–6 that are double, or forked and pointed apically, or have fingerlike projections (Figs. 17.22–17.24). The gills differ markedly from the lamellate, operculate, or fringed gills of other families. Larvae inhabit slow or fast currents in a variety of streams, where they are found among debris, on rocks or wood, or in gravel. While predominantly lotic, larvae of at least one species migrate upstream in spring and leave the stream habitat to enter adjacent ponds or marshes prior to emergence. Most species are univoltine, with eggs diapausing over winter or summer and adults emerging in spring, summer, or autumn; bivoltine and multivoltine life cycles occur in the South.

g. *Metretopodidae (2 Genera)*

Larvae reach a relatively large size and have lamellate gills, very long meso- and metatarsal claws, and shorter protarsal claws that are distinctively bifid (Figs. 17.18 and 17.19). They are excellent swimmers and may be found along banks and among vegetation in small to large streams and occasionally along shores of lakes. They occur in Canada, the upper Midwest, and in the East. Their life cycle is one year, with emergence in the spring or summer, depending on the species.

h. *Oligoneuriidae (3 Genera)*

The relatively large larvae, which are excellent swimmers, have two rows of long setae on the inner margin of the prothoracic legs (Fig. 17.21), which distinguish them from other mayflies. They use these setae to filter diatoms and other algae from the current; larvae of some species are at least partly carnivorous and feed also on other insects. Larvae are found either in riffles or in sandy bottoms of streams, depending on the genus. Depending on latitude and stream temperature, species are univoltine to multivoltine with emergence during the summer months.

i. *Palingeniidae (1 Genus)*

With their upcurved filamentous gills and outcurving mandibular tusks (Fig. 17.3), larvae closely resemble those of Ephemeridae, a family in which they were included until recently. They differ in having a distinct outer ridge of small spines on the tusks and prothoracic legs. Larvae occur in larger streams from central Canada south to Texas and the southeastern United States, where they burrow in clay and silt bottoms or banks. They are generally uncommon. They are probably univoltine, with adults emerging throughout the summer.

j. *Polymitarcyidae (3 Genera)*

Larvae in this small family of burrowing mayflies differ from other burrowing mayflies by having mandibular tusks that are curved inward and downward and covered with small spines. They inhabit medium to large streams and lakes, burrowing into silt, clay, or silt–gravel substrata; frequently they occur in riffles. Larvae circulate water through their burrows and filter out algae and detritus upon which they feed; they also graze on algae and detritus from the substrata. All species apparently are univoltine, with synchronized emergences during the summer months. In some species, there is a prolonged winter egg diapause.

k. *Potamanthidae (1 Genus)*

While having mandibular tusks and filamentous gills like other burrowing mayflies, gills of Potamanthidae larvae are horizontal instead of arched over the back and the legs also are spread more horizontally. They occur in medium to large streams where they most often sprawl on gravel and sand in shallow runs. They have a one-year life cycle with emergence during the summer months.

I. Siphlonuridae (7 Genera)

Most larvae are relatively large with lamellate gills. They most closely resemble larvae of Baetidae, but the antennae are short and most larvae have pronounced spines on the posterolateral corners of the ninth abdominal segment. Larval characteristics, feeding habits, larval habitat, and life histories vary so widely among the several genera that it impossible to produce an accurate characterization of this family. While most larvae are herbivore–detritivores like most other mayflies, a few species are carnivores. Species occur in all types of lotic and lentic habitats. Some are found in shifting sands of large rivers, others in fast riffles, others along banks of small- to medium-sized streams, and still others in small temporary streams. Several species are primarily lentic, occurring in vernal forest pools, marshes, or swamps. Still others leave lotic habitats and move into small pools just before emergence. Most, if not all species are probably univoltine, but there is a wide variety of slow and fast seasonal life cycles, with emergence from early spring to autumn, depending on the species.

B. Odonata—Dragonflies and Damselflies

This relatively small hemimetabolous order has two distinctive suborders in North America, Anisoptera (dragonflies) and Zygoptera (damselflies) (Figs. 17.27–17.40). Only about 415 species are known from North America. Larvae of all species are aquatic, with about two-thirds living in lentic habitats and one-third in lotic environments. Lotic species occur in all types of permanent stream habitats, including gravel and rock riffles, debris along banks, bank vegetation, soft sediments, and sand. Lentic species inhabit permanent and temporary ponds, marshes, swamps, and littoral and shoreline areas of lakes. Because all larvae attain a relatively large size and are predators, mostly on other insects, they are often important top predators in the invertebrate community. Adults are also predators, feeding on mosquitoes and other flying insects, and thus Odonata are regarded as highly beneficial insects.

Life cycles are relatively long, mostly one year in Zygoptera and one to four years in Anisoptera. Slow seasonal developmental cycles predominate, but fast seasonal cycles occur in both suborders. There are normally 11–12 larval instars, but the number may vary somewhat between species or even within a species. When larval development is complete, the larva crawls from the water onto emergent vegetation or onto the bank, where ecdysis occurs. Larvae of some species may crawl several meters from the shore to emerge. Adults may live for several weeks. Newly emerged "teneral" adults are feeble fliers for a day or two and vulnerable to predation. After the teneral stage has passed, adults often disperse widely and feed for a week or more on flying insects before returning to breeding sites. Here, males patrol selected areas, chasing away rival males and other species. Before mating, the male transfers sperm from primary genitalia at the tip of the abdomen to a secondary genital organ on the venter of abdominal segment 2. Using claspers on the terminal segment, a male will seize a female that flies into his territory, grasping her either by the pronotum (Zygoptera) or by the head (Anisoptera). The female then loops the tip of her abdomen forward and upward to receive sperm from the penis on the second abdominal sternum of the male.

Eggs are laid singly in plant tissues (endophytic oviposition) above, at, or below the surface of the water; or they may be laid singly or in masses on plants, debris, or other substrata, or randomly on the water surface (exophytic oviposition). Oviposition is endophytic in all Zygoptera and in Aeshnidae and Petaluridae; it is exophytic in other families. In many species, especially Zygoptera and Libellulidae, the male continues to grasp the female while she is ovipositing, protecting her from other males. Eggs hatch after an incubation period of one to several weeks, or after a long diapause. The first larval instar is a prolarval stage that lasts only a few minutes to several hours. Prolarvae lack the well-developed appendages of other larval instars.

Odonata larvae can be readily distinguished from all other aquatic insects by their elongate, hinged labium, which has been modified for seizing and grasping prey (Figs. 17.28–17.30). They have conspicuous compound eyes and relatively short filamentous antennae (Fig. 17.39). In Zygoptera, the abdomen is elongate and thin, terminating in three long, vertically oriented, caudal lamellae (Fig. 17.27). In Anisoptera, the abdomen is relatively broad and terminates in three triangular-shaped, pointed appendages (Fig. 17.40), the upper one being called the epiproct and the lower ones paraprocts. In between these are triangular-shaped cerci, which are usually distinctly shorter than the epiproct or paraprocts. Adults do not resemble larvae, being much more elongate, having conspicuous genitalia, two pairs of elongate net-veined wings, and much larger compound eyes. Like Ephemeroptera, the primitive wings of Odonata cannot be folded. Those of Anisoptera are held horizontally to the side when at rest; those of Zygoptera are held vertically just above or slightly to the side of the abdomen.

Odonata larvae move about on the substratum by crawling; some actively stalk prey. Zygoptera larvae may swim by moving their abdomen and caudal lamellae from side to side. Anisoptera larvae can move rapidly when disturbed by expelling water from a rectal chamber. There are tracheal gills within the rectal chamber, and movement of water in and out of the chamber aids in respiration of Anisoptera larvae. In Zygoptera, respiration is greatly enhanced by the caudal lamellae, enabling most species to live in water with relatively low levels of dissolved oxygen. Most larvae can be easily collected with an aquatic net using techniques described earlier. Adults are collected with large aerial nets after being sighted. Zygoptera and the Anisoptera that alight on vegetation are relatively easy to capture, but many large Anisoptera are extremely difficult to capture with a net.

Most larvae of North American Odonata can be identified to species by using keys and descriptions provided by Walker (1953, 1958), Walker and Corbet (1975), Needham and Westfall (1955), and Garman (1917, 1927). Most of these keys, however, are old and problems often arise when using them. Since their publication, new descriptions for many larvae have appeared in the literature, but unfortunately, revised larval keys are restricted to limited geo-

Table 17.2 Literature References for Identification of Odonate Species

---

Regional keys to larvae that include more than one
  family
  Canada and Alaska—(Zygoptera) Walker (1953);
    (Anisoptera) Walker (1958); (Anisoptera) Walker
    and Corbet (1975)
  California—Smith and Pritchard (1956)
  Connecticut—Garman (1927)
  Illinois—(Zygoptera) Garman (1917)
  North America—(Anisoptera) Needham and Westfall
    (1955)
  North and South Carolina—Huggins and Brigham
    (1981)
  Texas—Young and Bayer (1979)
  Utah—(Anisoptera) Musser (1962)

Keys to larvae listed by family
  Coenagrionidae
    Enallagma—(West Coast) Garrison (1984)
  Gomphidae
    Dromogomphus—Westfall and Tennessen (1979)
    Lanthus—Carle (1980)
  Libellulidae
    Ladona—Bennefield (1965)
    Sympetrum—Tai (1967)

---

graphic areas (Huggins and Brigham 1981) or a few genera. Because Odonata are relatively homogeneous, differences between families, genera, and species are not as obvious as in most other orders. Larvae of Corduliidae cannot be readily separated from those of Libellulidae without identifying them to genus. There is also considerable disagreement among odonatologists about the status of various genera as well as some disagreement about family status. But, at the species level, North American Odonata are better known as larvae and adults than aquatic insects in any other order. This is perhaps because they have attracted the interest of many amateur odonatologists who have made significant contributions to our knowledge. Books by Corbet (1962) and Miller (1987), and a review article by Corbet (1980) summarize the knowledge about the biology and ecology of dragonflies. Table 17.2 provides a list of references for the identification of Odonata species.

A brief coverage of the biology of dragonfly and damselfly families is presented next. See Section V.C for a taxonomic key to families of Odonata.

1. Families of Odonata—Suborder Anisoptera (Dragonflies)

a. *Aeshnidae (11 Genera)*

Larvae are large, with elongate tapered abdomens, and can be distinguished from other Anisoptera by their elongate shape, flat prementum, and thin six- or seven-segmented antennae (Fig. 17.33). Most inhabit lentic habitats, especially weedy permanent ponds, marshes, and littoral areas of lakes. Here, they move about on the vegetation and stalk prey composed of invertebrates and small vertebrates. A few species are lotic and inhabit either slower areas of streams or riffles. Life cycles of two to four years are likely for most species, especially in the North. One species, *Anax junius*, which does not overwinter in the North, migrates south in autumn and returns in early spring to oviposit. Larvae of this species develop more rapidly than other Aeshnidae, are bivoltine in the South, and may sometimes complete two generations in the northern United States. Flight periods of most species are during the summer months.

b. *Cordulegastridae (1 Genus)*

The large, hairy, somewhat elongate larvae with small anterolateral eyes can be easily distinguished from other Anisoptera by the jagged teeth on the palpal lobes of their spoon-shaped prementum (Fig. 17.35). They inhabit small streams, where they lie partially concealed in silt to ambush prey, mostly at the upstream edge of pools. A three- or four-year life cycle is likely for most species, with adults flying from spring into early summer. Eggs are laid in the substrata of shallow riffles.

c. *Corduliidae (8 Genera)*

Larvae are short and broad, with a spoon-shaped prementum, and cannot be readily separated from Libellulidae. In most, the palpal lobes are more deeply scalloped (Fig. 17.36) than in Libellulidae, but in a few genera, the lobes are rather shallowly scalloped (Fig. 17.37), and one genus of Libellulidae also has deeply scalloped palpal lobes. Larvae are mostly lentic, with two- to four-year life cycles probably predominating in the North and shorter cycles in the South. They commonly occur in marshes, swamps, cool ponds, and littoral areas of lakes. However, there are several lotic species, which inhabit debris in streams of all sizes. Synchronized emergences may lead to formation of exceptionally large swarms. Adult flight periods are mostly from May into early July.

d. *Gomphidae (15 Genera)*

Like Aeshnidae and Petaluridae, the prementum is flat, but the larvae have relatively shorter and broader abdomens. Their broad, four-segmented antennae are diagnostic (Fig. 17.32). Larvae of most species inhabit streams, where they may be found in riffles or partially buried in silt or sand in slower reaches. Other species inhabit lentic habitats, especially small permanent ponds and littoral areas of lakes, where they lie concealed in the substrata to ambush prey. Life cycles are long, usually two to four years, depending on water temperature and food. Most species have flight periods in late spring and summer; in some species eggs may diapause over the winter.

e. *Libellulidae (28 Genera)*

The short, broad larvae closely resemble those of Corduliidae, but in most, the crenulations on the palpal lobe of the mentum are extremely shallow (Fig. 17.38). Larvae occur in a variety of permanent and temporary lentic habitats where they crawl about in the vegetation and debris; occasionally they are found in vegetation along margins of streams. Littoral areas of lakes, permanent ponds, vernal ponds and marshes, cattail marshes, sphagnum swamps, and bogs all are habitats for the great variety of species. Many species are univoltine, with some having eggs that diapause. Other species may require two or more years to complete larval development. Flight periods range from spring through autumn, depending on the species.

f. *Macromiidae (2 Genera)*

Larvae are readily recognized by their relatively broad abdomens, extremely long legs, and the presence of a frontal horn that projects forward between the eyes (Fig. 17.39). They are found among debris in the currents of medium to large streams or in margins of lakes exposed to wave action. Life cycles of 2–4 years are likely. Flight periods extend from May through August, depending on the species.

g. *Petaluridae (2 Genera)*

The two North American species are rare and neither has been found in the central part of the continent. Larvae of one species probably inhabit swamps at high elevations in western mountains; those of the other species have been found in bogs and seeps in eastern forests. The relatively broad larvae have a flat prementum, relatively broad six- or seven-segmented antennae that are hairy (Fig. 17.34), and patches of black bristles on the posterior abdominal terga. Because of their cold habitat, up to six years may be required to complete larval development. The flight period ranges from early spring in Florida to summer farther north.

2. *Families of Odonata—Suborder Zygoptera (Damselflies)*

a. *Calopterygidae (2 Genera)*

The large larvae are readily distinguished from other Zygoptera by their very elongate first antennal segment (Fig. 17.27). All species are lotic, inhabiting permanent streams of all sizes, where they are found mostly among bank vegetation and accumulations of debris. They are univoltine or semivoltine, with flight periods from late May into October.

b. *Coenagrionidae (15 Genera)*

The relatively small larvae differ from Calopterygidae in having a short basal antennal segment and differ from Lestidae in not having the base of the prementum greatly narrowed (Fig. 17.30). They are predominantly lentic, although species in one genus (*Argia*) inhabit riffles of streams and a few species in other genera occur along banks of streams. Lentic species are found mostly in permanent ponds, marshes, swamps, and littoral areas of lakes; occasionally they occur among vegetation in parts of streams with little or no current. They crawl about on the vegetation and ambush or stalk prey, which consists mostly of small invertebrates. Most species are univoltine, some having eggs that diapause over the winter. A few species are bivoltine. Adults fly from spring to autumn, depending on the species.

c. *Lestidae (2 Genera)*

The long, thin larvae with their long, parallel-sided caudal lamellae are distinctive. They differ from all other Zygoptera in having a prementum that is greatly narrowed and elongate basally (Fig. 17.29). One western genus (*Archilestes*) with two species is lotic. Larvae of other species (*Lestes*) are lentic, only occasionally being found among emergent vegetation along slow streams. They commonly inhabit vegetation in both permanent and vernal ponds and marshes, and have a one-year life cycle. Eggs are laid in vegetation above the water and often diapause until autumn or the following spring. Eggs hatch when the now dead vegetation in which they were laid falls into the water, or in vernal ponds when water floods the dead vegetation. Adults fly from late spring through summer.

d. *Protoneuridae (2 Genera)*

This tropical family is found only in Texas, where larvae inhabit streams. Larvae are similar to Coenagrionidae, but their caudal lamellae are thick in the basal half and thin and leaflike in the distal half (Fig. 17.31).

C. Plecoptera—Stoneflies

In this relatively small hemimetabolous order, 550 species are known from North America (Stewart and Stark 1988) and many additional species will probably be described as the taxonomy of various families and genera is studied in greater detail and mountain streams are collected more thoroughly. The order is poorly represented in middle North America; species richness is greatest in mountain areas where fast, cold streams abound. Larvae of all species are aquatic. Almost all inhabit streams, but a few species have adapted to living in cold oligotrophic lakes. In the West, species in at least two genera inhabit the hyporheic zone of certain streams to depths of 10 m or more.

Stonefly larvae are an important part of the stream ecosystem. They provide food for fish and invertebrate predators and often are top predators in the invertebrate food chain. They are not sufficiently abundant to create nuisance problems; and because most species are intolerant of lowered levels of dissolved oxygen, they play an important role in biological monitoring of streams.

Most stoneflies are univoltine, but several larger species require two or three years to complete development. Species with two- or three-year life cycles in the North may be univoltine in the South. Univoltine species often have fast seasonal life cycles, with eggs or small larvae diapausing during the summer and rapidly developing in autumn, winter, or spring. Many species, however, have a slow seasonal life cycle with eggs hatching within a few weeks and larval development progressing slowly throughout the year. Species that have a slow seasonal life cycle in the North may have a fast seasonal life cycle in the South.

The number of larval instars varies with the species and even among individuals within a species, with 12–23 or more instars having been recorded for species that have been studied. After completing development, larvae crawl from the water onto a convenient substratum and ecdysis occurs. Adults then crawl or fly to nearby vegetation or other structures to find a mate. In most species, this is accomplished by drumming with their abdomen; males generally initiate the drumming signal and females respond. After repeated drumming and crawling, males find a female and mate. When eggs have matured, the female extrudes them into a mass on the tip of her abdomen. Females of some species fly over the water and land briefly on its surface to deposit their eggs; others jettison their egg mass while flying above the water. In still other species, females crawl into the water on a protruding substratum to oviposit beneath the surface. In most species, the female will lay multiple egg masses.

Stonefly larvae (Figs. 17.1 and 17.2) can be distinguished from most other aquatic insects by their two long filamentous cerci and generally elongate or flattened appearance (Figs. 17.41, 17.44–17.46, 17.50, 17.52, 17.54). They have relatively long antennae, distinct compound eyes, two or three ocelli, chewing mouthparts, two pairs of thoracic wing-pads, and three-segmented tarsi with two claws on each tarsus. They are likely to be confused only with Ephemeroptera larvae, which also have long cerci, but Ephemeroptera larvae usually have three caudal filaments instead of two, have only one tarsal claw, and have gills on the middle abdominal segments. Gills of Plecoptera, if present, are on the head, thorax (Fig. 17.2), and/or basal or apical segments of the abdomen. Adult stoneflies resemble their larvae, the most noticeable difference being the presence of two pairs of net-veined wings that fold flat over the abdomen; a few species are brachypterous or apterous. Genitalia are well developed in adults and gills are reduced to remnants in species that have them as larvae.

Stonefly larvae generally crawl on the substrata; when forced to swim, they do so weakly by moving their abdomen from side to side. Oxygen for respiration is obtained from the water through the cuticle, with thoracic and sometimes abdominal gills aiding oxygen uptake in most larger species. Because they generally lack extensive gills, most Plecoptera larvae occur only in oxygen-rich water. Feeding habits vary among families and also among species within a family or genus. Like many other aquatic insects, feeding habits within a species will vary with developmental stage and availability of food.

Adult stoneflies are relatively short-lived; most live a few days to two weeks, some a little longer. Several species feed as adults, mostly on algae, lichens, and riparian leaves or flowers. Larvae can be readily collected with an aquatic net by using standard sampling techniques. Adults of most species can be collected by using a net to sweep bank vegetation or by examining rocks, tree trunks, and bridges in the vicinity of streams. Some species, however, move high into trees and can be collected only by using nets with very long handles or sticky traps. Several species are attracted to lights, while others are rarely found at lights.

Through the years, there have been many changes in the higher classification of Plecoptera, but in the last decade it has remained stable. The order is divided into two suborders, Euholognatha and Systellognatha. The former includes four families (Capniidae, Leuctridae, Nemouridae, and Taeniopterygidae) that previously were in the family Nemouridae; larvae in all of these families are primarily herbivore–detritivores. Larvae of two families of Systellognatha (Perlidae and Chloroperlidae) are predators, in two other families (Peltoperlidae and Pteronarcyidae) they are herbivore–detritivores, and in the remaining family (Perlodidae), feeding habits depend on the species and larval age.

The biology of Plecoptera was reviewed by Hynes (1976) and a recent book by Stewart and Stark (1988) discusses classification, phylogeny, biogeography, ecology, and behavior of Plecoptera. This book includes keys and illustrations for larvae in all North American genera, notes on their biology, and the known distribution of all North American species. Species in some genera can be identified from mature larvae as well as from adults; in other genera, identification of larvae is not possible because larvae of several species remain unknown. About 55% of North American Plecoptera species are known in the larval stage (see references in Table 17.3). Information on individual families is presented as follows, and a key to stonefly families appears in Section V.D.

1. Families of Plecoptera— Suborder Euholognatha

a. *Capniidae (9 Genera)*

Capniidae are "winter stoneflies." Larvae are generally small, dark-colored, and elongate (Fig. 17.52). They closely resemble larvae of Leuctridae, but can

Table 17.3 Literature References for Identification of Stonefly Species

Regional keys to larvae that include more than one family
  California—Jewett (1956)
  Louisiana—Stewart *et al.* (1976)
  North and South Carolina—Unzicker and McCaskill (1981)
  Northeastern North America—Hitchcock (1974)
  Pennsylvania—Surdick and Kim (1976)
  Rocky Mountains—Baumann *et al.* (1977)
  Saskatchewan—Dosdall and Lehmkuhl (1979)
  Southwest British Columbia—Ricker (1943)

Keys to larvae listed by family
  Capniidae—(eastern Canada) Harper and Hynes (1971b)
  Leuctridae—(eastern Canada) Harper and Hynes (1971a)
  Nemouridae—(eastern Canada) Harper and Hynes (1971d)
  Peltoperlidae
    Soliperla—Stark (1983)
  Perlidae
    Agnetina—Stark (1986)
    Hesperoperla—Baumann and Stark (1980)
    Paragnetina—Stark and Szczytko (1981)
    Perlinella—Kondratieff *et al.* (1988)
  Perlodidae—(Wisconsin) Hilsenhoff and Billmyer (1973)
    Diploperla—Kondratieff *et al.* (1981)
    Helopicus—Stark and Ray (1983)
    Hydroperla—Ray and Stark (1981)
    Isogenoides—Ricker (1952)
    Isoperla—(western North America) Szczytko and Stewart (1979)
    Malirekus—Stark and Szczytko (1988)
    Setvena—Stewart and Stanger (1985)
  Pteronarcyidae—(West Virginia) Tarter *et al.* (1975)
  Taeniopterygidae—Harper and Hynes (1971c)
    Taeniopteryx—Fullington and Stewart (1980)

be recognized by their generally shorter and broader metathoracic wing-pads, which are nearly as widely separated as the mesothoracic wing-pads, and by posterior abdominal terga that are usually widened and fringed posteriorly. Larvae inhabit streams of all sizes. They are especially abundant in smaller streams and spring seeps; some even inhabit streams that become dry in the summer. Two genera occur in cold oligotrophic lakes. Life cycles are univoltine, with emergence in late winter or early spring. Eggs hatch a few weeks after being laid, and after some development, larvae of most species diapause. Larval development resumes in autumn, with some species developing most rapidly in late autumn and winter. A few species have a slow seasonal life cycle with steady development from late spring through autumn. Larvae are mostly detritivores that feed on allochthonous debris, especially decaying leaves. In the North, several species may emerge on warm days while streams are still ice-covered; adults often can be found crawling on the snow. Adults of some species are apterous or brachypterous.

b. *Leuctridae (7 Genera)*

The relatively small, elongate larvae are similar to those of Capniidae, but have metathoracic wing-pads that are similar in shape and distinctly closer together than those on the mesothorax (Fig. 17.54). Larvae also resemble Chloroperlidae, which have shorter cerci and meso- and metathoracic wing-pads that are equally far apart. Larvae are found mostly among gravel in small, cool, permanent streams where they feed on decaying allochthonous material; one genus occurs in temporary streams. Most species are univoltine with emergence from spring to autumn, depending on the species. A semivoltine life cycle occurs in at least one northern species.

c. *Nemouridae (12 Genera)*

The small, relatively dark-colored larvae are short and broad, with divergent metathoracic wing-pads in more mature larvae and metathoracic legs that reach or exceed the apex of the abdomen (Fig. 17.50). Most nemourids are univoltine, with emergence from spring to autumn, depending on the species. Fast seasonal life cycles predominate, with extended egg diapause in some species. At least one species is semivoltine. Larvae are detrital feeders that inhabit debris and soft sediments in smaller streams and spring seeps; they are often the only stonefly in spring seeps and runs. Larvae in at least one genus inhabit temporary streams, and one species has been collected from Arctic lakes.

d. *Taeniopterygidae (6 Genera)*

Like Capniidae, these are univoltine "winter stoneflies" that emerge in late winter or very early spring and probably spend the summer as diapausing larvae. In size and shape, larvae resemble Perlodidae, having divergent metathoracic wing-pads and elongate abdomens, but they are not as heavily patterned as most Perlodidae, they move much more slowly, and they swim very inefficiently. Larvae are herbivore–detritivores that occur in permanent streams of all sizes. They are frequently found among bank vegetation.

2. *Families of Plecoptera— Suborder Systellognatha*

a. *Chloroperlidae (13 Genera)*

Larvae are generally small, elongate, rounded in cross-section, and have short cerci (Fig. 17.45).

They are similar to larvae of Capniidae and Leuctridae and are difficult to distinguish from Leuctridae in the field. They are not likely to be confused with Capniidae, however, because late-instar larvae of Chloroperlidae are present in late spring and summer, while those of Capniidae occur in late autumn, winter, and early spring. Mouthparts differ greatly from those of Capniidae and Leuctridae, reflecting the carnivorous feeding habits of Chloroperlidae in the later larval instars. Larvae of Chloroperlidae inhabit mostly gravel bottoms of cold, small- to medium-sized streams. Life histories are poorly known. Univoltine and semivoltine fast seasonal life cycles have been documented, with emergence in late spring or early summer.

b. *Peltoperlidae (6 Genera)*

The broad, flat, roach-like appearance of the medium-sized larvae is distinctive (Fig. 17.41). They are often abundant in mountain streams of the eastern, southern, and western United States, but are absent from central North America. Larvae inhabit smaller streams, where they feed on decaying leaves, detritus, and associated microorganisms. Many species are semivoltine; some may be univoltine. Adults emerge in late spring or early summer.

c. *Perlidae (15 Genera)*

Larvae of these relatively large, broad stoneflies (Fig. 17.1) are readily recognized by the presence of branched filamentous gills at the base of each leg (Fig. 17.2) and the absence of gills on the first two visible abdominal sterna. They are widespread and common, inhabiting permanent streams of all sizes where they prey on other macroinvertebrates. They are often the dominant insect predator, although early-instar larvae may feed on detritus and algae. Most are found in fast water, especially among debris in riffles. Species in most genera have a two- or three-year slow seasonal life cycle in northern regions; univoltine life cycles may occur in the South. One common genus is univoltine with a fast seasonal life cycle. Emergence of most species is during the summer months.

d. *Perlodidae (29 Genera)*

Larvae are mostly small- to medium-sized stoneflies, which usually have distinctive dorsal patterns. They lack the thoracic gills of the similarly shaped and patterned Perlidae. The metathoracic wing-pads are divergent and the abdomen is rather elongate (Fig. 17.44). While larvae of some species of Perlodidae are herbivore–detritivores throughout their larval development, most species are predominantly predators. They develop in all sizes and types of streams; some even develop in cold oligotrophic lakes. All species are apparently univoltine, with emergence in spring. Eggs of many species diapause during the summer and hatch in autumn, but in other species, eggs hatch within a few weeks after being laid.

e. Pteronarcyidae (2 Genera)

These large stoneflies require one to four years to complete development. They have a slow seasonal life cycle, with univoltine life cycles occurring only in the South. Larvae are elongate, and can be readily recognized by the branched filamentous gills at the base of thoracic legs and on the first two or three abdominal sterna. They feed mostly on coarse particulate organic matter and inhabit wood or accumulations of debris in small- to medium-sized permanent streams with moderate to rapid currents. Emergence occurs in spring or early summer, being earliest in the South.

D. Trichoptera—Caddisflies

Trichoptera is a rather large holometabolous order (Figs. 17.55–17.80). Larvae and pupae of all species are aquatic, with one or two exceptions, and represent an important part of the insect fauna of most streams. More than 1340 species are known from North America. Most of them occur in lotic habitats, but there are some species in half of the families that inhabit lentic habitats. Adults are terrestrial, moth-like insects, which sometimes are attracted to lights in large numbers. Most have very long filiform antennae, and their wings, which they fold roof-like over their bodies are covered with hairlike setae. Eggs of most species are laid in masses in the water; a few species lay their eggs on vegetation above the water or in moist soil of temporary ponds that will eventually be flooded. Most species are univoltine, often with rather synchronized emergences; however, in some species different cohorts emerge throughout the warm weather period. A few species are bivoltine, with the tendency for bivoltinism increasing farther south; and several species, especially those that develop in cold-water streams, are known to be semivoltine, especially in the northern part of their range.

Adult caddisflies can be identified to species by using the many keys and descriptions that have been published, but larval taxonomy lags, with about two-thirds of the described species still unknown in the larval stage. This makes accurate identification of larvae to species impossible in most genera, and larvae of a few genera remain unknown. Wiggins (1977) published excellent keys, descriptions, and illustrations for larvae of known genera. Ross (1944) developed species keys for larvae in many genera

Table 17.4 Literature References for Identification of Caddisfly Species

Regional keys to larvae that include more than one family
    Illinois—(many incomplete) Ross (1944)
    North and South Carolina—Unzicker *et al.* (1981)

Keys to larvae listed by family
    Brachycentridae—(Wisconsin) Hilsenhoff (1985)
      Brachycentrus—Flint (1984)
      Micrasema—Chapin (1978)
    Goeridae—Wiggins (1973, 1977); (Eastern North America) Flint (1960)
    Hydropsychidae—(Wisconsin) Schmude and Hilsenhoff (1986)
    Arctopsychinae—Flint (1961)
      Ceratopsyche—(as *Hydropsyche*) Schefter and Wiggins (1986)
      Ceratopsyche and *Hydropsyche*—(eastern and central North America) Schuster and Etnier (1978)
    Lepidostomatidae—Weaver (1988)
    Leptoceridae
      Ceraclea—Resh (1976)
      Mystacides—Yamamoto and Wiggins (1964)
      Nectopsyche—Haddock (1977)
    Limnephilidae—(Eastern United States) Flint (1960)
      Desmona—Wiggins and Wisseman (1990)
      Dicosmoecus—Wiggins and Richardson (1982)
      Eocosmoecus—Wiggins and Richardson (1989)
      Hesperophylax—Parker and Wiggins (1985)
    Molannidae
      Molanna—Sherberger and Wallace (1971)
    Odontoceridae
      Psilotreta—Parker and Wiggins (1987)
    Phryganeidae—Wiggins (1960)
    Rhyacophilidae—(Eastern United States) Flint (1962)

that occur in the central United States as well as keys to adults; but in most of these genera, there are species for which larvae are unknown, so his larval keys must be used with caution. Readers should consult Table 17.4 for a list of references to regional keys.

Trichoptera is divided into three suborders, Annulipalpia, Integripalpia, and Spicipalpia (Wiggins and Wichard 1989), paralleling a previous division into three superfamilies, Hydropsychoidea, Rhyacophiloidea, and Limnephiloidea (Wiggins 1977). This modifies a classification proposed by Weaver and Morse (1986) by elevating one of their four infraorders (Spicipalpia) to a suborder. In the suborder Annulipalpia, larvae employ silk from their labial glands to construct retreats and nets, which they use to filter or gather food such as algae, detritus, or macroinvertebrates. In the suborder Spicipalpia, larvae in the four families are either free living or construct portable cases that are barrel-shaped, purse-like, or saddle-shaped. They also pupate in a completely closed cocoon of parchment-like silk, differing from most other caddisflies, which have openings in their pupal case to allow circulation of water. Those that are free living (Rhyacophilidae and Hydrobiosidae) are mostly predators on other arthropods. Those that build cases (Glossosomatidae and Hydroptilidae) are mostly herbivores that feed on periphyton. In the suborder Integripalpia, all larvae use silk from their labial glands to construct tubular cases, which vary widely in size and shape. The cases are often sufficiently distinctive to be employed in generic and species identification. Integripalpia larvae are mostly herbivores or detritivores, but in some families (Brachycentridae, Leptoceridae, Molannidae, Odontoceridae, and Phryganeidae) omnivores or predators can be found. Larvae in five families of Integripalpia (infraorder Plenitentoria of Weaver and Morse) have a prosternal horn, which larvae in other families lack.

Trichoptera larvae have a pair of simple eyes, chewing mouthparts, antennae that are so short they often are difficult to see, three pairs of thoracic legs with a single tarsal segment and tarsal claw, and a pair of fleshy prolegs on the last abdominal segment (Figs. 17.62 and 17.63). Many larvae have single or branched gills on the abdominal segments. Most species have five larval instars; after completion of larval development, a pupal case is constructed, either by sealing off the open end of the portable case or by constructing a case and cementing it to the substratum. The larva then ceases feeding and becomes quiescent with its appendages pressed against its body. This is the prepupal stage, which may last a few days to several weeks. The pupa is of the exarate type, with appendages free from the body. Pupae in most families have well-developed mandibles that the pharate adults (adult within the pupal integument) use to cut their way out of the case just prior to emergence. After a pharate adult leaves the pupal case, it swims to the surface of the water where ecdysis takes place, or more often, it swims to the surface and then rapidly to shore where it crawls out of the water onto some object to emerge. Ecdysis is rapid.

Trichoptera larvae have many adaptations to their aquatic environment. Respiration in caddisfly larvae is through the integument, and in many families this is aided by the presence of abdominal gills. Additionally, especially in Integripalpia, larvae are able to increase respiration by undulating their bodies within their tubular cases to increase the flow of water and dissolved oxygen through the cases. Larval cases and retreats also serve to protect larvae from predators, acting either as a physical barrier or serving to camouflage the larvae. Cases may also protect larvae from their environment. Hard cases protect those that live in sand and gravel from abrasion, and streamlined cases or nets allow others to inhabit substrata in very swift currents. Net-spinning larvae (Annulipalpia) use their nets to strain algae and other food material from the current in streams. Some caddisfly larvae (Leptoceridae) have fringes of setae on their legs that allow them to swim. Others (Brachycentridae) have fringes of setae on their legs to permit them to filter food particles from the current.

Most caddisfly larvae can be readily collected from lotic and lentic habitats with an aquatic net by using methods already described. In streams it is important to also examine large substrata such as wood and rocks, because many species attach themselves firmly to these substrata or inhabit moss or algae that grows on them. Species that burrow into decaying wood can be collected as they crawl from retreats when the wood is removed from a stream and allowed to dry. Species that inhabit sand or silt are best collected by sieving these substrata through a net or strainer, or by closely watching sand substrata and collecting larvae that expose themselves by moving. Extremely small species are often overlooked because they remain attached to the substratum and are too small to see. Microscopic examination of macrophytes and other substrata returned to the laboratory may yield many small larvae, especially larvae of Hydroptilidae. In the field, a hand lens is helpful in locating very small specimens.

Caddisflies are a very important part of the stream community; in some streams they dominate the insect biomass. Many species of fish feed on the larvae and emerging adults. Because of their general abundance and diversity in streams, and the wide variation in tolerance to pollution among various species, they are very important in biological monitoring. Occasionally, extremely large numbers emerge and because they are attracted to lights, they may create severe problems. These include contamination of products in factories, annoyance of people in restaurants and elsewhere, and allergic reactions in people who are sensitive to them.

Larvae of many species have never been associated with adults. Larvae of these species must be reared, associated with adults, and described before we will be able to develop reliable keys to larvae at the species level. Such keys are needed to enable us to complete needed studies of the life history, ecology, and behavior of the various species. Relatively little is presently known about the life history and ecology of most North American species. Most of what is known has resulted from studies within a single stream where the fauna is limited and thoroughly known because larvae have been associated with adults. Wiggins (1977) summarized knowledge of the biology, ecology, morphology, and distribution of larvae in each genus, and more recently Mackay and Wiggins (1979) reviewed the ecology.

Information on the biology of caddisfly families is presented here, and a taxonomic key to Trichoptera families appears in Section V.F.

1. Families of Trichoptera— Suborder Annulipalpia

a. Dipseudopsidae (1 Genus)

The only genus, *Phylocentropus,* is sometimes included in the generally predaceous family Polycentropodidae. Larvae differ from Polycentropodidae by having broad, flattened, setose tarsi (Fig. 17.68) and short, broad mandibles with mesal brushes. These are adaptations to their feeding on detritus, which they collect with a net placed within tubes that they construct in sand and silt along margins of sandy streams and lakes. The capture net and feeding habits have been described by Wallace *et al.* (1976). The five North American species all occur east of the Great Plains. The life cycle is probably one year, with cohorts emerging from May through August.

b. Ecnomidae (1 Genus)

Larvae of this Neotropical family were collected from a small, cold, rapid stream in the "hills" section of Texas. They represent the only record of this family north of Mexico (Waltz and McCafferty 1983). While previously considered a subfamily of Polycentropodidae, larvae most resemble those of Hydroptilidae, but the ninth tergum is not sclerotized, they have no case, and the abdomen is never enlarged.

c. Hydropsychidae (13 Genera)

Larvae in this very large and important family are easily recognized by the numerous, branched filamentous gills that occur ventrally on their abdomen. They frequently represent a large segment of the macroinvertebrate fauna in streams of all sizes, currents, and temperatures. Here, they build retreats on rocks and other larger substrata, in leafpacks, or in moss. Some species also tunnel into decaying wood and two others may occur on wave-swept lake shores. Most species are omnivorous, feeding primarily on algae, crustacea, and insects collected by their capture nets; a few species tend to eat mostly algae and diatoms and others are mostly predaceous, especially in later instars. Life cycles vary from bivoltine to semivoltine, depending very much on stream temperature, with the majority of species probably being univoltine. Because of their widespread abundance and the great variation in the tolerance of species to organic pollution, larvae are very important in biological monitoring of streams.

The subfamily Arctopsychinae is generally recognized in Europe as a separate family, and in North America this family has been recognized by Schmid (1968) and Nimmo (1987), but Wiggins (1981) presented compelling arguments for the retention of subfamily status. Capture nets of Arctopsychinae larvae are generally larger than those of other Hydropsychidae and larvae are predominantly predators in the later instars. Larvae differ from other Hydropsychidae by having an elongate gula that completely separates the genae.

d. Philopotamidae (3 Genera)

Larvae live in silken retreats, usually on the bottom of rocks in warm to cold streams of all sizes. Their unmarked head with an unsclerotized T-shaped labrum (Fig. 17.66) is distinctive and readily separates them from other net spinners with unsclerotized meso- and metanota. Larvae feed on algae and detritus that they capture in their silken nets, which have a smaller mesh size than those of other netspinning caddisflies. Most, if not all species are probably univoltine, except in the South where two or more generations may occur. Cohorts of most species emerge throughout the spring and summer months; one species also emerges in fall and winter.

e. Polycentropodidae (6 Genera)

While most species live in streams, larvae inhabit a wide variety of other habitats, including littoral areas of lakes and temporary ponds. Larvae can be recognized by their pointed protrochantin (Fig. 17.67), unmodified tarsi, and the presence of dark or light spots (muscle scars) on the heads of most species. Reliable keys to larvae do not exist because larvae of many species remain undescribed. Most larvae have a waxy cuticle that tends to repel water, giving the impression when collected that they are terrestrial. A few species build trumpet-shaped capture nets similar to other Annulipalpia and are primarily herbivores, but most species are predators and have tubular retreats that they sometimes use to detect prey. These tubular retreats also function as an aid to respiration, with larvae circulating water through them by undulating their body. This is especially important for species that live in lentic habitats that occasionally may be deficient in oxygen.

Life cycles are probably one year in most species, with emergence during the summer months.

f. *Psychomyiidae (4 Genera)*

Larvae live in streams, frequently on or in wood; they also occur on rocks, which they cover with silken retreats that incorporate sand or detritus. Larvae can be readily separated from other families without sclerites on the meso- and metanotum by their hatchet-shaped protrochantin (Fig. 17.65), but reliable keys to species do not exist. They graze food from the vicinity of their retreats, mostly consuming detritus, fungi, and periphyton. They are probably univoltine, with emergence in late spring and summer.

g. *Xiphocentronidae (1 Genus)*

Sometimes considered as part of Psychomyiidae, Xiphocentronidae larvae differ from this and other families by having their tarsi and tibia fused into a single segment on all legs (Fig. 17.64). Larvae are known in North America only from southern Texas, where they were collected from a spring run and described by Edwards (1961).

2. *Families of Trichoptera—Suborder Spicipalpia*

a. *Glossosomatidae (6 Genera)*

Larvae build unique turtle-like cases of sand and pebbles, but readily vacate them when collected. They can be separated from other Spicipalpia with unsclerotized meso- and metanota by their cases and their short anal prolegs with short claws (Fig. 17.69). Larvae can be identified to species only by associating them with adults. All species inhabit the upper surfaces of rocks or other large substrata in clean, cool streams or springs. Here they feed on diatoms and other algae that they scrape from the substrate. Most species probably have a one-year life cycle, often with many cohorts that emerge throughout the spring and summer.

b. *Hydrobiosidae (1 Genus)*

Although sometimes included in Rhyacophilidae, the free-living larvae are unique in having chelate prothoracic legs (Fig. 17.71), which are suited to their predatory habits. They occur in cool streams in Texas, Arizona, and Nevada. The larva of one species was described by Edwards and Arnold (1961).

c. *Hydroptilidae (14 Genera)*

Larvae are characterized by their completely sclerotized thoracic nota, lack of branched ventral gills, and very small size. They are frequently called mi-

crocaddisflies. The first four larval instars are free living, have no case, and have a normal-sized abdomen; in the last instar, the abdomen is greatly expanded and larvae have a case that is purse-shaped or barrel-shaped (Figs. 17.57 and 17.58) and sometimes attached to the substrate. Most species pass through the first four instars rapidly, and it is usually not possible to identify these early instars to genus. Larvae in the last instar can be identified to genus, but usually not to species. Larvae may be found in a wide variety of lotic and lentic habitats where they feed mostly on algae and other plant material. They are probably univoltine, or bivoltine in warmer habitats, but generally very little is known about life histories of the various species.

d. *Rhyacophilidae (2 Genera)*

The mostly predatory larvae, which live in cool or cold streams, have no case and do not build retreats. Most species occur in mountainous regions where cold, fast streams abound. They are recognized by their unsclerotized meso- and metanotum, sclerotized ninth abdominal tergum, and elongate anal prolegs (Fig. 17.70). Life cycles are probably univoltine for most species, but may be semivoltine in some.

3. *Families of Trichoptera— Suborder Integripalpia*

a. *Beraeidae (1 Genus)*

The gently curved and tapered case is made of fine sand. The larvae are small and can be recognized by a transverse ridge on the pronotum that curves forward laterally to form rounded lobes (Fig. 17.73). Larvae, which are known for only one of the North American species (Wiggins 1977), inhabit muck margins of spring seeps. Adults have been collected only in southeastern and northeastern North America, where they are rare. A species found in Ontario has been reported to be semivoltine.

b. *Brachycentridae (5 Genera)*

The larval case is tapered, of vegetable or sand grains, and square or round in cross-section. Larvae are distinctive, having their pronotum divided by a distinct furrow, with the area in front of the furrow depressed (Fig. 17.72). They have no dorsal or lateral humps on the first abdominal segment, which is unique among Integripalpia. They also lack a prosternal horn, which is present on all other Plenitentoria (Weaver and Morse 1986). Larvae in three of the genera are monotypic in North America; most others can be identified by the keys listed in Table 17.4. Most species are univoltine, but some that inhabit cold northern streams are semivoltine. All species inhabit streams, especially cold spring-fed streams, where they live in moss or other vegetation, or attach their cases to substrata in the current. The mostly herbivorous larvae graze on periphyton and other vegetation; some also consume small insects and crustaceans. Many are primarily filter feeders that use their elongate meso- and metathoracic legs to gather food from the current.

c. *Calamoceratidae (3 Genera)*

Cases are of leaf pieces or hollowed sticks. Larvae, which may reach a relatively large size, can be recognized by the projecting, divergent, anterolateral corners of the pronotum and a row of about 16 long setae dorsally on the labrum (Fig. 17.75). Most species can be identified as larvae by using descriptions and distributions reported in Wiggins (1977). Larvae inhabit pool areas of cool streams in the West or East, where they feed mostly on detritus. Many species probably have a two-year life cycle.

d. *Goeridae (5 Genera)*

Larvae resemble those of Limnephilidae and are often included in that family, but they differ in their unique modification of the mesepisternum, which is formed into an anteriorly projecting elongate process (Fig. 17.79) or a rounded spiny prominence. Since two genera are monotypic and the others have only a limited number of species, it is often possible to identify larvae to species using publications listed in Table 17.4. The sand cases are either cornucopiashaped or tubular with flanges of pebbles. Life cycles are probably univoltine or semivoltine. Larvae inhabit either cobbles and other substrata in the current or inhabit muck along the margins of springs. They feed on periphyton, vascular plants, or detritus.

e. *Helicopsychidae (1 Genus)*

The snail-like case (Fig. 17.55), which larvae never abandon, is unique. Only the larva of *Helicopsyche borealis,* which is widespread in North America, is known; larvae of the three southwestern species remain unknown. The normally lotic larvae may also be found along margins of lakes, often in rather deep water. Larvae inhabit rocks or wood and scrape periphyton from the substratum; early instars may burrow into the bed of sandy streams. They can tolerate very warm water, and have been collected from thermal springs with temperatures as high as 34°C. They are probably univoltine, with emergence of cohorts throughout the summer and a prolonged diapause of the egg.

f. *Lepidostomatidae (2 Genera)*

Cases are highly variable, being made of sand or bits of vegetation and square or round in cross-section. The location of the antenna very close to the eye (Fig. 17.74) readily identifies all larvae. Larvae are primarily detritivores and are found among detritus on a variety of substrata. While most species inhabit cold streams or springs, some occur in lakes and may be found a considerable distance from shore. Species that have been studied are univoltine, with spatial or temporal separation of different species that occur in the same habitat.

g. *Leptoceridae (7 Genera)*

Larvae construct a wide variety of cases. In some genera, they have elongate cases of vegetation, vegetation and sand, or pure silk, and are adapted for swimming by having fringes of long setae on their legs. Larvae in other genera have shorter, stouter cases that are often cornucopia-shaped and constructed of vegetable or mineral material; these larvae are unable to swim. Most larvae can be recognized by their relatively long antennae (Fig. 17.59), which are much longer than those of other caddisflies, and also by their elongate metathoracic legs, which are distinctly longer than their other legs. Larvae inhabit a variety of permanent lotic and lentic habitats; those in lotic habitats tend to be in slower water and those in lakes may be found far from shore as well as in the littoral zone. They are generally omnivorous in their feeding habits, but some are primarily predators and a few feed exclusively on freshwater sponges. Most species are probably univoltine.

h. *Limnephilidae (45 Genera)*

Case construction in this large and diverse family varies widely, with vegetable or mineral material incorporated into a variety of shapes and sizes. Larvae also vary greatly in size and structure, making them difficult to characterize. All have a prosternal horn and antennae that are located about halfway between each eye and mandible (Fig. 17.76). Most have numerous setae on the first abdominal segment, and sclerites at SA-1 of the metanotum. Many also have chloride epithelia ventrally or dorsally on abdominal segments. While species in some genera can be identified by keys in Table 17.4, accurate identification of larvae is not possible in most genera because larvae of too many species remain unknown. Most species are univoltine, but a few are semivoltine. The majority of species inhabit a wide range of lotic habitats, from springs and small streams to large rivers; others inhabit lakes, and several are found in permanent or temporary ponds. In these habitats, larvae can be found on vegetation, rocks, or detritus, or in gravel, sand, or soft sediments. The larvae are herbivores or detritivores, and are important in processing large particulate organic matter.

i. Molannidae (2 Genera)

Larvae construct characteristic sand cases with lateral flanges, and they can be distinguished from other caddisfly larvae by their very short metatarsal claws. Larvae live in sandy or silty substrata of lakes or streams where currents are reduced. Here they feed on algae, vascular plants, or even invertebrates. Most species are univoltine.

j. Odontoceridae (6 Genera)

Cases, which are typically cornucopia-shaped and made of sand grains, are hard and difficult to crush. Like Sericostomatidae, which they resemble, larvae lack a prosternal horn and have their eyes located close to the mandibles, but their protrochantin is small and not hook-shaped (Fig. 17.78). Species of larvae in the largest genus (Psilotreta) can be identified; other species are in monotypic genera and can be identified by referring to Wiggins (1977). Larvae inhabit small, cold streams or springs where they are found in sandy substrata. They are probably omnivorous, feeding on algae, vascular plants, and invertebrates. Most species are probably semivoltine.

k. Phryganeidae (8 Genera)

These large caddisflies have distinctive cases made mostly of pieces of vegetation that are spirally wound or of concentric rings. When disturbed, they readily abandon their cases, which they may reenter. Larvae are distinctive, generally having a boldly striped head (Fig. 17.61), which is more prognathous than other Integripalpia. They also have a prominent prosternal horn and lack significant sclerotization of the mesonotum. Identification of species is usually not possible. Larvae may be found among vegetation and detritus along streams of all sizes, in marshes, in temporary and permanent ponds, and even in lakes where they may occur far from shore. While many are mostly predators, vegetation is also consumed, especially by early instars. Life cycles are probably one year.

l. Sericostomatidae (3 Genera)

Cases are typically cornucopia-shaped and made of sand, but they are not hard as in Odontoceridae. Larvae can be recognized by the projecting anterolateral angles on their pronotum and a hook-shaped

protrochantin (Fig. 17.77). No key exists for separating larvae of Agarodes, the largest genus. Larvae live in sandy substrata of streams, springs, and even lake margins, where they feed primarily on detritus. Most species are probably univoltine.

m. Uenoidae (3 Genera)

Their cornucopia-shaped cases of sand grains are either extremely long and thin or short and stout, depending on the genus. Larvae occur in streams and springs, where they feed on detritus, diatoms, and algae. Until recently, they were considered part of Limnephilidae. Larvae differ from Limnephilidae by having an elongate pronotum and a mesally notched anterior of the mesonotum (Fig. 17.80). There are several species in each of the genera, and keys to larvae have not yet been developed. Two genera (Neothremma and Farula) occur in the western mountains; Neophylax, which is widespread in North America, was recently added to this family (Vineyard and Wiggins 1988).

E. Megaloptera—Fishflies and Alderflies

This very small holometabolous order contains two aquatic families that once were included as part of Neuroptera. The aquatic larvae are distinctive, having seven or eight pairs of lateral filaments and large, conspicuous mandibles (Figs. 17.81 and 17.82). They can be confused only with some aquatic Coleoptera larvae that have lateral filaments. Larvae often attain a very large size, and because all are predators, they may exert significant pressure on populations of macroinvertebrates that live in the same habitat. Studies indicate that there are usually ten or eleven larval instars, and that life cycles range from one to four years, depending on the species and climate. Larvae are aquatic; all other stages are terrestrial. Eggs are laid in masses on objects above the water, and pupation takes place in cells on land. The terrestrial adults are short-lived and generally secretive; however, because of their large size and often conspicuous mandibles, they attract attention when discovered. They are weak fliers, but are often attracted to lights. They fold their net-veined wings roof-like over their abdomens when at rest.

Larvae of most species occur only in relatively oxygen-rich environments because they usually obtain oxygen through their integument directly from the water in which they live. Their lateral filaments greatly increase surface area, allowing them to do this. Species that inhabit lentic habitats, which may at times be low in oxygen, have long caudal respiratory tubes, which they use to obtain oxygen from air at the surface. All larvae have spiracles and are

Table 17.5 Literature References for Identification of Megaloptera Species

Regional keys to larvae that include more than one
   family
      North and South Carolina—Brigham (1981a)

Keys to larvae listed by family
   Corydalidae
      Chauliodes—Cuyler (1958)
      Nigronia—Neunzig (1966)
   Sialidae
      Sialis—Canterbury (1978)

capable of living out of water in moist areas; some species survive in temporary streams.

Larvae can be readily collected with an aquatic net by using standard sampling procedures. However, greater numbers of stream species can often be collected by dislodging larger rocks. Rearing is difficult because of long life cycles, cannibalism, and the need to supply prey species and a terrestrial environment for pupation. Species of adults are well known; and as a result of recent studies, larvae of most species can also be identified (see references in Table 17.5). Information on the two families is presented as follows, and a key to Megaloptera families appears in Section V.F.

1. Families of Megaloptera

a. *Corydalidae (7 Genera)*

The very large size (20–90 mm), lateral filaments, and large mandibles of late instar "fishfly" or "Dobsonfly" larvae are distinctive (Fig. 17.82), but smaller larvae resemble those of Gyrinidae and two other Coleoptera genera. Larvae can be distinguished from Coleoptera by a pair of terminal prolegs, each with two hooks. Larvae in most genera inhabit streams or spring seeps, which have a relatively high dissolved oxygen content. Because these habitats are generally cold, life cycles of two or more years are likely for most species, with different cohorts and extended emergence periods in some species. The larval habitat of *Chauliodes* differs from that of other genera. They inhabit shallow lentic habitats or vegetated margins of streams and are univoltine, with emergence in late spring or early summer. They also are known to consume significant amounts of detritus in addition to prey, especially during the winter.

b. *Sialidae (1 Genus)*

The lateral filaments and long, single tail filament separate "alderfly" larvae from all other aquatic

insects (Fig. 17.81). Larvae live mostly in depositional sediments of streams, permanent ponds, and lakes. Species that live in lakes may inhabit areas that are several hundred meters from shore, necessitating a considerable migration to land for pupation. Most species are univoltine, but in northern regions species that live in streams or lakes are often semivoltine.

III. PARTIALLY AQUATIC ORDERS OF INSECTS

A. Aquatic Heteroptera—Aquatic and Semiaquatic Bugs

Because all aquatic and semiaquatic bugs are Heteroptera and none are Homoptera, I will follow the recent catalog by Henry and Froeschner (1988) and refer to this group of insects as the order Heteroptera rather than Hemiptera or "Hemiptera: suborder Heteroptera." The choice of names for this group of insects has created much controversy in recent years; pros and cons are discussed by Henry and Froeschner.

Heteroptera is a medium-sized paurometabolous order that is primarily terrestrial, but about 8.5% of the species are aquatic (Figs. 17.83–17.97), either living in water (217 species) or walking on its surface (107 species). It has been divided into seven suborders, only two of which have aquatic species. Those that walk on the surface of the water are in the suborder Gerromorpha, and I refer to them as semiaquatic Heteroptera. Species that live in the water belong to the suborder Nepomorpha. I refer to them as aquatic Heteroptera because they spend their entire lives under water, except for dispersal flights. Flightless species of Nepomorpha are the most aquatic of all insects because they rarely, if ever, leave the water. There are six families of Nepomorpha, in which all species are aquatic and five families of Gerromorpha, in which all species are semiaquatic or occasionally riparian. There are also four families of Heteroptera that are riparian; they are included in the key (Section V.G) because they occasionally are collected along with aquatic species. These are Gelastocoridae and Ochteridae (Nepomorpha), Macroveliidae (Gerromorpha), and Saldidae (Leptopodomorpha).

Most species of aquatic and semiaquatic Heteroptera breed in lentic habitats, but there are also several lotic species, and many of the lentic species fly to streams to overwinter. Most species are predators and their presence can significantly influence populations of other insects, especially in lentic habitats. They are known to have an impact on populations of mosquito larvae, and thus can benefit man.

Larger species, however, may prey on small fish and create problems in fish-rearing ponds. Some species also may bite people, and occasionally they become a nuisance when attracted by lights to swimming pools. The vast numbers of Corixidae that overwinter in streams are an important source of food for some fish, and dried corixids are sold as food for pet turtles and tropical fish. Most other Heteroptera are usually avoided by fish, probably because of secretions from their scent glands. Generally, however, aquatic and semiaquatic Heteroptera have little impact on man.

Heteroptera adults can be distinguished from other aquatic insects by their mesothoracic wings, or hemelytra (Figs. 17.83, 17.89–17.92), which are hardened in the basal half and membranous apically, and by sucking mouthparts that are formed into a tube or rostrum (Figs. 17.84 and 17.86). There are normally five nymphal instars; a few species have only four. Nymphs resemble adults, except that they lack wings and genitalia and are distinctly less sclerotized. They inhabit the same habitat as adults and are often associated with them.

Aquatic and semiaquatic Heteroptera have been rather thoroughly studied, and adults of almost all North American species can be readily identified. Families and genera are quite heterogeneous, making it possible to recognize families and many genera in the field. In some genera, even species can be readily identified in the field. The nymphs, however, have not been well studied except for Gerridae, and few can be identified to species except regionally or through association with adults. Some cannot even be identified to genus. Since the pioneering works of H.B. Hungerford and others 40–60 years ago, there have been relatively few additions to the known fauna in North America. For a list of literature references of species, see Table 17.6. Information on the biology of suborders and individual families is presented next, and a key to Heteroptera families appears in Section V.G.

1. Families of Aquatic Heteroptera— Suborder Nepomorpha

Most species are univoltine in northern North America; a few are bivoltine. Farther south, bivoltine life cycles probably predominate. Adults reach peak abundance in October or November, and overwinter in aquatic habitats. In areas where ponds and lakes freeze, species that typically breed in shallow lentic habitats usually fly to deep lentic habitats or larger streams to overwinter. In late April or early May, they return to their breeding sites to mate and oviposit. Eggs are laid on vegetation or other substrates, usually in the water. Some univoltine species diapause over the winter as eggs.

Table 17.6 Literature References for Identification of Heteroptera Species

Regional keys to larvae that include more than one family
  Alberta, Saskatchewan, and Manitoba—Brooks and Kelton (1967)
  California—Menke (1979a)
  Florida—Chapman (1958)
  Louisiana—Gonsoulin (1973a, 1973b, 1973c, 1974, 1975)
  Minnesota—Bennett and Cook (1981)
  Mississippi—Wilson (1958)
  Missouri—Froeschner (1949, 1962)
  North and South Carolina—Sanderson (1981)
  Virginia—Bobb (1974)
  Wisconsin—Hilsenhoff (1984, 1986)

Keys to larvae listed by family
  Belostomatidae—Menke (1979b)
  Corixidae—Hungerford (1948); (Oregon and Washington) Stonedahl and Lattin (1986)
    Cenocorixa—Jansson (1972)
    Hesperocorixa—Dunn (1979)
    Trichocorixa—Sailer (1948)
  Gerridae—(Arkansas) Kittle (1980); (Oregon and Washington) Stonedahl and Lattin (1982)
    Gerrinae—Drake and Harris (1934), Kuitert (1942)
    Gerris—(Connecticut) Calabrese (1974)
    Rheumatobates—Hungerford (1954)
    Trepobates—Kittle (1977)
  Hebridae—Porter (1950)
  Hydrometridae—Hungerford and Evans (1934); (Florida) Herring (1948)
  Mesoveliidae—Polhemus and Chapman (1979)
  Naucoridae
    Ambrysus—LaRivers (1951)
    Pelocoris—LaRivers (1948)
  Nepidae—Hungerford (1922)
  Notonectidae
    Buenoa—Truxal (1953)
    Notonecta—Hungerford (1933); (New England) Hutchinson (1945)
  Veliidae—Smith and Polhemus (1978); Smith (1980)

Most species are adapted for swimming by having fringes of long setae on their legs, especially the metathoracic legs. They obtain oxygen from air stores held under the hemelytra and from a bubble held ventrally on hydrofuge hairs, which may act as a physical gill or plastron. Air stores are renewed at the surface through air straps in Belostomatidae, air tubes in Nepidae, the pronotum in Corixidae, and the apex of the abdomen in Notonectidae and Naucoridae. In Corixidae, Notonectidae, Pleidae, and Naucoridae, the ventral bubble is large in relation to their size and acts as a physical gill, with oxygen from the water diffusing into the bubble and carbon dioxide from respiration diffusing out of it. This allows prolonged submergence in oxygen-rich water. Almost all species are carnivores as nymphs and adults. They feed mostly on small invertebrates and occasionally on vertebrates such as minnows and tadpoles. The largest family, Corixidae, is an exception, with most species being primarily detritivores.

Nepomorpha are best collected with an aquatic net and sorted by allowing them to crawl from the debris that has been spread on a 0.5-inch (13 mm) mesh screen over a large white pan. Many species feign death and will crawl from the debris only if left undisturbed for several minutes. Additional specimens in all families can often be collected with bottle traps (Hilsenhoff 1987). In northern regions, exceptionally large numbers of both lotic and lentic species can be collected in autumn from along banks of large streams where the current is slow, but some lentic species do not fly to streams and will be missed. Adults of many species are attracted to lights, and may be collected with light traps.

a. *Belostomatidae (3 Genera)*

"Giant water bugs" are easily recognized by their large size and a pair of short, strap-like, posterior respiratory appendages that they use to obtain air at the surface of the water (Fig. 17.88). Two genera (*Belostoma* and *Lethocerus*) inhabit permanent lentic habitats, especially weedy ponds, margins of lakes, and marshes, and are important as apex invertebrate predators. The third genus (*Abedus*) is found in streams among aquatic plants or under rocks in riffles. All are powerful predators that will capture and kill anything they can subdue, including fish, frogs, tadpoles, and other insects. Eggs are laid in large masses above the water on vegetation (*Lethocerus*) or on the backs of the males, which brood the eggs (*Abedus* and *Belostoma*). Belostomatids are univoltine or bivoltine, depending on the latitude, with adults of lentic species in northern regions flying in autumn to larger streams or deep lentic habitats where they overwinter. They return to their breeding sites in early May. Adults frequently fly to lights and are sometimes called "electric light bugs."

b. *Corixidae (17 Genera)*

As the most abundant aquatic Heteroptera, "water boatmen" are found in most permanent aquatic habitats and frequently invade temporary ones as well. Adults most closely resemble Notonectidae, but are dorsoventrally flattened (Fig. 17.83) and have a short, broad, unsegmented, triangular rostrum and widened, scoop-shaped protarsi (palae) (Figs. 17.84 and 17.85). Unlike other aquatic and semiaquatic Heteroptera, which are predators, corixids feed primarily on detritus, algae, protozoans, and other extremely small animals. A few species, however, will capture and eat larger insects such as mosquito larvae. Most species are lentic, inhabiting permanent

ponds and littoral areas of small lakes, but some species are strictly lotic and can be found only along margins of streams or in spring ponds and seeps. In northern latitudes, most lentic species fly to larger streams to overwinter, and in October and November, thousands can often be collected from slow, shallow areas of streams. Adults disperse widely, rapidly invading temporary ponds, and on warm nights they are attracted to lights. Eggs are laid on underwater structures in such great numbers that they are harvested for food in some tropical countries. Most northern species are probably univoltine, with overwintering adults returning to breeding sites in spring to mate and oviposit. In the South, two or more generations per year are likely. There is evidence that at least one species overwinters as an egg.

c. *Naucoridae (5 Genera)*

"Creeping water bugs" are most common in the southern United States and rarely are found as far north as Canada. They are dorsoventrally flattened, have greatly expanded profemora, and have a rounded appearance when viewed from above, with margins of the head, pronotum, and elytra being continuous (Fig. 17.89). They are usually associated with lotic habitats, living in streams, spring ponds, or impoundments. Eggs are glued to vegetation or mineral substrata, depending on the species. Most species probably are univoltine and overwinter as adults in deeper areas of their breeding habitat. Because they carry a large plastron and generally live in well-oxygenated water, they rarely have to surface for air. Although many species have fully developed wings, flight rarely has been observed.

d. *Nepidae (3 Genera)*

"Water scorpions" can be readily recognized by their long apical breathing tubes (Fig. 17.87). They breed in permanent lentic habitats, especially shallow areas with much vegetation. Here, they often remain with the breathing tube above the surface waiting to ambush invertebrates or small vertebrates, but they are capable of remaining completely submerged, using air stores under their hemelytra for respiration. Unlike other aquatic Heteroptera, they are very poor swimmers and usually cling to vegetation or other substrata. Most species are univoltine in northern latitudes and probably are bivoltine in the South. They overwinter as adults, and in the North usually fly to deeper lentic or lotic habitats in autumn and return to breeding sites in late April or early May. Eggs, which are laid on supports at the surface of the water, have respiratory horns that protrude from the water. When flooded, the eggs respire through a plastron.

e. Notonectidae (3 Genera)

"Backswimmers," as their name implies, swim ventral-side-up, but orient themselves dorsal-side-up when not in the water. They most resemble Corixidae, but are deep-bodied, not at all flattened (Fig. 17.91), and have an elongate, segmented rostrum (Fig. 17.86). Most species breed in littoral areas of lakes and permanent ponds, others in stream pools; some species are transients in temporary ponds. Two genera (*Beunoa* and *Notonecta*) can inhabit relatively deep water and remain submerged for long periods because they have ventral channels to retain a plastron. In *Notonecta*, adults and nymphs typically hang from the surface film, while *Buenoa* has hemoglobin, which enables adults and nymphs to maintain neutral buoyancy and remain planktonic. Species in both genera are active predators that pursue or ambush other invertebrates or small vertebrates. A third genus (*Martarega*) that occurs only in the extreme southwestern United States inhabits eddies of streams and ambushes prey carried by the current. Several species overwinter as diapausing eggs and have a univoltine life cycle with nymphal development in late spring and summer. Others have univoltine or bivoltine life cycles, with overwintering adults that mate and lay eggs in the spring. In the North, adults of a few species may overwinter in larger streams and return to lentic habitats in spring.

f. Pleidae (2 Genera)

"Pygmy backswimmers" are very small (< 3 mm) convex insects that swim upside-down (Fig. 17.90). Adults do not have typical hemelytra, but instead their mesothoracic wings are shell-like and resemble elytra of Coleoptera. Their metathoracic wings are vestigial and they cannot fly. They inhabit vegetation, mostly in permanent ponds, but also may be found in littoral areas of lakes, backwaters of streams, and swamps. If their habitat dries because of drought, they can survive for several months in the moist soil. Pleids are widespread and often abundant, except in the Pacific Coast region where they do not occur. Although they can swim quite well, they mostly crawl about on vegetation and feed on small invertebrates. Because of their small body size and relatively large plastron, they are able to remain submerged for long periods. They probably are bivoltine or multivoltine, but their life history is poorly known.

2. Families of Semiaquatic Heteroptera— Suborder Gerromorpha

Adults of semiaquatic species reach peak abundance in September, after which most of them find protected terrestrial sites in which to spend the winter (adults of species that overwinter as eggs are killed by freezing temperatures). Species that overwinter as adults are bivoltine or multivoltine; those that overwinter as eggs usually are univoltine. Eggs of semiaquatic species are laid on substrata at the surface of the water. Nymphs and adults feed mostly on invertebrates of an appropriate size in the surface film. Gerridae, and probably most other Gerromorpha, have sensillae on their femora and trochanters that detect vibrations in the surface film, enabling them to find their prey. In most semiaquatic Heteroptera, apterous and brachypterous adults are common. Apterous adults can be distinguished from nymphs by the presence of genitalia, by greater sclerotization of the body, and by having two tarsal segments instead of one on most legs.

Semiaquatic Heteroptera are adapted for walking on water by having hydrofuge hairs on their tarsi, and in families with larger and heavier species (Gerridae and Veliidae) by additionally having preapical claws. They are most readily collected by sighting them on the water surface and sweeping them from the surface film with an aquatic net.

a. Gerridae (8 Genera)

All of the largest species of semiaquatic Heteroptera are Gerridae, but the family has many diverse sizes, forms, and life histories. "Water striders" are distinguished by their relatively large size, preapical tarsal claws (Fig. 17.93), and elongate metafemora that extend well past the tip of the short abdomen. Apterous and brachypterous forms are common. Although found in all types of aquatic habitats, most species prefer quiet areas where there is minimal wave action and current. Lotic species inhabit streams of all sizes; lentic species inhabit swamps, marshes, permanent and temporary ponds, littoral areas of lakes, and even the ocean. However, most lotic species will occasionally inhabit lentic habitats and lentic species may occasionally be found in lotic habitats. Most species are usually associated with emergent vegetation. Univoltine or bivoltine life cycles prevail, with some species overwintering as adults and some univoltine species overwintering as eggs. All gerrids are predaceous. They feed on terrestrial insects and other invertebrates that become trapped in the surface film or aquatic insects that live just beneath the surface.

b. Hebridae (3 Genera)

Because of their very small size (< 2.5 mm), "velvet water bugs" are likely to be confused only with very small species of Veliidae. When on the water, Hebridae move much more slowly than similarly sized Veliidae, and upon close examination Hebridae have a rostral sulcus, which Veliidae lack. Hebridae inhabit very shallow, sheltered shoreline areas of lentic habitats, including swamps, marshes, ponds, lakes, and river sloughs. Some species are occasionally or even frequently riparian. Apterous forms are common in some species. Adults can be found from spring to autumn, suggesting that they are bivoltine or multivoltine and overwinter as adults.

c. *Hydrometridae (1 Genus)*

The small (8–11 mm), delicate, stick-like "marsh treaders" or "water measurers" with their elongate head are unlike any other aquatic insect (Fig. 17.95). Both macropterous and brachypterous adults are common. They walk on vegetation and the surface film of permanent habitats that lack wave action and current, including marshes, swamps, and vegetated margins of lakes, ponds, and slow streams. Here they feed by using their barbed rostrum to spear small macroinvertebrates in the surface film. They most often are found where fish are absent. They are multivoltine and overwinter as adults.

d. *Mesoveliidae (1 Genus)*

"Water treaders" become abundant in late summer in sheltered lentic habitats and margins of lotic habitats that have algae and duckweed along with emergent vegetation. Their relatively small size (2–4 mm) and the presence of black spines on the legs is distinctive. Apterous forms predominate, but winged forms are not uncommon, especially in autumn. They prey on small terrestrial and aquatic animals in the surface film. Adults and nymphs are killed by freezing temperatures, and in northern areas, mesoveliids overwinter as eggs. All species are probably multivoltine.

e. *Veliidae (5 Genera)*

"Broad-shouldered" or "short-legged water striders" are often abundant in a wide variety of aquatic habitats. Some inhabit streams, living in the vicinity of riffles or under the banks; others inhabit spring ponds, swamps, marshes, ponds, and margins of lakes. They can be distinguished from other semiaquatic Heteroptera, except Gerridae, by their preapical tarsal claws. They differ from Gerridae in their generally smaller size and relatively shorter legs, the metafemora not extending past the tip of the abdomen. Apterous adults predominate, with winged forms rare or absent in the various species. Most species are multivoltine and overwinter as adults, some along margins of streams or springs instead of in terrestrial habitats. A few species overwinter as eggs and are univoltine or bivoltine.

B. *Aquatic Neuroptera—Spongillaflies*

Neuroptera is a fairly large terrestrial order, but the only aquatic larvae are in the single family Sisyridae, with two genera. They live in and feed upon several genera of freshwater sponges (Spongillidae) that occur in permanent lotic or lentic habitats (see Chapter 4).

The small (< 8 mm) aquatic larvae are distinctive. They have sucking mouthparts in the form of a pair of long stylets and are covered with long, spinose setae (Fig. 17.98). Most species are probably multivoltine, with larvae having only three instars. Other life stages are terrestrial. Eggs are usually laid in very small masses, mostly in crevices of vegetation or substrata above aquatic habitats that support freshwater sponges. After an incubation period of about one week, the eggs hatch and larvae drop into the water where they swim or are carried by currents until they encounter a suitable species of sponge. After completing development on the sponge, they swim or crawl to shore, crawl out of the water, and find a suitable pupation site in a crack or crevice or on a flat surface that may be several meters from the water. Here, they spin a double-walled cocoon in which they pupate, the outer wall having a distinctive mesh-like construction. After several days, pharate adults use pupal mandibles to cut their way out of the cocoon and emerge. Adults feed on nectar and apparently live no more than a few weeks. Most species overwinter as larvae on sponges, but some may overwinter as prepupae in the pupal cocoon.

Larvae have no special adaptations for the aquatic environment. They obtain oxygen from the water through their cuticle and swim by repeatedly arching their abdomen forward while in a vertical position and snapping it back. Larvae are rarely collected by standard sampling techniques. Substantial numbers can be collected only by collecting sponges and removing larvae that crawl from them as they dry. Rearing has been accomplished, but a suitable colony of freshwater sponges must be maintained and a terrestrial habitat for pupation must be provided.

Taxonomic keys to larval species can be found in the following publications: Parfin and Gurney (1956), Poirrier and Arceneaux (1972), and Brigham (1981b) (for North and South Carolina).

C. *Aquatic Lepidoptera— Aquatic Caterpillars*

Larvae in several genera of this large, primarily terrestrial order, are associated with aquatic habitats. Some feed on emergent parts of aquatic macrophytes and others mine stems of these plants. Still others feed on submerged parts or plants attached to other substrata, and only these will be considered here. All are in the family Pyralidae (17 aquatic genera) and almost all are in the subfamily Nymphulinae.

Aquatic caterpillars have three pairs of thoracic legs and pairs of prolegs ringed with hook-like crochets on abdominal segments three to seven (Fig. 17.99). They are not likely to be confused with any other aquatic insect larvae, but often it is not possible to distinguish them from Lepidoptera that are inadvertently collected from emergent vegetation. Only those that have numerous filamentous gills covering their body and those that live in portable cases made from the aquatic vegetatation they inhabit can be recognized with certainty as being aquatic. Those that live in cases are so well camouflaged that they often escape notice, even after capture with a net. Some species may become sufficiently abundant such that they have an impact on the aquatic plant community. While adults have been well studied and larvae of many of the species have been reared and described, reliable keys to species of larvae have not been constructed because larvae of many species remain unknown.

Larvae inhabit a wide variety of permanent aquatic habitats. Some inhabit rocks in rapid water of larger streams, where they feed on diatoms and other algae. Others inhabit aquatic macrophytes or duckweed (*Lemna*) in ponds or along margins of streams and lakes. Here they feed on the plants they inhabit and make cases from them. Respiration is cutaneous, and is enhanced by numerous filamentous gills in species that inhabit rocks in streams and also in some lentic case-builders. At least one species has hydrofuge hairs that enable it to maintain a plastron that it uses as a physical gill. Pupation of aquatic species is in a cocoon, usually in the larval habitat. Emergence occurs after pupae swim or crawl to the surface of the water; legs of species that swim have long swimming hairs. Adults are generally nondescript small moths that hold their wings roof-like over their body and remain in the vicinity of their larval habitat. Stream species crawl under the water to oviposit on rocks, while most lentic species deposit rows of eggs just below the surface of the water on the preferred food plant. Most species are univoltine or bivoltine and have five larval instars.

D. Aquatic Coleoptera—Water Beetles

Only about 3% of Coleoptera species have an aquatic stage; but because this is the largest insect order, they represent a significant segment of the aquatic insect fauna. More than 1100 aquatic species in 18 families have been found in North America. There are also several species of riparian Coleoptera that occasionally may be collected while sampling margins of aquatic habitats. In eight families, both larvae and adults are aquatic in almost all species, and in six others either all larvae or all adults are aquatic. A few families with mostly terrestrial species also have some species with larvae that are adapted for living in the aquatic environment.

Three suborders of Coleoptera have aquatic representatives (Figs. 17.100–17.128). Adephaga is represented by five families in which all species are aquatic both as larvae and adults. These families are often referred to as the "Hydradephaga." The suborder Myxophaga is represented in North America by a single aquatic species in which both larvae and adults are aquatic. Several families of Polyphaga have some species with one or more aquatic stages, but only in Hydrophilidae, Hydraenidae, and five families of Dryopoidea do most species have a tie to the aquatic environment.

Most aquatic beetles can be collected with an aquatic net by using standard collecting techniques. However, larger Dytiscidae and Hydrophilidae are ineffectively sampled with nets, and bottle traps must be relied upon to capture substantial numbers of these larger beetles and also species that are nocturnal (Hilsenhoff 1987). Adult Gyrinidae, which mostly swim on the surface in deeper water, are seldom captured without special effort. In late summer, large schools often appear in shallow bays of lakes and streams, and by herding them toward shore and sweeping through the school, large numbers often can be captured. In northern regions where ponds and lakes freeze, lentic as well as lotic gyrinids congregate in streams in the fall to overwinter. Here, very large numbers of mixed species can be collected with a net from undercut banks where there is a current and the water is at least 0.5 m deep. Some overwintering lentic species of Dytiscidae, Haliplidae, and Hydrophilidae also may be found along stream banks in autumn and early spring.

Adult water beetles range from 1–40 mm in length and vary greatly in structure. All are characterized by chewing mouthparts and shell-like mesothoracic wings that are called elytra. They have functional spiracles and rely on atmospheric oxygen for respiration. Adult Adephaga carry a bubble of air under their elytra, which is renewed by swimming or crawling (Amphizoidae) to the surface and breaking the surface film with the tip of the abdomen. This air supply must be renewed frequently, except when water temperatures are cold and metabolism is slowed, or when the water is well oxygenated. Hydroscaphidae (Myxophaga) crawl to the surface to renew their air supply. Elmidae and Dryopidae adults (Polyphaga–Dryopoidea) usually live in oxygen-rich water and maintain a plastron of air that is carried on hydrofuge hairs. The plastron acts as a physical gill, with oxygen for respiration diffusing into it from the water, and carbon dioxide from respiration diffusing out into the water. Other Polyphaga adults also maintain an air supply on hydrofuge hairs, but must renew the air by swimming (most Hydrophilidae) or crawling (Hydraenidae and some Hydrophilidae) to the surface when the water contains too little oxygen for plastron respiration. Unlike Adephaga, adult Polyphaga break the surface film with their antennae.

Larvae of aquatic Coleoptera vary greatly in size and general morphology. All have three pairs of thoracic legs, chewing or biting mouthparts, and are relatively well sclerotized, especially in the later instars. Some have lateral filaments and resemble larvae of Megaloptera. Most larvae of aquatic Coleoptera rely on cutaneous respiration, which may be enhanced by gills, although some larger Dytiscidae and Hydrophilidae larvae obtain oxygen at the surface of the water through functional posterior spiracles. Larvae of Chrysomellidae and Noteridae can use oxygen from plant tissues, and by this mechanism are able to pupate under water. Most Coleoptera larvae lack spiracles in the early instars, but are equipped with a full complement of spiracles in the final instar, which allows them to leave the water and pupate in cells on land.

Adephaga adults (except Amphizoidae) and many Hydrophilidae are adapted for swimming by having long setae or swimming hairs on at least their metathoracic legs or by having greatly flattened legs (Gyrinidae). Some larvae of Dytiscidae are also adapted for swimming by having long setae on their legs, but larvae of most aquatic Coleoptera crawl on vegetation or other substrata. Gyrinidae larvae, which mostly crawl on vegetation, can swim rapidly with an undulating motion when disturbed.

Much of the taxonomy of North American water beetles was completed more than 50 years ago; but there are several genera that need revision, and species still remain to be discovered. Fortunately, adults of most species in all genera can be reliably identified (see references in Table 17.7); however, the distribution of species remains poorly defined because many regions of North America have not been thoroughly studied.

Unlike adults, very few larvae can be identified to species; some cannot even be identified to genus. There are almost no genera in which enough larvae are known at the species level to permit development of reliable keys. Fortunately, the European fauna is much better known and the European litera-

Table 17.7 Literature References for Identification of Coleoptera Species

Regional keys to adults that include more than one family
  California—Leech and Chandler (1956)
  Florida—Young (1954)
  Louisiana—(Dryopoidea) Barr and Chapin (1988)
  Maine—(Adephaga and Hydrophilidae) Malcolm (1971)
  North Dakota—Gordon and Post (1965)
  North and South Carolina—Brigham (1981c)
  Pacific Northwest—Hatch (1953, 1965)

Keys to adults listed by family
  Amphizoidae—Kavanaugh (1986)
  Dryopidae—(United States) Brown (1972)
  Dytiscidae—(Alberta) Larson (1975), (Virginia) Michael and Matta (1977)
    Acilius—Hilsenhoff (1975)
    Agabinus—Leech and Chandler (1956)
    Agabus—Fall (1922a), Leech (1938), Larson (1989)
    Bidessonotus—Young (1954)
    Celina—Young (1979a)
    Colymbetes—Zimmerman (1981)
    Copelatus—Young (1963a)
    Coptotomus—Hilsenhoff (1980)
    Cybister—Young (1954), Leech and Chandler (1956)
    Desmopachria—Young (1981a, 1981b)
    Dytiscus—Larson (1975)
    Graphoderus—Wallis (1939), Tracy and Hilsenhoff (1982)
    Heterosternuta—Matta and Wolfe (1981)
    Hydaticus—Roughley and Pengelly (1981)
    Hydroporus—Gordon (1969, 1981)
    Hydrovatus—Young (1956, 1963b)
    Hygrotus—Anderson (1970, 1975, 1983)
    Ilybius—Larson (1987)
    Laccophilus—Zimmerman (1970)
    Laccornis—Wolfe and Spangler (1985)
    Liodessus—Larson and Roughley (1990)
    Lioporeus—(as *Falloporus*) Wolfe and Matta (1981)
    Matus—Young (1953)
    Neoporus—(as *Hydroporus*) Fall (1923), Wolfe (1984)
    Neoscutopterus—Larson (1975)
    Oreodytes—Zimmerman (1985)
    Potamonectes—(as *Deronectes*) Zimmerman and A. H. Smith (1975)
    Rhantus—Zimmerman and R. L. Smith (1975)
    Sanfilippodytes—(as *Hydroporus*) Fall (1923)
    Uvarus—(as *Bidesus*) Young (1954)
  Elmidae—(United States) Brown (1972)
    Dubiraphia—Hilsenhoff (1973)
    Optioservus—White (1978)
  Gyrinidae—(Minnesota) Ferkinhoff and Gundersen (1983); (Quebec) Morrissette (1979)
    Dineutus—Hatch (1930)
    Gyrinus—Fall (1922b); Oygur (1988)
  Haliplidae—(Minnesota) Gundersen and Otremba (1988); (Virginia) Matta (1976); (Wisconsin) Hilsenhoff and Brigham (1978)
    Haliplus—Wallis (1933)
    Peltodytes—Roberts (1913)

(*continued*)

Table 17.7 (*Continued*)

Hydraenidae—Perkins (1980)
Hydrophilidae—(Canada and Alaska) Smetana (1988);
    (Illinois) Wooldridge (1967); (Pacific Northwest)
    McCorkle (1965), Miller (1965); (Virginia) Matta
    (1974)
    Ametor—Leech and Chandler (1956)
    Anacaena—(Europe) Berge Henegouwen (1986)
    Berosus—Van Tassell (1966)
    Crenitis—Winters (1926)
    Cymbiodyta—Smetana (1974)
    Dibolocelus—Young (1954)
    Enochrus—Gundersen (1978)
    Helophorus—Smetana (1985)
    Hydrobius—Wooldridge (1967)
    Hydrochara—Smetana (1980)
    Hydrochus—Smetana (1988)
    Hydrophilus—Young (1954)
    Laccobius—Cheary (1971)
    Paracymus—Wooldridge (1966)
    Tropisternus—Spangler (1960)
Noteridae
    Hydrocanthus—Young (1985)
    Suphisellus—Young (1979b)

Keys to larvae listed by family
Ptilodactylidae
    Anchytarsus—Stribling (1986)

ture gives us insight into probable life cycles, larval and adult habitats, overwintering sites, feeding habits, and larval taxonomy. Development of keys and descriptions for aquatic Coleoptera larvae remains a pressing need. To accomplish this, larva–adult associations must be obtained either by rearing larvae from eggs laid by identified adults or by rearing field-collected larvae to the better known adult stage. Rearing larvae to adults is difficult because almost all pupae are terrestrial and in any rearing endeavor a pupation site must be provided.

Because of our inability to identify larvae, life histories of most species remain poorly known. This includes our knowledge of adult habits, especially such things as feeding, dispersal, overwintering, and aestivation. Much research is needed before we will understand the bionomics of the many species of aquatic Coleoptera in North America.

Below you will find a brief, biologic discussion of the families of aquatic beetles. For taxonomic keys useful for identifying adult and larval aquatic beetles at the family level, see Sections V.H.1 and V.H.2, respectively.

1. Families of Aquatic Coleoptera— Suborder Adephaga

In the five aquatic families of this suborder, both larvae and adults of all species are aquatic. Eggs are laid singly or in masses, mostly on objects beneath the surface of the water. Larvae have three instars, and except for at least some Noteridae, pupation takes place in terrestrial cells near the larval development site. Almost all first and second instar larvae lack spiracles and rely on cutaneous respiration for oxygen. A few have functional posterior spiracles that they use to obtain oxygen at the surface of the water. Third instar larvae, most of which have functional spiracles on abdominal segments and the mesothorax, also rely primarily on cutaneous respiration until they leave the water to pupate. Life cycles are typically univoltine, although bivoltine and semivoltine life cycles have been recorded for some species. Some species complete larval development rapidly, requiring only a few weeks. Others require several weeks, and some with semivoltine life cycles may need almost two years to complete larval development. Adults are long-lived, with records of unmated beetles living for several years; but typically, most adults live several months and die after their final mating and oviposition, which is usually in spring.

a. *Amphizoidae (1 Genus)*

"Trout-stream beetles" are restricted to mountain streams in northwestern North America. Here both larvae and adults can be found near the edge of the water, crawling on plant roots, wood, or debris at or just beneath the surface of the water. Larvae are characterized by short, broad urogomphi and flattened lateral projections on each segment (Fig. 17.114). Adults are relatively large (12–14 mm long) and broad, with a narrow thorax and no swimming hairs on their legs (Fig. 17.104). Larvae cannot swim and adults swim very poorly; both usually remain in contact with the substratum. Third instar larvae frequently stay in the splash zone and enter the water only to capture prey. Both larvae and adults feed almost exclusively on Plecoptera larvae. Because their habitat is well-oxygenated, adults may remain submerged for prolonged periods, their air bubble acting as a physical gill. The life cycle is one year, with eggs laid on debris in the splash zone in late summer and pupation occurring the following summer, probably in cells that they build in the stream bank.

b. *Dytiscidae (44 Genera)*

The "predaceous diving beetles" are the largest family of water beetles, with more than 500 species distributed throughout North America. Adults range from 2–40 mm in length and are characterized by their ovate appearance, rounded sternum and dorsum, and by a usually spear-shaped prosternal process that projects back to the mesocoxae (Fig. 17.102). The beetles swim by moving both metathoracic legs backward simultaneously. Larvae are generally elongate (Fig. 17.116), and range from 3–60 mm in length at maturity. Most have paired, terminal appendages (urogomphi) of various lengths (Fig. 17.116). They have relatively elongate legs, which in several genera are modified by long hairs for swimming.

Most species are lentic, inhabiting all types of permanent and temporary habitats, and showing a preference for the shallow, vegetated margins of ponds, marshes, bogs, and swamps. They are most abundant in habitats that do not contain significant numbers of insectivorous fish. Most species breed in well-defined habitats, but adults may fly to other habitats to feed or overwinter. Lotic species are found mostly under stream banks or in shallow riffles and tend to remain in the habitat in which they breed.

Typically, adults overwinter, mate in the spring, oviposit, and die. Eggs are laid in or on vegetation beneath the water surface, or on other objects below or just above the surface. The larvae develop over a period of a few to several weeks, then leave the aquatic habitat to pupate in cells that they construct in the soil of protected areas nearby. The pupal stage lasts 5–14 days, after which the adult emerges and usually re-enters the aquatic habitat. Most species are univoltine, but bivoltine life cycles occur, especially in the South, and several northern species that breed in cold-water habitats are semivoltine. However, life histories of most species remain poorly known, and there are many variations in the life cycle just described. There is ample evidence for estivation of adults during dry periods in the summer and also evidence that several northern species overwinter as diapausing eggs. Many lotic species and a few lentic species overwinter as larvae, even in northern areas that freeze. While most overwintering adults are found in ponds, lakes, or streams, some pass the winter in terrestrial habitats and enter their breeding sites late in the spring. Both adults and larvae are predators, feeding primarily on other invertebrates and small vertebrates. Adults of most species are also scavengers.

c. *Gyrinidae (4 Genera)*

Adult "whirligig beetles" are small- to medium-sized beetles (3.5–14.0 mm long). They are widespread and often abundant. Most frequently, they are seen resting or swimming in groups of a few to several hundred on the surface of the water in areas protected from the wind, especially in late summer. However, they can and often do swim under the surface. Their greatly flattened legs and two pairs of eyes are distinctive. Larvae are also distinctive, having fringed projections on each abdominal seg-

ment and posteriorly (Fig. 17.113). They swim with a vertical undulating motion when disturbed.

While most species are lentic, there are several species that are lotic, and most lentic species frequent larger streams in autumn, or in late summer when it is dry. In streams, gyrinids will often escape the current by crawling onto vegetation just above the water. In northern regions, where ponds and lakes freeze, most lentic species fly to larger streams to overwinter. They return to breeding sites in early spring, as soon as air temperatures have warmed enough to permit flight. All species tend to inhabit deeper water than other Hydradephaga, with larvae being found mostly among submerged vegetation in large ponds or rivers, or in littoral zones of lakes or impoundments. Larvae are predators, feeding mostly on other invertebrates, while adults are scavengers on dead animals or predators of small invertebrates in the surface film. Most species are probably univoltine, with adults overwintering and larvae developing during the summer. Eggs are laid mostly on submerged aquatic vegetation. Pupation is in cocoons on emergent vegetation or on terrestrial vegetation adjacent to the larval habitat.

d. *Haliplidae (4 Genera)*

"Crawling water beetles" are small (2.5–5.0 mm long) yellowish to orangish beetles with black blotches and spots, apically narrowed elytra, and a narrowed thorax. They are often abundant in shallow lentic and lotic vegetation-choked habitats. In spite of their common name, they have fringed tarsi and tibiae and are able swimmers. The larvae, which crawl about on vegetation, are either elongate with a long terminal segment (Fig. 17.117) or have numerous long dorsal projections (Fig. 17.118). Life cycles of most species are probably univoltine, with eggs being deposited on or in filamentous algae or other vegetation. Both larvae and adults are herbivores, feeding on algae or macrophytes. Pupation takes place in cells in moist soil near the larval development site. Most species overwinter in the water as adults, but some adults and larvae are known to overwinter in terrestrial sites adjacent to the water.

e. *Noteridae (5 Genera)*

This small family, frequently referred to as "burrowing water beetles," is generally southern in its distribution and rarely found as far north as Canada or in the West. Adults are small (1.2–5.5 mm long) and resemble Dytiscidae, but are usually more attenuate apically than most Dytiscidae and have a truncate instead of a spear-shaped prosternal process (Fig. 17.107). Like Dytiscidae, adults are mostly predators that live in close association with aquatic vegetation. The larvae, which are poorly known, have short legs, short urogomphi, and telescoping body segments (Fig. 17.115). Their morphology suggests that they are omnivores, living among roots of aquatic vegetation. Larvae of a European species obtain oxygen from plant tissues and pupate in a cocoon on the plant roots; the pupae also obtain oxygen from the plant. Life cycles are probably one year with adults overwintering in the water.

2. Families of Aquatic Coleoptera— Suborder Myxophaga

a. Hydroscaphidae (1 Genus)

The only North American species occurs from the southwestern United States north to Idaho. The fusiform shape, truncate elytra, and very small size (1.5 mm) of adult "skiff beetles" is distinctive (Fig. 17.112). Larvae and adults are found most commonly on algae over which a thin film of water flows, but adults have also been collected in deeper water. Both adults and larvae are herbivores, feeding on algae. They tolerate a wide range of water temperatures, and have been found in thermal springs as warm as 46°C and in cold streams that frequently freeze. Females lay single, well-developed eggs on algae, and pupation also takes place on algae at the air–water interface. Larval respiration is aided by distinctive pairs of balloon-like spiracular gills on the prothorax, and first and eighth abdominal segments. Adults, which may remain submerged for long periods, retain an air bubble that acts as a physical gill.

3. Families of Aquatic Coleoptera— Suborder Polyphaga

All or most species in five families of the superfamily Dryopoidea (described in the following Sections 3.a–e) and two other families of the Polyphaga (Hydraenidae and Hydrophilidae; Sections 3.h–i) are aquatic as larvae or adults or both. In the remaining families of Polyphaga that have aquatic representatives (Sections 3.f–g, 3.j–k), the vast majority of species are strictly terrestrial. Additional families of Polyphaga have species that are associated with water, but not adapted for living in it, and they will not be treated here.

Aquatic members of the superfamily Dryopoidea differ from most aquatic beetles by generally having more larval instars, longer life cycles, and longer adult lives in species with aquatic adults. Aquatic adults and larvae are unable to swim, which confines them to well-oxygenated habitats. Aquatic larvae obtain oxygen from the water through their integument, usually with the aid of gills. Aquatic adults have a permanent air bubble (plastron) that is carried on hydrofuge hairs and acts as a physical gill.

Terrestrial adults of species with aquatic larvae typically inhabit the splash zone on rocks that project from their aquatic habitats, and they also use a plastron to remain submerged for relatively long periods when ovipositing. In Dryopoidea that have been studied, five to eight larval instars have been recorded, but there is disagreement about the actual number of larval instars, with six being suggested for many species. In Hydrophilidae and Hydraenidae, there are only three larval instars. Larval development probably takes about two years for most species of Dryopoidea, with one to three year development times having been suggested for various species. The shorter times are for species in warmer and more southern streams. Aquatic adults typically live a year or more, and some may live several years; terrestrial adults of species with aquatic larvae typically live only a few days to a week. To date, studies of life histories have been mostly on Elmidae and Psephenidae; additional studies are needed. Recently, Brown (1987) reviewed the biology of Dryopoidea.

Larvae of Dryopoidea are poorly known, and identification beyond genus is usually not possible. Brown (1972) summarized what was known about the distribution and taxonomy of aquatic Dryopoidea in the United States, and provided species keys for adults and generic keys for larvae. Since this publication, the aquatic Limnichidae have been recognized as a separate family, Lutrochidae. Several new species have been discovered recently, and revisions are needed for some genera.

a. Dryopidae (3 Genera)

In *Helichus,* the most common and widespread genus of "long-toed water beetles," only adults are aquatic, living among debris and on decaying wood in streams. The other two genera are rare, and while adults of both genera are probably aquatic, the aquatic status of their larvae is uncertain. *Pelonomus* occurs in woodland pools or swamps, especially in the South, and *Dryops* is found in streams in the extreme southwestern United States and in Quebec. Larvae and adults resemble those of other Dryopoidea. Adults are larger than the more abundant Elmidae and have a pectinate antenna (Fig. 17.110). Larvae are elongate and lack gills in the operculate caudal chamber. Little is known about life histories of species in this family.

b. Elmidae (26 Genera)

Commonly referred to as "riffle beetles," Elmidae are widespread and often abundant. Both larvae and adults are usually aquatic and often occur together; in a few species, adults are riparian. Adults are less than 4.5 mm long, smaller than Dryopidae, and have filiform or slightly clubbed antennae (Fig. 17.111). The sclerotized larvae are elongate, rounded in cross-section, and have a ventral caudal operculum that closes a chamber containing hooks and numerous filamentous gills (Fig. 17.126). Both adults and larvae are found mostly in streams, where they inhabit a variety of substrata, including gravel riffles, algae-laden rocks, aquatic macrophytes, and decaying wood. Some species may also occur on these same substrata in spring ponds or along wave-swept shores of lakes. Larvae and adults are herbivore–detritivores, feeding on algae, decaying wood, and detritus. When larvae complete their development, they leave the water and pupate in cells in protected areas on the adjacent shore. Upon emergence, adults disperse widely and frequently are captured in light traps. Once they find a suitable aquatic habitat, they rarely if ever fly again, but stream species may move downstream by drifting in the current. Most adults probably live a year or more, and a semivoltine life cycle seems probable for most species.

c. *Lutrochidae (1 Genus)*

Larvae of Lutrochidae occur on calcareous encrustations in hard-water streams in eastern, central, and southwestern United States. They closely resemble larvae of Elmidae, but the last abdominal tergum is rounded apically (Fig. 17.128) instead of being emarginate (Fig. 17.127), and thoracic sterna are membranous. They apparently feed on periphyton, but little is known about the bionomics of this family. Adults are riparian, inhabiting splash zones of larger rocks that project from the aquatic habitat.

d. *Psephenidae (6 Genera)*

"Water pennies" are flat, rounded larvae (Figs. 17.121 and 17.122) that occur on rocks and wood in streams from Texas and Arizona north to British Columbia, and east of the Great Plains; they may also be found on wave-swept shores of lakes. In two genera (*Eubrianax* and *Psephenus*), larvae (Fig. 17.122) lack the ventral caudal operculum of other aquatic dryopoid larvae, but have ventral abdominal gills that aid respiration. The remaining genera are in the subfamily Eubriinae, which is sometimes considered a separate family. Their larvae lack exposed ventral gills, possess a ventral caudal operculum, and have the lateral margins of each segment distinctly separated. Pupation of most species occurs above the water line on rocks near the larval habitat, but at least one species is known to pupate under water. Adults are riparian, typically being found in splash areas of projecting rocks. Most species probably have life cycles of one or two years.

e. *Ptilodactylidae (3 Aquatic Genera)*

Larvae have been collected from streams and springs in California and Nevada, and from Georgia to New York, but not in central North America. They are elongate, lack a ventral caudal operculum, and have gills on either the first seven abdominal sterna or the ninth sternum (Fig. 17.125). Very little is known about their life histories. Adults of aquatic species are riparian.

f. *Chrysomellidae (2 Genera)*

In this very large terrestrial family, there are several species in which larvae and adults feed on emergent aquatic vegetation, but there is only one subfamily in which aquatic larvae feed on submerged portions of aquatic plants. They are adapted for underwater existence by having a pair of caudal spiracular spines that they insert into plant tissues to obtain oxygen, and they can be recognized by these conspicuous spines (Fig. 17.124). Larvae pupate on underwater portions of plants, with pupae also obtaining oxygen from plants for respiration.

g. *Curculionidae (15 Genera)*

In this very large terrestrial family, adults of species in several genera feed on aquatic plants, both above and below the surface of the water, and there is no easy way to distinguish truly aquatic species. Adult Curculionidae are readily distinguished by their unique head, which is elongated into a snout, and by their geniculate antennae (Fig. 17.100). Larvae are grub-like, without distinct legs; most mine leaves and other plant tissues and are not aquatic. Aquatic adults crawl on underwater portions of plants and carry with them a plastron for respiration; a few species are capable of swimming. Most aquatic species are nocturnal and difficult to find during the day, when they apparently hide in the bottom substrata. Life histories are poorly known.

h. *Hydraenidae (4 Genera)*

Hydraenidae and Hydrophilidae are members of the superfamily Hydrophiloidea. Adult "minute moss beetles" (Hydraenidae) are aquatic, living mostly in extremely shallow water along the banks of streams, but also occurring along margins of lentic habitats. One genus (*Neochthebius*) is riparian and lives in intertidal areas. Adults resemble Hydrophilidae and once were considered part of that family. They are much smaller than most Hydrophilidae and have five segments in the antennal club instead of three. Larvae are riparian, developing and pupating in moist soil adjacent to the water. Adults cannot swim, and float upside-down in the surface film when dislodged from the substrata on which they live. Because of their small size (< 2 mm), they are easily overlooked. Their life history and feeding habits are poorly known.

i. *Hydrophilidae (23 Aquatic Genera)*

Adults and larvae of this large and abundant family mostly inhabit shallow lentic habitats, although a few species are found in deep water and others are strictly lotic. In some genera, only adults are aquatic; in one subfamily (Sphaeridiinae), none of the life stages are aquatic. While commonly called "water scavenger beetles," all aquatic larvae are predators and adults feed on algae and perhaps small animals, as well as scavenging. Adults range in size from 1.5–40 mm and somewhat resemble Dytiscidae, but are flat ventrally, have club-shaped antennae, and usually have very long maxillary palpi (Fig. 17.103). The predatory larvae have large, conspicuous mandibles and relatively long antennae and maxillary palpi (Fig. 17.119).

Adults of many genera mostly crawl on aquatic vegetation and have no adaptations for swimming. In genera with adults that swim, beetles have fringes of long setae on their meso- and metathoracic legs, which they move alternately instead of in unison as in the Dytiscidae. While most species can be readily collected with an aquatic net, larger species that swim rapidly are most effectively collected with bottle traps. Most species are univoltine, with adults being the normal overwintering stage. Adults of some species overwinter in terrestrial habitats, while others that inhabit shallow lentic habitats may fly to streams or larger lentic habitats in late summer and autumn to overwinter. Adults are often attracted to lights, and when they are dispersing on warm nights, large numbers may be collected in light traps.

j. *Lampyridae (1 Genus)*

Larvae of at least one species have been collected from shallow lentic and semilotic habitats. They are readily recognized by the flat, plate-like thoracic nota that mostly or completely conceal the head from above (Fig. 17.123). The larvae are predaceous, feeding on snails or other insects.

k. *Scirtidae (Helodidae; 7 Genera)*

Although adults are terrestrial, larvae of "marsh beetles" are aquatic and often common in marshes, swamps, margins of shallow ponds, tree holes, and a variety of other lentic habitats where they crawl about on the aquatic vegetation. Their elongate filiform antennae are unique among aquatic Coleoptera (Fig. 17.120). They are apparently detritivores, with little being known about their life history.

E. Aquatic Diptera—Flies and Midges

Diptera is a large, mostly terrestrial order, but larvae and pupae of numerous species are aquatic and it is the dominant order of insects in the aquatic environment (Figs. 17.129–17.153). Diptera inhabit all types of aquatic environments; because of very short development periods in many species, they are able to utilize ephemeral aquatic habitats as well as permanent ones. About 40% of all species of aquatic insects belong to this order, and at least a third of the aquatic Diptera species are in one family, Chironomidae. It is difficult to estimate the number of species of aquatic Diptera because in most families only the terrestrial adults can be identified, and in many families it is not possible to know which species have aquatic larvae. It is also difficult to estimate the relatively large number of aquatic species that probably remain undiscovered and undescribed.

The order is divided into two suborders, Nematocera and Brachycera (Agriculture Canada 1981), with Nematocera dominating the aquatic fauna. Formerly, a third suborder, Cyclorrhapha, was recognized; it is now included in Brachycera as the infraorder Muscomorpha. In Nematocera, there are several families in which all or almost all species have aquatic larvae and pupae. In Brachycera, most families that have species with aquatic larvae also have large numbers of terrestrial or semiaquatic species. Larvae of the infraorder Muscomorpha are often referred to as cyclorrhaphous Brachycera, with the remaining Brachycera being called orthorrhaphous Brachycera.

Dipteran larvae lack segmented thoracic legs, and thus can be readily distinguished from larvae of other aquatic insects. In Nematocera, a completely sclerotized, relatively round head capsule is present (except in most Tipulidae), and the mandibles, which usually have subapical teeth, move laterally. In Brachycera, the head is absent (cyclorrhaphous Brachycera) or poorly formed and not rounded (orthorrhaphous Brachycera), and the mandibles (mouth hooks) move vertically and lack subapical teeth.

Dipteran larvae have many adaptations that allow them to live in a wide variety of environments. Those that live in oxygen-rich environments usually rely on cutaneous respiration, which may be aided by gills that increase the surface area. In Chironomidae, some species have a hemoglobin-like pigment that aids in oxygen storage. Larvae in many families obtain oxygen at the surface of the water, aided by short or elongate caudal respiratory siphons that reach to the surface and may be retractile. Other larvae have fringes of hairs surrounding caudal spiracles that allow them to float in the surface film with their caudal spiracles exposed to the air. Some larvae repeatedly swim to the surface to obtain oxygen (Culicidae). Other larval adaptations include papillae for regulation of salt concentration, sucker disks to anchor lotic forms in rapid currents, and prolegs, pseudopods, or creeping welts on various segments to aid in locomotion. Adaptations of larvae in the various families are next discussed.

Pupae of Nematocera often have anterior respiratory horns that enable them to obtain oxygen at the surface; others rely on cutaneous respiration. A few swim actively to the surface (Culicidae) or pupate just above the water line (Dixidae). Most orthorrhaphous Brachycera pupate on land. Cyclorrhaphous Brachycera and Stratiomyidae pupate in a puparium, which may be terrestrial or float at the surface where oxygen is plentiful.

Aquatic Diptera larvae can be readily captured with an aquatic net as described earlier, and when placed in a pan with some water will usually swim or crawl free of the debris, allowing them to be collected. Active Nematocera larvae are difficult to capture with forceps, and the use of a small loop (30 mm diameter) of nylon mesh is often advantageous. Such a loop can be made by bending the end of a piece of thin wire into a loop at a right angle to the remainder of the wire (the handle) and gluing a piece of nylon stocking to the loop. Dipteran larvae are relatively easy to rear because most have short life cycles and pupate in the water. Lentic species, which require little or no aeration, are especially easy to rear.

Diptera are not only a very important part of almost all aquatic ecosystems, but adults in many aquatic families are important because they transmit disease or create severe nuisance problems. Culicidae (mosquitoes) are the most important group from a public health standpoint because they transmit diseases of humans, birds, and other animals. This family, along with Tabanidae (horse flies and deer flies), Simuliidae (black flies), and Ceratopogonidae (biting midges) may create severe nuisance problems because they bite humans and other animals. Nonbiting midges in the families Chironomidae and Chaoboridae may emerge in such tremendous numbers from larger lakes that they damage property and create public health problems by causing severe allergic reactions in some people.

Much research is needed on aquatic Diptera in North America. Many species remain to be described, and larvae of most described species remain unknown. Only in families that are of public health importance (Culicidae, Tabanidae, Simuliidae) or

useful as a biological control (Sciomyzidae) are larvae well enough known that reliable keys to most species have been developed. Even in these families there are taxonomic problems, especially with separation of sibling species. Great strides in larval tax-

Table 17.8 Literature References for Identification of Diptera Species

Regional keys to larvae that include more than one family
  California—Wirth and Stone (1956)
  North and South Carolina—Webb and Brigham (1981)

Keys to larvae listed by family
  Chaoboridae—Cook (1956)
    Chaoborus—Saether (1970)
    Chaoborus—*(Schadonophasma)* Borkent (1979)
  Chironomidae—References since Simpson (1982)
    Ablabesmyia—Roback (1985)
    Brillia—Oliver and Roussel (1983)
    Cricotopus—Simpson *et al.* (1983)
    Dicrotendipes—Epler (1988)
    Endochironomus—Grodhaus (1987)
    Endotribelos—Grodhaus (1987)
    Eukiefferiella and *Tvetenia*—Bode (1983)
    Guttipelopia—Bilyj (1988)
    Labrundinia—Roback (1987)
    Monopelopia—Roback (1986a)
    Nilotanypus—Roback (1986b)
  Orthocladiinae—Cranston *et al.* (1983)
    Pagastia—Oliver and Roussel (1982)
    Polypedilum—Boesel (1985)
    Stenochironomus—Borkent (1984)
    Synendotendipes—Grodhaus (1987)
    Tribelos—Grodhaus (1987)
  Culicidae—Carpenter and LaCasse (1955), Darsie and Ward (1981); (Canada) Wood *et al.* (1979); several regional keys published prior to 1981.
  Deuterophlebiidae—Courtney (1990)
  Sciomyzidae
    Dictya—Valley and Berg (1977)
    Elgiva—Knutson and Berg (1964)
    Hoplodictya—Neff and Berg (1962)
    Pherbella—Bratt *et al.* (1969)
    Sciomyza—Foote (1959)
    Sepedon—Neff and Berg (1966)
    Tetanocera—Foote (1961)
  Simuliidae—(Canada) Wood *et al.* (1963); (Michigan) Merritt *et al.* (1978); (New York) Stone and Jamnback (1955); (Southeastern United States) Snoddy and Noblet (1976)
    Gymnopais and *Twinnia*—Wood (1978)
    Prosimulium—Peterson (1970)
  Tabanidae—(Arizona) Burger (1977); (Eastern Nearctic) Goodwin (1987); (Eastern North America) Teskey (1969), Teskey and Burger (1976); (Illinois) Pechuman *et al.* (1983); (Louisiana) Tidwell (1973)
  Tipulidae
    Tipula—Gelhaus (1986)

Identification of species of Collembola
  Christianson and Bellinger (1980), Waltz and McCafferty (1979)

onomy have been made in recent years, especially in the largest family, Chironomidae, but until much more is accomplished it will not be possible for us to fully understand the role of Diptera in aquatic ecosystems. Recent publication of the two volume "Manual of Nearctic Diptera" by Agriculture Canada (1981, 1987) provides the most recent generic keys as well as information about morphology, biology, behavior, classification, and distribution. Consult Table 17.8 for a list of references to species keys. Information on Diptera families is given below, and a taxonomic key to families can be found in Section V.I.

1. Families of Aquatic Diptera— Suborder Nematocera

More than half of the families in this suborder have aquatic representatives, and in 12 families, all or almost all species have aquatic larvae and usually also aquatic pupae. Unless noted otherwise, larvae have four instars, respiration is through the integument, and pupation is in the larval habitat. Some generic names proposed by Meigen in 1800 and family names based on these genera were used in North America from about 1950 until 1963, when they were suppressed by the International Commission on Zoological Nomenclature. Family names based on Meigen [1800] genera are included in parentheses.

a. *Blephariceridae (5 Genera)*

The distinctive flattened larvae of "net-winged midges" have seven apparent body segments and a sucker disc ventrally on the first six segments (Fig. 17.129). They use their sucker discs to cling to rocks in very fast water of western and eastern mountain streams, and also streams in the northern United States and Canada. Here, they feed on diatoms and other algae. Larvae move forward slowly with an undulating motion, or sideways by alternately moving anterior and posterior sucker discs. Pupation occurs in cracks and depressions of rocks near the surface of the water. Adults live one to two weeks, with some females being predators on other midges. Eggs are laid just above the water on rocks, usually when water levels are receding. They hatch when flooded by higher water levels.

b. *Ceratopogonidae (Heleidae; 13 Aquatic Genera)*

Aquatic larvae of "biting midges" typically are small, elongate, and without prolegs (Fig. 17.145). While many species are riparian or live in moist terrestrial habitats, a large number of species occur in aquatic habitats that include tree holes, marshes, swamps, ponds, all areas of lakes, and streams. Most larvae are able swimmers and move about with a serpentine swimming motion. They are poorly known and most have not been associated with adults, making it difficult even to know which species develop in aquatic habitats. Reliable keys to North American genera have not been developed, although a key to species in Russia (Glukhova 1977) includes most genera found in North America. Most larvae are carnivores; others are herbivores or detritivores. Pupae of aquatic species hang in the surface film by their respiratory horns. Adults of a few aquatic species bite people and may become an annoying pest in some areas; most others feed on small insects. Kettle (1977) reviewed the biology and ecology of blood-sucking Ceratopogonidae.

c. *Chaoboridae (4 Genera)*

Most larvae are nearly transparent except for hydrostatic organs, causing them to be called "phantom midge larvae." They are unique in that their enlarged antennae have been modified for capturing prey such as insect larvae and small crustaceans (Fig. 17.142). They occur in a wide variety of lentic habitats, and are especially abundant in the profundal or sublittoral zones of some lakes. In lakes, third and fourth instar larvae rest in the bottom mud during daylight hours and prey on zooplankton in the limnetic area at night; earlier instars remain limnetic. They are the only insects frequently found in the limnetic area of lakes. Larvae also occur in permanent ponds, and may be found in spring ponds, vernal ponds, and margins of swamps. Adults do not feed, but their synchronized emergences may create severe nuisance problems around large lakes because adults are highly attracted to lights. Life cycles are univoltine to multivoltine, depending on species, climate, and habitat. Most larvae can be identified to species.

d. *Chironomidae (Tendipedidae; 161 Genera)*

Chironomidae is by far the largest family of aquatic insects, with the number of aquatic species surpassing that in most other orders. The larvae, which are recognized because they usually have anterior and posterior pairs of prolegs (Fig. 17.152), are diverse in form and size. They inhabit all types of permanent and temporary aquatic habitats and a few species inhabit semiaquatic or terrestrial habitats. Larvae are often the dominant insect in the profundal and sublittoral zones of lakes, and consequently adults are sometimes called "lake flies," although they are most often referred to as "midges." Species in larger lakes may emerge in such tremendous numbers that they create nuisance problems. The short-lived adults cause allergies in some people, invade factories, spot the paint on houses, and accumulate in large, odorous piles.

Larvae are an extremely important part of aquatic food chains, serving as prey for many other insects and food for most species of fish. Life cycles are highly variable, with many species being univoltine and many others being bivoltine or multivoltine. In the north, life cycles are longer, with some Arctic species requiring several years to complete development. Feeding habits also vary widely, with herbivore–detritivores and carnivores all commonly represented. Many larvae, especially predators, are free living, but most species construct loose cases of substrate cemented together with salivary secretions. Most herbivores and detritivores graze on fine particles on the substratum, but some are filter feeders that construct webs to filter water they circulate through their case or retreat. Larvae of most species are quite tolerant of lowered levels of dissolved oxygen; some can survive in areas where oxygen levels are so low that oxygen cannot be detected. Such species are usually red and contain a hemoglobin-like pigment that retains oxygen. These "blood worms" may become abundant in sewage lagoons or organically polluted areas of lakes or streams. The pupal stage is short, with pupae of most species being quiescent. After completing development, the pupa swims to the surface of the water where emergence occurs. Adults do not feed and generally live no more than two weeks. The biology of the family was recently reviewed by Pinder (1986).

In recent years, many new species have been named and great progress has been made in rearing and identifying larvae and pupae, but larvae of too many species remain unknown to permit reliable species keys to be constructed for larvae in almost all genera. A bibliography of taxonomic literature was compiled by Simpson (1982), and only keys to species or species groups that were not included in the bibliography by Simpson are listed in Table 17.8.

e. *Culicidae (13 Genera)*

Because they bite people and transmit disease, "mosquitoes" have been studied more thoroughly than any other family of aquatic Diptera. All species can be identified as adults or larvae, and the life histories and ecology of most species have been documented. Larvae are readily recognized by their swollen thoracic area (Fig. 17.141), the presence of a caudal respiratory tube (siphon) in most species, and their characteristic flip-flop swimming motion. Because of this swimming motion they are commonly called "wrigglers." Larvae occur in a variety of shallow lentic or semilotic habitats that include tree holes, artificial containers, catch basins, pitcher

plants, swamps, shallow temporary or permanent ponds and marshes, and heavily vegetated margins of lakes and streams. They are not found in moving water or water subjected to wave action. Larvae obtain oxygen at the surface of the water through a caudal siphon, which is very abbreviated in *Anopheles*. Those of *Coquillettidia* and *Mansonia* attach their spine-like siphon to aquatic plants to obtain oxygen. Most larvae are active swimmers and feed on detritus and microorganisms; larvae in two genera are predators, mostly on other mosquito larvae. Pupae are also active swimmers, and obtain oxygen at the surface through thoracic respiratory horns.

Many species are univoltine, with eggs that diapause over the winter and larval development and emergence in the spring. Many others are bivoltine or multivoltine. Although most species overwinter as eggs, a few overwinter as larvae and several species overwinter as adults. Most female adults obtain blood from mammals, birds, reptiles, or amphibians, depending on the species, and may live several weeks or months. Male mosquitoes do not feed on vertebrates, but instead obtain nectar from plants. Culicidae often dominate the insect fauna of temporary ponds and marshes, especially those that flood in spring and summer.

f. *Deuterophlebiidae (1 Genus)*

The unique larvae of "mountain midges" are less than 6 mm long and readily recognized by their elongate, branched antennae and the presence of seven pairs of stout ventrolateral prolegs on the abdomen (Fig. 17.147). They are found in rapid currents of western mountain streams. Here they inhabit the upper surface of light-colored, smooth rocks that have cracks or depressions, and are usually at or near the surface of the water. Pupation is in the same habitat, usually in depressions in darker-colored rocks. In cold water at higher elevations, there is one generation per year, while at lower elevations there may be several generations. Eggs within the pupa are mature, and when adults emerge, mating and oviposition occur almost immediately. Adults live only an hour or two.

g. *Dixidae (3 Genera)*

Larvae of "dixid midges" are found at or just below the surface of the water in lentic habitats or densely vegetated margins of lotic habitats. Here they characteristically lie bent in a U-shape, often resting on vegetation. They can be recognized by pairs of prolegs on the first one or two abdominal segments (Fig. 17.148). Larvae obtain oxygen from the air through caudal spiracles and feed on microorganisms and detritus in the surface film. Pupation occurs just above the water surface on vegetation. Life histories are poorly known; most species are probably multivoltine.

h. *Nymphomyiidae (1 Genus)*

The small, slender larvae are recognized by pairs of ventrally projecting, slender, two-segmented prolegs on abdominal segments 1–7 and 9 (Fig. 17.146). They are rare, and in North America have been found only in Quebec, New Brunswick, and Maine. Larvae occur mostly on moss-covered rocks in small, rapid, cold, spring-fed streams, where they apparently feed on diatoms and other algae. The only North American species is probably bivoltine, with emergences in spring and late summer.

i. *Psychodidae (4 Aquatic Genera)*

Most species of "moth flies" develop in semiaquatic or moist terrestrial habitats, but larvae of several species occur in aquatic habitats. Aquatic larvae can be recognized by their small size and secondary annulations on thoracic and abdominal segments, which often also have sclerotized dorsal plates on many annuli (Fig. 17.144). In one aquatic genus (*Maruina*), the larvae have ventral sucker-like discs. Aquatic larvae occur among vegetation and debris in shallow lentic habitats and along margins of streams, and are frequently associated with organically polluted water. They even occur in sink and floor drains. Larvae feed near the surface on microorganisms and detritus. Life histories of aquatic species are not well known; most are probably multivoltine, especially in warm-water habitats. Larvae cannot be identified to species.

j. *Ptychopteridae (Liriopeidae; 3 Genera)*

Larvae of "phantom crane flies" are easily recognized by their elongate caudal respiratory siphon and the presence of pairs of prolegs on the first three abdominal segments (Fig. 17.149). They live in very shallow water along margins of lentic or lotic habitats where there is an accumulation of detritus. Here they obtain oxygen at the surface through their caudal respiratory siphon and probably feed on detritus and associated microorganisms. Their life history is poorly known because they are generally uncommon.

k. *Simuliidae (11 Genera)*

"Black fly" larvae are uniquely shaped, having a swollen abdomen that they attach to the substratum with a caudal sucker (Fig. 17.150). They also have a relatively large head, and a single ventral proleg on the thorax. They inhabit a wide variety of lotic habitats where they are often abundant on rocks, submerged wood, or vegetation in fast to slow currents, each species usually having rather specific ecological requirements. Most larvae are filter feeders, using labral fans to filter diatoms and other food from the current. Larvae without labral fans feed by grazing on the substrata around them. Many species are univoltine, while others are bivoltine or multivoltine. Species overwinter as either eggs or larvae, with larvae of some species growing significantly during the winter months. Unlike most other Nematocera, which have only four larval instars, simuliid larvae have six or sometimes seven instars. Pupation occurs at the site of larval attachment, with pupae having a pair of highly branched thoracic gills that aid in respiration. Females usually require a blood meal for maturation of eggs and various species may be serious pests, biting humans, livestock, and poultry. They also transmit pathogenic organisms such as filarial nematodes, various viruses, and avian protozoan parasites.

l. *Tanyderidae (2 Genera)*

The elongate larvae of "primitive crane flies," which may reach a length of 18 mm, are rare in eastern and western North America and have not been found in central North America. They resemble chironomid larvae, but lack anterior prolegs and have three pairs of long caudal filaments (Fig. 17.151). Larvae inhabit the shallow sand, silt, or gravel margins of large streams but because they are rare, their life history and feeding habits remain virtually unknown.

m. *Thaumaleidae (2 Genera)*

The very rare "solitary midge" larvae resemble chironomid larvae and reach a length of 12 mm, but have a single, broad anterior and posterior proleg instead of paired prolegs (Fig. 17.153). They also have dorsally sclerotized segments and a pair of short, stalked spiracles dorsolaterally on the prothorax. Larvae inhabit vertical surfaces of rocks in cold, shady, mountain streams, where the water film is thin enough to not cover them completely. Here they feed mostly on diatoms. Almost nothing is known about their life history.

n. *Tipulidae (23 Aquatic Genera)*

Larvae of "crane flies" differ from those of other Nematocera by having the posterior portion of the head capsule incompletely sclerotized and retracted into the thorax (Figs. 17.130 and 17.131). In this, the largest family of Diptera, the vast majority of species are not aquatic, but there are several genera with species that have aquatic larvae, and these species are often widespread and abundant. Because larvae of fewer than 10% of the species have been described, it is difficult to know which species have aquatic larvae. While most aquatic larvae are found in lotic habitats, larvae in a few genera inhabit shallow lentic habitats. Most aquatic species have functional caudal spiracles, which they may utilize to obtain oxygen at the surface of the water, but those that live in well-aerated streams probably obtain most of their oxygen from the water through their cuticle. Larvae in some genera are very large. Feeding habits vary widely, with omnivores, herbivores, detritivores, and carnivores all represented. Most species are univoltine or bivoltine, but some that live in very cold water may be semivoltine. Pupation most often occurs in riparian habitats adjacent to the larval habitat, but some species pupate in shallow water and have thoracic respiratory horns through which they breathe. Adults are short-lived. The biology of Tipulidae was recently reviewed by Pritchard (1983).

2. Families of Aquatic Diptera—
Suborder Brachycera

In larvae of Brachycera, either the head and external head structures are replaced by an internal cephalopharyngeal skeleton (cyclorrhaphous Brachycera), or head structures are greatly reduced and not rounded and sclerotized as in Nematocera (orthorrhaphous Brachycera). Only in Stratiomyidae is there any significant head structure, but heads of Stratiomyidae larvae are angulate, partially retracted into the thorax, and quite unlike heads of Nematocera. Mandibles of Brachycera larvae move vertically, not horizontally as in Nematocera. They lack subapical teeth and are usually referred to as "mouth hooks."

The majority of species in all families except Athericidae and Sciomyzidae are terrestrial or semiaquatic, but it usually is not possible to know which species are aquatic because larvae of most species of Brachycera remain unknown. Furthermore, some genera that are known to have aquatic species also have terrestrial or semiaquatic species, so all larvae within a genus cannot be assumed to be aquatic because larvae of one or more species have been found to be aquatic. Because most larvae cannot be identified, life histories are poorly known.

a. *Athericidae (Formerly Included in Rhagionidae; 2 Genera)*

Aquatic larvae of *Atherix* are distinctive, with pairs of ventral abdominal prolegs on the first seven ab-dominal segments and with a single ventral proleg and a pair of fringed caudal projections on the eighth segment (Fig. 17.138). Adults of the other genus (*Suragina*) have been collected in Texas, but larvae are unknown in North America and possibly are not aquatic. *Atherix* larvae commonly occur in riffles or among vegetation in streams where they feed mostly on insect larvae. They pupate in moist soil along the edge of the stream. Adults are known to drink water and perhaps feed on nectar; adult females of *Suragina* suck blood from humans and cattle. Eggs of *Atherix* are laid on vegetation or supports above the stream, with several females often contributing eggs to a single mass. Larvae drop into the stream upon hatching. A one-year life cycle is likely for most species.

b. *Dolicopodidae (About 8 Aquatic Genera)*

The "long-legged flies" are mostly terrestrial, but larvae of a few species in eight genera are known to develop in a wide variety of lotic and lentic habitats. The taxonomy of larvae is so poorly known that identification to genus is not reliable and it is not possible to clearly define which species are aquatic. Aquatic larvae differ from other Brachycera by having a posterior spiracular pit that is surrounded by four lobes (Fig. 17.137). They are predaceous, feeding mostly on other insect larvae. Pupae differ from other orthorrhaphous Brachycera by having thoracic respiratory horns.

c. *Empididae (About 8 Aquatic Genera)*

The small (< 7 mm) larvae of "dance flies" are too poorly known to estimate the number of aquatic genera and species. Most species are terrestrial or semiaquatic. Most aquatic larvae can be recognized because they have seven or eight pairs of abdominal prolegs and a pair of caudal respiratory appendages (Fig. 17.139), which are not elongate and fringed as in Athericidae. Larvae are predaceous on smaller animals, and most aquatic species live in rapid water of streams. Because larval taxonomy is so poorly known, almost nothing is known about their life history.

d. *Ephydridae (About 65 Genera)*

There are about 65 North American genera, most of which are semiaquatic. Some, however, have aquatic species that live in vegetation or detritus in shallow habitats near shore, in algal mats, or in unusual habitats such as alkaline lakes and ponds, hot geyser pools, and ponds containing oil. Other species are terrestrial or mine leaves of aquatic plants. "Shore fly" larvae are less than 12 mm long and usually have a pair of very short to elongate caudal respiratory tubes that frequently have a black sclerotized tip (Fig. 17.134). Many also have abdominal prolegs. Most larvae feed on diatoms or other algae, others feed on detritus. Aquatic species pupate in a puparium that usually floats on the water surface. Eggs are laid singly at the surface of the aquatic habitat. Although the family has received a great deal of attention, larval taxonomy is still poorly Known and identification even at the generic level is difficult or impossible.

e. *Muscidae (About 8 Aquatic Genera)*

In this extremely large terrestrial family, larvae of several aquatic species live in streams, marshes, or ponds; they are mostly predators. Larvae are poorly known, but most have a pair of very short caudal respiratory tubes, a pair of ventral prolegs on the last abdominal segment that are longer than the respiratory tubes, and often creeping welts ventrally on other abdominal segments (Fig. 17.140). Eggs are laid in algae or debris, and larvae usually remain near the surface. Pupation is in a puparium, which remains at or near the surface. A pair of respiratory siphons that are thrust through the fourth segment of the puparium penetrate the surface of the water.

f. *Sciomyzidae (Tetanoceridae; About 16 Aquatic Genera)*

"Marsh fly" larvae are predators or parasitoids of snails, slugs, or fingernail clams, and only those species that attack land snails or slugs are not aquatic. They generally have a very wrinkled appearance, with girdles of pseudopods on abdominal segments. The abdomen terminates in a spiracular disc surrounded by eight to ten lobes and contains a pair of spiracles, each surrounded by numerous palmate hairs (Fig. 17.135). Larvae are found in lentic habitats and margins of lotic habitats inhabited by their prey. Here, they usually remain near the surface with their caudal spiracles at the surface. They swallow a large air bubble to keep themselves afloat, and are usually buoyant enough to keep their prey at the surface. Larvae pupate in a puparium that either remains in the snail shell or floats at the surface. Eggs are laid on emergent vegetation; newly hatched larvae drop into the water and actively search for prey. Adults of species that are parasitoids oviposit directly on the shell of a host snail. Species may be univoltine or multivoltine. Because snails are intermediate hosts of both liver flukes that infect cattle and of schistosomes that attack humans, sciomyzid

larvae have been employed as a biological control for snails. For this reason, they have been extensively studied in recent years and larvae in several genera can be identified to species. Berg and Knudson (1978) reviewed the biology and systematics.

g. *Stratiomyidae (About 10 Aquatic Genera)*

Although most "soldier fly" larvae are terrestrial or semiaquatic, there are many species that inhabit shallow, vegetated lentic habitats. While often quite abundant, they are so poorly known as larvae that identification at even the generic level is not always reliable. The relatively flat larvae are easily recognized by their mostly visible truncate head (Fig. 17.132), caudal spiracles surrounded by long hydrofuge hairs, and the calcium carbonate crystals that cover their integument. Larvae hang suspended in the surface film by hydrofuge hairs that surround the spiracular disc and use their palpi to move about. They may at times leave the water and inhabit moist shoreline debris or plants. They feed on detritus and also algae, differing from larvae of other orthorrhaphous Brachycera, which are predators. Pupation is in a puparium, which may float in the aquatic habitat. Adults feed mostly on flowers. Eggs are laid in masses, either on submerged or emergent aquatic plants, or on other objects in the water.

h. *Syrphidae (About 7 Aquatic Genera)*

Most "rat-tailed maggots" become fairly large and differ from other Brachycera by being broad and blunt anteriorly. Aquatic larvae are easily recognized by their very long, extensile, caudal breathing tube (siphon) (Fig. 17.133). Most also have ventral prolegs. Larvae inhabit shallow lentic habitats or margins of lotic habitats, especially areas high in decomposing organic matter. Here they feed on detritus and microorganisms. Some species occur in tree holes. Because they obtain oxygen from the air through their long respiratory siphon, they frequently inhabit very polluted areas such as sewage lagoons. They pupate in a puparium, which in some species has anterior respiratory horns. Life histories are poorly known, but most species probably have more than one generation per year. Adults are called "flower flies" because they feed on nectar from flowers. Eggs are laid in masses on the debris or vegetation that the larvae inhabit.

i. *Tabanidae (About 3 Aquatic Genera)*

Most "deer fly" or "horse fly" larvae develop in semiaquatic habitats that may be closely associated with aquatic habitats. Some larvae in three genera are known to be aquatic, and a few in other genera also may be aquatic. Many North American species have been reared and described, and larval keys have been developed, but identification is still risky because larvae of many species remain unknown. Larvae are recognized by the absence of distinct prolegs and the presence of a girdle of six or more pseudopods on most abdominal segments (Fig. 17.136). Aquatic larvae have been collected from stream riffles, shallow stream margins, and shallow vegetated lentic habitats. Most species are univoltine, but bivoltine and semivoltine species probably occur. Larvae are predators on other aquatic macroinvertebrates; pupation of aquatic species takes place in semiaquatic areas. Most adult females feed on blood of humans, livestock, and other animals, and are often a persistent nuisance. Eggs are laid in masses on vegetation above the larval habitat.

IV. SEMIAQUATIC COLLEMBOLA— SPRINGTAILS

Until recently, Collembola were regarded as insects, but most people no longer consider them to be in the class Insecta and place them in the class Entognatha or in a subclass of Hexopoda. The order Collembola is only partially aquatic, with the vast majority of the nearly 700 North American species inhabiting moist terrestrial habitats. Species that are collected from aquatic habitats are usually found on the surface of the water and are semiaquatic. Waltz and McCafferty (1979) reported that 96 species of springtails had been collected at least once from freshwater or coastal marine habitats in North America, but most of these records were believed to be incidental occurrences of terrestrial species. They regarded 10 species as semiaquatic and five as riparian with a propensity to venture onto the surface of the water. These 15 species exhibited various degrees of adaptation for the semiaquatic environment and are discussed in detail by Waltz and McCafferty, who also key species most likely to be found in freshwater habitats. Keys and descriptions for all North American species of Collembola are provided by Christiansen and Bellinger (1980).

Collembola are small, wingless arthropods, usually much less than 6 mm long. They resemble insects by having a distinct head with one pair of segmented antennae, a three-segmented thorax with three pairs of legs, and a segmented abdomen without legs (Figs. 17.154 and 17.155). They differ from insects in having only six abdominal segments and in having their maxillae and mandibles concealed by the wall of the head capsule with which their labium is fused. They also have the tibia and tarsus of each leg fused into a single segment. Except for their smaller size and lack of genitalia, young Collembola closely resemble adults.

Springtails can be readily recognized by a cylindrical colophore or ventral tube (Figs. 17.154 and 17.155) on the first abdominal segment and by the furcula, a midventral appendage on the fourth abdominal segment (Figs. 17.154 and 17.155) that is present in most species and in all that are known to be semiaquatic. The furcula consists of a basal piece, the manubrium, and two arms (Fig. 17.156). The basal segment of each arm is the dens and the apical segment is the mucro. The furcula can be folded forward under the abdomen where it is held in place by the tenaculum, a midventral clasp-like structure on the third abdominal segment. When the furcula is released, it springs backward and propels the Collembola several centimeters through the air, hence the common name "springtail." In most semiaquatic species, there is a widening of each mucro or lamellae on each mucro. This permits greater contact with the surface film, allowing individuals to spring farther when on the surface of the water.

The ecology and biology of Collembola remains very poorly known. Semiaquatic species are most often associated with lentic freshwater habitats; none occur on the ocean and reports from lotic habitats are rare. They feed primarily on algae, detritus, and other organic material in the surface film; some may eat bacteria. Adults do not copulate like insects. Instead, males deposit stalked or sessile spermatophores, which are gathered by the females; usually no contact between sexes is involved. Eggs are laid in moist habitats; in at least one species they are laid beneath the surface of the water and newly hatched young are truly aquatic, but once they pass through the surface film and come in contact with the air they remain semiaquatic. Eggs usually hatch within a few weeks, except those that diapause over the winter or during periods of dryness. The young reach sexual maturity within a few weeks and molt from three to seven times. In some species, the young may enter into diapause. Most adults live a few weeks to a few months and unlike insects, they continue to molt throughout their adult life, with more than fifty molts reported for one species. Some species are cold-hardy and may be active at temperatures near freezing; however, most species overwinter as diapausing eggs.

Because of their springing ability, Collembola are often difficult to capture. Semiaquatic species may be collected most easily by forcing them to jump into a white pan containing either 95% alcohol with 3% glacial acetic acid, or water with detergent to prevent them from jumping from the pan. Floating sticky traps may also be used to capture them. Christiansen and Bellinger recommend 70–95% isopropanol or 80–95% ethanol for preservation, with 1–2% glycerine added to prevent accidental desiccation.

V. IDENTIFICATION OF THE FRESHWATER INSECTS AND COLLEMBOLA

The following taxonomic keys are separated by class (Entognatha and Insecta) and orders (Insecta only). Collembola, which usually are much less than 6 mm long, can be distinguished from small larval insects by the possession ventrally of a colophore on the first abdominal segment and a furcula on the fourth (Figs. 17.154 and 17.155).

A. Taxonomic Key to Orders of Freshwater Insects

1a.	Thorax with three pairs of segmented legs	3
1b.	Thorax without segmented legs	2
2a(1b).	Mummy-like, with developing wings, legs, and other adult structures; in a case, which is often silk-cemented and contains vegetable or mineral matter	insect pupae (no keys)
2b.	Not mummy-like; not in a case, motile larvae, mostly with prolegs or pseudopods on one or more segments (Figs. 17.136–17.141, 17.144–17.153); larvae	Diptera
3a(1a).	With large functional wings	4
3b.	Wingless, or with developing wings (wing-pads) or brachypterous wings	6
4a(3a).	All wings completely membranous, with numerous veins	terrestrial insects or ovipositing adults of species with aquatic larvae
4b.	Mesothoracic wings hardened and shell-like, or leather-like in basal half (Figs. 17.83, 17.104, 17.111, 17.112)	5
5a(4b).	Mesothoracic wings hard, shell-like (Figs. 17.104, 17.111, 17.112); chewing mouthparts; adults	Coleoptera
5b.	Mesothoracic wings hardened in basal half (Fig. 17.83); sucking mouthparts formed into a broad or narrow tube (Figs. 17.84, 86); adults	Heteroptera (Hemiptera)
6a(3b).	With two or three long, filamentous terminal appendages (Figs. 17.9, 17.26, 17.41, 17.44, 17.45, 17.50, 17.52, 17.54)	7
6b.	Terminal appendages absent or not many-segmented	8
7a(6a).	Sides of abdomen with plate-like, featherlike, or leaflike gills (Figs. 17.4, 17.11–17.15, 17.22–17.26); usually with three tail filaments, occasionally only two; tarsi with one claw; larvae	Ephemeroptera
7b.	Gills absent from middle abdominal segments (Figs. 17.41, 17.44, 17.45, 17.50, 17.52, 17.54); two tail filaments; tarsi with two claws; larvae	Plecoptera
8a(6b).	Labium formed into an elbowed, extensile grasping organ (Figs. 17.28–17.30); abdomen terminating in three lamellae (Fig. 17.27) or five triangular points (Fig. 17.40); larvae	Odonata
8b.	Sucking or chewing mouthparts, not elbowed; abdomen not terminating in three lamellae or five triangular points	9
9a(8b).	Mouthparts sucking, formed into a broad or narrow tube (Figs. 17.84, 17.86) or a pair of long stylets (Fig. 17.98)	10
9b.	Mouthparts not sucking, not formed into a tube or pair of stylets	11
10a(9a).	Parasitic on sponges; mouthparts a pair of long stylets (Fig. 17.98); all tarsi with one claw (Fig. 17.98); larvae	Neuroptera, family Sisyridae
10b.	Free-living; mouthparts a broad or narrow tube (Figs. 17.84, 17.86); mesotarsi with at least two claws (Figs. 17.83, 17.89–17.92, 17.95); adults or nymphs	Heteroptera (Hemiptera)

11a(9b).	Ventral prolegs on middle abdominal segments, each with a ring of fine hooks (Fig. 17.99); larvae .... Lepidoptera, family Pyralidae	
11b.	Abdomen without ventral prolegs on middle segments, each with a ring of hooks 12	
12a(11b).	Antennae extremely small, inconspicuous, one-segmented (Figs. 17.59, 17.74, 17.76); larvae .... Trichoptera	
12b.	Antennae elongate, with three or more segments 13	
13a(12b).	A single claw on each tarsus (Figs. 17.117–17.120); larvae .... Coleoptera	
13b.	Each tarsus with two claws 14	
14a(13b).	Without conspicuous lateral filaments (Figs. 17.114–17.116); larvae .... Coleoptera	
14b.	With conspicuous lateral filaments (Figs. 17.81, 17.82, 17.113) 15	
15a(14b).	Abdomen terminating in two slender filaments or a median proleg with four hooks (Fig. 17.113); larvae .... Coleoptera	
15b.	Abdomen terminating in a single slender filament (Fig. 17.81) or in two prolegs, each with two hooks (Fig. 17.82); larvae .... Megaloptera	

B. Taxonomic Key to Families of Freshwater Ephemeroptera Larvae

1a.	Mandibles with large forward-projecting tusks (Fig. 17.3); gills on abdominal segments 2–7 with fringed margins (Fig. 17.4) and projecting laterally or dorsally over abdomen 2	
1b.	Mandibles without large tusks; conspicuous fringed gills absent from abdominal segments 2–7, if present, they project ventrolaterally 5	
2a(1a).	Gills dorsal, curving up over abdomen; protibia fossorial (Fig. 17.5) 3	
2b.	Gills lateral, projecting from sides of abdomen; protibiae slender, subcylindrical (Fig. 17.6) .... Potamanthidae	
3a(2a).	Apex of metatibiae rounded (Fig. 17.7); mandibular tusks curved inward and downward apically .... Polymitarcyidae	
3b.	Apex of metatibiae projected into an acute point ventrally (Fig. 17.8); mandibular tusks curved upward apically 4	
4a(3b).	Mandibular tusks with a toothed ridge along outer lateral margin (Fig. 17.3) .... Palingeniidae	
4b.	Mandibular tusks without a toothed ridge, more or less rounded .... Ephemeridae	
5a(1b).	Mesonotum modified into a carapace-like structure that covers the gills on abdominal segments 1–6 (Fig. 17.9) .... Baetiscidae	
5b.	Mesonotum not modified into a carapace, gills exposed 6	
6a(5b).	Head and pronotum with pads of long, dark setae on each side (Fig. 17.10); ventrolaterally projecting fringed gills; eastern United States .... Behningiidae	
6b.	Head and pronotum without pads of long, dark setae; gills not fringed and projecting ventrolaterally 7	
7a(6b).	Gills on abdominal segment 2 operculate or semioperculate, covering or partially covering gills on succeeding segments (Figs. 17.11–17.14) 8	
7b.	Gills on abdominal segment 2 similar to those on succeeding segments or absent 10	
8a(7a).	Operculate gills oval or subtriangular and well separated from each other mesally (Figs. 17.11 and 17.12); gills on segments 3–6 without fringed margins .... Tricorythidae	
8b.	Operculate gills quadrate and meeting along mesal edge (Figs. 17.13 and 17.14); gills on segments 3–6 with fringed margins 9	
9a(8b).	Operculate gills fused mesally (Fig. 17.13); Southeast and East .... Neoephemeridae	
9b.	Operculate gills not fused mesally, but overlapping (Fig. 17.14) .... Caenidae	

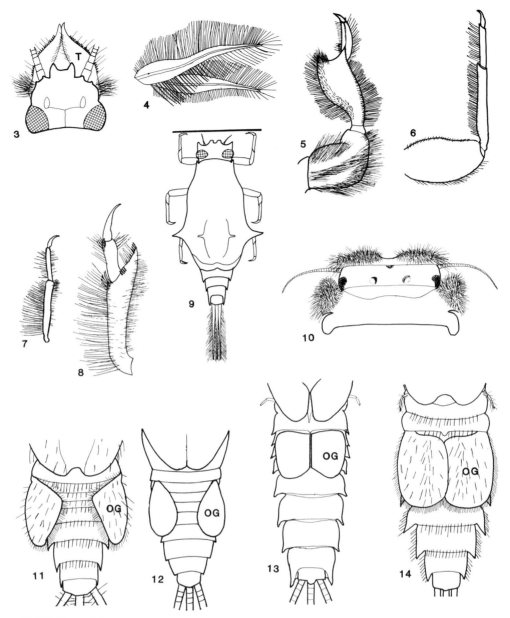

Figure 17.3 Palingeniidae: head of *Pentagenia* (dorsal view) showing mandibular tusk (T). **Figure 17.4** Ephemeridae; gills of *Hexagenia*. **Figure 17.5** Ephemeridae; prothoracic leg of *Hexagenia*. **Figure 17.6** Potamanthidae; prothoracic leg of *Potamanthus*. **Figure 17.7** Polymitarcyidae; metatibia and tarsus of *Ephoron*. **Figure 17.8** Ephemeridae; metatibia and tarsus of *Ephemera*. **Figure 17.9** Baetiscidae; *Baetisca* (dorsal view). **Figure 17.10** Behningiidae; head and pronotum of *Dolannia*. **Figure 17.11** Tricorythidae; abdomen of *Tricorythodes* (dorsal view) showing operculate gill (OG). **Figure 17.12** Tricorythidae; abdomen of *Leptohyphes* (dorsal view) showing operculate gill (OG). **Figure 17.13** Neoephemeridae; abdomen of *Neoephemera* (dorsal view) showing operculate gill (OG). **Figure 17.14** Caenidae; abdomen of *Caenis* (dorsal view) showing operculate gill (OG).

10a(7b).	Gills absent from abdominal segment 2 and sometimes also from 1 and 3, gills on segments 3 or 4 may be operculate (Fig. 17.15) .... Ephemerellidae
10b.	Gills present on abdominal segments 1 or 2 to 7 11
11a(10b).	Head and body dorsoventrally flattened; eyes and antennae dorsal (Fig. 17.16) .... Heptageniidae
11b.	Head and body not dorsoventrally flattened; eyes and usually also antennae along lateral margin of head (Fig. 17.17) 12

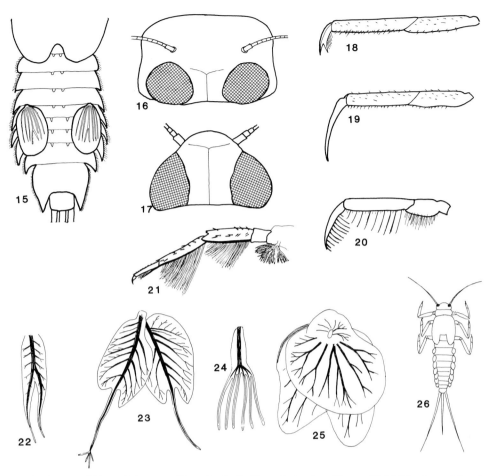

Figure 17.15 Ephemerellidae; abdomen of *Eurylophella* (dorsal view). **Figure 17.16** Heptageniidae; head of *Stenonema* (dorsal view). **Figure 17.17** Siphlonuridae; head of *Siphlonurus* (dorsal view). **Figure 17.18** Metretopodidae; protarsus and tibia of *Siphloplecton*. **Figure 17.19** Metretopodidae; metatarsus and tibia of *Siphloplecton*. **Figure 17.20** Ametropodidae; protarsus and tibia of *Ametropus*. **Figure 17.21** Oligoneuriidae; prothoracic leg of *Isonychia*. **Figure 17.22** Leptophlebiidae; gill on abdominal segment 3 of *Paraleptophlebia*. **Figure 17.23** Leptophlebiidae; gills on abdominal segment 3 of *Leptophlebia*. **Figure 17.24** Leptophlebiidae; gill on abdominal segment 3 of *Habrophlebia*. **Figure 17.25** Siphlonuridae; gills on abdominal segment 2 of *Siphlonurus*. **Figure 17.26** Baetidae; *Baetis* (dorsal view).

12a(11b).	Protarsal claws much shorter than those on other tarsi (Figs. 17.18 and 17.20), claws on meso- and metatarsi long and slender, about as long as tibiae (Fig. 17.19)	13
12b.	Claws on all tarsi similar in structure and length	14
13a(12a).	Protarsal claws simple, with long, slender denticles (Fig. 17.20); spinose pad present on procoxae	Ametropodidae
13b.	Protarsal claws bifid (Fig. 17.18); no spinose pad on procoxae	Metretopodidae
14a(12b).	Prothoracic legs with dense row of setae along inner margin (Fig. 17.21)	Oligoneuriidae
14b.	Prothoracic legs without dense row of setae along inner margin	15
15a(14b).	Gills forked (Fig. 17.22), bilamellate, and terminating in a point (Fig. 17.23), or terminating in filaments (Fig. 17.24)	Leptophlebiidae
15b.	Gills single or double lamellae (Figs. 17.25 and 17.26)	16
16a(15b).	Median caudal filament reduced or absent, if as long as cerci, antennae more than twice width of head (Fig. 17.26)	Baetidae
16b.	Median caudal filament as long as cerci, and antennae less than twice width of head	Siphlonuridae

C. Taxonomic Key to Families of Freshwater Odonata Larvae

1a.   Abdomen terminating in three caudal lamellae, longest more than half the length of abdomen (Fig. 17.27) .... suborder Zygoptera   2

1b.   Abdomen terminating in three stiff, pointed valves and two pointed cerci, longest less than 1/3 length of abdomen (Fig. 17.40) .... suborder Anisoptera   5

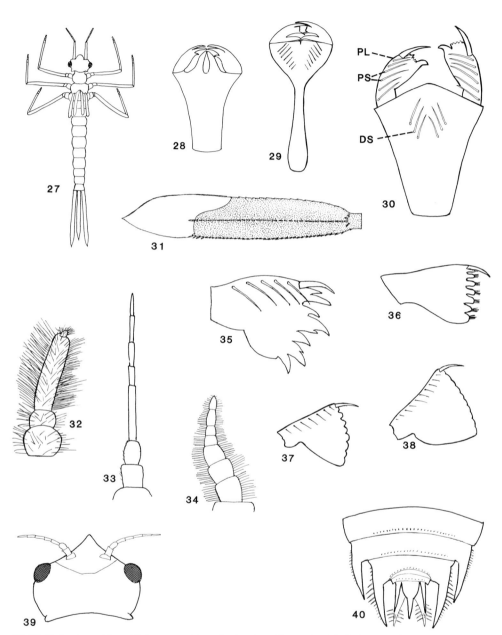

Figure 17.27 Calopterygidae; *Calopteryx* (dorsal view).   *Figure 17.28* Calopterygidae; prementum of *Hetaerina* (dorsal view).   *Figure 17.29* Lestidae; prementum of *Lestes* (dorsal view).   *Figure 17.30* Coenagrionidae; prementum of *Enallagma* (dorsal view) showing dorsal setae (DS), palpal lobe (PL), and palpal setae (PS).   *Figure 17.31* Protoneuridae; caudal lamella of *Protoneura* (lateral view).   *Figure 17.32* Gomphidae; antenna of *Gomphus*.   *Figure 17.33* Aeshnidae; antenna of *Anax*.   *Figure 17.34* Petaluridae; right antenna of *Tachopteryx* (dorsal view).   *Figure 17.35* Cordulegastridae; palpal lobe of *Cordulegaster* (dorsal view).   *Figure 17.36* Corduliidae; palpal lobe of *Neurocordulia* (dorsal view).   *Figure 17.37* Corduliidae; palpal lobe of *Cordulia* (dorsal view).   *Figure 17.38* Libellulidae; palpal lobe of *Tramea* (dorsal view).   *Figure 17.39* Macromiidae; head of *Macromia* (dorsal view).   *Figure 17.40* Libellulidae; abdominal segments 7–10 of *Pantala* (dorsal view).

2a(1a).   Zygoptera: first antennal segment as long as, or longer than remaining segments combined (Fig. 17.27); prementum with deep, median cleft (Fig. 17.28) .... Calopterygidae

2b.   First antennal segment much shorter than others combined; prementum with at most a very small median cleft (Figs. 17.29 and 17.30) 3

3a(2b).   Basal half of prementum greatly narrowed and elongate (Fig. 17.29); prementum in repose extends back to or past mesocoxae .... Lestidae

3b.   Basal half of prementum not greatly narrowed (Fig. 17.30); prementum in repose extends only to procoxae 4

4a(3b).   Caudal lamellae divided into a thick basal half and thin, lighter-colored distal half (Fig. 17.31); one dorsal seta on each side of prementum; Texas .... Protoneuridae

4b.   Caudal lamellae not distinctly divided as above; usually with two or more dorsal setae on each side of prementum .... Coenagrionidae

5a(1b).   Anisoptera: prementum flat or nearly so, without dorsal setae; palpal lobes also flat and usually without stout setae 6

5b.   Prementum rounded, spoon-shaped, and usually with dorsal setae; palpal lobes also rounded, always with stout setae, and covering face to base of antennae 8

6a(5a).   Antennae 4-segmented (Fig. 17.32); pro- and mesotarsi 2-segmented .... Gomphidae

6b.   Antennae 6- or 7-segmented (Figs. 17.33 and 17.34); pro- and mesotarsi 3-segmented 7

7a(6b).   Antennal segments slender, not hairy (Fig. 17.33) .... Aeshnidae

7b.   Antennal segments short, thick, and hairy (Fig. 17.34) .... Petaluridae

8a(5b).   Distal margin of palpal lobes with large, irregular teeth (Fig. 17.35) .... Cordulegastridae

8b.   Distal margin of palpal lobes with small, even crenulations (Figs. 17.36 and 17.37) or nearly flat (Fig. 17.38) 9

9a(8b).   Head with a prominent, almost erect, thick frontal process between bases of antennae (Fig. 17.39); legs very long, apex of each metafemur reaching to or beyond apex of abdominal segment 8; metasternum with broad, median tubercle .... Macromiidae

9b.   Head without a prominent frontal process (except in *Neurocordulia molesta*); legs shorter, apex of metafemur usually not reaching apex of abdominal segment 8; metasternum without a median tubercle 10

10a(9b).   Crenulations on palpal lobes very shallow, 1/6 to less than 1/10 as long as wide (Fig. 17.38); "or" with lateral spines on abdominal segment 8 longer than mid-dorsal length of segment 9 (Fig. 17.40) .... Libelluliidae

10b.   Crenulations on palpal lobes large, at least 1/4 as long as wide (Figs. 17.36 and 17.37); lateral spines on abdominal segment 8 shorter than mid-dorsal length of segment 9 .... Corduliidae

D. Taxonomic Key to Families of Freshwater Plecoptera Larvae

1a.   Conspicuous, finely branched gills present ventrally or laterally on all thoracic segments (Fig. 17.2) 2

1b.   Gills absent, confined to prosternum, or unbranched 3

2a(1a).   Finely branched gills on abdominal sterna 1, 2, and sometimes 3 .... Pteronarcyidae

2b.   Gills absent from first three visible abdominal sterna .... Perlidae

3a(1b).   Thoracic sterna produced into plates that overlap succeeding segment (Fig. 17.41); single, double, or forked gills behind meso- and metacoxae; form roachlike .... Peltoperlidae

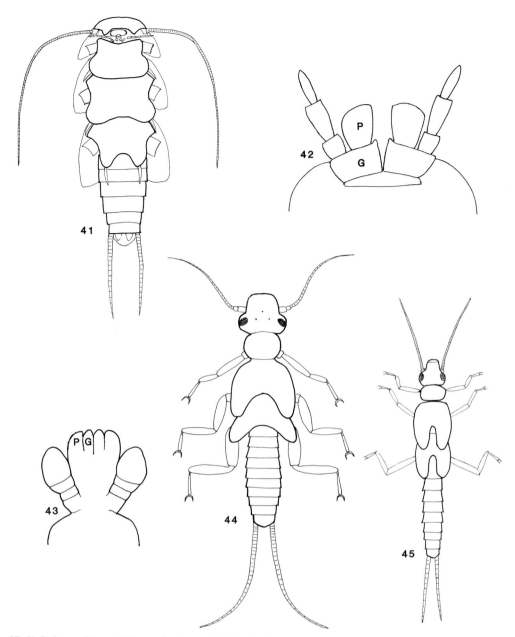

Figure 17.41 Peltoperlidae; *Peltoperla* (ventral view) with legs removed beyond coxae. **Figure 17.42** Chloroperlidae; labium of *Alloperla* (ventral view) showing glossae (G) and paraglossae (P). **Figure 17.43** Nemouridae; labium of *Nemoura* (ventral view) showing glossae (G) and paraglossae (P). **Figure 17.44** Perlodidae; *Isoperla* (dorsal view). **Figure 17.45** Chloroperlidae; *Haploperla* (dorsal view).

3b.   Thoracic sterna not produced into overlapping plates; coxal gills, if present, segmented and not apically pointed; form usually elongate, not roachlike 4

4a(3b).   Tips of glossae situated much behind tips of paraglossae (Fig. 17.42) 5

4b.   Tips of glossae produced nearly as far forward as tips of paraglossae (Fig. 17.43) 6

5a(4a).   Cerci almost as long or longer than abdomen; metathoracic wing-pads with inner and outer margins strongly diverging from axis of body (Fig. 17.44); head and thorax usually with a distinct pattern . . . . Perlodidae

5b.   Cerci distinctly shorter than abdomen; metathoracic wing-pads with inner and outer margins nearly parallel to axis of body (Fig. 17.45); head and thorax usually not patterned . . . . Chloroperlidae

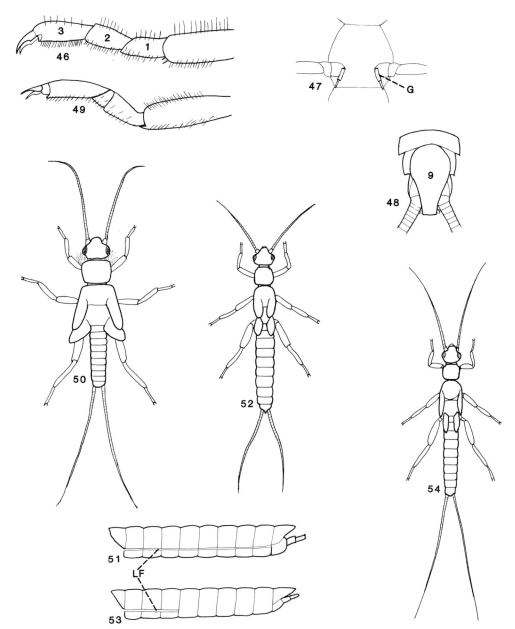

Figure 17.46 Taeniopterygidae; tarsal segments (1, 2, 3) of *Taeniopteryx* (lateral view). **Figure 17.47** Taeniopterygidae; mesosternum of *Taeniopteryx* showing gills (G) at base of coxae. **Figure 17.48** Taeniopterygidae; ninth abdominal sternum (9) of *Strophopteryx* (ventral view). **Figure 17.49** Nemouridae; tarsal segments of *Nemoura* (lateral view). **Figure 17.50** Nemouridae; *Amphinemura* (dorsal view). **Figure 17.51** Capniidae; abdomen of *Allocapnia* (lateral view) showing lateral fold (LF). **Figure 17.52** Capniidae; *Paracapnia* (dorsal view). **Figure 17.53** Leuctridae; abdomen of *Leuctra* (lateral view) showing lateral fold (LF). **Figure 17.54** Leuctridae; *Leuctra* (dorsal view).

6a(4b).	Second tarsal segment (lateral view) about as long as, or longer than first (Fig. 17.46); single segmented gills on inner side of each coxa (Fig. 17.47), "or" ninth abdominal sternum greatly produced (Fig. 17.48) .... Taeniopterygidae
6b.	Second tarsal segment much shorter than first (Fig. 17.49); gills absent from inner side of coxae and ninth sternum not produced 7
7a(6b).	Robust larvae, with extended metathoracic legs reaching to or beyond tip of abdomen; metathoracic wing-pads with inner and outer margins strongly diverging from axis of body (Fig. 17.50) .... Nemouridae

7b.	Elongate larvae, with extended metathoracic legs reaching well short of tip of abdomen; metathoracic wing-pads with inner and outer margins nearly parallel to axis of body (Figs. 17.52 and 17.54) 8
8a(7b).	Abdominal segments 1–9 divided by a membranous, ventrolateral fold (Fig. 17.51); metathoracic wing-pads, if present, about same distance apart as mesothoracic wing-pads and often short and broad (Fig. 17.52) .... Capniidae
8b.	Abdominal segments 7–9 not divided by a membranous fold (Fig. 17.53); metathoracic wing-pads similar in shape to mesothoracic wing-pads, but distinctly closer together (Fig. 17.54) .... Leuctridae

E. Taxonomic Key to Families of Freshwater Trichoptera Larvae

1a.	Larva in spiral case of sand grains or tiny stones that resembles a snail shell (Fig. 17.55); anal claws with many short teeth forming a comb (Fig. 17.56) .... Helicopsychidae
1b.	Case not like a snail shell or absent; anal claws not forming a comb with many teeth 2
2a(1b).	Each thoracic segment covered with a single dorsal plate, which may have a mesal or transverse fracture line 3
2b.	Metanotum mostly membranous, having only scattered hairs or small plates, or with two or more sclerites 5
3a(2a).	Abdomen with rows of branched gills ventrally; no portable case; often > 8 mm long .... Hydropsychidae
3b.	Abdomen without branched gills ventrally; often with a portable case; < 8 mm long 4
4a(3b).	Abdominal tergum 9 sclerotized; fifth instar with abdomen much enlarged and larva in barrel- or purse-like case (Figs. 17.57 and 17.58) .... Hydroptilidae
4b.	Abdominal tergum 9 entirely membranous; abdomen not enlarged and never in a case; south Texas .... Ecnomidae
5a(2b).	Antennae long, at least six times as long as wide, and arising near base of mandibles (Fig. 17.59); ''or'' mesonotum membranous, except for a pair of sclerotized, narrow, curved or angled bars (Fig. 17.60); larva in a case .... Leptoceridae
5b.	Antennae very short, not more than three times as long as wide, often inconspicuous and arising at various points; mesonotum never with a pair of sclerotized, narrow, curved or angled bars; with or without case 6
6a(5b).	Meso- and metanotum entirely membranous, or with only weak sclerites at SA-1 of mesonotum (Fig. 17.61) 7
6b.	Meso- and metanotum with some conspicuous sclerotized plates (Figs. 17.79 and 17.80); larva in a case (Fig. 17.62) 15
7a(6a).	Abdominal segment 9 with dorsum entirely membranous; no portable case (Fig. 17.63) 8
7b.	Abdominal segment 9 bearing a sclerotized dorsal plate; with or without portable case 12
8a(7a).	Tibia and tarsus fused to form a single segment on all legs; mesopleura with a forward projecting lobe (Fig. 17.64); Texas .... Xiphocentronidae
8b.	Tibia and tarsus distinct on all legs; mesopleura without a projecting lobe 9
9a(8b).	Protrochantins broad, hatchet-shaped (Fig. 17.65) .... Psychomyiidae
9b.	Protrochantins pointed or poorly developed 10
10a(9b).	Protrochantins poorly developed; head without markings; labrum membranous and T-shaped (Fig. 17.66) .... Philopotamidae
10b.	Protrochantins pointed anteriorly (Fig. 17.67); head usually with dark markings or dark or light muscle scars; labrum sclerotized and widest near base 11
11a(10b).	Tarsi broad and densely pilose (Fig. 17.68); mandibles short and triangular, each with a large, thick mesal brush; eastern and central United States .... Dipseudopsidae

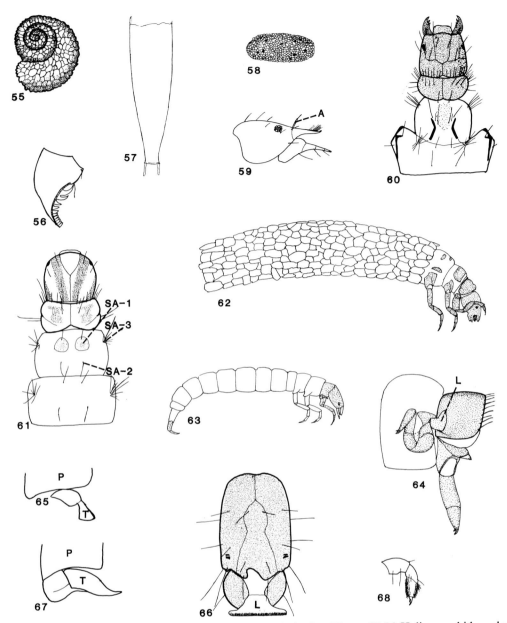

Figure 17.55 Helicopsychidae; case of *Helicopsyche* (dorsal view).   **Figure 17.56** Helicopsychidae; claw on anal proleg of *Helicopsyche*.   **Figure 17.57** Hydroptilidae; case of *Oxyethira* (lateral view).   **Figure 17.58** Hydroptilidae; case of *Hydroptila* (lateral view).   **Figure 17.59** Leptoceridae; head of *Oecetis* (lateral view) showing antenna (A).   **Figure 17.60** Leptoceridae; head and thoracic terga of *Ceraclea*.   **Figure 17.61** Phryganeidae; head and thoracic terga of *Oligostomis* showing location of setal areas (SA).   **Figure 17.62** Limnephilidae; *Hesperophylax* (lateral view).   **Figure 17.63** Polycentropodidae; *Polycentropus* (lateral view).   **Figure 17.64** Xiphocentronidae; pro- and mesothorax of *Xiphocentron* (lateral view) showing projecting lobe (L).   **Figure 17.65** Psychomyiidae; pronotum (P) and protrochantin (T) of *Psychomyia*.   **Figure 17.66** Philopotamidae; head of *Chimarra* (dorsal view) showing labrum (L).   **Figure 17.67** Polycentropodidae; pronotum (P) and protrochantin (T) of *Polycentropus*.   **Figure 17.68** Dipseudopsidae; tarsus of *Phylocentropus*.

11b.	Tarsi with little or no pile and with a large claw (Fig. 17.63); mandibles elongate	Polycentropodidae
12a(7b).	Prosternal horn present (Fig. 17.76); SA-3 on meso- and metanotum with a small sclerite and a cluster of setae (Fig. 17.61); case of vegetation, spirally wound or a series of rings and readily abandoned	Phryganeidae

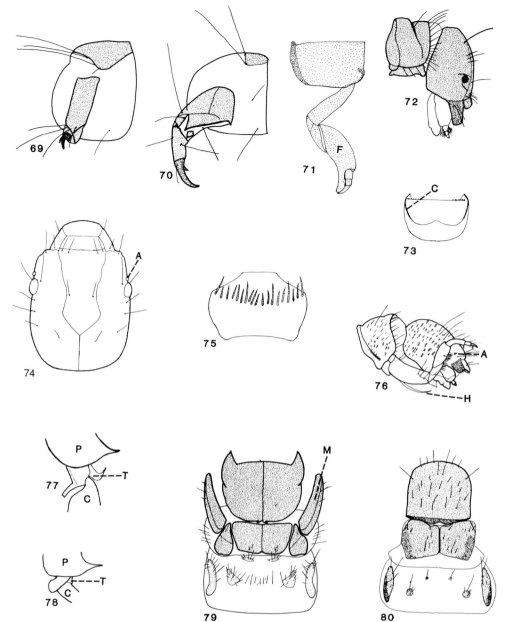

Figure 17.69 Glossosomatidae; last abdominal segment of *Glossosoma* (lateral view) showing anal proleg. **Figure 17.70** Rhyacophilidae; last abdominal segment of *Rhyacophila* (lateral view) showing anal proleg. **Figure 17.71** Hydrobiosidae; pronotum and prothoracic leg with chelate femur (F) of *Atopsyche* (posterolateral view). **Figure 17.72** Brachycentridae; pronotum and head of *Brachycentrus* (lateral view). **Figure 17.73** Beraeidae; pronotum of *Beraea* (dorsal view) showing carina (C). **Figure 17.74** Lepidostomatidae; head of *Lepidostoma* (dorsal view) showing antenna (A). **Figure 17.75** Calamoceratidae; labrum of *Heteroplectron* (dorsal view). **Figure 17.76** Limnephilidae; head and prothorax of *Platycentropus* (lateral view) showing antenna (A) and prosternal horn (H). **Figure 17.77** Sericostomatidae; pronotum (P), protrochantin (T), and coxa (C) of *Agarodes*. **Figure 17.78** Odontoceridae; pronotum (P), protrochantin (T), and coxa (C) of *Psilotreta*. **Figure 17.79** Goeridae; thoracic terga of *Goera* showing mesepisternum (M). **Figure 17.80** Uenoidae; thoracic terga of *Neophylax*.

12b.	Prosternal horn absent; SA-3 on meso- and metanotum without a sclerite and usually with a single seta; case, if present, of sand and pebbles	13
13a(12b).	Anal claws very small, much shorter than elongate sclerite on anal legs (Fig. 17.69); larva with turtle-like case of sand and pebbles, which is readily abandoned	Glossosomatidae

13b.	Anal claws large, as long as elongate sclerite on anal legs (Fig. 17.70); larva without a case	14
14a(13b).	Profemora with ventral projection to form chelate legs (Fig. 17.71); southwestern United States	Hydrobiosidae
14b.	Prothoracic legs not chelate	Rhyacophilidae
15a(6b).	Claws of metathoracic legs very small, those of other legs long; case of sand, usually with lateral flanges	Molannidae
15b.	Claws of metathoracic legs as long as those of mesothoracic legs; case never of sand, with lateral flanges	16
16a(15b).	Pronotum divided by a sharp furrow across middle, area in front of furrow depressed (Fig. 17.72); no dorsal or lateral humps on abdominal segment 1	Brachycentridae
16b.	Pronotum with at most a shallow furrow; lateral humps and usually dorsal hump present on abdominal segment 1	17
17a(16b).	Pronotum divided obliquely by a sharp carina that terminates as a rounded lobe anterolaterally (Fig. 17.73); East	Beraeidae
17b.	Pronotum without a distinct carina terminating in a rounded anterolateral lobe	18
18a(17b).	Antennae located extremely close to eyes (Fig. 17.74); median dorsal hump absent from abdominal segment 1	Lepidostomatidae
18b.	Antennae midway between eyes and base of mandibles or near base of mandibles; dorsal hump present on abdominal segment 1	19
19a(18b).	Anterolateral angles of pronotum produced and divergent; labrum with transverse dorsal row of about 18 long setae (Fig. 17.75); East and West	Calamoceratidae
19b.	Anterolateral angles of pronotum if produced, not divergent; labrum without a dorsal row of about 18 long setae	20
20a(19b).	Prosternal horn absent; antennae close to base of mandibles	21
20b.	Prosternal horn present (Fig. 17.76); antennae about midway between base of mandibles and eye (Fig. 17.76)	22
21a(20a).	Protrochantins large, hook-shaped (Fig. 17.77); anal prolegs each with about 30 long setae	Sericostomatidae
21b.	Protrochantins small, not hook-shaped (Fig. 17.78); anal prolegs each with about 5 long setae	Odontoceridae
22a(20b).	Mesepisternum formed anteriorly into a sharp, elongate process (Fig. 17.79) or a rounded spiny prominence	Goeridae
22b.	Mesepisternum not enlarged anteriorly as above	23
23a(22b).	Mesonotum with an anterior mesal notch (Fig. 17.80); SA-1 of metanotum unsclerotized and with only one or two setae (Fig. 17.80)	Uenoidae
23b.	Mesonotum without an anterior notch; SA-1 of metanotum with a sclerotized plate and/or more than two setae	Limnephilidae

F. Taxonomic Key to Families of Freshwater Megaloptera Larvae

1a.	Abdominal segments 1–7 with lateral filaments, last segment with a long median filament (Fig. 17.81)	Sialidae
1b.	Abdominal segments 1–8 with lateral filaments, last segment without a median filament, but with a pair of prolegs, each with a pair of claws (Fig. 17.82	Corydalidae

G. Taxonomic Key to Families of Aquatic, Semiaquatic, and Riparian Adults Heteroptera

1a.	Antennae shorter than head, inserted beneath eyes and (except Ochteridae) not visible from above (Figs. 17.83, 17.84, 17.89, 17.91) .... suborder Nepomorpha	2
1b.	Antennae longer than head, inserted in front of eyes and visible from above (Figs. 17.92, 17.95–17.97)	9

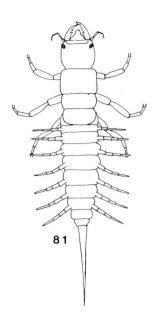

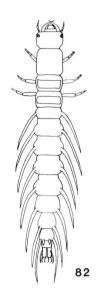

81

82

◀ **Figure 17.81** Sialidae; *Sialis* (dorsal view).
Figure 17.82 Corydalidae; *Nigronia* (dorsal view).

2a(1a).	Nepomorpha: rostrum broad, blunt, and triangular, not distinctly segmented (Fig. 17.84); each front tarsus a 1-segmented scoop (Fig. 17.85) .... Corixidae	
2b.	Rostrum cylindrical or cone-shaped, distinctly 3- or 4-segmented (Fig. 17.86); front tarsi not scoop-like 3	
3a(2b).	Apex of abdomen with a long, slender, tubular respiratory appendage (Fig. 17.87) .... Nepidae	
3b.	Apical respiratory appendages absent, if present, short and flat (Fig. 17.88) 4	
4a(3b).	Meso- and metathoracic legs with fringes of swimming hairs; ocelli absent; aquatic 5	
4b.	Meso- and metathoracic legs without fringes or swimming hairs; ocelli usually present; riparian 8	
5a(4a).	Dorsoventrally flattened, ovate insects; profemora broad, raptorial (Fig. 17.89) 6	
5b.	Elongate or hemispherical insects, not flattened dorsoventrally (Figs. 17.90 and 17.91); profemora slender, similar to other legs 7	
6a(5a).	Length > 18 mm; short, flat, strap-like apical respiratory appendages present (Fig. 17.88); eyes protrude from margin of head .... Belostomatidae	
6b.	Length < 16 mm; apical respiratory appendages absent; eyes do not protrude from margin of head (Fig. 17.89) .... Naucoridae	
7a(5b).	Hemispherical (Fig. 17.90); length < 3 mm .... Pleidae	
7b.	Elongate (Fig. 17.91); length > 5 mm .... Notonectidae	
8a(4b).	Profemora broad, raptorial; antennae concealed from above .... Gelastocoridae	
8b.	Profemora slender, similar to other legs; antennae visible from above .... Ochteridae	
9a(1b).	Membrane of hemelytra with four or five equal-sized cells (Fig. 17.92); metacoxae large, transverse; riparian .... Saldidae	
9b.	Membrane of hemelytra without veins or with dissimilar-sized cells; metacoxae small, conical; semiaquatic or riparian .... suborder Gerromorpha 10	
10a(9b).	Gerromorpha: claws of at least protarsi inserted before apex (Fig. 17.93) 11	
10b.	Claws of all tarsi inserted at apex (Fig. 17.94) 12	
11a(10a).	Metafemora very long, greatly surpassing apex of abdomen .... Gerridae	
11b.	Metafemora short, not or only slightly surpassing apex of abdomen .... Veliidae	
12a(10b).	Head as long as entire thorax, very slender, with eyes set about halfway to base (Fig. 17.95) .... Hydrometridae	
12b.	Head short and stout, eyes near posterior margin 13	
13a(12b).	Lower part of head grooved to receive rostrum; tarsi 2-segmented; < 2.5 mm long .... Hebridae	

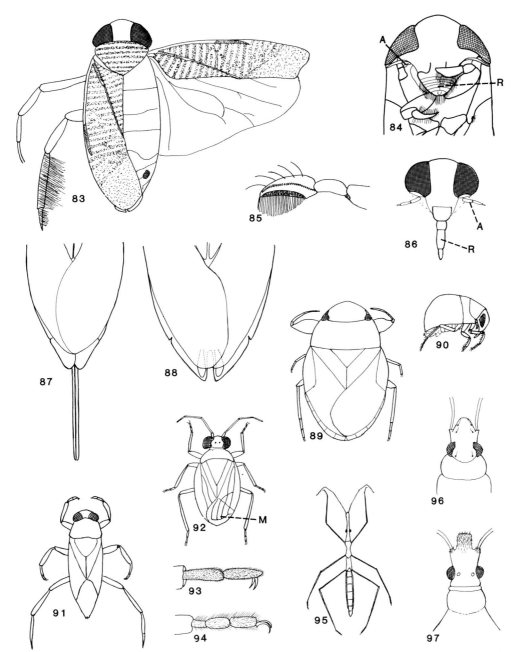

Figure 17.83 Corixidae; *Sigara* (dorsal view), with right wings extended laterally. **Figure 17.84** Corixidae; head and prothorax of *Sigara* (ventral view) showing rostrum (R) and antenna (A). **Figure 17.85** Corixidae; pala of male *Sigara*. **Figure 17.86** Notonectidae; head of *Notonecta* (ventral view) showing rostrum (R) and antenna (A). **Figure 17.87** Nepidae; apex of abdomen of *Nepa* (dorsal view). **Figure 17.88** Belostomatidae; apex of abdomen of *Belostoma* (dorsal view). **Figure 17.89** Naucoridae; *Pelocoris* (dorsal view). **Figure 17.90** Pleidae; *Neoplea* (lateral view). **Figure 17.91** Notonectidae; *Notonecta* (dorsal view). **Figure 17.92** Saldidae; *Salda* (dorsal view) showing membrane (M). **Figure 17.93** Gerridae; protarsus of *Gerris*. **Figure 17.94** Mesoveliidae; protarsus of *Mesovelia*. **Figure 17.95** Hydrometridae; *Hydrometra* (dorsal view). **Figure 17.96** Mesoveliidae; head of *Mesovelia* (dorsal view). **Figure 17.97** Macroveliidae; head of *Macrovelia* (dorsal view).

13b.		Lower part of head not grooved; tarsi 3-segmented; > 2.5 mm long 14
14a(13b).		Inner margins of eyes converge anteriorly (Fig. 17.96); femora with one or more dorsal black spines distally .... Mesoveliidae
14b.		Inner margins of eyes rounded (Fig. 17.97); femora without black spines; riparian, from Nebraska and North Dakota to west coast .... Macroveliidae

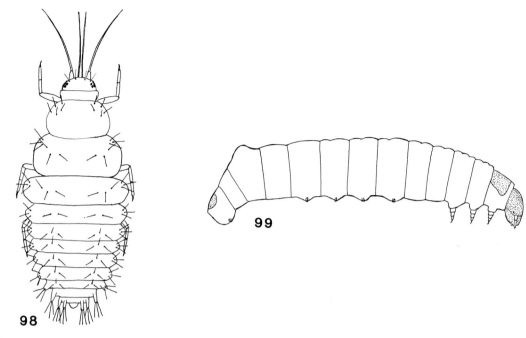

Figure 17.98 Neuroptera larva (Sisyridae); *Climacia* (dorsal view). **Figure 17.99** Lepidoptera larva (Pyralidae); *Nymphula* (lateral view).

H. Taxonomic Key to Families of Freshwater Coleoptera

1. Key to Adult Water Beetles

1a.	Head formed into a snout or beak anteriorly; antennae geniculate (Fig. 17.100) .... Curculionidae	
1b.	Head not formed into a beak or snout; antennae not geniculate 2	
2a(1b).	Two pairs of eyes, a dorsal and a ventral pair divided by sides of head; meso- and metathoracic legs short, extremely flat, with tarsi folding fan-like .... Gyrinidae	
2b.	One pair of eyes; meso- and metathoracic legs not extremely flat, with tarsi not folded fan-like 3	
3a(2b).	Metacoxae expanded into large plates that cover two or three abdominal sterna and base of metafemora (Fig. 17.101) .... Haliplidae	
3b.	Metacoxae not expanded into large plates 4	
4a(3b).	Prosternum with a postcoxal process that extends posteriorly to mesocoxae (Fig. 17.102); first visible abdominal sternum completely divided by metacoxal process (Fig. 17.102) 5	
4b.	Prosternum with postcoxal process absent or short; first visible abdominal sternum extending for its entire breadth behind metacoxae (Fig. 17.103) 8	
5a(4a).	Base of pronotum much narrower than base of elytra (Fig. 17.104); metatarsi rounded and not fringed with long setae, large, > 10 mm long; western mountains .... Amphizoidae	
5b.	Base of pronotum subequal in width to base of elytra; metatarsi flattened and fringed with long setae, "or" beetles < 2 mm long 6	
6a(5b).	Anterior of prosternum, its postcoxal process and metasternum in same plane (Fig. 17.105); pro- and mesotarsi distinctly 5-segmented, segment 4 as long as 3 7	
6b.	Anterior of prosternum greatly depressed and not in same plane as its postcoxal process and metasternum (Fig. 17.106); pro- and mesotarsi appear to be 4-segmented .... Dytiscidae (in part)	
7a(6a).	Prosternal process pointed or nearly so (Fig. 17.102); no curved spur or hooked apex on protibiae; > 4 mm long .... Dytiscidae (in part)	

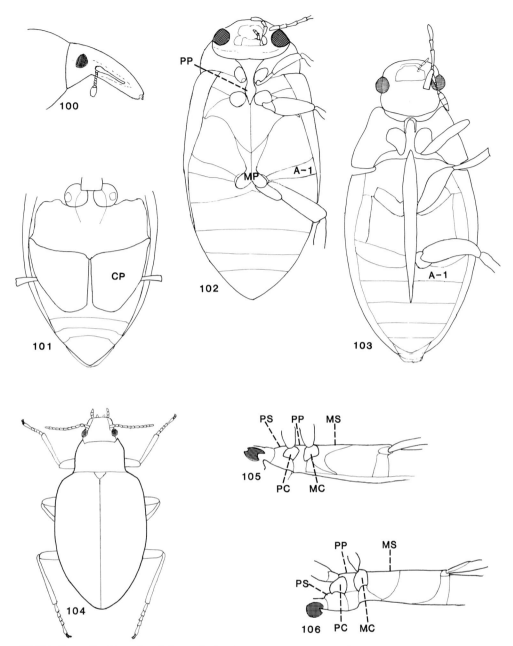

Figure 17.100 Curculionidae; head (lateral view). **Figure 17.101** Haliplidae; metathorax and abdomen of *Haliplus* (ventral view) showing metacoxal plates (CP). **Figure 17.102** Dytiscidae; *Agabus* (ventral view) showing prosternal process (PP), first abdominal sternum (A-1), and metacoxal process (MP). **Figure 17.103** Hydrophilidae; *Tropisternus* (ventral view) showing first abdominal sternum (A-1). **Figure 17.104** Amphizoidae; *Amphizoa* (dorsal view). **Figure 17.105** Dytiscidae; *Agabus* (lateral view) showing prosternum (PS), prosternal process (PP), metasternum (MS), procoxae (PC), and mesocoxae (MC). **Figure 17.106** Dytiscidae; *Hydroporus* (lateral view) showing prosternum (PS), prosternal process (PP), metasternum (MS), procoxae (PC), and mesocoxae (MC).

7b.	Prosternal process truncate or rounded apically (Fig. 17.107); protibiae with curved spur or hooked apex (Fig. 17.108), ''or'' beetles < 3 mm long .... Noteridae	
8a(4b).	Antennae short, club-shaped, with segment 4, 5, or 6 modified to form a cupule (Fig. 17.109); maxillary palpi usually longer than antennae	9
8b.	Antennae filiform or pectinate (Figs. 17.110 and 17.111), usually longer than maxillary palpi	10

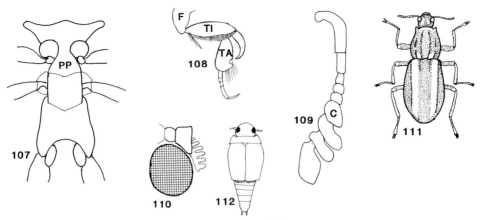

Figure 17.107 Noteridae; thorax of *Hydrocanthus* (ventral view) showing prosternal process (PP). **Figure 17.108** Noteridae; prothoracic leg of *Hydrocanthus* showing profemur (F), tibia (TI), and tarsus (TA). **Figure 17.109** Hydrophilidae; antenna of *Tropisternus* showing cupule (C). **Figure 17.110** Dryopidae; right antenna of *Helichus* (dorsal view). **Figure 17.111** Elmidae; *Stenelmis* (dorsal view). **Figure 17.112** Hydroscaphidae; *Hydroscapha* (dorsal view).

9a(8a).	Antennae with five segments past cupule; < 2.5 mm long	Hydraenidae
9b.	Antennae with three segments past cupule (Fig. 17.109); 1.5–40.0 mm long	Hydrophilidae
10a(8b).	Antennae short with pectinate club (Fig. 17.110); > 5.0 mm long	Dryopidae
10b.	Antennae slender, filiform; (Fig. 17.111); < 4.5 mm long	11
11a(10b).	Extremely small, < 1.5 mm long (Fig. 17.112); all tarsi 3-segmented; Southwest, north to Idaho	Hydroscaphidae
11b.	Larger, > 1.7 mm long; all tarsi 5-segmented	Elmidae

2. Key to Larval Water Beetles

1a.	Tarsi with two claws	2
1b.	Tarsi with one claw	5
2a(1a).	Abdomen with four conspicuous hooks on last segment; abdominal segments with at least eight pairs of lateral filaments (Fig. 17.113)	Gyrinidae
2b.	No hooks on last abdominal segment; if lateral abdominal filaments are present, there are only six pairs	3
3a(2b).	Abdominal and thoracic terga flattened and expanded laterally; (Fig. 17.114); western mountains	Amphizoidae
3b.	Abdominal and thoracic terga not flattened and expanded laterally	4
4a(3b).	Urogomphi shorter than last abdominal segment; legs short, stout, adapted for digging (Fig. 17.115); mandibles short, adapted for chewing	Noteridae
4b.	Urogomphi usually longer than last abdominal segment (Fig. 17.116); "if shorter," then legs are elongate with setal fringe for swimming; mandibles elongate, pointed, for piercing	Dytiscidae
5a(1b).	Legs distinctly 5-segmented; abdomen terminating in one or two long filaments (Figs. 17.117 and 17.118)	Haliplidae
5b.	Legs apparently 4-segmented; abdomen not terminating in long filaments	6
6a(5b).	Mandibles large, readily visible from above (Fig. 17.119)	Hydrophilidae
6b.	Mandibles not readily visible from above	7
7a(6b).	Antennae long, filiform, as long as head and thorax combined (Fig. 17.120)	Scirtidae
7b.	Antennae much shorter than head and thorax combined	8
8a(7b).	Body oval and extremely flat; head completely concealed from dorsal view (Figs. 17.121 and 17.122)	Psephenidae

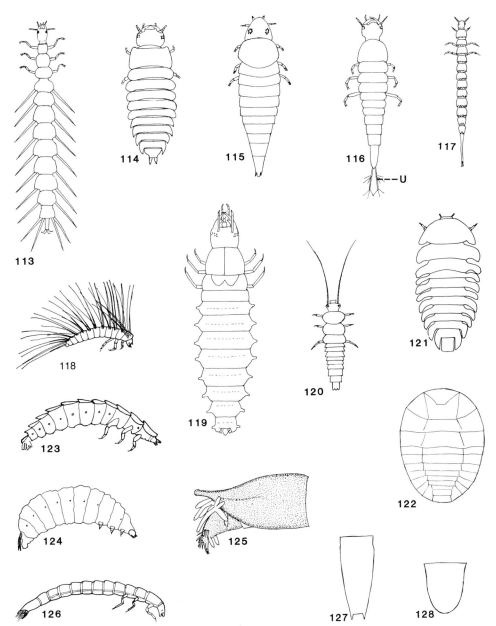

Figure 17.113 Gyrinidae; *Dineutus* (dorsal view).   *Figure 17.114* Amphizoidae; *Amphizoa* (dorsal view). *Figure 17.115* Noteridae; *Hydrocanthus* (dorsal view).   *Figure 17.116* Dytiscidae; *Agabus* (dorsal view) showing urogomphi (U).   *Figure 17.117* Haliplidae; *Haliplus* (dorsal view).   *Figure 17.118* Haliplidae; *Peltodytes* (lateral view).   *Figure 17.119* Hydrophilidae; *Tropisternus* (dorsal view).   *Figure 17.120* Scirtidae; *Cyphon* (dorsal view).   *Figure 17.121* Psephenidae; *Ectopria* (dorsal view).   *Figure 17.122* Psephenidae; *Psephenus* (dorsal view).   *Figure 17.123* Lampyridae (lateral view).   *Figure 17.124* Chrysomelidae; *Donacia* (lateral view).   *Figure 17.125* Ptilodactylidae; last abdominal segment of *Anchytarsus* (lateral view).   *Figure 17.126* Elmidae; *Stenelmis* (lateral view).   *Figure 17.127* Elmidae; last abdominal tergum of *Stenelmis*.   *Figure 17.128* Lutrochidae; last abdominal tergum of *Lutrochus*.

8b.	Body elongate and round or triangular in cross-section; head exposed, except mostly concealed in Lampyridae	9
9a(8b).	Each thoracic and abdominal segment covered by a flat, plate-like sclerite dorsally; prothoracic plate mostly or completely concealing head from above (Fig. 17.123)	Lampyridae
9b.	Thoracic and abdominal segments not covered by flat, plate-like sclerites; head visible from above	10

10a(9b).	All terga rounded and pale; grub-like larva with two spines on last abdominal tergum (Fig. 17.124) .... Chrysomelidae
10b.	Body elongate and sclerotized; no spines on last abdominal tergum (Figs. 17.125 and 17.126) 11
11a(10b).	Abdominal sterna with distinct tufts of gills, either on sterna 1–7 or on last sternum (Fig. 17.125) .... Ptilodactylidae
11b.	Abdominal gills, if present, on last abdominal segment and covered by a ventral operculum (Fig. 17.126) 12
12a(11b).	Last abdominal tergum bifid or notched apically (Fig. 17.127) .... Elmidae
12b.	Last abdominal tergum rounded apically (Fig. 17.128) .... Lutrochidae

I. Taxonomic Key to Families of Freshwater Diptera Larvae

1a.	Larvae apparently 7-segmented; first six segments each with a prominent ventral sucker (Fig. 17.129) .... Blephariceridae
1b.	Larvae with more than seven apparent segments; without six ventral suckers 2
2a(1b).	Head capsule completely sclerotized and fully visible; mandibles opposed, moving in a horizontal plane, and usually with two or more apical teeth; body never flattened and with a posterior spiracular chamber margined with long, soft hairs 12
2b.	Sclerotized head capsule absent, incomplete behind, or retracted at least partially into thorax; mandibles parallel, without secondary apical teeth, and moving in a vertical plane; ''or'' in a retracted head, may be opposed and moving in a horizontal plane 3
3a(2b).	Mandibles opposed, moving in a horizontal plane, and with two or more apical teeth; larva truncate anteriorly with head capsule retracted into thorax and incomplete posteriorly (Figs. 17.130 and 17.131) .... Tipulidae
3b.	Mandibles parallel, moving vertically, and without secondary apical teeth; larva narrowed anteriorly with head capsule lacking or poorly developed and incompletely sclerotized; ''or'' body terminating in a long respiratory tube (Fig. 17.133) or a posterior spiracular chamber margined with long, soft hairs 4
4a(3b).	Head mostly visible, truncate in shape (Fig. 17.132); body somewhat flattened; posterior spiracular chamber margined with long, soft hairs .... Stratiomyidae
4b.	Head mostly retracted into thorax and elongate, or indistinguishable; body nearly circular in cross-section; without a posterior spiracular chamber margined with long, soft hairs 5
5a(4b).	Larvae with a partially retractile caudal respiratory tube at least one-half as long as body (Fig. 17.133) .... Syrphidae
5b.	Larva without a long respiratory tube, if a short tube is present, it is divided apically 6
6a(5b).	Body terminating in a short tube that is divided apically (Fig. 17.134), or in a pair of spines .... Ephydridae
6b.	Body not terminating in a short tube or a pair of spines 7
7a(6b).	Caudal spiracular disc with palmate hairs and surrounded by 8–10 lobes, some of which may be very short (Fig. 17.135); body wrinkled .... Sciomyzidae
7b.	Caudal spiracular disc without palmate hairs, if surrounded by lobes, body not wrinkled 8
8a(7b).	Abdomen without distinct prolegs, paired ventral pseudopods, or terminal processes except a single spine, but with a girdle of six or more pseudopods on each segment (Fig. 17.136) .... Tabanidae
8b.	Distinct prolegs, paired pseudopods, or paired terminal processes present 9
9a(8b).	Body terminating in a spiracular pit surrounded by four pointed lobes (Fig. 17.137) .... Dolichopodidae
9b.	Body not terminating in a spiracular pit with four lobes 10
10a(9b).	Body terminating in a pair of ciliated, divergent processes; abdomen with eight pairs of ventral prolegs (Fig. 17.138) .... Athericidae
10b.	Terminal processes, if present, not ciliated; abdomen with or without prolegs 11

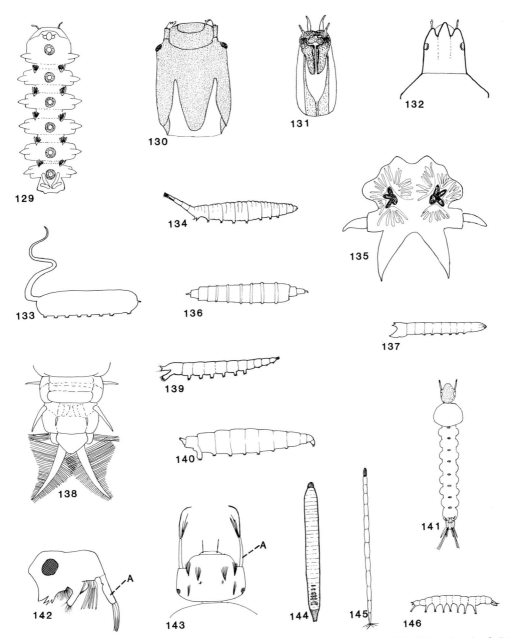

Figure 17.129 Blephariceridae; *Blepharicera* (ventral view). **Figure 17.130** Tipulidae; head of *Limonia* (dorsal view). **Figure 17.131** Tipulidae; head of *Hexatoma* (dorsal view). **Figure 17.132** Stratiomyidae; head of *Odontomyia* (dorsal view). **Figure 17.133** Syrphidae; *Eristalis* (lateral view). **Figure 17.134** Ephydridae (lateral view). **Figure 17.135** Sciomyzidae; spiracular disc of *Sepedon*. **Figure 17.136** Tabanidae; *Chrysops* (lateral view). **Figure 17.137** Dolicopodidae (lateral view). **Figure 17.138** Athericidae; terminal segments of *Atherix* (dorsal view). **Figure 17.139** Empididae (lateral view). **Figure 17.140** Muscidae; *Limnophora* (lateral view). **Figure 17.141** Culicidae; *Anopheles* (dorsal view). **Figure 17.142** Chaoboridae; head of *Chaoborus* (lateral view) showing antenna (A). **Figure 17.143** Culicidae; head of *Coquillettidia* (dorsal view) showing antenna (A). **Figure 17.144** Psychodidae; *Psychoda* (dorsal view). **Figure 17.145** Ceratopogonidae; *Palpomyia* (dorsal view). **Figure 17.146** Nymphomyiidae; *Palaeodipteron* (lateral view).

11a(10b).	Some external head structure visible, with palpi and antennae usually present; abdomen usually with ventral prolegs and elongate, paired, terminal appendages (Fig. 17.139) or a bulbous segment	Empididae
11b.	No visible external head structure; abdomen often with ventral pseudopods and usually short, paired, terminal appendages (Fig. 17.140)	Muscidae
12a(2a).	Prolegs absent	13

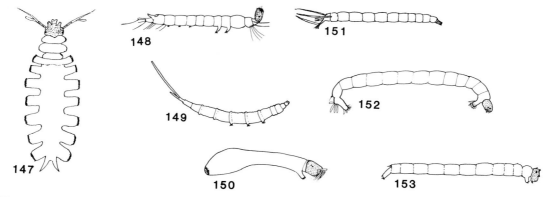

Figure 17.147 Deuterophlebiidae; *Deuterophlebia* (dorsal view). **Figure 17.148** Dixidae; *Dixella* (lateral view). **Figure 17.149** Ptychopteridae; *Ptychoptera* (lateral view). **Figure 17.150** Simuliidae; *Simulium* (lateral view). **Figure 17.151** Tanyderidae; *Protoplasa* (lateral view). **Figure 17.152** Chironomidae; *Chironomus* (lateral view). **Figure 17.153** Thaumaleidae; *Thaumalea* (lateral view).

12b.	Prolegs present at one or both ends of body or on abdominal segments (Figs. 17.146–17.153)	16
13a(12a).	Thoracic segments fused and distinctly thicker than abdomen (Fig. 17.141)	14
13b.	Thorax and abdomen about equal in diameter (Figs. 17.148 and 17.149)	15
14a(13a).	Antennae prehensile, with long, strong apical spines (Fig. 17.142)	Chaoboridae
14b.	Antennae not prehensile, lacking long apical spines (Fig. 17.143)	Culicidae
15a(13b).	Thoracic and abdominal segments each distinctly divided into two or three annuli, with sclerotized dorsal plates on some annuli (Fig. 17.144)	Psychodidae
15b.	No secondary annulations (Fig. 17.145)	Ceratopogonidae (in part)
16a(12b).	Prolegs on intermediate body segments (Figs. 17.146–17.149)	17
16b.	Prolegs on anterior and/or posterior ends of body only (Figs. 17.150–17.153)	20
17a(16a).	Seven or eight pairs of distinct prolegs on abdomen (Figs. 17.146 and 17.147)	18
17b.	Two or three pairs of weak prolegs on abdomen (Figs. 17.148 and 17.149)	19
18a(17a).	Eight pairs of slender, ventrally projecting prolegs (Fig. 17.146); Quebec, New Brunswick, Maine	Nymphomyiidae
18b.	Seven pairs of stout, ventrolaterally projecting prolegs (Fig. 17.147); western mountains	Deuterophlebiidae
19a(17b).	Paired ventral prolegs on abdominal segment 1 and usually also on segment 2; posterior end of body with two pairs of fringed processes (Fig. 17.148)	Dixidae
19b.	Paired ventral prolegs on abdominal segments 1, 2, and 3; body terminating in a long respiratory tube (Fig. 17.149)	Ptychopteridae
20a(16b).	A single proleg present only on prothorax; posterior of abdomen swollen and terminating in a ring of numerous small hooks (Fig. 17.150)	Simuliidae
20b.	Posterior prolegs usually present; posterior of abdomen not swollen and terminating in a ring of hooks	21
21a(20b).	Only posterior prolegs present (Fig. 17.151)	22
21b.	Anterior and usually posterior prolegs present (Figs. 17.152 and 17.153)	23
22a(21a).	Long filamentous processes arising from last two abdominal segments and from prolegs (Fig. 17.151); East and West	Tanyderidae
22b.	Last two abdominal segments without long filamentous processes	Ceratopogonidae (in part)
23a(21b).	Body covered with long, strong spines	Ceratopogonidae (in part)
23b.	Body covered at most with setae	24
24a(23b).	At least one pair of prolegs separated distally (Fig. 17.152); stalked spiracles lacking	Chironomidae
24b.	Prolegs unpaired (Fig. 17.153); short, stalked spiracles dorsolaterally on prothorax; mountainous regions	Thaumaleidae

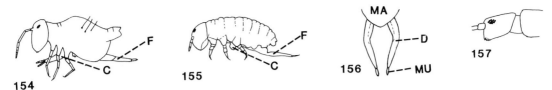

Figure 17.154 Sminthuridae (lateral view) showing furcula (F) and colophore (C). [Figs. 17.154–17.157 redrawn from Peckarsky *et al.* (1990).]   **Figure 17.155** Poduridae; *Podura* (lateral view) showing furcula (F) and colophore (C).   **Figure 17.156** Poduridae; furcula of *Podura* (dorsal view) showing manubrium (MA), dens (D), and mucro (MU).   **Figure 17.157** Isotomidae; anterior portion (lateral view).

J. Taxonomic Key to Families of Semiaquatic Collembola

1a.	Body somewhat globular; thorax and abdomen indistinctly segmented (Fig. 17.154)	Sminthuridae
1b.	Body elongate; thorax and abdomen distinctly segmented (Fig. 17.155)	2
2a(1b).	Mouthparts directed downward (Fig. 17.155); furcula distinctly convergent apically (Fig. 17.156)	Poduridae
2b.	Mouthparts directed forward (Fig. 17.157); furcula not distinctly convergent apically	Isotomidae

LITERATURE CITED

Agriculture Canada, Research Branch. 1981. Manual of Nearctic Diptera. Volume 1, Monograph No. 27. Canadian Government Publishing Centre, Supply and Services Canada, Hull, Quebec. 674 pp.

Agriculture Canada, Research Branch. 1987. Manual of Nearctic Diptera. Volume 2, Monograph No. 28. Canadian Government Publishing Centre, Supply and Services Canada, Hull, Quebec. 658 pp.

Allen, R. K. 1973. Generic revisions of mayfly nymphs. 1. *Traverella* in North and Central America (Leptophlebiidae). Annals of the Entomological Society of America 66:1287–1295.

Allen, R. K. 1978. The nymphs of North and Central American *Leptohyphes* (Ephemeroptera: Tricorythidae). Annals of the Entomological Society of America 71:537–558.

Allen, R. K., and G. F. Edmunds, Jr. 1959. A revision of the genus *Ephemerella* (Ephemeroptera: Ephemerellidae) I. The subgenus *Timpanoga*. Canadian Entomologist 91:51–58.

Allen, R. K., and G. F. Edmunds, Jr. 1961a. A revision of the genus *Ephemerella* (Ephemeroptera: Ephemerellidae) II. The subgenus *Caudatella*. Annals of the Entomological Society of America 54:603–612.

Allen, R. K., and G. F. Edmunds, Jr. 1961b. A revision of the genus *Ephemerella* (Ephemeroptera: Ephemerellidae) III. The subgenus *Attenuatella*. Journal of the Kansas Entomological Society 34:161–173.

Allen, R. K., and G. F. Edmunds, Jr. 1962a. A revision of the genus *Ephemerella* (Ephemeroptera: Ephemerellidae) IV. The subgenus *Dannella*. Journal of the Kansas Entomological Society 35:333–338.

Allen, R. K., and G. F. Edmunds, Jr. 1962b. A revision of the genus *Ephemerella* (Ephemeroptera: Ephemerellidae) V. The subgenus *Drunella* in North America. Miscellaneous Publications of the Entomological Society of America 3:147–187.

Allen, R. K., and G. F. Edmunds, Jr. 1963a. A revision of the genus *Ephemerella* (Ephemeroptera: Ephemerellidae) VI. The subgenus *Serratella* in North America. Annals of the Entomological Society of America 56:583–600.

Allen, R. K., and G. F. Edmunds, Jr. 1963b. A revision of the genus *Ephemerella* (Ephemeroptera: Ephemerellidae) VII. The subgenus *Eurylophella*. Canadian Entomologist 95:597–623.

Allen, R. K., and G. F. Edmunds, Jr. 1965. A revision of the genus *Ephemerella* (Ephemeroptera: Ephemerellidae) VIII. The subgenus *Ephemerella* in North America. Miscellaneous Publications of the Entomological Society of America 4:244–282.

Allen, R. K., and G. F. Edmunds, Jr. 1976. A revision of the genus *Ametropus* in North America (Ephemeroptera: Ametropodidae). Journal of the Kansas Entomological Society 49:625–635.

Allen, R. K., and C. M. Murvosh. 1987. Mayflies (Ephemeroptera: Tricorythidae) of the southwestern United States and northern Mexico. Annals of the Entomological Society of America 80:35–40.

Anderson, N. H., and J. R. Sedell. 1979. Detritus processing by macroinvertebrates in stream ecosystems. Annual Review of Entomology 24:351–377.

Anderson, R. D. 1970. A revision of the Nearctic representatives of *Hygrotus* (Coleoptera: Dytiscidae). Annals of the Entomological Society of America 64:503–512.

Anderson, R. D. 1975. A revision of the Nearctic species

of *Hygrotus* groups II and III (Coleoptera: Dytiscidae). Annals of the Entomological Society of America 69:577–584.

Anderson, R. D. 1983. Revision of the Nearctic species of *Hygrotus* groups IV, V, and VI (Coleoptera: Dytiscidae). Annals of the Entomological Society of America 76:173–196.

Barr, C. B., and J. B. Chapin. 1988. The aquatic Dryopoidea of Louisiana (Coleoptera: Psephenidae, Dryopidae, Elmidae). Tulane Studies in Zoology and Botany 26:89–164.

Baumann, R. W., and B. P. Stark. 1980. *Hesperoperla hoguei,* a new species of stonefly from California (Plecoptera: Perlidae). Great Basin Naturalist 40: 63–67.

Baumann, R. W., A. R. Gaufin, and R.F. Surdick. 1977. The Stoneflies (Plecoptera) of the Rocky Mountains. Memoirs of the American Entomological Society No. 31:208 pp.

Bay, E. C. 1974. Predator–prey relationships among aquatic insects. Annual Review of Entomology 19: 441–453.

Bednarik, A. F., and G. F. Edmunds, Jr. 1980. Descriptions of larval *Heptagenia* from the Rocky Mountain region (Ephemeroptera: Heptageniidae). Pan-Pacific Entomologist 56:51–62.

Bednarik, A. F., and W. P. McCafferty. 1979. Biosystematic revision of the genus *Stenonema* (Ephemeroptera: Heptageniidae). Canadian Bulletin of Fisheries and Aquatic Sciences No. 201:73 pp.

Bennefield, B. L. 1965. A taxonomic study of the subgenus *Ladona* (Odonata: Libellulidae). University of Kansas Science Bulletin 45:361–389.

Bennett, D. V., and E. F. Cook. 1981. The semiaquatic Hemiptera of Minnesota (Hemiptera: Heteroptera). Minnesota Agricultural Experiment Station Technical Bulletin No. 332:59 pp.

Berg, C. O., and L. Knudson. 1978. Biology and systematics of the Sciomyzidae. Annual Review of Entomology 23:239–258.

Berge Henegouwen, A. van. 1986. Revision of the European species of *Anacaena* Thomson (Coleoptera: Hydrophilidae). Entomologica Scandinavica 17:393–407.

Bergman, E. A., and W. L. Hilsenhoff. 1978. *Baetis* (Ephemeroptera: Baetidae) of Wisconsin. Great Lakes Entomologist 11:125–135.

Berner, L. 1956. The genus *Neoephemera* in North America (Ephemeroptera: Neoephemeridae). Annals of the Entomological Society of America 49:33–42.

Bilyj, B. 1988. A taxonomic review of *Guttipelopia* (Diptera: Chironomidae). Entomologica Scandinavica 19:1–26.

Bobb, M. L. 1974. The Insects of Virginia. No. 7: The aquatic and semi-aquatic Hemiptera of Virginia. Research Division Bulletin 87—Virginia Polytechnic Institute and State University, Blacksburg. 196 pp.

Bode, R. W. 1983. Larvae of North American *Eukiefferiella* and *Tvetenia* (Diptera: Chironomidae). New York State Museum Bulletin No. 452:40 pp.

Boesel, M. W. 1985. A brief review of the genus *Polypedilum* in Ohio, USA with keys to known stages of species occurring in northeastern USA. Ohio Journal of Science 85:245–262.

Borkent, A. 1979. Systematics and bionomics of the species of the subgenus *Schadonophasma* Dyar and Shannon (*Chaoborus,* Chaoboridae, Diptera). Quaestiones Entomologicae 15:122–255.

Borkent, A. 1984. The systematics and phylogeny of the *Stenochironomus* complex (*Xestochironomus, Harrisius,* and *Stenochironomus*) (Diptera: Chironomidae). Memoirs of the Entomological Society of Canada No. 128:269 pp.

Bratt, A. D., L. V. Knutson, B. A. Foote, and C. O. Berg. 1969. Biology of *Pherbellia* (Diptera: Sciomyzidae). Memoirs of the Cornell University Agricultural Experiment Station No. 404:247 pp.

Brigham, A. R., W. U. Brigham, and A. Gnilka, editors. 1981. The aquatic insects and oligochaetes of North and South Carolina. Midwest Aquatic Enterprises, Mahomet, Illinois. 837 pp.

Brigham, W. U. 1981a. Megaloptera. Pages 7.1–7.12 *in:* A. R. Brigham, W. U. Brigham, and A. Gnilka, editors. The aquatic insects and oligochaetes of North and South Carolina. Midwest Aquatic Enterprises, Mahomet, Illinois.

Brigham, W. U. 1981b. Aquatic Neuroptera. Pages 8.1–8.4 *in:* A. R. Brigham, W. U. Brigham, and A. Gnilka, editors. The aquatic insects and oligochaetes of North and South Carolina. Midwest Aquatic Enterprises, Mahomet, Illinois.

Brigham, W. U. 1981c. Aquatic Coleoptera. Pages 10.1–10.136 *in:* A. R. Brigham, W. U. Brigham, and A. Gnilka, editors. The aquatic insects and oligochaetes of North and South Carolina. Midwest Aquatic Enterprises, Mahomet, Illinois.

Brittain, J. E. 1982. Biology of mayflies. Annual Review of Entomology 27:119–147.

Brooks, A. R., and L. A. Kelton. 1967. Aquatic and semiaquatic Heteroptera of Alberta, Saskatchewan, and Manitoba (Hemiptera). Memoirs of the Entomological Society of Canada No. 51:92 pp.

Brown, H. P. 1972. Aquatic dryopoid beetles (Coleoptera) of the United States. Biota of Freshwater Ecosystems Identification Manual 6. U.S. Environmental Protection Agency Water Pollution Control Research Series 18050 ELDO4/72. 82 pp.

Brown, H. P. 1987. Biology of riffle beetles. Annual Review of Entomology 32:253–273.

Burger, J. F. 1977. The biosystematics of immature Arizona Tabanidae (Diptera). Transactions of the American Entomological Society 103:145–268.

Burks, B. D. 1953. The Mayflies or Ephemeroptera of Illinois. Bulletin of the Illinois Natural History Survey 26:1–216.

Calabrese, D. 1974. Keys to the adults and nymphs of the species of *Gerris* Fabricius occurring in Connecticut. Memoirs, Connecticut Entomological Society 1974:227–266.

Canterbury, L. E. 1978. Studies of the genus *Sialis* (Sialidae: Megaloptera) in eastern North America.

Ph.D. Thesis, University of Louisville, Louisville, Kentucky. 93 pp.

Carle, F. L. 1980. A new *Lanthus* (Odonata: Gomphidae) from eastern North America with adult and nymphal keys to American octogomphines. Annals of the Entomological Society of America 73:172–179.

Carpenter, S. J., and W. J. LaCasse. 1955. Mosquitoes of North America. University of California Press, Berkeley and Los Angeles. 360 pp.

Chapin, J. W. 1978. Systematics of Nearctic *Micrasema* (Trichoptera: Brachycentridae). Ph.D. Thesis, Clemson University, Clemson, South Carolina. 136 pp.

Chapman, H. C. 1958. Notes on the identity, habitat and distribution of some semiaquatic Hemiptera of Florida. Florida Entomologist 41:117–124.

Cheary, B. S. 1971. The biology, ecology, and systematics of the genus *Laccobius* (*Laccobius*) (Coleoptera: Hydrophilidae) of the new world. Ph.D. Thesis, University of California, Riverside. 178 pp. (Privately printed.)

Check, G. R. 1982. A revision of the North American species of *Callibaetis* (Ephemeroptera: Baetidae). Ph.D. Thesis, University of Minnesota, Minneapolis. 164 pp.

Christiansen, K., and P. Bellinger. 1980. The Collembola of North America north of the Rio Grande. A taxonomic analysis. Grinnell College, Grinnell, Iowa. 1322 pp.

Cook, E. F. 1956. The Nearctic Chaoborinae (Diptera: Culicidae). University of Minnesota Agricultural Experiment Station Technical Bulletin No. 218:102 pp.

Corbet, P. S. 1962. A biology of dragonflies. Witherby, London. 247 pp.

Corbet, P. S. 1980. Biology of Odonata. Annual Review of Entomology 25:189–217.

Courtney, G. W. 1990. Revision of Nearctic mountain midges (Diptera: Deuterophlebiidae). Journal of Natural History 24:81–118.

Cranston, P. S., D. R. Oliver, and O. A. Saether. 1983. The larvae of Orthocladiinae (Diptera: Chironomidae) of the Holarctic region—keys and diagnosis. Entomologica Scandinavica Supplement 9: 291 pp.

Cummins, K. W. 1973. Trophic relations of aquatic insects. Annual Review of Entomology 18:183–206.

Cuyler, R. D. 1958. The larvae of *Chauliodes* Latreille (Megaloptera: Corydalidae). Annals of the Entomological Society of America 51:582–586.

Darsie, R. F., and R. A. Ward. 1981. Identification and geographical distribution of the mosquitoes of North America, north of Mexico. Mosquito Systematics, Supplement 1, American Mosquito Control Association, Fresno, California. 313 pp.

Day, W. C. 1956. Ephemeroptera. Pages 79–105 *in:* R. L. Usinger, editor. Aquatic insects of California with keys to North American genera and California species. University of California Press, Berkeley and Los Angeles.

Dosdall, L., and D. M. Lehmkuhl. 1979. Stoneflies (Plecoptera) of Saskatchewan. Quaestiones Entomologicae 15:3–116.

Downing, J. A., and F. H. Rigler. 1984. A manual on methods for the assessment of secondary productivity in fresh waters. 2nd edition, IBP Handbook 17. Blackwell, Oxford. 501 pp.

Drake, C. J., and H. M. Harris. 1934. The Gerrinae of the Western Hemisphere (Hemiptera). Annals of the Carnegie Museum 23:179–240.

Dunn, C. E. 1979. A revision and phylogenetic study of the genus *Hesperocorixa* Kirkaldy (Hemiptera: Corixidae). Proceedings of the Academy of Natural Sciences of Philadelphia 131:158–190.

Edmunds, G. F., Jr. 1972. Biogeography and evolution of Ephemeroptera. Annual Review of Entomology 17: 21–42.

Edmunds, G. F., Jr., and J.R. Traver. 1959. The classification of the Ephemeroptera I. Ephemeroidea: Behningiidae. Annals of the Entomological Society of America 52:43–51.

Edmunds, G. F., Jr., S. L. Jensen, and L. Berner. 1976. The mayflies of North and Central America. University of Minnesota Press, Minneapolis. 330 pp.

Edwards, S. W. 1961. The immature stages of *Xiphocentron mexico* (Trichoptera). Texas Journal of Science 13:51–56.

Edwards, S. W., and C. R. Arnold. 1961. The caddis flies of the San Marcos River. Texas Journal of Science 13:398–415.

Epler, J. H. 1988. Biosystematics of the genus *Dicrotendipes* Kieffer, 1913 (Diptera: Chironomidae: Chironominae) of the world. Memoirs of the American Entomological Society Number 16. 214 p.

Fall, H. C. 1922a. A review of the North American species of *Agabus* together with a description of a new genus and species of the tribe Agabini. John D. Sherman, Jr., Mt. Vernon, New York. 36 pp.

Fall, H. C. 1922b. The North American species of *Gyrinus* (Coleoptera). Transactions of the American Entomological Society 47:269–307.

Fall, H. C. 1923. A revision of the North American species of *Hydroporus* and *Agaporus*. John D. Sherman, Jr., Mt. Vernon, New York. 238 pp.

Ferkinhoff, W. D., and R. W. Gundersen. 1983. A key to the whirligig beetles of Minnesota and adjacent states and Canadian provinces (Coleoptera: Gyrinidae). Scientific Publications of the Science Museum of Minnesota, New Series Vol. 5, No. 3. 53 pp.

Flint, O. S., Jr. 1960. Taxonomy and biology of Nearctic limnephelid larvae (Trichoptera), with special reference to species in eastern United States. Entomologica Americana 40:1–120.

Flint, O. S., Jr. 1961. The immature stages of the Arctopsychinae occurring in eastern North America (Trichoptera: Hydropsychidae). Annals of the Entomological Society of America 54:5–11.

Flint, O. S., Jr. 1962. Larvae of the caddis fly genus *Rhyacophila* in eastern North America (Trichoptera: Rhyacophilidae). Proceedings of the United States National Museum 113:465–493.

Flint, O. S., Jr. 1984. The genus *Brachycentrus* in North America, with a proposed phylogeny of the genera of

Brachycentridae (Trichoptera). Smithsonian Contributions to Zoology Number 398:58 pp.

Flowers, R. W. 1982. Review of the genus *Macdunnoa* (Ephemeroptera: Heptageniidae) with description of a new species from Florida. Great Lakes Entomologist 15:25–30.

Flowers, R. W., and W. L. Hilsenhoff. 1975. Heptageniidae (Ephemeroptera) of Wisconsin. Great Lakes Entomologist 8:201–218.

Foote, B. A. 1959. Biology and life history of the snail-killing flies belonging to the genus *Sciomyza* Fallen (Diptera: Sciomyzidae). Annals of the Entomological Society of America 52:31–43.

Foote, B. A. 1961. Biology and immature stages of the snail-killing flies belonging to the genus *Tetanocera* (Diptera: Sciomyzidae). Ph.D. Thesis, Cornell University, Ithaca, New York. 190 pp.

Froeschner, R. C. 1949. Contribution to a synopsis of the Hemiptera of Missouri. Part IV. Hebridae, Mesoveliidae, Cimicidae, Anthocoridae, Cryptostemmatidae, Isometopidae, Miridae. American Midland Naturalist 42:123–188.

Froeschner, R. C. 1962. Contribution to a synopsis of the Hemiptera of Missouri. Part V. Hydrometridae, Gerridae, Veliidae, Saldidae, Ochteridae, Gelastocoridae, Naucoridae, Belostomatidae, Nepidae, Notonectidae, Pleidae, Corixidae. American Midland Naturalist 67:208–240.

Fullington, K. E., and K. W. Stewart. 1980. Nymphs of the stonefly genus *Taeniopteryx* (Plecoptera: Taeniopterygidae) of North America. Journal of the Kansas Entomological Society 53:237–259.

Garman, P. 1917. The Zygoptera or damselflies of Illinois. Bulletin of the Illinois State Laboratory of Natural History 12:412–587.

Garman, P. 1927. Guide to the Insects of Connecticut. Part V: The Odonata or Dragonflies of Connecticut. State Geological and Natural History Survey Bulletin No. 39:331 pp.

Garrison, R. W. 1984. Revision of the genus *Enallagma* of the United States west of the Rocky Mountains and identification of certain larvae by discriminant analysis (Odonata: Coenagrionidae). University of California Publications in Entomology Volume 105:129 pp.

Gelhaus, J. K. 1986. Larvae of the cranefly genus *Tipula* in North America (Diptera: Tipulidae). University of Kansas Science Bulletin 53:121–182.

Glukhova, V. M. 1977. Ceratopogonidae. Pages 431–457 *in:* The identification of fresh water invertebrates of European USSR i.e. USSR west of the Urals (plankton and benthos). Zoological Institute., USSR Academy of Science, Moscow. (Translated from Russian by D.S. Kettle.)

Gonsoulin, G. J. 1973a. Seven families of aquatic and semiaquatic Hemiptera in Louisiana. Part I. Family Hydrometridae Billberg, 1820. "marsh treaders". Entomological News 84:9–16.

Gonsoulin, G. J. 1973b. Seven families of aquatic and semiaquatic Hemiptera in Louisiana. Part II. Family Naucoridae Fallen, 1814. "creeping water bugs". Entomological News 84:83–88.

Gonsoulin, G. J. 1973c. Seven families of aquatic and semiaquatic Hemiptera in Louisiana. Part III. Family Belostomatidae Leach, 1815. Entomological News 84:173–189.

Gonsoulin, G. J. 1974. Seven families of aquatic and semiaquatic Hemiptera in Louisiana. Part IV. Family Gerridae. Transactions of the American Entomological Society 100:513–546.

Gonsoulin, G. J. 1975. Seven families of aquatic and semiaquatic Hemiptera in Louisiana. Part V. Family Nepidae Latreille, 1802. Entomological News 86:23–32.

Goodwin, J. T. 1987. Immature stages of some eastern Nearctic Tabanidae (Diptera). VIII. Additional species of *Tabanus* Linnaeus and *Chrysops* Meigen. Florida Entomologist 70:268–277.

Gordon, R. D. 1969. A revision of the *niger-tenebrosus* group of *Hydroporus* (Coleoptera: Dytiscidae) in North America. Ph.D. Thesis, North Dakota State University, Fargo. 311 pp.

Gordon, R. D. 1981. New species of North American *Hydroporus, niger-tenebrosa* group (Coleoptera: Dytiscidae). Pan-Pacific Entomologist 57:105–123.

Gordon, R. D., and R. L. Post. 1965. North Dakota water beetles. North Dakota Insects, Publication No. 5. Department of Entomology, North Dakota State University, Fargo. 52 pp.

Grodhaus, G. 1987. *Endochironomus* Kieffer, *Tribelos* Townes, *Synendotendipes*, n. gen., and *Endotribelos*, n. gen. (Diptera: Chironomidae) of the Nearctic region. Journal of the Kansas Entomological Society 60:167–247.

Gundersen, R. W. 1978. Nearctic *Enochrus*. Biology, keys, descriptions and distribution (Coleoptera: Hydrophilidae). Department of Biological Science, St. Cloud State University, St. Cloud, Minnesota. 54 pp.

Gundersen, R. W., and C. Otremba. 1988. Haliplidae of Minnesota. Scientific Publications of the Science Museum of Minnesota, New Series, Vol. 6 No. 3. 43 pp.

Haddock, J. D. 1977. The biosystematics of the caddis fly genus *Nectopsyche* in North America with emphasis on the aquatic stages. American Midland Naturalist 98:382–421.

Harper, P. P., and H. B. N. Hynes. 1971a. The Leuctridae of eastern Canada (Insects; Plecoptera). Canadian Journal of Zoology 49:915–920.

Harper, P. P., and H. B. N. Hynes. 1971b. The Capniidae of eastern Canada (Insecta; Plecoptera). Canadian Journal of Zoology 49:921–940.

Harper, P. P., and H. B. N. Hynes. 1971c. The nymphs of the Taeniopterygidae of eastern Canada (Insecta; Plecoptera). Canadian Journal of Zoology 49:941–947.

Harper, P. P., and H. B. N. Hynes. 1971d. The nymphs of the Nemouridae of eastern Canada (Insecta: Plecoptera). Canadian Journal of Zoology 49:1129–1142.

Hatch, M. H. 1930. Records and new species of Coleoptera from Oklahoma and western Arkansas, with subsidiary studies. Publications of the University of Oklahoma Biological Survey 2:15–26.

Hatch, M. H. 1953. The Beetles of the Pacific Northwest. Part I: Introduction and Adephaga. University of Washington Publications in Biology Vol. 16:340 pp.

Hatch, M. H. 1965. The beetles of the Pacific Northwest. Part IV: Macrodactyles, Palpicornes, and Heteromera. University of Washington Publications in Biology Vol. 16:268 pp.

Henry, T. J., and R. C. Froeschner, editors. 1988. Catalog of the Heteroptera, or true bugs, of Canada and the continental United States. Brill, Leiden, Netherlands. 958 pp.

Herring, J. L. 1948. Taxonomic and distributional notes on the Hydrometridae of Florida (Hemiptera). Florida Entomologist 31:112–116.

Hilsenhoff, W. L. 1973. Notes on *Dubiraphia* (Coleoptera: Elmidae) with descriptions of five new species. Annals of the Entomological Society of America 66: 55–61.

Hilsenhoff, W. L. 1975. Notes on Nearctic *Acilius* (Dytiscidae), with the description of a new species. Annals of the Entomological Society of America 68:271–274.

Hilsenhoff, W. L. 1980. *Coptotomus* (Coleoptera: Dytiscidae) in eastern North America with descriptions of two new species. Transactions of the American Entomological Society 105:461–471.

Hilsenhoff, W. L. 1981. Aquatic insects of Wisconsin. Keys to Wisconsin genera and notes on biology, distribution and species. Publication of the Natural History Council, University of Wisconsin, Madison. 60 pp.

Hilsenhoff, W. L. 1984. Aquatic Hemiptera of Wisconsin. Great Lakes Entomologist 17:29–50.

Hilsenhoff, W. L. 1985. The Brachycentridae of Wisconsin (Trichoptera). Great Lakes Entomologist 18:149–154.

Hilsenhoff, W. L. 1986. Semiaquatic Hemiptera of Wisconsin. Great Lakes Entomologist 19:7–19.

Hilsenhoff, W. L. 1987. Effectiveness of bottle traps for collecting Dytiscidae (Coleoptera). Coleopterists Bulletin 41:377–380.

Hilsenhoff, W. L., and S. J. Billmyer. 1973. Perlodidae (Plecoptera) of Wisconsin. Great Lakes Entomologist 6:1–14.

Hilsenhoff, W. L., and W. U. Brigham. 1978. Crawling water beetles of Wisconsin (Coleoptera: Haliplidae). Great Lakes Entomologist 11:11–22.

Hitchcock, S. W. 1974. Guide to the insects of Connecticut. Part VII: The Plecoptera or stoneflies of Connecticut. State Geological and Natural History Survey of Connecticut Bulletin No. 107:262 pp.

Huggins, D. G., and W. U. Brigham. 1981. Odonata. Pages 4.1–4.100 *in:* A. R. Brigham, W. U. Brigham, and A. Gnilka, editors. 1981. The aquatic insects and oligochaetes of North and South Carolina. Midwest Aquatic Enterprises, Mahomet, Illinois.

Hungerford, H. B. 1922. The Nepidae of North America. Kansas University Science Bulletin 14:425–469.

Hungerford, H. B. 1933. The genus *Notonecta* of the world (Notonectidae–Hemiptera). University of Kansas Science Bulletin 21:5–193.

Hungerford, H. B. 1948. The Corixidae of the Western Hemisphere. University of Kansas Science Bulletin 32:1–288, 408–827.

Hungerford, H. B. 1954. The genus *Rheumatobates* Bergroth (Hemiptera–Gerridae). University of Kansas Science Bulletin 36:529–588.

Hungerford, H. B., and N. W. Evans. 1934. The Hydrometridae of the Hungarian National Museum and other studies in the family. Annales Historico-naturales Musei Nationalis Hungarici 28:31–112

Hutchinson, G. E. 1945. On the species of *Notonecta* (Hemiptera–Heteroptera) inhabiting New England. Transactions of the Connecticut Academy of Arts and Sciences 36:599–605.

Hynes, H. B. N. 1976. The biology of Plecoptera. Annual Review of Entomology 21:135–153.

Jansson, A. 1972. Systematic notes and new synonymy in the genus *Cenocorixa* (Hemiptera: Corixidae). Canadian Entomologist 104:449–459

Jewett, S. G., Jr. 1956. Plecoptera. Pages 155–181 *in:* R. L. Usinger, editor. Aquatic insects of California with keys to North American genera and California species. University of California Press, Berkeley and Los Angeles.

Kavanaugh, D. H. 1986. A systematic review of amphizoid beetles (Amphizoidae: Coleoptera) and their phylogenetic relationships to other Adephaga. Proceedings of the California Academy of Sciences 44: 67–109.

Kettle, D. S. 1977. Biology and bionomics of bloodsucking ceratopogonids. Annual Review of Entomology 22:33–51.

Kittle, P. D. 1977. A revision of the genus *Trepobates* Uhler (Hemiptera: Gerridae). Ph.D. Thesis, University of Arkansas, Fayetteville. 255 pp.

Kittle, P. D. 1980. The water striders (Hemiptera: Gerridae) of Arkansas. Arkansas Academy of Science Proceedings 34:68–71.

Knutson, L. V., and C. O. Berg. 1964. Biology and immature stages of snail-killing flies: the genus *Elgiva* (Diptera: Sciomyzidae). Annals of the Entomological Society of America 57:173–192.

Kondratieff, B. C., and J. R. Voshell, Jr. 1984. The North and Central American species of *Isonychia* (Ephemeroptera: Oligoneuriidae). Transactions of the American Entomological Society 110:129–244.

Kondratieff, B. C., R. F. Kirchner, and J. R. Voshell, Jr. 1981. Nymphs of *Diploperla* (Plecoptera: Perlidae). Annals of the Entomological Society of America 74:428–430.

Kondratieff, B. C., R. F. Kirchner, and K. W. Stewart. 1988. A review of *Perlinella* Banks (Plecoptera: Perlidae). Annals of the Entomological Society of America 81:19–27.

Kuitert, L. C. 1942. Gerrinae in the University of Kansas Collections. University of Kansas Science Bulletin 28:113–143.

LaRivers, I. 1948. A new species of *Pelocoris* from Nevada, with notes on the genus in the United States (Hemiptera: Naucoridae). Annals of the Entomological Society of America 41:371–376.

LaRivers, I. 1951. A revision of the genus *Ambrysus* in the United States. University of California Publications in Entomology 8:277–338.

Larson, D. J. 1975. The predaceous water beetles (Coleoptera: Dytiscidae) of Alberta: systematics, natural history and distribution. Quaestiones Entomologicae 11:245–498.

Larson, D. J. 1987. Revision of North American species of *Ilybius* Erichson (Coleoptera: Dytiscidae), with systematic notes on Palaearctic species. Journal of the New York Entomological Society 95:341–413.

Larson, D. J. 1989. Revision of North American *Aqabus* Leach (Coleoptera: Dytiscidae); introduction, key to species groups, and classification of the *Ambiquus-*, *Tristis-*, and *Arcticus-*groups. The Canadian Entomologist 121:861–919.

Larson, D. J., and R. E. Roughley. 1990. A review of the species of *Liodessus* Guignot of North America north of Mexico with the description of a new species (Coleoptera: Dytiscidae). Journal of the New York Entomological Society 98:233–245.

Leech, H. B. 1938. A study of the Pacific Coast species of *Agabus* Leach, with a key to the Nearctic species. Master's Thesis, University of California, Berkeley.

Leech, H. B., and H. P. Chandler. 1956. Aquatic Coleoptera. Pages 293–371 *in*: R. L. Usinger, editor. Aquatic insects of California with keys to North American genera and California species. University of California Press, Berkeley and Los Angeles.

Lehmkuhl, D. M., and N. H. Anderson. 1971. Contributions to the biology and taxonomy of the *Paraleptophlebia* of Oregon (Ephemeroptera: Leptophlebiidae). Pan-Pacific Entomologist 47:85–93.

Lewis, P. A. 1974. Taxonomy and ecology of *Stenonema* mayflies (Heptageniidae: Ephemeroptera). U.S Environmental Protection Agency Environmental Monitoring Series EPA-670/4-74-006:81 pp.

Mackay, R. J., and G. B. Wiggins. 1979. Ecological diversity in Trichoptera. Annual Review of Entomology 24:185–208.

Malcolm, S. E. 1971. The water beetles of Maine: including the families Gyrinidae, Haliplidae, Dytiscidae, Noteridae, and Hydrophilidae. Technical Bulletin 48—Life Sciences and Agricultural Experiment Station, University of Maine at Orono. 49 pp.

Matta, J. F. 1974. The Insects of Virginia. No. 8: The aquatic Hydrophilidae of Virginia (Coleoptera: Polyphaga). Research Division Bulletin 94—Virginia Polytechnic Institute and State University 44 pp.

Matta, J. F. 1976. The insects of Virginia. No. 10: The Haliplidae of Virginia (Coleoptera: Adephaga). Research Division Bulletin 109—Virginia Polytechnic Institute and State University, Blacksburg. 26 pp.

Matta, J. F., and Wolfe, G. W. 1981. A revision of the subgenus *Heterosternuta* Strand of *Hydroporus* Clairville (Coleoptera: Dytiscidae). Pan-Pacific Entomologist 57:176–219.

McCafferty, W. P. 1975. The burrowing mayflies (Ephemeroptera: Ephemeroidea) of the United States. Transactions of the American Entomological Society 101:447–504.

McCafferty, W. P. 1977. Biosystematics of *Dannella* and related subgenera of *Ephemerella* (Ephemeroptera: Ephemerellidae). Annals of the Entomological Society of America 70:881–889.

McCafferty, W. P. 1981. Aquatic entomology. The fishermen's and ecologists' illustrated guide to insects and their relatives. Science Books International, Boston, Massachusetts. 448 pp.

McCafferty, W. P., and A. V. Provonsha. 1988. Revisionary notes on predaceous Heptageniidae based on larval and adult associations (Ephemeroptera). Great Lakes Entomologist 21:15–17.

McCorkle, D. V. 1965. (Subfamily Elophorinae). Pages 23–38 *in*: M. H. Hatch, editor. The beetles of the Pacific Northwest. Part IV: Macrodactyles, Palpicornes, and Heteromera. University of Washington Publications in Biology Vol. 16.

Menke, A. S., editor. 1979a. The semiaquatic and aquatic Hemiptera of California (Heteroptera: Hemiptera). Bulletin of the California Insect Survey Volume 21: 166 pp.

Menke, A. S., editor. 1979b. Family Belostomatidae/giant water bugs, electric light bugs, toe biters. Pages 76–86 *in*: A. S. Menke, editor. The semiaquatic and aquatic Hemiptera of California (Heteroptera: Hemiptera). Bulletin of the California Insect Survey Volume 21.

Merritt, R. W., and K. W. Cummins, editors. 1984. An introduction to the aquatic insects of North America. 2nd Edition. Kendall/Hunt, Dubuque, Iowa. 722 pp.

Merritt, R. W., D. H. Ross, and B. V. Peterson. 1978. Larval ecology of some lower Michigan black flies (Diptera: Simuliidae) with keys to the immature stages. Great Lakes Entomologist 11;177–208.

Michael, A. G., and J. F. Matta. 1977. The insects of Virginia. No. 12: The Dytiscidae of Virginia (Coleoptera: Adephaga) (subfamilies: Laccophilinae, Colymbetinae, Dytiscinae, Hydaticinae and Cybistrinae). Research Division Bulletin 124—Virginia Polytechnic Institute and State University, Blacksburg. 53 pp.

Miller, D. C. 1965. Hydrophilidae, except Elophorinae and Sphaeridiinae. Pages 21–23, 38–46 *in*: M. H. Hatch, editor. The beetles of the Pacific Northwest. Part IV: Macrodactyles, Palpicornes, and Heteromera. University of Washington Publications in Biology Vol. 16.

Miller, P. L. 1987. Dragonflies. Naturalists' Handbooks, No. 7. Cambridge University Press, London. 84 pp.

Morihara, D. K., and W. P. McCafferty. 1979. The *Baetis* larvae of North America (Ephemeroptera: Baetidae). Transactions of the American Entomological Society 105:139–221.

Morrissette, R. 1979. Les Gyrinidae (Coleoptera) du Quebec. Fabreries 5:51–58.

Musser, R. J. 1962. Dragonfly nymphs of Utah (Odonata: Anisoptera). Biological Series 12(6). University of Utah, Salt Lake City. 66 pp.

Needham, J. G., and M. J. Westfall, Jr. 1955. A manual of the dragonflies of North America (Anisoptera). University of California Press, Berkeley and Los Angeles. 615 pp.

Needham, J. G., J. R. Traver, and Y.-C. Hsu. 1935. The biology of mayflies. Comstock, New York. 759 pp.

Neff, S. E., and C. O. Berg. 1962. Biology and immature stages of *Hoplodictya spinicornis* and *H. setosa* (Diptera: Sciomyzidae). Transactions of the American Entomological Society 88:77–93.

Neff, S. E., and C. O. Berg. 1966. Biology and immature stages of malacophagous Diptera of the genus *Sepedon* (Sciomyzidae). Bulletin of the Agricultural Experiment Station, Virginia Polytechnic Institute, Blacksburg. 566:5–113.

Neunzig, H. H. 1966. Larvae of the genus *Nigronia* Banks (Neuroptera: Corydalidae). Proceedings of the Entomological Society of Washington 68:11–16.

Nimmo, A. P. 1987. The adult Arctopsychidae and Hydropsychidae (Trichoptera) of Canada and adjacent United States. Quaestiones Entomologicae 23:1–189.

Oliver, D. R., and M. E. Roussel. 1982. The larvae of *Pagastia* Oliver (Diptera: Chironomidae) with descriptions of three Nearctic species. Canadian Entomologist 114:849–854.

Oliver, D. R., and M. E. Roussel. 1983. Redescription of *Brillia* Kieffer (Diptera: Chironomidae) with descriptions of Nearctic species. Canadian Entomologist 115:257–279.

Oygur, S. 1988. Taxonomy, distribution, and phylogeny of North American (north of Mexico) *Gyrinus* Geoffroy (Coleoptera: Gyrinidae). Ph.D. Thesis, Rutgers University, New Brunswick, New Jersey. 296 pp.

Parfin, S. I., and A. B. Gurney. 1956. The spongilla-flies, with species reference to those of the Western Hemisphere (Sisyridae: Neuroptera). Proceedings of the United States National Museum 105:421–529.

Parker, C. R., and G. B. Wiggins. 1985. The Nearctic caddisfly genus *Hesperophylax* Banks (Trichoptera: Limnephilidae). Canadian Journal of Zoology 63:2433–2472.

Parker, C. R., and G. B. Wiggins. 1987. Revision of the caddisfly genus *Psilotreta* (Trichoptera: Odontoceridae). Life Science Contributions Royal Ontario Museum 144:55 pp.

Pechuman, L. L., D. W. Webb, and H. J. Teskey. 1983. The Diptera or true flies of Illinois. I. Tabanidae. Bulletin of the Illinois Natural History Survey 33:122 pp.

Peckarsky, B. L., P. Fraissinet, M. A. Penton, and D. J. Conklin, Jr. 1990. Freshwater macroinvertebrates of northeastern North America. Cornell University Press, Ithaca, New York. 422 pp.

Perkins, P. D. 1980. Aquatic beetles of the family Hydraenidae in the western hemisphere: classification, biogeography and inferred phylogeny (Insects: Coleoptera). Quaestiones Entomologicae 16:1–554.

Pescador, M. L., and L. Berner. 1981. The mayfly family Baetiscidae (Ephemeroptera). Part II Biosystematics of the genus *Baetisca*. Transactions of the American Entomological Society 107:163–228.

Peterson, B. V. 1970. The *Prosimulium* of Canada and Alaska (Diptera: Simuliidae). Memoirs of the Entomological Society of Canada No. 69:216 pp.

Pinder, L. C. V. 1986. Biology of freshwater Chironomidae. Annual Review of Entomology 31:1–23.

Poirrier, M. A., and Y. M. Arceneaux. 1972. Studies on southern Sisyridae (spongillaflies) with a key to the third-instar larvae and additional sponge-host records. American Midland Naturalist 88:455–458.

Polhemus, J. T., and H. C. Chapman. 1979. Family Mesoveliidae/water treaders. Pages 39–42 *in:* A. S. Menke, editor. The semiaquatic and aquatic Hemiptera of California (Heteroptera: Hemiptera). Bulletin of the California Insect Survey Volume 21.

Porter, T. W. 1950. Taxonomy of the American Hebridae and the natural history of selected species. Ph.D. Thesis, University of Kansas, Lawrence. 185 pp.

Pritchard, G. 1983. Biology of Tipulidae. Annual Review of Entomology 28:1–22.

Ray, D. H., and B. P. Stark. 1981. The Nearctic species of *Hydroperla* (Plecoptera: Perlodidae). Florida Entomologist 64:385–395.

Resh, V. H. 1976. The biology and immature stages of the caddisfly genus *Ceraclea* in eastern North America (Trichoptera: Leptoceridae). Annals of the Entomological Society of America 69:1039–1061.

Resh, V. H., and D. M. Rosenberg, editors. 1984. The ecology of aquatic insects. Praeger, New York. 625 pp.

Ricker, W. E. 1943. Stoneflies of southwestern British Columbia. Indiana University Publications, Science Series 12:145 pp.

Ricker, W. E. 1952. Systematic studies in Plecoptera. Indiana University Science Publications, Science Series 18:200 pp.

Roback, S. S. 1985. The immature chironomids of the eastern United States. VI. Pentaneurini—genus *Ablabesmyia*. Proceedings of the Academy of Natural Sciences of Philadelphia 137:73–128.

Roback, S. S. 1986a. The immature chironomids of the eastern United States. VII. Pentaneurini—genus *Monopelopia* new record with redescription of the male adults and description of some Neotropical material. Proceedings of the Academy of Natural Sciences of Philadelphia 138:350–365.

Roback, S. S. 1986b. The immature chironomids of the eastern United States. VIII. Pentaneurini—genus *Nilotanypus* with the description of a new species from Kansas. Proceedings of the Academy of Natural Sciences of Philadelphia 138:443–465.

Roback, S. S. 1987. The immature chironomids of the eastern United States. IX. Pentaneurini—genus *Labrundinia* with the description of some Neotropical material. Proceedings of the Academy of Natural Sciences of Philadelphia 139:159–209.

Roberts, C. H. 1913. Critical notes on the species of Haliplidae of America north of Mexico with descriptions of new species. Journal of the New York Entomological Society 21:91–123.

Ross, H. H. 1944. The caddis flies or Trichoptera of Illinois. Bulletin of the Illinois Natural History Survey 23:326 pp.

Roughley, R. E., and D. H. Pengelly. 1981. Classification, phylogeny, and zoogeography of *Hydaticus* Leach (Coleoptera: Dytiscidae) of North America. Quaestiones Entomologicae 17:249–309.

Saether, O. A. 1970. Nearctic and Palaearctic *Chaoborus* (Diptera: Chaoboridae). Bulletin 174, Fisheries Research Board of Canada. 57 pp.

Sailer, R. I. 1948. The genus *Trichocorixa* (Corixidae,

Hemiptera). University of Kansas Science Bulletin 32:289–407.

Sanderson, M. W. 1981. Aquatic and semiaquatic Heteroptera. Pages 6.1–6.94 *in:* A. R. Brigham, W. U. Brigham, and A. Gnilka, editors. The aquatic insects and oligochaetes of North and South Carolina. Midwest Aquatic Enterprises, Mahomet, Illinois.

Schefter, P. W., and G. B. Wiggins. 1986. A systematic study of the nearctic larvae of the *Hydropsyche morosa* group (Trichoptera: Hydropsychidae). Life Sciences Miscellaneous Publication Royal Ontario Museum. 94 pp.

Schmid, F. 1968. La famille des Arctopsychides. Memoires de la Societe Entomologique du Quebec No. 1:84 pp.

Schmude, K. L., and W. L. Hilsenhoff. 1986. Biology, ecology, larval taxonomy, and distribution of Hydropsychidae (Trichoptera) in Wisconsin. Great Lakes Entomologist 19:123–145.

Schuster, G. A., and D. A. Etnier. 1978. A manual for the identification of the larvae of the caddisfly genera *Hydropsyche* Pictet and *Symphitopsyche* Ulmer in eastern and central North America (Trichoptera: Hydropsychidae). United States Environmental Protection Agency Environmental Monitoring and Support Laboratory 600/4-78-060: 129 pp.

Sherberger, F. F., and J. B. Wallace. 1971. Larvae of the southeastern species of *Molanna*. Journal of the Kansas Entomological Society 44:217–224.

Simpson, K. W. 1982. A guide to basic taxonomic literature for the genera of North American Chironomidae (Diptera)—adults, pupae, and larvae. Bulletin of the New York State Museum No. 447:43 pp.

Simpson, K. W., R. W. Bode, and P. Albu. 1983. Keys for the genus *Cricotopus* adapted from "Revision der Gattung *Cricotopus* van der Wulp und ihrer Verwandten (Diptera, Chironomidae)" by M. Hirvenoja. Bulletin No. 450 New York State Museum. 133 pp.

Smetana, A. 1974. Revision of the genus *Cymbiodyta* Bed. (Coleoptera: Hydrophilidae). Memoirs of the Entomological Society of Canada No. 93:113 pp.

Smetana, A. 1980. Revision of the genus *Hydrochara* Berth. (Coleoptera: Hydrophilidae). Memoirs of the Entomological Society of Canada No. 111:100 pp.

Smetana, A. 1985. Revision of the subfamily Helophorinae of the Nearctic region (Coleoptera: Hydrophilidae). Memoirs of the Entomological Society of Canada No. 131:154 pp.

Smetana, A. 1988. Review of the family Hydrophilidae of Canada and Alaska. Memoirs of the Entomological Society of Canada No. 142:316 pp.

Smith, C. L. 1980. A taxonomic revision of the genus *Microvelia* Westwood (Heteroptera: Veliidae) of North America including Mexico. Ph.D. Thesis, University of Georgia, Athens. 388 pp.

Smith, C. L., and J. T. Polhemus. 1978. The Veliidae (Heteroptera) of America north of Mexico—keys and check list. Proceedings of the Entomological Society of Washington 80:56–68.

Smith, R. F., and A. E. Pritchard. 1956. Odonata. Pages 106–153 *in:* R. L. Usinger, editor. Aquatic insects of California with keys to North American genera and California species. University of California Press, Berkeley and Los Angeles.

Snoddy, E. L., and R. Noblet. 1976. Identification of the immature black flies (Diptera: Simuliidae) of the southeastern United States with some aspects of the adult role in transmission of *Leucocytozoon smithi* to turkeys. South Carolina Agricultural Experiment Station Technical Bulletin No. 1057:58 pp.

Soldan, T. 1984. A revision of the Caenidae with ocellar tubercles in the nymphal stage (Ephemeroptera). Acta Universitatis Carolinae, Biologica 1982–1984:289–362.

Spangler, P. J. 1960. A revision of the genus *Tropisternus* (Coleoptera: Hydrophilidae). Ph.D. Thesis, University of Missouri, Columbia. 364 pp.

Stark, B. P. 1983. A review of the genus *Soliperla* (Plecoptera: Peltoperlidae). Great Basin Naturalist 43:30–44.

Stark, B. P. 1986. The Nearctic species of *Agnetina* (Plecoptera: Perlidae). Journal of the Kansas Entomological Society 59:437–445.

Stark, B. P., and D. H. Ray. 1983. A revision of the genus *Helopicus* (Plecoptera: Perlodidae). Freshwater Invertebrate Biology 2:16–27.

Stark, B. P., and S. W. Szczytko. 1981. Contributions to the systematics of *Paragnetina* (Plecoptera: Perlidae). Journal of the Kansas Entomological Society 54:625–648.

Stark, B. P., and S. W. Szczytko. 1988. A new *Malirekus* species from eastern North America (Plecoptera: Perlodidae). Journal of the Kansas Entomological Society. 61:195–199.

Stewart, K. W., and J. A. Stanger. 1985. The nymphs, and a new species, of North American *Setvena* Illies (Plecoptera: Perlodidae). The Pan-Pacific Entomologist 61:237–244.

Stewart, K. W., and B. P. Stark. 1988. Nymphs of North American stonefly genera (Plecoptera). Vol. 12. Thomas Say Foundation, 460 pp.

Stewart, K. W., B. P. Stark, and T.G. Huggins. 1976. The stoneflies (Plecoptera) of Louisiana. Great Basin Naturalist 36:366–384.

Stone, A., and H. A. Jamnback. 1955. The black flies of New York State (Diptera: Simuliidae). Bulletin of the New York State Museum 379:144 pp.

Stonedahl, G. M., and J. D. Lattin. 1982. The Gerridae or water striders of Oregon and Washington (Hemiptera: Heteroptera). Oregon State University Agricultural Experiment Station Technical Bulletin 144:36 pp.

Stonedahl, G. M., and J. D. Lattin. 1986. The Corixidae of Oregon and Washington (Hemiptera: Heteroptera). Oregon State University Agricultural Experiment Station Technical Bulletin 150:84 pp.

Stribling, J. B. 1986. Revision of *Anchytarsus* (Coleoptera: Dryopoidea) and a key to the new world genera of Ptilodactylidae. Annals of the Entomological Society of America 79:219–234.

Surdick, R. F., and K. C. Kim. 1976. Stoneflies (Plecoptera) of Pennsylvania: a synopsis. Bulletin 808 The Pennsylvania State University Agricultural Experiment Station, University Park. 73 pp.

Szczytko, S. W., and K. W. Stewart. 1979. The genus *Isoperla* (Plecoptera) of western North America; holomorphology and systematics, and a new stonefly genus *Cascadoperla*. Memoirs of the American Entomological Society Number 32:120 pp.

Tai, L. D. D. 1967. Biosystematic study of *Sympetrum* (Odonata: Libellulidae). Ph.D. Thesis, Purdue University, West Lafayette, Indiana. 234 pp.

Tarter, D. C., M. L. Little, R. F. Kirchner, W. D. Watkins, R. G. Farmer, and D. Steele. 1975. Distribution of pteronarcid stoneflies in West Virginia (Insecta: Plecoptera). Proceedings of the West Virginia Academy of Science 47:79–85.

Teskey, H. J. 1969. Larvae and pupae of some eastern North American Tabanidae (Diptera). Memoirs of the Entomological Society of Canada No. 63:147 pp.

Teskey, H. J., and J. F. Burger. 1976. Further larvae and pupae of eastern North American Tabanidae (Diptera). Canadian Entomologist 108:1085–1096.

Tidwell, M. A. 1973. The Tabanidae (Diptera) of Louisiana. Tulane Studies in Zoology and Botany 18:1–95.

Tracy, B. H., and W. L. Hilsenhoff. 1982. The female of *Graphoderus manitobensis* with notes on identification of female *Graphoderus* (Coleoptera: Dytiscidae). Great Lakes Entomologist 15:163–164.

Truxal, F. S. 1953. A revision of the genus *Buenoa* (Hemiptera: Notonectidae). University of Kansas Science Bulletin 35:1351–1517.

Unzicker, J. D., and P. H. Carlson. 1981. Ephemeroptera. Pages 3.1–3.97 *in:* A. R. Brigham, W. U. Brigham, and A. Gnilka, editors. The aquatic insects and oligochaetes of North and South Carolina. Midwest Aquatic Enterprises, Mahomet, Illinois.

Unzicker, J. D., and V.H. McCaskill. 1981. Plecoptera. Pages 5.1–5.50 *in:* A. R. Brigham, W. U. Brigham, and A. Gnilka, editors. The aquatic insects and oligochaetes of North and South Carolina. Midwest Aquatic Enterprises, Mahomet, Illinois.

Unzicker, J. D., V. H. Resh, and J. C. Morse. 1981. Trichoptera. Pages 9.1–9.138 *in:* A. R. Brigham, W. U. Brigham, and A. Gnilka, editors. 1981. The aquatic insects and oligochaetes of North and South Carolina. Midwest Aquatic Enterprises, Mahomet, Illinois.

Usinger, R. L., editor. 1956. Aquatic insects of California with keys to North American genera and California species. University of California Press, Berkeley and Los Angeles. 508 pp.

Valley, K. R., and C. O. Berg. 1977. Biology, immature stages, and new species of snail-killing Diptera of the genus *Dictya* (Sciomyzidae). Search 7(2):44 p.

Van Tassell, E. 1966. Taxonomy and biology of the subfamily Berosinae of North and Central America and the West Indies (Coleoptera: Hydrophilidae). Ph.D. Thesis, Catholic University, Washington, D.C. 329 pp.

Vineyard, R. N., and G. B. Wiggins. 1988. Further revision of the caddisfly family Uenoidae (Trichoptera): evidence for inclusion of Neophylacinae and Thremmatidae. Systematic Entomology 13:361–372.

Walker, E. M. 1953. The Odonata of Canada and Alaska. Vol. 1. Part I: General. Part II: The Zygoptera—damselflies. University of Toronto Press, Toronto. 292 pp.

Walker, E. M.. 1958. The Odonata of Canada and Alaska. Vol. 2. Part III: The Anisoptera—four families. University of Toronto Press, Toronto. 318 pp.

Walker, E. M., and P. S. Corbet. 1975. The Odonata of Canada and Alaska. Vol. 3. Part III: The Anisoptera—three families. University of Toronto Press, Toronto. 308 pp.

Wallace, J. B., and R. W. Merritt. 1980. Filter-feeding ecology of aquatic insects. Annual Review of Entomology 25:103–132.

Wallace, J. B., W. R. Woodall, and A. A. Staats. 1976. The larval dwelling-tube, capture net and food of *Phylocentropus placidus* (Trichoptera: Polycentropodidae). Annals of the Entomological Society of America 69:149–154.

Wallis, J. B. 1933. Revision of the North American species, (north of Mexico), of the genus *Haliplus*, Latreille. Transactions of the Royal Canadian Institute 19:1–76.

Wallis, J. B. 1939. The genus *Graphoderus* Aube in North America (north of Mexico) (Coleoptera). Canadian Entomologist 71:128–130.

Waltz, R. D., and W. P. McCafferty. 1979. Freshwater springtails (Hexopoda: Collembola) of North America. Purdue University Agricultural Experiment Station, West Lafayette, Indiana. Research Bulletin 960:32 pp.

Waltz, R. D., and W. P. McCafferty. 1983. *Austrotinodes* Schmid (Trichoptera: Psychomyiidae), a first U.S. record from Texas. Proceedings of the Entomological Society of Washington 85: 181–182.

Waltz, R. D., and W. P. McCafferty. 1987a. Revision of the genus *Cloeodes* Traver (Ephemeroptera: Baetidae). Annals of the Entomological Society of America 80:191–207.

Waltz, R. D., and W. P. McCafferty. 1987b. New genera of Baetidae for some Nearctic species previously included in *Baetis* Leach (Ephemeroptera). Annals of the Entomological Society of America 80:667–670.

Waltz, R. D. and W. P. McCafferty. 1987c. Systematics of *Pseudocloeon, Acentrella, Baetiella,* and *Liebebiella,* new genus (Ephemeroptera: Baetidae). Journal of the New York Entomological Society 95:553–568.

Waters, T. F. 1972. The drift of stream insects. Annual Review of Entomology 17:253–272.

Weaver, J. S. III. 1988. A synopsis of the North American Lepidostomatidae (Trichoptera). Contributions of the American Entomological Institute Volume 24. No. 2. 141 pp.

Weaver, J. S., III, and J. C. Morse. 1986. Evolution of feeding and case-making behavior in Trichoptera. Journal of the North American Benthological Society 5:150–158.

Webb, D. W., and W. U. Brigham. 1981. Aquatic Diptera. Pages 11.1–11.111 *in:* A. R. Brigham, W. U. Brigham, and A. Gnilka, editors. The aquatic insects and oligochaetes of North and South Carolina. Midwest Aquatic Enterprises, Mahomet, Illinois.

Westfall, M. J., Jr., and K. J. Tennessen. 1979. Taxonomic clarification within the genus *Dromogomphus*

Selys (Odonata: Gomphidae). Florida Entomologist 62:266–273.

White, D. S. 1978. A revision of the Nearctic *Optioservus* (Coleoptera: Elmidae), with descriptions of new species. Systematic Entomology 3:59–74.

Wiggins, G. B. 1960. A preliminary systematic study of the North American larvae of the caddisfly family Phryganeidae (Trichoptera). Canadian Journal of Zoology 38:1153–1170.

Wiggins, G. B. 1973. New systematic data for the North American caddisfly genera *Lepania, Goeracea* and *Goerita* (Trichoptera: Limnephilidae). Life Science Contributions Royal Ontario Museum No. 91:33 pp.

Wiggins, G. B. 1977. Larvae of the North American caddisfly genera (Trichoptera). University of Toronto Press, Toronto. 401 pp.

Wiggins, G. B. 1981. Considerations on the relevance of immature stages to the systematics of Trichoptera. Pages 395–407 *in:* G. P. Moretti, editor. Proceeding of the Third International Symposium on Trichoptera. Series Entomology, No. 20. Junk, The Hague.

Wiggins, G. B., and J. S. Richardson. 1982. Revision and synopsis of the caddisfly genus *Dicosmoecus* (Trichoptera: Limnephilidae; Dicosmoecinae). Aquatic Insects 4:181–217.

Wiggins, G. B., and J. S. Richardson. 1989. Biosystematics of *Eocomoecus,* a new Nearctic caddisfly genus (Trichoptera: Limnephilidae, Dicosmoecinae). Journal of the North American Benthological Society 8:355–369.

Wiggins, G. B., and W. Wichard. 1989. Phylogeny of pupation in Trichoptera, with proposals on the origin and higher classification of the order. Journal of the North American Benthological Society 8:260–276.

Wiggins, G. B., and R. W. Wisseman. 1990. Revision of the North American caddisfly genus *Desmona* (Trichoptera: Limnephilidae). Annals of the Entomological Society of America 83:155–161.

Wilson, C. A. 1958. Aquatic and semiaquatic Hemiptera of Mississippi. Tulane Studies in Zoology 6:116–170.

Winters, F. C. 1926. Notes on the Hydrobiini (Coleoptera-Hydrophilidae) of boreal America. Pan-Pacific Entomologist 3:49–58.

Wirth, W. W., and A. Stone. 1956. Aquatic Diptera. Pages 372–482 *in:* R. L. Usinger, editor. Aquatic insects of California with keys to North American genera and California species. University of California Press, Berkeley and Los Angeles.

Wolfe, G. W. 1984. A revision of the *Hydroporus vittatipennis* species group of the subgenus *Neoporus* (Coleoptera: Dytiscidae). Transactions of the American Entomological Society 110:389–434.

Wolfe, G. W., and J. F. Matta. 1981. Notes on nomenclature and classification of *Hydroporus* subgenera with the description of a new genus of Hydroporini (Coleoptera: Dytiscidae). Pan-Pacific Entomologist 57:149–175.

Wolfe, G. W., and P. J. Spangler. 1985. A synopsis of the *Laccornis difformis* species group with a revised key to North American species of *Laccornis* Des Gozis

(Coleoptera: Dytiscidae). Proceedings of the Biological Society of Washington 98:61–71.

Wood, D. M. 1978. Taxonomy of the Nearctic species of *Twinnia* and *Gymnopais* (Diptera: Simuliidae) and a discussion of the ancestry of the Simuliidae. Canadian Entomologist 110:1297–1337.

Wood, D. M., B. V. Peterson, D. M. Davies, and H. Gyorkos. 1963. The black flies (Diptera: Simuliidae) of Ontario. Part II. Larval identification, with descriptions and illustrations. Proceedings of the Entomological Society of Ontario 93:99–129.

Wood, D. M., P. T. Dang, and R. A. Ellis. 1979. The Insects and Arachnids of Canada. Part 6: The mosquitoes of Canada (Diptera: Culicidae). Agriculture Canada Publication 1686:390 pp.

Wooldridge, D. P. 1966. Notes on Nearctic *Paracymus* with descriptions of new species (Coleoptera: Hydrophilidae). Journal of the Kansas Entomological Society 39:712–725.

Wooldridge, D. P. 1967. The aquatic Hydrophilidae of Illinois. Illinois State Academy of Science 60:422–431.

Yamamoto, T., and G. B. Wiggins. 1964. A comparative study of the North American species in the caddisfly genus *Mystacides* (Trichoptera: Leptoceridae). Canadian Journal of Zoology 42:1105–1126.

Young, F. N. 1953. Two new species of *Matus,* with a key to the known species and subspecies of the genus (Coleoptera: Dytiscidae). Annals of the Entomological Society of America 46:49–55.

Young, F. N. 1954. The water beetles of Florida. University of Florida Press, Gainesville. 238 pp.

Young, F. N. 1956. A preliminary key to the species of *Hydrovatus* of the eastern United States (Coleoptera: Dytiscidae). Coleopterists Bulletin 10:53–54.

Young, F. N. 1963a. The Nearctic species of *Copelatus* Erichson (Coleoptera: Dytiscidae). Quarterly Journal of the Florida Academy of Science 26:56–77.

Young, F. N. 1963b. Two new North American species of *Hydrovatus,* with notes on other species (Coleoptera: Dytiscidae). Psyche 70:184–192.

Young, F. N. 1979a. A key to Nearctic species of *Celina* with descriptions of new species (Coleoptera: Dytiscidae). Journal of the Kansas Entomological Society 52:820–830.

Young, F. N. 1979b. Water beetles of the genus *Suphisellus* Crotch in the Americas north of Colombia (Coleoptera: Noteridae). Southwestern Naturalist 24:409–429.

Young, F. N. 1981a. Predaceous water beetles of the genus *Desmopachria:* the *convexa-grana* group (Coleoptera: Dytiscidae). Occasional Papers of the Florida Collection of Arthropods 2(III–IV):1–11.

Young, F. N. 1981b. Predaceous water beetles of the genus *Desmopachria* Babington: the *leechi-glabricula* group (Coleoptera: Dytiscidae). Pan-Pacific Entomologist 57:57–64.

Young, F. N. 1985. A key to the American species of *Hydrocanthus* Say, with descriptions of new taxa (Coleoptera: Noteridae). Proceedings of the Academy of Natural Sciences, Philadelphia 137:90–98.

Young, W. C., and C. W. Bayer. 1979. The dragonfly nymphs (Odonata: Anisoptera) of the Guadelupe River Basin, Texas. Texas Journal of Science 31:85–98.

Zimmerman, J. R. 1970. A taxonomic revision of the aquatic beetle genus *Laccophilus* (Dytiscidae) of North America. Memoirs of the American Entomological Society No. 26:275 pp.

Zimmerman, J. R. 1981. A revision of the *Colymbetes* of North America (Dytiscidae). Coleopterists Bulletin 35:1–52.

Zimmerman, J. R. 1985. A revision of the genus *Oreodytes* in North America (Coleoptera: Dytiscidae). Proceedings of the Academy of Natural Sciences, Philadelphia 137:99–127.

Zimmerman, J. R., and A. H. Smith. 1975. A survey of the *Deronectes* (Coleoptera: Dytiscidae) of Canada, the United States, and northern Mexico. Transactions of the American Entomological Society 101:651–722.

Zimmerman, J. R., and R. L. Smith. 1975. The genus *Rhantus* (Coleoptera: Dytiscidae) in North America. Part I. General account of the species. Transactions of the American Entomological Society 101:33–123.

Crustacea: Introduction and Peracarida

18

Alan P. Covich
Department of Zoology
University of Oklahoma
Norman, Oklahoma 73019

James H. Thorp
Department of Biology and Water Resources Laboratory
University of Louisville
Louisville, Kentucky 40292

Chapter Outline

I. INTRODUCTION TO THE SUBPHYLUM CRUSTACEA

II. ANATOMY AND PHYSIOLOGY OF CRUSTACEA
- **A. External Morphology**
  1. Skeleton
  2. Appendages
- **B. Organ System Function**
  1. Nutrition and Digestion
  2. Circulation and Respiration
  3. Fluid and Solute Balance
  4. Neural Systems
  5. Reproduction, Development, and Growth
- **C. Environmental Physiology**
  1. Salinity and pH Tolerances
  2. Photoreception
  3. Chemoreception
  4. Mechanoreception
  5. Thermoreception

III. ECOLOGY AND EVOLUTION OF SELECTED CRUSTACEA
- **A. General Relationships**
  1. Branchiura
  2. Peracarida
- **B. Ecological Distributions and Interactions**
  1. Habitats
  2. Food Resources
  3. Vertical Migrations and Feeding
  4. Responses to Water Quality
- **C. Life-History Traits**
  1. Longevity
  2. Patterns of Reproduction

- **D. Biogeography**
  1. Groundwaters
  2. Surface Waters

IV. COLLECTING, REARING, AND PREPARATION FOR IDENTIFICATION

V. CLASSIFICATION OF PERACARIDA AND BRANCHIURA
- **A. Taxonomic Key to Freshwater Genera of the Orders Mysidacea, Amphipoda, and Isopoda and the Subclass Branchiura**

Literature Cited

I. INTRODUCTION TO THE SUBPHYLUM CRUSTACEA

Crustaceans currently thrive in a wide array of habitats and adapted to freshwaters very early in their history (Abele 1982). The fossil record extends back to the lower Cambrian; throughout most of this long period, crustacean fossils are associated with a broad range of aquatic habitats (Schram 1982). Although only about 10% of the nearly 40,000 extant species in the subphylum Crustacea (phylum Arthropoda) occur in the inland pools, lakes, and streams of the world (Table 18.1) (Bowman and Abele 1982), freshwater crustaceans are extremely important in many ecosystem processes of inland waters.

Crustaceans have evolved into several major groups that are evolutionarily linked (Fig. 18.1) and represent distinct modifications for exploiting resources in many different habitats. The cladocerans and copepods together constitute the dominant com-

Table 18.1 Classification of Crustacea and Estimates of Their Numbers in North American Freshwaters[a, b]

Taxa	Worldwide Percentage Found in Freshwaters	Approximate Number of Species in the United States and Canada	Chapter of This Book
Subphylum Crustacea	< 10%	ca. 1500	18–22
Class Cephalocarida	0%	0	–
Class Branchiopoda [Eight orders; includes cladocerans]	> 95%	> 200	20
Class Remipedia	0%	0	–
Class Maxillopoda	< 15%		18, 21
Subclass Branchiura	90%	23	18
Subclass Copepoda	< 15%	> 200	21
Class Ostracoda	> 50%	420	19
Class Malacostraca	ca. 10%		18, 22
Superorder Pancarida [Order Thermosbaenacea]	ca. 50%	1	18
Superorder Peracarida [Orders Mysidacea, Amphipoda, and Isopoda]	< 10%	> 300	18
Superorder Eucarida [Order Decapoda]	10%	334	22

[a]North of Mexico.
[b]Classification based in part on Bowman and Abele (1982).

ponent of most freshwater planktonic assemblages (see Chapters 20 and 21, respectively). Crustacean zooplankton are major consumers of phytoplankton, transferring energy from these primary producers up the trophic pyramid to zooplanktivorous fish. Some crustacean zooplankton are predatory on rotifers, protozoans, other crustaceans, and even fish larvae. Within inland hypersaline environments, brine shrimp (Chapter 20) are sometimes the only metazoan capable of surviving hyperosmotic conditions. As salinity fluctuates, the crustacean community can rapidly change its species composition. For example, increased rainfall and decreased salinity in the Great Salt Lake basin of Utah caused major shifts in zooplankton (Stephens 1990, Wurtsbaugh and Berry 1990); declines in lake levels and higher salinity at Pyramid Lake, Nevada, altered the benthic community (Galat *et al.* 1988).

Crustaceans also comprise a significant portion of the benthic biomass in many lakes and streams. Some species are abundant in the littoral zone and throughout the oxygenated profundal zone of lakes where they feed on decaying vegetation or live algae (e.g., Hargrave 1970a, 1970b, Moore 1977, Gee 1982). As with the zooplankton, these small benthic consumers recycle nutrients that sustain high levels of microbial productivity. In streams, small crustaceans also occupy many different habitats ranging from quiet pools to fast-flowing riffles where they actively graze; many are abundant omnivores (Moore 1977, Marchant 1981, Marchant and Hynes 1981a, 1981b).

Some of these very productive benthic species live in extreme habitats, from cold arctic waters to hot springs. In fact, the aquatic metazoans with the highest known temperature tolerance are some crustaceans that live in hot springs, including the ostracode *Potamocypris*, which lives on algal-coated substrata that range from 30°–54°C (Wickstrom and Castenholz 1973), and the pancarids *Thermosbaena mirabilis*, which survives waters of 45°–48°C, and *Thermobatynella adami*, which lives in 55°C water (Capart 1951). Other benthic crustaceans, such as the amphipod *Hyallela azteca*, also live in very hot springs (Richard Wiegert, personal communication). Some isopods are found in warm springs of the southwestern United States; for example, *Thermosphaeroma thermophilum*, an endangered species, and *T. subequalum* thrive in 34°–35°C pools (Bowman 1981, Cole and Bane 1978, Schuster 1981, Manuel Molles, personal communication). Amphipods and isopods, as well as mysids and branchiura are discussed in this chapter; ostracodes are reviewed in Chapter 19.

One of the most diverse and abundant freshwater crustacean groups north of the Rio Grande are omnivorous crayfish (Decapoda; Chapter 22). These wide-ranging, benthic consumers are not only important in nutrient cycling but they also can change the life histories of some of their prey species that have evolved ways to avoid being eaten (Crowl and Covich 1990). Crayfish maintain very high densities in many different habitats where they feed on a wide range of living and dead plants and animals (Covich

1977, Goddard 1988, Rabeni 1985, Chambers *et al.* 1990, Hanson *et al.* 1990). Some of their prey are other crustaceans such as amphipods and ostracodes (Abrahamsson 1966, Tcherkashina 1977). Benthic amphipods, isopods, ostracodes, and various littoral cladocerans and copepods also consume a tremendous amount of living organisms and particulate organic matter. All these crustaceans, in turn, can be significant constituents in the diets of many invertebrate predators and bottom-feeding fishes. In addition to their crucial roles as consumers and secondary producers in natural aquatic ecosystems, some crustaceans are increasing in importance as a food source for humans (Huner 1988, Provenzano 1985). The highly nutritious crayfish have become widely available through aquacultural production as have the larger "river shrimp" (*Macrobrachium*) in warm temperate-zone locations (Brody *et al.* 1980).

The present chapter provides a brief introduction to crustacean anatomy and physiology. More de-

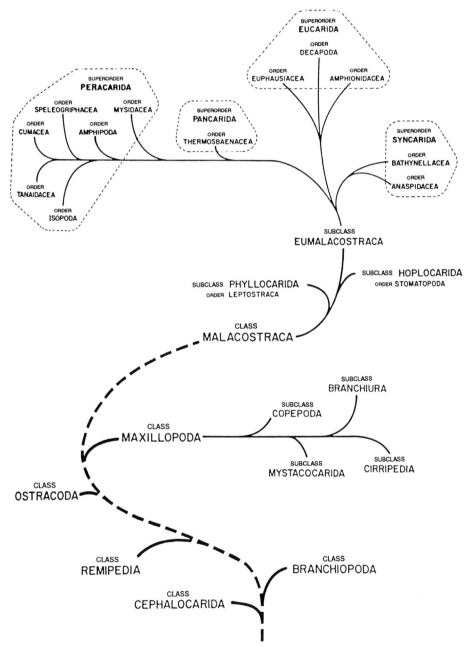

Figure 18.1 General evolutionary linkages of major groups of freshwater crustaceans from Bowman and Abele (1982) who stated, "This [classification] is not to indicate phylogenetic relationships and should not be so interpreted. The dashed line at the base emphasizes uncertainties concerning the origins of the five classes and their relationships to each other."

tailed information can be obtained from general invertebrate zoology texts (e.g., Barnes 1987, Thorp manuscript in preparation) and from specialty books on crustaceans, such as those by Schram (1986), Fitzpatrick (1983), McLaughlin (1980), Schmitt (1965), and the ten volume set edited by Bliss (1982–1985) and several co-editors. Following an introduction to anatomy and physiology (based primarily on information from these sources), this chapter summarizes the ecology and evolution of three orders of crustaceans: Mysidacea, Amphipoda, and Isopoda. A review of the subclass Branchiura is also included. This chapter concludes with a discussion of crustacean classification and a key to the genera of the major taxa listed above. Similar information is presented on other major crustacean groups in Chapters 18–22.

II. ANATOMY AND PHYSIOLOGY OF CRUSTACEA

A. External Morphology

As Schram (1986) has noted, the crustaceans are anatomically very diverse. Many other arthropod groups use a single basic body plan. Once you have seen one insect or spider taxon, you have a good idea of the general body form. The crustaceans, however, have evolved a high diversity of body forms by fusing various segments or by developing highly specialized body segments and appendages. Crustaceans are basically metameric, protostomate coelomates whose bodies are covered by a hard exoskeleton and often divided into three regions, or tagmata: the head (cephalic area), thorax, and abdomen (sometimes the first two regions are combined as a cephalothorax). The jointed appendages are primitively biramous and may be present in all three body regions. Mandibles, two pairs of maxillae, and two pairs of antennae are present in all species during at least one life stage. Crustaceans are distinguished from the Insecta, one of the other two large arthropod groups, by the presence of two pairs of antennae.

1. Skeleton

All crustaceans have a hard exoskeleton composed of chitin that varies in rigidity at various life-history stages. The name Crustacea is derived from the Latin word for shell and refers to this exoskeleton, which in many taxa is hardened with very small inclusions of calcium carbonate. Although not a true ''shell'' as is found among the Mollusca, this exoskeleton is very durable. Crustaceans then, like other arthropods, are characterized by a chitinized

exoskeleton consisting of a thin proteinaceous epicuticle stiffened by quinone tanning and a thick, multilayer procuticle strengthened by calcium carbonate. The skeleton attains its maximum thickness and rigidity in the decapods. Projecting inward are chitinized struts, known as apodemes, which serve as sites for internal attachment of the muscles; they also lend partial protection to some organs.

2. Appendages

Crustacean appendages are modified among species to serve a large variety of purposes, including locomotion (walking and swimming), feeding, grooming, respiration, sensory reception, reproduction, and defense. Consequently, the primitive, generally biramous appendages (terminal exopod and endopod) are often extensively modified with additional lateral and medial projections. Two extreme forms are recognized among adults (Fig. 18.2): the leaflike or lobed phyllopod appendage (as found among branchiopods) and the unbranched, segmented walking leg, or stenopod (typical of crayfish). The cephalic region contains five basic types of paired appendages: (1) antennules (first antennae), which are uniramous in all crustaceans except the malacostracans; (2) antennae (second antennae); (3) mandibles; (4) maxillules (first maxillae); and (5) maxillae (second maxillae). The first two pairs generally have a sensory function (aiding some taxa in food location and filtering), whereas the last three pairs normally function in food acquisition, handling, or processing. The number of appendages on the thorax and abdomen vary greatly among large taxonomic groups. Malacostracans (such as decapods and amphipods) generally possess 5–8 pairs of thoracic appendages (sometimes called pereiopods) and six pairs of abdominal appendages (pleopods and terminal uropods). Primary abdominal appendages are absent in all nonmalacostracans except Notostraca, although comparable structures may have secondarily evolved.

B. Organ System Function

1. Nutrition and Digestion

Among crustaceans as a whole, carnivory and scavenging are the dominant modes of obtaining food with filter feeding of algae and detritus next in importance. The picture for freshwater crustaceans, however, is slightly altered. These arthropods predominantly employ their antennae, maxillae, or thoracic appendages to filter or grab algae, bacteria, microzooplankton, or suspended dead organic matter (detrital seston) from the water column. In some habitats, a relatively large number of taxa are pri-

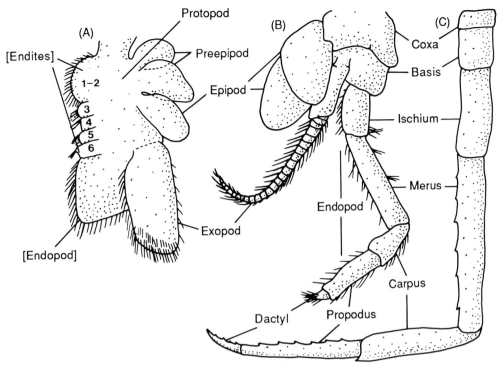

Figure 18.2 Representative crustacean appendages: (A) a phyllopod appendage of Anostraca; (B) biramous appendage of Anaspidacea (superorder Syncarida); (C) uniramous stenopod appendage of the Stenopodidea (Decapoda: Pleocyemata). (A–C from McLaughlin 1982.)

marily carnivores or scavengers (e.g., benthic decapods, certain copepods, and others). Parasitism is rare except among specialized copepods and branchiurans (Yamaguti 1963, Margolis and Kabata 1988). Cannibalism is only common at very high densities or when newly molted individuals are vulnerable, especially among the decapods (Goddard 1988).

Using their cephalic or thoracic appendages, crustaceans acquire food and pass it through a ventral mouth into a tripartite alimentary tract (Fig. 18.3). In most crustaceans, food is ground into small, digestible particles by the gastric mill (the posterior end of the cardiac stomach) of the muscular foregut (*e.g.,* Schmitz and Sherrey 1983). Enzymes (primarily from the hepatopancreas) empty into the short midgut, where food is digested and assimilated. Wastes are conveyed by peristalsis through the long hindgut and out the terminal anus.

2. Circulation and Respiration

Crustaceans generally have an open circulatory system consisting of a dorsal, neurogenic heart with a single chamber, one or more arteries (in malacostracans only), and several sinuses for returning blood to the heart. Many noncalanoid copepods (and marine barnacles) lack a heart altogether; fluids are circulated as a result of general body movements.

The heart is usually located in the thorax or cephalothorax but is present in the abdomen in species with abdominal gills (such as isopods). The blood contains low amounts of a viscous, copper-based respiratory pigment (hemocyanin), which is dissolved in the hemolymph.

Respiration takes place entirely across the body integument in nonmalacostracans, but thoracic and abdominal gills are typical of other freshwater crustaceans. Peracaridans (discussed later in this chapter) feature rudimentary epipodal gills (flattened outgrowths of the coxa), whereas freshwater decapods have trichobranchiate gills composed of a central axis (with afferent and efferent blood vessels) bearing numerous filamentous branches (Fig. 18.4).

3. Fluid and Solute Balance

Paired maxillary and antennal glands (also called green glands) are the principal excretory organs in crustaceans (Fig. 18.3). The "labyrinth" of the antennal gland is also involved in reabsorption of glucose, amino acids, and divalent ions from tubule fluids. Maxillary glands predominate within lower crustaceans, whereas the slightly more complex antennal glands characterize most malacostracans (Fig. 18.5). More than 90% of the nitrogenous wastes are voided as ammonia.

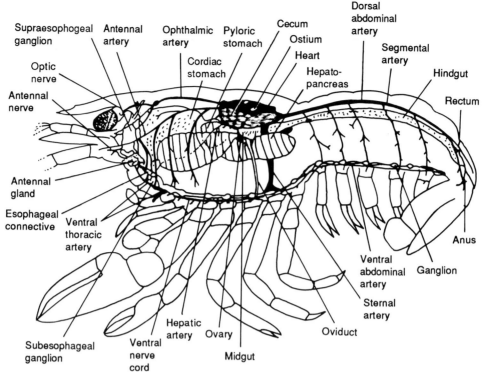

Figure 18.3 Internal anatomy of a generalized astacidean decapod. (From J. H. Thorp, in preparation, after McLaughlin 1980.)

Freshwater crustaceans are relatively competent osmoregulators. They face problems of water influx and salt loss (both primarily across the gills) and, therefore, must release copious quantities of urine that is isosmotic to the hemolymph (a few taxa produce urine that is hyposmotic to the blood).

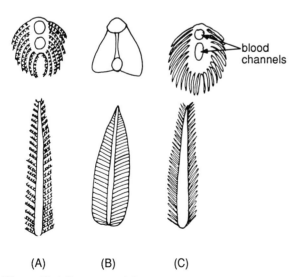

Figure 18.4 Structure of decapod gill branches, as shown in transverse (upper) and lateral (lower) views: (A) dendrobranchiate, (B) phyllobranchiate; (C) trichobranchiate. (From McLaughlin 1982.)

Freshwater branchiopods and some copepods maintain a dilute hemolymph equal to about 200 mOsm (20% of seawater), while the blood of crayfish is more salty (about 350 mOsm; see Mantel and Farmer 1983). An opposite situation is faced by hyporegulating crustaceans living in inland salt lakes. The brine shrimp *Artemia salina,* tolerates salinities ranging from 10% seawater to crystallizing brine; it is hypertonic to the medium in dilute waters but hypotonic in concentrated solutions.

4. Neural Systems

The crustacean neural system is comprised of a tripartite brain and paired ventral, ganglionated nerve cords linked by commissures in a ladderlike arrangement. The protocerebrum of the brain normally innervates the eye, sinus gland, frontal organs, and head muscles; the deutocerebrum controls the antennules; the tritocerebrum innervates the antennae and a portion of the alimentary tract. Crustaceans are generally sensitive to light, chemicals, temperature, touch, gravity, pressure, and sound. The eyes of adult malacostracans are compound and frequently mounted on a stalk, or peduncle; but the adult eye of many entomostracans (an older term inclusive of most nonmalacostracans) is not much advanced over the simple cluster of inverse pig-

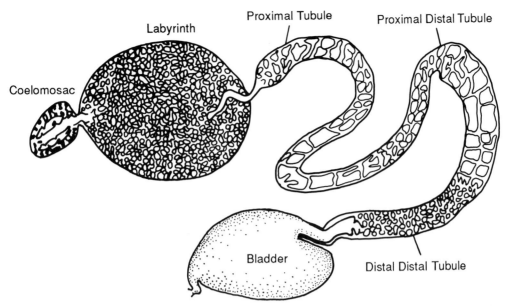

Figure 18.5 Structure of the antennal gland of a crayfish. (From Mantel and Farmer 1983.)

ment cup ocelli that characterizes the larval (naupliar) eye.

5. Reproduction, Development, and Growth

Crustaceans are primarily sexually reproducing, dioecious organisms, but hermaphroditism and parthenogenesis occur sporadically among entomostracans (especially in ostracodes and branchiopods) and a few marine malacostracans. Paired gonads lie above or lateral to the midgut in most species. In mysids, the ovaries are partially fused and linked by a cellular bridge. Paired, or rarely single, reproductive ducts open ventrally through simple gonopores (paired or single), elevated papillae, or elaborate copulatory structures (as in decapods). The location of the gonopore varies greatly in entomostracans, whereas in malacostracans, it opens on the sternite or coxae of the sixth thoracic somite in females and at the eighth somite in males. Internal fertilization is the general rule, with males transferring sperm through a penis or specialized trunk appendages known as gonopods (in crayfish these are highly modified pleopods).

Females of most species protect their embryos either by retaining them in an internal brood chamber (e.g., in cladocerans) or external ovisac (e.g., in copepods) or by gluing them to certain appendages (decapods employ their abdominal pleopods). Mysids, isopods, and amphipods safeguard their young until an advanced stage in a ventral brooding shelf, the marsupium.

The embryogeny of crustaceans includes modified spiral, total cleavage, and gastrulation by invagination. All taxa possess an initial nauplius stage in their development, but these larvae may be free swimming (e.g., in copepods) or enclosed in an egg (e.g., in decapods, cladocerans, and peracaridans). Direct development (i.e., without any external larval stages) characterizes taxa such as cladocerans and peracaridans. In contrast, indirect development with a free nauplius followed by either distinctive metamorphosis stages or gradual development to an adult is typical of most ostracodes, copepods, decapods (except crayfish), and Branchiopoda (other than cladocerans).

Growth throughout the larval, juvenile, or adult stages requires periodic shedding of the older, smaller exoskeleton in a process termed molting or ecdysis. Rapid expansion of the body occurs immediately after the crustacean extracts itself from its old exoskeleton and before the new shell hardens. The degree of expansion varies significantly among species, ranging from 8–9% in some mysids to 22% in many decapods to as high as 83% in certain cladocerans (Hartnoll 1982). Some taxa are characterized by indeterminate growth (e.g., the cladoceran *Daphnia* and apparently most crayfish species), while others reach a finite body size in a fixed number of molts (e.g., in ostracodes, copepods, and some isopods). The molt process is controlled by hormones released by several organs. With a decline in production of the molt-inhibiting hormones (normally secreted by the cephalic X-organ and sinus gland complex), the Y-organs are no longer suppressed and can begin secretion of ecdysone, a molt-stimulating hormone. The actual molt requires as little as a few minutes to as long as several hours in aquatic taxa. This period is extremely hazardous;

the animal may die from physiological stress or succumb when it is unable to extract a portion of its body (such as a crayfish cheliped) from the the old exoskeleton. The temporarily soft crayfish is also extremely susceptible to predators (including conspecifics) and may seek shelter prior to its molt. The larger and older the individual becomes, the longer it takes for the molting process to be completed, and the longer is the exposure to possible cannibalism and predation. Thus, indeterminate growth typically has limitations imposed by predatory pressures.

C. Environmental Physiology

1. Salinity and pH Tolerances

A large number of crustacean species have adapted to living completely in freshwaters (Table 18.1) while others spend only portions of their lives in these highly variable habitats (e.g., Mantel and Farmer 1983, Vernberg and Vernberg 1983). Branchiurans are distributed worldwide in both marine systems and freshwater habitats; some species can tolerate a rapid shifting of distributions from salt to freshwater and may even switch hosts from marine to nonmarine fishes. Among the Peracarida, as discussed later, some species are confined to coastal inland waters and apparently are not completely adapted to many of the relatively unbuffered, dilute freshwaters found farther inland (Sutcliffe 1971, 1974, Dormaar and Corey 1978).

The widespread acidification of lakes and streams has drawn attention to the importance of pH tolerances of many organisms, especially the crustaceans. Because calcification of the carapace immediately following molting is highly sensitive to low pH, changes in regional distributions may occur among crustaceans (e.g. Nero and Schindler 1983, France 1984, Berrill *et al.* 1985, Greenaway 1985, Davies 1989). There are both lethal and sublethal effects of increased acidity that alter ion regulation, especially by juveniles in soft-water lakes. Although many studies have emphasized lentic species, there is evidence that lotic species may have a narrower range of tolerance to low pH and that egg mortality may be important for some species.

Although crayfish are generally thought to require habitats with calcium concentrations in excess of 2 mg/liter, low population densities of some species may be maintained by extracting sufficient calcium from their food. In adult *Orconectes virilis*, calcium uptake is impaired in habitats below pH 5; survivorship is greatly influenced by age and molt stage (Davies 1989).

2. Photoreception

Sensitivity to small differences in light intensities is extremely well developed in crustaceans. Many groups are able to seek those microhabitats where they can optimize growth rates and reproduction (Hutchinson 1967). Crustaceans have evolved receptors to detect even slight differences in light intensities, especially in deep, thermally stratified waters. Marine and freshwater zooplankton exploit thermal gradients by migrating daily and seasonally in both vertical and horizontal patterns to obtain their preferred conditions. The adaptive value of this migration for zooplankton is thought to be a combination of: (1) reducing metabolic costs by remaining in cool waters; (2) maximizing growth through grazing at times when algae have the highest food value; and (3) avoiding exposure to visual predators. The relative importance of these variables can change seasonally and shift the exact timing of migration to maximize net energy gain under different conditions (Enright 1977, 1979). When algal food resources are scarce, zooplankton may ascend "early" from deep cool waters and begin grazing 1–2 hours before sunset, thereby gaining access to limited food supplies before competitors but at a higher metabolic cost and risk of predation (Zaret 1980, Gliwicz 1986). In arctic lakes and ponds, crustaceans are exposed to continuous daylight during most of their growing season and do not vertically migrate on a daily basis. However, they maintain phototactic reactions at approximately the same threshold values; their photosensitivity remains similar to temperate species (Buchanan and Haney 1980). Controversy exists regarding how to evaluate the adaptive significance of variable migratory responses to different light intensities in the presence of predators (see further discussion in Chapters 20 and 21).

3. Chemoreception

Crustaceans also have very well-developed chemosensory systems that allow individuals to locate food and mates while avoiding predators (Ache 1982, Atema 1988, Tierney and Atema 1988, Zimmer-Faust 1989). Amphipods and crayfish are two examples of crustaceans that demonstrate highly specific responses to pheromones (Dahl *et al.* 1970, Dunham 1978, Thorp 1984, Rose 1986, see Chapter 22). Hydrodynamics of stream flow and lake currents clearly mediate the effectiveness of diffuse chemicals but the persistent unidirectional nature of many signals does provide a basis for orientation and communication. Laboratory and field studies demonstrate that chemical communication between crustacean predators and their prey can influence prey life histories and may alter community structure (Crowl and Covich 1990). Within the boundary layer of stream flow, chemoreception can be highly effective as a means for two-dimensional orientation. For organisms drifting downstream or migrating upstream, chemical cues may be extremely important. The relative size of the organism to the depth of the boundary layer is an important ratio (Dodds 1990).

During periods of stable flow regimes in streams, crustaceans may depend on chemical cues as well as other sources of information to regulate daily and seasonal patterns of movement, to select foods or mates, and to avoid predators. As pointed out in Chapter 22, however, there are very few studies on the effectiveness of chemical communication in decapod crustaceans, and the situation is similar for other groups.

4. Mechanoreception

The ability to orient into currents (positive rheotaxis) is a common trait in many freshwater crustaceans. These physical, hydrodynamic cues are often associated with chemical cues that stimulate directed movement. Crustaceans have many types of setae, which function as mechanical and chemical sensors (Bush and Laverack 1982). These external receptors are typically positioned near the antennae and the mandibles. However, there is a wide range of other locations so that information can be obtained from several directions simultaneously. The functional anatomy of these mechanoreceptors is well studied in decapods. Behaviors associated with frequent cleaning activities are not as thoroughly studied in freshwater crustaceans as in marine taxa nor are there many studies on behavioral responses to changes in current speed (see Chapter 22 for further discussion). Behavioral studies of copepods have emphasized the importance of mechanoreceptors for avoidance of predators (see review in Chapter 21) but relatively little is known for other groups.

5. Thermoreception

Sensing differences in relative temperatures is an important ability for crustaceans migrating in search of optimal microhabitats. Many species can orient in thermal gradients and combine information on temperature with simultaneous inputs from photo- and chemoreceptors (Ache 1982, Cossins and Bowler 1987). Freshwater crustaceans have distinct thermal preferences (e.g., Cheper 1980, Taylor 1984, 1990, Peck 1985, Mundahl 1989) and considerable information is available for crayfish (as discussed in Chapter 22). As mentioned previously, the interac-

tions between thermal cues and light intensity are often associated with vertical migrations of zooplankton. In these behavioral responses to light and temperature gradients, there is growing evidence that zooplankton may use chemical cues to orient themselves and to avoid predators (see Chapter 20).

III. ECOLOGY AND EVOLUTION OF SELECTED CRUSTACEA

A. General Relationships

Evolution within the class Crustacea has resulted in a very large number of different species. The fossil record demonstrates that both freshwater and marine crustaceans existed in the early Paleozoic (Schram 1982). Almost all of the crustaceans that have adapted to freshwater are thought to have evolved directly from marine ancestors without having gone through a phase of terrestrial adaptation (see Hutchinson 1967 for a comprehensive review).

The class Remipedia is a recently recognized group that is reported to occur only in unique coastal (anchialine) cave environments (Yager 1981, 1987, Schram 1986, Schram *et al.* 1986). Because remipedes can tolerate saline waters with very little dissolved oxygen and they apparently feed in overlying freshwaters, they may have an advantage over some other crustaceans living in marine polyhaline coastal cave environments. Remipedes are considered to be very primitive crustaceans, and their evolution and Caribbean biogeography are currently under active study. Many other freshwater crustaceans also occur in hypogean habitats. Macrocrustaceans adapted to caves, artesian wells, and other groundwater habitats include a large number of amphipods, isopods, mysids, and decapods (Chapter 22) that inhabit these isolated freshwaters (Holsinger and Longley 1980, Abele 1982, Bousfield 1983, Barr and Holsinger 1985, Holsinger 1986).

1. Branchiura

The subclass Branchiura is widely distributed throughout the world, and yet relatively few taxa are found in freshwaters of the northern hemisphere (Yamaguti 1963, Cressey 1972, Margolis and Kabata 1988). Of the approximately 150 species and 4 genera that comprise this group, only *Argulus,* an ectoparasite on fishes, is typically found in North American freshwaters (McLaughlin 1980, Abele 1982, Fitzpatrick 1983, Schram 1986). Some 23 North American species are encountered; 3 are found in western and 19 in eastern North America. The largest number of species occurs along the Gulf of

Mexico. Only a few are very restricted in their geographic ranges, while others are widely distributed; for example, *A. maculosus* occurs from New York to Michigan and down the length of the Mississippi River to Louisiana.

For many years, the Branchiura, or "fish lice," were classified as a group of Copepoda (Wilson 1944) and earlier as Branchiopoda, but their separation as a distinct taxon is now widely accepted. Unlike copepods, the argulids have a pair of compound eyes. The Branchiura also differ from parasitic copepods in having several other unique structures that will be described. The branchiurans have undergone major modifications of the mandibles and associated feeding appendages. The mandibles typically form a pair of transversely toothed hooks between the labium and labrum, and in *Argulus*, the mandibles are incorporated into the specialized proboscis (McLaughlin 1982). The maxillules are large suckers that are characteristic of adult branchiurans (Fig. 18.6). In *Argulus*, a sheathed, hollow spine is used to pierce the skin of the host. Some species are blood suckers while others feed on extracellular fluids or mucous. Once engorged, they can wait 2–3 weeks between meals.

Some reproductive and larval behaviors are unique to Branchiura. In contrast to copepods, female Branchiura do not carry eggs in external ovisacs but, instead, fasten them in rows to rocks and other objects. The actively swimming larvae attach themselves within the gill chambers, mouth, or on the outer surfaces of fishes using small, specialized antennal hooks and maxillules. After 4–5 weeks of development as ectoparasites, they become adults that again may actively swim (actually more like

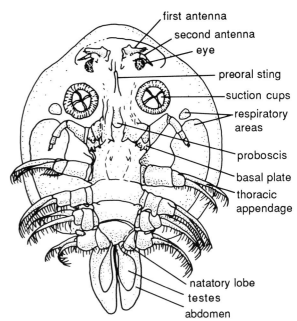

Figure 18.6 External anatomy of an ectoparasitic "fish lice" (Branchiura: *Argulus*).

somersaulting through the water) to find mates and other hosts. Adult females are highly motile in seeking egg-laying sites.

2. Peracarida

The superorder Peracarida has a relatively low percentage of taxa living in freshwaters. The three orders that comprise the Peracarida are the Amphipoda, Isopoda, and Mysidacea. These orders are commonly known as "scuds," "sow bugs," and "opossum shrimp," respectively (Figs. 18.7–18.9).

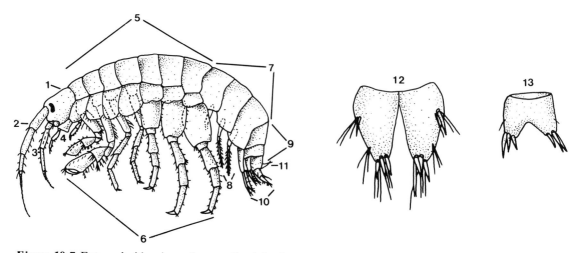

Figure 18.7 External side view of generalized freshwater gammarid amphipod: 1, head; 2, antenna 1; 3, antenna 2; 4, mouth parts; 5, pereonites 1–7; 6, pereopods 1–7 (including gnathopods 1 and 2); 7, pleonites 1–3; 8, pleopods 1–3; 9, uronites 1–3; 10, uropods 1–3; 11, telson (redrawn from Holsinger 1972). Top view of telsons of 12, *Gammarus*; 13, *Crangonyx*.

ISOPODA

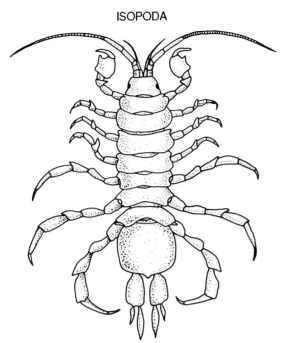

Figure 18.8 Diagram of top view of isopod *Caecidotea* (drawn without setation).

The largest order is the Amphipoda, and there are three large families, Gammaridae, Crangonyctidae and Hyalellidae. Worldwide, there are approximately 800 gammarid amphipods (Abele 1982). In North America, these peracarid crustaceans are represented by six families (Table 18.2) that commonly occupy surface waters. Another 10 families of amphipods (composed of 28 genera and 169 species) occupy subterranean freshwaters together with a limited diversity of isopods (Holsinger 1986). Most of the 780 species of mysids are marine; worldwide, only 25 species occur in freshwater, and 18 additional species live in freshwater caves (Abele 1982).

Members of a fourth order, the Thermosbaenacea, with some species living in thermal springs and others in fresh or saline groundwaters, have strong peracarid affinities but are generally considered to belong to the superorder Pancarida. Only one taxon, *Monodella texana*, lives in North American freshwaters (Maguire 1965, Stock and Longley 1981). Most carcinologists exclude the Thermosbaenacea from the Peracarida because they brood their eggs dorsally under the carapace rather than in a brood pouch, a unique feature of the peracarids (Schram 1982, 1986).

The Peracarida share several evolutionary similarities that demonstrate a common lineage. For example, all have modified the ancestral first thoracic leg into a mouth part (the maxilliped) and retain seven pairs of thoracic legs for movement. The first two pairs of these thoracic legs (the gnathopods) are specialized for grasping food while the remaining five pairs of legs (the pereiopods) lack any specialized grasping (chelate) appendages. As mentioned previously, peracarid females carry their larvae until a relatively advanced stage of development in a specialized ventral brooding chamber, the marsupium. Another character used for separating the peracarids from most other crustaceans is the highly developed *lacinia mobilis,* a small toothed process that articulates with the incisor process. A row of spines commonly separates the *lacinia mobilis* and the molar process (McLaughlin 1982). Several phylogenetic relationships have been proposed for the evolution of the Peracarida (see Schram 1982, 1986 for review) based on how the fossil record is interpreted relative to modern morphological criteria. Many workers view the Isopoda and Amphipoda as derived from a common mysid-like stock. Watling (1981) proposed a different arrangement that is more consistent with the known, but incomplete, fossil

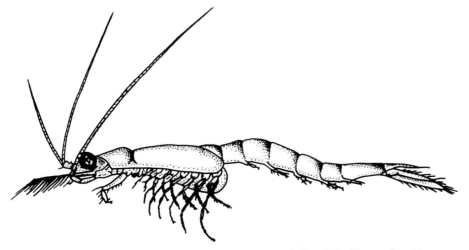

Figure 18.9 Diagram of side view of oppossum shrimp (Mysidacea: *Mysis*).

Table 18.2 Representative Taxa in North American Freshwaters

Superorder Peracarida
  Order Amphipoda
    Family Crangonyctidae
      Bactrurus, Crangonyx, Stygobromus, Synurella
    Family Gammaridae
      Gammarus
    Family Hyalellidae
      Hyalella azteca, H. montezuma
    Family Pontoporeiidae
      Monoporeia affinis
  Order Isopoda
    Family Asellidae
      Caecidotea, Licerus
  Order Mysidacea
    Family Mysidae
      *Mysis relicta, M. littoralis, Neomysis mercedis,
      Taphromysis louisianae*

(A)

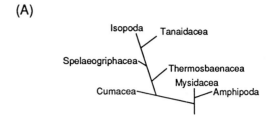

(B)

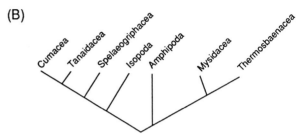

Figure 18.10 Alternative phylogenetic interpretations of the Peracarida: (A) widely accepted relationship of Mysidacea as the main stem from which two divergent lines evolved, one to Isopoda and the other to Amphipoda; (B) another view derived from Watling (1981) with independent lines of development for each major group. The orders Cumacea, Tanaidacea, and Spelaeogriphacea are marine peracarids, while the order Thermosbaenacea is often considered a pancarid group closely related to the peracarids (From Schram 1982.)

record (Fig. 18.10). Most workers agree that the amphipods and mysids are more closely related to each other than to the isopods (Schram 1982, 1986, Sieg 1983).

B. Ecological Distributions and Interactions

1. Habitats

The widespread distribution of Branchiura suggests a lack of any specific pH or salinity requirements, but detailed studies of these ecological relationships are lacking. Temperature influences the rate of hatching for eggs of different species with a range of 12–30 days being required (Yamaguti 1963). During the free-swimming stage *Argulus* may aggregate in a narrow zone of temperature and light conditions where encounters with host fishes can be increased. *Argulus* has no strict host specificity and is capable of attaching to both freshwater and marine telosts.

Most peracarid taxa that have fully adapted to the physiological and ecological conditions of inland waters share five ecological similarities in that they (1) are typically restricted to permanent bodies of water that are relatively cool, clean, and well oxygenated; (2) have distinct behavioral patterns of vertical migration (e.g., among mysids and two species of amphipods; these are similar to daily migration in other groups of crustacean zooplankton, as discussed previously and in Chapters 20–21); (3) have relatively limited ability to move upstream (positive rheotaxis is well studied in amphipods and isopods) or drift downstream in the current under specific ecological conditions (as do many of the lotic insects discussed in Chapter 17); (4) obtain much of their

energy while feeding on the bottom substrata (even if they also feed while in the nekton on suspended algae and smaller taxa of zooplankton); and (5) serve as important prey to a large number of predatory fishes. Some taxa also have various ways of inflicting damage to predatory populations either by being direct parasites on predators or by competing with young stages of those predators for the same prey (such as *Mysis* feeding on plankton) that would otherwise be potentially available to fish consumers.

In contrast to other crustaceans, most of these peracarid species are missing any distinct adaptation for avoiding desiccation, and they also typically lack a diapause phase. Thus, their chances for passive dispersal (e.g., being carried from one habitat to another by ducks or other migratory animals or being carried by the wind during storms) are low in comparison with cladocerans, copepods, or ostracodes. The peracarids also lack any strong active dispersal ability. They generally cannot move upstream against strong currents (Williams and Hynes 1976, Statzner 1988) so their migration throughout river-drainage networks is restricted, and they remain more or less in the same locality for long periods. Nonetheless, as discussed later, some taxa are very widely distributed and their patterns of distribution suggest a long history of slow dispersal and persistence. Many taxa of amphipods and some species of isopods are exceptionally well adapted to live in groundwater habitats and caves (Holsinger 1972, 1986, 1988, Culver 1982, Stanford and Ward 1988, Fong 1989, Williams 1989).

Ecologically, most of the amphipods and isopods are photonegative, positively rheotactic, thigmotactic, cold stenotherms (i.e., restricted to relatively constant, cold waters where they avoid bright light by moving into the current and into crevices or under leaves and roots). In complex substrata, they are less exposed to predators such as fish and crayfish. Typically, where refugia from predation exist, the peracarids occur at high densities in small, permanent, spring-fed streams, seeps, ponds, or sloughs (e.g., Allee 1914, 1929, Juday and Birge 1927). For instance, some species of *Gammarus* reach densities of thousands of individuals per square meter where detrital food and cover are abundant.

A relatively smaller number of isopod species have apparently adapted to live in intermittent or permanent surface waters but distributional data are scarce (Richardson 1905, Hubricht and Mackin 1949, Williams 1970, Ellis 1971). A similarly low diversity of described species is found in groundwater habitats and caves (Culver and Poulson 1971, Culver 1982, Holsinger 1988). Only a few known species tolerate the low oxygen concentrations found in polluted rivers (Williams 1970) or the stress of high temperatures (Dadswell 1974).

The order Mysidacea has only a single family and one very important species, *Mysis relicta* (Fig. 18.9), which occurs in cold, deep lakes as well as in shallow brackish ponds along the arctic coasts. A second species, *Neomysis mercedis,* has a more limited, coastal distribution but plays a significant role in some large lakes such as Lake Washington (Murtaugh 1989). *Mysis relicta,* the "opossum shrimp," is not a true shrimp (or decapod as defined in Chapter 22) but is an active swimmer and can make long daily trips up and down the water column of deep, well-oxygenated lakes (Beeton and Bowers 1982, Bowers and Vanderploeg 1982, Shea and Makarewicz 1989). A third species, *Taphromysis louisianae,* lives in roadside pools in Louisiana and Texas (Banner 1953, Pennak 1989).

Mysids have a holarctic distribution; they occur in oligotrophic lakes of northern regions of North America and in northern Europe but are known to tolerate relatively high temperatures and periods of low oxygen at least for short periods (Dadswell 1974, Sherman et al. 1987). As discussed later, their introduction into many western lakes has created several problems for fisheries managers (Martinez

and Bergersen 1989). Their natural limits of distribution apparently have been influenced by opportunities for migration during the Pleistocene and possibly by biotic interactions (Dadswell 1974).

2. Food Resources

All three peracaridian orders are similar in the range of foraging behaviors used in obtaining food resources. Juveniles are typically dependent on microbial foods such as algae and bacteria associated with either periphyton or aquatic plants in brightly illuminated habitats. They also consume dead organic matter in forested streams, ponds, and lakes where bacteria and fungi living on the detritus provide essential protein for amphipods and isopods (Barlocher and Kendrick 1973, 1975, Smock and Stoneburner 1980, Smock and Harlowe 1983, Gauvin et al. 1989). Adults are not limited to grazing or detritivory because they broaden their food niche to include larger food items. Some species become opportunistic scavengers, predators, and omnivores, depending on which foods are most available. In one unusual example, the amphipod *Hyalella montezuma* migrates to the surface waters of a large karst sink hole and filters out organic material entrapped on the surface film (Cole and Watkins 1977, Blinn and Johnson 1982, Blinn et al. 1987, Wagner and Blinn 1987).

Laboratory and field studies have focused on juvenile and adult growth rates that result from feeding on different types of foods (e.g., Hargrave 1970a, 1970b, Willoughby and Sutcliffe 1976, Marchant and Hynes 1981a, Sutcliffe et al. 1981). Only a few studies have considered the role of secondary plant chemicals as "defensive" compounds for aquatic plants to minimize effects of grazers. Concentrations of tannins and other inhibitory chemicals are known to differ among various macrophytes, and some compounds influence crustacean grazing preferences (e.g., Newman et al.1990). Plant foods that contain a high percentage of cellulose are often considered to be of relatively low quality for grazers and detritivores. However, some amphipods (Monk 1977) and mysids (Friesen et al. 1986) have digestive enzymes that hydrolyze cellulose *in vitro.* These cellulases, however, are apparently confined to degrading small particles in the gut rather than whole plant cell walls that are ingested during grazing. Microbial breakdown of ingested cellulose within the gut is generally the mechanism used for digesting this refractory material and requires development of a complex community of various microorganisms within the gut. Given that crustaceans molt periodically and would need to re-establish this microbial community frequently, the evolution of cellulases would be an important adaptation for expanding use of detrital food resources.

3. Vertical Migrations and Feeding

Amphipods and isopods are often confined to burrowing and feeding on the bottoms of streams and lakes, although species like *Monoporeia affinis* (*Pontoporeia affinis*) move from their benthic foraging areas during the day to nektonic, open-water foraging at night (Marzolf 1965, Winnell and White 1984, Donner *et al.* 1987). These actively swimming amphipods consume a wide range of particles while on the sediment surface, although the exact nature of the food is not completely known for those species found in the Laurentian Great Lakes or other large basins (Nalepa and Robertson 1981, Dermott and Corning 1988, Lopez and Elmgren 1989). As previously discussed, the amphipod *Hyalella montezuma* is an unusual forager in that it swims into the surface waters and filters out suspended nannoplankton and other very small particles of organic materials entrapped by surface tension (Blinn *et al.* 1987). Its twilight migration is followed by a predatory leech, which swims up to feed on the dense concentration of amphipods (Blinn *et al.* 1988); no fish predators occur in this sink hole lake. In contrast, *Hyalella azteca*, which coexists with *H. montezuma* in this lake, does not filter feed or migrate to the surface. Even in hot springs where all predatory fishes are absent, *H. azteca* does not migrate vertically (Strong 1972). The vertical movements of amphipods such as *Monoporeia affinis* in deep lakes overlap those of mysids with whom they may share food resources.

Mysids living in deep lakes provide an intriguing example of vertical migration and complex foraging dynamics. *Mysis relicta* regulates nocturnal feeding patterns by responding to changing light conditions. These very fast, active swimmers have well-developed eyes for precise photoreception. They rely on daily changes in intensity and quality of light (that penetrates through the photic zone) to determine when to begin swimming up through the water column from the dimly lit, deeper waters and to begin feeding on small zooplankton and phytoplankton (Morgan *et al.* 1978, Bowers and Vanderploeg 1982, Cooper and Goldman 1982, Folt *et al.* 1982).

By precisely regulating their travel times, migrating crustaceans reduce their exposure to fish predators and still obtain the food they need to grow and reproduce. Both *Mysis* and *Monoporeia* have potential fish predators that feed on benthic and open-water habitats so that minimizing exposures

to a wide range of predators is a complex process (McDonald *et al.* 1990). As mentioned previously, those zooplankton that seek deeper, cooler waters during the day not only avoid brightly lit waters and exposure to visual predators, but they also lower metabolic rates while digesting the food that they obtained the previous night. Depending on the temperatures of the upper strata, *Mysis* migrates faster and closer to the surface than does *Monoporeia* (Mundie 1959, Wells 1960). Because *Mysis* has less tolerance for low concentrations of dissolved oxygen in the deeper strata than does *Monoporeia*, it may not spend much time in those deep layers during the later portion of the growing season. Decomposition of organic matter in the deep hypolimnion depletes dissolved oxygen if the upper waters are highly productive and dead matter falls into these deep waters. Of considerable importance is the observation that *Mysis* can eat *Monoporeia* (Parker 1980). The spatial refuge from mysid predation for amphipods then is dependent upon the length of time that a lake remains thermally stratified, its level of productivity, and its bottom water temperature, because these factors determine both the concentration of dissolved oxygen in the deepest strata and the presence of various benthic predators.

Because mysids are consumed by a number of game fishes, there was a period of widespread, intentional introduction of *Mysis relicta* into several deep lakes in the western and northern United States (Martinez and Bergersen 1989) and in western Canada (Northcote 1991). As often happens with introductions of species into habitats where they do not naturally occur, the foodweb dynamics did not develop as expected. Of the 134 recorded introductions into the lakes of western North America, only 45 became directly established and 8 others became indirectly established through passive migrations to other lakes (Martinez and Bergersen 1989). In some lakes, instead of providing more food for predatory fishes such as lake trout and coregonids, the mysids ate many of the same zooplankton species that the juvenile fish had previously consumed. Thus, the mysids began competing with fish for zooplankton prey (Cooper and Goldman 1982, Spencer *et al.* 1991). In these lakes, the mysids did not migrate high enough into the surface waters where the visually oriented fish could capture them. In the spring and summer, mysids avoided the warmer, upper, dimly lit surface waters and thus escaped predation; furthermore, they intercepted the upwardly moving zooplankton at mid-water depths before the zooplankon could be consumed by fish. Even more unexpected was the predation by *Mysis relicta* on newly hatched larval fish (Seale and Binowski 1988).

4. Responses to Water Quality

The general requirement for high concentrations of dissolved oxygen by most peracaridian crustaceans usually limits them to clean, cold waters. Because they feed on a variety of materials and incorporate chemical constituents from their food into their bodies (often in the lipids they store internally), these organisms are important in research on water quality. Many species also move over relatively large areas and thus are exposed to a wide array of microhabitats. The ease of collecting amphipods and isopods from a broad range of habitats and their hardiness in laboratory mesocosms make these crustaceans ideal candidates for monitoring pollution. They, like many other freshwater species, are especially sensitive to copper (even from copper pipes in laboratory plumbing) and a number of other toxic heavy metals (Abel and Barlocher 1988, Thybaud and LeBrac 1988, Borgmann and Munawar 1989). They can be used as "sentinels" of chronic exposure to low concentrations of toxins or as indicators of acute episodes of toxic spills that might otherwise go unmeasured by routine sampling of streams and lakes (Aston and Milner 1980).

Accidental spills of toxins can have major impacts on survival of crustaceans and their ecological roles in maintaining high water quality. Experimental studies demonstrate the importance of detrital processing by benthic crustaceans through comparisons of reference "control" streams with "treated" or disturbed streams (e.g., Newman *et al.* 1987). Mysids also have become important organisms for monitoring the movement of toxins through foodwebs in large lakes (Evans *et al.* 1982).

C. Life-History Traits

1. Longevity

Although most small species appear to be "annuals" and may complete their life cycle within a single year, larger amphipod species such as *Monoporeia* are thought to live for two years or more. Troglobitic species may live even longer (5–6 yr) in stable habitats with low but relatively continuous inputs of organic detritus as food sources (Culver 1982). Long-term data from Swedish lakes demonstrate cyclic population oscillations (Johnson and Wiederholm 1989). Control mechanisms for the observed population growth cycles are not clear although effects of hydrographic cycles on pelagic primary productivity and related changes in food quality and quantity on amphipod fecundity seem likely, given the results of other studies (Hynes and Harper 1972, Moore 1977,

Siegfried 1985, Sarvala 1986). Isopods inhabiting freshwater and saline ponds typically have an annual life cycle in contrast to the longer lifespan of four years for many terrestrial isopods (Sastry 1983). *Caecidotea recurvata*, a cave-dwelling isopod, is known to live at least four years (D. C. Culver, personal communication).

2. Patterns of Reproduction

Generally, Peracarida mate at, or shortly after, the time of the female molt (Sastry 1983). In most amphipods and isopods, females produce only a single brood during the annual life cycle. However, some species, such as *Hyalella azteca,* produce multiple broods during an extended breeding season (Cooper 1965, Strong 1972). Adult females can reproduce at every stage of molting. They carry their eggs in the marsupium and after fertilization continue to bear their young until they release the fully developed juveniles (ranging from a few to more than 20 per brood). Some families of amphipods differ in how mating occurs. The Pontoporeiidae have a pelagic mating system while the Gammaridae have a benthic mode (Bousfield 1989).

Several types of broad life-history patterns have been documented for both North American and European species (e.g., Cooper 1965, Strong 1972, Gee 1988, Rosillon 1989, Tadini *et al.* 1988). In general, the cold-spring faunas have longer lifespans, produce fewer eggs, and lack marked seasonality in their reproduction compared to warm-water taxa, which only live one year and produce large numbers of young in the spring. A slowing down of reproduction during late summer is reported in some amphipod populations and may be related to food scarcity or shifts in breeding behavior controlled by other factors. In *Gammarus minus* the onset of amplexus is apparently induced by a rapid decline in temperature (D. C. Culver, personal communication). In widely distributed, warm-water species such as *Hyalella azteca,* day length has been ruled out as a controlling variable for triggering the onset of reproduction (Strong 1972). For cold-water species such as *Gammarus lacustris*, a period of short days and long nights (typical of winter) is needed to induce reproduction (DeMarch 1982). In *Monoporeia affinis,* the boreo-arctic amphipod, constant illumination inhibits gonadal development (Sergestrale 1970). Peracarid activity patterns are also known to be influenced by the presence of predators (Strong 1972, Andersson *et al.* 1986, Malmqvist and Sjostrom 1987, Holomuzki and Short 1988) and these behavioral changes may well influence life-history patterns.

D. Biogeography

The distribution and abundance of any species reflects a large number of variables that may have interacted over very long periods. Climatic changes have clearly influenced the stability of aquatic habitats and their availability for colonization by crustaceans for millions of years. The most recent glacial advances and retreats were associated with not only major fluctuations in global temperatures, but also with the water balances of lakes and river drainage ecosystems (Street-Perrott and Harrison 1985). When much of the earth's water was frozen in massive ice sheets, the levels of many lakes and rivers were much reduced in the temperate zones and throughout the neotropics (Covich 1988). As discussed later in this section, important refugia from thermal extremes and desiccation have persisted in below-ground habitats such as groundwater-fed caves. Global climate change may also alter species distributions in many freshwater habitats (for review see Molles and Dahm 1990).

The mechanisms of transport and available routes of dispersal for a species are primary factors in limiting access to new habitats and recolonization of old habitats that might have had major climatic disturbances. Physical and chemical barriers to entry can also limit range expansions, as may the presence of predators or strong competitors already established in adjacent waters. Among the Peracarida, there is a pattern of restricted distribution for many species (sometimes to a single lake or spring-fed stream) that has interested biologists, geologists, and geographers. A few species are widespread and seem to tolerate a broad range of ecological parameters while cooccuring with several other genera of amphipods and isopods. Consider these three amphipods as examples: *Gammarus lacustris* occurs throughout most of the northern and parts of the western United States; *Hyalella azteca* is present from Canada to South America in the littoral zone of glacial lakes as well as in small ponds and streams; species of *Crangonyx* are frequently found in the southeastern United States as well as in Canada (Bousfield and Holsinger 1989). Many species are considered to be relatively rare. The isopod, *Thermosphaeroma theromophilum*, has been declared legally "endangered;" it is restricted to a currently protected habitat near Socorro, New Mexico (Bowman 1981).

1. Groundwaters

The greatest number of amphipod species are found in subterranean interstitial habitats, and many are associated with caves or deep wells (Barr and Holsinger 1985). Bousfield (1983) suggested that more than 1,000 hypogean amphipod species will eventually be identified. J. R. Holsinger (personal communication) completed a world-wide survey and found that 117 genera have subterranean species; approximately 615 species are fully adapted to live in groundwaters. There are about 116 species that live in groundwaters in North America, north of Mexico. Holsinger (1972) estimated that 65–70% of the North American amphipod fauna occurs in these groundwater habitats; such species are known generally as "stygobionts." He pointed out that not all of the stygobiont species are restricted to cave waters *per se* (true cave-dwelling species are termed "troglobites"), yet they may have similar morphological specializations associated with cave living (e.g., loss of eyes and pigmentation). Groundwaters are insulated from many of the environmental fluctuations that characterize surface waters, but prolonged droughts or massive ice formations associated with glaciation might eliminate these otherwise "constant" habitats. Until recently, it was thought that the few subterranean amphipods and isopods known from glaciated regions represented post-glacial colonization from nonglaciated border regions to the south. Subglacial refugia are now thought likely to have persisted during the last glacial episode (known as the Wisconsin, which ended about 12,000 years ago). For example, Castleguard Cave in Alberta, Canada, is currently inhabited by the asellid isopod *Salmasellus steganothrix* and the crangonyctid amphipod *Stygobromus canadensis*. The amphipod is endemic to this single cave and was found about 2 km from the entrance. Eleven of the 100 species described in the genus *Stygobromus* are known to occur in various localities north of the southernmost boundary of the Wisconsin episode of Pleistocene glaciation (Holsinger *et al.* 1983). Thus, the action of glacial scour of surface habitats did act as a barrier to northern distributions of many species, but others were adapted to subsurface waters and found refuge.

The fluctuations of sea levels associated with Pleistocene glaciations and earlier Tertiary events also created barriers to distributions of Peracarida. Colonization of coastal streams and subterranean habitats by marine-derived species has resulted in a complex mosaic of biogeographic boundaries. Initially, those organisms that adapted to brackish waters were able to escape most of their marine-based fish predators and to obtain abundant food resources in freshwater habitats. As shallow Cenozoic marine waters receded, the isolated populations of amphipods and isopods developed distinct adaptations; subsequent speciation produced many new taxa (Holsinger 1986). In isolated refugia, these species diversified over a 70 million year period (possibly much longer) with apparently only slow and limited active dispersal. Because many of these underground habitats are deep aquifers, the faunal composition is difficult to sample. The Edwards Aquifer in south-central Texas has been studied intensively (Holsinger and Longley 1980), and a unique ancient fauna of at least 22 species, including 10 amphipod species, is known to live in this stable groundwater habitat. Abele (1982) concluded that this artesian well in San Marcos, Texas, contains ". . .probably the richest cave crustacean fauna in the world."

2. Surface Waters

Today, the peracarids occupy many types of surface waters that range widely in physical and chemical characteristics. Historically, the chemical barrier of dilute water was a major isolating mechanism (Vernberg and Vernberg 1983) in the evolution of crustaceans. Some species of isopods move through subterranean waters and into the surface waters for certain periods (e.g., Minckley 1961). The importance of hyporheic habitats as refugia for many types of benthic crustaceans and insects is now recognized as being of great importance, especially in previously glaciated regions where massive unconsolidated rock and sand deposits characterize the floodplains of large rivers (Stanford and Ward 1988). Neighboring populations can also be genetically isolated from one another by many types of physical barriers. Dispersal of individuals to distant headwaters of many fast-flowing streams is limited within a drainage area unless there are subsurface connections among aquifers or many short tributary branches of streams that connect. Gooch (1989) quantified seven levels of isolation in his studies of *Gammarus minus* in the Appalachian Mountains of West Virginia and Pennsylvania. The larger the barrier, the more isolated were the populations. He has shown that there is a cline of lower genetic variability along a sequence of northern populations. These populations have apparently recolonized habitats since the last glacial advance and now occupy what was once a periglacial zone or perimeter around the ice sheets (Gooch and Glazier 1986). The importance of the geomorphological setting in determining how gammarid populations are actively and passively distributed is also clear in other studies of karst populations (e.g., Gooch and Hetrick 1979, Culver *et al.* 1989).

Although few studies have considered the importance of habitat structure in the dispersal and population genetics of peracarid crustaceans (see Hedgecock *et al.* 1982 for a review), it is likely that the degree of genetic isolation among populations distributed within a drainage basin is related to topo-

graphic complexity for many types of slow-moving stream invertebrates. Long, high ridges dissect the landscape and form effective barriers to the mixing of genes. This isolation by topographic structure may also be important for benthic crustaceans in deeply dissected lake basins, as apparently has been the case in the ancient lakes such as Lake Titicaca in the Andes of Bolivia and Peru where some 11 species of *Hyalella* have evolved.

One of the most interesting cases of speciation and adaptive radiation among freshwater invertebrates is the high species richness of amphipods in Lake Baikal, the oldest and deepest lake in the world (Brooks 1950). Baikal is located in Siberia and is relatively isolated from other lakes; almost 25% of the total worldwide freshwater amphipod fauna known to exist today occurs in this ancient lake, and almost all of these species are restricted to this single lake (see Kozhov 1963). Because of the combination of low water temperatures and deep circulation, there is sufficient dissolved oxygen in this lake to allow benthic gammarids to live at all depths. There are at least 240 gammarid species in about 40 genera that have evolved specialized niches in this uniquely deep lake (maximal depth is over 1700 m). Some are armed with sharp cutaneous spines or ridges, and they range in size from 10–80 mm. Coloration is highly varied (red, green, yellow, violet, pink, and brown) with complex patterning, which is very unusual among freshwater invertebrates. Deep-water (1400 m) forms have evolved different modes of surviving high hydrostatic pressure (Brauer *et al.* 1980a, 1980b). Littoral forms are extremely abundant, reaching densities of 30,000 or more per square meter. In contrast, the isopods are represented by a single genus, *Asellus,* with only five species. Generally, the diversification of crustaceans reflects differences in the response of isolated populations to local and regional conditions of long-term habitat stability, plasticity of life-history characteristics, predator–prey interactions, and access to limited food resources.

IV. COLLECTING, REARING, AND PREPARATION FOR IDENTIFICATION

Most amphipods and isopods can be collected by sweeping a dip net through the littoral zone of lakes and ponds or in vegetated reaches of streams. Mysids require a large diameter plankton net that can be used for deep vertical tows. Being fast swimmers, mysids can avoid small nets and are more difficult to collect. Quantitative sampling of amphipods and isopods has relied on drift nets and Surber samplers in fast-flowing streams and Ekman grab sampler in shallow ponds and lakes. Drift nets can entrap organisms that may be moving upstream as well as those drifting downstream (if the net is not positioned off the bottom of the stream), and thus, interpretation of studies dealing with dispersal and migration can be complicated. The use of cores (e.g., Milstead and Threlkeld 1986), can be very effective for intensive study of microhabitats. Considerable time can be spent in sieving or sorting specimens; elutriators increase the efficiency of separating specimens from the inorganic and organic matrix (Magdych 1981). Baited traps with various mesh sizes also work well for quantitative sampling of isopods and gammarids that respond to current-dispersed chemicals given off by the bait (e.g., Fitzpatrick 1983, Allan and Malmqvist 1989). Sorting of specimens is relatively rapid, but construction of fine-meshed traps is necessary. Use of electrofishing equipment may also be useful in stunning larger decapod crustaceans so that they can be swept up more effectively with dip nets. Chemicals have been added to stun or poison animals, but these methods are not very effective and clearly can be very detrimental to the environment. As discussed previously, many species are extremely sensitive to even small concentrations of toxic chemicals and are used to monitor water quality.

Several papers provide details on care and feeding of laboratory stocks (e.g., Cooper 1965, Strong 1972, DeMarch 1981a, 1981b, Sutcliffe *et al.* 1981, Barlocher *et al.* 1989). Dead leaves conditioned with nutritious microbial growths of bacteria and fungi are sufficient food resources for most species of amphipods and isopods. These can be supplemented with cultures of green algae and diatoms or with boiled lettuce or spinach. As in rearing other taxa, the initial water used for setting up an aquarium must be well aerated and allowed to develop an adequate microbial community before the crustaceans are placed into the tank. Temperature fluctuations can be detrimental and some type of aquarium heater is useful to maintain optimal growing

conditions. Parasites and diseases are, of course, widespread in natural populations (e.g., Pixell Goodrich 1934, Brownell 1970, Laberge and McLaughlin 1989), and brood stock from previously reared laboratory cultures are less likely to suffer from these outbreaks as the complex life histories of many parasites require a variety of vertebrate and invertebrate hosts to maintain the life cycle.

Preparation of specimens for study usually requires rapid fixation and storage of field-collected organisms in vials of 60–70% ethyl alcohol. The alcohol will remove any pigment from the specimens, so careful notes or photographs will be useful if information on coloration or patterns of body markings are of interest. The long-term storage of specimens requires checking to ensure that the alcohol has not evaporated. Addition of a 5–20% solution of glycerin to the vials will protect the specimens from drying out. Glycerin jelly is a solid at typical room temperatures, and it is widely used as a permanent mounting medium. Thus, if the material is first stored in dilute glycerin, it is convenient to transfer specimens to slides for study and for preparing reference collections. These techniques are described in detail in several texts (e.g., Pennak 1989, Peckarsky *et al.* 1990).

V. CLASSIFICATION OF PERACARIDA AND BRANCHIURA

The taxonomic key for this chapter is intended to allow the reader to identify major taxa of amphipods, isopods, mysids, and Branchiura to the generic level. This key draws on characteristics used in previously published studies (Edmondson 1959, Williams 1970, Cressey 1972, Holsinger 1972, 1989, Fitzpatrick 1983, Pennak 1989, Peckarsky *et al.* 1990); some of these sources can be used to identify organisms to species level, others are useful for identifying crustaceans encountered in uncommon habitats.

A. Taxonomic Key to Freshwater Genera of the Orders Mysidacea, Amphipoda, and Isopoda and the Subclass Branchiura

1a.	Four pairs of legs and two prominent, highly modified suckers on upper ventral surface, specialized ectoparasites on fishes (Fig. 18.6) .... subclass Branchiura, order Arguloida, family Argulidae *Argulus*	
1b.	More than four pairs of walking legs (pereiopods), prominent second antennae, compound eyes	2
2a(1b).	Five pairs of legs, body flattened laterally (Fig. 18.7) .... order Amphipoda	4
2b.	Six or seven pairs of legs, body flattened dorsoventrally	3

3a(2b).   Seven pairs of legs, posterior legs longer than anterior legs, unstalked eyes (Fig.
18.8) .... order Isopoda   8
3b.   Six pairs of legs, prominent stalked eyes (Fig. 18.9) ...... order Mysidacea, family Mysidae   *Mysis*
4a(2a).   Amphipoda: antenna 1 shorter than antenna 2 (Fig. 18.7) 5
4b.   Antenna 1 longer than antenna 2 (Fig. 18.7) 6
5a(4a).   Accessory flagellum present on antenna 1 with 3–5 segments (Fig. 18.7), telson
deeply cleft but not to base .... family Pontoporeiidae   *Monoporeia*
5b.   Accessory flagellum absent on antenna 1, telson entire .... family Hyalellidae   *Hyalella*
6a(4b).   Accessory flagellum of antenna 1 with 2–7 segments (usually 3 or more), telson
typically cleft to base .... family Gammaridae   *Gammarus*
6b.   Accessory flagellum of antenna 1 with never more than 2 segments, telson usually not
deeply cleft but never to the base .... family Crangonyctidae   7
7a(6b).   Eyes may be present and pigmented, apical margin of telson distinctly
cleft .... *Crangonyx*
7b.   Eyes absent, apical margin of telson entire or with shallow cleft .... *Stygobromus*
8a(3a).   Isopoda: margin of head with a pointed median protuberance (carina) between bases
of first antennae (Fig. 18.8), eyes typically present .... family Asellidae   *Lirceus*
8b.   Anterior margin of head without carina, eyes present or absent .... family Asellidae   *Caecidotea*

LITERATURE CITED

Abele, L. G. 1982. Biogeography. Pages 242–304 *in:* (D. E. Bliss, editor). The biology of Crustacea, vol. 1. Systematics, the fossil record and biogeography. Academic Press, New York.

Abel, T. and F. Barlocher. 1988. Uptake of cadmium by *Gammarus fossarum* (Amphipoda) from food and water. Journal of Applied Ecology 25:223–231.

Abrahamsson, S. A. A. 1966. Dynamics of an isolated population of the crayfish *Astacus astacus*. Oikos 17:96–107.

Ache, B. W. 1982. Chemoreception and thermoreception. Pages 369–398 *in:* (H. L. Atwood and D. C. Sandeman, eds.), The biology of Crustacea, vol. 3. Academic Press, New York.

Allan, J. D. and B. Malmqvist 1989. Diel activity of *Gammarus pulex* (Crustacea) in a South Swedish stream: comparison of drift catches vs baited traps. Hydrobiologia 179:73–80.

Allee, W. C. 1914. The ecological importance of the rheotactic reaction of stream isopods. Biological Bulletin 27:52–66.

Allee, W. C. 1929. Studies in animal aggregations: natural aggregations of the isopod, *Asellus communis*. Ecology 10:14–36.

Andersson, K. A., C. Bronmark, J. Hermann, B. Malmqvist, C. Otto, and P. Sjostrom. 1986. Presence of sculpins (*Cottus gobio*) reduces drift and activity of *Gammarus pulex* (Amphipoda). Hydrobiologia 133:209–215.

Aston, R. J. and A. G. P. Milner. 1980. A comparison of populations of the isopod *Asellus aquaticus* above and below power stations in organically polluted reaches of the River Trent. Freshwater Biology 10:1–14.

Atema, J. 1988. Distribution of chemical stimuli. Pages 29–56 *in:* (J. Atema, editor) Sensory biology of aquatic animals. Springer-Verlag, New York.

Banner, A. H. 1953. On a new genus and species of mysid from southern Louisiana. Tulane Studies in Zoology 1:1–8.

Barlocher, F. and B. Kendrick. 1973. Fungi and food preferences of *Gammarus pseudolimnaeus*. Archiv fuer Hydrobiologie 72:501–516.

Barlocher, F. and B. Kendrick. 1975. Assimilation efficiency of *Gammarus pseudolimnaeus* (Amphipoda). Oikos 24:295–300.

Barlocher, F., P. G. Tibbo and S. H. Christie. 1989. Formation of phenol-protein complexes and their use by two stream invertebrates. Hydrobiologia 173:243–249.

Barnes, R. D. 1987. Invertebrate Zoology. Fifth edition. Saunders, New York. 893 pp.

Barr, T. C., Jr. and J. R. Holsinger. 1985. Speciation in cave faunas. Annual Review of Ecology and Systematics 16:313–337.

Beeton, A. M. and J. A. Bowers. 1982. Vertical migration of *Mysis relicta* Loven. Hydrobiologia 93:53–61.

Berrill, M., L. Hollet, A. Margosian and J. Hudson. 1985. Variation in tolerance to low environmental pH by the crayfish *Orconectes rusticus, O. propinquus,* and *Cambarus robustus*. Canadian Journal of Zoology 63:2586–2589.

Blinn, D. W. and D. B. Johnson. 1982. Filter-feeding of *Hyalella montezuma,* an unusual behavior for a freshwater amphipod. Freshwater Invertebrate Biology 1:48–52.

Blinn, D. W., N. E. Grossnickle and B. Dehdashti. 1987. Diel vertical migration of a pelagic amphipod in the absence of fish predation. Hydrobiologia 160:165–171.

Blinn, D. W., C. Pinney and V. T. Wagner. 1988. Intraspecific discrimination of amphipod prey by a freshwater leech through mechanoreception. Canadian Journal of Zoology 66:427–430.

Bliss, D. E., editor-in-chief, 1982–1985. The biology of Crustacea. Vol. 1–10. Academic Press, New York.

Borgmann, U. and M. Munawar. 1989. A new standardized sediment bioassay protocol using the amphipod *Hyalella azteca* (Saussure). Hydrobiologia 188/189:425–531.

Bousfield, E. L. 1983. An updated phyletic classification and paleohistory of the amphipods. pp.257–277 *in:* F. R. Schram, editor. Crustacean issues 1: Crustacean phylogeny. A. A. Balkema, Rotterdam.

Bousfield, E. L. 1989. Revised morphological relationships within the amphipod genera *Pontoporeia* and *Gammaracanthus* and the ''glacial relict'' significance of their postglacial distributions. Canadian Journal of Fisheries and Aquatic Sciences 46:1714–1725.

Bousfield, E. L. and J. R. Holsinger. 1989. A new crangonyctid amphipod crustacean from hypogean fresh waters of Oregon. Canadian Journal of Zoology 67:963–968.

Bowers, J. A. and H. A. Vanderploeg. 1982. In situ predatory behavior of *Mysis relicta* in Lake Michigan. Hydrobiologia 93:121–131.

Bowman, T. E. 1981. *Thermosphaeroma millerio* and *T. smithi,* new sphaeromatid isopod crustaceans from hot springs in Chihuahua, Mexico, with a review of the genus. Journal of Crustacean Biology 1:105–122.

Bowman, T. E. and L. G. Abele. 1982. Classification of the recent Crustacea. Pages 1–27 *in:* L. G. Abel (ed.), The biology of Crustacea. vol.1. Systematics, the fossil record, and biogeography. Academic Press, New York.

Brauer, R. W., M. Y. Bekman, J. B. Keyser, D. L. Nesbit, G. N. Sidelev and S. l. Wright. 1980a. Adaptation to high hydrostatic pressures of abyssal gammarids from Lake Baikal in eastern Siberia. Comparative Biochemistry and Physiology 65A:109–117.

Brauer, R. W., M. Y. Bekman, J. B. Keyser, D. L Nesbit, S. G. Shvetsov, G. N. Sidelev and S. L. Wright. 1980b. Comparative studies of sodium transport and its relation to hydrostatic pressure in deep and shallow water gammarid crustaceans from Lake Baikal. Comparative Biochemistry and Physiology 65A:119–127.

Brody, T., D. Cohen, A. Barnes, and A. Spector. 1980. Yield characteristics of the prawn *Macrobrachium rosenbergii* (de Man). Journal of World Mariculture Society 12:231–243.

Brooks, J. L. 1950. Speciation in ancient lakes. Quarterly Review of Biology 25:30–60.

Brownell, W. N. 1970. Comparison of *Mysis relicta* and *Pontoporeia affinis* as possible intermediate hosts for the acanthocephalan *Echinorhynchus salmonis*. Journal of the Fisheries Research Board of Canada 27:1864–1866.

Buchanan, C. and J. F. Haney. 1980. Vertical migrations of zooplankton in the arctic: a test of the environmental controls. Pages 69–79 *in:* (W. C. Kerfoot, ed.), Evolution and ecology of zooplankton communities. University Press of New England, Hanover.

Bush, B. M. H. and M. S. Laverack. 1982. Mechanoreception. Pages 399–468 *in:* ((H. L. Atwood and D. C. Sandeman, eds.), The biology of Crustacea, vol.3. Academic Press, New York.

Capart, A. 1951. *Thermobathynella adami,* gen. et spec. nov., Anaspidace du Congo Belge. Institut Royal des Sciences et Naturelles de Belgique, Bulletin 27:1–4.

Chambers, P. A., J. M. Hanson, J. M. Burke, and E. E. Prepas, 1990. The impact of the crayfish *orconectes virilis* on aquatic macrophytes. Freshwater Biology 24:81–91.

Cheper, N. J. 1980. Thermal tolerance of the isopod *Lirceus brachyurus* (Crustacea: Isopoda). American Midland Naturalist 104:312–318.

Cole, G. A. and C. A. Bane. 1978. *Thermosphaeroma subequalum,* n. gen., n. sp. (Crustacea: Isopoda) from Big Bend National Park, Texas. Hydrobiologia 59:223–228.

Cole, G. A. and R. L. Watkins. 1977. *Hyalella montezuma,* a new species (Crustacea: Amphipoda) from Montezuma Well, Arizona. Hydrobiologia 52:175–184.

Cooper, S. D. and C. R. Goldman. 1982. Environmental factors affecting predation rates of *Mysis relicta*. Canadian Journal of Fisheries and Aquatic Sciences 39:203–208.

Cooper, W. E. 1965. Dynamics and production of a natural population of a fresh-water amphipod, *Hyalella azteca*. Ecological Monographs 35:377–394.

Cossins, A. R. and K. Bowler. 1987. Temperature biology of animals. Chapman, London. 330 pp.

Covich, A. P. 1977. How do crayfish respond to plants and Mollusca as alternate food resources? Freshwater Crayfish 3:165–179.

Covich, A. P. 1988. Geographical and historical comparisons of neotropical streams: biotic diversity and detrital processing in highly variable habitats. Journal of the North American Benthological Society 7:361–386.

Cressey, R. F. 1972. The genus *Argulus* (Crustacea: Branchiura) of the United States. Biota of Freshwater Ecosystems, U.S. Environmental Protection Agency Identification Manual 2:1–14.

Crowl, T. A. and A. P. Covich. 1990. Predator-induced life-history shifts in a freshwater snail. Science 247:949–951.

Culver, D. C. 1982. Cave life. Harvard University Press, Cambridge.

Culver, D. C. and T. L. Poulson. 1971. Oxygen consumption and activity in closely related amphipod populations from cave surface habitats. American Midland Naturalist 85:74–84.

Culver, D. C., T. C. Kane, D. W. Fong, R. Jones, M. A. Taylor and S. C. Saureisen. 1989. Morphology of cave organisms—is it adaptive? Memoirs de Biospelogie 17:3–16.

Dadswell, M. J. 1974. Distribution, ecology, and postglacial dispersal of certain crustaceans and fishes in eastern North America. Publications in Zoology No.11, National Museum of Natural Sciences, National Museums of Canada, Ottawa.

Dahl, E., H. Emanuelsson and C. von Mecklenburg. 1970. Pheromone transport and reception in an amphipod. Science 170:739–740.

Davies, I. J. 1989. Population collapse of the crayfish *Orconectes virilis* in response to experimental whole-lake acidification. Canadian Journal of Fisheries and Aquatic Sciences 46:910–922.

DeMarch, B. G. E. 1981a. *Hyalella azteca* (Saussure). Pages 61–77 *in:* S. G. Lawrence (ed.), Manual for the culture of selected freshwater invertebrates. Canadian Special Publication in Fisheries and Aquatic Sciences No.54.

DeMarch, B. G. E. 1981b. *Gammarus lacustris lacustris* G. O. Sars. Pages 79–94 *in:* S. G. Lawrence (ed.), Manual for the culture of selected freshwater invertebrates. Canadian Special Publication in Fisheries and Aquatic Sciences No.54.

DeMarch, B. G. E. 1982. Decreased day length and light intensity as factors inducing reproduction in *Gammarus lacustris lacustris* Sars. Canadian Journal of Zoology 60:2962–2965.

Dermott, R. M. and K. Corning. 1988. Seasonal ingestion rates of *Pontoporeia hoyi* (Amphipoda) in Lake Ontario. Canadian Journal of Fisheries and Aquatic Sciences 45:1886–1895.

Dodds, W. K. 1990. Hydrodynamic constraints on evolution of chemically mediated interactions between aquatic organisms in unidirectional flows. Journal of Chemical Ecology 16:1417–1430.

Donner, K. O., A. Lindstrom and M. Lindstrom. 1987. Seasonal variation in the vertical migration of *Pontoporeia affinis* (Crustacea, Amphipoda). Annales Zoologica Fennica 24:305–313.

Dormaar, K. A. and S. Corey. 1978. Some aspects of osmoregulation in *Mysis relicta* Loven (Mysidacea). Crustaceana 34:90–93.

Dunham, P. 1978. Sex pheromones in Crustacea. Biological Review 53:555–583.

Edmondson, W. T., editor. 1959. Freshwater biology, Second Edition. Wiley, New York.

Ellis, R. J. 1971. Notes on the biology of the isopod *Asellus tomalensis* Harford in an intermittent pond. Transactions of the American Microscopical Society 90:51–61.

Enright, J. T. 1977. Diurnal vertical migration: adaptive significance and timing. Part 1. Selective advantage: a metabolic model. Limnology and Oceanography 22:856–872.

Enright, J. T. 1979. The why and when of up and down. Limnology and Oceanography 24:788–791.

Evans, M. S., R. W. Bathelt and C. P. Rice. 1982. Polychlorinated biphenyls and other toxicants in *Mysis relicta*. Hydrobiologia 93:205–215.

Fitzpatrick, J. F., Jr. 1983. How to know the freshwater Crustacea. W. C. Brown, Dubuque, Iowa. 227 pp.

Folt, C. L., J. T. Rybock and C. R. Goldman. 1982. The effect of prey composition and abundance on the predation rate and selectivity of *Mysis relicta*. Hydrobiologia 93:133–143.

Fong, D. W. 1989. Morphological evolution of the amphipod *Gammarus minus* in caves - quantitative genetic analysis. American Midland Naturalist 121:361–378.

France, R. L. 1984. Comparative tolerance to low pH of three life stages of the crayfish *Orconectes virilis*. Canadian Journal of Zoology 62:2360–2363.

Friesen, J. A., K. H. Mann and J. A. Novitsky. 1986. *Mysis* digests cellulose in the absence of a gut microflora. Canadian Journal of Zoology 64:442–446.

Galat, D. L., M. Coleman and R. Robinson. 1988. Experimental effects of elevated salinity on three benthic invertebrates in Pyramid Lake, Nevada. Hydrobiologia 158:133–144.

Gauvin, J. M., W. S. Gardner and M. A. Quigley. 1989. Effects of food removal on nutrient release rates and lipid content of Lake Michigan *Pontoporeia hoyi*. Canadian Journal of Fisheries and Aquatic Sciences 46:1125–1130.

Gee, J. H. R. 1982. Resource utilization by *Gammarus pulex* (Amphipoda) in a Cotswold stream; a microdistribution study. Journal of Animal Ecology 51:817–832.

Gee, J. H. R. 1988. Population dynamics and morphometrics of *Gammarus pulex* L.: evidence of seasonal food limitation in a freshwater detritivore. Freshwater Biology 19:333–343.

Gliwicz, M. Z. 1986. Predation and the evolution of vertical migration in zooplankton. Nature 320:746–748.

Goddard, J. S. 1988. Food and feeding. Pages 145–166 *in:* D. M. Holdich and R. S. Lowery (eds.), Freshwater crayfish. Biology, management and exploitation. Timber Press, Portland, Oregon.

Gooch, J. L. 1989. Genetic differentiation in relation to stream distance in *Gammarus minus* (Crustacea, Amphipoda) in Appalachian watersheds. Archiv fuer Hydrobiologie 114:505–519.

Gooch, J. L. and S. T. Hetrick 1979. The relation of genetic structure to environmental structure: *Gammarus minus* in a karst area. Evolution 33:192–206.

Gooch, J. L. and D. S. Glazier. 1986. Levels of heterozygosity in the amphipod *Gammarus minus* in an area affected by Pleistocene glaciation. American Midland Naturalist 116:57–63.

Greenaway, P. 1985. Calcium balance and molting in the Crustacea. Biological Review 60:425–454.

Hanson, J. M., P. A. Chambers, and E. E. Prepas. 1990. Selective foraging by the crayfish *Orconectes virilis* and its impact on macroinvertebrates. Freshwater Biology 24:69–80.

Hargrave, B. T. 1970a. The utilization of benthic microflora by *Hyalella azteca* (Amphipoda). Journal of Animal Ecology 39:427–437.

Hargrave, B. T. 1970b. Distribution, growth, and seasonal abundance of *Hyalella azteca* (Amphipoda) in relation to sediment microflora. Journal of Fisheries Research Board of Canada 27:685–699.

Hartnoll, R. G. 1982. Growth. Pages 111–196 *in:* L. G. Abele (ed.), The biology of Crustacea, vol. 2. Academic Press, New York.

Hedgecock, D., M. L. Tracey and K. Nelson. 1982. Genetics. Pages 284–403 *in:* L. G. Abele (ed.), The biology of Crustacea, vol. 2. Academic Press, New York.

Holomuzki, J. R. and T. M. Short. 1988. Habitat use and fish avoidance behaviors by the stream-dwelling isopod *Lirceus fontinalis*. Oikos 52:79–86.

Holsinger, J. R. 1972. The freshwater amphipod crustaceans (Gammaridae) of North America. Biota of

Freshwater Ecosystems, Identification Manual 5, U.S. Environmental Protection Agency.

Holsinger, J. R. 1986. Zoogeographic patterns of North American subterranean amphipod crustaceans. Pages 85–106 *in:* R. H. Gore and K. L. Heck (ed.), Crustacean biogeography. A. A. Balkema, Rotterdam.

Holsinger, J. R. 1988. Troglobites: the evolution of cave-dwelling organisms. American Scientist 76:146–153.

Holsinger, J. R. 1989. Allocrangonyctidae and Pseudo-crangonyctidae, two new families of Holarctic subterranean amphipod crustaceans (Gammaridae), with comments on their phylogenetic and zoogeographic relationships. Proceedings of the Biological Society of Washington 102:947–959.

Holsinger, J. R. and G. Longley. 1980. The subterranean amphipod crustacean fauna of an artesian well in Texas. Smithsonian Contributions to Zoology 308: 1–62.

Holsinger, J. R., J. S. Mort and A. D. Reckles. 1983. The subterranean crustacean fauna of Castleguard Cave, Columbia Icefields, Alberta, Canada, and its zoogeographical significance. Arctic and Alpine Research 15:543–549.

Hubricht, L. and J. G. Mackin. 1949. The American isopods of the genus *Lirceus* (Asellota, Asellidae). American Midland Naturalist 42:334–349.

Huner, J. V. 1988. *Procambarus* in North America and elsewhere. Pages 239–261, *in:* D. M. Holdich (eds.), Freshwater crayfish. Biology, management and exploitation. Timber Press, Portland, Oregon.

Hutchinson, G. E. 1967. A treatise on limnology, Vol. 2. John Wiley and Sons, New York. 1115 pp.

Hynes, H. B. N. and F. Harper. 1972. The life histories of *Gammarus lacustris* and *G. pseudolimnaeus* in southern Ontario. Crustaceana (Supplement) 3:329–341.

Johnson, R. K. and T. Wiederholm. 1989. Long-term growth oscillations of *Pontoporeia affinis* Lindstrom (Crustacea: Amphipoda) in Lake Malaren. Hydrobiologia 175:183–194.

Jones, R. and D. C. Culver. 1989. Evidence for selection on sensory structures in a cave population of *Gammarus minus* (Amphipoda). Evolution 43:688–693.

Juday, C. and E. A. Birge. 1927. *Pontoporeia* and *Mysis* in Wisconsin lakes. Ecology 7:445–452.

Kozhov, M. 1963. Lake Baikal and its life. Dr. W. Junk, The Hague. 344 p.

Laberge, R. J. A. and J. D. McLaughlin. 1989. *Hyalella azteca* (Amphipoda) as an intermediate host of the nematode *Streptocara crassicauda*. Canadian Journal of Zoology 67:2335–2340.

Lopez, G. and R. Elmgren. 1989. Feeding depths and organic absorption for the deposit-feeding benthic amphipods *Pontoporeia affinis* and *Pontoporeia femorata*. Limnology and Oceanography 34:982–991.

Magdych, W. P. 1981. An efficient, inexpensive elutriator design for separating benthos from sediment samples. Hydrobiologia 85:157–159.

Maguire, B. 1965. *Monodella texana,* an extension of the range of the crustacean order Thermosbaenacea to the western hemisphere. Crustaceana 9:149–154.

Malmqvist, B. and P. Sjostrom. 1987. Stream drift as a consequence of disturbance by invertebrate predators. Field and laboratory experiments. Oecologia 74:396–403.

Mantel, L. H. and L. L. Farmer 1983. Osmotic and ionic regulation. Pages 53–161 *in:* L. H. Mantel (ed.), The biology of Crustacea, vol.5 Academic Press, New York.

Marchant, R. 1981. The ecology of *Gammarus* in running water. Pages 225–249 *in:* M. A. Lock and D. D. Williams (eds.), Perspectives in running water ecology. Plenum Press, New York.

Marchant, R. and H. B. N. Hynes. 1981a. The distribution and production of *Gammarus pseudolimnaeus* (Crustacea: Amphipoda) along a reach of the Credit River, Ontario. Freshwater Biology 11:169–182.

Marchant, R. and H. B. N. Hynes. 1981b. Field estimates of feeding rate for *Gammarus pseudolimnaeus* (Crustacea: Amphipoda) in the Credit River, Ontario. Freshwater Biology 11:27–36.

Margolis, L. and Z. Kabata (editors). 1988. Guide to the parasites of fishes of Canada, Part II. Crustacea. Canadian Special Publication of Fisheries and Aquatic Sciences 101. Department of Fisheries and Oceans, Ottawa. 184 pp.

Martinez, P. J. and E. P. Bergersen. 1989. Proposed biological management of *Mysis relicta* in Colorado lakes and reservoirs. North American Journal of Fisheries Management 9:1–11.

Marzolf, G. R. 1965. Substrate relations of the burrowing amphipod *Pontoporeia affinis* in Lake Michigan. Ecology 46:579–592.

McDonald, M. E., L. B. Crowder and S. B. Brandt. 1990. Changes in *Mysis* and *Pontoporeia* populations in southeastern Lake Michigan: a response to shifts in the fish community. Limnology and Oceanography 35:220–227.

McLaughlin, P. A. 1980. Comparative morphology of recent Crustacea. W. H. Freeman and Company, San Francisco. 177 pp.

McLaughlin, P. A. 1982. Comparative morphology of crustacean appendages. Pages 197–256 *in:* L. G. Abele (ed.), The biology of crustacea, vol. 2. Academic Press, New York.

Milstead, B. and S. T. Threlkeld. 1986. An experimental analysis of darter predation on *Hyalella azteca* using semipermeable enclosures. Journal of the North American Benthological Society 5:311–318.

Minckley, W. L. 1961. Occurrence of subterranean isopods in the epigean environment. American Midland Naturalist 63:452–455.

Molles, M. C., Jr. and C. N. Dahm. 1990. A perspective on El Niño and La Niña: global implications for stream ecology. Journal of the North American Benthological Society 9:68–76.

Monk, D. C. 1977. The digestion of cellulose and other dietary components, and pH of the gut in the amphipod *Gammarus pulex* (L.). Freshwater Biology 7:431–440.

Moore, J. W. 1977. Importance of algae in the diet of subarctic populations of *Gammarus lacustris* and *Pontoporeia affinis*. Canadian Journal of Zoology 55:637–641.

Morgan, M. D., S. T. Threlkeld and C. R. Goldman. 1978. Impact if the introduction of kokanee (*Oncorhyncus nerka*) and opossum shrimp (*Mysis relicta*) on a subalpine lake. Journal of the Fisheries Research Board of Canada 35:1572–1579.

Mundahl, N. D. 1989. Seasonal and diel changes in thermal tolerance of the crayfish *Orconectes rusticus*, with evidence for behavioral thermoregulation. Journal of North American Benthological Society 8:173–179.

Mundie, J. H. 1959. The diurnal activity of the larger invertebrates at the surface of Lac la Ronge, Saskatchewan. Canadian Journal of Zoology 37:945–946.

Murtaugh, P. A. 1989. Fecundity of *Neopmysis mercedis* Holmes in Lake Washington (Mysidacea). Crustaceana 57:194–200.

Nalepa, T. F. and A. Robertson. 1981. Vertical distribution of the zoobenthos in southeastern Lake Michigan with evidence of seasonal variation. Freshwater Biology 11:87–96.

Nero, R. W. and D. W. Schindler. 1983. Decline of *Mysis relicta* during acidiification of Lake 223. Canadian Journal of Fisheries and Aquatic Sciences 40:1905–1911.

Newman, R. M., W. C. Kerfoot and Z. Hanscom, III. 1990. Watercress and amphipods: potential chemical defense in a spring stream macrophyte. Journal of Chemical Ecology 16:245–259.

Newman, R. M., J. A. Perry, E. Tam, R. L. Crawford. 1987. Effects of chronic chlorine exposure on litter processing in outdoor experimental streams. Freshwater Biology 18:415–428.

Northcote, T. G. 1991. Successes, problems, and control of introduced mysid populations in lakes and reservoirs. American Fisheries Society, Symposium Proceedings. (in press).

Parker, J. I. 1980. Predation by *Mysis relicta* on *Pontoporeia hoyi*: a food chain link of potential importance in the Great Lakes. Journal of Great Lakes Research 6:164–166.

Peck, S. K. 1985. Effects of aggressive interaction on temperature selection by the crayfish, *Orconectes virilis*. American Midland Naturalist 114:159–167.

Peckarsky, B. L., P. R. Frassinet, M. A. Penton and D. J. Conklin, Jr. 1990. Freshwater macroinvertebrates of northeastern North America. Cornell Univesity Press, Ithaca, New York. 422 pp.

Pennak, R. W. 1989. Fresh-water invertebrates of the United States: protozoa to mollusca. Third Edition. Wiley, New York. 628 pp.

Pixell Goodrich, H. P. 1934. Reactions of *Gammarus* to injury and disease, with notes on microsporidial and fungoid disease. Quarterly Journal of the Microscopy Science 72:325–353.

Provenzano, A. J., Jr. 1985. Commercial culture of decapod crustaceans. Pages 269–314, *in:* A. J. Provenzano (ed.), The biology of Crustacea, vol. 10. Academic Press, New York.

Rabeni, C. F. 1985. Resource partitioning by stream-dwelling crayfish: the influence of body size. American Midland Naturalist 113:20–29.

Richardson, H. 1905. A monograph of the isopods of North America. Bulletin of the U.S. National Museum 54:1–727.

Rose, R. D. 1986. Chemical detection of "self" and conspecifics by crayfish. Journal of Chemical Ecology 12:271–276.

Rosillon, D. 1989. The influence of abiotic factors and density-dependent mechanisms on between-year variations in a stream invertebrate community. Hydrobiologia 179:25–38.

Sarvala, J. 1986. Interannual variation of growth and recruitment in *Pontoporeia affinis* (Lindstrom) (Crustacea, Amphipoda) in relation to abundance fluctuations. Journal of Experimental Marine Ecology 101:41–59.

Sastry, A. N. 1983. Ecological aspects of reproduction. Pages 179–270 *in:* F. J. Vernberg and W. B. Vernberg (eds.), The biology of Crustacea, vol. 8. Academic Press, New York.

Schmitt, W. L. 1965. Crustaceans. University of Michigan Press, Ann Arbor, 204 pp.

Schmitz, E. H. and P. M. Scherrey. 1983. Digestive anatomy of *Hyalella azteca* (Crustacea, Amphipoda). Journal of Morphology 175:91–100.

Schram, F. R. 1982. The fossil record and evolution of Crustacea. Pages 93–174 *in:* L. G. Abele (ed.), The biology of Crustacea, vol. 1. Academic Press, New York.

Schram, F. R. 1986. Crustacea. Oxford University Press, New York. 606 pp.

Schram, F. R., J. Yager, and M. J. Emerson. 1986. Remipedia. Part 1. Systematics. Memoirs of the San Diego Society of Natural History 15:1–60.

Schuster, S. M. 1981. Life history characteristics of *Thermosphaeroma thermophilum*, the Socorro isopod (Crustacea: Peracarida). Biological Bulletin 161:291–302.

Seale, D. B. and F. P. Binowski. 1988. Vulnerability of early life intervals of *Coregonus hoyi* to predation by a freshwater mysid, *Mysis relicta*. Environmental Biology of Fishes 21:117–126.

Sergestrale, S. G. 1970. Light control of the reproductive cycle of *Pontoporeia affinis* Lindstrom (Crustacea: Amphipoda). Journal of Experimental Marine Biology and Ecology 5:272–275.

Shea, M. A. and J. C. Makarewicz. 1989. Production, biomass, and trophic interactions of *Mysis relicta* in Lake Ontario. Journal of Great Lakes Research 15:223–232.

Sherman, R. K., D. C. Lasenby and L. Hollet. 1987. Influence of oxygen concentration on the distribution of *Mysis relicta* Loven in a eutrophic temperate lake. Canadian Journal of Zoology 65:2646–2650.

Sieg, J. 1983. Evolution of the Tanaidacea. Pages 229–256 *in:* F. R. Schram (ed.), Crustacean phylogeny. A. A. Balkema, Rotterdam.

Siegfried, C. A. 1985. Life history, population dynamics and production of *Pontoporeia hoyi* (Crustacea, Amphipoda) in relation to the trophic gradient of Lake George, New York. Hydrobiologia 122:175–180.

Smock, L. A. and K. L. Harlowe. 1983. Utilization

and processing of freshwater wetland macrophytes by the detritivore *Asellus forbesi*. Ecology 64:1556–1565.

Smock, L. A. and D. L. Stoneburner. 1980. The response of macroinvertebrates to aquatic macrophyte decomposition. Oikos 33:397–403.

Spencer, C. N., B. R. McClelland and J. A. Stanford. 1991. Shrimp introduction, salmon collapse, and eagle displacement: cascading interactions in the food web of a large aquatic ecosystem. BioScience 41:14–21.

Stanford, J. A. and J. V. Ward. 1988. The hyporheic habitat of river ecosystems. Nature 335:64–66.

Statzner, B. 1988. Growth and Reynolds number of lotic macroinvertebrates: a problem for adaptation of shape to drag. Oikos 51:84–87.

Stephens, D. W. 1990. Changes in lake levels, salinity and the biological community of Great Salt Lake (Utah, USA), 1847–1987. Hydrobiologia 197:139–146.

Stock, J. H. and G. Longley. 1981. The generic status and distribution of *Monodella texana*, Maguire, the only known North American thermosbaenacean. Proceedings of the Biological Society of Washington 94:569–578.

Street-Perrott, F. A. and S. P. Harrison. 1985. Lake levels and climate reconstruction. Pages 291–340 *in:* A. D. Hecht (ed.), Paleoclimate analysis and modeling. Wiley, New York.

Strong, D. R. 1972. Life history variation among populations of an amphipod (*Hyalella azteca*). Ecology 53:1103–1111.

Sutcliffe, D. W. 1971. Regulation of water and some ions in gammarids (Amphipoda). I. *Gammarus duebeni* Lilljeborg from brackish water and fresh water. Journal of Experimental Biology 55:325–344.

Sutcliffe, D. W. 1974. Sodium regulation and adaptation to fresh water in the isopod genus *Asellus*. Journal of Experimental Biology 61:719–736.

Sutcliffe, D. W., T. R. Carrick and L. G. Willoughby. 1981. Effects of diet, body size, age and temperature on growth rates in the amphipod *Gammarus pulex*. Freshwater Biology 11:183–214.

Tadini, G. V., E. A. Tadini and M. Colangelo 1988. The life history of *Asellus aquaticus* (L.) explains its geographical distribution. Verhandlungen Internationale Vereinigung Fuer Theoretische und Angewandte Limnologie 23:2099–2106.

Taylor, R. C. 1984. Thermal preference and temporal distribution in three crayfish species. Comparative Biochemistry and Physiology 77A:513–517.

Taylor, R. C. 1990. Crayfish (*Procambarus spiculifer*) growth rate and final thermal preferendum. Journal of Thermal Biology 15:79–81.

Tcherkashina, N. Ya. 1977. Survival, growth, and feeding dynamics of juvenile crayfish (*Astacus leptodactylus cubanicus*) in ponds and the River Don. Freshwater Crayfish 3:95–100.

Thorp, J. H. 1984. Theory and practice in crayfish communication studies. Journal of Chemical Ecology 10:1283–1288.

Thorp, J. H. 1991. Invertebrate diversity and functional biology. John Wiley and Sons, New York. (in prep.).

Tierney, A. J. and J. Atema. 1988. Behavioral responses of crayfish (*Orconectes virilis* and *Orconectes rusticus*) to chemical feeding stimulants. Journal of Chemical Ecology 14:123–133.

Thybaud, E. and S. LeBrac. 1988. Absorption and elimination by *Asellus aquaticus* (Crustacea, Isopoda). Bulletin of Environmental Contamination and Toxicity 40:731–735.

Vernberg, W. B. and F. J. Vernberg 1983. Freshwater adaptations. Pages: 335–363 *in:* F. J. Vernberg and W. B. Vernberg (eds.), The biology of Crustacea, vol 8. Academic Press, New York.

Wagner, V. T. and D. W. Blinn. 1987. A comparative study of the maxillary setae for two coexisting species of *Hyalella* (Amphipoda), a filter feeder and a detritus feeder. Archiv fur Hydrobiologia 109:409–419.

Watling, L. 1981. An alternative phylogeny of peracarid crustaceans. Journal of Crustacean Biology 1:201–210.

Wells, L. 1960. Seasonal abundance and vertical movements of planktonic Crustacea in Lake Michigan. U.S. Fish and Wildlife Service Fishery Bulletin 60:343–369.

Wickstrom, C. E. and R. W. Castenholz. 1973. Thermophilic ostracod: aquatic metazoan with the highest known temperature tolerance. Science 181:1063–1064.

Williams, D. D. 1989. Towards a biological and chemical definition of the hyporheic zone in two Canadian rivers. Freshwater Biology 22:189–208.

Williams, D. D. and H. B. N. Hynes. 1976. The recolonization mechanism of stream benthos. Oikos 27:265–272.

Williams, D. D. and K. A. Moore. 1985. The role of semiochemicals to benthic community relationships of the lotic amphipod *Gammarus pseudolimneaus*: a laboratory analysis. Oikos 44:280–286.

Williams, W. D. 1970. A revision of North American epigean species of *Asellus* (Crustacea: Isopoda). Smithsonian Contributions to Zoology 49:1–80.

Willoughby, L. G. and D. W. Sutcliffe. 1976. Experiments on feeding and growth of the amphipod *Gammarus pulex* (L.) related to its distribution in the River Duddon. Freshwater Biology 6:577–586.

Wilson, C. B. 1944. Parasitic copepods of the United States National Museum. Proceedings of the U.S. National Museum 94:529–582.

Winnell, M. H. and D. S. White. 1984. Ecology of shallow and deep water populations of *Pontoporeia hoyi* (Smith) (Amphipoda) in Lake Michigan. Freshwater Invertebrate Biology 3:118–138.

Wurtsbaugh, W. A. and T. S. Berry. 1990. Cascading effects of decreased salinity on the plankton, chemistry, and physics of the Great Salt Lake (Utah). Canadian Journal of Fisheries and Aquatic Sciences 47:100–109.

Yager, J. 1981. Remipedia, a new class of Crustacea from a marine cave in the Bahamas. Journal of Crustacean Biology 1:328–333.

Yager, J. 1987. *Cryptocorynectes haptodiscus* and *Speleonectes benjamini* from anchialine caves in the Bahamas, with remarks on distribution and ecology.

Proceedings of the Biological Society of Washington 100:302–320.

Yamaguti. S. 1963. Parasitic Copepoda and Branchiura of fishes. Wiley Interscience Publishers, New York. 1104 pp.

Zaret, T. 1980. Predation and freshwater communities. Yale University Press, New Haven.

Zimmer-Faust, R.K. 1989. The relationship between chemoreception and foraging behavior in crustaceans. Limnology and Oceanography 34:1367–1374.

Ostracoda

19

L. Denis Delorme
National Water Research Institute
Canada Centre for Inland Waters
Burlington, Ontario L7R 4A6 Canada
and
Department of Earth Sciences
University of Waterloo
Waterloo, Ontario N2L 3G1 Canada

Chapter Outline

I. INTRODUCTION

II. ANATOMY AND PHYSIOLOGY
  A. External Shell Morphology
  B. Internal Shell Morphology
  C. External Body and Appendage Morphology
  D. Internal Anatomy and Physiology

III. LIFE HISTORY
  A. Egg
  B. Embryology
  C. Hatching and Seasonality
  D. Molting
  E. Competition, Coexistence, and Survival

IV. DISTRIBUTION
  A. Benthic Swimmers
  B. Benthic Nonswimmers

V. CHEMICAL HABITATS
  A. Chemical Water Types
  B. Salinity and Solute Composition
  C. Hydrogen Ion Concentration
  D. Dissolved Oxygen
  E. Carbonate

VI. PHYSICAL HABITAT
  A. Temperature
  B. Lakes, Ponds, and Streams
  C. Bromeliads
  D. Caves
  E. Sediment–Water Interface, Particle Size, Food

VII. PHYSIOLOGICAL AND MORPHOLOGICAL ADAPTATIONS
  A. Eggs
  B. Resting Stage and Torpidity

  C. Coloration
  D. Appendages

VIII. BEHAVIORAL ECOLOGY

IX. FORAGING RELATIONSHIPS

X. POPULATION REGULATION

XI. COLLECTING AND REARING TECHNIQUES
  A. Collecting and Preservation
  B. Rearing Techniques

XII. CURRENT AND FUTURE RESEARCH PROBLEMS

XIII. CLASSIFICATION OF FRESHWATER OSTRACODES
  A. Taxonomic Controversies
  B. Classification of the Order Podocopida
  C. Taxonomic Key to Genera of the Superfamily Cypridoidea
  D. Taxonomic Key to Genera of the Superfamily Cytheroidea
  Literature Cited

I. INTRODUCTION

The class Ostracoda (Latreille 1802) in the Crustacea contains both marine and freshwater forms. The subclass Podocopa (Müller 1894) has one order, Podocopida (Sars 1866), in which two of the suborders (Metacopina, Podocopina) contain all of the freshwater ostracodes (Bowman and Abele 1982, Schram 1986).

Ostracodes are the oldest known microfauna. Their preservation as fossils is attributed to the calcitic shell present in most species. The first fossil representatives are reported from Cambrian marine sediments (Moore 1961, Maddocks 1982). The earliest freshwater ostracode species come from coal-forming swamps, ponds, and streams of early Pennsylvanian age (Benson 1961). According to Moore (1961), the Podocopina evolved from Podocopida stock during the Ordovician period, and then branched out into the Cytheroidea and the Cypridoidea during the Devonian period (Maddocks 1982). The Darwinulidae evolved during the Carboniferous period (Maddocks 1982).

Freshwater ostracodes are free living except the Entocytheridae (Hoff 1942), which are commensal upon crayfish. This chapter will concentrate on freshwater ostracode genera found in North America. To date, there have been 54 genera and about 420 species identified for this continent.

II. ANATOMY AND PHYSIOLOGY

A. External Shell Morphology

Ostracodes have their body (visceral mass and appendages) suspended from the dorsal region in an elongate, chitinous pouch. The visceral mass is completely encased within a chitinous and calcitic exoskeleton (carapace). The only exception is the entocytherid group of ostracodes, which lacks calcium in the exoskeleton. The exterior surface of the exoskeleton (cuticle) is made up of chitin (Fig. 19.1) (Harding 1964). The outer chitinous coating of the epidermis is the only covering over the calcitic shell. The soft epidermal tissue lining the interior part of the carapace is composed of the inner and outer hypodermis. The hypodermal or epithelial cells of the outer lamella (Fig. 19.1) secrete the calcareous shells, and are larger and more irregular in shape than the inner hypodermal cells. Subdermal cells occur between the inner and outer hypodermal cells as do the liver, ovaries, and testes. The epidermis is contained within a thin chitinous lining. When the ostracode molts, the complete exoskeleton is shed, leaving only the visceral mass and the soft tissue or hypodermis. The hypodermis is continuous with the visceral mass in the dorsal region of the body.

When the ostracode molts, the space between the epidermal cells of the outer and inner layer is empty. The shells are secreted by the cells of the outer lamellae (Turpen and Angell 1971). The immediate source of calcium and magnesium carbonate is thought to be stored in the body and is not obtained directly from the water (Fassbinder 1912). Fassbinder has also shown that ostracodes fed with crushed calcium carbonate from snail shells developed thicker carapaces with more pronounced marginal ornamentation. He concluded that food is an

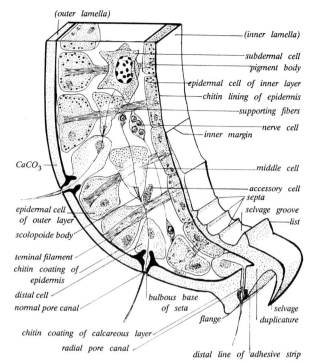

Figure 19.1 Diagrammatic cross-section of the shell and the enclosing dermal layers. (From Kesling 1951a,b. Figure redrawn by J. L. Delorme.)

important source of carbonate. Turpen and Angell (1971), however, found that calcium dissolved in water is the source of calcium for the shell. They used dissolved ^{45}Ca as a tracer (pH 7.4, 23°C) to also show that calcium is not resorbed from the shells prior to molting and that ostracodes do not store calcium in their bodies. Chivas *et al.* (1985) ran laboratory experiments on the Australian *Mytilocypris henricae* and found that temperature influenced the rate of shell calcification. Bodergat (1978) demonstrated that phosphorus was an integral part of the shells in all freshwater ostracodes investigated with a microprobe. She further noted a significant positive correlation between the amount of shell phosphorus and the habitat temperature of the ostracode.

The calcareous shell is the most obvious structure when the organism is viewed under a microscope. The shell is divided into two parts, the outer lamella and the duplicature (Fig. 19.1). The outer lamella is the major part of the shell; whereas the duplicature is the calcified part of the inner lamella bordering the posterior, ventral, and anterior margin. It forms a part of the free margin and projects toward the center of the carapace. The outer lamella may contain pores through which project setae (Fig. 19.2), which are sensitive to touch. The duplicature is welded to the outer lamella (Fig. 19.3) around the margins by a strip of chitin, which joins with the chitin coating and the chitin lining. This juncture is referred to as the line of concresence (Fig. 19.1). Within the chitinous strip, radial or marginal pore canals are found (Fig. 19.2). These are very fine tubes through which setae pass. The inner free margin of the duplicature is referred to as the inner margin (Figs. 19.1). When the duplicature extends beyond the line of concresence toward the center of the shell, a shelf is formed (Fig. 19.3). The space between the shelf and the outer lamella is referred to as the vestibule (Fig. 19.1). Several structures are found on the duplicature that are very important in taxonomy down to the species level (Sylvester-Bradley 1941).

Pustules, teeth, or crenulations (Fig. 19.3) may appear on the outer margin of the duplicature. A combination of septa, lists, and grooves are found on the duplicature. Some of these are shown in Fig. 19.1 (Kesling 1951a, 1951b). The shell may contain additional structures on the outer lamella such as lateral depressions called sulci (Fig. 19.27). Raised areas variously termed alae, pustules (Fig. 19.4), or knobs (Fig. 19.27), depending on their shape, size, and orientation, may also be found on the shell surface. The surface of the shell may also be pitted (Fig. 19.5), punctate, wrinkled (Fig. 19.6), or have a reticulate surface (Fig. 19.7). These ancillary structures of the shell are very often used for generic and specific identification (see Sylvester-Bradley and Benson 1971 for terminology). Some authors, notably Benson (1974 1981), have discussed the roles of ornamentation, form, function, and architecture of the ostracode shell.

Intraspecific morphological diversity of the shell may be pronounced. *Limnocythere itasca* has short recurved alae on its lateral surfaces. The individuals identified from Canada (Delorme 1971a) and northern Minnesota (Forester *et al.* 1987) appear to be characteristic of this species. R. M. Forester (personal communication) identified *L. itasca* from the Laguna de Texcoco, Mexico, which have very long and delicate recurved spines or alae. Intermediate forms have yet to be recovered. In another example, the caudal process of the posteroventral portion of the right valve of the female *Candona caudata* may be blunt (Delorme 1978) or sharp (Benson and MacDonald 1963). Both forms and intermediate morphs are found in Lake Erie.

B. Internal Shell Morphology

Structures on the internal surface of the shell may also be used in ostracode systematics. Many ostracode muscles are attached directly to the calcareous shells. The chitinous exoskeleton does not provide sufficient rigidity to anchor many of the muscles. The points of attachment of these muscles leave scars as raised or depressed areas on the interior of the shell (Figs. 19.8–19.11). The closing or adductor muscles form a distinctive pattern of scars as do the mandibular muscles (Benson 1967, Benson and MacDonald 1963, Smith 1965). Traces of the ovaries and testes can also be seen in the posterior of the shell (see Section II.D).

C. External Body and Appendage Morphology

The ostracode has a shortened body with a slight constriction in the midregion that separates the head from the thorax (Kesling 1951a). There is no abdomen. Instead, the posterior of the body tapers off bluntly and ends in a pair of preanal furcae. The head region contains four pairs of appendages that are used for swimming, walking, and feeding. The thoracic region features three pairs of appendages used or adapted for feeding, creeping, and cleaning of the shells. Ancillary cuticular structures such as setae, claws, and pseudochaetae, found on most limbs, are recognized as important in functional morphology and systematics (Broodbakker and Danielopol 1982). (Appendages are shown in Figs. 19.9–19.12.)

The head region is composed of the forehead, the upper lip, and the hypostome. The four pairs of appendages in the head region are the first antennae (antennules), second antennae (antennae), mandibles, and maxillae. When the distal four podomeres of the first antennae have very long, plumose setae, the ostracode is a good swimmer. If these setae are absent, as in the limnocytherids (Fig. 19.11), the ostracode can not swim. The first antennae are curved up and backward. The second antennae extend down and curve backward. These appendages are robustly constructed and have a strong claw for walking and climbing. Swimming setae may be present on the first podomere of the endopodite. These structures are used for rowing when the ostracode is swimming. In most species, well-developed mandibles occur between the upper lip and the hypostome. The dorsal tips of the mandibular palps are attached to the interior of the shell by muscles. The ventral portion of the palps terminate in heavily toothed mandibles; these teeth grind the food before it is ingested. A respiratory (branchial) plate extends from the mandibular palp. On the ventral portion of the head is located a keel-shaped structure called the hypostome, which forms the mouth. Rake-shaped organs are situated at the rear of the mouth. They consist of chitin shafts with terminal toothed structures and are used for straining and feeding. The fourth and final set of cephalic appendages are the maxillae. The maxillae, located on either side of the hypostome posterior of the mandibles, are made up

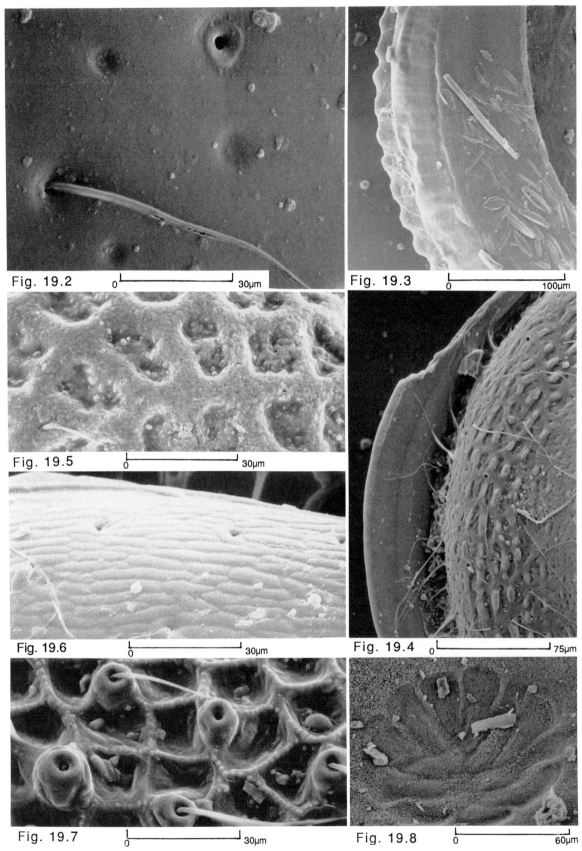

Fig. 19.2   0 ⊢———————⊣ 30µm

Fig. 19.3   0 ⊢———————⊣ 100µm

Fig. 19.5   0 ⊢———————⊣ 30µm

Fig. 19.6   0 ⊢———————⊣ 30µm

Fig. 19.4   0 ⊢———————⊣ 75µm

Fig. 19.7   0 ⊢———————⊣ 30µm

Fig. 19.8   0 ⊢———————⊣ 60µm

Figure 19.2 Shell surface of *Cyclocypris ampla* Furtos showing normal pore canals with funnels (without seta) and without funnels (with seta), and dimples.   ***Figure 19.3*** Anterior margin of *Cyprinotus glaucus* Furtos showing crenulations, duplicature with position of radial pore canals, and calcified inner lamella forming a shelf.   ***Figure 19.4*** Anterior margin of *Cyprinotus glaucus* Furtos with pustules on the right valve, left valve beneath.   ***Figure 19.5*** Reticulate exterior structure on *Limnocythere itasca* Cole.   ***Figure 19.6*** Anastomosing structure (lines) on the surface of *Cypria turneri;* note normal pore canals.   ***Figure 19.7*** Reticulate structure with superimposed mamillary pustules on *Paracandona euplectella* Brady and Norman.   ***Figure 19.8*** Distinctive muscle scar pattern of *Darwinula.*

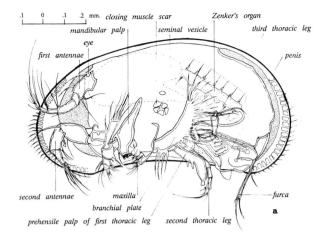

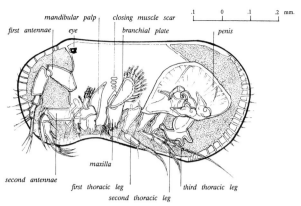

Figure 19.11 Sketch of the internal morphology of a male *Limnocythere sanctipatricii* (Brady and Robertson) (Cytheridae) from Kesling. (Figure redrawn by J. L. Delorme.)

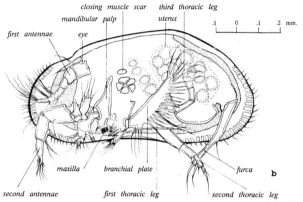

Figure 19.9 Sketch of the internal morphology of (a) male and (b) female *Candona suburbana* Hoff (Cyprididae) from Kesling. (Figure redrawn by J. L. Delorme.)

of several cylindrical masticatory processes. The mandibular exopodite is developed as a respiratory plate. The feathered setae of this plate are used in respiration. The maxillae pass the small particles toward the mouth.

There are three pairs of thoracic appendages. The first thoracic legs of most cyprids are modified for

feeding. In other groups, the first thoracic legs are modified in sexual ostracodes into prehensile palps (Fig. 19.8a) used in grasping the female shell during copulation. The second thoracic legs are robustly developed and used for walking, climbing, and clinging onto surfaces. The third thoracic legs, or cleaning legs, are the most flexible and well-muscled appendage. The ends of these legs, in most ostracodes, are modified pinchers used to grasp and remove foreign objects from between the shells. In the entocytherid group, the three pairs of thoracic legs (Fig. 19.12) are modified terminally as hooks to hold onto the gills of the host animal (Hart and Hart 1974).

Morphological diversity can be seen in the thoracic appendages of ostracodes. Most notable are the first thoracic legs. These can be modified for feeding, for use during copulation (prehensile

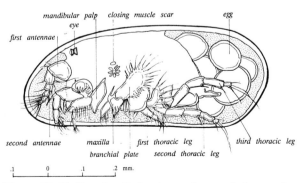

Figure 19.10 Sketch of the internal morphology of a female *Darwinula stevensoni* (Brady and Robertson) (Darwinulidae) from Kesling. (Figure redrawn by J. L. Delorme.)

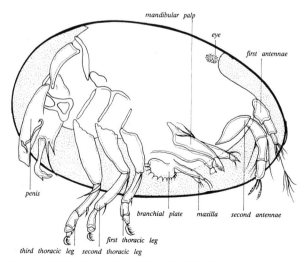

Figure 19.12 Generalized sketch of the internal morphology of a male entocytherid (Cytheridae) from Hart and Hart 1974. (Figure redrawn by J. L. Delorme.)

palps), or for walking. The commensal entocytherids have modified the three thoracic legs for grasping onto the gills of the host.

Paired hemipenes are found in syngamic species behind the third thoracic legs on the ventral side of the thorax. Paired uterine openings lie behind and inside the vaginal openings of sexual and asexual females. The distal part of the thorax terminates in paired furcae. In some species, these are whip-shaped structures, while in others they are well developed and armed at the end with long, serrated claws. Rome (1969) and Triebel (1953) consider the furcal attachment to be an important biocharacter to be used in systematics.

Sexual dimorphism is common in ostracodes. Determination of gender in most cases can be determined visually with a microscope. This is done by observing the presence/absence of testes, Zenker's organs, modification of the male first thoracic legs into prehensile palps, hemipenes, and ovaries. For many species, it is necessary to separate the valves to view these organs. Except in cyprinotids, the male is usually larger than the female. In candonids, only the male has an anteroventral notch on the shell.

D. Internal Anatomy and Physiology

The circulatory system of freshwater ostracodes lacks both heart and gills. Gaseous exchange is through the entire surface of the body and particularly the membranous inner lamella of the exoskeleton. The respiratory plates of some appendages move and renew oxygenated water past the inner surfaces of the organisms. Large respiratory cells are known to occur in the inner lamella, forming a respiratory epithelium over the valve cavity, which forms a blood sinus (Bernecker 1909).

The digestive system of ostracodes consists of the mouth, esophagus, stomach, intestine, rear gut, and anus. The mouth is a large opening with large teeth of the mandibular palps on either side. A gland, probably a salivary gland, opens into the mouth. The esophagus is very muscular and leads into the distended stomach. The hepatopancreas (liver) is found in the epidermal layers next to the shells and empties into the anterior part of the stomach. Most of the digestion takes place in the stomach, where the food is formed into balls, passes from the stomach through the rear gut for absorption of the nutrients, and exits the anus as fecal pellets.

Approximately eight excretory glands, including the liver, have been identified in freshwater ostracodes. McGregor (1967) showed that for *Chlamydotheca arcuata* some food particles pass through the hepatopancreas, aiding in the digestion of food.

Most glands have been studied in considerable detail (Claus 1899, Bergold 1910, Fassbinder 1912, Schreiber 1922, Cannon 1925, Rome 1947, Kesling 1951a, 1965).

The nervous system of the ostracode is composed of a cerebrum (protocerebrum, deutocerebrum, and tritocerebrum, Turner 1896, Hanström 1924, Weygoldt 1961), circumesophageal ganglion, and a ventral chain of fused ganglia (Kesling, 1951a). The first and second antennae, mandibular palps, maxillae, and first thoracic legs have small ganglia that connect to the central nervous system.

The ostracode has a single median eye made up of three optic cups, each containing lenses. The number of cells composing a lens is variable (Novikoff 1908, Rome 1947). The black opaque eye lies in the anterior part of the body just below the rim of the shells. When the shells are open, the animal can detect shapes and motion; however, when the shells are closed it can only distinguish light intensity. Subterranean-interstitial (hypogean) ostracodes are blind (Danielopol 1980a).

Reproduction in freshwater ostracodes may be either sexual or asexual. Depending on the gender, the ostracode has a pair of genital lobes or hemipenes. The ovaries are found in the hypodermis next to the shell as a continuous sequence of gametes. The ova are fairly large in the last instar of many species, forcing the chitin coating of the epidermis into the space normally occupied by the shell. When the new shell is secreted, a trace of the ovaries is commonly seen as part of the inner structure of the shell. The testes are similarly placed and appear as traces on the inner and posterior region of the shell. For some genera (e.g., *Cypricercus*), the testes form a spiral around the inner edge of the shell rather than being confined to the posterior area. The testes of Cyprididae change into the vasa deferentia as they leave the hypodermis and enter the body cavity. The vas deferens joins up with the ejaculation ducts or Zenker's organs (Fig. 19.9a). These large organs are made up of longitudinal muscles in the shape of a tube on which are found radiating bristles and are covered in a chitinous sheath. The alternate contraction and expansion of these paired organs force the sperm into the hemipenes during copulation. Zenker's organs lie in a nearly horizontal position in the posterior part of the body cavity and are visible in many live specimens. The hemipenes are very complex structurally. The ostracode must rotate them before copulation can take place (Kesling 1957, 1965, Danielopol 1969, McGregor and Kesling 1969a, 1969b). The testes and ovaries in Cytheridae are found next to the intestine; Zenker's organs are absent in this family. There are no males in the Darwinulidae.

Zissler (1969a, 1969b, 1970) studied in detail the formation of sperm in *Notodromas monacha*. A single nucleus is formed by the fusing of caryomeres; it is characterized by three granulated particles and by a double membrane-bounded region. The body then becomes spindle-shaped, which includes the nucleus and the "Nebenkern" derivatives. Further development of the spindle-shaped cells takes place by the extension of two wing-like structures, "Flüegelstruckturen," on either side of the nucleus. The nucleus and the wing-like structures extend throughout the sperm body, including the thread. The wing-like structures are interpreted as providing motility to the sperm. Further differentiation of the sperm occurs between the testes and the vas deferens.

Lowndes (1935), following the research of Zenker (1854), Stuhlmann (1886), and Müller (1889), reassessed the morphology and function of the sperm and concluded, "apparently the sperms are now functionless." Moore (1961) indicated that sperm do serve a function because females only lay eggs after copulation. However, Moore did not cite references to support his claim. Havel *et al.* (1990a) demonstrated that sperm have a function. Populations with males examined electrophoretically show genotypic frequencies in close agreement with those expected in a randomly mating sexual population. Ostracode sperm have the dubious distinction of being the longest of any sperm relative to the host animal (Lowndes 1935).

Based on earlier cytologic studies, Bell (1982) concluded that all ostracodes are diploid. White (1973) indicated the diploid chromosome number varies between 12 and 19. Electrophoretic studies by Havel *et al.* (1990b) showed that 10 of 12 asexual ostracodes included clones that were polyploid. They also found that clonal diversity varied among species. For instance, *Cyprinotus incongruens* averaged 1.2 clones, whereas *Cypridopsis vidua* had 8.3 clones per pond.

For further details on the anatomy and physiology, the reader is referred to Hoff (1942) and Kesling (1951a, 1965).

III. LIFE HISTORY

Ostracode eggs are constructed in a way that allows them to withstand physical and chemical extremes. The time of hatching may be soon after being laid or staggered over time. Once the egg is hatched, freshwater ostracodes will go through eight molt stages before becoming a mature adult. The length of time that the organism lives depends on the species.

A. Egg

The freshwater ostracode egg is a double-walled sphere of chitin impregnated with calcium carbonate. The space between the two spheres is occupied by a fluid (Wohlgemuth 1914). These two characteristics of the egg allow it to withstand desiccation and freezing. Ephemeral ponds in which the sediments stay frozen for up to six months during the winter produce many nauplii from eggs in the spring. Sohn and Kornicker (1979) freeze-dried eggs and found them to be viable after the process. Some workers have reported keeping dry lacustrine sediments in their laboratories for periods ranging from a few years to many decades and then hatching ostracodes (Sharpe 1918, Sars 1889, 1894, 1895, 1896, 1901). McLay (1978b) reported the eggs of *Cypricercus reticulatus* do not hatch until dried. The eggs of some species (*Cypridopsis vidua, Cyprinotus incongruens, Physocypria* sp., and *Potamocypris* sp.) can pass through the lower intestinal tract of both wild and domestic ducks and remain viable (Proctor 1964). Kornicker and Sohn (1971) reported that ostracode eggs were still viable after being egested by goldfish and swordtail fish.

Ostracodes develop from eggs that may or may not have been fertilized, depending on whether the species is bisexual or asexual. The eggs are laid singly or in masses, most often in the part of the habitat that will ensure an oxygen-rich environment when the eggs hatch (Wohlgemuth 1914). Most ostracode eggs are white, but some are green (*Cypridopsis vidua*) or bright orange (*Cyprinotus incongruens, Potamocypris smaragdina*). All North American freshwater ostracodes are oviparous except several species of *Darwinula* and the marine brackish water *Cyprideis*, which are ovoviviparous. These taxa brood their eggs in the posterior of the carapace from which the nauplii are released. Brood pouches are more common in marine ostracodes.

B. Embryology

Embryology of ostracodes has been studied by Woltereck (1898), Schleip (1909), Müller-Cale (1913), and Weygoldt (1960). The small size, double-walled sphere, and textured outer surface of the egg does not allow for direct observation of the cleavage process. It is known, however, that after seven divisions, a 128-celled blastula is developed (Kesling 1951a). Differentiation into ectoderm and endoderm occurs after the eighth division.

C. Hatching and Seasonality

Most species begin hatching in the spring in the temperate zone. The timing of this process is controlled in part by temperature and latitude. Species which live in temporary (vernal) ponds have a shorter life cycle. For example, *Cypridopsis vidua* develops in about one month (Kesling 1951a). Some of these same species can also exist in permanent ponds, where several generations per year may develop. For some species (e.g., *Candona caudata*, Delorme 1978, 1982; *Cyprinotus carolinensis*, McLay 1978c), hatching is delayed and spread over a much longer period. These apparent resting eggs allow the animal to sustain itself in the habitat in spite of stressful periods (such as anoxia and pollution, Delorme 1978) when the nauplii cannot survive. As early as 1915, Alm noted that temperatures affected the time of hatching (Alm 1915).

Most ostracodes are controlled seasonally in their habitat. Research on seasonality has concentrated on those species living in temporary ponds. For temporary ponds in the temperate zone, early spring collections made after the basin receives water yield few live ostracodes (Hoff 1943a [Illinois], Ferguson 1958a [South Carolina]). In these ponds, a regular succession of juveniles can be observed until the pond dries up, with some species exhibiting at least two generations per year. A temporary coastal habitat investigated by McLay (1978a, 1978b, 1978c) has ostracodes from the fall through to the late spring. Ostracodes were absent through the summer because of desiccation of the habitat. Freezing had only a marginal effect on a part of the habitat. McLay (1978b) stated that the maturation rate for *Herpetocypris reptans* was 190 days, and 45 days for *Cyprinotus carolinensis*. In permanent ponds, active species are absent for several months (Ferguson 1944 [Missouri]). Juveniles begin appearing in early spring and continue until late spring, while adults are present until late fall to early winter. This pattern was repeated for the three species studied by Ferguson: *Cypridopsis vidua*, *Potamocypris smaragdina*, and *Physocypria pustulosa*. Ostracodes that live in ponds or ephemeral bodies of water have a short life cycle. Martens *et al.* (1985) noted that an Australian species took longer (4–5 months) to reach sexual maturity in the winter than in the summer (2–3.5 months). *Candona subtriangulata*, *Cytherissa lacustris*, and *Darwinula stevensoni*, which inhabit large lakes where the habitat is considered to be stable, will have a life cycle of 6–24 months (McGregor 1969, Delorme 1978, Ranta 1979). The length of the life cycle depends on the species. Danielopol (1980b) observed the length of the life cycle for five species. These ranged from 126–489 days, the latter being for a hypogean ostracode.

D. Molting

During the eight molt stages, the appendages change in size, shape, and function. A summary of these developments is found in Table 19.1 (after Kesling 1951a).

Przibram (1931) determined that crustaceans doubled their weight from one molt to the next. He further concluded that because weight varies directly with volume, volume would also increase by a factor of two. This led him to believe that a given linear dimension would increase by about the cube root of two (1.26) for each molt. Based on this hypothesis, Kesling (1953) devised a circular slide rule that allows one to determine the variability of any linear dimension, area, or volume of any ostracode molt stage and adult. This slide rule is useful in determining which molt stage (shell or appendage) one is examining. Being able to identify an ostracode as an instar saves time because most instars can not be identified to the species level.

E. Competition, Coexistence, and Survival

The study of competition, coexistence, and survival by McLay (1978a, 1978b, 1978c) showed that, in a small temporary puddle, "competition is mediated via density-dependent mortality and maturation rates." He found that *Herpetocypris reptans* was a better competitor because of its longer maturation time (4419 degree days versus *Cyprinotus* with 2034 degree days) and was better able to withstand crowding and a shortage of food. His model shows that *Herpetocypris* was dominant over *Cyprinotus carolinensis* by a factor of 7 : 1. He also found that competitive exclusion of *Cyprinotus* by *Herpetocypris* was prevented by periodic freezing. The freezing action was also found to harm the population

Table 19.1 The Development of Appendages with Each Instar of the Freshwater Ostracode[a]

Instar	1	2	3	4	5	6	7	8	9
Antenna I	+	+	+	+	+	+	+	+	+
Antenna II	+	+	+	+	+	+	+	+	+
Mandible	+	+	+	+	+	+	+	+	+
Maxilla		−	+	+	+	+	+	+	+
Furca		−	−	+	+	+	+	+	+
Leg I				−	+	+	+	+	+
Leg II					−	+	+	+	+
Leg III						−	+	+	+
Ovaries							−	+	+
Testes								−	+

[a] (−) indicates the anlagen of a structure.

growth of *Herpetocypris* more than *Cyprinotus* because of the longer time required for maturation.

IV. DISTRIBUTION

Ostracodes are found in nearly every conceivable aquatic habitat. These range from temporary and permanent ponds, lakes, intermittent and permanent streams, to ditches and irrigation canals. Caves and the interstices of aquifers are additional habitats. Some are found in moist organic mats of fens (referred to as terrestrial ostracodes) and in axial cups of certain plants such as bromeliads. North American freshwater and brackish water ostracodes thrive in a full range of salinities from 10–74,800 mg/liter [high sulfate, low chloride waters (personal files)]. Brackish water forms are common in high sulfate waters (*Limnocythere staplini*), as well as in chloride waters mixed with freshwater (*Cytheromorpha fuscata*) in inland lakes (Neale and Delorme 1985). DeDeckker (1981) lists an Australian species found in saline waters of greater than 170 ppt. *Cyprideis* species are common in southern coastal areas of North America (Benson 1959).

A. Benthic Swimmers

Ostracodes are benthic organisms and are only rarely found in the plankton. They are divided into swimmers and nonswimmers. Most often they swim between aquatic plants. If they are disturbed while swimming, they often lose control and spiral down to the substratum. A few species of the genera *Cypria* and *Physocypria* are known to swim up into the water column, as they have been recovered from suspended sediment traps. Another swimmer, *Notodromas monacha,* has a peculiar habit of turning upside down when approaching the water surface. The venter of this species is flat, enabling the organism to adhere and hang beneath the water surface by surface tension.

B. Benthic Nonswimmers

The nonswimmers, or true benthic forms, use their well-developed antennae (without setae) and second thoracic legs to crawl on the sediment–water interface. Some species, such as the limnocytherids, are infaunal, crawling beneath the sediment–water interface between the sediment particles. Ostracodes live in the vents of cold-water springs where groundwater is discharging. Ostracodes may live in the fractures of bedrock adjacent to streams, lakes, and pools in caves. Hypogean (subterranean) ostracodes are being recovered more frequently from aquifers.

These blind animals live in the interstices of the sand aquifer. They are recovered by pumping the aquifer or trenching through it (Danielopol 1980a).

V. CHEMICAL HABITATS

A. Chemical Water Types

Physical and chemical characteristics of the habitat are important factors in ostracode distribution and abundance. From Canadian habitat-waters, four chemical types are defined based on the predominant equivalent anion or oxyanion. These are bicarbonate, sulfate, chloride, and carbonate (personal files). A high percentage of genera (61%) are found only in bicarbonate waters. Bicarbonate, after being converted to carbonate, is a primary constituent of the shell. Only 9% of the genera are found living in sulfate-rich waters. Several genera (30%) have species that have a preference for either of the two types of water. Chloride-dependent species, such as *Cytheromorpha fuscata* and species of *Cyprideis,* are more properly tolerant of marine brackish water. *C. fuscata* are occasionally found in inland lakes (Neale and Delorme 1985) where brine seepages occur. Such species as *Candona rectangulata* and *Cyprinotus salinus* are found in lake and pond habitats that may receive sea spray. These species are by no means restricted to chloride-enhanced waters.

B. Salinity and Solute Composition

Salinity and solute composition are of importance to freshwater ostracodes. Delorme (1989) has shown preliminary salinity tolerance distributions for some 43 species studied from Canadian habitats. The low end of the range, or pristine, low-salinity waters are inhabited by such species as *Candona subtriangulata, Limnocythere verrucosa* from the Great Lakes area, and *C. protzi* of the high arctic. The high end of the range or high-saline water bodies are inhabited by such species as *L. ceriotuberosa* and *L. staplini.* Species such as *Cyprinotus glaucus* are found in lakes where there is groundwater discharge of saline waters of intermediate flow systems (Delorme 1972). The sequence of species shown by Delorme (1989) for salinity tolerance shows a gradual increase in the mean salinity value. Furthermore, the range, as shown by minimum and maximum values of salinity tolerance, is a function of the species. For example, *Cypridopsis vidua* is a moderately saline-tolerant species able to survive in a very broad range of salt concentrations. On the other hand, species such as *Candona subtriangulata* and *Cytheromorpha fuscata* are much more restricted in their chemical and thus geographic range.

Forester (1983, 1986) and Forester and Brouwers (1985) have emphasized the importance of water composition over salinity. Examination of lakewater chemistry in North America, where *Limnocythere sappaensis* and *L. staplini* live, reveals that solute composition and not salinity is most important for their existence (Forester 1983). Forester found that *L. sappaensis* lives in water enriched in Na^+-$HCO_3CO_3^{2-}$ and depleted in Ca^{2+}. *L. staplini* lives in water enriched in various combinations of Na^{2+}-Mg^{2+}-Ca^{2+}-SO_4^{2-}-Cl^- and depleted in HCO_3. Forester (1987) and DeDeckker and Forester (1988) have expanded on the relationship between ostracode occurrences and major dissolved ion chemistry and salinity. DeDeckker and Forester (1988) further suggested that relationships between species diversity, carbonate saturation, and relative abundance of dissolved calcium and carbonate ions exist, based on limited data from the United States. Most ostracode species may be placed in one of six or seven hydrochemical groups. The groups have well-defined boundaries based on carbonate saturation, major dissolved ion composition, and salinity. The latter is the subject of a manuscript in preparation by Forester and Delorme.

C. Hydrogen Ion Concentration

The preservation of the calcareous exoskeleton for live ostracodes suggests that both the hydrogen ion concentration (pH) of the habitat must be circumneutral and a ready source of $CaCO_3$ must be available. Delorme (1989) indicated that the average pH of Canadian aquatic habitats for ostracodes is between 7.0 and 9.2. For carbonate-enriched lakes, common to groundwater discharge areas of the United States and Mexico, pH values are typically in the range of 9.6–10.5 (R. M. Forester personal communication). The species has an individualized pH range. Several of the species have a very broad pH range such as *Cypria ophthalmica* (5.2–13), *Candona candida* (5.4–13), *Cypridopsis vidua* (5.2–12), *Cyclocypris sharpei* (5.2–11), and *Cyclocypris ampla* (5–12). *Cytheromorpha fuscata*, with its limited distribution in the Manitoba Lakes, has a very restrictive range of 8.3–8.7. This may be an artifact of the lakes from which this species was collected. The majority of ostracodes have a larger pH range than *C. fuscata*. When the epidermal lining covering the shell is ruptured and the water and sediments are poorly buffered, the calcareous shell will dissolve, destroying the animal.

D. Dissolved Oxygen

Dissolved oxygen in the aquatic habitat is an important requirement for survival. The mean requisite for dissolved oxygen by ostracodes falls within a very narrow margin of 7.3–9.5 mg/liter (Fig. 19.13). In general, the concentration of dissolved oxygen in the water is broad. Most surprising in Figure 19.13 is the high number of species (34 out of 43 in Canadian habitats, L. D. Delorme unpublished) that can tolerate low dissolved oxygen concentrations. Those species that require a minimum of 3 mg/liter dissolved oxygen in the water are *Cytherissa lacustris*, *Potamocypris variegata*, *Cytheromorpha fuscata*, *Limnocyther herricki*, *Ilyocypris gibba*, *L. ceriotuberosa*, *L. verrucosa*, *Candona subtriangulata*, and *L. itasca*. Newrkla (1985) found *C. lacustris* could survive at 1 mg O_2/liter in the laboratory on a glass substrate and minimal food (20°C and 20 hr exposure). Nine of the 43 species listed in Fig. 19.13 can tolerate near zero dissolved oxygen (below the detection limit of the Winkler Azide Method, APHA 1965) for short periods. *Candona decora*, *Candona candida*, *Cyclocypris ampla*, *Cyclocypris sharpei*, and *Cypria ophthalmica* have been recovered from water with zero dissolved oxygen from a small river in early December (L. D. Delorme unpublished).

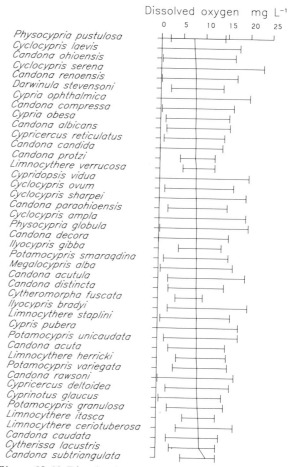

Figure 19.13 Dissolved oxygen range for some Canadian freshwater ostracodes. The solid line represents the mean, the bars represent the minimum and maximum values.

The river had been covered by one foot of ice, suggesting that these waters had been anoxic for some time. Fox and Taylor (1954, 1955) have found by experimentation that some ostracodes survive longer at oxygen levels below air saturation of 21%.

E. Carbonate

The effect of excess calcium carbonate and anoxia on *Cypridopsis vidua* was investigated by Reyment and Brännström (1962). Their results indicated that the morphology of the shell is adversely affected (smaller shell dimensions) by excess calcium carbonate and anoxia in the environment. They reiterate the common belief that the major contributor to the size and shape of an ostracode is the environment, with a secondary contributor being the genome of the animal.

VI. PHYSICAL HABITAT

A. Temperature

The water temperature of the aquatic habitat plays an important role in the habitat, seasonal, and geographical distribution of ostracodes. The temperatures in shallow water habitats can range from 0° to over 30°C, depending on the latitude and altitude. The exceptions to this are the shallow groundwater-fed streams, which may have a nearly constant temperature. Species living in this niche must have the ability to tolerate this broad range in one or more life stages. Such species as *Candona acutula, C. candida, C. ohioensis, Cyclocypris ampla, Cyclocypris sharpei, Cypria ophthalmica,* and *Limnocythere staplini* exhibit such a range for bottom water temperature (Fig. 19.14). *Candona subtriangulata* has a low mean value of 5.5°C, with a range of 2.6°–19.2°C (Delorme 1978). This species is common in Lakes Superior, Huron, and Ontario at considerable depth, where the bottom water temperature does not vary much. Korschelt (1915) ran several experiments with *Cypridopsis vidua* and *Cypria ophthalmica,* in which the species were frozen in ice for several hours. The ice was then allowed to melt; the majority of the specimens survived the freezing experiment.

At the other temperature extreme, *Cypris balnearia* and *C. thermalis* have been described from thermal springs (Moniez 1893) at temperatures of 45–50.5°C. Wickstrom and Castenholz (1973) have recovered *Potamocypris* from an algal–bacterial substratum of a hot spring at temperatures ranging between 30° and 54°C. The spring complex was located in Oregon.

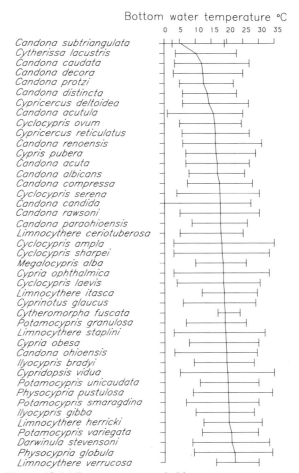

Figure 19.14 Bottom water or habitat temperature range (°C) for some Canadian freshwater ostracodes. The solid line represents the mean, the bars represent the minimum and maximum values.

Mean annual air temperature assists in explaining the geographic distribution of ostracodes in habitats shallower than 25 m. From Fig. 19.15, 20 out of 43 species can live in aquatic habitats north of latitude 60°. The minimum values of the range for mean annual air temperature for these species habitats are between −8.5° to −11°C. The species restricted to the arctic and low mean annual air temperatures are *Candona paralapponica, C. protzi, C. anceps, C. pedata, C. mülleri, C. ikpikpukensis, Cyclocypris globosa, Limnocythere liporeticulata,* and *Prionocypris glacialis.* The habitats for those species living in the mid-latitudes have a mean annual air temperature greater than −1.5°C.

B. Lakes, Ponds, and Streams

Most freshwater ostracode species may be found in more than one habitat. The pond–lake habitats can be viewed as a continuum from shallow to deep water. Some ponds are temporary in nature and, therefore, harbor a specific faunal association such

Mean annual temperature °C

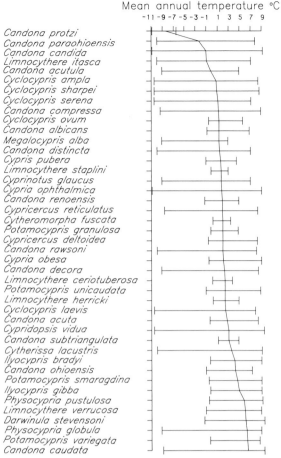

Figure 19.15 Mean annual temperature range (°C) for some Canadian freshwater ostracodes. The solid line represents the mean, the bars represent the minimum and maximum values.

as *Candona renoensis, Cypricercus deltoidea, Megalocypris alba,* and *Cyclocypris laevis* (Delorme 1989). At the opposite end of the scale are those species adapted or restricted to living in permanent lake habitats (*Candona subtriangulata, Cytherissa lacustris,* and *Cytheromorpha fuscata*). A faunal mix may occur between pond–lake and stream habitats. There may be a specific biotope developed at the mouth of a stream or river where it enters the lake or pond. The delta area is characterized by a combination of flowing and standing water. Many ostracode species are adapted but not restricted to living in this biotope such as *Candona acuta, Cyclocypris ampla, Cyclocypris sharpei, Potamocypris variegata, Ilyocypris bradyi,* and *Ilyocypris gibba.* Large lakes have bottom currents which accommodate such species as *Candona caudata* and *C. acuta.* None of the species listed by Delorme (1989) are completely restricted to flowing water. Those which are considered as lotic (fluvial) species are *Candona acuta, Potamocypris variegata, Ilyocypris bradyi, Ilyocypris gibba,* and *Cypria obesa.*

C. Bromeliads

Axial cups are formed by the spiral overlap of bromeliad leaves. These cups collect rain water and become inhabited by many kinds of invertebrates. Tressler (1956) described *Metacypris maracoensis,* recovered from the axial cups of Florida bromeliads. Ostracodes are more commonly found associated with bromeliads in Central and South America, being first discovered there by Müller (1880).

D. Caves

Ostracodes have been recovered from several caves. Klie (1931) reported *Entocythere donnaldsonensis* commensal on the crayfish *Cambarus* from the Donnelson Cave of Indiana. From the Marengo Cave of Indiana, Klie named two new species: *Candona marengoensis* and *C. jeanneli.* The Mammoth Cave of Kentucky yielded specimens of an unnamed candonid. Hart and Hobbs (1961) have described eight new troglobitic cave species from the eastern United States. Walton and Hobbs (1959) have also described troglobitic ostracodes from Florida caves.

Pools in deep sinkholes in limestone terrain (cenotes) of Yucatan, Mexico have a rich and varied ostracode fauna. Furtos (1936b) described 23 species from these habitats.

E. Sediment–Water Interface, Particle Size, Food

The distribution of ostracodes at the sediment–water interface is a function of the availability of food, whether the substratum surface is clearly or poorly defined, the particle size distribution of the top centimeters of the substratum, as well as the time of year. As detritivores and herbivores, ostracodes require a ready source of particulate organic matter that they can sweep into the mouth region. They also use their mandibular teeth to rasp or gnaw off small particles from large organic particles, living plants, or algae. The distribution of ostracodes is patchy. Where there is organic detritus, ostracodes will be abundant, particularly in shallow water. On predominantly mineral substrata, ostracode densities will be low. The presence of abundant floating organic debris, just above an ill-defined sediment–water interface, discourages ostracodes from living in this part of the habitat. As the sediment–water interface becomes deeper and further from the shoreline, the number of species decreases. In this niche, the only food available will be the resistant cellulose raining down through the water column, bacteria, and fungi. In the deep troughs located in the eastern end of Lake Superior, only *Candona subtriangulata* has been recovered from depths of 45–305 m. Tressler (1957) identified *Candona decora* at a depth of 600 m from Great Slave Lake, Canada.

For many limnocytherids, especially *Limnocythere ceriotuberosa*, and *L. staplini*, the particle size of the substratum is very important. These ostracodes crawl through the interstices of very fine sand (.0625 mm) to coarse sand (1.0 mm). In these niches, there must be sufficient food and oxygen for survival. Good porosity and permeability of these sediments allows a free flow of oxygenated waters. Most clay-sized, organic-rich sediments are anoxic and reducing and not suitable for ostracodes to live in the interstices.

VII. PHYSIOLOGICAL AND MORPHOLOGICAL ADAPTATIONS

Ostracodes live in a wide variety of habitats and, within a given habitat, physical and chemical conditions fluctuate. Physiological adaptations can be seen for some of these changes.

A. Eggs

Eggs produced by some species can be either spring or summer forms. The ostracodes hatched from eggs laid in the spring do not tolerate warm summer temperatures. Ostracodes hatched from eggs laid in the summer will live only in warm water and expire in cold water (Schreiber 1922). Martens *et al.* (1985) have noted that eggs of the Australian *Mytilocypris henricae* would not hatch below a certain temperature. The eggs can withstand freezing (Kesling 1951a, Sohn and Kornicker 1979).

B. Resting Stage and Torpidity

The primary resting stage for freshwater ostracodes is the egg. The structure of the egg permits the yolk to survive both desiccation and freezing. The other adaptation for surviving harsh conditions in ostracodes is torpidity (Barclay 1966, Delorme and Donald 1969, McLay 1978a). *Candona rawsoni* has been recovered in the seventh instar from frozen sediments of a pond that had gone dry the previous fall. Placing the sediments in water at room temperature caused rejuvenation of the seventh instar candonids. Survival rate was a function of the moisture content and temperature of the sediments holding the instar at the time of rejuvenation (Delorme and Donald 1969). In the fall or spring, hatching of the

eggs and development of the nauplii from the seventh instar may occur. In vernal ponds, torpidity allows sufficient time for the species to become sexually mature and lay eggs. In this way, the species is propagated in temporary bodies of water (Wiggins *et al.* 1980). The development of the torpid state allows ostracodes to survive the drying of vernal ponds.

C. Coloration

Coloration of ostracodes is a function of the pigment bodies located in epidermal cells of both the outer and inner layers. Ostracodes such as *Cypridopsis, Cypris, Cypricercus, Eucypris, Herpetocypris, Potamocypris,* and *Prionocypris* are normally green, although *Cypridopsis vidua* and *Cypricercus* may be found in shades of brown and purple. These genera are all prevalent in littoral areas of lakes and in ponds where there is abundant vegetation. Green (1962) investigated the green coloration of *Eucypris virens* and found it to be a bile pigment, biladienes, derived from blue-green algae. Other species, such as *Cypricercus horridus* and *Megalocypris alba,* are purplish to reddish-brown in color, which allows them to blend better with an organic substratum. The majority of the other species are a dirty white to a yellowish-buff color which blends in with the mineral substrata. Candonids, cyprinotids, and limnocytherids fall into the last category.

D. Appendages

In the male of some species, the first thoracic leg is modified as a prehensile palp. This changes the leg from a feeding function to a sexual function, used to grasp the female shell and spread it apart during copulation.

The first and second antennae are modified in swimming ostracodes. Finely feathered, swimming setae on the first antennae are in constant motion and offer resistance to the water when the ostracode is swimming. Setae on the second antennae are used in a rowing motion. When the swimming setae are absent, the second antennae are used exclusively for walking and climbing.

VIII. BEHAVIORAL ECOLOGY

Experiments were conducted by Towle (1900) on heliotropism using *Cypridopsis vidua*. She determined that ostracodes initially moved toward the light and then reversed directions and moved away.

Ostracodes that swim maintain a sustained motion. The swimming setae of the first antennae, used in propelling the animal forward, must be in constant motion or the animal will sink to the substrate. Observation of the swimming motion, using a binocular microscope, shows the swimming setae moving in a pattern similar to a handheld egg beater directed upward. Because swimming is energetically quite expensive, ostracodes can only swim short distances (usually between adjacent plants).

IX. FORAGING RELATIONSHIPS

The diet of most ostracodes is restricted to algae and organic detritus. Klugh (1927) reported certain algae as being part of the food supply. Kesling (1951a) indicated diatoms are a part of the ostracode diet, as did Strayer (1985) for *Cypria turneri*. Grant *et al.* (1983) reported feeding *Cyprinotus carolinensis* the blue-green algae *Nostoc* sp. When feeding, organic particles are swept into the mouth using the maxillae. The mandibular teeth grind large organic particles into a smaller size before they enter the mouth.

Although ostracodes are predominantly herbivores and detritivores, a few have shown carnivorous characteristics (Johansen 1912). Ostracodes have been observed attacking and eating the soft tissue of certain snails (Deschiens *et al.* 1953, Deschiens 1954, Sohn and Kornicker 1975). Some species (*Cypridopsis hartwigi, C. vidua, Cypretta kawatai,* and *Cyprinotus incongruens*) are known from experiments, by the above authors, to eat the soft tissue of snails such as the mantle, antennae, and crawl into the respiratory organs.

The trophic position of most ostracodes is that of a herbivore and detritivore. This group of benthic animals scavenges and consumes organic matter primarily from the sediment–water interface. Some of this food is fresh, while the remainder is in the process of decaying. During the feeding process, nutrients and the more resistant cellulose and silica (diatom frustules) are moved back into the food chain. Ostracodes are in turn consumed by higher animals. Predators include fish that dwell and feed on the bottom. There are several references to the presence of ostracodes in the gut of fishes. Bigelow (1924), Harding (1962), and Ferguson (1964) found ostracodes in the stomachs of suckers and trout. Stomach contents of brook stickleback contained *Cyclocypris ampla,* that of yellow perch had *Candona ohioensis,* and bullheads had *Cypridopsis vidua* and *Physocypria globula* (L. D. Delorme unpublished). Green (1954) noted that the oligochaete *Chaetogaster diaphanus* eats ostracodes as do cyclopoid copepods (Fryer 1957) and the tanypodine midge (Roback 1969). However, Swüste *et al.* (1973)

showed experimentally that the phantom midges *Chaoborus flavicans* released *Cypria* immediately after capturing it. This suggests that this ostracode either tastes bad or releases a noxious chemical.

X. POPULATION REGULATION

Abundance of ostracodes at the sediment–water interface is a function of substrata, availability of food, season, and depth of water. Strayer (1985) has provided a useful and current summary of abundance and biomass measurements of ostracodes in lake environments. The presence of food clearly has an impact on species richness and abundance. Warmer temperatures for long periods down to the bottom of the euphotic zone favor ostracode production. The number of species and specimens recovered from the profundal sediment–water interface is small. For example, the ostracode assemblage in the deep troughs (190–365 m) of eastern Lake Superior is limited to *Candona subtriangulata.* At these depths, during the last week in May and the first week in June, the bottom water temperature varied between 2.6° and 4.0°C. Based on Canadian habitats sampled for ostracodes, a maximum of nine genera have been counted from one sample. In these samples, between 13 and 19 species were represented, with a maximum of 10 being alive at the time of collection (L. D. Delorme, unpublished).

XI. COLLECTING AND REARING TECHNIQUES

A. Collecting and Preservation

Ostracodes have been collected using various kinds of gear such as a Birge cone net (Ferguson 1958b). Using a net has certain disadvantages when gathering benthic organisms. Many true benthic forms are not collected because the net usually does not thoroughly sample the substrata on and in which ostracodes live. The use of an Ekman dredge or similar device is recommended for a quantitative sample of the benthos (Delorme 1967). Care should be taken not to allow the dredge to settle too far into the substratum, 2 cm being optimal. The actual depth to which a sampling device should go may be checked by pushing a short core tube into the sediments to observe the depth to which ostracodes have penetrated. Any dredge that is used should have flaps on the top so water will flow through the dredge when it is being lowered. This will prevent a bow wave from preceding the dredge and disturbing the sediment–

water interface. When the dredge is retrieved, the flaps will close and prevent the interface and collected organisms from being disturbed and swept away by moving water.

The large quantity of sediment (>500 cm^3) retrieved using a dredge means that a method of concentrating the ostracodes must be used. The sample can be washed through a set of 8-inch sediment sieves (#20 mesh, 0.841 mm; #60 mesh, 0.25 mm; #100 mesh, 0.149 mm) using a gentle stream of water from a sprinkler head. This procedure removes particle size fractions (including instars) finer than fine sand (0.149 mm). The smallest freshwater ostracode is *Limnocythere friabilis* with a height of 0.23–0.31 mm. The adult shells of this species are readily retained by both the #60 and #100 meshes. Large cyprids, such as *Cypris pubera,* megalocyprids, herpetocyprids, and prionocyprids will be concentrated on the #20 mesh sieve. When the specimens are used to study appendages and other organs, ostracodes can be picked by hand from the washed residue by using a stereomicroscope. The specimens should then be placed in 5–10% buffered formalin solution. Concentrated formalin, ethanol, and methanol should not be used, as these preservatives will decalcify the shell. Specimens may be mounted in Hoyer's solution, glycerin, or polyvinyl lactophenol. However, with the use of the scanning electron microscope to study appendages, these mounting media are not necessary.

When the preservation of the visceral mass or the appendages is not critical, the suggested method to store and preserve the specimens is freeze-drying. The residue from the sieves is transferred to a beaker using a spatula and a small amount of water from a wash bottle. The sediment is then frozen. After freezing, filter paper or a suitable tissue is placed on the beaker with an elastic band and set into a freeze-drying apparatus. In the freeze-dryer, ice is sublimed by vacuum from the specimens and sediments. This process takes 24–48 hr depending on the volume of ice being removed, the number of samples, and the size of the freeze-dryer. The specimens are then hand-picked from the sample using a stereomicroscope and placed in a separate vial or a micropaleontologic slide. This method has the added advantage of not requiring labor and time curating wet samples. The dried specimens are also in excellent condition for further study with a scanning electron microscope (SEM) to view and photograph appendages.

When a large volume of sediment still remains in the sieves after washing, the ostracodes can be concentrated using acetone (Delorme 1967). The residue from the #60 and #100 mesh sieves is passed through a 3-inch filter sieve to ensure that the sedi-

ment particles are disaggregated. The sieved residue is then sprinkled into a funnel (4-inch) containing acetone. The sediment particles sink and the ostracode carapaces float to the surface. Normally, the carapace will trap an air bubble which makes the shell buoyant. The funnel is equipped with a latex tube on the lower tip of the funnel with a clamp. After the sediment has been sprinkled into the acetone, the clamp is carefully removed from the tube and the "sink" allowed to drain into another funnel lined with rapid-draining filter paper. Care must be taken not to drain all of the acetone from the top funnel, otherwise the ostracodes will be removed. When the sink has been removed, another funnel lined with filter paper is placed beneath, the clamp removed, and the top funnel rinsed with acetone from a wash bottle. In this way, all supernatant particles (the float), including the ostracodes, will be transferred to the lower filter paper. Individual shells will also be a part of the float except *Cytherissa, Darwinula, Cytheromorpha,* and *Limnocythere,* which do not have an anterior vestibule. The filter paper containing the ostracode residue is then air- or oven-dried, and the residue transferred to a vial. The shells and carapaces are picked with a very fine, wetted sable hair brush and mounted onto a micropaleontological slide. The slide is preglued with the water-soluble gum Tragacanth. The ostracodes are then ready to be identified and counted.

For an accurate count of adult ostracodes, the foregoing method is adequate. If, however, all live instars and adults are required for biomass studies then the complete sediment parcel will be required. Wet sieving would only assist in separating organisms and particles by size. For biomass studies, it might be preferable to use a short core tube pushed into the sediment some 10–15 cm. Nalepa and Robertson (1981) have recovered ostracodes using two mesh sizes, 0.595 and 0.106 mm, for biomass studies. The smallest mesh sized required to collect all the ostracodes including instars is not indicated by them, as 0.106 mm will not retain all the instars. Hummon (1981) recommends a sieve size of 0.062 mm to retain all of the ostracodes for quantitative analyses. This sieve size would miss the first three instars of *L. friabilis.*

B. Rearing Techniques

Cosmopolitan ostracodes, such as *Cypridopsis vidua* and *Cyprinotus incongruens,* are relatively easy to rear in an aquarium. Collection of water from the habitat associated with the species to be reared, along with some of the substrate, will allow the population to develop. Natural water from the habitat is preferred to distilled or tap water. Habitat water contains the proper mix of major and minor ions for an established solute composition. The pH of the water should be checked periodically and adjusted when necessary. This type of aquarium will probably go "wild" and anoxic after several months. A more pristine culture can be set up by filtering the water from the habitat, placing it in an aquarium with sterilized silt, sand, and clay as a substrata. Cultured algae (Klugh 1927, Kesling 1951a, Grant *et al.* 1983) have been used for food as well as grated boiled egg, which has been allowed to decompose bacterially for a month. Aquaria should not be placed on a window ledge where high temperatures may destroy the artificial habitat.

XII. CURRENT AND FUTURE RESEARCH PROBLEMS

Current research into ostracode systematics needs to be emphasized. This is particularly true of several generic pairs such as *Cypricercus* and *Eucypris*, *Cyprinotus* and *Heterocypris*, and *Megalocypris* and *Cypriconcha*. These groups require a detailed SEM study of shell and appendage morphology of species of the world (see Section XIII.A). Only in this way will we be able to resolve which biocharacters of a certain group should be used in defining a genus. The establishment of a universally acceptable systematic nomenclature is important for studies in ecology and paleolimnology.

Continuing studies are required in autecology of ostracodes. Data bases of chemical and physical parameters of ostracode habitats are being put together (R. M. Forester personal communication, L. D. Delorme unpublished). Analyses of these data bases are required to give a sense of the factors that control the species in its environment. The roles of solute composition and salinity of the habitat are an integral part of this analysis.

The role of ostracodes in the population dynamics of a pond or lake system is poorly known. Only a few studies have attempted to look at biomass, productivity, densities in communities, position in the aquatic food chain (as predators or prey), behavioral defense mechanisms, and host–parasite interactions. All of these factors and others may affect the success of ostracodes in their habitat and need to be studied in more detail.

Because ostracode shells are preserved as fossils and a partial autecological data base is available, these crustaceans can be used to reconstruct past chemical and physical habitats. With a suitable time frame, the limnology of a pond or lake may be reconstructed from the fossils extracted from a sediment core. Initially, subjective interpretations were made of past conditions of the lake or pond (Gutentag and Benson 1962, Benson and MacDonald 1963, Staplin 1963a, 1963b, Neale 1988). As autecological data bases were generated, paleointerpretive models were developed which used these data for objective reconstruction of the chemical and physical aspects of the habitat (Anderson *et al.* 1985, Cvancara *et al.* 1971, Delorme 1968, 1969b, 1971b, 1971c, 1982, Delorme *et al.* 1977, 1978, Delorme and Zoltai 1984, Forester *et al.* 1986, Karrow *et al.* 1975, Klassen *et al.* 1967, McAllister and Harrington 1969, Westgate *et al.* 1987, Westgate and Delorme, 1988). A summary of the use of North American ostracodes in paleolimnology and paleoclimatology is given by Delorme (1989).

Trace element analyses of Mg, Ca, and Sr in ostracode shells have been used for reconstructing past salinity changes in a variety of aquatic habitats (Chivas *et al.* 1983, 1985, 1986a, 1986b, DeDeckker 1983, 1988). Similar studies should be carried out on North American ostracode shells.

XIII. CLASSIFICATION OF OSTRACODES

Ostracodes are typically identified from adult specimens. Insufficient information is obtained from the instars (juveniles) to make a proper identification to the species level unless a pure culture has been obtained and adults are present. In some instances, identification of the instar can be taken to the generic level. The most useful biocharacter in taxonomy and systematics is the shell.

A. Taxonomic Controversies

A number of controversies exist around the naming of certain genera. These will each be discussed separately. In order not to confuse the issue any further, a conservative approach will be taken for the use of generic names. Detailed systematics using the scanning electron microscope will be required, using specimens from the global arena rather than from local areas, in order to resolve the issues.

The names of two genera, *Cypricercus* and *Eucypris*, have been commonly used. Broodbakker (1983) has provided a literature review of the status for *Cypricercus* and *Strandesia*. There is no mention of the genus *Eucypris*. Broodbakker (1983) erected the genus *Bradleystrandesia* to include those species of *Cypricercus* that occur in Europe and North America. This author states "further research is needed to prove if the genus *Bradleystrandesia* is confined to Europe and North America" (Broodbakker 1983, p. 329). Sars (1928) differentiated between *Eucypris* and *Cypricercus* on the absence and presence of males, respectively. He also noted the peculiar arrangement of the "spermatic vessels" (Sars 1928, p. 117) in the males of *Cypricercus*. Furtos (1933) maintained the two genera based on reproduction and length of the furcal ramus (<1/2 length of shell, *Eucypris*; >1/2 length of shell, *Cypricercus*). Hoff (1942) noted that the erection of genera on the basis of propagation was not an acceptable practice. He opted for one genus based on an extremely long furcal ramus for *Cypricercus*. The species *Eucypris crassa* was originally assigned to *Cypris* by O. F. Müller (1785) and later to *Eucypris* by G. W. Müller (1912). This species has always been referred to the genus *Eucypris*. *Cypricercus* and *Eucypris* will be used in the key.

The genera *Cyprinotus* and *Heterocypris* have had a confused nomenclature. Purper and Würdig-Macial (1974) did a literature review of these two genera as well as others. They concluded that the main difference between the two genera was the dorsal gibbosity of the right valve for *Cyprinotus* (Brady 1886). Both genera are reported as having the anteroventral margin of the right valve with denticles or crenulations. Sars (1928) did not consider the gibbosity of the right valve as a valid generic biocharacter and thus did not recognize *Heterocypris* (Claus 1893). This morphological biocharacter in *Physocypria* (Delorme 1970b) may or may not be present and is not used to split the genus. *Cyprinotus* is used in the key.

The genus *Cypriconcha* was erected by Sars (1926) to incorporate the large cyprinids of North America. Sars (1898) had already established *Megalocypris* for the large cyprinids of Africa. Delorme (1969a) reviewed the two genera and concluded that the similarities were many and the discrepancies few. As a result, he put the North American cypriconchids into the genus *Megalocypris*. Martens (1986) and McKenzie (1982, referenced in Martens 1986) argued for the retention of *Cypriconcha* for the large cyprinids of North America. *Megalocypris* is used in the key.

A number of genera may be found in the North American ostracode literature but are not included here. The following generic names are no longer considered valid.

Cyclocypria Dobbin 1941 = *Cyclocypris* Brady and Norman 1888

Cypriconcha Sars 1926 = *Megalocypris* Sars 1898 [see Delorme 1969a]

Cytherites Sars 1926 = *Entocythere* Marshall 1903 [see Hoff 1943b]

Pionocypris Brady and Norman 1896 = *Cypridopsis* Brady 1868

Pseudoilyocypris Ferguson 1967 = *Pelocypris* Klie 1939 [see Delorme 1970d]

Spirocypris Sharpe 1903 = *Cypricercus* Sars 1895 [see Hoff 1942]

Several species, listed as follows, have been assigned to an incorrect genus, therefore the generic names will not appear in the key.

Candonocypris deeveyi Tressler 1954 = *Megalocypris deeveyi* (Tressler) 1954

Candonocypris pugionis Furtos 1936 = *Megalocypris*[?] *pugionis* (Furtos) 1936 [see Furtos 1936b]

Candonocypris sarsi Danforth 1948 = *Megalocypris sarsi* (Danforth) 1948

Candonocypris serrato-marginata (Furtos) 1935 = *Eucypris serrato-marginata* (Furtos) 1935

Erpetocypris barbatus Turner 1899 = *Megalocypris* sp. (Turner) 1899

Pseudocandona ikpikpukensis Swain 1963 = *Candona ikpikpukensis* (Swain) 1963 [see Delorme 1970c]

Stenocypria longicomosa Furtos 1933 = *Isocypris longicomosa* (Furtos) 1933 [see Delorme 1970a]

B. Classification of the Order Podocopida

The Podocopida[1] is the only order that contains both marine and freshwater ostracodes (Table 19.2). The dorsal border of the valve is curved, or if straight, it is much shorter than the total length. No permanent anterior opening is present between the valves. There is no heart; there are two or three pairs of thoracic legs. Within Podocopida are two suborders: Metacopina and Podocopina. As will be described, the suborder Metacopina contains only one freshwater genus in North America, *Darwinula* (superfamily Darwinuloidea, family Darwinulidae; Fig. 19.9). Its characteristics are given as follows[2].

Shell surface smooth, vestibules absent, distinctive adductor muscle scars (Figs. 19.11 and 19.16); first antenna composed of six podomeres with strong spike-like setae, swimming setae lacking on second

---

[1] Modified after Kesling (1951a).

[2] Modified after Hoff (1942).

Table 19.2 Classification of the Class Ostracoda[a]

Class Ostracoda
Subclass Podocopa
Order Podocopida Sars 1866
  Suborder Metacopina Sylvester-Bradley 1961
    Superfamily Darwinuloidea Brady and Norman 1888
      Family Darwinulidae Brady and Norman 1888
        Genus *Darwinula* Brady and Norman
  Suborder Podocopina Sars 1866
    Superfamily Cypridoidea Baird 1845
      Family Candoniidae Kaufmann 1900
        Subfamily Candoninae Daday 1900
          Genus *Candocyprinotus* Delorme
          Genus *Candona* Baird
          Genus *Paracandona* Hartwig
      Family Cyprididae Baird 1845
        Subfamily Cypridinae Baird 1845
          Genus *Chlamydotheca* Sausseure
          Genus *Cypricercus* Sars
          Genus *Cyprinotus* Brady
          Genus *Cypris* Müller
          Genus *Dolerocypris* Kaufmann
          Genus *Eucypris* Vávra
          Genus *Herpetocypris* Brady and Norman
          Genus *Ilyodromus* Sars
          Genus *Isocypris* Müller
          Genus *Megalocypris* Sars
          Genus *Prionocypris* Brady and Norman
          Genus *Scottia* Brady and Norman
          Genus *Stenocypris* Sars 1889
          Genus *Strandesia* Stuhlmann 1888
        Subfamily Cyclocyprinae Hoff 1942
          Genus *Candocypria* Furtos
          Genus *Cyclocypris* Brady and Norman
          Genus *Cypria* Zenker
          Genus *Physocypria* Vávra
      Family Cypridopsidae Kaufmann 1910
        Subfamily Cypridopsinae Kaufmann 1900
          Genus *Cypretta* Vávra
          Genus *Cypridopsis* Brady
          Genus *Potamocypris* Brady 1870
      Family Ilyocyprididae Kaufmann 1900
        Subfamily Ilyocyprinae Kaufmann 1900
          Genus *Ilyocypris* Brady and Norman
          Genus *Pelocypris* Klie
      Family Notodromadidae Kaufmann 1900
        Subfamily Notodromadinae Kaufmann 1900
          Genus *Cyprois* Zenker
          Genus *Notodromas* Lilljeborg
    Superfamily Cytheroidea Baird 1850
      Family Cytheridae Baird 1850
        Subfamily Cytherideinae Sars 1928
          Genus *Cyprideis* Jones 1856
      Family Entocytheridae Hoff 1942
        Subfamily Entocytherinae Hoff 1942
          Genus *Ankylocythere* Hart
          Genus *Ascetocythere* Hart
          Genus *Cymocythere* Hart

(continues)

Table 19.2 *(Continued)*

          Genus *Dactylocythere* Hart
          Genus *Donnaldsoncythere* (Hart)
          Genus *Entocythere* Marshall
          Genus *Geocythere* Hart
          Genus *Harpagocythere* Hobbs III
          Genus *Hartocythere* Hobbs III
          Genus *Litocythere* Hobbs and Walton
          Genus *Lordocythere* Hobbs and Hobbs
          Genus *Okriocythere* Hart
          Genus *Ornithocythere* Hobbs
          Genus *Phymocythere* Hobbs and Hart
          Genus *Plectocythere* Hobbs III
          Genus *Rhadinocythere* Hart
          Genus *Saurocythere* Hobbs III
          Genus *Sagittocythere* Hart
          Genus *Thermastrocythere* Hobbs and Walton
          Genus *Uncinocythere* Hart
      Family Limnocytheridae Klie 1938
        Subfamily Limnocytherinae Klie 1938
          Genus *Limnocythere* Brady 1867
          Genus *Metacypris* Brady and Robertson 1870
      Family Loxoconchidae Sars 1928
        Subfamily Loxoconchinae Sars 1928
          Genus *Cytheromorpha* Hirschmann
      Family Neocytherideidae Puri 1957
        Subfamily Neocytherideidinae Puri 1957
          Genus *Cytherissa* Sars

[a] The systematic list is alphabetical (after Bowman and Abele 1982); however, the systematic treatment is not.

antenna, mandibular exopodite with short palp of three podomeres of which the first is wide and has row of long feathered setae, respiratory plate of mandible small, masticatory process of maxilla short and heavy, respiratory plate of maxilla with numerous feathered setae, first thoracic leg with strong masticatory structure and leg-like palp of three podomeres, second and third thoracic legs similar in structure and direction, furca lacking, ovaries do not originate between inner and outer hypodermis, female carries eggs in posterior of carapace .... *Darwinula*

Freshwater and some marine ostracodes are included in the suborder Podocopina. Members of this taxon have three pairs of thoracic legs. The exopodite of the second antenna is represented by at most a scale and a few setae. The furcae are rod-like rather than lamelliform; eyes are present in most forms. Closing muscle scars are discrete and arranged in a distinctive group. Taxonomic keys to genera within two superfamilies of this suborder are given as follows.

C. Taxonomic Key to Genera of the Superfamily Cypridoidea[3]

1a.	Furcal ramus greatly reduced, whip-shaped, without terminal claw (difficult to observe) .... Cypridopsinae[4]	2
1b.	Furcal ramus well developed, bar-shaped, with two terminal or subterminal claws; outer masticatory process of maxilla with six nearly equal setae modified to form toothed spines .... Notodromadinae[4]	4;
1c.	Swimming setae of antennae completely lacking, shell white; outer masticatory process with two or three setae modified as spines .... Candoninae[5]	5
1d.	Swimming setae of first antenna present; third thoracic leg distally modified as seizing apparatus, ultimate podomere beak-like with two well-developed, bristlelike setae, third setae lacking or hook-like .... Cypridinae[6]	7
1e.	Third thoracic leg with one long reflexed seta and two shorter unreflexed setae .... Cyclocyprinae[7]	20
1f.	Third thoracic leg not bearing chela, last podomere cylindrical and bearing three setae; shell oblong to subrectangular, dorsal margin straight, surface pitted, with one or more transverse sulci .... Ilyocyprinae[8]	23
2a(1a).	Shells tumid	3
2b.	Shells crescentic in shape, compressed, right valve higher than left and extending dorsally above left, usually hairy; swimming setae of second antenna well developed, usually extending to tips of claws or beyond; ultimate podomere of maxillary palp distally wider than long (Fig. 19.18) .... *Potamocypris*	
3a(2a).	Shells tumid, less than 1 mm in length, anterior margin of both valves with row of radiating septa; two maxillary spines may be smooth or toothed; ovaries or testes coiled in spiral pattern in posterior of carapace .... *Cypretta*	
3b.	Shells tumid, valves nearly equal; ultimate podomere of maxillary palp cylindrical, longer than wide (Fig. 19.17) .... *Cypridopsis*	
4a(1b).	Shell from side short and high, compressed; eyes not widely separated; first antenna of five podomeres; swimming setae almost reach tips of end claws; right and left prehensile palps of male first thoracic legs differing little, with well-developed respiratory plate; setae on distal end of third thoracic leg modified for grasping; furca with two claws and two setae (Fig. 19.19) .... *Cyprois*	
4b.	Shell short with height two-thirds of length, dorsal margin forming elevated arch, ventral surface flattened; eyes well separated; first antenna of six apparent podomeres, swimming setae extending beyond tips of terminal claws; palps of maxillae in males not similar, each formed with two podomeres; first thoracic leg without respiratory plate; distal three setae of third thoracic leg nearly equal, furca with terminal seta lacking .... *Notodromas*	
5a(1c).	Carapace with sides convex; outer masticatory process of maxilla smooth; apical claw of ultimate segment of third thoracic leg well developed, furcal claws weakly serrated .... *Candocyprinotus*	
5b.	Shell subrectangular to subtriangular in shape, two special sensory setae present at juncture of fourth and fifth podomeres of male first antenna, respiratory plate of first thoracic leg with two or three setae	6
6a(5b).	Male carapace shape and size usually different than female; third thoracic leg with three unequally long setae on last podomere; Zenker's organs usually with seven wreaths of chitinous spines, openings of ejaculatory duct funnel-shaped (Figs. 19.9a, b and 19.20) .... *Candona*	

---

[3] Modified after Hoff (1942) and Sars (1928).

[4] Modified after Hoff (1942).

[5] Modified after Hoff (1942) and Furtos (1933).

[6] Modified after Hoff (1942) and Delorme (1970a).

[7] Modified after Hoff (1942) and Furtos (1933).

[8] Modified after Hoff (1942) and Tressler (1949).

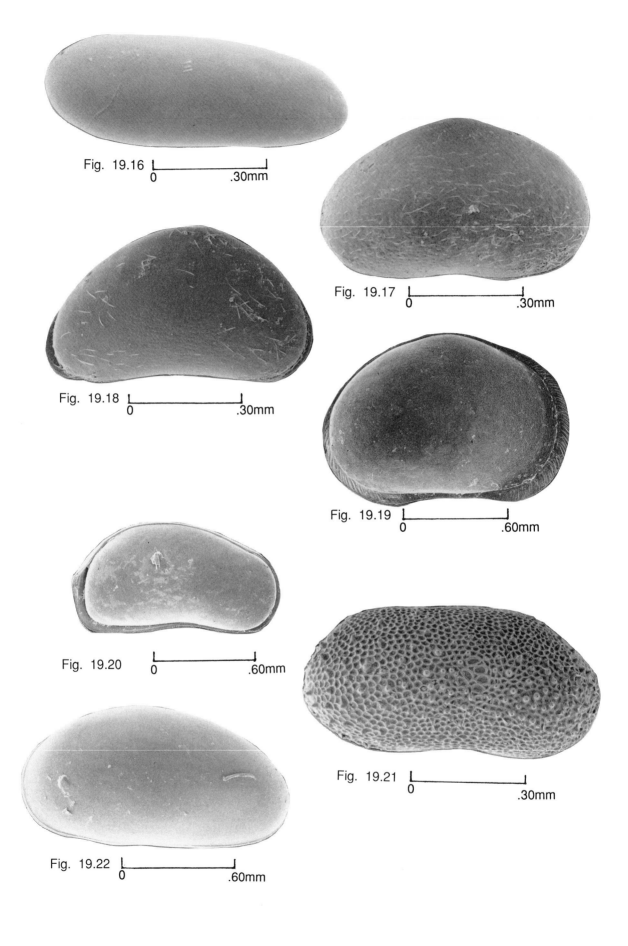

Fig. 19.16 ├─────────────┤
0   .30mm

Fig. 19.17 ├─────────────┤
0   .30mm

Fig. 19.18 ├─────────────┤
0   .30mm

Fig. 19.19 ├─────────────┤
0   .60mm

Fig. 19.20 ├─────────────┤
0   .60mm

Fig. 19.21 ├─────────────┤
0   .30mm

Fig. 19.22 ├─────────────┤
0   .60mm

6b.	Shell usually with small tubercles and reticulations on surface, shape subrectangular; penultimate segment of third thoracic leg divided, each division with strong distal seta; otherwise similar to *Candona* (Figs. 19.7 and 19.21)	*Paracandona*
7a(1d).	Margins of valve smooth	8
7b.	Margin of valve with spines or denticles	19
8a(7a).	Shells less than 2.2 mm long	9
8b.	Shells greater than 2.2 mm long	16
9a(8a).	Shell surface sculptured	10
9b.	Shell surface smooth	13
10a(9a).	Shell surface reticulate or scabrous especially in juveniles	11
10b.	Shell surface punctate or striated	12
11a(10a).	Outer masticatory process with two long, toothed spine-like setae; furcal ramus less than half of length of valve; testes coiled in anterior portion of each valve, males in some species rare (Fig. 19.22)	*Cypricercus*
11b.	Outer masticatory process with two long, toothed spine-like setae; furcal ramus greater than half of length of valve, males rare (Fig. 19.23)	*Eucypris*
12a(10b).	Shell elongate, narrow with scattered punctae; swimming seta of second antenna barely reaching tips of claws, third thoracic leg with well-developed apical claw; rami of furcae dissimilar in width, two end claws strongly pectinate	*Stenocypris*
12b.	Shells elongate and compressed, surface longitudinally striated; swimming setae of first and second antennae short, furcal ramus terminates in three short claws rather than two	*Ilyodromus*
13a(9b).	Swimming setae of second antenna rudimentary; shell elongate and moderately large, compressed and subrectangular; furcal ramus smooth	*Prionocypris*
13b.	Swimming setae of second antenna strongly developed	14
14a(13b).	Claws of furca finely denticulate; shell subrectangular, surface smooth and moderately hairy; characteristic marginal pore canals in anterior duplicature; outer masticatory process of maxilla with two smooth spines; third thoracic leg with well-developed apical claw	*Isocypris*
14b.	Claws of furca strongly developed	15
15a(14b).	Shell reniform, densely hairy; first thoracic leg bears well-developed respiratory plate; second thoracic leg with one long terminal seta and claw, furca with feathered dorsal seta	*Scottia*
15b.	Shells elongate, dorsum often with prominent dorsal flange, left valve always with conspicuous row of tubercle-like canals removed from the free margin; swimming setae of second antenna extend to tips of terminal claws, third masticatory process of maxilla with two strong spines, furcal ramus at least half as long as valve and with two claws and setae	*Strandesia*
16a(8b).	Shells with anterior flange-like projections which produce large vestibules; both antennae with well-developed swimming setae; outer masticatory process of maxilla with one toothed and one denticulate spine; third thoracic leg with well-developed apical claw; claws of furca finely denticulate	*Chlamydotheca*
16b.	Shells without anterior flange-like projection	17
17a(16b).	Shells narrow and spindle-shaped, vestibules well developed at both extremities; both antennae with well-developed swimming setae; maxilla with outer masticatory process with two smooth spines; third thoracic leg with well-developed apical claw, claws of furca coarsely denticulate	*Dolerocypris*
17b.	Shells subrectangular	18

Figure 19.16 External view of *Darwinula stevensoni* (Brady and Robertson). *Figure 19.17* External view of *Cypridopsis vidua* (Müller). *Figure 19.18* External view of *Potamocypris smaragdina* (Vávra). *Figure 19.19* External view of *Cyprois marginata* Straus. *Figure 19.20* External view of a female *Candona rawsoni* Tressler; note the overlap of the right valve by the left valve. *Figure 19.21* External view of *Paracandona euplectella* Brady and Norman, note the reticulate structure and the pustules. *Figure 19.22* External view of *Cypricercus reticulatus* (Zaddach).

18a(17b).	Shells elongate and large, compressed; swimming setae of second antenna not reaching midposition of terminal claw; spines of outer masticatory process of maxilla toothed; ventral edge of furca denticulate with combs of teeth .... *Herpetocypris*
18b.	Shells subrectangular to subtrapezoidal; swimming setae on first antenna extend to end of ultimate podomere and not to tips of claws of second antenna; outer masticatory process made up of short, smooth bristles; ultimate podomere of male first thoracic leg developed into prehensile palp; third thoracic leg with apical claw not well developed; claws of furca denticulate .... *Megalocypris*
19a(7b).	Shells less than 2 mm, margin of right or left valve tuberculate and margin of other valve smooth, dorsal margin may be arched or have an obtuse apex, valves commonly unequal, anterolateral surfaces pustulose; outer masticatory process of maxilla with two claw-like setae which may or may not be toothed; furcal claw longer than one-half of length of ventral margin of furcal ramus, ramus with length less than twenty times the least width (Figs. 19.3 and 19.4) .... *Cyprinotus*
19b.	Shells greater than 2 mm, tumid, highly arched, venter flattened, anterior margins denticulate, posteroventral margin of right valve with 2–4 sharp marginal spines, surface wrinkled to minutely crenelate to punctate; second antenna with well-developed swimming setae; outermost masticatory process of maxilla with two denticulate spines, third thoracic leg with well-developed apical claw, claws of furca strongly denticulate (Fig. 19.24) .... *Cypris*
20a(1e).	Swimming setae of second antenna lacking or rudimentary; shells ovoid, compressed; eyes well developed and fused; ultimate segment of third thoracic leg with two short and one long reflexed setae; tips of furcal claws serrate .... *Candocypria*
20b.	Swimming setae of second antenna well developed 21
21a(20b).	Shells tumid, height and width greater than one-half length, surface smooth and usually brown in color; eyes well developed; respiratory plate of first thoracic leg with six plumose setae; third thoracic leg consisting of four podomeres, ultimate podomere elongated and more than twice as wide and usually at least one-half as long as penultimate podomere; ultimate podomere of third thoracic leg with three distal setae, unequal, with outermost one very long and reflexed (Fig. 19.25) .... *Cyclocypris*
21b.	Shells compressed 22
22a(21b).	Shells short and high, occasionally elongate reniform, margins smooth not tuberculate; eyes well developed; second antenna of male with penultimate podomere divided and bearing specialized male setae; ultimate podomere of mandibular palp elongated three times as long as proximal width; palp of maxilla well developed, masticatory process weak; penultimate podomere of third thoracic leg undivided, ultimate podomere short, scarcely longer than wide, longest distal seta reflexed; terminal and subterminal claws of furca strong, dorsal seta may be rudimentary; Zenker's organ with seven whorls of chitinous rays, proximal end much inflated; hemipene with two terminal lobes only, outer lobe lacking .... *Cypria*
22b.	Shells commonly unequal in height or length or both, at least anterior margin of one of the valves tuberculate; otherwise as in genus *Cypria* (Fig. 19.26) .... *Physocypria*
23a(1f).	Shells with large, rounded hump-like projections, marginal spines and pustules; first antenna with some swimming setae shortened and claw-like; swimming setae of second antenna may be shortened, male without special setae; endopodite of male first thoracic leg transformed into prehensile palp, in female clearly leg-like; ultimate podomere in third thoracic leg cylindrical with three setae, longest may be reflexed, claws well developed; Zenker's organs with numerous chitinous rods and with spherical inflated openings at each end; vasa deferentia twisted .... *Ilyocypris*
23b.	Shells with pronounced mamillary-like projections on lateral surface, marginal spines and pustules; second antenna with swimming setae which extend considerably beyond tips of terminal claws; first thoracic leg with long curved terminal claw; second leg with one short and one long seta; furca well developed with dorsal seta longer than terminal claw (Fig. 19.27) .... *Pelocypris*

D. Taxonomic Key to Genera of the Superfamily Cytheroidea[9]

1a. Free margins of valves without conspicuous pore canals, carapaces noncalcareous, reniform or subelliptical in shape, furrows or protuberances may be present; respiratory plate of mandible reduced to two or three setae, masticatory lobes well developed, furca rudimentary, penis strongly curved; North American ostracodes commensal on crayfishes and a freshwater crab (Fig. 19.12) ...... subfamily Entocytherinae[10] 3

1b. Free margins of valves flattened, with many long marginal pore canals, subrectangular, often with protuberances or furrows, strongly reticulate; respiratory plate of mandible well developed, furca usually with two short setae .... subfamily Limnocytherinae[11] 22

1c. First thoracic leg of male prehensile, shell variable in shape and sculpturing, seldom smooth, usually with reticulations, spines, furrows, or tubercles, valves nearly equal, often with tooth-like projections along hinge, vestibules absent; eyes distinctly separated, antennae well developed, exopodite of second antenna in form of long hollow seta carrying secretions from a gland near base of second antenna, setae of first antenna short and stout, often claw-like, swimming setae on second antenna lacking, three pairs of thoracic legs similar, furca greatly reduced, hemipene well developed, Zenker's organs absent, testes lying in body lateral to intestine, in some species eggs retained in body cavity during development .... subfamily Cytherideinae[12]

[Shells distinctly sculptured, sometimes with lateral protuberances, right valve with spine in posteroventral corner] .... *Cyprideis*

1d. Thoracic legs successively increasing in length 2

2a(1d). Shell short and stout, hinge well developed; mandibular palp with terminal joint very small, respiratory plate well developed; furcal ramus edged at tip with two bristles .... subfamily Loxoconchinae[12]

[Shell short and stout, surface pitted, marginal zones thickened, duplicature narrow; eyes distinctly separated; ultimate segment of first antenna articulated and with four robust spines; second antenna with two claw-like spines inside penultimate joint; masticatory lobes of maxilla short] .... *Cytheromorpha*

2b. Shells plump, surface generally sculptured; hinge well defined, distinct closing teeth lacking in anterior and posterior of hinge line; mandibular palp slender, respiratory plate not well developed; furcal ramus extremely small, not well defined .... Neocytherideidinae[13]

[Shell club-shaped, surface rough and pitted; valves unequal with marginal zone thickened, duplicature narrow, vestibule absent, poorly developed teeth present on hinge; eyes well defined; antennae well developed, first antenna armed in front with three claw-like spines; respiratory plate of mandibular palp moderately well developed; maxilla with masticatory lobes short and stout; thoracic legs robustly developed; furcal ramus forming two oval thickened pieces placed vertically, each provided behind with two very small bristles (Fig. 19.28)] .... *Cytherissa*

3a(1a). Penis with prostatic and spermatic elements widely separated along much of their lengths 4

3b. Penis simple; or if two elements recognizable, contiguous along their entire lengths 6

4a(3a). Ventral portion of peniferum tapering with tip of penis reaching, or almost reaching apex .... *Plectocythere*

4b. Ventral portion of peniferum usually rounded or with one or more prominences, seldom tapering; if tapering, tip of penis never approaching apex 5

5a(4b). Ventral portion of peniferum rounded, without prominences .... *Phymocythere*

---

[9] Modified after Hoff (1942) and Sars (1928).

[10] Modified after Hoff (1942) and Hart and Hart (1974).

[11] Modified after Hoff (1942).

[12] Modified after Sars (1928).

[13] Modified after Sars (1928).

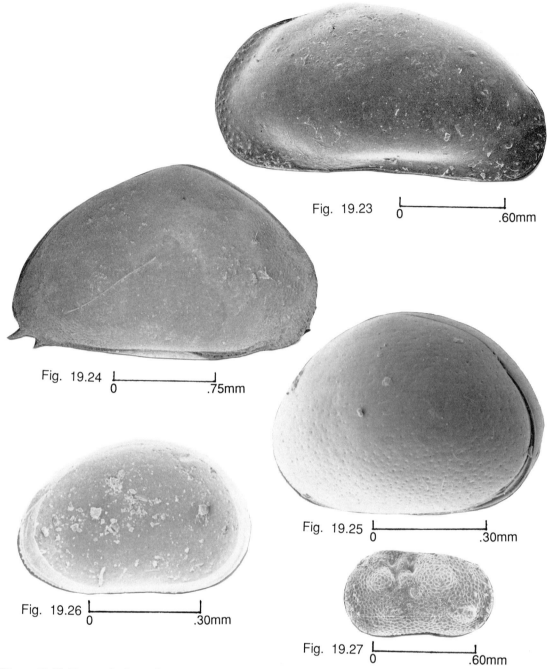

Fig. 19.23   0   .60mm

Fig. 19.24   0   .75mm

Fig. 19.25   0   .30mm

Fig. 19.26   0   .30mm

Fig. 19.27   0   .60mm

Figure 19.23 External view of *Eucypris crassa* (Müller).   ***Figure 19.24*** External view of *Cypris pubera* Müller, note the posteroventral spines on the left valve.   ***Figure 19.25*** External view of *Cyclocypris ampla* Furtos, note the pitted surface.   ***Figure 19.26*** Internal view of *Physocypria pustulosa* (Sharpe) showing the denticles on the anteroventral margin.   ***Figure 19.27*** Exterior view of *Pelocypris alatabulbosa* Delorme showing a reticulate surface, sulci, and nodes.   ***Figure 19.28*** Exterior view of *Cytherissa lacustris* (Sars) showing coarse reticulate surface and nodes.   ***Figure 19.29*** Exterior view of *Limnocythere itasca* Cole showing reticulate surface, sulci, and alae.

5b.	Ventral portion of peniferum with one or more prominences ventrally and/or anteriorly	*Ascetocythere*
6a(3b).	Penis directed posteroventrally from base	*Lordocythere*
6b.	Penis directed anteroventrally from base	7
7a(6b).	Finger guard absent	8
7b.	Finger guard present	18

Fig. 19.28 ├─────────┤ 0   .38mm

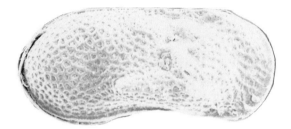

Fig. 19.29 ├─────────┤ 0   .30mm

8a(7a).	Anteroventral portion of peniferum with acute beak-like projection .... *Ornithocythere*
8b.	Anteroventral portion of peniferum never with beak-like projection 9
9a(8b).	External border of horizontal ramus of clasping apparatus with one or more excrescences 10
9b.	External border of horizontal ramus of clasping apparatus entire or with few shallow subapical grooves 14
10a(9a).	Anteroventral portion of peniferum produced ventrally in rounded lobe 11
10b.	Anteroventral portion of peniferum never produced ventrally in rounded lobe; or if produced, apex acute or truncate 12
11a(10a).	Spermatic loop horizontal, peniferum distal to dorsal margin of spermatic loop at least twice as long as portion dorsal to loop, clasping apparatus with external border bearing single tubercle and terminating in fan-like cluster of serrations .... *Saurocythere*
11b.	Spermatic loop vertical, peniferum distal to dorsal margin of spermatic loop much less than twice as long as portion dorsal to loop, clasping apparatus with external border broadly serrate and terminating in annulations .... *Okriocythere*
12a(11b).	Anteroventral portion of peniferum never with conspicuous anterodorsally directed projection .... *Ankylocythere*
12b.	Anteroventral portion of peniferum with conspicuous anterodorsally directed projection 13
13a(12b).	Anteroventral portion of peniferum entire .... *Geocythere*
13b.	Anteroventral portion of peniferum bifid .... *Hartocythere*
14a(9b).	Internal border of clasping apparatus with more than three teeth, apical cluster with more than two denticles, vertical ramus straight 15
14b.	Internal border of clasping apparatus usually with no more than three teeth, if more than three, with only two apical denticles or vertical ramus strongly convex posteriorly 16
15a(14a).	Ventral portion of peniferum tapering to slender tip .... *Rhadinocythere*
15b.	Ventral portion of peniferum never slender nor tapering .... *Entocythere*

16a(14b).   Clasping apparatus not clearly divisible into vertical and horizontal rami, extremities directed at angle of at least 100 degrees .... *Donnaldsoncythere*

16b.   Clasping apparatus usually divisible into vertical and horizontal rami, extremities directed at angle of no more than 90 degrees 17

17a(16b).   Penis large, S-shaped or sinuous and with curved posteroventral thickening of peniferum giving forcipate appearance to ventral portion of peniferum .... *Thermastrocythere*

17b.   Penis of moderate size, L-shaped, and never so disposed as to contribute forcipate appearance to ventral portion of peniferum .... *Uncinocythere*

18a(7b).   Peniferum with accessory groove (except in *Dactylocythere leptophylax* where finger guard always slender and trifid) 19

18b.   Peniferum without accessory groove; finger guard never slender and trifid 20

19a(18a).   Posteroventral portion of peniferum terminating in barbed point .... *Sagittocythere*

19b.   Posteroventral portion of peniferum variable, but never ending in barbed point .... *Dactylocythere*

20a(18b).   Ventral portion of peniferum bulbiform, clasping apparatus never extending as far ventrally as does peniferum .... *Cymocythere*

20b.   Ventral portion of peniferum slender or strongly flattened; clasping apparatus extending ventrally to or beyond ventral extremity of peniferum 21

21a(20b).   Ventral portion of peniferum slender, terminating in small recurved projection .... *Harpagocythere*

21b.   Ventral portion of peniferum flattened and with concave border .... *Litocythere*

22a(1b).   Shells subrectangular, reticulate, with sulci, alae, or pustules on surface, margins may be denticulate; males longer, more inflated in posterior region than females; second antenna with three apical claws; respiratory plate of mandibular palp well developed; masticatory lobes of maxilla short; thoracic legs moderately slender; furcal ramus well-defined, conical in shape with one terminal and one lateral seta; hemipene with basal part very large and protuberant in front (Figs. 19.11 and 19.29) .... *Limnocythere*

22b.   Shells short and broad, surface pitted, right valve with hinge line toothed; first antenna five or six segmented, second antenna four segmented, maxilla with three masticatory processes each longer than palp .... *Metacypris*

LITERATURE CITED

Alm, G. 1915. Monographie der Schwedischen Süsswasser-Ostracoden nebst systematischen Besprechungen der Tribus Podocopa. Zoologiska Bidrag fràn Uppsala 4:1–247.

American Public Health Association Inc. (APHA). 1965. Standard methods for the examination of water and wastewater, including bottom sediments and sludges. American Public Health Association, Inc., New York. 769p.

Anderson, T. W., R. J. Mott, and L. D. Delorme. 1985. Evidence for a pre-Champlain Sea glacial lake phase in Ottawa Valley, Ontario and its implications. Pages 239–245 *in:* Current Research. Part A. Geological Survey of Canada, Paper 85-1A.

Barclay, M. H. 1966. An ecological study of a temporary pond near Auckland, New Zealand. Australian Journal of Marine and Freshwater Research 17:239–258.

Bell, G. 1982. The masterpiece of nature, the evolution and genetics of sexuality. University of California Press, Berkeley. 600p.

Benson, R. H. 1959. Ecology of recent ostracodes of the Todos Santos Bay Region, Baja California, Mexico. University of Kansas Paleontological Contributions, Article 1:1–80.

Benson, R. H. 1961. Ecology of ostracode assemblages. Pages Q56–Q63 *in:* R. D. Moore, editor. Treatise on invertebrate paleontology. Part Q: Arthropoda 3, Crustacea, Ostracoda. Geological Society of America. University of Kansas Press, Lawrence.

Benson, R. H. 1967. Muscle-scar patterns of Pleistocene (Kansan) ostracodes; Essays in Paleontology and Stratigraphy Raymond C. Moore Commemorative Volume. University Kansas Department of Geology Special Publication 2:211–241.

Benson, R. H. 1974. The role of ornamentation in the design and function of the ostracode carapace. Geoscience and Man 6:47–57.

Benson, R. H. 1981. Form, function, and architecture of ostracode shells. Annual Review of Earth and Planetary Science 9:59–80.

Benson, R. H., and H. C. MacDonald. 1963. Postglacial (Holocene) ostracodes from Lake Erie. University of Kansas Paleontological Contributions 4:1–26.

Bergold, A. 1910. Beitäge zur Kenntnis des innern Baues der Süwasserostracoden. Zoologische Jahrbücher, Abteilung für Anatomie und Ontogenie der Tiere 30:1–41.

Bernecker, A. 1909. Zur Histologie der Respirationsorgane bei Crustaceen. Zoologischer Jahrbücher Abteilung der Anatomie und Ontogenie der Tiere 27: 583–630.

Bigelow, N. K. 1924. The food of young suckers in Lake Nipigon. University of Toronto Studies, Biological Series Number 24:81–116.

Bodergat, A. 1978. L'intensité lumineuse, son influence sur la teneur en phosphore des carapace d'ostracode. Géobios 11:715–735.

Bowman, T. E., and L. E. Abele. 1982. Classification of the recent Crustacea. Pages 221–239 *in:* L. G. Abele, editor. The biology of Crustacea. Vol. 1: systematics, the fossil record, and biogeography. Academic Press, New York.

Brady, G. S. 1868. A synopsis of the recent British Ostracoda. The Intellectual Observer 12:110–130.

Brady, G. S. 1886. Notes on Entomostraca collected by Mr. A. Haly in Ceylon. The Journal of the Linnean Society (Zoology) 19:293–316.

Brady, G. S., and A. M. Norman. 1888. A monograph of the marine and freshwater Ostracoda of the North Atlantic and of North-Western Europe. Transactions of the Royal Dublin Society of Science 4:63–270.

Brady, G. S., and A. M. Norman. 1896. A monograph of the marine and freshwater Ostracoda of the North Atlantic and of North-Western Europe. Transactions of the Royal Dublin Society of Science 5:621–784.

Broodbakker, N. W. 1983. The genus *Strandesia* and other cypricercini (Crustacea, Ostracoda) in the West Indies, part 1, taxonomy. Bijdragen tot de Dierkunde 53:327–368.

Broodbakker, N. W., and D. L. Danielopol. 1982. The chaetotaxy of Cypridacea (Crustacea, Ostracoda) limbs; proposals for a descriptive model. Bijdragen tot de Dierkunde 52:103–120.

Cannon, H. G. 1925. On the segmental excretory organs of certain freshwater ostracods. Philosophical Transactions of the Royal Society of London 214:1–27.

Chivas, A. R., P. DeDeckker, and J. M. G. Shelley. 1983. Magnesium, strontium and barium partitioning in non-marine ostracode shells and their use in paleoenvironmental reconstruction—a preliminary study. Pages 238–249 *in:* R. F. Maddocks, editor. Applications of Ostracoda. University of Houston Dept. of Geoscience, Houston, Texas.

Chivas, A. R., P. DeDeckker, and J. M. G. Shelley. 1985. Strontium content of ostracods indicate lacustrine paleosalinity. Nature (*London*) 316:251–253.

Chivas, A. R., P. DeDeckker, and J. M. G. Shelley. 1986a. Magnesium and strontium in non-marine ostracod shells as indicators of paleosalinity and palaeotemperature. Hydrobiologia 143:135–142.

Chivas, A. R., P. DeDeckker, and J. M. G. Shelley. 1986b. Magnesium content of non-marine ostracod shells: a new paleosalinometer and palaeothermometer. Palaeogeography, Palaeoclimatology, Palaeoecology 54:43–61.

Claus, C. 1893. Beiträge zur Kenntniss der Süsswasser-Ostracoden. Arbeiten aus dem Zoologischen Institute der Universität Wien und der Zoologischen Station in Triest 10:147–216.

Claus, C. 1899. Beiträge zur Kenntnis der Süsswasser-Ostracoden. Arbeiten aus den Zoologischen Instituten der Universität Wien und der Zoologischen Station in Triest 11:1-32.

Cvancara, A. M., L. Clayton, W. B. Bickley, Jr., A. F. Jacob, A. C. Ashworth, J. A. Brophy, C. I. Shay, L. D. Delorme, and G. E. Lammers. 1971. Paleolimnology of late Quaternary deposits, Seibold site, North Dakota. Science 171:172–174.

Danforth, W. A. 1948. A list of Iowa ostracods with descriptions of three new species. Proceedings of the Iowa Academy of Science 55:351–359.

Danielopol, D. L. 1969. Recherches sur la morphologie de l'organe copulateur male chez quelques ostracodes du genre *Candona* Baird (Fam. Cyprididae Baird). Pages 136–153 *in:* J. W. Neale, editor. The taxonomy, morphology and ecology of Recent Ostracoda. Oliver & Boyd, Edinburgh.

Danielopol, D. L. 1980a. An essay to assess the age of the freshwater interstitial ostracods of Europe. Bijdragen tot de Dierkunde 50:243–291.

Danielopol, D. L. 1980b. Sur la biologie de quelques ostracodes Candoninae épigés et hypogés d'Europe. Bulletin Museum National d'Histoire Naturelle (Paris) 2:471–506.

DeDeckker, P. 1981. Ostracods of athalassic saline lakes. Hydrobiologia 81:131–144.

DeDeckker, P. 1983. Australian Salt Lakes; their history, chemistry, and biota. Hydrobiologia 105:231–244.

DeDeckker, P. 1988. The use of ostracods in palaeolimnology in Australia. Palaeogeography, Palaeoclimatology, Palaeoecology 62:463–475.

DeDeckker, P., and R. M. Forester. 1988. The use of ostracods to reconstruct continental paleoenvironmental records. Pages 175–199 *in:* P. DeDeckker, J.-P. Collin, and J.-P. Peypouquet, editors. Ostracoda in the earth sciences. Elsevier, Amsterdam.

Delorme, L. D. 1967. Field key and methods of collecting freshwater ostracodes in Canada. Canadian Journal of Zoology 45:1275–1281.

Delorme, L. D. 1968. Pleistocene freshwater Ostracoda from Yukon, Canada. Canadian Journal of Zoology 46:859–876.

Delorme, L. D. 1969a. On the identity of the ostracode genera *Cypriconcha* and *Megalocypris*. Canadian Journal of Zoology 47:271–281.

Delorme, L. D. 1969b. Ostracodes as Quaternary paleoecological indicators. Canadian Journal of Earth Sciences 6:1471–1476.

Delorme, L. D. 1970a. Freshwater ostracodes of Canada, part I, Subfamily Cypridinae. Canadian Journal of Zoology 48:153–169.

Delorme, L. D. 1970b. Freshwater ostracodes of Canada, part II, Subfamilies Cypridopsinae, Herpetocyprindinae, and family Cyclocyprididae. Canadian Journal of Zoology 48:253–266.

Delorme, L. D. 1970c. Freshwater ostracodes of Canada, part III, Family Candonidae. Canadian Journal of Zoology 48:1099–1127.

Delorme, L. D. 1970d. Freshwater ostracodes of Canada, part IV, Families Ilyocyprididae, Notodromadidae,

Darwinulidae, Cytherideidae, and Entocytheridae. Canadian Journal of Zoology 48:1251–1259.

Delorme, L. D. 1971a. Freshwater ostracodes of Canada, part V, Families Limnocytheridae, Loxoconchidae. Canadian Journal of Zoology 49:43–64.

Delorme, L. D. 1971b. Paleoecological determinations using Pleistocene freshwater ostracodes. Bulletin du Centre de Recherche de Pau-SNPA 5:341–347.

Delorme, L. D. 1971c. Paleoecology of Holocene sediments from Manitoba using freshwater ostracodes. Geological Association of Canada Symposium, Special Paper No. 9:301–304.

Delorme, L. D. 1972. Groundwater flow systems, past and present. Pages 222–226 *in:* 24th International Geological Congress, Montreal. Sect. 11.

Delorme, L. D. 1978. Distribution of freshwater ostracodes in Lake Erie. Journal of Great Lakes Research 4:216–220.

Delorme, L. D. 1982. Lake Erie oxygen, the prehistoric record. Journal of Fisheries and Aquatic Sciences 39:1021–1029.

Delorme, L. D. 1989. Methods in Quaternary ecology. No. 7: Freshwater Ostracoda. Geosciences Canada 16:85–90.

Delorme, L. D., and D. Donald. 1969. Torpidity of freshwater ostracodes. Canadian Journal of Zoology 47:997–999.

Delorme, L. D., and S. C. Zoltai. 1984. Distribution of an arctic ostracod fauna in space and time. Quaternary Research (*N.Y.*) 21:65–73.

Delorme, L. D., S. C. Zoltai, and L. L. Kalas. 1977. Freshwater shelled invertebrate indicators of paleoclimate in Northwestern Canada during the late glacial. Canadian Journal of Earth Sciences 15:462–463.

Delorme, L. D., S. C. Zoltai, L. L. Kalas. 1978. Freshwater shelled invertebrate indicators of paleoclimate in Northwestern Canada during late glacial times; reply. Canadian Journal of Earth Sciences 14:2029–2046.

Deschiens, R. 1954. Mécansime de l'action léthale de *Cypridopsis hartwigi* sur les mollusques vecteurs des bilharzioses. Bulletin de la Société de Pathologie exotique 47:399–401.

Deschiens, R., L. Lamy, and H. Lamy. 1953. Sur un ostracode prédateur de bulins et de planorbes. Bulletin de la Société de Pathologie exotique 46:956–958.

Dobbin, C. N. 1941. Fresh-water Ostracoda from Washington and other western localities. University of Washington Publications in Biology 4:175–246.

Fassbinder, K. 1912. Beiträge zur Kenntnis der Süsswasserostracoden. Zoologische Jahrbücher, Abteilung für Anatomie und Ontogenie der Tiere 32:533–576.

Ferguson, E., Jr. 1944. Studies of the seasonal life history of three species of fresh-water ostracodes. American Midland Naturalist 32:713–727.

Ferguson, E., Jr. 1958a. Seasonal life history studies of two species of freshwater ostracods. Anatomical Records 131:549–550.

Ferguson, E., Jr. 1958b. Freshwater ostracods from South Carolina. The American Midland Naturalist 59:111–119

Ferguson, E., Jr. 1964. Stenocyprinae, a new subfamily of freshwater cyprid ostracods (Crustacea) with description of a new species from California. Proceedings of the Biological Society of Washington 77:17–24.

Ferguson, E., Jr. 1967. New ostracods from the playa lakes of eastern New Mexico and western Texas. Transactions of the American Microscopical Society 86:244–250.

Forester, R. M. 1983. The relationship of two lacustrine ostracode species to solute composition and salinity, implications for paleohydrochemistry. Geology 11:435–439.

Forester, R. M. 1986. Determination of the dissolved anion composition of ancient lakes from fossil ostracodes. Geology 14:796–799.

Forester, R. M. 1987. Paleoclimate records of deglaciation from lacustrine records. Pages 261–276 *in:* W. E. Ruddiman, and H. E. Wright, editors. North American and adjacent oceans during the last deglaciation. DNAG. Geological Society of America, Boulder, Colorado.

Forester, R. M., and E. M. Brouwers. 1985. Hydrochemical parameters governing the occurrence of estuarine and marginal estuarine ostracodes, an example from central Alaska. Journal of Paleontology 59:344–369.

Forester, R. M., L. D. Delorme, and J. P. Bradbury. 1987. Mid-Holocene climate in Northern Minnesota. Quaternary Research (*N.Y.*) 28:263–273.

Fox, H. M., and A. E. Taylor. 1954. Injurious effect of air-saturated water on certain invertebrates. Nature (*London*) 174:312.

Fox, H. M., and A. E. Taylor. 1955. The tolerance of oxygen by aquatic invertebrates. Proceedings of the Royal Society of London, Series B 143:214–225.

Fryer, G. 1957. The food of some freshwater cyclopoid copepods and its ecological significance. Journal of Animal Ecology 26:263–286.

Furtos, N. C. 1933. The Ostracoda of Ohio. Ohio Biological Survey, Bulletin 29:413–524.

Furtos, N. C. 1935. Fresh-water Ostracoda from Massachusetts. Journal of the Washington Academy of Science 25:530–544.

Furtos, N. C. 1936a. Freshwater ostracoda from Florida and North Carolina. American Midland Naturalist 17:491–522.

Furtos, N. C. 1936b. On the Ostracoda from the cenotes of Yucatan and vicinity. Carnegie Institution of Washington, Publ. Number 457:89–115.

Grant, I. F., E. A. Egan, and M. Alexander. 1983. Measurement of rate of grazing of the ostracod *Cyprinotus carolinensis* on blue-green algae. Hydrobiologia 106:199–208.

Green, J. 1954. A note on the food of *Chaetogaster diaphanus*. Annals of the Magazine of Natural History 12:842–844.

Green, J. 1962. Bile pigment in *Eucypris virens* (Jurine). Nature (*London*) 196:1318–1319.

Gutentag, E. D., and R. H. Benson. 1962. Neogene (Plio-Pleistocene) fresh-water ostracodes from the central high plains. Bulletin of the Geological Survey of Kansas. 157:60pp.

Hanström, B. 1924. Beitrage zur Kenntnis des zentralen Nervensystem der Ostracoden und Copepoden. Zoologischer Anzeiger 61:31–38.

Harding, J. P. 1962. *Mungava munda* and four other new species of ostracod crustaceans from fish stomachs. Natural History of the Rennell Island, British Solomon Islands 4:51–62.

Harding, J. P. 1964. Crustacean cuticle with reference to the ostracod carapace. Pubblicazione della Stazione Zoologica di Napoli 33:9–31.

Hart, C. W., Jr., and H. H. Hobbs, Jr. 1961. Eight new troglobitic ostracods of the genus *Entocythere* (Crustacea, Ostracoda) from the eastern United States. Proceedings of the Academy of Natural Sciences of Philadelphia 113:173–185.

Hart, D. G., and C. W. Hart, Jr. 1974. The ostracod family Entocytheridae. Monograph 18. Academy of Natural Sciences of Philadelphia, 239 pp.

Havel, J. E., P. D. N. Hebert, and L. D. Delorme. 1990a. Genetics of sexual Ostracoda from a low arctic site. Journal of Evolutionary Biology 3:65–84.

Havel, J. E., P. D. N. Hebert, and L. D. Delorme. 1990b. Genotypic diversity of asexual Ostracoda from a low arctic site. Journal of Evolutionary Biology 3:391–410.

Hoff, C. C. 1942. The ostracods of Illinois, their biology and taxonomy. Illinois Biological Monograph 19:196 pp.

Hoff, C. C. 1943a. Seasonal changes in ostracod fauna of temporary ponds. Ecology 24:116–118.

Hoff, C. C. 1943b. Two new ostracods of the genus *Entocythere* and records of previously described species. Journal of the Washington Academy of Science 33:276–286.

Hummon, W. D. 1981. Extraction by sieving: a biased procedure in studies of stream meibenthos. Transactions of the American Microscopical Society 100:278–284.

Johansen, F. 1912. Freshwater life in north-east Greenland. Meddelelser om Grønland 45:321–337.

Karrow, P. F., T. W. Anderson, L. D. Delorme, and A. J. Clarke, Jr. 1975. Stratigraphy, paleontology, and age of Lake Algonquin sediments in southwestern Ontario, Canada. Quaternary Research (*N.Y.*) 5:49–87.

Kesling, R. V. 1951a. The morphology of ostracod molt stages. Illinois Biological Monographs 21:1–324 pp.

Kesling, R. V. 1951b. Terminology of ostracod carapaces. Contribution from the Museum of Paleontology, University of Michigan 9:93–171.

Kesling, R. V. 1953. A slide rule for the determination of instars in ostracod species. Contributions from the Museum of Paleontology, University of Michigan 11:97–109.

Kesling, R. V. 1957. Notes on Zenker's organs in the ostracod *Candona*. American Midland Naturalist 57:175–182.

Kesling, R. V. 1965. Anatomy and dimorphism of adult *Candona suburbana* Hoff. Four Reports of ostracod Investigations. Report No. 1. University of Michigan Press, Ann Arbor. 56 pp.

Klassen, R. W., L. D. Delorme, and R. J. Mott. 1967. Geology and paleontology of Pleistocene deposits in southwestern Manitoba. Canadian Journal of Earth Sciences 4:433–447.

Klie, W. 1931. Campagne spéologique de C. Bolivar et R. Jeannel dan L'Amérique du Nord (1928), part 3, Crustacés Ostracodes. Archives Zoologie Expérimentale Générale 71:333–344.

Klie, W. 1939. Süsswasserostracoden aus Nordostbrasilien, part 1. Zoologischer Anzeiger 128:84–91.

Klugh, A. B. 1927. The ecology, food-relations and culture of fresh-water Entomostraca. Transactions of the Royal Canadian Institute 16:15–98.

Kornicker, L. S., and I. G. Sohn. 1971. Viability of ostracode eggs egested by fish and effect of digestive fluids on ostracode shells; ecologic and paleoecologic implications. Pages 125–135 *in:* H. J. Oertli, editor. Paleoecologie des ostracodes. Bulletin du Centre de Recherches de Pau-SNPA, Supplement 5:125–135.

Korschelt, E. 1915. Über das Verhalten verschiedener wirbelloser Tiere gegen niedere Temperaturen. Zoologischer Anzeiger 45:113–115.

Latreille, P. A. 1802. Histoire naturelle, generale et particuliere des crustaces et des insects. Paris, L'imprimerie Dufart. 3:17–18.

Lowndes, A. G. 1935. The sperms of fresh-water ostracods. Proceedings of the Zoological Society of London. Part 2:35–48.

Maddocks, R. F. 1982. Ostracoda. Pages 221–239 *in:* L. G. Abele, editor. The biology of Crustacea. Vol. 1: systematics, the fossil record, and biogeography. Academic Press, New York.

Marshall, W. S. 1903. *Entocythere cambaria* (nov. gen. et nov. spec.), a parasitic ostracod. Transactions of the Wisconsin Academy of Science, Arts, and Letters 14:117–144.

Martens, K. 1986. Taxonomic revision of the subfamily Megalocypridinae Rome, 1965. Verhandelingen van de Koninklijke Academie voor Wetenscappen, Letteren en schone Kunsten van Belgie 48:1–81.

Martens, K., P. DeDeckker, and T. G. Marples. 1985. Life history of *Mytilocypris henricae* (Chapman)(Crustacea: Ostracoda) in Lake Bathurst, New South Wales. Australian Journal of Marine and Freshwater Research 36:807–819.

McAllister, D. E., and C. R. Harrington. 1969. Pleistocene grayling, *Thymallus,* from Yukon, Canada. Canadian Journal of Earth Sciences 6:1185–1190.

McGregor, D. L. 1967. Rhythmic pulsation of the hepatopancreas in freshwater ostracods. Transactions of the American Microscopical Society 86:166–169.

McGregor, D. L. 1969. The reproductive potential, life history and parasitism of the freshwater ostracod, *Darwinula stevensoni* (Brady & Robertson). Pages 194–221 *in:* J. W. Neale, editor. Taxonomy, morphology and ecology of recent Ostracoda. Oliver & Boyd, Edinburgh.

McGregor, D. L., and R. V. Kesling. 1969a. Copulatory adaptations in ostracods, part 1, Hemipenes of *Can-*

dona. Contributions to the Museum of Paleontology University of Michigan 22:169–191.

McGregor, D. L., and R. V. Kesling. 1969b. Copulatory adaptations in ostracods, part 2, adaptations in living ostracods. Contributions to the Museum of Paleontology University of Michigan 22:221–239.

McLay, C. L. 1978a. Comparative observations on the ecology of four species of ostracods living in a temporary freshwater puddle. Canadian Journal of Zoology 56:663–675.

McLay, C. L. 1978b. The population biology of *Cyprinotus carolinensis* and *Herpetocypris reptans* (Crustacea, Ostracoda). Canadian Journal of Zoology 56:1170–1179.

McLay, C. L. 1978c. Competition, coexistence, and survival; a computer simulation study of ostracodes living in a temporary puddle. Canadian Journal of Zoology 56:1744–1758.

Moniez, R. 1893. Description d'une nouvelle èspece de *Cypris* vivents dans les eaux thermales du Hamman-Meskhoutine. Bulletin de la Société Zoologique de France 18:140–142.

Moore, R. C., editor. 1961. Treatise on invertebrate paleontology. Part Q: Arthropoda 3, Crustacea, Ostracoda. Geological Society of America. University of Kansas Press, Lawrence. 442 pp.

Müller, F. 1880. Wassertiere in Baumwipfeln, *Elpidium bromeliarum.* Kosmos 6:386–388.

Müller, G. W. 1889. Die Spermatogenese der Ostracoden. Zoologische Jahrbücher, Abteilung für Anatomie und Ontogenie der Tiere 3:677–726.

Müller, G. W. 1894. Die Ostracoden des Golfes von Neapel und der angrenzenden MeeresAbschnitte. Fauna und Flora des Golfes von Neaple 21:1–404.

Müller, G. W. 1912. Ostracoda. *In:* F. E. Schulze, editors. Das Tierreich. Friedlünder und Sohn, Lief, Berlin. 434 pp.

Müller, O. F. 1785. Entomostraca seu Insecta Testacea, Quae inaquis Daniae et Norvegiae reperit, descripsit et iconibus illustravit. Typis Thiele, Havniae. Pages 48–67.

Müller-Cale, K. 1913. Über die Entwicklung von *Cypris incongruens.* Zoologische Jahrbücher, Abteilund für Anatomie und Ontogenie der Tiere 36:1–56.

Nalepa, T. F., and A. Robertson. 1981. Screen mesh size affects estimates of macro- and meio-benthos abundance and biomass in the Great Lakes. Canadian Journal of Fisheries and Aquatic Sciences 38:1027–1034.

Neale, J. W. 1988. Ostracods and paleosalinity reconstruction. Pages 125–155 *in:* P. DeDeckker, J.-P. Colin, and J.-P. Peypouquet, editors. Ostracoda in the Earth Sciences. Elsevier, Amsterdam.

Neale, J. W., and L. D. Delorme. 1985. *Cytheromorpha fuscata,* relict Holocene marine ostracode from freshwater inland lakes of Manitoba, Canada. Revista Espanola de Micropaleontologia 17:41–64.

Newrkla, P. 1985. Respiration of *Cytherissa lacustris* (Ostracoda) at different temperatures and its tolerance towards temperature and oxygen concentration. Oecologia 67:250–254.

Novikoff, M. 1908. Über den Bau des Midiamauges der

Ostracoden. Zeitschrift für wissenschaftliche Zoologie 91:81–92.

Proctor, V. W. 1964. Viability of crustacean eggs recovered from ducks. Ecology 45:656–658.

Przibram, H. 1931. Connecting laws in animal morphology. Four lectures held at the University of London, March, 1929. University of London Press, London. 62 pp.

Purper, I., and N. I. Würdig-Macial. 1974. Occurrence of *Heterocypris incongruens* (Ramdohr), 1808—Ostracoda—in Rio Grande do Sul, Brazil; discussion on the allied genera *Cyprinotus, Hemicypris, Homocypris* and *Eucypris.* Pesquisas 3:69–91.

Ranta, E. 1979. Population biology of *Darwinula stevensoni* (Crustacea, Ostracoda) in an oligotrophic lake. Annales Zoologici Fennici 46:28–35.

Reyment, R. A., and B. Brännström. 1962. Certain aspects of the physiology of *Cypridopsis* (Ostracoda, Crustacea). Acta Universitatis Stockholmiensis 9:207–242.

Roback, S. S. 1969. Notes on the food of Tanypodine larvae. Entomological News 80:13–19.

Rome, D. R. 1947. *Herpetocypris reptans* (Ostracode) Étude morphologique et histologique I, morphologie externe et système nerveux. Cellule 51:51–152.

Rome, D. R. 1969. Morphologie de l'attache de la furca chez les Cyprididae et son utilisation en systematique. Pages 168–193 *in:* J. W. Neale, editor. The taxonomy, morphology and ecology of recent Ostracoda. Oliver & Boyd, Edinburgh.

Sars, G. O. 1866. Oversigt af Norges marine ostracoder. Norske Vidensdaps-Academi Forhandlinger. Pp. 1–130.

Sars, G. O. 1889. On some fresh-water Ostracoda and Copepoda raised from dried Australian mud. Christiania Videnskabs-Selakabs Forhandlinger 8:1–79.

Sars, G. O. 1894. Contributions to the knowledge of the fresh-water Entomostraca of New Zealand as shown by artificial hatching from dried mud. Videnskabs-Selskabets Skrifter. I. Mathematisk-naturv. Klasse 5:1–62.

Sars, G. O. 1895. On some South-African Entomostraca raised from dried mud. Videnskabs-Selskabets Skrifter. I. Mathematisk-naturv. Klasse 8:1–56.

Sars, G. O. 1896. On some west Australian Entomostraca raised from dried sand. Archiv for Mathematik og Naturvidenskab 19:3–35.

Sars, G. O. 1898. On *Megalocypris princeps,* a gigantic freshwater ostracod from South Africa. Archiv for Mathematik og Naturvidenskab 20:1–18.

Sars, G. O. 1901. Contributions to the knowledge of the fresh-water Entomostraca of South America, as shown by artificial hatching of dried material. Archiv for Mathematik og Naturvidenskab 24:16–52.

Sars, G. O. 1926. Freshwater Ostracoda from Canada and Alaska. Report of the Canadian Arctic Expedition 1913–1918 7:3–23.

Sars, G. O. 1928. An account of the Crustacea of Norway. Bergen Museum, Bergen, Norway. 277 pp.

Schleip, W. 1909. Vergleichende Untersuchung der Eireifung bei parthenogenetisch und bei geschlechtich sich fortpflanzendan Ostracoden. Archiv für Zellforschung 2:390–431.

Schram, F. R. 1986. Crustacea. Oxford University Press, New York. 606 pp.

Schreiber, E. 1922. Beiträge zur Kenntnis der Morphologie, Entwicklung und Lebensweise der Süsswasser-Ostracoden. Zoologische Jahrbücher, Abteilung für Anatomie und Ontogenie der Tiere 43:485–539.

Sharpe, R. W. 1903. Report of the fresh-water Ostracoda of the United States National Museum including a revision of the subfamilies and genera of the family Cyprididae. Proceedings of the U.S. National Museum 26:969–1001.

Sharpe, R. W. 1918. The Ostracoda. Pages 790–827 *in:* H. B. Ward, and G. C. Whipple, editors. Fresh-water biology. Wiley, New York.

Smith, R. N. 1965. Musculature and muscle scars of *Chlamydotheca arcuata* (Sars) and *Cypridopsis vidua* (O.F. Müller) (Ostracoda–Cyprididae). Four Reports on ostracode investigations. Vol. 3. Univ. of Michigan Press, Ann Arbor. Pages 1–40.

Sohn, I. G., and L. S. Kornicker. 1975. Variation in predation behavior of ostracode species on Schistosmoiasis vector snails. Bulletin of American Paleontology 65:217–223.

Sohn, I. G., and L. S. Kornicker. 1979. Viability of freeze-dried eggs of the freshwater *Heterocypris incongruens.* Pages 1–3 *in:* Proceedings of the Seventh International Symposium on Ostracoda, Belgrade.

Staplin, F. L. 1963a. Pleistocene Ostracoda of Illinois, part 1, Subfamilies Candoninae, Cyprinae, general ecology, morphology. Journal of Paleontology 37:758–797.

Staplin, F. L. 1963b. Pleistocene Ostracoda of Illinois, part 2, Subfamilies Cyclocyprinae, Cypridopsinae, Ilyocyprinae; families Darwinulidae and Cytheridae, stratigraphic ranges and assemblage patterns. Journal of Paleontology 37:1164–1203.

Strayer, D. 1985. The benthic micrometazoans of Mirror Lake, New Hampshire. Archiv für Hydrobiologie, Supplement 72:287–426.

Stuhlmann, F. 1886. Beitrage zur Anatomie der innern männlichen Geschlichts organe und zur Spermatogenese der Cypriden. Zeitschrift für Wissenschaftliche Zoologie 44:536–569.

Swain, F. M. 1963. Pleistocene Ostracoda from the Gubik formation, Arctic Coastal Plain, Alaska. Journal of Paleontology. 37:798–834.

Swüste, H. F. J., R. Cremer, and S. Parma. 1973. Selective predation by larvae of *Chaoborus flavicans* (Diptera, Chaoboridae). Internationale Vereinigung für Theoretische und Angewandte Limnologie. Verhundlungen 18:1559–1563.

Sylvester-Bradley, P. C. 1941. The shell structures of the Ostracoda and its application to their palaeontological investigation. The Annals and Magazine of Natural History Ser. 11:1–33.

Sylvester-Bradley, P. C., and R. H. Benson. 1971. Terminology for surface features in ornate ostracodes. Lethaia 4:249–286.

Towle, E. W. 1900. A study in the heliotropism of *Cypridopsis.* The American Journal of Physiology 3:345–365.

Tressler, W. L. 1949. Fresh-water Ostracoda from Brazil. United States National Museum 100:61–83.

Tressler, W. L. 1954. Fresh-water Ostracoda from Texas and Mexico. Journal of the Washington Academy of Science 44:138–149.

Tressler, W. L. 1956. Ostracoda from bromeliads in Jamaica and Florida. Journal of the Washington Academy of Science 16:333–336.

Tressler, W. L. 1957. The Ostracoda of Great Slave Lake. Journal of the Washington Academy of Science 47:415–423.

Triebel, E. 1953. Genotypus und Schalen-Merkmale der Ostracoden-Gattung *Stenocypris.* Senckenbergiana 34:5–14.

Turner, C. H. 1896. Morphology of the nervous system of *Cypris.* Journal of Comparative Neurology 6:20–44.

Turner, C. H. 1899. A male *Erpetocypris barbatus* Forbes. Zoological Bulletin 2:199–202.

Turpen, J. B., and R. W. Angell. 1971. Aspects of molting and calcification in the ostracod *Heterocypris.* Biological Bulletin (*Woods Hole, Mass.*) 140:331–338.

Walton, M., and H. H. Hobbs, Jr. 1959. Two new eyeless ostracods of the genus *Entocythere* from Florida. Quarterly Journal of the Florida Academy of Science 22:114–120.

Westgate, J. A., and L. D. Delorme. 1988. Lacustrine ostracodes in the late Pleistocene Sunnybrook diamicton of southern Ontario, Canada. Reply. Canadian Journal of Earth Sciences 25:1717–1720.

Westgate, J. A., F.-J. Chen, and L. D. Delorme. 1987. Lacustrine ostracodes in the late Pleistocene Sunnybrook diamicton of southern Ontario. Canadian Journal of Earth Sciences 24:2330–2335.

Weygoldt, P. 1960. Embryologische Untersuchungen an Ostrakoden; Die Entwicklung von *Cyprideis litoralis* (G. S. Brady) (Ostracoda, Podocopa, Cytheridae). Zoologische Jahrbücher Abteilund für Anatomie und Ontogenie der Tiere 78:369–426.

Weygoldt, P. 1961. Zur Kenntnis der Sekretion im Zentralnervensystem der Ostrakoden *Cyprideis litoralis* (G. S. Brady) (Podocopa Cytheridae) und *Cypris pubera* (O.F.M.) (Podocopa Cypridae), Neurosekretion und Sekretionzellen im Perineurium. Zoologischer Anzeiger 166:69–70.

White, M. J. D. 1973. Animal cytology and evolution. Cambridge University Press, London.

Wickstrom, C. E., and R. W. Castenholz. 1973. Thermophilic ostracod: aquatic metazoan with the highest known temperature tolerance. Science 181:1063–1064.

Wiggins, G. B., R. J. Mackay, and I, M. Smith. 1980. Evolutionary and ecological strategies of animals in annual temporary ponds. Archiv für Hydrobiologie Supplement 58:97–206.

Wohlgemuth, R. 1914. Beobactungen und Untersuchungen über die Biologie der Süsswasserostracoden; ihr Vorkommen in Sachsen und Böhmen, ihr Lebensweise und ihr Fortpflanzung. Biologisches Supplement zur Internationale Revue der Gesamten Hydrobiologie und Hydrobiologie 6: 1–72.

Woltereck, R. 1898. Zur Bildung und Entwickelung des Ostracoden-Eies. Zeitschrift für Wissenschaftliche Zoologie 64:596–620.

Zenker, W. 1854. Monographie der Ostracoden. Archiv für Naturgeschichte 20:1–87.

Zissler, D. 1969a. Die Spermiohistogenese des SüsswasserOstracoden; part 1, Die ovalen und spindelför-

migen Spermatiden. Zeitschrift für Zellforschung und Mikroscopische Anatomie 96:87–105.

Zissler, D. 1969b. Die Spermiohistogenese Süsswasser-Ostracoden *Notodromas monacha* O. F. Müller; part 2, Die spindelförmigen und schlauchürmigen Spermatiden. Zeitschrift für Zellforschung und Mikroscopische Anatomie 96:106–133.

Zissler, D. 1970. Zur Spermiohistogenese im Vas Deferens von Süsswasser-Ostracoden. Cytobiologie 2: 83–86.

Cladocera and Other Branchiopoda

20

Stanley I. Dodson
Department of Zoology
University of Wisconsin
Madison, Wisconsin 53706

David G. Frey
Department of Biology
Indiana University
Bloomington, Indiana 47405

Chapter Outline

CLADOCERA

I. ANATOMY AND PHYSIOLOGY
A. General External and Internal Anatomical Features
B. Relevant Physiological Information

II. ECOLOGY OF CLADOCERANS
A. Life History
B. Distribution
C. Physiological Adaptations
D. Behavioral Ecology
E. Foraging Relationships
F. Population Regulation
G. Functional Role in the Ecosystem
H. Paleolimnology

III. CURRENT AND FUTURE RESEARCH PROBLEMS

IV. COLLECTING AND REARING TECHNIQUES
A. Sampling Techniques
B. Culture Methods
  1. Cladoceran Food
  2. Culture Care

V. TAXONOMIC KEYS FOR CLADOCERAN GENERA
A. Available Keys and Specimen Preparation
B. Taxonomic Key to Families of Freshwater Cladocerans
C. Taxonomic Key to Genera of the Family Holopediidae
D. Taxonomic Key to Genera of the Family Sididae
E. Taxonomic Key to Genera of the Family Chydoridae (North of Mexico)
F. Taxonomic Key to Genera of the Family Daphniidae
G. Taxonomic Key to Genera of the Family Bosminidae
H. Taxonomic Key to Genera of the Family Macrothricidae: (North of Mexico)
I. Taxonomic Key to Genera of Freshwater Moinidae
J. Taxonomic Key to Genera of the Family Leptodoridae
K. Taxonomic Key to Genera of the Families Polyphemidae, Cercopagidae, and Podonidae

OTHER BRANCHIOPODS

VI. ANATOMY AND PHYSIOLOGY OF THE OTHER BRANCHIOPODS
A. General External and Internal Anatomical Features
B. Relevant Physiological Information

VII. ECOLOGY OF THE OTHER BRANCHIOPODS
A. Life History
B. Distribution and Biogeography
C. Physiological Adaptations
D. Behavioral Ecology
  1. Swimming Behavior
  2. Mating Behavior
E. Foraging Relationships
F. Population Regulation
G. Functional Role in the Ecosystem

VIII. CURRENT AND FUTURE RESEARCH PROBLEMS

IX. COLLECTING AND REARING TECHNIQUES

X. TAXONOMIC KEYS FOR NON-CLADOCERAN GENERA OF BRANCHIOPODS

  A. Available Keys and Specimen Preparation

  B. Taxonomic Key to Genera of Non-Cladoceran Freshwater Branchiopoda

Literature Cited

Branchiopods are small crustaceans that have flattened leaf-like legs. They occur in all freshwater habitats and can be abundant enough to form conspicuous swarms. Branchiopods are useful for studies of animal behavior, functional morphology, evolution of life history, speciation, and population and community ecology. They occupy a key position in aquatic communities, both as important herbivores eating algae and bacteria and as major prey items of fish, birds, or other predators. Their fossil remains open a window into the past climate and ecology of lakes.

The orders of branchiopods (Table 20.1), including eight living and two extinct orders, are related to each other only distantly (Fryer 1987, Kerfoot and Lynch 1987). The Branchiopoda are a heterogeneous group of crustaceans that share a few characteristics, principally similar thoracic legs called phyllopods, which are flat, edged with setae, not distinctly segmented, and usually appear to be unbranched. While this leg architecture resembles that of some of the earliest crustacean fossils, similarities among branchiopods are probably due at least as much to convergent evolution as to common ancestry. Branchiopods also have similar mouthparts. Mandibles are simple unsegmented rods with corrugated grinding inner surfaces. The first and second pairs of maxillae are either reduced to small scale-like structures or are absent. Branchiopods have on the last body segment a pair of spines or claws. A controversy exists concerning the evolutionary significance of these terminal spines (e.g., Bowman 1971, Schminke 1976).

The organisms commonly called cladocera (this term now has no taxonomic significance) are grouped today into 4 orders, 11 families, about 80 genera, and roughly 400 species (the first 4 orders in Table 20.1). The number of genera is only approximate, because new genera have been established recently in the families Sididae, Daphniidae, Macrothricidae, and Chydoridae, with certainly more to come.

Almost all members of the families Chydoridae and Macrothricidae and some members of the Sidi-

dae (Table 20.1) are benthic (bottom-dwelling), living on and in various surfaces, particularly macrophytes, coarse plant detritus (decaying plant material), and organic sediments and swim only short distances. They are called "meiobenthos" (Frey 1988a) because they live on the bottom or on surfaces and are small (mostly < 1 mm in total length). Chydorids are abundant (populations commonly have more than one million animals per square meter of bottom) and diverse (often with 20 or more species concurrent in the same small bit of habitat). Most species in the remaining eight cladoceran families are primarily or totally planktonic, where their facility for swimming makes them quite independent of surfaces.

Besides differing in habitat, planktonic and meiobenthic cladocerans differ also in ecological relationships and evolution. Hence, to imply, as stu-

Table 20.1 The Families and Present Number of Genera of Branchiopods[a,b]

Order	Family	Number of Genera in World
Anomopoda	Daphniidae	6
	Moinidae	2
	Bosminidae	2
	Macrothricidae	17
	Chydoridae	32
Ctenopoda	Sididae	8
	Holopediidae	1
Onychopoda	Polyphemidae	1
	Cercopagidae	2
	Podonidae	7
Haplopoda	Leptodoridae	1
Anostraca	Artemiidae	1
	Branchinectidae	1
	Branchipodidae	6
	Chirocephalidae	7
	Linderiellidae	2
	Polyartemiidae	2
	Streptocephalidae	1
	Thamnocephalidae	1
Spinicaudata	Cyclestheriidae	4
	Cyzicidae	4
	Leptestheriidae	3
	Limnadiidae	6
Laevicaudata	Lynceidae	3
Notostraca	Triopsidae	2

[a]Number of genera: for Onychopods is from Mordukhai-Boltovskoy and Riv'yer (1987); Macrothricidae from Smirnov (1976); Anostraca and Spinicaudata from Belk's annotated update of Belk (1982); Chydoridae from Smirnov (1971) and Smirnov and Timms (1983); Lynceidae from Martin and Belk (1988); Notostraca from Fryer (1988). Anostraca, Spinicaudata, Laevicaudata, and Notostraca are typically benthic, living in temporary ponds. Most of the species in the Macrothricidae and Chydoridae are meiobenthic, living on and within the sediments or aquatic vegetation. The others are mostly planktonic, living in open water.

[b]Arranged according to the eight orders established by Fryer (1987).

dents of plankton commonly do, that everything learned about *Daphnia* or *Bosmina* applies equally well to all the cladocera is often misleading or incorrect. The present chapter seeks to present an even treatment of all cladoceran families.

CLADOCERA

I. ANATOMY AND PHYSIOLOGY

A. General External and Internal Anatomical Features

Cladocerans (water fleas) are a group of small animals belonging to one of four crustacean orders (Table 20.1). Cladocerans have a single central compound eye as adults and possess a carapace that is used as a brood chamber. In most species, the carapace wraps around the entire body except for the head. Adults are 0.2–18.0 mm long. The four to six pairs of thoracic legs are generally covered by a transparent clear to yellow carapace. The large, paired appendages used for swimming are second antennae. [The first pair of antennae are often called antennules, and the second pair are termed antennae. Similarly, first and second pairs of maxillae are frequently termed maxillules and maxillae. To avoid confusion, in this chapter we will usually refer to appendages by their numerical rank (anterior to posterior).]

Most cladocerans are small transparent animals, whose general shape and jerky swimming accounts for their common name, "water fleas." Although abundant in most standing freshwater, they are small enough to be easily overlooked. Nevertheless, it is possible to observe their behavior and identify at least the larger species using the unaided eye. For a clear view of their structures, however, it is necessary to use a dissecting microscope for gross features and a compound microscope for finer details.

Anatomy is best observed using living animals. At first, it is easiest to focus on a large and transparent animal, such as an adult *Daphnia* or *Simocephalus*. Once structures are located in these large animals, they can also be found in small-sized taxa, such as chydorids and bosminids. Figure 20.1 indicates the appearance and relative arrangement of the major cladoceran structures.

One or more individuals can be put in a small volume of water in a petri dish. If most of the water is withdrawn, the animals will then be held down by the surface tension and be unable to dart out of the microscope's field of view. For more careful observation, an animal can be anchored to a small dab of grease. The grease can either be put on the bottom of a petri dish or on the tip of a small glass rod.

Cladocerans typically are discus-shaped. When dead or trapped by a surface film of water, they lie on their side. Anatomy is at first confusing, because the body in side view is rounded, the legs are obscured by a shell-like covering (the carapace), and the largest pigmented parts are likely to be eggs and not the eye. All species possess a single black compound eye in the head, except for the chydorids *Monospilus* and *Bryospilus*. The head smoothly joins the rest of the body, which is covered by the carapace. The carapace can be likened to an overcoat attached to the body only at the back of the neck. It is often ornamented with a faint geometric pattern in planktonic species or with striae, spines, pits, or hexagonal meshes in many littoral forms.

The body hangs within the overcoat-like carapace. The body ends in a pair of (postabdominal) claws which can reach out of the carapace. Several species have one or more stiff extensions of the carapace on the end opposite the head: teeth, spines, or one long tail spine. Attached to either side of the body just behind the head and partly or entirely protruding from the carapace are the second antennae (one to a side). These large branched appendages are the major swimming organs.

The thorax and abdomen of the animal can be seen moving within the carapace of living individuals. The thorax holds four to six pairs of legs, while the abdomen has no legs. The legs beat rhythmically back and forth at roughly five beats per second at room temperature. If a small amount of an algal suspension is placed with a pipette near the carapace margins, it is possible to follow the water current that brings the algae to the legs. Algae colleced on the legs are passed toward the mouth. The mouth is near the margin separating the carapace and the head. On either side of the mouth lies a pair of large but simple jaws (mandibles), each of which is a single rod, pointed toward the outer end. The inner dark chewing surfaces lie just in front of the mouth. The jaws are in nearly constant grinding motion.

Also near the front of the head and is a pair of usually short cigar-like appendages, the first antennae (antennules). These are usually shorter than the head and inconspicuous, but may be longer than the head as in genera such as *Moina* (Fig. 20.58) and most of the macrothricids (Figs. 20.44–20.57). First antennae are not used for swimming or feeding but are probably sensory organs. The head may curve toward and beyond the first antennae, forming a beak, as in *Daphnia* and the chydorids (Fig. 20.1), or the immobile first antennae can form a long tusk as in *Bosmina* (Fig. 20.42a).

The abdomen carries at its tip a pair of claws,

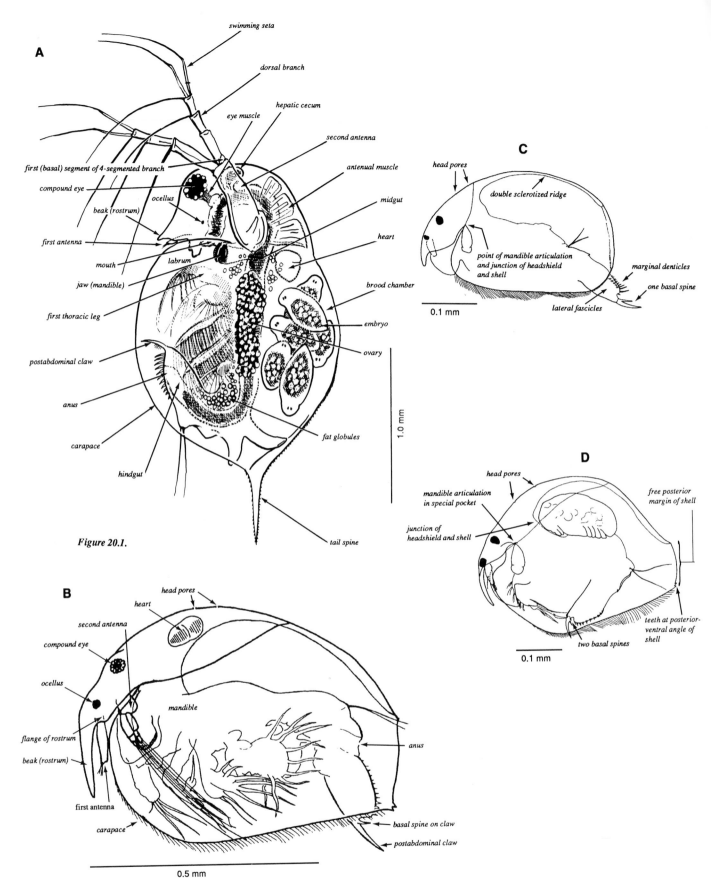

Figure 20.1 (A) The anatomy of a daphniid. *Daphnia pulicaria* Forbes, 1893, emend. Hrbáček, 1959. Nebish Lake, Vilas Co., Wisconsin. 6 June 1987. Figure drawn by Kandis Elliot (KE). The anatomy of a chydorid. *Pleuroxus trigonellus* (O. F. Mueller, 1776). Rybinsk Reservoir, USSR. 7 December 1962. Redrawn from Smirnov (1971) (KE); (B) whole body; (C) mandibular articulation in the chydorid subfamily Aloniae. *Oxyurella brevicaudis* Michael and Frey, 1983. D. G. Frey sample 5019. Clearwater Lake, Putnam Co., Florida. 3 March 1979; (D) Mandibular articulation in the chydorid subfamily Chydorinae. *Pleuroxus aduncus* (Jurine, 1820). D. G. Frey sample 3004. Langemosa, Sealand, Denmark. 23 October 1972.

which may appear to be single. These claws groom the thoracic legs and can be seen brushing over the thoracic legs as the abdomen is swung forward. Some smaller species, especially in the family Chydoridae, use the abdomen as a sort of foot to kick along surfaces. The part of the body near the end of the abdomen that bears the claws is called the postabdomen. It is anatomically ventral to the anus, which is considered the (theoretical) end of the abdomen.

Because most cladocerans reproduce asexually at least part of the time, most individuals will be females. Males (Figs. 20.2–20.3) resemble females, but are smaller. All cladoceran males have a "copulatory" hook (used for holding on to the female during mating) on the first thoracic leg, with the largest hooks in the Macrothricidae and Chydoridae and the smallest in the *Polyphemus* and *Leptodora* (Lilljeborg 1901). Some cladoceran males have longer first antennae than are present in females (especially *Sida, Diaphanosoma, Daphnia, Ceriodaphnia, Moina, Polyphemus,* and *Leptodora,* and many of the chydorids). *Daphnia* and *Ceriodaphnia* males have elongated setae on the anterior thoracic legs. Males of *Diaphanosoma, Latona, Bythotephes, Podon* (Lilljeborg 1901), and *Penilia,* (Della Croce and Gaino 1970) have a pair of penes on the postabdomen, as do several species of *Alona* and *Leydigia.*

In most females, the space between the body and the carapace is used as a brood chamber (Figs. 20.1). In the Onychopoda and Haplopoda, the carapace does not cover the body, but is reduced to a brood chamber on the back of the animal (e.g., Figs. 20.60 and 20.61).

Cladoceran eggs (e.g., Figs. 20.1, 20.7, 20.9b, 20.13, 20.27, 20.56, 20.61) develop into embryos in the brood chamber. The eggs have a central yolk droplet and are often pigmented yellow, blue, or green. As the embryos develop, they gradually assume the general appearance of the adult. After release from the brood chamber, the animals can grow only by molting, which occurs when enough energy has been stored. In preparation for molting, the inner part of the exoskeleton is reabsorbed and secretion of a new exoskeleton is begun below the remnants of the old exoskeleton. During molting, the animal absorbs water and pulls itself out of the remaining outer layers of the old exoskeleton, exiting at the back of the neck. The old exoskeleton provides an image of all of the outer surface of the animal, including all of the fine setae of the thoracic legs. This "ghost image" of the animal has been used to measure changes in size of various structures as the animals grow from neonates to adults.

Some planktonic cladocerans show variation in their external form depending on environmental stimuli (Jacobs 1987, Dodson 1989). Figure 20.4 is an example of a *Daphnia* with an elongated helmet. Animals as similar as clone mates have been described as distinct species (Krueger and Dodson 1981). The changes in form include elongated heads, tail spines, or (in *Bosmina*) elongated first antennae. These elongated body parts are sometimes accompanied by a reduced adult body size. Black and Slobodkin (1987) define cyclomorphosis as "temporal (seasonal or aseasonal) cyclic morphological changes that occur within a planktonic population." Animals generally lose elongation of body parts when cultured in the laboratory for one or two molts. Adults with extreme morphology, when taken from a natural population to the lab, produce offspring lacking the elongated parts. Changes in development leading to changes in morphology can be induced by either physical factors, such as high temperature or turbulence, or by chemical signals released by specific predator species (Dodson 1988a, 1988c, Dodson and Havel 1988). For example, *Daphnia galeata mendotae* and *D. retrocurva* both produce higher helmets in warm, turbulent laboratory environments. They also develop higher helmets in the presence of two invertebrate predators, *Chaoborus* and *Notonecta.* Similar responses to *Chaoborus* by *Daphnia pulex* and *D. ambigua* reduced the feeding rate of the predator (Havel 1987). The expression of anti-predator spines, which can be induced in the laboratory, also occurs in nature. *Daphnia pulicaria* produced neck teeth the first summer *Chaoborus* were present in Lake Lenore, Washington (Luecke and Litt 1987).

Internal anatomy can be observed through the carapace, especially in living animals. [The following discussion of anatomy is for *Daphnia* (Fig. 20.1); some details differ in other genera.] The dark black compound eye is perhaps the most conspicuous internal organ. The eye is usually in constant motion, twitching about and rotating slightly. The compound eye is single, the result of fusion of two eyes during embryonic development (Threlkeld 1979). Many species also possess a single simple eye (ocellus), which in *Daphnia,* lies in the head between the mouth and the compound eye. The ocellus usually appears as a black speck but in a few species, especially in the Chydoridae, it is larger than the compound eye. With a magnification of 50×, it is possible to see the nearly transparent muscles that are attached to the compound eye and are responsible for the motion. Once these muscles are seen, it is easier to find other muscles; for example, those attached between the basal segment of the swimming (second) antennae and the back of the thorax and head.

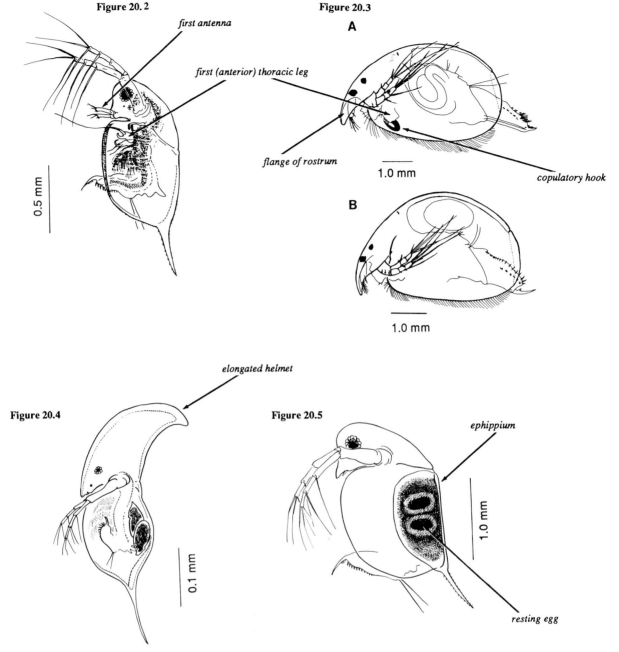

Figure 20.2 Male *Daphnia pulicaria* Forbes, 1893, emend. Hrbáček, 1959. Nebish Lake, Vilas Co., Wisconsin. 6 June 1987. (KE)   **Figure 20.3** Male and female chydorid. *Alona bicolor* Frey, 1965. (A) Male redrawn from Frey (1969); (B) female redrawn from Frey (1965). (KE)   **Figure 20.4** *Daphnia retrocurva* Forbes, 1882. With elongated helmet. Lake Wingra, Dane Col, Wisconsin. 5 October 1978. (KE)   **Figure 20.5** Ephippial female *Daphnia pulicaria* Forbes, 1893, emend. Hrbáček, 1959. Nebish Lake, Vilas Co., Wisconsin. 6 June 1987. (KE)

If the animals have been feeding, the gut will be easy to find. The gut runs from the mouth, loops through the head, continues along the body through the thorax and abdomen, and ends at the anus near the posterior end of the animal. The gut is divided into three regions. The foregut and hindgut are lined with cuticle. The midgut is lined with epithelium elaborated into microvilli, and is the site of absorption (Peters 1987a). In the head region are a couple of small sacs (hepatic secae) attached to the anterior midgut. In some species, especially in the chydorids, there is a sac attached to the gut in the abdominal region and none in the head region. The gut is filled with a green to brown, rather amorphous material (Tappa 1965) and undigested algal cells and small animals (Porter 1973, 1977). Waves of rhythmic (peristaltic) contraction move from the anus toward the midgut. By feeding a trapped animal two different kinds of food (yeast followed by algae), it is possible to measure the rate of movement of food particles through the gut.

The hindgut lies in the postabdomen and the anus opens near the tip of the postabdomen (Fig. 20.1). The anus lacks a muscle for closure (sphincter). In live animals, rhythmic (peristaltic) waves of contraction can be seen moving up the hindgut. Most, if not all cladocerans show this "anal drinking." The movements of the lower gut cause a current of water to flow into the anus; the water is excreted by the columnar cells of the anterior gut. Thus, organic compounds may stay in the gut much longer than the food from which they were derived (Fryer 1970).

Lying along the gut in the thorax region of well-fed adult females are fat globules and a pair of ovaries (Fig. 20.1). The fat globules are light yellow to orange and are spread throughout the body cavity in greater or lesser abundance. The ovaries resemble a mass of large fat globules, but the maturing eggs are more often slightly green in color. The number of eggs produced in a clutch depends on the nutritional state and size of the adult female. After the female releases the neonates that developed during the between-molt period, she molts, and then extrudes eggs into the brood chamber formed by the new carapace.

The heart, a clear muscular organ lying above the gut and just behind the head (Fig. 20.1), is obvious in living animals because of its rapid beating, approximately in phase with the thoracic leg beats. With correct lighting and 50X magnification, one can see the blood cells moving about, especially near the heart, in the bases of the second antennae, and in the beak of *Daphnia*.

B. Relevant Physiological Information

As small animals, cladocerans have relatively fast rates of metabolism and a high surface-to-volume ratio. The only reasonably complete picture available for branchiopod metabolic rates is for *Daphnia* (e.g., Peters 1987a). Respiration rate is influenced by a number of factors, including temperature, body size, and food concentration. Gas exchange probably takes place across the entire surface of the animal, but not at the branchial sacs and not particularly at the thoracic legs. Porter *et al.* (1982) and Peters (1987a) found no correlation between rate of leg movements and respiration rate.

Many entomostracans produce hemoglobin at low (1–2 mg O_2/liter) oxygen concentrations. (Some other crustaceans produce hemocyanin, or both.) The synthesis takes a week or so, and hemoglobin can be lost about as fast in well-oxygenated water (Waterman 1961, Engle 1985). The hemoglobin is dissolved in the hemolymph; if the blood is examined at high magnification, the blood cells will appear less pigmented than the surrounding fluid. The hemolymph is moved through the body by the action of the heart and other muscles. When the hemolymph is near the outer surface of the animal, oxygen combines with the hemoglobin and any CO_2 that had combined with hemoglobin dissolves passively in the external water (probably not actively, as it is in vertebrates). The oxyhemoglobin then releases the oxygen to the tissues, where the oxygen concentration is lower, carbon dioxide concentration is higher, and pH is lower than in the surrounding water. Oxygen diffuses passively across cell membranes, not by active transport.

We lack a good description of the relation between temperature and physiological rates in cladocerans, although the techniques now exist for studying the physiology of these small aquatic animals (Downing 1984a, Lampert 1984, 1987b). Egg development time and longevity are proportional to the inverse of temperature with an exponent of about 2.5 (Rigler and Downing 1984). Most cladocerans cannot live at temperatures much above 30°C, and animals native to cold waters may have much lower thermal limits. Sensitivity to high temperature increases if cladocerans feed on toxic blue-green algae (Threlkeld 1986a). Both *Ceriodaphnia* and *Diaphanosoma* live naturally in water of 27°–30°C. When blue-green algae are present, *Ceriodaphnia* shows a decline in population growth rate. For short periods of time (15 min), *Daphnia* can tolerate temperatures up to about 35°–38°C, depending on the temperature at which they were raised (MacIsaac *et al.* 1985).

Feeding rate is affected by food concentration, food quality, temperature, time of day, light intensity, oxygen concentration, pH, and crowding (Lampert 1987b). As food becomes more concentrated, the animals respire more rapidly, probably because of greater metabolic needs; assimilation rate also increases with food concentration (Peters 1984, 1987a, Lampert 1986a, 1987b). When food is superabundant, the grooming required to clean the filters on the thoracic legs may also increase the respiration rate (Porter *et al.* 1982). This phenomenon can lead to starvation at high food concentrations because the increased respiration requirements exceed the maximum rate at which energy can be ingested. In the absence of food, *Daphnia* continue to move their legs even though no food is being collected.

Richman (1958) reported that the respiration rate of *Daphnia* is proportional to body length raised to the exponent 2.14. Although this result is perhaps compromised by an unsatisfactory technique for measuring respiration, it is a value similar to the size-specific respiration rates reported for many invertebrates (Schmidt-Nielsen 1984).

Food caught on the thoracic legs is chewed by the mandibles and swallowed. The amount of food ingested depends on the algal concentration, until a concentration is reached at which the legs are operating at full capacity. Beyond this concentration, the amount of food ingested remains constant or even declines as the legs become clogged with food (Porter *et al.* 1982).

For most cladocerans, the blood is saltier than the water in which they swim. Thus, water tends to leak into the body and salts tend to leak out. *Daphnia* uses three strategies to regulate internal ion concentrations. First, a dilute hemolymph minimizes the rate of osmotic flow (Waterman 1961). Second, an active uptake of chloride through the thin cuticle of the branchial sacs at the base of each thoracic leg (Potts and Durning 1980, Peters 1987a) returns salts to the body. [Note the extremely long branchial sacs in *Holopedium* (Fig. 20.7).] Embryonic *Daphnia* and marine cladocerans also use the "neck organ" to take up cations. Third, water is excreted as urine, presumably along with ammonia and other waste products, although neither the site nor mechanism is mentioned in either Waterman (1961) or Mantel (1983). Potential sites for excretion include ductless antennal glands in *Daphnia,* glands attached to the hindgut diverticulum in some chydorids (Fryer 1969), or the maxillary gland (Fryer 1970). In the absence of a specific excretory organ in many cladocerans such as *Daphnia,* water and soluble wastes may be excreted across the general body surface (Peters 1987a).

Three cladoceran groups that live in saline waters—the polyphemoids, *Penilia,* and *Moina*—have the ability to close the brood sac. Adults in these groups can regulate the concentration of salts and possibly nutrients inside the brood sac using a "placenta" (Potts and Durning 1980).

The principal visual organ is a single compound eye lying in the middle of the head. Twenty-two ommatidia (multicellular units of light reception) comprise the *Daphnia* eye (Ringelberg 1987), while *Leptodora* has about 500 ommatidia (Wolken and Gallik 1983). *Polyphemus* shows the most morphologic specialization among its 108 ommatidia (Odselius and Nilsson 1983). The cladoceran eye is sensitive to the color, direction, and polarization of light and to point sources and light–dark boundaries. The eye of *Polyphemus pedicularis,* a visually

guided predator, has a fovea (region of high acuity). Odselius and Nilsson (1983) suggested that *Polyphemus* perceives its prey using cues provided by patterns of polarized light. Its prey are probably perceived as brightly glowing figures against a black background.

II. ECOLOGY OF CLADOCERANS

A. Life History

Detailed information exists for the life histories of many cladocerans (e.g., Shan 1969, Lynch 1980, Frey 1975, 1982c, 1987b, 1988c) and especially for *Daphnia* (Schwartz 1984, Lynch 1989, Threlkeld 1987a). Cladocerans typically reproduce parthenogenetically (Shan 1969, Hebert 1987a, 1988). This asexual reproduction can either be obligate or facultative (asexual alternating with bouts of sexual reproduction). Asexual (subitaneous) eggs develop into neonates inside the brood chamber of the mother. Egg developmental stages can be followed easily because the carapace is transparent (Threlkeld 1979). Neonates resemble the adult form, but are smaller. Neonates are released from the brood chamber just before the mother molts and extrudes another set of eggs into the brood chamber. The exception to this pattern is *Leptodora,* in which nauplii hatch from resting eggs (Warren 1901). *Leptodora* asexual reproduction is typical of that in other cladocerans.

Asexual eggs of smaller species are disproportionately large, producing relatively large neonates [compare the relative size of eggs in *Daphnia* (Fig. 20.1) with that of the much smaller *Bosminopsis* (Fig. 20.42)]. Except for *Eurycercus* and two other species, chydorids have a constant clutch size of two (Frey 1973, Robertson 1988). In other cladocerans, clutch size increases with body size. The smallest nonchydorid species produce one or two eggs per clutch, while the largest species can produce hundreds of eggs per clutch (Hann 1985).

Many cladocerans also can reproduce sexually (Shan 1969, Hebert 1978, 1987a, 1988). Males are produced asexually except in *Moina* and *Daphniopsis,* which produce males from sexually derived resting eggs. Asexual formation of males is induced by some deterioration of the environment, such as a change in food concentration, crowding, or decreasing photoperiod. The first step in sexual reproduction is asexual production of mixed clutches of males and females. The male-to-female ratio varies greatly between populations and in different years. Adult females also produce special haploid eggs and a modified carapace (except in Sididae and predaceous species). Upon fertilization by the male (Baird 1850), the resulting zygote undergoes several cell divisions (Banta 1939, Ojima 1958) and then enters diapause. These resting ''eggs'' (actually early embryos) are resistant to freezing and drying and are further protected by the modified part of the carapace, the ephippium (Fig. 20.5). Daphniid ephippia are designed to hold two resting eggs, while chydorids hold one [except for *Eurycercus* and some species of *Chydorus* and *Camptocercus* (Fryer and Frey 1981), and *Pleutoxus*-like species (Frey 1990), which hold two.]

The ephippium is thicker than normal carapace and often is darkly pigmented with melanin. As the female molts, the ephippium closes around the resting eggs and most of the remaining exoskeleton (except for the tail spine in *Daphnia*) detaches from the ephippium. Chydorid and macrothricid ephippia are usually attached to the substratum by either cement or trailing pieces of the carapace (Bretschko 1969, Fryer 1972, Fryer and Frey 1981). *Daphnia* ephippia tend to float because of their specific gravity and the hydrophobic character of the carapace. The dark pigmentation of ephippia may aid dispersal, because fish and bird predators can easily see the dark ephippium, which can pass undamaged through the gut of a predator (Mellors 1975). The tail spine and sometimes a strip of chitin located ventrally at the head end of the carapace (Fryer 1972) may aid dispersal by allowing the ephippium to stick to the fur or feathers of wading animals. Ephippia are small enough to blow about with the wind and resistant enough to remain alive on dry land for years (Moghraby 1977).

To continue developing, diapausing embryos in ephippia require stimuli such as water, long or increasing day lengths, proper temperature, and high oxygen concentrations (Schwartz and Hebert 1987a). Resting eggs usually develop into females, which produce more young asexually. Major exceptions are *Daphniopsis* and *Moina,* genera which specialize on short-lived temporary ponds and produce both males and females from ephippial eggs (Goulden 1968, Schwartz and Hebert 1987b).

Daphnia species often include both facultative and obligate asexual populations (Hebert 1987a, Innes and Hebert 1988). Occasionally, the obligate asexual populations produce a few males. These are functional and able to transfer genetic markers to offspring when mated with a sexual female. In the *Daphnia pulex* species complex, matings between males of obligate asexual clones and females of facultative asexual clones regularly produce viable resting eggs. These offspring may be either obligate or facultative parthenogens. However, few (about 5%) of the offspring can produce viable resting eggs themselves.

Two advantages of asexual reproduction are that the population growth rate is maximized (because all offspring are females) and genetic combinations particularly well adapted to present conditions are not lost, as they would be through sexual recombination (Banta 1939, Williams 1975). The disadvantage is that asexual clones are theoretically slow or unable to adjust to changing conditions by sexual recombination and so are restricted to specific habitat conditions. However, *Daphnia* clones accumulate new mutations at a surprisingly rapid rate and may therefore be able to adapt to changing conditions nearly as well as sexual lines (Banta 1939). The different adaptive advantages of the sexual and asexual eggs would seem to dictate that only the sexual eggs should diapause. The exception to this is the production of ''pseudosexual'' resting eggs in obligate asexual clones. These eggs are produced without genetic recombination but resemble sexual diapausing eggs and are enclosed in an ephippium.

Neonates tend to molt just two times in most chydorids (Shan 1969, Robertson 1988) and up to seven times in daphniids before producing eggs of their own. An ''instar'' is the developmental stage between molts. The rate of molting (instar duration) in *Daphnia pulex* depends mostly on temperature and less on food (Lynch 1989). If animals are poorly fed, they may even lose weight when they molt. An instar lasts from about a day to weeks, depending on the temperature. In laboratory conditions, cladocerans can live for months, even at warm temperatures. The chance of dying from ''physiological mortality'' is on the order of one in a hundred per instar.

Each time a well-fed juvenile female molts, it approximately doubles its volume, and its length increases by a factor of roughly 1.26. Growth slows as cladocerans mature. Successive adult instars grow slightly, but energy not allocated to maintenance is used to produce eggs (Frey and Hann 1985, Threlkeld 1987a). The energetic cost of molting may be a major constraint on the evolution of body size in *Daphnia* (Lynch 1989).

For nonchydorids, the number of juvenile instars is more or less species-specific and depends on adult body size, with the smaller species having the fewest molts. All chydorids except *Eurycercus* (Frey 1973) and *Monospilus* (Frey and Hann 1985) have just two juvenile instars. Female cladocerans can molt several times after becoming adults. Chydorid males become mature in the third instar and then stop molting, although *Scapholeberis* and *Daphnia* males continue to molt and grow after maturity.

Many life-history characteristics of *Daphnia* species show phenotypic plasticity. Instar duration, number of instars to first reproduction, time to first reproduction, primiparous and neonatal dimensions, and clutch size are all influenced by temperature, food supply, and chemical signals produced by predators (Hann 1984, Schwartz 1984, Havel and Dodson 1987, Dodson 1989). The patterns of phenotypic response (reaction norms) appear to be, in general, adaptive responses to size-selective predation and food limitation. Perhaps because of interactions among life-history characteristics, some of the individual responses to predation are not as predicted by the life history theory proposed by Stearns (1982). For example, *Daphnia pulex* exposed to *Chaoborus* larvae take significantly longer to reach maturity, even though *Chaoborus* feeds most heavily on the smaller (younger) animals. However, the longer time to maturity (and hence longer time at risk to the predator) is probably more than balanced by the protective "neck teeth" induced in small *Daphnia* by a chemical signal released from *Chaoborus* (Havel and Dodson 1987). In several cases, the induced ontogeny that produces neck teeth may also slow developmental rate.

B. Distribution

Cladocerans are a widespread group, occurring in all but the harshest freshwater habitats (Hutchinson 1967). While they are more abundant in lakes, ponds, and sluggish streams, they also occur in quiet water and in marginal vegetation in rushing streams. Several species probably inhabit groundwater, especially river gravels (Dumont 1987). One chydorid cladoceran has been found in water trapped in mosses that drape trees in the cloud forests of El Yunque, Puerto Rico, Venezuela, and New Zealand (Frey 1980b).

Cladocerans have a wide range of tolerance to salinity (Potts and Durning 1980); from near-distilled water (Dodson 1982) to the 3.2% salinity of oceans (Della Croce and Angelino 1987) to water more saline than seawater, such as *Moina* in Soap Lake, Washington, at salinities as high as 3.9% (Edmondson 1963) and *Daphniopsis* in Australia at salinities up to 5% or higher (Bayly and Edward 1969).

They are found from sea level to ponds above tree line and from arctic to tropical latitudes. In North America, the highest diversity of planktonic cladoceran species is found in the glaciated mid-temperate zone (Tappa 1965), whereas the highest diversity of meiobenthic taxa, especially small-bodied species, occurs in soft acidic waters of Nova Scotia and in the southeastern United States (Crisman 1980, Frey 1982c, Kerfoot and Lynch 1987).

Our understanding of the distribution of cladoceran species depends in part on how well we can distinguish among species. Cladoceran taxonomy is currently undergoing a re-evaluation. Because of an almost complete disregard for fine but significant details of morphology and a consequent impression that taxa on different continents were the same, investigators have claimed (Forbes 1925), and still do, that many species are cosmopolitan. But cosmopolitanism, at least among the Chydoridae, now seems to be mainly an illusory concept (Frey 1982a, 1987a).

Cosmopolitanism reflects an unquestioning acceptance of rapid passive dispersal of these organisms via their resting eggs by wind, water, birds, insects, and mammals (Frey 1982a). Studies in North America (Frey 1986b) demonstrate that the geographical ranges of chydorid species can be just as narrow and circumscribed as those of animals considered to be much less vagile.

The biological species concept (Mayr 1963) does not apply to species that are primarily parthenogenetic. Most cladoceran species are facultative parthenogens, but many daphniid species are obligate parthenogens, in which males have never been observed. Careful genetic analysis (Hebert 1987a, 1987b) has shown *Daphnia* species to be more diverse genetically and to have distributions restricted to small portions of North America. *Daphnia* species that are obligate parthenogens are genetically the most diverse asexual animals known (Hebert 1987b). Thus, cladoceran species are morphological species, defined by reproductive isolation where it has been measured, but more by morphological uniqueness as determined by taxonomists.

Traditional morphological species criteria may not be particularly useful, especially for the daphniids. For example, the well-known and abundant species in the *Daphnia pulex* species group cannot be distinguished morphologically (Dodson 1981) because populations of intermediate morphology link the described "species." It seems likely that electrophoretic techniques (Hebert 1987a) will provide a fresh perspective on the problem of defining morphological species. At this time, morphology appears to provide excellent criteria for defining species among the chydorids.

Hebert (1985) reviews reports of hybrids between cladoceran species. The evidence has until recently been either morphological or based on lab crosses. Recent studies based on proteins (allozymes) confirm that hybrids can indeed be formed between similar species of daphniids. Hebert (1985) used morphological and allozyme data to detect a group of hybrid clones intermediate between two species of *Daphnia* (*Ctenodaphnia*). He hypothesized that the persistence of clones in nature is due to their ability to reproduce asexually. Species remain distinct, in part, because the hybrids apparently cannot reproduce sexually in nature. Thus, while daphniid species are often difficult to distinguish using morphological characters (e.g., Dodson 1981), allozyme analyses sometimes show them to be clearly distinct (e.g., Hann and Hebert 1986, Hebert 1987b). Among the chydorids, *Pleuroxus denticulatus* and *P. procurvus* from the same lake were crossed in the laboratory, yielding one morphologically intermediate offspring that could reproduce successfully only asexually (Shan and Frey 1983). There is no evidence for the occurrence of natural hybrid populations in the Chydoridae.

The highest species richness values for cladocerans occur among the meiobenthic species, chiefly the chydorids. Frey (1982c) reported means of 13.7–14.0 chydorid species and 17.5–19.7 total cladoceran species in 17 bodies of water near Tallahassee, Florida, and four water bodies in Nova Scotia. The maximum species richness was 19 chydorids and 27 total cladocerans. Fifteen bodies of water near Gainsville, Florida, and in North Carolina had fewer species. All of the bodies of water have low electrolyte concentrations, generally low dissolved color, and a pH value that is usually less than 5.0. Fryer (1985), in a study of 207 English lakes, reported a range of chydorid species richness from 0–5 species in small ponds (< 5000 m^2) to 2–12 species in larger lakes ($> 15,000$ m^2).

Whether a species occurs in a body of water depends on a combination of factors: (1) zoogeographic region (that is, on the probability of colonizing the water body); (2) chemical and physical requirements of the species; (3) food conditions within the water; and (4) predators living in the water.

Although some cladocerans are thought to be capable of long-distance dispersal via resting eggs carried by the wind and by feathers and guts of water fowl, we still know little about actual rates of colonization. Inbreeding coefficients for small *Daphniopsis* populations suggest that populations receive an average of 0.3 migrants per generation (Schwartz and Hebert 1987b). Studies of *Daphnia* indicate slightly higher rates of migration, about 0.5 migrants per generation.

Once a species reaches a freshwater body, it may be excluded by chemical or physical features of the water. Carter *et al.* (1980) found that of 23 cladoceran species occurring in eastern Canada, 8 may have distributions restricted by water chemistry: *Ceriodaphnia* and a species of *Bosmina* tended to occur in hard water, while *Polyphemus*, *Sida*, two *Daphnia* species, and two other *Bosmina* species were found mostly in soft water. *Holopedium* may be restricted to soft waters with less than about 20 mg/liter of calcium (Stenson 1973). Small cla-

docerans, such as species of *Bosmina*, *Diaphanosoma*, small *Daphnia*, *Holopedium*, *Polyphemus*, and many chydorids appear to be particularly resistant to acidification, and are found in lakes with pH values between 5.0 and 3.8 (Sprules 1975, Frey 1982c).

There are extreme conditions such as high salinity that exclude most cladocerans, while providing a refuge for physiologically specialized genera such as some *Moina*. Most cladocerans are sensitive to salinity. The cladocerans of Pyramid Lake, Nevada, showed strong mortalities at salinities of 0.6% (*Diaphanosoma*), 0.7% (*Ceriodaphnia*), and 1.8% (*Moina*) (Galat and Robinson 1983). The marine cladocerans *Evadne* and *Penilia* are restricted to coastal and open waters, while *Podon* is found in less saline coastal and estuarine water (Della Croce and Angelino 1987).

A combination of laboratory life-table experiments and field observations suggested that *Moina* and *Diaphanosoma* are adapted to silt-laden water, while *Ceriodaphnia* and *Daphnia* (subgenus *Daphnia*) are not (Threlkeld 1986b). Some species of *Daphnia* (*Ctenodaphnia*) (Dodson 1985) and several chydorids (such as *Graptoleberis* and *Pleuroxus*) are found in muddy or very turbid water.

Cladocerans are sometimes food-limited (e.g., *Daphnia*, Haney and Buchanan 1987, Threlkeld 1987a) and may compete for food with other cladocerans and with other herbivores. In some cases, competition markedly reduces abundances or even leads to competitive exclusion. Vanni (1986) found that *Daphnia* could lower algae concentrations in a lake to a level that caused a decrease in *Bosmina* densities. He suggested that competition for food is a cause of the often observed scarcity of small species in lakes inhabited by large cladocerans such as *Daphnia*. This effect may be especially strong in nutrient-rich lakes, where *Daphnia* populations can increase rapidly enough to produce a sudden and major reduction in algal abundance. Fryer (1985) concluded that chydorid diversity within a lake depends more on microhabitat preferences and the presence of the appropriate species-specific food and less on dispersal, competition, or predation.

Size-selective predation sometimes affects the distribution of planktonic cladocerans (Brooks and Dodson 1965, Dodson 1974, Zaret 1980, Kerfoot and Sih 1987). In general, large-bodied species tend not to coexist with fish, and small-bodied species can be excluded by small predators such as the larvae of the midge *Chaoborus* or the backswimmer *Notonecta*. Results of 18 predator–prey community-level experiments with freshwater zooplankton suggest that adding predators often reduced the number of cladoceran individuals present, but few studies showed an effect on prey species number or relative abundance (Thorp 1986). Over the long term, predators may influence cladoceran diversity via coevolution, size-selective predation, and restrictions on prey distributions.

While food limitation and selective predation are components determining the distribution of zooplankton, it is clear that the two factors are intertwined. Predators influence competition and vice versa, via what are coming to be called "indirect effects" (Cooper and Smith 1982, Lampert 1987a, Kerfoot and Sih 1987, Carpenter 1988).

C. Physiological Adaptations

Although surrounded by water, many cladocerans are unwetted. If brought into contact with the surface film, they become trapped. Their hydrophobic carapace is probably an adaptation that discourages short-term growth of algae and protozoa on the carapace. Cladocerans cannot groom the outside of the carapace, and hence animals that have lived for a few weeks without molting often bear a thick load of encrusting organisms, which may reduce feeding and swimming efficiency, lower escape speed, and increase sinking rate.

Starvation is a problem faced by most cladoceran populations at some time during each year (Threlkeld 1976). Because smaller animals have a higher metabolic rate per unit body weight, neonates of any species are more prone to starvation than adults (Tessier and Goulden 1987). Eggs can receive a large store of energy-rich fat from the mother if the mother has had adequate food beforehand. Similarly, if food is temporarily absent, the large species can persist for a longer time than small species. Even for large adults, a single day without food results in a drop in energy stores, body weight, and respiration rate.

Body-size effects are complex at food concentrations above starvation levels. Large animals always have lower weight-specific respiration rates. However, smaller animals have greater weight-specific assimilation rates at low food concentrations, while large animals have greater rates when food is plentiful (Tessier and Goulden 1987). Thus, net growth, estimated from the difference between assimilation rate and respiration rate, is greatest for small animals at low food concentrations or for large animals at high food concentrations. In a study of three *Daphnia* species (Tillmann and Lampert 1984), large *Daphnia magna* out-reproduced smaller *Daphnia*. When food was marginally limiting, the three species had similar reproductive rates. Under severe food limitation, only the smallest species, *D. longispina* could still reproduce.

When a cladoceran is food-limited, it tends to mature at a smaller size and produce smaller (but as many) offspring. The main response of *Daphnia pulex* to low food levels was a reduction in size-specific food intake and egg size (Lynch 1989). Over a wide range, food concentration had no effect on the length–weight relationship, instar duration, and weight-specific investment of energy in reproduction.

Population growth rate is linked to body growth rate. Competition between various species of cladocerans may largely depend on which species can grow fastest at a given food level. Thus, small species such as *Ceriodaphnia* and *Bosmina* may be able to out-compete or at least out-grow large species such as *Daphnia pulex* at low but not at high food concentrations (Tessier and Goulden 1987).

Some important physiological adaptations are indicated by cladoceran pigmentation. Animals collected in oxygen-poor habitats (sewage lagoons or temporary pools with decaying vegetation) will often have a general pink to red color (Weider and Lampert 1985) due to hemoglobin dissolved in the blood. Oxyhemoglobin in *Daphnia* deoxygenates at about 1.18 mg O_2 per liter (Fox 1947). Compared to unpigmented animals, red *Daphnia* survive longer, move thoracic legs faster and gather more food, swim faster, produce more eggs, and eggs have a shorter development time (Waterman 1961). On the other hand, animals become clear in a day or so if grown in well-oxygenated water, suggesting an energetic cost to maintenance of hemoglobin (Landon and Stasiak 1983). Also, red pigmentation may have a high cost in the presence of visual predators such as fish, salamanders, or insects (Engle 1985).

Some cladocerans, such as *Scapholeberis* (Fig. 20.40) and some forms of *Daphnia* have deposits of black melanin over parts of the exoskeleton. Melanin is located in body surfaces habitually faced toward the sun, such as ventral carapace margins of *Scapholeberis* and the head and back of some *Daphnia*. Experiments have shown that the black pigment provides protection against photodamage (Luecke and O'Brien 1983a). It is possible that the induction of melanin is associated with a decrease in reproductive potential. Furthermore, the dark pigment may increase heating rate and, therefore, enhance reproductive rate in cold water. However, because cladocerans are small and water has a high thermal conductivity, this effect is probably insignificant. Finally, while melanin strengthens bird feathers and insect exoskeletons, it probably does not have this effect in cladocerans (Dodson 1984). Melanistic animals brought into the lab loose their pigmentation at the first molt, suggesting that stimulation of bright sunlight is needed to induce melanin. Melanization of the ephippia is often more intense than in the animals themselves. Because light stimulates the hatching of ephippial eggs, variability in ephippium pigmentation assures variability in hatching time, thus increasing the chance that some eggs will hatch under favorable conditions (Shan 1970).

Scapholeberis are the most darkly pigmented daphniid species. *S. kingii* is the only known distasteful cladoceran. It is not even ingested by *Hydra* (Schwartz *et al.* 1983). *Scapholeberis* homogenate causes *Hydra* to writhe and contract and induces concerts of tentacular spasms. It is not yet known if *Scapholeberis* is distasteful to visual predators. If so, the black pigmentation may be adaptive both as protection against photodamage and as a warning display.

The black color of the cladoceran eye is due to accessory screening pigments including melanins, ommochromes, pteridines, or purines (Shaw and Stowe 1982). Since the black color fades in several mounting media, such as Hoyer's, it is unlikely that much melanin is present. The actual photoreceptive pigment is a nearly colorless purple, masked by the accessory pigments.

Cladocerans can be colored a light yellow, blue, or green by carotenoids stored as globules of liquid fat (lipids) throughout the body cavity. The blue and green colors are due to carotenoids, derived from algae in the diet, which chemically combine with proteins. These colors are particularly evident along the gut, in the ovaries, and in the yolk of young eggs. Because carotenoids stain stored lipids, they make it easier to assess the relative amount of lipid carried by an animal and perhaps to make an estimate of the nutritional state and competitive ability of the animal (Tessier and Goulden 1982).

D. Behavioral Ecology

Cladocerans display a number of behaviors relating to mating, escape from predators, feeding, and migrations within lakes. Jurine (1820) provides us with one of the few descriptions of mating behavior in cladocera. In November of 1797, in the region of Geneva, he found a pond containing a dense population of *Daphnia* that included ardent males. He observed numerous couplings in the laboratory and reported (in French):

> The male jumps onto the back of the female, which sometimes escapes, but if he is able to seize her with the long filaments of his first legs and secure her with his elongated first antennae, he is able to catch hold of her firmly. He soon scuttles quickly over the surface of her carapace until he has attained the lower margin. When he finds the position so that the openings of the two carapaces are opposed, he

quickly introduces into her carapace the long setae of the first antennae and of the first thoracic legs, and thereby envelopes and binds, so to speak, the female's legs. When he is established in this position, he curves his postabdomen forward, putting it out far enough to get that of the female. When she feels that part, she is greatly agitated, and fleeing, carries the male so fast it is hard to follow the amorous couple in the vase that contains them. Finally this agitation ceases and the female in turn advances her postabdomen in order to touch that of the male. The postabdomens hardly touch before they separate. At the instant of this touch, the male is agitated by spasms which give his legs remarkable vibrations. It is during this contact that, in my opinion, the copulation occurs.

> The embrace is of variable duration; it rarely is sustained for more than eight or ten minutes. During this interval, the postabdomens were brought together more than once. Copulation terminated, the postabdomen of the male is slowly withdrawn into his carapace, but he is not yet able to withdraw the setae of his first antennae and first thoracic legs because of the spasms that persist in these parts. When these cease, the separation of the two animals takes place

> This is not the only mode of embrace in these animals. I have several times seen the male introduce only one first antenna, one first leg seta, and one swimming (second) antenna with which he envelops the setae of the females second and third legs. When the copulation is finished, he is not able to retrieve his swimming antenna before the spasms which he is experiencing are entirely gone. I noticed other animals that were embraced almost transversely, in such a way as to form a sort of cross.

> Sometimes other males arrived intending to unite with a female already in an embrace. One then sees them explore the female's body with a remarkable vivacity, trying, although futilely, to penetrate into the sanctuary of pleasure, elongating their postabdomen as much as they can into the opening of the carapace.

> . . . Males attacked indiscriminately all females they encountered. . . . Ephippial females pushed aside the males and rendered their attacks in vain. (pp. 108–110)

About sixty years later, Weismann (1880) gave cursory descriptions of mating behavior for *Moina*, *Daphnia*, and *Bythotrephes*, and conjectures on possible mating behavior, based on anatomy, for 16 other genera. These mid-Victorian descriptions lack the enthusiastic attention to behavioral details shown by the pre-Victorian Jurine (1820), but do include a wealth of detail concerning cellular and tissue structure and ontogeny. Weismann (1880) further argues that successful fertilization can occur in

Daphnia only with the partners face to face and parallel.

The veil of modesty is further drawn over mating behavior by monographers of functional morphology, species, and evolution who largely ignored the topic of mating behavior (e.g., Banta 1939, Brooks 1957, Fryer 1968, 1974, Goulden 1968). However, Shan (1969) and Smirnov (1971) have produced much more informative descriptions of chydorid mating behavior and functional morphology. The theme of female participation and even perhaps female choice, introduced by Jurine (1820), is developed further by Shan (1969) for the chydorid *Pleuroxus*:

> My observations on laboratory stocks of *P. denticulatus* do not agree with Weismann's (1880) description. At the beginning, the male swims vigorously here and there, using his first antennae, trying to find a female that is receptive, usually an ephippial female. A receptive female often cooperates in the copulation process. Sometimes, the ephippial female he approaches may not be receptive and she may reject him using her postabdomen as he approaches her. At times a male may hook onto a parthenogenetic female and try to copulate with her, or a male may hook onto another male. In one occasion, I saw a parthenogenetic female that had 3 males attached in a series, one after another!
>
> As the male hooks onto the edge of the female's carapace, he moves vigorously back and forth along the edge, often only on one side of the female. The contact points are the two copulatory hooks [on the first thoracic leg] and the tip of his rostrum. He jerks his postabdomen and moves his hooks vigorously while the female does not react to him in any way. This activity goes on for a while, usually a few minutes, possibly even longer than 10 minutes. The female may become excited when the male reaches a position where his hooks are at the top of the free posterior edge of the female's carapace. In this position, the male and the female are facing the same direction and have the same dorsal orientation. At this time the female may stop feeding, carry the male with her, and swim away. They swim together, both jerk their postabdomens and strike their antennae. They swim for some time, possibly 10 minutes or longer. Finally, they fall onto the bottom, still hooked together, and lie on one side. Both of them stretch and jerk their postabdomens back and forth. At this time the female's postabdomen is in maximum extension downward and backward, while the male is flexing his forward trying to get as close as possible to the brood pouch of the female. Presumably at this time the sperm are ejected. The mechanism that allows the sperm to get into the brood pouch and fertilize the egg is unclear. It is quite possible that the jerks of the female's postabdomen create an inward current that helps bring the sperm into the brood pouch. (pp. 518–519)

The spermatozoa are amoeboid, without a flagellum. In most species, they are simple ovals 5–20 μm in diameter. Spermatozoa of *Diaphanosoma* are ovals about 60 μm in diameter, and those of *Sida* and *Moina* are spiny and nearly 100 μm long (Weismann 1880, Wingstrand 1978)

The elongated first antennae of male daphniids and moinids may be used to feel the roughened carapace of a female (Goulden 1968, Kokkinn and Williams 1987). The female carapace ornamentation might tell the male whether she is the correct species and that she is producing haploid eggs. As with all aspects of behavior in *Daphnia*, we know few details of the individual behaviors that result in mating. Instead, cladoceran behavior has been studied mostly at the population level, using plankton nets in lakes.

One of the earliest observations of cladoceran behavior at the population level is Cuvier's 1817 description of *Daphnia* vertical migration (translated in Bayly 1986):

> In the morning and evening, and even during the day when the sky is overcast, the daphniids usually stay at the surface. But during very hot weather, and when the sun beats fiercely down on the pools of stagnant water where they live, they sink down in the water, and stay at a depth of six or eight feet or more; often not a single one can be seen at the surface. (p. 349)

Cladoceran populations have been observed to migrate, on a daily cycle, over great vertical distances in both fresh and marine waters (Hutchinson 1967, Bayly 1986). Distances migrated by pelagic species vary from less than a meter in a small rock pool to hundreds of meters in oceans. In freshwaters, the usual distance is between 5 and 20 meters. Because lake waters are vertically stratified, vertical migration takes the animals into very different habitats in terms of temperature, light, food, turbulence, oxygen and solute concentrations, and predators (Landon and Stasiak 1983). The greatest depth of the migration is often set by the bottom of the lake, by the depth to which light penetrates, or by the limit of sufficient oxygen (about 1.0–0.5 mg liter^{-1}, Weider and Lampert 1985). Littoral organisms also have a pronounced diel up-and-down migration. This behavior allows greater than 90% of a population to be trapped overnight in inverted funnels (Whiteside and Williams 1975).

The daily migration pattern often seems to have the greatest amplitude during the summer and fall months. Larger developmental stages typically migrate the farthest. Populations usually migrate downward at about dawn and do not return until after dark.

Physiological response to light intensity is a component of vertical migration behavior. Although an extensive literature exists on the light response, the behavior is notoriously variable between experiments (Ringelberg 1987). Nevertheless, it is felt that physiologic light response is in some way related to both diel vertical migration (Ringelberg 1987) and to shoreline avoidance (Siebeck 1980).

Directionality of light source affects the swimming position of *Daphnia*; if the light comes equally from every direction, the animal is unable to swim normally. Cladocerans probably orient in the water using polarized light (Ringelberg 1987). There is also some evidence that cladoceran swimming speed and direction are influenced differently by red and blue light.

Most ecologists have concluded that cladocerans may migrate to improve their habitat. In some cases, migration allows avoidance of dangerous levels of light, pH, or oxygen. Some estuarine and marine zooplankton may show diurnal (once daily) vertical migration to avoid horizontal movement caused by water currents. However, extreme changes in environmental conditions are not experienced on a daily basis in most freshwaters. Cladocerans probably migrate vertically in order to obtain food or avoid predators.

Vertical migration was once thought to allow cladocerans to feed near the surface, where algae are usually most dense, and then to digest the food more efficiently in colder waters. However, careful laboratory experiments have shown that just the opposite is true. Animals that migrate vertically show no increase in feeding efficiency; indeed, they must expend a small but significant portion of their daily energy budget for vertical migration (Lampert 1987a).

Instead, daily vertical migration seems to be a way of avoiding predators that need light to locate cladoceran prey. Because of the energy requirements, migration will probably be found only when food is at least moderately abundant. In most cases, vertical migration is between the food source in upper waters, which are safe only at night and the dark refuge of deeper waters during the day. Field and laboratory estimates of feeding rates suggest that the planktivores can cause high mortality in cladoceran populations that do not migrate vertically (Melville and Maly 1981, Bayly 1986, Lampert 1987a).

The vertical distribution of several species of *Daphnia* is, at least in the laboratory, an immediate response to feeding activity of predators (Dodson 1988b). When exposed to either feeding predators or even the water from cultures of predators, *Daphnia* will often swim up or down, depending on the specific predator and *Daphnia* species. Since the prey respond to water in which a predator had been living, the behavior is probably in response to a chemical released by the predator.

Cladocerans sometimes form swarms. Baird (1850) reported:

> I have, however, frequently seen large patches of water in different ponds assume a ruddy hue, like the red rust of iron, or as if blood had been mixed with it, and ascertained the cause to be an immense number of the *D. pulex*. The myriads necessary to produce this effect is really astonishing, and it is extremely interesting to watch their motions. On a sunshiny day, in a large pond, a streak of red, a foot broad, and ten or twelve yards in length, will suddenly appear in a particular spot, and this belt may be seen rapidly changing its position, and in a very short time wheel completely round the pond. Should the mass come near enough the edge to allow the shadow of the observer to fall upon them, or should a dark cloud suddenly obscure the sun, the whole body immediately disappears, rising to the surface again when they have reached beyond the shadow, or as soon as the cloud has passed over. (pp. 78–79)

Swarms and movements similar to those described by Baird (1850), if not as colorful, have been reported for other planktonic and littoral cladocerans (Weismann 1880). Tessier (1983) describes swarms of *Holopedium* and reviews reports of other cladocerans, including *Ceriodaphnia*, *Bosmina*, *Daphnia*, *Moina*, and *Sida*. *Daphnia* and *Polyphemus* sometimes form small swarms of a few hundred animals. Although earlier studies with *Ceriodaphnia*, *Daphnia*, and *Moina* suggested that females in swarms tended to produce ephippial eggs, Crease and Hebert (1983) failed to find any evidence for sexual pheromes produced by either sex of swarming *Daphnia*. Butorina (1986) observed *Polyphemus* swarms consisting of a few hundred animals in mushroom-shaped aggregations about 50 cm deep and 6–30 cm in diameter. Most individuals were in the top 5 cm of the swarm near the water surface. The swarms were simple aggregations, with no difference in age structure from the surrounding nonswarming *Polyphemus*. Swarms are dynamic, with animals arriving and leaving as long as the sun shines. The existence of the swarms appears to depend on proper light conditions. The adaptive significance of aggregations may be the result of animals seeking out optimal habitats. Weider (1985) found significant genetic heterogeneity in *Daphnia pulex* living in a small pond, with the different genotypes distributed nonrandomly both vertically and horizontally. Life-table data suggest temperature and food concentration affect genotype fitness, with no single genotype being most fit under all conditions.

Cladocerans show various behavioral responses to predators (e.g., Luecke and O'Brien 1983b, Havel and Dodson 1984, Dodson 1989). At least some of the planktonic prey species have a fast-swimming response, which reduces the encounter rate with predators. Small body size is probably the most effective defense against large predators, such as fish, that use vision to locate prey (O'Brien 1987). Large body size or spines are defenses against small predators, such as many invertebrates (Dodson 1974). Once encountered, prey may escape attack or capture. When caught by a small predator, such as a copepod or insect larva, cladocerans struggle to break free. *Bosmina*, once it escapes a nonvisual predator, closes up its carapace and sinks without making detectable hydrodynamic pressure waves (noise) (Kerfoot *et al.* 1980). Also, *Bosmina*, like chydorine chydorids, tucks the second antennae under the carapace for further protection when captured.

E. Foraging Relationships

Cladocerans are discriminating in their choice of food (Porter 1977). There are two major aspects of foraging relationships: what cladocerans eat and how they get it. They eat a variety of small particles, from bacteria less than a micrometer long (Porter *et al.* 1983, Lampert 1987b), through algae, up to ciliates and small rotifers 88 μm long (Porter 1973, Burns and Gilbert 1986), and copepod nauplii 100 μm long (Dodson 1975). The most important component of their diet is made up of small algae in the range of 1–25 μm (Lampert 1987b). Algae larger than about 50 μm or algae with spines or in colonies are usually rejected. Also, cyanobacteria (blue-green algae) in the preferred size range are often toxic and are not eaten if abundant (Porter and Orcutt 1980). The size of food particles is determined by more factors than the body size of the herbivore. For example, *Bosmina*, a small filter feeder, specializes on bigger particles than do the larger *Daphnia* (DeMott and Kerfoot 1982).

Leptodora, *Polyphemus*, and *Bythotrephes* are predaceous, eating protozoa, rotifers, planktonic stages of small chironomid larvae, and small crustaceans. *Leptodora* takes prey between about 0.5 and 1.5 mm and seems to prefer cladocerans (Karabin 1974). The smaller polyphemids consume smaller prey and prefer cladocerans, rotifers, and protozoans (Monakov 1972, Lehman 1987). Algae are a minor component of the diets of *Polyphemus* and *Bythotrephes*. Presumably, the metanauplii of *Leptodora* eat algae.

Herbivorous planktonic cladocerans obtain food by filtering water. The mechanism of this filtration is presently under investigation (Lampert 1987b), but it is clear that water is moved by the thoracic legs. Water enters the gap between the carapace margins, flows either over or through the filter combs of the thoracic legs, and then exits the carapace just above the postabdomen. Gerritson and Porter (1982) argued that electrostatic attraction is the major force attaching algae to the filter combs. Others are not ready to abandon the earlier concept of the filter combs as sieves whose mesh size (a function of the spacing of setae and setules) determines which algae are retained (Gophen and Geller 1984). It is not clear whether the filter combs, because of the viscous nature of water over small distances, are able to pass water. Brendelberger *et al.* (1986) claimed that the combs are good sieves, which require little energy to operate. Cheer and Koehl (1987) asserted that the sieves pass very little water in the normal feeding mode. Gerritsen *et al.* (1988) interpreted high-speed films of feeding *Daphnia* as showing a prefiltration sorting of algae in a vortex produced when water enters the carapace. Algae seem to enter the vortex and then move either toward the legs, directly toward the mandibles, or to remain in the feeding current. Observations also suggest that the right and left thoracic legs can be used independently to either manipulate or direct algae during the capture process. The apparent sorting of algal species suggests that cladocerans use chemical sensors, although DeMott (1986) found no evidence that *Daphnia* selects food by taste.

Whatever the mechanism, dense populations of filter-feeding cladocerans can remove significant fractions of the algae from a lake each day (Porter 1977). Each animal, depending on its size, can remove all of the algae in ≤ 4 ml per hour (Porter *et al.* 1982).

Chydorids typically feed by crawling along surfaces or through mud, where they scrape up or filter food (Fryer 1968, Smirnov 1971). Some chydorids employ the carapace as a suction cup. Negative pressure can be created inside the carapace if water is pumped out and if a good seal exists between the carapace and a surface. Second antennae, which are used for swimming by planktonic cladocerans, are used for crawling and jumping in some chydorids. Chydorids have a number of specialized feeding and related locomotory adaptations. For example, *Eurycercus*, the largest chydorid, climbs on upper surfaces of plants and is able to filter feed. *Alonopsis elongata* balances upright on its carapace margins on solid surfaces and scrambles along using its second antennae, scraping up food with its thoracic legs. *Peracantha truncata* keeps its second antennae inside the carapace when it moves. It can craw upside-down if the surface is rough enough to provide a hold for the claws on the first leg. Upside-down movement is aided by weak negative pressure created inside the carapace. *Alonella exigua* can hang upside-down on aquatic vegetation solely by means of negative pressure. *Graptoleberis testudinaria* has the most effective pressure chamber and exhibits snail-like behavior. *Leydigia leydigi* has neither pressure chamber nor scraping spines on the thoracic legs. It pushes itself into soft ooze using its first antennae and large spiny postabdomen. Food is sieved from the mud using setae along the margin of the carapace and on the shortened thoracic legs. In *Leydigia*, the respiratory current (which passes over the outside of the first two thoracic legs) is separated from the feeding current. While most chydorids are herbivorous, the scraping method of gathering food lends itself to carnivory. For instance, *Anchistropus* is a predator on *Hydra*, while *Pseudochydorus globosus* is a scavenger of dead animals.

Leptodora and the Onychopoda have a reputation of being "fluid feeders" (Cummins *et al.* 1969, Karabin 1974). They have been described as feeding on larger prey by grasping the prey, biting it with the mandibles, and then sucking up body fluids and perhaps small bits of tissue. Small prey, such as rotifers and protozoans are probably chewed and then swallowed. Edmondson and Litt (1987) reported large numbers of whole rotifers (*Conochilus*) and even whole cyclopoid copepods in the gut of *Leptodora*.

F. Population Regulation

Cladoceran populations, especially of planktonic species, are characterized by hundred-fold annual variations in population size. The peak population generally occurs during an algal bloom, which occurs when lakes are mixed in the spring in northern areas or when the rainy season begins in more tropical areas. Cladoceran populations are typically at low points during cold weather or when edible algae are scarce. Littoral species show similar population dynamics (Whiteside *et al.* 1978).

The abundance of planktonic cladocerans such as *Daphnia* is of great importance to the ecology of lakes, affecting the amount of algae via grazing and nutrient regeneration and providing food for young-of-the-year fish. What causes the large variations in *Daphnia* abundance? Until the work of Hall (1964), it was assumed that the only significant factor was the availability of edible algae. However, Hall showed that while the increase in *Daphnia* numbers was in response to an algal bloom, the decrease was due to predation rather than starvation. Recent studies suggest that the decline in abundance usually results from a combination of predation and food limitation (Lampert 1986b). As discussed previously (see Section II.C), *Daphnia* neonates are more susceptible to starvation than adults. Adults carry enough energy stores to carry them through short-term food shortages such as the "clear-water phase," which sometimes follows peak *Daphnia* abundances. However, fewer eggs per female during the summer, after the clear-water phase, indicates food limitation. Mortality due to predation is generally size-specific, depending on the size of the prey and the size and hunting behavior of the predators (Zaret 1980). Many studies (see Section II.B) have shown that predators can cause local apparent extinction of cladoceran populations (no adults are seen, although resting eggs remain in sediments). Cladocerans that coexist with predators tend to have specialized life histories, morphologies, and behaviors that provide defenses against predation (Dodson 1989). Even so, mortality from predation can be an important factor in determining the abundance of zooplankton.

Hall (1964) made a great contribution to aquatic ecology by showing that predation as well as food limitation restricts zooplankton populations. He did this by separating the population growth rate coefficient r into a component of birth or recruitment, b, and a component of mortality, d. He used a life-table technique that requires culturing individual animals in the lab to determine both rates of mortality in the absence of predation, and fecundity (number of young produced) at various food levels. This important process of separating mortality and recruitment can be achieved much more easily using Edmondson's (1960) egg-ratio technique. This technique requires that one know the number of eggs per female and the developmental time of an egg to estimate b. Then, the value of r depends on changes in population size and d is the difference between b and r. This technique provides good estimates of b when compared to life-table estimates (Lynch 1982), even when the assumptions of the egg-ratio technique (concerning age structure) are not met. The original Edmondson egg-ratio technique has been modified in various ways (Rigler and Downing 1984, Threlkeld 1987b). A comparison of these modifications with predictions based on life-table studies (Gabriel *et al.* 1987) suggests that different modifications are most accurate depending on the population age structure and the frequency with which the population is sampled. In general, a declining population with a high value of b indicates high predation, and a stable or declining population with a low value of d indicates food limitation (Threlkeld 1987b). The accuracy of r is influenced by sampling and counting techniques as well (Taylor 1988).

Studies of effects of food limitation and predation

suggests that both regulate populations. In a lake dominated by *Holopedium* and *Daphnia*, the *Holopedium* were mainly limited by the food supply, while *Daphnia* showed significant mortality from fish during the summer (Tessier 1986).

G. Functional Role in the Ecosystem

The concept of a functional role in an ecosystem implies that a species or a group of similar species has an effect on the number and kind of organisms living in a community and also has an effect on the way in which energy is passed from plants to herbivores to predators. Food chain dynamics may be the central hypothesis in ecology because of the perspective it provides and the potential it has for explaining the diversity of natural systems (Fretwell 1987). The power of this hypothesis results from its inclusion of both direct effects, such as predation of fish on zooplankton, and indirect effects, such as fish predation on the abundance of algae eaten by zooplankton. While members of each trophic level tend to be most strongly influenced by adjacent trophic levels (or nutrients for plants) (McQueen *et al.* 1986), taxa of nonadjacent trophic levels also have an effect (Carpenter *et al.* 1985). Cladocerans are major primary and secondary consumers in lake ecosystems (Downing 1984a, Kitchell and Crowder 1986, Edmondson 1987).

Hrbáček (1958) first pointed out the importance of certain size-selective fish in excluding cladocerans from communities. Concepts of size-selective predation and selective grazing by cladocerans have been combined to produce models of complex and dynamic food webs, in which the number and abundance of species in a community is the result of the interaction of competition, selective predation, and nutrients. Figure 20.6A shows a food-web model of four trophic levels. Cladocerans would occupy three boxes in this model: small and large crustacean zooplankton and invertebrate planktivore. Figure 20.6B is a more graphic representation of the flow of nutrients and biomass in a three-level trophic structure (note that *Mesocyclops* eats immature *Daphnia* and *Bosmina* as well as algae). The less complex food-chain model (Figure 20.6C) includes three consumer trophic levels and one decomposer level (the benthivorous fish).

Inorganic nutrients, especially phosphate, influence the kind and abundance of algae and hence the "water quality" of lake water. Over a wide range of concentrations, phosphate concentration is a particularly good predictor of water quality (Schindler 1977). However, for any given level of phosphate, there is considerable variation in water quality. Several recent studies suggest that food-chain dynamics

also influence water quality (Shapiro *et al.* 1982, Edmondson and Litt 1982, Carpenter *et al.* 1985, McQueen 1986, Carpenter 1988). Shapiro *et al.* (1982) suggested that water quality could be managed by food-chain manipulation, or "biomanipulation" (Table 20.2C). They reasoned that, at a given phosphate level, the kind and abundance of algae is influenced by the abundance and quality of herbivorus zooplankton, with large species of *Daphnia* having the greatest effect. In turn, cladoceran population dynamics are influenced by predation from populations of small fish, which are in turn controlled by predation from large predaceous fish. The authors recommended management of water quality through management of large "game" fish. In general, the goal of water-quality management is to reduce standing crops of algae, which can be accomplished by increasing the standing stocks of large predaceous fish. Biomanipulation is not seen as necessary in oligotrophic (nutrient- and algae-poor) lakes, and it may not work in eutrophic (nutrient- and algae-rich) lakes, but will probably be most effective in intermediate (mesotrophic) lakes (Carpenter *et al.* 1985). However, the planktonic food chain is complex and interconnected, and the effects of one trophic level are not straightforward. For example, Lynch (1979) found that trophic structure depended on both the intensity and the kind of predators present in a lake.

Large species of *Daphnia* appear to be the most efficient in cleaning up lakes made turbid by blue-green algae. For example, large *Daphnia pulicaria* was scarce in Lake Washington until 1976, when it became common (Edmondson and Litt 1982). At the same time, the mean summer transparency of the lake doubled, due to a decline in cyanobacteria. Improvement in water quality occurred even though phosphate concentration remained constant, and despite the continual presence of other herbivorous zooplankton. The size and high feeding rate of the large *Daphnia* make them especially efficient at cleaning up lakes (Edmondson 1987). Smaller common cladocerans, such as *Bosmina* and *Ceriodaphnia,* while they are less susceptible to fish predation, are also less efficient at cleaning up algae.

Meiobenthic cladocerans (chydorids, macrothricids, and most sidids) comprise a major part of the diverse littoral community of lakes, which also includes many other crustaceans, insects, and other invertebrates. In addition to this high diversity of animals, there are the various macrophytes, sessile and substratum-associated algae, and bacteria. Of all the animal taxa, cladocerans typically are most abundant, at times reaching densities greater than one million per m^2 of bottom, even under the ice in early winter (Smirnov 1971, Whiteside *et al.* 1978).

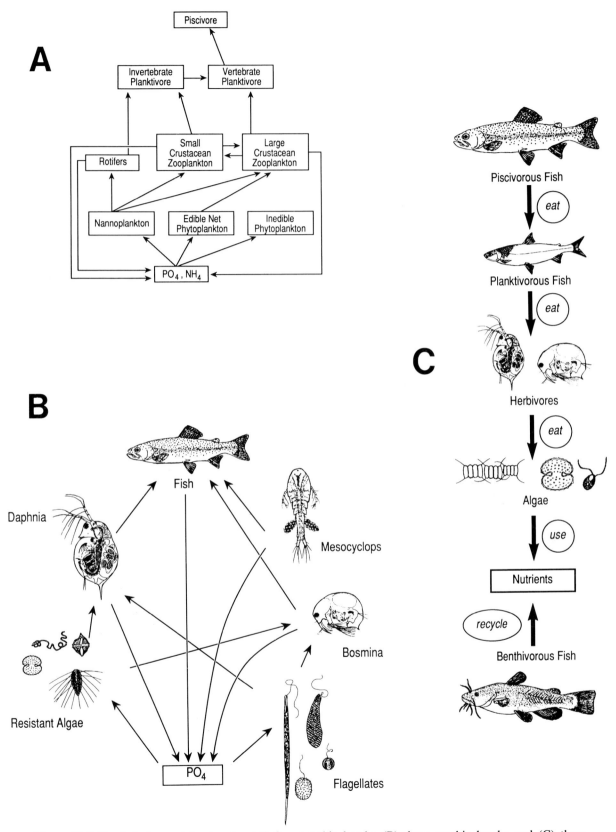

Figure 20.6 Food-chain models showing (A) four trophic levels; (B) three trophic levels; and (C) three consumer trophic levels and one decomposer level.

As many as 30 or more species may be present. The composition of this littoral community and its importance relative to the offshore planktonic and benthic associations vary with the geographic location of the lake and its size, depth, and transparency.

Our understanding of requirements and interrelationships of meiobenthic cladocerans is still very meager. Some species are associated only with sediments and others with macrophytes. Some taxa are restricted to acidic, soft-water habitats and others to quite strongly alkaline habitats. Fryer (1968) showed that the chydorid cladocerans vary greatly in what they eat. They include scavengers, ectoparasites, and substrate scrapers. Most of the specialization of the chydorids seems to be in obtaining food from slightly different microhabitats. The structure of the five pairs of trunk limbs is remarkably uniform among all these species, suggesting little variation in the basic feeding mechanism.

Meiobenthic cladocerans occur virtually wherever there is water: from water bodies of "normal" size down to marshes, swamps, pools, puddles, ditches, seasonal water bodies, groundwater, running water of all sizes and nearly all velocities, moist Sphagnum moss, and even in overgrowths and cushions of leafy liverworts in rain and cloud forests located several meters above the forest floor. Most species are sensitive to salinity, although a few occur at concentrations of up to 10 ppt or even higher.

Thus, the meiobenthic cladocerans are typically the most abundant metazoans in substrata-associated habitats, they have a high species diversity, and they occur in all kinds of aquatic and moist habitats. Yet they have long been neglected by limnologists in favor of planktonic cladocerans, even though in average-sized lakes the littoral community is more important than the open-water community in the annual production of a lake (Wetzel 1983).

Meiobenthic cladocerans constitute one of the prime converters of decaying organic material (detritus) with its associated microorganisms into particles, i.e., small cladocerans big enough to be handled by small fishes and predaceous invertebrates. Examination of the gut contents of minnows, sticklebacks, and small fingerlings of other fishes shows a heavy uptake of these cladocerans. The young of percid and centrarchid fishes moving through the macrophyte beds in the littoral zone find an abundance of these animals to feed upon. Small catfishes in streams near Bloomington, Indiana, seem to feed almost exclusively on chydorids. One of the possible causes of the midsummer population crash of chydorids in lakes of northern Minnesota is predation by early fingerling perch (Whiteside *et al.* 1978).

H. Paleolimnology

In paleolimnology, we seek to interpret how a lake functioned in past time (Frey 1969, Klekowski 1978, Löffler 1987, Meriläinen *et al.* 1983). The sediments accumulate chronologically. Therefore, from the changing representation of kinds of sediments and their sorting coefficients, organic and inorganic chemicals, and morphological remains of plants and animals, we attempt to interpret past conditions in the water bodies and their watersheds. Sometimes the record is exceptional. In some lakes, especially meromictic ones, sediments are deposited in distinct annual layers (varves), allowing year-by-year analysis. Remains of animals that are preserved are mostly either siliceous (sponges), calcareous (ostracods), or chitinous (e.g., cladoceran and midge exuviae). All groups of freshwater animals leave at least some recognizable fragments in sediments (Frey 1964), but cladocerans are usually the most abundant, represented by generally between a few thousand and several hundred thousand fragments per cubic centimeter of fresh sediment. Once the species pool of a region is known, virtually all fragments can be identified positively at the species level. Remains of chydorids and bosminids are preserved best and almost quantitatively. Remains of meiobenthic species become integrated over time (at least a year) and habitat before being incorporated into deep-water sediments. The quantity of these remains is greater toward shore than in the deepest part of the lake, but the percentage of composition by species is the same all over the bottom (Mueller 1964).

Close-interval stratigraphies are constructed of cladoceran remains using the same techniques employed for pollen or diatoms. The result is an estimate of changes in relative abundance of the various taxa over time (Frey 1986b). Interpretation of the changes depends on knowledge of the ecology of these forms. A few species, such as members of the *Chydorus sphaericus* complex, seem to respond positively to eutrophication (nutrient enrichment), sometimes comprising more than 90% of all remains. Other species appear associated with bog conditions or changes resulting from acidification. *Daphnia* and *Bosmina* remains, partly because of cyclomorphosis, record past predation regimes (Kerfoot 1981, Kitchell and Carpenter 1987).

Cotten (1985) studied cladoceran communities in 46 lakes in eastern Finland, using their remains in surficial sediments. She ordinated relative abundances of 70 taxa, using measured chemical and physical characters of the water bodies and their watersheds. She recognized five kinds or groups of lakes, which were almost completely concurrent with the five kinds revealed previously by diatom analysis. Thus, the cladocera and diatoms are equally good as indicators of chronologic changes that take place in a body of water in response to natural processes and human activities.

A study of the cladocera present in a lake in Denmark during the last interglacial (roughly 100,000 years ago) revealed 25 species of chydorids, all of which occur in Denmark today (Frey 1962). No real differences in morphology were apparent. Moreover, the relative order of abundance of these 25 species in the interglacial was the same as in Denmark today, indicating that the ecology of these forms had not changed either. Furthermore, evidence and suggestions from the meager remains of cladocera known from the present to the Tertiary, Cretaceous, and possibly even the Permian indicate that major evolution in the cladocera had occurred by the late Paleozoic or early Mesozoic. Consequently, species present today have been stable for millions of years, and as a result, we can employ our ecological knowledge of extant species to interpret community relationships within lakes of the distant past.

The presence of a vast record of cladocera in lake sediments is valuable for more than just paleolimnology. Sometimes the occurrence of a particular cladoceran species in a lake is suspected to be the result of recent transport (e.g., *Bythotrephes* in the Great Lakes). This can be checked by sedimentary analysis if desired.

The development of a cladoceran assemblage in a lake since late-glacial time, or even earlier in nonglacial lakes, is easy to follow (Goulden 1966). Species richness (number) increases over time and species diversity (evenness) tends toward a maximum as conditions stabilize (Goulden 1969). The distribution of species by their relative abundance in a stable community closely follows the MacArthur broken-stick model (Goulden 1969). Any major deviation in fit to the expected is considered to have arisen from changes in climate, agriculture, volcanism, or other large-scale controlling agents (Frey 1974, 1976). In lakes with varved sediments, one can follow changes in relative abundance of the total cladoceran community on a year-to-year basis and then relate these changes to potential causative agents (Boucherle and Züllig 1983).

III. CURRENT AND FUTURE RESEARCH PROBLEMS

Despite past emphasis on classification of cladoceran species, there is still a great deal of taxonomy that needs to be done. This is not just a matter of discovering and naming new species but, more fundamentally, of defining the species concept that makes the most sense for cladocerans. These animals are usually at least facultatively asexual, and in many groups, what initially seem to be two or more good morphologic species are made up of a combination of occasionally sexual and totally asexual populations. Descriptions of species before about 1950 are mostly very inadequate and, hence, cannot be used to compare populations on two different continents. To determine that any North American taxon is the same as or different from a taxon of the same name described from elsewhere is difficult. One first needs to compare both taxa closely, using large populations containing all instars and reproductive stages. Thus far, about 16 pairs of taxa with the same name from the New and Old Worlds, have been compared (partial summary in Frey 1986a), with the finding that almost all of them are separate species. Sometimes the same specific name has been applied to two or more related but different species in the same geographical region. Examples are the *Chydorus sphaericus* complex, *Eurycercus*, *Ephemeroporus*, the *Chydorus reticulatus* group, the *Chydorus faviformis* group, and *Pleuroxus laevis*. Consequently, for critical studies in ecology, one must be careful not to transfer the ecological requirements and indications of a species across an ocean or some other distributional barrier on the mere assumption that all taxa presently bearing the same name are actually the same species. This problem will be sorted out with a mixture of traditional morphological techniques, new molecular and genetic techniques, and some new ideas!

The study of food-web dynamics, using sets of whole lakes, is just getting under way, and shows promise for interesting research into population and community ecology. Applied food-web dynamics, or biomanipulation, may become an inexpensive and safe method for improving water quality.

The availability of great computer power means that studies can now be done that link cladoceran diversity in the sediments with knowledge of cladoceran biology. This will allow high-resolution reconstruction of past aquatic climates and aquatic communities.

Limnologists are just now finding that many zooplankton respond to chemical signals (Havel 1987). The chemicals are produced by both predators and food, and perhaps even competitors. The signals induce changes in development, morphology, reproduction, and behavior.

The field of behavior presents a wide-open field for future research in cladoceran biology. Almost nothing is known about the swimming, mating, or predator-escape behavior of these animals, except on the population level. An understanding of individual behavior provides an important perspective on the ecology and evolution of zooplankton (Sih 1987, Ohman 1988). Techniques have recently been developed that allow quantitative observations of these behaviors on individual animals (Strickler 1985, Ramcharan and Sprules 1989).

IV. COLLECTING AND REARING TECHNIQUES

A. Sampling Techniques

Cladocerans can be collected using a fine-mesh net pulled through the water, or a water sample can be poured through a sieve. A large number of specialized nets have been developed for specific applications (deBernardi 1984). For most purposes, the mesh should have a pore size of about 90–150 μm. Nets are most useful for towing behind a boat in open water. Cladocera can be collected from weedy habitats with a plankton net attached to a handle and with a wide-mesh (5 mm or so) cover over the opening of the net to reduce clogging by vegetation. Alternatively, one can use a bucket to collect water, which is poured through a series of nets or sieves (Downing 1984b). In any case, mud should be kept out of the sample. Filters can be made by cutting the bottom off of a plastic bottle. The center of the cap is cut out, and a piece of mesh glued over the hole. When the cap is screwed back onto the bottle, water can be poured into the inverted bottle, and cladocerans will be caught by the mesh. The captured cladocerans can then be washed into a jar of water previously filtered through the mesh.

Estimations of population dynamics or productivity require quantitative samples, which require special gear (deBernardi 1984, Downing 1984b), a careful experimental design, and proper statistical analyses of samples (McCauley 1984, Prepas 1984).

If the sample is to be killed, there are at least two ways to do it. For general studies, wash the animals off of a small sieve using a wash bottle filled with 95% ethanol or drop the sample into a large volume of 95% ethanol. This will kill the animals rapidly enough to prevent significant distortion or loss of eggs carried in the brood chamber and will preserve the sample indefinitely as long as the sample vials are tightly capped and 70% alcohol is added occasionally. Animals intended as museum specimens for taxonomic study should be killed and preserved by adding enough commercial formalin (containing 40 g sucrose per liter) to make a 5% formalin solution. For long-term preservation, a small amount of glycerol should be added to prevent an unintended complete desiccation.

B. Culture Methods

Animals will live in filtered pond water for at least a few days, especially if there are fewer than 50 animals per liter. Techniques have been developed for maintaining cladoceran cultures for years (e.g., Peters 1987b). These culture methods can be divided into two processes: (1) maintenance of food for the cladocerans and (2) care of the cultures themselves.

1. Cladoceran Food

The simplest to maintain and most reliable bulk food culture for cladocerans is a mixture of green algae and bacteria grown in a large aquarium.

Keep the algae in a glass aquarium as large as possible; a 40-gallon aquarium is sufficient. Place the aquarium in a south-facing window and choose a place that can be kept at about 18°–20°C. Fill the aquarium with lake or pond water that has been filtered through a fine-mesh (ca. 90 μm mesh) net or use one-day aged tap water with a little additional filtered lake water as a seed source for the phytoplankton. Add a dozen or so smallfish (goldfish are best) to this aerated tank and feed the fish regularly. Aerate the tank. Keep the glass walls scraped clean and occasionally siphon out bottom detritus. Regardless of the initial relative abundance of species, the culture will probably eventually stabilize with a mixture of small green algae (like *Chlorella*) and *Scenedesmus*, along with a few less common genera such as *Ankistrodesmus*. Keep the algae in a healthy state by removing at least 25% of the green algae culture every week or so with a large-diameter siphon.

2. Culture Care

First, bring the animals into the laboratory and let them equilibrate to the lab temperature. Many cladocerans tend to be trapped by the surface film. This happens more often for the smaller lake species, especially if they are initially in cold water with saturated oxygen (as the water warms, bubbles form on the carapace). If mortality is high due to surface entrapment, keep the animals in a narrow-mouth bottle filled to the top and plugged with cotton pushed below the water surface. After a couple of days it will be obvious which cultures are going to be vigorous. Label cultures (with tape on the jar) as to source and date of capture.

The cultures can be fed an algal suspension. If large objects are not visible through the culture jar, the algae are probably too concentrated and need to be diluted. The mixture should be green but not opaque. As the water in each jar clears, add algal suspension. Do not let the water become clear; even an overnight bout of clear water will cause some clones to starve. It is also possible to overfeed clones, but starvation seems a much more common occurrence. Continue adding algal suspension until the jar is full.

There are two ways to feed cultures, depending on the number of animals per jar (optimally 20–200, depending on their size). First, if there are too many animals in the jar, pour out half or more of the culture and refill with algal suspension. Beware of cultures in which the animals concentrate in the top few centimeters of water—it is possible to pour out the entire culture! Second, if too few cladocerans are present in the jar, siphon off about half of the water and add algal suspension. Use fine mesh (about 90 μm) on the end of the siphon in the jar or the animals will be lost. Wash and dry the siphon or dip it into hot water before using it on the next culture (to prevent cross-contamination). Once established, clones can be kept in a cold room or refrigerator at about 4°C. At this temperature, the animals need feeding only about once every two weeks. Beware: some species die when moved from 20° to 4°C, so avoid chilling all your cultures at once. When a thick layer of detritus accumulates on the bottom of the jar, decant off the culture and discard the detritus layer.

V. TAXONOMIC KEYS FOR CLADOCERAN GENERA

A. Available Keys and Specimen Preparation

Crustaceans are cladocerans if they have four to six pairs of legs, lack any paired eyes, swim with their second pair of antennae, and have at least the head not covered by a carapace.

There is currently a crisis in cladoceran taxonomy brought about by incomplete or no longer accurate keys, especially to the species level. Most details in the following keys to families and genera of cladocera are similar to those of Balcer *et al.* (1984) and Berner (1982). The keys to the families Macrothricidae and Chydoridae were written by Frey. To identify animals to the species level, the best choices are the two references just mentioned or the even more dated Brooks (1959) and Pennak (1978), both of which are based largely on Birge (1918). Also consult the more specific and recent references given for some genera in the following keys.

These keys are designed to allow identification of adult females. Dissection is seldom necessary, but it is a good idea to have several animals available for identification, because these small, delicate animals

are easily crushed out of recognizable shape. Animals shorter than about one millimeter should be mounted on a slide for high magnification viewing; the larger species can often be identified without being mounted, using a dissecting microscope at about 50× magnification or less.

While live animals trapped in a thin film of water can sometimes be identified, it will generally be necessary to kill them with alcohol and mount them on slides. The best mounting medium is one that is more or less permanent and is soluble in either water or alcohol solutions, so specimens can be added directly to the medium without tedious dehydration. If you have access to chloral hydrate (currently a controlled substance), a good medium is Hoyer's:

> Dissolve 30 gm of ground Gum Arabic in 150 ml of distilled water. Add 200 gm of chloral hydrate and 20 gm of glycerine.

> Divide the Hoyer's into two batches. Let one sit in a warm place until it is as thick as honey, and keep the second batch in a closed jar so it remains thin. You can always add more water to thin either batch. Keep at least some of the thick and thin Hoyer's in screw-top eye-dropper bottles for easy use.

To make a slide, use the eye dropper to put a small streak of dilute Hoyer's (about a quarter of a drop) toward one end of the slide. Arrange the specimens in this streak. It is wise to place four or so of what you think are the same kind of animal on a slide, to allow for comparison and viewing of difficult characters. Delicate specimens can be protected from compression by putting a few pieces of broken coverslips around the specimens. Let the Hoyer's streak dry on a slide warmer or in a warm place (below 100°C). When the streak is dry, add a drop or two of the thick Hoyer's and gently lower a cover glass over the Hoyer's. Put the slide back into a warm place to dry again. Label the slide as to the date and location of the collection! It is important to use thickened Hoyer's in the last step; otherwise, the Hoyer's will shrink as it dries and produce large bubbles in the final preparation. On the other hand, if you use thickened Hoyer's, small bubbles produced when you lower the coverslip will disappear as the medium dries. These slides are semipermanent. If the climate is humid, the Hoyer's will thin and cover glasses will slip off the slide. If the climate is arid, the Hoyer's will desiccate enough so that the cover glasses pop off the slide. These problems may be avoided by storing the slides horizontally in a climate that has moderate humidity, or you can ring the slide with nail polish.

An alternate method, especially useful for bosminids, chydorids, and macrothricids is to use polyvinyl lactophenol stained with lignin pink or glycerine jelly. Best of all for museum specimens is possibly Canada Balsam, although this mounting medium requires that the specimens be dehydrated through an alcohol series. For temporary mounting, glycerol is good (but messy). Specimens should be equilibrated to the glycerol through a series of dilutions in order to avoid distortion.

In the following taxonomic treatment, a key to families precedes the key to genera.

B. Taxonomic Key to Families of Freshwater Cladocerans

1a.	Thorax, abdomen, and thoracic legs covered by the carapace. The brood chamber is a space between body and carapace (Fig. 20.1)	2
1b.	Thorax, abdomen, and especially the thoracic legs not covered by the carapace; the carapace is present as a brood chamber only (Figs. 20.60–20.64)	8
2a(1a).	Swimming appendages (second antennae) unbranched in female (Fig. 20.7) .... family Holopediidae	
2b.	Swimming appendages branched (e.g., Fig. 20.1)	3
3a(2b).	Swimming appendages each with ten or fewer swimming setae (adding together the setae on both branches of the antenna, and not counting spines, which are shorter than the branches and unsegmented; e.g., Fig. 20.1)	4
3b.	Swimming appendages each with at least 15 setae on both branches of the antenna (Fig. 20.8) .... family Sididae	
4a(3a).	One branch of the second antenna with three segments, the other with four (the basal segment may be short); base of first antenna not covered by a flange on the side of the head e.g., (Fig. 20.1)	5
4b.	Both branches of the second antenna with three segments; first antennae at least partly covered by flanges on the side of the head (Fig. 20.2) .... family Chydoridae	
5a(4a).	First antennae less than three times as long as wide and attached to head near the carapace margin (e.g., Fig. 20.1) .... family Daphniidae	
5b.	First antennae more than five times as long as wide and attached to the head about halfway between the carapace margin and tip of the head (e.g., Figs. 20.44 and 20.58)	6
6a(5b).	First antennae in females rigidly fixed to head, tapered to a point and curved, resembling elephant tusks; first antennae with several aesthetascs about halfway between the tip and base, no aesthetascs at the tip (e.g., Fig. 20.42) .... family Bosminidae	
6b.	First antennae blunt and flexible; aesthetascs at the tip of the first antennae (e.g., Fig. 20.44)	7
7a(6b).	First antennae attached ahead of or just below the compound eye, or if attached posterior to compound eye, then the postabdomen without lateral feathered setae (lateral setae are perhaps present in *Ilyocryptus,* but they are just above the margin and are unfeathered); no distal bident tooth on postabdominal margin near base of postabdominal claw (e.g., Figs. 20.44 and 20.45) .... family Macrothricidae	
7b.	First antennae attached posterior to compound eye; postabdomen with a row of lateral feathered setae, and (in all but one species) a distal bident tooth (e.g., Fig. 20.58) .... family Moinidae	
8a(1b).	Body more than four times as long as high; second antenna with about 50 swimming setae, none longer than the branches of the antenna (Fig. 20.60) .... family Leptodoridae	
8b.	Body less than twice as long as high; second antennae with about 25 swimming setae, some as long as the branches (Figs. 20.61–20.64) .... families Polyphemidae, Cercopagidae, and Podonidae	

C. Taxonomic Key to Genera of the Family Holopediidae

John Havel (personal commmunication) feels that, based on morphologic and allozyme analyses, there is only one polymorphic species in North America, *Holopedium gibberum* Zaddach, 1855 (Fig. 20.7). This genus needs more study. Habitat: planktonic,

and abundant, especially in northern lakes and bogs. This species is remarkable in that each individual is surrounded (out to about an additional body diameter) by a mass of clear jelly attached to the carapace (easily removed by rough handling). The jelly probably defends *Holopedium* against an array of invertebrate predators. Adult females are about 0.6–2.0 mm long.

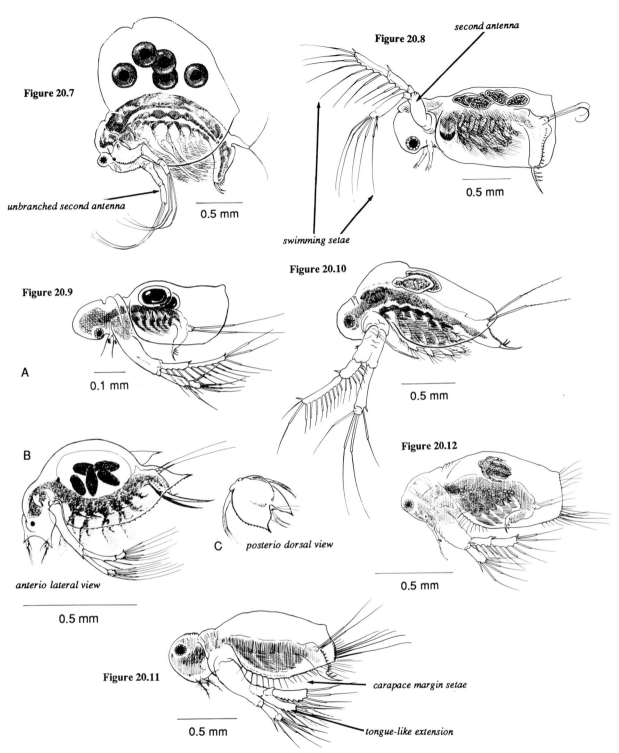

Figure 20.7 *Holopedium gibberum* Zaddach, 1855. Why Not Bog, Vilas Co., Wisconsin. 22 September 1973. (KE)   **Figure 20.8** *Sida crystallina* (O. F. Mueller, 1776). Lake Mendota, Dane Co., Wisconsin. 11 September 1987. (KE)   **Figure 20.9** (A) *Diaphanosoma birgei* (Kořínek, 1981. Lake Wingra, Dane Co., Wisconsin. 5 October 1978. (KE) (B) *Penilia avirostris* Dana, 1849. Anteriolateral view. Atlantic Ocean near Florida (29°55.8′ N and 81°07.4°W). 27 October 1987. (KE); (C) *P. avirostris.* Posteriolateral view. (KE)   ***Figure 20.10*** *Pseudosida bidentata* Herrick. 1884. Bull Frog Pond, S.R.P. area, South Carolina. 27 May 1985. (Coll. DGF #7483). (KE)   **Figure 20.11** *Latona setifera* (O. F. Mueller, 1776). Lake Mendota, Dane Co., Wisconsin. 30 June 1987. (KE)   ***Figure 20.12*** *Latonopsis occidentalis* Birge, 1891. Dick's Pond, S.R.P. area, South Carolina. 28 May 1985. (Coll. DGF #7496). (KE)

D. Taxonomic Key to Genera of the Family Sididae

1a.	Setae along the margin of the carapace shorter than 1/10 the height of the carapace	2
1b.	Setae along the margin of the carapace longer than 1/4 the maximum dorsal-ventral height of the carapace	4
2a(1a).	First antenna, including terminal setae, about half as long as the head height (Fig. 20.8)	3
2b.	First antenna, including terminal setae, about as long as the head height (Fig. 20.9A) .... *Diaphanosoma*	
3a(2a).	Posterior margin of carapace not flared laterally; posterior-ventral corners of carapace rounded; freshwater (Fig. 20.8) .... *Sida*	
3b.	Posterior margin of carapace flared out laterally, with a point on either posterior-lateral corner; marine (Figs. 20.9B,C) .... *Penilia*	
4a(1b).	Carapace margin with long setae only near the head (Fig. 20.10) .... *Pseudosida*	
4b.	Carapace with long setae along the whole margin	5
5a(4b).	A pointed tongue-like extension from the head, near the carapace (Fig. 20.11) .... *Latona*	
5b.	Without a pointed tongue-like extension from the head (Fig. 20.12) .... *Latonopsis*	

[Kořínek (1981) has the latest word on *Diaphanosoma* species in North America. *Diaphanosoma* and *Penilia* are planktonic, while the other genera are associated with aquatic vegetation. *Diaphanosoma* and *Penilia* adults are about 1.0 mm long, whereas adult females of the other genera tend to be larger (from 2–4 mm in length).]

E. Taxonomic Key to Genera of the Family Chydoridae (North of Mexico)

1a.	Postabdomen strongly flattened, broad, with a single row of 80–150 marginal denticles, resulting in a saw-like appearance; anus distal; subfamily Eurycercinae (Fig. 20.13) .... *Eurycercus*	

[Genus is Holarctic except for isolated populations in southern Brazil-northern Argentina and in South Africa. At least five species in North America, distributed among three subgenera: one species in the subgenus *Eurycercus* only within 300 km of the Gulf of Mexico and the Atlantic Ocean as far north as the Pinelands of New Jersey (maximum length ca. 2 mm); two very common species in subgenus *Bullatifrons,* one in the northern tier of states and northward, the other in the south (length to 2.4 mm); and two species in the subgenus *Teretifrons* in Newfoundland and the Arctic (length to 6 mm). See Frey (1971, 1975, 1978, 1982d) and Hann (1982).]

1b.	Postabdomen thicker, not so broad, and usually with two rows of marginal denticles, mostly fewer than 20 on each side; anus proximal	2
2a(1b).	Proximal tip of mandible articulates where headshield and shell touch each other (Fig. 20.1C); usually two or three major headpores on midline, which usually are connected by a double sclerotized ridge, resembling a channel (Fig. 20.1C); occasionally only one median pore (*Monospilus, Euryalona*) or none (*Notoalona*); minor headpores lateral to these; free posterior margin of shell not greatly less than maximum height of animal; typically one, occasionally zero, basal spine on postabdominal claw (Fig. 20.1C); postabdomen nearly always with lateral fascicles (Fig. 20.1C), sometimes without (*Monospilus*). .... subfamily Aloninae	3

[In North America contains the genera *Acroperus, Alona, Alonopsis, Bryospilus, Camptocercus, Euryalona, Graptoleberis, Kurzia, Leydigia, Monospilus, Notoalona, Oxyurella,* and *Rhynchotalona.*]

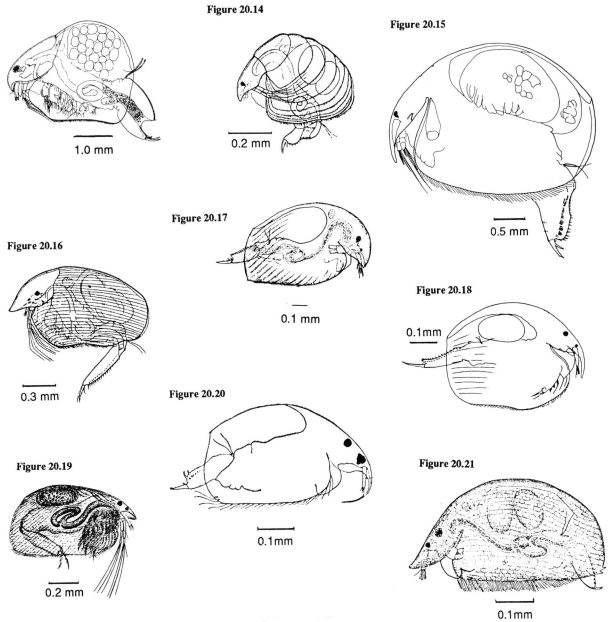

Figure 20.13 *Eurycercus (Eurycercus) lamellatus* (O. F. Mueller, 1776). Uppsala, Sweden. From Lilljeborg (1901). *Figure 20.14* *Monospilus dispar* Sars, 1862. Yddingesjon, Sweden. From Lilljeborg (1901). *Figure 20.15* *Bryospilus öepens* Frey, 1980. El Yungue, Puerto Rico. From Frey (1980b). *Figure 20.16* *Camptocercus rectirostris* Schoedler, 1862. Kristianstad, Sweden. From Lilljeborg (1901). *Figure 20.17* *Acroperus* cf. *harpae* (Baird, 1834). Eastern North America. From Birge (1918). *Figure 20.18* *Kurzia longirostris* (Daday, 1898). Tammanna wewa, Sri Lanka. From Rajapaksa (1986). *Figure 20.19* *Alonopsis elongata* Sars, 1862. England. From Norman and Brady (1867). *Figure 20.20* *Rhynchotalona falcata* (Sars, 1861). Northern Michigan. From Birge (1893). *Figure 20.21* *Graptoleberis testudinaria* (Fisher, 1851). Eastern North America. From Birge (1918).

2b.   Proximal tip of mandible articulates in a special pocket on the headshield, some
distance in a dorsal direction from the point of contact between the headshield and
shell (Fig. 20.1D); two major headpores on midline (Fig. 20.1B,D), completely
separated from one another and with no sclerotized ridge connecting them,
occasionally only one pore (*Dadaya*), or only one pore in first instar and none in later
instars (*Ephemeroporus*); minor pores most commonly near midline between major
pores; free posterior margin of shell usually considerably less than maximum height
of animal; typically two basal spines on postabdominal claw (Fig. 20.1D); articulated
lateral fascicles on postabdomen lacking, although *Pseudochydorus* has a row of
fascicle-like groups of setae . . . . subfamily Chydorinae   15
[In North America contains the genera *Alonella, Anchistropus, Chydorus, Dadaya,
Disparalona, Dunhevedia, Ephemeroporus, Pleuroxus,* and *Pseudochydorus.*]

3a(2a).   Only an ocellus present; no compound eye   4
3b.   Compound eye and ocellus both present   5

4a(3a).   Shells from previous instars firmly nested, so that instar number of specimen can be
counted; headshield transversely truncate posteriorly; single median headpore (Fig.
20.14) . . . . *Monospilus*
[Supposedly just one species in world, mainly Holarctic and Ethiopian in
distribution, with several records from New Zealand as well. Length to 0.5 mm.]
4b.   Shells lost on molting, so that no nesting of shells occurs; two separated median
pores without sclerotized connecting ridge; minor pores located far laterally; very
short antennae (Fig. 20.15) . . . . *Bryospilus*
[Two species in world, known from New Zealand, Venezuela, and Puerto Rico,
and hence may possibly occur on continent north of Mexico. Length to 0.35 mm.
Live among wet, leafy hepatics of rain forests and cloud forests (Frey 1980b,
1980d).]

5a(3b).   Body strongly compressed from side to side, at times almost wafer thin;
postabdominal claw with a secondary spine midway along concave margin, and
usually with a comb of fine setules decreasing in length proximally between it and the
basal spine (comb also occurs in *Euryalona*)   6
5b.   Body not so strongly compressed; postabdominal claw having only a basal spine of
variable length (except in *Euryalona*); sometimes none at all (one species of
Leydigia)   9

6a(5a).   Body strongly keeled, and head also usually strongly keeled   7
6b.   Body weakly keeled, and head without a keel   8

7a(6a).   Postabdomen narrow, elongate, tapered distally, with many (generally about 15 or
more) marginal denticles (Figs. 20.1C and 20.16) . . . . *Camptocercus*
[Nine species recognized in world in 1971, with lengths from 0.7–1.26 mm. Except
for *C. oklahomensis,* the species in North America are completely unknown and
undescribed, although certainly more than five species are present. *C. rectirostris,
C. lilljeborgi,* and *C. macrurus,* which were described from Europe, do not occur in
North America at all, even though reported by Brooks (1959).]
7b.   Postabdomen shorter and broader, parallel-sided, without any distinct marginal
denticles (Fig. 20.17) . . . . *Acroperus*
[Seven species in world in 1971, with maximal lengths from 0.59–0.85 mm. Like
Camptocercus, the species in *Acroperus* are not yet adequately described. There is
a marked conformity in body form and shape and in other morphologic characters,
making it very difficult to sort out the species satisfactorily.]

8a(6b).   Postabdomen rather slender, elongate, somewhat tapered distally; body high; shell
widely open behind (Fig. 20.18) . . . . *Kurzia*
[Three species in world and two in North America, *K. latissima* being widespread,
and *K. longirostris* occurring only in the southernmost Gulf states. Length to
0.6 mm.]
8b.   Postabdomen broader, elongate, parallel-sided; shell with parallel striae sloping
downward posteriorly, and with short longitudinal scratch marks in between (Fig.
20.19) . . . . *Alonopsis*
[Two species in world, the one in North America very similar to the one in Europe
(see Kubersky 1977). Occurs from northern New England into the adjacent
Canadian provinces and then into those farther East. Smirnov (1971) transferred
the European species to the genus *Acroperus,* which is not adequately justified.
Maximum length: American 0.91 mm, European 0.85 mm.]

9a(5b).   Rostrum long, attenuate, recurved, greatly exceeding antennules in length (Fig. 20.20) .... *Rhynchotalona*
[Only one species listed in world, although there is a second species from North America not yet described. Postabdomen with 2–4 stout marginal denticles distally and with an almost continuous row of long, hair-like setae laterally. Length to 0.5 mm. A mud dweller.]

9b.   Rostrum rounded, barely or only slightly exceeding antennules in length 10

10a(9b).   Seen from above, rostrum very broad, semicircular, wider than body; shell and head strongly reticulated (Fig. 20.21) .... *Graptoleberis*
[Possibly only one species in world, although the taxonomy is not yet clear. Reported maximum length 0.7 mm. Ventral margin of shell provided with a dense fringe of setae, and generally with two large, sharp, triangular teeth at posterior-ventral corner of shell; postabdomen tapered distally, marginal denticles small, lateral fascicles minute; postabdominal claws small, with a minute basal spine. This species functions much like a snail, scraping food off the substrate as it moves along. See Fryer (1968).]

10b.   Rostrum more narrowly rounded; shell may be weakly reticulated, but head never is 11

11a(10b).   Postabdomen elongate and rather narrow; marginal denticles well developed 12

11b.   Postabdomen much less elongate (except in *Leydigia*); marginal denticles usually well developed, but in some species greatly reduced 13

12a(11a).   Dorsal margin of postabdomen straight or slightly convex; marginal denticles very short proximally, increasing in length distally to three very long curved denticles; four median headpores (middle one is divided into two), usually completely separated from one another, or with middle pores joined by a sclerotized ridge (Figs. 20.1C and 20.22) .... *Oxyurella*
[Basal spine of postabdominal claw long, slender, attached some distance from base of claw; head and shell generally have a yellowish, hyaline appearance. Five species claimed to occur in world in 1971. Two species in North America, one being confined to the Gulf states, the other more widely distributed in the East. The common North American species is now separated from the formerly cognate species in Eurasia. See Michael and Frey (1983).]

12b.   Dorsal margin of postabdomen distinctly concave; marginal denticles increase in length distad, but never reach large size as in *Oxyurella;* one median headpore, sometimes completely absent (Fig. 20.23) .... *Euryalona*
[A tropical to subtropical genus. Three species in world, the one in America north of Mexico originally described from Sri Lanka. Head high; rostrum broadly rounded, with tip scarcely reaching halfway to ventral margin of shell; comb of setae on proximal half of concave margin of postabdominal claw, increasing in length distally; stout spine on inner distal lobe of trunklimb I, with coarse, rounded tubercles on concave edge. See Daday (1898) and Rajapaksa (1986).]

13a(11b).   Postabdomen large, broad, flattened, almost semicircular; armed laterally with long, spine-like setae in groups, those of distal groups projecting far beyond margin of postabdomen; margin of postabdomen with short, fine spinules; three median headpores close together, and with minor pores close-in laterally (Fig. 20.24) .... *Leydigia*
[Free posterior margin of shell very long; ocellus as large as compound eye or larger. Twelve species claimed for world in 1971, with maximal lengths from 0.8–1.15 mm. Two species are widely distributed in United States, and one tropical species barely gets into Florida.]

13b.   Postabdomen smaller, and with shorter setae in lateral fascicles; headpores variable 14

14a(13b).   Shell strongly sculptured, with longitudinal striae in posterior part and about five vertical striae anteriorly, roughly paralleling anterior margin; dorsal edge of postabdomen minutely serrate; about 14 fascicles laterally; no median headpores, but instead there are two comma-shaped thickenings, which presumably contain the minor pores (Fig. 20.25) .... *Notalona*
[A genus containing two tropical to subtropical species, only one of which gets into the United States. The headpore configuration is unique among the Chydoridae. Length to 0.44 mm. See Rajapaksa and Fernando (1987a).]

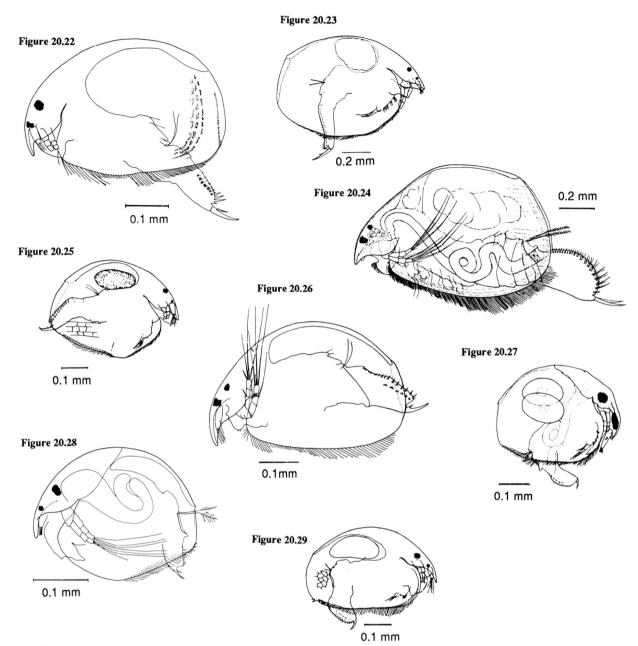

Figure 20.22 *Oxyurella brevicaudis* Michael and Frey, 1983. Skater's Pond, Monroe Co., Indiana. From Michael and Frey (1983).   **Figure 20.23** *Euryalona orientalis* (Daday, 1898). Ipiranga, Brazil. From Rajapaksa (1986).   **Figure 20.24** *Leydigia acanthocercoides* (Fisher, 1854). Uplandia, Sweden. From Lilljeborg (1901).   **Figure 20.25** *Notoalona freyi* Rajapaksa and Fernando, 1987. Homer Lake, Leon Co., Florida. From Rajapaksa and Fernando (1987a).   **Figure 20.26** *Alona bicolor* Frey, 1965. Goodwill Pond, Barnstable Co., Massachusetts. From Frey (1965).   **Figure 20.27.** *Dadaya macrops* (Daday, 1898). Polonnaruwa, Sri Lanka. From Rajapaksa and Fernando (1982).   **Figure 20.28** *Ephemeroporus acanthodes* Frey, 1987. Audubon Park, New Orleans. From Frey (1982b).   **Figure 20.29** *Dunhevedia americana* Rajapaksa and Fernando 1987. Glades Co., Florida. From Rajapaksa and Fernando (1987b).

14b.   Shell usually not sculptured at all, or if so, then rather weakly; postabdomen usually with well-developed marginal denticles and lateral fascicles; two or three median headpores, most commonly united by a sclerotized ridge, although sometimes completely separated (Fig. 20.26) .... *Alona*
   [This is the largest genus in the Chydoridae, with 52 species claimed in the world in 1971 and with more than 20 species in the United States and Canada. *Alona* intuitively consists of a number of genera, but these have not yet been defined satisfactorily. Smirnov (1971) placed the species having just two median headpores in a separate genus, *Biapertura*. Three of the species in North America, currently named *A. affinis*, *A. intermedia*, and *A. verrucosa*, belong here. Smirnov and Timms (1983) found a much greater proportion of two-pored species in Australia, which do not seem to be closely enough related to be included in the same genus. Length of species in America north of Mexico: two-pored species to 1.05 mm; three-pored species 0.4–0.9 mm.]

15a(2b).   Headshield not elongated posteriorly; in two-pored species, postpore distance usually considerably less than interpore distance 16
   [Includes the genera *Alonella, Dadaya, Disparalona, Dunhevedia,* and *Ephemeroporus.*]

15b.   Headshield much elongated posteriorly, extending well beyond the heart to the middle of the back; postpore distance usually greater than interpore distance, sometimes by several-fold (Fig. 20.1D) 19
   [Includes the genera *Anchistropus, Chydorus, Pleuroxus,* and *Pseudochydorus.*]

16a(15a).   Only one median headpore, or none 17

16b.   Two median headpores, well separated and not connected by a sclerotized ridge (Fig. 20.1D) 18

17a(16a).   One median headpore in all instars (Fig. 20.27) .... *Dadaya*
   [A single pantropical species occurring sparingly in the Gulf states. Color dark brown. Length to 0.41 mm.]

17b.   One median headpore only in first instar; none in later instars (Fig. 20.28) .... *Ephemeroporus*
   [Postabdomen with long proximal and distal marginal denticles, and with shorter ones in between; labral keel large, with 1–4 teeth on anterior margin. A tropical to subtropical genus with many species, few of which have been described to date. See Frey (1982b).]

18a(16b).   Postabdomen unique among chydorids, consisting of a much expanded postanal portion, having a series of short spines along dorsal margin and many small clusters of spinules laterally (Fig. 20.29) .... *Dunhevedia*
   [Four species claimed for world, two in the United States, one of which is strictly southern. Length from 0.46–0.8 mm.]

18b.   Postabdomen of more typical chydorid structure, with pre- and postanal angles and a well-developed series of postanal marginal denticles (Figs. 20.30 and 20.31) .... *Alonella* and *Disparalona*
   [The taxon *Disparalona rostrata* of Europe was formerly in the genus *Alonella*. Fryer (1968) removed it from this genus because it had a long "sweeping" seta on trunklimb III, which the other species of *Alonella* were claimed not to possess. Fryer (1971) subsequently transferred *Alonella acutirostris* to *Disparalona*, and Smirnov (1971) did the same for *Alonella dadayi*. Michael and Frey (1984) closely examined a number of species of *Alonella* and of *Disparalona* and concluded that the sweeper seta is not an additional seta in a few species but rather is the proximal member of the gnathobasic filter comb, and that species still in *Alonella* have the same seta variably developed and not always distinctly separable from the *Disparalona* species. I know of no characters that will clearly separate the two groups of species, suggesting that the validity of *Disparalona* is questionable. *Disparalona* presently contains the species *D. acutirostris*, *D. dadayi*, and *D. leei*, all occurring in North America, and *Alonella* the species *A. excisa*, *A. exigua*, *A. hamulata*, *A. nana*, and *A. pulchella*, out of a world total of 12. Maximal lengths of the North American taxa in *Disparalona* are 0.45–0.5 mm, and in *Alonella*, 0.25–0.6 (Michael and Frey 1984, Hann and Chengalath 1981).]

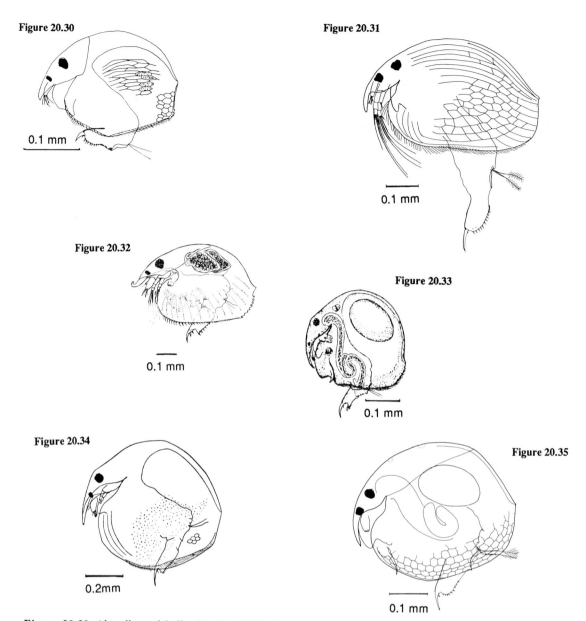

Figure 20.30 *Alonella pulchella* Herrick 1884. Kremer Lake, Minnesota. From Hann and Chengalath (1981). **Figure 20.31** *Disparalona leei* (Chien, 1970) Griffey Lake, Monroe Co., Indiana. From Michael and Frey (1984). **Figure 20.32** *Pleuroxus procurvus* Birge, 1878. Lake Mendota, Dane Co., Wisconsin. 11 September 1987. (KE) **Figure 20.33** *Anchistropus minor* Birge, 1893. Eastern North America. From Birge (1918). **Figure 20.34** *Pseudochydorus globosus* (Baird, 1843). River Sutka, USSR. From Smirnov (1971). **Figure 20.35** *Chydorus brevilabris* Frey, 1980. Salmon Lake, Montana. From Frey (1980a).

19a(15b).	Body elongate, somewhat flattened, usually with one or more teeth at the posterior-ventral angle of shell (Figs. 20.1D and 20.32) .... *Pleuroxus*
	[Fifteen species claimed in world, with maximal lengths from 0.4–0.8 mm. Seven species in America north of Mexico, of which *P. procurvus* and *P. denticulatus* are most frequent and most abundant (Frey 1988c). Lengths from 0.5–0.8 mm.]
19b.	Body globular, nearly round, seldom with any teeth at posterior-ventral angle 20
20a(19b).	Ventral margin of shell anteriorly with a hook-like development containing a groove, in which a stout seta with teeth on trunklimb I operates; postabdomen with a cluster of long, slender setae at distal angle (Fig. 20.33) .... *Anchistropus*
	[Surface of shell and head reticulated, with meshes having wavy edges. Three species in world, at least two of which are parasitic on *Hydra*. Lengths 0.35–0.46 mm, the North American species being the smallest. See Hyman (1926) and Smirnov (1985).]

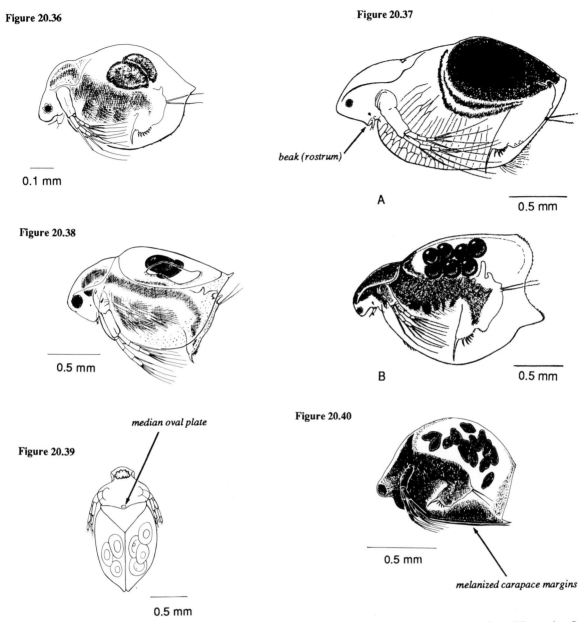

Figure 20.36

0.1 mm

Figure 20.37

beak (rostrum)

A

0.5 mm

Figure 20.38

0.5 mm

B

0.5 mm

median oval plate

Figure 20.39

Figure 20.40

0.5 mm

melanized carapace margins

0.5 mm

Figure 20.36 Ceridodaphnia cf. *quadrangula* (O. F. Mueller, 1785). Stewart Lake, Dane Co., Wisconsin. 5 September 1973. (KE) *Figure 20.37 Simocephalus* (A) *S. serrulatus* (Koch, 1841). Coliseum Pond, Madison, Dane Co., Wisconsin. 15 May 1987. (B) *S. vetulus* Schoedler, 1858. Gardner Pond, UW Arboretum, Dane Co., Wisconsin. 14 May 1987. (KE) *Figure 20.38 Daphniopsis ephemeralis* Schwartz and Hebert 1985. Baseline Pond, Ontario. 19 April 1988. (Coll. P.D.N. Hebert). (KE) *Figure 20.39 Megafenestra nasuta* (Birge, 1879). Dorsal view. Elk Island National Park, Alberta. 5 August 1978. (KE). Redrawn from Dumont and Pensaert (1983). *Figure 20.40 Scapholeberis rammneri* Dumont and Pensaert, 1983. Ojibwa, Ontario. 19 April 1988. (Coll. P.D.N. Hebert). (KE)

20b.	Ventral margin of shell without such a hook; marginal setae of shell arise submarginally posteriorly, forming a distinct duplicature (except in *Chydorus piger*) 21
21a(20b).	Postabdomen elongate, parallel-sided; postanal portion with about 15 slender marginal denticles; lateral surface with a row of fascicle-like clusters of setae (Fig. 20.34) .... *Pseudochydorus*

[*Pseudochydorus* presently contains a single species, distributed over much of the world, except possibly in South America. Length to 0.8 mm. It is a scavenger. See Fryer (1968).]

21b.   Postabdomen shorter, with a smaller number of marginal denticles, and laterally with crescentic clusters of short spinules, which do not resemble fascicles (Fig. 20.35) .... *Chydorus*

> [A highly complex genus, in which the species are difficult to separate, because they conform so closely to a common morphotype. At least ten species have been found in America north of Mexico. Smirnov (1971) recognized 18 species and many subspecies in the world, ranging in maximum length from 0.32–1 mm. The species occurring in North America range in length from 0.41–0.78 mm. The type species of the family, *Chydorus sphaericus* (sens. str.), quite possibly does not occur in North America at all. See Frey (1980a, 1982b, 1982c, 1987b).]

F. Taxonomic Key to Genera of the Family Daphniidae

1a.   Ventral margin of carapace rounded and not pigmented 2
1b.   Ventral margin of carapace straight and usually black 5

2a(1a).   Adults with a tail spine, at least four times as long as broad and pointed (Figs. 20.1, 20.2, 20.4, 20.5) .... *Daphnia*
2b.   Adults lack a tail spine, or if one is present, it is less than four times as long as broad and rounded 3

3a(2b).   Fourth (distal) segment longer than the third segment on the four-segment branch of the second (swimming) antenna (Fig. 20.38 .... *Daphniopsis*

> [Apparently only one species, *D. ephemeralis* in North America (Schwartz and Hebert 1987c); the genus is reviewed by Hann (1986).]

3b.   Fourth segment shorter than the third segment on the four-segment branch of the second antenna 4

4a(3b).   Second segment of the four-segment branch of the second antenna with an apical spine, the spine about 1/4 as long as the second segment (Fig. 20.37) .... *Simocephalus*

> [Allozyme studies of *Simocephalus* suggest there are at least five species in North America, two more than are recognized using morphologic characters (Hann and Hebert 1986).]

4b.   The second segment of the four-segment branch of the second antenna without an apical spine (Fig. 20.36) .... *Ceriodaphnia*

> [Species of *Ceriodaphnia* were discussed by Brandlova et al. (1972), but are still poorly known and in need of taxonomic attention.]

5a(1b).   Dorsum of headshield with a median oval plate about 20 μm in diameter, marked by slightly raised edges (Fig. 20.39) .... *Megafenestra*
5b.   Dorsum of headshield without such a plate (Fig. 20.40) .... *Scapholeberis*

> [With the exception of *Megafenestra* and *Scapholeberis* (Dumont and Pensaert 1983), and perhaps *Daphniopsis* (Hann 1986, Schwartz and Hebert 1987c), these genera are in need of revision.
>
> All of these genera tend toward the planktonic lifestyle. *Megafenestra* and *Scapholeberis* are the least planktonic; they are often found cruising along upside-down, at the surface of small pools. Their carapaces have flattened and turned-in margins (Dumont and Pensaert 1983), which may allow the carapace to serve as a suction cup, as in some chydorids. *Simocephalus* spends much of its time stuck to aquatic plants, using sticky mucous produced at the back of the neck. *Daphniopsis* has an extremely restricted habitat; it has been found in North America only in shallow temporary pools in maple forests. *Daphnia* and *Ceriodaphnia* are often restricted to open water of lakes and ponds.
> *Scapholeberis* and *Ceriodaphnia* adults are small, reaching about 1 mm, while the other genera produce adults in the range of 1–3 mm.]

G. Taxonomic Key to Genera of the Family Bosminidae

1a.   First antennae attached separately to the head, immobile in female, usually nearly vertical and usually curving somewhat posteriorly (Fig. 20.42A) .... *Bosmina*   2
1b.   First antennae fused together in basal half, separated and curved outward in distal half (resembling a mermaid), firmly attached to head (Fig. 20.41) .... *Bosminopsis*

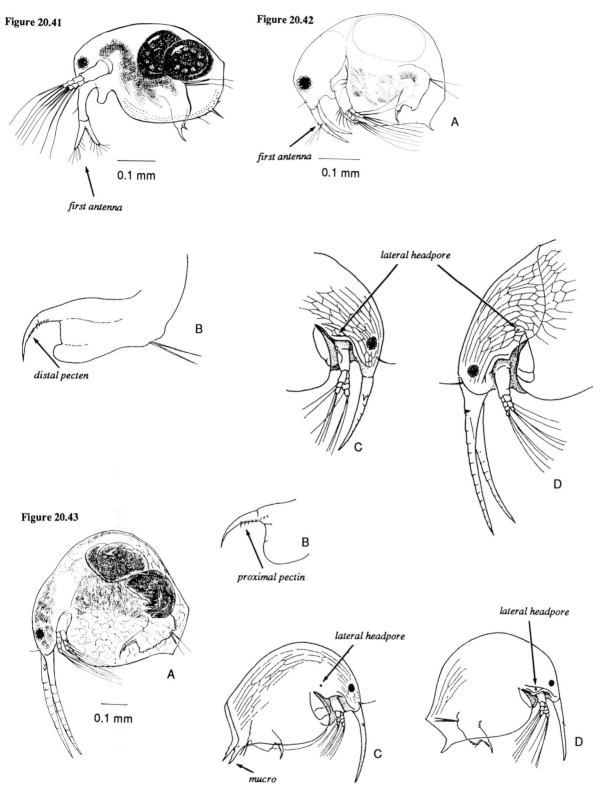

Figure 20.41

first antenna

0.1 mm

Figure 20.42

first antenna

0.1 mm

A

B

distal pecten

lateral headpore

C

D

Figure 20.43

A

0.1 mm

B

proximal pectin

lateral headpore

C

mucro

lateral headpore

D

Figure 20.41 *Bosminopsis deitersi* Richard 1895. Singletary Lake, North Carolina. 8 August 1948. (Coll. DGF #236. (KE)   **Figure 20.42** (A) *Bosmina(Bosmina)longirostris* (O. F. Mueller, 1776). Lake Mendota, Dane Co., Wisconsin. 11 September 1987. (KE); (B) Postabdomen of A. (KE); (C) Headshield of A, (KE); (D) Headshield of 20.43A. (KE)   **Figure 20.43** (A) *Bosmina (Eubosmina) coregoni* (Biard, 1850). Lake Waubesa, Dane Co., Wisconsin, 23 July 1988; (B) Postabdomen of A. (KE); (C) *Bosmina (Neobosmina) tubicen* Brehm, 1953. Presa Presidente Calles. Aquascalientes, Mexico. 8 October 1980. (KE); (D) *Bosmina (Sinobosmina) fatalis* Burchardt 1924. Redrawn from Kořínek (1971). (KE)

2a(1a).   Lateral headpore located within five pore diameters of the edge of the headshield just dorsal to the base of the second antenna and below the reticulated part of the headshield (Figs. 20.42C, 20.43D) 3

2b.   Lateral headpore located nearer the point of articulation of the mandibles than the base of the antenna, in the reticulated region of the headshield (Fig. 20.42D) 4

3a(2a).   Female postabdominal claw with a proximal comb (pecten) of 4–12 long slender setae, followed distally by a row of 7–10 minute spines (Fig. 20.42C); male postabdomen transversely truncated, with a deep anal embayment and with an incision in the dorsal margin; postabdominal claws unarmed, not set off from the claw peduncle. (Worldwide distribution, except possibly in Australia and New Zealand.).(Fig. 20.42A,B,C) .... *Bosmina (Bosmina) longirostris*

3b.   Female postabdominal claw without a distal pecten (as in Fig. 20.43B); male postabdominal claw of medium length, coarse and more or less blunt, with small short denticles on the claw and extending onto claw peduncle. (Reported only from eastern and southern Asia.) (Fig. 20.43D) .... *Bosmina (Sinobosmina)*

4a(2b).   Mucro absent in juvenile female (Fig. 20.43A), or if present, with minute incisions present only along the ventral margin; incisions may be lacking in the adult female. Lateral headpore near insertion of mandibles and at some distance from edge of headshield (Fig. 20.42D); male postabdomen strongly narrowed distally; postabdominal claws small, set off from claw peduncle, with 3–5 short spines. (Holarctic distribution.) (Figs. 20.42D, 20.43A, B) .... *Bosmina (Eubosmina)*

4b.   Mucro present in juvenile female, with minute incisions or teeth present only on the dorsal margin (Fig. 20.43C), incisions may be lacking in the adult female; lateral headpore position similar to that of *Eubosmina* (Fig. 20.43C); male postabdominal claw more or less elongate, slender, with six small denticles on the claw and no extension of the row onto the gibbous claw peduncle. (Reported mainly from the southern hemisphere, but occurs northward at least into Connecticut.) (Fig. 20.43C) .... *Bosmina (Neobosmina)*

   [There are just two genera in the family Bosminidae. The key to the subgenera of *Bosmina* is from Lieder (1983, based on his doctoral dissertation and on Lieder 1962). Goulden and Frey (1963) pointed out the different placement of lateral headpores that enables ready separation of *B. (Bosmina) longirostris* from *B. (Eubosmina) coregoni,* but were unable at that time to separate the species now in the subgenus *Neobosmina.* Deevey and Deevey (1971) accepted the latter suggestions, raised the subgenera *Bosmina* and *Eubosmina* to generic status, and combined *Neobosmina* with *Eubosmina. Sinobosmina* was not considered. These taxonomic decisions are not acceptable to *Bosmina* specialists (e.g., Lieder 1983). The four subgenera are distinct and should be retained. They can be separated chiefly by the placement of the lateral headpores (Kořínek 1971) and by the structure of the male postabdomen and postabdominal claws. *Bosmina* contains the single species *longirostris. Eubosmina* contains the species *longispina* and *coregoni* and the many species in the Baltic region of Europe. *Neobosmina* contains the species *tubicen* and *hagmanni* in North America. The taxonomy of *Bosmina* in North America is very unsatisfactory and desperately needs critical attention. Bosminids are more or less planktonic, with *Bosminopsis* the least planktonic. Adult *Bosmina* range in size from about 0.3–0.6 mm. Because of their small size, bosminids often are abundant when fish predation is intense.]

H. Taxonomic Key to Genera of the Family Macrothricidae (North of Mexico)

1a.   Antennule of two segments, the proximal one very short; postabdomen broad, dorsal margin broadly arched, with numerous long spines; posterior margin of shell armed to dorsal angle with stout, feathered setae (Fig. 20.44) .... subfamily Ilyocryptinae   *Ilyocryptus*

   [Smirnov (1976) lists nine species in the world, ranging in length from 0.6–0.9 mm, and with one species reaching 1.4 mm. At least five species in America north of Mexico. Almost all are mud inhabitants, although one species can swim weakly. See Williams (1978).]

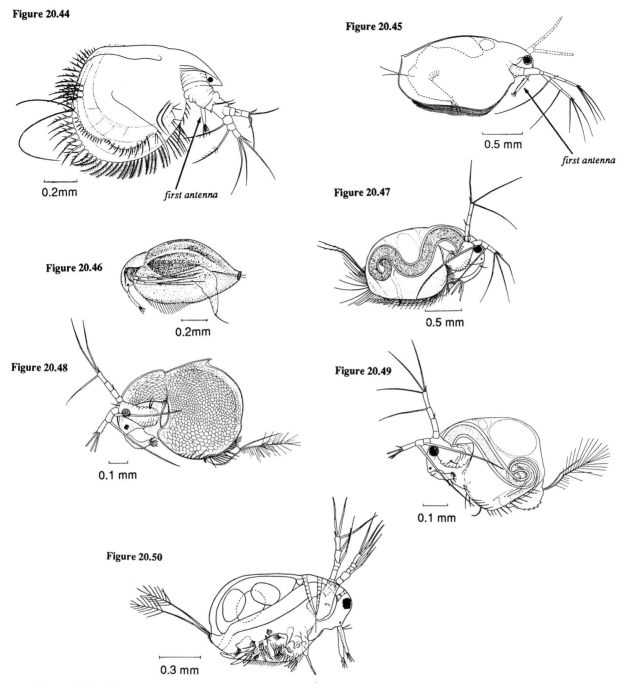

Figure 20.44

Figure 20.45

first antenna

0.5 mm

Figure 20.47

Figure 20.46

0.2mm

0.2mm

first antenna

Figure 20.48

Figure 20.49

0.1 mm

0.1 mm

Figure 20.50

0.3 mm

Figure 20.44 Ilyocryptus sordidus (Liévin 1848). English lake District. From Fryer (1974). *Figure 20.45* Ofryoxus gracilis Sars, 1861. English Lake District. From Fryer (1974). *Figure 20.46* Parophryoxus tubulatus Doolittle, 1909. Anonymous Pond, Maine. From Doolittle (1911). *Figure 20.47* Acantholeberis curvirostris (O. F. Mueller, 1776). English Lake District. From Fryer (1974). *Figure 20.48* Drepanothrix dentata (Eurén, 1861). English Lake District. From Fryer (1974). *Figure 20.49* Streblocerus serricaudatus (Fischer 1849). English Lake District. From Fryer (1974). *Figure 20.50* Lathonura rectirostris (O. F. Mueller 1776). Near Leningrad, USSR. From Smirnov (1976), after Sergeyev (1971).

1b.	Antennule consists of a single segment; dorsal margin of postabdomen either straight or if convex then without long spines; postabdominal claw short, with one or two, sometimes zero, short basal spines; posterior margin of shell not armed with stout feathered setae	2
2a(1b).	Anal opening in middle of postabdomen; two small hepatic caecae ...... subfamily Ofryoxinae	3

2b. Anal opening in distal part of postabdomen 4

3a(2a). Antennae with swimming setae 0-0-0-3/1-1-3 and with spines 1-1-0-1/0-0-1; shell with sharp, spine-like projection at posterior-dorsal angle (Fig. 20.45) .... *Ofryoxus*
[This is the official spelling, as it is what Sars used in his first paper, then a few months later changed the "f" to "ph." A single species in northern Eurasia and the eastern half of North America, including Florida. Length to 2.0 mm.]

3b. Antennae with swimming setae 0-0-0-3/0-0-3 and with spines 0-1-0-1/0-0-1; valves of shell narrowed behind, forming a tube-like structure; curved, horizontal ridge on each side of shell (Fig. 20.46) .... *Parophryoxus*
[A single species, described from Maine; also occurs in the Maritime Provinces. Length to 1.2 mm. See Doolittle (1911).]

4a(2b). Posterior margin of shell vertical, with distinct dorsal- and ventral-posterior angles, and many long, plumose setae projecting posteriorly; loop of intestine in postabdomen; six pairs of trunklimbs, numbers III-V with large exopodites (Fig. 20.47) .... subfamily Acantholeberinae *Acantholeberis*
[A single species in Europe and adjacent Asia and in northeastern North America plus northern Florida. Occurs in acidic bogs and other soft water bodies with much vegetation and organic detritus. Length to 2.0 mm.]

4b. Posterior margin of shell variable, generally much reduced in length, not vertical, and without long setae; five pairs of trunklimbs, all with small exopodites .... subfamily Macrothricinae 5

5a(4b). Intestine straight, without any loop 7
5b. Intestine with a loop 6

6a(5b). Shell crested dorsally, and with sharp tooth on crest; 4-segmented branch of antenna with three swimming setae (Fig. 20.48) .... *Drepanothrix*
[A single species in world, occupying much of the Holarctic region. In North America, it occurs from northern Florida north into the glaciated region. Length to 0.7 mm.]

6b. Shell without crest or tooth in dorsal margin; four swimming setae on 4-jointed branch of antenna; antennule of female bent outward distally (Fig. 20.49) .... *Streblocerus*
[Two species in world: one is mostly Holarctic, the other, smaller species ranges from South America into the southernmost United States. Lengths 0.25 and 0.6 mm.]

7a(5a). Ventral margin of shell with lancet-shaped or spatulate spines (Fig. 20.50) .... *Lathonura*
[Ten antennal setae, five on each branch, all plumose and similar in structure; postabdomen extending anteriorly as a conical process. One Holarctic species in world. Length to 1.15 mm.]

7b. Ventral margin of shell with more typical setae, not lancet-shaped 8

8a(7b). Postabdomen broadly oval, with a conspicuous notch in margin of preanal region 9
8b. Postabdomen with dorsal margin less strongly arched, or straight; no conspicuous notch in preanal region 10

9a(8a). Postabdomen with a group of stout setae lateral to anal opening, and with a longer and stouter seta immediately anterior to anal opening; antennules long and slender (Fig. 20.51) .... *Grimaldina*
[One species in world, pantropical; just gets into southernmost states bordering Gulf of Mexico. Length to 0.85 mm.]

9b. Postabdomen without any stout setae; antennule short and thickened, covered anteriorly with transverse rows of short ridges, resembling setae (Fig. 20.52) .... *Guernella*
[One species, pantropical; in the United States known thus far only from the Everglades. Length to 0.59 mm (Frey 1988b).]

10a(8b). Broad keel on shell, which extends posteriorly and forms a rounded posterior margin (Fig. 20.53) .... *Bunops*
[One species in world, possibly two. Holarctic region, scarce. Length to 1.3 mm. See Merrill (1893).]

10b. Shell without a broad keel dorsally 11

11a(10b). Antennule enlarged distally (Fig. 20.54) .... *Macrothrix*
[Smirnov (1976) lists 19 species in world, with maximal lengths mostly from 0.39–0.73 mm, but with 4 species greater than 1 mm and up to 2.0 mm. The 3 North American species range from 0.55–0.7 mm.]

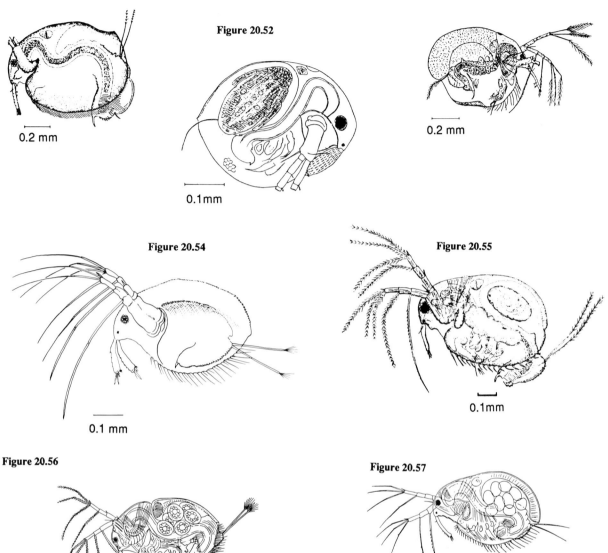

Figure 20.51 Grimaldina brazzai Richard, 1892. Louisiana. From Birge (1918). *Figure 20.52 Guernella raphaelis* Richard, 1892. Mayumba, Gabon. From Richard (1892). *Figure 20.53 Bunops acutifrons* Birge, 1893. Minocqua, Wisconsin. From Merrill (1893). *Figure 20.54 Macrothrix rosea* (Jurine, 1820). Hook Lake, Dane Co., Wisconsin. 25 October 1972. (KE) *Figure 20.55 Wlassicsia kinistinensis* Birge, 1910. Kinistino, Manitoba. From Birge (1910). *Figure 20.56 Iheringula paulensis* Sars, 1900. Sãn Paulo, Brazil. From Smirnov (1976), after Sars (1900). *Figure 20.57 Echinisca schauinslandi* (Sars, 1904). Lake Wakatipu, New Zealand. From Smirnov (1976), after Sars (1904).

11b.	Antennule not enlarged distally	12
12a(11b).	Exopodite of trunklimb IV with five setae (Fig. 20.55)	*Wlassiscia*
	[One European species and another known only from Manitoba. The latter reaches 0.8 mm in length.]	
12b.	Exopodite of trunklimb IV with only two or three setae	13
13a(12b).	Conspicuous deep depression or notch dorsally between head and shell; trunklimb V having a broad plate with three large, delicate, ciliated, rounded lobes (Fig. 20.56)	*Iheringula*
	[A single species in world, known from Brazil and the Everglades. See Sars (1900) and Frey (1988b).]	

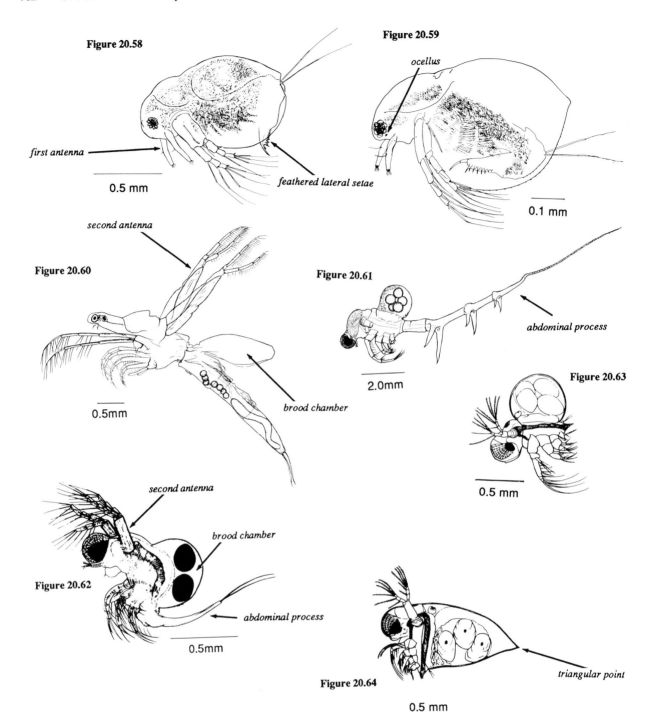

Figure 20.58

first antenna

0.5 mm

feathered lateral setae

Figure 20.59

ocellus

0.1 mm

second antenna

Figure 20.60

0.5mm

brood chamber

Figure 20.61

2.0mm

abdominal process

Figure 20.63

0.5 mm

second antenna

brood chamber

Figure 20.62

abdominal process

0.5mm

Figure 20.64

triangular point

0.5 mm

Figure 20.58 *Moina wierzejskii* Richard, 1895. 45 km NE of Las Cruces, New Mexico. 6 July 1984 (KE)   ***Figure 20.59*** *Moinodaphnia macleayii* (King, 1853). Roadside canal, Highlands Co., Florida. 20 July 1960. (Coll. DGF #96). (KE)   ***Figure 20.60*** *Leptodora kindti* (Focke, 1844). Lake Michigan, USA. From Balcer *et al.* (1984). Drawn by Nancy Korda.   ***Figure 20.61*** *Bythotrephes cederstroemii* Schodler, 1877. Summer female from Lake Karesuando in Norrbotten. From Lilljeborg (1901). Redrawn by KE.   ***Figure 20.62*** *Polyphemis pediculus* (Linnee, 1761). Lake Michigan, USA. From Balcer *et al.* (1984). Drawn by Nancy Korda.   ***Figure 20.63*** *Evadne nordmanni* (Lovén, 1836. Summer female from the sea near Dalaro. From Lilljeborg (1901). Redrawn by Cheryl Hughes.   ***Figure 20.64*** *Podon leuckartii* Sars, 1862. Summer female from the sea near Bergen, Norway. From Lilljeborg (1901). Redrawn by Cheryl Hughes.

13b.   No depression at all dorsally, or if present then very weak; trunklimb V without three ciliated lobes (Fig. 20.57) .... *Echinisca*
[Smirnov (1976) has 22 species of *Echinisca*, which in general tend to be larger in size than *Macrothrix* species. Of the species with lengths listed, 11 range from 0.35–0.85 mm, and 9 from 1.0–2.2 mm. The two taxa in North America have maximal lengths of 0.7 and 1.1 mm. *Macrothrix* and *Echinisca* are the macrothricid genera with the greatest number of species, making up more than half of all of the macrothricid species. They were formerly included in the single genus *Macrothrix*, but Sars (1916) pointed out that they have a different structure of the antennules. Smirnov (1976) has them separated also, which is one way of distributing the large number of species into two more manageable groups. Smirnov also placed *Iheringula paulensis* from Brazil into the genus *Echinisca*.]

I. Taxonomic Key to Genera of Freshwater Moinidae

1a.   Branch of second antenna with four segments appears to have four terminal setae; ocellus (a small black spot just back of the eye) absent in North American species (Fig. 20.58) .... *Moina*
1b.   Branch of second antenna with four segments has three setae and a short spine; an ocellus is present (Fig. 20.59) .... *Moinodaphnia*
[These two genera have been monographed recently by Goulden (1968). These animals can be planktonic, but the majority are restricted to small temporary ponds or saline or alkaline lakes. Adult females reach 1–2 mm, with one *Moina* species that is only about 0.5 mm long.]

J. Taxonomic Key to Genera of the Family Leptodoridae

[*Leptodora kindtii* (Fig. 20.60) is the sole species in this family. The adult females are up to 18 mm long. Despite their large size, the animals are so transparent as to be nearly invisible. They are common in the open water of large lakes, and occasionally are found in ponds. *L. kindtii* are voracious predators on smaller zooplankton.]

K. Taxonomic Key to Genera of the Families Polyphemidae, Cercopagidae, and Podonidae

1a.   Adult females 1–2 mm long, total length; abdominal process shorter than the body 2
1b.   Adult females longer than 2 mm, abdominal process longer than the rest of the body (Fig. 20.61) .... family Cercopagidae   *Bythotrephes*
2a(1a).   Abdominal process slender and about half as long as the body (Fig. 20.62) .... family Polyphemidae   *Polyphemus*
2b.   Abdominal process does not look like a slender tail and is shorter than half of the body length .... family Podonidae   2
3a(2b).   Body ends in a short extension of the abdomen, about twice as long as broad (Fig. 20.63) ...... *Evadne*
3b.   Body ends in a triangular point, but without a distinct abdominal process (Fig. 20.64) .... *Podon*
[*Polyphemus* and *Bythotrephes* are found in freshwater, the other two genera live in near-shore marine and occasionally estuarine water. *Bythotrephes*, a European native, has recently been introduced into the Great Lakes. All these genera are planktonic, although *Polyphemus* is the least so, and is typically found in small lakes and ponds, often along the shore or in aquatic vegetation. The Onychopoda are predators on smaller zooplankton.]

OTHER BRANCHIOPODS

VI. ANATOMY AND PHYSIOLOGY OF THE OTHER BRANCHIOPODS

A. General External and Internal Anatomic Features

The non-cladoceran branchiopods are composed of four extant orders (Table 20.1). They differ from cladocerans in a number of ways, including having more pairs of legs and paired compound eyes. The four orders are probably not closely related and are morphologically diverse (McLaughlin 1980, Fryer 1987).

Members of the first order, Anostraca (fairy shrimps), swim with their legs up (Fig. 20.65), have a pair of large compound eyes, and possess no shell-like covering. Anostracan morphology is described in detail by Linder (1941). Adults are 7–100 mm long and have 11, 17, or 19 pairs of thoracic legs along their rather delicate, transparent, and elongate body. The legs are all anterior to two fused genital segments (Fig. 20.65), and hence can be called thoracic legs. There are no abdominal legs, but rather a pair of terminal lobes (cercopods, Fig. 20.67) on the last abdominal segment (telson). The genital segments contain the gonads (which extend into abdominal and thoracic segments). Males have a pair of ventral penes (Figs. 20.70C and 20.71). Females possess a medial brood pouch (Fig. 20.70B). Within the brood pouch are lateral oviducal pouches, a single median ovisac, and shell glands. Males have large and sometimes complex second antennae (Fig. 20.69A) and sometimes outgrowths from the head (Fig. 20.66C) or antennae (Fig. 20.70A). The second antennae are used for grasping females (Fig. 20.73B) but not for swimming (as in cladocerans and conchostracans).

Laevicaudata and Spinicaudata (the two conchostracan orders of clam shrimps, Figs. 20.75 and 20.76) swim with their legs down or forward, have a pair of close-set compound eyes (separated, except in *Cyclestheria* where they are fused), and a carapace that wraps around the entire body (Figs. 20.75 and 20.76). Adults are 2–16 mm long, and have 10–32 pairs of thoracic legs covered by the transparent to brown carapace (Fig. 20.76). In species of Spinicaudata, the carapace closely resembles a clam shell, even having growth lines and an umbo (Fig. 20.81). In Laevicaudata, the carapace is more globular, and only one (Siberian) species has even a single growth line. The branched arm-like appendages that can be extended from the carapace are the second antennae (Fig. 20.81); they are used for swimming, as in the cladocera.

Notostraca, or tadpole shrimps, are sometimes confused with trilobites by the unwary. These crustaceans swim with their legs down, have a pair of dorsal compound eyes, and possess a broad and flat carapace that covers the head and thorax (Fig. 20.82). Adults are 10–58 mm long and have 35–70 pairs of trunk legs partially hidden by their dark green or brown carapace. Immature notostracans have a transparent carapace that is folded along the dorsal midline and encloses the body and legs, making them appear somewhat like a conchostracan. Long filaments (endites) protruding from either side of the carapace are attached to the first pair of thoracic legs (Fig. 20.83). Endites of the second pair of legs are less filamentous, and the succeeding legs lack filaments (Fryer 1988). The first pair of legs is used in swimming. Phyllopods posterior to the first pair are used for swimming and also walking, digging, and food handling (D. Belk personal communication, Fryer 1988).

Eggs are carried in brood pouches in the specialized eleventh pair of legs. The anterior lid of the concave brood pouch is derived from the exopodite of the leg and the bottom from the subapical lobe of the leg (Fryer 1988). The eleventh pair of thoracic legs of males resembles the tenth and twelfth pairs. The gross morphological difference between males and females is that males lack the brood pouch (and ovaries) and tend to be slightly larger than the hermaphroditic "females" (Beaton and Hebert 1988).

Maleless populations are usually (Akita 1971, Beaton and Hebert 1988, Longhurst 1954, Fryer 1988) but not always (Zaffagnini and Trentini 1980) made up of hermaphrodites. Zaffagini and Trentini (1980) present evidence that reproduction in maleless populations is by automictic parthenogenesis, not autogamic parthenogenesis (i.e., self-fertilizing hermaphrodites) as previously suggested (Longhurst 1954). Beaton and Hebert (1988) found no electrophoretic variation in *Lepidurus arcticus* and were not able to clarify the role of males in reproduction.

The notostracan trunk is not properly divided into thorax and abdomen, because the legs continue past the genital opening on the eleventh segment and there are many more legs than "segments." The legs gradually decrease in size toward the tail. The body ends in a pair of long thin cercopods (Fig. 20.82) and, in *Lepidurus*, a posterior extension of the telson (supra-anal plate, Fig. 20.82A).

B. Relevant Physiological Information

Branchiopods of the different orders are similar in regard to their physiological adaptations. Thus, the discussion of cladoceran physiology applies in general to the non-cladoceran orders as well. *Artemia* is a model animal for physiologists (e.g., Sorgeloos *et al.* 1987).

Conchostracans, notostracans, and anostracans probably depend more on gills for respiration than do the cladoceran orders (Eriksen and Brown 1980a). Notostracans and anostracans show the usual correlation between surface area and rate of oxygen consumption, although in conchostracans, small juveniles have an unusually high V_{O_2}, perhaps due to the well-developed gills on the trunk legs.

Osmoregulation (maintenance of proper salt concentration in the blood and tissues) is important to these animals, which often live in conditions of extreme or variable salinity. The general mechanisms of osmoregulation are probably similar in all branchiopods (Peters 1987a), but there are many variations on the common theme. Different species, even in the same genus, show differences in internal salt concentrations, degree of osmoregulation, and tolerance of extremes (e.g., Horne 1966, Broch 1988)

The hemoglobins of *Moina* (a cladoceran), *Cyzicus* (Spinicaudata), and *Triops* are electrophoretically similar (Horne and Beyenbach 1974).

Further physiological topics, including diapause, respiration, and ion balance are discussed in Section VII.C.

VII. ECOLOGY OF THE OTHER BRANCHIOPODS

A. Life History

Notostracans begin life as nauplii or metanauplii. There are a dozen or so molts before maturity, and the adults continue molting throughout their life (Fryer 1988).

Anostracans hatch from resting eggs as nauplii or metanauplii. Some species leave the egg shell as nauplii (*Artemia,* Anderson 1967; *Branchinecta,* Daborn 1977a), and some hatch as advanced metanauplii with as many as ten pairs of swimming legs (*Eubranchipus,* Daborn 1976). As the metanauplii molt, they add segments and appendages, gradually developing into adults (Fryer 1983). Adults molt several times. Fertilization normally takes place just after the adult female molts. The brood pouch has a pore through which sperm are injected by one or the other of the penes of a male. [It isn't clear how the eggs of *Artemiopsis,* which lacks a brood-pouch pore, are fertilized.] Fairy shrimp typically carry eggs until they molt. New eggs are released into the ovisac only after mating in all genera studied, except *Artemia* (Munuswamy and Subramoniam 1985). Anostracans probably do not store sperm (Wiman 1979).

Most conchostracans hatch from resting eggs as nauplii (Anderson 1967). The carapace appears in about the third naupliar stage. By successive molts, the juveniles gradually come to resemble adults.

The eggs of most species of large branchiopods enter diapause soon after development commences and become resistant resting eggs (Belk and Cole 1975). Exceptions include *Artemia,* in which development may continue in the ovisac and free-swimming nauplii may pass out of the brood pouch, and perhaps *Cyclestheria hislopi* (Spinicaudata), which may exhibit direct larval development, similar to cladocera (D. Belk personal communication). Eggs produced parthenogenetically or sexually have similar diapausing characteristics (*Eulimnadia antlei,* Belk 1972). Anostracans, conchostracans, and notostracans of temporary ponds typically have only one generation per wet episode and produce resting eggs. There is some evidence that diapausing eggs of different anostracan species specialize on hatching under specific rainfall patterns, allowing different species to take advantage of variations in climatic conditions from year to year in the same temporary pond (Donald 1982).

Thiel (1963) and Belk (1972) raised *Triops* (Notostraca) and *Eulimnadia* (Spinicaudata), respectively, in the laboratory without drying the eggs. The eggs of *Triops* overwinter best if exposed to low oxygen. The eggs of *Streptocephalus seali* show sensitivity to strong drying conditions. This may limit their distribution, excluding them from ephemeral (temporary) ponds in highly desiccating environments (Belk and Cole 1975). *Artemia* in permanent lakes can develop directly. *Artemia* of Mono Lake, a permanent saline lake in the Great Basin desert of mideastern California, hatch from resting eggs in the spring, resulting in adults that produce a second generation from eggs carried and matured in their brood sacs (Lenz 1984).

The large size achieved by most non-cladocerans requires a longer time to first reproduction, which is crucial in ephemeral ponds (Loring *et al.* 1988). Notostracans (1–2 cm long) require 2–3 weeks to develop from the egg to mature adults (Rzòska 1961). Anostracans (1–2 cm long) mature in 3 weeks at 15°C (Mossin 1986) and about 45 days at 4°–17°C (Modlin 1982). Successive generations of *Artemia* are separated by a month or so in permanent lakes (Lenz 1984). Depending on temperature, *Eulimnadia diversa* (3–4 mm) took 4–11 days to develop from the egg to an adult (Belk 1972). Cladocerans 1–2 mm long at maturity, which live in ephemeral ponds (e.g., *Daphnia, Moina*), could produce a clutch of eggs in as few as 2 days at high temperatures (Rzòska 1961). Thus, cladocerans can mature in the most temporary of ponds, while the larger non-cladocerans require ponds that retain water long enough to allow completion of their longer life cycles. These large branchiopods may then disappear as slower-developing predators and competitors become abundant (Sublette and Sublette 1967, Dodson 1987, Loring *et al.* 1988).

B. Distribution and Biogeography

Non-cladoceran branchiopods typically occur in the absence of fish (Hartland-Rowe 1972, Kerfoot and Lynch 1987). However, in each group, there are a few reports of occurrences in lakes with fish. Anderson (1974) gives examples of all three groups in lakes. Some anostracans co-occur with fish in large, deep freshwater lakes (*Branchinecta* in Canada, Anderson 1974; *Polyartemia* in Sweden, Nilsson and Pejler 1973).

Conchostracans are rare north of southern Canada, while anostracans and notostracans are found from arctic islands to Central America. All three groups appear to be especially diverse in the mid-temperate regions and are especially diverse in the arid southwestern United States. Anostracan diversity in Arizona depends mainly on two general factors: (1) chemical heterogeneity among (mostly temporary) habitats; and (2) thermal variation resulting both from ponds filling at different seasons and from altitudinal and latitudinal effects (Belk 1977a). It is uncommon to find congeneric assemblages within the same pond, although associations of different genera are common (Sublette and Sublette 1967, Dodson 1979, 1987, Donald 1982, Kerfoot and Lynch 1987); however, *Triops* and *Lepidurus* do not appear to co-occur.

Most of what we know about the distribution of these animals is a function of their physiological adaptions, discussed in the following section.

C. Physiological Adaptations

The four non-cladoceran orders are generally restricted to small (fishless) ponds, especially temporary systems. Some of these habitats are saline, and some species within the anostracan and conchostracan orders show a high tolerance to salinity (Hartland-Rowe 1966, 1972, D'Agostino 1980). *Artemia salina* and *Branchinecta campestris* are the most tolerant, surviving in water several times as saline as seawater. Most fairy shrimp, conchostracans, and notostracans tolerate a wide range of salinities; but a few taxa, probably most species of *Eubranchipus* and some of *Streptocephalus* and *Branchinecta*, are restricted to low-salinity water. On the other hand, *Artemia* are rarely found in water as dilute as seawater (D'Agostino 1980, Eriksen and Brown 1980b). The greater salt tolerance of

Branchinecta mackini (Broch 1988) may provide an occasional refuge in time or space from its co-occurring congeneric predator *B. gigas*.

The distributions of *Artemia* species and populations within species can be limited by their salinity tolerances (Hartland-Rowe 1972, Bowen *et al.* 1985, Abreu-Grobois 1987). The viability of several populations in the *A. franciscana* (Kellogg) superspecies depended on salinity, the concentration of bicarbonate, and the chloride/sulfate ratio. These differences in salinity requirements between populations may be a factor in the process of speciation. Populations in different types of saline lakes are genetically isolated in nature, because nauplii from one source will die in water from a different type of saline lake (Bowen *et al.* 1985).

Because temporary ponds can often get quite warm, these crustaceans must frequently tolerate high temperatures (Horne 1971, Hartland-Rowe 1972, Eriksen and Brown 1980a, 1980b, 1980c). Conchostracans do not perform well below 10°C, but they can live for at least a few hours at 30°–35°C. Fairy shrimp from desert areas tolerate for several hours temperatures as high as 40°C. Several species that tolerate the highest temperatures are eurythermal, in that they can also swim at temperatures near 0°C. Notostracans have thermal tolerances similar to those of fairy shrimp, but there is significant variation among taxa. For example, the higher temperature tolerance of *Branchinecta mackini* provides a temporal refuge from its predator *Lepidurus lemmoni* (Eriksen and Brown 1980b).

In warm climates, the combination of high temperatures and high productivities can lead to low oxygen concentrations during the night, or in stratified ponds at any time (Horne 1971, Loring *et al.* 1988). Functional hemoglobin has been observed in some notostracans (*Triops*), conchostracans (*Cyzicus*), and anostracans (*Artemia*, Horne and Beyenbach 1974; *Streptocephalus mackini*, D. Belk personal communication). All of the larger branchiopods can regulate their oxygen consumption and live at low oxygen concentrations. In the lab, adult conchostracans (*Cyzicus californicus*) showed the greatest tolerance to low oxygen, down to levels of 0.4–0.5 cm^3 liter^{-1} (roughly 3–4% saturation at 21°C) (Eriksen and Brown 1980a). Horne (1971) found a natural population of the conchostracan *Cyzicus setosa* able to tolerate oxygen concentrations of 0.1 cm^3 liter^{-1} for 2 hr with no significant mortality.) Anostracans (*Branchinecta mackini*) tolerated oxygen concentrations down to about 1–2 cm^3 liter^{-1} (roughly 10–20% saturation at 21°C) (Eriksen and Horne 1980b). Notostracans (*Lepidurus lemmoni*), despite their hemoglobin, have a minimum oxygen tolerance similar to anostracans, and show the least ability to regulate oxygen consumption at low oxygen concentrations (Eriksen and Brown 1980c). Oddly enough, a field study of lethal oxygen thresholds of these animals found that conchostracans (*Eulimnadia inflecta*) had a lower tolerance than two anostracan species (*Eubranchipus moorei* and *Streptocephalus seali*) (Moore and Burn 1968). The apparently greater tolerance of the anostracans may have been due to the behavioral tendency of anostracans (but not *E. inflecta*) to seek the oxygenated surface of the pond. All three groups show a maximum rate of weight-specific oxygen consumption at about 25xC (Eriksen and Brown 1980a, 1980b, 1980c).

The saturation level of oxygen concentration is reduced exponentially by salinity. Seawater holds about 20% less oxygen than freshwater at corresponding temperatures and pressures. More saline waters hold even less oxygen. Broch (1969) suggested that *Branchinecta campestris* may be limited to waters of lower salinity than *Artemia salina* both because of salt tolerances and because *Artemia* produces hemoglobin, which may be necessary at high concentrations of salt and the correlated low oxygen concentrations.

In Anostraca, Conchostraca, and Notostraca, the eggs are carried for at least a short period by the female. After a few cell divisions, the egg typically enters diapause. This resting egg (actually a developing embryo) has a dark covering (tertiary envelope or shell) and is prepared to survive drying, heat, freezing, and ingestion by birds (Belk and Cole 1975). Belk (1970) removed the covering of diapausing conchostracan embryos to demonstrate that the shell does not reduce desiccation effects but may reduce mortality due to abrasion or intense (ultraviolet) sunlight.

Signals that break diapause include temperature and oxygen concentration (Hartland-Rowe 1972, Belk and Cole 1975, Clegge and Conte 1980, Wiggins *et al.* 1980, Mossin 1986). In some anostracans, the final signal that breaks diapause may be a combination of high oxygen and an increase in CO_2 concentration (Mossin 1986). In fact, the fastest average hatching time of the European anostracan *Siphonophanes grubei* occurred at a pH of 5.5 (associated with high CO_2 concentration). At least one conchostracan, *Caenestheria setosa,* has a requirement for high temperatures to break diapause: its resting eggs hatch between 20° and 35°C. Hatching in *Eulimnadia antlei* (Spinicaudata) was inhibited by high salinity and darkness (Belk 1972).

Resistant eggs appear to be an adaptation to temporary habitats and to long-distance dispersal (Belk and Cole 1975). The habit of carrying diapausing eggs probably has two major advantages (Daborn 1977b). First, the eggs are protected from numerous egg predators, such as other crustaceans and chironomid larvae. Second, the eggs are more likely to be ingested by a bird for long-range dispersal. As with cladocerans and copepods, ingestion of diapausing eggs may be an important means of dispersal. Proctor (1964) raised *Artemia, Streptocephalus, Triops,* and *Cyzicus* from feces of mallard ducks (*Anas platyrhynchos*).

D. Behavioral Ecology

1. Swimming Behavior

Notostracans are detritus feeders and predators on anostracans and benthic invertebrates. They may be significant predators on resting eggs of other aquatic invertebrates such as fairy shrimp and clam shrimp (Daborn 1977b). Horne (1966) cultured *Triops* in the lab using live fairy shrimp. Notostracans capture mosquito larvae, and are occasionally pests in rice fields, eating the young plants off at mud level (Dodson 1987).

Most anostracans feed on their backs, filtering with rhythmic movements of their legs as they swim. *Artemiopsis* and *Branchinecta* also roll over and scrape surfaces with their thoracic legs (Daborn 1977b). *Branchinecta gigas* catches and holds smaller anostracans with its spiny phyllopods (Fryer 1966, Daborn 1975).

Conchostracans swim slowly, spending most of their time skimming along the pond bottom, in aquatic vegetation (Martin *et al.* 1986), or above algal mats (e.g., Sissom 1980). Notostracans and anostracans will swim away from a person arriving at the edge of a pond; conchostracans show no such response.

2. Mating Behavior

The immediate goal of notostracan mating behavior is the apposition of the eleventh pair of thoracic legs. Male and female gonopores are in the same position at the base of the first endopodite on these legs (Akita 1971), so some bodily contortion is required to deposit spermatozoa into the brood pouch. Notostracans have two mating positions (Mathias 1937). The two sexes either grasp each other in ventral apposition, or the male holds the carapace of the female and twists his body around and under the female.

Conchostracan sexes are morphologically distinct, with the male possessing specialized hooks on the first (Laevicaudata) or first and second (Spinicaudata) pairs of thoracic legs. These hooks are probably used in a similar manner to those of cladocerans; the hooks hold onto the female and position the animals for ventral apposition copulation (Mathias 1937, Martin *et al.* and 1986). In laboratory cultures and in nature, it is common to see a female swimming about with a male holding on to the lower (posterior) margins of her carapace.

Anostracans are sexually dimorphic. Males possess enlarged second antennae, often with long appendages. These antennae are used to grasp the female just anterior to her genital segments (just behind the last thoracic leg). Only the second antennae are prehensile. Belk (1984) offered evidence indicating no "holding" role for antennal appendages and suggested none for frontal appendages. Anostracans may swim in tandem for days, especially *Artemia* and *Artemiopsis* (Daborn 1977b).

Belk (1984) described the mating behavior of *Eubranchipus serratus*:

> Once a sexually active male *Eubranchipus* locates another fairy shrimp, he positions himself with his head below the dorsal surface of the other shrimp's genital segments. From this station-taking posture (Moore and Ogren 1962), the male may be able to assess suitability of mates (Wiman 1981). Assuming the station-taking male is below a female, his next action is to clasp her rapidly just anterior to her genital region with his second antennae. As he grabs her, he extends his antennal appendages, laying them along her back. Holding his body stiff with his phyllopods pressed against the ventral surface of his trunk and flexing only at the neck region, he bats at the female several times with quick ventral movements.

> As soon as she is grabbed, the female rolls the anterior half of her body into a ball with her head pressed ventrally against her genital region. Her abdomen remains straight. This position is usually held only a second or so. Then the female will either struggle violently and dislodge the male, or she will relax and begin swimming slowly. If the female accepts the male, he too relaxes and begins swimming with her while still clasping her with his second antennae. His antennal appendages remain stretched along her back. The male now attempts intromission, trying from first one side, then the other. Once one of his penes is inserted into the female's ovisac, the male assumes an S-posture. He ceases swimming and his phyllopods make only slight vibratory movements. The female may continue to swim or she may lie on the bottom. Disengagement is effected by the female with one or more rapid jerking movements. As soon as copulation terminates, the male resumes normal swimming. (p. 69)

Mating in *Streptocephalus mackini* was similar to that of *E. serratus*, except that males took station above rather than below females (Wiman 1981). Male *E. serratus* were successful in grasping a female in only about 20% of their attempts (Belk 1984), whereas *S. mackini* males were successful about 64% of the time. *S. mackini* males mate with "undiscriminating eagerness," approaching males as

well as females and also orienting toward fairy shrimp of different genera (Wiman 1981). Copulation is terminated by the female, possibly when she releases his penis from her gonopore (Wiman 1981). Males are sometimes observed being towed along by their penis, and one occasionally finds males with only one functional penis, the other being everted permanently. The spines on anostracan penes may be used to hold the penis inside the gonopore during swimming (Wiman 1981).

Eubranchipus bundyi occurs in April in roadside ditches near Wisconsin. Dodson (unpublished) watched four adult males and one adult female that were swimming in several adjacent depressions in a marshy area. Each depression was about 20 cm deep, with dead grasses at the bottom. One or two males swam around the edge of each depression, making circles about 20 cm in diameter and keeping at mid-depth, or about 10 cm most of the time. Although both sexes beat their legs 8–10 times sec^{-1}, males swam at about 1 cm sec^{-1}, females at 0.1 cm sec^{-1}. Females swam in long straight lines, changing direction every minute or so with a quick burst of speed. They emerged from grass cover for a few minutes every few hours. A female that came up to the mid-water level was followed by a male, who adjusted his swimming rate to hers, and swam with his head a mm or so below her uropods. He followed for about ten seconds, then moved closer. The female escaped with a quick burst of speed, about 10 cm sec^{-1} for 4 cm. The male showed no further interest. The impression these observations give is that males patrol bits of open water, and the females spend most of their time in concealment. When the females desire to mate, they swim upward, find a male, mate, and then return to the protection of the bottom. The lek-like pattern of male and female fairy shrimp mating behavior may result from their reproductive strategy (Wiman 1981). Because females produce large clutches of eggs while males spend little of their energy budget on gametes, life-history theory predicts that females should show discrimination in mating, while males should be more active and mate with as many females as possible. As a consequence of higher activity levels, males are more vulnerable than females to predators.

E. Foraging Relationships

In general, anostracans are free-swimming filter feeders, conchostracans process detritus from surfaces or collect plankton, and notostracans are benthic detritus feeders tending toward carnivory (Kaestner 1970).

There are many records of anostracans eating algae. The filtration mechanism of the thoracic legs is described by Fryer (1983). Cannon (1933) assumed anostracans live mainly on filtered algae. However, more recent work clearly indicates that anostracans are rather omnivorous, eating particles in the size range from algae to small crustaceans. Bernice (1971) reported that *Streptocephalus dichotomus* ate a variety of foods, including large algae (e.g., diatoms and blue-green bacteria), ciliates, small invertebrates (e.g., rotifers, nematodes, and cladocerans), and crustacean eggs. The diet of *Artemia* is well known both for natural populations and for lab cultures (D'Agostino 1980). *Artemia* feed on coccoid green algae in Mono Lake (Winkler 1977) and Baird (1850) relayed an account of *Artemia* catching and eating their own nauplii. *Branchinecta gigas*, the largest extant anostracan, catches and eats other smaller fairy shrimp, and possibly other small zooplankton and benthos (Fryer 1966, Daborn 1975). Several species of *Branchinecta* and *Artemiopsis* (Fryer 1966, Daborn 1977b) have the habit of rolling over and scraping the pond substrata with their thoracic legs. Several species of *Branchinecta* have been seen scooping up detritus and then sorting through it. *Branchinecta* species have a row of short curved spines along the inner distal edge of the legs which are probably used for scraping. Even congeneric species show differences in their feeding behavior. *Eubranchipus holmni* has spinier legs than *E. vernalis* and shows more scraping behavior (Modlin 1982). This scraping action seems a preadaption for feeding on benthos and catching larger planktonic prey.

Conchostracans feed somewhat like cladocerans, by drawing water into the carapace to remove food particles using the phyllopods (Kaestner 1970). They probably can remove large particles, such as large algal colonies, small crustaceans, and rotifers. They are able to feed in soft mud.

Notostracans feed by plowing along or through muddy sediments. They appear to pump mud posteriorly and probably extract any organic particles. Food includes algae, amphibian eggs, smaller crustaceans, insect larvae, anostracans, and tadpoles (Kaestner 1970, Dodson 1987). In the lab, notostracans have been observed to kill and eat goldfish (C. Sassaman personal communication).

F. Population Regulation

Cladoceran populations are often controlled by a combination of physical and biological factors, with food limitation and predation being important for several generations. Anostracans, conchostracans, and notostracans typically have one generation each time their habitat appears, in some cases briefly once a year (Belk and Cole 1975, Loring *et al.* 1988). A major strategy is to make as many small resistant eggs as possible in the shortest amount of time.

These eggs then hatch when the pond fills again, and the animals typically show rapid growth due to a combination of high temperatures and physiological capabilities specialized for rapid growth (Loring *et al.* 1988). Population regulation can be accomplished by physiologic factors (e.g., salinity, Broch 1988).

In the laboratory when food is not limiting, *Artemia* females produced the same total number of eggs with little regard as to whether they are mated early or late in life (Browne 1982). They accomplished this by having larger clutches closer together when mated late in life. Only at limiting food levels was there a trade-off between adult lifespan and number of eggs produced.

Populations of *Artemia* in permanent lakes can be reduced by bird predation. Mono Lake, a large saline lake in mideastern California, hosts a dense population of *Artemia*. The *Artemia* occupy the epilimnion (about 15 m deep) from early spring to December. Cooper *et al.* (1984) found that the *Artemia* population declined rapidly in response to feeding by eared grebes using the lake as a migratory stop in August and September. The *Artemia* in Mono Lake typically have two generations per year, with a significant fraction of the second generation being consumed by grebes. The *Artemia* population of Mono Lake lives at great enough depths to be protected from predation by most waterfowl, which are surface feeders or shallow divers. However, waterfowl can be important predators of large branchiopods in shallow ponds (Dodson and Egger 1980).

G. Functional Role in the Ecosystem

Anostracans, conchostracans, and notostracans play an important part in the community ecology of ephemeral ponds (Loring *et al.* 1988). Temporary ponds in southwestern North America, especially in physically simple playas or rock pools, provide community ecologists with simple, highly replicated, and easily manipulated natural communities (Dodson 1987, Loring *et al.* 1988). Also, artificial ponds can be used for community studies of ephemeral pond branchiopods (Tribbey 1965).

Ephemeral ponds in arid areas are closely linked to the surrounding terrestrial landscape (Loring *et al.* 1988). The ponds are an important water source for terrestrial animals, which then act as dispersal agents for crustaceans. Primary production both in the ponds and on the pond bottoms when they dry are an important food source for terrestrial animals. Anostracans are an important part of the diet of many waterfowl, especially dabbling ducks (Swanson *et al.* 1985) feeding in midwestern prairie ponds. *Artemia* in a permanent lake are a significant source of food for California gull chicks (Lenz 1984). The position of *Artemia* in the food web of a permanent saline lake is discussed by Winkler (1977).

VIII. CURRENT AND FUTURE RESEARCH PROBLEMS

Biologists living near known habitats who are willing to work around the vagaries of these animals have an almost unlimited field for genetic, ecological, and behavioral studies (e.g., Loring *et al.* 1988). Areas of great potential include application of electrophoresis to the species question, field studies of patterns of life-history evolution, grazing and predator–prey relationships, individual behavior, population dynamics, community structure, and landscape ecology.

IX. COLLECTING AND REARING TECHNIQUES

Collecting these animals is easy enough when they are in their adult forms. A large-mesh plankton net, a D-frame aquatic dip net, or even a large-bore pipette is sufficient to catch them. The *Eubranchipus* illustrated in this chapter were collected with a spoon. A traditional technique is to collect mud from a dried habitat, add water, and rear the animals that hatch from the resistant eggs.

Rearing is easy. Anostracans and conchostracans eat commercial tropical fish food. Notostracans can be fed anostracans, earthworms, or dried cladoceran fish food. *Artemia* diets are myriad, as are recipes for culture media. The most common culture medium is artificial seawater, although more specialized media may give better viability (Bowen *et al.* 1985).

X. TAXONOMIC KEYS FOR NON-CLADOCERAN GENERA OF BRANCHIOPODS

A. Available Keys and Specimen Preparation

This key applies to those crustaceans that have more than ten pairs of thoracic (or trunk) leaf-like legs and paired compound eyes. As in the cladoceran orders, the taxonomy of the larger branchiopods is undergoing revision based on laboratory studies of hybridization and studies of isozymes of natural populations (e.g., Wiman 1979, Browne and Sallee 1984). However, the genera of the larger branchiopods are probably well known and unlikely to change significantly.

The animals in this key can be identified to genus with a good dissecting microscope, without making dissections. Male or female conchostracans and notostracans may be identified to genus; males only are keyed for Anostraca.

B. Taxonomic Key to Genera of Non-Cladoceran Freshwater Branchiopoda

1a.	Body soft and flexible, not covered by a shell (order Anostraca) (for a key to species see Belk 1975)	3
1b.	Body covered at least in part by a hard shell (Figs. 20.75, and 20.82)	2
2a(1b).	Head and body covered with a clam-like shell that wraps around the legs; the abdomen is enclosed inside the shell (conchostracan orders Laevicaudata and Spinicaudata) (Figs. 20.75 and 20.76)	13
2b.	Head and body covered dorsally with a flattened shell; the abdomen is straight and trails out behind the shell (order Notostraca) (Fig. 20.82)	20
3a(1a).	Seventeen pairs of thoracic legs (Fig. 20.65) .... *Polyartemiella*	
3b.	Eleven pairs of thoracic legs	4
4a(3b).	Abdomen ends in and is bordered by a flattened blade (Fig. 20.66A, B) .... *Thamnocephalus*	
4b.	Abdomen ends in two blade-shaped appendages (cercopods) (Fig. 20.67B)	5
5a(4b).	Large branched "frontal appendage" growing from between the second antennae (similar to that of *Thamnocephalus*, Fig. 20.66C); when unrolled, frontal appendage is longer than the second antennae .... *Branchinella* (in part)	
5b.	Frontal appendage absent, or if present, unbranched and less than half as long as the basal segment of the second antennae	6

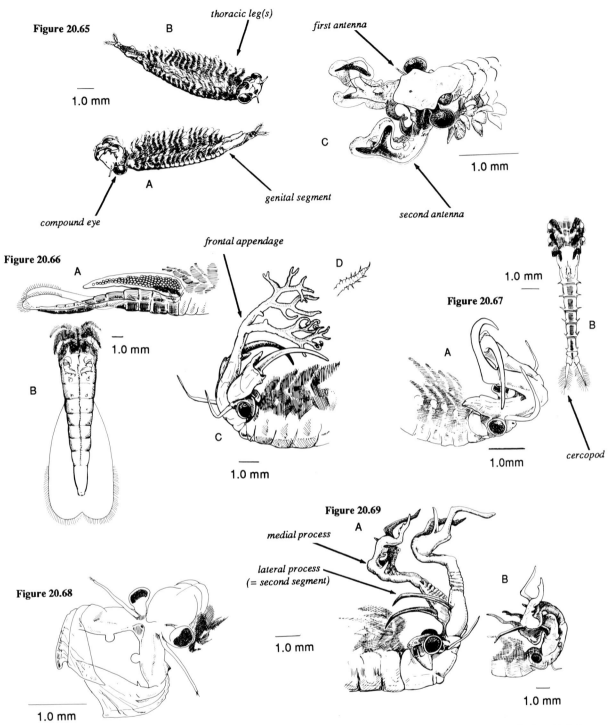

Figure 20.65 *Polyartemiella hazeni* (Murdoch, 1874). Pond #14, 18 July 1973, Point Barrow, Alaska. 18 July 1973. (A) female; (B) male; (C) head of male. (KE) ***Figure 20.66*** *Thamnocephalus platyurus* Packard, 1879. Pond near Ringold, McPherson Co., Nebraska. 28 June 1972. (Coll. S. Cooper). (KE) (A) Female abdomen from side; (B) male abdomen from below; (C) male head; (D) detail of spines on frontal process. ***Figure 20.67*** *Branchinella sublettei* Sissom, 1976. (Coll. D. Belk #717). (A) Head of male; (B) male abdomen. (KE) ***Figure 20.68*** *Artemia franchiscana* (Kellogg 1906). Male head. Alkali lake, Owen's Valley, California. 2 July 1974. (KE) ***Figure 20.69*** *Streptocephalus texanus* Packard, 1871. Texas. (Coll. F. Wiman). (A) Head of male, second antenna unrolled; (B) Head of male, second antenna coiled. (KE)

6a(5b).   Lateral spines at the posterior margin of each abdominal segment, spines about 1/3 as
            long as the width of the segment (Fig. 20.67B) .... *Branchinella* (in part)
6b.   Abdomen without lateral spines 7

7a(6b).   Distal (second) segment of the second antennae a flat triangular blade, the base of the
            triangle about half of the length (Fig. 20.68A) .... *Artemia*
7b.   Distal segment not triangular and thin, instead more or less rounded, but may be
            flattened toward the tip (do not confuse the distal segment with an antennal
            appendage, which is attached on the basal-medial margin of the second antenna and
            often carried folded or rolled up) (Fig. 20.70A) 8

8a(7b).   Antennal appendages present and longer than the basal (first) segment of the second
            antennae 9
8b.   Antennal appendages (projections) on the basal segment, if present, no longer than
            half of the length of the basal segment of the second antennae 10

9a(8a).   Complex (branched) medial process attached at the inner corner of the apex of the
            first segment of the second antennae. The second segment is also attached at the apex
            of the basal segment, but it is outside (lateral to) the medial process and unbranched
            (Fig. 20.69) .... *Streptocephalus*
                [Spicer (1985) gives a phylogenetic analysis of the nine North American species of
                Streptocephalus.]
9b.   Antennal appendage unbranched but with a series of lobes and teeth along at least
            one margin; antennal appendage attached near the proximal end of the basal segment
            of the second antenna 11

10a(8b).   Penis with a stout terminal spine about as long as the eversible (distal) part of the
            penis; several species have a second antenna with branch-like processes on the
            second segment (Fig. 20.70) .... *Eubranchipus*
10b.   Penis without a terminal spine, distal segment of second antenna without branch-like
            processes (Fig. 20.71) .... *Dexteria*

11a(9b).   Second segment of the second antenna with one or two sharp projections from the
            medial surface, about halfway along the segment, and approximately doubling the
            width of the segment at that point; tip of the second segment tapers to a blunt point
            (Fig. 20.72) .... *Artemiopsis*
11b.   Second segment of second antenna with no projections at middle, but sometimes
            flattened, curved, or with a knob at the tip 12

12a(11b).   First segment of second antenna with a dorsal-medial projection from the base of the
            segment, about half as long as segment; this is the only projection on the basal
            segment (Fig. 20.73) .... *Linderiella*
12b.   First segment of second antenna may be without ornamentation; however, usually
            there is a projection, bump, or patch or row of spines on the medial surface of the
            segment from about the middle of the segment, or a ventral-medial projection
            (curving upward) from the base of the segment, or some combination of the above;
            none of the projections more than half as long as the basal segment (Fig.
            20.74) .... *Branchinecta*

13a(2a).   Dorsal union of the carapace valves a true hinge with a groove .... (Fig. 20.75a) order
                                                                                        Laevicaudata)   19
                [Martin and Belk (1988) provide keys for the family Lynceidae in the Americas.]
13b.   Dorsal union of the valves a simple fold .... order Spinicaudata   14

14a(13b).   Dorsal margin of head with a stalked rounded organ (Fig. 20.76), and an occipital
            crest at the posterior end of the dorsal margin of the head 15
14b.   Dorsal margin of head with occipital crest only (Fig. 20.78), may be indistinct 16

15a(14a).   Ventral surface of telson at point of articulation of terminal claws with a spine (Fig.
            20.76) .... *Eulimnadia*
                [Belk (1989) reviews the North American species of *Eulimnadia*, recognizing five of
                the twelve described species as valid, based on traditional morphologic characters
                and new information on egg-shell morphology.]
15b.   Ventral surface of telson at point of articulation of terminal claws without a spine
            (Fig. 20.77) .... *Limnadia*

16a(14b).   Rostrum (snout) with a sharp spine (Fig. 20.78) .... *Leptistheria*
16b.   Rostrum without a spine 17

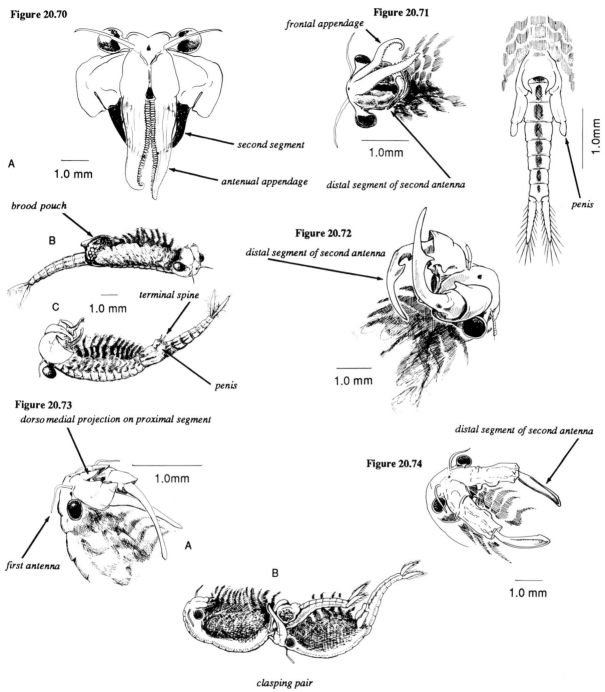

Figure 20.70

frontal appendage

Figure 20.71

second segment

antenual appendage

1.0 mm

A

1.0 mm

distal segment of second antenna

brood pouch

B

Figure 20.72

distal segment of second antenna

penis

1.0mm

terminal spine

C

1.0 mm

1.0 mm

penis

Figure 20.73

dorso medial projection on proximal segment

distal segment of second antenna

1.0mm

Figure 20.74

A

first antenna

B

1.0 mm

clasping pair

Figure 20.70 *Eubranchipus hundyi* Forbes, 1876. AB Woods Pond, Dane Co., Wisconsin. 27 April 1986. (Coll. K. Parejko). (A) Head of male, antennal appendage unrolled; (B) female; (C) male. (KE)   ***Figure 20.71*** *Dexteria floridanus* (Dexter 1953). Alachua Co., Florida. 7 March 1939. From paratypes in US National Museum. (Collection #93537). (KE)   ***Figure 20.72*** *Artemiopsis stephanssoni* (Johansen, 1922). Head of male. (Coll. D. Belk #506). (KE)   ***Figure 20.73*** *Linderiella occidentalis* (Dobbs, 1923). Sulfur Mtn. Pond, Venture Co., California. 2 March 1986. (Coll. S. Copper). (A) Head of male; (B) Male clasping female. (KE)   ***Figure 20.74*** *Branchinecta coloradensis* Packard, 1874. Head of male. Stock Pond in Rock Garden, Grand Co., Utah. 29 October 1986. (KE)

17a(16b).	Occipital crest acute, conspicuous	18
17b.	Occipital crest obtuse, inconspicuous (Fig. 20.79)	*Eocyzicus*
18a(17a).	Male and female rostrum terminates in an acute point (Fig. 20.80)	*Caenestheriella*
18b.	Male rostrum terminates bluntly, anterior margin almost as high as the length of the rostrum (Fig. 20.81), female rostrum acute	*Cyzicus*

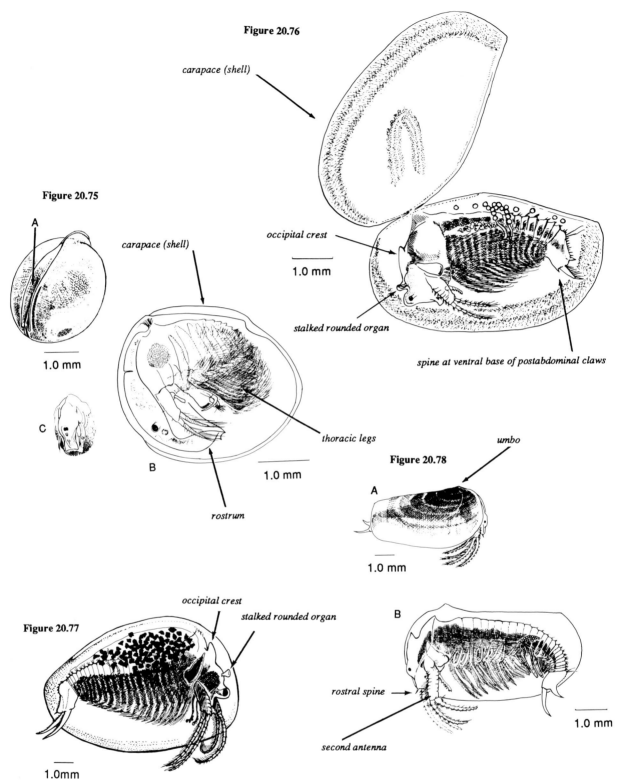

Figure 20.76

carapace (shell)

occipital crest

1.0 mm

stalked rounded organ

spine at ventral base of postabdominal claws

Figure 20.75

A

carapace (shell)

1.0 mm

C

thoracic legs

B

1.0 mm

rostrum

umbo

Figure 20.78

A

1.0 mm

occipital crest

stalked rounded organ

B

Figure 20.77

rostral spine

1.0 mm

1.0mm

second antenna

Figure 20.75 *Lynceus brachyurus* O. F. Mueller, 1785. Dane Co., Wisconsin. (A) Dorsal groove and hinge; (B) lateral view; (C) anterior view showing rostrum shape. (KE)   ***Figure 20.76*** *Eulimnadia* cf. *agassizii* Packard, 1874. Hwy 190, Oxaca, Mexico. 12 July 1976. (Coll. P. Mündel #235). (KE)   ***Figure 20.77*** *Limnadia lenticularis* (Linnacus, 1761). Female within shell. Woods Hole, Massachusetts. 27 August 1887. (USNM coll. Smith). (KE)   ***Figure 20.78*** *Leptisthera compleximanus* (Packard, 1877). Mexico? (A) Female within shell; (B) Female with one valve of shell removed. (KE)

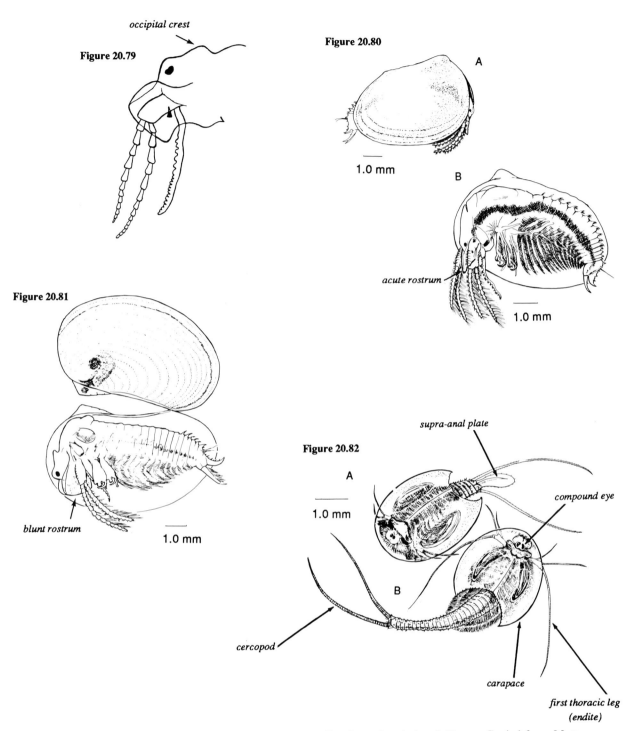

Figure 20.79 *Eocyzicus concavus* (Mackin, 1939). Profile view of male head. Texas. Copied from Mattox, 1959.   **Figure 20.80** *Caenestheriella* cf. *setosa* (Pearse, 1912). (A) Male within shell; (B) male with one valve removed. (KE)   **Figure 20.81** *Cyzicus californicus* (Packard, 1874) Male. Collected east of Lordsburg, Hidalgo Co., New Mexico. 13 August 1955. (Coll. Lynch). (KE)   **Figure 20.82** (A) *Lepidurus couessii* (Packard, 1875). Fish hatchery at Valentine, Cherry Co., Nebraska, Spring 1977. (Coll. D.C. Ashley). (KE) (B) *Triops longicaudatus* LeConte, 1846. Hwy 385, Philips Co., Colorado. 13 June 1972. (Coll. F. Wiman). (KE)

19a(13a).   Ridge running from top of the head down center of rostrum is unbranched; second
              thoracic leg of male is similar to the more posterior legs .... *Lynceus*
                   [Martin *et al.* (1986) provide comparative descriptions of the North American
                   species, and Martin and Belk (1988) review the family Lynceidae in the Americas.]

19b.	Head ridge branched toward the rostrum; second thoracic leg of male is modified, different from the more posterior legs .... *Paralimnetis*
20a(2b).	Last segment of abdomen (the telson) has a medial extension, the supra-anal plate, between the two long cercopods (Fig. 20.82A) .... *Lepidurus*
20b.	Telson is concave between the two cercopods (Fig. 20.82B) .... *Triops*

ACKNOWLEDGMENTS

Frey, a meiobenthic specialist, contributed the keys for the Bosminidae, Chydoridae, and Macrothricidae and discussion of these families, plus the section on paleolimnology. Dodson wrote the remaining sections and keys. Each author has modified the other's contribution. We are grateful for helpful comments from the reviewers, especially James Thorp, Alan Tessier, Brenda Hann, and Denton Belk. The section on non-cladoceran branchiopods could have been written only with the extensive advice of Denton Belk. Many of the figures for this chapter were drawn by Kandis Elliot ("KE" in the figure legends).

LITERATURE CITED

Abreu-Grobois, F. A. 1987. A review of the genetics of *Artemia*. Pages 61–73 *in:* P. Sorgeloos, D. A. Bengtson, W. Decleir, and E. Jaspers, editors. *Artemia* research and its applications. Vol. 1. Universa Press, Wetteren, Belgium.

Akita, M. 1971. On the reproduction of *Triops longicaudatus* (LeConte). Zoological Magazine (*Dobutsugaku Zasshi*) 80:242–250.

Anderson, D. T. 1967. Larval development and segment formation in the branchiopod crustaceans *Limnadia stanleyana* and *Artemia salina* (L.) (Anostraca). Australian Journal of Zoology. 15:47–91.

Anderson, R. S. 1974. Crustacean plankton communities of 340 lakes and ponds in and near the National Parks of the Canadian Rocky Mountains. Journal of the Fisheries Research Board of Canada 31:855–869.

Baird, W. 1850. The natural history of the British Entomostraca. Printed for the Ray Society, London. (Reprinted by Johnson Reprint, New York, 1968.)

Balcer, M. D., N. L. Korda, and S. I. Dodson. 1984. Zooplankton of the Great Lakes: a guide to the identification and ecology of the common crustacean species. Univ. of Wisconsin Press, Madison.

Banta, A. M. 1939. Studies on the physiology, genetics and evolution of some Cladocera. Department of Genetics, Paper No. 39. Carnegie Institution of Washington Publication No. 513.

Bayly, I. A. E. 1986. Aspects of diel vertical migration in zooplankton, and its enigma variations. Pages 349-368 *in:* P. De Deckker and W.D. Williams, editors. Limnology in Australia. CSIRO, Melbourne. Junk, Dordrecht, Netherlands.

Bayly, I. A. E., and D. H. Edward. 1969. *Daphniopsis pusilla* Serventy: a salt-tolerant cladoceran from Australia. Australian Journal of Science 32:21–22.

Beaton, M. J., and P. D. N. Hebert. 1988. Further evidence of hermaphroditism in *Lepidurus arcticus* (Crustacea, Notostraca) from the Melville Peninsula area, N.W.T. Pages 253–257 *in:* W.P. Adams and P.G. Johnson, editors. Student Research in Canada's North. Proceedings of the National Student's Conference on Northern Studies. Association of Canadian Universities for Northern Studies.

Belk, D. 1970. Functions of the conchostracan egg shell. Crustaceana 19:105–106.

Belk, D. 1972. The biology and ecology of *Eulimnadia antlei* Mackin (Conchostraca). The Southwestern Naturalist 16:297–305.

Belk, D. 1975. Key to the Anostraca (Fairy Shrimps) of North America. The Southwestern Naturalist 20:91-103.

Belk, D. 1977a. Zoogeography of the Arizona fairy shrimps (Crustacea: Anostraca). Arizona Academy of Science 12:70–78.

Belk, D. 1977b. Evolution of egg size strategies in fairy shrimps. The Southwestern Naturalist 22:99–105.

Belk, D. 1982. Branchiopoda. Pages 773–780 *in:* S. P. Parker, editor. Synopsis and Classification of Living Organisms. Vol. 2. McGraw-Hill, New York.

Belk, D. 1984. Antennal appendages and reproductive success in the Anostraca. Journal of Crustacean Biology 4:66–71.

Belk, D. 1989. Identification of species in the conchostracan genus *Eulimnadia* by egg shell morphology. Journal of Crustacean Biology 9:115–125.

Belk, D., and M. S. Belk. 1979. Hatching temperatures and new distributional records for *Caenestheriella setosa* (Crustacea: Conchostraca). The Southwestern Naturalist 20:409–411.

Belk, D., and G. A. Cole. 1975. Adaptational biology of desert temporary-pond inhabitants. Pages 207–226 *in:* N. F. Hadley, editor. Environmental physiology of desert organisms. Dowden, Hutchinson & Ross, Stroudsburg, Pennsylvania.

Berner, D. B. 1982. Key to the Cladocera of Par Pond on the Savannah River Plant. A publication of the Savannah River Plant. National Environment Research Park Program, U.S. Department of Energy NERP-SRO-11. Savannah River Ecology Laboratory, Aiken, South Carolina.

Bernice, R. 1971. Food, feeding, and digestion in *Streptocephalus dichotomus* Baird (Crustacea: Anostraca). Hydrobiologia 38:507–520.

Birge, E. A. 1893. Notes on Cladocera, III. Transactions of the Wisconsin Academy of Science, Arts, and Letters 9:275–317.

Birge, E. A. 1910. Notes on Cladocera, IV. Transactions of the Wisconsin Academy of Science, Arts, and Letters 16:1017–1066.

Birge, E. A. 1918. The water fleas (Cladocera), Pages 676–750 *in:* H.B. Ward and G. C. Whipple, editors. Freshwater Biology. Wiley, New York.

Black, R. W. III, and L. B. Slobodkin. 1987. What is cyclomorphosis? Freshwater Biology 18:373–378.

Boucherle, M. M., and H. Züllig. 1983. Cladoceran remains as evidence of change in trophic state in three Swiss lakes. Hydrobiologia 103:141–146.

Bowen, S. T., E. A. Fogarino, K. N. Hitchner, G. L. Dana, V. H. S Chow, M. R. Buoncristiani, and J. R. Carl. 1985. Ecological isolation in *Artemia;* population differences in tolerance of anion concentrations. Journal of Crustacean Biology 5:106–129.

Bowen, S. T., M. R. Buoncristiani, and J.R. Carl. 1988. *Artemia* habitats: ion concentrations tolerated by one superspecies. Hydrobiologia 158:201–214.

Bowman, T. E. 1971. The case of the nonubiquitous telson and the fraudulent furca. Crustaceana 21:165–175.

Brandlova, J., Z. Brandl, and C.H. Fernando. 1972. The cladocera of Ontario with remarks on some species and distribution. Canadian Journal of Zoology. 50:1373–1403.

Brendelberger, H., M. Herbeck, H. Lang, and W. Lampert. 1986. *Daphnia*'s filters are not solid walls. Archiv für Hydrobiologie 107:197–202.

Bretschko, G. 1969. Zur Ephippienablage bei Chydoridae (Crustacea, Cladocera). Zoologischer Anzeiger, Supplement No. 33, Verhandlungen der Zoologische Gesellschaft 1969:95–97.

Broch, E. S. 1969. The osmotic adaptation of the fairy shrimp *Branchinecta campestris* Lynch to saline astatic waters. Limnology and Oceanography 14:485–492.

Broch, E. S. 1988. Osmoregulatory patterns of adaptation to inland astatic waters by two species of fairy shrimps, *Branchinecta gigas* Lynch and *Branchinecta mackini* Dexter. Journal Crustacean Biology 8:383–391.

Brooks, J. L. 1957. The systematics of North American *Daphnia*. Memoirs of the Connecticut Academy of Arts and Sciences 13:1–180.

Brooks, J. L. 1959. Cladocera. Pages 587–656 *in:* W.T. Edmondson, editor. Freshwater Biology. 2nd Edition. Wiley, New York.

Brooks, J. L., and S. I. Dodson. 1965. Predation, body size, and composition of the plankton. Science 150:28–35.

Browne, R. A. 1982. The costs of reproduction in brine shrimp. Ecology 63:43–47.

Browne, R. A., and S. E. Sallee. 1984. Partitioning genetic and environmental components of reproduction and lifespan in *Artemia*. Ecology 65:949–960.

Burns, C. W., and J. J. Gilbert. 1986. Effects of daphnid size and density on interference between *Daphnia* and *Keratella cochlearis*. Limnology and Oceanography 31:848–858.

Butorina, L.G. 1986. On the problem of aggregation of planktonic crustaceans [*Polyphemus pediculus* (L.), Cladocera]. Archiv für Hydrobiologie 105:355–386.

Cannon, H. G. 1933. On the feeding mechanism of the branchiopods. Philosophical Transactions of the Royal Society of London 222:267–352

Carpenter, R., J. F. Kitchell, and J. R. Hodgson. 1985. Cascading trophic interactions and lake productivity. BioScience 35:634–639.

Carpenter, S., editor. 1988. Complex interactions in lake communities. Springer-Verlag, New York.

Carter, J. C. H., M. J. Dadswell, J. C. Roff, and W. G. Sprules. 1980. Distribution and zoogeography of planktonic crustaceans and dipterans in glaciated eastern North America. Canadian Journal of Zoology 58:1355–1387.

Cheer, A. Y. L., and M. A. R. Koehl. 1987. Paddles and rakes: fluid flow through bristled appendages of small organisms. Journal of Theoretical Biology 129:17–39.

Clegge, J. S., and F. P. Conte. 1980. A review of the cellular and developmental biology of *Artemia*. Pages 11–54 *in:* G. Persoone, P. Sorgeloos, O. Roels, and E. Jaspers, editors. The brine shrimp *Artemia*. Vol. 2: Physiology, Biochemistry, Molecular Biology. Universa Press, Wettern, Belgium.

Cooper, S. C., and D. W. Smith. 1982. Competition, predation and the relative abundances of two species of *Daphnia*. Journal of Plankton Research 4:859–879.

Cooper, S.D., D.W. Winkler, and P.H. Lenz. 1984. The effect of grebe predation on a brine shrimp population. Journal of Animal Ecology 53:51–64.

Cotten, C. A. 1985. Cladoceran assemblages related to lake conditions in eastern Finland. Ph.D. Thesis, Indiana University, Bloomington. 96 pp.

Crease, T. J., and P. D. N. Hebert. 1983. A test for the production of sexual pheromones by *Daphnia magna* (Crustacea: Cladocera). Freshwater Biology 13:491–496.

Crisman, T. L. 1980. Chydorid cladoceran assemblages from subtropical Florida. American Society of Limnology and Oceanography Special Symposium 3:657–668. New England.

Cummins, K. W., R. R. Costa, R. E. Rowe, G. A. Moshiri, R. M. Scanlon, and R. K. Zajdel. 1969 Ecological energetics of a natural population of the predaceous zooplankter *Leptodora kindtii* Focke (Cladocera). Oikos 20:189–223.

Daborn, G. R. 1975. Life history and energy relations of the giant fairy shrimp *Branchinecta gigas* Lynch 1937 (Crustacea: Anostraca). Ecology 56:1025–1039.

Daborn, G. R. 1976. The life cycle of *Eubranchipus bundyi* (Forbes) (Crustacea: Anostraca) in a temporary vernal pond of Alberta. Canadian Journal of Zoology 54:193–201.

Daborn, G. R. 1977a. The life history of *Branchinecta mackini* Dexter (Crustacea: Anostraca) in an agrillotrophic lake of Alberta. Canadian Journal of Zoology 55:161–168.

Daborn, G. R. 1977b. On the distribution and biology of an arctic fairy shrimp *Artemiopsis stefanssoni* Johansen, 1921 (Crustacea: Anostraca). Canadian Journal of Zoology. 55:280–287.

Daday, E. 1898. Mikroskopische Süsswasserthiere aus Ceylon. Termeszetrajzi Fuzetek, Anhangsheft 21:1–123.

D'Agusto, A. 1980. The vital requirements of *Artemia;* physiology and nutrition. Pages 56–82 *in:* G. Persoone,

P. Sorgeloos, O. Roels, and E. Jaspers, editors. The brine shrimp Artemia. Vol. 2: Physiology, biochemistry, molecular biology. Universa Press, Wetteren, Belgium.

deBernardi, R. 1984. Methods for the estimation of zooplankton abundance. Pages 59-86 *in:* J. A. Downing, and F. H. Rigler, editors. A manual on methods for the assessment of secondary productivity in fresh waters. 2nd Edition. Blackwell, Oxford.

Deevey, E. S., Jr., and G. B. Deevey. 1971. The American species of *Eubosmina* Seligo (Crustacea, Cladocera). Limnology and Oceanography 16:201–218.

Della Croce, N., and M. Angelino. 1987. Marine Cladocera in the Gulf of Mexico and the Caribbean Sea. Cahiers de Biologie Marine 28:263–268.

Della Croce, N., and E. Gaino. 1970. Osservazioni sulla biologia del maschio di *Penilia avirostris* Dana. Cahiers de Biologie Marine 11:361–365.

DeMott, W. R. 1986. The role of taste in food selection by freshwater zooplankton. Oecologia 69:334–340.

Demott, W. R., and W. C. Kerfoot. 1982. Competition among cladocerans: nature of the interaction between *Bosmina* and *Daphnia*. Ecology 63:1949–1966.

Dodson, S. I. 1974. Adaptive change in plankton morphology in response to size-selective predation: a new hypothesis of cyclomorphosis. Limnology and Oceanography 19:721–729.

Dodson, S. I. 1975. Predation rates of zooplankton in arctic ponds. Limnology and Oceanography 20:426–433.

Dodson, S. I. 1979. Body size patterns in arctic and temperate zooplankton. Limnology and Oceanography 24:940–949.

Dodson, S. I. 1981. Morphological variation of *Daphnia pulex* Leydig (Crustacea: Cladocera) and related species from North America. Hydrobiologia 83:101–114.

Dodson, S. I. 1982. Chemical and biological limnology of six west-central Colorado mountain ponds and their susceptibility to acid rain. American Midland Naturalist 107:173–179.

Dodson, S. I. 1984. Predation of *Heterocope septentrionalis* on two species of *Daphnia;* morphological defenses and their cost. Ecology 65:1249–1257.

Dodson, S. I. 1985. *Daphnia (Ctenodaphnia) brooksi* (Crustacea:Cladocera), a new species from eastern Utah. Hydrobiologia 126:75–79.

Dodson, S. I. 1987. Animal assemblages in temporary desert rock pools: aspects of the ecology of *Dasyhelea sublettei* (Diptera: Ceratopogonidae). Journal of the North American Benthological Society 6:65–71.

Dodson, S. I. 1988a. Cyclomorphosis in *Daphnia galeata mendotae* Birge and *D. retrocurva* Forbes as a predator-induced response. Freshwater Biology 19:109–114.

Dodson, S. I. 1988b. The ecological role of chemical stimuli for the zooplankton: predator-avoidance behavior in *Daphnia*. Limnology and Oceanography 33:1431–1439.

Dodson, S. I. 1988c. The ecological role of chemical stimuli for the zooplankton: predator-induced morphology in *Daphnia*. Oecologia 78:361–367.

Dodson, S. I. 1989. Predator-induced reaction norms. Bioscience 39:447–452.

Dodson, S. I., and D. L. Egger. 1980. Selective feeding of red phalaropes on zooplankton of arctic ponds. Ecology 61:755–763.

Dodson, S. I., and J. E. Havel. 1988. Indirect prey effects: some morphological and life history responses of *Daphnia pulex* exposed to *Notonecta undulata*. Limnology and Oceanography 33:1274–1285.

Donald, D. B. 1982. Erratic occurrence of anostracans in a temporary pond: colonization and extinction or adaptation to variations in annual weather. Canadian Journal of Zoology 61:1492–1498.

Doolittle, A. A. 1911. Descriptions of recently discovered Cladocera from New England. Proceedings of the United States National Museum 41:161–170.

Downing, J. A. 1984a. Assessment of secondary production: the first step. Pages 1–18 *in:* J. A. Downing, and F. H. Rigler, editors. A manual on methods for the assessment of secondary productivity in fresh waters. 2nd Edition. Blackwell, Oxford.

Downing, J. A. 1984b. Sampling the benthos of standing waters. Pages 87–130 *in:* J. A. Downing, and F. H. Rigler, editors. A manual on methods for the assessment of secondary productivity in fresh waters. 2nd Edition. Blackwell, Oxford.

Dumont, H. J., 1987. Groundwater Cladocera: a synopsis. Hydrobiologia 145:169–173.

Dumont, H. J., and J. Pensaert. 1983. A revision of the Scapholeberinae (Crustacea: Cladocera). Hydrobiologia 100:3–45.

Edmondson, W. T., editor. 1959. Freshwater biology. 2nd Edition. Wiley, New York.

Edmondson, W. T. 1960. Reproductive rates of rotifers in natural populations. Memorie dell'Istituto Italiano di Idrobiologia 12:21–77.

Edmondson, W. T. 1963. Pacific Coast and Great Basin. Pages 371–392 *in:* D.G. Frey, editor. Limnology in North America. University of Wisconsin Press, Madison.

Edmondson, W. T. 1987. *Daphnia* in experimental ecology: notes on historical perspectives. Pages 11–30 *in:* R. H. Peters and R. de Bernardi, editors. "*Daphnia.*" Memorie dell'Istituto Italiano di Idrobiologia Vol. 45.

Edmondson, W. T., and A. H. Litt. 1982. *Daphnia* in Lake Washington. Limnology and Oceanography 27:272–293.

Edmondson, W. T., and A. H. Litt. 1987. *Conochilus* in Lake Washington. Hydrobiologia 147:157–162.

Engle, D. L. 1985. The production of haemoglobin by small pond *Daphnia pulex;* intraspecific variation and its relation to habitat. Freshwater Biology 15:631–638.

Eriksen, C. H., and R. J. Brown. 1980a. Comparative respiratory physiology and ecology of phyllopod crustacea. I. Conchostraca. Crustaceana 39:1–10.

Ericksen, C. H., and R. J. Brown. 1980b. Comparative respiratory physiology and ecology of phyllopod crustacea. II. Anostraca. Crustaceana 39:11–21.

Eriksen, C. H., and R. J. Brown. 1980c. Comparative respiratory physiology and ecology of phyllopod crustacea. III. Notostraca. Crustaceana 39:22–32.

Forbes, S. A. 1925. The lake as a microcosm. Bulletin of the Illinois State Laboratory of Natural History (Survey) 15:537–550.

Fox, H. M. 1947. The haemoglobin of *Daphnia*. Proceedings of the Royal Society of London 135:195–212.

Fretwell, S. J. 1987. Food chain dynamics: the central theory of ecology? Oikos 50:291–301.

Frey, D. G. 1962. Cladocera from the Eemian Interglacial of Denmark. Journal of Paleontology 36:1133–1154.

Frey, D. G. 1964. Remains of animals in Quaternary lake and bog sediments and their interpretation. Archiv für Hydrobiologie. Supplement. Ergebnisse der Limnologie 2:1–116.

Frey, D. G. 1965. Differentiation of *Alona costata* Sars from two related species (Cladocera, Chydoridae). Crustaceana 8:159-173.

Frey, D. G. 1966. Phylogenetic relationships in the family Chydoridae. Pages 29–37 *in:* Marine Biological Association of India, Proceedings. Symposium on Crustacea. Part 1.

Frey, D. G. 1969. Symposium on paleolimnology. International Association of Limnology. Mitteilung 17:1–448.

Frey, D. G. 1971. Worldwide distribution and ecology of *Eurycercus* and *Saycia* (Cladocera). Limnology and Oceanography 16:254-308.

Frey, D. G. 1973. Comparative morphology and biology of three species of *Eurycercus* (Chydoridae: Cladocera) with a description of *Eurycercus macrocanthus* sp. nov. Internationale Revue der gesamten Hydrobiologie 58:221–267.

Frey, D. G. 1974. Paleolimnology. Pages 95–123 *in:* W. Rodhe, editor. Jubilee Symposium: 50 years of limnological research. International Association of Limnology. Mitteilung 20:1–402.

Frey, D. G. 1975. Subgeneric differentiation within *Eurycercus* (Cladocera, Chydoridae) and a new species from northern Sweden. Hydrobiologia 46:263–300.

Frey, D. G. 1976. Interpretation of Quaternary paleoecology from Cladocera and midges, and prognosis regarding usability of other organisms. Canadian Journal of Zoology 54:2208–2226.

Frey, D. G. 1978. A new species of *Eurycercus* (Cladocera, Chydoridae) from the southern United States. Tulane Studies in Zoology and Botany 20:1–26.

Frey, D. G. 1980a. On the plurality of *Chydorus sphaericus* (O. F. Muller) (Cladocera, Chydoridae) and designation of a neotype from Sjælsø, Denmark. Hydrobiologia 69:83–123.

Frey, D. G. 1980b. The non-swimming chydorid Cladocera of wet forests, with descriptions of a new genus and two new species. Internationale Revue der gesamten Hydrobiologie 65:613–641.

Frey, D. G. 1982a. Questions concerning cosmopolitanism in Cladocera. Archiv für Hydrobiologie 93:484–502.

Frey, D. G. 1982b. Relocation of *Chydorus barroisi* and related species (Cladocera, Chydoridae) to a new genus and descriptions of two new species. Hydrobiologia 86:231–269.

Frey, D. G. 1982c. The reticulated species of *Chydorus* (Cladocera, Chydoridae): two new species with suggestions of convergence. Hydrobiologia 93:255–279.

Frey, D. G. 1982d. Cladocera. Pages 177–185 *in:* S. H. Hulbert and A. Villalogos-Fiqueroa, editors. Aquatic Biota of Mexico, Central America, and the West Indies. San Diego State University Press, San Diego, California.

Frey, D. G. 1986a. The non-cosmopolitanism of chydorid Cladocera: implications for biogeography and evolution. Pages 237–256 *in:* R. H. Gore and K. L. Heck, editors. Crustacean biogeography. Balkema, Rotterdam.

Frey, D. G. 1986b. Cladocera analysis. Pages 667–692 *in:* B. E. Berglund, editor. Handbook of Holocene Palaeoecology and Palaeohydrology. Wiley, Chicester, England.

Frey, D. G. 1987a. The taxonomy and biogeography of the Cladocera. Hydrobiologia 145:5–17.

Frey, D. G. 1987b. The North American *Chydorus faviformis* (Cladocera, Chydoridae) and the honeycombed taxa of other continents. Philosophical Transactions of the Royal Society of London, Series B 315:353–402.

Frey, D. G. 1988a. Cladocera. Pages 365–369 *in:* R. P. Higgins and H. Thiel, editors. Introduction to the study of meiofauna. Smithsonian Institution Press, Washington, D.C.

Frey, D. G. 1988b. Are there tropicopolitan macrothricid Cladocera? Acta Limnologic Brasiliensia 2:513–525.

Frey, D. G. 1988c. Separation of *Pleuroxus laevis* Sars, 1961, from two resembling species in North America: *Pleuroxus straminius* Birge, 1879 and *P. chiangi* n. sp. (Cladocera, Chydoridae). Canadian Journal of Zoology 66:2534–2563.

Frey, D. G. 1991. The species of *Pleuroxus* and of three related genera (Cladocera, Chydoridae) in southern Australia and New Zealand. Records of the Australian Museum. in press.

Frey, D. G., and B. J. Hann. 1985. Growth in cladocera. Pages 315–335 *in:* A. W. Wenner, editor. Factors in adult growth. Balkema, Boston, Massachusetts.

Fryer, G. 1966. *Branchinecta gigas* Lynch, a non-filter-feeding raptatory anostracan, with notes on the feeding habits of certain other anostracans. Proceedings of the Linnean Society of London 177:19–34.

Fryer, G. 1968. Evolution and adaptive radiation in the Chydoridae (Crustacea: Cladocera): a study in comparative functional morphology and ecology. Philosophical Transactions of the Royal Society of London, Series B, Biological Sciences 254:221–385.

Fryer, G. 1969. Tubular and glandular organs in the Cladocera, Chydoridae. Zoological Journal of the Linnean Society 48:1–8.

Fryer, G. 1970. Defecation in some macrothricid and chydorid cladocerans. Zoological Journal of the Linnean Society 49:255–269.

Fryer, G. 1971. Allocation of *Alonella acutirostris* (Birge) (Cladocera, Chydoridae) to the genus *Disparalona*. Crustaceana 21:221–222.

Fryer, G. 1972. Observations on the ephippia of certain

macrothricid cladocerans. Zoological Journal of the Linnean Society 51:79–96.

Fryer, G. 1974. Evolution and adaptive radiation in the Macrothricidae (Crustacea: Cladocera): a study in comparative functional morphology and ecology. Philosophical Transactions of the Royal Society of London, Series B, Biological Sciences 269:137–274.

Fryer, G. 1983. Functional ontogenetic changes in *Branchinecta ferox* (Milne–Edwards) (Crustacea: Anostraca). Philosophical Transactions of the Royal Society of London, Series B. 303:229–343.

Fryer, G. 1985. Crustacean diversity in relation to the size of water bodies: some facts and problems. Freshwater Biology 15:347–361.

Fryer, G. 1987. A new classification of the branchiopod Crustacea. Zoological Journal of the Linnean Society 91:357–383.

Fryer, G. 1988. Studies on the functional morphology and biology of the Notostraca (Crustacea:Branchiopoda). Philosophical Transactions of the Royal Society of London, Series B. 321:27–124.

Fryer, G. and D. G. Frey. 1981. Two-egged ephippia in the chydorid Cladocera. Freshwater Biology 11:391–394.

Gabriel, W., B. E. Taylor, and S. Kirsch-Prokosch. 1987. Cladoceran birth and death rates estimates: experimental comparisons of egg-ratio methods. Freshwater Biology 18:361–372.

Galat, D.-L., and R. Robinson. 1983. Predicted effects of increasing salinity on the crustacean zooplankton community of Pyramid Lake, Nevada. Hydrobiologia 105:115–131.

Gerritsen, J., and K. G. Porter. 1982. The role of surface-chemistry in filter feeding by zooplankton. Science 216:1225–1227.

Gerritsen, J., K. G. Porter, and J. R. Strickler. 1988. Not by sieving alone: suspension feeding in *Daphnia*. Bulletin of Marine Science 43:366–376.

Gophen, M., and W. Geller. 1984. Filter mesh size and food particle uptake by *Daphnia*. Oecologia 64:408–412.

Goulden, C. E. 1966. The animal microfossils. Pages 84–120 *in*: U. M. Cogill, C. E. Goulden, G. E. Hutchinson, R. Patrick, A. A. Racek, and M. Tsukada, editors. The History of Laguna de Petenxil. Memoirs of the Connecticut Academy of Arts and Sciences Volume 17.

Goulden, C. E. 1968. The systematics and evolution of the Moinidae. Transactions of the American Philosophical Society, Volume 58, New Series, (Part 6):1–101.

Goulden, C. E. 1969. Developmental phases of the biocenosis. Proceedings of the National Academy of Sciences 62:1066–1073.

Goulden, C. E. 1971. Environmental control of the abundance and distribution of the chydorid Cladocera. Limnology and Oceanography 16:320–331.

Goulden, C. E., and D. G. Frey. 1963. The occurrence and significance of lateral head pores in the genus *Bosmina* (Cladocera). Internationale Revueder gesamten Hydrobiologie 48:513–522.

Hall, D. J. 1964. An experimental approach to the dynamics of a natural population of *Daphnia galeata mendotae*. Ecology 45:94–112.

Haney, J. F. and C. Buchanan. 1987. Distribution and biogeography of *Daphnia* in the arctic. Pages 77–105 *in*: R. H. Peters and R. de Bernardi, editors. "*Daphnia*." Memorie dell'Istituto Italiano di Idrobiologia Vol. 45.

Hann, B. J. 1982. Two new species of *Eurycercus* (*Bullatifrons*) from Eastern North America (Chydoridae, Cladocera). Taxonomy, ontogeny, and biology. Internationale Revue der gesamten Hydrobiologie 67:585–610.

Hann, B. J. 1984. Influence of temperature on life-history characteristics of two sibling species of *Eurycercus* (Cladocera, Chydoridae). Canadian Journal of Zoology 63:891–898.

Hann, B. J. 1985. Influence of temperature on life-history characteristics of two sibling species of *Eurycercus* (Cladocera, Chydoridae). Canadian Journal of Zoology 63:891–898.

Hann, B. J. 1986. Revision of the genus *Daphniopsis* Sars, 1903 (Cladocera: Daphniidae) and a description of *Daphniopsis chilensis*, new species, from South America. Journal of Crustacean Biology 6:246–263.

Hann, B. J., and R. Chengalath. 1981. Redescription of *Alonella pulchella* Herrick, 1884 (Cladocera, Chydoridae), and a description of the male. Crustaceana 41:249–262.

Hann, B. J., and P. D. N. Hebert. 1986. Genetic variation and population differentiation in species of *Simocephalus* (Cladocera, Daphniidae). Canadian Journal of Zoology 64:2246–2256.

Hartland-Rowe, R. 1966. The fauna and ecology of temporary pools in Western Canada. Verhandlungen—Internationale Vereinigung für theoretische und angewandte Limnologie 16:577–584.

Hartland-Rowe, R. 1972. The limnology of temporary waters and the ecology of Euphyllopoda. Pages 15–30 *in*: R. B. Clark and R. J. Wootton, editors. Essays in Hydrobiology presented to Leslie Harvey. University of Exeter, Exeter, England.

Havel, J. E. 1987. Predator-induced defenses: a review. Pages 263-278 *in*: W.C. Kerfoot and A. Sih, editors. Predation: direct and indirect impacts on aquatic communities. University Press of New England, Hanover, New Hampshire.

Havel, J. E., and S. I. Dodson. 1984. *Chaoborus* predation on typical and spined morphs of *Daphnia pulex*: behavioral observations. Limnology and Oceanography 29:487–494.

Havel, J. E. and S. I. Dodson. 1987. Reproductive costs of *Chaoborus*-induced polymorphism in *Daphnia pulex*. Hydrobiologia 150:273–281.

Hebert, P. D. N. 1978. The population biology of *Daphnia* (Crustacea, Daphnidae). Biological Review 53:387–426.

Hebert, P. D. N. 1985. Interspecific hybridization between cyclic parthenogens. Evolution 39:216–220.

Hebert, P. D. N. 1987a. Genetics of *Daphnia*. Pages 439–469 *in*: R. H. Peters and R. de Bernardi, editors. "*Daphnia*." Memorie dell'Istituto Italiano di Idrobiologia Vol. 45.

Hebert, P. D. N. 1987b. Genotypic characteristics of the Cladocera. Hydrobiologia 145:183–193.

Hebert, P. D. N. 1988. The comparative evidence. Pages 175–196 *in:* S. C. Stearns, editor. The evolution of sex and its consequences. Birkhauser, Boston, Massachusetts.

Horne, F. R. 1966. Some aspects of ionic regulation in the tadpole shrimp *Triops longicaudatus*. Comparative Biochemistry and Physiology 19:313–316.

Horne, F. R. 1971. Some effects of temperature and oxygen concentration on phyllopod ecology. Ecology 52:343–347.

Horne, F. R., and K. W. Beyenbach. 1974. Physiochemical features of hemoglobin of the Crustacean *Triops longicaudatus*. Archives of Biochemistry and Biophysics 161:369–374.

Hrbáček, J. 1958. Density of the fish population as a factor influencing the distribution and speciation of the species of *Daphnia*. Pages 794–796 *in:* Proceedings of the Fifteenth International Congress of Zoology. Section 10.

Hutchinson, G. E. 1967. A treatise on limnology. Vol. 2: Introduction to lake biology and the limnoplankton. Wiley, New York.

Hyman, L. H. 1926. Note on the destruction of *Hydra* by a chydorid cladoceran, *Anchistropus minor* Birge. Transactions of the American Microscopical Society 45:298–301.

Innes, D. J., and P. D. N. Hebert. 1988. The origin and genetic basis of obligate parthenogenesis in *Daphnia pulex*. Evolution 42:1024–1035.

Jacobs, J. 1987. Cyclomorphosis in *Daphnia*. Pages 325–352 *in:* R. H. Peters and R. de Bernardi, editors. "*Daphnia*." Memorie dell'Istituto Italiano di Idrobiologia Volume 45.

Jurine, L. 1820. Histoire des Monocles, qui se trouvent aux environs de Genève. J.J. Paschoud, Paris.

Kaestner, A. 1970. Invertebrate Zoology. Volume 3: Crustacea (English Edition). Wiley (Interscience), New York.

Karabin, A. 1974. Studies on the predatory role of the cladoceran, *Leptodora kindtii* (Focke), in secondary production of two lakes with different trophy. Ekologia Polska 22:295–310.

Kerfoot, W. C. 1981. Long-term replacement cycles in cladoceran communities: a history of predation. Ecology 62:216–233.

Kerfoot, W. C., and M. Lynch. 1987. Branchiopod communities: associations with planktivorous fish in time and space. Pages 367–378 *in:* W. C. Kerfoot and A. Sih, editors. Predation: direct and indirect impacts on aquatic communities. University Press of New England, Hanover, New Hampshire.

Kerfoot, W. K., and A. Sih, editors. 1987. Predation: direct and indirect impacts on aquatic communities. University Press of New England, Hanover, New Hampshire.

Kerfoot, W. C., D. L. Kellogg, Jr., and J. R. Strickler. 1980. Visual observations of live zooplankters: evasion, escape, and chemical defenses. American Society

of Limnology and Oceanography Special Symposium 3:10–27.

Kitchell, J. F., and S. A. Carpenter. 1987. Piscivores, Planktivores, Fossils, and Phorbins. Pages 132–146 *in:* W. C. Kerfoot and A. Sih, editors. Predation: direct and indirect impacts on aquatic communities. University Press of New England, Hanover, New Hampshire.

Kitchell, J. F., and L. B. Crowder. 1986. Predator–prey interactions in Lake Michigan: model predictions and recent dynamics. Environmental Biology of Fishes 16:205–211.

Klekowski, R. Z., editor. 1978. Proceedings Polskie Archiwum Hydrobiologii 25:1–501.

Kokkinn, M. J., and W. D. Williams. 1987. Is ephippial morphology a useful taxonomic descriptor in the Cladocera? An examination based on a study of *Daphniopsis* (Daphniidae) from Australian salt lakes. Hydrobiologia 145:67–73.

Kořínek, V. 1971. Comparative study of the head pores in the genus *Bosmina* Baird (Crustacea, Cladocera). Vestniu Ceskoslovenske Spolecnosti Zoologicke 35:275–296.

Kořínek, V. 1981. *Diaphanosoma birgei* n.sp. (Crustacea, Cladocera). A new species from America and its widely distributed subspecies *Diaphanosoma birgei* ssp. *lacustris* n.ssp. Canadian Journal Zoology 59:1115–1121.

Krueger, D. A., and S.I. Dodson. 1981. Embryological induction and predation ecology in *Daphnia pulex*. Limnology and Oceanography 26:219–223.

Kubersky, E. S. 1977. Worldwide distribution and ecology of *Alonopsis* (Cladocera: Chydoridae) with a description of *Alonopsis americana* sp. nov. Internationale Revue der gesamten Hydrobiologie 62:649–685.

Lampert, W. 1984. The measurement of respiration. Pages 413–468 *in:* J. A. Downing, and F. H. Rigler, editors. A Manual on Methods for the Assessment of Secondary Productivity in Fresh Waters. 2nd Edition. Blackwell, Oxford.

Lampert, W. 1986a. Response of the respiratory rate of *Daphnia magna* to changing food conditions. Oecologia 70:495–501.

Lampert, W. 1986b. Phytoplankton control of grazing zooplankton: a study on the spring clear-water phase. Limnology and Oceanography 31:478–490.

Lampert, W. 1987a. Vertical migration of freshwater zooplankton: indirect effects of vertebrate predators on algal communities. Pages 291–299 *in:* W. C. Kerfoot and A. Sih, editors. Predation: direct and indirect impacts on aquatic communities. University Press of New England, Hanover, New Hampshire.

Lampert, W. 1987b. Feeding and nutrition in *Daphnia*. Pages 143–192 *in:* R. H. Peters and R. deBernardi, editors. "*Daphnia*." Memorie dell'Istituto Italiano di Idrobiologia Volume 45.

Landon, M. S., and R. H. Stasiak. 1983. *Daphnia* hemoglobin concentration as a function of depth and oxygen availability in Arco Lake, Minnesota. Limnology and Oceanography 28:731–737.

Lehman, J. T. 1987. Palearctic predator invades North American Great Lakes. Oecologia 74:478–480.

Lenz, P. H. 1984. Life-history analysis of an *Artemia*

population in a changing environment. Journal of Plankton Research 6:967–983.

Levitan, C. 1987. Formal stability analysis of a planktonic freshwater community. Pages 71–100 *in:* W.C. Kerfoot and A. Sih, editors. Predation: direct and indirect impacts on aquatic communities. University Press of New England, Hanover, New Hampshire.

Lieder, U. 1962. Beschreibung einer neuen Bosminen-Art, *Neobosmina brehmi* n.sp., aus Äquatorialafrika und über die aus dem Orinoko veschreibene *Neobosmina tubicen* (Brehm), (Crustacea, Cladocera). Internationale Revue der Gesamnten Hydrobiologie 47: 313–320.

Lieder, U. 1983. Die Arten der Untergattung *Eubosmina* Seligo, 1900 (Crustacea: Cladocera, Bosminidae). Mitteilungen zoologisches Museum Berlin 59:195–292.

Lilljeborg, W. 1901. Cladocera Sueciae, oder Beitrage zur Kenntniss der in Schweden lebenden Krebsthiere von der Ordnung der Branchiopoden und der Unterordnung der Cladoceren. Nova Acta regiae Societatis Scientarum Upsaliensis. Series 3. 19:1–701.

Linder, F. 1941. Contributions to the morphology and the taxonomy of the Branchiopoda Anostraca. Zoologiska Bidrag fran Uppsala 10:101–302.

Löffler, H., editor. 1987. Paleolimnology IV. Proceedings of the Fourth International Symposium on Paleolimnology, Ossiach, Carinthia, Austria. Hydrobiologia 143:1–431.

Longhurst, A. R. 1954. Reproduction in Notostraca (Crustacea). Nature (London) 173:781–782.

Loring, S. J., W. P. MacKay, and W. G. Whitford. 1988. Ecology of small desert playas. Pages 89–113 *in:* J. L. Thame and C. D. Ziebel, editors. Small water impoundments in semi-arid regions. University of New Mexico Press, Albuquerque.

Luecke, C., and A. H. Litt. 1987. Effects of predation by *Chaoborus flavicans* of Lake Lenore, Washington. Freshwater Biology 18:185–192.

Luecke, C., and W. J. O'Brien. 1983a. Photoprotective pigments in a pond morph of *Daphnia middendorffiana*. Arctic 36:365–368.

Luecke, C., and W. J. O'Brien. 1983b. The effect of *Heterocope* predation on zooplankton communities in arctic ponds. Limnology and Oceanography 28:367–377.

Lynch, M. 1979. Predation, competition, and zooplankton community structure: an experimental study. Limnology and Oceanography 24:253–272.

Lynch, M. 1980. The evolution of cladoceran life histories. The Quarterly Review of Biology 55:23–42.

Lynch, M. 1982. How well does the Edmondson-Paloheimo model approximate instantaneous birth rates? Ecology 63:12–18.

Lynch, M. 1989. The life history consequences of resource depression in *Daphnia pulex*. Ecology 70:246–256.

MacIsaac, H. J., P. D. N. Hebert, and S. S. Schwartz. 1985. Inter- and intraspecific variation in acute thermal tolerance of *Daphnia*. Physiological Zoology 58:350–355.

Mantel, L. H., vol. editor. 1983. The biology of Crustacea Vol. 5: Internal anatomy and physiological regulation. Academic Press, New York.

Martin, J. W., and D. Belk. 1988. Review of the clam shrimp family Lynceidae Stebbing, 1902 (Branchiopoda: Conchostraca), in the Americas. Journal of Crustacean Biology 8:451–482.

Martin, J. W., B. E. Felgenhauer, and L. G. Abele. 1986. Redescription of the clam shrimp *Lynceus gracilicornis* (Packard) (Branchiopoda, Conchostraca, Lynceidae) from Florida, with notes on its biology. Zoologica Scripta 15:221–232.

Mathias, P. 1937. Biologie des Crustaces Phyllopodes. Actualites Scientifiques et Industrielles 447:1–107.

Mattox, N. T. 1959. Conchostraca. Pages 577–586 *in:* W. T. Edmondson, editor. Freshwater biology. Wiley, New York.

Mayr, E. 1963. Animal species and evolution. Belknap, Cambridge, Massachusetts.

McCauley, E. 1984. The estimation of the abundance and biomass of zooplankton in samples. Pages 228–265 *in:* J. A. Downing, and F. H. Rigler, editors. A Manual on methods for the assessment of secondary productivity in fresh waters. 2nd Edition. Blackwell, Oxford.

McLaughlin, P. A. 1980. Comparative morphology of recent Crustacea. Freeman, San Francisco, California.

McQueen, D. J, J. R. Post, and E. L. Mills. 1986. Trophic relationships in freshwater pelagic ecosystems. Canadian Journal of Fisheries and Aquatic Science. 43:1571–1581.

Mellors, W. K. 1975. Selective predation of ephippial *Daphnia* and the resistance of ephippial eggs to digestion. Ecology 56:974–980.

Melville, G. E., and E. J. Maly. 1981. Vertical distributions and zooplankton predation in a small temperate pond. Canadian Journal of Zoology 59:1720–1725.

Meriläinen, J., P. Huttunen, and R.W. Battarbee, editors. 1983. Paleolimnology. Proceedings of the Third International Symposium on Paleolimnology, Joensuu, Finland. Hydrobiologia 103:318 pp.

Merrill, H. B. 1893. The structure and affinities of *Bunops scutifrons*, Birge. Transactions of the Wisconsin Academy of Science, Arts, and Letters. 9:319–342.

Michael, R. G., and D. G. Frey. 1983. Assumed amphi-Atlantic distribution of *Oxyurella tenuicaudis* (Cladocera, Chydoridae) denied by a new species from North America. Hydrobiologia 106:3–35.

Michael, R. G. and D. G. Frey. 1984. Separation of *Disparalona leei* (Chien, 1970) in North America from *D. rostrata* (Koch, 1841) in Europe (Cladocera, Chydoridae). Hydrobiologia 114:81–108.

Mills, E. L., J. L. Forney, M. D. Clady, and W. R. Schaffner. 1978. Oneida Lake pp. 367–451. *in:* J. A. Bloomfield, editors. Lakes of New York State. Volume 11.

Modlin, R. F. 1982. A comparison of two *Eubranchipus* species (Crustacea: Anostraca). American Midland Naturalist 107:107–113.

Moghraby, A. E. el. 1977. A study on diapause of zooplankton in a tropical river—the Blue Nile. Freshwater Biology 7:207–212.

Monakov A. V. 1972. Review of studies on feeding of aquatic invertebrates conducted at the Institute of Biology of Inland Waters, Academy of Science, USSR.

Journal of the Fisheries Research Board of Canada 29:363–383.

Moore, W. G., and A. Burn. 1968. Lethal oxygen thresholds for certain temporary pond invertebrates and their applicability to field situations. Ecology 49:349–351.

Moore, W. G., and L. H. Ogren. 1962. Notes on the breeding behavior of *Eubranchipus holmani* (Ryder). Tulane Studies in Zoology 9:315–318.

Mordukhai-Boltovskoy, F. D., and I. K. Rivb'yer. 1987. Khishchnyye Vetvistousyye Podonidae, Polyphemidae, Cercopagidae i Leptodoridae Fauny Mira. Opredeliteli po Faunye SSSR, Izdavayemyye Zoologicheskim Institutom AN SSSR, No. 148.

Mossin, J. 1986. Physiochemical factors inducing embryonic development and spring hatching of the European fairy shrimp *Siphonophanes grubei* (Dybowsky) (Crustacea: Anostraca). Journal of Crustacean Biology 6:693–704.

Mueller, W. P. 1964. The distribution of cladoceran remains in surficial sediments from three northern Indiana lakes. Investigations of Indiana Lakes and Streams 6:1–63.

Munuswamy, N., and T. Subramoniam. 1985. Influence of mating on ovarian and shell gland activity in a freshwater fairy shrimp *Streptocephalus dichotomus* (Anostraca). Crustaceana 49:225–232.

Nilsson, N., and B. Pejler. 1973. On the relation between fish fauna and zooplankton composition in north Swedish lakes. Institute of Freshwater Research, Drottningholm. Report 53:51–77.

Norman, A. M., and G. S. Brady. 1867. A monograph of the British Entomostraca belonging to the families Bosminidae, Macrothricidae, and Lynceidae. Natural History Transactions of Northumberland & Durham 1:354–408.

O'Brien, J. E. 1987. Planktivory by freshwater fish: thrust and parry in the pelagia. Pages 3–16 *in:* W. C. Kerfoot and A. Sih, editors. Predation: direct and indirect impacts on aquatic communities. University Press of New England, Hanover, New Hampshire.

Odselius, R., and D.-E. Nilsson. 1983. Regionally different ommatidial structure in the compound eye of the water-flea *Polyphemus* (Cladocera, Crustacea). Proceedings of the Royal Society of London. Series B 217:177–189.

Ohman, M. D. 1988. Behavioral responses of zooplankton to predation. Bulletin of Marine Science 43:530–550.

Ojima, Y. 1958. A cytological study on the development and maturation of the parthenogenetic and sexual eggs of *Daphnia pulex* (Crustacea–Cladocera). Kwansei Gakuin University Annual Studies 6:123–176.

Pennak, R. W. 1978. Freshwater invertebrates of the United States. 2nd Edition. Wiley (Interscience), New York.

Peters, R. H. 1984. Methods for the study of feeding, grazing, and assimilation by zooplankton. Pages 336–412 *in:* J. A. Downing, and F. H. Rigler, editors. A manual on methods for the assessment of secondary productivity in fresh waters. 2nd Edition. Blackwell, Oxford.

Peters, R. H. 1987a. Metabolism in *Daphnia*. Pages 193–243 *in:* R. H. Peters and R. de Bernardi, editors.

"*Daphnia*." Memorie dell'Istituto Italiano di Idrobiologia Volume 45.

Peters, R. H. 1987b. *Daphnia* culture. Pages 483–495 *in:* R. H. Peters and R. de Bernardi, editors. "*Daphnia*." Memorie dell'Istituto Italiano di Idrobiologia Vol. 45.

Porter, K. G. 1973. Selective grazing and differential digestion of algae by zooplankton. Nature (London) 244:179–180.

Porter, K. G. 1977. The plant–animal interface in freshwater ecosystems. American Scientist 65:159–170.

Porter, K. G., and J. D. Orcutt. 1980. Nutritional adequacy, manageability, and toxicity as factors that determine the food quality of green and blue-green algae for *Daphnia*. American Society of Limnology and Oceanography Special Symposium 3:268–281.

Porter, K. G., J. Gerritsen, and J. D. Orcutt, Jr. 1982. The effect of food concentration on swimming patterns, feeding behavior, ingestion, assimilation, and respiration by *Daphnia*. Limnology and Oceanography 27:935–949.

Porter, K. G., Y. S. Feig, and E. F. Vetter. 1983. Morphology, flow regimes, and filtering rates of *Daphnia*, *Ceriodaphnia*, and *Bosmina* fed natural bacteria. Oecologia 58:156–163.

Potts, W. T. W., and C. T. Durning. 1980. Physiological evolution in the Branchiopods. Comparative Biochemistry and Physiology 67B:475–484.

Prepas, E. E. 1984. Some statistical methods for the design of experiments and analysis of samples. Pages 266–335 *in:* J. A. Downing, and F. H. Rigler, editors. A manual on methods for the assessment of secondary productivity in fresh waters, 2nd Edition. Blackwell, Oxford.

Proctor, V. W. 1964. Viability of crustacean eggs recovered from ducks. Ecology 45:656–658.

Rajapaksa, R. 1986. A contribution to the taxonomy, biogeography and gamogenesis of freshwater Cladocera, with special reference to the tropical region. Ph.D. Thesis, University of Waterloo, Waterloo, Ontario.

Rajapaksa, R., and C. H. Fernando. 1982. The first description of the male and ephippial female of *Dadaya macrops* (Daday, 1898) (Cladocera, Chydoridae), with additional notes on this common tropical species. Canadian Journal of Zoology 60:1841–1850.

Rajapaksa, R., and C. H. Fernando. 1987a. Redescription and assignment of *Alona globulosa* Daday, 1898 to a new genus *Notoalona* and a description of *Notoalona freyi* sp. nov. Hydrobiologia 144:131–153.

Rajapaksa, R., and C. H. Fernando. 1987b. Redescription of *Dunhevedia serrata* Daday, 1898 (Cladocera, Chydoridae) and a description of *Dunhevedia americana* sp. nov. from America. Canadian Journal of Zoology 65:432–440.

Ramcharan, C. W., and W. G. Sprules. 1989. Preliminary results from an inexpensive motion analyzer for free-swimming zooplankton. Limnology and Oceanography 34:457–462.

Richard, J. 1892. *Grimaldina Brazzai, Guernella Raphaelis, Moinodaphinia Macquerysi*. Cladòceres nouveaux du Congo. Memoires de la Societe Zoologique de France 5:213–226.

Richman, S. 1958. The transformation of energy by *Daphnia pulex*. Ecological Monographs 28:273–291.

Rigler, F. H., and J. A. Downing. 1984. The calculation of secondary productivity. Pages 19–58 *in*: J. A. Downing, and F. H. Rigler, editors. A Manual on Methods for the assessment of secondary productivity *in* fresh waters. 2nd Edition. Blackwell, Oxford.

Ringelberg, J. 1987. Light induced behavior in *Daphnia*. Pages 285–323 *in*: R. H. Peters and R. deBernardi, editors. "*Daphnia*." Memorie dell'Istituto Italiano di Idrobiologia Volume 45.

Robertson, A. L. 1988. Life histories of some species of Chydoridae (Cladocera: Crustacea). Freshwater Biology 20:75–84.

Rzòska, J. 1961. Observations on tropical rainpools and general remarks on temporary waters. Hydrobiologia 17:265–286.

Sars, G. O. 1900. Description of *Iheringula paulensis* G. O. Sars, a new generic type of Macrothricidae from Brazil. Archiv for Mathematikog Naturvidenskab 22:1–27.

Sars, G. O. 1904. Pacifische Plankton–Crustaceen. (Ergebnisse einer Reise nach dem Pacific. Schauinsland 1896/97.) Zoologische Jahrbücher. Abteilung für Systematik, Oekologie und Geographie der Tiere 19:629–646.

Sars, G. O. 1916. The fresh-water Entomostraca of Cape Province (Union of South Africa). Part I. Cladocera. Annals South African Museum 15:303–351.

Schindler, D. 1977. Evolution of phosphorus limitation in lakes. Science 195:260–262.

Schmidt-Nielsen, K. S. 1984. Scaling: why is animal size so important. Cambridge Univ. Press, London.

Schminke, H. K. 1976. The ubiquitous telson and the deceptive furca. Crustaceana 30:292–300.

Schwartz, S. S. 1984. Life history strategies in *Daphnia*: a review and predictions. Oikos 42:114–122.

Schwartz, S. S., and P. D. N. Hebert. 1987a. Methods for the activation of the resting eggs of *Daphnia*. Freshwater Biology 17:373–379.

Schwartz, S. S., and P. D. N. Hebert. 1987b. Breeding system of *Daphniopsis ephemeralis*: adaptations to a transient environment. Hydrobiologia 145:195–200.

Schwartz, S. S., and P. D. N. Hebert. 1987c. *Daphniopsis ephemeralis* sp. n. (Cladocera, Daphniidae): a new genus for North America. Canadian Journal of Zoology 63:2689–2693.

Schwartz, S. S., B. J. Hann, and P. D. N. Hebert. 1983. The feeding ecology of *Hydra* and possible implications in the structuring of pond zooplankton communities. Biological Bulletin 164:136–142.

Sergeyev, V. N. 1971. Povedeniye i mekhanizm pitaniya *Lathonura rectirostris* (Cladocera, Macrothricidae). Zoologicheskiy Zhurnal 50:1002–1010.

Shan, R. K. 1969. Life cycle of a chydorid cladoceran, *Pleuroxus denticulatus* Birge. Hydrobiologia 34:513–523.

Shan, R. K., 1970. Influence of light on hatching resting eggs of chydorids (Cladocera). Internationale Revue der gesamten Hydrobiologie 55:295–302.

Shan, R. K., and D. G. Frey. 1983. *Pleuroxus denticulatus* and *P. procurvus* (Cladocera, Chydoridae) in North America: distribution, experimental hybridization, and the possiblity of natural hybridization. Canadian Journal of Zoology 61:1605–1617.

Shapiro, J., B. Forsberg, V. Lamarra, M. Lynch, E. Smeltzer, and G. Zoto. 1982. Experiments and experiences in biomanipulation: studies of biological ways to reduce algal abundance and eliminate blue-greens. Limnological Research Center, Interim Report No. 19. University of Minnesota, Minneapolis.

Shaw, S. R., and S. Stone. 1982. Photoreception. Pages 291–367 *in*: H. L. Atwood and D. C. Sandeman, vol. editors. The biology of Crustacea. Vol. 3: Neurobiology: structure and function. Academic Press, New York.

Shen, C.-j., A.-y. Tai, and S.-c. Chiang. 1966. [On the cladoceran fauna of Hsi-song-pang-na and vicinity, Yunnan Province.] Acta Zootaxonomica Sinica 3:29–423. (In Chin.; Engl sum.)

Siebeck, H. O. 1980. Optical orientation of pelagic crustaceans and its consequence in the pelagic and littoral zones. American Society of Limnology and Oceanography Special Symposium 3:28–38.

Sih, A. 1987. Predators and prey lifestyles: an evolutionary and ecological overview. Pages 203–224 *in*: W. C. Kerfoot and A. Sih, editors. Predation: direct and indirect impacts on aquatic communities. University Press of New England, Hanover, New Hampshire.

Sissom, S. L. 1980. An occurrence of *Cycloestheria hislopi* in North America. Texas Journal of Science 32:175–176.

Smirnov, N. N. 1971. Chydoridae Fauny Mira. Fauna SSSR. Nov. Ser. No. 101. Rakoobraznyye. Volume 1. vyp. 2. (Available in English from Israel Program for Scientific Translations, Jerusalem, 1974.)

Smirnov, N. N. 1976. Macrothricidae i Moinidae Fauny Mira. Fauna SSSR. Rakoobraznyye. Volume 1, vyp. 3.

Smirnov, N. N. 1985. *Anchistropus ominosus* sp. n. (Cladocera, Chydoridae) iz Reki Shingy (pritok Amazonki). Zoologicheskiy Zhurnal 64:137–139.

Smirnov, N. N., and B. V. Timms. 1983. A revision of the Australian Cladocera (Crustacea). Records of the Australian Museum, Supplement 1:1–132.

Sorgeloos, P., D. A. Bengtson, W. Decleir, and E. Jaspers, editors. 1987. *Artemia* research and its applications. Vols. 1–3. Universa, Wetteren, Belgium.

Spicer, G. S. 1985. A new fairy shrimp of the genus *Streptocephalus* from Mexico with a phylogenetic analysis of the North American species (Anostraca). Journal of Crustacean Biology 5:168–174.

Sprules, W. G. 1975. Midsummer crustacean zooplankton communities in acid-stressed lakes. Journal of the Fisheries Research Board of Canada 32:389–385.

Stearns, S. C. 1982. The role of development in the evolution of life-histories. Pages 237–258 *in*: J. T. Bonner, editor. The role of development in evolution. Springer-Verlag, New York.

Stenson, J. A. E. 1973. On predation and *Holopedium gibberum* (Zaddach) distribution. Limnology and Oceanography 18:1005–1010.

Strickler, J. R. 1985. Feeding currents in calanoid copepods: two new hypotheses. Pages 459–485 *in*: M. S. Laverack, editor. Physiological Adaptions of Marine Animals. Symposia of the Society for Experimental Biology Volume 89.

Sublette, J. E., and M. S. Sublette. 1967. The limnology of playa lakes on the Llano Estacado, New Mexico and Texas. The Southwestern Naturalist 12:369–406.

Swanson, G. A., M. I. Meyer, and V. O. Adomaitis. 1985. Foods consumed by breeding mallards on wetlands of south-central North Dakota. Journal of Wildlife Management 49:197–203.

Tappa, D. W. 1965. The dynamics of the association of six limnetic species of *Daphnia* in Aziscoos lake, Maine. Ecological Monographs 35:395–423.

Taylor, B. E. 1988. Analyzing population dynamics of zooplankton. Limnology and Oceanography 33:1266–1273.

Tessier, A. J. 1983. Coherence and horizontal movements of patch of *Holopedium gibberum* (Cladocera). Oecologia 60:71–75.

Tessier, A. J. 1986. Comparative population regulation of two planktonic cladocera (*Holopedium gibberum* and *Daphnia catawba*). Ecology 67:285–302.

Tessier, A. J., and C. E. Goulden. 1982. Estimating food limitation in cladoceran populations. Limnology and Oceanography 27:707–717.

Tessier, A. J., and C. E. Goulden. 1987. Cladoceran juvenile growth: implications for competitive ability. Limnology and Oceanography 32:680–685.

Thiel, H. 1963. Zur Entwicklung von *Triops cancriformis* Bosc. Zoologischer Anzeiger 170:62–68.

Thorp, J. H. 1986. Two distinct roles for predators in freshwater assemblages. Oikos 47:75–82.

Threlkeld, S. T. 1979. Estimating cladoceran birth rates: the importance of egg mortality and the egg age distribution. Limnology and Oceanography 24:601–612.

Threlkeld, S. T. 1976. Starvation and the size structure of zooplankton communities. Freshwater Biology 6:489–496.

Threlkeld, S. T. 1986a. Differential temperature sensitivity of two cladoceran species to resource variation during a blue-green algal bloom. Canadian Journal of Zoology 64:1739–1744.

Threlkeld, S. T. 1986b. Life table responses and population dynamics of four cladoceran zooplankton during a reservoir flood. Journal of Plankton Research 8:639–647.

Threlkeld, S. T. 1987a. *Daphnia* life history strategies and resource allocation patterns. Pages 353–366 *in*: R. H. Peters and R. deBernardi, editors. "*Daphnia*." Memorie dell'Istituto Italiano di Idrobiologia Vol. 45.

Threlkeld, S. T. 1987b. *Daphnia* population fluctuations: patterns and mechanisms. Pages 367–388 *in*: R. H. Peters and R. deBernardi, editors. "*Daphnia*." Memorie dell'Istituto Italiano di Idrobiologia Vol. 45.

Tillmann, U., and W. Lampert. 1984. Competitive ability of differently sized *Daphnia* species: an experimental test. Journal of Freshwater Ecology 2:311–323.

Tribbey, B. A. 1965. A field and laboratory study of ecological succession in temporary ponds. Ph.D Thesis, Univ. of Texas, Austin.

Vanni, M. J. 1986. Competition in zooplankton communities: Suppression of small species by *Daphnia pulex*. Limnology and Oceanography 31:1039–1056.

Warren, E. 1901. A preliminary account of the development of the free-swimming nauplius of *Leptodora hyalina* (Lillj.). Proceedings of the Royal Society of London, Ser. B 68:210–218.

Waterman, T., editor. 1961. The physiology of Crustacea. Vol. 1. Academic Press, New York.

Weider, L. J. 1985. Spatial and temporal heterogeneity in a natural *Daphnia* population. Journal of Plankton Research 7:101–123.

Weider, L. J., and W. Lampert. 1985. Differential response of *Daphnia* genotypes to oxygen stress: respiration rates, hemoglobin content and low-oxygen tolerance. Oecologia 65:487–491.

Weismann, A. 1880. Beiträge zur Naturgeschichte der Daphnoiden. VI. Samen und Begattung der Daphnoiden. Zeitschrift für wissenschaftliche Zoologie 33:55–256.

Wetzel, R. G. 1983. Limnology. 2nd Edition. Saunders, Philadelphia, Pennsylvania.

Whiteside, M. C. 1970. Danish chydorid Cladocera: modern ecology and core studies. Ecological Monographs 40:79–118.

Whiteside, M. C., and J. B. Williams. 1975. A new sampling technique for aquatic ecologists. Verh. Int. Ver. Theor. Angew. Limnol. 19:1534–1539.

Whiteside, M. C., J. B. Williams, and C. P. White. 1978. Seasonal abundance and pattern of chydorid Cladocera in mud and vegetative habitats. Ecology 59:1177–1188.

Wiggins, G. B., R. J. Mackay, and I. M. Smith. 1980. Evolutionary and ecological strategies of animals in annual temporary pools. Archiv für Hydrobiologie. Supplement 58:97–206.

Williams, G. C. 1975. Sex and Evolution. Monographs in Population Biology, No. 8. Princeton Univ. Press, Princeton, New Jersey.

Williams, J. L. 1978. *Ilyocryptus gouldeni*, a new species of water flea, and the first American record of *I. agilis* Kurz (Crustacea: Cladocera: Macrothricidae). Proceedings of the Biological Society of Washington 91:666–680.

Wiman, F. H. 1979. Mating patterns and speciation in the fairy shrimp genus *Streptocephalus*. Evolution 33:172–181.

Wiman, F. H. 1981. Mating behavior in the *Streptocephalus* fairy shrimps (Crustacea: Anostraca). The Southwestern Naturalist 25:541–546.

Wingstrand, K. G. 1978. Comparative spermatology of the Crustacea Entomostraca. 1. Subclass Branchiopoda. Det Kongelige Danske Videnskabernes Selskab Biologiske Skrifter 22:1–67.

Winkler, D. W., editor. 1977. An ecological study of Mono Lake, California. Institute of Ecology, Publication Number 12. University of California, Davis.

Wolken, J. J., and G. J. Gallik. 1983. The compound eye of a crustacean. *Leptodora kindtii*. Journal of Cell Biology 26:968–973.

Zaddach, E. G. 1855. *Holopedium gibberum* ein neues

Crustacean aus der Familie Branchiopoden. Arch Naturgesch. 21:159–188.

Zaffagnini, F. and M. Trentini. 1980. The distribution and reproduction of *Triops cancriformis* (Bosc) in Europe (Crustacea Notostraca). Monitore zool. ital. (N.S.) 14:1–8.

Zaret, T. M. 1980. Predation and freshwater communities. Yale University Press, New Haven, Connecticut.

Copepoda

21

Craig E. Williamson
Department of Biology
Lehigh University
Bethlehem, Pennsylvania 18015

Chapter Outline

I. INTRODUCTION

II. ANATOMY AND PHYSIOLOGY
  A. External Morphology
  B. Internal Anatomy and Physiology

III. ECOLOGY AND EVOLUTION
  A. Reproduction and Life History
  B. Distribution
  C. Physiological Adaptations
  D. Behavioral Ecology
  E. Foraging Relationships
    1. Diet
    2. Feeding mechanisms
    3. Feeding rates
  F. Population Regulation
    1. Factors Controlling Population Density
    2. Methods for Analysis of Populations
  G. Functional Role in the Ecosystem

IV. CURRENT AND FUTURE RESEARCH PROBLEMS

V. COLLECTING AND REARING TECHNIQUES
  A. Collection and Preservation
  B. Rearing Techniques

VI. IDENTIFICATION TECHNIQUES
  A. Dissection Techniques for Identification
  B. Characteristics of the Three Major Orders

VII. IDENTIFICATION OF THE ORDER CALANOIDA
  A. General Morphology
  B. Taxonomic Key to Genera of Freshwater Calanoida

VIII. IDENTIFICATION OF THE ORDER CYCLOPOIDA
  A. General Morphology
  B. Taxonomic Key to Genera of Freshwater Cyclopoida

IX. IDENTIFICATION OF THE ORDER HARPACTICOIDA
  A. General Morphology
  B. Taxonomic Key to Genera of Freshwater Harpacticoida
Literature Cited

I. INTRODUCTION

The class Copepoda H. Milne Edwards, 1984, in the subphylum Crustacea, is the largest of the entomostracan classes. Copepods can be distinguished from other small aquatic invertebrates by a variety of morphological characteristics. They have a somewhat cylindrical, segmented body with numerous segmented appendages on the head and thorax, and two setose caudal rami on the posterior end of the abdomen. They possess an exoskeleton, conspicuous first antennae, and a single, simple, anterior eye.

Most authors recognize seven orders of copepods, four of which contain primarily parasitic species (Caligoida, Lernaeopodida, Monstrilloida, and Notodelphyoida). The Caligoida and Lernaeopodida are primarily parasites of freshwater and marine fishes; the Monstrilloida are parasitic in marine polychaetes; and the Notodelphyoida are commensals in tunicates. This chapter will focus on the three orders that contain primarily free-living copepods (Calanoida, Cyclopoida, and Harpacticoida), of which there are over 5500 species. The harpacticoids include approximately 2800 species, 10% of which inhabit freshwaters; calanoids include about 2300 species, 25% of which are freshwater; while there are about 450 marine and freshwater species of cyclopoids (Bowman and Abele 1982). Several cyclopoid species are known to be parasitic.

Free-living freshwater copepods generally range in size from < 0.5–2.0 mm in length, although some species such as the cyclopoids *Macrocyclops fuscus* and *Megacyclops gigas*, and calanoids in several genera including *Heterocope, Epischura, Limno-calanus,* and *Diaptomus* (subgenus *Hesperodiaptomus*), can reach lengths of 3–5 mm. While the vast majority of copepods are transparent or a pale gray or brown, colors may range from black to red, purple, or even blue. The brighter colors are caused by plant pigments such as carotenoids in oil droplets of the copepod.

Copepods are common in a variety of aquatic and semiaquatic habitats ranging from groundwaters and wetlands to lakes and open oceans. Most species are omnivorous to some extent, with foods ranging from detritus and pollen, to phytoplankton, other invertebrates, and even larval fish. Copepods frequently comprise a major portion of the consumer biomass in these habitats. They play a pivotal role in aquatic food webs both as primary and secondary consumers, and as a major source of food for many larger invertebrates and vertebrates.

II. ANATOMY AND PHYSIOLOGY

A. External Morphology

The generalized copepod body consists of an elongate, segmented body with an exoskeleton. There is usually a single major articulation that divides the body functionally into an anterior portion, the prosome, and a posterior portion, the urosome (Figs. 21.1 and 21.2). In cyclopoids and harpacticoids, the major body articulation is between the fifth and sixth thoracic segments (somites of the fourth and fifth legs); while in calanoids, it is between the sixth thoracic segment and the genital segment. The head is fused with the first and sometimes the second thoracic segment to form the cephalothorax. The thorax consists of seven segments, starting with the segment that bears the maxillipeds and ending with the genital segment. The number of segments in the abdomen is variable but is usually between three and five. The prosome thus includes the head and most of the thorax, while the urosome includes the last one or two segments of the thorax and the entire abdomen.

Five pairs of jointed appendages occur on the head: the first antennae (antennules), second antennae, mandibles, first maxillae (maxillules), and second maxillae (Figs. 21.2 and 21.3). The first antennae serve many important functions related to reproduction, locomotion, and feeding. They are armed with both chemoreceptors and mechanoreceptors that may aid in the discrimination between mates, prey, and potential predators. The first antennae of male copepods are geniculate (elbowed) and modified for grasping the female during copula-

tion. In male harpacticoids and cyclopoids, both first antennae are geniculate, while in male calanoids only the right antenna is geniculate. The second antennae of calanoids and harpacticoids are biramous, while those of cyclopoids lack an exopod and are thus uniramous.

Five or six pairs of well-developed thoracic appendages are present. The first thoracic segment bears the maxillipeds, while the second through the fifth bear four pairs of morphologically similar, biramous swimming legs. The swimming legs consist of two broad basal segments (the coxopod and basipod), to which are attached inner (endopod) and outer (exopod) rami of 1–3 segments each (Fig. 21.3). In cyclopoids and harpacticoids, the fifth pair of legs are highly reduced. In calanoids, the fifth legs are well developed and symmetrical in the female and highly asymmetrical and modified for grasping females during copulation in the male (Fig. 21.3). Some male cyclopoids and harpacticoids have vestigial sixth legs. Abdominal appendages are absent, although the abdomen terminates in two caudal rami that are variously armed with terminal setae and spines.

The terminology for defining the segmentation and various body parts of copepods varies. While arguments from homology divide the body into a head, thorax, and abdomen, the body is functionally divided into the prosome and urosome. Some authors refer to the prosome as the metasome, some believe that the maxillipeds are cephalic rather than thoracic appendages, and some consider the genital segment as part of the abdomen rather than as part of the thorax (see Dudley 1986 for further discussion).

B. Internal Anatomy and Physiology

The circulatory system of cyclopoids and harpacticoids consists of a hemocoel with no heart or blood vessels. Blood is circulated by body and gut movements, and respiration occurs through the general body surfaces. Copepods have no gills or other respiratory organs. Calanoids have a more developed circulatory system that includes a dorsal heart with one pair of lateral ostia and one ventral ostium. Blood is carried anteriorly by a short anterior aorta; no other blood vessels are present.

The digestive system of copepods includes a mouth, esophagus (foregut), midgut, hindgut, and anus. The midgut may have a sac-like diverticulum or caecum, and the anus is located at the posterior end of the urosome. The esophagus and hindgut are chitinized. A variable number of labral glands are located around the mouth. They function in a manner analogous to salivary glands by secreting mucus that mixes with food before entering the esophagus.

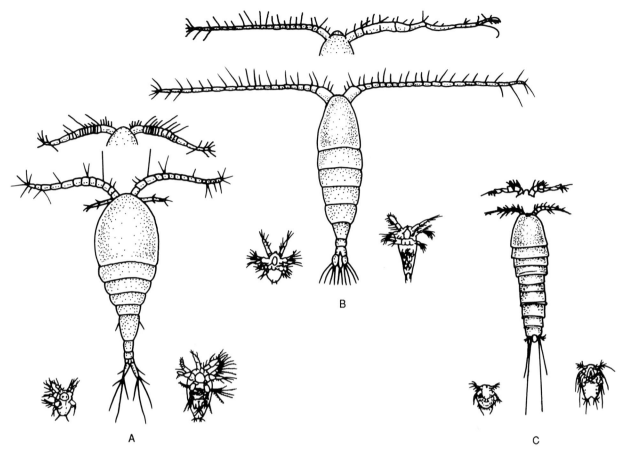

Figure 21.1 Diagrammatic representations of the three major orders of free-living copepods. Adult females are shown in full, while the adult male first antennae and early- and late-stage nauplii are shown for each group. Nauplii redrawn from Gurney (1932), male cyclopoid antennae from Wilson (1932), adult female cyclopoid and calanoid from Williamson (1980, 1987), adult harpacticoids from Coker (1934), calanoid male antennae from Gurney (1931).

The overall structure of the alimentary canal and the numbers and types of associated cells may vary with the feeding habits of the species (Musko 1983).

Nitrogenous excretion occurs through antennal glands in naupliar copepods, and through maxillary glands in adult copepods. Osmoregulation is accomplished largely by low osmolality of the hemolymph (< 200 mosmol, Mantel and Farmer 1983). Egested fecal material is surrounded by a peritrophic membrane secreted by cells in the midgut of at least some calanoid and harpacticoid copepods (Gould 1957, Fahrenbach 1962).

With the exception of a few species of harpacticoids (Sarvala 1979), reproduction in copepods is sexual. Copulation between male and female (followed by union of haploid gametes) is required for the formation of a fertile diploid zygote. Females have a paired or unpaired ovary dorsal to the midgut, with paired oviducts extending anteriorly. Diverticuli extend from the oviducts posteriorly to the gonopore on the genital segment. In calanoids, the diverticuli join just before entering an atrium.

Female calanoids also have a pair of seminal receptacles that open into the atrium. Female cyclopoids have a single seminal receptacle that is connected to the gonopores by fertilization ducts. Glandular cells along the terminal end of the oviducts secrete material for the egg shells. Skin glands adjacent to the gonopore secrete the egg-sac membrane.

Male copepods have a single dorsal gonad (testis). The vas deferens (one in calanoids, two in cyclopoids) is composed of the seminal vesicle, spermatophoric sac, and ejaculatory duct. Male calanoids have a single gonopore, female calanoids and both male and female cyclopoids and harpacticoids have paired gonopores. Spermatophores are elongate and cylindrical in calanoids and harpacticoids, and more compact and often kidney-shaped in cyclopoids.

The nervous system of copepods is composed of a supraesophageal ganglion (brain) connected to a ventral nerve cord by two large circumesophageal connectives. The ventral nerve cord has several thoracic ganglia. Copepods have a variety of sensory

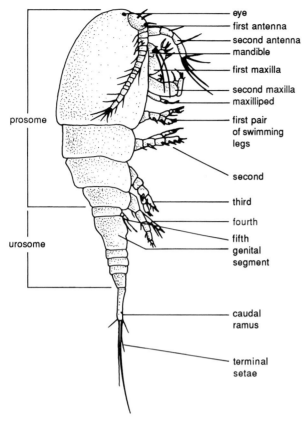

Figure 21.2 Generalized body parts of a copepod. (Redrawn from Smith and Fernando 1978.)

receptors. Freshwater copepods have a simple eye spot that is not capable of forming images; they do not exhibit the highly developed compound eyes with lenses characteristic of some marine pontellids (Park 1966, Wolken and Florida 1969, Gophen and Harris 1981). Most copepods have a frontal organ consisting of fine sensory hairs located on the anterior tip of the body. The organ is connected to the brain by paired nerves and probably functions in chemoreception (Elofsson 1971, Strickler and Bal 1973).

The structures of several other types of sensory receptors have been investigated in freshwater copepods, and their function established by analogy with other known arthropod receptors. The body surfaces of calanoids and cyclopoids are covered with two types of integumental sensilla: small pegs (sensilla basiconica) and fine hairs (sensilla trichodea). The pegs are thought to be chemoreceptors, while the hairs may serve as both chemoreceptors and mechanoreceptors (Strickler 1975a). Cyclopoids have rows of small bristles located at the joints between the segments of the abdomen, the swimming legs, and the caudal rami. These bristles are innervated by proprioceptors that sense the position of the relevant body parts (Strickler 1975b).

The first antennae of many cyclopoids and calanoids have setae that are modified ciliary processes that may serve in mechanoreception (Strickler and Bal 1973, Friedman 1980). The mouthparts of *Diaptomus* (mandibles, mandibular palps, first and second maxillae) have numerous contact chemoreceptors but no mechanoreceptors on their surfaces (Friedman and Strickler 1975; Friedman 1980). Experiments in which copepods were fed flavored and unflavored polystyrene beads confirm the chemoreceptive abilities of copepods; both *Diaptomus* and *Cyclops* selected flavored beads over unflavored beads (DeMott 1986).

For more information on the anatomy, cytology, genetics, and developmental biology of copepods, the reader is referred to the recent reviews in Bliss (1983), Wyngaard and Chinnappa (1982), and Blades-Eckelbarger (1986) and to the extensive monographs of Lang (1948) and Fahrenbach (1962).

III. ECOLOGY AND EVOLUTION

A. Reproduction and Life History

Copepods develop from fertilized eggs that hatch into a larval stage called a nauplius (Fig. 21.1). There are six naupliar stages (N1–N6), followed by six copepodite stages (C1–C6), the last of which is the adult. Growth is determinate, and adults do not continue to molt as in cladocerans. The nauplius larva starts out with a somewhat rounded body with three pairs of appendages: the first antennae, the second antennae, and the mandibles. During development, the body of the nauplius becomes more elongate, additional appendages appear, and existing appendages become more elongate and specialized (Fig. 21.1). A pronounced metamorphosis occurs between the last naupliar and the first copepodite instar. The copepodites are morphologically much more similar to the adults as the body is clearly divided into a prosome and a urosome with caudal rami (Fig. 21.1).

The sexes of adult copepods are dimorphic. The mechanism of sex determination is genetic in some species and not known in others (Wyngaard and Chinnappa 1982). Sexual dimorphism is characterized by differences in the structure of the first antennae and the fifth and sixth legs, as well as by the number of urosomal segments and the generally larger size of the females versus the males (see Section VI). Sexual dimorphism in body size is more pronounced in the cyclopoids than in the calanoids. Cyclopoids have a mean female:male body-length ratio of 1.44, while for calanoids the ratio is 1.13 (Gilbert and Williamson 1983). Morphological variation brought about by allometric growth of certain

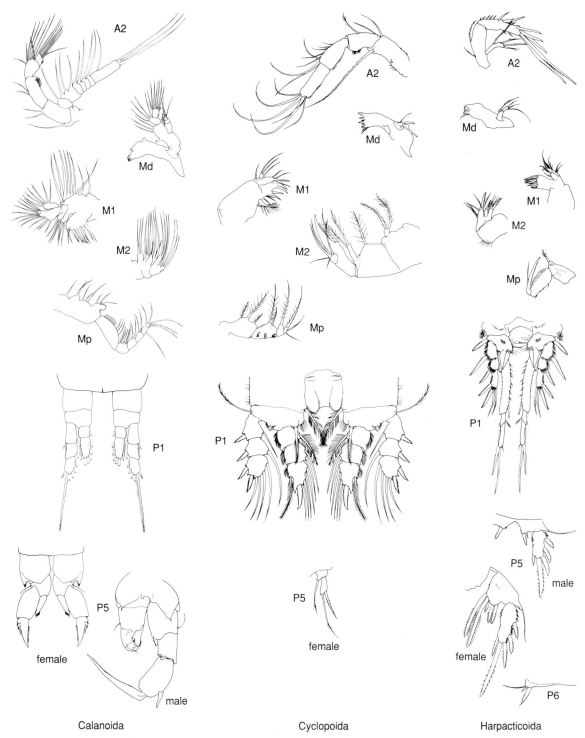

Figure 21.3 Appendages of three major orders of copepods (after Reid 1985, 1987, 1988). Second antenna (A2), mandibles (Md), first maxilla (M1), second maxilla (M2), maxilliped (Mp), and the first (P1), fifth (P5), and sixth (P6) legs.

body parts (cyclomorphosis) is absent or subtle in copepods, although seasonal variation in adult body size is common.

The development time from extrusion to hatching for subitaneous eggs is usually from 1–5 days. Development time from egg to adult is normally on the

order of 1–3 weeks, while the adult lifespan is generally from one to several months. Development times and lifespans are much longer at lower temperatures and may vary considerably with species. Some cyclopoids have periods of diapause that may extend their lifespans up to 1–2 years (Elgmork and Langeland 1980). Males generally mature more rapidly and have shorter lifespans than females.

After reaching maturity, copepods begin to produce gametes and can mate and produce viable embryos (usually called eggs) within a few days of maturation. Under favorable conditions, a single adult female produces a new clutch of eggs every few days and several hundred eggs over her lifetime. Clutches of 2–50 or more eggs may be carried laterally in two egg sacs (cyclopoids), ventrally in a single egg sac (most calanoids and harpacticoids), or scattered freely in the water (certain calanoids such as *Limnocalanus* and *Senecella*). Diel periodicities in egg laying have been reported for *Mesocyclops leuckarti* (Gophen 1978).

Calanoids and harpacticoids may produce either of two types of eggs. The normal mode of reproduction is through subitaneous eggs that hatch within a few days of being extruded. Resting eggs, capable of extended periods of dormancy, may also be produced. Some calanoids, such as *Limnocalanus macrurus,* may produce only resting eggs in some systems but both resting eggs and subitaneous eggs in other systems (Roff 1972). Cyclopoids produce only subitaneous eggs, but may enter diapause during late copepodite stages. These resting stages may vary from a simple developmental arrest to a fully encysted diapausing stage. Cyclopoid diapause is considered a true diapause similar to that observed in insects (Elgmork 1967, 1980, Elgmork and Nilssen 1978); it is more prevalent among planktonic than benthic species (Elgmork 1967, Tinson and Laybourn-Parry 1985). Harpacticoids may also encyst in response to adverse environmental conditions (Sarvala 1979, Frenzel 1980).

Diapause can be initiated during either the spring and summer or the autumn (Elgmork 1967, Vijverberg 1977). The proximate environmental cues that trigger the onset of diapause stages (eggs or copepodites) include overcrowding, photoperiod, and possibly age (Elgmork 1967, Watson and Smallman 1971, Cooley 1978, Walton 1985). The ultimate factors that confer a selective advantage on diapausing individuals may be related to the avoidance of adverse environmental conditions, including an increase in predation pressures (Strickler and Twombly 1975, Elgmork *et al.* 1978, Hairston and Olds 1984, Hairston and Walton 1986). The termination of diapause may be caused by either endogenous mechanisms, exogenous factors (such as low oxygen concentrations, changes in temperature, light, or physical disturbance), or some combination of these (Elgmork 1967).

Female cyclopoids and harpacticoids can store sperm and thus produce fertile clutches of eggs for extended periods of time in the absence of males

(Smyly 1970, Whitehouse and Lewis 1973, Hicks and Coull 1983). Most calanoid copepods on the other hand, must mate repeatedly in order to continue clutch production (Jacobs 1961, Parrish and Wilson 1978). *Diaptomus* must mate before each clutch is produced (Watras and Haney 1980, Williamson and Butler 1987), while *Epischura* can produce multiple clutches from a single mating (Chow-Fraser and Maly 1988). Under favorable environmental conditions, female diaptomids alternate between gravid (oviducts filled with mature ova) and nongravid reproductive phases. Females must mate when they are gravid in order to produce viable eggs (Watras and Haney 1980, Watras 1983b).

Mating behavior is similar for freshwater and marine calanoids (Blades and Youngbluth 1980, Watras 1983a). However, while male marine calanoids can locate females at distances of up to 20 cm with the aid of sex pheromones (Katona 1973, Griffiths and Frost 1976), diaptomids seem to respond to mates at distances of only a few body lengths (Watras 1983a). Watras (1983a) has described a behavioral sequence for mating in diaptomids wherein the male actively pursues the female (pursuit), grasps the female with his geniculate first antenna (precopula 1), grasps the urosome of the female with his chelate right fifth leg (precopula 2), and uses the left fifth leg to attach a cigar-shaped spermatophore to the genital segment of the female (copula). The spermatophore is affixed with the aid of a cement-like secretion. During both precopula 2 and copula, the male and female are aligned with their urosomes adjacent and their prosomes extending in opposite directions. It is likely that precopula is accompanied by stroking behavior as occurs in marine calanoids (Blades and Youngbluth 1980). The entire mating sequence lasts from one to several minutes. Eggs are fertilized upon extrusion from the genital segment. In diaptomids, eggs that are not fertilized are released into the environment and disintegrate rapidly (Watras and Haney 1980, Williamson and Butler 1987).

Mating behavior is similar in harpacticoid and cyclopoid copepods, but the behaviors of both groups differ somewhat from the reproductive activities of calanoids (Holmes 1909, Hill and Coker 1930, Coker 1935, Fahrenbach 1962, Gophen 1979). Male harpacticoids and cyclopoids are more active swimmers than females, and mates are encountered randomly, with no evidence of sex pheromones (as in calanoids). When encountering a female, a male grasps her with both of his geniculate antennae; usually her third or fourth legs or urosome are held, but any part of her body will do. A period of high activity follows, during which the pair jumps and turns through the water. While still grasping the female with his first antennae, the male works himself into the final mating position with his ventral prosome facing the ventral surface of the genital segment of the female. The male strokes this area of the female with his swimming legs and, with sharp flexing of the abdomen, deposits a compact spermatophore on her genital segment. Both individuals may remain passive for 5–10 sec after spermatophore transfer. The entire mating process may take from 5–20 min, after which the male releases the female. Fertilization occurs upon extrusion of the eggs.

Feifarek *et al.* (1983) have shown that there may be a cost to mating for cyclopoid copepods. In a laboratory experiment, female *Mesocyclops edax* that mated and reproduced throughout their lives exhibited lower survivorship than unmated females. Within the group of females that mated, however, there was no relationship between reproductive output and survival. This suggests that the cost is associated with the mating itself rather than with allocation of resources to reproduction versus survival. A similar phenomenon has been reported for fruitflies (Fowler and Partridge 1989).

Environmental factors have a great influence on the life-history characteristics of copepods. Temperature, food availability, and predation are the three factors most often implicated as causes of variations in life-history traits. The interactions between these factors and others are complex, and the relative importance of different factors undoubtedly varies between systems (Elmore 1983, Hicks and Coull 1983, Warren *et al.* 1986). In the field, adult body size and interclutch duration (time between successive clutches) are generally inversely related to temperature, while clutch size shows no consistent relationship to temperature. Clutch size is often positively correlated with female body size (Smyly 1968, Maly 1973, Hebert 1985). Food availability has also been demonstrated to be an important factor limiting rate of egg production and body size both in the laboratory and in the field (Edmondson *et al.* 1962, Edmondson 1964, Elmore 1982, 1983). Food limitation can decrease fecundity by either decreasing clutch size, increasing interclutch duration, or both. The highly transparent bodies of many copepods allow one to monitor the presence of ova in the oviducts as well as external eggs carried by females. The proportion of a copepod population that occurs in each of these reproductive phases at one point in time may be useful in estimating the relative importance of food and (in calanoids) mate limitation in nature (Fig. 21.4) (Williamson and Butler 1987). The presence or absence of ova in the oviducts alone is not by itself, however, a good indicator of reproductive activity (Watras and Haney 1980). This may be due at least in part to mate limitation and to an

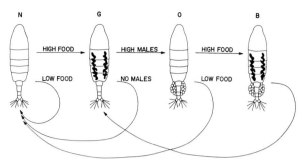

Figure 21.4 Four major reproductive phases of clutch-bearing female diaptomid copepods: G, gravid; O, ovigerous; N, neither; B, both. When abiotic environmental conditions are favorable for reproduction, transitions between phases are influenced by the availability of males and food. Phase B may be absent in species where embryos are released before oocyte maturation. (From Williamson and Butler 1987.)

increase in the percentage of gravid individuals caused by the need for diaptomids to remate before each clutch is produced.

Predation may influence either adult body size (Carter *et al.* 1983), sex ratio (Maly 1970, Vijverberg 1977, Hairston *et al.* 1983), body pigmentation (Hairston 1979a, 1979b, Luecke and O'Brien 1981), or the timing of diapause events in copepods (Strickler and Twombly 1975, Hairston and Olds 1984, Hairston and Walton 1986). Some of these life-history variations have a definite genetic component to them and, therefore, represent evolutionary adjustments rather than just simple acclimatory responses (Hairston 1979a, 1979b, Hairston and Olds 1984, Hairston and Walton 1986, Wynngaard 1986).

High light intensities may induce mortality in copepods (Hairston 1976). The presence of plant-derived carotenoid pigments in the body, low temperature conditions, or nutritional state may reduce these lethal light effects (Hairston 1976, 1979c, Ringelberg 1980, Ringelberg *et al.* 1984, Luecke and O'Brien 1981).

Life-history events of copepods may also be influenced by climate. Florida populations of *Mesocyclops edax* have shorter maturation times, larger bodies, and greater clutch sizes than populations of conspecifics in Michigan (Wyngaard 1986). These differences seem to be largely genotypic, as evidenced by their persistence through two generations in the laboratory under the same controlled environmental conditions.

B. Distribution

Copepods are found in a wide variety of aquatic environments, ranging from the benthic, littoral, and pelagic waters of lakes and oceans to the small volumes of water that collect in the various parts of plants (phytotelmata). Swamps, wetlands, marshes, temporary ponds, and small puddles may all have abundant copepod populations. Several species are common in interstitial and subterranean systems as well. Copepods are present but less abundant in the flowing waters of streams and rivers. Although most prevalent in aquatic habitats, copepods have been collected from many semiterrestrial habitats such as moist forest soil, leaf litter, and even arboreal mosses (Reid 1986). Several species, such as *Eurytemora affinis, Limnocalanus macrurus, Harpacticus gracilis,* and *Nitocra spinipes,* are known to be euryhaline, occurring in brackish and salt water as well as freshwater. *Halicyclops* spp. occur primarily in brackish waters, while *Megacyclops viridis* and *Diaptomus connexus* are common in inland saline lakes.

Calanoids are primarily planktonic, although a few species thrive in shallow ponds. Harpacticoids are generally associated with substrata in the littoral or benthic zones of lakes and ponds. Cyclopoids are also primarily littoral and benthic, although the few species that are planktonic are widespread and may contribute substantially to the zooplankton biomass in lakes and ponds. Benthic copepods generally do not have pelagic larvae, and dispersal may occur in all developmental stages (Hicks and Coull 1983).

The copepod component of the zooplankton is usually dominated by one or two species of calanoid or cyclopoid copepods. The most common and widely distributed planktonic cyclopoid species in North America are probably *Mesocyclops edax, Diacyclops bicuspidatus,* and *Acanthocyclops vernalis* (Hutchinson 1967, Confer *et al.* 1983, Patalas 1986). *Tropocyclops* is often found in association with *Mesocyclops. Mesocyclops edax* is frequently the dominant cyclopoid in subtropical regions and in the warmer months in temperate regions, but is rare or absent from the high arctic, with a northern limit of about 61 degrees north latitude (Wyngaard and Chinnappa 1982, Patalas 1986). In the North and in the cooler months in intermediate climates, *Mesocyclops* is often replaced by either *Diacyclops* or *Acanthocyclops* (Hutchinson 1967). In systems where *Mesocyclops* does not reach high densities (e.g., Lake Michigan), these other cyclopoids may persist through the summer.

The distribution of many copepods seems to be related to temperature. Other cold-water copepods include *Epischura lacustris, Heterocope septentrionalis, Diaptomus minutus, D. oregonensis, D. sicilis, Limnocalanus macrurus,* and *Senecella calanoides.* The three latter species are most common in large, deep lakes and are considered glacial relicts. The only calanoid genus endemic to North America is *Osphranticum.*

The pH of a system may also influence the distribution and abundance of copepods, but the effects vary with species. In a study of 20 small lakes in the northeastern United States ranging in pH from 4.5–7.2, *Mesocyclops edax* and *Diaptomus minutus* were both observed across the full range of pH values; but the former was more abundant at higher pH values while the latter was more abundant at lower pH values (Confer *et al.* 1983). In this same study, *Epischura lacustris* and *Cyclops scutifer* were found only at pH values of 5.3 or above. Bioassay experiments showed 100% mortality within 8.5 hr for *Mesocyclops leuckarti* at pH values of < 5 or > 9.2 (Ramalingam and Raghunathan 1982). The distribution and abundance of various copepod species may also vary with lake size, depth, water transparency, color, and the presence of vertebrate and invertebrate predators (Anderson 1971, 1974, 1980, Patalas 1971). The horizontal distribution of copepods may be influenced by aquatic macrophytes (Gehrs 1974) or fish (Carter and Goudie 1986).

Members of the family Diaptomidae are the most widely distributed and abundant calanoids in the plankton of North America. This group is thought to have undergone more extensive recent speciation than the other copepod groups due to the many species that are endemic to local regions.

It is not uncommon to observe two or more diaptomid species in the same plankton community (Hutchinson 1951, Cole 1961). This coexistence of similar diaptomid species has been the subject of much research due to its apparent contradiction of the competitive exclusion principle. Some mechanisms that have been proposed to explain the coexistence of similar species include food niche partitioning based on size differences and both temporal and spatial habitat partitioning (Fryer 1954, Sandercock 1967, Hammer and Sawchyn 1968). In some systems, differences in the morphology of the feeding appendages may be more important than body size (Maly and Maly 1974). One of the more intriguing hypotheses that has been proposed to explain the size difference between coexisting calanoids is the mating behavior hypothesis (Maly 1984). This hypothesis suggests that once one species of calanoid becomes established in the plankton, the successful invasion of another similarly sized species is highly unlikely due to the high frequency of accidental interspecific matings that lead to nonviable offspring in the invading species. The low initial densities of the invader increase the chances of this invader choosing the incorrect species as a mate. Recent evidence seems to support the importance of size in mating success (DeFrenza *et al.* 1986, Grad and Maly 1988).

In the benthos of lakes, copepods occur primarily in the top 1–2 cm of the sediments, although diapausing animals may be found to depths of 10–20 cm. The species richness of copepods in lake sediments is usually between 7 and 25 species, with about equal numbers of harpacticoids and cyclopoids (Strayer 1985). The vertical distribution of copepods within the sediments is primarily dependent upon redox potential, and only a few species can survive anaerobic sediments (Hicks and Coull 1983).

The abundance of harpacticoids in benthic communities tends to increase with larger sediment particle size (Hicks and Coull 1983). The most common species of harpacticoids are in the cosmopolitan family Canthocamptidae, with the most common genera being *Canthocamptus*, *Attheyella*, and *Bryocamptus*. Common littoral–benthic cyclopoids include *Eucyclops*, *Macrocyclops*, and *Paracyclops*.

C. Physiologic Adaptations

Copepods face a wide range of fluctuating environmental conditions in littoral, benthic, pelagic, and semiterrestrial habitats. Temperature, oxygen, light, pH, and even moisture may all vary tremendously over short distances and brief time periods in these habitats. Vertical gradients in environmental conditions are common in both benthic and pelagic habitats. Surface strata within these habitats are likely to be highly variable over a diel cycle, while deeper strata are generally more constant. In addition, the surface strata of both pelagic and benthic systems often have a higher quality and quantity of food, are warmer, and are better oxygenated than deeper strata. These conditions make the surface strata more conducive to rapid growth and reproduction of copepods. The deeper strata, on the other hand, often have lower predator densities and a more stable abiotic environment—conditions that tend to reduce mortality rates. Many copepods are exposed to further environmental variation during their diel vertical migrations from the surface to deeper strata in both the benthos and the pelagia. These daily migrations vary from lake to lake and season to season, and are thought to be induced by a variety of factors including light and predation. The exposure of copepods to such spatial and temporal variabilities requires that they adjust to diverse environmental conditions.

Perhaps the most widespread physiologic adaptation of copepods is their ability to respond to seasonally unfavorable environmental conditions by reducing their metabolic rate and entering diapause in either the egg or late copepodite stages (see Section III.A). Diapause may be induced by changes in temperature (Smyly 1962) or oxygen concentration (Wierzbicka 1962). These resting stages are highly resistant to temperature extremes and desiccation; they enable copepod populations to persist through periods of low food availability or high predation pressures.

Physiological responses to temperature may vary with the developmental stage and sex of the copepod. The longevity of adult male *Acanthocyclops viridis* may be extended under fluctuating versus constant temperature regimes, whereas female longevity is reduced under similar conditions (Smyly 1980). Nauplii of *Mesocyclops brasilianus* show a decrease in their metabolic rate over the temperature range of 26–28°C, while copepodites and adults exhibit no significant decrease (Epp and Lewis 1979). The variable responses of the different developmental stages is thought to permit the copepodites to maintain a constant metabolic rate during vertical migrations through strata that range from 26.5°–28.5°C in tropical lakes. The nauplii do not migrate.

In benthic harpacticoids, the ability to regulate respiration rates in response to temperature may be related to the fluctuation in environmental temperatures relative to food supply. Surface-dwelling harpacticoid species are exposed to a greater seasonal variability in food availability and temperature than are species that live deeper in the sediments. The surface dwellers are also better at regulating their respiration rates over the appropriate temperature ranges (Gee and Warwick 1984).

Copepods are frequently found in the benthic sediments of productive lakes and ponds that become hypoxic or anoxic. While low oxygen concentrations may induce diapause, many copepods still persist under these rigorous oxygen conditions. A few planktonic species (*Mesocyclops edax*, *Microcyclops varicans*) vertically migrate into anoxic hypolimnetic waters for several hours on a daily basis (Woodmansee and Grantham 1961, Chaston 1969, Williamson and Magnien 1982, Threlkeld and Dirnbirger 1986). Laboratory experiments have demonstrated that adult *M. edax* can tolerate anoxia for 6–12 hr at 24.6°C (Woodmansee and Grantham 1961), while adult *M. varicans* live for at least 36 hr. Cyclopoid nauplii may survive anoxia for less than 1 hr at unspecified temperatures (Chaston 1969). Benthic cyclopoids may persist for at least four days and probably indefinitely under hypoxic conditions (25% saturation, 8°C), but die within 2–5 hr under anoxic conditions at 10°C (Tinson and Laybourn-Parry 1985). Females are more tolerant of anoxia than males, and smaller species resist anoxia longer than larger species. A few species of harpacticoids are also known to tolerate anoxia (Hicks and Coull 1983).

In surface waters where oxygen is more plentiful, light radiation may have lethal effects on copepods due to photo-oxidation (Hairston 1976). The presence of carotenoid pigments such as astaxanthin in the bodies of copepods greatly reduces these lethal effects (Hairston 1976, Luecke and O'Brien 1981, Ringelberg *et al.* 1984). The darker color of the pigmented copepods may, however, make them more vulnerable to visually feeding vertebrate predators. Some species seem to have overcome these conflicting selective pressures by producing two separate morphs: one with dark red pigments that increase survivorship under high light conditions, and one with a carotenoid–protein complex that is a pale green in systems where vertebrate predation pressures are high (Hairston 1976, Luecke and O'Brien 1981).

D. Behavioral Ecology

Copepods are active and highly proficient swimmers that exhibit a diversity of behavioral responses to environmental stimuli ranging from the intensity and wavelength of light to the prospective reward or danger of other approaching organisms. The swimming mode of copepods varies with the species and habitat. Benthic harpacticoids and cyclopoids swim and crawl over substrates in the littoral, benthic, and interstitial habitats. The swimming behavior of planktonic copepods has been investigated in some detail (Rosenthal 1972, Strickler 1975b, 1977, 1982) and has been reviewed recently for cyclopoids (Williamson 1986).

Planktonic cyclopoids use their swimming legs, first antennae, and urosome to swim through the water, alternating between an active hop and a passive sink phase. The hop phase lasts about 20 msec and consists of a posteriorly directed thrust of the first antennae and a metachronal (sequential, one pair at a time, back to front) thrust of each of the pairs of swimming legs. The hops occur about once per second and during normal swimming will accelerate the copepod to a maximum velocity of about 80 mm s^{-1}. The sinking phase is the time between hops during which the copepod remains passive.

Many variations on this generalized swimming behavior occur. For example, the hops may have a reduced intensity and increased frequency and appear to be more of a shuttle behavior than a hop and sink. In addition, some cyclopoids swim right-side-up, while others swim upside-down. When startled, cyclopoids may exhibit escape responses that consist of more vigorous, continuous hops, which may accelerate the copepod up to a velocity of 350 mm s^{-1}.

The swimming behavior of calanoids is distinctly different from that of cyclopoids. Calanoids spend a majority of their time gliding slowly through the water, propelled by the rapid (20–80 Hz) vibration of their feeding appendages. The more carnivorous species, such as *Epischura* and *Limnocalanus,* may actively cruise through the water in an upright position (dorsal surface up), while the more herbivorous species tend to hang vertically or at a slight angle in the water with their anterior end up. Every few seconds the gliding motion may be interrupted by a hop that involves the same appendages and motions as observed in cyclopoids. Less frequently, calanoids stop vibrating their appendages and sink passively through the water.

Nauplii spend a larger proportion of their time in a stationary position than do adults and copepodites; calanoid nauplii vibrate their appendages for only short periods between passive phases, while cyclopoid nauplii hop less frequently than do copepodites or adults (Gerritsen 1978). Nauplii sink very slowly, if at all, but can exhibit escape responses that are the fastest recorded relative to body size for any metazoan (364 body lengths s^{-1} or 55 mm s^{-1}, Williamson and Vanderploeg 1988).

Copepods alter their swimming behavior and thus their distribution in response to many diverse environmental stimuli. Light seems to be a particularly important stimulus regulating the distribution of copepods. For example, *Heterocope septentrionalis* exhibits a swarming behavior in which it preferentially aggregates over light-colored substrata (Hebert *et al.* 1980). This behavior occurs only during the daylight and is most pronounced under conditions of high light and low wind. *Diaptomus tyrelli* may swarm to densities exceeding 11,000 per liter in response to contrasting substrates (Byron *et al.* 1983). Copepods may also respond to different wavelengths of light in different ways. For example, *Diaptomus nevadensis* swims faster in blue light than in red, and individuals that contain large amounts of red carotenoid pigments swim faster than those that do not (Hairston 1976).

Light is also responsible for the distinct "avoidance of shore" (Uferflucht) exhibited by pelagic copepods (Siebeck 1969, 1980). The elevation of the horizon affects the angular distribution of light entering the water column. Copepods orient their body position to this angular distribution of light in such a way that they swim away from the shore. This behavior is absent or less pronounced in littoral species.

The diel vertical migration of planktonic copepods through the water column also seems to be driven largely by responses to light (Hutchinson 1967, Strickler 1969). The most commonly observed migration patterns consist of downward migrations of 10 m or more into the deeper, darker strata at dawn, and a similar distance upward into the surface waters at dusk. Reverse migrations downward at dusk have also been noted (Hutchinson 1967). Adult females often migrate farther than do either males or juveniles.

The adaptive significance of these vertical migrations is thought to be related to the reduction of predation pressures by visually hunting vertebrate predators (Zaret and Suffern 1976, Wright *et al.* 1980). The damaging effects of visible radiation cannot, however, be ruled out as an additional ultimate cause of these migrations. Although ultraviolet radiation is absorbed in the top few meters of water, damaging blue light may penetrate more than 20 m in clear lakes and thus cause substantial mortality by photo-oxidation in copepods that remain in the surface waters throughout the day (Hairston 1976).

Copepods may also respond to changes in the density of food in the environment. Cyclopoid copepods exhibit two distinct types of looping behavior: horizontal and vertical looping (Williamson 1981). Horizontal loops consist of the copepod swimming in horizontal circles in a normal hop and sink swimming mode. When food densities are high, the frequency of horizontal looping behavior increases and serves to maintain the copepod in the vicinity of high food densities. Predatory cyclopoids may swim in rapid, tight, vertical loops after contacting individual prey organisms. This behavior aids the copepods in initially capturing prey and in recovering prey that attempt to escape after an initial attack (Williamson 1981). Calanoids also alter their swimming behavior in response to food. *Senecella calanoides* and *Epischura lacustris* both increase the frequency of their passive sinking phase in the presence of prey. This behavior may reduce the ''noise'' levels of the predators and thereby permit them to approach and capture their prey by surprise (Wong and Sprules 1986).

In addition to responding to food, copepods may also react to predators by altering their behavior. *Diaptomus tyrrelli* reduces its filtering rate (Folt and Goldman 1981), and *D. minutus* decreases the frequency of its hops (Wong *et al.* 1986) in the presence of predatory copepods. Similarly, *Cyclops vicinus* reduces its activity levels when the vertebrate predator *Abramis brama* (bream) is near (Winfield and Townsend 1983). These diminished activity levels decrease the ability of both types of predators to detect these small copepods.

E. Foraging Relationships

The foraging relationships of copepods have received a great deal of attention over the years due to the important intermediate position of copepods between the phytoplankton and fish in aquatic food chains and the fact that copepods are so abundant.

Copepods contribute a major portion of the biomass and secondary production in a wide variety of aquatic communities. The importance of copepods in these communities derives largely from their impact on energy and material processing, prey mortality rates, and depression of available resources resulting from foraging relationships.

1. Diet

Copepods feed on a wide variety of foods including algae, pollen, detritus, bacteria, rotifers, crustaceans, chironomids, and even larval fish (Fryer 1957b, Monakov 1976). Therefore, they occupy three of the four major trophic positions in the food chain: detritivore, herbivore, and carnivore. Most species of calanoids and cyclopoids are omnivorous and opportunistic but tend to be selective when possible. Calanoids ingest particles ranging in size from small ($< 1 \mu m$) bacteria to larger algae, microzooplankton, and macrozooplankton (> 1 mm). Cyclopoids are grasping feeders that generally eat larger foods than calanoids. Their diet may include algae, rotifers, copepods, crustaceans, oligochaetes, chironomids, larval fish, and a variety of other foods.

Harpacticoids consume microalgae, fungi, protozoa, bacteria, and detritus. The algae may include phytoflagellates, cyanobacteria, and both epipelic and epiphytic diatoms. Harpacticoids also appear to use dissolved organic matter as a food source (Hicks and Coull 1983). The microbial communities associated with dead foods may be more important than the foods themselves in the nutrition of harpacticoids. Although carnivory has been reported in harpacticoids, predation seems to be relatively unimportant as a mechanism for obtaining food (Hicks and Coull 1983).

The structure of the feeding appendages often correlates with dietary preferences (Maly and Maly 1974, Wong 1984). Both chemoreception and mechanoreception are important in selection of different types of food (Alcaraz *et al.* 1980, Strickler 1982, Andrews 1983, Price *et al.* 1983, DeMott 1986, Legier-Visser *et al.* 1986, Van Alstyne 1986).

Diet may vary with developmental stage, especially in the more carnivorous species. Nauplii and early copepodites are generally more herbivorous. The transition to more predatory feeding occurs in the late copepodite stages in calanoids (Maly and Maly 1974, Chow-Fraser and Wong 1986), while in cyclopoids even the earlier copepodite stages may be predatory (Jamieson 1980a, Brandl and Fernando 1986, Williamson 1986). Adult females are generally larger in size and correspondingly more voracious than adult males or juveniles. Many species are cannibalistic.

2. Feeding mechanisms

Our knowledge of the feeding mechanisms of marine calanoids has advanced rapidly in recent years with the experimental application of more sophisticated approaches such as high-speed filming, motion analyzers, and three-dimensional video (Alcaraz *et al.* 1980, Koehl and Strickler 1981, Strickler 1982, Price *et al.* 1983, Gill 1987). Some of these techniques have also been applied to freshwater calanoid species (Vanderploeg and Paffenhöfer 1985, Williamson and Vanderploeg 1988).

The feeding mechanisms of suspension-feeding calanoids appear to be similar for marine and freshwater species. Food is brought in toward the copepod on microcurrents created by rapid vibrations of the second antennae, mandibular palps, first maxillae, and maxillipeds (Fig. 21.5), (Koehl and Strickler 1981, Vanderplöeg and Paffenhöfer 1985). Smaller food particles are captured passively and are funneled into the mouth through the setae of the second maxillae (Vanderploeg and Paffenhöfer 1985, Price and Paffenhöfer 1986). During passive captures, the vibrations of the feeding appendages continue uninterrupted. Larger food particles are captured actively with an outward fling of the second maxillae followed by an inward squeeze to re-

move the excess water surrounding the food particle.

When microzooplankton such as rotifers and nauplii are brought in on the feeding currents of suspension-feeding diaptomids, the copepods will often attack these small animal prey from a distance with an active thrust response in which the antennae and swimming legs are used to orient and pounce toward the prey (Williamson and Butler 1986, Williamson 1987, Williamson and Vanderploeg 1988).

In diaptomids, particles less than 5 μm are generally captured passively; those greater than 50 μm are generally captured actively or by attack; while in intermediate size ranges, the frequency of active captures increases with increasing particle size (Vanderploeg and Paffenhöfer 1985, Williamson and Vanderploeg 1988).

Many of the larger calanoid species such as *Heterocope, Epischura, Limnocalanus,* and some of the large species of *Diaptomus* have predatory tendencies and feed on other zooplankton as well as algae. These predators cruise through the water, attack their prey with a pounce, and grasp them with their first and second maxillae. If a capture attempt fails, the copepod may swim in a vertical loop to try again to capture the prey (Kerfoot 1978).

Cyclopoid copepods do not create currents to aid

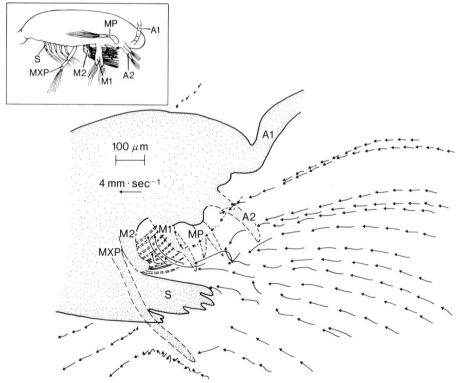

Figure 21.5 Water currents created by normal feeding activity of *Diaptomus*. Continuously moving appendages are indicated by dashed lines. First antenna (A1), second antenna (A2), mandibular palps (Mp), first maxilla (M1), second maxilla (M2), maxilliped (MXP), and swimming legs (S). Insert shows a marine calanoid for comparison. (From Vanderploeg and Paffenhöfer 1985.)

their feeding but instead grasp their food directly with their first maxillae or, to a lesser extent, with their second maxillae and maxillipeds. These appendages push food between the mandibles. By oscillating rapidly during feeding bouts, the mandibles tear prey into pieces and stuff them into the esophagus (Fryer 1957a).

Cyclopoids detect their prey with the help of mechanoreceptors on their first antennae. Prey that generate much disturbance as they swim can be detected at distances of several millimeters. Cyclopoids actively orient and attack prey with considerable precision (Kerfoot 1978). Larger prey such as cladocerans, copepods, and chironomids may be handled for 30 min or more before ingestion (Fryer 1957a, Gilbert and Williamson 1978, Williamson 1980). Small, generally less active prey such as algae, protozoans, rotifers, and nauplii may be detected only after contact; they are generally eaten whole. Cyclopoid nauplii use their mandibles and second antennae to capture and ingest food (Monakov 1976).

Harpacticoid copepods are primarily surface feeders that use their mouthparts for scraping food from a diversity of substrata. Suspension feeding has been reported in two primitive families, the Longipediidae and the Canuellidae (Hicks and Coull 1983). In surface feeding, the second maxillae and second antennae grasp the prey while the maxillipeds aid in anchoring the copepod to the substratum. Once the food is grasped, the first maxillae shred and stuff the food into the vibrating mandibles. The second maxillae and maxillipeds are not used in food handling. Harpacticoid nauplii have modified second antennae that are employed to grasp, masticate, and push food into the mouth (Fahrenbach 1962). The mandibles of nauplii serve only for locomotion and not for feeding.

Suspension-feeding harpacticoids feed by lying on their dorsal or lateral sides and vibrating their first and second maxillae and mandibles. These vibrations create feeding currents that bring water and food in toward the mouth and out posteriorly along the ventral surface of the copepod. Some harpacticoids can use both surface-feeding and suspension-feeding mechanisms (Hicks and Coull 1983).

3. Feeding rates

Feeding rates are usually expressed quantitatively as either filtering rates (equals clearance rate equals feeding rate coefficient, expressed as volume of water cleared of prey per unit time) or ingestion rates (food items ingested per copepod per unit time). Ingestion rates generally depend upon prey density below a saturation food density referred to as the incipient limiting concentration; above this density there is no further increase in ingestion rate. Clearance rates are density independent below this saturation food concentration and decrease above it. Feeding rates are calculated based on the exponential equations first applied to zooplankton feeding by Fuller and Clarke (1936) and later by Gauld (1951). The use of these exponential equations assumes that a constant proportion of the available food in an experimental vessel is consumed per unit of time, so that there is an exponential rather than a linear decrease in food concentration over time. This assumption is valid only below the incipient limiting concentration. The standard experimental design is to fill a number of experimental vessels with either food alone (controls) or food and copepods (experimentals) and then incubate them for a period of time on a rotating wheel to prevent settling and minimize patchiness of the food and copepods. Feeding rates can then be calculated as follows:

$$\text{Ingestion rate } (I) = \frac{D_{ct} - D_{et}}{TN} \qquad (1)$$

$$\text{Filtering rate } (F) = \frac{V(\ln D_{ct} - \ln D_{et})}{TN} \qquad (2)$$

where D_{ct} and D_{et} are the food densities (number per vessel) in the control and experimental vessels at the end of the incubation, T is the duration of the experiment, and N is the number of copepods per experimental vessel. The average food density to which the copepods are exposed during the experiment is then:

$$\bar{d} = \frac{D_{eo} - D_{et}}{V(\ln D_{eo} - \ln D_{et})} \qquad (3)$$

where $\bar{d}$ is the average food density, D_{et} and D_{eo} are the final and initial food densities in the experimental vessels, V is the volume of the experimental vessel, and $I = F\bar{d}$.

An alternative method for estimating feeding rates is to add different numbers of predators to the experimental vessels and use regression analysis to determine filtering rates ((Lehman 1980, Landry and Hassett 1982). This method assumes that changes in the experimental vessels are proportional to the number of predators per vessel and to the incubation time according to the following equation:

$$\ln D_{et} = D_{ct} + FNT \qquad (4)$$

This relationship is derived from the conventional population growth rate equation (see Equation 5). If D_{et}/T is plotted against N, the slope of the regression relationship is the filtering rate (F), the y intercept is an estimate of $\ln D_{eo}$, and standard regression statistics can be used to determine the statistical significance of the regression relationship as well as the percentage of the variation in prey densities that can be attributed to predation (or, how good was the experiment?). An assumption of both of the above methods of estimating feeding rates is that $D_{ct} = D_{eo}$, or, in other words, that prey birth rates as well as nonpredatory death rates were minimal.

Gut content analyses can also be helpful in distinguishing the diet of copepods. Copepods can be squashed between a coverslip and a microscope slide for a crude assessment of gut contents. A more quantitative analysis can be obtained by dissection. A fine pair of minuten needles (insect pins) secured to the tips of applicator sticks or glass pipets can be used to tease the prosome gut tube out of preserved copepods. The gut is placed in a very small drop of water, covered with a coverslip, and examined under a compound microscope before and after the addition of 5% sodium hypochlorite (commercial bleach) (Williamson 1984).

Filtering rates of predatory calanoids are generally in the range of 80–180 ml predator^{-1} day^{-1}, but values have been reported as high as 630 ml predator^{-1} day^{-1} for *Diaptomus shoshone* feeding on calanoid nauplii at 20°C (Maly 1976) and over 1000 ml predator^{-1} day^{-1} for *Heterocope septentrionalis* feeding on *Daphnia pulex* at 15°C (Leucke and O'Brien 1983). Ingestion rates averaging as high as 18.2 prey predator^{-1} day^{-1} have been reported for *Diaptomus shoshone* feeding on various crustacean prey (Anderson 1970); while rates of up to 54 prey predator^{-1} day^{-1} have been observed for *Heterocope septentrionalis* feeding on *Bosmina longirostris* at 15°C (O'Brien and Schmidt 1979).

The feeding rates of suspension-feeding calanoids vary widely and probably depend largely on dietary preferences and available food. Many investigators have obtained filtering rates of only a few ml copepod^{-1} day^{-1} or less on certain algae, and calanoids are thus often thought to have filtering rates that are much lower than similarly sized planktonic cladocerans (Wetzel 1983). However, when offered high quality algae as food, diaptomids filter 19–25 ml copepod^{-1} day^{-1}, rates comparable to or greater than those of cladocerans (Richman et al. 1980, Bogdan and Gilbert 1984, Vanderploeg *et al.* 1984, Williamson and Butler 1986). These filtering rates increase with increasing particle size; filtering and ingestion rates of diaptomids on rotifers may be five to six times greater than those for algae (Williamson and Butler 1986). The presence of algae reduces the predation rate of suspension-feeding diaptomids (Williamson and Butler 1986), but does not similarly affect more predatory species such as *Epischura* (Wong 1981). In general, *Diaptomus* feeds more selectively than *Daphnia* (Richman and Dodson 1983).

Cyclopoid ingestion rates are density dependent and may reach 42 prey predator^{-1} day^{-1} at prey densities approaching 1000 prey liter^{-1} (Williamson 1984, 1986). Clearance rates may exceed 150 ml predator^{-1} day^{-1} (Williamson 1983b). Ingestion rates, clearance rates, and prey selectivities may be influenced by hunger levels of the predator, water temperature, and the size, morphology, behavior, and density of the prey (McQueen 1969, Confer 1971, Li and Li 1979, Williamson 1980, 1986, Stemberger 1986, Roche 1987).

F. Population Regulation

1. Factors Controlling Population Density

Estimates of the per capita growth rates (r) for copepods vary from less than 0.1 day^{-1} to as high as 0.4 day^{-1} (Allan 1976, Hicks and Coull 1983). Doubling times (ln $2/r$) are generally on the order of 1–2 weeks but may be less than two days under optimal conditions. These high reproductive rates generate pronounced density oscillations in nature. These oscillations, in conjunction with 12 distinct life-history stages, small body size, and short generation times, make copepods good subjects for population studies (Rigler and Cooley 1974).

Several hypotheses have been proposed to explain the observed oscillations in copepod populations. The most compelling of these include food limitation, temperature, and predation. Mate limitation may also be important for diaptomids that must mate before the production of each clutch (Williamson and Butler 1987). The relative contributions of each of these factors vary between systems and are difficult to separate in nature.

The physiologic importance of temperature in controlling rates of growth and development has been demonstrated in numerous copepod species. Gamete maturation, egg development, clutch production, and instar development rates are all temperature dependent (Smyly 1974, Gophen 1976, Cooley 1978, Jamieson 1980b, Vijverberg 1980, Watras and Haney 1980, Watras 1983b, Hicks and Coull 1983). In addition, temperature may determine the seasonal timing of reproductive activity. Within the temperature range that permits reproductive activity, food availability may limit the rates of reproduction and development of copepods by prolonging interclutch duration and reducing clutch size (Weglenska 1971, Woodward and White 1981, Elmore 1982, 1983). Response to food limitation may be quite rapid in some species (Williamson *et al.* 1985).

The differential response of the reproductive phases of copepods to food density, mate density, and temperature may permit the analytic separation of food and mate limitation from temperature effects in some diaptomid copepods (Williamson and Butler 1987). The transitions between reproductive phases are largely regulated by food and mate availability and are independent of temperature; hence, the proportion of adult females in the different phases may reflect which factors are limiting (Fig. 21.4). Caution must be used in applying such indices to situations where either temperature extremes or other adverse environmental conditions may cause an arrest or change in the reproductive cycle.

Food limitation can be examined by looking at lipid storage in addition to reproductive condition (Tessier and Goulden 1982). Vanderploeg *et al.* (in preparation) have developed an index to analyze lipid storage and oviducal phase in copepods (Fig. 21.6). This index is similar to that developed for marine copepods (Marshall and Orr 1952) and freshwater cladocerans (Tessier and Goulden 1982). The response of copepods to food limitation will vary with species and environmental conditions, and these indices must therefore be applied only with an adequate understanding of the reproductive physiology of each species.

In many populations, the nauplii are the "bottleneck" stage, which exhibits the highest mortality rate in nature (Confer and Cooley 1977, Feller 1980). The small size of the nauplii relative to copepodites may result in an increased vulnerability to both starvation (due to higher weight-specific metabolic rates) and invertebrate predation, and consequently

higher mortality rates for the nauplii (Dodson 1974, Threlkeld 1976, Confer and Cooley 1977, Williamson *et al.* 1985).

The relative abundance of calanoid versus cyclopoid copepods may change with the trophic status of an ecosystem. Several investigators found that as system productivity rose, the abundance of cyclopoid copepods and cladocerans increased, while calanoid abundance either decreased or varied independently (Brooks 1969, McNaught 1975, Allan 1976, Byron *et al.* 1984, Pace 1986). Feeding and respiration-rate experiments with *Daphnia* and *Diaptomus* indicated that food quality may also play an important role in the relative abundance of calanoids versus cladocerans (Richman and Dodson 1983, Lampert and Muck 1985). Richman and Dodson (1983) suggested that calanoids will predominate when food quality is low or when food density is either extremely high or extremely low. Lampert and Muck (1985) suggested that *Diaptomus* is better adapted to low, fluctuating food densities than is *Daphnia*, because the former allocates more energy to storage for survival while the latter puts its resources into reproduction under these food-limiting conditions.

Predation may also influence copepod population densities. Cannibalism is common among the more carnivorous species (Gabriel 1985), and both vertebrate and invertebrate predators prey on copepods. The inconsistent relationship between calanoid abundance and lake trophic status discussed earlier may be due to invertebrate predation on *Diaptomus* (Edmondson 1985). Other invertebrate predators, including predatory copepods and larvae of the phantom midge *Chaoborus*, may inflict substantial mortality on copepod populations (Dodson 1975, Luecke and Litt 1987). Vertebrate predators, including fish, birds, and larval and adult salamanders, may influence the distribution and abundance of copepods in lakes (Dodson and Egger 1980, Zaret 1980, Carter and Goudie 1986). Vertebrate predators may also be instrumental in the induction of diapause (Hairston and Walton 1986) and may contribute to skewed sex ratios in copepod populations on which they prey (Maly 1970, Hairston *et al.* 1983).

2. *Methods for analysis of populations*

The instantaneous per capita birth rates, death rates, and growth rates of copepods can be estimated using the egg-ratio technique developed by Edmondson (1960) and Edmondson *et al.* (1962), as modified by Paloheimo (1974). The growth rate of the population (r) is estimated from the population densities of adult females at time zero (N_0) and time t (N_t) with the

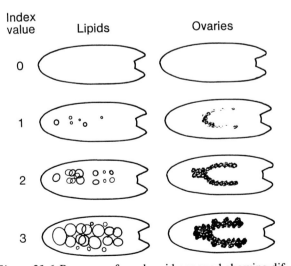

Figure 21.6 Prosome of a calanoid copepod showing different stages of accumulation of lipids, development of ovaries, and corresponding index values. Note that a copepod might have a lipid index value that is different from its ovary index value. [From Vanderploeg *et al.* (in preparation)].

following equation:

$$r = \frac{\ln N_t - \ln N_o}{t} \qquad (5)$$

The birth rate (*b*) can be estimated from the egg ratio (*E*) and the egg development time (*D*) as:

$$b = \frac{\ln (E + 1)}{D} \qquad (6)$$

where *E* = number of female eggs in the population (assume a 1:1 sex ratio) divided by the number of adult females in the population.

The death rate (*d*) can then be estimated from:

$$d = b - r \qquad (7)$$

These estimates of *r* are for the midpoint between the two sampling dates while the estimates of *b* are for the given sample date. Therefore, when estimating *d* from Equation (6), the values of *r* should be linearly interpolated for the given date (Wyngaard 1983). In addition, these methods assume a constant age structure and continuous egg laying in the populations. These assumptions may not always be valid in natural populations (e.g., Gophen 1978, Threlkeld 1979).

A careful analysis of the allocation of effort between and within sampling stations as well as between subsamples is critical in order to maximize the accuracy of population estimates (Nie and Vijverberg 1985). The variance associated with estimates of per capita birth, death, and growth rates can be obtained with either jackknife or bootstrap techniques (Meyer *et al.* 1986). Survivorship data can be analyzed with one of several statistical packages available on most mainframe computers (Pyke and Thompson 1986).

G. Functional Role in the Ecosystem

Copepods make up a major portion of the biomass and productivity of freshwater systems. In Mirror Lake, New Hampshire, copepods contributed about 55% of the total zooplankton biomass over a three-year period (Makarewicz and Likens 1979). Cyclopoid copepods may contribute up to 77% and calanoid copepods up to 49% of the macrozooplankton biomass in some lakes (Pace 1986). The dry weight biomass of calanoids is commonly 5–20 mg m^{-3} and may exceed 50 mg m^{-3}, while that of cyclopoids is commonly 10–50 mg m^{-3} and may exceed 90 mg m^{-3} (Pace 1986). Non-naupliar dry weight productivity estimates for planktonic cyclopoids are between 0.3 and 2.0 mg m^{-3} day^{-1} (Smyly 1973). The annual net productivity of *Diaptomus siciloides* and

Mesocyclops edax have been estimated at 1.39 and 0.37 cal cm^{-2} year^{-1}, respectively (Comita 1972). Copepods may also contribute to phosphorous regeneration in lakes (Bowers 1986).

In the benthic communities of lakes, it is not uncommon to find densities of copepods in excess of 10,000 and occasionally up to even 70,000 m^{-2}, with a biomass of 50–70 mg dry weight m^{-2} (Strayer 1985). Comparable data from marine systems suggest that harpacticoids may be more abundant in shallow-water habitats (400–3600 mg m^{-2}) than in deep-water habitats (4–36 mg m^{-2}) (Hicks and Coull 1983).

Benthic samples in a second-order stream in North Carolina revealed densities of harpacticoids of about 3000–18,000 m^{-2}, and cyclopoid densities of about 100–1600 m^{-2} (O'Doherty 1988). In this same study, the copepods were far more abundant in the sediments than in the associated leaf litter. The benthic harpacticoid *Atheyella* reduced bacterial densities in an Appalachian headwater stream by 9–58%; the greater reductions were observed under lower flow conditions (Perlmutter 1988).

In addition to their numerical contribution, copepods occupy an important intermediate position in aquatic food chains (Kerfoot and DeMott 1984). Their generally omnivorous habits make them important in energy transfer up the food chain; they also have the potential to regulate the structure of phytoplankton and zooplankton prey assemblages. Planktivorous vertebrate predators such as fish preferentially select larger prey items and are rather inefficient at ingesting small prey (O'Brien 1979). By packaging phytoplankton and microzooplankton prey into larger particles (their own bodies) copepods may actually increase food-chain efficiency in spite of the energy lost by the additional intermediate trophic level (Confer and Blades 1975).

As predators, copepods may have a substantial impact on their prey populations. Predatory calanoids and cyclopoids are intense, selective predators that can alter the distribution and morphology of their prey (Kerfoot 1977a,b, O'Brien and Schmidt 1979, Kerfoot and Peterson 1980). *Mesocyclops edax*, an important cyclopoid predator in the plankton, may crop up to 24% of its prey populations per day (Confer 1971, Brandl and Fernando 1979, Williamson 1984). Field correlation data suggest that predatory copepods may be a primary factor controlling the distribution and abundance of rotifers (Anderson 1977, 1980, Stemberger and Evans 1984). Populations of even some of the smaller suspension-feeding diaptomids have the potential to clear the rotifers from a volume of water equivalent to the entire volume of the lake in a single day (Williamson and Butler 1986).

Rotifers, the juvenile stages of copepods, and the smaller and more delicate cladocerans such as *Ceriodaphnia* and juvenile or recently molted *Daphnia* and *Bosmina* are generally the preferred prey of predatory calanoids and cyclopoids. The coincident vertical migration of the copepods with some of the larger cladocerans may, however, cause ingestion rates to be disproportionately greater on these otherwise undesirable prey types.

Many prey species exhibit elaborate morphologic and behavioral defense mechanisms that suggest a strong coevolutionary relationship between copepod predators and their prey. The hard loricas, spines, gelatinous sheaths, and rapid escape responses of many prey species are all effective at reducing predation by copepods (Gilbert and Williamson 1978, Kerfoot 1978, Li and Li 1979, O'Brien and Schmidt 1979, O'Brien *et al.*; 1979, Williamson 1983a, 1986, 1987, Stemberger 1985, Roche 1987). Some of the morphological defenses are nongenetic polymorphisms that are induced by chemical substances released by the copepods (Stemberger and Gilbert 1984). These defenses may come at some expense to the prey (Kerfoot 1977a, Dodson 1984).

Copepods are themselves important prey for other copepods (Confer 1971, Confer and Cooley 1977, Peacock and Smyly 1983), *Chaoborus* larvae (Swift and Fedorenko 1975, Peacock 1982, Wyngaard 1983), and fish. Adult copepods are preyed on by many planktivorous and benthic-feeding fish, and the nauplii are particularly important prey for larval fish (Siefert 1972, Guma'a 1978, Hicks and Coull 1983).

Cyclopoids and calanoids are also important intermediate hosts for several parasites (including flukes, nematodes, and tapeworms present in fish, amphibians, birds, and mammals). *Cyclops* is an intermediate host of the human parasite *Dracunculus medinensis* in the tropics.

IV. CURRENT AND FUTURE RESEARCH PROBLEMS

One of the most striking things about freshwater copepods is how little is known about them in comparison with either cladocerans or the marine copepods. Basic information on things as elementary as life cycles, reproductive biology, feeding ecology, physiology, and genetics is lacking for most species. This is particularly true for the harpacticoids.

In addition to expanding our knowledge of copepodology, research on copepods can contribute to our knowledge of more general ecological questions. The small size, amenability to culture, discrete instars, existence of diapause stages, and variability in life-history patterns permit many interesting questions in population biology to be addressed. The intense and selective feeding of copepods, as well as their importance as food for higher trophic levels, open the way for studies in foraging behavior, community ecology, and systems ecology.

The important position of copepods in aquatic food webs suggests that research on the trophic ecology of copepods will be particularly rewarding. Although a great deal has been learned about the feeding of copepods with recent advances in high-speed filming and video, we still do not understand the relative importance of chemoreception versus mechanoreception in the detection and discrimination of food particles by copepods. In addition, although many copepods are omnivorous, we know little about the relative value of plant versus animal foods in their diets. Good quantitative estimates of feeding rates and selectivities are available for only a very few species.

While we are aware that food limitation may limit copepod populations, we know very little about the relative importance of food in controlling the presence or abundance of different copepod species in nature. We are particularly naive about the interactive effects of environmental variables such as temperature, pH, the presence of alternate food, the hunger level of the predator, and predation risk on the feeding rates and selectivities of copepods.

Another interesting area of research to pursue concerns the relative abundance of cyclopoid versus calanoid copepods across lakes of different trophic status (see Section III.F.1). The densities of cyclopoid copepods and most other zooplankton groups (ciliates, rotifers, nauplii, cladocerans) are positively correlated with the primary productivity of lakes. This does not, however, seem to be universally true for calanoid copepods (Pace 1986). What are the characteristics of calanoids that permit them to respond differently from most other groups to variations in primary productivity?

Ecological genetics is another area that is advancing rapidly and is ripe for future study. Little is known about the endogenous versus exogenous control of the wide range of variations in life-history characteristics observed for copepods in different systems or through different seasons. Many interesting evolutionary questions can be asked about the timing and control of life-history events of copepods due to the genetic isolation of populations and variability in selective pressures acting on these populations between lakes.

V. COLLECTING AND REARING TECHNIQUES

A. Collection and Preservation

Planktonic copepods can be collected with a plankton net; if more quantitative estimates are required, a Schindler trap (Schindler 1969) or a Van Dorn bottle and rotifer cone (Likens and Gilbert 1970) should be used. Integrative samples can also be collected over 5 m of the water column with a device constructed of flexible dryer-vent hose (Lewis and Saunders 1979). Sampling devices with narrow openings and pump samplers may elicit avoidance responses in some of the larger copepods and should be tested for sampling efficiency. Quantitative samples must be taken at night for species that exhibit diel vertical migrations into the sediments. Benthic copepods and the resting stages of planktonic species can be collected with either an Ekman grab sampler or, in a somewhat more quantitative fashion, with a core sampler. Separation of the smaller copepods from the sediments presents difficulties. Sieving and sorting are tedious and not very efficient for copepods. Density-gradient centrifugation seems to be a more promising method (Strayer 1985).

Copepods can be preserved in a 10% solution of formalin (10 ml commercially available formaldehyde to 90 ml sample). Prior narcotization of copepods in carbonated water may enhance the retention of the gut contents of some species when they are placed in formalin (Gannon and Gannon 1975) but is not always necessary (Williamson 1984). The retention of clutches of eggs may be enhanced by prior narcotization in a 1 : 10 solution of chloroform:ethanol (Williamson and Butler 1987). The chilling and addition of 40 g of sucrose per liter of formalin commonly employed in the preservation of cladocerans (Prepas 1978) is not required, but it also has no adverse effect on copepods. The adult stages of copepods can be counted in a Bogorov chamber under a dissecting microscope (Gannon 1971), while the nauplii are more easily counted in a Sedgwick-Rafter chamber under a compound microscope.

Several techniques have been developed for the separation of copepods in samples. Many live copepods can be separated from cladocerans with siphon tubes that remove the slower cladocerans but not the evasive copepods; or one can take advantage of the greater resistance of copepods to pH shock (Bulkowski *et al.* 1985).

Live copepods can be separated from dead copepods by staining with neutral red (Dressel *et al.* 1972, Flemming and Coughlan 1978). A stock solution is made with 0.5 g of neutral red powder per liter

of distilled water. This stock solution is then added to the sample containing the copepods in a 2:100 ratio for a staining period of up to 1 hr before preservation in formalin. The prior addition of 4 g of NaOH per liter of formalin reduces the leaching of stain from the copepods for up to 28 weeks. The high pH makes the stain much less visible, however, and glacial acetic acid must be used to titrate the sample back down to an acid pH to regenerate the deep red color before sorting.

Techniques have also been developed for the critical examination of the anatomy of copepods. The external anatomy, including sensilla and other delicate integumental organs, can be prepared by critical-point drying and plating for examination under a scanning electron microscope, or cleared, stained, and examined under a light microscope (Fleminger 1973, Friedman and Strickler 1975). The gonads of some copepods can be selectively stained with fast green for easier visualization (Batchelder 1986), and techniques are available for the histopathological examination of copepods (Barszcz and Yevich 1976).

B. Rearing Techniques

Inorganic medium, spring water, or filtered lake water can be used in conjunction with a variety of foods to culture copepods. The utmost care should be taken to avoid glassware contaminated with soaps, acids, formaldehyde, or other toxins. Certain plastic containers may also be unsuitable. An easily rinsed detergent such as 7X (Flow Laboratories, McLean, Virginia 22102) should be used to clean all glassware thoroughly. Standard detergents may cause high mortalities (Jamieson 1980b). Dilute acids may also be satisfactory for washing glassware (Wyngaard and Chinnappa 1982), but thorough soaking or extra rinsing is suggested. Distilled water may have to be boiled or run through a mixed-bed resin to remove contaminants that remain after distillation. All glassware should be sterilized between uses to avoid contaminant buildup.

Harpacticoids have been cultured on a wide range of artificial and natural foods, including baker's yeast, Tetramin fish food, algae, bacteria, and dried, mashed, or rotting animals, eggs, and plants (Hicks and Coull 1983). Multiple foods are often more successful than a single food type.

Cyclopoid copepods are best reared on a combination of plant and animal foods. The nauplii will survive and grow well on a variety of algae including *Cryptomonas* and *Scenedesmus,* while animal food is usually necessary for the growth and survival of late copepodites and adults and for successful reproduction (Smyly 1970, Lewis *et al.* 1971, Gophen

1976, Jamieson 1980b, Wyngaard and Chinnappa 1982). The early copepodite stages may need a mixture of plant and animal foods. *Artemia* nauplii are the most common and convenient source of animal food for cyclopoids, but calanoid copepods, or soft-bodied cladocerans may also be used. Protozoans are a good food source, but they may lead to bacterial contamination unless care is taken to rinse and remove the organic medium or use an inorganic medium for culture (Lewis *et al.* 1971).

Although several marine calanoids have been successfully cultured through several generations, similar reports of techniques for the culture of freshwater calanoid copepods are less common. I have raised *Diaptomus pallidus* and *D. shoshone* through 22 and 3 generations, respectively, over a continuous two-year period on a food of *Cryptomonas reflexa* alone. The copepods were kept in 3 liters of 0.2 μm-filtered spring water in presterilized and covered 4-liter aquaria at 20°C (*D. pallidus*) or 10°C (*D. shoshone*) under a 12 : 12 L : D cycle. Vigorous reproduction was observed for both species.

Subtle nutritional differences between similar food types may be important to success when culturing copepods. The build up of bacteria associated with some of the dead foods as well as the addition of vitamins such as thiamine and pyridoxine may lead to contamination in some cases, but this may increase the chances of success with some detritivorous species. Isolation of food species from the same system as the copepods is likely to increase the chances of successful culture. When a constant supply of copepods is needed for experimentation and culturing is not possible, an alternative procedure is to collect sediments that contain diapausing stages from the benthos and keep the copepod-containing sediments under carefully controlled conditions (Nilssen 1982).

VI. IDENTIFICATION TECHNIQUES

A. Dissection Techniques for Identification

Dissection is usually required for positive identification of copepods. Small insect pins (minuten nadeln, available from entomological supply houses) can be mounted on the end of wood applicator sticks or melted glass pipets for use in dissection. The dissection should be performed on a glass microscope slide so that a coverslip can be placed over the final preparation for examination under a compound microscope. The dissection is most easily performed by placing the copepod on the microscope slide, ventral-side-up, and placing one pin in

the mouth region of the copepod to hold it down. A second insect pin is then used to tease off first the urosome, then each thoracic segment and its associated pair of legs—one segment at a time—and finally the first antennae. The dissection and handling of the pieces is easier when performed in a drop of glycerin or other viscous, water-miscible mounting medium such as Hoyer's. Placing a copepod directly in full-strength medium may cause osmotic stress and deformation. This problem can be avoided by placing the copepod in a dilute mixture of about 1 drop of medium to 5 drops of water and letting the water evaporate. Gentle warming will speed evaporation.

Semipermanent mounts can be made by sealing a glycerin coverslip with clear fingernail polish. One of several standard mounting media can be used for more permanent mounts. The water-miscible mounting medium Turtox CMCP-9 AF (Masters Chemical Co., Inc., Arlington Heights, Illinois) contains acid fuchsin that colors the dissected parts so that they are easier to locate after dissection. The dissection can be performed right in the viscous medium, the dissected appendages appropriately arranged in order, and the coverslip put in place. When using a mounting medium other than glycerin, it is useful to let the mounting medium dry somewhat after dissection before adding more medium and adding a coverslip. This helps the dissected parts stay put. If the medium is allowed to dry for a week and the edges sealed with fingernail polish, the mount will last indefinitely.

B. Characteristics of the Three Major Orders

Copepods are most easily identified from adult specimens, although information is available for the separation of the immature stages of some species (Comita and Tommerdahl 1960, Czaika and Robertson 1968, Katona 1971, Comita and McNett 1976, Shih and Maclellan 1977, Bourguet 1986, Pinel-Alloul and Lamoureux 1988a, 1988b). In freshwater, the three major copepod orders can be distinguished by a number of characteristics of the adults; the length of the first antennae and taper of the body are perhaps the most useful of these (Fig. 21.1). The first antennae of harpacticoids are very short (5–9 segments in the female) and rarely extend past the posterior end of the cephalothorax. In cyclopoids, the first antennae are intermediate in length (6–17 segments in the female) and rarely extend beyond the posterior end of the prosome. In calanoids, the first antennae have 23–25 segments in the female and generally extend well beyond the end of the prosome to the urosome or even to the caudal setae. The body of harpacticoids generally shows only a slight taper from the anterior to the posterior, so that the prosome and urosome are of similar widths. In cyclopoids and calanoids, the prosome is substantially wider than the urosome. In cyclopoids, there is a strong but gradual taper from front to back, while in calanoids, there is a more abrupt change in body width where the prosome joins the urosome. The body shape of harpacticoids is quite diverse and varies greatly with their habitat. For example, interstitial species are smaller and more elongate; burrowing species are larger and broader; while epibenthic species are large and variable in their body shape (Hicks and Coull 1983).

Other factors that distinguish the three free-living orders include general habitat preferences and the number of egg sacs. Calanoids are almost exclusively planktonic and either carry a single egg sac medially or release their eggs directly into the water. Cyclopoids are primarily benthic, although several planktonic species are widespread and abundant. Cyclopoids carry two egg sacs attached laterally to their genital segment. Harpacticoids are generally smaller than either calanoids or cyclopoids, almost exclusively benthic, and usually carry a single egg sac medially. Wilson and Yeatman (1959) provide a useful table that summarizes other differences between these three orders.

The larval stages within each order can be differentiated on the basis of body shape and the number and structure of the appendages. Adults are sexually dimorphic and can be distinguished from immatures on the basis of their larger size and certain secondary sexual characteristics. Adult male copepods have one (calanoids) or two (harpacticoids and cyclopoids) geniculate first antennae that are modified for grasping the females during mating (Fig. 21.1).

These modifications are generally not visible in stages C1–C4, and they are inconspicuous, if at all apparent, in stage C5.

VII. IDENTIFICATION OF THE ORDER CALANOIDA

A. General Morphology

The genera of North American calanoid copepods can be distinguished primarily by the characteristics of the caudal rami and fifth legs of adult males and females. Adult males can be distinguished from females and immatures by their right first antenna which is geniculate (Fig. 21.1b). Adult females are usually larger than males and immatures and often carry egg sacs. Characteristics of the genital segment such as a ventral swelling or more angular side profile in the adults may also be helpful in separating adult females from immatures. The fifth legs of living diaptomids are directed posteriorly in adult females but are absent or directed anteriorly in immatures (G. Grad, personal communication). Several adults should be compared to late copepodite stages to establish valid criteria for each species examined. In the following key, unless stated otherwise, the characteristics given are valid for either male or female specimens, and the identifications can be done with little or no dissection (see Section VI.A). The reader is referred to Wilson and Yeatman (1959) or Pennak (1978) for keys to the species of calanoids of North America, to Torke (1974), Smith and Fernando (1978), or Balcer *et al.* (1984) for keys covering more localized regions, or to Zo (1982) for an alternative key to the genera. A table of the genera and subgenera of freshwater calanoids of North America is given after the key.

B. Taxonomic Key to Genera of Freshwater Calanoida

1a.	Caudal ramus with three well-developed setae in addition to shorter inner and outer setae or spines; female with fifth leg	2
1b.	Caudal ramus with four well-developed setae in addition to slender inner and outer setae; female with no fifth leg; male with neither first antennae geniculate; body length 2.4–2.9 mm .... *Senecella calanoides* Juday	
1c.	Caudal ramus with five well-developed setae; female with fifth leg	3
2a(1a).	Outer seta of caudal ramus slender, setiform, and about the length of the ramus; male urosome symmetrical; female fifth leg and male left fifth leg with an apical spine that is as long or longer than the last segment .... *Heterocope septentrionalis* Juday and Muttkowski	
2b.	Caudal ramus with outer seta that is shorter than the length of the ramus, or spiniform; male urosome asymmetrical, with processes on the right side; female fifth leg and male left fifth leg without an elongate apical spine .... *Epischura* Forbes	
3a(1c).	Caudal ramus more than three times as long as wide	4
3b.	Caudal ramus length less than or equal to three times width	5

4a(3a). Maxillipeds shorter than body width in lateral view (Fig. 21.2); fifth leg with
no endopods .... *Eurytemora* Giesbrecht
4b. Maxillipeds about twice as long as body width in lateral view; fifth leg with
endopods .... *Limnocalanus* Sars
5a(3b). Caudal ramus, fourth seta from outside substantially larger than other four setae;
male right fifth leg without single terminal claw .... *Osphranticum labronectum* S. A. Forbes
5b. Setae of caudal ramus approximately equal in length (Fig. 21.1b); male right fifth leg
with single claw (Fig. 21.3a) 6
6a(5b). Male right first antenna, distal segment with a spine-like process .... *Acanthodiaptomus denticornis*
(Wierzjski)
6b. Male right first antenna, distal segment without spine-like process (Fig. 21.1b) ... *Diaptomus* Westwood

Table 21.1 Genera and Subgenera
of Freshwater Calanoids of
North America

Family Centropagidae (Sars)
  Limnocalanus
  Osphranticum
Family Diaptomidae (Sars)
  Acanthodiaptomus
  Diaptomus
    Aglaodiaptomus
    Arctodiaptomus
    Diaptomus
    Eudiaptomus
    Hesperodiaptomus
    Leptodiaptomus
    Mastigodiaptomus
    Microdiaptomus
    Mixodiaptomus
    Nordodiaptomus
    Onychodiaptomus
    Prionodiaptomus
    Sinodiaptomus
    Skistodiaptomus
Family Pseudocalanidae (Wilson)
  Senecella
Family Temoridae (Sars)
  Eurytemora
  Epischura
  Heterocope

particularly the fifth leg, first antennae, and caudal rami. These characteristics can usually be seen by placing a copepod in a drop of 1:5 mixture of glycerin:water, separating the urosome from the prosome with fine dissecting needles, and placing a coverslip over the preparation (see Section VI.A). If the urosome is oriented with its ventral-side-up, the small fifth legs can be seen by careful focusing under a compound microscope.

Females can be distinguished from males by their first antennae, which are not geniculate as in the males (Fig. 21.1a). After preservation, the first antennae of the males often become sharply bent. In addition, females are generally larger than males and frequently carry egg sacs. Adult males and females can be distinguished from immature copepodites by the last segment of the urosome which is shorter than the next to last (penultimate) segment. In immature cyclopoids, the last urosomal segment has not yet divided and is about twice as long as it is wide.

Body lengths given in the key are from the anterior end of the prosome to the posterior end of the caudal rami, not including the caudal setae. These lengths are given only as a general guide and are not strict taxonomic criteria. The reader is referred to Wilson and Yeatman (1959) or Pennak (1978) for keys to the species of cyclopoids of North America, to Torke (1974), Smith and Fernando (1978), or Balcer *et al.* (1984) for keys covering more localized regions, or to Zo (1982) for an alternative key to the genera. Due to the substantial recent changes in cyclopoid taxonomy at the genus level, a list of genera and species is presented after the key.

VIII. IDENTIFICATION OF THE ORDER CYCLOPOIDA

A. General Morphology

The following key to genera of North American cyclopoids is based on characteristics of adult females,

B. Taxonomic Key to Genera of Freshwater Cyclopoida

1a. Fifth leg a simple, unsegmented extension of fifth segment of prosome, with two inner
spines and one outer seta (Fig. 21.7a); first antenna with 9–11 (usually 11) segments;
body dorsoventrally flattened; caudal rami about twice as long as wide; body
length 0.7–1.3 mm .... *Ectocyclops phaleratus* (Koch)

1b.   Fifth leg a single broad segment with one inner spine and two outer setae (Fig. 21.7b) 2

1c.   Fifth leg of two segments (Fig. 21.7c), although in some species the basal segment of
the fifth leg is a simple, unsegmented extension of the fifth thoracic segment and has
an outer seta (Fig. 21.7d, g) 4

1d.   Fifth leg of 3 distinct segments (Fig. 21.7e); first antenna of 16 segments; body
length 0.7–1.3 mm .... *Orthocyclops modestus* (Herrick)

2a(1b).   First antenna of 11 or fewer (usually 8) segments; body length 0.6–0.9 mm .... *Paracyclops* Claus

2b.   First antenna of 12 segments; body length 0.5–1.6 mm 3

2c.   First antenna of 17 segments; body length 1.7–2.9 mm .... *Homocyclops ater* (Herrick)

3a(2b).   Caudal rami with spinules on outer margin and at least four times as long as wide;
body length 0.7–1.6 mm .... *Eucyclops* Claus

3b.   Caudal rami without spinules on outer margin and about three times as long as wide;
body length 0.5–0.9 mm .... *Tropocyclops* Kiefer

4a(1c).   Distal segment of fifth leg with one terminal seta and a highly reduced, often barely
visible internal spine located near the middle of the inner side of the distal segment
(Fig. 21.7d) 9

4b.   Distal segment of fifth leg with two well-developed setae or spines (Fig. 21.7c), or
rarely two short spines and one seta 5

4c.   Distal segment of fifth leg with two long spines and a median seta; first antenna of 17
segments (Fig. 21.7f); body length 1.5–4.0 mm .... *Macrocyclops* Claus

4d.   Distal segment of fifth leg with four or five setae or spines (Fig. 21.7g); first antenna
of 6 segments .... *Halicyclops* Norman

5a(4b).   Distal segment of fifth leg with two long setae or setose spines of roughly comparable
length (Fig. 21.7c) 6

5b.   Distal segment of fifth leg with one seta and one spine that is substantially shorter
than the seta, or rarely a single seta and two short, subapical, inner and outer
spines 7

6a(5a).   Distal segment of fifth leg with one terminal seta or spine and one seta or spine
inserted near the middle of the inner side of the segment (Fig. 21.7c); last two
segments of first antenna usually with hyaline membrane (difficult to see, Fig. 21.7h,
i); body length 0.68–1.5 mm .... *Mesocyclops* Sars

6b.   Distal segment of fifth leg with two terminal or subterminal setae or spines (Fig.
21.7j); body length 0.45–1.10 mm .... *Thermocyclops* Kiefer

7a(5b).   Distal segment of fifth leg with a short spine inserted near the middle of the inner side
of the segment (Fig. 21.7k); body length 1.07–5.0 mm .... *Cyclops* O. F. Muller

7b.   Distal segment of fifth leg with spine inserted distally or subdistally (Fig. 21.7l, m) 8

8a(7b).   Spine on distal segment of fifth leg less than or equal to 1/5 length of seta (Fig. 21.7l);
hairs present or absent on inner margins of caudal rami; body length
0.78–2.2 mm 10

8b.   Spine on distal segment of fifth leg more than 1/5 length of seta (Fig. 21.7m); body
length 0.45–1.57 mm .... *Diacyclops* Kiefer

9a(4a).   Distal segment of the endopod of fourth leg with two well-developed apical spines;
body length 0.5–1.0 mm .... *Microcyclops* varicans (Sars)

9b.   Distal segment of the endopod of fourth leg with the outer spine less than half the
length of the inner spine; body length 0.6–0.7 mm .... *Cryptocyclops bicolor* (Sars)

10a(8a).   Distal segment of fifth leg with an inner spine or spur located about halfway to the
end of the segment (Fig. 21.7n); caudal rami with hairs on inner
margin .... *Megacyclops* Kiefer

10b.   Distal segment of fifth leg with inner spine located more than halfway to the end of
the segment (Fig. 21.7l); caudal rami usually without hairs, or if present, hairs
distributed in tufts .... *Acanthocyclops* Kiefer

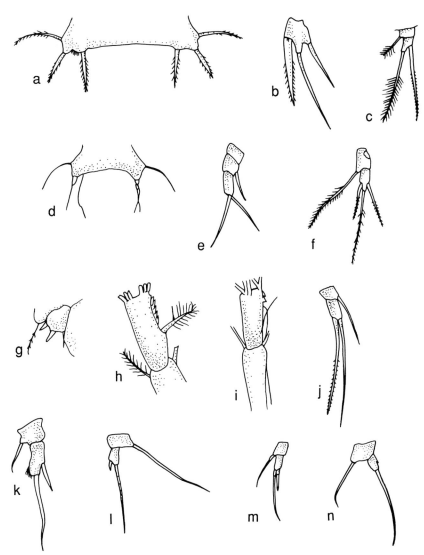

Figure 21.7 Cyclopoid appendages: (a) fifth leg and fifth metasomal segment of *Ectocyclops phaleratus;* (b) fifth leg of *Eucyclops agilis;* (c) fifth leg of *Mesocyclops edax;* (d) fifth leg of *Microcyclops varicans;* (e) fifth leg of *Orthocyclops modestus;* (f) fifth leg of *Macrocyclops albidus;* (g) fifth leg of *Halicyclops* sp.; (h) last two segments of first antenna of *Mesocyclops edax* showing hyaline membrane; (i) last two segments of first antenna of *Mesocyclops leuckarti* showing hyaline membrane; (j) fifth leg of *Thermocyclops tenuis;* (k) fifth leg of *Cyclops scutifer;* (l) fifth leg of *Acanthocyclops vernalis;* (m) fifth leg of *Diacyclops thomasi;* (n) fifth leg of *Megacyclops latipes.* [Redrawn from Yeatman (1959) except c and h are original, and i is from Smith and Fernando (1978).]

Table 21.2 North American Species in the Genera of the Family Cyclopidae Sars, with Range of Adult Female Body Lengths in mm

Subfamily Cyclopinae
    Acanthocyclops [*capillatus* (Sars) 1.80–2.20; *carolinianus* (Yeatman) 0.80–1.50; *exilis* (Coker) 0.78–0.88; *plattensis* Pennak and Ward; *venustoides* (Coker) 0.82–1.56; *venustus* (Norman and Scott) 1.00–1.30; *vernalis* (Fischer) 0.99–1.8]
    Cryptocyclops [*bicolor* (Sars) 0.6–0.7]
    Cyclops [*canadensis* Einsle; *columbianus* Lindberg; *insignis* Claus 2.50–5.00; *kolensis* Lilljeborg, (= *vicinus* Uljanin) (1.07–1.85); *scutifer* Sars 1.29–1.90; *strenuus* Fischer 1.42–2.35]
    Diacyclops [*bicuspidatus* (Claus) 0.95–1.57; *bisetosus* (Rehberg) 0.84–1.51; *crassicaudis brachycercus* (Kiefer) 0.72–1.10; *haueri* Kiefer 1.15–1.40; *jeanneli* Chappuis 0.90; *languidus* (Sars); *languidoides* (Lilljeborg) 0.51; *nanus* (Sars) 0.45–0.90; *navus* Herrick 0.90–1.16; *nearcticus* Kiefer 0.50–0.80; *palustris* Reid 1.13–1.14; *thomasi* (Forbes) 0.90–1.17; *yeatmani* Reid]
    Megacyclops [*donnaldsoni* Chappuis 1.45; *gigas* (Claus) 2.00–4.20; *latipes* (Lowndes) 1.85–2.50; *magnus* Marsh; *viridis* (Jurine) 1.50–3.00]
    Mesocyclops [*americanus* Dussart 1.0–1.2; *edax* (Forbes) 1.00–1.50]
    Microcyclops [*pumilis* Pennak and Ward; *varicans* (Sars) 0.50–1.0]
    Orthocyclops [*modestus* (Herrick) 0.7–1.3]
    Thermocyclops [*parvus* Reid 0.45–0.49; *tenuis* (Marsh) 1.10]
Subfamily Eucyclopinae
    Ectocyclops [*phaleratus* (Koch) 0.70–1.3]
    Eucyclops [*macruroides denticulatus* (Graeter) 0.87–0.94; *macrurus* (Sars) 1.10–1.40; *prionophorus* Kiefer 0.70–0.94; *agilis* (Koch) [= *serrulatus* (Fischer)] 0.80–1.50; *speratus* (Lilljeborg) 1.00–1.60]
    Homocyclops [*ater* (Herrick) 1.77–2.88]
    Macrocyclops [*albidus* (Jurine) 1.50–2.50; *distinctus* (Richard) 2.00–2.20; *fuscus* (Jurine) 1.80–4.00]
    Paracyclops [*affinis* (Sars) 0.60–0.85; *fimbriatus chiltoni* (Thomson) 0.72–0.96; *poppei* (Rehberg) 0.70–0.90; *yeatmani* Daggett and Davis 0.75–0.86]
    Tropocyclops [*extensus* (Kiefer); *prasinus* (Fischer) 0.50–0.90; *prasinus mexicanus* Kiefer 0.50–0.90]
Subfamily Halicyclopinae
    Halicyclops (*clarkei* Herbst; *coulli* Herbst; *fosteri* Wilson; *laminifer* Herbst)

IX. IDENTIFICATION OF THE ORDER HARPACTICOIDA

A. General Morphology

With no recent, comprehensive taxonomic work on the harpacticoid copepods of North America, the key of Wilson and Yeatman (1959), which is based primarily on the monographic efforts of Lang (1948), remains the most useful key to species. The current key follows these prior works to the generic level and is based on characteristics of the first to fifth legs of adults. Unless otherwise stated, the characters in the key apply to both males and females. Before an identification is attempted, an adult male and an adult female should be dissected and mounted in glycerin (see Section VI.A).

Most adult harpacticoids can be distinguished from immatures by the relative length of, the last and next to last segments of the urosome. In adults, the last two segments are similar in length while in the immatures the last segment is much longer. Males can be separated from females by their first antennae, which have fewer and broader segments and often more densely clumped setae. Males usually have ten body segments and females have nine due to the fusion of the genital and first abdominal segments. Other characteristics such as morphologic differences in the caudal rami and legs 2–4 can be used to distinguish males and females in many genera. In addition, females generally have a more highly developed fifth leg and no sixth leg as is found in many males.

B. Taxonomic Key to Genera of Freshwater Harpacticoida

1a.	First leg, distal segment of exopod with 3–5 short, clawlike spines, all shorter than segment itself	8
1b.	First leg, distal segment of exopod with some spines and/or setae longer than segment	2
2a(1b).	First leg, exopod of two segments	9
2b.	First leg, exopod of three segments	3
3a(2b).	First leg, distal segment of exopod with five or six spines and setae	15
3b.	First leg, distal segment of exopod with four spines and setae	4

4a(3b).	First leg, second segment of exopod with inner seta	7
4b.	First leg, second segment of exopod without inner seta	5
5a(4b).	First leg, second segment of exopod with outer spine	6
5b.	First leg, second segment of exopod without outer spine .... *Parastenocaris* Kessler	
6a(5a).	First leg, endopod of one segment	21
6b.	First leg, endopod of two segments	10
6c.	First leg, endopod of three segments	19
7a(4a).	First leg, endopod of two segments	23
7b.	First leg, endopod of three segments	25
8a(1a).	Female fifth leg, basal expansion with three or four setae; male fifth leg, basal portion not expanded inside, and without setae. Male second leg, second segment of endopod produced into outer marginal process .... *Harpacticus* Milne-Edwards	
8b.	Female fifth leg, basal expansion with five or six setae; male fifth leg, basal portion slightly expanded inside, with one seta. Male second leg, second segment of endopod produced into both inner and outer marginal processes .... *Tigriopus* Norman	
9a(2a).	First leg, endopod shorter than or equal to exopod	21
9b.	First leg, endopod about twice as long as exopod .... *Onychocamptus* Daday	
10a(6b).	First leg, endopod not reaching beyond second segment of exopod	20
10b.	First leg, endopod reaching from about the middle to a little beyond the end of the third segment of expopod	11
10c.	First leg, endopod reaching beyond third segment of exopod by half or more of the length of third segment	12
11a(10b).	Fifth leg with exopod segment distinctly separated from basal expansion	22
11b.	Fifth leg unsegmented .... *Stenocaris* Sars	
12a(10c).	First leg, apex of endopod with single stout spine; seta, if present, minute or hairlike	13
12b.	First leg, apex of endopod with stout spine and longer seta, or with two or more well-developed setae	14
13a(12a).	Female second, third, and fourth legs, distal segments with four, six, and four setae, respectively; male second and third legs, exopod directed inward with stout spines; male fourth leg, endopod of two segments; male fifth leg unsegmented .... *Heterolaophonte stromi* (Baird)	
13b.	Female second, third, and fourth legs, distal segments with four, five, and four setae, respectively; male second, third, and fourth legs, exopod directed inward with stout spines; male fourth leg, endopod of one segment; male fifth leg with exopod distinctly separated .... *Pseudonychocamptus proximus* (Sars)	
14a(12b).	Female second, third, and fourth legs, endopods of three segments. Male second leg, endopod of two segments, segment 2 with outer or apical spine expanded and enlarged as articulated spine or as enlarged process .... *Schizopera* Sars	
14b.	Female second, third, and fourth legs, endopods of one or two segments or lacking. Male second leg, endopod lacking or of two segments and normal setae .... *Paracamptus* Chappuis	
15a(3a).	First leg, second and third segments of exopod about same length	16
15b.	First leg, second segment of exopod about three times length of third segment .... *Paradactylopodia* Lang	
16a(15a).	Fifth leg unsegmented	17
16b.	Fifth leg with exopod distinctly separated from basal expansion	18
17a(16a).	First leg, third segment of exopod with total of six spines and setae .... *Microarthridion littorale* (Poppe)	
17b.	First leg, third segment of exopod with total of five spines and setae .... *Tachidius* Lilljeborg	
18a(16b).	Second and third legs, third segment of exopod with three outer marginal spines .... *Nitocra* Boeck	
18b.	Second and third legs, third segment of exopod with two outer marginal spines .... *Nitocrella* Chappuis	
19a(6c).	Second leg, endopod of two segments. Male second leg, endopod lacking outer apical modified spine or process .... *Bryocamptus* Chappuis	

19b.	Female second leg, endopod of three segments. Male second leg, endopod of two segments and with outer apical modified spine or process. Female fifth leg, exopod separated .... *Schizopera* Sars
19c.	Second leg, endopod of three segments. Female fifth leg, exopod not distinctly separated .... *Phyllognathopus* Mrazek
20a(10a).	Anterior body expanded, creating substantial taper in body posteriorly (Fig. 21.8a). Second, third, and fourth legs, endopods of three segments .... *Metis* Philippi
20b.	Body vermiform, slender throughout. Second and third legs, endopods usually of two segments; fourth leg of one segment .... *Epactophanes* Mrazek
21a(9a).	First leg, endopod of two segments and about as long as exopod .... *Maraenobiotus* Mrazek
21b.	First leg, endopod of one segment and considerably shorter than exopod .... *Huntemannia* Poppe
22a(11a).	Second, third, and fourth legs, second segment of exopod without inner seta .... *Moraria* T. Scott and A. Scott
22b.	Second, third, and fourth legs, second segment of exopod with inner seta .... *Bryocamptus* Chappuis
23a(7a).	First leg, endopod reaching to near end of exopod or beyond 24
23b.	First leg, endopod not reaching beyond second segment of exopod .... *Nannopus* Brady
24a(23a).	Fifth leg, exopod segment distinctly separated from basal expansion 25
24b.	Female fifth leg, and usually that of male, exopod segment separated from basal expansion only by notch or gap .... *Cletocamptus* Schmankewitsch
25a(24a,7b).	First leg, endopod of two segments and rostrum not extending to end of first segment of first antenna. Female first antenna of seven or (usually) eight segments; male third leg with hypophysis (stout spiniform process arising from inner margin or base of second segment or its equivalent and extending beyond apex of endopod, Fig. 21.8b) .... *Bryocamptus* Chappuis
25b.	First leg, endopod of three segments and rostrum not extending beyond first segment of first antennae. Female first antenna of 7–9 (usually 8) segments; male third leg with hypophysis 26
25c.	First leg, endopod of two or three segments; rostrum greatly enlarged, extending to end of first segment of first antenna or beyond (Fig. 21.8c). Female first antenna of six or seven segments; male third leg, second segment of endopod or its equivalent with simple, articulated spine on inner margin .... *Mesochra* Boeck
26a(25b).	Female fifth leg, second seta of basal expansion very reduced, not more than and usually less than 1/4 the length of first seta; male fourth leg, outer corner of apical endopod segment produced as spinous process (not an articulated spine) (Fig. 21.8d) .... *Canthocamptus* Westwood
26b.	Female fifth leg, second seta of basal expansion not very reduced, usually as long as or longer than first seta. Male fourth leg, outer corner of apical endopod segment with articulated spine or seta 27
27a(26b).	First leg, first segment of endopod reaching to middle of third segment of exopod or beyond .... *Attheyella* Brady
27b.	First leg, first segment of endopod segment not reaching beyond second segment of exopod 28
28a(27b).	Second, third, and fourth legs, third exopod segment with two outer marginal spines (total spines and setae: 5,6,6). Fifth leg, basal expansion with three or four setae in female, none in male .... *Elaphoidella* Chappuis
28b.	Second, third, and fourth legs, third segment of exopod with three outer marginal spines (total spines and setae: 6,7,7 or rarely 6 on leg 4). Fifth leg, basal expansion with five or six (rarely four) setae in female, usually with two setae in male 29
29a(28b).	Female second and third legs, endopods of three segments, or of two segments and female fifth leg with five setae on basal expansion. Male second leg, second segment of endopod with one or two spinous processes, notches or sclerotized knobs on outer margin, as in Fig. 21.8e .... *Bryocamptus* Chappuis
29b.	Female second and third legs, endopods of two segments and fifth leg with six setae on basal expansion. Male second leg, endopod not modified (outer margin continuous though it may have spinules) .... *Attheyella* Brady

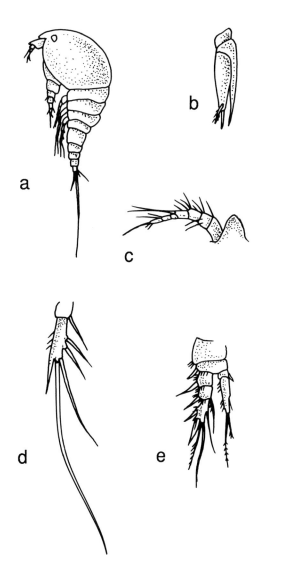

Figure 21.8 Harpacticoid characteristics. (a) lateral view of *Metis;* (b) endopod of third leg of male *Bryocamptus;* (c) dorsal view of rostrum of *Mesochra;* (d) endopod of fourth leg of male *Canthocamptus;* (e) second leg of male *Bryocamptus.* (Redrawn from Wilson and Yeatman 1959.)

Table 21.3 Genera of Freshwater Harpacticoids of North America

Family Ameiridae Monard
  Nitocra
  Nitocrella
Family Canthocamptidae Sars
  Attheyella
  Bryocamptus
  Canthocamptus
  Elaphoidella
  Epactophanes
  Maraenobiotus
  Mesochra
  Moraria
  Paracamptus
Family Cletodidae T. Scott
  Cletocamptus
  Huntemannia
  Nannopus
Family Cylindropsyllidae Sars
  Stenocaris
Family Diosaccidae Sars
  Schizopera
Family Harpacticidae Sars
  Harpacticus
  Tigriopus
Family Laophontidae T. Scott
  Heterolaophonte
  Onychocamptus
  Pseudonychocamptus
Family Metidae Sars
  Metis
Family Parastenocaridae Chappuis
  Parastenocaris
Family Phyllognathopodidae Gurney
  Phyllognathopus
Family Tachidiidae Sars
  Microarthridion
  Tachidius
Family Thalestridae Sars
  Paradactylopodia

LITERATURE CITED

Alcaraz, M., G.-A. Paffenhöfer, and J. R. Strickler. 1980. Catching the algae: A first account of visual observations on filter-feeding calanoids. Pages 241–248 *in*: W. C. Kerfoot, editor. Evolution and ecology of zooplankton communities. University Press of New England, Hanover, New Hampshire.

Allan, J. D. 1976. Life history patterns in zooplankton. American Naturalist 110:165–180.

Anderson, R. S. 1970. Predator–prey relationships and predation rates for crustacean zooplankters from some lakes in western Canada. Canadian Journal of Zoology 48:1229–1240.

Anderson, R. S. 1971. Crustacean plankton of 146 alpine and subalpine lakes and ponds in western Canada. Journal of the Fisheries Research Board of Canada 28:311–321.

Anderson, R. S. 1974. Crustacean plankton communities of 340 lakes and ponds in and near the National Parks of the Canadian Rocky Mountains. Journal of the Fisheries Research Board of Canada 31:855–869.

Anderson, R. S. 1977. Rotifer populations in mountain lakes relative to fish and species of copepods present. Archiv für Hydrobiologie Beihefte Ergebnisse der Limnologie 8:130–134.

Anderson, R. S. 1980. Relationships between trout and invertebrate species as predators and the structure of the crustacean and rotiferan plankton in mountain lakes. Pages 635–641 *in*: W. C. Kerfoot, editor. Evolution and ecology of zooplankton communities. University Press of New England, Hanover, New Hampshire.

Andrews, J. C. 1983. Deformation of the active space in

the low Reynolds number feeding current of calanoid copepods. Canadian Journal of Fisheries and Aquatic Sciences 40:1293–1302.

Balcer, M. D., N. L. Korda, and S. I. Dodson. 1984. Zooplankton of the Great Lakes: a guide to the identification and ecology of the common crustacean species. The University of Wisconsin Press, Madison. 174 p.

Barszcz, C. A., and P. P. Yevich. 1976. Preparation of copepods for histopathologic examinations. Transactions of the American Microscopical Society 95:104–108.

Batchelder, H. P. 1986. A staining technique for determining copepod gonad maturation: application to *Metridia pacifica* from the northeast Pacific Ocean. Journal of Crustacean Biology 6:227–231.

Blades, P. I., and M. J. Youngbluth. 1980. Morphological, physiological, and behavioral aspects of mating in calanoid copepods. Pages 39–51 *in*: W. C. Kerfoot, editor. Evolution and ecology of zooplankton communities. University Press of New England, Hanover, New Hampshire.

Blades-Eckelbarger, P. I. 1986. Aspects of internal anatomy and reproduction in the copepoda. *in*: G. Schriever, H. K. Schminke, and C.-t. Shih, editors. Proceedings of the second international conference on Copepoda. Syllogeus 58:26–50.

Bliss, D. E., series editor. 1983. The biology of Crustacea. 8 vols. Academic Press, New York.

Bogdan, K. G., and J. J. Gilbert. 1984. Body size and food size in freshwater zooplankton. Proceedings of the National Academy of Sciences 81:6427–6431.

Bourguet, J.-P. 1986. Contribution a l'étude de *Cletocamptus retrogressus* Schmankevitch, 1875 (Copepoda, Harpacticoida) II. Developpement larvaire—stades naupliens. Crustaceana 51:113–122.

Bowers, J. A. 1986. Phosphorus regeneration by the predatory copepod *Diacyclops thomasi*. Canadian Journal of Fisheries and Aquatic Sciences 43:361–365.

Bowman, T. E., and L. G. Abele. 1982. Classification of the recent Crustacea. Pages 1–25 *in*: L. G. Abele, editor. The biology of Crustacea Vol. 1: Systematics, the fossil record, and biogeography. Academic Press, New York.

Brandl, Z., and C. H. Fernando. 1979. The impact of predation by the copepod *Mesocyclops edax* (Forbes) on zooplankton in three lakes in Ontario, Canada. Canadian Journal of Zoology 57:940–942.

Brandl, Z., and C. H. Fernando. 1986. Feeding and food consumption by *Mesocyclops edax*. *in*: G. Schriever, H. K. Schminke, and C.-t. Shih, editors. Proceedings of the second international conference on Copepoda. Syllogeus No. 58:254–258.

Brooks, J. L. 1969. Eutrophication and changes in the composition of the zooplankton. Pages 236–255 in Eutrophication: Causes, consequences, correctives. National Academy of Sciences, Washington, D.C.

Bulkowski, L., W. F. Drise, and K. A. Kraus. 1985. Purification of *Cyclops* cultures by pH shock (Copepoda). Crustaceana 48:179–182.

Byron, E. R., P. T. Whitman, and C. R. Goldman. 1983. Observations of copepod swarms in Lake Tahoe. Limnology and Oceanography 28:378–382.

Byron, E. R., C. L. Folt, and C. R. Goldman. 1984. Copepod and cladoceran success in an oligotrophic lake. Journal of Plankton Research 6:45–65.

Carter, J. C. H., and K. A. Goudie. 1986. Diel vertical migrations and horizontal distributions of *Limnocalanus macrurus* and *Senecella calanoides* (Copepoda, Calanoida) in lakes of southern Ontario in relation to planktivorous fish. Canadian Journal of Fisheries and Aquatic Sciences 43:2508–2514.

Carter, J. C. H., W. G. Sprules, M. J. Dadswell, and J. C. Roff. 1983. Factors governing geographical variation in body size of *Diaptomus minutus* (Copepoda, Calanoida). Canadian Journal of Fisheries and Aquatic Sciences 40:1303–1307.

Chaston, I. 1969. Anaerobiosis in *Cyclops varicans*. Limnology and Oceanography 14:298–301.

Chow-Fraser, P., and E. J. Maly. 1988. Aspects of mating, reproduction, and co-occurrence in three freshwater calanoid copepods. Freshwater Biology 19:95–108.

Chow-Fraser, P., and C. K. Wong. 1986. Dietary change during development in the freshwater calanoid copepod *Epischura lacustris* Forbes. Canadian Journal of Fisheries and Aquatic Sciences 43:938–944.

Coker, R. E. 1934. Contribution to knowledge of North American freshwater harpacticoid copepod crustacea. Journal of the Elisha Mitchell Scientific Society 50:75–141.

Coker, R. E. 1935. Transfer of spermatophores by *Cyclops*. A special purpose served without a "specialized" structure. Archiv für Zoologie Experimentale et Generale 77:9–11.

Cole, G. A. 1961. Some calanoid copepods from Arizona with notes on congeneric occurrences of *Diaptomus* species. Limnology and Oceanography 6:432–442.

Comita, G. W. 1972. The seasonal zooplankton cycles, production and transformations of energy in Severson Lake, Minnesota. Archiv für Hydrobiologie 70:14–66.

Comita, G. W., and S. J. McNett. 1976. The postembryonic developmental instars of *Diaptomus oregonensis* Lilljeborg 1889 (Copepoda). Crustaceana 30:123–163.

Comita, G. W., and D. M. Tommerdahl. 1960. The postembryonic developmental instars of *Diaptomus siciloides* Lilljeborg. Journal of Morphology 107:297–355.

Confer, 1971. Intrazooplankton predation by *Mesocyclops edax* at natural prey densities. Limnology and Oceanography 16:663–666.

Confer, J. L., and P. I. Blades. 1975. Omnivorous zooplankton and planktivorous fish. Limnology and Oceanography 20:571–579.

Confer, J. L., and J. M. Cooley. 1977. Copepod instar survival and predation by zooplankton. Journal of the Fisheries Research Board of Canada 34:703–706.

Confer, J. L., T. Kaaret, and G. E. Likens. 1983. Zooplankton diversity and biomass in recently acidified lakes. Canadian Journal of Fisheries and Aquatic Sciences 40:36–42.

Cooley, J. M. 1978. Effect of temperature on development of freshwater copepod eggs. Crustaceana 35:27–34.

Czaika, S. C., and A. Robertson. 1968. Identification of

the copepodids of the Great Lakes species of *Diaptomus* (Calanoida, Copepoda). University of Michigan Great Lakes Research Division Contribution 17:39–60.

Daggett, R. F., and C. C. Davis. 1975. Distribution and occurrence of some littoral freshwater microcrustaceans in Newfoundland. Le Naturaliste Canadien 102:45–55.

DeFrenza, J., R. J. Kirner, E. J. Maly, and H. C. Van Leeuwen. 1986. The relationships of sex size ratio and season to mating intensity in some calanoid copepods. Limnology and Oceanography 31:491–496.

DeMott, W. R. 1986. The role of taste in food selection by freshwater zooplankton. Oecologia 69:334–340.

Dodson, S. I. 1974. Zooplankton competition and predation: an experimental test of the size-efficiency hypothesis. Ecology 55:605–613.

Dodson, S. I. 1975. Predation rates of zooplankton in arctic ponds. Limnology and Oceanography 20:426–433.

Dodson, S. I. 1984. Predation of *Heterocope septentrionalis* on two species of *Daphnia*: Morphological defenses and their cost. Ecology 65:1249–1257.

Dodson, S. I., and D. L. Egger. 1980. Selective feeding of red phalaropes on zooplankton of arctic ponds. Ecology 61:755–763.

Dressel, D. M., D. R. Heinle, and M. C. Grote. 1972. Vital staining to sort dead and live copepods. Chesapeake Science 13:156–159.

Dudley, P. L. 1986. Aspects of general body shape and development in Copepoda. *in*: G. Schriever, H. K. Schminke, and C.-t. Shih, editors. Proceedings of the Second International Conference on Copepoda. Syllogeus No. 58:7–25.

Dussart, B. 1967. Les copépodes des eaux continentales. Calanoides et Harpacticoides. N. Boubée, Paris. 500 p.

Dussart, B. 1969. Les copépodes des eaux continentales. Cyclopoides et Biologie. N. Boubée, Paris. 292 p.

Edmondson, W. T. 1960. Reproductive rates of rotifers in natural populations. Memoires Istituto Italiano di Idrobiologia 12:21–77.

Edmondson, W. T. 1964. The rate of egg production by rotifers and copepods in natural populations as controlled by food and temperature. Verhandlungen Internationale Vereinigung für Theoretische und Angewandte Limnologie 15:673–675.

Edmondson, W. T. 1985. Reciprocal changes in abundance of *Diaptomus* and *Daphnia* in Lake Washington. Archiv fur Hydrobiologie Beihefte Ergebnisse der Limnologie 21:475–481.

Edmondson, W. T., G. W. Comita, and G. C. Anderson. 1962. Reproductive rate of copepods in nature and its relation to phytoplankton population. Ecology 43:625–634.

Elgmork, K. 1967. Ecological aspects of diapause in copepods. Pages 647-954 *in*: Proceedings of the Symposium on Crustacea, Series 2, Part III, Ernakulam, Marine Biological Association of India.

Elgmork, K. 1980. Evolutionary aspects of diapause in freshwater copepods. Pages 411-417 *in*: W. C. Kerfoot, editor. Evolution and ecology of zooplankton communities. University Press of New England, Hanover, New Hampshire.

Elgmork, K., and A. Langeland. 1980. *Cyclops scutifer* Sars-one and two-year life cycles with diapause in the meromictic lake Blankvatn. Archiv für Hydrobiologia 88:178–201.

Elgmork, K., and J. P. Nilssen. 1978. Equivalence of copepod and insect diapause. Verhandlungen Internationale Vereinigung für Theoretische und Angewandte Limnologie 20:2511–2517.

Elgmork, K., J. P. Nilssen, and R. Ovrevik. 1978. Life cycle strategies in neighbouring populations of the copepod *Cyclops scutifer* Sars. Verhandlungen Internationale Vereinigung für Theoretische und Angewandte Limnologie 20:2518–2523.

Elmore, J. L. 1982. The influence of food concentration and container volume on life history parameters of *Diaptomus dorsalis* Marsh from subtropical Florida. Hydrobiologia 89:215–223.

Elmore, J. L. 1983. Factors influencing *Diaptomus* distributions: an experimental study in subtropical Florida. Limnology and Oceanography 28:522–532.

Elofsson, R. 1971. The ultrastructure of a chemoreceptor organ in the head of copepod crustaceans. Acta Zoologica 52:299–315.

Epp, R. W., and W. M. Lewis, Jr. 1979. Metabolic responses to temperature change in a tropical freshwater copepod (*Mesocyclops brasilianus*) and their adaptive significance. Oecologia 42:123–138.

Fahrenbach, W. H. 1962. The biology of a harpacticoid copepod. La Cellule 62:303–376.

Feifarek, B. P., G. A. Wyngaard, and J. D. Allan. 1983. The cost of reproduction in a freshwater copepod. Oecologia 56:166–168.

Feller, R. J. 1980. Development of the sand-dwelling meiobenthic harpacticoid copepod *Huntemannia jadensis* Poppe in the laboratory. Journal of Experimental Marine Biology and Ecology 46:1–15.

Fleminger, A. 1973. Pattern, number, variability, and taxonomic significance of integumental organs (sensilla and glandular pores) in the genus *Eucalanus* (Copepoda, Calanoida). Fisheries Bulletin 71:965–1010.

Flemming, J. M., and J. Coughlan. 1978. Preservation of vitally stained zooplankton for live/dead sorting. Estuaries 1:135–137.

Folt, C., and C. R. Goldman. 1981. Allelopathy between zooplankton: A mechanism for interference competition. Science 213:1133–1135.

Fowler, K., and L. Partridge. 1989. A cost of mating in female fruitflies. Nature (London) 338:760–761.

Frenzel, P. 1980. Die Populationsdynamik von *Canthocamptus staphylinus* (Jurine) (Copepoda, Harpacticoida) im Littoral des Bodensees. Crustaceana 39:282–286.

Friedman, M. M. 1980. Comparative morphology and functional significance of copepod receptors and oral structures. Pages 185–197 *in*: W. C. Kerfoot, editor. Evolution and ecology of zooplankton communities. University Press of New England, Hanover, New Hampshire.

Friedman, M. M., and J. R. Strickler. 1975. Chemoreceptors and feeding in calanoid copepods (Arthropoda: Crustacea). Proceedings of the National Academy of Sciences USA 72:4185–4188.

Fryer, G. 1954. Contributions to our knowledge of the biology and systematics of the freshwater Copepoda. Schweizerische Zeitschrift für Hydrologie 16:64–77.

Fryer, G. 1957a. The feeding mechanism of some freshwater cyclopoid copepods. Proceedings of the Zoological Society of London 129:1–25.

Fryer, G. 1957b. The food of some freshwater cyclopoid copepods and its ecological significance. Journal of Animal Ecology 26:263–286.

Fuller, J. L., and G. L. Clarke. 1936. Further experiments on the feeding of *Calanus finmarchicus*. Biological Bulletin 70:308–320.

Gabriel, W. 1985. Overcoming food limitation by cannibalism: a model study on cyclopoids. Archiv für Hydrobiologie Beihefte Ergebnisse der Limnologie 21:373–381.

Gannon, J. E. 1971. Two counting cells for the enumeration of zooplankton micro-crustacea. Transactions of the American Microscopical Society 90:486–490.

Gannon, J. E., and S. A. Gannon. 1975. Observations on the narcotization of crustacean zooplankton. Crustaceana 28:220–224.

Gauld, D. T. 1951. The grazing rate of planktonic copepods. Journal of the Marine Biological Association of the United Kingdom 29:695–706.

Gee, J. M., and R. M. Warwick. 1984. Preliminary observations on the metabolic and reproductive strategies of harpacticoid copepods from an intertidal sandflat. Hydrobiologia 118:29–37.

Gehrs, C. W. 1974. Horizontal distribution and abundance of *Diaptomus clavipes* Schacht in relation to *Potamogeton foliosus* in a pond and under experimental conditions. Limnology and Oceanography 19:100–104.

Gerritsen, J. 1978. Instar-specific swimming patterns and predation of planktonic copepods. Verhandlungen Internationale Vereinigung für Theoretische und Angewandte Limnologie 20:2531–2536.

Gilbert, J. J., and C. E. Williamson. 1978. Predator–prey behavior and its effect on rotifer survival in associations of *Mesocyclops edax*, *Asplanchna girodi*, *Polyarthra vulgaris*, and *Keratella cochlearis*. Oecologia 37:13–22.

Gilbert, J. J., and C. E. Williamson. 1983. Sexual dimorphism in zooplankton (Copepoda, Cladocera, and Rotifera). Annual Review of Ecology and Systematics 14: 1–33.

Gill, C. W. 1987. Recording the beat patterns of the second antennae of calanoid copepods with a microimpedance technique. Hydrobiologia 148:73–78.

Gophen, M. 1976. Temperature effect on life span, metabolism, and development time in *Mesocyclops leuckarti* (Claus). Oecologia (Berlin) 25:271–277.

Gophen, M. 1978. Errors in the estimation of recruitment of early stages of *Mesocyclops leuckarti* (Claus) caused by the diurnal periodicity of egg-production. Hydrobiologia 57:59–64.

Gophen M. 1979. Mating process in *Mesocyclops leuckarti* (Crustacea:Copepoda). Israel Journal of Zoology 28:163–166.

Gophen, M., and R. P. Harris. 1981. Visual predation by a marine cyclopoid copepod, *Corycaeus anglicus*. Journal of the Marine Biological Association of the United Kingdom 61:391–399.

Gould, D. T. 1957. A peritrophic membrane in calanoid copepods. Nature (London) 179:325–326.

Grad, G., and E. J. Maly. 1988. Sex size ratios and their influence on mating success in a calanoid copepod. Limnology and Oceanography 33:1629–1634.

Griffiths, A. M., and B. W. Frost. 1976. Chemical communication in the marine planktonic copepods *Calanus pacificus* and *Pseudocalanus* sp. Crustceana 30:1–8.

Guma'a, S. A. 1978. The food and feeding habits of young perch, *Perca fluviatilis* in Windermere. Freshwater Biology 8:177–187.

Gurney, R. 1931. British Freshwater Copepoda. Vol. 1. Ray Society, London.

Gurney, R. 1933. British Freshwater Copepoda. Vol. 3. Ray Society, London.

Hairston, N. G., Jr. 1976. Photoprotection by carotenoid pigments in the copepod *Diaptomus nevadensis*. Proceedings of the National Academy of Sciences, USA 73:971–974.

Hairston, N. G. 1979a. The adaptive significance of color polymorphism in two species of *Diaptomus* (Copepoda). Limnology and Oceanography 24:15–37.

Hairston, N. G. 1979b. The relationship between pigmentation and reproduction in two species of *Diaptomus* (Copepoda). Limnology and Oceanography 24:38–44.

Hairston, N. G. 1979c. The effect of temperature on carotenoid photoprotection in the copepod *Diaptomus nevadensis*. Comparative Biochemistry and Physiology 62:445–448.

Hairston, N. G., Jr., and E. J. Olds. 1984. Population differences in the timing of diapause: adaptation in a spatially heterogeneous environment. Oecologia 61:42–48.

Hairston, N. G., Jr., and W. E. Walton. 1986. Rapid evolution of a life history trait. Proceedings of the National Academy of Sciences USA 83:4831–4833.

Hairston, N. G., Jr., W. E. Walton, and K. T. Li. 1983. The causes and consequences of sex-specific mortality in a freshwater copepod. Limnology and Oceanography 28:935–947.

Hammer, U. T., and W. W. Sawchyn. 1968. Seasonal succession and congeneric associations of *Diaptomus* spp. (Copepoda) in some Saskatchewan ponds. Limnology and Oceanography 13:476–484.

Hebert, P. D. N. 1985. Ecology of the dominant copepod species at a low arctic site. Canadian Journal of Zoology 63:1138–1147.

Hebert, P. D. N., A. G. Good, and M. A. Mort. 1980. Induced swarming in the predatory copepod *Heterocope septentrionalis*. Limnology and Oceanography 25:747–750.

Herbst, H. V. 1977. Eine neuer *Halicyclops* aus Nord America. Gewasser Abwasser 62/63:121–126.

Herbst, H. V. 1982. Drei neue marine cyclopoida Gnathostoma (Crustacea, Copepoda) aus dem nordamerikanishen Kustenbereich. Gewasser Abwasser 68/69:107–124.

Hicks, G. R. F., and B. C. Coull. 1983. The ecology of marine meiobenthic harpacticoid copepods. Annual Review of Oceanography and Marine Biology 21:67–175.

Hill, L. L., and R. E. Coker. 1930. Observations on mating habits of *Cyclops*. Journal of the Elisha Mitchell Scientific Society 45:206–220.

Holmes, S. J. 1909. Sex recognition in *Cyclops*. Biological Bulletin 16:313–315.

Hutchinson, G. E. 1951. Copepodology for the ornithologist. Ecology 32:571–577.

Hutchinson, G. E. 1967. A treatise on limnology. Vol. 2. Wiley, New York.

Jacobs, J. 1961. Laboratory cultivation of the marine copepod *Pseudocoronatus* Williams. Limnology and Oceanography 6:443–446.

Jamieson, C. D. 1980a. The predatory feeding of copepod stages III to adult *Mesocyclops leuckarti* (Claus). Pages 518–537 *in*: W. C. Kerfoot, editor. Evolution and ecology of zooplankton communities. University Press of New England, Hanover, New Hampshire.

Jamieson, C. D. 1980b. Observations on the effect of diet and temperature on rate of development of *Mesocyclops leuckarti* (Claus) (Copepoda, Cyclopoida). Crustaceana 38:145–154.

Katona, S. K. 1971. The developmental stages of *Eurytemora affinis* (Poppe) (1880), (Copepoda, Calanoida) raised in laboratory cultures, including a comparison of the larvae of *Eurytemora americana* Williams, 1906, and *Eurytemora herdmani* Thompson and Scott, 1897. Crustaceana 21:5–20.

Katona, S. K. 1973. Evidence for sex pheromones in planktonic copepods. Limnology and Oceanography 18:574–583.

Kerfoot, W. C. 1977a. Competition in cladoceran communities: the cost of evolving defenses against copepod predation. Ecology 58:303–313.

Kerfoot, W. C. 1977b. Implications of copepod predation. Limnology and Oceanography 22:316–325.

Kerfoot, W. C. 1978. Combat between predatory copepods and their prey: *Cyclops*, *Epischura*, and *Bosmina*. Limnology and Oceanography 23:1089–1102.

Kerfoot, W. C., and W. R. DeMott. 1984. Food web dynamics: dependent chains and vaulting. Pages 347–382 *in*: D. G. Meyers and J. R. Strickler, editors. Trophic interactions within aquatic ecosystems. American Association for the Advancement of Science Selected Symposium 85.

Kerfoot, W. C., and C. Peterson. 1980. Predatory copepods and *Bosmina*: replacement cycles and further influences of predation upon prey reproduction. Ecology 61:417–431.

Koehl, M. A. R., and J. R. Strickler. 1981. Copepod feeding currents: Food capture at low Reynolds number. Limnology and Oceanography 26:1062–1073.

Lampert, W., and P. Muck. 1985. Multiple aspects of food limitation in zooplankton communities: the *Daphnia-Eudiaptomus* example. Archiv für Hydrobiologie Beihefte Ergebnisse der Limnologie 21:311–322.

Landry, M. R., and R. P. Hassett. 1982. Estimating the grazing impact of marine micro-zooplankton. Marine Biology 67:283–288.

Lang, K. 1948. Monographie der Harpacticiden. Vol. I and II. H. Ohlsson, Lund.

Légier-Visser, M. F., J. G. Mitchell, A. Okubo, and J. A. Fuhrman. 1986. Mechanoreception in calanoid copepods, a mechanism for prey detection. Marine Biology 90:529–535.

Lehman, J. T. 1980. Release and cycling of nutrients between planktonic algae and herbivores. Limnology and Oceanography 25:620–632.

Lewis, B. G., S. Luff, and J. W. Whitehouse. 1971. Laboratory culture of *Cyclops abyssorum* Sars. 1863. (Copepoda, Cyclopoida). Crustaceana 21:176–182.

Lewis, W. M., and J. F. Saunders, III. 1979. Two new integrating samplers for zooplankton, phytoplankton, and water chemistry. Archiv für Hydrobiologie 85:244–249.

Li, J. L., and H. W. Li. 1979. Species-specific factors affecting predator-prey interactions of the copepod *Acanthocyclops vernalis* with its natural prey. Limnology and Oceanography 24:613–626.

Likens, G. E., and J. J. Gilbert. 1970. Notes on quantitative sampling of natural populations of planktonic rotifers. Limnology and Oceanography 15:816–820.

Luecke, C., and A. H. Litt. 1987. Effects of predation by *Chaoborus flavicans* on crustacean zooplankton of Lake Lenore, Washington. Freshwater Biology 18:185–192.

Luecke, C., and W. J. O'Brien. 1981. Phototoxicity and fish predation: Selective factors in color morphs in *Heterocope*. Limnology and Oceanography 26:454–460.

Luecke, C., and W. J. O'Brien. 1983. The effect of *Heterocope* predation on zooplankton communities in arctic ponds. Limnology and Oceanography 28:367–377.

Makarewicz, J. C., and G. E. Likens. 1979. Structure and function of the zooplankton community of Mirror Lake, New Hampshire. Ecological Monographs 49:109–127.

Maly, E. J. 1970. The influence of predation on the adult sex ratios of two copepod species. Limnology and Oceanography 15:566–573.

Maly, E. J. 1973. Density, size, and clutch of two high altitude diaptomid copepods. Limnology and Oceanography 18:840–848.

Maly, E. J. 1976. Resource overlap between co-occurring copepods: effects of predation and environmental fluctuation. Canadian Journal of Zoology 54:933–940.

Maly, E. J. 1984. Interspecific copulation in and co-occurrence of similar-sized freshwater centropagid copepods. Austalian Journal of Marine and Freshwater Research 35:153–165.

Maly, E. J., and M. P. Maly. 1974. Dietary differences between two co-occurring calanoid copepod species. Oecologia 17:325–333.

Mantel, L. H., and L. L. Farmer. 1983. Osmotic and ionic regulation. Pages 53–162 *in*: L. H. Mantel, vol. editor. The biology of Crustacea, Vol. 5: Internal anatomy and physiological regulation. Academic Press, New York.

Marshall, S. M., and A. P. Orr. 1952. On the biology of *Calanus finmarchicus*. VII. Factors affecting egg production. Journal of the Marine Biological Association of the United Kingdom 30:527–547.

McNaught, D. C. 1975. A hypothesis to explain the succession from calanoids to cladocerans during eutrophication. Verhandlungen Internationale Vereinigung für Theoretische und Angewandte Limnologie 19:724–731.

McQueen, D. J. 1969. Reduction of zooplankton standing stocks by predaceous *Cyclops bicuspidatus* in Marion Lake, British Columbia. Journal of the Fisheries Research Board of Canada 26:1605–1618.

Meyer, J. S., C. G. Ingersoll, L. L. McDonald, and M. S. Boyce. 1986. Estimating uncertainty in population growth rates: jackknife vs. bootstrap techniques. Ecology 67:1156–1166.

Monakov, A. V. 1976. Feeding and food interrelationships in freshwater copepods. Nauka, Leningrad. 170 pp. (In Russ.)

Musko, I. B. 1983. The structure of the alimentary canal of two freshwater copepods of different feeding habits studied by light microscope. Crustaceana 45:38–47.

Nie, H. W. de, and J. Vijverberg. 1985. The accuracy of population density estimates of copepods and cladocerans, using data from Tjeukemeer (the Netherlands) as an example. Hydrobiologia 124:3–11.

Nilssen, J. P. 1982. A simple method for maintaining a laboratory supply of limnetic cyclopoid copepods. Hydrobiologia 94:213–216.

O'Brien, W. J. 1979. The predator-prey interaction of planktivorous fish and zooplankton. American Science 67:572–581.

O'Brien, W. J., and D. Schmidt. 1979. Arctic Bosmina morphology and copepod predation. Limnology and Oceanography 24:564–568.

O'Brien, W. J., D. Kettle, and H. Riessen. 1979. Helmets and invisible armor: structures reducing predation from tactile and visual planktivores. Ecology 60:287–294.

O'Doherty, E. C. 1988. The ecology of meiofauna in an appalachian headwater stream. Ph.D. Thesis, University of Georgia, Athens. 113 pp.

Pace, M. L. 1986. An empirical analysis of zooplankton community size structure across lake trophic gradients. Limnology and Oceanography 31:45–55.

Paloheimo, J. 1974. Calculation of instantaneous birth rate. Limnology and Oceanography 19:692–694.

Park, T. S. 1966. The biology of a calanoid copepod *Epilabidocera amphitrites* McMurrich. La Cellule 66:129–251.

Parrish, K. K., and D. F. Wilson. 1978. Fecundity studies on *Acartia tonsa* (Copepoda: Calanoida) in standardized culture. Marine Biology 46:65–81.

Patalas, K. 1971. Crustacean plankton communities in forty-five lakes in the experimental lakes area, Northwestern Ontario. Journal of the Fisheries Research Board of Canada 28:231–244.

Patalas, K. 1986. The geographical distribution of *Mesocyclops edax* (S.A. Forbes) in lakes of Canada. *In:* G. Schriever, H. K. Schminke, and C.-t. Shih, editors. Proceedings of the second international conference on Copepoda. Syllogeus No. 58:400–408.

Peacock, A. 1982. Responses of *Cyclops bicuspidatus*

thomasi to alterations in food and predators. Canadian Journal of Zoology 60:1446–1462.

Peacock, A., and W. J. P. Smyly. 1983. Experimental studies on the factors limiting *Tropocyclops prasinus* (Fischer) 1860 in an oligotrophic lake. Canadian Journal of Zoology 61:250–265.

Pennak, R. W. 1978. Freshwater invertebrates of the United States. 2nd Edition. Wiley, New York. 803 pp.

Perlmutter, D. G. 1988. Meiofauna in stream leaf litter: patterns of occurrence and ecological significance. Ph.D. Thesis, University of Georgia, Athens. 136 pp.

Pinel-Alloul, B., and J. Lamoureaux. 1988a. Developpement post-embryonnaire du copepode calanoide *Diaptomus* (*Aglaodiaptomus*) *leptopus* S. A. Forbes, 1882 I. Phase nauplienne. Crustaceana 54:69–84.

Pinel-Alloul, B., and J. Lamoureaux. 1988b. Developpement post-embryonnaire du copepode calanoide *Diaptomus* (*Aglaodiaptomus*) *leptopus* S. A. Forbes, 1882 II. Phase copepodite et adulte. Crustaceana 54:171–195.

Prepas, E. 1978. Sugar-frosted *Daphnia:* An improved fixation technique for Cladocera. Limnology and Oceanography 23:557–559.

Price, H. J., and G.-A. Paffenhöfer. 1986. Effects of concentration on the feeding of a marine copepod in algal monocultures and mixtures. Journal of Plankton Research 8:119–128.

Price, H. J., G.-A. Paffenhöfer, and J. R. Strickler. 1983. Modes of cell capture in calanoid copepods. Limnology and Oceanography 28:116–123.

Pyke, D. A., and J. N. Thompson. 1986. Statistical analysis of survival and removal rate experiments. Ecology 67:240–245.

Ramalingam, K., and M. B. Raghunathan. 1982. A study on the pH tolerance and survival of a cyclopoid copepod *Mesocyclops leucarti* (Claus). Comparative Physiology and Ecology 7:188–190.

Reid, J. W. 1985. Calanoid copepods (Diaptomidae) from coastal lakes, State of Rio de Janeiro, Brazil. Proceedings of the Biological Society of Washington 98:574–590.

Reid, J. W. 1986. Some usually overlooked cryptic copepod habitats. *In:* G. Schriever, H. K. Schminke, and C.-t. Shih, editors. Proceedings of the second international conference on Copepoda. Syllogeus No. 58:594–598.

Reid, J. W. 1987. *Attheyella* (*Mrazekiella*) *spinipes*, a new harpacticoid copepod (Crustacea) from Rock Creek Regional Park, Maryland. Proceedings of the Biological Society of Washington 100:694–699.

Reid, J. W. 1988. Copepoda (Crustacea) from a seasonally flooded marsh in Rock Creek Park, Maryland. Proceedings of the Biological Society of Washington 101:31–38.

Reid, J. W. 1990. The distribution of species of the genus *Thermocyclops* (Copepoda, Cyclopoida) in the western hemisphere, with description of *T. parvus*, new species. Hydrobiologia in press.

Richman, S., and S. I. Dodson. 1983. The effect of food quality on feeding and respiration by *Daphnia* and *Diaptomus*. Limnology and Oceanography 28: 948–956.

Richman, S., S. A. Bohon, and S. E. Robbins. 1980. Grazing interactions among freshwater calanoid copepods. Pages 219–233 *in:* W. C. Kerfoot, editor. Evolution and ecology of zooplankton communities. University Press of New England, Hanover, New Hampshire.

Rigler, F. H., and J. M. Cooley. 1974. The use of field data to derive population statistics of multivoltine copepods. Limnology and Oceanography 19:636–655.

Ringelberg, J. 1980. Aspects of red pigmentation in zooplankton, especially copepods. Pages 91–97 *In:* W. C. Kerfoot, editor. Evolution and ecology of zooplankton communities. University Press of New England, Hanover, New Hampshire.

Ringelberg, J., A. L. Keyser, and B. J. G. Flik. 1984. The mortality effect of ultraviolet radiation in a translucent and in a red morph of *Acanthodiaptomus denticornis* (Crustacea, Copepoda) and its possible ecological relevance. Hydrobiologia 112:217–222.

Roche, K. F. 1987. Post-encounter vulnerability of some rotifer prey types to predation by the copepod *Acanthocyclops robustus*. Hydrobiologia 147:229–233.

Roff, J. C. 1972. Aspects of the reproductive biology of the planktonic copepod *Limnocalanus macrurus* Sars, 1863. Crustaceana 22:155–160.

Rosenthal, H. 1972. Uber die Geschwindigkeit der Sprungbewegungen bei *Cyclops strenuus* (Copepoda). Internationale Revue der Gesamten Hydrobiologie 57:157–167.

Sandercock, G. A. 1967. A study of selected mechanisms for the coexistence of *Diaptomus* spp. in Clarke Lake, Ontario. Limnology and Oceanography 12:97–112.

Sarvala, J. 1979. A parthenogenetic life cycle in a population of *Canthocamptus staphylinus* (Copepoda, Harpacticoida). Hydrobiologia 62:113–129.

Schindler, D. W. 1969. Two useful devices for vertical plankton and water sampling. Journal of the Fisheries Research Board of Canada 26:1948–1955.

Shih, C.-T., and D. C. Maclellan. 1977. Description of copepodite stages of *Diaptomus (Leptodiaptomus) nudus* March 1904 (Crustacea: Copepoda). Canadian Journal of Zoology 55:912–921.

Siebeck, O. 1969. Spatial orientation of planktonic crustaceans. I. The swimming behaviour in a horizontal plane. Verhandlungen Internationale Vereinigung für Theoretische und Angewandte Limnologie 17:831–840.

Siebeck, O. 1980. Optical orientation of pelagic crustaceans and its consequences in the pelagic and littoral zones. Pages 28–38 *in:* W. C. Kerfoot, editor. Evolution and ecology of zooplankton communities. University Press of New England, Hanover, New Hampshire.

Siefert, R. E. 1972. First food of larval yellow perch, white sucker, bluegill, emerald shiner, and rainbow smelt. Transactions of the American Fisheries Society 101:219–225.

Smith, K., and C. H. Fernando. 1978. A guide to the freshwater calanoid and cyclopoid copepod crustacea of Ontario. Department of Biology, University of Waterloo, Waterloo, Ontario. 76 pp.

Smyly, W. J. P. 1962. Laboratory experiments with stage V copepodites of the freshwater copepod, *Cyclops leuckarti* Claus, from Windermere and Esthwaite Water. Crustaceana 4:273–280.

Smyly, W. J. P. 1968. Number of eggs and body-size in the freshwater copepod *Diaptomus gracilis* Sars in the English Lake District. Oikos 19:323–338.

Smyly, W. J. P. 1970. Observations on rate of development, longevity and fecundity of *Acanthocyclops viridis* (Jurine) (Copepoda, Cyclopoida), in relation to type of prey. Crustaceana 18:21–36.

Smyly, W. J. P. 1973. Bionomics of *Cyclops strenuus abyssorum* Sars (Copepoda: Cyclopoida). Oecologia 11:163–186.

Smyly, W. J. P. 1974. The effect of temperature on the development time of the eggs of three freshwater cyclopoid copepods from the English Lake District. Crustaceana 27:278–284.

Smyly, W. J. P. 1980. Effect of constant and alternating temperatures on adult longevity of the freshwater cyclopoid copepod, *Acanthocyclops viridis* (Jurine). Archiv für Hydrobiologie 89:353–362.

Stemberger, R. S. 1985. Prey selection by the copepod *Diacyclops thomasi*. Oecologia 65:492–497.

Stemberger, R. 1986. The effects of food deprivation, prey density and volume on clearance rates and ingestion rates of *Diacyclops thomasi*. Journal of Plankton Research 8:243–251.

Stemberger, R. S., and M. S. Evans. 1984. Rotifer seasonal succession and copepod predation in Lake Michigan. Journal of Great Lakes Research 10:417–428.

Stemberger, R. S., and J. J. Gilbert. 1984. Spine development in the rotifer *Keratella cochlearis:* induction by cyclopoid copepods and *Asplanchna*. Freshwater Biology 14:639–647.

Strayer, D. 1985. The benthic micrometazoans of Mirror Lake, New Hampshire. Archiv für Hydrobiologie Supplement 72:287–426.

Strickler, J. R. 1969. Über das Schwimmverhalten von Cyclopoiden bei Verminderungen der Bestrahlungsstarke. Schweizerische Zeitschrift für Hydrologie 31:150–180.

Strickler, J. R. 1975a. Intra- and interspecific information flow among planktonic copepods: Receptors. Verhandlungen Internationale Vereinigung für Theoretische und Angewandte Limnologie 19:2951–2958.

Strickler, J. R. 1975b. Swimming of planktonic *Cyclops* species (Copepoda, Crustacea): Pattern, movements and their control. Pages 599–613 *In:* T. Y. T. Wu, C. J. Brokaw, and C. Brennan, editors. Swimming and flying in nature. Vol. 2. Plenum Press, New York.

Strickler, J. R. 1977. Observation of swimming performances of planktonic copepods. Limnology and Oceanography 22:165–170.

Strickler, J. R. 1982. Calanoid copepods, feeding currents, and the role of gravity. Science 218:158–160.

Strickler, J. R., and A. K. Bal. 1973. Setae of the first antennae of the copepod *Cyclops scutifer* (Sars): Their structure and importance. Proceedings of the National Academy of Science U.S.A. 70:2656–2659.

Strickler, J. R., and S. Twombly. 1975. Reynolds number, diapause, and predatory copepods. Verhandlungen Internationale Vereinigung für Theoretische und Angewandte Limnologie 19:2943–2950.

Swift, M. C., and A. Y. Fedorenko. 1975. Some aspects of prey capture by *Chaoborus* larvae. Limnology and Oceanography 20:418–425.

Tessier, A. J., and C. E. Goulden. 1982. Estimating food limitation in cladoceran populations. Limnology and Oceanography 27:707–717.

Threlkeld, S. T. 1976. Starvation and the size structure of zooplankton communities. Freshwater Biology 6:489–496.

Threlkeld, S. T. 1979. Estimating cladoceran birth rates: the importance of egg mortality and the egg age distribution. Limnology and Oceanography 24:601–612.

Threlkeld, S. T., and J. M. Dirnberger. 1986. Benthic distributions of planktonic copepods, especially *Mesocyclops edax. In:* G. Schriever, H. K. Schminke, and C.-t. Shih, editors. Proceedings of the second international conference on Copepoda. Syllogeus No. 58:481–486.

Tinson, S., and J. Laybourn-Parry. 1985. The behavioural responses and tolerance of freshwater benthic cyclopoid copepods to hypoxia and anoxia. Hydrobiologia 127:257–263.

Torke, B. G. 1974. An illustrated guide to the identification of the planktonic crustacea of Lake Michigan with notes on their ecology. Spec. Rep. No. 17. Center for Great Lakes Study, University of Wisconsin, Milwaukee.

Van Alstyne, K. L. 1986. Effects of phytoplankton taste and smell on feeding behavior of the copepod *Centropages hamatus.* Marine Ecology Progress Series 34:187–190.

Vanderploeg, H. A., and G.-A. Paffenhöfer. 1985. Modes of algal capture by the freshwater copepod *Diaptomus sicilis* and their relation to food-size selection. Limnology and Oceanography 30:871–885.

Vanderploeg, H. A., D. Scavia, and J. R. Liebig. 1984. Feeding rate of *Diaptomus sicilis* and its relation to selectivity and effective food concentration in algal mixtures and in Lake Michigan. Journal of Plankton Research 6:919–941.

Vanderploeg, H. A., W. S. Gardner, C. C. Parrish, J. R. Liebig, and J. F. Chandler. Lipids and life cycle strategy of a cold-water calanoid copepod in Lake Michigan: are lipid levels high because low temperature is a bottleneck to reproducion? In preparation.

Vijverberg, J. 1977. Population structure, life histories and abundance of copepods in Tjeukemeer, the Netherlands. Freshwater Biology 7:579–597.

Vijverberg, J. 1980. Effect of temperature in laboratory studies on development and growth of Cladocera and Copepoda from Tjeukemeer, the Netherlands. Freshwater Biology 10:317–340.

Walton, W. E. 1985. Factors regulating the reproductive phenology of *Onychodiaptomus birgei* (Copepoda: Calanoida). Limnology and Oceanography 30:167–179.

Warren, G., M. S. Evans, D. J. Jude, and J. C. Ayers.

1986. Seasonal variations in copepod size: effects of temperature, food abundance, and vertebrate predation. Journal of Plankton Research 8:841–854.

Watras, C. J. 1983a. Mate location by diaptomid copepods. Journal of Plankton Research 5:417–423.

Watras, C. J. 1983b. Reproductive cycles in diaptomid copepods: Effects of temperature, photocycle, and species on reproductive potential. Canadian Journal of Fisheries and Aquatic Sciences 40:1607–1613.

Watras, C. J., and J. F. Haney. 1980. Oscillations in the reproductive condition of *Diaptomus leptopus* (Copepoda: Calanoida) and their relation to rates of egg-clutch production. Oecologia 45:94–103.

Watson, N. H. F., and B. N. Smallman. 1971. The role of photoperiod and temperature in the induction and termination of an arrested development in two species of freshwater cyclopoid copepods. Canadian Journal of Zoology 49:855–862.

Weglénska, T. 1971. The influence of various concentrations of natural food on the development, fecundity and production of planktonic crustacean filtrators. Ekologia Polska 19:427–473.

Wetzel, R. G. 1983. Limnology. 2nd Edition. Saunders, Philadelphia, Pennsylvania. 767 pp.

Whitehouse, J. W., and B. G. Lewis. 1973. The effect of diet and density on development, size, and egg production in *Cyclops abyssorum* Sars, 1863 (Copepoda, Cyclopoida). Crustaceana 25:225–236.

Wierzbicka, M. 1962. On the resting stage and mode of life of some species of Cyclopoida. Polish Archives of Hydrobiology 10:215–229.

Williamson, C. E. 1980. The predatory behavior of *Mesocyclops edax:* Predator preferences, prey defenses, and starvation-induced changes. Limnology and Oceanography 25:903–909.

Williamson, C. E. 1981. Foraging behavior of a freshwater copepod: Frequency changes in looping behavior at high and low prey densities. Oecologia 50: 332–336.

Williamson, C. E. 1983a. Behavioral interactions between a cyclopoid copepod predator and its prey. Journal of Plankton Research 5:701–711.

Williamson, C. E. 1983b. Invertebrate predation on planktonic rotifers. Hydrobiologia 104:385–396.

Williamson, C. E. 1984. Laboratory and field experiments on the feeding ecology of the freshwater cyclopoid copepod, *Mesocyclops edax.* Freshwater Biology 14:575–585.

Williamson, C. E. 1986. The swimming and feeding behavior of *Mesocyclops.* Hydrobiologia 134:11–19.

Williamson, C. E. 1987. Predator–prey interactions between omnivorous diaptomid copepods and rotifers: The role of prey morphology and behavior. Limnology and Oceanography 32:167–177.

Williamson, C. E., and N. M. Butler. 1986. Predation on rotifers by the suspension-feeding calanoid copepod *Diaptomus pallidus.* Limnology and Oceanography 31:393–402.

Williamson, C. E., and N. M. Butler. 1987. Temperature, food, and mate limitation of copepod reproductive rates: separating the effects of multiple hypotheses. Journal of Plankton Research 9:821–836.

Williamson, C. E., and R. E. Magnien. 1982. Diel vertical migration in *Mesocyclops edax*: Implications for predation rate estimates. Journal of Plankton Research 4:329–339.

Williamson, C. E., and H. A. Vanderploeg. 1988. Predatory suspension-feeding in *Diaptomus*: prey defenses and the avoidance of cannibalism. Bulletin of Marine Science 43:561–572.

Williamson, C. E., N. M. Butler, and L. Forcina. 1985. Food limitation in naupliar and adult *Diaptomus pallidus*. Limnology and Oceanography 30:1283–1290.

Wilson, C. B. 1932. The Copepods of the Woods Hole Region Massachusetts. Bulletin of the United States National Museum 158:635 pp.

Wilson, M. S., and H. C. Yeatman. 1959. Free-living Copepoda. Pages 735–868 *In:* W. T. Edmondson, editor. Fresh-water biology. Wiley, New York.

Winfield, I. J., and C. R. Townsend. 1983. The cost of copepod reproduction: increased susceptibility to fish predation. Oecologia 60:406–411.

Wolken, J. J., and R. G. Florida. 1969. The eye structure and optical system of the crustacean copepod, Copelia. The Journal of Cell Biology 40:279–285.

Wong, C. K. 1981. Predatory feeding behavior of *Epischura lacustris* (Copepoda, Calanoida) and prey defense. Canadian Journal of Fisheries and Aquatic Sciences 38:275–279.

Wong, C. K. 1984. A study of the relationships between the mouthparts and food habits in several species of freshwater calanoid copepods. Canadian Journal of Zoology 62:1588–1595.

Wong, C. K., and W. G. Sprules. 1986. The swimming behavior of the freshwater calanoid copepods *Limnocalanus macrurus* Sars, *Senecella calanoides* Juday and *Epischura lacustris* Forbes. Journal of Plankton Research 8:79–90.

Wong, C. K., C. W. Ramcharan, and W. G. Sprules. 1986. Behavioral responses of a herbivorous calanoid copepod to the presence of other zooplankton. Canadian Journal of Zoology 64:1422–1425.

Woodmansee, R. A., and B. J. Grantham. 1961. Diel vertical migrations of two zooplankters (*Mesocyclops* and *Chaoborus*) in a Mississippi Lake. Ecology 42:619–628.

Woodward, I. O., and R. W. G. White. 1981. Effects of temperature and food on the fecundity and egg development rates of *Boeckella symmetrica* Sars (Copepoda:Calanoida). Australian Journal of Marine and Freshwater Research 32:997–1002.

Wright, D., W. J. O'Brien, and G. L. Vinyard. 1980. Adaptive value of vertical migration: A simulation model argument for the predation hypothesis. Pages 138–147 *in:* W. C. Kerfoot, editor. Evolution and Ecology of Zooplankton Communities. University Press of New England, Hanover, New Hampshire.

Wyngaard, G. A. 1983. *In situ* life table of a subtropical copepod. Freshwater Biology 13:275–281.

Wyngaard, G. A. 1986. Genetic differentiation of life history traits in populations of *Mesocyclops edax* (Crustacea: Copepoda). Biological Bulletin 170:279–295.

Wyngaard, G. A., and C. C. Chinnappa. 1982. General biology and cytology of cyclopoids. Pages 485–533 *in:* F. W. Harrison and R. R. Cowden, editors. Developmental biology of freshwater invertebrates.

Yeatman, H. C. 1959. Cyclopoida. Pages 795–815 *in:* W. T. Edmondson, editor. Fresh-water biology. Wiley, New York.

Zaret, T. M. 1980. Predation in freshwater communities. Yale University Press, New Haven, Connecticut. 187 pp.

Zaret, T. M., and J. S. Suffern. 1976. Vertical migration in zooplankton as a predator avoidance mechanism. Limnology and Oceanography 21:804–813.

Zo, Z. 1982. The sequential taxonomic key: an application to some copepod genera. Hydrobiologia 96:9–13.

ADDITIONAL SUGGESTED READINGS

Anderson, R. S. 1971. A nomen novum to replace the junior homonym, *Diaptomus (Leptodiaptomus) intermedius*, Anderson and Fabris 1970. Canadian Journal of Zoology 49:133.

Anderson, R. S., and G. L. Fabris. 1970. A new species of diaptomid copepod from Saskatchewan with notes on the crustacean community of the pond. Canadian Journal of Zoology 48:49–54.

Dussart, B. 1985. Le genre *Mesocyclops* (Crustace, Copepode) en Amerique du Nord. Canadian Journal of Zoology 63:961–964.

Dussart, B., and D. Defaye. 1985. Repertoire mondial des copepodes cyclopoides. Editions du Centre National de la Recherche Scientifique, Bordeaux, Paris. 236 pp.

Dussart, B., and C. H. Fernando. 1986. The *Mesocyclops* species problem today. *In:* G. Schriever, H. K. Schminke, and C.-t. Shih, editors. Proceedings of the second international conference on Copepoda. Syllogeus No. 58:288–293.

Dussart, B. H., and C. H. Fernando. 1988. Sur quelques *Mesocyclops* (Crustacea, Copepoda). Hydrobiologia 157:241–264.

Einsle, U. 1975. Revision der Gattung *Cyclops* s. str. speziell der *abyssorum* Gruppe. Mem. Ist. Ital. Idrobiol. 32:57–219.

Einsle, U. 1985. A further criterion for the identification of species in the genus *Cyclops* s. str. (Copepoda, Cyclopoida). Crustaceana 49:299–309.

Gurney, R. 1932. British Freshwater Copepoda. Vol. 2. Ray Society, London.

Light, S. F. 1938. New subgenera and species of diaptomid copepods from the inland waters of California and Nevada. University of California Berkeley Publications in Zoology 43:67–78.

Light, S. 1939. New American subgenera of *Diaptomus* Westwood (Copepoda, Calanoida). Transactions of the American Microscopical Society 58:473–484.

Reed, E. B. 1986. Esteval phenology of an *Acanthocyclops* (Crustacea, Copepoda) in a Colorado tarn with remarks on the *vernalis-robustus* complex. Hydrobiologia 139:127–133.

Vervoot, W. 1986a. Bibliography of Copepoda up to and including 1980. Part I. (A–G). Crustaceana, Supplement 10:369 pp.

Vervoot, W. 1986b. Bibliography of Copepoda up to and including 1980. Part II. (H–R). Crustaceana, Supplement 10:371–845.

Vervoot, W. 1990. Bibliography of Copepoda up to and including 1980. Part III. S–Z. Addenda et Corrigenda; Supplement 1981–1985.

Wilson, M. S. 1958. The copepod genus *Halicyclops* in North America, with description of a new species from Lake Pontchartrain, Louisiana, and the Texas coast. Tulane Studies in Zoology 6:176–189.

Decapoda

22

H. H. Hobbs III
Department of Biology
Wittenburg University
Springfield, Ohio 45501-0720

Chapter Outline

I. INTRODUCTION

II. ANATOMY AND PHYSIOLOGY
  A. External Morphology
  B. Organ System Function
  C. Aspects of Decapod Physiology
    1. Respiration
    2. Ecdysis
    3. Reproductive Physiology
    4. Other Physiological Adaptations

III. ECOLOGY AND EVOLUTION
  A. Diversity, Distribution, and Evolutionary Relationships
    1. Shrimps
    2. Crayfishes
  B. Reproduction and Life History
    1. Shrimps
    2. Crayfishes
  C. Ecological Interactions
    1. Functional Role in the Ecosystem
    2. Foraging Relationships
    3. Behavioral Ecology
    4. Population Regulation

IV. CURRENT AND FUTURE RESEARCH PROBLEMS

V. COLLECTING AND REARING TECHNIQUES
  A. Shrimps
  B. Crayfishes: Astacidae and Cambaridae
VI. IDENTIFICATION
  A. Preparation of Specimens
  B. Taxonomic Key to Genera of Freshwater Decapoda
  Literature Cited

I. INTRODUCTION

The order Decapoda (Latreille) is one of many orders assigned to the class Malacostraca and encompasses a tremendous diversity of marine, freshwater, and semiterrestrial crustaceans, with nearly 10,000 species having been described. The two infraorders treated in this chapter, Caridea and Astacidea, are nearly worldwide in distribution and for purposes of this discussion are represented by freshwater shrimps and crayfishes (see Bowman and Abele 1982 for a detailed classification). These aquatic arthropods have successfully invaded a wide variety of aquatic and semiaquatic habitats and the crayfishes and one genus of shrimps, in particular, attain the greatest size among the freshwater crustaceans in North America.

This chapter summarizes the ecology and distribution of shrimps and crayfishes in North American freshwaters (generally excluding Mexico and Cuba). Species richness and niche diversification are particularly well demonstrated in aquatic ecosystems of the southeastern United States. That these crustaceans have successfully colonized such diverse habitats is reflected in various morphological and physiological adaptations discussed. Life-history patterns are reviewed, particularly with reference as to how they are influenced by environmental conditions. An examination is made of the role of these decapods in the functioning of various aquatic communities. As might be anticipated, a much larger body of data is available for crayfishes than for shrimps and is reflected in the treatment of these groups. Under each major section of the chapter, a discussion of shrimps appears first, followed by that of crayfishes. Where data are somewhat limited (particularly with reference to shrimps), both groups are discussed in the same section.

II. ANATOMY AND PHYSIOLOGY

A. External Morphology

Although the decapods are the largest order of crustaceans, all possess a number of common characteristics, including a carapace that encloses the branchial chamber. In addition, the first three pairs of thoracic appendages are modified in all species as maxillipeds. Carideans (Fig. 22.1) are easily distinguished from other shrimp groups (e.g., penaeids, stenopodids) by the large second abdominal pleura that overlap those of both the first and third somites; moreover, all carideans lack terminal chelae on the third pereiopods (Fig. 22.1a). Astacideans (Fig. 22.2) include the crayfishes of the northern and southern hemispheres and the marine lobsters; only North American crayfishes are treated here. In contrast to the carideans, crayfishes are rarely compressed laterally and their first three pairs of pereiopods are always chelate. Extensive body segmentation and the presence of jointed appendages on all metameres give most decapods a primitive appearance; yet the nervous, sensory, circulatory, and digestive systems are complex, a necessary requirement for animals of large body size. The body of these freshwater decapods is encased in an exoskeleton consisting of complex polysaccharides hardened with inorganic salts (except

at joints, where the cuticle is thin and pliable). The head and thoracic segments are fused to form a large cephalothorax covered by a single shield, the carapace (Figs. 22.1 and 22.2), but the six abdominal segments are individually distinct. The anterior portion of the cephalothorax contains a pair of large, stalked eyes, a median rostrum, two pairs of antennae, and a pair of mandibles. The body is composed of somites, each with a pair of ventral, jointed appendages that are serially homologous and basically biramous (Fig. 22.3), but are modified for various functions (e.g., sensory, food handling, cleaning, gill-bailing, pinching, walking, copulation, egg attachment and incubation, and swimming). The typical biramous appendage is Y-shaped and the base of the Y is attached to the somite. The base is the protopodite (consisting of two joints—coxopodite, basiopodite) that bears a mesial endopodite, typically of five podomeres, and a lateral exopodite, with a few to many segments.

B. Organ System Function

The digestive system includes the mouth, a short tubular esophagus, the stomach, two large hepatopancreatic glands, a short midgut, and a long tubular intestine extending dorsally through the abdomen to the anus (Fig. 22.4). The stomach functions as a

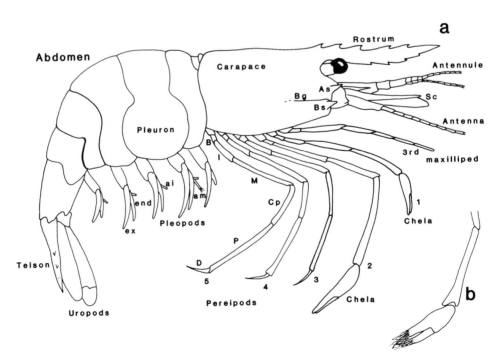

Figure 22.1 (a) Lateral view of generalized shrimp (after Hobbs and Jass 1988). Ai, appendix interna; Am, appendix masculina; As, antennal spine; B, basis; Bg, branchiostegal groove; Bs, branchiostegal spine; Cp, carpus; D, dactyl; end, endopod; ex, exopod; I, ischium; M, merus; P, propodus; Sc, scaphocerite. (b) Chela of second pereiopod with apical tufts of setae; *Palaemonias ganteri*. (After Hobbs *et al.* 1977.)

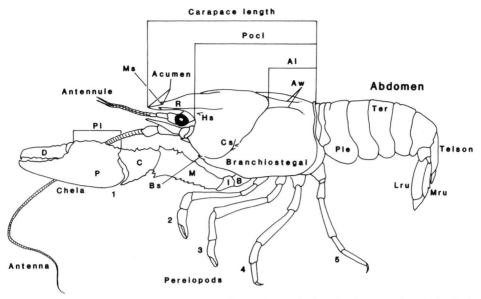

Figure 22.2 Dorsolateral view of generalized crayfish. Al, areola length; Aw, areola width; B, basis; Bs, branchiostegal spine; C, carpus; Cs, cervical spine; D, dactyl; Hs, hepatic spine; I, Ischium; Lru, lateral ramus of uropod; M, merus; Mru, mesial ramus of uropod; Ms, marginal spine; P, propodus; Pl, palm length; Ple, pleuron; Pocl, postorbital carapace length; R, rostrum; Ter, tergum; pleopods on ventral side of abdomen.

storage compartment (cardiac portion), as a masticating structure (gastric mill), and as a filter for gathering digestible material (pyloric portion). Morphological differences among species in the grinding surfaces of the gastric mill relate to diet (Caine 1975). For example, the surface is reduced in cave crayfishes feeding on organic silt and expanded in epigean species utilizing macromaterials.

The circulatory system is an open or lacunar system and consists of a heart, arteries (no veins), and sinuses. The heart is situated in a large mid-dorsal pericardial sinus, and blood in the sinus enters the heart via three pairs of ostia (valves). The nearly colorless blood is pumped into arteries that distribute it to various organs and afferent sinuses leading to the gills and the blood returns through a series of

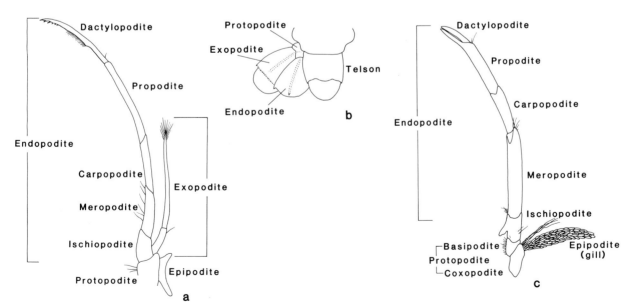

Figure 22.3 Biramous appendages: (a) *Palaemonias alabamae,* third maxilliped (after Smalley 1961); (b) *Cambarus strigosus,* dorsal view of telson and biramous uropod (after Hobbs 1981); (c) *Procambarus lunzi,* third pereiopod.

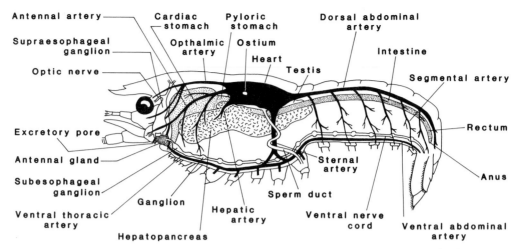

Figure 22.4 Generalized diagram of crayfish, demonstrating internal organization.

efferent sinuses that ultimately join to the pericardial sinus. Hemocyanin is the principal blood pigment transporting oxygen.

The gills are attached to the thoracic appendages (maxillipeds and pereiopods) and are situated along either side of the thorax in the branchial (gill) chamber. Lateral extensions of the carapace (branchiostegites) cover the gills (Figs. 22.1a and 22.2). In crayfishes, anterior to the gills, paddle-like second maxillae (scaphognathites, see Burggren and McMahon 1983) beat back and forth, drawing water over the gill filaments to ensure gas exchange. The gills of crayfishes (trichobranchs composed of unbranched, fingerlike filaments) and shrimps (phyllobranchs, consisting of platelike elements) are arranged in three series: the single podobranchia, borne on the coxa, the arthrobranchiae on the articular membrane between the coxa and basis, and the pleurobranchia borne on the pleura. The pleurobranch series is absent in all of the North American crayfishes except in members of the genus *Pacifastacus,* which have a single pair.

Considerable water constantly diffuses into the blood through the gill surfaces of freshwater decapods. Osmotic control and excretion are maintained by two large antennal (green) glands situated ventral to the subesophageal ganglia and anterior to the esophagus. These glands have an intimate association with the circulatory system and thus maintain water balance. The green gland consists of a complexly folded end sac, a convoluted labyrinth, a nephridial canal, a bladder, and an excretory pore situated at the coxae of the second antennae. Crayfishes excrete a quantity of hypotonic urine (some ammonia, urea, uric acid, and amines), which is generally isosmotic with the blood.

The central nervous system consists of paired supraesophageal and subesophageal ganglia, which

are joined, forming the "brain." This juncture is accomplished by a pair of connectives extending caudally from the subesophageal ganglia to the double ventral nerve cord which has segmental ganglia connected by commissures. The basal segment of each first antenna supports an organ of balance (statocyst). Chemoreceptors are abundant on mouth parts, pereiopods, and antennae; these last appendages also function as tactile organs. Lateral rami of the antennules bear chemoreceptive sensilla called aesthetascs that are innervated by sensory neurons. Tierney and Dunham (1984) and Tierney *et al.* (1984) suggested that these structures may be receptive to pheromones. Many setae on the body surface contain sensory neurons and are mechanoreceptors of one kind or another. A pair of large compound eyes characterizes most decapods (excluding troglobites); each eye is composed of discrete optical units called ommatidia.

Further details of the morphology of decapods in general can be examined in numerous invertebrate zoology textbooks and the external anatomy of crayfishes specifically is treated well by Crocker and Barr (1968).

C. Aspects of Decapod Physiology

1. Respiration

The adaptation of organisms to various habitats usually involves physiological, biochemical, and behavioral adjustments. Obviously a dynamic interaction exists between the external environment of an organism and the functioning of its internal machinery. For example, reduced metabolic rates in cave decapods may stem from biochemical and physiological adaptations, to greater environmental stability, to low number of predators, to reduced food concentrations (Huppop 1985), as demonstrated by whole animal respiration studies (see Weingartner 1977, Franz 1978). Also, the reduction of oxygen consumption and energy turnover of gill tissues reported by Dickson and Franz (1980) further substantiate the highly specialized nature of physiological and biochemical adaptations in troglobitic animals.

In order for any life form to survive, the variety of factors constituting the ambient environmental complex must not exceed genetically determined tolerance limits for that species. That crayfishes and shrimps have radiated into nearly every type of aquatic habitat suggests that these crustaceans have been tolerant and highly adaptable, particularly with regard to respiratory physiology. For example, Wiens and Armitage (1961) demonstrated that, for the crayfishes *Orconectes immunis* (Hagen) and *O. nais* (Faxon), the effect of body size on oxygen consumption and the effects of both temperature and oxygen saturation on oxygen consumption by them are highly significant (inverse relationship between body weight and metabolic rate; direct relationship between temperature and oxygen consumption). They also showed that the cumulative effect of increased temperature and lowered oxygen concentration produces greater stress than either parameter acting independently. *O. immunis* has a daily metabolic rate 10% lower than *O. nais,* and *O. nais* does not regulate as well as does *O. immunis* when under stressed conditions. Therefore, the inability of *O. nais* to maintain a higher and more nearly regulated metabolic rate (as does *O. immunis*) under such conditions, probably excludes it from roadside ditches where *O. immunis* is found. Hence, *O. immunis* is better adapted physiologically to live under environmental conditions of periodic high temperatures and low oxygen saturations. Nelson and Hooper (1982) showed that the shrimp, *Palaemonetes kadiakensis,* exhibited an acute thermal preference within a temperature range of 6°–10°C below their maximum tolerance limits of 37°–40°C. This shrimp was unable to tolerate sustained exposure to temperatures greater than 32°C. For additional treatment of metabolism, temperature acclimation, preference, and avoidance in these crustaceans, see McWhinnie and O'Connor (1967), Crawshaw (1974), Loring and Hill (1978), Mathur *et al.* (1982), and Layne *et al.* (1985).

Many species [e.g., the crayfish, *Procambarus s. simulans* (Faxon)] are oxygen regulators and increase their ventilation rates in response to reduced oxygen or increased carbon dioxide concentrations. Others [e.g., *Pacifastacus l. leniusculus* (Dana)] are oxygen conformers, lacking specific adaptations for regulation of oxygen uptake at low oxygen tensions (see Vernberg 1983). For further discussion of respiratory physiology in crayfishes and shrimps, see Moshiri *et al.* (1970), Becker *et al.* (1975), Thorp and Wineriter (1981), Cox and Beauchamp (1982), Nelson and Hooper (1982), and McMahon and Wilkes (1983).

2. Ecdysis

Molting (ecdysis), a process critical for growth and seasonal changes in form (dimorphic only in male cambarines), is a recurring crisis in the lives of shrimps and crayfishes. It involves far more than a simple splitting of the cuticle and secretion of a new exoskeleton. The life of a decapod consists of alternating periods of premolt, molt, postmolt, and intermolt; all of these phases are controlled by hormones (endocrines). Neurosecretory cells occur in the sensory papillae and in the X-organs of the medulla terminalis; both secretory masses are situated in the eye stalk. The sinus gland, also located within the eye stalk, receives secretions from the X-organs and, in turn, produces hormones that inhibit ecdysis. The Y-organs, a pair of glands in the maxillary somites, secrete hormones that stimulate molting (under the control of the X-organ). During postmolt and intermolt, the sinus gland liberates molt-inhibiting hormones, thus regulating the length of the intermolt period. Under appropriate external or internal conditions (light, temperature, loss of limbs, see Stoffel and Hubschman 1974), sinus gland hormones are not released and the Y-organ, no longer suppressed, secretes the molt-initiating hormone. This acts on the epidermis to initiate premolt activities which affect most body parts. Glycogen reserves are increased and minerals (e.g., calcium) are resorbed from the exoskeleton and stored. With the secretion of an enzyme that softens the cuticle at its base, it pulls away from the epidermal cells, stimulating the formation of a new epicuticle that is impervious to the molting enzyme. At the termination of premolt, the old cuticle splits and the soft body within retains the maximum volume due to the imbibing of water by the decapod. The tissues grow rapidly; and, during the short postmolt period, the new layer becomes rigid as the stored minerals are deposited in the hardening cuticle. After postmolt, the crustacean may enter intermolt; however, complete intermolt probably occurs only in adults and the rapidly growing (molting) juveniles generally have, at best, a very short period of stability.

3. Reproductive Physiology

The development and function of gonads and the ontogeny of secondary sexual characters in the dioecious decapods are regulated by hormones. The Y-organ influences the development and maturation of gonads in juveniles. In females, the sinus gland–X-organ complex is important in controlling the ovary. During the nonbreeding periods, the sinus gland produces a hormone that inhibits egg development. However, during the breeding season, the blood level of this hormone declines and egg development is subsequently initiated. Development of the testes and male sexual characteristics are controlled by hormones produced in the androgenic gland, a small mass of secretory tissue near the vas deferens. Additional information concerning the reproductive cycle for shrimps and crayfishes will be presented. Even though the source and nature of interspecific and intraspecific (pheromones) chemicals are unknown for crayfishes, studies have shown that these chemical signals play an integral role in the behavior of these decapods, although other studies have not demonstrated this (Itagaki and Thorp 1981, Thorp 1984). The role of chemoreception in mediating agonistic behavior, species recognition, and mate location in crayfishes has been documented by Hazlett (1985). Discrimination between male and female conspecifics using chemical cues has been reported by some authors; these pheromones may promote sex recognition and induce mating behavior (see Hazlett 1985). Tierney *et al.* (1984) determined that the site of pheromone reception in *Orconectes propinquus* (Girard) is the lateral antennular flagellum (exopod) bearing the chemoreceptive sensilla (aesthetascs).

4. Other Physiological Adaptations

As ectotherms, these decapods are subjected to temperature changes that directly affect metabolic rate, and most species demonstrate various thermal tolerances (most upper tolerances range from 32°–36°C) and preferences for their aquatic medium. Unfortunately, upper lethal temperatures are known for only a few species (see Cox and Beauchamp 1982 and references therein). Nelson and Hooper (1982) showed experimentally that *Palaemonetes kadiakensis* Rathbun exhibited an acute thermal preference (short-term response) within a temperature range of 6°–10°C below their maximum tolerance limits. Taylor (1984) determined for three southeastern United States crayfishes that both the acute preference and final preferendum primarily represented the temperature that they encountered during daily or seasonal activity cycles. He hypothesized that temperature choice does not function in temperature regulation except in the very broadest sense; instead, it serves as an "error detection system" and cues for behavioral arousal from either seasonal hibernation or from their daily refugia.

Biological rhythms are recognized in decapods and are reviewed by DeCoursey (1983). Not only are organismic cycles pronounced (e.g., crayfishes have daily locomotor rhythms) but cellular and physiological rhythms are prominently exhibited (e.g., circadian rhythmicity of heart rate, Pollard and Larimer 1977). Most seem to be in response to environmental cues (e.g., light, temperature); however, even where rhythmic environmental input may be lacking, circadian clock regulation can be maintained. For instance, Weingartner (1977) demonstrated a circadian activity pattern in the cave crayfish *Orconectes i. inermis* Cope from Indiana. In a similar way the Kentucky troglobite, *O. pellucidus* (Tellkampf), has retained a circadian clock regulation of locomotor activity and oxygen consumption in spite of isolation from day–night cycles.

Crayfishes and shrimps are faced with the problems of constant influx of water by osmosis and loss of salts by diffusion. Maintenance of constant body fluid composition is accomplished by active sodium uptake via the gills, production of a copious hypotonic urine, and by some reduction in boundary membrane permeability. As indicated earlier, the green gland is intimately associated with the circulatory system.

The chromatophore is an integument cell with branched, radiating, noncontractile processes that are associated with one or more types of pigment granules. These white, red, yellow, blue, brown, or black pigments can flow into the processes or may be confined to the center of the cell. A single chromatophore may possess any number of pigments and is accordingly classified as mono-, bi-, di-, or polychromatic (see Fig. 22.5). In shrimps, the number of pigments and the number of chromatophores can change (morphological color change). However, physiological color change (rapid color adaptation to background resulting from dispersal and concentration of pigments within the chromatophores) is the more common type of color alteration and is controlled by hormones and the central nervous system. Chromatophorotropins are released from both the sinus gland and the central nervous system. The latter elaborates several chromatophorotropins, resulting in pigments either dispersing or concentrating in the chromatophores (darkening or blanching of body color)(Rao and Fingerman 1983).

Increasing awareness of human impact on the environment has directed attention to the effects of various pollutants on the physiological ecology of crustaceans, including crayfishes and shrimps (Buikema and Benfield 1979). Accumulation of trace metals occurs through runoff or direct dumping of sewage or other effluents into aquatic systems. Lead has become particularly important because of its relative toxicity and increased environmental contamination via automobile exhaust and highway run-

Figure 22.5 Female *Palaemonetes kadiakensis* with eggs (note chromatophores).

off. Anderson (1978) demonstrated for *O. virilis* (Hagen) that exposure to lead resulted in damage to gills, which caused a decrease in the uptake of oxygen; ventilation rates increased as ambient concentrations increased. Hubschman (1967) noted that the toxic effect of copper on crayfishes is dependent not only upon concentration and duration of exposure but also on the age (or size) of the crustacean. At concentrations above 1 mg/liter, respiratory enzymes are inhibited quite rapidly as the mechanism of detoxification is apparently overwhelmed. At concentrations below 1 mg/liter, copper has a chronic effect on cell maintenance. For example, actively secreting cells such as those of the antennal gland apparently are destroyed because cell repair cannot keep pace with cellular activity. Cadmium, a highly toxic metal, has been implicated as both a carcinogen and a mutagen and causes a number of cardiovascular problems (all resulting from enzyme dysfunction). Cadmium is bioaccumulated and thus when incorporated in tissues of freshwater decapods can have far-reaching effects within aquatic ecosystems as well as in humans (e.g., Gillespie *et al.* 1977, Thorp *et al.* 1979, Dickson *et al.* 1982, Thorp and Gloss 1986). Vermeer (1972) noted high mercury levels in muscle tissue of *O. virilis* from several localities in Manitoba and Ontario and suggested that this species is a good indicator of mercury contamination in aquatic ecosystems. Insecticides (Albaugh 1972), herbicides (Naqvi and Leung 1983), lampricides, and numerous other chemicals (Buikema and Benfield 1979) have been shown to have various detrimental effects on freshwater decapods.

The bioaccumulation of metals and other toxic or persistent chemicals poses particular problems for long-lived cavernicoles (cave-dwellers). Dickson *et al.* (1979) showed higher concentrations of cadmium and lead in tissues of *Orconectes a. australis* (a troglobite) than in those of *Cambarus tenebrosus* (a troglophile), crayfishes occurring syntopically in a

Tennessee cave. It is possible that some troglobitic decapod populations could be extirpated not from a single, massive dose of a chemical (though this has occurred (Crunkilton 1982, Bechler 1983, Hobbs 1987), but slowly from long exposure to low levels of these metals or other toxic substances. Because the longevity of primary burrowing species is apparently greater than for nonburrowing taxa living in surface waters, a similar threat from chronic pollution exists for these crustaceans. This is particularly true for crayfishes living in roadside ditches that are periodically sprayed with herbicides (Hubschman 1967, Dimond *et al.* 1968).

III. ECOLOGY AND EVOLUTION

A. Diversity, Distribution, and Evolutionary Relationships

Most freshwater decapods of North America occur in one or more of the following habitats: shallow (1–2 m depths) lentic and lotic waters, epigean and hypogean habitats, oligotrophic to hypereutrophic lakes, ponds, marshes, ditches, steep-gradient headwater streams, low-gradient large rivers, springs, stygean streams and pools, and terrestrial burrows leading to groundwater. A few have been found in lakes or springs at depths of 30–90 m but most thrive in shallow, littoral areas. They are most abundant in the central, eastern, and southern regions of the United States while a few species are found in the Pacific slope drainages; they are rare or absent in the large area of the western Great Plains and the Rocky Mountains. This simply means that American crayfishes have become established in nearly every type of freshwater habitat, except glacial and thermal effluents, and some can tolerate brackish water. Considerable data suggest that habitat has a channelizing influence (convergence) on certain morphological aspects of crayfishes (Hobbs, Jr. 1976). For example, "troglobites" (a polyphyletic group) are characterized by an albino-sightless combination; whereas, "burrowers" (also polyphyletic) possess short, broad, flattened chelipeds, lack carapace spines, have a reduced abdomen and tail fan (length, width, and height), and possess an enlarged gill chamber. Water currents have certainly influenced the body form of crayfishes, with those species dwelling in fastest flowing water having depressed (most common) or compressed bodies (e.g., *Orconectes compressus* (Faxon) and *Procambarus tenuis* Hobbs); spination also is reduced. Those species inhabiting riffle areas of streams and residing in pebble and cobble substrata (rubble dwellers) are markedly depressed dorsoventrally (e.g., *Camba-*

rus friaufi Hobbs and *Cambarus brachydactylus* Hobbs).

Several decapods (infraorder Brachyura) occur sporadically (naturally or introduced) in coastal brackish or freshwater systems but are not included in the following discussion or key. They are represented by the portunid blue crab, *Callinectes sapidus* Rathbun, a xanthid mud crab, *Rhithropanopeus harrisii* (Gould) (found in streams of the Atlantic and Gulf coasts and introduced along the coasts of California and Oregon), and a grapsid river crab, *Platychirograpsus typicus* Rathbun, introduced from eastern Mexico into the Hillsboro River near Tampa, Florida (Marchand 1946). The giant Malaysian prawn, *Macrobrachium rosenbergii* (deMan) (family Palaemonidae), has been introduced into ponds of central Florida, South Carolina, and numerous other parts of the world (e.g., Israel, Jamaica) (Smith *et al.* 1978).

1. Shrimps

The atyid shrimps (fingers of chelate pereiopods with apical tufts of setae, Fig. 22.1b) of North America are restricted to two genera, one with two southeastern, obligate cave species and the other represented by two species found in coastal streams of California. Freshwater palaemonids (fingers of chelate pereiopods lacking apical tufts of setae, Fig. 22.1a) are a very successful family of shrimps; the genus *Palaemonetes* contains approximately 17 species in the western hemisphere, and *Macrobrachium* is assigned about 100 species worldwide (35 in the western hemisphere). These genera are represented in North America by 10 and 5 species, respectively. Shrimps are generally found in the macrophyte-rich littoral zone of lakes or in sluggish, vegetation-choked streams. Within these two families, five shrimps are troglobitic and are very restricted in distribution (three species known to occur only in a single karst locality).

a. Atyidae

North American freshwaters are inhabited by three federally endangered atyid species, *Palaemonias ganteri* Hay, *Palaemonias alabamae* Smalley, and *Syncaris pacifica* (Anonymous 1989). The first two taxa are troglobitic shrimps, the former restricted to Mammoth Cave in Kentucky, and the latter is known only from Bobcat and Shelta caves in northeastern Alabama. Cave systems occupied by these two species are approximately 280 km apart. Because most atyids currently inhabit warmer southern aquatic ecosystems, it is probable that these two troglobionts are relicts of a former, more widely distributed fauna. *Syncaris pacifica* (Holmes) and *S.*

pasadenae (Kingsley) are found in coastal streams of California (the former is reported from Marin, Napa, and Sonoma counties and the latter is previously known from Los Angeles, San Bernardino, and San Diego counties but may be extinct due to habitat destruction and over-collecting).

b. Palaemonidae

Those North American palaemonids that are strictly freshwater inhabitants are represented by nine species; three are troglobitic (two are candidate species, Fish and Wildlife Service, 1989) and six are inhabitants of lentic and sluggish lotic (including springs) habitats (Figs. 22.6 and 22.7 in part). In addition, five species are found in brackish waters along coastal regions of the Atlantic, Gulf of Mexico, and the Pacific.

Strenth (1976) offered hypothetical explanations for the origin and dispersal of *Palaemonetes* and *Macrobrachium;* he proposed that freshwater *Palaemonetes* may have arisen during the late Mesozoic or early Cenozoic and that the widely disjunct species of the genus are largely of monophyletic or of limited polyphyletic origin. He presented salinity tolerance data from laboratory experiments supporting the premise that worldwide dispersal of freshwater *Palaemonetes* over large expanses of open ocean would have been unlikely. Results of his studies on *P. kadiakensis,* coupled with those of Dobkin and Manning (1964) for *P. paludosus* (Gibbes), demonstrated that this group of shrimps cannot osmoregulate in salinities greater than 25 ppt. Thus, the distributional patterns of *Palaemonetes* and *Macrobrachium* appear consistent with the theories of plate tectonics and continental drift. He proposed that *Palaemonetes* preceded *Marcobra-*

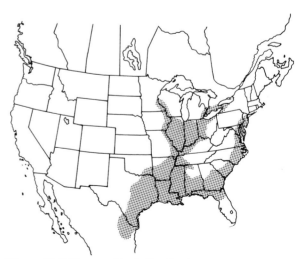

Figure 22.6 Geographical distribution of *Palaemonetes kadiakensis* and *P. paludosus.*

Figure 22.7 Geographical distribution of *Macrobrachium ohione*. Gross reduction in populations throughout its range is attributable to loss of suitable habitat.

chium into the Gulf of Mexico area and that *Macrobrachium* (primarily juveniles), through competitive exclusion, barred *Palaemonetes* from the middle latitudes. Most of the freshwater species of *Palaemonetes* are found above or below the 30° latitudes, or, if found between, always occur at higher elevations or are located at considerable distances from the coast. Parameters that appear to limit the further spread of *Macrobrachium* are current velocity, salinity, and temperature tolerances.

2. Crayfishes

The freshwater astacideans (superfamily Astacoidea) constitute a very successful group and currently are represented by 386 described species and subspecies (four extinct species), 319 species and subspecies which are restricted (introductions ignored) to North America, north of Mexico (see Hobbs, 1986 for lineages of the major groups in North America). These are assigned to twelve genera and two families. The families, Astacidae and Cambaridae, have historically occupied allopatric ranges, the former inhabiting the Pacific slope drainages [except for *Pacifastacus gambelii* (Girard) which has crossed the divide into the upper Missouri] and the Cambaridae occurring in some Hudson Bay watersheds, the Atlantic, and Gulf drainages from southern Canada to Honduras, certain Pacific slope basins in Mexico, and Cuba (Fig. 22.8). The three most successful genera of crayfishes are *Procambarus, Cambarus,* and *Orconectes.* The coastal plain of the southern states is dominated by *Procambarus,* whereas members of the genus *Cambarus* (Fig. 22.9) are major components of aquatic environments from the Cumberland Plateau to the North and East. West and northwest of the Cum-

berland Plateau, species of the genus *Orconectes* are the most commonly encountered crayfishes. Hobbs (1989) has produced a useful listing of the crayfish faunas by country and state and has included references treating species within each state. Two recent publications that discuss decapods in the northeastern United States are those of Smith (1988) and Peckarsky *et al.* (1990), and Page (1985) and Hobbs and Jass (1988) present data for decapods in Illinois and Wisconsin, respectively.

Many epigean crayfishes occur abundantly in areas that afford concealment, particularly during the daytime hours, such as weed beds, leaf litter, stones, brush piles, log jams, roots of riparian trees; and they often find shelter in cans, tires, or among other manmade debris. The genus *Cambarellus* in North America is distributed below the Fall Line zone in the Gulf Coastal Plain (Albaugh and Black 1973). Whereas a few species are known only from lotic habits, most members of this genus, as well as those of *Faxonella* and *Bouchardina,* frequent lentic habitats such as ponds, backwaters of streams, and roadside ditches; and these small crayfishes are successful in littoral waters subject to low oxygen concentrations and elevated temperatures. The members of the genus *Hobbseus* also occupy such habitats, but they are more common in very small, shallow streams. The monotypic *Bouchardina* probably lives in similar habitats because it has been collected from leaf litter and aquatic vegetation in a single stream backwater ditch. The monotypic *Barbicambarus* is found only under large limestone rocks in the swift currents of large streams in the Green and Barren river basins of the Highland Rim. Of note, introductions of non-native species into various parts of the world are leading to species displacements among crayfishes.

Hypogean species are represented by burrowing and cave-dwelling crayfishes. Primary burrowers, which are found in the genera *Cambarus, Distocambarus, Fallicambarus,* and *Procambarus,* construct tunnels in such habitats as seepage areas, along stream banks, in open fields, and in other low-lying areas. Among the troglobitic species are members of the genera *Cambarus, Orconectes* (Fig. 22.10), *Procambarus,* and *Troglocambarus;* some of these cavernicolous crayfishes occupy lotic habitats while others appear to have become adapted to lentic situations. Those species inhabiting caves with active streams generally are found in pooled areas where they have abandoned their preferred, and now less abundant, cover of cobble-gravel in favor of accumulated allochthonous, food-rich, silty substrata (Hobbs III 1976). Some troglobitic crayfishes living in sluggish lotic/lentic systems (e.g., caves developed in the porous Ocala limestone of north-

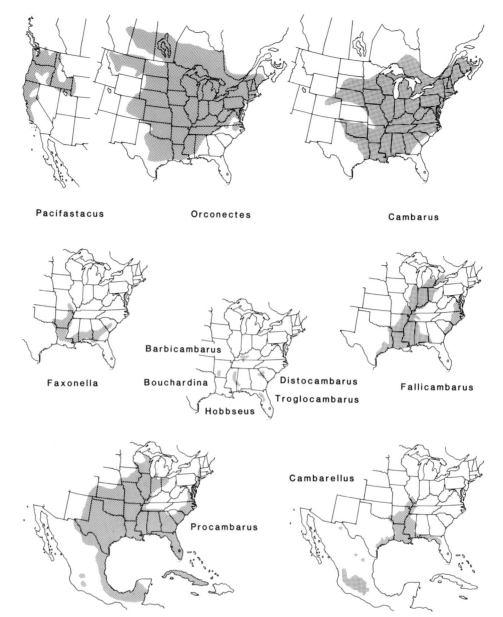

Figure 22.8 Geographical distribution of North American crayfish genera (not inclusive of introductions). (Modified from Hobbs 1988.)

central Florida) are found near cave entrances where organic debris is available; they are generally positioned on the walls and silty substrata of these submerged solution passages (Hobbs *et al.* 1977, Hobbs and Franz 1986). Franz and Lee (1982) showed that the geographical and ecological distributions of the cavernicolous crayfishes of Florida are correlated with low or high energy (food) cave systems. One species, *Troglocambarus maclanei* Hobbs, is often found clinging pendant to ceilings of submerged caves. The troglobitic crayfishes are derived from surface ancestors that entered caves during late Tertiary (Hobbs *et al.* 1977; see also Hobbs and Franz 1986, Holsinger 1988).

a. *Astacidae*

A single genus, *Pacifastacus,* is represented by eight species and subspecies in the Pacific drainages of western North America and headwaters of the Missouri River in Wyoming (Fig. 22.8). One and possibly two species are extinct and *P. fortis* (Faxon) is listed as a Federal Endangered Species. These crayfishes are found in lentic and lotic habitats ranging from alpine lakes and streams to streams crossing cold deserts. They occur in California, Idaho, Montana, Nevada, Oregon, Utah, Washington, Wyoming, and British Columbia (Fig. 22.8). Hobbs (1974) suggested that two evolutionary

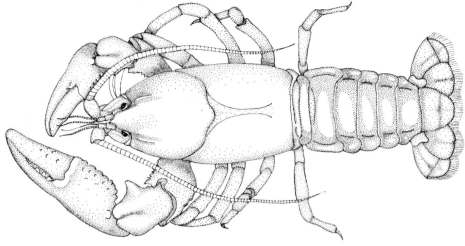

Figure 22.9 *Cambarus laevis* from epigean and hypogean lotic habitats in southern Illinois (?) and Indiana, and northern Kentucky.

lines arose from a Mesozoic marine nephropoid ancestral stock. One line (represented today by the genera *Astacus, Austropotamobius,* and *Pacifastacus*) moved into the freshwater habitat and maintained the primitive characteristic of absence of cyclic dimorphism in males, and females exhibited a sclerite (annular plate) lacking both a sinus and fossa. A second, less conservative stock also made the transition to freshwaters (represented by extant members of the family Cambaridae). These males evolved cyclic dimorphism and associated characters (e.g., hooks on ischia of certain pereiopods in males); females developed an annular sclerite with a sinus or fossa for sperm storage.

b. *Cambaridae*

As previously stated, the ancestral stock, in which cyclic dimorphism became established, moved into freshwater habitats and gave rise to the Cambaridae. This family consists of three subfamilies, two of which (Cambarellinae and Cambarinae) occur in North America (Fig. 22.8). The third, Cambaroidinae, is restricted to eastern Asia—Amur Basin, Korea, and Japan—and is treated only briefly in further discussions.

The following scenario is an abbreviation of arguments presented by Hobbs (1969, 1981, 1989) concerning the origin and dispersal of cambarid decapods. Whether or not the Asian Cambaroidinae and the American subfamilies shared a common freshwater ancestor or whether there were separate Asiatic and American invasions of freshwaters may never be known. Hobbs, however, suggested that the ancestral cambarine stock entered freshwaters of the southeastern portion of the North American continent during the late Cretaceous or early Cenozoic eras. Through adaptive radiation, this an-

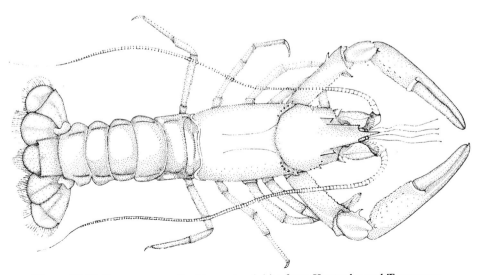

Figure 22.10 *Orconectes pellucidus,* a troglobite from Kentucky and Tennessee.

cestral *Procambarus* stock moved through estuarine situations and populated available habitats in many of the streams flowing from the southern and southeastern slopes of the existing continent. Except for rare passive dispersal resulting from stream piracy or from crayfishes moving during periods of low salinity from one stream mouth to another, stocks became isolated in their respective drainage systems.

Before the close of the Miocene (perhaps earlier), some stream-dwellers were able to exist in backwaters of lotic habitats; these forms gave rise to the ancestors of some of the lentic species of *Procambarus* (e.g., subgenera *Ortmannicus* and *Scapulicambarus*). These crayfishes moved into an ecological vacuum and were no longer confined to flowing water systems. Pools may have dried, or oxygen concentrations became low, or perhaps individuals simply moved across land from one aquatic habitat to another. Whatever the reason, additional ponds, lakes, ditches, streams, etc. were occupied and thus dispersal into lower elevation habitats occurred; also several stocks penetrated the water table by burrowing. *Distocambarus* occupied the groundwater niche and probably represents a remnant of a much more widely distributed Piedmont stock. *Troglocambarus* represents an offshoot of the *Procambarus* stock that moved into the karst groundwaters in northcentral Florida (see Hobbs *et al.* 1977 for further discussion of the evolution of troglobitic decapods). The genus *Cambarellus* probably arose from a primitive ancestral *Procambarus* stock in the Gulf coastal plain and occupied ephemeral swamps, ponds, and ditches.

One of the lotic ancestral procambarid stocks moved upstream, reaching the vicinity of the Cumberland Plateau, where ancestors to two very successful genera, *Cambarus* and *Orconectes,* became established. The major evolutionary lines were probably established by the Miocene Epoch, and the final stage was set for the refinement of evolutionary patterns begun during the early Cenozoic. Such major features as large rivers, the Appalachian and Ozark mountains, the coastal plain, the Cumberland Plateau, the Nashville Basin, and the Highland Rim influenced the dispersal of these stocks. Superimposed on these were the effects of Pleistocene glaciation.

Representatives of the ancestral *Cambarus* stock were successful in various habitats and spread through lowland areas, giving rise to the species currently assigned to *Fallicambarus*. This genus, which probably arose on the west Gulf coastal plain, developed into primary burrowers. *Barbicambarus* is another descendant of the early *Cambarus;* it moved into sizeable streams with large substrata,

rapid currents, and high oxygen concentrations. *Hobbseus* may also have evolved from the cambaroid line (Fitzpatrick 1977), but Hobbs (1969) suggested that it arose from the orconectoid line.

The genus *Orconectes* probably evolved in the Interior Lowland Plateau province (Kentucky/Tennessee) and radiated virtually in every direction. Some orconectoid stock successfully moved into the lowland regions where species arose that are now assigned to the genus *Faxonella*. This genus also seems to have evolved on the west Gulf coastal plain and today these small, tertiary burrowers are found in lentic situations (ponds, ditches). They occur in flood plains and, when the water level rises, are fairly common in the shallow flowing water. *Bouchardina* also is derived from the orconectoid stock and occupies backwater areas.

The profound effects of Pleistocene glaciation on the distribution of organisms should be mentioned. The obliteration of immense drainage systems (e.g., Teays) resulted in total habitat destruction for virtually all organisms. The principal zoogeographical effects were the elimination of populations and entire species from the northern portions of North America, with displacements of species farther south than they had occurred preglacially. Retreat of the glaciers resulted in the establishment of new drainage systems (e.g., the Ohio) which undoubtedly were rapidly invaded (reinvaded) primarily from the south and southeast, though some species may have entered from more than one refugium.

(1). Cambarellinae   The subfamily Cambarellinae is confined to North and middle America where its members are found in the coastal plain from northern Florida to southern Illinois and Texas (Fig. 22.8), and, disjunctly, on the Pacific slope and Central Plateau of Mexico. Seventeen species are assigned to three subgenera in the single genus *Cambarellus* (Fitzpatrick 1983); however, only eight of them are found north of the Rio Grande. In the United States, the eight species are known from the following states: Alabama, Arkansas, Florida, Illinois, Kentucky, Louisiana, Mississippi, Missouri, Tennessee, and Texas; they may have been introduced by man into Georgia.

(2). Cambarinae   The subfamily Cambarinae, which is comprised of 10 genera and 364 species, ranges from New Brunswick across southern Canada to Mexico, Guatemala, Honduras, and Cuba (Fig. 22.8). Ten genera represented by 319 species occur in North America, north of Mexico (see Hobbs 1989 for detailed presentation of genera, subgenera, species, and subspecies of American crayfishes). The following is a list of North American

genera, with a summary of subgenera and species (see Fig. 22.8 for distributions; refer also to Hobbs 1989):

Barbicambarus—monotypic;
Bouchardina—Monotypic;
Cambarus—10 subgenera and 82 species;
Distocambarus—2 subgenera and 4 species;
Fallicambarus—2 subgenera and 16 species;
Faxonella—4 species;
Hobbseus—6 species;
Orconectes—10 subgenera and 76 species and subspecies;
Procambarus—16 subgenera and 155 species and subspecies; only 11 subgenera and 114 species and subspecies in North America (*Procambarus primaevus* is an extinct species known only from Tertiary deposits in Wyoming and is not included in these figures);
Troglocambarus—monotypic.

B. Reproduction and Life History

1. Shrimps

The two common epigean, freshwater shrimps, *Palaemonetes kadiakensis* and *P. paludosus*, (Fig. 22.5) have similar life histories, both species living approximately one year (Fig. 22.11). Although many

individuals reproduce only once (semelparous), Nielsen and Reynolds (1977) observed ovigerous females of *P. kadiakensis* from Missouri also bearing mature eggs, indicating at least two broods of young per female (iteroparous); Beck and Cowell (1976) also noted that *P. paludosus* spawned twice. The length of the breeding season varies with latitude and is generally longer in southern localities. For instance, ovigerous females occur in populations of *P. kadiakensis* in Illinois and Michigan from April until August but are present in Louisiana populations from February to October; ovigerous females of *P. paludosus* occur throughout the year in the Florida Everglades. As a general rule, breeding does not begin until late spring or early summer in more northern areas. Only large females breed early in the season, and they are replaced by smaller females as the reproductive season progresses. Females are larger and usually more abundant than males. Mature (egg-bearing) females range from 20–49 mm in total length and produce one or two broods (two is particularly representative of populations of *P. paludosus* in Florida where the growing season is longer). The number of eggs produced per female is highly variable, ranging from 8–160, with fecundity generally a linear function of total length. The incubation period depends on water temperature and

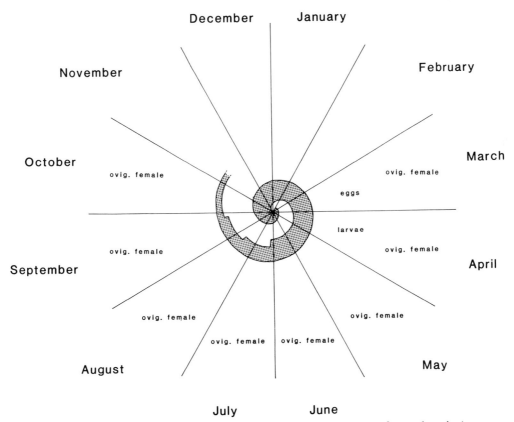

Figure 22.11 Generalized life history of *Palaemonetes* (see text for explanation).

varies from 12–24 days; zoeae larval lengths are approximately 4 mm at hatching and these free-swimming larvae pass through six stages (varies from 3–8) in about three weeks. Accounts of larval development and postembryonic growth in these two species are presented by Broad and Hubschman (1963), Hubschman and Rose (1969), Beck and Cowell (1976), Hobbs *et al.* (1976), Beck (1980), Kushlan and Kushlan (1980), Page (1985), and Hobbs and Jass (1988); shrimp usually mature when they attain a total length of 20 mm although this is not necessarily the case. Most shrimp die after reproducing, but some females molt within three days (generally within 24 hr) after young are released and can later produce another brood. Generally, adults disappear from the population in late summer or early fall, having a life span of approximately one year.

Marine and brackish water *Palaemonetes* (e.g., *P. intermedius, P. pugio* Holthuis) generally have smaller and more numerous eggs per female than do freshwater species. The production of fewer but larger eggs may enable the freshwater larvae to hatch at an advanced stage with a corresponding reduction in the number of free-swimming larval stages.

Dobkin (1971) described three larval stages in the troglobite, *P. cummingi* (Chace) and life-history studies of atyids consist of observations made on *Palaemonias ganteri* in Mammoth Cave, Kentucky (Hobbs *et al.* 1977, Leitheuser *et al.* 1985), the Alabama cave shrimp, *P. alabamae*, conducted by Cooper (1975), and *Syncaris pacifica* (Holmes) (Eng 1981). The study of the life history of *Macrobrachium ohione* (Smith) has not received as much attention as have the two taxa of *Palaemonetes* previously discussed. This species is larger (total length of ovigerous females ranging from 27–93 mm) and it lives a maximum of two years (Fig. 22.12). [Huner (1977) determined that *M. ohione* at Port Allen, Louisiana lived up to two years.] Fecundity is relatively enormous, the number of eggs varying from 6273–24,000 per female (Truesdale and Mermilliod 1979). Ovigerous females are known only during May in Illinois, from March through September in Louisiana, and during March in Texas. Generally, females are larger than males and only the largest individuals in the population bear eggs. Additional eological data can be obtained from McCormick (1934), Gunter (1937), Reimer *et al.* (1974), Truesdale and Mermilliod (1979), and Page (1985).

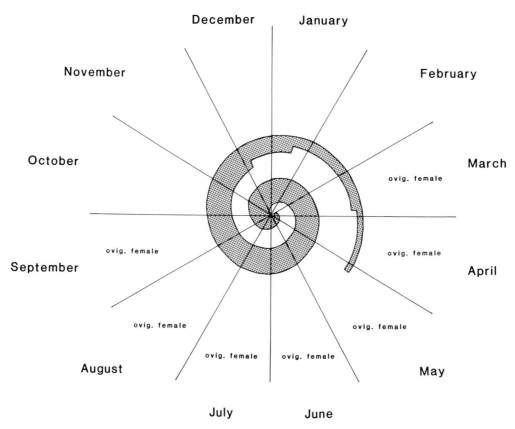

Figure 22.12 Generalized life history of *Macrobrachium ohione* (see text for explanation).

2. Crayfishes

Surprisingly few crayfishes have been studied carefully for extended periods. Although considerable life-history data are available for many species (e.g., Hart and Clark 1987), most are observations from isolated collections. Astacid life-history studies are limited to *Pacifastacus leniusculus leniusculus* and *Pacifastacus leniusculus trowbridgii* (Stimpson) (see references in Mason 1970a,b; McGriff 1983a); life-history studies of the more diverse cambarids have been conducted on only about 20 species.

In a general sense, the life history of a crayfish can be divided into several phases (Momot 1984), each having a slightly different goal important to the completion of the life history of a given cohort as well as to the long-term success of the species. Recently hatched juveniles search lake littoral and stream riffle areas for the best food and shelter available. Rapid growth of young-of-the-year (YOY) allows a partial escape from the devastating effects of predation (Gowing and Momot 1979). As juveniles grow larger, they abandon littoral and riffle areas for deeper waters, only to return after reaching adult size; this habitat segregation is probably related, in part, to such factors as competition with adults and vulnerability to predators. During postembryonic development, food conversion is expressed in maxi-

mum growth. Adulthood represents a conservative phase in which energy is shunted from rapid growth toward a periodic maximization of egg output among several cohorts per year. Rapid growth and shorter lifespan are reported to be more common in crayfishes in lower latitudes (Momot 1984, Hazlett and Rittschof 1985); however, many "northern" species demonstrate these same characteristics.

Unlike the cambarids, astacid males do not exhibit sexual dimorphism. Copulation in *Pacifastacus l. trowbridgii* coincides with the autumn decline of water temperature and is soon followed by oviposition (late October and early November) (Fig. 22.13). Females carry 90–251 eggs (number directly proportional to mass) on their abdomens, and hatching generally occurs after 7–8 months, during April to June. After the second molt, the young leave the mother as three- or four-week old juveniles. During the first year, individuals may undergo as many as 11 molts and nearly triple in length. Some individuals live for eight years, and a few become sexually mature at age three; the majority, however, become part of the breeding population during the fall of their fourth year. This certainly follows the trend suggested by Momot (1984) that crayfishes occurring at higher latitudes and in colder environments usually live longer and mature later. The smallest ovigerous female observed by

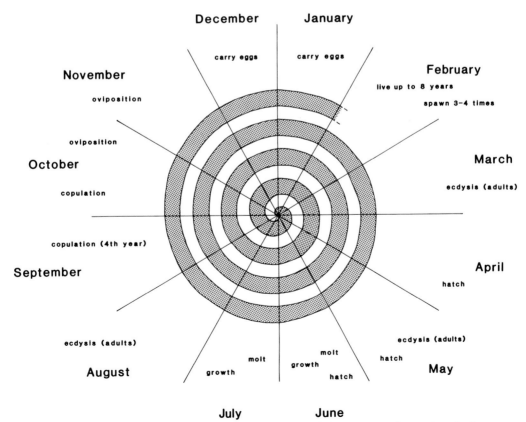

Figure 22.13 Generalized life history of astacid crayfishes (see text for explanation).

Mason (1974) was 60 mm total length and the largest 102 mm. This suggests that female astacids lay eggs more than once (iteroparous) and, indeed, may spawn as many as three or four times (Fig. 22.13). Refer to McGriff (1983a, 1983b) for life-history data for *Pacifastacus l. leniusculus*.

The life history of cambarid crayfishes (Fig. 22.14) is somewhat more variable and complex than is typical of the astacid crayfishes. The sexually dimorphic males exhibit two distinct and usually (Price and Payne 1980, Taylor 1985) alternating body forms. The change from one form to another occurs among mature males during the semi-yearly molts. The sexually competent stage (Form I) is first attained with the last juvenile molt. This more aggressive breeding form can be distinguished by the sclerotized, amber-colored, and lengthened condition of the terminal elements of at least one of the first pleopods. In addition, the ischial spines are more pronounced (Fig. 22.15), the first chelipeds are enlarged, and the sperm ducts are full of recently formed spermatids (Word and Hobbs 1958). In Form II males (the nonbreeding form), however, the terminal elements of the first pleopod are not as well differentiated (and never corneous), the ischial hooks are shorter and weaker, and the chelipeds are less robust.

Although Form I males may be present in a population year round (e.g., in *Cambarellus* and *Procambarus;* see also life-history data for numerous species in Hobbs 1981), generally they are more numerous during the fall and/or spring. Hence, many species demonstrate one or two peak periods of copulation (fall and/or spring). [Mason (1970a), Ingle and Thomas (1974), Ameyaw-Akumfi (1976, 1981), Bechler (1981), and Hobbs and Jass (1988) provide more detailed observations and literature review on the mating behavior in crayfishes.]

Because oviposition occurs usually in the spring or early summer, spermatophores (sperm plug) may be carried by the female for as long as six or seven months, but viable sperm may be carried for only a few days or weeks. During oviposition, the female secretes glair, a translucent, sticky, mucilaginous substance from "cement glands" located on the ventral side of the abdomen. After eggs are released from the genital pore and simultaneously fertilized, they attach to the abdominal pleopods by a short stalk. Amazingly, it is not known how nonmotile sperm inside a hard, waxy plug fertilize an egg mass! Once the glair hardens, the female is considered to be "in berry." Typically, larger females of a given species carry more eggs, yet the number of eggs is highly variable within as well as between species

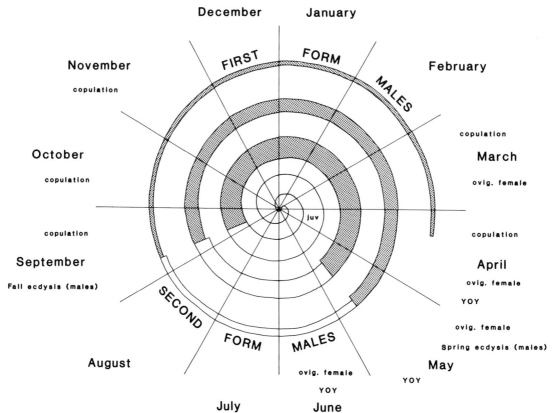

Figure 22.14 Generalized life history of cambarid crayfishes (see text for explanation).

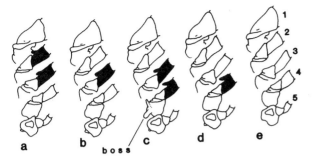

Figure 22.15 Ventral view of basal portions of left pereiopods with ischia bearing hooks darkened. (After Hobbs 1972.)

(epigean forms carry as many as 600–700 eggs). For example, Hobbs (1981, p. 415) reported that females of *Procambarus pubescens* (Faxon) carried between 89 and 443 eggs; he cited numerous other examples of variation in egg number for Georgia crayfishes. Little is known about the fecundity of troglobitic species because ovigerous females or those carrying young have not been observed in most species. For the few taxa studied, however, the eggs were larger (more yolk) but less numerous (27–80), reflecting the constraints of life for these K-strategists in such energy-poor systems.

Depending on the season of laying and the water temperature, eggs are carried by females for 2–20 weeks, during which time rhythmic movements of the pleopods effectively aerate the developing eggs. Corey (1987) noted an inverse relationship between the number of newly released eggs and the current velocity. Incomplete extrusion, failure to attach to pleopods, and fertilization failure are three other factors leading to egg loss at oviposition. Additional losses of eggs and young while being carried on the female result from abrasion, predation, high current velocity, low food availability, high density of females, and insufficient cover for females. After hatching, the first juvenile instar attaches to the mother with hooked chelae and a telson "thread" anchored by specialized hammate setae (Price and Payne 1984). The cephalothorax of this instar is disproportionately large, with immense eyes, a yolk-laden carapace, and an incompletely developed abdomen. Generally after 2–7 days of attachment, juveniles molt and grow to a form more closely approximating the adult appearance. This instar loses its posterior connection to the mother and uses only chelae to cling to her abdomen. An additional molt usually occurs within 4–12 days, producing a large, third instar crayfish. Although this juvenile hangs onto the mother with its chelae and pereiopods, it may leave the parental pleopods intermittently; subsequent instars are free living. [Refer to Price and

Payne (1984) for more details of postembryonic ontogeny and to Mason (1970b) for maternal–offspring behavior.] The adult female commonly molts 2–3 weeks after all young have left. By fall, 6–10 molts have occurred in the developing juvenile, resulting in a considerably larger and often sexually mature adult. Mating among mature yearlings frequently takes place; however, many individuals do not become sexually active until the following late summer or fall. Most adults can live 2.5–3 yr and the majority breed more than once. The longest lifespan reported for cambarids is 6 or 7 yr for the burrowing crayfish *Procambarus hagenianus* (Faxon) in eastern Mississippi and western Alabama (Lyle 1938). Longevity in troglobitic crayfishes has been demonstrated to be considerably greater than for epigean forms (Cooper and Cooper 1978, Hobbs 1980).

C. Ecological Interactions

1. Functional Role in the Ecosystem

Crayfishes have successfully invaded a wide variety of habitats and play important roles in processing organic matter and in the transformation and flow of energy. They are the largest and longest lived crustaceans in freshwater ecosystems of North America and process spatially and temporally large quantities of organic matter in most lotic and lentic systems. Although they can function as food specialists, most are opportunistic generalists, feeding virtually at all trophic levels (polytrophic). Shrimps, on the other hand, are generally restricted to large, sluggish backwaters of rivers where aquatic vegetation (cover) is dense. They function primarily as grazers and serve as food for a wide variety of aquatic organisms.

By feeding directly on carrion and organic debris of aquatic and terrestrial origin, these two decapods contribute to the processing of organic matter in aquatic ecosystems. They serve in part as decomposers by breaking down particulate matter, altering the chemical composition of detritus, and releasing fecal material back into the system. Budd *et al.* (1979) demonstrated that juvenile *Cambarus robustus* Girard and *Orconectes propinquus* function as collectors by filter feeding on algae. Lorman (1980) and Lorman and Magnuson (1978) showed that, in Wisconsin, *Orconectes rusticus* (Girard) males are more carnivorous than females yet are less carnivorous than juveniles. Others (Prins 1968, Momot *et al.* 1978) found that all age classes tend to utilize vegetation as a food source. Predation by crayfishes on both invertebrates and vertebrates is well documented (reviewed in Hobbs and Jass 1991), and it is clear that these crustaceans have ramifying impacts on the structure and dynamics of aquatic communities. That is, crayfishes prey upon organisms occupying various trophic levels and compete for food and other resources at the intra- and interspecific levels. Thus they are positioned high enough in the trophic structure to influence numerous interactions among several subwebs of species.

Crayfishes and shrimps are eaten by a myriad of predators (Hobbs and Jass 1991). Indeed, crayfishes are almost the sole source of food for some predators, such as the Florida water snake, *Regina alleni*. Naturally, shrimps do not play as important a role as do crayfishes in energy flow since they are generally smaller and much more limited ecologically (largely restricted to slowly moving streams or lakes and ponds with dense vegetation).

Rabeni (1985) determined that the sympatric species *O. luteus* (Creaser) and *O. punctimanus* (Creaser) in south-central Missouri coexist by partitioning the physical environment. *O. punctimanus* tends to be more restricted in its use of habitats, occurring as YOY in shallow water with macrophyte cover; both YOY and adults forage on larger-sized substrata; both life stages preferred slow current velocities. This species maintains a size advantage over *O. luteus* (larger individuals are dominant and occupy their preferred microhabitat). *O. luteus* tolerates a wider range of environmental conditions and exploits habitats unavailable or undesirable to its congener.

This discussion concerning natural range expansions and introductions of exotic species treats the competitive interaction of crayfishes. Displacement mechanisms explaining why and how the introduced species often replaces the native species are not well known; however, it is apparent that most introductions grossly alter the community composition and thus can have potentially severe effects on the flow of energy through the ecosystem.

Symbiotic and parasitic relationships involving crayfishes and shrimps have been summarized by Johnson (1977), Beck (1980), France and Graham (1985), Andolshek and Hobbs (1986), Lichtwardt (1986), and Corey (1987). These decapods serve as intermediate or definitive hosts and/or substrata to a wide variety of bacteria, algae, protozoa, fungi, worms, and crustaceans. For instance, the metacercariae of the lung fluke (*Paragonimus spp.*) invade crayfishes and are transmitted to another host when the crustacean is eaten. *Paragonimus westermani* is an important parasite of humans in Asia; fortunately for us, few human infestations have been noted for *P. kellicoti* in North America. Ecdysis provides periodic relief from the buildup of various ectocommensals.

Figure 22.16 Form I male of *Fallicambarus devastator*, a primary burrower from Texas.

Another functional role played by crayfishes is as a food resource as well as a pest for man. A multimillion dollar business centers around "crawfish" production in Louisiana and other parts of the world (Huner and Barr 1984). In addition, certain crayfishes are economically important as local seasonal nuisances for farming. Burrowing species can be extremely abundant: Hobbs and Whiteman (1987) report *Fallicambarus devastator* (Fig. 22.16) being responsible for constructing over 25,000 earthen mounds per acre! Farm equipment can obviously be damaged, crops destroyed, and earthen dams can be weakened by the activities of these fossorial decapods.

2. Foraging Relationships

Crayfishes play a polytrophic role in most ecosystems and, as such, contribute much to the complexities of energy flow in aquatic systems (see above; see also Lorman and Magnuson 1978, Momot *et al.* 1978, Momot 1984, Hobbs and Jass 1991). They are capable of switching roles from herbivore/carnivore to scavenger/detritivore merely in response to food availability. Although individuals of some species and sizes do forage selectively, many species are quite opportunistic and catholic in their diets. For example, Horns and Magnuson (1981) found that crayfishes consume significant numbers of Lake Trout eggs in northern Wisconsin when this seasonally restricted food item is available. Juvenile crayfishes are somewhat limited to detritivory and herbivory as they filter suspended particulates and grasp coarse particles. Adults actively prey and graze on larger items yet they too scrape microbes from hard substrata and shred vegetation. Because of this flexibility, crayfishes can maintain large population densities despite fluctuations in food resources. Although crayfishes forage nearly continuously, the peaks of this behavior occur shortly after sunset and during the night. This activity decreases as water temperatures begin to cool to around 14°C in Oklahoma (Covich 1978). In contrast, Price (1981), working in Arkansas, found that both diversity in the diet and length of foraging increased for *O. neglectus chaenodactylus* Williams as food abundance decreased during the winter months. Another factor significantly affecting foraging is the interaction of sympatric species, since dominance could be critical in determining both foraging and reproductive successes.

Crayfishes feed readily on macrophytes and they can significantly reduce species richness and biomass of aquatic vegetation by grazing selectively on plant species (Lorman and Magnuson 1978, Lodge and Lorman 1987). By moving through the shallow littoral zone of lakes and cutting stems near the bottom, they can wreak havoc even though they actually ingest very little of the "harvest" (an example of consumptive and nonconsumptive destruction). As predators (Lorman and Magnuson 1978, Covich *et al.* 1981, Covich 1982, Lodge *et al.* 1987), crayfishes apparently attempt to maximize energy gain with minimal energy expenditure. For instance, rates of predation by *Procambarus clarkii* (Girard) on thin-shelled snails (*Physa* sp.) were higher than those on thicker-shelled *Helisoma* sp. (Covich 1978).

Foraging behavior is altered in the presence of a predator. Stein (1977) showed that the susceptibilities of the crayfish *O. propinquus* to fish predators reflected the size and condition of the potential prey. Life stages with the greatest reproductive potential were those least vulnerable to small-mouth bass predation (berried females and Form I males). Young-of-the-year are particularly susceptible to predation, yet they, like other size classes, must balance predation risk with the benefits of foraging. Thus, juveniles, Form II males, females, and recently molted individuals modify their behaviors (reducing activity) and microdistribution (seeking substrata affording maximum protection) to minimize risk of predation. The degree of behavioral response appears to correlate positively with vulnerability.

Although several studies have determined the thresholds of detection by crayfishes for various stimulatory substances (Bauer *et al.* 1981, Hatt 1984, Tierney and Atema 1986), virtually no study has examined the role or influence of chemoreception on the foraging behavior of crayfishes. Studies of the hierarchial structure of food search and feeding in the spiny lobster (e.g., Zimmer-Faust and Case 1983) suggest that chemosensory-induced foraging in crayfishes may also be geared primarily to obtaining nearby rather than distant food sources. Johannes and Webb (1970) have demonstrated that

several species of invertebrates release free amino acids into the environment, and Rittschof (1980) has shown that amino acids diffuse rapidly from carrion. Thus, amino acids are ubiquitous in aquatic ecosystems and are readily available for use as chemical feeding cues for crayfishes. Of particular note, Hatt (1984) has shown that amino acid receptors are present on the pereiopods and antennules of crayfishes. Tierney and Atema (1988) suggest that the responsiveness of *Orconectes rusticus* and *O. virilis* to a wide diversity of chemicals (e.g., amino acids) may reflect the omnivorous foraging habits of these crustaceans.

The shrimp *Palaemonetes paludosus* feeds heavily on algae epiphytic on vascular plants. Grazing on diatoms (*Fragilaria, Navicula, Stephanodiscus, Gomphonema, Synedra,* and *Cymbella*) appears to be the primary method of food intake for populations in Hillsboro County, Florida (Beck and Cowell 1976); however, they also forage for immature insects.

Troglobitic crayfishes and shrimps live in heterotrophic environments where the sole, primary source of energy is from the surface in the form of allochthonous materials carried or washed in, such as vegetative debris and bat guano. In such resource-depressed systems, these crustaceans tend to congregate in regions of the cave where food is spatially or temporally abundant, and thus are commonly observed in pooled areas of streams among accumulated vegetation. Crayfish densities are particularly high in these microhabitats, and these crustaceans can often be observed foraging over the substratum and feeding on microbe-encrusted debris. Their particularly enhanced foraging efficiency reflects their improved sensory perception, reduced random movements, diminished body size, and increased metabolic efficiencies.

Barr and Kuehne (1971) reported that the troglobitic shrimp *Palaemonias ganteri* (Mammoth Cave, Kentucky) strained and filtered pool sediments with mouth parts, suggesting that its food consisted of microorganisms in the silt (sediments contained protozoa, Barr 1968). However, Lisowski (1983) observed this shrimp exploiting another resource; an individual was observed foraging upside-down on the surface film of a pool apparently feeding on floating material. This foraging behavior was similarly noted by Cooper (1975) for *Palaemonias alabamae* in Shelta Cave, Alabama (he also observed the sediment-straining behavior). The troglobitic crayfish, *Troglocambarus maclanei*, also is presumed to be a filter feeder; however, cannibalism by this troglobite has been noted (Hobbs *et al.* 1977, pp. 138–139).

Foraging behavior of burrowing crayfishes is

variable, but generally on warm, humid nights, individuals leave their burrows presumably in search of food (vegetative material). A peculiar phenomenon is noted regularly in Louisiana in which large numbers of crayfishes (primarily males) leave their marsh habitat, move over land, and die during the fall. They are generally in poor physical condition, and the cause of this mass "death-migration" is unknown (Penn 1943, p. 15). See the following references for additional information concerning burrowing crayfishes: Penn (1943); Hobbs (1981); Hobbs and Whiteman (1987); Hobbs and Jass (1988).

3. Behavioral Ecology

The behavioral and distributional responses of organisms to features of their physical and biotic environment is encompassed within the discipline of behavioral ecology. Behavioral responses to drought (burrowing), orientation to current (positive rheotaxis), light (negative phototrophism), reaction to cover, and substratum preference (cobble–pebble), are examples of responses by decapods to the physical environment; these factors, to a large degree, determine the nature of selected habitats. Aggressive behavior, maternal–offspring relationships, and behavioral mechanisms resulting in reproductive isolation of a shrimp or crayfish species are representative patterns related to the biotic features of the environment.

Hobbs (1981) recognized three categories of burrowing crayfishes. "Primary burrowers" (e.g., *Fallicambarus devastator*, Fig. 22.16) spend most of their lives in burrows, occasionally exiting to forage or to mate. Their burrows consist of complex tunnels that are often quite removed from open bodies of water (Fig. 22.17a, d). "Secondary burrowers" [e.g., *Fallicambarus hedgpethi* (Hobbs)] spend much of their life history in tunnels but frequent open water during rainy seasons (Fig. 22.17b). Their burrows usually feature a single subvertical tube that either slopes gently or descends in an irregular spiral (Hobbs 1981, p. 32). "Tertiary burrowers" [e.g., *Procambarus a. acutus* (Girard)] live in open waters and move into burrows only at various times (Fig. 22.17c, e). For example, they burrow to brood eggs, to move below the frost line during winter, and to avoid dessication; burrows also are refugia and can be defended. Nonburrowing crayfishes will occasionally construct burrows when surface waters disappear (Berrill and Chenoweth 1982). Burrowing crayfishes are often observed above the water table; many are capable of remaining out of water for extended periods in the humid environment of their tunnels (Hobbs 1981). The same is true for many cavernicolous species, which, in the saturated air, can move from one pool to another or remain out of the water for hours. Burrowers and surface-dwelling crayfishes are commonly noted at the air–water interface, aerating in response to low oxygen levels; the oxygen concentration in burrow waters is frequently below 2 mg/liter (Hobbs 1981) and pool and ditches often become oxygen-depleted. Kushlan and Kushlan (1980) observed *Palaemonetes paludosus* swimming at the surface using oxygen that diffused across it as a means of surviving low oxygen tension in the Everglades. [See Grow (1981) and

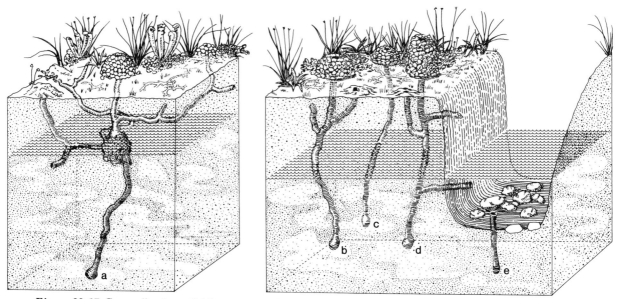

Figure 22.17 Generalized crayfish burrows: (a, d) those of primary burrowers; (b) that of secondary burrower; (c, e) those of tertiary burrowers. (After Hobbs 1981, p. 32.)

Rogers and Huner (1985) for further details concerning the burrowing behavior of crayfishes.]

Water currents certainly influence life processes of lotic organisms and bar many species from rapidly flowing streams. Maude and Williams (1983) demonstrated that when crayfishes are exposed to increases in current velocity, they alter their body posture to counteract the effects of drag and to maintain position. Not only is this a behavioral adaptation resulting in certain species optimally occurring in fast-flowing streams, but it may very well play a role in the widespread distribution of species as well as in the expansion of exotics, such as *O. rusticus,* into a variety of habitats following their introduction.

Many temperate crayfishes inhabit shallow littoral waters during early spring. In some regions, this time period coincides with the spring pulses of acidic water from melting snow. Because crayfishes are capable of extensive movement and can avoid streams acidified by industrial sulfuric acid or mine drainage (pH 4.0–4.5, France 1985b), they should be able to elude environments acidified by spring melt pulses. France (1985b) determined that *Orconectes virilis* can avoid lethally acidic waters, implying that during the spring melt, they could escape mass mortality by seeking refuge in the profundal zone of lakes. Yet, no avoidance was demonstrated to lake water with pH values as low as 5.6, levels that are reported to cause serious reproductive difficulties [from decreased aeration of eggs, leading to fungal infections and egg mortality (France 1985a)]. The failure of this crayfish to avoid sublethal acid waters could, therefore, interfere with recruitment; for this reason, France suggests that the prolonged survival of *O. virilis* in waters receiving acid runoff is doubtful. This may be the case for other crustaceans living in poorly buffered systems, but, surprisingly enough, some crayfishes [e.g., *Procambarus fallax* (Hagen) and *P. paeninsulanus* (Faxon)] reach maximal densities in acid waters.

Obviously important aspects of the ecology of crayfishes and shrimps are their intra- and interspecific behavioral responses. Although they are nonterritorial (Merkle 1969, Momot and Gowing 1972), they are aggressive (Bovbjerg 1953, Heckenlively 1970) and rely on agonism to procure shelter, food, and mates (Stein 1975). Most crayfishes tend to be solitary and occupy and defend crevices in a coarse substratum, both of which are of survival value to the species. Crevices not only provide shelter from pounding shore waves or from high stream velocities, but they also furnish a refuge that can be defended against predators and competitors. In the process, the crevice serves as a feature, which, by further partitioning the environment, permits a greater species richness in those habitats. Hayes (1975) demonstrated eight innate components of social interaction in the primary burrowing crayfish, *Procambarus gracilis* (Bundy): alert, approach, threat, combat, submission, avoidance, escape, and courtship. These have also been observed in other species and Ameyaw-Akumfi (1979) summarized the sequence of behaviors demonstrated by interacting crayfishes during aggressive encounters (Fig. 22.18). He proposed that movement of gnathal appendages and pereiopods as well as repeated antennal waving occurs prior to the termination of an agonistic encounter and this behavior may in fact constitute an appeasement display. He further stated that "the use of appeasement signals in mating behavior is, probably, a capitalization by the male as this behavior appears to reduce aggressive tendencies" (Ameyaw-Akumfi 1979, p. 41). Copulatory behavior (reviewed in Bechler 1981) is generally seasonal with the pair assuming several positions, the male mounting the female most commonly (Fig. 22.19) (mating duration ranging from 10 min to 10 hr, e.g., Andrews 1895, Mason 1970a). After oviposition, egg fanning, and hatching, maternal–offspring behavior is mediated by pheromones released into the water by the female. These cues are species specific but not brood specific, as juveniles are attracted to conspecific brooding females. Of concern, Tierney and Atema (1986) experimentally showed that acid exposure inhibits behavioral responses; this could cause acid-stressed populations to decline at pH levels well above those at which lethal physiological effects become apparent. Also, Peck (1985) showed that social interaction is an important factor affecting temperature selection by *O. virilis*. Individuals demonstrated behavioral mechanisms, not only to find favorable temperature ranges, but also to segregate according to social rank.

The presence of predators illicits particular behavioral response in crayfishes. Young-of-the-year are particularly susceptible to predation and must balance predation risks with foraging benefits (Butler and Stein 1985). Stein and Magnuson (1976) and Stein (1977) showed that in the presence of smallmouth bass (*Micropterus dolomieui*), *O. propinquus* foraged less, adopted defensive postures, and sought protective substrata. Ideally, crayfishes use crypsis to avoid detection by predators. However, when discovered, crayfishes typically backswim seeking shelter in vegetation, beneath rocks or debris, or in burrows. If escape is not possible, individuals assume and intensify a species-specific defensive posture [predator response postures (PRP)]. Hayes (1977) described PRP of crayfishes and discussed the evolutionary significance of these behavioral patterns in three species of

Figure 22.18 Agonistic behavior between two male *Orconectes propinquus* in the "body up" position; crayfish on right performing "antenna tap" (Bruski and Dunham 1987) (photograph courtesy of Ann Jane Tierney, Cornell University).

the genus *Procambarus*. The aquatic species examined have similar lateral merus displays, which result in a lateral presentation of the chelae, allowing for defensive coverage of the entire carapace and for quick flight by tail-flipping. Also, elongated chelae of these species enable them to sense tactilely predator approach and to keep the attacking predator at a greater distance. The PRP of the terrestrial (burrowing) species *P. gracilis* appears to have been modified for threat from terrestrial predators with attack coming primarily from above. Chelae are held shield-like in front of the carapace, and the tail is

Figure 22.19 Copulating pair of *Orconectes propinquus,* Form I male on top (photograph courtesy of Ann Jane Tierney, Cornell University).

folded tightly beneath the body allowing for rapid reorientation to the attacks of a shifting predator. This posture frees the tail from use as an escape organ on land. Stein and Magnuson (1976) also suggested that predators not only affect the distribution and behavior of crayfishes in natural communities, but also because of their impact on several trophic levels, the influence of predators goes beyond simple predator–prey interactions.

Many North American crayfishes occur sympatrically with close relatives, yet hybridization appears to occur relatively infrequently (if ever). One mechanism that may be operating to ensure reproductive isolation is chemical or visual recognition of species during mate selection (Smith 1988). Chemo-ethological isolating mechanisms could be effective for sympatric but not allopatric species. In consequence, hybridization of a resident and an exotic species might be more likely than between two distinct but geographically overlapping resident species.

Invasions of exotic species can elicit drastic changes in community structure (Hobbs *et al.* 1989). Of particular interest in recent years is the apparent replacement of native crayfishes in various parts of North America (and world) by introductions of nonnative species (Capelli 1982, Smith 1988). Displacement mechanisms are complex and the mechanisms regulating shifts and/or replacements of species are not well understood in most instances (Flynn and Hobbs 1984). However, field and laboratory results indicate that reproductive interference, acting synergistically with differences in aggressive dominance and YOY susceptibility to predation, probably serves as a major mechanism regulating replacement (Butler and Stein 1985). Based on data from Trout Lake in northern Wisconsin, Lodge *et al.* (1986) suggested that outcomes of such interactions are not necessarily predictable and are probably affected by a diversity of community-structuring forces such as parasitism, predation, competition, and disturbance.

4. Population Regulation

The population dynamics of shrimps and crayfishes are not well studied except for a few species. It should be emphasized, however, that traditional density-dependent and density-independent regulatory factors, while present, are too often superceded in importance by the effects on density of human activities (Hobbs and Hall 1974).

a. Shrimps

A one-year life history is characteristic of epigean shrimps, whereas cave forms probably live longer.

Food resources tend not to be limiting for epigean shrimps, but fish predation can have a significant impact on their densities (Nielsen and Reynolds 1977). The same investigators noted that fluctuating water conditions (seasonal drying and extended high water) also elicit population changes. Fecundity appears to be related to size (length) of female and to time of year.

In cave environments of the southeast, amblyopsid fishes (and crayfishes) are predators of troglobitic shrimps and crayfishes. In these food-limited systems, where shrimp and some crayfish populations are very small, loss of ovigerous females (and thus loss of recruitment of young) can be very significant to immediate and particularly long-term stability of population densities. Lisowski (1983) attributed the extreme decrease in density of *Palaemonias ganteri* from base-level streams in the Mammoth Cave System, Kentucky, to pollution of local groundwater. The sources of contaminants were human sewage and water from the Green River, which had received inputs of oil brines and hydrocarbons during the 1960s. He also indicated that modifications of habitat and flooding regimes caused by dams within the Green River drainage adversely affected shrimp by reducing the reproductive success and by increasing its predation pressure on them.

b. Crayfishes

Where birth rates, death rates, and immigration rates vary, population sizes will fluctuate, especially where birth recruitment is inadequate. As populations expand, they face potential resource depletion and the possibility of attracting more predators—both factors that can limit population size. Predation pressure may cause prey to modify their behavior (e.g., increased burrowing and chelae displays, suppressed grazing, and reduced overall activity). In turn, these antipredatory behavior patterns act as selective (coevolutionary) forces on the predator, thus operating to modify predator strategies. Stein and Magnuson (1976) and Stein (1977, 1979) have demonstrated that juvenile *O. propinquus* are the preferred crayfish prey of smallmouth bass and, as such, are rarely found on open, sandy substrata. Adult crayfish (least preferred prey) are abundant in these areas, demonstrating both that preferred prey make distributional shifts to protective habitats and that use of space by nonpreferred prey is regulated much less by predators.

Crayfishes commonly establish sole occupancy of protective crevices in coarse substrata (when available) which they defend agonistically. Considering the generalistic feeding habits of crayfishes and the apparent absence of food limitation in most systems, the principal resource bottleneck may be crevice availability. Caine (1978) observed that competition among Florida epigean crayfishes is primarily for shelter, and individuals unable to obtain cover succumb to predators. A specific home range has been demonstrated for a number of species (Hazlett *et al.* 1974, Hobbs 1980), and dispersion and dispersal often appear related not to population density but to water depth, number of crevices, predators, temperature, etc. A study conducted by Aiken (1968) in Alberta found that both male and female *Orconectes virilis* move to deeper water in late summer, females preceding males. Severe winter conditions would eliminate a population if it remained in the littoral zone where water freezes to the bottom. Rather than burrow to avoid ice, individuals escape by migrating to deeper water. Young-of-the-year appear to have higher mortality rates during the winter than adults. At spring thaw, both males and females return to shallow water. In streams, crayfishes generally burrow into the hyporheic zone to overwinter.

Length–frequency distributions have traditionally been used by investigators to show age and growth of individuals in populations. Growth of crayfishes is related directly to the number of molts during a growing season and the growth increment per molt; growth is stepwise rather than continuous. Average growth increments range from 0.41 mm for the dwarf crayfish *Faxonella clypeata* (Hay)(Black 1958) to 2.6 mm for *Procambarus clarkii* (Penn 1943). The difference in yearly size increment between males and females is due primarily to the frequency of molting and not necessarily to the growth increment at each molt (Hazlett and Rittschof 1985). However, the increment for individuals living in optimum conditions may well have as much impact as does the number of molts. Female growth is depressed, probably in response to the effects of carrying eggs and young, resulting overall in fewer molts.

Conditions that promote rapid juvenile growth rather than adjustments in egg production per se modify any stock recruitment interaction. In northern climates, the temperature of the water in the shallows, especially in "nursery areas," is an important determinant of the molting rate in juveniles. Flint (1975) found that a coastal stream population of *Pacifastacus leniusculus* had a slightly faster overall growth rate than a subalpine lake population; he ascribed this to higher water temperatures and a longer growing season. Because crayfishes molt more frequently as juveniles, any parameter reducing juvenile molt frequency also reduces growth rate; this, in turn, may determine size at maturity,

although there is no absolute size at which a crayfish becomes mature. Because fecundity is a function of female size, populations with smaller females should have lower recruitment of YOY. Obviously, a cohort that grows faster and reaches a larger overall size will have a better chance of ensuring the replacement of its population.

Momot and others, in a variety of population studies (Momot and Gowing 1977a, 1977b, Gowing and Momot 1979, France 1985a), have shown that various populations of *O. virilis* are regulated by several density-dependent and density-independent factors, including cannibalism, natural mortality at molting (physiological and mechanical problems), predation by snakes (Godley *et al.* 1984), fishes (e.g., trout, bass) and some insects (e.g., dragonfly) on yearlings, flooding, temperature, pH (Davies 1984), seiches (Emery 1970), and other factors. Crayfishes are generally characterized by high juvenile mortality and reduced adult mortality (concave survivorship curve). It has been suggested (Momot and Gowing 1977b) that overall, most changes in biomass and productivity of these populations result from internal density-dependent processes operating on fecundity and mortality rates rather than on growth. In several Michigan lakes during most years studied by Momot, the breeding success of *O. virilis* depended mainly on the survival and subsequent fecundity of the 1.5-year age group females. At 2.5 years, females are considered an important strategic reserve.

Periods of low water obviously affect all members of the aquatic community, yet crayfishes can escape desiccation by burrowing. Prior to complete loss of an aquatic habitat, other responses by crayfishes have been noted. Taylor (1983), studying a lotic population of *P. spiculifer* (LeConte) during a drought in northeastern Georgia, suggested that population shifts (decrease in adult population densities, change in reproductive timing, increase in number of juveniles) resulted from loss of preferred deep-water sites, emigration, and subsequent increase in predation of larger adults.

Population regulation among sympatric species has received considerable attention in the last 10–15 years (Lodge *et al.* 1986). Sequential invaders (natural or introduced) can elicit drastic changes in community structure (see earlier), including replacement of native species (Butler and Stein 1985). The outcome of interspecific interactions is not always predictable (e.g., interactions of *O. virilis*, *O. propinquus*, and *O. rusticus* in Wisconsin, Lodge *et al.* 1986) because a complex suite of factors such as competition for resources, predation, parasitism, and a variety of disturbances are involved in the "final" community structure.

IV. CURRENT AND FUTURE RESEARCH PROBLEMS

Even though considerable effort has been directed toward the study of these decapods, with particular emphasis placed historically on taxonomy, systematics, evolution, anatomy, and some areas of physiology, satisfying answers to only a few questions concerning them have been obtained. Barely a handful of functional morphology studies have been conducted. Recent realization of the impact of exotic species introductions (e.g., *Orconectes rusticus,* Capelli 1982, Magnuson and Beckel 1985, Hobbs and Jass 1988, and Hobbs *et al.* 1989) has initiated several studies to determine the underlying mechanisms controlling crayfish species replacements and community changes (Butler and Stein 1985).

Our knowledge of life histories of shrimps and crayfishes is incomplete, at best. Only about 20 species have been studied in any detail, and varying quantities of data exist for the remaining taxa. Although much time and energy are required for these research projects, the need is paramount. Also, much is lacking in our understanding of the dynamics of populations of virtually all species of freshwater decapods. Best known are factors influencing population dynamics in the crayfish *O. virilis.*

The Fish and Wildlife Service of the U.S. Department of Interior periodically issues a List of Endangered and Threatened Wildlife. Currently there are only six decapod species placed on this list (Anonymous 1989), three of which are troglobites. *Palaemonias alabamae* and *P. ganteri* are troglobitic shrimps in Alabama and Kentucky, respectively; *Cambarus zophonastes* is found in an Arkansas cave. The crayfishes *Orconectes shoupi* and *Pacifasticus fortis* (Faxon) are restricted to streams in western Tennessee and Shasta County, California, respectively, and the shrimp *Snycaris pacifica* inhabits streams in central California. Additionally, the Fish and Wildlife Service (1989) listed the following species as being considered for possible inclusion to the List of Endangered and Threatened Wildlife:

Shrimps (cave):
Palaemonetes antrorum (Texas),
P. cummingi (Florida); Crayfishes:
Cambarus aculabrum (Arkansas),
C. batchi (Kentucky),
C. bouchardi (Kentucky, Tennessee),
C. catagius (North Carolina),
C. chasmodactylus (North Carolina, Virginia, West Virginia),
C. extraneus (Georgia, Tennessee),
C. miltus (Alabama),
C. obeyensis (Tennessee),

C. spicatus (South Carolina),
C. tartarus (Oklahoma),
Distocambarus youngineri (South Carolina),
Fallicambarus jeanae (Arkansas),
Hobbseus orconectoides (Mississippi),
Orconectes deanae (New Mexico),
O. jeffersoni (Kentucky),
O. williamsi (Alabama),
Procambarus acherontis (Florida),
P. barbiger (Mississippi),
P. cometes (Mississippi),
P. connus (Mississippi),
P. echinatus (South Carolina),
P. lepidodactylus (North Carolina, South Carolina),
P. liberorum (Mississippi),
P. lylei (Mississippi),
P. pogum (Mississippi).

Obviously, these species should be closely monitored.

The ecology of decapods inhabiting resource-limited cave habitats is poorly understood. Further investigations should be conducted to determine the physiological mechanisms that allow the reduction of energy utilization in tissues of troglobites. These should be run in conjunction with studies delineating the levels of environmental flexibility of these forms.

Brown (1981), using electrophoretic analysis of 19 biochemical loci of six crayfishes, substantiated previous reports of low genetic variability in decapod crustaceans. She attributed the low heterozygosity in these crayfishes from the southeast United States to their limited geographical distribution in a low number of drainage systems and to their high mobility within those watersheds—both are factors allowing crayfishes to perceive their habitats as fine-grained. Nevo (1978) proposed that generalists are more genetically variable than habitat specialists among all major taxonomic groups and Brown, on the basis of her findings, suggested that crayfishes should be classified as habitat specialists. Biochemical investigations of other crayfishes occupying a variety of habitats and from different phylogenetic positions should be undertaken to test the latter hypothesis. Also, additional electrophoretic studies may aid in determining more precisely the relationships of currently recognized species groups, species, and subspecies.

Recently, considerable effort has been directed toward the study of pheromones and other chemical substances released into the environment by various decapods. Much is to be learned about the importance of chemical communication in the behavior of aquatic animals and about the impact of these substances on the structure of communities. Basic questions remain unanswered concerning the roles of pheromones and chemical cues in population and community dynamics. Obviously, a tremendous variety of behavioral studies await the ethologist.

Although James (1979, see his bibliography) has shown that benthic macroinvertebrates are good indicators of various forms of pollution and are useful as test organisms for determining the presence of toxic substances, the importance of the role of freshwater decapods has not been exhausted (France 1986). Recent technical advances and syntheses have brought together existing data on the effects and measurement of toxicity in decapods (Hobbs and Hall 1974, Buikema and Cairns 1980, Doughtie and Rao 1984, Roldman and Shivers 1987). Acidification due to acid precipitation and loss of buffering capability in freshwater systems in the northeastern United States and Canada have stimulated research on the effects of increased acidity on organisms (Schindler *et al.* 1985). Crayfishes are highly sensitive to an increase in the hydrogen ion concentration (Wood and Rogano 1986) and although there are inter- and intraspecific variations in the degree of tolerance (Berrill *et al.* 1985), France (1984, 1985a) has shown that *Orconectes virilis* juveniles are particularly sensitive. Additional studies are needed in order to determine the impact of toxic materials on spermatozoa, eggs, young, and individuals in various molt stages (premolt, postmolt, molt, intermolt).

As a result of increasing demands by humans for energy, aquatic ecosystems undoubtedly will be subjected to considerable additional waste heat. Because crayfishes and shrimps occupy important positions in complex food webs of lentic and lotic ecosystems, precise data concerning temperature preferences, heat tolerances, and thermal responses (e.g., McWhinnie and O'Connor 1967, Claussen 1980, Mathur *et al.* 1982, Taylor 1984) of these and other organisms will be invaluable in establishing temperature standards such that aquatic communities will be protected.

Much effort is and will continue to be required in order to recover disturbed lotic and lentic habitats and to conserve and preserve existing natural ecosystems; this includes surface and subsurface water resources. Aside from basic research conducted to recognize contamination and to treat perturbed areas, implementation of new laws, education, and stricter enforcement of existing legislature are necessary strategies of the future. In some parts of the region treated in this book (e.g., Louisiana), shrimp and crayfish production are important economic activities. Considerable work has been directed toward increasing the yields of crayfish farmers and this effort should be continued (Hudson and Fontenot 1970, Avault 1972). As marine shellfish populations are depleted, humans will undoubtedly turn to alternate food sources, including crayfishes. In various countries in Europe, these crustaceans are valued as food and extensive stocking of exotic species and a rehabilitation program for native species are ongoing. Most of the interest in North America is centered on development of an export trade for Europe, yet there is potential for an expanded domestic market. Those crayfish species with known economic importance in North America such as the burrowers *Cambarus diogenes* and *Fallicambarus devastator*, the stream and lake-dweller *Orconectes rusticus*, and the tasty *Procambarus clarkii* should particularly be studied in detail.

V. COLLECTING AND REARING TECHNIQUES

Because decapods are found in such a large diversity of habitats, a variety of techniques should be used in procuring representatives.

A. Shrimps

1. Atyidae

Collections should not be made for any of the atyid shrimps in North American freshwaters. Populations of the two species of *Syncaris* and the two cave shrimps assigned to *Palaemonias* are very small (two may already be extirpated) and should not be disturbed!

2. Palaemonidae

Freshwater shrimps occupy various sluggish lotic or lentic environments and are most readily collected with the aid of a sturdy, long-handled, fine-meshed dip net or a small-meshed seine. When a particular locality is choked with vegetation or if the water is greater than 1.5 m deep, the use of fine-meshed wire traps with inverted cones baited with meat can be employed. In clear to low turbidity waters, they may be collected at night with the aid of a headlight and dip net (the ruby-colored reflection of their eyes makes for easy location of individuals). Individuals can be maintained in well-aerated aquaria with considerable vegetation; the water should be changed weekly.

B. Crayfishes: Astacidae and Cambaridae

The North American astacids are known primarily from lakes and streams west of the Continental Divide, and methods for collecting these crayfishes are identical to those for the cambarid stream/lake-dwellers. In shallow streams that have little vegetation, a small-meshed seine (2–5 m long) can be most useful. Deep sections or pooled areas of streams as well as shallow ponds and the littoral zone of lakes may be sampled by pulling the seine (5–8 m length) through the water, taking care to ensure that the weighted margin of the seine remains in contact with the substrata. When electrofishing equipment (Fago 1982) is used, small-meshed seines can collect any organism that is stunned and carried downstream. Use of various wire traps is recommended for vegetation-choked streams, marshes, ponds, or lakes or when sampling deep waters of lakes. Inverted-cone minnow traps with enlarged openings (4–5 cm diameter) have proven to be quite effective. A residence time of 1–2 nights, using chicken and fish scraps, liver, canned cat food, etc. as bait, is an adequate and productive duration for this sampling technique.

A sturdy dip net is very helpful in sampling ditches, mud-bottomed pools, overhanging vegetation, and areas choked with vegetation. This type of net (delta net preferred) can be used to collect individuals from rocky substrata in lakes and streams. Probably most crayfishes burrow at least occasionally for one or more reasons. The diameter of the burrow tube and chimney pellet size provide good indicators of the size of the crayfish inhabiting the burrow (Hobbs and Jass 1988). One may examine the burrow aperture and assume that the smaller openings contain juveniles; to sample the adult population, concentrate on burrows with larger diameters. [This would not be an appropriate procedure for small burrowing species, such as *Cambarellus puer* Hobbs, *Faxonella clypeata,* and *Procambarus pygmaeus* Hobbs]. On warm, rainy, humid nights, burrowing crayfishes can be collected fairly easily. These crayfishes come to the mouths of their burrows and often leave them to forage for food; with the aid of a headlight, they can occasionally be collected in large numbers. Use of scuba equipment to observe and/or collect crayfishes in deep streams, pools, ponds, and lakes is also a very useful technique. Diver certification and various safety rules are requirements not to be ignored by the researcher!

Troglobitic decapods generally lack pigments and thus appear white or translucent. In subterranean streams, use of a small aquarium net is generally adequate to capture the readily visible organisms. Hand-grabbing is also a fairly productive means of collecting them once they have been disturbed and are "swimming" in open water. Wire minnow traps can be used if checked often. Note: cave decapod populations are usually small and because of reduced biotic potentials in troglobitic forms, yearly recruitment is very low. Thus, care must be taken not to collect many individuals from any cave locality—at any time!

Because crayfishes are aggressive, antisocial, and cannibalistic, rearing and maintaining them can be problematic. On the other hand, because they are scavengers and detritus feeders, they will feed on almost anything organic. Ideally, individuals should be maintained separately, each individual in a single container (e.g., 20 × 8 cm stackable glass bowl) and at room temperature. No aeration is required, provided the water does not become fouled with food, wastes, etc. The bowls should have small gravel substrata and the water should be sufficiently deep to cover the crayfish. If larger containers are used, small cobbles or broken pieces of flower pots or bricks can be added for cover. Individuals should be fed 2–3 times per week and the water should be changed approximately every other week. Food can range from various aquatic plants (*Potamogeton, Elodea*), scraps of meat, and earthworms, to dried cat food. Continuous illumination will ensure a growth of algae but is not necessary.

Shrimps can be raised in much the same manner as described for crayfishes except more vegetation and deeper water are required for these crustaceans. Females with eggs should definitely be kept isolated. When young hatch, they will remain attached to the pleopods of the female for several weeks. Most of the young will then leave the mother and the female should be separated from them to prevent cannibalism. As the young undergo molting and growth, some cannibalism will probably occur. By providing cover for the molting young or by dividing them among several containers, this can be reduced significantly.

On a much larger scale, large tanks or ponds can be utilized for raising crayfishes or shrimps. These should be designed for complete draining with sluice gates or standpipes and should be long and narrow rather than square. Water depth should be maintained at about 1–1.5 m and food can be variable, but catfish feed or poultry starter are good. For additional information on various aspects of crayfish aquaculture, see Avault (1972), Avault and Meyers (1976), Black and Huner (1979), Huner and Barr (1984), and Avault and Huner (1985).

Upon collection, crayfishes and shrimps should be killed in 5% neutral formalin and kept in the solution from 12 hours to a week, depending on the size and number of individuals. Specimens should then be washed in running tap water for several hours and transferred to 70–75% ethanol. If epizooites (e.g., protozoans, copepods, entocytherids, branchiobdellids) are to be saved, the formalin in which the decapods were killed should be filtered, the exoskeleton rinsed, and the rinse water poured through a fine seive. Any symbionts should be preserved in 75% ethanol and carefully labeled (Hart and Hart 1974, Sprague and Couch 1971). Where possible, crayfishes and shrimps should be stored in clamp-top glass jars with gaskets. The organisms should be placed in the jar anterior end down with room for adequate 70% ethanol preservative. The specimens should be accompanied by a 100% rag paper label on which data are clearly printed using insoluble black ink.

VI. IDENTIFICATION

A. Preparation of Specimens

Reliable identification of freshwater shrimps can be made only if appendages are removed, cleared, and mounted on a microscope slide. The second pleopod of adult males should be extirpated and then heated in a lactic acid–chlorazol Black E stain solution [3–5 drops in 1% ethanol (95%) stain solution per 20 ml of lactic acid—a 1% solution of fast green can be substituted] at 150°C for 15 min. The translucent appendage should be transferred to glycerine and then examined.

Crayfishes can be identified with the aid of a hand lens or a stereoscope. First-form males are required for positive identification, and a crayfish's left pleopod should be removed and used for working with the key.

B. Taxonomic Key to Genera of Freshwater Decapoda

Morphological characteristics used in the following key are illustrated in Figures 22.1–22.2, 22.15, 22.20–22.22. First-form males are required to identify crayfishes at the generic level.

1a.	Rostrum and abdomen compressed laterally; third pair of pereiopods never bearing chelae .... infraorder Caridea	2
1b.	Rostrum and abdomen compressed dorsoventrally; third pair of pereiopods bearing chelae .... infraorder Astacidea	5
2a(1a).	Fingers of chelae of first and second pereiopods with apical tufts of long setae (Fig. 22.1b); supraorbital spine on either side of base of rostrum; some pereiopods with exopods .... family Atyidae	3
2b.	Fingers of chelae of first and second pereiopods without tufts; no supraorbital spines; all pereiopods lacking exopods .... family Palaemonidae	4
3a(2a).	Eyes reduced, without pigment; all pereiopods with exopods; troglobitic .... *Palaemonias*	
3b.	Eyes well developed, pigmented; fifth pereiopod without exopod; epigean .... *Syncaris*	
4a(2b).	Second pereiopod distinctly longer than first; mandibular palp present; branchiostegal spine on carapace situated distant to anterior margin; epigean .... *Macrobrachium*	
4b.	Second pereiopod only slightly longer than first; mandibular palp absent; branchiostegal spine on carapace situated near or on anterior margin; epigean or hypogean .... *Palaemonetes*	
5a(1b).	Males lacking hooks on ischia of all pereiopods; first pleopod of male with distal portion rolled to form cylinder; male never demonstrating cyclic dimorphism; female lacking annulus ventralis .... family Astacidae *Pacifastacus*	
5b.	Males with ischial hooks on one or more of second through fourth pereiopods (Fig. 22.15); first pleopod of male complexly folded with sperm tube opening on one terminal element; male always demonstrating cyclic dimorphism; female with annulus ventralis .... family Cambaridae	6
6a(5b).	Males with hooks on ischia of second and third pereiopods; very small (15–33 mm total length) .... subfamily Cambarellinae *Cambarellus*	
6b.	Males lacking hooks on ischia of second pereiopods; hooks present on ischia of third, third and fourth, or fourth pereiopod .... subfamily Cambarinae	7
7a(6b).	Mesial surface of flagellum of antennae fringed .... *Barbicambarus*	
7b.	Mesial surface of flagellum of antennae never fringed	8
8a(7b).	Third maxilliped much enlarged; mesial margin of ischium without teeth .... *Troglocambarus*	
8b.	Third maxilliped not conspicuously large; mesial margin of ischium bearing teeth ..	9
9a(8b).	Coxa of fourth pereiopod with caudomesial boss (Fig. 22.15c)	10
9b.	Coxa of fourth pereiopod lacking caudomesial boss	15

Fig. 22.20

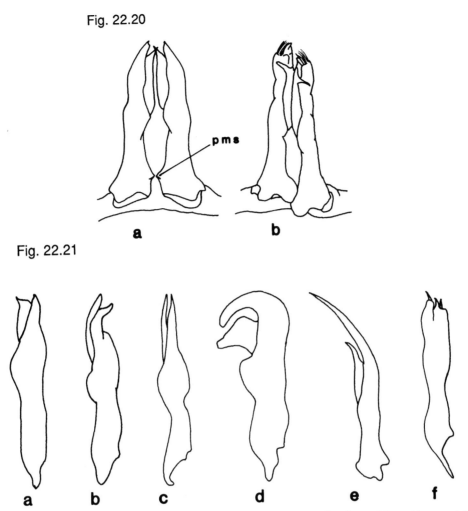

Fig. 22.21

Figure 22.20 Caudal view of first-form males (after Hobbs 1972). (a) *Bouchardina robinsoni* (symmetrical); (b) *Procambarus acutissimus* (asymmetrical); (pms, proximomesial spur). **Figure 22.21** (a–d, f) Lateral view of left first pleopods of first-form males (after Hobbs 1972); (e) caudal view of same. a, *Orconectes i. inermis;* b, *Procambarus pecki;* c, *Orconectes propinquus;* d, *Cambarus latimanus;* e, *Faxonella creaseri;* f, *Procambarus gracilis.*

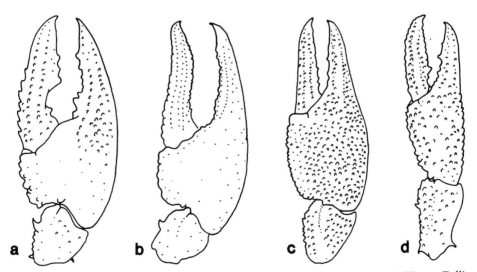

Figure 22.22 Dorsal view of chelae and carpus of first form males (after Hobbs 1972). a, *Fallicambarus fodiens;* b, *Orconectes rusticus;* c, *Hobbseus cristatus;* d, *Distocambarus crockeri.*

10a(9a).	First pair of pleopods asymmetrical (Fig. 22.20b) .... *Procambarus* (in part)	
10b.	First pair of pleopods symmetrical (Fig. 22.20a) 11	
11a(10b).	Well-developed hooks on ischia of third and pereiopods, or terminals of pleopods disposed cephalodistally 12	
11b.	Well-developed hooks on ischia of third pereiopods only, terminals of pleopods never disposed cephalodistally 13	
12a(11a).	Shaft of first pleopods straight and terminating in two very short elements directed caudodistally or distally (Fig. 22.21a) .... *Orconectes* (in part)	
12b.	Shaft of first pleopods straight or bent caudally; if straight, never terminating other than in two very short elements directed caudodistally or distally (Fig. 22.21b) ..... *Procambarus* (in part)	
13a(11b).	Opposable margin of dactyl of chela with angular excision in proximal half (Fig. 22.22a) .... *Fallicambarus*	
13b.	Opposable margin of dactyl of chela lacking angular excision in proximal half (Fig. 22.22b) 14	
14a(13b).	First pleopod with terminal elements directed distally (Fig. 22.21c) .... *Orconectes* (in part)	
14b.	First pleopod with terminal elements directed caudally or, rarely, caudodistally (Fig. 22.21d) .... *Cambarus*	
15a(9b).	First pleopod with two terminal elements, one of which at least twice as long as other (Fig. 22.21e) .... *Faxonella*	
15b.	First pleopod with two or more terminal elements, never two with one at least as long as other 16	
16a(15b).	First pleopod terminating in at least three distinct elements (Fig. 22.21f) ...... *Procambarus* (in part)	
16b.	First pleopod terminating in only two distinct elements 17	
17a(16b).	Dorsal surface of chela studded with crowded small tubercles (Fig. 22.22c) .... *Hobbseus*	
17b.	Dorsal surface of chela with tubercles over lateral half widely scattered at most 18	
18a(17b).	First pleopod with proximomesial spur (Fig. 22.20a) .... *Bouchardina*	
18b.	First pleopod lacking proximomesial spur 19	
19a(18b).	Carpus of cheliped slender and longer than mesial margin of palm of chela (Fig. 22.22d); rostrum without marginal spines .... *Distocambarus*	
19b.	Carpus of cheliped never conspicuously slender and seldom longer than mesial margin of palm (Fig. 22.22b); if so, rostrum with marginal spines .... *Orconectes* (in part)	

LITERATURE CITED

Aiken, D. E. 1968. The crayfish *Orconectes virilis:* survival in a region with severe winter conditions. Canadian Journal of Zoology 46:207–211.

Albaugh, D. W. 1972. Insecticide tolerances of two crayfish populations (*Procambarus acutus*) in south-central Texas. Bulletin of Environmental Contamination and Toxicology 8:334–338.

Albaugh, D. W., and J. B. Black. 1973. A new crayfish of the genus *Cambarellus* from Texas, with new Texas distributional records for the genus (Decapoda, Astacidae). Southwestern Natururalist 18:177–185.

Ameyaw-Akumfi, C. E. 1976. Some aspects of breeding biology of crayfish. Ph.D. Thesis, Univ. of Michigan, Ann Arbor. 252 pp.

Ameyaw-Akumfi, C. E. 1979. Appeasement displays in cambarid crayfish (Decapoda, Astacoidea). Crustaceana, Supplement 5:136–141.

Ameyaw-Akumfi, C. E. 1981. Courtship in the crayfish *Procambarus clarkii* (Girard) (Decapoda, Astacidea). Crustaceana 40:57–64.

Anderson, R. V. 1978. The effects of lead on oxygen uptake in the crayfish, *Orconectes virilis* (Hagen). Bulletin of Environmental Contamination and Toxicology 20:394–400.

Andolshek, M. D., and H. H. Hobbs, Jr. 1986. The entocytherid ostracod fauna of southeastern Georgia. Smithsonian Contributions to Zoology 424: 1–43.

Andrews, E. A. 1895. Conjugation in an American crayfish. American Naturalist 29:867–873.

Anonymous. 1989. Title 50. Wildlife and Fisheries. Part 17—Endangered and Threatened Wildlife and Plants. Subpart B—Lists. 50 CFR 17.11 & 17.12. U.S. Fish and Wildlife Service, Washington, D.C. 34 pp.

Avault, J. W., Jr. 1972. Crawfish farming in the United States. Sea Grant Reprint, Center for Wetland Research, Louisiana State University, Baton Rouge. 23 pp.

Avault, J. W., Jr., and J. V. Huner. 1985. Crawfish culture in the United States. *In:* J. V. Huner and E. E. Brown, editors. Crustacean and mollusk aquaculture in the United States. AVI, Westport, Connecticut. 476 pp.

Avault, J. W., Jr., and S. P. Meyers. 1976. Effects of feeding, fertilization, and vegetation on production of red swamp crayfish, *Procambarus clarkii*. Freshwater Crayfish 2:125–138.

Barr, T. E. 1968. Cave ecology and the evolution of troglobites. *In:* T. Dobzhansky, M. K. Hecht, and W. C. Steere, editors. Evolutionary Biology. 2:35–102.

Barr, T. E., and R. A. Kuehne. 1971. Ecological studies in the Mammoth Cave system of Kentucky. II. The ecosystem. Annales de Speleologie Revue Trimestrielle, 26:47–96.

Bauer, U., J. Dudel, and H. Hatt. 1981. Characteristics of single chemoreceptive units sensitive to amino acids and related substances in the crayfish leg. Journal of Comparative Physiology 144:67–74.

Bechler, D. L. 1981. Copulatory and maternal-offspring behavior in the hypogean crayfish, *Orconectes inermis inermis* Cope and *Orconectes pellucidus* (Tellkampf)(Decapoda, Astacidae). Crustaceana 40:136–143.

Bechler, D. L. 1983. Contamination of Maramec Spring. North American Biospeleology Newsletter 29:5–6.

Beck, J. T. 1980. Life history relationships between the bopyrid isopod *Probopyrus pandalicola* and one of its freshwater shrimp hosts *Palaemonetes paludosus*. American Midland Naturalist 104:135–154.

Beck, J. T., and B. C. Cowell. 1976. Life history and ecology of the freshwater caridean shrimp, *Palaemonetes paludosus* (Gibbes). American Midland Naturalist 96:52–65.

Becker, C. D., R. G. Genoway, and J. A. Merrill. 1975. Resistance of a northwestern crayfish, *Pacifastacus leniusculus* (Dana), to elevated temperatures. Transactions of the American Fisheries Society 104:374–387.

Berrill, M., and B. Chenoweth. 1982. The burrowing ability of nonburrowing crayfish. American Midland Naturalist 108:199–201.

Berrill, M., L. Hollett, A. Margosian, and J. Hudson. 1985. Variation in tolerance to low environment pH by the crayfish *Orconectes rusticus, O. propinquus,* and *Cambarus robustus*. Canadian Journal of Zoology 63:2586–2589.

Black, J. B. 1958. Ontogeny of the first and second pleopods of the male crayfish *Orconectes clypeatus* (Hay). Tulane Studies in Zoology 6:190–203.

Black, J. B., and J. V. Huner. 1979. Breeding crayfish. Carolina Tips, 42:13–14.

Bovbjerg, R. V. 1953. Dominance order in the crayfish *Orconectes virilis* (Hagen). Physiological Zoology 26:173–178.

Bowman, T. E., and L. G. Abele. 1982. Classification of the recent Crustacea. Pages 1–27 *In:* D. E. Bliss, editor. The biology of Crustacea. Vol. 1: Systematics, the fossil record, and biogeography. Academic Press, New York.

Broad, A. C., and J. H. Hubschman. 1963. The larval development of *Palaemonetes kadiakensis* M. J. Rathbun in the laboratory. Transactions of the American Microscopical Society 82:185–197.

Brown, K. 1981. Low genetic variability and high similarities in the crayfish genera *Cambarus* and *Procambarus*. American Midland Naturalist 105:225–232.

Bruski, C. A., and D. W. Dunham. 1987. The importance of vision in agonistic communication of the crayfish *Orconectes rusticus*. I: An analysis of bout dynamics. Behavior 103:83–107.

Budd, T. W., J. C. Lewis, and M. L. Tracey. 1979. Filtration feeding in *Orconectes propinquus* and *Cambarus robustus* (Decapoda, Cambaridae). Crustaceana, Supplement 5:131–134.

Buikema, A. L., Jr., and E. F. Benfield. 1979. Effects of pollution on freshwater invertebrates. Journal of the Water Pollution Control Federation 51:1708–1717.

Buikema, A. L., Jr., and J. Cairns. 1980. Aquatic invertebrate bioassays. American Society for Testing and Materials, Philadelphia, Pennsylvania. 209 pp.

Burggren, W. W., and B. R. McMahon. 1983. An analysis of scaphognathite pumping performance in the crayfish *Orconectes virilis:* compensatory changes to acute and chronic hypoxic exposure. Physiological Zoology 56:309–318.

Butler, M. J., and R. A. Stein. 1985. An analysis of the mechanisms governing species replacements in crayfish. Oecologia 66:168–177.

Caine, E. A. 1975. Feeding and masticatory structures of six species of the crayfish genus *Procambarus* (Decapoda, Astacidae). Forma et Functio 8:49–66.

Caine, E. A. 1978. Comparative ecology of epigean and hypogean crayfish (Crustacea:Cambaridae) from northwestern Florida. American Midland Naturalist 99:315–329.

Capelli, G. M. 1982. Displacement of northern Wisconsin crayfish by *Orconectes rusticus* (Girard). Limnology and Oceanography 27:741–745.

Claussen, D. L. 1980. Thermal acclimation in the crayfish, *Orconectes rusticus* and *O. virilis*. Comparative Biochemistry and Physiology 66A:377–384.

Cooper, J. E. 1975. Ecological and behavioral studies in Shelta Cave, Alabama, with emphasis on decapod crustaceans. Ph.D. Thesis, University of Kentucky, Lexington. 364 pp.

Cooper, J. E., and M. R. Cooper. 1978. Growth, longevity, and reproductive strategies in Shelta Cave crayfishes. National Speleological Society Bulletin 40:97.

Corey, S. 1987. Intraspecific differences in reproductive potential, realized reproduction and actual production in the crayfish *Orconectes propinquus* (Girard 1852) in Ontario. American Midland Naturalist 118:424–432.

Covich, A. P. 1978. Spatial and temporal patterns of foraging activity among radio-monitored crayfish. Page 13 *in:* 26th Annual Meeting of the North American Benthological Society, Titles and Abstracts.

Covich, A. P. 1982. Crayfish foraging and ecosystem control by selective omnivory. Page 14 *in:* J. F. Payne, editor. Crayfish Distribution Patterns. Symposium, American Society of Zoologists, Louisville, Kentucky.

Covich, A. P., L. L. Dye, and J. S. Mattice. 1981. Crayfish predation on *Corbicula* under laboratory conditions. American Midland Naturalist 105:181–188.

Cox, D. K., and J. J. Beauchamp. 1982. Thermal resistance of juvenile crayfish, *Cambarus bartoni* [sic] (Fabricius): experiment and model. American Midland Naturalist 108:187–193.

Crawshaw, L. I. 1974. Temperature selection and activity in the crayfish *Orconectes immunis*. Journal of Comparative Physiology 95:315–322.

Crocker, D. W., and D. W. Barr. 1968. Handbook of the crayfishes of Ontario. Miscellaneous Publication of the Royal Ontario Museum. University of Toronto Press, Toronto. 158 pp.

Crunkilton, R. 1982. Bitter harvest. Missouri Conservationist 43:4–7.

Davies, I. J. 1984. Effects of an experimental whole-lake acidification on a population of the crayfish *Orconectes virilis*. Page 35 *in:* North American Benthological Society Abstracts, 32nd Annual Meeting, North Carolina State University, Raleigh.

DeCoursey, P. J. 1983. Biological timing. Pages 107–162 *in:* D. E. Bliss, editor. The biology of Crustacea. Vol. 7: Behavior and Ecology. Academic Press, New York.

Dickson, G. W., and R. Franz. 1980. Respiration rates, ATP turnover and adenylate energy change in excised gills of surface and cave crayfish. Comparative Biochemistry and Physiology 65A:375–379.

Dickson, G. W., L. A. Briese, and J. P. Giesey, Jr. 1979. Tissue metal concentrations in two crayfish species cohabiting a Tennessee cave stream. Oecologia 44: 8–12.

Dickson, G. W., J. P. Giesy, Jr., and L. A. Briese. 1982. The effect of chronic cadmium exposure on phosphoadenylate concentrations and adenylate energy charge of gills and dorsal muscle tissue of crayfish. Environmental Toxicology and Chemistry 1:147–156.

Dimond, J. B., R. E. Kadunce, A. S. Gelchell, and J. A. Blease. 1968. Persistence of DDT in crayfish in a natural environment. Ecology 49:759–762.

Dobkin, S. 1971. The larval development of *Palaemonetes cummingi* Chace, 1954 (Decapoda, Palaemonidae), reared in the laboratory. Crustaceana 20:285–297.

Dobkin, S., and R. B. Manning. 1964. Osmoregulation in two species of *Palaemonetes* (Crustacea:Decapoda) from Florida. Bulletin of Marine Science of the Gulf and Caribbean 14:149–157.

Doughtie, D. G., and K. R. Rao. 1984. Histopathological and ultrastructural changes in the antennal gland, midgut, hepatopancreas, and gill of grass shrimp following exposure to hexavalent chromium. Journal of Invertebrate Pathology 43:89–108.

Emery, A. R. 1970. Fish and crayfish mortalities due to an internal seiche in Georgian Bay, Lake Huron. Journal of the Fisheries Research Board of Canada, 27:1165–1168.

Eng, L. L. 1981. Distribution, life history, and status of the California freshwater shrimp, *Syncaris pacifica* (Holmes). Inland Fish. Endangered Species Program, Special Publication 81- 1:1–27.

Fago, D. 1982. Distribution and relative abundance of fishes in Wisconsin. I. Greater Rock River Basin. Wisconsin Department of Natural Resources Technical Bulletin 136:1–120.

Fish and Wildlife Service, U.S. Department of the Interior. 1989. Endangered and threatened wildlife and plants; Annual Notice of Review. Federal Register 54(4):554–579.

Fitzpatrick, J. F. 1977. A new crawfish of the genus *Hobbseus* from northeast Mississippi, with notes on the origin of the genus (Decapoda, Cambaridae). Proceedings of the Biological Society of Washington 90:367–374.

Fitzpatrick, J. F. A revision of the dwarf crawfishes. Journal of Crustacean Biology 3:266–277.

Flint, R. W. 1975. Growth in a population of the crayfish *Pacifastacus leniusculus* from a subalpine lacustrine environment. Journal of the Fisheries Research Board of Canada 32:2433–2440.

Flynn, M. F., and H. H. Hobbs III. 1984. Parapatric crayfishes in southern Ohio: evidence of competitive exclusion? Journal of Crustacean Biology 4:382–389.

France, R. L. 1984. Comparative tolerance to low pH of three life stages of the crayfish *Orconectes virilis*. Canadian Journal of Zoology 62:2360–2363.

France, R. L. 1985a. Preliminary investigations of effects of sublethal acid exposure on maternal behavior in the crayfish *Orconectes virilis*. Bulletin of Environmental Contamination and Toxicology 35:641–645.

France, R. L. 1985b. Low pH avoidance by crayfish (*Orconectes virilis*): evidence for sensory conditioning. Canadian Journal Zoology 63:258–262.

France, R. L. 1986. Current status of methods of toxicological research on freshwater crayfish. Canadian Technical Report of Fisheries and Aquatic Sciences No. 1404:1–20.

France, R. L., and L. Graham. 1985. Increased microsporidian parasitism of the crayfish *Orconectes virilis* in an experimentally acidified lake. Water Air and Soil Pollution 26:129–136.

Franz, R. 1978. Ecological strategies of closely-related surface and troglobitic Florida crayfishes. Bulletin of the Ecological Society America 59:70.

Franz, R., and D. S. Lee. 1982. Distribution and evolution of Florida's troglobitic crayfishes. Bulletin of the Florida State Museum, Biological Sciences 28:53–78.

Gillespie, R., T. Reisine, and E. J. Massaro. 1977. Cadmium uptake by the crayfish, *Orconectes propinquus propinquus* (Girard). Environmental Research 13:364–368.

Godley, J. S., R. W. McDiarmid, and N. N. Rojas. 1984. Estimating prey size and number in crayfish-eating snakes, genus *Regina*. Herpetologica, 40:82–88.

Gowing, H., and W. T. Momot. 1979. Impact of brooktrout (*Salvelinus fontinalis*) predation on the crayfish *Orconectes virilis* in three Michigan lakes. Journal of the Fisheries Research Board of Canada 36:1191–1196.

Grow, L. 1981. Burrowing behaviour in the crayfish, *Cambarus diogenes diogenes* Girard. Animal Behaviour 29:351–356.

Gunter, G. 1937. Observations on the river shrimp, *Macrobrachium ohionis*, (Smith). American Midland Naturalist 18:1038–1042.

Hart, C. W., Jr., and J. Clark. 1987. An interdisciplinary bibliography of freshwater crayfishes (Astacoidea and Parastacoidea) from Aristotle through 1985. Smithsonian Contributions to Zoology No. 455:437 pp.

Hart, D. G., and C. W. Hart, Jr. 1974. The ostracod family Entocytheridae. Academy of Natural Sciences of Philadelphia, Monograph 18:239 pp.

Hatt, H. 1984. Structural requirements of amino acids and related compounds for stimulation of receptors in crayfish walking leg. Journal of Comparative Physiology 155A:219–231.

Hayes, W. A., II. 1975. Behavioral components of social interactions in the crayfish *Procambarus gracilis* (Bundy) (Decapoda, Cambaridae). Proceedings of the Oklahoma Academy of Science 55:1–5.

Hayes, W. A., II. 1977. Predator response postures of crayfish. I. The genus *Procambarus* (Decapoda, Cambaridae). Southwestern Naturalist 21:443–449.

Hazlett, B. A. 1985. Chemical detection of sex and condition in the crayfish *Orconectes virilis*. Journal of Chemical Ecology 2:181–189.

Hazlett, B. A., and D. Rittschof. 1985. Variation in rate of growth in the crayfish *Orconectes virilis*. Journal of Crustacean Biology 5:341–346.

Hazlett, B. A., D. Rittschof, and D. Rubenstein. 1974. Behavioral biology of the crayfish *Orconectes virilis*. I. Home range. American Midland Naturalist 92:301–319.

Heckenlively, D. B. 1970. Intensity of aggression in the crayfish *Orconectes virilis* (Hagen). Nature (London) 225:180–181.

Hobbs, H. H., Jr. 1969. On the distribution and phylogeny of the crayfish genus *Cambarus*. Pages 93–178 *in:* P. C. Holt., R. Hoffman, and C. W. Hart, Jr, editors. The distributional history of the biota of the southern Appalachians. Part I: Invertebrates. Research Division Monograph 1. Virginia Polytech. Inst., Blacksburg.

Hobbs, H. H., Jr. 1972. Crayfishes (Astacidae) of North and Middle America. Biota of Freshwater Ecosystems. U.S. Environmental Protection Agency, Water Pollution Control Research Service Identification Manual 9:173 pp.

Hobbs, H. H., Jr. 1974. Synopsis of the families and genera of crayfishes (Crustacea:Decapoda). Smithsonian Contributions to Zoology No. 164:32 pp.

Hobbs, H. H., Jr. 1976. Adaptations and convergence in North American crayfishes. Freshwater Crayfish 2:541–551.

Hobbs, H. H., Jr. 1981. The crayfishes of Georgia. Smithsonian Contributions to Zoology No. 318:549 pp.

Hobbs, H. H., Jr. 1986. Highlights of a half century of crayfishing. Freshwater Crayfish 6:12–23.

Hobbs, H. H., Jr. 1988. Crayfish distribution adaptive radiation and evolution. pages 52–82 *in:* D. M. Holdich and R. S. Lowery, editors. Freshwater crayfish biology: Management and exploitation. Croom Helm, London.

Hobbs, H. H., Jr. 1989. An illustrated checklist of American Crayfishes (Decapoda: Astacidae, Cambaridae, and Parastacidae). Smithsonian Contributions to Zoology No. 480 pp. 236.

Hobbs, H. H., Jr., and R. Franz. 1986. New troglobitic crayfish with comments on its relationship to epigean and other hypogean crayfishes of Florida. Journal of Crustacean Biology 6:509–519.

Hobbs, H. H., Jr., and E. T. Hall, Jr. 1974. Crayfishes (Decapoda:Astacidae). Pages 195–241 *in:* C. W. Hart, Jr. and S. L. H. Fuller, editors. Pollution ecology of freshwater invertebrates. Academic Press, New York.

Hobbs, H. H., Jr., and M. Whiteman. 1987. A new, economically important crayfish (Decapods:Cambaridae) from the Neches River Basin, Texas, with a key to the subgenus *Fallicambarus*. Proceedings of the Biological Society of Washington. 100:403–411.

Hobbs, H. H., Jr., H. H. Hobbs III, and M. A. Daniel. 1977. A review of the troglobitic decapod crustaceans of the Americas. Smithsonian Contributions to Zoology No. 244:183 pp.

Hobbs, H. H., III. 1976. Observations on the cave-dwelling crayfishes of Indiana. Freshwater Crayfish 2:405–414.

Hobbs, H. H., III. 1980. Studies of the cave crayfish, *Orconectes inermis inermis* Cope (Decapoda, Cambaridae). Part IV: Mark–recapture procedures for estimating population size and movements of individuals. International Journal of Speleology 10:303–322.

Hobbs, H. H., III. 1987. Gasoline pollution in an Indiana cave. Page 33 *in:* Program of the National Speleological Society Convention.

Hobbs, H. H., III, and J. P. Jass. 1988. The crayfishes and shrimp of Wisconsin (Decapoda: Palaemonidae, Cambaridae). Special Publications in Biology and Geology, No. 5. Milwaukee Public Museum. Milwaukee, Wisconsin. 177 pp.

Hobbs, H. H., III, and J. P. Jass. 1991. Trophic relationships of North American freshwater crayfishes and shrimps. Contrib. in Biol. and Geology, Milwaukee Public Museum. Milwaukee, Wisconsin. In press.

Hobbs, H. H., III., J. H. Thorp, and G. E. Anderson. 1976. The freshwater decapod crustaceans (Palaemonidae, Cambaridae) of the Savannah River Plant, South Carolina. Publication of the Savannah River Plant. National Environmental Research Park Program. 63 pp.

Hobbs. H. H., III, J. P. Jass, and J. V. Huner. 1989. A review of global crayfish introductions with particular emphasis on two North American species (Decapoda:Cambaridae). Crustaceana 56:299–316.

Holsinger, J. R. 1988. Troglobites: the evolution of cave-dwelling organisms. American Scientist 76:146–153.

Horns, W. H., and J. J. Magnuson. 1981. Crayfish predation on lake trout eggs in Trout Lake, Wisconsin. Rapports et Proces- Verbaux des Reunions, Conseil International pour Exploration de la Mer 178:299–303.

Hubschman, J. H. 1967. Effects of copper on the crayfish *Orconectes rusticus* (Girard). II. Mode of toxic action. Crustaceana 12:141–150.

Hubschman, J. H., and J. A. Rose. 1969. *Palaemonetes kadiakensis* Rathbun: post embryonic growth in the laboratory (Decapoda, Palaemonidae). Crustaceana 16:81–87.

Hudson, J. F., and W. F. Fontenot. 1970. Profitability of crawfish peeling plants in Louisiana. Louisiana Department of Agricultural Economics and Agribusiness, Research Report 408:57 pp.

Huner, J. V. 1977. Observations on the biology of the river shrimp from a commercial bait fishery near Port Allen,

Louisiana. Pages 380–386 *in:* 31st Annual Conference of the Southeastern Association of Fish and Wildlife Agencies, Proceedings.

Huner, J. V., and J. E. Barr. 1984. Red Swamp Crawfish: Biology and Exploitation. Louisiana Sea Grant College Program. Louisiana State University, Baton Rouge. 136 pp.

Huppop, K. 1985. The role of metabolism in the evolution of cave animals. National Speleological Society Bulletin 47:136–146.

Ingle, R. W., and W. Thomas. 1974. Mating and spawning of the crayfish *Austropotamobius palipes* (Crustacea:Astacidae). Journal of Zoology 173:525–538.

Itagaki, H., and J. H. Thorp. 1981. Laboratory experiments to determine if crayfish can communicate in a flow-through system. Journal of Chemical Ecology 7:115–126

James, A. 1979. The value of biological indicators in relation to other parameters of water quality. Pages 1-1–16 *in:* A. James and L. Evison, editors. Biological Indicators of water quality. Wiley, Chichester, England.

Johannes, R. E., and K. L. Webb. 1970. Release of dissolved organic compounds by marine and freshwater invertebrates. Pages 257–273 *in:* D. W. Hood, editor. Symposium on organic matter in natural waters. Institute of Marine Sciences. University of Alaska, Fairbanks.

Johnson, S. K. 1977. Crawfish and freshwater shrimp diseases. Texas A&M Univ., Agricultural Extension Service, College Station. 19 pp.

Kushlan, J. A., and M. S. Kushlan. 1980. Population fluctuations of the prawn, *Palaemonetes paludosus,* in the Everglades. American Midland Naturalist 103:401–403.

Layne, J. R., M. L. Manis, and D. L. Claussen. 1985. Seasonal variation in the time course of thermal acclimation in the crayfish *Orconectes rusticus.* Freshwater Invertebrate Biology 4:98–104.

Leitheuser, A. T., J. R. Holsinger, R. Olson, N. R. Pace, R. L. Whitman, and T. Whitmore. 1985. Ecological analysis of the Kentucky cave shrimp, *Palaemonias ganteri* Hay, at Mammoth Cave National Park (Phase V). Old Dominion University Research Foundation, Norfolk, 102 pp.

Lichtwardt, R. W. 1986. The Trichomycetes: Fungal Associates of arthropods. Springer-Verlag, New York. 343 pp.

Lisowski, E. A. 1983. Distribution, habitat, and behavior of the Kentucky cave shrimp, *Palaemonias ganteri* Hay. Journal of Crustacean Biology 3:88–92.

Lodge, D. M., and J. G. Lorman. 1987. Reductions in submersed macrophyte biomass and species richness by the crayfish *Orconectes rusticus.* Canadian Journal of Fisheries and Aquatic Sciences 44:591–597.

Lodge, D. M., T. K. Kratz, and G. M. Capelli. 1986. Long-term dynamics of three crayfish species in Trout Lake, Wisconsin. Canadian Journal of Fisheries and Aquatic Sciences 43:993–998.

Lodge, D. M., K. M. Brown, S. P. Klosiewski, R. A. Stein, A. P. Covich, B. K. Leathers, and C. Bronmark.

1987. Distribution of freshwater snails: spatial scale and the relative importance of physicochemical and biotic factors. American Malacological Bulletin 5:73–84.

Loring, M. W., and L. G. Hill. 1978. Temperature selection and shelter utilization of the crayfish *Orconectes causeyi.* Southwestern Naturalist 21:219–226.

Lorman, J. G. 1980. Ecology of the crayfish *Orconectes rusticus* in northern Wisconsin. Ph.D. Thesis, Univ. of Wisconsin, Madison. 227 pp.

Lorman, J. G., and J. J. Magnuson. 1978. The role of crayfishes in aquatic ecosystems. Fisheries, 3:8–10.

Lyle, C. 1938. The crawfishes of Mississippi, with special reference to the biology and control of destructive species. Iowa State College Journal of Science 13:75–77.

Magnuson, J. J., and A. L. Beckel. 1985. Exotic species: A case of biological pollution. Wisconsin Academy Review 32:8–10.

Marchand, L. J. 1946. The saber crab, *Platychirograpsus typicus* Rathbun, in Florida: a case of accidental dispersal. Quarterly Journal of the Florida Academy of Science 9:93–100.

Mason, J. C. 1970a. Copulatory behavior of the crayfish, *Pacifastacus trowbridgii* (Stimpson). Canadian Journal of Zoology 48:969–976.

Mason, J. C. 1970b. Maternal–offspring behavior of the crayfish, *Pacifastacus trowbridgi* (Stimpson). American Midland Naturalist 84:463–473.

Mason, J. C. 1974. Crayfish production in a small woodland stream. Department of Environmental Fisheries and Marine Sciences, Pacific Biological Station, Nanaimo, British Columbia. 30 pp.

Mathur, D., R. M. Schutsky, and E. J. Purdy, Jr. 1982. Temperature preference and avoidance responses of the crayfish, *Orconectes obscurus,* and associated statistical problems. Canadian Journal of Fisheries and Aquatic Sciences 39:548–553.

Maude, S. H., and D. D. Williams. 1983. Behavior of crayfish in water currents: hydrodynamics of eight species with reference to their distribution patterns in southern Ontario. Canadian Journal of Fisheries and Aquatic Sciences 40:68–77.

McCormick, R. N. 1934. *Macrobrachium ohionis,* the large freshwater shrimp. Proceedings of the Indiana Academy of Science 34:218–224.

McGriff, D. 1983a. The commercial fishery for *Pacifastacus leniusculus,* from the Sacramento–San Joaquin Delta. Freshwater Crayfish 5:403–417.

McGriff, D. 1983b. Growth, maturity, and fecundity of the crayfish *Pacifastacus leniusculus* from the Sacromento–San Joaquin Delta. California Fish and Game 69:227–242.

McMahon, B. R., and P. R. H. Wilkes. 1983. Emergence responses and aerial ventilation in normoxic and hypoxic crayfish *Orconectes rusticus.* Physiological Zoology 56:133–141.

McWhinnie, M. A., and J. D. O'Connor. 1967. Metabolism and low temperature acclimation in the temperate crayfish, *Orconectes virilis.* Comparative Biochemistry and Physiology 20:131–145.

Merkle, E. L. 1969. Home range of crayfish *Orconectes juvenalis*. American Midland Naturalist 81:228–235.

Momot, W. T. 1984. Crayfish production: a reflection of community energetics. Journal of Crustacean Biology 4:35–54.

Momot, W. T., and H. Gowing. 1972. Differential seasonal migration of the crayfish *Orconectes virilis* (Hagen), in marl lakes. Ecology 53:479–483.

Momot, W. T., and H. Gowing. 1977a. Response of the crayfish *Orconectes virilis* to exploitation. Journal of the Fisheries Research Board of Canada 34:1212–1219.

Momot, W. T., and H. Gowing. 1977b. Production and population dynamics of the crayfish *Orconectes virilis* in three Michigan lakes. Journal of the Fisheries Research Board of Canada 34:2041–2055.

Momot, W. T., H. Gowing, and P. D. Jones. 1978. The dynamics of crayfish and their role in ecosystems. American Midland Naturalist 99:10–35.

Moshiri, G. A., C. R. Goldman, G. L. Godshalk, and D. R. Mull. 1970. The effects of variations in oxygen tension on certain aspects of respiratory metabolism in *Pacifastacus leniusculus* (Dana)(Crustacea:Decapoda). Physiological Zoology 43:23–29.

Naqvi, S. M., and T. Leung. 1983. Trifluralin and Oryzalin herbicides toxicities to juvenile crawfish (*Procambarus clarkii*) and mosquitofish (*Gambusia affinis*). Bulletin of Environmental Contamination and Toxicology 31:304–308.

Nelson, D. H., and D. K. Hooper. 1982. Thermal tolerance and preference of the freshwater shrimp *Palaemonetes kadiakensis*. Journal Thermal Biology 7:183–187.

Nevo, E. 1978. Genetic variation in natural populations; pattern and theory. Theoretical Population Biology 13:121–177.

Nielsen, L. A., and J. B. Reynolds. 1977. Population characteristics of a freshwater shrimp, *Palaemonetes kadiakensis* Rathbun. Transactions of the Missouri Academy Science 10/11:44–57.

Page, L. M. 1985. The crayfishes and shrimps (Decapoda) of Illinois. Illinois Natutal History Survey Bulletin 33:335–448.

Peck, S. K. 1985. Effects of aggressive interactions on temperature selection by the crayfish, *Orconectes virilis*. American Midland Naturalist 114:159–167.

Peckarsky, B. L., P. R. Fraissinet, M. A. Penton, and D. J. Conklin, Jr. 1990. Freshwater macroinvertebrates of northeastern North America. Cornell Univ. Press, Ithaca, New York. 422 pp.

Penn, G. H., Jr. 1943. A study of the life history of the Louisiana red-crawfish, *Cambarus clarkii* Girard. Ecology 24:1–18.

Pollard, T. G., and J. L. Larimer. 1977. Circadian rhythmicity of heart rate in the crayfish, *Procambarus clarkii*. Comparative Biochemistry and Physiology 57A:221–226.

Price, J. O. 1981. Observations on the trophic ecology of the crayfish *Orconectes neglectus chaenodactylus* (Decapoda: Astacidae). ASB Bulletin 28:91–92.

Price, J. O., and J. F. Payne. 1980. Multiple summermolts in adult *Orconectes neglectus chaenodactylus* Williams (Crustacea, Decapoda). ASB Bulletin 27:57.

Price, J. O., and J. F. Payne. 1984. Postembryonic to adult growth and development in the crayfish *Orconectes neglectus chaenodactylus* Williams, 1952 (Decapoda, Astacidea). Crustaceana, 46:176–194.

Prins, R. 1968. Comparative ecology of the crayfishes *Orconectes rusticus* and *Cambarus tenebrosus* in Doe Run, Meade County, Kentucky. Internationale Revue der Gesamten Hydrobiologie 54:667–714.

Rabeni, C. F. 1985. Resource partitioning by stream-dwelling crayfish: The influence of body size. American Midland Naturalist 113:20–29.

Rao, K. R., and M. Fingerman. 1983. Regulation of release and mode of action of crustacean chromatophorotropins. American Zoologist 23:517–527.

Reimer, R., D. K. Strawn, and A. Dixon. 1974. Notes on the river shrimp, *Macrobrachium ohione* (Smith) 1874, in the Galveston Bay system of Texas. Transactions of the American Fisheries Society 103:120–126.

Rittschof, D. 1980. Chemical attraction of hermit crabs and other attendants to simulated gastropod predation sites. Journal of Chemical Ecology 6:103–118.

Rogers, R., and J. V. Huner. 1985. Comparison of burrows and burrowing behavior of five species of cambarid crawfish (Crustacea, Decapoda) from the Southern University campus, Baton Rouge, Louisiana. Proceedings of the Louisiana Academy of Science 48:23–29.

Roldman, B. M., and R. R. Shivers. 1987. The uptake and storage of iron and lead in cells of the crayfish (*Orconectes propinquus*) hepatopancreas and antennal gland. Comparative Biochemistry and Physiology 86C:201–214.

Schindler, D. W., K. H. Mills, D. F. Malley, D. L. Findlay, J. A. Shearer, I. J. Davies, M. A. Turner, G. A. Linsey, and D. R. Cruikshank. 1985. Long-term ecosystem stress: the effects of years of experimental acidification on a small lake. Science 228:1395–1401.

Smalley, A. E. 1961. A new cave shrimp from southeastern United States (Decapoda, Atyidae). Crustaceana 3:127–130.

Smith, D. G. 1988. Keys to the freshwater macroinvertebrates of Massachusetts. No. 3: Crustacea Malacostraca (Crayfish, isopods, amphipods). Publ. No. 15, 236-61-250-2-88-CR. Division of Water Pollution Control, Westboro, Massachusetts. 58 pp.

Smith, T. I. J., P. A. Sandifer, and M. H. Smith. 1978. Population structure of malaysian prawns, *Macrobrachium rosenbergii* (de Man), reared in earthen ponds in South Carolina, 1974–1976. Proceedings of the World Mariculture Society 9:21–38.

Sprague, V., and J. Couch. 1971. An annotated list of protozoa parasites, hyperparasites, and commensals of decapod crustacea. Journal of Protozoology 18:526–537.

Stein, R. A. 1975. Sexual dimorphism in crayfish chelae: functional significance linked to reproductive activities. Canadian Journal of Zoology 54:220–227.

Stein, R. A. 1977. Selective predation, optimal foraging,

and the predator–prey interaction between fish and crayfish. Ecology 58:1237–1253.

Stein, R. A. 1979. Behavioral response of prey to fish predators. Pages 343–353 *in:* R. H. Stroud and H. Clepper, editors. Predator–Prey Systems in Fisheries Management. Sport Fisheries Institute, Washington, D.C.

Stein, R. A., and J. J. Magnuson. 1976. Behavior response of crayfish to a fish predator. Ecology 57:751–761.

Stoffel, L. A., and J. H. Hubschman. 1974. Limb loss and the molt cycle in the freshwater shrimp, *Palaemonetes kadiakensis*. Biological Bulletin (*Woods Hole, Mass.*) 147:203–212.

Strenth, N. E. 1976. A review of the systematics and zoogeography of the freshwater species of *Palaemonetes* Heller of North America (Crustacea:Decapoda). Smithsonian Contributions to Zoology No. 228:27 pp.

Taylor, R. C. 1983. Drought-induced changes in crayfish populations along a stream continuum. American Midland Naturalist 110:286–298.

Taylor, R. C. 1984. Thermal preference and temporal distribution in three crayfish species. Comparative Biochemistry and Physiology 77A:513–517.

Taylor, R. C. 1985. Absence of Form I to Form II alternation in male *Procambarus spiculifer* (Cambaridae). American Midland Naturalist 114:145–151.

Thorp, J. H. 1984. Theory and practice in crayfish communication studies. Journal of Chemical Ecology 10:1283–1288.

Thorp, J. H., and S. P. Gloss. 1986. Field and laboratory tests on acute toxicity of cadmium to freshwater crayfish. Bulletin of Environmental Contamination and Toxicology 37:355–361.

Thorp, J. H., and S. A. Wineriter. 1981. Stress and growth response of juvenile crayfish to rhythmic and arrhythmic temperature fluctuations. Archives of Environmental Contamination and Toxicology 10:69–77.

Thorp, J. H., J. P. Giesy, Jr., and S. A. Wineriter. 1979. Effects of chronic cadmium exposure on crayfish survival, growth, and tolerance to elevated temperatures. Archives of Environmental Contamination and Toxicology 8:449–456.

Tierney, A. J., and J. Atema. 1986. Effects of acidification on the behavioral responses of crayfishes (*Orconectes virilis* and *Procambarus acutus*) to chemical stimuli. Aquatic Toxicology 9:1–11.

Tierney, A. J., and J. Atema. 1988. Behavioral responses of crayfish (*Orconectes virilis* and *Orconectes rusticus*) to chemical feeding stimulants. Journal of Chemical Ecology 14:123–133.

Tierney, A. J., and D. W. Dunham. 1984. Morphology of aesthetasc sensilla in surface and cave dwelling crayfishes. American Zoologist 24:67A.

Tierney, A. J., C. S. Thompson, and D. W. Dunham. 1984. Site of pheromone reception in the crayfish *Orconectes propinquus* (Decapoda, Cambaridae). Journal of Crustacean Biology 4:554–559.

Truesdale, F. M., and W. J. Mermilliod. 1979. The river shrimp *Macrobrachium ohione* (Smith)(Decapoda, Palaemonidae): its abundance, reproduction, and growth in the Atchafalaya River basin of Louisiana, USA. Crustaceana 36:61–73.

Vermeer, K. 1972. The crayfish, *Orconectes virilis*, as an indicator of mercury contamination. The Canadian Field-Naturalist 86:123–125.

Vernberg, F. J. 1983. Respiratory adaptations. Pages 1–42 *in:* D. E. Bliss, editor. The Biology of Crustacea. Vol. 8: Environmental Adaptations. Academic Press, New York.

Weingartner, D. L. 1977. Production and trophic ecology of two crayfish species cohabiting an Indiana cave. Ph.D. Thesis, Michigan State Univ., East Lansing, 323 pp.

Wiens, A. W., and K. B. Armitage. 1961. The oxygen consumption of the crayfish *Orconectes immunis* and *Orconectes nais* in response to temperature and to oxygen saturation. Physiological Zoology 34:39–54.

Wood, C. M., and M. Rogano. 1986. Physiological responses to acid stress in crayfish (*Orconectes*): haemolymph ions, acid-base status, and exchanges with the environment. Canadian Journal of Fisheries and Aquatic Sciences 43:1017–1026.

Word, B. H., Jr., and H. H. Hobbs, Jr. 1958. Observations on the testis of the crayfish *Cambarus montanus acuminatus* Faxon. Transactions of the American Microscopical Society 77:435–450.

Zimmer-Faust, R. K., and J. F. Case. 1983. A proposed dual role of odor in foraging by the California spiny lobster, *Panulirus interruptus* (Randall). Biological Bulletin (*Woods Hole, Mass.*) 164:341–353.

Glossary

acclimation the process of adjustment of a rate function (i.e., respiratory rate) to a change in ambient temperature.

acetabular plates sclerites associated with the genital field in adults (provisional genital field in deutonymphs) of certain mite taxa bearing the genital acetabula.

acetabulum (acetabula) knob-like structure with a porous cap associated with the genital field in deutonymph and adult mites, thought to function as an osmoregulatory chloride epithelium. Acetabula are strictly or roughly paired, and usually are borne on acetabular plates, but may lie in the gonopore or scattered in the integument flanking the gonopore.

acidosis a decline in the pH of the blood through accumulation of CO or anaerobic endproducts.

acrosome a cytoplasmic inclusion in the cell body of a sperm cell.

acute sharp angle at the end.

adductor muscles in bivalve molluscs, the muscles extending between shell valves, closing the valves on contraction.

adductor muscle scars raised areas or scars left on the interior of the ostracode shell by the adductor muscles, generally leave a distinctive pattern of scars.

adenal produced in the parenchyma (rhabdoids).

adenodactyl auxiliary glandular bulb in the male system of some turbellarian taxa.

adhesive strip a thin layer of chitin between the duplicature and the outer lamella of the ostracode valve.

adnate closely adherent, joined together.

adoral zone of membranelles orderly arrangement of three or more compound ciliary organelles (membranelles) that are serially arranged along the left side of the oral area of a ciliate.

afferent branchial vessel in Mollusca, the blood vessels bringing deoxygenated blood from the visceral mass, foot, and mantle to the distal ends of the gill filaments.

ala(e) a coarse or fine, wing-like spine, as in the posteriorly directed alae on the lateral surface of the ostracode carapace.

allelochemic a biochemical produced by one species and released into the environment that subsequently affects the behavior or physiology of another species.

allochthonous substances originating outside the immediate habitats (e.g., carbon produced by plants located upstream or in the riparian zone); opposite of autochthonous.

allorecruitive the type of growth of colonial rotifers whereby the young recruited into the colony come predominantly from other colonies so that as a consequence the genetic diversity of the colony is high. Originally defined as Type I colony formation (see also autorecruitive).

allozymes proteins produced by different forms of the same gene.

amictic (see also mictic and mixis) a phase in the life cycle of monogonont rotifers in which reproduction is parthenogenetic only.

amphidelphic with two opposite ovaries that normally join at the common vagina.

amphids paired sensory organs located on opposite lateral surfaces of nematodes; variable in position (lip or neck region) and shape (pore, slit, circular, spiral or cup-shaped).

amphimixis union of egg and sperm through ordinary sexual reproduction, with the gametes derived from separate individuals.

amphoteric type of female (e.g., rotifers of the class Monogononta) that is capable of producing, in the same clutch, two types of eggs, one developing into female (diploid) and the other into male (haploid) offspring.

anastomosing interconnecting.

ancestrula in ectoproct bryozoans, the original, single zooid that emerges from a statoblast, differing from subsequent zooids in the colony by its smaller number of tentacles and inability to form statoblasts.

anchoral processes posteriorly directed apodemes of the gnathosoma (capitulum) in deutonymphs and adult mites.

angulate having an angle.

anhydrobiosis ability of bdelloid rotifers to undergo desiccation and then be revived at some later time; a cryptobiotic (latent) state in tardigrades induced by loss of water through evaporation.

annual one generation per year; life-cycle length of one year.

annulate ringed; surrounded by a ring of a different color; formed into ring-like segments.

annulations transverse striations (trenches) in the cuticle of nematodes.

annulus ring; in ectoproct bryozoans, the ring of enlarged chambers, usually gas-filled, on the periphery of float-oblast valves.

annulus ventralis seminal (spermatophore) receptacle of female crayfishes.

anoxia lack of dissolved oxygen.

anoxybiosis cryptobiotic (latent) state in tardigrades induced by low oxygen tensions in the environmental water.

Note: Glossary terms were defined by individual authors in reference to their chapter. Consequently, terms may actually apply to other taxa and have broader definitions than indicated here.

antenna whiplike, paired, generally elongate sensory appendages arising from anterior area of cephalothorax.

antennal gland one of pair of complex excretory glands in many decapods with duct opening on antenna; green gland.

antennule paired appendage of first cephalic somite; sensory, often bearing aesthetascs.

aperture opening of the shell of a snail from which foot and body protrude.

apex that part of any structure opposite the base by which it is attached.

apical pertaining to the apex.

apodeme internal extension of thickened cuticle or exoskeleton in some arthropods, often functioning as muscle attachment site.

apomixis form of parthenogenesis in which meiosis is suppressed and the offspring are genetically identical to their mother.

apophyses see pharyngeal or buccal apophyses.

aposymbiotic describing an organism that would normally have symbionts but the symbionts are not present.

apotypical (apomorphic) derived, or modified, state of a biologic attribute or morphologic character.

apterous without wings.

areoles a segment of hairworm cuticle separated from adjacent segments (areoles) by longitudinal and transverse furrows; may bear granules, bristles, and contain pores.

articulation a joint.

ascus a glandular pocket in some turbellarians (Typhloplanidae).

atrium in flatworms: the space into which both the male and female systems open.

attenuate tapering rapidly or abruptly to a point.

aufwuchs a German term whose broader North American usage refers to all small, attached (except macrophytes) and free-moving organisms forming a living film on aquatic substrata such as rocks, snags, and plants; included are some microinvertebrates, fungi, algae, and bacteria.

autochthonous substances originating within the immediate habitat, for example, carbon produced by macrophytes in the surrounding river bottom; opposite from allochthonous.

autogamy in ciliates; a self-fertilization phenomenon (not true sexual reproduction) involving a single individual; haploid gametic nuclei are formed and two of them fuse with each other within a single individual; see cytogamy.

autorecruitive the type of growth of colonial rotifers whereby the young produced by a colony tend to remain in that colony so that as a consequence the genetic diversity of the colony is low. Originally defined as Type II colony formation (see also allorecruitive).

autotoky type of reproduction where progeny are produced by a single parent (i.e., hermaphroditism or parthenogenesis).

autotrophy synthesis of organic compounds from carbon dioxide using light as the energy source (in photosynthesis) or inorganic chemical compounds (in chemoautotrophy); contrasts with heterotrophy.

axe-head glochidium unionacean (Mollusca) glochidium larva characterized by a distinctly quadrate shell form.

axenic culture a bacteria-free culture containing only one species of organism.

axoneme a longitudinal bundle of microtubular fibers within flagella, cilia, and axopodia.

axopodium the characteristic feeding pseudopodium of Actinopoda, distinguished by needlelike shape, axonemes, and bidirectional streaming of cytoplasm.

baffles dorsal and/or transverse ridges posterior to the mucrones on the interior of the buccal cavity in some eutardigrades.

basal at or pertaining to the base or point of attachment to, or nearest the main body.

basal bulb the swollen, set-off, basal portion of the pharynx (esophagus) of a nematode.

basipod (basipodite) the second segment of an arthropod appendage.

basis the second segment from proximal end of segmented appendage.

beaks the raised portion of the dorsal margin of the bivalve shell; the oldest portion of the shell.

benthos animals living on or near the bottom of an aquatic habitat.

bident having two points.

bifid cleft, or divided into two parts; forked.

bilamellate divided into two lamellae or plates.

bivoltine having two broods or generations per year.

boss an expanded protuberance, as on caudomesial surface of coxa of fourth pereiopod of male crayfishes.

brachypterous with short or abbreviated wings.

bristles variously shaped cuticular projections arising from the cuticle in both nematodes and hairworms.

bromeliads the common name of plants belonging to the genus *Bromelia,* possessing water-filled axial cups in which live ostracodes and other invertebrates.

brosse a distinctive ''brush'' of cilia arising from specialized short kineties which is present on the anterodorsal surface of nondividing individuals of certain gymnostome ciliates.

buccal apparatus the anterior part of the tardigrade foregut, consisting of buccal tube, muscular pharynx, and a pair of piercing stylets.

buccal apophyses cuticular thickenings at the junction of the buccal tube for insertion of the stylet protractor muscles.

buccal ciliature compound ciliary organelles the bases of which are associated with the oral area of some ciliates.

buccal cirri in heterotardigrades, paired sensory appendages, usually short and hair-like, on either side of each cephalic papilla on the cephalic plate.

buccal tube support the median ventral lamina, also called a ''reinforcement rod,'' from the buccal ring to the midregion of the buccal tube in some eutardigrades.

bulbous see pharynx.

bursa folds of cuticle on the male nematode tail that assist in attaching the male tail to the female during mating.

bursa (copulatrix) copulatory bursa; the part of the female system into which donor spermatozoa are deposited.

byssus proteinaceous threads produced from a byssal gland at the base of the foot, used to attach juvenile or adult bivalves to hard surfaces.

calcium phosphate concretions small extracellular spherules found in the gills of unionacean mussels made up of calcium phosphate deposited in consecutive lamellae on a proteinaceous matrix.

calotte the head region of a Nematomorpha, usually lighter in color than the adjoining body.

capitulum the sclerotized base of the gnathosoma of a mite.

carapace the expanded, hard covering of a major region of the body (usually the head and/or thorax of some arthropods).

cardate a type of trophi (jaw structure) found in rotifers that functions by creating a sucking action.

cardinal teeth massive conical projections formed at the center of the shell-hinge plate that interdigitate to form the fulcrum on which the valves open and close.

carina sharp spiral edge or keel on the shell of a snail, usually along the center of the whorl.

carpus the fifth segment from proximal end of segmented appendage.

cauda a posterior extension of the idiosoma in male adults of certain mite taxa, functioning as an adaptation for copulation during spermatophore transfer.

caudal toward the posterior or rear end of an organism.

caudal glands glands found in the tail region (usually three behind the anus) of nematodes that secrete a mucous deposit.

caudal rami a bifurcated terminal extension of the abdomen of some arthropods.

cellulase an enzyme that degrades the cellulose of plant cell walls.

cement glands glands that secrete material to assist fixation of the egg capsules to a substrata, often near female gonopore in turbellarians.

centroplast central structure in some heliozoa from which axonemes radiate; synonym, central granule, axoplast; a type of microtubule organizing center.

cephalic pertaining to, or toward, the head.

cephalic papillae large papillae in the head region of nematodes; also, paired sensory appendages on the cephalic plate of heterotardigrades, located between the internal and external cirri.

cercus (cerci) a paired appendage of the last abdominal segment. In Anisoptera, one of two pairs of lateral, terminal, triangular sclerites.

cerebropleural ganglion a mass of nerve cell bodies in the region of the mouth.

chaetotaxy the number and arrangement of setae on the body and appendages.

chela a claw or pincer; in crayfish, two opposed distal podomeres (dactyl and propodus) of certain pereiopods.

chelate in mites, a condition of the pedipalp in deutonymphs and adults of certain taxa in which the tibia bears a dorsodistal seta, often associated with a projecting tubercle, that opposes the dorsal surface of the tarsus to form a grasping organ. In other invertebrates, the presence of pincer-like, exoskeletal or cuticular structures at the end of an appendage.

chelicera (chelicerae) in mites, a two-segmented, paired feeding appendage located dorsomedially on the gnathosoma in all instars; usually two-segmented, but one-segmented in deutonymphs and adults of Hydrachnidae; the distal segment, or "claw," functions in piercing and tearing the integument of hosts or prey.

chitin a long chain, polymeric glucosamine containing 6% nitrogen; used in the exoskeletons of many arthropods and often complexed with protein or infiltrated with calcium for strength.

chloride epithelia oval patches (as in insects) that differ from normal epithelia and are used for osmoregulation.

chrysolaminarin a liquid carbohydrate storage form, β-1:3-linked glucan; characteristic of flagellates in the order Chrysomonadida; usually in one or more large refractile vesicles at the posterior of the cell, optically bluish under strong light; synonym is leucosin.

cingulum (see also trochus); one of two ciliated rings present in the typical rotiferan corona.

circadian rhythm diurnal or daily rhythms of activity in animals.

cirrus A in heterotardigrades, paired filamentous sensory appendage at the anterior lateral margin of the scapular plate, situated slightly dorsal to the clava.

cirrus eversible, spiny male duct; may contain ejaculatory, prostatic, or accessory ducts and be enclosed in a bulb or special propulsory sac.

cladogram a classification, usually in the form of a branching diagram, of a group of organisms based on cladistic principles (i.e., tracing the evolution of shared, derived characters from ancestral to derived forms).

clasping apparatus in ostracodes, the heavily sclerotized, apparently movable sclerite articulating with the midportion of the periferum in a socket near the ventral cardo and in the vicinity of the loop of the spermatic tube of entocytherid hemipenes; includes the vertical and horizontal rami.

clava in heterotardigrades, a paired sensory appendage, usually short and broad, at the anterior lateral margin of the scapular plate, situated slightly ventral to lateral cirrus A.

claw true claws are paired, curved raptorial structures located terminally on the tarsi of the legs in all mite instars.

cocoon a structure formed by some species to enclose many egg capsules or embryos.

cohort a group of equal-aged individuals whose survivorship and fecundity are to be followed throughout the lifespan of the entire group.

columellar the strong central support around which a snail shell coils.

concentric in Mollusca, describing the growth lines of a snail operculum that lie entirely within each other (not forming a spiral).

conglutinate a mass of unionacean (Mollusca) glochidia larval bound together by mucous; often resembling the prey items of unionacean glochidial fish hosts (increasing chances of host contact).

conic shaped like a cone.

conjugation in ciliates, a reciprocal fertilization type of sexual phenomenon; it occurs only between members of differing mating types and involves temporary or total fusion of the pair.

contractile vacuole liquid-filled organelle which functions as an osmoregulator in the cytoplasm of many protozoan species.

copepodite the second larval stage of copepods (after the nauplius).

copulatory bulb the swollen muscular and connective tissue region containing the male copulatory organ and often also the seminal vesicle and/or prostate.

copulatory organ the terminal part of the male reproductive system which delivers spermatozoa into the body of a mate, i.e. an intromittent organ. When not armed with hard structures, yet capable of protrusion, it is simply called a penis (or penis papilla). The copulatory organ may be equipped with hard (sclerotized) structures such as a stylet or cirrus spines.

corneous slightly hardened but still pliable (cornified or "horny") proteinaceous material, as in the operculum of a snail or the modified terminal elements of first pleopods of first-form male crayfishes.

corona the apical region of rotifers; usually a ciliated band around the anterior end, characteristically composed of two ciliated rings called the trochus and cingulum; in some forms, the ciliation is absent and may be replaced by long setae which surround the rim of a funnel-shaped structure called the infundibulum.

costa a rib or ridge on a snail shell that lies transverse to (at right angles to) the coiling of the whorl.

coxa the proximal (closest to body) segment of a segmented appendage.

coxal plate in mites, a sclerite representing the modified basal leg segment, attached to the venter of the idiosoma in all instars. Coxal plates may be expanded and fused with one another to cover large areas of idiosomal integument.

coxoglandularia a series of paired glandularia associated with the coxal plates of mites.

coxopod (coxopodite) the proximal segment of an arthropod appendage.

creeping welt a slightly raised, often darkened structure on Diptera larvae.

crenulate evenly rounded and often rather deeply curved; having an irregularly wavy or serrate outline; possessing a scalloped margin or rounded tooth.

crescent (postanal crescent) a crescent-shaped fold between the cloaca and terminus in male hairworms; usually ornamented with setae or bristles.

cross fibers fine criss-cross fibers in the upper cuticular layers in mermithid nematodes.

cryobiosis a cryptobiotic (latent) state in tardigrades induced by low temperatures.

cryptobiosis a latent state in which metabolism comes to a reversible halt; also known as "anabiosis" or "abiosis."

crystalline style a mucopolysaccharide rod containing digestive enzymes projecting dorsally from the style sac in the floor of the bivalve stomach where it is secreted; the crystalline style is rotated by cilia lining the style sac against the gastric shield on the dorsal side of the stomach.

ctenidium a molluscan gill, in bivalves consisting of two demibranchs formed from the reflection of right and left gill filaments upon themselves.

cupule a cup-shaped segment at the base of the club on some antennae.

cuticular bar an external cuticular thickening on the leg, near or between the claws, in some eutardigrades.

cyclomorphosis a seasonal phenotypic change in body size, spine length, etc., found in successive generations of zooplankton within a single habitat.

cyrtos a basket-like structure supporting the cytopharynx of Ciliophora in the protistan subphylum Cyrtophora.

cytogamy occurrence of autogamy in each member of a pair of temporarily fused ciliates.

cytolosomes digestive organelles of protozoa, analogous with food vacuoles, but containing the organelles of the cell, which are being broken down under conditions of starvation; synonymous with autophagic vacuole or autophagic vesicle.

cytopharynx the cell pharynx, leading from the cytostome to the site of food vacuole formation in protozoa.

cytoproct the cell anus; a permanent and specialized site on the surface of some protozoa for ejection of the contents of food vacuoles.

cytostome the cell mouth; a permanent site for phagocytosis in some protozoa.

dactyl the most distal segment of the endopod of segmented appendages; smaller, mesially situated, movable finger of chela.

demibranch a portion of the bivalve ctenidium (gill) formed from a set of filaments reflecting upon themselves; in each ctenidium, the inner lateral filaments form the inner demibranch lying next to the visceral mass and the outer lateral filaments form the outer demibranch lying next to the mantle.

denticles small, delicate, spinelike projections, usually of shell material in numerous taxa; also the small teeth-like processes lining the stoma of some nematodes; small teeth.

depressed spire a spire that is not raised above the body whorl of a snail shell.

determinate growth describing an organism that ceases to grow in size after reaching a certain size or stage; contrast with indeterminate growth.

detritus decaying organic material.

deutocerebrum the portion of a crustacean's brain that controls the antennules and also innervates the antennae and a portion of the alimentary tract.

deutonymph the second nymphal instar in the generalized mite life cycle, and the only active, free-living nymphal instar in the water mite life cycle.

dextral snail shells with the aperture on right (when viewed with the apex pointed away from the observer and with the aperture fully visible).

diapause a period of arrested growth and development; often used to survive harsh environmental conditions.

digestive diverticulum in bivalve molluscs, a digestive organ surrounding the stomach, area in which the final intracellular phase of digestion takes place.

digitate many finger-like extensions of the mantle of a mollusc.

dioecious an individual possessing either a male or a female reproductive system; the reproductive systems are contained in separate individuals.

discobolocyst extrusome of some flagellated protozoa, especially Chrysomonadida.

distal toward the end of a structure, the opposite of proximal (near the base of a structure).

diurnal recurring every day; relating to the daytime.

diverticulum a sac-like structure branching from an internal organ such as the intestine.

doliiformis see pharynx.

dormancy a state of minimal metabolic activity when growth ceases, primarily enabling organisms to survive periods of adverse conditions.

dorsal furrow a strip of soft integument separating the dorsal and ventral shields in highly sclerotized adult mites.

dorsalia small platelets on the dorsum of the idiosoma in deutonymphs and adult mites that function as muscle attachment sites.

dorsal plate a large sclerite covering the prodorsal region of the idiosoma in any mite instar, often expanded to cover part of the opisthosomal dorsum as well.

dorsal shield a single large sclerite or a series of closely fitting platelets covering the entire dorsum of the idiosoma in deutonymphs and adults of certain mite taxa.

dorsocentralia in mites, a series of paired, medially located dorsalia.

dorsoglandularia in mites, a series of paired glandularia located on the dorsum of the idiosoma.

dorsolateralia in mites, a series of paired, laterally located dorsalia.

dorsum in ostracodes, a flattened area adjacent to the hinge line and set off from the lateral surface of the ostracode carapace.

ductus communis in turbellarians, the terminal (or common) part of the female canal distal to the joining of vitelline duct(s) and oviduct(s) in ectolecithal systems.

duplicature a narrow band of shell material around the outer edge of the proximal face of the epidermis, composed of the same three layers as the outer lamella; only calcified layer of the proximal covering, lining, or epidermis separated into the list strip, selvage strip, and flange strip of the ostracode shell.

ecdysis shedding or molting of an older, smaller exoskeleton in arthropods.

ecdysone the principal molt-stimulating hormone (in several forms) of arthropods.

ecophenotypic plasticity variation in a species induced by nongenetic environmental factors.

ectocyst in ectoproct bryozoans, the nonliving outer layer of the body wall; may be sclerotized or gelatinous.

ectolecithal pertains to those systems in which the yolk is not incorporated into the oocyte, i.e., part of the female gonad is a yolk gland (such as occurs in higher Turbellaria).

efferent branchial vessel a blood vessel in the bivalve gill axis carrying oxygenated hemolymph from the gill filaments to the longitudinal vessel of the kidney and eventually to the heart.

ejaculatory complex a series of membranous chambers and associated sclerotized framework located distally in the reproductive tract of male adult mites, functioning as a syringe-like organ for compacting masses of spermatozoa, assembling spermatophores, and expelling them from the genital tract through the gonopore.

ejaculatory duct (ductus ejaculatorius) in turbellarians, the terminal part of the male duct, sometimes eversible, which carries spermatozoa and usually prostatic secretions through the copulatory organ.

ejectisome an extrusome of some flagellated protozoa, especially in Cryptophyta.

electromorph an organism that possesses a set of enzymes that have a specific migration pattern as determined by electrophoretic techniques.

elytra hardened, shell-like mesothoracic wings of Coleoptera.

emarginate notched; with an obtuse, rounded, or quadrate section cut from the margin.

empodium medial unpaired, claw-like structure located terminally between the true claws on the tarsi of the legs in larvae of all mite taxa, and in deutonymphs and adults of Stygothrombidiidae.

encystment a latent state in aquatic tardigrades in which resistant cysts are produced under deteriorating environmental conditions.

endemic restricted to a particular geographic region.

endites projections or processes from the inner margin of the inner branch of a branched appendage.

endocyst in ectoproct bryozoans, the inner living tissues of the body wall, including epidermis, muscle layers, basement membrane, and peritoneum.

endocytosis the process by which minute food particles are engulfed by cells into food vacuoles for the final stages of digestion, such as occurs in the terminal tubules of the molluscan digestive diverticulum.

endogenous initiated from within an individual, such as nervous or hormonal cues controlling diurnal rhythmicity in the absence of external cues.

endolecithal (entolecithal) pertaining to those systems in which the yolk is incorporated into the oocyte (such as occurs in more primitive Turbellaria).

endopod (endopodite) the mesial, or inner branch of an appendage; the mesial ramus of biramous appendage, originating on second segment from base.

endosymbiotic describing an organism which normally, and currently, has internal symbionts.

epibranchial cavity in bivalve molluscs, the exhalant mantle cavity formed dorsal to the ctenidium, which carries water leaving the gill to the exhalant siphon; it also receives the openings of the gonad, kidney and anus.

epigean referring to surface (above ground) habitats as opposed to hypogean.

epilimnion portion of a lake above the thermocline and located in the limnetic zone.

epineuston organisms living on or slightly immersed in the air/water interface; see neuston.

estivation a general term for a physiologic process whereby animals reduce metabolic and activity levels in order to survive harsh conditions such as pond drying for freshwater bivalves.

eulaterofrontal cilia in bivalve molluscs, stiffened cilia extending laterally from the inhalant side of a gill filament to form a mesh with eulaterofrontal cilia of adjacent gill filaments on which particles are filtered.

eupathid a specialized, thickened seta of mites; eupathids are located on the distal segments of the legs in all active instars.

euryoecic pertaining to species tolerant of a wide range of habitats or environmental conditions.

exchange diffusion the transport of ions across epithelia such that diffusion of one ion species down its concentration gradient is coupled with the transport of a second ion species in the opposite direction.

excrescences teeth on the internal margins, denticles on the tips, and talons on the external margins of the horizontal ramus of the entocytherid ostracode hemipenes.

excretophore large vacuolated cells of the gut epithelium of turbellarians.

excretory pore plate in mites, a sclerite bearing the excretory pore in larvae, usually expanded to incorporate the bases of the setae associated with the pore.

exogenous initiated from outside an individual, such as a change in light intensity controlling the valve-gaping activity of a bivalve.

exopod (exopodite) the outer branch of a crustacean appendage; lateral ramus of biramous appendage, originating on the second segment from the base.

exploitative competition a form of competition in which the outcome is determined by differential abilities to harvest a resource and/or by differences in reproductive rates; competition in which one organism does not directly inhibit another's access to a limiting resource by behavioral means; compare with interference competition.

extrapallial fluid in bivalve molluscs, the fluid filling the space (extrapallial space) between the mantle and the shell.

extrapallial space the space between the mantle and shell in molluscs.

extrusome an organelle of the pellicle of protozoa, which extrudes a substance or structure, usually in response to a stimulus.

eye plate in mites, a sclerite bearing an eye or eyes.

facultative an optional process.

fascicles small tooth-like bundle of fibers.

fenestra in ectoproct bryozoans, the clear central portion of a floatoblast valve.

file row of cilia.

filiform thread-like; slender and of equal diameter.

filopodium long, filamentous pseudopodium of certain Rhizopoda (amoeba), sometimes branching out without anastomoses; may function in both food capture and locomotion.

finger guard an extension of the ventral cardo along the side of the dorsal and ventral fingers of the entocytherid ostracode hemipenes.

first-form male one of two morphological forms of cambarine crayfishes; sexually functional male; at least one terminal element of first pleopod usually corneous.

flagellum a multiarticulate, endopodite of antennae; whiplike locomotory structure of many protozoans.

flange the distal ridge of the contact margin of the duplicature between the list and groove, it may be a part of the outer lamella of the ostracode shell.

flange groove that part of the duplicature surface of the ostracode valve between the selvage and flange.

flange strip the part of the duplicature that forms the flange groove and the flange of the ostracode valve.

floatoblast in ectoproct bryozoans, a type of statoblast having a ring of gas-filled chambers; in some cases, the ring becomes buoyant only after being dried.

follicular an organ arranged in several small compartments (follicles), i.e., neither diffuse nor compact.

foot in Mollusca, a highly muscular spade-like structure which in bivalves can be projected a considerable distance anterioventrally from the valves through the pedal gape and is utilized both in burrowing and surface locomotion.

forcipate a type of trophi (jaw structure) found in rotifers having an action like forceps in which the trophi are projected from the mouth to grasp prey.

fossorial formed for or with the habit of digging or burrowing.

fragmentation spherules in Mollusca, the shed tips of digestive cells lining the terminal tubules of the digestive diverticulum which are released after intracellular digestion has been completed; they contain undigested material remaining in food vacuoles and are carried on ciliary rejection tracts in the digestive diverticulum back into the stomach where their breakdown releases the undigested contents of their food vacuoles to be egested.

frontal glands glands at anterior terminal region of turbellarians (frontal organ in Acoela).

fulcrate a type of aberrant trophi (jaw structure) found in rotifers in the order Seisonidae.

fulcrum one of the seven pieces that comprise the trophi of rotifers; together with the paired rami form the incus subunit.

funiculus in ectoproct bryozoans, a tubular strand of tissue joining the blind end of the gut to the inner colony wall; the site of statoblast and sperm formation.

furca the forked posterior end of some gastrotrichs. Also, an appendage-like structure that is the termination of the body proper in cypridid and cytherid ostracodes; it is articulated to the body and may assist in locomotion but is not considered a true thoracic leg.

fusiform spindle-shaped; tapering at each end.

gastric mill the posterior end of the cardiac stomach of a crustacean (and some other arthropods) where food is ground for later digestion.

gastric shield a chitinous plate on the dorsal side of the stomach of bivalves (Mollusca) against which the distal tip of the crystalline style rotates.

gena (genae) the cheek; the part of the head on each side below the eyes, extending ventrally to the gular suture.

genet a genetically distinct individual; often used to describe individuals that commonly grow as colonies in which the individual is not easily distinguished as a separate genetic entity; contrast with ramet.

geniculate jointed to permit bending at an abrupt angle; distinctly elbowed.

genital bay the area between the posterior coxal groups that accommodates the genital field in adult mites.

genital field the area of the idiosoma occupied by the gonopore and the genital acetabula in adult mites.

genital flaps paired movable flaps which close to cover the gonopore in adults of certain mite taxa; they typically also cover the genital acetabula, but in some derivative groups, the acetabula lie on the flaps.

genital papillae papillae situated on the surface of the cuticle or on elongate finger-like projections on the tail of male nematodes.

germarium an ovary.

germovitellarium an organ containing both oogonia and oocytes, i.e., ovary and vitelline cells.

gibbosity a hump, marked by convexity or swelling, protuberant structure normally found on the dorsal part of an ostracode carapace.

gill axis in Mollusca, the portion of the ctenidium from which the gill filaments extend and which is attached to the dorsal mantle wall.

gill filaments in bivalve molluscs, long, thin, fused lateral extensions from the gill axis which extend ventrally and then reflect dorsally to form the inner and outer demibranchs.

glandularium (glandularia) a specialized organ located in the integument of deutonymphs and adult mites, consisting of a small, goblet-shaped sac, or "gland," with a tiny external opening, and an associated seta. Glandularia have both a sensory and a secretory function.

globose globe-shaped.

glochidium the bivalved larval stage of unionacean bivalves (Mollusca), generally parasitic on fish.

glossa the median terminal (or subterminal) lobe of the labium.

glycocalyx the finely "hairy" secreted layer covering the outer cell membrane of many of the "naked" rhizopod amoebae.

gnathosoma in mites, the anterior region of the body comprising the mouth and associated appendages, the chelicerae and pedipalps.

gonochoristic two sexes; males and females in the same population reproducing by cross-fertilization; contrast with hermaphroditism.

gonopore the external opening of the reproductive system in adults.

green glands paired excretory glands in crustaceans; also known as antennal glands

gubernaculum the sclerotized structure that supports and guides the spicules out of the cloacal opening during mating.

gula the throat sclerite, forming the central part of the head beneath the genae.

gullet in flagellated protozoa, an anterior invagination from which the flagella emerge.

gyttja a watery, organic sediment that consists mainly of excretory material from benthic animals.

haptocyst an extrusome of some predatory Ciliophora (Sectoria) used to capture prey.

heliotropism the tendency of certain plants or other or-ganisms to turn or bend under the influence of light, especially sunlight.

helocrene a seepage area where spring water percolates through a substratum of mosses, leaf litter, and detritus.

hemelytra the mesothoracic wings of Heteroptera insects.

hemipenes paired penes.

hemocoel the internal body cavity formed by the expansion of the vascular system.

hemocyanin a copper-based respiratory pigment which is dissolved within the hemolymph of crustaceans.

hemolymph the circulatory fluid or blood of animals with open circulatory systems.

hemolymph sinuses or hemocoels open cavities in which the blood or hemolymph bathes the tissues directly, allowing exchange of gases, ions, nutrients, and wastes.

hermaphroditic an individual that contains both male and female reproductive organs.

heterotrophy consuming, rather than synthesizing, organic compounds for metabolic activities; contrast with autotrophy.

hibernaculum in gymnolaemate bryozoans, a thick-walled resting bud occurring in certain freshwater species.

hindgut portion of the intestine lying between the midgut and rectum, primarily involved with consolidation of undigested particles into feces in aquatic invertebrates.

hinge in bivalve molluscs, the section of the dorsal shell valve containing the teech and forming the fulcrum on which the valves open and close.

hinge ligament in bivalve molluscs, an elastic proteinaceous portion of the shell secreted by the mantle isthmus which connects the mineralized shell valves; it is flexed or compressed when the valves are closed and when adductor muscles relax, it functions to open the valves by returning to an unflexed state.

hinge teeth in bivalve molluscs, interdigitating mineralized projections of shell on the dorsal margins of the valves that hold the valves in position and serve as the fulcrum on which they open and close.

histophages protozoa that feed on the tissue of dead metazoa, as opposed to a parasite feeding on live hosts.

holophyletic group monophyletic clade including all species descended from a common ancestor.

homoplasy parallel adaptive modification of homologous attributes or structures in species of different clades due to similar selective pressures.

hooked glochidia in bivalve molluscs, glochidia larvae with spiny hooks on their ventral shell margins.

hyaline margin clear ectoplasmic material surrounding and/or preceding the granular body mass of some amoebae during locomotion.

hypersaline term generally referring to a water body whose salinity is higher than seawater.

hypodermal cords expansions of the hypodermis between the muscle fields of nematodes, best seen by cross-sectioning the nematode in the midsection and examining a thin body section.

hypogean beneath the surface of the earth; subterranean; burrow, cave, or groundwater environment.

hypolimnion portion of a lake below the thermocline and located in the limnetic zone.

hyponeuston organisms living just below the air/water interface of a water body; see neuston.

hyporheic stream bed where groundwater flows below or to the side of the flowing surface water.

hyporheic zone groundwater present below and to the side of a stream and located close enough to be affected by the movement of stream surface waters (unlike the phreatic zone).

hypoxia condition of environmental oxygen concentration below air oxygen saturation levels but not necessarily anoxic.

hysterosoma in mites, a region of idiosoma posterior to level of primitive sejugal furrow.

idiosoma the body proper of mites, comprising the fused cephalothorax and abdomen.

imagochrysalis the quiescent transitional instar between the deutonymph and adult in the water mite life cycle, representing the suppressed tritonymph.

incudate a type of trophi (jaw structure) found in rotifers, which functions by grasping prey with a forceps-like action.

incus the subunit of a rotifer trophi composed of the fulcrum and paired rami.

infauna sediment-dwelling animals; also, animals that burrow into and live inside another organism.

infraciliature in ciliates, the total assemblage of all kinetosomes and the associated subpellicular microfibrular and microtubular structures

infundibulum the corona of rotifers of the order Collothecacea.

inner lamella a thin layer of chitin, which, together with the duplicature, forms a proximal covering of the epidermis in ostracodes; at inner limit, it is folded back on itself to form the body wall of the pouch-shaped body of the animal.

instar any developmental stage between molts.

interareolar furrows a network of transverse and longitudinal trenches which separate the areoles in hairworms; may support bristles, warts, or bear pores.

interference competition a form of competition in which one organism directly or indirectly interferes with another organism (e.g., by aggresion), thus preventing it from obtaining a limiting resource; forms of interference include physical combat, the threat of combat, noxious chemical compounds, etc.

interlamellar space in Mollusca, a space formed between ascending and descending gill filaments; water flowing through the ostia of the gill filaments enters the interlamellar space to be carried dorsally to the epibranchial cavity; also called the water tube or water channel.

interstitial a microhabitat comprising the spaces between particles of submerged sand and gravel in groundwater and hyporheic habitats.

intracytoplasmic lamina a filament layer of varying thickness present within the syncytial integument of rotifers and acanthocephalans.

ischiopodite the third segment from the base of a segmented appendage.

isthmus that portion of the nematode pharynx between the metacorpus and basal bulb, usually surrounded by the nerve ring. In bivalve molluscs, the constricted dorsal section of the mantle separating the mantle into two lateral shell-secreting halves; the isthmus secretes the proteinaceous hinge ligament connecting the valves.

iteroparous reproducing repeatedly at multiple times during the life of an organism (cf., semelparous).

karst a terrain underlain by fractured and extensively dissolved carbonate rock (e.g., limestone) and typified by numerous sinkholes, springs, caves, and few surface streams.

kinetid the kinetosome and its associated pellicular structures in Ciliophora, including the cilium, the elementary repeating structure of the ciliate cortex; (polykinetid is an organellar complex composed of two or more kinetids).

kinetocyst the extrusome of Actinopoda, which secretes materials that probably serve to make the surface of an axopod sticky.

kinetoplast a DNA-rich organelle near the anterior of some flagellated protozoa (Kinetoplastida), associated with their unique large mitchondrion.

kinetosome a subpellicular structure that is the base of a cilium, but may be without a cilium; it is characterized by nine triplets of microtubules at right angles to the plasma membrane arranged in a circle.

kinety a single structurally and functionally integrated (somatic) row of kinetosomes plus their cilia (cilia not necessarily associated with each inetosome), i.e., a line of kinetids.

knob high rounded major protuberance with the sides joining the rest of the valve at a distinct line and at a steep angle.

K-selection a selection for life-history characteristics that increase fitness in stable environments where populations reach densities near environmental carrying capacity; associated with high levels of intraspecific competition.

labial palps a pair of thin flattened structures extending from either side of the mouth region of bivalve molluscs to lie against the outer surface of the outer demibranch and the inner surface of the inner demibranch; they function to receive filtered particles from the ctenidia and sort them into those accepted for ingestion and those rejected as pseudofeces.

lacustrine see lentic.

lamellae flattened, leaf-like structures in several phyla, such as the gills of bivalves and the buccal plate-like projections on the mouth (buccal) ring of some eutardigrades.

Lansing Effect putative negative effect of elevated calcium on survivorship in rotifers in which calcium accumulation in older mothers is passed on to their offspring (see also orthoclone).

lateral coxal apodeme conspicuous apodeme indicating the line of fusion between the second and third coxal plates in larvae of certain mite taxa.

lateral field track containing usually several longitudinal striations running on both sides of the nematode.

lateral teeth in bivalve molluscs, elongated lamellar hinge teeth forming anterior and posterior to the cardinal teeth or posterior to the pseudocardinal teeth in unionacean bivalves.

LC$_{50}$ the concentration of a toxic agent at which 50% of a cohort will perish within a certain predetermined period; also called the median lethal concentration and LD$_{50}$.

lentic standing water environments (e.g., lakes, ponds, and permanent and temporary pools).

limnetic open water, deeper areas of a lake or pond and away from the shoreline littoral zone.

line of concrescence proximal line of junction of the duplicature and the outer lamella of the ostracode valve, the inner border of the chitin adhesive strip.

lira rib or ridge running along the whorl of a snail shell in the direction of the coiling.

list proximal ridge on the contact margin of the duplicature of the ostracode valve, not always present.

list strip that part of the duplicature from the inner margin to and including the list of the ostracode valve.

littoral the zone or body of water shallow enough for growth of rooted aquatic vegetation.

lobopodium a blunt pseudopodium used in feeding and/or locomotion.

lophophore an organized structure of ciliated tentacles used for capturing suspended food particles from the water, and probably also providing an important surface for gas exchange; occurring in ectoprocts and certain other phyla.

lorica the thickened body wall of rotifers, organized as a syncytial integument possessing two keratin-like proteins which make up a layer called the intracytoplasmic lamina. A loose-fitting surface covering secreted and/or assembled by protozoa; generally synonomous with test.

lotic running water environments (e.g., rivers and springs).

lunule crescent or half-moon shaped structure at the base of each double claw in some eutardigrades.

lyrifissure (lyriform fissure) in mites slender sac-like structures opening through slits in the integument, apparently functioning as proprioceptors, in all active instars. Lyrifissures are arranged in paired series on both the idiosoma and appendages.

lysosome membrane-bound organelle containing hydrolytic enzymes; in protozoa, they are involved in intracellular digestion.

macroinvertebrate organisms larger than meiobenthos and which are retained on coarse sieves with a mesh size ≥ 2 mm.

macronucleus in ciliates, the nucleus active in transcription and thus responsible for the phenotype and regulation of metabolism; see micronucleus.

macrophagous in protozoa, a species that phagocytizes relatively large items singly.

macrophyte a macroscopic photosynthetic organism growing submersed, floating, or emergent in water; most are angiosperms but also includes some nonvascular plants and macroalgae.

macropterous with fully formed wings.

malleate a type of trophi (jaw structure) found in certain monogonont rotifers in which the rami are massive and may possess teeth along the inner margin.

malleoramate a type of trophi (jaw structure) found in monogonont rotifers of the order Flosculariacea, resembling the malleate form, except in the number of teeth on the unci.

malleus a subunit of rotifer trophi composed of an uncus and manubrium.

mandible hardened mouthparts (cephalic appendages) used to grasp, tear, and push food into the mouth; most anterior gnathal appendage.

mandibular palp endopodite segment of the mandible.

mantle thin extension of the dorsal body wall of molluscs underlying the shell and responsible for its secretion.

mantle cavity the space formed between the body and surrounding mantle tissues in molluscs; completely surrounds the body in bivalves.

manubrium one of the seven pieces that comprise the trophi of rotifers; a manubrium together with an uncus form the malleus subunit.

marsupium the space formed for incubation of embryos and larval stages.

mastax a muscular pharynx of rotifers that contains a set of jaws called trophi.

maxillae the first or second pair of mouthparts (cephalic appendages) posterior to the mandibles.

maxillary glands paired excretory glands of crustaceans.

maxilliped one of pair of three sets of gnathal appendages situated immediately posterior to second pair of maxillae.

maxillule the first maxilla.

medial coxal apodemes a series of short apodemes at the medial edges of the coxal plates in larvae of certain mite taxa.

medial eye plate (frontal plate) a sclerite bearing the medial eye in early derivative mite taxa; the medial eye may be absent in highly derived groups.

meiobenthos small benthic organisms that pass through a 2 mm mesh but are retained on a 40–200 μm mesh.

merus the fourth segment from proximal end of segmented appendage.

mesial (mesal) pertaining to the middle; toward the middle.

metachronal occurring one after another, in succession.

metacorpus the swollen median portion of the nematode pharynx, located between the procarpus and isthmus.

metamere homologous body segment, a somite.

metasome a portion of the body of copepods anterior to the major body articulation.

microaerophilic a protozoan with an aerobic metabolism that lives in microhabitats with very low oxygen concentrations.

microfauna organisms smaller than meiobenthos, that is, not retained on a 40 μm mesh sieve.

micronucleus in ciliates, the nuclei mediating sexual recombination and reproduction, typically much smaller than the macronucleus; diploid; often multiple.

microphagous in protozoa, species that phagocytize relatively small items, collecting many into a single food vacuole.

microplankton plankton that are 20–200 μm in size.

microsome a simple membrane-bound organelle containing oxidative enzymes; synonymous with microbody.

microvilli microscopic projections from a cell surface that increase the surface area of the cell.

mictic (mixis) a phase in the life cycle of monogonont rotifers in which reproduction is sexual, resulting in a diapausing embryo called a resting egg (see also amictic).

midden a refuse heap; native North American Indians left shells of unionacean clams which they utilized for food in middens now used as historic records of North American unionacean diversity.

middle piece a structure in sperm cells lying between the head and flagellum containing mitochondria.

midgut that portion of the intestine between the stomach and hindgut; in molluscs, it is characterized by the presence of the typhlosole, the ciliated surfaces of which sort particles into those returned to the stomach and those passed to the hindgut for consolidation into feces and eventual egestion.

mixotroph an organism deriving energy from both autotrophy and heterotrophy, including those whose autotrophy is via endosymbionts.

molluscivorous pertaining to predators specialized for feeding on molluscs.

molting the process of shedding the cuticle at periodic intervals.

monoecious an individual possessing both reproductive systems, at least during part of its life cycle.

monophyletic group a phylogenetic grouping of species descended from a common ancestor.

monopodial describing an unbranched, cylindrical type of pseudopodium.

morph a variant form within a species.

morphotype a morphologically distinct form present within a single rotifer population whose polymorphic feature(s) (e.g., spines, body size) undergo a seasonal phenotypic change.

mouth hook vertically oriented mandible-like structure in Diptera larvae.

muciferous body a variety of mucocyst found in some Phytomonadida.

mucocyst an extrusome of protozoa responsible for the secretion of mucous-like material onto the cell surface.

mucrones small cuticular teeth on the interior of the buccal cavity in some eutardigrades.

multispiral growth rings of an operculum of a snail comprised of many, slowly enlarging spirals.

multivoltine having more than two broods or generations per year.

muscle scars an area marking the former position of attachment of a muscle on the interior of a shell or other support structure, distinguished from the rest of the shell by being higher, depressed, or a surrounding groove.

mutualism an association between two organisms in which both benefit.

nacre or nacreous layer the inner shell layer of most molluscs, secreted by the mantle as successive layers of calcium carbonate crystals (absent in the Sphaeriidae).

nannoplankton plankton that range from 2–20 μm in size.

nauplius the earliest larval stage of many crustaceans.

net growth efficiency the percentage of assimilated energy represented by that allocated to tissue growth and gamete production.

neuston organisms living at or very near the air/water interface of a water body; see hyponeuston, epineuston, and pleuston.

node on the ostracode carapace, a protuberance smaller than a lobe or knob but larger than a tubercle, may have the form of a half sphere.

nonrespired assimilation that portion of assimilated energy or food not catabolized for maintenance energy and, therefore, available for production of new tissue or gametes.

nymphochrysalis a quiescent transitional instar between the larva and deutonymph in the water mite life cycle, representing the suppressed protonymph.

obtuse blunt angle at the end.

ocellus a simple eye consisting of a single, bead-like lens, occurring singly or in small groups.

omnivore consuming food at more than one trophic level, e.g., crayfish eat live animals (carnivory) and plants (herbivory) and will also consume dead animals and plants (detritivory).

open circulatory system a circulatory system characterized by blood not being continually maintained in vessels; rather, the blood bathes the tissues directly in large hemocoel sinuses before returning to the heart.

operculate formed as a cover.

operculum the horny or calcareous disk on the foot of a snail that covers the aperture when foot is withdrawn; used as protection for predators. Also, a lid or covering structure in some insects.

opsiblastic egg a type of resting egg produced by gastrotrichs; see tachyblastic egg.

orifice in ectoproct bryozoans, the terminal opening in a zooid through which the polypide is extended.

orthoclone a clone of a rotifer species established after many generations of culturing the ith offspring of every generation (see also Lansing Effect).

osmobiosis a cryptobiotic (latent) state in tardigrades induced by elevated osmotic pressures.

ostia in Mollusca, pores penetrating fused gill filaments in bivalves with eulamellibranch ctenidia which allow passage of water across the filaments into the interlamellar space.

outer lamella in ostracodes, a hard shell material that covers the outside of the epidermis that secreted it, composed of a thick layer of calcite enclosed between two layers of chitin.

ovate (ovoid) somewhat oval in shape.

oviduct a duct from the ovary, through which ova pass.

oviparous females that lay external eggs.

ovoviviparous females that lay eggs which hatch within the body of the female and are released as free-living offspring.

paddles feather-shaped locomotory structures attached at the anterior end of rotifers of the genus *Polyarthra*.

pallial pertaining to the pallium or mantle cavity of a mollusc.

pallial line line of muscle scars on the outer perimeter of the inside of the bivalve mollusc shell, marking the points at which pallial muscles attach the mantle to the shell.

pallial sinus a posterior indentation of the pallial line on the inside of the bivalve mollusc shell, marking the space into which the siphons are withdrawn on valve closure.

palmate like the palm of the hand, with fingerlike processes.

palpal lobes broad, paired, movable lobes on the distal end of the prementum in Odonata insects.

palustrine relating to wetland habitats.

papilla one of many small discrete protuberances (as on the ostracode shell surface) each with steep sides; smaller than tubercles.

papillate many lobe-like extensions.

paraglossa the lateral terminal lobe of the labium.

paramylon reserve food material characteristic of euglenid flagellates; occurs in a variety of shapes, usually with a laminated structure

paraphyletic group a monophyletic grouping including only some of the species descended from a common ancestor.

paratene repeated kinetidal patterns at right angles to the longitudinal axis of a ciliates body; superficially give the impression that the ciliary rows run circumferentially rather than longitudinally in the affected part of the body.

parietal the inside wall of a snail aperture.

paroral membrane synonomous with undulating membrane.

parthenogenesis asexual reproduction in which egg development occurs without fertilization. Amictic parthenogenesis is asexual reproduction without any recombination of the genetic material.

paucispiral growth lines of a snail operculum, present as a few, rapidly enlarging spirals.

pectinate having narrow parallel projections suggestive of the teeth of a comb.

pedal referring to the foot.

pedal feeding use of the foot in bivalve molluscs to feed on organic detrital deposits in or on sediments; generally involves ciliary tracts on the foot to bring detritus to the palps or gill food grooves.

pedal gland cement glands in the foot region of rotifers used to achieve temporary (in plankton or littoral forms) or permanent (in sessile forms) attachment to a surface.

pedicles lateral extensions of the corona of certain bdelloid rotifers.

pedipalp a feeding appendage; in mites, the pedipalp typically has five movable segments, located laterally on the gnathosoma in all instars. Pedipalps have both sensory and raptorial functions.

peduncle a pseudopodium-like cytoplasmic extension emerging from the sulcus in some dinoflagellates and used in feeding; also general name for a holdfast organelle in diverse groups of free-living protozoa.

pelagic see limnetic.

pellicle surface layer of protozoa, including the plasma membrane and associated membranous, microfibrillar and microtubular structures and organelles, but excluding nonliving external coverings such as tests or loricas.

peniferum the portion of the entocytherid ostracode copulatory apparatus that bears the penis (usually its distal portion) and its proximal portion forms a hinge on which the entire apparatus may be turned through an arc of 180 degrees about the zygum; includes the dorsal and ventral cardo, costa, spermatic tube, and penis.

penultimate next to last.

pereiopods thoracic appendages of many crustaceans; usually having a prime locomotory function (walking or swimming) and one or more secondary roles.

perennial living for several years.

periblast in ectoproct bryozoans, the part of a statoblast valve that excludes the capsule; in floatoblasts, the periblast includes both annulus and fenestra.

peribuccal lobes flattened cuticular projections surrounding the mouth of some eutardigrades.

peribuccal (oral) papillae elongated cuticular projections surrounding the mouth of some eutardigrades.

peribuccal papulae short, distinct cuticular projections surrounding the mouth of some eutardigrades.

pericardial cavity the remnant of the coelomic cavity surrounding the heart ventricle in molluscs.

pericardial gland specialized tissue in the wall of the heart auricles of molluscs through which hemolymph fluid is ultrafiltered into the pericardial space from which it enters the kidney.

pericardium an epithelial layer lining the pericardial cavity.

periostracum a proteinaceous outer layer of mollusc shells secreted by the mantle edge.

peristalsis waves of contraction that move along the gut.

peristome an area surrounding the cytostome of ciliates, especially where it is specialized relative to the rest of the body surface.

phagocytosis the process by which a cell engulfs a particle by extending its plasma membrane to form a vacuole or invagination.

phagotrophy nutrition by phagocytosis.

pharyngeal apophyses cuticular thickenings at the junction of the buccal tube and the lumen of the pharynx, alternating in position with the placoids.

pharyngeal intestinal junction; in nematodes, a cellular valve located between the pharynx and intestine; may protrude down into the intestine.

pharynx a portion of the alimentary tract located between the stoma and intestine, often called the esophagus. In turbellarians, "pharynx simplex" is a mostly short, unfolded tube; "pharynx plicatus" is a protrusible, more or less tubular fold enclosed in the pharynx cavity; "pharynx bulbosus" is a bulb separated from the surrounding tissue by a muscular septum. Pharynx bulbosus is represented by three types: rosulatus, doliiformis, and variabilis. Pharynx rosulatus is mostly round with a more or less vertical axis; pharynx doliiformis is barrelshaped, anteriorly situated and with a horizontal axis; pharynx variabilis is weakly muscular with variable shape and weakly differentiated septum.

phasmids paired sensory organs located laterally in the caudal area of nematodes.

pheromone a chemical messenger substance secreted by one individual into the external environment, which elicits a specific response in another individual of the same species.

phototaxis locomotion directed by light.

phreatic zone flowing groundwater located far enough from surface streams so that the movement of water in those streams has no influence on the movement of the phreatic waters; see also hyporheic zone.

phyllopod a leaflike or lobed, biramous appendage in some non-malacostracan crustaceans.

phytoplanktivorous pertaining to feeding on phytoplankton (i.e., suspended algae).

phytoplankton planktonic unicellular and multicellular algae and cyanophytic bacteria suspended in the water column.

picoplankton plankton 0.2–2 μm in size, primarily prokaryotic.

pinocytosis the process by which cells ingest dissolved or colloidal material, involving the attraction of these substances to the cell surface and then internalizing these substances through invagination of plasma membrane.

piptoblast in ectoproct bryozoans, a type of statoblast that has no annulus and is not cemented to the colony substratum; formed only in the family Fredericellidae.

placoids cuticular thickenings in the lumen of the tardigrade pharynx (pharyngeal bulb); large anterior ones are macroplacoids, small posterior row are microplacoids.

plankton eukaryotic and prokaryotic organisms living all or part of their lives in the water column of a freshwater or marine ecosystem.

plasma membrane a unit or cell that which is the outermost living layer of protozoa.

plasmodium multinucleate mass of protoplasm enclosed by a plasma membrane; nonhomologous trophic stage in the life cycle of some members of various protozoan taxa.

plastron a semipermanent bubble of air through which carbon dioxide from respiration is exchanged with oxygen from the water.

pleopod one of five pairs of appendages on first five segments of abdomen of a decapod; "swimmerets," or modified into male gonopod.

plesiotypical (plesiomorphic) conservative, or unmodified, state of a biologic attribute or morphologic character.

podomere a leg segment of an arthropod.

polykinetid see kinetid.

polykinety a compound ciliary organelle composed of several kineties, e.g., a cirrus.

polyphyletic group a taxonomic assemblage of species that are not all descended from a common ancestor.

polypide in bryozoans, the retractable portion of a zooid comprising the central ganglion and all structures related to feeding and digestion.

polytrophic see omnivore.

polyvoltine with several generations during the season as opposed to univoltine.

pore canals in nematomorphs, pores connected to fine tubes in the cuticle; in ostracodes, a passage through the entire valve in which are located sensory hairs to the nerves of the hypodermis.

postcloacal crescent see crescent.

posterior ridge an external ridge on unionacean (Mollusca) shells extending from the umbos posterioventrally to the shell margin.

postocularia a pair of prominent setae located on the dorsum of the idiosoma posterior to the level of the eyes in deutonymphs and adult mites.

prehensile fitted or adapted for grasping, holding, or seizing.

prehensile palp appendage modified to grasp, hold, or seize; generally formed of two podomeres, the proximal one referred to as the prodopus, distal one referred to as the dactylus of some crustaceans such as ostracodes.

prementum the distal segment of the labium in Odonata insects, which bears the palpal lobes.

preocularia a pair of prominent setae located on the dorsum of the idiosoma anterior to the level of the eyes in deutonymphs and adult mites.

prismatic layer a molluscan shell layer made up of elongated calcium carbonate crystals oriented at a 90 degree-angle to the surface plane of the shell and lying between the outer periostracum and inner nacreous layer.

procorpus the anterior portion of the nematode pharynx between the stoma and metacorpus.

prodelphic with a single anterior-directed ovary (in nematodes).

profundal deep zone of a lake below the littoral and open water photic zone; often restricted to benthic habitats but may include pelagic zone.

proleg any process or appendage that serves the purpose of a leg but is not a true leg.

promitosis form of nuclear division wherein the endosomal body in the nucleus divides and forms two polar masses with a spindle between them; characteristic of some naked amoeba of the class Heterolobosea.

propodosoma in mites, the region of idiosoma posterior to the level of the primitive sejugal furrow.

propodus the penultimate segment (sixth from base) of a segmented appendage.

proprioceptor a sensory receptor that responds to physical or chemical stimuli originating from the organism.

prostate cells, glands, or organs, part of the male reproductive system, releasing secretion that mixes with sperm.

prostaglandins local chemical mediators inducing short-lasting effects in adjacent tissues.

prostomium a region anterior to mouth in some worms.

protandry sequential hermaphroditism with the male stage preceding the female stage.

protocerebrum the part of a crustacean's brain that normally innervates the eyes, sinus gland, frontal organs, and head muscles.

protoconch the shell of newborn snail or "spat."

protonephridia excretory–osmoregulatory system of certain pseudocoelomates, including gastrotrichs and rotifers, comprised of flame cells and tubules.

protonymph the first nymphal instar in the generalized mite life cycle.

protopodite the basal part of a typical crustacean limb consisting of two more or less consolidated segments and bearing at its distal extremity an exopodite or endopodite or both.

proventriculus the large cavity in collothecid rotifers that is an extension of the mastax in which prey are held prior to being passed into the stomach.

provisional genital field a region of the idiosoma occupied by the rudimentary gonopore and the genital acetabula in deutonymphs of mites.

proximal toward the median plane of a body.

pseudobranch a "false gill" present in ancylids and planorbids (Mollusca, Gastropoda); a conical extension of epithelium used in respiration, analogous to ctenidium of prosobranchs.

pseudocardinal teeth in bivalve molluscs, compact shell lamellae on the anterior portion of the unionacean shell hinge plate performing the function of cardinal teeth.

pseudocoelom a body cavity that is not lined with mesodermal tissue, as is a true coelom

pseudofeces masses of mucous-bound particles rejected from the labial palps of bivalve molluscs before ingestion and released into the mantle cavity to be carried out the siphon on water currents induced by valve clapping (i.e., rapid valve adduction).

pseudopodium in protozoa, a dynamic extension of the cytoplasm used in feeding or locomotion; in insects, a soft, foot-like appendage that is less prominent than a proleg.

pseudosexual resting egg a diploid resting egg supposedly produced in rotifers by parthenogenesis in the absence of males (see also mixis).

pseudostome aperature of a test or shell of a testate amoeba; pseudopodia protrude from it during locomotion and/or feeding.

punctae a small pit-like opening of a pore canal on the ostracode shell; in bivalve molluscs, microscopic pores traversing the mineral portion of the shells of sphaeriids to open beneath the periostracum; contain extensions of the pyramidal cells.

pustule a protuberance with a pore in the middle, similar to a volcano with a crater, about the size of a tubercle on the ostracode carapace.

pyramidal cells in Mollusca, mantle epithelial cells that extend through shell punctae (pores) to lie just below the periostracum.

Q_{10} the difference in metabolic rate over a 10-degree temperature range (e.g., Q_{10} would equal 2 if the respiration rate doubled between 10 and 20°C).

Q_{10} (acc.) the factor by which a biologic rate function changes over a 10°C increase in acclimation temperature.

quadrate square or rectangular in shape.

radial pore canal a passage through the adhesive strip, between the duplicature and the outer lamella of the proximal valve edge of an ostracode.

ramate a type of trophi (jaw structure) found in bdelloid rotifers in which the rami are large and semicircular in shape and unci possess many teeth.

ramus one of the seven pieces that comprise the trophi of rotifers; paired rami together with the fulcrum form the incus subunit.

receptaculum seminis a part of the female reproductive system of turbellarians in which recipient spermatozoa are stored.

reniform suggesting a kidney in shape.

respiratory plate also referred to as branchial plate, modified from the exopodite and may contain numerous feathered setae.

resting egg an egg (actually 2N embryo) with a period of arrested development, usually resistant to extreme environmental conditions.

reticulopodium threadlike, repeatedly branching and anastomosing pseudopodium, functioning more often in food gathering than locomotion.

retrocerebral organ a structure of unknown function in rotifers composed of the unpaired retrocerebral sac and paired subcerebral glands together; possible function as an exocrine gland lubricating the anterior part of the body.

retrocerebral sac an unpaired glandular structure near the subcerebral gland in the head region of rotifers possessing a forked duct that opens into the corona (see also retrocerebral organ).

rhabdions cuticular segments of the nematode stoma, may be fused and produce an entire stomal wall or separate and exhibit a segmented wall.

rhabdos an organelle supporting the cytopharynx of some ciliates, especially in the subphylum Rhabdophora; although like a cyrtos, rhabdos is considered evolutionarily the more primitive organelle.

rheocrene a flowing spring with a substratum of sand and gravel.

rheotaxis orientation to current (can be positive or negative).

riparian living on the bank of a lake, pond, or stream.

rostrum any extended portion of the head end of an invertebrate.

r-selection a selection for those life-history traits that increase fitness in an unstable habitat where catastrophic population reductions prevent high levels of intraspecific competition.

sclerite a hardened area of the insect body wall bounded by sutures or membranes.

semivoltine having one brood or generation every two years.

sanguivorous feeding on blood.

scabrous having small, round, raised scales or cones.

second-form male one of two morphological forms of male cambarine crayfishes; sexually nonfunctional male lacking corneous terminal elements on first pleopod (gonopod).

sediment focusing uneven deposition of sediments in a basin so that certain portions of the basin accumulate sediments at a greater rate; usually this occurs in the deepest parts of the basin.

sejugal furrow a primitive line of demarcation on idiosoma at level between insertions of second and third pairs of legs in all mite instars, but expressed only in certain early derivative taxa.

selvage the middle and principal ridge of the contact margin of the duplicature of the ostracode valve.

selvage groove that part of the duplicature surface between the list and selvage of ostracode valve.

selvage strip the part of the duplicature that forms the selvage groove and the selvage of the ostracode valve.

semelparous breeding only once during life history; one mass reproductive effort.

seminal receptacle (spermatheca) a sac-like structure in the female reproductive system that receives and stores spermatozoa.

seminal vesicle a saclike structure in the male reproductive system that stores spermatozoa.

sensillum epithelial sense organ.

septulum a cuticular thickening in the lumen of the pharynx, posterior to the placoids, in the same plane as the pharyngeal apophyses in some eutardigrades.

septum a dividing wall or membrane; in ostracodes, a small ridge on the list strip on the duplicature of the valve.

serial homologies structures having the same embryonic and evolutionary origin but which differ segmentally (e.g., all appendages of a crayfish, except the first antennae; arose from similar embryonic primordia and from a series of similarly formed appendages in an ancient ancestral stock).

serotonin a chemical messenger stimulating active transport of sodium ions from the medium in freshwater bivalve molluscs.

serrate notched or toothed on the edge.

sessoblast in ectoproct bryozoans, a relatively large statoblast cemented firmly through the body wall to the substratum; formed only in the family Plumatellidae.

setae hair-like extensions of the epicuticle, produced by a trichogen.

sheath in nematodes, a cuticle retained from the last molt which may surround the entire body or head region.

simplex stage a molting stage in tardigrades characterized by the absence of the buccal apparatus.

sinistral snail shells with the aperture opening toward the left (when viewed with the apex pointed away from the observer and with the aperture fully visible).

siphon in bivalve molluscs, tubular structure formed by fusion of opposite mantle margins directing inhalant (inhalant siphon) and exhalant (exhalant siphon) respiratory and feeding water currents into and out of the mantle cavity; an elongated breathing tube in some insects.

solenidion (solenidia) in mites, a specialized setiform chemosensory structure; solenidia are located on the distal segments of the pedipalps and legs in all active instars; though seta-like in appearance, solenidia differ structurally and ontogenetically from setae and are usually given distinctive designations in chaetotactic formulae.

somite a homologous body segment, a metamere.

spear protrusible pointed, solid rod-like structure in the mouth region of some nematodes.

spermatogenesis the formation of sperm.

spermatophore a packet of spermatozoa enclosed in a

membranous capsule, produced by adult male mites and either deposited on the substratum or transferred to the gonopore of females.

sphincter a circular muscle that forms a ring around a tube, such as the gut; it can close the tube by contracting.

spicules internal structures of sponges composed principally of silica or calcium salts; also paired, sclerotized structures in the tail of the male nematode.

spindle a snail shell with a thick middle and a tapered point at both the apex and the aperture ends.

spinneret in nematodes, a terminal pore out of which passes secretions of the caudal glands.

spiracle a respiratory opening connected to the tracheal (respiratory) system.

standing crop biomass the amount of tissue produced per unit area at a set point in time, usually measured as g carbon or Kcal per m^2.

statoblast in phylactolaemate bryozoans, an asexually produced capsule composed of two convex valves enclosing a mass of yolky material and undifferentiated tissue; capable of germinating a new zooid after obligate dormancy.

statoconia in bivalve molluscs, many small mineral inclusions in the cavity of a statocyst (gravity orientation organ).

statocyst a gravity-orientation organ containing mineral inclusions (a statolith or statoconia) in a vesicle lined with sensory cilia.

statolith a mineral inclusion in the cavity of a statocyst (gravity-orientation organ).

stenopod an unbranched, segmented walking leg in some crustaceans, particularly decapods.

stenothermal having little variation in temperature.

stigmata paired external openings of the tracheal system, located between the bases of the chelicerae in all active instars of mites.

stolon in certain ectoproct bryozoans, a cylindrical stem-like structure from which individual zooids grow at intervals.

stoma that portion of the alimentary tract between the oral opening (mouth) and the pharynx.

stream order a method for classifying streams in which the smallest permanent streams is designated as a "first order" stream; two first orders streams combine to form a second order stream; a first and second in combination are still considered a second order, but when two second order tributaries join a third order stream results.

striations (striae) parallel grooves; also, the spiral, microscopic grooves along the whorls of a snail shell (sometimes incorrectly used to refer to spiral ridges or lirae).

style sac an evagination of the stomach floor of molluscs, which secretes the crystalline style.

stylet a sclerotized tube or channel of variable shape and complexity; a hollow, sclerotized, hard, rod-like structure in the mouth region of some nematodes.

stylet knobs enlarged protuberances (usually three) at the base of the stylet, usually serve for attachment of muscles that aid in stylet extension.

subcerebral glands paired glands near the retrocerebral

sac in the head region of rotifers with ducts that lead to the corona (see also retrocerebral organ).

subitaneous egg an egg that develops and hatches with no arrest in development.

sulcus a depression or furrow on the lateral surface of the organisms in several phyla (e.g., Ostracoda and Mollusca); in protozoa, a longitudinal groove on the surface of dinoflagellates from which the flagella and, if present, the peduncle or other feeding structures emerge.

suture a region where two structures (e.g., plates or valves) are joined.

symbiont one member of a symbiotic relationship.

sympatric species whose habitat distributions overlap.

syngamy fusion of haploid gametes or gametic cells in sexual reproduction to form a diploid zygote; phenomenon exibited by some members of all major protozoan taxa except the cilliates.

tachyblastic egg a type of egg produced by gastrotrichs in which development occurs immediately; see opsiblastic egg.

tagma or tagmata body regions (especially in arthropods) consisting of metameres that are grouped or fused to perform a similar function, e.g., the head, thorax, and abdomen of a crustacean or insect.

test loose surface covering of protozoa, created by the inhabitant with secreted and, sometimes, gathered materials; generally synonymous with lorica, but test is normally applied to Rhizopoda.

thermocline zone in a lake of relatively rapid temperature change over a small range of depth; commonly defined as 1°C change per meter of depth.

tocopherol (*a*-tocopherol or vitamin E) an organic compound that has been shown to induce mixis (sexual reproduction) in some species of rotifers (e.g., Asplanchna) and lengthen the lifespan in others.

torpidity the condition or quality of being dormant or inactive, a temporary loss of all or part of the power of sensation or motion.

transverse muscle attachment scar a conspicuous muscle attachment scar in the posterior region of the third coxal plates in larvae of certain mite taxa.

trichocyst a type of extrusome emitting a spindle-shaped, nontoxic projectile.

tritocerebrum the portion of a crustacean's brain that principally innervates the antennae and a section of the alimentary tract.

tritonymph the third nymphal instar in the generalized mite life cycle.

trochantin a small, forward projecting sclerite at the base of the trochanter.

trochus one of two ciliated rings of the typical rotiferan corona (see also cingulum).

troglobite an obligate cave-dweller, characterized by a reduction in eye structure and a lack of pigments; also, frequently demonstrating K-strategies.

troglobitic an animal living in a cave in an obligate relationship.

troglophile an organism living in a cave in a nonobligate relationship; such organisms usually complete their life history in the cave and commonly have conspecifics surviving outside the cave; see troglobite.

trophi a complex set of hard jaws characteristically found in the pharynx (mastax) of rotifers.

trophozoite mature, vegetative, feeding, adult stage in the life cycle of protozoa; used primarily with reference to parasitic species of nonciliate taxa; synonymous with trophont.

tubercles small, rounded knobs on the body surface of an animal, larger than papillae, smaller than nodes.

tun the barrel-shaped form of a tardigrade in a cryptobiotic (latent) state.

turnover time the amount of time necessary for the standing crop biomass to be replaced completely in a species.

type taxon (or specimen) the original individual of a species on which the species name is based; type specimens are kept in museums for comparative purposes and to reduce confusion about scientific nomenclature.

typhlosole an invaginated, cilated, particle-sorting structure running the length of the midgut and extending into the ciliated sorting surfaces of the floor of the stomach and digestive diverticulum in bivalve molluscs.

umbo in bivalve molluscs, that portion of the shell raised above the dorsal shell margin in the area of the hinge (also called the beak).

uncate a condition of the pedipalp, in deutonymphs and adults of certain mite taxa, in which the ventral surface of the tibia is expanded and produced distally to oppose the tarsus, forming an efficient grasping organ.

uncinate a type of trophi (jaw structure) found in rotifers characterized by unci possessing few teeth, usually with one large one and a few small ones.

uncus one of the seven pieces that comprise the trophi of rotifers; an uncus together with a manubrium form the malleus subunit.

undulating membrane in cilitates, a compound ciliary apparatus typically lying on the right side of the buccal cavity; synonomous with paroral membrane.

undulipodium a swimming organelle of eukaryotic cells, including protozoa, includes both the flagellum and the cilium.

univoltine having one brood or generation per year.

urogomphus a paired, unsegmented terminal appendage on larvae of Coleoptera.

uroid posterior part of moving "naked" lobose amoeba.

urosome the portion of the body of copepods posterior to the major body articulation.

urstigma (urstigmata) a knob-like structure with a porous cap located between the first and second coxal plates in larval mites. Urstigmata are paired, are apparently homologous with the genital acetabula of deutonymphs and adults, and probably function as an osmoregulatory chloride epithelium.

uterus the region in female system that stores egg capsules prior to release.

vagina that portion of the female reproductive tract connecting the vulva with the uterus or uteri; a female tract used in copulation, distal to entrance of the oviducts.

valve one of either the left or right shell of a seed shrimp

(Ostracoda) or mussel (Mollusca), generally referred to as left or right valve. In nematodes: disc or nipple-like structure which may occur in the metacorpus or basal bulb of the pharynx.

vas deferens a duct from the testes, through which spermatozoa pass.

veliger a free-swimming larval stage of molluscs characterized by the presence of a ciliated velum utilized for swimming and feeding; in North American freshwater bivalves, it occurs only in *Dreissena polymorpha*, the zebra mussel.

ventral shield a single large sclerite or series of closely fitting plates and platelets completely covering the venter of the idiosoma in deutonymphs and adults of certain mite taxa.

verge a penis.

vesicula seminalis an enlarged part of the sperm duct or ejaculatory duct holding sperm.

vestibule the space between the outer lamella and the duplicature in the proximal portion of an ostracode valve.

virgate a type of trophi (jaw structure) found in rotifers that are modified for piercing and pumping and generally can be recognized by the long fulcrum and manubria; some are asymmetrically shaped.

vitellarium a yolk gland associated with the ovary in bdelloid and monogonont rotifers.

vitelline membrane the most external membrane of a molluscan egg.

viviparous a type of reproduction in which the young are internally maintained and nourished by the mother before birth.

VO$_2$ an abbreviation for volume of oxygen consumed per unit time.

voltine refers to the number of generations during the year or during the reproductive season (e.g., univoltine, bivoltine, polyvoltine).

vulva ventrally located genital opening of the female nematode, usually located in the midbody region.

water tube in bivalve molluscs, the channels formed between the descending and asending limbs of lamellibranch bivalve gill filaments which carry water past gill ostia dorsally to the epibranchial cavity; also form gill marsupial chambers in which eggs and larval stages are incubated through development to juveniles or glochidia.

wetlands ecosystems whose soil is saturated for long periods seasonally or continuously; include marshes, swamps, ephemeral ponds, etc.

wing in bivalve molluscs, a thin, flattened, dorsally extended portion of the posterior slope occurring on the shell of some unionacean species.

YOY (young-of-the-year) juveniles during the period from the last larval stage to adulthood, or one year of age—whichever comes sooner.

Zenker's organ ejaculation ducts, made up of longitudinal muscles in the shape of a tube on which are found radiating bristles and are covered in a chitinous sheath in freshwater ostracodes.

zooid one of the physically connected, asexually replicated, morphologic units that comprise a colony; in bryozoans, the zooid includes the polypide and associated musculature as well as all parts of the adjacent body wall.

Taxonomic Index

Boldface page entries indicate figures; italic page entries indicate keys.

Abedus, 615
Ablabesmyia, 625
Abramis brama, 797
Abrochta, 227, **227**
Acalyptonotidae, 526, 531, 542, *549,* **560, 561,** *577, 578,* **586, 587**
Acalyptonotus, 549, 578
neoviolaceus, **560, 561, 586, 587**
Acanthamoeba, **67,** *68*
Acanthamoebidae, *68*
Acanthobdella, 414
peledina, 438, **439,** 444, 451, 454, *460*
Acanthobdellida, 7, *37–46,* 402, 410, **412,** 414, 444, *438–457*
peledina, 444
Acanthocephala, 221, 273
Acanthocyclops, 794, *808*
capillatus, 809
carolinianus, 809
exilis, 809
plattensis, 809
venustoides, 809
venustus, 809
vernalis, 794, 809
viridis, 795
Acanthocystis, 71
Acanthocystis
turfacea, **72**
Acanthodiaptomus, 807
denticornis, 807
Acantholeberis, 760
curvirostris, **759**
Acanthopodide, *68*
Acari, 9, *15,* 524
Acaridida, 523
Acarina, 501
Acella, 304
haldemani, **306**
Achromadora, 265, *258, 265*
Acilius, 619
Acineta, 73
limnetis, **74**
Acinetidae, *73*
Acipenser, 138
fulvesens, 363
guldenstadti, 138
ruthenus, 137, 138
Acipenseridae, 363
Aclitellata, 402, 405
Acoela, 146, *153,* **153**
Acolosomatidae, 405
Acoloxus, 309
Aconchulinida, 39, *70*
Aconopsinae, 539
Acrasida, 39, 47, *67*
Acrididae, 278

Acroloxidae, 291, *309*
Acroloxus, 291
Acroneuri, **595**
Acroperus, 748, *750*
harpae, **749**
Actinobdella, 456, *461*
annectens, 461
inequiannulata, 447, 456, *461,* **461**
Actinobolina, 44, *79*
radians, **80**
Actinobolinidae, *79*
Actinolaimus, 264
Actinomonas, **42**
Actinonaias, 388
carinata, **306–384**
Actinophryida, 39, *72*
Actinophrys, 72
sol, 54, **72**
Actinosphaerium, 48, 55, *72*
eichhorni, 72
Actinulida, 126
Acutogordius, 279
Acyclus, 226
Adephaga, 619, *620–622*
Adineta, 219, *226*
Adinetidae, **19,** *226,* 227
Adorybiotus, 503, **504, 506,** 516, *517*
Adropion, 518
Aeolosoma, **418**
Aeolosomatidae, 407, *417*
Aeromonas hydrophila, 443
Aeshnidae, 601, 602, 603, **636,** *637*
Agabinus, 619
Agabus, 619, **647,** *649*
Agarodes, 612, **642**
Agladiaptomus, 807
Agnetina, 605
Agriodrilidae, 438
Agriodriulus, 413
Alaimidae, *266*
Alaimus, 266
Alasmidonta, 384, 385
marginata, 385
raveneliana, 385
varicosa, **381-383,** *384,* 385
Albia, 536, *553, 578,* **587**
neogaea, **569**
Albiinae, 553, **569,** *587,* **587**
Alboglossiphonia, 456
heteroclita, 464, **464**
Alexandrovia, 424
Allocapnia, **639**
Allodero, 409, *423*
Allonais, 408, *423,* **423**
Alloperla, **638**

Alona, 727, 748, *753*
affinis, 753
bicolor, **728, 753**
intermedia, 753
verrucosa, 753
Alonella, 733
acutirostris, 753
dadayi, 753
excisa, 753
exigua, 739, *753*
hamulata, 753
nana, 753
Aloninae, 748
Alonopsis, 748, *750*
elongata, 738, **739**
Alosa
sapidissima, 363
Alturidae, 530
Alturinae, 530
Amblema, 383
neisleri, 383
plicata, 359, **380–382,** *383*
Ambleminae, *378*
Amblemini, *379, 380, 381, 382, 383*
Ambrysus, 614
Ametor, 620
Ametropodidae, 598, 599, 635, **635**
Ametropus, **635**
Amnicola, 296, *302*
Amnicolinae, *302,* **302**
Amoeba, **3,** 38, 47, 48, 58, 59, *68*
proteus, 47, **48,** *67*
Amoebidae, **47,** *68*
Ampharetidae, 458
Amphhizoa, **647**
Amphibia, 446, 528
Amphibolella, 160
Amphibolidae, 516
Amphibolus, 503, **504,** *506,* **506,** 510, 513, 516, *518*
weglarskae, 511
Amphichaeta, 406, 421, **422**
Amphidelus, 266
Amphileptidae, *81*
Amphileptus, 81
claparedi, **82**
Amphinemura, **639**
Amphionidacea, **667**
Amphipoda, 10, **11,** *15,* 449, **450,** 666, **667,** 668, 675, 676, **676,** *682*
Amphisiella, 78
oblonga, **77**
Amphisiellidae, *78*
Amphitremidae, *71*
Amphizoa, **649**
Amphizoidae, 619, 620, *646,* **647,** *648,* **649**

Amphoretidae, 458
Ampullaridae, 291
Ampullariidae, *304*
Anabaena
oscillarioides, 372
Anacaena, 620
Anas
platyrhynchos, 767
Anaspidacea, **667, 669**
Anatonchus, 266
Anatopynia, 151
Anax, **636**
junius, 602
Ancanthobdella
peledina, **438**
Anchistropus, 130, *750, 754, 739*
minor, **754**
Anchlidae, 286
Anchytarsus, 620, **649**
Ancylidae, 286, 291, 294, *309*
Ancylus
fluviatilis, 291, 294
Anheteromeyenia, *115*
argyrosperma, *115*, 116, **116**
ryderi, *115*, 116, **116**
Animalia, 38
Anisitsiellidae, 529, *549, 574*
Anisitsiellinae, 529, 538, *555, 552*, **559**,
563, *574*, **583**
Anisonyches, 503
Anisoptera, 601, 602, 602–603,
636
Ankistrodesmus, 372, 744
Ankylocythere, 715
Ankyrodrilus, 431
Annelida, 7–8, **8**, *14*, 401–479, **412,** 438,
459–460, 501
Annulipalpia, 607, 609–610
Anodonta, 31, 43, 324, **343**, 368, *384*
anatina, 345, 346, **346**, *384*
couperiana, *384*
cygnea, 323, 324, 329, 331, 332, 334,
335, **360**, 368
gibbosa, *384*
grandis, 327, 328, 329, 331, 334, 338,
354, **380–384**, *384*
imbecilis, 373
imbecilus, *384*
impicata, 335
implicata, **384**
kennerlyi, *384*
peggyae, *384*
simpsoniana, 364, *384*
suborbiculata, *384*
woodiana, 334
Anodontina, 341
Anodontini, *381*
Anodontoides, 382
ferussacianus, **378–380**, *382*
radiatus, 382
Anomopoda, 724
Anonchus, 266
Anopheles, 627, **651**
Anostraca, **669,** 724, 764, 767
Antechiniscus, 511
Anthozoa, 126
Antipodrilus, 413

Antrobia, *302*
Antroselates, *301*
Anura, 528
Anuraeopsis, 202
Anystoidea, 524, 527
Apatronemertes, 164
Aphanolaimus, 266
Aphanoneura, 405, *417*, *417–419*
Aphaostracon, *302*
Aphelenchida, *264*
Aphelenchidae, *264*
Aphelenchoides, 257, *264*, **269**
Aphelenchoididae, *264*, **269**
Aphelenchus, 257, *264*
Apidae, 279
Aplexa, 292, *306*
hypnorum, 294, **307**
Aplodinotus
grunniens, 363
Apochela, 503, 504, 516
Apodibius, 503, **506**, *517*
Apodidae, 278
Apophysis, **506**
Arachnida, 8, *15*
Araeolaimida, *265*
Araneae, 276, 278
Aranimermis, 271, *273*
Arcella, 59, *68*
vulgaris, **69**
Arcellidae, *68*
Arcellina, *68*
Arcellinida, 39, *68*
Archilestes, 604
Arcidens, 385
confragosus, **380–382**, *385*
Arcteonais, 408
lomondi, *422*
Arctodiaptomus, 807
Arctodrilus, 408
wulikensis, *408*, *428*
Arctopsychinae, 609
Arenohydracaridae, 531, *577*, **587**
Arenohydracarus, *577*, **587**
Argia, 603
Argulidae, *682*
Arguloida, *682*
Argulus, **667**, 674, **674,** 676,
682
maculosus, 674
Arhienemanniidae, *551*
Arhynchobdellida, *460*
Arkansia, 385
wheeleri, **380–382**, *385*
Armiger, 306
crista, **307**
Armophorida, *78, 79*
Arrenuridae, 531, 539, *550, 576*
Arrenurinae, 531, 537, 538, 540, 541, 542,
550, **561, 578**, *576*, **585, 586, 589**
Arrenuroidea, 524, 530, 533, 540, 535,
543, 538, 539, 541, 542, *549*
Arrenurus, 525, *550*, **578**, *576*, **591, 586,**
589
planus, **561**
Artemia, 34, 139, 140, 221, 764–770, *772*,
805
franchiscana, **771**

franciscana, 766
salina, 34, 132, 670, 766, 767
Artemiidae, 724
Artemiopsis, 765, 767, 768, *772*
stephanssoni, **773**
Arthropoda, 8–14, **10**, *14*, 501, 665
Arthrotardigrada, 503, 504, 513
Ascetocythere, 714
Aschelminthes, 173, 224
Ascomorpha, **194**, 196, *240*, **240**
Ascomorphella,
volvocicola, *237*, **238**
Ascophora, 163
Ascophorinae, *163*
Asellidae, 676, *618*
Asellus, 153, 681
Askenasia, 81
Askensia volvox, **80**
Aspidiophorus, 177, 181, *181*, **182**
Aspidisca, 75
Aspidisca costata, **76**
Aspidiscidae, 75
Asplanchna, 18, 193, **194,** 195, **195,** 196,
197, 200, 204, 206, 210, 216, 217,
218, 219, 222, *231*, **233**
brightwelli, 197, 199, 200, 208, 209,
210, 214
girodi, 214
intermedia, 199, 204
priodonta, 202, 213, **214**
sieboldi, 196, 199, 200, 207
Asplanchnidae, 193, *231*, **233**
Asplanchnopus, 193, 216, 218, *231*
Astacidae, 843, *850*
Astacidea, 823, *850*
Astacus, 833
Astylozoidae, 75
Astylozoon, 75
faurei, **81**
Athalamida, 39, *71*
Athericidae, 629, *650*, **651**
Atherix, 629, **651**
Athienemanniidae, 531, *551, 563*, **576**
Athienemanniinae, 531, *576*, **585**
Athlenchidae, *264*
Atopsyche, **642**
Atracides
grouti, **566**
Atractides, *552, 575, 577*, **584, 587**
Atrochus, 226
Attenella, 598
Attheyella, 795, *812*
Aturidae, 530, 541, 542, *553, 575, 577*
Aturinae, *553*, **567, 568, 578,** *577*, **587**
Aturus, 523, *553*, **567, 578,** *577*, **587**
deceptor, **568**
Atyidae, 830, 848, *850*
Atylenchus, *264*
Aulodrilus, 409, 413, **427**, *428*
pluriseta, *424*
Aulolaimoides, *264*
Aulophorus, 409, **423**, *423*
Austropotamobius, 833
Axonolaimidae, *266*
Axonopsinae, 530, 535, 539, 540, 542,
553, **568**, *577*
Axonopsis, *553, 577*, **587**

Bactrurus, 676
Baetidae, 278, 598, 599, *634*
Baetis, **10,** 598
Baetisca, **598**
Baetiscidae, 598, *633,* **634**
Balanonema, 85
   biceps, **82**
Bandakia, 552, *574*
   phreatica, **563, 583**
Bandakiopsis, 549, *574*
   fonticola, **559**
Barbicambarus, 831, 834, *850*
Barbidrilos, 417
Barbidrilus, 418, **418**
Barentsia, 490
Bastiania, 266
Bastianiidae, *266*
Bathynellacea, **667**
Batracobdella, **441,** 456, *462*
   cryptobranchii, 462
   michiganesis, 462
   paludosa, 456, *462*
   picta, 456, *462*
Bdellarogatis, 457
   plumbeus, 467
Bdellodrilus, 431
   illuminatus, **432**
Bdelloidea, 189, 196, 206, 221, 224, *225*
Beatogordius, 279
Beauchampi, **226, 228**
Beauchampia
   crucigera, 204
Behningiidae, *598,* 599 *633,* **634**
Belostoma, 614, **645**
Belostomatidae, **572,** 614, 615, *644,*
   645
Beraea, **642**
Beraeidae, 609, **642,** *643*
Berosus, 620
Beunoa, 616
Bharatalbia, 577
Biapertura, 753
Bicoeca, 66
   lacustris, **65**
Bidessonotus, 619
Bilateria, 146
Biomphalaria, 308
   glabrata, 295, **308**
Biomyxa, 71
   vagans, **70**
Biomyxidae, *71*
Birgea, **232**
   enanita, 232
Birgeidae, *232,* **232**
Birgella, 302
Bithynia, 304
   tentaculata, 295, **302**
Bithyniidae, 291, *304*
Bivalvia, 6, 6–7, **7,** *14,* 315–389
Blattidae, 278
Blepharicera, **651**
Blephariceridae, 626, *650,* **651**
Blepharisma, 55, 56, *79*
Blepharisma
   lateritium, **80**
Blepharismidae, *79*
Bodina, 46

Bodo, 46, 66
   caudatus, **65**
Bodonidae, *66*
Bogatiidae, 531, *576*
Borcobdella
   verrucata, 464
Bosmina, 724, 725, 727, 733, 734, 737,
   738, 740, 742, 756, **757,** 803
   coregoni, **757,** *758*
   fatalis, **757**
   hagmanni, 758
   longirostris, 216, **757,** *758,* 800
   tubicen, **757,** *758*
Bosminidae, 724, 746, *758, 758*
Bosminopsis, 730, 756, *758*
   deitersi, **757**
Bothrioneurum, 404, 405, **405,** 409, **427**
   americanum, 426
   vejdovskyanum, 426
Bothrioplana, **154,** *157,* **157**
   semperi, 149, **158**
Bothromesostoma, 150, **161,** *161*
Bouchardina, 831, *852*
   robinsoni, 851
Brachionidae, **236,** *236,* **237**
Brachionus, 195, 196, 197, 198, 202, 206,
   208, 210, 215, 218, **220,** 223, *236*
   angularis, 214, 223, **236**
   bidentata, 218
   calyciflorus, **195,** 199, 200, **207,** 209,
      210, 211, **211,** 213, **213, 214,** 214,
      215, 217, **217,** 218, **218,** 220, 223,
      236
   caudatus, 200
   patulus, 200
   plicatilis, 34, **188,** 190, 196, 197, 198,
      199, 208, **209,** 210, 211, 214, 220,
      223, *236*
   quadridentatus, 214, 215
   *rubens,*199, 216, 220, 223
   urceolaris, 211, **211,** 215, 218, **237**
Brachiopoda, 489
Brachycentridae, 278, 607, 608, 610, **642,**
   643
Brachycentrus, 607, **642**
Brachycera, 624, 625, 629–631
Brachycercus, 598
Brachypoda, 553, *577*
   setosicauda, **568**
Bradleystrandesia, 706
Bradyscela, 226
Branchinecta, 765, 766, 767, 769, *772*
   campestris, 766, 767
   coloradensis, **773**
   gigas, 767, 769
   mackini, 766
Branchinectidae, 724
Branchinella, 770
   sublettei, **771**
Branchiobdella
   astaci, 429
   hexodonta, 429
   kozarovi, 429
   parasita, 429
   pentodonta, 429
Branchiobdellida, 7, 404, 420, 428–433,
   430, *431–433*

Branchiobdellidae, 438, 439
Branchiopoda, 10, *15,* 666, **667,** 674, *770*
Branchipodidae, 724
Branchiura, **7,** 406, 408, 413, **425**
   sowerbyi, 409,414,**424**
Branchiura, 666, **667,** 668, 673-674,
   674, 676, 682, *682*
Bratislavia, 423
Bresslaua, 86
   vorax, **84**
Brillia, 625
Bromeliaceae, 150
Bryocamptus, 795, *811, 812,* **813**
Bryochoerus, 510, *516*
Bryodelphax, 510, *516*
   parvulus, **503**
Bryophyllum
   lieberkuhni, **80**
Bryospilus, 725, *748,* 750
   depens, **749**
Bryozoa, 8, **9,** *14,* 481–497, *492–587*
Buenoa, 614
Bulimnea, 305
   megasoma, **306**
Bullatiforns, 748
Bunops, 760
   acutifrons, **761**
Bursaria, 56, 79
   truncatella, **80**
Bursariidae, *79*
Bursariomorphida, *79*
Byrophuyllum, 81
Bythothephes, 727
Bythotrephes, 735, 738, 743, *763*
   cederstroemii, **764**

Cadona
   caudata, 693
Caecidotea, **11,** *638,* **676,** 675
   recurvata, 679
Caenestheria
   setosa, 767
Caenestheriella, 773
   setosa, **775**
Caenidae, 598, *633,* **634**
Caenis, **634**
Caenomorpha, 78
Caenomorpha medusula, **77**
Caenomorphidae, *78*
Caenorhabditis
   briggsae, 257
Calamoceratidae, 611, 642, *643*
Calanoida, 787, 806, *806*
Caligoida, 787
Callibaetis, 598
Callinectes
   sapidus, 830
Calliobdella
   vivida, 453
Calmoceratidae, *643*
Calohypsibiidae, 503, 515, 516
Calohypsibius, 503, **504,** 505, 506, **506,**
   515, 516, *517*
Calopterygidae, 604, **636,** *637*
Calopteryx, **636**
Calposoma, 4, 125, 137, 140, *142*
   dactyloptera, 137

Calyptostomatoidea, 524
Calyptotricha, 83
  pleuronemoides, **82**
Calyptotrichidae, *83*
Camacolaimidae, *266*
Cambarellinae, 833, 834, *850*
Cambarellus, 831, 834, 838, *850*
  puer, 849
Cambaridae, 849, *850*
Cambarinae, 833, 834, *850*
Cambarincola, 421, *431*
  fallax, **432**
Cambaroidinae, 833
Cambarus, 702, 831, 834, *852*
  aculabrum, 847
  batchi, 847
  bouchardi, 847
  brachydactylus, 830
  catagius, 847
  chasmodactylus, 847
  diogenes, 848
  extraneus, 847
  friaufi, 830
  laevis, **833**
  latimanus, **851**
  miltus, 847
  obeyensis, 847
  robustus, 839
  spicatus, 847
  strigosus, **825**
  tartarus, 847
  tenebrosus, 829
Campanella, 75
Campanella umbellaria, **76**
Campascua, 71
Campascus triqueter, **70**
Campbellonemertes, 164
Campeloma, 287, 288, *304*
  decisum, 293, 294, 296, **305**
Camptocercus, 731, *748, 750*
  lilljeborgi, 750
  marcrurus, 750
  rectirostris, 750
Campydoridae, *264*
Candocypria, 712
Candocyprinotus, 709
Candona, **11,** *709*
  acuta, 702
  acutula, 702
  albicans, 702
  anceps, 702
  candida, 700, 701, 702
  caudata, 698, 702
  compressa, 701, 702
  decora, 700, 701, 702, 703
  distincta, 701, 702
  ikpikpukensis, 701, 707
  jeanneli, 702
  marengoensis, 702
  mulleri, 701
  ohioensis, 701, 702, 704
  paralapponica, 702
  paraohioensis, 701, 702
  pedata, 701
  protzi, 699, 701, 702
  rawsoni, 701, 702, 703, **710**
  rectangulata, 699

  renoensis, 702
  subtriangulata, 698, 699, 701, 702, 703, 704
  suburbana, **695**
  thermalis, 701
Candoninae, *709*
Canthocamptidae, 795
Canthocamptus, 795, *812,* **813**
Canuellidae, 799
Capitomermis, 271
Capniidae, 605, 606, **639,** *640*
Carabidae, 279
Carchesium, 75
Carchesium polypinum, **76**
Caridae, 278
Caridea, 823, *850*
Carphania, 503, 504, **504,** 506, *522*
  fluviatilis, 513
Carphanidae, 503, 504
Carunculina, 387, **387**
  parva, **380–382,** *387*
  parva texasensis, 329
  pulla, 387
Caryoblastea, 39
Castrada, 152, **154,** *161*
  virginiana, **161**
Castrella, 163
  pinguis, **161**
Catenula, 147, 153, **154,** *155,* **162**
  lemnae, 148, **155**
  leptocephala, **155**
Catenulida, 146, 148, 150 *153,* **154,**
    155–156, **155**
Catenulidae, *155*
Catostomidae, 363
Caudatella, 598
Celina, 619
Cenocorixa, 614
Centrarchidae, 363
Centrohelida, 39, *70*
Centrolimnesia, 575
Centropagidae, 807
Cephaliophora, 220
Cephalobidae, 251, *264*
Cephalobus, 265
Cephalocarida, 666, **667**
Cephalodella, 190, **194,** 196, **238,** *239*
  cyclops, **238**
  forficula, 223
  gibba, **238**
  intuta, **238**
Cephalothamnion
  cyclopum, **65**
Cephalothamnium, 46, *66*
Ceraclea, 607, **641**
Ceratium hirudinella, **40**
Ceratodrilus, **7,** *431*
  thysanosomus, **432**
Ceratophylum
  demersum, 295
Ceratopogonidae, 538, 625, 626, **651,** *652*
Ceratopsyche, 607
Cercobrachys, 598
Cercomonadida, 39, 44, 46, *66*
Cercomonas, 46, **65,** *66*
Cercopagidae, 724, *746, 763*

Ceridodaphnia
  quandrangula, **755**
Ceriodaphnia, 727, 729, 733, 734, 737, 740, *756,* 803
Cestoidea, 146
Chactpmptis, 177
Chaena teres, **80**
Chaenea, 81
Chaetogaster, 408, 409, *423*
  diaphanus, 704
  limnaei, 362
Chaetonotida, 172, 179, *181*
Chaetonotidae, 174, *181*
Chaetonotus, **5,** 176, *181,* **182**
  maximum, **178**
Chaetospira, 77
  mulleri, 77
Chaetospiridae, 77
Chamydomonas
  reinhardtii, 491
Chamydotheca
  arcuata, 696
Chaoboridae, 59, 538, 625, 626, **651,** *652*
Chaoborus, 30, 197, 216, 626, **651,** 727, 732, 733, 801, 803
  flavicans, 704
Chaos, 68
Chaos illinoisense, **67**
Chappuisidees
  eremitus, **585**
Chappuisides, 575
Chappuisididae, 526, 531, *575,* **585**
Chappuisidinae, 531, *576*
Chauliodes, 613, *613*
Cheilostome, 490
Chelicerata, 8, *15*
Chelomideopsis, 551, *576*
  bessilingi, **585**
Chidastygacaridae, 531
Chiliodonellidae, *73*
Chilodonella, 73
Chilomonas, 66
Chilomonas paramecium, **64**
Chilopoda, 278
Chimarra, **641**
Chiorcephalidae, 724
Chironomidae, 30, 60, **258,** 449, **450,** 538, 539, 543, 599, 624, 625, 626, *652,* **652**
Chironomus, **10, 652**
Chlamydomonas, 61, 372
Chlamydomyxa, 71
  montana, **70**
Chlamydomyxidae, *71*
Chlamydophryidae, *71*
Chlamydophrys, 71
  minor, **69**
Chlamydotheca, 711
Chlorella, 4, 56, 133, 135, 178, 211, 372, 744
  fusca, 220
  pyrenoidosa, **211**
  vulgaris, 372
Chlorohydra, 4, 133, 140, 141
Chloroperlidae, 605, 606, *638,* **638**
Choanoflagellida, 39, 44, *66*
Chonotrichia, 39
Chordariidae, *155*

Chordarium europaeum, 155
   europaeum, 170
Chordodes, 275, 279
Chordodidae, *279*
Chordodiolus, 279
Choreotrichia, 39, 49, *78*
Chrodridae, *732*
Chromadora, 258
Chromadorida, *265*
Chromadoridae, *265*
Chromadorita, 265
Chromodorida, 258
Chromogaster, 239
Chronogaster, 266
Chrysomelidae, 279, **649,** *650*
Chrysomellidae, 619, 623
Chrysomonadida, 39, 46, *66*
Chrysops, **651**
Chydoridae, 724, 727, 732, 745, *746, 748,*
   750, 750–755
Chydorinae, *750*
Chydoris, 133
Chydorus, 731, *750, 753,* **754**
   brevilabris, **754**
   faviformis, 743
   piger, 755
   reticulatus, 743
   sphaericus, 742, 743, 756
Ciliophora, 3, 39, 48–50, 54, *63, 72–86*
Ciliophrida, 39
Ciliophrys, 72
   infusionum, **72**
Ciliophyrida, 72
Cincinnatia, 302
Cinetochilidae, *83*
Cinetochilum, 85
   margaritaceum, **82**
Cipangopaludina, 304
   japonica, **305**
Cirripedia, **667**
Cladocera, 10, 60, 449, **450,** 528, 529,
   530, 531, 723–776, *746*
Clappia, 302
Clathrosperchon, 572
   ornatus, **580**
Clathrosperchoninae, 527, 528, 541
Clathrosperchontinae, *572,* **580**
Clathrostomatidae, *85*
Clathrostoma, 86
   viminale, **84**
Clathrulina, 71
   elegans, **72**
Clenostomata, 481
Cleotocamptus, 812
Climacia, **646**
Climacostomidae, *79*
Climacostomum, 79
   virens, **80**
Clitellata, 402, 411, **412,** 430, 438
Cloeodes, 598
Clupeidae, 363
Cnidaria, 4, *13,* 61, 125–143, **127**
Cochliopina, 302
Cochliopodiidae, *68*
Cochliopodium, 68
   bilimbosum, **69**
Codonellidae, *78*

Codosiga, 66
   botrys, **63**
Coenagrionidae, 602, 603, **636,** *637*
Cohnilembidae, *83*
Cohnilembus, 83
   pusillus, **82**
Coleoptera, 277, 278, 298, 452, 528, 538,
   543, 594, 596, 612, 615, 616,
   618–624, *632, 633*
Colepidae, *79*
Coleps, 79
   hirtus, 60, **80**
Collembola, 9, **10,** *15,* 30, 528, 593–653,
   659
Collotheca, **192, 194,** 205, 215, *226*
   ferox, **227**
   gracilipes, 204, 205, **205**
   mutabilis, **192,** 221
   trilobata, **192**
Collothecacea, **192,** 193, 221, 225, *226*
Collothecidae, 191, 193, 204, 215, *226,*
   227
Colpidium, 85
   colpoda, 51, **84**
Colpoda, 56, 86
   steinii, **84**
Colpodea, 39, *79, 85,* 86
Colpodida, *85,* 86
Colpodidae, *86*
Colurella, 233, **233**
Colurellidae, *233,* **233**
Colymbetes, 619
Comptocercus
   rectirostris, **749**
Conchostraca, 767
Condylostoma, 79
   tardum, **80**
Condylostomatidae, *79*
Conochilidae, *227*
Conochiloides, 202, *227,* 228
Conochilus, 202, **203,** 206, 218, 219, *227,*
   228, 739
   hippocrepis, 204
   unicornis, **203,** 204
Cookacarus, 574
Cookidrilus, 413
Copelatus, 619
Copepoda, 11, **12,** *15,* 449, **450,** 529, 530,
   666, **667,** 674, 787–813
Coptotomus, 619
Coquillettidia, 627, **651**
Corbicula, 7, 334, **335, 358,** 364, 365,
   369, *374, 375,* **376**
   fluminea, 35, 316, 319, 322, 324, **325,**
   325, 327, 328, 329, 331, 332, 333,
   334, 335, 336, 337, 338, 339, 340,
   341, 342, **343,** 346, 347, 349–351,
   353, 354, **355,** 355, **357,** 359, 360,
   361, 362, 363, **363,** 364, 365, 366,
   367, 368, 369, 370, 371, 372, 373,
   381, *375,* 376
Corbiculacea, 316, 367, 373, *375,* 375
Corbiculidae, 316, **316, 320, 357,** *375*
Cordulegaster, **636**
Cordulegastridae, 602, **636**
Cordulesgastridae, *637*
Cordulia, **636**

Corduliidae, 603, **636**
Cordylophora, 126, 137–139, 140, *141,*
   lacustris, 130
Corixidae, **532,** 278, 543, 614, 615, 616,
   644, **645**
Cornechiniscus, 511, *516*
Corticacarus, 577
Coruduliidae, *637*
Corvomeyenia, 103, 108, *114*
   carolinensis, 108, 112, *114,* 116,
      116
   everetti, 105, 108, 112, *116,* **118**
Corvosponbilla becki, 114
Corvospongilla
   becki, 116
Corydalidae, 614, *643*
Cothurnia, 75
   imberbis, **76**
Cowichania, 575
   interstitialis, **581**
Cowichaniinae, 526, 528, *573,* **581**
Crangonyctidae, 675, 676, *683*
Crangonyx, **674,** 675, 680, *683*
Craspedacusta, 4, **4,** *13,* 125, 126, 129,
   130, 135–136, **136,** 139, 141, *144*
   sinensis, 136
   sowberii, **135**
   sowerbii, 135, **135,** 136,
Crenitis, 620
Crenodaphnia, 733
Cretaccous, 367
Criconematidae, *264*
Cricotopus, 34, 258, **258,** 625
Crimaldina
   brazzai, **761**
Cristatella
   magnifica, **496**
   mucedo, 487, 488, 489, **496,** *497*
Cristatellidae, 490, **496**
Cronodrilus, 431
   ogyguis, **432**
Crustacea, 8, 10, *15,* 446, 500, 665–683,
   691, 787
Crustipellis, 405, **419**
   tribranchiata, 418
Cryptobranchus
   alleganiensis, 463
Cryptocyclops
   bicolor, 808, 809
Cryptomonadida, 39, 46, *66*
Cryptomonads, 44
Cryptomonas, 804
   reflexa, 805
Cryptonchus, 266
Crytomonas, **45**
Crytonaias, 385
Crytonia, 235
Ctenidae, 278
Ctenodaphnia, 732
Ctenopoda, 724
Cubozoa, 126
Culex pipiens
   quinquefasciatus, 263
Culicidae, *273,* 538, 543, 625, 627, **651,**
   652
Culicimermis, 273
Cumacea, **667, 676**

Cumberlandia, 378
    monodonta, 378, **380–382**
Cumberlandiidae, *378*
Cupelopagis, **193,** 196, 215, *226*
    vorax, **227**
Cupeolpagis, 216
Cura, 158
    foremanii, 151
Curculionidae, 623, *646,* **642**
Cyanobacteria, 33, 57
Cyatholaimidae, *265*
Cybister, 619
Cyclestheria, 764
    hislopi, 765
Cyclestheriidae, 724
Cyclidiidae, *83*
Cyclidium, 83
    glaucoma, **82**
Cyclocypinae, *709*
Cyclocypris, 712
    ampla, **694,** 701, 702, 704, **714**
    globosa, 701
    laevis, 701, 702
    ovum, 701, 702
    serena, 701
    sharpei, 700, 701, 702
Cyclomomonia, 575
    andrewi, **584**
Cyclomomoniinae, 526, 530, *581,* **584**
Cyclonaias, 389
    tuberculata, **380–382,** *389*
Cyclopinae, 809
Cyclopoda, 60
Cyclopoida, 787, 790, *807–809*
Cyclops, 787, 803, *808*
    canadensis, 809
    columbianus, 809
    insignis, 809
    kolensis, 809
    scutifer, 794, 809, **809**
    strenuus, 809
    vicinus, 797, 809
Cyclorrhapha, 624
Cyclorrhaphous, 625
Cyclothyadinae, 526, 528, *573,* **581**
Cyclothyas, 573
    siskiyouensis, **581**
Cyldidium, 83
Cylindrolaimus, 266
Cylocypris
    ampla, 701
Cymbella, 841
Cymbiodyta, 620
Cymocythere, 716
Cyphoderia, 71
    ampulla, **70**
Cyphoderiidae, *71*
Cyphon, **649**
Cypretta, 709
    kawatai, 704
Cypria
    obesa, 701, 702
    ophthalmica, 700, 701, 702
    turneri, **694, 695,** 704
Cypricercus, 696, 703, 706, 707, 711
    deltoidea, 701, 702
    horridus, 703
    reticulatus, 697, 701, 702, **710**

Cyprichoncha, 706
Cypriconcha, 707
Cyprideis, 697, 698, *713*
Cyprididae, **695,** 697
Cypridinae, *709*
Cypridoidea, 692
Cypridopsinae, *709*
Cypridopsis, 703, 707, *709*
    hartwigi, 704
    vidua, 697, 698, 699, 701, 702, 703,
        704, 705, **710**
Cyprinidae, 363
Cyprinotus, 698, 706, 707, *712*
    carolinensis, 698, 698–699, 704
    glaucus, **694,** 700, 701, 702
    incongruens, 697, 698, 704, 705
    salinus, 699
Cyprinus
    carpio, 363
Cypriodoidea, *708, 711–713*
Cypris, 703, 707, *712*
    balnearia, 701
    pubera, 701, 702, 705, **714**
    thermalis, 701
Cyprogenia, 385
    aberti, 385
    alberti, **380–382**
    irrorata, 385
Cyprois, 709
    marginata, **710**
Cyrenoida
    floridana, 374
Cyrenoidacea, 373, *374*
Cyrinus
    carpio, 363
Cyrtolophosidae, *85*
Cyrtolophosidida, *85*
Cyrtolophosis, 85
    mucicola, **82**
Cyrtonaias
    tampicoensis, **380–382,** *385*
Cyrtophora, 39, 49
Cyrtophorida, 60, *73*
Cystobranchus, 461, *465*
    mammillatus, 465
    meyeri, **441,** *465*
    verrilli, 465
    virginicus, 465
Cystobrnachus
    verrilli, **441**
Cytheridae, **699, 695**
Cytherissa, 705, *713*
    lacustris, 698, 700, 701, **714**
Cytheroidea, 692, *713–716*
Cytheromorpha, 705, *713*
    fuscata, 699, 700, 701, 702
Cyzcidae, 724
Cyzicus, 765, 767, *773*
    californicus, 766, **775**

Dactylobiotus, 503, **504,** 505, **506, 507,**
    510, 513, 515, 516, *517*
    dispar, 508, 511
    macronyx, 510
Dactylocythere, 716
    leptophylax, 716

Dadaya, 753, 753
    macrops, **752**
Dalyellia, 151, 152, **157,** *163*
    viridis, 147, 148
Dalyelliida, 146
Dalyelliidae, 146, 153, **156**
Dalyellioda, *163*
Dalyellioida, 146, **154,** *154,* **161**
Dannella, 598
Daphnia, 10, **12,** 57, 69, 133, 197, 199,
    217, 671, 725, 727, 729, 730, 731,
    732, 733, 735, 736, 737, 738, 739,
    740, 742, *756, 763,* 765, 800, 801, 803
    ambigua, 727
    galeata
        mendotae, 727
    longispina, 733, 734
    magna, 133, 734
    pulex, 217, **217,** 727, 731, 732, 734,
        737, 800
    pulicaria, **726, 728,** 727, 740
    retrocurva, 727, **728**
    similis, 34
Daphniidae, 724, *746, 756*
Daphniopsis, 730, 731, 732, 733, *756*
    ephemeralis, **755,** *756*
Darwinula, **694, 697,** 698, 705, 707
    stevensoni, 698, 701, 702, **710**
Darwinulidae, 692, 707
Darwinuloidea, 707
Dasydytes, 177, **179,** *182*
    ornatus, **179**
Dasydytidae, 175, *181*
Dasyhormus, 156
Daubaylia, 264
Daubayliidae, *264*
Decapoda, *12,* **13,** *15,* 666, **667, 669,**
    823–852, *852*
Deltopylum, 60
Demospongiae, 111
Dendrocoelidae, 148
Dendrocoelopsis, **154,** *159*
Dendrocometes, 73
    paradoxus, **74**
Dendrocometidae, *73*
Dendrosoma, 73
    radians, **74**
Dendrosomatidae, *73*
Dero, 406, 408, 409, 416, **422,** *423, 424*
Desmarella, 66
    moniliformis, **65**
Desmona, 607
Desmopachris, 619
Desmothoracida, 39, *71*
Detritivores, 294
Deuterophelbiidae, 627
Deuterophlebia, **652**
Deuterophlebiidae, 625, *652,* **652**
Dexteria, 772
    floridanus, **773**
Diacyclops, 794, *808*
    bicuspidatus, 794, *810*
    biosetosus, 810
    crassicaudis brachycercus, *810*
    haueri, 810
    jeanneli, 810
    languidus, 810
    languidoides, 810

navus, 810
  nearcticus, 810
  palustris, 810
  thomasi, **809**, *810*
  yeatmani, 810
Diamphidaxona, 575, **585**
Diaphanosoma, 727, 729, 733, 736, *748*
  birgei, **747**
Diaptomidae, 794, 807
Diaptomus, 139, 788, 787, 790, **798**, 799,
    800, 801, 807, *807*
  connexus, 34, 794
  minutus, 794, 797
  nevadensis, 34, 796
  oregonensis, 794
  pallidus, 805
  shoshone, 800, 805
  sicilis, 794
  siciloides, 802
  tyrrelli, 796, 797
Dibolocelus, 620
Dichaetura, 176, 180, *181*, **182**
Dichaeturidae, 174, 176, *181*
Dicosmoecus, 607
Dicranophoridae, 194, *231*, **232**
Dicranophorus, **194**, **232**
  forcipatus, **232**
  uncinatus, **232**
Dicrotendipes, 625
Dictya, 625
Didiniidae, *81*
Didinium, **43**, 44, 56, *81*
  nasutum, 56, **80**
Difflugia, 59, 69
Difflugia corona, **69**
Difflugiidae, *69*
Difflugina, 68
Digononta, 224, *225*
Digordius, 279
Dileptus, 56, *81*
  anser, **82**
Dina, 457, **467**, 468
  anoculata, 468
  duba, 468
  parva, 468
Dina-Mooreobdella, 455, **456**
Dineutus, 619, **649**
Dinobryon, 66
Dinobryon sertularia, **64**
Dinoflagellida, 39, 46, 54, *63*
Dioctphyme
  renale, 430
Diphascon, 503, **504**, 506, **506**, 508, 513,
    515, 516, *518*
Diploeca, 66
  placita, **65**
Diplogasteridae, *264*
Diplois
  daviesiae, *234*
Diplomonadida, 39, 44, 46, 51, *66*
Diploperla, 605
Diplophyrys, 71
Diplophyrys archeri, **70**
Diplopoda, 278
Dipseudopsidae, 609, *640*, **641**
Diptera, 32, 30, 60, 298, 524, 528, 529,
    530, 531, 537, 538, 539, 543, 594,
    596, 624–631, *632*, *650–653*

Discomorphella, 78
  pectinata, **77**
Discomorphellidae, 78
Discophryidae, *74*
Discophyra, 74
  elongata, **74**
Disematostoma, 85
  butschlii, **84**
Disparalona, 750, 753, *753*
  acutirostris, 753
  dadayi, 753
  leei, 753, **754**
  rostrata, 753
Dissotrocha, **226**, *226*
Distocambarus, 831, 831, 834, *852*
  crockeri, **851**
  youngineri, 847
Dixella, **652**
Dixidae, 625, 627, *652*, **652**
Dochmiotrema, 163
Dolannia, **634**
Dolerocypris, 711
Dolichopodidae, *650*
Dolicopodidae, 629, **651**
Donacia, **649**
Donnaldsoncythere, 716
Dorylaimida, 258, *263*
Dorylaimidae, *264*, **267**, **270**
Dorylaimus, **252**, 258, *264*, **267**, **270**
  atratus, 254
  thermae, 257
Doryphoribius, 503, **504**, 513, 515, 516,
    517
Doryphoribus, **506**
Dosilia, 114
  palmeri, 116
  radiospiculata, 117, **117**
Dosilia palmeri, *114*
  radiospiculata, *114*
Dracunculus
  medinensis, 803
Dreissena, **7**, *374*
  polymorpha, 316, 318, 324, 325, 336,
      337, 339–342, **343**, 344, 345, 347,
      351, 351–355, 360, **360**, 362, 365,
      366, 369, 371, 372, *374*, **374**
Dreissenacea, 373, *374*
Drepanothrix, 760
  dentata, **759**
Drepanotrema, 307
  kermatoides, **307**
Drilomermis, 273
Dromogomphus, 602
Dromus, 386
  dromus, **380–382**, *386*
Drunella, 598
Dryopidae, 619, 619, 623, *648*,
    648
Dryopoidea, 619, 623
Dryops, 622
Dubiraphia, 619
Dugesia, **4**, 152, *158*
  dorotocephala, **158**
  polychroa, 148, **158**
  tigrina, 147, 148, **158**
Dugesiidae, 148
Dunhevedia, 750 753, 753
  americana, **752**

Dysnomia, 388
  triquetra, 373, **380–382**
Dytiscidae, 279, 543, 618, 619, 620–622,
    621, 624, *646*, **647**, *648*, **649**
Dytiscus, 619, 275

Echinamoeba, 68
  exudans, **67**
Echinamoebidae, 68
Echinisca, 763, *763*
  schauinslandi, **761**
Echiniscidae, 503, 504, 513, *516*,
Echiniscoidea, 503, 504, 513
Echiniscoides, 503
Echiniscoididae, 503
Echiniscus, 503, **504**, **506** 510, 513, *516*
  laterculus, 510
  oihonnae, 510
  spiniger, **502**
  testudo, 512
Echinisoides
  sigismundi, 509
Echinodermata, 2, 273
Echinursellus, 513
Echiursellus
  longiunguis, 513
Eclipidrilus, 420, **420**, *421*
  fontanus, 421
  frigidus, 421
  lacustris, 421
Ecnomidae, 609, *640*
Ectocyclops
  phaleratus, 807
  pharleratus, **809**, *810*
Ectopria, **649**
Ectoprocta, 8, **9**, *15*, 481, 489, 491
Edischura
  lacustris, 797
Eiseniella, 417
Elaphoidella, 812
Elgiva, 625
Elimia, 304 See also *Goniobasis*
  cohawbenses, 296
  clara 296
  livescens, **303**
  livescers, 297
Ellipsaria
  lineolata, **380–382**, *386*
Elliptio, 389
  complanata, 330, 331, 331, 338, 340,
      346, 359, 366, **380–382**
Elliptoideus, 381
  sloatianus, **380–382**, *381*
Ellisodrilus, 431
Ellissaria, 386
Elmidae, 618, 619, 622, 622–623, *648*,
    648, **649**, *650*
Elodea, 139, 204, 222, 849
  canadensis, 204
Eloranta, 57
Elosa, 237
Elpidoptera, *633*
Embafa
  parasitica, 221
Empididae, 629, *651*, **651**
Empidomermis, 273

Enallagma, 602, **636**
  clausum, 34
Encentrum, 223
  linnhei, 223
Enchelyidae, *81*
Enchelyodon, *81*
  elegans, **80**
Enchelys simplex, **80**
Enchytraeidae, **403**, 404, 405, **406**, 407, 412, 413, 417, *418*
Endochironomus, 625
Endogenida, *73*
Endotribelos, 625
Enghelys, *81*
Enochrus, 620
Enoplida, 258
Entamoebidae, **47**
Entocythere, 67, 707, *715*
  donaldsonensis, 702
Entocytheridae, 692
Entocytherinae, *713*
Entognatha, **10**, *15*, 594, 631
Entoprocta, 8, *492*
Entosiphon, *63*
  sulcatum, **64**
Entroprocta, 481
Eocosmoecus, 607
Eocyzicus, *773*
  concavus, **775**
Eohypsibiidae, 503
Eohypsibius, 503, **504**, 505, **506**, 516, *518*,
Eothinia, **194**
Epactophanes, 812
Epalxellidae, *78*
Ephemera, **634**
Ephemerella, 598
Ephemerellidae, 598, 599, *634*, **635**
Ephemeridae, 599, *634*, **634**
Ephemeroporus, 743, *750*, *753*, *753*
  acanthodes, **752**
Ephemeroptera, 594, 597–602, 604, *632*, 633–635,
Ephoron, **634**
Ephydatia, 113, *115*
  fluviatilis, **97**, 99, 103, 105, *115*, 117, **117**
  mackayi, **117**
  melleri, 117, **117**
  millsii, *115*, 117, **117**
  muelleri, **97**, 103, 104, 105, *115*, 117
Ephydridae, 629, *650*, **651**
Epiphanes, **194**, 196, 219, *235*, **235**
  senta, 213
Epiphanidae, *235*, **235**
Epischura, 218, 788, 792, 796, 798, 800, 806
  lacustris, 794
Epistylididae, *75*
Epistylis, *75*
  plicatilis, **76**
Erebaxonompsis, 577
Eristalis, **651**
Erpobdella, 454, 467, **467**
  montezuma, 437, 449, 449, 452, 453, 467
  octoculata, 452
  punctata, 445, **445**, 446, 449, **449**, 451, 452, 453, 454, *467*

Erpobdellidae, 284, **438, 441,** 441, **443,** 443, **444,** 444, 446, 449, 457, *461*, **461**
Erpobellidae, 278
Errantia, 458, *468*
Erythraeoidea, 524
Eschatigenae, *386*
Espejoia, 60
Estellacarus, *553*, *577*
Ethmolaimus, 256, 257, *265*
Euamoebida, 39, *68*
Euapius
  fragilis, 117
  mackayi, 117
Eubosimina, **757**
Eubosmina, **757**
Eubranchipus, **12**, 765, 766, 767, 770, 772
  bundyi, 768
  holmni, 769
  hundyi, **773**
  moorei, 771
  serratus, 768
  vernalis, 769
Eubrianax, 623
Eubriidae, **649**
Eubriinae, 623
Eucarida, 666, **667**
Eucephalobus, 265
Euchlanidae, *234*, 234, *235*
Euchlanis, **190**, 214, **234**, *235*
  dilatata, 202, 222
Euchordodes, 279
Euchromadora
  striata, 257
Eucyclops, 795, *808*
  agilis, **809**, *810*
  macruoides denticulatus 810
  macrurus, 810
  prionophorus, 810
  serrulatus, 810
  speratus, 810
Eucypris, 703, 706, 706, *711*
  crassa, 707, **714**
  serrato-marginata, 707
Eudiaptomus, 807
Eudorylaimus
  andrassyi, 257
Euglenida, 39, 47, *63*
Euglypha, 71
  tuberculata, **70**
Euglyphidae, 71
Euglyphina, 71
Euhirudinea, 438
Euholognatha, 605–606
Eukiefferiella, 625
Eulimnadia, 765, *772*
  agassizii, **774**
  antlei, 765, 767
  divers, 765
  inflecta, 767
Eumalacostraca, **667**
Eunapius, *115*, 120
  fragilis, 115
  igloviformis, 118
  mackayi, 115
Eunapius fragilis, 103
  mackayi, **96**, 103
Eupera, 376
  cubensis, 376, **377**

Euperinae, *376*
Euphausiacea, **667**
Euplotes, *75*
  patella, 52, **76**
Euplotida, *75*
Euplotidae, *75*
Eurotatoria, 224
Euryalona, 748, 751, *751*
  orientalis, **752**
Eurycercus, 730, 731, 738, 743, *748*
  lamellatus, **749**
Eurylophella, 598, **635**
Eurytemora, 807, *807*
  affinis, 794
Eutardigrada, 503, 504, **506**, 513, 516
Euteratocephalus, 266
Euthyas, 548, *573*
Evadne, 733, *763*
  nordmanni, **764**
Evaginogenida, *73*
Exogenida, *73*
Eylaidae, 528, 533, 534, 537–539, 541, 542, *548*, *554*, **557, 580**
Eylais, 548, *572*, **580**
  major, **557**
Eylaoidea, 524, 527, 528, 535–537, 539, 547, *554*

Fallicambarus, 831, 831, 834, *852*
  devastator, 840, **840**, 842, 848
  fodiens, **851**
  hedgpethi, 842
  jeanae, 847
Farula, 612
Faxonella, 831, 831, 834, *852*
  clypeata, 846, 849
  creaseri, **851**
Feltria, *553*, **567, 578,** *575*
  exilis, **584**
Feltriidae, 529, 540, *553*, **573, 578,** *575*, **584**
Ferrissia, **6,** *309*
  fragilis, 297
  rivularis, **308**
Ficopomatus
  enigmaticus, 468, **469**
  miamiensis, 469, **469**
Filamoeba, 68
  nolandi, **67**
Filinia, 197, 200, **201,** 219, **219,** *231*
  longiseta, 212
  longiseta passa, 218
Filiniidae, *231*
Filosea, 39, 48, 66
Fisherola, *309*
  nutalli, **308**
Floscularia, 205, 215, *228*, **228**
  conifera, 202, 204, 205, 210, **229**
  ringens, 202, 205
Flosculariacea, 193, 221, 225, *228*
Flosculariidae, 202, 204, *228*, **228, 229**
Fluminicola, 302
Fontelicella, 302
Fontigens, 303
Fontigentinae, **302**, *303*
Forelia, 554, 576, *584*
  floridensis, **586, 587**

onondaga, **570, 571**
ovalis, **570, 571**
Foreliinae, 530, 535, 540, 541, 542, 554,
   570, 571, *576, 578*, **586, 587**
Forficulidae, 278
Formicidae, 278
Fossaria, *305*
   humilis, **306**
Fragilaria, *841*
Fredericella, *482*, **483**, *487*, *489*, *490*, *492*
   australiensis, *493*
   indica, *487*, *488*, *489*, *492*, **495, 496**
   sultana, *487*, *488*, *493*
Fredericellidae, **495**
Frontipoda, *551*, *574*
Frontipodopsinae, 530, *575*, **584**
Frontipodopsis, *575*, 584
Frontonia, *59*, *85*
   leucas, **84**
Frontoniidae, *85*
Fusconaia, *383*
   flava, **380–382**, *383*

Gammaridae, 675, *675*, 679, *683*
Gammarus, **11**, 152, **674**, 676, 677, *683*
   lacustris, 679
   minus, 679, 681
Gastromermis, *271*
Gastrophoda, 449
Gastropoda, 6, **6**, *14*, *285–309*, *301–309*,
   449, **450**
Gastropodidae, *239*, **239**
Gastropus, 6, *239*, **239**
Gastrostyla, *75*
Gastrostyla steini, **77**
Gastrotricha, 5, **5**, *14*, *172–182*
Geayia, *550*, *576*, **585**
Gelastocoridae, 613, *644*
Geocentrophora, **154**, *156*, **156**
Geocythere, *715*
Gerridae, 30, 537, *616*, 614, 617, *644*, **645**
Gerrinae, 614
Gerris, 614, **645**
Gerromorha, 616–617
Gerromorpha, 613, 616, *644*
Giardia, 61
Giebula
   rotundata, **380–382**
Gieysztoria, **154**, *163*
Glaucoma, *85*
Glaucoma scintillans, **84**
Glaucomidae, *85*
Glebula, *386*
   rotundata, *386*, **386**
Glochidia, 341, 343, 369, *383*
Glomeridae, 278
Glossiphonia
   complanata, 445, 449, **449**, 451, 452,
      456, *464*, **464**
   heteroclita, 456
Glossiphoniidae, 278, **438**, 442, **442**, 443,
   444, **444**, 445, 446, 448, 449, 456,
   460, *461*
Glossosoma, **642**
Glossosomatidae, 608, 610, *642*, **642**
Gnaphosidae, 278
Gnesiotrocha, 225

Goera, **642**
Goeridae, 607, 611, **642**, *643*
Gomphidae, 602, 603, **636**, 637
Gomphonema, *841*
Gomphus, **636**
Gonidea, *378*
   angulata, *378*, **380–382**
Gonideini, *378*
Goniobasis, 296
Gonostomum, *78*
   affine, **77**
Gonostromatidae, *78*
Gordiacea, 273
Gordiidae, *279*
Gordioidea, *279*
Gordionus, *279*
Gordius, 273, **273**, *275*, 275, 277, *279*
   tenuifibrosus, 273
Granuloreticulosea, 39, *66*, *71*
Granuloreticulosida, 48
Graphoderus, 619
Graptoleberis, 733, *748*, 748
   testudinaria, 739, **749**
Gratrix, *163*
Grimaldina, 760
Gromiida, 39, *70*
Gromiina, *71*
Gruberellidae, *67*
Gryllacrididae, 278
Gryllidae, 278
Guernella, 760
   raphaelis, **761**
Guttipelopia, 625
Gymmolaemata, *492*
Gymnamoeba, 39, 54, *67*
Gymnodinium, *63*, **64**
Gymnolaemata, 481, 486, 490
Gymnopais, 625
Gyratrix, **154**
   hermaphroditus, 150, **158**
Gyraulus, *307*
   deflectus, **307**
Gyrinidae, 613, 618, 619, 620, 627, *646*,
   648, **649**
Gyrinus, 619
Gyrodinium, *63*, **64**
Gyrotoma, *303*
   excisum, **303**

Haber, **425**
   speciosus, *424*
Habrophelbia, **635**
Habrotrocha, **226**, *226*
   colaris, **236**
   constricta, **236**
   rosa, **199**, 222
Habrotrochidae, 225, **226**, *226*
Haementaria
   ghilianii, 442, 448, 453
Haementeria, 456
Haemonais
   waldvogeli, 423
Haemopis, 456
Haemopsis
   sanguisuga, 452
Halacaridae, 523
Halicyclopinae, *810*

Halicyclops, 794, *808*, **809**
   clarkei, *810*
   coulli, *810*
   fosteri, *810*
   laminifer, *810*
Haliplidae, 618, 619, 621, *646*, **647**, *654*,
   649
Haliplus, 619, **547, 649**
Halobiotus, 503, 512
Halteria, *79*
Halteria grandinella, **77**
Halteriidae, *79*
Haplomacrobiotus, 503, **504**, 506, **506**,
   507, 515, 516, *517*
Haploperla, **638**
Haplopoda, 724, 727
Haplotaxidae, 407, 411, 417, *417*
Haplotaxis, **406**, 407, 409, 411, *417*,
   418
   gordioides, *417*
Haptoglossa, **220**
Haptorida, *81*
Harpacticoida, 787, 810, *810–812*
Harpacticus, *811*
   gracilis, 794
Harpagocythere, 716
Harreolanus, *576*
Harringia, 232, **233**
Hartmanella veriformis, **67**
Hartmannella, 68
Hartmannellidae, *68*
Hartocythere, *715*
Hastatella, *75*
   radians, *76*
Hauffenia, *302*
Hebesuncus, 503, **504**, *518*
Hebetancylus, *309*
   excentricus, **308**
Hebridae, 616, *644*
Heleidae, 626
Heleidomermis, *271*
Helichus, 622, **648**
Helicopsyche, **641**
   borealis, 611
Helicopsychidae, 611, **640**, *641*
Heliophrya, *74*
Heliophryidae, *74*
Heliophyra reideri, *74*
Heliozoa, 44, **47**, 48, 54, 58, 59, *66*,
   71, *79*
Heliozoea, 39
Helisoma, **6**, 298, 299, *308*, **308**, *841*
   anceps, 293, **308**
   trivolvis, 293, 294, 295
Helobdella, **441**, *463*
   california, *463*, **464**
   elongata, *460*, *463*
   fusca, *463*
   papillata, *463*
   stagnalis, 445, 446, 449, 450, **450**, 451,
      453, *453*, *463*, **464**
   transversa, *463*
   triserialis, 451, *463*
Helodidae, 624
Helophorus, 620
Helopicus, 605
Hemicycliophora, **269**, *264*
Hemicycliophoridae, **269**

Hemiptera, 298, 452, 528, 537, 538, 543, 594, 613, **572,** *632*
Hemistena, 389
 lata, **380–382,** *389*
Heptagenia, 598
Heptageniidae, 278, 598, 600, *634,* **635**
Herbridae, 614
Herpetocypris, 699, 703, *712*
 reptans, 698, 698
Hesperocorixa, 614
Hesperodiaptomus, 787, 807
Hesperoperla, 605
Hesperophylax, 607, **641**
Hetaerina, **636**
Hetermeyenia
 tubisperma, **118**
Heterocope, 788, 798, 806
 septentrionalis, 794, 796, 800, *806*
Heterocypris, 706, 707
Heterogenae, *385*
Heterolaophonte, 811
 stromi, 811
Heterolepidoderma, 177, **179,** *181,* **182**
Heterolobosea, 39, *66*
Heterolobosia, 47
Heteromeyenia, 113, *114*
 baileyi, 114, 118
 latitenta, 114, 118, **118**
 tentasperma, 114, 118, **118**
 tubisperma,114, 118
Heterophrys, 71
 myriopoda, **72**
Heteroplectron, **642**
Heteroptera, 594, 595, 596, 613–617, *632,* *643–646*
Heterosternuta, 619
Heterotardigrada, 502, 503, **506,** 508, 513
Heterothricha, 49
Heterotrichia, 39, *78, 79*
Heterotrichida, *79*
Heterotrichina, *79*
Hetrophrys, **45**
Hexagenia, **634**
Hexamita, 51, *66*
 inflata, **65**
Hexapoda, **10,** 439
Hexapodibius, 503, **504, 506,** 515, 516, *517*
Hexarthra, 195, **195,** 197, 218, *228*
 fennica, 34, 212
Hexarthridae, *228*
Hexatoma, **651**
Hexopoda, 631
Himatismenida, 39, *68*
Hirschmanniella, 264
Hirudidae, 450
Hirudinea, 7, *14,* 402, 404, **413,** 414, 438, 439
Hirudinidae, 438, 441, **441, 442,** 443, **443,** 444, **444,** 446, 449, 450, 456, *461,* **461**
Hirudinoidea, 8, *14,* 438, 439, *460–468*
Hirudo
 medicinalis, 442, 443, **443,** 448, 453, 454, 467
Histobalantiidae, *83*
Histobalantium, 83
 natans, **82**

Hobbseus, 831, 831, 834, *852*
 cristatus, **851**
 orconectoides, 847
Hollphryidae, *81*
Holopediidae, 724, 746, *746*
Holopedium, 730, 733, 737, 740
 gibberum, 747, **747**
Holophrya simplex, **82**
Holophyra, 81
Holotrichous
 isorhiza, **140**
Homalozoon
 vermiculare, **80**
Homochaeta
 naidina, 421
Homocyclops
 ater, 818, 810
Homalozoon, 81
Hoplocarida, **667**
Hoplodictya, 625
Hoplonemertea, 164
Horatia, 302
Horella, 228
Horreolaninae, 526, 531, *576,* **583**
Horreolanus
 orphanus, **583**
Hoyia, 302
Htydryphantes, 573
Huitfeldtia, 554, *579*
 rectipes, **571, 588**
Huitfeldtiinae, 530, 542, *554,* **571, 588**
Huntemannia, 812
Hyalella, 679, *683*
 azteca, 453, 676, 679, 680
 montezuma, 449, 453, 676, 678
Hyalellidae, 676, *683*
Hyalinella, 490, *493*
 orbisperma, 487, 488, *494,* **495**
 punctata, 487, 488, 489, *494,* **494, 496**
 vaihiriae, 489, *494,* **494**
Hyallela
 azteca, 666
Hyalodiscidae, *68*
Hyalodiscus, 68
Hyalopyrgus, 302
Hyalosphenia, 69
 cunneata, 69
Hyalospheniidae, *69*
Hydaticus, 619
Hydra, **4,** *75,* 126, **127,** 130–134, **131,** 140–142, 735, 739, 754
 braueri, 141, *142*
 oligactis, 141, **141,** *142*
 viridis, **134,** *142*
 viridissima, 141
 vulgaris, 141, *142*
Hydraamoeba hydroxena, 130
Hydracarina, 523
Hydrachinidia, 523
Hydrachna, 549, **572, 579**
 magniscutata, **557, 558**
Hydrachnellae, 523
Hydrachnida, 8, 523
Hydrachnidae, 528, 533, 534, 537, 538, 539, 540, 542, *549,* *554,* **557, 558,** **572, 579**

Hydrachnoidea, 524, 527, 528, 535, 537, *549, 554*
Hydradephaga, 618
Hydraenidae, 618, 619, 620, 622, 623, 648
Hydrobiidae, 291, *302*
Hydrobiinae, *302,* **302**
Hydrobiosidae, 608, 610, **642,** *643*
Hydrobius, 620
Hydrocanthus, 620, **648, 649**
Hydrochara, 620
Hydrochoreutes, 578
 intermedius, **588**
 microporus, **570**
 minor, **570**
Hydrochoreutinae, 529, *554,* **570,** *578,* **588**
Hydrochorutes, 554
Hydrochus, 620
Hydrococorallina, 126
Hydrodroma, 548, *572,* **580**
 despiciens, 537, **556**
Hydrodromidae, 528, 537, 542, *548,* **556,** *572,* **580**
Hydroida, 126
Hydrolimax, 156
 grisea, 146, 147
Hydromermis, 271
Hydrometra, **645**
Hydrometridae, 537, 614, 617, *644,* **645**
Hydroperla, 605
Hydrophilidae, 618, 619, 620, 622, 623, 624, **647,** *648,* **648, 649**
Hydrophiloidea, 623
Hydrophilus, 620
Hydroporus, 619, **647**
Hydropsyche, 607
Hydropsychidae, 607, 609, *640*
Hydropsychoidea, 607
Hydroptila, **641**
Hydroptilidae, 608, 610, *640,* **641**
Hydroscapha, **648**
Hydroscaphidae, *648*
Hydrovatus, 619
Hydrovolzia, 547, *554*
 gerhardi, 555
 marshallae, **579**
Hydrovolziidae, 528, *547, 554*
Hydrovolziinae, 528, 537, 538, *547, 554,* **555,** *579*
Hydrovolzioidea, 524, 526, 527, 528, 534, *537, 547, 554*
Hydrozoa, 4, 126
Hydrscaphidae, *648,* **648**
Hydryphantes, 548, *572*
 ruber, **555, 556, 580, 581**
Hydryphantidae, 528, 537, *547, 548, 572*
Hydryphantinae, 527, 528, 532, 533, 534, 538, 539, 540, 542, *548,* **555, 556,** *572,* **580, 581**
Hydryphantoidea, 524, 527, 528, 534, 536, 537, 538, 539, *547, 552*
Hyalella
 azteca, 678
Hygrobates, 552, **566,** *578,* **580**
 neocalliger, **566**
Hygrobatidae, 529, 539, 541, 542, 543, *552,* **566,** *575,* *578,* **584, 585, 587**

Hygrobatoidea, 524, 527, 529, 534, 537, 538, 539, 540, 541, *552*
Hygrotus, 619
Hylodiscus rubicundus, **67**
Hymanella, 157, *159*
  retenuova, **158**
Hymenostomatia, 39, 49
Hymenostomatida, *85*
Hypaniola, **8**
  florida, 440, **440**, *468*
Hypechiniscus, 503, 511, 513, *516*
Hypotrichia, 39, 49, *75*
Hypotrichidium, *78*
  conicum, **77**
Hypsibiidae, 503, 515, 516
Hypsibius, 503, **504**, 505, **505**, 506, **506**, 510, 511, 512, 513, 515, 516, *518*
  convergens, 513
  dujardini, 511, 513

Ichthydium, 177, 181, *181*, **182**
Ichthyophthirius, 60
Ichtydium, 180, 185
Ictaluridae, 363
Ictalurus
  furcatus, 363
  punctatus, 363
Ictiobus
  bubalus, 363
  niger, 363
Iheringula, 761
  paulensis, **761**, *763*
Illinobdella, 440, 456, *460*, *465*
  alba, 465
  elongata, 466
  richardsoni, 465
Ilybius, 619
Ilyocyprinae, *709*
Ilyocryptus, **757**
  sordidus, **759**
Ilyocypris, 711
  bradyi, 701, 702
  gibba, 700, 701, 702
Ilyodrilus, 426, **427**
  frantzi, 426
  templetoni, 426
Ilyodrilus, 426
Ilyodromus, 711
Incertae sedis, 515
Infusoria, 501
Insecta, **10**
Integripalpia, 607, 608
Intergripalpia, 610–612
Io, *303*
  fluvialis, **303**
Ironidae, *266*
Ironus, 258, *266*
Isochaetides, *424*, **424**
Isocypris, 711
  longicomosa, 707
Isogenoides, 605
Isohypsibius, 503, **504**, 505, 506, **506**, **507**, 510, 511, 512, 513, 515, 516, *518*
  annulatus, 510
  granulifer, 510
Isomermis, 273

Isonychia, 598, **635**
Isoperla, 605, **638**
Isopoda, 10, **11**, *15*, 666, **667**, 668, 675, **675**, 676, **676**, 683
Isotomidae, *653*, **653**
Itaquascon, 503, **504**, **505**, 506, **506**, 507, 508, 515, 516, *518*
Itura, **194**

Juga, *304*
Julidae, 278

Kalyptorhychia, *154*
Kalyptorhynchia, 146, **154**, **158**, *163*, 163
Kamptozoa, 481
Karyoblastea, 47, *66*
Karyorelictea, 39, 49, *81*
Kawamuracarinae, 529, 541, *574*, **583**
Kawamuracarus, *574*, **583**
Kellicottia, 193, **236**, *237*
Kenkia, *159*
Keratella, **190**, 193, 197, 202, 217, 218, 237, 489
  cochlearis, 213, **214**, 216–218, 222, *243*, **243**
  crassa, **201**, **214**
  earlinae, **214**
  quadrata, 200, 212
  slacki, 218
  testudo, 197, 211, **211**, 218, 222, **237**
Kerona, 130
Khawkinea, **3**, *63*
  halli, **64**
Kijanebalola, *181*
Kincaidiana
  freidris, **419**, *421*
  hexatheca, *419*, **419**, *421*
Kinetoplastida, 39, 44, 46, 47, *66*
Kircheriella
  contorta, 220
Klattia, *163* See *Opisthocystis*
Koenikea, *552*, *575*, 577, **584**
  marshalli, **567**
Kongsbergia, 577
Krendowskia, *550*, *576*
  similis, **561**
Krendowskiidae, 527, 531, 538, *550*, **561**, 576, **585**
Krumbachia, *160*
  hiemalis, **161**
Kurzia, 748, *750*
  latissima, *750*
  longirostris, **749**, *750*

Labrundinia, 625
Laccobius, 620
Laccophilus, 619
Laccornis, 619
Lacinia
  mobilis, 675
Lacinularia, 202, 204, 205, 218, *228*, **228**
  elliptica, **229**
Lacrymaria, *81*
  olor, **80**

Lacrymariidae, *81*
Ladona, 602
Laenonereis
  culveri, 469
Laevapex, *309*
  fuscus, **308**
Laevicaudata, 724, 764
Lambornella, 60
Lambriculidae, 417
Lamellibranchia, 315
Lamellidens
  corrianus, 324, 329, **330**, 333, 335
  marginalis, 333
Lamnocythere
  staplini, 699
Lampsilini, *383*
Lampsilis, 324, **345**, *387*, **387**
  anodontes, 338
  corrianus, 329
  radiata, 327, 328, 329, 338
  radiata siliquoidea, 368, **380–382**
  ventricosa, **345**
Lampyridae, 624, *649*, **649**
Lanceimermis, 271
Lanthus, 602
Lanx, *309*
  patelloides, **308**
Lapedella, *233*, **233**
Lasmigona, 385
  complanata, **380–382**
Lathonura, 760
  rectirostris, **759**
Latona, 727, 748
  setifera, **747**
Latonopsis
  occidentalis, **747**
Latonpsis, 748
Latreille, 823
Laversia, *551*, 577
  berulophila, **562**, **563**, **586**
Laversiidae, 531, 526, 541, 542, *551*, **562**, **563**, 577, **586**
Lebertia, *551*, **564**, *574*, **583**
Lebertiidae, 529, 538, 539, *551*, **564**, *574*, **583**
Lebertioidea, 524, 529, 534, 535, 537, 538, 539, *549*, *551*, *551*, 573
Lecane, 202, **234**, *235*
  inermis, 222
  lunaris, **234**
Lecanidae, **234**, *235*
Lecithoepitheliata, 146, 150, *154*, **154**, 156
Lecythium, 71
  hyalinym, **69**
Lembadion, *83*
  magnum, **84**
Lembadionidae, *83*
Lemiox, 386
  rimosus, **380–382**, *386*
Lemmenius, 503
Lemna, 618
Lepadella, 202
Lepidodermella, 177, 181, **182**
  squamata, 175, **177**, 178
  trioba, **179**
Lepidoptera, 594, 617–618, 618, *633*

Lepidostoma, **642**
Lepidostomatidae, 607, 611, **642,** *643*
Lepidurus, 764, 766, *776*
    articus, 764
    couessii, **775**
    lemmoni, 766
Lepomis
    gibbosus, 298
    gulosus, 363
    macrochirus, 363
    microlophus, 298, 363
Leptestheriidae, 724
Leptisthera
    compleximanus, **774**
Leptistheria, 772
Leptoceridae, 607, 608, 611, *640,* **641**
Leptodea, 388
    fragilis, **380–382**
Leptodiaptomus, 807
Leptodora, 136, 727, 730, 738, 739
    kindtii, 30, **762,** *763*
Leptodoridae, 724, *746, 763*
Leptohyphes, 598, **634**
Leptolaimidae, *266*
Leptophelbia, **635**
Leptophlebiidae, 598, 600, *641,* **635**
Leptophrygidae, *85*
Leptopodomorpha, 613
Leptostraca, **667**
Leptoxis, 304
    carinata, **303**
Lepyrium, 302
Lernaeopodida, 787
Lesquereusia, 69
    spiralis, **69**
Lestes, 604, **636**
Lestidae, 604, **636,** 637
Lethaxona, 577
Lethaxonella, 577
Lethocerus, 615
Leuctra, **639**
Leuctridae, 605, 606, **639,** *640*
Lexingtonia, 389
    dolabelloides, **380–382,** *389*
    subplana, 389
Leydigia, 727, *748, 751, 751*
    acanthocercoides, **752**
    leydigi, 739
Libellulidae, 278, 543, 601, 602, 603,
    636
Libelluliidae, *637*
Licerus, 676
Lieberkuehnia, 71
Lieberkuehnia wagnerella, **70**
Lieberkuehnidae, *71*
Lieberkuhn, 60
Ligumia, 387
    nasuta, **380–382,** *387*
    recta, 359, *387*
    subrostrata, 329, 332, 333, 334, *387*
Limmenius, 503, **504, 506,** 508, 515, 516,
    517
Limnadia, 772
    lenticularis, **774**
Limnadiidae, 724
Limnephilidae, 607, 611, 612, **641, 642,**
    643

Limnephiloidea, 607
Limnesia, 552, 575, **579**
    marshalliana, **566**
Limnesiidae, 529, 538, *552, 574*
Limnesiinae, 529, 537, 539, 542, 543, *552,*
    566, 566, *575* **579**
Limnias, 202, 205, **228**
    melicerta, **229**
Limnichidae, 622
Limnocalanus, **57,** 788, 792, 896, 798,
    807, *807*
    macrurus, 794
Limnochares, 548, 572, **580**
    americana, **557, 558**
Limnocharidae, 528, 533, *548, 572*
Limnocharinae, 528, 537, 539, 540, 542,
    548, **557, 558,** *572*
Limnocnida, 126
Limnocodium, 125, 136
    victoria, 136, **135**
Limnocythere, 705, 716
    ceriotuberosa, 699, 701, 703, *703*
    friabilis, 705, *705*
    herricki, 700, *702*
    itasca, 693, **694,** 700, *700,* 702,
        714
    liporeticulata, 702
    protzi, 701
    sanctipatricii, **695**
    sappaensis, 700
    staplini, 699, 700, 703, *703*
    verrucosa, 699, 700, 702
Limnocytherinae, *713*
Limnodrilus, 409, 410, 415, *426,* **427**
    cervix, 409
    claparedeianus, 410
    hoffmeisteri, 407, 410
    profundicola, 408
Limnomedusae, 126
Limnomermis, 271
Limnophora, **651**
Limnoruanis, 160
    romanae, 149, **160**
Limonia, **651**
Limonia, 651
Linderiella, **12,** *772*
    occidentalis, **773**
Linderiellidae, 724
Lindia, 232, **232**
Lindiidae, 194, **194,** *232,* **232**
Liodessus, 619
Lioplax, 290, *304*
    subcarinata, **305**
Lioporeus, 619
Lipidodermella, 181
Lirceus, 683
Lithasia, 304
    geniculata, **303**
Lithobiidae, 278
Lithocolla, 71
Lithocolla globosa, **72**
Lithoglyphinae, *302,* **302**
Litocythere, 716
Litonotus, 81
Litonotus fasciola, **82**
Litostomatea, 39, 44, 49, 56, *79,* 81
Littoridininae, *302*
Littoridinops, 302

Ljania, 553, 577
    bipapillata, **568**
Lobosea, 39, 47, *66, 67*
Longenae, *385*
Longipediidae, 799
Lophocharis, 235
Lophopodella, **484,** 490
    carteri, 487, 489, 491, **496,** *497*
Lophopodidae, 490, **496**
Lophopus
    crystallinus, 487, 488, **496,** *497*
Lordocythere, 714
Loxocephalidae, *85*
Loxocephalus, 85
Loxocephalus plagius, **82**
Loxoconchinae, *713*
Loxodes, 49, 51, *81*
    magnus, **82**
Loxodida, *81*
Loxodidae, *81*
Loxophyllum, 44, *81*
    helus, **82**
Luectridae, 606
Lumbomerskiidae, 112
Lumbricidae, *418*
Lumbriculidae, 402, 404, 407, 409, 412,
    414, 417, *418,* **418,** *419–420*
Lumbriculus, 408, 409, 415, **420,** *420*
    variegatus, 408, 410, 415, *420*
Lutrochidae, 624, 625, **649,** *650*
Lutrochus, **649**
Lycastoides
    alticola, 469
Lycastopsis
    hummelincki, 469
Lydigia, 739
Lymnaea, 292, 305, **306**
    elodes, 292, 293, 295, 296, 298
    palustris, 295
    peregra, 293, 294
    reflexa, 294
    stagnalis, 34, **286,** 292, 293, **306**
Lymnaeidae, *286, 291, 294, 304, 309*
Lynceidae, 724, *775*
Lynceus, 775
    brachyurus, **774**
Lyogyrus, 302

Macdunnoa, 598
Macrobdella, 466
    decora, 442, *466,* **466**
    diplotertia, 466, **466**
    ditetra, 466, **466**
    sestertia, 466, **466**
Macrobiotidae, 503, 515, 516
Macrobiotus, 503, **504,** 505, 506, **506,**
    508, 510, **510,** 512, 515, 516, *517*
    areolatus, **510**
    hufelandi, **504,** 510, **510,** 512, 513
    joannae, 510
    tonollii, 502, **510**
Macrobrachium, 667, 830, 831, *850*
    ohione, **831,** 836, *836*
    rosenbergii, 830
Macrochaetus, 236, **236**
Macrocotyla, 159

Macrocyclops, 795, *808*
  albidus, **809**, *810*
    distinctus, *810*
    fuscus, 787, *810*
Macrodasyida, 170, 174, 179, *179*,
  181
Macromia, **636**
Macromiidae, 603, **636**, *637*
Macrostomida, 146, 148, *153*, *154*, **154**,
  156–157, **157**
Macrostomidae, 153, *156*
Macrostomum, **4**, 147, 150, 153, **154**, *156*,
  156
    gilberti, **157**
    tuba, **157**
Macrothricidae, 724, 727, 745, *746*,
  757–763, *763*
Macrothricinae, *760*
Macrothridae, 724
Macrothrix, *760*, *763*
    rosea, **761**
Macrotrachela, **226**
Macrotrachella, *226*
Macroturbellaria, 146
Macrovelia, **641**
Macroveliidae, 613, *645*, **645**
Mactracea, 373, *374*
Magmatodrilus, *431*
    obscurus, **432**
Malacostraca, 10, *10*, *15*, 666, **667**
Malirekus, 605
Mamersellides, *574*
Manayunkia
    speciosa, **440**, *458*, *468*
Mansonia, 627
Mantidae, 278
Maraenobiotus, *812*
Margaritifera, *378*
    falcata, *378*
    hembeli, *378*
    margaritifera, 332, 333, 338, 343, 344,
      345, 346, *378*, **380–382**
Margaritiferidae, *374*
Margaritiferinae, *374*
Marinellina, 174, 176, 180
    flagellata, 174, **174**
Marisa, 290, *304*
    cornuarietis, **305**
Marstonia, *302*
Martarega, 616
Maruina, 628
Marvinmeyeria, **441**
    lucida, *461*
Maryna, 85
    socialis, **84**
Marynidae, *85*
Mastigodiaptomus, 807
Mastigophora, 39, *63–66*
Matus, 619
Maxillopoda, 666, **667**
Mayorella bigemma, **67**
Mayorellidae, **47**
Mayroella, 68
Medionidus, *386*
    conradicus, **380–382**
Megacyclops, 808, *810*
    donnaldsoni, *810*

gigas, 787, *810*
    latipes, **809**, *810*
    magnus, *810*
    viridis, 794, *810*
Megadrili, 402, *417*
Megafenestra, *756*
    nasuta, **755**
Megalocypris, 716, 707, *707*, *712*
    alba, 702, *702*, *702*
    deeveyi, 707
    sarsi, 707
Megalocypris (?)
    pugionis, 707
Megalonaiadini, *383*
Megalonaias, 341, *383*
    boykiniana, *383*
    gigantea, **380–382**, *383*
Megaloptera, 594, 596, *612–613*, *633*,
  643
Megarili, 417
Melampus, 300
Melanoides, *303*
    tuberculata, **303**
Melosira, 491
Menetus, 307
    dilatatus, 307
Meramecia, 575
Mermithidae, 250, 251, 253, 255, 263,
  263, **270**, *271–273*
Mesobates, 578
Mesochra, 812, **813**
Mesocrista, 503, **504**, *518*
Mesocyclops, 218, 740, 794, *808*
    americanus, *810*
    brasilianus, 795
    edax, 793, 794, 795, 802, **809**, *810*
    leuckarti, 792, 794, **809**
Mesodiniidae, *81*
Mesodinium, *81*
    pulex, **80**
Mesodorylaimus, *264*
Mesogenae, *385*
Mesomermis, *271*
Mesorhabditis, *265*
Mesosotoma, 149
Mesostigmata, 523
Mesostoma, 147, 148, 150–152, **154**, *161*
    artica, 150
    californicum, 150
    craci, **158**, *162*
    ehrenbergi, 147, 150, 151
    lingua, 147, *149–151*
    vernale, 150, *161*
Mesotardigrada, 503
Mesovelia, **645**
Mesoveliidae, 614, 617, *645*, **645**
Metacineta, *73*
    mystacina, **74**
Metacopina, 691, 707
Metacypris, *711*
    maracoensis, 702
Metacystidae, *81*
Metacystis, *81*
    recurva, **80**
Metaniidae, 103, 112
Metis, *812*
Metopidae, 49, *79*

Metopus, *79*
    es, **77**
Metretopodidae, 600, *635*, **635**
Micorcometes, *71*
Micractinium, 216
Micrasema, 607
Microarthridion
    littorale, *811*
Microcodon
    clavus, *237*
Microcodonidae, *237*
Micrometes paludosa, **70**
Microcoryciidae, *68*
Microcyclops, *808*
    pumilis, *810*
    varicans, 795, *808*, **809**
Microdalyellia, **154**, *163*
    rossi, **157**
    tennesseensis, **157**
Microdiaptomus, 807
Microdrili, 402, 427
Microgromia, *71*
    haeckeliana **70**
Microgromiidae, *71*
Microhydra
    ryderi, 135, **135**
Microhypsibius, 503, **504**, **506**, 516,
  517
Microkalyptorhynchus, *163*
Microlaimidae, *265*
Microlaimus, *265*
Micromelaniidae, 291, *301*
Micropterus
    dolomieui, 843
Microsporidium, 219, **220**
Microstomidae, *154*, *156*
Microstomum, **154**, *156*, 258
    lineare, **157**
Microthoracida, 85
Microthoracidae, *85*
Microthorax, 85
    pusillus, **76**
Microturbellaria, 146, *170–171*
Midea, 551, 576
    expansa, **562**, *585*
Mideidae, 526, 530, 540, 541, 542, *551*,
  562, *576*, *585*
Mideopsidae, 531, *550*, *583*
Mideopsinae, 531, 535, 541, 542, *550*,
  561, *577*, **585**, *586*
Mideopsis, *550*, *577*, **585**, **586**
    borealis, **561**
Mikrocodides, *235*
Milnesiidae, 503, 504, 515, 516
Milnesium, 503, **504**, 506, **506**, **507**, 508,
  509, 510, 515, 517, *517*
    tartigradum, 513
Minibiotus, 503, **504**, 506, **506**, **507**, 515,
  516, *517*
Minytrema
    melanopus, 363
Mixodiaptomus, 807
Mobilida, 57, *74*
Moina, 725, 727, 730, 731, 732, 733, 735,
  736, 737, *763*, 765
    wierzejskii, **762**
Moinidae, 724, *746*, *763*

Moinodaphnia, 763
  macleayii, **764**
Molanna, 607
Molannidae, 607, 608, 612, *643*
Mollibdella, 456
  grandis, 442, *466*
Mollusca, 5–7, **7,** *14, 14,* 285–389, 530,
  668
Momonia, 575
  marciae, **561**
  projecta, **584**
Momoniidae, 530, *550, 575*
Momoniinae, 530, 542, *550,* **561,** *575,* **584**
Monas, 66
Monhystera, 256, 257, 258, *265*
  gerlachii, 257
  ocellata, 257
Monhysterida, *265*
Monhystrella, 265
Monochulus, 266
Monodella
  texana, 675
Monogononta, 189, 196, 206, **206,** 221,
  222, 224, 224, *225*
Monomia, 550
Mononchida, 258
Monochidae, *266*
Mononchus, 256, *266*
Monochulus, 266
Monopelopia, 625
Monoporeia, 678, *678, 683*
  affinis, 676, 678, 679
Monopylephorus, 407
Monosiga, **42,** 66
  ovata, **65**
Monospilus, 725, 731, *748, 750*
  dispar, **749**
Monostyla, **234,** *235*
Monothalamida, 39, *71*
Monstrilloida, 787
Mooreobdella, 42, 456, **468,** *468*
  bucera, 468
  fervida, 451, *468*
  melanostoma, 468
  microstoma, 453, *468*
  tetragon, 468
Mopsechiniscus, 511, *516*
Moraria, 812
Morimotacarinae, 526, 531, *577,* **586**
Morimotacarus, 577
Moxostoma
  carinatum, 363
Multifasciculatum, 73
  elegans, **74**
Multitubulata, 179
Muscidae, 630, **651,** *651*
Muscomorpha, 624
Musculium, 329, 333, 339, 342, 347, 360,
  367, *377, 378*
  lacustre, 328, 348, **377,** *378*
  partumeium, 325, 329, 332, 340, 348,
    359, 365, 373, *377,* **377**
  securis, 364, *377*
  transversum, **356,** 359, 360, 362, 364,
    366, *377*
Mya
  arenaria, 334

Myelostoma, 76
  flagellatum, **75**
Myelostomatidae, *76*
Myostenostomum, 156
  tauricum, **155**
Myriophyllum, 139, 222
Myriopoda, 284, 277, 439
Mysidacea, 10, **11,** *15,* 674, 666, **667,** 668,
  676, **675,** *683*
Mysidae, 676, *683*
Mysis, **11,** 676, **675,** 678, *683*
  littoralis, 676
  relicta, 30, 676, 677, 678, 678
Mystacides, 607
Mystacocarida, **677**
Mytilina, **233,** *235*
Mytilinidae, **233,** *235*
Mytilocypris
  henricae, 692
Mytilopsis, 369, *374*
Mytilposis
  leucophaeata, 374
*Mytilus edulis,*329
Myxophaga, 618, 622
Myzobdella, 456
  lugubris, **464,** *465*

Naegleria, 47, *67,* **67**
Naegleria fowleri, 47, 48
  gruberi, 47, **48**
Naididae, 60, 402, **403,** 404–406, **406,**
  407–410, 412, 417, **419,** *419–421*
Nais, 410, 416, *424*
  elinguis, 410
Najadicola, 362, *553, 578*
  ingens, **569, 570, 588**
Najadicolinae, 530, 542, *553,* **569, 570,**
  578, **588**
Namalycastis
  abiuma, 469
Namanereis
  hawaiiensis, 469
Nannochloris, **217**
  oculata, 217
Nannopus, 812
Nassophorea, 39, 49, 60, *73, 75, 83,*
  85, 86
Nassophoria, 39
Nassula, 59, **59,** *81*
  ornata, **76**
Nassulida, *73*
Nassulidae, *73*
Natricola, 302
Naucoridae, 614, 615, *644,* **645**
Nautarachna, 553, 576, 577, 579
  muskoka, **569**
Navicula, 60, 841
Nebela, 70
  collaris, **69**
Nebelidae, *70*
Necopinatidae, 503, 516
Necopinatum, 503, **504,** 506, 515, 516,
  517
Nectonema, 277, *279*
Nectonematoidea, *279*
Nectopsyche, 607

Necturus
  maculosus, 341, 362
Negossea, 177
Nemathelminthes, 170
Nematocera, 538, 624, 625, 626–629, 628,
  629
Nematoda, 5, **6,** *14,* 251–273, *263–270*
Nematomorpha, 5, **6,** *14,* 250, *263,*
  273–277, **274,** *277*
Nemertea, 4, **4,** *13,* 145–171, **162,**
  167–167
Nemoura, **638, 639**
Nemouridae, 605, 606, **638,** *639,* **639**
Neoacaridae, 526, 531, 542, *551,* **562,**
  576, **585**
Neoacarus, *551, 576*
  occidentalis, **562**
Neoatractides, 574, **582**
Neoatractidinae, 529, *574,* **582**
Neobosmina, **757,** *758*
Neobrachypoda, 553, 577
  ekmani, **569**
Neocacaridae, *576*
Neochordodes, **275,** 275, **276,** *279*
Neochthebius, 623
Neocytherideidinae, *713*
Neodasys, 179
Neoephemera, **634**
Neoephemeridae, 598, 599, *633,* **634**
Neogossea, 181, **182**
Neogosseidae, 175, *181*
Neolimnochares, 548, 572
Neomamersa, 575, **583**
Neomamersiinae, 575
Neomamersinae, 529, **583**
Neomideopsis, 576
Neomysis
  mercedis, 676, 677
Neophylax, 612, **642**
Neoplanorbis, 307
Neoplea, **645**
Neoporus, 619
Neoscutopterus, 619
Neothremma, 612
Neotiphys, 554, 578
  pionoidellus, **571**
Neotyrrellia, 575
Nepa, **645**
Nephelopsis, **447, 467**
  obscura, 298, 443, **443,** 445, **445,** 449,
    449, 450, 451, 452, 453, 454, 455,
    456, *468*
Nepidae, 614, 615, **644,** **645**
Neplmorpha, *644*
Nepomorpha, 614, 614–616, *643,*
  644
Nereidae, 438
Nereididae, 457, 459, *468*
Nereis
  limnicola, 459, *469*
  succinea, **439,** 459, *469*
Neritidae, 294
Neritina, 301
  reclivata, 290, **302**
Neritinidae, 290, 291, *301*
Neumania, 578, **588**
  punctata, **567**

Neurocordulia, **636**
  molesta, 637
Neuroptera, 594, 596, 617, *632,* **646**
Nigronia, 613
Nilotanypus, 625
Nitocra, 811
  spinipes, 794
Nitocrella, 811
Noacarus, 576
Nordodiaptomus, 807
Nostoc, 704
Notalona, 751
Noteridae, 619, 620, 621, *647, 648,* **648,**
  649
Notholca, 196, 202, *236,* **236**
  acuminata, 200
  squamula, **236**
Notoalona, 748
  freyi, **752**
Notodelphyoida, 787
Notodromadinae, *709*
Notodromas, 709
  monacha, 697, 699
Notogillia, 302
Notommata, 193, **194,** 206, *239,* **239**
  copeus, 215, 222
Notommatidae, **237,** *237,* **238**
Notonecta, 614, 616, **645,** 727, 733
Notonectidae, 614, 615, *644,* **645**
Notopanisus, 573
Notostraca, 724, 764, 765, 767
Nudomideopsidae, 531, 541, *551,* **562,**
  576, **586**
Nudomideopsis, 551, 576
  magnacetabula, **562**
Nudomideposis, 576
Nygolaimidae, *264*
Nygolaimus, 264
Nymphomyiidae, 628, **651,** *652*
Nymphophilinae, *302,* **302**
Nymphulinae, 618

Obliquaria, 385
  reflexa, **380–382,** *385*
Obovaria, 387
  subrotunda, **380–382**
Obvaria
  jacksoniana, 387
Ochromonas, **45,** *66*
  variabilis, **64**
Ochteridae, 613, *643,* **644**
Octomyomermis, 273
Octotrocha, **191, 228**
Odonata, 452, 528, 531, 537, 538, 543,
  572, 594, 596, 601–706, *632,* 636–
  637
Odontoceridae, 607, 608, 612, **642,** *643*
Odontolaimus, 265
Odontomyia, **651**
Odontostomatida, *78*
Oecetis, **641**
Oedipodrilus, 433
Ofryoxinae, *760*
Ofryoxus, 760
  gracilis, **759**
Okriocythere, 715

Oligobdella, **441,** *462*
  biannulata, 460, 462
Oligochaeta, 7, 8, 60, 401–429, *417,* 438,
  449, **450,** *460*
Oligohymenophora, *73, 83, 85*
Oligohymenophorea, 39, 49, *86*
Oligoneuriida, *635*
Oligoneuriidae, 598, 600, **635**
Oligophrmenophora, *73*
Oligostomis, **641**
Oligotrichida, *76*
Oliquaria
  reflexa, **358, 380–382**
Olisthanella, **154,** *161*
Omartacaridae, 529, *574*
Omartacarinae, 529, *580,* **583**
Omartacarus, *574,* **583**
Oncholaimidae, *266*
Oncholaimus, 266
Onychocamptus, *811*
Onychodiaptomus, 807
Onychophora, 439
Onychopoda, 724, 727, 739
Opercularia, 75
  nutans, **76**
Operculariidae, *75*
Ophidonais, 422, **422**
  serpentina, 421
Ophrydiidae, *75*
Ophrydium, 75
  eichhorni, **76**
Ophryoglena, **52,** 60
Opisthocystis, *163,* 163
  goettei, **158**
Opistocystidae, 402, 405, *417, 418,*
  419
Opistomum, 160
  pallidum, 147, **156**
Optioservus, 619
Orconectes, 829, 831, 832, 834, *852*
  compressus, 829
  deanae, 847
  immunis, 827
  inermis, 27
  inermis inermis, 828, **851**
  jeffersoni, 847
  luteus, 840
  nais, 827
  neglectus chaenodactylus, 841
  pellucidus, 828, **837**
  propinquus, 828, 839, 841, 843, **844,**
    845, 846, **851**
  punctimanus 840
  rusticus, 298, 339, 841, 843, 846, 848,
    851
  shoupi, 847
  testii, **28**
  virilis, 829, 839, 843, 846, 846, 848
  williamsi, 847
Oreela, 503, 511, *516*
Oreelidae, 503, 504
Oregonacarus, 574
Oreodytes, 619
Oribatida, 523
Orinthocythere, 715
Orthocladiinae, 625
Orthocladinae, 60

Orthocyclops
  *modestus,*808, **809,** *810*
Orthoptera, 275
Ortmannicus, 834
Orygocerus, 302
Oscillatoria, 59
Ospharanticum, 794
Osphranticum, 807
  auditivum, 149
Ostracoda, 10, **11,** *15,* 528, 531, 666, **667,**
  691–716
Otomesostoma, 147, *156*
  abronectum, 807
Oxidae, 529, 538, 542, *551,* **564,** *574,* **583**
Oxinae, 576
Oxtrema, 298
Oxus, *551, 574,* **583**
  elongatus, **564**
Oxyethira, **641**
Oxytricha, 75
  fallax, **77**
Oxytrichidae, 75
Oxyurella, 748, 751
  brevicaudis, **726, 752**

Pacifastacus, 826, 832, 833, *850*
  fortis, 832
  gambelii, 831
  leniusculus, 258, 837, 838
  leniusculus leniusculus, 837, 838
  leniusculus trowbridgii, 837
  spiculifer, 846
Paenecalyptonotus, 577
Pagastia, 625
Palaemonetes, 830, 831, **835,** 836,
  850
  antrorum, 847
  cummingi, 836, 847
  intermedius, 836
  kadiakensis, 828, **829,** 830, **830,**
    835
  paludosus, 830, **830,** 835, 841
  pugio, 836
Palaemonias, 850
  alabamae, **825,** 830, 836, 841, 847
  ganteri, **824,** 830, 836, 841, 845,
    847
Palaemonidae, 830–831, 848, *850*
Palaeodipteron, 651
Palingeniidae, 600, *633,* **634**
Palpomyia, 651
Palucidella
  articulata, 492
Paludicella
  articulata, 488, **494**
Pancarida, 666, **667,** 674
Panisopsis, 548, 573
Panisus, 548, 573
  condensatus, **581**
Pannota, 598–599
Pantala, **636**
Paracactylopodia, 811
Paracamptus, 811
Paracandona, 711
  euplectella, **694, 710**

Paracapnia, **639**
Parachela, 503, 504, 516
Parachordodes, 279
Paracineta, 73
  crenata, **74**
Paractinolaimus, 264
Paracyatholaimus, 265
Paracyclops, 795, *808, 810*
  affinis, 810
  fimbriatus chiltoni, 810
  poppei, 810
  yeatmani, 810
Paracymus, 620
Paradactylopodia, 811
Paradileptus, 81
  robusti, **82**
Paraeuglypha, 71
  reticulata, **70**
Paragnetina, 605
Paragonimus, 840
  kellicoti, 840
  westermani, 840
Paragordius, 279
Paraleptophelbia, **635**
Paraleptophlebia, 598
Paralimnetis, 776
Paramastix, 44, 63
Parameciidae, *86*
Paramecium, 38, 39, 43, **43,** 44, 49, 52,
  54, 56, 57, *86,* 137
  aurelia, 52, 56
  biaurelia, 52
  bursaria, 56
  caudatum, **84**
  monaurelia, 52
Paramideopsis, 551, 576
  susanae, **562, 586**
Paramoebidae, *68*
Paranais, 407, *421*
Paraphanolaimus, 259, *266*
Paraphysomonas, 59, 66
Paraphysomonas vestita, **64**
Paraquadrulidae, *70*
Parasitengona, 524, 527
Parastenocaris, 811
Parechiniscus, 510, *516*
Parhexapodibius, 515
  pilatoi, 511
Parophryoxus, 760
  tubulatus, **759**
Partnunia, 548, 573
  steinmanni, **581**
Partnuniella, 541
Paucitubulata, 179
Paulinella, 71
  chromataphora, **70**
Paulinellidae, *71*
Pectinatella
  magnifica, 484, 487, 488, 489, **496,** *497*
Pedalia, 228
Pediastrum
  duplex, 491
  simplex, 491
Pedicellinidae, 490
Pedinomonas, 372
Pellioditis, 265, **267, 269**
Pelobiontida, 39, *66*

Pelocoris, 614, **645**
Pelocypris, 707, *712*
  alatabulbosa, **714**
Pelomyxa, 59, 66
Pelomyxa palustris, 51
  palustrus, **69**
Pelomyxidae, **47,** *66*
Pelonomus, 624
Peltodytes, 631, **649**
Peltoperla, **638**
Peltoperlidae, 605, 606, *637,* **638**
Penardia granulosa, **69**
Penardochlamys, 68
  arcelloides, **69**
Peniculida, *83, 85, 86*
Penilia, 727, 730, 733, *748*
  avirostris, **747**
Pentagenia, **634**
Pepyrium, 302
Peracantha
  truncata, 738
Peracarida, 10, 665–683, **667,** 672, 674,
  674–676, 675, 676, **676,** 679, 679
Peranema, 56, 59, *63*
  trichophorum, **64**
Percichthyidae, 363
Percymoorensis, 456, *467*
  kingi, 467
  lateralis, 467
  lateromaculata, 467
  marmorata, 442, 451, *467*
  septagon, 467
Peridinium, 63, **64**
Peritrichia, 39, 50, *73*
Perlidae, **595,** 605, 606, *637*
Perlinella, 605
Perlodidae, 605, 606, *638,* **638**
Perutilimermis, 271
Petalomonas, 63
  abcissa, **64**
Petaluridae, 601, 603, **636,** *637*
Phaenocora, 147, 150, 152, *154, 161,* **161**
  beauchampi, 148
  typhlops, 147, 150, 150
  unipunctata, 147
Phagocata, 148, *159*
  velata, **159**
Phagodrilus, 407, 409, 413, 414
Phallodrilinae, 408
Phallodrilus
  hallae, 408, *426,* **427**
Pharyngophorida, *79, 81*
Pharyngostomoides
  procyonis, 430
Phasmatidae, 278
Pherbella, 625
Phermermis
  pachysoma, **272**
Pheromermis, 271, 271
  pachysoma, 255
Philaster, 60
Philasteridae, *83*
Philasterides, 83
  armata, **82**
Philobdella, 466
  floridana, 466
  gracilis, 466

Philodina, **189, 190,** 209, 219, **226**
  acticornis, 222
  citrina, 209
Philodinavidae, 225, **226,** *227*
Philodinidae, 225, **226,** *227*
Philopotamidae, 609, *641,* **641**
Phoronida, 489
Phreatobrachypoda, 577
Phreodrilidae, 412
Phryganeidae, 278, 608, 612, *641,* **641**
Phryganella, 68
  nidulus, **69**
Phryganellidae, *68*
Phryganellina, *68*
Phrynidae, 278
Phylactolaemata, 481, 486, 490, 496
Phyllocarida, **667**
Phyllognathopus, 812
Phyllopharyngea, 39, 44, 49, *73*
Phylocentropus, 611, **641**
Phymocythere, 713
Phyomastigophora, 39
Phyryganeidae, 607
Phrynidae, 278
Physa, **288,** 292, *306,* 841
  gyrina, 293, 294
  integra, 293
  vernalis, **289,** 294, 297
  virgata, 298
Physella, 306
  anatina, **307**
  gyrina, **307**
  integra, **307**
  virgata, **307**
  zionis, **307**
Physidae, 286, 291, 294, *305*
Physocypria, 697, 699, *712*
  globula, 701, *701,* 704
  pustulosa, 698, 700, 701, **714**
Phytomastigophora, 44–47, 54, 58, 63
Piectomerus
  dombeyanus, **380–382**
Pierisgia, 548
Piersigia, **557,** *572*
  limnophila, **580**
Piersigiidae, 528, *548,* **557, 580**
Piersigiinae, 528, 542, *548, 572*
Piguetiella, 422, 423
Pilgramilla, 163
Pilidae, *304*
Piona, 525, *553, 576,* 579, **586, 588**
  carnea, **568, 570**
  constricta, **570**
  interrupata, **569**
  interrupta, **570**
  mitchelli, **568, 570**
Pionacercus, 576, 578
Pionatacinae, 529, *553,* **567,** *575, 577,
  578,* **584, 588**
Pionidae, 529, 531, 541, *553, 576–579*
Pioninae, 530, 535, 537, 540, 542, 543,
  553, **568, 569, 570,** *576, 577, 579,*
  586, 588
Pionopsis, 554, 578
Piscicola, 460, *465*
  geometra, 453, 456, **457,** *465*
  milneri, 456, *465*

punctata, **457,** *465*
salmositica, **457,** *465*
Piscicolaria, 440, *460*
reducta, 460, **464,** *465*
Piscicolidae, **438,** 441, **441,** 443, 444, **444,**
    446, 450, 456–457, *460*
Pisidiinae, *376*
Pisidium, 325, 329, 338, 339, 342, 347,
    348, 349, 360, 366, 367, *376*
    amnicum, 329
    annandalie, 348
    casertanum, 349, 360, 361, 362, 366,
        373
    casetanum, 331
    clarkeanum, 348
    compressum, 327, 328, 377
    conventus, 360
    corneum, 366
    crassum, 366
    punctiferum, 337
    subtruncatum, 362
    variabile, 327
    walkeri, 327
Placobdella, **8,** **441,** 456, *462*
    costata, 453
    hollensis, **441,** 444, *461*
    montifera, 446, *460, 462,* **462**
    multilneata, 462
    nuchalis, 460, 462
    ornata, 444, *462*
    papillifera, 448, 451, 454, *462*
    parasitica, 444, *462*
    pediculata, 446, *462*
    phalera, 446, 456, *462*
    translucens, 462
Placozoa, 61
Plagiopyla, 86
    nasuta, **84**
Plagiopylida, *86*
Plagyophylidae, *86*
Plagiopyxidae, *70*
Plagiopyxis, *70*
    callida, **69**
Planaria, *159*
Planariidae, 148
Planolineus, 164
Planorbella, **6,** *309*
    companulata, **308**
    trivolvis, **308**
Planorbidae, 286, 291, *306*
Planorbis
    contortus, 297
    vortex, 294
Planorbula, *307*
    armigera, *307*
Planoribidae, 294
Platicrista, *518*
Platycentropus, **642**
Platychirograpsus
    typicus, 830
Platycola, *75*
    longicollis, **76**
Platycrista, 503, **504**
Platyhelminthes, 4, **4,** *13,* 146,
    221
Platyhydracarus, *551, 576*
    juliani, **563**

Plecoptera, 528, 538, 594, **595,** 604–607,
    620, *632, 637–640*
Plectidae, **250,** *266*
Plectocythere, *713*
Plectomerus, *382*
    dombeyanus, *382*
Plectus, **250,** 257, 258, *266*
Pleidae, 614, 616, *644,* **645**
Plenitentoria, 608, 610
Pleocyemata, **669**
Plethobasus, *389*
    cooperianus, 389
    cyphyus, **380–382,** *389*
Pleurobema
    cordatum, **380–382**
Pleurobemini, *383*
Pleuroblema, *389*
Pleurocera, *304*
    acuta, **303**
Pleurocerida, 294
Pleuroceridae, 291, *303*
Pleuromonas, 66
    jaculans, **65**
Pleuronema, 83
    coronatum, **82**
Pleuronematidae, *83*
Pleurostomatida, *81*
Pleurotrocha, *239*
Pleuroxus, 733, 736, *750, 757*
    aduncus, **726**
    denticulatus, 733, 736, *754*
    laevis, 743
    procurvus, 733, *754,* **754**
    trigonellus, **726**
Plistophora
    aerospora, 219
Ploesoma, **239,** *239*
Ploimida, 225, *228*
Ploltamanthus, **634**
Plumatella, **9,** 487, 489
    casmiana, 484, 488, 489, *495,* **495,**
        496
    corralloides, 495
    earginata, 488
    emarginata, 484, **485,** 487, 489, *494,*
        495
    fruticosa, 487, 488, *494,* **495**
    fungosa, 487, 488, *495*
    repens, **485,** 487, 488, 489, 491, *495,*
        495, 496
    reticulata, *494,* **495**
Plumatellidae, 490, *492,* **495**
Podiceps
    nigricollis, 34
Podocopa, 691
Podocopida, 692, 707–709
Podocopina, 691, 692, 694, 707
Podon, 727, 733, *762*
    leuckartii, **764**
Podonidae, 724, *746, 763*
Podophrya, *73*
Podophryidae, *73*
Podophyra fixa, **74**
Podura, **653**
Poduridae, *653,* **653**
Polomyxa, 61
Polyartemia, 766

Polyartemiella, *770*
    hazeni, **771**
Polyartemiidae, 724
Polyarthra, 190, 193, 197, 200, 214, 219,
    219, 223, *239*
    dolichoptera, 214
    major, 222
    remata, **214**
    vulgaris, 214, 222
Polycelis, 151, *158*
    coronata, **158**
    tenuis, 151
Polycentropodidae, 609, *641,* **641**
Polycentropus, **10,** *641*
Polychaeta, 7, 402, 404, **439, 458,** *459,*
    468–469
Polychaetes, 457–459
Polychaos timidum, **48**
Polygphaga, 624
Polymerurus, 177, 181, 182
Polymesoda, *374, 375*
    caroliniana, 375
Polymitarcyidae, 600, *633,* **634**
Polymorphus, 221
Polypedilum, 625
Polyphaga, 619, 619, 622–624
Polyphemidae, 724, *746, 763*
Polyphemis, 727
    pediculus, **764**
Polyphemus, 727, 730, 733, 737, 738, *763*
    peducularis, 730
Polypodium, 4, 125, 126, 137, **138,** 138,
    140, *141*
    hydriforme, 130, 137–138
Polytoma, 63
Polytomella, 63
    citri, **64**
Pomacea, **6,** 290, *304*
    paludosa, **305**
Pomatiopsidae, 291, *301*
Pomatiopsis, *301*
    lapidaria, **302**
Pompholyx, *231,* **231**
Pontoporeia
    affinis, 678
Pontoporeiidae, 676, 679, *683*
Popenaiadini, *383*
Popenaias, *383*
    buckleyi, 383
    popei, **380–382,** *383*
Porifera, 3–4, **3,** *13,* 46, 61, 95–121
Postciliodesmatophora, 39, 49
Potamanthidae, 600, *633,* **634**
Potamanibus, **634**
Potamlgeton, 849
Potamocypris, 666, 697, 701, 702, *709*
    granulosa, 701, *701*
    saragdina, **710**
    smaragdina, 697, 700, 701
    unicaudata, 700, 701
    variegata, 701, 702
Potamogeton, 452, 849
Potamolepidae, 112
Potamonectes, 619
Potamonemertes, 164
Potamopyrgus
    jenkinsi, 295

Potamothrix, 409, *424*
Pottsiella
 erecta, 429, **494**
Prinodiaptomus, 807
Prionchulus, 266
Prionobypris
 glacialis, 701
Prionocypris, 703, *711*
Prismatolaimus, 265
Pristina, 409, *422*
Pristinella, 409, *422*
Proales, **194, 235,** *235*
 daphnicola, 235
 gigantea, **194**
 sordida, 208
Proalidae, **235,** *235*
Probythinella, 302
Procambarus, 831, 832, 834, 838, 844,
  852, *852*
 acherontis, 847
 acutissimus, **851**
 acutus acutus, 842
 barbiger, 847
 clarkii, 845, 846, 848
 cometes, 847
 connus, 847
 echinatus, 847
 fallax, 843
 gracilis, 843, 844, **851**
 hagenianus, 839
 lepidodactylus, 847
 liberorum, 847
 lylei, 847
 paeninsulanus, 843
 pecki, **851**
 pogum, 847
 pubescens, 839
 pygmaeus, 844
 s.simulans, 827
 tenuis, 829
Procotyla, 159
 fluviatilis, **158**
Prodesmodora, 258, 265
Proechiniscus, 510
Progordius, 279
Proichthydidae, 174
Prolecithophora, 146, *154, 156–159*
Promenetus, 307
 exacuous, **308**
Promycetozoida, 39, *71*
Propeniculida, *85*
Proptera, 342, *383, 388*
 alata, **386,** *388*
 capax, *388*
 purpurata, *388*
Prorhynchella, *160*
 minuta, **160**
Prorhynchus, 150, **154,** *156,* **156**
 stagnalis, 150
Prorodon, 60, *81*
 teres, **82**
Prorodontida, *79, 81*
Prorodontidae, *81*
Proseriata, 146, *154,* **154,** *156–159,* **158**
Prosimulium, 625
Prosobranchia, 285, 291, 296, *301*
Prosotoma, **4**

Prostoma, 4, **162,** 164, **164,** 165, **165,** 166,
  258
 asensoriatum, 165
 canadiensis, 165, **165,** 167, *167*
 eilhardi, **165,** *167*
 graecense, **165**
 rubrum, 165
Prostomatea, 39, 44, 49, *79, 81*
Prostomatida, *81*
Protaiinae, 528
Protista, 3, *13*, 38, 61
Protoascus, *161*
Protoctista, 38
Protolimnesia, 575, **584**
Protolimnesiinae, 529, 575, **584**
Protonephridia, *153*
Protoneura, **636**
Protoneuridae, 604, **636,** *637*
Protoplanellinae, *160*
Protoplasa, **652**
Protozoa, 3, 30, 37–86
Protzia, *548,* **555, 556,** *573,* **581**
Protziinae, 528, *548,* **555, 556,** *573,* **581**
Psammoryctides, 425
 barbatus, 409
Psammoryctides **425**
Psedobiotus
 augusti, **513**
Psephenidae, 622, 623, *648,* **649**
Psephenus, 623, **649**
Pseudechiniscus, 503, 504, 510, 513, *516*
Pseudobiotus, 503, **504,** 505, 506, **506,**
  507, 509, 510, 511, **512, 513,** 515,
  516, *518*
 augusti, 509, **512**
 longiunguis, 513
 megalonyx, 509
Pseudocalanidae, 807
Pseudochordodes, 279
Pseudochydorus, 750, 755, *755*
 globosus, 739, **754**
Pseudodifflugia, 71
 gracilis, **69**
Pseudodifflugiidae, *71*
Pseudodiphascon, 503, **504,** 506, **506,**
  516, *517*
Pseudofeltria, 554, 576
 multipora, **571**
Pseudohydryphantes, 573
Pseudohydryphantinae, 528, 532, 533,
  573
Pseudomermis, 271
Pseudomicrothorax, 59, *85*
 agilis, **76**
Pseudonomas
 hirudinis, 442
Pseudonychocamptus
 proximus, 811
Pseudophaenocora, *161*
Pseudoploesoma, **239,** *239*
Pseudoprorodon, *81*
Pseudoprorodon
 ellipticus, **82**
Pseudosida, 748
 bidentata, **747**
Pseudosuccinea, 305
 columella, **289,** 294, 297, **306**

Pseudotorrenticola, 574
Psilotreta, 607, 612, **642**
Psilotricha, 76
 acuminata, **77**
Psilotrichidae, 76
Psychoda, **651**
Psychodidae, 628, **651,** *652*
Psycholyiidae, *640*
Psychomyia, **641**
Psychomyiidae, 610, 610, **641**
Pterodrilus, 431
 alcicornis, **432**
Pteronarcyidae, 605, 607, **637**
Ptilodactylidae, 620, 623, **649,** *650*
Ptychobranchus, 385
 fasciolare, **380–382**
Ptychogenae, *385*
Ptychoptera, **652**
Ptychopteridae, 628, **652,** *652*
Ptygura, **194,** 205, 215, *228,* **228**
 beauchampi, 197, 205, **229**
 crystallena, 204
 libera, 221
Pulmonata, 285, 291, 296, *301*
Punctodora, 265
Pyralidae, 618, *633*
Pyrgulopsis, 302
Pyrogophorus, 302
Pyxicola, 75
 affinis, **74**
Pyxidicula, 68
Pyxidicula operculata, **69**

Quadrula, **7,** 341, *379, 380, 383*
 cylindrica, *380*
 intermedia, *379*
 quadrula, **380–383**
 stapes, *379*
Quadrulella, 70
 symmetrica, **69**
Quincuncina, 383
 burkei, *383*
 guadalumensis, *383*
 infucata, **380–382,** *383*
Quistadrilus, 409, *424*

Radiolaria, 48
Radiospongilla, 113, *115*
 cerebellata, **115,** 119
 crateriformis, *115,* 119, **119**
Radix, 304
 auricularia, **306**
Ramazzottius, 503, 504, **506,** *518*
Rana temporaria, 275
Rangia
 cuneata, 363, **363,** *374*
Raphidiophrys, 71
 elegans, **72**
Rectocephala, 159
Redudasys, 176
 fornerise, *174,* **174**
Regina
 alleni, 840
Remipedia, 666, **667**
Renaudarctidae, 504

Reticulomyxa, 71
  filosa, **70**
Reticulomyxidae, *71*
Reticulosidae, *71*
Rhabditida, 258
Rhabditidae, 251, *265,* **267, 269**
Rhabditoidea, 258
Rhabditolaimus, 264
Rhabdocoela
Rhabdolaimus, 266
  brachyuris, 257
Rhabdophora, 39, 49
Rhabdophorine, 56
Rhabdostyla, 75
  pyriformis, **76**
Rhadinocythere, 715
Rhagionidae, 629
Rhantus, 619
Raphidiophrys
  elegasn, **72**
Rhapinema, 302
Rheumatobates, 614
Rhinoglena, 215, **235**
  fertoensis, 212
Rhithropanopeus
  harrisii, 830
Rhizodrilus, 406, 409
  lacteus, 425
Rhodacmea, 309
  rhodacme, **308**
Rhyacodrilinae, 406, 409
Rhyacodrilus, 409, *426*
  brevidentatus, **427**
  falciformis, 408, **427**
  sodialis, 410
Rhyacophila, **642**
Rhyacophilidae, 607, 608, 610, **642,**
  643
Rhyacophiloidea, 607
Rhynchelmis, 420
  brooksi, 419, **419**
Rhynchobdellida, *460*
Rhynchohydracaridae, 528, *572*
Rhyncholimnochares, 548, *572,* **580**
  kittatinniana, **558**
Rhyncholimnocharinae, 527, 528, 538,
  541, *548,* **558,** *572,* **580**
Rhynchomesostoma, 151, *160*
  rostratum, **160**
Rhynchomonas, 66
  nasuta, **65**
Rhynchoscolex, 147, 148, 151, *155,* **162**
  simplex, **155**
Rhynchotalona, 748, 751
  falcata, **749**
Rhyzodrilus
  lacteus, 424
Ripistes, 408, **422**
  parasita, 422
Roccus
  saxatilis, 363
Romanomermis, 271
  culicivorax, 263
Rotatoria, 187, **190, 226**
Rotifera, 5, *14,* 187–240, *199–240,* **221**
  splanchna, **190**
Rotosphaerida, 39, *71*

Rutripalpidae, 529, *573,* **582**
Rutripalpus, 573, **582**

Sabellidae, 458
Saccamoeba, 68
  lucens, **67**
Sagittocythere, 716
Salda, **645**
Saldidae, *644,* **645**
Salmasellus
  steganothrix, 680
Salpingoeca, 66
  fusiformis, **65**
Sanfilippodytes, 619
Saprodinium, 78
Saprodinium denatum, **77**
Sarcodina, 39, 47–48, *63,* 66, 72
Sarcomastigophora, 3, 39, **41,** 51, 61, *63*
Sathodrilus, 432
  attenuatus, **432**
  carolinensis, 432
  hortoni, **432**
Saturnidae, 278
Saurocythere, 715
Scapholeberis, 30, 217, 731, 734, 735, *756*
  kingii, 735
  rammneri, **755**
Scapholebris, 133
Scapulicambarus, 834
Scenedesmus, 216, 220, 372, 744, 804
  costatogranulatus, 220
Schadonophasma, 625
Schistonota, 599–601
Schizopera, 811, 812
Schizopyrenida, 39, 40, 44, 47
Schizopyreniida, 66
Sciaenidae, 363
Sciomyza, 625
Sciomyzidae, 625, 629, 630, *650,* **651**
Scirtidae, 624, *644,* **649**
Scolopendridae, 278
Scottia, 711
Scuticociliatida, *83,* 85
Scutolebertia, 574
Scyphidia, 75
  physarum, **76**
Scyphidiidae, 75
Scyphozoa, 126
Sedentaria, 458, *468*
Seinura, 264
Seisonidea, 189, 194, 196, 209, 224
Selenastrum, 372
Semigordiunus, 279
Senecella, 792, 807
  calanoides, 794, 797, *806*
Sepedon, 625, **651**
Sericostomatidae, 612, **642,** *643*
Serpulidae, 458, *468*
Serratella, 598
Sessilida, *75*
Setvena, 605
Sialidae, 278, 613, *643*
Sialis, 613, *644*
Sida, 727, 736, 737, *748*
  crystallina, **747**
Sididae, 724, 730, 746, *748*

Sigara, **645**
Silphidae, 279
Simocephalus, 133, 725, *756*
  serrulatus, **755**
  vetulus, **755**
Simpsoniconcha, 385
  ambigua, 341, **380–382,** *385*
Simuliidae, 625, 628, *652,* **652**
Simulium, 23, **652**
Sinantherina, **6,** 202, **203,** 206, 215, 218,
  228, **228**
  socialis, 204, 205, 208, 223, **229**
  solicalis, 215
  sponosa, 218
Sinobosmina, **757,** *758*
Sinodiaptomus, 807
Siolineus, 164
Siphlonuridae, 601, *635,* **635**
Siphlonurus, **635**
Siphloplecton, **635**
Siphonophanes
  grubei, 767
Siphonophora, 126
Sisyridae, 617, *632,* **646**
Skistodiaptomus, 807
Slavina
  appendiculata, 423
Sminthuridae, *653,* **653**
Snycaris
  pacifica, 830, 847
Solenopsis
  invincta, 354
Soliperla, 605
Somatogyrus, 302
Sparganophilidae, *417*
Sparganophilus, 417
Spathidiidae, 73, 81
Spathidium, 81
  spathula, **80**
Specaria, 423
Spelaecogriphacea, **676**
Spelaedrilus, **420**
  ultiporus, 421
Speleographacea, 667
Sperchon, 525, *549,* 574, **581**
  glandulosus, **559**
Sperchonopsis, 549, 574
  ecphyma, **532, 533**
Sperchontidae, 529, **532, 533, 559,** *574*
Sperchontinae, 529, 538, 542, *574,* **581**
Spermatozoa, 164
Sphaeridiinae, 624
Sphaeriidae, **6,** 342, 347–349 **356, 358,**
  367, 374, 375, *375,* **377**
Sphaeriinae, *376*
Sphaerium, 325, 329, 338, 339, 342, 347,
  360, 367, *377, 378*
  corneum, 329, 354, 377, **377,** *378*
  fabale, 377
  fluminea, 335
  nitidum, 377
  occidentale, 332, 333, *377, 378*
  simile, 331, *377,* **377**
  straiatinum, 359
  striatinum, 327, 328, 329, 348, 359,
    361, 366, *377*
  striatnum, **377**

Sphaeroeca, 66
  volvox, 65
Sphaerophrya, 73
Sphaerophyra magna, **74**
Sphalloplana, 159
Sphenoderia, 71
  lenta, **70**
Spicipalpia, 607, 610
Spilochlamys, 302
Spinicaudata, 724, 764, 765, 772
Spirobolidae, 278
Spirofilidae, *78*
Spirogyra, 294
Spirosperma, 405, 424, **424**
Spirostomidae, *79*
Spirostomum, 49, 79
  minus, **80**
Spirotrichea, 39, 49, *75, 78, 79*
Spirotrichs, 49
Spongeilidae, 112
Spongilla, 110, 114, 120
  alba, 103, 105, *114,* 119, **120**
  aspinosa, 114, 120
  cenota, 115, 120, **120**
  heterosclerifera, 114, 120
  lacustris, **96, 101,** 102, **102,** 103, 104,
    105, 106–109, 110, 111, 114, *115,*
    120, **120**
Spongillidae, 103, 113, 617
Spumella (=Monas) vivapara, **64**
Squalorophyra, 74
  macrostyla, **74**
Squatinella, **234**
Stachyamoeba, 67
  lipophora, **67**
Stagnicola, 305
  elodes, **306**
Stenelmis, **10, 648, 649**
Stenocaris, 811
Stenochironomus, 625
Stenocypris, 711
Stenonema, 598, **635**
Stenoninereis
  martini, 469, **469**
Stenopelmatidae, 278
Stenophylax, 275
Stenophysa, 306
Stenopodidea, **669**
Stenostomidae, *155*
Stenostomum, 147, 148, 150, 152, **154,**
  156
  beauchampi, **156**
  brevipharyngium, **156**
  leucops, 148, 151, **156**
  pegephilum, 149
  predatorium, 150, 151
Stenotele, **141**
Stensotomum, 152
  glandulosum, **156**
Stentor, **3,** 49, 59, *79*
  polymorphus, **80**
Stentoridae, *79*
Stephanella
  hina, 487, 488, *493,* **495**
Stephanoceros, 205, 215, **226**
Stephanodiscus, 841
Stephensoniana, 422

Stichotricha aculeata, **77**
Stichotrichia, 39, *75*
Stichotrichida, *75*
Stokesia, 85
Stokesia vernalis, **84**
Stokesiidae, *85*
Stolella, 493
  evelinae, 488, 493, **495**
  indica, 493
Stomatopoda, **667**
Stombidium, 57
Stophoteryx, **639**
Strandesia, 706, 711
Stratiomyidae, 625, 629, 630, *650,* **651**
Stratospongilla
  penneyi, 121
Stratospongilla penneyi, 114
Streblocerus, 760
  serricaudatus, **759**
Strelkovimermis, 273
Strepocephalus, 767, 768
  mackini, 768
Streptocephalidae, 724
Streptocephalus, 766, 772
  dichotomus, 769
  mackini, 766, 768
  seali, 765, 767
  texanus, **771**
Strichotricha, 78
Strichotrichina, *8*
Striobia, 302
Strobilidiidae, *79*
Strobilidium, 79
Strombidiidae, *79*
Strombidinopsidae, *78*
Strombidinopsina, *78*
Strombidinopsis, 78
  setigera, **77**
Strombidium, 79
  viride, 57, **77**
Strombilidium gyrans, **77**
Strongylidium crassum, **77**
Strongylidiidae, *78*
Strongylidium, 78
Strongylostoma, 160, **162**
  simplex, **160**
Strophitus, 383, 384, 385
  subvexus, 385
  undulatus, **380–382,** *384*
Stygabiella, 577
Stygameracarinae, 526, 531, *576,* **585,**
  586
Stygameracarus, 576
  cooki, **585, 586**
Stygarctidae, 504
Stygobromus, 676, 680, **683**
  canadensis, 680
Stygomomonia, 550, 575
  mitchelli, **560, 561**
  riparia, **584**
Stygomomoniinae, 526, 530, *550,* **560,**
  561, *575,* **584**
Stygothrombidiidae, 528, 537, 538, 541,
  554, **558, 579**
Stygothrombidioidae, 548
Stygothrombidioidea, 524, 526, 528, *548,*
  554

Stygothrombium, 548, 554, **558, 579**
Stylaria, 408
  lacustris, 422, **422**
Stylarinae, 408
Stylochacia, 177
Stylochaeta, **5, 182**
Stylocometes, 73
  digitalis, **74**
Stylodrilus, 413
  absoloni, 421
  beatiei, 421
  heringiansu, 410, 411, 421
  heringianus, *421*
Stylodrlus
  sovaliki, 421
Stylonychia, 75
  mytilus, **77**
Styloscolex, 421
  opisthothecus, 421
Styraconyx, 513
  hallasi, 503, 513
Submiraxona, 577
Suctoria, 39, 40, 49, 56, 57, *72*
Suomina, 155
  turgida, **154**
Suphisellus, 620
Suragina, 629
Sychaeta, **239**
Sympetrum, 602
Syncarida, **667, 669**
Syncaris, 850
  pacifica, 836
  pasadenae, 830
Synchaeta, 189, **190,** 193, **194,** 202, 214,
  219, *239*
  cecelia, 222
  oblonga, 208, 214, 217
  pectinata, 202, 213, 219
Synchaetidae, **239,** *239*
Synedra, 841
Synendotendipes, 625
Synurella, 676
Syrphidae, 630, *650,* **651**
Systellognatha, 605, 606–607
Sytherideinae, 713

Tabanidae, 625, 630, *650,* **651**
Tachidius, 811
Tachopteryx, **636**
Taeniopterygidae, 605, 606, *639,* **639**
Taeniopteryx, 605, **639**
Tanaidacea, **667, 676**
Tanyderidae, 628, *652,* **652**
Taphromysis
  louisianae, 676, 677
Tardigrada, 8, **9,** *15, 15,* 501–518,
  516–518
Tartarothyadinae, 528, 532, 541, *548,* **556,**
  573, **581**
Tartarothyas, 548, **556,** *573,* **581**
Tasserkidrilus, 426
Telmatodrilus, 424, **427**
  onegensis, 424
  vejdovskyi, 406, 426
Temnocephalida, 146
Temoridae, 807

Tendipedidae, 626
Tendipes
  pectinatellae, 489
Tenebrionidae, 279
Teneridrilus
  mastix, 426, **427**
Teratocephalidae, *265*
Teratocephalus, 266
Teretifrons, 748
Testacealobosia, 39, 47, *67*
Testechiniscus, 510
Testudacarinae, 529, *552,* **564, 565,** *574,*
  582
Testudacarus, 552, **564, 565,** *574*
  americanus, **582**
Testudinella, 231, **231**
Testudinellidae, *231,* **231**
Tetanocera, 625
Tetanoceridae, 630
Tetrahymena, 38, 39, **44,** 51, 55, 56, 59
  60, *85*
  pyriformis, **40,** 52, 55, **84**
Tetrahymenidae, *85*
Tetrahymenina, *85*
Tetramermis, 271
Tettigonidae, 278
Tettodrilus, 433
Teutonia, 549, 574, **583**
  lunata, **559, 560**
Teutoniidae, 526, 529, 538, 542, *549,* **559,**
  560, *574,* **583**
Teutoniinae, *574*
Thamnocephalidae, 724
Thamnocephalus, 770
  platyurus, **771**
Thaumalea, **652**
Thaumaleidae, 628, **652,** *652*
Thecacineta, 73
Thecacineta cothurniodes, **74**
Thecacinetidae, *73*
Thecamoeba, 47, 59, *68*
  sphaeronucleolus, **67**
Thecamoebidae, *68*
Theristus, 258
  pertenuis, 257
Thermacaridae, 528, 541, *547,* **555,** *573,*
  581
Thermacarus, 547, 573
  nevadensis, **555, 581**
Thermastrocythere, 716
Thermobatynella
  adami, 666
Thermocyclops, 808
  parvus, 810
  tenuis, **809,** *810*
Thermosbaena
  mirabilis, 666
Thermosbaenacea, 666, **667,** 675, **676**
Thermosphaeroma
  subequalum, 666
  thermophilum, 666, 680
Thermozodiidae, 503
Thermozodium, 503
  esakii, 503, 513
Thermozoida, 503
Theromyzon, 463
  biannulatum, 456, *463*

maculosum, 456
  rude, 446, 448, **448,** 450, 451, 453, 455,
    463, *464*
  tessulatum, 448, 450, 453, 456, *464*
Thiara, 303
  granifera, **303**
Thiaridae, 291, *303*
Thinodrilus, 421
Thulinia, 503, **504,** 506, **506,** 513, *518*
Thuricola, 75
  folliculata, **76**
Thyadinae, 527, 528, 534, 539, 541, 542,
  548, **555,** 556, *573,* **581**
Thyas, 548, 573
  stolli, **555, 556, 581**
Thyasides, 548, 573
Thyopsella, 573
Thyopsis, 573
Thysanoptera, 528
Tigriopus, 811
Tiguassidae, **413**
Tiguassu, **413**
Tillina, 86
  magna, **84**
Timpanoga, 598
Tintinnididae, *78*
Tintinnidium, 78
  fluviatile, **77**
Tintinnina, *78*
Tintinnopsis, 78
  cylindricum, **77**
Tiphyinae, 530, 540, 541, 542, *554,* **562,**
  571, *578*
Tiphys, 554, 578
  americanus, **571**
  ornatus, **571**
  weaveri, **580**
Tipula, 625
Tipulidae, 624, 625, 628, *650,*
  651
Tjadikothyas, 573
Tobrilus, 258, *266,* **270**
  gracilus, 257
  grandipapellatus, 256
Tokophrya, 57, *73*
Tokophrya quadripartita, **57**
Tokophryidae, *73*
Tokophyra quadripartita, **74**
Torrenticola, 552, 565, 574, 582
Torrenticolidae, 529, *551, 573*
Torrenticolinae, 529, 542, *552,* **565,** *574,*
  582
Tracheliidae, *81*
Trachelius, 81
  ovum, **82**
Trachelophyllidae, *81*
Trachelophyllum, 81
  apiculatum, **80**
Tramea, **636**
Traverella, 598
Trematoda, 146
Trepobates, 614
Trepomonas, 51
Triaculus, **220**
Triannulata, 431
  magna, **432**
Tribelos, 625

Trichamoeba, 68
  cloaca, **67**
Trichocerea, 237, **238**
  rattus, **194,** 215
Trichocericidae, *237*
Trichocorixa, 614
Trichodina, 75
  pediculis, **76**
Trichodinidae, *75*
Trichodrilus, 408, 413, *421*
  allegheniensis, 420
Trichophyra, 73
Trichophyra epistylidis, **74**
Trichoptera, **272,** 528, 529, 530, 537, 538,
  594, 607–612, *633, 640–643*
Trichothyas, 573
Trichotria, 236, **236**
Trichotriidae, *236,* **236**
Tricladida, 146, 149, *154,* **154,** *156–159,*
  157
Tricorythidae, 598, 599, *633,* **634**
Tricorythodes, **634**
Trilobus, 258
Trinema, 71
  enchelys, **70**
Triogonia, 379
Triops, 765, 766, 767, 776
  longicaudatus, **775**
Triopsidae, 724
Tripyla, 258, *266*
Tripylidae, *266*
Tritogonia
  verrucosa, 379, **380–382**
Trochospaera, 228, **231**
Trochosphaeridae, 228
Trochospongilla, 115
  horrida, 103, *115,* **120,** 121
  leidii, 115, **120,** 121
  pennsylvanica, 103, 105, *115,* 121, **121**
Troglocambarus, 831, 832, 834, *850*
  maclanei, 832, 841
Trombidioidea, 524
Tropisternus, 620, **647, 648, 649**
Tropocyclops, 218, 794, *808*
  extensus, 810
  parsinus, 810
  parsinus mexicanus, 810
Truncilla, 388
  donaciformis, **380–382,** *388*
  macrodon, 388
  truncilla, 388
Tryonia, 302
Trypanosoma, 46
Tubifex, 152, 409, *426*
  tubifex, 407, 408, 409, **427**
Tubificidae, 401, **402, 403,** 405, 406, **406,**
  408, 409, 412, 413, 414, 417, *419,* **419**
Tubificinae, 404, 406, 409
Tubificoides
  benedeni, 407
  benedii, 407, 409
Tubifididae, *424–428*
Tulellaria, **162**
Tulotoma, 290, *304*
  magnifica, **305**
Turaniella, 85
  vitrea, **84**

Turaniellidae, *85*
Turbanella
  hyalina, 178
Turbellaria, 4, *13,* 49, 145–171
Tvetenia, 625
Twinnia, 625
Tyhasella, 573
Tylenchida, *264*
Tylenchidae, 257, *264*
Tylenchus, 264
Tylocephalus, 257
Typhloplana, 152, **157,** *161*
  viridata, 148, 149
Typhloplanella, 161
Typhloplanida, 146, 149, 150, **161**
Typhloplanidae, 153
Typhloplanoida, 146, *154, 159–162,* **159,**
  160
Typholoplanoida, **154**
Tyrrellia, 552, **566,** *575*
  ovalis, **584**
Tyrrelliinae, 529, 542, *552,* **566,** *575,*
  584

Uchidastygacaridae, 531, **575,** *577*
Uchidastygacarinae, 526, 531, *575,* **585**
Uchidastygacarus, 575
  ovalis, **585**
Uenoidae, 612, **642,** *643*
Uglukodrilus, 431
Uncinais, 422
  uncinata, 422
Uncinocythere, 716
Unio, 341
  pictorum, 324, **346, 360**
Uniomerus, 389
  tetralasmus, 354, **380–382,** *389*
Unionacea, 7, 316, **316, 318,** 341–347,
  342, 367, 367, 373, *374,* 378, **380–**
  382
Unionicola, 362, 553, **566,** *578,* **588**
Unionicolidae, 529, *553, 578*
Unionicolinae, 529, 537, 540, 542, *553,*
  567, *578,* **588**
Unionidae, **358,** *378*
Unioninae, *378*
Unionoida, 367, 368
Uniramia, 7, 8, *15,* 438
Urccolariidae, *75*
Urceolaria, 75
  mitra, **76**
Urceolus, 63
  cyclostomas, **64**

Urnatella, **9,** 490
  gracilis, 8, 490, 491, *492,* **493**
Urnulidae, *73*
Urocentridae, *83*
Urocentrum, 83
  turbo, **84**
Uroglena, 66
  americana, **65**
Uroleptus, 77
  piscis, **77**
Uronema, 85
  griseolum, **82**
Uronematidae, *85*
Urostyla, 77
  grandis, 77
Urostylidae, *77*
Urotricha, 81
  farcta, **80**
Urotrichidae, *81*
Urozona, 83
  butschlii, **82**
Urozonidae, *83*
Utaxatax, 549, 574
  ovalis, **583**
Utricularia, 222
Uvarus, 619

Vaejovidae, 278
Vaginicola, 75
Vaginicola ingenita, **76**
Vaginicolidae, *75*
Vahlkampfiidae, *67*
Valkampfia, 67
  avaria, 67
Valvata, 296, *301*
  sincera, **302**
  tricarinata, **302**
Valvatidae, 291, *301*
Vampryella lateritia, **70**
Vampyrella, 70
Vampyrellidae, *70*
Vannella, 68
  miroides, **67**
Vannellidae, *68*
*Varichaetadrilus,*425, *426*
Varsoviella kozminski, 148
Vasicola, 81
  ciliata, **80**
Vejdovskyella, 408, **423,** *423*
Veliidae, 614, 616, 617, *644*
Veneroida, 367
Vexillifera, 68
  telemathalassa, **67**

Vexilliferidae, 68
Victorella
  pavida, 488, *492,* **494**
Villosa, **387,** *388*
  iris, **380–382**
Viviparidae, 291, 294, *304*
Viviparus, 290, 296, *304*
  ater, 292
  geogianus, **305**
  georgianus, 292, 294
  subpurpureus, 294, 296
Volsellacarus, 576
  sabulonus, **585**
Volvocida, 39, *63*
Volvox, 237, **238**
Vorticella, 50, 57, 75
  campanula, **76**
Vorticellidae, 75
Vorticifex, 308
  effusa, **308**

Wandesia, 557, **555,** *573,* **580**
Wandesiinae, 528, 538, 541 *547,* **555,** *573,*
  580
Wettina, 553, 578
  octopora, **587**
  ontario, **568**
Wettininae, 529, 542, *553,* **568,** *578,* **587**
Wlassiscia, 761
  kinistinensis, **761**
Woolastookia, 553, *577*
  setosipes, **568**

Xiphocentron, **641**
Xiphocentronidae, 610, *640,* **641**
Xironodrilus, 431
Xironogiton, 431
  instabilis, **432**

Yachatsia, 577
  mideopsoides, **586**

Zoomastigophora, 39, 44–46, 54, 59, 66
Zoothamniidae, *75*
Zoothamnium, 53, 75
  arbuscula, **76**
Zschokkea, 548, 573
Zygoptera, **564,** 601, 602, 603–604, *636,*
  637

Subject Index

Boldface page entries indicate figures; italic page entries indicate keys.

Acanthobdellidae. *See* Hirudinoidea
  (leeches) and Acanthobdellidae
Acanthocephala, 190, 221, **221**
acetylcholine, 196
acid mine drainage, effects on bivalves,
  361
acidosis, 332
acid precipitation, 23, 848
active ion transport, in bivalves, 321–322
adaptive plasticity, 285
adhesive organ, in flatworms, 147
adrenergic system, 196
aerial gas exchange, in bivalves, 332
aging, of rotifers, 208–209
agonistic behavior, in crayfishes,
  843, **844**
Alabama River, 379, 382, 386, 388
Alaska, 408
albuminotrophic cryptolarva, 430
alderflies, 612
algae, 288, 289, 293, 294, 295, 297, 299
algal–invertebrate associations, 107–108
algivory, 48, 49, 58, 294, 299
alkalinity, effects on bivalve distribution,
  340, 361–362
allelochemic, 218
allochthonous carbon, 27, 30, 31, 841
allochthonous substrata, 831
allometric growth, 790
allorecruitive colony, 204
allozymes, 213, 732–733
alpha-tocopherol (vitamin E), 199–200
Altamaha River, 384
ameba, as predators of hydra, 130
amictic females, 206, **206,** 210, 211, **211,**
  213
amino acids, 333
  chemoreceptors to, 841
ammonia, 198, 220, 253, 290, 330, 333,
  353, 444
amphimixis, 254, 274
Amphipoda (scuds), 10, **11,** 666, 667, 671,
  672, 674, 677, 678, 679, 680, 681,
  682. *See also* Peracarida, Crustacea
amphoteric females, **206,** 206–208
anabiosis, 254. *See also* anhydrobiosis,
  254
anaerobic conditions, 147, 253, 257, 445
anaerobic metabolism
  in bivalves, 329, 332, 354
  in gastrotrichs, 176
anal drinking, 729
anemones, 125
anhydrobiosis
  in rotifers, 199, **199**
  in tardigrades, 508, 509

Annelida, 401–479. *See also*
  Hirudinoidea (leeches) and
  Acanthobdellidae, Polychaeta,
  Oligochaeta, Branchiobdellida
  synopsis of, 7, **8**
anoxia, 253, 257, 299, 329, 331, 333, 698,
  700, 701, 703, 706
anoxybiosis, in tardigrades, 508, 509
antennal gland, 669, **670, 671**
anticoagulant, 438, 441
Apalachicola River, 381, 383
aphotic zone, 29, **29,** 31
apophyses, 506, **506**
aposymbiotic hydra, 133, 134
aptera, generations, in rotifers, 200
apterous species, 604, 616, 617
aquaculture
  of crayfishes, 849
  of rotifers, 219–220
aquatic bugs, 613–617
aquatic caterpillars, 617–618
aquatic insects, 594–631, 632–653. *See
  also* Hexapoda gills, 595, 597, 598,
  599, 600, 604, 606, 608, 609, 610,
  614, 618, 619, 620, 622, 623, 624
architomy, in macroturbellarians, 148
arctic–alpine distribution, 526
Artemia (brine shrimp), 34, **34,** 132, 139,
  220, 666, 670. *See also* Cladocera
  (water fleas) and other Branchiopoda
Arthropoda
  as hosts for nematomorphs, 277
  synopsis of, 8–13
asbestos, effects on bivalves, 362
Aschelminthes, 173, 223–224. *See also*
  Gastrotricha, Nematoda,
  Nematomorpha, Rotifera,
  Acanthocephala
asexual reproduction
  in branchiopods, 727, 730, 731, 732,
    733, 743
  in coelenterates, 128
  in hydra, 130, 131
  in ostracods, 696, 697
  in rotifers, 196, **206,** 206–207
  in sponges, 101, 101–102, 104–106
  in turbellarians, 148
assimilation, 216, 323–324, 349, 351,
  366–367, 407, 729, 734
Astacidea (crayfishes), 666, 667, 668, 670,
  671, 672, 673, 677, 823, 824, **825.** *See
  also* Decapoda
  "death migration" of, 842
  distribution and evolution of , 831–835
  feeding on bivalves, 362
  as food resources, 848

geographic distribution, 831–835, **832**
morphological effects of habitat on,
  829, 830
predation by, 258
reproduction and life history of,
  837–839
ATP, 318, 332
aufwuchs, 31, 57, 60, 62, 489
  in rivers, 24, 26
autochthonous carbon, 27
autoecology, research on, 51
autogamy, 54
autorecruitive colony, 204
autotoky, 254
autotrophy, 46, 60, 107
"a" values, 328
"avoidance of the shore" behavior,
  200–202, 796–797
axenic culture, 62, 223

backswimmers, 616
bacteria, 484, 486
  filtering by bivalves, 323, 366
bacterivory, 48, 49, 50, 55
  in gastropods, 294
  in protozoa, 58, 60
Barren River, 831
Bayou Teche, 385
beetles. *See* water beetles
benthos, **29,** 30, 31, 172
  littoral, 29, 31, 31
  profundal, 31
bile pigment, 703
binary fission, 53, 54
biological rhythms, in decapods, 828
biomagnification, 51
biomass, 189, 216, 328, 345, 351, 352, 491
birds, feeding on bivalves, 362, 363
biting midges, 625, 626
Bivalvia (mussels, clams), 315–399
  adductor muscles, 315
  anaerobic metabolism , 329, 332, 354
  anoxia in, 329, 331–332
  assimilation, 323–324, 349, 351,
    366–367
  bacteria filtered by, 323, 366
  biomass, 328, 345, 351, 352, 364
  as biomonitors, 367, **368**
  brood chambers, 324, 325
  burrowing by, 319, 327, 353
  byssus, 319, 336, 351, 367, 368
  calcium carbonate in, 315, 316, 317,
    318, 366
  calcium phosphate concretions in gills,
    332, 333–334

calcium utilization, 317–318, 321, 361–362

caloric content of flesh, 366

carbon to nitrogen ratios, 366

cardinal teeth, **316,** 318–319

caruncle, 387, **387**

cerebropleural ganglion, 324, 326, **326**

circadian activity, 354

circulatory system, 319–320

competition , 364

conglutinates, 344, 346

crawling behavior, 319, 354

crystalline style, 323

demibranchs, **317,** 320, 321, **321,** 334, 356, **356, 357,** 361

deposit feeding in, 360

desiccation resistance, 332, 350, 354

detritus feeding, 323, 360

digestive system , 323–324

dispersal of, 335–337, 348, 349, 351, 353, 355, 369

diurnal cycles in, 325, 326, 329

diversity, 316, 335–341

eggs, 324, 325, 350, 351

emersion response, 354

endangered species, 316

epibranchial cavity, 323–325

estivation in, 367

eulamellibranch gills, 367, 369

evolution of, 367–370

excretion, 321–323, **322**

extrapallial fluid, 317, 318

fecundity, 326, 341, 342, 344, 345, 350, 351, 352, 369

fertilization, 351, 369

filibranch gills, 367

filter feeding, 315, 316, 355, 359–361, 364–365, 372

filtration rate, 349, 359

foot, 316, **317,** 319, **320, 321,** 326, **326,** 360–361

fossil history, 337, 367–370

gaping of shell valves, 332–333

gas exchange in, 332

gill ciliation, 355, **356–358,** 359, **360**

gill (ctenidium), **317,** 320, **320,** 321, 323, 332

gill filament, 320, **321**

glochidium, 324, 335–336, 338, 341–347, **343,** 353

glycogenesis in, 324, 329

gonads, 324, **324,** 350

growth rate, 341, 342, 348, 349, 350, 351, 352, 367, 369

haemolymph, 317, 319–320, 321, 322, 323, 332, 359

hermaphroditism, 324, 325, 341, 342, 350, 369

hinge, **316,** 318

interpopulation variation, 349, 350, 364

interstitial feeding, 360

intrapopulation variation, 349, 364

ion depletion, 322, 323

iteroparity in, 342, 344, 346, 348, 350, 351

life history traits, 341–353

life span, 342, 348, 350, 351, 352, 367

locomotion, 319, 327, 353, 361

mantle, 315, 316, 317, **317, 318,** 321, 322, 334

maturation time, 341, 350, 351, 369

mechanoreceptors, 326, 327

metabolic rates, 320, 327–332, 360

nacreous shell layer, 317, **318**

net growth efficiency, 367

nitrogen excretion, 333, 351

osmoregulation in, 321–322, 334–335

parasites of, 362

pedal feeding, 360–361

pericardium, 319, **320,** 322, **322**

periostracum, 316, 318, **318**

population regulation, 361–364

predators of, 349, 362–363

prismatic layer, 316, 317, **318**

productivity of, 349, 350, 353, 364, 366

reproduction in, 324–326, 349, 350

seasonal changes in biochemical composition, 329

semelparity, 348, 349, 351

sense organs, 316, 326, 354

sexual dimorphism, 325

shell growth, 344, **345, 352**

shell morphology, 316, **316,** 317–319

shell organic (proteinaceous) matrix, 315, 316–317, 318

siphons, **323,** 320, 320–321

sperm, 324, 325, 350, 351

statocysts, **326,** 326–327

statolith, 327

synopsis of, 6–7

turn-over times in, 342, 346, 349, 353

typhlosole, 323

umbo, **316, 317,** 318, 319

veliger larva, 324, 325, **343,** 351, 369

visceral mass, 326, **326**

water depth effect on, 339

bivoltine reproduction in bivalves, 348, 350

blackflies, 625, 628

bladder, **188,** 196

blue-green algae, 295, 729, 738, 740

Bobcat Cave, Alabama, 830

body size, effect of temperature on, 132, **132**

bogs, 112, 177, 234, 542, 544, 546

boreal distribution, 526

brachypterous species, 604, 616, 617

brain, 147, **164**
    of rotifers, **188, 189,** 195

Branchiobdellida, 428–433
    attachment organ, 428
    central nervous system, 429
    chromosome numbers, 429
    clitellar gland, 429
    cocoon, 429
    commensal, 429
    decapod behavior in cleaning off, 430
    digestive system, 428, **428**
    distribution, 429
    ectosymbiotic, 428, 429–430
    embryo, 430
    evolution of, 430
    feeding, 429
    hatching, 430

    jaws, 428, **428**
    life history, 429
    locomotion, 428
    muscle ultrastructure, 428
    peristomium, 428, **428**
    prostate gland, 429
    reproduction, 429
    segmental numbering, 428
    spermatheca, 428, 429, **428**
    spermatogenesis, 429
    spermatophore, 429
    sperm ultrastructure, 429
    sulcus, **428**
    synopsis of, 7–**8**
    temperature effect on, 429
    temperature tolerance, 429
    troglobitic hosts for, 430
    vascular system, 429

Branchiopoda. *See* Cladocera (water fleas) and other Branchiopoda, *Artemia*

Branchiura (fish lice), 673–674, **674**

brine shrimp. *See* Artemia (brine shrimp)

broad-shouldered water striders, 617

bromeliad plants, 699, 702

brooding. *See* brood chamber under Cladocera (water fleas) and other Branchiopoda, and under Bivalvia

Bryozoa (Ectoprocta), 481–490, 491–499
    See also Entoprocta
    adventitious bud, **483,** 486
    ancestrula, 488
    annulus, 482, **485,** 492
    bacteriophagous, 486
    body wall, **483,** 484, 486
    budding, 486, 488–489, 490
    caecum, 482, **483,** 485, 486
    cardia, **483,** 485
    coelom, 483, **483,** 484, 486, 489, 490
    dispersal, 487, 489
    distribution, 488
    duplicate bud, 483, **483,** 486
    ectocyst, 483, 484
    eggs, 486, 489
    endocyst, **483,** 484
    epistome, 482, **483,** 484, 485, 490
    esophagus, **483,** 485, 486
    evolution of, 489–490
    fecal pellets, 485, 491
    feeding mechanics, 485
    fenestra, 482, 485, 492
    floatoblast, 482, **483,** 484, **485,** 487, 488, 490, 492
    funiculus, 482, **483,** 484, 485
    gametogenesis, 487, 489
    hibernaculum, 482, 487
    intertentacular groove, 482, 485
    intertentacular membrane, 482, **483**
    larvae, 486, 488, 489, 490
    lophophore, 482, 483, **483,** 484, **484,** 485, 490, 492
    main bud, 486
    metamorphosis, 486
    nerve ganglion, 483, **483**
    ovary, 483
    periblast, 482, 492
    peritoneal cilia, 483, **483,** 484

peritoneum, **483,** 484, 486, 489
pharynx, **483,** 485
pH tolerance, 487
piptoblast, 482, 487
polypide, 482, 486, 490
predation by, 489
predation on, 489
pylorus, **483,** 485
reproductive system, 486
retractor muscles, 482, **483,** 490
self-fertilization, 486
sessoblast, 482, **483, 485,** 487, 490
sexual reproduction, 486, 489
sperm, 484, 486, 489
statoblast, 482, 484, **485,** 486, 487, 488–489, 490, 491, 492
substrate selectivity, 488
synopsis of, 8–**9**
temperature tolerance, 487
tentacle, 482, 483, **483,** 485, 489
tentacle sheath, 482, **483,** 484
tentacular cilia, 482, 485, 490
testes, **483**
vestibule, **483,** 484
zooid, 487, **488,** 483, 484, 486, 487, 488, 489, 491
budding, 49, 53, 129, 486, 488–489, 490, 491
bugs. *See* aquatic bugs
Burgess Shale, 125
burrowers (crayfishes), **842,** 842–843
primary, 829, 831, 834, **839,** 842
secondary, 842
tertiary, 842
burrowing water beetles, 621
"b" values, 328–329

caddisflies, 538, 607, 608, 609, 611, 612
cadmium, effects on decapods, 829
calcareous shell, 692, 693, 700
calcium, 209, 290, 296, 297, 299, 317–318, 321, 361–362
calcium carbonate, 287, 315, 316, 317, 318, 366
calcium phosphate, 332, 333–364
Cambrian, 692
cannibalism, 55, 842, 846, 849
capacity adaptation, 289–290
carbon, 257
carbonate saturation, 700
carbonic anhydrase, 317, 334
carbon to nitrogen ratios, 293, 295, 366
carcinogen, 829
Caridea (freshwater shrimps), 823, 824, **824, 830,** 830–831, **831.** *See also* Decapoda
carnivores, 626, 627, 629, 705, 840–841
Carolina bays, 32, **33**
carotenoids, 735
caryomere, 697
catecholaminergic system, of rotifers, 196
caterpillars, 617–618
cave-dwelling species. *See* troglobite
cave fauna. *See* troglobite
cavernicole. *See* troglobite

caves, **27,** 27–28, **28,** 673, 675, 677, 677, 679, 680, 680
cell–cell recognition studies sponges as models for, 109
cellulase, 288, **289,** 294, 297, 299
Cenozoic, 833
channelization, effects on bivalves, 338, 348
chemical patterns, in rivers, 22, 23
chemoreception, 147, 151, 174, 195, 208, 218, 294–295, 453, 457, 527, 530, 547, 788, 790, 797, 803, 826, 828
chemoreceptors, 195, 208, 218, 294–295, 452, 457, 527, 536, 547, 788, 790, 797, 803
chemostat, 217, 223
Chipola River, 383
chironomid larvae, 489
chironomids, 538, 539
chitin, 254, 482, 484
chloramine, 198
Chlorella
as endosymbiont in coelenterates, 133, 134
expelled from or digested by hydra, 134
in green hydra, 133–135, **134**
chlorine
effects on bivalves, 354
effects on rotifers, 198
chlorophyll, in algal symbionts, 107–108
chlorophyll a, 295
reduction by bivalves, 365
chloroplast, 57, 61
choanocyte, 46
cholinergic system, 196
Chowan River, 365
chromatophores, in decapods, 828, **829**
chromatophorotropins, in decapods, 828
cilia, in protozoa, 41, 42, **42**
ciliate protozoa, as ectosymbionts on branchiobdellidans, 430
circadian clock, 828. *See also* biological rhythms
cladistics, 402, 411, 524
Cladocera (water fleas) and other Branchiopoda, 216–217, 218, 236, 723–786
asexual reproduction, 727, 730, 731, 732, 733, 743
birth rate, 739
brood chamber, 725, **726,** 727, 729, 730, 736, 744, **762,** 765, 767
carapace, 725, **726,** 727, 729, 730, 731, 734, 735, 736, 738, 739, 744, 745, 746, 765, 767, 769
clutch size, 729, 730, 732, 769
detritus feeders, 742
dispersal, 732, 733
egg ratio, 730, 739
ephippium, 731, 735, 736,
eye, 725, **726,** 727, 730, 735, 764
feeding rate, 727, 740
filter feeders, 738, 768
fluid feeding, 739
growth rate, 729, 731, 739
inbreeding, 733

instar, 731, 732, 734
mating behavior, 735, 736
metabolic rate, 729
mortality rate, 731, 733, 739
nauplii, 730, 765, 769
neck organ, 730
ommatidia, 730
placenta, 730
predation of protozoa by, 60
recruitment rate, 739
reproductive rate, 730, 732, 734
respiration rate, 729, 730, 734
resting eggs, **728,** 730, 731, 732, 733, 739, 765, 767
scraping behavior, 738, 739, 742, 767, 769
sexual reproduction, 730–731, 743
synopsis of, 10, **12**
clams. *See* Bivalvia
clarification of water, by bivalves, 366
clearance rates, of rotifers vs.other zooplankton, 216
Clear Fork of the Trinity River, 365
climatic effects, 526, 535
cloacal chamber, 251, 254
clonal development, of sponges, 101
Cnidaria (Coelenterata), 125–142
body plan, **126,** 126–127, **127**
budding in, 129
classification of, 140–142
cnidoblasts, 127
coelenteron, 125, 126, **126, 127**
coelomate prey of, 130
collecting techniques for, 138
desmonemes, 127–128
ectoderm, 126, **126**
endoderm, 126, **126, 127**
feeding in, 128
food vacuoles in, 128
gonophores, 137
gonozooids, 130
hydrostatic skeleton, 126
as laboratory animals, 130
larva, 125, 129, **129,** 135, **135,** 137, 138–139
life cycle of *Cordylophora,* 137
life cycle of *Craspedacusta,* 135
life cycle of *Polypodium,* 138–139
maintenance procedures for, 139–140
manubrium, 135, 137
medusa, 126, 127, **127,** 135, 137, 139
mesoglea, 126, **126, 127**
microhydra larvae, 129, 135, 139
natural habitat of *Craspedacusta,* 137
nematocysts, **126,** 126–128, **127, 128**
penetrants, 127, 128
planula, 137–138
polyp, 125, 126
stenoteles, 127, 128
stolon, 138
synopsis of, 4, **4**
tentacles, 126–127, **127**
tentaculocyte, 137
vellum, 136
volvonts, 127–128
cocoon, 404, 410, 429
Coelenterata. *See* Cnidaria

coelenterates. *See* Cnidaria
coevolution, 296, 524, 538, 803
cohort, 841
coldwater spring, 699
collagen, 100, 101, 102, 274, 274
collecting techniques
 for bivalves, 370–373
 for branchiobdellians, 430–431
 for crayfishes, 848–850
 for decapods, 848–849
 for gastrotrichs, 180
 for leeches and acanthobdellids, 454
 for nemerteans, 166
 for oligochaetes, 414–415
 for polychaetes, 459
 for protozoa, 61–62
 for rotifers, 222
 for shrimps, 848
 for sponges, 112
 for tardigrades, 514
 for turbellaria, 152
 for water mites, 544–546
collectors, 24–25, **25,** 839
Collembola (springtails), **9,** 10, 631–632,
 653
colonial forms, of rotifers, 189, 202–204,
 203
colonization, 53, 55, 56
 by microturbellarians, 148
colony, ecological criterion for, 94
colony formation, by coelenterates, 129
Colorado River, 385
commensalism, 110, 149, 402, 429, 542,
 692, 695
competition, 38, 55–56, 150, 453, 733
 among crayfishes, 846
 among sponges, 109–110
 between sponges and other organisms,
 109, 110
 of rotifers vs. other zooplankton, 216,
 217, **217**
conglutinates, of bivalve glochidia, 343,
 346
conjugation, 49, 54, 55
Connecticut River, 336
Copepoda, 216, 217, 787–822
 caudal rami, 787, 788, 790, **790,** 806,
 807, 810
 cephalothorax, 788, 805
 chemoreceptors, 788, 790, 797, 803
 clearance rate of, 799, 800
 copepodite, 730, 792, 795
 feeding rate of, 799, 800, 801, 803
 frontal organ, 790
 labral glands, 788
 mating behavior of, 792–793, 794, 800,
 801
 mechanoreceptors, 788, 790, 797, 799,
 803
 metasome, 788, 806, 807
 mortality rate, 793, 794, 795, 801, 802
 nauplius, **789,** 790, 795, 796, 797, 798,
 799, 800, 801, 802, 803, 804, 805
 parasitic forms of, 787
 peritrophic membrane, 789
 photo-response (light response), 793,
 796, 797

population growth rate, 800, 802
 productivity, 801, 802, 803
 proprioceptors, 790
 selective predation of protozoa, 60
 sensilla basiconica, 790
 sensilla trichodea, 790
 sex pheromones, 792
 sex ratio, 801
 sexual dimorphism, 790
 spermatophore, 792, 793
 surface feeders, 799
 suspension feeders, 798, 799, 800, 803
 swarming behavior, 796
 synopsis of, 11–12, **12**
 urosome, 788, 790, **790,** 792, 793, 796,
 805, 806, 807, 810
 vertical migration, 795, 797, 803, 804
copepodite, 730, 792, 795
copper, effects on crayfishes, 829
copulating behavior, in crayfishes,
 838–839, **844**
copulation, 208, **208,** 788, 789, 837–838
corals, 125
cosmopolitanism, 732
crane flies, 628–629
crawling water beetles, 621
crayfishes. *See* Astacidea, Decapoda
creeping water bugs, 615
Cretaceous, 367, 526, 833
Crustacea, 665–689. *See also*
 Ostracoda, Cladocera (water fleas),
 Amphipoda, Isopoda, Mysidacea,
 Branchiopoda, Copepoda,
 Decapoda
 adaptive radiation, 681
 antennal gland (green gland), 669, **670,**
 671
 endangered species, 680
 evolution of, 673–675
 indeterminate growth, 671, 672
 parasitic, 669, 676
 pheromones, 672
 predation by, 666, 673
 predation on, 667, 672, 673, 676, 677,
 678, 679, 680, 681
 as prey for hydra, 133
 stygobionts, 680
 synopsis of, 10–13
 thermal tolerance, 678
 troglobite, 678
 vertical migration, 672, 676, 678
cryobiosis, in tardigrades, 508, 509
crypsis, 847
cryptobiosis
 in rotifers, 199
 in tardigrades, 508, 510
cryptolarva, 429
culturing techniques
 for bivalves, 373
 for bryozoans, 492
 for decapods, 849
 for gastrotrichs, 180
 for hydra, 138–139
 for leeches and acanthobdellids,
 454–455
 for nemerteans, 166
 for oligochaetes, 414–415

for protozoa, 62
 for rotifers, 219–220, 222
 for sponges, 112
 for turbellaria, 152–153
 for water mites, 546
Cumberland Plateau, 834
current velocity, 20–21
 adaptations to, 20, 21
 determination of, 20
 effects on bivalve distribution, 338–339
 patterns along stream continuum, 20,
 20, 21
 variations in, 20–21
cyanobacteria, 59. *See also*
 blue-green algae
cyclic AMP, 335
cyclic dimorphism, in crayfishes, 833,
 837, 838
cyclomorphosis, 199, 202, 791
cysts, 57, 60, 408

damselflies, 601–604
dance flies, 629
"death-migration," of crayfishes, 842
Decapoda (crayfishes, freshwater
 shrimps), 823–858
 abdominal pleura, 824, **824, 825**
 aesthetasc, 826, 828
 androgenic gland, 828
 antenna, 824, **824**
 antennal gland (green gland), 826, **826,**
 829
 antennule, **824**
 basiopodite, 824, **825**
 basis, **825**
 biramous appendage, 824, **825**
 branchial chamber, 824
 branchiostegite, **824, 825,** 826
 carapace, 824, **824**
 cardiac stomach, 825, **826**
 carpopodite, **825**
 carpus, **824, 825**
 cement gland, 838–839
 central nervous system, 826
 cephalothorax, 824
 chela, **824, 825**
 cheliped, 838
 chromatophore, 828, **829**
 circulatory system, 825, 826, **826**
 coxa, 826
 coxopodite, 824, **825**
 cuticle, 824, 827
 dactyl, **824, 825**
 dactylopodite, **825**
 dimorphism in, 833, 837, 838
 endopodite, 824, **825**
 epipodite, **825**
 exopodite, 823, **825**
 exoskeleton, 824, 827
 eye stalk, 827, 828
 fecundity in, 835, 836–837, 839, 846
 Form I male, 838, **839, 840,** 850, **851**
 Form II male, 838
 fossa, 833
 gastric mill, 825
 gill, 826

hepatopancreas, 824, **826**
ischiopodite, **825**
ischium, **824, 825**
mandible, 824
maxilla, 826
maxilliped, 824, **824**
medulla terminalis, 827
meropodite, **825**
merus, **824, 825,** 844
metabolic rate of, 827
ommatidium, 826
ostium, 825, 826, **826**
ovary, 828
pereiopod, **824, 825**
pericardial sinus, 825–826
pleopod, **824, 825**
pleurobranch, 826
pleuron, 824, **824, 825**
podobranch, 826
predation by, 298
propodite, **825**
propodus, **824, 825**
protopodite, 824, **825**
pyloric stomach, 825, **826**
rostrum, 824, **824**
scaphocerite, **824**
scaphognathite, 826
sclerite, 833
sensory papilla, 827
seta, **824, 825,** 826
sinus, 825, 833
sinus gland, 827
somite, 824, **824**
statocyst, 826
subesophageal ganglion, 826, **826**
supraesophageal ganglion, 826, **826**
synopsis of, 12, **13**
telson, **824, 825**
testis, 828
trichobranch, 826
uropod, **824, 825**
vas deferens, 828
X-organ, 827, 828
Y-organ, 827, 828
decomposer communities, 60, 257, 366
deer flies, 625, 630
definitive hosts, for nematomorphs, 294
density dependence, in sponges, 105–106
density independence, in sponges, 105,
  106
deposit feeding, 60, 360
desiccation, 255, 408, 487
  of rotifers, 199, **199**
  of turbellarian eggs, 148
determinate growth, 675
detritivore, 285, 288, **289,** 293, 294, 299,
  300, 597, 599, 605, 606, 608, 611,
  614, 624, 626, 627, 629, 702, 704,
  742, 840–841
developmental rate, in rotifers, 209, 213
Devonian, 692
diapause, 488, 489, 731, 765, 767, 792,
  793, 795, 801, 803, 805
  in rotifers, 206
  in sponges, 105
  in turbellarians, 148
diatoms, 293, 297, 299, 489, 491

didelphic females, 254
diel periodicity, 792, 797, 804
diet-induced polymorphisms, in rotifers,
  199–200
dinoflagellates, 489
diorchic males, 254
diploid chromosome, 697
Diptera, 538, 539, 624–631
dispersal
  of bivalves, 335–337, 348, 349, 350,
    351, 352, 369
  of branchiopods, 732, 733
  of bryozoans, 487, 489
  of gastropods, 297, 299, 300
  of leeches and acanthobdellids,
    451–452
  of microturbellarians, 148
  of nematodes, 255
  of sponges, 103, 105
distribution
  of bivalves, 335–341
  of branchiobdellidans, 429
  of bryozoans, 487–488
  of cladocerans, 732–734
  of copepods, 794–795
  of decapods (crayfishes, shrimps), 829,
    835
  of gastropods, 290–291
  of gastrotrichs, 170–176
  of leeches and acanthobdellids, 446,
    448, 476–478
  of nematodes, 256–258
  of nemerteans, 165
  of non-cladoceran branchiopods, 766
  of oligochaetes, 408–410
  of ostracodes, 699
  of Peracarida (amphipods, isopods,
    mysids), 674–676
  of polychaetes, 458–459, 478–479
  of protozoa, 52–53
  of rotifers, 200–202, **201**
  of sponges, 102–103
  of tardigrades, 514
  of turbellarians, 148–150
  of water mites, 524–527, 541–542
diversity, 149, 172, 177, 296, 297. *See
  also* distribution
dixid midges, 627
dobsonflies, 613
dolines, 27
dormancy, 103–104, 792
dragonflies, 602, 603
dun, 597
dwarfism
  in rotifers, 200

earthworms, 402, 411, 413
ecdysis. *See* molting
ecological role
  of copepods, 802–803
  of decapods, 839–846
  of nemerteans, 166
  of rotifers, 216–217
  of sponges, 111
  of turbellarians, 151–152
ectoparasitism, 149, 437, 446, 447

ectoplasm, 48
Ectoprocta. *See* Bryozoa, Entoprocta
ectosymbiosis, 56, 428, 429–430
Edwards aquifer, 681
egg ratio, 213, 730, 739
emergence, 607, 613
emersion responses, in bivalves, 354
encephalitis, 47–48
encystment
  of glochidia in fish, 343
  in macroturbellarians, 148
  in nematomorphs, 275
  in protozoa, 54
  in tardigrades, 508–509
endangered species
  of bivalves, 316
  of crayfishes, 832–833, 847
  of isopods, 680
  of shrimps, 830, 847
endemism, 149, 202, 526
endoparasitic coelenterate, in sturgeon
  eggs, 125, 137
endoparasitism, by fungi in rotifers, 219
endosymbiont, 51, 53, 55, 56–57, 108,
  152
endosymbiotic bacteria, 442, 443
energetics, of protozoa, 52, 61
energy flow, through bivalves to higher
  trophic levels, 366–367
Entoprocta, 481, 490–491. *See also*
  Bryozoa (Ectoprocta)
  atrium (vestibule), 491
  basal plate, 491
  biomass, 491
  calyx, 491
  flame bulb, 491
  larvae, 491
  medial ganglion, 491
  nephridipore, 491
  nephridium, 491
  phylogeny, 491
  pseudocoel, 491
  substrate selectivity, 491
  tentacles, 491
  tentacular membrane, 491
  zooid, 497
environmental physiology
  of protozoa, 50–52
Eocene, **252,** 256, 273
ephemeral habitats, 765, 766, 769
ephemeral ponds, 31, **33,** 348
ephemeral streams, 25
ephippium, 731, 735, 736
epibenthos, 62
epibionts, 429
epidermal cells, 692, **692,** 700, 703
epigean species, 149
  of crayfishes, 831
epilimnion, **29,** 30
epiphytic protozoa, 57
epirhithron zone, 149–150, 151
epithelial cells, 692, 696
epizooic protozoa, 57
Escambia River, 383, 384
eutrophic waters, 198, 216, *228,* 295, 488
Everglades, 835
eversion, 44

evolution
    of bivalves, 367–370
    of crustacea, 673–675
    of gastropods, 299–300
    of gastrotrichs, 179–180
    of protozoa, 61
    of rotifers, **221,** 221–222
    of sponges, 112
    of water mites, 524, 534–535, 538
exocrine gland, of rotifers, 196
exotic species, effects of invasion or
        introduction, 845, 847
extinction, 724
extinction rates, 297, 294
extrapallial fluid, in bivalves, 317, 318

fairy shrimps. *See* Cladocera (water fleas)
        and other Branchiopoda
fast seasonal life cycle, 601, 606
fecundity, 209, 210, **210,** 211, **211,** 213,
        292, 835, 836–837, 839, 846
filter feeding, 24–25, **25,** 60, 315, 316,
        355, 359–361, 364–365, 372, 738, 768
filtration rates, 216
fire ants, feeding on bivalves, 354, 362
fish
    feeding on bivalves, 363–364
    larvae, 666, 678
    predation by, 489, 676, 677, 678, 678,
        680
fishflies, 612–613
fish hosts, for bivalve glochidia, 341–344,
        363
fish lice. *See* Branchiura
fission
    equal binary, 53
    multiple, 53, 54
    rate of, 54
    unequal binary, 49, 53
flame cell, of rotifers, **189,** 196
Flathead River, 27
flatworms. *See* Turbellaria
flies, 624, 625, 628–631
Flint River, 383
flocculation, of bacteria, 44, 58
fluvial geomorphology. *See* rivers
food chain. *See* food web
food limitation, 732, 733, 734, 739, 769,
        793, 800, 801, 803. *See also*
        competition
food resources, in rivers, 24–25
food vacuoles, in coelenterates, 128
food web, 56, 58, 60, 111, 733, 740, 743,
        770
foraging
    by crayfishes, 840–842
    by rotifers, 215–216
fouling, by bivalves, 341
fragmentation (clonal development), in
        sponges, 101–102, 103, **103,** 105
freshwater shrimps. *See* Caridea
functional feeding groups
    in rivers, 24–25, **25**
fungi, 219, **220**

gametogamy, 54
gametogenesis, 205, 487, 489
gamontogamy, 54
Gastropoda (snails), 285–314
    albumen gland, 287, **287, 288**
    aperture, 286, **286,** 287
    apex, 286, **286**
    biomass, 296
    carina, 290, *301, 307, 308*
    coelom, 289
    coelomoduct, 289
    columella, 286, **286,** 287
    concentric opercula, **286,** 287
    conical shell, 286, **286**
    ctenidium (gill), 286, 287, **287,** 288–289,
        290, 294, 300
    dextral shell, 286, 290, 291
    dioecious, 287, 291
    dispersal of, 297, 299, 300
    distribution, 290–291
    epiphragm, 290
    esophagus, 288
    evolution of, 299–300
    eye, 287, **287**
    feeding preferences, 293–294, 299
    foot, 285, 287, **287**
    globose shell, *303, 304*
    hepatopancreas, 288, 289
    hermaphroditic duct, 287, 288, **288**
    hermaphroditic (monoecious), 287, 288,
        288, 290, 291
    iteroparous, 291, 292, **292**
    mantle, 285, 286, 287, **287**
    mantle cavity, 286, 290
    monoecious (hermaphroditic), 287, 288,
        288, 290, 291
    multispiral opercula, **286,** 287
    operculum, 286, 287, **287,** 280
    oviduct, 288, **288**
    oviparous, 291
    ovotestis, 287, **287, 288**
    ovoviviparous, 291, 292
    parthenogenetic, 287, 291
    paucispiral opercula, **286,** 287
    perennial, 291, 292
    perforate shell, 287
    pericardium, 289
    periostracum, 287
    planospiral shell, 286, **286**
    pneumostome, 289
    productivity, 296, 299
    prosobranch, 285–314
    protoconch, 286, **286**
    pseudobranch, 287, *301, 305, 309*
    pulmonate, 285–314
    radula, 285, 288, **289,** 294, 301
    semelparous, 291, 292, **292**
    sexual dimorphism, 292
    shell morphology, 285, **286,** 286–287,
        290, 291, 299, 301
    sinistral shell, 286, 291
    spiral shell, **286,** 286–287
    substratum selection by, 293, 297
    suture, 286, **286,** 287
    synopsis of, 6–7
    temperature tolerance of, 289–290

    tentacle, 287, **287**
    turnover time, 296
    umbilicus, **286,** 287
    verge (penis), 287, **287,** 288, 291, 301
    visceral mass, 285, 287, 288
    whorl, **286,** 286–287
Gastrotricha, 172–182, 489
    adhesive tube, 172, 174, **174,** 175, **175**
    brain, 175, **175**
    buccal capsule, 174
    ciliated pits, 175
    furca, 174, **174,** 174, **175**
    life history, 172, 176–178, **177, 178**
    parthenogenetic egg, 175, **175,** 176,
        177, **177,** 178
    pharyngeal pore, 172, **173**
    pharynx, 174, **174**
    plaque-bearing egg, 178
    protonephridia, 175, **175**
    sexual reproduction in, 178
    sperm sac, 174, **175**
    synopsis of, **5**
    tachyblastic eggs, 176–178, **177, 178**
    X-organ, 174, **175,** 177, 178
gel electrophoresis, 293, 299, 349
generation time, of protozoa, 53
genetic polymorphism, 293, 299
genetic variation
    in decapods, 847
    in rotifers, 213–215
genome, of protozoa, 53, 54
geomorphology of streams. *See* rivers
geotaxis, 51
giant nerve fiber, 408
giant water bugs, 615
gills, 595, 597, 598, 599, 600, 604, 606,
        608, 609, 610, 614, 618, 619, 620,
        622, 623, 624
glacial relict, 794, 804
glaciations, 526
    effects on crustaceans, 680
glair, 838–839
glochidium larvae, 324, 335–336, 338,
        341–344, **343,** 345–347
glossary, 859–874
Gordiacea. *See* Nematomorpha
grazers, 24–25, **24,** 110, 285, 288, 293,
        294, 295, 297, 429
green algae, 107–108
green gland, 669, **670, 671.** *See also*
        antennal gland
Green River, 831, 845
Green River sediments, **252,** 256
groundwater, 174–176, 408, 409, 673,
        676, 677, 680, 681, 732, 742
growth rates
    in crayfishes, 846
    in cultured gastrotrichs, 178
    effect of temperature on, 132
    of protozoa, 50, 52, 58
    of sponges, 105–106
gyttja, 256

habitats
    of lake plankton, 38

of nematodes, 256–258
of protozoa, 38
of stream aufwuchs, 38
stream sediments, 38
of turbellarians, 148
of wetlands, 38
habitat stability, reproductive adaptations to, 348
hairworms. *See* Nematomorpha
halophile, 190, 198
haploid egg, 730, 736
headwater streams, **22**, 24, 25
heated effluents, effects on bivalves, 361
heavy metals
    effects on bivalves, 330, 354
    response of nematodes to, 258
    response of protozoa to, 52
    toxicity to rotifers, 198
heliotropism, 708
helocrene habitats, 534
hemimetabolous life cycle, 594
hemocyanin, 729
    in decapods, 826
hemoglobin, 289, 291, 300, 729, 734, 766, 767
herbicide, 829
herbivore, 215, 597, 599, 605, 606, 608, 609, 611, 621, 622, 623, 626, 627, 629, 703, 704, 840–841
hermaphroditism, 254, **412**, 443, 444, 449, 764
    in bivalves, 324, 325, 341, 342, 349, 369
    in gastropods, 287, 288, **288**, 290, 291
    in gastrotrichs, 174, **175**, 177, **177**
    in nemerteans, 164
    in tardigrades, 509, 511
    in turbellarians, 148
heterotrophy, 24–25, **25**, 38, 44, 46, 47, 50, 56, 58, 107
Hexapoda (aquatic insects), synopsis of, **8**, 9
Highland Rim, 831, 834
Hirudinea. *See* Hirudinoidea (leeches)
Hirudinoidea (leeches) and
        Acanthobdellidae, 437–439, 439–457
    annulus, 438, **439**, 443, 455, 456
    buccal cavity, 441, **442**
    caecae, 442, **442**, 443
    chaetae, 438, **439**
    clitellum, 444, 450
    cocoon, 443, 446, 450, 451, 452, 454
    coelom, 438, 444
    copulation, 450
    cosexuality, 450
    crop, 442, **442**, 443
    digestive enzymes, 442
    dispersal, 451–452
    distribution, 446, 453
    ectoparasitic, 437, 446, 448
    eggs, 443, 450, 451
    eyes, 440, **440**, **441**, 453, 455
    feeding behavior, 448–449
    fertilization, 450
    fish hosts for, 446
    foraging behavior, 448–449, 453
    gametogenesis, 443

gonopores, 438, **439**, 443, **443**, 443, 455, 456
growth, **448**, 451
haemocoel, 444
haemaphroditic, 443, 443, 450
jaws, 441, **442**, 448, 455
life history, 449–451, **451**
mechanoreceptors, 437, 453
metamere, 438, 444
metanephridia, 444
nerve cord, 440, 443, 443, **444**
oculiform spots (eye spots), 440, 441, **441**, 456, 457
ovisacs, 443, **443**, 450
penis, 443, **443**
pharynx, 441, **441**, 443
predation by, 437, 442, 442, 446, **449**, 449, **449**, 454–455
predation on, 451–452
proboscis, 441, 442, 448, 449
prostomium, 438
protandry, 443, 443, 449
pulsatile vesicles, 440, **441**, 444, 455
salivary glands, 441
sanguivorous, 437, 441, 442, 446, 448–449, 450, 451, 454
sensillae, **438**, 440, 453
septa, 438
sexual reproduction, 443, 449–450
spermatophore, 449
sucker, 438, **438**, **439**, 440, **441**, 455, 455, **457**
swimming, 444, 444, 452
synopsis of, 7–8
testisacs, 443, **443**
trachelsome, 440, **441**
tubercules, 438, **441**, 455
urosome, 440, **441**
vagina, **443**, 450
vascular system, 444
velum, 441, **442**, 455
ventilation (respiration), 444
histophage, 50, 53, **53**
holometabolous life cycle, 594, 607
homozygosity, 214
hormones, in decapods, 827, 828
horse flies, 625, 630
horsehair worms. *See* Nematomorpha
host associations, 528–531, 537, 539
hot springs, 257, 666, 678
Hudson Bay, 389
hybridization, 214, 845
hybrids, 732–733
hydra, 130–135
    algal symbionts in, 133–135, **134**
    Artemia as food for, 139
    asexual reproduction, 130, 131, 132
    bisexual forms, 131
    body size, 131, 132, **132**
    body size vs. food supply in, 132
    brown (asymbiotic), 130, 131, 132, 133
    buds, 131, 132
    capitulum, 135, 137
    cosmopolitan distribution of, 131
    culture media for, 139
    effect of heavy metals on, 131

feeding in, 132–133
food requirements, 132
gametes, 130, 131
gonads, 131, 132
green, 130, 131, 132, 133
growth rate, 132
hormonal control of algal symbiont mitosis, 134–135
hormonal control of size, 134–135
immortal polyps of, 132
immunity of potential prey to, 133
as laboratory animals, 138–139
lack of evidence for senescence in, 132
lysozymes, 133
nematocysts, 132–133
oxygen requirements of, 131
pedal disk, 131
planktonic, 131
planula, 135
predation by, 133
regeneration in, 133
removal of algae from, 133
sexual reproduction, 130, 131, **131**, 132
species groups of, 139–140
stability of endosymbiosis in, 133–134
substrates preferred by, 130
tentacles, 130, 132–133
theca, 131
thermal tolerance of, 131
unisexual forms, 131
hydrocarbons, response of protozoa to, 51
hydrophilous species, 501, 512
hydrostatic skeleton, 195, 218
hygrophilous species, 512–513
hyperoxia, 437–438, 445, 445, **447**
hypersaline lakes, 32–35
hypodermal cell, 692
hypodermis, 692, 696, 696, 698, 708
hypogean habitat, 695, 696, 697, 699
hypogean species
    of crayfishes, 831–832
    of decapods, 831
hypolimnion, **29**, 30, 51, 257
hyporheic zone, 25, 26, 173, 174, 534, 541, 542, 551, 610
hyporheos. *See* hyporheic zone
hyporhithron zone, 150
hypoxia, 290, 297, 299, 331, 339–340, 348, 350, 445

immigration rates, 297, 300
impoundments, effects on bivalves, 338, 369
inbreeding depression, in gastropods, 288
indeterminate growth, 671, 672
Indian middens, bivalve assemblages in, 338
insecticides, effects on decapods, 833
instars, 594, 597, 601, 604, 606, 608, 609, 610, 611, 612, 613, 617, 620, 622, 626, 628, 629, 731, 732, 734, 839
Interior Lowland Plateau (Kentucky/Tennessee), 834
intermittent streams, 25

intermolt phase, 827
International Commission on Zoological
  Nomenclature, 2
interspecific competition, 297–298
interstitial habitats, 149, 176, 526, 535,
  538, 541
island biogeography, 297
isogamy, 54
Isopoda (sow bugs), 10, **11,** 666, 671,
  674, 676, 677, 678, 679, 680, 681,
  682. *See also* Peracarida, Crustacea
iteroparity, 291, 292, **292,** 437, 450, 835,
  838

jellyfish, 125
junior homonym, in taxonomic
  classification, 2

karsts, 27, **27**
keratin-like proteins, 190
Kiamichi River, 385
Kleiner Wasser Bar, 501
K-selected traits
  of bivalves, 348
K-selection, 55–56

lacustrine ecosystems. *See* lentic
  ecosystems
Lake Baikal, 413, 413, 681
Lake Champlain, 335, 377, 378, 388
Lake Erie, 316, 337, **337,** 366, 374, 377,
  378, 388, 389
lake flies, 626
Lake Huron, 337, **337,** 374, 385, 388
Lake Michigan, 337, **337,** 338, 374
Lake Ontario, 337, **337,** 374
lakes. *See also* lentic ecosystems
  geologic age of, 106
Lake St. Clair, 316, 337, **337,** 374, 388,
  389
Lake Superior, 337, **337,** 374
Lake Tahoe, 408
Lake Tanganyika, 130, 136
lake turnover, 30
Lake Winnepeg, 383
lampricide, 829
latency (latent stages). *See* cryptobiosis
Laurasia, 526
LC, 46, 198
lead, 828, 829
learning, in turbellarians, 150
lecithotrophic embryo, 429
leeches. *See* Hirudinoidea (leeches) and
  Acanthobdellidae
lentic ecosystems, 29–35, 368, 369
  abiotic zonation of, 29, **29**
  biotic zonation of, 30–35
  geomorphology of, 29
  psammon, 26
  species diversity in, 149
life history
  of bivalves, 341–353
  of branchiobdellidans, 429
  of bryozoans, 488–489

of cladocerans, 730–732
of cnidarians (coelenterates), 129–130
of copepods, 790–793
of crayfishes, 837, **837**
of gastropods, 291–293
of gastrotrichs, 172, 176–178, **177, 178**
of leeches, 449–451
of nematodes, 254–255
of nematomorphs, 215–277
of nemerteans, 164–165
of oligochaetes, 410
of ostracodes, 697–699
of Peracarida (amphipods, isopods,
  mysids), 679–680
of polychaetes, 459
of protozoa, 53–55
of rotifers, 205–215
of shrimps, **835,** 835–837
of sponges, **103,** 103–106
of tardigrades, 511–512
of turbellarians, 148
of water mites, 537–541
life span, of crayfishes, 839
limnetic zone, 29, **29**
lipid storage, importance to population
  dynamics, 148
littoral zone, 29, **29,** 30, 31, 149, *231, 232,*
  235, 235, 237, 237, 736, 739, 740, 742
locomotion, 528–531, 534–535
long-legged flies, 629
long-toed water beetles, 622
lotic ecosystems. *See* rivers
Lotka-Volterra equations, 55–56

macrofauna, 31, 256
macrophytes, 285, 293, 294, 295, 297, 299
mammals, feeding on bivalves, 364
Mammoth Cave, Kentucky, 830, 836, 845
marine species, 733
marsh, 32–33
marsh beetles, 624
marsh flies, 630
marsh treaders, 617
''mate-killer'' ciliates, 55, 57
mating behavior, 208, **208,** 541, 601, 620,
  735, 736, 792–793, 794, 800, 801
mayflies, 597–601
mechanoreceptors, 195, 326, 327, 788,
  790, 797, 799, 807
meiobenthos, 31
meiofauna, 31, 256
meiosis, 205
melanin, 731, 734, 735
mercury, effects on decapods, 829
Mesozoic, 833
metalimnion, **29,** 30
metamere, 438, 444, 457
metamerism, 189
metamorphosis, 255, 352, 486, 524, 540,
  790
metanauplii, 738, 765
metarhithron zone, 150
microaerophilic protozoa, 51, 56–57
microbenthos, 31
microbial loop, 56, 60
microfauna, 31, 256

microsporidian spores, 258
microzooplankton, 38
mictic females, 206, **206,** 210, 211, **211**
midges, 624–627
migration, 733, 735
  vertical, 51
mine drainage, 24, 843
minute moss beetles, 623
Miocene, 838
Mirror Lake, 253, 256
Mississippi River, 18, 335, 378, 379, 382,
  383, 384, 385, 387, 388, 389
Missouri River, 832, **832**
mites. *See* water mites
mixotrophy, 46, 57, 108
Mollusca, 285–399. *See also* Gastropoda,
  Bivalvia
  synopsis of, 6–7
molting, 251, 255, 511–512, 537, 538, 597,
  727, 729, 730, 731, 734, 765, 827,
  837, 838, 840
molt-inhibiting hormone, 827
molt-initiating hormone, 827
monodelphic females, 254
Mono Lake, 34, **34**
monorchic males, 254
monoxenic culture, 62
mortality rate, 793, 794, 795, 797, 801,
  802
mosquitoes, 625, 627
moth flies, 628
mountain midges, 627
mucus, 150, 165
multiple fission, 53–54
muskrats, feeding on bivalves, 364
mussels. *See* Bivalvia
mussel shrimp. *See* Ostracoda
mutagen, 829
mutualism, in sponges, 108–109
Mysidacea (opossum shrimp), 10, **11,**
  674, 677. *See also* Peracarida,
  Crustacea

naiad, 594
nanoplankton, 49, 51, 58
Nashville Basin, 834
nauplius, 730, 765, 769, **789,** 790, 795,
  796, 797, 798, 799, 800, 801, 802,
  803, 804, 805
nekton, 30
nektoplankton, 30
Nelson River, 383
Nematoda (roundworms), 249–273
  alae, 251
  amphids, 251, 253, 254, **268, 270**
  basal bulb, **250,** 251, 253, **269, 270**
  biomass of, 256
  bristles, 251, **268**
  bryophilic, 257
  bursa, 251, 254, **267**
  caudal glands, 250, **250, 252, 267**
  circumpharyngeal commissure, **250,**
    253, 254
  cloacal chamber, 251, 254
  collecting techniques, **259,** 259–260
  corpus, 251, **269**

culturing techniques of, 262–263
cuticle, 251, 254, 255, **267, 268, 269, 270**
deirids, 254
development of, 254–255
digestive system, **250,** 251, 252, **268, 269, 270**
dispersal, 255
egg hatching of, 255
endobenthic, 258, **258**
epibenthic, 258, **258**
euhygrophilic, 257
eurytopic, 258
excretory pore, **250,** 251, 253
excretory system, 250, 253
extraction of, **259,** 260–262, **261**
food requirements of, 255, 256
fossil record, 255–256
free-living, 256
genital papillae, **250,** 254
gubernaculum, **250,** 254, **267**
habitats of, 256–258
haptobenthic, 258, **258**
hypodermal cords, 251, 271
hypodermal glands, 250, **250**
hypodermis, 251, 253, 255, **270**
as indicators of pollution, 268
isthmus, 251
juveniles, 249, **252,** 254, 255, 256, 262, 263, **270, 272**
larvae, 255
larval parasites on branchiobdellidans, 430
life history of, 254–255
locomotion, 251, **253**
longitudinal muscles, 250, 251
longitudinal striations, 251, **267**
mating behavior of, 251, 254, **267**
mermithid, 249, 250, 251, 253, 254, 255, 256, 260, 262, 263, **270,** *271,* **272,** *273*
metacorpus, 251, **269**
microbotrophic, 249, 251, 255
microvillae, 253
molts, 251, 255
muscular system, 251
nerve ring, **250,** 251, 253, 254
nervous system, 250, 253
ocellus, 254
omnivorous, 255
ovary, **256,** 254, **269**
as parasites on branchiobdellidans, 430
parasitic, 249, 256, 362
periphytic, 257
pharyngeal glands, 250
pharynx, **250,** 251, **252,** 253
photoreceptors, 254
planktonic, 258, **258**
postdeirids, 254
predaceous, 249, 251, 255, **268**
predation by, 258
predation on, 258, 489
productivity of, 256
protozoan diseases of, 258
punctations, 251
rectal glands, 253
reproductive system, 254

respiration in, 253
rhabdions, 251
seminal vesicle, 254
sense organs, 253–254, **267, 268**
setae, 251, **268**
sexual dimorphism in, 253–254
specimen preparation of, 262
spermatheca, 254
spicule, **250,** 254, **267**
spinneret, 250, **250, 252**
stenohygrophilic, 257, 258
stenotopic, 258
stoma, **250,** 251, **268, 269**
stylet, 251, **269, 270**
synopsis, 5–6
trophosome, 253
uterus, 254
vagina, 254
vas deferens, 254
Nematomorpha (hairworms, horsehair worms), 273–280, 281, 282, 283
antrum, 276
areoles, 274, 275, 280
calotte, 276, **277**
cloaca, 276
collecting techniques for, 277
cuticle, 273, 274, **274, 275,** 276, 277
cuticular fibers, 274–275, **275**
cuticular spines, 275, **276**
cysts, 275
definitive hosts for, 275, 277
development of, 275, 276
direct life cycle, 275
eggs, 275, 277
epicuticle, 274, **274, 275,** 277
epidermis, **274, 275,** 274, 277
evolution of, 273
fossil record, 273
free-living adult, 275
host invertebrate families for, 277
indirect-free-living cycle, 275
indirect-paratenic cycle, 275
larva, 275–276
life history of, 275, 276
mesenchyme, **274,** 275
microvilli, 275
mouth, 275, **276**
ovary, 275
oviduct, 275
parasitic larva, 275
paratenic (transport) hosts for, 275, 277
preparasitic larva, 275, **276**
presoma, 275, **276**
proboscis, 275
pseudocoel, **280,** 275
setae (bristles), 274
stylets, **276**
synopsis of, 5–6
testes, 275
tubercles, 274
Nemertea (ribbon worms), 145, 164–166
brain, **164**
cerebral organ, 164
coelom, 164, **164**
epidermis, 164, **165**
esophagus, 165
eye, 168, **164**

frontal organ, 164, **164, 165**
gonad, 164, **164**
lateral organ, 164
mucus, 165
osmoregulation, 165
predation by, 258
proboscis, 164, **164, 165,** 166
protonephridia, 164
rhynchocoel, 164, **164, 165**
rhyncodaeum, **165**
stylet, **164**
synopsis of, **4**
net-winged midges, 626
neurosecretory cells, in decapods, 827
neurotransmitter, 196
neuston, 30
epineuston, 30
hyponeuston, 30
niche, 454
niche boundaries, for rotifers, 198
niche partitioning, 297
nitrogen, 290, 299, 411
nonovigerous females, 200
normoxia, 445, **447**
North River, 389
nuclear dualism, 49, 54
nutrient cycling, by protozoa, 60–61

ocellus, 536
Ochlockonee River, 381, 383
Ohio River, 21, 26, 338, 378, 380, 385, 388
oil, response of protozoa to, 51
Old River, 385
Oligochaeta (oligochaete worms), 401–428
assimilation efficiency, 407
chaetae, 402, **403,** 404, 405, 406, 409, 414, 416
clitellum, 402
cocoon, 404, 408, 410
cysts, 408
distribution, 408–410
diversity, 408–410
endemic species, 408
evolution of, 411–414
feces, 407, 410–411
giant nerve fiber, 408
gills, 406, 409
parasitic, 362
peristomium, 405
photoreceptors (eye spots), **403,** 407, 408, 412
predation by, 489
proboscis, 405, 416
prostomial pit, 405, **405**
prostomium, **403,** 405, **405,** 414, 416
respiration, 412–413
spermatheca, **403,** 404, 406
spermatophore, 404
spermatozeugmata, 404, 406
swimming, 408
symbionts, 402, 409
synopsis of, **7–8**
oligotrophic waters, 58, 198, 402

omnivory, 59, 60, 788, 802
opossum shrimp. *See* Mysidacea
Ordovician, 692
organic shell matrix. *See* proteinaceous
shell matrix
orthoclones, 209, **209**
osmobiosis, 199, **199,** 508, 509
osmoconformer, 198
osmoregulation
in cladocerans, 730, 765
in decapods, 826, 828
in leeches, 444, 446
in nematodes, 253
in nemerteans, 165
in rotifers, **188, 189,** 196, 198
in shrimp, 830–831
in sponges, 98
in water mites, 535, 536
Ostracoda (mussel and seed shrimps),
691–722
alae, 693
antennae, 693, **695,** 696, 708
asexual reproduction, 696, 697
calcareous shell, 692, 693, 700
caryomere, 697
cerebrum, 696
circumesophageal ganglion, 696
coloration, 703
deutocerebrum, 696
duplicature, 693, **694**
eggs, 697, 698, 703, 704, 706, 708
food, 696, 699, 703, 704, 706
forehead, 693
furca, 693, **695,** 696, 707, 708
hemipenis, 695, 696, 697
hepatopancreas, 696
hypodermis, 692, 696, 697, 708
hypostome, 693, 695
knob, 693
lamellae, 692, **692,** 693, **694,** 696, 708
life cycle, 697–698
mandible, 693, 695, 708
mandibular palp, 693, **695,** 696, 703,
704, 708
marginal pore canal, 693
maxillae, 693, 695, **695,** 704, 708
prehensile palp, 695, **695**
protocerebrum, 696
pustules, 693, **694**
radial pore canal, **692,** 693, **694**
respiratory plate, 693, 696, 704, 708
salivary gland, 696
septa, 693
sexual dimorphism, 696
sexual reproduction, 695, 696, 704
species diversity, 706
sperm, 697, 707
subdermal cell, **692**
sulci, 693
synopsis of, 10, **11**
thoracic legs, 693, 695, **695,** 696, 699,
704, 708
tritocerebrum, 696
upper lip, 693
vas deferens, 697
vestibule, 693
Zenker's organ, **695,** 696, 697

Ottawa River, 385
Ouachita Mountains, 385
ovigerous females, 200, 835, **835, 836,**
836–837, 838–839
oviparity, 196, 291
oviposition, 537, 541, 597, 601, 620, 627,
837, **837,** 838–839
ovoviviparity, 148, 196, 291, 292, 324
oxygen, 22, 50–51, 147, 165, 198, 220,
253, 256, 257, 262, 288, 289, 290,
320, 329, 328, 330–331, **331,** 332,
332–333, 353, 354, 362, 406, 409,
429, 444, 445, 446, 454, 458, 487,
700–701
oxygen conformers, 290, 445, 827
oxygen regulators, 827

Pacific Rim, 408
paleolimnology, 742–743
sponge spicules as research tools for,
111–112
Paleozoic, 273
palustrine habitats, 32–33
Pangea, 526, 534
parasite load, 543
parasites
in eggs of sturgeon, 125, 137
of protozoa, 56, 57
of sponges, 110
parasitism
by branchiobdellids, 429
by copepods, 787
by nematodes, 249, 256
by nematomorphs, 276–277, **276**
on copepods, 803
on fish, **674,** 675, 676
on rotifers, 219
paratenic host, for nematomorphs, 275,
277
paratomy, 148, 410
parthenogenesis, 410, 730, 731, 732, 736,
764, 765
in cladocerans, 730, 731, 732, 736
in gastropods, 287, 291
in gastrotrichs, 176–178, **177, 178**
in nematodes, 254
in non-cladoceran branchiopods, 764,
765
in oligochaetes, 410
in rotifers, 189, 196, 200, 206, **206,** 207,
210, 213, 225
in tardigrades, 509, 510–511
pearling industry, 338, 340
Pearl River, 338, 340
pelagic zone, 29, **29**
Peracarida (scuds, sow bugs, opossum
shrimps), 10, **11,** 674–675, 682. *See
also*
Amphipoda (scuds), Isopoda (sow
bugs),
Mysidacea (opossum shrimps)
periphyton, 31, 172, 285, 292, 293, 294,
295, 299, 300
peristaltic contraction, 729
peritrophic membrane, 789

pesticides
response of protoza to, 51
toxicity to rotifers, 198
pH, 22, 24, 34, 40, 197–198, 220, 254,
296, 297, 340, 350, 361, 404, 486,
487, 729, 733, 767, 794, 795, 803,
804, 843, 846, 848
phagocytosis, 48
in protozoa, 39, **40,** 59
in sponges, 99
phagotrophy, 38, 46, 58, 60
phantom crane flies, 628
phantom midge larvae, 626
phenotypic plasticity, 731, 732
phenotypic variation, in rotifers, 199–200
pheromones, 792
crustacean responses to, 672
released by decapods, 828, 847
phosphorus, 411
photic zone, 29, **29**
photodamage, 734, 735
photoperiod, 730
photoreceptors, **189, 192,** 195, 221, **403,**
407, 408, 412, 440, 441, **441,** 456, 457
photo-response (light response), 793, 796,
797
phototaxis, 56–57, 150, 175, 453
phototropism, 842
phreatic zone, 27
phytoplankton productivity, bivalve
consumption of, 349, 364–366, 367
picoplankton, 38, 46, 56, 58, 60
pigmentation, 731, 734, 735
pigmy backswimmers, 616
pinocytosis, 48
in coelenterates, 128
in protozoa, 58
planaria. *See* Turbellaria (flatworms)
plankton, 60, 175, 189, 200, 215, 216,
217, 218, 220, 222, *235, 236, 239,*
258, **258,** 542. *See also* zooplankton
plate tectonics, 526
playas, 769
Pleistocene, 368, 369, 526
Pleistocene glaciation, effects on
distribution, 834
pleuston. *See* neuston
pocosins, 32, **33**
polarized light, 730
pollutants, effects on decapods
cadmium, 829
copper, 829
herbicides, 829
insecticides, 829
lead, 828–829
mercury, 829
sewage, 828
pollution
bioassays with protozoa, 52
effects on bivalves, 330, 338, 346, 350,
362, 367, 368
effects on bryozoans, 482, 487–488
effects on gastrotrichs, 176
by metals, 52
nematodes as indicators of, 258
organic, 50
response of protozoa to, 50, 51–52

Polychaeta (polychaete worms), 439–440, 457–459
  aciculum, **439**, 457
  antennae, 457, 458
  chaetae, 457, 458, **458**
  cirri, 439, **439**, 457
  deposit feeders, 458
  eggs, 459
  epitokes, 459
  eyes, 439, **439**, 457
  feeding behavior, 458–459
  filter feeders, 458
  gills (branchia), 457, 458
  jaws, **439**, 457, 458
  metamere, 457
  neuropodium, 457, 458
  notopodium, 457, 458
  ommnivorous, 458
  osmoregulation in, 439
  palps, 439, **439**, 457, 458
  parapodia, 439, **439**, 457, 458
  peristomium, 439, **439**, 458
  pharynx, 439
  proboscis, 458
  prostomium, 439, **439**, 457, 458
  setae, **439**
  sexual reproduction, 459
  synopsis of, **8**
polykinety, 42, **42**
polymorphism (non-genetic), 803
polymorphisms, in rotifers, 199–200
polyps, 125
  ambulatory, 125
  predaceous, 125
ponds. *See* lentic ecosystems
population density, 151, 152, 172, 176, **178**, 179, **179**, 188–189, 200, 201, **214**, 216, 218, 222, 293
population dynamics. *See also* fecundity, population growth rate, population regulation
  of copepods, 800–802
  of crayfishes, 845–847
  of field populations, 212–213
  of gastrotrichs, 178–179
  life table analysis of, 212–213
  of rotifers, 198, 210–215
  of shrimps, 845
  of sponges, 105–106
population growth rate, 800, 802
population regulation, 295–296, 800–802
Porifera (sponges), 95–124
  archaeocytes, **98**, 100, 101
  atrium, **98**, 99, 100
  choanocyte chamber, **98**, 99, **99**, 100, 101
  choanocytes, **98**, 99, 100, 101
  cirrous projections, 113
  "colony vs. individual" debate for, 95
  development of spongillaflies on, 617
  excurrent canal, **98**, **99**, 100
  feeding canal, 98, **98**
  fertilization in, 101
  filtering rates of, 107, 111
  flagellum, 99
  foraminal aperture, 102, 113
  fragmentation, 101–102, 103, **103**, 106

gemmoscleres, 100, 102, 112, 113–114
gemmules, 102, 103, **104**, 104, 106, 112, 113
  growth rate of, 105–106
  incurrent canal, **98**, 99, **99**, 100
  life history pattern of, **104**, 104–106
  megascleres, 100, **100**, **101**, 112, 113
  mesohyl, 98, **98**, 102
  micropyle, 102, 113
  microscleres, 100, **100**, 112, 113
  microvilli, 99, 100
  oocytes, 101
  osculum, **98**, **99**, 100
  ostia, 98, **98**, 100
  parenchymula, 101, **104**, 106
  pinacocytes, 98, **98**, 100, 101
  population dynamics of, 106–107
  porocytes, 100
  sclerocytes, 100, 101
  selective feeding by, 100
  spicules, 100, 100, **100**, 109, 110, 111, 112–114, 115–121
  subdermal cavity, **98**, 99, **99**
  synopsis of, 3–4
  thesocytes, 101
postmolt phase, 827
potamon zone, 150
prawns, predation by, 489
Precambrian, 255, 256
predaceous diving beetles, 620
predation
  on branchiopods, 732, 733, 734, 736, 738, 739, 740, 743, 746, 766, 767, 768, 769, 770
  with chemical detection, 151, 178
  chemical deterrence to, 110, 732, 735
  by coelenterates, 126, 130, 133
  by copepods, 802–803
  by crustaceans, 666, 673
  by decapods, 840–841
  by dragonflies and damselflies, 601
  by nemerteans, 166
  on coelenterates, 126, 130
  on copepods, 801, 802, 803
  on crustaceans, 667, 672, 673, 676, 677, 678, 678, 678, 681
  on dragonflies and damselflies, 601
  on gastrotrichs, 178
  on protozoa by metazoans, 60
  on rotifers, 217, **218**, 219, **219**, 219
  on sponges, 109, 110–111
  on turbellarians, 151
  size-selective, 732, 733, 734, 739
predator-induced polymorphisms, in rotifers, 199, 200
predator-prey studies, of protozoa, 56
predator response postures (PRP), of crayfishes, 842–845
premolt phase, 827
primitive crane flies, 628
priority designation, in taxonomic classification, 2
prismatic shell layer, 316, 317, **318**
production efficiency, in bivalves, 349
production to biomass ratio, of protozoa, 52

productivity
  of copepods, 801, 802, 803
  of nematodes, 256
profundal zone, 29, **29**, 31, 542, 843
propagule, 53
proprioceptors, 790
prostaglandins, in bivalves, 335
protandry
  in nemerteans, 164
  in turbellarians, 148
protein, 287
proteinaceous shell matrix, 315, 316, 317, 318
Protista. *See* protozoa
Protoctista (=Protista). *See* protozoa
protonymph, 539
protozoa, 37–86, 489
  acidosome, 40
  autophagic vacuole, **40**
  autotrophic, 46
  axoneme, 48
  axopod, 44, **45**
  axopodia, **47**, 48
  brosse, 41, 49
  buccal cavity, **40**
  cilia, 41, 42, **42**
  cirri, 48–49
  contractile vacuole, 40–41, **41**, 46
  cyrtos, 49, **59**, 59–60
  cytolosome, 39
  cytoproct, 40
  cytostome, 39–40, 49, 50
  definition of, 38
  discobolocyst, 44, **45**
  dispersal of, 53
  distribution of, 50
  ejectisome, 44, **45**, 46
  extrusome, 42–44, **47**, 48, 58
  filopodia, **47**, 48
  flagella, 41, 42, **42**
  food vacuole, 39–40, **40**
  gas vacuole, 47
  genome, 53, 54
  growth efficiency, 52
  growth rate of, 50
  haptocyst, 44, **45**, 49
  heterotrophic, 46, 47, 50, 60
  histophagous, 50, 60
  kinetocyst, 44
  kinetoplast, 46
  kinetosome, **40**, 49
  kinety, 42, **42**, 49
  Lieberkuhnia, 60
  lobopodia, 47
  lorica, 46
  macrophagous, 40
  mastigoneme, 41–42, **42**, 46
  microaerophilic, 51, 56–57
  microphagous, 40
  mixotrophic, 46
  as model cells, 38
  as model organisms, 37–38
  mucocyst, 44, 58
  Muller's vesicles, 51
  peduncle, 46
  pexicyst, **43**
  pharopodia, **47**

pinocytic vacuole, **40**
polykinety, 42, **42**, 49–50
predators of, 56, 60
as prey, 60
proboscis, 49
pseudopodia, 44, **45**, 46, 47, **47**
pusule, 41
reticulopodia, 48
rhabdos, 49
selective feeding of, 56, 58, 59
sex in, 52, 53, 54–55
spongiome, 41, **41**
swimming speed of, 42
synopsis of, 3, **3**
tests of, 47, **69**
toxicyst, 40, **43**, 44, 49
trichocyst, **43**, 43–44, 46
undulating membrane, **40**, 49
psammon, 26, 31, 174
pseudocoelom, in rotifers, **188**, 191, 195, 199, 218, 219, 220, **220**, 221
pseudosexual resting eggs, 731
pupation, 607, 613, 617, 620, 621, 622, 623, 626, 627, 628, 629, 630

Q$_{10}$, 50, 197, 289–290, 327–329

raccoons, feeding on bivalves, 364
radioisotopes, 215
R and K theory, 293
rat-tailed maggots, 630
rearing techniques. *See* culturing techniques
Red River of the North, 383, 387, 388
refugia, 361, 369, 526
relict distributions, 526
reproduction. *See* life history
reproductive isolating mechanisms, 214–215
research problems
of decapods, 847–848
of gastropods, 299
of nemerteans, 166
of turbellarians, 145, 152
resistance adaptation, 290
resistant stages, 255
respiration, 147–148, 197, 198, 253, 406–407, 602, 608, 609, 614, 615, 618, 618, 620, 622, 623, 625, 626, 628, 824–825
respiration rate
of decapods, 827
of flatworms, 147
relationship to body size, 147–148
of rotifers, 197, 198
resting eggs
of cladocerans, **728**, 730, 731, 732, 733, 739
of copepods, 792
of non-cladoceran branchiopods, 765, 767
of rotifers, 206, **206**, 207, **207**, 208, 214
resting stage, of ostracodes, 703
rheocrene habitats, 534, 544
rheotaxis, 293, 842

ribbon worms. *See* Nemertea
ribosomal RNA analyses, of protozoa, 44, 61
riffle beetles, 622
riffle habitats, 541, 545
Rio Grande River, 389
riparian zone, 24, 25
river continuum concept, 24–25, **25**, 26, 150
rivers, 16–27
acidic ecosystems in, 24
bars in, 19, **19**
braiding of, 18
current velocity, 19–21, **20**, **21**
Flathead River, 27
fluvial geomorphology, **18**, 18–19, **19**
hyporheic zone, 25, 26–27
hyporhithron, 151
large, 26
lentic refugia in, 369
meandering of, 18–19, **19**
metarhithron zone, 151
Mississippi River, 18, 335, 378, 379, 382, 383, 384, 385, 387, 388, 389
Ohio River, 21, 26
osmoregulatory requirements in, 23–24
oxygen in, 22–23
pools in, 19, **19**
riffles in, **19**
riparian zone, 24, 25
river continuum concept, 24–25
salinity of, 22, 23, 24
serial discontinuity concept, 22
species distribution in, 149
species diversity, 149
stream discharge data on, 20
stream ordering system, 18, **18**
temperature patterns in, 22
Tennessee River, 26
thalweg in, **19**
river shrimp, 667
r/K selection hypothesis, 55–56
rock pools, 769
Rotifera, 187–248
aging of, 209–210
alpha-tocopherol (vitamin E) effect on, 199–200
amictic females, 197, 206, **206**, 210, 211, **211**, 213
amphoteric females, **206**, 206–207
aptera generations of, 200
areolations, *239*
auricle, *239, 239*
bioassay for toxicity testing, 199
biogeography of, 202
biomass of, 189, 216
birth rate, 213
bladder, **188**, 196
brain, **188**, **189**, 195
buccal field, **188**, 215, 216
cauda, 193
cingulum, 187, **188**, 191, **191**
clearance rate of, 216
cloaca, **188**, **189**, 191, 196
colonial, 189, 202–204, **203**
corona, 187, **188**, 189, **189**, **191**, 191–192, **192**, 195, 199, **199**, 205,

208, **208**, 215, 216, 218, **218**, 219, 221, **221**, 224
death rate, 212
developmental rate, 209, 212
epiphytic, 189, 204–205, **205**, *226, 234*
epizooic, *235*
esophagus, **188**, 193, 218
feeding rate, 215
flame cell, **189**, 196
foraging behavior, 215, 216
fulcrum, 193, **193**, **194**
generation time, 210
genetic discontinuity in, 214
hypopharyngeal muscle, 193
incus, 193, **193**, **194**
as indicators in toxicity testing, 198
infundibulum, 191, **192**, 215
inbreeding depression in, 213–214
integument, **188**, 190, 191, 195, 218, 219, 221
intracytoplasmic lamina, 190, 221, **221**
larva, 189, 204, 205
life history of, 205–215
lorica, **188**, 190, 191, 195, 218, 219, 221
malleus, 193, **193**, **194**
manubrium, 193, **193**, **194**
mastax, 187, **188**, 189, **192**, 192–193, 195, 216, **221**
mating behavior of, 208, **208**
mictic, 197, 211
mictic females, 197, 206, **206**, 210, 211, **211**
mucus sheaths, 218–219
net reproductive rate, 210
ovary, **188**, **189**, 196
parasites of, 219–220
parasitic, **188**, *237, 238*
pedal gland, **188**, **189**, 190, 205, 224
pharynx, 192, 218
photoreceptors (eye spots), **189**, **192**, 195, 221
prostate gland, 196
protonephridium, **188**, **189**, 193, 195, 196, 221
proventriculus, **192**, 193, 215–216
pseudocoelom, **188**, 191, 195, 199, 218, 219, 219, **219**, 219
pseudotrochal cirri, 192, 215
ramus, 193, **193**, **194**
respiration rate of, 197, 198
resting egg, 206, **206**, 207, **207**, 208, 213
retrocerebral organ, 195–196
retrocerebral sac, **188**, 195
sessile, 189, **191**, **192**, 204–205, 215, 220, 225
setae, 191–192, **192**, 215, *239, 240*
spines of, 197, 217–219
subcerebral sac, **188**, 195
substratum selection by, 204–205
summer egg, 206
survivorship, **209**, 210, **210**, 211
synopsis of, 5–**6**
temperature tolerance in, 197, **210**, 210–211
threshold food concentration for, 211, **211**
trochus, 187, **188**, 191, **191**, 220

trophi, 187, 188, **188, 189,** 192–193,
  193, 198, 216, **220,** 221, 223, 225
uncus, 193, **193, 194**
vas deferens, 196
vertical migration of, 200
vitellarium, **188, 191,** 196, **220,** 221, 224
winter egg, 206, **206,** 207, **207,** 208, 214
rotifers, 489
roundworms. *See* Nematoda
r-selected traits, in bivalves, 348

St. Croix River, 338
St. Lawrence River, 374, 383, 384, 385,
  387, 388
saline lakes and ponds, 33–35, 674, 683
salinity, 22, 23–24, 32–35, 198, 407, 437,
  446, 459, 482, 699, 700, 706, 732,
  733, 765, 766, 767
salinity tolerance, 23, 32–35, 732, 733,
  765, 766, 767
salivary gland, 696
saprobe, 50
Savannaha River, 385
scavenger, 840–841
scrapers, 24–25, **25,** 60
scraping behavior
  in cladocerans, 738, 739, 742
  in non-cladoceran branchiopods, 767,
    769
scuds. *See* Amphipoda
secondary production, of gastrotrichs,
  179
seed shrimp. *See* Ostracoda
seepage area, 534, 541, 544
segmentation, 189
selective feeding, by gastrotrichs, 178,
  179
self-fertilization, 325, 350, 369, 486
semelparity, 291, 292, **292,** 348, 348, 351,
  437, 450, 835
senescence, 54–55, 209–210
serial discontinuity concept, 22
serotonin, in bivalves, 335
seston, consumption by bivalves, 359,
  365
sewage effluents
  effects on bivalves, 338, 340, 362
  effects on decapods, 845
sex attractants, 254
sex pheromones, 792
sex ratio, 801
sexual dimorphism
  in bivalves, 325
  in copepods, 790
  in crayfishes, 833, 837, 838
  in gastropods, 292
  in nematodes, 253–254
  in ostracodes, 696
  in rotifers, 208
sexual reproduction
  in bryozoans, 486, 489, 491
  in cladocerans, 730–731, 743
  in coelenterates, 129, **129**
  in gastrotrichs, 178
  in hydra, 130, 131, **131,** 132

in leeches and acanthobdellids, **443,**
  443–444, 449
in nemerteans, 164–165
in ostracodes, 695, 696, 704
in polychaetes, 459
in rotifers, 205–206, **206,** 207, 208, **208,**
  210, 211, **211,** 213, 214–215
in sponges, 100, 101, 103, 105
in turbellarians, 148
shell length, at birth in bivalves, **348**
shell organic matrix. *See* proteinaceous
  shell matrix
shell strength, in bivalves, 363, **363**
Shelta Cave, Alabama, 830
shore flies, 629–630
short-legged water striders, 617
shredders, in rivers, 24–25, **25**
shrimps. *See* Caridea
silica
  effect on sponge growth, 109
  influence on sponge populations, 107
siltation, effects on bivalves, 338–339,
  341, 359–361
skeleton
  hydrostatic, 126
  of sponges, 100–101
skiff beetles, 622
slow seasonal life cycle, 607
sludge worms. *See* Tubificidae
snails. *See* Gastropoda
snapping behavior, in glochidia, 342–343
sodium, 321, 334
sodium uptake, 828
soldier flies, 630
solitary midges, 628
sow bugs. *See* Isopoda
species introductions, 408
species invasions, 409
specimen preparation
  of bivalves, 371–372
  of decapods, 849–850
  of leeches and acanthobdellids, 455
  of oligochaetes, 415
  of polychaetes, 459
  of rotifers, 222–224
  of sponges, 112–113
  of tardigrades, 515
spermatheca, **403,** 404, 406, 428, 429, **430**
spermatic cysts, 101
spermatogenesis, 101
spermatophore, 404, 429, 792, 793, 838,
  839
spermatozoa, 736
spermiogenesis, 350
sponges. *See* Porifera
spongillaflies, 110, 617
sporozoan parasites, 219
springs, 25, 26, 534, 541
springtails, 631–632. *See also* Collembola
statocyst, 147
stenothermic pools, 526, 542
stoneflies, 597, 604, 605, 606, 607
stream discharge, calculation of, 20
streams. *See* rivers
stygobiont, 680
subimago, 597, 598, 599
subitaneous egg, 792

substrata
  competition for, 109–110
  importance of, 21–22, 149
substrata availability
  control over sponge populations, 107
  limiting factor for bryozoan
    populations, 488
substrate selection, 293, 297, 488, 491
substratum selection, 293, 297
succession, of protozoan species, 51
sulfide ciliates, 51
sulfuric acid, 843
surface feeding, 799
surface film, 537
suspension feeding, 215, 216, 481, 489,
  798, 799, 800, 803. *See also* filter
  feeding
Suwannee River, 383
swamps, 32, **33**
swarming, 724, 796
swimming speed
  accelerated bursts ("jumps") in, 197
  of protozoa, 42
  of rotifers, 196–197
symbiosis, 61
  in decapods, 840
  in hydra, 133–134, **134**
  involving oligochaetes and
    branchiobdellidans, 402, 409, 428,
    429–430
  in protozoa, 56–57
  in sponges, 107
  in turbellarians, 147, 148, 150
symbiotic algae, 147, 148
syncytial tissue. *See* syncytium
syncytium, 191, 196, 221, 251
syngen, 52

Tangipahoa River, 339
Tardigrada (water bears), 501–521
  anhydrobiosis in, 509, 509
  anoxybiosis in, 508, 509
  apophyses, 506, **506**
  buccal apparatus, 505–508, **506,** 509,
    511, 512, 515, 516
  buccal ring, 506, **506**
  buccal tube, **503,** 505, 506, **506,** 512
  cirri, **502,** 503
  clavae, **502,** 503
  claws, 501, **504,** 504–505, **505,** 509,
    511, 512, 515, 516
  cloaca, **504,** 505, 508, 510
  cryobiosis in, 508, 509
  cryptobiosis in, 508, 510
  cuticle, 503–504, 505, 506, 508, 509,
    511, 512, 515
  cuticular band, 505
  cuticular bar, 505, **505**
  desiccation of, 509
  development, 511
  digestive system, **503, 504,** 505–508
  dispersal, 515
  distribution, 501, 514
  eggs, **503, 504,** 509, 510, **510,** 511, 514,
    515
  encystment, 508–509

esophagus, **503, 504,** 505, 508, 512
habitats, 503, 512–513
hermaphroditism, 509, 511
hydrophilous, 501, 512
hygrophilous, 512–513
hypodermis, 502
lamellae, 506, **506, 507**
latency (latent states), 508–509, 511
life history, 512
life span, 512
lunule, **510,** 505
molting in, 511–512
osmobiosis in, 508, 509
osmoregulation in, 508, 509
ovary, **503, 504,** 510
papillae, **502,** 503, 505, **506, 507**
parthenogenesis, 509, 510–511
pharyngeal bulb (sucking pharynx),
    503, 504, 505, 506, **506, 507,** 508
photoreceptors (eye spots), **503, 504,**
    508
placoids, **506,** 507, 508
population density, 513–514
predation on, 513
septulum, **506,** 507
sexual reproduction in, 509, 511
simplex stage, 508, 512
stylets, **504,** 505, 506, **506,** 507, 511
synopsis of, 8, **9**
tuns, 509
ventral lamina, 506, **506**
taxonomic classification, approaches to,
    1–2
taxonomic keys
    Amphipoda, 683
    Annelida (higher taxa), *459–460*
    Bivalvia (Corbiculacea), 375–378
    Bivalvia (Superfamilies), 373–374
    Bivalvia (Unionacea), 378–389
    Branchiobdellida, *431–433*
    Branchiopoda (non-cladoceran),
        770–776
    Branchiura, 682–683
    Bryozoa (Ectoprocta and Endoprocta),
        492–497
    Ciliophora, 72–86
    Cladocera, 746–763
    Cnidaria (Coelenterata), 140
    Collembola, 653
    Copepoda, *806–813*
    Decapoda (crayfishes, shrimps),
        850–852
    freshwater insects and larvae, 632–653
    Gastropoda, 301–309
    Gastrotricha, 181–182
    Hirudinoidea, *460–468*
    Isopoda, 682
    "major taxa" of freshwater
        invertebrates, 13–15
    Mastigophora, 63–66
    Mermithidae (family in Nematoda),
        271–273
    Mysidacea, 682
    Nematoda (ex-Mermithidae), *263, 264,
        265, 266*
    Nematomorpha, *279*
    Nemertea, 167
    Oligochaeta, *416–428*
    Ostracoda, 709–716
    Peracarida, 682
    Polychaeta, *468–469*
    Porifera, 114–115
    protozoa, 63
    Rotifera, 225–240
    Sarcodina, 66–72
    Tardigrada, *516–518*
    Turbellaria, 153–163
    water mites, 547–589
temperature acclimation, in decapods,
    827
temperature patterns, in rivers, 22
temperature tolerance
    in bryozoans, 487
    in cladocerans, 729
    in decapods, 827, 828
    in gastropods, 289–290
    in hydra, 131
    in non-cladoceran branchiopods, 766
    in protozoa, 47–48, 50
    in rotifers, 197, **210,** 210–211
temporary pools. *See* ephemeral habitats,
    ephemeral ponds
Tennessee River, 26, 338, 378, 379, 380,
    385, 386, 388, 389
Tertiary, 524, 526, 830
test organisms, for toxic substances, 198,
    848
thermal spring, 702
thermocline, **29,** 30
threshold food concentration, for rotifers,
    211, **211**
tolerance adaptation, 289
tolerance curves, 197–198
Tombigbee River, 379, 383, 387
torpidity, 703
toxicity, of pollutants in decapods,
    828–829, 848
trace metals, effects on decapods,
    828–829
tree frogs, 402
trematode mesocercaria, 430
trematode parasites, in bivalves, 362
Triassic, 367
tritonymph, 524
troglobites, 26, 27–28, **28,** 149, 408, 430,
    680, 702, 829, 830, 832, **833,** 836,
    839, 841–842, 847, 849
troglophile, 27
Trout Lake, Wisconsin, 840, 845
trout-stream beetles, 620
Tubificidae (sludge worms), 401, **402,**
    408, 409
Turbellaria (flatworms), 145–164
    adhesive organ, 147
    ascus, **160,** *163*
    atrium, *158, 157, 159, 161,* **161**
    average body size of, 146
    brain, 147
    bursa, *158,* **157, 159**
    chemoreceptors, 147
    ciliated pits, 147
    copulatory organ, *154,* **154, 156, 160,**
        161
    ductus ejaculatorius, 160, **159, 160**
    entolecithal, *153*
    epidermis, 147
    eye, 147
    life cycle of, 148
    macroturbellaria, 146
    microturbellaria, 146
    mucus, 147, 150, 152
    neurotoxins in, 150
    oviduct, *153, 158,* **157, 159**
    papillae, 147
    parenchyma, 147
    pharynx, 150
    phototaxis, 147
    predation by, 258
    prostomium, *155*
    refractive body, 147
    rhabdoid, 152
    rhabdoid (adenal), 147
    rhabdoid (dermal), 147
    selected characteristics of, 146
    seminal vesicle, **159, 160**
    statocyst, 147
    stylet, 150, *156,* **158, 160,** 163, *163*
    synopsis, 4, **4**
    testicle, *154,* **162**
    thigmotaxis, 147
    vitellarium, *154*
    yolk, *153, 154,* **154**
    zooids, 147
Type I colony, 204
Type II colony, 204
type taxa, in taxonomic classification, 2
typhlosole, 323

underground aquatic habitats, 26–28
    caves, **27,** 27–28, **28**
    hyporheic zone in, 26–27
    phreatic zone in, 27
underground water species, 149
univoltine reproduction, in bivalves, 348,
    351
Ural River, 353
urea, 253, 290
urine, 730

veliger larvae, 324, 325, **343,** 351, 369
velvet water bugs, 616
vertical migration, 200
    of cladocerans, 736–737
    of copepods, 795, 797, 803, 804
    of crustaceans, 672, 676, 678
    of gastropods, 293
vicariance, 526
viviparity, 101
$\dot{V}_{O2}$, 327, 328–329, 330–331, 332

Waldsea Lake, 34
waste heat, effects on decapods, 848
water balance, in bivalves, 334–335, 354
water bears. *See* Tardigrada
water beetles, 618–624
water boatmen, 615
water fleas. *See* Cladocera

water level, effects on bivalves, 340,
   353–354, 362
water measurers, 617
water mites, 489, 523–592
   adult, **525,** 524, 528–531, 534, 536, 539,
      540, 541, 543, 544, 546
   aposematic coloration, 543
   attachment sites, 537, 538
   behavioral flexiblity, 542
   as bioindicators, 543–544
   capitulum, **532**
   cauda, 540
   chaetotaxy, 524, 527, 534, 539, 546,
      547
   chelicera, 527, **532,** 533, 536, 537, 540
   circulation in, 535
   color patterns of, 535
   commensal, 542
   copulation in, 540–541
   crawling forms, 534, 542
   deutonymph, 524, **525,** 528–531, 534,
      536, 539–540, 543, 544, 546
   digestive system, 535
   dispersal, 526, 538, 539, 541
   disruptive camouflage, 535
   distribution, 528–531
   dorsal plate, 527, **532**
   dorsal shield, 535, 546
   ecological characteristics, 528–531
   ectoparasitic larva, 524, **525,** 528–531,
      532, **532, 533,** 537, 537–539, 541,
      546, 547
   eggs, 524, **525,** 537, 538, 541, 546
   ejaculatory complex, 536
   engorgement in, 524, 539
   esophagus, 527, 535, 536
   evolution of, 524, 534–535, 538
   excretory pore, 527, **532,** 535, 536
   excretory system, 535–536
   excretory tubule, 535, 536
   exoskeleton, 534, 535
   eye, 536, 541
   eye spots, 536
   fossil record, 524
   genital acetabulum, 534, 535, 536, 539
   genital field, 534, 536, 539
   genital valve, 534
   glandularium, 534, 535, 536
   gnathosoma, 527, **532,** 533
   gonopore, 534, 535, 536, 540–541
   in groundwater habitats, 534
   haemocoel, 535
   haemolymph, 535, 536
   hindgut, 536
   hosts, 524, 528–531, 541
   host-seeking behavior, 537–538
   idiosoma, 527, **532,** 533, 536
   imagochrysalis, 524, **525,** 539–540
   integument, 527, 534, 535, 536, 540
   in lentic habitats, 525, 542, 545
   life history, 537–541
   in lotic habitats, 525, 534, 535, 542
   lyrifissure, 527, 534, 535, 536
   midgut, 535, 536
   mouthparts, 533, 536, 546
   nearctic species, 525, 526, 537
   nerve trunk, 536
   nymphochrysalis, 524, **525,** 539
   origin of, 524, 528–531, 538
   osmoregulatory chloride epithelium,
      535, 536
   ovary, 536
   oviduct, 536
   palearctic species, 526, 537
   as parasites of bivalves, 362
   pedipalp, 527, 532, **532,** 533, 536, 546
   petiole, 540
   population dynamics, 543
   as predators, 543
   as prey, 543
   proprioreceptors, 536
   sclerotization, 524, 534, 535, 536
   setae, 527, 532, **532,** 534, 535, 536, 542,
      546, 547
   site selection on host of, 538–539
   solenidia, 527, 536, 547
   spermatheca, 536
   spermatophore, 528–531, 536, 540–541
   stigmata, 536
   swimming forms, 527, 534, 535, 542
   tactile receptors, 536
   testes, 536
   trachea, 536
   urstigmata, 527, **532,** 536
   vas deferens, 536
   ventral shield, 546
   walking forms, 534, 542
   zoogeographic characteristics,
      526–527, 528–531
water pennies, 623
water scavenger beetles, 624
water scorpions, 615
water striders, 616
water treaders, 617
water velocity, effects on bivalve
   distribution, 338–339
wells, 409
wetlands, 32–33
whirligig beetles, 621
Winnipeg River, 338
winter eggs. *See* resting eggs
Wisconsin River, 338
wrigglers, 627

Yang-tse River (China), 135
yellow-green algae, 107–108

zebra mussel, 316, 319, 324, 336–337,
   337, 343, 351–352
zoeae larva, of shrimp, 836
zoobenthic community, 256
zoochlorellae, 56–57, 107–108
zoogeography, 526–527, 733
zooplankton, 30–31, 200, 216, 217, 218,
   219, 666, 668, 672, 673, 676, 678, 679